HANDBOOK OF
NANOPHYSICS

Handbook of Nanophysics

Handbook of Nanophysics: Principles and Methods

Handbook of Nanophysics: Clusters and Fullerenes

Handbook of Nanophysics: Nanoparticles and Quantum Dots

Handbook of Nanophysics: Nanotubes and Nanowires

Handbook of Nanophysics: Functional Nanomaterials

Handbook of Nanophysics: Nanoelectronics and Nanophotonics

Handbook of Nanophysics: Nanomedicine and Nanorobotics

HANDBOOK OF NANOPHYSICS

Nanotubes and Nanowires

Edited by

Klaus D. Sattler

CRC Press
Taylor & Francis Group
Boca Raton London New York

CRC Press is an imprint of the
Taylor & Francis Group, an **informa** business

CRC Press
Taylor & Francis Group
6000 Broken Sound Parkway NW, Suite 300
Boca Raton, FL 33487-2742

First issued in paperback 2020

ISBN 13: 978-0-367-38361-9 (pbk)
ISBN 13: 978-1-4200-7542-7 (hbk)

Library of Congress Cataloging-in-Publication Data

Handbook of nanophysics. Nanotubes and nanowires / editor, Klaus D. Sattler.
    p. cm.
  Includes bibliographical references and index.
  ISBN 978-1-4200-7542-7 (alk. paper)
  1. Nanotubes. 2. Nanowires. I. Sattler, Klaus D. II. Title: Nanotubes and nanowires.

TA418.9.N35N35777 2009
620'.5--dc22
                                                 2009038058

Visit the Taylor & Francis Web site at
http://www.taylorandfrancis.com

and the CRC Press Web site at
http://www.crcpress.com

Contents

PART I   Carbon Nanotubes

PART II Inorganic Nanotubes

PART III Types of Nanowires

PART IV Nanowire Arrays

PART V Nanowire Properties

Preface

The *Handbook of Nanophysics* is the first comprehensive reference to consider both fundamental and applied aspects of nanophysics. As a unique feature of this work, we requested contributions to be submitted in a tutorial style, which means that state-of-the-art scientific content is enriched with fundamental equations and illustrations in order to facilitate wider access to the material. In this way, the handbook should be of value to a broad readership, from scientifically interested general readers to students and professionals in materials science, solid-state physics, electrical engineering, mechanical engineering, computer science, chemistry, pharmaceutical science, biotechnology, molecular biology, biomedicine, metallurgy, and environmental engineering.

What Is Nanophysics?

Modern physical methods whose fundamentals are developed in physics laboratories have become critically important in nanoscience. Nanophysics brings together multiple disciplines, using theoretical and experimental methods to determine the physical properties of materials in the nanoscale size range (measured by millionths of a millimeter). Interesting properties include the structural, electronic, optical, and thermal behavior of nanomaterials; electrical and thermal conductivity; the forces between nanoscale objects; and the transition between classical and quantum behavior. Nanophysics has now become an independent branch of physics, simultaneously expanding into many new areas and playing a vital role in fields that were once the domain of engineering, chemical, or life sciences.

This handbook was initiated based on the idea that breakthroughs in nanotechnology require a firm grounding in the principles of nanophysics. It is intended to fulfill a dual purpose. On the one hand, it is designed to give an introduction to established fundamentals in the field of nanophysics. On the other hand, it leads the reader to the most significant recent developments in research. It provides a broad and in-depth coverage of the physics of nanoscale materials and applications. In each chapter, the aim is to offer a didactic treatment of the physics underlying the applications alongside detailed experimental results, rather than focusing on particular applications themselves.

The handbook also encourages communication across borders, aiming to connect scientists with disparate interests to begin interdisciplinary projects and incorporate the theory and methodology of other fields into their work. It is intended for readers from diverse backgrounds, from math and physics to chemistry, biology, and engineering.

The introduction to each chapter should be comprehensible to general readers. However, further reading may require familiarity with basic classical, atomic, and quantum physics. For students, there is no getting around the mathematical background necessary to learn nanophysics. You should know calculus, how to solve ordinary and partial differential equations, and have some exposure to matrices/linear algebra, complex variables, and vectors.

External Review

All chapters were extensively peer reviewed by senior scientists working in nanophysics and related areas of nanoscience. Specialists reviewed the scientific content and nonspecialists ensured that the contributions were at an appropriate technical level. For example, a physicist may have been asked to review a chapter on a biological application and a biochemist to review one on nanoelectronics.

Organization

The *Handbook of Nanophysics* consists of seven books. Chapters in the first four books (*Principles and Methods*, *Clusters and Fullerenes*, *Nanoparticles and Quantum Dots*, and *Nanotubes and Nanowires*) describe theory and methods as well as the fundamental physics of nanoscale materials and structures. Although some topics may appear somewhat specialized, they have been included given their potential to lead to better technologies. The last three books (*Functional Nanomaterials*, *Nanoelectronics and Nanophotonics*, and *Nanomedicine and Nanorobotics*) deal with the technological applications of nanophysics. The chapters are written by authors from various fields of nanoscience in order to encourage new ideas for future fundamental research.

After the first book, which covers the general principles of theory and measurements of nanoscale systems, the organization roughly follows the historical development of nanoscience. *Cluster* scientists pioneered the field in the 1980s, followed by extensive work on *fullerenes*, *nanoparticles*, and *quantum dots* in the 1990s. Research on *nanotubes* and *nanowires* intensified in subsequent years. After much basic research, the interest in applications such as the *functions of nanomaterials* has grown. Many bottom-up and top-down techniques for nanomaterial and nanostructure generation were developed and made possible the development of *nanoelectronics* and *nanophotonics*. In recent years, real applications for *nanomedicine* and *nanorobotics* have been discovered.

Acknowledgments

Many people have contributed to this book. I would like to thank the authors whose research results and ideas are presented here. I am indebted to them for many fruitful and stimulating discussions. I would also like to thank individuals and publishers who have allowed the reproduction of their figures. For their critical reading, suggestions, and constructive criticism, I thank the referees. Many people have shared their expertise and have commented on the manuscript at various stages. I consider myself very fortunate to have been supported by Luna Han, senior editor of the Taylor & Francis Group, in the setup and progress of this work. I am also grateful to Jessica Vakili, Jill Jurgensen, Joette Lynch, and Glenon Butler for their patience and skill with handling technical issues related to publication. Finally, I would like to thank the many unnamed editorial and production staff members of Taylor & Francis for their expert work.

Klaus D. Sattler
Honolulu, Hawaii

Editor

Klaus D. Sattler pursued his undergraduate and master's courses at the University of Karlsruhe in Germany. He received his PhD under the guidance of Professors G. Busch and H.C. Siegmann at the Swiss Federal Institute of Technology (ETH) in Zurich, where he was among the first to study spin-polarized photoelectron emission. In 1976, he began a group for atomic cluster research at the University of Konstanz in Germany, where he built the first source for atomic clusters and led his team to pioneering discoveries such as "magic numbers" and "Coulomb explosion." He was at the University of California, Berkeley, for three years as a Heisenberg Fellow, where he initiated the first studies of atomic clusters on surfaces with a scanning tunneling microscope.

Dr. Sattler accepted a position as professor of physics at the University of Hawaii, Honolulu, in 1988. There, he initiated a research group for nanophysics, which, using scanning probe microscopy, obtained the first atomic-scale images of carbon nanotubes directly confirming the graphene network. In 1994, his group produced the first carbon nanocones. He has also studied the formation of polycyclic aromatic hydrocarbons (PAHs) and nanoparticles in hydrocarbon flames in collaboration with ETH Zurich. Other research has involved the nanopatterning of nanoparticle films, charge density waves on rotated graphene sheets, band gap studies of quantum dots, and graphene foldings. His current work focuses on novel nanomaterials and solar photocatalysis with nanoparticles for the purification of water.

Among his many accomplishments, Dr. Sattler was awarded the prestigious Walter Schottky Prize from the German Physical Society in 1983. At the University of Hawaii, he teaches courses in general physics, solid-state physics, and quantum mechanics.

In his private time, he has worked as a musical director at an avant-garde theater in Zurich, composed music for theatrical plays, and conducted several critically acclaimed musicals. He has also studied the philosophy of Vedanta. He loves to play the piano (classical, rock, and jazz) and enjoys spending time at the ocean, and with his family.

Contributors

Adekunle O. Adeyeye
Information Storage Materials
    Laboratory
Department of Electrical and Computer
    Engineering
National University of Singapore
Singapore, Singapore

Lihi Adler-Abramovich
Department of Molecular Microbiology
    and Biotechnology
George S. Wise Faculty
    of Life Sciences
Tel Aviv University
Tel Aviv, Israel

Nicolás Agraït
Departamento de Física de la Materia
    Condensada
Universidad Autónoma de Madrid

and

Instituto Madrileño de Estudios
    Avanzados en Nanociencia
Madrid, Spain

A. Sasha Alexandrov
Department of Physics
Loughborough University
Loughborough, United Kingdom

Jeffery R. Alston
Nanoscale Science Program
University of North Carolina
    at Charlotte
Charlotte, North Carolina

Dora Altbir
Department of Physics
Universidad de Santiago de Chile
Santiago, Chile

Ioan Bâldea
Theoretische Chemie
Physikalisch-Chemisches Institut
Universität Heidelberg
Heidelberg, Germany

Pietro Ballone
Atomistic Simulation Centre
Queen's University Belfast
Belfast, United Kingdom

Frank Balzer
Mads Clausen Institute
Syddansk Universitet
Sønderborg, Denmark

Prabhakar R. Bandaru
Department of Mechanical
    and Aerospace Engineering
University of California, San Diego
La Jolla, California

Damien J. Carter
School of Physics
The University of Sydney

and

Curtin University of Technology
Sydney, New South Wales, Australia

Lorenz S. Cederbaum
Theoretische Chemie
Physikalisch-Chemisches Institut
Universität Heidelberg
Heidelberg, Germany

Kwok Sum Chan
Department of Physics and Materials
    Science
City University of Hong Kong
Hong Kong, People's Republic
    of China

Harsh Chaturvedi
Optical Science and Engineering
    Program
University of North Carolina
    at Charlotte
Charlotte, North Carolina

Robinson Cortes-Huerto
Atomistic Simulation Centre
Queen's University Belfast
Belfast, United Kingdom

Gene Dresselhaus
Francis Bitter Magnet Laboratory
Massachusetts Institute of Technology
Cambridge, Massachusetts

Mildred S. Dresselhaus
Departments of Electrical Engineering
    and Computer Science and Physics
Massachusetts Institute of Technology
Cambridge, Massachusetts

Peter C. Eklund
Departments of Physics and Materials
    Science and Engineering
The Pennsylvania State University
University Park, Pennsylvania

Andrey Enyashin
Physikalische Chemie
Technische Universität Dresden
Dresden, Germany

and

Institute of Solid State Chemistry
Ural Branch of Russian Academy
    of Science
Ekaterinburg, Russia

Juan Escrig
Department of Physics
Universidad de Santiago de Chile
Santiago, Chile

Muthusamy Eswaramoorthy
Nanomaterials and Catalysis Laboratory
Chemistry and Physics of Materials Unit
    and DST Unit on Nanoscience
Jawaharlal Nehru Centre for Advanced
    Scientific Research
Bangalore, India

Hong Jin Fan
Division of Physics and Applied Physics
School of Physical and Mathematical
    Sciences
Nanyang Technological University
Singapore, Singapore

Adalberto Fazzio
Instituto de Física
Universidade de São Paulo

and

Centro de Ciências Naturais e Humanas
Universidade Federal do ABC
São Paulo, Brazil

Michael W. Forney
Nanoscale Science Program
University of North Carolina
    at Charlotte
Charlotte, North Carolina

Takeo Fujiwara
Center for Research and Development
    of Higher Education
The University of Tokyo
Tokyo, Japan

and

Core Research for Evolutional Science
    and Technology
Japan Science and Technology Agency
Kawaguchi, Japan

Fei Gao
Pacific Northwest National
    Laboratory
Richland, Washington

Ehud Gazit
Department of Molecular Microbiology
    and Biotechnology
George S. Wise Faculty of Life Sciences
Tel Aviv University
Tel Aviv, Israel

Sarjoosing Goolaup
Information Storage Materials
    Laboratory
Department of Electrical and Computer
    Engineering
National University of Singapore
Singapore, Singapore

Zhennan Gu
College of Chemistry and Molecular
    Engineering
Peking University
Beijing, People's Republic of China

Stephan Haas
Department of Physics and Astronomy
University of Southern California
Los Angeles, California

Natalie Herring
Department of Chemistry
University of North Carolina
    at Charlotte
Charlotte, North Carolina

Takeo Hoshi
Department of Applied Mathematics
    and Physics
Tottori University
Tottori, Japan

and

Core Research for Evolutional Science
    and Technology
Japan Science and Technology Agency
Kawaguchi, Japan

David Hughes
Atomistic Simulation Centre
Queen's University Belfast
Belfast, United Kingdom

Yusuke Iguchi
Department of Applied Physics
The University of Tokyo
Tokyo, Japan

Siarhei Ihnatsenka
Department of Physics
Simon Fraser University
Burnaby, British Columbia, Canada

Ado Jorio
Departamento de Física
Universidade Federal de Minas Gerais
Belo Horizonte, Brazil

Swetlana Jungblut
Fakultät für Physik
Universität Wien
Vienna, Austria

Victor V. Kabanov
Department of Complex Matter
Josef Stefan Institute
Ljubljana, Slovenia

Stefan Kettemann
School of Engineering and Science
Jacobs University Bremen
Bremen, Germany

and

Division of Advanced Materials Science
Pohang University of Science
    and Technology
Pohang, South Korea

Sanjay V. Khare
Department of Physics and Astronomy
The University of Toledo
Toledo, Ohio

Oleg V. Kibis
Department of Applied and Theoretical
    Physics
Novosibirsk State Technical University
Novosibirsk, Russia

Na Young Kim
Edward L. Ginzton Laboratory
Stanford University
Stanford, California

Jakob Kjelstrup-Hansen
Mads Clausen Institute
Syddansk Universitet
Sønderborg, Denmark

Suneel Kodambaka
Department of Materials Science
    and Engineering
University of California
Los Angeles, California

Katla Sai Krishna
Nanomaterials and Catalysis Laboratory
Chemistry and Physics of Materials Unit
    and DST Unit on Nanoscience
Jawaharlal Nehru Centre for Advanced
    Scientific Research
Bangalore, India

Meng Hau Kuok
Department of Physics
National University of Singapore
Singapore, Singapore

Jingbo Li
State Key Laboratory for Superlattices
    and Microstructures
Institute of Semiconductors
Chinese Academy of Sciences
Beijing, People's Republic of China

Chwee Teck Lim
Department of Mechanical
    Engineering
National University of Singapore
Singapore, Singapore

Hock Siah Lim
Department of Physics
National University of Singapore
Singapore, Singapore

Wentao Trent Lu
Department of Physics
Electronic Materials Research Institute
Northeastern University
Boston, Massachusetts

Morten Madsen
Mads Clausen Institute
Syddansk Universitet
Sønderborg, Denmark

Sergey A. Maksimenko
Institute for Nuclear Problems
Belarus State University
Minsk, Belarus

Mark A. Miller
University Chemical Laboratory
Cambridge, United Kingdom

Joel E. Moore
Department of Physics
University of California

and

Materials Sciences Division
Lawrence Berkeley National
    Laboratory
Berkeley, California

Padraig Murphy
Department of Critical Studies
California College of the Arts
San Francisco, California

and

Department of Physics
University of California
Berkeley, California

Andrei M. Nemilentsau
Institute for Nuclear Problems
Belarus State University
Minsk, Belarus

Sunil Kumar R. Patil
Department of Mechanical
    Engineering
The University of Toledo
Toledo, Ohio

Jordan C. Poler
Department of Chemistry
University of North Carolina
    at Charlotte
Charlotte, North Carolina

Mikhail E. Portnoi
School of Physics
University of Exeter
Exeter, United Kingdom

Nick Quirke
School of Physics
University College Dublin
Dublin, Ireland

Atikur Rahman
Surface Physics Division
Saha Institute of Nuclear Physics
Kolkata, India

Horst-Günter Rubahn
Mads Clausen Institute
Syddansk Universitet
Sønderborg, Denmark

Jan M. van Ruitenbeek
Kamerlingh Onnes Laboratorium
Universiteit Leiden
Leiden, the Netherlands

Milan K. Sanyal
Surface Physics Division
Saha Institute of Nuclear Physics
Kolkata, India

Elke Scheer
Department of Physics
University of Konstanz
Konstanz, Germany

Manuela Schiek
Mads Clausen Institute
Syddansk Universitet
Sønderborg, Denmark

Tanja Schilling
Bâtiment des Sciences
Université du Luxembourg
Luxembourg

Gotthard Seifert
Physikalische Chemie
Technische Universität Dresden
Dresden, Germany

Li Shi
Department of Mechanical
    Engineering
Texas Materials Institute
The University of Texas at Austin
Austin, Texas

Zujin Shi
College of Chemistry and Molecular
    Engineering
Peking University
Beijing, People's Republic of China

Antônio J. R. da Silva
Instituto de Física
Universidade de São Paulo
São Paulo, Brazil

and

Laboratório Nacional de Luz Síncrotron
Campinas
São Paulo, Brazil

Edison Z. da Silva
Gleb Wataghin Institute of Physics
University of Campinas
São Paulo, Brazil

Ferenc Simon
Institute of Physics
Budapest University of Technology
    and Economics
Budapest, Hungary

and

Fakultät für Physik
Universität Wien
Wien, Austria

Gregory Ya. Slepyan
Institute for Nuclear Problems
Belarus State University
Minsk, Belarus

Roel H. M. Smit
Kamerlingh Onnes Laboratorium
Universiteit Leiden
Leiden, the Netherlands

Chorng Haur Sow
Department of Physics
Faculty of Science
National University of Singapore
Singapore, Singapore

Srinivas Sridhar
Department of Physics
Electronic Materials Research Institute
Northeastern University
Boston, Massachusetts

Catherine Stampfl
School of Physics
The University of Sydney
Sydney, New South Wales, Australia

Oleg A. Starykh
Department of Physics
University of Utah
Salt Lake City, Utah

Iorwerth O. Thomas
Department of Physics
Loughborough University
Loughborough, United Kingdom

Kálmán Varga
Department of Physics and Astronomy
Vanderbilt University
Nashville, Tennessee

Patricio Vargas
Department of Physics
Universidad Técnica Federico Santa
    María
Valparaíso, Chile

Binni Varghese
Department of Physics
National University of Singapore
Singapore, Singapore

Igor Vasiliev
Department of Physics
New Mexico State University
Las Cruces, New Mexico

Eugenio E. Vogel
Department of Physics
Universidad de La Frontera
Temuco, Chile

Zhiguo Wang
Department of Applied Physics
University of Electronic Science
    and Technology of China
Chengdu, People's Republic of China

and

State Key Laboratory for Superlattices
    and Microstructures
Institute of Semiconductors
Chinese Academy of Sciences
Beijing, People's Republic of China

Zhiyong Wang
College of Chemistry and Molecular
    Engineering
Peking University
Beijing, People's Republic of China

William J. Weber
Pacific Northwest National
    Laboratory
Richland, Washington

Paul Wenk
School of Engineering and Science
Jacobs University Bremen
Bremen, Germany

Max Whitby
Department of Chemistry
Imperial College London
and
RGB Research Ltd
London, United Kingdom

Jian Wu
Departments of Physics
The Pennsylvania State University
University Park, Pennsylvania

Andrei D. Zaikin
Institute for Nanotechnology
Karlsruhe Institute of Technology
Karlsruhe, Germany

and

I.E. Tamm Department of Theoretical
    Physics
P.N. Lebedev Physics Institute
Moscow, Russia

Wen Zhang
Department of Physics and Astronomy
University of Southern California
Los Angeles, California

Igor V. Zozoulenko
Department of Science and Technology
Linköping University
Linköping, Sweden

Xiaotao Zu
Department of Applied Physics
University of Electronic Science
    and Technology of China
Chengdu, People's Republic of China

I

Carbon Nanotubes

Pristine and Filled Double-Walled Carbon Nanotubes

Zujin Shi
Peking University

Zhiyong Wang
Peking University

Zhennan Gu
Peking University

1.1 Introduction

As one of the most important materials in the nano area, carbon nanotubes have generated broad and interdisciplinary attention in the last two decades (Dresselhaus and Dai 2004). Their outstanding properties have been studied extensively and much effort has been devoted to their applications in areas of energy storage, electronics, sensors, and more (Baughman et al. 2002; Pengfei et al. 2003; Avouris and Chen 2006; Ajayan and Tour 2007). Research on carbon nanotubes has been primarily focused on multi-walled carbon nanotubes (MWNTs) (Iijima 1991) and single-walled carbon nanotubes (SWNTs) (Bethune et al. 1993; Iijima and Ichihashi 1993). Ever since the breakthrough of the macroscale selective synthesis of double-walled carbon nanotubes (DWNTs) (Hutchison et al. 2001), they have increasingly drawn scientific interest due to their attractive structures and properties. Strictly speaking, DWNTs are one kind of MWNTs. However, DWNTs' properties are remarkably different from those of MWNTs with three or more graphitic shells. As the intermedium between SWNTs and MWNTs, DWNTs possess the advantages of both MWNTs and SWNTs, i.e., excellent mechanical and electrical properties. More importantly, DWNTs offer lots of unique characteristics over SWNTs and MWNTs. For example, DWNTs provide simple models for the investigation of intertube interaction (Saito et al. 1993a; Zolyomi et al. 2006; Tison et al. 2008). Impressive results have been attained regarding the effects of intertube interaction on the properties of DWNTs. Furthermore, the outer tube of a DWNT can serve as a protector for the inner tube (Iakoubovskii et al. 2008).

When the outer tubes are covalently functionalized, the inner tubes still retain their electronic and optical properties (Hayashi et al. 2008). Thus, the inner and outer tubes can play different roles simultaneously in electronic and optical devices.

Due to the hollow structure of carbon nanotubes, they can be filled with molecules in their interior nanometer-sized space, thus providing a new class of hybrid materials with novel structures and properties (Monthioux 2002; Kitaura and Shinohara 2006). The filling of carbon nanotubes is an effective way to modify the properties of carbon nanotubes. Interactions between carbon nanotubes and the encapsulated materials including van der Waals force, electron transfer, and orbital mixing have been shown to alter the properties of carbon nanotubes markedly. On the other hand, the spatial confinement of carbon nanotubes is expected to impart novel and distinct physical and chemical properties to the encapsulated species from their corresponding bulk samples. Carbon nanotubes are transparent to light and electron beams so that they can be used as nano test tubes or nano vessels, which also serve as a protector for the molecules that are otherwise unstable in air or could be used to study the physical and chemical properties in situ, e.g., the structure, phase transition, or chemical reactions in the nanospace, using high-resolution transmission electron microscopy (HRTEM) and optical techniques.

The filling of carbon nanotubes was primarily performed on MWNTs with the aim of fabricating nanowires using MWNTs as templates. The encapsulated materials include metals, oxides, halides, and carbides (Monthioux et al. 2006). After the macroscale synthesis of SWNTs was achieved, they attracted much more attention because of their smaller diameters and more uniform and defect-free structures compared to MWNTs. The encapsulation of $RuCl_3$ into SWNTs by Sloan et al. (1998) and the discovery of $C_{60}@SWNTs$ in the SWNT sample synthesized by a laser-ablation method (Smith et al. 1998) stimulated the research subject of filling of SWNTs. A large variety of molecules have been encapsulated into SWNTs, ranging from fullerenes, metallofullerenes, metal and nonmetal elements, inorganic compounds, and organic molecules (Monthioux et al. 2006). The encapsulated molecules exhibit unique properties in terms of structure, phase transition, motion behavior, and chemical properties. Meanwhile, it has been found that the electronic structure of SWNTs can be modified by dopant insertion, which makes it possible to tune the electronic properties and mechanical properties of the SWNTs. For example, arrays of C_{60} molecules nested inside SWNTs can change the local electronic structure of the SWNTs to give it a hybrid electronic band (Hornbaker et al. 2002); integrating organic molecules of electron donors or acceptors into SWNTs provides stable and controllable doped SWNTs for fabricating molecular electronic devices (Takenobu et al. 2003). Doped endohedral metallofullerenes $Gd@C_{82}$ can divide a semiconducting SWNT into multiple quantum dots, where the band gap is narrowed from 0.5 eV down to 0.1 eV (Lee et al. 2002).

With regard to the encapsulation of guest molecules, DWNTs possess unique characteristics compared to SWNTs and MWNTs. On one hand, larger inner diameters of DWNTs compared to SWNTs impart the ability of accommodating large-size molecules to DWNTs. On the other hand, inner diameters of DWNTs are smaller than those of MWNTs; thus the quantum effects of the encapsulated materials are expected to be more notable in the former case. Nevertheless, there are fewer reports on the filling of DWNTs than both SWNTs and MWNTs, because the synthesis of pure DWNTs is more difficult than the synthesis of SWNTs and MWNTs. Since the successful macroscale synthesis of DWNTs (Hutchison et al. 2001), the filling of DWNTs has been drawing increasing attention.

This chapter focuses on the synthesis and properties of pristine and filled DWNTs. Section 1.2 introduces the synthesis, and structural and electronic properties of pristine DWNTs. As an important method for characterizing carbon nanotubes, features of the Raman spectra of DWNTs are also given. Section 1.3 describes DWNTs filled with fullerenes, and inorganic and organic materials. The structure, phase transition, and chemical reactions of the encapsulated species as well as the doping effects on DWNTs are discussed in detail.

1.2 Pristine Double-Walled Carbon Nanotubes

1.2.1 Synthesis of Double-Walled Carbon Nanotubes

The selective synthesis of DWNTs was first achieved by a hydrogen arc discharge method (Hutchison et al. 2001). The critical factor of this method is the use of sulfur and metals as a catalyst

in the hydrogen atmosphere. From then on, large numbers of reports have emerged on the synthesis of DWNTs, which can be divided into three categories, i.e., the arc discharge method (Huang et al. 2003; Saito et al. 2003; Sugai et al. 2003; Chen et al. 2006; Qiu et al. 2006a, 2007b), chemical vapor deposition (CVD) (Ci et al. 2002, 2007; Bacsa et al. 2003; Flahaut et al. 2003; Hiraoka et al. 2003; Lyu et al. 2003; Wei et al. 2003; Zhou et al. 2003; Zhu et al. 2003; Wei et al. 2004; Li et al. 2005; Liu et al. 2005, 2007; Ramesh et al. 2005; Yamada et al. 2006; Bachmatiuk et al. 2007; Gunjishima et al. 2007; Qi et al. 2007), and the SWNT-template method (Bandow et al. 2001, 2004; Fujita et al. 2005; Guan et al. 2005b, 2008; Kalbac et al. 2005; Pfeiffer et al. 2007; Kuzmany et al. 2008; Shiozawa et al. 2008).

The arc discharge method is an effective way of producing high structural quality carbon nanotubes. MWNTs observed by Iijima using HRTEM in 1991 were by-products in an arc discharge process toward the synthesis of fullerenes. In the following years, the synthesis of carbon nanotubes using the arc discharge method was developed (Ebbesen and Ajayan 1992; Bethune et al. 1993; Iijima and Ichihashi 1993; Saito et al. 1993b; Seraphin and Zhou 1994; Journet et al. 1997). The commonly used conditions were inert atmosphere and metals (Fe, Co, Ni, etc.) used as a catalyst. In the arc discharge process, the evaporation of graphite and metals at high temperature leads to the formation of carbon nanotubes. As for the synthesis of the DWNTs, the atmosphere of hydrogen was frequently used (Saito et al. 2003; Chen et al. 2006; Qiu et al. 2006a). Another important aspect of the synthesis condition for DWNTs is the adding of sulfur to the metal catalysts or using sulfide (Huang et al. 2003; Saito et al. 2003; Chen et al. 2006; Qiu et al. 2006a). In addition, KCl was reported to behave as a promoter for the growing of DWNTs (Qiu et al. 2006a); however, the mechanism is not clear yet.

The production of carbon nanotubes by CVD is a process from gaseous carbon sources to nanotube structures catalyzed by nanoparticles. Generally, the growing conditions of SWNTs, DWNTs, and MWNTs are similar in the CVD process. In many cases, the products consist of all these kinds of carbon nanotubes. Fine control of the conditions is necessary for selectively obtaining one kind of carbon nanotubes. For the synthesis of DWNTs, catalysts reported in the literature include Fe, Co, Ni, Mo, ferrocene, sulfide, etc. Carbon sources include hydrocarbon (methane, ethene, acetylene, benzene), ethanol, etc. (Ci et al. 2002, 2007; Bacsa et al. 2003; Flahaut et al. 2003; Hiraoka et al. 2003; Lyu et al. 2003; Wei et al. 2003; Zhou et al. 2003; Zhu et al. 2003; Wei et al. 2004; Li et al. 2005; Liu et al. 2005, 2007; Ramesh et al. 2005; Yamada et al. 2006; Bachmatiuk et al. 2007; Gunjishima et al. 2007; Qi et al. 2007). One advantage of the CVD method is the facility of preparing a special assembly of DWNTs. For example, a synthesis of vertical arrays of DWNTs on flat substrates has been achieved by using catalysts with a high density (Yamada et al. 2006).

Besides the above two traditional methods, another ingenious route is taking advantage of chemical reactions of guest molecules encapsulated inside SWNTs. Bandow et al. (2001) discovered that C_{60} molecules inside SWNTs coalesce to dimers, trimers,

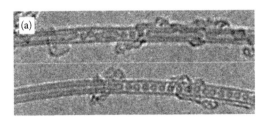

FIGURE 1.1 HRTEM images of C_{60}@SWNTs after heat treatment in vacuo ($<10^{-6}$ Torr) at (a) 1000°C and (b) 1200°C for 14 h. (Reproduced from Bandow, S. et al., *Chem. Phys. Lett.*, 337(1–3), 48, 2001. With permission.)

and so on, and transform into tubes eventually upon heating at temperatures higher than ~800°C (Figure 1.1). Detailed Raman scattering analysis revealed that the diameters of the inner tubes derived from encapsulated C_{60} molecules are close to that of C_{60} molecules at an early stage, whereas the inner tubes turn into wider tubes that match the size of the parent SWNTs upon long-time heat treatment (Bandow et al. 2004). Kalbac et al. (2005) found that the irradiation of C_{60}@SWNTs and C_{70}@SWNTs by ultraviolet light also yields DWNTs. Molecules that can transform into carbon nanotubes inside SWNTs are not limited to fullerenes. As the research on the filling of SWNTs develops rapidly, some other kinds of molecules were found to behave similarly, including ferrocene (Guan et al. 2005b, 2008), PTCDA (Fujita et al. 2005), and $GdCp_3$ (Shiozawa et al. 2008). It was found that the diameter distributions of the inner tubes derived from C_{60} and ferrocene are different even for the same set of parent SWNTs (Pfeiffer et al. 2007).

1.2.2 Electronic Properties of Double-Walled Carbon Nanotubes

Generally, DWNTs are composed of two coaxial tubes interacting through van der Waals forces (DWNTs with an uncoaxial structure were occasionally observed [Hashimoto et al. 2005]). The electronic properties of DWNTs are determined basically by the electronic properties of the two constituent tubes. However, the interaction between the two tubes has significant effects on the electronic properties of DWNTs, which has been demonstrated by theoretical calculations (Saito et al. 1993a; Liang 2004; Song et al. 2005; Zolyomi et al. 2006, 2008; Lu and Wang 2007).

Saito et al. (1993a) investigated the effect of intertube interaction on the energy dispersion relations of DWNTs, which consist of metal–metal, metal–insulator, and insulator–metal constituents. The splitting of some energy bands induced by intertube interactions was observed (Figure 1.2). Moreover, it was predicted that metallic@metallic (M@M)-type DWNTs are still metallic and the metallic tube retains its metallic state in the case of metallic@semiconducting (M@S) and semiconducting@metallic (S@M)-type DWNTs. Liang (2004) performed calculations considering the intertube coupling strength of commensurate DWNTs as a function of the radius difference of the inner and

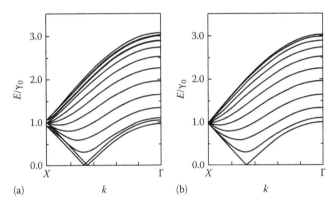

FIGURE 1.2 Energy dispersion relations of a (5, 5)@(10, 10) DWNT (a) with and (b) without considering the intertube interaction. (Reproduced from Saito, R. et al., *J. Appl. Phys.*, 73(2), 494, 1993a. With permission.)

outer tubes, i.e., $t_p = (d_g/d)t_g$, where t_p is the intertube coupling strength of DWNTs with a radius difference of d, and t_g is the interwall coupling strength of graphite with an interlayer distance of $d_g = 0.344$ nm. (If a (n_1, m_1)@(n_2, m_2) DWNT satisfies the relation $n_2/n_1 = m_2/m_1 = \lambda$, where λ is an integer, the carbon atoms in the inner tube match those in the outer tube perfectly. These kinds of DWNTs are commensurate DWNTs.) It was found that the electronic structure of semiconducting@semiconducting (S@S)-type DWNTs depends strongly on the intertube coupling strength. Specifically, strong intertube coupling tends to turn S@S type DWNTs into metallic states. Similar results were found by Moradian et al. (2007) that S@S type DWNTs become metallic when the intertube distance is small, which is attributed to the overlap of the $2p_z$ orbitals of the walls. In addition, it was revealed that the diameter of the tubes also affects the electronic properties of S@S type DWNTs (Liang 2004).

The metallic state of S@S-type DWNTs has been confirmed by experimental studies, including HRTEMs combined with $I(V)$ measurements (Kociak et al. 2002), NMR spectroscopy (Singer et al. 2005), and scanning tunneling microscopy-scanning tunneling spectroscopy (STM–STS) measurements (Tison et al. 2008). STS measurements showed evidence of the presence of a finite density of states for an individual S@S-type DWNT. Apart from the Van Hove singularities of both the inner and the outer tubes, some additional bands were observed during

STS measurements, which might arise from local defects or the occurrence of the new Van Hove singularities due to the intertube interaction (Tison et al. 2008). The Raman spectra of DWNTs also imply the occurrence of new Van Hove singularities compared to SWNTs (Ren et al. 2006), which is discussed in detail in Section 1.2.4.

The inner tube diameter of DWNTs synthesized can be down to ~0.4 nm (Guan et al. 2008). It is known that small-diameter carbon nanotubes exhibit different properties compared with large-diameter ones, arising from their highly curved structure (Blase et al. 1994). For DWNTs with small-diameter inner tubes, a curvature induced metallization was revealed (Okada and Oshiyama 2003). As shown in Figure 1.3, (7, 0)@ (16, 0) and (7, 0)@(17, 0) DWNTs consisting of semiconducting tubes exhibit a metallic nature, resulting from the overlap between the conduction band of the inner nanotube and the valence band of the outer nanotube. The band overlap is a consequence of larger downward shifts of the inner tube rather than those of the outer tubes, because of stronger σ-π rehybridization. It was also demonstrated that electron transfer occurs from the π orbitals of both inner and outer tubes to the intertube region. On the other hand, calculations by Zolyomi et al. (2008) revealed that electron transfer from outer tubes to inner tubes is a universal character for all DWNTs. The transferred electron is in the range of 0.005–0.035 e/Å, depending on the diameters and chiralities of the tubes. The photoemission spectra of DWNTs also support that the inner tubes are negatively charged (Shiozawa et al. 2008).

Due to the nested structure of DWNTs and weak van der Waals interaction between the constituent tubes, displacement of the inner tube with respect to the outer tubes occurs spontaneously (Charlier and Michenaud 1993). Interestingly, it was predicted that rotational displacement gives rise to periodical pseudogaps near the Fermi level (Kwon and Tomanek 1998). Figure 1.4 shows the density of states of a (5, 5)@(10, 10) DWNT with mutual rotation angles between the constituents of 0° and 3°, respectively. It can be seen that four pseudogaps open and close periodically when the rotation angle is 3°. This striking feature was attributed to symmetry lowering during the rotational displacement.

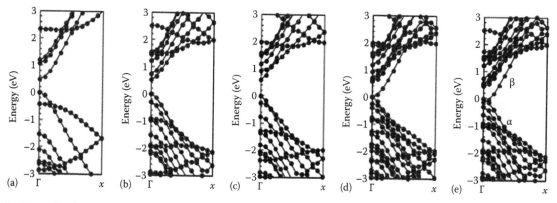

FIGURE 1.3 Energy band structures of (a) (7, 0) SWNT, (b) (16, 0) SWNT, (c) (17, 0) SWNT, (d) (7, 0)@(16, 0) DWNT, and (e) (7, 0)@(17, 0) DWNT. The α and β denote the electron states of the highest branch of the π band and the lowest branch of the π* band, respectively. (Reproduced from Okada, S. and Oshiyama, A., *Phys. Rev. Lett.*, 91(21), 216801, 2003. With permission.)

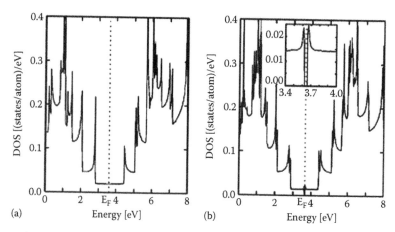

FIGURE 1.4 Density of states of a (5, 5)@(10, 10) DWNT with rotation angles between the two constituents of (a) 0° and (b) 3°.

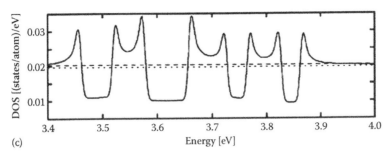

FIGURE 1.4 (continued)　(c) Amplified density of states showing the pseudogaps. (Reproduced from Kwon, Y.K. and Tomanek, D., *Phys. Rev. B*, 58(24), R16001, 1998. With permission.)

1.2.3 Structure Determination of Double-Walled Carbon Nanotubes

Owing to the structure-dependent properties of carbon nanotubes, it is important to determine their structural parameters, including diameters, chiral angles, and handedness. A number of techniques can be used to characterize the structure of carbon nanotubes, such as HRTEM (Hashimoto et al. 2005), STM (Giusca et al. 2007), electron diffraction (Kociak et al. 2003; Zuo et al. 2003), x-ray diffraction (Abe et al. 2003), Raman (Pfeiffer et al. 2003), and fluorescence spectroscopy (Kishi et al. 2006). These techniques allow for the determination of the indices of DWNTs at the individual tube level.

A method for identifying the two independent chiral indices of an individual DWNT based on HRTEM imaging was developed by Hashimoto et al. (2005). The detailed process is as follows. Firstly, the apparent chiral angles (α) of the inner and outer tubes were estimated from the fast Fourier transform (FFT) of an HRTEM image of the nanotube. α is the inclination angle of the hexagon spots to the tube axis in the FFT pattern. Secondly, simulation was performed to obtain the diameters of tubes. It is notable that direct measurements of the diameters from TEM images will get values smaller than the true diameters, because the maximum atomic density on the projection is biased toward the inside of the diameter due to the tube curvature (Hirahara et al. 2006). Lastly, some possible chiral indices were obtained from the combination of chiral angles and diameters, and the accurate chiral indices can be achieved by iterative image simulations. Figure 1.5 shows a DWNT with indices of (12, 1)@(13, 13) identified by this method.

Hirahara et al. (2006) developed an alternative method using electron diffraction with a procedure similar to the above method. They examined the structures of more than 100 DWNTs and revealed the structural characteristics of the DWNTs synthesized with the arc discharge method. They found that DWNTs with diameters larger than 3 nm exhibit a chiral angle distribution over the whole chiral map (0°–30°), while smaller DWNTs display a tendency toward high chiral angles for both inner and outer tubes. In addition, they proposed that the interlayer distance of carbon nanotubes depends on the number of graphitic layers of nanotubes. The interlayer distance of DWNTs were found to be in the range

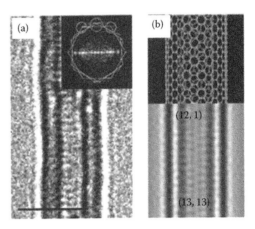

FIGURE 1.5　(a) HRTEM image and the fast Fourier transform of the image for a (12, 1)@(13, 13) DWNT. The scale bar is 2 nm. (b) An atomic model (upper) and simulated image (bottom) of the (12, 1)@(13, 13) DWNT. (Reproduced from Hashimoto, A. et al., *Phys. Rev. Lett.*, 94(4), 045504, 2005. With permission.)

of 0.34–0.38 nm, with an average of 0.358 ± 0.001 nm, which is 5% greater than that of MWNTs with several layers (0.34 nm) (Iijima 1992).

In addition to the chiral indices of carbon nanotubes, the chirality is another structural parameter of carbon nanotubes. The interaction between adjacent graphene layers in carbon nanotubes is associated with the chirality of constituent nanotubes. For a chiral DWNT with indices of $(n_1, m_1)@(n_2, m_2)$, there are four optical isomers: $(n_1, m_1)@(n_2, m_2)$, $(m_1, n_1)@(n_2, m_2)$, $(n_1, m_1)@(m_2, n_2)$, and $(m_1, n_1)@(m_2, n_2)$. These isomers cannot be distinguished by the above methods based on an analysis of FFT or the electron diffraction pattern. Liu et al. (2005) addressed this issue by analyzing HRTEM images on a series of specimen tilts. The principle of this method is schematically shown in Figure 1.6. When the zigzag chains in the skeleton of a carbon nanotube is parallel to the incident electron beam, asymmetric lattice fringes appear at one side of the nanotube (denoted by horizontal arrows). The lattice spacing is 0.216 nm, which corresponds to the distance between two neighboring zigzag chains (Friedrichs et al. 2001). Based on the chirality analysis of 18 chiral DWNTs, a conclusion was drawn that right-handed and left-handed nanotubes are

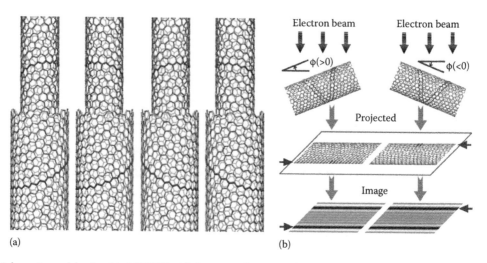

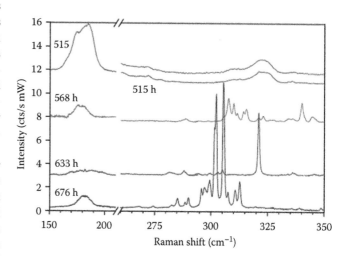

FIGURE 1.6 (a) Schematic models of a chiral DWNT with four optical isomers (from left to right): (14, 3)@(17, 10), (3, 14)@(17, 10), (14, 3)@ (10, 17), and (3, 14)@(10, 17). (b) A schematic sketch of imaging chiral carbon nanotubes with an electron beam. (Reproduced from Liu, Z. et al., *Phys. Rev. Lett.*, 95, 187406, 2005. With permission.)

equally distributed for both the inner and outer nanotubes, whereas the same chirality is preferable for a couple of nested inner and outer nanotubes.

1.2.4 Raman Spectra of Double-Walled Carbon Nanotubes

Raman spectroscopy is a key analytical technique for characterizing the properties of carbon nanotubes (Dresselhaus and Eklund 2000). Compared to electron microscopy, one advantage of Raman spectroscopy is that it is nondestructive to the samples. The spot size of the focused laser used for Raman scattering is usually similar to 1 μm, and thus the overall average properties of the samples can be easily obtained. For carbon nanotubes with a low distribution density, Raman scattering can be performed at the individual nanotube level (Dresselhaus et al. 2002). It is well known that not all the carbon nanotubes are responsive for a laser with a specific wavelength. Raman intensity is enhanced when the energy of the incident photons matches one of the optical transition energy of carbon nanotubes, which is known as resonance Raman (Dresselhaus and Eklund 2000). Generally, the carbon nanotubes out of the resonance window are nondetectable. Thus, multiple lasers or a laser with a tunable wavelength are necessary for the general evaluation of carbon nanotubes.

Raman spectra of carbon nanotubes usually contains four dominant bands (Dresselhaus and Eklund 2000). The first one is radial breathing modes (RBM) corresponding to symmetric in-phase displacements of all the carbon atoms in the radial direction. The frequencies of RBM are usually in the range of ~ 100–$450\,cm^{-1}$. For large-diameter carbon nanotubes, the frequencies of RBM are too low to be detected. The second band is the D band centered around $1300\,cm^{-1}$, which arises from dispersive double resonance processes related to defects.

The third one is tangential modes (G band) arising from vibrations of two adjacent carbon atoms along the tube or in the circumferential direction. G bands are usually in the range of ~ 1500–$1600\,cm^{-1}$. The fourth one is overtones of the D band called G′ band at $\sim 2700\,cm^{-1}$. The vibrational properties of DWNTs are affected by the interaction between the two constituent tubes. Thus, in the following part, we stress special emphasis on the unique characteristics of DWNTs that differ from SWNTs.

The inner tubes of DWNTs derived from C_{60}@SWNTs exhibit two striking features with respect to the RBM band: (1) unusual narrow line widths and (2) a splitting into a large number of lines (Pfeiffer et al. 2003). As shown in Figure 1.7, the peaks in the range of 265–$350\,cm^{-1}$ correspond to the inner tubes. Some

FIGURE 1.7 Raman spectra of DWNTs derived from C_{60}@SWNTs excited by lasers of 515, 568, 633, and 676 nm. (Reproduced from Pfeiffer, R. et al., *Phys. Rev. Lett.*, 90(22), 225501, 2003. With permission.)

of the line widths are down to 0.35 cm⁻¹, which is several times smaller than that of SWNTs (Jorio et al. 2001). The extremely narrow lines are attributed to the highly defect-free and unperturbed environment in the interior of the outer tubes. On the other hand, the splitting of the RBM lines is a consequence of the combination of one inner tube with different outer tubes. The interaction between the inner tubes and outer tubes results in shifts of the RBM frequency.

It is well established that the RBM frequencies of carbon nanotubes scale inversely with diameters of carbon nanotubes as $\omega(RBM) = C_1/d + C_2$, where C_1 and C_2 are constants related to the environments (Dresselhaus et al. 2007). For a specific (n, m) type carbon nanotube, its Raman intensity is enhanced when the energy of the incident photons matches one of its optical transition energy, as stated before. Based on these rules, the assignment of the indices of inner tubes can be made. Figure 1.8 gives an example (Pfeiffer et al. 2005b). Two RBM clusters in 300–370 cm⁻¹ excited by lasers of 585 and 600 nm are assigned to (6, 5) and (6, 4) tubes, respectively. The origin of the splitting of RBM was demonstrated by theoretical calculations. Figure 1.8 shows the calculated RBM frequency of a (6, 4) inner tube in

various outer tubes with different diameters. The trend is clear that the RBM frequency of the inner tube upshifts as the diameter difference to the outer tubes decreases. By comparing the experimental RBM frequencies of the inner tubes with the calculated values, the corresponding outer tubes can be tentatively assigned.

The physical origin of the D band in the Raman spectra of carbon nanotubes is based on the double resonance mechanism, which involves a resonance with both the incident or scattered photons and an intermediate intraband scattering process (Ren et al. 2006). It is known that the D band features are very sensitive to the electronic structure of the carbon nanotubes. For SWNTs with a Gaussian diameter distribution, only one main peak is observed in the D band. In the case of DWNTs, four peaks with narrow line width are observed (Figure 1.9) (Ren et al. 2006). A comparative experiment with a mixture of two samples of SWNTs with diameters similar to the outer and inner tubes of the DWNTs exhibits a D band composed of two peaks, which are close in frequency to the two inner peaks of the D band for the DWNTs. This indicates that the two inner peaks of the D band are from DWNTs with a weak intertube

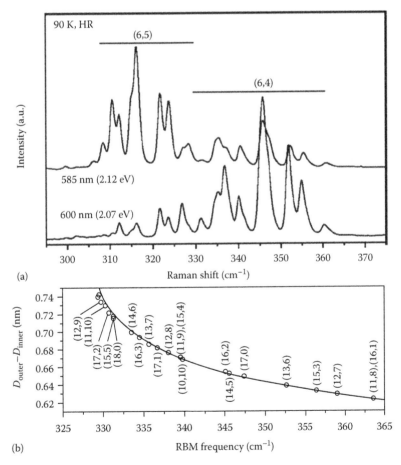

FIGURE 1.8 (a) RBM bands of (6, 4) and (6, 5) inner tubes of DWNTs derived from C₆₀@SWNTs excited by lasers of 585 and 600 nm. (b) Calculated RBM frequency of a (6, 4) inner tube as a function of diameter difference to various outer tubes. (Reproduced from Pfeiffer, R. et al., *Phys. Rev. B*, 72(16), 161404, 2005b. With permission.)

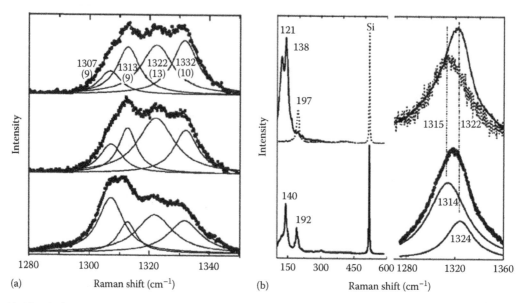

FIGURE 1.9 (a) D band of DWNTs obtained at different positions of an identical sample. (b) Raman spectra of DWNTs (bottom) and superposition of Raman lines of two SWNTs samples with different diameter distributions (upper). (Reproduced from Ren, W.C. et al., *Phys. Rev. B*, 73(11), 115430, 2006. With permission.)

interaction, i.e., incommensurate DWNTs. The two outer peaks are attributed to commensurate DWNTs that have strong coupling between the two constituent tubes. This finding suggests the influence of the atomic correlation between the two constituent tubes on their electronic structures and the occurrence of additional Van Hove singularities in the density of states of commensurate DWNTs compared with those of independent SWNTs.

The G bands of DWNTs derived from C_{60}@SWNTs are investigated by Bandow et al. (2002) using different excitation wavelengths. From a comparison of the G bands of DWNTs with those of the parent SWNTs, the frequency range of G bands of the inner tubes were identified. As shown in Figure 1.10, the position of G bands for the inner and outer tubes are 1587 and 1994 cm^{-1}, respectively. This frequency downshift of the G band for the inner tubes with respect to the outer tubes is a consequence of more sp^3 components in the C–C interaction in the former case due to their high curvature.

The G′ band of double-wall carbon nanotubes exhibits four main features (Figure 1.11) (Pfeiffer et al. 2005a). First, the dispersion for the response from the inner tubes (85.4 cm^{-1}) is lower than that from the outer tubes (99.1 cm^{-1}), which is a consequence of the high curvature of the inner tubes and suggests a flattening of the phonon dispersion at the K point in the Brillouin zone. Second, the frequency of inner tubes is lower than that of carbon nanotubes with larger diameters that are obeyed to the relation $\omega(G') = 2708.1 - 35.4/d$ (Souza et al. 2003). This deviation is also associated with the strong curvature of the small-diameter inner tubes. Third, the G′ band cross sections reflect the resonance with the Van Hove singularities. The peak and the humps in Figure 1.11 correspond to resonance transitions that are marked with arrows.

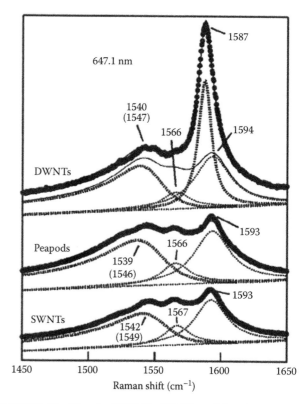

FIGURE 1.10 G bands of DWNTs, SWNTs, and C_{60}@SWNTs excited by a laser of 647.1 nm. (Reproduced from Bandow, S. et al., *Phys. Rev. B*, 66(7), 075416, 2002. With permission.)

Fourth, there is an enhancement of the cross section from the inner tubes compared with the outer tubes, indicating a strongly enhanced electron–phonon coupling for smaller-diameter tubes.

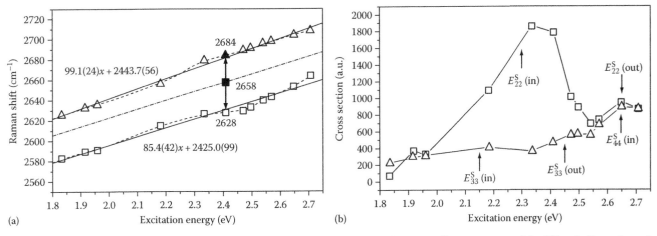

FIGURE 1.11 (a) Position of the G′ band of inner (bottom) and outer tubes (upper) of DWNTs. (b) Cross section of the G′ band of inner (upper) and outer tubes (bottom) of DWNTs. (Reproduced from Pfeiffer, R. et al., *Phys. Rev. B*, 71(15), 155409, 2005a. With permission.)

1.3 Filled Double-Walled Carbon Nanotubes

Materials confined in nanospace are of great interest because of their unique properties arising from low-dimensionality and spatial confinement. The hollow structure of carbon nanotubes enables them to serve as templates for encapsulation of guest molecules. In general, the interior space of carbon nanotubes are smooth and have a wide diameter distribution from less than 1 nm to several tens of nanometers; thus, carbon nanotubes are ideal candidates for the investigation of various encapsulated molecules. In this section, we introduce the progress of filling of DWNTs with fullerenes, inorganic and organic materials, as well as chemical reactions inside DWNTs. Due to the similarity of endohedral doping and exohedral doping effect on DWNTs, we also discuss exohedral doping of DWNTs in this section.

1.3.1 Fullerene-Filled Double-Walled Carbon Nanotubes

Fullerenes received the most interest regarding the filling of carbon nanotubes because of their high stability and multiple phases inside carbon nanotubes. Theoretical calculations showed that the packing modes of C_{60} molecules inside carbon nanotubes are strongly dependent on the diameters of the latter and up to 10 crystalline phases could be formed in a wide range of nanospace (Hodak and Girifalco 2003), which was proved by experiments (Khlobystov et al. 2004; Qiu et al. 2006b). Two phases of C_{60} in DWNTs, i.e., linear chains and achiral zigzags, observed by HRTEM and their relevant schematic illustrations are shown in Figure 1.12a and b, respectively (Qiu et al. 2006b). The upper image of Figure 1.12a shows a linear chain of C_{60} molecules encapsulated inside a DWNT with an inner diameter of ~1.2 nm and a model of C_{60} accommodated [(10, 10)@(16, 16)] DWNTs. Figure 1.12b shows an HRTEM image of an achiral zigzag packing phase of C_{60} in a DWNT with an inner diameter

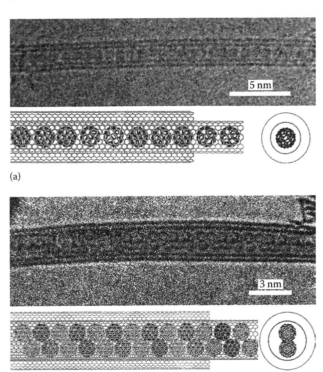

FIGURE 1.12 Typical HRTEM images of different phases of C_{60} molecules inside individual DWNTs and their corresponding schematic models: (a) C_{60}@DWNT with a single chain phase and its schematic illustration of the model C_{60}@[(10,10)@(16,16)]; (b) C_{60}@DWNT with a zigzag phase and the schematic model of C_{60}@[(15,15)@(22,22)]. (Reproduced from Qiu, H.X. et al., *Carbon*, 44(3), 516, 2006b. With permission.)

of ~2.0 nm and its schematic C_{60}@[(15, 15)@(22, 22)]. As most of the DWNTs have the outer diameters from 2.5 to 3.5 nm (typically 1.7–2.7 nm in inner diameter), C_{60}@DWNTs with a ~2.2 nm inner diameter should be dominant. Therefore, the main stacking phases of C_{60} molecules should be the zigzag phase, double-helix phase, and even the phase of two-molecule layers as predicted by

Hodak and Girifalco (2003). Figure 1.13 is an HRTEM image of C_{60}@DWNTs in bundles in which a three-nanotube bundle can be clearly seen in the lower part. As these tubes are filled with C_{60} in a relatively high yield, it is hard to work out whether the C_{60} molecules are well arranged or not, which is hindered by the superposition of fullerene molecules.

Figure 1.14 shows the room temperature Raman spectra of C_{60}@DWNTs and the corresponding parent empty DWNTs. From the spectra, two additional peaks can be seen clearly in the former. A strong peak attributable to the intramolecular Raman active frequency of C_{60} (pentagonal pinch mode) is at 1464 cm^{-1},

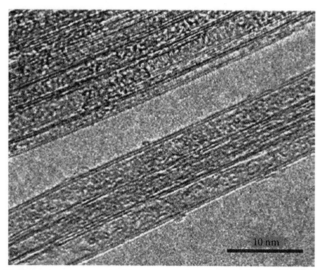

FIGURE 1.13 An HRTEM image of a bundle of C_{60}@DWNTs. (Reproduced from Qiu, H.X. et al., *Carbon*, 44(3), 516, 2006b. With permission.)

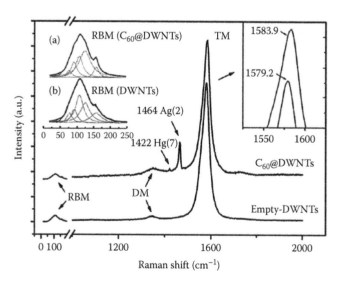

FIGURE 1.14 Raman spectra of C_{60}@DWNTs and empty DWNTs (room temperature, 514.5 nm excitation). The RBM observed either in (a) or (b) is simulated by a linear combination of five Lorentzian lines. Inset on the top-right corner is the zoomed tangential modes (TM). (Reproduced from Qiu, H.X. et al., *Carbon*, 44(3), 516, 2006b. With permission.)

Ag(2), and another weak peak at 1422 cm^{-1}, Hg(7). Ag(2) is obviously down-shifted by about 5 cm^{-1} comparing to that of the bulk C_{60} (Pokhodnia et al. 1996). It is well known that the high-frequency Ag(2) mode at 1469 cm^{-1} for C_{60} is sensitive to the polymerization degree, especially in the case of dimerization (Rao et al. 1993). The softening of the Ag(2) mode suggests that a certain degree of polymerization among fullerene molecules has indeed taken place inside the tubes. Moreover, it is known that the charge partially transferred between SWNTs and C_{60}, typically about 6 cm^{-1} per electron per C_{60} cage (Pichler et al. 2001) can also contribute to the softening of the Ag(2) mode. So, 5 cm^{-1} softening of the Ag(2) mode proves the charge transfer between DWNTs and C_{60}. In addition, the tangential mode (TM) is slightly upshifted from 1579.2 to 1583.9 cm^{-1}, which further supports the charge transfer between C_{60} and DWNTs.

Ning et al. (2007) found that C_{60}@DWNTs exhibit a higher thermal stability in air than pristine DWNTs. Thermogravimetric analysis revealed that the burning temperatures of C_{60}@DWNTs and pristine DWNTs are 636.2°C and 581.8°C, respectively. The enhanced thermal stability of C_{60}@DWNTs was attributed to charge redistribution in this composite. Theoretical calculations demonstrated that the rearrangement of the electronic structure of the C_{60}@DWNTs system gives rise to a slightly negatively charged area between fullerenes and inner tubes and a positively charged area on the surface of outer tubes (Okada et al. 2001). Consequently, the reducing ability of the outer tubes decreases, and the oxidation of C_{60}@DWNTs needs a higher temperature than pristine DWNTs.

It is comprehensible that the stability of C_{60}-filled carbon nanotubes is dependent on the inner diameter of the nanotubes. When the inner diameter of a carbon nanotube matches the size of C_{60} well, the insertion of C_{60} into the carbon nanotube is energetically favorable (Okada et al. 2001). Scipioni et al. (2008) reported a somewhat surprising result that encapsulation energies of C_{60} in an SWNT and in an DWNT with a same inner diameter are different. Specifically, the encapsulation energy of C_{60} in a (15, 0) SWNT is +1.7 eV, while the encapsulation energy is −0.2 eV in the case of a (15, 0)@(24, 0) DWNT. The reduction of the encapsulation energy for a DWNT originates from the charge transfer from the area between the fullerene and the inner tube to the area between the outer and the inner tube, which reduces the electrostatic repulsion. Therefore, besides the inner diameter of carbon nanotubes, the number of carbon walls is also an important parameter that affects the stability of C_{60}-filled carbon nanotubes.

Apart from spherical shape C_{60} molecules, the fullerene-family includes many ellipsoidal shape molecules (e.g., C_{70}). When the latter ones are inserted into carbon nanotubes, their molecular orientations are determined by the inner diameters of carbon nanotubes. For instance, standing and lying orientations of C_{70} molecules in SWNTs of different diameters have been predicted theoretically and have been proved by experimental studies (Guan et al. 2005a). On the other hand, in the case of DWNTs, the internal space is big enough for C_{70} molecules to rotate freely. TEM images in Figure 1.15 show the rotation of C_{70} molecules

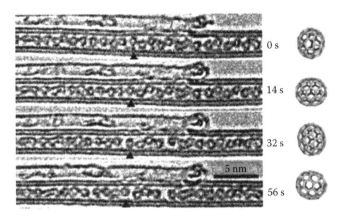

FIGURE 1.15 TEM images of C_{70}@DWNT arranged in a time sequence showing the rotation of C_{70} molecules inside a DWNT. (Reproduced from Ning, G. et al., *Chem. Phys. Lett.*, 441(1–3), 94, 2007. With permission.)

inside a DWNT (Ning et al. 2007). One of the rotating C_{70} molecules is labeled by a black triangle. It can also be seen from the TEM images that C_{70} molecules coalesce gradually into capsuliform and short-tubular structures under electron irradiation, which is a general characteristic of fullerenes.

1.3.2 Inorganic-Material-Filled Double-Walled Carbon Nanotubes

DWNTs are ideal templates for producing nanowires of ionic compounds. Some halides (KI, CsI, AgCl) have been encapsulated into DWNTs using a melt-phase procedure, which involves heating the mixture of DWNTs and the melt encapsulated materials (Costa et al. 2005; Chen et al. 2008). For better contact between DWNTs and the encapsulated materials, the mixture was ground before the heat treatment. For encapsulation of AgCl into DWNTs, a three-step process was employed (Chen et al. 2008). Firstly, a KCl aqueous solution containing dispersed DWNTs was evaporated to form KCl-coated DWNTs. Then the KCl on DWNTs was transformed to AgCl through reaction with $AgNO_3$ in nitric acid solution. Finally, AgCl was melted and encapsulated into DWNTs by capillarity. One merit of this method is that the compact contact between AgCl and DWNTs facilitates the diffusion of AgCl into DWNTs. Figure

1.16 shows several typical HRTEM images of AgCl nanowires inside DWNTs. The interplanar distance of these AgCl crystals is about 0.28 nm, which is close to the (200) lattice spacing of bulk AgCl crystals. Some structural defects such as small twists, vacancy, and interstitial defects are observed in the AgCl crystal nanowires. Such kinds of defects also exist in KI- and CsI-filled DWNTs (Costa et al. 2005). Another interesting issue is that the lattice spacing of KI and CsI are almost identical to their counterparts in bulk phases, which is contrary to the case of KI@ SWNTs in that the lattice expansions are up to 14% (Sloan et al. 2000). This discrepancy originates from the diameter difference between SWNTs and DWNTs. For SWNTs with inner diameters of around 1 nm, only two or three layers of KI along the tube axis are allowed to locate inside SWNTs, giving rise to a high proportion of surface atoms. By contrast, larger diameters of DWNTs result in structures much closer to the bulk phases.

Jorge et al. (2008) reported the ferromagnetic properties of encapsulated α-Fe nanowires inside DWNTs. The encapsulation of α-Fe nanowires was achieved by encapsulating $FeCl_3$ with a solution method and subsequently decomposing the $FeCl_3$. Mösbauer spectra revealed that most of the encapsulated phase is α-Fe. Due to the confinement of DWNTs, the structure of α-Fe is relaxed, leading to an increase in the hyperfine field and permission of some Raman modes that are absent in bulk α-Fe.

The structure and phase transition of water confined in nanometer-sized spaces have been investigated by theoretical simulations and experiments, and novel properties of water were revealed (Koga et al. 2001; Kolesnikov et al. 2004). Chu et al. (2007) investigated water confined in DWNTs with inner diameters of ~1.6 nm through quasielastic neutron scattering (QENS) and observed a fragile to strong liquid transition at 190 K. This dynamic crossover temperature of water confined in a hydrophobic substrate (carbon nanotubes) is lower than in the case of a hydrophilic substrate (e.g., MCM-41).

DWNTs with inner diameters of 0.6–0.8 nm are suitable for the encapsulation of individual atomic metallic chains. Choi et al. (2002) discovered that SWNTs can reduce metal ions that have redox potentials higher than that of SWNTs at room temperature. The resultant metal particles deposit on the surface of SWNTs. This method was improved by Muramatsu et al. (2008) for synthesizing metal-filled DWNTs. Specifically, DWNTs and hexaammonium heptamolybdate tetrahydrate

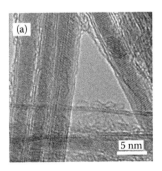
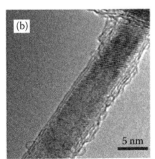
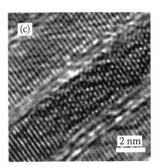

FIGURE 1.16 HRTEM images of AgCl@DWNTs. (Reproduced from Chen, G. et al., *Scr. Mater.*, 58(6), 457, 2008. With permission.)

were treated in hydrochloric acid (18 wt%) solution at 100°C for 24 h and then oxidized in air at 500°C for 30 min. In this process, molybdenum chloride ions are reduced to Mo atoms by carbon nanotubes and then the Mo atoms are inserted into the interior spaces of the nanotubes and form the wires. An HRTEM examination showed that the length of encapsulated Mo chains ranges from several to several tens of nanometers, and the interatomic distance between Mo atoms ranges from 0.32 to 0.38 nm, which is slightly larger than the interatomic distance in bulk Mo crystals (0.315 nm). Fluorescence spectra provides evidence of a strong interaction between Mo atoms and the inner tubes. The PL intensity from the inner tubes with chiralities of (10, 2), (9, 4), (8, 6), and (7, 6) were highly quenched, indicating that the electronic structure of the inner tubes was modified due to the interaction between the Mo atomic chains and the inner tubes.

Cs-filled DWNTs were synthesized by Li et al. (2006c) using plasma-ion irradiation. In that process, Cs atoms were ionized and accelerated by a plasma sheath to bombard DWNTs, generating Cs@DWNTs. Raman scattering measurements revealed that both the outer tubes and inner tubes are doped by Cs. Both the G band of the outer tubes (1591 cm^{-1}) and the inner tubes (1582 and 1556 cm^{-1}) downshift, indicating an electron transfer from Cs to carbon nanotubes.

One important application of filled carbon nanotubes is their use for nano-electronic devices. The transport properties of Cs@DWNTs were investigated by Li et al. (2006a). They found that Cs@DWNT-based field-effect transistor devices exhibit excellent n-type characteristics. Figure 1.17a depicts the current versus voltage (I_{DS}–V_G) characteristics of a Cs@DWNT-FET device measured for different V_{DS}. The n-type semiconducting characteristics are clear and the I_{on}/I_{off} ratio is up to ~5 × 10^5. In the case of Cs@SWNT-FET devices, the I_{on}/I_{off} ratio is ~10^3–10^4. The higher I_{on}/I_{off} ratio of the Cs@DWNT-FET device is associated with better structural properties of DWNTs than SWNTs. It has been observed that much less structural and chemical defects are generated by the Cs encapsulation in the case of DWNTs compared with SWNTs. Figure 1.17b shows the room-temperature output characteristic of I_{DS}–V_{DS} curves with V_{DS} ranging from –3 to 3 V by applying different gate voltages, further confirming the

n-type properties. One striking feature of the Cs@DWNT-FET device is the Coulomb oscillation phenomena at low temperatures. Figure 1.17c shows the conductance versus the gate voltage characteristic measured at 11.5 K for bias voltage V_{DS}=10 mV, from which we can see a series of sharp peaks, indicating that Cs@DWNT consists of a number of Coulomb islands connected in a series.

1.3.3 Organic-Material-Filled Double-Walled Carbon Nanotubes

DWNTs provide templates for the investigation of the arrangement manner of organic molecules in a one-dimensional nanospace. One example is cobalt phthalocyanine (CoPc) encapsulated in DWNTs (Schulte et al. 2007). CoPc is a square-shape molecule possessing useful optical properties. The properties of CoPc assemblies are dependent on their dimensions and their arrangement manner. Schulte et al. investigated the assembly of CoPc that stacks inside DWNTs as well as few-walled carbon nanotubes with inner diameters larger than that of DWNTs. Figure 1.18a shows the HRTEM images of individual carbon nanotubes filled with CoPc. Although it is clear that the tubes are filled, no obvious ordering of the encapsulated molecules could be observed. The researchers attribute this to the fact that the intense electron beam used in TEM disturbs or even damages the CoPc molecules while imaging. The molecular orientations of CoPc were then determined by a near-edge x-ray absorption fine structure (NEXAFS). The N K-edge NEXAFS spectra of CoPc-filled MWNTs for θ = 90° and 28° are shown in Figure 1.18 (θ is the angle between the surface normal and the dominant electric field vector component). The value for the angle ν between the vector perpendicular to the CoPc molecular plane and the surface normal was obtained as ν = 54° ± 5° by comparing the normalized NEXAFS intensity at θ = 90° and 28° with the theoretical curves. Then the stacking angle of CoPc inside the nanotubes was determined to be 36° ± 5°. For the bulk phases of CoPc, i.e., α-CoPc and β-CoPc, the stacking angles are 26° and 45°, respectively. Therefore, it can be concluded that a mixture of the two phases coexist inside carbon nanotubes.

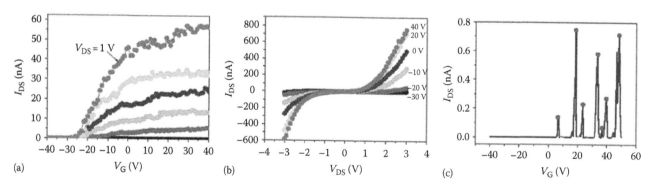

FIGURE 1.17 (a) Drain current versus gate voltage (I_{DS}–V_G) characteristics of a Cs@DWNT-FET device measured for different V_{DS}. (b) I_{DS}–V_{DS} curves with V_{DS} ranging from –3 to 3 V by applying different gate voltages. (c) Conductance versus gate voltage characteristics measured at 11.5 K for bias voltage V_{DS}=10 mV. (Reproduced from Li, Y.F. et al., *Appl. Phys. Lett.*, 89(9), 093110, 2006a. With permission.)

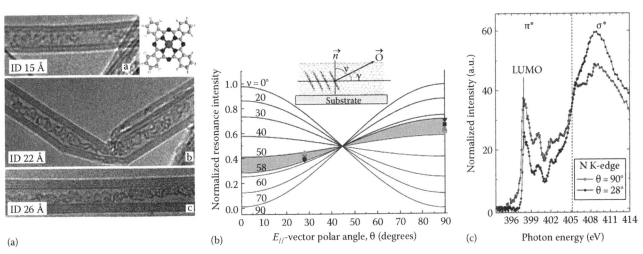

(a)   (b)   (c)

FIGURE 1.18 (a) HRTEM images of carbon nanotubes filled with CoPc. Inset is the schematic drawing of a CoPc molecule. (b) Theoretical normalized NEXAFS intensity curves and the experimental intensity at $\theta = 90°$ and 28°. Top: a schematic drawing of encapsulated CoPc viewed side-on, defining the measured polar angle ν and the internal stacking angle γ inside the nanotube. (c) Nitrogen K-edge NEXAFS spectra of encapsulated CoPc for $\theta = 90°$ and 28°. (Reproduced from Schulte, K. et al., *Adv., Mater.*, 19(20), 3312, 2007. With permission.)

Ferrocene is a frequently used charge-transfer mediator. The catalytic properties of ferrocene@SWNTs have been demonstrated by Sun et al. (2006). Ferrocene was also encapsulated into DWNTs and was shown to modify the transport properties of DWNTs (Li et al. 2006b). Pure DWNT-based FET

exhibits ambipolar characteristics. Figure 1.19a shows the drain current versus the gate voltage (I_{DS}–V_{GS}) curves of a DWNT-FET measured with various V_{DS} ranging from 0 to 1 V in steps of 0.1 V. p-Type conduction is observed when $V_{GS} < -27\,V$ and n-type conductance is observed when $V_{GS} > -20\,V$. In the case of

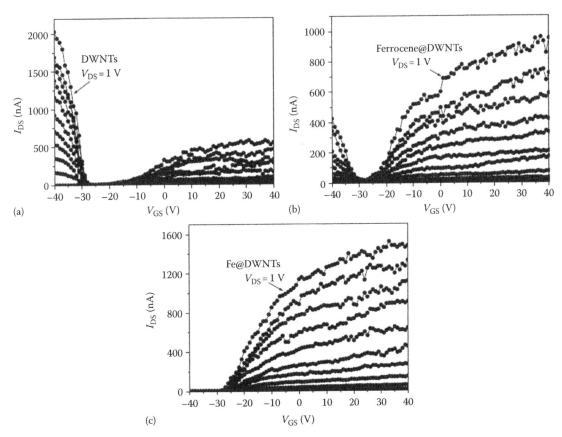

FIGURE 1.19 Drain current versus gate voltage (I_{DS}–V_{GS}) curves of FETs based on (a) DWNT, (b) ferrocene@DWNTs, and (c) Fe@DWNTs measured with $V_{DS} = 0$–1 V. (Reproduced from Li, Y.F. et al., *Nanotechnology*, 17(16), 4143, 2006b. With permission.)

ferrocene@DWNTs, the conductance in the n-channel increases and the conductance in the p-channel is significantly suppressed (Figure 1.19b). This change arises from the transfer of electrons from the encapsulated ferrocene molecules to the DWNTs. Decomposition of ferrocene inside DWNTs leads to the formation of Fe@DWNTs. Fe@DWNTs exhibit a unipolar n-type semiconducting behavior (Figure 1.19c). The different properties between ferrocene@DWNTs and Fe@DWNTs are attributed to their different electron donating abilities. A higher concentration of electrons is injected into the DWNTs in the case of Fe@DWNTs. These results demonstrate that FET devices with controllable characteristics can be fabricated using DWNTs filled with appropriate molecules.

1.3.4 Chemical Reactions inside Double-Walled Carbon Nanotubes

Carbon nanotubes can serve as nano test tubes for reactions in their interior space. Due to the spatial confinement, the reaction process and the products exhibit distinct features compared with reactions free of confinement. In general, the polymerization of encapsulated molecules inside carbon nanotubes generates products with one-dimensional structures. Examples include a transformation from C_{60} into nanotubes (Bandow et al. 2001, 2004; Kalbac et al. 2005) and from $C_{60}O$ into linear polymers (Britz et al. 2005), etc.

Qiu et al. (2007a) studied the pyrolysis of ferrocene inside DWNTs. They found that triple-walled carbon nanotubes (TWNTs) formed after high temperature annealing of ferrocene@ DWNTs, and the carbon nanotubes' base-growth mechanism was proved. Figure 1.20 shows HRTEM images of TWNT bundles and an individual TWNT (inset), revealing clearly that each

nanotube consists of three concentric graphene sheets. On average, the TWNTs have outer diameters ranging from 3 to 5 nm. For the individual TWNT shown in the inset of Figure 1.20, its outer diameter is 3 nm with an interlayer spacing of ~0.4 nm. The HRTEM examinations reveal the purity or the ratio of TWNTs to all nanotubes is over 50%, which is strongly dependent on the filling rate and the inner diameters of DWNTs. In general, for DWNTs with outer diameters larger than 3 nm, high quality TWNTs are favored and formed more easily inside the tubes. Otherwise, few TWNTs are obtained.

Figure 1.21 shows the room temperature Raman spectra of parent DWNTs and TWNTs, respectively. Several new RBM

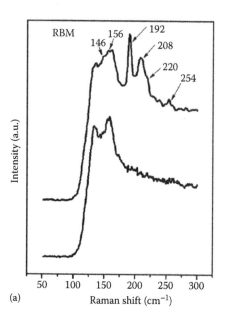

(a)

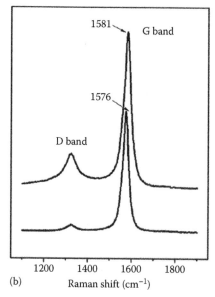

(b)

FIGURE 1.20 An HRTEM image of TWNT bundles derived from ferrocene@DWNTs. Top-right inset: an individual TWNT with clearly resolved graphitic layers. (Reproduced from Qiu, H.X. et al., *Chem. Commun.*, (10), 1092, 2007a. With permission.)

FIGURE 1.21 Raman spectra (633 nm excitation) for empty DWNTs (bottom) and TWNTs (upper) showing (a) RBM and (b) the D band and G band. (Reproduced from Qiu, H.X. et al., *Chem. Commun.*, (10), 1092, 2007a. With permission.)

peaks are clearly seen in the case of TWNTs (top curve in Figure 1.21a), which is evidence that novel carbon nanotubes with smaller diameters have been formed after the annealing of ferrocene inside DWNTs. From these new Raman peaks, it can be calculated that the smallest tertiary tube has a diameter of 0.9 nm, implying that the template DWNTs should have an inner diameter of about 1.6 nm. In other words, DWNTs with inner diameters larger than 1.6 nm would favor the growth of the tertiary carbon nanotubes. Although the largest inner tube of TWNTs is 1.6 nm, this is unlikely the largest tertiary tube because the Raman peaks lower than 146 cm^{-1}, corresponding to tubes with a diameter of 1.6 nm, are hard to detect. In comparison with the Raman spectra from DWNTs, the I_G/I_D ratio of TWNTs sample is lower than that of DWNTs, implying that the new nanotubes formed from ferrocene have a higher defect density than the original template DWNTs prepared by the arc discharge method.

Structure details of TWNTs may render some information on their growth mechanism. It has been found that both capped and open-ended nanotubes are found inside the inner cavities of DWNTs. There are no catalyst particles at the closed-tip, see Figure 1.22a, but something black is packed right at the open-end of the tube (indicated by the lower arrow, in Figure 1.22b). Energy Dispersive x-ray (EDX) analysis reveals that black substances are small iron particles formed during the decomposition of encapsulated ferrocene when they are heated to 900°C inside the central cavity of DWNTs. This information

leads one to believe that the formation of inner carbon nanotubes follows the base-growth mechanism, as schematically illustrated in Figure 1.22c.

The growth of CNTs conforms to the vapor–liquid–solid (VLS) mechanism (Gavillet et al. 2001). Firstly, ferrocene undergoes a decomposition to release a series of hydrocarbons and Fe species at temperatures over 500°C. Theoretical calculations and molecular dynamics simulations revealed that the hydrogen atoms left the ferrocene first due to the weaker C–H bond (492 kJ/mol), and then the cyclopentadienyl ring with higher C–C bond energy (602 kJ/mol) started to break up, followed by the breakage of the C–Fe bonds (1480 kJ/mol) in combination with the remaining C–C bonds (Elihn and Larsson 2004). Hence, the fragments from the decomposition of ferrocene are sequent C–H, C, C_2 and C_3, and finally Fe. During the pyrolysis process, the Fe species serves as a catalyst and those hydrocarbons work as carbon sources. By increasing the temperature to 900°C, these hydrocarbon species and C clusters dissolve into the ion nanoparticles, diffuse through them and precipitate and thus react with each other, promoting the growth of carbon tubular structures under the confinement of DWNTs. The diameter of the newly formed tube is strictly controlled by the diameter of the template, i.e., by the inner tube diameter of DWNTs, and usually smaller by 0.7–0.8 nm. Moreover, the formation of new nanotubes is strongly dependent on the amount of ferrocene encapsulated inside DWNTs and on the pyrolysis conditions. DWNTs with large diameters contain a large amount of ferrocene, accordingly favoring the growth of nanotubes with longer lengths and good structures, while the small template is apt to produce nanotubes with finite length and imperfect structures.

The transformation from ferrocene-filled DWNTs to TWNTs proves that the pyrolysis of molecules inside carbon nanotubes is an effective route for increasing the wall number of carbon nanotubes. An appropriate selection of the encapsulated species enables the alteration of the composition of carbon nanotubes. For instance, nitrogen-doped TWNTs were prepared by encapsulation and pyrolysis of iron(II) phthalocyanine in the interior space of DWNTs (Wang et al. 2008).

Figure 1.23 shows HRTEM images of the as-prepared TWNTs, in which three main features of the TWNTs are explicit. First, the innermost walls are corrugated. In contrast, most of the innermost walls formed from ferrocene are straight. Second, there are bamboo structures in the innermost wall. Third, fullerene structures are present inside the TWNTs. In addition, it was found that the newly formed innermost wall has more defects and is less stable under electron irradiation than the parent walls. Some of the innermost walls collapsed and transformed into an amorphous structure after a long-time electron irradiation. Theoretical calculations and experimental studies have demonstrated that the presence of nitrogen atoms in the carbon nanotubes would lead to the formation of bamboo structures and fullerene structures (Sumpter et al. 2007). In addition, pyridine-like nitrogen atoms in the skeleton of carbon nanotubes result in structural defects (Terrones et al.

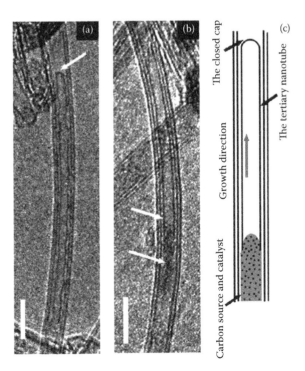

FIGURE 1.22 (a,b) HRTEM images of an individual TWNT derived from ferrocene@DWNT, showing how the tertiary carbon layer is formed. (c) A schematic illustration of the growth mechanism of the TWNTs. The scale bars are 5 nm. (Reproduced from Qiu, H.X. et al., *Chem. Commun.*, (10), 1092, 2007a. With permission.)

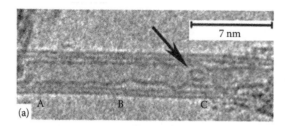 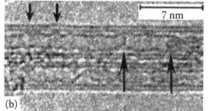

FIGURE 1.23 HRTEM images of TWNTs derived from iron(II) phthalocyanine@DWNTs. (Reproduced from Wang, Z.Y. et al., *Chin. J. Inorg. Chem.*, 24(8), 1237, 2008. With permission.)

2002). Therefore, the structural characteristics of TWNTs are probably related to the nitrogen atoms in the innermost wall of TWNTs.

1.3.5 Doping of Double-Walled Carbon Nanotubes

The modulation of DWNTs' properties is a crucial step toward their applications. A great deal of interest is focused on the doping of DWNTs. p-Type doping of DWNTs has been achieved with electron acceptors such as Br_2 (Chen et al. 2003; Souza et al. 2006), I_2 (Cambedouzou et al. 2004), H_2SO_4 (Barros et al. 2007), etc. On the other hand, n-type doping of DWNTs has been achieved with electron donors including Cs (Li et al. 2006c), K (Rauf et al. 2006; Chun and Lee 2008; Desai et al. 2008), Fe (Li et al. 2006b; Jorge et al. 2008), etc. For DWNTs doped with Cs and Fe, as shown above, the dopants are encapsulated inside the DWNTs. Whereas for other dopants listed here, both endohedral doping and exohedral doping are possible, depending on the experimental conditions.

The radial charge distribution on Br_2-doped DWNTs was studied as a cylindrical molecular capacitor (Chen et al. 2003). Raman scattering measurements on Br_2-doped DWNTs reveal that most of the charge resides on the outer tubes. Figure 1.24 shows that the RBM of the outer tubes disappears after Br_2-doping, while the RBM of the inner tubes are not affected. Meanwhile, the G band of inner tubes remains at $1581\,cm^{-1}$, whereas the G band of the outer tubes vanishes. These results demonstrate that the charge transfer between Br_2 and DWNTs only involves the outer tubes. This system was modeled as a three-layer cylindrical capacitor with the bromine anions forming a shell around the outer nanotube. The calculated number of holes on the inner tube versus the total number of holes on the double-walled tubes is consistent with the experimental results. For M@S type DWNTs, the inner tube dopes slightly at a low doping level. As the doping level increases, the charge distribution rapidly begins to favor the outer tube because of the increase in the electrostatic effects. Similar results were obtained in the case of iodine-doped DWNTs (Cambedouzou et al. 2004). However, the doping effect of H_2SO_4 on the inner tubes of DWNTs (Barros et al. 2007) is different from that of Br_2 doping. In the case of H_2SO_4-doped DWNTs, only the metallic inner tubes are affected, while there is no obvious change for

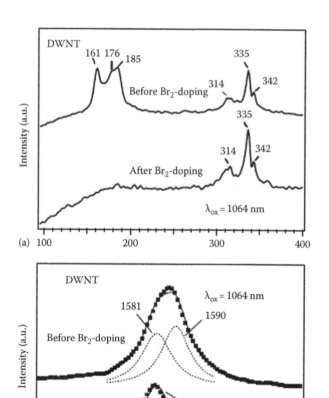

FIGURE 1.24 Raman spectra of DWNTs and Br_2-doped DWNTs: (a) RBM band and (b) G band. (Reproduced from Chen, G.G. et al., *Phys. Rev. Lett.*, 90(25), 257403, 2003. With permission.)

the RBM peaks of the inner semiconducting tubes in both frequency and intensity.

On the other hand, n-type doping of DWNTs by potassium was investigated in detail (Rauf et al. 2006). The behavior of the RBM response of DWNTs upon potassium-doping is similar to the p-type doping of DWNTs. For the excitation of 568 nm (Figure 1.25a), the RBM of the outer tubes and the metallic inner tubes decreases significantly at a low doping level, whereas the response of the semiconducting inner tubes response is hardly

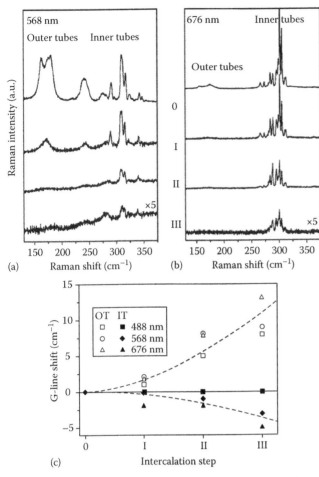

FIGURE 1.25 RBM of potassium-doped DWNTs for different doping steps (0, I, II, and III) excited by lasers of (a) 568 nm and (b) 676 nm. (c) G band shift of outer tubes and inner tubes of potassium-doped DWNTs for different doping steps excited with three different lasers. (Reproduced from Rauf, H. et al., *Phys. Rev. B*, 74(23), 235419, 2006. With permission.)

changed. At a higher doping level, the RBM of the semiconducting inner tubes starts to decrease. For the excitation of 676 nm (Figure 1.25b), the RBM peaks at the frequency region of 250–320 cm^{-1} corresponding to (7, 5) and (8, 3) inner tubes. An identical inner tube exhibits a number of RBM peaks when the outer tubes are different, so each peak in Figure 1.25b represents an inner/outer tube pair. Based on the assignment of these peaks, the influence of the diameter difference between the inner and outer tubes on the charge transfer to the inner tubes was investigated. For instance, the intensity of an RBM peak from an (8, 3) inner tube with a smaller diameter difference [(8, 3)@ (14, 1) tubes] decreases faster than that with a larger diameter difference [(8, 3)@(15, 6) tubes]. Figure 1.25c depicts the G band shift of outer tubes and inner tubes of potassium-doped DWNTs excited with three different lasers. The G band of outer tubes upshifts rapidly with an increase in doping. The maximum shift values are 8, 9, and 13 cm^{-1} for the excitation of 488, 568, and 676 nm, respectively. In contrast, the G band of inner tubes

remains unchanged for the excitation of 488 nm and downshifts for the excitation of 568 and 676 nm. A comparison of charge concentration between outer tubes and inner tubes cannot be made from the degree of the G band shift because the relationship of the G band shift of carbon nanotubes with the doping level is complicated. A nonmonotonic behavior of the G band of SWNTs upon potassium-doping has been observed (Chen et al. 2005). The frequency shift in Figure 1.25c is explained by the Coulomb interaction in a system composed of negatively charged carbon ions and positively charged potassium ions.

1.4 Summary

DWNTs possess distinct characteristics compared with SWNTs and MWNTs. The interaction between the two constituent tubes of DWNTs not only modifies the properties of DWNTs, including electronic band splitting, the occurrence of new Van Hove singularities, and semiconducting-metal transitions, but also affects the stability of encapsulated molecules. The inner and outer tubes of DWNTs exhibit different responses to endohedral or exohedral doping, which is related to the nested structure and the metallic or semiconducting nature of the constituents. Taking advantage of the different properties of the inner and outer tubes and imparting different functions to them are research issues that need to be addressed. Materials encapsulated inside DWNTs including fullerenes, inorganic and organic molecules display novel structures and properties due to the spatial confinement and the interaction with DWNTs. The new phenomena observed in the nanospace inside carbon nanotubes have promoted the understanding of the low-dimensional system.

Acknowledgments

This work was supported by the National Natural Science Foundation of China (Nos. 90206048, 20371004, and 20771010) and the Ministry of Science and Technology of China (Grant 2006CB932701 and 2007AA03Z311).

References

Abe, M., Kataura, H., Kira, H., Kodama, T., Suzuki, S., Achiba, Y., Kato, K., Takata, M., Fujiwara, A., Matsuda, K., and Maniwa, Y. (2003). Structural transformation from single-wall to double-wall carbon nanotube bundles. *Physical Review B* 68(4): 041405.

Ajayan, P. M. and Tour, J. M. (2007). Materials science—Nanotube composites. *Nature* 447(7148): 1066–1068.

Avouris, P. and Chen, J. (2006). Nanotube electronics and optoelectronics. *Materials Today* 9(10): 46–54.

Bachmatiuk, A., Borowiak-Palen, E., Rummeli, M. H., Kramberger, C., Hubers, H. W., Gemming, T., Pichler, T., and Kalenczuk, R. J. (2007). Facilitating the CVD synthesis of seamless double-walled carbon nanotubes. *Nanotechnology* 18(27): 275610.

Bacsa, R. R., Flahaut, E., Laurent, C., Peigney, A., Aloni, S., Puech, P., and Bacsa, W. S. (2003). Narrow diameter double-wall carbon nanotubes: Synthesis, electron microscopy and inelastic light scattering. *New Journal of Physics* 5: 131.

Bandow, S., Takizawa, M., Hirahara, K., Yudasaka, M., and Iijima, S. (2001). Raman scattering study of double-wall carbon nanotubes derived from the chains of fullerenes in single-wall carbon nanotubes. *Chemical Physics Letters* 337(1–3): 48–54.

Bandow, S., Chen, G., Sumanasekera, G. U., Gupta, R., Yudasaka, M., Iijima, S., and Eklund, P. C. (2002). Diameter-selective resonant Raman scattering in double-wall carbon nanotubes. *Physical Review B* 66(7): 075416.

Bandow, S., Hiraoka, T., Yumura, T., Hirahara, K., Shinohara, H., and Iijima, S. (2004). Raman scattering study on fullerene derived intermediates formed within single-wall carbon nanotube: From peapod to double-wall carbon nanotube. *Chemical Physics Letters* 384(4–6): 320–325.

Barros, E. B., Son, H., Samsonidze, G. G., Souza, A. G., Saito, R., Kim, Y. A., Muramatsu, H., Hayashi, T., Endo, M., Kong, J., and Dresselhaus, M. S. (2007). Raman spectroscopy of double-walled carbon nanotubes treated with H_2SO_4. *Physical Review B* 76(4): 045425.

Baughman, R. H., Zakhidov, A. A., and de Heer, W. A. (2002). Carbon nanotubes—The route toward applications. *Science* 297(5582): 787–792.

Bethune, D. S., Kiang, C. H., Devries, M. S., Gorman, G., Savoy, R., Vazquez, J., and Beyers, R. (1993). Cobalt-catalyzed growth of carbon nanotubes with single-atomic-layer-walls. *Nature* 363(6430): 605–607.

Blase, X., Benedict, L. X., Shirley, E. L., and Louie, S. G. (1994). Hybridization effects and metallicity in small radius carbon nanotubes. *Physical Review Letters* 72(12): 1878–1881.

Britz, D. A., Khlobystov, A. N., Porfyrakis, K., Ardavan, A., and Briggs, G. A. D. (2005). Chemical reactions inside single-walled carbon nano test-tubes. *Chemical Communications* (1): 37–39.

Cambedouzou, J., Sauvajol, J. L., Rahmani, A., Flahaut, E., Peigney, A., and Laurent, C. (2004). Raman spectroscopy of iodine-doped double-walled carbon nanotubes. *Physical Review B* 69(23): 235422.

Charlier, J. C. and Michenaud, J. P. (1993). Energetics of multilayered carbon tubules. *Physical Review Letters* 70(12): 1858–1861.

Chen, G., Qiu, J., and Qiu, H. X. (2008). Filling double-walled carbon nanotubes with AgCl nanowires. *Scripta Materialia* 58(6): 457–460.

Chen, G. G., Bandow, S., Margine, E. R., Nisoli, C., Kolmogorov, A. N., Crespi, V. H., Gupta, R., Sumanasekera, G. U., Iijima, S., and Eklund, P. C. (2003). Chemically doped double-walled carbon nanotubes: Cylindrical molecular capacitors. *Physical Review Letters* 90(25): 257403.

Chen, G. G., Furtado, C. A., Bandow, S., Iijima, S., and Eklund, P. C. (2005). Anomalous contraction of the C-C bond length in semiconducting carbon nanotubes observed during Cs doping. *Physical Review B* 71(4): 045408.

Chen, Z. G., Li, F., Ren, W. C., Cong, H. T., Liu, C., Lu, G. Q., and Cheng, H. M. (2006). Double-walled carbon nanotubes synthesized using carbon black as the dot carbon source. *Nanotechnology* 17(13): 3100–3104.

Choi, H. C., Shim, M., Bangsaruntip, S., and Dai, H. J. (2002). Spontaneous reduction of metal ions on the sidewalls of carbon nanotubes. *Journal of the American Chemical Society* 124(31): 9058–9059.

Chu, X. Q., Kolesnikov, A. I., Moravsky, A. P., Garcia-Sakai, V., and Chen, S. H. (2007). Observation of a dynamic crossover in water confined in double-wall carbon nanotubes. *Physical Review E* 76(2): 021505.

Chun, K. Y. and Lee, C. J. (2008). Potassium doping in the double-walled carbon nanotubes at room temperature. *Journal of Physical Chemistry C* 112(12): 4492–4497.

Ci, L. J., Rao, Z. L., Zhou, Z. P., Tang, D. S., Yan, Y. Q., Liang, Y. X., Liu, D. F., Yuan, H. J., Zhou, W. Y., Wang, G., Liu, W., and Xie, S. S. (2002). Double wall carbon nanotubes promoted by sulfur in a floating iron catalyst CVD system. *Chemical Physics Letters* 359(1–2): 63–67.

Ci, L., Vajtai, R., and Ajayan, P. M. (2007). Vertically aligned large-diameter double-walled carbon nanotube arrays having ultralow density. *Journal of Physical Chemistry C* 111(26): 9077–9080.

Costa, P., Friedrichs, S., Sloan, J., and Green, M. L. H. (2005). Imaging lattice defects and distortions in alkali-metal iodides encapsulated within double-walled carbon nanotubes. *Chemistry of Materials* 17(12): 3122–3129.

Desai, S., Rivera, J., Jalilian, R., Hewaparakrama, K., and Sumanasekera, G. U. (2008). Studies of electronic distribution in potassium-doped mats of single-walled carbon nanotubes, double-walled carbon nanotubes, and peapods. *Journal of Applied Physics* 104(1): 013707.

Dresselhaus, M. S. and Dai, H. (2004). Carbon nanotubes: Continued innovations and challenges. *MRS Bulletin* 29(4): 237–239.

Dresselhaus, M. S. and Eklund, P. C. (2000). Phonons in carbon nanotubes. *Advances in Physics* 49(6): 705–814.

Dresselhaus, M. S., Dresselhaus, G., Jorio, A., Souza, A. G., Pimenta, M. A., and Saito, R. (2002). Single nanotube Raman spectroscopy. *Accounts of Chemical Research* 35(12): 1070–1078.

Dresselhaus, M. S., Dresselhaus, G., and Jorio, A. (2007). Raman spectroscopy of carbon nanotubes in 1997 and 2007. *Journal of Physical Chemistry C* 111(48): 17887–17893.

Ebbesen, T. W. and Ajayan, P. M. (1992). Large-scale synthesis of carbon nanotubes. *Nature* 358(6383): 220–222.

Elihn, K. and Larsson, K. (2004). A theoretical study of the thermal fragmentation of ferrocene. *Thin Solid Films* 458(1–2): 325–329.

Flahaut, E., Bacsa, R., Peigney, A., and Laurent, C. (2003). Gram-scale CCVD synthesis of double-walled carbon nanotubes. *Chemical Communications* (12): 1442–1443.

Friedrichs, S., Sloan, J., Green, M. L. H., Hutchison, J. L., Meyer, R. R., and Kirkland, A. I. (2001). Simultaneous determination of inclusion crystallography and nanotube conformation for a Sb_2O_3/single-walled nanotube composite. *Physical Review B* 6404(4): 045406.

Fujita, Y., Bandow, S., and Iijima, S. (2005). Formation of small-diameter carbon nanotubes from PTCDA arranged inside the single-wall carbon nanotubes. *Chemical Physics Letters* 413(4-6): 410-414.

Gavillet, J., Loiseau, A., Journet, C., Willaime, F., Ducastelle, F., and Charlier, J. C. (2001). Root-growth mechanism for single-wall carbon nanotubes. *Physical Review Letters* 87(27): 275504.

Giusca, C. E., Tison, Y., Stolojan, V., Borowiak-Palen, E., and Silva, S. R. P. (2007). Inner-tube chirality determination for double-walled carbon nanotubes by scanning tunneling microscopy. *Nano Letters* 7(5): 1232-1239.

Guan, L. H., Li, H. J., Shi, Z. J., You, L. P., and Gu, Z. N. (2005a). Standing or lying C(70)s encapsulated in carbon nanotubes with different diameters. *Solid State Communications* 133(5): 333-336.

Guan, L. H., Shi, Z. J., Li, M. X., and Gu, Z. N. (2005b). Ferrocene-filled single-walled carbon nanotubes. *Carbon* 43(13): 2780-2785.

Guan, L. H., Suenaga, K., and Iijima, S. (2008). Smallest carbon nanotube assigned with atomic resolution accuracy. *Nano Letters* 8(2): 459-462.

Gunjishima, I., Inoue, T., Yamamuro, S., Sumiyama, K., and Okamoto, A. (2007). Synthesis of vertically aligned, double-walled carbon nanotubes from highly active Fe-V-O nanoparticles. *Carbon* 45(6): 1193-1199.

Hashimoto, A., Suenaga, K., Urita, K., Shimada, T., Sugai, T., Bandow, S., Shinohara, H., and Iijima, S. (2005). Atomic correlation between adjacent graphene layers in double-wall carbon nanotubes. *Physical Review Letters* 94(4): 045504.

Hayashi, T., Shimamoto, D., Kim, Y. A., Muramatsu, H., Okino, F., Touhara, H., Shimada, T., Miyauchi, Y., Maruyama, S., Terrones, M., Dresselhaus, M. S., and Endo, M. (2008). Selective optical property modification of double-walled carbon nanotubes by fluorination. *ACS Nano* 2(3): 485-488.

Hirahara, K., Kociak, M., Bandow, S., Nakahira, T., Itoh, K., Saito, Y., and Iijima, S. (2006). Chirality correlation in double-wall carbon nanotubes as studied by electron diffraction. *Physical Review B* 73(19): 195420.

Hiraoka, T., Kawakubo, T., Kimura, J., Taniguchi, R., Okamoto, A., Okazaki, T., Sugai, T., Ozeki, Y., Yoshikawa, M., and Shinohara, H. (2003). Selective synthesis of double-wall carbon nanotubes by CCVD of acetylene using zeolite supports. *Chemical Physics Letters* 382(5-6): 679-685.

Hodak, M. and Girifalco, L. A. (2003). Ordered phases of fullerene molecules formed inside carbon nanotubes. *Physical Review B* 67(7): 075419.

Hornbaker, D. J., Kahng, S. J., Misra, S., Smith, B. W., Johnson, A. T., Mele, E. J., Luzzi, D. E., and Yazdani, A. (2002). Mapping the one-dimensional electronic states of nanotube peapod structures. *Science* 295(5556): 828-831.

Huang, H. J., Kajiura, H., Tsutsui, S., Murakami, Y., and Ata, M. (2003). High-quality double-walled carbon nanotube super bundles grown in a hydrogen-free atmosphere. *Journal of Physical Chemistry B* 107(34): 8794-8798.

Hutchison, J. L., Kiselev, N. A., Krinichnaya, E. P., Krestinin, A. V., Loutfy, R. O., Morawsky, A. P., Muradyan, V. E., Obraztsova, E. D., Sloan, J., Terekhov, S. V., and Zakharov, D. N. (2001). Double-walled carbon nanotubes fabricated by a hydrogen arc discharge method. *Carbon* 39(5): 761-770.

Iakoubovskii, K., Minami, N., Ueno, T., Kazaoui, S., and Kataura, H. (2008). Optical characterization of double-wall carbon nanotubes: Evidence for inner tube shielding. *Journal of Physical Chemistry C* 112(30): 11194-11198.

Iijima, S. (1991). Helical microtubules of graphitic carbon. *Nature* 354(6348): 56-58.

Iijima, S. (1992). Growth of carbon nanotubes. *4th Symposium on Fundamental Approaches to New Material Phases, Physics and Chemistry of Nanometer Scale Materials*, Karuizawa, Japan, pp. 172-180.

Iijima, S. and Ichihashi, T. (1993). Single-shell carbon nanotubes of 1-nm diameter. *Nature* 363(6430): 603-605.

Jorge, J., Flahaut, E., Gonzalez-Jimenez, F., Gonzalez, G., Gonzalez, J., Belandria, E., Broto, J. M., and Raquet, B. (2008). Preparation and characterization of alpha-Fe nanowires located inside double wall carbon nanotubes. *Chemical Physics Letters* 457(4-6): 347-351.

Jorio, A., Saito, R., Hafner, J. H., Lieber, C. M., Hunter, M., McClure, T., Dresselhaus, G., and Dresselhaus, M. S. (2001). Structural (n, m) determination of isolated single-wall carbon nanotubes by resonant Raman scattering. *Physical Review Letters* 86(6): 1118-1121.

Journet, C., Maser, W. K., Bernier, P., Loiseau, A., delaChapelle, M. L., Lefrant, S., Deniard, P., Lee, R., and Fischer, J. E. (1997). Large-scale production of single-walled carbon nanotubes by the electric-arc technique. *Nature* 388(6644): 756-758.

Kalbac, M., Kavan, L., Juha, L., Civis, S., Zukalova, M., Bittner, M., Kubat, P., Vorlicek, V., and Dunsch, L. (2005). Transformation of fullerene peapods to double-walled carbon nanotubes induced by UV radiation. *Carbon* 43(8): 1610-1616.

Khlobystov, A. N., Britz, D. A., Ardavan, A., and Briggs, G. A. D. (2004). Observation of ordered phases of fullerenes in carbon nanotubes. *Physical Review Letters* 92(24): 245507.

Kishi, N., Kikuchi, S., Ramesh, P., Sugai, T., Watanabe, Y., and Shinohara, H. (2006). Enhanced photoluminescence from very thin double-wall carbon nanotubes synthesized by the zeolite-CCVD method. *Journal of Physical Chemistry B* 110(49): 24816-24821.

Kitaura, R. and Shinohara, H. (2006). Carbon-nanotube-based hybrid materials: Nanopeapods. *Chemistry—An Asian Journal* 1(5): 646-655.

Kociak, M., Suenaga, K., Hirahara, K., Saito, Y., Nakahira, T., and Iijima, S. (2002). Linking chiral indices and transport properties of double-walled carbon nanotubes. *Physical Review Letters* 89(15): 155501.

Kociak, M., Hirahara, K., Suenaga, K., and Iijima, S. (2003). How accurate can the determination of chiral indices of carbon nanotubes be? An experimental investigation of chiral indices determination on DWNT by-electron diffraction. *European Physical Journal B* 32(4): 457–469.

Koga, K., Gao, G. T., Tanaka, H., and Zeng, X. C. (2001). Formation of ordered ice nanotubes inside carbon nanotubes. *Nature* 412(6849): 802–805.

Kolesnikov, A. I., Zanotti, J. M., Loong, C. K., Thiyagarajan, P., Moravsky, A. P., Loutfy, R. O., and Burnham, C. J. (2004). Anomalously soft dynamics of water in a nanotube: A revelation of nanoscale confinement. *Physical Review Letters* 93(3): 035503.

Kuzmany, H., Plank, W., Pfeiffer, R., and Simon, F. (2008). Raman scattering from double-walled carbon nanotubes. *Journal of Raman Spectroscopy* 39(2): 134–140.

Kwon, Y. K. and Tomanek, D. (1998). Electronic and structural properties of multiwall carbon nanotubes. *Physical Review B* 58(24): R16001–R16004.

Lee, J., Kim, H., Kahng, S. J., Kim, G., Son, Y. W., Ihm, J., Kato, H., Wang, Z. W., Okazaki, T., Shinohara, H., and Kuk, Y. (2002). Bandgap modulation of carbon nanotubes by encapsulated metallofullerenes. *Nature* 415(6875): 1005–1008.

Li, L. X., Li, F., Liu, C., and Cheng, H. M. (2005). Synthesis and characterization of double-walled carbon nanotubes from multi-walled carbon nanotubes by hydrogen-arc discharge. *Carbon* 43(3): 623–629.

Li, Y. F., Hatakeyama, R., Kaneko, T., Izumida, T., Okada, T., and Kato, T. (2006a). Electronic transport properties of Cs-encapsulated double-walled carbon nanotubes. *Applied Physics Letters* 89(9): 093110.

Li, Y. F., Hatakeyama, R., Kaneko, T., Izumida, T., Okada, T., and Kato, T. (2006b). Synthesis and electronic properties of ferrocene-filled double-walled carbon nanotubes. *Nanotechnology* 17(16): 4143–4147.

Li, Y. F., Hatakeyama, R., Okada, T., Kato, T., Izumida, T., Hirata, T., and Qiu, J. S. (2006c). Synthesis of Cs-filled double-walled carbon nanotubes by a plasma process. *Carbon* 44(8): 1586–1589.

Liang, S. D. (2004). Intrinsic properties of electronic structure in commensurate double-wall carbon nanotubes. *Physica B-Condensed Matter* 352(1–4): 305–311.

Liu, B. C., Yu, B., and Zhang, M. X. (2005). Catalytic CVD synthesis of double-walled carbon nanotubes with a narrow distribution of diameters over Fe-Co/MgO catalyst. *Chemical Physics Letters* 407(1–3): 232–235.

Liu, Q. F., Ren, W. C., Li, F., Cong, H. T., and Cheng, H. M. (2007). Synthesis and high thermal stability of double-walled carbon nanotubes using nickel formate dihydrate as catalyst precursor. *Journal of Physical Chemistry C* 111(13): 5006–5013.

Lu, J. and Wang, S. D. (2007). Tight-binding investigation of the metallic proximity effect of semiconductor-metal double-wall carbon nanotubes. *Physical Review B* 76(23): 233103.

Lyu, S. C., Lee, T. J., Yang, C. W., and Lee, C. J. (2003). Synthesis and characterization of high-quality double-walled carbon nanotubes by catalytic decomposition of alcohol. *Chemical Communications* (12): 1404–1405.

Monthioux, M. (2002). Filling single-wall carbon nanotubes. *Carbon* 40(10): 1809–1823.

Monthioux, M., Flahaut, E., and Cleuziou, J. P. (2006). Hybrid carbon nanotubes: Strategy, progress, and perspectives. *Journal of Materials Research* 21(11): 2774–2793.

Moradian, R., Azadi, S., Refii-Tabar, H. (2007). When double-wall carbon nanotubes can become metallic or semiconducting. *Journal of Physics-Condensed Matter* 19(17): 176209.

Muramatsu, H., Hayashi, T., Kim, Y. A., Shimamoto, D., Endo, M., Terrones, M., and Dresselhaus, M. S. (2008). Synthesis and isolation of molybdenum atomic wires. *Nano Letters* 8(1): 237–240.

Ning, G., Kishi, N., Okimoto, H., Shiraishi, M., Kato, Y., Kitaura, R., Sugai, T., Aoyagi, S., Nishibori, E., Sakata, M., and Shinohara, H. (2007). Synthesis, enhanced stability and structural imaging of C-60 and C-70 double-wall carbon nanotube peapods. *Chemical Physics Letters* 441(1–3): 94–99.

Okada, S. and Oshiyama, A. (2003). Curvature-induced metallization of double-walled semiconducting zigzag carbon nanotubes. *Physical Review Letters* 91(21): 216801.

Okada, S., Saito, S., and Oshiyama, A. (2001). Energetics and electronic structures of encapsulated C-60 in a carbon nanotube. *Physical Review Letters* 86(17): 3835–3838.

Pengfei, Q. F., Vermesh, O., Grecu, M., Javey, A., Wang, O., Dai, H. J., Peng, S., and Cho, K. J. (2003). Toward large arrays of multiplex functionalized carbon nanotube sensors for highly sensitive and selective molecular detection. *Nano Letters* 3(3): 347–351.

Pfeiffer, R., Kuzmany, H., Kramberger, C., Schaman, C., Pichler, T., Kataura, H., Achiba, Y., Kurti, J., and Zolyomi, V. (2003). Unusual high degree of unperturbed environment in the interior of single-wall carbon nanotubes. *Physical Review Letters* 90(22): 225501.

Pfeiffer, R., Kuzmany, H., Simon, F., Bokova, S. N., and Obraztsova, E. (2005a). Resonance Raman scattering from phonon overtones in double-wall carbon nanotubes. *Physical Review B* 71(15): 155409.

Pfeiffer, R., Simon, F., Kuzmany, H., and Popov, V. N. (2005b). Fine structure of the radial breathing mode of double-wall carbon nanotubes. *Physical Review B* 72(16): 161404.

Pfeiffer, R., Peterlik, H., Kuzmany, H., Shiozawa, H., Gruneis, A., Pichler, T., and Kataura, H. (2007). Growth mechanisms of inner-shell tubes in double-wall carbon nanotubes. *Physica Status Solidi B-Basic Solid State Physics* 244(11): 4097–4101.

Pichler, T., Kuzmany, H., Kataura, H., and Achiba, Y. (2001). Metallic polymers of C-60 inside single-walled carbon nanotubes. *Physical Review Letters* 87(26): 267401.

Pokhodnia, K., Demsar, J., Omerzu, A., Mihailovic, D., and Kuzmany, H. (1996). Low temperature Raman spectra on TDAE-C-60 single crystals. *International Conference on the Science and Technology of Synthetic Metals*, Snowbird, UT, pp. 1749–1750.

Qi, H., Qian, C., and Liu, J. (2007). Synthesis of uniform double-walled carbon nanotubes using iron disilicide as catalyst. *Nano Letters* 7(8): 2417–2421.

Qiu, H. X., Shi, Z. J., Guan, L. H., You, L. P., Gao, M., Zhang, S. L., Qiu, J. S., and Gu, Z. N. (2006a). High-efficient synthesis of double-walled carbon nanotubes by arc discharge method using chloride as a promoter. *Carbon* 44(3): 516–521.

Qiu, H. X., Shi, Z. J., Zhang, S. L., Gu, Z. N., and Qiu, J. S. (2006b). Synthesis and Raman scattering study of double-walled carbon nanotube peapods. *Solid State Communications* 137(12): 654–657.

Qiu, H. X., Shi, Z. J., Gu, Z. N., and Qiu, J. S. (2007a). Controllable preparation of triple-walled carbon nanotubes and their growth mechanism. *Chemical Communications* (10): 1092–1094.

Qiu, J. S., Wang, Z. Y., Zhao, Z. B., and Wang, T. H. (2007b). Synthesis of double-walled carbon nanotubes from coal in hydrogen-free atmosphere. *Fuel* 86(1–2): 282–286.

Ramesh, P., Okazaki, T., Taniguchi, R., Kimura, J., Sugai, T., Sato, K., Ozeki, Y., and Shinohara, H. (2005). Selective chemical vapor deposition synthesis of double-wall carbon nanotubes on mesoporous silica. *Journal of Physical Chemistry B* 109(3): 1141–1147.

Rao, A. M., Zhou, P., Wang, K. A., Hager, G. T., Holden, J. M., Wang, Y., Lee, W. T., Bi, X. X., Eklund, P. C., Cornett, D. S., Duncan, M. A., and Amster, I. J. (1993). Photoinduced polymerization of solid C-60 films. *Science* 259(5097): 955–957.

Rauf, H., Pichler, T., Pfeiffer, R., Simon, F., Kuzmany, H., and Popov, V. N. (2006). Detailed analysis of the Raman response of n-doped double-wall carbon nanotubes. *Physical Review B* 74(23): 235419.

Ren, W. C., Li, F., Tan, P. H., and Cheng, H. M. (2006). Raman evidence for atomic correlation between the two constituent tubes in double-walled carbon nanotubes. *Physical Review B* 73(11): 115430.

Saito, R., Dresselhaus, G., and Dresselhaus, M. S. (1993a). Electronic-structure of double-layer graphene tubules. *Journal of Applied Physics* 73(2): 494–500.

Saito, Y., Yoshikawa, T., Inagaki, M., Tomita, M., and Hayashi, T. (1993b). Growth and structure of graphitic tubules and polyhedral particles in arc-discharge. *Chemical Physics Letters* 204(3–4): 277–282.

Saito, Y., Nakahira, T., and Uemura, S. (2003). Growth conditions of double-walled carbon nanotubes in arc discharge. *Journal of Physical Chemistry B* 107(4): 931–934.

Schulte, K., Swarbrick, J. C., Smith, N. A., Bondino, F., Magnano, E., and Khlobystov, A. N. (2007). Assembly of cobalt phthalocyanine stacks inside carbon nanotubes. *Advanced Materials* 19(20): 3312–3316.

Scipioni, R., Oshiyama, A., and Ohno, T. (2008). Increased stability of C-60 encapsulated in double walled carbon nanotubes. *Chemical Physics Letters* 455(1–3): 88–92.

Seraphin, S. and Zhou, D. (1994). Single-walled carbon nanotubes produced at high-yield by mixed catalysts. *Applied Physics Letters* 64(16): 2087–2089.

Shiozawa, H., Pichler, T., Gruneis, A., Pfeiffer, R., Kuzmany, H., Liu, Z., Suenaga, K., and Kataura, H. (2008). A catalytic reaction inside a single-walled carbon nanotube. *Advanced Materials* 20(8): 1443–1449.

Singer, P. M., Wzietek, P., Alloul, H., Simon, F., and Kuzmany, H. (2005). NMR evidence for gapped spin excitations in metallic carbon nanotubes. *Physical Review Letters* 95(23): 236403.

Sloan, J., Hammer, J., Zwiefka-Sibley, M., and Green, M. L. H. (1998). The opening and filling of single walled carbon nanotubes (SWNTs). *Chemical Communications* (3): 347–348.

Sloan, J., Novotny, M. C., Bailey, S. R., Brown, G., Xu, C., Williams, V. C., Friedrichs, S., Flahaut, E., Callender, R. L., York, A. P. E., Coleman, K. S., Green, M. L. H., Dunin-Borkowski, R. E., and Hutchison, J. L. (2000). Two layer 4: 4 co-ordinated KI crystals grown within single walled carbon nanotubes. *Chemical Physics Letters* 329(1–2): 61–65.

Smith, B. W., Monthioux, M., and Luzzi, D. E. (1998). Encapsulated C-60 in carbon nanotubes. *Nature* 396(6709): 323–324.

Song, W., Ni, M., Lu, J., Gao, Z. X., Nagase, S., Yu, D. P., Ye, H. Q., and Zhang, X. W. (2005). Electronic structures of semiconducting double-walled carbon nanotubes: Important effect of interlay interaction. *Chemical Physics Letters* 414(4–6): 429–433.

Souza, A. G., Jorio, A., Samsonidze, G. G., Dresselhaus, G., Pimenta, M. A., Dresselhaus, M. S., Swan, A. K., Unlu, M. S., Goldberg, B. B., and Saito, R. (2003). Competing spring constant versus double resonance effects on the properties of dispersive modes in isolated single-wall carbon nanotubes. *Physical Review B* 67(3): 035427.

Souza, A. G., Endo, M., Muramatsu, H., Hayashi, T., Kim, Y. A., Barros, E. B., Akuzawa, N., Samsonidze, G. G., Saito, R., and Dresselhaus, M. S. (2006). Resonance Raman scattering studies in Br-2-adsorbed double-wall carbon nanotubes. *Physical Review B* 73(23): 235413.

Sugai, T., Yoshida, H., Shimada, T., Okazaki, T., and Shinohara, H. (2003). New synthesis of high-quality double-walled carbon nanotubes by high-temperature pulsed arc discharge. *Nano Letters* 3(6): 769–773.

Sumpter, B. G., Meunier, V., Romo-Herrera, J. M., Cruz-Silva, E., Cullen, D. A., Terrones, H., Smith, D. J., and Terrones, M. (2007). Nitrogen-mediated carbon nanotube growth: Diameter reduction, metallicity, bundle dispersability, and bamboo-like structure formation. *ACS Nano* 1(4): 369–375.

Sun, N. J., Guan, L. H., Shi, Z. J., Li, N. Q., Gu, Z. N., Zhu, Z. W., Li, M. X., and Shao, Y. H. (2006). Ferrocene peapod modified electrodes: Preparation, characterization, and mediation of H_2O_2. *Analytical Chemistry* 78(17): 6050–6057.

Takenobu, T., Takano, T., Shiraishi, M., Murakami, Y., Ata, M., Kataura, H., Achiba, Y., and Iwasa, Y. (2003). Stable and controlled amphoteric doping by encapsulation of organic molecules inside carbon nanotubes. *Nature Materials* 2(10): 683–688.

Terrones, M., Ajayan, P. M., Banhart, F., Blase, X., Carroll, D. L., Charlier, J. C., Czerw, R., Foley, B., Grobert, N., Kamalakaran, R., Kohler-Redlich, P., Ruhle, M., Seeger, T., and Terrones, H. (2002). N-doping and coalescence of carbon nanotubes: Synthesis and electronic properties. *Applied Physics A—Materials Science & Processing* 74(3): 355–361.

Tison, Y., Giusca, C. E., Stolojan, V., Hayashi, Y., and Silva, S. R. P. (2008). The inner shell influence on the electronic structure of double-walled carbon nanotubes. *Advanced Materials* 20(1): 189–194.

Wang, Z. Y., Zhao, K. K., Shi, Z. J., Gu, Z. N., and Jin, Z. X. (2008). Preparation of nitrogen-doped triple-walled carbon nanotubes. *Chinese Journal of Inorganic Chemistry* 24(8): 1237–1241.

Wei, J. Q., Ci, L. J., Jiang, B., Li, Y. H., Zhang, X. F., Zhu, H. W., Xu, C. L., and Wu, D. H. (2003). Preparation of highly pure double-walled carbon nanotubes. *Journal of Materials Chemistry* 13(6): 1340–1344.

Wei, J. Q., Jiang, B., Wu, D. H., and Wei, B. Q. (2004). Large-scale synthesis of long double-walled carbon nanotubes. *Journal of Physical Chemistry B* 108(26): 8844–8847.

Yamada, T., Namai, T., Hata, K., Futaba, D. N., Mizuno, K., Fan, J., Yudasaka, M., Yumura, M., and Iijima, S. (2006). Size-selective growth of double-walled carbon nanotube forests from engineered iron catalysts. *Nature Nanotechnology* 1(2): 131–136.

Zhou, Z. P., Ci, L. J., Chen, X. H., Tang, D. S., Yan, X. Q., Liu, D. F., Liang, Y. X., Yuan, H. J., Zhou, W. Y., Wang, G., and Xie, S. S. (2003). Controllable growth of double wall carbon nanotubes in a floating catalytic system. *Carbon* 41(2): 337–342.

Zhu, J., Yudasaka, M., and Iijima, S. (2003). A catalytic chemical vapor deposition synthesis of double-walled carbon nanotubes over metal catalysts supported on a mesoporous material. *Chemical Physics Letters* 380(5–6): 496–502.

Zolyomi, V., Rusznyak, A., Kurti, J., Gali, A., Simon, F., Kuzmany, H., Szabados, A., and Surjan, P. R. (2006). Semiconductor-to-metal transition of double walled carbon nanotubes induced by inter-shell interaction. *Physica Status Solidi B—Basic Solid State Physics* 243(13): 3476–3479.

Zolyomi, V., Koltai, J., Rusznyak, A., Kuerti, J., Gali, A., Simon, F., Kuzmany, H., Szabados, A., and Surjan, P. R. (2008). Intershell interaction in double walled carbon nanotubes: Charge transfer and orbital mixing. *Physical Review B* 77(24): 245403.

Zuo, J. M., Vartanyants, I., Gao, M., Zhang, R., and Nagahara, L. A. (2003). Atomic resolution imaging of a carbon nanotube from diffraction intensities. *Science* 300(5624): 1419–1421.

2

Quantum Transport in Carbon Nanotubes

Kálmán Varga
Vanderbilt University

This chapter summarizes the transport properties of carbon nanotubes. Due to their unusual electronic and structural physical properties, carbon nanotubes are promising candidates for a wide range of nanoscience and nanotechnology applications. We review the most important experimental and theoretical research exploring future carbon nanotube devices.

2.1 Introduction

Carbon nanotubes (CNs) are thin hollow cylinders made entirely out of carbon atoms (see Figure 2.1). There are many types of CNs and CN-like structures. The most basic ones are multiwall nanotubes (with diameters, d_t, of order ~10 nm) and single-wall nanotubes (SWNT) (d_t ~1 nm). Multiwall carbon nanotubes were discovered by the Japanese scientist, Sumio Iijima, in 1991 [1] and, 2 years later, individual single-wall carbon nanotubes (see Figure 2.2) were reported [2,3]. Immediately after their discovery, it became clear that these tiny objects would have very remarkable electronic properties [4,5]. Still, it was not until 1997 that the first electronic transport measurements on CNs were performed [6,7], thanks largely to a new growth method developed by the group of R. Smalley, which enabled the production of large amounts of CN material. Currently, the physical properties of CNs are still being discovered and disputed. These studies are interesting and challenging due to the fact that CNs have a very broad range of electronic, thermal, and structural properties that change depending on the different kinds of nanotubes (defined by their diameter, length, chirality, and twist).

The most appealing feature of CNs, for nanoelectronics, is their near-ballistic transport due to a limited carrier–phonon interaction [8]. When electrons travel through a conventional metal-based wire, they encounter resistance as they bump into atoms, defects, and impurities. CNs can conduct electricity better than metals due to the ballistic transport. Ballistic transport refers to the motion of charge carriers driven by electric fields in a conducting or semiconducting material without scattering. It is a highly desirable phenomenon for a wide range of applications needing high currents, high speeds, and low power dissipations. Several groups have observed ballistic transport in CNs. In metallic tubes, the mean-free paths for phonon scattering have been found to be about 1 μm at low field and in the range of 10–100 nm at high field. For semiconducting tubes, values close to 300–500 nm and 10–100 nm have been obtained at low and high field, respectively [9–15]. The high electron mobilities (10^4–10^5 cm^2/V s) in CNs [9,16] indicate that semiconducting nanotubes should be an excellent material for a number of semiconductor applications, especially in high-speed transistors where mobility is crucial. The high mobility in CNs is partly due to the fact that graphite itself is a good conductor of electricity (about 20,000 cm^2/V s at room temperature) and also partly to its one-dimensional structure. In addition, CNs can withstand about three orders of magnitude larger current densities [17,18] than a typical metal such as copper or aluminum, making them excellent nanomaterials for nanodevice applications. Moreover, due to their cylindrical shape, CNs may be an ideal element of a coaxial field-effect transistor, which is expected to be the ultimate geometry due to its short-channel behavior [19–21]. With the unique electrical properties of CNs, one could envision all-carbon electronics with metallic nanotube leads connected to semiconducting nanotube devices [22–24].

CNs can be synthesized by various methods [25,26]. The arc-evaporation method, which produces the best-quality

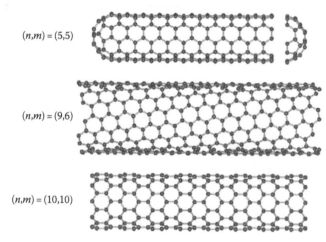

$(n,m) = (5,5)$

$(n,m) = (9,6)$

$(n,m) = (10,10)$

FIGURE 2.1 CNs with different chiralities. See text for further explanation and the definition of the chiral indices (n,m).

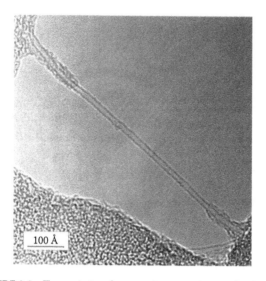

FIGURE 2.2 Transmission electron microscope image of an SWNT.

nanotubes, involves passing a current of about 50 A between two graphite electrodes in an atmosphere of helium. This causes the graphite to vaporize, some of it condensing on the walls of the reaction vessel and some of it on the cathode. It is the deposit on the cathode that contains the CNs. SWNT are produced when Co and Ni or some other metal is added to the anode. It has long been known that CNs can also be made by passing a carbon-containing gas, such as a hydrocarbon, over a catalyst. The catalyst consists of nano-sized particles of metal, usually Fe, Co, or Ni. These particles catalyze the breakdown of the gaseous molecules into carbon, and a tube then begins to grow with a metal particle at the tip. It was shown that SWNT can also be produced catalytically. The perfection of CNs produced in this way has generally been poorer than those made by arc-evaporation, but great improvements in this technique have been made in recent years. The big advantage of catalytic synthesis over arc-evaporation is that it can be scaled up for volume production. The third important method for making CNs involves using a

powerful laser to vaporize a metal-graphite target. This can be used to produce single-walled tubes with high yield.

2.2 Electronic Structure of Carbon Nanotubes

2.2.1 Band Structure of Graphene

Since CNs can be thought of as rolled-up ribbons of graphene sheets, one can start from the band structure of graphene to understand that of CNs. Many properties of CNs can be understood by a simple tight-binding model description [27–30]. This model is based on the assumption (besides the approximations present in the tight-binding description) that the effect of the curvature of the tube can be neglected and the band structure of the nanotube can be derived from the tight-binding band structure of a graphene sheet. This assumption is good for tubes of sufficiently large diameter, that is, ones whose diameter is much larger than the nearest-neighbor distance.

A graphene sheet consists of carbon atoms arranged in a hexagonal lattice illustrated in Figure 2.3. The unit vectors in real and reciprocal space are defined as (see Figure 2.4)

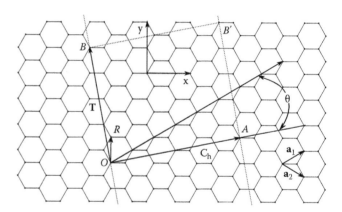

FIGURE 2.3 Honeycomb lattice of graphene. See the text for the definition of the lattice vectors.

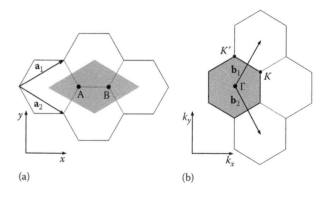

(a)   (b)

FIGURE 2.4 Real- and reciprocal-space lattice vectors of graphene. The shaded areas mark the unit cell and the Brillouin zone. $\mathbf{a}_1$ and $\mathbf{a}_2$ are the real-space lattice vectors, and $\mathbf{b}_1$ and $\mathbf{b}_2$ are the reciprocal-space lattice vectors.

$$\mathbf{a}_1 = \left(\frac{\sqrt{3}}{2}a, \frac{a}{2}\right), \quad \mathbf{a}_2 = \left(\frac{\sqrt{3}}{2}a, -\frac{a}{2}\right), \tag{2.1}$$

$$\mathbf{b}_1 = \left(\frac{2\pi}{\sqrt{3}a}, \frac{2\pi}{a}\right), \quad \mathbf{b}_2 = \left(\frac{2\pi}{\sqrt{3}a}, -\frac{2\pi}{a}\right), \tag{2.2}$$

where $a = \sqrt{3}a_{C\text{-}C}$ is the lattice constant of the graphene and $a_{C\text{-}C} = 1.41\,\text{Å}$ is the carbon–carbon bond length. The first Brillouin zone is marked by shaded areas (see Figure 2.4). Carbon has four valence electrons per atom, three of which are used to form sp^2 bonds with neighboring atoms in σ orbitals. The corresponding energy bands lie far below the Fermi level and do not contribute to the electrical conduction. The transport properties are determined by the remaining π electrons, which occupy the bonding and antibonding band resulting from the superposition of the $2p_z$ orbitals. The valence orbital is thus the $\pi(2p_z)$ orbital and there is no interaction between the π and the $\sigma(2s$ and $2p_{x,y})$ orbitals because of their different symmetries. The mixing of π and σ orbitals due to the curvature is neglected as indicated above. Since the carbon atoms in a graphene plane can be divided into two sublattices A and B (bipartite lattice, see Figure 2.4), the π bands of the two-dimensional graphene is derived from the following 2 × 2 Hamiltonian matrix [31]:

$$H = \begin{pmatrix} 0 & h^* \\ h & 0 \end{pmatrix}. \tag{2.3}$$

Here h is the nearest-neighbor interaction between the A and the B sublattices expressed as

$$h = t\left(e^{ik_x a/\sqrt{3}} + 2e^{-ik_x a/2\sqrt{3}}\cos\left(\frac{k_y a}{2}\right)\right), \tag{2.4}$$

where t is the nearest-neighbor transfer integral. Diagonalizing the Hamiltonian, we obtain the two-dimensional energy dispersion relation of the graphene, $E_{2D}(k_x, k_y)$, as follows:

$$E_{2D}(\mathbf{k}) = \pm t\left(1 + 4\cos\left(\frac{\sqrt{3}k_x a}{2}\right)\cos\left(\frac{k_y a}{2}\right) + 4\cos^2\left(\frac{k_y a}{2}\right)\right)^{1/2}. \tag{2.5}$$

This equation describes the two bands resulting from bonding (Figure 2.5) and antibonding (Figure 2.6) states. Figure 2.7 shows a surface plot of the energy dispersion, E_{2D}. The bonding and antibonding bands touch at six K points at the corners of the first Brillouin zone. There are two triplets of Fermi points, K and K', that are inequivalent under translations of the reciprocal lattice. On the other hand, the hexagonal lattice symmetry provides an equivalence under 60° rotations. Because of this symmetry, the K and K' points are energetically degenerate and lead to the peculiar touching of the conduction and the valence band. At zero temperature, the bonding bands are completely

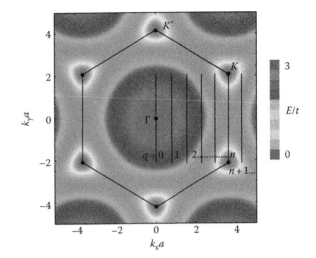

FIGURE 2.5 **(See color insert following page 20-16.)** Plot of the bonding π band of graphene.

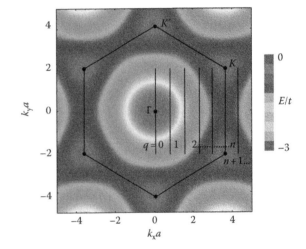

FIGURE 2.6 **(See color insert following page 20-16.)** Plot of the antibonding π band of graphene.

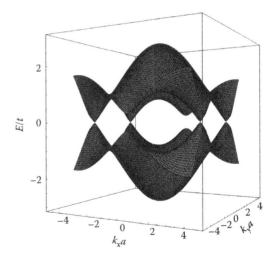

FIGURE 2.7 **(See color insert following page 20-16.)** Band structure of graphene. The figure shows the various energy bands in the Brillouin zone.

filled and the antibonding bands are empty. Thus, undoped graphene is a zero-gap semiconductor. In practice, there is always some doping shifting the Fermi energy and leading to a small density of states (DOS).

This band structure (see Figures 2.5 through 2.7), which determines how electrons scatter from the atoms in the crystal lattice, is quite unusual. It is not like that of a metal, which has many states that freely propagate through the crystal at the Fermi energy. This is not the band structure of a semiconductor either; there is no energy gap with no electronic states near the Fermi energy. The band structure of graphene is instead somewhere in between these two extremes. In most directions, electrons moving at the Fermi energy are backscattered by atoms in the lattice, which gives the material an energy band gap like that of a semiconductor. However, in other directions, the electrons that scatter from different atoms in the lattice interfere destructively, which suppresses the backscattering and leads to metallic behavior. Graphene is therefore called a semimetal, since it is metallic in special directions and semiconducting in the others. Looking more closely at Figure 2.7, the band structure of the low-energy states appears to be a series of cones. At low energies, graphene resembles a two-dimensional world populated by massless fermions [32].

In passing we note that, although this model has become very popular due to its simplicity, it only gives good description close to the K point of the Brillouin zone. First-principles and more elaborated tight-binding calculations have to be used to obtain better dispersion relations [33].

2.2.2 Band Structure of Carbon Nanotubes

The geometry of a CN is described by a wrapping vector. The wrapping vector encircles the waist of a CN so that the tip of the vector meets its own tail (see Figure 2.3). The wrapping vector can be any

$$\mathbf{C}_h = n\mathbf{a}_1 + m\mathbf{a}_2, \qquad (2.6)$$

where

    n and m are integers
    $\mathbf{a}_1$ and $\mathbf{a}_2$ are the unit vectors of the graphene lattice (see Figures 2.3 and 2.4)

The angle between the wrapping vector and the lattice vector $\mathbf{a}_1$ is called the chiral angle. The pair of indexes (n,m) identifies the nanotube, and each (n,m) pair corresponds to diameter d_t:

$$d_t = \frac{L}{\pi} \quad L = |\mathbf{C}_h| = a\sqrt{n^2 + m^2 + nm} \qquad (2.7)$$

and to a specific chiral angle, θ, (the angle between $\mathbf{a}_1$ and $\mathbf{C}_h$)

$$\theta = \arctan\left(\frac{\sqrt{3}m}{m+2n}\right). \qquad (2.8)$$

The translation vector, $\mathbf{T}$, is parallel to the axis of the tube (see Figure 2.3)

$$\mathbf{T} = t_1\mathbf{a}_1 + t_2\mathbf{a}_2 \quad t_1 = \frac{2m+n}{d_r} \quad t_2 = -\frac{2n+m}{d_r}, \qquad (2.9)$$

where

$$d_r = \begin{cases} d & \text{if } n-m \text{ is not multiple of } 3d \\ 3d & \text{otherwise} \end{cases} \quad d = \gcd(n,m) \qquad (2.10)$$

(gcd is the greatest common divisor).

The length of $\mathbf{T}$ is $|\mathbf{T}| = \frac{\sqrt{3}L}{d_r}$. Using $\mathbf{T}$ and $\mathbf{C}_h$, we can define the unit cell of the CN with an area given, $|\mathbf{C}_h \times \mathbf{T}|$, and the number of atoms in a unit cell is

$$N = \frac{|\mathbf{C}_h \times \mathbf{T}|}{|\mathbf{a}_1 \times \mathbf{a}_2|} = \frac{2L^2}{a^2 d_r} \qquad (2.11)$$

The reciprocal lattice vectors corresponding to $\mathbf{C}_h$ and $\mathbf{T}$ are

$$\mathbf{K}_1 = \frac{1}{N}\left(-t_2\mathbf{b}_1 + t_1\mathbf{b}_2\right) \quad \mathbf{K}_2 = \frac{1}{N}\left(m\mathbf{b}_1 - n\mathbf{b}_2\right). \qquad (2.12)$$

Now, by zone folding of the two-dimensional dispersion relation of the graphene, the dispersion relation of the CN in the tight-binding approximation is given by

$$E_{\text{nanotube}} = E_{2D}\left(k\frac{\mathbf{K}_1}{|\mathbf{K}_1|} + \mu\mathbf{K}_2\right), \qquad (2.13)$$

where

    k is a continuous variable

$$-\frac{\pi}{|\mathbf{T}|} < k < \frac{\pi}{|\mathbf{T}|}$$

    μ labels the discrete bands

If $n-m$ is a multiple of 3, the nanotube is metallic.

There is an energy gap in the graphene band structure along the K–K' line (Figure 2.5) of the first Brillouin zone. It has the form of

$$\Delta = 2t(2\cos(\sqrt{3}k_y a) - 1). \qquad (2.14)$$

The gap reaches its maximum at the center of the edge (Figure 2.5) and decreases to zero at the corners (points K and K'), where the bonding and antibonding bands connect with each other. The same gap exists in a CN. However, because of the periodic boundary condition around the circumference of a nanotube, the allowed k values are quantized in the direction perpendicular to the tube axis, so that the allowed states are lines parallel to the tube axis (Figure 2.5), with a momentum separation $\Delta k = 1/R$ ($R = d_t/2$ is the tube radius). Each line corresponds to a one-dimensional subband, and hence the conduction channel. The energy separation between these subbands is characterized by $\Delta E = \hbar v_F/d_t$, where v_F is the Fermi velocity. An undoped nanotube at temperatures below this energy scale can be either metallic or semiconducting, depending on its diameter and chirality. However, higher or lower subbands above or below the gap will contribute to the conductance if there is charge transfer caused by defects, impurities, or lead attachment.

There are two special chiral angles, i.e., $\theta = 0°$ and $\theta = 30°$, which correspond to highly symmetric non-chiral nanotubes. These are termed "zigzag" and "armchair", having chiral indexes $(n, 0)$ and (m, m), respectively. The labels zigzag and armchair refer to the pattern of the carbon bonds along the circumference (see Figure 2.3). These two special cases are described in detail below.

Armchair nanotubes: Armchair nanotubes are described by (m, m) chiral vectors, therefore

$$\mathbf{C}_h = m(\mathbf{a}_1 + \mathbf{a}_2) \quad \mathbf{T} = \mathbf{a}_1 - \mathbf{a}_2 \tag{2.15}$$

and the dispersion relation is

$$E_{armchair}(k) = \pm t \left(1 \pm 4\cos\left(\frac{q\pi}{m}\right)\cos\left(\frac{ka}{2}\right) + 4\cos^2\left(\frac{ka}{2}\right) \right)^{1/2} \tag{2.16}$$

with $-\pi < ka < \pi$ and $q = 1,\ldots, 2m$. The result is plotted in Figure 2.8. Note the crossing of the bands at the Fermi energy at $k = 2\pi/3a$, two-thirds of the way to the Brillouin zone boundary. The crossing bands are nondegenerate and are the highest occupied valence band and lowest unoccupied conduction band. Since $m - m = 0$ is a multiple of 3, all armchair nanotubes are zero band gap semiconductors.

Zigzag nanotubes: Zigzag nanotubes are described by $(n, 0)$ chiral vectors and the dispersion relation is

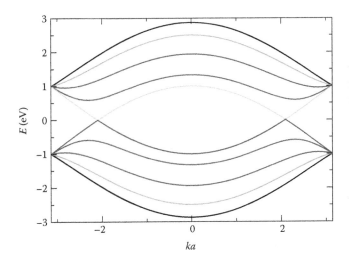

FIGURE 2.8 Band structure of a (5,5) armchair nanotube.

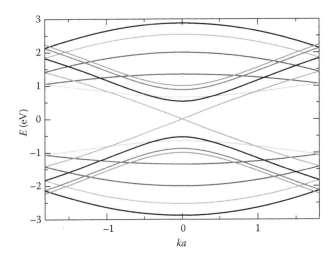

FIGURE 2.9 Band structure of a metallic (9,0) zigzag nanotube.

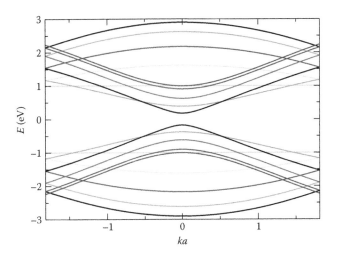

FIGURE 2.10 Band structure of a nonmetallic (10,0) zigzag nanotube.

$$E_{zigzag}(k) = \pm t \left(1 \pm 4\cos\left(\frac{\sqrt{3}ka}{2}\right)\cos\left(\frac{q\pi a}{n}\right) + 4\cos^2\left(\frac{q\pi}{n}\right) \right)^{1/2} \tag{2.17}$$

with $(-\pi/\sqrt{3}) < ka(\pi/\sqrt{3})$ and $q = 1,\ldots, 2n$. Examples are plotted in Figures 2.9 and 2.10. Zigzag nanotubes are metallic if n is an integer and a multiple of 3 (see Figure 2.9).

All other nanotubes are called chiral nanotubes. Figure 2.11 shows an example for the band structure of a metallic chiral nanotube.

2.3 Transport in Nanotubes

2.3.1 Background

To measure the conducting properties of nanotubes, one has to wire them up by contacting them to metallic electrodes (see Figure 2.12). The electrodes, which can be connected to either a

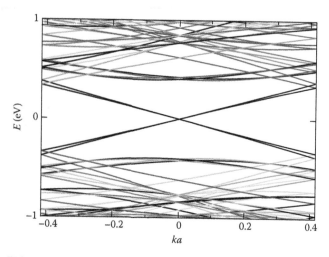

FIGURE 2.11 Band structure of a metallic chiral (9,6) nanotube.

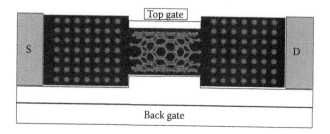

FIGURE 2.12 Schematic picture of a CN field-effect transistor.

single tube or a bundle of up to several hundred tubes, are usually made using electron-beam lithography. The tubes can be attached to the electrodes in different ways. One way is to make the electrodes and then drop the tubes on top of them. Another is to deposit the tubes on a substrate, locate them with a scanning electron microscope or an atomic force microscope, and then attach leads to the tubes using lithography. More advanced techniques are also being developed to make device fabrication more reproducible and controlled. These include the possibility of growing the tubes between electrodes or by attaching the tubes to the surface in a controllable fashion using either electrostatic or chemical forces. The "source" and "drain" electrodes (so named in analogy to standard semiconducting devices) allow the conducting properties of the nanotube to be measured. In addition, a third terminal—called a "gate"—is often used. The gate and the tube act like the two plates of a capacitor, which means that the gate can be used to electrostatically induce negative or positive carriers onto the tube.

As we already mentioned, CNs are promising candidates for ballistic one-dimensional conductors. For ideal, i.e., highly transparent contacts and perfect tubes, the one-dimensional subbands as described in the previous section form one-dimensional conduction channels. Each of these channels contributes $2e^2/h$ to the conductance if the different spin orientations are degenerate. Because of the double degeneracy of the states at the corner points of the Brillouin zone, two spin degenerate channels contribute to the transport, resulting in an estimated conductance of $G = 4e^2/h = 6.4 \text{ k}\Omega^{-1}$.

2.3.2 Transport Calculations

An accurate prediction of the transport properties of nanoscale atomic or molecular systems (nanotubes, in particular), including their current–voltage (*I–V*) characteristics, is essential for the realization of a broad spectrum of device applications. Theoretical investigations of the transport properties of these nanoscale systems are based on calculations with various levels of sophistication, including phenomenological, semiclassical, and quantum mechanical *ab initio* methods. Semiempirical methods based on parametrized Hamiltonians, such as tight-binding or extended Huckel models, have been widely used. Since these parametrized Hamiltonians are generally derived from the bulk or isolated molecular systems, the effects such as external bias and gate potential are usually not accounted for. Moreover, these models do not allow for structural relaxations to be performed. Semiempirical methods are not typically self-consistent but they are relatively simple and easy to implement. These methods provide some quantitative insight especially in cases where the charge transfer effects are relatively small.

Ab initio transport calculations are rapidly developing but they still have significant limitations. The Lippmann–Schwinger method [34] is based on a plane-wave basis set, which allows systematic convergence studies, but has so far been implemented only for model "jellium" electrodes, which fails to capture the essence of real molecule-electrode contacts that ultimately control the total current. The nonequilibrium Green's function (NEGF) method accommodates real electrodes, [35–49] but has so far been implemented only with localized basis sets, for which convergence tests are not practical. Such basis sets are generally *optimized for ground-state properties* for which a variational principle is applicable and both experimental and theoretical benchmarks are available. In the absence of both experimental benchmarks and manifestly converged theoretical benchmarks, and in the absence of a variational principle for currents, optimization of localized basis sets for current calculations presents difficulties. In Ref. [50], it was demonstrated that different choices of localized basis sets yield different current–voltage characteristics, but there is no way to tell which is a better set. The numerical challenge and the need for manifestly *converged* as opposed to *optimized* basis sets can be appreciated if one recognizes that ground-state properties, for which a variational principle is operative, generally require the convergence of the ground-state electron density, not of individual wave functions. For currents, without a variational principle, we need *convergence of the derivatives of individual wave functions in a narrow energy range.* The presence of a strongly slanting electrostatic potential that destroys point-group symmetries about atoms adds further to the need for a fully converged basis, unencumbered by ground-state point-group symmetry constraints.

To calculate the conductance of CN, we will use Green's function approach [35–49]. In the following, we give a brief introduction to this formalism. In a suitable chosen basis representation the Hamiltonian and the overlap matrix of the lead-device-lead system (see Figure 2.13), under the assumption that there is no interaction between the leads, takes the form

$$H = \begin{pmatrix} H_L & H_{LC} & 0 \\ H_{LC}^\dagger & H_C & H_{RC}^\dagger \\ 0 & H_{RC} & H_R \end{pmatrix} \quad O = \begin{pmatrix} O_L & O_{LC} & 0 \\ O_{LC}^\dagger & O_C & O_{RC}^\dagger \\ 0 & O_{RC} & O_R \end{pmatrix}, \quad (2.18)$$

where

$H_L(O_L)$, $H_C(O_C)$, and $H_R(O_R)$ are the Hamiltonian (overlap) matrices of the leads and the device

$H_{LC}(O_{LC})$ and $H_{RC}(O_{RC})$ are the coupling matrices between the central region and the leads

By defining the self-energies of the leads $(X = L, R)$ as

$$\Gamma_X(E) = i\left(\Sigma_X(E) - \Sigma_X^\dagger(E)\right), \quad (2.19)$$

$$\Sigma_X(E) = (EO_{XC} - H_{XC})g_X(E)(EO_{XC} - H_{XC}), \quad (2.20)$$

where

$$g_X(E) = (EO_X - H_X)^{-1} \quad (2.21)$$

is Green's function of the semi-infinite leads, and defining Green's function of the central region

$$G_C(E) = (EO_C - H_C - \Sigma_L(E) - \Sigma_R(E))^{-1} \quad (2.22)$$

the transmission probability is given by

$$T(E) = \text{Tr}[G_C(E)\Gamma_L(E)G_C^\dagger(E)\Gamma_R(E)]. \quad (2.23)$$

To calculate the transmission probability, one first has to calculate Green's functions of the leads $g_L(E)$ and $g_R(E)$ for a given set of energy values and then determine the self-energies and Green's function of the central region, $G(E)$. The self-energy of the lead is usually calculated by assuming that the lead is made of periodically repeated cells (see Figure 2.13) and that the basis functions in these cells only overlap between neighboring cells. This is a computationally intensive part of the calculation.

The simplest method to calculate Green's function of the lead is the iterative method [51]. It starts with the calculation of Green's function for a single cell of the periodic lead and adds more and more cells until convergence. The convergence, however, is extremely slow because the termination of the infinite lead at a finite distance generates reflections from the artificial boundary, and this effect will only be negligible if the boundary is far away. The decimation method [52] is a clever variant of the iterative approach which greatly speeds up the convergence by increasing the lead in a recursive way. The decimation method is a very efficient way to calculate Green's function of the lead and is used in many quantum transport codes. Other approaches calculate Green's function by solving quadratic eigenvalue problems [49,53,54]. The review paper [51] gives a pedagogical introduction to these methods, shows the equivalence of the different approaches, and is an extensive source of references of self-energy calculations.

Recently, we have developed a simpler and faster method to evaluate the self-energies. The essence of the approach is to add a complex absorbing potential (CAP) [55–59] to the Hamiltonian of the leads (Figure 2.14). This potential absorbs the outgoing waves without reflection, transforming the infinite open system into a closed finite system. By adding the CAP to the Hamiltonian of the leads one obtains

$$H_L' = H_L - iW_L(x) \quad H_R' = H_R - iW_R(x), \quad (2.24)$$

where W_L and W_R are the matrix elements of the complex potential on the left and the right. Assuming that the basis states only connect the neighboring cells in the lead, these matrices will have the same block tridiagonal structure as the leads Hamiltonian. These are finite dimensional Hamiltonians; beyond the range of the complex potential, the lead is effectively cut off.

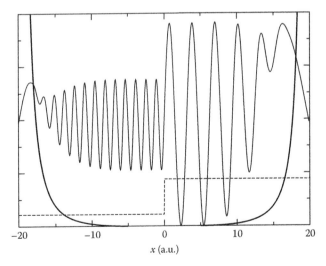

FIGURE 2.14 Scattering on a potential step. The bold line shows the complex absorbing potential, the dashed line is the potential step, and the thin solid line is the square of the scattering wave function. In the middle region where the CAP is zero, the wave function is in perfect agreement with the exact wave function (not shown in the figure).

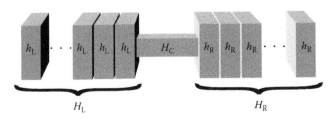

FIGURE 2.13 Schematic diagram of a two-probe device. The device is modeled by two semi-infinite electrodes (L and R) and a central region (C). The electrodes are divided into principal layers (blocks) that interact only with the nearest-neighbor layers.

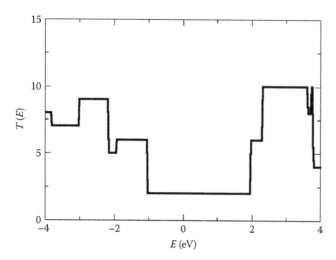

FIGURE 2.15 Conductance of a (10,10) armchair nanotube. The conductance was calculated by using the decimation method and by the CAP approach. The two results are in complete agreement and the results are not distinguishable on the figure.

The simplest way to calculate Green's functions of the leads now is to diagonalize the complex Hamiltonians H'_L and H'_R:

$$H'_X C_X = E_X O_X C_X \quad (X = L, R). \tag{2.25}$$

Green's function of the leads now can be calculated by the spectral representation

$$(g_L)_{ij} = \sum_k \frac{C_{Xik} C_{Xjk}}{E - E_{Xk}}, \tag{2.26}$$

where C_{Xik} is the ith component of the kth eigenvector belonging to eigenvalue E_{Xk}. As the Hamiltonian matrix of the lead is complex symmetric, the left and right eigenvalues are equal and the left and right eigenvectors are complex conjugates of each other. For a given lead, this diagonalization only has to be done once in the beginning of the calculations and the self-energies are then available for any desired energy.

To show the accuracy of the method, we have calculated the conductance of carbon nanotubes and compared the transmission obtained from CAP and decimation. The agreement between the two approaches (Figure 2.15) is excellent.

2.3.3 Conductance of Carbon Nanotubes

Using the method described in the previous section we have calculated the DOS and transmission probability of nanotubes of different chirality. The conductance is given by the transmission probability at Fermi energy:

$$G = G_0 T(E_F), \tag{2.27}$$

where $G_0 = e^2/h$ is the unit of the quantum conductance. Figure 2.16 shows the electron transmission probability of an (5,5)

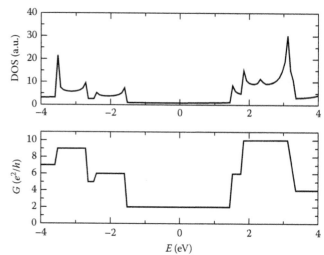

FIGURE 2.16 DOS and transmission probability of a (5,5) nanotube.

armchair nanotube. This nanotube is metallic, the conductance is equal to $2G_0$. The next two examples show two semiconducting CNs (10,0) (Figure 2.17) and (13,0) (Figure 2.18). As the conductance increases the resistance decreases. The conductance increases if there are more conduction channels available, i.e., in multiwall nanotubes, nanotubes of larger diameter, or bundles of CNs.

Figure 2.19 shows the effect of Na encapsulation on the transmission probability. In the device region, the (4,4) CN encapsulates an Na atom chain placed on the axis of the nanotube. The Na chain substantially changes the transmission probability, showing the possibility of sensing application of CN devices.

A recent theoretical simulation concluded that the CN bundles boasted a much smaller electrical resistance than the copper nanowires [60]. The resistivity of copper interconnects, with cross-sectional dimensions of the order of the mean free path of electrons (~40 nm in Cu at room temperature) in current and

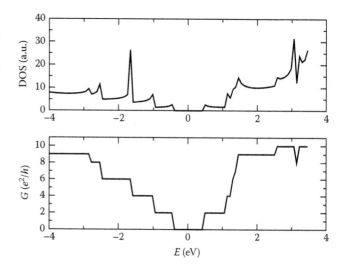

FIGURE 2.17 DOS and transmission probability of a (10,0) nanotube.

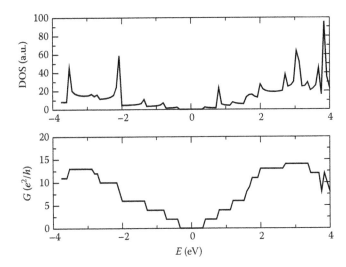

FIGURE 2.18 DOS and transmission probability of a (13,0) nanotube.

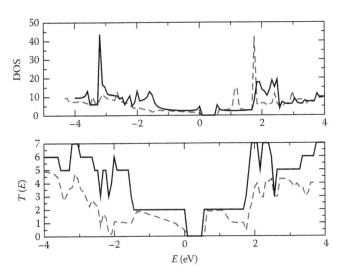

FIGURE 2.19 DOS and transmission probability of a (4,4) nanotube (solid line). The dashed line shows the effect of Na encapsulation.

imminent technologies is increasing rapidly under the combined effects of enhanced grain boundary scattering and surface scattering. The rising Cu resistivity also poses a reliability concern due to Joule heating induced significant metal temperature rise. Metallic CNs have been suggested as an interconnect material due to their high current-carrying capacity and mechanical stability [61]. Several groups have reported on fabrication of CN interconnects. Recently, performance and thermal analysis of CN interconnects reported by the authors in [62] have shown that CN-based interconnects can potentially offer significant advantages over copper.

2.3.4 Carbon Nanotube Transistors

Semiconducting nanotubes can work as transistors [63–67]. The electrical conductivity of semiconducting tubes can be effectively modulated by an external electric field, which makes them suitable for use as high-performance field-effect transistors (FET). Such devices may operate with either Schottky or Ohmic source and drain contacts. Complementray metal-oxide-semiconductor (CMOS)-like logic circuits using n-type and p-type MOS-like FETs have been demonstrated.

The tube can be turned on, i.e., made to conduct by applying a negative bias to the gate and turned off with a positive bias. A negative bias induces holes, increasing conductance. The positive bias works in the opposite way, while it depletes the holes and decreases conductance. The resistance of the off state can be more than a million times greater than the on state. This behavior is similar to the behavior of a p-type metal-oxide-silicon field-effect transistor (MOSFET). The nanotube becomes p-type because the metal electrodes remove electrons from the tube leaving the remaining holes responsible for the conduction. This charge transfer is due to the different work functions between the metallic contacts and the CN, leading to the appearance of Schottky barriers, and consequently contact resistances. A Schottky barrier is a potential barrier formed at a metal–semiconductor junction that has rectifying characteristics, suitable for use as a diode. In the simplest picture, the electrical resistance of a CN device is the series connection of the two contact resistances and the channel resistance. Both the contact and channel resistance is affected by the gate voltage, and their relative importance determines the device characteristics. The tubes can even be doped n-type using elements such as potassium that donate electrons to the tube.

Experimentally, most of the CN transistors realized at this moment have metal source and drain contacts with the Schottky barrier dominating their behavior [68–77]. The largest difference between a Schottky barrier and a p–n junction is its typically lower junction voltage and decreased (almost nonexistent) depletion width in the metal. The transistor action observed in CN-FET can be understood on a basis of transport across a Schottky barrier at the metal–CN contact. The gate induces an electric field at the contact, which controls the width of the barrier and hence the current. Changes in work function affect the Schottky barrier and hence the device characteristic.

As a prototypical example, we show a transport calculation of the current–voltage characteristics of a semiconducting CN with a perpendicular electric field acting as a gate (see Figure 2.12) [78]. The response of nanotubes to external electric fields is of interest both for transport devices [79] and electromechanical systems. The CN-FET [79,80] is a particularly promising candidate as an element of future electronic devices. Calculations so far have been based on semiempirical Hamiltonians [14,81]. We have found that, in contrast to Si-based devices, where a large transverse electric field merely "bends" the bulk energy bands, the CN band structure is modified significantly, with new band crossings, an ultimate collapse of the energy gap, and substantial changes in the DOS around the Fermi energy and the resulting transport properties.

We used a prototypical (10,0) semiconducting CN (see Figure 2.12). We placed a 4.9 nm long CN (12 unit cells) between Al(100) electrodes. The CN-electrode separation is 1.5 Å, providing

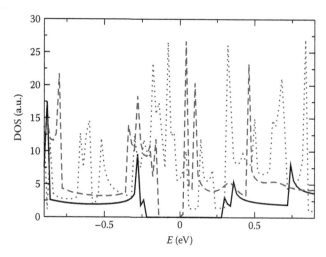

FIGURE 2.20 DOS of the CN-FET device for two different electric fields (solid line $F = 0.0$ V/Å, dashed line $F = 0.2$ V/Å, and dotted line $F = 0.4$ V/Å).

strong coupling at the interface. The DOS for $E = 0.0$, 0.2, and 0.4 V/Å transverse electric fields are shown in Figure 2.20. The energy gap of this Al–CN–Al system is ~0.5 eV when no electric field is present (the value of the energy gap depends on the CN length); it vanishes for CNs shorter than 8 unit cells because of the coupling to the metal leads and approaches the value of the infinite (periodically repeated unit cells) system for longer CNs. An increasing electric field results in a decreasing energy gap that vanishes at ~0.4 V/Å. The transmission probability, $T(E)$, shows similar characteristics: it has a gap around the Fermi level when no electric field is present and the gap gradually decreases and disappears with increasing the electric field.

The current–voltage characteristic as a function of the external field is shown in Figure 2.21. When the external field is zero or small, the current is very small due to the energy gap where the transmission is zero. By increasing the external

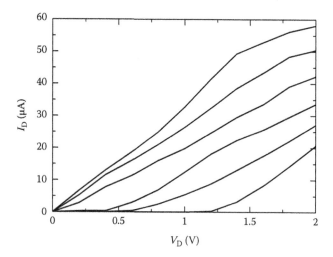

FIGURE 2.21 Drain current versus drain voltage of the CN-FET device for different gate voltages from $F = 0$ V/Å (lowest curve) to $F = 1$ V/Å in 0.2 V/Å steps.

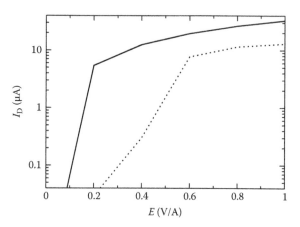

FIGURE 2.22 Drain current versus gate voltage of the CN-FET device at a fixed $V_{SD} = 0.5$ V (lower curve) and $V_{SD} = 1.0$ V (upper curve) source-drain voltages.

electric field, the energy gap decreases and the current becomes substantially larger. This characteristic is in good agreement with the experimental data on semiconducting CNs [82]. The calculated current is larger than the experimentally measured current, most likely because of differences in the contacts. As a test, we calculated the current for different CN-electrode distances and found significant changes in the absolute magnitude but not the overall behavior (in the experiments, the CN is either buried in or laid on the leads so that the precise contact geometry is not known).

The source-drain current is plotted versus gate voltage for two different source-drain voltages in Figure 2.22. This behavior is the same as found in experiments [82,83] (our curves are symmetric about the zero gate voltage because of our symmetric geometry; in experiments, the gate is asymmetric). We note definite saturation of the source-drain current at high gate voltage. The plateaus can be used to define the "on" current. With the "off" current defined at zero gate voltage, we find the on–off ratio to be about 10^2. The experimentally observed on–off ratio varies between 10^2 and 10^5 [80,83]. We repeated the calculation for a 9.8 nm long CN-FET (24 unit cell) and the on–off ratio increased to 10^3. The increase came entirely from a reduction in the off current (smaller tunneling current in a longer CN). The saturation current remained unchanged, which means that, for a certain CN diameter and a certain contact, there exists an intrinsic maximum saturation current (scattering mechanisms in longer CNs would reduce the maximum current).

2.3.5 Functionalized Carbon Nanotube Devices

Physical and chemical properties of an SWNT can also be modified by the adsorption of foreign atoms or molecules. This process is usually named functionalization and carries great potential in tailoring new nanostructures for engineering them according to a desired application. For example, depending on the pattern of hydrogen atom coverage, while a metallic armchair SWNT can be transformed to a wide band gap semiconductor, a semiconducting zigzag tube may become a metal with very high state density. A free SWNT, which is normally nonmagnetic, becomes magnetic with unpaired spins upon the adsorption of oxygen molecules or specific transition metal atoms. A recent study demonstrates that a semiconducting zigzag tube becomes both a magnetic and a high conductance wire as a result of Ti coating. A selectable functionalization of the (5,5) SWNT resulting from CH_n ($n = 13$) adsorption and decoration gives rise to a substantial change in the DOS [84]. Suitably doped CNs can be functionalized by selectively forming chemical bonds with ligands at the chemically active impurity site [85].

2.3.6 Spin Transport in Carbon Nanotubes

The quantum mechanical spin degree of freedom is now widely exploited to control current transport in electronic devices. For example, the readout of magnetic hard disc devices is based on the spin-valve effect, i.e., the tunability of the conductance with the relative orientation of ferromagnetic polarizations. Spin transport in a nonmagnetic lateral channel between a spin-polarized source and a spin-polarized drain is at the basis of several concepts of logic devices or spin transistors. The spin lifetimes in graphenes and CNs are predicted to be relatively long [86,87]. Therefore, CNs could become a very good material for future spintronic devices. There are two reasons why spin relaxation time in CNs is expected to be long. First, the weak hyperfine interaction between the electron spins and the nuclear spins. One of the reasons is that only about 1% of the carbon atoms consists of the isotope C13, which carries a nuclear spin. Second, a weak spin–orbit interaction. Spin orbit interaction is a relativistic effect where the electric fields around the nuclei are seen by the moving electrons as a magnetic field, which can therefore give rise to a randomization of their spin directions, and thus effectively cause spin relaxation. Because of the low atomic number of carbon ($Z = 6$) this spin–orbit interaction should be weak, since its strength scales with Z.

The first organic spintronic device was reported by Tsukagoshi et al. in 1999 [88], and remarkably, it consisted of a multiwalled CN contacted by Co contacts. Many other spintronic experiments on multi- and single-walled CNs followed up. Unfortunately, the experiment of Tsukagoshi and all other experiments performed after his work have made use of the conventional two-terminal spin-valve geometry [89–93]. The use of this geometry makes it difficult to separate spin transport from other effects, such as Hall effects, anisotropic magnetoresistance, interference effects, and magneto-coulomb effects [94]. These may obscure and even mimic the spin accumulation signal. Recent report of oscillations in the magnitude and sign of magnetoresistence with gate voltage provide the strongest evidence that the magnetoresistence is primarily due to spin-polarized transport [95].

The spin transport through CNs contacted by ferromagnetic electrodes was studied in [96]. It was found that the resistance of a ferromagnetically contacted multiwalled nanotube switches hysteretically as a function of applied magnetic field, with a maximum resistance change of 9% at 4.2 K.

2.4 Summary

Many potential applications of CNs have been put forward, including field-emission devices, high-strength composites, energy storage and energy conversion devices, hydrogen storage media, probes, sensors, interconnects, and transistors. Some of these applications are already realized as products (e.g., CN sheets as large as 3 ft by 6 ft have been produced and will be used as lightweight conductors in planes and satellites), others are in the developmental stage.

In this work, we mainly concentrated on the electron-transport properties of nanotubes. We have seen that a fully conjugated π-electron system can be used as a metallic wire at lengths of hundreds of nanometers showing that molecular systems can be good electrical conductors and can become a major part of nanoelectronic devices.

New CN growth techniques are playing a key role in future applications. To demonstrate the growing capabilities, nanotube transistor radios have been fabricated [97], in which nanotube devices provided all of the key functions. The radios were based on a heterodyne receiver design consisting of four capacitively coupled stages: an active resonant antenna, two radio-frequency amplifiers, and an audio amplifier, all based on nanotube devices.

In summary, the future of nanotubes looks very bright: Nanotubes are interesting model systems for fundamental studies of one-dimensional systems, but they are equally (or even more) attractive for applied researches and industry due to the wide variety of their potential applications. They offer lot of creativity in material preparation. Furthermore, since nanotubes are very user friendly and very robust, they can also act as excellent model systems for learning manipulation at a nanometer scale, which is the scale of biological macromolecules like DNA, microtubules, and proteins.

References

1. Iijima, S., 1991. Helical microtubules of graphitic carbon, *Nature*, 354 (6348), 56–58.

2. Iijima, S. and Ichihashi, T., 1993. Single-shell carbon nanotubes of 1-nm diameter, *Nature*, 363 (6430), 603–605.

3. Bethune, D. S., Klang, C. H., de Vries, M. S., Gorman, G., Savoy, R., Vazquez, J., and Beyers, R., 1993. Cobalt-catalysed growth of carbon nanotubes with single-atomic-layer walls, *Nature*, 363 (6430), 605–607.

4. Hamada, N., Sawada, S., and Oshiyama, A., 1992. New one-dimensional conductors: Graphitic microtubules, *Physical Review Letters*, 68 (10), 1579.

5. Saito, R., Fujita, M., Dresselhaus, G., and Dresselhaus, M. S., 1992. Electronic structure of chiral graphene tubules, *Applied Physics Letters*, 60 (18), 2204–2206.

6. Tans, S. J., Devoret, M. H., Dai, H., Thess, A., Smalley, R. E., Geerligs, L. J., and Dekker, C., 1997. Individual single-wall carbon nanotubes as quantum wires, *Nature*, 386 (6624), 474–477.

7. Bockrath, M., Cobden, D. H., McEuen, P. L., Chopra, N. G., Zettl, A., Thess, A., and Smalley, R. E., 1997. Single-electron transport in ropes of carbon nanotubes, *Science*, 275 (5308), 1922–1925.

8. McEuen, P., Fuhrer, M., and Park, H., 2002. Single-walled carbon nanotube electronics, *IEEE Transactions on Nanotechnology*, 1 (1), 78–85.

9. Dürkop, T., Cobas, E., and Fuhrer, M. S., 2003. High-mobility semiconducting nanotubes, in *Molecular Nanostructures: XVII International Winterschool Euroconference on Electronic Properties of Novel Materials*, H. Kuzmany, J. Fink, M. Mehring, and S. Roth, (eds.), AIP, Melville, NY, vol. 685, pp. 524–527.

10. Javey, A., Guo, J., Paulsson, M., Wang, Q., Mann, D., Lundstrom, M., and Dai, H., 2004. High-field quasiballistic transport in short carbon nanotubes, *Physical Review Letters*, 92 (10), 106804.

11. Lin, Y., Appenzeller, J., Chen, Z., Chen, Z., Cheng, H., and Avouris, P., 2005. High-performance dual-gate carbon nanotube FETs with 40-nm gate length, *IEEE Electron Device Letters*, 26 (11), 823–825.

12. Guo, J., 2005. A quantum-mechanical treatment of phonon scattering in carbon nanotube transistors, *Journal of Applied Physics*, 98 (6), 063519-(6).

13. Akturk, A., Pennington, G., and Goldsman, N., 2005. Quantum modeling and proposed designs of CNT-embedded nanoscale MOSFETs, *IEEE Transactions on Electron Devices*, 52 (4), 577–584.

14. Alam, K. and Lake, R. K., 2005. Leakage and performance of zero-Schottky-barrier carbon nanotube transistors, *Journal of Applied Physics*, 98 (6), 064307-(8).

15. Perebeinos, V., Tersoff, J., and Avouris, P., 2006. Mobility in semiconducting carbon nanotubes at finite carrier density, *Nano Letters*, 6 (2), 205–208.

16. Durkop, T., Getty, S. A., Cobas, E., and Fuhrer, M. S., 2004. Extraordinary mobility in semiconducting carbon nanotubes, *Nano Letters*, 4 (1), 35–39.

17. Dekker, C., 1999. Carbon nanotubes as molecular quantum wires, *Physics Today*, 52 (5), 22.

18. Yao, Z., Kane, C. L., and Dekker, C., 2000. High-field electrical transport in single-wall carbon nanotubes, *Physical Review Letters*, 84 (13), 2941.

19. Oh, S., Monroe, D., and Hergenrother, J., 2000. Analytic description of short-channel effects in fully-depleted double-gate and cylindrical, surrounding-gate MOSFETs, *IEEE, Electron Device Letters*, 21 (9), 445–447.

20. Winstead, B. and Ravaioli, U., 2000. Simulation of Schottky barrier MOSFETs with a coupled quantum injection/Monte carlo technique, *IEEE Transactions on Electron Devices*, 47 (6), 1241–1246.

21. Guo, J., Goasguen, S., Lundstrom, M., and Datta, S., 2002. Metal–insulator–semiconductor electrostatics of carbon nanotubes, *Applied Physics Letters*, 81 (8), 1486–1488.

22. Collins, P. G. and Avouris, P., 2001. Nanotubes for electronics., *Scientific American*, 283 (6), 62–69.

23. Gutierrez, R., Fagas, G., Cuniberti, G., Grossmann, F., Schmidt, R., and Richter, K., 2002. Theory of an all-carbon molecular switch, *Physical Review B*, 65 (11), 113410.

24. Martel, R., Schmidt, T., Shea, H. R., Hertel, T., and Avouris, P., 1998. Single- and multi-wall carbon nanotube field-effect transistors, *Applied Physics Letters*, 73 (17), 2447–2449.

25. Dai, H., 2001. *Nanotube Growth and Characterization*, Springer, Berlin, Germany, pp. 29–53.

26. Dai, H., 2002. Carbon nanotubes: Synthesis, integration, and properties, *Accounts of Chemical Research*, 35 (12), 1035–1044.

27. Odom, T. W., Huang, J., Kim, P., and Lieber, C. M., 2000. Structure and electronic properties of carbon nanotubes, *The Journal of Physical Chemistry B*, 104 (13), 2794–2809.

28. Jorio, A., Saito, R., Hafner, J. H., Lieber, C. M., Hunter, M., McClure, T., Dresselhaus, G., and Dresselhaus, M. S., 2001. Structural (n, m) determination of isolated single-wall carbon nanotubes by resonant Raman scattering, *Physical Review Letters*, 86 (6), 1118.

29. Reich, S. and Thomsen, C., 2000. Chirality dependence of the density-of-states singularities in carbon nanotubes, *Physical Review B*, 62 (7), 4273.

30. Saito, R., Dresselhaus, G., and Dresselhaus, M. S., 1998. *Physical Properties of Carbon Nanotubes*, Imperial College Press, London, U.K.

31. Wallace, P. R., 1947. The band theory of graphite, *Physical Review*, 71 (9), 622.

32. Geim, A. K. and Novoselov, K. S., 2007. The rise of graphene, *Nature Mater*, 6 (3), 183–191.

33. Reich, S., Maultzsch, J., Thomsen, C., and Ordejn, P., 2002. Tight-binding description of graphene, *Physical Review B*, 66 (3), 035412.

34. Ventra, M. D., Pantelides, S. T., and Lang, N. D., 2000. First-principles calculation of transport properties of a molecular device, *Physical Review Letters*, 84 (5), 979.

35. Taylor, J., Guo, H., and Wang, J., 2001. Ab initio modeling of quantum transport properties of molecular electronic devices, *Physical Review B*, 63 (24), 245407.

36. Brandbyge, M., Mozos, J., Ordejn, P., Taylor, J., and Stokbro, K., 2002. Density-functional method for nonequilibrium electron transport, *Physical Review B*, 65 (16), 165401.

37. Datta, S., 1997. *Electronic Transport in Mesoscopic Systems*, Cambridge University Press, Cambridge, U.K.

38. Faleev, S. V., Lonard, F., Stewart, D. A., and van Schilfgaarde, M., 2005. Ab initio tight-binding LMTO method for nonequilibrium electron transport in nanosystems, *Physical Review B*, 71 (19), 195422.

39. Thygesen, K. and Jacobsen, K., 2005. Molecular transport calculations with Wannier functions, *Chemical Physics*, 319 (1–3), 111–125.

40. Derosa, P. A. and Seminario, J. M., 2001. Electron transport through single molecules: Scattering treatment using density functional and green function theories, *The Journal of Physical Chemistry B*, 105 (2), 471–481.

41. Tomfohr, J. and Sankey, O. F., 2004. Theoretical analysis of electron transport through organic molecules, *The Journal of Chemical Physics*, 120 (3), 1542–1554.

42. Zhang, X., Fonseca, L., and Demkov, A., 2002. The application of density functional, local orbitals, and scattering theory to quantum transport, *Physica Status Solidi (b)*, 233 (1), 70–82.

43. Palacios, J. J., Prez-Jimnez, A. J., Louis, E., SanFabin, E., and Vergs, J. A., 2003. First principles phase-coherent transport in metallic nanotubes with realistic contacts, *Physical Review Letters*, 90 (10), 106801.

44. Stokbro, K., Taylor, J., Brandbyge, M., Mozos, J. L., and Ordejn, P., 2003. Theoretical study of the nonlinear conductance of di-thiol benzene coupled to Au(1 1 1) surfaces via thiol and thiolate bonds, *Computational Materials Science*, 27 (1–2), 151–160.

45. Emberly, E. G. and Kirczenow, G., 2001. Models of electron transport through organic molecular monolayers self-assembled on nanoscale metallic contacts, *Physical Review B*, 64 (23), 235412.

46. Nardelli, M. B., Fattebert, J., and Bernholc, J., 2001. O(N) real-space method for ab initio quantum transport calculations: Application to carbon nanotubemetal contacts, *Physical Review B*, 64 (24), 245423.

47. Xue, Y., Datta, S., and Ratner, M. A., 2001. Charge transfer and "band lineup" in molecular electronic devices: A chemical and numerical interpretation, *The Journal of Chemical Physics*, 115 (9), 4292–4299.

48. Ke, S., Baranger, H. U., and Yang, W., 2004. Electron transport through molecules: Self-consistent and non-self-consistent approaches, *Physical Review B*, 70 (8), 085410.

49. Sanvito, S., Lambert, C. J., Jefferson, J. H., and Bratkovsky, A. M., 1999. General green's-function formalism for transport calculations with spd Hamiltonians and giant magnetoresistance in Co- and Ni-based magnetic multilayers, *Physical Review B*, 59 (18), 11936.

50. Bauschlicher, C. W., Lawson, J. W., Ricca, A., Xue, Y., and Ratner, M. A., 2004. Current-voltage curves for molecular junctions: The effect of Cl substituents and basis set composition, *Chemical Physics Letters*, 388 (4–6), 427–429.

51. Velev, J. and Butler, W., 2004. Topical review: On the equivalence of different techniques for evaluating the green function for a semi-infinite system using a localized basis, *Journal of Physics Condensed Matter*, 16, 637.

52. Sancho, M. P. L., Sancho, J. M. L., Sancho, J. M. L., and Rubio, J., 1985. Highly convergent schemes for the calculation of bulk and surface green functions, *Journal of Physics F Metal Physics*, 15, 851–858.

53. Ando, T., 1991. Quantum point contacts in magnetic fields, *Physical Review B*, 44 (15), 8017.

54. Krsti, P. S., Zhang, X., and Butler, W. H., 2002. Generalized conductance formula for the multiband tight-binding model, *Physical Review B*, 66 (20), 205319.

55. Kosloff, R. and Kosloff, D., 1986. Absorbing boundaries for wave propagation problems, *Journal of Computational Physics*, 63 (2), 363–376.

56. Seideman, T. and Miller, W. H., 1992. Calculation of the cumulative reaction probability via a discrete variable representation with absorbing boundary conditions, *Journal of Chemical Physics (United States)*, 96(6), 4412–4422.

57. Muga, J. G., Palao, J. P., Navarro, B., and Egusquiza, I. L., 2004. Complex absorbing potentials, *Physics Reports*, 395, 357–426.

58. Manolopoulos, D. E., 2002. Derivation and reflection properties of a transmission-free absorbing potential, *The Journal of Chemical Physics*, 117 (21), 9552–9559.

59. Zhang, X., Varga, K., and Pantelides, S. T., 2007. Generalized Bloch theorem for complex periodic potentials: A powerful application to quantum transport calculations, *Physical Review B (Condensed Matter and Materials Physics)*, 76 (3), 035108-(9).

60. Zhou, Y., Sreekala, S., Ajayan, P. M., and Nayak, S. K., 2008. Resistance of copper nanowires and comparison with carbon nanotube bundles for interconnect applications using first principles calculations, *Journal of Physics Condensed Matter*, 20, 5209.

61. Kreupl, F., Graham, A. P., Duesberg, G. S., Steinhgl, W., Liebau, M., Unger, E., and Hnlein, W., 2002. Carbon nanotubes in interconnect applications, *Microelectronic Engineering*, 64 (1–4), 399–408.

62. Srivastava, N., Joshi, R., and Banerjee, K., 2005. Carbon nanotube interconnects: Implications for performance, power dissipation and thermal management, in *Electron Devices Meeting, 2005. IEDM Technical Digest. IEEE International*, Washington, DC, pp. 249–252.

63. Lin, Y., Appenzeller, J., and Avouris, P., 2004. Novel carbon nanotube FET design with tunable polarity, in *Electron Devices Meeting, 2004. IEDM Technical Digest. IEEE International*, San Francisco, CA, pp. 687–690.

64. Javey, A., Tu, R., Farmer, D. B., Guo, J., Gordon, R. G., and Dai, H., 2005. High performance n-type carbon nanotube field-effect transistors with chemically doped contacts, *Nano Letters*, 5 (2), 345–348, PMID: 15794623.

65. Appenzeller, J., Lin, Y.-M., Knoch, J., and Avouris, P., 2004. Band-to-band tunneling in carbon nanotube field-effect transistors, *Physical Review Letters*, 93 (19), 196805.

66. Appenzeller, J., Lin, Y., Knoch, J., Chen, Z., and Avouris, P., 2005. Comparing carbon nanotube transistors - the ideal choice: A novel tunneling device design, *IEEE Transactions on Electron Devices*, 52 (12), 2568–2576.

67. Clifford, J., John, D., and Pulfrey, D., 2003. Bipolar conduction and drain-induced barrier thinning in carbon nanotube FETs, *IEEE Transactions on Nanotechnology*, 2 (3), 181–185.

68. Appenzeller, J., Knoch, J., Martel, R., Derycke, V., Wind, S., and Avouris, P., 2002. Short-channel like effects in Schottky barrier carbon nanotube field-effect transistors, in *Electron Devices Meeting, 2002 (IEDM'02). Digest. International*, pp. 285–288. Available at: http://ieeexplore.ieee.org/xpls/absall.jsp?arnumber=1175834&tag=1

69. Avouris, P., Martel, R., Heinze, S., Radosavljevic, M., Wind, S., Derycke, V., Appen-zeller, J., Terso, J., Kuzmany, H., Fink, J., Mehring, M., and Roth, S., 2002. The role of

Schottky barriers on the behavior of carbon nanotube field-effect transistors, in *Structural and Electronic Properties of Molecular Nanostructures: XVI International Winterschool on Electronic Properties of Novel Materials*, Kirchberg, Tirol, Austria, AIP, vol. 633, pp. 508–512.

70. Heinze, S., Tersoff, J., Martel, R., Derycke, V., Appenzeller, J., and Avouris, P., 2002. Carbon nanotubes as Schottky barrier transistors, *Physical Review Letters*, 89 (10), 106801.

71. Knoch, J. and Appenzeller, J., 2002. Impact of the channel thickness on the performance of Schottky barrier metal-oxide–semiconductor field-effect transistors, *Applied Physics Letters*, 81 (16), 3082–3084.

72. Appenzeller, J., Knoch, J., and Avouris, P., 2003. Carbon nanotube field-effect transistors-an example of an ultra-thin body Schottky barrier device, in *Device Research Conference, 2003*, Salt Lake City, UT, pp. 167–170.

73. Heinze, S., Radosavljevi, M., Tersoff, J., and Avouris, P., 2003. Unexpected scaling of the performance of carbon nanotube Schottky-barrier transistors, *Physical Review B*, 68 (23), 235418.

74. Castro, L., John, D., and Pulfrey, D., 2002. Towards a compact model for Schottky-barrier nanotube FETs, in *2002 Conference on Optoelectronic and Microelectronic Materials and Devices*, Sydney, Australia, pp. 303–306.

75. Guo, J., Lundstrom, M., and Datta, S., 2002. Performance projections for ballistic carbon nanotube field-effect transistors, *Applied Physics Letters*, 80 (17), 3192–3194.

76. Anantram, M. P., Datta, S., and Xue, Y., 2000. Coupling of carbon nanotubes to metallic contacts, *Physical Review B*, 61 (20), 14219.

77. John, D. L. and Pulfrey, D. L., 2006. Switching-speed calculations for Schottky-barrier carbon nanotube field-effect transistors, *The Journal of Vacuum Science and Technology. A*, AVS, 24, 708–712.

78. Varga, K. and Pantelides, S. T., 2007. Quantum transport in molecules and nanotube devices, *Physical Review Letters*, 98 (7), 076804.

79. Appenzeller, J., Knoch, J., Derycke, V., Martel, R., Wind, S., and Avouris, P., 2002. Field-modulated carrier transport in carbon nanotube transistors, *Physical Review Letters*, 89 (12), 126801.

80. Tans, S. J., Verschueren, A. R. M., and Dekker, C., 1998. Room-temperature transistor based on a single carbon nanotube, *Nature*, 393 (6680), 49–52.

81. Xia, T., Register, L. F., and Banerjee, S. K., 2004. Calculations and applications of the complex band structure for carbon nanotube field-effect transistors, *Physical Review B*, 70 (4), 045322.

82. Chen, Y. and Fuhrer, M. S., 2005. Electric-field-dependent charge-carrier velocity in semiconducting carbon nanotubes, *Physical Review Letters*, 95 (23), 236803.

83. Javey, A., Guo, J., Farmer, D. B., Wang, Q., Yenilmez, E., Gordon, R. G., Lundstrom, M., and Dai, H., 2004. Self-aligned ballistic molecular transistors and electrically parallel nanotube arrays, *Nano Letters*, 4 (7), 1319–1322.

84. Li, F., Xia, Y., Zhao, M., Liu, X., Huang, B., Tan, Z., and Ji, Y., 2004. Selectable functionalization of single-walled carbon nanotubes resulting from CHn (n = 1–3) adsorption, *Physical Review B*, 69 (16), 165415.

85. Nevidomskyy, A. H., Csanyi, G., and Payne, M. C., 2003. Chemically active substitutional nitrogen impurity in carbon nanotubes, *Physical Review Letters*, 91 (10), 105502.

86. Kane, C. L. and Mele, E. J., 2005. Quantum spin hall effect in graphene, *Physical Review Letters*, 95 (22), 226801.

87. Trauzettel, B., Bulaev, D. V., Loss, D., and Burkard, G., 2007. Spin qubits in grapheme quantum dots, *Nature Physics*, 3 (3), 192–196.

88. Tsukagoshi, K., Alphenaar, B. W., and Ago, H., 1999. Coherent transport of electron spin in a ferromagnetically contacted carbon nanotube, *Nature*, 401 (6753), 572–574.

89. Kim, J., So, H. M., Kim, J., and Kim, J., 2002. Spin-dependent transport properties in a single-walled carbon nanotube with mesoscopic co contacts, *Physical Review B*, 66 (23), 233401.

90. Chakraborty, S., Walsh, K. M., Alphenaar, B. W., Liu, L., and Tsukagoshi, K., 2003. Temperature-mediated switching of magnetoresistance in co-contacted multiwall carbon nanotubes, *Applied Physics Letters*, 83 (5), 1008–1010.

91. Orgassa, D., Mankey, G. J., and Fujiwara, H., 2001. Spin injection into carbon nanotubes and a possible application in spin-resolved scanning tunnelling microscopy, *Nanotechnology*, 12, 281–284.

92. Zhao, B., Monch, I., Muhl, T., Vinzelberg, H., and Schneider, C. M., 2002. Spin dependent transport in multiwalled carbon nanotubes, *Journal of Applied Physics*, AIP, 91, 7026–7028.

93. Sahoo, S., Kontos, T., Furer, J., Hoffmann, C., Graber, M., Cottet, A., and Schonenberger, C., 2005. Electric field control of spin transport, *Nature Physics*, 1 (2), 99–102.

94. Jedema, F. J., Filip, A. T., and van Wees, B. J., 2001. Electrical spin injection and accumulation at room temperature in an all-metal mesoscopic spin valve, *Nature*, 410 (6826), 345–348.

95. Sahoo, S., Kontos, T., Schonenberger, C., and Surgers, C., 2005. Electrical spin injection in multiwall carbon nanotubes with transparent ferromagnetic contacts, *Applied Physics Letters*, 86 (11), 112109-(3).

96. Alphenaar, B. W., Tsukagoshi, K., and Wagner, M., 2001. Spin transport in nanotubes (invited), *Journal of Applied Physics*, AIP, 89, 6863–6867.

97. Kocabas, C., Kim, H. S., Banks, T., Rogers, J. A., Pesetski, A. A., Baumgardner, J. E., Krishnaswamy, S. V., and Zhang, H., 2008. Radio frequency analog electronics based on carbon nanotube transistors, *Proceedings of the National Academy of Sciences*, 105 (5), 1405–1409.

3

Electron Transport in Carbon Nanotubes

Na Young Kim
Stanford University

3.1 Introduction

Carbon ($_6$C) exists ubiquitously in nature. It is the fourth most abundant element by mass after hydrogen, helium, and oxygen. Carbon has four valence electrons in 2*s* and 2*p* orbitals that can form three hybridization arrangements: *sp*, *sp*², and *sp*³. Via these hybrid orbitals, carbon can bond to themselves and to other elements. Even with the carbon element only, there are many systems with distinct properties: diamond, graphene, and buckyball fullerene C_{60} (Figure 3.1c). It is mysterious that carbon bonding in diamond can be as strong as to scribe other crystals, but in graphite as weak as to scribble on paper.

In 1991, graphitic needle-like carbon structures were synthesized fortuitously in a furnace that was used to produce C_{60} and other fullerene structures (Figure 3.1d, Iijima 1991). Since such concentric cylindrical structures were unprecedented, in early literature they were denoted as "microtubules" (Iijima 1991), "fullerene tubules" (Mintmire et al. 1992), or "graphene tubules" (Saito et al. 1992), before the term carbon nanotube (CNT) was coined. These early terms reflected the different perspectives of newly discovered structures. One school of thought believed that CNTs could be elongated from fullerene as mutations, while another school of thought believed that they could be formed by folding several graphene layers seamlessly. Although the starting material to make needle-like carbon structures from these two thoughts was not the same, both perspectives commonly envisioned CNTs as one-dimensional (1D) objects. A high geometric ratio of length to diameter (10⁵–10⁶) indeed supports this 1D picture from experimental statistical analyses that diameters are typically up to a few tens of nanometers and lengths are in microns and even in millimeters.

One-dimensional systems are very special in comparison to their two-dimensional or three-dimensional counterparts in that there is only one particular direction along which particles can move freely while they are strongly confined along two other directions. Hence, scientists and engineers have longed for a physical medium to explore unique 1D properties and have searched for alternative structures to attain miniaturization. Shortly after their discovery, CNTs have been one of the hot research topics in investigating low-dimensional properties in a variety of disciplines in science and engineering.

Many researchers have been intrigued by the CNTs' extraordinary mechanical, chemical, optical, and electrical properties and their fabrication advantages (low cost and defect-free crystalline structure) (Saito et al. 1998, Harris 1999, Jorio et al. 2008). Indeed, CNTs have already been influential in widespread areas as sensitive chemical sensors, high-resolution imaging probes, field emission displays, supercapacitors, conductive flexible electrodes, and more. In particular, CNTs have been regarded as potentially promising for future molecular electronics based on their phenomenal electrical characteristics and naturally small sizes (Dekker 1999). For example, high-performance nanotube field effect transistors have been demonstrated with semiconducting tubes (Bockrath et al. 1997, Javey et al. 2003), and the interplay of electrons and phonons in CNT transistors was thoroughly examined in order to develop practical room-temperature electronics (Yao et al. 2000, Javey et al. 2004, Park et al. 2004). Due to the spatial confinement in 1D where electrons can move only back and forth in principle, several questions naturally arise: How do electrons carry electrical information in 1D conductors like CNTs? What are the unique electrical phenomena occurring in 1D conductors?

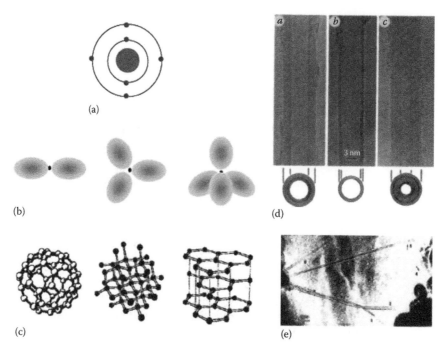

FIGURE 3.1 (a) A cartoon of the $_6$C atomic shell structure in a semiclassical picture. (b) The diagram of *s*- and *p*-orbital hybridization: linear *sp* (left), planer *sp²* (middle), and tetrahedral *sp³* (right). (c) Structures of carbon-based material: C_{60} (left), diamond (middle), and graphite (right). (d) Electron micrograph of multi-walled carbon nanotubes. (Reprinted from Iijima, S., *Nature*, 354, 56, 1991. With permission.) (e) An electron micrograph of single-walled carbon nanotubes. (Reprinted from Iijima, S. and Ichihashi, T., *Nature*, 363, 603, 1993. With permission.)

This chapter describes the fundamental concepts and recent progress in CNT electron transport fields. Electron transport measurements are convenient methods for characterizing materials in order to study the underlying physical principles. By applying external stimulations, the corresponding response of materials provides an answer to the question of how electrons flow through them. Section 3.2 provides an overview of CNT electron transport research areas and presents several basic concepts and terminologies of quantum electron transport and two theoretical frameworks pertinent for interpreting CNT transport properties. Based on these fundamentals, Section 3.3 develops the details of quantum electron transport measurements using metallic single-walled carbon nanotubes in terms of differential conductance and shot noise. A summary of the discussion and future perspectives are given in Sections 3.4 and 3.5, respectively.

3.2 Fundamental Concepts

3.2.1 Overview of Carbon Nanotube Transport

CNTs are regarded as one of the ideal 1D systems and a good candidate for exploring low dimensional physics. The basic electronic understanding of CNTs starts from their band structure. Shortly after the discovery of concentric multi-layered carbon tubules in 1991, condensed matter theorists immediately established the theoretical framework to compute the band structure

of CNTs by extending the two-dimensional graphene knowledge. They first simplified a target system as one single tube that was conceptually folded from one graphene layer even before such tubes were physically identified (Mintmire et al. 1992, Saito et al. 1992).

Figure 3.2 depicts how to form CNTs from a sheet of graphene by rolling up along a certain direction with a different diameter. A (*n, m*) configuration classifies possible nanotubes in terms of the integer coefficients of the two-dimensional roll-up vector ($\vec{C}_h$, Figure 3.2b, Saito et al. 1992). This (*n, m*) configuration is a compact representation to specify diameters and chiralities of CNTs. A spatial confinement along the circumferential direction in CNTs requires satisfying a periodic boundary condition. Thus, by including this additional boundary condition into the well-established graphite band structure calculation (Wallace 1947), the CNT band structure was readily computed (Mintmire et al. 1992, Saito et al. 1992).

Condensed matter theorists conjectured from the band structure calculation that such tubules would exhibit either metallic or semiconducting behavior depending on the values of *n, m* (Mintmire et al. 1992, Saito et al. 1992). When *n-m* is an integer in multiples of 3, the (*n, m*) nanotubes will be metallic. Specific terms became available according to theoretical insights: single-walled nanotubes (SWNTs) and multi-walled nanotubes (MWNTs) are distinguished by the number of constituent layers; armchair (*n = m*) and zigzag tubes (*n = 0* or *m = 0*) indicate a hexagonal arrangement around the circumference.

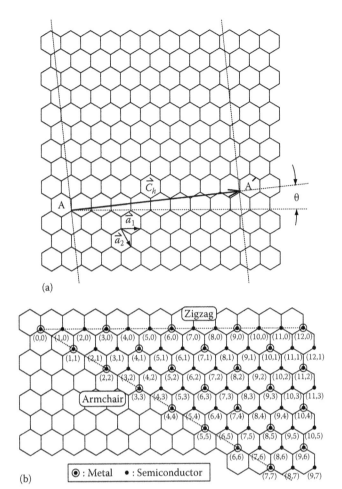

(a)

(b)

⊙ : Metal   • : Semiconductor

FIGURE 3.2 (a) Theoretical visualization to form a nanotube by rolling up from A to A' in a two-dimensional graphene sheet with a roll-up vector $\vec{C}_h = n\vec{a}_1 + m\vec{a}_2 \equiv (n, m)$ where $\vec{a}_1$ and $\vec{a}_2$ are two unit vectors. (b) Possible (n, m) configuration of SWNTs with various diameters with the denotation of metallic ($|n-m| = 3l$, where l is an integer) and semiconducting properties. (Reprinted from Saito, R. et al., *Appl. Phys. Lett.*, 60, 2204, 1992. With permission.)

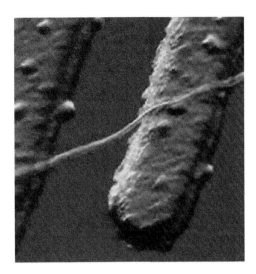

FIGURE 3.3 Atomic force microscope image of a SWNT between Pt electrodes spaced by 50 nm. (Reprinted from Tans, S.J. et al., *Nature*, 386, 474, 1997. With permission.)

Furthermore, these theoretical conjectures were initiated to search for SWNTs and to investigate mysterious properties. In the year 1993, another landmark in the history of the CNT research field emerged—the discovery of theoretically imagined SWNTs (Iijima and Ichihashi 1993). Finally, physical systems are now available to verify the theoretical predictions.

Early experiments using CNTs were rather limited in material investigations such as synthesis (Iijima 1991, Ebbesen and Ajayan 1992, Iijima et al. 1992, Iijima and Ichihashi 1993), chemical treatments for purifications (Ebbesen et al. 1994, Tsang et al. 1994, Tohji et al. 1996), surface imaging by scanning electron microscopy (Iijima 1991, Iijima and Ichihashi 1993), atomic force microscopy (Figure 3.3), and scanning tunneling microscopy (Ge and Sattler 1993, Wilder et al. 1998). The aforementioned experimental works confirmed unequivocally the theoretical predictions based on the band structure calculation

of CNTs. However, in order to perform electrical transport measurements, it was critical to access individual nanotubes and to apply voltages to them. Ebbesen et al. (1996) and Tans et al. (1997) accomplished this goal and reported their first electrical access to individual SWNTs in the late 1990s.

To enhance experimental efficiency and reproducibility in electrical transport measurements, two primary issues needed to be solved. One was to synthesize nanotubes at designated locations and the other was to improve the coupling between the tube and the electrodes. The first target was achieved by using a chemical vapor deposition technique to produce high-yield, high-quality SWNTs nearby catalyst islands (Kong et al. 1998). The second target has been continuously attempted by using different metal electrodes and annealing, and recently ohmic contacted nanotube devices were fabricated (Kong et al. 2001, Liang et al. 2001, Javey et al. 2003). In addition, when integrated nanotube circuits were achieved in Si-substrates (Soh et al. 1999), the CNT transport field made rapid progress and demonstrated numerous functional devices including transistors, oscillators, and sensors by utilizing advanced semiconductor fabrication techniques.

Originating from phase-coherent electrons and strong interactions among electrons in 1D structures, quantum transport phenomena have also been continuously revealed in SWNTs coupled with electron reservoirs: Coulomb blockade (Bockrath et al. 1997, Tans et al. 1998, Postma et al. 2001), Tomonaga–Luttinger liquid behavior (Bockrath et al. 1999, Yao et al. 1999), quantum ballistic interference (Kong et al. 2001, Liang et al. 2001), Kondo (Liang et al. 2002, Nygard et al. 2000) and orbital Kondo phenomena (Jarillo-Herrero et al. 2005), Aharonov-Bohm interference (Cao et al. 2004), magnetic orbital moment determination (Minot et al. 2004), and supercurrent behavior (Jarillo-Herrero et al. 2006). Some basic concepts of quantum electron transport are introduced in the following subsection.

3.2.2 Electron Transport

Mesoscopic systems refer to materials whose physical dimensions typically span from millimeters to nanometers. A SWNT is a representative mesoscopic system. Rigorously, a SWNT is an ideal 1D system since its Fermi wavelength λ_F is shorter than a longitudinal SWNT length, L. λ_F is a wavelength of carriers at Fermi energy (E_F) level about 0.8 nm, and it is an important quantity since the majority of carriers are electrons near E_F. In such low-dimensional systems, subtle and sophisticated features appear that are deeply rooted in quantum mechanics and tightly associated with many-body interactions.

Discerned from their classical counterpart, quantum transport properties reflect unique traits of coherence and quantization in the system. Quantum transport is classified into several regimes primarily by relevant lengthscale comparison. These transport divisions are crucial in understanding transport properties in each regime. Besides λ_F, other relevant characteristic lengths are defined for classified transport regimes: (1) mean free path, l_{mfp}, (2) thermal diffusion length, l_T, and (3) phase coherence length, l_ϕ.

3.2.2.1 Mean Free Path l_{mfp}

The mean free path, as the name indicates, is the average distance in which particles can move freely. Major sources to inhibit free motions of electrons are scatterers such as defects, impurities, or grain boundaries. Elastic scattering does not conserve momentum but energy, while inelastic scattering changes both momentum and energy of incident particles. Thus, mean free paths due to elastic and inelastic scattering should be differentiated accordingly, although generally, l_{mfp} refers to the elastic mean free path. In semiconductors, l_{mfp} is closely related to the mobility of carriers, and in metals l_{mfp} is much longer than λ_F. As l_{mfp} becomes comparable to λ_F, systems with such l_{mfp} are called in the dirty limit.

3.2.2.2 Thermal Diffusion Length l_T

At nonzero temperatures, electron wavepackets have an energy width of about $k_B T$ where k_B is the Boltzmann's constant and T is the temperature. This energy uncertainty induces diffusion in time. l_T is a characteristic length of diffusion process due to thermal energy.

3.2.2.3 Phase Coherence Length l_ϕ

Within l_ϕ, particles preserve their phase. Dynamical interactions including mutual Coulomb interactions among electrons and electron-phonon interactions disturb phase coherence. Therefore, this length is important to determine whether quantum interference effects from phase coherent sources can be detectable or not in systems.

Comparisons of such scales define three distinct transport regimes summarized in Table 3.1. Varying the physical length of mesoscopic conductors, all enlisted transport regimes are indeed within practical reach. Both in dissipative and in

TABLE 3.1 Classified Quantum Electron Transport Regimes. L_i ($i = x, y, z$) Represent a System Length along Each Direction

Regime	Condition
Ballistic	$L_x, L_y, L_z < l_{mfp}, l_T, l_\phi$
Diffusive	$l_{mfp}, l_T \ll L_x, L_y, L_z$
Dissipative	$l_\phi < L_x, L_y, L_z$

diffusive regimes, scattering processes determine transport quantities similar to the classical case. Elaborately, electrons in dissipative conductors suffer from inelastic as well as elastic scattering, losing the history of momentum and energy trajectories. Electrons in diffusive conductors encounter elastic scatterers so that they preserve energy but not momentum information. In the ballistic regime, on the other hand, all dimensions of ballistic conductors are much smaller than all aforementioned length scales; therefore, electrons participating in the conduction process do not encounter any kind of scattering sources, consequently, both momentum and energy are conserved in this ballistic regime.

In principle, ballistic conductors have zero resistance, which is caused by scattering processes; however, when we measure their two-terminal resistance, it is not zero. The finite resistance occurs at the interface of a ballistic mesoscopic conductor and two metal electrodes, which are unavoidable to access the conductor (Datta 1995). This resistance does not come from scattering processes; instead it comes from the fact that the number of electron modes has to be matched between an electrode and a conductor. Metal electrodes can support infinite numbers of electron modes, whereas a mesoscopic conductor can only allow several discrete electron modes determined by spatial confinement. Hence, most of the electron modes are reflected at the interface of the metal electrode and the conductor and only a few modes are able to flow through the conductor. The reflection of electron modes yields such finite resistance. There needs to be an alternative way to express such resistance beyond Ohm's law. Landauer captured the significance of the wave nature of charge carriers in mesoscopic conductors, and he developed a theory to estimate resistance or conductance in terms of transmission probabilities of propagating electron modes analogous to electromagnetic photon modes. This "Landauer–Büttiker formalism" is one of fundamental theory to express a system conductance in terms of its quantum mechanical transmission probabilities.

3.2.2.4 Landauer–Büttiker Formalism

Suppose a simple 1D ballistic conductor coupled with bulk (3D) electron reservoirs. An adiabatic transition from bulk reservoirs to the device and zero temperature is assumed. Due to the spatial confinement, the allowed modes in the conductor are discretized whereas the modes in the bulk are relatively dense. Therefore, not all modes below the Fermi energy can propagate into the conductor due to energy and momentum conservation, yielding that only certain modes can exist in both regions. Mode reflection at the interface of two dissimilar materials causes finite conductance even with a ballistic conductor. Sometimes this finite resistance is called "contact resistance." In the simplest case, only one channel in the conductor exists. The current I across the conductor with the applied bias voltage V is derived by definition from $I = \int_{E_F}^{E_F+eV} e\rho(E)v_g(E)\mathrm{d}E$ with energy-dependent density of states ρ, group velocity v_g, and Fermi energy E_F. The density of states ρ in 1D is written as $\rho = \dfrac{1}{2\pi\hbar v_g(E)}$. Note that in 1D, there is a magic cancellation of the velocity component, yielding the product of ρ and v_g is constant $1/h$. Hence, the current including spin degeneracy (a factor of 2) is

$$I = \int_{E_F}^{E_F+eV} e\frac{2}{h}\mathrm{d}E = 2\frac{e^2}{h}V \equiv G_Q V$$

where G_Q is denoted as a spin-degenerate quantum unit of conductance when the mode is completely transmitting. If partial transmission occurs with a probability τ, the conductance becomes $G_Q\tau$. Straightforwardly, this scheme can be extended to multi-mode channels whose individual transmission probability is τ_i, where the conductance G is summed over all modes, $G = G_Q \sum_i \tau_i$, known as the "Landauer formula" (Datta 1995, Imry and Landauer 1999). Since the hexagonal Brillouin zone of SWNTs contains two inequivalent K and K' points, four bands degenerate at the same energy reflecting the orbital and spin degeneracy. Therefore, the ideal resistance of SWNTs is $2G_Q = 4e^2/h \sim (6.45\,\mathrm{k\Omega})^{-1}$.

Besides conductance, noise (current fluctuations) has also been actively studied in mesoscopic conductors (Blanter and Büttiker 2000). There are several sources to generate current fluctuations that are pronounced in mesoscopic conductors: internal microscopic random processes caused by thermal fluctuations, scattering and tunneling, and quantum effects (Buckingham 1983, Blanter and Büttiker 2000). Noise theories have been developed intensively since its properties disclose correlations of charge carriers, scattering mechanisms, and quantum coherence, which are essential for understanding the electron transport properties especially in low-dimensional conductors. Recent endeavors in this direction are put into gaining a complete statistical analysis of charge transport under the name of "full counting statistics" (Levitov and Lesovik 1993, Levitov et al. 1996, Kindermann and Nazarov 2002, Nazarov et al. 2002). In the stochastic transport of quantized charged carriers, nonequilibrium current fluctuations are denoted as shot noise, the second moment of characteristic functions in the full counting statistics. What stimulates the advent of full counting statistics is the fact that higher moments provide additional information of systems beyond the first moment, conductance. The focus of this chapter lies on two moments: conductance (the first moment) and shot noise (the second moment) from which SWNTs' electronic properties are quantified.

3.2.3 Fermi-Liquid vs. Non-Fermi-Liquid Theories

Often, a single-particle picture describes macroscopic conductors and their properties sufficiently well with a valid justification of effective screening. However, in lower dimensional systems as devices shrink down, screening among particles becomes insufficient. Consequently, the many-body picture assumes a significant role. In this section, two theoretical models are introduced: the Fermi-liquid (FL) theory and the Tomonaga–Luttinger liquid (TLL) theory. The former explains nominal features within a single particle picture, whereas the latter explains the unique behaviors arising from many-body interactions particularly in 1D conductors.

3.2.3.1 Fermi-Liquid Theory

The FL theory is one of the successful solid-state frameworks to describe the physical properties of weakly interacting many-body condensed matter systems, such as the liquid state of ^{3}He and conductivity in metals and semiconductors (Mahan 2007). In the FL system, there are still non-negligible interactions among particles. However, Landau ingeniously approached the FL system with a hypothesis that interactions are adiabatically switched on. Furthermore, he considered the long wavelength limit, namely, low energy excitations near the Fermi energy. He captured the idea that the interactions would modify the energy dispersion relation, consequently changing the mass of electrons in the system. By introducing the effective mass m^*, which manifests the strength of mutual interactions, Landau established the FL theory within the single particle picture.

The essence of the FL theory is the existence of quasi-particles with the effective mass. Quasi-particles are low-lying elementary excitations consisting of electrons whose density fluctuations arise from the particle interactions. Due to the fact that quasi-particles are formed from electrons, they possess fermionic nature such as obeying Pauli Exclusion Principle. The validity of the FL theory in higher dimensions relies on effective charge screening, which reduces the long-range Coulomb interactions among electrons. Therefore, the FL theory works well to describe transport processes in systems whose interactions are short ranged and isotropic such as metals, semiconductors, and liquid ^{3}He.

3.2.3.2 Tomonaga–Luttinger Liquid Theory

The successful FL theory fails in 1D. The breakdown of the FL theory in 1D conductors can be understood intuitively in terms of inefficient charge screening. The long-range Coulomb interactions survive among strongly correlated electrons. In 1D, any excitation at a particular site spreads over the whole lattice similar to the domino effect. This collectiveness is unique in 1D excitations, and Landau's quasi-particles do not exist in 1D (Giamarchi 2004). A rigorous attempt to describe 1D electron gas systems is formulated as the TLL theory. Tomonaga and Luttinger came up with an exactly solvable model in 1D with insights that collective modes are bosonic in nature. By linearizing the dispersion relation near the Fermi level, low energy properties of system can be extracted (Voit 1994, Giamarchi 2004).

The Fermi surface of 1D consists of two points at $\pm k_F$, where k_F is the Fermi wavenumber. Particle-hole excitations in 1D are only possible near momentum $q = 0$ or $q = 2k_F$ near the Fermi points, whereas any q values below $2k_F$ are allowed for particle-hole excitations in higher dimensions by conserving the energy and momentum. In the limit of $q \to 0$ and low energy, the excitation spectrum is linear, resembling a phonon mode. This resemblance hints that the Hamiltonian of 1D electron gas system can be derived by boson-like phonon displacements as a rather intuitive approach. The positive (negative) slope at $k_F(-k_F)$ corresponds to right (left) moving channels. When interactions are renormalized in a system, the dimensionless quantity, the TLL parameter g, is defined as $g = \left(1 + \dfrac{V_0}{\pi \hbar v_F}\right)^{-\frac{1}{2}}$, where V_0 is the interaction potential. It is a measure of competition between the interaction potential energy and the kinetic energy. In the absence of V_0, g becomes 1, recovering the noninteracting Fermi gas system. On the other hand, $V_0 > \hbar v_F > 0$ for a repulsive Coulomb interaction leads to $g < 1$. The stronger the interactions V_0, the smaller the value of g. Note that g can also be greater than 1 if attractive Coulomb interactions are dominant among the particles. The TLL parameter g emerges in various 1D properties such as the fractional charge ge, the charge mode velocity v_F/g, and the power-exponents of correlation functions.

As one specific example of 1D conductors, metallic SWNTs have been predicted as the TLL system (Egger and Gogolin 1997, Kane et al. 1997). The transport properties in the tunneling regime, where tubes are isolated from metal reservoirs, exhibited the TLL features as the power-scaling conductance by means of the bias voltage and the temperatures (Bockrath et al. 1999). This nonlinear behavior certainly cannot be explained by a noninteracting single-particle picture. Recently, the spectral function from SWNT mats was obtained from angle-integrated photoemission measurements, which was claimed as the direct observation of the TLL features in SWNTs (Ishii et al. 2003).

The TLL parameter g for the SWNT is $g = \left(1 + \dfrac{8e^2}{\pi \hbar v_F} \ln\left(\dfrac{R_S}{R}\right)\right)^{-\frac{1}{2}}$, where R_S is the screening length and R is the radius of the SWNT (Kane et al. 1997). The logarithmic dependence on R_S/R explains that the value of g is rather insensitive to the actual value of R_S (Kane et al. 1997), and the value falls between 0.2 and 0.3 with $v_F = 8 \times 10^7$ cm/s as $R_S/R > 4$. The search of the TLL behavior in SWNTs is actively pursued since the strongly correlated SWNTs serve as a basic ingredient of quantum electron entanglers (Bena et al. 2002, Recher and Loss 2002, Bouchiat et al. 2003, Crépieux et al. 2003).

3.3 Electron Transport in Carbon Nanotubes

3.3.1 Synthesis and Device Fabrication

3.3.1.1 Synthesis

The discovery of MWNTs seems fortuitous in a carbon arc-discharge chamber that was designed to produce fullerenes

(Iijima 1991). Two years later, SWNTs were found by the same arc-discharge method except that catalytic components had been added into the chamber (Iijima and Ichihashi 1993). For a systematic characterization of new materials toward functional device fabrication and quantum nature investigation, efficient synthesis methods have been on demand that aim to isolate individual nanotubes and grow specific types of SWNTs in a controllable way. There are three major synthetic methods for growing SWNTs using catalytic nanoparticles: electric arc-discharge, laser ablation, and chemical vapor deposition (CVD). Among them, the CVD method has been superior in producing high-quality SWNTs.

Typically, the CVD chamber for growing SWNTs in the laboratory consists of a 1 in. diameter tube vessel inserted into a furnace, gas sources of CH_4, H_2 and Ar, a Si-substrate containing catalyst islands, and an exhaust system. Catalysts are essential for designating the location of SWNTs during growth. In 1999, Professor Dai's group managed to synthesize a high-yield of SWNTs near the catalyst islands. Iron-based alumina-supported catalysts were under a carbon feedstock: 99.999% CH_4 and H_2 at the right concentration for 5–7 min at 900°C–1000°C followed by an Ar flush and a cool-down to room temperature (Kong et al. 1998). This work has advanced the SWNT research field in which ballistic transport studies could be performed and prototypes of nanotube-electronics can be built.

The synthesis mechanism of SWNTs in the catalytic CVD method is associated with the details of nanoparticles. Recently, Li et al. (2001) have attempted to assess the role of catalysts and have shown that the diameter of SWNTs indeed closely links to the nanoparticle size based on statistical analysis. The report presents that the synthesis can be understood in three stages: First, nanoparticles as catalysts absorb decomposed carbon atoms from CH_4 or other carbon feedstock in the CVD process. Second, the absorption of carbon atoms to nanoparticles continues until saturation. Once it reaches the saturation point, carbon atoms start to grow outward from the catalysts with a closed-end. Third, an excess carbon supply adds to the carbon precipitation on the surface, yielding finite-length nanotubes in the end. It is reasonable, therefore, that the SWNT diameter would be determined by the nanoparticle size as the initial basis. Although Li et al. (2001) provided valuable information as to the microscopic level of understanding of the synthesis in the catalytic CVD process, the complete controllability to produce tailor-made SWNTs with an expected diameter, chirality (roll-up vector direction), length, position, and orientation on demand is yet to be acquired, which is the present SWNT fabrication challenge. Once this goal is achieved, it is not difficult to imagine that SWNTs would become widely utilized in various applications as electrical, chemical, mechanical, and optical components.

3.3.1.2 Device Fabrication

The configuration of SWNT devices for electron transport measurements resembles conventional semiconductor field-effect transistors, which have three terminals: source, drain, and gate. The fabrication goal is to produce three-terminal isolated SWNT nanotube devices on top of a Si-wafer. Figure 3.4 shows

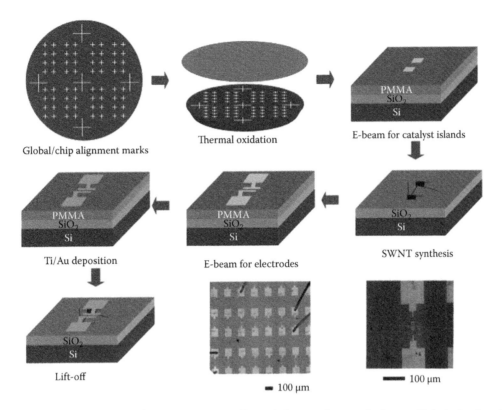

FIGURE 3.4 Schematics of the SWNT device fabrication processes. Two right bottom photographs show multiple devices in a single chip and an individual SWNT device. (From Bockrath, M. et al., *Nature*, 397, 598, 1999. With permission.)

the steps of conventional fabrication processes. Combinations of standard photolithography and electron-beam lithography (EBL) techniques enable one to produce about one hundred devices in a 1 cm × 1 cm Si substrate chip during a short period of time.

The starting material is a 4 in. heavily doped Si-wafer, which serves as a backgate to control the electron density or the Fermi energy of carbon nanotubes. First, global wafer marks and chip marks should be patterned in the blank wafer. These marks are very useful primarily in that the overlapping processes can be performed within the lithographic resolution limit and also in that they draw boundaries of chips in the whole wafer along which the wafer can be cleaved into each chip for further steps. Those alignment marks are patterned by a photolithography recipe: photoresist spin-coating followed by exposure and development, etching process, and removal of the resist. The second step is the thermal oxidation on top of the marked Si-wafer. It is a very critical step to avoid any possible impurities on wafers during this process, since any dirt on the wafers would lead to a current leakage when devices are biased. Therefore, before inserting the wafer to a diffusion furnace, the wafer should be cleaned thoroughly and properly through the diffusion wet bench process. The next task is to pattern catalyst islands at intended locations using the EBL method consisting of polymethylmethacrylate EBL resist coating, exposure, and development. Nanotubes are then synthesized by the aforementioned CVD method with methane and hydrogen gas. The second EBL is processed for

patterning metal electrodes followed by thin metal deposition and a lift-off in acetone.

Once devices are prepared, at room temperature several preliminary characterizations are performed. Atomic force microscopy imaging measurements give the number of SWNTs between the electrodes and differentiate SWNTs from MWNTs based on the tube height. Nanotubes, whose diameters are 1.5–3.5 nm from atomic force microscopy (AFM) images, are presumably considered to be SWNTs according to the statistical analysis at a given recipe. AFM images cannot identify a SWNT, MWNT, or a double-walled nanotube with certainty unlike transmission electron microscopy (TEM). Since TEM requires conducting substrates, transport devices on top of Si-substrates are not quite adequate to TEM at this time. Thus, the determination of SWNTs or MWNTs relies heavily on the statistics of TEM results with synthesized nanotubes on conducting substrates by the same growth recipe.

The current (I) vs. drain-source voltage (V_{ds}) characteristics are measured by varying gate voltages (V_g). This room-temperature electrical characterization has a particular purpose to tell metallic tubes from semiconducting tubes. Metallic tubes have weak or no dependence on V_g in principle since there are always free electrons no matter where the Fermi energy lies. On the other hand, semiconducting tubes, where there is an energy gap between conduction and valence bands, exhibit strong V_g dependent I–V characteristics. When the Fermi energy lies in the energy gap regime, current values through tubes are suppressed to zero. In this way, no or very weak $-V_g$-dependent tubes can be selected, and room-temperature characterization has also another purpose to identify the best and good ohmic contacted devices for low-temperature measurements.

The subsequent two sections describe low-temperature electron transport properties: differential conductance and low-frequency shot noise. The main discussion focuses on three-terminal metallic SWNT devices that are well contacted to electrodes. The device dimension is fixed between 200 and 600 nm by a distance between two electrodes. The Ti/Au, Ti-only, and Pd metal electrodes are used, which feature low-resistance contacts. Metallic SWNTs are considered to have both the elastic and the inelastic mean free path at least on the order of microns at low temperatures. Therefore, the electron transport within 200–600 nm-long SWNTs is believed to be ballistic, where quantum coherence is preserved inside (Kong et al. 2001, Liang et al. 2001).

3.3.2 Differential Conductance

At low temperatures, an interference pattern in differential conductance was observed in well-contacted SWNTs to Ti/Au metals with finite reflection coefficients, as shown in Figure 3.5. This diamond interference pattern arises from quantum coherence along a finite SWNT length (longitudinal confinement) due to the potential barriers at the interfaces with two metal electrodes. The spatial confinement quantizes energy levels and the energy spacing between maxima corresponds to $\Delta E = \hbar v_F / L$, where L is the SWNT length. The inset in Figure 3.5 shows that the diamond structure size is inversely proportional to L.

Liang et al. modeled this system as an electronic analog Fabry-Perot (FP) cavity as a two-channel double-barrier problem using the Landauer–Büttiker formalism within the context of the FL theory. They captured the wave nature of electrons through an isolated nanotube as an electron waveguide. Two interfaces at metal and tubes have a one-to-one correspondence with partially reflecting mirrors in the FP interferometer (Liang et al. 2001). Similar to photons in the FP cavity, electrons would experience multiple reflections between two barriers separating the metal reservoirs from the SWNT before escaping. The approach is to establish three 4×4 scattering matrices at the left and right interfaces and inside the tube. The scattering matrices contain energy-dependent components, satisfying unitary property by Born approximation (Liang et al. 2001). In metallic infinite SWNTs, forward scatterings are dominant in comparison to backscattering and interbranch scattering. Backscattering and interbranch scattering require a big momentum transfer of $2k_F$ between two K points in the Brillouin zone and satisfy the symmetry selection rules between two orthogonal π and π^* bondings among p_z-orbitals. Thus, in principle, such scattering processes inside tubes are prohibited. However, backscattering can occur at the interfaces between metal electrodes and the tube that increases the overall resistance. Meanwhile, inside the ballistic tube, the phase is accumulated over multiple reflections. The differential conductance is calculated by a simple expression,

$$\frac{dI}{dV} = \frac{2e^2}{h} \sum_{i=1}^{2} \text{Tr}(S^{T*}S),$$

taking a trace of the total scattering matrix, S. Indeed, this noninteracting theoretical model reproduces quantum interference diamond patterns within a ±10 mV bias voltage range, supporting that the observation of quantum interference is the evidence of the ballistic transport. Note that the diamond structure is persistent regardless of the bias voltage window in this FL model.

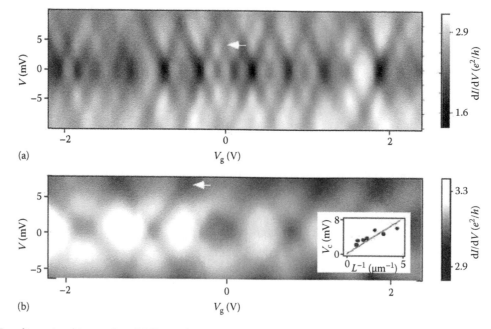

(a)

(b)

FIGURE 3.5 Two-dimensional image plot of differential conductance (dI/dV) as a function of drain-source voltage (V) and gate voltage (V_g). dI/dV is renormalized by e^2/h. There is a clear interference pattern whose feature is dependent on the SWNT length, (a) a 530 nm long SWNT device and (b) a 220 nm long SWNT device. (Reprinted from Liang, W. et al., *Nature*, 411, 665, 2001.)

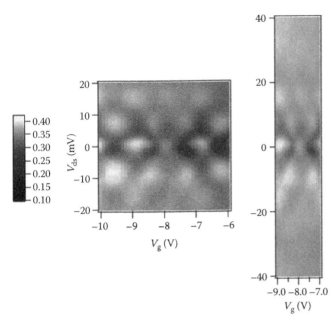

FIGURE 3.6 Density plot of differential conductance in units of $2G_Q$ against two voltages, V_{ds} and V_g. (From Liang, W., *Nature*, 411, 665, 2001. With permission.)

However, FP quantum interference pattern fringe contrast becomes reduced in magnitude at high drain-source voltages (V_{ds}). Figure 3.6 clearly exhibits that the diamond structure disappears as V_{ds} increases above 20 mV. This feature cannot be explained by the standard FL theory, which predicts constant oscillation amplitude regardless of the bias voltage values.

An alternative attempt to explain this experimental data is to model the SWNT device within the TLL theory. It is motivated by the fact that SWNTs have exhibited the features of strong correlations among charge carriers in experiments (Bockrath et al. 1999) and in theories (Egger and Gogolin 1997, Kane et al. 1997, Peça et al. 2003) owing to intrinsic many-body interactions in 1D systems. The device consisting of a ballistic SWNT and two metal electrodes is theoretically simplified as one infinite 1D conductor with two different values of the TLL parameter g for the SWNT and metal electrodes. The interaction is assumed to be strong in

the SWNT ($0 < g < 1$) and weak in the higher dimensional metal reservoirs ($g = 1$) for metals (Peça et al. 2003, Recher et al. 2006). The four conducting transverse channels of the SWNTs in the FL theory are transformed to four collective excitations in the TLL theory: one interacting collective mode of the total charge and three neutral noninteracting collective modes including spin. There are two distinct propagating velocities, $v_c = v_F/g$ (the total charge mode) and v_F (the rest of three modes).

The inter-channel and intra-channel scattering processes are assumed to be allowed to reflect channels partially only at the two barriers. The application of nonzero bias voltages is treated within Keldysh formalism, a powerful method to study nonequilibrium many-body condensed-matter systems in terms of Green's functions (Mahan 2007). The transport properties are computed from correlation and retarded Green's functions (Recher et al. 2006). Three noninteracting modes encounter backscattering at the physical barrier, whereas the interacting mode encounters the momentum-conserving backscattering due to g mismatch at the interfaces in addition to the physical barrier backscattering. It was found that one interacting mode exhibits the power-law behavior, whereas three noninteracting modes still show oscillatory behavior. The overall behavior combining two effects more closely resembles the real experiments. In Figure 3.7, experimental data (left) are compared with theoretical graphs (right) at three V_g voltages (blue, green, red). Note that the tendency of amplitude reduction in experimental data cannot be reproduced by the reservoir heating model (Henny et al. 1999), which asserts that the dissipated power $V_{ds}^2(dI/dV_{ds})$ leads to a bias-voltage-dependent electron temperature (Liang et al. 2001).

Figure 3.7, furthermore, presents the following pronounced features: the period of the oscillations at low V_{ds} depends on the value of V_g, and it becomes elongated at high V_{ds}. The TLL model suggests that the elongated period appears when there is nonzero contribution to the overall oscillation by only interacting mode (propagating at a slower group velocity v_c) while those from the three noninteracting modes (propagating at Fermi velocity v_F) are completely canceled by destructive interference. This V_g-dependent oscillation period in differential conductance has been interpreted as a signature of spin-charge separation

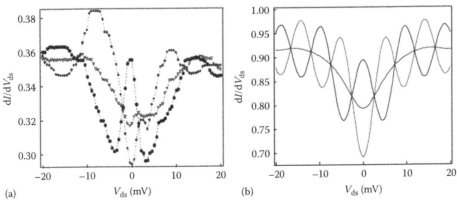

FIGURE 3.7 (a) Experimental and (b) theoretical differential conductance traces at $V_g = -9$ V (diamond in (a), line with peak at V = 0 V in (b)), -8.3 V (triangle in (a), line without oscillation in (b)), and -7.7 V (circle in (a), line with dip at V = 0 V in (b)) at 4 K.

in the SWNT (Peça et al. 2003). A comparison of the primary periods at different V_g values yields the TLL parameter $g \sim 0.22$, which is consistent with the predicted g values in SWNT. Although it seems to be compelling evidence, further experiments focusing on the periodicity with V_{ds} should be performed to be conclusive.

3.3.3 Low-Frequency Shot Noise

The low-frequency shot noise probes the second-order temporal correlation of electron current in the nonequilibrium condition. It often manifests certain microscopic physical mechanisms of the conduction process. When Poisson statistics govern the emission of electrons from a reservoir electrode, in other words, the propagation of an electron has no relationship with the previous or the successive electron, the spectral density of the current fluctuations reaches its full shot noise spectral density, $S_I = 2eI$, where I is the average current. In a mesoscopic conductor, nonequilibrium shot noise occurs due to the random partitioning of electrons by a scatterer, and it may be further modified as a consequence of the quantum statistics and interactions among charged carriers (Blanter and Büttiker 2000). A conventional measure for characterizing the shot noise level in mesoscopic conductors is the Fano factor $F \equiv S_{I,m}/2eI$, the ratio of the measured noise power spectral density $S_{I,m}$ to the full shot noise value ($2eI$). In statistics, the Fano factor is related to be a variance-to-mean ratio, so a Poisson process whose variance equals to mean yields $F = 1$. Despite growing interest in the shot noise properties of the TLLs, current noise measurements in nanotubes have only recently been executed due to the difficulty of achieving highly transparent ohmic contacts and a high signal-to-noise ratio between the weak excess-noise signal and the prevalent background noise (Roche et al. 2002), although the shot noise properties of SWNTs in the tunneling (strong barriers between metals and SWNTs) regime with the TLL features have been reported recently (Onac et al. 2006).

In the ballistic regime, the shot noise properties reveal the TLL features in the SWNT device unambiguously. Two-terminal shot noise measurements were implemented at 4 K. It is critical to calibrate an experimental apparatus to reach an accurate data acquisition and analysis. For this purpose, two current noise sources are placed in parallel: a SWNT device and a full shot noise generator. The well-characterized full shot noise source quantifies complicated cryogenic transfer functions and filters. The standard of such sources is a weakly coupled light emitting diode (LED) and photodiode (PD) pair. At 4 K, the overall coupling efficiency from the LED input current to the PD output current was about 0.1%, which eliminated completely the shot noise squeezing effect due to the constant current operation (Kim and Yamamoto 1997). In order to recover the weak shot noise of the SWNT embedded in the background thermal noise, an AC modulation lock-in technique is implemented and a resonant tank-circuit together with a home-built cryogenic low-noise preamplifier is incorporated (Reznikov et al. 1995, Liu et al. 1998, Oliver et al. 1999). The specific shot noise measurement techniques for the SWNT devices are described in detail by Kim et al. (2007).

Figure 3.8a presents a typical log–log plot (base 10) of S_{SWNT} in V_{ds} at a particular V_g. S_{SWNT} (dot) is clearly suppressed to values below full shot noise S_{PD} (triangle), and it suggests that the relevant backscattering for shot noise is indeed weak. Note that S_{SWNT} and S_{PD} have clearly different scaling slopes vs. V_{ds}. The deviation from the full shot noise in the SWNT device is beyond the scope of the FL theory. Hence, the previous TLL theory for differential conductance is extended to compute the low-frequency shot noise spectral density, $S_{SWNT} = \int dt e^{i\omega t} \langle \{\delta\hat{I}(t), \delta\hat{I}(0)\} \rangle$ with $\delta\hat{I}(t) = \hat{I}(t) - \bar{I}$, the current fluctuation operator, and the anticommutator relation $\{\delta\hat{I}(t), \delta\hat{I}(0)\} = \delta\hat{I}(t)\delta\hat{I}(0) + \delta\hat{I}(0)\delta\hat{I}(t)$ (Recher et al. 2006).

The SWNT noise in the zero-frequency limit is expressed as

$$S_{SWNT} = 2e \coth\left(\frac{eV_{ds}}{2k_BT}\right)I_B + 4k_BT(dI/dV_{ds} - dI_B/dV_{ds})$$

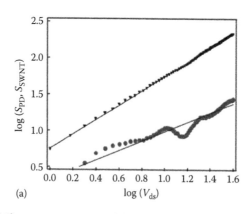

(a)

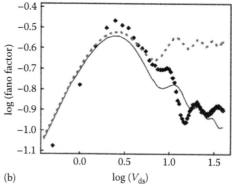

(b)

FIGURE 3.8 (a) Shot noise power spectral density vs. V_{ds} for the LED/PD pair (S_{PD}, triangle) and the SWNT (S_{SWNT}, dot) at $V_g = -7.9$ V. The slopes of S_{PD}, S_{SWNT} are 1 and 0.64, respectively. The inferred g value for the SWNT is 0.16. (b) Fano factor vs. V_{ds} on a log–log scale. The theoretical Fano factor curves where thermal noise $4k_BT(dI/dV_{ds})$ is subtracted are drawn for $g = 1$ (dotted line) and $g = 0.25$ (straight line) at $T = 4$ K. The power exponent α is -0.35 for the measured Fano factor (diamond) at $V_g = -7.9$ V., and the inferred g value is 0.18. The theoretical $g = 1$ (dotted line) plot gives $\alpha \sim 0$ as expected.

where I_B is the backscattered current. In the limit of $eV_{ds} > k_B T$, the SWNT noise spectral power density becomes simplified as $S_{SWNT} = 2eI_B$. The asymptotic behavior of I_B follows the power-law scaling $I_B \sim V_{ds}^{1+\alpha}$ with $\alpha = -(1/2)(1-g)/(1+g)$. The power exponent α is uniquely determined by the TLL parameter g.

The experimental Fano factor $F(V_{ds})$ is displayed on a log–log (base 10) scale in Figure 3.8b. The TLL model predicts that at low bias voltages $eV_{ds} < k_B T < \hbar/2gt_F$, experimental Fano factor F_{exp} is proportional to V_{ds} if we subtract the thermal noise component. In addition, the slope between F_{exp} and V_{ds} is insensitive to g-values in the region of $\log(V_{ds}) < \log(\hbar/2gt_F) \sim 0.47$. On the other hand, if $eV_{ds} > \hbar/2gt_F$, a power-law $F \sim V_{ds}^\alpha$ is expected by assumption that the backscattered current is smaller than the ideal current $2G_Q V_{ds}$. A linear regression analysis of the Fano factor F with V_{ds} in this region, therefore, is another means to obtain the g value. The Fano factors F for $g = 0.25$ (red) and $g = 1$ (yellow) are displayed on a log–log scale in Figure 3.8b. The experimental data (diamonds) agree well with the theoretical Fano factor of $g = 0.25$. The stiffer slope (α) corresponds to a stronger electron–electron interaction. The measured exponent α and inferred g values from the spectral density and the Fano factor from four different devices with various metal electrodes (Ti/Au, Ti-only, Pd) show similar statistics of $\alpha \sim -0.31 \pm 0.047$ and $g \sim 0.26 \pm 0.071$ as derived from several V_g values for each sample. Many-body TLL behavior in the ballistic SWNT is clearly probed in the shot noise properties.

3.4 Summary

The electron transport properties of SWNTs are discussed based on experimental results and theoretical models. The discussion started from the discovery of carbon nanotubes and the history of their research field with a focus on the electron transport area. The noninteracting FL and interacting TLL theories, which have been used widely for bulk and 1D systems respectively, are briefly introduced as background knowledge. The ballistic transport regime is clearly defined, in which no inelastic and elastic scattering occurs, consequently quantum coherence is preserved.

A metallic SWNT is a model system to investigate ballistic transport properties in 1D owing to quantum many-body interactions. The differential conductance and the shot noise have been measured, and their experimental signatures are examined by the FL and the TLL theoretical frameworks. It is clear that quantum coherent properties manifest as an interference pattern in differential conductance, and unique power-law scaling behavior quantifies correlations among charge carriers. A noninteracting picture may describe conductance data in the low bias regime; however, it fails to explain the high-energy regions of experimental data. The TLL theory has explained the qualitative trend of conductance as a function of the drain-source voltage. In addition, it has captured the quantitative information of the strong electron–electron interactions both in conductance and the shot noise quantities. The strength of the interactions is parameterized by the TLL parameter, g, which has been obtained

from the conductance period at various V_g and power-law scaling exponents from both shot noise and the Fano factor. The search of many-body collective phenomena in SWNT devices would provide fundamental physical knowledge in 1D electron transport properties.

3.5 Future Perspective

CNTs have been greatly influential in numerous areas based on extraordinary properties in physics, chemistry, chemical engineering, mechanical engineering, electrical engineering, and more. In particular, conductance and shot noise properties are important because they would ultimately provide the limiting performance of electronic devices. This chapter focuses on the early work of this property on a particular type of carbon nanotubes: metallic SWNTs. However, there has been growing interest with regard to these current fluctuations of diverse SWNT device structures. The knowledge acquired from SWNTs can certainly be transferred to other 1D systems and also carbon nanotube mother material including fullerene and graphene (Bréchignac et al. 2007, Dupas et al. 2007). At present, an experimental investigation on graphene is indeed very exciting and progressively moving forward in comparison with carbon nanotube properties. Therefore, this field will continue to deepen low-dimensional physical knowledge and to build pragmatic devices and systems that can impact everyday lives of human beings (Baughman et al. 2002).

References

Baughman, R. H., Zakhidov, A. A., and de Heer, W. A. 2002. Carbon nanotubes—The route toward applications. *Science* 297: 787–792.

Bena, C., Vishveshware, S., Balents, L., and Fisher, M. P. A. 2002. Quantum entanglement in carbon nanotubes. *Phys. Rev. Lett.* 89: 037901.

Blanter, Ya. M. and Büttiker, M. 2000. Shot noise in mesoscopic conductors. *Phys. Rep.* 336: 1–166.

Bockrath, M., Cobden, D. H., McEuen, P. L. et al. 1997. Single-electron transport in ropes of carbon nanotubes. *Science* 275: 1922–1925.

Bockrath, M., Cobden, D. H., Lu, J. et al. 1999. Luttinger-liquid behavior in carbon nanotubes. *Nature* 397: 598–601.

Bouchiat, V., Chtchelkatchev, N., Feinberg, D., Lesovik, G. B., Martin, T., and Torres, J. 2003. Single-walled carbon nanotube-superconductor entangler: Noise correlations and Einstein-Podolsky-Rosen states. *Nanotechnology* 14: 77–85.

Bréchignac, C., Houdy, P., and Lahmani, M. 2007. *Nanomaterials and Nanochemistry*. Berlin/Heidelberg, Germany: Springer.

Buckingham, M. J. 1983. *Noise in Electronic Devices and Systems*. New York: John Wiley & Sons.

Cao, J., Wang, Q., Rolandi, M., and Dai, H. 2004. Aharonov-Bohm interference and beating in single-walled carbon-nanotube interferometers. *Phys. Rev. Lett.* 93: 216803.

Crépieux, A., Guyon, R., Devillard, P., and Martin, T. 2003. Electron injection in a nanotube: Noise correlations and entanglement. *Phys. Rev. B* 67: 205408.

Datta, S. 1995. *Electronic Transport in Mesoscopic Systems.* Cambridge, U.K.: Cambridge University Press.

Dekker, C. 1999. Carbon nanotubes as molecular quantum wires. *Phys. Today* 5: 22–28.

Dupas, C., Houdy, P., and Lahmani, M. 2007. *Nanoscience— Nanotechnologies and Nanophysics.* Berlin/Heidelberg, Germany: Springer.

Ebbesen, T. W. and Ajayan, P. M. 1992. Large-scale synthesis of carbon nanotubes. *Nature* 358: 220–222.

Ebbesen, T. W., Ajayan, P. M., Hiura, H., and Tanigaki, K. 1994. Purification of nanotubes. *Nature* 367: 519.

Ebbesen, T. W., Lezec, T. H., Hiura, H., Bennett, J. W., Ghaemi, H. F., and Thio, T. 1996. Electrical conductivity of individual carbon nanotubes. *Nature* 382: 54–56.

Egger, R. and Gogolin, A. O. 1997. Effective low-energy theory for correlated carbon nanotubes. *Phys. Rev. Lett.* 79: 5082–5085.

Ge, M. and Sattler, K. 1993. Vapor-condensation generation and {STM} analysis of fullerene tubes. *Science* 260: 515–518.

Giamarchi, T. 2004. *Quantum Physics in One Dimension.* Oxford, U.K.: Oxford University Press.

Harris, P. J. F. 1999. *Carbon Nanotubes and Related Structures: New Materials for the Twenty-First Century.* Cambridge, U.K.: Cambridge University Press.

Henny, M., Oberholzer, S., Strunk, C., and Schonenberger, C. 1999. 1/3-Shot-noise suppression in diffusive nanowires. *Phys. Rev. B* 59: 2871–2880.

Iijima, S. 1991. Helical microtubules of graphitic carbon. *Nature* 354: 56–58.

Iijima, S. and Ichihashi, T. 1993. Single-shell carbon nanotubes of 1-nm diameter. *Nature* 363: 603–605.

Iijima, S., Ichihashi, T., and Ando, Y. 1992. Pentagons, heptagons and negative curvature in graphite microtubule growth. *Nature* 356: 776–778.

Imry, Y. and Landauer, R. 1999. Conductance viewed as transmission. *Rev. Mod. Phys.* 71: S306–S312.

Ishii, H., Kataura, H., Shiozawa, H. et al. 2003. Direct observation of Tomonaga-Luttinger-liquid state in carbon nanotubes at low temperatures. *Nature* 426: 540–544.

Jarillo-Herrero, P., Kong, J., Van der Zant, H. S. J. et al. 2005. Orbital Kondo effect in carbon nanotubes. *Nature* 434: 484–488.

Jarillo-Herrero, P., van Dam, J., and Kouwenhoven, L. 2006. Quantum supercurrent transistors in carbon nanotubes. *Nature* 439: 953–956.

Javey, A., Guo, J., Paulsson, M. et al. 2004. High-field quasi ballistic transport in short carbon nanotubes. *Phys. Rev. Lett.* 92: 106804.

Javey, A., Guo, J., Wang, Q., Lundstrom, M., and Dai, H. 2003. Ballistic carbon nanotube field-effect transistors. *Nature* 424: 654–657.

Jorio, A., Dresselhaus, G., and Dresselhaus, M. S. 2008. *Carbon Nanotubes: Advanced Topics in the Synthesis, Structure, Properties and Applications.* Berlin/Heidelberg, Germany: Springer.

Kane, C., Balents, L., and Fisher, M. P. A. 1997. Coulomb interactions and mesoscopic effects in carbon nanotubes. *Phys. Rev. Lett.* 79: 5086–5089.

Kim, J. and Yamamoto, Y. 1997. Theory of noise in P-N junction light emitters. *Phys. Rev. B* 55: 9949–9959.

Kim, N. Y., Recher, P., Oliver, W. D., Yamamoto, Y., Kong, J., and Dai, H. 2007. Tomonaga-Luttinger liquid features in ballistic single-walled carbon nanotubes: Conductance and shot noise. *Phys. Rev. Lett.* 99: 036802.

Kindermann, M. and Nazarov, Yu. V. 2002 Full counting statistics in electric circuits. In *Quantum Noise in Mesoscopic Physics,* eds. Yu. V. Nazarov and Ya. M. Blanter, pp. 403–429. Dordrecht, the Netherlands: Kluwer Academic Publishers.

Kong, J., Soh, H. T., Cassell, A. M., Quate, C. F., and Dai, H. 1998. Synthesis of individual single-walled carbon nanotubes on patterned silicon wafers. *Nature* 395: 878–881.

Kong, J., Yenilmez, E., Tombler, T. W. et al. 2001. Quantum interference and ballistic transmission in nanotube electron waveguides. *Phys. Rev. Lett.* 87: 106801.

Levitov, L. S., Lee, H., and Lesovik, G. B. 1996. Electron counting statistics and coherent states of electric current. *J. Math. Phys.* 37: 4845–4866.

Levitov, L. S. and Lesovik, G. B. 1993. Charge distribution in quantum shot noise. *JETP Lett.* 58: 230–235.

Li, Y., Kim, W., Zhang, Y., Rolandi, M., Wang, D., and Dai, H. 2001. Growth of single-walled carbon nanotubes from discrete catalytic nanoparticles of various sizes. *J. Phys. Chem. B* 105: 11424–11431.

Liang, W., Bockrath, M., Bozovic, D., Hafner, J. H., Tinkham, M., and Park, H. 2001. Fabry-Perot interference in a nanotube electron waveguide. *Nature* 411: 665–669.

Liang, W., Bockrath, M., and Park, H. 2002. Shell filling and exchange coupling in metallic single-walled carbon nanotubes. *Phys. Rev. Lett.* 88: 126801.

Liu, R. C., Odom, B., Yamamoto, Y., and Tarucha, S. 1998. Quantum interference in electron collision. *Nature* 391: 263–265.

Mahan, G. D. 2007. *Many-Particle Physics.* New York: Springer.

Minot, E. D., Yaish, Y., Sazonova, V., and McEuen, P. L. 2004. Determination of electron orbital magnetic moments in carbon nanotubes. *Nature* 428: 536–539.

Mintmire, J. W., Dunlap, B. I., and White, C. T. 1992. Are fullerene tubules metallic? *Phys. Rev. Lett.* 68: 631–634.

Nazarov, Yu. V. and Bagrets, D. A. 2002. Circuit theory for full counting statistics in multiterminal circuits. *Phys. Rev. Lett.* 88: 196801.

Nygard, J., Cobden, D. H., and Lindelof, P. E. 2000. Kondo physics in carbon nanotubes. *Nature* 408: 342–346.

Oliver, W. D., Kim, J., Liu, R. C., and Yamamoto, Y. 1999. Hanbury Brown and Twiss-type experiment with electrons. *Science* 284: 299–302.

Onac, E., Balestro, F., Trauzettel, B., Lodewijk, C. F. J., and Kouwenhoven, L. P. 2006. Shot-noise detection in a carbon nanotube quantum dot. *Phys. Rev. Lett.* 96: 026803.

Park, J., Rosenblatt, S., Yaish, Y. et al. 2004. Electron-phonon scattering in metallic single-walled carbon nanotubes. *Nano Lett.* 4: 517–520.

Peça, C. S., Balents, L., and Wiese, K. J. 2003. Fabry-Perot interference and spin filtering in carbon nanotubes. *Phys. Rev. B* 68: 205423.

Postma, H. W. Ch., Teepen, T, Yao, Z., Grifoni, M., and Dekker, C. 2001. Carbon nanotube single-electron transistors at room temperature. *Science* 293: 76–79.

Recher, P., Kim, N. Y., and Yamamoto, Y. 2006. Tomonaga-Luttinger liquid correlations and Fabry-Perot interference in conductance and finite-frequency shot noise in a single-walled carbon nanotube. *Phys. Rev. B* 74: 235438.

Recher, P. and Loss, D. 2002. Superconductor coupled to two Luttinger liquids as an Entangler for electron spins. *Phys. Rev. B* 65: 165327.

Roche, P.-E., Kociak, M., Gueron, S., Kasumov, A., Reulet, B., and Bouchiart, H. 2002. Very low shot noise in carbon nanotubes. *Euro. Phys. J. B* 28: 217–222.

Reznikov, M., Heiblum, H., Shtrikman, H., and Mahalu, D. 1995. Temporal correlation of electrons: Suppression of shot noise in a ballistic quantum point contact. *Phys. Rev. Lett.* 75: 3340–3343.

Saito, R., Dresselhaus, G., and Dresselhaus, M. S. 1998. *Physical Properties of Carbon Nanotubes*. London, U.K.: Imperial College Press.

Saito, R., Fujita, M., Dresselhaus, G., and Dresselhaus, M. S. 1992. Electronic structure of chiral graphene tubules. *Appl. Phys. Lett.* 60: 2204–2206.

Soh, H. T., Quate, C. F., Morpurgo, A. F., Marcus, C. M., Kong, J., and Dai, H. 1999. Integrated nanotube circuits: Controlled growth and ohmic contacting of single-walled carbon nanotubes. *Appl. Phys. Lett.* 75: 627–629.

Tans, S. J., Devoret, M. H., Dai, H. et al. 1997. Individual single-wall carbon nanotubes as quantum wires. *Nature* 386: 474–477.

Tans, S. J., Verschueren, A. R. M., and Dekker, C. 1998. Room-temperature transistor based on a single carbon nanotube. *Nature* 393: 49–52.

Tohji, K., Goto, T., Takahashi, H., Shinoda, Y., and Shimizu, N. 1996. Purifying single-walled nanotubes. *Nature* 383: 679.

Tsang, S. C., Chen, Y. K., Harris, P. J. F., and Green, M. L. H. 1994. A simple chemical method of opening and filling carbon nanotubes. *Nature* 372: 159–162.

Voit, J. 1994. One-dimensional Fermi liquids. *Rep. Prog. Phys.* 57: 977–1116.

Wallace, P. R. 1947. The band theory of graphite. *Phys. Rev.* 71: 622–634.

Wilder, J. W. G., Venema, L. C., Rinzler, A. G., Smalley, R. E., and Dekker, C. 1998. Electronic structure of atomically resolved carbon nanotubes. *Nature* 391: 59–52.

Yao, Z., Kane, C. L., and Dekker, C. 2000. High-field electrical transport in single-wall carbon nanotubes. *Phys. Rev. Lett.* 84: 2941–2944.

Yao, Z., Postma, H. W. Ch., Balents, L., and Dekker, C. 1999. Carbon nanotube intramolecular junctions. *Nature* 402: 273–276.

4

Thermal Conductance of Carbon Nanotubes

Li Shi
The University of Texas at Austin

4.1 Introduction

Carbon nanotubes are a class of unique nanostructures that are being explored for applications in nanoelectronics, interconnects, sensors, biomedicine, and energy applications. They also provide an ideal low-dimensional system for investigating nanoscale thermal transport physics that often dictates the performance and reliability of functional devices made of carbon nanotubes. For example, the current-carrying capability of nanotubes depends on the temperature rise during self-heating, which is in turn limited by the thermal property of nanotubes. This chapter starts with a review of the fundamentals of thermal conduction in solids and a discussion of the theoretical descriptions of the unique features of thermal conduction in carbon nanotubes.

4.1.1 Fundamentals of Thermal Conduction in a Solid

The rate of heat conduction (Q) along a solid bar can be obtained by using the phenomenological Fourier's law

$$Q = G\Delta T \tag{4.1}$$

where G and ΔT are the thermal conductance and the temperature difference between the two ends of the solid bar, respectively. The thermal conductance can be calculated from the thermal conductivity (κ), cross-sectional area (A), and length (L) of the solid bar according to

$$G = \frac{\kappa A}{L} \tag{4.2}$$

The thermal conductance and thermal conductivity consist of contributions from different energy carriers including electrons and phonons, the latter of which are the energy quanta of crystal vibration waves, i.e.,

$$G = G_e + G_{ph} \tag{4.3a}$$

and

$$\kappa = \kappa_e + \kappa_{ph} \tag{4.3b}$$

where the subscripts e and ph denote the electron contribution and phonon contribution, respectively.

4.1.2 Effects of Size Confinement on Heat Conduction

Both thermal conductance and thermal conductivity depend on the crystal structure and the temperature. While the thermal conductance depends on the sample dimension, the thermal conductivity is independent of the sample size until the size is scaled down to the fundamental length scales such as the wavelengths and scattering mean free paths of electrons and phonons, which are both waves and particles.

The electron wavelength can be calculated according to

$$\lambda_e = \frac{h}{\sqrt{2m^* E}} \tag{4.4}$$

where
  h is the Planck's constant
  m^* is the electron effective mass
  E is the electron energy

The as-calculated electron wavelength varies from picometer (pm) for high energy free electrons to the order of 100 nm for electrons in some semimetals and semiconductors such as bismuth and indium antimonide that possess a very small effective mass.

In comparison, the phonon wavelength (λ_{ph}) can range in size from as small as two times the primitive unit cell size (a) of the lattice to as large as the size of a crystal. At low temperatures (T), the dominant wavelength ($\lambda_{ph,max}$) of phonons that carries the maximum spectral energy density is given by an expression that resembles Wien's displacement law for photons, i.e.,

$$\lambda_{ph,max} T \approx \frac{h v_g}{2.8 k_B} \tag{4.5}$$

where
  v_g is the phonon group velocity
  k_B is Boltzmann's constant

For a group velocity of about 3000 m/s, Equation 4.5 is reduced to $\lambda_{ph,max} T \approx 50$ nm-K. Equation 4.5 fails at sufficiently high temperatures, where $\lambda_{ph,max}$ approaches the minimum allowable wavelength on the order of $2a$, corresponding to the edge of the first Brillouin zone of the reciprocal lattice.

As particles, both electrons and phonons are scattered by electrons and phonons, crystal defects, and boundaries. The scattering mean free path (l) is the average distance that electrons or phonons can travel before being scattered, and it ranges from atomic spacing in amorphous materials to over micrometers in high-quality crystals at low temperatures.

In the diffusive transport limit where the length scale is much longer than the mean free path, the thermal conductivity contribution of each energy carrier type can be calculated using a simple result from the kinetic theory (Chen 2005)

$$\kappa \approx C v_x^2 \tau \tag{4.6}$$

where
  C is the specific heat
  v_x is the average velocity component of the energy carrier along the transport direction
  τ is the scattering mean free time

When the cross-sectional dimension (d) of the solid bar is reduced to the order of mean free paths, the scattering mean free path can be modified from the bulk value (l_{bulk}). According to the Matthiessen's rule,

$$l^{-1} = l_{bulk}^{-1} + l_s^{-1} \tag{4.7}$$

where l_s is the surface scattering mean free path, which can be estimated with the use of the specularity parameter (p) of the surface according to (Ziman 1962, Dames and Chen 2004, Moore et al., 2008)

$$l_s = \frac{1+p}{1-p} d \tag{4.8}$$

Depending on the relative magnitude of the surface roughness and the wavelength of the energy carrier, the specularity parameter of a surface with random surface roughness ranges between 0 and 1. If the surface is smooth and specular, $p = 1$ and l_s approaches infinity. Specular surface scattering does not alter the mean free path or thermal conductivity. If the surface is rough and diffuse, $p = 0$ and l_s approaches d. Diffuse surface scattering can greatly suppress the mean free path and the thermal conductivity.

When the length of the solid bar is reduced to be shorter than the scattering mean free path, energy carriers can travel from one end to the other end in a ballistic manner without encountering scattering along the path. In the ballistic thermal transport regime, the thermal conductance is finite because the energy carriers are scattered at the two end contacts when they enter or exit the short solid bars from or into the two thermal reservoirs at the two ends. The maximum but finite thermal conductance found in the ballistic limit is the ballistic thermal conductance, which is in effect the reciprocal of the minimum thermal resistance achievable at the two contacts to the solid bar. In the ballistic transport regime, the thermal conductivity is not well defined because thermal conductivity is a property inherently associated with diffusive transport and is based on the assumption of a local thermal equilibrium among the energy carriers via scattering. Nevertheless, an effective thermal conductivity can still be calculated from the ballistic thermal conductance according to Equation 4.2. Apparently, the as-calculated thermal conductivity is length dependent in the ballistic transport regime because the ballistic thermal conductance is the contact thermal conductance that is length independent to the first order.

When the lateral dimension (d) of the solid is reduced to be comparable with the carrier wavelength of the energy carriers, only few discrete transverse wavevector states can be supported in the solid that acts as a waveguide of the energy carrier. On the other hand, the longitudinal wavevector states can be quasi-continuous if the length of the solid is much longer than the carrier wavelength. Consequently, thermal energy is carried by few one-dimensional (1D) waveguide modes that result in well-separated energy subbands in the solid. The energy quantization in the transverse direction modifies the electronic band structure or phonon dispersion. Consequently, the density of states of electrons or phonons as well as the group velocity can be different from those in the bulk, leading to modified thermal property in the quantum transport regime.

4.2 Theory of Thermal Conduction in Carbon Nanotubes

4.2.1 Structure of Carbon Nanotubes

Carbon nanotubes are made of graphitic cylinders (Dresselhaus et al., 2001). One distinguishes between multi-walled carbon nanotubes (MWNTs), consisting of a series of coaxial graphite cylinders, and single-walled carbon nanotubes (SWNTs) with a one atom thick wall, usually a small number (20–40) of carbon atoms along the circumference, and microns along the cylinder axis. The nanotube is specified by the chiral vector $\mathbf{C_h}$

$$\mathbf{C_h} = n\mathbf{a_1} + m\mathbf{a_2} \equiv (n,m) \qquad (4.9)$$

which is often described by the pair of indices (n, m) that denote the number of unit vectors $n\mathbf{a_1}$ and $m\mathbf{a_2}$ in the hexagonal honeycomb lattice contained in the vector $\mathbf{C_h}$. The chiral vector $\mathbf{C_h}$ makes an angle θ, called the chiral angle, with the zigzag or $\mathbf{a_1}$ direction. A nanotube with $m = 0$ is called a zigzag tube; while one with $n = m$ is called an armchair tube. Other nanotubes are called chiral tubes. The axis of the zigzag nanotube corresponds to $\theta = 0°$, while the axis of an armchair nanotube corresponds to $\theta = 30°$, and the general chiral nanotube axis corresponds to $0° \leq \theta \leq 30°$. The diameter of the SWNT depends on the indices n and m and is usually in the range of 1–2 nm. For comparison, a MWNT has a typical diameter of about 10 nm and consists of a few tens of graphitic layers with an inter-layer spacing of about 0.34 nm.

4.2.2 Electronic Thermal Conductance of Carbon Nanotubes in the Ballistic Regime

When a graphene sheet is rolled up into a tube, the allowed electron wavevector components perpendicular to the tube axis become quantized, resulting in 1D subbands in the electronic band structure of a SWNT (Dresselhaus et al., 2001). The energy (E) band structure in the wavevector (k) space for the subband near the Fermi energy E_f is shown in Figure 4.1 for a metallic and a semiconducting nanotube, with the dark lines representing filled electron states. The subband structures depend on the indices n and m and are different for different tubes, resulting in both metallic SWNTs with a vanishing bandgap and semiconducting SWNTs. It has been shown that the metallic bandstructure in a (n, m) SWNT is achieved when

$$2n + m = 3q \qquad (4.10)$$

where q is an integer (Dresselhaus and Eklund 2000). All armchair carbon nanotubes with $n = m$ are metallic, satisfying Equation 4.10. While SWNTs can either be metals or semiconductors, MWNTs are mostly metallic.

At room temperature, electrons near the Fermi level in a metallic SWNT can have a long electron mean free path (Bockrath et al., 1997, Ando and Nakanishi 1998, Ando et al., 1998, Tans et al., 1998, McEuen et al., 1999, Javey et al., 2003, Park et al., 2004, Gao et al., 2005, Purewal et al., 2007). If the length of a SWNT is shorter than the mean free path, electrons can flow from one end of the tube to the other end without scattering with phonons, defects, and boundaries. The ballistic electron transport phenomenon is illustrated in Figure 4.2a. On the other hand, transport measurements of MWNTs (Collins et al., 2001b) and SWNTs (Yao et al., 2000, Collins et al., 2001a, Park et al., 2004, Pop et al., 2005, Foa Torres and Roche 2006, Lazzeri and Mauri 2006, Sundqvist et al., 2007) under high bias voltages have suggested the scattering of the energetic electrons by optical and zone boundary phonons, giving rise to diffusive transport, as illustrated in Figure 4.2b.

The unique transport property of electrons near the Fermi level in a metallic SWNT arises from its unusual lattice structure.

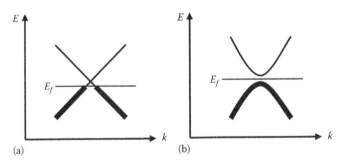

(a) (b)

FIGURE 4.1 The electron subbands near Fermi energy for a (a) metallic and (b) semiconducting SWNT.

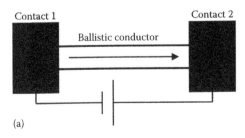 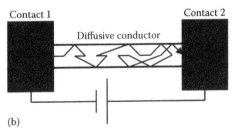

(a) (b)

FIGURE 4.2 Electron trajectory in a (a) ballistic and (b) diffusive conductor between two contacts.

The SWNT has a periodic structure in the circumferential direction with an atomically smooth surface, resulting in the absence of diffuse electron-boundary scattering in the bulk of the tube. Furthermore, although the metallic and semiconducting SWNTs are nearly identical in structure with the amount of disorder likely very similar, a long-length scale disorder breaks the semiconducting tubes into a series of dots with large barriers and a dramatically reduced conductance; whereas, metallic tubes are insensitive to disorder and remain near-perfect 1D conductors (McEuen et al., 1999).

Ballistic electron transport in a short metallic SWNT at a low electric field gives rise to finite electrical conductance or resistance. For the perfect ballistic conductor shown in Figure 4.2a, the current is carried in the contacts by infinitely many modes, but inside the size-confined ballistic conductors, the current is carried by only a few 1D energy subbands due to quantization. This requires a redistribution of the current among the current-carrying modes at the contacts, leading to contact resistance. According to the Landauer formula, for a conductor with currents carried by 1D subbands, the electric current can be calculated as (Datta 1997)

$$I = \sum_m 2 \int_0^\infty \frac{dk}{2\pi} e v_m(k)[f(E_m(k),\mu_1) - f(E_m(k),\mu_2)]T_{r,m}(k) \quad (4.11)$$

where
the factor of 2 accounts for the spin degeneracy
e is the elemental charge
m is an 1D electron subband
$v_m(k) = dE_m(k)/\hbar dk$ is the electron group velocity
$f(E,\mu) = (\exp((E-\mu)/k_B T) + 1)^{-1}$ is the Fermi-Dirac distribution function
μ is the electrochemical potential, the subscripts 1 and 2 denote the two contacts
T_r is the transmission coefficient that represents the average probability that an electron injected at one end of the conductor will transmit to the other end. For a ballistic conductor, $T_r = 1$.

After changing the integration variable in Equation 4.11 from wavevector k to energy E, Equation 4.11 becomes

$$I = \sum_m 2 \int_0^\infty dE \frac{e}{h} [f(E_m,\mu_1) - f(E_m,\mu_2)]T_{r,m}(E) \quad (4.12)$$

In the low-bias, linear-response regime

$$f(E_m,\mu_1) - f(E_m,\mu_2) = \frac{\partial f}{\partial \mu}(\mu_1 - \mu_2) = -\frac{\partial f}{\partial E}(\mu_1 - \mu_2) \quad (4.13)$$

Hence, the low-bias electrical conductance that would be measured between two perfect contacts is

$$g = \frac{I}{(\mu_1 - \mu_2)/e} = -\frac{2e^2}{h} \sum_m \int_0^\infty dE \frac{\partial f}{\partial E} T_{r,m}(E) \quad (4.14)$$

If the transmission coefficient T_r is constant and the number of 1D subbands crossing the Fermi level is M, Equation 4.14 is simplified to the following for the metallic case of $\mu \gg k_B T$:

$$g = 2g_0 M T_r \quad (4.15)$$

where $g_0 = e^2/h$ is the universal quantum of electrical conductance. The corresponding low-bias electrical resistance is

$$R = \frac{1}{g} = \frac{h}{2e^2 M T_r} \approx \frac{12.9 \text{ k}\Omega}{M T_r} \quad (4.16)$$

In a metallic SWNT, currents are carried by two 1D subbands, i.e., $M = 2$, leading to a ballistic resistance of $6.5\,\text{k}\Omega$ that is in effect the minimum contact resistance to the SWNT for perfect contacts.

The electronic thermal conductance of a 1D conductor in the ballistic regime can be obtained in a similar manner based on the Landauer formulism. The heat current carried by the 1D subbands is obtained by replacing the charge e in Equation 4.12 with thermal energy $(E - \mu)$ carried by an electron (Chen 2005):

$$Q = \sum_m 2 \int_0^\infty dE \frac{(E-\mu)}{h} [f(E_m,T_1) - f(E_m,T_2)]T_{r,m}(E) \quad (4.17)$$

In the linear-response regime where the temperature difference between the two ends of the conductor is small,

$$f(E_m,T_1) - f(E_m,T_2) = \frac{df}{dT}(T_1 - T_2) \quad (4.18)$$

The electronic thermal conductance is given by

$$G_e = \frac{Q}{T_1 - T_2} = \sum_m 2 \int_0^\infty dE \frac{(E-\mu)}{h} \frac{df}{dT} T_{r,m}(E) \quad (4.19)$$

After changing the integration variable to the dimensionless form of $x = (E - \mu)/k_B T$ and assuming $\mu \gg k_B T$ for the metallic limit and a constant transmission coefficient T_r, Equation 4.19 becomes

$$G_e = \frac{2k_B^2 T}{h} T_r \sum_m \int_{-\infty}^\infty dx \frac{x^2 e^x}{(1+e^x)^2} \quad (4.20)$$

When the number of subbands crossing the Fermi level is M, the quantized electronic thermal conductance is

$$G_e = 2G_0 M T_r \quad (4.21)$$

where $G_0 \equiv \pi^2 k_B^2 T/3h$ is the universal quantum of thermal conductance. The ratio of the thermal conductance quantum to the electrical conductance quantum satisfies the Wiedemann-Franz law.

For a metallic SWNT, two linear subbands crossing the Fermi level contribute to an electronic thermal conductance $G_e = 4T_rG_0$. For an intrinsic semiconducting SWNT with an energy gap of the order of 0.1 eV, the electronic thermal conductance is expected to vanish roughly exponentially when the temperature decreases to zero (Yamamoto et al., 2004).

4.2.3 Phonon Transport in Carbon Nanotubes

In addition to the unique electronic structures, nanotubes also possess an unusual phonon dispersion relationship between frequency ω and wavevector k. Among 66 distinct phonon branches for a (10, 10) SWNT (Dresselhaus and Eklund 2000), only four acoustic modes with a linear dispersion have a zero frequency at the Γ point ($k = 0$). These are a longitudinal acoustic mode, doubly degenerate transverse acoustic modes, and a twisting acoustic mode. Among other higher energy phonon branches, the lowest phonon mode is an optical mode that is doubly degenerate and has a nonzero energy of $\hbar\omega_{op} = 2.1$ meV at the Γ point.

The phonon thermal conductance of a short SWNT in the ballistic phonon transport regime can be obtained in a similar manner as the electronic thermal conductance based on the Landauer formulism. After replacing the Fermi-Dirac distribution function in Equation 4.17 with the Bose-Einstein distribution, i.e., $< n(\omega,T) > = [\exp(\hbar\omega/k_BT) - 1]^{-1}$, the heat current carried by all the phonon branches can be written as (Yamamoto et al., 2004)

$$Q = \sum_m \int_{\omega_m^{\min}}^{\omega_m^{\max}} \frac{d\omega_m}{2\pi} \hbar\omega_m [< n(\omega_m, T_1) > - < n(\omega_m, T_2) >] T_{r,m}(\omega) \quad (4.22)$$

where the integration is carried out between the lower and upper frequency limits of the phonon branch. A procedure similar to Equations 4.18 through 4.20 can be used to obtain the phonon thermal conductance

$$G_{ph} = \frac{k_B^2 T}{h} T_{r,ph} \sum_m \int_{x_m^{\min}}^{x_m^{\max}} dx \frac{x^2 e^x}{(e^x - 1)^2} \quad (4.23)$$

where a constant phonon transmission coefficient $T_{r,ph}$ is assumed and $x = \hbar\omega/k_BT$. At a sufficiently low temperature, only the four acoustic branches are occupied. For these four acoustic branches, $x_m^{\min} = 0$ and $x_m^{\max}$ approaches infinity at low temperatures. Under this condition, the quantized phonon thermal conductance becomes

$$G_{ph} = \frac{4k_B^2 T}{h} T_r \int_0^\infty dx \frac{x^2 e^x}{(e^x - 1)^2} = 4G_0 T_{r,ph} \quad (4.24)$$

Therefore, at the limit of ballistic electron and phonon transport where the transmission coefficient for electrons and phonons are unity, electrons and phonons each contribute to $4G_0$ for a metallic SWNT at low temperatures. As the temperature increases,

higher energy phonon branches are occupied, making the phonon thermal conductance much higher than the electronic thermal conductance.

The above discussion is focused on the ballistic transport regime relevant for a short nanotube at low temperatures. For a long nanotube in the diffusive transport regime, the thermal conductivity of the nanotube is dominated by the phonon contribution at a sufficiently high temperature. Diffusive phonon transport in a nanotube has been studied using molecular dynamics (MD) simulation (Berber et al., 2000). The calculated thermal conductivity of nanotubes has been compared with diamond and graphite. Because of the stiff SP^3 bonds and the high speed of sound, monocrystalline diamond is one of the best thermal conductors, with the highest room temperature thermal conductivity of 3320 W/m-K reported for isotopically enriched ^{12}C diamond (Anthony et al., 1990). Because nanotubes are held by even stronger SP^2 bonds as in the basal plane of bulk graphite, they are expected to be very efficient thermal conductors. Because of the atomically smooth surface and the absence of interlayer phonon scattering or coupling to soft phonon modes of a medium, a free-standing SWNT may give rise to higher thermal conductivity than the room-temperature in-plane value of about 1950 W/m-K for graphite (Holland et al., 1966), which is also held by SP^2 bonds. An equilibrium MD prediction states that the room-temperature thermal conductivity exceeds 6000 W/m-K for a free-standing (10, 10) nanotube (Berber et al., 2000). However, when nanotubes are entangled in a mat, the thermal conductivity could be suppressed by intertube phonon coupling, similar to the case of graphite, where interlayer phonon coupling quenches its thermal conductivity.

4.3 Measurements of Thermal Conductance of Carbon Nanotubes

4.3.1 Thermal Measurement of Carbon Nanotube Bundles

The intriguing thermal transport phenomena in nanotubes have motivated intense experimental investigations. Hone et al. first measured the thermal conductivity of a nanotube rope with dimensions of 5 mm × 2 mm × 2 mm using a steady-state comparative method. Based on an estimated filling fraction of the nanotube rod, they obtained a thermal conductivity of only 35 W/m-K for the densely packed mat of the nanotube rod (Hone et al., 1999). Subsequently, a higher room-temperature thermal conductivity value exceeding 200 W/m-K was measured on a dense nanotube rope aligned in high magnetic fields (Hone et al., 2000). The low thermal conductivity can be attributed to contact thermal resistance between interconnected nanotubes in the mat and coupling of phonon modes in adjacent nanotubes in the mat. Interestingly, a linear temperature dependence of the thermal conductivity was observed in the nonaligned sample at temperatures below 30 K. The linear dependence was attributed to the occupation of only the acoustic phonon modes in a SWNT at low temperatures and a constant phonon mean free path dominated by boundary scattering. As the temperature increases, the thermal conductivity was found to increase faster than the linear temperature dependence because higher energy phonon modes in the nanotube were occupied.

4.3.2 Direct Thermal Conductance Measurement of Individual Carbon Nanotubes

For direct thermal measurement of individual nanotubes, a suspended micro-device was developed (Kim et al., 2001, Shi 2001, Shi et al., 2003). Figure 4.3 shows scanning electron microscopy (SEM) images of one of the various designs of the microdevice. This design consisted of two adjacent, low-stress silicon nitride (SiN_x) membranes suspended with six $0.5\,\mu m$-thick, $420\,\mu m$-long, and $2\,\mu m$-wide silicon nitride beams. The length and number of the supporting beams were varied in different designs. One $50\,nm$-thick and $200\,nm$-wide platinum resistance thermometer (PRT) serpentine was patterned on each membrane. The PRT was connected to $200\,\mu m \times 200\,\mu m$ Pt bonding pads on the substrate via $2\,\mu m$ wide Pt leads on the long SiN_x beams. Depending on the number of supporting beams for each membrane, up to two additional $2\,\mu m$ wide Pt electrodes were patterned on each membrane, providing electrical contact to a nanotube or nanowire sample bridging between the two membranes. In the design shown in Figure 4.3, a through-wafer hole was etched under the suspended structure to allow transmission electron microscopy (TEM) characterization of the nanostructure sample bridging the two membranes.

An individual nanotube or nanowire sample can be placed between the two suspended membranes by several methods. In one method, a sharp probe was used to pick up an individual nanotube. The probe was manipulated to place the nanostructure between the two membranes. The process requires a high-resolution optical microscope or nanomanipulator in a scanning electron microscope. This method was employed for placing MWNT bundles and individual MWNTs between the two membranes (Kim et al., 2001), as shown in Figure 4.4a.

Alternatively, a suspension of the nanostructures in isopropanol (IPA) was dropped on a wafer piece containing many suspended devices. After the IPA evaporated, occasionally a nanostructure was left bridging the two suspended membranes. The yield could be improved by using a micro pipette to place a micro droplet of the suspension on the suspended device.

In another approach, a chemical vapor deposition (CVD) method was employed to grow individual SWNTs bridging the two suspended membranes (Yu et al., 2005). In this approach, catalytic nanoparticles made of Fe, Mo, and Al_2O_3 were delivered to the suspended membranes using a sharp probe tip. Alternatively, a nanometer thick Fe film was patterned on the two suspended membranes. The suspended device was then placed in a 900°C CVD tube with flowing methane. The Fe film was annealed into nanoparticles at a high temperature. The catalytic Fe particles seeded the growth of SWNTs, which occasionally bridged the two suspended membranes. Figure 4.4b shows a SWNT grown between the two suspended membranes using the CVD method.

Figure 4.5 shows the schematic diagram of the experimental setup for measuring the thermal conductance of the nanotube sample using the suspended device. During the measurement, the sample was placed in an evacuated cryostat. The two suspended membranes are denoted as the heating membrane and sensing membrane, respectively. When a dc current (I) flows to one of the two PRTs, a Joule heat $Q_h = I^2R_h$ is generated in this heating PRT that has a resistance of R_h. The PRT on each membrane is connected to the contact pads by four Pt leads, allowing a four-probe resistance measurement. The resistance of each Pt lead is R_L. A Joule heat of $2Q_L = 2I^2R_L$ is dissipated in the two Pt leads that supply the dc current to the heating PRT. The temperature of the heating membrane is raised to a relatively uniform temperature, T_h, because the internal thermal resistance of the membrane is much smaller than the thermal resistance of the nanostructure sample or the thermal resistance of the long narrow beams thermally connecting the membrane to the silicon chip at the ambient temperature T_0. The temperature uniformity has been verified by a numerical simulation (Yu et al., 2006).

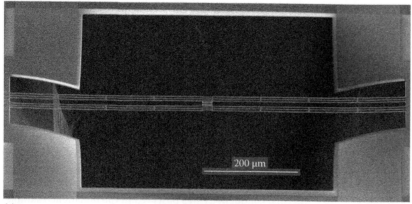

(a)

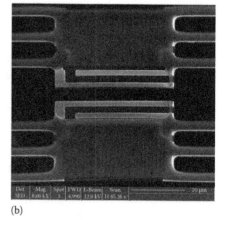

(b)

FIGURE 4.3 SEM images of a suspended device for measuring the thermal and thermoelectric properties of an individual nanotube or nanowire assembled between the two central membranes of the device: (a) low magnification SEM of the device and (b) high magnification SEM of the two central membranes of the device.

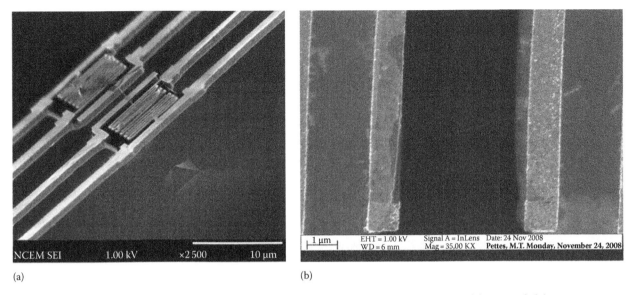

FIGURE 4.4 SEM of (a) a MWNT assembled and (b) a SWNT and grown between the two membranes of the suspended device.

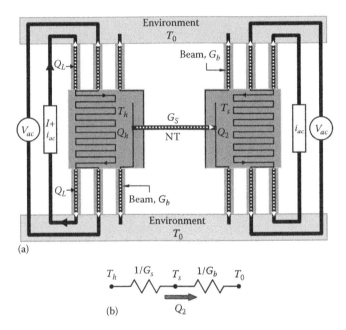

FIGURE 4.5 (a) Schematic diagram and (b) thermal resistance circuit of the measurement device. The left and right membranes are the heating and sensing membranes, respectively. The black arrow indicates the heating current flow direction. The white arrows indicate heat flow direction. The four-probe electrical resistances of the two serpentines were measured with a small ac current.

A fraction of the Joule heat, Q_2, is conducted through the nanostructure sample from the heating membrane to the sensing one, raising the temperature of the latter to T_s. The sensing membrane temperature rise consists of background contributions from heat conduction due to the residual molecules in the evacuated cryostat, radiation and heating of the Si chip because of the finite spreading thermal resistance of the Si chip, and the packaging thermal resistance. When measured using a blank suspended device without a sample bridging the two membranes, the background contribution was usually found to be much smaller than the heat conduction through the nanostructure sample.

The heat flow in the amount of Q_2 is further conducted to the environment through the beams supporting the sensing membrane. The rest of the heat, i.e., $Q_1 = Q_h + 2Q_L - Q_2$, is conducted to the environment through the supporting beams connected to the heating membrane.

The several beams supporting each membrane were designed to be identical. It was found by calculation that the radiation and residual molecular conduction heat losses from the membrane and the supporting beams to the environment are negligible compared to the conduction heat transfer through the supporting beams. Hence, the total thermal conductance of the supporting beams can be simplified as $G_b = n\kappa_l A/L$, where n is the number of the supporting beams for each membrane, κ_l, A, and L are the thermal conductivity, cross-sectional area, and the length of the supporting beams, respectively. We can obtain the following equation from the thermal resistance circuit shown in Figure 4.5

$$Q_2 = G_b(T_s - T_0) = G_s(T_h - T_s) \tag{4.25}$$

where G_s is the thermal conductance of the sample and consists of two components, i.e.,

$$G_s = (G_n^{-1} + G_c^{-1})^{-1} \tag{4.26}$$

where

$G_n = \kappa_n A_n/L_n$ is the intrinsic thermal conductance of the nanostructure

κ_n, A_n, and L_n are the effective thermal conductivity, cross-sectional area, and length of the free-standing segment of the sample between the two membranes, respectively

G_c is the contact thermal conductance between the tube and the two membranes

Considering 1D heat diffusion, one can obtain a temperature profile in the supporting beams. A Joule heat of Q_L is generated uniformly in each of the two Pt leads supplying the heating current, yielding a parabolic temperature distribution along the two beams; while linear temperature distribution is obtained for the other beams without DC Joule heating. The heat conduction to the environment from the two Joule-heated beams can be derived as $Q_{h,1} = 2(G_b \Delta T_h/n + Q_L/2)$; while that from the other supporting beams connected to the heating membrane is $Q_{h,2} = (n-2)G_b \Delta T_h/n$, and that from all the beams connected to the sensing membrane is $Q_s = G_b \Delta T_s$, where $\Delta T_s \equiv T_s - T_0$. The energy conservation requirement, i.e., $Q_{h,1} + Q_{h,2} + Q_s = Q_h + 2Q_L$, is used to obtain

$$G_b = \frac{Q_h + Q_L}{\Delta T_h + \Delta T_s} \qquad (4.27)$$

and

$$G_s = G_b \frac{\Delta T_s}{\Delta T_h - \Delta T_s} \qquad (4.28)$$

Q_h and Q_L can be obtained from the measured dc current and the voltage drops across the heating PRT and the Pt leads. ΔT_h and ΔT_s are obtained from the measured resistance increase of the two PRTs and their temperature coefficient of resistance (TCR).

Based on this direct measurement method, the thermal conductance of individual MWNTs and SWNTs have been measured. One reported result of a MWNT is shown in Figure 4.6 (Kim et al., 2001). For the MWNT, the measured thermal conductance shows a $T^{2.5}$ dependence at temperatures between 8 K and

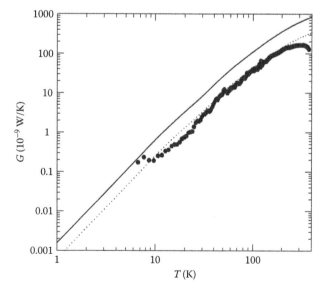

FIGURE 4.6 Measured thermal conductance (circles) of a 14 nm diameter MWNT as a function of temperature. The solid and dotted lines are 1-times and 0.4-times the calculated ballistic thermal conductance of a 14 nm diameter MWNT, respectively. (From Mingo, N. and Broido, D.A., *Phys. Rev. Lett.*, 95, 096105, 2005.)

below 50 K, and a T^2 dependence at the intermediate temperature between 50 and 150 K.

The measurement results can be better understood by reviewing the temperature dependence of the specific heat of graphite (DeSorbo and Tyler 1953, Kelly 1981), from which the MWNT is derived. At very low temperatures, the inter-layer phonon modes in graphite combined with the 2D in-plane phonon modes are expected to give rise to a T^3 dependence of the specific heat. At an intermediate temperature range, the soft inter-layer phonon modes of relatively low energy are mostly occupied and do not contribute to the specific heat. Consequently, a T^2 dependence of the specific heat has been observed as the signature of the 2D in-plane phone modes. As temperature increases further, the specific heat starts to saturate to the classical limit given as $3k_b$ per atom.

The observed quadratic temperature dependence of the thermal conductance of the MWNT in the intermediate temperature range can be attributed to a quadratic-specific heat in combination with a temperature independent mean free path, which is dominated by boundary and defect scattering. On the other hand, the $T^{2.5}$ dependence at the lower temperature range reveals a transition from the T^3 dependence to T^2 dependence of the specific heat.

The diameter of the MWNT was determined to be about 14 nm using a high-resolution SEM. The measured thermal conductance below a temperature of 200 K was about 0.4 times the calculated phonon ballistic conductance of a 14 nm diameter MWNT (Mingo and Broido 2005). The 0.4 factor can be attributed to the thermal contact resistance between the two thermal reservoirs and the MWNT, as well as the static scattering processes of phonons that reduce the phonon transmission coefficient to be below unity. At temperatures higher than 200 K, the measured thermal conductance is lower than 0.4 times the ballistic conductance because phonon-phonon umklapp scattering processes reduce the phonon mean free path, resulting in a peak thermal conductance at about 320 K. The relatively high temperature corresponding to the thermal conductance peak suggests that the umklapp scattering mean free path is shorter than that for other static scattering process or the nanotube length of about 2.5 μm until at near room temperature.

This method has also been used to measure the thermal conductance of individual SWNTs. One reported measurement result is shown in Figure 4.7 (Yu et al., 2005). The diameter of this SWNT was believed to be smaller than 3 nm based on a high-resolution SEM. However, the diameter determination was not accurate because of the limited resolution of the SEM. The thermal conductance was measured between a temperature of 100 and 300 K. At temperatures below 100 K, the signal to noise ratio becomes low. The observed thermal conductance was about 0.6 times that of the calculated ballistic thermal conductance of a (22, 0) SWNT 1.73 nm in diameter in the 100–300 K temperature range, without showing signatures of umklapp phonon scattering that would reduce the thermal conductance with increasing temperature.

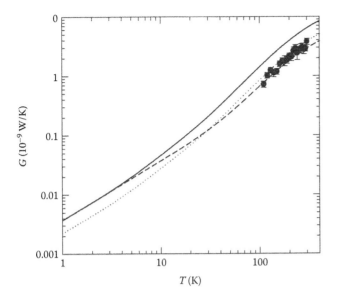

FIGURE 4.7 Measured thermal conductance (circles) of a SWNT as a function of temperature. The solid and dotted lines are 1-times and 0.6-times the calculated ballistic thermal conductance of a (22, 0) SWNT of 1.72 nm, respectively, and the dashed line is the calculated ballistic thermal conductance of a (10, 0) SWNT of 0.78 nm. (From Mingo, N. and Broido, D.A., *Phys. Rev. Lett.*, 95, 096105, 2005.)

Because of the uncertainty in the diameter and the contact thermal resistance, the intrinsic thermal conductivity cannot be obtained for both the MWNT and SWNT. Figure 4.8 shows the effective thermal conductivity calculated using Equation 4.2 with the contact thermal resistance ignored, the diameter of the SWNT assumed to be the upper bound of 3 nm, and the MWNT assumed to be a 14 m solid cylinder. The intrinsic thermal conductivity is expected to be higher than the as-obtained values of ~3000 W/m-K for the SWNT and MWNT at room temperature. Nevertheless, the room-temperature thermal conductivity of the

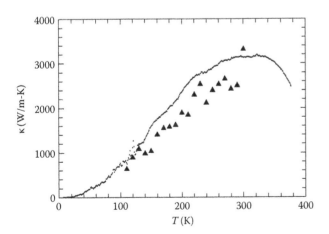

FIGURE 4.8 Calculated effective thermal conductivity of the 14 nm diameter MWNT (dots) and the SWNT (triangles) if the diameter of the SWNT is 3 nm.

nanotube is already about 50% higher than that of the pyrolytic graphite (Kelly 1981). Since the room-temperature thermal conductivity is limited by umklapp scattering, this finding suggests that umklapp scattering in nanotubes could be weaker than in graphite.

Besides the suspended device, a T-shape junction sensor has been developed for the direct measurement of the thermal conductance of an individual MWNT (Fujii et al., 2005). The sensor contains a suspended pattered Pt nanofilm that serves as both a heater and thermometer. A MWNT was placed between the center of the suspended nanofilm and the Si substrate, and formed a T-shape junction with the suspended nanofilm. The average temperature of the nanofilm was calculated by measuring its electrical resistance at different electrical currents. The thermal conductivity of the nanofilm was measured using a self-heating method when no NT was placed on the nanofilm. The self-heating measurement was repeated after the NT was placed on the nanofilm. The latter measurement was used to determine the thermal conductance and thermal conductivity of the MWNT. Using this method, the thermal conductivity of MWNTs of three different diameters was measured. It was found that the effective thermal conductivity increases as the MWNT diameter decreases, and exceeds 2000 W/m-K for a diameter of 9.8 nm.

A common issue with these two direct thermal conductance measurement methods is that the contact thermal resistance was not eliminated from the measured thermal resistance or conductance. It is possible that the observed diameter dependence of the MWNTs (Fujii et al., 2005) was caused by contact thermal resistance, the effect of which is expected to be more apparent for larger diameter NTs because of a smaller surface-to-volume ratio (Prasher 2008).

4.3.3 Self-Heating Measurement of Thermal Transport in Carbon Nanotubes

Besides the two direct thermal conductance measurement methods, other methods based on self electrical heating and resistance thermometry of individual nanotubes and wires have been reported. One reported method is based on the extension of the 3-ω technique to individual metallic micro-wires and MWNT bundles (Yi et al., 1999, Lu et al., 2001). In this method, the micro-wire or MWNT bundle was suspended across a trench and heated by a sinusoidal current at a frequency of ω. The temperature and consequently the electrical resistance of the suspended sample contained an oscillatory component at a frequency of 2ω. Thus, the voltage drop along the suspended sample contained an oscillatory component at a frequency of 3ω. The amplitude of the voltage oscillation at 3ω is given by

$$V_{3\omega} \approx \frac{4I^3 L R R'}{\pi^4 \kappa A \sqrt{1 + (2\omega\gamma)^2}} \tag{4.29}$$

where

> I is the root mean squared amplitude of the heating current
>
> R is the electrical resistance of the suspended sample, $R' = dR/dT$
>
> κ and A are the thermal conductivity and cross-section area of the sample
>
> γ is the thermal time constant of the sample given by $\gamma \equiv L^2/\pi^2\alpha$, where L is the suspended length
>
> α is the thermal diffusivity of the sample

Apparently, this self-heating resistance thermometry method requires that the sample has a sufficiently large TCR or R'. In addition, the phase lag (ϕ) between the voltage oscillation and the temperature oscillation is given as

$$\tan\phi \approx \omega\gamma \qquad (4.30)$$

In this method, the κ and γ can be extracted by fitting the measured $V_{3\omega}$ and ϕ at different ω values. The specific heat is obtained as

$$C = \pi^2\gamma\kappa/\rho L^2 \qquad (4.31)$$

where ρ is the density.

The κ measurement is usually conducted in the low frequency range where $\omega\gamma \ll 1$. On the other hand, for measuring γ and C, the $\omega\gamma$ product needs to be sufficiently large, e.g., $\omega\gamma > 0.1$, which is equivalent to $\omega > 0.1\pi^2\alpha/L^2$. For an individual NT with α estimated to be of the order of 2×10^{-3} m^2/s and L of about $5\,\mu$m, ω needs to be larger than about 80 MHz. This very high frequency and especially its third harmonic component exceeds the frequency range of a lock-in amplifier that is needed for measuring $V_{3\omega}$ and ϕ. In addition, capacitive coupling can lead to a large leakage of current at this high frequency, further complicating the analysis of the measurement result.

The self-heating 3ω method was used to measure the specific heat and thermal conductivity of a millimeter long suspended defective MWCN tube bundle (Yi et al., 1999). A linear temperature dependence was found in the measured specific heat from 10 to 300 K, and accompanied by a quadratic temperature dependence of the thermal conductivity at temperatures below 120 K. It was speculated that the linear-specific heat was caused by a dominant out-of-plane acoustic phonon mode in the MWNT bundle. The quadratic temperature dependence in the thermal conductivity was attributed to the linear-specific heat and a phonon mean free path that increases linearly with temperature. These findings are rather intriguing. In addition, the measured thermal conductivity was about 20 W/m-K at room temperature. The low value can be attributed to inter-tube phonon coupling, the thermal resistance at tube–tube junctions, and defects in the bundle.

In two other reports (Choi et al., 2006, Wang et al., 2007), the 3ω method was used to obtain the lattice thermal conductivity of individually suspended MWNTs and individual SWNTs on a SiO$_2$ substrate. In these self-heating methods, the electric resistance increase with increasing bias voltage was assumed to be caused by an increase in the lattice temperature alone. However, electron transport in nanotubes has been known to be highly nonlinear. Electron transport is characterized by ballistic transport in short SWNTs at low electric fields and optical phonon emissions at high electric fields (Yao et al., 2000, Javey et al., 2003, Pop et al., 2005). Transport in MWNTs can also be nonlinear with increasing bias, which has been attributed to an increase in the number of current-carrying shells and/or to electrons subjected to Coulomb interaction that tunnel across the MWNT-electrode interface (Bourlon et al., 2006). If the nonlinear current–voltage (I–V) characteristic of a NT is fitted using a polynomial, the obtained V expression can contain an I^3 term caused not only by simply an increased lattice temperature, but also by other nonlinear processes including optical phonon emissions. Consequently, the 3ω voltage $U_{3\omega}$ is not entirely caused by a rise in lattice temperature, especially at a high field or in short SWNTs where nonequilibrium temperatures of electrons, acoustic phonons, and optical phonons need to be taken into account (Pop et al., 2005, Shi 2008).

For addressing this issue, the electric current–voltage (I–V) characteristics of suspended SWNTs were measured and fitted to a coupled electron-phonon transport model that takes into account the nonequilibrium between different energy carriers (Pop et al., 2006). The lattice temperature rise and the thermal conductivity of the NT were extracted by adjusting several fitting parameters. The obtained thermal conductance of a 2 nm diameter SWNT shows inverse temperature ($1/T$) dependence at above room temperature, revealing the effect of the umklapp process. The effective thermal conductivity was determined to be about 3600 W/m-K at room temperature.

In general, these self-heating methods provide a simpler approach for thermal measurement of nanotubes than the direct thermal conductance methods that require additional prefabricated suspended thermal sensors. Nevertheless, the inconvenience of the self-heating-based approaches are that the results obtained on the thermal properties depend on the models employed as well as on several parameters that are difficult to characterize, such as the coupling between the optical and acoustic phonons and the contact electrical and thermal resistances at the nanotube-electrode interfaces.

4.3.4 Optical Measurement of Thermal Transport in Carbon Nanotubes

In addition to the aforementioned electrical thermometry methods, an optical heating and Raman thermometry technique has been developed for measuring the thermal conductance of nanotubes (Hsu et al., 2009). In this method, a laser beam was focused on a suspended nanotube and the frequency of the Raman G band was measured. Based on the calibrated Raman G band frequency shifts as a function of the nanotube temperature, the nanotube temperature at the laser spot was obtained. If the nanotube was suspended between two suspended membranes with integrated resistance thermometers, as illustrated in Figure 4.9a, the heat absorption by the NT from the laser can be obtained as

$$Q = Q_L + Q_R \qquad (4.32)$$

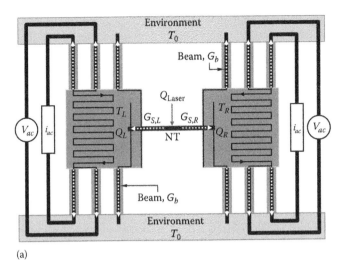

(a)

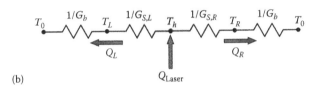

(b)

FIGURE 4.9 (a) Schematic diagram and (b) thermal resistance circuit for measuring the optical absorption and thermal conductance of a nanotube suspended between two suspended membranes with integrated resistance thermometers. The white dashed lines indicate heat flow direction. The electrical resistances of the two serpentines were measured with a small ac current.

where Q_L and Q_R are the heat conduction rates to the left and right membranes and subsequently through the supporting beams of the membranes to the environment. Based on the thermal resistance circuit of Figure 4.9b,

$$Q_L = G_{S,L}(T_h - T_L) = G_b(T_L - T_0) \qquad (4.33)$$

and

$$Q_R = G_{S,R}(T_h - T_R) = G_b(T_R - T_0) \qquad (4.34)$$

where

$G_{S,L}$ and $G_{S,R}$ are the thermal conductance of the nanotube segment to the left and to the right of the laser spot
G_b is the total thermal conductance of the supporting beams for each membrane and can be measured using Equation 4.27
T_h is the nanotube temperature at the laser spot measured using Raman thermometry
T_L and T_R are the temperatures of the two membranes measured by using resistance thermometry

The thermal conductance of the nanotube is obtained as

$$G_S = \left(G_{S,L}^{-1} + G_{S,R}^{-1}\right)^{-1} = G_b \left(\frac{T_h - T_L}{T_L - T_0} + \frac{T_h - T_R}{T_R - T_0}\right)^{-1} \qquad (4.35)$$

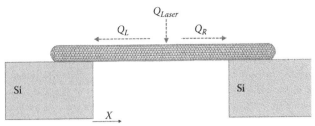

FIGURE 4.10 Schematic diagram of the contact thermal resistance measurement using laser heating and Raman thermometry.

Compared to the method shown in Figure 4.5 based solely on resistance thermometry, the potential advantage of this method employing Raman thermometry and resistance thermometry is that the contact thermal resistance may be obtained by scanning the laser spot along the length of the NT. A similar contact resistance measurement approach has been demonstrated on a nanotube suspended across a trench etched in a Si substrate (Hsu et al., 2008), as illustrated in Figure 4.10.

When the laser spot was focused at different locations of the suspended nanotube, the nanotube temperature rise (ΔT) at the laser spot was obtained using Raman thermometry. If the phonon transport in a nanotube is diffusive, one can solve the Fourier heat equation for 1D heat conduction. For this experiment, the total heat generation rate (Q) by the incident laser is equal to the sum of the heat flow rates to the right and left hand sides of the nanotube, which is

$$Q = Q_L + Q_R = \frac{\Delta T}{\dfrac{x}{\kappa A} + R_{c,\text{left}}} + \frac{\Delta T}{\dfrac{L-x}{\kappa A} + R_{c,\text{right}}} \qquad (4.36)$$

where

κ is the nanotube thermal conductivity
A and L are the geometrical cross section and length of the suspended nanotube
x is the distance of the laser spot from the left edge of the trench
$R_{c,\text{left}}$ and $R_{c,\text{right}}$ are the contact thermal resistances at the two ends of the nanotube

The measured temperature at the laser spot moved along the nanotube is given by

$$\Delta T(x) = \frac{Q\left[-\dfrac{x^2}{(\kappa A)^2} + x\left(\dfrac{L}{(\kappa A)^2} + \dfrac{R_{c,\text{right}} - R_{c,\text{left}}}{\kappa A}\right) + R_{c,\text{right}}R_{c,\text{left}} + \dfrac{L}{\kappa A}R_{c,\text{left}}\right]}{\dfrac{L}{\kappa A} + R_{c,\text{left}} + R_{c,\text{right}}}$$

$$(4.37)$$

When the two contact thermal resistances are negligible compared to the intrinsic thermal resistance of the nanotube $R_{NT} = L/\kappa A$, Equation 4.38 is reduced to

$$\Delta T(x) = \frac{Q}{\kappa AL}(-x^2 + Lx) \qquad (4.38)$$

While the amount of heat generated in the nanotubes by the laser (Q) was not determined in the experiment based on the configuration of Figure 4.10, one can still determine the ratios between the contact and nanotube thermal resistances, $r_l \equiv R_{c,\text{left}}/R_{NT}$ and $r_r \equiv R_{c,\text{right}}/R_{NT}$. Equation 4.38 can be expressed in the general form of $\Delta T(x) = -ax^2 + bx + c$. One can define the following coefficient ratios $\alpha = b/a = L(1 - r_l + r_r)$ and $\beta = c/a = L^2 r_l(1 + r_r)$. The ratios between the contact and nanotube thermal resistances are then given by

$$r_l \equiv \frac{R_{c,\text{left}}}{R_{NT}} = \frac{-\alpha + \sqrt{\alpha^2 + 4\beta}}{2L} \tag{4.39}$$

and

$$r_r \equiv \frac{R_{c,\text{right}}}{R_{NT}} = -1 + \frac{\alpha + \sqrt{\alpha^2 + 4\beta}}{2L} \tag{4.40}$$

Using this method, the ratios of thermal contact resistance to the thermal resistance of the nanotube were found to span the range from smaller than 0.02 to larger than 17 for four 2.6–5.0 µm long suspended SWNTs. In addition, by spatially resolving the temperature rise profile along the length of an optically heated nanotube, this technique can distinguish between diffusive and ballistic phonon transport, the latter of which would result in a constant temperature rise in the nanotube independent of the location of the heating laser spot. The results obtained on four nanotubes indicate that phonon transport in these nanotubes is diffusive or the phonon mean free path is shorter than the 2.6–5 µm suspended segment of the nanotubes. As such, thermal transport in the four nanotubes can be explained with the simple Fourier heat transport model.

4.4 Conclusion and Outlook

This chapter provided an introduction to thermal conductance in carbon nanotubes and reviewed recent research progress in this topic, which is still being actively investigated. As such, it is not possible for this chapter to cover all the interesting works that have been reported on thermal transport in nanotubes. While breakthroughs have been made on the theoretical understanding and experimental characterization of thermal transport in nanotubes, the intrinsic thermal conductance and conductivity of nanotubes are yet to be found. On the theoretical front, different approaches, including molecular dynamics simulations, have been employed to calculate the thermal property of nanotubes. A common problem is that the computation capability is currently limited to short nanotubes. Consequently, long wavelength phonon modes were cut off in the simulation, leading to length-dependent thermal conductivity values obtained by these numerical methods. On the experimental side, the reported thermal conductance values all contain errors caused by contact thermal resistance, which is difficult to measure for

nanotubes. In addition to the optical method (Hsu et al., 2008), recent demonstration of a new method for measuring the contact thermal resistance to nanowires provides some promise in solving this difficult problem (Mavrokefalos et al., 2007, Zhou et al., 2007). Moreover, the crystal structure including the diameter, chirality, and defects of the nanotubes being measured are often not well characterized. Hence, there is a need to establish the thermal property-structure relationship of nanotubes. Given the increased interest and the importance of the thermal property of nanotubes on their wide range of possible applications, many of these challenges will likely be overcome in the coming years.

Acknowledgments

The author's research on thermal transport in carbon nanotubes has benefited from collaborations with Philip Kim, Arun Majumdar, Paul L. McEuen, Choongho Yu, Deyu Li, Michael Pettes, Stephen B. Cronin, and I-Kai Hsu, and has been supported by the Department of Energy award DE-FG02-07ER46377, the National Science Foundation Thermal Transport Processes Program, and the University of Texas System.

References

Ando, T. and Nakanishi, T. (1998) Impurity scattering in carbon nanotubes—Absence of back scattering. *Journal of the Physical Society of Japan*, 67, 1704–1713.

Ando, T., Nakanishi, T., and Saito, R. (1998) Berry's phase and absence of back scattering in carbon nanotubes. *Journal of the Physical Society of Japan*, 67, 2857–2862.

Anthony, T. R., Banholzer, W. F., Fleischer, J. F. et al. (1990) Thermal diffusivity of isotopically enriched C12 diamond. *Physical Review B*, 42, 1104.

Berber, S., Kwon, Y.-K., and Tománek, D. (2000) Unusually high thermal conductivity of carbon nanotubes. *Physical Review Letters*, 84, 4613.

Bockrath, M., Cobden, D. H., McEuen, P. L. et al. (1997) Single-electron transport in ropes of carbon nanotubes. *Science*, 275, 1922–1925.

Bourlon, B., Miko, C., Forro, L., Glattli, D. C., and Bachtold, A. (2006) Beyond the linearity of current-voltage characteristics in multiwalled carbon nanotubes. *Semiconductor Science and Technology*, 21, S33–S37.

Chen, G. (2005) *Nanoscale Energy Transport and Conversion: A Parallel Treatment of Electrons, Molecules, Phonons, and Photons*, Oxford, U.K./New York: Oxford University Press.

Choi, T. Y., Poulikakos, D., Tharian, J., and Sennhauser, U. (2006) Measurement of the thermal conductivity of individual carbon nanotubes by the four-point three-omega method. *Nano Letters*, 6, 1589–1593.

Collins, P. C., Arnold, M. S., and Avouris, P. (2001a) Engineering carbon nanotubes and nanotube circuits using electrical breakdown. *Science*, 292, 706–709.

Collins, P. G., Hersam, M., Arnold, M., Martel, R., and Avouris, P. (2001b) Current saturation and electrical breakdown in multiwalled carbon nanotubes. *Physical Review Letters*, 86, 3128–3131.

Dames, C. and Chen, G. (2004) Theoretical phonon thermal conductivity of Si/Ge superlattice nanowires. *Journal of Applied Physics*, 95, 682–693.

Datta, S. (1997) *Electronic Transport in Mesoscopic Systems*, Cambridge, U.K./New York: Cambridge University Press.

Desorbo, W. and Tyler, W. W. (1953) The specific heat of graphite from 13[degree] to 300[degree] K. *The Journal of Chemical Physics*, 21, 1660–1663.

Dresselhaus, M. S. and Eklund, P. C. (2000) Phonons in carbon nanotubes. *Advances in Physics*, 49, 705–814.

Dresselhaus, M. S., Dresselhaus, G., and Avouris, P. (2001) *Carbon Nanotubes: Synthesis, Structure, Properties, and Applications*, Berlin, Germany/New York: Springer.

Foa Torres, L. E. F. and Roche, S. (2006) Inelastic quantum transport and peierls-like mechanism in carbon nanotubes. *Physical Review Letters*, 97, 076804.

Fujii, M., Zhang, X., Xie, H. Q. et al. (2005) Measuring the thermal conductivity of a single carbon nanotube. *Physical Review Letters*, 95, 065502.

Gao, B., Chen, Y. F., Fuhrer, M. S., Glattli, D. C., and Bachtold, A. (2005) Four-point resistance of individual single-wall carbon nanotubes. *Physical Review Letters*, 95, 196802.

Holland, M. G., Klein, C. A., and Straub, W. D. (1966) The Lorenz number of graphite at very low temperatures. *Journal of Physics and Chemistry of Solids*, 27, 903–906.

Hone, J., Whitney, M., Piskoti, C., and Zettl, A. (1999) Thermal conductivity of single-walled carbon nanotubes. *Physical Review B*, 59, R2514.

Hone, J., Llaguno, M. C., Nemes, N. M. et al. (2000) Electrical and thermal transport properties of magnetically aligned single wall carbon nanotube films. *Applied Physics Letters*, 77, 666–668.

Hsu, I. K., Kumar, R., Bushmaker, A. et al. (2008) Optical measurement of thermal transport in suspended carbon nanotubes. *Applied Physics Letters*, 92, 063119.

Hsu, I. K., Pows, M. T., Bushmaker, A. et al. (2009) Optical absorption and thermal transport of individual suspended carbon nanotube bundles. *Nano Letters*, 9, 590–594.

Javey, A., Guo, J., Wang, Q., Lundstrom, M., and Dai, H. J. (2003) Ballistic carbon nanotube field-effect transistors. *Nature*, 424, 654–657.

Kelly, B. T. (1981) *Physics of Graphite*, London, U.K./Englewood, NJ: Applied Science Publishers.

Kim, P., Shi, L., Majumdar, A., and McEuen, P. L. (2001) Thermal transport measurements of individual multiwalled nanotubes. *Physical Review Letters*, 8721, 215502.

Lazzeri, M. and Mauri, F. (2006) Coupled dynamics of electrons and phonons in metallic nanotubes: Current saturation from hot-phonon generation. *Physical Review B*, 73, 165419.

Lu, L., Yi, W., and Zhang, D. L. (2001) 3 omega method for specific heat and thermal conductivity measurements. *Review of Scientific Instruments*, 72, 2996–3003.

Mavrokefalos, A., Pettes, M. T., Zhou, F., and Shi, L. (2007) Four-probe measurements of the in-plane thermoelectric properties of nanofilms. *Review of Scientific Instruments*, 78, 034901.

McEuen, P. L., Bockrath, M., Cobden, D. H., Yoon, Y. G., and Louie, S. G. (1999) Disorder, pseudospins, and backscattering in carbon nanotubes. *Physical Review Letters*, 83, 5098–5101.

Mingo, N. and Broido, D. A. (2005) Carbon nanotube ballistic thermal conductance and its limits. *Physical Review Letters*, 95, 096105.

Moore, A. L., Saha, S. K., Prasher, R. S., and Shi, L. (2008) Phonon backscattering and thermal conductivity suppression in sawtooth nanowires. *Applied Physics Letters*, 93, 083112.

Park, J. Y., Rosenblatt, S., Yaish, Y. et al. (2004) Electron-phonon scattering in metallic single-walled carbon nanotubes. *Nano Letters*, 4, 517–520.

Pop, E., Mann, D., Cao, J. et al. (2005) Negative differential conductance and hot phonons in suspended nanotube molecular wires. *Physical Review Letters*, 95, 155505.

Pop, E., Mann, D., Wang, Q., Goodson, K., and Dai, H. J. (2006) Thermal conductance of an individual single-wall carbon nanotube above room temperature. *Nano Letters*, 6, 96–100.

Prasher, R. (2008) Thermal boundary resistance and thermal conductivity of multiwalled carbon nanotubes. *Physical Review B (Condensed Matter and Materials Physics)*, 77, 075424.

Purewal, M. S., Hong, B. H., Ravi, A. et al. (2007) Scaling of resistance and electron mean free path of single-walled carbon nanotubes. *Physical Review Letters*, 98, 186808.

Shi, L. (2001) Mesoscopic thermophysical measurements of microstructures and carbon nanotubes, PhD thesis, University of California, Berkeley, CA.

Shi, L. (2008) Comment on "Length-dependant thermal conductivity of an individual single-wall carbon nanotube" [*Applied Physics Letters*, 91, 123119 (2007)]. *Applied Physics Letters*, 92, 206103.

Shi, L., Li, D. Y., Yu, C. H. et al. (2003) Measuring thermal and thermoelectric properties of one-dimensional nanostructures using a microfabricated device. *Journal of Heat Transfer-Transactions of the ASME*, 125, 881–888.

Sundqvist, P., Garcia-Vidal, F. J., Flores, F. et al. (2007) Voltage and length-dependent phase diagram of the electronic transport in carbon nanotubes. *Nano Letters*, 7, 2568–2573.

Tans, S. J., Devoret, M. H., Groeneveld, R. J. A., and Dekker, C. (1998) Electron-electron correlations in carbon nanotubes. *Nature*, 394, 761–764.

Wang, Z. L., Tang, D. W., Li, X. B. et al. (2007) Length-dependent thermal conductivity of an individual single-wall carbon nanotube. *Applied Physics Letters*, 91, 123119.

Yamamoto, T., Watanabe, S., and Watanabe, K. (2004) Universal features of quantized thermal conductance of carbon nanotubes. *Physical Review Letters*, 92, 075502.

Yao, Z., Kane, C. L., and Dekker, C. (2000) High-field electrical transport in single-wall carbon nanotubes. *Physical Review Letters*, 84, 2941–2944.

Yi, W., Lu, L., Zhang, D. L., Pan, Z. W., and Xie, S. S. (1999) Linear specific heat of carbon nanotubes. *Physical Review B*, 59, R9015–R9018.

Yu, C. H., Shi, L., Yao, Z., Li, D. Y., and Majumdar, A. (2005) Thermal conductance and thermopower of an individual single-wall carbon nanotube. *Nano Letters*, 5, 1842–1846.

Yu, C. H., Saha, S., Zhou, J. H. et al. (2006) Thermal contact resistance and thermal conductivity of a carbon nanofiber. *Journal of Heat Transfer-Transactions of the ASME*, 128, 234–239.

Zhou, F., Szczech, J., Pettes, M. T. et al. (2007) Determination of transport properties in chromium disilicide nanowires via combined thermoelectric and structural characterizations. *Nano Letters*, 7, 1649–1654.

Ziman, J. M. (1962) *Electrons and Phonons: The Theory of Transport Phenomena in Solids*, Oxford, U.K.: Clarendon Press.

5

Terahertz Radiation from Carbon Nanotubes

Andrei M. Nemilentsau
Belarus State University

Gregory Ya. Slepyan
Belarus State University

Sergey A. Maksimenko
Belarus State University

Oleg V. Kibis
Novosibirsk State Technical University

Mikhail E. Portnoi
University of Exeter

5.1 Introduction

Creating a compact, reliable source of terahertz (THz) radiation is one of the most challenging problems in contemporary applied physics (Lee and Wanke, 2007). Despite the fact that THz technology is at the boundaries of microwave and photonic technologies, it is quite underdeveloped compared to the achievements in the microwave or the photonic technology. There are very few commercially available instruments for the THz frequency region, and most of them lack the precision required to perform accurate measurements. There are also no miniaturized and low-cost THz sources. One of the latest trends in THz technology (Dragoman and Dragoman, 2004a) is to use single-walled carbon nanotubes (SWNTs) as building blocks of novel high-frequency devices.

An SWNT is a hollow cylindrical molecule made up of carbon atoms (Saito et al., 1998). We can formally consider the SWNT as a graphene sheet rolled up into a cylinder along the vector $\mathbf{R}_h$ connecting to crystallographically equivalent sites of the graphene lattice (see Figure 5.1). This vector is called the chiral vector and is usually defined in terms of the basic vectors, $\mathbf{a}_1$ and $\mathbf{a}_2$, of the graphene lattice: $\mathbf{R}_h = m\mathbf{a}_1 + n\mathbf{a}_2$, where m, n are integers. The dual index (m, n) is usually used to characterize SWNT type. Three different SWNT types are defined: $(m, 0)$ zigzag SWNTs, (m, m) armchair SWNTs, and (m, n) $(0 < n \neq m)$ chiral SWNTs. The SWNT radius, R_{cn}, and chiral angle, θ (the angle between the $\mathbf{R}_h$ and $\mathbf{a}_1$), are defined as follows:

$$R_{cn} = \frac{|\mathbf{R}_h|}{2\pi} = \frac{\sqrt{3}b}{2\pi}\sqrt{m^2 + mn + n^2}, \tag{5.1}$$

$$\cos\theta = \frac{\mathbf{R}_h \cdot \mathbf{a}_1}{|\mathbf{R}_h||\mathbf{a}_1|} = \frac{2n + m}{2\sqrt{n^2 + nm + m^2}}, \tag{5.2}$$

where $b = 0.142$ nm is the C–C bond length. Typically, SWNTs are 0.1–10 μm in length; their cross-sectional radius varies within the range 1–10 nm, while their chiral angle is $0 \leq \theta_{cn} \leq 30°$.

There are several promising proposals of using carbon nanotubes for THz applications including a nanoklystron using extremely efficient high-field electron emission from nanotubes (Dragoman and Dragoman, 2004a; Manohara et al., 2005; Di Carlo et al., 2006); devices based on negative differential conductivity (NDC) in large-diameter semiconducting SWNTs (Maksimenko and Slepyan, 2000; Pennington and Goldsman, 2003); high-frequency resonant-tunneling diodes (Dragoman and Dragoman, 2004b) and Schottky diodes (Léonard and Tersoff, 2000; Odintsov, 2000; Yang et al., 2005; Lu et al., 2006); as well as electric-field-controlled carbon nanotube superlattices (Kibis et al., 2005a,b), frequency multipliers (Slepyan et al., 1999, 2001), THz amplifiers (Dragoman and Dragoman, 2005), and switches (Dragoman et al., 2006). Among others, the idea of SWNT-based optical devices enabling the control and enhancement of radiation efficiency on the nanoscale, i.e., nanoscale antennas for THz, infrared, and visible light, is actively

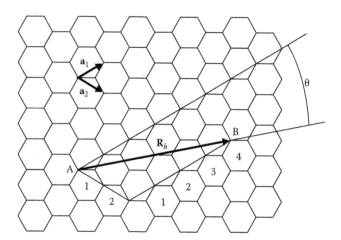

FIGURE 5.1 Graphene crystalline lattice. Each lattice node contains a carbon atom.

discussed (Dresselhaus, 2004; Hanson, 2005; Burke et al., 2006; Slepyan et al., 2006). Noise properties and operational limits of such antennas are substantially determined by the thermal fluctuations of the electromagnetic field.

In this chapter, several novel schemes are discussed (Kibis and Portnoi, 2005; Portnoi et al., 2006; Kibis et al., 2007, 2008; Nemilentsau et al., 2007; Portnoi et al., 2008) to utilize the physical properties of SWNTs for the generation and detection of THz radiation.

5.2 Electronic Properties of SWNTs

Electrodynamic processes in any medium are dictated by its electronic properties, although they may be missing in an explicit form of the macroscopic electrodynamics equations. In that sense, SWNTs are not an exception. Many researches (Charlier et al., 2007) have been devoted to the development of the theory of electronic properties of the SWNT. Both the sophisticated methods of modern solid-state physics and first-principles simulations are among them. In this section, we give only an elementary introduction for later use in the analysis of the THz radiation from SWNTs.

Each carbon atom in graphene and SWNT has four valence orbitals ($2s$, $2p_x$, $2p_y$, and $2p_z$). Three orbitals (s, p_x, and p_y) combine to form in-plane σ orbitals. The σ bonds are strong covalent bonds responsible for most of the binding energy and elastic properties of the graphene sheet and SWNT. The remaining p_z orbital, pointing out of the graphene sheet, cannot couple with σ orbitals. The lateral interaction with the neighboring p_z orbitals creates delocalized π orbitals. The π bonds are perpendicular to the surface of the SWNT and are responsible for the weak interaction between SWNTs in a bundle, similar to the weak interaction between graphene layers in pure graphite. The energy levels associated with the in-plane bonds are known to be far away from the Fermi energy in graphene, and thus do not play a key role in its electronic properties. In contrast, the bonding and antibonding π bands cross the Fermi level at high-symmetry

points in the Brillouin zone of graphene (Wallace, 1947). Thus, we restrict our consideration to the π electrons, assuming that their movement can be described in the framework of the tight-binding approximation (Saito et al., 1998); the overlapping of wave functions of only the nearest atoms is taken into account. In the beginning, we apply this approach to the plane mono-atomic graphite layer, and then show how the model must be modified to analyze an SWNT.

To describe graphene π bands we use the 2×2 Hamiltonian matrix (Wallace, 1947):

$$\hat{\mathbf{H}}_0 = \begin{pmatrix} 0 & H_{12}(p_x, p_y) \\ H_{12}^*(p_x, p_y) & 0 \end{pmatrix}, \tag{5.3}$$

where

$$H_{12}(p_x, p_y) = -\gamma_1 \exp\left\{ i \frac{b}{\hbar} p_x \right\} - \gamma_2 \exp\left\{ i \frac{b}{\hbar} \left(\frac{1}{2} p_x - \sqrt{3} p_y \right) \right\}$$
$$- \gamma_3 \exp\left\{ -i \frac{b}{\hbar} \left(\frac{1}{2} p_x - \sqrt{3} p_y \right) \right\}. \tag{5.4}$$

Here, $\gamma_{1,2,3}$ are the overlapping integrals; $p_{x,y}$ are the projections of the quasi momentum of electrons, $\mathbf{p}$, on the corresponding axes; and $\hbar$ is the Planck constant. As the electronic properties of graphene are isotropic in the in-plane, we set $\gamma_1 = \gamma_2 = \gamma_3 = \gamma_0$ in Equation 5.4, where $\gamma_0 \simeq 3\,\mathrm{eV}$ is the phenomenological parameter, which can be determined experimentally (see, e.g., Saito et al., 1998). The electron energy values are found as the eigenvalues of the matrix on the right side of (5.3) as

$$\varepsilon_{c,v}(\mathbf{p}) = \pm \gamma_0 \sqrt{1 + 4\cos\left(\frac{3bp_x}{2\hbar}\right)\cos\left(\frac{\sqrt{3}bp_y}{2\hbar}\right) + 4\cos^2\left(\frac{\sqrt{3}bp_y}{2\hbar}\right)}. \tag{5.5}$$

The plus and minus signs in this equation correspond to conduction (c) and valence (v) bands, respectively. The range of quasi momentum variation (the first Brillouin zone) is the hexagons shown in Figure 5.2. The vertices are the Fermi points where $\varepsilon = 0$, which is indicative of the absence of the forbidden zone for π electrons in graphene.

The dispersion properties of electrons in SWNTs are quite different from those in graphene, as a plane monolayer is transformed into a cylinder. In a cylindrical structure, an electron located at the origin and an electron located at the position defined by the vector $\mathbf{R}_h = m\mathbf{a}_1 + n\mathbf{a}_2$ are identical. Hence, we should impose the periodic boundary conditions along the tube circumference on the wave functions of π electrons in SWNTs:

$$\Psi(\mathbf{r} + \mathbf{R}_h) = e^{i\mathbf{p}\mathbf{R}_h/\hbar}\Psi(\mathbf{r}) = \Psi(\mathbf{r}). \tag{5.6}$$

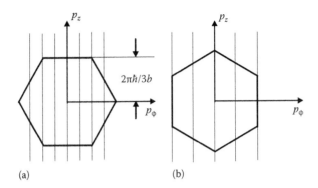

FIGURE 5.2 First Brillouin zone for (a) zigzag and (b) armchair SWNTs.

The second equality here is due to the Bloch theorem. This leads to the quantization of the transverse quasi momentum of electrons:

$$p_\phi = \hbar s / R_{cn}, \qquad (5.7)$$

where s is an integer. The cylindrical coordinate system with the z-axis oriented along the SWNT axis is used here. The axial projection, p_z, of the quasi momentum is continuous. In order to derive the dispersion equation for zigzag SWNTs from Equation 5.5, one must perform the substitutions $\{p_x \to p_z, p_y \to p_\phi\}$, which yields

$$\varepsilon_{c,v}(p_z, s) = \pm \gamma_0 \sqrt{1 + 4\cos\left(\frac{3bp_z}{2\hbar}\right)\cos\left(\frac{\pi s}{m}\right) + 4\cos^2\left(\frac{\pi s}{m}\right)},$$
$$s = 1, 2, \ldots, m. \qquad (5.8)$$

For armchair SWNTs, the dispersion law is obtained from Equation 5.5 by means of the substitutions $\{p_x \to p_\phi, p_y \to p_z\}$. For chiral SWNTs, the analogous procedure is specified by $\{p_x \to p_z \cos\theta + p_\phi \sin\theta, p_x \to p_z \sin\theta - p_\phi \cos\theta\}$.

It follows from Equation 5.7 that the first Brillouin zone in SWNTs is transformed from a hexagon to a family of one-dimensional zones defined by segments of straight lines confined to the interior of the hexagon. Depending on the dual index (m,n), these segments can be oriented differently either by bypassing or crossing the Fermi points, as shown in Figure 5.2. Correspondingly, the forbidden zone either appears or disappears in the electron spectrum of an SWNT. In the absence of the forbidden zone, a material is a metal; otherwise, it is a semiconductor. The condition for the forbidden zone to appear is (Saito et al., 1998)

$$m - n \neq 3q, \qquad (5.9)$$

where q is an integer. For armchair SWNTs, this condition is not valid at any m, and the forbidden zone is always absent, thus proving that the armchair SWNTs are always metallic. For zigzag SWNTs, the zone appears when $m \neq 3q$, and, thus, zigzag SWNTs can be either metallic or semiconducting, depending on R_{cn}.

Strictly speaking, the curvature of the SWNT surface breaks the isotropic symmetry of electronic properties so that the overlapping integrals, $\gamma_{1,2,3}$, in Equation 5.4 turn out to be different from one another. For the zigzag SWNTs, these integrals are as follows (Kane and Mele, 1997; Lin and Chuu, 1998): $\gamma_1 = \gamma_0, \gamma_2 = \gamma_3 = (1 - 3b^2/32R_{cn}^2)\gamma_0$. Then, instead of Equation 5.8, we have the dispersion equation:

$$\varepsilon_{c,v}(p_z, s) = \pm \sqrt{\gamma_0^2 + 4\gamma_0\gamma_2 \cos\left(\frac{3bp_z}{2\hbar}\right)\cos\left(\frac{\pi s}{m}\right) + 4\gamma_2^2 \cos^2\left(\frac{\pi s}{m}\right)},$$
$$(5.10)$$

which shows the presence of the forbidden zone even for $m = 3q$. However, this zone is much narrower compared to that for $m \neq 3q$.

The nontrivial electronic structure of SWNTs dictates their response to the electromagnetic field. Due to the quasi one-dimensional nature of SWNTs, their optical response is strongly anisotropic. The optical response to the axially polarized incident electric field significantly exceeds the optical response to the electric field polarized transversely to the CNT (carbon nanotube) axis (Tasaki et al., 1998; Milošević et al., 2003; Murakami et al., 2005). Due to the quantization of the transverse quasi momentum of electrons (Equation 5.7), divergences arise in the electronic density of states (DOS) of SWNTs (Saito et al., 1998). These divergences, which are known as Van Hove singularities, produce discrete energy levels or "subbands," the energy of which is determined solely by the chirality of SWNTs (Saito et al., 1998). As the inter-subband gap corresponds to the energy of infrared to visible light, the spectra of optical conductivity of an SWNT demonstrate the number of resonant lines in the region.

In the spectral range of 1–100 THz, the nonmonotonic frequency dependence of the reflectance and transmittance of CNT-based composite media that does not follow from the standard Drude theory has been observed (Ugawa et al., 1999; Ruzicka et al., 2000). Ugawa et al. (1999) found empirically that the effective permittivity of a CNT-based composite medium can be represented as a superposition of Drudian and Lorentzian functions. The spectral width of the resonance is of the order of the resonant frequency. The origin of this resonance could be attributed the inhomogeneously broadened geometric resonance in an isolated CNT (Slepyan et al., 2006).

5.3 Thermal Radiation from SWNTs

In this section, we investigate the thermal electromagnetic field radiated by an SWNT at temperature T placed in cold environment and show that the thermal radiation from metallic SWNTs can serve as an efficient source of the THz radiation. Our consideration is based on the method developed by Rytov (1958), which is known as fluctuational electrodynamics (see details in Lifshitz and Pitaevskii, 1980; Rytov et al., 1989; Joulain et al., 2005). The key idea of this method is that the thermal radiation sources in a material are the fluctuation currents, which are due to the random thermal motion of charged carriers the material consists. To determine the statistical properties of the electromagnetic field, we have to know the statistical properties of random currents and the radiation of the elemental volume of the material. The first information is given by the fluctuation-dissipative theorem while the second information is given by the Green tensor of the system. It should be noted that the application of the equilibrium laws, such as the fluctuation-dissipative theorem, is not very rigorous in this case, but it is justified when the role of the heat transport phenomena (such as thermal conductivity) is negligible. Hence, we will not consider them further.

The thermal radiation from SWNTs is of interest not only because of possible applications of THz device. Fundamental interest to the thermal radiation is dictated by the ability of nanostructures to change the *photonic local density of states* (LDOS), i.e., the electromagnetic vacuum energy (Agarwal, 1975; Joulain et al., 2003; Novotny and Hecht, 2006). The effect has been observed in microcavities, photonic crystals, and nanoparticles in the vicinity of surface-plasmon resonances (Novotny and Hecht, 2006). Thus, as the electromagnetic fluctuations are defined by photonic LDOS, the investigation of the thermal radiation is expected to bring new opportunities for the reconstruction of photonic LDOS in the presence of nanostructures. The apertureless scanning near-field optical microscopy provides a possibility for the experimental detection of LDOS (Joulain et al., 2003). In turn, the photonic LDOS is a key physical factor defining a set of well-known quantum electrodynamic effects: the Purcell effect and the electromagnetic friction (Novotny and Hecht, 2006), the Casimir–Lifshitz forces (Lifshitz and Pitaevskii, 1980), etc.

Thermal radiation in systems with surface plasmons is known to be considerably different from blackbody radiation (Carminati and Greffet, 1999; Henkel et al., 2000; Schegrov et al., 2000). Earlier theoretical studies of SWNTs showed the existence of low-frequency plasmon branches (Lin and Shung, 1993) and the formation of strongly slowed-down electromagnetic surface waves in SWNTs (Slepyan et al., 1999). Such waves define a pronounced Purcell effect in SWNTs (Bondarev et al., 2002) and the potentiality of SWNTs in the development of Cherenkov-type nano-emitters (Batrakov et al., 2006). Geometrical resonances—standing surface waves excited due to the strong reflection from the SWNT tips—qualitatively distinguish SWNTs from the planar structures investigated in Carminati and Greffet (1999), Henkel et al. (2000), and Schegrov et al. (2000). One can expect an essential role of these resonances in the formation of SWNTs' thermal radiation.

5.3.1 Fluctuation-Dissipative Theorem

The fluctuation-dissipative theorem relates the fluctuations of physical quantities to the dissipative properties of the system when it is subjected to an external action. We are interested in the space–time correlation function of the electromagnetic field fluctuations $\langle A_n(\mathbf{r},t)A_m(\mathbf{r}',t')\rangle$, where $\mathbf{A}(\mathbf{r},t)$ is the vector potential of the electromagnetic field. We use the Hamiltonian

gauge, which implies the scalar potential to be equal to zero for the electromagnetic field. For a stationary field, the correlation function depends on the time difference, $t - t'$, only. The Fourier transform of the correlation function is called the cross-spectral density (Joulain et al., 2005):

$$\left\langle A_n(\mathbf{r})A_m^*(\mathbf{r}')\right\rangle_\omega = \int_{-\infty}^{\infty} \left\langle A_n(\mathbf{r},t)A_m(\mathbf{r}',t')\right\rangle e^{i\omega(t-t')}d(t-t'). \quad (5.11)$$

Then the fluctuation-dissipative theorem for the electromagnetic field vector potential is formulated as follows (Lifshitz and Pitaevskii, 1980):

$$\left\langle A_n(\mathbf{r}_1)A_m^*(\mathbf{r}_2)\right\rangle_\omega = \left[\hbar + \frac{2\Theta(\omega,T)}{\omega}\right]\text{Im}\left[G_{nm}(\mathbf{r}_1,\mathbf{r}_2,\omega)\right], \quad (5.12)$$

where

$\underline{\mathbf{G}}(\mathbf{r}_1,\mathbf{r}_2,\omega)$ is the retarded Green tensor
$n,m = x,y,z$ designates the Cartesian coordinate system axis
$\Theta(\omega,T) = \hbar\omega/[\exp(\hbar\omega/k_B T)-1]$, $\hbar$ and k_B are the Planck and Boltzmann constants, respectively

The first term in square brackets is due to the zero vacuum fluctuations, and will be omitted further. Thus, to calculate the intensity of thermal radiation emitted by an SWNT, we elaborate the method of calculation of the electromagnetic field Green tensor in the vicinity of a CNT.

5.3.2 Free-Space Green Tensor

The electromagnetic field Green tensor is defined by the equation

$$(\nabla_{\mathbf{r}_1} \times \nabla_{\mathbf{r}_1} \times -k^2)\underline{\mathbf{G}}(\mathbf{r}_1,\mathbf{r}_2,\omega) = 4\pi\underline{\mathbf{I}}\delta(\mathbf{r}_1 - \mathbf{r}_2), \quad (5.13)$$

where

$\nabla_{\mathbf{r}_1}$ indicates that operator ∇ acts only on the variable $\mathbf{r}_1$ of the Green tensor
$\underline{\mathbf{I}}$ is the unit tensor
$k = \omega/c$, ω is the electromagnetic field frequency
c is the speed of light in vacuum

In general, this equation should be supplemented by boundary conditions.

In the Cartesian coordinate system, Equation 5.13 takes the following index form:

$$\left(\varepsilon_{iln}\varepsilon_{nkj}\frac{\partial^2}{\partial x_{1l}\partial x_{1k}} - k^2\delta_{ij}\right)G_{jm}(\mathbf{r}_1,\mathbf{r}_2,\omega) = 4\pi\delta_{im}\delta(\mathbf{r}_1 - \mathbf{r}_2), \quad (5.14)$$

where $x_{1l,k} = x_1, y_1, z_1$ and summation over the repeated indices is assumed. For each index m, Equation 5.14 gives us an independent equation that describes the evolution of the mth column of the Green tensor. Thus, for a given m, the column $G_{nm}(\mathbf{r}_1,\mathbf{r}_2,\omega)$ can formally be considered as a field vector, $\mathbf{G}^{(m)}(\mathbf{r}_1;\mathbf{r}_2)$, induced at point $\mathbf{r}_1$ by a delta source located at point $\mathbf{r}_2$; here m and $\mathbf{r}_2$ are parameters. Let us introduce the Hertz vector, $\Pi^{(m)}(\mathbf{r}_1;\mathbf{r}_2)$:

$$\mathbf{G}^{(m)}(\mathbf{r}_1;\mathbf{r}_2) = (k^2 + \nabla\nabla\cdot)\Pi^{(m)}(\mathbf{r}_1;\mathbf{r}_2) \qquad (5.15)$$

Then we obtain three independent equations instead of Equation 5.14:

$$(\Delta + k^2)\Pi^{(m)}(\mathbf{r}_1;\mathbf{r}_2) = -\frac{4\pi}{k^2}\mathbf{e}_m\delta(r_1 - r_2), \qquad (5.16)$$

where $\mathbf{e}_m = (\delta_{xm},\delta_{ym},\delta_{zm})$ is the basis vector of the Cartesian coordinate system. In the free-space case, the solution of Equation 5.16 is straightforward, (Jackson, 1999) and the free-space Hertz vector has the following form:

$$\Pi^{(0m)}(\mathbf{r}_1;\mathbf{r}_2) = \frac{1}{k^2}G^{(0)}(\mathbf{r}_1,\mathbf{r}_2,\omega) = \frac{1}{k^2}\frac{e^{ik|\mathbf{r}_1-\mathbf{r}_2|}}{|\mathbf{r}_1-\mathbf{r}_2|}. \qquad (5.17)$$

where

$$G^{(0)}(\mathbf{r}_1,\mathbf{r}_2,\omega) = \frac{e^{ik|\mathbf{r}_1-\mathbf{r}_2|}}{|\mathbf{r}_1-\mathbf{r}_2|} \qquad (5.18)$$

is the free-space Green function. Thus, we obtain the standard expression for the free-space Green tensor (Lifshitz and Pitaevskii, 1980):

$$\underline{\mathbf{G}}^{(0)}(\mathbf{r}_1,\mathbf{r}_2,\omega) = (\underline{\mathbf{I}} + k^{-2}\nabla_{\mathbf{r}_1}\otimes\nabla_{\mathbf{r}_1})G^{(0)}(\mathbf{r}_1,\mathbf{r}_2,\omega) \qquad (5.19)$$

with $\nabla_{\mathbf{r}_1}\otimes\nabla_{\mathbf{r}_1}$ as the operator dyadic acting on variables $\mathbf{r}_1$.

5.3.3 Green Tensor in the Vicinity of SWNT

Consider an isolated single-walled CNT of cross-sectional radius R_{cn} and length L, aligned along the z axis of the Cartesian coordinate basis (x, y, z) with the origin in the geometrical center of the CNT (see Figure 5.3). We restrict our consideration to the case $R_{cn} \ll 2\pi/k$, which implies that the incident field should be slowly varied within the CNT cross section.

To calculate the electromagnetic field Green tensor, we should solve Equation 5.13 with the effective boundary conditions (Slepyan et al., 1999) imposed on the SWNT surface. The general solution of the problem can be presented as follows:

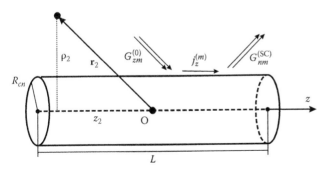

FIGURE 5.3 Free-space Green tensor scattering by an SWNT.

$$\underline{\mathbf{G}}(\mathbf{r}_1,\mathbf{r}_2,\omega) = \underline{\mathbf{G}}^{(0)}(\mathbf{r}_1,\mathbf{r}_2,\omega) + \underline{\mathbf{G}}^{(SC)}(\mathbf{r}_1,\mathbf{r}_2,\omega), \qquad (5.20)$$

where $\underline{\mathbf{G}}^{(0)}(\mathbf{r}_1,\mathbf{r}_2,\omega)$ is the solution of the inhomogeneous Equation 5.13, for the free-space case and $\underline{\mathbf{G}}^{(SC)}$ satisfies the equation

$$(\nabla_{\mathbf{r}_1}\times\nabla_{\mathbf{r}_1}\times -k^2)\underline{\mathbf{G}}^{(SC)}(\mathbf{r}_1,\mathbf{r}_2,\omega) = 0 \qquad (5.21)$$

and boundary conditions on the SWNT surface. From a formal point of view, $\underline{\mathbf{G}}^{(SC)}$ can be considered as the free-space Green tensor scattered by the SWNT (see Figure 5.3). To calculate the scattered Green tensor, we use the method developed in Lifshitz and Pitaevskii (1980, see problem 1 after paragraph 77). Each column of the free-space Green tensor induces current density $j_z^{(m)}$ on the SWNT surface, which generates the mth column of the scattered Green tensor, $m = x, y, z$. We take into account only the axial component of the induced current due to the fact that the SWNT length is much greater than the SWNT radius. By analogy with Section 5.3.2, we introduce three independent scattered Hertz vectors, $\Pi^{SC(m)}(\mathbf{r}_1;\mathbf{r}_2) = \mathbf{e}_z\Pi^{SC(m)}(\mathbf{r}_1;\mathbf{r}_2)$. Each Hertz vector has only z nonzero component. Further, we omit parameter $\mathbf{r}_2$ in the notation of the Hertz vector to simplify the designations. These Hertz vectors satisfy scalar equations,

$$(\Delta + k^2)\Pi^{SC(m)}(\mathbf{r}) = 0, \qquad (5.22)$$

and effective boundary conditions on the CNT surface. Thus, we have to solve three equations (Equation 5.22) to calculate scalar quantities $\Pi^{SC(m)}$. We could do this in the arbitrary coordinate system. We use the cylindrical coordinate system (ρ,ϕ,z) in which the effective boundary conditions (Slepyan et al., 1999, 2006) have the simplest form:

$$\left.\frac{\partial\Pi^{SC(m)}(\rho,z)}{\partial\rho}\right|_{\rho=R_{cn}+0} - \left.\frac{\partial\Pi^{SC(m)}(\rho,z)}{\partial\rho}\right|_{\rho=R_{cn}-0}$$

$$= \frac{4\pi}{i\omega}j_z^{(m)}(z;\mathbf{r}_2), \quad -L/2 \le z \le L/2 \qquad (5.23)$$

$$\left.\frac{\partial\Pi^{SC(m)}(\rho,z)}{\partial\rho}\right|_{\rho=R_{cn}+0} - \left.\frac{\partial\Pi^{SC(m)}(\rho,z)}{\partial\rho}\right|_{\rho=R_{cn}-0} = 0, \quad |z| > L/2, \quad (5.24)$$

where

$$j_z^{(m)} = \sigma_{zz}\left[\frac{\partial^2\Pi^{SC(m)}(R_{cn},z)}{\partial z^2} + k^2\Pi^{SC(m)}(R_{cn},z) + G_{zm}^{(0)}(\mathbf{R},\mathbf{r}_2,\omega)\right]. \qquad (5.25)$$

Vector $\mathbf{R}$ designates the point on the CNT surface, and in the cylindrical coordinate system, has the following form: $\{R_{cn},\phi,z\}$. Due to the cylindrical symmetry of the system, the scattered Hertz potential does not depend on the azimuthal variable, ϕ. Here,

$$\sigma_{zz}(\omega) = -\frac{2e^2}{\sqrt{3}\pi\hbar mb(\nu - i\omega)}\sum_{s=1}^{m}\int\frac{\partial\varepsilon_c(p_z,s)}{\partial p_z}\frac{\partial f(p_z,s)}{\partial p_z}dp_z \qquad (5.26)$$

is the axial conductivity of the zigzag SWNT (Slepyan et al., 1999), $f(p_z, s)$ is the equilibrium Fermi distribution, and $v = (1/3) \times 10^{12}$ s^{-1} is the relaxation frequency. Applying the Green theorem (Jackson, 1999) to Equations 5.22 through 5.24, we obtain the integral equations for the normalized axial current density, $j_z^{(m)}(z; \mathbf{r}_2)$, induced on the CNT surface by the incident electric field, $G_{zm}^{(0)}(\mathbf{R}, \mathbf{r}_2, \omega)$ (Nemilentsau et al., 2007):

$$\int_{-L/2}^{L/2} j_z^{(m)}(z'; \mathbf{r}_2) K(z - z') dz' + C_1 e^{-ikz} + C_2 e^{ikz}$$

$$= \frac{1}{2\pi} \int_{-L/2}^{L/2} \frac{e^{ik|z-z'|}}{2ik} \int_0^{2\pi} G_{zm}^{(0)}(\mathbf{R}', \mathbf{r}_2, \omega) d\phi' dz', \quad (5.27)$$

where $C_{1,2}$ are constants determined by the edge conditions, $j_z^{(m)}(\pm L/2; \mathbf{r}_2) = 0$ (Slepyan et al., 2006),

$$K(z) = \frac{\exp(ik|z|)}{2ik\sigma_{zz}(\omega)} - \frac{2iR_{cn}}{\omega} \int_0^{\pi} \frac{e^{ikr}}{r} d\phi, \quad (5.28)$$

and $r = \sqrt{z^2 + 4R_{cn}^2 \sin^2(\phi/2)}$. After three independent integral equations have been solved (Equation 5.27) (for three different values of $m = x, y, z$) and three current density values have been calculated, we again return to the Cartesian coordinate system.

Finally, we can present the solution of the scattering problem in the form of the simple layer potential (Colton and Kress, 1983; Nemilentsau et al., 2007):

$$G_{nm}(\mathbf{r}_1, \mathbf{r}_2, \omega) = G_{nm}^{(0)}(\mathbf{r}_1, \mathbf{r}_2, \omega) + \frac{i\omega R_{cn}}{c^2}$$

$$\times \int_{-L/2}^{L/2} j_z^{(m)}(z; \mathbf{r}_2) \int_0^{2\pi} G_{nz}^{(0)}(\mathbf{r}_1, \mathbf{R}, \omega) d\phi dz. \quad (5.29)$$

While deriving Equation 5.29, we assumed the incident field source distance from the CNT farther than its radius; therefore, we can neglect the dependence of the current $j_z^{(m)}$ on the azimuthal variable, ϕ.

Equation 5.29 with an arbitrary $j_z^{(m)}$ satisfies the aforementioned equation for the retarded Green tensor and the radiation condition at $|\mathbf{r}_1 - \mathbf{r}_2| \to \infty$. Peculiar electronic properties of CNTs (Dresselhaus et al., 2000) influence the Green tensor through the axial conductivity $\sigma_{zz}(\omega)$ (for details, see Slepyan et al. (1999)). The index m and the variable $\mathbf{r}_2$ appear in Equations 5.29 and 5.27 only as parameters. Note that these equations, as they couple the Green tensor of the system considered and the free-space Green tensor, play the role of the Dyson equation for CNTs.

It is important that the role of scattering by CNTs *is not reduced* to a small correction to the free-space Green tensor. This means that the Born approximation conventionally used for solving the Dyson equation (Lifshitz and Pitaevskii, 1980) becomes inapplicable to our case. Because of this, the direct

numerical integration of Equation 5.27 has been performed with integral operators approximated by a quadrature formula and subsequent transition to a matrix equation.

5.3.4 Thermal Radiation Calculation

Let us calculate the thermal radiation of a hot SWNT placed into an optically transparent cold environment. As the fluctuation-dissipative theorem (Equation 5.12) is applicable only at thermal equilibrium, we cannot directly apply it in this case.

To solve the problem, let us consider in more detail the case when the SWNT is in thermal equilibrium with the environment. At thermal equilibrium, the thermal fluctuation field in the system is the superposition of three different fields: thermal electromagnetic field radiated by the SWNT itself; blackbody radiation of the surrounding medium in the absence of the CNT, $\mathbf{A}^{(0)}$; and the field $\mathbf{A}^{(s)}$ resulting from the scattering of radiation of the medium by the SWNT. In the case of the cold medium, only the electromagnetic field radiated by the SWNT remains. Thus, to calculate the thermal radiation of the hot SWNT placed in the cold medium, we should calculate the total thermal electromagnetic field radiated in the equilibrium and separate the blackbody contribution.

The thermal radiation intensity in equilibrium is easily calculated by substituting Equation 5.29 for the electromagnetic field Green tensor to Equation 5.12 for the fluctuation-dissipative theorem. To separate the blackbody radiation contribution, we use the method developed in Lifshitz and Pitaevskii (1980, see problems after Sect. 77). We introduce the blackbody radiation vector potential,

$$\mathbf{A}^{(B)}(\mathbf{r}) = \mathbf{A}^{(0)}(\mathbf{r}) + \mathbf{A}^{(s)}(\mathbf{r}), \quad (5.30)$$

and calculate the correlator:

$$\left\langle A_n^{(B)}(\mathbf{r}_1) A_m^{(B)*}(\mathbf{r}_2) \right\rangle_\omega \equiv D_{nm}^{(B)}(\mathbf{r}_1, \mathbf{r}_2, \omega). \quad (5.31)$$

Then, the electric field intensity of the SWNT thermal radiation in the case when the SWNT temperature is much higher that the temperature of the surrounding medium, $I_\omega(\mathbf{r}_0) = |\mathbf{E}(\mathbf{r}_0)|^2$ is given, in view of the relation $E_n = -ikA_n$, by

$$I_\omega(\mathbf{r}_0) = k^2 \sum_{n=1}^{3} \left[\left\langle |A_n(\mathbf{r}_0)|^2 \right\rangle_\omega - D_{nn}^{(B)}(\mathbf{r}_0, \mathbf{r}_0, \omega) \right]. \quad (5.32)$$

To calculate the blackbody radiation correlator, we should calculate the scattered vector potential, $A^{(s)}$. To do this, we should solve Equations 5.22 through 5.25 with $A^{(0)}$ instead of the free-space Green tensor, $G_{zm}^{(0)}$, in Equation 5.25. By analogy with Equation 5.29, the vector $A_n^{(B)}(\mathbf{r}_1)$, potential, is written as

$$A_n^{(B)}(\mathbf{r}_1) = A_n^{(0)}(\mathbf{r}_1) + \frac{R_{cn}}{c} \int_{-L/2}^{L/2} j(z) \int_0^{2\pi} G_{nz}^{(0)}(\mathbf{r}_1, \mathbf{R}, \omega) d\phi dz, \quad (5.33)$$

where the current density, $j(z)$, induced on the SWNT surface by the free-space fluctuation electromagnetic field, $A^{(0)}$, is the solution of the integral equation

$$\int_{-L/2}^{L/2} K(z-z')j(z')dz' + C_1 e^{-ikz} + C_2 e^{ikz} = \frac{1}{2}\int_{-L/2}^{L/2} A_z^{(0)}(\mathbf{R}')e^{ik|z-z'|}dz'.$$

(5.34)

The second term in Equation 5.33 describes scattering of the free-space blackbody radiation by the SWNT. To calculate $D_{nm}^{(B)}$, we utilize Equation 5.33 and take into account that the correlator, $\left\langle A_n^{(0)}(\mathbf{r}_1)A_m^{(0)*}(\mathbf{r}_2)\right\rangle_\omega$, is defined by Equation 5.12 with the free-space Green tensor, $\mathbf{G}_{nm}^{(0)}(\mathbf{r}_1,\mathbf{r}_2,\omega)$, on the right-hand side.

5.3.5 Numerical Results

In this section, we present the results of the numerical calculations of the thermal radiation emitted by metallic (15,0) (see Figure 5.4) and semiconducting (23,0) SWNTs (see Figure 5.5). The following parametrization of the radius vector is used throughout this section:

$$\mathbf{r} = \{\rho,\phi,z\} = \mathbf{e}_x\rho\cos\phi + \mathbf{e}_y\rho\sin\phi + \mathbf{e}_z z.$$

(5.35)

The spectra of the thermal radiation from the SWNT (15,0) at different distances from the SWNT axis are presented in Figure 5.4a and b presents one of the spectra in the logarithmic scale. The spectrum depicted in Figure 5.4a demonstrates a number of equidistant discrete spectral lines with decreasing intensities superimposed by the continuous background. Such a structure

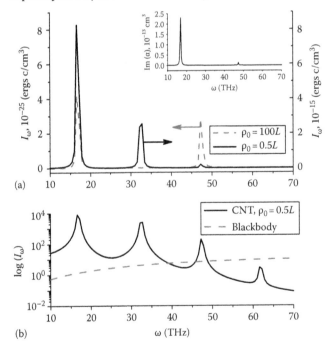

FIGURE 5.4 (a) Thermal radiation spectra of a metallic (15,0) SWNT. $\mathbf{r}_0 = \{100L,\phi_0,0\}$ (dashed line, left ordinate axis) and $\mathbf{r}_0 = \{0.5L,\phi_0,0\}$ (solid line, right ordinate axis); ϕ_0 is arbitrary; $T = 300\,\mathrm{K}$; and $L = 1\,\mu\mathrm{m}$. The inset presents the CNT's polarizability. (b) Thermal radiation from CNT in the near-field zone (solid line) compared to blackbody radiation, $I_\omega^{(B)}(\mathbf{r}_0) = 4\omega^2\Theta(\omega,T)/c^3$ (dashed line). (From Nemilentsau, A.M. et al., *Phys. Rev. Lett.*, 99, 147403, 2007.)

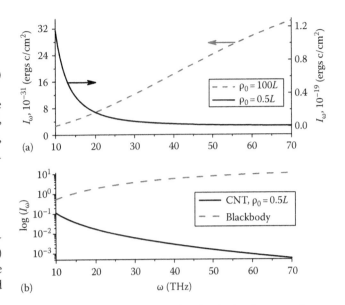

FIGURE 5.5 Same as Figure 5.4, but for the (23,0) semiconducting SWNT. (From Nemilentsau, A.M. et al., *Phys. Rev. Lett.*, 99, 147403, 2007.)

is inherent to spectra both in the far-field (dashed line) and near-field (solid line) zones. The peculiarity of the near-field zone is the presence of additional spectral lines absent in the far-field zone. Thus, the thermal radiation spectra presented in the figure qualitatively differ from both blackbody radiation (Lifshitz and Pitaevskii, 1980) and radiation of semi-infinite SiC samples (Schegrov et al., 2000). In the latter case, the discrete spectrum is observed only in the near-field zone (Schegrov et al., 2000).

The comparison of the thermal radiation and the SWNT's polarizability spectra (Slepyan et al., 2006) depicted in Figure 5.4a reveals the coincidence in the far-field zone of the thermal radiation resonances and the polarizability resonances. The latest are the *dipole* geometrical resonances of surface plasmons (Slepyan et al., 2006) defined by the condition $\mathrm{Re}[\kappa(\omega)]L \cong \pi(2s-1)$, with $\kappa(\omega)$ as the plasmon wavenumber; s is a positive integer. It should be noted that the polarizability (and the thermal radiation) resonances are found to be significantly shifted to the red as compared to the perfectly conducting wire of the same length, because of the strong slowing-down of surface plasmons in SWNTs: $\mathrm{Re}[\kappa(\omega)]/k \approx 100$ (Slepyan et al., 1999). In particular, for $L = 1\,\mu\mathrm{m}$, the geometrical resonances fall into the THz frequency range. The attenuation is small in a wide frequency range below the interband transitions. Additional spectral lines in the near-field zone are described by the condition $\mathrm{Re}[\kappa(\omega)]L \cong 2\pi s$. We refer to these resonances as *quadrupole* geometric resonances because the current density distribution for these modes is antisymmetrical with respect to $z = 0$ and, consequently, the dipole component of their field is identically zero. Thus, the resonant structure of the thermal radiation spectra is determined by the finite-length effects and also depends on the peculiar conductivity of SWNTs. Note that a similar structure of the thermal radiation spectra is predicted for the two-dimensional electron gas (Richter et al., 2007).

Resonances in the article by Richter et al. (2007) are due to the excitations of other physical nature—optical phonon modes of the barrier material.

The presence of singled out resonances illustrated in Figure 5.4a allows us to propose metallic SWNTs as far-field and near-field *thermal antennas* for the THz range (optical thermal antennas based on photonic crystals have recently been considered in the article by Laroche et al. (2006) and Florescu et al. (2007). Taking into account the high temperature stability of SWNTs, the SWNT thermal antennas can be excited by Joule heating from the direct electric current. Low-frequency alteration of the current allows the amplitude modulation of thermal emission and, consequently, allows the use of the thermal emission for information transmission (similar to modulated RF fields in present-day radioengineering). The scattering pattern of the thermal antenna can be calculated using the approaches developed in Hanson (2005) and Slepyan et al. (2006) and is found to be partially polarized and directional. A polarization of the thermal radiation from bundles of multi-walled CNTs has been observed experimentally in Li et al. (2003). The blackbody spectrum reported in Li et al. (2003) is due to the inhomogeneous broadening originated from the SWNT length and radius dispersion and multi-walled effects. Moreover, the observation was made above the frequency range of surface plasmons.

According to the article by Hanson (2005), Slepyan et al. (2006), and Burke et al. (2006) the maximal efficiency of vibrator SWNT antennas is reached at frequencies of the surface-plasmon dipole resonances. Figure 5.4a shows that the intensities of spectral lines of the thermal radiation go down with the resonance number much slowly than the polarizability peaks. This means that the signal–noise ratio for the SWNT-based antennas is maximal for the first resonance and decreases fast with the resonance number.

As different from metallic SWNTs, semiconducting ones do not reveal isolated resonances in both far-field and near-field zones (see Figure 5.5 as an illustration). Such a peculiarity can easily be understood by accounting for the strong attenuation of surface plasmons in semiconducting SWNTs, whereas the slowing down remains of the same order. That is why in this case the Q factor of geometrical resonances turns out to be substantially smaller, and the resonances do not manifest themselves as separated spectral lines. In the same way, the thermal radiation intensity of semiconducting SWNTs is substantially smaller than that of metallic ones and displays qualitatively different spectral properties in the near-field zone: monotonous growth of the intensity with frequency inherent to the far-field zone changes into monotonous declining (see Figure 5.5a). As the thermal spectra are strongly dependent on the SWNT conductivity type and length, the near-field thermal radiation spectroscopy proposed in Schegrov et al. (2000) for testing the surface-plasmon structures can be expanded to SWNTs.

Figure 5.5b demonstrates that in the frequency range considered, the blackbody radiation intensity considerably exceeds the thermal radiation of semiconducting SWNTs: $I_\omega \ll I_\omega^{(B)}$. In the regions between geometrical resonances, the same property is

inherent to metallic SWNTs (see Figure 5.4b). This means that CNTs as building blocks for nanoelectronics and nanosensorics possess uniquely low thermal noise and, thus, provide *high electromagnetic compatibility on the nanoscale*: Their contribution to the electromagnetic fluctuations in nanocircuits is negligibly small as compared to the contribution of dielectric substrate. More generally, the latter example illustrates the peculiarity of the electromagnetic compatibility problem on the nanoscale, motivating future research investments into the problem.

Next, we have studied the spatial structure of the electromagnetic fluctuations near SWNTs, characterized by the normalized first-order correlation tensor:

$$g_{nm}^{(1)}(\mathbf{r}_1, \mathbf{r}_2, \omega) = \frac{\left\langle A_n(\mathbf{r}_1) A_m^*(\mathbf{r}_2) \right\rangle_\omega}{\sqrt{\left\langle |A_n(\mathbf{r}_1)|^2 \right\rangle_\omega \left\langle |A_m(\mathbf{r}_2)|^2 \right\rangle_\omega}}. \tag{5.36}$$

The axial–axial component of this tensor is depicted in Figure 5.6. The figure clearly displays the distinctive behavior of the correlation in the far- and near-field zones. In the vicinity of geometrical resonances, where $\mathrm{Re}(\kappa)L \sim 1$, the near-field zone is defined by the condition $\rho \lesssim L$. Because of strong slowing down of surface plasmons in SWNTs (Slepyan et al., 1999), the latter condition corresponds to $k\rho \ll 1$; for the first geometrical resonance in the 1 pm length SWNT, depicted in Figure 5.6, $k\rho \lesssim 0.06$. Thus, the figure demonstrates a strong correlation between points inside the near-field zone (curves 1–3) and its fast falling down as ρ increases, indicating a weak correlation between near- and far-field zones. Physically, this is related to the fact that the dominant field component in the near-field zone is a nonradiative surface plasmon while radiative modes dominate in the far-field zone. The latter condition also explains that in the far-field zone, the correlation function is well approximated by the blackbody radiation law, $\sin(k\rho)/k\rho$, with the radial correlation length $\sim 1/k$. Note,

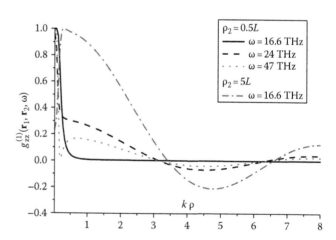

FIGURE 5.6 First-order correlation function, $g_{zz}^{(1)}(\mathbf{r}_0, \mathbf{r}_2, \omega)$, for a metallic (15,0) SWNT in radial direction. $\mathbf{r}_1 = \{\rho, \phi, 0\}$ with an arbitrary ϕ. Curves 1–3 depict correlation for $\mathbf{r}_2 = \{0.5L, \phi, 0\}$, while curve 4 presents the case $\mathbf{r}_2 = \{5L, \phi, 0\}$. (From Nemilentsau, A.M. et al., *Phys. Rev. Lett.*, 99, 147403, 2007.)

that the long-range spatial correlation characteristics for surface-plasmon planar structures (Carminati and Greffet, 1999; Henkel et al., 2000) are found to be absent in SWNTs. The reason is that SWNTs in the vicinity of geometrical resonances are electrically small oscillators ($kL \ll 1$) with wide scattering patterns (Slepyan et al., 2006).

5.4 Quasi-Metallic Carbon Nanotubes as Terahertz Emitters

The next scheme of the THz generation (Kibis and Portnoi, 2005; Kibis et al., 2007) is based on the electric-field induced heating of electron gas in an SWNT, resulting in the inversion of population of optically active states with the energy difference within the THz spectrum range. It is well known that the elastic backscattering processes in metallic SWNTs are strongly suppressed (Ando et al., 1997), and in a high-enough electric field, charge carriers can be accelerated up to the energy allowing emission of optical/zone-boundary phonons. At this energy, corresponding to the frequency of about 40 THz, the major scattering mechanism switches on abruptly, resulting in current saturation(Yao et al., 2000; Freitag et al., 2004; Javey et al., 2004; Park et al., 2004; Perebeinos et al., 2005). As will be shown hereafter, for certain types of carbon nanotubes, the heating of electrons to the energies below the phonon-emission threshold results in a spontaneous THz emission with the peak frequency controlled by an applied voltage.

The electron energy spectrum, $\varepsilon(k)$ (Equation 5.8), of a metallic SWNT in the vicinity of the Fermi energy linearly depends on the electron wave vector, k, and has the form $\varepsilon(k) = \pm \hbar v_F |k - k_0|$, where $v_F \approx 9.8 \times 10^5$ m/s is the Fermi velocity of graphene, which corresponds to the commonly used tight-binding matrix element, $\gamma_0 = 3.033$ eV (Saito et al., 1998; Reich et al., 2004). Here and in what follows, the zero of energy is defined as the Fermi

energy position in the absence of an external field. When the voltage, V, is applied between the SWNT ends, the electron distribution is shifted in the way shown by the thick lines in Figure 5.7a, corresponding to the filled electron states.

This shift results in inversion of population and, correspondingly, in optical transitions between filled states in the conduction band and empty states in the valence band. The spectrum of optical transitions is determined by the distribution function for hot carriers that, in turn, depends on the applied voltage and scattering processes in the SWNT. It is well known that the major scattering mechanism in SWNTs is due to the electron–phonon interaction (Yao et al., 2000; Javey et al., 2004; Park et al., 2004; Perebeinos et al., 2005). Since the scattering processes erode the inversion of electron population, an optimal condition for observing the discussed optical transitions takes place when the length of the SWNT $L < l_{ac}$, where the electron mean-free path for acoustic phonon scattering is $l_{ac} \approx 2\,\mu m$ (Park et al., 2004). Further, only such short SWNTs with ideal Ohmic contacts (Javey et al., 2004) are considered in the ballistic transport regime, when the energy acquired by the electron on the whole length of the tube, $\Delta \varepsilon = eV$, does not exceed the value of $\hbar \Omega = 0.16$ eV at which a fast emission of high-energy phonons begins (Park et al., 2004). In this so-called low-bias regime (Yao et al., 2000; Javey et al., 2004; Park et al., 2004), in which the current in the nanotube is given by the Büttiker–Landauer-type formula, $I \approx (4e^2/h)V$, the distribution function of hot electrons is

$$ f_e(k) = \begin{cases} 1, & 0 < k - k_0 < \Delta \varepsilon / 2\hbar v_F \\ 0, & k - k_0 > \Delta \varepsilon / 2\hbar v_F \end{cases}. \qquad (5.37) $$

The distribution function for hot holes, $f_h(k)$, has the same form as $f_e(k)$.

Let us select an SWNT with a crystal structure most suitable for the observation of the discussed effect. First, the required nanotube should have metallic conductivity, and, second, the

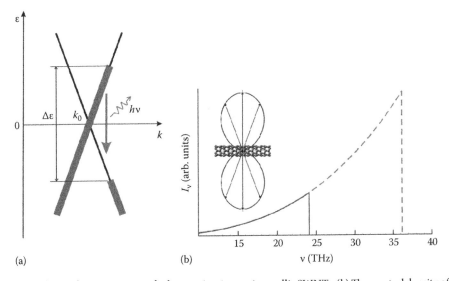

FIGURE 5.7 (a) The scheme of THz photon generation by hot carriers in quasi-metallic SWNTs. (b) The spectral density of spontaneous emission as a function of frequency for two values of applied voltage: solid line for $V = 0.1$ V and dashed line for $V = 0.15$ V. The inset shows the directional radiation pattern of the THz emission with respect to the nanotube axis. (From Portnoi, M.E. et al., *Superlattices Microstruct.*, 43, 399, 2008.)

optical transitions between the lowest conduction subband and the top valence subband should be allowed. SWNTs having a true metallic energy band structure, for which the energy gap is absent for any SWNT radius, are armchair (*n,n*) SWNTs only (Kane and Mele, 1997; Ouyang et al., 2001; Reich et al., 2004; Gunlycke et al., 2006; Li et al., 2006). However, for armchair SWNTs, the optical transitions between the first conduction and valence subbands are forbidden (Milošević et al., 2003; Jiang et al., 2004). So, for the observation of THz generation, it is suitable to use the so-called quasi-metallic (*n, m*) SWNTs with $n - m = 3p$, where *p* is a nonzero integer. These nanotubes, which are gapless within the frame of a simple zone-folding model of the π-electron graphene spectrum (Saito et al., 1998), are in fact narrow-gap semiconductors due to curvature effects. Their bandgap is given by $\varepsilon_g = \hbar v_F b \cos 3\theta/(8R_{cn}^2)$ (Kane and Mele, 1997; Gunlycke et al., 2006). It can be seen from the expression for ε_g that the gap is decreasing rapidly with the increasing nanotube radius. For large values of R_{cn}, this gap can be neglected even in the case of moderate applied voltages due to Zener tunneling of electrons across the gap. It is easy to show in a fashion similar to the original Zener's work (Zener, 1934) that the tunneling probability in quasi-metallic SWNTs is given by $\exp(-\alpha \varepsilon_g^2/eE\hbar v_F)$, where α is a numerical factor close to unity.* For example, for a zigzag (30,0) SWNT the gap is $\varepsilon_g \approx 6\,\text{meV}$, and the Zener breakdown takes place for the electric field $E \sim 10^{-1}\,\text{V}/\mu\text{m}$. Since almost the whole voltage drop in the ballistic regime occurs within the few-nanometer regions near the contacts (Svizhenko and Anantram, 2005), a typical bias voltage of 0.1 V corresponds to an electric field, which is more than sufficient to achieve a complete breakdown. In what follows, all calculations are performed for a zigzag (3*p*, 0) SWNT of large enough radius, R_{cn}, and for applied voltages exceeding the Zener breakdown, so that the finite-gap effects can be neglected. The obtained results can be easily generalized for any quasi-metallic large-radius SWNT.

Optical transitions in SWNTs have been a subject of extensive research (see, e.g., Grüneis et al., 2003; Milošević et al., 2003; Jiang et al., 2004; Popov and Henrard, 2004; Saito et al., 2004; Goupalov, 2005; Oyama et al., 2006). Let us treat these transitions using the results of the nearest-neighbor orthogonal π-electron tight-binding model (Saito et al., 1998). Despite its apparent simplicity and well-known limitations, this model has been extremely successful in describing low-energy optical spectra and electronic properties of SWNTs (see, e.g., Sfeir et al. (2006) for one of the most recent manifestations of this model's success). The main goal is to calculate the spectral density of spontaneous emission, I_v, which is the probability of optical transitions per unit time for the photon frequencies in the interval (v, v + *d*v) divided by *d*v. In the dipole approximation (Berestetskii et al., 1997), this spectral density is given by

---

* For the energy spectrum near the band edge given by $\varepsilon = \pm \left[\varepsilon_g^2/4 + \hbar^2 v_F^2 (k - k_0)^2 \right]^{1/2}$, it can be shown that $\alpha = \pi/4$.

$$I_v = \frac{8\pi e^2 v}{3c^3} \sum_{i,f} f_e(k_i) f_h(k_f) \left| \left\langle \Psi_f \,|\, \hat{v}_z \,|\, \Psi_i \right\rangle \right|^2 \delta(\varepsilon_i - \varepsilon_f - hv). \quad (5.38)$$

Equation 5.38 contains the matrix element of the electron velocity operator. In the frame of the tight-binding model, this matrix element for optical transitions between the lowest conduction and the highest valence subbands of the (3*p*, 0) zigzag SWNT can be written as (Jiang et al., 2004; Grüneis et al., 2003)

$$\left\langle \Psi_f \,|\, \hat{v}_z \,|\, \Psi_i \right\rangle = \frac{b\omega_{if}}{8} \delta_{k_f, k_i}, \quad (5.39)$$

where $b\omega_{if} = \varepsilon_i - \varepsilon_f$ is the energy difference between the initial (*i*) and the final (*f*) state. These transitions are associated with the light polarized along the nanotube axis *z*, in agreement with the general selection rules for SWNTs (Milošević et al., 2003). Substituting Equation 5.39 in Equation 5.38 and performing necessary summation, we get

$$I_v = Lf_e(\pi v / v_F) f_h(\pi v / v_F) \frac{\pi^2 e^2 b^2 v^3}{6c^3 \hbar v_F}. \quad (5.40)$$

Equation 5.40 has broader applicability limits than the considered case of $L < l_{ac}$ and $eV < \hbar\Omega$, in which the distribution functions for electrons and holes are given by Equation 5.37. In the general case, there is a strong dependence of I_v on the distribution functions, which have to be calculated taking into account all the relevant scattering mechanisms (Yao et al., 2000; Javey et al., 2004; Park et al., 2004; Perebeinos et al., 2005). In the discussed ballistic regime, the spectral density has a universal dependence on the applied voltage and photon frequency for all quasi-metallic SWNTs. In Figure 5.7b, the spectral density is shown for two values of the voltage. It is clearly seen that the maximum of the spectral densities of emission has a strong voltage dependence and lies in the THz frequency range for experimentally attainable voltages. The directional radiation pattern, shown in the inset of Figure 5.7b, reflects the fact that the emission of light polarized normally to the nanotube axis is forbidden by the selection rules for the optical transitions between the lowest conduction subband and the top valence subband.

For some device applications, it might be desirable to emit photons propagating along the nanotube axis, which is possible in optical transitions between the SWNT subbands characterized by angular momenta differing by one (Milošević et al., 2003; Reich et al., 2004). To achieve the emission of these photons by electron heating, it is necessary to have an intersection of such subbands within the energy range accessible to electrons accelerated by attainable voltages. From the analysis of different types of SWNTs, it follows that the intersection is possible, for example, for the lowest conduction subbands in several semi-conducting zigzag nanotubes and in all armchair nanotubes. However, for an effective THz emission from these nanotubes, it is necessary to move the Fermi level very close to the subband intersection point (Kibis and Portnoi, 2005). Therefore, obtaining the THz emission propagating along the nanotube axis is a more difficult technological problem than generating the emission shown in Figure 5.7b.

5.5 Chiral Carbon Nanotubes as Frequency Multipliers

Another proposal for using SWNTs for THz applications (Kibis et al., 2008; Portnoi et al., 2008) is based on chiral nanotubes, which represent natural super-lattices. For example, a (10, 9) SWNT has a radius that differs from the radius of the most commonly studied (10,10) nanotube by less than 5%, whereas a translational period, T, along the axis of the (10,9) SWNT is almost 30 times larger than the period of the (10, 10) nanotube. Correspondingly, the first Brillouin zone of the (10, 9) nanotube is 30 times smaller than the first zone for the (10,10) tube. However, such a Brillouin zone reduction cannot influence electronic transport unless there is a gap opening between the energy subbands resulting from the folding of the graphene spectrum. It can be shown that an electric field normal to the nanotube axis opens noticeable gaps at the edge of the reduced Brillouin zone, thus turning a long-period nanotube of certain chirality into a "real" superlattice. This gap opening is a general property of chiral nanostructures exposed to a transverse electric field (Kibis et al., 2005a,b; Kibis and Portnoi, 2007, 2008). The field-induced gaps are most pronounced in $(n, 1)$ SWNTs (Kibis et al., 2005a,b; Portnoi et al., 2006, 2008).

Figure 5.8a shows the opening of an electric-field induced gap near the edge of the Brillouin zone of a (6,1) SWNT. This gap opening results in the appearance of a negative effective-mass region in the nanotube energy spectrum. The typical electron energy in this part of the spectrum of 15 meV is well below the optical phonon energy $\hbar\Omega \approx 160$ meV, so that it can easily be accessed in moderate heating electric fields. The negative effective mass results in NDC, as can be seen from Figure 5.8b. The NDC characteristic presented in Figure 5.8b is calculated assuming the energy-independent scattering time $\tau = 1$ ps. However, when the carrier energy reaches the optical or edge-phonon energy, the scattering time, τ, increases abruptly. This results in more pronounced NDC, which can be used for generating electromagnetic radiation in the

THz range. In fact, recent Monte Carlo simulations (Akturk et al., 2007a,b) show that the phonon-induced effects alone might result in THz current oscillations in SWNTs.

The effect of the negative effective mass in chiral nanotubes (Portnoi et al., 2008) not only results in NDC but also leads to an efficient frequency multiplication in the THz range. The results of calculations of the electron velocity in the presence of the time-dependent longitudinal electric field are presented in Figure 5.9. One of the advantages of a frequency multiplier based on chiral SWNTs, in comparison with the conventional superlattices (Alekseev et al., 2006), is that the dispersion relation in such a system can be controlled by the transverse electric field, $E_\perp$.

5.6 Armchair Nanotubes in a Magnetic Field as Tunable THz Detectors and Emitters

The problem of detecting THz radiation is known to be at least as challenging as creating reliable THz sources. The proposal of a novel detector (Kibis et al., 2008; Portnoi et al., 2008) is based on several features of truly gapless (armchair) SWNTs. The main property to be utilized is opening of a bandgap in these SWNTs in a magnetic field along the nanotube axis (Saito et al., 1998; Reich et al., 2004). For a (10,10) SWNT, this gap corresponds to approximately 1.6 THz in the field of 10 T. For attainable magnetic fields, the gap grows linearly with increasing both the magnetic field and the nanotube radius. It can be shown (Portnoi et al., 2008) that the same magnetic field also allows dipole optical transitions between the top valence subband and the lowest conduction subband, which are strictly forbidden in armchair SWNTs without the field (Milošević et al., 2003).

In Figure 5.10, it is shown how the energy spectrum and matrix elements of the dipole optical transitions polarized along the nanotube axis are modified in the presence of a longitudinal magnetic field. In the frame of the nearest-neighbor tight-binding model, one can show that for a (n,n) armchair nanotube the squared matrix element of the velocity operator between the states at the edge of the gap opened by the magnetic field is given by a simple analytic expression

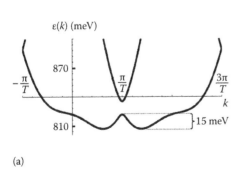

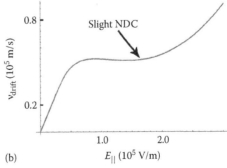

(a)   (b)

FIGURE 5.8 (a) Energy spectrum of the (6,1) SWNT in a transverse electric field, $E_\perp = 4$ V/nm. (b) The electron drift velocity in the lowest conduction subband of a (6, 1) SWNT as a function of the longitudinal electric field in the presence of acoustic phonon scattering. (From Portnoi, M.E. et al., *Superlattices Microstruct.*, 43, 399, 2008.)

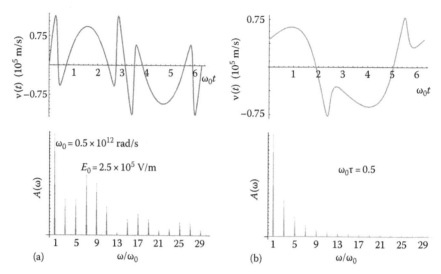

FIGURE 5.9 Time dependence of the electron velocity in the lowest conduction subband of a (6,1) SWNT under the influence of a pump harmonic longitudinal electric field, $E_\parallel(t) = E_0\sin(\omega_0 t)$, and its correspondent spectral distribution, $A(\omega)$: (a) in the ballistic transport regime and (b) in the presence of scattering with the relaxation time $\tau = 10^{-12}$ s. (From Portnoi, M.E. et al., *Superlattices Microstruct.*, 43, 399, 2008.)

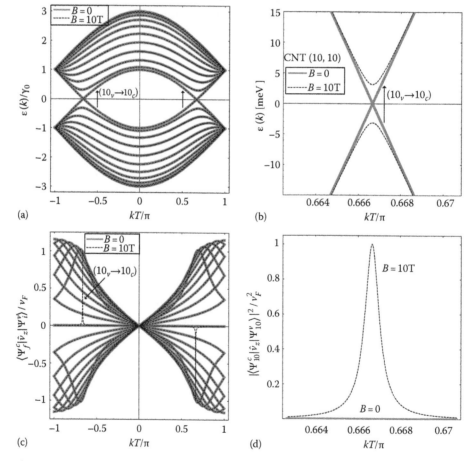

FIGURE 5.10 (a) Band structure of a (10, 10) nanotube, with and without an external magnetic field along the nanotube axis. (b) Detailed view of the gap, which is opened between the top valence subband and the lowest conduction subband in an external field, $B = 10$ T. (c) The change in the matrix elements of the dipole optical transitions, for the light polarized along the SWNT axis, due to the introduction of the external magnetic field. The only appreciable change is in the appearance of a high, narrow peak associated with the transition $(10_v \rightarrow 10_c)$, which is not allowed in the absence of the magnetic field. Here and in what follows, the energy subbands are numbered in the same way as in Saito et al. (1998). (d) Dependence of the squared dipole matrix element for the transition $(10_v \rightarrow 10_c)$ on the 1D wave vector, k, with and without an external magnetic field. (From Portnoi, M.E. et al., *Superlattices Microstruct.*, 43, 399, 2008.)

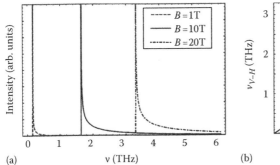
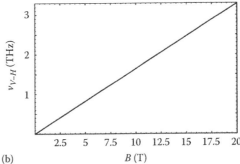

FIGURE 5.11 (a) Calculated photon absorption spectra for a (10,10) SWNT for three different magnetic field values. The absorption intensity is proportional to the product of $\left|\left\langle \Psi_{10}^{v} \left| \hat{v}_z \right| \Psi_{10}^{c} \right\rangle\right|^2$ and the joint DOS. (b) Dependence of the position of the peak in the absorption intensity associated with the Van Hove singularity on the magnetic field. (From Portnoi, M.E. et al., *Superlattices Microstruct.*, 43, 399, 2008.)

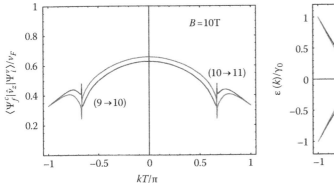
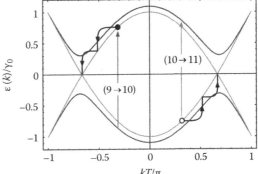

FIGURE 5.12 A scheme for creating a population inversion between the lowest conduction subband and the top valence subband of an armchair SWNT in a magnetic field. The left plot shows the calculated matrix elements of the relevant dipole optical transitions polarized normally to the axis of a (10, 10) SWNT. The right plot shows several energy subbands closest to the Fermi level and illustrates the creation of photoexcited carriers and their nonradiative thermalization. (From Portnoi, M.E. et al., *Superlattices Microstruct.*, 43, 399, 2008.)

$$\left|\left\langle \Psi_n^{v} \left| \hat{v}_z \right| \Psi_i^{c} \right\rangle\right|^2 = \frac{4}{3}\left[1 - \frac{1}{4}\cos^2\left(\frac{f}{n}\pi\right)\right]v_F^2, \qquad (5.41)$$

where $f = eBR^2/(2\hbar)$. For experimentally attainable magnetic fields, when the magnetic flux through the SWNT is much smaller than the flux quantum, the absolute value of the velocity operator is close to v_F. Equation 5.41 is relevant to the transitions between the highest valence subband and the lowest conduction subband only for $f \leq 1/2$, since for the higher values, the order of the nanotube subbands is changed. Notably, the same equation allows to obtain the maximum value of the velocity operator in any armchair SWNT for the transitions polarized along its axis: this value cannot exceed $2v_F/\sqrt{3}$ (see Figure 5.10c).

The electron (hole) energy spectrum near the bottom (top) of the bandgap produced by the magnetic field is parabolic as a function of a carrier momentum along the nanotube axis. This dispersion results in a Van Hove singularity in the reduced DOS that, in turn, leads to a very sharp absorption maximum near the band edge and, correspondingly, to a very

high sensitivity of the photocurrent to the photon frequency (see Figure 5.11).

Notably, the same effect can be used for the generation of a very narrow emission line having the peak frequency tunable by the applied magnetic field. A population inversion can be achieved, for example, by optical pumping with the light polarized normally to the nanotube axis, as shown in Figure 5.12.

5.7 Conclusion

In this chapter we have demonstrated several novel schemes for the emission and detection of the THz radiation by the SWNTs.

We have demonstrated that intensity spectra of the thermal electromagnetic field emitted by metallic SWNTs of micron length reveal resonances in the THz range. These resonances are the geometrical resonances of the surface plasmons in SWNTs. It is quite important that we could vary the resonance frequency by changing the SWNT length or conductivity. The predicted effect allows to formulate the conception of metallic SWNT as a thermal antenna in the THz range and is of importance for the SWNT spectroscopy, the nanoantenna design, the high-resolution near-field optical microscopy, and the thermal noise control in nanocircuits.

We also have demonstrated that a quasi-metallic carbon nanotube can emit the THz radiation when potential difference is applied to its ends. The typically required voltages and nanotube parameters are similar to those available in the state-of-the-art transport experiments. The maximum of the spectral density of emission is shown to have a strong voltage dependence, which is universal for all quasi-metallic carbon nanotubes in the ballistic regime. Therefore, the discussed effect can be used for creating a THz source with frequency controlled by the applied voltage. Appropriately arranged arrays of nanotubes should be considered as promising candidates for active elements of amplifiers and generators of coherent THz radiation.

We have also shown that an electric field, which is applied normally to the axis of long-period chiral nanotubes, significantly modifies their band structure near the edge of the Brillouin zone. This results in the negative effective-mass region at the energy scale below the high-energy phonon-emission threshold. This effect can be used for an efficient frequency multiplication in the THz range. Finally, we have discussed the feasibility of using the effect of the magnetic field, which opens energy gaps and allows optical transitions in armchair nanotubes, for creating tunable THz detectors and emitters.

Acknowledgments

This work was supported by INTAS (project 05-1000008-7801) and the EU FP7 TerACaN (project FP7-230778), Royal Society (U.K.); BRFFR (Belarus) projects F07F-013 and F08R-009, RFBR (Russia) project 08-02-90004; and "Development of Scientific Potential of Russian Higher Education" program (project 2.1.2/2115). RFBR (Russia) projects 10-02-00077 and 10-02-90002, and Russian Federal goal-oriented program "Scientific and Scientific-Educational Personnel of Innovative Russia."

References

Agarwal, G. S. (1975). Quantum electrodynamics in the presence of dielectrics and conductors. I. Electromagnetic-field response functions and black-body fluctuations in finite geometries. *Phys. Rev. A 11*, 230–242.

Akturk, A., N. Goldsman, and G. Pennington (2007a). Self-consistent ensemble Monte Carlo simulations show terahertz oscillations in single-walled carbon nanotubes. *J. Appl. Phys. 102*, 073720.

Akturk, A., N. Goldsman, and G. Pennington (2007b). Terahertz current oscillations in single-walled zigzag carbon nanotubes. *Phys. Rev. Lett. 98*, 166803.

Alekseev, K. N., M. V. Gorkunov, N. V. Demarina, T. Hyart, N. V. Alexeeva, and A. V. Shorokhov (2006). Suppressed absolute negative conductance and generation of high-frequency radiation in semiconductor superlattices. *Europhys. Lett. 73*, 934–940.

Ando, T., T. Nakanishi, and R. Saito (1997). Impurity scattering in carbon nanotubes—Absence of back scattering. *J. Phys. Soc. Jpn. 67*, 1704–1713.

Batrakov, K. G., P. P. Kuzhir, and S. A. Maksimenko (2006). Radiative instability of electron beam in carbon nanotubes. *Proc. SPIE 6328*, 63280Z.

Berestetskii, V. B., E. M. Lifshitz, and L. P. Pitaevskii (1997). *Quantum Electrodynamics*. Oxford, U.K.: Butterworth-Heinemann.

Bondarev, I. V., G. Y. Slepyan, and S. A. Maksimenko (2002). Spontaneous decay of excited atomic states near a carbon nanotube. *Phys. Rev. Lett. 89*, 115504.

Burke, P. J., S. Li, and Z. Yu (2006). Quantitative theory of nanowire and nanotube antenna performance. *IEEE Trans. Nanotechnol. 5*, 314–334.

Carminati, R. and J.-J. Greffet (1999). Near-field effects in spatial coherence of thermal sources. *Phys. Rev. Lett. 82*, 1660–1663.

Charlier, J.-C., X. Blase, and S. Roche (2007). Electronic and transport properties of nanotubes. *Rev. Mod. Phys. 70*, 677–732.

Colton, D. and R. Kress (1983). *Integral Equation Methods in Scattering Theory*. New York: Wiley.

Di Carlo, A., A. Pecchia, E. Petrolati, and C. Paoloni (2006). Modelling of carbon nanotube-based devices: From nanofets to THz emitters. *Proc. SPIE 6328*, 632808.

Dragoman, D. and M. Dragoman (2004a). Terahertz fields and applications. *Prog. Quantum Electron. 28*, 1–66.

Dragoman, D. and M. Dragoman (2004b). Terahertz oscillations in semiconducting carbon nanotube resonant-tunnelling diodes. *Physica E 24*, 282–289.

Dragoman, D. and M. Dragoman (2005). Terahertz continuous wave amplification in semiconductor carbon nanotubes. *Physica E 25*, 492–496.

Dragoman, M., A. Cismaru, H. Hartnagel, and R. Plana (2006). Reversible metal-semiconductor transitions for microwave switching applications. *Appl. Phys. Lett. 88*, 073503.

Dresselhaus, M. (2004). Nanotube antennas. *Nature (London) 432*, 959.

Dresselhaus, M. S., G. Dresselhaus, and P. Avouris (Eds.) (2000). *Carbon Nanotubes*. Berlin, Germany: Springer.

Florescu, M., K. Busch, and J. P. Dowling (2007). Thermal radiation in photonic crystals. *Phys. Rev. B 75*, 201101(R).

Freitag, M., V. Pereibenos, J. Chen et al. (2004). Hot carrier electroluminescence from a single carbon nanotube. *Nano Lett. 4*, 1063–1066.

Goupalov, S. V. (2005). Optical transitions in carbon nanotubes. *Phys. Rev. B 72*, 195403.

Grüneis, A., R. Saito, G. G. Samsonidze et al. (2003). Inhomogeneous optical absorption around the k point in graphite and carbon nanotubes. *Phys. Rev. B 67*, 165402.

Gunlycke, D., C. J. Lambert, S. W. D. Bailey, D. G. Pettifor, G. A. D. Briggs, and J. H. Jefferson (2006). Bandgap modulation of narrow-gap carbon nanotubes in a transverse electric field. *Europhys. Lett. 73*, 759–764.

Hanson, G. W. (2005). Fundamental transmitting properties of carbon nanotube antennas. *IEEE Trans. Antennas Propag. 53*, 3426–3435.

Henkel, C., K. Joulain, R. Carminati, and J.-J. Greffet (2000). Spatial coherence of thermal near-fields. *Opt. Commun. 186*, 57–67.

Jackson, J. D. (1999). *Classical Electrodynamics*. New York: Wiley.

Javey, A., J. Guo, M. Paulsson et al. (2004). High-field quasiballistic transport in short carbon nanotubes. *Phys. Rev. Lett. 92*, 106804.

Jiang, J., R. Saito, A. Gruneis, G. Dresselhaus, and M. S. Dresselhaus (2004). Optical absorption matrix elements in single-wall carbon nanotubes. *Carbon 42*, 3169–3176.

Joulain, K., R. Carminati, J.-P. Mulet, and J.-J. Greffet (2003). Definition and measurement of the local density of electromagnetic states close to an interface. *Phys. Rev. B 68*, 245405.

Joulain, K., J.-P. Mulet, F. Marquier, R. Carminati, and J.-J. Greffet (2005). Surface electromagnetic waves thermally excited: Radiative heat transfer, coherence properties and casimir forces revisited in the near field. *Surf. Sci. Rep. 57*, 59–112.

Kane, C. L. and E. J. Mele (1997). Size, shape, and low energy electronic structure of carbon nanotubes. *Phys. Rev. Lett. 78*, 1932–1935.

Kibis, O. V. and M. E. Portnoi (2005). Carbon nanotubes: A new type of emitter in the terahertz range. *Tech. Phys. Lett. 31*, 671–672.

Kibis, O. V. and M. E. Portnoi (2007). Semiconductor nanohelix in electric field: A superlattice of the new type. *Tech. Phys. Lett. 33*, 878–880.

Kibis, O. V. and M. E. Portnoi (2008). Superlattice properties of semiconductor nanohelices in a transverse electric field. *Physica E 40*, 1899–1901.

Kibis, O. V., S. V. Malevannyy, L. Hugget, D. G. W. Parfitt, and M. E. Portnoi (2005a). Superlattice properties of helical nanostructures in a transverse electric field. *Electromagnetics 25*, 425–435.

Kibis, O. V., D. G. W. Parfitt, and M. Portnoi (2005b). Superlattice properties of carbon nanotubes in a transverse electric field. *Phys. Rev. B 71*, 035411.

Kibis, O. V., M. R. da Costa, and M. E. Portnoi (2007). Generation of terahertz radiation by hot electrons in carbon nanotubes. *Nano Lett. 7*, 3414–3417.

Kibis, O. V., M. R. da Costa, and M. E. Portnoi (2008). Carbon nanotubes as a basis for novel terahertz devices. *Physica E 40*, 1766–1768.

Laroche, M., R. Carminati, and J.-J. Greffet (2006). Coherent thermal antenna using the photonic slab. *Phys. Rev. Lett. 96*, 123903.

Lee, M. and M. C. Wanke (2007). Searching for a solid-state terahertz technology. *Science 316*, 64–65.

Léonard, F. and J. Tersoff (2000). Negative differential resistance in nanotube devices. *Phys. Rev. Lett. 85*, 4767–4780.

Li, P., P. Li, K. Jiang, M. Liu, Q. Li, S. Fana, and J. Sun (2003). Polarized incandescent light emission from carbon nanotubes. *Appl. Phys. Lett. 82*, 1763–1765.

Li, Y., U. Ravaioli, and S. V. Rotkin (2006). Metal-semiconductor transition and fermi velocity renormalization in metallic carbon nanotubes. *Phys. Rev. B 73*, 034415.

Lifshitz, E. M. and L. P. Pitaevskii (1980). *Statistical Physics: Theory of the Condensed State*. Oxford, U.K.: Pergamon Press.

Lin, M. F. and D. S. Chuu (1998). Electronic states of toroidal carbon nanotubes. *J. Phys. Soc. Jpn. 67*, 259–263.

Lin, M. F. and W.-K. Shung (1993). Elementary excitations in cylindrical tubules. *Phys. Rev. B 47*, 6617–6624.

Lu, C., L. An, Q. Fu, J. Liu, H. Zhang, and J. Murduck (2006). Schottky diodes from asymmetric metal-nanotube contacts. *Appl. Phys. Lett. 88*, 133501.

Maksimenko, A. S. and G. Y. Slepyan (2000). Negative differential conductivity in carbon nanotubes. *Phys. Rev. Lett. 84*, 362–365.

Manohara, H. M., M. J. Bronikowski, M. Hoenk, B. D. Hunt, and P. H. Siegel (2005). High-current-density field emitters based on arrays of carbon nanotube bundles. *J. Vac. Sci. Technol. B 23*, 157–161.

Milošević, I., T. Vuković, S. Dmitrović, and M. Damnjanović (2003). Polarized optical absorption in carbon nanotubes: A symmetry-based approach. *Phys. Rev. B 67*, 165418.

Murakami, Y., S. M. E. Einarsson, and T. Edamura (2005). Polarization dependent optical absorption properties of single-walled carbon nanotubes and methodology for the evaluation of their morphology. *Carbon 43*, 2664–2667.

Nemilentsau, A. M., G. Y. Slepyan, and S. A. Maksimenko (2007). Thermal radiation from carbon nanotubes in the terahertz range. *Phys. Rev. Lett. 99*, 147403.

Novotny, L. and B. Hecht (2006). *Principles of Nano-Optics*. Cambridge, U.K.: Cambridge University Press.

Odintsov, A. A. (2000). Schottky barriers in carbon nanotube heterojunctions. *Phys. Rev. Lett. 85*, 150–153.

Ouyang, M., J.-L. Huang, C. L. Cheung, and C. M. Lieber (2001). Energy gaps in "metallic" single-walled carbon nanotubes. *Science 292*, 702–705.

Oyama, Y., R. Saito, K. Sato et al. (2006). Photoluminescence intensity of single-wall carbon nanotubes. *Carbon 44*, 873–879.

Park, J.-Y., S. Resenblatt, Y. Yaish et al. (2004). Electron-phonon scattering in metallic single-wall carbon nanotubes. *Nano Lett. 4*, 517–520.

Pennington, G. and N. Goldsman (2003). Semiclassical transport and phonon scattering of electrons in semiconducting carbon nanotubes. *Phys. Rev. B 68*, 045426.

Perebeinos, V., J. Tersoff, and P. Avouris (2005). Electron-phonon interaction and transport in semiconducting carbon nanotubes. *Phys. Rev. Lett. 94*, 086802.

Popov, V. N. and L. Henrard (2004). Comparative study of the optical properties of single-walled carbon nanotubes within orthogonal and nonorthogonal tight-binding models. *Phys. Rev. B 70*, 115407.

Portnoi, M. E., O. V. Kibis, and M. R. da Costa (2006). Terahertz emitters and detectors based on carbon nanotubes. *Proc. SPIE 6328*, 632805.

Portnoi, M. E., O. V. Kibis, and M. R. da Costa (2008). Terahertz applications of carbon nanotubes. *Superlattices Microstruct. 43*, 399–407.

Reich, S., C. Thomsen, and J. Maultzsch (2004). *Carbon Nanotubes: Basic Concepts and Physical Properties*. Berlin, Germany: Wiley.

Richter, M., S. Butscher, M. Schaarschmidt, and A. Knorr (2007). Model of thermal terahertz light emission from a two-dimensional electron gas. *Phys. Rev. B 75*, 115331.

Ruzicka, B., L. Degiorgi, R. Gaal, L. Thien-Nga, R. Bacsa, J.-P. Salvetat, and L. Forro (2000). Optical and dc conductivity study of potassium-doped single-walled carbon nanotube films. *Phys. Rev. B 61*, R2468–R2471.

Rytov, S. (1958). Correlation theory of thermal fluctuations in isotropic medium. *Sov. Phys. JETP 6*, 130–140.

Rytov, S., Y. Kravtsov, and V. Tatarskii (1989). *Principles of Statistical Radiophysics*, Volume 3. Berlin, Germany: Springer-Verlag.

Saito, R., G. Dresselhaus, and M. S. Dresselhaus (1998). *Physical Properties of Carbon Nanotubes*. London, U.K.: Imperial College Press.

Saito, R., A. Grüneis, G. G. Samsonidze et al. (2004). Optical absorption of graphite and single-wall carbon nanotube. *Appl. Phys. A 78*, 1099–1105.

Schegrov, A. V., K. Joulain, R. Carminati, and G.-G. Greffet (2000). Near-field spectral effects due to electromagnetic surface excitations. *Phys. Rev. Lett. 85*, 1548–1551.

Sfeir, M. Y., T. Beetz, F. Wang et al. (2006). Optical spectroscopy of individual single-walled carbon nanotube of defined chiral structure. *Science 312*, 554–556.

Slepyan, G. Y., S. A. Maksimenko, V. P. Kalosha, J. Herrmann, E. E. B. Campbell, and I. V. Hertel (1999). Highly efficient high-order harmonic generation by metallic carbon nanotubes. *Phys. Rev. A 60*, R777–R780.

Slepyan, G. Y., S. A. Maksimenko, V. P. Kalosha, A. V. Gusakov, and J. Herrmann (2001). High-order harmonic generation by conduction electrons in carbon nanotubes ropes. *Phys. Rev. A 63*, 053808.

Slepyan, G. Y., M. Shuba, S. Maksimenko, and A. Lakhtakia (2006). Theory of optical scattering by achiral carbon nanotubes and their potential as optical nanoantennas. *Phys. Rev. B 73*, 195416.

Svizhenko, A. and M. P. Anantram (2005). Effect of scattering and contacts on current and electrostatics in carbon nanotubes. *Phys. Rev. B 72*, 085340.

Tasaki, S., K. Maekawa, and T. Yamabe (1998). π-Band contribution to the optical properties of carbon nanotubes: Effects of chirality. *Phys. Rev. B 57*, 9301–9318.

Ugawa, A., A. G. Rinzler, and D. B. Tanner (1999). Far-infrared gaps in single-wall carbon nanotubes. *Phys. Rev. B 60*, R11305–R110308.

Wallace, P. R. (1947). The band theory of graphite. *Phys. Rev. 71*, 622–634.

Yang, M. H., K. B. K. Teo, W. I. Milne, and D. G. Hasko (2005). Carbon nanotube schottky diode and directionally dependent field-effect transistor using asymmetrical contacts. *Appl. Phys. Lett. 87*, 253116.

Yao, Z., C. Kane, and C. Dekker (2000). High-field electrical transport in single-wall carbon nanotubes. *Phys. Rev. Lett. 84*, 2941–2944.

Zener, C. (1934). A theory of the optical breakdown of solid dielectrics. *Proc. R. Soc. (Lond.) 145*, 523–529.

6

Isotope Engineering in Nanotube Research

Ferenc Simon
*Budapest University of
Technology and Economics*

and

Universität Wien

6.1 Introduction

The fields of nanosciences received an enormous boost with the discovery of carbon nanotubes (CNTs) by Sumio Iijima in 1991 (Iijima 1991). Before 1991, nanoscience and nanotechnology usually meant small clusters of atoms or molecules of seemingly fundamental interest only. The discovery of fullerenes in 1985 (Kroto et al. 1985) revolutionized several fields in chemistry, physics, and also in biomedical sciences. Fullerenes gave material scientists a fresh look at carbonaceous systems: it suggested that there may exist a number of other forms of carbon that await discovery. The discovery of Iijima fulfilled this expectation and brought the attention to nanosciences, even though the discovery of nanotubes had been reported before (Monthioux and Kuznetsov 2006). In fact, he was using the same apparatus that was used for the production of fullerenes. The originally discovered multiwall carbon nanotubes (MWCNTs) were soon followed by the discovery of single-wall carbon nanotubes (SWCNTs) (Bethune et al. 1993, Iijima and Ichihashi 1993), which can be grown in similar conditions as the fullerenes and MWCNTs but exclusively in the presence of metal catalysts.

SWCNTs are the one-dimensional allotropes of carbon, completing the list of zero- (the fullerenes (Kroto et al. 1985)), two- (the graphene (Novoselov et al. 2004)), and three-dimensional (graphite and diamond) carbon allotropes. An interesting property of SWCNTs is that all constituent carbons are equivalent and closely sp^2 bound, like in graphite, which provides unique mechanical and transport properties. This, combined with their huge, >1000, aspect ratio (the diameters being 1–20 nm and their lengths over 1 micron), endows them with an enormous

application potential and a range of unique and exotic physical properties.

The not exhaustive list of applications includes field-emission displays due to their sharp tips (Obraztsov et al. 2000), cathode emitters for small-sized x-ray tubes for medical applications (Yue et al. 2002), reinforcing elements for CNT–metal composites, tips for scanning probe microscopy (Hafner et al. 1999), high-current-transmitting wires, cables for a future space elevator, elements of nanotransistors (Bachtold et al. 2001), and elements for quantum information processing (Harneit et al. 2002).

However, several fundamental questions need to be answered before the benefits of these novel nanostructures can be fully exploited. Recent theoretical and experimental efforts focused on the understanding of the electronic and optical properties of SWCNTs. It has been long thought that the one-dimensional structure of SWCNTs renders their electronic properties inherently one dimensional (Hamada et al. 1992, Saito et al. 1998). This was suggested to result in a range of exotic correlated phenomena such as the Tomonaga–Luttinger-liquid (TLL) state (Egger and Gogolin 1997), the Peierls transition (Bohnen et al. 2004, Connétable et al. 2005), ballistic transport, and bound excitons (Kane and Mele 2003, Perebeinos et al. 2004, Spataru et al. 2004). The presence of the TLL state is established (Bockrath et al. 1999, Ishii et al. 2003, Rauf et al. 2004), there is evidence for the ballistic transport properties (Tans et al. 1997) and there is growing experimental evidence for the presence of excitonic effects (Maultzsch et al. 2005, Wang et al. 2005).

Isotope engineering is a useful tool in material science. It can be defined as isotope enrichment of materials with a control over the isotope allocation. The fundamental property of isotope engineering lies in the fact that to a good approximation (this is the so-called Born–Oppenheimer approximation) a different isotope leaves the electronic properties unaffected while changing the energy of the vibrations only. Therefore, isotope engineering provides an extra degree of freedom for studies of physical phenomena wherever the energy of vibrations (or phonons in a solid) are involved. Such physical phenomena includes, e.g., phonon-mediated superconductivity (Bardeen et al. 1957) and phonon-dominated heat conduction (Capinski et al. 1997). Isotope engineering facilitates the interpretation of vibrational spectra such as, e.g., in infrared and Raman spectroscopies (Cardona 2005).

An additional benefit of isotope engineering is the ability to change the nucleus itself. Different nuclei (1) have different nuclear spin, I, which allows to perform nuclear magnetic resonance (NMR) and (2) allow nuclear reactions with well-defined end-products when an external irradiation is used. Several stable and most abundant isotopes such as, e.g., the ^{16}O and ^{12}C have $I = 0$, which are NMR silent, but their isotopes, ^{17}O and ^{13}C, have $I = \frac{1}{2}$, which make them suitable NMR probes. A particularly compelling use of isotope engineering is the layer-selective phosphorus doping of isotope engineered and ready-prepared Si heterostructures by means of neutron irradiation (Meese 1979). A recent proposal suggested the use of isotope engineering to provide the basic architecture for spintronics and quantum computing (Shlimak 2004). In biomedical sciences, isotope engineering is commonly used to allow, e.g., molecule site-specific NMR experiments and to trace in vivo reactions with radioactive agents such as tritium, ^{3}H.

The most abundant isotope of carbon is ^{12}C; however, 1.1% of natural carbon is the stable ^{13}C isotope. ^{13}C-enriched carbon is available commercially in the form of enriched organic solvents such as ^{13}C-benzene, fullerenes, or graphite. ^{13}C-enriched SWCNTs have only been prepared for scientific research. Here, we show how ^{13}C isotope engineering can be used to study the properties of SWCNTs. The examples include the study of SWCNT growth, the vibrational properties including the identification of previously unknown vibrational modes, and the NMR investigations aimed at understanding the nature of the correlated ground state in SWCNTs. This chapter is organized as follows: first, the fundamentals of the physics of SWCNTs are described; the applied experimental tools and the synthesis methods for isotope engineering are presented next. Finally, the knowledge gained using isotope engineering on the vibrational properties and on the correlated ground state in SWCNTs is discussed.

6.2 State of the Art of SWCNT Research

6.2.1 Geometry and Electronic Properties of SWCNTs

CNTs can be represented as rolled-up graphene sheets, i.e., single layers of graphite. Depending on the number of coaxial CNTs, they are usually classified into MWCNTs and SWCNTs.

Some general considerations have been clarified in the last 19 years of nanomaterial research related to these structures. MWCNTs are more homogeneous in their physical properties as the large number of coaxial tubes smears out individual tube properties. This makes MWCNTs suitable candidates for applications where their nanometer size and the conducting properties can be exploited as, e.g., nanometer-sized wires. In contrast, SWCNT materials are grown as an ensemble of weakly interacting tubes with different diameters. The physical properties of similar diameter SWCNTs can change dramatically as the electronic structure is very sensitive to the rolling-up direction, the so-called chiral vector (Hamada et al. 1992, Saito et al. 1998). We show the geometry of a graphene sheet and the folding/chiral vector in Figure 6.1.

The chiral vector is characterized by the (n,m) vector components that denote the direction along which a graphene sheet is rolled up to form a nanotube.

The (n,m) indices determine the diameter, d, of the SWCNTs:

$$d = \frac{a_0\sqrt{n^2 + m^2 + n \cdot m}}{\pi}, \qquad (6.1)$$

where $a_0 = 0.2461$ nm is the lattice constant of graphene (Saito et al. 1998). It turned out that for a given sample, the diameters of the SWCNTs follow a Gaussian distribution, which is characterized by the mean diameter, d_{mean}, and its variance, σ. In Figure 6.2, we show the diameter distribution for a typical sample with $d_{mean} = 1.4$ nm and $\sigma = 0.1$ nm. In addition to their local needle-like structure, SWCNTs arrange themselves in a three-dimensional, closely packed hexagonal lattice due to van der Waals forces, which are the so-called bundles. The tube wall-to-tube wall distance is close to the 3.35 Å found in graphite (Saito et al. 1998).

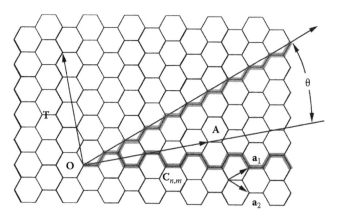

FIGURE 6.1 Geometry of a graphene sheet. a_1 and a_2 are the primitive lattice vectors. **A** is the vector that joins two carbons which become identical after the rolling-up of a stripe cut out along the vector **T**. θ is measured between a_1 and **A** and is called the chiral angle. The Hamada vectors for an armchair (lower solid curve) and a zigzag (upper solid curve) are indicated.

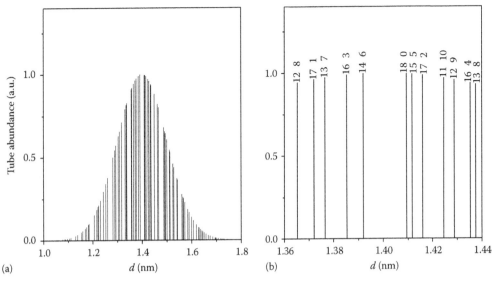

FIGURE 6.2 Diameter distribution for a typical SWCNT (a) and diameter distribution on a magnified scale showing the chiral indices as well (b). Vertical lines represent a geometrically allowed chirality.

The quasi one-dimensional structure of SWCNTs is reflected in their electronic properties: there is a quasi-continuous electron dispersion along the k direction that corresponds to the tube axis, and we find discrete states along the other two directions. The SWCNT band structure can be derived from that of graphene. The latter is a zero band metal where valence and conduction bands touch at the edges of the Brillouin zone at 6 points (the K points). Upon rolling up the graphene sheet in real space, the SWCNT valence and conduction bands can meet at the K points (giving a metallic SWCNT) or miss each other (making a semiconducting or insulating SWCNT). Simple geometry rules summarize the metallicity versus chiral indices dependence: an SWCNT is metallic if $(n − m)$ mod 3 = 0 and it is semiconducting if $(n − m)$ mod 3 ≠ 0. These rules apply to SWCNTs with $d \geq 1.5$ nm as for smaller diameters curvature effects, i.e., the deviation of the local bonding from the planar sp^2 arrangement, play an important role (Zólyomi and Kürti 2004). According to the above rules, an SWCNT sample contains bundles with mixed semiconducting and metallic SWCNTs with a 2:1 abundance. Clearly, this property severely limits applicability of the tubes in, e.g., nanoelectronics, where well-defined metallicity is desired.

We find Van Hove singularities in the density of states (DOS) of SWCNTs due to their quasi one-dimensionality. The singularities are symmetric to the Fermi energy when calculated in the tight-binding (TB) model (Dresselhaus et al. 2001), and we show the DOS for two SWCNTs in Figure 6.3, which are representative for the metallic (the 10,10 tube) and for the semiconducting (the 11,9 tube) SWCNTs. These tubes have very similar diameters according to Equation 6.1, while their electronic property is very different. The Van Hove singularities were first detected using scanning tunneling spectroscopy (Wildör et al. 1998). The selection rules allow optical transitions between symmetric Van Hove singularity pairs only. The optical-transition energies of the metallic and semiconducting SWCNTs are also different. Denoting

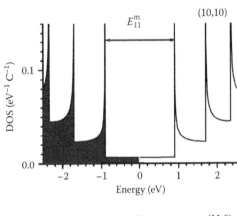

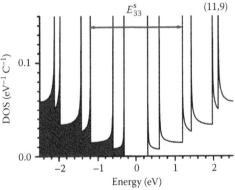

FIGURE 6.3 Density of states for the metallic (10,10) and the semiconducting (11,9) SWCNTs near the Fermi energy calculated with the tight-binding model. Filled and open areas indicate filled and empty states. We give as an example the first (E_{11}^{m}) optical transition for the metallic, and the third optical transition (E_{33}^{s}) for the semiconducting nanotube.

the optical transitions for metallic (m) and semiconducting (s) SWCNTs with $E_{ii}^{m/s}$, where ii is the index of the singularity pair, the TB calculation yields (Kataura et al. 1999)

$$E_{ii}^{s} \approx i \cdot \frac{0.85 \,[\text{eV nm}]}{d \,[\text{nm}]},$$

$$E_{ii}^{m} \approx i \cdot \frac{2.55 \,[\text{eV nm}]}{d \,[\text{nm}]}.$$

(6.2)

The TB result shows that the optical-transition energies are inversely proportional to the diameter, and that they are very different for metallic and semiconducting tubes. The presence of Van Hove singularities dominate the optical properties of SWCNTs and have been experimentally confirmed using Raman spectroscopy (Kuzmany et al. 2001, Fantini et al. 2004, Telg et al. 2004) and band-gap fluorescence (Bachilo et al. 2002). More recent "symmetry-adapted TB" calculations by Popov (2004) have refined Equation 6.2 and have shown that characteristic deviations arise from the simplest $E_{ii} \propto 1/d$ rule for tubes in the same "families", i.e., for which $2n + m = $ const.

As mentioned, the presence of locally mixed semiconducting and metallic nanotubes significantly limits the range of applications. To date, neither the chirality controlled growth nor the selection of SWCNTs with a well-defined chiral vector has been performed successfully. Correspondingly, current research is focused on the post-synthesis separation of SWCNTs with a narrow range of chiralities (Chattopadhyay et al. 2003, Chen et al. 2003, Krupke et al. 2003, Zheng et al. 2003) or with separated metallic and semiconducting nanotubes (Arnold et al. 2006). Additionally, methods that yield information that are specific to SWCNTs with different chiralities are important. Examples for the latter are the observation of chirality-selective band-gap fluorescence in semiconducting SWCNTs (Bachilo et al. 2002) and chirality-assigned resonant Raman scattering (Fantini et al. 2004, Telg et al. 2004).

6.2.2 Synthesis of SWCNTs

Synthesis methods of SWCNTs share the common ingredient that they all employ a catalyst, unlike the synthesis of MWCNTs or fullerenes. The catalysts are usually transition metal elements with 3d (such as Ni, Co, and Fe), 4d (e.g., Y, Mo, Ru, Rh, Pd), or with 5d electron shells (e.g., Pt). There are two major synthesis methods: one is based on the evaporation of a graphite source and other kinds are the chemical vapor deposition (CVD) methods. Graphite evaporation–based methods are the laser ablation (i.e., evaporating a graphitized carbon + catalyst target by a powerful laser) and the arc-discharge method (two graphite rods are evaporated by a high-current arc in between), which are performed in helium atmosphere. For the CVD methods, carbon source is some small organic molecule such as, e.g., ethanol, while a powder of the catalyst is placed in the reactor that is kept at a few hundred degrees centigrade.

The commonly accepted model of SWCNT growth (Reich et al. 2004) states that the small catalyst particles absorb carbon and form metastable metal carbides. Upon further absorption, they become saturated with carbon, and the spontaneous growth of SWCNTs starts in a "hedgehog"-like fashion. The end

of the nanotubes are closed with caps. As the reaction proceeds, further absorbed carbons continue the growth of the SWCNTs. The SWCNT phase has a larger surface energy compared to the MWCNTs due to the larger curvature. Therefore, the effect of the metals is to lower this surface energy, thus allowing the formation of SWCNTs. The growth of SWCNTs is essentially a random process with no preferred chirality of a nanotube, however, experimental conditions such as the temperature and the type of catalyst selects a nominal diameter (Kataura et al. 1999). This results in the already discussed Gaussian distribution of tube diameters.

6.2.3 Raman Spectroscopy of SWCNTs

We discuss Raman scattering as it is one of the most important experimental methods of the SWCNT research. Raman spectroscopy is an inelastic light-scattering method. The Raman process involves the absorption and reemission of light quanta whose energy differs by the energy of a vibration (a vibration for a molecule and an optical phonon for a solid). The process is called, correspondingly, Stokes or anti-Stokes if a vibration is induced or carried away by the light. The Raman process is of a very low efficiency, and laser sources in combination with highly dispersive optics are used to enable the efficient discrimination of the inelastically scattered light from the elastically scattered stray light.

An important property of Raman scattering, which is particularly pronounced for SWCNTs, is the so-called resonant Raman enhancement. It turns out that the Raman process can be described by the creation of a short living quasi-particle that carries the energy and momentum of the incoming light. The probability of creation and decay of the quasi-particle according to the Raman process is orders of magnitude enhanced if either the incoming laser energy (the so-called incoming resonance) or the outgoing light energy (the so-called outgoing resonance) matches an optical eigen-transition of the system. This situation is depicted in Figure 6.4, and the corresponding Raman cross section (or Raman intensity) is given by (Kuzmany 1998)

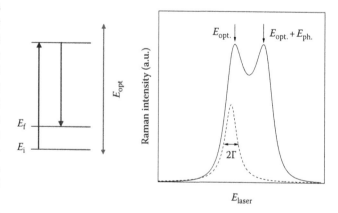

FIGURE 6.4 Left panel: Schematics of the Stokes Raman process, E_i and E_f are the initial and final states, respectively. Right panel: Schematics of the resonant Raman intensity as a function of the exciting laser energy. Dashed curve indicates the incoming resonant curve alone and the linewidth, 2Γ, as defined in the text.

$$I(E_1) \propto M_{\text{eff}}^4 \left| \int \frac{(E_1 - E_{\text{ph}})^2 \cdot g_{\text{JDS}}(\varepsilon) d\varepsilon}{(E_1 - \varepsilon - i\Gamma)(E_1 - E_{\text{ph}} - \varepsilon - i\Gamma)} \right|^2, \quad (6.3)$$

where $I(E_1)$ is the Raman intensity for a given exciting laser energy, E_1. The Raman intensity gives the proportion between the number of incoming and scattered photons for a solid angle of unity. M_{eff} is the effective electron–phonon coupling constant that determines the probability of the quasi-particle creation and its decay to the outcoming photon and a phonon. E_{ph} is the energy of the investigated vibration or phonon and Γ is the so-called damping parameter and $\Gamma = \hbar/\tau$, where τ is the lifetime of the excited quasi-particle. $g_{\text{JDS}}(\varepsilon)$ is the strength or density of the optical transition that depends on the occupation of the starting and final states of the optical transitions (the *Joint Density of States*). $g_{\text{JDS}}(\varepsilon)$ is directly related to the strength of the Van Hove singularities.

As mentioned, SWCNTs have particularly strong optical transitions due to the Van Hove singularities in the DOS. This means that with a particular choice, Raman spectroscopy can be tuned to particular SWCNT chiralities that are in resonance. This photoselective property is exploited extensively for the characterization of SWCNT samples with unknown chirality distributions.

The energy difference between the outgoing and incoming light is negative for the Stokes process which is by definition a positive Raman shift. It directly gives the energy of Raman-allowed vibrations. In Figure 6.5, we show a typical Raman spectrum. We also give the labeling of the most important Raman-active modes. As SWCNTs are macromolecules, their vibrational states can be referred to as both vibrations in the molecular language

or as phonons in the solid-state physics terminology. The strongest *G* (or graphitic) mode is related to the tangential motion of carbon, the *D* mode is related to a non $k \neq 0$ phonon and is defect induced (Thomsen and Reich 2000, Kürti et al. 2002, Zólyomi et al. 2003), and the *G'* mode is its overtone, i.e., when two phonons of the same energy are involved. The most important and unique vibration of SWCNTs is the so-called radial breathing mode or RBM. The motion of carbons is what the name indicates, and it is depicted in the inset of Figure 6.5. The principal interest in this mode comes from the fact that its energy is inversely proportional to the diameter of an SWCNT (Kürti et al. 1998):

$$\nu_{\text{RBM}} = C_1/d + C_2, \quad (6.4)$$

where $C_1 \sim 230\,\text{cm}^{-1}\text{nm}$ and $C_2 \sim 10\,\text{cm}^{-1}$ are constants. C_2 accounts for the tube–tube interactions in the tube bundles. Equation 6.4 allows to determine the diameter of the SWCNTs in a sample. For example, for the Raman spectrum in Figure 6.5, the RBM is around $\nu_{\text{RBM}} \approx 170\,\text{cm}^{-1}$, thus it is a sample with a mean diameter of $d_{\text{mean}} \approx 1.4\,\text{nm}$.

6.2.4 Modified SWCNTs

As mentioned in the introduction, SWCNTs and fullerenes are both carbon allotropes with different dimensionality. A significant event that further connects nanotubes and fullerenes was the discovery of fullerene *peapods*. In 1998, Smith, Monthioux, and Luzzi reported the observation of fullerenes encapsulated inside SWCNTs using high-resolution transmission electron microscopy (HR-TEM) (Smith et al. 1998). The schematics of the structure is shown in Figure 6.6. It was suggested that fullerenes that are coproduced with the SWCNTs enter the tubes through openings and remain inside as it is energetically preferred as it was shown by first principles calculations (Berber et al. 2002, Melle-Franco et al.

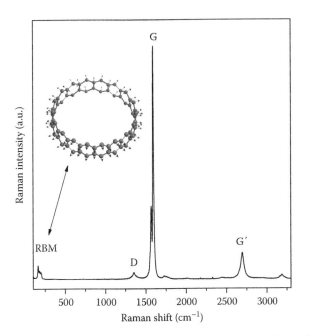

FIGURE 6.5 Raman spectrum of a typical SWCNT sample taken with $\lambda = 488\,\text{nm}$ (2.54 eV) laser excitation at room temperature. Labeling of the most important Raman-active vibrational modes are given. Inset indicates the motion of carbons for the radial breathing mode.

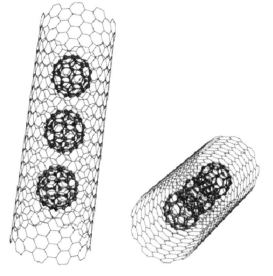

FIGURE 6.6 Schematics of the peapod structure from two viewpoints.

2003, Otani et al. 2003, Rochefort 2003, Dubay and Kresse 2004). The stability of this structure can be understood if we consider that the nominal diameter of the C_{60}s is 0.7 nm. Therefore an SWCNT with $d \approx 1.4$ can accommodate them so that the fullerene–tube wall distance is close to the optimal van der Waals distance of 0.335 nm (Dresselhaus et al. 1996).

It was also shown that macroscopic or close-packed filling with the fullerenes can be achieved (Smith et al. 1999, Smith and Luzzi 2000, Kataura et al. 2001). For such peapods, the presence of C_{60}s can be observed by scanning tunneling spectroscopy through the influence on the electronic properties of the tubes (Hornbaker et al. 2002). The high filling opens the way for magnetic resonance studies through encapsulation of isotope engineered fullerenes and the results are discussed herein.

The high-yield synthesis of peapods is achieved by opening the SWCNTs by oxidation in air at 400°C–500°C for about 0.5 h duration. The opened SWCNTs are sealed in a glass tube together with abundant fullerene powder under vacuum. The glass tube is then heated to 650°C for 2 h (Kataura et al. 2001). Fullerenes sublime above ~350°C, thus encapsulation proceeds due to the high fullerene vapor pressure inside the glass tube. Non-encapsulated fullerenes can be removed by washing in toluene or by heating the peapod material to 800°C while being continuously evacuated. Peapods of functionalized fullerenes were also effectively encapsulated and their removal was suggested to be suitable for drug delivery (Simon et al. 2007).

The fate of fullerenes and nanotubes were further linked by the observation that upon intensive electron irradiation (Smith et al. 1999) or upon heating to 1200°C (Smith and Luzzi 2000, Bandow et al. 2001) the encapsulated fullerenes are transformed to inner tubes. These inner tubes are SWCNTs such as the host outer tubes, and the whole structure forms the so-called double-wall carbon nanotube (DWCNT) structure, which was extensively studied using Raman spectroscopy (Bandow et al. 2001, Kramberger et al. 2003, Pfeiffer et al. 2003). In Figure 6.7, we show the changes of the Raman spectra of the tubes upon fullerene encapsulation and

transformation to DWCNTs. A number of narrow lines develop in the 220–370 cm^{-1} Raman shift range, which indicates the presence of smaller-diameter SWCNTs with $d \approx 0.6..1$ nm according to Equation 6.4. These modes were identified as the RBMs of the inner tubes (Bandow et al. 2001, Pfeiffer et al. 2003).

The inner tubes grown inside SWCNTs from peapods turned out to be a particularly interesting system as they are remarkably defect free and are isolated from the environment, which results in very long phonon lifetimes, i.e., very narrow vibrational modes (Pfeiffer et al. 2003). In addition, their smaller diameter results in a larger spectral splitting for diameter-dependent phonon modes such as, e.g., the radial breathing mode. These two effects make the inner tubes very suitable to study diameter-dependent physics of the small-diameter tubes with precision.

Here, we show that using ^{13}C-enriched fullerenes as starting materials for the DWCNT synthesis, DWCNTs can be synthesized where the outer shell consists of natural carbon whereas the inner shell is ^{13}C enriched. This allows a nanotube-specific enrichment. Usual ^{13}C enrichment of nanotubes involves their synthesis from enriched graphite (Tang et al. 2000, Rümmeli et al. 2007). However, the resulting material contains other carbonaceous phases, such as graphite nanoparticles that are also isotope enriched. The use of the isotope engineered DWCNTs is twofold: on one hand they facilitate the identification of inner tube Raman modes, on the other hand they are excellent probes for NMR, which requires the presence of a sizeable amount of ^{13}C isotope.

Alternatively, DWCNTs can be produced with usual synthesis methods such as arc discharge (Hutchison et al. 2001) or chemical vapor deposition, CVD, (Ren et al. 2002) under special conditions. However, such methods do not allow SWCNT-specific isotope enrichment and the side-product carbon phases are also isotope enriched. Therefore, such methods are not considered as isotope engineering.

6.3 Isotope Engineering of SWCNTs

6.3.1 SWCNT-Specific Isotope Engineering

The ability to grow inner tubes from encapsulated fullerenes provided the idea to grow inner tubes from isotope-enriched fullerenes. Such fullerenes are available commercially (MER Corp., Tucson, Arizona) and are produced by the standard Krätschmer–Huffmann process (Krätschmer et al. 1990), i.e., by an arc-discharge synthesis from ^{13}C graphite rods. Two supplier-specified grades of ^{13}C-enriched fullerene mixtures were used: 25% and 89%. These are mean values of the enrichment, and the number of ^{13}C nuclei for a given fullerene follows a binomial (or with a good approximation a Poisson) distribution. These enrichment values were refined using Raman spectroscopy. The above-detailed standard routes were followed for the peapod and DWCNT synthesis: SWCNTs were opened by oxidation in air, were placed inside glass tubes with excess fullerenes, and were finally subject to a 1250°C heat treatment for 1 h for the DWCNT transformation (Simon et al. 2005a–b). As we discuss here, this results in a compelling isotope engineered system: DWCNTs

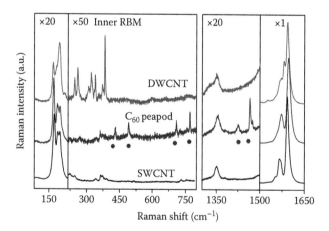

FIGURE 6.7 Transformation of peapods to DWCNTs followed by Raman spectroscopy. Note the disappearance of the fullerene modes (solid circles) and the development of narrow RBM modes in the 220–370 cm^{-1} spectral range.

with ^{13}C isotope-enriched inner walls and outer walls containing natural carbon (Simon et al. 2005a).

In Figure 6.8, we show the inner tube RBM-range Raman spectra for a natural DWCNT and two DWCNTs with differently enriched inner walls, 25% and 89%. These two latter samples are denoted as ^{13}C$_{25}$- and ^{13}C$_{89}$-DWCNT, respectively. The inner-wall enrichment is taken from the nominal enrichment of the fullerenes used for the peapod production, whose value is slightly refined based on the Raman data. An overall downshift of the inner tube RBMs is observed for the ^{13}C-enriched materials accompanied by a broadening of the lines. The downshift is clear evidence for the effective ^{13}C enrichment of inner tubes. The magnitude of the enrichment and the origin of the broadening are discussed below.

The RBM lines are well separated for inner and outer tubes due to the $\nu_{RBM} \propto 1/d$ relation and a mean inner tube diameter of $d \sim 0.7$ nm (Abe et al. 2003, Simon et al. 2005b). However, other vibrational modes such as the defect-induced D and the tangential G modes strongly overlap for inner and outer tubes. Arrows in Figure 6.8. indicate a gradually downshifting component of the observed D and G modes. These components are assigned to the D and G modes of the inner tubes. The sharper appearance of the inner tube G mode, as compared to the response from the outer tubes, is related to the excitation of semiconducting inner tubes and metallic outer tubes (Pfeiffer et al. 2003, Simon et al. 2005c).

The shifts for the RBM, D, and G modes can be analyzed for the two grades of enrichment. The average value of the relative shift for these modes was found to be $(\nu_0 - \nu)/\nu_0 = 0.0109(3)$ and $0.0322(3)$ for the ^{13}C$_{0.25}$- and ^{13}C$_{0.89}$-DWCNT samples, respectively. Here, ν_0 and ν are the Raman shifts of the same inner tube mode in the natural carbon and enriched materials, respectively.

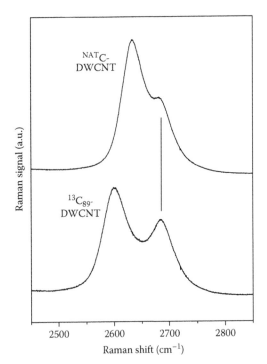

FIGURE 6.9 G′ spectral range of DWCNTs with natural carbon and ^{13}C enriched inner walls with 515 nm laser excitation. Note the unchanged position of the outer tube G′ mode indicated by a vertical line.

In the simplest continuum model, the shift originates from the increased mass of the inner tube walls. This gives

$$(\nu_0 - \nu)/\nu_0 = 1 - \sqrt{\frac{12 + c_0}{12 + c}}, \qquad (6.5)$$

where

> c is the concentration of the ^{13}C enrichment on the inner tube
> $c_0 = 0.011$ is the natural abundance of ^{13}C in carbon

The resulting values of c are 0.277(7) and 0.824(8) for the 25% and 89% samples, respectively.

The growth of isotope-labeled inner tubes allows to address whether carbon exchange between the two walls occurs during the inner tube growth. In Figure 6.9, we show the G′ spectral range for DWCNTs with natural carbon and ^{13}C-enriched inner walls with 515 nm laser excitation. The G′ mode of DWCNTs corresponds to the two-phonon process of the defect-induced D mode. The upper G′ mode component corresponds to the outer tubes and the lower to the inner tubes. The outer tube G′ components are unaffected by the ^{13}C enrichment within the 1 cm^{-1} experimental accuracy. This gives an upper limit of 14% to the extra ^{13}C in the outer wall. This proves that there is no sizeable carbon exchange between the two walls as this would result in a measurable ^{13}C content on the outer wall, too. We show the schematics of the isotope engineered DWCNTs in Figure 6.10.

This result is important for the contrast of the NMR signal between the two walls as it is discussed further below: were the

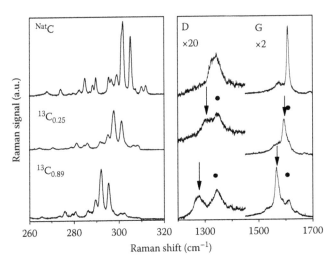

FIGURE 6.8 Raman spectra of DWCNTs with natural carbon and ^{13}C enriched inner tubes at 676 nm laser excitation and 90 K. The inner tube RBM and D and G mode spectral ranges are shown. Arrows and filled circles indicate the D and G modes corresponding to the inner and outer tubes, respectively. (Reprinted from Simon, F. et al., *Phys. Rev. Lett.*, 95, 017401, 2005a. With permission.)

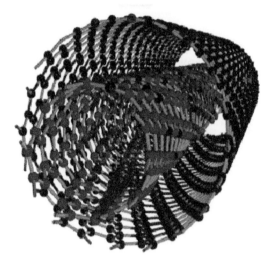

FIGURE 6.10 Schematic structure of an isotope engineered DWCNT with (14,6) outer and (6,4) inner tubes. ^{12}C and ^{13}C are shown in black and grey, respectively. The inner tube is 89% ^{13}C enriched and the outer contains natural carbon (1.1% ^{13}C abundance), which are randomly distributed for both shells.

outer shell also enriched, one would not obtain a reliable information about the DOS on the inner shell alone.

The narrow RBMs of inner tubes and the freedom to control their isotope enrichment allows to precisely compare the isotope-related phonon energy changes in the experiment and in *ab initio* calculations. This was performed by J. Kürti and V. Zólyomi (Simon et al. 2005a). The validity of the above simple continuum model in Equation 6.5 for the RBM frequencies was verified by performing first principles calculations on the (5,5) tube as an example. In the calculation, the Hessian matrix was determined by DFT using the Vienna Ab Initio Simulation Package (Kresse and Joubert 1999). Then, a large number of random ^{13}C distributions were generated and the RBM vibrational frequencies were determined from the diagonalization of the corresponding dynamical matrix for each individual distribution. The distribution of the resulting RBM frequencies can be approximated by a Gaussian where center and variance determine the isotope-shifted RBM frequency and the spread in these frequencies, respectively. The difference between the shift determined from the continuum model and from the *ab initio* calculations is below 1%.

6.3.2 Growth Mechanism of Inner Tubes

The growth of inner tubes from fullerenes raises the question, whether the fullerene geometry plays an important role in the inner tube growth or it acts as a carbon source only. According to the first scenario, inner tube growth starts with the bond formation between adjacent fullerenes, and a low-energy "bond-jumping" Stone–Wales transformation (Stone and Wales 1986) proceeds the reaction until a tube-like structure is attained. The second scenario suggests that fullerene geometry plays no particular role and it acts as a carbon source only, fullerenes fully disintegrate into small units such as, e.g., gaseous C_2 inside the host outer tubes and fuse together to form the inner tubes.

Theoretical results favor the first possibility (Zhao et al. 2002, Han et al. 2004) as it was found that the Stone–Wales transformations require little energy.

An earlier experimental work found that inner tube growth starts with the formation of inner tubes with $d \approx 0.7\,nm$, which change their diameter for longer heat treatments (Bandow et al. 2004). This finding also favors the first possibility as the diameter of C_{60} is 0.7 nm, which is not optimal when the host outer tube has $d = 1.4\,nm$ concerning that the tube shell-to-tube shell distance is the van der Waals separation of 0.335 nm. The growth of the inner tubes from the isotope-enriched fullerenes allows to distinguish between the two scenarios as we present it here.

The presence of ^{13}C isotopes on the inner tubes shifts the energy of the vibrational modes as we discussed it above. However, in addition to the shift, inhomogeneous broadening of the vibrational modes also occurs as the ^{13}C isotope distribution is statistically random along the tube axis. This effect is best observed on the inner tube RBM lines, which are very narrow; their half width at half maximum (HWHM) can be as small as $0.5\,cm^{-1}$. Regions that are somewhat richer in ^{13}C give more downshifted modes than those poorer in ^{13}C. To study the effect of isotope inhomogeneity, we consider two types of experiments. In the first one, we encapsulate ^{13}C fullerenes inside the host outer tubes and grow the inner tubes. We call this system $^{13}C_{0.28}$-DWCNT, and we refer to this material as having a uniform isotope distribution. In the second experiment, we prepare a 1:1 mixture of $^{13}C_{0.28}$-C_{60} with C_{60} of natural carbon. We call this second system as having a mixed isotope distribution. Clearly, for the second experiment the enriched and natural fullerenes enter the host outer tubes in a random fashion, and we call the resulting DWCNTs as $^{13}C_{0.15-M}$-DWCNT. The resulting RBM modes for the two kinds of experiments is shown in Figure 6.11. along with the data on the inner tubes based on natural C_{60}.

We observe that the inner tube RBMs are narrowest for the natural carbon sample, they are significantly broadened for the $^{13}C_{0.15-M}$-DWCNT sample, and somewhat less broadened for the $^{13}C_{0.28}$-DWCNT material. The resulting inhomogeneous broadenings are summarized in Figure 6.12. V. Zolyomi et al. attempted to reproduce the observed data with first principles calculations (Zolyomi et al. 2007). A (5,5) inner tube that has a 20 atom unit cell was considered. For the uniform isotope enrichment, it was assumed that 60 atoms, i.e., 3 nanotube unit cells contain an amount of ^{13}C isotopes that correspond to the nominal enrichment, r, but the exact amount of the isotopes follows the binomial distribution: $p(n_{13} = k) = \binom{60}{k} r^k (1 - r)^{60-k}$ where $p(n_{13} = k)$ is the probability of finding k pieces of ^{13}C nuclei in the 3 unit cells.

A large number of such configurations were generated and the resulting RBM frequencies were calculated whose distribution gave the inhomogeneous broadening. The result is shown as a solid curve in Figure 6.12. For the mixed distribution, it was assumed that 3 unit cell entities are randomly distributed, where a given 3 unit cell entity contains enriched carbon and another 3 unit cell entity contains natural carbon. The distribution of these 3 unit cell entities was also considered as random, which reflects

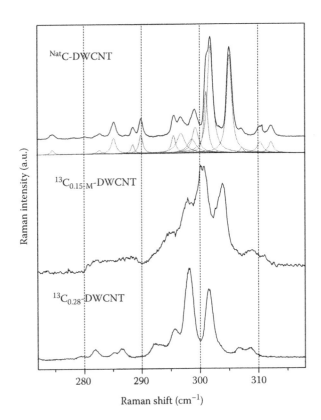

FIGURE 6.11 Raman spectra of DWCNTs with ^{Nat}C, $^{13}C_{28}$, and $^{13}C_{15-M}$ enriched inner tubes at $\lambda = 676$ nm laser excitation and 90 K measured with high resolution. We show deconvolution of the DWCNT inner tube RBMs into components for the natural carbon sample. Vertical dashed lines are intended to guide the eye. (Reprinted from Zólyomi, V. et al., *Phys. Rev. B*, 75, 195419-1, 2007. With permission.)

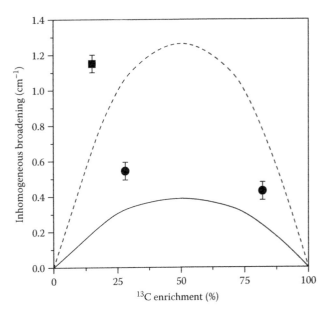

FIGURE 6.12 Measured (symbols) and calculated (solid and dashed curves) inhomogeneous broadening of isotope engineered inner tubes as a function of the ^{13}C enrichment. (■) $^{13}C_{0.28}$- and $^{13}C_{0.82}$-DWCNT, (●) $^{13}C_{0.15}$-M-DWCNT. Solid and dashed curves are calculated inhomogeneous broadening for a (5,5) inner tube for the uniform and mixed distributions, respectively. (Reprinted from Zólyomi, V. et al., *Phys. Rev. B*, 75, 195419-1, 2007. With permission.)

the real situation that natural and enriched fullerenes are located in a random fashion along the outer host tube axis. The result of this calculation is shown with a dashed curve in Figure 6.12.

Some general properties can be drawn from the result. First, both kinds of broadenings are symmetric for the 50% enrichment and are monotonously increasing for the 1%–50% enrichment. Second, the mixed distribution calculation gives about a factor 3 larger inhomogeneous broadening than the uniform one. For both kinds of calculations the resulting inhomogeneous broadening is about a factor two smaller than the experimental broadening. However, if both calculated curves are multiplied by the same amount (by 1.65) a close agreement between the calculation and the experiment is found (not shown). This means that the above models of uniform and mixed distribution properly account for the experimentally observed inhomogeneous broadening.

This agreement has an important consequence for the inner tube growth. The mixed model assumed that carbon from the fullerenes are fixed to the locations of their fullerene "mother" compound. If carbons were diffusing during the inner tube growth, one would expect a significantly smaller inhomogeneous broadening for the mixed sample, which is clearly not the case. The broadening for the $^{13}C_{15-M}$-DWCNT sample is about 3.5 times larger than for a uniformly enriched $^{13}C_{15}$-DWCNT material.

This means that the data and its theoretical explanation rules out any carbon diffusion during the inner tube growth. In turn, this supports the above model of inner tube growth due to fullerene fusion and Stone–Wales transformations rather than due to a complete disintegration into small carbon units.

6.3.3 NMR Studies on Isotope Engineered Heteronuclear Nanotubes

The growth of the "isotope engineered" nanotubes, i.e., DWCNTs with a highly enriched inner wall allows to study these samples with NMR with an unprecedented specificity for the small-diameter CNTs. For normal SWCNTs, either grown from natural or ^{13}C-enriched carbon, the NMR signal originates from all kinds of carbon-like amorphous or graphitic carbon.

NMR is usually an excellent technique for probing the electronic properties at the Fermi level of metallic systems. The examples include conducting polymers, fullerenes, and high-temperature superconductors. However, the 1.1% natural abundance of ^{13}C with nuclear spin $I = 1/2$ limits the sensitivity of such experiments. As a result, meaningful NMR experiments have to be performed on ^{13}C isotope-enriched samples. NMR data were taken with the samples sealed in quartz tubes filled with a low-pressure, high-purity Helium gas for thermal exchange (Simon et al. 2005a).

NMR allows to determine the amount of enriched tubes in our sample as it is sensitive to the number of ^{13}C nuclei. In Figure 6.13, we show the static and magic angle spinning, MAS, spectra of ^{13}C-enriched DWCNTs, and the static spectrum for

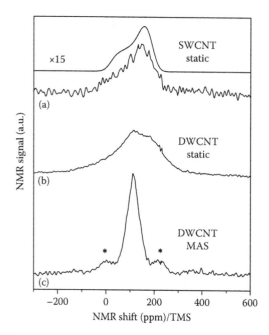

FIGURE 6.13 NMR spectra normalized by the total sample mass, taken with respect to the tetramethylsilane (TMS) shift. (a) Static spectrum for nonenriched SWCNT enlarged by 15. Smooth solid line is a chemical shift anisotropy powder pattern simulation with parameters published in the literature (Tang et al. 2000). (b) Static and (c) MAS spectra of $^{13}C_{0.89}$-DWCNT, respectively. Asterisks show the sidebands at the 8 kHz spinning frequency. (Reprinted from Simon, F. et al., *Phys. Rev. Lett.*, 95, 017401, 2005a. With permission.)

the SWCNT material. The mass fraction which belongs to the highly enriched phase can be calculated from the integrated signal intensity by comparing it to the signal intensity of the 89% ^{13}C enriched fullerene material. We found that the mass fraction of the highly enriched phase relative to the total sample mass is 13(4)%. The expected mass ratio of inner tubes as compared to the total sample mass is 15%, which is obtained from the SWCNT purity (50%), the ~70% volume filling for peapod samples (Liu et al. 2002), and the mass ratio of encapsulated fullerenes to the mass of the SWCNTs. Thus, the measured mass fraction of the highly enriched phase is very similar to that of the calculated mass fraction of inner tubes. This proves that the NMR signal comes nominally from the inner tubes.

The typical chemical shift anisotropy (CSA) powder pattern is observed for the SWCNT sample in agreement with previous reports (Tang et al. 2000, Goze-Bac et al. 2002). However, the static DWCNT spectrum cannot be explained with a simple CSA powder pattern even if the spectrum is dominated by the inner tube signal. The complicated spectral structure suggests that the chemical shift tensor parameters are distributed for the inner tubes. It is the result of the higher curvature of inner tubes as compared to the outer ones: the variance of the diameter distribution is the same for the inner and outer tubes (Simon et al. 2005b) but the corresponding bonding angles show a larger variation (Kurti et al. 2003). In addition, the residual linewidth

in the MAS experiment, which is a measure of the sample inhomogeneity, is 60(3) ppm, i.e., about twice as large as the ~35 ppm found previously for SWCNT samples (Tang et al. 2000, Goze-Bac et al. 2002). The isotropic line position, determined from the MAS measurement, is 111(2) ppm. This value is significantly smaller than the isotropic shift of the SWCNT samples of 125 ppm (Tang et al. 2000, Goze-Bac et al. 2002). However, recent theoretical *ab initio* calculations successfully explained this anomalous isotropic chemical shift (Marques et al. 2006). It was found that diamagnetic demagnetizing currents on the outer walls cause the diamagnetic shift of the inner tube NMR signal.

The dynamics of the nuclear relaxation is a sensitive probe of the local electronic properties (Slichter 1989). As NMR is a low-energy ($\hbar\omega \approx 0.3 \mu eV$ for ^{13}C in 7 T field) method, it is only sensitive to the immediate vicinity of the Fermi surface. It can be probed using the spin-lattice relaxation time, T_1, defined as the characteristic time it takes the ^{13}C nuclear magnetization to recover after saturation (Singer et al. 2005). The signal intensity after saturation, $S(t)$, is deduced by integrating the fast Fourier transform of half the spin echo for different delay times, t. The value of T_1 can be obtained by fitting the t dependence of $S(t)$ to the form $S(t) = S_a - S_b \cdot M(t)$, where $S_a \simeq S_b$ (>0) are arbitrary signal amplitudes, and

$$M(t) = \exp\left[-(t/T_1^e)^\beta\right] \tag{6.6}$$

is the reduced magnetization recovery of the ^{13}C nuclear spins. Figure 6.14. shows the results of $M(t)$ for the inner tubes as a function of the scaled delay time, t/T_1^e, under various experimental conditions listed in the figure. $M(t)$ does not follow the single exponential form with $\beta = 1$ (dashed line), but instead fits well to a stretched exponential form with $\beta \simeq 0.65(5)$, implying a distribution in the relaxation times, T_1.

The data in Figure 6.14. is displayed on a semilog scale for the time axis in order to emphasize the data for earlier decay times and to illustrate the collapse of the data set for the upper 90% of the NMR signal. For a broad range of experimental conditions, the upper 90% of the $M(t)$ data is consistent with constant $\beta \simeq 0.65(5)$ (see inset), implying a field- and temperature-independent underlying distribution in T_1. The lower 10% of the $M(t)$ data, corresponding to longer delay times, comes from the nonenriched outer walls which have much longer relaxation times under similar experimental conditions (Tang et al. 2000, Goze-Bac et al. 2002).

The collapse of the data set in Figure 6.14 to Equation 6.6 with constant $\beta = 0.65(5)$ is a remarkable experimental observation. From an experimental point of view, it implies that all one needs to characterize the T and H dependence of the underlying T_1 distribution is the bulk (or average) value, T_1^e (Equation 6.6). From an interpretational point of view, it implies that each inner tube in the powder sample has a different value of T_1, yet *all* the T_1 components and therefore all the inner tubes follow the same T and H dependence within experimental uncertainty. This finding is in contrast to earlier reports in SWCNTs where $M(t)$ fits well to a biexponential distribution, 1/3 of which had

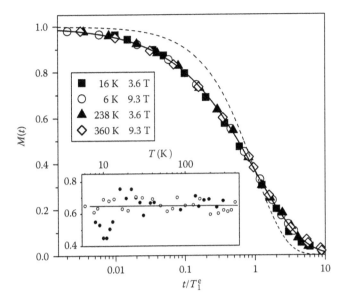

FIGURE 6.14 Reduced nuclear magnetization recovery, $M(t)$, as a function of the scaled delay time, t/T_1^e, for various temperature and magnetic field values. Both axes are dimensionless. Solid curve shows stretched exponential fit with $\beta = 0.65$ and dashed curve shows single exponential with $\beta = 1$. Inset shows temperature dependence of the best fit values of β at 3.6 Tesla (•) and 9.3 Tesla (◦), and average value of the data set $\beta = 0.65$ (solid line). (Reprinted from Singer, P.M. et al., *Phys. Rev. Lett.*, 95, 236403-1, 2005. With permission.)

a short T_1 value characteristic of fast relaxation from metallic tubes, and the remaining 2/3 had long T_1 corresponding to the semiconducting tubes (Tang et al. 2000, Goze-Bac et al. 2002, Shimoda et al. 2002, Kleinhammes et al. 2003), as expected from a macroscopic sample of SWCNTs with random chiralities. The data for the inner tubes in DWCNTs differ in that a similar biexponential fit to $M(t)$ is inconsistent with the shape of the recovery in Figure 6.14. Furthermore, if there were 1/3 metallic and 2/3 truly semiconducting inner tubes in the DWCNT sample, one would expect the ratio of T_1 between semiconducting and metallic tubes to increase exponentially with decreasing T below the semiconducting gap (~5000 K). As a consequence, one expects an increasingly large change in the underlying distribution in T_1 with decreasing T. This change would manifest itself as a large change in the shape of the $M(t)$; however, this is not the case as shown in Figure 6.14. The possibility of two components in T_1 with different T dependence can therefore be ruled out, and instead it could be concluded that all T_1 components (corresponding to distinct inner tubes) exhibit the same T and H dependence within experimental scattering.

The experimentally observed uniform metallicity of inner tubes is a surprising observation. This is suggested to be caused by the shifting of the inner tube Fermi levels due to charge transfer between the two tube walls. Indeed, *ab initio* calculations found that charge transfer and hybridization can render an otherwise semiconducting tube metallic (Okada and Oshiyama 2003, Zólyomi et al. 2008).

With these arguments, the bulk average, T_1^e, defined in Equation 6.6 is considered and its uniform T and H dependence can be followed. The $M(t)$ data can be fitted with the constant exponent $\beta = 0.65(5)$, which reduces unnecessary experimental scattering in T_1^e. In Figure 6.15, we show the temperature dependence of $1/T_1^e T$ for two different values of the external magnetic field, H. The data can be separated into two temperature regimes; the high-temperature regime ≥ 150 K, and the low T regime ≤ 150 K. At high temperatures, $1/T_1^e T$ is independent of T, which indicates a metallic state (Slichter 1989) for all of the inner tubes.

The simplest explanation for the experimental data is a noninteracting electron model of a 1D semiconductor with a small secondary gap (SG). The SG may be a result of the finite inner-wall curvature (Hamada et al. 1992, Kane and Mele 1997, Mintmire and White 1998, Zólyomi and Kürti 2004). The $1/T_1^e T$ data can be fitted using this noninteracting model with only one free parameter, the homogeneous SG, 2Δ. The normalized form of the gapped 1D DOS, $n(E)$

$$n(E) = \begin{cases} \dfrac{E}{\sqrt{E^2 - \Delta^2}} & \text{for } |E| > \Delta, \\ 0 & \text{otherwise,} \end{cases} \tag{6.7}$$

here, E is taken with respect to the Fermi energy. Equation 6.7 is used to calculate $1/T_1^e T$ (Moriya 1963) as such

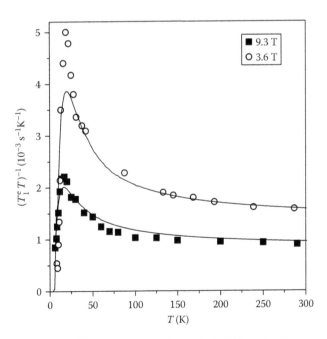

FIGURE 6.15 Temperature dependence of spin-lattice relaxation rate divided by temperature, $1/T_1^e T$, in units of ($10^3 \times$ s^{-1} K^{-1}). Solid curves are best fits to Equation 6.8 with $2\Delta = 46.8(40.2)$ K for $H = 3.6(9.3)$ Tesla, respectively. (Reprinted from Singer, P.M. et al., *Phys. Rev. Lett.*, 95, 236403-1, 2005. With permission.)

$$\frac{1}{T_1^e T} = \alpha(\omega) \int_{-\infty}^{\infty} n(E) n(E+\omega) \left(-\frac{\delta f}{\delta E} \right) dE, \qquad (6.8)$$

where

E and ω are in temperature units for clarity

f is the Fermi function, $f = [\exp(E/T) + 1]^{-1}$

the amplitude factor $\alpha(\omega)$ is the high temperature value for $1/T_1^e T$.

The results of the best fit of the data to Equation 6.8 are presented in Figure 6.15, where $2\Delta = 43(3)$ K ($\equiv 3.7$ meV) is H independent within experimental scattering between 9.3 and 3.6 Tesla.

The explanation in the noninteracting electron picture has two shortcomings: (1) calculations show that SG is of the order of a few 100 meV, which is two orders of magnitude larger than the experimental value; (2) the SG is expected to be strongly chirality dependent. In fact, a gap induced by electron–electron or electron–phonon correlations could be of the right order of magnitude (Bohnen et al. 2004, Connétable et al. 2005) and it could be uniform, i.e., chirality independent. The problem with such a correlation-induced gap is its magnitude: experimentally, the gap is open above 300 K, thus the critical transition of the correlation is $T_c > 300$ K. However, a mean-field expression between the gap and the critical temperature usually satisfies that $2\Delta/T_C > 3.52$ (Bardeen et al. 1957) but the current value is 0.13 or smaller. To avoid this contradiction, the NMR data was reinterpreted in the framework of the TLL theory by B. Dora and coworkers (Dóra et al. 2007), which we outline here.

The TLL state occurs in one-dimensional systems with strong electron–electron correlation. Formally, the interacting electrons can be treated with a noninteracting bosonic Hamiltonian that contains two TLL parameters $K_c \approx 0.2$ for the charge and $K_s \approx 1$ for the spin degree of freedom. The physically relevant correlation functions, such as, e.g., the spin–spin or current–current correlation functions follow power-law dependencies with exponents related to the TLL parameters. This leads to power-law behavior in the experimental measurables as, e.g., $(T_1 T)^{-1} \propto T^{-1}$. The latter result can be qualitatively explained by the localization of electrons in the TLL state, which leads to a paramagnetic-like fluctuating fields, giving the increase of $(T_1 T)^{-1}$ with decreasing temperature (Abragam 1961). The apparent presence of the gap at low temperatures was explained by the formation of the so-called Luther–Emery liquid (Dóra et al. 2007), which is a ground state that contains a gap in the excitation spectrum and competes with the TLL state.

We show the calculated $(T_1 T)^{-1}$ values in Figure 6.16. Although, the calculation relies on essentially three parameters only (the two TLL and a vertical scaling parameter, the gap being fixed to the temperature where the gap opens) we observe a much better agreement between the data and the calculation as compared to the gapped Fermi-liquid explanation shown in Figure 6.15. This shows that the TLL description is indeed relevant in describing the NMR data.

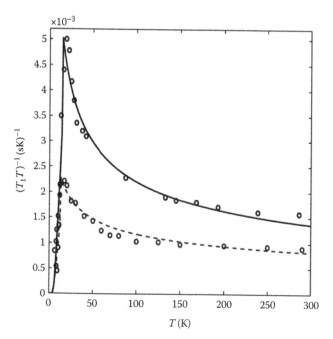

FIGURE 6.16 Calculated $(T_1 T)^{-1}$ assuming a TLL ground state of the electrons in the inner tube SWCNTs. The magnetic field dependence is associated to the slight field dependence of the K_s TLL parameter. (Reprinted from Dóra, B. et al., *Phys. Rev. Lett.*, 99, 166402-1, 2007. With permission.)

Summarizing the NMR studies on DWCNTs, it was shown that T_1 has a similar T and H dependence for all the inner tubes with no indication of a metallic/semiconducting separation due to chirality distributions. Below ~150 K, $1/T_1^e T$ increases dramatically with decreasing T and a gap in the spin excitation spectrum is found below $\Delta \simeq 20$ K. The result can be understood if electrons on the inner tubes are in the TLL state above 20 K and in the Luther–Emery liquid state below. This makes SWCNTs the only known example of materials where the TLL state is observed using NMR.

6.4 Summary

In summary, we reviewed recent advances in the isotope engineering of SWCNTs. We showed how nanotube-specific ${}^{13}C$ isotope enrichment can be achieved by encapsulating ${}^{13}C$ enriched fullerenes inside host SWCNTs and by transforming them into a smaller inner nanotube. The process produces highly ${}^{13}C$ enriched inner tubes while the host outer tube and other carbonaceous side-products in the SWCNT sample consists of natural carbon. This material allows to identify Raman modes of the DWCNTs. The use of mixtures of natural and ${}^{13}C$ enriched fullerenes allows to prove that no diffusion of carbon happens along the nanotube axis during inner nanotube synthesis, which supports the fullerene fusion model for their growth. The isotope-enriched inner tubes are excellent for NMR studies. Measurement of the ${}^{13}C$ NMR T_1 relaxation time allows to identify a non-Fermi-liquid behavior above 20 K and a low-energy, correlation-related gap providing direct experimental evidence for the TLL state in SWCNTs.

Acknowledgments

This work was supported by the Hungarian State Grants No. F61733 and NK60984, and by the Bolyai postdoctoral fellowship of the Hungarian Academy of Sciences. H. Kuzmany, R. Pfeiffer, T. Pichler, M. Rümmeli, C. Kramberger, H. Alloul, P. M. Singer, P. Wzietek, B. Dóra, M. Gulácsi, J. Bernardi, F. Fülöp, A. Rockenbauer, A. Jánossy, L. Forró, V. N. Popov, J. Koltai, V. Zólyomi, and J. Kürti are acknowledged for their participation in the reviewed works. Dario Quintavalle is acknowledged for preparing some of the figures.

References

Abe, M., Kataura, H., Kira, H., Kodama, T., Suzuki, S., Achiba, Y., Kato, K.-I., Takata, M., Fujiwara, A., Matsuda, K., and Maniwa, Y. (2003). Structural transformation from single-wall to double-wall carbon nanotube bundles, *Phys. Rev. B* **68**: 041405(R).

Abragam, A. (1961). *Principles of Nuclear Magnetism*, Oxford University Press, Oxford, U.K.

Arnold, M., Green, A., Hulvat, J., Stupp, S., and Hersam, M. (2006). Sorting carbon nanotubes by electronic structure via density differentiation, *Nat. Nanotechnol.* **1**: 60–65.

Bachilo, S. M., Strano, M. S., Kittrell, C., Hauge, R. H., Smalley, R. E., and Weisman, R. B. (2002). Structure-assigned optical spectra of single-walled carbon nanotubes, *Science* **298**: 2361–2366.

Bachtold, A., Hadley, P., Nakanishi, T., and Dekker, C. (2001). Logic circuits with carbon nanotube transistors, *Science* **294**: 1317–1320.

Bandow, S., Takizawa, M., Hirahara, K., Yudasaka, M., and Iijima, S. (2001). Raman scattering study of double-wall carbon nanotubes derived from the chains of fullerenes in single-wall carbon nanotubes, *Chem. Phys. Lett.* **337**: 48–54.

Bandow, S., Hiraoka, T., Yumura, T., Hirahara, K., Shinohara, H., and Iijima, S. (2004). Raman scattering study on fullerene derived intermediates formed within single-wall carbon nanotube, *Chem. Phys. Lett.* **384**: 320.

Bardeen, J., Cooper, L. N., and Schrieffer, J. R. (1957). Theory of superconductivity, *Phys. Rev.* **108**: 1175–1204.

Berber, S., Kwon, Y.-K., and Tománek, D. (2002). Microscopic formation mechanism of nanotube peapods, *Phys. Rev. Lett.* **88**: 185502.

Bethune, D. S., Kiang, C. H., DeVries, M. S., Gorman, G., Savoy, R., and Beyers, R. (1993). Cobalt-catalysed growth of carbon nanotubes with single-atomic-layer walls, *Nature* **363**: 605.

Bockrath, M., Cobden, D. H., Lu, J., Rinzler, A. G., Smalley, R. E., Balents, L., and McEuen, P. L. (1999). Luttinger-liquid behaviour in carbon nanotubes, *Nature* **397**: 598–601.

Bohnen, K. P., Heid, R., Liu, H. J., and Chan, C. T. (2004). Lattice dynamics and electron-phonon interaction in (3,3) carbon nanotubes, *Phys. Rev. Lett.* **93**: 245501-1–4.

Capinski, W. S., Maris, H. J., Bauser, E., Silier, I., Asen-Palmer, M., Ruf, T., Cardona, M., and Gmelin, E. (1997). Thermal conductivity of isotopically enriched Si, *Appl. Phys. Lett.* **71**: 2109–2111.

Cardona, M. and Thewalt, M. L. W. (2005). Isotope effects on the optical spectra of semiconductors, *Rev. Mod. Phys.* **77**: 1173–1224.

Chattopadhyay, D., Galeska, L., and Papadimitrakopoulos, F. (2003). A route for bulk separation of semiconducting from metallic single-wall carbon nanotubes, *J. Am. Chem. Soc.* **125**: 3370–3375.

Chen, Z. H., Du, X., Du, M. H., Rancken, C. D., Cheng, H. P., and Rinzler, A. G. (2003). Bulk separative enrichment in metallic or semiconducting single wall carbon nanotubes, *Nano Lett.* **3**: 1245–1259.

Connétable, D., Rignanese, G.-M., Charlier, J.-C., and Blase, X. (2005). Room temperature peierls distortion in small diameter nanotubes, *Phys. Rev. Lett.* **94**: 015503-1–015503-4.

Dóra, B., Gulácsi, M., Simon, F., and Kuzmany, H. (2007). Spin gap and Luttinger liquid description of the NMR relaxation in carbon nanotubes, *Phys. Rev. Lett.* **99**: 166402-1–166402-4.

Dresselhaus, M., Dresselhaus, G., and Ecklund, P. C. (1996). *Science of Fullerenes and Carbon Nanotubes*, Academic Press, New York.

Dresselhaus, M. S., Dresselhaus, G., and Avouris, P. (2001). *Carbon Nanotubes: Synthesis, Structure, Properties, and Applications*, Springer, Berlin, Germany.

Dubay, O. and Kresse, G. (2004). Density functional calculations for C_{60} peapods, *Phys. Rev. B* **70**: 165424.

Egger, R. and Gogolin, A. O. (1997). Effective low-energy theory for correlated carbon nanotubes, *Nature* **79**: 50825085.

Fantini, C., Jorio, A., Souza, M., Strano, M. S., Dresselhaus, M. S., and Pimenta, M. A. (2004). Optical transition energies for carbon nanotubes from resonant Raman spectroscopy: Environment and temperature effects, *Phys. Rev. Lett.* **93**: 147406.

Goze-Bac, C., Latil, S., Lauginie, P., Jourdain, V., Conard, J., Duclaux, L., Rubio, A., and Bernier, P. (2002). Magnetic interactions in carbon nanostructures, *Carbon* **40**: 1825–1842.

Hafner, J. H., Cheung, C. L., and Lieber, C. M. (1999). Growth of nanotubes for probe microscopy tips, *Nature* **398**: 761.

Hamada, N., Sawada, S., and Oshiyama, A. (1992). New one-dimensional conductors: Graphitic microtubules, *Phys. Rev. Lett.* **68**: 1579–1581.

Han, S. W., Yoon, M., Berber, S., Park, N., Osawa, E., Ihm, J., and Tománek, D. (2004). Microscopic mechanism of fullerene fusion, *Phys. Rev. B* **70**: 113402-1–4.

Harneit, W., Meyer, C., Weidinger, A., Suter, D., and Twamley, J. (2002). Architectures for a spin quantum computer based on endohedral fullerenes, *Phys. Status Solidi B* **233**: 453–461.

Hornbaker, D. J., Kahng, S. J., Misra, S., Smith, B. W., Johnson, A. T., Mele, E., Luzzi, D. E., and Yazdani, A. (2002). Mapping the one-dimensional electronic states of nanotube peapod structures, *Science* **295**: 828–831.

Hutchison, J. L., Kiselev, N. A., Krinichnaya, E. P., Krestinin, A. V., Loutfy, R. O., Morawsky, A. P., Muradyan, V. E., Obraztsova, E. D., Sloan, J., Terekhov, S. V., and Zakharov, D. N. (2001). Double-walled carbon nanotubes fabricated by a hydrogen arc discharge method, *Carbon* **39**: 761–770.

Iijima, S. (1991). Helical microtubules of graphitic carbon, *Nature* **354**: 56–58.

Iijima, S. and Ichihashi, T. (1993). Single-shell carbon nanotubes of 1-nm diameter, *Nature* **363**: 603–605.

Ishii, H., Kataura, H., Shiozawa, H., Yoshioka, H., Otsubo, H., Takayama, Y., Miyahara, T., Suzuki, S., Achiba, Y., Nakatake, M., Narimura, T., Higashiguchi, M., Shimada, K., Namatame, H., and Taniguchi, M. (2003). Direct observation of Tomonaga-Luttinger-liquid state in carbon nanotubes at low temperatures, *Nature* **426**: 540–544.

Kane, C. and Mele, E. (1997). Size, shape, and low energy electronic structure of carbon nanotubes, *Phys. Rev. Lett.* **78**: 1932–1935.

Kane, C. and Mele, E. (2003). Ratio problem in single carbon nanotube fluorescence spectroscopy, *Phys. Rev. Lett.* **90**: 207401-1–207401-4.

Kataura, H., Kumazawa, Y., Maniwa, Y., Umezu, I., Suzuki, S., Ohtsuka, Y., and Achiba, Y. (1999). Optical properties of single-wall carbon nanotubes, *Synth. Met.* **103**: 2555–2558.

Kataura, H., Maniwa, Y., Kodama, T., Kikuchi, K., Hirahara, K., Suenaga, K., Iijima, S., Suzuki, S., Achiba, Y., and Krätschmer, W. (2001). High-yield fullerene encapsulation in single-wall carbon nanotubes, *Synth. Met.* **121**: 1195–1196.

Kleinhammes, A., Mao, S.-H., Yang, X.-J., Tang, X.-P., Shimoda, H., Lu, J. P., Zhou, O., and Wu, Y. (2003). Gas adsorption in single-walled carbon nanotubes studied by NMR, *Phys. Rev. B* **68**: 75418-1–75418-6.

Kramberger, C., Pfeiffer, R., Kuzmany, H., Zólyomi, V., and Kürti, J. (2003). Assignment of chiral vectors in carbon nanotubes, *Phys. Rev. B* **68**: 235404.

Krätschmer, W., Lamb, L. D., Fostiropoulos, K., and Huffmann, D. R. (1990). Solid C_{60}: A new form of carbon, *Nature* **347**: 354.

Kresse, G. and Joubert, D. (1999). From ultrasoft pseudopotentials to the projector augmented-wave method, *Phys. Rev. B* **59**: 1758–1775.

Kroto, H. W., Heath, J. R., O'Brien, S. C., Curl, R. F., and Smalley, R. E. (1985). C_{60}: Buckminsterfullerene, *Nature* **318**: 162–163.

Krupke, R., Hennrich, F., von Lohneysen, H., and Kappes, M. M. (2003). Separation of metallic from semiconducting single-walled carbon nanotubes, *Science* **301**: 344–347.

Kürti, J., Kresse, G., and Kuzmany, H. (1998). First-principles calculations of the radial breathing mode of single-wall carbon nanotubes, *Phys. Rev. B* **58**: R8869–R8872.

Kürti, J., Zólyomi, V., Grüneis, A., and Kuzmany, H. (2002). Double resonant Raman phenomena enhanced by van Hove singularities in single-wall carbon nanotubes, *Phys. Rev. B* **65**: 165433.

Kürti, J., Zólyomi, V., Kertész, M., and Guangyu, S. (2003). The geometry and the radial breathing mode of carbon nanotubes: Beyond the ideal behaviour, *New J. Phys.* **5**: 125.

Kuzmany, H. (1998). *Solid-State Spectroscopy, An Introduction*, Springer Verlag, Berlin, Germany.

Kuzmany, H., Plank, W., Hulman, M., Kramberger, C., Grüneis, A., Pichler, T., Peterlik, H., Kataura, H., and Achiba, Y. (2001). Determination of SWCNT diameters from the Raman response of the radial breathing mode, *Eur. Phys. J. B* **22**(3): 307–320.

Liu, X., Pichler, T., Knupfer, M., Golden, M. S., Fink, J., Kataura, H., Achiba, Y., Hirahara, K., and Iijima, S. (2002). Filling factors, structural, and electronic properties of C_{60} molecules in single-wall carbon nanotubes, *Phys. Rev. B* **65**: 045419-1–045419-6.

Marques, M. A. L., d'Avezac, M., and Mauri, F. (2006). Magnetic response of carbon nanotubes from ab initio calculations, *Phys. Rev. B* **73**: 125433-1–125433-6.

Maultzsch, J., Pomraenke, R., Reich, S., Chang, E., Prezzi, D., Ruini, A., Molinari, E., Strano, M. S., Thomsen, C., and Lienau, C. (2005). Exciton binding energies in carbon nanotubes from two-photon photoluminescence, *Phys. Rev. B* **72**: 241402.

Meese, J. (1979). *Neutron Transmutation Doping of Semiconductors*, Plenum Press, New York.

Melle-Franco, M., Kuzmany, H., and Zerbetto, F. (2003). Mechanical interaction in all-carbon peapods, *J. Phys. Chem. B* **109**: 6986–6990.

Mintmire, J. and White, C. (1998). Universal density of states for carbon nanotubes, *Phys. Rev. Lett.* **81**: 2506.

Monthioux, M. and Kuznetsov, V. L. (2006). Who should be given the credit for the discovery of carbon nanotubes? *Carbon* **44**: 1621–1624.

Moriya, T. (1963). The effect of electron-electron interaction on the nuclear spin relaxation in metals, *J. Phys. Soc. Jpn.* **18**: 516.

Novoselov, K. S., Geim, A. K., Morozov, S. V., Jiang, D., Zhang, Y., Dubonos, S. V., Grigorieva, I. V., and Firsov, A. A. (2004). Electric field effect in atomically thin carbon films, *Science* **306**: 666–669.

Obraztsov, A. N., Pavlovsky, I., Volkov, A. P., Obraztsova, E. D., Chuvilin, A. L., and Kuznetsov, V. L. (2000). Aligned carbon nanotube films for cold cathode applications, *J. Vac. Sci. Technol. B* **18**: 1059–1063.

Okada, S. and Oshiyama, A. (2003). Curvature-induced metallization of double-walled semiconducting zigzag carbon nanotubes, *Phys. Rev. Lett.* **91**: 216801-1–216801-4.

Otani, M., Okada, S., and Oshiyama, A. (2003). Energetics and electronic structures of one-dimensional fullerene chains encapsulated in zigzag nanotubes, *Phys. Rev. B* **68**: 125424-1–8.

Perebeinos, V., Tersoff, J., and Avouris, P. (2004). Scaling of excitons in carbon nanotubes, *Phys. Rev. Lett.* **92**: 257402-1–257402-4.

Pfeiffer, R., Kuzmany, H., Kramberger, C., Schaman, C., Pichler, T., Kataura, H., Achiba, Y., Kürti, J., and Zólyomi, V. (2003). Unusual high degree of unperturbed environment in the interior of single-wall carbon nanotubes, *Phys. Rev. Lett.* **90**: 225501-1–225501-4.

Popov, V. N. (2004). Curvature effects on the structural, electronic and optical properties of isolated single-walled carbon nanotubes within a symmetry-adapted non-orthogonal tight-binding model, *New J. Phys.* **6**: 17.

Rauf, H., Pichler, T., Knupfer, M., Fink, J., and Kataura, H. (2004). Transition from a tomonaga-luttinger liquid to a fermi liquid in potassium-intercalated bundles of single-wall carbon nanotubes, *Phys. Rev. Lett.* **93**: 096805-1–096805-4.

Reich, S., Thomsen, C., and Maultzsch, J. (2004). *Carbon Nanotubes*, Wiley-VCH, Weinheim, Germany.

Ren, W., Li, F., Chen, J., Bai, S., and Cheng, H.-M. (2002). Morphology, diameter distribution and Raman scattering measurements of double-walled carbon nanotubes synthesized by catalytic decomposition of methane, *Chem. Phys. Lett.* **359**: 196–202.

Rochefort, A. (2003). Electronic and transport properties of carbon nanotube peapods, *Appl. Magn. Reson.* **67**: 11540117.

Rümmeli, M. H., Löffler, M., Kramberger, C., Simon, F., Fülöp, F., Jost, O., Schönfelder, R., Grüneis, R., Gemming, T., Pompe, W., Büchner, B., and Pichler, T. (2007). Isotope-engineered single-wall carbon nanotubes; A key material for magnetic studies, *J. Phys. Chem. C* **111**: 4094.

Saito, R., Dresselhaus, G., and Dresselhaus, M. (1998). *Physical Properties of Carbon Nanotubes*, Imperial College Press, London, U.K.

Shimoda, H., Gao, B., Tang, X. P., Kleinhammes, A., Fleming, L., Wu, Y., and Zhou, O. (2002). Lithium intercalation into opened single-wall carbon nanotubes: Storage capacity and electronic properties, *Phys. Rev. Lett.* **88**: 15502.

Shlimak, I. (2004). Isotopically engineered silicon nanostructures in quantum computation and communication, *HAIT J. Sci. Eng.* **1**: 196–206.

Simon, F., Kramberger, C., Pfeiffer, R., Kuzmany, H., Zólyomi, V., Kürti, J., Singer, P. M., and Alloul, H. (2005a). Isotope engineering of carbon nanotube systems, *Phys. Rev. Lett.* **95**: 017401.

Simon, F., Kukovecz, A., Kramberger, C., Pfeiffer, R., Hasi, F., Kuzmany, H., and Kataura, H. (2005b). Diameter selective characterization of single-wall carbon nanotubes, *Phys. Rev. B* **71**: 100–122.

Simon, F., Pfeiffer, R., Kramberger, C., Holzweber, M., and Kuzmany, H. (2005c). The Raman response of double wall carbon nanotubes, in *Applied Physics of Carbon Nanotubes*, S. V. Rotkin and S. Subramoney (eds.), Springer, New York, pp. 203–224.

Simon, F., Peterlik, H., Pfeiffer, R., and Kuzmany, H. (2007). Fullerene release from the inside of carbon nanotubes: A possible route toward drug delivery, *Chem. Phys. Lett.* **445**: 288–292.

Singer, P. M., Wzietek, P., Alloul, H., Simon, F., and Kuzmany, H. (2005). NMR evidence for gapped spin excitations in metallic carbon nanotubes, *Phys. Rev. Lett.* **95**: 236403-1–236403-4.

Slichter, C. P. (1989). *Principles of Magnetic Resonance*, 3rd edn. 1996 edn, Spinger-Verlag, New York.

Smith, B. W. and Luzzi, D. (2000). Formation mechanism of fullerene peapods and coaxial tubes: A path to large scale synthesis, *Chem. Phys. Lett.* **321**: 169–174.

Smith, B. W., Monthioux, M., and Luzzi, D. E. (1998). Encapsulated C_{60} in carbon nanotubes, *Nature* **396**: 323–324.

Smith, B. W., Monthioux, M., and Luzzi, D. (1999). Carbon nanotube encapsulated fullerenes: A unique class of hybrid materials, *Chem. Phys. Lett.* **315**: 31–36.

Spataru, C. D., Ismail-Beigi, S., Benedict, L. X., and Louie, S. G. (2004). Excitonic effects and optical spectra of single-walled carbon nanotubes, *Phys. Rev. Lett.* **92**: 077402-1–077402-4.

Stone, A. J. and Wales, D. J. (1986). Theoretical-studies of icosahedral C_{60} and some related species, *Chem. Phys. Lett.* **128**: 501–503.

Tang, X.-P., Kleinhammes, A., Shimoda, H., Fleming, L., Bennoune, K. Y., Sinha, S., Bower, C., Zhou, O., and Wu, Y. (2000). Electronic structures of single-walled carbon nanotubes determined by NMR, *Science* **288**: 492.

Tans, S. J., Devoret, M. H., Dai, H., Thess, A., Smalley, R. E., Geerligs, L. J., and Dekker, C. (1997). Individual single-wall carbon nanotubes as quantum wires, *Nature* **386**: 474–477.

Telg, H., Maultzsch, J., Reich, S., Hennrich, F., and Thomsen, C. (2004). Chirality distribution and transition energies of carbon nanotubes, *Phys. Rev. Lett.* **93**: 177401.

Thomsen, C. and Reich, S. (2000). Double resonant Raman scattering in graphite, *Phys. Rev. Lett.* **85**: 5214–5217.

Wang, F., Dukovic, G., Brus, L. E., and Heinz, T. F. (2005). The optical resonances in carbon nanotubes arise from excitons, *Science* **308**: 838–841.

Wildör, J. W. G., Venema, L. C., Rinzler, A. G., Smalley, R. E., and Dekker, C. (1998). Electronic structure of atomically resolved carbon nanotubes, *Nature* **391**: 59–62.

Yue, G. Z., Qiu, Q., Gao, B., Cheng, Y., Zhang, J., Shimoda, H., Chang, S., Lu, J. P., and Zhou, O. (2002). Generation of continuous and pulsed diagnostic imaging x-ray radiation using a carbon-nanotube-based field-emission cathode, *Appl. Phys. Lett.* **81**: 355–368.

Zhao, Y., Yakobson, B. I., and Smalley, R. E. (2002). Dynamic topology of fullerene coalescence, *Phys. Rev. Lett.* **88**: 18550-1–185501-4.

Zheng, M., Jagota, A., Strano, M. S., Santos, A. P., Barone, P., Chou, S. G., Diner, G., B. A. Dresselhaus, M. S., McLean, R. S., Onoa, G. B., Sam-sonidze, G. G., Semke, E. D., Usrey, M., and Walls, D. J. (2003). Structure-based carbon nanotube sorting by sequence-dependent DNA assembly, *Science* **302**: 1545–1548.

Zólyomi, V., Koltai, J., Rusznyák, A., Kürti, J., Gali, A., Simon, F., Kuzmany, H., Szabados, A., and Surján, P. R. (2008). Intershell interaction in double walled carbon nanotubes: Charge transfer and orbital mixing, *Phys. Rev. B* **77**: 245403-1–10.

Zólyomi, V. and Kürti, J. (2004). First-principles calculations for the electronic band structures of small diameter single-wall carbon nanotubes, *Phys. Rev. B* **70**: 085403-1–085403-8.

Zólyomi, V., Kürti, J., Grüneis, A., and Kuzmany, H. (2003). Origin of the fine structure of the Raman *D* band in single-wall carbon nanotubes, *Phys. Rev. Lett.* **90**: 157401.

Zólyomi, V., Simon, F., Rusznyák, A., Pfeiffer, R., Peterlik, H., Kuzmany, H., and Kürti, J. (2007). Inhomogeneity of ^{13}C isotope distribution in isotope engineered carbon nanotubes: Experiment and theory, *Phys. Rev. B* **75**: 195419-1–195419-8.

Raman Spectroscopy of sp² Nano-Carbons

Mildred S. Dresselhaus
*Massachusetts Institute
of Technology*

Gene Dresselhaus
*Massachusetts Institute
of Technology*

Ado Jorio
*Universidade Federal
de Minas Gerais*

7.1 Introduction

Different from scanning probe and electron microscopy-related techniques, optics is one of the oldest characterization techniques for materials science, being largely used long before nanoscience could even be imagined. In the age of nano, optics still sustains its kingdom. The advantages of optics for nanoscience relate to both experimental and fundamental aspects. Experimentally, the techniques are readily available, relatively simple to perform, possible at room temperature and under ambient pressure, and require relatively simple or no sample preparation. Fundamentally, the optical techniques (normally using infrared and visible wavelengths) are nondestructive and noninvasive because they use the photon, a massless and charge-less particle, as a probe.

The sp² carbon materials and Raman spectroscopy have a special place in the nanoworld. It is possible to observe Raman scattering from one single sheet of sp²–hybridized carbon atoms, the two-dimensional (2D) graphene sheet, and from a narrow strip of graphene sheet rolled up into a 1 nm-diameter cylinder—the one-dimensional (1D) single-wall carbon nanotube (SWNT). These observations are possible just by shining a light on the nanostructure focused through a regular microscope. This chapter focuses on both the basic concepts of sp² carbon nanomaterials and Raman spectroscopy, together with their interaction.

7.1.1 Nanoscience, Nanotechnology, and sp² Carbon

To fully understand the importance of Raman spectroscopy of sp² carbons in the nano-carbon context, it is important to understand that nano-carbons are part of the nanoworld and address structures with sizes between the molecular and the macroscopic. The Technical Committee (TC-229) for nano-technologies standardization of the International Organization for Standardization (ISO) defines the field of nanotechnologies as the application of scientific knowledge to control and utilize matter at the nanoscale, where size-related properties and phenomena can emerge. The nanoscale is the size range from approximately 1 to 100 nm. It is not possible to clearly envisage the future results of nanotechnology, or even the limit for the potential of nanomaterials, but clearly, serious fundamental challenges have to be overcome, such as

- Constructing nanoscale building blocks precisely and reproducibly
- Finding and controlling the rules for assembling these objects into complex systems
- Predicting and probing the emergent properties of these systems

These challenges are not only technological, but also conceptual: how do you treat a system that is too big to be solved by present-day first principles calculations, and too small for statistical methods? Although these challenges punctuate nanoscience and nanotechnology, the success here will represent a revolution in larger-scale scientific challenges in the fields of emergent phenomena and information technology. The answers to questions like "How do complex phenomena emerge from simple ingredients?" and "How will the information technology revolution be extended?" will probably come from nanotechnology.

It is exactly in this context that nano-carbon is playing a very important role. From one side, nature shows that it is possible to manipulate matter and energy the way integrated circuits manipulate electrons, by assembling complex self-replicating carbon-based structures able to sustain life. From another side, carbon is the upstairs neighbor of silicon in the periodic table, with more flexible bonding and unique physical, chemical, and biological properties, which nevertheless hold promise of a revolution in electronics at some future time. Three important aspects make sp² carbon materials special for facing the nano-challenges listed in the previous paragraph: first, the unusually strong covalent bonding between neighboring atoms; second, the extended π-electron clouds; and third, the simplicity of the system. We elaborate shortly on these aspects in the following paragraphs.

Carbon has six electrons: two are core 1s states, and four occupy the 2s and 2p orbitals. In the sp² configuration, the 2s, p_x, and p_y orbitals mix to form three covalent bonds, 120° from each other in the *xy* plane (see Figure 7.1). Each carbon atom has three neighbors, forming a hexagonal network. These nearest-neighbor sp² bonds are stronger than the nearest-neighbor sp³ bonds in diamond, making graphene (a single sheet of sp² atoms) stronger than diamond in tensile strength. This added strength is advantageous for sp² carbons as a prototype material for the development of nanoscience and nanotechnology, since different interesting nanostructures (sheets, ribbons, tubes, horns, fullerenes, etc.) are stable and strong enough for exposure to many different types of characterization and processing steps.

The p_z electrons that remain perpendicular to the hexagonal network (see Figure 7.1) form the delocalized π electron states. For this reason, sp² carbons, which include graphene, graphite, carbon nanotubes, fullerenes, and other carbonaceous materials, are also called π-electron materials. These delocalized

electronic states are highly unusual, because they behave like relativistic Dirac Fermions, i.e., they exhibit a massless-like linear momentum–energy relation (like a photon), and are responsible for unique optical and transport (both thermal and electronic) properties.

These two properties accompany a very important aspect of sp² carbons—the simplicity of a system formed by only one type of atom in a periodic hexagonal structure. Therefore, different from most materials, sp² nano-carbons allow us to access their special properties from both experimental and theoretical approaches. Being able to model the structure is crucial for the development of our methodologies and knowledge.

7.1.2 Importance of Graphite, Carbon Nanotubes, and Graphene

The ideal concept of sp² nano-carbons starts with the graphene sheet (see Figure 7.2). Adding one or two layers produces the bilayer and trilayer graphene. Roll up a narrow strip of graphene into a seamless cylinder and you have a SWNT. Add one-layer or two-layer concentric cylinders and you have double-wall and triple-wall carbon nanotubes. Many rolled-up cylinders would make a multi-wall carbon nanotube (MWNT), and many flat layers give graphite. This ideal concept is didactic, but historically these materials came into human knowledge in the opposite order.

Three-dimensional (3D) graphite is one of the longest-known forms of pure carbon, formed by graphene planes usually in an ABAB stacking (Wyckoff 1981). Of all materials, graphite has the highest melting point (4,200 K), the highest thermal conductivity (3,000 W/mK), and a high room temperature electron

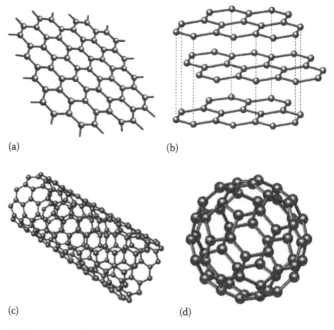

(a)   (b)

(c)   (d)

FIGURE 7.2 The sp² carbon materials, including (a) single-layer graphene, (b) triple-layer graphene, (c) SWNT, and (d) a C_{60} fullerene, which includes pentagons in the structure. (From Castro Neto, A.H. et al., *Rev. Mod. Phys.*, 81, 109, 2008. With permission.)

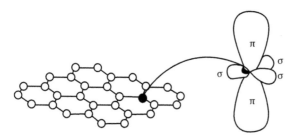

FIGURE 7.1 The carbon atomic orbitals in the sp² honeycomb lattice. (From Pfeiffer, R. et al., in *Carbon Nanotubes: Advanced Topics in the Synthesis, Structure, Properties and Applications, Springer Series in Topics in Applied Physics*, Vol. 111, Jorio, A. et al. (eds.), Springer-Verlag, Berlin, Germany, 2008, 495–530. With permission.)

mobility (30,000 cm²/Vs; Dresselhaus et al. 1988). Graphite and its related carbon fibers (Dresselhaus et al. 1988) have been used commercially for decades. Their applications range from use as conductive fillers and mechanical structural reinforcements in composites (e.g., in the aerospace industry) to their use in electrode materials utilizing their resiliency (e.g., in batteries; Endo et al. 2008).

In 1985, a unique discovery in another sp² carbon system took place: the C_{60} fullerene molecule (see Figure 7.2; Kroto et al. 1985). The fullerenes stimulated and motivated a large scientific community from the time of its discovery up to the end of the century, but their applications remain sparse to date. Carbon nanotubes arrived on the scene following in the footsteps of the C_{60} fullerene molecule. Carbon nanotubes have evolved into one of the most intensively studied materials, and are held responsible for co-triggering the nanotechnology revolution.

The big rush on carbon nanotube science started after the observation of MWNTs on the cathode of a carbon arc used to produce fullerenes (Iijima 1991), even though they were identified in the core structure of vapor grown carbon fibers as very small carbon fibers in the 1970s (Oberlin et al. 1976) and in the 1950s in Russian literature (Radushkevich and Lukyanovich 1952; see Figure 7.3). SWNTs were first synthesized in 1993 (Bethune et al. 1993, Iijima and Ichihashi 1993). The interest in the fundamental properties of carbon nanotubes and their exploitation through a wide range of applications is due to their unique structural, chemical, mechanical, thermal, optical, optoelectronic, and electronic properties (Saito et al. 1998, Reich et al. 2003). The growth of a single SWNT at a specific location and pointing in a given direction (Zhang et al. 2001, Huang et al. 2003), and the growth of a huge amount of millimeter-long tubes with nearly 100% purity (Hata et al. 2004) have been achieved (Joselevich et al. 2008). Substantial success with the separation of nanotubes (Arnold

et al. 2006) by metallicity and length has been achieved and advances have been made with doping nanotubes for the modification of their properties (Terrones et al. 2008). Studies on nanotube optics, magnetic properties, transport, and electrochemistry have exploded, revealing many rich and complex fundamental excitonic and other collective phenomena (Jorio et al. 2008a). Quantum transport phenomena, including quantum information, spintronics, and superconducting effects have also been explored (Biercuk et al. 2008). After a decade and a half of intense activity in carbon-nanotube research, more and more attention is now focusing on the practical applications of the many unique and special properties of carbon nanotubes (Endo et al. 2008).

In the meantime, the study of nano-graphite was under development, and in 2004, Novoselov et al. discovered a simple method to transfer a single atomic layer of carbon from the c-face of graphite to a substrate suitable for the measurement of its electrical and optical properties (Novoselov et al. 2004). This finding led to a renewed interest in what was considered to be a prototypical, yet theoretical, 2D system, providing a basis for the structure of graphite, fullerenes, carbon nanotubes, and other nano-carbons. Surprisingly, this very basic graphene system first prepared by Boehm (Boehm 1962) in monolayer form, which had been studied for many decades, suddenly appeared with many novel physical properties that were not even imagined previously (Geim and Novoselov 2007, Castro Neto et al. 2008). In one or two years, the rush on graphene science started. The scientific interest was stimulated by the report of the relativistic properties of the conduction electrons (and holes) in a single graphene layer less than 1 nm thick, which is responsible for the unusual properties in this system (see Figure 7.4) (Novoselov et al. 2005, Zhang et al. 2005). Many groups are now making devices using graphene and also graphene ribbons, which have a long length and a small width, and where the ribbon edges play an important role.

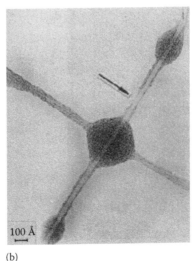

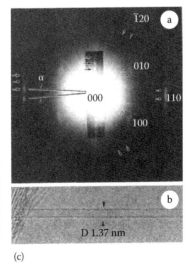

(a)   (b)   (c)

FIGURE 7.3 Transmission electron microscopy images of carbon nanotubes. The early reported observations (a) in 1952 (From Radushkevich, L.V. and Lukyanovich, V.M., *Zum. Fisc. Chim.*, 26, 88, 1952.), (b) in 1976 (From Oberlin, A. et al., *J. Cryst. Growth*, 32, 335, 1976. With permission.), and (c) in 1993 the observation of SWNTs that launched the field (From Iijima, S. and Ichihashi, T. et al., *Nature*, 363, 603, 1993. With permission.)

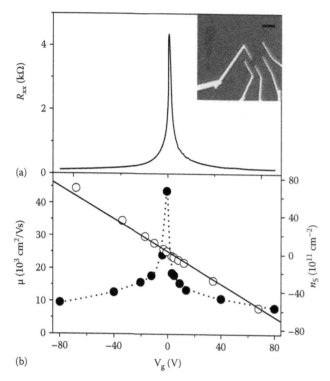

FIGURE 7.4 Resistivity, mobility, and carrier density as a function of gate voltage V_g (Vg in the figure) in a single-layer graphene field effect transistor device (Charlier et al. 2008). (a) V_g-dependent R_{xx} showing a finite value at the Dirac point, where the valence and conduction bands are degenerate and the carriers are massless. The resistivity ρ_{xx} can be calculated from R_{xx} using the geometry of the device. The inset is an image of a graphene device on a Si:SiO$_2$ substrate. The Si is the bottom gate; five top electrodes that formed via e-beam lithography are shown in the inset. The scale bar is 5 μm. (b) Mobility μ and carrier density n$_S$ as a function of V_g (for holes $V_g < 0$ and for electrons $V_g > 0$). The mobility (dotted curve) diverges at the Dirac point due to the finite resistivity. (From Pfeiffer, R. et al., in *Carbon Nanotubes: Advanced Topics in the Synthesis, Structure, Properties and Applications, Springer Series in Topics in Applied Physics*, Vol. 111, Jorio, A. et al. (eds.), Springer-Verlag, Berlin, Germany, 2008, 495–530. With permission.)

Having introduced nano-carbons in the nanoworld, we now focus on the Raman spectroscopy of nano-carbons. Section 7.2 discusses the basic concepts of light–matter interaction and Raman spectroscopy. A more detailed presentation of state-of-the-art Raman spectroscopy of sp² nano-carbons is given in Section 7.3, followed by a critical discussion in Section 7.4. Section 7.5 summarizes the main aspects of this paper, and Section 7.6 presents future perspectives.

7.2 Background

7.2.1 Light–Matter Interaction and Raman Spectroscopy

When shining light into a material, part of the energy can just pass through (transmission), while the remaining energy interacts with the system through several different mechanisms.

From the light that interacts with the system, many different effects might occur: (1) a photon, which is the quantum unit of light, can be absorbed and transformed into atomic vibrations, i.e., heat, which can be represented by phonons, the quantum units of lattice vibrations; (2) a photon can be absorbed and transformed into a photon with a lower energy that is emitted. This process is called photoluminescence and it happens in semiconducting materials when the energy of the incident photon exceeds the energy gap between the valence and conduction bands; (3) a photon may not be absorbed, but it just shakes the electrons, which will scatter that energy back into another photon with the same energy as the incident one. This is an elastic scattering process named Rayleigh scattering; or (4) the photon may shake the electrons causing oscillations of the atoms according to their natural vibrational frequencies, thereby changing the electronic configuration of the atoms. In this case, when the electrons scatter the energy back into another photon, this photon will have lost or gained energy to or from the atoms. This is an inelastic scattering process called Raman scattering. Many other processes may occur but they are usually less important for the energetic balance of the light–matter interaction. The amount of light that will be transmitted, as well as the details for all the light–matter interactions will be determined by the electronic and vibrational properties of the material, thus making light a very powerful characterization tool for materials science studies, while gently perturbing the material.

7.2.2 Basic Concepts of Raman Spectroscopy

This section gives the basic definitions for the terms that are used in describing Raman spectroscopy.

7.2.2.1 Raman Scattering

The Raman effect refers to the inelastic scattering of light. An incident photon with energy $E_i = E_{laser}$ and momentum $k_i = k_{laser}$ reaches the sample and is scattered, resulting in a photon with a different energy E_S and momentum k_S. For energy and momentum conservation,

$$E_S = E_i \pm E_q, \tag{7.1}$$

$$k_S = k_i \pm k_q, \tag{7.2}$$

where E_q and k_q are the energy and momentum change during the scattering event, mediated by an excitation of the medium. Although electronic excitations can result from Raman scattering, the most usual scattering outcome is the excitation of atomic normal mode vibrations. These vibrational modes are related to the chemical and structural properties of materials, and since every material has a unique set of such normal modes, Raman spectroscopy can be used to probe materials properties in detail and to provide an accurate characterization of specific materials. Raman spectroscopy in particular provides a rich variety of characterization information regarding carbon nanostructures.

7.2.2.2 Stokes and Anti-Stokes Raman Processes

In the inelastic scattering process, the incident photon can decrease or increase its energy by destroying (anti-Stokes) or creating (Stokes) a quantum of normal mode vibration (i.e., a phonon) in the medium. The plus and minus signs in Equations 7.1 and 7.2 apply when energy has been received from or transferred to the medium, respectively. The probability for the two types of events depends on the excitation photon energy E_i, and this dependence can be explored for an accurate determination of electronic transition energies. Furthermore, the probability to destroy a phonon depends on the phonon population given by the Bose–Einstein distribution and, consequently, the anti-Stokes event also depends on the temperature, according to

$$I_S / I_{AS} = \exp(E_q / k_B T), \qquad (7.3)$$

where

   I_S and I_{AS} denote the measured intensity for the Stokes and anti-Stokes peaks, respectively

   k_B is the Boltzmann constant

   T is the temperature

Because the anti-Stokes process is usually less probable than the Stokes process, it is usual that people only care about the Stokes spectra. In this work, when not referring explicitly to the type of process, it is the Stokes process that is being addressed.

7.2.2.3 Energy Conservation

A Raman spectrum is a plot of the scattering intensity as a function of $E_S - E_{laser}$. Therefore, the energy conservation relation given by Equation 7.1 is the most important aspect of Raman spectroscopy. Although the anti-Stokes process has a positive net energy, it is the Stokes spectra that are most usually measured, and for simplicity, the Stokes process is designated as being positive. The Raman spectra will show peaks at $\pm E_q$, where E_q is the energy of the excitation associated with the Raman effect. The quantum of excitation denoting the normal vibrational modes is named the phonon, and is regularly used to describe the lattice vibrations in crystals. We will use the term "phonon" frequently in this chapter. The phonon excitation energies are found by decomposing the atomic vibrations into the vibrational normal modes of the material.

7.2.2.4 Energy Units

The energy axis in the Raman spectra is usually displayed in units of cm⁻¹. Lasers are usually described by the wavelength of the light, i.e., in nanometers, but the phonon energies are usually too small a number when displayed in nm, which is not a comfortable unit for denoting Raman shifts. The accuracy of a common Raman spectrometer is on the order of 1 cm⁻¹, which is

equivalent to 10^{-7} nm. The energy conversion factors are: 1 eV = 8065.5 cm⁻¹ = 2.418×10^{14} Hz = 11,600 K. Also 1 eV corresponds to a wavelength of 1.2398 µm.

7.2.2.5 Shape of the Raman Peak

The response of a forced damped harmonic oscillator, in the limit that the peak frequency ω_q is much larger than the peak width Γ_q, is a Lorentzian curve. Therefore, the Raman peaks usually have a Lorentzian shape. The center of the Lorentzian gives the natural vibration frequency, and the full width at half maximum (FWHM) is related to the damping, which gives the phonon lifetime. However, in specific cases, the Raman feature can deviate from the simple Lorentzian shape. One obvious case is when the feature is actually composed of many phonon contributions. Then the Raman peak will be a convolution of several Lorentzian peaks, giving rise to Gaussian, Voigt, or more complex lineshapes. Another case is when the lattice vibration couples with electrons. In this case, additional line broadening and even an asymmetric lineshape can result. These effects are observed in certain metallic sp² carbon materials, including SWNTs (Dresselhaus et al. 2005).

7.2.2.6 Resonance Raman Effect

The laser energies are usually much higher than the phonon energies. Therefore, although the exchange in energy between the light and the medium is transferred to the atomic vibrations, the light–matter interaction is mediated by electrons. Usually, the photon energy is not large enough to achieve a real electronic transition, and the electron that absorbs the light is said to be excited to a "virtual state," from where it couples to the lattice, generating the Raman scattering. However, when the excitation laser energy matches the energy gap between the valence and conduction bands in a semiconducting medium (or between an occupied initial state and an unoccupied final state more generally), the probability for the scattering event to occur increases by many orders of magnitude, and the process is then called a resonance Raman process (non-resonant otherwise). The same happens if the scattered light matches such an electronic transition. The resonance effect is extremely important in systems where electronic transitions are in the visible range, which includes sp² carbon systems.

7.2.2.7 First and Higher-Order Raman Processes

The order of the Raman process is given by the number of scattering events involved in the scattering process. The most usual case is the first-order Stokes Raman scattering process, where the photon energy exchange creates one phonon in the medium with a very small momentum ($q \approx 0$). If 2, 3, or more phonons are involved in the same scattering event, the process is of a second, third, or higher-order, respectively. The first-order Raman process gives the basic quanta of vibration, while higher-order processes give very interesting information about harmonics and combination modes (Dresselhaus et al. 2005).

7.2.2.8 Vibrational Structure

Describing the vibrations of small molecules is simple. The number of vibrational modes is given by the number of degrees of freedom for atomic motion (i.e., the number of atoms N multiplied by three dimensions, minus six; the six coming from translations along x, y, z, and rotations around these axes). The number of vibrational energy levels may be smaller than $3N - 6$ since some levels can be energy degenerate. Finding the normal modes of large and complex molecules, such as proteins, is not an easy task because of the large number of degrees of freedom. In cases like that, it is common to find the spectral features identified with local bondings (e.g., C–C, C=C, C=O, etc.) rather than the actual molecular normal modes. Crystals have a large number of atoms (ideally infinite), but periodic systems are, again, quite simple to describe, although such descriptions require an understanding of the *phonon dispersion relations*.

Since ideal crystalline structures have an infinite number of atoms, the number of vibrational levels is infinite. The vibrational structures of these systems are displayed in a plot of the phonon energy (ω_q) vs. phonon wave vector q, i.e., the phonon dispersion relation for each distinct normal mode phonon. The ω_q vs. q plots are composed of continuous curves called phonon branches. Being continuous, they account for the infinity of vibrational levels. The number of phonon branches depends on the number of degrees of freedom for the atomic motion for a unit cell for this material. The wave vector q is defined by the magnitude and phase for a normal mode vibration that can involve more than one consecutive unit cell.

7.2.2.9 Momentum Conservation and Back-Scattering

The q vector carries information about the wavelength of the vibration ($q = 2\pi/\lambda$) and the direction along which the oscillation occurs. For $\lambda \to \infty$, we have $q \to 0$, and we usually denote the center of the phonon dispersion relation as the Γ point. The phonon wavelength λ cannot be smaller than the unit cell vector $\mathbf{a}$, which defines an upper limit for the relevant values of q, namely, the unit cell of the crystal in the reciprocal space, which is called the first Brillouin zone. This zone provides a bounding polygon that confines the phonon dispersion. For $q > |\pm\pi/\mathbf{a}|$, the phonon structure will repeat itself.

For the momentum conservation given by Equation 7.2, different scattering geometries are possible, and specific choices may be used to select different phonons due to the symmetry selection rules (Dresselhaus et al. 2005). The back-scattering configuration, for which k_i and k_s have the same direction and opposite signs, is the most common when working with nanomaterials, because a microscope is usually needed to focus the light onto small samples. Furthermore, in the first-order Raman process, the momentum transfer is usually neglected, i.e., $k_s - k_i \sim 0$. Phonon momentum $k_s \neq 0$ becomes important in defect-induced or higher-order Raman scattering processes, and the theoretical background for such processes can be found in the article by Saito et al. (2003).

This discussion gives an explanation for why the first-order Raman process can only access phonons at $q \to 0$, i.e., at the Γ point. The momenta associated with the first-order light scattering process are on the order of $k_i = 2\pi/\lambda_{\text{light}}$, where λ_{light} is in the visible range (800–400 nm). Therefore, k_i is a very small number when compared to the dimensions of the first Brillouin zone, which is limited to vectors no longer than $q = 2\pi/\mathbf{a}$, where the unit cell vector $\mathbf{a}$ in real space is on the order of tenths of nm.

7.2.2.10 Coherence

It is not trivial to define whether a real system is big enough to be considered as effectively infinite and to exhibit a quasi-continuous phonon (or electron) energy dispersion relation. Whether or not a dispersion relation can be defined indeed depends on the process that is under evaluation. In the Raman process, how long does it take for an electron excited by the incident photon to decay? Considering this scattering time, what is the distance felt by an electron? These problems are described in solid–state physics textbooks by the concept of coherence. The coherence time is the time the electron takes to suffer an event such as scattering that changes its state. Thus, the coherence length is the size over which the electron maintains its integrity, its coherence, and it is defined by the electron speed and the coherence time, which can be measured experimentally. The Raman process is an extremely fast process, in the range of femtoseconds (10^{-15} s). Considering the speed of electrons in graphite (10^6 m/s), this gives a coherence length on the order of nm. Interestingly, this number is much smaller than the wavelength of visible light. Actually, this is a particle picture for the scattering process and playing with these concepts is actually quite interesting.

7.3 Presentation of State-of-the-Art Raman Spectroscopy of sp² Nano-Carbons

Figure 7.5 shows the Raman spectra from different crystalline and disordered sp² carbon nanostructures. The first-order Raman bands go up to 1620 cm⁻¹, and the spectra above this value are composed of overtone and combination modes. We focus here on the Raman spectra from crystalline structures (Sections 7.3.1 through 7.3.4) and on the disorder caused by defects in the sp² structure (Section 7.3.5). The bottom spectrum in Figure 7.5 is from amorphous carbon that exhibits a considerable amount of sp³ bonds and some hydrogen satisfying dangling bonds. This is a rich field with important applications for industry, but outside the scope of this chapter. Discussions on amorphous carbon can be found in the article by Ferrari and Robertson (2004).

7.3.1 The sp² Model System: Graphene

Among the sp² carbon systems, monolayer graphene is the simplest and has, consequently, the simplest Raman spectra (see Figure 7.6). The ideal graphene is a 2D crystalline sheet, one atom thick, with two C atoms in a unit cell that repeats itself to infinity.

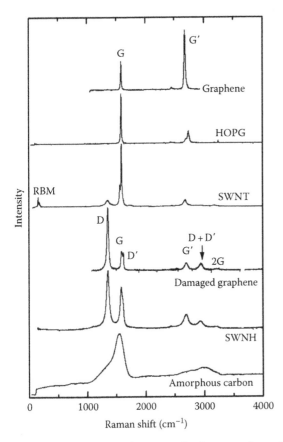

FIGURE 7.5 Raman spectra from several sp² nano-carbons. From top to bottom: crystalline graphene, highly oriented pyrolytic graphite (HOPG), SWNT bundles, damaged graphene, single-wall carbon nano-horns (SWNH), and hydrogenated amorphous carbon. The most intense Raman peaks are labeled in a few of the spectra.

Its Raman spectrum is marked by two strong features, named the G and G′ bands (G from graphite), as discussed below.

7.3.1.1 The G Band

The most usual Raman peak observed in the Raman spectra of any sp² carbon system has historically been named the G band. The G mode is related to the stretching of the bonds between the nearest neighbor A and B carbon atoms in the unit cell. This feature appears around 1585 cm⁻¹ (see Figure 7.6), but exhibits some specificities according to the particular nano-carbon sp² system under discussion and according to ambient conditions of temperature, pressure, doping, and disorder. Because of the peculiar electronic dispersion of graphene, being a zero gap semiconductor with a linear $E(k)$ dispersion relation, the G band phonons (energy of 0.2 eV) can promote electrons from the valence to the conduction band. For this reason, the electron-phonon coupling in this system is quite strong, and gives rise to a renormalization of the electronic and phonon energies, including a sensitive dependence on electron or hole doping (Das et al. 2008). The FWHM intensity observed for graphene on top of a Si/SiO₂ substrate usually ranges from 6 to 16 cm⁻¹, depending on the graphene to substrate interactions.

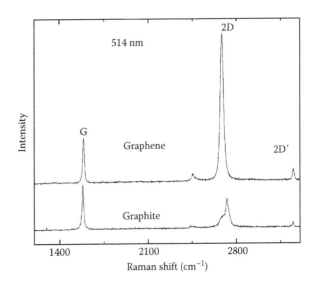

FIGURE 7.6 Raman spectrum of single-layer graphene and graphite. The two most intense features are the Raman allowed first order G band and the second-order G′ band (labeled 2D in the figure making reference to its origin as a second-order process related to the disorder-induced D peak). The spectrum of pristine single-layer graphene is unique in sp² carbons for exhibiting a very intense G′ band as compared to the G band. (From Ferrari, A.C. et al., *Phys. Rev. Lett.*, 97, 187401, 2006; Pfeiffer, R. et al., in *Carbon Nanotubes: Advanced Topics in the Synthesis, Structure, Properties and Applications, Springer Series in Topics in Applied Physics*, Vol. 111, Jorio, A. et al. (eds.), Springer-Verlag, Berlin, Germany, 2008, 495–530. With permission.)

Furthermore, interesting confinement and polarization effects can be observed in the G band of a graphene nanoribbon, as shown by Cancado et al. (2004). The G₁ band from the nanoribbon on top of a highly oriented pyrolytic graphite (HOPG) substrate can be separated from the substrate G₂ band by use of laser heating (see Figure 7.7). The temperature rise of the ribbon due to laser heating is greater than that of the substrate and ω_{G1} for the ribbon decreases more than for the substrate because of the higher thermal conductivity of the substrate relative to the graphene ribbon. The ribbon G₁ band shows a clear antenna effect, where the Raman signal disappears when crossing the light polarization direction with respect to the ribbon axis, in accordance with theoretical predictions (Dresselhaus et al. 2005).

7.3.1.2 The G′ Band

The second most usual Raman peak observed in the Raman spectra of any sp² carbon system has been historically named the G′ band. It is a second-order peak involving two phonons with opposite wave vectors q and $-q$; the atomic motion related to the phonon looks like the hexagon rings are breathing. This feature appears at around 2700 cm⁻¹ (see Figure 7.6) for laser excitation energy of 2.41 eV (514 nm), while the highest phonon frequency in graphene is around 1620 cm⁻¹. The scattering by the related one-phonon process is not allowed by symmetry and has been named the D-band, denoting the dominant disorder-induced band (see Section 7.3.5.1). Besides a rich dependence of the G′ band on the ambient conditions (temperature, pressure, doping), this band exhibits a

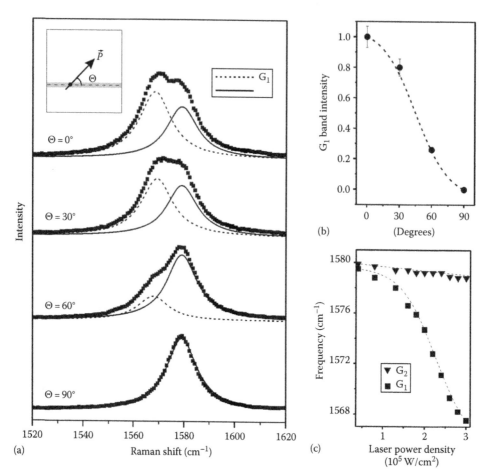

FIGURE 7.7 (a) The G band Raman spectra from the graphene nanoribbon (G_1, dashed line) and from the graphite substrate (G_2, solid line). (b) The G_1 dependence on the light polarization direction with respect to the ribbon axis including experimental points and theoretical predictions in the dashed curve. (c) Frequency of the G peaks as a function of incident laser power for the graphene ribbon (G_2) and the HOPG substrate (G_1). (From Cancado, L.G. et al., *Phys. Rev. Lett.*, 93, 047403, 2004. With permission.)

very interesting resonance phenomena related to the laser excitation energy. By increasing (decreasing) E_{laser}, the G′ peak frequency $\omega_{G'}$ will increase (decrease), as is also the case of the D band at $\omega_D \approx \omega_{G'}/2 \approx 1350\,cm^{-1}$ for $E_{laser} = 2.41\,eV$ (see Section 7.3.5.3).

7.3.2 Adding Graphene Layers: From a Single-Layer Graphene to Graphite

When increasing the number of graphene layers from one to two, two atoms are added to the unit cell (see Figure 7.8), thus increasing the number of phonon branches. Here we consider trilayer graphene as having an ABA Bernal stacking, like HOPG. Rhombohedral graphite has ABC stacking, but it will not be considered here (Wyckoff 1981). We now discuss the main changes in the Raman spectra when adding layers to the graphene system, i.e., bilayer, trilayer, many-layer (graphite) (Malard et al. 2009a).

7.3.2.1 The G′ Band

While the first-order G band spectrum is approximately independent of the number of layers, the G′ spectrum shows the most characteristic changes and for this reason the G′ band can

be used to determine the number of layers (Ferrari et al. 2006) and the stacking order (Pimenta et al. 2007, Ni et al. 2008) of few-layer graphene. The G′ spectrum from single-layer graphene has one peak, while from the bilayer graphene with an AB stacking order, four distinct peaks can be observed (see Figure 7.8d). If the double layer has no stacking order, there is a small upshift in frequency, but only one peak is observed (Ni et al. 2008). The number of peaks increases with the increasing number of layers. However, it is not possible to clearly distinguish all of these peaks since the splitting among them is not larger than their linewidths. Thus, for graphite (the limit of a semi-infinite number of layers), the G′ band is found to contain only two well-defined peaks when the out-of-plane AB stacking is perfect. Turbostratic graphite (no stacking order) has, again, a one peak G′ band lineshape (Pimenta et al. 2007).

7.3.2.2 Other Raman Features

Other features are usually observed in graphite, and are identified as combination modes and overtones (other than the G′ band). Some examples are the features observed above 2000$\,cm^{-1}$ in Figure 7.5. The mode assignment upon

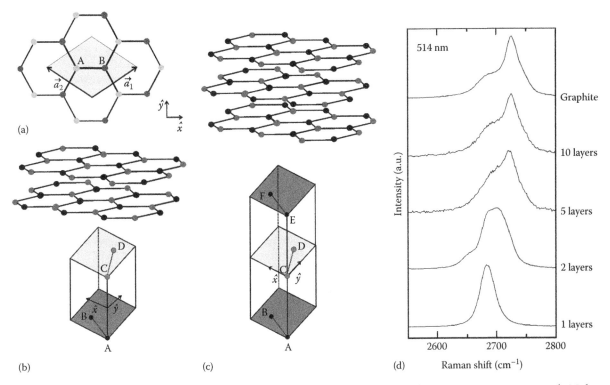

FIGURE 7.8 Schematics for (a) mono-, (b) bi-, and (c) trilayer graphene (From Malard, L.M. et al., *Phys. Rev. B*, 79, 125426, 2009b. With permission.). (d) Evolution of the G' band spectra at 514 nm ($E_{laser} = 2.41$ eV) with the number of graphene layers. (From Ferrari, A.C. et al., *Phys. Rev. Lett.*, 97, 187401, 2006; Pfeiffer, R. et al., in *Carbon Nanotubes: Advanced Topics in the Synthesis, Structure, Properties and Applications, Springer Series in Topics in Applied Physics*, Vol. 111, Jorio, A. et al. (eds.), Springer-Verlag, Berlin, Germany, 2008, 495–530. With permission.)

which these features are based is discussed in the article by Dresselhaus et al. (2005). Furthermore, many other Raman features are observed when disorder is present in nano-carbons, as discussed in Section 7.3.5.

7.3.3 Rolling Up One Graphene Layer: The Single-Wall Carbon Nanotube

The SWNT can be obtained by rolling up a strip of graphene into a cylinder (see Figure 7.9). This procedure can generate tubes with different diameters (d_t) by changing the width of the graphene strip, by using different chiral angles (θ), and by changing the angle between the carbon bonds and the tube axis. These two characteristic parameters (d_t, θ) define the SWNT structure and are usually represented by the (n, m) indices, which describe the number of $\mathbf{a}_1$ and $\mathbf{a}_2$ graphene lattice vectors needed to build the so called chiral vector $\mathbf{C}_h = n\mathbf{a}_1 + m\mathbf{a}_2$. The chiral vector $\mathbf{C}_h$ spans the circumference of the tube cylinder (see Figure 7.9).

Of course, rolling up a graphene strip is an idealized picture. The carbon nanotubes are actually formed by carbon atoms in the vapor phase self-organizing themselves to form a tube structure when growing from a nanometer size catalyst particle. Different growth methods generate samples with very different aspects, like isolated tubes on different substrates or suspended tubes over trenches, tubes in bundles, or tubes (isolated or in bundles) that are aligned along a specific direction. Furthermore, tube

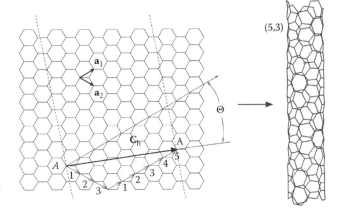

FIGURE 7.9 Schematic diagram showing a possible rolling up of a strip of 2D graphene sheet into a tubular form. In this example, a (5, 3) nanotube is under construction and the resulting tube is illustrated on the right. (From Saito, R. et al., *Physical Properties of Carbon Nanotubes*, Imperial College Press, London, U.K., 1998.)

processing can generate samples in different environments, such as liquids or wrapped by DNA molecules (Joselevich et al. 2008).

The most spectacular differences when comparing SWNTs and graphene are observed due to quantum confinement effects, which are also completely different from the effects on graphite ribbons due to satisfying the carbon bond requirements at the edges of the ribbon. As a very simple example, the zero-energy translation of the graphene generates, when rolling it up into a tube, the so-called radial breathing mode (RBM), which is discussed in Section 7.3.3.1. Furthermore, due to the spatial confinement in this 1D system, their optical properties are discrete, similar to molecules. The physics behind it is related to the formation of spikes, named van Hove singularities, in the density of electronic states (Saito et al. 1998, Dresselhaus et al. 2005) and the formation of excitons (bounded electron-hole pairs) for the optical transitions, as is discussed in the article by Dresselhaus et al. (2006). These van Hove singularities and the exciton formation generate an absorption picture for SWNTs that, although very rich in different phenomena (Dresselhaus et al. 2006), is mostly dominated by strong discrete optical transition levels usually denoted by E_{ii}, where $i = 1, 2, 3...$ enumerates the optical transition energy levels (Saito et al. 1998, Dresselhaus et al. 2005). The E_{ii} optical transition energies depend on (d_t, θ), or alternatively on (n, m), and many of them lie in the visible range, thus generating strong resonance Raman effects, as discussed in the following sections.

7.3.3.1 The Radial Breathing Mode

The RBM Raman feature corresponds to the coherent vibration of all the C atoms of the SWNT in the radial direction, as if the tube were "breathing." This feature is unique to carbon nanotubes and occurs with frequencies ω_{RBM} typically between 55 and 350 cm^{-1} for SWNTs with diameters in the range 0.7 nm < d_t < 4 nm. These RBM frequencies are therefore very useful for identifying whether a given carbon material contains SWNTs, through the presence of the Raman-active RBM modes (see Figure 7.5). When the excitation laser energy is in resonance with the optical transition energy E_{ii} from one isolated SWNT, its RBM can be seen, as shown in the left panel of Figure 7.10. The (n, m) assignment can be made based

on the RBM measurement, as discussed in the next section. The natural linewidth (FWHM) for isolated SWNTs on a SiO$_2$ substrate is $\Gamma = 3$ cm^{-1} (Dresselhaus et al. 2005), but much narrower linewidths (down to ~0.25 cm^{-1}) have been observed for the inner tubes of double-wall carbon nanotubes (DWNTs) at low temperatures (Pfeiffer et al. 2008).

The RBM gives the nanotube diameter through the use of the relation $\omega_{RBM} = (A/d_t)(1 + Ce/d_t^2)^{1/2}$. Therefore, from the ω_{RBM} measurement of an individual isolated SWNT (see left panel in Figure 7.10), it is possible to obtain its d_t value. The parameter $A = 227$ cm^{-1} nm has been obtained experimentally for a single type of sample (Araujo et al. 2008), and it is in agreement with theoretical predictions based on the elastic constant of graphite (Mahan 2002). Most of the samples, however, exhibit an upshift in ω_{RBM} (see Figure 7.10b) due to the tube-environment vander Waals type interactions, and this interaction can be taken into account by making the parameter $Ce \neq 0$ for each specific sample (Araujo et al. 2008). The RBM spectra for SWNT bundles contain RBM contributions from different SWNTs in resonance with the laser excitation line (see Figure 7.10), and a careful spectral analysis gives the tube diameter distribution in the bundles.

7.3.3.2 The Resonance Raman Effect in the RBM and the (n, m) Assignment

When the excitation laser energy matches one of the discrete optical transition energies E_{ii} for a given SWNT, there is large enhancement in the Raman intensity (resonance Raman effect). Therefore, for interpreting the Raman spectra of SWNTs, a very useful guide is the so-called Kataura plot, where the transition energies E_{ii} are plotted as a function of nanotube diameter d_t (see the left panel in Figure 7.11). Such a plot can be directly related to the RBM spectra (see the right panel in Figure 7.11).

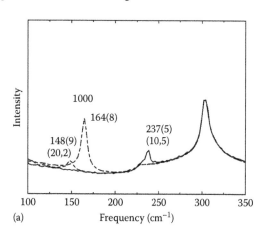
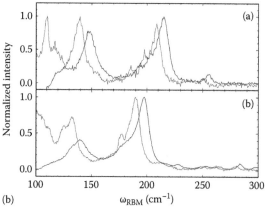

FIGURE 7.10 (a) Raman spectra of a Si/SiO$_2$ substrate containing isolated SWNTs grown by the CVD method. (From Jorio, A. et al., *Phys. Rev. Lett.*, 86, 1118, 2001. With permission.) The spectra are taken at three different spots on the substrate where the RBM Raman signals from resonant SWNTs are found. The RBM frequencies (linewidths) are displayed in cm^{-1}. Also shown are the (n, m) indices assigned from the Raman spectra for each resonant tube. The step at 225 cm^{-1} and the peak at 303 cm^{-1} come from the Si/SiO$_2$ substrate. (b) The RBM Raman spectra for "super-growth" SWNTs (gray) and for "alcohol CVD" SWNTs (black). The samples are different and exhibit different E_{ii} resonance values. The four spectra are obtained using different excitation laser lines: (a) 590 nm (gray) and 600 nm (black); (b) 636 nm (gray) and 650 nm (black), so that in (a) and (b) the same (n, m) SWNTs are in resonance. The gray and black spectra are shifted from each other due to the different environments seen by the SWNTs in the two types of samples. (From Araujo, P.T. et al., *Phys. Rev. B*, 77, 241403(R), 2008. With permission.)

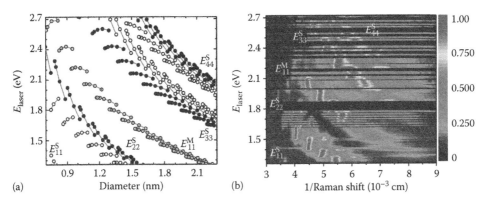

FIGURE 7.11 **(See color insert following page 20-16.)** (a) The optical transition energies (dots) of SWNTs as a function of tube diameter. (From Jiang, J. et al., *Phys. Rev. B*, 75, 035407, 2007a. With permission.) The superscripts S and M for the E_{ii} labels indicate the optical transitions for semiconducting and metallic SWNTs, respectively. (b) 2D color map showing the SWNT RBM spectral evolution as a function of laser excitation energy. The intensity of each spectrum is normalized to the strongest peak, and we plot the RBM results are plotted in terms of the inverse Raman shift. (a) and (b) show coincident axes and can be directly correlated. (From Araujo, P.T. et al., *Phys. Rev. Lett.*, 98, 067401, 2007. With permission.)

If resonance with a single tube is achieved ($E_{laser} = E_{ii}$) and its RBM is observed, it is possible to assign its specific (n, m), since (E_{ii}, ω_{RBM}) can be related to (d_t, θ) (Jorio et al. 2001). For a more general diameter characterization of a bundled SWNT sample based on the RBMs, it is necessary to work with the Kataura plot (Dresselhaus et al. 2005). A single Raman measurement gives the RBM for the specific tubes that are in resonance with that laser line, which may not give a complete characterization of the diameter distribution of the sample. By taking the Raman spectra with many excitation laser lines, a good characterization of the diameter distribution in the sample can be obtained (Dresselhaus et al. 2005, Araujo et al. 2007). Since semiconducting (S) and metallic (M) tubes of similar diameters do not occur at similar E_{ii} values, ω_{RBM} measurements using several laser energies (E_{laser}) can also be used to characterize the ratio of metallic to semiconducting SWNTs in a given sample (Samsonidze et al. 2004, Miyata et al. 2008).

A careful analysis of the resonance Raman intensities in the right panel of Figure 7.11 shows that the RBM intensity has a strong (n, m) dependence, as explained in the article by Jiang et al. (2007b). This effect is mostly due to a chiral angle dependence of the electron-phonon coupling, plus a diameter dependence of the electron-radiation interaction due to excitonic effects (Dresselhaus et al. 2006, Jiang et al. 2007b).

7.3.3.3 The G Band

In contrast to the graphite Raman G band, which exhibits one single Lorentzian peak at 1584 cm⁻¹ related to the tangential mode vibrations of the C atoms, the SWNT G-band is composed of multi-peaks due to the phonon wave vector confinement along the SWNT circumferential direction and due to symmetry-breaking effects associated with the SWNT curvature (see Figure 7.12). The G-band frequency in SWNTs can be used for (1) diameter characterization, the lower frequency G⁻ peak exhibiting a frequency dependence on diameter (see Figure 7.12b and c); (2) distinguishing between metallic and semiconducting SWNTs, through major differences in their Raman lineshapes

(see Figure 7.12a); (3) probing the charge transfer arising from doping a SWNT (Terrones et al. 2008); and (4) studying the selection rules in the various Raman scattering processes and scattering geometries (Dresselhaus et al. 2005).

Elaborating a bit more on item (2) above, the difference between the G band spectra from metal and semconducting tubes is the strong coupling between electrons and G band phonons in metals. Although up to six modes of Raman are allowed in the G band of SWNTs (Dresselhaus et al. 2005), the G band is dominated by two strong peaks that can be represented by the C–C stretching along the circumferential direction or along the axis, usually named the transverse optical (TO) and longitudinal optical (LO) modes, respectively. The LO mode strongly couples with electrons in metals, being largely downshifted in frequency and broadened (see Figure 7.12a). The physics related to this coupling have been discussed in the articles by Piscanec et al. (2004) and Ando (2006).

7.3.3.4 The G′ Band

Like for the other sp² carbons, the G′ Raman spectra provide unique information about the electronic structure of both semiconducting and metallic SWNTs. The G′-band sometimes appears (at the individual nanotube level) in the form of unusual two-peak structures for both semiconducting and metallic nanotubes (Samsonidze et al. 2003, Dresselhaus et al. 2005), even if there is no interlayer coupling, like in bilayer graphene and graphite. The two-peak G′-band Raman features observed from semiconducting and metallic isolated nanotubes are shown in Figure 7.13a and b, respectively. The presence of two peaks in the G′-band Raman feature from semiconducting SWNTs indicates resonance with two different van Hove singularities of the same nanotube, occurring independently for both the incident E_{laser} and scattered $E_{laser} - E_{G'}$ photons.

7.3.3.5 Other Features

Carbon nanotubes also exhibit several combination modes and overtones. Basically, all the Raman features observed in graphite can also be seen in carbon nanotubes. Furthermore,

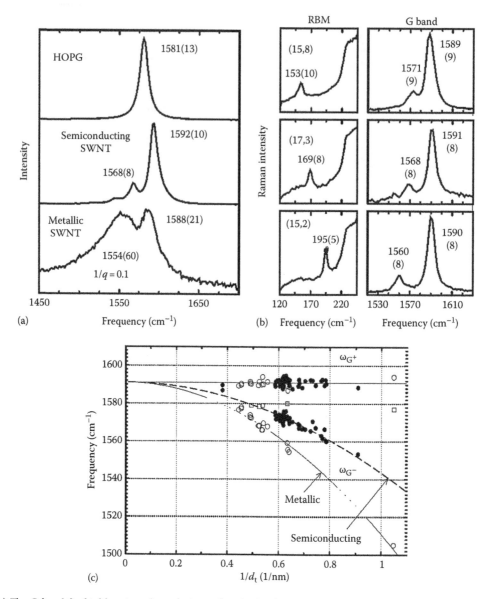

FIGURE 7.12 (a) The G-band for highly oriented pyrolytic graphite (HOPG), one semiconducting SWNT, and one metallic SWNT. For C_{60} fullerenes, a peak is observed at $1469\,cm^{-1}$, but it is not considered a G band. (b) The RBM and G-band Raman spectra for three semiconducting isolated SWNTs of the indicated (n, m) values. (c) Frequency vs. $1/d_t$ for the two most intense G-band features (ω_G- and ω_G+) from isolated SWNTs. (From Dresselhaus, M.S. et al., *Phys. Rep.*, 409, 47, 2005. With permission.)

carbon nanotubes exhibit several Raman features in the spectral region between 400 and $1200\,cm^{-1}$, which are called intermediate frequency modes (IFMs), since their frequencies lie between the common ω_{RBM} and ω_G modes (Dresselhaus et al. 2005). Some of the IFMs are fixed in energy and some of the IFMs are "dispersive" (Raman shift changes when changing the excitation energy) and are attributed to combination modes. It is not yet clear whether these modes are related to disorder or not. Theory relates their observation with confinement along the tube length (Saito et al. 1998), and some supporting experimental evidence has been found for such an effect (Chou et al. 2007). However, the IFM picture is not yet fully understood.

7.3.4 Adding Tube Layers: Double- and Multi-Wall Carbon Nanotubes

The differences between single-, double-, and many-wall carbon nanotubes are often classified by the diameter distribution in the tubes. MWNTs, for example, usually have tubes with very large diameters d_t, with inner diameters over 10 nm, and outer diameters rising up to about 100 nm. Their spectra approach that of graphite, since no confinement effects can be observed. Only general line broadening differentiates large MWNTs from graphite (see Figure 7.14).

For DWNTs, however, very interesting outer vs. inner tube effects have been observed. The most striking is the observation of many

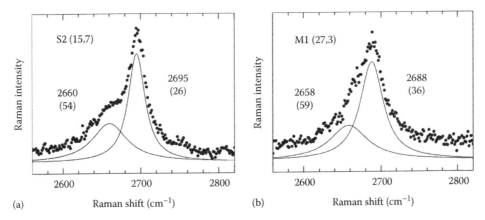

FIGURE 7.13 The G′-band Raman features for (a) semiconducting (15, 7) and (b) metallic (27, 3) nanotubes showing unusual two-peak structures. (From Samsonidze, Ge. G. et al., *Appl. Phys. Lett.*, 85, 1006, 2004. With permission.)

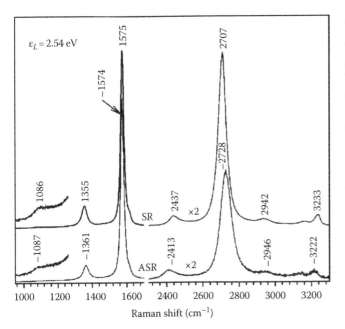

FIGURE 7.14 Stokes (SR) and anti-Stokes (ASR) Raman spectra from MWNTs. The small difference in frequencies when comparing the same peak in the SR and ASR spectra are due to the double resonance effect, as discussed in detail in Cancado et al. (2002) and Tan et al. (2002). (From Tan, P. et al., *Phys. Rev. B*, 66, 245410, 2002. With permission.)

different ω_{RBM} for the same (*n*, *m*) inner SWNT, due to different possible outer tubes (see Figure 7.15). Furthermore, charge transfer effects depending on the metal vs. semiconducting outer vs. inner configuration have been observed (Souza Filho et al. 2007). While most of these results have been obtained from DWNTs in bundles, measurements of isolated DWNTs are expected to be very informative (Villalpando-Paez et al. 2008).

7.3.5 Disorder in sp² Systems

When the size of a graphite system is reduced, disorder effects start to be seen. Disorders can also be seen at defect locations or at graphene/graphite edges.

7.3.5.1 The D Band

The so-called D band (D comes from disorder) is observed in any sp² carbon system when the lattice periodicity is broken by a disorder mechanism, such as defects. This feature appears around 1350 cm⁻¹ (for E_{laser} = 2.41 eV) and the phonon-related step of the D band double resonance process corresponds to one of the steps in the G′ band process. The D band itself is a second-order process in which the elastic scattering of the defect allows for the conservation of momentum in the phonon creation, and the defect breaks the Raman scattering selection rules. The D band has been largely used to characterize disorder in carbon materials (see Figure 7.16). The D band is dispersive, like the G′ band, and shows interesting phenomena in SWNTs due to the electron and phonon confinement (Dresselhaus et al. 2005, Pimenta et al. 2007).

7.3.5.2 The D′ Band

The so-called D′ band, appearing around 1620 cm⁻¹, is another feature commonly observed in many of the sp² disordered carbon systems. This feature is usually much weaker than the D band, and for this reason the D band is more often used for disorder characterization. Whereas the D band is connected with an intervalley scattering process from the K point to the K′ point, in the Brillouin zone, the D′ band is connected with an intravalley scattering process around the K point or the K′ point. The D and D′ bands tend to be sensitive to defects of a different physical origin, but these differences require further study.

7.3.5.3 The G′ Band

The G′ band is not disorder induced but it can be used to study changes in the electronic and vibrational structure related to disorder. The 2D vs. 3D stacking order of graphene layers is one example. Highly crystalline 3D graphite shows two G′ peaks (see the top spectra of Figure 7.8d). When the interlayer stacking order is lost, a one-peak feature starts to develop, identified with 2D graphite, and the peak is centered near the middle of the two peaks in the G′ lineshape from ordered graphite (Pimenta et al. 2007).

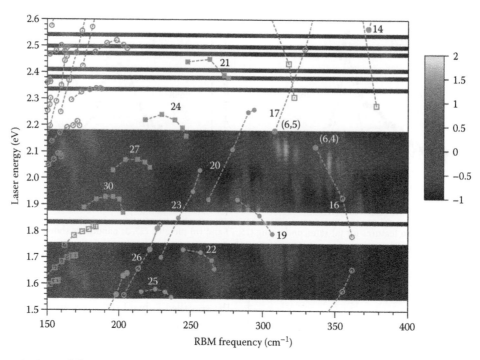

FIGURE 7.15   (See color insert following page 20-16.) The 2D RBM Raman map for DWNTs. The E_{ii} points from the Kataura plot are superimposed (green bullets). It is clear there are many more RBM features than (n, m) related E_{ii} values. (From Pfeiffer, R. et al., in *Carbon Nanotubes: Advanced Topics in the Synthesis, Structure, Properties and Applications, Springer Series in Topics in Applied Physics*, Vol. 111, Jorio, A. et al. (eds.), Springer-Verlag, Berlin, Germany, 2008, 495–530. With permission.)

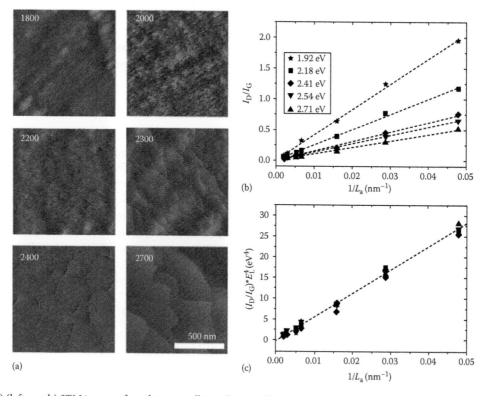

FIGURE 7.16 (a) (left panels) STM images of graphite crystallites. The crystallite size L_a varies with the annealing temperature in degree Celsius (displayed in the images). (right panels) The evolution of the D/G intensity ratio (I_D/I_G) as a function of crystallite size (a) for different laser excitation energies. All the curves in (b) collapse into one curve (c) when considering the dependence of the G band intensity on the fourth power of the incident laser excitation energy, E_{laser}^4. (From Pimenta, M.A. et al., *Phys. Chem. Chem. Phys.*, 9, 1276, 2007. With permission.)

Furthermore, localized emission of a red-shifted G′ band was observed and related to the local distortion of the nanotube lattice by a negatively charged defect. The opposite occurs for *p* doping and this effect can be used to study SWNT doping (see Figure 7.17; Maciel et al. 2008).

7.3.6 Other Raman Modes and Other sp² Carbon Structures

In general, the observation of overtones and combination modes in condensed matter systems is rare because of dispersion effects that make these features too weak and too broad to pick out from the noisy background. The double resonance process (Saito et al. 2003), however, allows such overtones and combination modes to be quite clearly observed (Dresselhaus et al. 2005), thereby providing new information about SWNT properties. Nanowiskers, nanobuds, nanorods, and nanohorns exhibit a large number of

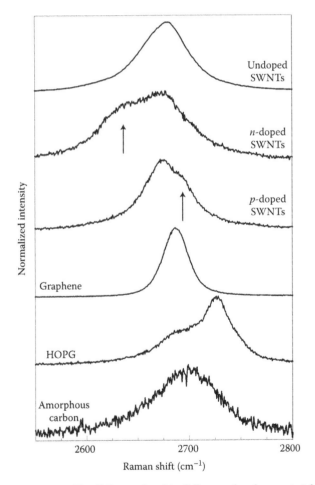

FIGURE 7.17 The G′ Raman band in different sp² carbon materials measured at room temperature with E_{laser} = 2.41 eV (514 nm). The arrows point to defect-induced peaks in the G′ band for doped SWNTs. The *p/n* doping comes from substitutional boron/nitrogen atoms (Terrones et al. 2008), the nearest neighbors of carbon in the periodic table. The spectra of graphene, HOPG, and amorphous carbon are shown for comparison. (From Maciel, I.O. et al., *Nat. Mat.*, 7, 878, 2008. With permission.)

peaks (see Figure 7.18), always related to and assignable from the phonon structure of graphene.

Although we do not cover fullerenes here, when talking about overtones, it is especially interesting to mention their Raman spectra. As a molecular carbon structure, a fullerene has an especially rich overtone spectra that can be seen in both Raman and infrared spectra. A detailed discussion on this topic can be found in the book by Dresselhaus et al. (1996).

7.4 Critical Discussions

As discussed in Section 7.3, Resonance Raman spectroscopy (RRS) was shown to provide a powerful metrological tool for distinguishing among the different sp² nano-carbons, the number of layers of a graphene sample, the AB stacking order in many-layers graphene of graphite and the metallic (M) from semiconducting (S) tubes; and for determining the diameter distribution of SWNTs in a given sample, the (*n*, *m*) values for specific tubes, the doping, and many other important properties. Recent RRS measurements done on the $E_{22}^S, E_{11}^M, E_{33}^S, E_{44}^S, E_{22}^M, E_{55}^S,$ and E_{66}^S transitions on water-assisted chemical vapor deposition (CVD)-grown SWNTs (Araujo et al. 2008) suggest that tubes in the interior of the forest of aligned SWNTs seem to be well-shielded from environmental effects and may provide a standard reference material for SWNTs that show minimal environmental effects. This suggestion needs further experimental confirmation, but if this interpretation is correct, such a reference material might represent the first nano-carbon standard reference material, which could be useful for quantitative determinations of environmental effects in nanotubes. Such effects are important for understanding the current nanotube photo-physics studies where samples routinely experience environmental effects due to substrates, wrapping agents, functionalization, strain or molecules adsorbed in suspended nanotubes, and similar problems also arise for graphene and the other nano-carbons. Controlling environmental effects are vital for a variety of uses of nano-carbons for sensors and biomedical applications.

Calculations of Raman frequencies and RRS matrix elements are now at an advanced stage for SWNTs, but for understanding lifetime effects, dark singlet and triplet states are still incomplete. The characterization of MWNTs as well as many layers of graphene by RRS is at an early stage, though a good start has been made on the characterization of DWNTs and bilayer graphene that are, respectively, the simplest examples of a MWNT or of graphite. The effect of edge states in graphene is expected to be a rich field and has been, basically until now, weakly explored by spectroscopy.

7.5 Summary

In summary, we presented here the basic concepts of Raman spectroscopy and the Raman signatures of sp² carbon materials. The G band at ~1585 cm⁻¹ has a single-Loretzian peak structure for 2D and 3D materials, and shows a special lineshape for their 1D counterparts, carbon nanotubes, with a doublet rather than a single Lorentzian feature. Raman spectroscopy

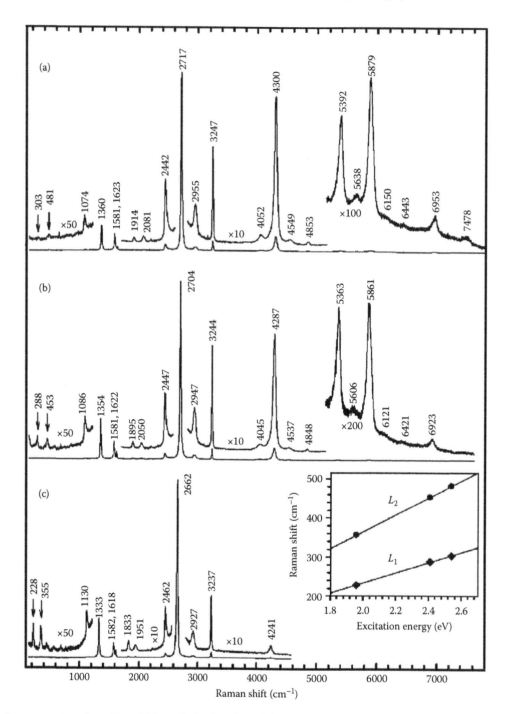

FIGURE 7.18 Raman spectra of graphite whiskers obtained at three different laser excitation energies: (a) 488.0 nm, (b) 514.5 nm, and (c) 632.8 nm. Note that some phonon frequencies vary with E_{laser}, some do not. The inset to (c) shows details of some peaks that are dispersive, and are explained theoretically by the double-resonance process. (From Tan, P. et al., *Phys. Rev. B*, 64, 214301, 2001. With permission.)

can also be used to differentiate between metallic and semiconducting carbon nanotubes, as well as to obtain information on tube diameter and the tube diameter distribution of a bundled sample. Another important feature is the G′ band, which appears at ~2700 cm^{-1} for 2.41 eV (514 nm) excitation. This feature is highly sensitive to the electronic structure, changing shape when comparing single-layer graphene, bilayer graphene, trilayer graphene, stack-ordered (3D) graphite, and

pyrolytic (2D) graphite and when nano-carbons are doped. The cylindrical carbon nanotubes are the only nano-carbon material showing the low frequency radial breathing modes that are strongly dependent on diameter, and usually appear in the range 50–350 cm^{-1}. Finally, at ~1350 cm^{-1} ($\approx \omega_{G'}/2$), the D peak can be observed when there is disorder in the material, i.e., any perturbation that breaks the sp^2 periodic structure, and the D-band can be used for characterizing the structural quality of carbon-based materials. Finally, many other features related to disorder, overtone, and combination modes are observed, most of them being weak, but sometimes they are strong. Although the assignment and frequency behavior has been quite well understood, the variation in intensity of these features still remains an open issue.

7.6 Future Perspectives

It is clear that Raman spectroscopy of sp² carbons remains a fruitful field of research even though many advances have already been achieved. A big step forward in the field will occur when Raman spectroscopy starts to be combined in a single instrument with microscopy techniques, either high resolution electron transmission microscopy (HRTEM) or scanning probe microscopies, like atomic force microscopy (AFM) or scanning tunneling spectroscopy (STM). Combining the information from real space (microscopy) with information from momentum space (spectroscopy) should bring in new insights, including a new understanding of different types of disorders. In this context, near-field studies should also bring in an important new understanding, since Raman spectroscopy with spatial resolution down to 10 nm can be achieved, which will provide important information on the science of defects (Hartschuh et al. 2003, Maciel et al. 2008).

Furthermore, a major focus is now devoted to directing the nano-carbon field to applications, which urges studies on metrology, standardization, and industrial quality control (Jorio and Dresselhaus 2007, Jorio et al. 2008b). The development of protocols for the definition of sample parameters like structural metrics (carbon–carbon distance, surface area, tube diameter, chiral angle, ribbon width), physical properties (optical, thermal, mechanical), composition (impurity content, spatial homogeneity), and stability (dispersability, bio-compatibility and health effects) are important for both research and applications of nano-carbons. These metrological protocols are expected to be applicable not only to nanocarbon materials, but also to the exploding field of nanomaterials, where metrology issues will drive technological growth and innovation.

Acknowledgments

M.S.D. and G.D. acknowledge NSF-DMR 07-04197. A.J. acknowledges financial support from CNPq, CAPES, and FAPEMIG. The authors thank Mario Hofmann for help with the preparation of the manuscript.

References

Ando, T., 2006. Anomaly of optical phonon in monolayer graphene. *J. Phys. Soc. Jpn.* 75: 124701.

Araujo, P. T., Doorn, S. K., Kilina, S. et al., 2007. Third and fourth optical transitions in semiconducting carbon nanotubes. *Phys. Rev. Lett.* 98: 067401.

Araujo, P. T., Maciel, I. O., Pesce, P. B. C. et al., 2008. Nature of the constant factor in the relation between radial breathing mode frequency and tube diameter for single-wall carbon nanotubes. *Phys. Rev. B* 77: 241403(R).

Arnold, M. S., Green, A. A., Hulvat, J. F. et al., 2006. Sorting carbon nanotubes by electronic structure using density differentiation. *Nat. Nanotechnol.* 1: 60–65.

Bethune, D. S., Kiang, C. H., deVries, M. S. et al., 1993. Cobalt catalysed growth of carbon nanotubes with single atomic layer wells. *Nature* 363: 605–607.

Biercuk, M. J., Ilani, S., Marcus, C. M. et al., 2008. Electrical transport in single-wall carbon nanotubes. In *Carbon Nanotubes: Advanced Topics in the Synthesis, Structure, Properties and Applications*, eds. A. Jorio, M. S. Dresselhaus, and G. Dresselhaus, pp. 63–100. *Springer Series in Topics in Applied Physics*, Springer-Verlag, Berlin, Germany, Vol. 111.

Boehm, H. P., Clauss, A., Fischer, G. O., and Hofman, U., 1962. Thin carbon leaves. *Zeitschrift fur Naturforschung.* 176: 150–153.

Cancado, L. G., Pimenta, M. A., Neves, B. R. A. et al., 2004. Anisotropy of the Raman spectra of nanographite ribbons. *Phys. Rev. Lett.* 93: 047403.

Cancado, L. G., Pimenta, M. A., Saito, R. et al., 2002. Stokes and anti-Stokes double resonance Raman scattering in two-dimensional graphite. *Phys. Rev. B* 66: 035415.

Castro Neto, A. H., Guinea, F., Peres, N. M. R. et al., 2008. The electronic properties of graphene. *Rev. Mod. Phys.* 81: 109.

Charlier, J.-C., Eklund, P. C., Zhu, J. et al., 2008. Electron and phonon properties of graphene: Their relationship with carbon nanotubes. In *Carbon Nanotubes: Advanced Topics in the Synthesis, Structure, Properties and Applications*, eds. A. Jorio, M. S. Dresselhaus, and G. Dresselhaus, pp. 673–708. *Springer Series in Topics in Applied Physics*, Springer-Verlag, Berlin, Germany, Vol. 111.

Chou, S. G., Son, H., Zheng, M. et al., 2007. Finite length effects in DNA-wrapped carbon nanotubes. *Chem. Phys. Lett.* 443: 328–332.

Das, A., Pisana, S., Charkraborty, B. et al., 2008. Monitoring dopants by Raman scattering in an electrochemically top-gated graphene transistor. *Nat. Nanotechnol.* 3: 210–215.

Dresselhaus, M. S., Dresselhaus, G., and Eklund, P., 1996. *Science of Fullerenes and Carbon Nanotubes*. Academic Press, New York.

Dresselhaus, M. S., Dresselhaus, G., Saito, R. et al., 2005. Raman spectroscopy of carbon nanotubes. *Phys. Rep.* 409: 47–99.

Dresselhaus, M. S., Dresselhaus, G., Saito, R. et al., 2006 Exciton photophysics of carbon nanotubes. *Ann. Rev. Phys. Chem.* 58: 719–747.

Dresselhaus, M. S., Dresselhaus, G., Sugihara, K. et al., 1988. *Graphite Fibers and Filaments. Springer Series in Materials Science*, Springer-Verlag, Berlin, Germany, Vol. 5.

Endo, M., Strano, M. S., Ajayan, P. M., 2008. Potential Applications of Carbon Nanotubes. In *Carbon Nanotubes: Advanced Topics in the Synthesis, Structure, Properties and Applications*, eds. A. Jorio, M. S. Dresselhaus, and G. Dresselhaus, pp. 13–61. *Springer Series in Topics in Applied Physics*, Springer-Verlag, Berlin, Germany, Vol. 111.

Ferrari, A. C., Meyer, J. C., Scardaci, V. et al., 2006. Raman spectrum of graphene and graphene layers. *Phys. Rev. Lett.* 97: 187401.

Ferrari, A. C. and Robertson, J., 2004. Raman spectroscopy in carbons: From nanotubes to diamond. *Philos. Trans. R. Soc. Lond. A* 362: 2267–2565.

Geim, A. K. and Novoselov, K. S., 2007. The rise of graphene. *Nat. Mat.* 6(3): 183–191.

Hartschuh, A., Sanchez, E. J., Xie, X. S. et al., 2003. High-resolution near-field Raman microscopy of single-walled carbon nanotubes. *Phys. Rev. Lett.* 90: 095503–095506.

Hata, K., Futaba, D. N., Mizuno, K. et al., 2004. Water-assisted highly efficient synthesis of impurity-free single-walled carbon nanotubes. *Science* 306: 1362–1364.

Huang, S. M., Cai, X. Y., and Liu, J., 2003. Growth of millimeter-long and horizontally aligned single-walled carbon nanotubes on flat substrates. *J. Am. Chem. Soc.* 125: 5636–5637.

Iijima, S., 1991. Helical microtubules of graphitic carbon. *Nature* 354: 56.

Iijima, S. and Ichihashi, T., 1993. Single shell carbon nanotubes of 1-nm diameter. *Nature* 363: 603–605.

Jiang, J., Saito, R., Samsonidze, Ge. G. et al., 2007a. Chirality dependence of exciton effects in single-wall carbon nanotubes: Tight-binding model. *Phys. Rev. B* 75: 035407.

Jiang, J., Saito, R., Sato, K. et al., 2007b. Exciton-photon, exciton-phonon matrix elements, and resonant Raman intensity of single-wall carbon nanotubes. *Phys. Rev. B* 75: 035405.

Jorio, A. and Dresselhaus, M. S., 2007. Nanometrology links state-of-the-art academic research and ultimate industry needs for technological innovation. *MRS Bull.* 34(12): 988–993.

Jorio, A., Dresselhaus, M. S., and Dresselhaus, G., 2008a. *Carbon Nanotubes: Advanced Topics in the Synthesis, Structure, Properties and Applications. Springer Series in Topics in Applied Physics*, Springer-Verlag, Berlin, Germany, Vol. 111.

Jorio, A., Kauppinen, E., and Hassanien, A., 2008b. Carbon-nanotube metrology. In *Carbon Nanotubes: Advanced Topics in the Synthesis, Structure, Properties and Applications*, eds. A. Jorio, Dresselhaus, M. S., and G. Dresselhaus, pp. 63–100. *Springer Series in Topics in Applied Physics*, Springer-Verlag, Berlin, Germany, Vol. 111.

Jorio, A., Saito, R., Hafner, J. H. et al., 2001. Structural (n, m) determination of isolated single-wall carbon nanotubes by resonant Raman scattering. *Phys. Rev. Lett.* 86: 1118–1121.

Joselevich, E., Dai, H., Liu, J., and Hata, K., 2008. Carbon nanotube synthesis and organization. In *Carbon Nanotubes: Advanced Topics in the Synthesis, Structure, Properties and Applications*, eds. A. Jorio, M. S. Dresselhaus, and G. Dresselhaus, pp. 101–164. *Springer Series in Topics in Applied Physics*, Springer-Verlag, Berlin, Germany, Vol. 111.

Kroto, H. W., Heath, J. R., O'Brien, S. C., Curl, R. F., and Smalley, R. E., 1985. C60: Buckminsterfullerene. *Nature* 318: 162–163.

Maciel, I. O., Anderson, N., Pimenta, M. A. et al., 2008. Electron and phonon renormalization near charged defects in carbon nanotubes. *Nat. Mat.* 7: 878.

Mahan, G. D., 2002. Oscillations of a thin hollow cylinder: Carbon nanotubes. *Phys. Rev. B* 65: 235402.

Malard, L. M., Pimenta, M. A., Dresselhaus, G., and Dresselhaus, M. S., 2009a. Raman spectroscopy in graphene. *Phys. Rep.* 473: 51–87.

Malard, L. M., Mafra, D. L., Guimaraes, M. H. D. et al., 2009b. Group theory analysis of optical absorption and electron scattering by phonons in mono- and multi-layer graphene. *Phys. Rev. B* 79: 125426.

Miyata, Y., Yanagi, K., and Kataura, H., 2008. Evaluation of the metal-to-semiconductor ratio of single–wall carbon nanotubes using optical absorption spectroscopy. *Ninth International Conference on the Science and Applications of Nanotubes*, Montpellier, France, p. 56, T10.

Ni, Z., Wang, Y., Yu, T. et al., 2008. Reduction of Fermi velocity in folded graphene observed by resonance Raman spectroscopy. *Phys. Rev. B* 77: 235403.

Novoselov, K. S., Geim, A. K., Morozov, S. V. et al., 2004. Electric field effect in atomically thin carbon films. *Science* 306: 666.

Novoselov, K. S., Geim, A. K., Morozov, S. V. et al., 2005. Two-dimensional gas of massless Dirac fermions in graphene. *Nature* 438: 197.

Oberlin, A., Endo, M., and Koyama, T., 1976. Filamentous growth of carbon through benzene decomposition. *J. Cryst. Growth* 32: 335.

Pfeiffer, R., Pichler, T., Kim, Y. A. et al., 2008. Double-wall carbon nanotubes. In *Carbon Nanotubes: Advanced Topics in the Synthesis, Structure, Properties and Applications*, eds. A. Jorio, M. S. Dresselhaus, and G. Dresselhaus, pp. 495–530. *Springer Series in Topics in Applied Physics*, Springer-Verlag, Berlin, Germany, Vol. 111.

Pimenta, M. A., Dresselhaus, G., Dresselhaus, M. S. et al., 2007. Studying disorder in graphite-based systems by Raman spectroscopy. *Phys. Chem. Chem. Phys.* 9: 1276–1291.

Piscanec, S., Lazzeri, M., Mauri, F. et al., 2004. Kohn anomalies and electron-phonon interactions in graphite. *Phys. Rev. Lett.* 93: 185503.

Radushkevich, L. V. and Lukyanovich, V. M., 1952. O strukture ugleroda, obrazujucegosja pri termiceskom razlozenii okisi ugleroda na zeleznom konarte. *Zurn. Fisc. Chim.* 26: 88.

Reich, S., Thomsen, C., and Maultzsch, J., 2003. *Carbon Nanotubes: Basic Concepts and Physical Properties*. Wiley-VCH, Weinheim, Germany.

Saito, R., Dresselhaus, G., and Dresselhaus M. S., 1998. *Physical Properties of Carbon Nanotubes*. Imperial College Press, London, U.K.

Saito, R., Gruneis, A., Samsonidze, Ge. G. et al., 2003. Double resonance Raman spectroscopy of single-wall carbon nanotubes. *New J. Phys.* 5: 157.1–157.15.

Samsonidze, Ge. G., Chou, S. G., Santos, A. P. et al., 2004. Quantitative evaluation of the octadecylamine-assisted bulk separation of semiconducting and metallic single wall carbon nanotubes by resonance Raman spectroscopy. *Appl. Phys. Lett.* 85: 1006–1008.

Samsonidze, Ge. G., Saito, R., Jorio, A. et al., 2003. The concept of cutting lines in carbon nanotube science. *J. Nanosci. Nanotechnol.* 3(6): 431–458.

Souza Filho, A. G., Endo, M., Muramatsu, H. et al., 2007. Resonance Raman scattering studies in Br_2-adsorbed double-wall carbon nanotubes. *Phys. Rev. B* 73: 235413.

Tan, P., An, L., Liu, L. et al., 2002. Probing the phonon dispersion relations of graphite from the double-resonance process of Stokes and anti-Stokes Raman scatterings in multiwalled carbon nanotubes. *Phys. Rev. B* 66: 245410.

Tan, P., Hu, C., Dong, J. et al., 2001. Polarization properties, high-order Raman spectra, and frequency asymmetry between Stokes and anti-Stokes scattering of Raman modes in a graphite whisker, *Phys. Rev. B* 64: 214301.

Terrones, M., Souza Filho, A. G., and Rao, A. M., 2008. Doped carbon nanotubes: Synthesis, characterization and applications. In *Carbon Nanotubes: Advanced Topics in the Synthesis, Structure, Properties and Applications*, eds. A. Jorio, M. S. Dresselhaus, and G. Dresselhaus, pp. 531–566. *Springer Series in Topics in Applied Physics*, Springer-Verlag, Berlin, Germany, Vol. 111.

Villalpando-Paez, F., Son, H., Nezich, D. et al., 2008. Raman spectroscopy study of isolated double-walled carbon nanotubes with different metallic and semiconducting configurations. *Nano Lett.* 8: 3879.

Wyckoff, R. W. G., 1981. *Crystal Structures*, 2nd edn. Krieger, New York.

Zhang, Y. G., Chang, A. L., Cao, J. et al., 2001. Electric-field-directed growth of aligned single-walled carbon nanotubes. *Appl. Phys. Lett.* 79: 3155–3157.

Zhang, Y., Tan, Y. W., Stormer, H. L. et al., 2005. Experimental observation of the quantum Hall effect and Berry's phase in graphene. *Nature* 438: 201.

Dispersions and Aggregation of Carbon Nanotubes

Jeffery R. Alston
University of North Carolina, Charlotte

Harsh Chaturvedi
University of North Carolina, Charlotte

Michael W. Forney
University of North Carolina, Charlotte

Natalie Herring
University of North Carolina, Charlotte

Jordan C. Poler
University of North Carolina, Charlotte

8.1 Introduction

Nanostructured carbon has shown great utility in modifying and enhancing the physiochemical properties of composite materials (Lau et al., 2006). Specifically, single-walled carbon nanotubes (SWNTs) are unique in their ability to enhance the electrical and mechanical properties of materials. The science and technology of formulating stable dispersions of SWNTs in various media has been well reviewed in the literature. This chapter is intended to be an encapsulated tutorial on the formation, stability, and properties of carbon nanotube (CNT) dispersions. That said, the reader is advised to also access several well-cited reviews in this important area. Bundles of tubes aligned parallel with each other and interacting through extensive van der Waals (vdW) forces is the thermodynamically stable state of SWNTs. If we can provide enough energy to overcome this binding energy, it is possible to disperse the SWNTs as individual tubes into various solvents and polymeric matrices. Fu and Sun (2003) review some of the earlier work on forming SWNT dispersions in various solvents and for various surface modifications of the tubes. This is a good place to get started for those unfamiliar with this field. Regardless of the solvent one wants to disperse nanotubes into, one must use some form of mechanical mixing to separate the individual tubes from the stable bundle. Mechanically grinding nanotube solids into a dispersant is effective, but often damages

the tubes too much (Chen et al., 2001a). In order to intentionally shorten and damage SWNTs, high-impact mixing techniques such as Ball–Milling are used (Pierard et al., 2004). When one's intentions are to put as much nanostructured carbon into the matrix regardless of its final properties, these first two mechanical methods are straightforward to apply. Another mechanical method that has shown good utility in dispersing and separating SWNTs is high shear mixing. This technique forces the nanotube bundles through small pores or plates that pull individual tubes from the bundles through extrusion. The article by Hilding provides an excellent review of the mechanical mixing methods and also includes an introduction to the structure and properties of CNTs (Hilding et al., 2003).

Since we want to take advantage of the superior properties of nanotubes for various applications, we must minimize damage as they are brought into dispersion. For most nanotube applications, the dispersion medium is aqueous, organic, or polymeric. The kinetic stability of nanotube dispersions depends on several factors. The nanotube–solvent interaction dominates the kinetics. One cannot achieve a kinetically stable SWNT dispersion in water without first functionalizing the nanotube surface or adding surfactants. Either way, aqueous nanotubes are no longer pristine. There are many organic solvents that can support kinetically stable dispersions of pristine SWNTs. Most of these solvents are good Lewis bases, which do not have significant hydrogen-bonding capabilities, such as *N,N*-dimethylformamide (DMF). Dispersions in these solvents will be detailed below. For composite materials, nanotubes must be dispersed into polymer precursors or directly into the polymer matrix. For these systems, the higher viscosity of the matrix makes mechanical mixing more difficult. Moreover, if the polymer matrix does not interact strongly with the nanotube walls, then the nanotubes will aggregate back to their bundled form and behave like defects within the condensed polymer matrix. Many of the challenges have been overcome and are illustrated in the well-cited review by Xie et al. (2005). To effectively stabilize SWNTs in a polymer, the tubes are chemically modified, then added to polymer precursors and then the matrix is polymerized while it is being extruded, cast, or spun. To further enhance the properties of these polymer composites, the nanotubes are aligned. Nanotube orientation can be established *in situ* by extrusion (Fischer, 2002), by magnetic force (Kimura et al., 2002), or dielectrophoresis (Wang et al., 2008a). Aligned nanotube composites can also be formed *ex situ* by chemical vapor deposition (CVD) growth or self-assembly, functionalized, and finally set into polymer (Feng et al., 2003). Regardless of how the composite is formulated, the stability and utility of the material depends on the nanostructured carbon staying dispersed. Therefore, one must also design a system that inhibits the aggregation of the nanoscale materials.

This chapter addresses the fundamental processes involved in the dispersion and aggregation of SWNTs. We address the structure and properties of CNTs and the intermolecular and interparticle attractive forces that result in nanotube bundles. The thermodynamics of solute-particle interactions leads to a dispersion limit or loading of SWNTs into a matrix. Once dispersed, the nanotubes are kinetically stabilized and this stability is discussed in terms of classical colloidal stability arguments. While a stable dispersion is typically desired, the process of aggregation under various chemical, physical, and optical stimuli is discussed. After an extended background section, we explain in detail some common methods for producing dispersions and measuring their properties. The end result of this work is to make something useful. From these stable dispersions, we aim to move the nanotubes into a more ordered and functional form. Procedures to make SWNT mats, thin films, and devices for electronic, optical, and electrochemical sensing applications will be detailed.

8.2 Background

8.2.1 Structure and Properties of CNTs

Carbon is the sixth element of the periodic table and is the element with the lowest atomic number in column IV. Each carbon atom has six electrons, which occupy 1s, 2s, and 2p atomic orbitals. The 1s orbital contains two strongly bound core electrons. Four more weakly bound electrons occupy the 2s and 2p valence orbitals. In the crystalline phase, the valence electrons give rise to 2s, $2p_x$, $2p_y$, and $2p_z$ orbitals, which are important in forming covalent bonds in carbon materials. Since the energy difference

between the upper 2p energy levels and the lower 2s level in carbon is small compared with the binding energy of the chemical bonds, the electronic wave functions for these four electrons can readily mix with each other. Consequently, this changes the occupation of the 2s and three 2p atomic orbitals so as to enhance the binding energy of the carbon atom with its neighboring atoms. Carbon can form stable morphologies in various dimensions due to several possible hybridizations of the carbon 2s and 2p orbitals, e.g., diamond (three-dimensional), graphite (two-dimensional), SWNT (one-dimensional graphene tubules), and fullerenes (C_{60} zero-dimensional). The various bonding states correspond to certain structural arrangements; e.g., sp bonding gives rise to chain structures, sp^2 bonding to planar structures, and sp^3 bonding to tetrahedral structures. CNTs are theoretically considered as a graphene sheet appropriately rolled into a cylinder with essentially sp^2 atomic bonding between the nearest neighbors.

Electronic properties of SWNTs vary from semiconducting to metallic depending on the tubular structure; i.e., the diameter and chirality of the SWNTs. From the real space lattice of the two-dimensional graphene, several different tubular arrangements differing in diameter and helical arrangement of the carbon hexagons are possible. As SWNTs can be considered a rolled-up graphene sheet, the tubular arrangement of the carbon atoms can be explained in terms of the two-dimensional hexagonal lattice structure of graphene. From the origin of the lattice structure, subsequent atomic arrangements can be referred to using the coordinates of the lattice points, (n, m). As the sheet is rolled into a tube, the origin is superimposed onto itself to get the tubular structure. Due to periodic boundary conditions, only certain wave vectors are allowed. Thus, momentum vectors are allowed only for certain energies as represented by lines in the Brillouin zones (BZ). The properties of electrons in a periodic potential are calculated using band structure analyses. The collection of energy eigenstates in the first BZ is called the band structure. The momentum of an electron in the infinite periodic potential is defined by the crystalline lattice point. Since lattice points are periodic in a crystalline material, only the first BZ is considered. For a given wave vector and potential, there are a number of distinct solutions for Schrodinger's equation of Bloch electrons in the first BZ. These solutions are denoted as different bands.

In a simple one-dimensional infinite square well, energy levels are labeled by a single quantum number n and the energies are given by

$$E_n = (\hbar^2 \pi^2 / 2mL^2)n^2.$$

Now consider N electrons instead of just one, with either spin in the box. Due to Pauli's Exclusion Principle, there are two states for each energy level. The Fermi level is determined at $E = 0$ where states of lower energy are fully occupied, while higher energy states are completely empty such that the Fermi energy is

$$E_F = E_{N/2} = (\hbar^2 \pi^2 / 2mL^2)(N/2)^2.$$

The unique electronic properties of CNTs are due to the quantum confinement of the electrons with a wave vector perpendicular to the nanotube's axis. In the radial direction, electrons are confined by the monolayer thickness of the graphene sheet. Around the circumference of the nanotube, periodic boundary conditions come into play. For example, if a zigzag or armchair nanotube has 10 hexagons around its circumference, the 11th hexagon will coincide with the first. Because of this quantum confinement, electrons can only propagate along the nanotube axis, and so their wave vectors point in this axial direction. The resulting number of one-dimensional conduction and valence bands effectively depends on the standing waves that are set up around the circumference of the nanotube. These simple ideas can be used to calculate the dispersion relations of the one-dimensional bands, which relate momentum wave vector to energy, using a well-known dispersion relation in graphene. Dispersion relations show how the kinetic energy in different SWNTs varies with the wave vector. Each curve corresponds to a single semiconductor. Bands lower and higher than the Fermi level are denoted as bonding and antibonding π molecular orbitals.

An armchair (5, 5) nanotube and a zigzag (9, 0) nanotube exhibit metallic properties. An electron in the highest occupied molecular orbital (HOMO) requires only an infinitesimally small amount of energy to excite it into the lowest unoccupied molecular orbital (LUMO). For a zigzag (10, 0) nanotube, there is a finite band gap between the HOMO and LUMO states (An et al., 2003), so this type of nanotube exhibits semiconductor properties. Significant variations in the absorption and Raman spectra are observed for slight changes in the diameter of the tubes. Thus, a small change in diameter has a major impact on the electronic and vibrational properties of SWNTs. The electronic and chiral structure of the SWNTs has been atomically resolved using various scanning tunneling technologies. Structural information like chirality and diameter, along with the band structure have been experimentally verified. In general, an (n, m) CNT will be metallic when $n - m = 3q$, where q is an integer. All armchair nanotubes are metallic, as are one-third of zigzag nanotubes.

Plotting the density of states (DOS) with respect to the Fermi energy (see Figure 8.9) shows a number of prominent peaks called van Hove singularities (vHs). We expect the nanotube to absorb at energies corresponding to the inter-peak energy gap separations. We observe these singularities in the absorption spectra of SWNTs. This model considers independent electron–hole pairs and provides us with a basic phenomenological understanding. Experimental observations correlate better with the excitonic model. In this model, the electron in the conduction band and the hole in the valance band are found to act as quasi particles. The difference between the calculated energy and the observed energy in the absorption spectra of SWNTs is better explained by the binding energy of the exciton.

8.2.2 Interparticle Forces in Nanotube Bundles

SWNTs are typically found as aggregated bundles, as illustrated in Figure 8.1. The force of adhesion between individual SWNTs in a bundle is a result of the π–π stacking and dispersion vdW forces present throughout the length of SWNT walls. Using a "universal

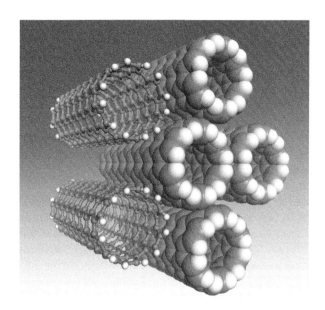

FIGURE 8.1 Model of (6, 6) SWNTs assembled into a closest packed bundle. The nanotubes on the right are shown with their van der Waals contact surfaces.

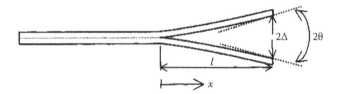

FIGURE 8.2 Model of two SWNTs bound to catalyst particles at the right (particles not shown) and subsequently adhered together from the left due to π–π stacking and other van der Waals forces. (From Chen, B. et al., *Appl. Phys. Lett.*, 83, 3570, 2003. With permission.)

graphitic potential," Girifalco calculated a Lennard–Jones type potential energy well depth of 95.16 meV/Å for a (10, 10) SWNT bound to another (10, 10) SWNT (Girifalco et al., 2000). As nanotubes are growing near each other, they bind together and sometimes twist about each other, forming a rope-like structure. A transmission electron microscopy (TEM) image of two such nanotubes was modeled by Chen et al. (2003) as illustrated in Figure 8.2. Based on the experimentally measured mechanical properties of SWNTs, and the equilibrium geometry shown here, they calculated a 0.36 nN/(unit length) binding energy between the nanotubes. Dispersing SWNT bundles and ropes into a solvent requires a significant amount of energy. Moreover, these dispersed nanotubes will re-form bundles unless they are stabilized by the solvent or other molecular and ionic species destablize in the solution.

8.2.3 Thermodynamics of Nanotube Dispersions

There is no evidence that SWNTs are thermodynamically stable in a solvent, therefore we refer to dispersions rather than solutions of nanotubes. Using ultrasonication or other mechanical mixing techniques, SWNTs can be dispersed into solvent up to some dispersion limit, D_L. Keep in mind that the D_L depends on many factors that are often not well controlled. Many of the amide organic solvents such as DMF or *N*-methyl-2-pyrrolidone (NMP) used to disperse SWNTs rapidly absorb water from the air. The amount of water in these dispersions affects the D_L. Moreover, since the dispersions are only kinetically stable, the D_L is time-dependent and also depends on sample manipulation. Simple mixing, vortexing, or centrifugation can initiate nanotube aggregation from a seemingly stable dispersion.

To maintain the integrity of pristine SWNTs, they are dispersed into organic solvents. Properties of solvents that enable high D_L (>1 mg/L) are good Lewis-based electron pair donation and low hydrogen bond donating character (Ausman et al., 2000; Bahr et al., 2001a). The best solvents also have large solvatochromic parameters such that the molecular properties of the solvent have a high polarity-polarizability (Kolling, 1981).

As discussed below, SWNTs can also be dispersed into aqueous solutions. Acid oxidized SWNTs disperse into water with a high D_L value; however, the ends and sidewalls of the nanotubes are highly defected. Addition of surfactants, such as sodium dodecyl sulfate, to water increase the surface energy of the nanotubes. The non-polar tails of the surfactant wrap around the nanotubes to maximize their attractive dispersion interactions. The polar head group then interacts more effectively with the water. Another successful strategy is to use single-strand DNA as a surfactant. The electrostatic energy of the DNA strand can be minimized by wrapping around the nanotube. Careful selection of the DNA strand length and sequence can effectively select specific nanotube chiralities from the dispersion (Zheng et al., 2003b). Essentially, the thermodynamics of SWNT dispersions is driven by the free energy interactions of the solvent with the nanotube. Both intermolecular interactions and entropic considerations are important.

8.2.4 Theory of Colloid Stability

Molecular aggregation and self-assembly is a well-studied field and will not be reviewed here (Hill, 1964; Israelachvili, 1992). The study of the aggregation of colloidal systems is also a mature field that is well described by the DLVO theory named after Derjaguin, Landau, Verwey, and Overbeek (Verwey and Overbeek, 1948). Often, the stability of a dispersion is characterized by the Schulze–Hardy (SH) rule, where the critical coagulation concentration (CCC) is related to the valence of the charged coagulant: $(CCC) \alpha Z_+^{-6}$. This behavior is predicted by the DLVO theory for certain particle geometries under a limited set of experimental conditions. Sano et al. (2001) have studied the rapid coagulation of CNTs in aqueous media as a function of valence on several inorganic coagulants. Their analysis shows good agreement with the SH rule. This result is surprising, given the assumptions made in the DLVO theory and the extraordinary properties of SWNTs (Syue et al., 2006; Tao, 2006; Yanagi et al., 2006).

The kinetics of coagulation depend on the surface functionality of the pristine or modified CNTs and on the nature of the

solvent. CNT dispersions can collapse in minutes or be "stable" for months. The physical state of dispersed CNTs is still not clear. Several solution phase scattering experiments indicate that dispersions of nanotubes consist of intertwined particles forming fractal geometries (Chen et al., 2004; Saltiel et al., 2005). Only under intense ultrasonication do these light scattering results imply a more rod-like morphology consistent with dispersed and isolated SWNTs (Schaefer et al., 2003). We have shown that deposited CNTs are found in mats of bundles and ropes, and as isolated and aligned SWNTs (Chaturvedi and Poler, 2006). It is unlikely that CNTs with fractal geometry in solution would separate into isolated CNTs upon deposition. We believe an accurate description of dispersed CNTs is still needed.

It is likely that a dispersion of rod-like isolated CNTs will eventually aggregate into more complex morphologies, form a floc, and segregate from the liquid phase. While interacting plates and spheres are accurately described by the DLVO theory, the geometry and surface properties of CNTs are not. The well-accepted method of treating interacting particles in solution is to sum the attractive dispersion forces between the particles and the repulsive electrical double-layer forces. By linearizing the Poisson–Boltzmann equation and integrating, the net potential energy of interacting spheres is (Evans and Wennerstrom, 1999; Verwey and Overbeek, 1948)

$$V(x)_{sphere-sphere} = \pi r \left[\frac{-H_{121}}{12\pi x} + \frac{64 k_B T N_0 \Gamma_0^2}{\kappa^2} \exp[-\kappa x] \right]$$

where

r is the radius of the sphere

H is the Hamaker constant for the system

Γ is a result of linearization and depends on the zeta-potential of the particles and the valence charge, Z, of the solvated ions

$1/\kappa$ is the Debye length, which depends on the number density of ions in the solution, N_0 and the dielectric strength of the solvent,

$$1/\kappa = \left[\frac{\varepsilon_r \varepsilon_0 k_B T}{\sum_i (Z_i e)^2 N_{i0}} \right]^{1/2}$$

The sphere–sphere and slab–slab $V(x)_{slab-slab}$ potential energies are both results of several approximations. The Debye–Huckle approximation, used in linearizing the electrostatics, requires the zeta-potential on the particles to be less than 25 mV ($k_B T$). CNTs in aqueous solvent have zeta potentials in the range of −15 to −65 mV. In order to integrate these equations in closed form, the second assumption is that the radius of the particle be much larger than the Debye length, $\kappa d > 5$. However, the diameter of a CNT is much smaller than the Debye length in the dispersions we are using. This requires a numerical integration to model the system properly. Obviously, the geometry of interacting CNTs is neither spherical nor planar. Recent work has derived the electrical double-layer repulsion for a sphere interacting with a cylinder (Gu, 2000). Much progress has been made on the vdW attraction between CNTs using a universal graphite potential (Girifalco et al., 2000), resulting in a general attractive vdW potential between SWNTs (Sun et al., 2005, 2006). While progress on theoretical descriptions of CNT aggregation is being made, further experimental work is required. Recent molecular dynamics simulations of CNTs in water show an interesting solvent-induced nanotube–nanotube repulsive interaction that is not taken into account in other models (Li et al., 2006). This interaction should be found for other nonaqueous solvents (Giordano et al., 2007).

8.2.5 Theory of Particle Aggregation

8.2.5.1 Effects of Solvent and Solute on Particle Interactions

Particles dispersed in a medium behave differently than in free space. Solvent may change the properties of the solute molecules, thereby changing the solution's stability. Solute molecules are in constant motion and move by displacing the solvent. This means that if the work required to displace the solvent is greater than the energy gained by the approaching solute molecules, then the molecules will repel one another. Therefore, the interactions between the solvent and solute molecules play a vital role in the stability of a solution and a dispersion.

The polarizability (α) of solvent molecules is affected by the medium if the solvent and solute have different dielectric constants. A continuum approach models the solute molecules, i, as dielectric spheres with a radius, a_i, and dielectric constants, ε_i (Israelachvili, 1995; Landau and Lifshitz, 1960). The spheres gain an excess dipole moment and the solvent, with a dielectric constant ε, feels this as a polarizability defined by

$$\alpha_i = 4\pi\varepsilon_0\varepsilon\left(\frac{\varepsilon_i - \varepsilon}{\varepsilon_i + 2\varepsilon}\right)a_i^3.$$

The values of the dielectric constants are important in determining the attractive/repulsive force between dissolved particles. These changes imply that molecules with high dielectric constants attract ions and molecules with low dielectric constants repel ions. The importance of a solvent's ionic strength is discussed in greater detail in the following section.

Solvent–solute attraction is a stabilizing interaction (Belloni, 2000). Solvent molecules will form an adsorbed layer around each solute molecule. Therefore, as solute particles approach one another, the adsorbed solvent layer must first overlap; thus, preventing the two molecules from coming into true contact. Unless the solvent molecules can stick to two surfaces at the same time, a bridging attraction, it is more difficult for the solute molecules to stick together. Furthermore, the additional layer provides a steric repulsion at short distances. An example of solute–solvent attraction is hydrophilic interactions. Water molecules bind strongly to hydrophilic surfaces and hydration repulsion stabilizes the solute molecules. Overall, dispersions with solvent–solute attraction are more stable.

Conversely, solvent–solute repulsion is a destabilizing interaction (Belloni, 2000). When solute molecules repel solvent molecules, a depletion of solvent is created around the solute molecules. This leads to a strong attraction between solute molecules. An example of this destabilization is exemplified by the formation of micelles when molecules are dispersed in incompatible solvents.

Since the attractive forces between solute particles are dependent upon the solvent, it is critical to incorporate all of these forces to determine the stability of a solution. Dissolving a solute in a solvent will change the behavior from that predicted by placing the molecules in free space.

8.2.5.2 Effects of Ionic Strength and Charge Transfer Reagents on Dispersion Stability

Interactions between particles are strongly dependent on the electrolyte concentration of the surrounding medium. Charged surfaces in weak electrolyte solutions will experience a strong, long-range repulsion, shown by line 1 in Figure 8.3. Particles are unable to overcome this high energy barrier. Increasing the electrolyte concentration will create a secondary minimum that appears before the primary minimum. Line 2 in Figure 8.3 shows a secondary minimum. At low electrolyte concentrations, particles are unable to overcome the energy barrier and will either aggregate in the secondary minimum or remain dispersed in the solution. These suspensions are considered kinetically stable; adhesion in the secondary minimum is weak and easily reversible. Further increasing the electrolyte concentration will lead to slow aggregation until the electrolyte concentration reaches the CCC. At the CCC, the peak of the energy barrier crosses zero and particles can fall into the primary minimum and flocculate rapidly. This potential energy is shown as line 2 in Figure 8.3. The condition required for rapid aggregation is

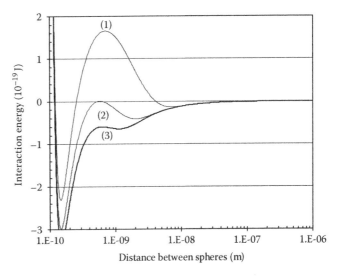

FIGURE 8.3 Graph showing the DLVO potential energy as a function of distance for various electrolyte concentrations. Line 1 shows an potential when the electrolyte concentration value is below the CCC. Line 2 is when the electrolyte concentration is at the CCC. Line 3 is the potential when the concentration is above the CCC.

$$V(x) = 0; \quad \left. \frac{\partial V(x)}{\partial x} \right|_{x=1/\kappa} = 0.$$

This aggregation is irreversible and the suspension is considered kinetically unstable. The energy barrier will continue to fall below zero, as shown by line 3 in Figure 8.3, as the electrolyte concentration is increased above the CCC; particles will irreversibly coagulate.

Electronic, vibrational, and structural properties of the nanotubes are characterized using absorption or Raman spectroscopy. Characteristic and significant changes in the UV-Vis-NIR absorption spectrum and in the Raman active vibrational modes are observed due to charge transfer, selective functionalization, or induced stress on SWNTs (Alvarez et al., 2000; Bahr and Tour, 2002; Lin et al., 2003; O'Connell et al., 2001; Paiva et al., 2004; Pompeo and Resasco, 2002; Rao et al., 1997; Yudasaka et al., 2002).

Studies of the covalent and noncovalent interactions of porphyrins with CNTs are directed toward the design of novel hybrid nanomaterials combining unique electronic and optical properties of the two components.

CNTs are metallic or semiconducting, based upon delocalized electrons occupying a one-dimensional density of states. However, any covalent bond on SWNT sidewalls causes localization of these electrons. Water-soluble diazonium salts (Bahr et al., 2001b) react with CNTs forming a stable covalent aryl bond. These salts are found to preferentially interact with metallic SWNTs, forming covalent bonds. It is believed this is due to greater electron density in metallic SWNTs. The reactant first forms a charge-transfer complex at the nanotube surface by noncovalent physisorption of salt onto the nanotube surface, forming a charge-transfer complex. This is considered to be a rapid, selective noncovalent adsorption of salts on the sidewall of SWNTs. During the second step, the charge transfer complex decomposes to form a covalent bond with the nanotube surface, thus forming more defect sites. Upon thermal treatment, this process is found to be reversible, where the salts can be washed away from the extracted metallic SWNTs. Characteristic features observed in the absorption and Raman spectra of extracted SWNTs are similar to pristine SWNTs.

The noncovalent functionalization of SWNTs with molecules like porphyrins is based on π–π interactions between the two components, and thus does not disrupt the intrinsic electronic structure of CNTs, which is important for electronic applications. Variations in the characteristic absorption and Raman spectra are observed in functionalized SWNTs, due to noncovalent interaction and charge transfer.

8.2.5.3 Kinetics of Aggregation

The process of colloid aggregation is well understood for monodisperse samples, where all the particles have the same size, shape, and mass. In a solution, particles move randomly due to the Brownian motion. When this motion causes two particles to collide and irreversible stick to one another, clusters begin to form; clusters collide and stick together forming a polydisperse solution of aggregates. This is known as cluster–cluster aggregation. Recall, for particles to irreversibly aggregate, the energy of the interactions must be greater than the thermal energy $k_B T$. This type of aggregation exhibits universal aggregation kinetics; therefore, the kinetics are independent of the chemical properties of the colloids. There exist two limiting regimes for irreversible aggregation: reaction limited colloid aggregation (RLCA) and diffusion limited colloid aggregation (DLCA) (Ball et al., 1987; Lin et al., 1990b; Lin et al., 1989; Sandkuhler et al., 2005). Each regime is characterized by unique aggregation kinetics and aggregate structures (Weitz and Oliveria, 1984).

Analytical approaches for describing kinetics use dynamic light scattering (DLS) to follow changes in the effective hydrodynamic radius of the particles. These data are modeled using the Smoluchowski rate equation (Lin, Lindsay, Weitz, Ball et al., 1990; Lin, Lindsay, Weitz, Klein et al., 1990) to describe the number of clusters of mass, M:

$$N(M) = M_n^{-2} \psi(M/M_n).$$

Here, M_n is defined as the nth moment of distribution

$$M_n(t_a) = \frac{\sum_M N(M) \cdot M^n}{\sum_M N(M) \cdot M^{n-1}}$$

where t_a is the time that has elapsed since the initiation of the aggregation. The Smoluchowski rate equation exhibits dynamic scaling where $\psi(x)$ is a scaling function that reflects the shape of the cluster mass. Furthermore, the mass of a cluster is related to its radius of gyration, R_g, fractal dimension, d_f, and radius of individual particles, a, by

$$M = (R_g/a)^{d_f}.$$

Together, these basic equations create a foundation to describe the aggregation kinetics for both limiting regimes.

When the repulsive energy barrier (E_B) is slightly greater than or equal to $k_B T$, the particles must overcome the barrier to stick together. In this case, several collisions must occur before particles stick, which is defined as RLCA or slow aggregation. The rate of aggregation is limited by the probability of overcoming E_B (Lin et al., 1990a), written as

$$P \sim \exp(-E_B/k_B T).$$

The probability of two clusters sticking to one another is also proportional to the number of binding sites. Therefore, as particles stick together and form clusters, the probability changes. Since larger clusters have more binding sites, they grow more rapidly than smaller clusters and individual particles. This leads to an exponential rate increase in the average cluster size over time resulting in a highly polydisperse sample. Most samples formed in the slow regime contain large aggregates and many individual, non-aggregated particles.

Cluster growth has been studied and analytical models have been developed to describe $N(M)$ (Lin et al., 1990a). Experimental results for the shape of the cluster mass distribution are well described by a power law equation with an exponential cutoff:

$$N(M) = AM^{-\tau} \exp\left(\frac{-M}{M_C}\right)$$

Assuming the conservation of total mass, normalization is determined by

$$N_0 = \sum_M N(M) \cdot M$$

where N_0 is the initial number of colloid particles. Here,

$$A = \frac{N_0 M_C^{\tau-2}}{\Gamma(2-\tau)}$$

where $\Gamma(\upsilon)$ is the gamma function, which is determined by DLS and is inversely proportional to the effective diameter. The mass of the cluster increases as

$$M_C \sim \exp(t_a/t_0)$$

where t_0 is a sample-dependent time constant. Computer simulations and experimental results suggest a value of $\tau \sim 1.5$ for these systems. DLS studies following the change in the average hydrodynamic radius as a function of aggregation time show linear behavior on a semi-logarithmic plot, shown in Figure 8.4 (Lin et al., 1990a; Sandkuhler et al., 2005). This confirms that the absolute rate of RLCA is exponential.

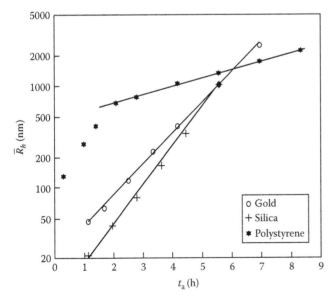

FIGURE 8.4 Graph showing the average hydrodynamic radius, $\bar{R}_h$, as a function of aggregation time, t_a during RLCA. $\bar{R}_h$ is logarithmic to show exponential growth kinetics. (From Lin, M.Y. et al., *Phys. Rev. A*, 41, 2005, 1990a.)

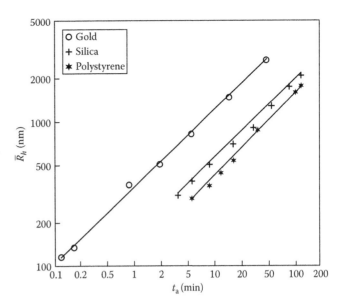

FIGURE 8.5 Graph showing the change in the average hydrodynamic radius, $\bar{R}_h$, as a function of aggregation time, t_a, during DLCA. Both axes are logarithmic to show power-law kinetics. (From Lin, M.Y. et al., *J. Phys.-Condens. Matter*, 2, 3093, 1990b. With permission.)

Reducing the repulsive energy barrier to much less than $k_B T$ changes the regime to fast aggregation or DLCA. In this regime, every collision results in particles sticking to one another and the rate is limited by the time it takes for the particles to encounter one another. For DLCA, the average cluster mass (Lin et al., 1990b), $\bar{M}$, is defined by $\bar{M} = M_0/N_t$. M_0 is the total mass of the system and N_t is the time-dependent number of clusters at time, t, given by $N_t(t_a) = \sum N(M)$. A dynamic exponent, z, describes the time dependence of $\bar{M}$ by $\bar{M} \sim t_a^z$. Studies following the change in an average hydrodynamic radius as a function of aggregation time show linear behavior on a log–log plot, shown in Figure 8.5, which is indicative of the power law dependence (Lin et al., 1990b; Sandkuhler et al., 2005).

Additionally, most colloid aggregates form highly disordered structures that are described by fractals. The fractal dimension of each aggregate describes how completely it fills space and depends on the kinetics of its creation. Aggregates formed during RLCA have fractal dimensions greater than 2 (Lin et al., 1989; Weitz and Oliveria, 1984). This is because the particles and clusters are able to sample all possible binding sites before choosing the site with the most contact. Thus, the structures created are denser. DLCA aggregates are characterized by fractal dimensions less than 2 (Lin et al., 1989; Weitz and Oliveria, 1984). In this case, every collision results in sticking and less dense aggregates are formed. Therefore, studying the resultant aggregate structure reveals information about the kinetics of its formation.

8.2.5.4 Photon-Induced Aggregation

One important concern when studying colloidal dispersions is photon-induced aggregation. When metal clusters are illuminated at their plasmon wavelength, light-induced aggregation may occur. Studies show that when slow-aggregating metal colloids are irradiated by light, the aggregation rate is drastically accelerated (Eckstein and Kreibig, 1993; Karpov et al., 2002; Kimura, 1994; Kreibig and Vollmer, 1995; Satoh et al., 1994). There are several explanations that account for the increase in aggregation kinetics. One explanation for this acceleration is photoelectron emission changing the electric charge of either the double layers or the clusters thereby changing the Coulomb forces between the clusters. Another explanation is that light induces forces similar to vdW forces. Due to the acceleration of aggregation rates when clusters are irradiated by light, care must be taken when using optical methods to collect aggregation data.

The first explanation assumes the diffuse region of the EDL experiences light-induced compression; and the DLVO forces are altered (Karpov et al., 2002; Kreibig and Vollmer, 1995; Satoh et al., 1994). The negative charge on the particles decreases due to the emission of photoelectrons. This decrease leads to a smaller surface potential, and consequently a decrease in the Coulombic repulsion between the two particles. These changes are accompanied by an accumulation of positive charge at the metallic core, which leads to an increase in adsorption potential at the surface. Counter-ions will increase at the dense region of EDL, leading to the compression of the diffuse region. Reducing the EDL surrounding the colloids accelerates the aggregation rate.

The second explanation suggests the photon-induced enhancement of vdW forces (Eckstein and Kreibig, 1993; Karpov et al., 2002; Kimura, 1994). Recall, the Hamaker constant describes the attractive interaction between particles and is a result of zero point fluctuations of electronic polarizations. These fluctuations are predominately determined by conduction band electrons, which exist as surface plasmons for metal colloids. Excitation of the surface plasmons by light irradiation induces electromagnetic multipolar interactions. Therefore, the vdW attraction between particles increases, which leads to an increase in the aggregation kinetics.

The interaction of charge transfer reagents and SWNTs in solution is essentially driven by dispersion forces and other vdW forces. These forces between molecules are essentially electromagnetic in nature and, along with the surface charges in the solution, drive the molecules into supramolecular assemblies. The aggregation of nanoscale particles from the solution depends on the type and charge of the coagulant and also on the surface charge or zeta-potential on the nanotubes themselves. The zeta-potential of the CNTs in the aqueous solvent is in the range of $\zeta = -15$ to $-65\,mV$, depending on the solution pH and nanotube preparation. Along with using the charge and concentration of coagulants, photons can be used to reduce the repulsive potential barrier between the SWNTs in stable dispersion causing enhanced rapid flocculation. The double-layer repulsion term essentially depends on the counter-ion concentration and Debye length. The counter-ion concentration interacting with the SWNT surface can be changed by charge donation into the SWNT. Optically active molecules can absorb light and donate charges into the SWNT, thereby affecting the surface potential

and counter-ion concentration. It is shown that when SWNT solutions include photoactive metallodendrimers and are optically illuminated at the metal to ligand charge transfer (MLCT) absorption band, photon enhanced aggregation occurs. This rate of photon enhanced aggregation rate depends linearly on the illumination power (Chaturvedi and Poler, 2007).

8.3 State-of-the-Art Procedures and Techniques

8.3.1 How to Make a Dispersion of CNTs

Dispersing CNTs in liquids is important for many reasons. CNTs are typically synthesized by laser ablation, electric arc discharge, CVD, electrolysis, or sonochemical methods (Hilding et al., 2003). Regardless of the method and growth conditions, CNTs in a condensed phase are bundled together due to π–π stacking along their length. Whether one purchases CNTs in powder form or grows them, it is necessary to disperse them in a liquid for processing. CNTs do not easily disperse into solution due to their strong tube–tube interactions and their length (hundreds of nanometers to microns).

Sonication is the most common method used to de-bundle CNTs, allowing them to be dispersed in a solvent. Depending on one's needs, three different types of sonicators are used: bath, tip, and cup-horn sonicators. Bath sonication is the gentlest form of sonication and does not cause as many sidewall defects, whereas tip or cup-horn sonicators can be far more forceful at de-bundling the CNTs. Sonication can shorten CNTs and damage the sidewalls, so sonication power and time must be chosen carefully. Furthermore, the choice of solvent also affects the sonication power and time, which can influence the dispersion quality and stability.

8.3.1.1 Aqueous versus Nonaqueous Dispersions

8.3.1.1.1 Aqueous Dispersions

CNTs are completely insoluble in water when pristine. CNTs must be modified through acid treatment, a surfactant coating, or covalent functionalization. Surfactants such as sodium dodecyl sulfate (SDS) (Cardenas and Glerup, 2006; Islam et al., 2003; Moore et al., 2003; O'Connell et al., 2002; Strano et al., 2003), Triton X-100 (Hilding et al., 2003; Islam et al., 2003; Liu et al., 1998; Moore et al., 2003; Tan and Resasco, 2005; Wang et al., 2004), or sodium dodecylbenzene sulfonate (SDBS) (Cardenas and Glerup, 2006; Islam et al., 2003; Matarredona et al., 2003; Moore et al., 2003; Yamamoto et al., 2008) are used most frequently. However, other surfactants have also been explored, such as Gum Arabic (Bandyopadhyaya et al., 2002), 4-(10-hydroxy)decyl benzoate (Mitchell et al., 2002), trimethyl-(2-oxo-2-pyren-1-yl-ethyl)-ammonium bromide (Nakashima et al., 2002), PmPV-based polymer (Star and Stoddart, 2002), or even single-stranded DNA (Zheng et al., 2003a).

Typical preparation methods use deionized water or D_2O with 1% surfactant by weight (O'Connell et al., 2002), as this solution is above the critical micelle concentration for SDS as shown

FIGURE 8.6 SWNT dispersions of varying concentrations 54, 27, 13.5, 6.75, and 3.38 mg/L in filtered, 1% w/w aqueous SDS.

in Figure 8.6. However, lower surfactant concentrations can still exfoliate and de-bundle CNTs sufficiently. An alternative method to the 1% by weight method is a gradual addition of surfactant at intervals during sonication (Yamamoto et al., 2008). There is evidence that the gradual addition method may do a better job of dispersing individual tubes without a large excess of surfactant left in the solution. Once a surfactant solution has been prepared, powdered CNTs are added and one of the sonication methods is used to accelerate the de-bundling process.

An alternative to using surfactants is to covalently functionalize the CNT sidewalls. Many types of covalent functionalization have been employed, including amidation, thiolation, halogenation, bromination, esterification, chlorination, fluorination, and hydrogenation (Hirsch and Vostrowsky, 2005). Covalent sidewall functionalization will significantly affect the electronic and mechanical properties, because the change from sp^2 to sp^3 hybridization of the sidewall carbon atoms causes irregularities in the hexagonal lattice (Hirsch and Vostrowsky, 2005). Consequently, this approach to CNT dispersion is not acceptable when studying properties of pristine CNTs.

The last major method for rendering CNTs dispersible in water is to use an acid treatment. Acid treatments severely damage the integrity of the CNT sidewalls by causing many defect sites. Moreover, when the acid/CNT mixture is sonicated, CNTs will be cut at those defect sites because of the collapse of sonication-induced cavitation bubbles (Liu et al., 1998). Acid treatment is an effective way to make CNTs dispersible or rapidly reduce the mean length of CNTs in a dispersion. This method should be avoided if features of pristine CNTs are to be studied.

8.3.1.1.2 Nonaqueous Dispersions

To study the properties of pristine CNTs, nonaqueous solvents must be used since they do not require any type of surfactant, defect creation, or covalent functionalization, all of which modify CNT properties (Ausman et al., 2000). As with aqueous dispersions, sonication is used to disperse CNTs in nonaqueous solvents. A wide variety of nonaqueous solvents have been tested and the concentration of CNTs that can be dispersed in each solvent varies significantly (Ausman et al., 2000; Bahr et al., 2001a; Landi et al., 2004).

8.3.1.2 Techniques to Determine SWNT Dispersion Properties

8.3.1.2.1 UV-Vis-NIR Spectroscopy

SWNT samples prepared according to the methods described above can be analyzed by several techniques. UV-Vis-NIR spectrometry is a low-cost, convenient, and accurate method for ascertaining the quality and concentration of a SWNT dispersion. SWNTs are strong optical absorbers and have a distinctive optical absorption spectrum with several salient features. Figure 8.7 illustrates a typical UV-Vis-NIR absorption spectrum of a SWNT dispersion.

The shape of a SWNT absorption spectrum is dominated by the background π–plasmon resonance peak at 260 nm, which comes from the sp^2 hybridized bonds. If low quality, highly defected SWNTs have been dispersed, the π–plasmon resonance is the only visible feature.

When high-quality, pristine SWNTs are dispersed, vHs can be easily detected in the spectrum. vHs are a consequence of the one-dimensional electronic density of states in SWNTs (Dresselhaus et al., 2004). vHs manifest themselves as small peaks along the π–plasmon background and the magnitude of those peaks, with respect to the background π–plasmon resonance, can be used as a qualitative way to determine whether or not a good dispersion has been produced. In general, the ratio of the absorption from the vHs to the π–plasmon resonance should be greater than 1 for a good dispersion. As shown in Figure 8.8, there are three distinct regions of vHs in a SWNT optical absorption spectrum (Kataura et al., 1999). In the S$_{11}$ and S$_{22}$ regions, 1–1 and 2–2 transitions for semiconducting SWNTs are observed. The M$_{11}$ region is where the vHs are due to 1–1 transition in metallic SWNTs.

To quantify the concentration of SWNT dispersions, Smalley's group developed a Beer's Law approach to determine the

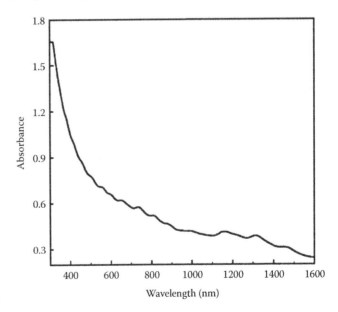

FIGURE 8.7 UV-Vis-NIR spectrum for SWNTs has the right tail of the π-plasmon resonance, with vHs peaks clearly visible. (Reproduced from Bahr et al., *Chem. Commun.*, 193, 2001. With permission.)

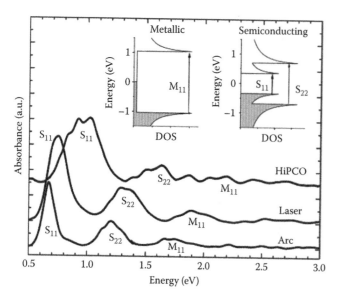

FIGURE 8.8 UV-Vis-NIR absorption spectrum, indicating the S_{11}, S_{22}, and M_{22} ranges. In this figure, the π-plasmon background has been removed. The insets schematically show the transitions in energy vs. density of states plots. (Reprinted from Niyogi, S. et al., *Acc. Chem. Res.*, 35, 1105, 2002. With permission.)

concentration of SWNT dispersions based on optical absorption at 500 nm (Bahr et al., 2001a). Samples of pristine SWNTs in 16 organic solvents with varying concentrations were prepared by sonication and the optical absorption at 500 nm was measured. The solvent was then removed by oven drying and the SWNT residue was weighed, allowing Smalley et al. to determine an absorption coefficient. This approach allows the rapid quantification of the concentration of a SWNT dispersion based on optical absorption.

Based on the simple tight-binding theory, the semiconducting SWNTs give rise to a series of electronic transitions between the principal mirror spikes in the electronic density of states (DOS). Electronic transition energy between corresponding levels in the valance and conduction band can be generally formulated as $S_n = (2na_{C-C}\beta/d_t)$, where n is an integer value corresponding to the transition, i.e., $n = 1, 2,...$ for S_{11}, S_{22} respectively, a_{C-C} is the carbon–carbon bond length (0.142 nm), β is defined as the transfer integral between π–orbitals (β ≈ 2.9 eV), and d is the SWNT diameter (nm), e.g., $S_{11} = 2na_{C-C}\beta/d$ and $S_{22} = 4a_{C-C}\beta/d$, while the metallic SWNTs show their first transition at $M_{11} = 6a_{C-C}\beta/d$. These vHs in the absorption spectra are observed over a broad monotonic background absorption due to π–plasmon resonance. In a bulk characterization technique such as absorption spectroscopy S_{11}, S_{22} transitions from different diameter semiconductor tubes and M_{11} transitions from different metallic tubes are simultaneously observed. π–plasmon absorption from both the SWNTs and carbonaceous impurities are observed in UV. These spectral characteristics provide an opportunity to distinguish between the SWNTs and impurities present in the sample using absorption spectroscopy. While the S_{11} transition is the most prominent, the second semiconducting transition (S_{22}) is used as a reference for evaluation purity because the S_{22} transition is less susceptible to doping

as compared to S_{11}. The ratio $A(S_{22})/A(T)$ can be used to ascertain SWNT purity, where $A(S_{22})$ is the area of the S_{22} interband transition after linear baseline subtraction and $A(T)$ is the total area under the spectral curve. This ratio was then normalized by dividing by 0.141, which is experimentally determined using the ratio of $A(S_{22})/A(T)$ in the reference sample of SWNTs. This procedure is used to calculate the relative purity of SWNTs (Haddon et al., 2004). An absolute determination of the purity of SWNT is not yet possible due to the lack of reference SWNTs.

8.3.1.2.2 Fluorescence Spectroscopy

Band-gap fluorescence provides another method by which SWNT dispersions may be characterized. Fluorescence is especially useful when analyzing SWNT dispersions because it can reveal the chirality of SWNTs present in the dispersion as well as the relative abundance of each chirality. SWNTs have characteristic excitation and emission energies depending on the chirality of the SWNT, as shown in Figure 8.9.

Sample preparation for measuring SWNT fluorescence is more challenging than for UV-Vis-NIR absorption. Bundled SWNTs disturb each other's electronic properties and quench the already weak fluorescence signal. The predominant method for preparing SWNT samples for fluorescence measurements was developed by Smalley's group (O'Connell et al., 2002). HiPCO grown SWNTs are first dispersed into D_2O/SDS by sonication. The surfactant-based method was chosen to minimize the re-aggregation of SWNTs once the dispersion has been made. Next, ultracentrifugation is employed to remove as many of the small (or large) bundles that remain partially dispersed. Ultracentrifugation fields on

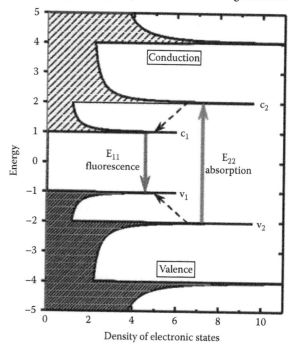

FIGURE 8.9 Schematic representation of SWNT fluorescence, where a photon of energy E_{22} is absorbed and a photon of energy E_{11} is emitted. (Reprinted from Bachilo, S.M. et al., *Science*, 298, 2361, 2002. With permission.)

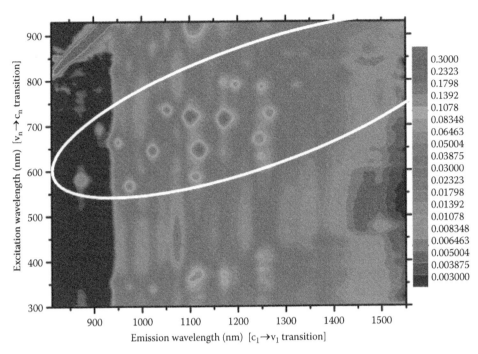

FIGURE 8.10 **(See color insert following page 20-16.)** Three-dimensional fluorescence scan of HiPCO grown SWNTs. Circled in white are some of the well defined fluorescence peaks that are characteristic of the chirality distribution found in SWNTs produced by the HiPCO method. (Reprinted from Bachilo, S.M. et al., *Science*, 298, 2361, 2002. With permission.)

the order of 100,000 g for several hours ensure that only the well-dispersed, individual SWNTs remain in the supernatant, which is then used for fluorescence spectroscopy.

SWNT excitation/emission peaks in the near infrared are used to uniquely identify SWNT chirality. Consequently, three-dimensional fluorescence scans are commonly used to observe the emission over a range of excitation wavelengths. From these scans, a contour plot is generated, as shown in Figure 8.10.

In order to correctly interpret the fluorescence spectra from SWNT dispersions, it was necessary to use both fluorescence and Raman spectroscopy to assign specific chirality SWNTs to the appropriate peaks. Over 30 assignments were made by Weisman et al. (Bachilo et al., 2002) and can be used to rapidly identify some of the SWNTs that are typically found in SWNT dispersions. A shortcoming of fluorescence is that some SWNT chirality do not fluoresce strongly enough to be detected and metallic SWNTs, of course, do not fluoresce at all.

8.3.1.2.3 Raman Spectroscopy

Raman spectroscopy is a widely used tool to study the vibrational properties of materials. It is an important tool for analyzing nanomaterials, especially SWNTs. It provides ease in sample preparation and is a nondestructive, efficient tool for sample characterization. Incident photons interact with an electron that makes a transition to a higher energy virtual state where the electron interacts with phonons or Raman active vibrational modes before making a transition back to the electronic ground state. In this instantaneous process, both the energy and momenta are conserved, hence we can write that $E_{scattered} = E_{laser} \pm h\Omega$ for

Stokes (−) and anti-Stokes (+) scattering respectively, with Ω being the frequency of the particular phonon mode. The energy of the inelastically scattered light is measured with respect to the applied laser energy, and by convention, the Stokes shift is typically plotted as a negative shift. Typically, laser energy in the visible or near infrared is used. Laser energy does not affect the Raman shift, but if the laser energy is resonant with the electronic transition of SWNTs, it results in an exceptional increase in the observed Raman intensity. For CNTs that have a one-dimensional density of states, resonant Raman scattering dominates over nonresonant contributions. Raman scattering from nanotubes gives information about the vibrational modes of SWNTs. Raman scattering can be used for quantitative measurements to provide details about the diameter and chirality distribution of SWNTs, functionalization, stress, and charge transfer properties in SWNTs. However, it is commonly used in conjunction with other tools as Raman modes of SWNTs are found to be exceptionally sensitive to changes in the SWNT environment and aggregation states.

Raman scattering is caused by the interaction of light with matter. The vibrations (phonons) of the nanotubes exhibit unique properties due to their one-dimensional tubular nature. Novel radial breathing modes (RBMs) are observed in SWNT that are not observed in any other carbon allotrope. In Raman scattering (inelastic scattering of light), photons excite or absorb phonons in the nanotubes and their frequency change is observed in the spectrum of the scattered light. Raman scattering thus enables us to understand the vibrational and electronic structure of the SWNTs.

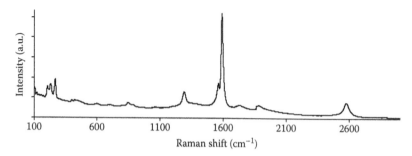

FIGURE 8.11 Typical Raman spectra of pristine SWNT. Peaks in range (150–300 cm⁻¹) are denoted as RBM. Peaks between the wavenumber 1550–1600 cm⁻¹ are the G-band. And peaks about 1250 and 2570 cm⁻¹ are the D-band and G′-band. Weak bands in the range 650–1000 cm⁻¹ are identified as the intermediate frequency band.

Raman spectroscopy of SWNTs has been well studied and is broadly disseminated (Dresselhaus et al., 2004). It provides us with detailed information about the electronic, vibrational, and structural properties of SWNTs. Raman spectra of SWNTs have been found to be remarkably distinct from graphite due to spatial confinement at the nanoscale and their one-dimensional behavior. The most important features in typical Raman spectra of SWNTs are shown in Figure 8.11.

Typical SWNT spectra have two distinct bands of transitions. At small Raman shifts ~150–300 cm⁻¹, RBMs are observed and at higher Raman shifts ~1590 cm⁻¹, the G-band is observed. There is a less intense but significant band at ~1250 cm⁻¹ called the D-band. The origin of other bands, including the weak intermediate frequency bands and a fairly intense peak at 2570 cm⁻¹ called the G′-band are explained by the processes such as overtones, double resonance, or combinational modes.

8.3.1.2.3.1 Radial Breathing Modes   RBMs are due to the radial motion of the atoms perpendicular to the tube axis. These modes are found due to isotropic radial vibrations of the tube, which are a manifestation of the one-dimensional tubular structure of SWNTs. The RBM frequency is inversely proportional to the diameter of the tube, making it an important feature for determining the diameter distribution in a sample, as shown in Figure 8.12. The frequency of these transitions have strong diameter dependence, $\omega_{RBM} = \alpha_{RBM}/d_t$. RBMs are also strongly

affected by the neighboring tubes and sample environment due to vdW forces. For SWNT aggregates and bundles, a constant α_{bundle} is added to the expression for ω_{RBM}. The modified expression is given as $\omega_{RBM} = \alpha_{RBM}/d_t + \alpha_{bundle}$. Values of α_{RBM} and α_{bundle} are found to be 227 and 14 cm⁻¹, respectively. The peaks exhibit a Lorentzian line shape described by $I(\omega) = I_0 + (2A/\pi)(\tau/(4(\omega - \omega_c)^2 + \tau^2))$. $I(\omega)$ is the intensity at the frequency ω and I_0, ω_c, A, and τ represent the constant shift, center frequency, area under the curve, and full width half maximum, respectively.

8.3.1.2.3.2 Tangential G-Band   The G-band consists of several peaks in the frequency range of 1500–1600 cm⁻¹. These Raman active modes are theoretically explained by zone folding of the graphite modes, which makes Raman inactive graphene modes become Raman active G⁺ and G⁻ modes for SWNTs (Brown et al., 2001). The G-band is characterized by a tangential shear vibrational mode of the carbon atoms in the nanotube. Two prominent peaks at higher (~1590 cm⁻¹) and lower (~1560 cm⁻¹) frequencies are denoted as G⁺ and G⁻, respectively. The frequency of the G⁺ mode depends on the laser excitation energy. G⁺ is expected to be diameter independent and G⁻ is expected to be diameter dependent. The G-band is the most intense peak in the spectrum. The physical origin of the G-band is unclear, because it depends on the combination of several processes such as the resonance, polarization effects, and electron-phonon coupling. The G-band forms a distinctive, asymmetric line shape at a lower frequency described by a Breit–Wigner–Fano (BWF) line shape. The origin of this line shape is attributed to the metallic nature of the tubes. This line shape is assumed to emerge because of plasmon–phonon coupling in metallic tubes. The ratio between the intensity of G⁺ and G⁻ can be used for quantifying the amount of semiconducting and metallic SWNTs in the dispersion or in the sample. Metallic tubes are easily recognized from the broad and asymmetric BWF line shape of the G⁻-band. The frequency down shift of the G⁻ is particularly strong for metallic nanotubes, with down shifts of ~100 cm⁻¹ for small diameter tubes.

8.3.1.2.3.3 D-Band   The D-band is a common feature found in all defected sp² carbon hybridized carbon materials. Defects in graphene result in more sp³ type bonding. The intensity of the D-band has been found to increase with the defects introduced. The most convincing explanation for the formation of

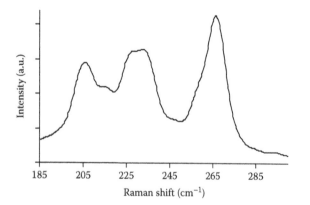

FIGURE 8.12 RBM region of the Raman spectra of pristine SWNTs using a 785 nm laser.

the D-band can be explained by the double resonant process. Incoming photons create an electron–hole pair in the nanotube. The electron is then scattered resonantly to another point in the Brillouin zone by a phonon with momentum $k \neq 0$. This electron is then scattered by a defect back to a virtual state having the same momentum as before it was scattered by the phonon. Consequently, it recombines with the hole and creates a photon. Since the electrons are scattered back by the defect, the D-band is an important tool for characterizing the concentration of defects in the samples. The G'-band is explained by a two phonon double resonance process. The degree of disorder in the system can be quantified using the G'-band since the intensity, $I_{G'}$, does not depend on the disorder, and thus the ratio, $I_D/I_{G'}$, is the measure of the disorder scattering. Since, both of these bands involve the same phonons, the ratio $I_D/I_{G'}$ should provide a good estimate for the disorder in the tubes due to defects. Moreover, the width of the G' Lorentzian peak is inversely correlated with the degree of the nanotube dispersion (Cardenas, 2008).

8.3.2 How to Make CNT Thin Films and Devices

The synthesis and proper characterization of SWNTs in 1991 (Iijima and Ichihashi, 1993), inspired many studies and encouraged the suggestion of countless applications using these unique allotropes of carbon. The properties that make CNTs unique also could enhance the performance of well-established devices and produce new technology. Devices based on classical physics, such as textile, building, or polymer materials would gain enhanced tensile strength and durability. Moreover, electronic components benefit from both the enhanced physical and tunable electrical properties of CNTs. New devices based on quantum physics and confinement phenomena are also possible when SWNTs are used as the building blocks.

As with all nanofabrication, there are two schools of thought when building a device formed from nanoscale building blocks. The two methods are typically referred to as the bottom-up approach and the top-down approach. With respect to CNTs, the bottom-up approach would translate as building a device containing one to several CNTs with nanoscale dimensions. Bottom-up nanoscale devices would exploit the properties and phenomena of individual CNTs and could then be combined to form arrays or integrated into larger devices. An example of such a device is the CNT-FET, which is discussed below. Top-down device fabrication uses CNTs formed together as a quasi-bulk material then manipulated and incorporated into a device. A device made this way could utilize well-established fabrication techniques and still exhibit many of the unique phenomena afforded by the individual CNTs that are interspersed throughout the device. These devices can utilize agglomerations of CNTs such as CNT bundles, ropes, fibers, or as-grown SWNT forests (Pint et al., 2008). Or they can be deposited as films of CNTs (Behnam et al., 2007; de Andrade et al., 2007; Gonnet et al., 2006; Gupta et al., 2004; Kavan et al., 2008; Kazaoui

et al., 2005; Lima et al., 2008; Liu et al., 1999; Merchant and Markovic, 2008; Ng et al., 2008; Pint et al., 2008; Song et al., 2008; Takenobu et al., 2006; Wang et al., 2008b; Wu et al., 2004; Zhang et al., 2004b; Zhu and Wei, 2008), or as mats (Deck et al., 2007; Gupta et al., 2004; Sun et al., 2008) or buckypaper (Rinzler et al., 1998; Whitby et al., 2008).

8.3.2.1 Fabrication of CNT Films and Substrate Adhesion

Regardless of the naming conventions, almost all CNT film fabrication processes consist of two major steps: organization of the CNTs into a film and attachment of said film to a substrate. An important distinction needs to be made between CNT film fabrication and mats formed during CNT growth. Some of the most common CNT synthesis methods, like arc discharge and pulsed laser vaporization, produce complex structures of CNT ropes that are intertwined and entangled into mats (Gupta et al., 2004). These mats are highly disordered and there is very little control of their final morphology and properties. CNT film fabrication techniques can employ many different methods that affect the properties of a CNT film. Currently, researchers fabricating CNT films use a variation of one of the following five fabrication methods depicted in Figure 8.13.

8.3.2.1.1 Vacuum Filtration

Vacuum filtration of CNT dispersions to form a film seems to be the most widely used method (de Andrade et al., 2007; Liu et al., 2006; Whitby et al., 2008; Wu et al., 2004), probably due to the ease of setup and the many adjustable parameters. Figure 8.14 clearly illustrates the wide variability of results one can produce from vacuum filtration, from transparent 100 nm thickness flexible films to buckypaper so thick and rigid they are referred to as buckydiscs (Whitby et al., 2008). There are typically two steps in the vacuum filtration of CNT dispersions. First, a dispersion of CNTs is filtered through a porous membrane or filter. Since filtration is intrinsically self-limiting, the uniformity of the film on the filter membrane is therefore self-leveling. The second step usually involves transferring the deposited CNT film from the membrane or filter paper to a more favorable substrate. In the case of a buckypaper thick enough to be self-supporting, transfer to a substrate may not be necessary at all, as evidenced by Figure 8.14b.

The filtration step is simple in concept, but the method can vary enormously depending upon the desired product. While a CNT is dispersed and unbundled, it can be modified either chemically or physically, changing the inherent properties of the dispersed tubes and ultimately modifying the properties of the deposited film. Changing the amount of CNTs dispersed in a solution is the most effective way to adjust film properties such as thickness, opacity, rigidity, and conductivity. Sometimes it is necessary to increase the volume of the dispersion while increasing the amount of CNTs in a solution to remain below the CCC and avoid coagulation and bundling of the dispersed CNTs before they can be deposited into a film.

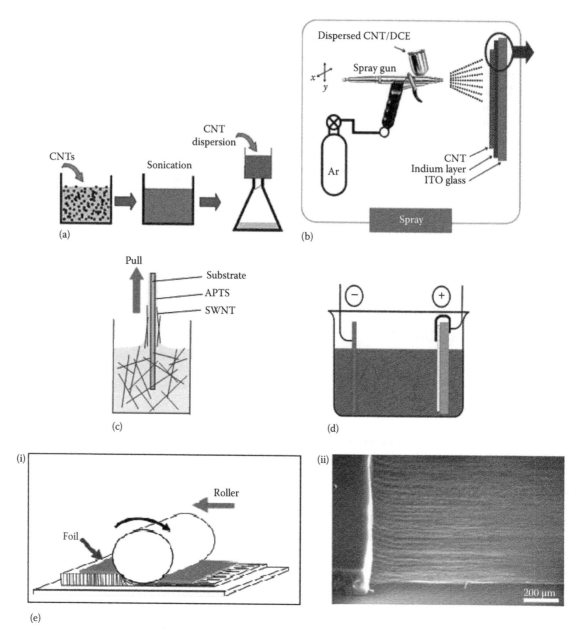

FIGURE 8.13 CNT film preparation by: (a) Vacuum Filtration, (b) Spray-Coating (Jeong et al., 2006), (c) Dip-Coating (Ng et al., 2008), (d) Electrophoretic Deposition (Lima et al., 2008) Reprinted by permission of the Royal Society of Chemistry, and (e) i. and ii. Post-Growth Extrusion. (Reprinted from Pint, C.L. et al., *ACS Nano*, 2, 1871, 2008; Zhang, M. et al., *Science*, 309, 1215, 2005. With permission.)

CNT dispersions are also affected by the choice of solvent and the possible requirement of a surfactant to stabilize the CNTs in the dispersion. Ideally, surfactant is present only in the solution phase of the process and is removed from the CNT film during filtration. However, it has been shown that vigorous washing of a deposited film is required to completely remove a surfactant, possibly due to an intercalating process that traps surfactant molecules between the intermingled and woven CNTs within a film. The most apparent result of inadequate surfactant removal is the significantly reduced integrity of the film. Intercalated surfactant will inhibit the vdW interactions between the CNTs and diminishes the intrinsic electrical properties of the film due to

a reduction of π-conjugation, resulting in an apparent increase of surface resistance (de Andrade et al., 2007). To reduce the risk of having excess surfactant, the minimum amount of surfactant required to maintain a stable dispersion should be used. Many publications report a surfactant concentration of 0.1%–1% by weight (de Andrade et al., 2007; Whitby et al., 2008; Wu et al., 2004), which seems to be appropriate for maintaining dispersions on the order of a few milligrams per milliliter. However, it has been difficult to remove surfactant from films made from filtrate containing 1% by weight surfactant. An effective way to reduce the amount of surfactant in the filtrate is to create two separate dispersions; a stock dispersion with a large

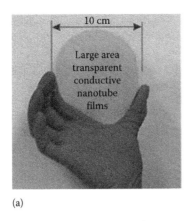

(a)   (b)

FIGURE 8.14 (a) Examples of ultrathin (From Wu, Z.C. et al., *Science*, 305, 1273, 2004.) and (b) ultra-thick (From Whitby, R.L.D. et al., *Carbon*, 46, 949, 2008.) CNT films and mats.

concentration of CNTs with 0.1%–1% surfactant and a second dispersion, made from an aliquot of the first and a large amount of solvent. This allows the surfactant molecules that are interacting strongly with CNTs to remain associated and maintain stability while dispersing excess surfactant in large amounts of solvent. This seems to reduce the required washings of a CNT film to a minimum, maintaining film uniformity and integrity.

A chemical treatment of the CNTs before or during their dispersion can also be implemented. The most common chemical treatments use strong acids, such as concentrated sulfuric acid mixed with concentrated nitric acid to encourage defect sites along the CNTs and at their ends. This strong acid solution can release $NO_{2(g)}$ and should only be done under a vented chemical hood. These kinds of treatments typically affect the electrical properties of the film, changing the conductivity and photoconductivity of the end product. In a similar process, it may also be possible to physically bind CNTs with substances that would enhance or amplify desirable properties such as absorption and conductivity.

After a film is deposited, it must be removed from the filter and adhered to a substrate. Removal of the film from the filter is a fairly straightforward process. Once the film is intimately attached to a new substrate, the filter membrane can be removed either by peeling the film from the filter or dissolving the filter. The latter tends to be the least destructive method for thinner films. A common problem that occurs during this process is that adhesion of the film to the new substrate does not occur. This method assumes that a wetted filter attached CNT film can be pressed onto a substrate and, as the wetting agent (usually water) dries, capillary forces will pull the CNT film into intimate contact with the substrate, where vdW forces hold them together firmly. This method works well and is widely used, however little or no information is presented on the preparation of the new substrates.

New substrates must be sufficiently cleaned. Oxide coated substrates are cleaned by means of piranha etchant, where the substrate is placed into concentrated sulfuric acid: 30% hydrogen peroxide at a 3:1 ratio in a bath sonicator for 30 min. and rinsed copiously with deionized water. This wet process is followed by an oxygen plasma scrubbing to remove any surface contaminants. Typical plasma conditions of 200 W under 200 mTorr $O_{2(g)}$ for 30 min are sufficient. Polymer substrates, such as polycarbonate, are also cleaned using similar methods. Leaving the substrate in water or methanol immediately after plasma cleaning helps preserve the surface until one is ready to transfer the film.

8.3.2.1.2 Spray-Coating

Spray-coating is widely used to deposit films, adhesives, and pigments. It is also capable of producing very thin and uniform CNT films. Spray-coating to deposit CNT films is the most precise method of deposition. Figure 8.13b illustrates the spray-coating method using an air brushing technique. Another variation of this film fabrication method is inkjet printing. Inkjet printing has been used by many groups to print patterned layers of nanoparticles on substrates just like a printer lays down ink when printing on paper. These two variations can print highly resolved images using either masks or a computer controlled printing mechanism patterning CNT films onto a substrate. The initial preparation of the CNTs is similar to that of vacuum filtration. The dispersion is made following established techniques, with the stipulation that no surfactant is used. *N*-methyl-2-pyrrolidone (NMP) has been observed as being a good solvent for this purpose (Beecher et al., 2007). Then, using either an air gun or printing mechanism, the CNT solution is sprayed as an aerosol onto the substrate (Artukovic et al., 2005; Beecher et al., 2007; de Andrade et al., 2007; Merchant and Markovic, 2008). Figure 8.15 shows an image of an electrode pattern in which the gaps are printed with CNTs (Beecher et al., 2007). Problems arise in the form of continuity while using this method. The CNT film morphology depends on the temperature of the substrate, droplet size, and solvent type. Typically, the capillary forces induced by the drying droplets will ball up the CNTs or bundle and separate CNTs within the droplet creating an incongruent layer. This effect will manifest as a higher film resistivity.

8.3.2.1.3 Dip-Coating

Dip coating has the distinction of being the simplest of the CNT film deposition techniques. Using previously described

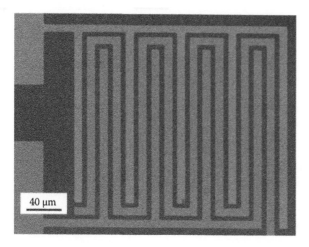

FIGURE 8.15 Interdigitated CNT electrodes deposited by spray coating. (From Beecher, P. et al., *J. Appl. Phys.*, 102, 7, 2007.)

methods, CNT dispersions are created with CNT concentrations of the order of milligram per milliliter (Ng et al., 2008; Song et al., 2008). Then, much like one would dip a candle, the substrate is immersed into the dispersion in a vertical orientation. The dipping process is repeated until a coating of CNTs of the desired thickness has formed. The thickness can usually be estimated by a calibrated optical absorption curve correlated with AFM measurements. As one would imagine, this process is highly dependent on the attraction and adhesion of CNTs to the substrate surface, so a clean substrate is always required and in many cases a coating of polyethylene terephthalate (PET) is used as the substrate coated on glass (Song et al., 2008). In other cases, when a conductive substrate is required, a binding agent can be utilized (Ng et al., 2008). Figure 8.13c illustrates dip-coating with a glass substrate utilizing 1,2-aminopropyltriethoxysilane (APTS) as an adhesion promoter.

8.3.2.1.4 Electrophoretic Deposition

Electrophoretic deposition (EPD) utilizes an electric field to deposit charged particles onto a surface (Boccaccini et al., 2006; Guo et al., 2007; Lima et al., 2008; Poulin et al., 2002; Wang et al., 2007; Zhao et al., 2005). EPD is particularly attractive to those wanting to incorporate CNTs with electrodes. Typically, a conductive material acts as both an electrode and a substrate, attracting CNTs towards an oppositely charged electrode and building up a densely and uniformly packed/bonded layer. Until recently, this method produced CNT films attached to opaque conductive substrates. New methods have shown that a substrate coated with a very thin layer of metal will serve as an electrode for the deposition of CNTs. The ultrathin metal layer is then oxidized to form an optically transparent metal-oxide such as Al_2O_3 or TiO_2 (Lima et al., 2008).

8.3.2.1.5 Post-Growth Extrusion

Post-growth extrusion (PGE) is a noteworthy recent development for forming films from CNTs. PGE is a term that suggests extruding a film from an as-grown CNT material,

typically a CNT forest. Two different methods have been utilized for this technique (Pint et al., 2008; Zhang et al., 2005). The first method uses an adhesive tape to adhere to the exterior edge of a multi-walled carbon nanotube (MWNT) forest. Subsequently, the edge of the MWNT forest is extended from the substrate with a constant pulling force. This method results in an extended continuous transparent sheet of unraveled MWNTs (Zhang et al., 2005). The second technique uses a foil laid over the top of a CNT forest, followed by a roller pressing the foil to compress and bend the CNTs over each other, forming them into a highly aligned dense film (Pint et al., 2008). Figure 8.13e (i) and (ii) illustrate these techniques. Either chemical or physical means can be used to attach these films to new substrates.

8.3.2.2 Devices Made from CNT Dispersions

8.3.2.2.1 CNT-Based Field Effect Transistors

The exceptional mechanical, electronic, and opto-electronic characteristics of SWNTs make them important materials for novel sensors and nanodevices. The field effect transistor (FET) response of SWNT devices has been fairly well documented (Appenzeller et al., 2002; Avouris, 2002; Guo et al., 2004; Snow et al., 2003). These carbon nanotube field effect transistors (CNT-FETs) are used to probe the electronic properties of functionalized SWNTs.

8.3.2.2.1.1 Theory
As discussed above, carbon has four valence electrons. In SWNTs, these electrons in trigonal planar sp^2 orbitals form strong σ covalent bonds with the neighboring carbon atoms. The robust mechanical properties of SWNTs are mainly caused by strong covalent bonds between the tightly-bound σ-orbitals. The electron in a $2p_z$ orbital forms a weaker π bond with the $2p_z$ orbital of the neighboring C atoms. The resulting π-band structure defines the Fermi surface, and hence these orbitals are responsible for the electronic transport properties.

8.3.2.2.1.2 Basics of One-Dimensional Transport
A detailed discussion of the conductance of CNTs as one-dimensional conductors can be found in various reviews and textbooks (e.g., Datta, 1995; Landauer, 1989). Mesoscopic systems such as SWNTs are defined by two characteristic lengths. The first is the mean free path L_m, which is the average length that an electron travels before it is scattered by a scattering center. Impurities, lattice mismatch, and any potential variation can act as scattering centers. The second characteristic length is the Fermi wavelength λ_F, defined by $\lambda_F = 2\pi/k_F$, which is the de Broglie wavelength for electrons at the Fermi energy.

8.3.2.2.1.3 Resistance of a Ballistic Conductor
Figure 8.16 illustrates a narrow two-dimensional conductor with width W and length L ($W \ll L$) between two large contacts. According to Ohm's law, its conductance is given by $G = W/\sigma L$, here the conductivity σ is a material parameter independent of the sample dimensions. However, this Ohmic behavior is true only for large dimensions of the conductor.

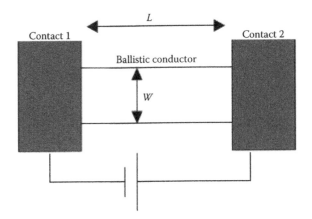

FIGURE 8.16 A narrow two-dimensional conductor with width W and length L ($W \ll L$) sandwiched between two contacts across which an external bias is applied.

When W is large, the associated wave vectors are approximately continuous and there are a large number of conducting bands available. The conductance of this narrow conductor is proportional to W. However, when W is comparable to λ_F, the associated wave vectors become discrete and separated by $2\pi/W$. Each wave vector corresponds to a conducting band. Therefore, only a limited number of bands are available. Then the conductance is not proportional to W; instead it depends on the number of available bands.

When W is small and the L is reduced, following Ohm's law leads one to expect the conductance of this narrow conductor to increase without limit. However, this has not been observed experimentally. It was found that the measured conductance approaches a limiting value G_c, when the length of the conductor becomes shorter than the mean free path ($L < L_m$). The mean free path in this context refers to the momentum scattering length L_m, which includes any process that alters the electronic momentum and hence affects the resistance. In this limit, the conductor is ballistic (an ideal conductor without scatters), and so it should have zero scattering resistance. The observed resistance $R_c = G_c^{-1}$ arises not from scattering in the conductor but rather from the interface between the conductor and the contact. Contacts with large width have more conducting bands available, but the narrow conductor with a small width has only a few available bands. This requires a redistribution of the current at the interface leading to the interface resistance. The contact resistance R_C is ultimately determined from quantum mechanical considerations discussed using the Landauer formulism.

8.3.2.2.1.4 Landauer Formulism Using Landauer's formulism, conductivity can be written as $G = G_0 NT$. The factor T represents the average probability that an electron injected at one end of the conductor will be transmitted to the other end. This transmission is determined by the properties of the conductor and also on the potential at the interfacial contacts. N is the number of available conducting sub-bands inside the conductor. The intrinsic quantum conductance depends on the fundamental constants of charge and energy (Plank's) and is calculated to be $G_0 = 2e^2/h$. This model has proven very useful in describing mesoscopic transport in one-dimensional conductors like SWNTs.

8.3.2.2.1.5 Transport Properties of Carbon Nanotubes A metallic SWNT has two conducting sub-bands as described above. According to the Landauer equation, the conductance of the system is $G = 2G_0$. However, for a SWNT with perfect contacts in a scattering medium, we assume the transmission coefficients of both the conducting channels are TI ($0 \leq$ TI ≤ 1), then $G = 2T_1 G_0$. From these results, we see that the conductance of a ballistic one-dimensional channel depends only on fundamental constants. In SWNTs, the one-dimensional channels come in degenerate pairs, due to the clockwise/anticlockwise symmetry of the wrapping modes around a cylinder, where a metallic SWNT at least has one pair of channels with energies that overlap the Fermi level. At low bias, only this pair of channels can be unequally populated with left- and right-moving electrons; all other wrapping modes remain completely filled or completely empty. Therefore, a perfect metallic SWNT, with a length smaller than $l_{scatter}$, is expected to have a quantized conductance $2(2e^2/h) = (6.5\,\text{k}\Omega)^{-1}$. In fact, conductance of nearly $4e^2/h$ has been shown by Kong in cases of well-contacted metallic SWNTs with a length of ~200 nm (Kong et al., 1999). In semiconducting SWNTs, there is an energy gap in which no electron states exist. Because of this energy gap interrupting the one-dimensional channels of a semiconducting SWNT, the conductance can be turned on and off by applying an electric field (Tans et al., 1998). This FET behavior has various technological applications as both biological and gas sensors. Moreover, these SWNT devices can be used in densely packed integrated circuits or as optical sensors.

Typical SWNT transistor geometry is shown in Figure 8.17a. The SWNT and the gate are two sides of a capacitor where a voltage difference will cause opposite charges to accumulate on the SWNT and the gate. The number of electrons flowing through a SWNT can be controlled by applying the appropriate potential to the gate. Conductance is turned on when the number of electrons on the SWNT is such that a one-dimensional channel, either above or below the energy gap, is partially occupied. Conduction is turned off when the one-dimensional channels above the energy gap are empty or when the one-dimensional channels below the energy gap are filled. A semiconducting SWNT in the "on" state, like a metallic SWNT, is expected to have a maximum conductance of $4e^2/h$. Semiconducting SWNTs with ohmic electrical contacts have been observed with on-state conductance ~$4e^2/h$. The energy gap separating these states in a typical semiconducting SWNT is E_{gap} ~0.5 eV. The Fermi level is aligned below the valence band edge when semiconducting SWNTs are in contact with certain metals like Pt, Au, or Ti. This causes band bending at the junctions. When the gate voltage on the device, V_g, equals zero, there are unoccupied valence states on the tube, and a current can pass through the SWNT. Because conductance is through the unoccupied valence states, the SWNT is a conductor with positive-type charge carriers (p-type). As the gate voltage is increased ($V_g > 0$),

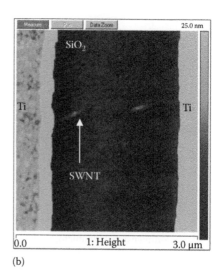

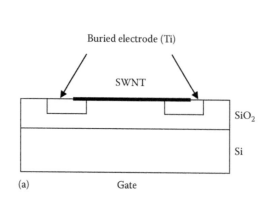

FIGURE 8.17 (a) Schematic diagram of back gated SWNT-FET. (b) AFM image of back gated SWNT-FET fabricated by standard "lift off" technique.

valence states become completely filled in the middle of the SWNT. The SWNT is in the "off" state and conductance drops. Changes in electrostatic energy are greatest in the middle section of the SWNT because the electric field is screened close to the electrodes. As V_g is increased further, the potential energy of states in the middle of the SWNT is lowered far enough so that conduction states become occupied. The SWNT is insulated from the gate by a dielectric material, typically SiO_2. The conductance of a SWNT device, G, changes as a function of gate voltage, V_g.

8.3.2.2.1.6 Fabrication

Devices were fabricated using a standard photolithographic "lift off" technique. In the first step, an insulating oxide layer is grown over the commercially purchased silicon wafer. Care should be taken that the oxide layer grown is thick enough to prevent electrical shorts. Oxide layers may be grown thermally or by sputtering over the Si wafer. AFM characterization of thermally grown oxide layers are found to be more uniform and hence are preferred.

A clean silicon wafer is a critical requirement for the process. Impurities on the surface can act as traps and shorts. The following steps were used to prepare the samples:

1. Rinse wafers in acetone with ultrasound for 180 s
2. Rinse in running DI water for 60 s
3. Rinse in methanol with ultrasound for 60 s
4. Rinse in DI water for 120 s
5. Soak in Buffered Oxide Etch for 30 s
6. Rinse in running DI water for 10 min

Wafers were then transferred while in methanol. The wafers were then loaded into a tube furnace, where they were oxidized. To thermally grow ~1200 Å of oxide layer on a silicon wafer, the highly P+ doped Si wafers were first heated to 1200°C in $N_{2(g)}$. Oxide can be grown either by using dry $O_{2(g)}$ for 135 min or by using wet O_2 where $O_{2(g)}$ flow was turned on for 5 min before $H_{2(g)}$ was introduced for another 45 min.

8.3.2.2.1.7 Photolithography Process Recipe

1. Place the wafer on a spinner; set the time for 45 s and the speed to 4000 rpm
2. With a clean pipette, cover the entire wafer with S1813 photoresist
3. Spin the wafer @ 4000 rpm for 60 s
4. Soft bake @ 115°C for 60 s, then let cool for 60 s
5. Expose the pattern for 27 s based on a 200 W lamp power
6. Develop in 1:1 Microposit Developer:DI for 75 s

Rinsing well with DI water after development is extremely important in order to wash away the residual solvents to provide better contact between the metal and the oxide on the wafer. The developed pattern is then checked by optical microscopy. If the features were found to be underdeveloped or overdeveloped (missing pattern/features), the resist may be stripped off by placing the sample into acetone and then methanol washes for 2 min each. Bath sonication may be used to make the rinsing and stripping processes more effective.

8.3.2.2.1.8 Metal Deposition

Before the metal contacts were deposited, the oxide was etched back in the areas where the resist was developed. The etch rates were carefully controlled so that the thickness of the removed oxide was equal to the thickness of the deposited metal film. These buried contacts enable a more conformal coating of SWNTs. A thin film of titanium (~500 Å) was deposited onto the samples using an AJA International sputter tool at 300 W and 3×10^{-3} Torr for 5 min. The wafers were then sonicated for ~1 min in a wash bath of acetone. The resist, along with metal deposited on it, are lifted off leaving the metal features behind. The samples are then rinsed in methanol and DI water. The samples are checked using optical microscopy.

SWNTs were deposited by drop casting them onto the Source-Gate-Drain structures. The samples were left to dry in a laminar flow hood. In Figure 8.17b, the SWNTs are shown to be aligned across the pads. Control of the nanotube concentration helps to control the number of nanotubes crossing the pads. Electrical characteristics such as: I_{ds} vs. V_{ds}, I_{ds} vs. V_g, and the conductivity of the SWNT-FET devices can be measured using a three terminal probe station.

8.3.2.2.2 Photoconductivity of CNT Thin Films

Referring back to the beginning of this chapter, we learned that SWNTs, because of their one-dimensional tubular structure, have band gaps that vary from semiconducting to metallic. SWNTs are unique in that their electronic properties are tunable based on chirality. Photoconductive properties of SWNTs were first reported by Chen et al. (2001a). Like other semiconducting materials, it is possible to excite an electron from the valence band up to the conduction band of the material if a photon with energy equal to or greater than the band gap is absorbed. Unique transport properties of SWNTs are related to their band structure and, because of their morphology, SWNTs have low carrier scattering rates (Newson et al., 2008). Electron–hole pairs generated by photonic excitation cause the electrons to become ballistic charge carriers, free to move along the length of a nanotube and also through π-conjugation onto other CNTs within a film. When a bias voltage is applied to a film being excited by light, those ballistic electrons can be directed towards an electrode and into a circuit (Newson et al., 2008). The use of SWNT thin films in photoelectronic devices is still in its infancy. However, if SWNTs are aggregated into structures such as films or mats, the photoconductive properties of the SWNT fabricated structure could become the basic building blocks for future electronic and photonic devices.

8.3.2.2.3 CNT-Modified Electrodes for Electrochemical Sensing

Carbon has been used for electrodes in electrochemical systems for many years because of its high conductivity, chemical robustness, and the adjustable surface area to volume ratio of different forms of carbon. CNT electrodes are the next logical step in the evolution of the carbon electrode. CNTs used for electrochemical sensing was first reported in 2003 (Cai et al., 2003; Ye et al., 2003; Zhao et al., 2003), where their superior effectiveness was demonstrated by detecting uric acid (Ye et al., 2003) and nitrates (Zhao et al., 2003). It is apparent that the electronic properties of CNTs make them ideal candidates for electrode and sensing materials. Electrodes can be as small as an individual SWNT or as large as arrays of MWNTs. The procedure to attach CNT films to substrates, as discussed above, has been shown to produce effective electrochemical sensing electrodes. Pristine CNTs can serve as the chemical sensing material or the CNTs can be chemically modified to enhance their sensitivity or specificity. Alternatively, they could simply be used as conductive scaffolding, supporting a sensing agent, or as an aggregate in an epoxy (Pumera et al., 2006). It has been noted by Rocha et al. that "one remarkable feature of CNTs is that their conductance is affected by the interaction with certain foreign objects" (Rocha et al., 2008). Other benefits from using CNTs as electrodes are their high chemical stability and their mechanical strength. A review of this application

is given by Merkoci et al., (2005), and the reader of this tutorial is referred there for a discussion of the current state of CNT electrochemical sensing. It is important to note that for all the benefits CNTs bring to the area of electrodes and sensing, there are still fundamental issues with all phases of CNT production and use that must be researched and developed before the full potential of this approach can be realized.

8.4 Summary

At the crux of the problem in developing SWNT-based devices and materials is our ability to bring these nanoparticles into dispersion and back out again. Forming a kinetically stable dispersion of SWNTs that are isolated from each other is challenging due to their strong interparticle attractions. Care must be taken when dispersing the nanotubes so that they are not cut or damaged unintentionally. Through various chemical and mechanical mechanisms it is now straightforward to make these dispersions with technologically acceptable dispersion limits. The more challenging task is to get the nanotubes out of the dispersion without forming random aggregates that preclude the intended functionality. This is critically important when making SWNT composite materials. Aggregates of bundles and ropes will simply act as defects within the matrix and possibly yield a weaker material then the matrix itself. Colloid science is a good place to start with regard to SWNT stability and aggregation. However, that theory poorly describes the geometry and chemical environment of SWNTs in aqueous, nonaqueous, or polymeric media. Further theoretical analysis of SWNT dispersions is required.

The most technologically viable SWNT materials consist of random arrangements of the nanotubes in a polydisperse morphology. Thin films and mats have been used for optical switching and sensing. Fibers and ropes of MWNTs have been spun, and exhibit extraordinary strength and utility (Zhang et al., 2004a). Dry adhesives have been constructed from vertically aligned arrays of nanotubes that can hold an object onto a smooth surface better then a gecko's sticky feet (Qu and Dai, 2007). Further advances in optical and electronic applications of CNTs are predicated on our ability to isolate SWNTs with chiral purity and length homogeneity. Understanding the fundamental aspects of the dispersed and aggregated state of SWNTs is central to this process and the commercial viability of these materials.

References

Alvarez, L., Righi, A., Guillard, T., Rols, S., Anglaret, E., Laplaze, D., and Sauvajol, J. L. (2000). Resonant Raman study of the structure and electronic properties of single-wall carbon nanotubes. *Chemical Physics Letters*, **316**, 186–190.

An, K. H., Park, J. S., Yang, C. M., Jeong, S. Y., Lim, S. C., Kang, C., Son, J. H., Jeong, M. S., and Lee, Y. H. (2003). A diameter-selective attack of metallic carbon nanotubes by nitronium ions. *Science*, **301**, 344–347.

Appenzeller, J., Knoch, J., Derycke, V., Martel, R., Wind, S., and Avouris, P. (2002). Field-modulated carrier transport in carbon nanotube transistors. *Physical Review Letters*, **89**, 126801.

Artukovic, E., Kaempgen, M., Hecht, D. S., Roth, S., and GrUner, G. (2005). Transparent and flexible carbon nanotube transistors. *Nano Letters*, **5**, 757–760.

Ausman, K. D., Piner, R., Lourie, O., Ruoff, R. S., and Korobov, M. (2000). Organic solvent dispersions of single-walled carbon nanotubes: Toward solutions of pristine nanotubes. *Journal of Physical Chemistry B*, **104**, 8911–8915.

Avouris, P. (2002). Molecular electronics with carbon nanotubes. *Accounts of Chemical Research*, **35**, 1026–1034.

Bachilo, S. M., Strano, M. S., Kittrell, C., Hauge, R. H., Smalley, R. E., and Weisman, R. B. (2002). Structure-assigned optical spectra of single-walled carbon nanotubes. *Science*, **298**, 2361–2366.

Bahr, J. L., Mickelson, E. T., Bronikowski, M. J., Smalley, R. E., and Tour, J. M. (2001a). Dissolution of small diameter single-wall carbon nanotubes in organic solvents? *Chemical Communications*, **2**, 193–194.

Bahr, J. L. and Tour, J. M. (2002). Covalent chemistry of single-wall carbon nanotubes. *Journal of Materials Chemistry*, **12**, 1952–1958.

Bahr, J. L., Yang, J. P., Kosynkin, D. V., Bronikowski, M. J., Smalley, R. E., and Tour, J. M. (2001b). Functionalization of carbon nanotubes by electrochemical reduction of aryl diazonium salts: A bucky paper electrode. *Journal of the American Chemical Society*, **123**, 6536–6542.

Ball, R. C., Weitz, D. A., Witten, T. A., and Leyvraz, F. (1987). Universal kinetics in reaction-limited aggregation. *Physical Review Letters*, **58**, 274–277.

Bandyopadhyaya, R., Nativ-Roth, E., Regev, O., and Yerushalmi-Rozen, R. (2002). Stabilization of individual carbon nanotubes in aqueous solutions. *Nano Letters*, **2**, 25–28.

Beecher, P., Servati, P., Rozhin, A., Colli, A., Scardaci, V., Pisana, S., Hasan, T., Flewitt, A. J., Robertson, J., Hsieh, G. W., Li, F. M., Nathan, A., Ferrari, A. C., and Milne, W. I. (2007). Ink-jet printing of carbon nanotube thin film transistors. *Journal of Applied Physics*, **102**, 7.

Behnam, A., Guo, J., and Ural, A. (2007). Effects of nanotube alignment and measurement direction on percolation resistivity in single-walled carbon nanotube films. *Journal of Applied Physics*, **102**, 7.

Belloni, L. (2000). Colloidal interactions. *Journal of Physics-Condensed Matter*, **12**, R549–R587.

Boccaccini, A. R., Cho, J., Roether, J. A., Thomas, B. J. C., Minay, E. J., and Shaffer, M. S. P. (2006). Electrophoretic deposition of carbon nanotubes. *Carbon*, **44**, 3149–3160.

Brown, S. D. M., Jorio, A., Corio, P., Dresselhaus, M. S., Dresselhaus, G., Saito, R., and Kneipp, K. (2001). Origin of the Breit-Wigner-Fano lineshape of the tangential G-band feature of metallic carbon nanotubes. *Physical Review B*, **63**, 155414.

Cai, H., Cao, X. N., Jiang, Y., He, P. G., and Fang, Y. Z. (2003). Carbon nanotube-enhanced electrochemical DNA biosensor for DNA hybridization detection. *Analytical and Bioanalytical Chemistry*, **375**, 287–293.

Cardenas, J. F. (2008). Protonation and sonication effects on aggregation sensitive Raman features of single wall carbon nanotubes. *Carbon*, **46**, 1327–1330.

Cardenas, J. F. and Glerup, M. (2006). The influence of surfactants on the distribution of the radial breathing modes of single walled carbon nanotubes. *Nanotechnology*, **17**, 5212–5215.

Chaturvedi, H. and Poler, J. C. (2006). Binding of rigid dendritic ruthenium complexes to carbon nanotubes. *Journal of Physical Chemistry B*, **110**, 22387–22393.

Chaturvedi, H. and Poler, J. C. (2007). Photon enhanced aggregation of single walled carbon nanotube dispersions. *Applied Physics Letters*, **90**, 223109.1–223109.3.

Chen, J., Dyer, M. J., and Yu, M. F. (2001a). Cyclodextrin-mediated soft cutting of single-walled carbon nanotubes. *Journal of the American Chemical Society*, **123**, 6201–6202.

Chen, R. J., Franklin, N. R., Kong, J., Cao, J., Tombler, T. W., Zhang, Y. G., and Dai, H. J. (2001b). Molecular photodesorption from single-walled carbon nanotubes. *Applied Physics Letters*, **79**, 2258–2260.

Chen, B., Gao, M., Zuo, J. M., Qu, S., Liu, B., and Huang, Y. (2003). Binding energy of parallel carbon nanotubes. *Applied Physics Letters*, **83**, 3570–3571.

Chen, Q., Saltiel, C., Manickavasagam, S., Schadler, L. S., Siegel, R. W., and Yang, H. C. (2004). Aggregation behavior of single-walled carbon nanotubes in dilute aqueous suspension. *Journal of Colloid and Interface Science*, **280**, 91–97.

Datta, S., Klimeck, G., Lake, R. K., and Anantram, M. P. (1995). In *Compound Semiconductors 1994* (Eds., Goronkin, H. and Mishra, U.) IOP Publishing Ltd, Bristol, U.K., pp. 775–780.

de Andrade, M. J., Lima, M. D., Skakalova, V., Bergmann, C. P., and Roth, S. (2007). Electrical properties of transparent carbon nanotube networks prepared through different techniques. *Physica Status Solidi-Rapid Research Letters*, **1**, 178–180.

Deck, C. P., Flowers, J., McKee, G. S. B., and Vecchio, K. (2007). Mechanical behavior of ultralong multiwalled carbon nanotube mats. *Journal of Applied Physics*, **101**, 9.

Dresselhaus, M. S., Dresselhaus, G., and Jorio, A. (2004). Unusual properties and structure of carbon nanotubes. *Annual Review of Materials Research*, **34**, 247–278.

Eckstein, H. and Kreibig, U. (1993). Light-induced aggregation of metal-clusters. *Zeitschrift Fur Physik D-Atoms Molecules and Clusters*, **26**, 239–241.

Evans, D. and Wennerstrom, H. (1999). *The Colloidal Domain where Physics, Chemistry, Biology and Technology Meet*, Wiley-VCH, New York.

Feng, W., Bai, X. D., Lian, Y. Q., Liang, J., Wang, X. G., and Yoshino, K. (2003). Well-aligned polyaniline/carbon-nanotube composite films grown by in-situ aniline polymerization. *Carbon*, **41**, 1551–1557.

Fischer, J. E. (2002). Chemical doping of single-wall carbon nanotubes. *Accounts of Chemical Research*, **35**, 1079–1086.

Fu, K. F. and Sun, Y. P. (2003). Dispersion and solubilization of carbon nanotubes. *Journal of Nanoscience and Nanotechnology*, **3**, 351–364.

Giordano, A. N., Chaturvedi, H., and Poler, J. C. (2007). Critical coagulation concentrations for carbon nanotubes in non-aqueous solvent. *Journal of Physical Chemistry C*, **111**, 11583–11589.

Girifalco, L. A., Hodak, M., and Lee, R. S. (2000). Carbon nanotubes, buckyballs, ropes, and a universal graphitic potential. *Physical Review B*, **62**, 13104–13110.

Gonnet, P., Liang, S. Y., Choi, E. S., Kadambala, R. S., Zhang, C., Brooks, J. S., Wang, B., and Kramer, L. (2006). Thermal conductivity of magnetically aligned carbon nanotube buckypapers and nanocomposites. *Current Applied Physics*, **6**, 119–122.

Gu, Y. G. (2000). The electrical double-layer interaction between a spherical particle and a cylinder. *Journal of Colloid and Interface Science*, **231**, 199–203.

Guo, J., Datta, S., and Lundstrom, M. (2004). A numerical study of scaling issues for Schottky-barrier carbon nanotube transistors. *IEEE Transactions on Electron Devices*, **51**, 172–177.

Guo, Z. H., Wood, J. A., Huszarik, K. L., Yan, X. H., and Docoslis, A. (2007). AC electric field-induced alignment and long-range assembly of multi-wall carbon nanotubes inside aqueous media. *Journal of Nanoscience and Nanotechnology*, **7**, 4322–4332.

Gupta, S., Hughes, M., Windle, A. H., and Robertson, J. (2004). In situ Raman spectro-electrochemistry study of single-wall carbon nanotube mat. *Diamond and Related Materials*, **13**, 1314–1321.

Haddon, R. C., Sippel, J., Rinzler, A. G., and Papadimitrakopoulos, F. (2004). Purification and separation of carbon nanotubes. *MRS Bulletin*, **29**, 252–259.

Hilding, J., Grulke, E. A., Zhang, Z. G., and Lockwood, F. (2003). Dispersion of carbon nanotubes in liquids. *Journal of Dispersion Science and Technology*, **24**, 1–41.

Hill, T. L. (1964). *Thermodynamics of Small Systems*, Benjamin, New York.

Hirsch, A. and Vostrowsky, O. (2005). Functional molecular nanostructures. *Top. Curr. Chem.*, **245**, 193–237.

Iijima, S. and Ichihashi, T. (1993). Single-shell carbon nanotubes of 1-nm diameter. *Nature*, **363**, 603–605.

Islam, M. F., Rojas, E., Bergey, D. M., Johnson, A. T., and Yodh, A. G. (2003). High weight fraction surfactant solubilization of single-wall carbon nanotubes in water. *Nano Letters*, **3**, 269–273.

Israelachvili, J. (1992). *Intermolecular & Surface Forces*, Academic Press, London, U.K.

Israelachvili, J. N. (Ed.) (1995). *Intermolecular and Surface Forces*, Academic Press, London, U.K.

Jeong, H. J., Choi, H. K., Kim, G. Y., Il Song, Y., Tong, Y., Lim, S. C., and Lee, Y. H. (2006). Fabrication of efficient field emitters with thin multiwalled carbon nanotubes using spray method. *Carbon*, **44**, 2689–2693.

Karpov, S. V., Slabko, V. V., and Chiganova, G. A. (2002). Physical principles of the photostimulated aggregation of metal sols. *Colloid Journal*, **64**, 425–441.

Kataura, H., Kumazawa, Y., Maniwa, Y., Umezu, I., Suzuki, S., Ohtsuka, Y., and Achiba, Y. (1999). Optical properties of single-wall carbon nanotubes. *Synthetic Metals*, **103**, 2555–2558.

Kavan, L., Frank, O., Green, A. A., Hersam, M. C., Koltai, J., Zolyomi, V., Kurti, J., and Dunsch, L. (2008). In situ Raman spectroelectrochemistry of single-walled carbon nanotubes: Investigation of materials enriched with (6,5) tubes. *Journal of Physical Chemistry C*, **112**, 14179–14187.

Kazaoui, S., Minami, N., Nalini, B., Kim, Y., and Hara, K. (2005). Near-infrared photoconductive and photovoltaic devices using single-wall carbon nanotubes in conductive polymer films. *Journal of Applied Physics*, **98**, 6.

Kimura, K. (1994). Photoenhanced Van-der-Waals attractive force of small metallic particles. *Journal of Physical Chemistry*, **98**, 11997–12002.

Kimura, T., Ago, H., Tobita, M., Ohshima, S., Kyotani, M., and Yumura, M. (2002). Polymer composites of carbon nanotubes aligned by a magnetic field. *Advanced Materials*, **14**, 1380–1383.

Kolling, O. W. (1981). Commentary on the Kamlet-Taft scale of solvent polarity for Aprotic Lewis bases. *Transactions of the Kansas Academy of Science*, **8**, 32–38.

Kong, J., Zhou, C., Morpurgo, A., Soh, H. T., Quate, C. F., Marcus, C., and Dai, H. (1999). Synthesis, integration, and electrical properties of individual single-walled carbon nanotubes. *Applied Physics A: Materials Science & Processing*, **69**, 305–308.

Kreibig, U. and Vollmer, M. (1995). *Optical Properties of Metal Clusters*, Springer-Verlag, Berlin, Germany.

Landau, L. D. and Lifshitz, E. M. (1960). *Electrodynamics of Continuous Media*, Pergamon Press, Oxford, U.K.

Landauer, R. (1989). Conductance determined by transmission-probes and quantized constriction resistance. *Journal of Physics-Condensed Matter*, **1**, 8099–8110.

Landi, B. J., Ruf, H. J., Worman, J. J., and Raffaelle, R. P. (2004). Effects of alkyl amide solvents on the dispersion of single-wall carbon nanotubes. *Journal of Physical Chemistry B*, **108**, 17089–17095.

Lau, K. T., Gu, C., and Hui, D. (2006). A critical review on nanotube and nanotube/nanoclay related polymer composite materials. *Composites Part B-Engineering*, **37**, 425–436.

Li, L. W., Bedrov, D., and Smith, G. D. (2006). Water-induced interactions between carbon nanoparticles. *Journal of Physical Chemistry B*, **110**, 10509–10513.

Lima, M. D., de Andrade, M. J., Bergmann, C. P., and Roth, S. (2008). Thin, conductive, carbon nanotube networks over transparent substrates by electrophoretic deposition. *Journal of Materials Chemistry*, **18**, 776–779.

Lin, M. Y., Lindsay, H. M., Weitz, D. A., Ball, R. C., Klein, R., and Meakin, P. (1989). Universality in colloid aggregation. *Nature*, **339**, 360–362.

Lin, M. Y., Lindsay, H. M., Weitz, D. A., Ball, R. C., Klein, R., and Meakin, P. (1990a). Universal reaction-limited colloid aggregation. *Physical Review A*, **41**, 2005–2020.

Lin, M. Y., Lindsay, H. M., Weitz, D. A., Klein, R., Ball, R. C., and Meakin, P. (1990b). Universal diffusion-limited colloid aggregation. *Journal of Physics-Condensed Matter*, **2**, 3093–3113.

Lin, Y., Zhou, B., Fernando, K. A. S., Liu, P., Allard, L. F., and Sun, Y. P. (2003). Polymeric carbon nanocomposites from carbon nanotubes functionalized with matrix polymer. *Macromolecules*, **36**, 7199–7204.

Liu, C. Y., Bard, A. J., Wudl, F., Weitz, I., and Heath, J. R. (1999). Electrochemical characterization of films of single-walled carbon nanotubes and their possible application in supercapacitors. *Electrochemical and Solid State Letters*, **2**, 577–578.

Liu, H., Zhai, J., and Jiang, L. (2006). Wetting and anti-wetting on aligned carbon nanotube films. *Soft Matter*, **2**, 811–821.

Liu, J., Rinzler, A. G., Dai, H. J., Hafner, J. H., Bradley, R. K., Boul, P. J., Lu, A., Iverson, T., Shelimov, K., Huffman, C. B., Rodriguez-Macias, F., Shon, Y. S., Lee, T. R., Colbert, D. T., and Smalley, R. E. (1998). Fullerene pipes. *Science*, **280**, 1253–1256.

Matarredona, O., Rhoads, H., Li, Z. R., Harwell, J. H., Balzano, L., and Resasco, D. E. (2003). Dispersion of single-walled carbon nanotubes in aqueous solutions of the anionic surfactant NaDDBS. *Journal of Physical Chemistry B*, **107**, 13357–13367.

Merchant, C. A. and Markovic, N. (2008). Effects of diffusion on photocurrent generation in single-walled carbon nanotube films. *Applied Physics Letters*, **92**, 3.

Merkoci, A., Pumera, M., Llopis, X., Perez, B., del Valle, M., and Alegret, S. (2005). New materials for electrochemical sensing VI: Carbon nanotubes. *Trac-Trends in Analytical Chemistry*, **24**, 826–838.

Mitchell, C. A., Bahr, J. L., Arepalli, S., Tour, J. M., and Krishnamoorti, R. (2002). Dispersion of functionalized carbon nanotubes in polystyrene. *Macromolecules*, **35**, 8825–8830.

Moore, V. C., Strano, M. S., Haroz, E. H., Hauge, R. H., Smalley, R. E., Schmidt, J., and Talmon, Y. (2003). Individually suspended single-walled carbon nanotubes in various surfactants. *Nano Letters*, **3**, 1379–1382.

Nakashima, N., Tomonari, Y., and Murakami, H. (2002). Water-soluble single-walled carbon nanotubes via noncovalent sidewall-functionalization with a pyrene-carrying ammonium ion. *Chemistry Letters*, **6**, 638–639.

Newson, R. W., Menard, J. M., Sames, C., Betz, M., and van Driel, H. M. (2008). Coherently controlled ballistic charge currents injected in single-walled carbon nanotubes and graphite. *Nano Letters*, **8**, 1586–1589.

Ng, M. H. A., Hartadi, L. T., Tan, H., and Poa, C. H. P. (2008). Efficient coating of transparent and conductive carbon nanotube thin films on plastic substrates. *Nanotechnology*, **19**, 5.

Niyogi, S., Hamon, M. A., Hu, H., Zhao, B., Bhowmik, P., Sen, R., Itkis, M. E., and Haddon, R. C. (2002). Chemistry of single-walled carbon nanotubes. *Accounts of Chemical Research*, **35**, 1105–1113.

O'Connell, M. J., Bachilo, S. M., Huffman, C. B., Moore, V. C., Strano, M. S., Haroz, E. H., Rialon, K. L., Boul, P. J., Noon, W. H., Kittrell, C., Ma, J. P., Hauge, R. H., Weisman, R. B., and Smalley, R. E. (2002). Band gap fluorescence from individual single-walled carbon nanotubes. *Science*, **297**, 593–596.

O'Connell, M. J., Boul, P., Ericson, L. M., Huffman, C., Wang, Y., Haroz, E., Kuper, C., Tour, J., Ausman, K. D., and Smalley, R. E. (2001). Reversible water-solubilization of single-walled carbon nanotubes by polymer wrapping. *Chemical Physics Letters*, **342**, 265–271.

Paiva, M. C., Zhou, B., Fernando, K. A. S., Lin, Y., Kennedy, J. M., and Sun, Y. P. (2004). Mechanical and morphological characterization of polymer–carbon nanocomposites from functionalized carbon nanotubes. *Carbon*, **42**, 2849–2854.

Pierard, N., Fonseca, A., Colomer, J. F., Bossuot, C., Benoit, J. M., Van Tendeloo, G., Pirard, J. P., and Nagy, J. B. (2004). Ball milling effect on the structure of single-wall carbon nanotubes. *Carbon*, **42**, 1691–1697.

Pint, C. L., Xu, Y.-Q., Pasquali, M., and Hauge, R. H. (2008). Formation of highly dense aligned ribbons and transparent films of single-walled carbon nanotubes directly from carpets. *ACS Nano*, **2**, 1871–1878.

Pompeo, F. and Resasco, D. E. (2002). Water solubilization of single-walled carbon nanotubes by functionalization with glucosamine. *Nano Letters*, **2**, 369–373.

Poulin, P., Vigolo, B., and Launois, P. (2002). Films and fibers of oriented single wall nanotubes. *Carbon*, **40**, 1741–1749.

Pumera, M., Merkoci, A., and Alegret, S. (2006). Carbon nanotube-epoxy composites for electrochemical sensing. *Sensors and Actuators B-Chemical*, **113**, 617–622.

Qu, L. and Dai, L. (2007). Gecko-foot-mimetic aligned single-walled carbon nanotube dry adhesives with unique electrical and thermal properties. *Advanced Materials*, **19**, 3844–3849.

Rao, A. M., Eklund, P. C., Bandow, S., Thess, A., and Smalley, R. E. (1997). Evidence for charge transfer in doped carbon nanotube bundles from Raman scattering. *Nature*, **388**, 257–259.

Rinzler, A. G., Liu, J., Dai, H., Nikolaev, P., Huffman, C. B., Rodriguez-Macias, F. J., Boul, P. J., Lu, A. H., Heymann, D., Colbert, D. T., Lee, R. S., Fischer, J. E., Rao, A. M., Eklund, P. C., and Smalley, R. E. (1998). Large-scale purification of single-wall carbon nanotubes: process, product, and characterization. *Applied Physics A-Materials Science & Processing*, **67**, 29–37.

Rocha, A. R., Rossi, M., Fazzio, A., and da Silva, A. J. R. (2008). Designing real nanotube-based gas sensors. *Physical Review Letters*, **100**, 4.

Saltiel, C., Manickavasagam, S., Menguc, M. P., and Andrews, R. (2005). Light-scattering and dispersion behavior of multiwalled carbon nanotubes. *Journal of the Optical Society of America A-Optics Image Science and Vision*, **22**, 1546–1554.

Sandkuhler, P., Lattuada, M., Wu, H., Sefcik, J., and Morbidelli, M. (2005). Further insights into the universality of colloidal aggregation. *Advances in Colloid and Interface Science*, **113**, 65–83.

Sano, M., Okamura, J., and Shinkai, S. (2001). Colloidal nature of single-walled carbon nanotubes in electrolyte solution: The Schulze-Hardy rule. *Langmuir*, **17**, 7172–7173.

Satoh, N., Hasegawa, H., Tsujii, K., and Kimura, K. (1994). Photoinduced coagulation of Au nanocolloids. *Journal of Physical Chemistry*, **98**, 2143–2147.

Schaefer, D. W., Zhao, J., Brown, J. M., Anderson, D. P., and Tomlin, D. W. (2003). Morphology of dispersed carbon single-walled nanotubes. *Chemical Physics Letters*, **375**, 369–375.

Snow, E. S., Novak, J. P., Campbell, P. M., and Park, D. (2003). Random networks of carbon nanotubes as an electronic material. *Applied Physics Letters*, **82**, 2145.

Song, Y. I., Yang, C. M., Kim, D. Y., Kanoh, H., and Kaneko, K. (2008). Flexible transparent conducting single-wall carbon nanotube film with network bridging method. *Journal of Colloid and Interface Science*, **318**, 365–371.

Star, A. and Stoddart, J. F. (2002). Dispersion and solubilization of single-walled carbon nanotubes with a hyperbranched polymer. *Macromolecules*, **35**, 7516–7520.

Strano, M. S., Moore, V. C., Miller, M. K., Allen, M. J., Haroz, E. H., Kittrell, C., Hauge, R. H., and Smalley, R. E. (2003). The role of surfactant adsorption during ultrasonication in the dispersion of single-walled carbon nanotubes. *Journal of Nanoscience and Nanotechnology*, **3**, 81–86.

Sun, C. H., Lu, G. Q., and Cheng, H. M. (2006). Simple approach to estimating the van der Waals interaction between carbon nanotubes. *Physical Review B*, **73**, 195414.

Sun, C. H., Yin, L. C., Li, F., Lu, G. Q., and Cheng, H. M. (2005). Van der Waals interactions between two parallel infinitely long single-walled nanotubes. *Chemical Physics Letters*, **403**, 343–346.

Sun, J. L., Xu, J., Zhu, J. L., and Li, B. L. (2008). Disordered multi-walled carbon nanotube mat for light spot position detecting. *Applied Physics A-Materials Science & Processing*, **91**, 229–233.

Syue, S.-H., Lu, S.-Y., Hsu, W.-K., and Shih, H.-C. (2006). Internanotube friction. *Applied Physics Letters*, **89**, 163115.

Takenobu, T., Takahashi, T., Kanbara, T., Tsukagoshi, K., Aoyagi, Y., and Iwasa, Y. (2006). High-performance transparent flexible transistors using carbon nanotube films. *Applied Physics Letters*, **88**, 033511.

Tan, Y. Q. and Resasco, D. E. (2005). Dispersion of single-walled carbon nanotubes of narrow diameter distribution. *Journal of Physical Chemistry B*, **109**, 14454–14460.

Tans, S. J., Verschueren, A. R. M. and Dekker, C. (1998). Room-temperature transistor based on a single carbon nanotube. *Nature*, **393**, 49–52.

Tao, N. J. (2006). Electron transport in molecular junctions. *Nature Nanotechnology*, **1**, 173–181.

Verwey, E. J. W. and Overbeek, J. T. G. (1948). *Theory of the Stability of Lyophobic Colloids*, Elsevier, Amsterdam, the Netherlands.

Wang, L. L., Chen, Y. W., Chen, T., Que, W. X., and Sun, Z. (2007). Optimization of field emission properties of carbon nanotubes cathodes by electrophoretic deposition. *Materials Letters*, **61**, 1265–1269.

Wang, F., Dukovic, G., Brus, L. E., and Heinz, T. F. (2004). Time-resolved fluorescence of carbon nanotubes and its implication for radiative lifetimes. *Physical Review Letters*, **92**, 177401–177404.

Wang, M. W., Hsu, T. C., and Weng, C. H. (2008a). Alignment of MWCNTs in polymer composites by dielectrophoresis. *European Physical Journal-Applied Physics*, **42**, 241–246.

Wang, D., Song, P. C., Liu, C. H., Wu, W., and Fan, S. S. (2008b). Highly oriented carbon nanotube papers made of aligned carbon nanotubes. *Nanotechnology*, **19**, 6.

Weitz, D. A. and Oliveria, M. (1984). Fractal structures formed by kinetic aggregation of aqueous gold colloids. *Physical Review Letters*, **52**, 1433–1436.

Whitby, R. L. D., Fukuda, T., Maekawa, T., James, S. L., and Mikhalovsky, S. V. (2008). Geometric control and tuneable pore size distribution of buckypaper and buckydiscs. *Carbon*, **46**, 949–956.

Wu, Z. C., Chen, Z. H., Du, X., Logan, J. M., Sippel, J., Nikolou, M., Kamaras, K., Reynolds, J. R., Tanner, D. B., Hebard, A. F., and Rinzler, A. G. (2004). Transparent, conductive carbon nanotube films. *Science*, **305**, 1273–1276.

Xie, X. L., Mai, Y. W., and Zhou, X. P. (2005). Dispersion and alignment of carbon nanotubes in polymer matrix: A review. *Materials Science & Engineering R-Reports*, **49**, 89–112.

Yamamoto, T., Miyauchi, Y., Motoyanagi, J., Fukushima, T., Aida, T., Kato, M., and Maruyama, S. (2008). Improved bath sonication method for dispersion of individual single-walled carbon nanotubes using new triphenylene-based surfactant. *Japanese Journal of Applied Physics*, **47**, 2000–2004.

Yanagi, K., Iakoubovskii, K., Kazaoui, S., Minami, N., Maniwa, Y., Miyata, Y., and Kataura, H. (2006). Light-harvesting function of beta-carotene inside carbon nanotubes. *Physical Review B*, **74**, 155420.

Ye, J. S., Wen, Y., De Zhang, W., Gan, L. M., Xu, G. Q., and Sheu, F. S. (2003). Selective voltammetric detection of uric acid in the presence of ascorbic acid at well-aligned carbon nanotube electrode. *Electroanalysis*, **15**, 1693–1698.

Yudasaka, M., Kasuya, Y., Kokai, F., Takahashi, K., Takizawa, M., Bandow, S., and Iijima, S. (2002). Causes of different catalytic activities of metals in formation of single-wall carbon nanotubes. *Applied Physics A: Materials Science & Processing*, **74**, 377–385.

Zhang, M., Atkinson, K. R., and Baughman, R. H. (2004a). Multifunctional carbon nanotube yarns by downsizing an ancient technology. *Science*, **306**, 1358–1361.

Zhang, M., Fang, S. L., Zakhidov, A. A., Lee, S. B., Aliev, A. E., Williams, C. D., Atkinson, K. R., and Baughman, R. H. (2005). Strong, transparent, multifunctional, carbon nanotube sheets. *Science*, **309**, 1215–1219.

Zhang, X. F., Sreekumar, T. V., Liu, T., and Kumar, S. (2004b). Properties and structure of nitric acid oxidized single wall carbon nanotube films. *Journal of Physical Chemistry B*, **108**, 16435–16440.

Zhao, H. F., Song, H., Li, Z. M., Yuan, G., and Jin, Y. X. (2005). Electrophoretic deposition and field emission properties of patterned carbon nanotubes. *Applied Surface Science*, **251**, 242–244.

Zhao, Y. D., Zhang, W. D., Luo, Q. M., and Li, S. F. Y. (2003). The oxidation and reduction behavior of nitrite at carbon nanotube powder microelectrodes. *Microchemical Journal*, **75**, 189–198.

Zheng, M., Jagota, A., Semke, E. D., Diner, B. A., McLean, R. S., Lustig, S. R., Richardson, R. E., and Tassi, N. G. (2003a). DNA-assisted dispersion and separation of carbon nanotubes. *Nature Materials*, **2**, 338–342.

Zheng, M., Jagota, A., Semke, E. D., Diner, B. A., McLean, R. S., Lustig, S. R., Richardson, R. E., and Tassi, N. G. (2003b). DNA-assisted dispersion and separation of carbon nanotubes. *Nature Materials*, **2**, 338–342.

Zhu, H. W. and Wei, B. Q. (2008). Assembly and applications of carbon nanotube thin films. *Journal of Materials Science & Technology*, **24**, 447–456.

9

Functionalization of Carbon Nanotubes for Assembly

Igor Vasiliev
New Mexico State University

9.1 Introduction

Carbon nanotubes (CNTs) are microscopic cylindrical structures obtained by "rolling up" two-dimensional graphene sheets. Depending on the number of concentric graphene layers, CNTs can be divided into single-walled carbon nanotubes (SWCNTs) and multiwalled carbon nanotubes (MWCNTs). CNTs exhibit a variety of remarkable physical characteristics resulting from their small size and one-dimensional periodicity (Ajayan 1999; Rao et al. 2001; Bernholc et al. 2002). The unusual structural, electronic, and mechanical properties of CNTs make them attractive for a wide range of technological applications. Within the rapidly growing field of nanotechnology, a major research effort is dedicated to the use of CNTs as building blocks for nanopolymer, nano-organic, and organoceramic composite materials (Dai and Mau 2001; Dresselhaus et al. 2001). Among the key properties of CNTs are their ability to show semiconducting or metallic behavior depending on the structural parameters, remarkable tensile strength, high stability in air, and the ability to retain conductivity without chemical doping (Mureau et al. 2008; Naito et al. 2008). The unique characteristics of CNTs open the way for the development of a new class of electronic and engineering materials. CNTs can play a role of reinforcing agents in polymer and epoxy composites leading to the production of lightweight materials that possess superior mechanical strength, chemical stability, and high electric and thermal conductivity (Blake et al. 2004; Cadek et al. 2004; Byrne et al. 2008). Composite materials containing CNTs can also be used in nanoscale electronic devices such as single-electron transistors (Tans et al. 1998; Stokes and

Khondaker 2008), molecular diodes (Wei et al. 2006), memory elements (Meunier et al. 2007), logic gates (Derycke et al. 2001), optoelectronics (Avouris et al. 2008), and chemical sensors (Sun and Sun 2008).

9.2 Functionalization of CNTs

One of the challenging problems in the design of nanocomposite materials is the creation of a sufficiently strong interface connecting CNTs to each other and the surrounding polymer network. Interfacial bonding between CNTs and polymers can be strengthened by the functionalization of CNT sidewalls (de la Torre et al. 2003; Byrne et al. 2008). Functionalization increases the extent of cross-linking in nanocomposites and provides a method for self-assembly of CNTs to organic matrices and substrates. In recent years, several approaches to the functionalization of CNTs have been developed. They include noncovalent and covalent sidewall functionalization, as well as the functionalization of the ends and defects of oxidatively etched CNTs (de la Torre et al. 2003).

Noncovalent sidewall functionalization is one of the most common types of chemical interface that can be created between CNTs and surrounding materials. In most cases, these interactions are limited to relatively weak van der Waals and electrostatic forces. In experiments, noncovalent interactions have been observed between SWCNTs and various organic molecules such as amines (Kong and Dai 2001), amides (Sun et al. 2001), and a monoclonal antibody (Erlanger et al. 2001). SWCNTs have also been shown to interact with sodium dodecylsulfate, forming CNT-containing micelles surrounded by hydrophobic moieties

(Bandow et al. 1997; Duesberg et al. 1998). Somewhat stronger π–π interactions have been detected between the walls of CNTs and amphiphiles containing aromatic groups (Chen et al. 2001). A separate class of supramolecular systems with CNTs can be obtained by using polymers. Several research groups have reported the formation of CNT–polymer complexes, in which polymer molecules show a selective interaction toward CNTs (Curran et al. 1998; Dalton et al. 2000; Star et al. 2001). The selectivity of surface interaction in these experiments is demonstrated by the fact that the polymer molecules wrap around the CNTs, while amorphous graphite precipitates out of solution (Curran et al. 1998; Dalton et al. 2000). The properties of CNT–polymer composite materials are different from those of single polymers. The addition of CNTs can substantially increase electrical conductivity of the polymer, as well as change its photoconductivity and fluorescence behavior, thus creating a new class of nanostructured materials with potential application to sensors, light emitters, detectors, and other electronic devices (Star et al. 2001). At the same time, due to the relatively weak nature of the van der Waals interaction, the noncovalent bonding has only a limited effect on the mechanical strength of CNT–polymer composite materials. This suggests that the other mechanisms of interaction, including surface intercalation and covalent bonding, should be considered for mechanical reinforcement of polymer composites (Coleman et al. 2003; Byrne et al. 2008).

End-group and defect functionalization of CNTs is most commonly produced by treatment with oxidizing acids, such as concentrated sulfuric and nitric acids (Liu et al. 1998). Unlike the noncovalent sidewall functionalization, this treatment results in much stronger covalent bonding between carboxylate or oxygen-containing ether groups and the surface of CNTs. The covalent attachment normally occurs at surface defect sites introduced by the acid treatment. Once the defects are created along the CNT surface, carboxylic acids can be used to covalently attach various organic groups, leading to highly soluble materials (de la Torre et al. 2003). The carboxylate groups can be derivatized to acid chlorides and coupled to alkylamines (Chen et al. 1998; Hamon et al. 1999), monoamine-terminated polyethylene oxide (Sano et al. 2001), amine groups (Shen et al. 2007), or imine-containing polymers (Czerw et al. 2001). The resulting functionalized CNTs have shown substantial solubility in organic chlorinated and aromatic solvents (Hamon et al. 1999, 2001), which makes these structures suitable for use in nanocomposite materials. The presence of carboxylate groups can also be used to covalently bind small metal particles, such as gold nanoclusters, to defect sites on the surface of CNTs (Azamian et al. 2002). Covalent functionalization increases the strength of the interaction between CNTs and the surrounding polymer. This broadens the range of potential applications for composite materials containing CNTs.

Covalent sidewall functionalization represents a promising way to obtain soluble CNT-based materials. However, in the absence of surface defects, CNTs do not readily interact with other chemical compounds. The high stability and low reactivity of CNT sidewalls limits the range of experimental methods available for the assembly of atoms and functional groups on the surface of CNTs. Generally, these methods require exposure of pristine CNTs to aggressive chemical environments. The application of aggressive chemical agents to CNTs must be carefully controlled to ensure that the electronic structure of CNTs remains mostly unchanged. One of the early approaches to covalent sidewall functionalization of CNTs involved reactions with fluorine-containing agents (Mickelson et al. 1998, 1999). Experiments demonstrated that this technique can produce a variety of sidewall functionalizations. Unfortunately, treatment with fluorine often led to highly modified structures and resulted in significant changes of the electronic properties of fluorinated CNTs (Mickelson et al. 1999). Other types of sidewall functionalization can be achieved by hydrogenation (Pekker et al. 2001); interaction with carbenes (Chen et al. 1998b), nitrines, and radicals (de la Torre et al. 2003); electrochemical reduction of aryl diazonium salts (Bahr et al. 2001); electrochemical modification under oxidative conditions (Kooi et al. 2002); and by thermal treatment (Bahr and Tour 2001). Organic covalent functionalization of CNTs can be performed through the addition of nitrenes and carbenes (Holzinger et al. 2001; Georgakilas et al. 2002). Compared to noncovalent interaction, covalent sidewall functionalization produces stronger bonds between CNTs and the surrounding polymer matrix.

9.3 Theoretical Modeling of CNTs

Theoretical studies of nanoscale systems, such as functionalized CNTs and CNT-based organic complexes, present major challenges to computational methods employed in quantum chemistry and condensed matter physics. The challenges are mainly related to the structural complexity and the lack of three-dimensional space periodicity in these systems. The complex structure and composition of functionalized CNTs necessitates the use of efficient numerical techniques in conjunction with massively parallel computing. Low-cost classical and semiclassical computational methods based on empirical force fields or interatomic potentials usually do not work well for these structures. For this reason, accurate numerical calculations for functionalized CNTs and CNT–polymer composites require a direct quantum mechanical approach.

An *ab initio* method based on density-functional theory (DFT) (Hohenberg and Kohn 1964; Kohn and Sham 1965) combined with the pseudopotential approximation (Troullier and Martins 1991) represents a quantum mechanical computational technique ideally suited for medium and large-scale modeling of functionalized CNTs. The pseudopotential method effectively reduces the total number of particles in the system by solving the quantum mechanical problem for the valence electrons only (Troullier and Martins 1991). The density-functional formalism transforms the many-body Schrödinger equation into a set of single-electron Kohn–Sham equations (Kohn and Sham 1965) given by*

---

* Atomic units ($\hbar = e = m_e = 1$) are used throughout this chapter.

$$\left(-\frac{\nabla^2}{2} + \sum_a v_{\text{ion}}(\mathbf{r} - \mathbf{R}_a) + v_{\text{H}}[\rho](\mathbf{r}) + v_{xc}[\rho](\mathbf{r})\right)\psi_i(\mathbf{r}) = \varepsilon_i\psi_i(\mathbf{r}).$$

$$(9.1)$$

In Equation 9.1, the true potential of each ion at $\mathbf{R}_a$ is replaced by a pseudopotential, $v_{\text{ion}}(\mathbf{r} - \mathbf{R}_a)$, accounting for the interaction of valence electrons with core electrons and nuclei; the Hartree potential, $v_{\text{H}}[\rho](\mathbf{r})$, describes the electrostatic interactions among valence electrons; the exchange-correlation potential, $v_{xc}[\rho](\mathbf{r})$, represents the nonclassical part of the Hamiltonian, and $\rho(\mathbf{r})$ is the charge density. The single-electron Kohn–Sham eigenvalues, ε_i, and eigen wave functions, $\psi_i(\mathbf{r})$, in Equation 9.1 pertain to valence electrons only.

The nonlocal ionic pseudopotential simulates the angular-momentum-dependent interaction between the valence and core electrons. In practical computational schemes, the Kleinman–Bylander (Kleinman and Bylander 1982) form of the nonlocal pseudopotential is usually employed:

$$v_{\text{ion}}(\mathbf{r} - \mathbf{R}_a)\psi_i(\mathbf{r}) = v_{\text{local}}(\mathbf{r} - \mathbf{R}_a)\psi_i(\mathbf{r}) + \sum_{l,m} G_{lm}\Delta v_l(\mathbf{r} - \mathbf{R}_a)\phi_{lm}(\mathbf{r} - \mathbf{R}_a),$$

$$(9.2)$$

where
v_{local} is the local ionic pseudopotential
$\Delta v_l = v_l - v_{\text{local}}$ is the difference between the local potential and the potential component with the angular momentum l
ϕ_{lm} are the atomic pseudo-wave functions

The projection coefficients G_{lm} are calculated as

$$G_{lm} = \frac{\left\langle \phi_{lm} \middle| \Delta v_l \middle| \psi_i \right\rangle}{\left\langle \phi_{lm} \middle| \Delta v_l \middle| \phi_{lm} \right\rangle}.$$

$$(9.3)$$

The exchange-correlation potential is approximated by a parametrized analytical expression of the charge density. Among the common approximations for the exchange-correlation functional are the local-density approximation (LDA) (Kohn and Sham 1965; Perdew and Zunger 1981) and the generalized-gradient approximation (GGA) (Perdew and Wang 1986; Perdew et al. 1996), although new types of hybrid functionals, such as B3LYP (Becke 1993), are becoming increasingly popular. Due to the nonlinear nature of the exchange-correlation functional, the accuracy of the approximation can be improved by correcting the analytical formula to account for the core electronic density. The exchange-correlation potential is then evaluated as a functional of the core-corrected charge density (Louie et al. 1982):

$$\rho(\mathbf{r}) = \rho_v(\mathbf{r}) + \sum_a \rho_{\text{core}}(|\mathbf{r} - \mathbf{R}_a|),$$

$$(9.4)$$

where
$\rho_{\text{core}}(|\mathbf{r} - \mathbf{R}_a|)$ is a fixed partial correction for the charge density of core electrons
$\rho_v(\mathbf{r})$ is the charge density of valence electrons calculated as

$$\rho_v(\mathbf{r}) = \sum_i n_i |\psi_i(\mathbf{r})|^2,$$

$$(9.5)$$

where
$\psi_i(\mathbf{r})$ are single-electron wave functions
n_i are occupation numbers

The combination of the Kohn–Sham scheme, exchange-correlation approximation, and pseudopotential method reduces the overall computational cost without significant loss of accuracy. Furthermore, calculations performed in the framework of the DFT method do not require any adjustable external parameters (Parr and Wang 1989). The absence of structure and material-dependent parameters and a relatively low computational cost makes the *ab initio* DFT approach applicable to complex nanosystems containing hundreds or, in some cases, thousands of atoms (Soler et al. 2002; Kronik et al. 2006). Unlike classical or semiclassical methods based on empirical interatomic potentials, the DFT approach provides the complete information about electronic wave functions of the system under study and can predict a wide variety of electronic, physical, and photochemical properties. The first-principles density-functional computational method has been successfully applied to pristine and functionalized CNTs. This approach have been used to compute the structural and electronic properties of isolated and bundled CNTs (Gulseren et al. 2002; Reich et al. 2002), physical and electronic properties of intercalated CNT ropes (Zhao et al. 2000), oxidation (Jhi et al. 2000), atomic and molecular adsorption and chemisorption on the surface of CNTs (Froudakis 2002; Krasheninnikov et al. 2004; Ricca and Bauschlicher 2006), and CNT doping (Jhi et al. 2002; Nevidomskyy et al. 2003; Bai and Zhou 2007). Other theoretical studies based on the first-principles density-functional formalism include the interaction of CNTs with organic molecules (Fagan et al. 2003; Kang 2004; Cho et al. 2008), electronic transport (Palacios et al. 2003; Biel et al. 2008), and electromechanical effects in CNTs (Verissimo-Alves et al. 2003).

The central theorem of DFT states that the external potential and the ground-state energy of a system of interacting electrons are uniquely determined by the ground-state charge density (Hohenberg and Kohn 1964). However, the original formulation of the DFT formalism has been restricted to the *time-independent* case only. A proper treatment of electronic excitations is not possible within the time-independent framework. This limitation has led to the development of *time-dependent* density-functional theory (TDDFT), which extends the central DFT theorem to time-dependent phenomena (Bartolotti 1982; Deb and Ghosh 1982; Runge and Gross 1984; Gross and Kohn 1985). The TDDFT formalism can be used to calculate absorption spectra and predict optical properties of nanostructured materials. Similarly to time-independent DFT developed by Kohn and Sham, TDDFT reduces the many-electron problem to a set of self-consistent single-electron equations:

$$\left(-\frac{\nabla^2}{2} + v_{\text{eff}}[\rho](\mathbf{r}, t)\right)\psi_i(\mathbf{r}, t) = i\frac{\partial}{\partial t}\psi_i(\mathbf{r}, t). \qquad (9.6)$$

The single-particle wave functions, $\psi_i(\mathbf{r}, t)$, and the effective potential, $v_{\text{eff}}[\rho](\mathbf{r}, t)$, in Equation 9.6 explicitly depend on time. The effective potential is given by

$$v_{\text{eff}}[\rho](\mathbf{r}, t) = \sum_a v_{\text{ion}}(\mathbf{r} - \mathbf{R}_a) + \int \frac{\rho(\mathbf{r}', t)}{|\mathbf{r} - \mathbf{r}'|}d\mathbf{r}' + v_{xc}[\rho](\mathbf{r}, t). \qquad (9.7)$$

The three terms on the right side of Equation 9.7 describe the external ionic potential, Hartree potential, and the exchange-correlation potential, respectively. The time-dependent charge density is defined as $\rho(\mathbf{r}, t) = \sum_i n_i |\psi_i(\mathbf{r}, t)|^2$, where n_i are occupation numbers.

Electronic excitations can be computed in the framework of TDDFT by considering a linear response to an external periodic perturbation (Casida 1995; Gross et al. 1996). In this approach, the linear response formalism is used to derive a density-functional expression for the dynamic polarizability. The excitation energies, Ω_I, which correspond to the poles of the dynamic polarizability, are obtained from the solution of an eigenvalue problem:

$$\mathbf{QF}_I = \Omega_I^2 \mathbf{F}_I, \qquad (9.8)$$

where the matrix $\mathbf{Q}$ is given by

$$\mathbf{Q}_{ij\sigma,kl\tau} = \delta_{i,k}\delta_{j,l}\delta_{\sigma,\tau}\omega_{kl\tau}^2 + 2\sqrt{\lambda_{ij\sigma}\omega_{ij\sigma}}K_{ij\sigma,kl\tau}\sqrt{\lambda_{kl\tau}\omega_{kl\tau}}. \qquad (9.9)$$

In this equation, the indices i, j, and σ (k, l, and τ) refer to the space and spin components, respectively, of the unperturbed static Kohn–Sham orbitals $\phi_{i\sigma}(\mathbf{r})$, $\omega_{ij\sigma} = \varepsilon_{j\sigma} - \varepsilon_{i\sigma}$ are the differences between the eigenvalues of the single-particle states, $\lambda_{ij\sigma} = n_{i\sigma} - n_{j\sigma}$ are the difference between their occupation numbers, and the coupling matrix, $\mathbf{K}$, in the adiabatic approximation is given by

$$K_{ij\sigma,kl\tau} = \iint \phi_{i\sigma}^*(\mathbf{r})\phi_{j\sigma}(\mathbf{r})\left(\frac{1}{|\mathbf{r} - \mathbf{r}'|} + \frac{\delta^2 E_{xc}[\rho]}{\delta\rho_\sigma(\mathbf{r})\,\delta\rho_\tau(\mathbf{r}')}\right)\phi_{k\tau}(\mathbf{r}')\phi_{l\tau}^*(\mathbf{r}')d\mathbf{r}\,d\mathbf{r}' \qquad (9.10)$$

where E_{xc} is the exchange-correlation energy of the system. The oscillator strengths f_I, which correspond to the residues of the dynamic polarizability, are given by

$$f_I = \frac{2}{3}\sum_{\beta = \{x, y, z\}} |\hat{\beta}\mathbf{R}^{1/2}\mathbf{F}_I|^2, \qquad (9.11)$$

where

$\mathbf{F}_I$ are the eigenvectors of Equation 9.8
$R_{ij\sigma,kl\tau} = \delta_{i,k}\delta_{j,l}\delta_{\sigma,\tau}\lambda_{kl\tau}\omega_{kl\tau}$
$\hat{\beta}$ is the dipole matrix element, $\hat{\beta}_{ij\sigma} = \int \phi_{i\sigma}(\mathbf{r})\beta\phi_{j\sigma}(\mathbf{r})d\mathbf{r}$, $\beta = \{x, y, z\}$

The static Kohn–Sham orbitals, $\phi_i(\mathbf{r})$, and their eigenvalues, ε_i, used in Equations 9.8 through 9.11 are obtained from the solution of the *time-independent* Kohn–Sham equations given by Equation 9.1.

Compared to other *ab initio* methods for excited states, such as the configuration-interaction method (Saunders and van Lenthe 1983), quantum Monte Carlo simulations (Ceperley and Bernu 1988), or the Green's function method based on the GW approximation (Hedin 1965; Sham and Rice 1966), the TDDFT approach is less computationally demanding and can be applied to nanosystems containing up to several hundreds of atoms (Vasiliev et al. 2001, 2002). The accuracy of the TDDFT formalism strongly depends on the quality of analytical approximation for the exchange-correlation functional. The agreement between experimental and theoretical optical spectra, especially in the range of high excitation energies, can be improved by using asymptotically correct exchange-correlation potentials (Vasiliev and Martin 2004), such as the Leeuwen–Baerends potential (van Leeuwen and Baerends 1994), or the asymptotically corrected LDA potential introduced by Casida and Salahub (2000).

The TDDFT formalism together with other first principle time-dependent computational methods have been used to calculate excitation energies, optical absorption, electron energy loss spectra, and excited carrier dynamics in CNTs (Machon et al. 2002; Marinopoulos et al. 2003, 2004; Miyamoto et al. 2006). These calculations have demonstrated that the TDDFT approach can be applied to functionalized CNTs and CNT-based complexes.

9.4 Carboxylated CNTs

Pristine CNTs are virtually insoluble in water and most organic solvents. The poor solubility of pristine CNTs makes it difficult to disperse them in liquids and polymers, complicating the manufacture of CNT-based composite materials. Chemical functionalization of CNT sidewalls improves the solubility of CNTs in organic solvents and makes them more easily dispersible in polymer matrices.

One of the common approaches to chemical functionalization of CNTs is the carboxylation of CNT edges and outer sidewalls. Carboxylation is conducted via oxidation by treating CNTs with a mixture of concentrated sulfuric and nitric acids. Acid treatment creates dangling bonds on the CNT surface that are progressively oxidized to hydroxyl (–OH), carbonyl (=CO), and carboxyl (–COOH) functional groups (Lordi et al. 2001; Li et al. 2002; Curran et al. 2004). The mechanism of sidewall carboxylation of CNTs has been investigated in the framework of *ab initio* DFT computational methods. The binding energies of carboxyl groups attached to CNTs were calculated using a parallel version of the PARSEC (Pseudopotential Algorithms for Real Space Electronic Calculations) electronic structure code (Kronik et al. 2006). Within PARSEC, the system of Kohn–Sham equations for electronic states is solved self-consistently on a real-space three-dimensional Cartesian grid without the use of explicit basis functions. The real-space approach does not produce an artificial periodicity and does not impose any restrictions on the net charge of the system, which makes this method particularly suitable for studying functionalization of CNTs and self-assembly of molecules on CNTs. Norm-conserving Troullier–Martins nonlocal pseudopotentials (Troullier and Martins 1991) in the Kleinman–Bylander form (Kleinman and Bylander 1982) were employed in these calculations. The exchange-correlation energy was evaluated using the GGA functional implemented in the form developed by Perdew, Burke, and Ernzerhof (Perdew et al. 1996). Spin polarization was explicitly included in the computational formalism. The sidewalls of CNTs were modeled by surface fragments passivated with hydrogen atoms along the perimeter. Similar theoretical models have been used in the past to study the adsorption of potassium on CNTs and graphene (Lou et al. 2000; Lugo-Solis and Vasiliev 2007). The surface fragments were sufficiently large to ensure that the attached carboxyl group did not interact with the fragment edges. The structures and binding energies of the carboxyl group attached to the CNT sidewall were examined in the three cases shown in Figure 9.1: a CNT sidewall containing no defects, sidewall containing a Stone–Wales (SW) defect (Stone and Wales 1986), and sidewall

containing a vacancy. Vacancies and SW defects represent common point defects naturally occurring in CNTs. SW defects are produced by rotating a C–C bond by 90° about its center, which converts a group of four adjacent hexagonal rings into a (5,7,7,5) ring cluster (Stone and Wales 1986). Vacancies are created by the removal of a single carbon atom from the CNT sidewall (Rossato et al. 2005).

The optimized lowest-energy structures of carboxylated defect-free and defective CNT fragments are shown in Figure 9.2. The DFT–GGA calculations predict site 1 to be the most energetically favorable for the attachment of the carboxyl group to a SW-defective CNT. The electronic configuration of the carbon atom at the COOH attachment site of defect-free and SW-defective CNTs suggests a partial transition from sp^2 to sp^3 hybridization for this atom. The computed binding energies, E_{bind}, and the equilibrium bond lengths, d_{min}, for the carboxyl group attached to a CNT are summarized in Table 9.1. The comparison of the calculated binding energies reveals the important role of surface defects in sidewall functionalization of CNTs. In the absence of surface defects, the interaction between the carboxyl group and the CNT is relatively weak. The binding energy between the COOH group and the CNT surface in this case is estimated to be less than 1 eV. The introduction of a SW defect to the surface of a CNT raises the binding energy to 1.88 eV, while in the presence of a vacancy on the CNT surface the binding energy is increased to 3.41 eV. These energies

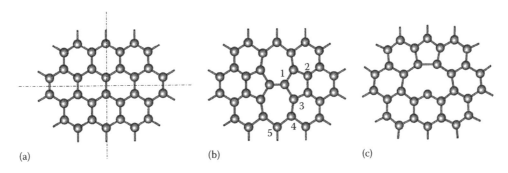

(a)　　　　　　　　　　　(b)　　　　　　　　　　　(c)

FIGURE 9.1　Point defects in CNTs. (a) A defect-free CNT sidewall. The dash–dot lines show the directions of the principal axes of the arm-chair and zig-zag CNTs. (b) A sidewall containing a Stone–Wales defect. The numbers 1 through 5 indicate different functionalization points. (c) A sidewall containing a vacancy.

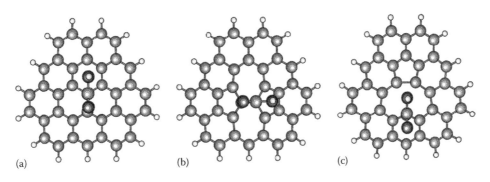

(a)　　　　　　　　　　　(b)　　　　　　　　　　　(c)

FIGURE 9.2　Lowest-energy structures of carboxylated CNT fragments (a) containing no defects, (b) containing a Stone–Wales defect, and (c) containing a vacancy.

TABLE 9.1 Binding Energies, E_{bind}, and Equilibrium Bond Lengths, d_{min}, between the Carboxyl Group and the CNT Sidewall Containing No Surface Defects, a Stone–Wales Defect, and a Vacancy

CNT	d_{min} (Å)	E_{bind} (eV)
No defects	1.55	0.94
SW defect	1.50	1.88
Vacancy	1.48	3.41

Note: Calculations were carried out using a real-space DFT formalism combined with the GGA exchange-correlation functional.

are in good agreement with the values obtained in localized-orbital DFT–GGA calculations, which predicted the binding energies of 0.92, 2.00, and approximately 3.3 eV for carboxylated (10,0) SWCNTs with no defects, with a SW defect, and with a vacancy, respectively (Wang et al. 2006a,b). The length of the CNT–COOH bond for the CNTs containing surface defects is about 3%–4% shorter than that for the defect-free CNT. The difference in the bond length can be explained by stronger bonding between the carboxyl group and the defective CNTs. At the same time, the CNT–COOH bonds are longer than the σ_{C-C} bonds in CNTs and graphite. This result is consistent with the fact that the CNT–COOH binding energies are weaker than the covalent C–C binding energies within the CNT sidewall.

9.5 Thiolated CNTs

Organic nanocomposite materials with good electrical conductivity can be obtained by covalently linking CNTs dispersed in a polymer. Conductive nanocomposites can be used as an active phase of electrochemical sensors, actuators, and photovoltaic devices (Long et al. 2004; Qi et al. 2007). The formation of covalent links between CNTs has been observed experimentally after S-thiolation of acid-treated MWCNTs with phosphorus pentasulfide (P_4S_{10}) (Curran et al. 2006a,b). The thiolation of MWCNTs was conducted in two steps: During the first step, pristine MWCNTs were treated with a mixture of concentrated acids to produce dangling bonds oxidized to carboxyl functional groups (Curran et al. 2004). In the second step, the functionalized MWCNTs were refluxed with P_4S_{10} in toluene, producing thiols, thioesters, and finally thiocarboxylic and dithiocarboxylic ester links connecting MWCNTs to each other. A similar method of catalytic thiolation leading to the formation of thiocarboxylic and dithiocarboxylic esters has been previously described for the interaction of carboxylic acids with thiols and alcohols (Sudalai et al. 2000). The presence of cross-linked MWCNTs in experimental samples was confirmed by atomic force microscopy, field emission scanning electron microscopy, and high resolution transmission electron microscopy (Curran et al. 2006a,b). An active role of dithioesters in the formation of intertube bonds was demonstrated by the existence of S=C–S stretching and bending modes in the Raman spectra of interconnected MWCNTs

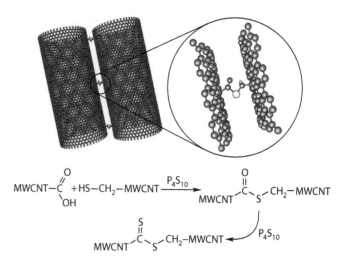

FIGURE 9.3 Schematic illustration of two MWCNTs linked by thiocarboxylic or dithiocarboxylic ester. Only the external layers of the MWCNTs are shown. (From Vasiliev, I. and Curran, S.A., *J. Appl. Phys.*, 102, 024317-1, 2007. With permission.)

(Curran et al. 2006a). A schematic illustration of MWCNTs tethered by thiocarboxylic or dithiocarboxylic ester is shown in Figure 9.3. The bottom portion of Figure 9.3 displays the proposed mechanism of interaction between functionalized MWCNTs (Curran et al. 2006b).

The mechanism of cross-linking between thiolated MWCNTs has been investigated using the *ab initio* DFT computational approach (Vasiliev and Curran 2007). The goal of this theoretical study was to verify the validity of the proposed reaction scheme and evaluate the influence of surface defects on the strength and stability of intertube bonds. Calculations were carried out using the PARSEC electronic structure code (Kronik et al. 2006) combined with Troullier–Martins nonlocal pseudopotentials (Troullier and Martins 1991). The exchange-correlation energy was evaluated within the LDA and GGA approximations. The LDA correlation term was described using the Perdew–Zunger parametrization (Perdew and Zunger 1981) of the Ceperley–Alder functional (Ceperley and Alder 1980). The GGA exchange-correlation functional was implemented in the form developed by Perdew, Burke, and Ernzerhof (Perdew et al. 1996). Interacting MWCNTs were modeled by surface fragments. Dangling bonds along the fragment perimeter were passivated with hydrogen atoms. Due to the complexity of the modeled system and a large number of possible bonding configurations, the study was focused on the analysis of single isolated thiocarboxylic and dithiocarboxylic ester molecules bridging two MWCNTs. To determine the most probable mechanism of intertube bond formation, calculations were performed for MWCNTs without surface defects, a MWCNT containing a SW defect, and a MWCNT containing a vacancy. The equilibrium structures of cross-linked MWCNTs were obtained by minimizing the total energy of the system with respect to the mutual separation and orientation of the MWCNT fragments.

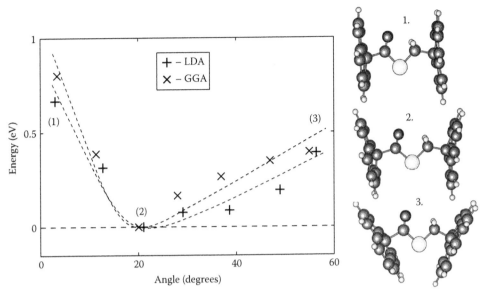

FIGURE 9.4 Potential energy profiles of cross-linked MWCNTs as a function of the tilt angle between MWCNT surfaces. Calculations were conducted using a DFT formalism combined with the LDA and the GGA exchange-correlation functionals. (From Vasiliev, I. and Curran, S.A., *J. Appl. Phys.*, 102, 024317-1, 2007. With permission.)

Figure 9.4 shows the variation of the total energy of defect-free MWCNTs linked by a thiocarboxylic ester molecule as a function of the tilt angle between MWCNT surfaces. The tilt angle was altered by adjusting the relative positions of the MWCNT fragments. The structures of cross-linked MWCNTs at each angle were optimized by fixing the coordinates of hydrogen atoms along the perimeters of the MWCNT fragments and relaxing the positions of all other atoms in the system. The optimized structures of cross-linked MWCNTs at different tilts are shown

on the right side of Figure 9.4. The DFT–LDA and DFT–GGA computational approaches predict a minimum in the potential energy of linked MWCNTs at an angle of 21° and 20°, respectively. When the tilt angle between MWCNTs deviates from the optimum value, the total energy of the system increases at a rate of approximately 0.01–0.02 eV per degree.

Figure 9.5 displays the dependence of the total energy of defect-free MWCNTs linked by thiocarboxylic ester on the distance between MWCNT surfaces. As before, the geometries of

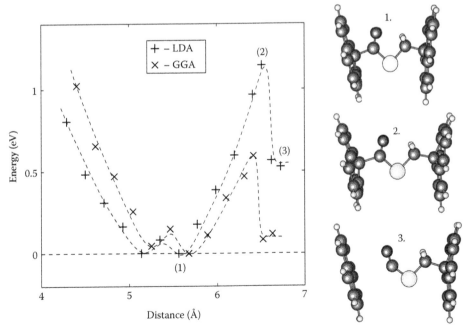

FIGURE 9.5 Potential energy profiles of cross-linked MWCNTs as a function of the distance between MWCNT surfaces. The images on the right side show the structures of connected MWCNT fragments (1) at the lowest-energy position, (2) at the point of maximum link extension, and (3) after link breaking. (From Vasiliev, I. and Curran, S.A., *J. Appl. Phys.*, 102, 024317-1, 2007. With permission.)

cross-linked MWCNT fragments at each distance were structurally optimized by fixing the positions of the perimeter hydrogen atoms and relaxing all other atomic coordinates. The DFT calculations based on the LDA and GGA functionals predict the lowest potential energy of linked MWCNTs at a distance of 5.56 and 5.68 Å, respectively. As expected, the GGA equilibrium distance is about 2% greater than the corresponding LDA

value. When the MWCNT fragments are pulled apart, the total energy of the system raises at a rate of approximately 1.2 eV/Å (LDA) and 0.8 eV/Å (GGA), respectively. The link between the MWCNTs breaks when the intertube separation exceeded 6.51 Å for LDA and 6.42 Å for GGA. The energy barrier for the link dissociation computed within the LDA and GGA methods is 1.1 and 0.6 eV, respectively. The structures of bound MWCNT fragments at the lowest-energy position, at the point of maximum link extension, and after link breaking are shown on the right side of Figure 9.5.

The potential energy profiles of cross-linked MWCNTs without surface defects, with a SW defect, and with a vacancy are compared in Figure 9.6. Structural optimization of interconnected MWCNT fragments was carried out in the framework of the DFT–GGA formalism. For the MWCNT containing a SW surface defect, the link was attached to the lowest-energy position marked in Figure 9.1b as site 1. Surprisingly, the theoretical study indicates that the presence of SW defects does not enhance the stability of intertube bonds. The computed potential barrier for the dissociation of a thioester link attached to the SW defect site is nearly the same as that for a defect-free MWCNT. In contrast, the calculations show that the addition of a vacancy to the MWCNT surface significantly increases the strength of intertube bonding and raises the height of the dissociation barrier to more than 3 eV.

The calculated binding energies and dissociation barriers of MWCNTs linked by thiocarboxylic and dithiocarboxylic ethers are summarized in Table 9.2. The binding energies, E_{bind}, are computed as the differences between the total energies of cross-linked MWCNTs and the sum of the total energies of the MWCNTs fragments after the dissolution of the intertube bond. Positive values of E_{bind} correspond to thermodynamically stable structures. The potential barriers for link dissociation, $E_{barrier}$, are evaluated as the differences in the total energies of bound

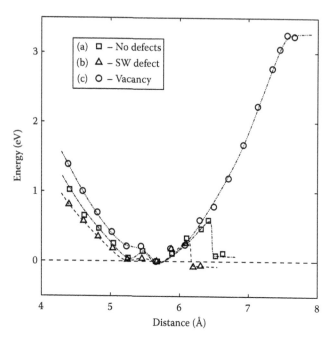

FIGURE 9.6 Energy profiles of cross-linked MWCNTs (a) without surface defects, (b) with a Stone–Wales defect, and (c) with a vacancy as a function of the distance between the MWCNTs. Calculations were performed using a DFT formalism with the GGA exchange-correlation functional. (From Vasiliev, I. and Curran, S.A., *J. Appl. Phys.*, 102, 024317-1, 2007. With permission.)

TABLE 9.2 Equilibrium Tilt Angles, α_{min}, Intertube Distances, d_{min}, Binding Energies, E_{bind}, and Dissociation Barriers, $E_{barrier}$, for MWCNTs Linked by Thiocarboxylic and Dithiocarboxylic Ethers

CNT	XC functional	α_{min} (degrees)	d_{min} (Å)	E_{bind} (eV)	$E_{barrier}$ (eV)
MWCNT–CO–S–CH$_2$–MWCNT					
No defects	LDA	21.1	5.56	0.9 (0.8)	1.1
No defects	GGA	20.1	5.68	0.2 (0.1)	0.6
SW defect	GGA	23.2	5.67	0.1 (0.0)	0.4
Vacancy	GGA	21.7	5.65	3.2 (3.0)	3.2
MWCNT–CS–S–CH$_2$–MWCNT					
No defects	LDA	22.9	5.57	0.1 (0.1)	0.3
No defects	GGA	23.8	5.69	−0.3 (−0.3)	0.1
SW defect	GGA	28.5	5.64	−0.2 (−0.3)	0.1
Vacancy	GGA	24.3	5.61	2.7 (2.6)	2.8

Source: Vasiliev, I. and Curran, S.A., *J. Appl. Phys.*, 102, 024317-1, 2007. With permission.

Note: Positive values of E_{bind} correspond to thermodynamically stable structures. The geometries of interacting MWCNTs were optimized by fixing the positions of hydrogen atoms along the perimeters of the MWCNT fragments and relaxing all other atomic coordinates. The values of binding energies in parentheses were obtained by further relaxing the coordinates of the perimeter hydrogen atoms after link breaking.

MWCNT fragments at minimum energy and at the point of maximum link extension. The equilibrium tilt angle, α_{min}, represents the minimum-energy angle between the surfaces of MWCNT fragments. The equilibrium length of the intertube link, d_{min}, is measured as the minimum-energy distance between two carbon atoms at the opposite ends of the linker molecule. The values in Table 9.2 show that the DFT calculations based on the GGA exchange-correlation functional predict close to zero or negative binding energies for defect-free and SW-defective MWCNTs. The link dissociation barriers for these structures are found to be in the range of 0.1–0.6 eV. At the same time, the study indicates that the introduction of a MWCNT surface vacancy increases the binding energy of cross-linked MWCNTs to 2.7–3.0 eV and raises the dissociation barrier to 2.8–3.2 eV. These observations imply that the strength and stability of intertube bonding could be enhanced by the presence of defect sites that change the coordination number of carbon atoms within the MWCNT. These findings are consistent with the results of other theoretical studies that have demonstrated an important role of vacancies and interstitial atoms in covalent functionalization and cross-linking of graphitic structures and CNTs (Telling et al. 2003; Rossato et al. 2005; Wang et al. 2006b).

A comparison of the values presented in Table 9.2 reveal significant differences between the binding energies and dissociation barriers for cross-linked MWCNTs obtained using the LDA and GGA exchange-correlation functionals. The DFT–LDA calculations predict a binding energy of 0.9 eV for MWCNTs linked by thiocarboxylic ether, while the corresponding GGA value is only 0.2 eV. It is well known that the DFT formalism based on the local-density approximation tend to overestimate interatomic binding energies by approximately 10%–30% (Becke 1992; Hammer et al. 1993; Perdew et al. 1996). However, disagreements between the LDA and the GGA binding energies and dissociation barriers for MWCNTs linked by thioesters appear to be greater than a typical difference between the LDA and GGA values. A large discrepancy between the LDA and GGA results observed in this study could be attributed to the fact that the electronic orbitals forming the σ_{C-C} bonds in CNTs are strongly localized. The localization of interatomic bonds leads to nonuniform distribution of electronic charge density in CNTs, which emphasizes the importance of the gradient correction term in the exchange-correlation functional for this system. Furthermore, the relative differences between the LDA and GGA energies obtained in these calculations may also appear to be larger than usual due to the small magnitudes of E_{bind} and $E_{barrier}$ for defect-free MWCNTs. These considerations suggest that the use of gradient-corrected functionals is essential for an accurate *ab initio* theoretical description of functionalized and cross-linked CNTs.

9.6 Assembly of Phenosafranin to CNTs

Organic light-emitting diodes (OLEDs), photodiodes, and field-effect transistors made from semiconducting polymers are poised to become a key technology for the next generation of electrically pumped solid-state lasers, high efficiency solar cells, and flexible flat-panel displays (Kelley et al. 2004; Sheats 2004). The fluorescent emission in OLEDs is produced by the radiative recombination of singlet excitons (Kulkarni et al. 2004). The performance of OLEDs can be improved by using a buffer layer and doping the emissive material, which leads to the reduction of the hole injection potential barrier, a more even charge distribution with a larger contact area at the anode and the organic interfaces, and an excitation migration from host to guest molecules (Tang et al. 1989; Gustafsson et al. 1992; Mori et al. 1995; Kim et al. 1999).

Dispersion of nanostructures within emissive polymers has been suggested as a way of increasing the efficiency of the recombination process without modification of the polymer backbone (Curran et al. 1998). Experimental studies show that the blending of CNTs in a host-emissive polymer matrix improves the OLEDs performance in terms of the turn-on voltage and efficiency. The addition of CNTs creates a polymer-nanostructure matrix with cooperative behavior between the host and the additive, thus modifying the electronic properties of the polymer (Ago et al. 1999). The fluorescent emission of polymer nanocomposite OLEDs can be enhanced by assembling certain organic molecules on the surface of CNTs (Curran et al. 2004). The enhancement mechanism is based on the fact that the molecules serve as fillers producing the effects of nanoscale optical and electrical antennae. The efficiency of light emission can be further amplified by selecting luminescent organic molecules that possess greater charge-carrying capabilities within the polymeric structure and reduce the aspect of charge trapping and quenching.

The possibility of self-assembly of luminescent organic molecules on CNTs has been confirmed experimentally (Curran et al. 2004). In the experiment, phenosafranin (PSF), a cationic dye (PS+, 3,7-diamino-5-phenylphenazinium chloride), was adsorbed on the surface of MWCNTs treated with a mixture of concentrated sulfuric and nitric acids. Atomic force microscopy images showed numerous "bright spots" indicating the positions of luminescent PSF molecules attached to the surface of acid-treated MWCNTs. An interaction between MWCNTs and PSF was also detected through changes in the Raman and UV-visible absorption spectra. Spectroscopic measurements showed that the absorption maximum of PSF shifted from 520 nm for pure PSF dye to 562 nm for PSF molecules attached to MWCNTs. A substantial bathochromic shift of 0.17 eV in the absorption spectrum of PSF indicated the formation of a PSF–MWCNT complex. It has been suggested that the active sites produced by acid treatment of MWCNTs may interact with PSF dye through a charge transfer process, in which MWCNTs and PSF molecules act as electron acceptors and donors, respectively (Curran et al. 2004).

The mechanism of interaction between PSF molecules and CNTs has been studied in the framework of *ab initio* DFT computational methods with the purpose of assessing the potential impact of the PSF–MWCNT self-assembly on the light output of nanocomposite OLEDs (Vasiliev and Curran 2006). The structures of functionalized MWCNTs and PSF were optimized

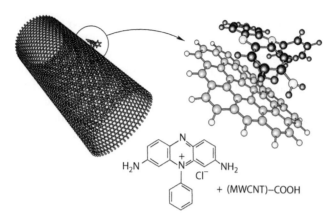

H_2N — [ring] — NH_2

Cl^-

+ (MWCNT)–COOH

FIGURE 9.7 Interaction between carboxylated MWCNTs and phenosafranin (PSF). Only the external layer of the MWCNT is shown. (From Vasiliev, I. and Curran, S.A., *Phys. Rev. B*, 73, 165420-1, 2006. With permission.)

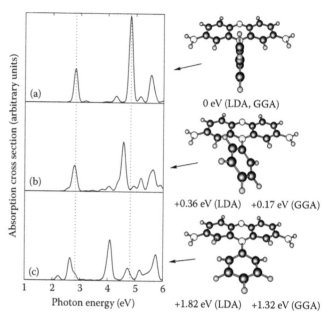

0 eV (LDA, GGA)

+0.36 eV (LDA)   +0.17 eV (GGA)

+1.82 eV (LDA)   +1.32 eV (GGA)

FIGURE 9.8 Optimized geometries, energies, and calculated absorption spectra of the three lowest-energy structures of PSF⁺. (From Vasiliev, I. and Curran, S.A., *Phys. Rev. B*, 73, 165420-1, 2006. With permission.)

using the PARSEC electronic structure code (Kronik et al. 2006). MWCNTs were modeled by hydrogen-passivated surface fragments. The size of the fragments was selected to be sufficiently large to ensure that the attached PSF molecule did not interact with the fragment edges. Figure 9.7 shows the proposed structural model for a single PSF molecule attached to the surface of an acid-treated MWCNT. UV-visible absorption spectra of isolated PSF molecules and complexes formed between PSF and MWCNTs were calculated using the linear response formalism based on TDDFT (Vasiliev et al. 1999, 2002). Due to computational complexity, these calculations were carried out using a parallel version of the TDDFT algorithm (Burdick et al. 2003).

Before analyzing the mechanism of interaction between PSF and MWCNTs, the DFT and TDDFT methods were applied to compute the ground-state energy, equilibrium geometry, and optical absorption spectrum of a single PSF⁺ molecule. The calculations revealed the existence of several isomeric structures for PSF⁺. The optimized geometries and absorption spectra of the three low-energy isomers are shown in Figure 9.8. The structures mainly differ by the orientation of a benzene ring with respect to the plane of the molecule. In the lowest-energy structure shown in Figure 9.8a, the benzene ring is orthogonal to the molecular plane. In the isomer shown in Figure 9.8b the ring is tilted with respect to the plane of the molecule. Calculations based on the LDA and the GGA exchange-correlation functionals predict the energy of this isomer to be higher than that of the previous structure by 0.36 and 0.17 eV, respectively. In the isomer shown in Figure 9.8c the benzene ring is parallel to the molecular plane. Depending on the type of the exchange-correlation functional used in DFT calculations, the energy of this structure is found to be higher than that of the lowest-energy isomer by 1.82 eV (LDA) or 1.32 eV (GGA). The comparison of TDDFT spectra for the three isomers demonstrates that the rotation of the benzene ring results in a red shift of the calculated absorption peaks and creates a new absorption band in the low-energy part of the spectrum. The intensity of this additional absorption band increases with decreasing angle between the benzene ring

and the molecular plane. Compared to the other two isomers, the structure shown in Figure 9.8c has the lowest absorption edge. The calculated energy of the main visible optical transition for the most stable PSF isomer (Figure 9.8a) is higher than the experimental value by approximately 0.4 eV. The discrepancy can most likely be attributed to solvent effects not accounted for by the computational approach.

A theoretical model of the interaction between PSF and acid-treated MWCNTs is based on the assumptions that (i) acid treatment produces carboxylated MWCNTs and (ii) PSF molecules interact primarily with the surface defect states produced by acid treatment of MWCNTs. The first assumption is justified by the fact that the reaction of carboxylation has been described by many authors as the most probable outcome of an acid treatment of CNTs (Lordi et al. 2001; Li et al. 2002; Curran et al. 2004). The second assumption is based on the observation that the change in the absorption maximum of PSF was detected only when the PSF dye was mixed with the acid-treated MWCNTs, while no spectral change was recorded after mixing PSF with pristine MWCNTs (Curran et al. 2004). The surfaces of carboxylated MWCNTs in the DFT calculations were simulated using two different models: one model was based on a fragment of a flat graphite surface corresponding to a "MWCNT" of infinite radius, and the other was based on a fragment of curved carbon surface representing a MWCNT of 2 nm in diameter. Both fragments consisted of 42 carbon atoms and were passivated with 16 hydrogen atoms along the perimeter.

Structural optimization of complexes formed between PSF and MWCNTs was performed in several steps. First, hydrogen-passivated MWCNT surface fragments were prepared and relaxed. Next, defect states were introduced into the carbon surface by functionalizing fragments with the carboxyl groups. After that, the positions of the perimeter hydrogen atoms were fixed and the structures of carboxylated fragments were optimized. In the last step, PSF molecules were added to carboxylated MWCNTs and the final structural optimization of the PSF–MWCNT systems was performed. The lowest-energy configurations of the PSF–MWCNT complexes were obtained by selecting a number of different initial geometries, including structures where the main symmetry axis of the PSF molecule was rotated by 0°, 30°, 60°, and 90° with respect to the axis of the MWCNT. Each initial structure was optimized using molecular dynamics simulations, followed by a minimization of interatomic forces at "zero" temperature. The final geometries of the PSF–MWCNT complexes were obtained by selecting the structures with the lowest total energies.

Figure 9.9a and b shows the optimized lowest-energy configurations of the complexes built on the flat and curved carbon surface fragments, respectively. The binding energy between the PSF molecule and the carboxylated MWCNT is estimated to be approximately 1.2 eV in DFT–LDA calculations and 0.8 eV in DFT calculations based on the GGA exchange-correlation functional. In both cases, interaction energies are essentially the same for the flat and curved carbon surfaces. It is interesting to note that the optimized structures of PSF–MWCNT complexes no longer favor the lowest-energy structure of PSF shown in Figure 9.8a. Instead, the geometry of the PSF molecule attached to the carboxylated MWCNT most closely resembles the isomeric structure shown in Figure 9.8b, in which the benzene ring is tilted with respect to the molecular plane. This effect can be attributed to an interference between the PSF benzene ring and MWCNT surface.

The three-dimensional contour plots in Figure 9.9a and b show the spatial distributions of electron densities for the lowest unoccupied molecular orbital (LUMO) and the highest occupied molecular orbital (HOMO) of the PSF–MWCNT complexes. The plots demonstrate that the wave function of the HOMO state is located on the MWCNT surface and has a configuration typical for a defect state. In contrast, the wave function of the LUMO state is almost entirely localized on the PSF molecule. The characters of the HOMO and LUMO states are consistent with the formation of a charge transfer complex between PSF and the carboxylated MWCNT. A partial transfer of electronic charge from PSF to MWCNTs is confirmed by the calculated values of electric dipole moments of the PSF–MWCNT complexes. The projection of the dipole moment on the axis orthogonal to the surface of the MWCNT changes from 0.4 to 0.6 D for isolated carboxylated MWCNT fragments to 5.3–5.9 D for the PSF–MWCNT complexes.

The calculated UV-visible absorption spectra of an isolated PSF molecule and PSF molecules attached to MWCNTs are compared in Figure 9.10. The upper plot (Figure 9.10a) shows the optical absorption spectrum of a single PSF$^+$ molecule. The spectra of complexes formed between PSF and MWCNTs are shown in the insets of Figure 9.10b and c. Shaded areas outline the change in absorption which occurred after an attachment of the PSF molecule to the surface of the carboxylated MWCNT fragment. The main panels of Figure 9.10b and c show the differential absorption spectra calculated as the difference in the optical absorption of the same carboxylated MWCNT fragment with and without the PSF molecule. Positive values of differential photoabsorption correspond to the new absorption peaks that appear after the attachment of PSF to the fragment surface. The comparison of the differential spectra of PSF–MWCNT complexes with the optical absorption of a single PSF molecule indicates a bathochromic shift of the absorption maximum of PSF molecules attached to carboxylated MWCNTs. The shift is equal to 0.14 eV for the model based on the flat carbon surface (Figure 9.10b) and 0.10 eV for the curved surface corresponding to a MWCNT of 2 nm in diameter (Figure 9.10c). The calculated spectral shift is in good agreement with the experimental value of 0.17 eV (Curran et al. 2004). The calculations do not rule out the possibility that the bathochromic shift of the absorption maximum of PSF is related to charge transfer. However, it appears more likely that the shift could be explained, at least in part, by isomerization of PSF molecules attached to the surface of carboxylated MWCNTs. The interaction between PSF and MWCNTs changes the orientation of the benzene ring in

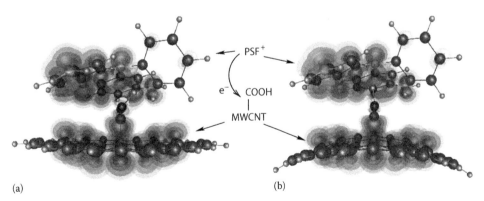

(a)   (b)

FIGURE 9.9 Structures of charge transfer complexes formed between carboxylated MWCNTs and PSF. MWCNTs were modeled by fragments of (a) a flat graphite surface and (b) a curved carbon surface corresponding to a MWCNT of 2 nm in diameter. Contour plots show the spatial distributions of electron densities in the HOMO and LUMO orbitals. (From Vasiliev, I. and Curran, S.A., *Phys. Rev. B*, 73, 165420-1, 2006. With permission.)

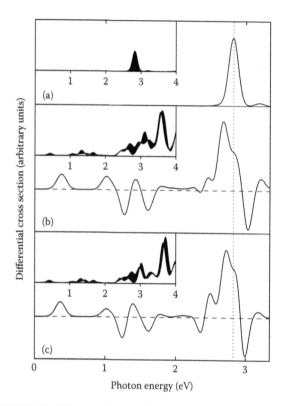

FIGURE 9.10 Calculated UV-visible absorption spectra of (a) an isolated PSF molecule and differential absorption spectra of PSF molecules attached to carboxylated MWCNT fragments modeled by the (b) flat and (c) curved carbon surfaces. Shaded areas in the inline plots show the difference in optical absorption of MWCNT fragments with and without the PSF molecule. (From Vasiliev, I. and Curran, S.A., *Phys. Rev. B*, 73, 165420-1, 2006. With permission.)

the PSF molecule, as shown in Figure 9.9. A comparison of the absorption spectra for the three lowest-energy isomers of PSF (Figure 9.8) indicates that rotation of the benzene ring could be responsible for the observed red shift of the main visible absorption band of PSF.

9.7 Summary

Carbon nanotubes exhibit a variety of unusual physical characteristics resulting from their extremely small size and one-dimensional periodicity. The unusual mechanical and electronic properties of CNTs make them attractive for a wide range of technological applications. Functionalization of CNTs provides a method for connecting CNTs together, as well as to organic matrices and substrates. First principles computational techniques based on DFT and TDDFT are well suited for the modeling of functionalized CNTs. These techniques have been applied to study the covalent sidewall functionalization of CNTs, the cross-linking of thiolated CNTs, and the assembly of organic molecules on functionalized CNTs. The results of density-functional calculations show that *ab initio* theoretical methods can successfully describe the mechanisms of functionalization, cross-linking, and self-assembly of CNTs.

Acknowledgments

The author would like to thank Professor Seamus A. Curran and Nabil Al-Aqtash for their valuable contributions to this chapter. The author also acknowledges support from the National Science Foundation, the U.S. Department of Energy, and the Donors of the American Chemical Society Petroleum Research Fund.

References

Ago, H., K. Petritsch, M. S. P. Shaffer, A. H. Windle, and R. Friend. 1999. Composites of carbon nanotubes and conjugated polymers for photovoltaic devices. *Adv. Mater.* 11: 1281–1285.

Ajayan, P. M. 1999. Nanotubes from carbon. *Chem. Rev.* 99: 1787–1799.

Avouris, P., M. Freitag, and V. Perebeinos. 2008. Carbon-nanotube photonics and optoelectronics. *Nat. Photonics* 2: 341–350.

Azamian, B. R., K. S. Coleman, J. J. Davis, N. Hanson, and M. L. H. Green. 2002. Directly observed covalent coupling of quantum dots to single-wall carbon nanotubes. *Chem. Commun.* 4: 366–367.

Bahr, J. L. and J. M. Tour. 2001. Highly functionalized carbon nanotubes using in situ generated diazonium compounds. *Chem. Mater.* 13: 3823–3824.

Bahr, J. L., J. Yang, D. V. Kosynkin, M. J. Bronikowski, R. E. Smalley, and J. M. Tour. 2001. Functionalization of carbon nanotubes by electrochemical reduction of aryl diazonium salts: A Bucky paper electrode. *J. Am. Chem. Soc.* 123: 6536–6542.

Bai, L. and Z. Zhou. 2007. Computational study of B- or N-doped single-walled carbon nanotubes as NH_3 and NO_2 sensors. *Carbon* 10: 2105–2110.

Bandow, S., A. M. Rao, K. A. Williams, A. Thess, R. E. Smalley, and P. C. Eklund. 1997. Purification of single-wall carbon nanotubes by microfiltration. *J. Phys. Chem. B* 101: 8839–8842.

Bartolotti, L. J. 1982. Time-dependent Kohn-Sham density-functional theory. *Phys. Rev. A* 26: 2243–2244.

Becke, A. D. 1992. Density-functional thermochemistry. I. The effect of the exchange-only gradient correction. *J. Chem. Phys.* 96: 2155–2160.

Becke, A. D. 1993. Density-functional thermochemistry. III. The role of exact exchange. *J. Chem. Phys.* 98: 5648–5652.

Bernholc, J., D. Brenner, M. B. Nardelli, V. Meunier, and C. Roland. 2002. Mechanical and electrical properties of nanotubes. *Annu. Rev. Mater. Res.* 32: 347–375.

Biel, B., F. J. Garcia-Vidal, A. Rubio, and F. Flores. 2008. *Ab initio* study of transport properties in defected carbon nanotubes: An O(N) approach. *J. Phys.: Condens. Matter* 20: 294214-1–294214-8.

Blake, R., Y. K. Gun'ko, J. N. Coleman, M. Cadek, A. Fonseca, J. B. Nagy, and W. J. Blau. 2004. A generic organometallic approach toward ultra-strong carbon nanotube polymer composites. *J. Am. Chem. Soc.* 126: 10226–10227.

Burdick, W. R., Y. Saad, L. Kronik, I. Vasiliev, M. Jain, and J. R. Chelikowsky. 2003. Parallel implementation of time-dependent density functional theory. *Comp. Phys. Commun.* 156: 22–42.

Byrne, M. T., W. P. McNamee, and Y. K. Gun'ko. 2008. Chemical functionalization of carbon nanotubes for the mechanical reinforcement of polystyrene composites. *Nanotechnology* 19: 415707-1–415707-8.

Cadek, M., J. N. Coleman, K. P. Ryan, V. Nicolosi, G. Bister, A. Fonseca, J. B. Nagy, K. Szostak, F. Beguin, and W. J. Blau. 2004. Reinforcement of polymers with carbon nanotubes: The role of nanotube surface area. *Nano Lett.* 4: 353–356.

Casida, M. E. 1995. Time-dependent density functional response theory for molecules. In *Recent Advances in Density-Functional Methods, Part I*, D. P. Chong (Ed.), pp. 155–188. Singapore: World Scientific.

Casida, M. E. and D. R. Salahub. 2000. Asymptotic correction approach to improving approximate exchanger-correlation potentials: Time-dependent density-functional theory calculations of molecular excitation spectra. *J. Chem. Phys.* 113: 8918–8935.

Ceperley, D. M. and B. J. Alder. 1980. Ground state of the electron gas by a stochastic method. *Phys. Rev. Lett.* 45: 566–569.

Ceperley, D. M. and B. Bernu. 1988. The calculation of excited state properties with quantum Monte Carlo. *J. Chem. Phys.* 89: 6316–6328.

Chen, J., M. A. Hamon, H. Hu, Y. S. Chen, A. M. Rao, P. C. Eklund, and R. C. Haddon. 1998a. Solution properties of single-walled carbon nanotubes. *Science* 282: 95–98.

Chen, Y., R. C. Haddon, S. Fang, A. M. Rao, P. C. Eklund, W. H. Lee, E. C. Dickey et al. 1998b. Chemical attachment of organic functional groups to a single-walled carbon nanotube material. *J. Mater. Res.* 13: 2423–2431.

Chen, R. J., Y. Zhang, D. Wang, and H. Dai. 2001. Noncovalent sidewall functionalization of single-walled carbon nanotubes for protein immobilization. *J. Am. Chem. Soc.* 123: 3838–3839.

Cho, E., S. Shin, and Y. G. Yoon. 2008. First-principles studies on carbon nanotubes functionalized with azomethine ylides. *J. Phys. Chem. C* 112: 11667–11672.

Coleman, J. N., W. J. Blau, A. B. Dalton, E. Munoz, S. Collins, B. G. Kim, J. Razal, M. Selvidge, G. Vieiro, and R. H. Baughman. 2003. Improving the mechanical properties of single-walled carbon nanotube sheets by intercalation of polymeric adhesives. *Appl. Phys. Lett.* 82: 1682–1684.

Curran, S. A., P. M. Ajayan, W. J. Blau, D. L. Carroll, J. N. Coleman, A. B. Dalton, A. P. Davey et al. 1998. A composite from poly(m-phenylenevinylene-co-2,5-dioctoxy-p-phenylenevinylene) and carbon nanotubes: A novel material for molecular optoelectronics. *Adv. Mater.* 10: 1091–1093.

Curran, S. A., A. V. Ellis, A. Vijayaraghavan, and P. M. Ajayan. 2004. Functionalization of carbon nanotubes using phenosafranin. *J. Chem. Phys.* 120: 4886–4889.

Curran, S. A., J. Cech, D. Zhang, J. L. Dewald, A. Avadhanula, M. Kandadai, and S. Roth. 2006a. Thiolation of carbon nanotubes and sidewall functionalization. *J. Mater. Res.* 21: 1012–1018.

Curran, S. A., D. Zhang, W. T. Wondmaqegn, A. V. Ellis, J. Cech, S. Roth, and D. L. Carroll. 2006b. Dynamic electrical properties of polymer-carbon nanotube composites: Enhancement through covalent bonding. *J. Mater. Res.* 21: 1071–1077.

Czerw, R., Z. Guo, P. M. Ajayan, Y. P. Sun, and D. L. Carroll. 2001. Organization of polymers onto carbon nanotubes: A route to nanoscale assembly. *Nano Lett.* 1: 423–427.

Dai, L. and A. W. H. Mau. 2001. Controlled synthesis and modification of carbon nanotubes and C_{60}: Carbon nanostructures for advanced polymeric composite materials. *Adv. Mater.* 13: 899–913.

Dalton, A. B., C. Stephan, J. N. Coleman, B. McCarthy, P. M. Ajayan, S. Lefrant, P. Bernier, W. J. Blau, and H. J. Byrne. 2000. Selective interaction of a semiconjugated organic polymer with single-wall nanotubes. *J. Phys. Chem.* 104: 10012–10016.

Deb, B. M. and S. K. Ghosh. 1982. Schrodinger fluid dynamics of many-electron systems in a time-dependent density-functional framework. *J. Chem. Phys.* 77: 342–348.

de la Torre, G., W. Blau, and T. Torres. 2003. A survey on the functionalization of single-walled nanotubes. The chemical attachment of phthalocyanide moieties. *Nanotechnology* 14: 765–771.

Derycke, V., R. Martel, J. Appenzeller, and Ph. Avouris. 2001. Carbon nanotube inter- and intramolecular logic gates. *Nano Lett.* 1: 453–456.

Dresselhaus, M. S., G. Dresselhaus, and Ph. Avouris. 2001. *Carbon Nanotubes: Synthesis, Structure, Properties, and Applications.* Berlin, Germany: Springer-Verlag.

Duesberg, G. S., M. Burghard, J. Muster, G. Philipp, and S. Roth. 1998. Separation of carbon nanotubes by size exclusion chromatography. *Chem. Commun.* 3: 435–436.

Erlanger, B. F., B. X. Chen, M. Zhu, and L. Brus. 2001. Binding of an anti-fullerene IgG monoclonal antibody to single wall carbon nanotubes. *Nano Lett.* 1: 465–467.

Fagan, S. B., R. Mota, R. J. Baierle, A. J. R. da Silva, and A. Fazzio. 2003. *Ab initio* study of an organic molecule interacting with a silicon-doped carbon nanotube. *Diam. Relat. Mater.* 12: 861–863.

Froudakis, G. E. 2002. Hydrogen interaction with carbon nanotubes: A review of *ab initio* studies. *J. Phys.: Condens. Matter* 14: R453–R465.

Georgakilas, V., K. Kordatos, M. Prato, D. M. Guldi, M. Holzinger, and A. Hirsch. 2002. Organic functionalization of carbon nanotubes. *J. Am. Chem. Soc.* 124: 760–761.

Gross, E. K. U. and W. Kohn. 1985. Local density-functional theory of frequency-dependent linear response. *Phys. Rev. Lett.* 55: 2850–2852.

Gross, E. K. U., J. F. Dobson, and M. Petersilka. 1996. Density functional theory of time-dependent phenomena. In *Density Functional Theory*, R. F. Nalewajski (Ed.), pp. 81–172. Berlin, Germany: Springer-Verlag.

Gulseren, O., T. Yildirim, and S. Ciraci. 2002. Systematic *ab initio* study of curvature effects in carbon nanotubes. *Phys. Rev. B* 65: 153405-1–153405-4.

Gustafsson, G., Y. Cao, G. M. Treacy, F. Klavetter, N. Colaneri, and A. J. Heeger. 1992. Flexible light-emitting diodes made from soluble conducting polymers. *Nature* 357: 477–479.

Hammer, B., K. W. Jacobsen, and J. K. Norskov. 1993. Role of nonlocal exchange correlation in activated adsorption. *Phys. Rev. Lett.* 70: 3971–3974.

Hamon, M. A., J. Chen, H. Hu, Y. Chen, M. E. Itkis, A. M. Rao, P. C. Eklund, and R. C. Haddon. 1999. Dissolution of single-walled carbon nanotubes. *Adv. Mater.* 11: 834–840.

Hamon, M. A., H. Hu, P. Bhowmik, S. Niyogi, B. Zhao, M. E. Itkis, and R. C. Haddon. 2001. End-group and defect analysis of soluble single-walled carbon nanotubes. *Chem. Phys. Lett.* 347: 8–12.

Hedin, L. 1965. New method for calculating the one-particle Green's function with application to the electron gas problem. *Phys. Rev.* 139: A796–A823.

Hohenberg, P. and W. Kohn. 1964. Inhomogeneous electron gas. *Phys. Rev.* 136: B864–B871.

Holzinger, M., O. Vostrowsky, A. Hirsch, F. Hennrich, M. Kappes, R. Weiss, and F. Jellen. 2001. Sidewall functionalization of carbon nanotubes. *Angew. Chem. Int. Ed. Engl.* 40: 4002–4005.

Jhi, S. H., S. G. Louie, and M. L. Cohen. 2000. Electronic properties of oxidized carbon nanotubes. *Phys. Rev. Lett.* 85: 1710–1713.

Jhi, S. H., S. G. Louie, and M. L. Cohen. 2002. Electronic properties of bromine-doped carbon nanotubes. *Solid State Commun.* 123: 495–499.

Kang, H. S. 2004. Organic functionalization of sidewall of carbon nanotubes. *J. Chem. Phys.* 121: 6967–6971.

Kelley, T. W., P.F. Baude, C. Gerlach, D. E. Ender, D. Muyres, M. A. Haase, D. E. Vogel, and S. D. Theiss. 2004. Recent progress in organic electronics: Materials, devices, and processes. *Chem. Mater.* 16: 4413–4422.

Kim, J. S., R. H. Friend, F. Cacialli, R. Daik, and W. J. Feast. 1999. Built-in field electroabsorption spectroscopy of polymer light-emitting diodes incorporating a doped poly(3,4-ethylene dioxythiophene) hole injection layer. *Appl. Phys. Lett.* 75: 1679–1681.

Kleinman, L. and D. M. Bylander. 1982. Efficacious form for model pseudopotentials. *Phys. Rev. Lett.* 48: 1425–1428.

Kohn, W. and L. J. Sham. 1965. Self-consistent equations including exchange and correlation effects. *Phys. Rev.* 140: A1133–A1138.

Kong, J. and H. Dai, 2001. Full and modulated chemical gating of individual carbon nanotubes by organic amine compounds. *J. Phys. Chem. B* 105: 2890–2893.

Kooi, S. E., U. Schlecht, M. Burghard, and K. Kern. 2002. Electrochemical modification of single carbon nanotubes. *Angew. Chem. Int. Ed. Engl.* 41: 1353–1355.

Krasheninnikov, A. V., K. Nordlund, P. O. Lehtinen, A. S. Foster, A. Ayuela, and R. M. Nieminen. 2004. Adsorption and migration of carbon adatoms on carbon nanotubes: Density-functional *ab initio* and tight-binding studies. *Phys. Rev. B* 69: 073402-1–073402-4.

Kronik, L., A. Makmal, M. L. Tiago, M. M. G. Alemany, M. Jain, X. Y. Huang, Y. Saad, and J. R. Chelikowsky. 2006. PARSEC – the pseudopotential algorithm for real-space electronic structure calculations: Recent advances and novel applications to nano-structures. *Phys. Stat. Solidi (B)* 243: 1063–1079.

Kulkarni, A. P., C. J. Tonzola, A. Babel, and S. A. Jenekhe. 2004. Electron transport materials for organic light-emitting diodes. *Chem. Mater.* 16: 4556–4573.

Li, Y. H., C. Xu, B. Wei, X. Zhang, M. Zheng, D. Wu, and P. M. Ajayan. 2002. Self-organized ribbons of aligned carbon nanotubes. *Chem. Mater.* 14: 483–485.

Liu, J., A. G. Rinzler, H. J. Dai, J. H. Hafner, R. K. Bradley, P. J. Boul, A. Lu et al. 1998. Fullerene pipes. *Science* 280: 1253–1256.

Long, Y. Z., Z. J. Chen, X. T. Zhang, J. Zhang, and Z. F. Liu. 2004. Electrical properties of multi-walled carbon nanotube/polypyrrole nanocables: Percolation-dominated conductivity. *J. Phys. D* 37: 1965–1969.

Lordi, V., N. Yao, and J. Wei. 2001. Method for supporting platinum on single-walled carbon nanotubes for a selective hydrogenation catalyst. *Chem. Mater.* 13: 733–737.

Lou, L., L. Osterlung, and B. Hellsing. 2000. Electronic structure and kinetics of K on graphite. *J. Chem. Phys.* 112: 4788–4796.

Louie, S. G., S. Froyen, and M. L. Cohen. 1982. Nonlinear ionic pseudopotentials in spin-density-functional calculations. *Phys. Rev. B* 26: 1738–1742.

Lugo-Solis, A. and I. Vasiliev. 2007. *Ab initio* study of K adsorption on graphene and carbon nanotubes: Role of long-range ionic forces. *Phys. Rev. B* 76: 235431-1–235431-8.

Machon, M., S. Reich, C. Thomsen, D. Sanchez-Portal, and P. Ordejon. 2002. *Ab initio* calculations of the optical properties of 4-angstrom-diameter single-walled nanotubes. *Phys. Rev. B* 66: 155410-1–155410-5.

Marinopoulos, A. G., L. Reining, A. Rubio, and N. Vast. 2003. Optical and loss spectra of carbon nanotubes: Depolarization effects and intertube interactions. *Phys. Rev. Lett.* 91: 046402-1–046402-4.

Marinopoulos, A. G., L. Wirtz, A. Marini, V. Olevano, A. Rubio, and L. Reining. 2004. Optical absorption and electron energy loss spectra of carbon and boron nitride nanotubes: A first-principles approach. *Appl. Phys. A* 78: 1157–1167.

Meunier, V., S. V. Kalinin, and B. G. Sumpter. 2007. Nonvolatile memory elements based on the intercalation of organic molecules inside carbon nanotubes. *Phys. Rev. Lett.* 98: 056401-1–056401-4.

Mickelson, E. T., C. B. Huffman, A. G. Rinzler, R. E. Smalley, R. H. Hauge, and J. L. Margrave. 1998. Fluorination of single-wall carbon nanotubes. *Chem. Phys. Lett.* 296: 188–194.

Mickelson, E. T., I. W. Chiang, J. L. Zimmerman, P. J. Boul, J. Lozano, J. Liu, R. E. Smalley, R. H. Hauge, and J. L. Margrave. 1999. Solvation of fluorinated single-wall carbon nanotubes in alcohol solvents. *J. Phys. Chem. B* 103: 4318–4322.

Miyamoto, Y., A. Rubio, and D. Tomanek. 2006. Real-time *ab initio* simulations of excited carrier dynamics in carbon nanotubes. *Phys. Rev. Lett.* 97: 126104-1–126104-4.

Mori, T., K. Miyachi, and T. Mizutani. 1995. A study of the electroluminescence process of an organic electroluminescence diode with an Alq3 emission layer using a dye-doping method. *J. Phys. D* 28: 1461–1467.

Mureau, N., P. C. P. Watts, Y. Tison, and S. R. P. Silva. 2008. Bulk electrical properties of single-walled carbon nanotubes immobilized by dielectrophoresis: Evidence of metallic or semiconductor behavior. *Electrophoresis* 29: 2266–2271.

Naito, K., J. M. Yang, Y. Tanaka, and Y. Kagawa. 2008. Tensile properties of carbon nanotubes grown on ultrahigh strength polyacrylonitrile-based and ultrahigh modulus pitch-based carbon fibers. *Appl. Phys. Lett.* 92: 231912-1-231912-3.

Nevidomskyy, A. H., G. Csanyi, and M. C. Payne. 2003. Chemically active substitutional nitrogen impurity in carbon nanotubes. *Phys. Rev. Lett.* 91: 105502-1–105502-4.

Palacios, J. J., A. J. Perez-Jimenez, E. Louis, E. San Fabian, and J. A. Verges. 2003. First-principles phase-coherent transport in metallic nanotubes with realistic contacts. *Phys. Rev. Lett.* 90: 106801-1–106801-4.

Parr, R. G. and W. Wang. 1989. *Density-Functional Theory of Atoms and Molecules*. New York: Oxford University Press.

Pekker, S., J. P. Salvetat, E. Jakab, J. M. Bonard, and L. Forro. 2001. Hydrogenation of carbon nanotubes and graphite in liquid ammonia. *J. Phys. Chem. B* 105: 7938–7943.

Perdew, J. P. and A. Zunger. 1981. Self-Interaction correction to density-functional approximations for many-electron systems. *Phys. Rev. B* 23: 5048–5079.

Perdew, J. P., K. Burke, and M. Ernzerhof. 1996. Generalized gradient approximation made simple. *Phys. Rev. Lett.* 77: 3865–3868.

Perdew, J. P. and Y. Wang. 1986. Accurate and simple density functional for the electronic exchange energy: Generalized gradient approximation. *Phys. Rev. B* 33: 8800–8802.

Qi, Z. M., M. Wei, I. Honma, and H. Zhou. 2007. Thin films composed of multiwalled carbon nanotubes, gold nanoparticles and myoglobin for humidity detection at room temperature. *Chem. Phys. Chem.* 8: 264–269.

Rao, C. N. R., B. C. Satishkumar, A. Govindaraj, and M. Nath. 2001. Nanotubes. *Chem. Phys. Chem.* 2: 78–105.

Reich, S., C. Thomsen, and P. Ordejon. 2002. Electronic band structure of isolated and bundled carbon nanotubes. *Phys. Rev. B* 65: 155411-1–155411-11.

Ricca, A. and C. W. Bauschlicher. 2006. The adsorption of NO_2 on (9,0) and (10,0) carbon nanotubes. *Chem. Phys.* 323: 511–518.

Rossato, J., R. J. Baierle, A. Fazzio, and R. Mota. 2005. Vacancy formation process in carbon nanotubes: First-principles approach. *Nano Lett.* 5: 197–200.

Runge, E. and E. K. U. Gross. 1984. Density-functional theory for time-dependent systems. *Phys. Rev. Lett.* 52: 997–1000.

Sano, M., A. Kamino, J. Okamura, and S. Shinkai. 2001. Self-organization of PEO-graft-single-walled carbon nanotubes in solutions and Langmuir-Blodgett films. *Langmuir* 17: 5125–5128.

Saunders, V. R. and J. H. van Lenthe. 1983. The direct CI method. A detailed analysis. *Mol. Phys.* 48: 923–954.

Sham, L. J. and T. M. Rice. 1966. Many-particle derivation of the effective-mass equation for the Wannier exciton. *Phys. Rev.* 144: 708–714.

Sheats, J. R. 2004. Manufacturing and commercialization issues in organic electronics. *J. Mater. Res.* 19: 1974–1989.

Shen, J. D., W. S. Huang, L. P. Wu, Y. Z. Hu, and M. X. Ye. 2007. Study on amino-functionalized multiwalled carbon nanotubes. *Mater. Sci. Eng. A* 464: 151–156.

Soler, J. M., E. Artacho, J. D. Gale, A. Garca, J. Junquera, P. Ordejon, and D. Sanchez-Portal. 2002. The Siesta method for *ab initio* order-N materials simulation. *J. Phys.: Condens. Matter* 14: 2745–2779.

Star, A., J. F. Stoddart, D. Steuerman, M. Diehl, A. Boukai, E. W. Wong, X. Yang, S. W. Chung, H. Choi, and J. R. Heath. 2001. Preparation and properties of polymer-wrapped single-walled carbon nanotubes. *Angew. Chem. Int. Ed. Engl.* 40: 1721–1725.

Stokes, P. and S. I. Khondaker. 2008. Controlled fabrication of single electron transistors from single-walled carbon nanotubes. *Appl. Phys. Lett.* 92: 262107-1–262107-3.

Stone, A. J. and D. J. Wales. 1986. Theoretical studies of icosahedral C_{60} and some related species. *Chem Phys. Lett.* 128: 501–503.

Sudalai, A., S. Kanagasabapathy, and B. C. Benicewicz. 2000. Phosphorus pentasulfide: A mild and versatile catalyst/reagent for the preparation of dithiocarboxylic esters. *Org. Lett.* 2: 3213–3216.

Sun, X. Y. and Y. G. Sun. 2008. Single-walled carbon nanotubes for flexible electronics and sensors. *J. Mater. Sci. Technol.* 24: 569–577.

Sun, Y., S. R. Wilson, and D. I. Schuster. 2001. High dissolution and strong light emission of carbon nanotubes in aromatic amine solvents. *J. Am. Chem. Soc.* 123: 5348–5349.

Tang, C. W., S. A. van Slyke, and C. H. Chen. 1989. Electroluminescence of doped organic thin films. *J. Appl. Phys.* 65: 3610–3616.

Tans, S. J., A. R. M. Verschueren, and C. Dekker. 1998. Room-temperature transistor based on a single carbon nanotube. *Nature* 393: 49–52.

Telling, R. H., C. P. Ewels, A. A. El-Barbary, and M. I. Heggie. 2003. Wigner defects bridge the graphite gap. *Nat. Mater.* 2: 333–337.

Troullier, N. and J. L. Martins. 1991. Efficient pseudopotentials for plane-wave calculations. *Phys. Rev. B* 43: 1993–2006.

van Leeuwen, R. and E. J. Baerends. 1994. Exchange-correlation potential with correct asymptotic behavior. *Phys. Rev. A* 49: 2421–2431.

Vasiliev, I. and S. A. Curran. 2006. *Ab initio* study of the self-assembly of phenosafranin to carbon nanotubes. *Phys. Rev. B* 73: 165420-1–165420-5.

Vasiliev, I. and S. A. Curran. 2007. Cross-linking of thiolated carbon nanotubes: An *ab initio* study. *J. Appl. Phys.* 102: 024317-1–024317-5.

Vasiliev, I. and R. M. Martin. 2004. Time-dependent density-functional calculations with asymptotically correct exchange-correlation potentials. *Phys. Rev. A* 69: 052508-1–052508-10.

Vasiliev, I., S. Öğüt, and J. R. Chelikowsky. 1999. *Ab initio* excitation spectra and collective electronic response in atoms and clusters. *Phys. Rev. Lett.* 82: 1919–1922.

Vasiliev, I., S. Öğüt, and J. R. Chelikowsky. 2001. *Ab initio* absorption spectra and optical gaps in nanocrystalline silicon. *Phys. Rev. Lett.* 86: 1813–1816.

Vasiliev, I., S. Öğüt, and J. R. Chelikowsky. 2002. First-principles density-functional calculations for optical spectra of clusters and nanocrystals. *Phys. Rev. B* 65: 115416-1–115416-18.

Verissimo-Alves, M., B. Koiller, H. Chacham, and R. B. Capaz. 2003. Electromechanical effects in carbon nanotubes: *Ab initio* and analytical tight-binding calculations. *Phys. Rev. B* 67: 161401-1–161401-4.

Wang, C., G. Zhou, H. Liu, J. Wu, Y. Qiu, B. L. Gu, and W. Duan. 2006a. Chemical functionalization of carbon nanotubes by carboxyl groups on Stone-Wales defects: A density functional theory study. *J. Phys. Chem. B* 110: 10266–10271.

Wang, C., G. Zhou, J. Wu, B. L. Gu, and W. Duan. 2006b. Effects of vacancy-carboxyl pair functionalization on electronic properties of carbon nanotubes. *Appl. Phys. Lett.* 89: 173130-1–173130-3.

Wei, Z., M. Kondratenko, L. H. Dao, and D. F. Perepichka. 2006. Rectifying diodes from asymmetrically functionalized single-wall carbon nanotubes. *J. Am. Chem. Soc.* 128: 3134–3135.

Zhao, J., A. Buldum, J. Han, and J. P. Lu. 2000. First-principles study of Li-intercalated carbon nanotube ropes. *Phys. Rev. Lett.* 85: 1706–1709.

10

Carbon Nanotube Y-Junctions

Prabhakar R. Bandaru
University of California, San Diego

10.1 Introduction

In recent years, carbon nanotubes (CNTs) have emerged as one of the foremost manifestations of nanotechnology, and extensive research has been expended in probing their various properties. While many desirable attributes in terms of electrical, mechanical, and biological properties have been attributed to CNTs, many obstacles remain before their widespread, practical application (Baughman et al. 2002) becomes feasible. Some of the foremost hindrances are (1) the variation of properties from one nanotube to another, partly due to the unpredictability in synthesis and the random occurrence of defects, and (2) the lack of a tangible method for wide-scale synthesis. Generally, the variation of properties is a natural consequence of nanoscale structures and could be difficult to solve, at least in the short term. It would then seem that fundamentally new ideas might be needed. Some interesting viewpoints are also being considered, where defect manipulation could be used on purpose (Nichols et al. 2007). Wide-scale synthesis methods, for example, by aligning the nanotubes with the underlying crystal orientation (Kang et al. 2007) have recently proved successful, but it is still not clear as to whether such methods would allow for practical implementation, say on the scale of silicon microelectronics.

In this context, it would be pertinent to pause and consider the rationale for the use of nanotubes, especially in the context of electronic characteristics and devices. Carbon-based nanoelectronic technologies (McEuen 1998) promise greater flexibility compared to conventional silicon electronics, one example being the extraordinarily large variety of carbon-based organic

structures. It would then be interesting to look into the possible implications of this "large variety," particularly with respect to the morphology and its associated properties. Such an outlook gives rise to the possibility of examining nonlinear forms, some examples of which are depicted in Figure 10.1. While Y-junction constituted CNTs can be used as three-terminal switching devices or diodes, alternate forms such as helical nanostructures can give rise to nanoscale inductors or, more interestingly, a sequence of metallic and semiconducting junctions (Castrucci et al. 2004). It is to be noted at the outset that we now seek to explore completely novel forms of electronics as laid out, for example, in the International Technology Roadmap for Semiconductors (ITRS) recommendations on *Emerging Research Devices*. We quote "The dimensional scaling of CMOS devices and process technology, as it is known today, will become much more difficult as the industry approaches 16 nm (6 nm physical channel length) around the year 2019 and will eventually approach an asymptotic end. Beyond this period of traditional CMOS scaling, it may be possible to continue functional scaling by integrating alternative electronic devices onto a silicon platform. These alternative electronic devices include 1D structures such as *CNTs....*" The purpose of investigating novel nanotube morphologies is to demonstrate many of the "compelling attributes" as laid out in the ITRS roadmap, which include (1) "room temperature operation," (2) "functionally scalable by orders of magnitude," and (3) "energy dissipation per functional operation substantially less than CMOS." We will show specifically how the exploration of Y-junction topologies would help in laying the foundation for an entirely new class of electronic and optical devices.

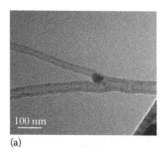 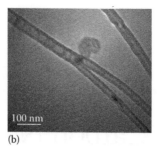 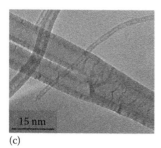

(a)   (b)   (c)

FIGURE 10.1 The nano-engineering of CNTs to produce nonlinear structures is manifested most clearly through the Y-junction morphology. Such structures can be prepared by adding carbide-forming elements, such as Ti, Zr, and Hf to ferrocene-based thermal CVD at different branching angles [i.e., (a) vs. (b)] and spatial locations of catalyst particles [i.e., (a) vs. (c)] through varying the growth conditions.

10.1.1 Carbon Nanotubes

We commence with an overview of the underlying constituents of Y-shaped structures—linear nanotubes. Several comprehensive expositions (Dai 2002) of the fundamental aspects are extant in literature (Ajayan 1999, Dai 2002, McEuen et al. 2002, Dresselhaus et al. 2004, Bandaru 2007). Nanotubes are essentially graphene sheets rolled up into varying diameters (Saito et al. 1998) and are attractive from both a scientific and a technological perspective, as they are extremely robust (elastic modulus approaching 1 TPa) and, at least in pristine forms, chemically inert. By varying the nature of wrapping of a planar graphene sheet and consequently their diameter, nanotubes can be constructed to be either semiconductors or metals (Yao et al. 1999), which can be used in electronics (Collins and Avouris 2000). In the literature, there are a variety of tubular structures composed of carbon that are referred to as nanotubes (single-walled nanotubes [SWNTs] and multi-walled nanotubes [MWNTs]) when the graphene walls are parallel to the axis of the tube, and as nanofibers for other configurations, e.g., where the graphene sheets are at an angle to the tube axis.

The electrical and thermal conductivity (Hone et al. 2000) properties of both SWNTs (Tans et al. 1997) and MWNTs have been well explored. While SWNTs (diameter ~1 nm) can be described as quantum wires due to the ballistic nature of electron transport (White and Todorov 1998), the transport in MWNTs (with a diameter in the range 10–100 nm) is found to be diffusive/quasi-ballistic (Delaney et al. 1999, Buitelaar et al. 2002). Quantum dots can be formed in both SWNTs (Bockrath et al. 1997) and MWNTs (Buitelaar et al. 2002) and the Coulomb blockade and the quantization of the electron states can be used to fabricate single-electron transistors (Tans et al. 1998). Several electronic components, based on CNTs, such as single-electron transistors (Tans et al. 1998, Freitag et al. 2001, Postma et al. 2001), nonvolatile random-access memory (Rueckes et al. 2000, Radosavljevic et al. 2002), field-effect transistors (FET) (Radosavljevic et al. 2002), and logic circuits (Bachtold et al. 2001, Martel et al. 2002, Javey et al. 2003), have also been fabricated. However, most of these devices use conventional lithography schemes and electronics principles, either using nanotubes as conducting wires or modifying them along their length, say

through atomic force microscopy (AFM)-based techniques (Postma et al. 2001). While extremely important in elucidating fundamental properties, the above experiments have used external electrodes, made through conventional lithographic processes, to contact the nanotubes and do not represent truly nanoelectronic circuits. Additionally, the well-known metal oxide semiconductor field-effect transistor (MOSFET) architecture is used, where the nanotube serves as the channel between the electrodes (source and drain), and an SiO_2/Si-based gate modulates the channel conductance. In other demonstrations, cumbersome AFM manipulations (Postma et al. 2001) were needed.

It would, therefore, be more attractive to propose new nanoelectronic elements to harness new functionalities peculiar to novel CNT forms such as nanotubes with bends, Y-junctions (Bandaru et al. 2005), etc. One can also envision a more ambitious scheme and circuit topology where both interconnect and circuit elements are all based on nanotubes, realizing true nanoelectronics (Figure 10.2). For example, the nanotube-based interconnect does not suffer from the problems of electro-migration that plague copper-based lines, due to the strong carbon–carbon bonds, and can support higher current densities (Collins et al. 2001b) (~10 μA/nm^2 or 10^9 A/cm^2 vs. 10 nA/nm^2 or 10^6 A/cm^2 for noble metals such as Ag). Additionally, the predicted large thermal conductivity (Kim et al. 2001) (~3000 W/mK at 300 K), up to an order of a magnitude higher than copper, could help alleviate the problem of heat dissipation in ever-shrinking devices. Developing nanotube-based devices, besides miniaturization and lower power consumption, could also allow us to exploit the advantages of inherently quantum mechanical systems for practical devices, such as ballistic transport and low switching voltages (Wesstrom 1999) (~26 mV at room temperature $\equiv k_BT/e$).

10.1.2 Branched Carbon Nanostructures—Initial Work

At the very outset, any deviation from linearity, say in a branched Y-junction, must be accompanied by the disruption of the regular hexagonal motif. This can be accomplished, in the simplest case, by the introduction of pentagons and heptagons to account for the curvature (Iijima et al. 1992) (Figure 10.3a). Since the charge

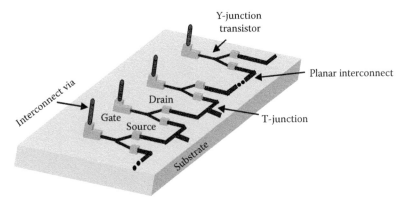

FIGURE 10.2 A conceptual view of a possible CNT technology platform, including Y-junction devices, interconnect vias, and directed nanotube growth. The overall objective is to create nano devices with novel functionalities that go beyond existing technologies.

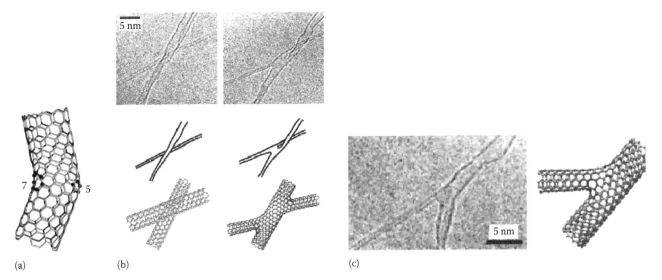

FIGURE 10.3 (a) A bend in a nanotube, say in a Y-junction, introduces regions of positive and negative curvature. The associated heptagons and pentagons can have local excess/deficit of charge and can be used as scattering centers for nanoelectronics. (From Yao, Z. et al., *Nature*, 402, 273, 1999. With permission.) (b) X-shaped and (c) Y-shaped nanotube molecular junctions can be fabricated by irradiating crossed SWNT junctions with high energy (~1.25 MeV) and beam intensity (10 A/cm²) electron beams. (From Terrones, M. et al., *Phys. Rev. Lett.*, 89, 075505, 2002. With permission.)

distribution is likely to be nonuniform in these regions, the interesting possibility of localized scattering centers can be introduced. For example, rectification behavior was posited due to different work functions of contacts with respect to metal (M) and semiconductor (S) nanotubes (electrostatic doping) on either side of the bend (Yao et al. 1999). Later in this chapter, we will discuss how this can be exploited for more interesting device electronics.

Nanotube junctions were formed through the use of high-energy (1.25 MeV) electron beam exposure (in a transmission electron microscope)–based welding of linear SWNTs, at high temperature (800°C) to form X-, Y-, or T-junctions as illustrated in Figure 10.3b and c (Terrones et al. 2002). The underlying mechanism invoked was primarily the "knock-off" of carbon atoms and *in situ* annealing. Molecular dynamics simulations intimate that vacancies and interstitials play a role. However, the purposeful synthesis of branched morphologies can be accomplished through more conventional chemical vapor deposition (CVD) methods.

Preliminary work on individual Y-junctions, grown through CVD, in branched nano-channel alumina templates (Li et al. 2001) resulted in the observation of nonlinear *I–V* characteristics at room temperature through Ohmic contact (Papadapoulos et al. 2000) and tunneling conductance (Satishkumar et al. 2000) measurements. From an innate synthesis point of view, nanotubes with T-, Y-, L-junctions, and more complex junctions (resembling those in Figure 10.1), were initially observed in arc-discharge produced nanotubes (Zhou and Seraphin 1995). Beginning in 2000, there was a spate of publications reporting on the synthesis of Y-junctions through the use of organometallic precursors, such as nickelocene and thiophene, in CVD (Papadapoulos et al. 2000, Satishkumar et al. 2000). Y-junction (Li et al. 2001) and multi-junction carbon nanotube networks (Ting and Chang 2002) were also synthesized through the pyrolysis of methane over cocatalysts and through the growth on roughened Si substrates. However, the mechanism of growth was not probed into adequately.

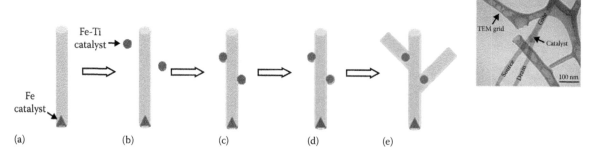

FIGURE 10.4 The postulated growth sequence of a Y-junction nanotube (Gothard et al. 2004) involves (a) initial seeding of a straight nanotube through conventional catalytic synthesis (Teo et al. 2004), (b) Ti-doped Fe catalyst particles (from ferrocene and $C_{10}H_{10}N_4Ti$) attach (c) to the sidewalls and nucleate (d) the side branches (e).

10.2 Controlled Carbon Nanotube Y-Junction Synthesis

It is to be noted at the very outset that the Y-junctions synthesized are quite different in form and structure compared to crossed nanotube junctions (Fuhrer et al. 2000, Terrones et al. 2002), where the nanotubes are individually placed and where the junctions are produced through electron irradiation (Terrones et al. 2002). Significant control in the growth of Y-junction nanotubes (Gothard et al. 2004) on bare quartz or SiO_2/Si substrates through thermal CVD was accomplished through the addition of Ti-containing precursor gases to the usual nanotube growth mixture. In one instance, a mixture of ferrocene ($C_{10}H_{10}Fe$), xylene ($C_{10}H_{10}$), and a Ti-containing precursor gas-$C_{10}H_{10}N_4Ti$ was decomposed at 750°C in the presence of flowing argon (~600 sccm) and hydrogen (75 sccm) carrier gases. The two-stage CVD reactor consisted of (1) a low temperature (~200°C) preheating chamber for the liquid mixture vaporization followed by (2) a high temperature (~750°C) main reactor. A yield of 90% MWNT Y-junction nanotubes, which grew spontaneously on quartz substrates in the main reactor, was obtained. The mechanism for the Y-junction growth was hypothesized to depend on the carbide-forming ability of Ti as measured by its large heat of formation (ΔH_f of −22 Kcal/g-atom). The Ti-containing Fe catalyst particles seed nanotube nucleation by a *root growth* method, in which carbon was absorbed at the root and then ejected to form vertically aligned MWNTs (Figure 10.4a). As the supply of Ti-containing Fe catalyst particles continues (Figure 10.4b), some of the particles (Fe-Ti) attach onto the sidewalls of the growing nanotubes (Figure 10.4c). The catalysts on the side then promote the growth of a side branch (Figure 10.4d), which when further enhanced forms a full-fledged Y-junction. The correlation of the carbide-forming ability to branch formation was also supported by Y-junction synthesis in Hf-, Zr-, and Mo-doped (Choi and Choi 2005) Fe catalyst particles (Gothard et al. 2004), which also have large ΔH_f (HfC: −26 Kcal/g-atom and ZrC: −23 Kcal/g-atom). It was generally found that the use of Zr and Hf catalysts yields larger diameter Y-junctions.

The ratio of the Ti-precursor gas and the feedstock gases could be adjusted to determine the growth of the side-branches at specific positions (Figure 10.5). For example, a decreased flow of the xylene gas, at a point in time, would halt the growth of the nanotube while preponderance of the Fe-Ti precursor gas/catalyst particles would nucleate the branch. The Y-junction formation has also been found to be sensitive to temperature, time, and catalyst concentration. The optimal temperature range is

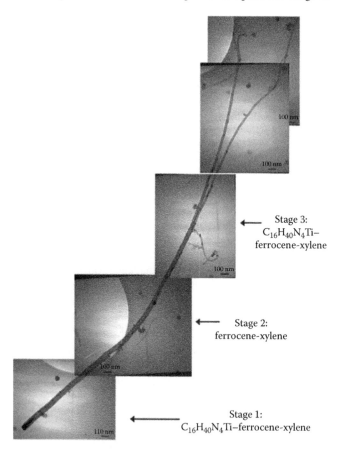

FIGURE 10.5 Controlled addition of Ti (see Stage 3) can induce branching in linear nanotubes. (From Gothard, N. et al., *Nano Lett.*, 4, 213, 2004. With permission.)

between 750°C and 850°C; below 750°C, the yield is very low and temperatures greater than 850°C produce V-shaped nanotube junctions. Y-junction CNTs with minimal defects, at the junction region, were obtained when the atomic compositions of Fe:Ti:C were in the ratio of 1:3:96.

The growth of the Y-junctions essentially seems to be a nonequilibrium phenomenon and various other methods have been found to be successful in proliferating branches, such as the sudden reduction of temperature during a normal tip growth process (Teo et al. 2004), where over-saturation by the carbon feedstock gas causes a surface-energy driven splitting of the catalyst particle and branch nucleation. Other catalyst particles, such as Ca and Si, have also been found to nucleate side branches (Li et al. 2001). The location of the junction is in any case a point of structural variation (Ting and Chang 2002), the control of which seems to determine the formation of Y-junctions and its subsequent properties. It is also possible to undertake a rigorous thermodynamic analysis (Bandaru et al. 2007) to rationalize the growth of nonlinear forms.

10.3 Electrical Characterization of Y-Junction Morphologies

The initial research has focused mainly on the electrical characterization of the nanostructures. In a large part, it was motivated by the possible influence of topology on electrical transport, and supported through theoretical predictions of SWNT junctions. Current theoretical explanations of electrical behavior in Y-junctions are mainly based on SWNT Y-junctions, and the experimental demonstrations detailed in this chapter were made on both SWNT and MWNT Y-junctions (Papadapoulos et al. 2000, Satishkumar et al. 2000). It is speculated that relatively low temperature CVD methods that have been used to date may not be adequate to reliably produce SWNT-based Y-junctions synthesized through high energy electron beam welding (Terrones et al. 2002).

While SWNTs have been extensively studied theoretically (Ajayan 1999, Dai 2001), MWNTs have been relatively less scrutinized. An extensive characterization of their properties is found in literature (Forro and Schonenberger 2001, Bandaru 2007). MWNTs are generally found to be metal-like (Forro and Schonenberger 2001) with possibly different chiralities for the constituent nanotubes. Currently, there is some understanding of transport in *straight* MWNTs, where it has been shown that electronic conduction mostly occurs through the outermost wall, (Bachtold et al. 1999) and inter-layer charge transport in the MWNT is dominated by thermally excited carriers (Tsukagoshi et al. 2004). While the outer wall dominates in the low-bias regime (<50 mV), at a higher bias, many shells can contribute to the conductance with an average current carrying capacity of 12 μA/shell at room temperature (Collins et al. 2001b). In contrast to SWNTs with μm coherence lengths, the transport in MWNTs is quasi-ballistic (Buitelaar et al. 2002) with mean free paths <100 nm. Based on the above survey of properties in *straight* MWNTs, we could extend the hypothesis in that noncoherent electronic transport dominates the Y-junctions and other branched morphologies.

10.4 Carrier Transport in Y-Junction–Electron Momentum Engineering

The progenitor of a Y-junction topology, for electronic applications, was basically derived from an electron-wave Y-branch switch (YBS; Palm and Thylen 1992) where a refractive index change of either branch through an electric field modulation can affect switching. This device, demonstrated in the GaAs/AlGaAs (Worschech et al. 2001) and InP/InGaAs (Hieke and Ulfward 2000, Lewen et al. 2002)-based two-dimensional electron gas (2-DEG) system, relies on ballistic transport and was proposed for low power, ultra-fast (THz) signal processing. It was derived theoretically (Wesstrom 1999) and proven experimentally (Shorubalko et al. 2003) that based on the ballistic electron transport, nonlinear and diode-like *I–V* characteristics were possible. These devices based on III-V materials, while providing proof of concept, were fabricated through conventional lithography. It was also shown in 2-DEG geometry (Song et al. 1998) with artificially constructed defects/barriers, that the defect topology can affect the electron momentum and guide the current to a predetermined spatial location independent of input current direction. This type of rectification involves a new principle of *electron momentum engineering* in contrast to the well-known *band engineering*. Nanotubes provide a more natural avenue to explore such rectification behavior. It was theoretically postulated (Andriotis et al. 2001) that switching and rectification (Figure 10.6) could be observed in symmetric (e.g., no change in chirality from stem to branch) Y-junction SWNTs, assuming quantum conductivity of electrons where the rectification could be determined (Andriotis et al. 2003) by (1) formation of a quantum dot/asymmetric scattering center (Song et al. 1998) at the location of the Y-junction, (2) finite length of the stem and branches connected to metallic leads, (3) asymmetry of the bias applied/the potential profile (Tian et al. 1998) across the nanotube, (4) strength of the nanotube–metal lead interactions, and influence of the interface (Meunier et al. 2002). Some of the possibilities, where the nature of the individual Y-junction branches determines the electrical transport, are illustrated in Figure 10.7.

10.5 Applications of Y-CNTs to Novel Electronic Functionality

Generally, the motivation for use of new CNT morphologies, such as Y-junctions, based on either SWNTs or MWNTs, in addition to the miniaturization of electronic circuits, is the possible exploration of new devices and technologies through new physical principles. The existence of negative-curvature fullerene-based units (Scuseria 1992), and branching in nanotubes necessitates

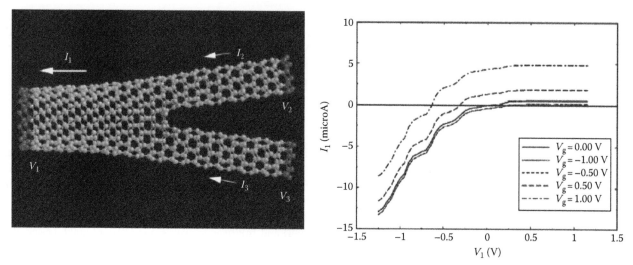

FIGURE 10.6 Asymmetry, and rectification like behavior, in the *I–V* characteristics of a single walled Y-junction nanotube is indicated, through quantum conductivity calculations. (From Andriotis, A.N. et al., *Phys. Rev. Lett.*, 87, 066802, 2001. With permission.)

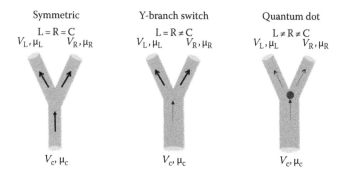

FIGURE 10.7 The CNT Y-junction as a prototypical *structural* element for a variety of functions such as switching or as a quantum dot, depending on the characteristics of the individual branches. L, R, and C, refer to the left, right, and central/stem branches of a Y-junction.

the presence of topological defects—in the form of pentagons, heptagons, and octagons—at the junction regions for maintaining a low energy sp^2 configuration (Andriotis et al. 2002). These *intrinsic* defects are natural scattering centers that could affect/modulate the electrical transport characteristics of a nanotube.

At the nanometer scale, the dimensions of the device are also comparable with the electron wavelength (λ_F) and the electron travel/current must be considered in terms of wave propagation (Davies 1998), analogous to the propagation of light down an optical fiber. Wave phenomena, such as interference and phase shifting, can now be used to construct new types of devices. For example, constructive and destructive interferences can be used to cause transmission and reflection of current leading to switching and transistor-like applications with the added advantage of very low power dissipation. Novel applications have been proposed theoretically (Xu 2002b, Csontos and Xu 2003) for ballistic nano-junctions, of which the Y-junction is only one example. Several of these have been demonstrated in preliminary experiments and will be elucidated later in the chapter. A brief overview follows.

10.5.1 Switching and Transistor Applications

In a basic Y-junction switch, an electric field can direct electrons into either of two branches, while the other branch is cut off (Wesstrom 1999). It has been shown, in computer simulations (Palm and Thylen 1992), that a sufficient lateral field for electron deflection is created by applying a very small voltage of the order of millivolts. The specific advantage of a Y-junction switch is that it does not need single-mode electron waveguides for its operation and can function over a wide range of electron velocities and energies, the reason being that the electrons are not stopped by a barrier but only deflected. An operational advantage over a conventional FET could be that the current is switched between two outputs rather than completely turned on/off (Palm and Thylen 1996), leading to higher efficiency of operation.

An electrical asymmetry can also be induced through structural or chemical means across the two branches in a nanostructured junction. The Y-junction region, for instance, can possess a positive charge (Andriotis et al. 2001) due to two reasons, viz., the presence of (1) topological defects, due to the formation of non-hexagonal polygons at the junction to satisfy the local bond order (Crespi 1998), where delocalization of the electrons over an extended area leads to a net positive charge, and (2) catalyst particles, which are inevitably present during synthesis (Gothard et al. 2004, Teo et al. 2004). This positive charge and the induced asymmetry is analogous to a "gating" action that could be responsible for rectification. While the presence of defects at the junction seems to assist switching, there is also a possibility that such defects may not be needed as some instances of novel switching behavior in Y-junctions are observed in the noticeable absence of catalyst particles. Additional studies are necessary to elucidate this aspect, but such an observation is significant in that a three-dimensional array of Y-junction devices based on CNTs would be much easier to fabricate if a particle is not always required at the junction region.

10.5.2 Rectification and Logic Function

It is possible to design logic circuitry, based on electron wave guiding in Y-junction nanotubes (Xu 2002b), to perform operations similar to and exceeding the performance of conventional electronic devices (Palm and Thylen 1996). When finite voltages are applied to the left and the right branches of a Y-junction, in a push-pull fashion (i.e., $V_{left} = -V_{right}$ or vice versa), the voltage output at the stem would have the same sign as the terminal with the lower voltage. This dependence follows from the principle of continuity of electro-chemical potential ($\mu = -eV$) in electron transport through a Y-junction and forms the basis for the realization of an AND logic gate, i.e., when either of the branch voltages is negative (say, corresponding to a logic state of 0), the voltage at the stem is negative and positive voltage (logic state of 1) at the stem is obtained only when both the branches are at positive biases. The change of μ is also not completely balanced out due to the scattering at the junction, and results in nonlinear interaction of the currents from the left and the right sides (Shorubalko et al. 2003). To compensate, the resultant center branch voltage (V_S) is always negative and varies quadratically (as V^2) with the applied voltage.

10.5.3 Harmonic Generation/Frequency Mixing

The nonlinear interaction of the currents and the V^2 dependence of the output voltage at the junction region also suggest the possibility of higher frequency/harmonic generation. When an AC signal of frequency ω, $V_{L-R} = A \cos[\omega t]$, is applied between the left (L) and right (R) branches of the Y-junction, the output signal from the stem (V_S) would be of the form:

$$V_S = \mathbf{a} + \mathbf{b} \cos[2\omega t] + \mathbf{c} \cos[4\omega t]$$

where **a**, **b**, and **c** are constants.

The Y-junction can then be used for second and higher harmonic generation or for frequency mixing (Lewen et al. 2002). The second harmonic (2ω) output is orthogonal to the input voltage and can be easily separated out. These devices can also be used for an ultrasensitive power meter, as the output is linearly proportional to V^2

to very small values of *V*. A planar CNT Y-junction, with contacts present only at the terminals, suffers from less parasitic effects than a vertical transistor structure and high frequency operation—up to 50 GHz at room temperature (Song et al. 2001) is possible. It can be seen from the brief discussion above that several novel devices can be constructed on CNT Y-junction technology, which could be the forerunner of a new paradigm in nanoelectronics.

10.6 Experimental Work on Electrical Characterization

Compared to the large body of work on electrical transport through linear nanotubes, the characterization of nonlinear nanotubes is still in its infancy. The samples for electrical measurements are typically prepared by suspending nanotube Y-junctions, say, in isopropanol and depositing them on an SiO_2/Si substrate with patterned Au pads. Y-junctions, in proximity to the Au contact pads, are then located at low voltages (<5 kV) using a scanning electron microscope (SEM) and contacts are patterned to each branch of the Y-junction, either through electron-beam lithography (Kim et al. 2006) or focused ion beam induced metal deposition (Gopal et al. 2004), see Figure 10.8. In the latter case, special care needs to be taken not to expose the nanotube to the ion-beam, to prevent radiation damage. The early measurements in Y-junction nanotubes explored the theoretical idea of rectification between any two branches of the Y-junction.

10.6.1 Rectification Characteristics

The initial research in the measurement of the current–voltage (I–V) characteristics of the Y-junctions was accomplished through two-terminal measurements using the stem as one terminal and the two branches, connected together, as the other terminal. With a stem/branch diameter ratio of approximately 60:40 nm, diode-like behavior was observed (Papadapoulos et al. 2000). The authors ascribing SWNT-like p-type semiconductor characteristics, modeled the behavior on a p–p isotype heterojunction where the concentration (N) of carriers (holes) varied inversely as the fourth power of the diameter (D) i.e.,

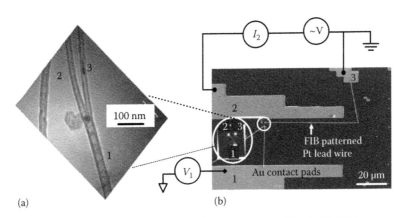

FIGURE 10.8 The (a) MWNT Y-junction electrical measurement configuration as imaged in the (b) SEM.

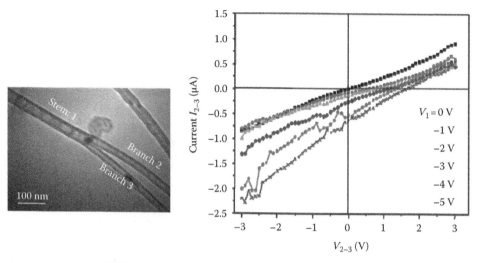

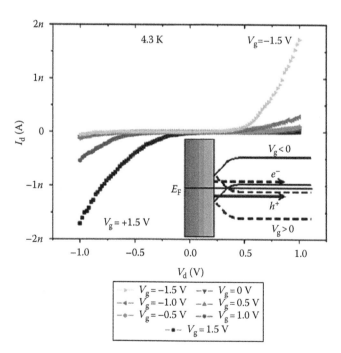

FIGURE 10.9 *I–V* characteristics of an MWNT Y-junction. A constant DC voltage is applied on the stem, while the *I–V* behavior across branches 2 and 3 are monitored. The gating action of the stem voltage (V_1) and the asymmetric response are to be noted.

$N_{stem}/N_{branch} \sim (D_{branch}/D_{stem})^4$. The doping mismatch would then presumably account for the rectifying behavior. In the absence of further characterization, it is difficult to see how semiconducting characteristics could be assigned to >40 nm MWNTs. A subsequent publication from the same group (Perkins et al. 2005) ascribed the electrical characteristics to be dominated by activated conduction, presumably through carrier hopping. More complex nonlinear, quasi-diode-like behavior has been seen (Bandaru et al. 2005) in MWNT Y-junctions (Figure 10.9), which corresponds to a saturation of current at positive bias polarities. This is predicted from theoretical considerations (Xu 2001) where the current cannot decrease beyond a certain value, but saturates at a value (~0.5 μA at positive V_{1-2}) corresponding to the intrinsic potential of the junction region itself. Such behavior has also been seen in Y-junctions fabricated from 2–DEG systems (Shorubalko et al. 2003, Wallin et al. 2006). The underlying rationale, in more detail, is as follows: the necessity of maintaining a uniform electrochemical potential in the overall structure gives rise to nonuniform/nonlinear interactions (Xu 2002a). For example, when V_2 (the voltage on branch 2) decreases, μ_2 (= $-eV_2$) increases, and an excess electron current flows towards the central junction. The balance between the *incoming* current and the *outgoing* currents, at the junction itself, is achieved by increasing μ_1 (decreasing V_1). On the other hand, when μ_2 decreases, μ_1 decreases also, but cannot decrease past a certain critical point, viz., the fixed electrochemical potential dictated by the geometry/defects of the junction.

Work on SWNT-based Y-junctions has revealed the diode-like behavior in richer detail. It was established through Raman spectroscopy analyses (Choi and Choi 2005) that constituent Y-junction branches could be either metallic/semiconducting. This gives rise to the possibility of forming intrinsic metal (M)-semiconductor (S) junctions within/at the Y-junction region. Experiments carried out on SWNTs in the 2–5 nm diameter range (Choi and Choi 2005) (which could, depending on the

chirality, correspond to either metallic or semiconducting nanotubes with energy gaps in the range of 0.37–0.17 eV (Ding et al. 2002)) reveal ambipolar behavior where carrier transport due to both electrons and holes could be important. Considering, for example, that the Fermi level (E_F) for the metallic branch of the Y-junction is midway through the semiconductor branch band gap, positive or negative biases on the semiconductor branch can induce electron/hole tunneling from the metal (Figure 10.10).

FIGURE 10.10 *I–V* measurements on single walled Y-CNTs, where a metallic CNT interfaces with a semiconducting CNT (see band diagram in inset) indicates ambipolar behavior, as a function of applied gate voltage (on the semiconducting nanotube). (From Kim, D.-H. et al., *Nano Lett.*, 6, 2821, 2006. With permission.)

Such temperature-independent tunneling behavior was invoked through modeling the *I–V* characteristics to be of the Fowler–Nordheim type. However, at higher temperatures (>100 K), thermionic emission corresponding to barrier heights of 0.11 eV could better explain the electrical transport results.

On an interesting note, it should be mentioned that the contact resistance at a branch–metal contact could also play a major role and contribute to the relatively low currents observed in experiments (Meunier et al. 2002). While MWNTs should theoretically have a resistance smaller than h/e^2 (~26 kΩ), ideal, and reproducible Ohmic contacts, through metal evaporation, have been difficult to achieve.

10.6.2 Electrical Switching Behavior

Intriguing experiments have brought forth the possibility of using the CNT Y-junctions for switching applications as an electrical inverter analogous to earlier (Palm and Thylen 1992, Hieke and Ulfward 2000, Shorubalko et al. 2003) Y-switch studies in 2–DEG systems. In this measurement (Bandaru et al. 2005), a DC voltage was applied on one branch of the Y-junction while the current through the other two-branches was probed under a small AC bias voltage (<0.1 V). As the DC bias voltage is increased, at a certain point, the Y-junction goes from nominally conducting to a "pinched-off" state. This switching behavior was observed for all three branches of the Y-junction, at different DC bias voltages. The absolute value of the voltage at which the channel is pinched off is similar for two branches (~2.7 V, as seen in Figure 10.11a and b) and is different for the third stem branch (~5.8 V, as seen in Figure 10.11c). The switching behavior was seen over a wide range of frequencies, up to 50 kHz. The upper limit was set by the capacitive response of the Y-junction when the branch current tends to zero.

The detailed nature of the electrical switching behavior is currently not understood. The presence of catalyst nano-particles (Figure 10.1) in the conduction paths could blockade current flow, and their charging could account for the abrupt drop-off of the current. The exact magnitude of the switching voltage would then be related to the exact size of the nanoparticle, which suggests the possibility of nano-engineering the Y-junction to get a variety of switching behaviors. However, it was deduced (Andriotis and Menon 2006) through tight binding molecular dynamics simulations on SWNTs, that interference effects could be solely responsible for the switching behavior, even in the absence of catalyst particles. An associated possibility is that there is inter-mixing of the currents in the Y-junction, where the

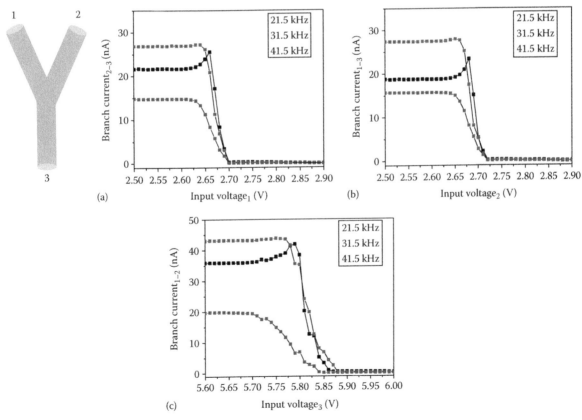

FIGURE 10.11 An abrupt modulation of the current through two branches of the Y-junction, indicative of electrical switching, is seen on varying the voltage on the third branch. The voltage, at which the switching action occurs, on the two branches (1 & 2) is similar and smaller (~2.7 V, see a and b) compared to the turn-off voltage (~5.8 V) on the stem (3) in (c). Such abrupt switching characteristics are seen up to 50 kHz, the upper limit arising from the capacitive response of the Y-junction. (From Bandaru, P.R. et al., *Nature Mater.*, 4, 663, 2005. With permission.)

electron transmission is abruptly cut off due to the compensation of currents, for example, the current through branches 2 and 3 is canceled by current leakage through stem 1. The simultaneous presence of an AC voltage on the source–drain channel and a DC voltage on the control/gate terminal could also result in an abrupt turn-off, due to defect mediated negative capacitance effects (Beale and Mackay 1992). Further research is needed to clarify the exact mechanisms in these interesting phenomena.

A suggestion was also made that AND logic gate behavior could be observed (Bandaru et al. 2005) in a Y-junction geometry. The continuity of the electro-chemical potential from one branch of the CNT Y-junction to another is the basis for this behavior (see Section 10.5.2 for a more detailed explanation).

10.6.3 Current Blocking Behavior

Other interesting characteristics were seen when the CNT Y-morphologies were *in situ* annealed in the ambient, in a range of temperatures 20°C–400°C and *I–V* curves measured for various configurations of the Y-CNT. The observations are summarized in Figure 10.12 and are fascinating from the point of view of the tunability of electrical characteristics. As the annealing is

continued, the onset of nonlinearity in the *I–V* characteristics is observed (Figure 10.12b). With increasing times, the nonlinearity increases, but is *limited* to one polarity of the voltage. This is reminiscent of diode-like behavior and can be modeled as such (Figure 10.12c and inset). With progressive annealing, it was seen that the current is completely cut-off (<1 pA) and perfect rectification seems to be obtained (Figure 10.12d). A current rectification ratio (I_{ON}/I_{OFF}) of >10⁴ has been calculated, where I_{ON} denotes the current through the Y-device at −3 V while I_{OFF} is the current at +3 V. The transition to linear characteristics was more rapid at elevated temperatures of annealing (>200°C). In yet another set of Y-junction samples, reversible behavior from Ohmic to blocking type was observed (Figure 10.13). Interestingly, it was observed that the current through one set of branches (S1–B3 in Figure 10.13b) was not affected by the annealing, while the blocking voltages differ for the other two configurations (viz., ~0.1 V for S1–B2 and ~2.6 V for B2–B3).

The time-dependent behavior and rectification seems to be both a function of temperature and of time of annealing. The sharp cut-off of the current at positive voltages is also remarkable. (The cut-off of the currents at the positive polarity of the voltage seems to be indicative of hole transport in the Y-junctions, through

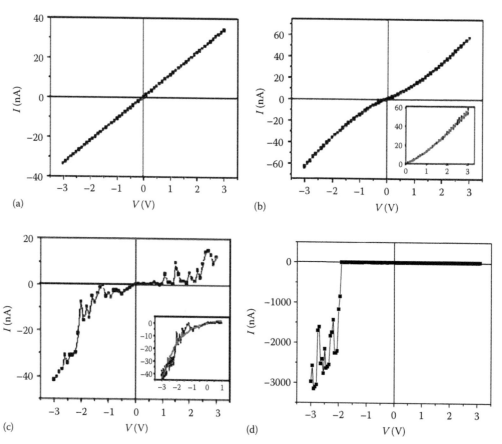

FIGURE 10.12 The *I–V* characteristics for a CNT Y-junction subject to high temperature annealing (~150°C) as a function of annealing time, after being cooled down to room temperature. (a) Prior to annealing Ohmic behavior is observed. At increasing times, nonlinearity is introduced (b) which can be modeled in terms of space charge currents (see inset) as $I = A V + B V^{3/2}$. Further annealing results in a (c) diode-like behavior (the inset shows a fit: $I = I_0(e^{(eV/k_BT)} − 1)$ and finally in a (d) current-blocking/rectification behavior where a 10⁴ fold suppression in current is seen.

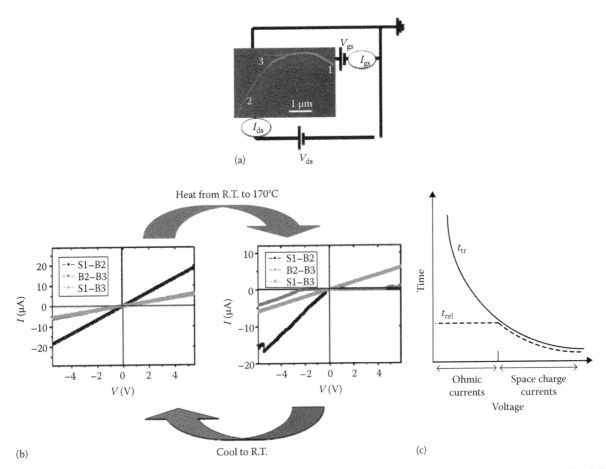

FIGURE 10.13 (a) Circuit arrangement used to probe the *I–V* characteristics of Y-CNT (SEM micrograph). (b) Reversible current blocking behavior induced in the CNT Y-junction. The individual segments' electrical transport characteristics exhibit different blocking/linear characteristics and are geometry dependent. (c) The transition from Ohmic behavior to space charge behavior is a function of voltage and can be accelerated at higher temperatures. The ratio of the transit time (t_{tr})—solid line, to the dielectric relaxation time (t_{rel})—dotted line, determines the dynamics of the carrier transport.

electrostatic doping, due to the work function of Pt, $\phi_{Pt} \sim 5.7\,eV$ being larger than ϕ_{CNT}.) While it is well known that contacts to p-type CNTs through high work-function metals, where $\phi_M > \phi_{CNT}$ ($\sim 4.9\,eV$) result in Ohmic conduction (Javey et al. 2003, Yang et al. 2005), the transition from conducting to blocking behavior is usually more gradual. A continuous change in the current was also seen when a gate voltage modulates the p-CNT channel conduction and in CNT-based p–n junctions (Lee et al. 2004) with diode-like behavior (Lee et al. 2004, Manohara et al. 2005, Yang et al. 2005). Consequently, it is thought that the observed behavior is unique to the CNT branched topologies examined here.

The annealing-induced rectification behavior in CNT Y-junctions could be intrinsic to the Y-nanotube form or be related to the nature of the contacts. It was reported that SWNTs contacted by Pt (Javey et al. 2003) could yield nonmetallic behavior, arising from a discontinuous (Zhang et al. 2000) contact layer to the nanotubes. It is also possible that the outer contacting walls of the MWNT Y-junction are affected by the annealing procedure, resulting in a modification of the Schottky barrier (Collins et al. 2001a). Several CNT device characteristics, such as transistors (Appenzeller et al. 2002) and photodetectors (Freitag et al. 2003), are Schottky barrier mediated. Exposure to

oxygen is also known to affect the density of states of CNTs and the *I–V* characteristics (Collins et al. 2000). However, the low temperatures (<300°C) employed in the experiment preclude oxidation (Ajayan et al. 1993, Collins et al. 2001b), and the blocking behavior that was observed in these experiments cannot be justified on the above principles.

A hint for explaining this intriguing *I–V* behavior is obtained by modeling the *I–V* characteristics of Figure 10.12b, an intermediate stage in the annealing process. A supra-linear behavior, viz., **I** proportional to $A\mathbf{V} + B\mathbf{V}^{3/2}$, can be fitted (Figure 10.12b inset), which could be indicative of space-charge-limited currents (A and B are numerical constants). To further understand the transport behavior, it becomes necessary to examine the role of the contacts in detail. Prior to annealing, linear behavior was observed (Figure 10.12a), i.e., the Ohmic contact is a reservoir of free holes. Generally, the ratio of the hole transit time (t_{tr}) to the dielectric relaxation time (t_{rel}) in the CNT determines the carrier dynamics and currents. A large (t_{tr}/t_{rel}) ratio obtained, say at smaller voltages, would result in Ohmic currents while a lower (t_{tr}/t_{rel}) ratio, at increased voltages, would imply space charge currents (Figure 10.13c). As t_{rel} is inversely proportional (Muller and Kamins 1986) to the electrical conductivity (σ),

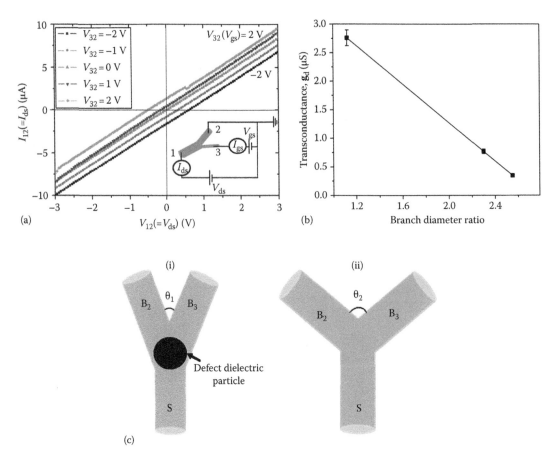

(a)

(b)

(c)

FIGURE 10.14 (a) *I–V* characteristics of a Y-CNT consisting of metallic branches. The circuit diagram is indicated in the inset. (b) The conductance ($\mathbf{g}_d$) through two nanotubes of different diameters in the Y-CNT is inversely proportional to their ratio (= $d_>/d_<$). (c) Proposed Y-junction CNT-based switching devices, (i) where a smaller angle (θ_1) between the branches (B_2 and B_3) can result in a higher gating efficiency for the stem (S). (ii) A Y-CNT with uniform gating/electrical switching characteristics can be fabricated by synthesizing all the constituent nanotubes to be of the same diameter and with $\theta_2 = 120°$. (From Park, J. et al., *Appl. Phys. Lett.*, 88, 243113, 2006. With permission.)

such effects would be more important at elevated temperatures for metallic Y-CNTs. It was then suggested that the saturation of the current at positive voltages could be due to space charge effects. In other words, at sufficiently high voltages, further hole generation is choked due to the preponderance of space charge and negligible current is observed. Note that such a scenario can also be predicted from the band structure diagrams at the metal contact-CNT interface. At forward bias, there is a significant barrier to hole transport and space charge processes could be important. However, at negative voltages, no hole barriers exist and Ohmic conduction is obtained. The space charge hypothesis is also supported by the observation that over a period of time, the blocking behavior reverts to Ohmic conduction. In an earlier study (Antonov and Johnson 1999) of a straight SWNT diode containing a charged impurity, similar space-charge-based arguments were invoked to explain the observed current rectification. Experiments are in progress to probe the time constants and dynamics of current flow through the Y-CNTs.

It is speculated that the different diameters of the branches of the Y-junction (Figure 10.1 or Figure 10.13a) could be playing a

role in the current blocking behavior. The basis is the observation (Figure 10.13b) where rectification is not observed for a particular (S1–B3) two-terminal configuration. It was found that the magnitude of the conductance ($\mathbf{g}_d$) between any two terminals of the Y-junction is *inversely* proportional to the ratio of the diameters through which the current is flowing (see Figure 10.13b), i.e., a larger discrepancy between the branch diameters lowers the conductance. In an analogy to fluid flow through a pipe, the hole current can similarly be modulated by the Y-junction geometry. It would be instructive to characterize the blocking behavior systematically as a function of branch diameters in the Y-junction geometry.

The phenomena of current being low when the voltage was high in terms of space charge distribution, was noted by Shockley (1950) who remarked on its presence in the collector region of a junction transistor amplifier. The *I–V* characteristics observed in these CNT Y-junctions could be compared to that of a transit-time device such as a barrier injection and transit time (BARITT) diode (Streetman and Banerjee 2000). Such devices consider space charge limited currents for creating *negative resistance* regions, which can be exploited for switching, oscillation, amplification, and other functions in high-speed circuitry (Sze 1981). The time delay associated with space charge dissipation is useful for generating microwaves and the BARITT devices are consequently used as local oscillators in microwave receivers.

10.6.4 Metallic Y-Junctions

It is also possible that the constituent nanotubes in the Y-junction can exhibit metallic conduction, in which case the behavior is analogous to that of three interconnected metal lines! Such behavior has indeed been observed (Park et al. 2006) (Figure 10.14a). A control voltage (V) is applied on one of the three terminals (gate) of the Y-CNT while the current (I) through the other two terminals (source and drain) is monitored. The I–V curves of the Y-CNTs, as a function of gate (V_{gs})/bias (V_{ds}) voltage, are shown in Figure 10.14a. The notable features are: (a) Ohmic conduction, (b) a proportional displacement of the current with V_{gs} and V_{ds}, and (c) a geometry dependent conductance. The devices are also seen to have both current and voltage gain. In line with an earlier study (Perkins et al. 2005), the differential current gain (g_{diff}) is defined as the ratio of $(\partial I_{gs}/\partial V_{gs})$ to $(\partial I_{ds}/\partial V_{gs})$ at a constant source–drain voltage (V_{ds}). In the Y-CNT devices, a g_{diff} of up to 10 is obtained at room temperature. The voltage gain was calculated (Tans et al. 1998) by considering the voltage change (ΔV_{ds}) for a given increment in gate voltage (ΔV_g), at a constant value of the current. The voltage gain when the stem (1 in Figure 10.13a) was used for electrical gating is ~6 while the value drops to ~0.3 when the branch (3 in Figure 10.13a) was used for a gate.

Other quantities of importance in quantifying the electrical characteristics are the *transconductance*, $\mathbf{g_m} = (\partial I_{ds}/\partial V_{gs})_{V_{ds}}$, the ratio of the output current (I_{ds}) to the modulating/gating voltage (V_{gs}) at a constant bias voltage (V_{ds}), and the *output conductance*, $\mathbf{g_d} = (\partial I_{ds}/\partial V_{ds})_{V_{gs}}$, the ratio of the output current to the bias voltage at a constant gate voltage. It was found that the magnitude of the $\mathbf{g_d}$ is *inversely* proportional to the ratio of the diameters of the branches through which the current is flowing (Figure 10.14b), i.e., a larger discrepancy between the branch diameters lowers the conductance. On the other hand, the $\mathbf{g_m}$, which denotes the gating efficiency for current modulation, seems to be *directly* proportional to the diameter of the gating nanotube (Park et al. 2006).

From the Y-CNT electrical characteristics, it was seen that the mode of operation of the device is quite different from that of a conventional FET and more akin to a Schottky barrier type FET (Muller and Kamins 1986). The current modulation is controlled by the electric field (= $\nabla_r \cdot \mathbf{V}$) at the gating terminal and not by the carrier density in the channel. In this case, the device performance can be improved by increasing the diameter of the gate so as to have a greater modulating effect on the source–drain current. The assumption of electric field control can also be tested (Park et al. 2006) by varying the angle (θ) between the branches of the Y-CNT. From elementary electrostatics (Jackson 1999), the surface-charge density

at a distance, ρ, from the intersection of the gate with the junction varies as $\rho^{(\pi/\theta-1)}$, which implies a greater modulating effect at a sharper angle. As an example, for the Y-CNT in Figure 10.14c(i), the stem would have a higher gating efficiency than for Figure 10.14c(ii). The manipulation of the junction region to be dielectric/insulating (say, by localized ion-implantation) can also be used to change the electrochemical potential and vary device characteristics. It is interesting to note an analogy of Figure 10.14c to a MOSFET, where the stem is analogous to a *gate*, the dielectric particle plays the role of the *gate oxide*, and the *channel* is formed by conduction through the branches. On the other hand, uniform electrical gating/switching characteristics could be obtained by making all the nanotubes the same diameter and length, at an angle of 120° to each other (Figure 10.14c(ii)).

It is also suggested that even more interesting applications are facilitated by such experimental results and observations. For example, a single Y-CNT switching device can be made to have diverse operating characteristics through the use of different gating terminals. Technology based on Y-junctions can be used for devices (1) with multiple programmable characteristics and (2) as components in Field Programmable Gate Arrays (FPGAs) for reconfigurable computing (Rabaey 1996), where it would be possible to dynamically alter circuit paths. The large scale integration density and high speed of signal propagation intrinsic to nanotubes would be advantageous in this regard.

10.7 Topics for Further Investigation

For the novel applications of Y-CNTs to be practicable, it would be important and necessary to gain better control of the geometry, through synthesis or processing, of Y-CNTs. It is plausible that the remarkable features of Y-junction-based transistors such as the abrupt switching (Bandaru et al. 2005) and differential gain (Perkins et al. 2005) observed in earlier studies could be related to the presence and location of defects. *In situ* engineering of CNT morphology, e.g., exposure to intense electron-beam radiation (Terrones et al. 2002) could also be utilized to tailor individual Y-CNT characteristics. Other aspects that merit further study include the following:

a. The characterization of the detailed morphology of the Y-junctions and their effect on the electrical properties. Catalyst particles have been found both at the junction, along the length of the Y-junctions and at the tips (Li et al. 2001) (Figure 10.1). The effect of the location, type, and composition of these particles and their scattering characteristics on the electrical transport would yield insight into their influence.

b. A detailed study of the growth mechanisms of multi-junction carbon nanotubes vis-à-vis the influence of the catalyst particles. It has been contended (Teo et al. 2004) that a perturbation during growth could promote the formation of branches/junctions. For example, if the temperature is reduced during the growth process, a catalyst particle over-saturated with carbon can be induced to nucleate another branch. On the other hand, if the Ti catalyst does play a role, the importance of carbide formers in inducing side-branch growth is interesting from a basic thermodynamic point of view, i.e., is a high negative heat of formation (ΔH_f) of the carbide formers necessary? It is worth noting that Y-junctions and other morphologies, such as H-junctions and T-junctions (Ting and Chang 2002) are also formed when methane (which could yield positive ΔH_f) is used as a precursor. The composition of the catalyst particle, which nucleates the nanotube branches (Teo et al. 2004) and the effects of stress generated at the growing nanotube tips could also be playing important roles (Ting and Chang 2002) in Y-junction growth.

c. The issue of electrical conductivity through nanotubes with bends, junctions, and catalyst particles is also important for the future viability of nanotube-based electronics. While the MWNT has an intrinsic resistance, say due to ballistic transport, a capacitive and inductive component will also have to be considered due to the presence of particles, inter-tube transport etc. Such a study will also yield insight into the speed of operation of the nanotube-based devices and affecting factors, in terms of the RC delay. It has been theoretically proposed (Burke 2003) that *straight* SWNTs are capable of THz operation and can even be used as nano-antennae (Burke 2002) for radiation and detection in these very high frequency ranges. Whether SWNT or MWNT Y-junctions are capable of being operated at such high frequencies due to the presence of defects, and determining an ultimate limit (Guo et al. 2002) to their performance is worth investigating.

d. It has been proposed (Xu 2005) that novel CNT-based circuits based on Y-junctions and branched morphologies can be created. These circuits could be fabricated so as to construct universal logic gates such as NAND/X-NOR. Such a demonstration together with the possibility of multi-functionality, e.g., where the catalyst particle could be used for photon detection (Kosaka et al. 2002) will be important. A variety of such novel circuits can be fabricated leading to a potentially new paradigm for nanoelectronics that goes well beyond traditional FET architecture, yielding shape-controlled logic elements.

e. Assembly into a viable circuit topology and large scale fabrication are issues that are of paramount importance. Currently, the Y-junctions are grown in mats/bundles and individually isolated and measured. A scheme where each junction is assembled in this way is not viable practically. Ideas into self-assembly and controlled placement (Huang et al. 2005) of nanotubes will have to be addressed. While some measure of success has been achieved in coordinating the placement of loose nanotubes, e.g., through the use of chemically functionalized substrates (Liu et al. 1999)

employing dip-pen lithography (Rao et al. 2003), through orienting electric fields (Joselevich and Lieber 2002) and magnetic fields (Hone et al. 2000), using microfluidic arrangements (Huang et al. 2001) etc., such techniques are difficult to scale up and need to be resolved. An array of Y-junctions can be prepared on the same nanotube stem, as was found in preliminary studies, by exposing only periodic locations (Qin et al. 2005) along the length of a nanotube, sputter depositing catalyst, and growing parallel Y-junction branches from this linear array of catalysts using electric field-induced direction control during a subsequent CVD process, e.g., at any angle from the main stem. The above growth technique could be used to make multiple Y-junction devices in series/parallel. This proof of growth will go a long way in demonstrating the feasibility of large scale nanoelectronic device assembly—a question of the highest importance in recent times.

10.8 Conclusions

The study of Y-junctions is still in its infancy. Their synthesis, while reasonably reproducible, is still challenging in terms of the precise placement on a large scale. It is worth mentioning that this is an issue even with linear nanotubes and might well determine the feasibility of CNT applications, in general. Presently, both single-walled and multi-walled Y-CNTs, the constituent branches of which could have different diameters or semiconducting/metallic character, have been synthesized. Such internal diversity gives rise to novel phenomena such as (1) rectification/current blocking behavior, (2) electrical switching, and (3) logic gate characteristics. It is also interesting to investigate whether the three terminals of the Y-CNT can be interfaced into a transistor-like paradigm. Future investigations should correlate the detailed physical structure of the nanostructure morphologies with the electrical measurements to gain a better understanding of the conduction processes vis-à-vis the role of defects and geometry in nonlinear structures. Such a comprehensive and correlated study would be useful to the nanotube/nanowire community and could pave the way to the realization of *shape controlled* nanoelectronic devices exclusive to the nanoscale.

References

Ajayan, P.M., 1999. Nanotubes from carbon. *Chemical Reviews*, 99, 1787–1799.

Ajayan, P.M., Ebbesen, T.W., Ichihashi, T. et al., 1993. Opening carbon nanotubes with oxygen and implications for filling. *Nature*, 362, 522–525.

Andriotis, A.N. and Menon, M., 2006. Are electrical switching and rectification inherent properties of carbon nanotube Y-junctions? *Applied Physics Letters*, 89, 132116.

Andriotis, A.N., Menon, M., Srivastava, D. et al., 2001. Rectification properties of carbon nanotube "Y-junctions". *Physical Review Letters*, 87(6), 066802.

Andriotis, A.N., Menon, M., Srivastava, D. et al., 2002. Transport properties of single-wall carbon nanotube Y-junctions. *Physical Review B*, 65, 165416.

Andriotis, A.N., Srivastava, D., and Menon, M., 2003. Comment on Intrinsic electron transport properties of carbon nanotube Y-junctions. *Applied Physics Letters*, 83, 1674–1675.

Antonov, R.D. and Johnson, A.T., 1999. Subband population in a single-wall carbon nanotube diode. *Physical Review Letters*, 83, 3274–3276.

Appenzeller, J., Knoch, J., Martel, R. et al., 2002. Carbon nanotube electronics. *IEEE Transactions on Nanotechnology*, 1, 184–189.

Bachtold, A., Hadley, P., Nakanishi, T. et al., 2001. Logic circuits with carbon nanotube transistors. *Science*, 294, 1317–1320.

Bachtold, A., Strunk, C., Salvetat, J.-P. et al., 1999. Aharonov-Bohm oscillations in carbon nanotubes. *Nature*, 397, 673–675.

Bandaru, P.R., 2007. Electrical properties and applications of carbon nanotube structures. *Journal of Nanoscience and Nanotechnology*, 7, 1239–1267.

Bandaru, P.R., Daraio, C., Jin, S. et al., 2005. Novel electrical switching behavior and logic in carbon nanotube Y-junctions. *Nature Materials*, 4, 663–666.

Bandaru, P.R., Daraio, C., Yang, K. et al., 2007. A plausible mechanism for the evolution of helical forms in nanostructure growth. *Journal of Applied Physics*, 101, 094307.

Baughman, R.H., Zakhidov, A.A., and De Heer, W.A., 2002. Carbon nanotubes-the route toward applications. *Science*, 297, 787.

Beale, M. and Mackay, P., 1992. The origins and characteristics of negative capacitance in metal-insulator-metal devices. *Philosophical Magazine B*, 65, 47–64.

Bockrath, M., Cobden, D.H., Mceuen, P.L. et al., 1997. Single-electron transport in ropes of carbon nanotubes. *Science*, 275, 1922–1924.

Buitelaar, M.R., Bachtold, A., Nussbaumer, T. et al., 2002. Multiwall carbon nanotubes as quantum dots. *Physical Review Letters*, 88, 156801.

Burke, P.J., 2002. Luttinger liquid theory as a model of the gigahertz electrical properties of carbon nanotubes. *IEEE Transactions on Nanotechnology*, 1, 129–144.

Burke, P.J., 2003. An RF circuit model for carbon nanotubes. *IEEE Transactions on Nanotechnology*, 2(1), 55–58.

Castrucci, P., Scarselli, M., De Crescenzi, M. et al., 2004. Effect of coiling on the electronic properties along single-wall carbon nanotubes. *Applied Physics Letters*, 85, 3857–3859.

Choi, Y.C. and Choi, W., 2005. Synthesis of Y-junction single-wall carbon nanotubes. *Carbon*, 43, 2737–2741.

Collins, P.G., Arnold, M.S., and Avouris, P., 2001a. Engineering carbon nanotubes and nanotube circuits using electrical breakdown. *Science*, 292, 706–709.

Collins, P.G. and Avouris, P., 2000. Nanotubes for electronics. *Scientific American*, December, 283(6), 62–69.

Collins, P.G., Bradley, K., Ishigami, M. et al., 2000. Extreme oxygen sensitivity of electronic properties of carbon nanotubes. *Science*, 287, 180–1804.

Collins, P.G., Hersam, M., Arnold, M. et al., 2001b. Current saturation and electrical breakdown in mutiwalled carbon nanotubes. *Physical Review Letters*, 86, 3128–3131.

Crespi, V.H., 1998. Relations between global and local topology in multiple nanotube junctions. *Physical Review B*, 58, 12671.

Csontos, D. and Xu, H.Q., 2003. Quantum effects in the transport properties of nanoelectronic three-terminal Y-junction devices. *Physical Review B*, 67, 235322.

Dai, H., 2001. Nanotube growth and characterization. In Dresselhaus, M.S., Dresselhaus, G., and Avouris, P. eds., *Topics in Applied Physics*. Berlin: Springer-Verlag, pp. 29–53.

Dai, H., 2002. Carbon nanotubes: From synthesis to integration and properties. *Accounts of Chemical Research*, 35, 1035–1044.

Davies, J.H., 1998. *The Physics of Low-Dimensional Semiconductors*. New York: Cambridge University Press.

Delaney, P., Di Ventra, M., and Pantelides, S., 1999. Quantized conductance of multiwalled carbon nanotubes. *Applied Physics Letters*, 75, 3787–3789.

Ding, J.W., Yan, X.H., and Cao, J.X., 2002. Analytical relation of band gaps to both chirality and diameter of single-wall carbon nanotubes. *Physical Review B: Condensed Matter*, 66, 073401.

Dresselhaus, M.S., Dresselhaus, G., and Jorio, A., 2004. Unusual properties and structure of carbon nanotubes. *Annual Review of Materials Research*, 34, 247–278.

Forro, L. and Schonenberger, C., 2001. Physical properties of multi-wall nanotubes. In Dresselhaus, M.S., Dresselhaus, G., and Avouris, P. eds., *Carbon Nanotubes—Topics in Applied Physics*. Heidelberg: Springer-Verlag.

Freitag, M., Martin, Y., Misewich, J.A. et al., 2003. Photoconductivity of single carbon nanotubes. *Nano Letters*, 3, 1067–1071.

Freitag, M., Radosavljevic, M., Zhou, Y. et al., 2001. Controlled creation of a carbon nanotube diode by a scanned gate. *Applied Physics Letters*, 79, 3326–3328.

Fuhrer, M.S., Nygard, J., Shih, L. et al., 2000. Crossed nanotube junctions. *Science*, 288, 494.

Gopal, V., Radmilovic, V.R., Daraio, C. et al., 2004. Rapid prototyping of site-specific nanocontacts by electron and ion beam assisted direct-write nanolithography. *Nano Letters*, 4, 2059–2063.

Gothard, N., Daraio, C., Gaillard, J. et al., 2004. Controlled growth of Y-junction nanotubes using ti-doped vapor catalyst. *Nano Letters*, 4(2), 213–217.

Guo, J., Datta, S., Lundstrom, M. et al., 2002. Assessment of silicon mos and carbon nanotube fet performance limits using a general theory of ballistic transistors. *International Electron Devices Meeting (IEDM)*, San Francisco, CA, pp. 711–714.

Hieke, K. and Ulfward, M., 2000. Nonlinear operation of the Y-branch switch: Ballistic switching mode at room temperature. *Physical Review B*, 62, 16727–16730.

Hone, J., Laguno, M.C., Nemes, N.M. et al., 2000. Electrical and thermal transport properties of magnetically aligned single wall carbon nanotube films. *Applied Physics Letters*, 77, 666.

Huang, X.M.H., Caldwell, R., Huang, L. et al., 2005. Controlled placement of individual carbon nanotubes. *Nano Letters*, 5, 1515–1518.

Huang, Y., Duan, X., Wei, Q. et al., 2001. Directed assembly of one-dimensional nanostructures into functional networks. *Science*, 291, 630–633.

Iijima, S., Ichihashi, T., and Ando, Y., 1992. Pentagons, heptagons and negative curvature in graphite microtubule growth. *Nature*, 356, 776–778.

Jackson, J.D., 1999. *Classical Electrodynamics*. New York: John Wiley & Sons.

Javey, A., Guo, J., Wang, Q. et al., 2003. Ballistic carbon nanotube field-effect transistors. *Nature*, 424, 654–657.

Joselevich, E. and Lieber, C.M., 2002. Vectorial growth of metallic and semiconducting single-wall carbon nanotubes. *Nano Letters*, 2, 1137–1141.

Kang, S.J., Kocabas, C., Ozel, T. et al., 2007. High-performance electronics using dense, perfectly aligned arrays of single-walled carbon nanotubes. *Nature Nanotechnology*, 2, 230–236.

Kim, D.-H., Huang, J., Shin, H.-K. et al., 2006. Transport phenomena and conduction mechanisms of single walled carbon nanotubes (SWNTs) at Y- and crossed junctions. *Nano Letters*, 6, 2821–2825.

Kim, P., Shi, L., Majumdar, A. et al., 2001. Thermal transport measurements of individual multiwalled nanotubes. *Physical Review Letters*, 87, 215502.

Kosaka, H., Rao, D.S., Robinson, H. et al., 2002. Photoconductance quantization in a single-photon detector. *Physical Review B (Rapid Communications)*, 65, 201307.

Lee, J.U., Gipp, P.P., and Heller, C.M., 2004. Carbon nanotube p-n junction diodes. *Applied Physics Letters*, 85, 145–147.

Lewen, R., Maximov, I., Shorubalko, I. et al., 2002. High frequency characterization of gainas/inp electronic waveguide TBS switch. *Journal of Applied Physics*, 91, 2398–2402.

Li, W.Z., Wen, J.G., and Ren, Z.F., 2001. Straight carbon nanotube Y junctions. *Applied Physics Letters*, 79, 1879–1881.

Liu, J., Casavant, M.J., Cox, M. et al., 1999. Controlled deposition of individual single-walled carbon nanotubes on chemically functionalized templates. *Chemical Physics Letters*, 303, 125–129.

Manohara, H.M., Wong, E.W., Schlecht, E. et al., 2005. Carbon nanotube Schottky diodes using Ti-schottky and Pt-ohmic contacts for high frequency applications. *Nano Letters*, 5, 1469–1474.

Martel, R., Derycke, V., Appenzeller, J. et al., 2002. Carbon nanotube field-effect transistors and logic circuits. *Design Automation Conference, ACM*, New Orleans, LA.

Mceuen, P.L., 1998. Carbon-based electronics. *Nature*, 393, 15–16.

Mceuen, P.L., Fuhrer, M.S., and Park, H., 2002. Single-walled carbon nanotube electronics. *IEEE Transactions on Nanotechnology*, 1, 78–85.

Meunier, V., Nardelli, M.B., Bernholc, J. et al., 2002. Intrinsic electron transport properties of carbon nanotube Y-junctions. *Applied Physics Letters*, 81, 5234–5236.

Muller, R.S. and Kamins, T.I., 1986. *Device Electronics for Integrated Circuits*, 2nd edn. New York: John Wiley & Sons.

Nichols, J.A., Saito, H., Deck, C. et al., 2007. Artificial introduction of defects into vertically aligned multiwall carbon nanotube ensembles: Application to electrochemical sensors. *Journal of Applied Physics*, 102, 064306.

Palm, T. and Thylen, L., 1992. Analysis of an electron-wave Y-branch switch. *Applied Physics Letters*, 60, 237–239.

Palm, T. and Thylen, L., 1996. Designing logic functions using an electron waveguide Y-branch switch. *Journal of Applied Physics*, 79, 8076–8081.

Papadapoulos, C., Rakitin, A., Li, J. et al., 2000. Electronic transport in Y-junction carbon nanotubes. *Physical Review Letters*, 85, 3476–3479.

Park, J., Daraio, C., Jin, S. et al., 2006. Three-way electrical gating characteristics of metallic Y-junction carbon nanotubes. *Applied Physics Letters*, 88, 243113.

Perkins, B.R., Wang, D.P., Soltman, D. et al., 2005. Differential current amplification in three-terminal Y-junction carbon nanotube devices. *Applied Physics Letters*, 87, 123504.

Postma, H.W.C., Teepen, T., Yao, Z. et al., 2001. Carbon nanotube single-electron transistors at room temperature. *Science*, 293, 76–79.

Qin, L., Park, S., Huang, L. et al., 2005. On-wire lithography. *Science*, 309, 113–115.

Rabaey, J., 1996. *Digital Integrated Circuits: A Design Perspective*. New York: Prentice Hall.

Radosavljevic, M., Freitag, M., Thadani, K.V., and Johnson, A.T., 2002. Non-volatile molecular memory elements based on ambipolar nanotube field effect transistors. *Nano Letters*, 2(7), 761–764.

Rao, S.G., Huang, L., Setyawan, W. et al., 2003. Large-scale assembly of carbon nanotubes. *Nature*, 425, 36–37.

Rueckes, T., Kim, K., Joselevich, E. et al., 2000. Carbon nanotube-based nonvolatile random access memory for molecular computing. *Science*, 289, 94–97.

Saito, R., Dresselhaus, G., and Dresselhaus, M.S., 1998. *Physical Properties of Carbon Nanotubes*. London U.K.: Imperial College Press.

Satishkumar, B.C., Thomas, P.J., Govindaraj, A. et al., 2000. Y-junction carbon nanotubes. *Applied Physics Letters*, 77, 2530–2532.

Scuseria, G.E., 1992. Negative curvature and hyperfullerenes. *Chemical Physics Letters*, 195, 534–536.

Shockley, W., 1950. *Electrons and Holes in Semiconductors*. New York: D. Van. Nostrand Company.

Shorubalko, I., Xu, H.Q., Omling, P. et al., 2003. Tunable nonlinear current-voltage characteristics of three-terminal ballistic nanojunctions. *Applied Physics Letters*, 83, 2369–2371.

Song, A.M., Lorke, A., Kriele, A. et al., 1998. Nonlinear electron transport in an asymmetric microjunction: A ballistic rectifier. *Physical Review Letters*, 80, 3831–3834.

Song, A.M., Omling, P., Samuelson, L. et al., 2001. Room-temperature and 50 GHz operation of a functional nanomaterial. *Applied Physics Letters*, 79, 1357–1359.

Streetman, B.G. and Banerjee, S., 2000. *Solid State Electronic Devices*, 5th edn., Upper Saddle River, NJ: Prentice Hall.

Sze, S.M., 1981. *Physics of Semiconductor Devices*, 2nd edn. New York: John Wiley & Sons.

Tans, S.J., Devoret, M.H., Dai, H. et al., 1997. Individual single-wall carbon nanotubes as quantum wires. *Nature*, 386, 474–477.

Tans, S.J., Verschueren, A.R.M., and Dekker, C., 1998. Room-temperature transistor based on a single carbon nanotube. *Nature*, 393, 49–52.

Teo, K.B.K., Singh, C., Chhowalla, M. et al., 2004. Catalytic synthesis of carbon nanotubes and nanofibers. In Nalwa, H.S. ed., *Encyclopedia of Nanoscience and Nanotechnology*. Stevenson Ranch, CA: American Scientific Publishers.

Terrones, M., Banhart, F., Grobert, N. et al., 2002. Molecular junctions by joining single-walled carbon nanotubes. *Physical Review Letters*, 89, 075505.

Tian, W., Datta, S., Hong, S. et al., 1998. Conductance spectra of molecular wires. *Journal of Chemical Physics*, 109, 2874–2882.

Ting, J.-M. and Chang, C.-C., 2002. Multijunction carbon nanotube network. *Applied Physics Letters*, 80, 324–325.

Tsukagoshi, K., Watanabe, E., Yagi, I. et al., 2004. Multiple-layer conduction and scattering property in multi-walled carbon nanotubes. *New Journal of Physics*, 6, 1–13.

Wallin, D., Shorubalko, I., Xu, H.Q., and Cappy, A., 2006. Nonlinear electrical properties of three-terminal junctions. *Applied Physics Letters* 89, 092124.

Wesstrom, J.O., 1999. Self-gating effect in the electron Y-branch switch. *Physical Review Letters*, 82, 2564–2567.

White, C.T. and Todorov, T.N., 1998. Carbon nanotubes as long ballistic conductors. *Nature*, 393, 240.

Worschech, L., Xu, H.Q., Forchel, A. et al., 2001. Bias-voltage-induced asymmetry in nanoelectronic y-branches. *Applied Physics Letters*, 79, 3287–3289.

Xu, H.Q., 2001. Electrical properties of three-terminal ballistic junctions. *Applied Physics Letters*, 78(14), 2064–2066.

Xu, H.Q., 2002a. Diode and transistor behaviors of three-terminal ballistic junctions. *Applied Physics Letters*, 80, 853–855.

Xu, H.Q., 2002b. A novel electrical property of three-terminal ballistic junctions and its applications in nanoelectronics. *Physica E*, 13, 942–945.

Xu, H.Q., 2005. The logical choice for electronics? *Nature Materials*, 4, 649–650.

Yang, M.H., Teo, K.B.K., Milne, W.I. et al., 2005. Carbon nanotube schottky diode and directionally dependent field-effect transistor using asymmetrical contacts. *Applied Physics Letters*, 87, 253116.

Yao, Z., Postma, H.W.C., Balents, L. et al., 1999. Carbon nanotube intramolecular junctions. *Nature*, 402, 273–276.

Zhang, Y., Franklin, N.W., Chen, R.J. et al., 2000. Metal coating on suspended carbon nanotubes and its implication to metal-tube interaction. *Chemical Physics Letters*, 331, 35–41.

Zhou, D. and Seraphin, S., 1995. Complex branching phenomena in the growth of carbon nanotubes. *Chemical Physics Letters*, 238, 286–289.

11

Fluid Flow in Carbon Nanotubes

Max Whitby
Imperial College

and

RGB Research Ltd

Nick Quirke
University College Dublin

11.1 Introduction

Nanofluidics involves the scientific investigation and technical application of fluid flow in systems where a relevant dimension lies between one and a hundred nanometers. This is the conventional definition of the field. In practice, relevant work encompasses flows through somewhat larger channels, up to a few hundred nanometers in a cross section. Research into fluid flow in this size regime has been greatly facilitated in recent years by the availability of nanoscale tubular carbon structures with intrinsically open central pores. Over 100 studies investigating fluid transport involving these materials have been published in the past decade. A number of reviews covering closely related aspects of nanofluidics have recently appeared (Bau et al., 2005; Eijkel, 2005; Nicholson and Quirke, 2006; Noy et al., 2007; Whitby and Quirke, 2007; Abgrall and Nguyen, 2008; Mattia and Gogotsi, 2008; Rasaiah et al., 2008; Schoch et al., 2008). Here, we review selected papers with an emphasis on fluid transport through the central pores of carbon tubes.

Understanding and controlling how fluids behave as they flow through carbon pores is important from multiple perspectives. First, carbon nanotubes and nanopipes provide a versatile platform for conducting experiments at the nanoscale to investigate whether novel and potentially useful physical and physicochemical

phenomena may occur. Techniques exist for fabricating highly aligned nanopipe arrays and carbon membranes with nanoscale pores that facilitate the measurement of the parameters, such as flow rates, by aggregating large numbers of parallel channels. This opens up the possibility of industrial scale applications. It also allows nanofluidic processes to be investigated using standard instruments and benchtop laboratory equipment. In this chapter, we consider the still-emerging evidence from such studies that theoretical models of fluid flow, long established to apply in continuum fluids, require modification when applied to some nanofluidic systems.

Second, carbon nanotubes and nanopipes offer many attractive features for the fabrication of practical nanofluidic devices. Well-developed processes exist for the deterministic placement of tubular structures with morphologies that make them adaptable for fluid transport experimentation (Bau et al., 2004a; Abgrall and Nguyen, 2008). Carbon nanotubes have useful electronic properties that offer potential for sensing and signaling (Martin and Kohli, 2003; Jain, 2005; Mauter and Elimelech, 2008). Ready chemical modification allows for functionalization, leading to the control of surface properties (Rakov, 2006; Hirsch and Vostrowsky, 2007; Herrero and Prato, 2008). And low-friction transport, due to the well-ordered graphitic inner surfaces and the highly curved geometry of molecular nanotubes, offers the prospect of enhanced throughput for materials separation (Supple and Quirke, 2005; Hinds, 2006; Yuan et al., 2007).

Third, carbon nanomaterials have promising characteristics for investigating and intervening in biological systems. In other words, they have important applications in the emerging field of nanomedicine. The biocompatibility of bulk elemental carbon is in principle acceptable, although recently concerns about the potential toxicity of nanocarbons have been raised (Dowling, 2004; Lacerda et al., 2006; Smart et al., 2006; Jain et al., 2007; Bianco et al., 2008; Yu et al., 2008). These concerns include, by analogy, the possibility of cell damage similar to that caused by asbestos fibers, which present a similar size and aspect ratio. Bearing this important caveat in mind, carbon nanotubes and nanopipes are promising tools for direct interaction with living systems at the cellular level. Their diameter is considerably smaller than typical animal cells and so nanopipes are potentially suitable for delivering therapeutic drugs to, and extracting analytic samples from, the living interior cytoplasm (Martin and Kohli, 2003; Gardeniers and Van den Berg, 2004; Sinha et al., 2004; Tegenfeldt et al., 2004; Jain, 2005; Staufer et al., 2007; Liang et al., 2008).

Biology has frequently provided inspiration for engineers at the human scale. Now with the advent of nanoscience and nanoengineering, scientists are examining how nature has evolved nanoscale mechanisms to drive its remarkable processes (Preece and Stoddart, 1994; Jones, 2006). The 2003 Nobel Prize for Chemistry was awarded to Peter Agre and Roderick MacKinnon for the discovery of water channels and for structural and mechanistic studies of ion channels in cell membranes (Agre, 2004). This work is closely related to nanofluidics where the cross-disciplinary collaboration of physicists, chemists, and biologists is leading to a rapid advance in our understanding.

We start this chapter with a detailed consideration of the materials, specifically the different types of carbon nanotubes and nanopipes that can be used to transport fluids at the nanoscale. Next, we reprise the classical theory of fluid flow through pipes and pores and consider the phenomenon of slip flow that becomes increasingly important as the influence of surface interaction becomes dominant at the nanoscale. We also consider the rich literature on the simulation of nanofluidic systems, using techniques such as molecular dynamics (MD), which has pointed the way for the experimentalists. The central sections of this chapter then deal with the key experimental results of recent years: for capillary filling, for wetting at the nanoscale, for fluid flow, and for mechanisms to switch and regulate transport. We consider the controversial evidence for "super flows" through carbon nanotubes and we look at methods that have been proposed to achieve pumping, which is a key requirement for nanofluidic applications. Finally, we examine the current state of nanofluidic device fabrication, looking particularly at nanomedical applications in diagnostics and therapy.

11.2 Materials

Two distinct classes of nanoscale carbon tubes have been used experimentally in nanofluidics research, each with significantly different properties and methods of synthesis. The first type are the well-known carbon nanotubes, both single- and multi-walled, first described in 1952 (Radushkevich and Lukyanovich, 1952; Monthioux and Kuznetsov, 2006) and subsequently identified as having a cylindrical fullerene molecular structure in 1991 (Iijima, 1991). The second class are carbon nanopipes (see Figure 11.1) produced by the chemical vapor deposition (CVD) of amorphous carbon in alumina templates with a honeycomb morphology (Martin, 1994; Kyotani et al., 1995; Masuda and Fukuda, 1995). In the literature, there is a tendency to not always clearly differentiate between the two classes. This can lead to confusion as their physical and chemical characteristics are significantly different.

In this review, the term "nanotube" is used to refer to the molecular carbon structures, i.e., single-walled carbon nanotubes (SWNT) and multi-walled carbon nanotubes (MWNT). The term "nanopipe" is used to describe the carbon pipes produced in templates using CVD. The latter are generally larger and are composed mainly of amorphous rather than well-ordered graphitic carbon (Mattia et al., 2006b).

11.2.1 Carbon Nanotubes

Numerous excellent reviews of molecular carbon nanotubes are available (Dresselhaus et al., 1996; Saito et al., 1998; Harris, 1999; Awasthi et al., 2005; Dai, 2006; Gogotsi, 2006). The key characteristic of carbon nanotubes as compared to carbon nanopipes is that they are composed of rolled cylinders of graphene with each carbon atom sp^2 bonded to three others in a planar hexagonal network. In the case of SWNTs, only a single layer of carbon atoms is involved. They can be conducting (metallic) or semi-conducting, depending on their chirality, i.e., the exact folding vector whereby the graphene network is rolled up to make a seamless closed tube. The outer diameter of SWNTs can be as small as 0.6 nm and as large as ~5 nm with a typical range of 1.0–1.4 nm (Rakov, 2006). Large single-walled tubes tend to kink and collapse making then difficult to use for nanofluidic applications (Galanov et al., 2002). MWNTs have many concentric layers usually in a "Russian Doll" configuration but can also possibly be rolled up like a parchment (Lavin et al., 2002). Arc- and CVD-produced MWNTs are larger than SWNTs with greater rigidity and central pores as wide as 20 nm (Govindaraj and Rao, 2006). Double-walled nanotubes (DWNT) are also known and have an outer diameter in the range of 1.9–5 nm and an inner pore diameter in the range of 1.1–4.2 nm (Hutchison et al., 2001).

From a nanofluidics perspective, an interesting type of MWNT is a less commonly encountered variation produced using hydrothermal synthesis (Gogotsi et al., 2000; Libera and Gogotsi, 2001). Production takes place at a high pressure (60–100 MPa) in polyethylene/water mixtures in the presence of a nickel catalyst. The resulting carbon nanotubes have well-ordered and relatively defect-free graphitic walls, which are typically only 10% of the overall tube width. The hydrothermal method can produce tubes with diameters as large as 800 nm, and the resulting carbon pipes have very wide central pores and are thus suitable for high-throughput nanofluidic applications. They also have the unique characteristic that water is sometimes trapped inside closed tubes during synthesis and its hydrodynamic behavior can then

Mag = 21.53 KX — 2 μm — EHT = 5.00 kV — Signal A = InLens — Date: 30 Jan 2006
WD = 8 mm — Photo No. = 9129 — Time: 16:43:45

FIGURE 11.1 A bundle of amorphous carbon nanopipes produced using chemical vapor deposition from commercially available anodic aluminum oxide (AAO) templates with nominal pore size 200 nm. The template has been etched with NaOH solution to release and expose the nanopipes. Note the branching morphology towards the ends of the nanopipes that reduces their effective inner diameter in this terminating region. This irregularity is an artifact of the electrochemical etching process used to manufacture the AAO template. Caution must be exercised when calculating flow rates based on the diameter of the main pipe sections.

be studied in the transmission electron microscope (TEM) as described in the experimental Section 11.3.3. Although methods for aligning nanotubes post-synthesis have been demonstrated (Kouklin et al., 2005; Li et al., 2006), it is much less straightforward to create aligned arrays of hydrothermal pipes in contrast to the similarly dimensioned templated nanopipes produced using CVD as described in Section 11.2.2.

In practice, several factors may prevent or limit the transport of fluids through the central pores of carbon nanotubes. Structural defects, including dislocations, disclinations, and stacking faults, frequently lead to blocking of the central pore. In SWNTs and smaller MWNTs, an accumulation of minor defects may affect transport even when the channel itself remains partially open (Zimmerli et al., 2005; Rivera et al., 2007; Li et al., 2008). MWNTs sometimes show a so-called bamboo structure in which the inner carbon cylinder is closed with a cap, forming periodic compartments between which fluid cannot be exchanged (Endo et al., 1993).

Similar capping can occur at the ends of both SWNTs and MWNTs, effectively sealing them and again rendering the untreated nanopipes incapable of transport. A number of procedures exist for "opening" carbon nanotubes by selective oxidation, taking advantage of the relatively strained bonds in the 5 and/or 7 member carbon rings that are necessary to achieve closure of the highly curved tube ends (Tsang et al., 1993; Ugarte et al., 1998; Harris, 1999). Simply heating a sample of carbon nanotubes in air until oxidation begins (above ~600°C) can be effective in this regard (Ajayan et al., 1993). So too is prolonged boiling in a concentrated acid with reflux

(Tsang et al., 1994). This latter procedure has the further merit of helping to remove metal catalyst particles (typically nickel, cobalt, and iron) left over from carbon nanotube synthesis. These are often located at the end of carbon nanotubes that are extruded from the molten catalyst particle during the vapor–liquid–solid (VLS) growth process. Hence, the metal particles are liable to block the central pores unless they are removed.

Even with the ends of the SWNTs and MWNTs opened to expose continuous central pores, a number of factors can prevent or restrict fluid transport through nanotubes. An important consideration is the extent to which a given fluid wets the surface of the pore. This has been investigated experimentally and is also the subject of extensive theoretical analysis as further discussed below.

11.2.2 Carbon Nanopipes

The second broad class of nanoscale carbon pipes that are relevant to nanofluidics are produced using CVD in anodic aluminum oxide (AAO) templates. This CVD technique was developed following a broader interest in the use of nanopatterned materials for producing templated nanostructures (Martin, 1994). The method takes advantage of the self-organization of highly ordered nanopores arranged in a dense hexagonal pattern and produced when aluminum foil is double etched in an acid solution as the anode in an electrochemical cell (Itaya et al., 1984; Masuda and Fukuda, 1995; Routkevitch et al., 1996; Masuda et al., 1997). As the metal converts to the oxide, its density decreases, thus creating stress forces that direct the etching centers into an energy-minimizing close-packed hexagonal configuration. The pore diameter and pore separation can be controlled by selecting different acids, by varying the applied potential, and by controlling the temperature and duration of the reaction. AAO templates with monodisperse pore diameters in the range 10–300 nm can be made with densities as high as 10^{11} cm^{-2} (Miller et al., 2001). The length of the pores depends on the thickness of the starting aluminum foil. A typical value is 50 μm with longer pores up to 200 μm having been achieved. A slight tapering of the channel walls due to etching of the side walls during pore formation can be a problematic feature in the experimental context (L. Cagnon, personal communication, 2008).

To line the walls of the AAO nanopores with carbon, the alumina template is placed in a furnace and heated under an inert atmosphere to a temperature between 650°C and 900°C (Kyotani et al., 1995, 1996; Hulteen and Martin, 1997; Che et al., 1998a). Prior annealing of the templates between flat plates of quartz or alumina can be used to prevent curling. A hydrocarbon gas, such as ethylene, is then flowed over the template for several hours, whereupon carbon is catalytically decomposed by the hot alumina and deposited on the surface of the template. Carbon tubes with an outside diameter as small as 9 nm have been formed in similar silica templates (Joo et al., 2001). Using alumina, larger 200 nm nanopipes are easily produced with commercially available AAO filters (Whatman Anodisc). Due to the way they are manufactured, these templates have an irregular branching morphology on one side (see Figure 11.1), this morphology must be considered when interpreting the results of the flow through the tubes produced using them. This is particularly the case with commercial templates having nominal pore sizes <200 nm (Whitby et al., 2008a). Regardless of which templates are used, the deposited carbon wall thickness depends in part on the duration of the CVD process as well as the gas feedstock selected, with 10–20 nm being typical for a 200 nm diameter template and 5–10 nm being typical for sub-200 nm diameter nanopipes (Whitby et al., 2008a).

It is important to appreciate that, as synthesized, the inner walls of carbon pipes produced by the basic AAO templating method just described do not consist of a well-ordered graphene tube as is the case with carbon nanotubes, but rather will be at least partly amorphous. Electron diffraction studies in a TEM produce a pattern characteristic of nonaligned graphite fragments (Che et al., 1998b). There is evidence that high temperature annealing can improve the order of the tube walls through graphitization (Kyotani et al., 1996), which, as discussed later in this chapter, may enhance transport properties. In a recent paper, 200–300 nm CVD carbon nanopipes originally synthesized at a temperature of 670°C were annealed in an inert atmosphere at 2000°C. Contact angles with water were reported to increase from 44° to 77° (Mattia et al., 2006b). In a separate paper, the same authors reported contact angles for a range of polar and nonpolar fluids with amorphous carbon films, produced in the same manner as CVD carbon nanopipes (Mattia et al., 2006a). As produced, templated CVD nanopipes are more hydrophilic than carbon nanotubes. Exposure to the air at room temperature

immediately following CVD leads to the partial oxidation of the carbon surface and by analogy with similar carbon materials and can be expected to yield acidic sites on the inner nanopipe walls and pore entrances (Panzer and Elving, 1975; Garcia et al., 1997; Bismarck et al., 1999).

Carbon nanopipes produced by templates have several important advantages for nanofluidics applications. Pore size, distribution, and nanopipe length can all be well controlled at the template production stage. Wall thickness can be determined by the appropriate adjustment of CVD parameters (Cooper et al., 2004). There are no catalyst particles to block channels. Their central pores are large compared to their outer diameter, giving similar throughput benefits as with hydrothermal MWNTs. And their amorphous composition, while resulting in reduced mechanical strength and lower electrical conductivity, also means it is easier for them to be broken down, which may assist in elimination from the body in a nanomedical context. There is preliminary evidence that carbon nanopipes are less toxic to human cells in vitro compared to MWNTs (Whitby et al., 2008b).

11.2.3 Carbon Nanomembranes and Filters

Nanopipes also possess an immediate advantage that follows from the method of their synthesis: they are highly aligned, replicating precisely the morphology of their parent AAO templates. Furthermore, their pores span this impermeable alumina matrix with fully open entrances on both sides. So, as produced, they constitute a ready-made nanoporous membrane with carbon-lined pores. This makes templated carbon nanopipes highly suitable for investigating fluid flow through nanoscale channels, and several studies using these materials are discussed later in this chapter.

Membranes based on carbon nanotubes have also been produced. The relatively complex steps required for their manufacture are summarized in Section 11.4.2 that discusses the fluid flow experiments that employ them. Also relevant to note in this context are carbon nanopipe filters that have been produced by spray pyrolysis of a ferrocene/benzene mixture in at tube furnace at 900°C (Srivastava et al., 2004). However, in this case, the fluid to be filtered is much more likely to pass *between* the outside surfaces of closely packed aligned nanotubes, rather than *through* their central pores. The same caveat applies to several other recent reports of filtration using carbon nanotubes (Lu et al., 2005; Yan et al., 2006b; Brady-Estevez et al., 2008). A comprehensive review covering the aligned carbon nanotube membranes, in which fluids do flow exclusively through the central pores, is available (Hinds, 2006).

11.2.4 Releasing Carbon Nanopipes

For some experiments and applications, such as delivering drugs and interacting with cells, it is necessary to separate individual carbon nanopipes from their alumina template. Fortunately, this can be achieved very simply by wet chemical etching using an alkali solution such as 2N NaOH, which dissolves Al_2O_3 but leaves the carbon intact (Che et al., 1998a). Mild sonication may be used to speed up the process, which can thereby be accomplished in a few minutes. With sonication at higher power levels, the nanopipes tend to become fragmented into segments, which may be useful for some applications such as drug delivery (Lin, 2007; Whitby et al., 2008b). It should be noted here that treatment with NaOH and other etchants does modify the surface chemistry of the carbon nanopipes, causing them to become more hydrophilic (Mattia et al., 2006a; Yan et al., 2006a). Also worth appreciating is the common observation that isolated carbon nanopipes appear to have a slightly greater diameter, as measured using SEM, than the AAO template channels in which they are formed. This may be due to van der Waals contact forces acting between the carbon nanopipe and the substrate on which it rests during SEM characterization. The adhesion causes an ovoid deformation that leads to an overestimation of the true circular diameter (Sinha et al., 2007).

11.2.5 Chemical Surface Modification

Finally, with regard to controlling the surface properties of both nanotubes and nanopipes, a range of techniques have been developed for their functionalization and chemical modification. Comprehensive reviews of the field are available (Rakov, 2006; Hirsch and Vostrowsky, 2007; Herrero and Prato, 2008). The possibilities include the attachment of oxidic groups ($-COOH$, $-C=O$, and $-OH$) using aqueous solution chemistry, fluorination by direct gaseous exposure, and amidation including processes that involve microwave heating. Reaction with diazonium salts has been shown to be a versatile approach offering a high level of selectivity that can even distinguish metallic from semiconducting SWNTs (Strano et al., 2003). Sonication-assisted reactions and surface modifications with reactive gas plasmas have also been demonstrated (Chirila et al., 2005; Felten et al., 2005). Non-covalent bonding of surfactants to carbon nanotubes has been widely used to improve solubility, although this is already good in the case of alkali-exposed carbon nanopipes (Mattia and Gogotsi, 2006; Lin, 2007).

There is considerable interest in improving the biocompatibility of carbon nanotubes and nanopipes by chemical modification of their surface. The attachment of polyethylene glycol (PEG) groups is one widely used technique (Howard et al., 2008). Success has been reported recently with biomimetic polymers designed to mimic cell surface mucin glycoproteins (Chen et al., 2006).

11.3 Theory of Nanoscale Flow

The pioneers who studied *static* properties of liquids in macroscopic channels were the nineteenth century investigators Laplace, Poisson, and Young (Young, 1805; Poisson, 1831; Laplace, 1878, Rowlinson and Widom, 1982). Steady-state *flow* in a channel with a width, h, of simple incompressible fluids, driven for example by gravity, ρg, or a pressure gradient, dP/dy,

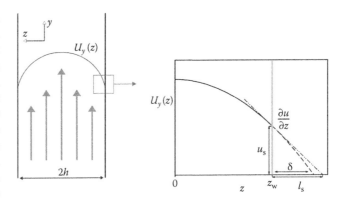

FIGURE 11.2 (Left) Parabolic velocity profile $U_y(z)$ for Poiseuille flow in a slit pore of width $2h$. (Right) Enlarged diagram of fluid-wall interface region as indicated by the square. Solid line is fluid velocity profile for the plane Poiseuille flow with slip at the solid surface; dashed line shows the extrapolated velocity profile with δ the slip length; dashed-dotted line denotes the velocity gradient at the wall. Also shown are slip velocity u_s and slip coefficient l_s. Note some authors (Thomas and MacGaughey, 2008) call the slip coefficient as defined here the slip length and our slip length the excess radius.

can be described by the Navier–Stokes equation (Navier, 1823; Palmer and Rogalski, 1996). The solution is parabolic for velocity in the direction of flow, y, as a function of distance from the wall z (see Figure 11.2):

$$U_y(z) = (\rho g / 2\eta)((\delta + h)^2 - z^2) \tag{11.1}$$

where δ, the slip length, is the distance into the wall at which the velocity extrapolates to zero. Conventionally, the slip length is assumed to be zero. If it is not then we are said to have slip boundary conditions at the wall. Integrating over the flow profile gives the Hagen–Poiseuille law for the flux through a pore.

11.3.1 Hagen–Poiseuille Flow

The equation of flow for incompressible fluids at a constant density and viscosity is the (momentum) Navier–Stokes equation (Navier, 1823; Palmer and Rogalski, 1996) amounting to Newton's second law:

$$\rho dV/dt = \rho F - \nabla P + \eta \nabla^2 V \tag{11.2}$$

where

the velocity V of the fluid is a vector quantity

F is any external field

P is the (scalar) pressure

At the steady-state for no external fields $F = 0$, and for a slit geometry we have (defining the direction of flow as the y direction, there is no pressure gradient in x or z) $V_x = V_z = 0$ and

$$dP/dy = \eta(\partial^2 V_y / \partial x^2 + (\partial^2 V_y \partial z^2)) \tag{11.3}$$

since $\partial^2 V_y / \partial x^2 = 0$, we obtain

$$dP/dy = \eta(\partial^2 V_y / \partial z^2) \tag{11.4}$$

integrating twice with respect to z

$$V_y(z) = (1/\eta)\int (dP/dy)\,dz = (1/\eta)(dP/dy)z^2/2 + cz + d \tag{11.5}$$

where c and d are constants of integration. At $z = \pm h$ (lower wall, upper wall, $H = 2h$), we have $V_y(\pm h) = V_0$ and

$$V_y(z) = (1/(2\eta))(dP/dy)[z^2 - h^2] + V_0 \tag{11.6}$$

For a cylindrical geometry, the problem is solved in cylindrical polar coordinates to obtain (where R is the radius of the pipe)

$$V_y(r) = (1/(4\eta))(dP/dy)[r^2 - R^2] + V_0 \tag{11.7}$$

To calculate the flow rate, we calculate a mean velocity across the pore. Assuming a uniform density over a cross section i.e., $\rho(z) = \rho$, the mean velocity is

$$\overline{V_y} = 1/\pi R^2 \int_0^R V_y(r)2\pi r\,dr = -R^2/(8\eta)(dP/dy) + V_0 \tag{11.8}$$

To obtain the flow rate (Q = the volume of fluid flowing per unit of time), we multiply the mean velocity by the cross-sectional area

$$Q = (\pi/(8\eta))\,|(dP/dy)|[R^4] + \pi R^2 V_0 \tag{11.9}$$

And we know

$$V_0 = (\partial V_y(r)/\partial r)_R l_s \tag{11.10}$$

Hence, if Q_O is the volumetric flow rate with no slip and Q_E is the enhanced volumetric flow rate with slip,

$$Q_E = Q_O\left(1 + \frac{4l_s}{R}\right) \tag{11.11}$$

This is the Hagen–Poiseuille equation describing the pressure-driven flow along a pipe with a circular cross-section modified to include a slip-length term describing non-zero flow rates at the interface between the transported fluid and the wall of the pipe.

The dynamics of capillary non steady-state filling were described by Washburn using the Hagen–Poiseuille law with the driving force for flow given by the Laplace equation for the pressure difference across the invading liquid meniscus (Washburn, 1921; de Gennes et al., 2004). Thus, the penetration length L at time t in a capillary of radius R is given by

$$L^2 = (R\gamma/2\eta)\,t \tag{11.12}$$

where
   γ is the liquid/vapor surface tension
   η is the shear viscosity of the invading liquid

In deriving the Washburn equation, it is assumed that there is no fluid motion at the wall (i.e., non-slip boundary conditions). The Washburn equation is very successful on macroscopic time scales. It breaks down for short times due to the neglect of inertia. A more general solution can be found for non-steady, viscous, incompressible flow in the form of the Bosanquet equation, which tends to the Washburn equation at long times (Bosanquet, 1923).

For nanoscale capillaries, the flow behavior is dominated not by bulk properties (γ, η) but by the interaction of the fluid with the capillary walls. In this regard, it is important to include the role of surface friction due to molecular corrugation. A convenient parameter here is the Maxwell coefficient α (Maxwell, 1879), which represents the fraction of molecule-wall collisions undergoing diffuse scattering and/or trapping–desorption; the remainder being specularly reflected (no energy loss). The Maxwell coefficient has been shown to provide a useful description of molecular friction (Sokhan et al., 2001) for a variety of materials (Nicholson and Quirke, 2006) and for liquid imbibition (Supple and Quirke, 2004) and is of particular significance for carbon, which has an exceptionally low value of alpha for smaller nanotubes. Although α can be directly related to the slip coefficient l_s for a single wall at a low density (Maxwell, 1879) (and clearly the smaller α, the greater the slip), there was no general relation relevant to nanopores until 2008 (Sokhan and Quirke, 2008). For a gas at a *single* wall using the kinetic theory, Maxwell derived a microscopic expression for the slip coefficient, which can be written as

$$l_s = \lambda\left(\frac{2}{\alpha} - 1\right) \tag{11.13}$$

From a balance of forces in a steady-state liquid flow in a slit pore of width h, Sokhan and Quirke obtained the expression

$$l_s = \frac{\tau\eta}{\rho h} - \frac{h}{3} \tag{11.14}$$

where τ is the relaxation time of the center of mass velocity autocorrelation function in the pore. This expression, for the first time, relates the slip coefficient to wall friction (through τ), the shear viscosity in the pore, and the pore width. It is valid for gas and liquid densities. In what follows, we understand the flow enhancement Q_E in terms of the slip coefficient. As is clear from Equation 11.14, the slip depends, among other things, explicitly on the shear viscosity in the pore, which is itself a function of pore width.

A comparison with Maxwell's expression for a single wall at low fluid density can be made by taking the kinetic theory expressions for the frequency of wall collisions and the Chapman–Cowling expression for the viscosity (Cercignani, 1969), giving

$$l_s = \lambda\frac{2}{\alpha} - \frac{h}{3} \tag{11.15}$$

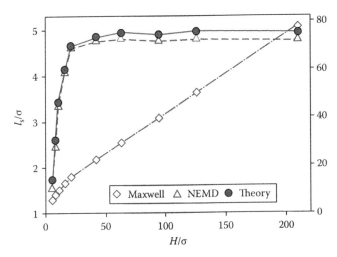

FIGURE 11.3 Slip coefficients (left scale) calculated directly in NEMD of fluid methane slit pores, triangles; and estimated from the relaxation time, circles, as a function of pore width. For comparison, slip coefficients estimated from the Maxwell's theory; diamonds, are also shown (right scale). Lines are drawn to guide the eye. (Reproduced from Sokhan, V.P. and Quirke, N., *Phys. Rev. E*, 78, 015301, 2008. With permission.)

where λ is the mean free path and now the slip length depends both explicitly and implicitly (through α) on the pore width (see Figure 11.3) and differs from the single wall case in the second term.

This result (Equation 11.14, Figure 11.3) is consistent with MD simulations at gas *and liquid* densities and shows that the slip coefficient depends strongly on the diameter of smaller pores, tending to a constant value for pores with a width $>h_s$ where h_s depends on, e.g., the Knudsen number in contrast to the linear scaling predicted by Maxwell's theory of slip. There is a reduction in slip for *nanopores* with respect to larger pores, but note that this result is for slit pores. For cylindrical geometry, η is replaced by 2η in Equation 11.14 with h replaced by R; Equation 11.15 is unchanged except again h is replaced by R. Note that in the case of cylinders, such as carbon nanotubes, α will become very small as the radius decreases.

Several extended discussions of slip flow in the context of nanofluidics have been recently published to which the interested reader is referred (Neto et al., 2005; Eijkel, 2007; Martini et al., 2008).

11.3.2 Theory Underlying Simulation

A central question in nanofluidics concerns the extent to which the classical equations describing the way fluids interact with nanomaterials hold at the nanoscale (i.e., nanometer dimensions or nanosecond time scales). Molecular simulation is ideally suited to shed light on this problem and has shown that static properties, such as surface tensions and contact angles of simple fluids, obey classical relations down to almost single nanometer dimensions (Powell et al., 2002). Steady-state Poiseuille flows maintain a parabolic velocity profile down to widths of several nanometers (Travis et al., 1997). Deviations arise due to the confining walls and are associated with density inhomogeneities.

In performing simulations relevant to experimental work on flow in nanopores, a major challenge has been to choose realistic potentials. Although it has proved possible to be reasonably confident about say alkane/carbon interactions, for fluids such as water, a considerable amount of uncertainty has arisen. For a graphite surface, there are a number of different reported values of the water contact angle (Werder et al., 2003) and depending on the choice of potential, a model nanotube can be filled by water (Hummer et al., 2001) or not. A detailed review of the literature on water-graphite contact angles is available (Mattia and Gogotsi, 2008), which includes a histogram showing the frequency of contact angles reported in the literature. The authors conclude that the interaction is highly dependent on surface chemistry, including the extent of chemisorption and hydrogen termination of edge defects. Both characteristics can be influenced by the particular history of the sample being tested including surface contamination and artifacts resulting from the manufacturing process. It is also the case that even if a reliable water–graphene potential was available, SWNTs have extremely high curvature, which may well make the potential very sensitive to the pore radius for say $r \leq 1$ nm, or indeed to the tube symmetry and the consequent metallic, semiconductor, or insulating behavior.

Recent work has shown how it may be possible to calibrate diameter-dependent solvent-nanotube potentials using the radial breathing mode Raman shifts in solution (Longhurst and Quirke, 2006a) but that accurate water/CNT potentials are still not available. The uncertainty in water potentials notwithstanding, it has been possible to determine some general features of nanoscale flow from MD using simple fluids relevant to the experimental work. For example, Sokhan et al. were the first to show that fluids flowing through (slit) carbon nanopores experience very low surface friction, i.e., very small Maxwell coefficients ($\alpha \sim 0.01$) or equivalently large slip boundary conditions, while pores made from other materials show higher values (~ 0.5) and stick to the boundary conditions (Sokhan et al., 2001, 2002).

Subsequent work showed that cylindrical geometry reduced the Maxwell coefficients by orders of magnitude ($\alpha \sim 0.000002$ for decane in a 7,7 CNT, with a diameter of 0.75 nm) and that *carbon* nanotubes are predicted (Supple and Quirke, 2003, 2004) to be essentially frictionless pipes. Skoulidas et al. compared transport diffusivities for CNTs to zeolites using MD and found exceptionally high transport rates for CNTs due to the smooth carbon surface (Skoulidas et al., 2002). We should expect therefore that crystalline carbon channels (as opposed perhaps to amorphous surfaces found in carbon nanopipes) show a significantly increased flow. Another consequence is the extremely rapid imbibition of wetting fluids observed in non-equilibrium molecular dynamics. Decane, for example, is predicted to fill 7,7 nanotubes, at 800 m/s (Supple and Quirke, 2004). Note that the same model predicts external wetting to be much slower by factors of e.g., 30 for a 13,13 tube with a diameter of 1.765 nm.

From a Bosanquet-like equation, where we approximate the surface friction by an expression involving the Maxwell coefficient and the surface tension driving force is represented by the wall-wetting liquid tension, it is possible to obtain a solution

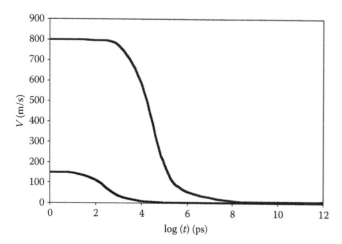

FIGURE 11.4 Decane imbibition velocity at room temperature as a function of log observation time obtained from an analytic model using parameters derived from nanoscale simulation, the upper line shows results for a (7,7) carbon nanotube (diameter = 0.951 nm), the bottom line, a non-carbon tube of similar diameter (From Whitby, M. and Quirke, N., *Nat. Nanotechnol.*, 2, 87, 2007. With permission.).

for the imbibition velocity and penetration length L at all times (Supple and Quirke, 2004). There is an $L \propto t^1$ dependence for short times tending to $t^{1/2}$ at long times (cf. Washburn equation)—the timescale for which depends on the value of the Maxwell coefficient. The following plot shows how the imbibition speed for decane in a SWNT (configuration on left) varies as a function of time with an ultrafast uptake for $t < \mu s$ falling to cm/s at longer times (see Figure 11.4).

11.3.3 Filling Experiments with Carbon Nanotubes

The early confirmation that the inner pores of carbon nanotubes are indeed hollow came from experiments in which these channels were filled with a variety of fluids including molten metals, salts, and oxides (Ajayan and Iijima, 1993; Lago et al., 1995; Chen et al., 1997; Ugarte et al., 1998). Subsequent examinations using TEM revealed the presence of the inserted materials as revealed by contrast differences in the images corresponding to the filled pores. In 1994, Dujardin and coworkers studied the wetting and filling of carbon nanotubes and proposed an empirical rule for SWNTs that liquids must have a surface tension below 100–200 mN/m in order to enter the inner pore by capillary filling (Dujardin et al., 1994). This rule has generally been found to hold, but with a few apparent exceptions, such as the common observation that metal from catalyst particles sometimes penetrates arc-produced nanotubes. Presumably, this occurs as a result of capillary uptake when the metal is molten during the synthesis conditions. A recent molecular dynamics study of molten palladium in carbon nanopipes has shown that this filling can be explained by taking into consideration the Laplace pressure exerted by small droplets of molten metal attached to the ends of SWNTs (Schebarchov and Hendy,

2008). The authors found large slip lengths for the ensuing flow of liquids along the nanotubes of up to 10 nm (i.e., several times the channel diameter).

The surface tension depends on the relative polarizability of the liquid and solid in a system, and so another way of expressing the filling rule is to say that a liquid must be less polar than the carbon surface for capillary uptake to occur. What about the case of water and carbon nanotubes: the most important fluid in the context of many applications? The question of the contact angle between water and carbon arises, which is a complex subject with a history of debate (Mattia and Gogotsi, 2008). The surface tension of water (72 mN/m) is below Dujardin's empirical threshold. However, cleaved graphite is generally considered not to be wettable by polar fluids including water, forming instead droplets with a high contact angle. Early simulations based on such contact angles predicted that water would not wet or readily enter carbon nanotubes (Werder et al., 2001). However, other research has shown that the situation is more complex and determined both by the dimensions of the nanopore (Hummer et al., 2001; Kotsalis et al., 2004) and crucially by the chemical properties both of the pore entrance and the inner carbon walls of the nanopipe.

The group led by Y. Gogotsi based at the Materials Science and Engineering Department at Drexel University in Philadelphia, Pennsylvania, has made a major contribution to this area of research, contributing to numerous papers over the past decade that illuminate the interactions between water and the inner pores of carbon nanopipes. Their first paper describing the synthesis of hydrothermal MWNTs included TEM observations of liquid (water with polyethylene residues following pyrolysis synthesis) trapped in closed nanotubes (Gogotsi et al., 2000). The pressure of the trapped liquid was estimated to be as high as 30 MPa and the gaseous component consisted of a mixture of water vapor, CO_2, and CH_4 (Yarin et al., 2005b). Distinct liquid/vapor menisci were visible and the contact angle between the fluid and the carbon inner walls was measured at 5°: clearly the water was wetting these hydrothermal nanotubes.

In the following few years, the Drexel group, in collaboration with other teams at the University of Philadelphia and at the University of Illinois at Chicago, published a series of papers reporting a detailed investigation of water trapped in large ~100 nm pore diameter hydrothermal carbon nanotubes (Gogotsi et al., 2001; Libera and Gogotsi, 2001; Megaridis et al., 2002; Yarin et al., 2005a; Yazicioglu et al., 2005). Their key technique involved the use of the TEM electron beam to apply local heating to selected areas of the system, allowing vaporization, condensation, and flows resulting from the pressure changes to be observed. Stage heating was also used to apply a global thermal stimulus to the system. The Drexel group proposes an explanation for the small contact angle observed for water in hydrothermal nanotubes based on hydrogen bonding occurring between the water molecules and carboxyl groups attached to edge defects in the carbon layers of the inner nanotube walls (Gogotsi et al., 2002). These carboxyl groups resulted from the hydrothermal synthesis conditions.

In 2004, the work was extended to MWNTs with much smaller internal pores (2–5 nm in diameter). These were filled by autoclaving so that water infiltrated the closed nanotubes under high pressure. A TEM showed that the vapor/liquid interface in such nanotubes was much less ordered than had been seen with the hydrothermal MWNTs. The authors speculate that this is likely to have been due to both more hydrophobic carbon walls and to the increased confinement.

Their research into the dynamic behavior of water trapped inside larger closed carbon nanotubes led to the publication of a joint paper from the Drexel University and the University of Illinois at Chicago groups presenting a theoretical model of experimentally observed fluid behavior in these systems (Yarin et al., 2005b). The model uses a continuum approach combining thermally driven mass transport with Lennard-Jones intermolecular interactions between the fluid and the nanotube walls. A variety of simulated configurations were compared with good agreement to experimental observations including phase changes, meniscus movement, water membrane pinchoff on heating, fluid relocation due to heating, and jetting.

Subsequent work by members of the Drexel group (Mattia et al., 2007) presented TEM images of water highly confined inside much smaller SWNTs and DWNTs (~2 nm pore diameter) forming straight and helical chains with quasi-crystalline structures. These findings provide some support for numerical simulations showing similar ice-like phase transitions for water in very small nanopores (Koga et al., 2001a,b; Noon et al., 2002; Bai et al., 2003). Other experimental techniques have been used to probe the characteristics of water highly confined inside SWNTs including x-ray diffraction (Maniwa et al., 2005), neutron scattering (Kolesnikov et al., 2004), infrared spectroscopy (Byl et al., 2006), and nuclear magnetic resonance (NMR; Mao et al., 2006; Matsuda et al., 2006; Sekhaneh et al., 2006). Interesting novel phenomena have been observed in nanotubes where the confinement is sufficient to force water molecules into a single file. New solid-state ice phases are also formed in slightly larger tubes, particularly at lower temperatures. A comprehensive review of this topic is available in the article by Rasaiah et al. (2008).

11.3.4 Filling Carbon Nanopipes with Fluids

Closed nanopipes suffer from the limitation that the study of dynamic interactions with fluids is restricted by spring-like back-pressure from trapped gas and vapor bubbles that hinder significant transport. To overcome this obstacle, the Drexel group moved on to study open carbon nanopipes with large (~200–300 nm) diameters produced using the CVD template-assisted process. They reported a series of observations of liquid/carbon pipe interactions made in an environmental scanning electron microscope (ESEM). The advantage of using an ESEM is that *in situ* experiments involving exposed fluids can be undertaken. (This is also possible in a conventional SEM using nonvolatile ionic liquids.) Furthermore, by carefully varying the pressure in the ESEM sample chamber, and by controlling the

temperature using a peltier cooling stage, it is possible to cause condensation and evaporation may occur in a controlled manner on the sample (Rossi et al., 2004).

Using this method, the Drexel team captured the behavior of liquid menisci inside CVD-grown carbon pipes with diameters from 200 to 300 nm that were freed from the template by chemical etching as previously described. The thin walls of these tubes (12–15 nm) are transparent to electrons and also to visible light. This made it possible to observe fluid behavior without unduly heating the sample and driving off the water (see Figure 11.5).

A key result was that water condensed on the internal wall of the carbon tubes in preference to the external steel surface of the stage. The authors conclude that the disordered walls of the AAO template grown tubes are hydrophilic and they measured contact angles with water between 5° and 20°. They also report a slight deformation of the tubes containing plugs of water, indicating negative pressure inside the channel due to tensile capillary forces. They conclude that carbon tubes of this type may be

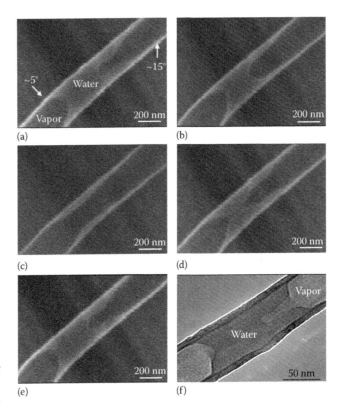

FIGURE 11.5 ESEM micrographs showing the dynamic behavior of a water plug close to the open end of a carbon nanopipe. The meniscus shape changes when, at a constant stage temperature, the vapor pressure of water in the chamber is changed (a) 5.5 Torr; (b) 5.8 Torr; (c) 6.0 Torr; (d) 5.8 Torr; and (e) 5.7 Torr, where the meniscus returns to the shape seen in (a). The asymmetrical shape of the meniscus, especially the complex shape of the meniscus on the right side in (a, e), is a result of the difference in the vapor pressure caused by the open left end and closed right end of the tube. (f) TEM image showing a similar plug shape in a closed CNT under pressure. (From Rossi, M.P. et al., *Nano Lett.*, 9, 989, 2004.)

useful for guiding aqueous fluids to and from specific locations at the nanoscale. They may also be useful for collecting attoliter and picoliter amounts of liquids.

Also in 2004, a team from the University of Pennsylvania reported work on liquid interactions with similar carbon pipes that were studied using an optical microscope (Kim et al., 2004). The optical resolution however, even when using an oil immersion objective lens, is at least two orders of magnitude lower than with an ESEM. Kim and his coworkers therefore used even larger carbon nanopipes with diameters from 300 to 800 nm. A novel aspect was the use of an applied potential to a suspension of the isolated tubes to cause their alignment and size segregation by means of dielectrophoresis.

The interaction with five liquids was studied: water (distilled), ethylene glycol, 2-propanol, acetone, and 96% glycerine. Two principal techniques were used, both using carbon tubes immobilized by drying the electrophoretically aligned suspension on a glass cover slip. The first technique involved bringing a 300 μm glass pipette filled with the test liquid close to the carbon tube being observed. An electrical current was then passed through a fine wire passing through the tip of the pipette to heat the liquid. This caused a small amount to evaporate, some of which condensed in the vicinity of the target carbon tube.

All liquids were seen to condense on or inside the tubes, confirming their hydrophilic character. In the latter case, menisci formed as a plug of fluid built up inside the pipe. The size of the plug grew as long as vapor production continued. When the current to the heater was switched off, the menisci would retreat as the fluid evaporated. Sometimes both menisci would move. In other tubes studied, one end of the plug was seen to be pinned at a fixed location. The authors speculate that this may have been due to defects in the tube walls at these locations. The observations were generally repeatable, with menisci moving backwards and forward in response to vapor pressure in a consistent manner (Bau et al., 2004a).

The second technique for initiating interaction was to bring a droplet of liquid on the end of a micromanipulator into direct contact with an open carbon tube. The results are reported with 96% glycerol, which is sufficiently viscous so that the progress of the resulting meniscus can be followed along the carbon tube at the video frame rates employed (14 fps). With other liquids, imbibition was too rapid to follow.

The University of Pennsylvania team plotted filling length against time and found a good fit with $x(t) = At^{1/2}$ (Washburn equation) as expected for the long (ms) observation.

11.3.5 Filling Nanopipes with Nanoparticles

A year later, the same group published a follow-on paper (Kim et al., 2005) reporting on imbibition experiments that involved introducing fluorescent polystyrene nanoparticles (diameter ~50 nm) suspended in a variety of liquids. The suspended particles were taken up by 500 nm diameter carbon tubes through a combination of capillary action and evaporation forces. The introduced particles could be observed with an optical microscope moving along the tubes since their fluorescence was visible through the

tube walls. Kim et al. speculate that this technique may be useful for observing the interactions of biological macromolecules in the vacuum environment of both SEMs and TEMs.

The authors of this chapter have similarly filled smaller amorphous carbon nanopipes with an inside diameter of 200 nm with ~40 nm silica nanoparticles using both capillary uptake and simple diffusion (see Figure 11.6).

The same year the Drexel group also reported results from experiments filling CVD carbon nanopipes with nanoparticles (Korneva et al., 2005). In this case, the inserted material was a commercially available magnetic ferrofluid. The nanopipes were filled using simple capillary filling of both the intact membrane (i.e., nanopipes still in their AAP template) and isolated nanopipes removed in the standard manner with alkali etching. In the former case, a magnetic field was applied to assist filling although the authors determined that it was not necessary. They did, however, convincingly demonstrate that the resulting filled nanopipes were magnetic and capable of highly controlled manipulation using appropriate external fields.

Remaining briefly on the topic of filling nanopipes and nanotubes with nanoparticles, which may be important for drug delivery as discussed later in this chapter, two recent studies are worth mentioning. The first describes a novel method for loading both nanoparticles and viscous fluids using centrifugation of a suspension of nanotubes in an alginate solution (Nadarajan et al., 2007). This causes the nanopipes to be forced through the viscous medium with a natural tendency to travel in the direction of their major axis. A consequence is that their open ends are presented to the nanoparticle-containing medium that flows through them to fill the tubes. The method was found to be effective for nanoparticles in the size range of 10–40 nm. In another paper, the methods are described for capping filled carbon nanotubes, carbon nanopipes, and ultra-fine glass capillaries using low molecular weight polymers in a self-sustaining diffusion process (Bazilevsky et al., 2008a). The technique is important because often it is necessary to retain material in a filled nanotube, for example for the purpose of drug delivery, at least for a period prior to release.

11.3.6 Wetting

A number of experimental studies have been published reporting results for the wetting of carbon nanotubes, nanopipes, and amorphous carbon films similar to the material forming nanopipe walls. In an elegant series of experiments, a team from the Weizmann Institute of Science in Rehovot, Israel investigated contact angles between carbon MWNTs and a variety of fluids using an atomic force microscope (AFM) to measure contact forces (Barber et al., 2004, 2005).

Their data shows firstly that the AFM can be used to measure contact forces between carbon nanotubes and liquids such as polyethylene glycol and glycerol that are partially wetting. Following contact, an equilibrium state is established and measurement of the forces pulling the nanotube into the liquid allows the contact angle to be determined using the equation

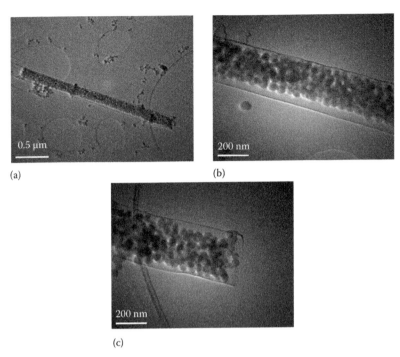

(a) (b)

(c)

FIGURE 11.6 A carbon nanopipe ~5 μm long and ~250 nm in outside diameter filled with ~40 nm silica nanoparticles. The nanopipe was produced by the authors using CVD in an AAO template and subsequently fragmented into 3–5 μm lengths by means of high power sonication. Filling was achieved by incubating the segmented nanopipes in an aqueous suspension of the nanoparticles and subsequently washing in distilled water with centrifugation in cuvette incorporating a mesoporous membrane. The nanoparticles are assumed to enter the nanopipes by diffusion.

$$F_{out} = \gamma_{\ell}\pi d_{out}\cos\theta_{out} \qquad (11.16)$$

where

γ_{ℓ} is the liquid surface tension

d_{out} is the outer nanotube diameter

θ_{out} is the contact angle of the liquid with the outer nanotube surface

As can be seen from Figure 11.7, the force on the nanotube due to the wetting by the two organic liquids increases with the diameter of the tube in accordance with this equation. The situation in the case of water is complicated by the fact that this liquid is observed to completely wet at least some nanotubes.

The authors attempt to extend this experimental work to consider open carbon nanotubes and developed the following

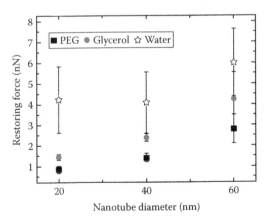

Probe Liquid	Nanotube External Diameter (nm)	Contact Angle (°)	
		External, θ_{out}	Internal, θ_{in}
PEG	20	73 ± 2	
	40	53 ± 3	
	60	49 ± 12	
Glycerol	20	77 ± 3	
	40	74 ± 3	
	60	57 ± 10	
Water	20	>77*	0
	40	>74*	0–66
	60	>57*	0–104

FIGURE 11.7 Plot of the equilibrium force acting on a carbon nanotube during partial immersion in three different liquids measured using an AFM as a force balance. The right-hand plot shows derived contact angles. The asterisked values for water should be treated with caution since the contact angle may be equal to or greater than 1 (implying complete wetting) within the measurement error. The authors' derivation of internal contact angle is discussed in the text. (From Barber, A.H. et al., *Phys. Rev. B*, 71, 115443, 2005. With permission.)

equation to describe the wetting of both the outside and the inside of a nanotube:

$$F_r = \gamma_\ell \pi \left(d_{out} \cos\theta_{out} + d_{in} \cos\theta_{in} \right) \qquad (11.17)$$

where $F_r = F_{out} + F_{in}$ (i.e., the restoring force is the sum of the wetting forces on both the inside and the outside of the nanotube). The ratio of d_{out} to d_{in} was taken to be 2.5:1 from the TEM.

There were a number of difficulties noted by the authors that make the interpretation of the imbibition experiments problematic. No internal wetting was observed for the two organic liquids. This may be due to tubes that are not properly open (the TEM images included in the paper are not definitive) or it may be due to defects in the nanotubes, which were synthesized using CVD. Notwithstanding these problems, the authors interpret the greatly increased pull-in forces observed for the open nanotubes in water as being due to the influence of imbibition.

As discussed previously in Section 11.2.2 on materials, the Drexel group has published a study of the wetting of amorphous carbon films by a range of polar and nonpolar liquids (Mattia et al., 2006a). These films were produced using CVD on alumina substrates and are assumed to have surface properties similar to templated carbon nanopipes. This assumption was supported by Raman and the infrared spectroscopy of the two sets of materials. CVD amorphous carbon surfaces were generally found to be more hydrophilic than carbon nanotubes. Contact angles, particularly for polar liquids, decreased substantially following exposure to alkali solutions, which are commonly used for etching during the manufacture of nanopipes. Conversely, annealing amorphous carbon at high temperatures in an inert atmosphere promotes graphitization and leads to a sharp increase in contact angles and reduced wettability of nanopipes by water (Mattia et al., 2006b).

As also discussed earlier, Raman spectroscopy offers a unique opportunity to probe the intermolecular forces between nanotubes and fluids, especially as peak shifts are of a level that is easily detectable by modern Raman instruments. Figure 11.8 shows the predicted solvent shift of the radial breathing mode for a [22,0] nanotube in water obtained by molecular dynamics simulation and theory (based on treating the nanotube and its solvent shell as elastic membranes) for a range of possible water-carbon nanotube Lennard-Jones interaction strengths ε. The figure shows the separate contributions from the internal and external wetting of the tube and a combination of both. The solid line is the experimental upshift for diameter corresponding to a [22,0] SWNT taken from Izard et al. (2005). The intersection between this line and the triangles indicates a [22,0] CNT–water interaction of around 0.4 kJ/mol. This is weaker than predicted from external wetting only. Detailed predictions of water carbon interactions will require more accurate Raman data for pure solvent; however, the approach seems promising.

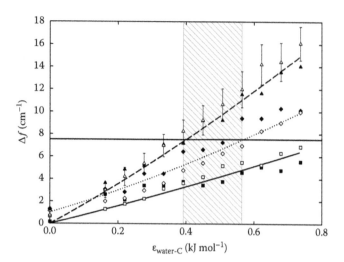

FIGURE 11.8 Upshift in the RBM; theory (lines) against simulation (symbols) for [22,0] nanotubes as a function of the nanotube-water interaction strength $\varepsilon_{water\text{-}C}$. Solid symbols correspond to atomistic simulations, open symbols correspond to mean field shell approximation, for water present on the inner surface (squares), outer surface (diamonds) or both surfaces (triangles). Errors are of the order of ±1 wave number. The hatched region corresponds to intermediate values of the nanotube-water interaction strength, which result in contact angles of water on graphite between 42° and 86°. (From Longhurst, M.J. and Quirke, N., *J. Chem. Phys.*, 125, 184705, 2006b.)

Finally, with regard to dynamic wetting, a comprehensive review of the experimental results in this field, including consideration of nanoscale phenomena, has recently appeared (Ralston et al., 2008)

11.4 Experimental Investigation of Nanoscale Fluid Flow

We turn now to consider a series of recent experiments that directly investigate the flow of fluids through the central pores of carbon nanotubes and nanopipes. Two of these experiments report dramatically enhanced flow rates in small (<10 nm) diameter channels with correspondingly huge slip lengths (Majumder et al., 2005a; Holt et al., 2006). This phenomenon is sometimes described as "super flow." In the first paper, reported flow enhancements were as much as 61,404 times those predicted using conventional theory. In the second paper, the enhancement measured was up to 8400 times. These results have attracted widespread attention, having been published respectively in the journals *Nature and Science*. If confirmed, the discovery would have major implications both for our theoretical understanding of nanoscale fluid flow and for practical nanofluidic devices particularly in the field of materials separation. However, a number of papers have subsequently appeared that cast doubt on the existence or at least the magnitude of "super flows" through carbon nanotubes. At the time of writing, the authors of the two main papers have yet to publish detailed responses to these criticisms or further data supporting their original findings.

11.4.1 Early Fluid Flow Experiments

The first indications of unexpectedly rapid fluid flow through very small channels in an experimental system were published by a group from the University of Pennsylvania in 1990 (Pfahler et al., 1990). Working with rectangular channels etched in a silicon substrate, Pfahler et al. found an *n*-propanol flux approximately three times greater than expected in the smallest channels (height = 0.8 μm and width = 100 μm).

Following up on this observation, a team working at Purdue University in Indiana reported measurements for Newtonian fluids flowing under pressure through nanometer-scale channels produced lithographically by building up layers of photoresist on a glass substrate (Cheng and Giordano, 2002). The height of the channels varied from 2.7 μm down to 40 nm. A simple optical observation of the motion of a meniscus along the channel was used to determine the flow rate for water, silicone oil, decane, and hexadecane.

While not directly involving carbon materials, both studies are relevant in so far as the dimensions of the channels investigated the approach of the central pores in nanotubes and nanopipes. The results showed a marked and progressive departure from the Poiseuille flow as the channel height falls below 200 nm. This effect was observed most strongly for hexadecane and is also evident for decane and for silicone oil. Slip lengths were estimated at 25–30 nm and ~9 and 14 nm, respectively. The effect was not seen for water in silicon or photoresist channels at any of the length scales investigated.

Of historical note—particularly given the current interest in single molecule detection by the passage through nanopores (Dekker, 2007)—was the first experimental study of mass transport through the central pore of a carbon nanotube (Sun and Crooks, 2000). This team from Texas A&M University created a single nanopore membrane by embedding a MWNT in epoxy and then using a diamond knife to cut an ultra-thin section containing a short 660 nm length of the nanotube crossing the polymer slice. The central pore of this fragment provided a unique channel between the two sides. This membrane was sealed in a dual compartment cell and a Coulter counter was used to detect the passage of negatively charged probe particles including 60 ± 10 nm and 100 ± 10 nm COOH-modified polystyrene spheres under an applied electric field. The effect of opposing this electrophoretic flow with a pressure gradient was also investigated. At the relatively large pore size studied (a diameter of 153 nm), the authors found no evidence of departure from conventional continuum models.

The following year, in 2001, a group from the University of Florida reported results of electro-osmotic flow (EOF) experiments through templated CVD carbon nanopipes (Miller et al., 2001). As far as the authors of this chapter are aware, this is the first report of experiments involving transport through these materials. The outside diameter of the carbon pipes used in this study were ~300 nm with an average wall thickness of ~40 nm of amorphous carbon. The as-produced membrane, traversed by large numbers of these well-aligned nanopipes, was used to separate two sides of a U-tube permeation cell. The concentration of a probe molecule (phenol) in the permeate side of the half-cell was continuously measured using High Performance Liquid Chromatography (HPLC). The phenol flowed as an ionic current from a reservoir on the feed-side through the carbon membrane under an applied potential. For as-synthesized carbon membranes, the Florida group reported a linear relationship between electro-osmotic flow velocity and applied current density. This study is discussed further below in Section 11.5 dealing with methods of modulating transport through nanopipes.

In 2004, a team from NASA's Ames Research Center in California and a group from Kettering University in Michigan joined forces to publish a paper on gas transport through carbon nanopipes (Cooper et al., 2004). Again they used intact AAO templates covered in a layer amorphous carbon by means of CVD with nominal pore diameters of 200 nm. They measured the pressure driven flow of argon, nitrogen, and oxygen through a series of membranes with progressively thicker coatings of carbon deposited at increasing CVD temperatures. With no carbon coating, they found non-slip flow and concluded that the transport was well-described by diffuse molecular reflection at the wall. With the thickest 20–30 nm coating of amorphous carbon, they measured a tangential-momentum accommodation coefficient (i.e., the Maxwell coefficient α) of 0.52 ± 0.1 indicating the onset of slip flow in carbon tubes of these dimensions.

11.4.2 Super Flow in Carbon Nanotubes

We turn now to the studies mentioned at the start of this section, which both report dramatically enhanced flow rates and long slip lengths for fluid transport through carbon nanotube membranes with sub-10 nm pore diameters. A group from the University of Kentucky previously published an account of experiments involving transport through a membrane composed of true molecular nanotubes (Hinds et al., 2004). They used dense, well-aligned arrays of MWNTs grown using CVD (ferrocene-xylene-argon-hydrogen) on a flat quartz substrate. One advantage of this approach is that the pore diameters are much smaller (4.3 ± 2.3 nm) than with the AAO template method. Another is that the crystallinity of the carbon nanotube walls is greater, with likely benefits for low friction fluid flow (see Figure 11.9).

As synthesized, MWNT arrays are permeable not only through the central pores of the nanotubes, but also extensively through the gaps between the individual tubes. Furthermore, many nanotubes will be closed by end caps or plugged by a residual iron catalyst. To overcome these problems, a polystyrene solution was spin-coated onto the array to fill the gaps. Then, plasma etching was used to remove the top layer of the composite and to open the carbon nanotube central pores. SEM, TEM, and electrical conductivity characterization demonstrated that ~70% of the nanotube tips had been opened by the plasma oxidation process. Pore density was estimated at 6 (±3) × 10^10 cm^{-2}.

Using a gas flow apparatus, a 3.1 cm² surface area and 5 μm thick MWNT composite array showed a linear relationship between the flow rate and the pressure drop across the membrane.

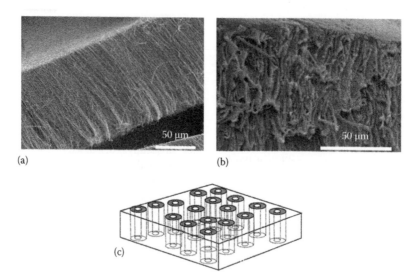

FIGURE 11.9 Composite picture shows (a) an as-grown, dense, multiwalled CNT array produced with an Fe-catalyzed chemical vapor deposition process; (b) the cleaved edge of the CNT-polystyrene membrane after exposure to H$_2$O plasma oxidation (the PS matrix is slightly removed to contrast the alignment of the CNTs across the membrane); (c) schematic of the target membrane structure with a polymer embedded between the CNTs allowing a viable membrane structure to be readily produced. (From Hinds, B.J. et al., *Science*, 303, 62–65, 2004.)

The slope of the line indicated a permeance of 2.6 μmol/(m^2 s Pa). The diffusion transport of aqueous ionic species (Ru(NH$_3$)$_6^{3+}$ diameter = 0.55 nm) across the MWNT composite membrane was measured using cyclic voltametry on the permeation side. The authors noted that the enhanced diffusion coefficient was near the bulk aqueous-solution diffusion for the Ru cation and concluded that there was only limited interaction between the ion and the carbon nanotube tip and tube walls (Hinds et al., 2004).

In a subsequent report published in *Nature*, the Kentucky group presented results of further experiments on fluid transport through the MWNT composite membranes (Majumder et al., 2005a). A simple pressure-driven flow apparatus was used with the accumulated mass of transported fluids including water, ethanol, and alkanes

being measured by weighing after a fixed time period. Their results suggest a dramatically enhanced flow through ~7 nm diameter MWNT cores compared to conventional fluid flow theory. The observed flow rates were four to five orders of magnitude greater than standard hydrodynamics predicts. These results are summarized in Table 11.1.

In the studies reviewed so far, increasingly interesting fluid flow results were found at progressively smaller length scales. In 2006, a team working at the Lawrence Livermore National Laboratory and at the University of California achieved a milestone in measuring water flow through the central pores of double-walled carbon nanotubes (DWNTs; Holt et al., 2006). These DWNTs had inner diameters less than 2 nm with relatively

TABLE 11.1 Observed and Expected Flow Rates of Five Different Fluids through ~7 nm Diameter Carbon Nanopores Together with a Slip Length Calculated from the Experimental Data

Liquid	Initial Permeability[a]	Observed Flow Velocity[b]	Expected Flow Velocity[b]	Slip Length (μm)
Water	0.58	25	0.00057	54
	0.01	43.9	0.00057	68
	0.72	9.5	0.00015	39
Ethanol	0.35	4.5	0.00014	28
iso-Propanol	0.088	1.12	0.00077	13
Hexane	0.44	5.6	0.00052	9.5
Decane	0.053	0.67	0.00017	3.4

Source: Majumder, M. et al., *Nature*, 438, 44, 2005a. With permission.

Note: These results indicate liquid flow rates 4–5 orders of magnitude faster than would be predicted from conventional fluid-flow theory. The authors conclude that the implied slip lengths (3–70 μm) are consistent with a nearly frictionless interface.

[a] Initial permeability in cm^3/cm^2 at minimum pressure.

[b] Flow velocities in cm/s at 1 bar.

defect-free graphitic walls. Simulations (see Section 11.3.2) suggest that these smooth carbon channels should present a low friction surface to transported fluids (Hummer et al., 2001).

The practical challenges involved in making a membrane composed of such very small carbon nanotubes are formidable. The approach used by the California team was to use fabrication techniques adapted from the semiconductor industry. A dense array of carbon nanotubes was first grown on a silicon substrate onto which metal catalyst particles had been deposited. The spaces between the tubes were then completely filled with silicon nitride using low-pressure CVD. Finally, the closed ends of the tubes with catalyst particles and the substrate were removed by a series of chemical etching, ion milling, and reactive ion etching steps. The resulting membranes had a thickness in the range 2.0–3.0 µm and pore densities ≤0.25 × 10^{12} cm^{-2} as characterized by TEM. Pore diameter, as determined by size exclusion tests using gold nanoparticles of known dimensions, were in the range 1.3–2.0 nm (Holt et al., 2004).

Gas flow and water flow measurements were both undertaken in o-ring sealed flow cells. A total of five hydrocarbon and eight non-hydrocarbon gases were tested to determine flow rates and to demonstrate the molecular weight selectivity compared to helium. The water flow was pressure driven at 0.82 atm and measured by following the level of a meniscus in a feed tube. The results for both gas and liquid showed dramatic enhancements over flux rates predicted with continuum flow models. Gas flow rates were between 16 and 120 times that expected according to the Knudsen diffusion model. Water flow rates were 560–8400 times that calculated according to the Hagen–Poiseuille equation. Minimum slip lengths were estimated in the range of 140–1400 nm. To express these findings in a practical context, the nanotube membranes showed flow rates several orders of magnitude greater than those of conventional polycarbonate membranes, despite having pore sizes an order of magnitude smaller. The authors attribute the observed flow enhancement either to the effect of the confinement of the water molecules and/or to the near frictionless surface of the nanotube pores. They point out the significance of this finding for separation applications (Holt et al., 2006).

In the supporting information published alongside both the Majumder et al. and the Holt et al. papers are details of control experiments performed to verify that the observed flow was occurring through the central pores of the nanotubes and not, for example, through cracks in the membrane. The Kentucky group used 10 nm gold nanocyrstals suspended in a solution of UV active small-molecular sized citrate. Both passive diffusion and pressure flow control experiments were conducted with their ~7 nm diameter MWNT membranes. They reported detection of the citrate in the permeate, but no detectable Au nanoparticles in the transported fluid. In the case of the flow control experiment, rapid fouling occurred, presumably as a result of pore blocking by the gold. Majumder et al. did state that the metal support used in their flow cell had a tendency to absorb the Au nanoparticles, thus "making this pressure flow experiment inconclusive." They did not state whether the metal support was also present

in the diffusion case. Holt et al. presented a similar size exclusion control, using 2 nm Au nanoparticles. They report a ten-fold reduction in the flow rate when the nanoparticles were present (presumably due to fouling) and estimated that less than 0.1% of their observed water flux can be attributed to pores larger than 2 nm. Holt et al. did not mention whether metal supports were used in their system, leaving open any determination on the concern raised about possible gold absorption affecting the control.

Two surprising aspects of the main results presented in these high profile papers are worth highlighting. The first is simply that the magnitude of the reported enhancements were in the opposite relationship to that which might reasonably be predicted from the size of the respective pores. In the case of water, Majumder et al. reported flow enhancement of 61,404 times through their 7 nm diameter MWNT. Holt et al. reported flow enhancements in the range of 560–8400 times for their sub-2 nm DWNT. If nanoscale confinement plays a part in the reported "super flows," then one would expect to see greater enhancement with the smaller nanotubes. In fact, the opposite seems to be the case. It should perhaps be remarked that flow rates are proportional to the fourth power of the pore radius and hence, small errors in the estimated pore radius could introduce large errors in calculated flow enhancement.

The second surprising observation in the Majumder et al. paper is that the reported flow enhancement for decane (3,941 times) is so much less than for water (61,404). In macroscale pipes ($d > 100$ µm), where viscosity of the bulk fluid dominates, decane flows faster than water (Cheng and Giordano, 2002). Majumder et al. attributed the relationship reversal they observed as being due to the interaction between the water molecules and the hydrophobic carbon walls, with formation of a hydrogen bond network leading to very low friction.

Further theoretical support for the "super-flow" papers appeared in 2008 with the publication of a molecular dynamics simulation of water flowing through carbon SWNTs with diameters in the range of 2.1–2.5 nm, i.e., similar to the DWNTs used in the Holt et al. study (Joseph and Aluru, 2008b). These authors predicted a flow enhancement of 2052 times compared to the non-slip Poiseuille flow, which they attributed to the occurrence of a pronounced depletion layer in the water adjacent to the carbon walls. According to their model, the density of water molecules is just 5% of the bulk value in this region. They postulate that this low-density layer acts as a kind of lubricant to facilitate super flow in smooth walled carbon nanopipes. Far less enhancement (<5 times) was seen in the case of polymer tubes of similar dimensions but with rough walls, which were also modeled.

11.4.3 Reassessing Super Flows

Two further experimental papers, both published since the Nature and Science reports of "super flows," throw further light on fluid transport through nanoscale carbon pipes. In 2008, the authors of this chapter working at Imperial College in London published results for the flow of water, ethanol, and decane through $43 \pm 3\,\text{nm}$ nanopipes produced using the templated CVD method. This study provides data for nanoscale fluid flow from a new size regime intermediate between MWNT/DWNT membranes and the larger 200–300 nm diameter nanopipes previously discussed. The Imperial College team found evidence of only modest flow enhancement compared to the sub-10 nm nanotubes investigated by Majumder et al. and by Holt et al. Furthermore, observed flow rates for decane were greater than for water, in line with the relationship of the bulk viscosities and hydrodynamic behavior in macroscale channels.

The experimental method was similar to the earlier papers. A syringe pump was used to force fluids through a membrane composed of CVD templated nanopipes in their intact alumina template. A novel method of mounting the membranes in brass adapter rings using epoxy was employed to ensure a fluid-tight seal. The pressure was measured at a series of flow rates and adjusted for the effective area of each membrane under test. The fluid passing through the nanopipes was collected and periodically weighed to verify flow rates. In all the cases, a good linear fit was found for plots of flow rate against pressure standardized for the membrane area. The inner diameter of the nanopipe was determined using TEM and SEM and pore density was determined by SEMs of the two membrane surfaces. Flow enhancement and slip lengths were calculated using the slip-adjusted Hagen–Poiseuille equation from these parameters and the known viscosity of the three liquids. Results are summarized in Table 11.2 with data from the earlier studies for comparison. Whitby et al. quote a range rather than absolute values for observed enhancement and slip lengths. This is due to uncertainty over the possible pore-blocking influence of a supporting metal grid used to prevent the brittle carbon nanopipe arrays from cracking under hydraulic pressure.

In 2007, the Drexel/University of Philadelphia teams published another joint paper investigating fluid flow through templated CVD nanopipes (Sinha et al., 2007). This time the focus was on measuring flows through an individual carbon nanopipe rather than large numbers arrayed to span a membrane. They used an elegant technique that takes advantage of the small pressure difference between differently sized fluid droplets. The difference arises from the fact that the Laplace pressure due to surface tension varies inversely with the radius of the drop according to the formula

$$P = \frac{2\sigma \sin\theta}{R} \qquad (11.18)$$

where

P is the pressure inside the droplet (above atmospheric)

σ is the surface tension of the liquid

θ is the contact angle with the substrate (90° in the case of the gold film on glass used in the experiment)

Using an optical microscope and a very fine pump-driven pulled-glass capillary on a micromanipulator, drops of fluids of unequal size were attached to either open end of an isolated carbon nanopipe of ~300 nm in diameter. The nanopipe rested on the substrate, thus the droplets assumed a flattened shape with a contact angle of 90° in the case of glycerin on the gold film used in the experiment.

Fluids are known to travel between droplets along the outer surfaces of carbon MWNTs (Regan et al., 2004). To preclude this possible mode of transport, Sinah et al. placed a barrier of photoresist at right-angles mid-way along the nanopipe, ensuring that fluid could flow between the drops only through the inner pore. This fluid transport was monitored by video recording the changing size of the droplets as they equalized. Direct observation of flow was also made by including fluorescent nanoparticles in the fluid, the light from which could pass through the thin carbon tube walls. From the observation of the flow rate, the diameter of the nanopipes could be calculated assuming a non-slip flow. Alternatively, by independently determining the pore size from SEM characterization, slip rates could be derived. Using the later method, Sinah et al. found that the flow rates they observed were close to theoretical predictions based on non-slip flow. Extrapolating the slip lengths reported by Majumder et al. and by Holt et al. to take account of the much larger nanopipes in the

TABLE 11.2 A Summary of Flow Enhancement and Implied Slip Lengths Observed in Three Recent Studies for Three Fluids through Carbon Nanopipes

	Decane	Ethanol	Water
43 nm diameter nanopipes (Whitby et al., 2008a, 2009)			
Flow enhancement factor	28–45	16–25	22–34
Calculated slip length[a] (nm)	145–237	81–129	113–177
7 nm diameter nanotubes (Majumder et al., 2005a)			
Flow enhancement factor	3,941	32,143	61,404
Calculated slip length (nm)	3,448	28,124	53,728
<2 nm diameter nanotubes (Holt et al., 2006)			
Flow enhancement factor			560–8,400
Calculated slip length (nm)			140–1,400

[a] The slip lengths reported here are those based on Equation 11.11 and differ from the originally published values, see erratum (Whitby et al., 2009).

Sinha et al. study, one should still expect a flow enhancement in the range of 3000–4000 times. The authors saw nothing like such flow rates and concluded that their observations did not support the earlier studies. It should, however, be noted that due to their experimental method, Sinha et al. were not able to study water or decane, which are both transported much too rapidly through the nanopipes for the changing size of droplets to be measured.

In 2008, a further molecular dynamics study appeared from a team at Carnegie Mellon University provocatively entitled "Reassessing fast water transport through carbon nanotubes" and challenging the Majumder et al. and Holt et al. findings of super flow (Thomas and McGaughey, 2008). The authors reviewed the experimental evidence as well as the results of their own MD simulations, which found that flow enhancement decreased from 433 down to 47 times as the nanotube diameter increased from 1.66 to 4.99 nm. In their calculations, they allowed for variations in water viscosity and slip length as a function of the nanotube diameter. The authors concluded that their results can be fully explained in the context of continuum fluid mechanics. Extrapolating to larger pores, they predicted an enhancement of 35 times for the ~7 nm nanotubes used in the Majumder et al. study and just 2 times for larger 250 nm diameter nanopipes. The 22–34 times flow enhancement found by Whitby et al. for water through 43 nm diameter nanopipes appears to be more in line with Thomas and McGaughey's predictions. Note that slip is very sensitive to the molecular structure of the pore wall, different forms of carbon, for instance, would be expected to have quite different slips. The results of Whitby et al. imply that the "amorphous" carbon nanopipe inner surfaces may well have a locally crystalline nature.

Thomas and McGaughey also pointed out a possible error in the Joseph and Aluru study, relating to the assumed diameter of the nanotube under consideration. By using their proposed correction, which is based on using the full inner diameter of the pore in the calculations, the enhancement predicted from the data of Joseph and Aluru is reduced from 2052 times to 459 times. The enhancements measured during both simulations are then within 10% of one another.

11.5 Controlling Fluid Flow through Carbon Nanopipes

Regardless of the rate at which fluids flow through carbon nanotubes and nanopipes, the fact that they do so at all opens up a rich set of possibilities for the design and fabrication of nanofluidic devices. These channels offer scope for precise delivery, manipulation, and extraction of fluids. Further possibilities include size selection of transported species and surface chemical interactions with functionalized pore entrances and walls, which may also provide a useful route to selective materials processing. This can be combined with the sensitive detection and measurement of target analytes, perhaps taking advantage of carbon's excellent electrical conductivity. Other basic facilities include pumping and gating, allowing nanoflows to be switched on and off on demand. All these possibilities have been explored

in recent experimental and simulation papers, a selection of which we will now review.

In the previously discussed study by Miller et al., electrochemical derivatization was used to change the surface functionalization of carbon nanopipes (Miller et al., 2001, Figure 8). Using this technique, both the magnitude and the direction of EOF could be modified. A simple application of a potential across the membrane can also act as an immediate switch for molecular transport as reported by several researchers (Melechko et al., 2003; Hinds et al., 2004; Miller and Martin, 2004).

Miller and Martin describe a more sophisticated redox modulation of ionic transport, which they used to switch both the rate and direction of EOF (Miller and Martin, 2004). Working again with carbon nanopipes, they coated the carbon surfaces with the redox polymer poly (vinylferrocene) (PVFc). They used an electrochemical technique that takes advantage of the electrical conductivity of the carbon film. Following deposition of the polymer, a variable potential was used to totally oxidize, totally reduce, or to set the redox state of the PVFc to an intermediate value. Once set, the membrane would remember its state until altered. The flow of an electrically neutral probe molecule (phenol) revealed the direction and magnitude of EOF under a driving cross membrane potential. The authors compare their approach to the field-effect concept in semiconductors, although in the nanofluidic case, the switching potential needed is very much lower: ~1 V as compared to >50 V.

Similar findings were reported by the University of Kentucky team using their finer MWNT composite membranes described in Section 11.4.2 on super flows (Majumder et al., 2005b). Instead of functionalizing the entire length of the nanopores, the authors selectively modified just the tip region of the nanotubes. They showed that different chemical end groups (straight chain alkanes, anionically charged dye molecules, and an aliphatic amine elongated by polypeptide spacers) attached near the entrance to the channels could selectively modulate the ionic flux. This method takes advantage of the differential rate of plasma etching of carbon nanotubes and the polymer matrix, which conveniently acts both to expose the tips and to introduce carboxylic acid groups onto them. The tips were then functionalized using a carbodiimide mediated coupling.

Babu et al., from the active nanofluidics group at Drexel University, reported a technique for guiding the entry of water into carbon nanopipes that were selectively modified by the electrodeposition of polypyrrole (PPy) on their tips (Babu et al., 2005). PPy is a hydrophilic polymer, more strongly hydrophilic than the AAO template synthesized carbon nanopipes (length = 5–60 μm and diameter ~200 nm) used in these experiments. Selective deposition is possible using bipolar electrochemistry in which a potential is set up across an object in an electric field by virtue of its greater relative conductivity. An ohmic contact is not required and the location of deposition is determined by field polarity.

The Drexel group used an ESEM to observe the dynamic filling of the PPy-treated tips with condensing water as previously described. Carbon nanopipes that were free of PPy, on which

PPy had been selectively deposited on one end, and on which a larger quantity of PPy had been electrodeposited to completely block the pore entrance were studied. They found that droplets of liquid water nucleated preferentially on the PPy coated tips and then entered the adjacent pore by capillary action. Having filled the nanopipe, the level of water in the tube could be controlled by adjusting the ESEM chamber pressure so that menisci visible inside the nanopipes moved back and forth. Where excess PPy had been deposited, water uptake was blocked. The authors concluded that this technique for selectively depositing conducting polymers may be useful for constructing a wide range of devices capable of controlling liquid flows at the nanoscale.

Several studies have investigated the control of nanofluidic transport using electrostatics. In 2005, a team from Berkeley, California demonstrated the transistor-like gating of ionic transport in etched silica nanoscale channels (Fan et al., 2005b; Karnik et al., 2005). Modeling the single-file transport of water molecules through a small carbon SWNT, Li et al. found that a single external charge with a value of +1.0e was sufficient to stop water permeation through the tube when the charge approached closer than 0.85 Å to the nanotube wall (Li et al., 2007). The onset of gating is pronounced, with negligible effect on flow found when the charge distance is >0.85 Å or half the diameter of a water molecule. The authors suggest that the gating effect is due to the perturbation of the water–wall interaction in this highly confined regime and may have similarities to the way that pores in biological membranes switch on and off.

In the same year, the Kentucky group again published the results of experiments in which the flow of ionic species is controlled by voltage applied to a functionalized carbon nanotube membrane (Majumder et al., 2007). They used long diazonium-based actuator molecules tethered both near the ends of the MWNT and along the inner pore walls. The former positioning was achieved by undertaking key steps of the reaction while rapidly flowing an inert solvent through the core during electrochemical functionalization. By applying a voltage in the range of ±200 mV, Majumder et al. found a potential at which the selective transport of large and small probe molecules was achieved. They postulate, in the case of the actuator molecules tethered near the nanopipe tip, that an applied positive charge has the effect of drawing the negative tails of the actuators into the pore entrance, thus partially blocking it. In the case of actuator molecules tethered inside the pore, negative charging of the nanopipes causes the tails to be repelled from the walls into the center of the channel, again partially blocking it. Selectivity between the two isovalent but differently sized probe molecules was as high as 23 times when the nanopipe channels were switched closed.

A different electrical method of controlling transport is to modify the wetting behavior of a fluid be applying a potential to the contact surface, a phenomenon known as electrowetting. Using this technique, mercury can be persuaded to enter the interior of carbon MWNTs by capillary uptake, which normally it will not (Chen et al., 2005). A subsequent MD simulation of nanoscale wetting shows that the contact angle of water nanodroplets on a graphite surface is remarkably sensitive to an applied potential (Daub et al., 2007). The authors explain the effect as being due to the modulation of interfacial hydrogen bonding in the nanodrop, which in turn affects the interfacial tensions.

11.6 Pumping

Flow switching and control is just one of several basic facilities required for the realization of practical nanofluidic devices. No less fundamental are the methods for inducing flow through nanochannels by pumping. A variety of pumping methods involving carbon nanotubes have been proposed in the literature and explored using MD simulations. These include: using two-beam coherent lasers to induce an electric field gradient along the axis of the nanotube such that the resulting electron transport in the carbon walls induces motion in intercalated ions (Kral and Tomanek, 1999); using ultrasound or pulsed laser heating to induce surface acoustic waves that travel along the carbon nanotube, propelling gas molecules inside the central pore by a form of nanoperistalsis (Insepov et al., 2006); placing a sequence of precisely positioned external charges near the pore entrances of nanotubes to induce water molecules to enter or exit by dipole coupling (Gong et al., 2007; Hinds, 2007); locally heating fluid imbibed by a SWNT from a cooler reservoir to generate high internal pressures and jetting (Longhurst and Quirke, 2007); use of thermal gradients (Shiomi and Maruyama, 2009); and the alignment of water molecules inside a carbon nanopipe by an external electric field to promote unidirectional transport due to asymmetrical coupling between rotational and translational motions (Joseph and Aluru, 2008a).

A recent theoretical paper has examined whether simple mechanical propellers, as widely deployed in macroscale pumps, can be used at the nanoscale (Wang and Kral, 2007). The authors conclude that performance is highly sensitive to the surface interactions between the driven fluid and the material of the propeller surface. Micron scale motors are of course known in the form of flagella used by spermatozoa and many unicellular organisms to achieve mobility. As suggested in the introduction to this chapter, the sophisticated biological mechanisms found in specialized transport proteins embedded in cell membranes can also be considered as nanofluidic pumps. A recently published review of transport mechanisms involving the confinement of water in nanopores provides extended background (Rasaiah et al., 2008).

11.7 Interfacing, Interconnections, and Nanofluidic Device Fabrication

Also necessary for the fabrication of practical nanofluidic devices incorporating nanotubes and nanopipes are general architectural facilities for forming and accurately positioning channels, for interconnecting fluid conduits, for forming branches, and for interfacing to macroscale inputs and outputs. Microfabrication techniques, originally developed for use in the semiconductor industry, offer a broad range of relevant capabilities (Harnett et al., 2001; Madou, 2002; Lorenz et al., 2004; Flachsbart et al., 2006; Han et al., 2006). These use "top-down" methods such as lithography and etching to form patterns and structures such as those typically found in integrated circuits. Features as small as a few tens of nanometers can be defined in this way using the latest methods. For fabrication below this size limit, self-assembly or "bottom-up" techniques are appropriate. These take advantage of natural chemical and physical processes to rationally arrange and organize nanoscale objects, for example, through the formation of self-assembling molecular monolayers and the natural segregation of block co-polymers (Liang et al., 2004). The self-ordering of pores in AAO templates, discussed above in the contexts of nanopipe synthesis, is another instance. A combination of top-down and bottom-up methodologies is a promising strategy for fabricating practical devices with nanoscale components (Mendes and Preece, 2004; Mijatovic et al., 2005; Riegelman et al., 2006).

A related suite of resources, already tailored to handling fluids, has been developed for applications in the rapidly advancing field of microfluidics, where great progress has been made in recent years (Whitesides and Stroock, 2001; deMello, 2006; Whitesides, 2006; Abgrall and Gue, 2007). This technology platform provides a solid foundation for nanofluidic device design. Using microfluidic technology, the "lab on a chip" concept has been realized with the launch of commercial miniaturized devices for sensing, diagnosis, and synthesis. Designers take advantage of the unique characteristics of microfluidic systems including laminar flow, very small sample volumes, rapid and highly efficient mixing schemes, and the possibility of ultra-sensitive (ultimately single molecule) detection (deMello and deMello, 2004; Dekker, 2007; Horsman et al., 2007; Kuswandi et al., 2007; Hansen and Miro, 2008; Mansur et al., 2008; Winkle et al., 2008).

There are a number of reports of experimental work addressing the challenge of microfluidic to nanofluidic interfacing. Liquid samples for characterization or processing using nanoscale components are likely to start out as macroscale droplets. These must be guided into progressively smaller channels for ultimate delivery to the nanoscale device components. A number of potential problems arise. One is channel blocking due to large macromolecules or insoluble debris (Stone et al., 2004). This can already be a limiting hazard with microscale lab-on-a-chip systems. Work is currently taking place in many laboratories to understand the dynamics of particulate transport at the nanoscale and to find low friction coatings or channel materials that help to reduce blocking (Sharp and Adrian, 2005).

Another problem is more fundamental: the large size of polymers, including biologically relevant polymeric molecules such as DNA, which are often tangled and tightly folded *in vivo*. A typical DNA molecule from a virus has a length of 100–200 kb and will form a random coil with a radius of some 700 nm in aqueous solution at 20°C (Cao et al., 2002). This is several times greater than the pore diameter of even large carbon pipes and two orders of magnitude greater than the diameter of a SWNT, such as might be functionalized to detect the presence of specific base sequences (Heller et al., 2006). DNA molecules will fit into the central channel of even small nanotubes, but only when unraveled and fed into the pore opening lengthwise. The entropic barrier to achieve this from the disordered state is very high and therefore such long molecules are normally excluded.

To overcome this problem, investigators at Princeton University used optical lithography to fabricate an array of microchannels forming a gradient fluidic device to interface the microscale to the nanoscale (Cao et al., 2002).

They used a novel modified form of diffraction gradient lithography involving a photosensitive blocking mask resist on a silicon wafer substrate. The technique is inherently parallel and both faster and more efficient than using e-beam lithography. The end result is a massive array of microposts, with a continuous reduction in the gaps that form fluidic channels as the chip is traversed from one side to the other.

To test the device, long DNA strands stained with a fluorescent dye were introduced on the microscale side. Diffusion of the molecules was then observed under a UV light and captured on video. Still frames from the recording showed individual DNA molecules straightening out and moving through the interface in an extended configuration towards the nanoscale region (see Figure 11.10). The authors reported the transport of the stretched DNA molecules with significantly greater efficiency compared

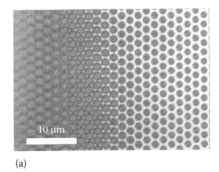
(a)

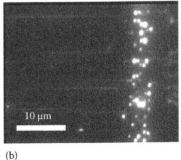
(b)

FIGURE 11.10 In the above pair of optical micrographs (reproduced with permission of the authors) the left image shows the chip after development of the photoresist. A continuous reduction in the gaps between the posts can be seen moving in a horizontal direction from right to left. The right-hand image shows integrated video recordings of fluorescent-labeled DNA molecules entering the gradient zone and becoming elongated as they enter the nanochannels on the left side of the picture. (From Cao, H. et al., *Appl. Phys. Lett.*, 81, 3058, 2002.)

to random diffusion through comparable nanoscale pores in the absence of any gradient interface (Cao et al., 2002).

Wang et al. have created a nanoscale preconcentration device using standard photolithography and etching methods, which has achieved concentration factors in the range of 10^6–10^8 (Wang et al., 2005). They exploit the electrokinetic trapping effect found in nanofluidic filters. A team from Oak Ridge National Laboratory in Tennessee reports a general method for creating patterned arrays of silica nanopipes precisely positioned over pores in a silicon nitride membrane on a silicon substrate (Melechko et al., 2003). The method requires expensive equipment and top-down fabrication techniques, but allows a high-degree of control over the device architecture. Critically, precise control of the nanopipe location is achieved by deterministic positioning of catalyst particles for CVD growth. Bau et al. describe another method for constructing a nanofluidic device comprising a single carbon nanopipe (diameter 250 nm) connecting two fluid reservoirs (Bau et al., 2005). Dielectrophoresis was used to orient and manipulate the nanopipe into position prior to depositing a dividing wall on top. Other workers have fabricated nanofluidic devices with microfluidic interconnects using interferometric lithography (O'Brien et al., 2003) and the creation of multiple polymer layers involving contact printing with a thermally cured adhesive (Flachsbart et al., 2006).

Finally, there have been efforts to realize Y-junctions and other interconnect topologies involving carbon nanotubes (Ho et al., 2001; Wang et al., 2006). The motivation for such research has primarily been driven by interest in the electronic properties of junction structures, where transistor-like behavior can be expected (Kim et al., 2006; Choi et al., 2007). The integrity of the inner fluid channel is not critical for such applications and so the techniques developed are not immediately applicable to nanofluidic devices. Some preliminary investigation of mechanical stability, fluid flow dynamics, and possible applications of Y-junctions, such as for ion separation, have been made using MD simulation (Hanasaki et al., 2004; Meng et al., 2006; Park et al., 2006).

The varied approaches described above currently enable prototyping of functional nanofluidic devices for research and testing. Once effective designs have been demonstrated to the proof-of-concept stage, less costly and highly repeatable fabrication methods will need to be developed to make large scale, affordable manufacturing possible. This will require considerable effort and investment, which can only be justified if there are clear applications for nanofluidic devices incorporating carbon nanotubes.

11.8 Applications for Fluid Flow through Nanopipes

The size scale of nanopores discussed in this chapter is essentially the same as the dimension of many important biological entities (antibodies, enzymes, viruses, DNA molecules). Thus, carbon nanotubes and nanopipes are potential conduits, concentrators, detectors, containers, and probes for biomedical applications. Doubtlessly, many challenges remain before such devices become practical including: ensuring the mechanical strength and biochemical compatibility of nanocomponents in proximity with living cells and tissue, developing methods for assembling huge numbers of nanoscale components, using precautions to avoid fouling of the channels and surfaces, controlling defects in the components, and managing information flow from nanoscale sensors to the outside world.

11.8.1 Filtering and Purification

An obvious nanofluidics application involving carbon nanotubes is materials separation, purification, and processing; taking advantage of the small pore size to selectively transport target molecules while excluding larger contaminants. Desalination of seawater, which is likely to require nanotubes with very small pore diameters, is an exciting example with enormous global demand. Physical filtering can be supplemented with electrochemical and surface modification strategies (Schoch et al., 2008). Selectivity has been demonstrated for gas transport through membranes incorporating nanotubes (Hinds, 2006; Pietrass, 2006; Sholl and Johnson, 2006). Promising results have been demonstrated using EOF, specifically taking advantage of the electric double layer on electrokinetic transport (Yuan et al., 2007). A number of studies were mentioned in Section 11.2.2 involving filtration through tangled mats of SWNTs and MWNTs with fluids passing between rather than through the nanopipes (Srivastava et al., 2004; Lu et al., 2005; Yan et al., 2006b; Brady-Estevez et al., 2008). An interesting study, which fully exploits the interior pores of carbon nanotubes, presents evidence to show that viruses can be destroyed by disassembly induced by confinement inside the pore (Fan et al., 2008).

A review of the environmental applications of carbon nanomaterials, including the use of carbon nanotubes for filtering and purification, is available in the article by Mauter and Elimelech (2008).

11.8.2 DNA Sensing

Rapid progress has been made during the past several decades in sequencing DNA and in analyzing the resulting genetic information. The field of bioinformatics accelerates our understanding of living systems. This knowledge is enabling improvements in medical care, particularly by customizing therapies to the specific needs of the individual. To fully realize this promise, it will be necessary to achieve a dramatic reduction in the time and cost associated with DNA analysis. Ultimately, the goal is make sequencing of a person's genome in a few hours using equipment that costs just tens, rather than tens of millions, of dollars possible (Hood, 2004). Researchers in this swiftly developing field are looking to nanofluidics to provide key components of the necessary technology (Tegenfeldt et al., 2004).

An MD simulation study in 2003 indicated that short DNA molecules spontaneously enter carbon nanotubes in an aqueous environment (Gao et al., 2003). Similar studies have been carried out for RNA using an aligned SWNT membrane with a pore diameter of 1.5 nm (Yeh and Hummer, 2004). In order to fit, the nucleic acid chains must stretch out into a linear configuration that may be a useful feature for subsequent sequencing. An electric field is used to drive the charged macromolecules through the carbon nanochannels. Soon after these simulation papers appeared, the first experimental results were published confirming that DNA can enter 50 nm gold nanotubes (Kohli et al., 2004) and that 5–8 μm long DNA sequences can be transported through 50 nm wide silica nanochannels (Fan et al., 2005a). Monitoring of the ionic current was used to detect the passage of the DNA molecules. More recent studies have investigated ways of slowing the transit times with the ultimate aim of identifying the individual nucleotide bases as they pass through the nanopore (Kim et al., 2007). An optical readout from parallel arrays of nanopores has also been demonstrated (Mulero and Kim, 2008).

11.8.3 Drug Delivery

The application of nanoscale principles and engineering to the challenge of drug delivery is a rapidly advancing field with a fast-growing literature and high levels of commercial and government investment (Martin and Kohli, 2003; Sinha et al., 2004; Wagner et al., 2006; Emerich et al., 2007; Sahoo et al., 2007; Hervella et al., 2008; Singh et al., 2008; Venugopal et al., 2008). A growing capability to control both physical and chemical characteristics at very small scales (<100 nm) is enabling the development of materials, pharmacological agents, and biomedical devices with useful novel properties. Multiple perceived benefits motivate this effort including: the precise targeting of therapeutic agents only to diseased cells; the delivery of drugs to otherwise inaccessible locations; improved control of doses at the precise site of the disease; extended release mechanisms reducing the frequency of medication; the lowering of systematic side effects, and the widening of the scope of compounds available for therapy. In particular, research into the use of nanoparticulates and nanosuspensions is yielding promising results (Rabinow, 2004; Torchilin, 2006; Emerich and Thanos, 2007). In the midst of this enthusiasm for nanomedicine, it is, however, important to keep in mind well-founded concerns that some nanomaterials may involve new risks to health (Dowling, 2004; Lam et al., 2006; Maysinger et al., 2006; Jain et al., 2007; Lacerda et al., 2008; Vega-Villa et al., 2008; Yu et al., 2008)

In parallel with the general interest in nanoparticles for drug delivery, a number of researchers have recognized the potential of nanotubes and nanopipes in biomedical applications (Martin and Kohli, 2003; Klumpp et al., 2006; Son et al., 2007a,b; Hilder and Hill, 2008). Son et al. have employed templated synthesis in AAO to produce composite multifunctional nanotubes containing iron and silica nanoparticles for controlled release and Magnetic Resonance Imaging (MRI) applications (Son et al., 2006). This group, based at the University of Maryland, has concentrated on the use of wet chemistry (sol gel) techniques for growing the nanotubes, rather than CVD as previously described for making carbon nanopipes. The authors stress the benefits of

being able to differentially functionalize the interior and exterior surfaces of the nanopipes that they have produced. The outside can be optimized for targeting and solubility; the interior can be configured for the transport and controlled release of a drug payload. Successful functionalization strategies have already been developed for liposomes and nanoparticles (Emerich and Thanos, 2007; Torchilin, 2007).

Gong et al. (2003) from Pennsylvania State University have demonstrated the use of nanoporous alumina capsules, with pores in the range of 25–55 nm, for controlled drug release. The authors created a cylindrical membrane by direct anodization of an aluminum metal tube. They also employed a technique for creating a branching pore structure to control pore size. In related work, Orosz et al. (2004) at Ohio State University used alumina nanoporous membranes with 20 nm diameter pores to deliver antiangiogenic and antioxidant drugs to human retinal endothelial cells in culture. They report improved (i.e., slower and more linear) release kinetics and better selectivity than conventional membranes. In another study of release kinetics, investigating silicon pores with diameters down to 7 nm and including in vivo experiments, zero-order diffusion was observed indicating non-Fickian behavior as the nanopore width approaches the hydrodynamic diameter of the solute (Martin et al., 2005). This phenomenon could be useful in achieving constant drug levels in the body over extended periods of time. A recent review of nanopipe loading and unloading for drug delivery is available in an article by Hilder and Hill (2008).

11.8.4 Cellular Probes and Nanoneedles

An exciting biomedical application for carbon nanotubes and nanopipes is to interact directly with living cells, both to deliver drugs by penetrating the cell membrane and to take samples for analysis from the cytoplasm. Animal cells are typically 5–30 μm in size, which means that carbon probes with diameters up to a few hundred nm can be expected to cross the membrane in the same manner as a hypodermic needle without necessarily destroying the cell (Vereb et al., 2003). This offers a way to deliver therapeutic drugs, for example macromolecules such as DNA or RNA sequences, that may otherwise be difficult to introduce inside living cells.

In one of the first such studies, individual MWNTs (diameter <100 nm) and bundles of aligned SWNTs were attached to the ends of fine tungsten needles with the help of dielectrophoretic alignment (Kouklin et al., 2005). Van der Waals forces were sufficient to maintain attachment and high aspect ratio probes with lengths >100 μm were constructed. These probes were brought into contact with the human epithelial cells using an optical microscope. It was shown that they were strong enough to penetrate the cell membranes. Furthermore, the cells withstood the contact and continued to be viable after penetration.

Progress in this direction was reported 2 years later in a study involving the attachment of single MWNTs to an AFM tip (Chen et al., 2007). Although transport through the central pore of the nanotubes was not attempted with this configuration, streptavidin-coated quantum dots were attached to the outer surface of the nanotubes and successfully introduced into individually selected target cells. Penetration of the membrane was achieved by means of the AFM, which allowed controlled contact between the loaded MWNT and the cell membrane.

Further progress was made by the nanofluidics group at Drexel University, Philadelphia with the construction of a carbon nanopipe tipped glass pipette (Freedman et al., 2007). Templated CVD was used to produce nanopipes with a ~200 nm diameter and 60 μm length. These were filled with a suspension of magnetite nanoparticles (ferrofluid) and introduced into a pulled glass capillary that tapered down to ~900 nm. Using a magnetic field, the nanopipes were aligned along the axis of the glass capillary and manipulated so that around half the length of a single pipe protruded. An ultraviolet polymerizable adhesive was then locally cured at the pipette entrance by focusing the objective of a UV microscope in that region. This resulted in sealing the carbon nanopipe in place without blocking its inner pore; 100 nm fluorescent polystyrene beads were shown to pass down the glass pipette and out exclusively through the carbon nanopipe tip, confirming the integrity of the seal. Preliminary experiments were then undertaken demonstrating the successful penetration and withdrawal from dog kidney cells without apparent damage to either the cell or the nanopipette. Similar results were achieved using a different technique to manufacture the nanopipettes in which carbon was deposited inside a pulled quartz capillary using CVD (Schrlau et al., 2008). The outer layer of quartz was then subsequently removed by etching. Somewhat larger fluidic probes have also been fabricated from electrospun carbon nanopipe bundles (Bazilevsky et al., 2008b). All three approaches allow for electrical connectivity to the nanoprobe tip in addition to fluid transport and make a range of novel experiments on single living cells possible.

11.9 Conclusions and Future Directions

Research involving fluid flow through carbon nanotubes and nanopipes has seen rapid growth in the past 5 years with the publication of more than 100 experimental and theoretical papers bearing on the field, many of which have been reviewed here. Evidence has been presented for non-classical fluid behavior involving slip flow as the width of carbon channels approaches a few nanometers. Results showing several orders of magnitude increase for flow velocities of water through carbon nanotubes have been claimed independently by two groups. While there is a continuing debate over the actual magnitude of this effect, there is an emerging consensus that some degree of flow enhancement over classical predictions does occur for nanoscale flows (Mattia and Gogotsi, 2008; Thomas and McGaughey, 2008). Theoretical models of nanoscale flow, particularly taking into account the interaction between fluid molecules and pore surfaces, need refinement to fully account for observed phenomena.

From a practical perspective, a broad range of techniques have been established to create a variety of tubular carbon nanofluidic channels with controllable diameters from 1 to several 100 nm.

Methods have been established to modulate fluid and ionic transport through these conduits and preliminary progress has been made towards implementing pumping systems. Applications being actively developed include filtration, drug delivery, and nanofluidic interaction with single cells. The control of surface properties, through chemical modification and functionalization, is recognized as critical for future progress.

Acknowledgments

The authors wish to thank EPSRC, United Kingdom for funding under grant number P09017 "Experimental Nanofluidics" and for a Doctoral Training Award to M. Whitby. We are grateful to M. Thanou, A. MacGaughey, and J. Thomas for helpful comments on the manuscript.

References

Abgrall, P. and Gue, A. M. (2007) Lab-on-chip technologies: making a microfluidic network and coupling it into a complete microsystem—A review. *Journal of Micromechanics and Microengineering*, 17, R15–R49.

Abgrall, P. and Nguyen, N. T. (2008) Nanofluidic devices and their applications. *Analytical Chemistry*, 80, 2326–2341.

Agre, P. (2004) Aquaporin water channels (Nobel lecture). *Angewandte Chemie-International Edition*, 43, 4278–4290.

Ajayan, P. M. and Iijima, S. (1993) Capillarity-induced filling of carbon nanotubes. *Nature*, 361, 333–334.

Ajayan, P. M., Ebbesen, T. W., Ichihashi, T., Iijima, S., Tanigaki, K., and Hiura, H. (1993) Opening carbon nanotubes with oxygen and implications for filling. *Nature*, 362, 522–525.

Awasthi, K., Srivastava, A., and Srivastava, O. N. (2005) Synthesis of carbon nanotubes. *Journal of Nanoscience and Nanotechnology*, 5, 1616–1636.

Babu, S., Ndungu, P., Bradley, J.-C., Rossi, M. A. P. A., and Gogotsi, Y. (2005) Guiding water into carbon nanopipes with the aid of bipolar electrochemistry. *Microfluidics and Nanofluidics*, 1, 284–288.

Bai, J., Su, C. R., Parra, R. D., Zeng, X. C., Tanaka, H., Koga, K., and Li, J. M. (2003) Ab initio studies of quasi-one-dimensional pentagon and hexagon ice nanotubes. *Journal of Chemical Physics*, 118, 3913–3916.

Barber, A. H., Cohen, S. R., and Wagner, H. D. (2004) Static and dynamic wetting with various liquids. *Physical Review Letters*, 92, 186103.

Barber, A. H., Cohen, S. R., and Wagner, H. D. (2005) External and internal wetting of carbon nanotubes with organic liquids. *Physical Review B*, 71, 115443-1–115443-5.

Bau, H. H., Sinha, S., Kim, B. M., and Riegelman, M. (2005) Fabrication of nanofluidic devices and the study of fluid transport through them. *Proceedings of SPIE*, 5592, 201–213.

Bazilevsky, A. V., Sun, K., Yarin, A. L., and Megaridis, C. M. (2008a) Room-temperature, open-air, wet intercalation of liquids, surfactants, polymers and nanoparticles within nanotubes and microchannels. *Journal of Materials Chemistry*, 18, 696–702.

Bazilevsky, A. V., Yarin, A. L., and Megaridis, C. M. (2008b) Pressure-driven fluidic delivery through carbon tube bundles. *Lab on a Chip*, 8, 152–160.

Bianco, A., Kostarelos, K., and Prato, M. (2008) Opportunities and challenges of carbon-based nanomaterials for cancer therapy. *Expert Opinion on Drug Delivery*, 5, 331–342.

Bismarck, A., Wuertz, C., and Springer, J. (1999) Basic surface oxides on carbon fibers. *Carbon*, 37, 1019–1027.

Bosanquet, C. H. (1923) The flow of liquids into capillary tubes. *Philosophical Magazine*, 6, 525–531.

Brady-Estevez, A. S., Kang, S., and Elimelech, M. (2008) A single-walled-carbon-nanotube filter for removal of viral and bacterial pathogens. *Small*, 4, 481–484.

Byl, O., Liu, J. C., Wang, Y., Yim, W. L., Johnson, J. K., and Yates, J. T. (2006) Unusual hydrogen bonding in water-filled carbon nanotubes. *Journal of the American Chemical Society*, 128, 12090–12097.

Cao, H., Tegenfeldt, J. O., Austin, R. H., and Chou, S. Y. (2002) Gradient nanostructures for interfacing microfluidics and nanofluidics. *Applied Physics Letters*, 81, 3058–3060.

Cercignani, C. (1969) *Mathematical Methods in Kinetic Theory*, New York, Plenum.

Che, G., Lakshmi, B. B., Martin, C. R., Fisher, E. R., and Ruoff, R. S. (1998a) Chemical vapor deposition based synthesis of carbon nanotubes and nanofibers using a template method. *Chemistry of Materials*, 10, 260–267.

Che, G. L., Lakshmi, B. B., Fisher, E. R., and Martin, C. R. (1998b) Carbon nanotubule membranes for electrochemical energy storage and production. *Nature*, 393, 346–349.

Chen, J. Y., Kutana, A., Collier, C. P., and Giapis, K. P. (2005) Electrowetting in carbon nanotubes. *Science*, 310, 1480–1483.

Chen, X., Tam, U. C., Czlapinski, J. L., Lee, G. S., Rabuka, D., Zettl, A., and Bertozzi, C. R. (2006) Interfacing carbon nanotubes with living cells. *Journal of the American Chemical Society*, 128, 6292–6293.

Chen, X., Kis, A., Zettl, A., and Bertozzi, C. R. (2007) A cell nano-injector based on carbon nanotubes. *Proceedings of the National Academy of Sciences of the United States of America*, 104, 8218–8222.

Chen, Y. K., Chu, A., Cook, J., Green, M. L. H., Harris, P. J. F., Heesom, R., Humphries, M., Sloan, J., Tsang, S. C., and Turner, J. F. C. (1997) Synthesis of carbon nanotubes containing metal oxides and metals of the d-block and f-block transition metals and related studies. *Journal of Materials Chemistry*, 7, 545–549.

Cheng, J. T. and Giordano, N. (2002) Fluid flow through nanometer-scale channels. *Physical Review E*, 65, 031206.

Chirila, V., Marginean, G., and Brandl, W. (2005) Effect of the oxygen plasma treatment parameters on the carbon nanotubes surface properties. *Surface & Coatings Technology*, 200, 548–551.

Choi, W. B., Kim, D. H., Choi, Y. C., and Huang, J. (2007) Y-junction single-wall carbon nanotube electronics. *JOM*, 59, 44–49.

Cooper, S. M., Cruden, B. A., Meyyappan, M., Raju, R., and Roy, S. (2004) Gas transport characteristics through a carbon nanotubule. *Nano Letters*, 4, 377–381.

Dai, L. (2006) *Carbon Nanotechnology*, Amsterdam, the Netherlands, Elsevier.

Daub, C. D., Bratko, D., Leung, K., and Luzar, A. (2007) Electrowetting at the nanoscale. *Journal of Physical Chemistry C*, 111, 505–509.

de Gennes, P.-G., Brochard-Wyart, F., and Quere, D. (2004) *Capillarity and Wetting Phenomena*, New York, Springer.

Dekker, C. (2007) Solid-state nanopores. *Nature Nanotechnology*, 2, 209–215.

deMello, A. (2006) Control and detection of chemical reactions in microfluidic systems. *Nature*, 442, 5062.

deMello, J. and deMello, A. (2004) Microscale reactors: Nanoscale products. *Lab on a Chip*, 4, 11N–15N.

Dowling, A. (2004) *Nanoscience and Nanotechnologies: Opportunities and Uncertainties*, London, U.K., The Royal Society and The Royal Academy of Engineering.

Dresselhaus, G., Dresselhaus, M. S., and Eklund, P. C. (1996) *Science of Fullerenes and Carbon Nanotubes*, New York, Academic Press.

Dujardin, E., Ebbesen, T. W., Hiura, H., and Tanigaki, K. (1994) Capillarity and wetting of carbon nanotubes. *Science*, 265, 1850–1852.

Eijkel, J. C. T. (2005) Nanofluidics: What is it and what can we expect from it? *Microfluidics and Nanofluidics*, 1, 249–267.

Eijkel, J. (2007) Liquid slip in micro- and nanofluidics: recent research and its possible implications. *Lab on a Chip*, 7, 299–301.

Emerich, D. F. and Thanos, C. G. (2007) Targeted nanoparticle-based drug delivery and diagnosis. *Journal of Drug Targeting*, 15, 163–183.

Emerich, D. F., Halberstadt, C., and Thanos, C. (2007) Role of nanobiotechnology in cell-based nanomedicine: A concise review. *Journal of Biomedical Nanotechnology*, 3, 235–244.

Endo, M., Takeuchi, K., Igarashi, S., Kobori, K., Shiraishi, M., and Kroto, H. W. (1993) The production and structure of pyrolytic carbon nanotubes (PCNTS). *Journal of Physics and Chemistry of Solids*, 54, 1841–1848.

Fan, R., Karnik, R., Yue, M., Li, D. Y., Majumdar, A., and Yang, P. D. (2005a) DNA translocation in inorganic nanotubes. *Nano Letters*, 5, 1633–1637.

Fan, R., Yue, M., Karnik, R., Majumdar, A., and Yang, P. D. (2005b) Polarity switching and transient responses in single nanotube nanofluidic transistors. *Physical Review Letters*, 95, 086607.

Fan, X. B., Barclay, J. E., Peng, W. C., Li, Y., Li, X. Y., Zhang, G. L., Evans, D. J., and Zhang, F. B. (2008) Capillarity-induced disassembly of virions in carbon nanotubes. *Nanotechnology*, 19, 165702.

Felten, A., Bittencourt, C., Pireaux, J. J., Van Lier, G., and Charlier, J. C. (2005) Radio-frequency plasma functionalization of carbon nanotubes surface O_2, NH_3, and CF_4 treatments. *Journal of Applied Physics*, 98, 074308.

Flachsbart, B. R., Wong, K., Iannacone, J. M., Abante, E. N., Vlach, R. L., Rauchfuss, P. A., Bohn, P. W., Sweedler, J. V., and Shannon, M. A. (2006) Design and fabrication of a multilayered polymer microfluidic chip with nanofluidic interconnects via adhesive contact printing. *Lab on a Chip*, 6, 667–674.

Freedman, J. R., Mattia, D., Korneva, G., Gogotsi, Y., Friedman, G., and Fontecchio, A. K. (2007) Magnetically assembled carbon nanotube tipped pipettes. *Applied Physics Letters*, 90, 3.

Galanov, B. A., Galanov, S. B., and Gogotsi, Y. (2002) Stress-strain state of multiwall carbon nanotube under internal pressure. *Journal of Nanoparticle Research*, 4, 207–214.

Gao, H. J., Kong, Y., Cui, D. X., and Ozkan, C. S. (2003) Spontaneous insertion of DNA oligonucleotides into carbon nanotubes. *Nano Letters*, 3, 471–473.

Garcia, A. B., Cuesta, A., Montesmoran, M. A., Martinezalonso, A., and Tascon, J. M. D. (1997) Zeta potential as a tool to characterize plasma oxidation of carbon fibers. *Journal of Colloid and Interface Science*, 192, 363–367.

Gardeniers, H. and Van den Berg, A. (2004) Micro- and nanofluidic devices for environmental and biomedical applications. *International Journal of Environmental Analytical Chemistry*, 84, 809–819.

Gogotsi, Y. (2006) *Carbon Nanomaterials*, Boca Raton, FL, Taylor & Francis.

Gogotsi, Y., Libera, J. A., and Yoshimura, M. (2000) Hydrothermal synthesis of multiwall carbon nanotubes. *Journal of Materials Research*, 15, 2591–2594.

Gogotsi, Y., Libera, J. A., Guvenc-Yazicioglu, A., and Megaridis, C. M. (2001) In situ multiphase fluid experiments in hydrothermal carbon nanotubes. *Applied Physics Letters*, 79, 1021–1023.

Gogotsi, Y., Naguib, N., and Libera, J. A. (2002) In situ chemical experiments in carbon nanotubes. *Chemical Physics Letters*, 365, 354–360.

Gong, D. W., Yadavalli, V., Paulose, M., Pishko, M., and Grimes, C. A. (2003) Controlled molecular release using nanoporous alumina capsules. *Biomedical Microdevices*, 5, 75–80.

Gong, X. J., Li, J. Y., Lu, H. J., Wan, R. Z., Li, J. C., Hu, J., and Fang, H. P. (2007) A charge-driven molecular water pump. *Nature Nanotechnology*, 2, 709–712.

Govindaraj, A. and Rao, C. N. R. (2006) Synthesis, growth mechanism and processing of carbon nanotubes. In Dai, L. M. (Ed.) *Carbon Nanotechnology*, Amsterdam, the Netherlands, Elsevier.

Han, A. P., de Rooij, N. F., and Staufer, U. (2006) Design and fabrication of nanofluidic devices by surface micromachining. *Nanotechnology*, 17, 2498–2503.

Hanasaki, I., Nakatani, A., and Kitagawa, H. (2004) Molecular dynamics study of Ar flow and He flow inside carbon nanotube junction as a molecular nozzle and diffuser. *Science and Technology of Advanced Materials*, 5, 107–113.

Hansen, E. H. and Miro, M. (2008) Interfacing microfluidic handling with spectroscopic detection for real-life applications via the lab-on-valve platform: A review. *Applied Spectroscopy Reviews*, 43, 335–357.

Harnett, C. K., Coates, G. W., and Craighead, H. G. (2001) Heat-depolymerizable polycarbonates as electron beam patternable sacrificial layers for nanofluidics. *Journal of Vacuum Science & Technology B*, 19, 2842–2845.

Harris, P. J. F. (1999) *Carbon Nanotubes and Related Structures*, Cambridge, U.K., Cambridge University Press.

Heller, D. A., Jeng, E. S., Yeung, T. K., Martinez, B. M., Moll, A. E., Gastala, J. B., and Strano, M. S. (2006) Optical detection of DNA conformational polymorphism on single-walled carbon nanotubes. *Science*, 311, 508–511.

Hervella, P., Lozano, V., and Garcia-Fuentes, M. (2008) Nanomedicine: New challenges and opportunities in cancer therapy. *Journal of Biomedical Nanotechnology*, 4, 276–292.

Herrero, M. A. and Prato, M. (2008) Recent advances in the covalent functionalization of carbon nanotubes. *Molecular Crystals and Liquid Crystals*, 483, 21–32.

Hilder, T. A. and Hill, J. M. (2008) Modeling the loading and unloading of drugs into nanotubes. *Small*, 5, 300–308.

Hinds, B. (2007) A blueprint for a nanoscale pump. *Nature Nanotechnology*, 2, 673–674.

Hinds, B. J. (2006) Aligned carbon nanotube membranes. In Dai, L. (Ed.) *Carbon Nanotechnology*, Amsterdam, the Netherlands, Elsevier.

Hinds, B. J., Chopra, N., Rantell, T., Andrews, R., Gavalas, V., and Bachas, L. G. (2004) Aligned multiwalled carbon nanotube membranes. *Science*, 303, 62–65.

Hirsch, A. and Vostrowsky, O. (2007) Functionalization of carbon nanotubes. In *Functional Organic Materials*. (Eds., Prof. Dr. Bunz, U. H. F. and Prof. Dr. Müller, T. J. J.) Wiley-VCH Verlag, Weinheim, Germany, pp. 1–57.

Ho, G. W., Wee, A. T. S., and Lin, J. (2001) Electric field-induced carbon nanotube junction formation. *Applied Physics Letters*, 79, 260–262.

Holt, J. K., Noy, A., Huser, T., Eaglesham, D., and Bakajin, O. (2004) Fabrication of a carbon nanotube-embedded silicon nitride membrane for studies of nanometer-scale mass transport. *Nano Letters*, 4, 2245–2250.

Holt, J. K., Park, H. G., Wang, Y. M., Stadermann, M., Artyukhin, A. B., Grigoropoulos, C. P., Noy, A., and Bakajin, O. (2006) Fast mass transport through sub-2-nanometer carbon nanotubes. *Science*, 312, 1034–1037.

Hood, L. (2004) *Lecture on Bioinformatics*, London, U.K., Imperial College.

Horsman, K. M., Bienvenue, J. M., Blasier, K. R., and Landers, J. P. (2007) Forensic DNA analysis on microfluidic devices: A review. *Journal of Forensic Sciences*, 52, 784–799.

Howard, M. D., Jay, M., Dziublal, T. D., and Lu, X. L. (2008) PEGylation of nanocarrier drug delivery systems: State of the art. *Journal of Biomedical Nanotechnology*, 4, 133–148.

Hulteen, J. C. and Martin, C. R. (1997) A general template-based method for the preparation of nanomaterials. *Journal of Materials Chemistry*, 7, 1075–1087.

Hummer, G., Rasaiah, J. C., and Noworyta, J. P. (2001) Water conduction through the hydrophobic channel of a carbon nanotube. *Nature*, 414, 188–190.

Hutchison, J. L., Kiselev, N. A., Krinichnaya, E. P., Krestinin, A. V., Loutfy, R. O., Morawsky, A. P., Muradyan, V. E., Obraztsova, E. D., Sloan, J., Terekhov, S. V., and Zakharov, D. N. (2001) Double-walled carbon nanotubes fabricated by a hydrogen arc discharge method. *Carbon*, 39, 761–770.

Iijima, S. (1991) Helical microtubules of graphitic carbon. *Nature*, 354, 56–58.

Insepov, Z., Wolf, D. and Hassanein, A. (2006) Nanopumping using carbon nanotubes. *Nano Letters*, 6, 1893–1895.

Itaya, K., Sugawara, S., Arai, K., and Saito, S. (1984) Properties of porous anodic aluminum-oxide films as membranes. *Journal of Chemical Engineering of Japan*, 17, 514–520.

Izard, N., Riehl, D., and Anglaret, E. (2005) Exfoliation of single-wall carbon nanotubes in aqueous surfactant suspensions: A Raman study. *Physical Review B*, 71, 195417.

Jain, A. K., Mehra, N. K., Lodhi, N., Dubey, V., Mishra, D. K., Jain, P. K., and Jain, N. K. (2007) Carbon nanotubes and their toxicity. *Nanotoxicology*, 1, 167–197.

Jain, K. K. (2005) Nanotechnology in clinical laboratory diagnostics. *Clinica Chimica Acta*, 358, 37–54.

Jones, R. (2006) What can biology teach us? *Nature Nanotechnology*, 1, 85–86.

Joo, S. H., Choi, S. J., Oh, I., Kwak, J., Liu, Z., Terasaki, O., and Ryoo, R. (2001) Ordered nanoporous arrays of carbon supporting high dispersions of platinum nanoparticles. *Nature*, 412, 169–172.

Joseph, S. and Aluru, N. R. (2008a) Pumping of confined water in carbon nanotubes by rotation-translation coupling. *Physical Review Letters*, 101, 064502.

Joseph, S. and Aluru, N. R. (2008b) Why are carbon nanotubes fast transporters of water? *Nano Letters*, 8, 452–458.

Karnik, R., Fan, R., Yue, M., Li, D. Y., Yang, P. D., and Majumdar, A. (2005) Electrostatic control of ions and molecules in nano-fluidic transistors. *Nano Letters*, 5, 943–948.

Kim, B. M., Sinha, S., and Bau, H. H. (2004) Optical microscope study of liquid transport in carbon nanotubes. *Nano Letters*, 4, 2203–2208.

Kim, B. M., Murray, T., and Bau, H. H. (2005) The fabrication of integrated carbon pipes with sub-micron diameters. *Nanotechnology*, 16, 1317–1320.

Kim, D. H., Huang, J., Rao, B. K., and Choi, W. B. (2006) Pseudo Y-junction single-walled carbon nanotube based ambipolar transistor operating at room temperature. *IEEE Transactions on Nanotechnology*, 5, 731–736.

Kim, Y. R., Min, J., Lee, I. H., Kim, S., Kim, A. G., Kim, K., Namkoong, K., and Ko, C. (2007) Nanopore sensor for fast label-free detection of short double-stranded DNAs. *Biosensors & Bioelectronics*, 22, 2926–2931.

Klumpp, C., Kostarelos, K., Prato, M., and Bianco, A. (2006) Functionalized carbon nanotubes as emerging nanovectors for the delivery of therapeutics. *Biochimica et Biophysica Acta-Biomembranes*, 1758, 404–412.

Koga, K., Gao, G. T., Tanaka, H., and Zeng, X. C. (2001a) Formation of ordered ice nanotubes inside carbon nanotubes. *Nature*, 412, 802–805.

Koga, K., Gao, G. T., Tanaka, H., and Zeng, X. C. (2001b) How does water freeze inside carbon nanotubes? *Meeting on Horizons in Complex Systems*, Messina, Italy, Elsevier Science BV.

Kohli, P., Harrell, C. C., Cao, Z. H., Gasparac, R., Tan, W. H., and Martin, C. R. (2004) DNA-functionalized nanotube membranes with single-base mismatch selectivity. *Science*, 305, 984–986.

Kolesnikov, A. I., Zanotti, J. M., Loong, C. K., Thiyagarajan, P., Moravsky, A. P., Loutfy, R. O., and Burnham, C. J. (2004) Anomalously soft dynamics of water in a nanotube: A revelation of nanoscale confinement. *Physical Review Letters*, 93, 035503.

Korneva, G., Ye, H. H., Gogotsi, Y., Halverson, D., Friedman, G., Bradley, J. C., and Kornev, K. G. (2005) Carbon nanotubes loaded with magnetic particles. *Nano Letters*, 5, 879–884.

Kotsalis, E. M., Walther, J. H., and Koumoutsakos, P. (2004) Multiphase water flow inside carbon nanotubes. *International Journal of Multiphase Flow*, 30, 995–1010.

Kouklin, N. A., Kim, W. E., Lazareck, A. D., and Xu, J. M. (2005) Carbon nanotube probes for single-cell experimentation and assays. *Applied Physics Letters*, 87, 3.

Kral, P. and Tomanek, D. (1999) Laser-driven atomic pump. *Physical Review Letters*, 82, 5373–5376.

Kuswandi, B., Nuriman, Huskens, J., and Verboom, W. (2007) Optical sensing systems for microfluidic devices: A review. *Analytica Chimica Acta*, 601, 141–155.

Kyotani, T., Tsai, L. F., and Tomita, A. (1995) Formation of ultra-fine carbon tubes by using an anodic aluminum-oxide film as a template. *Chemistry of Materials*, 7, 1427–1428.

Kyotani, T., Tsai, L. F., and Tomita, A. (1996) Preparation of ultra-fine carbon tubes in nanochannels of an anodic aluminum oxide film. *Chemistry of Materials*, 8, 2109–2113.

Lacerda, L., Bianco, A., Prato, M., and Kostarelos, K. (2006) Carbon nanotubes as nanomedicines: From toxicology to pharmacology. *Advanced Drug Delivery Reviews*, 58, 1460–1470.

Lacerda, L., Ali-Boucetta, H., Herrero, M. A., Pastorin, G., Bianco, A., Prato, M., and Kostarelos, K. (2008) Tissue histology and physiology following intravenous administration of different types of functionalized multiwalled carbon nanotubes. *Nanomedicine*, 3, 149–161.

Lago, R. M., Tsang, S. C., Lu, K. L., Chen, Y. K., and Green, M. L. H. (1995) Filling Carbon nanotubes with small palladium metal crystallites—The effect of surface acid groups. *Journal of the Chemical Society-Chemical Communications*, 13, 1355–1356.

Lam, C. W., James, J. T., Mccluskey, R., Arepalli, S., and Hunter, R. L. (2006) A review of carbon nanotube toxicity and assessment of potential occupational and environmental health risks. *Critical Reviews in Toxicology*, 36, 189–217.

Laplace, P. S. (1878) *Oeuvres Complètes de Laplace*, Paris, France, Gauthier-Villars.

Lavin, J. G., Subramoney, S., Ruoff, R. S., Berber, S., and Tomanek, D. (2002) Scrolls and nested tubes in multiwall carbon nanotubes. *Carbon*, 40, 1123–1130.

Li, J. Y., Gong, X. J., Lu, H. J., Li, D., Fang, H. P., and Zhou, R. H. (2007) Electrostatic gating of a nanometer water channel. *Proceedings of the National Academy of Sciences of the United States of America*, 104, 3687–3692.

Li, Q. W., Zhu, Y. T. T., Kinloch, I. A., and Windle, A. H. (2006) Self-organization of carbon nanotubes in evaporating droplets. *Journal of Physical Chemistry B*, 110, 13926–13930.

Li, S. Y., Xiu, P., Lu, H. J., Gong, X. J., Wu, K. F., Wan, R. Z., and Fang, H. P. (2008) Water permeation across nanochannels with defects. *Nanotechnology*, 19, 105711.

Liang, C. D., Hong, K. L., Guiochon, G. A., Mays, J. W., and Dai, S. (2004) Synthesis of a large-scale highly ordered porous carbon film by self-assembly of block copolymers. *Angewandte Chemie-International Edition*, 43, 5785–5789.

Liang, X. J., Chen, C. Y., Zhao, Y. L., Jia, L., and Wang, P. C. (2008) Biopharmaceutics and therapeutic potential of engineered nanomaterials. *Current Drug Metabolism*, 9, 697–709.

Libera, J. and Gogotsi, Y. (2001) Hydrothermal synthesis of graphite tubes using Ni catalyst. *Carbon*, 39, 1307–1318.

Lin, J. (2007) Drug delivery systems using carbon nanopipes. *Chemistry*, London, U.K., Imperial College.

Longhurst, M. J. and Quirke, N. (2006a) Environmental effects on the radial breathing modes of carbon nanotubes in water. *Journal of Chemical Physics*, 124, 234708.

Longhurst, M. J. and Quirke, N. (2006b) The environmental effect on the radial breathing mode of carbon nanotubes. II. Shell model approximation for internally and externally adsorbed fluids. *Journal of Chemical Physics*, 125, 184705.

Longhurst, M. J. and Quirke, N. (2007) Temperature-driven pumping of fluid through single-walled carbon nanotubes. *Nano Letters*, 7, 3324–3328.

Lorenz, R. M., Kuyper, C. L., Allen, P. B., Lee, L. P., and Chiu, D. T. (2004) Direct laser writing on electrolessly deposited thin metal films for applications in micro- and nanofluidics. *Langmuir*, 20, 1833–1837.

Lu, C., Chung, Y.-L., and Chang, K.-F. (2005) Adsorption of trihalomethanes from water with carbon nanotubes. *Water Research*, 39, 1183–1189.

Madou, M. J. (2002) *Fundamentals of Microfabrication*, Boca Raton, FL, CRC Press.

Majumder, M., Chopra, N., Andrews, R., and Hinds, B. J. (2005a) Nanoscale hydrodynamics—Enhanced flow in carbon nanotubes. *Nature*, 438, 44.

Majumder, M., Chopra, N., and Hinds, B. J. (2005b) Effect of tip functionalization on transport through vertically oriented carbon nanotube membranes. *Journal of the American Chemical Society*, 127, 9062–9070.

Majumder, M., Zhan, X., Andrews, R., and Hinds, B. J. (2007) Voltage gated carbon nanotube membranes. *Langmuir*, 23, 8624–8631.

Maniwa, Y., Kataura, H., Abe, M., Udaka, A., Suzuki, S., Achiba, Y., Kira, H., Matsuda, K., Kadowaki, H., and Okabe, Y. (2005) Ordered water inside carbon nanotubes: Formation of pentagonal to octagonal ice-nanotubes. *Chemical Physics Letters*, 401, 534–538.

Mansur, E. A., Ye, M. X., Wang, Y. D., and Dai, Y. Y. (2008) A state-of-the-art review of mixing in microfluidic mixers. *Chinese Journal of Chemical Engineering*, 16, 503–516.

Mao, S. H., Kleinhammes, A., and Wu, Y. (2006) NMR study of water adsorption in single-walled carbon nanotubes. *Chemical Physics Letters*, 421, 513–517.

Martin, C. R. (1994) Nanomaterials—A membrane-based synthetic approach. *Science*, 266, 1961–1966.

Martin, C. R. and Kohli, P. (2003) The emerging field of nanotube biotechnology. *Nature Reviews Drug Discovery*, 2, 29–37.

Martin, F., Walczak, R., Boiarski, A., Cohen, M., West, T., Cosentino, C., and Ferrari, M. (2005) Tailoring width of microfabricated nanochannels to solute size can be used to control diffusion kinetics. *Journal of Controlled Release*, 102, 123–133.

Martini, A., Roxin, A., Snurr, R. Q., Wang, Q., and Lichter, S. (2008) Molecular mechanisms of liquid slip. *Journal of Fluid Mechanics*, 600, 257–269.

Masuda, H. and Fukuda, K. (1995) Ordered metal nanohole arrays made by a 2-step replication of honeycomb structures of anodic alumina. *Science*, 268, 1466–1468.

Masuda, H., Hasegwa, F., and Ono, S. (1997) Self-ordering of cell arrangement of anodic porous alumina formed in sulfuric acid solution. *Journal of the Electrochemical Society*, 144, L127–L130.

Matsuda, K., Hibi, T., Kadowaki, H., Kataura, H., and Maniwa, Y. (2006) Water dynamics inside single-wall carbon nanotubes: NMR observations. *Physical Review B*, 74, 073415.

Mattia, D. and Gogotsi, Y. (2006) Surface functionalization to control the wetting behavior of nanostructured carbons. *Abstracts of Papers of the American Chemical Society*, 231, 1789–1794.

Mattia, D. and Gogotsi, Y. (2008) Review: Static and dynamic behavior of liquids inside carbon nanotubes. *Microfluidics and Nanofluidics*, 5, 289–305.

Mattia, D., Ban, H. H., and Gogotsi, Y. (2006a) Wetting of CVD carbon films by polar and nonpolar liquids and implications for carbon nanopipes. *Langmuir*, 22, 1789–1794.

Mattia, D., Rossi, M. P., Kim, B. M., Korneva, G., Bau, H. H., and Gogotsi, Y. (2006b) Effect of graphitization on the wettability and electrical conductivity of CVD-carbon nanotubes and films. *Journal of Physical Chemistry B*, 110, 9850–9855.

Mattia, D., Rossi, M. P., Ye, H. H., and Gogotsi, Y. (2007) In situ fluid studies in carbon nanotubes with diameters ranging from 1 to 500 nm. *Proceedings of the 5th IASME/WSEAS International Conference on Fluid Mechanics and Aerodynamics (FMA '07)*, Athens, Greece, pp. 297–299.

Mauter, M. S. and Elimelech, M. (2008) Environmental applications of carbon-based nanomaterials. *Environmental Science & Technology*, 42, 5843–5859.

Maxwell, J. C. (1879) On stresses in rarified gases arising from inequalities of temperature. *Philosophical Transactions of the Royal Society of London*, 170, 231–256.

Maysinger, D., Behrendt, M., and Przybytkowski, E. (2006) Death by nanoparticles. *NanoPharmaceuticals Online*, 1, http://www.nanopharmaceuticals.org/files/OCT2006.htm

Megaridis, C. M., Yazicioglu, A. G., Libera, J. A., and Gogotsi, Y. (2002) Attoliter fluid experiments in individual closed-end carbon nanotubes: Liquid film and fluid interface dynamics. *Physics of Fluids*, 14, L5–L8.

Melechko, A. V., Mcknight, T. E., Guillorn, M. A., Merkulov, V. I., Ilic, B., Doktycz, M. J., Lowndes, D. H., and Simpson, M. L. (2003) Vertically aligned carbon nanofibers as sacrificial templates for nanofluidic structures. *Applied Physics Letters*, 82, 976–978.

Mendes, P. M. and Preece, J. A. (2004) Precision chemical engineering: Integrating nanolithography and nanoassembly. *Current Opinion in Colloid & Interface Science*, 9, 236–248.

Meng, F. Y., Shi, S. Q., Xu, D. S., and Chan, C. T. (2006) Surface reconstructions and stability of X-shaped carbon nanotube junction. *Journal of Chemical Physics*, 124, 024711.

Mijatovic, D., Eijkel, J. C. T., and Van den Berg, A. (2005) Technologies for nanofluidic systems: top-down vs. bottom-up—A review. *Lab on a Chip*, 5, 492–500.

Miller, S. A. and Martin, C. R. (2004) Redox modulation of electroosmotic flow in a carbon nanotube membrane. *Journal of the American Chemical Society*, 126, 6226–6227.

Miller, S. A., Young, V. Y., and Martin, C. R. (2001) Electroosmotic flow in template-prepared carbon nanotube membranes. *Journal of the American Chemical Society*, 123, 12335–12342.

Monthioux, M. and Kuznetsov, V. L. (2006) Who should be given the credit for the discovery of carbon nanotubes? *Carbon*, 44, 1621–1623.

Mulero, R. and Kim, M. J. (2008) An integrated nanoporous chip for detecting single DNA molecules. *Biochip Journal*, 2, 73–77.

Nadarajan, S. B., Katsikis, P. D., and Papazoglou, E. S. (2007) Loading carbon nanotubes with viscous fluids and nanoparticles—A simpler approach. *Applied Physics A-Materials Science & Processing*, 89, 437–442.

Navier, C. L. M. H. (1823) Memoire sur les lois du mouvement des fluids. *Mémoires de l'Académie Royale des Sciences de l'Institut de France*, 6, 389–436.

Neto, C., Evans, D. R., Bonaccurso, E., Butt, H. J., and Craig, V. S. J. (2005) Boundary slip in Newtonian liquids: A review of experimental studies. *Reports on Progress in Physics*, 68, 2859–2897.

Nicholson, D. and Quirke, N. (2006) Adsorption and transport at the nanoscale (Chapter 1). In Quirke, N. (Ed.) *Adsorption and Transport at the Nanoscale*, Boca Raton, FL, CRC Press.

Noon, W. H., Ausman, K. D., Smalley, R. E., and Ma, J. P. (2002) Helical ice-sheets inside carbon nanotubes in the physiological condition. *Chemical Physics Letters*, 355, 445–448.

Noy, A., Park, H. G., Fornasiero, F., Holt, J. K., Grigoropoulos, C. P., and Bakajin, O. (2007) Nanofluidics in carbon nanotubes. *NanoToday*, 2, 22–29.

O'Brien, M. J., Bisong, P., Ista, L. K., Rabinovich, E. M., Garcia, A. L., Sibbett, S. S., Lopez, G. P., and Brueck, S. R. J. (2003) Fabrication of an integrated nanofluidic chip using interferometric lithography. *Journal of Vacuum Science & Technology B*, 21, 2941–2945.

Orosz, K., Gupta, S., Hassink, M., Abdel-Rahman, M., Moldovan, L., Davidorf, F. H., and Moldovan, N. I. (2004) Delivery of antiangiogenic and antioxidant drugs of ophthalmic interest through a nanoporous inorganic filter. *Molecular Vision*, 10, 555–565.

Palmer, S. B. and Rogalski, M. S. (1996) *Advanced University Physics*, New York, Gordon and Breach Publishers.

Panzer, R. E. and Elving, P. J. (1975) Nature of surface compounds and reactions observed on graphite electrodes. *Electrochimica Acta*, 20, 635–647.

Park, J. H., Sinnott, S. B., and Aluru, N. R. (2006) Ion separation using a Y-junction carbon nanotube. *Nanotechnology*, 17, 895–900.

Pfahler, J., Harley, J., Bau, H., and Zemel, J. (1990) Liquid transport in micron and submicron channels. *Sensors and Actuators A-Physical*, 22, 431–434.

Pietrass, T. (2006) Carbon-based membranes. *MRS Bulletin*, 31, 765–769.

Poisson, S. D. (1831) *Nouvelle Théorie de l'action capillaire*. Paris: Bachelier.

Powell, C., Fenwick, N., Bresme, F., and Quirke, N. (2002) Wetting of nanoparticles and nanoparticle arrays. *Colloids and Surfaces A: Physicochemical and Engineering Aspects* 206, 241.

Preece, J. A. and Stoddart, J. F. (1994) Concept transfer from biology to materials. *Nanobiology*, 3, 149–166.

Rabinow, B. E. (2004) Nanosuspensions in drug delivery. *Nature Reviews Drug Delivery*, 3, 785–796.

Radushkevich, L. V. and Lukyanovich, V. M. (1952) O strukture ugleroda, obrazujucegosja pri termiceskom razlozenii okisi ugleroda na zeleznom kontakte. *Zurn Fisic Chim*, 26, 88–95.

Rakov, E. G. (2006) Chemistry of carbon nanotubes. In Gogotsi, Y. (Ed.) *Nanomaterials Handbook*, Boca Raton, FL, CRC Press.

Ralston, J., Popescu, M., and Sedev, R. (2008) Dynamics of wetting from an experimental point of view. *Annual Review of Materials Research*, 38, 23–43.

Rasaiah, J. C., Garde, S., and Hummer, G. (2008) Water in nonpolar confinement: From nanotubes to proteins and beyond. *Annual Review of Physical Chemistry*, 59, 713–740.

Regan, B. C., Aloni, S., Ritchie, R. O., Dahmen, U., and Zettl, A. (2004) Carbon nanotubes as nanoscale mass conveyors. *Nature*, 428, 924–927.

Riegelman, M., Liu, H., and Bau, H. (2006) Controlled nanoassembly and construction of nanofluidic devices. *Journal of Fluids Engineering (Transactions of the ASME)*, 128, 6–13.

Rivera, J. L., Rico, J. L., and Starr, F. W. (2007) Interaction of water with cap-ended defective and nondefective small carbon nanotubes. *Journal of Physical Chemistry C*, 111, 18899–18905.

Rossi, M. P., Ye, H. H., Gogotsi, Y., Babu, S., Ndungu, P., and Bradley, J. C. (2004) Environmental scanning electron microscopy study of water in carbon nanopipes. *Nano Letters*, 4, 989–993.

Routkevitch, D., Bigioni, T., Moskovits, M., and Xu, J. M. (1996) Electrochemical fabrication of CdS nanowire arrays in porous anodic aluminum oxide templates. *Journal of Physical Chemistry*, 100, 14037–14047.

Rowlinson, J. S. and Widom, B. (1982). *Molecular Theory of Capillarity*, Oxford, U.K., Oxford University Press, vol. Section 1.3.

Sahoo, S. K., Parveen, S., and Panda, J. J. (2007) The present and future of nanotechnology in human health care. *Nanomedicine-Nanotechnology Biology and Medicine*, 3, 20–31.

Saito, R., Dresselhaus, G., and Dresselhaus, M. S. (1998) *Physical Properties of Carbon Nanotubes*, London, U.K., Imperial College.

Schebarchov, D. and Hendy, S. C. (2008) Dynamics of capillary absorption of droplets by carbon nanotubes. *Physical Review E*, 78, 046309.

Schoch, R. B., Han, J. Y., and Renaud, P. (2008) Transport phenomena in nanofluidics. *Reviews of Modern Physics*, 80, 839–883.

Schrlau, M. G., Falls, E. M., Ziober, B. L., and Bau, H. H. (2008) Carbon nanopipettes for cell probes and intracellular injection. *Nanotechnology*, 19, 015101.

Sekhaneh, W., Kotecha, M., Dettlaff-Weglikowska, U., and Veeman, W. S. (2006) High resolution NMR of water absorbed in single-wall carbon nanotubes. *Chemical Physics Letters*, 428, 143–147.

Sharp, K. and Adrian, R. (2005) On flow-blocking particle structures in microtubes. *Microfluidics and Nanofluidics*, 1, 376–380.

Shiomi, J. and Maruyama, S. (2009) Water transport inside a single-walled carbon nanotube driven by temperature gradient, *Nanotechnology*, 20, 055708.

Sholl, D. S. and Johnson, J. K. (2006) Making high-flux membranes with carbon nanotubes. *Science*, 312, 1003–1004.

Singh, A. K., Pandey, A., Rai, R., Tewari, M., Pandey, H. P., and Shukla, H. S. (2008) Nanomaterials as emerging tool in cancer diagnosis and treatment. *Digest Journal of Nanomaterials and Biostructures*, 3, 135–140.

Sinha, P. M., Valco, G., Sharma, S., Liu, X. W., and Ferrari, M. (2004) Nanoengineered device for drug delivery application. *Nanotechnology*, 15, S585–S589.

Sinha, S., Rossi, M. P., Mattia, D., Gogotsi, Y., and Bau, H. H. (2007) Induction and measurement of minute flow rates through nanopipes. *Physics of Fluids*, 19, 013603.

Skoulidas, A. I., Ackerman, D. M., Johnson, J. K., and Sholl, D. S. (2002) Rapid transport of gases in carbon nanotubes. *Physical Review Letters*, 89, 185901.

Smart, S. K., Cassady, A. I., Lu, G. Q., and Martin, D. J. (2006) The biocompatibility of carbon nanotubes. *Carbon*, 44, 1034–1047.

Sokhan, V. P. and Quirke, N. (2008) Slip coefficient in nanoscale pore flow. *Physical Review E*, 78, 015301.

Sokhan, V. P., Nicholson, D., and Quirke, N. (2001) Fluid flow in nanopores: An examination of hydrodynamic boundary conditions. *Journal of Chemical Physics*, 115, 3878.

Sokhan, V. P., Nicholson, D., and Quirke, N. (2002) Fluid flow in nanopores: Accurate boundary conditions for carbon nanotubes. *Journal of Chemical Physics*, 117, 8531.

Son, S. J., Bai, X., Nan, A., Ghandehari, H., and Lee, S. B. (2006) Template synthesis of multifunctional nanotubes for controlled release. *Journal of Controlled Release*, 114, 143–152.

Son, S. J., Bai, X., and Lee, S. (2007a) Inorganic hollow nanoparticles and nanotubes in nanomedicine. Part 2: Imaging, diagnostic, and therapeutic applications. *Drug Discovery Today*, 12, 657–663.

Son, S. J., Bai, X., and Lee, S. B. (2007b) Inorganic hollow nanoparticles and nanotubes in nanomedicine. Part 1. Drug/gene delivery applications. *Drug Discovery Today*, 12, 650–656.

Srivastava, A., Srivastava, O. N., Talapatra, S., Vajtai, R., and Ajayan, P. M. (2004) Carbon nanotube filters. *Nature Materials*, 3, 610–614.

Staufer, U., Akiyama, T., Gullo, M. R., Han, A., Imer, R., de Rooij, N. F., Aebi, U., Engel, A., Frederix, P. L. T. M., Stolz, M., Friederich, N. F., and Wirz, D. (2007) Micro- and nanosystems for biology and medicine. *Microelectronic Engineering*, 84, 1681–1684.

Stone, H. A., Stroock, A. D., and Ajdari, A. (2004) Engineering flows in small devices: Microfluidics toward a lab-on-a-chip. *Annual Review of Fluid Mechanics*, 36, 381–411.

Strano, M. S., Dyke, C. A., Usrey, M. L., Barone, P. W., Allen, M. J., Shan, H. W., Kittrell, C., Hauge, R. H., Tour, J. M., and Smalley, R. E. (2003) Electronic structure control of single-walled carbon nanotube functionalization. *Science*, 301, 1519–1522.

Sun, L. and Crooks, R. M. (2000) Single carbon nanotube membranes: A well-defined model for studying mass transport through nanoporous materials. *Journal of the American Chemical Society*, 122, 12340–12345.

Supple, S. and Quirke, N. (2003) Rapid imbibition of fluids in carbon nanotubes. *Physical Review Letters*, 90, 214501.

Supple, S. and Quirke, N. (2004) Molecular dynamics of transient oil flows in nanopores I: Imbibition speeds for single wall carbon nanotubes. *Journal of Chemical Physics*, 121, 8571–8579.

Supple, S. and Quirke, N. (2005) Molecular dynamics of transient oil flows in nanopores II: Density profiles and molecular structure for decane in carbon nanotubes. *Journal of Chemical Physics*, 122, 104706-1–104706-6.

Tegenfeldt, J. O., Prinz, C., Cao, H., Huang, R. L., Austin, R. H., Chou, S. Y., Cox, E. C., and Sturm, J. C. (2004) Micro- and nanofluidics for DNA analysis. *Analytical and Bioanalytical Chemistry*, 378, 1678–1692.

Thomas, J. A. and McGaughey, A. J. H. (2008) Reassessing fast water transport through carbon nanotubes. *Nano Letters*, 8, 2788–2793.

Torchilin, V. P. (Ed.) (2006) *Nanoparticulates as Drug Carriers*, London, U.K., Imperial College Press.

Torchilin, V. P. (2007) Targeted pharmaceutical nanocarriers for cancer therapy and Imaging. *AAPS Journal*, 9, E128–E147.

Travis, K. P., Todd, B. D., and Evans, D. J. (1997) Departure from Navier-Stokes hydrodynamics in confined liquids. *Physical Review E*, 55, 4288.

Tsang, S. C., Harris, P. J. F., and Green, M. L. H. (1993) Thinning and opening of carbon nanotubes by oxidation using carbon-dioxide. *Nature*, 362, 520–522.

Tsang, S. C., Chen, Y. K., Harris, P. J. F., and Green, M. L. H. (1994) A simple chemical method of opening and filling carbon nanotubes. *Nature*, 372, 159–162.

Ugarte, D., Stockli, T., Bonard, J. M., Chatelain, A., and de Heer, W. A. (1998) Filling carbon nanotubes. *Applied Physics a-Materials Science & Processing*, 67, 101–105.

Vega-Villa, K. R., Takemoto, J. K., Yanez, J. A., Remsberg, C. M., Forrest, M. L., and Davies, N. M. (2008) Clinical toxicities of nanocarrier systems. *Advanced Drug Delivery Reviews*, 60, 929–938.

Venugopal, J., Prabhakaran, M. P., Low, S., Choon, A. T., Zhang, Y. Z., Deepika, G., and Ramakrishna, S. (2008) Nanotechnology for nanomedicine and delivery of drugs. *Current Pharmaceutical Design*, 14, 2184–2200.

Vereb, G., Szollosi, J., Matko, J., Nagy, P., Farkas, T., Vigh, L., Matyus, L., Waldmann, T. A., and Damjanovich, S. (2003) Dynamic, yet structured: The cell membrane three decades after the Singer-Nicolson model. *Proceedings of the National Academy of Sciences of the United States of America*, 100, 8053–8058.

Wagner, V., Dullaart, A., Bock, A. K., and Zweck, A. (2006) The emerging nanomedicine landscape. *Nature Biotechnology*, 24, 1211–1217.

Wang, B. Y. and Kral, P. (2007) Chemically tunable nanoscale propellers of liquids. *Physical Review Letters*, 98, 266102.

Wang, M. S., Wang, J. Y., Chen, Q., and Peng, L. M. (2005) Fabrication and electrical and mechanical properties of carbon nanotube interconnections. *Advanced Functional Materials*, 15, 1825–1831.

Wang, Z., Ba, D. C., Yu, C. H., and Liang, J. (2006) Synthesis of carbon nanotube junction by ECR-CVD. *Journal of Inorganic Materials*, 21, 1244–1248.

Washburn, E. W. (1921) The dynamics of capillary flow. *Physical Review*, 17, 273–283.

Werder, T., Walther, J. H., Jaffe, R. L., Halicioglu, T., Noca, F., and Koumoutsakos, P. (2001) Molecular dynamics simulation of contact angles of water droplets in carbon nanotubes. *Nano Letters*, 1, 697–702.

Werder, T., Walther, J. H., Jaffe, R. L., Halicioglu, T., and Koumoutsakos, P. (2003) On the water-carbon interaction for use in molecular dynamics simulations of graphite and carbon nanotubes. *Journal of Physical Chemistry B*, 107, 1345.

Whitby, M. and Quirke, N. (2007) Fluid flow in carbon nanotubes and nanopipes. *Nature Nanotechnology*, 2, 87–94.

Whitby, M., Cagnon, L., Thanou, M., and Quirke, N. (2008a) Enhanced fluid flow through nanoscale carbon pipes. *Nano Letters*, 8, 2632–2637.

Whitby, M., Lin, J., Quirke, N., and Thanou, M. (2008b) Carbon nanopipe dispersions in aqueous solutions and their effect on cell viability. *Nanotech 2008*, Boston, MA.

Whitby, M., Cagnon, L., Thanou, M., and Quirke, N. (2009) Enhanced fluid flow through nanoscale carbon pipes (erratum). *Nano Letters*, 8, 2632–2637.

Whitesides, G. M. (2006) The origins and the future of microfluidics. *Nature*, 442, 368–373.

Whitesides, G. M. and Stroock, A. D. (2001) Flexible methods for microfluidics. *Physics Today*, 54, 42–48.

Winkle, R. F., Nagy, J. M., Cass, A. E. G., and Sharma, S. (2008) Towards microfluidic technology-based Maldi-MS platforms for drug discovery: A review. *Expert Opinion on Drug Discovery*, 3, 1281–1292.

Yan, A. H., Xiao, X. C., Kulaots, I., Sheldon, B. W., and Hurt, R. H. (2006a) Controlling water contact angle on carbon surfaces from 5 degrees to 167 degrees. *Carbon*, 44, 3116–3120.

Yan, H., Gong, A., He, H., Zhou, J., Wei, Y., and Lv, L. (2006b) Adsorption of microcystins by carbon nanotubes. *Chemosphere*, 62, 142–148.

Yarin, A. L., Yazicioglu, A. G., and Megaridis, C. M. (2005a) Thermal stimulation of aqueous volumes contained in carbon nanotubes: Experiment and modeling. *Applied Physics Letters*, 86, 013109.

Yarin, A. L., Yazicioglu, A. G., Megaridis, C. M., Rossi, M. P., and Gogotsi, Y. (2005b) Theoretical and experimental investigation of aqueous liquids contained in carbon nanotubes. *Journal of Applied Physics*, 97, 124309.

Yazicioglu, A. G., Megaridis, C. M., Nicholls, A., and Gogotsi, Y. (2005) Electron microscope visualization of multiphase fluids contained in closed carbon nanotubes. *Journal of Visualization*, 8, 137–144.

Yeh, I. C. and Hummer, G. (2004) Nucleic acid transport through carbon nanotube membranes. *Proceedings of the National Academy of Sciences of the United States of America*, 101, 12177–12182.

Young, T. (1805) Cohesion of fluids. *Philosophical Transactions of the Royal Society of London*, 95, 65–87.

Yu, Y. M., Zhang, Q., Mu, Q. X., Zhang, B., and Yan, B. (2008) Exploring the immunotoxicity of carbon nanotubes. *Nanoscale Research Letters*, 3, 271–277.

Yuan, Z., Garcia, A. L., Lopez, G. P., and Petsev, D. N. (2007) Electrokinetic transport and separations in fluidic nanochannels. *Electrophoresis*, 28, 595–610.

Zimmerli, U., Gonnet, P. G., Walther, J. H., and Koumoutsakos, P. (2005) Curvature induced L-defects in water conduction in carbon nanotubes. *Nano Letters*, 5, 1017–1022.

II

Inorganic Nanotubes

Inorganic Fullerenes and Nanotubes

Andrey Enyashin
Technische Universität Dresden

and

Institute of Solid State
Chemistry, UBRAS

Gotthard Seifert
Technische Universität Dresden

12.1 Introduction

Nanosized allotropes of inorganic compounds demonstrate a high variety of their morphology from small nanoplatelets, nanostripes, and nanorods, which may be understood as cutouts of the bulk modifications up to larger hollow nanotubes, fullerene-like particles, and fullerenes with a polymorphic atomistic structure, which deviates from that of the bulk state. The small particles are characterized by a large amount of chemically active atoms with dangling bonds, whereas the atoms in the fullerenic and tubular species possessing a cavity are coordinatively saturated and are more inert.

Both types of objects have the potential to enhance the performance of bulk materials in their typical industrial applications in catalysis tribology, electronics, and electrochemistry. For this purpose, the knowledge of the basic trends determining the stability, mechanical resistance, reactivity, and electronic properties of these nano-objects came into the focus of extensive theoretical and experimental investigations. Numerous studies, review papers and books about hollow nanoparticles like nanotubes and fullerenes of inorganic compounds have been published (see Tenne 1996, 2006; Patzke et al. 2002; Rao and Nath 2003; Remškar 2004; Tenne and Rao 2004; Bar-Sadan et al. 2007; Enyashin et al. 2007a; Kaplan-Ashiri and Tenne 2007; Deepak and Tenne 2008; Tenne et al. 2008). Taking into account the incredibly large research field for such nano-objects, this chapter will focus on their basic characteristics in comparison to the corresponding bulk materials and illustrate the most recent and important results of both experimental and theoretical studies.

12.2 Models of Hollow Inorganic Nanostructures

Most of the inorganic nanotubes (INTs) currently known were synthesized from layered—quasi two-dimensional (2D)—bulk compounds as carbon ones from graphite. One characteristic property of these compounds is the clearly expressed anisotropy of the strong (covalent) and weak (van der Waals') bondings within and between the layers, respectively. Experimentally obtained nanotubes usually have complex structures: they can be composed of various numbers of coaxial cylindrical layers and the bonds within the layers exhibit different orientations relative to the tube axis. Nanotubes can be open or closed at the ends. They can have not only cylindrical, but also a scroll-like morphology. Nevertheless, many tubular forms of matter may be structurally characterized by using the classification developed for the single-walled nanotubes (SWNTs), which can be constructed by rolling up of the monolayer.

The structure of the crystalline monolayer can be characterized using the terms of a 2D lattice spanned by the translation vectors $(\vec{a}_1, \vec{a}_2)$. Five types of 2D Bravais lattices exist (Figure 12.1): oblique $(|\vec{a}_1| \neq |\vec{a}_2|, \varphi \neq 90°)$, square $(|\vec{a}_1| = |\vec{a}_2|, \varphi = 90°)$, hexagonal $(|\vec{a}_1| = |\vec{a}_2|, \varphi = 120°)$, primitive rectangular $(|\vec{a}_1| \neq |\vec{a}_2|, \varphi = 90°)$,

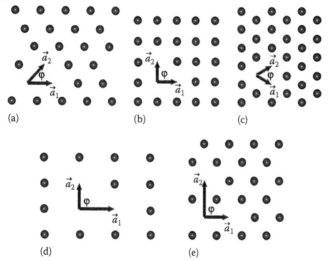

FIGURE 12.1 The five 2D Bravais lattices: (a) oblique; (b) square; (c) hexagonal; (d) primitive rectangular; (e) centered rectangular.

discovered, where the corresponding bulk structures exhibit layers with a hexagonal atom arrangement (Figures 12.2a through d). Therefore, the classification of the carbon nanotubes can often be used for the classification of the structure of INTs as well. All cylindrical nanotubes are characterized by a radius R and by the type of the helicoidal atomic arrangement (chirality) determined by a chiral angle θ. Using the basis vectors of the 2D hexagonal lattice $|\vec{a}_1| = |\vec{a}_2|$ and the chiral vector $\vec{c} = n\vec{a}_1 + m\vec{a}_2$, it is possible to describe the basic geometry parameters of the tubulene, R and θ, which is produced by the rolling of a ribbon cut from a monolayer:

$$\theta = \arctan \frac{\sqrt{3}m}{m + 2n} \qquad (12.1)$$

$$R = \frac{|\vec{c}|}{2\pi} = \frac{|\vec{a}|}{2\pi}\sqrt{3(n^2 + m^2 + mn)} \qquad (12.2)$$

and centered rectangular $\left(|\vec{a}_1| \neq |\vec{a}_2|, \varphi = 90°\right)$. Cylindrical surfaces may be constructed by rolling up these lattices.

The walls of carbon nanotubes are composed of carbon hexagons with the same ordering as in the hexagonal graphenic layer. The basic principles of the geometry specification of the "ideal" carbon nanotubes are described in detail in the article by Dresselhaus and Avouris (2001). Numerous INTs were

Since R and θ are associated uniquely with $\vec{a}_1$ and $\vec{a}_2$, using the integer indexes n and m, these indices may be used for the structural classification of the INTs in the same way as for carbon nanotubes—(n, m). Hence, depending on the values of n and m, all such nanotubes can be subdivided into two groups: chiral tubes with $0 < \theta < 30°$ and the nonchiral ones, the so-called zigzag and armchair tubes with $\theta = 0°$ and $30°$, respectively.

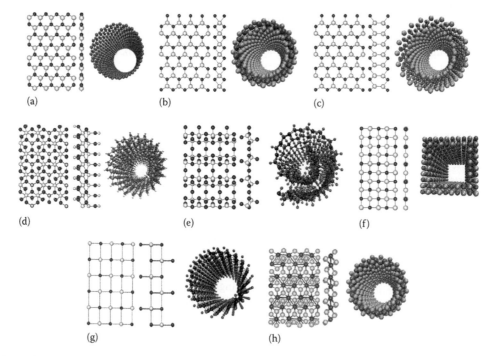

FIGURE 12.2 **(See color insert following page 20-16.)** Structures of the monolayers of some inorganic compounds and respective nanotubes: (a) hexagonal BN and (20,0) nanotube; (b) hexagonal MoS_2 and (20,0) nanotube; (c) hexagonal GaS and (20,0) nanotube; (d) hexagonal imogolite $Al_2O_3(OH)_3Si(OH)$ and (12,0) nanotube; (e) rectangular V_2O_5 and (20,0) nanoroll; (f) square MgO and (7@9) nanotube; (g) centered rectangular lepidocrocite TiO_2 and (20,0) nanotube; (h) oblique ReS_2 and (20,0) nanotube.

The same scheme may be applied for the classification of nanostripes and nanorolls.

Single-walled tubular structures based on nonhexagonal layers may analogously be described by the primitive vectors of the respective lattice (Figures 12.2e through h). In these cases, R and θ will be associated with n and m by other relations than those given above. A detailed description of the classification and stability of nanotubes based on square lattice was given recently by Wilson et al. (Bishop and Wilson 2008) for nanotubes of MX compounds (M = alkali-metal, X = halogen). The radius of such nanotubes can be written as a function of an integer number n and the length of the M–X bond a:

$$R = \frac{a}{2\sin(\pi/2n)} \quad (12.3)$$

The symmetry of the nanotubes determines many of their physical properties connected particularly with electronic and phonon states. A detailed review for tubes based on hexagonal layers, such as MS_2, (M = Mo, W), XN (X = B, Ga), C, BC_3, or BC_2N, is given in the articles by Milošević et al. (2000) and Damnjanović et al. (2001).

It is important to note that single-walled carbon nanotubes (SWNTs) are based on a planar 2D graphitic monolayer. In contrast, most of the layered inorganic compounds have a more complex atomic arrangement and consist mostly of multilayer sheets. Therefore, inorganic SWNTs are composed of several concentric and interconnected cylinders of atoms (Figure 12.2). This fact suggests *a priori* that INTs might be less stable than carbon ones with the same radii, since rolling up a thicker layer to a cylinder is energetically less preferable.

The layers of the inorganic compounds described so far represent only a small portion of the vast possible structural variety. As carbon nanotubes, the INTs are often closed at the ends by "caps." Structural models of these "caps" were described for tubes based on the hexagonal layers of the graphitic-like boron nitride (BN), metal dichalcogenides, and dihalides with prismatic or octahedral coordination of metallic atoms (Saito and Maida 1999; Seifert et al. 2000b; Enyashin et al. 2003) (Figure 12.3). Similarly to the carbon nanostructures, the "caps" are fragments of the corresponding hollow molecules, the so-called inorganic fullerenes (IF).

IFs are closed nano-objects with regular polyhedral shapes and are based on the pieces of the same monolayers of inorganic compounds that are able to form nanotubes. However, the atomic models of fullerenes are more complicated because a simple rolling of a monolayer is not sufficient for the construction of their cage-like structure. The construction principle of a fullerenic cage depends on the lattice type of the corresponding layered compound. The concept can be illustrated exemplarily at the hexagonal lattice. For this case, the construction of the polyhedra relies on the insertion of point defects into a hexagonal layer. In an elemental compound like carbon based on

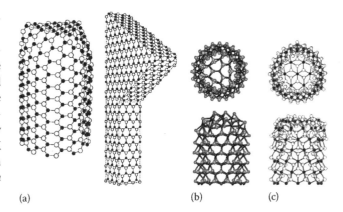

(a)   (b) (c)

FIGURE 12.3 The "capped" single-walled inorganic nanotubes with structure of: (a) BN (From Saito, Y. and Maida, M., *J. Phys. Chem. A*, 103, 1291, 1999. With permission.); (b) MoS_2 (From Seifert, G. et al., *Phys. Rev. Lett.*, 85, 146, 2000b. With permission.); (c) ZrS_2. (From Enyashin, A.N. et al., *IEJMD*, 1, 499, 2003. With permission.)

graphene layers or hexagonal P, such a defect can be realized by the replacement of the hexagons by pentagons, squares, or triangles that provide a positive curvature to the layer and facilitate the formation of a closed cage. The most known examples are carbon fullerenes, which can be constructed by introducing 12 pentagons or onion-like carbon nanoparticles with their sphericity provided by numerous heptagons and pentagons (Terrones et al. 2004).

In a binary compound like BN or MoS_2 with its own sublattice for each atomic species, point defects are only low in energy if they maintain the overall connectivity of the parent compound. Therefore, the most easily accessible point defects are those that are related with squares, octagons, and dodecagons (Enyashin et al. 2004a). Among them, only the square-like defects can provide a surface with a positive curvature and can act as corners of a polyhedral particle. Facets of fullerene-like cages can be cut from a hexagonal layer without any defects. Octagons or dodecagons create negative curvatures and can also build square and pentagonal facets of the polyhedra (Figure 12.4). All such IFs retain a bonding environment of atoms similar to those seen in bulk compounds. Thus, using these simple principles, fullerenes with the shapes of a hexagonal bipyramid, a trigonal prism, or Platonic or Archimedean solids can be designed (Figure 12.4).

The atomistic models of INTs and fullerenes were supported by the data from electron or x-ray diffraction (Wada and Yoshinaga 1969; Amelinckx et al. 1996; Margulis et al. 1996; Hsu et al. 2000). The development of the ultra-high-resolution electron microscopy allows a direct imaging of inorganic nanostructures (Bar-Sadan et al. 2008a,b) with an atomic resolution, confirming the theoretical atomistic models. This was demonstrated very recently for fullerene-like MoS_2 and WS_2 particles (Bar-Sadan et al. 2008a,b).

The understanding of atomic arrangement in nanostructures can lead to new insights into their growth mechanism or physical

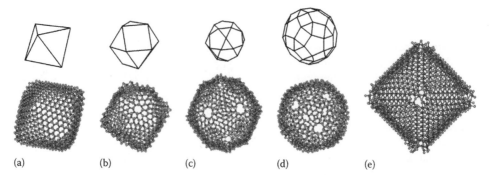

FIGURE 12.4   Ideal structures of stoichiometric MoS_2 fullerenes with regular or semiregular shapes: (a) octahedron $(MoS_2)_{432}$; (b) cuboctahedron $(MoS_2)_{384}$; (c) icosadodecahedron $(MoS_2)_{540}$; (d) rhomboicosadodecahedron $(MoS_2)_{480}$, and (e) the structure of octahedron $(MoS_2)_{576}$ distorted after molecular dynamics DFTB simulations. (From Enyashin, A.N. et al., *Angew. Chem. Int. Ed.*, 46, 623, 2007. With permission.)

properties. Control over shape, size, and atomic architecture is a key issue in synthesis and further design of functional nanoparticles for applications.

12.3 General Criteria Describing the Stability of Inorganic Nanotubes and Fullerenes

For the characterization of the stability of a tubular or a fullerenic structure, one can compare the energy of such a structure with that of the corresponding layer structure. Usually the formation of a tube or a cage out of a layer is connected with an increase in the energy due to the strain of the bending. Therefore, one can define the strain energy (E_{str}) as the difference of the energies (E_{tot}) of a nanoparticle and the corresponding infinite layer. On the other hand, the formation of a closed structure (nanotube, cage) from a finite layered structure (stripe) can decrease the energy due to the saturation of dangling bonds at the edges of the flat layered stripe. As it will be shown below, the competition between these two effects will determine the occurrence of certain morphologies of the nanostructure.

For the estimation of the stability, one can apply phenomenological models based on the principles of the classical theory of elasticity (Seifert and Frauenheim 2000). In the framework of this approach, the atomic structure of the nanoparticle is not taken into account. In such a model, the dependence of E_{str} on the radius R for a nanotube is given by

$$E_{str} = \frac{\pi Y L h^3}{12R}, \qquad (12.4)$$

where
- h is the thickness of the layer
- L is the length of the nanotube
- Y is the elastic module of the layer

The number of atoms in the nanotube wall is given as $N = 2\pi R L \rho_a$, where ρ_a is the number of atoms on the surface unit of the monolayer. Then the strain energy per atom is equal to

$$\frac{E_{str}}{N} = \frac{Y h^3}{24 \rho_a R^2}. \qquad (12.5)$$

In atomistic approaches, the quantitative estimates of the energetic parameters are realized in the framework of the molecular mechanics or the quantum-mechanics methods. The validity of Equation 12.5 was successfully proven by the quantum-mechanical calculations for the nanotubes of carbon and different inorganic compounds: BN (Hernandez et al. 1999), MoS_2 (Seifert et al. 2000b), TiS_2 (Ivanovskaya and Seifert 2004), GaS (Köhler et al. 2004), $MoTe_2$ (Wu et al. 2007), TiO_2 (Enyashin and Seifert 2005), AlOOH (Enyashin et al. 2006a), etc.

The calculated energies per atom follow the $1/R^2$ trend also observed for carbon nanotubes (Hernandez et al. 1999). However, except for BN nanotubes, the strain energy for the smallest inorganic tubes is usually at least one order of magnitude larger than that of carbon nanotubes with a similar diameter. This result can be easily understood, since many INTs have a larger thickness of the walls. The rolling of a layer into a narrow tube leads to increased steric hindering, which results in a higher strain energy than the one of a thin (single-atomic) graphite monolayer (Figure 12.5a through d).

The atomistic approach allows to derive energetic models of stability of the various nanostructures—nanotubes, nanostrips, nanorolls, and fullerenes (Seifert et al. 2002; Enyashin and Seifert 2005; Enyashin et al. 2007a,b). The strain energy per atom of a multi-walled nanotube (MWNT) may be written as

$$\frac{E_{str}}{N} = \frac{\beta}{N} \sum_{i=1}^{k} \frac{1}{N_i} + \frac{k-1}{k} \varepsilon_{vdW}, \quad N = \sum_{i=1}^{k} N_i \qquad (12.6)$$

The first term reflects the curvature energies, whereas the second term takes into account the contribution of the van der Waals' interaction between the neighboring layers in the nanotube. The integer k is the number of walls, N_i is the number of atoms within the unit cell of a SWNT, N is the total number of atoms, and β and ε_{vdW} are the parameters, which determine the resistance against bending and van der Waals' interaction between the monolayers.

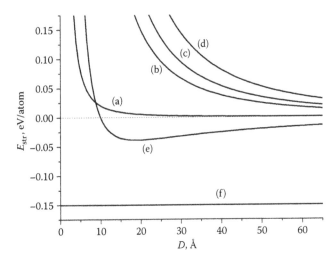

FIGURE 12.5 Comparison of the calculated strain energies E_{str} vs. diameter for (a) carbon nanotubes (From Hernandez, E. et al., *Appl. Phys. A Mater. Sci. Process.*, 68, 287, 1999. With permission.), and those of the (b) inorganic nanotubes TiS_2 (From Ivanovskaya, V.V. and Seifert, G., *Solid State Commun.*, 130, 175, 2004. With permission.), (c) boehmite AlOOH (From Enyashin, A.N. et al., *Mendeleev Commun.*, 6, 292, 2006a. With permission.), (d) GaS (From Köhler, T. et al., *Phys. Rev. B*, 69, 193403, 2004. With permission.), (e) imogolite $Al_2O_3(OH)_3Si(OH)$ (From Guimaraes, L. et al., *ACS Nano*, 1, 362, 2007. With permission.), and (f) TiC (From Enyashin, A.N. and Ivanovskii, A.L., *Phys. E Low-Dim. Syst. Nanostruct.*, 30(1–2), 164, 2005b.).

A similar expression can be derived for layered nanostripes of a finite width:

$$\frac{E_{str}}{N} = 3k\frac{\varepsilon_x - \varepsilon_\infty}{N} + \frac{k-1}{k}\varepsilon_{vdW} \qquad (12.7)$$

The parameters ε_x and ε_∞ reflect the dangling bonds at the edges of the stripe and the energy of the infinite planar monolayer.

All these parameters can be determined by atomistic calculations of model structures (Seifert et al. 2002).

Atomistic descriptions are limited with respect to the size of the system that can be studied. Relations as equations (12.5 and 12.6) are very useful for the systematic study of the size evolution of the stability of tubular nanostructures. In particular, the energetic model of the cylindrical MoS_2 nanotubes and the plain MoS_2 nanostripes shows that the nanotubes are more stable than the nanostripes if $R > 6$ nm, which is in agreement with the experimental findings (Seifert et al. 2002) (Figure 12.6). In accordance with experimental data, multilayered nanosystems are more stable than the monolayered ones, owing to the attractive interlayer van der Waals' interactions. Moreover, inorganic SWNTs have not been synthesized yet.

A similar approach was used to compare the relative stability of SWNTs and the corresponding nanorolls of anatase TiO_2 (Enyashin and Seifert 2005). The study has shown that the strain energy for a nanoroll with a cross-section of an Archimedian spiral depends on the distance L between the coils and the starting angle φ_1:

$$\frac{E_{str}}{N} = \frac{4\pi^2\beta}{NL^2}\sum_{i=1}^{k}\frac{\varphi_i^2 + 2}{(\varphi_i^2 + 1)^3} + \frac{6}{N}\varepsilon_x,$$

$$\varphi_i^2 = \varphi_1^2 - \varphi_{i-1}^2 + 2\varphi_i\varphi_{i-1}\cos(\varphi_i - \varphi_{i-1}) \qquad (12.8)$$

An analysis of Equations 12.6 and 12.8 shows that cylindrical nanotubes are always more stable than the corresponding nanorolls with the same perimeter. This result is confirmed by the numerous investigations of the INTs, which appear mainly as a system of coaxial cylinders. However, there are some exceptions like nanotubes of vanadium oxides VO_x and titanic acid $H_2Ti_3O_7$ that are in fact scroll-like (Krumeich et al. 1999; Chen et al. 2002b). The possible mechanisms of their formation are different from those of the cylindrical nanotubes and

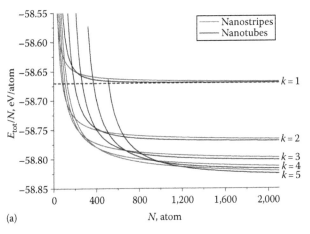

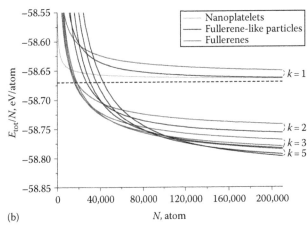

FIGURE 12.6 Total energies of multilayered nanostructures of MoS_2: (a) 1D nanostripes and nanotubes (From Seifert, G. et al., *J. Phys. Chem. B*, 106, 2497, 2002. With permission.) (b) zero-dimensional nanoplatelets, octahedral fullerenes and spherical fullerene-like nanoparticles depending on the number of atoms within a unit cell and number of walls k. Dashed line, energy of a MoS_2 monolayer. (From Bar-Sadan, M. et al., *J. Phys. Chem. B*, 110, 25399, 2006a. With permission.)

can be explained by the difference in surface tensions of the two sides of the layer (Zhang et al. 2005). For instance, titanate nanotubes are delivered from the proton-containing layers of $H_2Ti_3O_7$, which are able to exchange the protons by other cations like Na^+. The difference in the concentration of Na^+ ions deposited on both sides leads to the difference in surface tension, which can exceed bending forces and lead to the self-rolling of the layers. Evidently, a similar mechanism is involved in the formation of VO_x nanorolls using the adsorption of alkylamines or other organic molecules initially on one side of the monolayer (Patzke et al. 2002).

The formation of such INTs attracted much interest, since they can be formed in a natural environment, as has been shown for nanotubes and nanorolls of alumina- and magnesiasilicates, which bundle into minerals species like imogolite (Guimaraes et al. 2007) and chrysotile (Piperno et al. 2007), respectively. These nanotubes represent curved aluminum $Al(OH)_3$ or magnesium $Mg(OH)_2$ hydroxide sheets, where at one side the hydroxyl groups are completely substituted by silicate anions. In these cases, the incommensurabilty in the lattice vectors of hydroxide and silica sublattices results in negative values of energy for the rolling of the planar sheet into a cylinder (Figure 12.5e) i.e., in contrast to most of the nanotubes, these tubes are not metastable. The description of the stability for such nanotubes implies a second contribution in the strain energy (Equation 12.5) (Guimaraes et al. 2007), which is responsible for the difference in the surface tensions, $\Delta\sigma < 0$, between the outer and inner tube surfaces (internal and external surface energies):

$$\frac{E_{str}}{N} = \frac{Yh^3}{24\rho_a R^2} + \frac{\Delta\sigma h}{R} \qquad (12.9)$$

The interaction between the walls of coaxial nanotubes may also play an important role in the formation of these nanostructures, especially if they are composed of layers of different chemical compounds. Mendelev et al. proposed a thermodynamic model for two-phase multilayered films and nanotubes (Mendelev et al. 2002). They showed that the dominance of one or another nanoform is determined by the balance between the surface and interfacial energies, the energies of interactions between the interfaces and interfaces and surfaces, and the bending energies.

The stability of IFs can be also evaluated using the classic theory of elasticity or an atomistic approach. Such an approach may be refined on the basis of quantum-mechanical calculations for representative smaller entities of the whole structure. As in the corresponding carbon-based nanostructures, a spherical shape can not be built from strained, but otherwise unperturbed fragments of a hexagonal crystalline sheet, but structural defects are required to provide the curvature of the fullerene-like shape.

As shown in the framework of a continuum model (Srolovitz et al. 1995), line defects such as dislocations and more extended defects such as grain boundaries and stacking faults are intrinsic features of spherical fullerene particles, once the thickness exceeds a critical value. Such defects are essential for relieving the large inherent strains in quasi-spherical fullerene-like structures. It has also been established that the ratio of the wall thickness to the radius of a multi-walled particle is determined by the ratio of the surface energy to the energies of curvature and dislocation (or grain boundary).

The number and the distribution of the defects within a fullerene-like particle are not known. However, one may assume that the number of defects is quite small in comparison with the total number of atoms in a nanoparticle, as each defect introduces a significant curvature. Assuming the absence of defects within the shell of a spherical nanoparticle, the strain energies per atom for this morphological type can be derived using the theory of elasticity as (Enyashin et al. 2007b)

$$\frac{E_{str}}{N} = \frac{Yh^3(1+\sigma)}{12\rho_a R^2}, \qquad (12.10)$$

where σ is the Poisson ratio. The difference between Equations 12.5 and 12.10 by $2(1+\sigma)$ reflects the fact of a higher curvature of a sphere compared with a cylinder and, accordingly, a lower stability of fullerenic forms at a given radius R (Figure 12.6).

The nanoparticles with regular shapes have a countable number of defects. Thus, the exact expression for their strain energies can be derived. It may be assumed that the energy of a polyhedral fullerene can be subdivided into contributions from the atoms of the facets, the edges, and the corners with the corresponding energies ε_∞, ε_e, and ε_c, respectively (Bar-Sadan et al. 2006a). For example, the strain energy of an octahedron, based on a layered MX_2 compound, with k-layers can be written as

$$\frac{E_{str}}{N} = \frac{6\varepsilon_e \sum_{i=1}^{k}(\sqrt{N_i}-6)-6\varepsilon_\infty \sum_{i=1}^{k}\sqrt{N_i}+36k\varepsilon_c}{\sum_{i=1}^{k}N_i} + \frac{k-1}{k}\varepsilon_{vdW} \qquad (12.11)$$

The energies as a function of size (number of atoms N) for octahedral MoS_2-based fullerenes obtained using Equation 12.11 are drawn in Figure 12.6 in comparison with the corresponding energies for spherical fullerene-like particles and nanoplatelets. All single-walled fullerenes are less stable than a flat monolayer and even the nanoplatelets with their unsaturated edges. Obviously, the stability of fullerene-like particles and octahedral particles increases with the number of van der Waals bound shells. However, the most striking result is the occurrence of "crossover points" between the energy curves of the nano-octahedra and the nanospheres at the values of N of the order of a few 10^5 atoms. Thus, several phase transitions between the different morphologies can be predicted: first, for particles with less than 1.5×10^4 atoms, the flat platelet structure is the most stable one. Second, in the interval between 1.5×10^4 and 1.5×10^5 atoms, the most stable structures are the octahedral particles (nano-octahedra) with a small number of dangling bonds at the corners and flat facets. Third, above 10^5 atoms, the most stable "zero-dimensional" modifications are quasi-spherical, fullerene-like particles. It turns out, therefore, that the nano-octahedra are the smallest hollow clusters of MoS_2, i.e., they can be considered as the genuine IFs of this compound. Obviously, this observation is not limited to MoS_2 and can probably be extended to various other layered compounds.

Equations 12.9 and 12.10 can also explain the preference for the formation of uncapped inorganic tubes with open ends, because the strain energies for spherical or octahedral fullerene-like caps are too large for the typical radii of the nanotubes. This conclusion is supported by the quantum-mechanical density-functional tight-binding (DFTB) calculations of the MoS_2 octahedral fullerenes (Bar-Sadan et al. 2006a; Enyashin et al. 2007a,b). Geometry optimizations of hollow $(MoS_2)_x$ particles up to $x = 576$ reveal that the initially assigned octahedral shape of the small stoichiometric fullerenes (with $x < 100$) was unstable. The hollow structures of the larger nano-octahedra were found to be stable. While the facets remain unchanged, changes occur at the corners: two sulfur atoms split off from each corner and the Mo:S composition degrades to Mo_xS_{2x-12}. Additional molecular dynamics simulations showed a considerable distortion of the initial structure around the corners already at ambient conditions, but the fullerenic facets and edges preserve their integrity (Figure 12.4e). This can be attributed to the high strain energy in the corners of the nano-octahedra.

The numerous observations of nanotubes based on "nonlayered" compounds like MgO (Li et al. 2003b), ZnO (Xing et al. 2003), and Mn_5Si_3 (Yang et al. 2004), which will be described below, evoke the question of their stability. These nanotubes have preferentially prismatic faceted morphology and, actually, can be represented as hollow monocrystals fashioned from the corresponding bulk compounds. Atomistic simulations using a force-field method on MgO nanotubes (Enyashin et al. 2006b; Enyashin and Ivanovskii 2007) and quantum-chemical calculations on $TiSi_2$ and TiC nanotubes (Enyashin and Ivanovskii 2005a,b; Enyashin and Gemming 2007) show that the energy of such nanotubes weakly depends on their perimeters. Though, by increasing the wall thickness, the energy of a nanotube based on a nonlayered compound will approach the energy of the bulk material.

12.4 Synthesis of Inorganic Nanotubes and Fullerenes

There is a broad range of synthetic strategies that have been developed for the preparation of INTs and fullerene-like nanoparticles. While the structure and shape control of inorganic nanoparticles is not as effective as in the case of carbon fullerenes and nanotubes, some of the successful methods demonstrated partial controllability. Furthermore, heuristic arguments derived from a combination of theory and experiment allowed one to draw some far-reaching conclusions as to the propensity of a

particular technique to produce certain nanostructures. Despite some exceptional cases, unlike their carbon counterparts, INTs and fullerenes appear preferentially in multiwall structures. Quantum-mechanical calculations provide a clue to this phenomenon, but far more work is needed on both the experimental and theoretical fronts to clarify the full complexity of these inorganic nanostructures. Remarkably though, chemistry provided some neat ways to produce various fullerene-like nanoparticles and nanotubes, other than carbon, in large amounts (a few kg/day and more) and at affordable costs (<100 dollars/kg).

12.4.1 Physical Techniques

Laser ablation and arc-discharge techniques, which played such a pivotal role in the discovery and further developments of the large-scale synthesis of carbon nanotubes and fullerenes, have been embraced only to a limited extent for the synthesis of INTs and fullerenes. The arc discharge of a MoS_2 target in water was shown to produce closed-cage nanoparticles consisting of 2–3 layers with a polyhedral topology and sizes between 5 and 15 nm (Sano et al. 2003). The role of water is not fully clear but it seems to mend the violent reaction near the anode (temperatures reaching 3000 K), which would otherwise lead to excess sulfur evaporation. In yet another set of experiments, arc discharge of MoS_2 nanoparticles with a Mo core were reported (Hu and Zabinski 2005). Films of these nanoparticles were prepared and were found to provide superior tribological behavior as compared with films of $2H\text{-}MoS_2$ that were prepared by pulsed laser ablation (see Section 12.5.1).

Using a Nd:YAG laser, multi-wall polyhedral structures of 5–15 nm in size of both MoS_2 and WS_2 were obtained (Sen et al. 2001). Short multi-wall WS_2 nanotubes were obtained, though at low yields. Nano-octahedra of MoS_2 were prepared by laser ablation (Parilla et al. 2004; Bar-Sadan et al. 2006a). Figure 12.7 shows a transmission electron microscope (TEM) image of a typical product of the laser ablation (above) and their structural models. Nano-octahedra can be considered to be the ultimate closed-cage structure of MoS_2 (as well as many other layered compounds)—somewhat analogous to C_{60} in the carbon fullerene series.

Nanostructured titanium oxide nanoparticles were produced in a pulsed microplasma cluster source in the presence of a helium–oxygen mixture as carrier gas (Ducati et al. 2005). Using a high resolution transmission electron (HRTEM), isolated TiO_x cages that closely resemble carbon fullerenes were found in the deposit. The diameter of the cages ranges from about 0.9 to 2.7 nm. A fraction of the cages have irregular shapes, possibly induced by oxygen vacancies. It was proposed that the TiO_x fullerenoids grow in the gas phase, in a narrow temperature/pressure range within the cluster source and that they are preserved through low-energy deposition. The authors have also attributed the stoichiometry Ti:O = 1:3 to these nanoparticles and have offered their structural models as the analogues of carbon-based fullerenes. However, later quantum-mechanical calculations (Enyashin and Seifert 2007) have shown that such

(a)   (b)

FIGURE 12.7 (a) Experimental TEM and modeled images of nanooctahedra of MoS₂ prepared by laser ablation (From Bar-Sadan, M. et al., *J. Phys. Chem. B*, 110, 25399, 2006a. With permission.) and (b) TEM image of a NiCl₂ nanotube also prepared by laser ablation (From Hacohen, Y. et al., *Adv. Mater.*, 14, 1075, 2002. With permission.).

composition is not stable and such compounds should disintegrate into TiO_2 and O_2. The models of TiO_2 nano-octahedra were found to be more stable.

In a series of experiments, ablation of the layered compound Cs_2O (with an anti-cadmium chloride structure) using a Nd:YAG laser (Albu-Yaron et al. 2005), and subsequently with a focused solar beam (Albu-Yaron et al. 2006), led to the formation of closed-cage fullerene-like nanoparticles. In contrast to the bulk material that comes in a platelet form and is extremely unstable in the ambient atmosphere, the closed cage Cs_2O nanoparticles were found to be stable in the ambient for some time (<1 h).

Electron-beam (e-beam) irradiation was found to produce closed-cage structures from various inorganic compounds and particularly those belonging to the layered metal-halide series. Early on, fullerene-like structures of MoS_2 were obtained by the e-beam irradiation of MoS_2 within the TEM (JoseYacaman et al. 1996). More recently, closed-cage polyhedral structures of layered $CdCl_2$ (Popovitz-Biro et al. 2001) were obtained by e-beam irradiation. Unlike the bulk structure, the fullerene-like $CdCl_2$ was found to have no bound water molecules in their lattices. Furthermore, the nanoparticles were found to be stable in the ambient atmosphere for some time, before water started to dissolve the reactive corners, which suffer from excess strain. This observation provides more evidence of the kinetic stabilization of the closed-cage fullerene-like structure. Faceted polyhedra and short nanotubes of CdI_2 (Popovitz-Biro et al. 2003) were also obtained by the e-beam irradiation of the bulk materials.

Fullerenes of $NiBr_2$ were produced (Bar-Sadan et al. 2006b) by sublimation-condensation in a furnace at 980°C. The closed-cage nanoparticles were found to engulf a nonhollow core consisting of amorphous Ni. Alternatively, a $NiBr_2$-hydrate powder was first dried, and then laser ablated. The soot obtained by the laser ablation was placed on a TEM grid and was irradiated by focusing the e-beam on a spot for a few seconds, which also resulted in fullerenic $NiBr_2$ nanoparticles. In a few cases, nano-octahedra about 6 nm in size and consisting of three layers were observed.

12.4.2 Wet Chemistry

These methods are used for the low-temperature synthesis of materials, including among others hydrothermal (solvothermal) synthesis, that is carried out in a high-pressure vessel (bomb reactor): the intercalation–exfoliation method (of layered crystals); electrochemical synthesis, which in general yields polycrystalline nanotubes; the solution–liquid solution (SLS), which is the low-temperature solution analog of the high-temperature vapor–liquid–solid (VLS) process, sol-gel, sonochemical methods, etc. Various kinds of nanotubes have been prepared using soft-chemistry methods, most of them being oxides. A number of examples of the synthesis of inorganic nanoparticles through soft-chemistry methods were reviewed (Tenne and Zettl 2001).

Such synthetic approaches are very common for the preparation of oxide structures due to the usage of wide-spread water-based solvents. The variety of the reaction conditions leads to a wide range of morphologies and properties of the produced structures. Therefore, further comparison of their stability and properties, and the study of the impurity influence are very important for the purposeful control of their functionalization.

One of the first successful uses of solution chemistry in the production of nanostructured matter is the synthesis of vanadium oxide nanotubes VO_x involving an amine with long alkyl chains as a molecular, structure directing template (Spahr et al. 1998; Patzke et al. 2002). These nanotubes have a mostly scroll-like structure. The tube diameters can be tuned from a few up to several hundreds of nm. VO_x nanotubes are easily accessible in high yield by treating a vanadium (V) precursor with an amine ($C_nH_{2n-1}NH_2$ with $4 < n < 22$) or an α,ω-diaminoalkane ($H_2N(CH_2)_nNH_2$ with $14 < n < 20$), followed by hydrolyzation, aging of the gel, and a hydrothermal reaction. The possibility of using V_2O_5, $VOCl_3$, or HVO_3 as the vanadium source instead of a vanadium (V) alkoxide provides a low-cost alternative (Niederberger et al. 2000). VO_x nanotubes containing aromatic amine were also obtained with phenylpropylamine (Bieri et al. 2001).

Perhaps the most outstanding example of the success of soft-chemistry methods is the synthesis of the titanate nanoscrolls and the arrays of polycrystalline titania nanotubes. In 1998, a new method for the synthesis of titania nanotubes first using the sol-gel process and subsequently treating the product in an NaOH solution at 110°C for 20 h was reported (Kasuga et al. 1998). A more detailed structural analysis of the product revealed that the titania nanotubes are not of the anatase polytype, but are rather nanoscrolls obtained by the self-winding of molecular sheets of the hydrated layered titanate ($H_2Ti_3O_7$) compound (Du et al. 2001). Further study (Zhang et al. 2003) of the synthetic process produced a plausible growth mechanism for the titanate nanotubes, which has been described above.

Nanotubes of very few other ternary compounds with layered structures have so far been reported. One such case is the hydrothermal synthesis of nanotubes of the layered compound $Bi_{24}O_{31}Br_{10}$ (Deng et al. 2005). In a different kind of experiment, crystalline $PbCrO_4$ nanotubes were prepared by mixing lead acetate and sodium chromate in a mixture of aqueous and aprotic solvents in the presence of a surfactant and at room temperature (Chen et al. 2005). This compound possesses a monoclinic unit cell and not a layered structure. Not surprisingly therefore, the rectangular cross-section of the nanotubes reflected the unit-cell structure. Furthermore, the large diameter of the nanotubes (>100 nm) far exceeds the typical value for nanotubes from layered compounds, which can wind around themselves, producing nanotubes with a characteristic diameter of 5 nm and above.

In another remarkable experiment (Zwilling et al. 1999; Gong et al. 2001), arrays of polycrystalline titania (anatase polytype) nanotubes were prepared by anodization of Ti foil in a fluoride ion-containing solution. Anodization of Si, Al, and various II–VI and III–V semiconductors is known to lead to the formation of a network of microporous holes with aspect ratios exceeding 10^3 for some three decades. These arrays and most particularly the alumina porous membrane serve to prepare numerous one-dimensional (1D) nanostructures, which is mostly out of the scope of the present review.

Template techniques for the fabrication of polycrystalline nanotube arrays have been used extensively in recent years. Thus, $Ni(OH)_2$ nanotubes were prepared by immersing a porous alumina membrane in hydrous nickel chloride and subsequently in ammonia solutions (Chou et al. 2005). This study was motivated by the need for a high surface area material for the positive electrode of rechargeable Ni/Cd, Ni/Fe, Ni/metal–hydride, and Ni/Zn batteries.

Some other examples of the use of anodized (porous) alumina (AAO) membranes as templates for the synthesis of polycrystalline nanotubes are, for example, silica nanotubes (Son and Lee 2006), which were gold functionalized at their (open) tip. Such nanotubes can be used to bind specifically various biomolecules through the strong thiol–gold interaction. Another example is that of hydrous-RuO_2 nanotubes that can potentially be used for supercapacitors (Hu et al. 2006).

In another kind of study, hollow MoS_2 nanoparticles were obtained by the sonochemical reaction of $Mo(CO)_6$ and sulfur in isodurene with silica nanospheres used as a template (Dhas and Suslick 2005). After the sonochemical reaction was completed, the silica nanospheres were removed by light hydrofluoric (HF) etching. The MoS_2 hollow nanoparticles showed a high reactivity and selectivity towards the decomposition of thiophene in a strongly reducing atmosphere, indicating that these nanomaterials could be very useful for the hydrodesulfurization of sulfur-rich gasoline. Elements of the pnictide series (P, As, Sb, and Bi) are able to form a quasilayered puckered structure, with the lone pair of adjacent atoms oriented in opposite directions. A remarkably simple synthesis of Bi nanotubes by the room-temperature reduction of the $BiCl_3$ solution with metallic zinc was proposed (Yang et al. 2003).

It is remarkable that the fullerene-like particles and fullerenes of oxides are quite rare cases. It can be explained by a high ionicity of oxides, which does not promote a layered structure, and the higher strain energies of closed nanoparticles compared to the nanotubes. Though, together with Cs_2O and TiO_x synthesis, which were described above, the soft chemistry also gives the possibility of observing onion-like nanoparticles of layered Tl_2O (Avivi et al. 2000). Such particles can be prepared by the sonochemical treatment of $TlCl_3$ in water solution. About 40% of the as-prepared Tl_2O material is composed of nanoparticles with closed structure and the number of layers in such structures is typically 15–20. A similar procedure for a $GaCl_3$ water solution produces a scroll-like nanostructure of GaOOH (Avivi et al. 1999).

12.4.3 High-Temperature Reactions

This category of methods includes IFs, fullerene-like particles, and nanotubes produced by reactions at ~450°C and above (Figure 12.8). The products of these reactions are, in general, of higher crystalline quality. Starting with the report on WS_2 nanotubes and fullerene-like structures obtained by reacting thin films of WO_3 in a reducing atmosphere and under an H_2S flow at 850°C (Tenne et al. 1992), a wealth of reports of new nanotubular structures have appeared in the literature. Viewing the substantial potential for applications, extensive work has been devoted to the synthesis of nanotubes from transition-metal dichalcogenides.

One of the most important developments in recent years involves reacting volatile transition-metal dihalides or carbonyls and H_2S or other sulfur-containing gases as precursors. The rationale for developing this reaction is that the synthesis of, e.g., fullerene-like MoS_2 from the molybdenum oxide nanoparticles is a slow diffusion-controlled reaction, where the oxide core is gradually consumed by the reaction with the sulfur. Furthermore, some oxides like TiO_2 are very stable making the reaction with H_2S at <1200°C practically unfeasible. Not least important is the fact that instead of a temperature between 750°C and 950°C used for the oxide to sulfide conversion, here, lower temperatures (below 750°C) can be employed for the synthesis, though short annealing at higher temperatures is often needed. The first report of this new kind of reaction path was demonstrated for the synthesis of fullerene-like NbS_2, which was obtained by reacting $NbCl_5$ vapors with H_2S in the range of temperatures between 400°C and 550°C (Schuffenhauer et al. 2002). Clear evidence was obtained that the reaction is proceeded by a nucleation and growth mechanism, i.e., nucleation and then a layer-by-layer growth mode. When higher temperatures were used, the yield of the product became smaller, probably due to the fast self-decomposition of the niobium chloride vapors. This study was followed by a series of similar studies in which the metal-halide vapors were reacted with H_2S to yield fullerene-like nanoparticles and nanotubes (Li et al. 2004; Deepak et al. 2006).

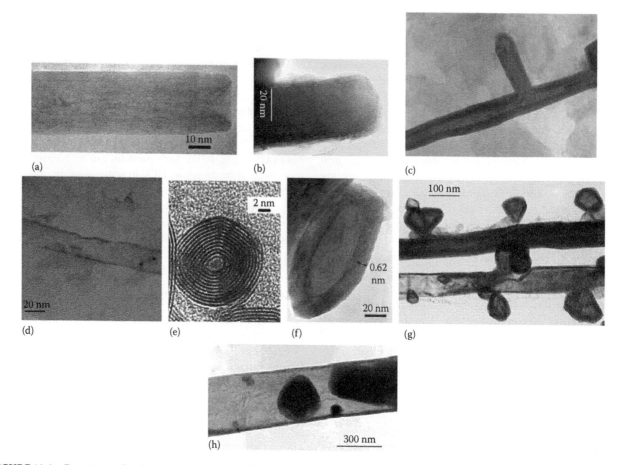

FIGURE 12.8 Experimentally observed morphologies of MoS$_2$ nanostructures: (a) Open-ended nanotube (From Hsu, W.K. et al., *J. Am. Chem. Soc.*, 122, 10155, 2000. With permission.) (b) capped nanotube (From Mastai, Y. et al., *J. Am. Chem. Soc.*, 121, 10047, 1999. With permission.) (c) T-shaped nanotube (From Tenne, R. et al., *Chem. Mater.*, 10, 3225, 1998. With permission.) and (d) twisted nanotube (From Santiago, P. et al., *Appl. Phys. A–Mater.*, 78, 513, 2004. With permission.) (e, f) typical fullerene-like MoS$_2$ particles (From Zak, A. et al., *J. Am. Chem. Soc.*, 122, 11108, 2000; Li, X.L. and Li, Y.D., *J. Phys. Chem. B*, 108, 13893, 2003. With permission.). (g) Nanobuds of WS$_2$ (From Remškar, M. et al., *Adv. Mater.*, 19, 4276, 2007. With permission.) and (h) nanopods of MoS$_2$ are first inorganic hybrid nanostructures. (From Remškar, M. et al., *Nano Lett.*, 8, 76, 2008. With permission.)

In another series of studies, vapors of metal carbonyls were used as the volatile precursor to react with H$_2$S, leading to fullerene-like MX$_2$ (M = Mo, W; X = S, Se) nanoparticles (Etzkorn et al. 2005). Similarly, TiS$_2$ nanotubes were obtained by reacting TiCl$_4$ vapors with H$_2$S at 450°C (Chen et al. 2003a). The nanotubes were subsequently used to demonstrate a rechargeable Li-intercalation battery (Chen et al. 2003b).

A great perspective of this method is the possibility of obtaining nanoparticles from alloyed chalcogenides, i.e., containing at least two types of d-metals. Recently, fullerene-like Mo$_{1-x}$Nb$_x$S$_2$ nanoparticles have been synthesized (Deepak et al. 2007) by a vapor-phase reaction involving the respective metal halides with H$_2$S. These nanoparticles, containing up to 25% Nb, were characterized by a variety of experimental techniques. An analysis of the x-ray powder diffraction, x-ray photoelectron spectroscopy, and different electron microscopy techniques shows that the majority of Nb atoms are organized as nanosheets of NbS$_2$ within the MoS$_2$ host lattice. Most of the remaining Nb atoms (3%) are interspersed individually

and randomly in the MoS$_2$ host lattice. Very few Nb atoms, if any, are intercalated between the MoS$_2$ layers. In another set of experiments, fullerene-like Mo$_{1-x}$Re$_x$S$_2$ and W$_{1-x}$Re$_x$S$_2$ nanoparticles and nanotubes have been prepared by the same method (Deepak et al. 2008). Furthermore, it was shown that the addition of small amounts of Nb (Re) into the MoS$_2$ lattice of fullerene-like particles shifts the Fermi level endowing a p-(n-type) character to the pristine nanoparticles.

The application of different kinds of d-metal containing precursors in high-temperature reactions allows the synthesis of new hybrid nanostructures, which can be related to similar carbon nanostructures—peapods (fullerenes inside nanotubes) and nanobuds (coalescence of outside fullerenes and nanotubes; Figure 12.8g and h). First-known inorganic peapods or so-called mama-tubes were observed recently (Remškar et al. 2007) on a sample of MoS$_2$ by sulfurization of Mo$_6$S$_2$I$_8$ nanowires in the flow of Ar with 1% of H$_2$S and 1% of H$_2$ at 1100 K. The peapod formation appears to originate from the following sequence of events: (a) the formation of the MoS$_2$ nanotube as a kind of envelope, by decomposition of the topmost layers of a $Mo_6S_2I_8$ nanowire (it is possible that the transformation takes place through the local appearance of other $Mo_xS_yI_z$ phases with low iodine content, including different Chevrel phases), (b) the subsequent decomposition of the inner $Mo_6S_2I_8$ phase causing a large reduction of mass after the complete removal of the released iodine, and (c) further sulphurization of the remaining internal material and formation of MoS_2 fullerenes by diffusion of the molecules along the inner nanotube surface. Due to very thin walls, which break under short ultra sound agitation, the fullerene-like particles can be released in a controlled way. Using similar conditions and W_5O_{14} nanowires as precursors, the first inorganic nanobuds (nanotubes decorated by nanoparticles) of WS_2 were prepared (Remškar et al. 2008). The fullerene-like particles nucleate in surface corrugations of the nanowires and grow by a diffusion process simultaneously with the transformation of nanowires into hollow MWNTs.

The pyrolysis of ammonium thio-metallate was used on various occasions to synthesize nanotubes and fullerene-like nanoparticles of metal dichalcogenides compounds. In a remarkable experiment, WS_2 nested nanoparticles exhibiting a cuboid form of a rectangular parallelepiped (rather than the usual spherical morphology) were obtained by spray pyrolysis of $(NH_4)_2WS_4$ ethanolic solution (Bastide et al. 2006). A microwave-induced plasma reduction of WO_3, ZrS_3, and HfS_3 at an effective plasma temperature >1000 K yielded fullerene-like MS_2 (M = W, Zr, Hf) nanoparticles and nanotubes/nanorods (Brooks et al. 2006). The latter two were found to oxidize very rapidly, which was attributed to the imperfect structure of the nanotubes and fullerene-like nanoparticles. Templated ReS_2 nanotubes were synthesized by reacting $ReCl_3$-coated carbon nanotubes with H_2S gas at 1000°C (Brorson et al. 2002).

A most remarkable development in recent years is the synthesis of nanotubes with high crystalline order from compounds with a 3D (quasi-isotropic) lattice. Frequently, the synthesis of such nanotubes involves a templating nano-object. The first example was provided by the synthesis of an array of GaN nanotubes obtained by reacting trimethyl gallium and ammonia on the surface of ZnO nanowires that served as a template and the subsequent removal of the template (Goldberger et al. 2003). The nanotubes exhibited a hexagonal cross section following the contour of the ZnO nanowires, which grow along the [0001] growth axis with (100) and (110) facets. Likewise, In_2O_3 nanotubes with high crystalline order were obtained by heating a mixture of In and the above compound in an induction furnace to 1300°C (Li et al. 2003a). More recently, nanotubes and microtubes of 3D compounds, like InN were synthesized by a number of methods, like the carbothermal synthesis (Yin et al. 2004). Furthermore, crystalline nanotubes of the ternary compound indium germanate ($In_2Ge_2O_7$) were synthesized by mixing In_2O_3, GeO_2 powders with active carbon and heating to 1000°C in a sealed quartz ampoule (Zhan et al. 2006). The active carbon reduced the indium oxide into the volatile intermediate InO, which led to a 1D axial growth of the nanotube. This, and similar germanate compounds, are being used as scintillating materials, e.g., for detecting solar neutrinos and as catalysts. $ZnAl_2O_4$

spinel nanotubes were obtained by first forming core (ZnO) shell (Al_2O_3) 1D nanostructures. Subsequent annealing of the sample led to excessive outdiffusion of the ZnO core (Kirkendall effect) and spontaneous formation of the spinel nanotubes (Fan et al. 2006a,b). Similarly, $MgAl_2O_4$ nanotubes were obtained by conformal deposition of alumina on MgO nanowires. Following the annealing at 700°C, the excess MgO core was removed by chemical etching, leaving behind the spinel structured nanotubes (Fan et al. 2006a,b). Heating $MnCl_2$ and Mg_2Si in excess of the former reactant at 650°C led to a reaction of a molten-flux type that produced Mn_5Si_3 nanocages and bamboo-like nanotubes (Yang et al. 2004).

12.5 Physical Properties of Inorganic Nanostructures

The availability of a new toolbox for the nanomanipulation of individual nanotubes and nanowires offers new opportunities for physical measurements and opens a whole range of issues to be addressed. The most interesting is the fact that INTs can be synthesized free of defects thereby providing a safe ground for quantitative comparisons with theoretical calculations, which was hitherto very difficult. Moreover, many physical properties show a remarkable size dependence in the transition from a bulk state to that of a single state or few molecules. These changes can be systematically analyzed with the current technology. However, much of this physics is yet to be unraveled. While relatively a lot of work has been done on the mechanical properties of INTs, the number of studies on the electronic, optical, and magnetic properties of such nanotubes is very small and should be extended.

12.5.1 Mechanical and Tribological Properties

The mechanical properties of INTs are interesting not merely for academic reasons. Many of these nanotubes show extensive potential for becoming part of ultrahigh strength nanocomposite technology.

The most systematic and comprehensive study of the mechanical properties of INTs was undertaken with WS_2 using scanning electron microcopy (SEM) and scanning probe microscopy (SPM) (Kaplan-Ashiri et al. 2004, 2006, 2007). Several techniques for the handling and mounting of individual nanotubes were used for these studies. Conceptually, performing similar experiments within the TEM could provide accurate atomistic pictures during the mechanical measurements, which could then be compared with quantitative computer simulations. However, technologically this is the most demanding and the least mature technique of all. Further developments in this direction are in progress. Another area where progress is slow and much further work is needed, is the coupling between mechanical and electrical, thermal, or optical effects in a single nanotube.

Figure 12.9 displays a series of micrographs taken during the tensile test of an individual WS_2 nanotube within the SEM. These experiments provided the full strain–stress curve of a

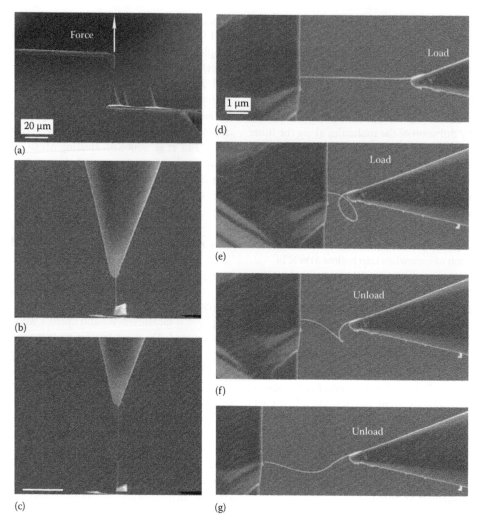

FIGURE 12.9 SEM image of a WS$_2$ nanotube with both its edges glued to the Si cantilevers in a typical tensile test (a, b) before and (c) after it has been broken (From Kaplan-Ashiri, I. et al., *Proc. Natl. Acad. Sci. U. S. A.*, 103, 523, 2006. With permission.), and in a buckling test during the (d, e) loading and the (f, g) unloading processes. (From Sheng Wang, M., *Nano Res.*, 1, 22, 2008. With permission.)

nanotube. Repeating these experiments many times provides statistically averaged meaningful values for the Young's modulus (Y), the strength, and the elongation to failure. In a limited series of about 10 nanotubes, the values of these parameters were found to be 150 GPa, 16 GPa, and 12%, respectively. The experimentally observed Young's modulus agrees very well with that of bulk WS$_2$. The strength of the nanotube is 11% of the Young's modulus, which is rarely observed in bulk materials.

The quantum-mechanical calculations for single-wall MoS$_2$ nanotubes were performed and were generally in agreement with the experimental findings. The calculated stress and strain are 40 GPa and 17% for the *zigzag* (22,0) tube and 34 GPa and 19% for the *armchair* (14,14) tube. The Young's modulus was found to be 230 GPa, which is the same as that of bulk MoS$_2$. The calculated strength results are equivalent to 17.4% and 14.7% (for the *zigzag* and *armchair* nanotubes, respectively) of the Young's modulus of MoS$_2$. Notwithstanding the differences between the Young's modulus of MoS$_2$ and WS$_2$, the agreement between the

experimentally observed and the calculated values is very good. When the nanotube reaches its ultimate elongation, a single chemical bond in the middle of the nanotube breaks. This failure then leads to a stress concentration in the adjacent chemical bonds that becomes overstrained and consequently fails, leading to the immediate destruction of the nanotube. The experimentally determined strength of the WS$_2$ nanotubes is 11% of its Young's modulus, which is an exceedingly high value in comparison to high-strength materials. It is important to realize that the onset of failure for the nanotubes emerges from the excessive distortion of a chemical bond, while the role of macroscopic failure mechanisms, like dislocation diffusion and propagation of cracks along grain boundaries, seems not to be applicable here. A recent tensile test of a multi-wall WS$_2$ nanotube within the TEM shows a telescopic failure, which confirms the notion that the strain is taken by the outermost layer of the nanotube (Kaplan-Ashiri and Tenne 2007). It should also be noted that the mechanical behavior of bulk (crystalline) MoS$_2$ was measured in the basal plain, i.e., along the $< hk0 >$ direction. The mechanical properties in this direction are determined by the tight Mo–S chemical bond. On the other hand, the mechanical parameters of bulk MoS_2 crystals along the c-axis < 001 > exhibit appreciably inferior values. Here, the properties are determined by the weak van der Waals interactions between the layers.

The shear modulus of a beam is expressed as: $G = Y/2(1 + v)$, with v = the Poisson ratio. This expression shows that the three important mechanical parameters are not independent. In particular, if one assumes the value of 0.3 for the Poisson ratio of a WS_2 nanotube, G is found to be 57 GPa. DFTB calculations of the intralayer shear modulus yielded the values of 53 and 81.7 GPa for *zigzag* and *armchair* single-wall MoS_2 nanotubes, respectively. This value is significantly different from C_{44} values that were previously determined by the neutron and x-ray scattering of the linear compressibilities for bulk $2H$–MoS_2 (15 GPa; Feldman 1976). This data indicated that the interlayer shear of multi-wall MoS_2, which is affected by the weak van der Waals interactions, is appreciably smaller than the one obtained within a layer. To prove this issue quantitatively, a bending test of an individual WS_2 nanotube was carried out (Kaplan-Ashiri and Tenne 2007), which allowed to determine a shear of 2 GPa reflecting the slippage between the adjacent WS_2 layers of the nanotubes and designated as a sliding modulus. The DFTB calculation of the interlayer shear of two adjacent layers in $2H$–MoS_2 resulted in a modulus of 4.09 GPa, which is in reasonable agreement with the experimental data for the multi-wall WS_2 nanotubes (the van der Waals interaction between adjacent layers in the two materials is not likely to be very different).

Due to the strong C–C bond in sp^2 (graphitic) hybridization, the Young's and the shear moduli of carbon nanotubes are appreciably higher than those of most INTs. Furthermore, being made of carbon, these nanotubes are very light. Nonetheless, carbon nanotubes, and especially the multi-wall ones, which are very important, for example, in ultra-high strength nanocomposites, suffer from a number of disadvantages making the presently available inorganic (WS_2 or MoS_2) nanotubes suitable for a variety of mechanical applications. In particular, the C–C bond is unstable under compression and transforms easily into the sp^3 (diamond) bond. Contrarily, WS_2 or MoS_2 do not have a high-pressure phase, and under high pressure, they eventually break down making their inorganic fullerene-like and tubular nanoparticles much more robust under compression. Furthermore, the narrow scattering in the strength and elongation data of the WS_2 nanotubes (Kaplan-Ashiri et al. 2004, 2006; Kaplan-Ashiri and Tenne 2007) indicates that they are almost free of critical defects, permitting a predictable assessment of their mechanical behavior in a variety of media.

The mechanical properties of WS_2 nanotubes under loading were also studied by in situ TEM and SEM experiments (Sheng Wang et al. 2008). Centers of deformation in the form of kinks occurred while the nanotubes were loaded until complete or partial fracture occurred. These kinks can be correlated with the nonlinear elastic deformations of the nanotubes, which were observed before. A hint for the nanotubes' remarkable resistance against fracture was also demonstrated in the deformation experiments. Here a rip occurred in the nanotube but it has not propagated. The incredible capability of the tubular morphology to highly deform without failure was demonstrated as well (Figure 12.9). The deformation mechanism of the WS_2 nanotubes is somehow different from the one observed for carbon nanotubes because no ripples were observed. This difference can be related to the more complex atomic structure of the WS_2 layers, which makes it stiffer than carbon nanotubes.

The mechanical properties of INTs of other compounds also attracted attention and they were studied by means of both theoretical and experimental methods. The Young's moduli were found to be ~80 GPa for $MoTe_2$ (Wu et al. 2007), ~290 GPa for GaS (Köhler et al. 2004), ~240 GPa for imogolite $(HO)_3Al_2O_3SiOH$ (Guimaraes et al. 2007), and 159±125 GPa for chrysotile $Mg_3Si_2O_5(OH)_4$ (Piperno et al. 2007) nanotubes. The stretching of the mentioned INTs leads mainly to the deformation of the valence angles. These values are a bit lower than for carbon nanotubes having Y ~ 1–2 TPa. Here, Y is determined by the strong C–C covalent bonds along a tube axis, which suffers the strain. Evidently, INTs with similar mechanisms of deformation should also be more resistant to the strain. It is nicely illustrated, for example, by BN nanotubes, which have a graphite-like structure of walls and Y ~ 0.5 –1 TPa (Hernandez et al. 1999). Quite recently, rhenium disulfide ReS_2 nanotubes with Y ~ 0.4 TPa were claimed to be the most rigid among the INTs with a nongraphitic structure of walls (Enyashin et al. 2009). This uniqueness may be explained by the presence of intralayer covalent bonding between the metal atoms within ReS_2, which is nearly absent in other dichalcogenides.

For a long time, d-transition metal dichalcogenides (MX_2, M = Mo, W, Nb, Ta, X = S, Se) were known as compounds with excellent antifrictional properties (Kalikhman and Umanskii 1972). Not only did they reduce the friction coefficient, they were also found to lubricate the reciprocating metal contacts at higher loads than traditional lubricants such as, for example, grease. These results were obtained in the last few years through a long series of experiments, and a realization was found in the preparation of fluid or dry solid lubricants and self-lubricating metal coatings with nanostructured fullerene-like WS_2 (MoS_2) additives (Spalvins 1971; Moser and Levy 1993). The general dependency of the friction coefficient on the load, however, is that of a standard grease-based lubricant. Early on it was hypothesized that the spherical fullerene-like MS_2 nanoparticles would behave like nanoball bearings thereby providing superior solid lubrication as compared to the existing technology. Further work suggested that under mechanical stress the nanoparticles would slowly deform and exfoliate, transferring MS_2 nanosheets onto the underlying surfaces (third-body effect), and continue to provide effective lubrication until they are completely gone, or oxidized (Figure 12.10). The beneficial effect of the powder of fullerene-like nanoparticles as an additive to lubricating fluids has been studied in quite some detail and this phenomenon has been summarized (Hu and Zabinski 2005; Joly-Pottuz et al. 2005; Rapoport et al. 2005). This effect is particularly important when the clearance (gap) between the two mating surfaces and

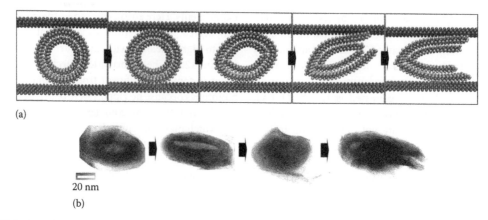

(a)

20 nm

(b)

FIGURE 12.10 Tribological properties of inorganic nanoparticles are the subject of both experimental and theoretical research and find industrial application. (a) Snapshots of molecular dynamics DFTB simulation are shown for the structural evolution of a double walled MoS_2 nanotube under squeezing between two Mo(001) planar gripes. (From Stefanov, M. et al., *J. Phys. Chem. B*, 112, 17764, 2008. With permission.) (b) Damage of fullerene-like WS_2 particle under friction and wear. (From Rapoport, L. et al., *Wear*, 229, 975, 1999. With permission.)

the surface roughness are approximately of the same order of magnitude as the nanoparticles themselves, i.e., 30–300 nm.

More recently, fullerene-like WS_2 nanoparticles were impregnated into metal and polymer films, endowing them with a self-lubricating character (Chen et al. 2002a; Katz et al. 2006; Friedman et al. 2007) and offering them a variety of applications. Clearly, the rolling and sliding friction of the nanoparticles is not possible in this case unless they are gradually released from the metal/polymer/ceramic matrix onto the surface. Here too, some of the beneficial effects of such nanoparticles can be attributed to their gradual exfoliation and the transfer of WS_2 nanosheets onto the asperities of the mating metal surface (third body effect). Furthermore, the bare metal surface is shown to oxidize during the test, leading to a gradual increase in the friction coefficient to very high values (0.3–0.6). In contrast to this, the metal surface impregnated with fullerene-like nanoparticles does not seem to oxidize during the tribological test, although the coverage of the metal surface by the nanoparticles does not appear to exceed 20%–30%. This observation suggests that the temperature of the WS_2-impregnated interface is lower than that of the pure metal surface during the tribological test. It, furthermore, suggests that the fullerene-like nanoparticles may act as a kind of "cathodic protection" against the oxidation of the metal surface, which prevents the oxidation of the metal surface. This technology offers numerous applications, among them various medical devices, like improved orthodontic practice (Katz et al. 2006). In order to capitalize on these applications, NanoMaterials, Ltd. recently constructed a manufacturing pilot plant with a production capacity of about 75 kg/batch and sales of their product under the title "NanoLub" have been launched.

The friction process at high loads was studied also at the atomistic level using quantum-mechanical DFTB simulations (Stefanov et al. 2008). The effect of high loads was performed on cylindrical, defect-free MoS_2 nanotubes of different diameters, chirality, and number of tube walls (Figure 12.10). Two external Mo grips apply a mechanical pressure and gradually squeeze the nanotubes, while chemical interactions between the

grips and the nanotubes were deliberately excluded. The strain–stress curves of single-, double-, and triple-walled nanotubes have shown several remarkable trends: (1) the strain–stress relation is steeper for smaller nanotubes, which is in agreement with the well-known inverse proportionality of the tube energy with respect to its diameter (Seifert et al. 2000a,b, 2002); and (2) the strain–stress relation is essentially independent of the tube's chirality. In contrast to tensile strain–stress curves, the systems deviate from linearity towards higher stress values, that is, if the nanotubes are deformed it becomes harder to deform them any further. The reason for this behavior is that the tubes form planar surface segments close to the grips that are connected with half-tubes, so further compression is essentially equivalent with the compression of tubes with smaller diameters. Further compression leads to irreversible deformation of the nanotubes, and the final product of this process is, however, MoS_2 sheets attached to the two grips, at least by van der Waals interactions, in a face-to-face position, which is ideal for lubrication. It might be meaningful to interpret this result as local coating of the grips. This study has also shown that the strain–stress relationship of the MWNTs is determined by the smallest, innermost tube. In all simulations, the innermost tube determines the breaking process of the whole nanostructure. The tubes are being broken from the inside to the outside: when the innermost tube bursts and unbends under the load, much of the original stress is transferred to the next-largest tube, which breaks down and so forth. The result of the process is, however, the same as for the SWNTs: at the end, MoS_2 platelets are formed partially attached to the grips, which will provide good lubrication at the position of the closest contact of the grips. The results of this study, which are also supported by the existing experimental data, suggest that the ball-bearing effect does not seem to play a major effect in the lubrication process provided by the fullerene-like nanoparticles. This is supported by the fact that the nanostructures break easily under mechanical pressure, but also because the friction coefficient of MoS_2 platelets and nanostructures is identical for smaller loads. The excellent lubrication of nanostructures is hence interpreted as "nano-coating," by attaching lubricating platelets to those parts of the material that are exposed most closely to each other i.e., the asperities.

12.5.2 Electronic Properties

The electronic band structure of many INTs has been studied with many semi-empirical and density-functional-based methods. In the latter, the Kohn–Sham orbital energies were used for the representation of the band structures. Simple zone-folding schemes for the estimation of the nanotube band structure from the band structure of the corresponding layered structure are less meaningful for INTs than for carbon nanotubes. This is mainly due to the much more significant structural relaxation in the case of the rolled layer for inorganic tubes as compared to carbon nanotubes. This conjecture was nicely shown for the nanotubes of MoS_2 and WS_2. In an analogy to carbon nanotubes, the qualitative picture of the band structure of *armchair* and *zigzag* MoS_2 can be derived from the band structure of the molecular MoS_2 layer with its hexagonal structure (Seifert et al. 2000a,b)—see Figure 12.11. But there is a strong reduction in the gap size with the decreasing radius of the tube. This reduction is caused by the compression of the inner sulfur "shell" in the S–Mo–S triple layer in the tubular structure as compared with the flat undistorted triple layer in MoS_2. This calculated bandgap reduction is consistent with experimental observations of the optical-absorption spectra of MoS_2 nanotubes, inorganic fullerene-like structures, and scanning tunneling microscopy (STM) studies of WS_2 nanotubes (Scheffer et al. 2002). A similar size dependence on the electronic gap was also predicted for GaS and GaSe nanotubes (Côté et al. 1998; Köhler et al. 2004).

Many INTs investigated up to now are semiconductors or insulators and they show a considerable dependence on the gap size of the tube diameter. This holds for hypothetical SiH- and GeH-based nanotubes (Seifert et al. 2001a,b), hypothetical phosphorus tubes (Seifert and Hernandez 2000), Bi tubes (Su et al. 2002), and many others. The diameter dependence of the gap size is in all cases nearly independent of the chirality of the nanotube. However, in many cases, the *zigzag* nanotubes have

a direct bandgap, whereas for the *armchair* nanotubes, the gap is an indirect one. Also, the electronic properties of nanotubes based on compounds with insulating character (wide band semiconductors) remain insulating and weakly depend on their chirality, which follows from DFTB calculations of TiO_2 (Enyashin and Seifert 2005), AlOOH (Enyashin et al. 2006a), $Al(OH)_3$ (Enyashin and Ivanovskii 2008), and imogolite nanotubes (Guimaraes et al. 2007).

For several tubular structures, it has been shown that the semiconducting nanotubes can be transformed into metallic ones by intercalation or substitution. This has been demonstrated theoretically, e.g., for Si-based nanotubes in terms of silicide nanotubes $CaSi_2$ (Gemming and Seifert 2003), for BC-based nanotubes by intercalation with Li (Ponomarenko et al. 2003), and by partial substitution of Mo by Nb in MoS_2 nanotubes (Ivanovskaya et al. 2006). Doping of semiconducting MoS_2 nanotubes by Re results in the change of the character of conductivity to *n*-type (Deepak et al. 2008). NbS_2 nanotubes (Seifert et al. 2000a) should be metallic with the Fermi energy in the Nb electronic *d*-band, which is related to a rather high density of states at the Fermi energy. Boron-based nanotubes (Kunstmann and Quandt 2005; Quandt and Boustani 2005) and metal-boride nanotubes (Quandt et al. 2001; Guerini and Piquini 2003; Ivanovskaya et al. 2003) should also be metallic.

An analysis of the optical properties of various kinds of INTs was undertaken in recent years. Many of the measurements were taken from an ensemble of nanoparticles with a few studies dedicated only to individual nanotubes. Raman and IR spectroscopies were used to follow the structural transformation of TiO_2 powder into titanate nanotubes (Qian et al. 2005a). The conversion of the anatase/rutile nanoparticles to sodium titanate during the NaOH reflux was confirmed by the loss of the B_{2g} and the second order of the B_{1g} phonon mode at 398 cm^{-1} of anatase that individually peaked at 516 and 784 cm^{-1}. Instead, a new peak appears at 906 cm^{-1}, which is typical for the short Ti–O bond stretching in the layered sodium titanate and the shoulder at 3208 cm^{-1}, which was assigned to the Ti–OH bonds. This study confirmed the existence of the Ti–OH bonds in the nanotubes.

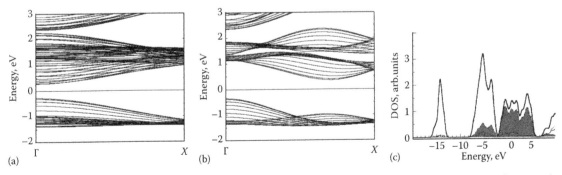

FIGURE 12.11 The band structures of the MoS_2 nanotubes: (a), (22,0) *zigzag*; (b), (14,14) *armchair* configurations. (From Seifert, G. et al., *Phys. Rev. Lett.*, 85(1), 149, 2000b. With permission.); and (c), the total (solid line) and Mo4d- (painted area) density of states for octahedral fullerene $(MoS_2)_{576}$. (From Enyashin, A.N. et al., *Angew. Chem. Int. Ed.*, 46, 623, 2007. With permission.)

Optical absorption, photoluminescence (PL), and luminescence excitation of titanate nanotubes were undertaken (Bavykin et al. 2005). The bandgap of the nanotubes (3.87 eV) is close to that of the layered sodium titanate (3.84 eV), but is appreciably higher than that of the anatase phase of titania (3.2 eV). Studies have shown that changing the internal diameter of TiO_2 nanotubes in the tube diameter range of 2.5–5 nm does not lead to any changes in the position of absorption and emission bands, indicating small quantum size effects in this size range. It was concluded that the electronic structure of TiO_2 nanotubes is very close to that of TiO_2 nanosheets (Bavykin et al. 2005). In yet another study, a strong and broad sub-bandgap PL with a peak at 570 nm was observed in samples consisting of titanate nanotubes. The PL was associated with the Ti–OH complex within the tubular structure (Qian et al. 2005b).

The electronic structure and optical properties of VO_x–alkylamine nanotubes were investigated by absorption, photoelectron, and electron energy-loss spectroscopies (Liu et al. 2005). The photoemission and core-level electron energy-loss spectroscopies confirmed the mixed-valence character of VO_x–alkylamine nanotubes. Indeed, the vanadium ion was found to have an average valency of +4.4 in these nanotubes. In another study, the temperature dependence of the optical gap of VO_x–alkylamine nanotubes was determined (Cao et al. 2004). The optical gap at 0.56 eV was found to be insensitive to the tube diameter. The Raman spectrum of VO_x–alkylamine nanotubes was measured and the optical transitions were assigned to the various modes (Chen et al. 2004; Souza et al. 2004). In latter work (Souza et al. 2004), Cu ions were exchanged successfully with the alkylamine moiety (dodecylamine), which was followed by both Raman and IR spectroscopies.

The optical properties of MoS_2 nanotubes and fullerene-like nanoparticles have been reported in some detail before. More recently, the optical–limiting (OL) properties of MoS_2 nanotubes in aqueous suspensions were investigated (Loh et al. 2006). The OL performance of MoS_2 nanotubes at 1064 and 532 nm was found to surpass that of the carbon-nanotube sample. The resonance Raman spectrum of individual WS_2 nanotubes was recorded (Rafailov et al. 2005). The Raman spectra of agglomerated WS_2 nanotubes were measured under hydrostatic pressure as well. The 2D (in-plane) Grüneisen parameter value found for the WS_2 nanotubes is 0.45, compared to the value of 2 for carbon nanotubes, which shows the softness of the WS_2 nanotube material. Another interesting observation is the blocking of the rigid-layer (E_{2g}) mode (33 cm^{-1}) in the fullerene-like nanoparticles. This observation indicates that the shear of the two layers with respect to each other in the unit cell is damped in the closed fullerene-like structure.

In a separate study (Luttrell et al. 2006), the IR reflectivity at room temperature and at 10 K of bulk (2H) and fullerene-like WS_2 nanoparticles was studied. It was found that the oscillator strength of the E_{1u} transition is appreciably stronger in the former (2H) material at both room temperature and 10 K. By analyzing the two modes at both the 2H (bulk) and fullerene-like WS_2 nanoparticles, it was concluded that the interlayer

charge polarization, i.e., electron transfer from the metal to the sulfur atom, is appreciably smaller in the 2H as compared to the fullerene-like nanoparticles. On the other hand, the interlayer charge polarization between the layers is slightly larger in the fullerenic nanoparticles as compared to the 2H, indicative of the somewhat larger interlayer interaction as compared to the 2H–WS_2 particles.

The electronic and optical properties of octahedral IFs in spite of quasispherical fullerene-like particles have still not been investigated experimentally because of the low stability of such nanostructures. However, the first theoretical studies (Enyashin et al. 2004a,b) using the semi-empirical Extended Hückel Theory (EHT) method have shown that small hollow fullerenes of dichalcogenides MX_2 (M = Mo, Nb, Ti, Zr, Sn, X = S, Se) should be metalloid irrespective of the electronic character of the bulk material. The results of an *ab initio* study on the same sulfide particles and chlorides, like $NiCl_2$, $FeCl_2$, and $CdCl_2$ also yielded a very low HOMO–LUMO gap and also predicted a noticeable spin-polarization of *d*-metal atoms in the case of MoS_2, $NiCl_2$, and $FeCl_2$ fullerenes (Enyashin and Ivanovskii 2004, 2005a; Enyashin et al. 2005). A bond population analysis demonstrated on both semi-empirical and *ab initio* levels of theory that the M–S bonds on the inner surface of the particle were weaker than on the outer surface, and indirectly suggested a propensity to form nonstoichiometric structures.

The fully quantum-mechanical DFTB studies (Bar-Sadan et al. 2006a; Enyashin et al. 2007a,b) support the conclusions about the metalloid character of MoS_2 nano-octahedra drawn from the semi-empirical modeling (Figure 12.11). The densities-of-states are quite similar in profile to those of semiconducting nanotubular and bulk MoS_2 (Seifert et al. 2000b). The general features of the electronic spectra are the same in all cases. However, in spite of this great similarity and in sharp contrast to the nanotubular and bulk MoS_2, the HOMO–LUMO gap in all investigated MoS_2 fullerenes does not exceed a few 0.01 eV, nearly irrespective of the nanocluster size (Figure 12.11). The valence band is composed of mixed Mo4d–S3p-states. As in the bulk, the states around the HOMO and LUMO levels are mainly Mo4d-states. A Mulliken charge distribution analysis shows a charge transfer from the Mo to the S atoms with average charges of −0.41 e and −0.47 e for the internal and external S atoms and +0.91 e for the Mo atoms, hence there is a tendency to form a surface dipole that points along the facet normals. Referring to the activity of nanoplatelets, such an enlarged electronic density on the external sulfur atoms of a fullerenic wall can be associated with a high reactivity. This finding opens up a perspective for the MoS_2 nano-octahedra as a catalyst similar to the nanoplatelets that await an experimental verification.

DFTB calculations (Enyashin and Seifert 2007) reveal that in spite of the above mentioned (MoS_2) IFs with metal-like character, the fullerenes of wide-band gap semiconductors like TiO_2 should be also semiconducting like their bulk or nanotubular forms (Enyashin and Seifert 2005). The calculated densities of states for TiO_2 fullerenes are similar to those of the anatase bulk phase and various 1D titania nanostructures, which were discussed above, do not differ for particles of different sizes. The valence band of the fullerenes is composed of O2p states, whereas the lower part of the conduction band is formed by Ti3d states. All titania fullerenes have large HOMO–LUMO gaps of about 4.3–4.7 eV.

12.6 Chemical Modification

In addition to the structure-property relationship of pure inorganic compounds in nanostructured and bulk allotropes, which was described in detail in the previous sections, the chemical properties of nanosized allotropes can also be modified in comparison to their bulk materials. The most promising approaches are doping, defect formation, intercalation, surface functionalization, and incorporation into composites. The nanotubes, fullerene-like particles, and fullerenes (nano-octahedra) are especially well-suited candidates for these purposes, because being hollow they provide additional ways for functionalization. Four possibilities may be distinguished by the placement of an admixture into the hollow nanostructures: the substitution within the walls (intra-layer doping), the intercalation in the space between the walls (interlayer doping), an adsorption on the outer surfaces of the walls (exohedral functionalization), or an injection into the inner cavities (endohedral functionalization). The first attempts in these directions have already been accomplished and were described above by the substitutional doping in wall-doped $Mo_{1-x}W_xS_2$, $Mo_{1-x}Nb_xS_2$ and $Mo_{1-x}Re_xS_2$ nanotubes (Deepak and Tenne 2008). Ti-doped MoS_2 nanostructures were produced by pyrolysing H_2S over an oxidized Ti–Mo alloy powder (Hsu et al. 2001). However, such chemical modification is possible only during the synthesis because direct substitutional doping is a very energy-expensive process due to the quite strong covalent bonding within a monolayer. Crystalline Fe-doped trititanate nanotubes were synthesized via a wet-chemistry method (Han et al. 2007). These nanotubes exhibited noticeable catalytic activity in the water-gas-shift reaction. Magnetic measurements indicated that the Fe-doped trititanate nanotubes comprised a mixture of ferromagnetic and paramagnetic phases.

The second alternative has more perspectives: the processes of intercalation and de-intercalation by insertion of alkali-metal atoms into MoS_2 and WS_2 fullerene-like particles (Zak et al. 2002; Kopnov et al. 2008). A small increase of the a-axis was observed and the XPS analysis of the rubidium intercalated material showed a rise in the Fermi energy as a result of the intercalation, endowing the originally p-type nanoparticles an n-type character. Mg^{2+} cations were electrochemically injected in various MoS_2 nanostructures including nanotubes and fullerene-like particles (Li and Li 2004). It was shown that even larger particles, such as organic surfactant molecules or even fragments of graphite can be incorporated into MoS_2 nanotubes (Mirabal et al. 2004; German et al. 2005). MoS_2 nanotubes alloyed with gold or silver and with encapsulated nickel sulfide $Ni_{17}S_{18}$ have been synthesized (Remškar et al. 2000; Hofmann et al. 2002).

Third, results of the covalent surface functionalization of fullerene-like MoS_2 particles by organic ligands have been reported (Tahir et al. 2006). Surface functionalization was achieved using an nitrilotriacetic ligand functionalized with a fluorescent 7-nitrobenzofurazan unit and using a polymeric ligand carrying a nitrilotriacetic group and a catechol type ligand, which has been used as the anchor groups for the functionalization of metal oxides. It provides the basis of a toolbox to construct supramolecular assemblies of organic–inorganic hybrid nanomaterials. Multi-walled BN nanotubes grown by a chemical vapor deposition (CVD) method were used for functionalization by naphthoyl chloride $C_{10}H_7COCl$, butyryl chloride $CH_3(CH_2)_2COCl$, and stearoyl chloride $CH_3(CH_2)_{16}COCl$ to fabricate the respective covalent adducts of alkylcarbonyl groups and BN walls, which showed a dramatic change in the electronic structure of BN nanotubes (Zhi et al. 2005, 2006). These studies demonstrated that INTs and fullerene-like particles may be functionalized in the same manner as the analogous carbon nanostructures.

Finally, such a functionalization (by organic molecules or by metal and semiconductor nanoparticles) may even facilitate their dispersion in composites. One recent example concerns the integration of MoS_2 with carbon nanostructures: carbon nanotubes covered by the MoS_2 sheets (Song et al. 2004) and coaxial MoS_2@carbon nanotubes, which demonstrate improved MoS_2 properties regarding lithium storage (Wang and Li 2007).

Already these initial and still not numerous studies on modified nanostructures underline the necessity for more profound experimental and theoretical investigations. In this way, the great potential of these versatile nanomaterials due to their wealth of polymorphic structures may be explored and exploited in large-scale applications.

12.7 Conclusions

Currently, many layered compounds have shown the ability to be prepared in nanostructured allotropes—nanotubes and fullerenes. The first known to be artificially curved into hollow nanostructures and the most explored type of such compounds are d-metal chalcogenides, especially, MoS_2 and WS_2. An extensive effort was especially devoted to the development of advanced synthetic routes and the characterization of the nanotubes and fullerene-like particles of tungsten and molybdenum disulfides. This is not surprising in view of their excellent tribological and mechanical properties, which was recently capitalized in a series of commercial products. In the meantime, analogous structures of BN, VO_x, $H_2Ti_3O_7$, and TiO_2 were discovered suggesting numerous applications for these nanoparticles. Potential applications for such nanostructures in ultrastrong nanocomposites, as biosensors, catalysts for green chemistry, and renewable energy devices are being intensively explored.

However, the materials science of the INTs and fullerenes is currently at an incipient stage of development. The investigation of the properties of INTs and fullerenes is still lacking in many points and needs an evaluation both from experimental and theoretical perspectives. No doubt, the results of further investigations of INTs and fullerenes promote not only the development of our ideas about matter but also will find their incarnation in advanced technological decisions.

Acknowledgments

The authors would like to acknowledge Prof. Reshef Tenne from the Weizmann Institute of Science (Rehovot, Israel) for the helpful discussions and valuable comments made during the preparation of the manuscript.

References

Albu-Yaron, A. et al. (2005). Preparation and structural characterization of stable Cs$_2$O closed-cage structures. *Angewandte Chemie-International Edition* **44**(27): 4169–4172.

Albu-Yaron, A. et al. (2006). Synthesis of fullerene-like Cs$_2$O nanoparticles by concentrated sunlight. *Advanced Materials* **18**(22): 2993–2996.

Amelinckx, S. et al. (1996). Geometrical aspects of the diffraction space of serpentine rolled microstructures: Their study by means of electron diffraction and microscopy. *Acta Crystallographica Section A* **52**: 850–878.

Avivi, S. et al. (1999). Sonochemical hydrolysis of Ga^{3+} ions: Synthesis of scroll-like cylindrical nanoparticles of gallium oxide hydroxide. *Journal of the American Chemical Society* **121**(17): 4196–4199.

Avivi, S. et al. (2000). A new fullerene-like inorganic compound fabricated by the sonolysis of an aqueous solution of TlCl$_3$. *Journal of the American Chemical Society* **122**(18): 4331–4334.

Bar-Sadan, M. et al. (2006a). Structure and stability of molybdenum sulfide fullerenes. *Journal of Physical Chemistry B* **110**(50): 25399–25410.

Bar-Sadan, M. et al. (2006b). Closed-cage (fullerene-like) structures of NiBr$_2$. *Materials Research Bulletin* **41**(11): 2137–2146.

Bar-Sadan, M. et al. (2007). Inorganic fullerenes and nanotubes: Wealth of materials and morphologies. *European Physical Journal-Special Topics* **149**: 71–101.

Bar-Sadan, M. et al. (2008a). Insights into inorganic nanotubes and fullerene-like structures using ultra-high-resolution electron microscopy. *Proceedings of the National Academy of Sciences of the United States of America* **105**(41): 15643–15648.

Bar-Sadan, M. et al. (2008b). Toward atomic-scale bright-field electron tomography for the study of fullerene-like nanostructures. *Nano Letters* **8**(3): 891–896.

Bastide, S. et al. (2006). WS$_2$ closed nanoboxes synthesized by spray pyrolysis. *Advanced Materials* **18**(1): 106–109.

Bavykin, D. V. et al. (2005). Apparent two-dimensional behavior of TiO$_2$ nanotubes revealed by light absorption and luminescence. *Journal of Physical Chemistry B* **109**(18): 8565–8569.

Bieri, F. et al. (2001). The first vanadium oxide nanotubes containing an aromatic amine as template. *Helvetica Chimica Acta* **84**(10): 3015–3022.

Bishop, C. L. and M. Wilson (2008). The energetics of inorganic nanotubes. *Molecular Physics* **106**(12–13): 1665–1674.

Brooks, D. J. et al. (2006). Synthesis of inorganic fullerene (MS$_2$, M = Zr, Hf and W) phases using H$_2$S and N$_2$/H$_2$ microwave-induced plasmas. *Nanotechnology* **17**(5): 1245–1250.

Brorson, M. et al. (2002). Rhenium(IV) sulfide nanotubes. *Journal of the American Chemical Society* **124**(39): 11582–11583.

Cao, J. et al. (2004). Effect of sheet distance on the optical properties of vanadate nanotubes. *Chemistry of Materials* **16**(4): 731–736.

Chen, W. X. et al. (2002a). Wear and friction of Ni-P electroless composite coating including inorganic fullerene-WS$_2$ nanoparticles. *Advanced Engineering Materials* **4**(9): 686–690.

Chen, Q. et al. (2002b). Trititanate nanotubes made via a single alkali treatment. *Advanced Materials* **14**(17): 1208–1211.

Chen, J. et al. (2003a). Low-temperature synthesis of titanium disulfide nanotubes. *Chemical Communications* **8**: 980–981.

Chen, J. et al. (2003b). Lithium intercalation in open-ended TiS$_2$ nanotubes. *Angewandte Chemie-International Edition* **42**(19): 2147–2151.

Chen, W. et al. (2004). Raman spectroscopic study of vanadium oxide nanotubes. *Journal of Solid State Chemistry* **177**(1): 377–379.

Chen, D. et al. (2005). Fabrication of PbCrO$_4$ nanostructures: From nanotubes to nanorods. *Nanotechnology* **16**(11): 2619–2624.

Chou, S. L. et al. (2005). Electrochemical deposition of Ni(OH)$_2$ and Fe-doped Ni(OH)$_2$ tubes. *European Journal of Inorganic Chemistry* **20**: 4035–4039.

Côté, M. et al. (1998). Theoretical study of the structural and electronic properties of GaSe nanotubes. *Physical Review B* **58**(8): R4277–R4280.

Damnjanović, M. et al. (2001). Symmetry of single-wall nanotubes. *Acta Crystallographica Section A* **57**: 304–310.

Deepak, F. L. et al. (2006). MoS$_2$ fullerene-like nanoparticles and nanotubes using gas-phase reaction with MoCl$_5$. *Nano* **1**(2): 167–180.

Deepak, F. L. et al. (2007). Fullerene-like (IF) Nb$_x$Mo$_{1-x}$S$_2$ nanoparticles. *Journal of the American Chemical Society* **129**(41): 12549–12562.

Deepak, F. L. and R. Tenne (2008). Gas-phase synthesis of inorganic fullerene-like structures and inorganic nanotubes. *Central European Journal of Chemistry* **6**(3): 373–389.

Deepak, F. L. et al. (2008). Fullerene-like Mo(W)$_{1-x}$Re$_x$S$_2$ Nanoparticles. *Chemistry—An Asian Journal* **3**(8): 1568–1574.

Deng, H. et al. (2005). Controlled hydrothermal synthesis of bismuth oxyhalide nanobelts and nanotubes. *Chemistry—A European Journal* **11**(22): 6519–6524.

Dhas, N. A. and K. S. Suslick (2005). Sonochemical preparation of hollow nanospheres and hollow nanocrystals. *Journal of the American Chemical Society* **127**(8): 2368–2369.

Dresselhaus, M. S. and P. Avouris (2001). Introduction to carbon materials research. *Carbon Nanotubes* **80**: 1–9.

Du, G. H. et al. (2001). Preparation and structure analysis of titanium oxide nanotubes. *Applied Physics Letters* **79**(22): 3702–3704.

Ducati, C. et al. (2005). Titanium fullerenoid oxides. *Applied Physics Letters* **87**(20): 201906.

Enyashin, A. N. and S. Gemming (2007). TiSi$_2$ nanostructures—enhanced conductivity at nanoscale? *Physica Status Solidi B-Basic Solid State Physics* **244**(10): 3593–3600.

Enyashin, A. N. and A. L. Ivanovskii (2004). Electronic structure of fullerene-like titanium, zirconium, niobium, and molybdenum disulfide nanoparticles from ab initio calculations. *Russian Journal of Inorganic Chemistry* **49**(10): 1531–1535.

Enyashin, A. N. and A. L. Ivanovskii (2005a). Calculating the atomic and electronic structure and magnetic properties of inorganic fullerenes. *Russian Journal of Physical Chemistry* **79**(6): 940–945.

Enyashin, A. N. and A. L. Ivanovskii (2005b). Structural and electronic properties of the TiC nanotubes: Density functional-based tight binding calculations. *Physica E-Low-Dimensional Systems and Nanostructures* **30**(1–2): 164–168.

Enyashin, A. N. and A. L. Ivanovskii (2007). The mechanically induced tuning of structural properties for MgO tubes under uniaxial tension, torsion and bending: Computer molecular modelling. *Nanotechnology* **18**(20): 205707.

Enyashin, A. N. and A. L. Ivanovskii (2008). Theoretical prediction of Al(OH)$_3$ nanotubes and their properties. *Physica E-Low-Dimensional Systems and Nanostructures* **41**(2): 320–323.

Enyashin, A. N. and G. Seifert (2005). Structure, stability and electronic properties of TiO2 nanostructures. *Physica Status Solidi B-Basic Solid State Physics* **242**(7): 1361–1370.

Enyashin, A. N. and G. Seifert (2007). Titanium oxide fullerenes: Electronic structure and basic trends in their stability. *Physical Chemistry Chemical Physics* **9**(43): 5772–5775.

Enyashin, A. N. et al. (2003). Computational studies of electronic properties of ZrS$_2$ nanotubes. *Internet Electronic Journal of Molecular Design* **1**(2): 499–510.

Enyashin, A. N. et al. (2004a). Structure and electronic spectrum of fullerene-like nanoclusters based on Mo, Nb, Zr, and Sn disulfides. *Inorganic Materials* **40**(4): 395–399.

Enyashin, A. N. et al. (2004b). Electronic structure of nanotubes and fullerene-like molecules of superconducting niobium diselenide. *Russian Journal of Inorganic Chemistry* **49**(8): 1204–1216.

Enyashin, A. N. et al. (2005). Electronic structure and magnetic states of crystalline and fullerene-like forms of nickel dichloride NiCl$_2$. *Physics of the Solid State* **47**(3): 527–530.

Enyashin, A. N. et al. (2006a). Stability and electronic properties of single-walled gamma-AlO(OH) nanotubes. *Mendeleev Communications* **6**: 292–294.

Enyashin, A. N. et al. (2006b). Simulation of the structural and thermal properties of tubular nanocrystallites of magnesium oxide. *Physics of the Solid State* **48**(4): 801–805.

Enyashin, A. et al. (2007a). Nanosized allotropes of molybdenum disulfide. *European Physical Journal-Special Topics* **149**: 103–125.

Enyashin, A. N. et al. (2007b). Structure and stability of molybdenum sulfide fullerenes. *Angewandte Chemie-International Edition* **46**(4): 623–627.

Enyashin, A. N. et al. (2009). Stability and electronic properties of rhenium sulfide nanotubes. *Physica Status Solidi B-Basic Solid State Physics* **246**(1): 114–118.

Etzkorn, J. et al. (2005). Metal-organic chemical vapor deposition synthesis of hollow inorganic-fullerene-type MoS$_2$ and MoSe$_2$ nanoparticles. *Advanced Materials* **17**(19): 2372–2375.

Fan, H. J. et al. (2006a). Single-crystalline MgAl$_2$O$_4$ spinel nanotubes using a reactive and removable MgO nanowire template. *Nanotechnology* **17**(20): 5157–5162.

Fan, H. J. et al. (2006b). Monocrystalline spinel nanotube fabrication based on the Kirkendall effect. *Nature Materials* **5**(8): 627–631.

Feldman, J. L. (1976). Elastic-constants of 2H-MoS$_2$ and 2H-NbSe$_2$ extracted from measured dispersion curves and linear compressibilities. *Journal of Physics and Chemistry of Solids* **37**(12): 1141–1144.

Friedman, H. et al. (2007). Fabrication of self-lubricating cobalt coatings on metal surfaces. *Nanotechnology* **18**(11): 115703.

Gemming, S. and G. Seifert (2003). Nanotube bundles from calcium disilicide: A density functional theory study. *Physical Review B* **68**(7): 075416.

German, C. R. et al. (2005). Graphite-incorporated MoS$_2$ nanotubes: A new coaxial binary system. *Journal of Physical Chemistry B* **109**(37): 17488–17495.

Goldberger, J. et al. (2003). Single-crystal gallium nitride nanotubes. *Nature* **422**(6932): 599–602.

Gong, D. et al. (2001). Titanium oxide nanotube arrays prepared by anodic oxidation. *Journal of Materials Research* **16**(12): 3331–3334.

Guerini, S. and P. Piquini (2003). Theoretical investigation of TiB$_2$ nanotubes. *Microelectronics Journal* **34**(5–8): 495–497.

Guimaraes, L. et al. (2007). Imogolite nanotubes: Stability, electronic, and mechanical properties. *ACS Nano* **1**(4): 362–368.

Hacohen, Y. R. et al. (2002). Vapor-liquid-solid growth of NiCl$_2$ nanotubes via reactive gas laser ablation. *Advanced Materials* **14**(15): 1075–1078.

Han, W. Q. et al. (2007). Fe-doped trititanate nanotubes: Formation, optical and magnetic properties, and catalytic applications. *Journal of Physical Chemistry C* **111**(39): 14339–14342.

Hernandez, E. et al. (1999). Elastic properties of single-wall nanotubes. *Applied Physics A-Materials Science & Processing* **68**(3): 287–292.

Hofmann, S. et al. (2002). Low-temperature self-assembly of novel encapsulated compound nanowires. *Advanced Materials* **14**(24): 1821–1824.

Hsu, W. K. et al. (2000). An alternative route to molybdenum disulfide nanotubes. *Journal of the American Chemical Society* **122**(41): 10155–10158.

Hsu, W. K. et al. (2001). Titanium-doped molybdenum disulfide nanostructures. *Advanced Functional Materials* **11**(1): 69–74.

Hu, J. J. and J. S. Zabinski (2005). Nanotribology and lubrication mechanisms of inorganic fullerene-like MoS$_2$ nanoparticles investigated using lateral force microscopy (LFM). *Tribology Letters* **18**(2): 173–180.

Hu, C. C. et al. (2006). Design and tailoring of the nanotubular arrayed architecture of hydrous RuO$_2$ for next generation supercapacitors. *Nano Letters* **6**(12): 2690–2695.

Ivanovskaya, V. V. and G. Seifert (2004). Tubular structures of titanium disulfide TiS$_2$. *Solid State Communications* **130**(3–4): 175–180.

Ivanovskaya, V. et al. (2003). Quantum chemical simulation of the electronic structure and chemical bonding in (6,6), (11,11) and (20,0)-like metal-boron nanotubes. *Journal of Molecular Structure-Theochem* **625**: 9–16.

Ivanovskaya, V. V. et al. (2006). Structure, stability and electronic properties of composite Mo$_{1-x}$Nb$_x$S$_2$ nanotubes. *Physica Status Solidi B-Basic Solid State Physics* **243**(8): 1757–1764.

Joly-Pottuz, L. et al. (2005). Ultralow-friction and wear properties of IF-WS$_2$ under boundary lubrication. *Tribology Letters* **18**(4): 477–485.

JoseYacaman, M. et al. (1996). Studies of MoS$_2$ structures produced by electron irradiation. *Applied Physics Letters* **69**(8): 1065–1067.

Kalikhman, V. L. and Y. S. Umanskii (1972). Chalcogenides of transition-metals with layered structure and peculiarities of filling of their brillouin zones. *Uspekhi Fizicheskikh Nauk* **108**(3): 503–528.

Kaplan-Ashiri, I. and R. Tenne (2007). Mechanical properties of WS$_2$ nanotubes. *Journal of Cluster Science* **18**(3): 549–563.

Kaplan-Ashiri, I. et al. (2004). Mechanical behavior of individual WS$_2$ nanotubes. *Journal of Materials Research* **19**(2): 454–459.

Kaplan-Ashiri, I. et al. (2006). On the mechanical behavior of WS$_2$ nanotubes under axial tension and compression. *Proceedings of the National Academy of Sciences of the United States of America* **103**(3): 523–528.

Kaplan-Ashiri, I. et al. (2007). Microscopic investigation of shear in multiwalled nanotube deformation. *Journal of Physical Chemistry C* **111**(24): 8432–8436.

Kasuga, T. et al. (1998). Formation of titanium oxide nanotube. *Langmuir* **14**(12): 3160–3163.

Katz, A. et al. (2006). Self-lubricating coatings containing fullerene-like WS$_2$ nanoparticles for orthodontic wires and other possible medical applications. *Tribology Letters* **21**(2): 135–139.

Köhler, T. et al. (2004). Tubular structures of GaS. *Physical Review B* **69**(19): 193403.

Kopnov, F. et al. (2008). Intercalation of alkali metal in WS$_2$ nanoparticles, revisited. *Chemistry of Materials* **20**(12): 4099–4105.

Krumeich, F. et al. (1999). Morphology and topochemical reactions of novel vanadium oxide nanotubes. *Journal of the American Chemical Society* **121**(36): 8324–8331.

Kunstmann, J. and A. Quandt (2005). Constricted boron nanotubes. *Chemical Physics Letters* **402**(1–3): 21–26.

Li, X. L. and Y. D. Li (2003). Formation MoS$_2$ inorganic fullerenes (IFs) by the reaction of MoO$_3$ nanobelts and S. *Chemistry—A European Journal* **9**(12): 2726–2731.

Li, X. L. and Y. D. Li (2004). MoS$_2$ nanostructures: Synthesis and electrochemical Mg^{2+} intercalation. *Journal of Physical Chemistry B* **108**(37): 13893–13900.

Li, Y. B. et al. (2003a). Single-crystalline In$_2$O$_3$ nanotubes filled with In. *Advanced Materials* **15**(7–8): 581–585.

Li, Y. B. et al. (2003b). Ga-filled single-crystalline MgO nanotube: Wide-temperature range nanothermometer. *Applied Physics Letters* **83**(5): 999–1001.

Li, X. L. et al. (2004). Atmospheric pressure chemical vapor deposition: An alternative route to large-scale MoS$_2$ and WS$_2$ inorganic fullerene-like nanostructures and nanoflowers. *Chemistry-A European Journal* **10**(23): 6163–6171.

Liu, X. et al. (2005). Structural, optical, and electronic properties of vanadium oxide nanotubes. *Physical Review B* **72**(11): 115407.

Loh, K. P. et al. (2006). Templated deposition of MoS$_2$ nanotubules using single source precursor and studies of their optical limiting properties. *Journal of Physical Chemistry B* **110**(3): 1235–1239.

Luttrell, R. D. et al. (2006). Dynamics of bulk versus nanoscale WS$_2$: Local strain and charging effects. *Physical Review B* **73**(3): 035410.

Margulis, L. et al. (1996). TEM study of chirality in MoS$_2$ nanotubes. *Journal of Microscopy-Oxford* **181**: 68–71.

Mastai, Y. et al. (1999). Pulsed sonoelectrochemical synthesis of cadmium selenide nanoparticles. *Journal of the American Chemical Society* **121**(43): 10047–10052.

Mendelev, M. I. et al. (2002). Equilibrium structure of multilayer van der Waals films and nanotubes. *Physical Review B* **65**(7): 075402.

Milošević, I. et al. (2000). Symmetry based properties of the transition metal dichalcogenide nanotubes. *European Physical Journal B* **17**(4): 707–712.

Mirabal, N. et al. (2004). Synthesis, functionalization, and properties of intercalation compounds. *Microelectronics Journal* **35**(1): 37–40.

Moser, J. and F. Levy (1993). MoS$_{2-x}$ Lubricating films—Structure and wear mechanisms investigated by cross-sectional transmission electron-microscopy. *Thin Solid Films* **228**(1–2): 257–260.

Niederberger, M. et al. (2000). Low-cost synthesis of vanadium oxide nanotubes via two novel non-alkoxide routes. *Chemistry of Materials* **12**(7): 1995–2000.

Parilla, P. A. et al. (2004). Formation of nanooctahedra in molybdenum disulfide and molybdenum diselenide using pulsed laser vaporization. *Journal of Physical Chemistry B* **108**(20): 6197–6207.

Patzke, G. R. et al. (2002). Oxidic nanotubes and nanorods—Anisotropic modules for a future nanotechnology. *Angewandte Chemie-International Edition* **41**(14): 2446–2461.

Piperno, S. et al. (2007). Characterization of geoinspired and synthetic chrysotile nanotubes by atomic force microscopy and transmission electron microscopy. *Advanced Functional Materials* **17**(16): 3332–3338.

Ponomarenko, O. et al. (2003). Properties of boron carbide nanotubes: Density-functional-based tight-binding calculations. *Physical Review B* **67**(12): 125401.

Popovitz-Biro, R. et al. (2001). Nanoparticles of $CdCl_2$ with closed cage structures. *Israel Journal of Chemistry* **41**(1): 7–14.

Popovitz-Biro, R. et al. (2003). CdI_2 nanoparticles with closed-cage (fullerene-like) structures. *Journal of Materials Chemistry* **13**(7): 1631–1634.

Qian, L. et al. (2005a). Raman study of titania nanotube by soft chemical process. *Journal of Molecular Structure* **749**(1–3): 103–107.

Qian, L. et al. (2005b). Bright visible photoluminescence from nanotube titania grown by soft chemical process. *Chemistry of Materials* **17**(21): 5334–5338.

Quandt, A. and I. Boustani (2005). Boron nanotubes. *Chemphyschem* **6**(10): 2001–2008.

Quandt, A. et al. (2001). Density-functional calculations for prototype metal-boron nanotubes. *Physical Review B* **6412**(12): 125422.

Rafailov, P. M. et al. (2005). Orientation dependence of the polarizability of an individual WS_2 nanotube by resonant Raman spectroscopy. *Physical Review B* **72**(20): 205436.

Rao, C. N. R. and M. Nath (2003). Inorganic nanotubes. *Dalton Transactions* **1**: 1–24.

Rapoport, L. et al. (1999). Inorganic fullerene-like material as additives to lubricants: Structure-function relationship. *Wear* **229**: 975–982.

Rapoport, L. et al. (2005). Applications of WS_2 (MoS_2) inorganic nanotubes and fullerene-like nanoparticles for solid lubrication and for structural nanocomposites. *Journal of Materials Chemistry* **15**(18): 1782–1788.

Remškar, M. (2004). Inorganic nanotubes. *Advanced Materials* **16**(17): 1497–1504.

Remškar, M. et al. (2000). Structural stabilization of new compounds: MoS_2 and WS_2 micro- and nanotubes alloyed with gold and silver. *Advanced Materials* **12**(11): 814–818.

Remškar, M. et al. (2007). Inorganic nanotubes as nanoreactors: The first MoS_2 nanopods. *Advanced Materials* **19**(23): 4276–4278.

Remškar, M. et al. (2008). WS_2 nanobuds as a new hybrid nanomaterial. *Nano Letters* **8**(1): 76–80.

Saito, Y. and M. Maida (1999). Square, pentagon, and heptagon rings at BN nanotube tips. *Journal of Physical Chemistry A* **103**(10): 1291–1293.

Sano, N. et al. (2003). Fabrication of inorganic molybdenum disulfide fullerenes by arc in water. *Chemical Physics Letters* **368**(3–4): 331–337.

Santiago, P. et al. (2004). Synthesis and structural determination of twisted MoS_2 nanotubes. *Applied Physics A–Materials Science & Processing* **78**(4): 513–518.

Scheffer, L. et al. (2002). Scanning tunneling microscopy study of WS_2 nanotubes. *Physical Chemistry Chemical Physics* **4**(11): 2095–2098.

Schuffenhauer, C. et al. (2002). Synthesis of NbS_2 nanoparticles with (nested) fullerene-like structure (IF). *Journal of Materials Chemistry* **12**(5): 1587–1591.

Seifert, G. and T. Frauenheim (2000). On the stability of non carbon nanotubes. *Journal of the Korean Physical Society* **37**(2): 89–92.

Seifert, G. and E. Hernandez (2000). Theoretical prediction of phosphorus nanotubes. *Chemical Physics Letters* **318**(4–5): 355–360.

Seifert, G. et al. (2000a). Novel NbS_2 metallic nanotubes. *Solid State Communications* **115**(12): 635–638.

Seifert, G. et al. (2000b). Structure and electronic properties of MoS_2 nanotubes. *Physical Review Letters* **85**(1): 146–149.

Seifert, G. et al. (2001a). Tubular structures of germanium. *Solid State Communications* **119**(12): 653–657.

Seifert, G. et al. (2001b). Tubular structures of silicon. *Physical Review B* **6319**(19): 193409.

Seifert, G. et al. (2002). Stability of metal chalcogenide nanotubes. *Journal of Physical Chemistry B* **106**(10): 2497–2501.

Sen, R. et al. (2001). Encapsulated and hollow closed-cage structures of WS_2 and MoS_2 prepared by laser ablation at 450–1050 degrees C. *Chemical Physics Letters* **340**(3–4): 242–248.

Sheng Wang, M. et al. (2008). In Situ TEM measurements of the mechanical properties and behavior of WS_2 nanotubes. *Nano Research* **1**(1): 22–31.

Son, S. J. and S. B. Lee (2006). Controlled gold nanoparticle diffusion in nanotubes: Platform of partial functionalization and gold capping. *Journal of the American Chemical Society* **128**(50): 15974–15975.

Song, X. C., Xu, Z. D., Zheng, Y. F., Han, G., Liu, B., and Chen W. X. (2004). Molybdenum disulfide sheathed carbon nanotubes. *Chinese Chemical Letters* **15**(5): 623–626.

Souza, A. G. et al. (2004). Raman spectra in vanadate nanotubes revisited. *Nano Letters* **4**(11): 2099–2104.

Spahr, M. E. et al. (1998). Redox-active nanotubes of vanadium oxide. *Angewandte Chemie-International Edition* **37**(9): 1263–1265.

Spalvins, T. (1971). Lubrication with sputtered MoS_2 films. *Tribology Transactions* **14**(4): 267.

Srolovitz, D. J. et al. (1995). Morphology of nested fullerenes. *Physical Review Letters* **74**(10): 1779–1782.

Stefanov, M. et al. (2008). Nanolubrication. How do MoS_2-based nanostructures lubricate? *Journal of Physical Chemistry B* **112**(46): 17764–17767.

Su, C. R. et al. (2002). Bismuth nanotubes: Potential semiconducting nanomaterials. *Nanotechnology* **13**(6): 746–749.

Tahir, M. N. et al. (2006). Overcoming the insolubility of molybdenum disulfide nanoparticles through a high degree of sidewall functionalization using polymeric chelating ligands. *Angewandte Chemie-International Edition* **45**(29): 4809–4815.

Tenne, R. (1996). Fullerene-like structures and nanotubes from inorganic compounds. *Endeavour* **20**(3): 97–104.

Tenne, R. (2006). Inorganic nanotubes and fullerene-like nanoparticles. *Nature Nanotechnology* **1**(2): 103–111.

Tenne, R. and C. N. R. Rao (2004). Inorganic nanotubes. *Philosophical Transactions of the Royal Society of London Series A–Mathematical Physical and Engineering Sciences* **362**(1823): 2099–2125.

Tenne, R. and A. K. Zettl (2001). Nanotubes from inorganic materials. *Carbon Nanotubes* **80**: 81–112.

Tenne, R. et al. (1992). Polyhedral and cylindrical structures of tungsten disulfide. *Nature* **360**(6403): 444–446.

Tenne, R. et al. (1998). Nanoparticles of layered compounds with hollow cage structures (inorganic fullerene-like structures). *Chemistry of Materials* **10**(11): 3225–3238.

Tenne, R. et al. (2008). Inorganic nanotubes and fullerene-like structures (IF). *Carbon Nanotubes* **111**: 631–671.

Terrones, H. et al. (2004). Shape and complexity at the atomic scale: The case of layered nanomaterials. *Philosophical Transactions of the Royal Society of London Series A–Mathematical Physical and Engineering Sciences* **362**(1823): 2039–2063.

Wada, K. and N. Yoshinaga (1969). Structure of imogolite. *American Mineralogist* **54**(1–2): 50–71.

Wang, Q. and J. H. Li (2007). Facilitated lithium storage in MoS_2 overlayers supported on coaxial carbon nanotubes. *Journal of Physical Chemistry C* **111**(4): 1675–1682.

Wu, X. J. et al. (2007). Single-walled $MoTe_2$ nanotubes. *Nano Letters* **7**(10): 2987–2992.

Xing, Y. J. et al. (2003). Optical properties of the ZnO nanotubes synthesized via vapor phase growth. *Applied Physics Letters* **83**(9): 1689–1691.

Yang, B. J. et al. (2003). A room-temperature route to bismuth nanotube arrays. *European Journal of Inorganic Chemistry* **20**: 3699–3702.

Yang, Z. H. et al. (2004). Preparation of Mn_5Si_3 nanocages and nanotubes by molten salt flux. *Solid State Communications* **130**(5): 347–351.

Yin, L. W. et al. (2004). Growth of single-crystal indium nitride nanotubes and nanowires by a controlled-carbonitridation reaction route. *Advanced Materials* **16**(20): 1833–1838.

Zak, A. et al. (2000). Growth mechanism of MoS_2 fullerene-like nanoparticles by gas-phase synthesis. *Journal of the American Chemical Society* **122**(45): 11108–11116.

Zak, A. et al. (2002). Alkali metal intercalated fullerene-like MS_2 (M = W, Mo) nanoparticles and their properties. *Journal of the American Chemical Society* **124**(17): 4747–4758.

Zhan, J. H. et al. (2006). Hollow and polygonous microtubes of monocrystalline indium germanate. *Angewandte Chemie-International Edition* **45**(2): 228–231.

Zhang, S. et al. (2003). Formation mechanism of $H_2Ti_3O_7$ nanotubes. *Physical Review Letters* **91**(25): 256103.

Zhang, S. et al. (2005). Structure and formation of $H_2Ti_3O_7$ nanotubes in an alkali environment. *Physical Review B* **71**(1): 014104.

Zhi, C. Y. et al. (2005). Covalent functionalization: Towards soluble multiwalled boron nitride nanotubes. *Angewandte Chemie-International Edition* **44**(48): 7932–7935.

Zhi, C. Y. et al. (2006). Engineering of electronic structure of boron-nitride nanotubes by covalent functionalization. *Physical Review B* **74**(15): 153413.

Zwilling, V. et al. (1999). Structure and physicochemistry of anodic oxide films on titanium and TA6V alloy. *Surface and Interface Analysis* **27**(7): 629–637.

Spinel Oxide Nanotubes and Nanowires

Hong Jin Fan
Nanyang Technological University

13.1 Introduction

Spinel oxides have been one of the most important oxides in material science. In the form of ceramics, spinels have applications such as gas sensors, pigment materials, phosphors, catalysts, battery electrodes, infrared windows, and transparent electric conductors. However, ceramics are made up of highly compressed grain particles whose sizes range from nano- to micrometers. With nanotechnology, it is possible to make nanoparticles with more uniform sizes and finer grain structures. If traditional ceramic spinels are substituted with such nanostructured spinel nanoparticles, it is expected that some of the application performance can be enhanced. Driven by this, the preparation and characterization of nanoscale spinel oxides, in the form of nanoparticles, nanotubes, and nanowires (dimensions below 100 nm), have been recently receiving much attention (Song and Zhang 2004, Zeng et al. 2004a,b, Wang et al. 2005: 2928, Tirosh et al. 2006). Because of a large surface-to-volume ratio and symmetry breaking on the surface, nanoscale spinel oxides have shown physical properties different from their bulk counterparts. This chapter discusses only quasi one-dimensional (1D) nanostructures (e.g., tubes and wires).

We will first provide an introduction to the fabrication of nanowires and nanotubes, followed by some fundamentals of spinel oxides. In the main body, a material-by-material summary will be given of the examples of spinel-type 1D nanostructures that have been demonstrated so far. As there are relatively few physical characterizations of these nanomaterials, our focus is on the fabrication and associated structure properties. We will discuss the merits and drawbacks of different fabrication routes, analyze the common features of the structures (particularly the twinned nanowires), and also point out the potential spinel-forming interface reaction problem that is still not so well known to the metal-oxide nanowire community and hence may not have been addressed. Some well-known binary spinel-type oxides, such as Fe$_3$O$_4$ and Co$_3$O$_4$, will not be included here.

13.1.1 General Fabrication Methods for Nanowires and Nanotubes

The fabrication of solid nanowires can be realized through various strategies, as elaborated in several review articles (Fan et al. 2006c, Kuchibhatla et al. 2007, Comini et al. 2009). For example, molecules/atoms physically fill up a hollow channel to form nanowires, or they self-organize in a 1D manner due to an anisotropic elongation, or they grow with the assistance of a metal catalyst, the way that carbon nanotubes grow. The most commonly applied techniques are the vapor–transport–deposition and hydrothermal growth. Spinel nanowires have been synthesized following these two methods. However, the difficulty for ternary spinel nanostructures is the control of morphology and composition. The thermal evaporation of the powder mixture of two constituent oxides (or metal plus one oxide) usually results in more than one type of nanostructure in terms of their morphology and phase. A relatively new fabrication method is the solid-state reaction of core–shell nanowire, where the core and shell are the two constituent materials. Compared to thermal evaporation, this method allows for better control.

Unlike nanowires, which can grow directly from the precursor molecules, the formation of nanotubes usually needs the usage of templates. Templates can be of a negative type (such as porous anodic alumina and polycarbonate, which possess 1D parallel channels) or of a positive type (such as solid nanowires). For negative templates of porous alumina, in most cases, the pores are a pure morphology-defining agent, which means that the template does not react with the depositing material; but in some cases, the template itself reacts with the depositing precursor. In other words, the templates can be both reactive and nonreactive. The same holds true for positive templates.

In general, the template-based fabrication methods for spinel nanotubes can be classified into the following five categories (references can be found in Table 13.1):

1. Thermal annealing of a core–shell nanowire to form a spinel nanoshell, followed by eliminating the core via etching or dissolution
2. Reaction of a nanotube of one oxide with the other oxide in a gas or liquid

3. Reaction at pore surfaces of a porous alumina with the other oxide material
4. Infiltration of liquid precursors inside nanochannels followed by solidification
5. Transformation of a nanowire through a solid–state reaction directly to a nanotube through the Kirkendall effect

As we can see, the common feature of these methods is that they are based on templates. Compared to the growth of solid nanowires, such template-based processes for nanotubes allow more control in the tube wall thickness, total diameter, and composition stoichiometry.

13.1.2 Basics of Spinels

The AB_2O_4 type spinel has a cubic lattice structure, which at low temperatures consists of an fcc sublattice of O with the divalent cation (A^{2+}) and the trivalent cations (B^{3+}) occupying one-eighth of the tetrahedral interstices and one half of the octahedral interstices, respectively (see Figure 13.1). One conventional unit cell comprises 8 formula units: 8 A metal cations, 16 B metal cations,

TABLE 13.1   A Summary of Recently Reported One-Dimensional Spinel Oxide Nanomaterials

Material	Nano Morphology	Fabrication Method	Refs.
$MgAl_2O_4$	Wires	Reaction of Mg vapor with ceramic alumina substrate	Wu et al. (2003)
	Tubes	Solid-state reaction of core–shell nanowires	Fan et al. (2006a: 5157)
$ZnAl_2O_4$	Tubes	Reaction of porous alumina with Zn precursor	Wang and Wu (2005), Wang et al. (2006), Zhao et al. (2006)
	Tubes	Solid–state reaction of core–shell nanowires	Fan et al. (2006b: 627)
$ZnGa_2O_4$	Wires	Thermal evaporation of ZnO/Ga powders, with Au catalyst	Bae et al. (2004)
	Helical wires and springs	Thermal evaporation of ZnO/Ga powder	Bae et al. (2005)
	Shells with Ga_2O_3 core	Reaction of Ga_2O_3 nanowires with Zn vapor	Chang and Wu (2006)
	Tubes	Reaction of ZnO nanowires with Ga–O vapor and removal of ZnO core	Li et al. (2006)
$ZnFe_2O_4$	Chainlike wires	Infiltration of nanoparticles into porous alumina and ripening	Jung et al. (2005)
	Wires	Solution infiltration in silica template	Liu et al. (2006)
$ZnCr_2O_4$	Tubes	Reaction of Zn nanowires with CrO_2Cl_2 vapor	Raidongia and Rao (2008)
Zn_2TiO_4	Twinned zigzag wires	Solid-state reaction of ZnO–Ti core–shell nanowires	Yang et al. (2007, 2009)
Zn_2SnO_4	Ribbons	Thermal evaporation of powder mixture and vapor–solid	Wang et al. (2004b: 435, 2005)
	Rings	Thermal evaporation of powder mixture and vapor–solid	Wang et al. (2004b: 435)
	Chainlike zigzag wires	Thermal evaporation of ZnO/SnO powder	Wang et al. (2004a: 177, 2008a,b: 707), Kim et al. (2008)
	Twinned zigzag wires	Thermal evaporation of ZnO/SnO powder	Wang et al. (2004a: 177, 2008a,b: 707), Chen et al. (2005), Jeedigunta et al. (2007)
	Thin rods	Hydrothermal synthesis	Zhu et al. (2006)
$ZnSb_2O_4$	Wires and belts	Thermal evaporation of Sb_2O_3/ZnO powders	Zeng et al. (2004a)
Mn_2SnO_4	Wires	Thermal evaporation of $MnCl_2$/Sn powder, with Au catalyst	Na et al. (2006)
$CrAl_2O_4$	Tubes	Reaction of porous alumina with Cr	Liu et al. (2008)
Mg_2TiO_4	Shells with MgO core	Pulsed laser deposition and in situ reaction with MgO nanowires	Nagashima et al. (2008)
$CoAl_2O_4$	Tubes with peapod structure	Electrodeposition of Co within porous alumina and annealing	Liu et al. (2008)
$LiMn_2O_4$	Wires	Hydrothermal reaction	Hosono et al. (2009)
Fe_3O_4	Tubes	Epitaxial deposition on MgO nanowires	Liu et al. (2005)
	Tubes	Atomic layer deposition inside porous alumina	Bachmann et al. (2007), Escrig et al. (2008)

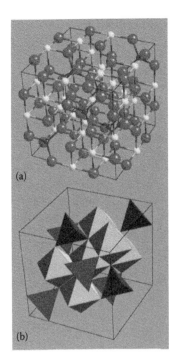

FIGURE 13.1 (a) A ball-and-stick and (b) polyhedral model of the AB_2O_4 spinel structure. Cations A locate at tetrahedral interstices and surrounded by four oxygen; whereas cations B at octahedral interstices and surrounded by six oxygen. The oxygen atoms have an fcc-type close pack sublattice occupying the corners of the polyhedra. (Reproduced from Fan, H.J. et al., *J. Mater. Chem.*, 19, 885, 2009. With permission.)

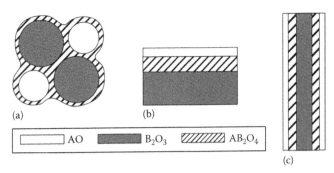

FIGURE 13.2 Interface solid-state reaction for the growth of spinels. (a) Conventional powder reaction where two powders are mixed thoroughly and calcinated at high temperatures (near 1000°C). Spinels form at the contact face in which the material transport methods involve volume and surface diffusion, grain boundary diffusion, and evaporation and deposition. (b) Film-substrate interface reaction to grow spinel films. One of the oxide materials can also be vapors which react with the other oxide due to high substrate temperatures. (c) Nanostructures (tubes or wires) of spinels formed via solid-state reactions of a solid nanowire core with a shell or with vapors of the other oxide.

and 32 O anions. The divalent, trivalent, or quadrivalent metal cations can be Mg, Zn, Fe, Mn, Al, Cr, Ti, and Si. Some crystals have an inverse spinel structure, which contains only half A cations at tetrahedral sites while the other half A and B are at the octahedral sites. A review article on spinel crystal structures is available in Sickafus et al. (1999).

Solid-state reactions of the type $AO + B_2O_3 \rightarrow AB_2O_4$ are a common method used for the fabrication of spinel-oxide films or single crystals. Traditional studies on spinel-formation reactions are usually conducted at planar interfaces or in the form of a powder mixture by bringing two solid binary oxides, or a solid oxide and a vapor or liquid phase, into contact at temperatures near or higher than 1000°C (Schmalzried 1974, Bolt et al. 1998). Figure 13.2a and b show this process schematically. In addition, this sol–gel method is also commonly used for the fabrication of spinel thin films (e.g., see Lee et al. 1998, Meyer et al. 1999).

The name spinel originally refers to $MgAl_2O_4$. It has been a model material for the study of spinel-forming solid-state reaction thermodynamics and kinetics, as well as cation disordering (Irifune et al.1991, Askarpour et al. 1993, Redfern et al. 1999). MgO is cubic with a lattice constant (0.421 nm) about half of most cubic spinels. Moreover it has the same fcc-type oxygen sublattice as that in cubic spinels. The lattice misfit, defined as $2(a_f - a_s)/(a_f + 2a_s)$, where a_f and a_s are, respectively, the lattice constants for the film and substrate, is around 4%. Hence, single-crystal MgO substrates have been frequently used as one of the reactants for the formation of a variety of spinels through solid-state reactions. Besides $MgAl_2O_4$, spinels of Mg_2TiO_4, $MgFe_2O_4$, $MgIn_2O_4$, $MgCr_2O_4$, $MgCo_2O_4$, etc., can be formed through reactions between the corresponding oxide and a MgO(100) substrate (Hesse and Bethge 1981, Hesse 1987, Sieber et al. 1996, 1997). A cubic-to-cubic orientation of (001) $[100]_{spinel} \parallel$ (001) $[100]_{MgO}$ has been established. A detailed electron microscopy analysis of the lattice misfits between MgO and various MgO-based spinels was available by Sieber et al. (1997). In other materials, when the two oxides are non-cubic, the O sublattice has to rearrange its type, e.g, from hcp in sapphire to fcc in cubic, which easily results in the grain growth of different orientations and a rough interface.

The growth process of classical spinel oxides involves Wagner's cation counterdiffusion mechanism (Carter 1961, Rigby and Cutler 1965), namely, cations migrating through the reaction interface in opposite directions and the oxygen sublattice remaining essentially fixed (see Figure 13.3a). This mechanism applies to many types of spinels, for example, $MgAl_2O_4$, $ZnFe_2O_4$, and Mg_2TiO_4.

The reaction of ZnO (wurzite structure; lattice constants: $a = 0.3250$ nm, $c = 0.5207$ nm) with Al_2O_3 into $ZnAl_2O_4$ spinel (cubic structure; lattice constants: $a = 0.880$ nm) is, however, unique: The growth mechanism involves the diffusion of both Zn and O and an effective unilateral transfer of ZnO into the spinel (see Figure 13.3b). This was first pointed out by Bengtson and Jagitsch (1947) and later readdressed by Navias (1961), Branson (1965), and Keller et al. (1988). This means that an inert marker plane placed at the initial interface will be found at the ZnO/spinel interface for the ZnO–Al_2O_3 reaction, whereas in the case of the

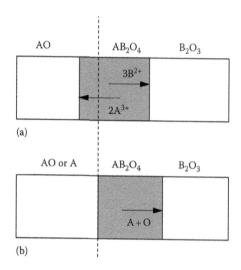

FIGURE 13.3 Schematics of the two main diffusion mechanisms during the spinel-forming solid-state reactions. (a) Wagner counterdiffusion. Typical reaction is $MgO–Al_2O_3$. (b) Uni-directional diffusion. Typical reaction is $ZnO–Al_2O_3$. The dashed line indicates the initial interface. An inert marker plane placed at the initial interface will be found at the ZnO/spinel interface in (b), whereas within the spinel layer in (a) dividing this layer in a ratio of 1:3 as a result of charge neutrality of the overall diffusion flux.

$MgO–Al_2O_3$ reaction, the marker plane is within the spinel layer (see Figure 13.3).

When the nanoscale templates are the materials of one of the oxides, the same reaction process and diffusion mechanism can occur, resulting in the formation of spinel nanowires or tubes (see Figure 13.2c). This is a straightforward and controllable fabrication route. In addition, in order to grow spinel 1D nanostructures of high crystalline quality, one of the oxides should be a single-crystal substrate. This is essentially the same as the film-substrate reactions in conventional studies. On the other hand, there are also reports of the direct formation of spinel oxides nanowires by the thermal evaporation of powder mixture of the two oxides or two metals. Table 13.1 lists the examples that have been published so far of spinel-type ternary oxide nanotubes and nanowires. It should be mentioned that Fe_3O_4, in its form of $(Fe^{3+})(Fe^{3+} Fe^{2+})O_4$, is a prototype inverse spinel ferrite. There are a number of publications of Fe_3O_4 nanotubes and nanowires, but they will not be covered in this chapter.

In bulk solid-state reactions, the diffusion of atoms through the growing product phase generally represents the rate-controlling step; therefore, a high annealing temperature and long time are usually needed for a complete sintering reaction. But a temperature higher than 1300°C brings a stoichiometry problem because of the volatility of the reactants, e.g., ZnO. In a nanoscale system, the diffusion path is shortened so that the reaction steps at the interface may become rate controlling. Furthermore, during the growth of hollow nanoparticles or nanotubes, surface diffu-

sion and grain boundary diffusion might become the dominating transport process over the volume diffusion.

13.2 $MgAl_2O_4$

Conventional interests in the magnesium aluminate spinel are the growth kinetics and mechanism order–disorder phase transitions (Irifune et al. 1991, Askarpour et al. 1993, Redfern et al. 1999). From the application point of view, $MgAl_2O_4$ is a material of interest for optical windows in the visible to mid-wavelength infrared ranges, and had been expected to be a low-cost substitute for sapphire or other optical ceramics. Therefore, most of the synthesis experiments of $MgAl_2O_4$ spinel focus on the modification of its microstructure, in terms of a more uniform grain size and finer grain microstructure, to improve its ceramic strength. From this content, nanostructures (wires and tubes) of $MgAl_2O_4$ spinel might be advantageous.

The first report on $MgAl_2O_4$ spinel nanowires was made by Wu et al. (2003), who obtained spinel nanowires (sometimes the product is spinel/Al_2O_3 core–sheath wires) through the reaction of Mg vapors with a ceramic alumina substrate at 1350°C. The growth direction of the nanowires was along [011]. Generation of the Mg vapor by carbothermal reduction, and subsequent formation of eutectic droplets was the keypoint here to make the reaction process different from the traditional powder-mixture solid–solid reaction. In our own experiments of the vapor-phase synthesis of MgO nanowires by evaporating Mg_3N_2 powder at 950°C inside an alumina tube, a spinel layer and possibly nanostructures might have also formed.

The implementation of the solid-state reaction using nanowires as one reactant provides a generic method for the synthesis of spinel nanotubes. Following this concept, $MgAl_2O_4$ nanotubes have been fabricated by using single-crystalline MgO nanowires as the template (Fan et al. 2006a: 5157). The nanowires were first coated with a conformal layer of alumina via atomic layer deposition (ALD). The interfacial solid–solid reaction took place upon annealing the core–shell nanowires at 700°C–800°C under ambient conditions. During the reaction, metal cation pairs (Mg^{2+} and Al^{3+}) diffuse in opposite directions while the oxygen sublattice is stationary. This is the well-accepted Wagner counterdiffusion mechanism for $MgAl_2O_4$. As the core nanowire has a larger diameter than necessary for a complete spinel reaction with the alumina shell, part of the core remains and an MgO/spinel core–shell nanowire structure was formed (see Figure 13.4a). In order to have a pure phase of spinel nanotube, the remaining MgO core was dissolved in an $(NH_4)_2SO_4$ solution. Figure 13.4b shows a TEM image of a particular web-like $MgAl_2O_4$ spinel nanotube, which is a perfect single crystalline with a cubic cross section (indicated by the arrow). This cubic structure was inherited from an excellent interface lattice matching (the lattice misfit equals to −4.1%) with an MgO core whose side faces are also {100} planes.

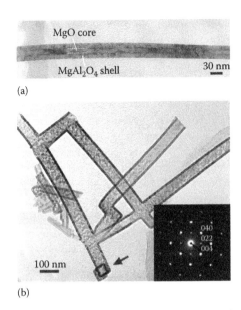

(a)

(b)

FIGURE 13.4 Spinel $MgAl_2O_4$ nanotubes based on the reaction and Wagner cation counterdiffusion. (a) Schematics of the fabrication process. (b) TEM image of one MgO-$MgAl_2O_4$ core–shell nanowire. (c) TEM image of one complicated nanotubes web of spinel $MgAl_2O_4$. The arrow indicates the square cross section of the nanotubes. Inset is the corresponding electron diffraction pattern revealing the single crystallinity. (Reproduced from Fan, H.J. et al., *Nanotechnology*, 17, 5157, 2006a. With permission.)

13.3 ZnO-Based Spinels

Among all semiconductor 1D nanomaterials, ZnO is the most studied material compared to Si, Ge, and III–V. This is driven partly by the simple and cheap growth setup and high throughput of ZnO nanostructures. ZnO itself is a multifunctional material because of its intrinsic interesting physical properties. Research on the nanostructure ZnO has boomed since 2001 when the Wang group (Pan et al. 2001) and the Yang group (Huang et al. 2001) reported ZnO nanobelts and nanowires, respectively. Worldwide there are now numerous groups dedicated to the basic research and device application of ZnO in its various morphologies from thin films to quantum wells to nanostructures. Most of the attention on 1D ZnO nanostructures is

focused on their structural, optical, and optoelectronic properties. Also, due to their easy availability, ZnO nanowires have been widely used as a physical (nonreactive) template for the fabrication of nanotubes like single-crystalline GaN (Goldberger et al. 2003), amorphous alumina (Shin et al. 2004) and silica (Chen et al. 2005a,b), and Pt–$BaTiO_3$–Pt double-electroded ferroelectrics (Alexe et al. 2006).

In most cases, compounds based on ZnO are technologically more important than its pure phase. For example, alloying with Cd or Mg is an effective route for tuning the optical bandgap of ZnO; doping with lithium or vanadium has been shown to enhance the piezoelectricity because the spontaneous polarization (due to the non-centrosymmetric crystal structure $P6_3mc$) is enlarged by replacing the Zn ions with smaller Li or V ions that cause lattice distortion (Joseph et al. 1999, Yang et al. 2008a: 012907); and alloying ZnO with In_2O_3 gives thermoelectric applications (Kaga et al. 2004). Most of the work is still on bulk materials or films for their potential applications like SAW (surface acoustic waves) devices, piezoelectric actuators, and transparent conductors.

ZnO can form spinel-type ternary compounds with many other oxides. They have a formula of either $ZnM_2^{[3+]}O_4$ or $Zn_2M^{[4+]}O_4$ (M = Al, Ga, Fe, In, Sn, Sb, Ti, Mn, V, Cr). The formation process can be either solid-state reactions by thermal annealing, or hydrothermal growth, or in-situ vapor–liquid–solid (VLS) growth by co-evaporating powders of the constituent oxides. There is now increasing interest in ZnO-based spinel 1D nanostructures within the domain of ZnO research (see Table 13.1). Such nanoscale materials show a large diversity of morphologies, such as tubes, wires, periodically zigzagged and/or twinned wires, which might reflect the uniqueness of the nanoscale solid-state reaction (versus bulk reactions) and VLS growth of ternary compound nanowires (versus binary compounds). Most of the ZnO-based spinels are wide bandgap semiconductors and phosphorous materials.

13.3.1 $ZnAl_2O_4$

Porous aluminum oxide (see Figure 13.5a) is the most widely used template for the fabrication of 1D nanostructures, and generally it is used as a morphology-defining agent. When in contact

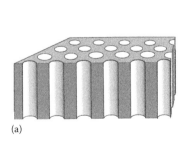

(a)

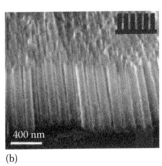

(b)

(c)

FIGURE 13.5 $ZnAl_2O_4$ by reacting Zn precursor with porous alumina template. (a) Schematics of the alumina template. (b) Single-crystal $ZnAl_2O_4$ nanonet. (Reprinted from Wang, Y. and Wu, K., *J. Am. Chem. Soc.*, 127, 9687, 2005. With permission.) (c) Polycrystalline $ZnAl_2O_4$ nanotubes. (Reprinted from Zhao, L. et al., *Angew. Chem. Int. Ed*, 45, 8042, 2006. With permission.)

with ZnO, however, the alumina is also one of the reactants used to form zinc aluminate spinel (ZAO). Wang et al. (2005, 2006) and Zhao et al. (2006) obtained ZAO nanotubes by reacting a Zn-containing precursor with the porous alumina template (see Figure 13.5). In Wang's experiment, the Zn vapor was generated by an H_2 reduction of ZnO or ZnS powder, whereas no spinel was formed using the pristine ZnO solid source because of an insufficient contact. As shown in Figure 13.5b, the overall single-crystalline ZAO nanonets were obtained after annealing at 680°C for 1000 min (Wang and Wu 2005) or 650°C–660°C for 800 min (Wang et al. 2006). A variation of the tube wall thickness was realized by varying the temperature and reaction time, but was restrained below 20 nm because of the limited diffusion depth of Zn vapor into alumina. Note that a multilayer of ZAO/$Zn_4Al_{22}O_{37}$/ZnO was observed on the surfaces of the template. The thin (<4 nm) layer of the Zn-deficient phase $Zn_4Al_{22}O_{37}$ is a transition phase of the eventual ZAO. The existence of such a multilayer supports that the reaction is diffusion limited and based on the bulk diffusion mechanism. Interestingly, after longer sintering (viz., >800 min at 660°C), a ZnO nanonet can also be formed epitaxially on the ZAO nanonet surface. Such an epitaxial growth might explain the vertical alignment of ZnO nanowires that we obtained using a ZAO covered sapphire as the substrate.

In Zhao's experiment, the Zn precursor was provided by the [Zn(TePh)$_2$(tmeda)] solution that was infiltrated into the alumina pores (Zhao et al. 2006). After annealing in air for 24 h at 500°C, nanowires of the single-crystalline Te core surrounded by a polycrystalline ZAO shell was formed. It was suggested that the reaction underwent an intermediate phase of ZnTe that was then oxidized to ZnO plus elemental Te. Such freshly formed ZnO then reacted with the alumina pore wall to form the spinel. After the removal of the Te core, ZAO nanotubes with a thin (8 nm) and polycrystalline wall were obtained (see Figure 13.5c).

We have demonstrated a different fabrication route to ZAO spinel nanotubes based on the Kirkendall effect (see Figure 13.6; Fan et al. 2006b: 627; Fan et al. 2007). In this strategy, pre-synthesized ZnO nanowires (10–30 nm thick, up to 20 μm long) were coated with a 10 nm thick conformal layer of alumina via ALD. Thus, the formed ZnO–Al_2O_3 core–shell nanowires were annealed in air at 700°C for 3 h, causing an interfacial solid-state reaction and diffusion. Because the reaction is effectively a one-way transfer of ZnO into the alumina, it represents an extreme Kirkendall effect (Aldinger 1974). It is noted that this nano Kirkendall mechanism also accounts for the formation of other spinel nanotubes of $CoAl_2O_4$ (Liu et al. 2008) and $ZnCr_2O_4$ (Raidongia and Rao 2008), $ZnCo_2O_4$ hollow cubes (Tian et al. 2008), as well as the conversion of Zn wires into ZnO nanotubes (Qiu and Yang 2008). Upon a suitable matching of the thickness of the core and the shell, highly crystalline single-phase spinel nanotubes are obtained, as shown in Figure 13.6a and b. Figure 13.6b shows a TEM image of one nanotube together with the remaining gold particle atop the pristine ZnO nanowire. The spinel nanotube wall thickness can be precisely controlled through the variation of the alumina shell thickness i.e., the number of ALD cycles.

In addition, a suitable annealing temperature is very important. Figure 13.6c illustrates how the morphology of the final product depends on the annealing temperature (Yang et al. 2008c: 4068): Tiny voids are generated at the ZnO/Al_2O_3 interface near 600°C, but do not grow to large voids due to kinetic

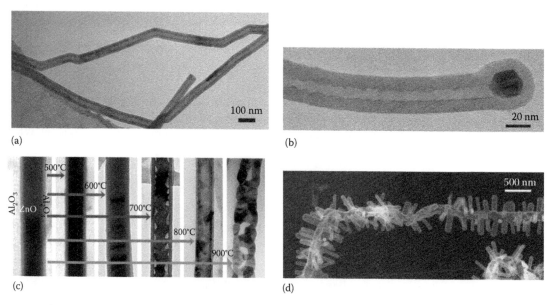

(a)   (b)

(c)   (d)

FIGURE 13.6 $ZnAl_2O_4$ tubular nanostructures by solid-state reaction of ZnO/Al_2O_3 core–shell nanowires. (a,b) TEM images of straight nanotubes. (Reprinted from Fan, H.J. et al., *Nat. Mater.*, 5, 627, 2006b. With permission.) (c) Effect of the annealing temperature to the morphology. (Reprinted from Yang, Y. et al., *J. Phys. Chem. C*, 112, 4068, 2008c. With permission.) (d) Complex tubular structures using hierarchical ZnO nanowires as the starting material. (Reprinted from Yang, Y. et al., *Chem. Mater.*, 20, 3487, 2008b. With permission.)

reasons. The reaction at 700°C results in good crystalline nanotubes as long as the thickness of the wires and shells matches well to each other. Interestingly, a higher annealing temperature near 800°C results in hollow nanotubes even if the initial ZnO nanowire is thicker than necessary for a complete spinel formation of the former shell. However, when the temperature is further increased, the tube wall collapses driven by the thermodynamic instability. Therefore, the temperature window for an optimal solid-state reaction of the core–shell nanowires is 700°C–800°C.

The conformity and uniformity characteristics of ALD are essential to the formation of smooth nanotubes. Particularly, the synthesis of complex tubular ZAO nanostructures by such a shape-preserving transformation is possible if one starts with hierarchical three-dimensional (3D) ZnO nanowires (Yang et al. 2008b: 3487). Recent literature shows an abundant variety of ZnO 3D nanostructures including bridges, nails, springs, and stars. Figure 13.6d gives one example of a Chinese firecracker-like 3D hollow ZAO spinel. Such uniform hollow 3D structures cannot be obtained by other coating methods like pulsed laser deposition, sputtering, or physical vapor deposition, which have poor step coverage characteristics.

Because of the above mentioned reaction, we expect that the ZnO nanowires grown inside the porous alumina template, especially those by Zn electrodeposition plus post oxidation (Li et al. 2000, Liu et al. 2003, Fan et al. 2006d: 213110) are not pure ZnO, but a mixture of ZnO with a layer of ZAO or $Zn_4Al_{22}O_{37}$, since their growth or annealing temperatures were high enough for the spinel-forming solid-state reaction to occur. Unfortunately, none of these papers have shown a high-resolution TEM of the nanowires. Likewise, the growth of ZnO nanowires directly on the lattice-matched sapphire (single crystal Al_2O_3) might also end up with the formation of a ZAO spinel, as pointed out recently by Grabowska et al. (2008).

13.3.2 $ZnGa_2O_4$

Zinc gallate (ZGO) is most intensively studied among all spinel nanostructures, because of its interesting luminescent properties. It is a promising candidate for a blue light source, for a vacuum fluorescent display (Itoh et al. 1991), and for gas sensing (based on surface states-related electric property). ZGO has a wide bandgap of 4.4 eV, thus it is transparent from the violet to near ultraviolet region. It is proposed that ZGO can be a host for full color emitting materials when doped with various activators: Mn^{2+} for green (Shea et al. 1994) and Eu^{3+} or Cr^{3+} for red (Rack et al. 2001).

The demonstrated fabrication routes of 1D ZGO nanostructures can be divided into two main categories. The first type of fabrication is a high-temperature (around 1000°C) co-evaporation of a mixture of ZnO–Ga powders and deposition substrates. The growth is driven by the well-known vapor–liquid–solid mechanism with gold nanoparticles acting as the catalyst. One drawback of this method is the difficulty in controlling the Ga/Zn ratio so that a ZnO phase might co-exist with the ZGO

spinel even if the source material is mixed with a defined molar ratio. Nevertheless, Bae et al. (2004) fabricated pure cubic phase spinel nanowires vertically standing on an Si substrate using a gold nanoparticle catalyst for the vapor–liquid–solid growth by employing this method. All the nanowires grow along the [111] direction. Using the same technique, Feng et al. (2007) obtained ultralong nanowires on the Au covered Si substrate, which were mainly ZGO but with a small amount of ZnO. A particularly interesting discovery was the helical ZGO nanowires wrapped around the straight ZnSe nanowire support, as well as free-standing ZGO nanowires self-coiled into a spring-like structure (Bae et al. 2005). These zigzag wires elongate periodically along the four equivalent <011> directions, with an oblique angle of 45° (so as to maintain a coherent lattice at the kinks). According to the authors (Bae et al. 2005), this unique growth behavior might be related to the ZnSe nanowires, which provide the Zn source at the beginning and simultaneously an epitaxial substrate for ZGO. Note that the evaporation temperature used by the authors was as low as 600°C.

The second type of fabrication route is a two-step process: synthesis of single-crystalline nanowires followed by their reaction with the other oxide. The latter step is a typical solid-state reaction on the nano scale, which results in an outer ZGO layer surrounding the nanowire core. Pure ZGO spinel can be obtained by dissolving the remaining core material. The spinel nanoshell or nanotubes can be single crystalline as a result of the epitaxial relationship between the growing ZGO and the core material. Chang and Wu (2006) made a systematic study of the formation of the ZGO layer on top of β-Ga_2O_3 nanowires. Nanowires of a single-crystalline Ga_2O_3 core and a chemical vapor deposition (CVD)-grown polycrystalline ZnO shell were annealed at 1000°C for 1 h. Depending on the thickness of the ZnO layer, the final product after the solid-state reaction was either Ga_2O_3/ZGO core–shell nanowires, ZGO nanowires, or ZGO nanowires capped with ZnO nanocrystals. In any case, the ZGO were single crystalline.

Li et al. (2006) did a similar experiment but in the opposite way: they started with ZnO nanowire array, which in situ reacted at 500°C with a Ga_xO_y surface layer deposited from the vapor, forming a ZGO spinel layer. Because of the epitaxial interface, $ZnO[1\bar{1}0]\|ZGO[1\bar{1}2]$; $ZnO[110]\| ZGO[4\bar{4}0]$, the ZGO layer as well as the subsequent ZGO nanotubes were single crystalline. Figure 13.7a shows the nanotubes together with an electron diffraction pattern that proves its single-crystallinity. Here, the phase was controlled by the deposition temperature, unlike Chang and Wu's case where the phase was controlled by the shell thickness.

Another special way to obtain ZGO spinel nanotubes was reported by Gautam et al. (2008). In their experiment, Ga-doped ZnS nanowires were heated up in an oxygen gas flow at 500°C–850°C for 200 min, during which the Ga:ZnS nanowires were converted into nanotubes whose walls are comprised of ZnO + ZGO composite fragments (see Figure 13.7c). While the overall 1D structure and morphological homogeneity of the initial nanowire were preserved after the reaction, the nanotube wall became

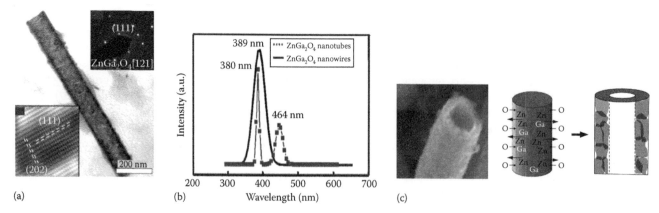

FIGURE 13.7 ZnGa$_2$O$_4$. (a) ZnGa$_2$O$_4$ nanotubes by reaction of initial ZnO nanowire with Ga-O vapor and removal of the remaining ZnO core. Insets are the corresponding electron diffraction pattern and high-resolution TEM image showing the single-crystallinity of the nanotube. (From Li, Y. J. et al., *Appl. Phys. Lett.*, 88, 143102, 2006. With permission.) (b) Room temperature cathodoluminescence spectra showing sharp a UV and a blue peak of ZnGa$_2$O$_4$ nanotubes relative to the spectrum from solid nanowires. (Reprinted from Li, Y.J. et al., *Appl. Phys. Lett.*, 88, 143102, 2006. With permission.) (c) ZnGa$_2$O$_4$–ZnO compound nanotubes and the proposed formation mechanism which involves a faster outward diffusion of Zn/Ga than the inward diffusion of O. (Reprinted from Gautam, U.K. et al., *Adv. Mater.*, 20, 810, 2008. With permission.)

polycrystalline and rugged. The possible formation mechanism involves unequal diffusion rates for the outward Ga/Zn and inward oxygen during the oxidation of the ZnS nanowires at the interface plane. This is similar to the Kirkendall effect in the case of the ZnO + Al$_2$O$_3$ reaction discussed in Section 13.3.1. As the Ga concentration in the initial ZnS nanowires is only on doping level, it is not surprising that the nanotube walls are not pure ZGO. Annealing of a Ga-doped ZnO nanowire will certainly not give similar nanotubes as above.

All the so-far demonstrated ZGO nanowires or nanotubes are undoped, so the observed emissions in their luminescence spectra are related mainly to the self-activated Ga–O emission or composition stoichiometry (e.g., O vacancies, Zn interstitials). Two main peaks have been commonly observed: one near 380 nm and the other near 430 nm. An example from Li et al. (2006) is shown in Figure 13.7b, but there seems to be little point in comparing data from different authors since the peak position is highly sensitive to the preparation condition and can shift by more than 200 nm depending on the Zn/Ga stoichiometry. For the application in nanoscale light source, doping with rare-earth elements into the 1D ZGO nanostructures is needed.

ZGO spinels are also known to have applications as solid sensors. Like ZnO, the large surface-to-volume ratio makes the surface states play a major role in the optoelectronic properties of 1D nanostructures. The optically driven oxygen and temperature sensing behavior of ZGO nanowires was recently demonstrated (Feng et al. 2007). The current across individual nanowires was nearly zero at ambient conditions, whereas the current increases drastically under illumination by 254 nm UV light (the bandgap of ZGO is around 270 nm). Such enhancement was attributed to the increased charge carrier concentration and decreased contact resistance. The current level is also dependent on the temperature and oxygen pressure, giving the ability to construct sensing devices.

13.3.3 ZnFe$_2$O$_4$

Nanostructured ferrite spinels are of special interest because of their size- and morphology-related magnetic properties. While most ferrite spinels have inverse spinel structure and are ferrimagnetic, ZnFe$_2$O$_4$ (ZFO) have a normal spinel structure and show long-range antiferromagnetic ordering at its equilibrium state below the Néel temperature $T_N \approx 9$–11 K (Ho et al. 1995, Schiessl et al. 1996). At room temperature, it is paramagnetic because of the weak magnetic exchange interaction of the B site Fe^{3+} ions. Nanoscale ZFO, however, can have a higher T_N or show ferromagnetic with the magnetization generally increasing with grain size reduction. There are a number of reasons for this feature, such as the inversion of the cation distribution (Fe^{3+} into tetrahedral interstices and Zn^{2+} into octahedral interstices), small size effect (a large surface area gives uncompensated magnetization moment), and nonstoichiometry (Oliver et al. 2000, Grasset et al. 2002, Hofmann et al. 2004, Yao et al. 2007). For example, the ZFO nanoparticles with diameters of ≈ 10 nm have shown a saturated magnetization of $M_s = 44.9$ emu/g at room temperature (Yao et al. 2007). The magnetic behavior of the nanocrystals was related to the ferromagnetic coupling of Fe ions at A–B sites in the $(Zn_{1-x}Fe_x)[Zn_xFe_{2-x}]O_4$ particles and surface spin canting.

While there are quite a number of reports on ZFO nanoparticles, there are strangely only a few publications on ZFO nanotube or wires. Jung et al. (2005) fabricated ZFO nanowires by annealing the densely packed ZFO nanoparticles infiltrated into porous alumina templates. The nanowires are superparamegnetic and show evidence of an increase in the coercive field and M_r/M_s value compared to the unannealed nanoparticles. Liu et al. (2006) adopted a strategy similar to the standard solution synthesis of perovskite nanotubes/wires (e.g., Hernandez et al. 2002). In their experiment, Zn(NO$_3$)$_2 \cdot 6H_2O$ and Fe(NO$_3$)$_3 \cdot 9H_2O$ with 1:2 molar ratio of Zn/Fe were dissolved into mesopores of silica SBA-15 template. After drying, decomposition, and annealing, polycrystalline ZFO nanowire bundles were obtained, which also shows enhanced paramagnetism compared to the bulk phase.

Different from the above negative template methods, ZFO nanoshells using ZnO nanowires as the template were prepared. Single-crystalline ZnO nanowires were coated with a uniform amorphous layer of ALD Fe_2O_3 using a $Fe_2(O^tBu)_6$ precursor (Bachmann et al. 2007). After annealing the core–shell nanowires at 800°C for 4 h, an outer layer of ZFO surrounding the remaining ZnO was formed. It is expected that the solid-state reaction of $ZnO–Fe_2O_3$ occurs by a Wagner counterdiffussion mechanism. Unlike the $ZnO–Al_2O_3$, which is characterized by a very smooth inner surface of the nanoshells, the interface after the solid-state reaction is rugged but the ZFO layer is single-crystalline. The crystallinity can be attributed to an orientation relationship $ZnFe_2O_4(111)$ [110]//$ZnO(0001)$ [11$\bar{2}$0], as established by Zhou et al. (2007). Since these nanoshells have a large aspect ratio of up to 100 and tunable shell thickness on an atomic scale, it would be interesting to examine its anisotropy magnetic properties. Systematic characterization is still underway.

13.3.4 $ZnCr_2O_4$

There is a recent report on the conversion of elemental metal nanowires into the corresponding nanotubes, among which is the spinel $ZnCr_2O_4$ (Raidongia and Rao 2008). The reaction is an unconventional solid–vapor type: solid Zn nanowires reacted with a CrO_2Cl_2 vapor in an oxygen environment at 400°C for 3 h in a resistance furnace. The nanotube walls are granular and polycrystalline and possibly contain pores; nevertheless, the XRD spectrum does correspond to a cubic, normal, spinel-type structure. Note that the reaction temperature was much lower than 900°C, temperatures above which are used for the conventional solid–solid reaction of the $ZnO–Cr_2O_3$ powder mixture. As was learned from the properties of bulk or thin films, it would be of great interest to investigate the antiferromagnetic ordering (Martinho et al. 2001) and gas-sensing properties (Zhuiyko et al. 2002) of the $ZnCr_2O_4$ nanotubes.

13.3.5 Zn_2TiO_4

Zinc titanate is a useful material for low-temperature sintering dielectrics. Usually, there are three modifications for the ZnO–TiO_2 system: Zn_2TiO_4 (inverse spinel, cubic), $ZnTiO_3$ (hexagonal), and $Zn_2Ti_3O_8$ (cubic). ZTO has been widely used as a regenerable catalyst as well as an important pigment in the industry. It is also a good sorbent for removing sulfur-related compounds at high temperatures. Since the nanoscaled ZTO is expected to achieve low-temperature sintering, a desirable property for microwave dielectrics, increasing attention has been paid to the preparation of ZTO in a nanoscaled form in recent years.

Conventional synthesis methods such as high-energy ball milling and a high-temperature solid reaction have been reported to prepare ZTO nanocrystallites (Manik et al. 2003). However, the reports on 1D ZTO nanostructure growth are scarce. Only recently, Yang et al. (2007) reported the synthesis of twinned ZTO nanowires by using ZnO nanowires as a reactive template. In the experiments, ZnO nanowires with an axis direction along [01$\bar{1}$0] were coated with amorphous Ti by magnetron sputter deposition to form ZnO/Ti core–shell nanowires. After thermal annealing at 800°C for 8 h at low vacuum, the solid-state reaction led to a phase transformation from wurtzite ZnO to ZTO. The final product contained a large amount of (111)-twinned ZTO nanowires elongated along the [111] direction. It was discussed that two types of nonuniformities have contributed to the formation of ZnTO subcrystallites, eventually causing the zigzag morphology of the nanowires: first, the sputter-deposited Ti layer was not conformal and second, the electrons and ions bombardment causes content fluctuation in the ZnO nanowires.

We synthesized 1D twinned ZTO nanowires via a solid-state reaction approach using ZnO nanowires as the template (Yang et al. 2009). Different from above, a very thin (5 nm) layer of TiO_2 was deposited via ALD. The TiO_2 layer was amorphous and surrounded the ZnO nanowires with a uniform thickness. The ZnO/TiO_2 core–shell nanowires transformed into a zigzag structure, too, after annealing at 900°C for 4 h in an open furnace (see Figure 13.8a and b). The twinned nanowire is composed of large parallelogram-shaped subcrystallites, but does not present periodic stacking (see Figure 13.8b). The interplanar spacings of 0.49 nm measured from the high-resolution TEM image perfectly match the d_{111} lattice distance of ZTO crystal, demonstrating the [111]-oriented growth direction of each individual grain. The fast Fourier transform (FFT) pattern of the twin boundary (TB) (inset of Figure 13.8c) reveals clearly the (111) twin structure: two (1$\bar{1}$1) mirror planes sharing a common (111) face. The twinning angle across the boundary is measured to be about 141°, and the relative rotational angle is about 70.5° (see Figure 13.8c and d). These results are in agreement with other cubic nanowires like SiC and InP (Section 13.4), and Zn_2SnO_4 (Section 13.3.6). No misfit dislocations were observed at the interface.

It is proposed that the formation of twinned ZTO nanowires in our case includes multiple stages, which differs from the one by Yang et al. (2007) Since amorphous TiO_2 easily crystallizes at a temperature of 500°C or higher, the initially continuous TiO_2 shell on the surface of ZnO nanowires are expected to transform into anatase TiO_2 islands first. The islands are textured as a result of the volume shrinkage of the amorphous phase. During the annealing at 900°C, the TiO_2 phase could be incorporated in the ZnO lattice as a segregation at early stages because of the faster diffusion speed of Ti^{4+} as compared with that of Zn^{2+}. Subsequently, ZTO spinel nanocrystallites are formed through the lattice rearrangement and were attached to the surface of the unconsumed ZnO core (note that the TiO_2 layer was only 5 nm thick). No voids or tubular structures appeared at the ZTO/ZnO interface, indicating that the inner-diffusion of Ti^{4+} is prominent for the TiO_2/ZnO couple. With prolonged annealing at 900°C, the unconsumed ZnO core is desorbed or evaporated at 900°C through the gaps of the nanocrystallites. Subsequently,

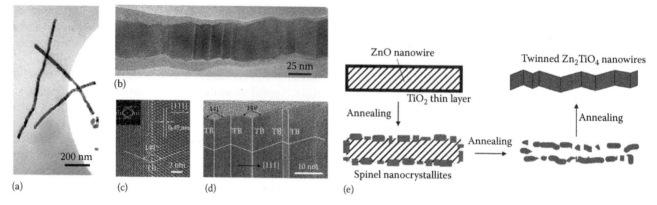

FIGURE 13.8 Zn_2TiO_4 zigzag twinned nanowires by solid-state reaction of ZnO/TiO_2 core–shell nanowires. A uniform shell of TiO_2 was coated using atomic layer deposition and the annealing condition is: 900°C, 4h. (a,b) TEM images at different magnifications showing the zigzag morphology of the nanowires. Note that the twinning is non periodic. (c) A closer view at one twin boundary and the corresponding FFT pattern. (d) Several twinning boundaries with a fixed zigzag angle of 141°. (Reproduced from Fan, H.J. et al., *J. Mater. Chem.*, 19, 885, 2009. With permission.) (e) Schematics of the proposed formation process of the Zn_2TiO_4 zigzag nanowires.

the loosely interconnected nanocrystallites were [111]-orient attached, coalesced, and finally evolved into twinned nanowires seen in Figure 13.8b.

By comparing the above two solid-state reaction experiments (Yang et al. 2007, 2009), we are able to draw the following conclusions: first, the preferred [111] growth direction of the twinned ZTO nanowires is independent on the orientation of the starting ZnO nanowires ([01$\bar{1}$0] or [0001]). During sintering, individual ZTO nanocrystallites may rotate to constitute an energetically favorable assembly along the [111] direction, since the surface energy of [111] facets is the lowest for an fcc structure. Secondly, the diffusion species in the ZnO/TiO_2 couple is unambiguously Ti^{4+}, not the metallic Ti as described in the case of the ZnO/Ti couple. Lastly, a fluctuation of the ZnO content is not a necessity for forming the subcrystallites.

13.3.6 Zn_2SnO_4

As in the case of Zn_2TiO_4, Zn_2SnO_4 (ZSO) crystals also have an inverse spinel structure. The interest in ZSO has been in the electrical circuit as a transparent conducting electrode as a replacement of the expensive ITO (In-doped SnO_2). This is because of its high electron mobility, high electrical conductivity, and low visible absorption. Like other metal–oxide semiconductors, ZSO can also have application potential in photovoltaic devices and Li-ion batteries (Rong et al. 2006, Tan et al. 2007). While ZSO nanoparticles have been demonstrated to be useful as electrodes of dye-sensitized solar cells with a light-to-electricity efficiency of 3.8% (Rong et al. 2006), it is envisaged that 1D nanostructures could enhance the performance due to a continuous electrical conduction path.

One-dimensional ZSO nanostructures are fabricated mainly by thermal evaporation of a powder mixture of $ZnO + SnO_2$ or Zn + Sn metal (and some by hydrothermal reaction). No reports have been published so far on the solid-state reaction of

core–shell nanowires, although a controlled growth of ZnO and SnO_2 nanowires has become mature in many research labs. The high-temperature evaporation and deposition method involves a complicated thermodynamic and kinetic process. The partial pressures of oxygen and ZnO/SnO vapor are very inhomogeneous, which causes disturbances of the Au–Zn–Sn ternary droplet size in the VLS growth and hence diameter oscillations (Jie et al. 2004). Moreover, such an evaporation method usually ends up with a mixture of different phases and structures in one growth set. For example, ZnO, ZSO nanowires, and chainlike ZTO nanowires were identified in a single VLS growth (Jie et al. 2004). The ZSO composition was off stoichiometric with an element ratio of Zn:Sn:O = 1.75:1:(2.6–3.5), indicating a deficiency of zinc and oxygen. Reports on a series of ZSO nanostructures with very different morphologies have been published by the Xie group, who conducted similar evaporation experiments (Wang et al. 2004a: 177,b: 435, 2008b: 707, Chen et al. 2005). The so-far demonstrated ZSO quasi-1D nanostructures are listed in Table 13.1, including smooth belts (Wang et al. 2004b: 435, 2005: 2928), rings (Wang et al. 2004b: 435), chainlike single-crystal wires (Wang et al. 2004a: 177, 2008b: 707, Kim et al. 2008), twinned wires (Wang et al. 2004a: 177, 2008b: 707, Chen et al. 2005, Jeedigunta et al. 2007), and short rods (Zhu et al. 2006). In the following section, we selectively discuss only two types of rather extraordinary nanostructures.

The first is the twinned nanowire (see Figure 13.9a). These 1D ZSO nanostructures have a twinning morphology similar to the aforementioned ZTO. Wang et al. (2004a: 177,b: 435, 2008b: 707) reported VLS-grown twinned ZSO nanowires by evaporating powder mixtures at 1000°C with two types of weight ratios. In one case, a ratio of ZnO:SnO = 2:1 gave twinned nanowires (among many other shaped wires), whereas in the other case, a ratio of ZnO:SnO = 1:4 was needed for the production of similar twined wires. All these ZSO wires have a [1$\bar{1}$1] growth direction, with the twin planes being (1$\bar{1}$1) and the twinning direction

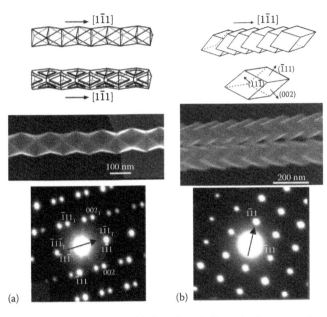

FIGURE 13.9 Zn_2SnO_4. (a) Pseudoperiodic twinning nanowire. (b) Single-crystal nanowires by periodic stacking of rhombohedral nanocrystals. Top: schematics of the structural; middle: SEM images; bottom: the corresponding electron diffraction patterns. The arrows in the diffraction patterns indicate the nanowire growth directions. (Reprinted from Wang, J.X. et al., *Cryst. Growth Des.*, 8, 707, 2008b. With permission.)

perpendicular to the wire axes, same as was observed for ZTO nanowires.

The second is the single-crystal chainlike nanowire (see Figure 13.9b). Structurally, this type of wire is formed by a sequential stacking of rhombohedral nanocrystals along the [1$\bar{1}$1] direction. The stacked nanocrystals constitute a single-crystal wire on the whole, as revealed by the electron diffraction pattern (e.g., the bottom of Figure 13.9b). This means lattice coherence at the boundary of the nanocrystals. Jie et al. (2004) observed the so-called diameter-modulated ZSO nanowires composed of linked ellipses. Wang et al. (2005: 2928) reported diamond-like wires composed of intersected rhombohedra (see Figure 13.9b). But according to the careful electron tomography study by Kim et al. (2008), who also obtained both types of chainlike ZSO nanowires, the above two structures are essentially the same, that is, the observed zigzag angles appear different (125° or 120°) under the electron microscopy simply because of a rotation of the wire around its axis.

Hydrothermal synthesis as a low-cost, high-yield method for 1D nanostructures has been applied to single-crystalline ternary compound nanowires like $BaTiO_3$ and $PbTiO_3$, which are important ferroelectric materials (Rorvik et al. 2008), and recently to ZSO nanowires. A hydrothermal process was applied with the use of hydrazine hydrate as the alkaline mineralizer for the growth of ultrathin (2–4 nm in diameter) ZSO nanowires (Zhu et al. 2006). The optical bandgap was determined to be 3.87 eV by diffuse UV-vis reflectance measurement, which is a blueshift of 0.27 eV from bulk ZSO (3.6 eV), indicative of the quantum confinement effect.

13.4 Twinning of Spinel Nanowires

Twinning as planar defects in nanowires is becoming a very interesting topic and is widely studied for cubic crystals including not only the above ZTO and ZSO but also binary compounds like GaAs, GaP, InP, ZnSe, and SiC (Xiong et al. 2006, Davidson et al. 2007, Mattila et al. 2007, Wang et al. 2008a: 215602). These nanowires are formed through either solid-state reactions or metal-catalyzed unidirectional growth. Quasiperiodic twinnings have been observed intersecting the entire wire cross-section. It is known that the (111) twined crystals have a relative rotational angle of 70.5° and the zigzag angle between the two twinning nanounits is about 141°. This is in accordance with all the so-far demonstrated nanowires of the cubic phase.

Figure 13.10 schematically illustrates the two typical growth processes of twinned nanowires. The first (Figure 13.10a) represents the spinel nanowires by solid-state reactions, for example, the aforementioned ZTO (Figure 13.8). This is mainly a self-assembling process where subcrystallites rearrange to minimize the total energy: two adjacent (111) faces meet via a shear movement along the [21$\bar{1}$] direction of subcrystallites that are driven by a dislocation-induced strain, and then the mirror planes arrange symmetrically to reduce the total energy. Alternatively, the twinned structure can also be formed when new crystallites start to grow in a limited space between two existing grains via an Ostwald ripening process during the sintering stage.

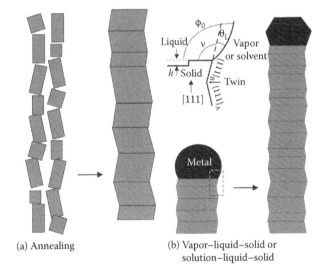

(a) Annealing   (b) Vapor–liquid–solid or solution–liquid–solid

FIGURE 13.10 Two main formation processes proposed for twinned nanowires of cubic crystals including spinels. (a) Nanocrystallites are formed first through solid-state reactions, and then transformed into twinned nanowires during annealing. This process is mainly for ternary alloyed nanowires like Zn_2TiO_4 and Zn_2SnO_4. (b) Twinned nanowires formed by vapor–liquid–solid or solution–liquid–solid growth processes. It is for both ternary compound spinel nanowires and binary nanowires like GaP, InP, GaAs, and ZnSe. Inset of (b) shows the three-phase boundary where a fluctuation in the droplet/nanowire contact angle causes nucleation of a twinned plane. (Reproduced from Fan, H.J. et al., *J. Mater. Chem.*, 19, 885, 2009. With permission.)

In the second route (see Figure 13.10b), the nanowires grow through a VLS or solution–liquid–solid process with a ternary-alloy droplet at the growth front as the catalyst. The ZSO twinned nanowires (see Figure 13.9) and most of the binary compound zinc blende nanowires (Xiong et al. 2006, Davidson et al. 2007, Mattila et al. 2007, Wang et al. 2008a: 215602) fall into this category. In this case, the formation of twins is strongly related to the contact angle fluctuation at the three-phase boundary (inset of Figure 13.10b), as elaborated by Davidson et al. (2007). A smooth untwined nanowire has (112) sidewall faces. But twinning eliminates the (112) sidewall surface by converting it to two lower energy (111) mirror planes. The sidewall surface alternates between the two (111) mirror planes with subsequent twins in order to maintain a straight nanowire growth along the [111] direction. In the Davidson model, fluctuations in the droplet/nanowire contact angle for Au-seeded nanowires are most likely necessary for twinning to occur. This requirement (the contact angle at the three-phase boundary must fluctuate sufficiently) indicates that even when twin planes might be expected based on their formation in bulk crystals, they may not form in nanowires. Interestingly, this model also suggests that nanowires with smaller diameters (<10 nm) exhibit less twinning because of their higher tendency to exhibit their lowest energy (111) and (110) planes.

Overall, twinning and zigzag shape is a common phenomenon for spinel nanowires. Similar twins can form from different growth processes (solid-state reactions, VLS growth) and/or different materials; the twinning mechanism also appears to be different but a universal mechanism might exist. Possibly, an in situ reaction annealing experiment inside a TEM might provide valuable hints. Also, little is known about the electrical, mechanical, and especially the light-emitting properties of these twinned nanostructures. Mechanical studies indicate that twin boundaries act commonly as dislocation obstacles to affect the deformation behavior of an fcc system. Hence, twins are expected to have important influence on the electronic and mechanical properties.

13.5 CoAl$_2$O$_4$

Cobalt aluminate has a normal spinel structure with a lattice constant of 0.81 nm (Madelung et al. 2006). The fabrication of high-surface area powders of CoAl$_2$O$_4$ has been made by the sol-gel process (Areán et al. 1999, Meyer et al. 1999). So far there is only one publication on 1D nanostructure CoAl$_2$O$_4$ spinel nanotubes (Liu et al. 2008).

In Liu's experiment, metallic Co reacted with porous alumina into spinel CoAl$_2$O$_4$ nanotubes, essentially the same way as the spinel-forming reaction between ZnO and the alumina pore wall we previously discussed in Section 13.3.1 (Wang and Wu 2005, Wang et al. 2006, Zhao et al. 2006). The fabrication process is shown schematically in Figure 13.11a. Specifically, the alumina channels were filled up with Co/Pt multisegments by pulsed electrodeposition, so that multilayered nanowires formed. Subsequent thermal annealing at 700°C transferred the

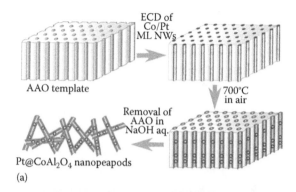

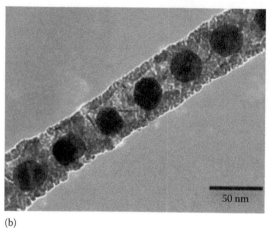

(b)

FIGURE 13.11 CoAl$_2$O$_4$ nanotubes with peapods, formed due to Kirkendall effect and Rayleigh instability. (a) Schematics of the fabrication process. Multi-segmented Cr/Pt nanorods were electrodeposited inside the channel of anodic alumina, which then reacted with the alumina during a thermal annealing process. Free-standing nanotubes were obtained after chemical etching of the alumina. (b) TEM image of one of the nanotubes with peapods. (Reprinted from Liu, L.F. et al., *Angew. Chem. Int. Ed.*, 47, 7004, 2008. With permission.)

nanowires into CoAl$_2$O$_4$ nanotubes whose interiors are encapsulated with Pt nanopeapods. The former is a solid–solid reaction Co + Al$_2$O$_3$ → CoAl$_2$O$_4$, whereas the later is a result of the rearrangement of the nonreactive Pt, during which Rayleigh instability occurs. Interestingly, the reaction also involves the aforementioned nanoscale Kirkendall effect, that is, Co metals diffuse into the alumina lattice leaving voids behind. Such hybrid nanostructure is of high interest for their electronic transport properties, or even optoelectronic properties, which have been investigated previously.

13.6 LiMn$_2$O$_4$

The spinel LiMn$_2$O$_4$ has been regarded as a promising candidate as cathode material for Li-ion batteries. 1D nanostructures are now receiving increasing attention in the application of lithium ion battery anodes or cathodes. Single-crystalline nanowire arrays have several advantages over bulk or powder counterparts: first, the contact area between the electrodes and electrolytes is increased; second, a continuous charge transport path is

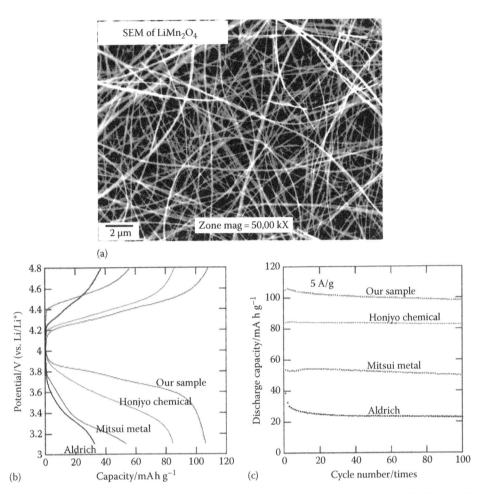

FIGURE 13.12 Spinel $LiMn_2O_4$ nanowires and their application as lithium ion battery cathode material. (a) SEM image of the single-crystalline nanowires. (b) The charge-discharge curves at the second cycle at a 5 A/g rate. (c) Cycle performance at a 5 A/g rate. The performances of other $LiMn_2O_4$ samples from various commercial suppliers are also included for comparison (Reprinted from Hosono, E. et al., *Nano Lett.*, 9, 1045, 2009. With permission.)

established due to the contact of the nanowires on the substrate (which is the charge collector); and third, the enhanced mechanical robustness as the nanowires can sustain larger volume expansion than bulk materials (Chan et al. 2008). While scientists are searching new cathode/anode materials, it is equally valuable to test the performance of conventional electrode materials of the 1D nanostructure.

Spinel $LiMn_2O_4$ nanobelts and nanowires have been fabricated in hydrothermal growth (Zhang et al. 2003, Hosono et al. 2009). The former is a self-seeded one-step growth process while the latter is a template-based two-step process. The work by Hosono et al. (2009) represents a significant step forward, because the fabrication method is simple but with large throughput, and the obtained spinel $LiMn_2O_4$ nanowires are long, smooth, and single crystalline. Specifically, single crystalline ultralong $Na_{0.44}MnO_2$ nanowires were first obtained by the hydrothermal reaction of Mn_3O_4 powders with a NaOH solution. Such nanowires were then transformed into $LiMn_2O_4$ nanowires in the subsequent ion exchange reaction, during which the single-crystallinity characteristics were preserved.

Figure 13.12a shows a SEM image of the spinel nanowires that are indeed long and smooth. The charge–discharge performance of the $LiMn_2O_4$ nanowires was tested in comparison to various commercial $LiMn_2O_4$ materials (Figure 13.12b and c). Evidently, the capacity can reach up to and be maintained at 100 mA h/g after 100 cycles at a relatively high current density of 5 A/g. Such performance is competitive to other mainstream Li-metal oxide electrodes, such as $LiCoO_2$, $LiFePO_4$, and $Li_4Ti_5O_{12}$ (e.g., see the review by Cheng et al. 2008). In addition, the capacity and reversible stability are enhanced compared to commercial samples (see Figure 13.12b and c), which is ascribed to the higher crystalline quality and thermal stability of the ultralong spinel nanowires.

13.7 Summary and Conclusions

In this chapter, the fabrication and structural properties of spinel oxide nanotubes and nanowires are reviewed and discussed case by case. One-dimensional spinel nanostructures demonstrated in literature are mainly MgO-based and ZnO-based spinels;

14

Magnetic Nanotubes

Eugenio E. Vogel
Universidad de La Frontera

Patricio Vargas
*Universidad Técnica Federico
Santa María*

Dora Altbir
Universidad de Santiago de Chile

Juan Escrig
Universidad de Santiago de Chile

14.1 Introduction

Nanotubes are special nanoparticles in several senses. Opposed to most tiny objects that tend to adopt a spherical shape with only one geometrical parameter (the average radius), hollow cylindrical objects have an immediate advantage over spheres as they possess three independent geometrical parameters: length L, external radius R, and internal radius a, as illustrated in Figure 14.1. Moreover, in some cases the inner and outer surfaces are electrically polarized in the opposite sense so they have a different affinity toward foreign atoms, allowing for diverse functionalizations (Mitchell et al. 2002). Magnetic nanotubes have still another advantage over the rest of the nanotubes: they can be driven to the point of interest by means of externally applied magnetic fields. This is true whether the application of the magnetic nanotube is on a semiconductor chip or is intended to reach an ill organ of a living body for drug delivery.

In this chapter, we will concentrate on nanotubes made out of magnetic materials, leaving out of consideration the nonintrinsically magnetic nanotubes that can later adsorb magnetic atoms. This is the case with magnetically doped carbon nanotubes for instance, which can be filled in by magnetic materials. The reader interested in this line of thought is referred to the abundant literature of which we mention just one paper leading to further references (Korneva et al. 2005).

Most of the existing magnetic nanotubes are thin, namely, $d_w \ll a$, which leads to the most interesting case as they lack the inner magnetic core (Escrig et al. 2007b), which is characteristic of magnetic nanowires. Although similar in external shape and some fabrication methods, magnetic nanowires are generally cheaper to produce and have their own interesting properties

(Nielsch and Stadler 2007). However, the advantages of nanotubes over nanowires are easy to grasp: they have one more surface, their magnetization is easily reversed due to the lack of an inner core, and they can float in most liquids. This last property is a result of a combination of facts: surface tension, bubbles tend to remain naturally in the inner space, and this property can be increased by hydrophobic functionalization.

What kind of special physics is possible in magnetic nanotubes? In short, we can say that the nanotubes produced this far are small enough so they usually present one dominant magnetic domain but they are large enough so quantum effects still do not show up in their magnetic properties. These points are discussed in detail before getting into the specific properties of these artifacts.

Magnetic nanotubes have diameters of the same order of magnitude as the so-called magnetostatic exchange length, which is defined as $L_x = (2A/\mu_0 M_0^2)^{1/2}$, where A is the exchange stiffness constant (approximately 10^{-11} J/m) and M_0 is the saturation magnetization. Typical values for this length are in the range of 2–10 nm. This length determines the order of magnitude where the existence of a single magnetic domain is energetically possible. (i.e., all magnetic moments are aligned along a preferential direction). Therefore, existing magnetic nanotubes are special because their thickness is comparable to the typical exchange length parameters of many ferromagnetic materials. On the contrary, the spatial dimensions of micro-size tubes or nanowires are large enough, thus favoring the formation of several magnetic domains separated by Bloch walls.

On the other hand, to observe quantum phenomena, a smaller spatial dimension (hence, a smaller total magnetic moment) is needed. Quantum commutation relationships for the magnetic moments allow the possibility of quantum tunneling of

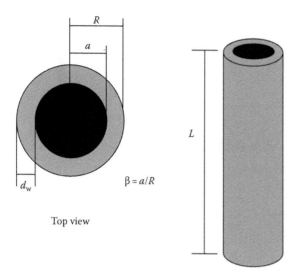

FIGURE 14.1 Geometrical parameters defining a nanotube: Length L, external radius R, internal radius a, thickness $d_w = R - a$.

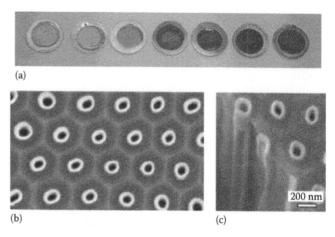

FIGURE 14.2 (**See color insert following page 20-16.**) Several morphological aspects of magnetic nanotubes. (a) Macroscopic view of the samples, consisting of circular porous anodic alumina membranes containing embedded tubes of different thickness. Each circular sample is surrounded by an Al circle whose outer diameter is 2 cm. The first sample on the left corresponds to alumina without magnetic material: the other samples to the right progressively increase their thickness going over 1, 2, 4, 8, 12, and 16 nm approximately. Upon illumination with daylight they vary their color according to the wall thickness as it can be readily seen from the illustration. (b,c) Scanning electron microscope images of nanotubes embedded in porous alumina, observed from a top view (b) and at an angle at a break in the sample (c). These micrographs give an idea of shapes, diameters, thicknesses, and distances for the systems under consideration in the present paper. (Reprinted from Escrig, J. et al., *Phys Rev. B*, 77, 214421, 2008b. With permission.)

the magnetic moment, as is the case in smaller systems such as magnetic molecules.

For very small systems, the magnetic moment (magnetic content of the system) is measured in units of the so-called Bohr magneton μ_B, which is approximately the magnetic moment of a small molecule. However, for a typical nanotube made out of any ferromagnetic material, the total magnetic moment of a magnetic domain is already in the order of a few thousands to a million μ_B, therefore, it can be considered classical in the sense that the orientation of the three components of the magnetic moment can be determined by a macroscopic measurement. In that sense, a nanometer-size ferromagnetic particle is qualitatively similar to the arrow of a magnetic compass.

For those readers somewhat familiar with quantum mechanical procedures, previous physical discussions can be put into mathematical terms in the following way. The quantum commutation relationship among the components of the magnetic moments is given by: $M_i M_j - M_j M_i = 2i\mu_B \epsilon_{ijk} M_k$. Since μ_B is very tiny, the right-hand side of this equation is negligible unless the magnetization (and its components) is of the same order of magnitudes. So, for classical magnetic systems, this commutator vanishes in practical terms and quantum effects (such as tunneling) are fully suppressed.

The wall thickness can reach a few nm, which makes these devices unique in many respects beyond their magnetic properties discussed here. Thus, for instance, optical properties can change noticeably at this scale as shown in Figure 14.2a, where the optical response to the natural light of the magnetite nanotubes of several different geometries is illustrated (Escrig et al. 2008b). It can be noticed that the color of the top end of the tubes changes by varying only the wall thickness in the nanoscopic range. Parts (b) and (c) of this figure should also provide an idea of the way nanotubes are produced in porous alumina membranes and also the typical sizes reached by magnetic nanotubes so far.

There are basically two kinds of magnetic nanotubes: metallic or oxide. In each case, there are appropriate fabrication techniques that we will review in the next section where we also review the most important characterization methods applied to magnetic nanotubes. Toward the end of Section 14.2, we give a summary of the best known magnetic nanotubes, indicating their preparation methods and main characterizations. Section 14.3 is devoted to the theory behind the magnetic properties of nanotubes, where the basic theoretical aspects are presented and applied to some cases. Then we go on to Section 14.4 to briefly discuss some actual and potential applications of magnetic nanotubes; this area is rapidly changing so several potential applications are still to be found and revealed. We then summarize with an overview of the possible future perspectives of the field.

14.2 Synthesis and Characterization

In general, low-dimensional magnetic nanostructures attract attention both due to interesting fundamental properties and potential applications. The interest is stimulated by continuum improvements in both synthesis techniques and characterization methods, which are precisely the subjects to be covered in this section. We first review the methods used to fabricate magnetic nanotubes, then we go on to discuss the characterization methods, and we end by summarizing the discussion with a table merging the fabrication and characterization for known magnetic nanotubes so far.

14.2.1 Fabrication Methods

There are many ways to fabricate magnetic nanotubes. Among these, self-ordered porous alumina (Al_2O_3) is a suitable template system for the fabrication of arrays of nanotubes. A two-step anodization process commonly produces templates of porous alumina. This method can be traced back to the report of Masuda and Fukuda (1995). In this method, oxalic acid and phosphoric acid are used as electrolytes. The porous alumina layer builds a perfect hexagonal structure over the micrometer range under well defined conditions. The hexagonally ordered pores have diameters of the order of 30–450 nm, and an inter-pore (center to center) distance of the order of 50–500 nm, and lengths ranging from 1 to 120 μm. Such an array with a high aspect ratio (length over external radius) exhibits a so-called 2D polycrystalline arrangement with a hexagonal structure. These templates are known as anodic aluminum oxide (AAO) templates. A general idea of the morphology attained by these templates and the nanotubes produced by them can be obtained from Figure 14.3.

There are also polycarbonate porous membranes with specified pore diameters in the range of 50 nm to 2 μm, pore lengths of the order of 6–10 μm, and a pore density of ~10^8 pores/cm². These membranes also constitute suitable templates for growing tubular nanostructures (Leyva et al. 2004). Silicon templates have also been used to produce arrays with large diameters (Nielsch et al. 2005a). Moreover, by using certain block copolymers it is also possible to fabricate nanoscale templates. In block copolymers, nanoscale structures are created from phase separation due to incompatibilities between the different blocks. Copolymers could potentially be used for creating devices for use in computer memory, nanoscale-templating, and nanoscale separations. There are already reports of the fabrication of magnetic nanotubes using triblock copolymers as templates (Tao et al. 2006).

Once the templates are prepared, two methods are used to deposit the actual magnetic materials on the surface of the porous cylinders: atomic layer deposition (ALD) and chemical vapor deposition (CVD). The ALD is a self-limited gas–solid chemical reaction (Puurunen 2005). Two thermally stable gaseous precursors are pulsed alternatively into the reaction chamber, whereby direct contact of both precursors in the gas phase is prevented. Each precursor specifically reacts with the chemical functional groups present on the surface of the substrate (as opposed to nonspecific thermal decomposition), which ensures that one monolayer of the precursor adsorbs onto the surface during each pulse despite an excess of it in the gas phase. This peculiarity of ALD makes it suitable for coating substrates of complex geometry (in particular, highly porous ones). Moreover, it is possible to attain an outstanding thickness control (Lim et al. 2003).

CVD is a chemical process used to produce high-purity, high-performance solid materials. The process is often used in the semiconductor industry to produce thin films. In a typical CVD process, the wafer (substrate) is exposed to one or more

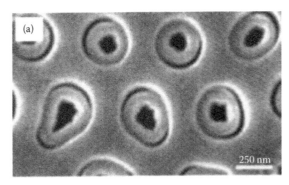

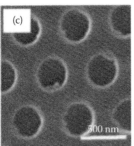

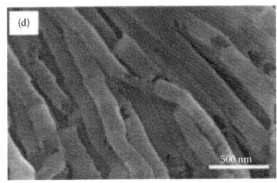

FIGURE 14.3 Some morphological aspects of cobalt nanotubes produced using AAO templates. Images were obtained by the technique known as scanning electron microscopy (SEM) discussed below. (Reprinted from Nielsch, K. et al., *J. Appl. Phys.*, 98, 034318, 2005b. With permission.)

volatile precursors, which react and/or decompose on the substrate surface to produce the desired deposit. Frequently, volatile by-products are also produced that are removed by gas flow through the reaction chamber. Microfabrication processes widely use CVD to deposit materials in various forms, including in the monocrystalline, polycrystalline, amorphous, and epitaxial forms. These materials include silicon, carbon fiber, carbon nanofibers, filaments, carbon nanotubes, SiO_2, silicon-germanium, tungsten, silicon carbide, silicon nitride, silicon oxynitride, titanium nitride, and various high-k dielectrics. The CVD process is also used to produce synthetic diamonds. The main difference between ALD and CVD is that in the latter, precursors react simultaneously on the surface in a way that they decompose in a high temperature process. To reach uniformity, several conditions must be fulfilled: uniform flux of precursors, precursor dosing control, and temperature control.

Another process used in the fabrication of magnetic nanotubes in conjunction with templates is the so called sol–gel process. It consists of a wet-chemical technique (chemical solution deposition) for the fabrication of materials (typically a metal oxide) starting either from a chemical solution (*sol* short for solution) or colloidal particles (*sol* for nanoscale particle) to produce an integrated network (*gel*). Typical precursors are metal alkoxides and metal chlorides, which undergo hydrolysis and polycondensation reactions to form a colloid, a system composed of solid particles (sizes ranging from 1 nm to 1 μm) dispersed in a solvent. The sol then evolves toward the formation of an inorganic continuous network containing a liquid phase (*gel*). The formation of a metal oxide involves connecting the metal centers (M) with oxo (M–O–M) or hydroxo (M–OH–M) bridges, therefore generating metal-oxo or metal-hydroxo polymers in the solution. The *drying* process serves to remove the liquid phase from the gel thus forming a porous material. Then, a thermal treatment (*firing*) may be performed in order to favor further polycondensation and enhance mechanical properties. The precursor sol can be either deposited on a substrate to form a film (e.g., by dip-coating or spin-coating), cast into a suitable container with the desired shape (e.g., to obtain monolithic ceramics, glasses, fibers, membranes, aerogels), or used to synthesize powders (e.g., microspheres, nanospheres). The sol–gel approach is interesting in that it is a cheap and low-temperature technique that allows for the fine control of the product's chemical composition, as even small quantities of dopants, such as organic dyes and rare earth metals, can be introduced in the sol and end up in the final product finely dispersed. It can be used in ceramics manufacturing processes, as an investment casting material, or as a means of producing very thin films of metal oxides for various purposes. Sol–gel derived materials have diverse applications in optics, electronics, energy, space, (bio) sensors, medicine (e.g., controlled drug release), and separation (e.g., chromatography) technology (Krumeich et al. 1999, Krusin-Elbaum et al. 2004, Saleta et al. 2007).

14.2.2 Characterization Techniques

Morphological characterizations and studies on the material composition of magnetic nanotubes have been made mainly by means of scanning electron microscope (SEM) and transmission electron microscope (TEM).

SEM generates a continuous beam of electrons in vacuum. The beam is collimated by electromagnetic condenser lenses, focused by an objective lens, and scanned across the surface of the sample by electromagnetic deflection coils. The primary imaging method is by collecting the secondary electrons that are released by the sample. The secondary electrons are detected by a scintillation material, which produces photons from the electrons. These photons are then detected and amplified by a photomultiplier tube. By correlating the sample scan position to the resulting signal, an image can be formed that is similar to what would be seen trough an optical microscope. The result is a digital file presenting illumination and shadows looking quite similar to a regular photography, thus reproducing the surface topography. Typical SEM magnification is of the order of 200,000 times, which translates to a theoretical resolution of about 1 nm as it can be seen, for instance, in Figure 14.3 (Nielsh et al. 2005b).

On the other hand, TEM is an image technique whereby a beam of electrons is focused onto a thin specimen so electrons are transmitted through the sample and then gathered on a fluorescent screen or layer of photographic paper producing an enlarged image. The wavelength of the electrons depends on their energy, and so it can be much smaller than that of light, which provides a correspondingly improved resolution. Other than this image mode, TEM can be used in the diffraction mode, which is also called selected area electron diffraction (SAED). The diffraction mode is a complementary tool to analyze the crystalline structure of the material. A typical TEM operating at 100 kV has a theoretical resolution of 0.2 nm, which is the case in the example shown in Figure 14.4.

Magnetic characterization of the nanotubes can be obtained using a vibrating sample magnetometer (VSM) and/or superconductor quantum interference device (SQUID) techniques. These setups have been mainly used on nanotubes to measure

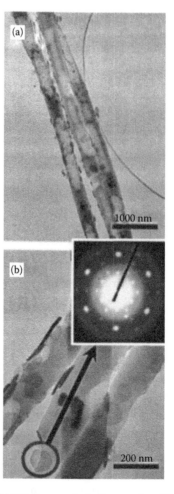

FIGURE 14.4 (a) TEM image of cobalt nanotubes. (b) SAED showing the local hexagon produced by a hcp-Co crystallite. (Reprinted from Nielsch, K. et al., *J. Appl. Phys.*, 98, 034318, 2005b. With permission.)

their magnetic properties, such as saturation magnetization, magnetic anisotropy, and coercive fields. In iron oxide magnetic nanotubes, for instance, it has been observed that the coercive field has a maximum value when measured as a function of wall thickness. This effect can be attributed to a crossover in the reversion process going from a vortex mode to a transverse mode as the wall thickness increases (Escrig et al. 2008b) as it will be discussed in Section 14.3.3.

The main components of a current magnetometer are the following: a temperature control system, which controls with precision the sample temperature in a wide range; a magnet control system, which provides magnetic fields from zero to several tesla through current provided by a power supply; a magnetic detector such as a SQUID; a sample handling system, which gives the ability to step or rotate the sample smoothly through the detection coils; and a computer operating system, which automatically controls all the features of the magnetometer. The SQUID amplifier system is the main component of a magnetic moment detection system in which the magnetic sample in motion induces an electrical current that can be amplified and measured.

A measurement of a sample containing magnetic nanotubes is performed in the magnetometer by moving a sample through the superconducting detection coils. The sample is usually mounted in a straw oriented with the tube axes parallel or perpendicular to the external magnetic field. As the sample moves through the coils, the magnetic moment of the sample induces an electric current in the detection coils. Variations of the current in the detection coils produce corresponding variations in the SQUID output voltage, which are proportional to the magnetic moment of the sample. In this way, hysteresis loops can be measured under a controlled temperature. A scheme illustrating the oriented measurement is shown in Figure 14.5.

Characterization of the samples is easily done in the inert media in which they are produced. Namely, magnetization curves can be obtained for the set of magnetic nanotubes in porous alumina for instance. Later on, the supporting material can be chemically dissolved freeing the nanotubes, which will tend to pack following their magnetic polarity. If single nanotubes are needed, special methods must be used to separate the tubes in the bundle one by one.

14.2.3 Summary of Reported Magnetic Nanotubes

To summarize, Table 14.1 lists the best-known tubular cylindrical nanoparticles characterized since 2000. The summary includes magnetic materials (first column), the fabrication method (second column), properties already characterized (third column), and references (fourth column). A small glossary of the nomenclature used in the table is provided at the end.

14.3 Modeling Magnetic Properties

14.3.1 Analytical and Numerical Methods

The different properties exhibited by magnetic nanotubes can be understood using both analytical and numerical methods. The first ones are based on the continuum theory of magnetism, in which the ultimate details of the atomic structure have been neglected, and then the material is considered as a continuum in accordance with the discussion in the introduction. In this way, our methods will find a first limitation when investigating nanotubes with walls thinner than the typical magnetic lengths involved in the problem, like the exchange length for instance. Also, analytical calculations require the previous knowledge of the possible stationary remnant states a particular system can achieve. Eventually, dynamic processes are not well covered in this way. However, this method provides analytical expressions that are useful for investigating static properties like magnetic configurations, coercivity, and reversal modes for nanotubes with a wall thickness greater than 10 nm, generally speaking.

The numerical methods are mainly based on the Monte Carlo (MC) approach, which is easily implemented in computers. However, under the frame of this method, the main limitation is that only a small number of particles can be investigated with no approximations. When studying nanotubes like those reported in the literature discussed above, we must deal with over 10^8 magnetic moments. Then it is necessary to use some approximations that lead to replacing the system by an equivalent one with fewer magnetic centers. However, in spite of the use of approximations, this method is useful for investigating the ground state and aspects of the dynamical process such as the detailed microscopic states of the system during reversal.

These two techniques are thus complementary and the simultaneous use of both of them leads to a better understanding of the magnetic properties of magnetic nanotubes. They are presented and discussed separately in the following sections.

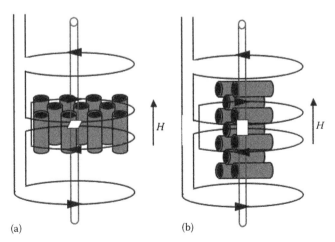

(a)   (b)

FIGURE 14.5 Schematic illustration of the parallel (a) and perpendicular (b) orientations of the SQUID measurement for arrays of magnetic nanotubes.

TABLE 14.1 Summary of Fabrication Methods, Characterizations Techniques and Reported Properties for Some of the Most Popular Magnetic Nanotubes Reported After 2000

Materials	Methods of Synthesis	Characterizations	References
FePt, Fe_3O_4	Filling of AAO templates	TEM, XRD, hysteresis loops	Sui et al. (2004a,b)
$CoFe_2O_4/Pb(Zr_{0.52}Ti_{0.48})O_3$	Sol–gel on AAO template	TEM, SEM, magnetic and electric hysteresis loops	Hua et al. (2008)
$La_{0.67}Ca_{0.33}MnO_3$	Filling of polycarbonate templates	Hysteresis loops[a]	Dolz et al. (2008), Leyva et al. (2004)
$BiFeO_3$	Sol–gel on AAO template	Hysteresis loops and photoabsorption	Wei et al. (2008)
FeCo	Sol–gel on AAO template	Moesbauer spectra and hysteresis loops	Li et al. (2007)
$Ni-Ni_3P$	Deposition on carbon nanofibers	SEM, TEM, XRD	Xie et al. (2007)
Fe_3O_4	ALD, template AAO	SEM, TEM XPS, hysteresis loops, XRD, Moesbauer	Bachmann et al. (2007), Wang et al. (2006), Son et al. (2005)
$La_{0.67}Sr_{0.33}MnO_3$	Filling of polycarbonate templates	IRM, DCD	Curiale et al. (2007)
$La_{0.67}Ca_{0.33}MnO_3$			
$VO_x/PAni$	Sol–gel	TEM, EPR	Saleta et al. (2007)
$(PAH9.3/PSS9.3)_{20.5}$	Filling of polycarbonate templates	TEM, SEM, hysteresis loops	Lee et al. (2007a)
$(Fe_3O_4/PAH9.3)_3$			
Ni, Co, Fe_2O_3	ALD on AAO, electrodeposition on AAO	SEM,TEM, hysteresis loop	Daub et al. (2007a,b), Tao et al. (2006), Nielsch et al. (2005a)
$Fe_{32}Ni_{68}$	Filling of polycarbonate templates	SEM, TEM, XRD, EDS, hysteresis loops	Xue et al. (2005, 2007)
Fe oxide	Filling of polycarbonate templates	TEM, SEM	Chen et al. (2006)
FeB, CoB, NiB	Filling of polycarbonate templates	TEM, SEM, XRD, EDS	Zhu et al. (2006)
Co/polymer	Filling of AAO template	SEM, TEM, hysteresis loops, XRD	Nielsch et al. (2005b)
Ni–Au (multisegmented)	Filling of AAO template	SEM, hysteresis loops	Lee et al. (2005)
VO_x (doped)	Sol–gel	SEM, TEM, XRD, hysteresis loops	Krusin-Elbaum et al. (2004), Krumeich et al. (1999)
$Ni_{45}Fe_{55}$	FIB trimming of magnetic heads	MFM	Khizroev et al. (2002)

[a] Electric polarization vs. electric field loops were also measured besides standard magnetization loops.

AAO, anodic aluminum oxide template; IRM, isothermal remanent magnetization; DCD, dc-demagnetization; PAni, polyaniline; EPS, electronic paramagnetic resonance; PAH, poly(allylamine hydrochloride); PSS, poly(sodium 4-styrenesulfonate); XRD, x-ray diffraction; XPS, x-ray photoelectron spectroscopy; EDS, energy dispersive spectrometry; MFM, magnetic force microscopy; FIB, focused ion beam.

14.3.1.1 Magnetostatic Continuum Theory of Ferromagnetism

Analytical calculations are based on the well-known theory of ferromagnetism (Aharoni 1996, Bertotti 1998), where a discrete distribution of magnetic moments is replaced by a continuous one characterized by $M(r)$, such that $M(r)\delta v$ gives the total magnetic moment within the element of the volume δv centered at r. Contributions to the magnetic energy come from three different mechanisms: exchange coupling, dipolar interactions, and anisotropy. The exchange term is given by $E_{ex} = A \int \sum (\vec{\nabla} m_i)^2 dv$, with $m_i = M_i/M_0$ ($i = x, y, z$); the components of the magnetization normalized to the saturation value M_0 and A is the stiffness constant of the particular material. The dipolar contribution is $E_d = (\mu_0/2)\int \vec{M} \cdot \vec{\nabla} U\, dv$, with U representing the magnetostatic potential. Finally, the cubic anisotropy energy can be added by means of $E_c = K_c \int (m_x^2 m_y^2 + m_y^2 m_z^2 + m_z^2 m_x^2) dv$, while the uniaxial anisotropy energy is given by $E_u = -K_u \int m_z^2 dv$. Besides, the exchange length defined as $L_x = \sqrt{2A/\mu_0 M_0^2}$ will also prove to be useful.

14.3.1.2 Monte Carlo Simulations

When analytic methods fail (or as a complement to analytic studies), properties can be simulated by means of numerical approaches among which MC simulations are frequently used (Landau and Binder 2005). The starting point in magnetism is the magnetic moment μ as a unit, which is appropriately defined for each system. They are led to interact among themselves and with any external field. The magnetostatic energy E of a nanotube, due to its N magnetic moments, is written as $E = \sum_{j>i}^{N} E_{ij}^d - J \sum_{ij \in \{nn\}} \vec{\mu}_i \cdot \vec{\mu}_j + E_a$, where E_{ij}^d is the dipolar energy given by $E_{ij}^d = \left[\vec{\mu}_i \cdot \vec{\mu}_j - 3(\vec{\mu}_i \cdot \hat{n}_{ij})(\vec{\mu}_j \cdot \hat{n}_{ij}) \right]/r_{ij}^3$, with $\hat{n}_{ij}$ the unit vector along the direction from the dipole moment $\vec{\mu}_i$ to the dipole moment $\vec{\mu}_j$; they are separated by a distance r_{ij}. The nearest neighbor exchange coupling constant is J and the interaction with an external field is written as $E_a = -\sum_{i=1}^{N} \vec{\mu}_i \cdot \vec{H}_a$.

Although quite nanoscopic in thickness, the known nanotubes listed in Table 14.1 contain more than 10^8 atoms, which is out of reach for a MC simulation with dipolar interactions considering the available computational resources. In order to reduce the number of interacting centers, we made use of a scaling technique (d'Albuquerque e Castro et al. 2002) originally formulated to investigate the equilibrium phase diagram of cylindrical particles of length L and diameter d. The authors showed that this diagram is equivalent to the one for a smaller particle with $d' = d\chi^\eta$ and $L' = L\chi^\eta$, with $\chi < 1$ and $\eta \approx 0.56$, if the exchange constant is also scaled as $J' = \chi J$.

14.3.2 Magnetic Configurations and Phase Diagram

14.3.2.1 Magnetic Configurations in Single Nanotubes

The geometry of the tubes (see Figure 14.1) is characterized by their external and internal radii, R and a, respectively, and length L. It is convenient to define the ratios $\beta = a/R$ and $\gamma = L/R$. β can be also written as a function of the thickness of the wall, d_w, and is given by $\beta = 1 - (d_w/R)$. Notice that the volume of such a tube is given by $V = \pi R^2 L(1 - \beta^2)$. Thus, $\beta = 0$ represents a solid cylinder (or wire) and β close to 1.0 corresponds to a tube with very thin walls. In this section, we consider an isolated magnetic nanotube, which in the limit case of an array corresponds to a sufficiently large separation between the tubes. In this situation, the interaction between them can be ignored. These particles present two characteristic ideal internal configurations according to their magnetization: F, with magnetic moments parallel to the tube axis, and V, with concentric magnetic moments on a plane perpendicular to the tube axis (Escrig et al. 2007b,c). The latter is the vortex state, where magnetic moments have no component along the tube axis and they orientate themselves tangentially circling around the tube axis. The existence of a core region in tubular structures has been analyzed recently by Kravchuk and collaborators (Kravchuk et al. 2007). They have shown that the core vanishes for the inner radius $a \approx L_x$, or safely $\beta > 0.3$. These results have been corroborated by using the public OOMMF computer program package (OOMMF year unknown) (Escrig et al. 2007b). Thus, we focus the following description on core-free magnetic nanotubes.

For the F configuration, the magnetization $M(r)$ can be approximated by $M_0 z$, where M_0 is the saturation magnetization and z is the unit vector parallel to the axis of the nanotube. In this configuration, the exchange contribution is nil, and the total energy is given by the dipolar contribution and the uniaxial anisotropy (Escrig et al. 2007b,c):

$$E^F = \frac{1}{2}\mu_0 M_0^2 V N_z(L) - \pi K_u R^2 L(1 - \beta^2) \quad (14.1)$$

where K_u is the uniaxial anisotropy constant and $N_z(l)$ corresponds to the demagnetizing factor along z, which adopts an integral form in terms of the Bessel functions $J_1(z)$:

$$N_z(l) = \frac{2R}{(1-\beta^2)l}\int_0^\infty \frac{dq}{q^2}\left[J_1(q) - \beta J_1(q\beta)\right]^2\left(1 - e^{-q(l/R)}\right) \quad (14.2)$$

For the V configuration, the magnetization can be approximated by $M_0\,\phi$ in terms of the unit vector in the sense of increasing the polar angle. In such a case, the dipolar contribution and the

uniaxial anisotropy are both nil. Thus, only exchange and cubic anisotropy characterized by the constant K_c contributes to the total energy (Escrig et al. 2007b,c):

$$E^V = -2\pi LA\ln\beta + \frac{K_c}{8}\pi LR^2(1 - \beta^2) \quad (14.3)$$

The equilibrium state for nanotubes of a given geometry has been obtained by means of numerical computer simulations (Lee et al. 2007b) using a three-dimensional hybrid finite element/boundary element micromagnetic code (Fidler and Schrefl 2000). This state can be obtained by minimizing the total magnetic energy from the randomly magnetized initial state, or by finding a remnant state from the hysteresis curve with an external field applied parallel to the tube axis. They found opposite directed vortices separated by a domain wall, which seems to be due to the relatively short aspect ratio γ of the nanotubes used in the simulations.

14.3.2.2 Magnetic Phase Diagrams

Phase diagrams in the R–L plane present a critical line that separates the V and F phases (Escrig et al. 2007b,c). This property is highly sensitive to the thickness of the nanotube through parameter β, leading to the condition

$$\frac{L}{L_x} = \frac{\frac{4}{3\pi}(1+\beta^3) - \beta^2 F_{21}[\beta]}{\ln(1/\beta) + \frac{R^2}{L_x^2}(1-\beta^2)\left(\frac{\kappa_c}{16} + \frac{\kappa_u}{2}\right)}\frac{R^3}{L_x^3}, \quad (14.4)$$

where $F_{21}[\beta] = F_{21}[-1/2,1/2,2,\beta^2]$ is an hypergeometric function, $\kappa_u = 2K_u/\mu_0 M_0^2$, $\kappa_c = 2K_c/\mu_0 M_0^2$, and the exchange length L_x was already defined above. For further details, we refer the reader to the original paper (Escrig et al. 2007c). The principal aspects of this analysis are presented in Figure 14.6 for three materials in

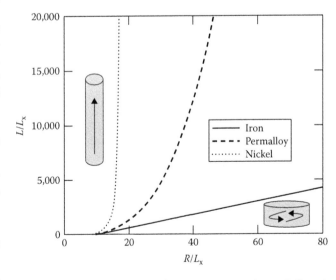

FIGURE 14.6 Phase diagrams for magnetic nanotubes with $\beta = 0.9$ giving the regions in the RL plane where one of the configurations has lower energy: F phase to the upper left, V phase to the lower right.

a representative range for the geometrical parameters. Regions where each phase dominates are graphically indicated. However, it is clear that most of the phase diagram for long iron thin tubes is dominated by the F phase, namely, with the magnetic moment along the axis with two possible and equivalent orientations. This property is what makes such magnetic nanotubes suitable for binary information storage.

14.3.3 Reversal Modes in Magnetic Nanotubes

The polarization of the magnetic moment along one of the two possible orientations in state F can be changed by means of external magnetic fields. This is precisely what is done in the process of magnetic recording in a bunch of magnetic nanotubes. There are some studies concerning the way this process is achieved, which we summarize this section.

The properties of virtually all magnetic materials are controlled by domains—extended regions where the spins of individual magnetic centers are tightly locked together and point in the same direction. A domain wall forms where two domains meet. Measurements on elongated magnetic nanostructures highlighted the importance of nucleation and the propagation of a domain wall going from the extreme of one domain to the extreme of the other domain with opposite magnetization; this is called the magnetization reversal process (Wernsdorfer et al. 1996, Atkinson et al. 2003, Thomas et al. 2006).

For one isolated magnetic nanotube, magnetization reversal (e.g., the change in the magnetization from one of its energy minima $M(r) = M_0 z$ to the other $M(r) = -M_0 z$) can occur by one of the following three idealized mechanisms (Landeros et al. 2007): coherent mode (C) in which all spins are supposed to reverse their spins coherently and simultaneously; vortex mode (V) with zero net magnetization in a segment of the tube; or transverse mode (T) with a net magnetization component in the (x, y) plane in a segment of the tube as depicted in Figure 14.7. Instabilities arise at any of the ends where surface magnetic moments are weekly coordinated to the rest of the material and propagates from one end to another (or from both ends to the center). In the case of any of the last two mechanisms (T or V), a domain wall appears at any end of the tube and propagates toward the other end.

According to the available simulations, the actual mechanism followed to achieve the reversal is strongly dependent on the geometry of the nanotube. The energy cost associated with the reversal for each of the three modes (C, V, and T) can be calculated in a way similar to the energy calculations done to find the phase diagram above (Landeros et al. 2007, Usov et al. 2007). It turns out that the C mechanism is extremely costly in energy and can be left out of the analysis. In the case of the other two mechanisms (V and T), a domain wall is tentatively defined and continuity of the magnetic quantities at the borders is imposed (Chen et al. 2007, Landeros et al. 2007, Usov et al. 2007). Energy for each configuration is found. Analytic studies have been complemented by numerical simulations to decide upon which mechanism requires less energy to switch the magnetization

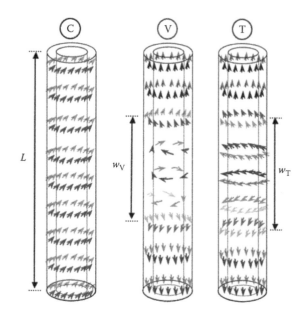

FIGURE 14.7 **(See color insert following page 20-16.)** Magnetization reversal modes in nanotubes. Arrows represent the orientation of magnetic moments within the tube. Left: Coherent-mode rotation, C. Center: Vortex-mode rotation, V, with a domain wall of thickness w_V. Right: Transverse-mode rotation, T, with a domain wall of thickness w_T.

completely from one end to the other. The reader who is interested in the details of this study is referred to the already quoted literature. Now we go directly to a discussion on the conditions for the prevalence of one mechanism over the other.

The results of the simulations indicate that the preferred reversal mode depends on the actual material and on the geometry mainly through the internal and external radius (Landeros et al. 2007, Usov et al. 2007). For each β value, there exists a critical radius, $R_c(\beta)$, at which the cost in energy is the same for V and T mechanisms. Thus, for $R < R_c(\beta)$, the tube reverses its magnetization creating a transverse domain wall (T), while for $R > R_c(\beta)$, a vortex domain wall (V) appears. The locus for the equilibrium curve $R_c(\beta)$ can be found looking like the lower curve in Figure 14.9.

It can be seen that the formation of the vortex domain walls is favored for the dimensions at which magnetic nanotubes are presently obtained. Actually, nanotubes with a radius smaller than a few exchange lengths are very difficult to produce currently (Nielsch et al. 2005a,b, Bachmann et al. 2007, Daub et al. 2007a,b). Thus, the propagation of vortex-like domain walls is the expected reversal process. However, for some particular nanotubes with values of R and β lying close to the boundary line separating the T and V mechanisms (see Figure 14.9), instabilities may appear giving chances to the T mechanism. MC simulations have even shown that in such cases the reversion can begin as a T process, changing to the V mechanism as the domain wall propagates to the center of the tube (Landeros et al. 2007).

In any case, this is a field where experimental progress is needed as simulations depend strongly on the assumptions. Thus, for instance, ideal long nanotubes are considered to have anisotropy with the magnetization aligned along the nanotube axis. In such models, a coherent reversal of the magnetization would be needed to invert the magnetic polarization. However, such mechanisms are extremely costly in terms of energy. Contrary to such an ideal picture (Stoner and Wohlfarth 1948), magnetic nanotubes tend to exhibit inhomogeneous reversal processes. Micromagnetic and MC simulations provide a more realistic picture to deal with the switching process for real nanotubes, which can still be put in agreement with analytic models (Escrig et al. 2007a, 2008b, Allende et al. 2008). We now proceed to study each reversion mode independently.

14.3.3.1 Transverse Mode

In the simple model by Escrig and collaborators (Escrig et al. 2007a, 2008b), the coercive field H_n^T can be approximated by an adapted Stoner–Wohlfarth model (Stoner and Wohlfarth 1948) in which the length of the coherent rotation is replaced by the width of the transverse domain wall, w. Following this approach,

$$\frac{H_n^T}{M_0} = \frac{2K(w)}{\mu_0 M_0^2} \tag{14.5}$$

where

$$K(l) = \frac{1}{4}\mu_0 M_0^2 \left[1 - 3N_z(l)\right] \tag{14.6}$$

The calculated switching field for an isolated nanotube based on the model of Escrig et al. (2007a, 2008b) is always lower in the T mode than in the Stoner–Wohlfarth approximation (Stoner and Wohlfarth 1948).

14.3.3.2 Curling Mode

The curling mode (Aharoni 1996, 1997) is a noncoherent calculation that minimizes the total magnetic energy. When magnetic reversal occurs via curling, a magnetic vortex structure is formed at the nanotube ends and for magnetic tubes with an infinite length, an analytical solution can be calculated (Chang et al. 1994, Escrig et al. 2007a, 2008b). For an infinite tube, the nucleation field for the V mode, H_n^V, is given by (Chang et al. 1994, Escrig et al. 2007a)

$$\frac{H_n^V}{M_0} = \alpha(\beta)\frac{L_x^2}{R^2} \tag{14.7}$$

The function $\alpha(\beta)$ in previous expressions has been obtained by means of two different methods. Escrig and collaborators (Escrig et al. 2007a) obtained an analytical approach using a Ritz model, which leads to

$$\alpha(\beta) = \frac{8}{3}\frac{(14 - 13\beta^2 + 5\beta^4)}{(11 + 11\beta^2 - 7\beta^4 + \beta^6)} \tag{14.8}$$

Equation 14.7 had been previously obtained by Chang and collaborators (Chang et al. 1994) starting from Brown's equations. They obtained $\alpha(\beta) = q^2$, where q satisfies the condition

$$\frac{qJ_0(q) - J_1(q)}{qY_0(q) - Y_1(q)} - \frac{\beta qJ_0(\beta q) - J_1(\beta q)}{\beta qY_0(\beta q) - Y_1(\beta q)} = 0 \tag{14.9}$$

Here $J_p(z)$ and $Y_p(z)$ are Bessel functions of the first and second kind, respectively. Equation 14.8 has an infinite number of mathematical solutions, out of which only the one with the smallest nucleation field has to be considered (Aharoni 1996) for practical purposes. Therefore, the nucleation field depends on $\alpha(\beta)$, which is related to the internal and external radii of the tube.

A crossover between the two reversal modes T and V (that have been previously presented) has been reported (Escrig et al. 2008b). It turns out that the curling mode (V) is more stable for thinner tubes, whereas thicker tube walls favor the transverse mode (T). However, the absolute values are computed for the coercivity by means of equations. Values obtained by means of Equations 14.5 and 14.7 are greater than the experimental data, a discrepancy which can be solved by considering the weak magnetostatic interactions among nanotubes that are present in the experiment but they were left out of the theoretical calculations based on a single nanotube.

14.3.4 Magnetostatic Interactions among Nanotubes

A couple of isolated magnetic nanotubes can attract or repel each other upon approach depending on their relative orientations. Based on the nanotubes reported in the literature, the interacting force has been estimated to be of the order of tens to hundreds of microdynes (Suarez et al. 2009). The performance of such an experiment could provide the basis for testing the different theoretical models at the nanoscale.

In the case of nanotubes produced by porous membranes, they are trapped in the array and the interaction in such a triangular lattice (with an overall hexagonal appearance) tends to produce all kinds of orientations, frustrating many local fields and tending to give a null effective magnetization in the absence of external fields (Figure 14.8). In a way, each tube interacts with

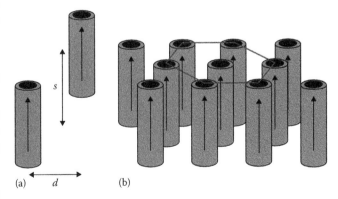

FIGURE 14.8 (a) Relative position of interacting tubes: d is the interaxial distance and s is the vertical separation. (b) Hexagonal array of magnetic nanotubes on a triangular lattice where contributions from the 6 nearest neighbors, the 12 second-degree neighbors, and the tubes situated farthest away from the probe tube combine to form the so-called stray field.

the stray fields produced by the array, so an effective antiferromagnetic coupling between neighboring tubes is expected, thus reducing the coercive field. In these interacting systems, the process of magnetization reversal can be viewed as the overcoming of a single energy ΔE.

In an array with all the nanotubes initially magnetized in the same direction, the magnetostatic interaction between neighboring tubes favors the magnetization reversal of some of them. A reversing field aligned opposite to the magnetization direction lowers the energy barrier, thereby increasing the probability of switching. Escrig and collaborators (Escrig et al. 2008a) have calculated this magnetostatic interaction using a simple mathematical expression. They start by calculating the magnetic field generated by a ferromagnetic tube assuming that it is a continuum material (see Section 14.3.1.1). From the expression for the field, the researchers can calculate if the field is attractive or repulsive at any point in space. Next, they put another magnetic tube near the first one and calculate the interaction energy between them. The expression for the energy they obtained is usually quite complicated and has to be numerically solved. However, assuming that the tubes investigated satisfy $R/L \ll 1$, they obtain a simple analytical expression for the magnetic interaction, which provides an excellent tool for the understanding of interactions between those nanoelements (Escrig et al. 2008a).

The angle θ formed by the externally applied magnetic field and the axis of the nanotube play an important role in defining the coercivity of these objects; such angles also determine the preferred reversal mode (Allende et al. 2008, Escrig et al. 2008b). Under a critical angle, the nanotube reverses by means of the T mechanism, while over that critical angle, the preferred

mechanism is V. Phase diagrams similar to the one shown in Figure 14.9 can be obtained for the combination of the geometrical parameters according to the case of interest.

14.4 Applications

Since the discovery of the first nanotubes by Iijima in 1991 (Iijima 1991), their applications have been mainly focused on the development of microelectronic devices; however, their biomedical and biotechnological applications remain in the early stages. In these areas, the focus has been put on spherical nanoparticles, which exhibit many advantages related to the synthesis process. However, spherical particles still need to be improved in terms of surface modification and environmental compatibility, especially when multifunctionality is required.

In the last few years, improvements in the synthesis of magnetic nanotubes have opened new possibilities in the field due to the fabrication of magnetic nanoscopic objects of different geometrical shapes with important advantages with respect to spherical nanoparticles. Nanotubes have an inner section that can be filled with species ranking from large proteins to small molecules and their open ends can serve as a gate to regulate the interaction of the inner agent with the medium. Also they have inner and outer surfaces, which can be functionalized differently depending on the specific application deseeded. Above all, the main advantage of magnetic nanotubes used as drug carriers is precisely their magnetic behavior that allows them to be directed to specific target sites by means of externally applied magnetic fields. Complementarily, it is also possible to use magnetic resonance imaging to track the drug delivery process. The combination of their geometrical variations based on their tubular shape and their magnetic properties have made magnetic nanotubes the most promising candidates for clinical diagnosis and therapeutic applications. In the following sections, we describe some of the recent applications of these particles.

Silica nanotubes embedded with magnetic nanoparticles have been synthesized (Son et al. 2005). Since they have an external silica surface, they have a hydrophilic behavior and can be suspended on aqueous phases. These particles were used to remove dye molecules from an aqueous solution. The dye molecule enters the nanotubes where hydrophobic interactions hold them in place within the tube structure. Also, these nanotubes were labeled with anti-human IgG and were separated specifically from human IgG.

Son and collaborators later investigated the cytotoxicity of magnetic silica nanotubes against the human metastatic breast cancer cell line MDA-MB-231 (Son et al. 2006). They looked for the effect of varying size and surface functionalization of these nanotubes on cytotoxicity and concluded that the toxicity was strongly dependent on the concentration of nanotubes rather than on the size or surface functionalization. They also observed that positively functionalized magnetic nanotubes were more toxic than nonfunctionalized ones.

Techniques used to fabricate multilayer nanoparticles have been successfully used to produce nanotubes composed of

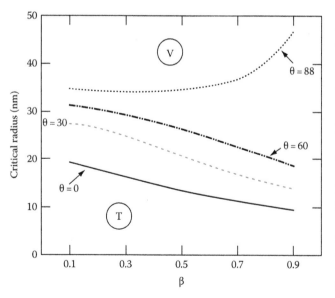

FIGURE 14.9 Critical radius as a function of thickness β for different values θ, which is the angle between the direction of the externally applied magnetic field and the tube axis. This picture can be interpreted as a phase diagram in which each line separates the T mode of magnetization reversal, which prevails in the lower region of the (β, R) space, from the V mode, present in the upper area.

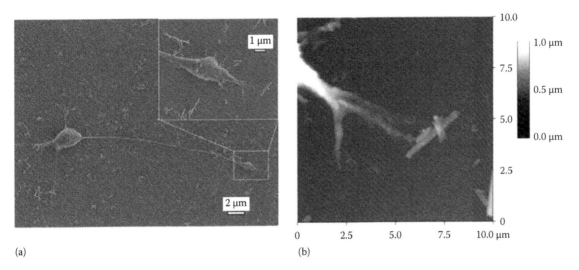

FIGURE 14.10 (a) SEM micrograph of PC12 cell culture. Inset: magnified image of the growth cone area (b) AFM image of the growth cone with filopodia rolling an NGF-incorporated magnetic nanotube. (Reproduced from Xie, J. et al., *Nanotechnology*, 19, 1051, 2008. With permission.)

alternating layers of cationic poly (allylamine hydrochloride) and anionic poly (styrene sulfonate) substances (Lee et al. 2007a). One of the inner layers is magnetic material (usually Fe_3O_4, known as magnetite), which makes the device easy to functionalize and drive to a designated target. Such magnetic tubes could be successfully used to separate (or remove) a high concentration of dye molecules from the solution by activating the nanotubes in acidic solution. The release of the anionic molecules in a physiologically relevant buffer solution showed that whereas bulky molecules (e.g., rose bengal) release slowly, small molecules (i.e., ibuprofen) immediately release from the multilayers.

Functional magnetic nanotubes have been recently used in *in vitro* experiments using rat tissue to load growth factors and induce cell differentiation (Xie et al. 2008). It was observed that magnetic nanotubes loaded with nerve growth factor can induce the differentiation of rat pheochromocytoma (PC12) cells into neurons exhibiting growth cones and neurite outgrowth. This effect is illustrated in Figure 14.10. The functionalization of the external surface of the magnetic nanotube makes it appealing to these nerve cells thus allowing its differentiation from others.

14.5 Summary and Perspectives

Magnetic nanotubes are just at the onset of their development. Improvements in the ALD method will make possible a controlled fabrication of some desired compositions at least in the succession of different layers. Just one of them needs to be magnetic to make the object directed by means of applied magnetic fields. Outer and inner layers can be functionalized for different purposes thus making the whole nanotube quite versatile.

Applications may range from microelectronics or nanoelectronics to fundamental biology. Additionally, information storage and drug delivery are also open possibilities. Traditional spherical nanoparticles are poor competitors to nanotubes due to their only free parameter subject to variations (the radius) and just one contact surface to the external world. On the other hand, nanowires are still good competitors when simple and direct applications are needed. In particular, nanowires are still cheaper to produce than nanotubes, so under similar responses, nanowires could be preferred. However, nanotubes are unbeatable in the long run when light multifunctional magnetic particles are needed.

A magnetic nanotube can be taken to the region of applicability by means of noninvasive externally applied magnetic fields. If necessary, its magnetization can be reversed on the spot. On the other hand, the functionalization of the external surface can make such nanotubes attractive to a certain desired surrounding (a particular kind of biological cell or a particular chemical surrounding in an inorganic device). In this way, the nanotube can reach a well-determined target. This is enough if the magnetic nanotube is to be used as a tracer for certain surroundings, which can then be visualized by appropriate imaging techniques. However, still further applications are possible if the magnetic nanotube is a vehicle transporting drugs or any other chemical agent, which can be accommodated in the hollow space by functionalizing the inner surface appropriately. Techniques must then be used to release the needed dose of the chemical agent where and when needed. Independently of this chemical (drug) delivery process, the metallic nature of the nanotube can be used to produce local heat by electromagnetic induction for burning undesired tissue, as in the treatment of cancer. In inorganic applications, local heating can be used to produce soldering fixing contacts or to produce any other kind of unions. This short list of possible applications in different areas will probably prove to be even shorter as more advances are made in the production, characterization, and functionalization of magnetic nanotubes.

In the most fundamental area, extremely thin magnetic nanotubes (with thicknesses close to or under the correlation length) could pose entirely new basic problems as classical magnetism could be no longer valid and quantum magnetism should be invoked. Eventually, this could open entirely new areas of research with eventual links to molecular magnetism at a certain scale.

Acknowledgments

The authors are grateful to Millennium Scientific Initiative (Chile) under contract P06-022-F and FONDECYT (Chile) under codes 1060317, 1070224, 1080300, and 11070010 for partial support. Complementary support from Financiamiento Basal Para Centros Científicos y Tecnológicos de Excelencia (Chile) through the Center for Development of Nanoscience and Nanotechnology (CEDENNA) is also acknowledged.

References

Aharoni, A. 1996. *Introduction to the Theory of Ferromagnetism*, Clarendon Press, Oxford, U.K.

Aharoni, A. 1997. Angular dependence of nucleation by curling in a prolate spheroid. *J. Appl. Phys.* 82: 1281–1287.

Allende, S., Escrig, J., Altbir, D., Salcedo, E., Bahiana, M. 2008. Angular dependence of the reversal modes in small diameter magnetic nanotubes. *Eur. Phys. J. B* 66: 37. DOI:10.1140/e2008-00385-4.

Atkinson, D., Allwood, D.A., Xiong, G., Cooke, M.D., Faulkner, C.C., Cowburn, R.P. 2003. Magnetic domain-wall dynamics in a submicrometre ferromagnetic structure. *Nat. Mater.* 2: 85–87.

Bachmann, J., Jing, J., Knez, M., Barth, S., Shen, H., Mathur, S., Gösele, U., Nielsch, K. 2007. Ordered iron oxide nanotube arrays of controlled geometry and tunable magnetism by atomic layer deposition. *J. Am. Chem. Soc.* 129: 9554–9555.

Bertotti, G. 1998. *Hysteresis in Magnetism*, Academic Press, San Diego, CA.

Chang, C.-R., Lee, C.M., Yang, J.-S. 1994. Magnetization curling reversal for an infinite hollow cylinder. *Phys. Rev. B* 50: 6461–6464.

Chen, L., Xie, J., Srivatsan, M., Varadan, V.K. 2006. Magnetic nanotubes and their potential use in neuroscience applications. *Smart Electronics, MMES, BIOMEMS and Nanotechnology, SPIE*, 6172: 61720J1–61720J10.

Chen, A.P., Usov, N.A., Blanco, J.M., Gonzalez, J. 2007. Equilibrium magnetization states in magnetic nanotubes and their evolution in external magnetic field. *J. Magn. Magn. Mater.* 316: e317–e319.

Curiale, J., Sánchez, R.D., Troiani, H.E., Leyva, A.G., Levy, P. 2007. Magnetic interactions in ferromagnetic manganite nanotubes of different diameters. *Appl. Surf. Sci.* 254: 368–370.

d'Albuquerque e Castro, J., Altbir, D., Retamal, J.C., Vargas, P. 2002. Scaling approach to the magnetic phase diagram of nanosized systems. *Phys. Rev. Lett.* 88: 237202-(4).

Daub, M., Bachmann, J., Jing, J., Knez, M., Gösele, U., Barth, S., Marthur, S., Escrig, J., Altbir, D., Nielsch, K. 2007a. Ferromagnetic nanostructures by atomic layer deposition: From thin films towards core-shell nanotubes. *ECS Trans.* 11: 139–148.

Daub, M., Knez, M., Gösele, U., Nielsch, K. 2007b. Ferromagnetic nanotubes by atomic layer deposition in anodic alumina membranes. *J. Appl. Phys.* 101: 09J111-(3).

Dolz, M.I., Bast, W., Antonio, D., Pastoriza, H., Curiale, J., Sánchez, R.D., Leyva, A.G. 2008. Magnetic behavior of single $La_{0.67}Ca_{0.33}MnO_3$ nanotubes: Surface and shape effects. *J. Appl. Phys.* 103: 083909-(5).

Escrig, J., Daub, M., Landeros, P., Nielsch, K., Altbir, D. 2007a. Angular dependence of coercivity in magnetic nanotubes. *Nanotechnology* 18: 445706-(5).

Escrig, J., Landeros, P., Altbir, D., Vogel, E.E., Vargas, P. 2007b. Phase diagrams of magnetic nanotubes. *J. Magn. Magn. Mater.* 308: 233–237.

Escrig, J., Landeros, P., Altbir, D., Vogel E.E. 2007c. Effect of anisotropy in magnetic nanotubes. *J. Magn. Magn. Mater.* 310: 2448–2450.

Escrig, J., Allende, S., Altbir, D., Bahiana, M. 2008a. Magnetostatic interactions between magnetic nanotubes. *Appl. Phys. Lett.* 93: 023101-(3).

Escrig, J., Bachmann, J., Jing, J., Daub, M., Altbir, D., Nielsch, K. 2008b. Crossover between two different magnetization reversal modes in arrays of iron oxide nanotubes. *Phys. Rev. B* 77: 214421-(7).

Fidler, J., Schrefl, T., 2000. Micromagnetic modeling—The current state of the art. *J. Phys. D: Appl. Phys.* 33: R135–R156.

Hua, Z.H., Yang, P., Huang, H.B., Wan, J.G., Yu, Z.Z., Yang, S.G., Lu, M., Gu, B.X., Du, Y.W. 2008. Sol-gel template synthesis and characterization of magnetoelectric $CoFe_2O_4/Pb(Zr_{0.52}Ti_{0.48})O_3$ nanotubes. *Mater. Chem. Phys.* 107: 541–546.

Iijima, S. 1991. Helical microtubules of graphitic carbon. *Nature* 354: 56–58.

Khizroev, S., Kryder, M.H., Litvinov, D., Thompson, D.A. 2002. Direct observation of magnetization switching in focused-ion-beam-fabricated magnetic nanotubes. *Appl. Phys. Lett.* 81: 2256–2257.

Korneva, G., Ye, H., Gogotsi, Y., Halverson, D., Friedman, G., Bradley, J.-C., Kornev, K.G. 2005. Carbon nanotubes loaded with magnetic particles. *Nano Lett.* 5: 879–884.

Kravchuk, V.P., Sheka, D.D., Gaididei, Y.B. 2007. Equilibrium magnetization structures in ferromagnetic nanorings. *J. Magn. Magn. Mater.* 310: 116–125.

Krumeich, F., Muhr, H.-J., Niederberber, M., Bieri, F., Schnyder, B., Nesper, R. 1999. Morphology and topochemical reactions of novel vanadium oxide nanotubes. *J. Am. Chem. Soc.* 121: 8324–8331.

Krusin-Elbaum, L., Newns, D.M., Zeng, H., Derycke, V., Sun, J.Z., Sandstrom, R. 2004. Room-temperature ferromagnetic nanotubes controlled by electron or hole doping. *Nature* 431: 672–676.

Landau, D.P., Binder, K. 2005. *A Guide to Monte Carlo Simulations in Statistical Physics*, 2nd edn., Cambridge University Press, Cambridge, U.K.

Landeros, P., Allende, S., Escrig, J., Salcedo, E., Altbir, D., Vogel, E.E. 2007. Reversal modes in magnetic nanotubes. *Appl. Phys. Lett.* 90: 102501-(3).

Lee, W., Scholz, R., Nielsch, K., Gösele, U. 2005. A template-based electrochemical method for the synthesis of multi-segmented metallic nanotubes. *Angew. Chem. Int. Ed.* 44: 6050–6054.

Lee, D., Cohen, R.E., Rubner, M.F. 2007a. Heterostructured magnetic nanotubes. *Langmuir* 23: 123–129.

Lee, J., Suess, D., Schrefl, T., Oh, K.H., Fidler, J. 2007b. Magnetic characteristics of ferromagnetic nanotube. *J. Magn. Magn. Mater.* 310: 2445–2447.

Leyva, A.G., Stoliar, P., Rosenbusch, M., Lorenzo, V., Levy, P., Albonetti, C., Cavallini, M., Biscarini, F., Troiani, H.E., Curiale, J., Sanchez, R.D. 2004. Microwave assisted synthesis of manganese mixed oxide nanostructures using plastic templates. *J. Solid State Chem.* 177: 3949–3953.

Li, F.S., Zhou, D., Wang, T., Wang, Y., Song, L.J., Xu, C.T. 2007. Fabrication and magnetic properties of FeCo alloy nanotube array. *J. Appl. Phys.* 101: 014309-(3).

Lim, B.S., Rahtu, A., Gordon, R.G. 2003. Atomic layer deposition of transition metals. *Nature Mat.* 2: 749–754.

Masuda, H., Fukuda, K. 1995. Ordered metal nanohole arrays made by a two-step replication of honeycomb structures of anodic alumina. *Science*, New Series, 268: 1466–1468.

Mitchell, D.T., Lee, S.B., Trofin, L., Li, N., Nevanen, T.K., Soderlund, H., Martin, C.R. 2002. Smart nanotubes for bioseparations and biocatalysis. *J. Am. Chem. Soc.* 124: 11864–11865.

Nielsch, K., Stadler, B.J.H. 2007. Template-based synthesis and characterization of high-density ferromagnetic nanowire arrays. In *Handbook of Magnetism and Advanced Magnetic Materials Vol. 4 Novel Materials*, Eds. H. Kronmüller and S. Parkin, John Wiley & Sons Ltd., Chichester, U.K., ppi-ppf.

Nielsch, K., Castaño, F.J., Matthias, S., Lee, W., Ross, C.A. 2005a. Synthesis of cobalt/polymer multilayer nanotubes. *Adv. Eng. Mater.* 7: 217–221.

Nielsch, K., Castaño, F.J., Ross, C.A., Krishnan, R. 2005b. Magnetic properties of template-synthesized cobalt/polymer composite nanotubes. *J. Appl. Phys.* 98: 034318-(6).

OOMMF. The public domain package is available at (http://math.nist.gov/oommf/).

Puurunen, R.L. 2005. Surface chemistry of atomic layer deposition: A case study for the trimethylaluminum/water process. *J. Appl. Phys.* 97: 121301–121352 and references therein.

Saleta, M.E., Curiale, J., Troiani, H.E., Ribeiro-Guevara, S., Sánchez, R.D., Malta, M., Torresi, R.M. 2007. Magnetic characterization of vanadium oxide/polyaniline nanotubes. *Appl. Surf. Sci.* 254: 371–374.

Son, S.J., Reichel, J., He, B., Schuchman, M., Lee, S.B. 2005. Magnetic nanotubes for magnetic-field-assisted bioseparation, biointeraction, and drug delivery. *J. Am. Chem. Soc.* 127: 7316–7317.

Son, S.J., Bai, X., Nan, A., Ghandehari, H., Lee, S.B. 2006. Template synthesis of multifunctional nanotubes for controlled release. *J. Control Release* 114: 143–152.

Stoner, E.C., Wohlfarth, E. P. 1948. A mechanism of magnetic hysteresis in heterogeneous alloys. *Philos. Trans. R Cos. A* 240: 599. Reprinted in 1991. *IEEE Trans. Magn.* 27: 3475–3518.

Suarez, O.J., Vargas, P., Vogel, E.E. 2009. Energy and force between two magnetic nanotubes. *J. Magn. Magn. Mater.* 321: 3658–3664.

Sui, Y.C., Skomski, R., Sorge, K.D., Sellmyer, D.J. 2004a. Nanotube magnetism. *Appl. Phys. Lett.* 84: 1525–1527.

Sui, Y.C., Skomski, R., Sorge, K.D., Sellmyer, D.J. 2004b. Magnetic nanotubes produced by hydrogen reduction. *J. Appl. Phys.* 95: 7151–7153.

Tao, F., Guan, M., Jiang, Y., Zhu, J., Xu, Z., Xue, Z. 2006. An easy way to construct an ordered array of nickel nanotubes: The triblock-copolymer-assisted hard-template method. *Adv. Mater.* 18: 2161–2164.

Thomas, L., Hayashi, M., Jiang, X., Moriya, R., Rettner, C., Parkin, S. S. P. 2006. Oscillatory dependence of current-driven magnetic domain wall motion on current pulse length. *Nature* 443: 197–200.

Usov, N.A., Zhukov, A., Gonzalez, J. 2007. Domain walls and magnetization reversal process in soft magnetic nanowires and nanotubes. *J. Magn. Magn. Mater.* 316: 255–261.

Wang, T., Wang, Y., Li, F., Xu, C., Zhou, D. 2006. Morphology and magnetic behaviour of an Fe_3O_4 nanotube array. *J. Phys. Condens. Matter* 18: 10545–10551.

Wei, J., Xue, D., Xu, Y. 2008. Photoabsorption characterization and magnetic property of multiferroic $BiFeO_3$ nanotubes synthesized by a facile sol-gel template process. *Scr. Mater.* 58: 45–48.

Wernsdorfer, W., Doubin, B., Mailly, D., Hasselbach, K., Benoit, A., Meier, J., Ansermet, J.-Ph., Barbara, B. 1996. Nucleation of magnetization reversal in individual nanosized nickel wires. *Phys. Rev. Lett.* 77: 1873–1876.

Xie, G., Wang, Z., Li, G., Shi, Y., Cui, Z., Zhang, Z. 2007. Templated synthesis of metal nanotubes via electroless deposition. *Mater. Lett.* 61: 2641–2643.

Xie, J., Chen, L., Varadan, V.K., Yancey, J., Srivatsan, M. 2008. The effects of functional magnetic nanotubes with incorporated nerve growth factor in neuronal differentiation of PC12 cells. *Nanotechnology* 19: 1051.

Xue, S., Cao, C., Wang, D., Zhu, H. 2005. Synthesis and magnetic properties of $Fe_{0.32}Ni_{0.68}$ alloy nanotubes. *Nanotechnology* 16: 1495–1499.

Xue, S., Li, M., Wang, Y., Xu, X. 2007. Electrochemically synthesized binary alloy FeNi nanorod and nanotube arrays in polycarbonate membranes. *Thin Solid Films* 517: 5922–5926.

Zhu, Y., Liu, F., Ding, W., Guo, X., Chen, Y. 2006. Noncrystalline metal-boron nanotubes: Synthesis, characterization, and catalytic-hydrogenation properties. *Angew. Chem. Int. Ed.* 45: 7211–7214.

15

Self-Assembled Peptide Nanostructures

Lihi Adler-Abramovich
Tel Aviv University

Ehud Gazit
Tel Aviv University

15.1 Molecular Self-Assembly

The spontaneous formation of ordered structures at the nanoscale is a key issue in nanotechnology (Whitesides et al. 1991; Zhang 2003). In a "bottom-up" process, simple building blocks self-assemble to form large and more complex supramolecular assemblies (Figure 15.1). In the molecular self-assembly process, the molecules spontaneously interact with each other through noncovalent bonds to form well-ordered ultrastructures; therefore, the structure of the molecular building blocks determines the architecture of the assembly. This assembly process is mediated through weak intermolecular interactions, such as van der Waals bonds, hydrogen bonds, aromatic interactions, and electrostatic interactions. The overall coordinated combination of the various molecular forces, which are quite weak individually, results in the process of the self-organization from simple blocks into elaborate and ordered structures. Several natural building blocks such as nucleic acids, phospholipids, and polypeptides self-assemble to form novel materials. In recent years, there has been a great interest in the fabrication of new materials using natural building blocks and combining them in artificial systems. Therefore, the biomolecular self-assembly mechanism is extensively being studied by many research groups to obtain a better understanding.

15.2 Self-Assembly of Proteins and Peptides

Proteins and peptides serve as the major molecular scaffold material of the biological world. An example of nanoscale elements is the self-assembled actin cytoskeleton, the molecular structures that give the cell its physical rigidity, and the self-assembled microtubules that serve as nanoscopic protein railways that enable the transport of "cargo" within the cell using nanoscale protein motors. Proteins also serve as the building blocks for macroscopic structural elements, for instance, the collagen proteins in the skin and the keratin proteins in nails and hair. In addition, proteins serve as the building blocks for elaborate structures possessing unique physical properties such as silk, whose tensile strength-to-density ratio is about five times higher than steel. The formation of inorganic biological structures such as bones, teeth, and marine animal shells is directed by protein templates via the specific interaction of proteins and peptides with calcium or silicon (Naik et al. 2002).

Besides their structure role, proteins and peptides facilitate biological recognition. The specific binding to various molecules is mediated by protein antibodies and receptors, whereas messages in the body are delivered by polypeptide hormones such as insulin, vasopressin, and luteinizing. In addition, almost all the enzymatic activities in every biological system are carried out by protein enzymes ranging from simple reaction enzymes to multicomponent molecular synthesizers (Aggeli et al. 1997). Also, the mechanical components in biological systems, such as molecular motors at the nano-scale and muscles at the macroscale, are composed of proteins. Thus, taken together, proteins serve both as the building scaffold as well as the functional entities in the biological world. Hence, researchers envision proteins and their peptide fragments as potential sources of engineered "smart function materials."

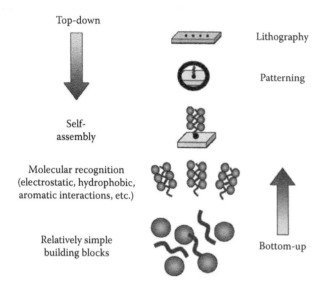

FIGURE 15.1 The process of "top-down" as compared to "bottom-up" self-assembly. The top-down process as developed to superb efficiency by the microelectronics industry is based on the patterning of assemblies by lithographic definition. The "bottom-up" approach is based on the interaction of simple building blocks to form a well-ordered assembly by means of molecular recognition and self-assembly. (Reproduced from Gazit, E., *Chem. Soc. Rev.*, 36, 1263, 2007. With permission.)

15.3 Peptide-Based Nanostructures

The self-assembled protein- and peptide-based nanostructures are envisioned to serve as important building blocks in future nanotechnological devices. Such assemblies offer the advantages of chemical diversity deriving from the great structure range of the building block, which not only includes the 20 natural occurring amino acids but hundreds of modified residues. Furthermore, amino acid motifs can enable the binding of metals to the biological structures, for example, the high-affinity binding of nickel or cobalt to histidine stretches and the affinity of gold to the thiol group in cysteine. The protein and peptide structures could also be decorated through the coupling of functional and protective groups. Moreover, protein and peptides are biocompatible and have the ability to be biodegradable. Another major advantage is their spontaneous formation through the process of self-organization based on efficient molecular recognition modules. Additionally, short peptides are known to self-assemble into various forms at the nano-scale, including tapes, fibrils, tubes, and spheres (Ghadiri et al. 1993; Aggeli et al. 1997; Bong et al. 2001; Hartgerink et al. 2001; Vauthey et al. 2002; Banerjee et al. 2003; Zhang 2003; Djalali et al. 2004).

15.4 The First Peptide Nanotubes

In 1993, Ghadiri and coworkers presented a pioneered work by defining a new class of organic nanotubes. They demonstrated the formation of hollow tubular nanostructures by the self-assembly of cyclic peptides designed with an even number of alternating D- and L-amino acids (Figure 15.2) (Ghadiri et al. 1993;

Hartgerink et al. 1996; Fernandez-Lopez et al. 2001). This unique architecture results in flat ring-shaped subunits that are stacked together through intermolecular hydrogen bonds in a β-sheet conformation. The closed cycle and the alternating D- and L-conformations direct the side chains outward of the ring and the backbone amides approximately perpendicular to the ring's plane. The internal diameter of the nanotubes ranges between 7 and 8 Å and can be controlled by changing the number of the amino acids in the cyclic peptide sequence. A notable feature of cyclic D,L-peptide nanotubes is the ability to alternate the external surface properties through rationally changing the amino acid side chains. The diversity in the side group of the peptide nanotubes enabled the design of transmembrane channels. Cyclic D- and L-peptides bearing appropriate hydrophobic side chains could partition into nonpolar lipid bilayers, which undergo self-assembly, then they serve as ion channels and display transport activities for potassium and sodium (Figure 15.2) (Ghadiri et al. 1994). Other potential applications include drug delivery as these structures can serve as nano-containers, and potential applications in material sciences since new composite material can be formed by the nucleation of inorganic materials onto the peptide structures (Ghadiri et al. 1994). A recent study dealing with the electronic properties of these nanotubes suggested a wide highest occupied molecular orbital–lowest unoccupied molecular orbital (HOMO-LUMO) gap for the nanotubes of interest to bioelectronic device applications. Unmodified peptides are electronically insulating as are most biomaterials made of natural amino acids. Cyclic peptides have been designed bearing four cationic 1,4,5,8-naphthalenetetracarboxylic groups (NDI). These molecules undergo redox-triggered self-assembly in aqueous solution into long peptide nanotubes that possess highly delocalized electronic states (Ashkenasy et al. 2006). Thus, the cyclic peptide assembly is used as a scaffold to promote the stacking of NDI groups and charge transfer between them and to provide ordered electronically active biomaterials with potential utility in optical and electronic devices.

15.5 Charge-Complementary and Surfactant-Like Peptide Nanostructures

Peptide-based tubular structures can also self-assemble by linear peptides based on self-complementary ionic peptides. Zhang and coworkers demonstrated a new direction for using these peptide nanostructures as a hydrogel scaffold for tissue engineering (Zhang et al. 1993; Holmes et al. 2000). The constituents of the hydrogel are amphiphilic oligopeptides that have alternatively repeating units of positively charged lysine or arginine and negatively charged aspartate and glutamate. The hydrogel scaffold, consisting of more than 99% water content, was used as a template for tissue-cell attachment, extensive neurite outgrowth, and the formation of active nerve connections (Holmes et al. 2000) (Figure 15.3c). A later work by Zhang and coworkers discovered that surfactant-like peptides undergo self-assembly to form nanotubes and nanovesicles. The peptide monomer contains 7–8

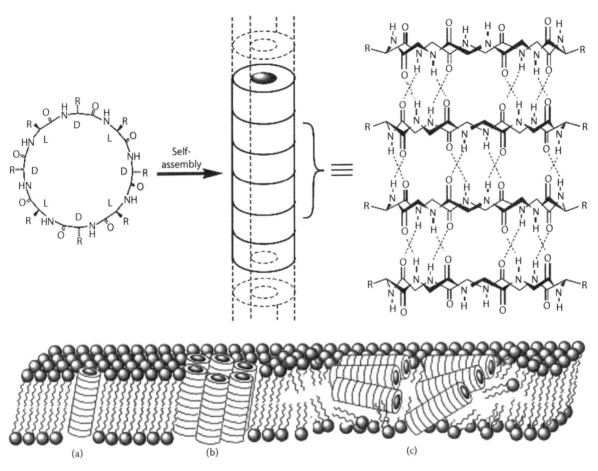

FIGURE 15.2 Cyclic peptide nanotubes described by Ghadiri and coworkers. A typical chemical structure of a cyclic peptide and schematic illustrations of the self-assembly of such peptides into nanotubes and nanotube arrays. Bottom, cyclic D,L-peptide nanotubes can display sequence-dependent modes of membrane permeation: (a) intramolecular pore; (b) barrel stave; and (c) carpet-like (cyclic peptides are depicted as ring structures). (Reprinted from Fernandez-Lopez, S. et al., *Nature*, 412, 452, 2001. With permission.)

residues and has a hydrophilic head composed of aspartic acid and a tail composed of hydrophobic amino acids, such as alanine, valine, or leucine. The surfactant-like peptides formed a network of open-ended nanotubes with remarkable size uniformity (Vauthey et al. 2002) (Figure 15.3a).

15.6 Amphiphile Peptide Nanotubes

Matsui and coworkers presented another group of linear peptides, bolaamphiphile peptides. The bolaamphiphiles have two amide head groups connected by a hydrocarbon tail group resulting in two hydrophilic heads that are conjugated through a hydrophobic linker. The driving force for the formation of these structures is, most likely, the intermolecular association of the hydrophobic moieties in the aqueous solution to form ordered structures similar to a micellization process (Banerjee et al. 2003). The self-assembly process is pH dependent; for example, the heptane bolaamphiphiles, bis (*N*-R-amido-glycylglycine)-1,7-heptane dicarboxylate, displays a sensitivity to the acidity of a solution. At pH 4, the heptane bolaamphiphile grows to a crystalline tubule in 2 weeks. At pH 8, a helical ribbon structure is

formed in 1 week (Banerjee et al. 2003). These peptide nanotubes were immobilized on Au substrates functionalized with self-assembled monolayers of 4-mercaptobenzoic acid via hydrogen bonds between the peptide nanotubes and the monolayers. Subsequently, the immobilized nanotubes were metallized by nickel via the electroless deposition coating process (Matsui et al. 2001). Moreover, Matsui and coworkers presented a new biological approach to fabricate Au nanowires by using sequenced histidine-rich peptide nanowires as templates. They used peptide nanotubes that were self-assembled from bolaamphiphile monomers, and then immobilized a histidine-rich peptide, A-H-H-A-H-H-A-A-D, on the amide group of the nanotubes by hydrogen binding. This histidine-rich peptide, the biological recognition motif, was previously reported to mineralize Au with the aid of a reducing agent. The addition of Au ions and a reducing agent led to efficient uniform gold coating on the nanotubes without contamination of gold precipitations. This is an essential feature in order to use the nanowires for the fabrication of electronic and sensor devices (Djalali et al. 2002). Further work demonstrated that similar histidine-rich peptide nanowires sequenced could serve as a template for cooper coating as well (Figure 15.4).

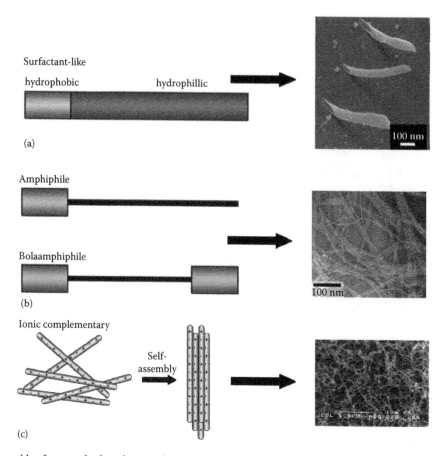

FIGURE 15.3 Self-assembly of nanoscale object by peptides and peptide-conjugates. (a) Schematic representation of surfactant like peptide and quick-freeze deep-etch TEM image of self assembled tubular structures formed by the surfactant like peptide, A6D, and V6D. Arrows point to the hollow opening at the ends. (Reproduced from Vauthey, S. et al., *Proc. Natl. Acad. Sci. U. S. A.*, 16, 5355, 2002. With permission.) (b) Schematic representation of two families of amphipile peptide conjugates, the bolaamphiphiles peptides and peptide conjugated with hydrophobic alkyl tail. On the right, TEM image of bolaamphiphile tubule negatively stained with phosphotungstic acid. (Reproduced from Hartgerink, J.D. et al., *Proc. Natl. Acad. Sci. U. S. A.*, 99, 5133, 2002. With permission.) (c) Schematic representation of ionic self-complementary peptides and scanning electron micrograph of self assembled fibrillar scaffold form by ionic self-complementary peptides. (Reproduced from Holmes, T.C. et al., *Proc. Natl. Acad. Sci. U. S. A.*, 97, 6728, 2000. With permission.)

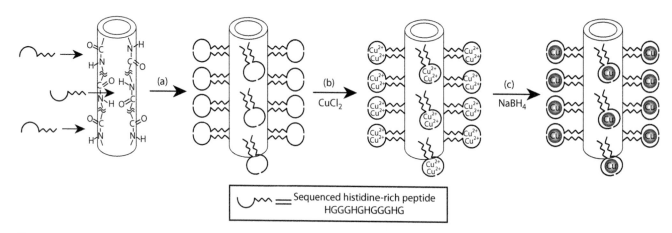

FIGURE 15.4 Cu nanocrystal growth on bolaamphiphile peptide nanotubes. Scheme of the Cu nanotube fabrication. (a) Immobilization of the sequenced histidine-rich peptide at the amide-binding sites of the template nanotubes. (b) The Cu ion-histidine-rich peptide complex on the nanotube surfaces. (c) Cu nanocrystal growth on the nanotubes nucleated at Cu ion-binding sites after reducing trapped Cu ions. (Reproduced from Banerjee, I.A. et al., *Proc. Natl. Acad. Sci. U. S. A.*, 100, 14678, 2003. With permission.)

The biological recognition of the specific sequence toward copper led to efficient and uniform copper coating on the nanotubes. Those nanotubes showed a significant change in electronic structure by varying the nanocrystal diameter, which ranges between 10 and 30 nm; therefore, this system may be developed to a conductivity-tunable building block for microelectronics and biological sensors (Banerjee 2003). In a similar procedure, gold nanocrystals were grown inside the cavities of doughnut-shaped peptide nano-assemblies. The Au nanocrystals inside the nano-doughnuts were extracted by destroying the nano-doughnuts with long UV irradiation (Djalali 2004).

A parallel molecular direction for the assembly of nanoscale peptide structures that was developed independently by Stupp and coworkers is based on the assembly of nanoscale structures from peptide amphiphiles (PAs) that are composed by the conjugation of a hydrophilic peptide chain head with a long hydrophobic alkyl tail (Figure 15.3b) (Hartgerink et al. 2001, 2002). The structure of the formed nano-fiber is a cylindrical micelle in which the PA is perpendicular to the fiber axis with the hydrophobic tail in the fiber center and the hydrophilic head at the external part of the fiber (Figure 15.5). The self-assembled peptide amphiphile fibers served as a scaffold reminiscent of an extracellular matrix. After the nanofibers were formed, they underwent different manipulations, such as cross-linking and mineralization. These were achieved by the design of PAs that presented several active groups, such as cysteine residues, which are oxidized to form intermolecular disulfide bonds. The cross-links can be reversed by the reduction of the disulfides back to free thiol groups. After cross-linking, the fibers can nucleate calcium ions resulting in the mineralization of hydroxyapatite (HA) through phosphorylated serine residue to form a composite

material. The phosphor-serine residues that were incorporated into the PAs are closely associated with the collagen extracellular matrix and are known to play an important role in HA mineralization. Therefore, following the self-assembly process, a highly phosphorylated surface was formed. The crystallographic *c* axes of HA were aligned with the long axes of the fibers. This alignment is the same as that observed between collagen fibrils and HA crystals in bone (Hartgerink et al. 2001). Moreover, the designed PAs presented the RGD peptide motif, which is part of the fibronectin, a collagen-associated protein. This sequence has been found to play an important role in integrin-mediated cell adhesion and could promote adhesion and growth of cells on the nanofibers' surface (Hartgerink et al. 2001). Similar nanofibers were also formed by peptide bolaamphiphiles. In this case, the tube's internal wall is also hydrophilic as well as the external surface (Claussen et al. 2003). A possible application of PAs is to use them as artificial three-dimensional (3D) scaffolds for cell growth since the self-assembled nanofibers form a fibrillar network and produce a gel-like solid. Furthermore, in order to encourage cell differentiation, an adequate peptide sequence was incorporated into the PAs. The pentapeptide epitope, IKVAV, which is known to promote neurite sprouting and to direct neurite growth displayed on the artificial nanofiber scaffold and induced very rapid differentiation of cells into neurons, while discouraging the development of astrocytes (Silva et al. 2004).

In another attempt to exploit the intrinsic self-assembly of peptides as an avenue to emerging materials, Aggeli et al. have designed different short peptides that self-assemble in nonaqueous solvent into long, semi-flexible, polymeric β-sheet peptide nanotapes. These systems were designed rationally to provide strong cross-strand-attractive forces between the side chains such as electrostatic, hydrophobic, or hydrogen-bonding interactions (Aggeli et al. 1997, 2001).

15.7 Amyloid Fibrils as Natural Nanoscale Supramolecular Structures and Their Nanotechnological Applications

The process of the self-assembly of proteins is widespread in nature and has a pathogenic role in the case of amyloid fibrils. This process is associated with a large number of unrelated human diseases. A partial list includes Alzheimer's disease, Type II diabetes, Prion disease, and Parkinson's disease (Harper and Lansbury 1997; Dobson 1999; Rochet and Lansbury 2000; Sipe and Cohen 2000; Wickner et al. 2000; Soto 2001; Gazit 2002a,b). All of these diseases are characterized by the transformation of soluble proteins into aggregated fibrillar deposits in different organs and tissues. At the chemical level, the amyloid fibrils are large and highly ordered self-organized structures with a clear x-ray diffraction pattern with a 4.6–4.8 Å diffraction on the meridian and an average diameter of 7–10 nm (Sunde and Blake 1998; Dobson 1999; Rochet and Lansbury 2000; Soto 2001).

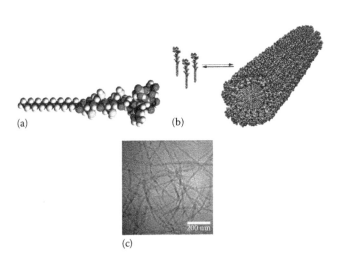

FIGURE 15.5 **(See color insert following page 20-16.)** PAs molecules form cylindrical micelle. (a) Molecular model of the PA showing the overall conical shape of the molecule going from the narrow hydrophobic tail to the bulkier peptide region. Color scheme: C, black; H, white; O, red; N, blue; P, cyan; S, yellow. (b) Schematic presentation of the self-assembly of PA molecules into a cylindrical micelle. (c) Cryo-TEM image of the fibers formed by PA molecules. (From Hartgerink, J.D. et al., *Science*, 294, 1684, 2001. With permission.)

They are rigid and organized in bundles. The secondary structure of the fibrils is also very uniform and consists mainly of cross-β-sheet elements. In a cross-section, amyloid fibrils appear as hollow cylinders or ribbons (Shirahama and Cohen 1967; Kirschner et al. 1987; Serpell et al. 1995). Hence, they have also been referred to as water-filled nanotubes by Perutz et al. (2002).

Amyloid fibrils may actually represent a much more fundamental structural state of protein, since some disease-unrelated proteins can form typical amyloid fibrils under various conditions. Furthermore, typical amyloid fibrils can be found in bacterial biofilm and other bacterial structural proteins (Chapman et al. 2002; Claessen et al. 2003; Cherny et al. 2005). Upon examination, amyloid fibrils were found to have a very strong physical rigidity (Kenney et al. 2002; Smith et al. 2006). This suggests that indeed amyloids may represent a fundamental scaffold that supports the physical structures with a nanoscale order. As will be further discussed, amyloid-derived peptides serve as major elements in peptide nanotechnology.

The use of yeast amyloid fibrils for nanotechnological applications has already been demonstrated. Their small diameter together with high stability under a wide variety of harsh physical conditions made these fibrils favorable for use as a template for conducting nanowires. The lengths of the fibril could be roughly controlled by assembly conditions in the range of 60 nm to several hundred microns. A genetically modified variant of the yeast protein that presents reactive, surface-accessible, cysteine residues was used to covalently bind gold particles. An enhancement protocol was later performed in order to get continuous 100 nm conductive silver and gold nanowire (Scheibel et al. 2003).

Amyloid fibrils are usually formed by polypeptides of 30–40 amino acids, but they can also be formed by larger proteins. Yet, recent studies have demonstrated the ability of much shorter peptides, namely, tetra- to hexapeptides, to form typical amyloid fibrils that exhibit all the typical biophysical and ultrastructural properties of amyloid fibrils (Tenidis et al. 2000; Reches et al. 2002). The change from larger polypeptides or proteins into short amyloid fragments now enables the large-scale synthesis of fibrils and their application in various nanotechnological settings. A peptide derived from the amphiphilic core Aβ(16–22) of the Alzheimer peptide can assemble into parallel β-sheets that produce bilayer structures at low pH levels. These stack on top of each other in large numbers to give rise to helical ribbons that fuse at the edges to produce highly homogeneous nanotubes (Lu et al. 2003, 2007).

15.8 Aromatic Nanostructures

Another class of peptide nanostructures is based on the use of short aromatic peptides that form well-ordered nanostructures. In the path to discover the shortest peptide fragment that can form amyloid fibrils, the diphenylalanine aromatic module, which is a peptide fragment corresponding to the core recognition motif of the Alzheimer's β-amyloid polypeptide, was examined. It was discovered that the diphenylalanine self-assembles

into ordered and discrete peptide nanotubes with a remarkable persistence length. The biocompatible and water soluble tubes are formed under mild conditions and are inexpensive and easy to manufacture (Reches and Gazit 2003). Although the aromatic dipeptide nanotubes (ADNT) are bio-inspired materials, they are remarkably stable at the presence of organic solvent and have extraordinary thermal stability properties (Adler-Abramovich et al. 2006; Sedman et al. 2006). ADNTs were suggested to be among the stiffest bio-inspired materials presently known in view of their unique mechanical strength (Kol et al. 2005; Niu et al. 2007). This is most likely due the aromatic interactions that stabilize the structures as observed with aromatic polyamides such as Kevlar®.

Nanospherical structures can self-assemble from a simpler analogue, the diphenylglycine peptide (Reches and Gazit 2004) and by the self assembly of the end termini analogue, Boc-diphenylalanine (Adler-Abramovich and Gazit 2008). Tubular structures can assemble from the noncharged aromatic dipeptide analogue Ac-Phe–Phe-NH_2, in which the N-terminal amine is acetylated and the C-terminal carboxyl is amidated, as well as the assembly of other amine-modified analogues (Reches and Gazit 2005). Other end-termini analogs, the Fmoc-diphenylalanine peptide, forms biocompatible macroscopic hydrogels with nanoscale order (Mahler et al. 2006) (Figure 15.6a).

A limiting factor in the utilization of the ADNT system was the ability to temporally control the assembly process. This was resolved by the use of a self-immolative dendritic system as a platform for the controlled assembly of peptide nanotubes that was enzymatically activated. The extremely short length of the peptide building blocks and their ability to self-assemble enable the controlled assembly applications (Adler-Abramovich et al. 2007). Various methodologies were also developed for the horizontal and vertical alignment of the ADNT. A vertically aligned nanoforest was formed by the axial unidirectional growth of a dense array of these peptide tubes (Reches and Gazit 2006a) (Figure 15.6b). Furthermore, the horizontal alignment of the tubes was achieved through noncovalent coating of the tubes with a ferrofluid solution and the application of an external magnetic field (Reches and Gazit 2006a). In addition, the alignment of the ADNT without any additional coating in an external strong magnetic field was demonstrated. The alignment was attributed to the effect of the magnetic torque associated with the diamagnetic anisotropy of the aromatic rings of phenylalanine (Hill et al. 2007). By dissolving the diphenylalanine peptide in N-methyl-2-pyrrolidone (NMP), two-dimensional (2D) ordered films of nanotubes ~1 μm in thickness have been created, comprising closely arranged spherulites of multiple nanotube bundles (Hendler et al. 2007). Recently, inkjet technology has been applied for the patterning of peptide nanostructures on nonbiological surfaces. The ADNTs were used as "ink" and patterned on transparent foil and indium tin oxide (ITO)-coated plastic surfaces by a modified commercial inkjet printer (Adler-Abramovich and Gazit 2008). ADNTs can be integrated into micro-fabricated devices using specially tailored photo-lithography methods. The propensity of ADNTs

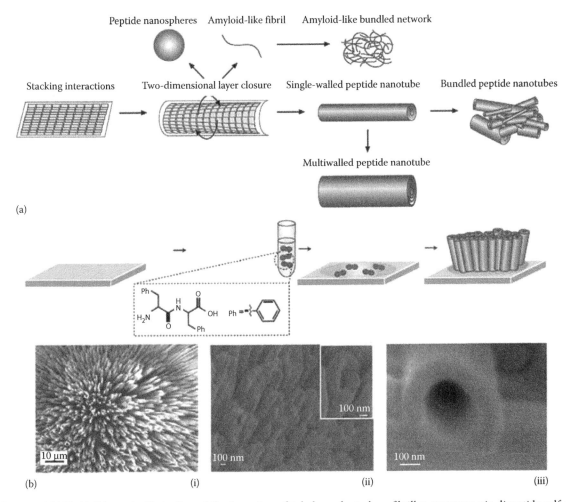

FIGURE 15.6 ADNT. (a) Schematic illustration of the formation of tubular, spherical, or fibrillar structures via dipeptide self-assembly. (Reproduced from Reches, M. and Gazit, E., *Phys. Biol.*, 3, S10, 2006b. With permission.) (b) Proposed model for the formation of aligned peptide nanotube arrays. (i) Scanning electron micrograph of the vertically aligned peptide nanotubes. (ii) Cold field-emission gun high resolution scanning electron (CFEG-HRSEM) micrograph of the nanotube arrays. (iii) High-magnification micrograph of an individual nanotube obtained by HRSEM. (Reprinted from Reches, M. and Gazit, E., *Nat. Nanotechnol.*, 1, 195, 2006a. With permission.)

to be destroyed by some of the standard photo-lithography developers, as well as their inherent inability to stick to several standard surfaces, was overcome. Two fundamental types of devices include the attaching of electrical contacts to individual ADNTs and integrating ADNTs into nano-fluidic devices were demonstrated (Sopher et al. 2007).

Several research groups have used the simple diphenylalanine peptide nanotubes. The ADNTs delivered oligonucleotides into the interior of the cells as a proof of the concept of the potential applications of the system in gene and drug delivery. This was done by using the diphenylalanine peptide properties in which diphenylalanine peptide self-assembles into nanotubes at a neutral pH level and rearranges into spherical structures upon dilution. These tubes can be absorbed by cells through endocytosis upon spontaneous conversion into vesicles (Yan et al. 2007). Moreover, the diphenylalanine peptide can self-assemble into long nanofibrils in organic solvents and entangle further to form gels. Such gels can be readily

used to encapsulate quantum dots (QDs) and gold nanoparticles through gelating the organic solution of nanocrystals. The obtained gels with the incorporated QDs display an obvious photoluminescence (PL). Encapsulation using the diphenylalanine peptide gelator provides an effective method to protect QDs from oxidation and can improve the stability of the QDs (Yan et al. 2008).

Kern and coworkers have demonstrated the possibility of fabricating extended surface-supported dipeptide architectures by the 2D co-crystallization method. The organic linker terephthalic acid (TPA) was used as a molecular "glue" to bridge isolated diphenylalanine peptide chains and improve their regularity (Wang et al. 2007). Furthermore, Kern and coworkers presented electrospinning from concentrated diphenylalanine solutions in a low-boiling point solvent, resulting in tubes that are chemically identical to self-assembled tubes, but show different morphologies, such as extreme length. The electrospinning of tubes offers more possibilities for manipulation, for example, bridging electrodes in parallel orientation; a possible patterning strategy for electrospun material (Singh et al. 2008).

15.9 Additional Applications of Peptide Nanostructures

Applications of peptide nanostructures are diverse and include unrelated fields such as tissue engineering and nanoelectronics. Several of the nanostructure applications were already mentioned in this chapter; here we highlight a few more interesting fields. One of the earlier applications of peptide nanotubes was the development of novel antibacterial agents (Ghadiri et al. 1994; Fernandez-Lopez et al. 2001). The concept underlying this use was envisioned from naturally occurring antibacterial peptides that disintegrate the bacterial membrane, for example, the insect antibacterial Cecropins that are produced by various insects including the silk moth *Bombyx mori*. Interestingly, the nanotubes were able to form nanoscale channels in the membranes of bacteria, leading to their death by osmotic collapse. The interaction of these supramolecular structures with biological membranes is highly dependent upon the amino acid composition of the D- and L-peptides and the chemical properties of the residues that are in contact with the components of the cell membrane. Peptide nanotubes formed from amphipathic cyclic peptides adopt an orientation parallel to the membrane plane, where the hydrophobic side chains are inserted into the lipidic components of the membrane and the hydrophilic residues remain exposed to the hydrophilic components of the cell membrane. In this format, peptide nanotubes are believed to permeate membranes through a carpet-like mechanism, collapse transmembrane potential and/or gradient, and cause rapid cell death (Figure 15.2) (Fernandez-Lopez et al. 2001; Dartois et al. 2005).

Peptide nanotubes were used to fabricate 20 nm silver nanowires by their use as a degradable casting mold at the nano-scale (Reches and Gazit 2003). In this study, silver ions were reduced to metallic silver in the lumen of the tube and the peptide template was removed by enzymatic degradation. Moreover, the nanotubes were used as a template for the formation of coaxial nano-cables (Carny et al. 2006). As mentioned previously, bolaamphiphiles peptide nanotubes were coated with various metals (Djalali et al. 2002, 2004; Banerjee et al. 2003) as well as gold binding to amyloid fibrils (Scheibel et al. 2003). These fabrications may have applications in molecular electronics, such as small nano-wires, which could not be made by conventional lithography.

Another application of nano-order peptide structures, from quite a different angle, is in the field of neurological regeneration. Proteins and peptides can form unique materials at macroscopic as well as nanoscopic scales. An important group of materials are nanoscale ordered hydrogels (Holmes et al. 2000; Kisiday et al. 2002; Schneider et al. 2002; Aggeli et al. 2003). These hydrogels are of great interest as a class of materials for tissue engineering and regeneration as they offer 3D scaffolds to support the growth of cultured cells and support neural growth in damaged optic nerves, leading to the recovery of visual functions in model animals (Holmes et al. 2000; Kisiday et al. 2002; Schneider et al. 2002; Silva et al. 2004; Almany and Seliktar 2005; Ellis-Behnke et al. 2006). Conventionally, hydrogels are made of either synthetic building blocks including poly-(ethylene oxide) and poly-(vinyl alcohol), or natural building blocks including agarose, collagen, fibrin, alginate, gelatin, and hyaluronic acid (HA) (Almany and Seliktar 2005). Peptide-based hydrogels have various advantages: they are biocompatible, easy to manufacture in large quantities, and can also be easily decorated chemically and biologically. Such decoration gives the ability to design an ultrastructure that presents ligands, as well as other functional groups, hence promoting cell adhesion and growth (Holmes et al. 2000; Silva et al. 2004). Two examples of hydrogel-based peptides were discussed above: the self-assembled peptide-amphiphile fibrous scaffold (Hartgerink et al. 2001, 2002) and a hydrogel-based on self-complementary ionic peptides (Zhang et al. 1993; Holmes et al. 2000). Another study by Xu and coworkers was the first to report the formation of fibrous scaffolds with fluorenylmethoxycarbonyl (Fmoc)-protected amino acids and dipeptides (Yang et al. 2004). This Fmoc group, which is widely used as a protecting group in peptide synthesis, also showed remarkably anti-inflammatory properties for a number of Fmoc-amino acids (Burch et al. 1991). Based on these results, Ulijn and coworkers studied Fmoc dipeptides composed of the combination of diverse amino acids, thus covering a range of hydrophobicities. They demonstrated the spontaneous assembly, under physiological conditions, of Fmoc dipeptides into fibrous hydrogels, while the architecture and physical properties of these gels were found to be dictated by the nature of the amino acid sequence (Toledano et al. 2006). Furthermore, the time spatial control of the self-assembly process was demonstrated by selectively connecting non gelling Fmoc amino acids and dipeptides to form amphiphilic Fmoc tripeptides. This unique connection was achieved by providing the proteases, enzymes that normally hydrolyze peptide bonds, with an appropriate environment to favor thermodynamically the reverse direction to form reversed hydrolysis i.e., peptide synthesis (Toledano et al. 2006).

Peptide nanostructures were also shown to have applications in the field of diagnosis and biosensors. Peptide nanostructures were modified with antibodies to allow highly sensitive electrical detection of viruses with an extremely low detection limit (de la Rica et al. 2008). Other directions in the field of biosensors involve the modification of electrodes with native peptide nanostructures or with enzyme-modified ones to significantly increase the sensitivity of these devices (Yemini et al. 2005).

Self-assembled peptide structures could be used for various biomaterials applications. One direction is the formation of macroscopic fibrils with nanoscale order (Mitraki et al. 2006). The order and self-assembly process are important for the ease of fabrication while the biological nature is a key for biocompatibility. Such fibrils could be based on natural proteins or peptide fragments that could be useful for large-scale production. Among many directions, the fibers can be used for the fabrication of bandages, medical fabric that can allow the slow release of various drugs, and degradable medical materials.

Taken together, the peptide nanostructure study conducted in the last two decades has revealed a novel direction for their utilizations. Here, we presented only a brief overview of the selected applications of peptide nanostructures. In the years to come, we expect many more applications in the fields of micro-electromechanical systems, electronics, diagnostics, and medicine. It is important to keep in mind that most of these building blocks were developed in an academic setting and their translation into practical application may take a few more years.

References

Adler-Abramovich, L. and Gazit, E. 2008. Controlled patterning of peptide nanotubes and nanospheres using inkjet printing technology. *J. Pept. Sci.* 14: 217–223.

Adler-Abramovich, L., Perry, R., Sagi, A., Gazit, E., and Shabat, D. 2007. Controlled assembly of peptide nanotubes triggered by enzymatic activation of self-immolative dendrimers. *ChemBioChem* 8: 859–862.

Adler-Abramovich, L., Reches, M., Sedman, V. L. et al. 2006. Thermal and chemical stability of diphenylalanine peptide nanotubes: Implications for nanotechnological applications. *Langmuir* 22: 1313–1320.

Aggeli, A., Bell, M., Boden, N., Carrick, L. M., and Strong, A. E. 2003. Self-assembling peptide polyelectrolyte beta-sheet complexes form nematic hydrogels. *Angew. Chem. Int. Ed.* 42: 5603–5606.

Aggeli, A., Bell, M., Boden, N. et al. 1997. Responsive gels formed by the spontaneous self-assembly of peptides into polymeric beta-sheet tapes. *Nature* 386: 259–262.

Aggeli, A., Nyrkova, I. A., Bell, M. et al. 2001. Hierarchical self-assembly of chiral rod-like molecules as a model for peptide beta-sheet tapes, ribbons, fibrils, and fibers. *Proc. Natl. Acad. Sci. U. S. A.* 98: 11857–11862.

Almany, L. and Seliktar, D. 2005. Biosynthetic hydrogel scaffolds made from fibrinogen and polyethylene glycol for 3d cell cultures. *Biomaterials* 26: 2467–2477.

Ashkenasy, N., Horne, S., and Ghadiri, M.R. 2006. Design of self-assembling peptide nanotubes with delocalized electronic states. *Small* 2: 99–102.

Banerjee, I. A., Yu, L., and Matsui, H. 2003. Cu nanocrystal growth on peptide nanotubes by biomineralization: Size control of cu nanocrystals by tuning peptide conformation. *Proc. Natl. Acad. Sci. U. S. A.* 100: 14678–14682.

Bong, D. T., Clark, T. D., Granja, J. R., and Ghadiri, M. R. 2001. Self-assembling organic nanotubes. *Angew. Chem. Int. Ed.* 40: 988–1011.

Burch, R. M., Weitzberg, M., Blok, N., Muhlhauser, R., Martin, D., Farmer, S. G., Bator, J. M., Connor, J. R., Green, M., and Ko, C. 1991. N-(fluorenyl-9-methoxycarbonyl) amino acids, a class of antiinflammatory agents with a different mechanism of action. *Proc. Natl. Acad. Sci. U. S. A.* 88: 355–359.

Carny, O., Shalev, D. E., and Gazit, E. 2006. Fabrication of coaxial metal nanocables using a self-assembled peptide nanotube scaffold. *Nano Lett.* 6: 1594–1597.

Chapman, M. R., Robinson, L. S., Pinkner, J. S. et al. 2002. Role of *Escherichia coli* curli operons in directing amyloid fiber formation. *Science* 295: 851–855.

Cherny, I., Rockah, L., Levy-Nissenbaum, O. et al. 2005. The formation of *Escherichia coli* curli amyloid fibrils is mediated by prion-like peptide repeats. *J. Mol. Biol.* 352: 245–252.

Claessen, D., Rink, R., de Jong, W. et al. 2003. A novel class of secreted hydrophobic proteins is involved in aerial hyphae formation in *Streptomyces coelicolor* by forming amyloid-like fibrils. *Genes Dev.* 17: 1714–1726.

Claussen, R. C., Rabatic, B. M., and Stupp, S. I. 2003. Aqueous self-assembly of unsymmetric peptide bolaamphiphiles into nanofibers with hydrophilic cores and surfaces. *J. Am. Chem. Soc.* 125: 12680–12681.

Dartois, V., Sanchez-Quesada, J., Cabezas, E., Chi, E., Dubbelde, C., Dunn, C., Granja, J., Gritzen, C., Weinberger, D., Ghadiri, R., and Parr, T. R. 2005. Systemic antibacterial activity of novel synthetic cyclic peptides. *Antimicrob. Agents Chemother.* 49: 3302–3310.

de la Rica, R., Mendoza, E., Lechuga, L. M., and Matsui, H. 2008. Label-free pathogen detection with sensor chips assembled from peptide nanotubes. *Angew. Chem. Int. Ed.* 47: 9752–9755.

Djalali, R., Chen, Y. F., and Matsui, H. 2002. Au nanowire fabrication from sequenced histidine-rich peptide. *J. Am. Chem. Soc.* 124: 13660–13661.

Djalali, R., Samson, J., and Matsui, H. 2004. Doughnut-shaped peptide nano-assemblies and their applications as nanoreactors. *J. Am. Chem. Soc.* 126: 7935–7939.

Dobson, C. M. 1999. Protein misfolding, evolution and disease. *Trends Biochem. Sci.* 24: 329–332.

Ellis-Behnke, R. G., Liang, Y. X., You, S. W. et al. 2006. Nano neuro knitting: Peptide nanofiber scaffold for brain repair and axon regeneration with functional return of vision. *Proc. Natl. Acad. Sci. U. S. A.* 103: 5054–5059.

Fernandez-Lopez, S., Kim, H. S., Choi, E. C. et al. 2001. Antibacterial agents based on the cyclic D, L-α-peptide architecture. *Nature* 412: 452–455.

Gazit, E. 2002a. The correctly folded state of proteins: Is it a metastable state? *Angew. Chem. Int. Ed.* 41: 257–259.

Gazit, E. 2002b. A possible role for π-stacking in the self-assembly of amyloid fibrils. *FASEB J.* 16: 77–83.

Gazit, E. 2007. Self-assembled peptide nanostructures: The design of molecular building blocks and their technological utilization. *Chem. Soc. Rev.* 36: 1263–1269.

Ghadiri, M. R., Granja, J. R., and Buehler, L. K. 1994. Artificial transmembrane ion channels from self-assembling peptide nanotubes. *Nature* 369: 301–304.

Ghadiri, M. R., Granja, J. R., Milligan, R. A., McRee, D. E., and Hazanovich, N. 1993. Self-assembling organic nanotubes based on a cyclic peptide architecture. *Nature* 366: 324–327.

Harper, J. D. and Lansbury, P. T. 1997. Models of amyloid seeding in alzheimer's disease and scrapie: Mechanistic truths and physiological consequences of the time-dependent solubility of amyloid proteins. *Annu. Rev. Biochem.* 66: 385–407.

Hartgerink, J. D., Beniash, E., and Stupp, S. I., 2001. Self-assembly and mineralization of peptideamphiphile nanofibers. *Science* 294: 1684–1688.

Hartgerink, J. D., Beniash, E., and Stupp, S. I. 2002. Peptide-amphiphile nanofibers: A versatile scaffold for the preparation of self-assembling materials. *Proc. Natl. Acad. Sci. U. S. A.* 99(8): 5133–5138.

Hartgerink, J. D., Granja, J. R., Milligan, R. A., and Ghadiri, M. R. 1996. Self-assembling peptide nanotubes. *J. Am. Chem. Soc.* 118: 43–50.

Hendler, N., Sidelman, N., Reches, M. et al. 2007. Formation of well-organized self-assembled films from peptide nanotubes. *Adv. Mater.* 19: 1485–1488.

Hill, R. J. A., Sedman, V. L., Allen, S. et al. 2007. Alignment of aromatic peptide tubes in strong magnetic fields. *Adv. Mater.* 19: 4474–4479.

Holmes, T. C., de Lacalle, S., Su, X., Liu, G. S., Rich, A., and Zhang, S. 2000. Extensive neurite outgrowth and active synapse formation on self-assembling peptide scaffolds. *Proc. Natl. Acad. Sci. U. S. A.* 97: 6728–6733.

Kenney, J. M., Knight, D., Wise, M. J., and Vollrath, F. 2002. Amyloidogenic nature of spider silk. *Eur. J. Biochem.* 269: 4159–4163.

Kirschner, D. A., Inouye, H., Duffy, L., Sinclair, A., Lind, M., and Sekoe, D. A. 1987. Synthetic peptide homologous to beta protein from Alzheimer disease forms amyloid-like fibrils in vitro. *Proc. Natl. Acad. Sci. U. S. A.* 84: 6953–6957.

Kisiday, J., Jin, M., Kurz, B. et al. 2002. Self-assembling peptide hydrogel fosters chondrocyte extracellular matrix production and cell division: Implications for cartilage tissue repair. *Proc. Natl. Acad. Sci. U. S. A.* 99: 9996–10001.

Kol, N., Adler-Abramovich, L., Barlam, D. et al. 2005. Self-assembled peptide nanotubes are uniquely rigid bioinspired supramolecular structures. *Nano Lett.* 5: 1343–1346.

Lu, K., Guo, L., Mehta, A. K. et al. 2007. Macroscale assembly of peptide nanotubes. *Chem. Commun.* 2729–2731.

Lu, K., Jacob, J., Thiyagarajan, P., Conticello, V. P., and Lynn, D. G. 2003. Exploiting amyloid fibril lamination for nanotube self-assembly. *J. Am. Chem. Soc.* 125: 6391–6393.

Mahler, A., Reches, M., Rechter, M., Cohen, S., and Gazit, E. 2006. Rigid, self-assembled hydrogel composed of a modified aromatic dipeptide. *Adv. Mater.* 18: 1365–1370.

Matsui, H., Gologan, B., Pan, S., and Douberly Jr, G. E. 2001. Controlled immobilization of peptide nanotube-templated metallic wires on au surfaces. *Eur. Phys. J. D* 16: 403–406.

Mitraki, A., Papanikolopoulou, K., Van Raaij, M. J. et al. 2006. Natural triple b-stranded fibrous folds. *Adv. Protein Chem.* 73: 97–124.

Naik, R. R., Brott, L. L., Clarson, S. J., and Stone, O. M. 2002. Silica-precipitating peptides isolated from a combinatorial phage display peptide library. *J. Nanosci. Nanotechnol.* 2: 95–100.

Niu, L., Chen, X., Allen, S., and Tendler, S. J. B. 2007. Using the bending beam model to estimate the elasticity of diphenylalanine nanotubes. *Langmuir* 23: 7443–7446.

Perutz, M. F., Finch, J. T., Berriman, J., and Lesk, A. 2002. Amyloid fibers are water-filled nanotubes. *Proc. Natl. Acad. Sci. U. S. A.* 99: 5591–5595.

Reches, M. and Gazit, E. 2003. Casting metal nanowires within discrete self-assembled peptide nanotubes. *Science* 300: 625–627.

Reches, M. and Gazit, E. 2004. Formation of closed-cage nanostructures by self-assembly of aromatic dipeptides. *Nano Lett.* 4: 581–585.

Reches, M. and Gazit, E. 2005. Self-assembly of peptide nanotubes and amyloid-like structures by charged-termini capped diphenylalanine peptide analogues. *Israel J. Chem.* 45: 363–371.

Reches, M. and Gazit, E. 2006a. Controlled patterning of aligned self-assembled peptide nanotubes. *Nat. Nanotechnol.* 1: 195–200.

Reches, M. and Gazit, E. 2006b. Designed aromatic homo-dipeptides: Formation of ordered nanostructures and potential nanotechnological applications. *Phys. Biol.* 3: S10–S19.

Reches, M., Porat, Y., and Gazit, E. 2002. Amyloid fibril formation by pentapeptide and tetrapeptide fragments of human calcitonin. *J. Biol. Chem.* 277: 35475–35480.

Rochet, J. C. and Lansbury, P. T. Jr. 2000. Amyloid fibrillogenesis: Themes and variations. *Curr. Opin. Struct. Biol.* 10: 60–68.

Scheibel, T., Parthasarathy, R., Sawicki, G., Lin, X., Jaeger, H., and Lindquist, S. L. 2003. Conducting nanowires built by controlled self-assembly of amyloid fibers and selective metal deposition. *Proc. Natl. Acad. Sci. U. S. A.* 100: 4527–4532.

Schneider, J. P., Pochan, D. J., Ozbas, B. et al. 2002. Responsive hydrogels from the intramolecular folding and self-assembly of a designed peptide. *J. Am. Chem. Soc.* 124: 15030–15037.

Sedman, V. L., Adler-Abramovich, L., Allen, S., Gazit, E., and Tendler, S. J. 2006. Direct observation of the release of phenylalanine from diphenylalanine nanotubes. *J. Am. Chem. Soc.* 128: 6903–6908.

Serpell, L. C., Sunde, M., Fraser, P. E., Luther, P. K., Morris, E. P., Sangren, O., Lundgren, E., and Blake, C. C. 1995. Examination of the structure of the transthyretin amyloid fibril by image reconstruction from electron micrographs. *J. Mol. Biol.* 254: 113–118.

Shirahama, T. and Cohen, A. S. 1967. High resolution electron microscopic analysis of the amyloid fibrils. *J. Cell. Biol.* 33: 679–706.

Silva, G. A., Czeisler, C., Niece, K. L. et al. 2004. Selective differentiation of neural progenitor cells by high-epitope density nanofibers. *Science* 303: 1352–1355.

Singh, G., Bittner, A. M., Loscher, S., Malinowski, N., and Kern, K., 2008. Electrospinning of diphenylalanine nanotubes. *Adv. Mater.* 20: 2332–2336.

Sipe, J. D. and Cohen, A. S. 2000. Review: History of the amyloid fibril. *J. Struct. Biol.* 130: 88–98.

Smith, J. F., Knowles, T. P. J., Dobson, C. M., MacPhee, C. E., and Welland, M. E. 2006. Characterization of the nanoscale properties of individual amyloid fibrils. *Proc. Natl. Acad. Sci. U. S. A.* 103: 15806–15811.

Sopher, N. B., Abrams, Z. R., Reches, M., Gazit, E., and Hanein, Y. 2007. Integrating peptide nanotubes in micro-fabrication processes. *J. Micromech. Microeng.* 17: 2360–2365.

Soto, C. 2001. Protein misfolding and diseases; protein refolding and therapy. *FEBS lett.* 498: 204–207.

Sunde, M. and Blake, C. C. 1998. From the globular to the fibrous state: Protein structure and structural conversion in amyloid formation. *Q. Rev. Biophys.* 31: 1–39.

Tenidis, K., Waldner, M., Bernhagen, J. et al. 2000. Identification of a penta- and hexapeptide of islet amyloid polypeptide (IAPP) with amyloidogenic and cytotoxic properties. *J. Mol. Biol.* 295: 1055–1071.

Toledano, S., Williams, R. J., Jayawarna, V., and Ulijn, R. V. 2006. Enzyme-triggered self-assembly of peptide hydrogels via reversed hydrolysis. *J. Am. Chem. Soc.* 128: 1070–1071.

Vauthey, S., Santoso, S., Gong, H., Watson, N., and Zhang, S. 2002. Molecular self-assembly of surfactant-like peptides to form nanotubes and nanovesicles. *Proc. Natl. Acad. Sci. U. S. A.* 16: 5355–5360.

Wang, Y., Lingenfelder, M., Classen, T., Costantini, G., and Kern, K. 2007. Ordering of dipeptide chains on cu surfaces through 2d cocrystallization. *J. Am. Chem. Soc.* 129: 15742–15743.

Whitesides, G. M., Mathias, J. P., and Seto, C. T. 1991. Molecular self-assembly and nanochemistry: A chemical strategy for the synthesis of nanostructures. *Science* 254: 1312–1319.

Wickner, R. B., Taylor, K. L., Edskes, H. K. et al. 2000. Prions of yeast as heritable amyloidoses. *J. Struct. Biol.* 130: 310–322.

Yan, X., Cui, Y., He, Q., Wang, K., and Li, J. 2008. Organogels based on self-assembly of diphenylalanine peptide and their application to immobilize quantum dots. *Chem. Mater.* 20: 1522–1526.

Yan, X., He, Q., Wang, K., Duan, L., Cui, Y., and Li, J. 2007. Transition of cationic dipeptide nanotubes into vesicles and oligonucleotide delivery. *Angew. Chem. Int. Ed.* 46: 2431–2434.

Yang, Z. M., Gu, H. W., Zhang, Y., Wang, L., and Xu, B. 2004. Small molecule hydrogels based on a class of antiinflammatory agents. *Chem. Commun.* 2: 208–209.

Yemini, M., Reches, M., Rishpon, J., and Gazit, E. 2005. Novel electrochemical biosensing platform using self-assembled peptide nanotubes. *Nano Lett.* 5: 183–186.

Zhang, S. 2003. Fabrication of novel biomaterials through molecular self-assembly. *Nat. Biotech.* 21: 1171–1178.

Zhang, S., Holmes, T., Lockshin, C., and Rich, A. 1993. Spontaneous assembly of a self-complementary oligopeptide to form a stable macroscopic membrane. *Proc. Natl. Acad. Sci. U. S. A.* 90.

III

Types of Nanowires

16

Germanium Nanowires

Sanjay V. Khare
The University of Toledo

Sunil Kumar R. Patil
The University of Toledo

Suneel Kodambaka
University of California

16.1 Introduction

In a nanowire (NW), the momentum of an electron is confined in two directions, thus allowing for electron motion only in one direction (a NW is a one degree of freedom structure and is often called a one-dimensional nanostructure if its diameter is less than 100 nm). This reduction in dimensionality results in dramatic quantum effects dependent on wire material, axis orientation, length, and diameter. These quantum effects in NWs change the electrical, chemical, and mechanical properties to name but a few. Thus, NWs exhibit properties and applications very different from their bulk form. Therefore, they have been assiduously studied recently by experimentalists and theorists for their potential applications in electronic devices and sensors. Investigations for a thorough theoretical understanding of the structure–property relationship for many NWs and NW devices are currently in progress across scientific and engineering disciplines. This research is being carried out on the theoretical side by a multi-scale approach to the atomic simulation of these materials. On the experimental side, a careful study of growth, characterization, and device assembly is ongoing. This chapter gives a brief introduction to this twofold approach. Ample references for further study are also provided. This chapter is divided into five sections. Section 16.2 briefly reviews an elementary NW model derived from the steady-state (time-independent) Schrödinger equation. The aim of this section is to show the profound effect dimensionality has on the electronic properties of a nanostructure. Section 16.3 reviews some Ge NW growth experiments and measurements of Ge NW properties. The electrical, optical, mechanical, and surface properties of Ge NW along with their applications are dealt with briefly in Section 16.4. Section

16.5 introduces some of the simulation methods used to study NWs and describes their capabilities and limits and Section 16.6 forms the conclusion.

16.2 Conceptual Theoretical Model for Nanowires

In this section, we will work with a fundamental equation of quantum mechanics—the time-independent Schrödinger equation (TISE), for the energy eignenfunction $\psi(x, y, z)$ of the electron and its solutions in different dimensions. Effects of these solutions on the electronic properties of NWs will be elucidated. This section is heavily influenced by the presentations in the articles by Harrison (Harrison 1999) and Chen (Chen and Shakouri 2002), which are valuable for exploring this topic further. The three-dimensional (3D) TISE for a constant effective mass is

$$-\frac{\hbar^2}{2m^\star}\nabla^2\psi(x, y, z) + V(x, y, z)\psi(x, y, z) = E\,\psi(x, y, z)$$

$$\text{with } \hbar \equiv h/(2\pi), \tag{16.1}$$

where

h is the universal Planck's constant
E is the energy
$m^\star$ is the effective mass of the electron

Assuming in the plane dispersion and axis of the wire along the x axis that the total potential $V(x, y, z)$ can be written as a sum of a two-dimensional (2D) confinement potential and a potential along the axis of the wire

$$V(x, y, z) = V(x) + V(y, z), \quad (16.2)$$

and eigenfunction can be written as

$$\psi(x, y, z) = \psi(x)\psi(y, z). \quad (16.3)$$

By substituting Equations 16.2 and 16.3 in Equation 16.1, we get

$$-\frac{\hbar^2}{2m^*}\left[\psi(y, z)\frac{\partial^2\psi(x)}{\partial x^2}\psi(x)\frac{\partial^2\psi(y, z)}{\partial y^2} + \psi(x)\frac{\partial^2\psi(y, z)}{\partial z^2}\right]$$
$$+ \psi(x)V(y, z)\psi(y, z) + \psi(x)V(x)\psi(y, z)$$
$$= \psi(x)(E_x + E_{y,z})\,\psi(y, z). \quad (16.4)$$

Associating the kinetic and potential energies on the left-hand side of Equation 16.4 to the corresponding energies on the right-hand side and realizing that $V(x) = 0$ (meaning there is no potential gradient along the wire axis), yields two decoupled equations:

$$-\frac{\hbar^2}{2m^*}\frac{\partial^2\psi(x)}{\partial x^2} = E_x\psi(x), \quad (16.5)$$

and

$$-\frac{\hbar^2}{2m^*}\left[\frac{\partial^2\psi(y, z)}{\partial y^2} + \frac{\partial^2\psi(y, z)}{\partial z^2}\right] + V(y, z)\psi(y, z) = E_{y,z}\psi(y, z). \quad (16.6)$$

The solution to Equation 16.5 is

$$E_x = \frac{\hbar^2 k_x^2}{2m^*}. \quad (16.7)$$

Here k_x is any real number.

Equation 16.6 is the Schrodinger equation for a 2D confinement potential in NWs. The solution to this for different wire shapes is more involved but could be simplified by making some assumptions. Assuming an infinitely deep rectangular NW as shown in Figure 16.1, and taking the potential inside the wire to be zero and the potential outside to be infinity changes Equation 16.6 inside the wire to

$$-\frac{\hbar^2}{2m^*}\left[\frac{\partial^2\psi(y, z)}{\partial y^2} + \frac{\partial^2\psi(y, z)}{\partial z^2}\right] = E_{y,z}\psi(y, z). \quad (16.8)$$

Outside the wire, the eigenfunction $\psi(x, y, z) = 0$, since the potential is infinite. Using separation of variables

$$\psi(y, z) = \psi(y)\psi(z), \quad (16.9)$$

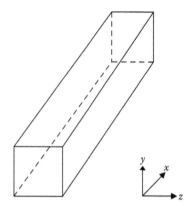

FIGURE 16.1 Infinitely long rectangular nanowire with the potential inside taken to be $V = 0$ and outside as infinity. (Adapted from Harrison, P., *Quantum Wells, Wires and Dots*, John Wiley & Sons, West Sussex, U.K., 1999.)

and substituting Equation 16.9 in Equation 16.8 and assigning the proper energies along the different axes, yields two decoupled equations again:

$$-\frac{\hbar^2}{2m^*}\frac{\partial^2\psi(y)}{\partial y^2} = E_y\psi(y), \quad (16.10)$$

$$-\frac{\hbar^2}{2m^*}\frac{\partial^2\psi(z)}{\partial z^2} = E_y\psi(z). \quad (16.11)$$

The solutions for Equations 16.10 and 16.11 with the origin at a corner and the wire dimensions shown in Figure 16.1 is

$$\psi(y) = \sqrt{\frac{2}{L_y}}\sin\left(\frac{\pi n_y y}{L_y}\right), \quad \psi(z) = \sqrt{\frac{2}{L_z}}\sin\left(\frac{\pi n_z z}{L_z}\right), \quad (16.12\text{a})$$

where n_x and n_y are restricted to be positive integers with components of energy as

$$E_y = \frac{\hbar^2\pi^2 n_y^2}{2m^* L_y^2}, \quad E_z = \frac{\hbar^2\pi^2 n_z^2}{2m^* L_z^2}. \quad (16.12\text{b})$$

The condition that the eigenfunction be zero at the boundary of the wire leads to the quantization of these energy values through the integers n_x and n_y. So spatial charge density distributions that are proportional to $|\psi(y, z)|^2$ are described by two principal quantum numbers, n_y and n_z, as seen in Equation 16.12a. Thus, the confinement energy decreases as the size of the NW increases, due to the presence of the L terms in the denominators in Equation 16.12b. We emphasize that these energies are quantized as a consequence of the integer values n_y and n_z.

16.2.1 Density of States

The electronic density of states (DOS) $D(E)$ is the number of states per unit of energy per unit of volume of real space and is an important quantity. This quantity primarily determines many physical and chemical properties of materials. It gives us the number of possible energy states, like the ones shown in Equation 16.12, which are allowed to be occupied by electrons. Thus

$$D(E) = \frac{dN}{dE}. \tag{16.13}$$

N is sometimes called the cumulative or total density of states upto an energy E. For bulk material, the three degrees of freedom for electron momentum maps out a sphere in k-space. In quantum wells with two degrees of freedom, the electron momenta fill successively larger circles. For a NW with just one degree of freedom, the electron momenta then fill states along a line. So in one dimension (1D), the total number of states, N^{1D}, is equal to the ratio of the length of the line in k-space ($2k$, for positive and negative values of k) to the length occupied by one state divided by the length in real space, which is L. The factor 2 in the equation accounts for spin degeneracy (i.e., the up and down spin allowed for each electronic state).

$$N^{1D} = 2(2k)\left(\frac{L}{2\pi}\right)\left(\frac{1}{L}\right) = \frac{2k}{\pi}. \tag{16.14}$$

The density of states for NWs can then be derived as

$$D(E)^{1D} = \frac{dN^{1D}}{dE} = \frac{dN^{1D}}{dk}\frac{dk}{dE}. \tag{16.15}$$

From Equation 16.7,

$$\frac{dk}{dE} = \left(\frac{2m^{\star}}{\hbar^2}\right)^{\frac{1}{2}}\frac{E^{\frac{-1}{2}}}{2}. \tag{16.16}$$

Differentiating Equation 16.14 with respect to k and substituting the result and Equation 16.16 in Equation 16.15 yields

$$D(E)^{1D} = \left(\frac{2m^{\star}}{\hbar^2}\right)^{\frac{1}{2}}\frac{E^{\frac{-1}{2}}}{\pi}.$$

Similar calculations may be carried out in 0 dimension (0D; i.e., for a quantum dot), 2D, and 3D. Table 16.1 shows the dependence

TABLE 16.1 Density of States for 1D, 2D, and 3D Nanostructures

Dimensionality	Density of States, $D(E)$
3D	$\left(\dfrac{2m^{\star}}{\hbar^2}\right)^{\frac{3}{2}}\dfrac{E^{\frac{1}{2}}}{2\pi^2}$
2D	$\left(\dfrac{2m^{\star}}{\hbar^2}\right)^{1}\dfrac{E^0}{2\pi}$
1D	$\left(\dfrac{2m^{\star}}{\hbar^2}\right)^{\frac{1}{2}}\dfrac{E^{\frac{-1}{2}}}{\pi}$

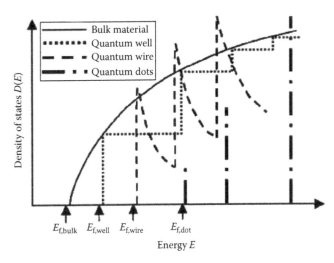

FIGURE 16.2 Density of states of different nanostructures compared to bulk material. (Adapted from Chen, G. and Shakouri, A., *J. Heat Transfer T ASME*, 124, 242, 2002.)

on energy of the DOS for different dimensionalities. Similar data showing how the density of states changes from bulk (3D), to 2D, 1D, and 0D nanostructures are shown in Figure 16.2. The profound effect of dimensionality is clearly seen in Figure 16.2 and is one of the main reasons for the interest in nanostructures in general and nanowires in particular for electronic and optical device applications. While the DOS is continuous for 3D structures, for 1D or NWs it has sharp spikes at different band edges. This leads to strong signals in many optical and electronic measurements at these energies. Similar effects influence the chemical bonding and mechanical properties of NWs due to the confinement of electrons in 2D.

Having briefly explored one of the motivations for NW applications, we now turn our attention to their synthesis in a laboratory. We now focus our attention mostly on germanium NWs for the remainder of this chapter.

16.3 Growth of Ge Nanowires

NWs are most commonly grown via the vapor–liquid–solid (VLS) process (Wagner and Ellis 1964). The VLS process, first proposed in the early 1960s for the growth of Si wires using Au (Wagner et al. 1964b, Wagner and Doherty 1966, Wagner 1967a,b), involves the dissociative adsorption of material from the vapor phase via a low-melting alloy liquid as a catalyst. The preferential incorporation of the material at the solid substrate-catalyst liquid interface leads to growth in the form of cylindrical pillars or "wires." A schematic of the wire growth process is shown in Figure 16.3. In this process, the wire length is determined by the growth flux and the deposition time while the wire diameter is controlled by the catalyst size. Using VLS, wires as long as a few millimeters and as narrow as a few nanometers wide, have been grown (Cui et al. 2001, Park et al. 2008). To date, a wide variety of materials (elemental metals, semiconductors as well as compound arsenides, borides, carbides, nitrides, oxides,

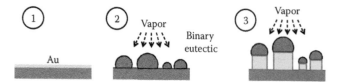

FIGURE 16.3 Schematic of vapor–liquid–solid process. Au is shown in panel (1), half-circles in panel (2), and (3) at the top of the wires are Au catalysts. Bottom region is the substrate. (Adapted from Kodambaka, S., Tersoff, J., Ross, M.C., and Ross, F.M., *Proc. SPIE*, 7224, 72240C, 2009.)

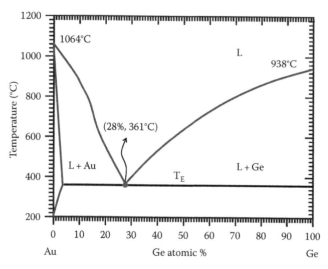

FIGURE 16.4 Au-Ge phase diagram. (Adapted from Okamoto, H. and Massalski, T.B., *Binary Alloy Phase Diagrams* I, ASM International, Gaithersburg, MD, 1986.)

phosphides, selenides, sulphides, and tellurides) have been successfully grown in the form of whiskers or NWs using this technique (Givargizov 1978). Readers are encouraged to refer to the article by Givargizov (1987) for an excellent review of the entire list of materials grown, methods employed, catalysts, and growth parameters used.

Interest in the semiconducting NWs stems from their potential for applications in a wide variety of areas ranging from opto- and nano-electronics to chemical and biological sensors (Haraguchi et al. 1992, Duan et al. 2000, Cui and Liueber 2001, Bjork et al. 2002, Gudiksen et al. 2002, DeFranceschi et al. 2003). More recently, Si and Ge NWs have attracted considerable attention owing to their significantly enhanced energy storage and energy generation properties (Tian et al. 2007, Boukai et al. 2008, Chan et al. 2008a,b, Hochbaum et al. 2008). Here, we focus on the Au-catalyzed growth of Ge NWs and the influence of growth parameters such as substrate temperature, source flux, and catalyst state (solid or liquid) on the morphological and structural evolution of the NWs.

The growth of Ge using Au was first reported four decades ago (Wagner et al. 1964a) and more recently using other metals such as Ti (Wan et al. 2003), Ni (Tuan et al. 2005), and Mn (Lensch-Falk et al. 2007). It is interesting to note that the Au-catalyzed VLS growth of Si occurs at temperatures T that are above the bulk Au–Si eutectic temperature $T_E = 363°C$ (Okamoto and Massalski 1986) while Au-catalyzed Ge wire growth has been reported to occur at T as low as 260°C (Miyamoto and Hirata 1975, Wang and Dai 2002, Kamins et al. 2004, Adhikari et al. 2006, Jin et al. 2006) i.e., ~100 K below the bulk Au–Ge eutectic temperature $T_E = 361°C$ (see Figure 16.4) (Okamoto and Massalski 1986).

However, until recently, the exact growth mechanisms were not clear. This is because most existing studies focused on postgrowth characterization to determine the growth processes. While *ex situ* characterization is essential for the determination of chemical composition, interface structure, and wire morphologies, inference of the growth mechanisms from such data is difficult. *In situ* observations permit the direct determination of the growth rates and hence the kinetics and help identify several aspects of the VLS growth process that have not been previously discussed.

In this section, we review the recent *in situ* transmission electron microscopy (TEM) studies of Ge NW growth kinetics as a means of developing a fundamental understanding of their growth

mechanisms. TEM images of the wires and the catalyst droplets are collected at video rate as a function of the growth pressure, temperature, and gas environment. From the images, the wire growth rates were measured, the catalyst states were identified, and the rate-limiting steps were determined. Ge wire growth, as shown in the following sections, occurs in the presence of a liquid catalyst at temperatures below the bulk AuGe alloy eutectic temperature. This liquid phase, key to the successful growth of Ge wires, is found to be stable only at high Ge concentrations.

All the experimental results pertaining to Ge NW growth experiments were carried out at the IBM T. J. Watson Research Center in a multi-chamber UHV (base pressure 2×10^{-10} Torr) transmission electron microscope (300 kV Hitachi H-9000 TEM) equipped with *in situ* physical and chemical vapor deposition (CVD) facilities. First, Si(111) wafers (miscut < 0.5°) were sliced into $1.5 \times 4 \times 0.5$ mm sections, cleaned chemically, and mounted in the TEM with the polished surface vertical. The samples were degassed in the UHV at 600°C for 2 h followed by annealing at 1250°C for 30–60 s. Au thin films, 2–3 nm thick, were deposited at room temperature by thermal evaporation from a Knudsen cell at a rate of 3×10^{-3} nm/s onto the samples in a preparation chamber at a base pressure of 2×10^{-8} Torr during the deposition. These Au-covered Si(111) samples were then transferred under the UHV to the TEM chamber and wire growth was initiated by resistively heating the samples in a gas mixture (purity 99.999%) containing 80% He and 20% of digermane (Ge_2H_6) for the growth of Ge NWs, respectively. Prior to deposition, the samples were first annealed in vacuum at ~400°C for 5 min and then cooled to the growth temperatures $T < 350°C$. As soon as the samples are heated, eutectic droplets form and act as the catalysts for the formation of individual wires. NW growth is initiated beneath these eutectic droplets and the wires grow away from the surface, most of which are perpendicular to the substrate and hence are imaged with the electron beam perpendicular to the wire axis. Individual wires can be observed in bright or dark fields in transmission mode, and the effects of the pressure, temperature, and gas environment can be observed directly.

Ge wires are observed to grow at Ge_2H_6 pressures $P > 9 \times 10^{-7}$ Torr and temperatures between 300°C and 380°C. The maximum pressure in these experiments is limited to 10^{-5} Torr by the design of the TEM. The gases are leaked continuously into the microscope column to ensure a constant pressure during wire growth. Under these conditions, $\langle 111 \rangle$-oriented Ge wires that are several hundreds of nanometers long can be grown for times between 1 and 6 h. TEM images are acquired at video rate (30 frames/s). Wires grown under continuous, as well as intermittent, electron beam irradiation exhibit similar growth rates indicating that the electron beam does not affect the wire growth kinetics. Substrate temperatures are measured before and after the deposition using an infrared pyrometer. After growth, the surface can be cleaned by heating to 1250°C, so that a series of growth experiments can be carried out on the same sample in one area. For such a series, the relative temperature can be measured to within 20 K, while for different samples the measurement uncertainties in absolute temperature are ~50 K.

In these experiments, the Ge_2H_6 pressure range accessible for growth was between 1×10^{-7} and 1×10^{-5} Torr. Sustained epitaxial growth of the $\langle 111 \rangle$-oriented single-crystalline Ge wires was observed at $250°C < T < 400°C$. In contrast to Si, Ge wires are bounded by smooth sidewalls. Figure 16.5 is a typical bright field TEM image of Ge NWs obtained during deposition at $T = 330°C$ using 4.8×10^{-6} Torr Ge_2H_6.

Note that the wire tips show smoothly curved catalyst particles indicative of the liquid phase and suggest that the Ge wires grow via the VLS process. This is an important observation since $T_E = 361°C$ for AuGe alloy (see Figure 16.4) while liquid droplets are observed at $T = 330°C$.

In order to understand the Ge wire growth kinetics, a series of experiments were carried out while systematically varying the substrate temperature and Ge_2H_6 pressure during growth.

In one such experiment, the effect of the substrate temperature on the AuGe catalyst state was studied during wire growth at a constant Ge_2H_6 pressure. Solidification of the droplets occurred at temperatures far below (~100 K) T_E and required significantly higher temperatures (>400°C) to re-establish the liquid phase. This hysteresis in the solid–liquid phase transformation is seen in all Ge growth experiments and for wires with a range of diameters (20–140 nm). Interestingly, the wires continue to grow even after the catalyst particle has solidified, i.e., via the vapor–solid–solid (VSS) process. Measurements made on several wires showed that VSS growth is 10–100 times slower than VLS growth at the same Ge_2H_6 pressure and temperature, presumably due to weaker surface reactivity and/or lower diffusivity through the solid. Both VLS and VSS growth were observed to occur simultaneously on neighboring wires in some instances. All the wires, irrespective of the growth mode, are crystalline and the only obvious difference between the growth modes is that VSS process yields more tapered wires owing to their relatively slower growth rates. This demonstration of dual growth modes may be relevant to the controversy regarding the role of VSS and VLS growth in other systems (Persson et al. 2004, Harmand et al. 2005).

The growth experiments carried out while varying Ge_2H_6 pressure at a constant substrate temperature showed that a significant Ge_2H_6 pressure appears to be essential for stabilizing the liquid state below T_E. Whenever the Ge_2H_6 pressure is reduced during VLS growth, the solidification of catalyst droplets was observed (see Figure 16.6).

Here, the first image shows a typical VLS-grown Ge wire. In this experiment, growth was initiated using 4.6×10^{-6} Torr Ge_2H_6 and continued at pressures $\geq 1.1 \times 10^{-6}$ Torr for ~78 min. The Ge_2H_6 pressure was then reduced to 2.8×10^{-7} Torr while maintaining a constant temperature. Within 681 s, the droplet abruptly solidifies. The fact that the droplets can solidify confirms that the temperature is definitely below the bulk T_E, independent of any uncertainties in temperature calibration.

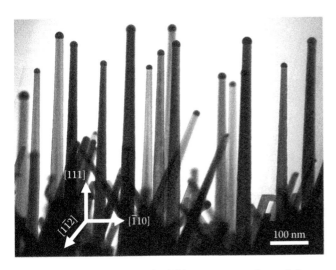

FIGURE 16.5 A typical bright-field TEM image obtained from a Si(111) sample during the VLS growth of Ge nanowires. Most wires grow epitaxially in the $\langle 111 \rangle$ direction. (Adapted from Kodambaka, S., Tersoff, J., Ross, M.C., and Ross, F.M., *Proc. SPIE*, 7224, 72240C, 2009.)

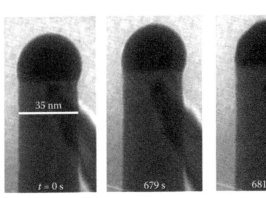

FIGURE 16.6 Representative bright-field TEM image series showing the solidification of AuGe catalyst on top of a Ge wire when the Ge_2H_6 pressure is reduced during growth at a constant temperature $T = 340°C$. In this experiment, Ge_2H_6 pressure was dropped from 1.1×10^{-6} Torr to 2.8×10^{-7} Torr at $t = 0$. (Adapted from Kodambaka, S., Tersoff, J., Ross, M.C., and Ross, F.M., *Proc. SPIE*, 7224, 72240C, 2009.)

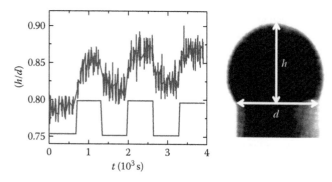

FIGURE 16.7 Plots of aspect ratio (top curve) of a AuGe eutectic droplet on top of a Ge wire (diameter = 59 nm) and Ge_2H_6 pressure (bottom curve) vs. *t* at *T* = 355°C. The data is acquired from a TEM video sequence as the pressure is cycled repeatedly between 1.9×10^{-6} Torr and 8.4×10^{-6} Torr. Aspect ratio (*h/d*) is defined as the ratio of the height *h* and the base width *d* of the droplet as labeled on the TEM image. (Adapted from Kodambaka, S. et al., *Science*, 316, 729, 2007.)

This behavior is typical of all droplets observed, although the exact time delay between smaller and larger droplets depends on the growth history and in some cases can be as long as several tens of minutes.

In order to understand the role of Ge_2H_6 pressure on the droplet state, shapes of AuGe droplets were measured during wire growth as a function of Ge_2H_6 pressure at a constant *T*. Figure 16.7 is a typical plot of the aspect ratio of a droplet as the pressure is varied repeatedly between higher and lower values.

Although the changes are small, note that when the Ge_2H_6 pressure is decreased, the aspect ratio decreases and when the Ge_2H_6 pressure is increased, the aspect ratio increases. Clearly, the droplet shape is varying with Ge_2H_6 pressure, suggesting that there are observable changes in surface energy with pressure (Kodambaka et al. 2007). All the above observations suggest that the liquid phase may be effectively stabilized against solidification by Ge supersaturation, which arises from the growth process (Kodambaka et al. 2007). For more details, the reader is encouraged to refer to the article by Kodambaka et al. (2007).

In conclusion, *in situ* TEM experiments enable the quantitative determination of the NW growth mechanisms. In case of Au-catalyzed growth of Ge NWs, it was found that the AuGe catalyst state may be either solid or liquid below the bulk eutectic temperature, with the state depending not just on temperature but also on the Ge_2H_6 pressure and history. Remarkably, both VLS and VSS processes can operate under the same conditions to grow Ge wires. Most surprisingly, a significant Ge_2H_6 pressure is essential for growth via the VLS process below the eutectic temperature. Clearly, *in situ* observations provide valuable insights into the physical processes controlling the morphological and structural evolution of NWs and are expected to be general and applicable to other material systems.

Large-scale fabrication of NW-based devices for the above-mentioned applications requires precise control over NW morphology (shape, length, and size), crystalline structure, chemical composition, and interfacial abruptness. This is an

extremely challenging task governed by a complex interplay of the thermodynamics of the materials and the kinetics of nucleation and growth processes. For example, growth orientations of Ge wires can be varied between ⟨111⟩, ⟨110⟩, or ⟨112⟩ by changing the growth temperature, precursor, and the growth technique (Hanrath and Korgel 2005). However, very little is known concerning the factors affecting the wire orientation (Tan et al. 2002, Borgström et al. 2004, Wu et al. 2004a, Schmidt et al. 2005, Wang et al. 2006). Despite several years of research in this area, none of the following has been achieved, even for relatively simple elemental Si and Ge NWs: (1) ⟨100⟩-oriented Si or Ge wires, (2) Si/Ge heterostructures with atomically-abrupt interfaces (Clark et al. 2008), or (3) large-scale synthesis of sub-5 nm diameter wires. Success in the rational synthesis of NWs with desired architecture can only be achieved through a fundamental understanding of all the processes influencing the nucleation and growth.

16.4 Properties of Ge Nanowires

Having looked at the VLS growth of Ge NWs, we turn to their properties and applications. NWs could be the building blocks for the post-CMOS era bottom-up assembly of nano devices. The primary requirements for this new technological era are compatibility with traditional silicon manufacturing processes and integration for cost reduction. For this reason, Ge is the front runner to replace silicon. Ge has a lighter effective mass for the electron and hole charge carriers than Si implying higher carrier mobilities and thus high-performance transistors. Figure 16.8, adapted from the article by Yu (Yu et al. 2006), shows that bulk Ge has direct (~0.88 eV) and indirect (~0.66 eV) band gaps making it an attractive material for electronic and photonic circuitry applications. These band gaps can be further tuned for specific electronic/photonic applications by controlling the wire

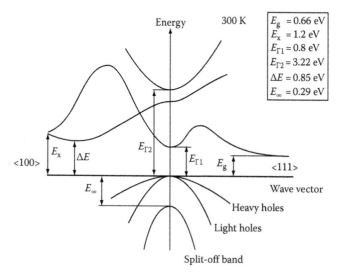

FIGURE 16.8 Band gaps in bulk Ge. (Adapted from Yu, B. et al., *J. Clust. Sci.*, 17, 579, 2006.)

dimensions thus giving rise to a myriad of design possibilities. Some of the quantum effects leading to various properties and applications for Ge NWs are discussed here.

16.4.1 Structural Properties

For NWs, the surface to volume ratio is very large. This leads to the surface reconstruction of the atoms in the NW as seen in Figure 16.9, which shows the simulated Ge NWs with H termination of the structures adapted from Medabonia et al. (2007). After analyzing bond lengths in these structures, Medaboina et al. observed that bond lengths between Ge atoms at the surface were relaxed by ~1%, whereas no relaxation was observed for atoms in the interior region away from the surface. When the NWs were allowed to dimerize, it was observed that the atoms relaxed by ~50% as expected. Cross-sections of wires along [110] were found to have cylindrical structures for all diameters. Along the [001] and [111] axes, the smaller wires with diameters $d < dc$ have circular cross-sections. As seen from Figure 16.9, the critical diameter dc, above which the cross-section acquires a faceted shape, lies in the range of 2.0 nm < dc < 3.0 nm for [001] and [111] axes. The value of dc is determined by the competition to lower the total energy between the energies of the different exposed surfaces of H terminated Ge under the constraint of a fixed volume. Wires along the [001] direction appear in Figure 16.9 to have a rectangular bonding geometry rather than the expected square surface arrangement of atoms in a single [001] surface layer. This expected square arrangement of atoms of a diamond lattice is not made of the nearest

neighbor atoms. The rectangular structure seen in Figure 16.9 arises because the middle atoms, along the longer side of the rectangle, are the nearest neighbors of the top layer atoms and are one layer below the top layer. For larger diameters, d > dc = 2.15 nm wires along [001], cross-sections were found to be octagonal-shaped with facets of the [001] and [110] type, which were normal to [001]. For larger diameters, d > dc = 2.11 nm wires along [111], they were hexagonal-shaped with facets of the [110] type, which were normal to [111]. Thus, we see that the arrangement of atoms of a NW cut along a particular direction depends on its diameter, its axis orientation, and the termination of its surface. We will see in the following discussion that the properties of a NW are intimately related to its crystal structure.

16.4.2 Electronic Properties and Applications

As seen from Equation 16.12, the quantum confinement of electrons (in the NW cross-section) leads to the difference between successive energy states to increase as NW diameter decreases. This leads to an increase in the band gap between the filled and unoccupied electronic energy states, which correspondingly increases with a decreasing diameter. For example, in InP NWs, it was found that the band gap E_g varies with diameter d of the wires as $E_g \sim 1/d^{1.45}$. Simple particle in box type explanations (as shown in Section 16.2) though qualitatively adequate, do not account for this scaling quantitatively (Yu et al. 2003). As expected from previous theoretical analysis, Figure 16.10 shows the dependence of the band gap of hydrogen passivated Ge NWs

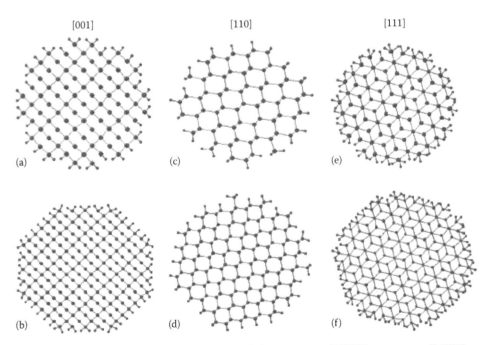

FIGURE 16.9 Cross-sectional views of the relaxed Ge NWs. (a) $NW_{[001]}^{(Ge-89,H-44)}$ (2.03), (b) $NW_{[001]}^{(Ge-185,H-60)}$ (3.03), (c) $NW_{[110]}^{(Ge-69,H-32)}$ (2.12), (d) $NW_{[111]}^{(Ge-133,H-40)}$ (3.3), (e) $NW_{[111]}^{(Ge-170,H-66)}$ (2.11), (f) $NW_{[111]}^{(Ge-326,H-90)}$ (3.03). Larger circles represent Ge atoms and the outer smaller ones represent the H atoms used to saturate dangling bonds. Local geometrical patterns corresponding to the axis of the wires are evident: square shapes for [001] axis, hexagonal for [110], and parallelograms for [111]. (Adapted from Medaboina, D. et al., *Phys. Rev. B*, 76, 205327, 2007.)

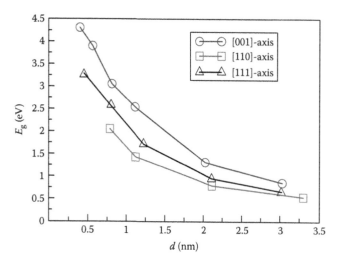

FIGURE 16.10 Dependence of band gap on the wire diameter and orientation. (Adapted from Medaboina, D. et al., *Phys. Rev. B*, 76, 205327, 2007.)

on their diameter and orientation (Medaboina et al. 2007). Band structures for different Ge NWs (shown in Figure 16.9) as calculated by Medaboina et al. are shown in Figure 16.11. As expected, the band structure varies depending on the wire orientation, diameter, and passivation material. The dispersion of the valence band for wires with approximately the same diameter is greatest for wires along [110] and least for wires along [111]. As expected from a quantum size effect, Figure 16.11 shows the absolute

value of the valence band maximum decreases in energy and the absolute value of the conduction band minimum increases in energy as the thickness of the wire decreases. This leads to an increase in the electronic band gap with a decreasing NW diameter. The band gap could be either direct or indirect, depending on the crystallographic orientation. NWs along [110] and thin ($d < 1.3$ nm) ones along [001] have direct band gaps occurring at the gamma point. This is different from bulk Ge band structure as shown in Figure 16.8. Such wires, due to their direct gaps, would be suitable for applications in optics. Wires along [001] were found to transit from direct to indirect band gaps as the diameter increased above 1.3 nm, while all wires along [111] had indirect band gaps. Another important factor used to control the electronic properties of NWs is doping. Dopants of *n*- and *p*-type in NWs are required for faster and low power consuming logic devices. Doping has been studied in several NWs and has been used to design inverters, LEDs, and bipolar transistors. Obtaining good electronic properties for *n*- and *p*-channel FETs is more complicated (Greytak et al. 2004). Band bending caused by doping Ge NWs has been demonstrated experimentally (Wang and Dai 2006). Figure 16.12 shows the simulation results of band structures of doped ~2 nm diameter Ge NWs for a single wire along each crystallographic direction [001], [110], and [111]. A high level of doping (0.5%–1.5%) obviously has an impact on the electronic structures of Ge NWs. As shown in Figure 16.12, adding a *p*-type (*n*-type) dopant moves the Fermi energy toward the valence band (conduction band). The maximum of the valence band and the minimum of the conduction band increase

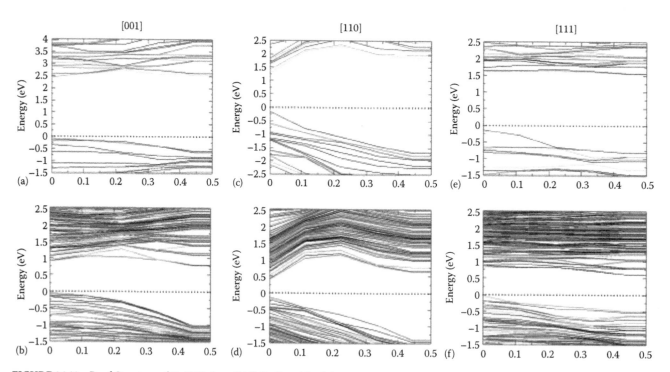

FIGURE 16.11 Band Structure of Ge NW along [001], [110], and [111] directions. The Fermi level in each panel is set to zero and is shown by the dotted line. (a) $NW_{[001]}^{(Ge-25,H-20)}$ (1.12) and (b) $NW_{[001]}^{(Ge-185,H-60)}$ (3.03) represent the band structure along the [001] direction. (c) $NW_{[110]}^{(Ge-17,H-12)}$ (1.12) and (d) $NW_{[110]}^{(Ge-133,H-40)}$ (3.3) represent the band structure along the [110] direction. (e) $NW_{[111]}^{(Ge-62,H-42)}$ (1.23) and (f) $NW_{[111]}^{(Ge-326,H-90)}$ (3.03) represent the band structure along the [111] direction. (Adapted from Medaboina, D. et al., *Phys. Rev. B*, 76, 205327, 2007.)

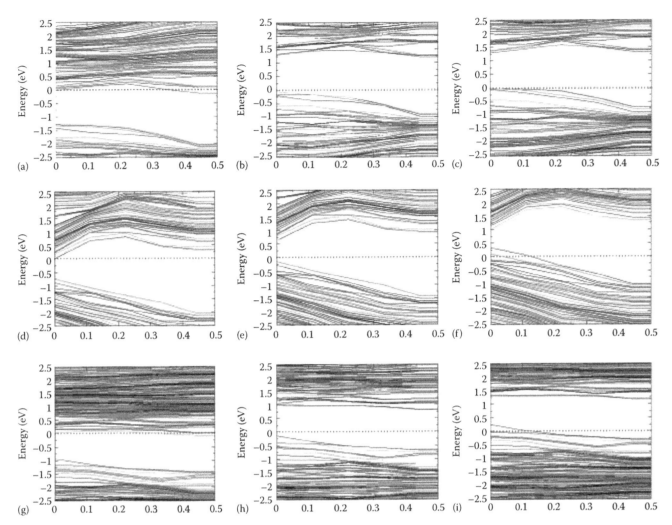

FIGURE 16.12 Comparison of band structures of Ge NWs with and without doping. The Fermi level in each panel is set to zero and is shown by the dotted line. (a) $NW_{[001]}^{(Ge-88,H-44,P-1)}$ (2.03), (d) $NW_{[110]}^{(Ge-68,H-32,P-1)}$ (2.12), and (g) $NW_{[111]}^{(Ge-169,H-66,P-1)}$ (2.11) are *n*-type doped wires. (b) $NW_{[001]}^{(Ge-89,H-44)}$ (2.03), (e) $NW_{[110]}^{(Ge-69,H-32)}$ (2.12), and (h) $NW_{[111]}^{(Ge-170,H-66)}$ (2.11) are undoped wires. (c) $NW_{[001]}^{(Ge-88,H-44,B-1)}$ (2.03), (f) $NW_{[110]}^{(Ge-68,H-32,B-1)}$ (2.12), and (i) $NW_{[111]}^{(Ge-169,H-66,B-1)}$ (2.11) are *p*-type doped wires. (Adapted from Medaboina, D. et al., *Phys. Rev. B*, 76, 205327, 2007.)

(decrease) in energy with the addition of *p*-type (*n*-type) dopant when measured relative to the Fermi level. Figure 16.12 also shows that the doping of wires does not have a significant effect on the dispersion of valence and conduction bands. Thus, the effect of doping on the band structure of hydrogenated Ge NWs is similar to that in doped bulk Ge. This property would make Ge NWs suitable for applications that require the tuning of the NW conductivity by doping.

Ge NW-FETs are ideal for very-low-power circuit applications. The majority of the published work (Yu et al. 2006) shows that Ge NW-FETs function at nano- to micro-amperes on-state current with a large on-to-off current ratio (10^4–10^6). The switching energy of a single-NW-FET is 3 to 6 orders of magnitude lower than that of a typical top-down conventional FET. The standby power is almost negligible due to the pA-level off-state leakage per NW. These excellent properties make Ge NW-FETs an ideal nano electric device for micro or nano power

chips with significantly improved power performance tradeoff (Yu et al. 2006).

16.4.3 Optical Properties and Applications

Germanium's compatibility with Group III–V materials and germanium oxide's optical properties allow for the realization of radical integrated optoelectronic circuitry designs. Photoluminescence (Canham 1990, Duan et al. 2000, Katz et al. 2002) data revealed a substantial blue shift with a decreasing size of NWs. It has also been shown recently that Ge NWs could be used in optoelectronic components fabricated within silicon-based technology (Halsall et al. 2002). Optical properties of NWs also mainly depend on the size of the wire, orientation of the wire (Bruno et al. 2007), and passivation material and doping utilized (Wu et al. 2004b). As explained earlier, the band gap and band structure change with wire size and orientation leading to numerous ways of obtaining a particular optoelectronic functionality. The simulation of optical properties of NWs becomes expensive if the proper incorporation of many body effects like self energy, local field, and excitonic effects are taken into account correctly (Bruno et al. 2005). Ge NWs of diameters in the range of 0.8 nm have the main absorption peak in the visible range (Bruno et al. 2007), which could lead to an efficient application of Ge NWs in optoelectronic devices when compared with Si NWs. Already, Ge NW-based nano devices like solar cells, magnets (Alguno et al. 2003), and FETs (Wang et al. 2003) have been characterized.

16.4.4 Mechanical Properties and Applications

Mechanical stability: The high surface to volume ratio of NWs gives them interesting mechanical properties, as well. For example, simulations on Au NWs show that a spontaneous transition from face centered cubic (fcc) to body centered tetragonal (bct) structure occurs in ⟨001⟩ oriented gold (Au) NWs or cross-sections less than 4 nm². The simulations showed no transitions when the wires were oriented along ⟨111⟩ or ⟨110⟩ directions (Diao et al. 2003). It may be speculated that similar phenomena driven by surface reconstruction may cause structural phase transitions into the interior of Ge NWs. However, there has been no such observation to date.

Mechanical strength: If NWs are to be used as building blocks in nano devices, their mechanical properties need to be characterized for avoiding mechanical failure. With current NW production techniques, NWs can be grown as single crystals with the absence of any structural defects such as point or line defects. This sometimes leads to the high mechanical strength and stiffness of NWs close to their theoretical single crystal limits (Wong et al. 1997, Diao et al. 2003, Kis et al. 2003). This property makes them attractive for use in composites and nano-electromechanical devices. Many test set-ups have been used to characterize the mechanical properties of NWs. Figure 16.13 shows the most commonly used clamped beam experimental set-up for lateral loading of the Ge NW bending test using a atomic force microscope (AFM). Despite the widespread use of the clamped beam configuration, a comprehensive model that accounts for the detailed shape of these force-displacement (*F-d*) curves over the entire elastic region is yet to be described and validated for NW systems. Heidelberg et al. (2006) provides a method for the complete description of the elastic properties in a double-clamped beam configuration over the entire elastic regime for diverse wire systems. This method can be used to perform a comprehensive analysis of *F-d* curves using a single closed-form analytical description. It can be applied to extract linear material constants such as the Young's modulus (*E*) and to describe the entire elastic range and hence to identify the yield points for dramatically different systems of NWs. The same set-up and theory have been used for tests on Ge NWs (Ngo et al. 2006) of sizes between 20 and 80 nm. These tests revealed that the bulk moduli of NWs is comparable with that of bulk Ge but their mechanical strength

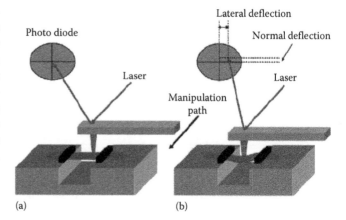

FIGURE 16.13 Schematic of the lateral nanowire manipulation experiment setup: (a) before manipulation, (b) during manipulation. The nanowire under investigation is suspended over a trench in the substrate and fixed by Pt deposits at the trench edges. During the manipulation the lateral and normal cantilever deflection signals are simultaneously recorded. (Adapted from Heidelberg, A. et al., *Nano Lett.*, 6, 1101, 2006.)

is closer to the theoretical value for pure bulk Ge and is substantially greater than the mechanical strength of any whisker semiconductor. Simple bending stress analysis on the clamped–clamped beam system shows a linear variation of displacement with force (Heidelberg et al. 2006). This cannot explain the nonlinear behavior seen in Figure 16.13. As the NW is displaced, an axial tensile force is inherently induced due to the stretching of the NW. This force affects the total stress experienced by the NW, leading to an enhancement of its rigidity and thus non-linearity of the *F-d* curve. The curve fit to the elastic deformation plot using the closed-form solution (Heidelberg et al. 2006) gives a Young's modulus of 137.6 GPa and a fracture strength of 17.1 GPa. An important observation is that the Ge NW breaks without plastic deformation, revealing its brittle nature.

16.4.5 Surface Chemistry and Applications

Two commonly occurring adsorbents for Si and Ge surfaces are hydrogen and oxygen. An adsorbent layer on a surface can significantly affect the surface reconstruction and also lead to faceting or de-faceting transitions of the surface. Density functional theory (DFT) computations (Mingwei et al. 2006) show that the passivation of Ge NWs with more stable ethine reduces the quantum confinement effects when compared with that of Ge NWs passivated with hydrogen. NWs have been thought to be the active elements in sensing devices because the large surface area is believed to cause the sensitive dependence of electronic properties on changes in surface adsorption by different chemical species (Wang and Dai 2006). Ge NWs along [111], which have facets of [110] and [100] types have been studied (Medaboina et al. 2007) by changing the surface hydrogen concentration. This orientation has the tendency of the unhydrogenated facets to dimerize. Hydrogen atoms were removed from the surface of relaxed wires. Band structures of these wires were found to have electronic states with energies lying in the middle of band gap due to the occurrence of unpaired electrons on the surface of the NW after the removal of H atoms. This change in the electronic properties with a change in hydrogen concentration for wires along [111] could be potentially exploited in sensing applications, due to the disappearance of the band gap, which would cause a significant increase in electronic conduction. Synthesized Ge NWs could be oxidized when exposed to air forming Ge/GeO_2 surfaces leading to the high density of surface states causing appreciable Fermi level pinning (Kingston 1957). This oxidation phenomenon is more severe for small wires. So Ge NWs have to be chemically passivated with appropriate adsorbents to avoid the degradation of their electronic properties. Band bending could occur in Ge NWs due to the pinning of Fermi level at surface states (Wang et al. 2004, 2005). Band bending due to surface oxidation of Ge NWs has been explored by Wang and Dai (Wang and Dai 2006).

16.5 Simulation Methods

The grand challenge of producing complex nanodevices by elementary nano building blocks still faces several hurdles. NWs due to the nano-length scale along one axis also show properties intermediate between bulk and single atoms. This broken symmetry means quantum mechanical computations are expensive and a multi-scale approach is necessary. A thorough theoretical understanding of the properties of nano structures involves understanding the issues of their symmetry, lower dimensionality, size, and chemical composition. Computational techniques are very useful to deal with the multiplicity of these parameters. Table 16.2 lists some of the simulation methods that have been commonly used to characterize nanostructures. All these theoretical techniques attempt to solve complex body problems arising due to the interaction between different atoms. Moving down the table increases assumptions employed in the theory thus simplifying the model and decreasing the fidelity of the result, the advantage being reduced computational cost. The simulation method that would yield the best results in a reasonable amount of time differs from case to case and is entirely at the discretion of the user. For the interested reader, Martin (2004) gives additional information of these simulation methods. Quantum Monte Carlo (QMC) simulations approximate the many electron solutions by a

TABLE 16.2 Hierarchy of Computational Simulation Methods Used for Studying Nanostructures

Quantum Monte Carlo (QMC)
GW
Time dependent density functional theory
Time independent density functional theory (DFT)
Tight binding (TB)
Classical molecular dynamics (MD)

Note: Going down the column increases the physical assumptions in the input parameters, but yields computation-time savings leading to large scale simulations.

wave function obtained from Monte Carlo methods. Many types of QMC (Hammond et al. 1992) techniques are available for various applications. These simulations are very expensive and so are limited to few tens of atoms. In DFT, the many-electron system is approximated by functionals that depend on spatial electron density. DFT (Thomas 1927, Hohenberg and Kohn 1964, Kohn and Sham 1965, Parr and Yang 1989, Fiolhais et al. 2003) has been a workhorse in condensed matter physics with comparatively less computational expense than QMC. After the exchange and correlation interaction effects were added to DFT (Kohn and Sham 1965), it has also been used in computational chemistry with improved results. Some of the disadvantages of DFT include the difficulty of accounting for the exchange and correlation effects in strongly correlated materials like high temperature superconductors. Specifically as related to NWs, DFT methods do not account correctly for electron–hole interactions that could become appreciable for small NWs (Bruno et al. 2007). In tight binding (TB) methods (Turchi et al. 1997, Slater and Koster 1954, Ashcroft and Mermin 1976, Goringe et al. 1997), the Hamiltonian of the many-body system is approximated as the Hamiltonian of an isolated center at each lattice point and the resultant atomic orbitals are assumed to become localized within a lattice constant. These assumptions make the computational solution to larger systems with many atoms less expensive. In classical molecular dynamics (MD) simulations, the many atom problem is approximated by a potential or force field between atoms or molecules. MD (McCammon and Harvey 1987, Haile 2001, Oren et al. 2001, Schlick 2002) is generally used when interaction effects between thousands of atoms or molecules are of interest. It is very commonly used in large material systems and bio-molecule (or large protein) simulations.

16.6 Conclusions

We have shown a rudimentary theory connecting the physical dimensions of a NW to its properties. We have described state-of-the-art Ge NW growth techniques. TEM characterization of the growth process was reported. A brief overview of the effect of Ge NW diameter, orientation, passivation, and doping on surface reconstruction, band gap, and band structure has been provided. The resulting electrical, optical, mechanical, and surface properties have been explained and their applications have been cited. Different nanoscale simulation techniques used for the numerical analysis of NW were briefly listed. Ge NWs have a considerable potential for next generation microelectronics, photonics, and nanodevices owing to their high carrier mobilities, tenability of their band gaps, and recently discovered high mechanical strength at NW scales. This chapter should serve as a quick reference for Ge NW growth methods, properties, and applications.

Acknowledgments

SVK thanks the NSF, DARPA, Wright Center for PVIC from the State of Ohio, Wright Patterson Air Force Base, and Kirtland Air Force Base for funding this work.

References

Adhikari, H., Marshall, A.F., Chidsey, C.E.D., and McIntyre, P.C. 2006. Germanium nanowire epitaxy: Shape and orientation control. *Nano Lett.* 6: 318–323.

Alguno, A., Usami, N., Ujihara, T., Fujiwara, K., Sazaki, G., and Nakajima, K. 2003. Enhanced quantum efficiency of solar cells with self-assembled Ge dots stacked in multilayer structure. *Appl. Phys. Lett.* 83: 1258.

Ashcroft, W. and Mermin, N.D. 1976. *Solid State Physics*, Brooks Cole, Belmont, CA.

Bjork, M.T., Ohlsson, B.J., Thelander, C. et al. 2002. Nanowire resonant tunneling diodes. *Appl. Phys. Lett.* 81: 4458–4460.

Borgström, M., Deppert, K., Samuelson, L., and Seifert, W.J. 2004. Size- and shape-controlled GaAs nano-whiskers grown by MOVPE: A growth study. *J. Cryst. Growth* 260: 18–22.

Boukai, A.I., Bunimovich, Y., Tahir-Kheli, J., Yu, J.-K., Goddard III, W.A., and Heath, J.R. 2008. Silicon nanowires as efficient thermoelectric materials. *Nature* 451: 168–171.

Bruno, M., Palummo, M., Marini, A. et al. 2005. Excitons in germanium nanowires: Quantum confinement, orientation, and anisotropy effects within a first-principles approach. *Phys. Rev. B* 72: 153310.

Bruno, M., Palummo, M. Ossicinic, S., and Del Sole, R. 2007. First-principles optical properties of silicon and germanium nanowires. *Surf. Sci.* 601: 2707.

Canham, L.T. 1990. Silicon quantum wire array fabrication by electrochemical and chemical dissolution of wafers. *Appl. Phys. Lett.* 57: 01046.

Chan, C.K., Peng, H., Liu, G. et al. 2008a. High-performance lithium battery anodes using silicon nanowires. *Nat. Nanotechnol.* 3: 31–35.

Chan, C.K., Zhang, X.F., and Cui, Y. 2008b. High capacity Li ion battery anodes using Ge nanowires. *Nano Lett.* 8: 307–309.

Chen, G. and Shakouri, A. 2002. Heat transfer in nanostructures for solid-state energy conversion. *J. Heat Transfer–T ASME* 124: 242–252.

Clark, T.E., Nimmatoori, P., Lew, K.-K., Pan, L., Redwing, J.M., and Dickey, E.C. 2008. Diameter dependent growth rate and interfacial abruptness in vapor–liquid–solid Si/Si$_{1-x}$Ge$_x$ heterostructure nanowires. *Nano Lett.* 8: 1246–1252.

Cui, Y., Lauhon, L.J., Gudiksen, M.S., Wang, J., and Lieber, C.M. 2001. Diameter-controlled synthesis of single-crystal silicon nanowires. *Appl. Phys. Lett.* 78: 2214–2216.

Cui, Y. and Lieber, C.M. 2001. Functional nanoscale electronic devices assembled using silicon nanowire building blocks. *Science* 291: 851–853.

De Franceschi, S., J van Dam, A., Bakkers, E.P.A.M., Feiner, L.F., Gurevich, L., and Kouwenhoven, L.P. 2003. Single-electron tunneling in InP nanowires. *Appl. Phys. Lett.* 83: 344–346.

Diao, J.K., Gall, K., and Dunn, M.L. 2003. Surface-stress-induced phase transformation in metal nanowires. *Nat. Mater.* 2: 656.

Duan, X., Wang, J., and Lieber, C.M. 2000. Synthesis and optical properties of gallium arsenide nanowires. *Appl. Phys. Lett.* 76: 01116.

Fiolhais, C., Nogueira, F., and Marques, M. 2003. *A Primer in Density Functional Theory*, Springer-Verlag, Berlin, Germany.

Givargizov, E.I. 1978. Growth of whiskers by VLS mechanism. *Curr. Top. Mater. Sci.* 1: 82–145.

Givargizov, E.I. 1987. *Highly Anisotropic Crystals*. Springer-Verlag, Berlin, Germany.

Goringe, C.M., Bowler, D.R., and Hernández, E. 1997. Tight-binding modelling of materials. *Rep. Prog. Phys.* 60: 1447.

Greytak, A.B., Lauhon, L.J., Gudiksen, M.S., and Lieber, C.M. 2004. Growth and transport properties of complementary germanium nanowire field-effect transistors. *Appl. Phys. Lett.* 84: 4176.

Gudiksen, M.S., Lauhon, L.J., Wang, J., Smith, D.C. and Lieber C.M. 2002. Growth of nanowire superlattice structures for nanoscale photonics and electronics. *Nature* 415: 617–620.

Haile, M. 2001. *Molecular Dynamics Simulation: Elementary Methods*, Wiley Professional, New York.

Halsall, P., Omi, H. and Ogino. T. 2002. Optical properties of self-assembled Ge wires grown on Si(113). *Appl. Phys. Lett.* 81: 2448.

Hammond, B.J., Lester W.A., and Reynolds, P.J. 1992. Monte Carlo methods in Ab Initio. *Int. J. Quantum Chem.* 42: 837.

Hanrath, T. and Korgel, B.A. 2005. Crystallography and surface faceting of germanium nanowires. *Small* 1: 717–721.

Haraguchi, K., Katsuyama, T., Hiruma, K., and Ogawa, K. 1992. GaAs p-n-junction formed in quantum wire crystals. *Appl. Phys. Lett.* 60: 745–747.

Harmand, J.C., Patriarche, G., Pere-Laperne, N., Merat-Combes, M.N., Travers, L., and Glas, F. 2005. Analysis of vapor-liquid-solid mechanism in Au-assisted GaAs nanowire growth. *Appl. Phys. Lett.* 87: 203101.

Harrison, P. 1999. *Quantum Wells, Wires and Dots*. John Wiley & Sons, West Sussex, U.K.

Heidelberg, A., Ngo, L.T., Wu, B. et al. 2006. A generalized description of the elastic properties of nanowires. *Nano Lett.* 6: 1101.

Hochbaum, I., Chen, R., Delgado, R.D. et al. 2008. Enhanced thermoelectric performance of rough silicon nanowires. *Nature* 451: 163–167.

Hohenberg, P. and Kohn, W. 1964. Inhomogeneous electron gas. *Phys. Rev. B* 864: 136.

Jin, C.-B., Yang, J.-E., and Jo, M.-H. 2006. Shape-controlled growth of single-crystalline Ge nanostructures. *Appl. Phys. Lett.* 88: 193105.

Kamins, T.I., Li, X., and Williams, R.S. 2004. Growth and structure of chemically vapor deposited Ge nanowires on Si substrates. *Nano Lett.* 4: 503–506.

Katz, D., Wizansky, T., Millo, O., Rothenberg, E., Mokari, T., and Banin, U. 2002. Size-dependent tunneling and optical spectroscopy of CdSe quantum rods. *Phys. Rev. Lett.* 89: 86801.

Kingston, R.H. (ed.) 1957. *Semiconductor Surface Physics*, University of Pennsylvania Press, Philadelphia, PA.

Kis, A., Mihailovic, D., Remskar, M. et al. 2003. Shear and Young's moduli of MoS2 nanotube ropes. *Adv. Mater.* 15: 733.

Kodambaka, S., Tersoff, J., Reuter, M.C. and Ross, F.M. 2007. Germanium nanowire growth below the eutectic temperature. *Science* 316: 729–732.

Kodambaka, S., Tersoff, J., Ross, M.C., and Ross, F.M. 2009. Growth Kinetics of Si and Ge nanowires. *Proc. SPIE* 7224: 72240C.

Kohn, W. and Sham, L.J. 1965. Self-consistent equations including exchange and correlation effects. *Phys. Rev. A* 140:1133.

Lensch-Falk, J.L., Hemesath, E.R., Lopez, F.J., and Lauhon, L.J. 2007. Vapor-solid-solid synthesis of Ge nanowires from vapor-phase-deposited manganese germanide seeds. *J. Am. Chem. Soc.* 129: 10670–10671.

Martin, R.M. 2004. *Electronic Structure: Basic Theory and Practical Methods*, Cambridge University Press, Cambridge, U.K.

McCammon, J.A. and Harvey S.C. 1987. *Dynamics of Proteins and Nucleic Acids*, Cambridge University Press, Cambridge, U.K.

Medaboina, D., Gade, V., Patil, S.K.R., and Khare, S.V. 2007. Effect of structure, surface passivation, and doping on the electronic properties of Ge nanowires: A first principles study. *Phys. Rev. B* 76: 205327.

Mingwei, J., Ming, N., Wei, S. et al. 2006. Anisotropic and passivation-dependent quantum confinement effects in germanium nanowires: A comparison with silicon nanowires. *J. Phys. Chem. B* 110: 18332–18337.

Miyamoto, Y. and Hirata, M. 1975. Growth of new form Germanium whiskers. *Jpn. J. Appl. Phys.* 14: 1419–1420.

Ngo, L.T., Almecija, D., Sader, J.E. et al. 2006. Ultimate-strength germanium nanowires. *Nano Lett.* 6: 2964.

Okamoto, H. and Massalski, T.B. 1986. *Binary Alloy Phase Diagrams* I. ASM International, Gaithersburg, MD.

Oren, M., Alexander, D.M., Benoît, R., and Masakatsu, W. 2001. *Computational Biochemistry and Biophysics*, Marcel Dekker, New York.

Park, W.I., Zheng, G., Jiang, X., Tian, B., and Lieber, C.M. 2008. Controlled synthesis of millimeter-long silicon nanowires with uniform electronic properties. *Nano Lett.* 8: 3004–3009.

Parr, R.G. and Yang, W. 1989. *Density-Functional Theory of Atoms and Molecules*, Oxford University Press, New York.

Persson, A.I., Larsson, M.W., Stenstrom, S., Ohlsson, B.J., Samuelson, L., and Wallenberg L.R. 2004. Solid-phase diffusion mechanism for GaAs nanowire growth. *Nat. Mater.* 3: 677–681.

Schlick, T. 2002. *Molecular Modeling and Simulation*. Springer-Verlag, New York.

Slater, J.C. and Koster, G.F. 1954. Simplified LCAO method for the periodic potential problem, *Phys. Rev.* 94, 1498.

Schmidt, V., Senz, S., and Gösele, U. 2005. Diameter-dependent growth direction of epitaxial silicon nanowires. *Nano Lett.* 5: 931–935.

Tan, T.Y., Lee, S.T., and Gösele, U. 2002. A model for growth directional features in silicon nanowires. *Appl. Phys. A* 74: 423–432.

Tian, B., Zheng, X., Kempa, T.J. et al. 2007. Coaxial silicon nanowires as solar cells and nanoelectronic power sources. *Nature* 449: 885–890.

Thomas, L.H. 1927. The calculation of atomic fields. *Proc. Cam. Philos. Soc* 23: 542.

Tuan, H.Y., Lee, D.C., Hanrath, T. and Korgel, B.A. 2005. Germanium nanowire synthesis: An example of solid-phase seeded growth with nickel nanocrystals. *Chem. Mater.* 17: 5705–5711.

Turchi, P.E., Gonis, A., and Colombo, L. 1997. Tight-binding approach to computational materials science. *Materials Research Society Symposia Proceedings*, vol. 491, Boston, MA.

Wagner, R.S. 1967a. *Crystal Growth*. Pergamon Press, Oxford & New York, pp. 45–119.

Wagner, R.S. 1967b. Defects in silicon crystals grown by VLS technique. *J. Appl. Phys.* 38: 1554–1560.

Wagner, R.S. and Doherty, C.J. 1966. Controlled vapor-liquid-solid growth of silicon crystals. *J. Electrochem. Soc.* 113: 1300–1305.

Wagner, R.S. and Ellis, W.C. 1964. Vapor-liquid-solid mechanism of single crystal growth. *Appl. Phys. Lett.* 4: 89–90.

Wagner, R.S., Doherty, C.J., and Ellis, W.C. 1964a. Preparation + Morphology of crystals of Silicon + Germanium grown by vapor-liquid-solid mechanism. *J. Met.* 16: 761.

Wagner, R.S., Ellis, W.C., Arnold, S.M., and Jackson, K.A. 1964b. Study of filamentary growth of silicon crystals from vapor. *J. Appl. Phys.* 35: 2993–3000.

Wan, Q., Li, G., Wang, T.H., and Lin, C.L. 2003. Titanium-induced germanium nanocones synthesized by vacuum electron-beam evaporation: Growth mechanism and morphology evolution. *Solid State Commun.* 125: 503–507.

Wang, C.X., Hirano, M., and Hosono, H. 2006. Origin of diameter-dependent growth direction of silicon nanowires. *Nano Lett.* 6: 1552–1555.

Wang, D. and Dai, H. 2002. Low-temperature synthesis of single-crystal germanium nanowires by chemical vapor deposition. *Angew. Chem. Int. Ed.* 41: 4783–4786.

Wang, D. and Dai, H. 2006. Germanium nanowires: From synthesis, surface chemistry, and assembly to devices. *Appl. Phys. A: Mater. Sci. Process.* 85: 217.

Wang, D., Chang, Y.L., Wang, Q. et al. 2004. A novel method for preparing carbon-coated germanium nanowires. *J. Am. Chem. Soc.* 126: 11602.

Wang, D.W., Tu, R., Zhang, L., Dai, H.J. 2005. Deterministic one-to-one synthesis of germanium nanowires and individual gold nanoseed patterning for aligned arrays. *Angew. Chem. Int. Ed.* 44: 2925.

Wang, D., Wang, W., Javey, A., Tu, R., and Dai, H. 2003. Germanium nanowire field-effect transistors with SiO2 and high-kappa HfO2 gate dielectrics. *Appl. Phys. Lett.* 83: 2432.

Wong, W.W., Sheehan, P.E., and Lieber, C.M. 1997. Nanobeam mechanics: Elasticity, strength, and toughness of nanorods and nanotubes. *Science* 277: 1971.

Wu, Y. Cui, Y., Huynh, L., Barrelet, C.J., Bell, D.C., and Lieber, C.M. 2004a. Controlled growth and structures of molecular-scale silicon nanowires. *Nano Lett.* 4: 433–436.

Wu, J., Punchaipetch, P., Wallace, R., and Coffer, J. 2004b. Fabrication and optical properties of erbium-doped germanium nanowires. *Adv. Mater.* 16: 1444.

Yu, B., Sun, X.H., Calebotta, G.A., Dholakia, G.R., and Meyyappan, M. 2006. One-dimensional germanium nanowires for future electronics. *J. Clust. Sci.* 17: 579.

Yu, H., Li, J., Loomis, R.A., Wang, L.W., and Buhro, W.E. 2003. Two- versus three-dimensional quantum confinement in indium phosphide wires and dots. *Nat. Mater.* 2: 517.

One-Dimensional Metal Oxide Nanostructures

Binni Varghese
National University of Singapore

Chorng Haur Sow
National University of Singapore

Chwee Teck Lim
National University of Singapore

17.1 Introduction

Metal oxides have numerous technological applications and provide an excellent platform to study various fundamental physical processes and phenomena existing in the material systems. Metal oxides crystallize in a multitude of crystal structures and exhibit diverse properties. For example, several metal oxides have the ability to undergo reversible surface oxidation and reduction processes due to the adsorption of certain specific gases (Henrich and Cox 1994). The adsorption of gas molecules results in band bending at the surface and effectively modifies the surface conductivity. Such a change in the conductivity due to gas adsorption is readily detectable in most of the metal oxides, including SnO_2, In_2O_3, and ZnO. Deviations in the properties of metal oxides due to the adsorption of specific gases render them as potential gas sensors. Transition metal oxides, in particular, are attractive for their range of properties (Rao 1999). This is partly due to the partially filled d-orbital and the mixed valency of the constituent transition metal atoms and the defect-induced self-doping capability. Transition metal oxides are potentially useful in a variety of applications including as a catalysis in the petroleum industry, magnetic data storage in information technology, and gas sensing.

Surface processes play key roles in various applications of metal oxides. Nanostructures of metal oxides with huge surface areas are therefore valuable for many potential applications. A large surface area of nanostructures could result in the improvement of material functionalities. In addition to the properties that originate from the large surface area, confinement effects in low-dimensional systems are expected to provide additional properties that can be tuned by varying the physical size or shape. The confinement effect occurs when the size of the nanostructures are comparable to the characteristic length scale of the physical properties of interest (typically in the sub-ten nanometer range). Nanostructures hold great promise in device applications where small size, faster operation, and high density integration is of great importance. In addition to the technological applications, studies on nanometric metal oxide structures may aid in the improvement of our understanding of various fundamental physical phenomena associated with metal oxides.

High quality nanostructures of metal oxides with a tailored geometrical size and shape are needed for studying their behavior at the nanometric regime. In general, nanostructures can be fabricated either by lithography based "top down" approaches or self assembly based "bottom up" approaches. For metal oxides, "bottom up" approaches were found to be more effective in crafting structures at the nanometric regime. Over the years, a variety of metal oxide nanostructures with varied dimensionality and morphologies including nanoparticles (Jun et al. 2006), nanowires (NWs) (Lu et al. 2006, Wang 2007b), nanobelts (Rao and Nath 2003, Wang 2004a, 2007b), nanorods (Patzke et al. 2002), nanotubes (NTs) (Patzke et al. 2002, Bae et al. 2008), core-shell, and other complex hierarchical structures (Rao and Nath 2003, Wang 2004a, Jun et al. 2006, Lu et al. 2006, Wang 2007b) were synthesized by adopting various "bottom up" methods. Once appropriate nanostructures are obtained, one can explore various properties and nanoscale phenomena using these structures. Due to the size-dependent properties, there is a need for the characterization of individual nanostructures. Techniques that allow the manipulation and investigation of properties of individual metal oxide nanostructures are in the forefront of low-dimensional material research.

In this chapter, an overview of the research activities on one-dimensional (1D) metal oxide nanostructures is presented. The chapter is organized as follows. After this short introduction, various available methods for the synthesis of 1D metal oxide nanostructures are described. The relative merits and demerits of each synthesis approach to create nanostructured metal oxides are highlighted. Following this, selected physical properties of 1D metal oxide nanostructures are discussed. The techniques adopted for the characterization of individual metal oxide nanostructures are detailed. In addition, properties of metal oxide nanostructures that led to the discovery of various prototype nanodevices are emphasized. This chapter ends with some remarks on the future perspective of 1D metal oxide nanostructures.

17.2 Controlled Synthesis of 1D Metal Oxide Nanostructures

The discovery of carbon nanotubes (CNTs) in 1991 (Ijima 1991) and the realization of their amazing physical properties stimulated interest in inorganic nanomaterials as well. Over the years, efficient methods have been established to synthesize metal oxide nanostructures with fine control over their chemical composition, crystal structure, dimensionality, size, and shape. Depending on the medium in which nanostructures are formed, the growth techniques are broadly classified as (1) liquid phase growth and (2) vapor phase growth. In the following section, various synthesis techniques are described with a special emphasis on the new developments in the field.

17.2.1 Vapor Phase Growth

In vapor phase growth, nanostructures are formed from gaseous state precursor reactants. Using vapor phase techniques, highly crystalline, contamination-free nanostructures can be synthesized. The major advantage of vapor phase growth is the feasibility of manipulating and organizing nanostructures during their growth. In addition, hybrid and complicated nanostructures with multiple functionalities can be synthesized by vapor growth techniques. The much needed impurity doping, which is essential for constructing various nanodevices, can be realized using vapor phase growth methods. It is customary to further divide the vapor phase growth techniques in terms of the governing mechanisms. Different vapor phase growth strategies used for the growth of metal oxide nanostructures are elaborated in the following section.

17.2.1.1 Vapor–Liquid–Solid Growth

The growth of micron sized whiskers from gas phase reactants on substrates covered with metal impurities was developed more than 40 years ago (Wagner and Treuting 1961, Wagner and Ellis 1964). When metal coated substrates are annealed above a certain temperature, the metal film melts and forms droplets. Due to the high sticking coefficient of the liquid as compared with the solid substrate, the reactant gases adsorb on the metal

droplet surfaces. Such adsorbed gas molecules undergo surface and bulk diffusion in the metal droplet and form a eutectic mixture (liquid). As the metal droplet supersaturates with the precursor atoms or compounds, phase segregation occurs, leading to the formation of nuclei at the droplet–substrate interface. Subsequent growth occurs as more and more atoms joined to the nuclei at the liquid–solid interface. The metal droplet functions as a virtual template by promoting crystal growth at the liquid–solid interface and restricting growth in other directions. The metal droplet remains at the tip of the resultant nanostructure and solidifies in the post-growth cooling phase to form a nanoparticle. The appearance of such nanoparticles at the tip of the nanostructures indicates the vapor–liquid–solid (VLS) growth mechanism. VLS routes often promote anisotropic growth leading to the formation of 1D nanostructures. The use of a metal catalyst and the formation of an eutectic mixture largely reduces the activation energy required for the growth of nanostructures via the VLS route compared with noncatalytic growth. Moreover, the growth conditions can be retrieved from the binary phase diagram of the metal component of the targeted nanostructure and the catalyst metal.

The diameter of the as-grown structures is largely determined by the size of the metal droplet. For a sustained growth via the VLS route, the stability of the catalyst liquid droplet is essential. Using thermodynamic considerations, the minimum equilibrium size of a metal droplet can be expressed as (Tan et al. 2003, Li et al. 2007b)

$$r_{\mathrm{m}} = \frac{2\Omega_{\mathrm{l}}\sigma_{\mathrm{lv}}}{k_{\mathrm{B}}T \ln S}, \tag{17.1}$$

where

Ω_{l} is the volume of an atom in the liquid
σ_{lv} is the liquid–vapor surface energy
k_{B} is the Boltzmann constant
T is the temperature
S is the degree of supersaturation

This sets a limit for the smallest achievable size of a nanostructure by VLS growth. In addition, the morphology, density, and size of the 1D structure formed by VLS growth are dependent on the nature of the metal catalysts (Song et al. 2005b). Studies showed that metals that have a low meting point and are oxidation resistant are found to have a better catalyzing capability. For example, Figure 17.1 shows the SEM images of SnO_2 NWs grown via the VLS route by using different catalyst metals (Nguyen et al. 2005).

Thermal chemical vapor deposition (CVD) technique can be employed to grow nanostructures through the VLS route. In a typical setup, a tube furnace with a vacuum-sealed ceramic tube is used. One end of the ceramic tube is connected to a vacuum pump and the other end is connected to gas cylinders through mass flow controllers. Substrates coated with a catalyst metal thin film are placed inside the tube furnace and heated to form

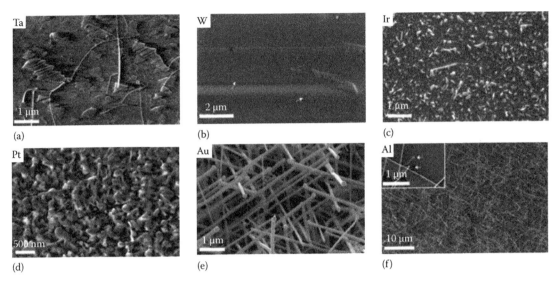

FIGURE 17.1 SEM images of SnO$_2$ NWs grown on α-sapphire substrate via VLS mode using different metal catalysts (as indicated in the top left corner). (From Nguyen, P. et al., *Adv. Mater.*, 17, 1773, 2005. With permission.)

metal droplets. The gas phase precursors are introduced at an optimal flow rate into the tube furnace. The pressure inside the ceramic tube is regulated and controlled.

The availability of precursor gases is an issue in the VLS-based nanostructure synthesis. The reactant gases can be produced by evaporating respective metals or metal nitrides in the presence of oxygen (Choi et al. 2000, Guha et al. 2004, Johnson et al. 2006). Yang et al. initiated a method to create the reactant gases using carbothermal reduction of metal oxide powder and successfully synthesized ZnO NW using the VLS route by using gold as the catalyst metal (Huang et al. 2001b, Yang et al. 2002). The ZnO was first reduced by carbon into Zn and CO/CO$_2$ in the high temperature zone of the tube furnace. The Zn metal evaporated and transported to the substrates placed at the low temperature zone. This is followed by a metal catalyst–assisted growth of ZnO NWs. The density and diameter of the as-synthesized NWs is controlled by the thickness of the gold catalyst film. NWs with diameters as small as 40 nm can be synthesized using gold as the catalyst. A number of researchers have utilized the carbothermal reduction–assisted VLS method to synthesize 1D nanostructures of SnO$_2$, Ga$_2$O$_3$, In$_2$O$_3$, Al$_2$O$_3$, ZnO, and V$_2$O$_5$ (Kam et al. 2004, Rao et al. 2004, Nguyen et al. 2005, Song et al. 2005, Zhang et al. 2005).

The laser ablation–assisted VLS growth developed by Lieber's group was found to be effective for the growth of many 1D semiconductor nanostructures (Morales and Lieber 1998). In laser-assisted VLS growth, a high-intensity laser beam evaporates the target containing NW material and condensates on a substrate with catalyst metal clusters. The laser ablation–assisted VLS growth was used for growing metal oxide nanostructures including In$_2$O$_3$ NWs (Stern et al. 2006) and ZnO NWs (Son et al. 2007). Pulsed laser deposition (PLD) was also used for vaporizing respective bulk materials and subsequent growth of nanostructures by the VLS route (Morber et al. 2006, Son et al. 2007).

In some cases, the VLS growth of nanostructures can take a different route in which the constituent metal of the targeted oxide nanostructure itself functions as the catalyst. The governing mechanism of such growth is usually denoted as a self-catalytic VLS mechanism (Mohammad 2006). In addition to its simplicity, the self catalytic growth avoids the unintentional doping of the nanostructures due to the use of a foreign metal catalyst. Many metal oxide nanostructures such as dentritic ZnO NWs (Fan et al. 2004a), SnO$_2$ NWs (Chen et al. 2003, 2004a), CuO nanofibers (Hsieh et al. 2003a), indium doped tin oxide NWs (Chen et al. 2004c), and Al$_4$B$_2$O$_9$ NWs (Liu et al. 2003b) were synthesized following the self-catalytic growth.

Nanostructures of mixed metal oxides or impurity doping can be achieved by choosing a mixture of appropriate source materials or gas phase components (Wan et al. 2006). By a one step evaporation method using a mixture of In and Sn as the source for reactant vapor production, SnO$_2$–In$_2$O$_3$ heterostructured NWs have been produced (Kim et al. 2007). The SnO$_2$ NWs covered with a In$_2$O$_3$ shell were formed most likely due to the difference in the bulk and surface diffusion coefficients of the InO$_x$ and SnO$_x$ species in the catalyst droplet.

The VLS approach has the great advantage of yielding high-quality single-crystalline nanostructures. In most occasions, the VLS grown nanostructures are dislocation free. The morphology of the nanostructures formed by the VLS route depends on the selection of catalyst particles, source material, thickness of the catalyst layer, and growth duration (Yang et al. 2002, Ng et al. 2003, Nguyen et al. 2005, Zhang et al. 2007a). By precisely adjusting the catalyst layer thickness, Ng et al. demonstrated the possibility of creating 1D and two-dimensional (2D) ZnO structures on different substrates (Figure 17.2; Ng et al. 2003).

The VLS route has the feasibility of manipulating and positioning the NWs during growth (Fan et al. 2006c). For many applications, the proper alignment and precise positioning of nanomaterials is necessary. In the VLS route, aligned

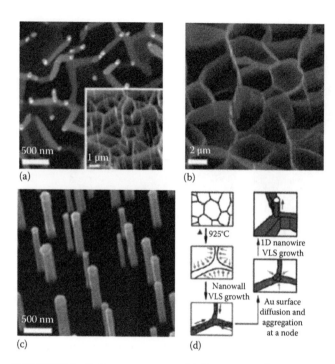

FIGURE 17.2 ZnO nanostructures of different morphologies synthesis via VLS route by choosing appropriate catalyst layer thickness. (a) SEM image of quasi-3D ZnO nanostructures grown on a sapphire using ~40–50 Å Au thin film as the catalyst. The inset shows a SEM perspective view. (b) ZnO nanowalls on a sapphire substrate. (c) ZnO NWs on a highly ordered pyrolytic graphite substrate using ~15 Å thick Au film as the catalyst. (d) Schematic illustration showing the growth mechanism of ZnO nanowalls and NWs. (From Ng, H.T. et al., *Science*, 300, 1249, 2003. With permission.)

nanostructures can be produced by using lattice matching substrates. For example, vertically aligned ZnO nanostructures can be obtained with substrates like sapphire (Huang et al. 2001, Yang et al. 2002), GaN (Fan et al. 2006a), and SiC (Ng et al. 2004). Positioning of the nanostructures can be accurately achieved by the VLS route using various catalyst patterning strategies (Wang et al. 2004).

The actual growth mechanism of metal oxide nanostructures using the VLS method is complicated due to the presence of oxygen. The mechanism that governs the growth of metal oxide nanostructures through the VLS route still remains controversial. Nanostructures of ZnO, for example, can form via the VLS route for a broad range of temperatures. The state of the catalyst alloy particle (solid or liquid) during the growth over this entire range of temperature is unclear. Campos et al. proposed a vapor–solid–solid (VSS) growth mechanism instead of the VLS mechanism for the growth of ZnO using gold as the catalyst at low temperatures (Campos et al. 2008).

17.2.1.2 Vapor–Solid Growth

The growth of nanostructures from gas phase reactants could be possible even in the absence of any metal catalyst. Gas phase precursor reactants of the targeted nanomaterial are directly adsorbed on the substrates, followed by nucleation and the

subsequent growth of nanostructures. Since the gaseous reactants directly condense into solid structures, the governing mechanism is known as the vapor–solid (VS) mechanism. Probability in the formation of nuclei via the vapor–solid process can be expressed as (Blakely and Jackson 1962, Dai et al. 2003)

$$P_n = A \exp\left(\frac{-\pi\sigma^2}{k_B T^2 \ln\alpha}\right), \qquad (17.2)$$

where

> A is a constant
> σ is the surface energy
> α is the supersaturation ratio
> T is the temperature in Kelvin
> k_B is the Boltzmann constant

The supersaturation ratio is given by, $\alpha = p/p_0$, with p as the vapor pressure and p_0 as the equilibrium vapor pressure of the condensed phase at the same temperature.

Similar to the VLS method, the thermal CVD technique can be used for growing nanostructures via the VS route. The source material is normally placed at a high temperature zone of the furnace. The substrates to support the nanostructures were located at a lower temperature zone. The reactant gases were first formed by using techniques such as thermal evaporation (Zhang et al. 1999, Pan et al. 2001, Wang 2003, Zhou et al. 2003a,b, Lilach et al. 2005, Zhou et al. 2005a, Chueh et al. 2006, Zhao et al. 2006) of the respective source materials. Reactants were then transported by carrier gas to the substrate kept at a favorable temperature. The resultant morphology of the nanostructures largely depends on the substrate temperature, processing pressure, carrier gas flow rate, and source material (Wang 2003). Pan et al. reported a versatile approach to create metal oxides in a unique nanobelt morphology by direct evaporation of the respective metal oxide powders without using any metal catalysts (Pan et al. 2001). Despite the crystallographic structure diversity among binary oxides including ZnO, SnO_2, In_2O_3, CdO, and Ga_2O_3, nanobelts are readily formed via the VS route (Figure 17.3). Using the VS route, nanostructures including ZnO nanotubes (Mensah et al. 2007), ZnO NWs (Umar et al. 2005), and nitrogen doped tungsten oxide NWs (Chang et al. 2007) were also synthesized.

The versatility of creating complex hierarchical nanostructures via the VS route has been established. Such capability will facilitate our efforts to achieve high density integration of nanostructure assembly. Following the VS route, ZnO comb-like structures (Wang et al. 2003) and three-dimensional (3D) WO_{3-x} NW networks (Zhou et al. 2005b) have been reported. Lao et al. prepared hierarchical ZnO nanostructures on In_2O_3 NWs by using ZnO, with In_2O_3 and graphite powders as source materials (Lao et al. 2002). The InOx vapors first evaporated and formed In_2O_3 NWs on the collector substrates. Then ZnOx vaporized and the secondary growth produced branches on the already existing In_2O_3 NW sidewalls. The radial In_2O_3–SnO_2 heterostructure was also reported by

FIGURE 17.3 (a) Metal oxide nanobelts via VS route. (b) SEM and TEM image of the In_2O_3 nanobelts, respectively. (c) A nanobelt with an abruptly reduced width. (d) SEM image of CdO nanobelts and sheets. (e, f) TEM images and a corresponding electron diffraction pattern of the CdO nanobelts. (From Pan, Z.W. et al., *Science*, 291, 1947, 2001. With permission.)

Vomiero et al. using the VS approach (Vomiero et al. 2007). In another example, Sun et al., produced SnO_2 hierarchical nanostructures in a multi-step thermal evaporation method (Sun et al. 2007).

To fabricate vertically aligned structures through the VS route, one can either choose a lattice matching foreign substrate to promote heterogeneous epitaxial growth or a seed layer for homogeneous epitaxial growth (Fang et al. 2007, Li et al. 2007). Figure 17.4 displays the SEM images of vertically well-aligned ZnO NWs on substrates with a ZnO seed layer. The structural characterizations reveal the high crystalline quality of such synthesized NWs. In addition, with modification of the surface roughness of the substrates with nonmatching lattices, one can also effectively improve the alignment of nanostructures via the VS route (Ho et al. 2007).

The exact physical mechanism that governs the anisotropic growth of nanostructures via the VS route is not clear. The morphology of the resultant nanostructures is found to be largely determined by the anisotropy in the growth rates of different crystallographic surfaces. Certain crystal surfaces have relatively higher surface energy and tend to grow faster to minimize the total energy of the system resulting in anisotropic crystal growth. In addition, the presence of defects like screw dislocations also facilitates the growth in the VS process.

17.2.1.3 Template-Assisted Growth

The use of proper templates to direct crystal growth is a versatile technique for producing monodisperse metal oxide nanostructures. By using the template-assisted approaches, various compositions of materials can be crafted at the nanometric regime. Either negative templates with nanosized pores (e.g., anodic alumina (AAO), track etched polycarbonate films) or positive templates (e.g., NWs and CNTs) can be used as scaffolds to confine crystal growth. The use of templates to create oxide nanostructures was first reported in the early 1990s. Early efforts on the template-assisted synthesis of metal oxide nanostructures were focused on CNTs as positive templates (Ajayan et al. 1995). In the CNT templating method, the surface of the CNT is first coated with desired metal oxide. This is followed by the removal of the CNT templates either by thermal heating or by chemical means.

Recently, many other 1D nanostructures were employed as a template for creating various metal oxide nanostructures. The epitaxial deposition of technologically important mixed oxides such as superconducting YBCO, magnetic LCMO, ferroelectric PZT, and Fe_3O_4 on vertically oriented MgO NWs by pulsed laser deposition has been reported (Han et al. 2004). Figure 17.5 describes the experimental procedure and the results of the morphological and structural characterizations of as-obtained MgO/YBCO core-shell structures. A similar approach was used for the creation of MgO/titanate hetrostructures (Nagashima et al. 2008).

The NW templating method can be effectively used for the realization of nanostructures of structurally complicated multinary metal oxides by various solid–state reaction mechanisms (Chang and Wu 2007, Yang 2005, Fan et al. 2006b). Well-aligned β-Ga_2O_3 NWs were coated with ZnO using the metal organic CVD technique and subsequent annealing at 1000°C in an O_2 atmosphere produced Ga_2O_3/$ZnGa_2O_4$ core-shell NWs, single-crystalline $ZnGa_2O_4$ NWs, and $ZnGa_2O_4$ NWs inlaid with ZnO nanocrystals (Chang and Wu 2005). Spinel Zn_2TiO_4 NWs were synthesized by coating ZnO NWs with Ti and subsequent annealing at 800°C in a low vacuum condition. Heating causes a solid–state reaction via diffusion of Ti atoms into the ZnO leading to the phase transformation from wurtzite ZnO to spinel Zn_2TiO_4 (Yang et al. 2007). In another example, depositing Al_2O_3 on ZnO NWs using the atomic layer deposition (ALD) technique and subsequent annealing of the resultant ZnO–Al_2O_3 core-shell structures produced spinel $ZnAl_2O_4$ nanotubes by using the nanoscale kirkendall effect (Fan et al. 2006b).

17.2.1.4 Direct Growth by Solid–Vapor Interaction

Whisker or needle-shaped metal oxide structures have drawn the attention of the scientific community as early as the 1950s (Cowley 1954, Takagi 1957). Microscopic studies on pure metal pieces thermally oxidized in air or in an oxygen environment revealed the presence of whiskers grown perpendicular to the metal surface. The anisotropic growth of oxides along a certain crystal axis is believed to be due to the presence of screw

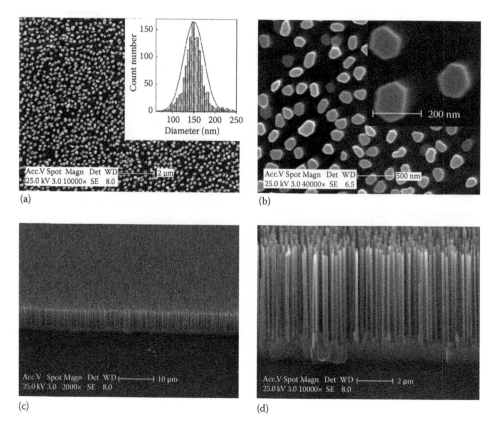

(a)

(b)

(c)

(d)

FIGURE 17.4 SEM images of vertically aligned ZnO nanorod arrays using ZnO seed layer. (a and b) Top view. (c and d) 50° tilted view. (From Li, C. et al., *J. Phys. Chem. C*, 111, 12566, 2007a. With permission.)

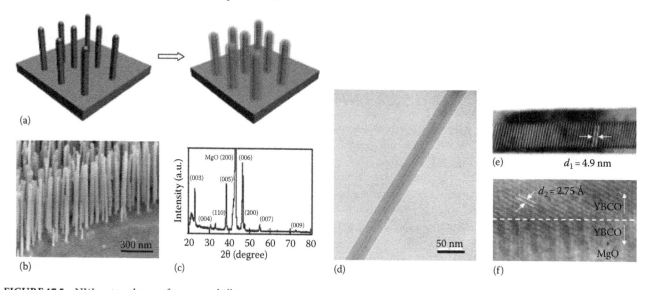

(a)

(b)

(c)

(d)

(e) $d_1 = 4.9$ nm

(f)

FIGURE 17.5 NWs as templates to form core-shell nanostructures. (a) Schematic illustration of vertically aligned MgO NWs and core-shell NWs consisting of MgO NWs coated with the desired material by PLD process. (b) SEM image of YBCO NWs after the PLD process. (c) XRD data of as-grown YBCO NWs on a MgO (100) substrate. (d) Low magnification TEM image showing the MgO/YBCO coreshell structure. (e) TEM image of MgO/YBCO NWs. (f) HRTEM image of the MgO/YBCO NW. (From Han, S. et al., *Nano Lett.*, 4, 1241, 2004. With permission.)

dislocations (Cowley 1954). Another type of whiskers with pores along their axes was grown when beryllium metal was heated in a silica furnace tube in hydrogen with a trace of water vapor (Edwards and Happel 1962). In this particular growth mode, a metal ball always appeared at the tip of the whisker. By heating a W foil that is partly covered by a SiO_2 plate in the Ar atmosphere at ~1600°C, Zhu et al. observed the formation of tungsten oxide tree-like microstructures with nanoneedle branches (Zhu et al. 1999). Gu et al. reported the formation of tungsten oxide NWs on W wires/foil by heating in an Ar atmosphere (Gu et al. 2002).

They suggested that the mechanism of formation of NWs on the clean metal surface may be governed by the VS mechanism. Recently, a number of reports presented the direct growth of metal oxides on the respective metal surfaces heated at the right conditions (Dang et al. 2003, Fu et al. 2003, Xu et al. 2004. Wen et al. 2005, Rao et al. 2006, Varghese et al. 2008b).

17.2.1.4.1 Hotplate Method

We have developed a simple yet efficient method of growing metal oxide nanostructures in large quantities by heating metal foils in ambient conditions on a thermal hotplate. Using this technique, vertically oriented α-Fe$_2$O$_3$ nanoflakes (Zhu et al. 2005, Yu et al. 2006), Co$_3$O$_4$ nanowalls (Yu et al. 2005c), CuO NWs (Yu et al. 2005b), and CuO–ZnO hybrid nanostructures (Zhu et al. 2006) have been synthesized. The coverage and size of the hotplate grown nanostructures can be controlled by varying the growth time as shown in Figure 17.6 (exemplified by the growth of α-Fe$_2$O$_3$ nanoflakes). Surprisingly, the hotplate method produced metal oxide nanostructures with high crystalline quality. Being at a low temperature (200°C–550°C) and a catalyst-free method, the hotplate technique is particularly attractive. The morphology and size of the nanostructures can be controlled by simply varying the growth duration (Yu et al. 2005). The hotplate method for the direct growth of nanostructures on metal foils has advantages of low cost and large scale production.

The mechanism that governs the direct growth of oxide nanostructures on respective metal substrates is not well understood. Yu et al. proposed a solid–liquid–solid (SLS) mechanism for the direct growth (Yu et al. 2005c). Due to heating, surface melting occurs even at low temperatures. The adsorbed oxygen atoms from the surroundings react with the melt forming various sub-oxides of the metal. This is followed by the nucleation of the most stable oxide phase. Surface diffusion of the constituent atoms towards the nucleated crystals fueled the growth of the nanostructures. The morphology of the resultant nanostructures is most likely controlled by kinetic factors.

17.2.1.4.2 Plasma-Assisted Direct Growth

Plasma-assisted direct growth is another feasible direct growth method for producing nanostructures of low melting metal (for example, gallium) oxides in large quantities. Sharma et al. synthesized β-gallium oxide tubes, NWs, and nano-paintbrushes by heating molten gallium in microwave plasma containing mono-atomic oxygen and hydrogen mixture (Sharma and Sunkara 2002). Using the plasma-assisted direct growth technique, Varghese et al. have demonstrated the feasibility of tailoring morphology of metal oxide nanostructures (Varghese et al. 2007, 2008a). Here, the respective metal foils were heated in the presence of oxygen at low pressure conditions in a vacuum chamber. The chamber is equipped with a RF plasma generator. By varying the plasma power, vertically oriented Co$_3$O$_4$ NWs, nanowalls, or a mixture of these two structures were successfully synthesized (Figure 17.7). Similar techniques can be used for the creation of NiO nanowalls and nanoflake-like morphologies. Such control over the morphology of the nanostructures is believed to be due to the variation in the rate of oxide formation with plasma power.

17.2.2 Liquid Phase Growth

Metal oxide nanostructures with controlled size, shape, and structure can be synthesized by solution-based methods using

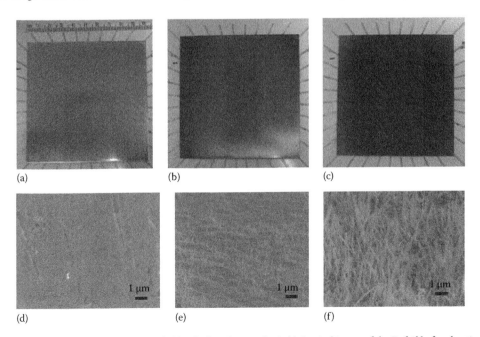

FIGURE 17.6 Synthesis of α-Fe$_2$O$_3$ nanoflakes on Fe foil by the hotplate method. (a) Optical image of the Fe foil before heating. (b and c) Optical images of the Fe foil after heating at 300°C for 10 min and 24 h, respectively. (d–f) Corresponding SEM images of the foil surfaces shown in (a–c). (From Yu, T. et al., *Small*, 2, 80, 2006. With permission.)

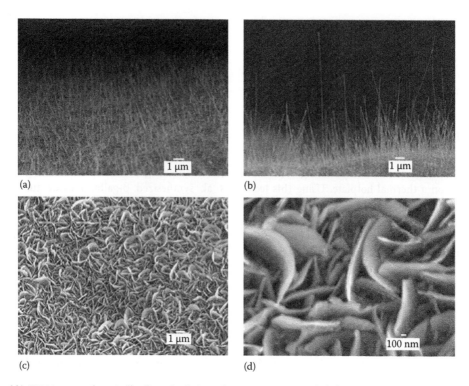

(a)

(b)

(c)

(d)

FIGURE 17.7 (a and b) SEM images of vertically aligned cobalt oxide NWs grown on cobalt foil, by heating Co foils in the presence of oxygen. (c and d) SEM images of cobalt oxide nanowalls grown by heating the Co foils in an oxygen plasma. (From Varghese, B. et al., *Adv. Funct. Mater.*, 17, 1932, 2007. With permission.)

relatively simple laboratory equipments. In solution-based methods, metal precursors are dissolved in appropriate solvents and the nucleation and growth of the nanostructures are controlled by the degree of supersaturation, temperature, pH value, etc. Due to the large surface energy associated with the nano entities, suitable surfactants are employed to stabilize and/or direct the growth of nanostructures. The solution-based growth can be broadly classified as (1) aqueous, (2) nonaqueous, and (3) template-assisted solution routes. In the aqueous solution process, appropriate metal salts are dissolved in water and the oxide is formed when the mixture is heated at certain optimum temperatures. On the other hand, the nonaqueous solvothermal process involves mostly organic solvents as the growth medium. The shape and morphology control is achieved by properly selecting the growth parameters or using appropriate surfactants/ligands to direct the growth. In the template-assisted method, the 1D growth control is achieved by confining the growth inside nanopores or channels. A brief note on these techniques is provided in this chapter.

17.2.2.1 Aqueous Solution Route

The most common solution-based synthesis of metal oxide nanostructures is the aqueous solution method in which the chemical reaction leading to the formation of nanostructures takes place in the presence of water. Nanostructured metal oxides, particularly transition metal oxides, in the form of spherical or faceted nanoparticles (Seshadri 2004) to highly anisotropic NWs or nanotubes (Vayssieres et al. 2001a,b, Vayssieres 2003) have

been synthesized via the aqueous solution method. Normally, metal alcoxides or metal halides are used as the metal precursors. The purity as well as crystal quality of nanostructures can be improved by carrying out the synthesis at elevated temperatures. Such reactions can be conducted in a closed container like autoclave at high pressure. The high pressure allows the reaction temperature to be higher than the boiling point of water. The reaction in such a closed system at high temperatures above the boiling point of solvents is known as the solvothermal process. If the solvent is water, it is called a hydrothermal reaction. For a more detailed description on the hydrothermal synthesis of metal oxide nanostructures, please refer to recent reviews by Mao et al. (2007).

17.2.2.2 Nonaqueous Solution Route

Recently, the nonaqueous solution route to synthesize crystalline metal oxide nanostructures has attracted much attention (Niederberger 2007). In the nonaqueous solution route, the growth medium is usually organic solvents. Park et al. demonstrated the feasibility of using the nonaqueous method to synthesize ultra large scale metal oxide nanocrystals (Park et al. 2004).

Nanostructures of mixed metal oxides, which are otherwise hard to synthesize, can be controllably produced using nonaqueous solvothermal routes. O'Brien et al. reported a generalized method for synthesizing complex oxides like $BaTiO_3$ nanoparticles using an "injection-hydrolysis" protocol (O'Brien et al. 2001). In a typical experiment, barium titanium ethyl hexano-isopropoxide is injected into a mixture of biphenyl ether and stabilizing agent oleic acid at 140°C under argon or nitrogen. The mixture is cooled to 100°C and a 30 wt% hydrogen peroxide solution is injected through the septum (vigorous exothermic reaction). The solution is maintained in a close system and stirred at 100°C over 48 h to promote further hydrolysis and crystallization of the product in an inverse micelle condition. Size control is achieved by varying the reagent concentration. Using this technique, ferroelectric $BaTiO_3$ nanoparticles 6–12 nm in size were synthesized. Single crystalline perovskite nanorods of $BaTiO_3$ and $SrTiO_3$ were synthesized by the decomposition of bimetallic alkoxide in the presence of coordinating ligants (Jeffrey 2002). The reaction carried out at a temperature of ~100°C in a mixture of heptadecane, H_2O_2, and oleic acid. The anisotropic growth is attributed to the precursor decomposition and crystallization in a structured inverse micelle medium formed by precursors and oleic acid under these reaction conditions.

Size and shape control in solution routes can be achieved by employing various strategies. An efficient means of obtaining nanostructures with a uniform size is through the Ostwald ripening process (Zeng 2007). The Ostwald ripening process occurs during the aging of the nanomaterial suspension by which the growth of bigger structures is facilitated at the cost of smaller ones due to size-dependent dissolution. Considerably narrow size distribution on the nano-products can be achieved by the separation of the nucleation and growth process. A principally different approach to form anisotropic nanostructures including metal oxides is the so-called oriented attachment of nanoparticles during aging (Penn and Banfield 1998, Pacholski et al. 2002). The oriented attachment refers to the process in which adjacent particles spontaneously self-organized to share a common crystallographic orientation (Penn and Banfield 1998). Metal oxide nanostructures with complex morphologies can be produced using the oriented attachment mechanism (Zitoun et al. 2005).

17.2.2.3 Template-Assisted Liquid Phase Growth

The use of nano-porous materials as a host for the synthesis of nanostructures was pioneered by Martin's group (Martin 1994, Parthasarathy and Martin 1994). Early attempts were focused on the synthesis of metals and conducting polymer structures via the template-assisted method. The filling of nanopores of the negative templates by means of solution-based techniques is a feasible way to synthesize metal oxide nanostructures. The subsequent removal of the template by selective etching yields nanostructures. As mentioned earlier, the most common negative templates used for the growth of nanostructures are AAO and track-etched polycarbonate. The AAO templates can be produced by anodizing pure Al foils in various acids. AAOs have high chemical, thermal, and mechanical stability, which makes them an ideal template for nanofabrication. Porous AAO templates with high nanopore density, in various pore sizes are now available. The solution-based filling of the template pores is either achieved by sol–gel chemistry or electrochemical route.

17.2.2.3.1 Sol–Gel Processing

The template-assisted sol–gel chemistry route is a viable method for producing nanostructures of many chemical compositions. In the sol–gel technique, a suspension of the colloidal sol of the materials was first prepared by the hydrolysis and polymerization of precursor molecules. Either inorganic metal salts or organic metal alkoxides can be used as precursors. The subsequent condensation of as-prepared sol yields the gel. The pores of the templates can be filled by the as-prepared sol by direct infiltration due to capillary action or the electrophorectic method (Shankar and Raychaudhuri 2005). The template-assisted sol–gel technique is particularly useful for many materials to be sculptured into 1D nanostructures (NWs or nanotubes) using the appropriate processing conditions. Lakshmi et al. first extended the template-assisted sol–gel method to produce an array of 1D metal oxides (Lakshmi et al. 1997). They demonstrated the feasibility of using the AAO template to create 1D nanostructures of TiO_2, ZnO, and WO_3 by the direct immersion of the template in respective sols prepared using the sol–gel chemistry approach. Due to the capillary action, the sols fill the pores of the AAO template. Heat treatment and the subsequent removal of the template by dissolution in aqueous NaOH solution yields the respective 1D nanostructures. The end-products can be nanotubes or NWs depending on the immersion time and sol temperature. Following this pioneering work, numerous materials were engineered into nanometric structures using the template-assisted sol–gel route. One of the advantages of using the well-developed sol–gel chemistry method is the possibility to control the stoichiometry of complex multi-component oxides that are not straightforward or impossible to achieve via vapor phase techniques (Zhou and Li 2002, Jian et al. 2004, Yang et al. 2006b). Recently, Kim et al. reported the preparation and ferroelectric properties of ultra-thin walled Pb(Zr,Ti)O_3 (PZT) nanotube arrays by using the AAO template-assisted sol–gel process (Kim et al. 2008). In their work, the infiltration of AAO nanopores was facilitated by spin coating. The schematic of the processing steps, images of the template, and products are shown in Figure 17.8.

17.2.2.3.2 Electrochemical Deposition

The electrochemical deposition in conjunction with templates is a viable low temperature method for the production of various metal nanostructures. The deposition is normally carried out in a conventional three electrode electrochemical bath with the template to be deposited configured as the cathode. Since most of the templates are insulating, a metal coating on one of the surfaces is essential in order to use them as electrodes. The salt solution of the metal to be deposited was used as the electrolyte. The production of metal oxide nanostructures by the electrochemical deposition route can be realized by either direct oxide deposition (Hoyer 1996, Zheng et al. 2002, Oh et al. 2004, Takahashi et al. 2004) or by adopting a post-oxidation protocol on the electrochemically deposited metal nanostructures (Yi et al. 2008).

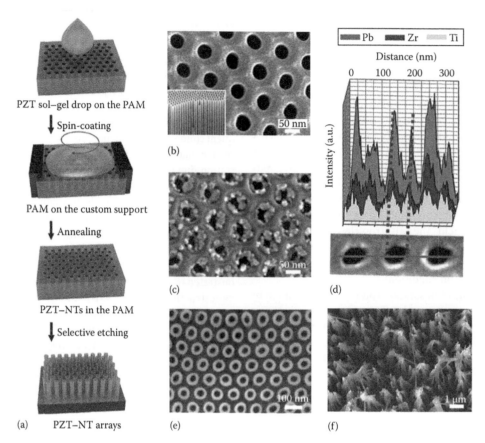

FIGURE 17.8 (a) A schematic illustration of the PZT–NT arrays synthesis procedure using template-assisted sol–gel route. (b) SEM image of a typical porous alumina membrane with a corresponding cross-sectional image shown in the inset. (c) SEM image of a alumina template after spin-coating with a PZT sol–gel solution. (d) The lower part of this figure is a STEM image and the upper part is the EDS line profile along the green line in the STEM image. The dotted red line highlights the periodic intensity of Pb, Zr, and Ti in the sectioned PZT–NTs. (e and f) SEM image of PZT–NTs after wet chemical etching for 15–30 min, respectively. (From Kim, J. et al., *Nano Lett.*, 8, 1813, 2008. With permission.)

17.3 Physical Properties of 1D Metal Oxide Nanostructures

17.3.1 Electrical Properties

Electrical properties of low-dimensional nanostructures show deviation from their bulk form. The variation in the electronic properties of materials with dimensionality can be explained on the basis of the difference in the electronic density of states (Yoffe 2002). In general, the density of states, $\rho \propto E^{D/2-1}$, where E is the energy and D is the dimensionality (3, 2, or 1 depending on whether it is 3D, 2D, or 1D). In addition, the spatial confinement in nanostructures causes a blueshift in their band gap with a reduction in size. The shift in band gap of nanostructures, $\Delta E_g \propto 1/d^2$, where d is the characteristic size of the nanostructure. Due to the varied degree of confinement, the band gap shift evolves differently with size in nanostructures of different dimensionality (Yoffe 2002, Yu et al. 2003). The evolution of band gap with size and dimension is displayed in Figure 17.9 (Yu et al. 2003). Here, the band gap energy was estimated from the energy states of electron and holes calculated using the simple particle-in-a-box model. Thus, the

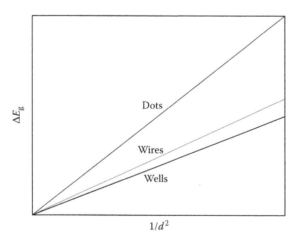

FIGURE 17.9 Predictions of band gap variation with size for 2D (wells), 1D (wires), and 0D (dots) materials. (From Yu, H. et al., *Nat. Mater.*, 2, 517, 2003. With permission.)

electric transport properties of nanostructures are expected to be dependent on the characteristic size and shape.

The electrical transport properties of nanostructures can be sensitively affected by the large electron scattering at the boundaries (Zhang et al. 2000). The large surface scattering results in an increase in the resistivity of the nanostructures compared with the bulk materials. This makes the electrical properties of the nanostructures sensitive and dependent on the surface and surrounding medium (Nanda et al. 2001).

Metal oxides exhibit the whole spectrum of conductivity, which ranges from metallic through semiconductor to insulators. The electrical properties of metal oxide nanostructures are particularly interesting as one can follow how the myriad of electrical phenomena observed in bulk evolve with size and shape at the nanometric regime. Over the years, research efforts have been focused on the transport properties of metal oxide nanostructures. In particular, the transport properties of 1D metal oxides have attracted tremendous attention owing to their possible dual role as functional electronic components as well as interconnects. An overview of the electric transport properties of metal oxide 1D nanostructures is given in the following sections.

17.3.1.1 Electrical Properties of 1D Metal Oxides

The electrical transport properties of 1D structures are usually determined by performing two-probe or four-probe measurements by laying them across metal electrodes with a few micron gap. The nanostructures are first dispersed on a substrate and then suitable metal electrodes are deposited using appropriate masking techniques. Alternatively, one can deposit the nanostructures on substrates that are pre-patterned with metal electrodes. In some cases, aligning strategies were employed to precisely place the nanostructures across the electrodes (Smith et al. 2000, Hu et al. 2006).

Studies suggested that for most of the nanostructured metal oxides investigated, the electrical properties resemble their bulk material (semiconducting, metallic, or insulating). However, significant quantitative variations from the bulk material properties are often observed. The *I–V* characteristics of CdO nanoneedles (Liu et al. 2003a), RuO_2 NWs (Liu et al. 2007b), and ITO NWs (Wan et al. 2006) showed metallic behavior. Whereas ZnO NWs (Kang et al. 2007), SnO_2 NWs (Ramirez et al. 2007), VO_2 nanobelts (Liu et al. 2004), V_2O_5 NWs (Muster et al. 2000), $W_{18}O_{49}$ NWs (Shi et al. 2008), and Nb_2O_5 NWs (Varghese et al. 2009) exhibited semiconducting *I–V* characteristics.

A semiconductor NW with metal pads on either side can be modeled as a metal–semiconductor–metal circuit. Due to the work function difference between the NW material and the metal pads, the two contacts form Schottky-type barriers. This circuit is equivalent to two Schottky diodes connected back to back through a semiconductor (Zhang et al. 2006a). The *I–V* characteristics of such circuits at an intermediate bias condition are determined by the reverse biased Schottky junction. Assuming the thermionic field emission theory for the reverse biased Schottky junction, the current–voltage relationship for such circuits at intermediate bias conditions can be expressed as (Zhang et al. 2006, 2007b)

$$\ln I = \ln(S) + V\left(\frac{e}{kT} - \frac{1}{E_0}\right) + \ln J_s, \qquad (17.3)$$

where

S is an area factor

e is an electronic charge

k is the Boltzmann constant (1.38×10^{-23} m^2 kg s^{-2} K^{-1})

T is the temperature

J_s is a slowly varying function of the applied voltage

E_0 is a function of the carrier density (n) and is given by the equation

$E_0 = E_{00} \coth(eE_{00}/kT)$, where $E_{00} = (\hbar/2)(n/(m^\star \, \varepsilon_s \, \varepsilon_0))^{1/2}$, with $m^\star$ as the effective mass of electrons in the NW material, ε_s is its relative permittivity, ε_0 is the permittivity of free space, and $\hbar$ is the Planks constant.

This implies that the slope of $\ln(I)$ versus the voltage curve can be approximated to $e/kT - 1/E_0$. Through this analogous, we can retrieve the various electronic transport parameters of the NWs from their respective *I–V* curves. According to this theoretical formulation, Zhang et al. estimated the carrier density in ZnO NW to be ~1.0×10^{17} cm^{-3}, which is of the same order of magnitude as the bulk ZnO (Zhang et al. 2006a).

17.3.1.2 Nanowire Field Effect Transistors

Quantitative information regarding the carrier type (electron or holes), carrier density, and carrier mobility could be retrieved from measurements on a NW designed in a three terminal device configuration. The working principle is analogous to the field effect transistor (FET) in the micro electronic industry. Typically, NWs are dispersed on a degenerately doped Si substrate with an SiO_2 over-layer. This is followed by patterned electrode fabrication using lithography techniques. More advanced metal deposition techniques using electron beam lithography or focused ion beam (FIB) techniques can be used to selectively deposit metals to perform single NW level measurements. The metal contacts on either side of the NW could function as source (S) and drain (D) electrodes and the bottom Si bulk can be used as the gate electrode. At the moderate doping level, the Debye screening length (λ_d) of most of the metal oxides is in the range of 10–100 nm (Kolmakov and Moskovits 2004). This implies that one can control the current through the NW by varying the gate voltage (V_g). In such configurations, the total charge on the NW can be expressed as (Martel et al. 1998) $Q = CV_{gT}$, where C is the NW capacitance with respect to the back gate and V_{gT} is the threshold gate voltage required to completely deplete the carriers from the channel. By assuming the NW as a metallic cylinder,

$$C = \frac{2\pi\varepsilon\varepsilon_0 L}{\ln(2h/r)}, \qquad (17.4)$$

where

L is the NW length

r is its radius

h is the thickness of the SiO_2 layer

ε is the average dielectric constant of the NW material

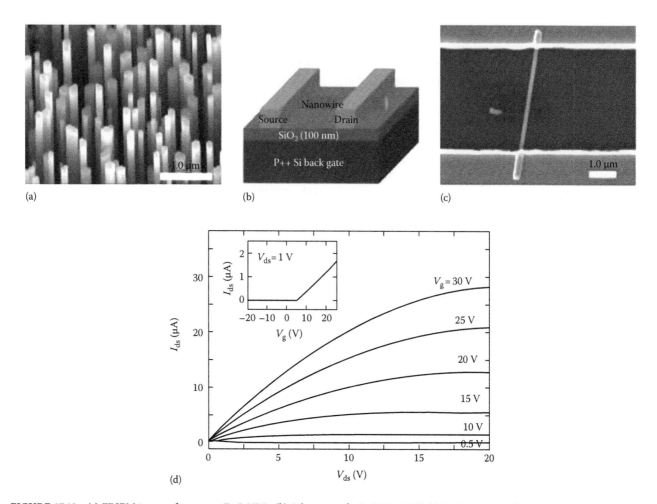

FIGURE 17.10 (a) FESEM image of as grown ZnO NWs. (b) Schematic of a ZnO NW FET. (c) FESEM image of a ZnO NW across metal electrodes. (d) Typical I_{ds}–V_{ds} curves for different gate biases from 0 to 30 V with 5 V step. The inset shows I_{ds}–V_g curve measured at V_{ds} = 1 V. (From Maeng, J. et al., *Appl. Phys. Lett.*, 92, 233120, 2008. With permission.)

The carrier density is given by $n = Q/eL$ cm^{-1}. The carrier mobility (μ) can be determined according to $dI/dV_g = \mu(C/L^2)V$, where I and V are the source–drain current and voltage, respectively.

The FET performance of many semiconducting metal oxide nanostructures has been reported recently. A single ZnO NW FET, for example, was reported by many researchers (Fan et al. 2004b, Chang et al. 2006). By appropriately doping the NWs, the FET characteristics of ZnO NWs can be tuned (Yuan et al. 2008a,b). Recent studies revealed that the performance of single ZnO NWs largely depends on their surface microstructures (Hong et al. 2008). Maeng et al. fabricated a FET using ZnO NWs and its performance was evaluated under different environments (Maeng et al. 2008). The SEM image of the ZnO NW arrays, the schematic of the single NW FET design, the SEM image of a typical NW FET, and the performance of the as fabricated device are shown in Figure 17.10.

Improved device performance was achieved by fabricating the vertical surrounded-gate FETs ZnO NWs (Ng et al. 2004). This design facilitates high density integration and eliminates the alignment and lithographic issues associated with the horizontal single NW-based FETs.

17.3.1.3 Conductometric Nano Sensors

Most of the commercially available gas sensors contain doped or pristine metal oxides. This is due to the selective adsorption of specific analyte molecules on certain metal oxide surfaces (Henrich and Cox 1994). As a result, its properties modify and provide quantitative information on the presence of the analyte molecules. If the sensor functions on the basis of the electrical conductivity changes of the active material, it is called a conductometric type sensor. Presumably, the huge surface fraction of the nanostructures enhances the sensing capability of the metal oxides compared with the coarse-grained polycrystalline bulk materials (Kolmakov and Moskovits 2004). In addition, nanostructures having reduced defects and free of dislocations will improve the stability and performance in sensing applications (Comini 2006). The conductivity of the ZnO NWs was found to be highly sensitive to UV light (Kind et al. 2002, Li et al. 2004). The ultraviolet photoconductivity of ZnO was found to be significantly enhanced when the size was reduced to the nano regime (Kind et al. 2002). Such variations in the electrical conductivity are attributed to the desorption of the adsorbed oxygen species from the surface of the NW. This effect can be utilized for the fabrication of ultrafast optical switches and photodetectors.

17.3.1.4 Nanowire FET Sensors

The FET characteristics of metal oxide nanostructures are found to be highly dependent on the surrounding medium. ZnO NW FETs are sensitive to oxygen partial pressures (Maeng et al. 2008). Individual and multiple In_2O_3 NWs in the FET configuration are found to be sensitive to NO_2 gases at the ppb level (Zhang et al. 2004). The sensing properties of NW FETs depend on the doping level of the NWs as well. Zhang et al. reported the sensing capability of single In_2O_3 NW transistors for the detection of NH_3 gas (Zhang et al. 2003). Sysoev et al. demonstrated an electronic nose device based on conductivity measurement on an array of three kinds of metal oxide (Ni surface doped and pristine SnO_2, TiO_2, and In_2O_3) NWs in a single chip to selectively detect the presence of H_2 and CO gases in an oxygen environment (Sysoev et al. 2006).

The electrical properties of metal oxide nanostructures are dependent on various factors such as size, defects and microstructures, surface properties, and environment in which the measurements are carried out. Due to these multiple factors, electrical properties of the same kind of NWs reported by different researchers showed large inconsistencies (Schlenker et al. 2008).

17.3.2 Mechanical Properties

The theoretical calculations and subsequent experimental verification of the ultrahigh strength of CNTs have stimulated intensive research on the mechanical properties of nanosized structures (Ebbesen 1994, Treacy et al. 1996, Wong et al. 1997). Different from bulk materials, the mechanical properties of many nanostructures vary as a function of their characteristic size. Such size effect has great importance from both a fundamental as well as a technological point of view. Particularly, studies on the mechanical properties of nanostructures will provide greater insight into the fundamental mechanism of material deformation and failure.

Due to their small size, the experimental characterization of the mechanical properties of nanosized structures proves to be challenging. The challenges include their manipulation, application of force, and the measurement of the corresponding deformation. Accuracy in the range of nano-Newton in force and nanometer in deflection measurements are required to extract elastic constants of the nanostructures. Recently, direct bending or indentation techniques using an atomic force microscope (AFM) were developed to characterize the mechanical properties of NWs/NTs (Wong et al. 1997, Salvetat et al. 1999, Kis et al. 2003, Zhu et al. 2007). Another way to extract the elastic constants of NWs/NTs is to excite them into mechanical resonance vibration inside electron microscopes (Treacy et al. 1996, Poncharal et al. 1999, Yu et al. 2000).

17.3.2.1 AFM-Based Techniques

AFM provides exceptionally high precision in force and deflection measurement at the nano-Newton and nanometer level, respectively. The nanoscale three-point bend test, lateral force microscopy, or the nanoindentation test can be performed using an AFM on a nanostructure to extract its elastic constants. A brief overview on the experimental strategies developed using an AFM for mechanical testing is discussed below and the obtained results on various metal oxide nanostructures are highlighted.

17.3.2.1.1 Nanoscale Three-Point Bend Test

The nanoscale three-point bend test can be performed using an AFM on suspended NWs across the trenches fabricated on hard substrates like Si. An AFM cantilever of an accurately calibrated force constant is used for applying a normal force on the midpoint of the suspended NW. From the recorded force–distance curve (vertical deflection of the cantilever versus Z-piezo position), the force and the corresponding deflection on the NW can be estimated (Tombler et al. 2000). By assuming the suspended NW as an end clamped cantilever beam, the Young's modulus is given by

$$Y = \frac{FL^3}{192\delta I},\tag{17.5}$$

where
  F is the force
  L is the suspended length of the NW
  δ is the deflection
  I is the second moment of area of the NW

For a NW with a cylindrical cross-section, $I = \pi d^4/64$, with d as the diameter.

Tan et al. reported the elastic properties of CuO NWs by the nanoscale three-point bend test using AFM (Tan et al. 2007). The effects of crystallinity, surface properties, and size of the CuO NW on its elastic constant were discussed. Following similar methodologies, Cheong et al. observed a size-dependent Young's modulus of WO_x NWs (Cheong et al. 2007).

17.3.2.1.2 Lateral Force Microscopy

One of the early works on the mechanical characterization of 1D structures was based on quantifying the lateral force signal obtained while an AFM cantilever was used for deflecting a one-end pinned NW (Wong et al. 1997). Song et al. demonstrated the feasibility of AFM lateral force microscopy technique to characterize the mechanical properties of vertically aligned ZnO NWs avoiding the tedious manipulation and assembly steps (Song et al. 2005a).

17.3.3.1.3 Nano-Indentation

In nano-indentation, a sharp tip made of a hard material and with a known geometry and elastic properties is used to make an indent on the material to be tested. From the details of

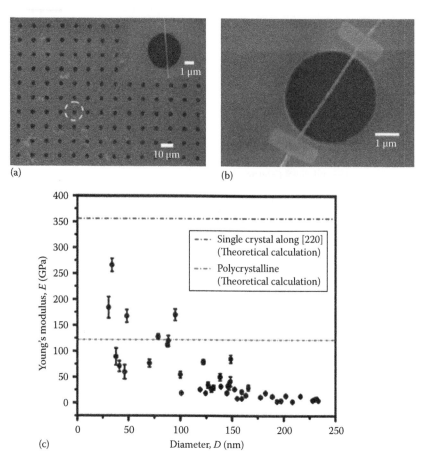

(a)

(b)

(c)

FIGURE 17.11 (a) SEM image of a SiN TEM grid with circular holes with dispersed Co_3O_4 NWs. Inset shows a closeup view of the highlighted region (white circle) showing a suspended NW. (b) SEM image of a suspended NW after securing the ends with Pt deposition. (c) Plot of Young's modulus of the Co_3O_4 NWs against diameter. (From Varghese, B. et al., *Nano Lett.*, 8, 3226, 2008c. With permission.)

applied load and penetration depth, the mechanical properties of materials can be extracted. For nano-indentation tests, the nanostructures are dispersed on a hard substrate. It is recommended that the nanostructures are secured by using the FIB technique or the like, to prevent sliding during indentation. The nano-indenter is often used in conjunction with an AFM, so that the testing and subsequent imaging of the indent region can be carried out (Tao et al. 2007, Tao and Li 2008). Lucas et al. investigated the size-dependent elastic modulus of the ZnO nanobelts using a modified nano-indentation technique (Lucas et al. 2007). The observed aspect ratio dependence on the elastic constant of ZnO nanobelts was attributed to the growth direction-dependent aspect ratio and variation in defects.

AFM-based techniques provide a platform for acquiring force–deflection curves simultaneously during the application of force. However, AFM techniques lack the *in situ* structural characterization and imaging during the test. To circumvent some of these drawbacks, we have developed a combinatory approach that permits both visualization of the microscopic details as well as characterization of its mechanical properties (Varghese et al. 2008). In this approach, suspended NW configurations are constructed on a SiN TEM grid with circular

holes. The position of each suspended NW can be noted and this facilitates multiple experiments on the same NW. The elastic constants were obtained from the nanoscale three-point bend test using AFM and the microstructure of the same NW was examined using TEM. Figure 17.11a shows SEM images of the SiN TEM grid with Co_3O_4 NWs. The ends of the suspended NWs are secured using Pt deposition (Figure 17.11b). Figure 17.11c shows a plot of Young's modulus versus the size of the Co_3O_4 NWs.

17.3.2.2 Resonance Method

The resonance tests developed to obtain the mechanical properties of 1D nanostructures are normally conducted inside electron microscopes for visualization purposes. First, experimental calculations of Young's modulus of the CNTs were performed by setting a freestanding CNT into thermal vibration inside a TEM and measuring the resonance frequency (Treacy et al. 1996). Later, mechanical vibrations induced by an electric field emerged as a versatile tool for nano-mechanical characterization (Poncharal et al. 1999). When a static potential is applied to the projected NW/NT, its end is electrically charged and attracted to the counter electrode. If the NW/NT is not perpendicular to the counter electrode, it bends towards the counter

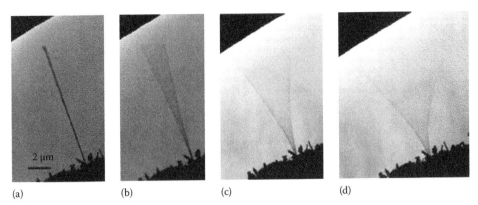

(a)　　　　　　　　(b)　　　　　　　　(c)　　　　　　　　(d)

FIGURE 17.12 Images of ZnO NWs of different length and diameter captured during resonance vibration arising from the application of alternating electric field. (From Huang, Y. et al., *J. Phys.: Condens. Matter.*, 18, L179, 2006. With permission.)

electrode. On the other hand, applications of alternating field to such protruding NWs/NTs results in dynamic deflections. If the frequency of the applied field is varied, the NW/NT can be excited into the resonance vibrations. The resonance frequency of such cantilevered beams is given by

$$f_i = \frac{\beta_i^2}{L^2}\sqrt{\frac{YI}{\rho A}}, \tag{17.6}$$

where
β_i is a constant for the ith harmonic with the values $\beta_1 = 1.875$ and $\beta_2 = 4.694$
L is the length of the cantilever beam
Y is the elastic modulus
I is moment of inertia
ρ is the mass density
A is the cross-sectional area of the nanostructures

For a NW with a circular cross-section with diameter d, the expression can be written as

$$f_i = \frac{\beta_i^2}{2\pi}\frac{d}{L^2}\sqrt{\frac{Y}{16\rho}}. \tag{17.7}$$

For nanotubes with an outer diameter d and inner diameter d_i, the expression is

$$f_i = \frac{\beta_i^2}{8\pi}\frac{1}{L^2}\sqrt{(d^2 + d_i^2)}\sqrt{\frac{Y}{\rho}}. \tag{17.8}$$

For a NW with an equilateral triangular cross-section,

$$f_i = \frac{\beta_i^2}{2\pi}\frac{1}{L^2}\sqrt{\frac{Yd^2}{24\rho}}. \tag{17.9}$$

Chen et al. studied the size-dependent Young's modulus of the ZnO NWs by resonance technique inside SEM (Chen et al. 2006). Young's modulus of other metal oxide nanostructures including,

ZnO nanobelts (Bai et al. 2003), WO$_x$ NWs (Liu et al. 2006), and β-Ga$_2$O$_3$ NWs (Yu et al. 2005a) were also determined using the resonance method. Figure 17.12 shows the resonance oscillation of ZnO NWs of different lengths and diameters (Huang et al. 2006).

Table 17.1 shows a summary of the mechanical properties of the various metal oxide nanostructures estimated using the techniques described above. Many of the metal oxide nanostructures exhibited a size-dependent elastic behavior. The origin of the size-dependent elastic modulus is a topic of fundamental research. As size shrinks to nanoscale, the surface atoms contribute to the material properties significantly. The surface atoms are in a state of strain due to the uneven bonding compared with the atoms in the interior of the nanostructures. The effects of surface stress and reduced defects are considered to be the causes of the size-dependent mechanical properties of the nanosized materials.

Recently, theoretical modeling on the mechanical properties of nanostructures by considering a combined effect of surface and bulk properties was reported. Miller et al. explained the size-dependent elastic properties of nanostructures by taking into account the surface elasticity (Miller and Shenoy 2000). Their model predicts that the deviation of an elastic property D from that of conventional continuum mechanics D_c can be expressed as

$$\frac{D - D_c}{D_c} = \alpha\frac{S}{Yh} = \alpha\frac{h_0}{h}, \tag{17.10}$$

where
α is a dimensionless constant that depends on the geometry of the structural element
h is a length defining the size of the structure
$h_0 \equiv S/Y$ is a material length that sets the scale at which the effect of the free surfaces become significant

The quantity S is a surface elastic constant relevant to the structural element being considered and Y is the corresponding elastic modulus of the bulk material.

Similarly, Chen et al. modeled the NW as composite material with a bulk core and a surface with different elastic properties

TABLE 17.1   Mechanical Properties of Metal Oxide Nanostructures

Method	Material	Young's Modulus	Hardness	Size Dependent	Reference
Bending test	CuO nanowires	70–300 GPa	—	Yes	Tan et al. (2007)
	WO_x nanowires	10–110 GPa	—	Yes	Cheong et al. (2007)
	Co_3O_4 nanowires	4–250 GPa	—	Yes	Varghese et al. (2008c)
	Nb_2O_5 nanowires	5–30 GPa	—	Yes	Varghese et al. (2009)
Lateral force microscopy	ZnO nanowires	29 ± 8 GPa	—	—	Song et al. (2005a)
Nano indentation	ZnO nanowires	—	3.4 ± 0.9 GPa	—	Feng et al. (2006)
	ZnO nanobelts	31.1 ± 1.3 Gpa	—	—	Ni and Li (2006)
	$Mg_2B_2O_5$ nanowires	125.8 ± 3.6 GPa	15 ± 0.7 GPa	—	Tao and Li (2008)
	$Al_4B_2O_9$ nanowires	80–120 GPa	8–15 Gpa	—	Tao et al. (2007)
	$Al_{18}B_4O_{33}$ nanowires	80–200 GPa	10–20 GPa	—	Tao et al. (2007)
Resonance	ZnO nanowires	130–250 GPa	—	Yes	Chen et al. (2006)
	ZnO nanobelts	35–50 GPa	—	Yes	Bai et al. (2003)
	WO_x nanowires	100–300 GPa	—	Yes	Liu et al. (2006)
	Ga_2O_3 nanowires	~300 GPa	—	—	Yu et al. (2005a)

(Chen et al. 2006). The flexural rigidity of such composite material can be written as

$$YI = Y_b I_b + S I_s, \tag{17.11}$$

where

I_b and I_s are the moment of inertia of the cross-section of the core and the shell, respectively
Y_b is the Young's modulus of the bulk core

By substituting I_b and I_s, we get

$$Y = Y_b \left[1 + 8 \left(\frac{S}{E_b} - 1 \right) \left(\frac{r_s}{D} - 3 \frac{r_s^2}{D^2} + 4 \frac{r_s^3}{D^3} - 2 \frac{r_s^4}{D^4} \right) \right], \tag{17.12}$$

where

r_s is the depth of the shell
D is the total NW diameter

The experimental data obtained from the resonance method fit well in the above equation. However, the applicability of this model is limited by the arbitrary nature of the constants S and r_s.

On the other hand, adding the surface stress contribution to the total energy of a bend NW, Cuenot et al. derived an expression for the Young's modulus as (Cuenot et al. 2004)

$$Y = Y_b + \frac{8}{5}\tau(1-\nu)\frac{L^2}{d^3}, \tag{17.13}$$

where

τ is the surface stress
ν is the Poisson's ratio
L is the length of the NW
d is the diameter of the NW

Thus, the elastic modulus of the NWs can be larger or lower than the corresponding bulk material depending on the positive or negative surface stresses.

Jing et al. take into account both the surface stress and the surface elastic constant to calculate the elastic constant of the nanostructure (Jing et al. 2006). The total surface stress of the nanostructure due to an applied strain of ε can be expressed as

$$\tau = \tau_0 + S\varepsilon, \tag{17.14}$$

where τ_0 is the surface stress on the NW at $\varepsilon = 0$ and the elastic constant of NW can be expressed as

$$\frac{Y - Y_b}{Y_b} \approx \frac{8}{d}\frac{S}{Y_b} + \frac{8L^2}{5d^3}\frac{\tau_0}{Y_b}. \tag{17.15}$$

Although large amounts of contributions have been made to this topic during the last few years, a satisfactory theoretical framework to elucidate the size dependence on the mechanical properties of nanostructures is still lacking. This is fundamentally due to many contributing factors that determine the overall mechanics of nanostructured materials. Some of these factors are surface properties, internal microstructure (e.g., anisotropic growth direction), defects, and geometry of the nanostructure.

Utilizing the mechanical properties of metal oxide nanostructures, many prototype devices are proposed and some of them have been demonstrated recently. Wang et al. demonstrated a direct-current nanogenerator making use of the piezoelectric properties of ZnO (Wang and Song 2006, Wang et al. 2007a). In their work, either an AFM tip or a micro fabricated zigzag electrode was used to deflect the free-end of vertically oriented ZnO NWs. The strain field induces charge separation on the piezoelectric ZnO and generates continuous direct current that would be adequate enough for powering various nanodevices.

17.3.3 Optical Properties

The energy states near the band edges of low-dimensional materials are densely packed. This enhances the probability for optical transition to occur. Metal oxide nanostructures, in particular those having a direct band gap, are found to have attractive optical properties and could function as various nano-photonic components

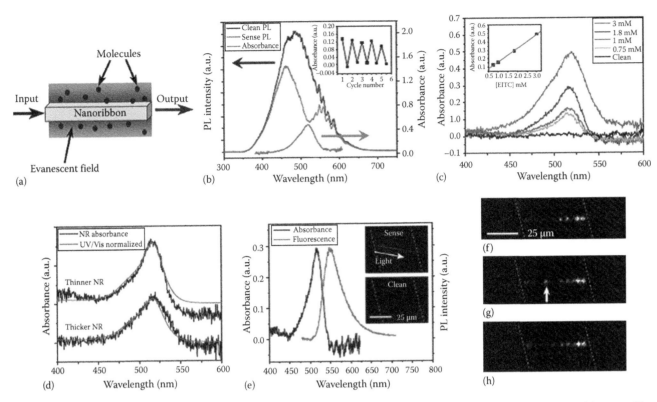

FIGURE 17.13 (a) Schematic illustration of the SnO$_2$ nanoribbon absorption sensor. (b–d) PL and optical absorption spectra of the nanoribbon waveguides immersed in solutions containing the dye molecules in different concentrations. (e) Absorption and fluorescence of the dye molecules containing solution using the nanoribbon waveguide and its photoluminescence images. (f–h) Photoluminescence images of λ-DNA flowing past the nanoribbon sensor. (From Sirbuly, D.J. et al., *Adv. Mater.*, 19, 61, 2007. With permission.)

(Agarwal and Lieber 2006, Djurišić and Leung 2006, Pauzauskie and Yang 2006). Research activities on nano-photonics have been stimulated by the discovery of room temperature ultraviolet lasing in vertically oriented ZnO NWs (Huang et al. 2001a). Well-aligned ZnO NWs on sapphire substrates could function as natural laser cavities when excited with the fourth harmonic of the Nd:YAG laser. The excitation light was incident at an angle to the aligned NWs and the emission spectra were collected parallel to the long axis of the NWs. Stimulated emission from the NWs was observed when the excitation intensity was greater than a threshold power of ~40 kW/cm^2. The lasing is likely to be caused by an exciton recombination in ZnO. The threshold power for crystalline NW lasers are found to be significantly lower than that required for lasing for ZnO thin films (>300 kW/cm^2). Later, the same group demonstrated the lasing action in an isolated ZnO nanobelt dispersed on a sapphire substrate (Yan et al. 2003). The low threshold lasing from ZnO nanorods synthesized via the solution growth technique was also reported recently (Hirano et al. 2005). Optical lasing action is observed in other metal oxide systems like SnO$_2$ NWs (Liu et al. 2007a).

One dimensional nanostructures of SnO$_2$, ZnO, and Ga$_2$O$_3$ are found to have sub-wavelength optical wave-guiding capability (Yan et al. 2003, Law et al. 2004, Sirbuly et al. 2005, Zhang et al. 2007c). This could be useful for the realization of integrated nano-optoelectronic devices, computing, and sensing. Recently, the wave-guiding ability of the SnO$_2$ nanoribbons was

utilized for the demonstration of a prototype multifunctional optical sensor (Sirbuly et al. 2007). The sensor element consists of a SnO$_2$ nanoribbon positioned across a microfluidic channel on a PDMS matrix. This geometry allows the measurement of the absorption/fluorescence spectra of the solution containing the analyte molecules excited by the evanescent wave of the guided light through the nanoribbon. A schematic illustration of the experimental technique and optical spectroscopic results obtained from the SnO$_2$ nanoribbon evanescent sensor is displayed in Figure 17.13.

17.3.4 Field Emission Properties

Field emission is the process of electron emission from a condensed phase to a vacuum by electron tunneling caused by the application of a high-intensity electric field (Gomer 1961, Fursey 2005). The applied field modifies and narrows the potential barrier of electrons at the solid–vacuum interface. This enables the electrons to tunnel through the barrier. Field emitters have applications in diverse fields including flat panel displays, x-ray sources, electron microscopes, etc.

The field emission from metals is explained by the quantum mechanical tunneling theory developed by Fowler and Nordheim (Fowler and Nordheim 1928). Figure 17.14 shows a schematic of the energy diagram of electrons at the surface of a metal at 0 K. In the absence of the field, the potential energy of

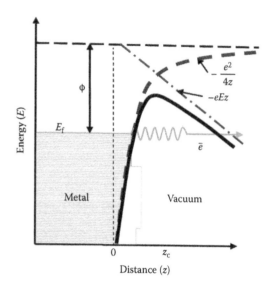

FIGURE 17.14 Potential energy diagram of electrons at the surface of a metal.

the electron at a distance x from the surface is determined by the image potential $(-e^2/4x)$. After the application of an electric field, the potential energy of the electron is represented by the solid curve and is given by

$$U(x) = -\frac{e^2}{4x} - eFx, \qquad (17.16)$$

where

  e is electronic charge
  F is the local electric field at the metal surface

The application of electric field reduced the work function of the electron by

$$\Delta\varphi = \sqrt{\frac{e^3 F}{4\pi\varepsilon_0}}. \qquad (17.17)$$

Assuming a planar metal surface, the current density can be written as

$$j = e \int_0^\infty n(E_\perp) D(E_\perp, F) \mathrm{d}E_\perp, \qquad (17.18)$$

where

  $E_\perp$ is part of the electron energy due to the momentum perpendicular to the metal surface
  $n(E_\perp)$ is the number of electrons with energies in between E and $E_\perp + \mathrm{d}E_\perp$ incident on the unit area of the potential barrier surface from the metal
  $D(E_\perp, F)$ is the tunneling probability of the incident electron with the energy component $E_\perp$ due to momentum perpendicular to the metal surface

The above integral can be simplified into the form of the Fowler–Nordheim (F–N) equation:

$$j = \frac{A}{\phi} F^2 \exp\left(-\frac{B\phi^{3/2}}{F}\right). \qquad (17.19)$$

According to Equation 17.19, the plot between $\ln(J/F^2)$ versus $1/F$ (the Fowler–Nordheim (F–N) plot) gives a straight line. Although Equation 17.19 is derived for metal field emitters, it is often employed for most of the semiconducting field emitters as well.

A useful field emitter should need to have attributes such as a low turn-on field (defined as the applied field at which the emission current density reaches $10\,\mu A/cm^2$), high current emission capability, long-term emission stability, high mechanical and chemical stability, and the availability of cost effective fabrication techniques. Conventional field emitters are micro-fabricated sharp tips made of metals or semiconductors (Brodie 1994, Temple 1999, Xu and Huq 2005). The fabrication involves expensive multi-step lithography techniques including thin film deposition, photolithography, etching, and lift-off procedures. Two types of design normally used in vacuum microelectronics are the gated and un-gated Spindt-type field emitters (Brodie 1994, Temple 1999, Xu and Huq 2005). In gated Spindt field emitters, the electric field is applied using a gate electrode that is fabricated ~1 μm away from the emitter tip and an additional electrode is used for the collection of the emitted electrons. In such a design, a field of 100 V/μm is necessary to obtain appreciable current density. Greater efforts have been devoted for the realization of field emitting structures that can be fabricated by simple means and can be operated at low voltage conditions.

The discovery of efficient field emission from CNT arrays (de Heer et al. 1995, Fan et al. 1999, Jonge and Bonard 2004) has stimulated intense efforts in the investigation of the field emission properties of other nanostructures. The CNT emitters exhibit a low turn-on field and long-term stability. However, the lack of structure-controlled growth strategies for CNTs is a challenging issue in device integration. Alternatively, FE emitters based on metal oxide nanostructures have attracted attention due to the feasibility of precise structural control during the growth process. In addition, the oxides emitters can be operated in the presence of oxygen. These factors guarantee a predictable voltage–current characteristic from nanostructured metal oxide field emitters.

Typically, the field emission properties of nanostructures are characterized in vacuum conditions with a base pressure better than ~10^{-6} Torr. Table 17.2 summarizes the field emission properties of various metal oxide nanostructure arrays. Nanostructures of zinc oxides, molybdenum oxides, tungsten oxides, and niobium oxides are found to exhibit efficient electron field emission capability. The field emission characteristics are sensitively dependent on the geometrical shape and areal density of nanostructures.

TABLE 17.2 Field Emission Properties of Various Metal Oxide Nanostructure Arrays

Material	Geometry	Turn-On field (V/μm)	Threshold Field (V/μm)	Stability	Reference
ZnO	Nanowires	6.0 ($0.1 \mu A/cm^2$)	11.0 ($1 mA/cm^2$)	—	Lee et al. (2002)
	Nanoneedles	2.4 ($0.1 \mu A/cm^2$)	6.5	—	Zhu et al. (2003)
	Nanopencils	3.7	<4.6 ($1 mA/cm^2$)	—	Wang et al. (2005)
	Nanopins	1.9 ($0.1 \mu A/cm^2$)	5.9 ($1 mA/cm^2$)	—	Xu and Sun (2003)
MoO_3	Nanobelts	8.7	12.9	Good	Li et al. (2002)
	Nanowires	3.5	7.65	—	Zhou et al. (2003a)
MoO_2	Nanowires	2.4	5.6	—	Zhou et al. (2003b)
$WO_{2.9}$	Nanorodes	1.2	—	Good	Liu et al. (2005)
WO_2	Nanorodes	>7	—	—	Liu et al. (2005)
$W_{18}O_{49}$	Nanowires/nanotubes	2.6	6.2	—	Li et al. (2003)
	Nanotips	2.0	4.37	Good	Zhou et al. (2005c)
CuO	Nanowires	3.5–4.5	—	—	Zhu et al. (2005)
In_2O_3	Nanowires	>7	11.3 ($1 mA/cm^2$)	—	Li et al. (2005)
	Nanopyramids	2.7	6.0 ($1 mA/cm^2$)	—	Jia et al. (2003)
RuO_2	Nanowires	10.3	—	Good	Cheng et al. (2005)
	Nanorodes				
NiO	Nanorods	11.5	—	—	Zhang et al. (2006b)
	Nanowalls	7.4	—	—	Varghese et al. (2008a)
Co_3O_4	Nanowires	6.4	—	—	Varghese et al. (2007)
	Nanowalls				
Nb_2O_5	Nanowires	~6.7	9.2 ($1 mA/cm^2$)	Good	Varghese et al. (2008b)
α-Fe_2O_3	Nanowires	6.3	10	—	Chueh et al. (2006)
TiO_2	Nanowires	5.7	—	—	Wu et al. (2005)
$V_2O_5 \cdot nH_2O$	Nanotubes	6.35	—	—	Zhou et al. (2007)
SnO2	Nanobelts	2.3–4.5 ($1 \mu A/cm^2$)	4.4 ($2 mA/cm^2$)	Good	Chen et al. (2004b)
	Nanowhiskers	1.4 ($0.1 \mu A/cm^2$)	8.1 ($1 mA/cm^2$)	—	Luo et al. (2004)
AlZnO	Nanowires	2.9 ($1 \mu A/cm^2$)	3.7 ($1 mA/cm^2$)	Good	Xue et al. (2006)

Note: Turn-on field is defined as the applied field at which the emission current density reaches $10 \mu A/cm^2$ (unless otherwise indicated). The threshold field is defined as the field at which the emission current reaches $10 mA/cm^2$ (unless otherwise indicated).

17.3.4 1 Effect of Morphology

The field emission from nanostructures is found to be largely dependent on their morphological shape (Hsieh et al. 2003b, Zhao et al. 2005, Korotcov et al. 2006, Marathe et al. 2006, Shen et al. 2006, Yang et al. 2006a, Varghese et al. 2007). Among different morphological manifestations of the same material, one with a sharp tip and high aspect ratio was found to exhibit better field emitting characteristics. The effect of morphology on field emission of nanostructures predominantly arises from the variations in the local field enhancement at the emitter tip. Zhao et al. reported the field emission characteristics of three different morphological variations of 1D ZnO nanostructures (Figure 17.15). The efficient field emission current density obtained from the nanoneedles with a sharp tip was explained by the basis of its large field enhancement factor. The enhancement factor can be expressed as $\beta = 1 + s(d/r)$, where d is the inter-electrode separation and r is the radius of the emitter. The screening factor, s, is dependent on the areal density of the emitters in the sample. Evidently, the enhancement factor increases with a decrease in the radius of the emitter.

17.3.4.2 Effect of Areal Density

The local electric field at the emitter tips of the field emitter is affected by its neighboring emitters. The screening factor of the field emitter arrays is a determining factor of the emission characteristics. Thus, the screening factor ranges from 0 for a densely packed emitter array to 1 for a single emitter (Jo et al. 2003). For this reason, the coverage of nanostructures on the sample affects its field emission characteristics significantly. Wang et al. studied the field emission properties of vertically well-aligned ZnO NW arrays of different areal densities (Figure 17.16). Their results showed that an areal density of ~60–80 μm^{-2} and an average length of ~1 μm exhibits a better field emission performance with low turn-on and threshold fields (Figure 17.16) (Wang et al. 2007b).

In view of future applications as electron sources in electron microscopes, the field emission characteristics of single metal oxide nanostructures are investigated by many researchers. Typically, the measurements were carried out inside a SEM or TEM to facilitate alignment through direct visualization. For FE characterization, one end of the NW can be attached to the tip of

Hirano, S., Takeuchi, N., Shimada, S. et al., *J. Appl. Phys.*, 2005, 98, 094305.

Ho, S. T., Chen, K. C., Chen, H. A. et al., *Chem. Mater.*, 2007, 19, 4083.

Hong, W. K., Sohn, J. I., Hwang, D. K. et al., *Nano Lett.*, 2008, 8, 950.

Hoyer, P., *Langmuir*, 1996, 12, 1411.

Hsieh, C. T., Chen, J. M., Lin, H. H., Shih, H. C., *Appl. Phys. Lett.*, 2003a, 82, 3316.

Hsieh, C. T., Chen, J. M., Lin, H. H., Shih, H. C., *Appl. Phys. Lett.*, 2003b, 83, 3383.

Hu, Z., Fischbein, M. D., Querner, C., Drndić, M., *Nano Lett.*, 2006, 6, 2585.

Huang, M. H., Mao, S., Feick, H. et al., *Science*, 2001a, 292, 1897.

Huang, M. H., Wu, Y., Feick, H. et al., *Adv. Mater.*, 2001b, 13,113.

Huang, Y., Bai, X., Zhang, Y., *J. Phys.: Condens. Matter.*, 2006, 18, L179.

Huang, Y., Zhang, Y., Gu, Y. et al., *J. Phys. Chem. C*, 2007, 111, 9039.

Ijima, I, *Nature*, 1991, 354, 56.

Jefferey, J. U., Wan, S. Y., Qian, G., Hongkun, P., *J. Am. Chem. Soc.*, 2002, 124, 1186.

Jia, H., Zhang, Y., Chen, X. et al., *Appl. Phys. Lett.*, 2003, 82, 4146.

Jian, X., Xiaohe, L., Yadong, L., *Mater. Chem. Phys.*, 2004, 86, 409.

Jing, G. Y., Duan, H. L., Sun, X. M. et al., *Phys. Rev. B*, 2006, 73, 235409.

Jo, S. H., Lao, J. Y., Ren, Z. F. et al., *Appl. Phys Lett.*, 2003, 83, 4821.

Johnson, M. C., Aloni, S., McCready, D. E., Courchesne, E. D. B., *Cryst. Growth Des.*, 2006, 6,1936.

Jonge, N. D., Bonard, J. M., *Phil. Trans. R. Soc. Lond. A*, 2004, 362, 2239.

Jun, Y. W., Choi, J. S., Cheon, J., *Angew. Chem. Int. Ed.*, 2006, 45, 3414.

Kam, K. C., Deepak, F. L., Cheetham, A. K., Rao, C. N. R., *Chem. Phys. Lett.*, 2004, 397, 329.

Kang, J., Keem, K., Jeong, D. Y., Kim, S., *Jpn. J. Phys.*, 2007, 46, 6227.

Kim, D. W., Hwang, I. S., Kwon, S. J. et al., *Nano Lett.*, 2007, 7, 3041.

Kim, J., Yang, S. A., Choi, Y. C. et al., *Nano Lett.*, 2008, 8, 1813.

Kind, H., Yan, H., Messer, B. et al., *Adv. Mater.*, 2002, 14, 158.

Kis A., Mihailovic D., Remskar M. et al., *Adv. Mater.*, 2003, 15, 733.

Kolmakov, A., Moskovits, M., *Annu. Rev. Mater. Res.*, 2004, 34, 151.

Korotcov, A., Huang, Y. S., Tsai, T. Y. et al., *Nanotechnology*, 2006, 17, 3149.

Lakshmi, B. B., Dorhout, P. K., Martin, C. R., *Chem. Mater.*, 1997, 9, 587.

Lao, J. Y., Wen, J. G., Ren, Z. F., *Nano Lett.*, 2002, 2, 1287.

Law, M., Sirbuly, D. J., Johnson, J. C. et al., *Science*, 2004, 305, 1269.

Lee, C. J., Lee, T. J., Lyu, S. C. et al., *Appl. Phys. Lett.*, 2002, 81, 3648.

Li, C., Fang, G., Liu, N. et al., *J. Phys. Chem. C*, 2007a, 111, 12566.

Li, N., Tan, T. Y., Gösele, U., *Appl. Phys. A*, 2007b, 86, 433.

Li, Q. H., Wan, Q., Liang, Y. X., Wang, T. H., *Appl. Phys. Lett.*, 2004, 84, 4556.

Li, S. Q., Liang, Y. X., Wang, T. H., *Appl. Phys. Lett.*, 2005, 87, 143104.

Li, Y., Bando, Y., Golberg, D., *Adv. Mater.*, 2003, 15, 1294.

Li, Y. B., Bando, Y., Golberg, D., Kurashima, K., *Appl. Phys. Lett.*, 2002, 81, 5048.

Lilach, Y., Zhang, J. P., Moskovits, M., Kolmakov, A., *Nano Lett.*, 2005, 5, 2019.

Liu, J., Li, Q., Wang, T. et al., *Angew. Chem. Int. Ed.*, 2004, 43, 5048.

Liu, J., Zhang, Z., Zhao, Y. et al., *Small*, 2005, 1, 310.

Liu, K. H., Wang, W. L., Xu, Z. et al., *Appl. Phys. Lett.*, 2006, 89, 221908.

Liu, R. B., Chen, Y. J., Wang, F. F. et al., *Physica E*, 2007a, 39, 223.

Liu, X., Li, C., Han, J., Zhou, C., *Appl. Phys. Lett.*, 2003a, 82, 1950.

Liu, Y., Li, Q., Fan S., *Chem. Phys. Lett.*, 2003b, 375, 632.

Liu, Y. L., Wu, Z. Y., Lin, K. J. et al., *Appl. Phys. Lett.*, 2007b, 90, 013105.

Lu, J. G., Chang, P., Fan, Z., *Mater. Sci. Eng. R*, 2006, 52, 49.

Lucas, M., Mai, W., Yang, R. et al., *Nano Lett.*, 2007, 7, 1314.

Luo, S. H., Wan, Q., Liu, W. L. et al., *Nanotechnology*, 2004, 15, 1424.

Maeng, J., Jo, G., Kwon, S. S. et al., *Appl. Phys. Lett.*, 2008, 92, 233120.

Mao, Y., Park, T. J., Zhang, F. et al., *Small*, 2007, 3, 1122.

Marathe, S. K., Koinkar, P. M., Ashtaputr, S. S. et al., *Nanotechnology*, 2006, 17, 1932.

Martel, R., Schmidt, T., Shea, H. R. et al., *Appl. Phys. Lett.*, 1998, 73, 2447.

Martin, C. R., *Science*, 1994, 266, 1961.

Mensah, S. L., Kayastha, V. K., Ivanov, I. N. et al., *Appl. Phys. Lett.*, 2007, 90, 113108.

Miller, R. E., Shenoy, V. B., *Nanotechnology*, 2000, 11, 139.

Mohammad, S. N., *J. Chem. Phys.*, 2006, 125, 094705.

Morales, A. M., Lieber, C. M., *Science*, 1998, 279, 208.

Morber, J. R., Ding, Y., Haluska, M. S. et al., *J. Phys. Chem. B*, 2006, 110, 21672.

Muster, J., Kim, G. T., Krstić, V. et al., *Adv. Mater.*, 2000, 12, 420.

Nagashima, K., Yanagida, T., Tanaka, H. et al., *J. Am. Chem., Soc.*, 2008, 130, 5378.

Nanda, K. K., Kruis, F. E., Fissan, H., *Nano Lett.*, 2001, 1, 605.

Ng, H. T., Li, J., Smith, M. K. et al., *Science*, 2003, 300, 1249.

Ng, H. T., Han, J., Yamada, T. et al., *Nano Lett.*, 2004, 4, 1247.

Nguyen, P., Ng, H. T., Meyyappan, M., *Adv. Mater.*, 2005, 17, 1773.

Ni, H., Li, X., *Nanotechnology*, 2006, 17, 3591.

Niederberger, M., *Acc. Chem. Res.*, 2007, 40, 793.

Oh, J., Tak, Y., Lee, J., *Electrochem. Solid State Lett.*, 2004, 7, C27.

O'Brien, S., Brus, L., Murray, C. B., *J. Am. Chem. Soc.*, 2001, 123, 12085.

Pacholski, C., Kornowski, A., Weller, H., *Angew. Chem. Int. Ed.*, 2002, *41*, 1188.

Pan, Z. W., Dai, Z. R., Wang, Z. L., *Science*, 2001, *291*, 1947.

Park, J., An, K., Hwang, Y. et al., *Nat. Mater.*, 2004, *3*, 891.

Parthasarathy, R. V., Martin, C. R., *Nature*, 1994, *369*, 298.

Patzke, G. R., Krumeich, F., Nesper, R., *Angew. Chem. Int. Ed.*, 2002, *41*, 2446.

Pauzauskie, P. J., Yang, P., *Mater. Today*, 2006, *9*, 36.

Penn, R. L., Banfield, J. F., *Science*, 1998, *281*, 969.

Poncharal, P., Wang, Z. L., Ugarte, D., de Heer, W. A., *Science*, 1999, *283*, 1513.

Ramirez, F. H., Tarancon, A., Casals, O. et al., *Phys. Rev. B*, 2007, *76*, 085429.

Rao, C. N. R., *J. Mater. Chem.*, 1999, *9*, 1.

Rao, C. N. R., Nath, M., *Dalton Trans.*, 2003, 1–24.

Rao, C. N. R., Gundiah, G., Deepak, F. L. et al., *J. Mater. Chem.*, 2004, *14*, 440.

Rao, R., Chandrasekaran, H., Gubbala, S. et al., *J. Electron. Mater.*, 2006, *35*, 941.

Salvetat, J. P., Briggs, A. D., Bonard, J. M. et al., *Phy. Rev. Lett.*, 1999, *82*, 944.

Schlenker, E., Bakin, A., Weimann, T. et al., *Nanotechnology*, 2008, *19*, 365707.

Seshadri, R., Oxide nanoparticle, Chapter 5, *The Chemistry of Nanomaterials: Synthesis, Properties and Applications*, Volume I, C. N. R. Rao, A. Müller, A. K. Cheetham (Eds.), Wiley-VCH Verlag GmbH & Co., Weinheim, Germany, 2004.

Shankar, K. S., Raychaudhuri, A. K., *Mater. Sci. Eng. C*, 2005, *25*, 738.

Sharma, S., Sunkara, M. K., *J. Am. Chem. Soc.*, 2002, *124*, 12288.

Shen, G., Bando, Y., Liu, B. et al., *Adv. Funct. Mater.*, 2006, *16*, 410.

Shi, S., Xue, X., Feng, P. et al., *J. Cryst. Growth*, 2008, *310*, 462.

Sirbuly, D. J., Law, M., Pauzauskie, P. et al., *Proc. Natl. Acad. Sci.*, 2005, *102*, 7800.

Sirbuly, D. J., Tao, A., Law, M. et al., *Adv. Mater.*, 2007, *19*, 61.

Smith, P. A., Nordquist, C. D., Jackson, T. N. et al., *Appl. Phys. Lett.*, 2000, *77*, 1399.

Son, H. J., Jeon, K. A., Kim, C. E. et al., *Appl. Surf. Science*, 2007, *253*, 7848.

Song, J., Wang, X., Riedo, E., Wang, Z. L., *Nano Lett.*, 2005a, 5, 1954.

Song, J., Wang, X., Riedo, E., Wang, Z. L., *J. Phys. Chem. B*, 2005b, *109*, 9869.

Stern, E., Cheng, G., Guthrie, S. et al., *Nanotechnology*, 2006, *17*, S246.

Sun, S., Meng, G., Zhang, G. et al., *Chem. Eur. J.*, 2007, *13*, 9087.

Sysoev, V. V., Button, B. K., Wepsiec, K. et al., *Nano Lett.*, 2006, *6*, 1584.

Takahashi, K., Limmer, S. J., Wang, Y., Cao, G., *J. Phys. Chem. B*, 2004, *108*, 9795.

Takagi, R., *J. Phys. Soc. Jpn.*, 1957, *12*, 1212.

Tan, E. P.S., Zhu, Y., Yu, T. et al., *Appl. Phys. Lett.*, 2007, *90*, 163112.

Tan, T. Y., Li, N., Gösele, U., *Appl. Phys. Lett.*, 2003, *83*, 1199.

Tao, X., Li, X., *Nano Lett.*, 2008, *8*, 505.

Tao, X., Wang, X., Li, X., *Nano Lett.*, 2007, *7*, 3172.

Temple, D., *Mater. Sci. Eng. R*, 1999, *24*, 185.

Tombler, T. W., Zhou, C., Alexseyev, L. et al., *Nature*, 2000, *405*, 769.

Treacy, M. M. J., Ebbesen, T. W., Gibson, J. M., *Nature*, 1996, *381*, 678.

Umar, A., Kim, S. H., Lee, Y. S., Nahm, K. S., Hahn, Y. B., *J. Cryst. Growth*, 2005, *282*, 131.

Urban, J. J., Yun, W. S., Qian Gu, Q., Park, H., *J. Am. Chem. Soc.*, 2002, *124*, 1186.

Varghese, B., Teo, C. H., Yanwu, Z. et al., *Adv. Funct. Mater.*, 2007, *17*, 1932.

Varghese, B., Reddy, M. V., Yanwu, Z. et al., *Chem. Mater.*, 2008a, *20*, 3360.

Varghese, B., Sow, C. H., Lim C. T., *J. Phys. Chem. C*, 2008b, *112*, 10008.

Varghese, B., Zhang, Y., Dai, L. et al., *Nano Lett.*, 2008c, 8, 3226.

Varghese, B., Zhang, Y., Feng, Y. P. et al., *Phys. Rev. B*, 2009, *79*, 115419.

Vayssieres, L., *Adv. Mater.*, 2003, *15*, 464.

Vayssieres, L., Beermann, N., Lindquist, S. E., Hagfeldt, A., *Chem. Mater.*, 2001a, *13*, 233.

Vayssieres, L., Keis, K., Lindquist, S. E., Hagfedt, A., *J. Phys. Chem. B*, 2001b, *105*, 3350.

Vomiero, A., Ferroni, M., Comini, E. et al., *Nano Lett.*, 2007, *7*, 3553.

Wagner, R. S., Ellis, W. C., *Appl. Phys. Lett.*, 1964, *4*, 89.

Wagner, R. S., Treuting R. G., *J. Appl. Phys.*, 1961, *32*, 2490.

Wan, Q., Dattoli, E. N., Fung, W. Y. et al., *Nano Lett.*, 2006, *6*, 2909.

Wang, R. C., Liu, C. P., Huang, J. L. et al., *Appl. Phys. Lett.*, 2005, *87*, 013110.

Wang, X., Summers, C. J., Wang, Z. L., *Nano Lett.*, 2004, *4*, 423.

Wang, X., Song, J., Liu, J., Wang, Z. L., *Science*, 2007a, *316*, 102.

Wang, X., Zhou, J., Lao, C. et al., *Adv. Mater.*, 2007b, *19*, 1627.

Wang, Z. L., *Adv. Mater.*, 2003, *15*, 432.

Wang, Z. L., Kong, X. Y., Zuo, J. M., *Phys. Rev. Lett.*, 2003, *91*, 185502.

Wang, Z. L., *Annu. Rev. Phys. Chem.*, 2004a, *55*, 159.

Wang, Z. L., *Dekker Encyclopedia of Nanoscience and Nanotechnology*, 2004b, 1773.

Wang, Z. L., Song, J., *Science*, 2006, *312*, 242.

Wang, Z. L., *Adv. Mater.*, 2007a, *19*, 889.

Wang, Z. L., *J. Nanosci. Nanotech.*, 2007b, *8*, 27.

Wen, X., Wang, S., Ding, Y. et al., *J. Phys. Chem. B*, 2005, *109*, 215.

Wong, E. W., Sheehan P. E., Lieber, C. M., *Science*, 1997, *277*, 1971.

Wu, J. M., Shih, H. C., Wu, W. T., *Chem. Phys. Lett.*, 2005, *413*, 490.

Xu, C. H., Woo, C. H., Shi, S. Q., *Chem. Phys. Lett.*, 2004, *399*, 62.

Xu, C. X., Sun, X. W., *Appl. Phys. Lett.*, 2003, *83*, 3806.

Xu, N. S., Huq, S. E., *Mater. Sci. Eng. R*, 2005, *48*, 47.

Xue, X. Y., Li, L. M., Yu, H. C. et al., *Appl. Phys. Lett.*, 2006, *89*, 043118.

Yan, H., Johnson, J., Law, M. et al., *Adv. Mater.*, 2003, *15*, 1907.

Yang, P., Yan, H., Mao, S. et al., *Adv. Funct. Mater*, 2002, *12*, 323.

Yang, Y., Sun, X. W., Tay, B. K. et al., *Adv. Mater.*, 2007, *19*, 1839.

Yang, Y. H., Wang, B., Xu, N. S., Yang, G. W., *Appl. Phys. Lett.*, 2006a, *89*, 043108.

Yang, Z., Huang, Y., Dong, B., Li, H. L., *Mater. Res. Bull.*, 2006b, *41*, 274.

Yeong, K. S., Maung, K. H., Thong, J. T. L., *Nanotechnology*, 2007, *18*, 185608.

Yeong, K. S., Thong, J. T. L., *J. Vac. Sci. Technol. B*, 2008, *26*, 983.

Yi, J. B., Pan, H., Lin, J. Y. et al., *Adv. Mater.*, 2008, *20*, 1170.

Yoffe, A. D., *Adv. Phys.*, 2002, *51*, 799.

Yu, H., Li, J., Loomis, R. A. et al., *Nat. Mater.*, 2003, *2*, 517.

Yu, M. F., Lourie, O., Dyer, M. J. et al., *Science*, 2000, *287*, 637.

Yu, M. F., Atashbar, M. Z., Chen, X., *IEEE Sen. J.*, 2005a, *5*, 20.

Yu, T., Sow, C. H., Gantimahapatruni, A. et al., *Nanotechnology*, 2005b, *16*, 1238.

Yu, T., Zhu, Y., Xu, X. et al., *Adv. Mater.*, 2005c, *17*, 1595.

Yu, T., Zhu, Y., Xu, X. et al., *Small*, 2006, *2*, 80.

Yuan, G. D., Zhang, W. J., Jie, J. S. et al., *Adv. Mater.*, 2008a, *20*, 168.

Yuan, G. D., Zhang, W. J., Jie, J. S. et al., *Nano Lett.*, 2008b, *8*, 2591.

Zeng, H. C., *Curr. Nanosci.*, 2007, *3*, 177.

Zhang, D., Li, C., Liu, X. et al., *Appl. Phys. Lett.*, 2003, *83*, 1845.

Zhang, D., Liu, Z., Li, C. et al., *Nano Lett.*, 2004, *4*, 1919.

Zhang, H. Z., Kong, Y. C., Wang, Y. Z. et al., *Solid State Commun.*, 1999, *109*, 677.

Zhang, J., Jiang, F., Yang, Y., Li, J., *J. Phys. Chem. B*, 2005, *109*, 13143.

Zhang, Z., Sun, X., Dresselhaus, M. S. et al., *Phys. Rev. B*, 2000, *61*, 4850.

Zhang, Z., Zhao, Y., Zhu, M., *Appl. Phys. Lett.*, 2006b, *88*, 033101.

Zhang, Z., Wang, S. J., Yu, T., Wu, T., *J. Phys. Chem., C*, 2007a, *111*, 17500.

Zhang, Z., Yao, K., Liu, Y. et al., *Adv. Funct. Mater.*, 2007b, *17*, 2478.

Zhang, Z., Yuan, H., Gao, Y. et al., *Appl. Phys. Lett.*, 2007c, *90*, 153116.

Zhang, Z. Y., Jin, C. H., Liang, X. L. et al., *Appl. Phys. Lett.*, 2006a, *88*, 073102.

Zhao, Q., Zhang, H. Z., Zhu, Y. W. et al., *Appl. Phys. Lett.*, 2005, *86*, 203115.

Zhao, Y. M., Li, Y. H., Ahmad, I., McCartney, D. G. et al., *Appl. Phys. Lett.*, 2006, *89*, 133116.

Zheng, M. J., Zhang, L. D., Li, G. H., Shen, W. Z., *Chem. Phys. Lett.*, 2002, *363*, 123.

Zhou, C., Mai, L., Liu, Y. et al., *J. Phys. Chem.*, 2007, *111*, 8202.

Zhou, J., Deng, S. Z., Xu, N. S. et al., *Appl. Phys. Lett.*, 2003a, *83*, 2653.

Zhou, J., Xu, N. S., Deng, S. Z. et al., *Adv. Mater.*, 2003b, *15*, 1835.

Zhou, J., Ding, Y., Deng, S. Z. et al., *Adv. Mater.*, 2005a, *17*, 2107.

Zhou, J., Gong, L., Deng, S. Z. et al., *Appl. Phys. Lett.*, 2005b, *87*, 223108.

Zhou, Y., Li, H., *J. Solid State Chem.*, 2002, *165*, 247.

Zhu, Y., Ke, C., Espinosa, H. D., *Exp. Mech.*, 2007, *47*, 7.

Zhu, Y. Q., Hu, W., Hsu, W. K. et al., *Chem. Phys. Lett.*, 1999, *309*, 327.

Zhu, Y. W., Zhang, H. Z., Sun, X. C. et al., *Appl. Phys. Lett.*, 2003, *83*, 144.

Zhu, Y. W., Yu, T., Cheong, F. C. et al., *Nanotechnology*, 2005a, *16*, 88.

Zhu, Y. W., Yu, T., Sow, C. H. et al., *Appl. Phys. Lett.*, 2005b, *87*, 023103.

Zhu, Y. W, Sow C. H., Yu, T. et al., *Adv. Funct. Mater.*, 2006, *16*, 2415.

Zitoun, D., Pinna, N., Frolet, N., Belin, C., *J. Am. Chem. Soc.*, 2005, *127*, 15034.

Žumer, M., Nemanić, V., Zajec, B. et al., *J. Phys. Chem. C*, 2008, *112*, 5250.

18

Gallium Nitride Nanowires

Catherine Stampfl
The University of Sydney

Damien J. Carter
The University of Sydney

and

Curtin University of Technology

18.1 Introduction

Semiconductor nanostructures, which have one dimension of the order of 10^{-9} m, have attracted huge interest from the scientific community due to the unique quantum confinement effects that become important on this scale, and the resulting potential for size-tunable nanodevices (Cui and Lieber 2001; Duan et al. 2003). Such nanostructures include (1) quantum wells, e.g., a heterostructure or superlattice, which are two-dimensional structures and confined in one dimension; (2) quantum wires, a one-dimensional structure confined in two dimensions; and (3) quantum dots, a zero-dimensional structure confined in three dimensions. These systems are illustrated in Figure 18.1, as well as the three-dimensional bulk system. The terminology of "quantum" indicates that the electronic properties become quantized as the size of a dimension(s) of the structure diminishes, that is, they change from being continuous to discrete values, as depicted in the electronic density of states shown in Figure 18.1.

One of the most identifiable aspects of quantum confinement in semiconductors is the nanostructure size dependence of the band gap; namely, the band gap increases as the size decreases (Yoffe 1993, 2001; Efros and Rosen 2000). The variation in the band gap with nanostructures size for one-, two-, and three-dimensionally confined wells, wires, and dots, respectively, will be different, as illustrated in Figure 18.2. These theoretical results are obtained from simple effective-mass approximation, particle-in-a-box models (EMA-PIB) for confinement in planar wells, cylindrical wires, and spherical dots, as described by Dong et al. (2008). In these models, the change in band gap compared to the bulk value (ΔE_g) depends linearly on $1/d^2$ (where d is the inverse square of the diameter or thickness) with the slope of each curve varying

for wells, wires, and dots. Such a simple description should be regarded as providing only a first approximation of quantum confinement in semiconductors; as the relevant dimension(s) of the nanostructure becomes very small, this relationship can be expected to break down. It is however useful for showing how quantum confinement should depend on the geometry of confinement. Interestingly, recent results for quantum wires, wells, and dots demonstrate that the band gap scales largely according to this prediction (Yu et al. 2003; Dong et al. 2008).

Since the band gap is of fundamental importance for the properties of a solid, and much of a material's behavior depends on it (e.g., conductivity, optical transitions, and electronic transitions), any change of the band gap may significantly alter the material's physics and chemistry. From the relationship between band gap and nanostructure size and dimensionality as explained above, it can therefore be understood that having control of these size aspects of the nanostructure can lead to a tailoring of the system's properties and functionality according to desire. To give an indication of the exciting prospects nanostructures hold for future high-technology applications, some examples from recent research into nanowires, in particular, are described below.

Recently, silicon nanowires have been reported to produces 10 times the amount of electricity of existing lithium-ion batteries, which is related to how much lithium can be held in the battery's anode. The nanowires evidently inflate four times their normal size as they absorb lithium, but they do not fracture (Chan et al. 2008). Silicon nanowires, therefore, could have a high potential for future use in rechargeable lithium-ion batteries that power laptops, iPods, video cameras, cell phones, and countless other devices (Chan et al. 2008). In other potential applications, indium phosphide nanowires, through their found reduction in electron-hole

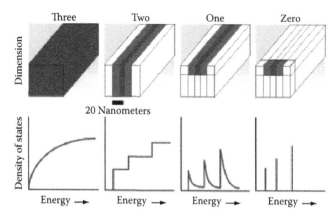

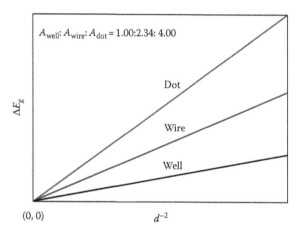

FIGURE 18.1 Schematic illustration of (upper panel) three-, two-, one-, and zero-dimensional bulk, superlattice (or heterostructure), wire, and dot systems. Corresponding electronic density of states versus energy (lower panels), showing the effect of quantization. (From Corcoran, E., *Sci. Am.*, 263, 122, 1990.)

FIGURE 18.2 Size dependence of the change in the effective band gaps, ΔE_g, compared to the bulk semiconductor for quantum wells, wires, and dots, derived from a simple effective-mass approximation particle-in-a-box (EMA-PIB) model. d^{-2} denotes the inverse square of the diameter or thickness of the quantum nanostructure. The slope ratios between the quantum well line, the wire line, and the dot line are predicted to be A_{well}: A_{wire}: A_{dot} = 1.00:2.34:4.00. (Reprinted from Dong, A. et al., *J. Am. Chem. Soc.*, 130, 5954, 2008. With permission.)

recombination, could lead to highly efficient thin-film solar cells (Novotny et al. 2008). Furthermore, nanowires made of the transparent semiconductor indium oxide can be used to create transparent transistors and circuits, which could one day be used in applications such as electronic paper, flexible monitors, and displays in car windshields (Ju et al. 2008). As a final example, recently, a mesh composed of potassium manganese oxide nanowires has been created that can absorb up to 20 times its weight in oil. The membrane selectively absorbs hydrophobic liquids from water and could have applications in the cleanup of oil and other organic pollutants (Yuan et al. 2008). With regard to gallium nitride nanowires, GaN nanowire lasers with improved performance have recently

been fabricated by surrounding the nanowire in a shell made of alternating layers of GaN and InGaN (Qian et al. 2008). InGaN has a smaller band gap than GaN and acts to confine the electrons.

Gallium nitride is an important wide band-gap semiconductor for micro- and optoelectronic applications, such as blue light-emitting diodes and lasers. It is probably the most important semiconductor material since silicon. As well as emitting brilliant light, it is a key material for the next-generation high-frequency, high-power transistors capable of operating at high temperatures. Significantly, it has been reported that a new gallium nitride-based power device has been developed that could deliver cost-effective performance that is at least 10 times better than existing silicon devices (International Rectifier 2009). The key to this structure is the successful growth of GaN on silicon; typically, GaN has been grown on expensive sapphire substrates. This new growth development would enable dramatic reductions in energy consumption in applications like computing and communications, consumer appliances, lighting systems, and automotive. The recent ability to use cheap silicon substrates was also reported recently in the creation of light-emitting diodes (LED) (Humphreys 2009). Such an advancement reportedly could reduce household lighting bills by up to 75% within 5 years.

Future research in GaN nanostructures, in particular on nanowires, used as building blocks for devices, holds great promise for future revolutionary developments. For example, transistors made of nanowires represent one potential way to continue Moore's law (an unofficial rule stating that the number of transistors on a computer chip doubles about every 18 months), which in 5–10 years time will see silicon transistor dimensions scaled to their limit. Nanowires of silicon and, e.g., gallium arsenide, gallium nitride, or indium arsenide, are actively being investigated as a step toward continuing to scale electronics down. For nanowires to be utilized to their full potential, a microscopic understanding of their properties, and of how they differ compared to the bulk, is crucial. Likewise, since native point defects control many aspects of a semiconductor's function, a detailed understanding of defects and the difference to the behavior in bulk, is mandatory.

As described in the review paper on semiconductor nanowires and nanotubes (Law et al. 2004), the mechanisms of creating one-dimensional nanostructures include (1) the growth of an intrinsically anisotropic crystallographic structure, (2) the use of various templates with one-dimensional morphologies to direct the formation of one-dimensional structures, (3) the introduction of a liquid–solid interface to reduce the symmetry of a seed, (4) the use of an appropriate capping reagent to kinetically control the growth of various facets of a seed, and (5) the self-assembly of 0D nanostructures. Further details and information on experimental techniques, and results, can be found in recent reviews on growth, characterization, assembly, and integration of semiconductor nanowires (Lu and Lieber 2006), and on the energy band gap of clusters, nanoparticles, and quantum dots (Sattler 2002).

In this chapter, on the basis of first-principles calculations, the physical properties of GaN nanowires are discussed, as well as how they change as a function of wire size and shape, and whether the dangling bonds at the wire surfaces are saturated or not. Also the behavior of single and multiple gallium and nitrogen vacancies in the nanowires are described. Finally, a brief comparison of the properties of nanowires and nanodots are mentioned, highlighting the important effect of the dimensionality of the nanostructure system on the electronic structure in particular.

18.2 Geometry and Diameter Dependence of the Electronic and Physical Properties of GaN Nanowires

Gallium nitride nanowires typically form in the wurtzite structure (Kim et al. 2002; Nam et al. 2004; Kipshidze et al. 2005; Byeun et al. 2006; Simpkins et al. 2006; Xu et al. 2006), although they have also been reported in the zinc blende structure (Dhara et al. 2004). The diameter of wires synthesized experimentally typically range from approximately 5 to 100 nm. A number of different growth directions have been reported, including the [0001] (Kim et al. 2002; Byeun et al. 2006; Simpkins et al. 2006; Xu et al. 2006), [10$\bar{1}$0] (Peng et al. 2002;

Kuykendall et al. 2004), and [11$\bar{2}$0] (Kuykendall et al. 2004; Byeun et al. 2006) directions.

Since nanowires are not created under ultrahigh vacuum conditions experimentally, atoms or molecules from the environment (e.g., hydrogen or oxygen) may adsorb on the surfaces and saturate the dangling bonds. For this reason, it is valuable for theoretical studies to consider both the case of pristine nanowires with no adsorbed impurity species, as well as those with atoms adsorbed on the nanowires surface in such a way as to saturate dangling bonds, thus simulating the situation that may occur in experiment.

In this section, the dependence of the electronic and physical properties, of both saturated and unsaturated GaN nanowires in the [0001] growth direction, on the shape and size of the nanowire will be discussed. All the theoretical results were obtained using density-functional theory (DFT) as implemented in the DMol³ (Delley 1990, 2000) and SIESTA (Ordejon et al. 1996; Soler et al. 2002) codes. For more details of the calculations for nanowires, the original papers are referred to (Carter et al. 2008; Carter and Stampfl 2009).

18.2.1 Atomic and Electronic Structure of Unsaturated and Saturated Nanowires

The GaN nanowires considered are shown in Figure 18.3. They have both hexagonal and triangular cross sections. The

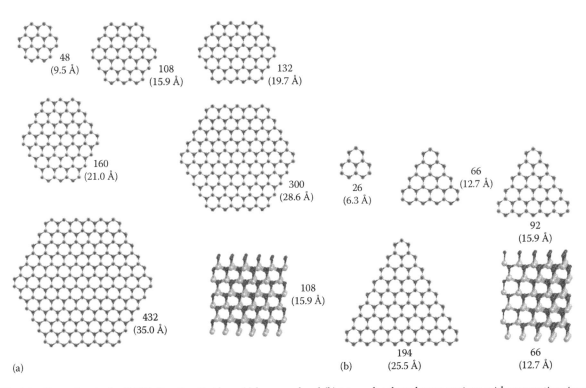

FIGURE 18.3 Nanowires in the [0001] direction that have (a) hexagonal and (b) triangular shaped cross sections, with perspective views of the 108- and 66-atom nanowires included. The number of atoms used in the calculations for each nanowire, and the diameter (in parenthesis) are also labeled. Nitrogen and gallium atoms are indicated by the dark and light spheres, respectively. (Reprinted from Carter, D.J. et al., *Phys. Rev. B*, 77, 115349, 2008. With permission.)

diameters vary from approximately 8–35 Å. The nanowire diameter is defined here as the maximum distance between edge atoms on opposite sides of the nanowire. All the nanowires considered have (10$\bar{1}$0) facets since the surface energy of the (10$\bar{1}$0) surface of GaN is lower than that of the (11$\bar{2}$0) surface. Thus, nanowires composed of (10$\bar{1}$0) facets are expected to be more stable than those composed of (11$\bar{2}$0) facets. Nanowires both with unsaturated dangling bonds (pristine or "pure" GaN wires) and those with hydrogen atoms saturating dangling bonds are investigated.

18.2.1.1 Atomic Structure

The surfaces of all the unsaturated nanowires, hexagonal and triangular, and for all diameters considered exhibit a similar atomic relaxation (change in atomic positions). In particular, the outermost bond lengths along the [0001] direction are contracted by around 6.0%–7.4% compared to that of bulk GaN (Carter et al. 2008). This behavior is very similar to that exhibited by the GaN (10$\bar{1}$0) surface, where a value of 6% was reported (Northrup and Neugebauer 1996). It is also similar to the findings of Tsai et al. (2006), who, using DFT, examined

GaN nanowires in the [0001] growth direction with diameters of 10, 15, and 18 Å, and found that this bond length contracts by 6.2%–6.4%. On saturating the dangling bonds of the nanowires with hydrogen, only a very small atomic relaxation occurs, namely, the bonds contract by 0.5%–1.7% (Carter et al. 2008). This is understandable since the atoms at the surface of the wires experience an environment very similar to that which they have in the bulk material, and hence the bond lengths are very close to the bulk value.

18.2.1.2 Electronic Structure

In Figure 18.4, the band structures for unsaturated hexagonal and triangular nanowires are shown for large and small diameter wires. The results are very similar for all sized wires of both shapes. Figure 18.5 shows the corresponding band structure for the saturated nanowires. By comparing Figures 18.4 and 18.5, it can be noticed that the effective "band gaps" for the unsaturated nanowires are smaller than for the saturated nanowires; this is due to dangling bonds at the surface of the unsaturated nanowires. These dangling bonds induce states (bands) in the band gaps, located above the valence band maximum (VBM)

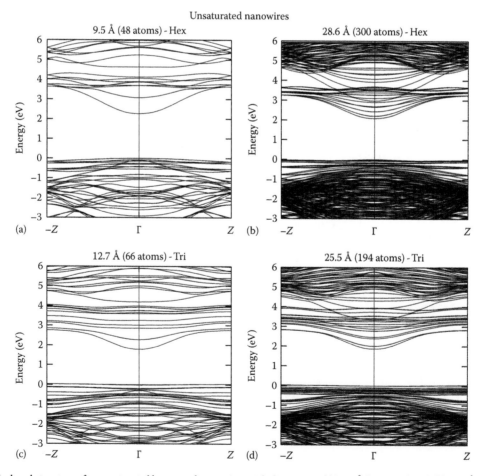

Unsaturated nanowires

FIGURE 18.4 The band structures for unsaturated hexagonal nanowires with diameters of (a) 9.5 Å (48 atoms) and (b) 28.6 Å (300 atoms) and triangular nanowires with diameters of (c) 12.7 Å (66 atoms) and (d) 25.5 Å (194 atoms). The energy zero is set at the highest occupied level. "Hex" and "Tri" represent "hexagonal" and "triangular," respectively. (Reprinted from Carter, D.J. et al., *Phys. Rev. B*, 77, 115349, 2008. With permission.)

Saturated nanowires

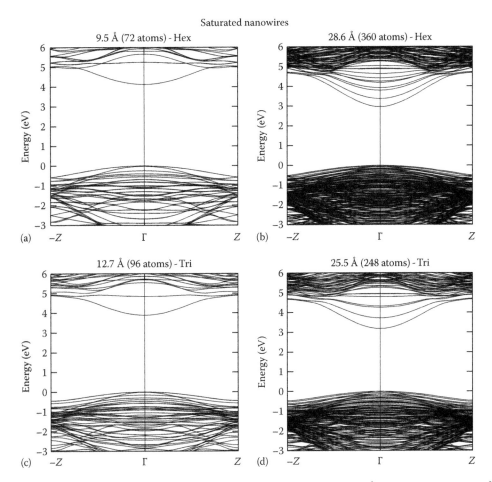

FIGURE 18.5 The band structures for saturated hexagonal nanowires with diameters of (a) 9.5 Å (72 atoms) and (b) 28.6 Å (360 atoms) and triangular nanowires with diameters of (c) 12.7 Å (96 atoms) and (d) 25.5 Å (248 atoms). The energy zero is set at the highest occupied level. "Hex" and "Tri" represent "hexagonal" and "triangular," respectively. (Reprinted from Carter, D.J. et al., *Phys. Rev. B*, 77, 115349, 2008. With permission.)

and below the conduction band minimum (CBM) (Carter et al. 2008). The "effective" band gap of unsaturated nanowires, i.e., the gap between the edge-induced states, does not change significantly with nanowire diameter since they are quite localized. When the nanowires are saturated with hydrogen, these dangling bond bands are removed from the band gap.

The nature of the edge-induced states for the unsaturated nanowires can be seen by plotting their spatial distribution (at the Γ-point). This is shown in Figure 18.6 for both the large and the small hexagonal and triangular wires, for the highest occupied orbital (HOMO) and the lowest unoccupied orbital (LUMO). It can be seen that the states have a significant weight at the edge of the nanowires (Carter et al. 2008). This character is similar for all the nanowire diameters considered and for both shapes. For the smaller diameter nanowire, there is still a visible contribution at the center of the nanowire due to the very small diameter. Through close inspection of the HOMO, it can be determined that the orbital contributions are mainly due to the nitrogen atoms, with p character, while the LUMO states are mainly due to gallium atoms, also with p character. For the saturated nanowires shown in Figure 18.7, the HOMO and LUMO states are distributed across the center of the nanowires.

18.2.2 Dependence of the Band Gap and Formation Energy on Nanowire Diameter

18.2.2.1 Band Gap Dependence on Nanowire Size and Shape

The dependencies of the band gaps of the hexagonal and triangular nanowires on diameter are shown in Figure 18.8. All saturated nanowires exhibit a decrease in the band gap with increasing diameter, tending toward the bulk band-gap values. For unsaturated nanowires, there is little change in the band gaps, illustrating the influence of the localized edge-like dangling bond states in the band gap, mentioned above. The unsaturated triangular nanowires possess a smaller gap than the hexagonal wires. This is understandable since the triangular wires are less stable, and the occupied edge-induced states are at a higher energy compared to the hexagonal wires (and for a given diameter, there are more dangling bonds for the triangular wires), and thus the formation energy is higher, i.e., less stable (as seen also from Figure 18.10).

In Figure 18.9, the change in energy band gap relative to the bulk value is plotted versus $1/d^2$ (with d the diameter of the wire).

Unsaturated nanowires

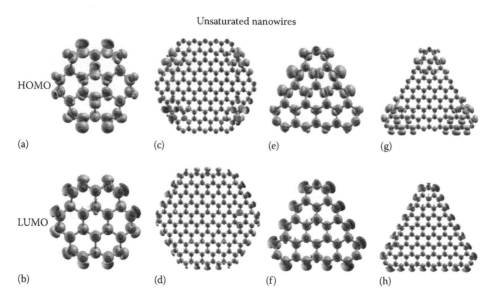

FIGURE 18.6 Spatial distribution of the HOMO and LUMO states at the Γ-point for unsaturated nanowires. The HOMO states (upper panel) are shown for hexagonal wires with diameters of (a) 9.5 Å (48 atoms) and (c) 28.6 Å (300 atoms) and for triangular nanowires with diameters of (e) 12.7 Å (66 atoms) and (g) 25.5 Å (194 atoms). The LUMO states (lower panel) are shown for hexagonal wires with diameters of (b) 9.5 Å and (d) 28.6 Å and for triangular nanowires with diameters of (f) 12.7 and (h) 25.5 Å. Nitrogen and gallium atoms are indicated by dark, and light spheres, respectively, and the orbitals are pale gray. (Reprinted from Carter, D.J. et al., *Phys. Rev. B*, 77, 115349, 2008. With permission.)

Saturated nanowires

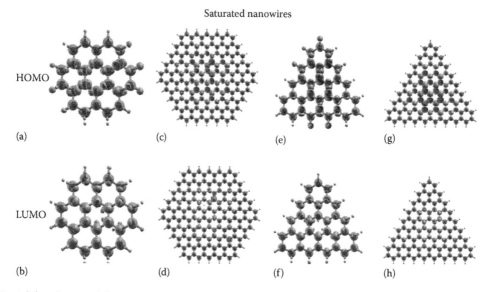

FIGURE 18.7 Spatial distribution of the HOMO and LUMO states at the Γ-point for saturated nanowires. The HOMO states (upper panel) are shown for hexagonal wires with diameters of (a) 9.5 Å (72 atoms) and (c) 28.6 Å (360 atoms) and triangular nanowires with diameters of (e) 12.7 Å (96 atoms) and (g) 25.5 Å (248 atoms). The LUMO states (lower panel) are shown for hexagonal wires with diameters of (b) 9.5 and (d) 28.6 Å, and triangular nanowires with diameters of (f) 12.7 and (h) 25.5 Å. Nitrogen and gallium atoms are dark and light spheres, respectively, hydrogen atoms are represented by very small light gray spheres, and the orbitals are pale gray. (Reprinted from Carter, D.J. et al., *Phys. Rev. B*, 77, 115349, 2008. With permission.)

It is found that hexagonal nanowires with diameters ranging from 19.7 to 35.0 Å and triangular nanowires with diameters ranging from 12.7 to 25.5 Å appear to follow closely the $1/d^2$ proportionality. Nanowires with smaller diameters deviate from this proportionality relationship. Schmidt et al. (2005) report a similar behavior for InP nanowires, where nanowires with diameters ranging from 18.0 to 21.3 Å were studied. They

suggest that for smaller diameter nanowires, the contribution of the nanowire surface to the electronic properties is not negligible, compared to the contribution for larger nanowires, leading to the deviation from the proportionality of the EMA-PIB model. It should be noted of course that band gaps calculated using DFT are systematically underestimated. Thus, in order to obtain band gaps closer to experimental values, more accurate

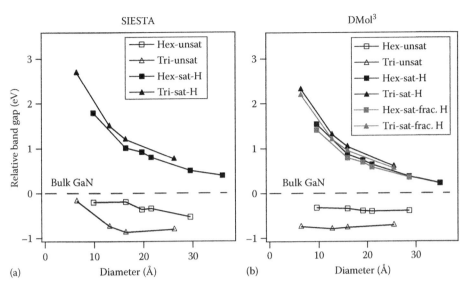

FIGURE 18.8 Relative band gap as a function of nanowire diameter as obtained using the (a) SIESTA and (b) DMol^3 codes. Band gaps are relative to the calculated bulk GaN band gap. "Hex" and "Tri" represent "hexagonal" and "triangular," respectively, and "unsat," "sat," and "sat-frac. H" represent "unsaturated," "saturated," and "saturated with fractional charge hydrogen," respectively. (Reprinted from Carter, D.J. et al., *Phys. Rev. B*, 77, 115349, 2008. With permission.)

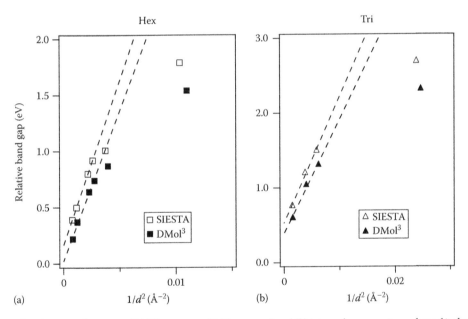

FIGURE 18.9 Relative band gap as a function of $1/d^2$ for saturated (a) hexagonal and (b) triangular nanowires, where *d* is the nanowire diameter. The dashed lines indicate the linear regions for each of the data sets. "Hex" and "Tri" represent "hexagonal" and "triangular," respectively. Band gaps are relative to the calculated bulk GaN band gap. (Reprinted from Carter, D.J. et al., *Phys. Rev. B*, 77, 115349, 2008. With permission.)

descriptions are required. Recently Carter et al. (2009a) have carried out DFT calculations employing self-interaction corrected (SIC) pseudopotentials, to determine the band gaps of saturated hexagonal nanowires, in comparison to the results obtained from standard DFT-GGA pseudopotential methods. The value of the bulk band gap was significantly improved in the SIC approach (4.17 eV) over that obtained by standard DFT-GGA (using SIESTA with Troullier–Martin pseudopotentials, 1.44 eV) in comparison to the experimental value (3.50 eV). For the nanowires with diameters ranging from 9.5 to 21 Å,

the SIC approach resulted in systematically larger band gaps (by ≈2.7 eV) compared to standard DFT-GGA, but the relative band gap change was very similar to that obtained by standard DFT-GGA. This dependence on the band gap with nanowire size clearly offers potential to design or tailor a wire to yield the desired band gap for particular device applications.

Another interesting trend with (saturated) nanowire size is the behavior of the effective mass. In particular, it has been found (Carter et al. 2008) that the smaller the nanowire, the larger the effective mass. For example, for bulk, the effective mass, m_c^*,

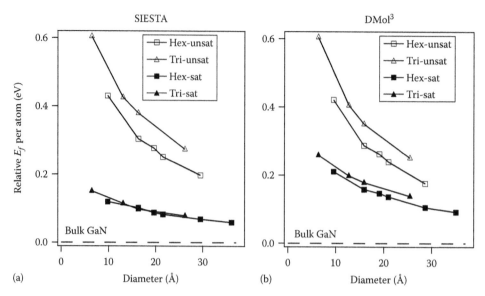

FIGURE 18.10 Relative heat of formation (per atom) as a function of nanowire diameter for unsaturated and saturated nanowires in the [0001] direction, from (a) SIESTA and (b) DMol³. Energies are relative to the heat of formation (per atom) of bulk GaN. "Hex" and "Tri" represent "hexagonal" and "triangular," respectively, and "unsat" and "sat" represent "unsaturated" and "saturated," respectively. (Reprinted from Carter, D.J. et al., *Phys. Rev. B*, 77, 115349, 2008. With permission.)

around the CBM with respect to the free electron mass (m_c^*) is calculated to be $0.16\,m_0$ and $0.27\,m_0$ using DMol³ and SIESTA, respectively, while for a hexagonal nanowire of small diameter (9.5 Å) it is $0.48\,m_0$ (SIESTA) and $0.41\,m_0$ (DMol³). For a larger diameter of (28.6 Å) it is $0.22\,m_0$ (SIESTA) and $0.25\,m_0$ (DMol³). This trend is understandable because as the diameter of the nanowire increases, the band gap becomes smaller, which leads to an increase in the curvature of the conduction band minimum about the Γ-point (and correspondingly, a larger effective mass).

18.2.2.2 Formation Energy Dependence on Nanowire Size and Shape

It could be expected that as the size of a nanowire increases, the relative affect of its surface on the physical properties of the nanowire becomes less. Indeed, this is the case for the formation energy, where it exhibits a pronounced decrease for increasing diameter of the nanowire as shown in Figure 18.10. In other words, the relative stability of a nanowire increases as its size increases, gradually approaching the bulk situation. Due to the unsaturated dangling bonds of clean nanowires, the formation energy compared to a saturated wire is considerably higher (less stable), and the dependence with size greater. In Figure 18.10, the relative formation energy of saturated and unsaturated nanowires for varying diameters is shown, for both hexagonal and triangular wires, as calculated with SIESTA and DMol³. The formation energy, E_f, per atom is defined as the total energy of the nanowire minus that of a bulk Ga atom (times the number of Ga atoms that it contains), minus half that of the N_2 molecule (times the number of N atoms that it contains), divided by the total number of atoms in the wire. For the saturated wires, there is in addition, a term subtracting half the total energy of an H_2

molecule (times the number of **H** atoms that the wire contains). From Figure 18.10, it can furthermore be noticed that for the saturated wires, the results for hexagonal and triangular wires are very similar, which is because once the dangling bonds are saturated, the local bonding environment is very similar for both shaped wires. For the unsaturated wires, the triangular ones are somewhat less stable since for a given diameter, they contain a greater number of dangling bonds; and, as mentioned above, the occupied edge-induced states are at a higher energy compared to the hexagonal wires.

In relation to this, earlier DFT studies of nanowires in the [0001] direction of approximately 20 Å diameter (Gulans and Tale 2007) found, that out of nanowires with cross-sectional shapes ranging from hexagonal to essentially circular, the hexagonal-shaped nanowire is the most stable.

18.3 Atomic and Electronic Structure of Single and Multiple Vacancies in GaN Nanowires

For nanowires to be utilized to their full potential, an improved understanding of their properties and of the behavior of defects, which control many aspects of a semiconductor's function, and of how these properties differ compared to the bulk is crucial.

The properties of bulk III-nitride compounds containing defects and dopants from first-principles calculations has been reviewed by Van de Walle and Neugebauer (2004). For GaN in particular, there have been numerous first-principles studies of native defects (e.g., Neugebauer and Van de Walle 1994; Park and Chadi 1997; Limpijumnong and Van de Walle 2004; Ganchenkova and Nieminen 2006). However to date, the structure, relative stability, and electronic properties of defects in gallium nitride nanowires, has only very recently been investigated (Carter and Stampfl 2009).

It is useful to first review the behavior of the most abundant defects in bulk wurtzite GaN, which are the nitrogen and gallium vacancies. First-principles studies are consistent in predicting that gallium vacancies act as triple acceptors in more *n*-type material. For the nitrogen vacancy, while there is agreement that it acts as a single donor for a considerable range of the Fermi level, some theoretical studies predict that in more *p*-type material it also acts as a triple donor, while other studies predict it only acts as a single donor. Furthermore, very recent calculations, in contrast to earlier studies, report that in more *n*-type material, the nitrogen vacancy can also act as an acceptor in single, double, and triple charge states (Ganchenkova and Nieminen 2006). Recent calculations by Carter and Stampfl report that the nitrogen vacancy is stable as a single acceptor and as a donor in the single and triple charge states (Carter and Stampfl 2009).

For the study of vacancies in nanowires, two sized hexagonal wires are considered, namely, 96- and 216-atom wires. These wires have double the periodicity along the *c*-axis in order to separate the vacancies, and thus the cross sections of the nanowires appear as for the 48- and 108-atom nanowires, as shown in Figure 18.3.

18.3.1 Nitrogen Vacancies

Let us consider first, single nitrogen vacancies in unsaturated 96-atom (9.5 Å diameter) and 216-atom (15.9 Å diameter) nanowires. To determine the most stable configurations, all possible locations in the wire are considered. It is found that for both sized wires, the most stable position is where the vacancy is located at the edge of the nanowire (Carter and Stampfl 2009) and coordinated to three Ga atoms (instead of four as in the bulk). The next

most favorable site is 1.1 eV less stable and is also located at the edge of the nanowire and is fourfold coordinated to Ga atoms.

It is interesting to consider the energy cost of creating a vacancy, the so-called formation energy. It is defined as the total energy of the nanowire containing the vacancy, minus that of the corresponding pure nanowire, plus the total energy of a nitrogen atom in a free nitrogen molecule (resulting in so-called nitrogen-rich conditions), or instead, plus the total energy of a nitrogen atom in bulk GaN (resulting in so-called gallium-rich conditions). The formation energies for the nitrogen vacancy in the 96-atom nanowire are in the range of 0.7–2.0 eV, and 0.6–2.2 eV for the 216-atom nanowire (for Ga-rich conditions). The largest values correspond to the vacancy located in the center of the wires. The notably smaller value of the formation energy at the edge of the wires is related to fewer Ga dangling bonds that the defect has at the wire edge (three versus four), and also to the greater space the neighboring atoms have to relax (move) in response to the creation of the vacancy. This lower formation energy indicates that nitrogen vacancies may be more abundant in nanowires compared to in bulk, where the formation energy is calculated to be 2.5 eV (using the same calculation procedure) (Carter and Stampfl 2009).

The single nitrogen vacancy induces several defect states below the edge states (ES) region near the CBM, as illustrated in the band structure in Figure 18.11a and b for the 96-atom and 216-atom nanowires, respectively. The character of the states is determined by investigating the spatial distribution of the corresponding wave functions at the Γ-point. Those that are defect related are indicated by "D" in the figure. There are three singlet states just below the top of the ES region near the CBM, with the lowest of these states occupied by one electron, while the other two defect states are unoccupied. This is illustrated schematically in Figure 18.11c. Compared to the single nitrogen vacancy in the bulk (Carter and Stampfl 2009), there is one less singlet state for

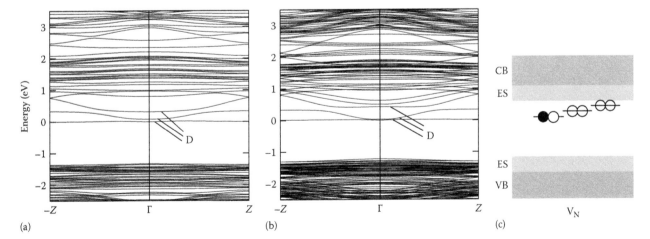

FIGURE 18.11 Band structure of the most stable configurations for the single nitrogen vacancy in the (a) 96-atom and (b) 216-atom nanowires. The energy zero is set at the highest occupied state, and the defect states are labeled with a "D." (c) Corresponding schematic representations of the defect-induced levels for both size nanowires, where filled and open circles denote electrons and holes, respectively. "CB," "VB," and "ES" represent the conduction band, the valence band, and the edge states, respectively. (Reprinted from Carter, D.J. and Stampfl, C., *Phys. Rev. B*, 79, 195302, 2009. With permission.)

a vacancy in the nanowires, i.e., there is no fully occupied singlet state near the VBM/ES region.

Comparing the position of the defect-induced states for the small and larger wire, there is only a small difference in the position of the states where those of the larger wire are slightly closer to the CBM. In Figure 18.12a, the spatial distribution of the occupied defect state at the Γ-point is shown, where it can be seen that it is localized in the region of the vacancy. Figure 18.12b shows (in the plane of the vacancy) the difference electron-density distribution, which is the difference between the total electron density of the wire and the superimposed sum of atomic densities of the Ga and N atoms in the positions they have in the wire. The increase in electron density at the vacancy site can be clearly seen.

Having investigated the behavior of single N vacancies in GaN nanowires, it is interesting to consider multiple vacancies, namely, the behavior of two and three N vacancies. For these investigations the large 216-atom nanowires are used. Through considering 50 different configurations for two nitrogen vacancies, and 20 for three vacancies, the energetically most favorable geometries are found to be as follows: The most stable two-vacancy configuration has both vacancies at the edge of the nanowire, with the vacancies as close as possible to each other, but in different nitrogen (0001) planes (referred to as "out-of-plane"). The total formation energy of the most stable configuration is 1.0 eV more favorable than the next most favorable configuration, which also has the vacancies at the edge of the nanowire, but "in-plane" (same nitrogen (0001) plane) with each other, and as close as possible to each other, i.e., nearest-neighbor (like-species) sites (Carter and Stampfl 2009). When the two vacancies are well separated, but still at the edge of the nanowire, the configuration is 1.8 eV less stable than the most favorable configuration, showing that the vacancies prefer to be clustered. The reason for the attraction between the vacancies is attributed to the significant atomic relaxations of the surrounding Ga

atoms. In particular, there is a reduced separation between two gallium atoms around the vacancy, resulting in a Ga–Ga distance of 2.65 Å. This separation is noticeably smaller than the Ga–Ga separation of 3.28 Å in bulk GaN, and more similar to the Ga–Ga separation of 2.55 Å in bulk α-Ga.

For three-nitrogen vacancies, the most stable configuration is where all vacancies are located at the edge of the nanowires, and are clustered together. The geometry is similar to the most stable configuration for two nitrogen vacancies, with two of the vacancies out-of-plane to each other (different nitrogen (0001) planes), while the third nitrogen vacancy is in-plane with one of the vacancies. The atomic structure shows that again there are significant relaxations of the surrounding Ga atoms, such that they move toward each other, resulting in Ga–Ga distances of 3.03, 3.08, and 3.11 Å.

The two- and three-nitrogen vacancy complexes induce additional states in the band gap, acting as double and triple donors respectively, as seen in Figure 18.13, where the results for the single nitrogen vacancy are shown again for comparison.

Given that the nitrogen vacancy is expected to be the major native point defect in bulk GaN (Ganchenkova and Nieminen 2006), the lower formation energies found for this defect in the nanowires indicates it will be an even more dominant defect in wires and will contribute to an *n*-type conductivity of the wires. For the calculations with one- and three-nitrogen vacancies, it was seen that there is an unpaired electron in singlet states below the ES region near the CBM. This occupancy and position appears consistent with recent electrical transport measurements of GaN nanowires (Calarco et al. 2005) that suggest that the Fermi level is pinned at the surface of the nanowires at about 0.5–0.6 eV below the CBM.

18.3.2 Gallium Vacancies

Single gallium vacancies have also been investigated using 96- (9.5 Å diameter) and 216-atom (15.9 Å diameter) unsaturated nanowires. For both nanowires, the most stable gallium vacancy sites are the same as for the nitrogen vacancy, namely, at the edge of the nanowire and coordinated to three neighboring atoms (instead of four as in bulk) (Carter and Stampfl 2009). The next most favorable site is 0.3–0.4 eV less stable and located in the more bulk-like center region of the wire.

The formation energies for Ga vacancies are in the range of 4.5–5.5 eV for the 96-atom wire, and in the range 5.9–7.1 eV in the large 216-atom nanowire. Thus, the energy cost to create a (neutral) Ga vacancy is considerably more than to create a (neutral) N vacancy. In bulk GaN, the corresponding formation energy is 7.1 eV (Carter and Stampfl 2009); thus, that of the vacancy in the large wire has already converged to the bulk value.

The band structures for a single gallium vacancy in the 96-atom and 216-atom nanowires are shown in Figure 18.14a and b, respectively. The induced defect states are shown schematically in Figure 18.14c. There are two defect-induced states just above the ES region near the VBM, with the lower of these two states occupied by one electron, while the higher of the two

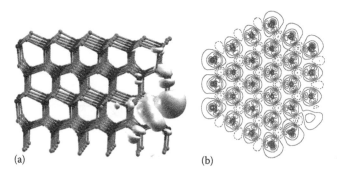

(a) (b)

FIGURE 18.12 (a) The spatial distribution of the singly occupied defect state (see Figure 18.11) at the Γ-point for the single nitrogen vacancy in the 216-atom nanowire. Nitrogen and gallium atoms are indicated by dark and light spheres, respectively, and the orbitals are pale gray. (b) The difference electron-density distribution in the plane of the vacancy for the single nitrogen vacancy in the 216-atom nanowire. Solid lines indicate charge accumulation and dashed lines indicate charge depletion. (Reprinted from Carter, D.J. and Stampfl, C., *Phys. Rev. B*, 79, 195302, 2009. With permission.)

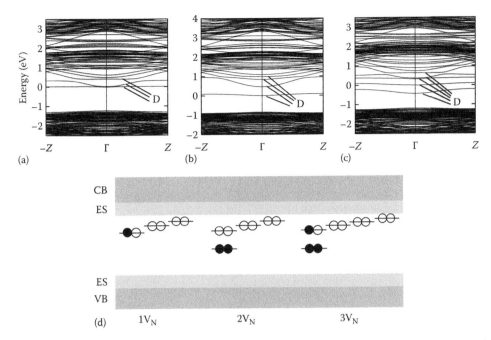

FIGURE 18.13 The band structure of the most stable nitrogen (a) single-, (b) two-, and (c) three-vacancy configurations in the 216-atoms nanowire. The energy zero is set at the highest occupied state, and the defect states are labeled with a "D." (d) Corresponding schematic representation of the defect-induced levels for one, two, and three nitrogen vacancies, where filled and open circles denote electrons and holes, respectively. "CB," "VB," and "ES" represent the conduction band, the valence band, and the edge states, respectively. (Reprinted from Carter, D.J. and Stampfl, C., *Phys. Rev. B*, 79, 195302, 2009. With permission.)

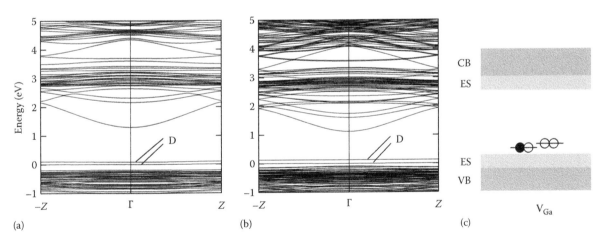

FIGURE 18.14 Band structure of the most stable configurations for the single gallium vacancy in the (a) 96-atom and (b) 216-atom nanowires. The energy zero is set at the highest occupied state, and the defect states are labeled with a "D." (c) Corresponding schematic representations of the defect-induced levels for both size nanowires, where filled and open circles denote electrons and holes, respectively. "CB," "VB," and "ES" represent the conduction band, the valence band, and the edge states, respectively. (Reprinted from Carter, D.J. and Stampfl, C., *Phys. Rev. B*, 79, 195302, 2009. With permission.)

is unoccupied. For a gallium vacancy in the bulk, there are three singlet states above the VBM; thus, for the wires there is one less state. In Figure 18.15a, the spatial distribution of the occupied defect state at the Γ-point is shown, illustrating how it is localized in the region of the vacancy. Figure 18.15b illustrates (in the plane of the vacancy) the difference electron-density distribution, clearly showing an increase in the electron density at the vacancy site.

When two gallium vacancies are considered in the 216-atom nanowire, it is found that the energetically most favorable configuration (out of the 20 investigated) is when both vacancies are located close to each other at the edge of the nanowire, but out-of-plane to each other (Carter and Stampfl 2009). This configuration is the same as the most stable nitrogen two-vacancy configuration and is 1.2 eV lower in energy than the next most favorable two-Ga-vacancy configuration, which also has the

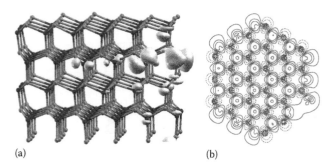

(a)   (b)

FIGURE 18.15 (a) The spatial distribution of the singly occupied defect state (see Figure 18.14) at the Γ-point for the single gallium vacancy in the 216-atom nanowire. Nitrogen and gallium atoms are indicated by dark and light spheres, respectively, and the orbitals are pale gray. (b) The difference electron-density distribution in the plane of the vacancy for the single gallium vacancy in the 216-atom nanowire. Solid lines indicate charge accumulation and dashed lines indicate charge depletion. (Reprinted from Carter, D.J. and Stampfl, C., *Phys. Rev. B*, 79, 195302, 2009. With permission.)

vacancies next to each other, but in the same plane. The reason for the notably lower energy of the favored structure is due to a considerable atomic relaxation, which results in the surrounding nitrogen atoms forming an N_3-like geometry. The N–N separations are 1.30 and 1.41 Å, significantly shorter than the N–N separation in bulk GaN of 3.28 Å, and more comparable to that of the N_2 molecule (1.12 Å) or the azide ion (1.16 Å).

The band structure of the most stable two-Ga-vacancy configuration is shown in Figure 18.16b (that of the most stable single-vacancy configuration is also shown in Figure 18.16a). The positions of the defect-induced states and the occupations are shown schematically in Figure 18.16c. The two Ga-vacancy configuration induces seven defect singlet states spread out across the region of the band gap between the ES states near the VBM and the CBM. The four lower singlet states are fully occupied while the remaining three are unoccupied. These induced states are quite different to those resulting from a single gallium vacancy, which is related to the formation of the N_3 trimer-like

structure; this gives rise to extra dangling bonds around the vacancies that contribute the extra defect-induced states.

It is interesting to consider the effect of saturation of the dangling bonds at the nanowire surfaces on the nature and position of defect-induced states. In Figure 18.17a and b, the band structures of the unsaturated and saturated wires with a single gallium vacancy in the center of the wire are shown. In Figure 18.17c and e, the corresponding results for a single nitrogen vacancy are shown. The positions and occupations of the defect-induced states are shown schematically in Figure 18.17e (Carter and Stampfl 2009). For the gallium vacancy there are three singlet states located just above the ES region near the VBM, with the lowest state fully occupied, the second lowest containing one electron, and the highest singlet state unoccupied. This can be compared to the result for the gallium vacancy at the edge of the wire (Figure 18.14) where there are only two defect-induced singlet states. At the center of the wire, the defect is fully coordinated and more similar to in bulk GaN. In bulk GaN, there are three singlet states just above the VBM, with the lowest of the singlet states fully occupied, the second lowest singlet containing one electron, and the highest singlet state unoccupied, which is qualitatively similar to what we find for the unsaturated nanowire with the defect in the centre (Carter and Stampfl 2009). For the case of the saturated nanowire, the nature of the induced states are qualitatively similar to the unsaturated wires, except that the band gap is larger due to the removal of the edge-induced states.

For the nitrogen vacancy at the centre of the unsaturated nanowire (Figure 18.17c) there are four defect-induced singlet states, with one singlet state fully occupied below the ES region near the VBM, and three below the ES region near the CBM, with the lowest of these half occupied, and the other two unoccupied. Compared to the case when the vacancy is in the most stable configuration at the edge, there is one additional state just below the ES region near the VBM. For the saturated nanowire, there are similarly to the saturated case four defect-induced states. Interestingly, the fully occupied singlet state is located below the

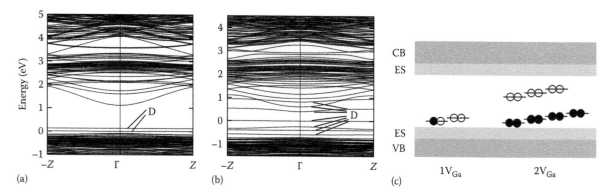

(a)   (b)   (c)

FIGURE 18.16 Band structures of the most stable gallium (a) single-vacancy and (b) two-vacancy configurations, with the energy zero set at the highest occupied state, and the defect states are labeled with a "D." (c) Corresponding schematic representation of the defect-induced levels for one and two gallium vacancies, where filled and open circles denote electrons and holes, respectively. "CB," "VB," and "ES" represent the conduction band, the valence band, and the edge states, respectively. (Reprinted from Carter, D.J. and Stampfl, C., *Phys. Rev. B*, 79, 195302, 2009. With permission.)

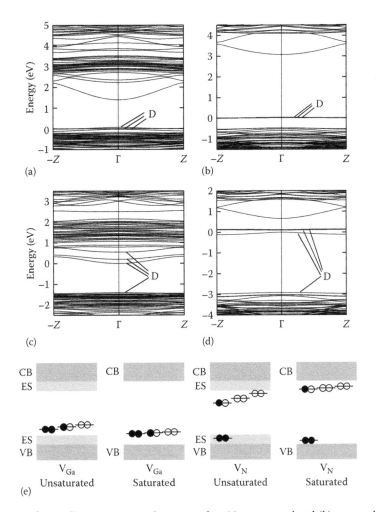

FIGURE 18.17 The band structure for a gallium vacancy in the center of an (a) unsaturated and (b) saturated 96-atom nanowire, and for a nitrogen vacancy in the center of (c) an unsaturated and (d) saturated 96-atom nanowire. The energy zero is set at the highest occupied state, and the defect states are labeled with a "D." (e) Schematic representation of the defect-induced levels, where filled and open circles denote electrons and holes, respectively. "CB," "VB," and "ES" represent the conduction band, the valence band, and the edge states, respectively. (Reprinted from Carter, D.J. and Stampfl, C., *Phys. Rev. B*, 79, 195302, 2009. With permission.)

VBM in bulk GaN, while in the saturated nanowire it is located just above the VBM. This suggests that potentially the nitrogen vacancy in the saturated nanowire can act as a triple donor.

18.4 Comparison of Properties of GaN Nanowires and Nanodots

In this section, recent results of DFT calculations for saturated GaN nanodots are described (Carter et al. 2009b), in comparison to those of nanowires. Nanodots with diameters ranging from 11.0 to 22.8 Å are investigated (see Figure 18.18) where all structures are fully atomically relaxed, as for the wires described in the previous sections. The calculations are carried out using DMol³ (Delley 1990, 2000). In Figure 18.19, the change in energy gap with diameter is shown as indicated by the position of the HOMO and LUMO states at the Γ-point (Figure 18.19a), and relative to the bulk band gap (Figure 18.19c), and as a function of $1/d^2$ (Figure 18.19d). In Figure 18.19b, the

formation energy (per atom) of the nanodots are shown versus the diameter. For each quantity, comparison with the results for the wires is included. It can be seen that the LUMO exhibits a greater variation than the HOMO over the nanodot sizes examined, as is the case for the nanowires (Carter and Stampfl 2009). In particular, as the nanodot diameter increases from 11.0 to 22.8 Å the shifts of the HOMO and LUMO values are +0.60 and −1.12 eV, respectively.

From Figure 18.19c and d, it can be seen, as expected from the simple EMA-PIB model described in the introduction, the band gap of the nanodot is larger than that of the nanowire. Also, its variation with diameter is greater with the ratio of the slopes of the lines of best fit to the calculated points in Figure 18.19d for wires to dots being 1.29:2.72 = 0.474. This value can be compared to that obtained from the EMA-PIB of 0.585, and to that found experimentally for InP quantum dots and wires of 0.62 (Yu et al. 2003).

If instead the theoretical results are fitted to an equation of the form $\Delta E_g = A/d^x$ (see Figure 18.20), we find values of x for

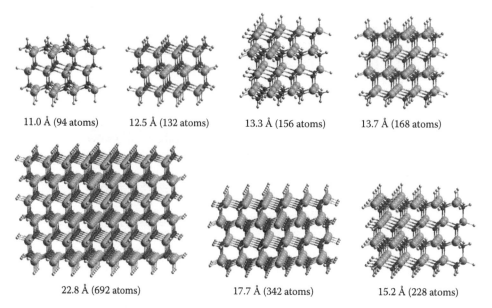

11.0 Å (94 atoms)   12.5 Å (132 atoms)   13.3 Å (156 atoms)   13.7 Å (168 atoms)

22.8 Å (692 atoms)   17.7 Å (342 atoms)   15.2 Å (228 atoms)

FIGURE 18.18 Illustrations of the wurtzite GaN nanodots considered. The diameter of each nanodot and the number of atoms (in parenthesis) is also labeled. Nitrogen, gallium and fractional charge hydrogen atoms are indicated by the dark, large light, and small light gray spheres, respectively. (Reprinted from Carter D.J. et al., *Nanotechnol.*, 20, 425401, 2009b. With permission.)

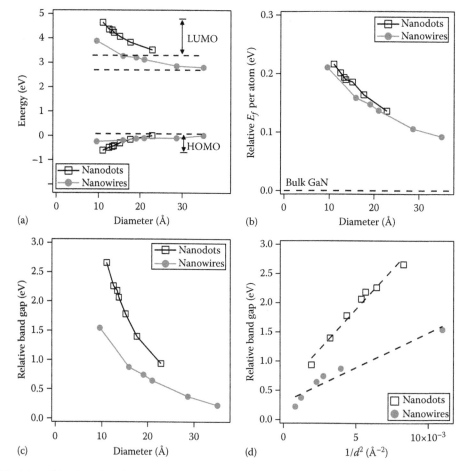

FIGURE 18.19 (a) Position of the HOMO and LUMO energies as a function of the nanodot and nanowire diameter. (b) Heat of formation (per atom) as a function of diameter. (c) Band gap of the nanodots and nanowires as a function of diameter. (d) Energy gap (relative to that of bulk GaN) of the nanodots and nanowires as a function of $1/d^2$, where d is the diameter and the dashed lines are the best fit to the calculated points. (Reprinted from Carter D.J. et al., *Nanotechnol.*, 20, 425401, 2009b. With permission.)

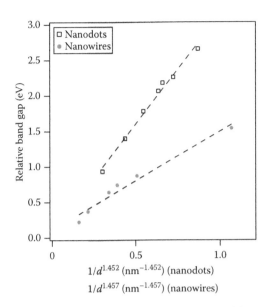

FIGURE 18.20 Relative band gaps (to the bulk value) of the nanodots and nanowires as a function of $1/d^x$, where d is the diameter. The dashed lines are the best fit to the calculated points using an expression for the change in band gap of A/d^x. The obtained values of x for the dots and wires are 1.451 and 1.460, respectively. (Reprinted from Carter D.J. et al., *Nanotechnol.*, 20, 425401, 2009b. With permission.)

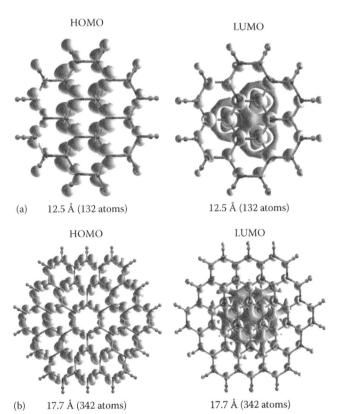

FIGURE 18.21 Spatial distribution of the HOMO and LUMO states at the Γ-point for nanodots with diameters of (a) 12.5 Å (132 atoms) and (b) 17.7 Å (342 atoms). Nitrogen and gallium atoms are indicated by dark and light spheres, respectively, and the orbitals are light gray. (Reprinted from Carter D.J. et al., *Nanotechnol.*, 20, 425401, 2009b. With permission.)

the wires and dots of 1.457 and 1.452, respectively, and for the gradients we find 1.784 and 3.179, respectively, yielding a ratio of the gradients equal to 0.561. This can be compared to the results obtained for InP quantum dots and wires, using semiempirical pseudopotential calculations, of values for x of 1.45 and 1.35 for wires and dots, respectively, and gradients of 2.49 and 3.30 yielding a ratio of gradients wires:dots of 0.75 (Yu et al. 2003).

Figure 18.21 shows the spatial distribution of the HOMO and LUMO at the Γ-point for a "small" (12.5 Å) and a "large" (17.7 Å) diameter nanodot. For the "small" nanodot, the HOMO states mainly form localized regions centered on atoms, while the LUMO states have some extended delocalized regions where orbital regions appear to connect. Similar behavior is observed for the "large" diameter nanodot in Figure 18.21b. Such a localized/delocalized behavior of the HOMO and LUMO states has been reported for hydrogen-saturated diamond clusters (McIntosh et al. 2004).

18.5 Summary

The dependence of the physical properties of unsaturated and saturated hexagonal and triangular cross-section GaN nanowires on wire diameter are found to exhibit the following trends: for saturated nanowires, with increasing diameter, the band gap decreases approaching the bulk value; for unsaturated nanowires, dangling bonds at the wire surface induce electronic states in the band gap that are localized at the wire surfaces and their position in the band gap remain relatively unchanged with varying wire diameter. The occupied edge-induced states are of nitrogen p character and the unoccupied edge-induced states are of gallium p character. The relative average stability of the wires increases with increasing diameter, where the hexagonal cross-section wires are more stable than triangular ones. For the saturated nanowires, the effective electron mass around the conduction band minimum decreases with increasing diameter, approaching that of the bulk value. Comparing to the behavior of nanowires, nanodots exhibit a larger band gap for a given diameter, and the variation with diameter is greater as expected on the basis of EMA-PIB, which indicates how quantum confinement should depend on the geometry of confinement. Interestingly, for saturated nanodots, in contrast to nanowires, the lowest unoccupied molecular orbital is found to exhibit a delocalized character, being spread across the center region of the quantum dot.

Single and multiple gallium and nitrogen vacancies in hexagonal GaN nanowires are found to prefer to be located at the edge of the nanowires. For two- and three-nitrogen vacancies, the most stable configurations are those where the vacancies are located at the edge of the nanowires, and on nearest-neighbor sites, clustering together. Similar behavior is found for two gallium vacancies, where in this case there is a large reconstruction of the nitrogen atoms around the vacancy leading to the formation of an N_3 trimer-like species. The nitrogen vacancies (single and multiple) are predicted to act as both donors and acceptors, while the single gallium vacancy to act as an acceptor. For two gallium vacancies, the significant reconstruction leads to multiple singlet states in the band gap with four being fully occupied and three unoccupied. The formation energies of nitrogen and gallium vacancies in the (neutral charge state) are considerably lower than in bulk GaN, suggesting they could exist in greater concentrations in nanowires.

References

Byeun, Y.-K., K.-S. Han, and S.-C. Choi 2006. Single crystal growth of one-dimensional GaN nanostructures by halide vapor-phase epitaxy. *J. Electroceram.* 17: 903–907.

Calarco, R., M. Marso, T. Richter, A. I. Aykanat, R. Meijers, A. V. D. Hart, T. Stoica, and H. Lüth 2005. Size-dependent photoconductivity in MBE-grown GaN-nanowires. *Nano Lett.* 5: 981–984.

Carter, D. J. and C. Stampfl 2009. Atomic and electronic structure of single and multiple vacancies in GaN nanowires from first-principles. *Phys. Rev. B* 79: 195302.

Carter, D. J., J. D. Gale, B. Delley, and C. Stampfl 2008. Geometry and diameter dependence of the electronic and physical properties of GaN nanowires from first-principles. *Phys. Rev. B* 77: 115349.

Carter, D. J., J. D. Gale, B. Delley, M. Fuchs, and C. Stampfl 2009a. Atomic and electronic structure of single and multiple vacancies in GaN nanowires from first-principles (to be published).

Carter, D. J., M. Puckeridge, B. Delley, and C. Stampfl 2009b. Quantum confinement effects in gallium nitride nanostructures: ab initio investigations. *Nanotechnology* 20: 425401.

Chan, C. K., H. Peng, G. Liu, K. McIlwrath, X. F. Zhang, R. A. Huggins, and Y. Cui 2008. High-performance lithium battery anodes using silicon nanowires. *Nat. Nanotechnol.* 3: 31–35.

Corcoran, E. 1990. Diminishing dimensions. *Sci. Am.* 263: 122–131.

Cui, X. and C. M. Lieber 2001. Functional nanoscale electronic devices assembled using silicon nanowire building blocks. *Science* 291: 851–853.

Delley, B. 1990. An all-electron numerical method for solving the local density functional for polyatomic molecules. *J. Chem. Phys.* 92: 508–517.

Delley, B. 2000. From molecules to solids with the DMol³ approach. *J. Chem. Phys.* 113: 7756–7764.

Dhara, S., A. Datta, C. T. Wu, Z. H. Lan, K. H. Chen, Y. L. Wang, C. W. Hsu, C. H. Shen, L. C. Chen, and C. C. Chen 2004. Hexagonal-to-cubic phase transformation in GaN nanowires by Ga+ implantation. *Appl. Phys. Lett.* 84: 5473–5475.

Dong, A., H. Yu, F. Wang, and W. E. Buhro 2008. Colloidal GaAs quantum wires: Solution-liquid-solid synthesis and quantum-confinement studies. *J. Am. Chem. Soc.* 130: 5954–5961.

Duan, X., Y. Huang, R. Agarwal, and C. M. Lieber 2003. Single-nanowire electrically driven lasers. *Nature* 421:241–245.

Efros, A. L. and M. Rosen 2000. The electronic structure of semiconductor nanocrystals. *Annu. Rev. Mater. Sci.* 30: 475–521.

Ganchenkova, M. G. and R. M. Nieminen 2006. Nitrogen vacancies as major point defects in gallium nitride. *Phys. Rev. Lett.* 96: 196402.

Gulans, A. and I. Tale 2007. Ab initio calculation of wurtzite-type GaN nanowires. *Phys. Status Solidi C* 4: 1197–1200.

Humphreys, C. 2009. Low-cost LEDs to slash household electric bills. Press Release of the Engineering and Physical Sciences Research Council, U.K.

International Rectifier, 2009. Proprietary GaN-on-silicon epitaxy and power device technology heralds new era for power conversion. News announcement http://www.irf.com/whats-new/nr080909.html.

Ju, S., J. Li, J. Liu, P.-C. Chen, Y. Ha, F. Ishikawa, H. Chang, C. Zhou, A. Facchetti, D. B. Janes, and T. J. Marks 2008. Transparent active matrix organic light-emitting diode displays driven by nanowire transistor circuitry. *Nano Lett.* 8: 997–1004.

Kim, J.-R., H. M. So, J. W. Park, J.-J. Kim, J. Kim, C. J. Lee, and S. C. Lyu 2002. Electrical transport properties of individual gallium nitride nanowires synthesized by chemical-vapor-deposition. *Appl. Phys. Lett.* 80: 3548–3550.

Kipshidze, G., B. Yavich, A. Chandolu, J. Yun, V. Kuryatkov, I. Ahmad, D. Aurongzeb, M. Holtz, and H. Temkin 2005. Controlled growth of GaN nanowires by pulsed metalorganic chemical vapor deposition. *Appl. Phys. Lett.* 86: 033104.

Kuykendall, T., P. J. Pauzauskie, Y. Zhang, J. Goldberger, D. Sirbuly, J. Denlinger, and P. Yang 2004. Crystallographic alignment of high-density gallium nitride nanowire arrays. *Nat. Mater.* 3: 524–528.

Law, M., J. Goldberger, and P. Yang 2004. Semiconductor nanowires and nanotubes. *Annu. Rev. Mater. Res.* 34: 83–122.

Limpijumnong, S. and C. G. Van de Walle 2004. Diffusivity of native defects in GaN. *Phys. Rev. B* 69: 035207.

Lu, W. and C. M. Lieber 2006. Semiconductor nanowires. *J. Phys. D: Appl. Phys.* 39: R387–R406.

McIntosh, G. C., M. Yoon, S. Berber, and D. Tomanek 2004. Diamond fragments as building blocks of functional nanostructures. *Phys. Rev. B* 70: 045401.

Nam, C. Y., D. Tham, and J. E. Fischer 2004. Effect of the polar surface on GaN nanostructure morphology and growth orientation. *Appl. Phys. Lett.* 85: 5676–5678.

Neugebauer, J. and C. G. Van de Walle 1994. Atomic geometry and electronic structure of native defects in GaN. *Phys. Rev. B* 50: 8067–8070.

Northrup, J. E. and J. Neugebauer 1996. Theory of GaN($10\bar{1}0$) and ($11\bar{2}0$) surfaces. *Phys. Rev. B* 53: R10477–R10480.

Novotny, C. J., E. T. Yu, and P. K. Y. Yu 2008. InP nanowire/polymer hybrid photodiode. *Nano Lett.* 8: 775–779.

Ordejon, P., E. Artacho, and J. M. Soler 1996. Self-consistent order-N density-functional calculations for very large systems. *Phys. Rev. B* 53: R10441–R10444.

Park, C. H. and D. J. Chadi 1997. Stability of deep donor and acceptor centers in GaN, AlN, and BN. *Phys. Rev. B* 55: 12995–13001.

Peng, H. Y., N. Wang, X. T. Zhou, Y. F. Zheng, C. S. Lee, and S. T. Lee 2002. Control of growth orientation of GaN nanowires. *Chem. Phys. Lett.* 359: 241–245.

Qian, F., Y. Li, S. Gradeak, H.-G. Park, Y. Dong, Y. Ding, Z. L. Wang, and C. M. Lieber 2008. Multi-quantum-well nanowire heterostructures for wavelength-controlled lasers. *Nat. Mater.* 7: 701–706.

Sattler, K. 2002. The energy gap of clusters, nanoparticles, and quantum dots. In *Handbook of Thin Film Materials: Nanomaterials and Magnetic Thin Films*, Vol. 5, ed. H. S. Nalwa, pp. 61–97. Academic Press, San Diego, CA.

Schmidt, T. M., R. H. Miwa, P. Venezuela, and A. Fazzio 2005. Stability and electronic confinement of free-standing InP nanowires: Ab initio calculations. *Phys. Rev. B* 72: 193404.

Simpkins, B. S., L. M. Ericson, R. M. Stroud, K. A. Pettigrew, and P. E. Pehrsson 2006. Gallium-based catalysts for growth of GaN nanowires. *J. Cryst. Growth* 290: 115–120.

Soler, J. M., E. Artacho, J. D. Gale, A. Garcia, J. Junquera, P. Ordejon, and D. Sanchez-Portal 2002. The SIESTA method for ab initio order-N materials simulation. *J. Phys.: Condens. Matter* 14: 2745–2779.

Tsai, M.-H., Z.-F. Jhang, J.-Y. Jiang, Y.-H. Tang, and L. W. Tu 2006. Electrostatic and structural properties of GaN nanorods/nanowires from first principles. *Appl. Phys. Lett.* 89: 203101.

Van de Walle, C. G. and J. Neugebauer 2004. First-principles calculations for defects and impurities: Applications to III-nitrides. *J. Appl. Phys.* 95: 3851–3879.

Xu, B.-S., L.-Y. Zhai, J. Liang, S.-F. Ma, H.-S. Jia, and X.-G. Liu 2006. Synthesis and characterization of high purity GaN nanowires. *J. Cryst. Growth* 291: 34–39.

Yoffe, A. D. 1993. Low-dimensional systems: Quantum size effects and electronic properties of semiconductor microcrystallites (zero-dimensional systems) and some quasi-two-dimensional systems. *Adv. Phys.* 42: 173–266.

Yoffe, A. D. 2001. Semiconductor quantum dots and related systems: Electronic, optical, luminescence and related properties of low dimensional systems. *Adv. Phys.* 50: 1–208.

Yu, H., J. Li, R. A. Loomis, L.-W. Wang, and W. E. Buhro 2003. Two- versus three-dimensional quantum confinement in indium phosphide wires and dots. *Nat. Mater.* 2: 517–520.

Yuan, J., X. Liu, O. Akbulut, J. Hu, S. L. Suib, J. Kong, and F. Stellacci 2008. Superwetting nanowire membranes for selective absorption. *Nat. Nanotechnol.* 3: 332–336.

Gold Nanowires

Edison Z. da Silva
University of Campinas

Antônio J. R. da Silva
Universidade de São Paulo

and

Laboratório Nacional
de Luz Síncrotron

Adalberto Fazzio
Universidade de São Paulo

and

Universidade Federal do ABC

19.1 Introduction

Gold, a noble metal, which is also considered to be very precious, has been the desire of people for millennia. There is a popular saying about it: "Who has the gold has the power." Due to its remarkable properties, gold stands out as a very important metal used as money and in sculpturing for very many centuries, and continues today as a favorite for jewelry, such as the wedding rings that some of us use. Legend has it that gold was the goal of alchemists who worked very hard in their attempt to transmute other lesser-valued metals, such as lead, into gold. All this has to do with the important properties of bulk gold: it is inert and does not, unlike other metals, react easily with other elements, in particular with oxygen, which prevents its oxidation. It is the most malleable and ductile metal known and, therefore, is easy to work with to produce the desired object, a jewel or a segment of wire.

Its conductance (capacity to transmit electrical current) is one of the best among metals, which, together with the other properties, makes it a high-quality electrical wiring component.

It is frequently used, for example, to connect electronic components in modern laptops as well as in modern high-definition multimedia interface (HDMI) high-quality cables to interconnect the new high-resolution digital television (HDTV) screens and blue ray disk players.

In the final decade of the twentieth century, new forms of gold were produced, bringing new possibilities to science and new hope for future technology applications. Small gold clusters and, more importantly, very thin gold nanowires (NWs) were produced, which were as thin as a line of atoms. Some of these NWs presented new forms not observed previously in bulk gold. It is well

known that the electrical resistance of gold wires with sizes ranging from millimeters to kilometers obey Ohm's Law, i.e., the resistance increases linearly as the wire's length is increased, whereas it decreases as the diameter of the wire increases. The reason for this behavior is that the electrons evolve diffusively in the metal. This behavior, however, is not observed in gold or other metallic NWs where the conductance becomes quantized, changing discontinuously in steps as the diameter of the wire changes.

The reasons for the discovery of these new forms of gold, in particular of gold NWs, the subject of this chapter, are intimately related with the important development of new experimental techniques and tools to study condensed matter at the atomic scale, unveiling a new field of research, now called nanoscience. In fact, the three important tools were the development of the scanning tunneling microscope (STM), the atomic force microscope (AFM), and the high-resolution transmission electron microscope (HRTEM). The STM was developed by Gerd Binning and Heinrich Roher at IBM, a research that led to the Nobel Prize in physics in 1986. Following the ideas of the STM, many new probing tools [1] were developed, and one of the most important was the AFM, which can be used in samples that are not good conductors and can be employed to determine atomic forces as well as in the imaging of small objects, such as atoms and molecules on surfaces. These new probe tools have an important new ingredient that, apart from imaging structures at the atomic scale, can also manipulate particles as an atomic LEGO, as was shown by D. Eigler group at IBM. Among other interesting examples of atomic manipulation, they wrote the IBM logo with 35 Xenon atoms on a Ni surface [2]. Regarding the third important tool, the HRTEM, although TEM was known for a long time (Huska, together with Binning and Roher, was awarded the Nobel Prize in physics in 1986 for its invention), high resolution at the atomic level was achieved only in the 1990s, making it possible to image structures such as the single metal atoms that are discussed here.

Another important motivation to study gold NWs comes from charge transport experiments. Many important results associated with the understanding of transport came from mesoscopic physics. It was for these kinds of systems that the Landauer [3] picture, i.e., viewing electrical conductance as transmission probabilities for quantum mechanical waves, became the natural framework to think about transport. One important difference, though, in using the results from mesoscopic physics to nanophysics is that while in the intermediary situation of a mesoscopic scale, which is much smaller than bulk but larger than atomic distances, it is possible to simply concentrate on the universal features of the problem, at the nanometer scale, the particular composition, the detailed atomic distribution, as well as the material properties play a fundamental role. As will be shown later on, the problem of conductance quantization and how it is coupled to the structural details is of fundamental importance for gold NWs.

Besides experimental advances, another fundamental tool in the study of nanoscience, and in particular gold NWs, are computer simulations and modeling. The capability of experiments to go into the nanometer scale meets computer simulations that, dealing with a small number of atoms, can make interesting and new predictions as well as help in the understanding of mechanisms that explain experimental results. In nanoscience, the interplay of simulations and experiments has been very fruitful. The new experimental and theoretical tools were applied to the study of the formation of thin NWs in a fundamental work in 1990 [4]. They used computer simulations to show that the contact of a thin metal STM tip on a flat metal surface produces, upon stretching, a series of successive stages of elastic deformation with atomic rearrangements that eventually led to the formation of atomically thin NWs. The measured forces were compared successfully with the computer simulations. Following these developments, a new technique to study atomic-sized junctions, known as mechanically controllable breaking junction (MCBJ), was introduced in 1992 by Muller et al. [5]. The scanning electron microscope image of a gold MCBJ is shown in Figure 19.1. The first results of this technique were for Nb and Pt contacts. They [7] observed steps in the conductance of the order of unit in terms of the conductance quantum $G_0 = 2e^2/h$. These experiments triggered an intense research by many groups [6,8–11]. Later on, it was shown that the conductance quantization could be achieved at room temperature by touching two bulk metals. Many more results were published for different metals and under different conditions. In 1996, a paper by Rubio et al. [12] established the link between atomic rearrangements and conductance steps using a combined technique of conductance and force measurement to show that jumps in conductance were associated with jumps in the force.

Another fundamental contribution to the study of gold NWs was the discovery of the unusual behavior of gold contacts, which

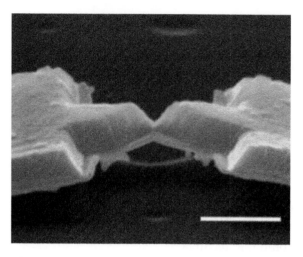

FIGURE 19.1 A scanning electron microscope image of a fabricated gold MCBJ. The contact is formed by a thin 20 nm gold layer (white) in close contact with a 400 nm aluminum layer on top (gray). The bridge is anchored to the substrate at the left and right. Bending the substrate breaks the narrow bridge forming gold single atom contacts that break and reconnect depending on the bending. (From Scheer, E., *Nature*, 394, 154, 1998. With permission.)

upon stretching evolved into one-atom linear atomic chains (LACs). This result, initially inferred from conductance measurements [13], was confirmed unambiguously by the seminal work of Ohnishi et al. [14], which combined STM measurements at the focal point of a HRTEM to show that the one-dimensional atomic gold chains have a conductance that is close to one quantum G_0.

While STM, AFM, and HRTEM techniques, conductance measurements, and computer simulations continued to be used, many more developments on gold NWs occurred. A nice review on these issues was published in 2003 by Agrait et al. [15]. In the present work we focus more on computer simulations. The structure of this chapter is as follows: In Section 19.2, we briefly discuss some experiments, how NWs are fabricated, their mechanical properties, as well as the important problem of conductance quantization and how it is related to the atomic structure. In Section 19.3, we discuss how theories and simulations helped in the understanding of the experiments, and the predictions that can be inferred from them. We discuss molecular dynamics (MD) simulations—tight-binding molecular dynamics (TBMD) and also *ab initio* DFT static and dynamic calculations. Finally, we discuss new forms of gold NWs, in particular helical NWs.

19.2 Experiments

Fabrication techniques used in the production of metal NWs and, in particular, gold NWs have improved over the years since the first experiments were reported. Early developments, long before nanofabrication, were experiments by Yanson [16] followed by those of Jansen et al. [17]. The techniques used a metal needle in contact with a metal surface, controlled by a differential-screw mechanism. Ballistic transport in metallic contacts was established by this technique. However, the metallic contact diameters obtained by this technique ($d \simeq 10$ and $100\,nm$) were not suitable to study the quantum regime.

The advent of the new tools, STM and AFM, made possible the study of much smaller tip-surface contacts in metals, allowing experiments in the regime of a very few atoms, making possible atomic resolution. In these experiments, the tip-sample separation is controlled by the current that flows between them due to tunneling. Atomic resolution is achieved because of the exponential dependence of the tunneling current on the tip-sample separation. This permits such a fine control that only the foremost tip atoms will be in contact with the surface. Thus, distances in STM and AFM experiments are very small. In fact this possibility can be used to our advantage. The tip-sample contact can be used to modify the surface. In this way, the first atomic contacts were produced [18]. This work used the tip to touch the surface and observed the transition from the tunneling to the metallic regime. Since then, many groups have performed STM experiments with atomic-sized contacts in very diverse settings, from cryogenic temperature [8,19] to room temperature [9,20,21] and UHV conditions [11,22]. Therefore, the STM used to study conductance became an important tool for NW studies, establishing clearly the conductance quantization. An illustrative example of the conductance quantization at room temperatures and UHV [22] is shown in Figure 19.2. These experiments triggered an intense research by many groups [6,8–11].

In 1992, a new technique was introduced by Muller et al. [5], the mechanically controllable breaking junction (MCBJ), to study atomic-sized junctions. The image from a scanning electron microscope shows a fabricated gold MCBJ (Figure 19.1). The gold metal to be studied has the form of a notched wire of a size around $0.1\,mm$ in diameter, fixed onto an insulating elastic substrate by glue. By bending this support using a piezo element with a three-point configuration, it is possible to break and reconnect the wire at the notch with great control, allowing conductance measurements. These experiments are in UHV and can be performed at cryogenic temperatures [23] as well as at room temperature [24]. MCBJ is the other technique that allows a controlled way of

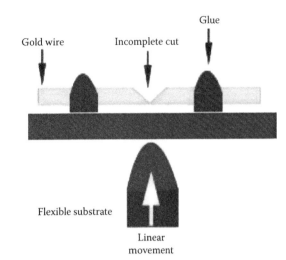

FIGURE 19.3 Illustration of a MCBJ setup for measurements of conductance on atomic-scale NWs.

measuring conductance. By connecting and breaking the metallic contact many times and measuring the conductance, histograms could be produced that showed clearly the conductance quantization. The principle is illustrated in Figure 19.3. A comparison of MCBJ- and STM-based techniques shows that in the former, there is no information on the geometry of the NWs formed.

One solution to the problem of gathering more information on the nature of the formation of atomic thin gold NWs was the use of combined techniques. One interesting example was the work of Rubio et al. [12], where they used a conventional AFM to measure the deflection of a cantilever beam attached to the STM gold sample. This experiment allowed simultaneous conductance and force measurements at room temperature, and showed that in the conductance plateaus, the force had an elastic behavior, while the jumps in conductance occurred associated to jumps in the force, indicative of atomic rearrangements, as can be seen in Figure 19.4. This setup was used later to clearly relate the simultaneous force and conductance measurements of an atomic chain of gold atoms by Rubio-Bollinger et al. [25] (Figure 19.5). In this experiment, a histogram of the breaking forces was produced, giving an experimental value of the breaking force for a linear chain of gold atoms of $1.5 \pm 02\,nN$.

All these ways of producing gold NWs as well as the experiments measuring forces and currents, although important, still needed further confirmation regarding the structural nature of the systems. This was only possible via the use of imaging techniques, which was essential to obtain structural information. Very important experiments were performed using the HRTEM. Kizuka [26,27] used a piezo-driven STM, placed at the focal point of a microscope, to show images of the atomic formation and evolution of a tip as it retracted from the contact with a gold surface (Figure 19.6).

Ohnishi et al. [14] placed an STM at the focal point of a microscope and measured conductance while acquiring the TEM image of the gold structure. It is possible to see the tip scanning the surface and also indenting it. When the tip is retracted, the tip-sample connection is thinned down forming linear chains.

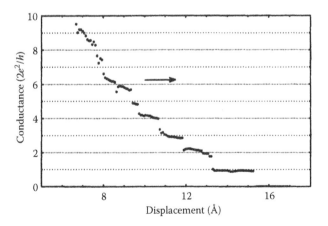

FIGURE 19.2 Conductance curve for a gold contact measured at room temperature in UHV by pressing an STM tip into a clean gold surface and recording the conductance while retracting the tip. (Reprinted from Brandbyge, M. et al., *Phys. Rev. B*, 52, 8499, 1995. With permission.)

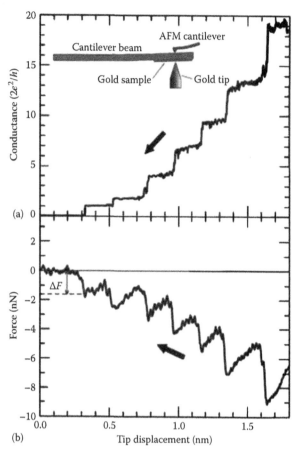

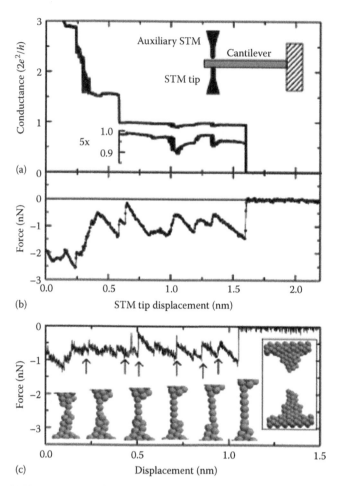

FIGURE 19.4 Simultaneous recording of the measured conductance and force during the elongation of an atomic-sized constriction at 300 K. The inset shows the experimental setup. (Reprinted from Rubio, G. et al., *Phys. Rev. Lett.*, 76, 2302, 1996. With permission.)

FIGURE 19.5 High-resolution simultaneous conductance (a) and force (b) measurements during chain fabrication and breaking. The conductance in the last plateau has been zoomed in to show detailed variations. Inset: schematic drawing of the experimental setup. (c) Calculated force during an MD simulation. Arrows indicate where a new atom pops into the chain and snapshots of the structures at these positions are shown. (Reprinted from Rubio-Bollinger, G. et al., *Phys. Rev. Lett.*, 87, 026101, 2001. With permission.)

This seminal experiment showed clearly that in the case of gold, a conductance plateau of $2G_0$ occurs for a two stranded wire, whereas the plateau of $1G_0$ occurs for a single atomic chain. They further showed an image of a gold NW with a four-atom linear chain with atomic distances of 0.36–0.40 nm. These experiments were conducted in ultrahigh vacuum and at ambient temperature (Figure 19.7). After these experiments, many more were performed involving imaging of gold NWs.

In subsequent studies, to understand the evolution of thin gold NWs, yet another method to produce gold NWs was developed. It used the electron beam of the TEM to make holes in a thin gold foil. This melting of nearby holes (contrary to the cold welding of STM indentation) formed a neck that evolved into atomically thick chains [28]. This technique has been used by different authors since then [24,29,69]. In particular, this method has been used [24] to correlate the NW crystallographic direction, atomic structure, and conductance before rupture (see Figure 19.8), as well as to image NWs close to the rupture point, allowing the measurement of interatomic Au–Au distances [29] (see Figure 19.9).

More recently, Kizuka [30], who did the pioneering work in images of gold NWs back in 1997, now presented very nice experiments, this time simultaneously combining HRTEM, force,

and conductance measurements on Au NWs. He used a gold tip attached to a cantilever. This tip touched a gold plate, itself attached to another plate. This arrangement was inserted into an *in situ* HRTEM focus. This setup allowed him to make images of the evolution of Au NWs under controlled tension while recording the conductance simultaneously. These experiments confirmed some of the previous findings in a more detailed way. They showed that NWs with LACs of up to five atoms are stable and show conductance quantization of $1G_0$. These experiments have control of the stretching force applied to the NWs force, and it was observed that a force of 1.6 ± 0.7 nN was needed to form the LACs. NWs as long as 10 atoms were presented in the HRTEM images, the longest presented so far. The conductance of these NWs showed a value of $1G_0$ for LACs up to 5 atoms long, and decreased to $0.1G_0$ for LACs of 6 to 10 atoms. Considering (1) the results of previous works on the role of light impurities, such as H, as a possible explanation for the observed long Au–Au

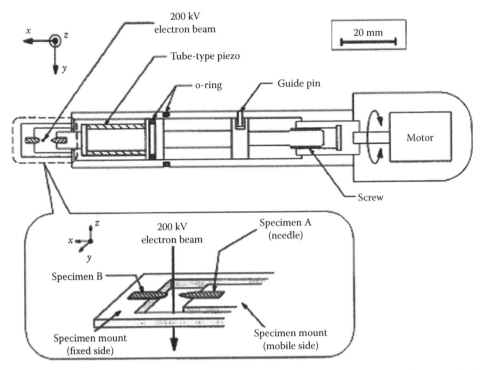

FIGURE 19.6 Illustration of a specimen holder of a HRTEM for atomic-scale surface scanning. (Reprinted from Kizuka, T. et al., *Phys. Rev. B*, 55, R7398, 1997. With permission.)

interatomic distances, and (2) the possibility of using impurities such as oxygen to make longer LACs, he concluded that the formation of these LACs with more than 5 atoms, with longer interatomic Au–Au distances, was attributed to the effect of impurities. This work provided a great deal of new results and left some more questions to be addressed in this very exciting field of metal NW research.

19.3 Theory, Computer Simulations, and Modeling

In the previous sections, we discussed the experimental aspects of the production of gold NWs and the different techniques to observe them or their properties, for example, the conductance. These NWs are produced using different techniques and can be made very thin, as thin as one-atom thick wires. One way to use theory to understand this formation, evolution, and breaking is to use computer simulations and modeling. The next sections discuss the evolution of the computer simulation studies done over the past 20 years.

19.3.1 Molecular Dynamics of Pure Nanowires

Attempts to theoretically understand experimental results on the behavior of gold NWs started almost at the same time as the first experiments were presented. Many studies using effective potentials addressed this question. In particular, the seminal work of Landman et al. [4] showed using MD that the indentation of a Ni STM tip on a gold surface produced very thin NWs

upon retraction of the tip, as shown in Figure 19.10. In later years, more calculations using MD with effective potentials were done, as well as other studies using more accurate approaches, ranging from tight-binding to first-principle DFT calculations. It is desirable that a method would include the electronic structure explicitly. Certainly, *ab initio* DFT dynamical methods would be the best choice, but a full evolution under these conditions would be too hard on computer time. A reasonable alternative was the use of a tight-binding formulation to perform MD in order to study the formation and evolution of gold NWs. This was done using the tight-binding MD method, TBMD [31–33]. The TBMD method is a good compromise, since it lies in between first principles and empirical methods: It is more accurate than empirical potential methods because it explicitly includes the electronic structure, and is much faster than first principles methods. Of course, this gain in speed comes with the cost of losing some of the flexibility of fully *ab initio* methods. The TBMD basically divides the problem of the dynamical evolution of a system into two: (a) The TB accurate parametrization for the system of interest [31,34]. This is previously done, and a variety of parametrizations based on precise fitting to DFT are tabulated [31,34]. (b) Use of this basis set to calculate the quantum forces to be used in the MD calculation [32,33]. Since the basis sets used are usually much smaller than in full *ab initio* calculations, the required matrix diagonalizations are performed much faster.

In order to understand the evolution and breaking of a gold NW, simulations using TBMD were performed [31,34]. Details of the used procedure can be found in Refs. [35,36,38]. Very briefly, the electronic structure of gold is described using a TB

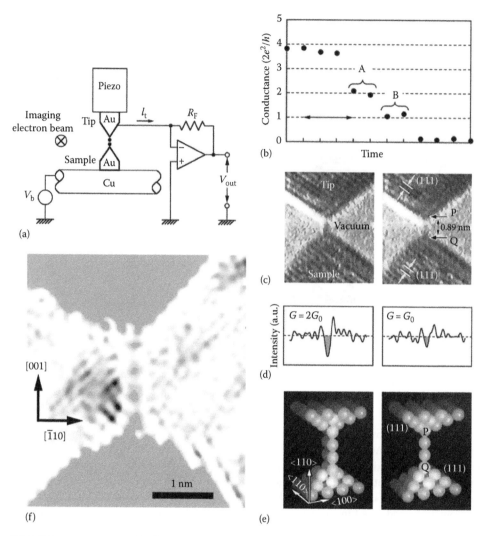

(a)

(b)

(c)

(d)

(e)

(f)

FIGURE 19.7 The STM–HRTEM experiment. (a) Illustration of the experimental setup: STM gold tips in the focal point of a HRTEM. (b) Plot of the conductance of a gold NW showing steps of (A) $2G_0$ and (B) $1G_0$. (c) The two electron microscope images of a gold NW at the two stages of the evolution marked (A) and (B) in Figure (b). (d) Intensity profiles of the left and right bridges shown in (c). (e) Models of the left and right bridges. The bridge at step (A) has two rows of atoms; the bridge at step (B) has only one row of atoms. (f) Electron microscope image of a linear strand of gold atoms (four gray dots) forming a bridge between two gold films (gray areas). The spacings between the four gold atoms are 0.35–0.40 nm. (Reprinted from Ohnishi, H. et al., *Nature*, 395, 780, 1998. With permission.)

fit developed by Mehl and Papaconstantopoulos [34], which gave very good results [33] when applied to bulk solid and liquid gold, for both static as well as dynamic properties. The periodic super-cells used in all calculations had dimensions (20 Å, 20 Å, L_W), where L_W is the length of the NW.

A gold NW consisting of a stack of ten (111) gold planes with seven atoms each, under stress, was simulated using the TBMD. The evolution of the wire can be followed in Figure 19.11, where six snapshots along the dynamical evolution of the wire are presented. As can be seen from Figure 19.11a, the wire tends to become hollow as it is pulled, which is caused by the motion of atoms from the center of the wire toward its surface. As a consequence, the seven-atom planes are transformed into six-atom rings, stacked along the tube axis, as can be seen in Figure 19.11a ($L_W = 25.5$ Å). In this configuration, the Au atoms form a tube

that is essentially a folded (111) sheet, similar to what happens in carbon nanotubes [39]. Hollow NWs, with six-atom rings, have been reported recently in Pt NWs [40]. This is consistent with the fact that the Au atoms always try to expose a close-packed surface. We observe that a new ring of six atoms is inserted when the increase in the length of the wire is of the order of the (111) interplanar distance in bulk Au (~2.3 Å) [35]. The evolution from the hollow tube of Figure 19.11a to the one-atom constriction in Figure 19.11b happens due to a very interesting mechanism of defect formation that has been discussed by da Silva et al. [35]. The further evolution with inclusion of more atoms into the chain is due to a competition between elastic forces and atomic displace-ments. At one side the elastic forces grow, and if the tips are not very symmetric, atoms move to relax these forces [36,37]. When the tips attain a very symmetric and stable configuration and, as

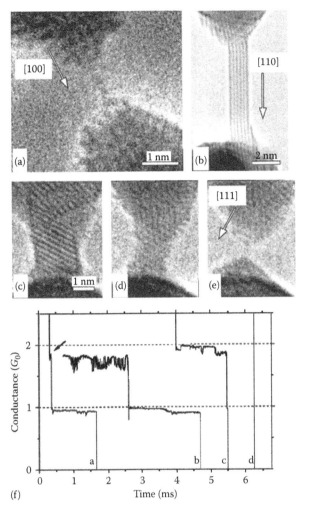

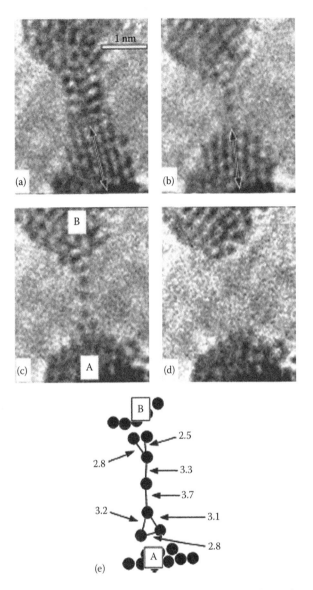

FIGURE 19.8 Images of gold NWs (atomic positions appear dark) obtained via HRTEM. (a) [100] NW; (b) [110] NW showing a rodlike structure; (c)–(e) time evolution of a NW where the apexes are sliding. The bottom image shows the conductance behavior of gold NW obtained via UHV-MCBJ measurements. Curve (a) shows a single step at $1G_0$; (b) shows a staircase with plateaus at approximately $2G_0$ and $1G_0$; in (c) the last plateau lies close to $2G_0$; finally, in (d) there is an abrupt rupture of the NW. (Reprinted from Rodrigues, V. et al., *Phys. Rev. Lett.*, 85, 4124, 2000. With permission.)

a consequence, the force required to extract any atom from them becomes too large, then the wire will withhold the pulling forces up to a critical breaking value when rupture occurs.

19.3.2 First-Principles Simulations

First-principles simulations of gold NWs are also interesting avenues to understand these systems. Fully *ab initio* MD (AIMD) calculations are computationally demanding, and *ab initio* calculations tend to be done statically, only relaxing forces at zero temperature, for example, using the conjugated gradient form of minimization of forces and energies. Noteworthy exceptions are the AIMD simulations by Anglada et al. [41] and Hobi et al. [42]. For this reason, most of the *ab initio* calculations considered a

FIGURE 19.9 Temporal evolution of a gold NW, obtained via HRTEM. The images show the sequence of formation, elongation, and fracture of a chain of Au atoms: (a) 0 s; (b) 0.64 s; (c) 1.12 s; and (d) 3.72 s. In (e), the atomic chain is schematically represented. The distances are in Å (error bar is 0.1 Å). The letters A and B indicate the apex position. (Reprinted from Rodrigues, V. and Ugarte, D., *Phys. Rev. B*, 63, 073405, 2001. With permission.)

limited number of possible configurations. One of the first such calculations was done by Sanchez-Portal et al. [43]. In that work, they studied a one-dimensional atomic chain of gold atoms in an attempt to explain the large Au–Au distances reported previously [13,14] and found that the optimized geometry was a zigzag chain that upon stretching became a linear chain. Hakkinen et al. [44] also studied the evolution of two parallel two-atom strands connected to gold tips that under tension evolved into a one-atom thick chain, four atoms long. The study of the dynamical evolution of NWs using TBMD previously discussed provided very good and novel insights into the mechanisms of formation and also breaking of Au NWs, but one question

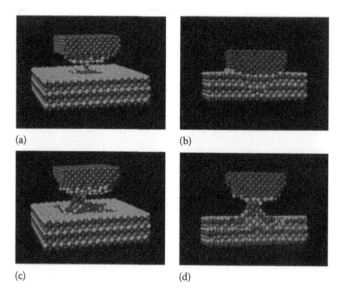

FIGURE 19.10 Atomic configurations generated by the MD simulations. (a) Separation after contact illustrating adherence of the top Au layer to the Ni tip and the formation of an atomically thin connective neck. (b) A cut through the system at the point of maximum indentation, illustrating deformation of the Au substrate. (c) Separation after indentation, illustrating wetting of the Ni tip by Au atoms, and formation of an extensive connective neck between the tip and the substrate. (d) A cut through the system shown in (c) illustrating the crystalline structure of the neck and the extent of structural deformations of the substrate. (Reprinted from Landman, U. et al., *Science*, 248, 454, 1990. With permission.)

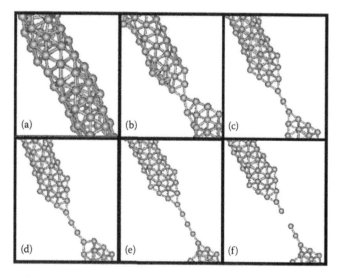

FIGURE 19.11 Evolution, thinning, and breaking of an Au NW under a TBMD simulation. Configurations at selected L_w elongations of (a) 25.5 Å; (b) 33.0 Å; (c) 37.0 Å; (d) 38.0 Å; (e) 40.5 Å; (f) 41.0 Å. (Reprinted from da Silva, E.Z. et al., *Phys. Rev. Lett.*, 86, 256102, 2001.)

remained unanswered, viz., how different would these results be from those obtained via an *ab initio* calculation, in particular regarding the final stages of rupture. This question was answered by *ab initio* total-energy DFT [45,46] calculations for selected structures from the TBMD simulations. These calculations used

a description based on a localized basis set, with the calculations performed using the SIESTA [47] code, which is a fully self-consistent procedure for solving the Kohn–Sham [48] equations.

Calculations were performed both at the local density approximation (LDA) [49] and the generalized gradient approximation (GGA) [50] for the exchange-correlation functional. A series of tests for both bulk gold and gold dimer gave confidence that a good description was achieved. Calculations using the GGA and LDA approximations were used to examine the wire's rupture. The calculations showed that the GGA breaking forces are smaller than the LDA ones and in better agreement with the experimental results (for the LDA, it was obtained 2.4 nN; for the GGA, 1.9 nN; and the experimental result [25] was 1.5 ± 0.3 nN). It should be mentioned that, even though the GGA Au–Au equilibrium distances tend to be larger than the LDA ones, the maximum Au–Au rupture distances were very similar in both cases, indicating that the differences between the GGA and LDA geometrical parameters for the equilibrium structures do not reflect their behavior under stress. As discussed in Ref. [36], the *ab initio* calculations gave very similar results to the previous TBMD ones.

Besides the DFT calculations using a localized basis set (SIESTA code), some calculations using a plane-wave (PW) basis set using the VASP code [51,52] were performed to compare the influence of basis sets in the final results. All other approximations were the same as in the SIESTA calculations. These PW calculations provided results that were very similar to the localized basis-set calculations [36]. Overall, all these studies of pure Au NWs confirmed the conclusions of the TBMD calculations. They served to provide more insight into the mechanism of bonding as well as into the electronic structure of these wires [36], and have shown that indeed the breaking distances in pure Au NWs are not bigger than 3.0–3.1 Å.

Figure 19.12a shows a structure obtained from the TBMD much before its breaking point, but already with a five-atoms neck. The rupture was also studied using DFT total-energy *ab initio* calculations, using both a localized basis (SIESTA code) as well as a PW basis set (VASP code), as mentioned before. Both procedures provided similar overall conclusions. The TBMD structure in Figure 19.12a was simply relaxed (all the atoms) using SIESTA (GGA approximation), and the final result is presented in Figure 19.12b. As can be seen from Figure 19.12 and Table 19.1, the relaxation with SIESTA did not change the overall distribution of Au atoms, but caused mainly a slight change of the bond lengths, thus resulting in a decrease of the total length of the neck region. This caused the structure at the neck to attain a zigzag configuration, as can be clearly seen from Figure 19.12b. From this configuration, the wire was quasistatically pulled, at the *ab initio* level using SIESTA, all the way up to its rupture. The structure in Figure 19.12c shows the NW just before its rupture. The bond distances of the atoms in the neck are displayed in Table 19.1 (the bond that breaks is displayed in boldface). Calculations using PW (VASP code) [38] provided distances that were somewhat smaller than in the localized basis-set calculation, except for the Au–Au bond that broke, which had a maximum distance

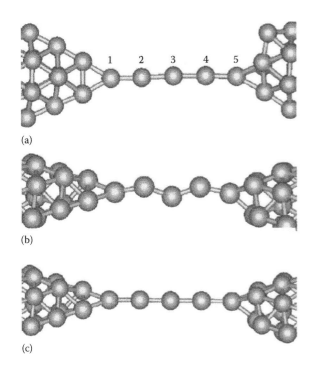

(a)

(b)

(c)

FIGURE 19.12 Final stages of evolution and breaking of a NW using *ab initio* calculations. Structure (a) shows the result from TBMD. Structure (b) is the same structure after *ab initio* relaxation of the forces, using the localized orbitals (LO) description. Structure (c) shows the wire prior to breaking using the LO–*ab initio* description (distances are given in Table 19.1). Atoms at the neck are labeled from 1 to 5.

TABLE 19.1 Interatomic Distances (i, j) (in Å) between the Atoms Au_i and Au_j in the Neck of the Structures of Figure 19.12, and the Breaking Force Obtained with the *ab Initio* (Structures (b) and (c)) Calculation

Structure	(1,2)	(2,3)	(3,4)	(4,5)	F(nN)
a	2.67	2.75	2.77	2.66	—
b	2.64	2.62	2.62	2.63	—
c	2.94	2.93	**2.96**	2.89	1.9

Note: The boldface distance in structure (c) marks the Au–Au bond where the wire will break.

of approximately 3.1 Å. All these results strongly indicate that, in pure Au NWs under tension, the limit for the Au–Au breaking distance is somewhere around 3.0–3.1 Å. Moreover, a comparison [38] between the breaking force for the two calculations (PW and LO) has shown that the PW result was smaller and closer to the experimental value. Nevertheless, the overall behavior was well described in both cases (and also in the TB calculations), especially the maximum bond distances before the NW rupture.

19.3.3 The Role of Light Impurities

In the Introduction, we discussed some aspects of the gold metal. One of the attractive features of bulk gold is that it is a rather inert metal. This property, however, changes with dimensionality.

In small clusters and very thin wires, in particular, Au becomes much more reactive, and these chemically active structures can form a new class of nanomaterials. For this reason, it is interesting to study the interaction of light impurities with Au NWs with the focus on how these impurities may alter the mechanical properties of these NWs. In fact, it is an interesting result that a single atom, like oxygen, for instance, when inserted in the NW neck can produce a drastic effect on the properties of the pure system [57,64]. Many groups studied the effect of impurities in gold NWs [38,53–57]. Novaes et al. [53] studied this problem by first-principles calculations using the two basis-set expansions mentioned previously for pure NWs. To generate structures for the contaminated NWs, structures previously obtained from *ab initio* calculations for the pure NW prior to the rupture were used. The impurities were inserted in the neck, after which all the atoms were relaxed. See Ref. [53] for more details. Many different impurities [38,53] were studied: C, CH, CH_2, H, H_2, O, N, B, and S. Some of them are discussed in detail below.

Out of the many impurities that could contaminate gold NWs, C was one of the first to be considered. The main reason was the possibility that C could be responsible for the large observed Au–Au distance of $\simeq 3.6$ Å. Novaes et al. [53] studied C in Au NWs and considered the impurity in a few different configurations; a wire with five and four Au atoms in the neck, and for this latter geometry the C was taken either in a symmetrical or an asymmetrical position. For all the configurations studied, the wire never ruptured at an Au–C bond. In fact, this is a general property of all the impurities that have been investigated.

In all cases, the behavior of C as a contaminant was to make the Au–C–Au bond, right before the wire's rupture, of the order of 3.85–3.9 Å (see Table 19.2), which is significantly larger than the experimentally reported values of $\simeq 3.5$–3.6 Å. The Au–C distances remain almost constant during the stretch of the wires, with values close to 1.9 Å. Figure 19.13a shows a gold NW with a C atom inserted between the Au atoms labeled 1 and 2, therefore in an asymmetric position in the neck. As already mentioned, the Au–C–Au is a very stiff bond that results in an Au–C–Au distance of 3.85 Å and the breaking of the 2–4 bond, as shown in Figure 19.13b. Table 19.2 gives the values of bond distances between the Au atoms in the neck. The Au–Au bond that breaks attains a maximum interatomic distance of 3.05 Å with a pulling force prior to rupture of 1.36 nN. Both these values are similar to

TABLE 19.2 Interatomic Distances (i, j) (in Å) between the Atoms Au_i and Au_j in the Neck of the Structures with inserted C, CH, and CH_2 Impurities Shown in Figure 19.13a, c, and e, respectively

NW	(1,2)	(2,3)	(3,4)	(Au–C)	(1-C-2)	F(nN)
a	3.85	2.67	**3.05**	1.92	177.8°	1.36
c	3.87	2.75	**3.08**	2.01	148.8°	1.64
e	3.77	**3.01**	2.81	2.10	127.5°	1.75

Source: Novaes, F.D. et al., *Appl. Phys. A*, 81, 1551, 2005.
Note: The average Au–C bonds (in Å), the angle (1-C-2) between the Au_1–C–Au_2 atoms, and the breaking forces are also presented. The boldface distances mark the Au–Au bond where the wires will break.

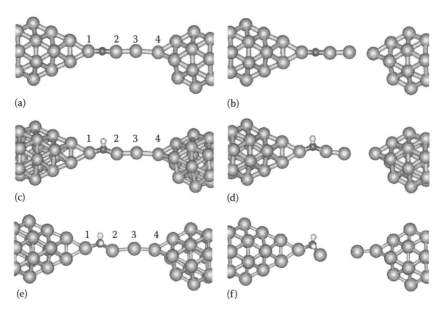

FIGURE 19.13 Final stages of evolution of Au NWs with C, CH, and CH$_2$ impurities. The studies were performed using *ab initio* DFT calculations (plane-waves basis set was used—VASP code). Numbers refer to the Au atoms in the neck, and bond distances and angles given in Table 19.2. (a), (c), and (e) are the structures with C, CH, and CH$_2$, respectively, just prior to rupture, whereas (b), (d), and (f) are the same wires after the breaking. (Reprinted from Novaes, F.D. et al., *Appl. Phys. A*, 81, 1551, 2005.)

what is obtained for pure NWs, which is not completely surprising since the rupture occurs in a pure Au–Au bond in both cases.

Given the experimental conditions where these wires are usually fabricated, how hydrocarbons such as CH and CH$_2$ alter the properties of the NWs is also an interesting question. Figure 19.13c and d shows structures for a CH inserted in the same neck position where the C atom discussed above was inserted, whereas in Figure 19.13e and f, similar structures for a CH$_2$ are presented. Figure 19.13c and e shows configurations prior to rupture, and Figure 19.13d and f after the breaking. All relevant interatomic distances are shown in Table 19.2.

As in the C impurity case, the rupture occurred in an Au–Au bond when this bond reached a distance of approximately 3.0–3.1 Å. Some clear trends can be observed as one considers the sequence C, CH, and CH$_2$. First, the Au–C average bonds have an almost regular increase from 1.92 Å for C, to 2.01 Å for CH, and finally to 2.10 Å for CH$_2$. This almost constant increase of 0.1 Å per added H atom is also accompanied by a decrease in the angle (1-C-2) between the Au$_1$–C–Au$_2$ atoms, from a rather linear configuration for C with (1-C-2) = 177.8°, to bent structures with (1-C-2) = 148.8° for CH and (1-C-2) = 127.5° for CH$_2$. Note that if the C in the CH configuration had a tendency toward an *sp^2* hybridization whereas the C in CH$_2$ configuration had a tendency toward an *sp^3* hybridization, the "optimum" (nonstressed) Au–C–Au angles would be 120° and 109.47°, respectively. Another trend is the steady increase in the maximum pulling force necessary to rupture the wire. It goes from 1.36 nN for a single C atom, to 1.64 nN for CH, to finally 1.75 nN for CH$_2$. Therefore, it seems that the added H atoms to the C impurity are making the overall neck stronger. Finally, the maximum Au–X–Au distance, when X = CH, is very similar to

the Au–C–Au distance (3.87 Å for the CH versus 3.85 Å for pure C), and it is ≃ 0.1 Å smaller when X = CH$_2$. These values suggest that under a quasistatic pulling setup, and if the system is not perturbed by external conditions, it seems unlikely that either one of the impurities C, CH, and CH$_2$ could be responsible for the large Au–Au distances in the range of 3.6 Å [38].

Hydrogen is a very difficult impurity to get rid of, even under ultrahigh vacuum conditions. Therefore, it is a possible candidate to be found in Au NWs. Many groups studied the effect of H in gold NWs [38,53,56]. Novaes et al. [53] investigated the effect of a variety of impurities (using *ab initio* DFT calculations), coming to the conclusion that, as far as the large Au–Au distances of approximately 3.6 Å found in the experiments are concerned, only H presented an Au–X–Au distance (with X representing an impurity) with a similar value. This work was later extended [38] to study a variety of structures with different number of H atoms inserted in the neck of a NW, as shown in Figure 19.14. The authors initially considered a contamination by one single H in the NW. Figure 19.14a depicts the structure prior to the rupture, and Figure 19.14b the structure after the breaking. The relevant interatomic distances are presented in Table 19.3. As can be seen, the Au–H–Au distance prior to breaking has a value of approximately 3.6 Å (a similar result was also obtained in Ref. [53]). This indicates that unless the experimental setup, such as electron bombardment in a HRTEM experiment, disrupts the structure of the wire, it seems that H would be the most likely impurity responsible for the measured Au–Au distances in the range of 3.5–3.6 Å.

Finally, it should be mentioned that all these studies were performed under a quasistatic pulling condition. Thus, other effects, such as the thermally activated process, may alter some of these conclusions.

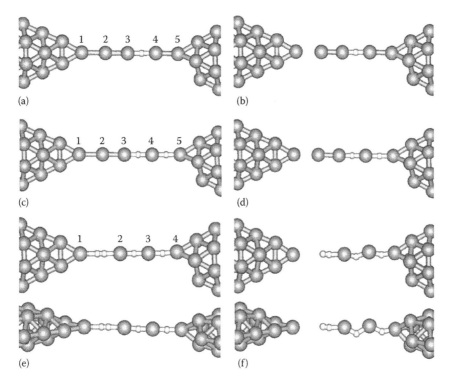

FIGURE 19.14 Final stages of evolution of Au NWs with different numbers of inserted H impurities (using *ab initio* DFT calculations). Numbers label the atoms that are in the neck and bond distances are given in Table 19.3. (a) and (b) show the wire with only one H atom just before and after the rupture, respectively. (c) and (d) show similar configurations for two H atoms, whereas (e) and (f) show two views of similar structures with a H_2 molecule and two H atoms in the NW. (Reprinted from Novaes, F.D. et al., *Appl. Phys. A*, 81, 1551, 2005.)

TABLE 19.3 Bond Distances (i, j) (in Å) between the Atoms Au_i and Au_j in the Neck of the Structures with Inserted H Impurities Shown in Figure 19.14a, c, and e

NW	(1,2)	(2,3)	(3,4)	(4,5)	F(nN)
a	**3.04**	2.84	3.60	2.84	1.71
c	**3.05**	2.81	3.58	3.63	1.58
e	**4.93**	3.48	3.53	—	0.93

Note: The breaking forces are also presented. The bold-face distances mark the bond where the wires will break.

To further investigate the possibility that H impurities inserted in Au NW could be responsible for the large Au–Au distances, the effect of more H atoms in the NW's neck was considered. Figure 19.14c and d presents configurations with two inserted H atoms in neighboring Au–Au bonds, one of them just before (Figure 19.14c) and the other just after (Figure 19.14d) the breaking. Similar configurations are presented in Figure 19.14e and f for structures with a H_2 molecule and two H atoms in the NW. When the two H atoms contaminate the NW with an arrangement Au–H–Au–H–Au, two similar distances of 3.6 Å were obtained. Such a structure closely resembles the experimental result of Rodrigues et al. shown in Figure 3 of Ref. [29]. The case of a further H_2 molecule inserted in the neck resulted in a configuration like Au–H–H–Au–H–Au–H–Au. The final Au–X–Au distances for such a structure, as shown in Table 19.3,

also compare well with the HRTEM image shown in Figure 2 of the paper by Legoas et al. [55].

Even though there are still no experiments that can directly probe the NWs in order to identify which impurities are present causing the large Au–Au distances, an experiment sheds some light into this problem. Zahai et al. [58] have done experiments with Au clusters obtained from bulk Au, and they have shown that the only clusters that incorporate H are Au dimers. Furthermore, they have shown that a linear structure Au–H–Au is stable and gives an Au–Au distance of 3.44 Å, which compares rather well with the results for H in Au NWs if one considers the fact that the dimer is not under stress. If dimers such as those produced by Zahai et al. were part of a NW under tension, it is reasonable to imagine that they would give the observed experimental values. These experiments point to H as a possible contaminant in Au NWs, a very likely candidate responsible for the large Au–Au distances. The maximum pulling forces for the H-contaminated wires are also shown in Table 19.3. The force decreased as more H atoms were added to the NW (see Table 19.3). In particular, for the configuration with an inserted H_2 molecule, the pulling force is significantly smaller, since the wire now breaks at an Au–H–H bond. In this way, measurements of the required force to cause the rupture of the NW could be used to identify how many atoms are present in the linear chain. Another possibility to identify if H atoms are inserted in the neck, and their possible configuration, is the use of inelastic conductance measurements [59].

Further work provided detailed studies of an H_2 molecule inserted in the neck of a gold NW. Barnett et al. [60], on the basis of computer simulations, argued that "welding" and restoration of electric conductance of a broken gold wire could be achieved through the incorporation of an H_2 molecule, and an electric switching action would result from structural fluctuations of an adsorbed molecule caused by mechanical forces applied to the wire. Experiments using MCBJ later argued that their results of conductance measurements of gold NWs in an H_2-saturated chamber were in agreement with those computer simulations [61].

As mentioned previously, an important question is also to understand the effect of temperature on the stability of the NWs, both with and without impurities. Some early AIMD simulations using small linear structures [62] argued that inserted H atoms in the chain would not be stable at temperatures between 300–500 K. However, further work [63] questioned these results indicating that they would be mostly related to particular choices of the simulation protocol. Another work used AIMD simulations [41] in order to investigate the formation of gold NWs with impurities. They studied H, C, O, and S impurities, concluding that sulfur was the most likely impurity to be incorporated in the chain. However, in order to observe the required processes in the timescale of the simulations, very high temperatures (of the order of 2000 K) had to be used, which casts some doubts in the conclusions. Finally, a detailed AIMD investigation of both pure as well as H- and C-doped NWs was performed with the objective of understanding the stability and rupture mechanism of these NWs [42]. This work clearly showed that both H- and C-doped wires are stable at ambient temperature, at least within the simulation times considered (around 20 ps). Moreover, similar to the quasistatic simulations, the Au–H–Au distances were the only ones providing distances close to 3.5–3.6 Å, with the Au–C–Au distances being still around 3.9 Å. Other important findings of this work were as follows: (1) Quantum effects might be important to understand the dynamics of H atoms in these wires, due to the light mass of hydrogen. (2) Triplets of atoms are the important units to look at in order to understand the rupture process. (3) There is a change in the potential energy profile for the position of central atom in these triplets from a single minimum at the center of the triplet to a double-well type of curve, as schematically illustrated in Figure 19.15. Quasistatically, one has to wait for this change to occur in order to observe the rupture, whereas at finite temperature thermally activated processes might happen that cause the rupture. (4) The Au–Au–Au triplets have "softer" bonds when compared to Au–X–Au ones, where X stands for an impurity such as H, C, or O. Thus, the instability happens first for the pure gold bonds, and this is the reason why the wire never breaks at a bond where an impurity is located.

Oxygen, which is not reactive to Au even in Au surfaces, can become significantly more reactive when gold forms small clusters or NWs under tension [54]. Novaes et al. [38,57] considered the effect of the contamination of one O atom to a NW with four atoms in the neck. Results are displayed in Figure 19.16 for a

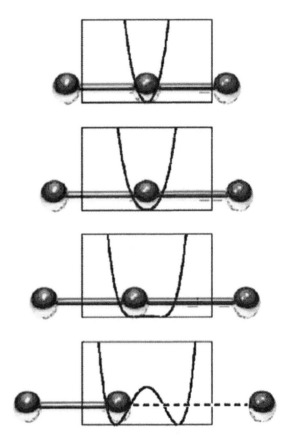

FIGURE 19.15 Schematic evolution of the breaking process of a triplet of gold atoms, where an increase of the triplet length causes a potential energy surface instability (third panel) resulting in the rupture of the wire.

sequence of configurations along the pulling of the wire all the way up to its rupture.

One distinguished property of oxygen as an impurity is the fact that atoms were extracted from the tip, as shown in Figure 19.16a through c. Figure 19.16a shows the NW after O contamination in a symmetrical configuration, in the middle of the four Au neck atoms. In Figure 19.16b, one can see that the bond Au_1–Au_3 starts to break, and in Figure 19.16c, after the tip rearrangements, the Au_2 atom is also extracted from the tip to the neck. These tips have a rather stable configuration, as already discussed above. Also, all the other impurities studied so far simply evolved in such a way that one of the Au–Au bonds in the neck broke, but the structures of the tips were never modified. These results concluded that oxygen is in some sense a special type of impurity, since it stabilizes the neck in such a way that upon application of stress the system favors the removal of atoms from the tip rather than rupturing [38,57].

As a result of this neck-tip reconstruction, there is a stress release, characterized by a sudden drop in the pulling force, as shown in Figure 19.17. We observe that after this stress release, as the neck became larger, it formed a zigzag structure. This straightening of the structure continues until the breaking of the wire. Once more, the rupture occurs at an Au–Au bond with the force around 1.7 nN, a value within the experimental window

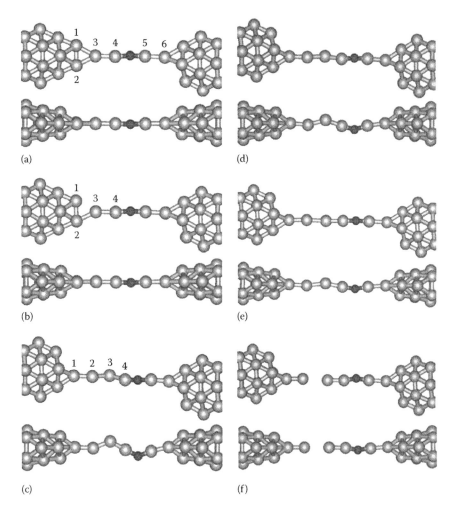

FIGURE 19.16 Final stages, from (a) through (f), of the evolution of an Au NW with an oxygen impurity all the way up to its rupture. Numbers refer to the atoms that are involved in major rearrangements in the NW's neck and tip. (Reprinted from Novaes, F.D. et al., *Phys. Rev. Lett.*, 96, 016104, 2006.)

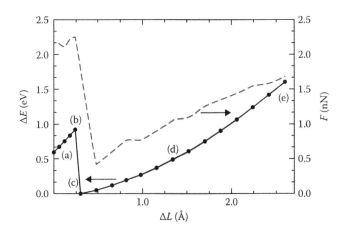

FIGURE 19.17 Total energy (lower curve; left axis) and calculated pulling force acting on the NW (upper curve; right axis) along the stages of the simulation. The total-energy values are displayed relative to the lowest energy structure, shown in Figure 19.16c. The results are presented as a function of the displacement, ΔL, of the wire. The configurations presented in Figure 19.16 are marked accordingly in the graph. (Reprinted from Novaes, F.D. et al., *Phys. Rev. Lett.*, 96, 016104, 2006.)

of $1.5 \pm 0.3\,$nN, as expected for a pure Au–Au bond breaking. Experiments performed almost simultaneously and independently came to similar conclusions that O helps the production of longer gold atomic chains [64]. Recently Kizuka [30] presented simultaneous HRTEM images, force and conductance measurements on the longest LACs experimentally produced so far (LACs of up to 10 Au atoms), and attributed these longer LACs to the effect of impurities, such as oxygen. Further theoretical work on oxygen, addressing the influence of an O_2 molecule in the conductance of Au NWs, provided evidence of spin effects caused by the presence of O_2 [65].

Kruger et al. [66], in a recent calculation of organic molecules attached to an Au surface, have also observed a behavior similar to what was obtained for the O impurity. This simulation showed that when the molecular structure is pulled out of the surface, it is transformed into a composite system, namely, a molecule attached to an Au NW being extracted out of the Au surface. Similar to the NWs simulated in the present work, the molecule, instead of breaking when pulled out from the surface, draws out from it a one-atom thick NW that upon stretching will eventually break.

19.3.4 Magnetic Impurities

A recent work [67] investigated the effect that transition metal impurities, such as a Co atom, might have in the transmittance of Au NWs when they are inserted at different positions in the neck. In Figure 19.18, the transmittance of a pure gold NW, with three atoms in the neck, is presented. As can be seen, the total transmittance at the Fermi level, i.e., the sum of the transmittances for the up and down spin channels, is close to $1G_0$, indicating the presence of only one channel, which is related to Au s-orbitals. Moreover, there is no spin polarization, meaning that the up and down transmittances are identical. However, when a Co atom is inserted in the neck, there are now strong spin-dependent transport properties (Figure 19.19).

In particular, the local symmetry of the NW can dramatically change these transport properties. When such symmetry permits the mixing between the wire s-orbitals with the transition metal d-states, there are interference effects that resemble Fano-like resonances (Figure 19.19, left panel). On the other hand, if this symmetry decouples such states, the result is simply a sum of independent transmission channels (Figure 19.19, right panel). This opens up new vistas to the field of nanospintronics. With the present manipulation techniques, it is possible to design wires connected to systems with the appropriate local symmetry in such a way that spin anisotropy effects are enhanced. For example, connecting the wire discussed in the present work to magnetic leads can lead to a spin filter with the ultimate size limit, which can be used to explore spintronic devices at a new scale [67].

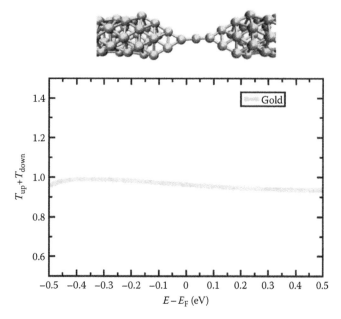

FIGURE 19.18 Top panel shows the structure of a gold NW with an LAC of three atoms. Bottom panel displays the calculated conductance of this NW metal. The curve displays the sum of transmittances for the up and down spin channels, in units of $G_0 = 2e^2/h$.

19.3.5 Novel NW Structures

One of the very interesting aspects of research in gold NWs is the possibility of new forms of atomic arrangements beyond bulk face-centered cubic (FCC)-derived structures. Novel types of

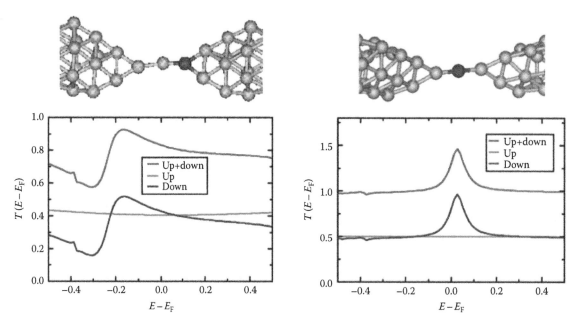

FIGURE 19.19 Panels at the left show in the top figure the structure of a gold NW with a nonsymmetrical Co impurity. Bottom figure displays the calculated conductance of this NW metal. As in this configuration, the local symmetry of the Co atom permits the mixing between the NW s-orbital with the Co atom d-orbitals, there appear structures in the transmittance that resemble Fano-like resonances. Panels at the right show in the top figure the structure of a gold NW with a symmetrical Co impurity. Bottom figure displays the calculated conductance of this NW metal. For this configuration, the symmetry is such that the NW s-orbitals are decoupled from the Co atom d-states. Thus, the total transmittance simply reflects the sum of these channels. (Reprinted from Pontes, R.B. et al., *J. Am. Chem. Soc.*, 130(30), 9897, 2008.)

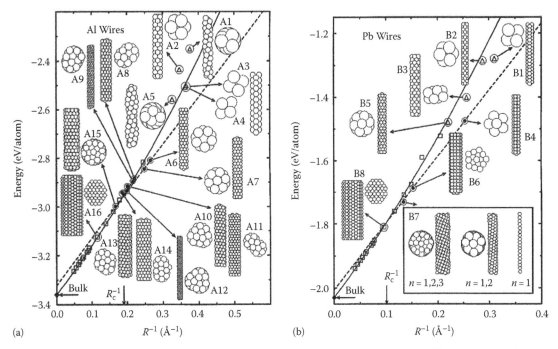

FIGURE 19.20 Total energy per atom, E, vs. inverse wire radius, $1/R$, for the relaxed structures obtained by optimization for (a) Al and (b) Pb wires. FCC wires are represented by open squares and weird wires by full circles. Weird structures become favored for $R \mathrel{i} R_c$. The inset in (b) shows the structure of the helical Pb wire B7: complete wire $n = 1, 2, 3$ with outer shell; with outer shell removed $n = 1, 2$; and the inner strand $= 1$. (Reprinted from Gulseren, O. et al., *Phys. Rev. Lett.*, 80, 3775, 1998. With permission.)

atomic packing including helical NWs have been theoretically predicted in MD simulations for unsupported lead and aluminum NWs by Gulseren et al. [68]. In that work, many forms of these metal NWs were studied, and they showed that below a characteristic radius R_c, of the order of three interatomic distances, new structures, that they called "weird wires," were formed in order to minimize the surface energy. The calculated structures can be seen in Figure 19.20. This important work stimulated both experimental and further theoretical work. This predicted effect was experimentally observed in gold NWs by Kondo and Takayanagi [69]. They showed that NWs formed along the [110] direction, obtained in experiments of the electron-beam irradiation technique, the same technique that years before demonstrated the existence of NWs with chains of atoms, become helical, exposing a (111) outermost shell, when these NWs are sufficiently thin. In that article it was argued that when very thin helical NWs were formed they produced structures with magic shell numbers, each shell having 7 more atoms than the previous one. The observed structures are illustrated in Figure 19.21. The magic structures observed in the experiments were 7-1, 11-4, 13-6, 14-7-1, and 15-8-1. As the NWs gets thicker there is a crossover to bulk behavior. Platinum NWs were also reported to produce helical structures [40].

The computer simulations of Gulseren et al. and the experiments of Kondo and Takayanagi stimulated further research. Extending their previous studies, Tosatti et al. [70] used total-energy density-functional theory calculations to show that when wires formed with concentric cylindrical sheets of atoms are under tension, the helical NWs with magic numbers were stable.

Similarly to the considerations of Kondo et al., they discussed the simulated wires as folding of a (111) Au plane. Depending on how the folding is done, several coaxial NWs appear. This is graphically displayed in Figure 19.22. They argued that the appropriate thermodynamic function to be considered was the wire tension rather than the free energy. This work found good agreement with the experiments of Kondo and Takayanagi [69].

Besides these works on the stability of helical NWs, few other contributions on the question of mechanisms for the helical formation in these NWs are available at present. Recently Iguchi et al. [71] proposed a mechanism for helical NW formation. It is a two-stage model that uses an additional line of atoms attached to the otherwise perfect [110] NW, that upon reconstruction shows the helicity being formed (they used TBMD).

More recently, Amorim and da Silva [72] using TBMD computer simulations presented an explanation for the formation of helical structures in gold NWs. Using computer MD simulations, they showed that an intrinsic mechanism is responsible for the helical formation in [110] gold NWs under stress. It was also shown why this helical NW under tension produces one-atom LACs that are longer than LACs formed from other nonhelical NWs. They also performed *ab initio* calculations to study the NWs obtained from the TBMD simulations, at stages close to rupture, and compared LAC interatomic distances obtained with both methods. Details of the electronic structure of the NWs close to rupture were also investigated.

This work used the same methodology previously employed by da Silva et al. [35] to study a [110] gold NW under tension. It found that NWs grown along the [111] direction form straight

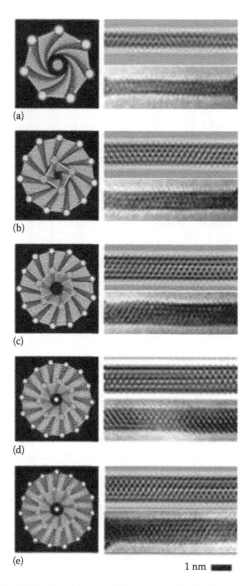

(a)

(b)

(c)

(d)

(e)

1 nm ▬▬

FIGURE 19.21 Frontal view of the proposed helical structures are displayed in Panels A–E. For each panel, the upper right figure shows the simulated image whereas the lower figure displays the experimental image. The atomic row numbers for the structures are 7-1 (a); 11-4 (b); 13-6 (c); 14-7-1 (d); and 15-8-1 (e). (Reprinted from Kondo, Y. and Takayanagi, K., *Science*, 289, 606, 2000. With permission.)

wires. The reason being that they are composed by stacks of (111) planes perpendicular to the growth direction. The surface atoms from these planes reconstruct forming rings that compose the rounded surface of the cylindrical NWs, which is a {111} surface displaying hexagons with no chirality. On the other hand, NWs grown along the [110] direction have (111) planes that are at an angle with the NW direction. These planes are very compact, with shorter bonds than those from (110) planes, along the growth direction or perpendicular to it. When the NW is under tension along the [110] direction, the compact (111) planes relax to form rings, keeping registry of their initial angular arrangement. As a consequence, the outermost shell

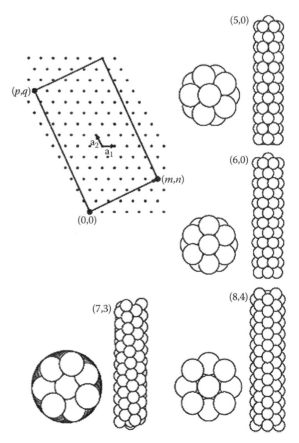

FIGURE 19.22 Cylindrical folding of a triangular lattice for an (m, n) tube, with views of several coaxial tube NWs. Each atom is pictured as a sphere of atomic radius. The (7,3) gold NW (note its chirality) was reported to be magic in Ref. [70]. (From Tosatti, E. et al., *Science*, 291, 288, 2001. With permission.)

that would otherwise expose facets reconstructs into a helical rounded {111} surface, which has the lowest surface energy for gold. The first stage of the evolution of the NW is shown in Figure 19.23, which presents the evolution of an FCC NW formed in the [110] direction. Right panels show the structure before relaxation, and the left panels present the helical structure formed under MD evolution. Before relaxation the structure exhibits {001} and {110} facets, which after relaxation under MD become a rounded surface.

Figure 19.23 presents three views of the NW's structure to explain the mechanism of the formation of the helical structure. Figure 19.23a presents the initial structure with the (111) planes dark gray so we can follow their evolution. They are responsible for the formation of the hexagons that form the {111} outer surface, and in trying to keep registry of this direction they produce the helical line around the NW. Note that in the {001} facets the dark gray atoms from the (111) planes are lined perpendicular to the NW's axis, as can be seen in Figure 19.23a. Therefore these lines (black and dark gray lines in Figure 19.23b) rearrange by slipping to accommodate the hexagonal formation to become {111} rounded surfaces that match the tilted (111) rings

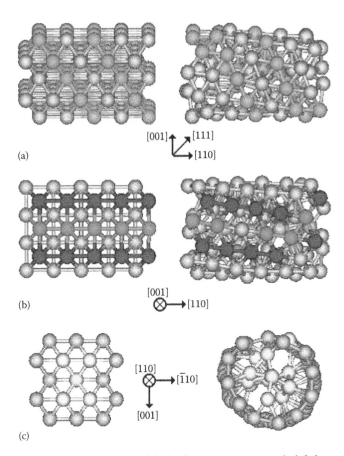

(a)

[001] [111]
[110]

(b)

[001] [110]

(c)

[110] [1̄10]

[001]

FIGURE 19.23 Evolution of the [110] FCC NW: Figures at the left show the FCC structure that, after relaxing under MD evolution, evolves to the structure shown in the right panels. Three views of this evolution are shown: (a) shows the (111) planes, (b) shows the {001} facet, and (c) shows a frontal view of the ring formation. (From Amorim, E.P.M. and da Silva, E.Z., *Phys. Rev. Lett.*, 101, 125502, 2008.)

completing the helical formation. The process of helical formation also involves rounding the planar facets into ring-type structures (Figure 19.23c) with atoms from the interior migrating to the outer surface to complement the formation of the {111} external surface but accompanied with formation of defects. The overall process of relaxation is the reconstruction toward a NW presenting a tilted {111} helical surface.

This work showed that [110] gold NWs are intrinsically helical and this behavior is caused by the compact (111) planes forming an angle with the NW's direction. They also showed that this helical formation of the NW is sustained even when it starts to further thin to make necks. The neck under tension evolves into a LAC that grows longer than NWs grown in other directions. This is again due to the helical structure of the tips that lack symmetry, so the bonds near the tip break more easily than they would in a more symmetrical NW. LACs as long as 8 atoms were obtained in these simulations. Using *ab initio* calculations, they corroborate the reliability of TBMD calculations with a discussion of the electronic states in the final stages of the evolution of the NW studied.

19.4 Conclusions

In this chapter, we have tried to convey some of the important historical landmarks in the development of the research in gold NWs. We started with the early experiments, which not only depended on the development, but at the same time were also nice examples of the application of the new tools that made possible nanoscience as is practiced today, such as the STM and the HRTEM. Along with these developments, we have tried to show how theoretical tools have also been used to help understand these experiments, elucidate mechanisms, and propose new phenomena.

In these roughly 20 years since the first experiments on the Au NWs, it is clear that they have provided a wealth of exciting results regarding the properties of atomic arrangement at the nanoscale. They are examples of the thinnest wires that can be produced in nature, with the diameter of only a single atom. Mechanical and charge transport properties have been investigated in detail, and they have demonstrated high stability considering the timescale of atomic motion. Even though much has been learned, there are still many open questions that will stimulate further work. One of the areas that needs to be explored is the interaction of foreign atoms with the NWs. It is clear that it is important to fully understand how a variety of impurities affect the intrinsic properties of NWs, such as their strength and charge transport, how to control the insertion of these impurities, how to experimentally study them, or even how to use them, in connection to the increased reactivity of NWs, to perform catalytic tasks, for instance.

This chapter started with the gold metal and its importance through time, and discussed the new form of Au, namely, Au NWs. One of the important motivations of most of these endeavors was to produce these new structures with the hope that they will be part of new devices in the future. Whether the dream of reinventing in nanoscience a device as important as the transistor invented by Brattain, Bardeen, and Shockley, which ruled microelectronics for the past 50 years, is achievable can only be answered in the future.

Acknowledgments

The simulations were performed at the National Center for High Performance Computing in São Paulo (CENAPAD-SP). We acknowledge support from FAPESP and CNPq.

References

1. Wiesendanger, R., *Scanning Probe Microscopy and Spectroscopy*, Cambridge University Press, Cambridge, U.K., 1994.
2. Eigler, D.M. and Schweizer, E.K., Positioning single atoms with a scanning tunneling microscope. *Nature*, 344, 524–526, 1990.
3. Landauer, R., Spatial variation of currents and fields due to localized scatterers in metallic conduction. *IBM J. Res. Dev.*, 344, 524–526, 1990.

4. Landman, U., Luedtke, W.D., Burnham, N.A., and Colton, R.J., Atomistic mechanisms and dynamics of adhesion, nanoindentation, and fracture. *Science*, 248, 454, 1990.

5. Muller, C.J., van Ruitenbeek, J.M., and de Jong, L.J., Experimental observation of the transition from week link to tunnel junction. *Phys. C*, 191, 455–504, 1992.

6. Scheer, E., Agrait, N., Cuevas, J.C., Levy Yeyati, A., Ludoph, B., Martn-Rodero, A., Rubio-Bollinger, G., van Ruitenbeek, J.M., and Urbina, C., The signature of chemical valence in the electrical conduction through a single-atom contact. *Nature*, 394, 154–157, 1998.

7. Muller, C.J., van Ruitenbeek, J.M., and de Jong, L.J., Conductance and supercurrent discontinuities in atomic-scale metallic constrictions of a variable width. *Phys. Rev. Lett.*, 69, 140–143, 1992.

8. Agrait, N., Rodrigo, J.G., and Vieira., S., Conductance steps and quantization in atomic contacts. *Phys. Rev. B* 47, 12345–12348, 1993.

9. Pascual, J.I., Mendez, J., Gomez-Herrero, J., Baro, A.M., Garcia, N., and Thien Binh, V., Quantum contact in gold nanostructures by scanning tunneling microscopy. *Phys. Rev. Lett.*, 71, 1852–1855, 1993.

10. Krans, J.M., Muller, C.J., Yanson, I.K., Govaert, Th.C.M., Hesper, R., and van Ruitenbeek, J.M., One-atom point contacts. *Phys. Rev. B* 48, 14721–14724, 1993.

11. Olesen, L., Laeggaard, E., Stensgaard, I., Besenbacher, F., Schiotz, J., Stolze, P., Jacobsen, K.W., and Norskov, J.K., Quantized conductance in an atom-sized point contact. *Phys. Rev. Lett.*, 72, 2251–2254, 1994.

12. Rubio, G., Agrait, N., and Vieira., S., Atomic-sized metallic contacts: Mechanical properties and electronic transport. *Phys. Rev. Lett.*, 76, 2302–2305, 1996.

13. Yanson, A.I., Rubio-Bollinger,G., van den Brom, H.E., Agrait, N., and van Ruitenbeek, J.M., Formation and manipulation of metallic wire of single gold atoms. *Nature*, 395, 783–785, 1998.

14. Ohnishi, H., Kondo, Y., and Takayanagi, K., Quantized conduction through individual rows of suspended gold atoms. *Nature*, 395, 780–782, 1998.

15. Agrait, N., Yeyati, A.L., and van Ruitenbeek, J.M. Quantum properties of atomic-sized conductors. *Phys. Rep.*, 377, 81–279, 2003.

16. Yanson, I.K., Nonlinear effects in the electric conductivity of point junctions and electron-phonon interaction in metals. *Zh. Eksp. Teor. Fiz.*, 66, 1035-1–1035-50, 1974. (*Sov. Phys., JETP* 39, 506–513, 1974).

17. Jansen, A.G.M., van Gelder, A.P., and Wyder, P., Point contact spectroscopy in metals. *J. Phys. C: Solid State Phys.*, 13, 6073–6118, 1980.

18. Gimzewski, J.K. and Moller, R., Transition from the tunneling regime to point contact studied using scanning tunneling microscopy. *Phys. B*, 36, 1284–1287, 1987.

19. Agrait, N., Rodrigo, J.G., Sirvent, C., and Vieira, S., Atomic-scale connective neck formation and characterization. *Phys. Rev. B*, 48, 8499–8501, 1993.

20. Pascual, J.I., Mendez, J., Gomez-Herrero, J., Baro, A.M., Garcia, N., Landman, U., Luedtke, W.D., Bogachek, E.N., and Cheng, H.P., Properties of metallic nanowires: From conductance quantization to localization. *Science*, 267, 1793–1795, 1995.

21. Landman, U., Luedtke, W.D., Salesbury, B.E., and Wetten, R.L., Reversible manipulations of room temperature mechanical and quantum transport properties in nanowire junctions. *Phys. Rev. Lett.*, 77, 1362–1365, 1996.

22. Brandbyge, M., Schiotz, J., Sorensen, M.R., Stoltze, P., Jacobsen, K.W., Norskov, J.K., Olesen, L., Laegsgaard, E., Stensgaard, I., and Besenbacher, F., Quantized conductance in atomic-sized wires between two metals. *Phys. Rev. B*, 52, 8499–8514, 1995.

23. van der Post, N. and van Ruitenbeek, J.M., High stability STM made of a break junction. *Czech. J. Phys.*, 46(Supp. S5), 2853–2854, 1996.

24. Rodrigues, V., Fuhrer, T., and Ugarte, D., Signature of atomic structure in the quantum conductance of gold nanowires. *Phys. Rev. Lett.*, 85, 4124–4127, 2000.

25. Rubio-Bollinger, G., Bahn, S.R., Agrait, N., Jacobsen, K.W., and Vieira, S., Mechanical properties and formation mechanisms of a wire of single gold atoms. *Phys. Rev. Lett.*, 87, 026101, 2001.

26. Kizuka, T., Yamada, K., Degushi, S., Naruse, M., and Tanaka, T., Cross-sectional atomic time-resolved high-resolution transmission electron microscopy of atomic scale contact-type scannings on gold surfaces. *Phys. Rev. B*, 55, R7398–R7401, 1997.

27. Kizuka, T., Atomic process of contact in gold studied by time-resolved high-resolution transmission electron microscopy. *Phys. Rev. Lett.*, 81, 4448–4451, 1998.

28. Kondo, Y. and Takayanagi, K., Gold nanobridge stabilized by surface structure. *Phys. Rev. Lett.*, 79, 3456–3459, 1997.

29. Rodrigues, V. and Ugarte, D., Real time imaging of atomistic process in one-atom-thick metal junctions. *Phys. Rev. B*, 63, 073405, 2001.

30. Kizuka, T., Atomic configuration and mechanical and electrical properties of stable gold wires of single-atom width. *Phys. Rev. B*, 77, 155401, 2008.

31. The NRL TB parameters are available on the World Wide Web at http://cst-www.nrl.navy.mil/bind.

32. The NRL TB-MD software is available on the World Wide Web at http://cst-www.nrl.navy.mil/bind/dodtb.

33. Kirchhoff, F., Mehl, M.J., Papanicolaou, N.I., Papaconstantopoulos, D.A., and Khan, F.S., Dynamical properties of Au from tight-binding molecular-dynamics simulations. *Phys. Rev. B*, 63, 195101, 2001.

34. Mehl M.J. and Papaconstantopoulos, D.A. Applications of a tight-binding total-energy method for transition and noble metals: Elastic constants, vacancies, and surfaces of monatomic metals. *Phys. Rev. B*, 54, 4519, 1996.

35. da Silva, E.Z., da Silva, A.J.R., and Fazzio, A., How do gold nanowires break? *Phys. Rev. Lett.*, 86, 256102, 2001.

36. da Silva, E.Z., Novaes, F.D., da Silva, A.J.R., and Fazzio, A., Theoretical study of formation evolution and breaking of gold nanowires. *Phys. Rev. B*, 69, 115411-1–115411-11, 2004.

37. da Silva, E.Z., da Silva, A.J.R., and Fazzio, A., Breaking of gold nanowires. *Comput. Mat. Sci.*, 30(1–2), 73–76, 2004.

38. Novaes, F.D., da Silva, A.J.R., Fazzio, A., and da Silva, E.Z., Computer simulations in the study of gold nanowires: The effect of impurities. *Appl. Phys. A*, 81, 1551–1558, 2005.

39. Hamada, N., Sawada, S., and Oshiyama, A., New one-dimensional conductors, graphitic microtubes. *Phys. Rev. Lett.*, 68, 1579, 1992.

40. Oshima, Y., Koizimi, H., Mouri, K., Hirayama, H., Takayanagi, K., and Kondo, Y., Evidence of single-wall platinum nanotube. *Phys. Rev. B*, 65, 121401, 2002.

41. Anglada, E., Torres, J.A., Yndurain, F., and Soler, J.M., Formation of gold nanowires with impurities: A first-principles molecular dynamics simulation. *Phys. Rev. Lett.*, 98, 096102, 2007.

42. Hobi, E., Fazzio, A., and da Silva, A.J.R., Temperature and quantum effects in the stability of pure and doped gold nanowires. *Phys. Rev. Lett.*, 100, 056104, 2008.

43. Sanchez-Portal, D., Artacho, E., Junquera, J., Ordejon, P., Garcia, A., and Soler, J.M., Stiff monoatomic gold wires with a spinning zigzag geometry. *Phys. Rev. Lett.*, 83, 3884–3887, 1999.

44. Hakkinen, H., Barret, R.N., Scherbakov, A.G., and Landman, U., Nanowire gold chains: Formation mechanisms and conductance. *J. Phys. Chem. B*, 104 9063–9066, 2000.

45. Hohenberg, P. and Kohn, W., Inhomogeneous electron gas. *Phys. Rev.*, 136, B864, 1964.

46. Kohn, W., Nobel Lecture: Electronic structure of matter-wave functions and density functionals. *Rev. Mod. Phys.*, 71, 1253–1266, 1999.

47. Ordejon, P., Artacho, E., and Soler, J.M., Self-consistent order-N density-functional calculations for very large systems. *Phys. Rev. B*, 53, 10441–10444, 1996.

48. Kohn, W. and Sham, L.J., Self-consistent equations including exchange and correlation effects. *Phys. Rev.*, 140, 1133, 1965.

49. Ziesche, P., Kurth, S., and Perdew, J.P., Density functionals from LDA to GGA. *Comput. Mat. Sci.*, 11, 122, 1998.

50. Perdew, J.P., Burke, J.K., and Ernzerhof, M., Generalized gradient approximation made simple. *Phys. Rev. Lett.*, 77, 3865–3868, 1996.

51. Kresse, G. and Hafner, J., Ab initio molecular-dynamics for liquid-metals. *Phys. Rev. B*, 47, 558–561, 1993.

52. Kresse, G. and Furthmüller, J., Efficient iterative schemes for *ab initio* total-energy calculations using a plane-wave basis set. *Phys. Rev. B*, 54, 11169–11186, 1996.

53. Novaes, F.D., da Silva, E.Z., da Silva, A.J.R., and Fazzio, A., Effect of impurities in the large Au-Au distances in gold nanowires. *Phys. Rev. Lett.*, 90, 036101, 2003.

54. Bahn, S.R., Lopez, N., Norskov, J.K., and Jacobsen, K.W., Adsorption-induced restructuring of gold nanochains. *Phys. Rev. B*, 66, 081405(R), 2002.

55. Legoas, S.B., Galvão, D.S., Rodrigues, V., and Ugarte, D., On the origin of long interatomic distances in suspended gold nanowires. *Phys. Rev. Lett.*, 88, 076105, 2002.

56. Skorodumova, N.V. and Simak, S.I., Stability of gold nanowires at large Au-Au separations. *Phys. Rev. B*, 67, 121404, 2003.

57. Novaes, F.D., da Silva, A.J.R., da Silva, E.Z., and Fazzio, A., Oxygen clamps in gold nanowires. *Phys. Rev. Lett.*, 96, 016104, 2006.

58. Zahai, H., Boggavarapu, B., and Wang, J., Observation of Au_2H^- impurity in pure gold clusters and implications for the anomalous Au-Au distances in gold nanowires. *J. Chem. Phys.*, 121, 8231, 2004.

59. Frederiksen, T., Paulsson, M., and Brandbyge, M., Inelastic fingerprints of hydrogen contamination in atomic gold wire systems. *J. Phys.: Conf. Series*, 61, 312, 2006.

60. Barnett, R.N., Hakkinen, H., Scherbakov, A.G., and Landman, U., Hydrogen welding and hydrogen switches in a monatomic gold nanowire. *Nano. Lett.*, 4, 1845–1852, 2004.

61. Csonka, Sz. Halbritter, A., and Mihaly G., Pulling gold nanowires with a hydrogen clamp: Strong interactions of hydrogen molecules with gold nanojunctions. *Phys. Rev., B*, 73, 075405-1–075405-6, 2006.

62. Legoas, S.B., Rodrigues, V., Ugarte, D., and Galvão, D.S. Contaminants in suspended gold chains: An *ab initio* molecular dynamics study. *Phys. Rev. Lett.*, 93, 216103, 2004.

63. Hobi, E., da Silva, A.J.R., Novaes, F.D., da Silva, E.Z., and Fazzio, A., Comment on "contaminants in suspended gold chains: An *ab initio* molecular dynamics study." *Phys. Rev. Lett.*, 95, 169601, 2005.

64. Thijssen, W.H.T., Marjenburgh, D., Bremmer, R.M., and van Ruitenbeek, J.M., Oxygen-enhanced atomic chain formation. *Phys. Rev. Lett.*, 96, 026806, 2006.

65. Zhang, C., Barnett, R.N., and Landman, U., Bonding, conductance, and magnetization of oxygenated Au nanowires. *Phys. Rev. Lett.*, 100, 046801, 2006.

66. Kruger, D., Fucks, H., Rousseau, R., Markx, D., and Parrinello, M., Pulling monatomic gold wires with single molecules: An *ab initio* simulation. *Phys. Rev. Lett.*, 89, 186402, 2002.

67. Pontes, R.B., da Silva, E.Z., Fazzio, A., and da Silva, A.J.R., Symmetry controlled spin polarized conductance in Au nanowires. *J. Am. Chem. Soc.*, 130(30), 9897–9903, 2008.

68. Gulseren, O., Ercolessi, F., and Tosatti, E., Noncrystalline structures of ultrathin unsupported nanowires. *Phys. Rev. Lett.*, 80, 3775, 1998.

69. Kondo, Y. and Takayanagi, K., Synthesis and characterization of helical multi-shell gold nanowires. *Science*, 289, 606, 2000.

70. Tosatti, E., Prestipino, S., Kostlmeier, S., Dal Corso, A., Di Tolla, F.D., String tension and stability of magic tip-suspended nanowires, *Science*, 291, 288–290, 2001.

71. Iguchi, Y., Hoshi, T., and Fujiwara, T., Two-stage formation model and helicity of gold nanowires. *Phys. Rev. Lett.*, 99, 125507, 2007.

72. Amorim, E.P.M. and da Silva, E.Z., Helical [110] gold nanowires make longer linear atomic chains. *Phys. Rev. Lett.*, 101, 125502, 2008.

Polymer Nanowires

Atikur Rahman
Saha Institute of Nuclear Physics

Milan K. Sanyal
Saha Institute of Nuclear Physics

20.1 Introduction

A polymer molecule is composed of several repetitive units called monomers; the word "poly" means many and "mer" means unit. The term monomer means a chemical repeat unit of small molecules that are polymerized to produce a polymer. For example, a useful polymer polyethylene $-CH_2-CH_2-CH_2-CH_2-CH_2-$ or $(-CH_2-CH_2-)_n$ is obtained by a polymerization reaction of n number of ethylene ($CH_2=CH_2$) molecules. It is interesting to note that for n of the order of 20,000, a straight polymer chain will have a length of a few microns and a width of about 1 nm. Unlike conventional molecules, polymers will have a distribution of n, hence, the molecular weight of a polymer will be around the mean value—for this particular example, the mean molecular weight will be 280,000 g mol^{-1}.

The concept of such a "macromolecule" (or Makromolekül) was given by Staudinger on the basis of his experimental research on the structure of rubber in the 1920s. The rapid development of various synthetic polymer materials that started as a substitute for natural rubber occurred between 1940 and 1970. Now the volume of the world production of various polymer materials has become comparable to that of steel production, and polymers are used in various application areas like packaging, building and construction, electrical insulators, and the electronic and transport industries.

The history of polymer materials and the basics of polymer are beyond the scope of this chapter and one can refer to several textbooks (Bower, 2002). Here we shall discuss nanowires of synthetic polymers that are produced in the chemical industry, in general, and of conducting polymer, in particular. Polymer materials produced by living systems, known as biopolymers, will not be discussed here, though one can make nice "wires" like silk, wool, and linen out of these biopolymers.

The synthetic polymers can be classified in various groups either based on their structures or on their properties. It is the combination of monomer structure, length, and the flexibility of macromolecules that generates such versatile properties of polymers. Three broad types of structures are generally used for the classification of polymers, namely, linear-like polyethylene mentioned before, branched, or network combinations of monomers. Again, one can have three classifications of polymers based on their properties: (thermo) plastics, rubbers (elastomers), and thermozets.

The (thermo) plastics are made of either linear or branched macromolecules and plastics soften or melt above a transition temperature. These polymer materials are easy to mold and are widely used for technological applications. On cooling, these plastics generally form a noncrystalline structure below a transition temperature, known as the glass transition temperature, T_g, from a molten state. Some plastic materials may form partially crystalline structures or liquid crystalline phases on cooling. Rubbers or elastomers are generally made of macromolecules that are lightly cross-linked to allow reversible stretching and to prevent melting in the conventional sense—these materials do not flow even in a molten state. Thermozets are made of heavily cross-linked macromolecules and are rigid materials that do not soften even at higher temperatures, for example, epoxy resin materials like Araldites.

Polymer synthesis and industrial production has come a long way since the production of the first synthetic polymer, cellulose nitrate, or celluloid from cotton in the late 1800s. The present day production techniques involve precise control over the monomer chain length, control over monomer sequences in copolymers that use more than one monomer, and control over the precise molecular weight. It is also important to note that right from the beginning "wire" and "fiber" were important products of polymer industries not only for the textile industry but also for other applications like filter materials, etc., that require a high surface-to-volume ratio.

In the subsequent sections, we shall briefly discuss the fabrication methods of nanofibers or nanowires of various polymers that can be efficiently used in varieties of applications like filtration technology, life science, and sensor development. Apart from all these applications that demand chemical sensitivity of the polymers, two prime physical properties make polymer materials technologically unique and motivated large amounts of research activities for both basic and applied physics. One is mechanical property, for example, Young's modulus of two syntectic polymer fibers made of elastomer and plastic can be 10 MPa and 350 GPa, respectively (Bower, 2002)—that is, 35,000 times higher. Even more dramatic is the electrical conductivity of polymers—it can vary more than 20 orders of magnitude from a good insulator having a conductivity of 10^{-18} Ω^{-1} cm^{-1} to a good conductor with a conductivity of 10^4 Ω^{-1} cm^{-1}. In this chapter, we shall primarily concentrate on the conductivity of polymers especially when it takes the form of a nanowire. Fascinating new properties are obtained in these research activities.

Polymers are, in general, electrically insulating materials. It was only after the discovery by Heeger, McDiarmid, Shirakawa, and coworkers that plastic could be made to conduct electricity (Heeger, 2001) and this discovery was awarded the Nobel Prize

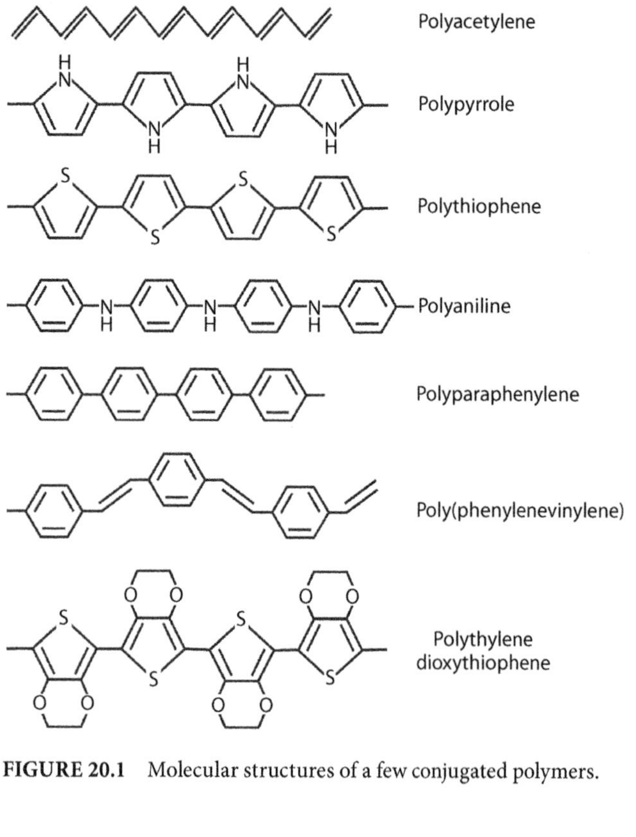

for chemistry in 2000. Conducting polymers can be both organic (such as polyacetylene, polyphenylene, polyaniline, polypyrrole, poly(p-phenylenevinylene), polythiophene, poly(3,4-ethylene-dioxythiophene), etc., refer to Figure 20.1) or inorganic (such as polysulfur nitride $(SN)_x$, alkali fulleride compounds $(AC60)_n$ [with A = K, Rb, or Cs], etc.) in nature. Conducting polymers are generally of two types: conjugated polymers and redox polymers. In conjugated polymers, alternate double bonds and single bonds (conjugated bonds) occur along the polymer chain. In a redox polymer, a redox-active transition metal-based pendant group is covalently bound to the polymer backbone. We will confine our discussion to conjugated polymers.

20.2 Conjugated Polymers

In conjugated polymers, three out of the four valence electrons of carbon form strong σ bonds (through sp^2 hybridization) where electrons are localized and these help to form the polymer backbone (Salaneck, 1991). The remaining unpaired (π) electron of each carbon atom stays in a p-orbital, which is nonlocalized. The π electrons of the successive carbon atoms overlap to form a π bond and this leads to the electron delocalization along the polymer backbone. As each carbon atom contributes only one electron, a half-filled valence band should have been formed by the delocalized π orbitals and one should have always observed a metallic behavior. However, due to the electron-lattice coupling, such long-chain one-dimensional (1D) metal is unstable (Peierls instability) and the polymer lowers its energy by bond alteration (alternating short and long bonds). This conjugated structure increases the π electron density between alternate pairs of carbon atoms (due to the formation of a double bond, the separation between the pair of atoms decreases, which in turn increases the electron density) and the π band is then divided into π (bonding) and π^* (antibonding) bands. Due to the Pauli exclusion principle, each band can accommodate only two electrons per atom (spin up and spin down) leaving a filled π band and an empty π^* band. An energy gap, E_g, appears between the highest occupied state in the π band and the lowest unoccupied state in the π^* band. Due to the presence of a large band gap ($E_g \sim$ eV) and the absence of a partially filled band, conjugated polymers behave as a high band gap semiconductor (Chiang et al., 1977) with negligible room temperature conductivity. To increase the conductivity, one needs to create sufficient charge carriers in the conduction band; this has been done by the doping of conducting polymers (Chiang et al., 1977, 1978). Conducting polymer doping can be done by redox reaction, which converts the insulating neutral polymer into an ionic complex consisting of a polymeric cation (or anion) and a counterion that comes from the oxidizing agent (or from the reducing agent). This oxidation or reduction process creates p-type or n-type charge carriers, respectively, in the conjugated polymer. The doping process in conducting polymer is markedly different from that of conventional inorganic semiconductors (like silicon, germanium, etc.). In inorganic semiconductors, the concentration of dopant atoms is very low ($\sim$1 in 10^6 host atoms) and they are substituted directly

FIGURE 20.1 Molecular structures of a few conjugated polymers.

in the host lattice; but in conducting polymers, the dopant ions reside outside the polymer chain and are present in several percent with respect to the carbon atoms of the polymer. The doping of polymers can be done using chemical, electrochemical, or photoabsorption processes and also by charge injection at a metal-semiconducting polymer interface (Heeger, 2001).

Depending on the doping level, conducting polymers can be insulating or semiconducting and one can even make a metallic conducting polymer by proper doping (Heeger et al., 1988; Heeger, 2001; Kaiser, 2001; Lee et al., 2006a,b). Several functional groups can be easily attached to conducting polymers using chemical methods to make sensors for chemical and biological applications. Metal decorated conducting polymers can act as hydrogen storage materials and these composite materials find important applications in fuel cell technology (Bashyam and Zelenay, 2006). Conducting polymer has shown its potential for organic electronic memory devices (Möller et al., 2003; Jianyong et al., 2004; Smith and Forrest, 2004). Conducting polymers find several applications in LEDs (Burroughes et al., 1990; Sheats et al., 1996; Friend et al., 1999), electrochromic devices (Mortimer, 1999; Dimitrakopoulos and Malenfant, 2002; Hayes and Feenstra, 2003; Sonmez et al., 2004), solar cells (Tang, 1986; Sariciftci et al., 1992; Yang et al., 2005; Na et al., 2008), rechargeable batteries (Kitani et al., 1986; Gurunathan et al., 1999), actuators (Smela et al., 1995; Baughman, 1996; Jager and Edwin et al., 2000; Berdichevsky and Lo, 2005; Lee et al., 2008), and biological and chemical sensors (Parthasarathy et al., 1997; Milella and Penza, 1998; Albert et al., 2000; McQuade et al., 2000; Hwang et al., 2001; Gerard et al., 2002; Janata and Josowicz, 2003).

The electrochemical synthesis of conducting polymers has been very versatile and in this technique, counterions are incorporated into the polymer chains to maintain charge neutrality. The doping level of electrochemically synthesized conducting polymers can be varied by changing the electrochemical potential between the polymer and a reference electrode, which drives ions into (or out of) the polymer. This change in the amount of ion intercalation between polymer chains causes various changes in the physical and chemical properties. This method is used to build electrochemically gated transistors, chemical and biological sensors, actuators, capacitors, etc.

20.2.1 Charge Excitations: Solitons, Polarons, and Bipolarons

Solitons, polarons, and bipolarons are the self-localized fundamental excitations of conjugated polymers (Bredas and Street, 1985; Roth and Bleier, 1987; Heeger et al., 1988) and here we shall discuss the basics of these processes.

Solitons: The ground state of long-chain quasi-one-dimensional (quasi-1D) polymers like trans-polyacetylene and polythiophene, having alternate double and single bonds is doubly degenerate. One can easily understand this with a simple example (Heeger et al., 1988): consider a polymer chain where atoms are marked by 1, 2, and 3, then the polymer structure can have two phases A → (1=2–3=4–5=6–) or B → (1–2=3–4=5–6=),

where – and = represent single and double bonds, respectively. It is easy to understand that if all the atoms are identical (as in polyacetylene) and the system is of infinite extent then the energy of the two phases is the same, i.e., the ground state is doubly degenerated (see Figure 20.3; Su et al., 1979). If we remove the charges from the polymer chain, a double bond will be replaced by a single bond and an atom will have single bonds on both sides (refer to Figure 20.2); this broken pattern of alternate double and single bonds represents a soliton. So a soliton can be viewed as a domain boundary between two ground state configurations, "A" and "B" phases (Heeger et al., 1988) and the misfit between the "A" and "B" phases is known as "kink" (see Figure 20.3). With the formation of a soliton, a localized electronic state is created in the middle of the π–π* band gap (see Figure 20.4). Due to the presence of a structural kink, the mid gap state can accommodate up to 2 electrons. If the state is singly occupied, then it is a neutral soliton and if it is empty (doubly occupied) then one

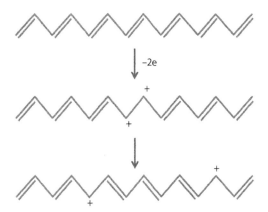

FIGURE 20.2 Creation of charged soliton by removing electrons from the polymer backbone of *trans*-polyacetylene.

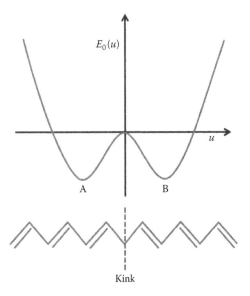

FIGURE 20.3 Total energy as a function of mean amplitude of distortion. The double minima indicates the presence of twofold-degenerate ground state.

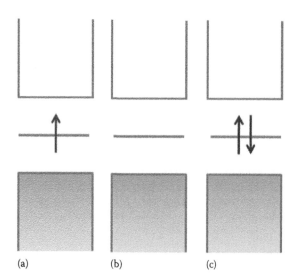

FIGURE 20.4 Band structure of (a) neutral soliton, (b) positively charged solitons, and (c) negatively charged soliton.

gets a positively (negatively) charged soliton (see Figure 20.4). So unlike other excitation processes, soliton poses an unusual spin–charge relationship; a neutral soliton is a spin 1/2 radical whereas a charged soliton is spin-less (Su et al., 1979).

Polarons and bipolarons: In conjugated polymers where ground state degeneracy has been lifted due to structural reasons, polarons and bipolarons are the elementary charge excitations (refer to Figure 20.5). A polaron is a quasi particle composed of a charge carrier and a lattice distortion bounded with it. A charge carrier when trapped in a localized state distorts its surrounding to some extent to lower its energy. A charge moving in a polar crystal attracts the oppositely charged ions and repels similar type charges and results in polarization of the medium, which in turn affects the motion of charge carriers. The polarization energy (W_P) depends on the deformation potential (V_D), localization volume (a^3) of the charge carrier, and bulk modulus (κ) of the lattice (Mott and Davis, 1979).

$$W_P = \frac{(1/2)V_D^2}{\kappa a^3} \qquad (20.1)$$

It is clear that polaron formation becomes more probable with increasing polarization energy, i.e., with decreasing bulk modulus and localization volume. In soft (low κ value) materials like conjugated polymer, polaron formation is a manifestation of a strong electron–phonon interaction. Polarons have both charge and spin and show a usual spin–charge relationship ($q = \pm e$, $s = 1/2$) like Fermions. If a second electron is removed from the polymer chain, it is energetically favorable to take out the electron from the polaron (which has already formed during the removal of the first electron) rather than to create another polaron in the polymer chain. This bound state of two polarons, which consists of a pair of like charges having an opposite spin (so that the net spin is zero) combined with strong lattice distortion is called bipolaron (see Figure 20.5). The formation of bipolaron demands that the energy gained by

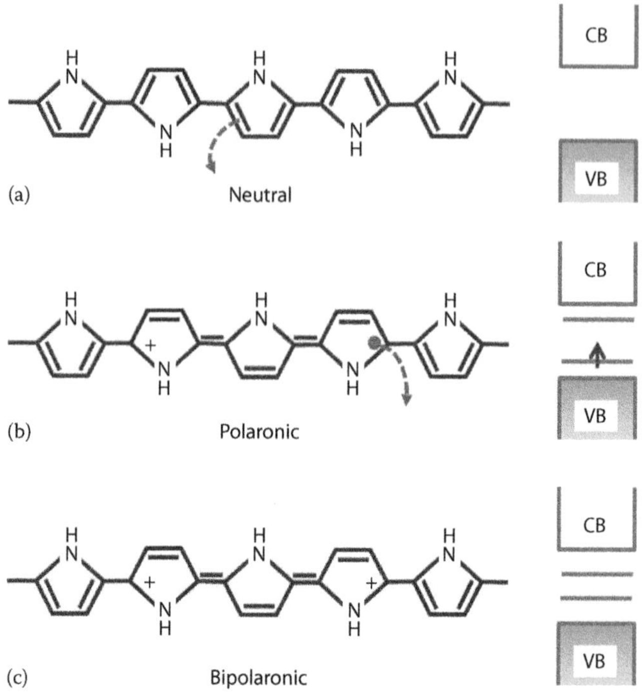

FIGURE 20.5 Formation of polarons and bipolarons in polypyrrole has been shown. Band structures for (a) neutral, (b) polaronic, and (c) bipolaronic polymer has been shown in the right side.

lattice distortion is larger than the Columbic repulsion between the two similar type charges (Anderson, 1975). As the bipolarons are spinless, its movement contributes to charge transport without any spin transport. Though solitons, polarons, and bipolarons are the fundamental excitations of conjugated polymers, their contribution to transport mechanisms are largely different: the movement of (1) neutral soliton along a polymer chain does not contribute to charge transport, (2) bipolaron contributes only to charge transport, while (3) polaron carries both charge and spin.

20.2.2 Electrical Conductivity of Bulk Conducting Polymer

At very large doping, one could expect a metallic behavior of the conducting polymer due to a finite density of states at the Fermi level. However, the metallic behavior of the conducting polymer is generally obscured due to the unavoidable presence of disorder. Though the resistivity (ρ) of the doped conducting polymer at room temperature can be as low as that of metal ($\sim 10^{-4}$–10^{-5} Ω cm), in most of the cases, the temperature coefficient of resistivity ($d\rho/dT$) is negative, i.e., like conventional semiconductors, with decreasing temperature, the resistivity increases. Several models have been proposed to explain the electrical conductivity of bulk polymers (Heeger et al., 1988; Heeger, 2001; Kaiser, 2001). However, the exact mechanism still remains controversial and here we will discuss some of the models used to explain the electrical conductivity of bulk conducting polymers.

20.2.2.1 Hopping Model

When a conducting polymer is doped with some dopant ion, it introduces a localized state in the band gap. Depending on the nature of the dopant, the localized state may be created near the conduction band (n-type doping) or near the valence band (p-type doping). It has been observed that irrespective of the nature of charge excitations (solitons, polarons, or bipolarons), electron transport in such disordered materials takes place from one localized site to another by phonon-assisted hopping. Due to the presence of covalent bonding along a chain and only weak bonding between adjacent chains, conducting polymers are highly anisotropic materials. However, the quasi-1D nature is often obscured due to a lack of chain alignment and disorder. The nature of disorder and localization of the wave function is also not very clear; it may arise due to the passage of 1D chains through disordered regions (inhomogeneous disorder model) (Epstein et al., 1994; Joo et al., 1994; Kohlman et al., 1995, 1996) or three-dimensional (3D) homogeneous disorder (Reghu et al., 1993; Väkiparta et al., 1993).

If the electron transport takes place by hopping only through the nearest neighbor localized sites, then the temperature dependence of resistivity shows a Miller–Abrahams type of behavior (Miller and Abrahams, 1960)

$$\rho(T) = \rho_1 \exp\left(\frac{T_0}{T}\right) \tag{20.2}$$

where $T_0 = \Delta/k_B$ is a characteristic temperature that depends on the energy difference (Δ) between two neighboring sites. However, with decreasing temperatures, hopping over large barriers to the nearest neighbor sites become energetically unfavorable than hopping to a distant site with a low barrier height, i.e., variable range hopping (VRH) dominates. According to Mott, in d-dimension VRH-type hopping exhibits the following temperature dependence of resistivity

$$\rho(T) = \rho_2 \exp\left(\frac{T_M}{T}\right)^{1/(1+d)} \tag{20.3}$$

where

$$T_M = \frac{\eta}{k_B g(\varepsilon_F) a^3} \tag{20.4}$$

T_M is Mott's characteristic temperature
$g(\varepsilon_F)$ is the density of states at the Fermi level
a is the localization length of charge carriers
η is a numerical coefficient (Mott and Davis, 1979)

The prefactor ρ_2 is also temperature-dependent but its dependence can be neglected compared to the stronger temperature dependence of the exponential term. If electron hopping along the chain length dominates over the interchain hopping, then one observes a 1D VRH with ln ρ showing a $T^{-1/2}$-type temperature dependence and for considerable interchain hopping it shows a $T^{-1/4}$ dependence.

So far in our discussion we have not included the effect of interactions between electrons of various localized sites. Efros and Shklovskii (ES) (Shklovskii and Efros, 1984; Pollak and Ortuno, 1985) showed that in the presence of Coulombic interaction, Mott's VRH law is modified to

$$\rho(T) = \rho_3 \exp\left(\frac{T_{ES}}{T}\right)^{1/2} \tag{20.5}$$

for all dimensions and a gap opens (Coulomb gap) at the Fermi level. $T_{ES} = (\eta_1 e^2)/(\varepsilon_0 k_B a)$ is the characteristic temperature for ES-type hopping, where η_1 is a numerical coefficient, ε_0 is the static dielectric constant, and a is the localization length. In case of hopping transport in 1D, both the Mott and ES models give the same result.

20.2.2.2 Granular Metallic Model

In this model, conducting polymer is considered to be composed of regions having partially aligned chains (crystalline region-metallic), where electrons are delocalized and embedded in a disordered matrix (insulating region). The conduction of electrons take place by tunneling from one mesoscopic metallic island to another through nonconducting material, which acts as a barrier (Sheng et al., 1973; Sheng and Klafter, 1983; Zuo et al., 1987; Li et al., 1993; Pelster et al., 1994) (see Figure 20.6). Due to the small size of the metallic island, Coulomb charging energy ($\propto$ diameter of the metallic island) plays a dominant role here and at low electric filed ρ shows an $\exp(-A/T^\gamma)$-type temperature dependence, where γ ranges from 1/4 to 1 (Sheng and Klafter, 1983). Over a large temperature range, $\gamma = 1/2$ is generally observed with possible crossovers to $\gamma = 1/4$ at low temperatures and to $\gamma > 1/2$ at high temperatures.

For the large size of the metallic islands, the Coulomb charging energy is much smaller and can be neglected compared to the

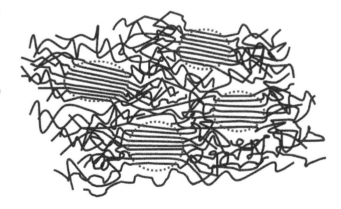

FIGURE 20.6 Granular metallic model of conducting polymer. Aligned portion of the polymer chains formed the crystalline region (wave functions are extended in this region—metallic) which are separated by disordered polymer chains (form the insulating barriers between the crystalline regions).

thermal energy ($k_B T$) and then tunneling can take place between the same energy states of the metallic islands without thermal activation. Fluctuation in voltage across the barrier greatly increases the tunneling probability with increasing temperature; this is known as fluctuation induced tunneling (FIT) (Sheng et al., 1978; Sheng, 1980). Considering a parabolic barrier, the temperature dependence of conductivity through a single tunnel junction (for a small applied field) for FIT can be expressed as

$$\rho(T) = \rho_\infty \exp\left(\frac{T_1}{T + T_s}\right) \qquad (20.6)$$

$$T_1 = \frac{8A\varepsilon_0 V_0^2}{w k_B e^2} \qquad (20.7)$$

$$T_s = \frac{16\hbar A\varepsilon_0 V_0^{3/2}}{(2m)^{1/2}\pi k_B e^2 w^2} \qquad (20.8)$$

In the above expression T_1 is the measure of the energy required for an electron to cross over the top of the barrier and T_s represents the temperature below which thermal fluctuation become insignificant. The ratio T_1/T_s represents the resistivity at $T = 0$ and ρ_∞ is a weak temperature-dependent parameter. V_0 is the barrier height, w is the barrier width, and A is the junction area. At sufficiently high temperatures, $T \gg T_s$ resistivity becomes thermally activated.

20.3 Growth of Polymer Nanowires

The fascinating world of nanoscience and nanotechnology has evolved due to the tunable properties of nanomaterials. The properties of nanomaterials change with shape and size, as the nanomaterial size becomes comparable to the molecular size. As the macromolecules are large in size, this effect becomes prominent in polymer materials even when the size in any of the three directions becomes 50 nm. For polymers, the size at which properties deviate from the bulk properties depends on the "radius of gyration," r_g, of the macromolecules. Long-chain macromolecules form randomly coiled structures in polymer materials and r_g is defined as the root-mean-square distance of atoms to the center of gravity of the chain of length nl and can be given as $\sqrt{n/6l}$. Even for thin polymer films when the thickness of the films becomes comparable to r_g, several anomalous properties are obtained (Sanyal et al., 1996) like layering of molecules and the reduction of the glass transition temperature (Bhattacharya et al., 2005).

With the advancement of nanotechnology, it has now been possible to fabricate quasi-1D systems such as nanowires and nanotubes that are one of the most promising candidates for the future nano-electronic device. It is generally agreed now that nanowires will not only be used as inter-connectors of nano-devices, but they will also be used as active circuit elements. Among nanowires and nanotubes of various materials, polymer nanowires and nanotubes will have advantages in several

nanotechnology applications as polymers can be used in printable and wearable electronic devices.

20.3.1 Electrospinning Technique

The synthesis of polymer nanofibers or nanowires is done mainly by template-free and template-based synthesis techniques. Among template-free techniques, the electrospinning technique is inexpensive and the most popular method for making nanofibers or nanowires from various types of polymers. The production techniques and applications of electrospun polymer nanofibers have been reviewed recently (Huang et al., 2003a,b; Ramakrishna et al., 2006). In this technique, the polymer melts/solutions come out through a small diameter tube under the influence of an electric field kept between this tube and a metal collector. The electrically charged jet of polymer solution/melt gets deposited on the metal collector and takes the shape of a nanowire (or nanofiber) as the jet becomes long and thin due to instability and elongation processes. It is also possible to form core-shell nanowires by this electrospinning technique having two different types of co-axial polymers. One can also form polymer composite nanofibers using this technique, for example, 70 nm fibers of poly(vinylidene fluoride) (PVDF) and single-walled carbon nanotubes (CNTs) have been produced using this technique for electrical applications (Seoul et al., 2003). It is difficult to control fiber orientation during electrospinning and also the doping, because the polymerization reaction takes place before the nanofiber formation.

20.3.2 Template-Based Synthesis

Template-based synthesis is one of the popular bottom-up techniques of nanostructure fabrication. Depending on the requirements and material properties, different types of templates have been used, such as the Stepped substrate (Barth et al., 2005), the Grooved substrate (Kapon et al., 1989), self-assembly using organic surfactant or a block copolymer (Thurn-Albrecht et al., 2000), biological macromolecules such as DNA or rod shaped viruses (Quake and Scherer, 2000; Yan et al., 2003; Ma et al., 2004; Mao et al., 2004; Nam et al., 2006), and porous membranes/materials (Martin, 1994; Wu and Bein, 1994; Hong et al., 2001; Wang et al., 2003). The main merits of the template-based synthesis are (a) fine control on the shape and size, (b) easy processing, (c) high yield cost-effectiveness, and (d) doping control.

Conducting polymer nanowires have been synthesized both by template-based (Martin, 1994; Wu and Bein, 1994; Hong et al., 2001; Wang et al., 2003) and template-free (Huang et al., 2003a,b) methods. However, the template-based synthesis provides nearly monodisperse nanowires of a desired size and high-aspect ratio. After the synthesis, the nanowires can be easily separated from the template and the individual nanowires can be isolated and manipulated for further applications because they are not interconnected with each other, which is generally found in the template-free method. A synthesis of various kinds of nanowires of polymers (polypyrrole, polyaniline, and poly[3,4-ethylenedioxythiophene]) using this technique by chemical and electrochemical routes has been done.

A chemical synthesis of conducting polymer nanowires has been done inside the pores of several templates like polycarbonate membrane, alumina membrane, aluminosilicate MCM-41, etc. (Martin, 1994; Wu and Bein, 1994; Hong et al., 2001; Wang et al., 2003). Commercially available polycarbonate membranes are of various pore diameters ranging from 10 nm to more than a few μm. One can prepare conducting polymer nanowires with a diameter as low as ~10 nm using these polycarbonate templates. The thickness of the templates are generally 6–20 μm and this determines the maximum length of the nanowires so the aspect ratio can be as large as ~1000. The pore density in the polycarbonate membranes ranges from ~10^7 to 10^9 pores cm^{-2} depending on the pore diameter. The pores inside the polycarbonate membrane were prepared using the track-etch method (Fleischer et al., 1975; Ferain and Legras, 1994, 1997) where a polycarbonate sheet of a few micron thickness is bombarded with energetic (~MeV) heavy ions that lead to linear narrow paths of radiation damage called tracks. The tracks can be revealed by using a suitable chemical agent (like NaOH, HF, etc.) that selectively etches the latent track to create a hollow channel keeping the remaining part unaltered (Price and Walker, 1962). The pores created by this method are hydrophobic and this helps the attachment of the polymer chains on the pore walls.

Alumina membrane is among other widely used templates that are commercially available. The pore diameter of the alumina membranes ranges from 20 nm to a few hundred nm and the pore density ranges from ~10^{10} to 10^{12} pores cm^{-2}. The thickness of the membranes is ~60 μm, hence, alumina templates are useful for synthesizing relatively long nanowires. However, the brittle nature of the alumina template deserves careful handling. The nanopores in the alumina are fabricated using the anodization process. Commercially, nanoporous alumina templates were manufactured by the sulfuric acid (H_2SO_4)-based hard anodization (HA) process of oxide films. Using this method, pores having a diameter ranging from 5 to 200 nm and a density of 10^{10}–10^{12} pores cm^{-2} can be fabricated. The nanopores prepared in this method are disorderly organized. For the fabrication of well-ordered pores, one generally uses oxalic acid ($H_2C_2O_4$) and the potential for anodization is kept between 100 and 150 V (Masuda and Fukuda, 1995; Nielsch et al., 2002; Lee et al., 2006a,b).

The mechanical rigidity of the polycarbonate membrane is low compared to alumina membrane but alumina membranes are brittle. The pore density of alumina membranes is much higher than that of polycarbonate. It is also possible to fabricate well ordered pores in the alumina membrane. After the nanowire synthesis inside the porus membrane, it is necessary to remove the template in order to get the individual nanowires. A polycarbonate membrane can be dissolved using chloroform, *N*-mehyl-2 pyrrolidone, methylene chloride, etc. These solvents have no apparent effect on the doping level of conducting polymer nanowires so after removing the template, the properties of nanowires remain intact. On the other hand, the alumina membrane can be dissolved in concentrated (1–6 M) NaOH solution, HF, etc. The concentrated NaOH solution can alter the doping of the conducting polymer nanowires and may change the intrinsic properties of the nanowires.

20.3.2.1 Chemical Synthesis

A chemical synthesis of conducting polymer nanowires of various doping concentrations can be done using the oxidative polymerization technique, where the monomers (pyrrole, aniline, etc.) get polymerized by an oxidizing agent (like ferric chloride, ammonium persulfate, potassium persulfate, etc.). For the synthesis, a porus membrane (polycarbonate or alumina) of a particular pore diameter is placed between a two-compartment glass cell with a rubber o-ring and clips (refer to Figure 20.7a and b). Before using the monomer, it is better to distill it under reduced pressure. An aqueous (a solvent other than water like acetonitrile, water/alcohol mixture can also be used) solution of monomer should be added in one compartment and the other compartment should have the solution of the oxidizing agent. Polymerization is done by the oxidizing agent (like $FeCl_3$) that can also provide dopant counter anion (Cl^-). Polymerization takes place within each pore as the oxidizing agent starts diffusing through the pores toward the compartment containing the monomer. This diffusion process of the oxidizing agent may create a profile of doping concentration (the amount of the dopant can vary depending on the nanowire diameter or along the length of the nanowires) that helps one to fabricate conducting polymer nanowires with various average doping levels, and a doping profile at one end of the nanowires (Rahman et al., 2006). As the pore walls of the polycarbonate are hydrophobic, the polymer chains prefer to be deposited on the walls and remain

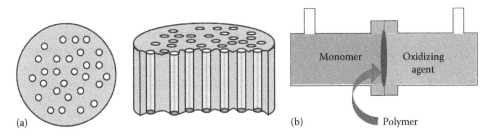

FIGURE 20.7 (a) Schematic of the porus membrane used for the nanowire synthesis. (b) Schematic of cell used for chemical synthesis; membrane is place in between the two cell and monomer is added in one compartment while oxidizing agent is added in another.

extended, which is in contrast to the polymerization reaction in the hydrophilic substrate where they form spherical shapes due to larger covalent interactions among themselves (Wang et al., 1999). So polymer chains remain well aligned in nanowires synthesized inside the nanopores of membranes.

The doping of the conducting polymer nanowires prepared by this technique can be very low. The doping concentration can be determined by the relative amount of the counterions, for example polypyrrole nanowires synthesized using an oxidising agent $FeCl_3$, the atomic ratio of Cl to N determines the degree of doping. One can obtain two to three orders of magnitude lower doping (Rahman et al., 2006) as compared to a doping concentration (c) = 0.33 for fully doped polypyrrole (Armes, 1987; Wu and Chen, 1997). It is to be noted that lower doping concentrations are obtained for membranes with lower diameter nanopores.

20.3.2.2 Electrochemical Synthesis

For the electrochemical deposition of nanowires inside the porus membrane, a metal layer is deposited on the back side of the membrane that acts as a working electrode (anode) in a three-electrode electrochemical cell. Ag/AgCl or saturated calomel (SCE) is used as a reference electrode and a platinum or gold foil is used as a counter electrode (see Figure 20.8). For the electropolymerization, lithium perchlorate, sodium perchlorate, sodium polystyrenesulfonate, dodecylbenzene sulfonic acid, tetrabutylammonium hexafluorophosphate, etc., are used as supporting electrolytes, and perchlorate, sulfonate, hexafluorophosphate, etc., act as dopant ions. Electrochemical deposition was done in constant potential (chronoamperometric) or in constant current (chronopotentiometric) mode using a potentiostat/galvanostat. In potentiostatic mode, the electrodeposition of various conducting polymers initiates at different voltages (with respect to the reference electrode); as an example, the electropolymerization of

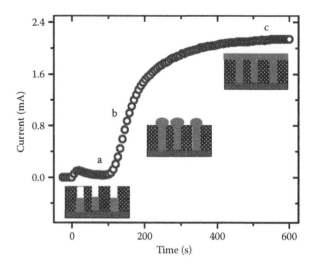

FIGURE 20.9 Current vs. time (I–t) plot for electrochemical deposition of nanowires inside 15 nm pore diameter membrane showing the different regions of deposition.

pyrrole was done at +0.8 V with respect to the SCE reference. By monitoring the current vs. time (I–t) data (see Figure 20.9) during the potentiostatic deposition mode, one can get information about the growth process of the nanowires. As long as the nanowires are growing inside the pore, the current remains constant (region a) because the deposition area remains the same. When the pore fills up and the capping layer starts to form, the current increases due to an increase in the area (region b) and finally when they are all connected, the surface deposition area remains constant and the current also becomes constant (region c). The steady nature of the region (a) indicates a homogeneous deposition inside the pores.

During the deposition of metal film (gold or platinum) on the back side of the membrane, one should be very careful about the penetration of metal inside the pores of the membrane. This problem could be solved by keeping the membrane at a certain angle with respect to the deposition direction. In Figure 20.10, the secondary ion mass spectroscopic (SIMS) measurement on the gold-coated polycarbonate membrane has been presented to show the penetration of metal during the deposition process (done using magnetron sputtering). It has been observed that if the membranes were not rotated (static mode) and the membrane surface is held normal (i.e., the pores are parallel) to the deposition direction, then the gold goes through the pore to the other side and is deposited on the sample holder (see Figure 20.10a). In dynamic mode (the sample holder is rotated at maximum speed), if the membrane surface is not tilted up to a certain angle with respect to the target, then gold also enters up to a considerable depth inside the pores (see Figure 20.10b). This unwanted penetration of gold causes serious problems to the physical properties of the sample deposited inside the pore.

It is easy to understand that for a fixed pore diameter membrane the extent of region (a) in Figure 20.9 (i.e., time needed to fill the pores) indicates the length of the nanowires. It has been seen that region (a) is smaller, i.e., the length of the nanowires

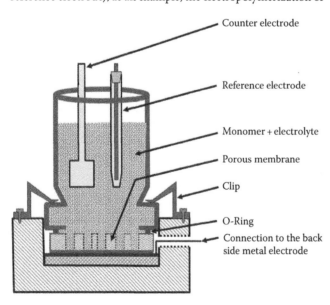

FIGURE 20.8 Schematic of electrochemical cell used the electrodeposition of conducting polymer nanowires.

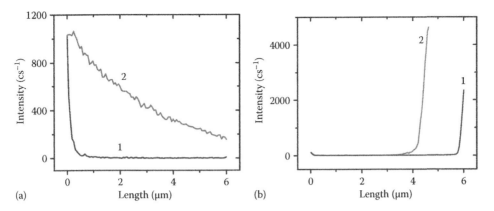

FIGURE 20.10 (a) SIMS measurements show that (1) gold deposited on 30 nm pore diameter membrane does not enter deep inside the pores when membrane was held at ~30° angle with respect to the target and rotated at maximum speed. (2) Gold enters deep inside the pores and reaches on the other side when it was not rotated and not tilted. (b) SIMS profile of gold, deposited at ~30° angle with respect to the target on 10 nm diameter membrane. During deposition if the membrane was not rotated then gold enters up to a large extent inside the nanopores (2) compared to the deposition done at maximum speed (1).

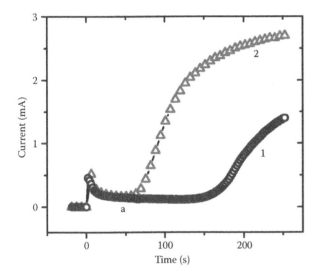

FIGURE 20.11 *I–t* plot for electrochemical deposition of nanowires inside two 10 nm pore diameter membrane. Length of nanowires are smaller (corresponds to extent of region (a)) in (2) where gold has entered deep inside the pore compared to (1).

is smaller, for those membranes where gold has entered up to a considerable length inside the pores (during sputtering deposition). An *I–t* curve shown in Figure 20.11 for nanowires synthesized inside a 10-nm pore diameter membrane supports this argument. Similar behavior is observed for higher pore diameter membranes.

20.4 Properties of Polymer Nanowires

In the nanostructured form, conducting polymers show wide versatility and enhanced efficiency. Due to a large surface-to-volume ratio, a conducting polymer nanowire shows better sensitivity when used as an actuator or sensor (Smela et al., 1995; Swager, 1998; Jager and Edwin et al., 2000; Huang et al., 2003a,b; Hernndez et al., 2004; Ramanathan et al., 2004; Berdichevsky

and Lo, 2005; Ramanathan et al., 2005; Lee et al., 2008). Due to better alignment of the polymer chains and enhanced conjugation length, a conducting polymer nanowire shows higher electrical conductivity compared to bulk polymer (Cai and Martin, 1989; Cai et al., 1991; Wu and Bein, 1994; Demoustier-Champagne and Stavaux, 1999; Choi and Park, 2000). Using electrochemical gating, a conducting polymer nanowire serves as a better filed effect transistor with large transconductance and on/off current ratio (Wanekaya et al., 2007). Functionalized polymer nanotubes show better chemical selectivity (Ramaseshan et al., 2006; Wang et al., 2006; Savariar et al., 2008). Also, an enhancement in the surface-to-volume ratio in this nanowire has made this material more effective in filter application as efficiency in this process is closely associated with the fineness of the fiber. The enhancement of the surface area has also made polymer nanowires ideal low-mass materials for several applications. The mechanical properties of polymer nanowires or nanotubes change dramatically in comparison to that of bulk (Bergshoef and Vancso, 1999; Cuenot et al., 2000; Qian et al., 2000; Ge et al., 2004; Sreekumar et al., 2004; Ye et al., 2004; Tan and Lim, 2005; Arinstein et al., 2007; Liu et al., 2007). It has been observed that the elastic modulus of the polypyrrole nanotube increases strongly when the thickness or outer diameter of the nanotubes decreases (Cuenot et al., 2000). The increase of mechanical strength is due to the better alignment of polymer chains and reduced voids in the nanowires.

20.4.1 Applications of Polymer Nanowires

Several materials have been incorporated with polymer nanofiber to make composites that show better mechanical strength. CNTs have been widely used as reinforced materials in polymer nanofibers (Bergshoef and Vancso, 1999; Qian et al., 2000; Ge et al., 2004; Sreekumar et al., 2004; Ye et al., 2004; Liu et al., 2007). CNTs and poly(acrylonitrile) (PAN) composite nanofiber has been used for mechanical reinforcement (Ge et al., 2004; Sreekumar et al., 2004; Ye et al., 2004). The formation of CNTs and nanocomposite wires through electrospinning is an ongoing research field and it is known that the enhancement of the physical properties like mechanical strength and electrical conductivity will depend on the even distribution and will control the alignment of CNTs in the nanowires of the polymer.

Electrochemical supercapacitors are high-power density charge-storage devices. Several polymers have been used for making supercapacitors. Nanocomposites, nanofiber, or nanowires of polymers such as polypyrrole (Hughes et al., 2002; Li et al., 2002a), *p*-phenylenevinylene (Deng et al., 2002), polyaniline (Zhou et al., 2004) and polymethyl methacrylate (Sun et al., 2001), poly(3,4-ethylenedioxythiophene) (Cho and Lee, 2008; Liu et al., 2008), etc., show enhanced energy storage capacity and are suitable for fabrication of high-power density and long-life supercapacitors.

Polymer nanofibers produced by this electrospun technique are finding wide applications in biomedical applications and tissue engineering (Deitzel et al., 2002; Li et al., 2002b; Khil et al., 2004; Smitha and Ma, 2004; Xu et al., 2004; Buttafoco et al., 2006). Biocompatible nanofibers of poly(D,L-lactide-*co*-glycolide) (PLGA) play an important role in modulating tissue growth. The nanofibers are capable of supporting cell attachment and guide cell growth (Li et al., 2002b). Poly(*E*-caprolactone) (PCL) nanofibers are used for the guiding and proliferation of fibroblasts and myoblasts cells (Williamson and Coombes, 2004). Collagen nanofibers and their composites have been found to be ideal for cell attachment and proliferation (Huang et al., 2001; Matthews et al., 2002).

20.4.2 Electronic Transport Properties of Conducting Polymer Nanowires

Conducting polymer nanowires are quasi-1D systems composed of aligned polymer chains where charge carriers are created by doping. The electronic transport properties of such systems are strongly dependent on the detailed nature of interaction among charge carriers and with the environment and disorder present in the systems. Due to the presence of disorder, the charge carriers are localized and electron transport can take place by hopping from one localized site to another. In the absence of interaction, if the interchain electron hopping is negligible compared to intrachain, then one observes $\ln \rho \propto T^{-1/2}$, which is a signature of 1D hopping. On the other hand, if interchain hopping plays a significant role, then 3D VRH behavior is observed ($\ln \rho \propto T^{-1/4}$). In the presence of electron–electron interaction (EEI), the ES-type behavior ($\ln \rho \propto T^{-1/2}$) is observed in all dimensions. However, recent studies on polymer nanowires at low temperatures have revealed exciting features of the 1D transport properties of interacting electrons. We have given a brief overview on the effect of interaction and disorder in 1D in Appendix 20.A.

20.4.2.1 Lüttinger Liquid Behavior in Polymer Nanowires

The low temperature electronic transport study of helical polyacetylene (PA) fibers doped with iodine shows characteristics of LL behavior (Aleshin et al., 2004; Aleshin, 2006, 2007). The fibers are nearly 10 μm long, have a thickness of a few tens of nanometers, and are composed of several polyacetylene chains. The low temperature *I–V* characteristics of these nanofibers are highly asymmetric and the asymmetry increases with decreasing temperature. The current–voltage characteristics show power-law behavior ($I \propto V^{1+\beta}$) in the temperature range 30 K < *T* < 300 K. The β value for R-helical-PA fibrils was found to be ~1.0–4.7 depending on the diameter and the temperature. Power-law dependence was also observed in the conductance vs. temperature ($G \propto T^{\alpha}$) and the α value increases from ~2.2 to ~7.2 with a decreasing cross-section of the nanofibers. Different temperature *I–V* characteristics show expected scaling to a master curve by plotting $I/T^{1+\alpha}$ versus $eV/(k_{B}T)$. The value of $\alpha \neq \beta$ and the value of β is always less than α, however, up to a certain extent, the results indicate toward a LL-like behavior of helical polyacetylene nanofibers above ~30 K (Aleshin et al., 2004; Aleshin, 2006, 2007). Below 30 K, the electronic transport characteristics of R-helical-PA nanofibers show Coulomb blockade behavior (Aleshin et al., 2005).

20.4.2.2 Wigner Crystal-Like Behavior of Polymer Nanowires

Conducting polymer nanowires are disordered quasi-1D system where formation of Wigner crystal (WC) may be favored at a low carrier concentration. The disorder restricts the zero-point motion of the electron, which in turn reduces the quantum fluctuation, and the quasi-1D nature restricts the increase of quantum fluctuation (due to the statistical averaging of fluctuations) even in the absence of disorder. Confinement (in 1D) also enhances the EEI; all these conditions favor the WC formation.

It has been observed that conducting polymer nanowires synthesized by the template-based technique using the chemical method have very low charge carrier density compared to the electrochemically synthesized nanowires or bulk polymer (Rahman et al., 2006, 2007; Rahman and Sanyal, 2007b). Also, the polymer chains are well aligned in template-based synthesized nanowires, which helps to explore its quasi-1D nature. Due to the presence of the unavoidable defect, the charge carriers are localized and the carrier concentrations are further decreased with decreasing temperature; hence, the screening of interactions by charge carriers becomes less effective and the Coulomb interaction stars to play a significant role. For very low electron density (as in the case of chemically synthesized nanowires), the EEI becomes long ranged which favors the Wigner crystal formation. The low temperature *I–V* characteristics of these nanowires show characteristic features of a charge density wave (CDW) system that can arise due to the formation of 1D WC in these structurally disordered materials. At low temperatures, the system shows a gap in the d*I*/d*V* vs. *V* data, a power-law-dependent *I–V* was observed up to certain bias and above a threshold voltage the nanowires show switching transition and hysteresis. Current driven measurement shows the presence of negative differential resistance (NDR) and a huge enhancement of noise near the switching transition. The rapid decrease of the gap with increasing temperature suggests the presence of strong EEI and a correlated nature of the system. The switching transition, NDR, and noise enhancement suggest the sliding motion of the WC. Below we will give a brief detail about the possible formation of WC in polymer nanowires.

Low doped polymer nanowires with a quasi-1D nature and low electron density ($r_s \gg 1$) are a potential candidate to form Wigner crystals that can exhibit characteristics of a charge density wave state (Heeger et al., 1988; Schulz, 1993; Lee et al., 2000; Fogler et al., 2004; Aleshin et al., 2004; Aleshin, 2006, 2007). For weakly pinned Wigner crystals, the tunneling density of states shows a power law behavior with the applied bias (Maurey and Giamarchi, 1995) and the exponent ranges from ~3 to 6 (Jeon et al., 1996; Lee, 2002).

In the chemically synthesized nanowires, a gap (V_G) was observed in the low temperature I–V characteristics. The gap shows strong temperature dependence and it vanishes at relatively high temperatures (see Figure 20.12), also V_G decreases with the increasing diameter of nanowires (Rahman and Sanyal, 2007b). Strong temperature dependence of the gap suggests the presence of EEI and the collective nature of the charge carriers.

I–V characteristics of all the nanowires show power law behavior ($I \propto V^{1+\beta}$) above the gap. In Figure 20.13, one such bit of representative data has been shown for various diameter nanowires. The value of β increases with an increasing diameter (refer to the inset of Figure 20.13) and decreases with increasing temperature for all the nanowires (Rahman and Sanyal, 2007b; Rahman et al., 2007). It has been shown (Jeon et al., 1996) that for 1DWC with increasing pinning strength, β should decrease. So the decrease of V_G with an increasing diameter of nanowires clearly indicates that pinning strength increases (Middleton and Wingreen, 1993) with a decreasing diameter. Hence, the reduction in β value with a decreasing diameter of nanowires is consistent with the 1DWC model.

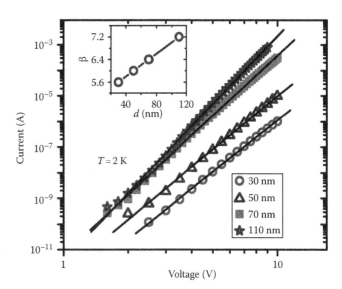

FIGURE 20.13 I–V data for various diameter nanowires at $T = 2$ K is plotted in a double logarithmic scale. The data is fitted (solid line) by the power law $I \propto V^{1+\beta}$. β value for various diameter nanowire has been shown in the inset.

Above a certain temperature (~30 K), low bias resistance vs. temperature data shows a 3D VRH behavior (Rahman et al., 2006). However, power law gives a better fit for the data taken with a higher bias. A clean LL state predicts $\alpha = \beta$ and scaling of I–V curves of different temperatures to a master curve (Balents, 1999). Most of the nanowires do not show such collapse, also the exponents are of different values. It has been shown previously that above a certain temperature (>30 K), electronic transport properties of polymer nanowires can show LL-type behavior (Aleshin et al., 2004; Aleshin, 2006, 2007). The absence of a single master curve and unequal exponents in these wires show that the LL theory is not applicable to describe the electronic properties of these nanowires in the low temperature (<30 K) regime. Moreover, higher values of the exponents (compared to the LL theory) observed here also indicate that 1DWC has formed in these nanowires.

Under the application of high bias voltage, the nanowires show a sharp transition to a highly conducting state above a certain threshold voltage $V_{Th}(>V_G)$ and the change in conductance is found to be more than three orders of magnitude (Rahman and Sanyal, 2007a). One such piece of representative data has been shown in Figure 20.14 for a nanowire with a diameter of 450 nm measured at 2.1 K. After switching, the current does not follow the same path with the reduction of voltage and the system returns to its low conducting state only below a certain threshold voltage V_{Re} ($|V_{Th}| > |V_{Re}| > |V_G|$) (see Figure 20.14). With increasing temperature, the hysteresis, defined as $(E_{Th} - E_{Re})/E_{Re}$ ($E_{Th,Re}$ is the field corresponding to $V_{Th,Re}$), decreases. The area under the hysteresis loop is independent of the scan speed of bias voltage (current) used in the experiment. The observed switching transition can be explained by considering the sliding motion of the WC that depinnes above the threshold voltage. This sliding motion of WC exhibits characteristics of depinning of the

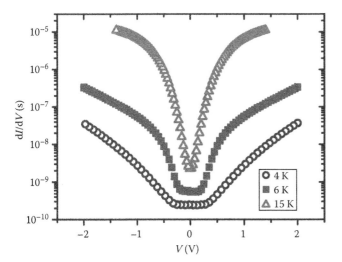

FIGURE 20.12 Differential conductance (dI/dV) is plotted as a function of bias voltage (V) for 450 nm diameter nanowire at various temperatures.

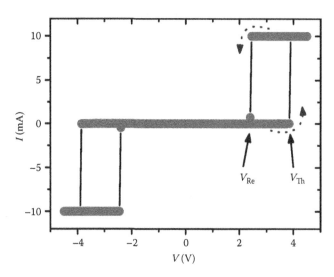

FIGURE 20.14 *I–V* characteristics of 450 nm diameter nanowires measured at 2.1 K showing the threshold voltage and switching transition. Arrow indicates the direction of voltage scan.

pinned CDW state (Zettl and Grüner, 1982; Coleman et al., 1987; Grüner, 1988; Maeda et al., 1990; Levy et al., 1992).

As the resistance of the nanowires in the highly conducting state is very less (a few Ohm), one cannot get information about the switched state from the voltage bias measurements (due to the presence of input impedance or the current limit of the source meters used). Current biased measurements show the existence of NDR in the switched state (refer to Figure 20.15). The switching and NDR were observed for all the nanowires of various diameters and the data were highly reproducible (Rahman and Sanyal, 2008). The sharp threshold and its time independence confirm that switching is not due to field heating. The reproducibility of observed switching and NDR rules out the burning of nanowires. The scan speed independence of the switching and zero crossing (zero current at zero voltage) rules out any capacitive effect. The possibility of any interface-dependent effect has

been ruled out by taking different contact materials. The observed switching is obviously not due to dielectric breakdown because the length of higher diameter (>110 nm) nanowires are larger where low threshold voltages have been observed (Rahman and Sanyal, 2007a).

In conclusion, all the observations like the existence of gap, the power-law behavior of *I–V* and *R–T* characteristics, switching transition, NDR, noise enhancement, etc., suggest the formation of Wigner crystal in these nanowires.

20.5 Conclusion

Nanowires of polymers provide us unique systems to investigate various ideas of low dimensional physics. Moreover, the nanowires and nanofibers of the polymers have remarkable application potential in the emerging field of nanotechnology. It is known that the physical properties of materials change drastically as one of the three dimensions of a material becomes comparable to the molecular size. For polymers, which are macromolecules, the interesting changes in physical properties start happening even when the diameter of the nanowires is only 100 nm (0.1 μm). This provides us great technological potential as drastic changes in mechanical and electrical properties in these structurally disordered and easy-to-form nanomaterials can be exploited. We have presented a brief summary of growth, properties, and applications of the nanowires of polymers. We have also elaborated novel electronic transport properties of conducting polymer nanowires at the end of this chapter.

Appendix 20.A: Physics in One Dimension: Effect of Interactions and Disorder

The electronic transport properties of 1D systems are very interesting as interactions and disorder play a very important role in determining them. In the absence of interactions, properties of the many-Fermion system can be described by the fermi gas model, which is a quantum mechanical version of an ideal gas model. To explain the properties of interacting fermions in higher (more than 1) dimensions, Landau gave a successful theory—the Fermi liquid theory, where elementary excitations can be considered as a collection of free quasi-particles obeying Fermi statistics (Abrikosov et al., 1963; Pines and Noziéres, 1966). Fermi liquid theory breaks down in 1D systems in the presence of interactions (Voit, 1994; Giamarchi, 2004). For the 1D electronic system, the ground state is strongly correlated in the presence of the EEI and the low energy excitations are bosonic sound-like density waves (plasmons). Depending on the range of the EEI, the properties of these systems can be described by Lüttinger liquid (for short-range interactions) or Wigner crystal (for long-range interactions).

Lüttinger liquid theory: In the presence of short-range EEI, the electronic properties of a 1D system are described by the Lüttinger liquid theory. The most striking behavior, which

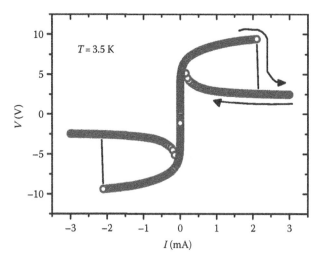

FIGURE 20.15 Current driven *I–V* characteristics of 450 nm diameter nanowire measured at 3.5 K showing presence of NDR (arrows indicates the direction of current scan).

clearly distinguishes the Lüttinger liquid from the Fermi liquid, is the anomalous power-law dependence of various correlation functions in the low-energy region. For example, the density–density correlation function between two distant positions x and x exhibits the power-law dependence in the asymptotic region

$$< \rho_e(x)\rho_e(x') > \sim e^{2ik_F(x-x')}(x-x')^{-\alpha} \qquad (20.9)$$

In contrast to the Fermi liquid, which has the finite density of states at the Fermi level, the tunneling density of states of these systems shows power-law energy dependence $n(\varepsilon) \propto \varepsilon^{\alpha}$. The differential conductance shows power-law dependence on either bias or temperature (Voit, 1994; Giamarchi, 2004)

$$dI/dv \propto T^{\alpha} \quad \text{for low bias } (V \ll k_BT/e) \qquad (20.10)$$

$$dI/dV \propto V^{\beta} \quad \text{for low bias } (V \gg k_BT/e) \qquad (20.11)$$

Pure (in the absence of disorder) Lüttinger (LL) state predicts $\alpha = \beta$ and the bias (V) and temperature (T) dependence of the I–V curve can be expressed as (Balents, 1999; Giamarchi, 2004)

$$I = I_0 T^{1+\alpha} \sinh\left(\frac{eV}{2k_BT}\right)\left|\Gamma\left(1 + \frac{\alpha}{2} + \frac{ieV}{2\pi k_BT}\right)\right|^2 \qquad (20.12)$$

where

I_0 is a constant
$\Gamma(z)$ is the gamma function

For LL, the I–V curves measured at different temperatures can be scaled to a master curve by plotting $I/T^{1+\alpha}$ versus $eV/(k_BT)$ (Balents, 1999). The value of the exponent α depends on the number of channels and on the strength parameter (g) of the LL. Depending on whether the electron tunnels into the bulk or the end of the LL, α can be expressed as (Matveev and Glazman, 1993)

$$\alpha_{\text{end}} = \frac{(g^{-1}-1)}{4} \qquad (20.13)$$

$$\alpha_{\text{bulk}} = \frac{(g^{-1}+g-2)}{8} \qquad (20.14)$$

Since $g < 1$, the exponent α is larger near the end, which is due to the fact that the spread away of an added charge at the end is small compared with the bulk. If $g \ll 1$ then $\alpha_{\text{end}} = 2\alpha_{\text{bulk}}$.

For a strictly 1D system, the power-law behavior may be washed out by quantum fluctuations. However, in quasi-1D systems (systems composed of several parallel chains) where the different channels are statistically independent, the power-law behavior may be observed clearly due to the averaging of fluctuations. In quasi-1D systems, the exponents depend on the number

of the 1D channels (Matveev and Glazman, 1993). However, in these systems, if inter-chain electron hopping plays a significant role, then the system looses its one-dimensionality and β becomes equal to zero. This could be the case for quasi-1D systems at very low temperatures where arbitrarily weak interchain hopping can drive the system toward a 3D system (Firsov et al., 1985; Schulz, 1991; Arrigoni, 1999). In disordered 1D systems, impurity acts as an infinite tunneling barrier (Kane and Fisher, 1992; Matveev and Glazman, 1993). However, the presence of disorder in 1D systems can stabilize the LL state, where the LL state is preserved between two consecutive impurities, thus a disordered quasi-1D system can be considered to be a collection of LL stubs (Artemenko and Remizov, 2005). For such systems, weak interchain hopping does not destroy the LL state at low enough temperatures (compared to the minimum excitation energy of the LL collective modes), and low bias conductivity shows VRH behavior while high bias conductivity shows power law behavior. Low bias R–T behavior of such systems follows Mott's 3D VRH law for short-range interaction and Efros–Shklovskii law for long-range interactions between electrons of different segments (Artemenko and Remizov, 2005).

Wigner Crystal: In the presence of long-range EEI, in very low electron density materials, if Coulomb repulsion dominates over the kinetic energy of the electrons, then electrons get arranged periodically in space forming an electron crystal (Wigner, 1934) known as Wigner crystal (WC). To quantify the condition of WC formation, a parameter, strength of interaction, is defined as $r_s = a/2a_B$ where a is the average distance between electrons and a_B is the effective Bohr radius. At a very low electron density (n_s), the separation between electrons becomes large ($r \propto 1/n_s$) so $r_s \gg 1$ and this is the condition for WC formation. It has been found that (a) $r_s \simeq 36$ for clean unbound system, (b) $r_s \simeq 7.5$ for disordered unbound systems and (c) $r_s \simeq 7.4$ for confined systems (Tanatar and Ceperley, 1989; Chui and Esfarjani, 1991; Chui and Tanatar, 1995; Reimann et al., 2000). The decrease of r_s in disordered systems is due to the breaking of continuous translational invariance that stabilizes the crystalline state.

It has been predicted theoretically that one dimensional WC exhibits the characteristic of a charge density wave (CDW) (Schulz, 1993). The charge–charge correlation function in 1D for long-range EEI can be written as

$$\langle \rho(x)\rho(0)\rangle = \frac{A_1\cos(2k_Fx)\exp\left(-c_2\sqrt{\ln(x)}\right)}{x}$$
$$+ A_2\cos(4k_Fx)\exp\left(-4c_2\sqrt{\ln(x)}\right) + \cdots \qquad (20.15)$$

where A_1 and A_2 are interaction-dependent constants. In the expression, other fast decaying Fourier components have been omitted. It is interesting to note that the $4k_F$ component decays very slowly (much slower than any power law), showing an incipient charge density wave having a wave vector of $4k_F$ (Schulz, 1993) (in a CDW system formed due to Peierls transition, one observes a $2k_F$ periodicity). The decay of the $4k_F$ component observed in long-range interaction is much slower compared to short-range interactions where $2k_F$ and $4k_F$ components show power-law decay. As the $4k_F$ oscillation period is the same as interparticle spacing, the structure is the same as 1D WC but a true long-range order does not exist in such systems because of its 1D nature. It should be mentioned that $4k_F$ oscillation arises only due to the long-range nature of the EEI and can exist even for its extremely small value.

It has been reported that the WC state may occur in quasi-1D systems (Wen, 1992; Boies et al., 1995; Arrigoni, 2000) or even in nanowires of structurally disordered materials (Fogler et al., 2004). This electron crystal is pinned by the impurities present in the system. Theoretically, it has been shown that if the impurities act as strong pinning centers then quasi-1D WC shows VRH conductivity with various exponents that depend on the impurity concentration (Fogler et al., 2004). At large impurity concentrations, conductivity follows Efros–Shklovskii law and at very low impurity concentrations, Mott's 3D VRH-type behavior is predicted. For weakly pinned Wigner crystals, the tunneling density of states shows a power law behavior with the applied bias (Maurey and Giamarchi, 1995) and the exponent ranges from ~3 to 6 (Glazman et al., 1992; Maurey and Giamarchi, 1995; Jeon et al., 1996; Lee, 2002). The impurities destroy the long-range order; however, for weak impurity strength, quasi-long-range orders may exist (Cha and Fertig, 1994) and a Wigner glass (Andrei et al., 1988; Li et al., 1995; Chakravarty et al., 1999; Fogler and Huse, 2000; Slutskin et al., 2003; Akhanjee and Rudnick, 2007) may form. Depending upon the pinning strength, a pinned WC becomes nonconducting below a certain threshold field. When an applied field is strong enough to overcome the pinning energy, the WC depins and the sliding motion starts giving rise to a switching transition similar to CDW systems (Zettl and Grüner, 1982; Grüner, 1988; Maeda et al., 1990; Levy et al., 1992).

References

Abrikosov A. A., Gorkov L. P., and Dzyaloshinski I. E., *Methods of Quantum Field Theory in Statistical Mechanics*, Dover, New York (1963).

Akhanjee S. and Rudnick J., *Phys. Rev. Lett.*, **99**, 236403 (2007).

Albert K. J. et al., *Chem. Rev.*, **100**, 2595 (2000).

Aleshin A. N., *Adv. Mater.*, **18**, 17 (2006).

Aleshin A. N., *Phys. Solid State*, **49**, 2015 (2007).

Aleshin A. N., Lee H. J., Park Y. W., and Akagi K., *Phys. Rev. Lett.*, **93**, 196601 (2004).

Aleshin A. N., Lee H. J., Jhang S. H., Kim H. S., Akagi K., and Park Y. W., *Phys. Rev. B*, **72**, 153202 (2005).

Anderson P. W., *Phys. Rev. Lett.*, **34**, 953 (1975).

Andrei E. Y., Deville G., Glattli D. C., Williams F. I. B., Paris E., and Etienne B., *Phys. Rev. Lett.*, **60**, 2765 (1988).

Arinstein A., Burman M., Gendelman O., and Zussman E., *Nat. Nanotechnol.*, **2**, 59 (2007).

Armes S. P., *Synth. Met.*, **365**, 20 (1987).

Arrigoni E., *Phys. Rev. Lett.*, **3**, 128 (1999).

Arrigoni E., *Phys. Rev. B*, **61**, 7909 (2000).

Artemenko S. N. and Remizov S. V., *Phys. Rev. B*, **72**, 125118 (2005).

Balents L., *Proceedings of the Moriond Les Arcs Conference*, Les Arcs, France, (1999) (unpublished).

Barth J. V., Costantini G., and Kern K., *Nature*, **437**, 671 (2005).

Bashyam R. and Zelenay P., *Nature*, **443**, 63 (2006).

Baughman R. H., *Synth. Met.*, **78**, 339 (1996).

Berdichevsky Y. and Lo Y.-H., *Adv. Mater.*, **18**, 122 (2005).

Bergshoef M. M. and Vancso G. J., *Adv. Mater.*, **11**, 1362 (1999).

Bhattacharya M., Sanyal M. K., Geue T., and Pietsch U., *Phys. Rev. E*, **71**, 041801 (2005).

Boies D., Bourbonnais C., and Tremblay A.-M. S., *Phys. Rev. Lett.*, **74**, 968 (1995).

Bower D. I., *An Introduction to Polymer Physics*. Cambridge University Press, Cambridge, U.K. (2002).

Bredas J. L. and Street G. B., *Acc. Chem. Res.*, **18**, 309 (1985).

Burroughes J. H., Bradley D. D. C., Brown A. R., Marks R. N., Mackay K., Friend R. H., Burns P. L., and Holmes A. B., *Nature* **347**, 539 (1990).

Buttafoco L., Kolkman N. G., Engbers-Buijtenhuijs P., Poot A. A., Dijkstra P. J., Vermes I., and Feijen J., *Biomaterials*, **27**, 724 (2006).

Cai Z. and Martin C. R., *J. Am. Chem. Soc.*, 111, 4138 (1989).

Cai Z., Lei J., Liang W., Menon V., and Martin C. R., *Chem. Mater.*, **3**, 960 (1991).

Cha, M. C. and Fertig H. A., *Phys. Rev. Lett.*, **73**, 870 (1994).

Chakravarty S., Kivelson S., Nayak C., and Voelker K., *Philos. Mag. B*, **79**, 859 (1999).

Chiang C. K., Fincher C. R., Park Y. W., Heeger A. J., Shirakawa H., Louis E. J., Gau S. C., and MacDiarmid A. G., *Phys. Rev. Lett.*, **39**, 1098 (1977).

Chiang C. K., Gau S. C., Fincher C. R. Jr., Park Y. W., and MacDiarmid A. G., *Appl. Phys. Lett.*, **33**, 18 (1978).

Cho S. I. and Lee S. B., *Acc. Chem. Res.*, **41**, 699 (2008).

Choi S.-J. and Park S.-M., *Adv. Mater.*, **20**, 1547 (2000).

Chui S. T. and Esfarjani K., *Europhys. Lett.*, **14**, 361 (1991).

Chui S. T. and Tanatar B., *Phys. Rev. Lett.*, **74**, 458 (1995).

Coleman R. V., Everson M. P., Eiserman G., Johnson A., and Lu H.-A., *Synth. Met.*, **19**, 795 (1987).

Cuenot S., Demoustier-Champagne S., and Nysten B., *Phys. Rev. Lett.*, **85**, 1690 (2000).

Deitzel J. M., Kosik W., McKnight S. H., Tan N. C. B., DeSimone J. M., and Crette S., *Polymer*, **43**, 1025 (2002).

Demoustier-Champagne S. and Stavaux P.-Y., *Chem. Mater.*, **11**, 829 (1999).

Deng J., Ding X., Zhang W., Peng Y., Wang J., Long X., Li P., and Chan A. S. C., *Eur. Polym. J.*, **38**, 2497 (2002).

Dimitrakopoulos Ch. D. and Malenfant P. R. L., *Adv. Mater.*, **14**, 99 (2002).

Epstein A. J., Joo J., Kohlman R. S., Du G., MacDiarmid A. G., Oh E. J., Min Y., Tsukamoto J., Kaneko H., and Pouget J. P., *Synth. Met.*, **65**, 149 (1994).

Ferain E. and Legras R., *Nucl. Instrum. Method B*, **84**, 331 (1994); Ferain E. and Legras R., *Nucl. Instrum. Method B*, **131** 97 (1997).

Firsov Y. A., Prigodin V. N., and Seidel Chr., *Phys. Rep.*, **126**, 245 (1985).

Fleischer R. L., Price P. B., and Walker R. M., *Nuclear Tracks in Solids*, University of California Press, Berkeley, CA (1975).

Fogler M. M. and Huse D. A., *Phys. Rev. B*, **62**, 7553 (2000).

Fogler M. M., Teber S., and Shklovskii B. I., *Phys. Rev. B*, **69**, 035413 (2004).

Friend R. H. et. al., *Nature,* **397**, 121 (1999).

Ge J. J., Hou H., Li Q., Graham M. J., Greiner A., Reneker D. H., Harris F. W., and Cheng S. Z. D., *J. Am. Chem. Soc.*, **126**, 15754 (2004).

Gerard M., Chaubey A., and Malhotra B. D., *Biosens. Bioelectron.*, **17**, 345 (2002).

Giamarchi T., *Quantum Physics in One Dimension*, Clarendon Press, Oxford, U.K. (2004).

Glazman L. I., Ruzin I. M., and Shklovskii B. I., *Phys. Rev. B*, **45**, 8454 (1992).

Grüner G., *Rev. Mod. Phys.*, **60**, 1129 (1988).

Gurunathan K., Murugan A. V., Marimuthu R., Mulik U. P., and Amalnerkar D. P., *Mater. Chem. Phys.*, **63**, 173 (1999).

Hayes R. A. and Feenstra B. J., *Nature*, **425**, 383 (2003).

Heeger A. J., *Rev. Mod. Phys.*, **73**, 681 (2001).

Heeger A. J., Kivelson S., Schrieffer J. R., and Su W.-P., *Rev. Mod. Phys.*, **60**, 781 (1988).

Hernndez R. M., Richter L., Semancik S., Stranick S., and Mallouk T. E., *Chem. Mater.*, **16**, 3431 (2004).

Hong B. H., Bae S. C., Lee C.-W., Jeong S., and Kim K. S., *Science*, **294**, 348 (2001).

Huang L., Nagapudi K., Apkarian R. P., and Chaikof E. L., *J. Biomater. Sci. Polym. Ed.*, **12**, 979 (2001).

Huang J., Virji S., Weiller B. H., and Kaner R. B., *J. Am. Chem. Soc.*, **125**, 314 (2003a).

Huang Z. M., Zhang Y. Z., Kotaki M., and Ramakrishna S., *Comput. Sci. Technol.*, **63**, 2223 (2003b).

Hughes M., Chen G. Z., Shaffer M. S. P., Fray D. J., and Windle A. H., *Chem. Mater.*, **14**, 1610 (2002).

Hwang B. J., Yang J. Y., and Lin C. W., *Sens. Actuators B*, **75**, 67 (2001).

Jager Edwin W. H., Smela E., and Inganäs O., *Science*, **290**, 1540 (2000).

Janata J. and Josowicz M., *Nat. Mater.*, **2**, 19 (2003).

Jeon G. S., Choi M. Y., and Eric Y. S.-R., *Phys. Rev. B*, **54**, R8341 (1996).

Jianyong O., Chih-Wei C., Charles R. S., Liping M., and Yang Y., *Nat. Mater.*, **3**, 918 (2004).

Joo J., Prigodin V. N., Min Y. G., MacDiarmid A. G., and Epstein A. J., *Phys. Rev. B*, **50**, 12226 (1994).

Kaiser A. B., *Rep. Prog. Phys.*, **64**, 1 (2001).

Kane C. L. and Fisher M. P. A., *Phys. Rev. Lett.*, **68**, 1220 (1992).

Kapon E., Hwang D. M., and Bhat R., *Phys. Rev. Lett.*, **63**, 430 (1989).

Khil M. S., Kim H. Y., Kim M. S., Park S. Y., and Lee D. R., *Polymer*, **45** 295 (2004).

Kitani A., Kaya M., and Sasaki K., *J. Electrochem. Soc.*, **133**, 1069 (1986).

Kohlman R. S., Joo J., Wang Y. Z., Pouget J. P., Kaneko H., Ishiguro T., and Epstein A. J., *Phys. Rev. Lett.*, **74**, 773 (1995).

Kohlman R. S., Joo J., Min Y. G., MacDiarmid A. G., and Epstein A. J., *Phys. Rev. Lett.*, **77**, 2766 (1996).

Lee H. C., *Phys. Rev. B*, **66**, 052202 (2002).

Lee K., Menon R., Heeger A. J., Kim K. H., Kim Y. H., Schwartz A., Dressel M., and Grüner G., *Phys. Rev. B*, **61**, 1635 (2000).

Lee K., Cho S., Park S. H., Heeger A. J., Lee C.-W., Lee S.-H., *Nature*, **441**, 65 (2006a).

Lee W., Ji R., Gösele U., and Nielsch K., *Nat. Mater.*, **5**, 741 (2006b).

Lee A. S., Peteu S. F., Ly J. V., Requicha A. A. G., Thompson M. E., and Zhou C., *Nanotechnology*, **19**, 165501 (2008).

Levy J., Sherwin M. S., Abraham F. F., and Wiesenfeld K., *Phys. Rev. Lett.*, **68**, 2968 (1992).

Li Q., Cruz L., and Phillips P., *Phys. Rev. B*, **47**, 1840 (1993).

Li Q. W., Yan H., Cheng Y., Zhang J., and Liu Z. F., *J. Mater. Chem.*, **12**, 1179 (2002a).

Li W. J., Laurencin C. T., Caterson E. J., Tuan R. S., and Ko F. K., *J. Biomed. Mater. Res.*, **60**, 613 (2002b).

Li Y. P., Tsui D. C., Sajoto T., Engel L. W., Santos M., and Shayegan M., *Solid State Commun.*, **95**, 619 (1995).

Liu R., Cho S. I., and Lee S. B., *Nanotechnology*, **19**, 215710 (2008).

Liu L.-Q., Tasis D., Prato M., and Wagner H. D., *Adv. Mater.*, **19**, 1228 (2007).

Ma Y., Zhang J., Zhang G., and He H., *J. Am. Chem. Soc.*, **126**, 7097 (2004).

Maeda A., Notomi M., and Uchinokura K., *Phys. Rev. B*, **42**, 3290 (1990).

Mao C. et al., *Science*, **303**, 213 (2004).

Martin C. R., *Science*, **266**, 1961 (1994).

Masuda H. and Fukuda K., *Science*, **268**, 1466 (1995).

Matthews J. A., Wnek G. E., Simpson D. G., and Bowlin G. L., *Biomacromolecules*, **3**, 232 (2002).

Matveev K. A. and Glazman L. I., *Phys. Rev. Lett.*, **70**, 990 (1993).

Maurey H. and Giamarchi T., *Phys. Rev. B*, **51**, 10833 (1995).

McQuade D. T., Pullen A. E., and Swager T. M., *Chem. Rev.*, **100**, 2537 (2000).

Middleton A. A. and Wingreen N. S., *Phys. Rev. Lett.* **71**, 3198 (1993).

Milella E. and Penza M., *Thin Solid Films*, **329**, 694 (1998).

Miller A. and Abrahams E., *Phys. Rev.*, **120**, 745 (1960).

Möller S., Perlov C., Jackson W., Taussig C., and Forrest S. R., *Nature*, **426**, 166 (2003).

Mortimer R. J., *Electrochim. Acta*, **44**, 2971 (1999).

Mott N. F. and Davis E. A., *Electron Processes in Non-Crystalline Materials*, Clarendon Press, Oxford, U.K. (1979).

Na S.-I., Kim S.-S., Jo J., and Kim D.-Y., *Adv. Mater.*, **20**, 4061 (2008).

Nam K. T. et al., *Science*, **312**, 855 (2006).

Nielsch K. et al., *Nano Lett.*, **2**, 677 (2002).

Parthasarathy R. V., Menon V. P., and Martin C. R., *Chem. Mater.*, **9**, 560 (1997).

Pelster R., Nimtz G., and Wessling B., *Phys. Rev. B*, **49**, 12718 (1994).

Pines D. and Noziéres P., *The Theory of Quantum Liquids*, AddisonWesley, Menlo Park, CA (1966).

Pollak M. and Ortuno M., *Electron-Electron Interaction in Disordered Systems*, Eds. Efros A. L. and Pollak M., North-Holland, Amsterdam, the Netherlands (1985).

Price P. B. and Walker R. M., *J. Appl. Phys.*, **33**, 3407 (1962).

Qian D., Dickeya E. C., Andrews R., and Rantell T., *Appl. Phys. Lett.*, **76**, 2868 (2000).

Quake S. R. and Scherer A., *Science*, **290**, 1535 (2000).

Rahman A. and Sanyal M. K., *Adv. Mater.*, **19** 3956 (2007a).

Rahman A. and Sanyal M. K., *Phys. Rev. B*, **76**, 045110 (2007b).

Rahman A. and Sanyal M. K., *Nanotechnology*, **19**, 395203 (2008).

Rahman A., Sanyal M. K., Gangopadhayy R., De A., and Das I., *Phys. Rev. B*, **73**, 125313 (2006).

Rahman A., Sanyal M. K., Gangopadhayy R., and De A., *Chem. Phys. Lett.*, **447**, 268 (2007).

Ramanathan K., Bangar M. A., Yun M., Chen W., Mulchandani A., and Myung N. V., *Nano Lett.*, **4**, 1237 (2004).

Ramanathan K., Bangar M. A., Yun M., Chen W., Myung N. V., and Mulchan-dani A., *J. Am. Chem. Soc.*, **127**, 496 (2005).

Ramakrishna S., Fujihara K., Ganesh V. K., Teo W. E., and Lim T. C., *Functional Nanomaterials*, Eds. Geckeler K. E. and Rosenberg E., p. 113, American Scientific Publishers, Stevenson Ranch, CA (2006).

Ramaseshan R., Sundarrajan S., Liu Y., Barhate R. S., Lala N. L., and Ramakrishna S., *Nanotechnology*, **17**, 2947 (2006).

Reghu M., Yoon C. O., Moses D., Heeger A. J., and Cao Y., *Phys. Rev. B*, **48**, 17685 (1993).

Reimann S. M., Koskinen M., and Manninen M., *Phys. Rev.*, **62**, 8108 (2000).

Roth S. and Bleier H., *Adv. Phys.*, **36**, 385 (1987).

Salaneck W. R., *Rep. Prog. Phys.*, **54**, 1215 (1991).

Sanyal M. K., Basu. J. K., Datta A., and Banerjee S., *Europhys. Lett.*, **36**, 265 (1996).

Sariciftci N. S., Smilowitz L., Heeger A. J., and Wudl F., *Science*, **258**, 1474 (1992).

Savariar E. N., Krishnamoorthy K., and Thayumanavan S., *Nat. Nano*, **3**, 112 (2008).

Schulz H. J., *Int. J. Mod. Phys. B*, **5**, 57 (1991).

Schulz H. J., *Phys. Rev. Lett.*, **71**, 1864 (1993).

Seoul C., Kim Y. T., and Baek C. K., *J. Polym. Sci. Pt. B: Polym. Phys.*, **41**, 1572 (2003).

Sheats J. R. et al., *Science*, **273**, 884 (1996).

Sheng P., *Phys. Rev. B*, **21**, 2180 (1980).

Sheng P. and Klafter J., *Phys. Rev. B*, **27**, 2583 (1983).

Sheng P., Abeles B., and Arie Y., *Phys. Rev. Lett.*, **31**, 44 (1973).

Sheng P., Sichel E. K., and Gittleman J. I., *Phys. Rev. Lett.*, **40**, 1197 (1978).

Shklovskii B. I. and Efros A. L., *Electronic Properties of Doped Semiconductor*, Springer-Verlag, Berlin, Germany (1984).

Slutskin A. A., Peeper M., and Kovtun H. A., *Europhys. Lett.*, **62**, 705 (2003).

Smela E., Inganas O., and Lundstrom I., *Science*, **268**, 1735 (1995).

Smith S. and Forrest S. R., *Appl. Phys. Lett.*, **84**, 5019 (2004).

Smitha L. A. and Ma P. X., *Colloids Surf. B: Biointerfaces*, **39**, 125 (2004).

Sonmez G., Meng H., and Wudl F., *Chem. Mater.*, **16**, 574 (2004).

Sreekumar T.V., Liu T., Min B.G., Guo H., Kumar S., Hauge R. H., and Smalley R. E., *Adv. Mater.*, **16**, 58 (2004).

Su W. P., Schrieffer J. R., and Heeger A. J., *Phys. Rev. Lett.*, **42**, 1698 (1979).

Sun Y., Wilson S. R., and Schuster D. I., *J. Am. Chem. Soc.*, **123**, 5348 (2001).

Swager T. M., *Acc. Chem. Res.*, **31**, 201 (1998).

Tan E. P. S. and Lim C. T., *Appl. Phys. Lett.*, **87**, 123106 (2005).

Tanatar B. and Ceperley D. M., *Phys. Rev. B*, **39**, 5005 (1989).

Tang C. W., *Appl. Phys. Lett.*, **48**, 183 (1986).

Thurn-Albrecht T. et.al., *Science*, **290**, 2126 (2000).

Väkiparta K., Reghu M., Andersson M. R., Cao Y., Moses D., and Heeger A. J., *Phys. Rev. B*, **47**, 9977 (1993).

Voit J., *Rep. Prog. Phys.*, **57**, 977 (1994).

Wanekaya A. K. et al., *J. Phys. Chem. C*, **111**, 5218 (2007).

Wang D. et al., *Adv. Mater.*, **15**, 130 (2003).

Wang H.-J., Zhou W.-H., Yin X.-F., Zhuang Z.-X., Yang H.-H., and Wang X.-R., *J. Am. Chem. Soc.*, **128**, 15954 (2006).

Wang P.-C., Haung Z., and MacDiarmid A. G., *Synth. Met.*, **101**, 852 (1999).

Wen X. G., *Phys. Rev. B*, **42**, 6623 (1992).

Wigner E., *Phys. Rev.*, **46**, 1002 (1934).

Williamson M. R. and Coombes A. G. A., *Biomaterials*, **25**, 459 (2004).

Wu C.-G. and Bein T., *Science*, **264**, 1757 (1994).

Wu C.-G. and Chen C.-Y., *J. Mater. Chem.*, **1409**, 7 (1997).

Xu C. Y., Inai R., Kotaki M., and Ramakrishna S., *Biomaterials*, **25**, 877 (2004).

Yan H., Park S. H., Finkelstein G., Reif J. H., and LaBean T. H., *Science*, **301**, 1882 (2003).

Yang W. Ma, C., Gong X., Lee K., and Heeger A. J., *Adv. Mater.*, **15**, 1617 (2005).

Ye H., Lam H., Titchenal N., Gogotsi Y., and Ko F., *Appl. Phys. Lett.*, **85**, 1775 (2004).

Zettl A. and Grüner G., *Phys. Rev. B*, **26**, 2298 (1982).

Zhou Y. K., He B. L., Zhou W. J., Huang J., Li X. H., Wu B., and Li H. L., *Electrochim. Acta*, **49**, 257 (2004).

Zuo F., Angelopoulos M., MacDiarmid A. G., and Epstein A. J., *Phys. Rev. B*, **36**, 3475 (1987).

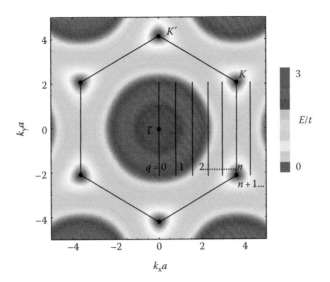

FIGURE 2.5   Plot of the bonding π band of graphene.

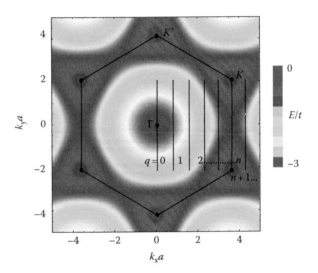

FIGURE 2.6   Plot of the antibonding π band of graphene.

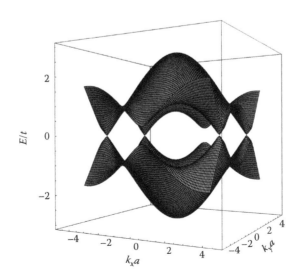

FIGURE 2.7   Band structure of graphene. The figure shows the various energy bands in the Brillouin zone.

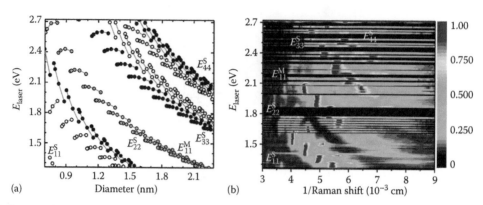

(a)   (b)

FIGURE 7.11 (a) The optical transition energies (dots) of SWNTs as a function of tube diameter. (From Jiang, J. et al., *Phys. Rev. B*, 75, 035407, 2007a. With permission.) The superscripts S and M for the E_{ii} labels indicate the optical transitions for semiconducting and metallic SWNTs, respectively. (b) 2D color map showing the SWNT RBM spectral evolution as a function of laser excitation energy. The intensity of each spectrum is normalized to the strongest peak, and we plot the RBM results are plotted in terms of the inverse Raman shift. (a) and (b) show coincident axes and can be directly correlated. (From Araujo, P.T. et al., *Phys. Rev. Lett.*, 98, 067401, 2007. With permission.)

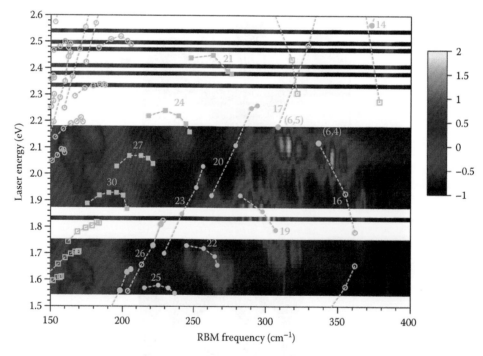

FIGURE 7.15 The 2D RBM Raman map for DWNTs. The E_{ii} points from the Kataura plot are superimposed (green bullets). It is clear there are many more RBM features than (n, m) related E_{ii} values. (From Pfeiffer, R. et al., in *Carbon Nanotubes: Advanced Topics in the Synthesis, Structure, Properties and Applications, Springer Series in Topics in Applied Physics*, Vol.111, Jorio, A. et al. (eds.), Springer-Verlag, Berlin, Germany, 2008, 495–530. With permission.)

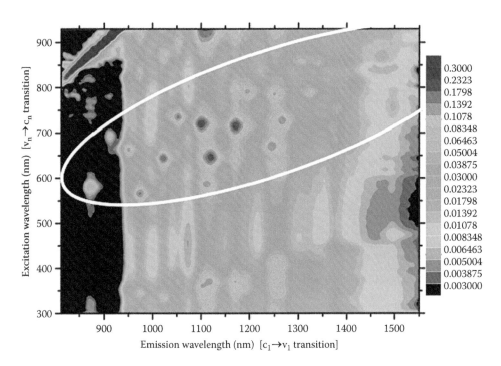

FIGURE 8.10 Three-dimensional fluorescence scan of HiPCO grown SWNTs. Circled in white are some of the well defined fluorescence peaks that are characteristic of the chirality distribution found in SWNTs produced by the HiPCO method. (Reprinted from Bachilo, S.M. et al., *Science*, 298, 2361, 2002. With permission.)

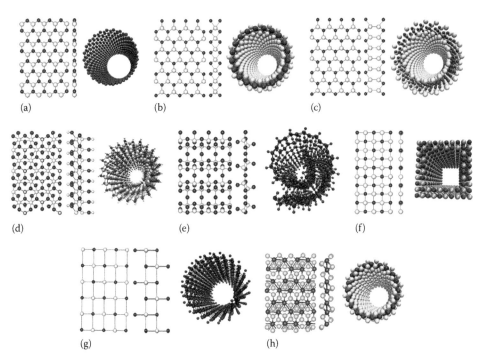

FIGURE 12.2 Structures of the monolayers of some inorganic compounds and respective nanotubes: (a) hexagonal BN and (20,0) nanotube; (b) hexagonal MoS_2 and (20,0) nanotube; (c) hexagonal GaS and (20,0) nanotube; (d) hexagonal imogolite $Al_2O_3(OH)_3Si(OH)$ and (12,0) nanotube; (e) rectangular V_2O_5 and (20,0) nanoroll; (f) square MgO and (7@9) nanotube; (g) centered rectangular lepidocrocite TiO_2 and (20,0) nanotube; (h) oblique ReS_2 and (20,0) nanotube.

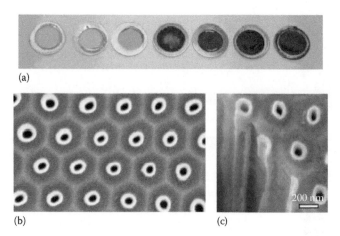

(a)

(b)   (c)

FIGURE 14.2 Several morphological aspects of magnetic nanotubes. (a) Macroscopic view of the samples, consisting of circular porous anodic alumina membranes containing embedded tubes of different thickness. Each circular sample is surrounded by an Al circle whose outer diameter is 2 cm. The first sample on the left corresponds to alumina without magnetic material: the other samples to the right progressively increase their thickness going over 1, 2, 4, 8, 12, and 16 nm approximately. Upon illumination with daylight they vary their color according to the wall thickness as it can be readily seen from the illustration. (b,c) Scanning electron microscope images of nanotubes embedded in porous alumina, observed from a top view (b) and at an angle at a break in the sample (c). These micrographs give an idea of shapes, diameters, thicknesses, and distances for the systems under consideration in the present paper. (Reprinted from Escrig, J. et al., *Phys Rev. B*, 77, 214421, 2008b. With permission.)

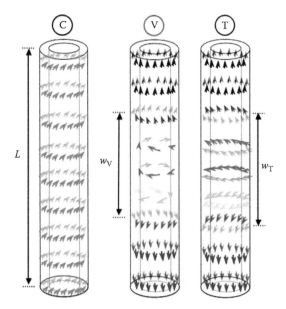

FIGURE 14.7 Magnetization reversal modes in nanotubes. Arrows represent the orientation of magnetic moments within the tube. Left: Coherent-mode rotation, C. Center: Vortex-mode rotation, V, with a domain wall of thickness w_V. Right: Transverse-mode rotation, T, with a domain wall of thickness w_T.

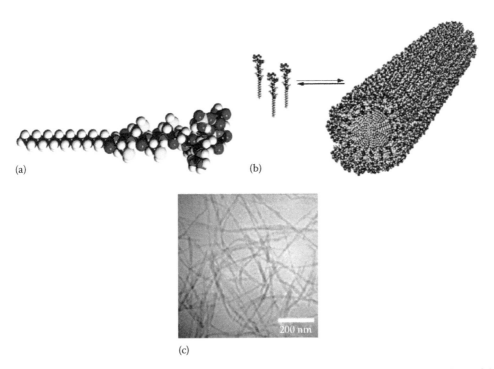

FIGURE 15.5 PAs molecules form cylindrical micelle. (a) Molecular model of the PA showing the overall conical shape of the molecule going from the narrow hydrophobic tail to the bulkier peptide region. Color scheme: C, black; H, white; O, red; N, blue; P, cyan; S, yellow. (b) Schematic presentation of the self-assembly of PA molecules into a cylindrical micelle. (c) Cryo-TEM image of the fibers formed by PA molecules. (From Hartgerink, J.D. et al., *Science*, 294, 1684, 2001. With permission.)

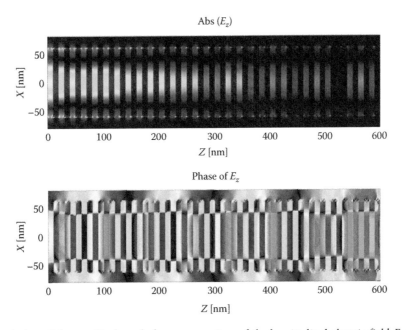

FIGURE 25.9 FDTD simulation of the amplitude and phase propagation of the longitudinal electric field E_z along the nanowire with radius $a = 60$ nm at $\lambda = 488$ nm. The metamaterial nanowire consists of alternative disks of silver and glass disks of thickness 10 nm. (From Huang, Y.J. et al., *Phys. Rev. A*, 77, 063836, 2008. With permission.)

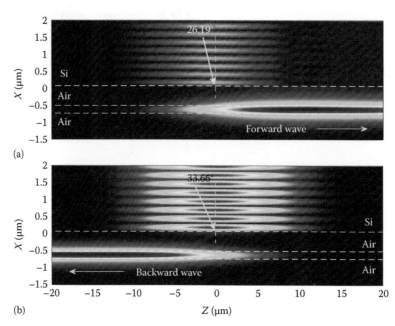

FIGURE 25.13 Gaussian beam excitation through prism coupling of the forward-wave (a) and backward-wave mode (b) at incident angle 26.19° and 33.66°, respectively. The air gap between the prism and the waveguide (d = 234 nm) is 600 nm. Plotted is the absolute value of the magnetic field H_y.

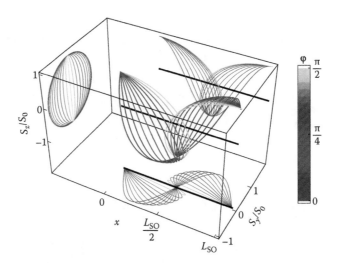

FIGURE 28.8 Persistent spin helix solution of the spin diffusion equation in a quantum wire whose width, W, is smaller than the spin precession length, L_{SO}, for varying ratio of linear Rashba, $\alpha_2 = \alpha \sin \varphi$, and linear Dresselhaus coupling, $\alpha_1 = \alpha \cos \varphi$ (Equation 28.46) for fixed α and $L_{SO} = \pi/m^*\alpha$.

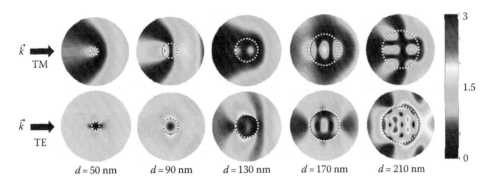

FIGURE 32.5 Scattered internal electric field *amplitude* maps calculated for 488 nm waves incident from the left and for $L = \infty$ GaP NWs with diameters from 50 to 210 nm. The amplitude scale is color coded and appears to the right. TM and TE refer, respectively, to transverse magnetic ($\theta = 0°$) and transverse electric ($\theta = 90°$) excitation with the light incident at right angles to the NW axis. The dashed circles indicate the boundary of the NW. (From Chen, G. et al., *Nano Lett.*, 8, 1341, 2008. With permission.)

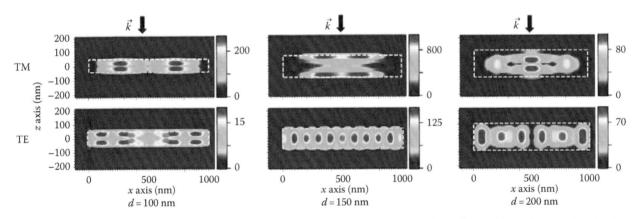

FIGURE 32.7 Calculated internal electric field *amplitude* maps for the axial cross section. Results are from DDA simulation code using 488 nm excitation incident at right angles to the axis of $L = 1\,\mu m$ GaP NWs of diameter $d = 100$, 150, and 200 nm. Color-coded scales appear to the right correspond to differing maximum amplitude. The dashed lines indicate the boundary of the NW. (From Chen, G. et al., *Nano Lett.*, 8, 1341, 2008. With permission.)

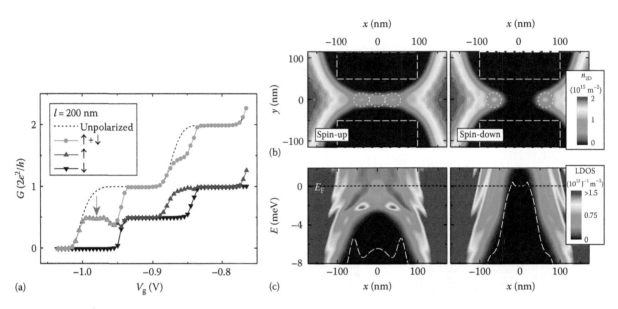

FIGURE 38.18 (a) Conductance of the quantum point contact calculated within the spin DFT approach as a function of the gate voltage V_g. The geometrical width of the constriction is $w = 100\,\text{nm}$; the geometrical length is $l = 200\,\text{nm}$. Dashed line corresponds to the spin-unpolarized solution. (b, c) Formation of quasibound states in the quantum point contact. (b) The charge density and (c) the local density of states are shown for the regime of one transmitted spin-up and totally blocked spin-down channel. The left and right columns correspond to the spin-up and spin-down electrons. White dashed lines in (c) indicate the self-consistent Kohn-Sham potential in the center of the QPC along the transport direction. The geometrical length and width of the QPC are $l = 200\,\text{nm}$ and $w = 100\,\text{nm}$, respectively; the gate voltage $V_g = -0.98\,\text{V}$. (Adapted from Ihnatsenka, S. and Zozoulenko, I.V., *Phys. Rev. B*, 76, 045338, 2007.)

21

Organic Nanowires

Frank Balzer
Syddansk Universitet

Morten Madsen
Syddansk Universitet

Jakob Kjelstrup-Hansen
Syddansk Universitet

Manuela Schiek
Syddansk Universitet

Horst-Günter Rubahn
Syddansk Universitet

21.1 Introduction

Fiber-like, light-emitting nanoaggregates from small organic molecules, nanorods, nanowires, or nanofibers have evolved as a very active research field during the last years [1–3]. Depending on the material they are made of, the nanosizing in one or two dimensions leads to interesting new properties. To zeroth order, the surface-to-volume ratio is greatly enhanced and electrons and photons are much better confined as compared to micron-sized components, while the "long" axis allows easy connection to the macroscopic world. Among the properties that have found special interest are waveguiding [4], lasing [5], electrical transport [6,7], mechanical properties [8], and nonlinear optical properties [9–11]. The organic aggregates have either been grown directly on surfaces by organic molecular beam deposition (OMBD) [12], by hot-wall epitaxy [13], or by solvent vapor annealing [14], or they have been assembled in solution and then deposited onto a surface [15–18]. The formation of upright nanowires on a substrate has been facilitated by, e.g., filling of mesoporous substrates [19,20] or even by simple vapor phase deposition [21,22].

As for the organic nanofibers that are oriented parallel to the surface plane, the most detailed experimental data exists for the growth of nanofibers made of *para*-hexaphenylene (*p*-6P) molecules, Figure 21.1 [23,24]. Their growth is investigated in most detail on dielectric and metallic surfaces such as muscovite and phlogopite mica [23,25,26], KCl and NaCl [27–29], TiO_2 [30], thin Au films, Au foil and Au single crystals [31–33], and GaAs [34].

Micas as substrate surfaces are known to promote uniaxial growth. Examples include mesochannels in a mesoporous silica film [35], needles from organic molecules such as ph-thalocyanines [36,37] or anthraquinone [38], protein microfibrils [39], and cationic surfactants grown from solution [40]. On muscovite mica, e.g., *p*-6P forms clusters as well as mutually parallel nanofibers from lying molecules, the fibers growing mainly by cluster aggregation. The first step is the formation of a wetting layer from lying molecules [41]. The wetting layer is crystalline, resulting in a clear low-energy electron diffraction (LEED) pattern [26]. On substrates such as KCl similar needles form, but without a wetting layer [42]. For another class of conjugated molecules, the α-thiophenes, α-quaterthiophene and α-sexithiophene, no wetting layer on muscovite has been detected by LEED yet [43].

Because of their fortunate optical and electrical properties phenylene/thiophene, cooligomers represent another class of interesting molecules [44–46]. Field-effect transistors [45], amplified spontaneous emission [47], and lasing [48,49] have, e.g., been demonstrated. Therefore, both from a fundamental point of view as well as from the application side, it is of interest to study the formation of nanofibers from thiophene/phenylene cooligomers on mica and on other dielectric surfaces such as KCl or NaCl. Choosing different sequences and numbers of thiophene and phenylene rings, more thiophene- or phenylene-like molecules can be tested, with different (e.g., zigzag or banana-like) shapes [50]. Reports have been published about the overall morphology of the cooligomers 5,5′-di-4-biphenyl-2,2′-bithiophene (PPTTPP) and 4,4′-di-2,2′-bithienyl-biphenyl (TTPPTT) [51] as well as 2,5-di-4-biphenyl-thiophene (PPTPP) thin films on muscovite and on phlogopite mica.

21.2 Growth via Organic Molecular Beam Deposition

One of the most prominent ways for the formation of organic nanofibers is their growth by OMBD on crystalline or rough substrates in high (1×10^{-7} mbar) or even in ultrahigh vacuum

FIGURE 21.1 Organic molecules used for producing nanofibers. From top to bottom, the molecules are *para*-hexaphenylene (*p*-6P); 4,4′-di-2,2′-bithienyl-biphenyl (TTPPTT); 5,5′-di-4-biphenyl-2,2′-bithiophene (PPTTPP); 2,5-di-4-biphenyl-thiophene (PPTPP); α-quaterthiophene (α-4T); and α-sexithiophene (α-6T).

(1 × 10^{-10} mbar). The proper cleaning of the substrates as well as of the organic molecules ensures well-defined chemical and morphological conditions during the deposition process for thermally and chemically stable molecules. Furthermore, many of the organic semiconducting molecules such as *para*-hexaphenylene possess only a poor solubility, making growth from solution impractical, or even impossible. The resulting growth morphology and the question if the molecules form organic nanowires at all, however, depends on details of the molecule–substrate interaction together with surface

energies of adsorbate, substrate, and interface, and on details of the deposition process such as the level of supersaturation [52,53] or on the substrate temperature [29]. On the micas, e.g., a large deposition rate and/or a small adsorbate/substrate interaction does not lead to fibers, but to islands from upright molecules.

In this section, we will exemplarily present growth by OMBD of needle-like aggregates from three different types of organic molecules: *para*-phenylenes, α-thiophenes, and thiophene/phenylene cooligomers, Figure 21.1. As substrates, the KCl (0 0 1) surface together with the two micas phlogopite and muscovite [54] are used. Their surface symmetries are different, resulting in different simultaneous growth directions. Typical fluorescence microscope images of grown samples from *para*-hexaphenylene on these three substrates are shown in Figure 21.2. All fibers fluoresce after normal incidence UV irradiation due to their buildup from laying molecules on the substrate surface. Mean lengths of the fibers vary between a few hundred nanometers and a few hundred micrometers, depending on the deposition conditions. The widths and heights are typically in the range of a few hundred and a few ten nanometers, respectively.

Differences in the mean length of the aggregates are due to different deposition conditions, but also due to details in the growth mechanism. Obviously, on KCl two simultaneous needle orientations evolve. On phlogopite mica, three needle orientations are present, whereas on muscovite only a single needle orientation evolves. Due to the quasi-singly crystalline nature of the fibers [24], the emitted light is polarized either along two (KCl), three (phlogopite), or one (muscovite) direction.

Another example—the deposition of the thiophene/phenylene PPTPP on the same substrates as in Figure 21.2—is presented in Figure 21.3. Very similar to *p*-6P on KCl two needle directions evolve, whereas on muscovite only a single direction is present. On phlogopite, three growth directions are realized, but fibers start bending after they reach a certain length. In this manner, rings, loops, and ridgets with a few micrometers diameter evolve.

On KCl, the growth directions are the ⟨110⟩ directions. This is a rather typical growth direction, in which fibers from many other molecules grow [44,55]. On phlogopite mica, the growth

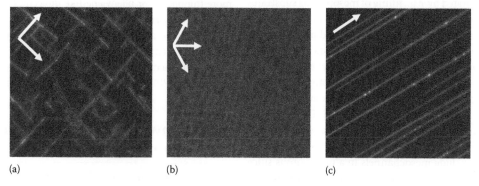

(a)   (b)   (c)

FIGURE 21.2   100 × 100 μm² fluorescence microscope images of *p*-6P on (a) KCl, (b) phlogopite mica, and (c) muscovite mica. White arrows mark the ⟨110⟩ directions for KCl, [100] and ⟨110⟩ for phlogopite, and a single ⟨110⟩ direction on muscovite.

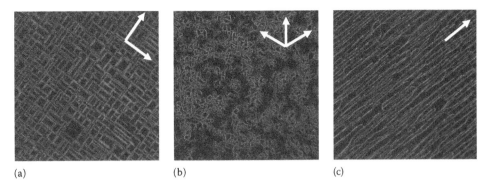

FIGURE 21.3 $150 \times 150\,\mu m^2$ fluorescence microscope images of PPTPP (a) KCl, (b) phlogopite mica, and (c) muscovite mica. Arrows mark the same directions, as in Figure 21.2.

directions are the three high-symmetry directions [100] and $\langle 110 \rangle$, and on muscovite it is one of the $\langle 110 \rangle$ directions. Especially on muscovite the $\langle 110 \rangle$ is not necessarily a typical growth direction as demonstrated in Figure 21.4. For example, for the thiophenes α-4T and α-6T roughly three simultaneous growth directions are realized, whereas for PPTTPP two directions are present with one of the $\langle 110 \rangle$ substrate directions now serving as the bisecting line of the two needle growth directions. Note that for all three molecules, only *two* polarization directions of the fluorescence light are observed.

The reason for the rather different morphologies on muscovite lies in the subtle interplay between epitaxial growth of the molecules, their bulk packing within the fibers (which might even be dictated by the substrate [56]), and energetically favorable molecule orientations on the substrate due to other-than-epitaxial interactions. The muscovite mica (001) plane possesses an uniaxial symmetry, with a single mirror plane along one of the $\langle 110 \rangle$ directions. Along this direction, grooves exist due to the stacking of the underlying crystal sheets [57]. Both the α-thiophenes as well as PPTTPP have their long molecular axes on the surface oriented along one out of two of the substrate high-symmetry directions, which is not the direction of the mirror axis. From this, the needle directions are determined by the packing of the molecules within the fibers. For PPTTPP, the angle between the molecules long axis and the fiber long axis is close to 90°, while for the thiophenes, it is close to 70°. This way, either roughly three or two needle directions are realized. The *para*-phenylene *p*-6P as well as PPTPP grow with the short unit cell axis along the special $\langle 110 \rangle$ directions, which then results only in a single needle direction.

For the micas, needle formation is not necessarily the initial growth stage. A number of AFM images (JPK NanoWizard in intermittent contact mode) for different growth stages of *p*-6P needles on phlogopite mica are shown in Figure 21.5. The first step is the formation of clusters, not of fibers. These clusters are also made of lying molecules, emitting light after normal incidence UV irradiation. Only after a critical coverage is reached, the clusters assemble into fibers, which then continue growing. Note that needle formation can also be induced from such a cluster film by moving the fibers by an AFM tip. The fiber formation also shows a sensitive substrate temperature dependence. At room temperature, only a closed, packed film of very short fibers forms. At elevated temperatures of about 450 K, the fibers are rather long and isolated, but again, only clusters grow at larger temperatures. The other two morphological parameters, that is, the mean width and height, only change marginally with deposition parameters. The width is increased from about 100 to 600 nm, whereas the mean height almost stays constant, because

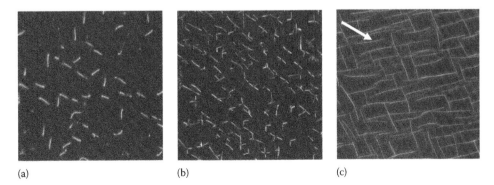

FIGURE 21.4 $100 \times 100\,\mu m^2$ fluorescence microscope images of (a) α-4T, (b) α-6T, and of (c) PPTTPP on muscovite mica. Whereas α-4T and α-6T grow roughly along the three muscovite high-symmetry directions, one of the muscovite $\langle 110 \rangle$ directions (white arrow) for PPTTPP serves as the bisecting line of the two needle orientations.

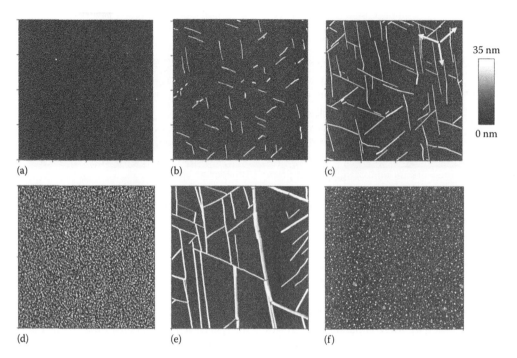

FIGURE 21.5 AFM images of *p*-6P growth on phlogopite mica as a function of the nominal thickness (upper panel) and of the deposition temperature (lower panel). Nominal thicknesses are 0.25, 0.5, and 1 nm at 440 K for the $20 \times 20\,\mu m^2$ images in (a), (b), and (c). For the $10 \times 10\,\mu m^2$ AFM images (d), (e), and (f), the deposition temperatures have been 313, 458, and 503 K, respectively. The arrows in (c) denote the three phlogopite high-symmetry directions.

of the growth by cluster agglomeration. This way, varying the deposition conditions allows an easy control over the fiber dimensions.

The functionalization of the building blocks is a means to tune the fiber morphology, their linear and nonlinear optical properties, their chemical reactivity, electronic properties, etc. An overview over symmetrically functionalized, nonsymmetrically functionalized, and monofunctionalized *para*-quaterphenylenes in *para* positions are shown in Figure 21.6 [58,59]. Details of the growth vary between the molecules, but almost all of them emit blue light after UV excitation and form mutually parallel nanofibers on muscovite.

Figure 21.7 demonstrates the variety in morphology, which can be achieved using different molecules and different growth conditions. The needle in Figure 21.7a from a di-chloro functionalized *para*-quaterphenylene (CLP4) is 700 nm long (rather short), only 25 nm tall, and 115 nm wide. The cross section is triangular. The fiber from α-6T in Figure 21.7b on the opposite is 16 μm long and 250 nm tall, with a width of 330 nm. In Figure 21.7c for a 3.5 μm long and 45 nm tall PPTPP fiber, single clusters as building blocks are still visible.

21.3 Growth on Microstructured Templates

In most cases, fabrication of nanowires and their device integration are two separate processes, and that imposes strong constraints on the usefulness of the nanoaggregates. The most

obvious way of integrating fragile nanofibers into device platforms is to grow them on purpose directly at the place where they are supposed to act as new optoelectronic components. One way to achieve that goal is the selected deposition of growth catalysts, from which the nanowires would grow. This works sufficiently well in the case of carbon nanotubes. However, oriented growth of nanofibers from arbitrary organic molecules cannot be tailored in this manner.

Since gold (Au) is an interesting surface for devices (bottom contact), let us begin with the direct growth of nanofibers on Au. Nanofibers from *para*-hexaphenylene molecules have been successfully grown on Au surfaces, but they did not show the strict alignment that has been observed on muscovite mica as the growth template. Instead, on Au (1 1 1), for example, the nanofibers have preferred growth directions along one of the three crystalline high-symmetry directions for deposition at elevated temperatures [33]. The reason for the oriented growth of organic nanofibers on muscovite mica is a combination of epitaxy and electric field-induced alignment of the initially deposited organic molecules. Such a combination is *solely* possible on a specific, single-crystalline growth substrate.

Thus, until now, the mostly implemented way to generate oriented organic nanofibers on substrates that are interesting for device applications is via growth on a template and subsequent soft transfer [60,61]; see the below section. Such methodology has obvious drawbacks such as the fact that remnants from the transfer liquid perturb the functionality of the nanofibers on the surface; that the parallelism of the nanofibers over large areas is usually not retained; and that it is

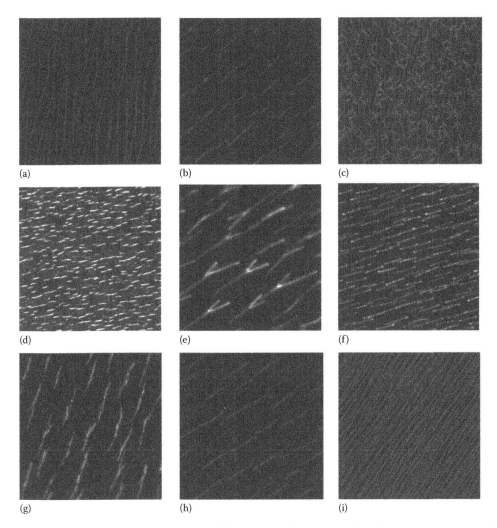

FIGURE 21.6 Fluorescence microscope images, 85 × 85 μm², of functionalized *para*-quaterphenylenes on muscovite mica. In the upper panel fibers from symmetrically di-functionalized *p*-4P by O—CH₃ (a), Cl (b), and CN (c) are shown, in the middle panel fibers from nonsymmetrically functionalized ones by O—CH₃ on one side, by NH₂ (d) and Cl (e), (f) on the other side. In the lower panel fibers from monofunctionalized *p*-4P are displayed, functionalized by O—CH₃ (g), Cl (h), and CN (i). (From Schiek, M., *Tomorrow's Chemistry Today: Concepts in Nanoscience, Organic Materials and Environmental Chemistry*, Pignatoro, B. (ed.), Wiley-VCH, Weinheim, Germany 2008.)

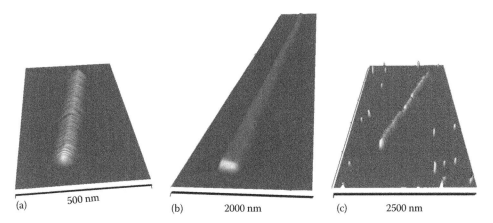

FIGURE 21.7 AFM images of organic nanofibers from (a) a di-chloro functionalized quater-phenylene, (b) from α-sexithiophene, and (c) from PPTPP. Heights, widths, and lengths vary between 25 and 250 nm, 100 and 350 nm, and 700 nm and 16 μm, respectively.

extremely difficult to place them specifically on micrometer-sized regions defined by the device substrate. In order to meet the demands needed for devices, another method that relies on a combination between top-down fabrication of microstructured substrates and bottom-up growth of the organic nanofibers has been developed recently and is discussed here [62]. We note that the other possibilities for specific transfer such as scatter onto structured surfaces [16] or the alignment by electric fields between electrodes [63] might be useful for other classes of nanowires, but are in general not compatible with the specific material properties of organic nanowires as described in this chapter.

As template, microstructured silicon is most conveniently applied. Silicon is structured by optical lithography and subsequent reactive ion etching. Afterward, a thin Au layer (55 nm) is deposited on the substrate. The resulting microstructured Au surface is then used as a template for growth of *p*-6P nanofibers. A nominal thickness of 5 nm *p*-6P is deposited at different substrate temperatures. Figure 21.8 shows a fluorescence microscopy image (Figure 21.8a) and a scanning electron microscopy image (Figure 21.8b) of *p*-6P nanofibers grown on a 5 μm wide Au-coated ridge at a substrate temperature of 388 K.

As observed from the figure, the nanofibers are grown both on top of the Au-coated ridge and on the bottom of the substrate, since the whole template is coated with gold. Although this allows us to grow the nanofibers directly on prefabricated microstructures, there is at this substrate temperature no preferred growth direction of the nanofibers. However, when increasing the substrate temperature during growth, it is possible to grow the nanofibers almost perpendicular to the long axis of the Au-coated ridges. This is demonstrated in Figure 21.9, which shows a fluorescence microscopy image (Figure 21.9a) and a scanning electron microscopy image (Figure 21.9b) of *p*-6P nanofibers grown on a 5 μm wide Au-coated ridge at a substrate temperature of 435 K. Note also that the nanofiber length is increased with the increasing temperature as it is the case for nanofibers grown on mica [64].

One reason for this oriented growth on the microstructured Au ridges is temperature gradients near the edges of the microstructures that guide the nanofiber growth in preferred

(a)   (b)

FIGURE 21.9   (a) Fluorescence microscopy image and (b) tilted scanning electron microscopy image of *p*-6P nanofibers grown on a 5 μm wide Au-coated ridge at a substrate temperature of 435 K.

directions. We have investigated this effect for several different substrate temperatures during growth, and it is seen that there is a strong correlation between substrate temperature and orientation of the nanofibers. At low substrate temperatures the nanofibers grow almost randomly on the Au-coated ridges (Figure 21.8), whereas nanofibers grown at high substrate temperatures result in growth nearly perpendicular to the long axis of the ridges (Figure 21.9). Since growth perpendicular to the ridge starts from the edges, the clusters that the nanofibers assemble from have to diffuse to an edge. At low substrate temperatures the diffusion length of the clusters is smaller than at high temperatures, and growth in the center part of the ridges can take place. At high substrate temperatures the clusters can diffuse to an edge and thus perpendicular growth takes place. Therefore, it is possible by controlling the substrate temperature during growth to achieve oriented organic nanofibers directly on prefabricated microstructures.

21.4 Embedding and Integration

After growth on structured Au substrates (as detailed in Section 21.3), the fibers might be further contacted with a top electrode (see Section 21.6) or coated by a dielectric layer that protects them against oxygen and thus reduces problems arising from optical bleaching [65].

In general, growth on structured surfaces is still restricted by the fact that an ultrathin Au film on the structured surface is needed. On silver or palladium, for example, oriented nanofibers cannot be grown. Therefore, there is still a need for transferring nanofibers individually or in arrays from the growth substrate to an arbitrary target substrate. Individual fibers can conveniently be transferred into a liquid and then drop casted or transferred into thin hollow fibers by applying capillary forces. Figure 21.10 shows an example.

Alternatively, arrays of oriented nanofibers can be stamped using a delicate mixture of pressure and humidity. Figure 21.11 shows such an array of nanofibers, stamped from the single crystalline growth substrate to a glass plate. The transfer process does not change the photonic properties, here demonstrated via a conserved strong dichroism of the emitted blue light.

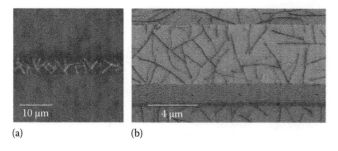

(a)   (b)

FIGURE 21.8   (a) Fluorescence microscopy image and (b) tilted scanning electron microscopy image of *p*-6P nanofibers grown on a 5 μm wide Au-coated ridge at a substrate temperature of 388 K.

(a)

(b)

FIGURE 21.10 (a) *p*-6P nanofibers inside a hollow fiber, excited by UV light. The nanofibers are aligned along the long axis of the hollow fiber. (b) *p*-6P nanofibers inside a glass micropipette. No alignment is visible.

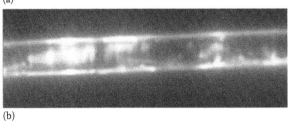

FIGURE 21.11 Array of blue-light-emitting nanofibers, stamped from the original growth substrate to a glass substrate. The image on the left-hand side has been taken with a polarization analyzer set parallel to the dipole emitter axis, the image on the right-hand side with the polarizer set perpendicular to that axis. The large dichroic ratio proofs that the transfer process does not damage the optical properties of the nanofibers.

21.5 Linear and Nonlinear Optical Properties

An understanding of the optical properties of nanofibers is of importance for optical applications which include, e.g., super-radiance [66], lasing, or electroluminescence [67,68].

21.5.1 Temperature-Dependent Spectroscopy

In the case of the phenylenes, thiophenes, or phenylene/thiophene cooligomers, the UV excitation-induced fluorescence spectra are dominated by several excitonic transitions between the electronic ground state, S_0, and the first excited singlet state, S_1, [69], that is, display vibronic progression series. For example, for *p*-6P, vibronic peaks from the $S_1 \rightarrow S_0$ transition with energetic differences of 160 ± 10 meV are clearly resolved, the energetic distance between the maximum in absorption and the first maximum in emission being approximately 0.7 eV. Samples of nanofibers emit at 3.09 eV (00), 2.94 eV (01), 2.77 eV (02), and 2.62 eV (03). These are transitions between the vibrational ground state of the S_1 electronic state and the different vibrational states of the electronic ground state [69,71].

A detailed investigation as a function of surface temperature from 300 to 30 K reveals three classes of spectra for slightly different molecule–substrate systems: (a) well-resolved excitonic peaks, which shift to the blue up to 35 meV with decreasing temperature, (b) a similar spectrum with an additional intermediate broadening around 150 K, and (c) an excitonic spectrum similar to (b), but with an additional green defect emission band. Quantitative fitting of type (a) results in an exciton–phonon coupling factor of 80 ± 10 meV and an average phonon temperature of $\Theta = 670 \pm 70$ K. The Huang–Rhys factor decreases linearly from 1.2 to 1.0 with decreasing temperature. Fitting of type (b) spectra reveals that the apparent intermediate temperature broadening is due to the additional fluorescence peaks, the relative importance of which increases monotonically with decreasing temperature.

In Figure 21.12, contour plots of the three types of spectra as a function of temperature are shown [70]. With decreasing temperature the overall intensity for all types of fluorescence spectra

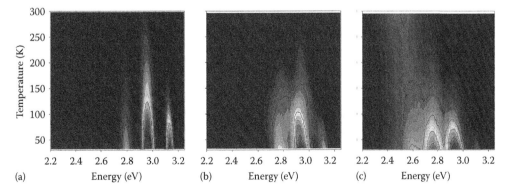

FIGURE 21.12 Contour plots of the three types of temperature-dependent fluorescence spectra. The overall fluorescence intensity increases with decreasing temperature, whereas the green emission band at 2.6 eV from (c) shows only a very weak temperature dependence. (From Balzer, F. et al., *J. Lumin.*, 129, 784, 2009.)

increases strongly, that is, an increase in the fluorescence yield is observed. However, the intensity of the green emission band from series (c) decreases slightly in intensity with decreasing temperature. Such a behavior has been observed before and hints, together with different decay constants from time-resolved measurements, to a different origin as compared to the regular excitonic peaks such as aggregate states [72–74]. Contamination of the material by, e.g., other organic molecules might be another possible reason.

Besides investigations of arrays of nanofibers, also individual nanofibers have been spectroscopically investigated. Extended spectroscopic measurements along a single nanofiber have shown that the spectra depend on the morphology of the aggregates, and that they become significantly narrower if the nanofiber width decreases, e.g., at the tip of the nanofiber [75].

As a general result of the spectroscopic investigations, it is observed that the optical emission of nanofibers delicately depends on the growth conditions—even though the orientation and packing of the individual molecular emitters in the nanoscaled crystals is rather well defined. This—among with other findings—points to the importance of well-defined growth conditions especially for nanoaggregates that are supposed to serve in future devices.

21.5.2 Determination of Molecular Orientations

As noted above, due to the quasi-singly crystalline nature of the fibers [24], the emitted light of arrays of p-6P nanofibers is polarized either along two (KCl), three (phlogopite), or one (muscovite) directions. For a single fiber the polarization vector of the emitted light is directed under a well-defined angle with respect to the characteristic axes of the nanofibers such as their long axes. Since in many cases the optical transition dipole moment of the molecules is along the long molecular axes, local measurements of the polarization of the emitted light will provide direct information about the local orientation of the molecules that build the nanofibers. The resolution of this method in a far-field optical approach is limited by the focus diameter of the excitation light, that is, of the order of $\lambda/2 \approx 200\,\text{nm}$. If one detects the light polarization in the near field by the use of, e.g., a SNOM (scanning near-field optical microscope), the resolution is eventually given by the effective diameter of the SNOM fiber tip. That could in principle be less than 50 nm. In experimental praxis, 400 nm is a more realistic value.

An experimental implementation of the method uses two two-dimensional scanning measurements of emitted light intensity, I_{xy}, behind a polarization analyzer set to a specific angle, and two cross-polarized excitation geometries. Here, the subscripts x and y stand for polarization orientations parallel to the transition dipole moment, p, and perpendicular to it, s. One finds, for example, that

$$\frac{I_{sp}}{I_{pp}} = \tan^4\theta, \qquad (21.1)$$

where θ is the molecular orientation angle with respect to the chosen coordinate system of the nanofiber on the surface. Besides the solution $+\theta$, this equation allows also the solution $-\theta$. One of the solutions can be removed by either rotating the analyzer (and obtaining a continuous polarization dependence) or by setting the sample at two different, fixed angles with respect to the incoming and outgoing electric field vectors.

Some recent polarized second-harmonic SNOM measurements [76,77] have shown that for *para*-hexaphenylene nanofibers on mica, the orientation angle, θ, is fairly constant along the fiber with small-scale deviations of less than 10°. These deviations are attributed to inhomogeneities of individual fibers as being confirmed by polarized linear fluorescence measurements.

Since at present, crystallographic measurements with local molecular resolution are nearly impossible for single nanofibers, the discussed optical measurements represent still the only direct way of obtaining information about local molecular order in nanofibers.

21.6 Devices

21.6.1 On the Way to Nanofiber Light Sources

Thin films from *para*-hexaphenylene molecules have successfully been used as the light-emitting layers in blue-light-emitting devices (LEDs) [78]. It must therefore be expected that a nanoscale LED could be made using p-6P nanofibers as the light emitters.

Electroluminescence in organic semiconductors originates from the radiative decay of an exciton, which is formed by the binding of an electron and a hole polaron. This process requires injection of both charge species from the cathode and anode, respectively, and their transport to the recombination zone by an applied electric field [79]. Optimizing the light generation process therefore involves balancing the electron and the hole current, which requires a detailed study of the charge injection and transport properties.

In order to investigate this, a technique for electric contacting of the organic nanofibers was devised recently [80]: Nanofibers are transferred to a prefabricated silicon dioxide platform on a silicon chip either by the spreading of a small volume of water with dispersed nanofibers or by a stamping technique. A suitable nanofiber is located and a rigid silicon wire with a diameter of a few hundred nanometers is positioned on top of and perpendicular to the organic nanofiber. This silicon wire acts as a local shadow mask during the subsequent deposition of electrode material (typically gold) by electron beam evaporation. Upon removal of the silicon wire shadow mask, two electrodes are formed with a gap in between, the size of which is determined by the shadow mask wire diameter. This electrode gap is spanned only by the organic nanofiber, which can then be probed electrically.

Figure 21.13a shows a scanning electron microscope image of a p-6P nanofiber with two gold contacts. Two-point measurements were performed by applying a DC voltage and recording

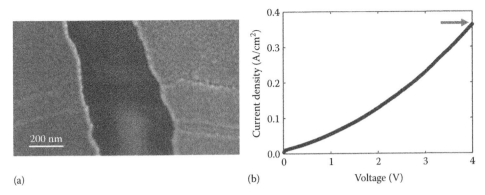

(a)　　　　　　　　　　　　(b)

FIGURE 21.13 (a) SEM image of a *p*-6P nanofiber with two gold electrodes. (b) Typical current density vs. voltage characteristic for a gold contacted *p*-6P nanofiber.

the resulting current. Figure 21.13b shows typical results. Here, the measured current has been scaled with the nanofiber cross-sectional dimensions to provide the current density.

In order to investigate this curve in more detail, the measured electrical characteristics were analyzed with the Mott–Guerney theory. This approach assumes that the current is bulk-limited and that the interface effects are not important. However, in a real sample such effects will be present. This theory can therefore only provide an intrinsic "theoretical" upper limit to the current flow. Contact effects, defects, traps, or any other influencing factors will always cause the current to be smaller than the prediction [81]. However, if the permittivity, $\varepsilon_r\varepsilon_0$, and device length, L, are known, an estimate of the *minimum* mobility, μ_{min}, can be extracted from the current density, J, vs. voltage, V, characteristics by

$$\mu_{min} = \frac{8JL^3}{9\varepsilon_r\varepsilon_0 V^2}. \tag{21.2}$$

Similar to the method used by de Boer et al. [81], the current measured at the maximum bias voltage was used in the calculation of μ_{min} as indicated with an arrow in Figure 21.13b. Several nanofiber samples were investigated [82], which showed values with a significant spread over four orders of magnitude between 3×10^{-5} and 3×10^{-1} cm²/V s. This indicates that for the majority of the investigated samples interface effects play a significant role. The conclusion to be drawn from these results is therefore that the carrier mobility for *para*-hexaphenylene nanofibers is at least 3×10^{-1} cm²/V s. In addition, it was found that gold was a suitable electrode material for hole injection [82]. The next step toward realizing a nanofiber LED is then to locate a suitable electron injector material and to develop an efficient method capable of depositing *two different* electrode materials for the anode and the cathode. The shadow mask method described above can be extended to facilitate the deposition of different anode and cathode material through the use of two shadow masks and two metal deposition steps [80].

Figure 21.14 shows an example of a nanofiber contacted with two different contact materials: gold and titanium. Even though gold injects holes, titanium is apparently not an efficient

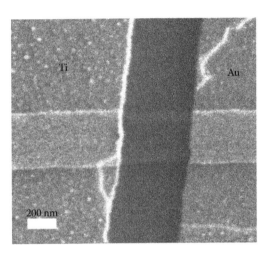

FIGURE 21.14 SEM image of a *p*-6P nanofiber contacted with Au and Ti for the anode and cathode, respectively.

electron-injecting material since no electroluminescence was observed from such devices. Thus, a more efficient electron injecting material is required.

It is concluded that it is well possible to contact selected organic nanofibers with electrodes made from different materials. The current through the fibers is, however, limited by the injection through the contacts. If a too high voltage is applied, the breakdown threshold will be reached and the material will desorb. Thus, in contrast to thin films made of the same material (where electroluminescence has easily been achieved) the very fact that the material is nanoscaled makes it necessary to develop other integration and contacting methods. In fact, as detailed above, a direct, sophisticated growth of nanoscaled material under appropriately clean conditions might result in a significantly lower threshold for electroluminescence.

21.6.2 Waveguides

Nanofibers act as waveguides if the index of refraction of the underlying substrate or surrounding medium is smaller than their index of refraction $\left(\sqrt{\varepsilon_{iso}} = 1.7\right)$ and if a critical width is overcome. Note that the cutoff wavelength for the guided modes in the nanofibers is [28]

$$\lambda_c = \frac{2\sqrt{\varepsilon_\perp}\, a}{m} \qquad (21.3)$$

and the number of possible modes, $m = 1, 2, 3\dots$, is restricted by the condition

$$m < \frac{2a}{\lambda}\sqrt{\frac{\varepsilon_\perp}{\varepsilon_\parallel}}\sqrt{\varepsilon_\perp - \varepsilon_s}, \qquad (21.4)$$

with λ the wavelength, ε_s the dielectric constant of the substrate, $\varepsilon_\parallel$ and $\varepsilon_\perp$ the components of the dielectric permittivity tensors, and a the fiber width. Hence, waveguiding occurs roughly for $a \approx \lambda/2$.

In the case of negligible scattering from surface irregularities, low absorption in the fiber, and if one assumes negligible damping of the evanescent field in the substrate, propagation losses in the fiber are very small. However, it is difficult to couple light to a nanofiber. Thus, the experiments performed so far have relied on the fact that the light to be guided can be generated inside the fiber via UV excitation. It then propagates along the fiber and is scattered at breaks, where it can be detected. Such a method, of course, suffers from reabsorption of the light that is generated in the nanofiber.

As a result, the complex dielectric functions of nanofibers have been determined and a very good agreement with simple electromagnetic theory has been found [28,83]. Thus nanofibers can be treated and used as dielectric waveguides or connecting elements in opto-chips. Since the dielectric function of the nanofibers can be changed by changing the molecular basis units, future applications might even involve active waveguides, which alter the light that is propagating along the fibers. This might prove especially useful in the context of combined photonic and plasmonic circuits.

21.6.3 Nanofiber Frequency Doublers

Recent developments of blue-light-based photonic devices such as fully integrated opto-chips have renewed strong interest in the development of small and efficient units for light conversion. A nanoscaled, integrated frequency doubler would allow one to use cheap, power-saving, and brilliant near-infrared laser sources as the main source for intense, coherent blue light sources. Such a frequency-doubling unit could rely on easily integrable nanofibers.

Optimization of frequency-doubling efficiency asks for optimized hyperpolarizabilities of the involved molecular building blocks. *para*-Hexaphenylene molecules can be easily grown into nearly single crystalline nanoaggregates, and thus resemble the optimum configuration for the given molecular constitutes. They also show the optimum optoelectronic properties, just limited by the molecular building blocks themselves. However, their hyperpolarizabilities are very low due to a lack of donor

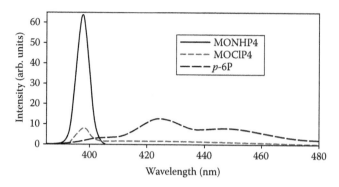

FIGURE 21.15 Optical two-photon spectra of nanofibers made from different molecules as a result of near-infrared (790 nm) femtosecond laser excitation. The nonsymmetrically functionalized nanofibers result in strong second-harmonic generation.

and acceptor groups. A way out of this dilemma bases on the nonsymmetrical chemical functionalization of a *para*-quaterphenylene (*p*-4P) block with electron push and pull groups, for example, methoxy, amino, or cyano groups. As shown above, these molecules also form oriented nanofibers.

The generation of a strong second-harmonic signal intensity asks for an in-phase adding of the second-harmonic signals of the individual molecular emitters. So the phases of the fundamental and the second-harmonic light have to be matched within the nonlinear crystal (phase-matching condition). Fortunately, the use of doubling elements with dimensions significantly smaller than the wavelength of the doubled light makes such considerations of phase matching obsolete. However, aggregation of the molecules into nano-elements still imposes restrictions on doubling efficiency related to the aggregation state (single-crystalline, poly-crystalline, amorphous).

In Figure 21.15, we demonstrate how the optical emission spectrum upon infrared excitation changes as one uses differently functionalized molecules. For *p*-6P, the spectrum consists solely of two-photon luminescence (TPL), while it consists solely of second-harmonic generated (SHG) light in the case of methoxy- and amino-functionalized *p*-4P (MONHP4). If one uses chlorine instead of the amino groups (MOCLP4), one obtains both SHG and TPL. Obviously, molecular tailoring provides one with high optoelectronic flexibility.

21.7 Conclusions

Besides the above-mentioned device applications (lighting, waveguiding, and frequency doubling), there are many other areas where nanofibers might prove useful as nanoscaled integrated elements. For example, nanosensing is an obvious choice since the molecules that make up the nanofibers can easily be made sensitive to specific binding agents. Upon binding, the waveguiding and fluorescence properties of the nanofibers might be changed, which can readily be read out once the fibers are integrated into simple circuits. Obstacles here are low renewability of the sensing material and possibly insignificant surface-to-volume ratio; that is, the optical properties might be dominated by the volume rather than the surface of the nanofiber, which reduces the sensitivity and points to the use of very small nanofibers.

In terms of photonic applications, nonlinear optical response and lasing [5,84] seem to be the most interesting. Both take advantage of the fact that nanofibers are made of active photonic material and do not simply resemble a passive dielectric slab. Optical gain and nonlinear response can be tailored by modifying the molecular units [10], which opens up a broad perspective for future application areas.

Acknowledgments

We thank the Danish research foundations FNU and FTP as well as the Danish national advanced technology trust HTF for financial support.

References

1. Schiek, M., Balzer, F., Al-Shamery, K., Brewer, J., Lützen, A., and Rubahn, H.-G. *Small* 4, 176–181 (2008).
2. Schiek, M., Balzer, F., Al-Shamery, K., Lützen, A., and Rubahn, H.-G. *Soft Matter* 4, 277–285 (2008).
3. Al-Shamery, K., Rubahn, H.-G., and Sitter, H., editors. *Organic Nanostructures for Next Generation Devices*, volume 101 of *Springer Series in Materials Science*, Springer, Berlin, Germany (2008).
4. Zhao, Y., Xu, J., Peng, A., Fu, H., Ma, Y., Jiang, L., and Yao, J. *Angew. Chem. Int. Ed.* 47, 7301–7305 (2008).
5. Quochi, F., Cordella, F., Mura, A., Bongiovanni, G., Balzer, F., and Rubahn, H.-G. *J. Phys. Chem. B* 109, 21690–21693 (2005).
6. Kjelstrup-Hansen, J., Henrichsen, H., Bøgild, P., and Rubahn, H.-G. *Thin Solid Films* 515, 827–830 (2006).
7. Briseno, A., Mannsfeld, S., Reese, C., Hancock, J., Xiong, Y., Jenekhe, S., Bao, Z., and Xia, Y. *Nano Lett.* 7, 2847–2853 (2007).
8. Kjelstrup-Hansen, J., Hansen, O., Rubahn, H.-G., and Bøggild, P. *Small* 2, 660–666 (2006).
9. Brewer, J., Schiek, M., Lützen, A., Al-Shamery, K., and Rubahn, H.-G. *Nano Lett.* 6, 2656–2659 (2006).
10. Brewer, J., Schiek, M., Wallmann, I., and Rubahn, H.-G. *Opt. Commun.* 281, 3892–3896 (2008).
11. Quochi, F., Saba, M., Cordella, F., Gocalinska, A., Corpino, R., Marceddu, M., Anedda, A., Andreev, A., Sitter, H., Sriciftci, N., Mura, A., and Bongiovanni, G. *Adv. Mater.* 20, 3017–3021 (2008).
12. Balzer, F. In *Organic Nanofibers for Next Generation Devices*, Al-Shamery, K., Rubahn, H.-G., and Sitter, H., editors, volume 101 of *Springer Series in Materials Science*, Chapter 3, pp. 31–65. Springer, Berlin, Germany (2008).
13. Sitter, H. In *Organic Nanostructures for Next Generation Devices*, Al-Shamery, K., Rubahn, H.-G., and Sitter, H., editors, volume 101 of *Springer Series in Materials Science*, Chapter 5, pp. 89–117. Springer, Berlin, Germany (2008).
14. Mascaro, D., Thompson, M., Smith, H., and Bulovic, V. *Org. Electron.* 6, 211–220 (2005).
15. Wang, Z., Ho, K., Medforth, C., and Shelnutt, J. *Adv. Mater.* 18, 2557–2560 (2006).
16. Briseno, A., Mannsfeld, S., Lu, X., Xiong, Y., Jenekhe, S., Bao, Z., and Xia, Y. *Nano Lett.* 7, 668–675 (2007).
17. Kim, D., Lee, D., Lee, H., Lee, W., Kim, Y., Han, J., and Cho, K. *Adv. Mater.* 19, 678–682 (2007).
18. Mille, M., Lamere, J.-F., Rodrigues, F., and Fery-Forgues, S. *Langmuir* 24, 2671–2679 (2008).
19. O'Carroll, D., Lieberwirth, I., and Redmond, G. *Small* 3, 1178–1183 (2007).
20. Moynihan, S., Iacopino, D., O'Carroll, D., Lovera, P., and Redmond, G. *Chem. Mater.* 20, 996–1003 (2008).
21. Zhao, Y., Xiao, D., Yang, W., Peng, A., and Yao, J. *Chem. Mater.* 18, 2302–2306 (2006).
22. Chung, J., An, B.-K., Kim, J., Kim, J.-J., and Park, S. *Chem. Commun.*, 2998–3000 (2008).
23. Kankate, L., Balzer, F., Niehus, H., and Rubahn, H.-G. *J. Chem. Phys.* 128, 084709 (2008).
24. Resel, R. *J. Phys.: Condens. Matter* 20, 184009 (2008).
25. Andreev, A., Matt, G., Brabec, C., Sitter, H., Badt, D., Seyringer, H., and Sariciftci, N. *Adv. Mater.* 12, 629–633 (2000).
26. Balzer, F. and Rubahn, H.-G. *Appl. Phys. Lett.* 79, 3860–3862 (2001).
27. Yanagi, H. and Morikawa, T. *Appl. Phys. Lett.* 75, 187–189 (1999).
28. Balzer, F., Bordo, V., Simonsen, A., and Rubahn, H.-G. *Phys. Rev. B* 67, 115408 (2003).
29. Balzer, F. and Rubahn, H.-G. *Surf. Sci.* 548, 170–182 (2004).
30. Koller, G., Berkebile, S., Krenn, J., Tzvetkov, G., Hlawacek, G., Lengyel, O., Netzer, F., Teichert, C., Resel, R., and Ramsey, M. *Adv. Mater.* 16, 2159–2162 (2004).
31. Balzer, F., Kankate, L., Niehus, H., Frese, R., Maibohm, C., and Rubahn, H.-G. *Nanotechnology* 17, 984–991 (2006).
32. Müllegger, S., Mitsche, S., Pölt, P., Hänel, K., Birkner, A., Wöll, C., and Winkler, A. *Thin Solid Films* 484, 408–414 (2005).
33. Müllegger, S., Hlawacek, G., Haber, T., Frank, P., Teichert, C., Resel, R., and Winkler, A. *Appl. Phys. A* 87, 103–111 (2007).
34. Erlacher, K., Resel, R., Hampel, S., Kuhlmann, T., Lischka, K., Müller, B., Thierry, A., Lotz, B., and Leising, G. *Surf. Sci.* 437, 191–197 (1999).
35. Suzuki, T., Kanno, Y., Morioka, Y., and Kuroda, K. *Chem. Commun.*, 3284–3286 (2008).
36. Uyeda, N., Ashida, M., and Suito, E. *J. Appl. Phys.* 36, 1453–1460 (1965).
37. Ashida, M. *Bull. Chem. Soc. Jpn.* 39, 2625–2631 (1966).
38. Kobzareva, S. and Distler, G. *J. Cryst. Growth* 10, 269–275 (1971).
39. Sun, M., Stetco, A., and Merschrod S. E. *Langmuir* 24, 5418–5421 (2008).

40. Hou, Y., Cao, M., Deng, M., and Wang, Y. *Langmuir* 24, 10572–10574 (2008).

41. Frank, P., Hlawacek, G., Lengyel, O., Satka, A., Teichert, C., Resel, R., and Winkler, A. *Surf. Sci.* 601, 2152–2160 (2007).

42. Frank, P., Hernandez-Sosa, G., Sitter, H., and Winkler, A. *Thin Solid Films* 516, 2939–2942 (2008).

43. Kankate, L., Balzer, F., Niehus, H., and Rubahn, H.-G. *Thin Solid Films* 518, 130–137 (2009).

44. Yanagi, H., Morikawa, T., Hotta, S., and Yase, K. *Adv. Mater.* 13, 313–317 (2001).

45. Ichikawa, M., Yanagi, H., Shimizu, Y., Hotta, S., Sugunuma, N., Koyama, T., and Taniguchi, Y. *Adv. Mater.* 14, 1272–1275 (2002).

46. Yamao, T., Taniguchi, Y., Yamamoto, K., Miki, T., Ohira, T., and Hotta, S. *Jap. J. Appl. Phys.* 47, 4719–4723 (2008).

47. Bando, K., Nakamura, T., Masumoto, Y., Sasaki, F., Kobayashi, S., and Hotta, S. *J. Appl. Phys.* 99, 013518 (2006).

48. Sasaki, F., Kobayashi, S., Haraichi, S., Fujiwara, S., Bando, K., Masumoto, Y., and Hotta, S. *Adv. Mater.* 19, 3653–3655 (2007).

49. Fujiwara, S., Bando, K., Masumoto, Y., Sasaki, F., Kobayashi, S., Haraichi, S., and Hotta, S. *Appl. Phys. Lett.* 91, 021104 (2007).

50. Hotta, S., Goto, M., Azumi, R., Inoue, M., Ichikawa, M., and Taniguchi, Y. *Chem. Mater.* 16, 237–41 (2004).

51. Balzer, F., Schiek, M., Al-Shamery, K., Lützen, A., and Rubahn, H.-G. *J. Vac. Sci. Technol. B* 26, 1619–1623 (2008).

52. Verlaak, S., Steudel, S., Heremans, P., Janssen, D., and Deleuze, M. *Phys. Rev. B* 68, 195409 (2003).

53. Campione, M., Sassella, A., Moret, M., Marcon, V., and Raos, G. *J. Phys. Chem. B* 109, 7859–7864 (2005).

54. Griffen, D. *Silicate Crystal Chemistry.* Oxford University Press, New York (1992).

55. Yamada, Y. and Yanagi, H. *Appl. Phys. Lett.* 76, 3406–3408 (2000).

56. Hoshino, A., Isoda, S., and Kobayashi, T. *J. Cryst. Growth* 115, 826–830 (1991).

57. Kuwahara, Y. *Phys. Chem. Miner.* 28, 1–8 (2001).

58. Schiek, M., Al-Shamery, K., and Lützen, A. *Synthesis* 4, 613–621 (2007).

59. Schiek, M., Light-emitting organic nanoaggregates from functionalized para-quaterphenylenes. In B. Pignatoro (ed.), *Tomorrow's Chemistry Today: Concepts in Nanoscience, Organic Materials and Environmental Chemistry*, Wiley-VCH, Weinheim, Germany (2008).

60. Brewer, J., Henrichsen, H., Balzer, F., Bagatolli, L., Simonsen, A., and Rubahn, H.-G. *Proc. SPIE* 5931, 250–257 (2005).

61. Javey, A., Nam, S., Friedman, R., Yan, H., and Lieber, C. *Nano Lett.* 7, 773–777 (2007).

62. Madsen, M., Kjelstrup-Hansen, J., and Rubahn, H.-G. *Nanotechnology* 20, 115601 (2009).

63. Talapin, D., Black, C., Kagan, C., Shevchenko, E., Afzali, A., and Murray, C. *J. Phys. Chem. C* 111, 13244–13249 (2007).

64. Balzer, F. and Rubahn, H.-G. *Adv. Funct. Mater.* 15, 17–24 (2005).

65. Maibohm, C., Brewer, J., Sturm, H., Balzer, F., and Rubahn, H.-G. *J. Appl. Phys.* 100, 054304 (2006).

66. Zhao, Z. and Spano, F. *J. Phys. Chem. C* 111, 6113–6123 (2007).

67. Stampfl, J., Tasch, S., Leising, G., and Scherf, U. *Synth. Met.* 71, 2125–2128 (1995).

68. Meghdadi, F., Tasch, S., Winkler, B., Fischer, W., Stelzer, F., and Leising, G. *Synth. Met.* 85, 1441–1442 (1997).

69. Guha, S. and Chandrasekhar, M. *Phys. Stat. Solidi (B)* 241, 3318–3327 (2004).

70. Balzer, F., Pogantsch, A., and Rubahn, H.-G. *J. Lumin.* 129, 784 (2009).

71. Graupner, W., Meghdadi, F., Leising, G., Lanzani, G., Nisoli, M., De Silvestri, S., Fischer, W., and Stelzer, F. *Phys. Rev. B* 56, 10128–10132 (1997).

72. Kadashchuk, A., Andreev, A., Sitter, H., and Sariciftci, N. *Synth. Met.* 139, 937–940 (2003).

73. Kadashchuk, A., Andreev, A., Sitter, H., Sariciftci, N., Skryshevski, Y., Piryatinski, Y., Blonsky, I., and Meissner, D. *Adv. Funct. Mater.* 14, 970–978 (2004).

74. Faulques, E., Wéry, J., Lefrant, S., Ivanov, V., and Jonussauskas, G. *Phys. Rev. B* 65, 212202 (2002).

75. Simonsen, A. and Rubahn, H.-G. *Nano Lett.* 2, 1379–1382 (2002).

76. Beermann, J., Bozhevolnyi, S., Bordo, V., and Rubahn, H.-G. *Opt. Commun.* 237, 423–429 (2004).

77. Beermann, J., Marquart, C., and Bozhevolnyi, S. *Laser Phys. Lett.* 1, 264–268 (2004).

78. Klemenc, M., Meghdadi, F., Voss, S., and Leising, G. *Synth. Met.* 85, 1243–1244 (1997).

79. Parker, I. *J. Appl. Phys.* 75, 1656–1666 (1994).

80. Kjelstrup-Hansen, J., Dohn, S., Madsen, D. N., Mølhave, K., and Bøggild, P. *J. Nanosci. Nanotechnol.* 6, 1995–1999 (2006).

81. de Boer, R., Jochemsen, M., Klapwijk, T., Morpurgo, A., Niemax, J., Tripathi, A., and Pflaum, J. *J. Appl. Phys.* 95, 1196 (2004).

82. Henrichsen, H., Kjelstrup-Hansen, J., Engstrøm, D., Clausen, C., Bøggild, P., and Rubahn, H.-G. *Org. Electron.* 8, 540–544 (2007).

83. Volkov, V., Bozhevolnyi, S., Bordo, V., and Rubahn, H.-G. *J. Microscopy* 215, 241–244 (2004).

84. Quochi, F., Cordella, F., Mura, A., Bongiovanni, G., Balzer, F., and Rubahn, H.-G. *Appl. Phys. Lett.* 88, 041106 (2006).

IV

Nanowire Arrays

22

Magnetic Nanowire Arrays

Adekunle O. Adeyeye
National University of Singapore

Sarjoosing Goolaup
National University of Singapore

22.1 Introduction

Magnetic nanostructured materials are of scientific interest both from a fundamental point of view and because of their potential in a wide range of emerging applications. Nanomagnets, by virtue of their extremely small size, possess both static and dynamic properties that are quantitatively and qualitatively very different from their parent bulk material. The magnetization reversal mechanism can therefore be drastically modified in nanomagnets confined to sizes that preclude the formation of domain walls. Topical reviews on the magnetic properties of nanostructures can be found in articles by Bader (2006), Srajer et al. (2006), Bader et al. (2007), Adeyeye et al. (2008), and Adeyeye and Singh (2008).

Magnetic nanostructures are the basic building blocks of various spintronic applications. In data storage for example, as the recording media rapidly approaches the superparamagnetic limit (whereby stored information is unstable due to thermal fluctuations), patterned magnetic media consisting of arrays of single domain nanomagnets have been proposed as a candidate for recording density up to 1 Tb/in.^{2} (Ross, 2001; Martin et al., 2003; Terris et al., 2007).

Ferromagnetic (FM) nanowires are attracting considerable interest due to their unique and tunable magnetic properties and their potential in a wide range of applications. The shape-induced magnetic anisotropy of planar magnetic nanowires creates a relatively simple magnetization structure that is being exploited for scientific and technological studies. The interplay between electronic transport and magnetic structure in nanowires has been the focus of many recent experiments. Arrays of giant magnetoresistive nanowires offer attractive potential to serve diverse applications such as high-density magnetic recording devices and magnetic field sensors. Understanding the magnetic and transport properties of nanowires is important for the design and optimization of miniature magnetoresistance (MR) heads for ultra-high-density data storage (Yuan and Bertram, 1993). The movement of domain walls (DW) in planar magnetic nanowires forms the basis of several recently proposed technological applications from magnetic logic to magnetic memory devices (Allwood et al., 2002, 2004, 2005). In magnetic-field-driven nanowire devices, it is necessary to control the direction of DW motion. The time taken to change the direction of magnetization of a nanowire device is directly related to the writing and reading of such a device and therefore an understanding of both its static and dynamic properties is very important.

In the field of biomagnetism, magnetic nanoparticles are used in a broad range of applications, including cell separation (Moore et al., 1998; Safarik and Safarikova, 1999), bio-sensing (Baselt et al., 1998), studies of cellular functions (MacKintosh and Schmidt, 1999; Alenghat et al., 2000), as well as a variety of potential medical and therapeutic uses. Most of the applications to date have used spherical magnetic nanoparticles consisting of a single magnetic species and a suitable coating to allow functionalization with bioactive ligands. It has been proposed that by using electrodeposited magnetic nanowires, a variety of functions would be possible (Fert and Piraux, 1999). Due to their large aspect ratios, FM nanowires have large remanent magnetizations and, hence, can be used in low-field environments where the superparamagnetic beads are unsuitable. Hultgren et al. (2003) have demonstrated the use of FM Ni nanowires in cell-sorting applications. For example, by precisely modulating the composition along the length of the nanowires and carefully selecting the ligands that bind selectively to different segments of a multicomponent wire, it is possible to introduce spatially modulated multiple functionalization in these nanowires.

Arrays of magnetic nanowires are being exploited in microrheology, an emerging technique for investigating the viscoelastic properties of complex fluids. In microrheology, nanometer or micrometer scale particles suspended in the fluid are used to probe the local mechanical environment. Recently, Anguelouch et al. (2006) have shown a microrheological approach that applies FM nanowires to the study of interfacial rheology and have described the application of this technique to the measurements of thin viscous oil films on aqueous subphases. Magnetic nanowires are also being explored as artificial cilia to sense acoustic signals. Cilia are employed in nature for sensing sound, fluid flow, touch, and other stimuli (McGary et al., 2006). A better understanding of the fundamental properties of nanowires is important for the realization of these practical applications and novel devices. The ability to characterize and extract quantitative information about the magnetic properties and the reversal mechanisms of nanowire arrays is very crucial in the design of the various magnetoelectronic devices mentioned above.

This chapter is organized as follows. Section 22.2 is devoted to a review of the various techniques for synthesizing ordered magnetic nanowire arrays. This includes the electrodeposition technique, the electron beam lithography, the nanoimprint lithography, the interference lithography, and the deep ultraviolet lithography. In Section 22.3, we focus on the magnetic properties of the nanowire arrays as a function of the various geometrical parameters. The chapter ends with a summary in Section 22.4.

22.2 Synthesis of Magnetic Nanowires

The fabrication of uniformly distributed high-quality magnetic nanowires over a very large area is a major challenge. The key issues to be considered when developing a nanofabrication technique are critical dimension control, resolution, patterned area, size, and shape homogeneity. In the last decade, various techniques for synthesizing ordered magnetic nanowire arrays have been developed.

22.2.1 Electrodeposition

Template synthesis using electrodeposition, first reported by Possin in 1970 (Possin, 1970), is a low-cost, high-yield technique for producing large arrays of nanowires in the fabrication of tin nanowires in tracked mica films. This technique has been used to synthesize a variety of metal (Liu et al., 1998; Sun et al., 1999) and semiconductor (Routkevitch et al., 1996; Ohgai et al., 2005) nanowires. It has been shown that it is possible to pattern wires of various materials with diameters as small as 5 nm (Zeng et al., 2002).

A schematic illustration of the electro-chemical deposition technique is shown in Figure 22.1. A metallic film serving as a cathode is normally evaporated on one side of the membrane

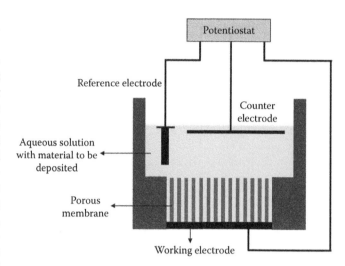

FIGURE 22.1 A schematic illustration of the experimental setup for the electrodeposition of nanowires in membrane pores.

prior to the electrodeposition process. The growth of the material within the pores is monitored from the current response at a constant potential (Whitney et al., 1993). Nanowires form in the pores of the template as they are filled with the electrodeposited material. This method can also be used to make multilayer wires. Two materials can be deposited from a single bath by switching between the deposition potentials of the two constituents materials (Blondel et al., 1994; Piraux et al., 1994; Doudin et al., 1996; Heydon et al., 1997; Schwarzacher et al., 1997). The electrodeposition process is stopped when the wire emerges from the surface, leading to a sudden increase in the plating current.

Several factors should be taken into account in choosing the template material: the uniformity in pore size and shape, pore density, pore orientation, surface roughness of the pores, and the template thickness. The two most commonly used templates are: nuclear track-etched polycarbonate membranes (Bean et al., 1970; Guillot and Rondelez, 1981; Fischer and Spohr, 1983) and self-ordered anodized aluminum oxide films (Masuda et al., 1997; Ba and Li, 2000). This technique has been used to fabricate magnetic nanowire arrays from different materials (Ferre et al., 1997; Aranda and Garcia, 2002; Chien et al., 2002).

Track-etched polymer membranes have been shown to possess good properties with respect to pore shape, size, and the parallel alignment of the pores (Ferain and Legras, 1997). The template is made using the nuclear track-etched technology, where a high-energy heavy ion accelerated in a cyclotron is used to bombard a film of polymer polycarbonate. The irradiated template is then etched in an adequate solution. The etch rate of the damaged tracks is much higher as compared with that of bulk film, leading to the formation of the pores. The diameters of the pores are dependent on the etching time and the etch selectivity between the bulk material and the tracks. The main limitation of this method is the random distribution of the pores on the membrane.

The anodic anodized oxide templates are prepared by the anodization of high-purity aluminum in an acid electrolyte under a constant voltage. The templates have a packed array of columnar hexagonal cells with central, cylindrical, uniformly sized pores. Anodized aluminum oxide (AAO) film templates are stable at high temperatures in organic solvent and the pore channels in AAO films are uniform, parallel, and perpendicular to the membrane surface. Long periods of anodization have been shown to improve the pore arrangement, with an almost ideal honeycomb lattice structure being possible (Masuda and Fukuda, 1995; Friedman and Menon, 2007). The pore sizes in the AAO templates are controlled by a further process that involves dipping the template into phosphoric acid. AAO film is ideal for the electrodeposition of nanowire arrays. The pore sizes typically range from 4 to 200 nm. The method has been used by various research groups to fabricate nanowire arrays of various materials (Metzger et al., 2000; Nielsch et al., 2001; Sellmyer et al., 2001; Khan and Petrikowski, 2002; Chiriac et al., 2003; Chun-Guey et al., 2006; Vazquez et al., 2006; Wu et al., 2006; Napolskii et al., 2007).

This template synthesis technique is however limited by the distribution in pore size and orientation making it difficult to control the period and uniformity of the nanostructures (Meier et al., 1996). Also, nanowires grown via electrodeposition processes exhibit the so-called skyscraper effect associated with a lack of length uniformity and control (Yin et al., 2001).

22.2.2 Electron Beam Lithography

Electron beam lithography (EBL) is a versatile high-resolution technique for patterning magnetic nanostructures. The principle of the EBL technique is the direct writing of desired structures on a thin resist layer with a focused beam of electrons as shown in Figure 22.2. The e-beam can create extremely fine patterns due to the small spot size of the electron. The substrate to be patterned is coated with an electron beam sensitive polymeric resist film, normally polymethylmethacrylate (PMMA). The PMMA resist is then exposed to high-beam energy with a small spot size. Following the exposure, the sample is usually developed in a methylisobutylketone (MIBK): isopropyl alcohol (IPA) (1:3) to form a resist template on the substrate. PMMA is a positive resist, so the regions that are exposed to the e-beam are removed after development. Following the development of the resist, the patterns in the resist may be transferred to the substrate by using an etching process (Shearwood et al., 1993; Adeyeye et al., 1996; Fraune et al., 2000; Katine et al., 2000; Remhof et al., 2008) with the PMMA acting as an etching mask. The developed patterns can also be converted into magnetic nanostructures by metal deposition and the lift-off process (Kirk et al., 1997; Cowburn, 2000; Castano et al., 2003; Tsoi et al., 2003; Vavassori et al., 2003; Adeyeye and White, 2004; Miyawaki et al., 2006) or electroplating (Chou et al., 1994; Obarr et al., 1997; Martin et al., 2002). For electroplating, a thin gold layer is deposited onto the substrate prior to the resist coating. After the development, the sample is immersed into a plating bath and the magnetic material is electrodeposited into the openings created in the resist. Adeyeye et al. (Adeyeye et al., 1997b) have shown that it is possible to combine

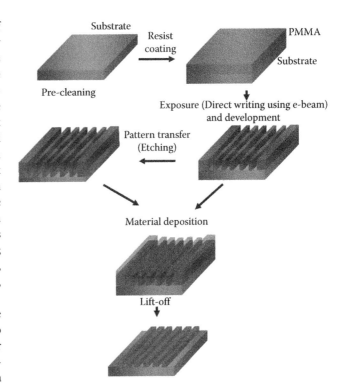

FIGURE 22.2 A schematic illustration of the fabrication processes of nanowires via electron beam lithographic combined with lift-off techniques.

the wet etching selectivity of different materials to enable the patterning of epitaxial magnetic films.

There are limitations, however, with the use of EBL in fabricating large area magnetic nanostructures. The writing process in EBL is serial and very slow, thus making large area fabrication extremely difficult, although it can be used in the preparation of masks for optical lithography. It is also very difficult to fabricate closely packed high aspect ratio nanostructure arrays due to proximity effects.

22.2.3 Interference Lithography

Interference lithography is another method used by researchers to fabricate large area magnetic nanostructures. In this method, a resist layer is exposed by an interference pattern generated by two obliquely incident laser beams without the use of a mask, as shown in Figure 22.3. The interference pattern, due to the two laser beams, consists of standing waves whose intensity vary with a period of $p = \lambda/(2 \sin \theta)$, where λ is the wavelength of the light and θ is the half angle at which the two beams intersect (Schattenburg et al., 1995; Spallas et al., 1996). This method of patterning allows for the fabrication of regular arrays of fine features, without the use of complex optical systems.

This technique is particularly useful for patterning parallel arrays of lines. In order to pattern complex structures, the sample has to go through successive exposures. Arrays of rectangular dots may be patterned by rotating the sample by 90° followed by a second exposure. The patterned area is determined by the diameter of the

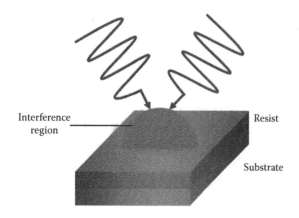

FIGURE 22.3   A simplified illustration of the interference of two laser beams on a sample surface.

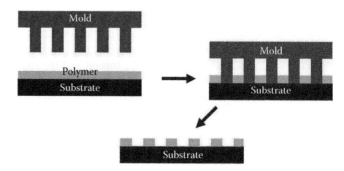

FIGURE 22.4   A schematic illustration of the NIL process.

two laser beams. For processes where no alignment is required, the technique is relatively cheap and simple. Several groups have used this method to fabricate large arrays of magnetic nanostructures (Wassermann et al., 1998; Farhoud et al., 1999; Vogeli et al., 2001; Zheng et al., 2001; Heyderman et al., 2004). The minimum period that can be achieved in interference lithography has a lower limit of λ/2. For commonly used Argon-ion lasers, with a wavelength of 350 nm, the lower limit on the period is about 200 nm. Shorter wavelength lasers tend to generate a broad range of wavelengths to enable the formation of sufficiently sharp interference patterns (Savas et al., 1999). Achromatic interferometric lithography has been used to overcome this limitation, allowing for the patterning of structures with a period down to 100 nm (Savas et al., 1996). The main drawback of interference lithography is that it is limited to patterning arrays of highly symmetrical elements only. Hence, for fabricating patterns that are arbitrarily shaped, other lithography techniques have to be explored.

22.2.4 Nanoimprint Lithography

Nanoimprint lithography (NIL) is one of the most promising low-cost, high-throughput technologies for patterning magnetic nanowires over a large area. NIL is based on a different fundamental principle compared with the conventional lithography process where the local chemical properties are changed using radiation. The patterned nanostructures are formed by physical deformation of a deformable material using a mold. This creates a thickness contrast in the materials. The best way to fabricate the mold is to select a mold substrate deposited with a suitable mold surface, and then use electron beam lithography to pattern a resist, followed by etching to transfer the patterns in the resist into the mold (Chou, 1997). In order to deform a resist on a substrate by a mold, the mold and the substrate must be pressed toward each other to have direct contact. To achieve good uniformity, the pressure should be uniform across the imprint area and should be sufficiently high to create the necessary deformation in either the mold or the substrate or both in order to make their surfaces conform. A good mold separation from the imprinted patterns is achieved by using good mold release

agents, which reduce the bonding of the mold and the resist and also by reducing the stress between the mold and the resist. A typical NIL process is shown in Figure 22.4. Features as small as 10 nm have been replicated onto the resist layer.

NIL can be implemented using various methods. In the *thermal imprint* process, a thermoplastic material starts as a solid film, then becomes a viscous liquid when its temperature is raised higher than the glass transition temperature (T_g) and returns to a solid when its temperature is reduced to below T_g (Chou et al., 1995; Chou, 2001). *Photo-NIL* uses a photocurable material as a resist (Haisma et al., 1996; Chou, 2001). A photo-curable material is initially in liquid form but is cured photochemically using photons rather than heat. Like the thermal process, this is also an irreversible process. Step and flash imprint lithography is a photo-NIL in which drops of a resist liquid are dispensed and imprinted on one single die at a time. The process is repeated as the imprint mold is stepped from die to die across the wafer repeating the resist drop and imprint cycle (Bailey et al., 2001) This method of patterning allows for higher alignment accuracy, as only a small area is patterned at one time and more importantly, a small mold can be used to fabricate the nanostructures over a large area. Magnetic nanostructures can be fabricated using the resist as a deposition template or etch mask. This technique has been used to fabricate large area magnetic nanostructures from different materials.

22.2.5 Deep Ultraviolet Lithography

An optical lithography system consists of a light source, a condenser lens, mask, an objective lens, and finally the resist-coated wafer. The mask image is projected onto the resist-coated wafer. In the process of imaging, the light source uniformly illuminates the mask through the condenser lens. The radiation passing through the transparent regions of the mask is partially diffracted before reaching the objective lens. The low spatial frequencies corresponding to the larger patterns appear closer to the lens center, whereas high frequencies corresponding to the smaller patterns and pattern corners fall toward the periphery of the lens pupil. The objective lens, being of finite size, cannot collect all of the light in the diffraction pattern. The diffracted radiation accepted by the pupil is collimated by the objective lens and interferes at the wafer plane to constitute the image. The loss of diffraction information is the ultimate limiter of the image quality and resolution. The smallest features that can be printed, that is, the resolution is given by the relation

$$R = K_1 \frac{\lambda}{\text{NA}}$$

where

K_1 is a process-dependent parameter usually ~0.6 for conventional lithography with a theoretical limit of 0.25

λ is the wavelength of the exposure tool

NA is the numerical aperture

In our fabrication technique, we have used a KrF exposing wavelength of 248 nm, and a DUV lithography scanner with a maximum NA of 0.68. The K_1 factor is a measure of the degree of difficulty for printing a particular feature. There has been tremendous progress to reduce K_1 through the use of resolution enhancement technology such as various phase shift mask approaches, off-axis illumination, optical proximity correction methods, and other approaches (Brunner, 2003). A process with K_1 of 0.8 is considered easy; a process with K_1 smaller than 0.5 is extremely difficult to achieve without any resolution enhancement techniques. For a general review of the application of deep ultraviolet lithography in the fabrication of magnetic nanostructures; the reader is referred to Adeyeye and Singh (Adeyeye and Singh, 2008). In order to pattern nanostructures below the conventional resolution limit of the optical exposure tool, the use of an aggressive resolution enhancement technique is necessary. Arrays of FM nanowires with lateral dimensions below the conventional resolution limit have been fabricated using resolution enhancement techniques such as alternating phase shift and chromeless phase shift masks (Singh et al., 2004). All the magnetic nanowire arrays presented in this chapter were fabricated using this technique.

22.3 Magnetic Properties

Magnetic nanowires have attracted a lot of interest both from a fundamental viewpoint and because of their potential for use in various magnetoelectronic applications. A lot of research has focused on understanding the static and dynamic properties of homogeneous width FM nanowire arrays. This section will focus on some of the recent works. It will be shown that the reversal process evolved from a DW-dominated process to coherent spin rotation when the wire width is reduced to submicron.

22.3.1 Demagnetizing Fields

Polycrystalline FM films consisting of small randomly distributed single crystalline grains will possess no crystal anisotropy. If such a film is patterned into spherical shaped magnets, the applied field will magnetize the sample to the same extent in any direction because there is no preferred direction of magnetization.

However, if the film is patterned into nanowires, magnetization prefers to lie along the length of the wire due to shape anisotropy. Shape anisotropy occurs because the magnetization vector prefers to lie along the long axis where the demagnetizing field is minimum and the magnetostatic energy is lowest. Demagnetizing fields hold the key to an understanding of the magnetic properties of nanowire arrays. When a magnetic wire of finite size is magnetized, the field experienced (H_{exp}) is different from the applied field (H_{appl}). The difference that is usually known as the demagnetizing field (H_{dem}) is due to the presence of magnetic poles at the magnetized surface, which gives rise to a magnetic field that counteracts the applied field.

$$H_{exp} = H_{appl} - H_{dem}$$

The magnitude of the demagnetizing field is a function of the magnetization in the material (i.e., pole strength) and the pole separation determined by the sample geometry. The demagnetizing field is both opposite and proportional to the magnetization. The constant of proportionality is known as the demagnetization factor, or more generally as the demagnetization tensor.

$$H_{dem} = -\left|N_d\right| M$$

The demagnetization factor for a normally magnetized disk is 1, while for an infinite cylinder magnetized along the long direction is 0. If the magnetization has components along more than one of the three axes of the sample, then a tensor relation is required. For a magnetic nanowire array, the strength of the demagnetizing field is determined by the ratio of the lateral dimension to the film thickness. The computation of the demagnetizing field is rather complicated for nonellipsoidal shaped nanomagnets because the magnetization is nonuniform.

There is a general theorem in micromagnetics that assumes that under rather relaxed conditions, a body of arbitrary shape is equivalent to an ellipsoid, both of them uniformly magnetized to saturation and of equal volume (Brown and Morrish, 1957; Brown, 1960). The concept of the *equivalent ellipsoid* is intuitively very appealing, but it is not straightforward to find the correct shape parameters of an ellipsoid in order to establish this equivalence (Beleggia et al., 2006b). It may be misleading, however, in some cases to assume that some symmetrical FM bodies would closely resemble an actual ellipsoid.

The justification for the substitution of one shape (ellipsoid of revolution) for the other cylinder, for example, is due to the fact that until recently (Beleggia and De Graef, 2003; Millev et al., 2003), the demagnetizing factors for the cylinder have only been given in terms of elliptic integrals (Rhodes et al., 1962; Joseph, 1966) or numerically (Brown, 1962), while the counterpart expressions for the spheroid have been known for quite a while in terms of elementary functions (Osborn, 1945; Stoner, 1945). In addition, the difference in the demagnetization factor due to this swap of shapes has been considered to be negligible. However,

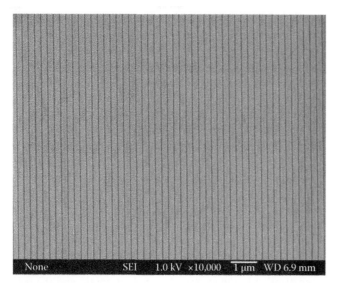

FIGURE 22.5 Scanning electron micrograph of 20 nm thick $Ni_{80}Fe_{20}$ nanowire array with width 185 nm and edge-to-edge spacing of 35 nm.

it has been shown in the case of very thin disks that the actual difference can be quite significant (Vedmedenko et al., 2003). Beleggia et al. (Beleggia et al., 2006a) have presented a self-contained method for the calculation of the demagnetization tensor for a uniformly magnetized ellipsoid based on a Fourier-space approach to the micromagnetics of magnetized bodies.

22.3.2 Shape Anisotropy

The concept of shape anisotropy in nanowire arrays can be illustrated by considering an array of 20 nm thick $Ni_{80}Fe_{20}$ nanowires with a width of 185 nm and edge-to-edge spacing of 35 nm, as shown in Figure 22.5. This nanowire array was fabricated using DUV lithography and the lift-off process.

Experimentally, by measuring the magnetic properties of the nanowire arrays with the applied field parallel ($\theta = 0°$) and perpendicular ($\theta = 90°$) to the wire axis, it is possible to investigate the effects of shape anisotropy. Typical normalized hysteresis loops for fields applied along $\theta = 0°$ (long axis) and $\theta = 90°$ (short axis) are shown in Figure 22.6. For fields applied along $\theta = 0°$ (corresponding to the easy axis), the nanowire array displays a near rectangular hysteresis loop as expected with a large jump, as seen in Figure 22.6a. The nanowire array exhibits a high squareness ratio of 0.93, a coercive field of 220 Oe, and a saturation field of 350 Oe. For fields applied perpendicular to the wire axis, $\theta = 90°$ (the hard axis), however, a sheared $M–H$ loop with a very small coercive field, 2 Oe and squareness ratio of 0.02 is obtained, as shown in Figure 22.6b. A significant increase in the saturation field of the $M–H$ loop is obtained in comparison with the loops obtained for fields applied along the wire axis, with a field of 800 Oe. The observed changes in the magnetic properties for fields applied along long and short axes can be attributed to the shape anisotropy effects being the dominant anisotropy for

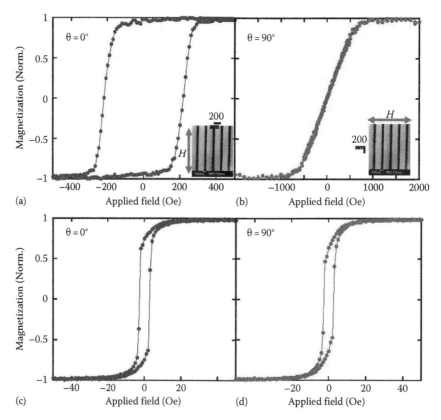

FIGURE 22.6 $M–H$ loops for 20 nm thick $Ni_{80}Fe_{20}$ nanowire arrays, $w = 185$ nm and $s = 35$ nm, with fields applied along (a) $\theta = 0°$ and (b) $\theta = 90°$, with respect to the long axis. The corresponding $M–H$ loops for the 20 nm $Ni_{80}Fe_{20}$ reference film with fields applied along (c) $\theta = 0°$ and (d) $\theta = 90°$.

the nanowire array, with the easy axis being along the long axis of the nanowire (Goolaup et al., 2005b). This is also in qualitative agreement with results of earlier published works (Adeyeye et al., 1997a).

The corresponding normalized magnetization loops for 20 nm thick $Ni_{80}Fe_{20}$ reference (unpatterned) samples with fields applied along the $\theta = 0°$ and $\theta = 90°$ axes are shown in Figure 22.6c and d. This reference sample was deposited at the same time and under the same conditions as the nanowire arrays. The M–H loops are almost identical for both orientations of the applied field and display a very low coercivity of 2 Oe and a small saturation field of 20 Oe. This implies that, as expected, the reference sample has negligible intrinsic magnetic anisotropy as compared with the shape anisotropy of the nanowire. The difference between the M–H loops of the reference film and the nanowire array can be attributed to different mechanisms mediating the reversal process. For the continuous film, the reversal process is dominated by DW propagation.

22.3.3 Effects of Nanowires Thickness

As shown earlier, the magnetic properties of nanowire arrays are strongly dependent on the spatially varying demagnetizing field, which is related to the geometric parameters. In order to understand the effects of the thickness of the $Ni_{80}Fe_{20}$ nanowire array on its magnetic properties, a series of experiments were performed. In order to investigate the effect of film thickness, the lateral dimensions of the wire arrays were fixed while the $Ni_{80}Fe_{20}$ wire thickness (t) was varied from 10 to 150 nm.

22.3.3.1 Easy Axis Behavior

The magnetization loops for fields applied along the easy axis ($\theta = 0°$) for the nanowires of $w = 185$ nm and $s = 35$ nm, as a function of the $Ni_{80}Fe_{20}$ wire thickness, is shown in Figure 22.7. As expected, the M–H loops are markedly sensitive to the wire thickness. In general, for all the wire thicknesses, near rectangular M–H loops were obtained. For thin films ($t = 10$ nm), as the field is increased from negative saturation, the maximum moment is retained until the applied field reaches zero, as shown in Figure 22.7a, followed by a gradual increase in the magnetization as the field is increased toward positive saturation. A further increase in the applied field, to 120 Oe, leads to an abrupt rise in the magnetization resulting in positive saturation. This field corresponds to the coercivity of the nanowire array. As the nanowire thickness is increased to 40 nm, there is a corresponding increase of the coercivity to 410 Oe, accompanied by a slight tilt, within the region where the magnetization changes direction, as shown in Figure 22.7b. The slight tilting in the M–H loops observed has been attributed to the switching field distribution and the dipolar coupling among the nanowires (Castano et al., 2001a,b; Gubbiotti et al., 2005; Vavassori et al., 2007). A further increase in the wire thickness to 80 nm results in a slight increase in the coercivity to 475 Oe, as shown in Figure 22.7c. The tilt in the M–H loop, first seen for $t = 40$ nm, becomes more pronounced. As the wire thickness is further increased to 150 nm, a drastic

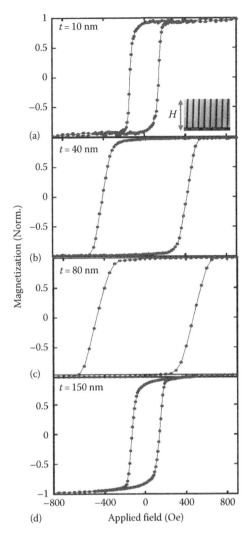

FIGURE 22.7 Representative M–H loops for $Ni_{80}Fe_{20}$ nanowire arrays, of width = 185 nm and edge-to-edge spacing = 35 nm, with field applied along the easy axis for different wire thicknesses.

reduction in the coercivity to 140 Oe with a slight shearing in the M–H loop is observed, as shown in Figure 22.7d.

The results clearly show that the easy axis coercive field of the $Ni_{80}Fe_{20}$ nanowire array of fixed width is strongly dependent on the film thickness. To gain an insight into the trend of the easy axis coercive field, as a function of $Ni_{80}Fe_{20}$ film thickness, the coercivity was extracted from the M–H loops for all the thicknesses studied. A plot of the easy axis coercivity as a function of the thickness to width (t/w) ratio of the nanowires is shown in Figure 22.8. The coercive field follows a nonlinear increase with a t/w ratio, reaching a peak of 520 Oe when the t/w ratio = 0.54. This t/w ratio corresponds to a film thickness of 100 nm. A further increase in the t/w ratio leads to a rapid reduction in the coercivity of the nanowires. This non-monotonic t/w ratio dependence suggests that there is a crossover in the magnetization reversal mechanism in the nanowires, as the thickness of the film is increased. This has also been confirmed by others (Goolaup et al., 2005a; Vavassori et al., 2007).

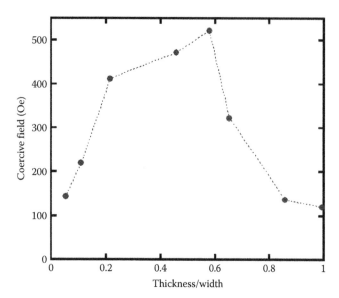

FIGURE 22.8 Easy axis coercive field of the $Ni_{80}Fe_{20}$ nanowire array of width = 185 nm and s = 35 nm as a function of thickness/width ratio. The dotted line is a visual guide.

22.3.3.2 Hard Axis Behavior

The magnetization loops for fields applied along the hard axis (θ = 90°) of the nanowire array as a function of the $Ni_{80}Fe_{20}$ film thickness are shown in Figure 22.9. Again, an evolution in the hysteresis curve as a function of the $Ni_{80}Fe_{20}$ wire thickness is observed. The shape of the M–H loop changes from an almost sharp switching to a highly sheared curve as the wire thickness is increased. For t = 10 nm, a linear reversible M–H curve with a steep slope is obtained, as seen in Figure 22.9a. The M–H loop exhibits a small hard axis saturation field of 345 Oe. When t = 40 nm, there is a slight shearing of the M–H loop, resulting in an increase in the saturation field to 1160 Oe, as seen in Figure 22.9b. Interestingly, when t is increased to 80 nm, the M–H curve displays a small coercive field of 140 Oe, as shown in Figure 22.9c. Also, the hysteresis curve for t = 80 nm is less steep as compared with t = 40 nm, thus exhibiting a much larger saturation field of 1670 Oe. As t is increased to 150 nm, the M–H loop becomes highly sheared displaying an almost "S" shaped curve with a saturation field of 3500 Oe, as shown in Figure 22.9d.

In order to investigate the effect of wire thickness on the hard axis behavior, we have extracted the saturation field of the nanowires from the M–H loops shown in Figure 22.9. A plot of the saturation field of the nanowires as a function of the $Ni_{80}Fe_{20}$ wire thickness is shown in Figure 22.10. It can be seen that the saturation field of the nanowire array increases monotonically as the $Ni_{80}Fe_{20}$ thickness is increased. This can be explained by the strong influence of the demagnetizing field across the wire width. The hard axis saturation field, H_s, is given by (Bajorek et al., 1974)

$$H_s \approx H_k + \frac{3}{2}H_d$$

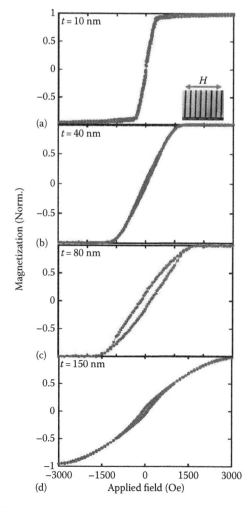

FIGURE 22.9 Representative M–H loops for $Ni_{80}Fe_{20}$ nanowire arrays of w = 185 nm and s = 35 nm, with field applied along the hard axis for different wire thicknesses.

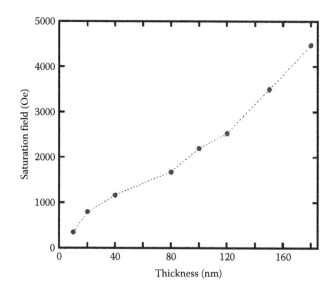

FIGURE 22.10 Hard axis saturation field as a function of film thickness for $Ni_{80}Fe_{20}$ nanowires with w = 185 nm and s = 35 nm.

where

 H_k is the magnetic anisotropy constant, which is negligible in our structures

 H_d represents the average demagnetizing field

For wire arrays, the demagnetizing field is given by (Pant, 1996):

$$\mu_o H_d = M_s \frac{t}{w} \alpha(r)$$

where $\alpha(r)$ is a function of the ratio of inter-wire spacing to the wire width (s/w). In the limit of (s/w) → 0, the factor $\alpha(r)$ → 0; in the opposite limit of (s/w) → ∞, $\alpha(r)$ → 1. The first limit corresponds to a continuous film, where the wires are in physical contact and the second limit corresponds to the isolated wires. For the wire arrays investigated, $s = 35$ nm, $w = 185$ nm, and $\alpha(r) \approx 0.34$, a constant value. Thus, the demagnetizing field varies as t/w. As the thickness of the nanowire increases, the demagnetizing field across the wire increases leading to an increase in the saturation field, consistent with the experimental results obtained.

22.3.4 Coercivity Variations as a Function of Field Orientations

The angular dependence of coercivity is known to provide information on the magnetization reversal mode in nanowires

(Lederman et al., 1995; Adeyeye et al., 1997a; Hao et al., 2001; Han et al., 2003; Perez-Junquera et al., 2003). To better understand the evolution of the magnetization reversal modes, the values of coercive fields were extracted from the hysteresis loops as a function of the orientation of the applied field for different $Ni_{80}Fe_{20}$ film thicknesses. The variation of the coercivity (H_c) with the applied field orientation (θ) for various wire thicknesses is shown in Figure 22.11. Generally, for all the thickness ranges investigated, the coercivity is markedly sensitive to the orientation of the applied field and is symmetrical about field orientation of 180°. Several mechanisms may be responsible for magnetization reversal: coherent rotation, magnetization curling, magnetization buckling, or DW motion. For nanowires, two modes are considered as being important; coherent rotation (Stoner and Wohlfarth, 1991) and curling magnetization (Brown, 1957; Frei et al., 1957; Aharoni and Shtrikman, 1958; Shtrikman and Treves, 1959; Ishii, 1991).

It is well-known that $Ni_{80}Fe_{20}$ is a soft magnetic material with low intrinsic magnetic anisotropy. Since all the wire dimensions are larger than the exchange length (Ha et al., 2003), we can expect deviations from uniform magnetization and as a result, the occurrence of DW nucleation at lower fields than the curling field or anisotropy field, for fields applied along the hard-axis. For $t = 10$ nm, the maximum coercive field occurs along the nanowire axis ($\theta = 0°$), as shown in Figure 22.11a. As the field orientation (θ) increases, the coercive field decreases, reaching a minimum for fields applied along the hard-axis ($\theta = 90°$). A bell-shape angular variation is observed as field

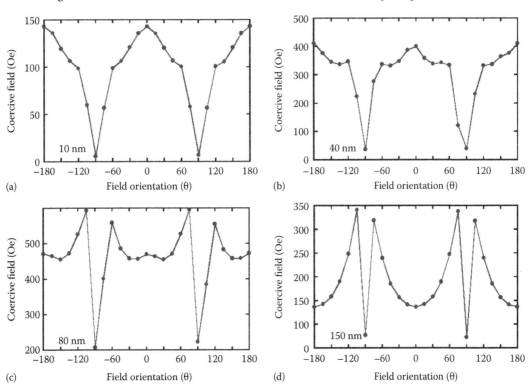

(a) (b) (c) (d)

FIGURE 22.11 Coercive field as a function of the field orientation with respect to the nanowire ($w = 185$ nm and $s = 35$ nm) axis, for $Ni_{80}Fe_{20}$ wire thickness ranging from 10 to 150 nm.

orientation is varied from θ = −90° to θ = +90°. This angular variation of the coercive field is consistent with the coherent rotation reversal mode based on the Stoner–Wohlfarth model (Stoner and Wohlfarth, 1991). As the wire thickness is increased to 40 nm, a slight departure from coherent rotation is observed for the field orientation of 60° and 120°, as shown in Figure 22.11b. This departure from coherent rotation can be attributed to the onset of a curling mode of reversal. The coercive field vs. the θ curve for a nanowire array with t = 80 nm is shown in Figure 22.11c. A completely different field orientation dependence of the coercive field is observed when compared with the coherent mode for t = 10 nm. As the field orientation increases from θ = 0° to θ = 30°, a slight decrease in the coercive field is first observed. With a further increase in field orientation, an increase in the coercive field with a peak at θ = 75° is seen. For θ ≥ 75°, the coercive field decreases reaching a minimum along the hard-axis (θ = 90°). This field orientation dependence of the coercive field suggests that the reversal process is dominated by a combination of coherent rotation and the curling mode of the magnetization reversal process. The region where there is an increase in coercivity with field orientation (i.e., 30° ≤ θ ≤ 75°) may be attributed to curling mode reversal (Brown, 1957; Frei et al., 1957; Aharoni and Shtrikman, 1958; Shtrikman and Treves, 1959). The coherent rotation reversal mode is present for fields applied close to the easy and hard axis.

For t = 150 nm, the coercive field increases as the field orientation increases with respect to the wire axis, reaching a peak for field orientation θ = 75°, as shown in Figure 22.11d. Beyond this field orientation, θ > 75°, the coercive field decreases with increasing field orientation, reaching a minimum at θ = 90°. The curve displays a U-shape within the field orientation θ = −90° and θ = +90°. This angular dependence of the coercive field suggests that the magnetization reversal mechanism is dominated by the curling mode of the reversal process.

To aid in the understanding of the reversal process of nanowire arrays with t = 150 nm, the experimental results were modeled using the theoretical prediction proposed in the article by Meier et al. (1996). For an infinite cylinder with a curling mode of reversal, the coercive field is given as (Aharoni and Shtrikman, 1958)

$$\mu_o H_c = \frac{M_s}{2} \frac{a(1+a)}{\sqrt{a^2 + (1+2a)\cos^2\theta}}$$

where $a = -1.08\,(d_0/d)^2$. The exchange length $d_0 = 2(\sqrt{A}/M_s)$, where A is the exchange constant. This model is valid for cylindrical wires; however, the structures investigated in this work have rectangular geometry, which is not analytically solvable. Hence, the experimental results were fitted with this theoretical prediction. A plot of the experimental data and the theoretical prediction is shown in Figure 22.12; the dotted line is the predicted coercive field curve for the curling mode of magnetization reversal and the dots are the experimental points.

There is a very good agreement between the measured coercivity and the theoretical prediction except for field orientation

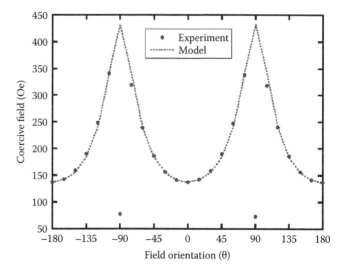

FIGURE 22.12 Angular variation of coercive field together with theoretical prediction based on curling magnetization for 150 nm thick Ni$_{80}$Fe$_{20}$ nanowire array (w = 185 nm and s = 35 nm).

when θ = −90° and θ = +90°. For fields applied along the hard axis, a minimum is obtained as opposed to the theoretical prediction of the maximum coercive field.

This is consistent with the results reported by Han et al. (2003), where curling magnetization is present for small θ angles and the coherent rotation occurs at larger θ values. Thus, for thick nanowires, t = 150 nm, the magnetization reversal is dominated by the curling mode except at fields applied along the hard axis, θ = 90°, where coherent rotation is responsible for the reversal.

22.3.5 Pseudo-Spin Valve Nanowires

The effect of magnetostatic interactions in patterned heterostructures consisting of FM layers of different materials or two FM layers of the same material of different thicknesses, separated by a spacer layer is of fundamental interest. Patterned layered nanomagnets exhibit interesting magnetic properties because of interlayer magnetostatic coupling that can lead to a relative anti-parallel alignment of magnetization at remanence (Castano et al., 2001a). In addition, magnetotransport properties of patterned layered nanostructures have been shown to yield a higher giant magnetoresistance (GMR) when compared with the unpatterned film (Hylton et al., 1995). Recent works on patterned heterostructures have typically been aimed at understanding the magnetization reversal process and transport properties of isolated pseudo-spin-valve (PSV) nanoelements (Castano et al., 2001b; Ross et al., 2005) and nanowires (Kume et al., 1996; Katine et al., 1999; Castano et al., 2002b; Morecroft et al., 2005).

A systematic study of the magnetic properties of closely packed and isolated homogeneous width PSV nanowire arrays is presented in this section. PSV nanowires were fabricated by exploiting the differential thickness coercivity of single film nanowires, shown in Figure 22.7. Nanowire arrays with a width of 185 nm and spacing (s) of 185 nm (isolated) and 35 nm (closely packed) were patterned using DUV lithography and the lift-off process. The PSV structures consisting of (bottom to top) $Ni_{80}Fe_{20}$ (10 nm)/Cu(t_{Cu} nm)/$Ni_{80}Fe_{20}$ (80 nm) were fabricated using the electron beam deposition technique. As the inter-element spacing is reduced in arrays of magnetic nanostructures, the stray fields generated during the magnetization reversal process may influence the internal magnetic domain structure and reversal mechanisms of neighboring elements. The effect of the inter-element interaction is complicated by the fact that the dipolar fields depend on the magnetization state of each element, which in turn depends on the fields due to the adjacent elements. The thickness of the Cu spacer layer, t_{Cu}, was varied from 2 to 35 nm in order to investigate the effect of interlayer coupling on the overall magnetization reversal process. Scanning electron micrographs (SEM) of the closely packed and isolated PSV nanowires are shown in Figure 22.13a and b, respectively. A schematic of the PSV structure is shown in Figure 22.13c. The respective insets in Figure 22.13 show the tilted cross-sectional view of the PSV nanowire arrays with t_{Cu} = 35 nm. The nanowire arrays have uniform width and inter-wire spacing as can be clearly seen from the SEM micrographs.

The normalized hysteresis loops for $Ni_{80}Fe_{20}$ (10 nm)/Cu(10 nm)/$Ni_{80}Fe_{20}$(80 nm) nanowire arrays for fields applied along the easy axis, $\theta = 0°$, of the nanowire arrays are shown in Figure 22.14. The respective arrows correspond to the possible magnetization alignment of the FM layers comprising the PSV nanowire arrays. Both the closely packed and isolated nanowire arrays display a double-step reversal *M–H* loop. For the array with s = 35 nm, as the field is reduced from positive saturation, the first drop in magnetization occurs at an external field of −5 Oe, followed by a quasi-stable plateau-like region within the field range of −20 Oe to −190 Oe, as seen in Figure 22.14a. Further reduction in the external field leads to a gradual decrease in magnetization resulting in negative saturation at an external field of −720 Oe. For the PSV nanowire array with s = 185 nm, the first drop in magnetization occurs at the same external field as the wire array with s = 35 nm, as shown in Figure 14b. The quasi-stable plateau, however, occupies a larger field range of −30 Oe to −245 Oe. The magnetization then decreases to lead to negative saturation at an external field of −500 Oe. For both sets of wire arrays, the first drop in magnetization corresponds to the switching of the 10 nm thin layers, followed by the switching of the 80 nm thick layers leading to negative saturation. This is

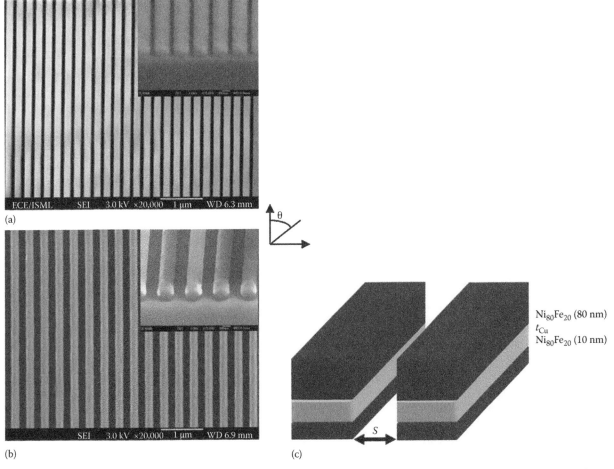

FIGURE 22.13 Scanning electron micrograph of $Ni_{80}Fe_{20}$(10 nm)/Cu(35 nm)/$Ni_{80}Fe_{20}$(80 nm) spin valve nanowire arrays with width 185 nm, (a) edge-to-edge spacing = 35 nm and (b) edge-to-edge spacing = 185 nm; (c) schematic representation of the spin valve nanowires. (Reproduced from Goolaup, S. et al., *J. Appl. Phys.*, 100, 114301, 2006. With permission.)

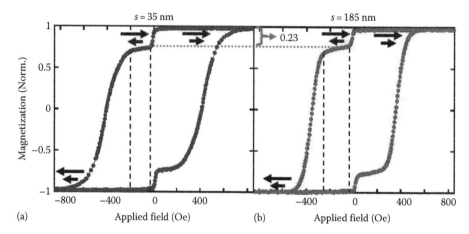

FIGURE 22.14 *M–H* loops for $Ni_{80}Fe_{20}(10\,nm)/Cu(10\,nm)/Ni_{80}Fe_{20}(80\,nm)$ nanowire arrays with fields applied along the wire axis for (a) $s = 35\,nm$ and (b) $s = 185\,nm$.

evidenced by the percentage drop in magnetization following the reversal of each layer comprising the PSV nanowire. For the reversal of the 10 nm thin layers, there is a drop of 0.23 in the magnetization. This is consistent with the percentage of magnetic moment of the 10 nm $Ni_{80}Fe_{20}$ thin layer as compared with the total magnetic moment of layers comprising the PSV nanowire.

The quasi-stable plateau observed for both sets of PSV nanowire arrays correspond to the region of anti-parallel alignment of the magnetization of the two layers comprising the PSV nanowires. Interestingly, for both sets of PSV nanowires, the switching fields for the thin (10 nm) and thick (80 nm) layer do not coincide with the coercive field of the individual single-layer films. This difference is due to the interlayer coupling between the two FM layers in the nanowires. Interlayer coupling between the FM layers results in a change of the effective field at which the magnetization of each layer reverses its orientation. As the exchange length of $Ni_{80}Fe_{20}$ is 5.4 nm (Ha et al., 2003), for $t_{Cu} = 10\,nm$, the interlayer magnetostatic coupling is expected to dominate the reversal process in the PSV nanowires. The interlayer magnetostatic coupling tends to promote the anti-parallel alignment of the magnetization of the two layers to reduce the magnetostatic energy of the system during the reversal process.

For $s = 35\,nm$, following the plateau-like region, the reversal of the 80 nm $Ni_{80}Fe_{20}$ thick layer occurs over a field range extending from −190 Oe to −720 Oe. The 80 nm $Ni_{80}Fe_{20}$ thick layer for the PSV nanowire array with $s = 185\,nm$ switches over a smaller field range of −245 Oe to −500 Oe. This is attributed to the much larger switching field distribution (SFD) in the closely packed nanowire array, due to the dipolar coupling effect between neighboring nanowires. It has been shown that the dipolar contribution for a homogeneously magnetized cylindrical nanowire, with a density of magnetic poles +M and −M at the two edges of the wire, is given by (Velazquez et al., 2003)

$$H(r,z) = \frac{-MR^2L}{4}\left[\frac{s^2 - 2z^2}{(s^2 + z^2)^{\frac{5}{2}}}\right]$$

where

s corresponds to the edge-to-edge spacing between the wires
z is the distance along the wire axis

The radius and length of the cylindrical nanowire is denoted by R and L, respectively. The dipolar field varies both as a function of the edge-to-edge spacing, s, and along the wire length, z. The dipolar field has been shown by Velázquez et al. (2003) to decay rapidly as the distance between the wires is increased. As the equation above is not solvable for rectangular geometry, the dipolar field can be approximated by computing the ratio for different edge-to-edge wire spacing of cylindrical wires. By substituting s with 35 and 185 nm, the dipolar coupling is found to be reduced by a factor of 150 as the edge-to-edge spacing is increased from 35 to 185 nm.

22.3.5.1 Effects of Cu Spacer Layer Thickness

To understand the effect of the interlayer coupling in the PSV nanowires, a systematic investigation of the effects of Cu spacer layer thickness has been conducted (Goolaup et al., 2006). The representative hysteresis loops for PSV nanowire arrays with $s = 35\,nm$ and $s = 185\,nm$ as a function of the spacer layer thickness, t_{Cu}, for fields applied along the easy axis of the wire are shown in Figure 22.15. The hysteresis loops are markedly sensitive to both the edge-to-edge spacing and the Cu spacer layer thickness. The evolution in the magnetization reversal process as a function of the spacer layer thickness can be attributed to the different coupling mechanisms between the two magnetic layers comprising the PSV nanowire arrays (Parkin et al., 1990; Barnas, 1992; Bruno, 1993; Bloemen et al., 1994; Castano et al., 2002a; Zhu et al., 2003).

For $t_{Cu} = 2\,nm$, the FM exchange coupling between the two FM layers dominates the reversal process in the PSV nanowire arrays. In FM coupling, the parallel alignment of the magnetization of the FM layers is favored and the field at which the magnetizations of the FM layers are anti-parallel aligned is reduced. When the FM coupling is very strong, the magnetization of the layers reverses their orientation simultaneously and a single rectangular loop is obtained. For exchange coupled FM layers, where each

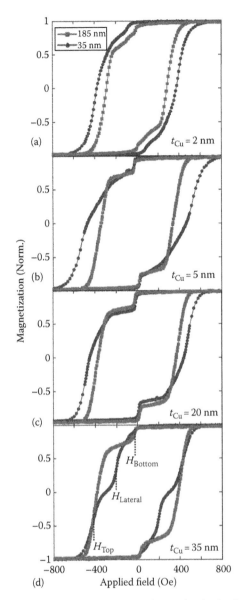

(a) $t_{Cu} = 2\,nm$

(b) $t_{Cu} = 5\,nm$

(c) $t_{Cu} = 20\,nm$

(d) H_{Bottom}, $H_{Lateral}$, H_{Top}, $t_{Cu} = 35\,nm$

Applied field (Oe)

Magnetization (Norm.)

185 nm, 35 nm

FIGURE 22.15 Representative *M–H* loops for both the closely packed and isolated nanowire arrays with $Ni_{80}Fe_{20}(10\,nm)/Cu(t_{Cu}\,nm)/Ni_{80}Fe_{20}(80\,nm)$ film as a function of the Cu spacer layer thickness, t_{Cu}. (Reproduced from Goolaup, S. et al., *J. Appl. Phys.*, 100, 114301, 2006. With permission.)

layer is in the single domain state, the reversal field, H_r, is given by (Yelon, 1971)

$$H_r = \frac{t^i M_S^i H_{r0}^i + t^j M_S^j H_{r0}^j}{t^i M_S^i + t^j M_S^j}$$

where the i and j represent the two layers with small and large coercive fields, respectively. The thickness and saturation magnetization of the magnetic layers are denoted by t^z and M_S^z, respectively, with $z = i, j$. H_{r0}^z denotes the field at which the magnetization of the respective single film magnetic

layer switches. For the structures: $t^i = 10\,nm$, $t^j = 80\,nm$, and $M_S^i = M_S^j$. The computed reversal fields are 292 Oe and 424 Oe for the PSV nanowire arrays with $s = 35\,nm$ and $s = 185\,nm$, respectively. Due to the FM exchange coupling between the two FM layers, the switching of the 10 nm $Ni_{80}Fe_{20}$ layer exerts an additional field on the top 80 nm $Ni_{80}Fe_{20}$ layer. This field, coupled with the external applied field, causes the magnetic moment of the 80 nm $Ni_{80}Fe_{20}$ (thick) layer to rotate prior to switching, leading to a gradual decrease in magnetization as observed in Figure 22.15a. The slight deviation between the computed and experimental values may be due to the presence of additional FM coupling mechanisms such as pin holes, due to a break in the spacer layer and "orange peel" coupling (Neel, 1962). This will lead to a stronger FM coupling in the layers, resulting in a smaller reversal field.

When $t_{Cu} = 5\,nm$, a slight change in the *M–H* loop of the PSV nanowires is noted, due to the weakening of the FM exchange coupling between the two FM layers. For $s = 35\,nm$, the switching of the 10 nm $Ni_{80}Fe_{20}$ layer is characterized by an almost abrupt drop in the magnetization. This is followed by the gradual decrease of magnetization until the 80 nm $Ni_{80}Fe_{20}$ layer reverses. For $s = 185\,nm$, however, the switching of the 10 nm $Ni_{80}Fe_{20}$ (thin) layer is followed by a stable plateau-like region in the *M–H* loop, as shown in Figure 22.15b. As t_{Cu} is increased to 20 nm, both sets of nanowire arrays clearly exhibit two-step switching with the nanowire array with $s = 185\,nm$ displaying a slightly larger region of anti-parallel alignment. Interestingly, when the spacer layer thickness becomes equal to the wire edge-to-edge spacing of the closely packed PSV nanowire arrays, $t_{Cu} = s = 35\,nm$, a totally different *M–H* loop is obtained, as shown in Figure 22.15d. This significant change in the magnetization behavior is attributed to the competition between the dipolar and inter-layer magnetostatic coupling in the nanowires. For the closely packed PSV nanowires, both the dipolar coupling between neighboring nanowires and the interlayer coupling between the thick and thin $Ni_{80}Fe_{20}$ layers are comparable in strength. The 80 nm thick $Ni_{80}Fe_{20}$ layer of the neighboring wire arrays may adopt an anti-parallel configuration due to dipolar coupling (Sampaio et al., 2000; Velazquez et al., 2003; Vavassori et al., 2007). This is evidenced by the appearance of an intermediate switching field $H_{Lateral}$, as indicated in Figure 22.15d. For PSV nanowire array with $s = 185\,nm$ however, no significant change in the *M–H* loop is observed as the spacer layer thickness is increased to 35 nm. The thin and thick layer reversal field, H_{Thin} and H_{Thick}, are consistent with the reversal field obtained for smaller t_{Cu} thicknesses, suggesting that the neighboring wires are decoupled. These results clearly show that it is possible to laterally engineer the magnetic properties of nanowire arrays by carefully controlling the various geometrical parameters.

To further elucidate our understanding of the different switching processes in the PSV system, the differential *M–H* loops, as a function of the applied field (d*M*/d*H*), were calculated for both sets of nanowire arrays from the measured *M–H* loops for Cu spacer layer thicknesses of 20 and 35 nm.

The curves for the 20 nm Cu spacer layer are used as a reference. The representative dM/dH curves for both sets of PSV nanowire arrays for Cu spacer layer thickness of 20 and 35 nm are shown in Figure 22.16. The peaks in the dM/dH curve correspond to the different switching processes occurring as the field is swept from positive saturation to negative saturation. For t_{Cu} = 20 nm, as shown in Figure 22.16a, both the PSV nanowire arrays display two distinct peaks corresponding to the switching of the two FM layers, 10 and 80 nm, respectively. For the isolated PSV nanowire array, the peaks occur at an external field of ~20 Oe and −395 Oe, whereas for the closely packed nanowire array, the peaks occur at −25 Oe and −482 Oe, respectively. The switching of the thin 10 nm $Ni_{80}Fe_{20}$

layer and thick 80 nm $Ni_{80}Fe_{20}$ layer correspond to peak position B and T, respectively. The differences in the peak values for the two wire geometries can be attributed to the effect of dipolar coupling.

When t_{Cu} is increased to 35 nm, the dM/dH curve of s = 185 nm displays the same two peaks response as observed for t_{Cu} = 20 nm, as shown in Figure 22.16b. The PSV nanowire array with s = 35 nm, however, displays four peaks. Two intermediate peaks, at position B' and T', are sandwiched between the peaks B and T, indicating the presence of two additional magnetic states in the PSV nanowire array. For the PSV nanowire array with s = 35 nm, as the spacer layer is comparable to the edge-to-edge spacing of the closely packed nanowires, t_{Cu} = s = 35 nm, the field acting on each FM layer is a result of the competition between the dipolar coupling from the FM layers in neighboring wires and the interlayer magnetostatic field. Due to the strong effect of the dipolar coupling, FM layers in neighboring wires may adopt an anti-parallel alignment.

22.3.6 Alternating Width Nanowire Arrays

A lot of research has focused on understanding both the static (Goolaup et al., 2005b; Wegrowe et al., 1999; Wernsdorfer et al., 1997., Adeyeye et al., 1996, 1997a) and dynamic properties (Gubbiotti et al., 2004, 2005; Bayer et al., 2006) of homogeneous width FM nanowire arrays. It has been observed that the magnetic properties of nanowires are strongly dependent on the lateral size due to the spatially varying demagnetizing field (Adeyeye et al., 1996, 1997c). While most of the research has focused on vertically stacked multilayer nanowires with a view for application in miniaturized advanced read head sensor and nonvolatile magnetic random access memories, few works have exploited the lateral engineering of microwire arrays (Adeyeye et al., 2002; Husain and Adeyeye, 2003). The switching field of the wire arrays of a fixed film thickness is highly sensitive to the wire width and it increases as the wire width is reduced. By exploiting the width dependence of the coercivity, we have fabricated alternating nanowire arrays with unique magnetic properties.

Alternating $Ni_{80}Fe_{20}$ nanowires with differential width, Δw = 200 nm consisting of nanowires with a width of w_1 = 330 nm; w_2 = 530 m, and Δw = 570 nm constituting of nanowires with a width of w_1 = 330 nm; w_2 = 900 nm alternated in an array were fabricated on silicon substrate using deep ultraviolet lithography. A control experiment (reference nanowire) consisting of homogeneous nanowire arrays with a width of w = 330 nm was also patterned using the same technique. For all the geometry patterned, the length of all the nanowire arrays was maintained at 4 mm. In order to ensure that the nanowires are magnetostatically coupled, the edge-to-edge spacing for all the nanowire arrays patterned was maintained at 70 nm. To minimize the formation of end-domains during the reversal process, the nanowires were patterned with rounded edges. Polycrystalline $Ni_{80}Fe_{20}$ of thickness (t) in the range from 20 to 100 nm was deposited by DC magnetron sputtering at room temperature. The SEM images of the alternating and homogeneous width nanowire arrays are

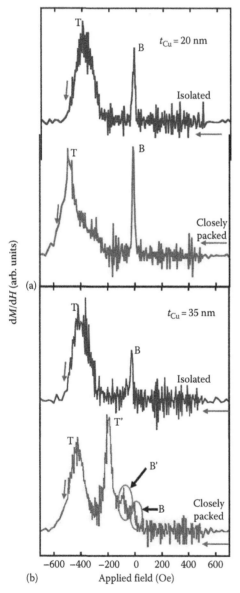

FIGURE 22.16 Differentiated *M–H* loops for the isolated and closely packed PSV nanowire arrays with (a) t_{Cu} = 20 nm and (b) t_{Cu} = 35 nm. (Reproduced from Goolaup, S. et al., *J. Appl. Phys.*, 100, 114301, 2006. With permission.)

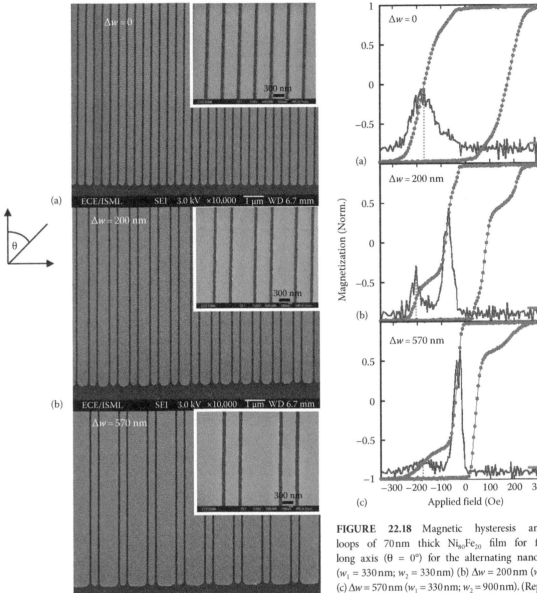

(a)

(b)

(c)

FIGURE 22.17 Scanning electron micrographs of 40 nm $Ni_{80}Fe_{20}$ thick; alternating nanowire arrays with (a) $\Delta w = 200$ nm constituting of wires $w_1 = 330$ nm; $w_2 = 530$ nm, (b) $\Delta w = 570$ nm consisting of wires $w_1 = 330$ nm; $w_2 = 900$ nm, (c) homogeneous width nanowire array with width = 330 nm. The edge-to-edge spacing for all the nanowire arrays is maintained at 70 nm. (Reproduced from Goolaup, S. et al., *Phys. Rev. B*, 75, 144430, 2007. With permission.)

FIGURE 22.18 Magnetic hysteresis and differentiated M–H loops of 70 nm thick $Ni_{80}Fe_{20}$ film for fields applied along the long axis ($\theta = 0°$) for the alternating nanowire arrays, (a) $\Delta w = 0$ ($w_1 = 330$ nm; $w_2 = 330$ nm) (b) $\Delta w = 200$ nm ($w_1 = 330$ nm; $w_2 = 530$ nm), (c) $\Delta w = 570$ nm ($w_1 = 330$ nm; $w_2 = 900$ nm). (Reproduced from Goolaup, S. et al., *Phys. Rev. B*, 75, 144430, 2007. With permission.)

shown in Figure 22.17. The large area view shows well-defined wires with uniform wire spacing and good edge definition.

The representative M–H loops for 70 nm thick $Ni_{80}Fe_{20}$ nanowire arrays for fields applied along the long (easy) axis of the wires are shown in Figure 22.18a. Both alternating width nanowire arrays display a totally different M–H behavior as compared with the homogeneous width nanowire array. The homogeneous nanowire array with $w_1 = w_2 = 330$ nm, as expected, displays an

almost rectangular M–H loop with a coercivity of 170 Oe. For alternating width nanowire arrays with $\Delta w = 200$ nm, however, we observed a double-step hysteresis loop. As the applied field is reduced from positive saturation, a sharp drop in magnetization within the field range of -30 Oe to -100 Oe was observed. Beyond this field, a gradual decrease in magnetization is observed until an external field of -190 Oe. This is followed by an abrupt drop in magnetization leading to negative saturation. A similar trend was observed for alternating width nanowires with $\Delta w = 570$ nm, although, the switching fields were shifted to lower external fields due to the contribution from the larger width wire.

The corresponding differentiated half M–H loop, for fields applied from positive to negative saturation, for the 70 nm $Ni_{80}Fe_{20}$ thick nanowire arrays is shown in Figure 22.18b. For the homogeneous wire, a broad base peak with the maximum at an external field of $-175\,$Oe was observed. This value is consistent with the coercivity obtained from the M–H loop.

As expected for $\Delta w = 200$ and $570\,$nm, two peaks, corresponding to the switching of the two sets of wires, w_1 and w_2, comprising the array are seen. The first peak corresponds to the low field switching of the larger width wire w_2, while the second peak corresponds to the high field switching of the smaller wire width, w_1. For $\Delta w = 200\,$nm, the switching of w_1 and w_2 occurs at an external field of $-205\,$Oe and $-70\,$Oe, respectively. For $\Delta w = 570\,$nm, however, the switching of w_1 and w_2 are at $-190\,$Oe and $-25\,$Oe, respectively. As w_1 is the same for both alternating width nanowire arrays, the difference in the switching fields may be attributed to the effects of the magnetostatic coupling, due to the small inter-wire spacing $s = 70\,$nm, between the wires in the array, which greatly influences the reversal process. The clear and distinct differences between the two peaks in the M–H loops imply that a region of anti-parallel alignment in the magnetization of neighboring wires in the array exists.

A schematic representation of the possible magnetic states in the 70 nm thick $Ni_{80}Fe_{20}$ alternating width nanowire arrays as the field is swept from positive to negative saturation are shown in Figure 22.19. At positive saturation, the magnetizations of all the wires are aligned along the field direction, as seen in Figure 22.20a. At the occurrence of the first peak in the dM/dH, in Figure 18b and c, the larger width (w_2) wire switches magnetization direction, as shown in Figure 22.19b. A further

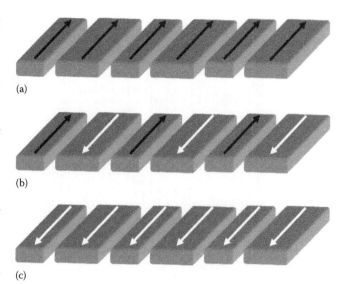

(a)

(b)

(c)

FIGURE 22.19 Schematic representation of the different states of the 70 nm thick $Ni_{80}Fe_{20}$ alternating width nanowires.

decrease of the field leads to the switching of the smaller width wire, w_1, as shown in the schematic in Figure 22.19c.

22.3.6.1 Effect of Thickness

The effect of film thickness on the magnetic properties of alternating width nanowire arrays has also been investigated. The $Ni_{80}Fe_{20}$ film thickness was varied from 20 to 100 nm, while

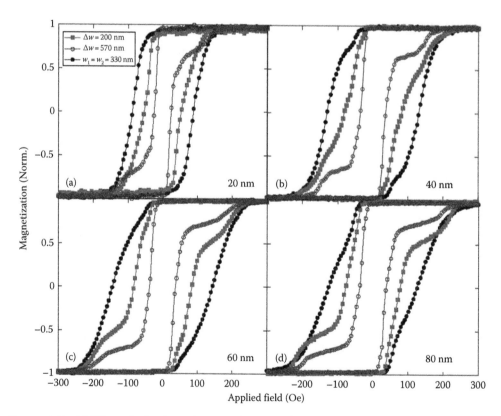

FIGURE 22.20 Representative M–H loops for the alternating nanowire arrays, $\Delta w = 200\,$nm, and $\Delta w = 570\,$nm and reference nanowire arrays as a function of the $Ni_{80}Fe_{20}$ film thickness. (Reproduced from Goolaup, S. et al., *Phys. Rev. B*, 75, 144430, 2007. With permission.)

all other geometric parameters of the wire arrays were kept fixed. The representative *M–H* loops for fields applied along the long (easy) axis of the nanowire arrays, as a function of the $Ni_{80}Fe_{20}$ film thickness are shown in Figure 22.20. The evolution of the magnetization as a function of the $Ni_{80}Fe_{20}$ film thickness can be attributed to the spatially varying demagnetizing field along the width of the alternating nanowire array. When the $Ni_{80}Fe_{20}$ film thickness is increased, the thickness to width ratio of w_1 and w_2, constituting the alternating nanowire array, will change at a different rate. Thus, as t is increased, the switching of the two sets of nanowires, w_1 and w_2, constituting the array becomes more distinct, resulting in a double-step reversal. The switching field distribution (SFD) was estimated by differentiating one branch of the hysteresis loop for all the thickness rang investigated. The peaks represent the switching of the respective wires, w_1 and w_2, whereas the base reflects the SFD. It was observed that the homogeneous nanowire array and alternating width nanowire with $\Delta w = 200$ nm, exhibit a much larger SFD as compared with $\Delta w = 570$ nm.

The SFD during the reversal of nanostructures is attributed to the process variation and the dipolar coupling between the elements. From the SEM image, the shape homogeneity of the nanowire array is confirmed, thus, the broadening of the slope in the *M–H* curve can be attributed to the dipolar coupling between the nanowires in the array. For the larger width wires, we expect the formation of edge domains at the end of the wires as the field is relaxed along the wire axis. The edge domains inhibit the accumulation of magnetic charges at the wire end, thus reducing the effective coupling field between alternate wires. This is evidenced by the abrupt drop in the magnetic moment for $\Delta w = 570$ nm, for all thicknesses investigated. Collective spin-wave modes have also been observed in these structures using brillouin light scattering (Gubbiotti et al., 2007; Kostylev et al., 2008).

22.4 Summary

In this chapter, the magnetization reversal mechanism in $Ni_{80}Fe_{20}$ nanowire arrays as a function of wire thickness has been presented. For fields applied along the nanowire array easy axis, a non-monotonic variation of the coercive field was observed due to the different mechanisms of magnetization reversal dominating the switching process in the nanowire arrays. The angular dependence of coercivity was used to map the reversal mechanism in the nanowires. A cross-over from coherent rotation to the curling mode of reversal was observed for a t/w ratio > 0.5. The question of how the magnetostatic interaction affects the reversal process in PSV nanowire arrays has been addressed. Closely packed and isolated homogeneous width $Ni_{80}Fe_{20}(10\,nm)/Cu(t_{Cu})/Ni_{80}Fe_{20}(80\,nm)$ PSV nanowire arrays with varied Cu spacer layer thicknesses were studied. The magnetization reversal process is strongly sensitive to the Cu spacer layer thickness. When the spacer layer thickness becomes comparable to the edge-to-edge spacing of the nanowire array, a drastic change in the reversal mechanism was observed due to the competition between the dipolar coupling in the neighboring nanowires and the interlayer magnetostatic coupling between the two FM layers. By exploiting the width dependence of the coercive field in nanowires, we have shown that complex nanowires with unique magnetic properties can be engineered. Alternating width nanowires consisting of two sets of $Ni_{80}Fe_{20}$ nanowires differentiated by their width, which are alternated in an array, were fabricated and systematically studied. The magnetization reversal process in the alternating width nanowire arrays was found to be markedly sensitive to the $Ni_{80}Fe_{20}$ wire thickness and differential width between the two sets of nanowire arrays.

Acknowledgments

The authors would like to thank Dr. N. Singh from the A*Star Institute of Microelectronics, Singapore, for his contributions to the fabrication of magnetic nanowires. This work was supported by the Ministry of Education, Singapore, under Grant No. R-263-000-437-112. The authors are grateful to Dr. D. Tripathy for proofreading the manuscript.

References

Adeyeye, A. O. and Singh, N. (2008) Large area patterned magnetic nanostructures. *Journal of Physics D—Applied Physics*, 41, 153001–153028.

Adeyeye, A. O. and White, R. L. (2004) Magnetoresistance behavior of single castellated Ni80Fe20 nanowires. *Journal of Applied Physics*, 95, 2025–2028.

Adeyeye, A. O., Bland, J. A. C., Daboo, C., Lee, J., Ebels, U., and Ahmed, H. (1996) Size dependence of the magnetoresistance in submicron FeNi wires. *Journal of Applied Physics*, 79, 6120–6122.

Adeyeye, A. O., Bland, J. A. C., Daboo, C., and Hasko, D. G. (1997a) Magnetostatic interactions and magnetization reversal in ferromagnetic wires. *Physical Review B*, 56, 3265–3270.

Adeyeye, A. O., Bland, J. A. C., Daboo, C., Hasko, D. G., and Ahmed, H. (1997b) Optimized process for the fabrication of mesoscopic magnetic structures. *Journal of Applied Physics*, 82, 469–473.

Adeyeye, A. O., Lauhoff, G., Bland, J. A. C., Daboo, C., Hasko, D. G., and Ahmed, H. (1997c) Magnetoresistance behavior of submicron Ni80Fe20 wires. *Applied Physics Letters*, 70, 1046–1048.

Adeyeye, A. O., Husain, M. K., and NG, V. (2002) Magnetic properties of lithographically defined lateral Co/Ni80Fe20 wires. *Journal of Magnetism and Magnetic Materials*, 248, L355–L359.

Adeyeye, A. O., Goolaup, S., Singh, N., Jun, W., Wang, C. C., Jain, S., and Tripathy, D. (2008) Reversal mechanisms in ferromagnetic nanostructures. *IEEE Transactions on Magnetics*, 44, 1935–1940.

Aharoni, A. and Shtrikman, S. (1958) Magnetization curve of the infinite cylinder. *Physical Review*, 109, 1522.

Alenghat, F. J., Fabry, B., Tsai, K. Y., Goldmann, W. H., and Ingber, D. E. (2000) Analysis of cell mechanics in single vinculin-deficient cells using a magnetic tweezer. *Biochemical and Biophysical Research Communications*, 277, 93–99.

Allwood, D. A., Xiong, G., Cooke, M. D., Faulkner, C. C., Atkinson, D., Vernier, N., and Cowburn, R. P. (2002) Submicrometer ferromagnetic NOT gate and shift register. *Science*, 296, 2003–2006.

Allwood, D. A., Xiong, G., and Cowburn, R. P. (2004) Domain wall diodes in ferromagnetic planar nanowires. *Applied Physics Letters*, 85, 2848–2850.

Allwood, D. A., Xiong, G., Faulkner, C. C., Atkinson, D., Petit, D., and Cowburn, R. P. (2005) Magnetic domain-wall logic. *Science*, 309, 1688–1692.

Anguelouch, A., Leheny, R. L., and Reich, D. H. (2006) Application of ferromagnetic nanowires to interfacial microrheology. *Applied Physics Letters*, 89, 3.

Aranda, P. and Garcia, J. M. (2002) Porous membranes for the preparation of magnetic nanostructures. *Journal of Magnetism and Magnetic Materials*, 249, 214–219.

Ba, L. and Li, W. S. (2000) Influence of anodizing conditions on the ordered pore formation in anodic alumina. *Journal of Physics D—Applied Physics*, 33, 2527–2531.

Bader, S. D. (2006) Colloquium: Opportunities in nanomagnetism. *Reviews of Modern Physics*, 78, 1–15.

Bader, S. D., Buchanan, K. S., Chung, S. H., Guslienko, K. Y., Hoffmann, A., Ji, Y., and Novosad, V. (2007) Issues in nanomagnetism. *Superlattices and Microstructures*, 41, 72–80.

Bailey, T., Smith, B., Choi, B. J., Colburn, M., Meissl, M., Sreenivasan, S. V., Ekerdt, J. G., and Willson, C. G. (2001) Step and flash imprint lithography: Defect analysis. *Journal of Vacuum Science & Technology B*, 19, 2806–2810.

Bajorek, C. H., Coker, C., Romankiw, L. T., and Thompson, D. A. (1974) Hand-held magnetoresistive transducer. *IBM Journal of Research and Development*, 8, 541.

Barnas, J. (1992) Coupling between two ferromagnetic films through a non-magnetic metallic layer. *Journal of Magnetism and Magnetic Materials*, 111, L215–L219.

Baselt, D. R., Lee, G. U., Natesan, M., Metzger, S. W., Sheehan, P. E., and Colton, R. J. (1998) A biosensor based on magnetoresistance technology. *Biosensors & Bioelectronics*, 13, 731–739.

Bayer, C., Jorzick, J., Demokritov, S. O., Slavin, A. N., Guslienko, K. Y., Berkov, D. V., Gorn, N. L., Kostylev, M. P., and Hillebrands, B. (2006) Spin-wave excitations in finite rectangular elements. *Spin Dynamics in Confined Magnetic Structures III*. Berlin: Springer. pp. 57–103.

Bean, C. P., Doyle, M. V., and Entine, G. (1970) Etching of submicron pores in irradiated mica. *Journal of Applied Physics*, 41, 1454.

Beleggia, M. and De Graef, M. (2003) On the computation of the demagnetization tensor field for an arbitrary particle shape using a Fourier space approach. *Journal of Magnetism and Magnetic Materials*, 263, L1–L9.

Beleggia, M., De Graef, M., and Millev, Y. (2006a) Demagnetization factors of the general ellipsoid: An alternative to the Maxwell approach. *Philosophical Magazine*, 86, 2451–2466.

Beleggia, M., De Graef, M., and Millev, Y. T. (2006b) The equivalent ellipsoid of a magnetized body. *Journal of Physics D—Applied Physics*, 39, 891–899.

Bloemen, P. J. H., Johnson, M. T., Van De Vorst, M. T. H., Coehoorn, R., De Vries, J. J., Jungblut, R., Aan De Stegge, J., Reinders, A., and De Jonge, W. J. M. (1994) Magnetic layer thickness dependence of the interlayer exchange coupling in (001) Co/Cu/Co. *Physical Review Letters*, 72, 764.

Blondel, A., Meier, J. P., Doudin, B., and Ansermet, J. P. (1994) Giant magnetoresistance of nanowires of multilayers. *Applied Physics Letters*, 65, 3019–3021.

Brown, W. F. (1957) Criterion for uniform micromagnetization. *Physical Review*, 105, 1479.

Brown, J. W. F. (1960) Single-domain particles: New uses of old theorems. *American Journal of Physics*, 28, 542–551.

Brown, W. F. (1962) *Magnetostatic Principles in Ferromagnetism*. Amsterdam, the Netherlands: Elsevier/North-Holland.

Brown, W. F. and Morrish, A. H. (1957) Effect of a cavity on a single-domain magnetic particle. *Physical Review*, 105, 1198.

Brunner, T. A. (2003) Why optical lithography will live forever. *Journal of Vacuum Science & Technology B*, 21, 2632–2637.

Bruno, P. (1993) Oscillations of interlayer exchange coupling vs ferromagnetic-layers thickness. *Europhysics Letters*, 23, 615–620.

Castano, F. J., Hao, Y., Haratani, S., Ross, C. A., Vogeli, B., Walsh, M., and Smith, H. I. (2001a) Magnetic switching in 100 nm patterned pseudo spin valves. *IEEE Transactions on Magnetics*, 37, 2073–2075.

Castano, F. J., Hao, Y., Hwang, M., Ross, C. A., Vogeli, B., Smith, H. I., and Haratani, S. (2001b) Magnetization reversal in sub-100 nm pseudo-spin-valve element arrays. *Applied Physics Letters*, 79, 1504–1506.

Castano, F. J., Hao, Y., Ross, C. A., Vogeli, B., Smith, H. I., and Haratani, S. (2002a) Switching field trends in pseudo spin valve nanoelement arrays. *Journal of Applied Physics*, 91, 7317–7319.

Castano, F. J., Haratani, S., Hao, Y., Ross, C. A., and Smith, H. I. (2002b) Giant magnetoresistance in 60–150-nm-wide pseudo-spin-valve nanowires. *Applied Physics Letters*, 81, 2809–2811.

Castano, F. J., Ross, C. A., and Eilez, A. (2003) Magnetization reversal in elliptical-ring nanomagnets. *Journal of Physics D—Applied Physics*, 36, 2031–2035.

Chien, C. L., Sun, L., Tanase, M., Bauer, L. A., Hultgren, A., Silevitch, D. M., Meyer, G. J., Searson, P. C., and Reich, D. H. (2002) Electrodeposited magnetic nanowires: Arrays, field-induced assembly, and surface functionalization. *Journal of Magnetism and Magnetic Materials*, 249, 146–155.

Chiriac, H., Moga, A. E., Urse, M., and Ovari, T. A. (2003) Preparation and magnetic properties of electrodeposited magnetic nanowires. *Sensors and Actuators A—Physical*, 106, 348–351.

Chou, S. Y. (1997) Patterned magnetic nanostructures and quantized magnetic disks. *Proceedings of the IEEE*, 85, 652–671.

Chou, S. Y. (2001) Nanoimprint lithography and lithographically induced self-assembly. *MRS Bulletin*, 26, 512–517.

Chou, S. Y., Wei, M. S., Krauss, P. R., and Fischer, P. B. (1994) Single-domain magnetic pillar array of 35 nm diameter and 65 Gbits/in. density for ultrahigh density quantum magnetic storage. *Journal of Applied Physics*, 76 (10), 6679.

Chou, S. Y., Krauss, P. R., and Renstrom, P. J. (1995) Imprint of Sub-25 Nm vias and trenches in polymers. *Applied Physics Letters*, 67, 3114–3116.

Chun-Guey, W., Hu Leng, L., and Nai-Ling, S. (2006) Magnetic nanowires via template electrodeposition. *Journal of Solid State Electrochemistry*, 10, 198–202.

Cowburn, R. P. (2000) Property variation with shape in magnetic nanoelements. *Journal of Physics D—Applied Physics*, 33, R1–R16.

Doudin, B., Blondel, A., and Ansermet, J. P. (1996) Arrays of multilayered nanowires. *Journal of Applied Physics*, 79, 6090–6094.

Farhoud, M., Ferrera, J., Lochtefeld, A. J., Murphy, T. E., Schattenburg, M. L., Carter, J., Ross, C. A., and Smith, H. I. (1999) Fabrication of 200 nm period nanomagnet arrays using interference lithography and a negative resist. *Journal of Vacuum Science & Technology B*, 17, 3182–3185.

Ferain, E. and Legras, R. (1997) Characterisation of nanoporous particle track etched membrane. *Nuclear Instruments & Methods in Physics Research Section B—Beam Interactions with Materials and Atoms*, 131, 97–102.

Ferre, R., Ounadjela, K., George, J. M., Piraux, L., and Dubois, S. (1997) Magnetization processes in nickel and cobalt electrodeposited nanowires. *Physical Review B—Condensed Matter*, 56, 14066–14075.

Fert, A. and Piraux, L. (1999) Magnetic nanowires. *Journal of Magnetism and Magnetic Materials*, 200, 338–358.

Fischer, B. E. and Spohr, R. (1983) Production and use of nuclear tracks: Imprinting structure on solids. *Reviews of Modern Physics*, 55, 907.

Fraune, M., Rudiger, U., Guntherodt, G., Cardoso, S., and Freitas, P. (2000) Size dependence of the exchange bias field in NiO/Ni nanostructures. *Applied Physics Letters*, 77, 3815–3817.

Frei, E. H., Shtrikman, S., and Treves, D. (1957) Critical size and nucleation field of ideal ferromagnetic particles. *Physical Review*, 106, 446.

Friedman, A. L. and Menon, L. (2007) Optimal parameters for synthesis of magnetic nanowires in porous alumina templates–Electrodeposition study. *Journal of the Electrochemical Society*, 154, E68–E70.

Goolaup, S., Singh, N., and Adeyeye, A. O. (2005a) Coercivity variation in $Ni_{80}Fe_{20}$ ferromagnetic nanowires. *IEEE Transactions on Nanotechnology*, 4, 523–526.

Goolaup, S., Singh, N., Adeyeye, A. O., Ng, V., and Jalil, M. B. A. (2005b) Transition from coherent rotation to curling mode reversal process in ferromagnetic nanowires. *European Physical Journal B*, 44, 259–264.

Goolaup, S., Adeyeye, A. O., and Singh, N. (2006) Dipolar coupling in closely packed pseudo-spin-valve nanowire arrays. *Journal of Applied Physics*, 100, 114301.

Goolaup, S., Adeyeye, A. O., Singh, N., and Gubbiotti, G. (2007) Magnetization switching in alternating width nanowire arrays. *Physical Review B—Condensed Matter*, 75, 144430–144431.

Gubbiotti, G., Kostylev, M., Sergeeva, N., Conti, M., Carlotti, G., Ono, T., Slavin, A. N., and Stashkevich, A. (2004) Brillouin light scattering investigation of magnetostatic modes in symmetric and asymmetric NiFe/Cu/NiFe trilayered wires. *Physical Review B*, 70, 224422.

Gubbiotti, G., Tacchi, S., Carlotti, G., Vavassori, P., Singh, N., Goolaup, S., Adeyeye, A. O., Stashkevich, A., and Kostylev, M. (2005) Magnetostatic interaction in arrays of nanometric permalloy wires: A magneto-optic Kerr effect and a Brillouin light scattering study. *Physical Review B*, 72, 224413.

Gubbiotti, G., Tacchi, S., Carlotti, G., Singh, N., Goolaup, S., Adeyeye, A. O., and Kostylev, M. (2007) Collective spin modes in monodimensional magnonic crystals consisting of dipolarly coupled nanowires. *Applied Physics Letters*, 90, 092503.

Guillot, G. and Rondelez, F. (1981) Characteristics of submicron pores obtained by chemical etching of nuclear tracks in polycarbonate films. *Journal of Applied Physics*, 52, 7155.

Ha, J. K., Hertel, R., and Kirschner, J. (2003) Configurational stability and magnetization processes in submicron permalloy disks. *Physical Review B*, 67, 224432.

Haisma, J., Verheijen, M., Vandenheuvel, K., and Vandenberg, J. (1996) Mold-assisted nanolithography: A process for reliable pattern replication. *Journal of Vacuum Science & Technology B*, 14, 4124–4128.

Han, G. C., Zong, B. Y., Luo, P., and Wu, Y. H. (2003) Angular dependence of the coercivity and remanence of ferromagnetic nanowire arrays. *Journal of Applied Physics*, 93, 9202–9207.

Hao, Z., Shaoguang, Y., Gang, N., Shalogn, T., and Youwei, D. (2001) Fabrication and magnetic properties of $Fe_{14}Ni_{86}$ alloy nanowire array. *Journal of Physics—Condensed Matter*, 13, 1727–1731.

Heyderman, L. J., Solak, H. H., David, C., Atkinson, D., Cowburn, R. P., and Nolting, F. (2004) Arrays of nanoscale magnetic dots: Fabrication by x-ray interference lithography and characterization. *Applied Physics Letters*, 85, 4989–4991.

Heydon, G. P., Hoon, S. R., Farley, A. N., Tomlinson, S. L., Valera, M. S., Attenborough, K., and Schwarzacher, W. (1997) Magnetic properties of electrodeposited nanowires. *Journal of Physics D—Applied Physics*, 30, 1083–1093.

Hultgren, A., Tanase, M., Chen, C. S., Meyer, G. J., and Reich, D. H. (2003) Cell manipulation using magnetic nanowires. *Journal of Applied Physics*, 93, 7554–7556.

Husain, M. K. and Adeyeye, A. O. (2003) Magnetotransport properties of lithographically defined lateral Co/Ni80Fe20 wires. *Journal of Applied Physics*, 93, 7610–7612.

Hylton, T. L., Parker, M. A., Coffey, K. R., Howard, J. K., Fontana, R., and Tsang, C. (1995) Magnetostatically induced giant magnetoresistance in patterned NIFE/Ag multilayer thin-films. *Applied Physics Letters*, 67, 1154–1156.

Ishii, Y. (1991) Magnetization curling in an infinite cylinder with a uniaxial magnetocrystalline anisotropy. *Journal of Applied Physics*, 70, 3765–3769.

Joseph, R. I. (1966) Ballistic demagnetizing factor in uniformly magnetized cylinders. *Journal of Applied Physics*, 37, 4639–4643.

Katine, J. A., Palanisami, A., and Buhrman, R. A. (1999) Width dependence of giant magnetoresistance in Cu/Co multilayer nanowires. *Applied Physics Letters*, 74, 1883–1885.

Katine, J. A., Albert, F. J., and Buhrman, R. A. (2000) Current-induced realignment of magnetic domains in nanostructured Cu/Co multilayer pillars. *Applied Physics Letters*, 76, 354–356.

Khan, H. R. and Petrikowski, K. (2002) Synthesis and properties of the arrays of magnetic nanowires of Co and CoFe. *Materials Science & Engineering C—Biomimetic and Supramolecular Systems*, 19, 345–348.

Kirk, K. J., Chapman, J. N., and Wilkinson, C. D. W. (1997) Switching fields and magnetostatic interactions of thin film magnetic nanoelements. *Applied Physics Letters*, 71, 539–541.

Kostylev, M., Schrader, P., Stamps, R. L., Gubbiotti, G., Carlotti, G., Adeyeye, A. O., Goolaup, S., and Singh, N. (2008) Partial frequency band gap in one-dimensional magnonic crystals. *Applied Physics Letters*, 92.

Kume, M., Maeda, A., Tanuma, T., and Kuroki, K. (1996) Giant magnetoresistance effect in multilayered wire arrays. *Journal of Applied Physics*, 79, 6402–6404.

Lederman, M., Obarr, R., and Schulz, S. (1995) Experimental-study of individual ferromagnetic submicron cylinders. *IEEE Transactions on Magnetics*, 31, 3793–3795.

Liu, K., Chien, C. L., Searson, P. C., and Kui, Y. Z. (1998) Structural and magneto-transport properties of electrodeposited bismuth nanowires. *Applied Physics Letters*, 73, 1436–1438.

Mackintosh, F. C. and Schmidt, C. F. (1999) Microrheology. *Current Opinion in Colloid & Interface Science*, 4, 300–307.

Martin, J. I., Velez, M., Morales, R., Alameda, J. M., Anguita, J. V., Briones, F., and Vicent, J. L. (2002) Fabrication and magnetic properties of arrays of amorphous and polycrystalline ferromagnetic nanowires obtained by electron beam lithography. *Journal of Magnetism and Magnetic Materials*, 249, 156–162.

Martin, J. I., Nogues, J., Liu, K., Vicent, J. L., and Schuller, I. K. (2003) Ordered magnetic nanostructures: Fabrication and properties. *Journal of Magnetism and Magnetic Materials*, 256, 449–501.

Masuda, H. and Fukuda, K. (1995) Ordered metal nanohole arrays made by a 2-Step replication of honeycomb structures of anodic alumina. *Science*, 268, 1466–1468.

Masuda, H., Yamada, H., Satoh, M., Asoh, H., Nakao, M., and Tamamura, T. (1997) Highly ordered nanochannel-array architecture in anodic alumina. *Applied Physics Letters*, 71, 2770–2772.

Mcgary, P. D., Liwen, T., Jia, Z., Stadler, B. J. H., Downey, P. R., and Flatau, A. B. (2006) Magnetic nanowires for acoustic sensors (invited). *Journal of Applied Physics*, 99, 8–310.

Meier, J., Doudin, B., and Ansermet, J. P. (1996) Magnetic properties of nanosized wires. *Journal of Applied Physics*, 79, 6010–6012.

Metzger, R. M., Konovalov, V. V., Sun, M., Xu, T., Zangari, G., Xu, B., Benakli, M., and Doyle, W. D. (2000) Magnetic nanowires in hexagonally ordered pores of alumina. *IEEE Transactions on Magnetics*, 36, 30–35.

Millev, Y. T., Vedmedenko, E., and Oepen, H. P. (2003) Dipolar magnetic anisotropy energy of laterally confined ultrathin ferromagnets: Multiplicative separation of discrete and continuum contributions. *Journal of Physics D—Applied Physics*, 36, 2945–2949.

Miyawaki, T., Toyoda, K., Kohda, M., Fujita, A., and Nitta, J. (2006) Magnetic interaction of submicron-sized ferromagnetic rings in one-dimensional array. *Applied Physics Letters*, 89, 122508.

Moore, L. R., Zborowski, M., Sun, L. P., and Chalmers, J. J. (1998) Lymphocyte fractionation using immunomagnetic colloid and a dipole magnet flow cell sorter. *Journal of Biochemical and Biophysical Methods*, 37, 11–33.

Morecroft, D., Van Aken, B. B., Prieto, J. L., Kang, D. J., Burnell, G., and Blamire, M. G. (2005) In situ magnetoresistance measurements during nanopatterning of pseudo-spin-valve structures. *Journal of Applied Physics*, 97, 054302.

Napolskii, K. S., Eliseev, A. A., Yesin, N. V., Lukashin, A. V., Tretyakov, Y. D., Grigorieva, N. A., Grigoriev, S. V., and Eckerlebe, H. (2007) Ordered arrays of Ni magnetic nanowires: Synthesis and investigation. *Physica E—Low-Dimensional Systems & Nanostructures*, 37, 178–183.

Neel, L. (1962) A magnetostatic problem concerning ferromagnetic films. *Comptes rendus de l'Académie des Sciences*, 255, 1545.

Nielsch, K., Wehrspohn, R. B., Barthel, J., Kirschner, J., Gosele, U., Fischer, S. F., and Kronmuller, H. (2001) Hexagonally ordered 100 nm period nickel nanowire arrays. *Applied Physics Letters*, 79, 1360–1362.

Obarr, R., Yamamoto, S. Y., Schultz, S., Xu, W. H., and Scherer, A. (1997) Fabrication and characterization of nanoscale arrays of nickel columns. *Journal of Applied Physics*, 81, 4730–4732.

Ohgai, T., Gravier, L., Hoffer, X., and Ansermet, J. P. (2005) CdTe semiconductor nanowires and NiFe ferro-magnetic metal nanowires electrodeposited into cylindrical nano-pores on the surface of anodized aluminum. *Journal of Applied Electrochemistry*, 35, 479–485.

Osborn, J. A. (1945) Demagnetizing factors of the general ellipsoid. *Physical Review*, 67, 351.

Pant, B. B. (1996) Effect of interstrip gap on the sensitivity of high sensitivity magnetoresistive transducers. *Journal of Applied Physics*, 79, 6123–6125.

Parkin, S. S. P., More, N., and Roche, K. P. (1990) Oscillations in exchange coupling and magnetoresistance in metallic superlattice structures: Co/Ru, Co/Cr, and Fe/Cr. *Physical Review Letters*, 64, 2304.

Perez-Junquera, A., Martin, J. I., Velez, M., Alameda, J. M., and Vicent, J. L. (2003) Temperature dependence of the magnetization reversal process in patterned Ni nanowires. *Nanotechnology*, 14, 294–298.

Piraux, L., George, J. M., Despres, J. F., Leroy, C., Ferain, E., Legras, R., Ounadjela, K., and Fert, A. (1994) Giant magnetoresistance in magnetic multilayered nanowires. *Applied Physics Letters*, 65, 2484–2486.

Possin, G. E. (1970) A method for forming very small diameter wires. *Review of Scientific Instruments*, 41, 772.

Remhof, A., Schumann, A., Westphalen, A., Zabel, H., Mikuszeit, N., Vedmedenko, E. Y., Last, T., and Kunze, U. (2008) Magnetostatic interactions on a square lattice. *Physical Review B*, 77, 134409.

Rhodes, P., Rowlands, G., and Birchall, D. R. (1962) Magnetization energies and distributions in ferromagnetics. *Journal of the Physical Society of Japan*, 17, 543–547.

Ross, C. (2001) Patterned magnetic recording media. *Annual Review of Materials Research*, 31, 203–235.

Ross, C. A., Castano, F. J., Rodriguez, E., Haratani, S., Vogeli, B., and Smith, H. I. (2005) Size-dependent switching of multilayer magnetic elements. *Journal of Applied Physics*, 97, 053902.

Routkevitch, D., Bigioni, T., Moskovits, M., and Xu, J. M. (1996) Electrochemical fabrication of CdS nanowire arrays in porous anodic aluminum oxide templates. *Journal of Physical Chemistry*, 100, 14037–14047.

Safarik, I. and Safarikova, M. (1999) Use of magnetic techniques for the isolation of cells. *Journal of Chromatography B*, 722, 33–53.

Sampaio, L. C., Sinnecker, E., Cernicchiaro, G. R. C., Knobel, M., Vazquez, M., and Velazquez, J. (2000) Magnetic microwires as macrospins in a long-range dipole-dipole interaction. *Physical Review B*, 61, 8976–8983.

Savas, T. A., Schattenburg, M. L., Carter, J. M., and Smith, H. I. (1996) Large-area achromatic interferometric lithography for 100 nm period gratings and grids. *Journal of Vacuum Science & Technology B*, 14, 4167–4170.

Savas, T. A., Farhoud, M., Smith, H. I., Hwang, M., and Ross, C. A. (1999) Properties of large-area nanomagnet arrays with 100 nm period made by interferometric lithography. *Journal of Applied Physics*, 85, 6160–6162.

Schattenburg, M. L., Aucoin, R. J., and Fleming, R. C. (1995) Optically matched trilevel resist process for nanostructure fabrication. *Journal of Vacuum Science & Technology B*, 13, 3007–3011.

Schwarzacher, W., Attenborough, K., Michel, A., Nabiyouni, G., and Meier, J. P. (1997) Electrodeposited nanostructures. *Journal of Magnetism and Magnetic Materials*, 165, 23–29.

Sellmyer, D. J., Zheng, M., and Skomski, R. (2001) Magnetism of Fe, Co, and Ni nanowires in self-assembled arrays. *Journal of Physics—Condensed Matter*, 13, R433–R460.

Shearwood, C., Ahmed, H., Nicholson, L. M., Bland, J. A. C., Baird, M. J., Patel, M., and Hughes, H. P. (1993) Fabrication and magnetization measurements of variable-pitch gratings of cobalt on GaAs. *Microelectronic Engineering*, 21, 431–434.

Shtrikman, S. and Treves, D. (1959) The coercive force and rotational hysteresis of elongated ferromagnetic particles. *Journal de Physique et Le Radium*, 20, 286.

Singh, N., Goolaup, S., and Adeyeye, A. O. (2004) Fabrication of large area nanomagnets. *Nanotechnology*, 15, 1539–1544.

Spallas, J. P., Boyd, R. D., Britten, J. A., Fernandez, A., Hawryluk, A. M., Perry, M. D., and Kania, D. R. (1996) Fabrication of sub-0.5 mu m diameter cobalt dots on silicon substrates and photoresist pedestals on 50 cm × 50 cm glass substrates using laser interference lithography. *Journal of Vacuum Science & Technology B*, 14, 2005–2007.

Srajer, G., Lewis, L. H., Bader, S. D., Epstein, A. J., Fadley, C. S., Fullerton, E. E., Hoffmann, A., Kortright, J. B., Krishnan, K. M., Majetich, S. A., Rahman, T. S., Ross, C. A., Salamon, M. B., Schuller, I. K., Schulthess, T. C., and Sun, J. Z. (2006) Advances in nanomagnetism via X-ray techniques. *Journal of Magnetism and Magnetic Materials*, 307, 1–31.

Stoner, E. C. (1945) The demagnetizing factors for ellipsoids. *Philosophical Magazine*, 36, 803–820.

Stoner, E. C. and Wohlfarth, E. P. (1991) A mechanism of magnetic hysteresis in heterogeneous alloys (Reprinted from *Philosophical Transaction Royal Society-London*, 240, 599–642, 1948). *IEEE Transactions on Magnetics*, 27, 3475–3518.

Sun, L., Searson, P. C., and Chien, C. L. (1999) Electrochemical deposition of nickel nanowire arrays in single-crystal mica films. *Applied Physics Letters*, 74, 2803–2805.

Terris, B. D., Thomson, T., and Hu, G. (2007) Patterned media for future magnetic data storage. *Microsystem Technologies—Micro- and Nanosystems—Information Storage and Processing Systems*, 13, 189–196.

Tsoi, M., Fontana, R. E., and Parkin, S. S. P. (2003) Magnetic domain wall motion triggered by an electric current. *Applied Physics Letters*, 83, 2617–2619.

Vavassori, P., Grimsditch, M., Novosad, V., Metlushko, V., and Ilic, B. (2003) Metastable states during magnetization reversal in square permalloy rings. *Physical Review B*, 67, 134429.

Vavassori, P., Bonanni, V., Gubbiotti, G., Adeyeye, A. O., Goolaup, S., and Singh, N. (2007) Cross-over from coherent rotation to inhomogeneous reversal mode in interacting ferromagnetic nanowires. *Journal of Magnetism and Magnetic Materials*, 316, E31–E34.

Vazquez, M., Hernandez-Velez, M., Asenjo, A., Navas, D., Pirota, K., Prida, V., Sanchez, O., and Baldonedo, J. L. (2006) Preparation and properties of novel magnetic composite nanostructures: Arrays of nanowires in porous membranes. *Physica B—Condensed Matter*, 384, 36–40.

Vedmedenko, E. Y., Oepen, H. P., and Kirschner, J. (2003) Size-dependent magnetic properties in nanoplatelets. *Journal of Magnetism and Magnetic Materials*, 256, 237–242.

Velazquez, J., Pirota, K. R., and Vazquez, M. (2003) About the dipolar approach in magnetostatically coupled bistable magnetic micro and nanowires. *IEEE Transactions on Magnetics*, 39, 3049–3051.

Vogeli, B., Smith, H. I., Castano, F. J., Haratani, S., Hao, Y. W., and Ross, C. A. (2001) Patterning processes for fabricating sub-100 nm pseudo-spin valve structures. *Journal of Vacuum Science & Technology B*, 19, 2753–2756.

Wassermann, E. F., Thielen, M., Kirsch, S., Pollmann, A., Weinforth, H., and Carl, A. (1998) Fabrication of large scale periodic magnetic nanostructures. *Journal of Applied Physics*, 83, 1753–1757.

Wegrowe, J. E., Meier, J. P., Doudin, B., Ansermet, J. P., Wernsdorfer, W., Barbara, B., Coffey, W. T., Kalmykov, Y. P., and Dejardin, J. L. (1999) Magnetic relaxation of nanowires: Beyond the Néel-Brown activation process. *Europhysics Letters*, 38, 329–334.

Wernsdorfer, W., Orozco, E. B., Hasselbach, K., Benoit, A., Barbara, B., Demoncy, N., Loiseau, A., Pascard, H., and Mailly, D. (1997) Experimental evidence of the Néel-Brown model of magnetization reversal. *Physical Review Letters*, 78, 1791–1794.

Whitney, T. M., Jiang, J. S., Searson, P. C., and Chien, C. L. (1993) Fabrication and magnetic-properties of arrays of metallic nanowires. *Science*, 261, 1316–1319.

Wu, C.-G., Lin, H. L., and Shau, N.-L. (2006) Magnetic nanowires via template electrodeposition. *Journal of Solid State Electrochemistry*, 10, 198–202.

Yelon, A. (1971) Interactions in multilayer magnetic films. In: Hoffman, R. W. and M. H. Francombe, (Eds.) *Physics of Thin Films*. New York: Academic Press.

Yin, A. J., Li, J., Jian, W., Bennett, A. J., and Xu, J. M. (2001) Fabrication of highly ordered metallic nanowire arrays by electrodeposition. *Applied Physics Letters*, 79, 1039–1041.

Yuan, S. W. and Bertram, H. N. (1993) Micromagnetics of small unshielded MR elements. *Journal of Applied Physics*, 73, 6235–6237.

Zeng, H., Skomski, R., Menon, L., Liu, Y., Bandyopadhyay, S., and Sellmyer, D. J. (2002) Structure and magnetic properties of ferromagnetic nanowires in self-assembled arrays. *Physical Review B*, 65, 8.

Zheng, M., Yu, M., Liu, Y., Skomski, R., Liou, S. H., Sellmyer, D. J., Petryakov, V. N., Verevkin, Y. K., Polushkin, N. I., and Salashchenko, N. N. (2001) Magnetic nanodot arrays produced by direct laser interference lithography. *Applied Physics Letters*, 79, 2606–2608.

Zhu, X. B., Grutter, P., Metlushko, V., Hao, Y., Castano, F. J., Ross, C. A., Ilic, B., and Smith, H. I. (2003) Construction of hysteresis loops of single domain elements and coupled permalloy ring arrays by magnetic force microscopy. *Journal of Applied Physics*, 93, 8540–8542.

23

Networks of Nanorods

Tanja Schilling
Université du Luxembourg

Swetlana Jungblut
Universität Wien

Mark A. Miller
University Chemical Laboratory

23.1 Introduction

The term nanorod may be applied to an unbranched macromolecule or supramolecular assembly whose diameter is a few nanometers or tens of nanometers and whose length is much longer than its diameter, giving a high aspect ratio. Additionally, nanorods are distinguished from long chain-like polymers by their stiffness; a nanorod has a clear orientation in space, and although it may be somewhat flexible, it does not readily fold or form coils.

This chapter provides an introduction to the physics of interconnected assemblies of nanorods. Although such networks are encountered in areas of science as apparently disparate as materials science and cell biology, their physics is based on some unifying underlying concepts that can be explained with a generic description of rod-like particles. Many of these concepts are geometrical in origin. One of the best known examples is Onsager's explanation in 1949 of the spontaneous orientational ordering of long rods into a "nematic" phase, on the basis that the sacrificed orientational entropy is more than repaid by the gain in translational entropy (Onsager, 1949). This general theory explains results in a diverse range of specific experimental systems, including some that are apparently rather complex, such as suspensions of rod-like viruses (Bawden et al., 1936). In this chapter, the main underlying theme will be percolation theory, which describes the transition from isolated finite clusters of particles to a system-spanning network that produces connectivity on a macroscopic scale. Because of the generality of percolation theory, many inferences can be made about specific systems composed of nanorods based on knowledge gained from simple geometrical models. Indeed, the appealing simplicity of rod-like particles has led to simple models of rods being used as a forum for exploring fundamental physical properties of anisotropic objects such as phase transitions and scaling laws.

Following an outline of relevant aspects of percolation theory in Section 23.2, we turn in Section 23.3 to a survey of applications of carbon nanotube (CNT) networks in composite materials. In this field, the remarkable properties of individual CNTs are exploited to develop novel and useful materials. Although nanorods have many applications in materials science, such as liquid crystals, we have focused on cases where the involvement of a network is crucial. Most of our interest will lie in equilibrium properties, though it is important to bear in mind that nonequilibrium phenomena such as flow and gelation come into play, especially during the preparation of the composites.

In Section 23.4, we address the importance of nanorod networks in cell biology, concentrating on the physics of cross-linked assemblies of actin filaments, as found in the cytoskeleton. The cytoskeleton itself is a dynamic system, and its constant consumption of energy keeps it permanently out of equilibrium. However, in order to understand the rich physics of this complex network, simplified experimental systems have been extracted from it and explored under more controlled conditions *in vitro*. The analysis will bring us into contact with the highly developed field of semiflexible polymers, and we refer readers to the book by Doi and Edwards (1986) for further detail on this topic.

23.2 Percolation

23.2.1 Connectivity Percolation

The physical networks that we are considering are collections of connected nanoparticles and can therefore be regarded as clusters that have grown to a very large size. In a system of finite clusters, the distribution of cluster sizes is governed by many factors, such as the strength and type of the interactions, the temperature, and the density. At certain combinations of these conditions, the particles form a cluster large enough to span the system. The locus of conditions where spanning first occurs is known as the percolation threshold and it demarcates the divergence of the average cluster size, where the system is transformed from an ensemble of finite, disconnected clusters into a macroscopic network.

The point at which a connected pathway first appears is important in a wide range of applications. An important example in the context of this chapter is a suspension of electrically conducting particles in an insulating medium, which is an insulator below the percolation density but suddenly becomes a conductor above the threshold. Similar considerations arise in quite different fields, such as porous materials, which only become permeable to fluids when the pores form a connected network on the scale of the sample.

The connectivity percolation transition is one of the archetypal transitions in statistical physics. Although it is a purely geometrical problem, it shares certain properties with other transitions. It therefore serves as a useful model, in particular, for the mathematical treatment of critical phenomena, which is concerned with how properties change or diverge as a transition is approached. Introductions to percolation often start by considering bonds on a lattice for simplicity, but the theory and results (including the values of critical exponents) are universal for a given dimensionality of space and transfer directly from regular lattices to the continuum. Here, we will outline some continuum methods that are in current use and can be applied to networks of nanorods, referring the reader to the standard texts (Stauffer and Aharony, 1994, Grimmett, 1999, Bollobás and Riordan, 2006) for a more general treatment.

For simplicity, we will start with a system of spherical particles to illustrate percolation theory, noting that the concepts transfer directly to the case of particles with anisotropic shape, such as rods. Consider N freely interpenetrable spheres of diameter σ. We place these spheres randomly in a cube of length L and volume $V = L^3$ and apply periodic boundary conditions, as illustrated schematically in two dimensions in Figure 23.1. Periodic boundary conditions are useful for two reasons. First of all, they mitigate the effect of studying a finite system by removing the exposed surfaces of an isolated box. This point is particularly important in computer simulations, since the number of particles that can be treated is usually much smaller than that encountered in an experiment. Secondly, as the system size L is increased, the remaining finite-size effects, associated with the fact that fluctuations larger than L are missing, scale in a well-understood way that allows the percolation threshold in the thermodynamic limit to be obtained relatively easily (Škvor et al., 2007). The most noticeable effect of studying a finite system is that percolation does not set in at a sharp threshold, but spanning clusters appear more gradually (i.e., with a smoothly increasing probability) over a narrow range of densities.

For the purposes of defining clusters, we now consider two spheres to be connected if they overlap, i.e., if the distance between their centers is less than σ. We may then ask above what density threshold $\rho_c = N_c/V$ a cluster of spheres is first observed to be connected to its own periodic image via the boundaries. At this point, the periodically repeated system appears to contain an infinite cluster. Below ρ_c, disconnected clusters with a distribution of sizes are found, and we may want to know the mean cluster size, defined as the mean number of particles in the cluster to which a randomly chosen particle belongs:

$$S = \frac{\sum_s s^2 n_s}{\sum_s s n_s},$$

where n_s is the number of clusters that contain s spheres. Later in this chapter, specific properties such as conductivity and mechanical response will be explored. Certain aspects of the internal structure of clusters are then of interest in addition to the cluster size distribution, and these will depend on the nature of the interactions between the particles even for clusters of a given size.

These properties can be determined in various ways. For a mathematical treatment see, for example, the book by Meester and Roy (1996). Much progress can be made numerically by a direct implementation of the model in a computer simulation, generating many random configurations to improve statistics. It can be enlightening to follow a theoretical approach, and here we will describe one that makes use of a formal correspondence between the percolation theory and the statistical theory of fluids. A detailed understanding of the following analysis is not necessary for the subsequent discussion or for the other sections of this chapter.

In order to show the connection between the percolation theory and the theory of fluids, we start with a brief summary of the key concepts in the latter (Hansen and McDonald, 2006). Consider the equation of state of an ideal gas

$$\beta P = \rho,$$

where

P is the pressure

ρ is the particle number density

$1/\beta$ is the product of the Boltzmann constant k_B with the temperature T

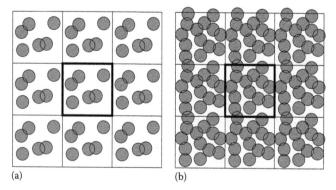

(a)   (b)

FIGURE 23.1 Percolation in a system of freely interpenetrable spheres. The bold square marks the volume V, while the surrounding squares are the periodic images. (a) $\rho < \rho_c$, below the percolation threshold. (b) $\rho > \rho_c$, a cluster has formed that is connected to its own periodic images.

The equation of state of a real gas deviates from this simple relation because particles interact with each other. The higher the density is, the stronger the interactions and hence the greater the deviation from ideal gas behavior. This fact is expressed by writing the equation of state of an interacting gas as a power series in the density

$$\frac{\beta P}{\rho} = 1 + \sum_{i=2}^{\infty} B_i(T)\rho^{i-1},$$

which is known as the virial expansion with the virial coefficients, $B_i(T)$.

Computing virial coefficients is in general a difficult task. In order to do so systematically, one defines the Mayer f-function

$$f(r) = \exp[-\beta \upsilon(r)] - 1,$$

where $\upsilon(r)$ is the potential energy due to the interaction of two particles separated by a distance r (assuming an isotropic potential here for simplicity). Using the f-function, the virial coefficients can be mapped to graphs that represent multidimensional integrals, with vertices standing for particle coordinates and edges for integration over the associated f-function. For short-ranged potentials, the f-function decays to zero rapidly with increasing distance, making a systematic evaluation of the graphs feasible. In this scheme the second virial coefficient, for example, is simply given by

$$B_2(T) = -\frac{1}{2} \int f(r)\,\mathrm{d}r.$$

While the virial expansion approach is rather intuitive, much more information about the liquid is contained in its structure. Hence, one usually approaches a liquid state theory problem using the pair-distribution function,

$$g^{(2)}(r) = \frac{1}{N^2} \sum_{i,j=1}^{N} \left\langle \delta\!\left(\left|\mathbf{r}_i - \mathbf{r}_j\right| - r\right) \right\rangle,$$

where $\mathbf{r}_i$ is the position of particle i and the angle brackets denote a thermal average. $g^{(2)}(r)$ describes the probability of finding a particle at a given distance r from another particle, normalized with respect to the overall density. For short distances, $g^{(2)}(r)$ reflects the structure of the fluid that is due to interactions between the particles. In the limit of large distances $g^{(2)}(r)$ goes to 1, because there are no long-range correlations in a fluid. In order to focus on the structure, which is contained in the deviations from the overall density, one therefore often uses the total pair-correlation function instead

$$h^{(2)}(r) = g^{(2)}(r) - 1.$$

Ornstein and Zernike suggested splitting $h^{(2)}(r)$ into two parts: the direct correlation function $c^{(2)}(r)$, which accounts for the direct interactions between the two particles; and the remaining indirect contributions that are mediated by other particles,

$$h^{(2)}(r) = c^{(2)}(r) + \rho \int c\left(\left|\mathbf{r} - \mathbf{r}'\right|\right) h^{(2)}(r')\,\mathrm{d}\mathbf{r}'. \tag{23.1}$$

The idea behind this strategy is to separate the part of the correlations that stays short-ranged even at criticality (Ornstein and Zernike, 1914). Equation 23.1 defines the direct correlation function, but it does not yet provide a route to the computation of, for example, an equation of state for a given system. In order to solve the Ornstein–Zernike Equation 23.1, one needs a closure relation, i.e., one further condition that relates $h^{(2)}(r)$ and $c^{(2)}(r)$. A variety of closures are available, and their suitability depends on the specific interactions between the particles and the relative difficulty of solving the resulting equations. One relation that is fairly robust for many types of interactions is the Percus–Yevick approximation,

$$c^{(2)}(r) = g^{(2)}(r)\left[1 - \exp\left(\beta \upsilon(r)\right)\right],$$

which sets the direct correlation function to zero where the potential is zero. Finally, a density expansion similar to that described for the equation of state can also be made for the correlation functions.

These concepts can now be transferred to the percolation problem. The idea goes back to work by Hill (1955) and more detailed subsequent work by Coniglio and coworkers (Coniglio et al., 1977), who suggested decomposing the partition function of the system into clusters of connected particles and unconnected particles and then applying a density expansion to a "pair-connectedness function" in analogy to the expansion of the pair-correlation function.

First, we define an effective interaction that distinguishes between connected and unconnected particles. In the case of the freely interpenetrable spheres introduced above, this could be

$$u^+(\mathbf{r}) = \begin{cases} 0 & \text{if } r < \sigma \\ \infty & \text{otherwise} \end{cases}$$

$$u^\star(\mathbf{r}) = \begin{cases} \infty & \text{if } r < \sigma \\ 0 & \text{otherwise} \end{cases}$$

and

$$f^+(\mathbf{r}) = \exp[-\beta u^+(\mathbf{r})], \qquad f^\star(\mathbf{r}) = \exp[-\beta u^\star(\mathbf{r})] - 1,$$

known as the Mayer cluster functions (Bug et al., 1986, DeSimone et al., 1986). We then introduce the pair-connectedness function, $H^+(\mathbf{r}_1, \mathbf{r}_2)$, by analogy with the pair-correlation function, where

$$\rho^2 H^+(\mathbf{r}_1, \mathbf{r}_2)\,\mathrm{d}\mathbf{r}_1\mathrm{d}\mathbf{r}_2$$

is the probability that sphere 1 is in $d\mathbf{r}_1$, sphere 2 is in $d\mathbf{r}_2$, and the spheres overlap. Analogously to the Ornstein–Zernike approach, we then define the direct pair-connectedness function, $C^+(1, 2)$:

$$C^+(\mathbf{r}_1, \mathbf{r}_2) = H^+(\mathbf{r}_1, \mathbf{r}_2) - \rho \int C^+(\mathbf{r}_1, \mathbf{r}_3) H^+(\mathbf{r}_2, \mathbf{r}_3) d\mathbf{r}_3.$$

Given a closure condition we can now compute the percolation threshold as follows. We first note that C^+ and H^+ are translationally invariant, i.e., they can be written as functions of the difference $\mathbf{r}_{12} = \mathbf{r}_1 - \mathbf{r}_2$, so that $C^+(\mathbf{r}_1, \mathbf{r}_2) \equiv C^+(\mathbf{r}_{12})$ and similarly for H^+. We may then Fourier transform the Ornstein–Zernike equation, yielding the reciprocal space equation

$$\tilde{H}^+(k) = \frac{\tilde{C}^+(k)}{1 - \rho \tilde{C}^+(k)},$$

where k is the wave number. At the percolation threshold, the mean cluster size S diverges. The mean cluster size is related to $\tilde{C}^+$ by

$$S = 1 + \rho \int H^+(\mathbf{r}_{12}) d\mathbf{r}_{12} = 1 + p\tilde{H}^+(0) = \frac{1}{1 - \rho\tilde{C}^+(0)}.$$

Hence, S diverges at

$$\rho_c \tilde{C}^+(0) = 1,$$

where we must note carefully that $\tilde{C}^+(0)$ is dependent on the density.

If we consider freely overlapping particles, the interpretation of $C^+(\mathbf{r}_1, \mathbf{r}_2)$ is particularly graphic. Within the second virial approximation,

$$C^+(\mathbf{r}_1, \mathbf{r}_2) = f^+(\mathbf{r}_1, \mathbf{r}_2). \tag{23.2}$$

Hence, the percolation threshold, ρ_c, is inversely proportional to the so-called excluded volume, i.e., the volume around the center of mass of one particle into which a second particle's center of mass has to enter in order to produce a connection (see Figure 23.2).

The general approach described above for an ideal gas of spheres can also be applied to systems of interpenetrable non-spherical particles as well as particles that interact with each other. The particle shape and interaction potential enter in the cluster Mayer functions, f^+ and f^*, leading to more complicated expressions, but the analysis proceeds analogously.

The result of Coniglio theory in Equation 23.2 begins to explain a key observation in the percolation of rod-shaped objects, that the number density of particles at percolation is proportional to the inverse square of the rod aspect ratio, or equivalently, the fraction of the system volume occupied by the rods is inversely proportional to the length of the rods, since the volume of a single rod is proportional to its length (Balberg et al., 1984, Bug et al., 1985, 1986). The excluded volume of two

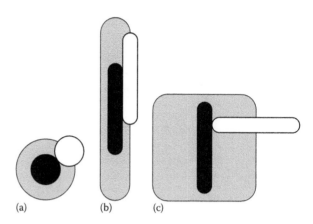

(a)   (b)   (c)

FIGURE 23.2 Illustration of the excluded volume (gray area) into which the center of the white particle may not enter if it is to avoid overlap with the black particle. (a) The excluded volume between two spheres is a sphere with twice the diameter. (b, c) The excluded volume between two rods depends on the relative orientation and is largest when the rods are perpendicular.

rods depends on their mutual orientations, as illustrated in Figure 23.2. For long rods, the average excluded volume over all orientations increases as the square of the rod length, producing a corresponding decrease in the percolation density.

In order to account more thoroughly for the excluded volume of long rod-like polymers that do not freely interpenetrate due to a hard core, the approach by Coniglio has been combined with the reference interaction site model (RISM) (Leung and Chandler, 1991, Chatterjee, 2000, Wang and Chatterjee, 2003). Here, the pair-connectedness function takes into account the fact that the particles are joined to form molecules, using

$$\rho^2 H_{\alpha\beta}^+(\mathbf{r}_1, \mathbf{r}_2) \, d\mathbf{r}_1 d\mathbf{r}_2$$

as the probability that site α of one molecule is in $d\mathbf{r}_1$, site β of another molecule is in $d\mathbf{r}_2$, and the sites are connected. In order to compute $H_{\alpha\beta}^+$ the relevant Ornstein–Zernike-like equation is now a matrix equation containing both the intramolecular correlations and the site–site correlations (Chandler and Andersen, 1972). Using RISM, Leung and Chandler showed that, just as for ideal rods, the percolation threshold of rods with a hard core scales as the inverse aspect ratio. However, the larger the size of the hard core, the longer the rods have to be before the scaling regime is entered (Leung and Chandler, 1991).

In a series of articles, Wang and Chatterjee extended the RISM approach to incorporate attractive interactions between the rods. The attraction can be a direct result of site–site interactions in the one-component fluid of rods (Wang and Chatterjee, 2001), or the indirect result of explicitly adding a second component to the suspension of rods (Wang and Chatterjee, 2002, 2003). In particular, these authors showed that the addition of coiled polymers (which can be regarded as small spheres) to the suspension lowers the percolation threshold of the rods. The polymers produce an effective attraction between the rods because closing the gaps between the rods increases the free volume (and thereby the entropy) of the polymers, as illustrated

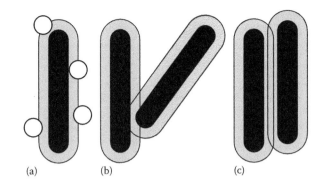

FIGURE 23.3 (a) In a mixture of impenetrable rods and spheres, each rod is surrounded by an excluded volume (gray) into which the center of the spheres cannot enter. (b) When two rods approach, their excluded volumes overlap, increasing the free volume available to the spheres. (c) The greatest increase in free volume occurs when the rods are aligned as they approach.

in Figure 23.3. This depletion effect has also been studied in explicit Monte Carlo simulations using mixtures of hard rods and spheres (Schilling et al., 2007). The simulations are computationally challenging because they must equilibrate the configurations thoroughly, taking care to respect detailed balance (Frenkel and Smit, 2002), in order to capture the structural correlations introduced by the impenetrability of the particles.

In a recent piece of work (Kyrylyuk and van der Schoot, 2008), a different theoretical approach was taken, by combining Coniglio theory with liquid state theory for rods on the level of the second virial approximation (which is exact for infinitely long rods interacting by a short-ranged, repulsive potential), fully including translation–rotation coupling. It was found that polydispersity in rod length decreases the percolation threshold—a counter-intuitive effect that has considerable advantages for the production of composite materials on a large scale. The same study also predicted that weak attractions decrease the percolation threshold and that flexibility of the rods increases it, in agreement with independent theoretical studies (Wang and Chatterjee, 2001). An increase in the percolation threshold has also been predicted for "wavy" nanorods, using a model where the lack of straightness is captured by considering rigid sinusoidal or helical rods (Berhan and Sastry, 2007). The inverse proportionality between the percolation threshold and the rod length transfers from straight to wavy rods, but the constant of proportionality becomes a function of the nanorod waviness.

23.2.2 Rigidity Percolation

At the connectivity percolation threshold itself, the percolating cluster is only transient if the bonding is reversible, since the breaking of one bond may be enough to disconnect the cluster into two pieces. Further into the percolated regime, however, the greater number of connections leads to multiple independent paths through the cluster, so that the cluster continues to percolate even though its structure is changing dynamically.

Even if the bonds are permanent, a network at the connectivity threshold will not be able to support stress if rotation at the contacts is possible. The network is said to have floppy modes because it may deform without energetic penalty. If the number of connections is increased, some degrees of freedom will become constrained and floppy modes will be lost, leaving a rigid cluster that cannot be deformed without bending or stretching a rod. The point at which rigidity occurs on a macroscopic scale defines the rigidity percolation threshold, whose physics is independent of connectivity percolation (Jacobs and Thorpe, 1995).

The concept and theory of rigidity percolation were introduced by Thorpe in the context of polymeric glasses (Thorpe, 1983). In such systems, isolated rigid regions may develop, but it is only when the average coordination number is increased that these regions join to make a true amorphous solid. The changes in mechanical properties at the rigidity threshold are also important in random fiber networks such as paper (Latva-Kokko et al., 2001) and the cytoskeleton in living cells (Head et al., 2003a).

To illustrate the problem, imagine a network of rods in d dimensions and connected at M sites. The structure of the network is described by the positions of these sites, which collectively have Md degrees of freedom. Each rod represents a distance constraint between two sites and reduces the number of modes available for motion to the network by one, unless the sites were already constrained directly or indirectly by other rods, in which case the constraint in question is redundant. If all constraints were independent, simple counting would establish when no floppy modes remain. The difficulty comes in establishing the number of redundant constraints, since these do not reduce the number of floppy modes.

Floppy modes can always, in principle, be identified by a standard normal mode analysis. A spring constant and equilibrium length are assigned to each rod, such that the network is at mechanical equilibrium. The $Md \times Md$ Hessian matrix of second derivatives of the potential energy with respect to site displacements is then assembled. Diagonalization of the Hessian leads to the normal mode frequencies, with each zero eigenvalue revealing a floppy mode. This method rapidly becomes impractical as M increases because of the time taken to diagonalize the Hessian (which typically increases as M^3) and due to the numerical problem of obtaining unambiguously zero eigenvalues for the floppy modes.

A less costly numerical method for detecting floppy modes that is also exact in principle involves relaxing the network after a perturbation (Chubynsky and Thorpe, 2007). Spring constants are assigned to the constraints as in the normal mode analysis, but then the connection sites are given small but finite random displacements. A local minimization of the energy is then performed. The network will be returned to equilibrium, but the position of equilibrium is not unique if floppy modes are present, since displacements along such modes do not affect the energy. Hence, the initial and relaxed configurations generally differ by a combination of displacements along the floppy modes, leaving the distances between mutually rigid sites unaffected. Hence, the mutual rigidity of sites can be tested systematically, and if there is any ambiguity due to numerical precision, an alternative set of random displacements can be tried.

A more rational approach to identifying rigid clusters and redundant constraints in two-dimensional networks, involving only integer arithmetic and avoiding artificial spring constants, has been developed by Thorpe and coworkers (Jacobs and Thorpe, 1995). This graph-theoretic method, called the pebble game, represents degrees of freedom by pebbles that may either be free or account for a constraint. The pebble game starts by assigning two free pebbles to each site (connection between rods). Each constraint (rod) is then accounted for by covering it at one of its ends using a free pebble from the site at that end. An example of such an arrangement is shown in Figure 23.4. To test whether an additional rod represents an independent or a redundant constraint, it is added to the network and the pebbles are shuffled around in an attempt to free two pebbles at each of the two sites connected by the test constraint while keeping all the existing constraints covered at one or other end. Each pebble may only be moved between being free at its original site on the one hand or covering an adjacent constraint on the other, but shuffling is possible by choosing to cover a constraint at the opposite end from the pebble currently covering it, thereby freeing the latter pebble. For example, if the test constraint labeled A in Figure 23.4a is added, the sequence of pebble movements indicated by the curved arrows produces two free pebbles at each end of A, while keeping all existing constraints covered. Rod A is therefore not redundant, and indeed rigidifies the rest of the network. In contrast, no sequence of pebble movements can free two pebbles at both ends of the test constraint B, and it is easy to see that B would be redundant due to the indirect rigidity of the two sites it connects. Redundant constraints are left uncovered, and testing of further constraints proceeds in the same way.

The two-dimensional pebble game can be extended to "bond-bending" networks in three dimensions (Jacobs, 1998). In such networks, not only do the rods generate distance constraints between the sites at which they are connected, but the angle between rods meeting at any given site is also constrained. Constraining the bond angles is equivalent to inserting an implicit distance constraint between all next-nearest neighbor sites. The extended pebble game proceeds by freeing three pebbles at each end of a test constraint and then attempting to free a further pebble at each neighbor of both ends of the test constraint (Chubynsky and Thorpe, 2007).

Three-dimensional bond-bending networks share an important property with the two-dimensional networks described above: rigid clusters are always contiguous, so that two mutually constrained sites always belong to the same rigid cluster. This is not true of non-bond-bending networks in three dimensions, where it is possible for the rigidity between nodes in the network to arise from connections to nodes that are not part of the cluster. An example is shown in Figure 23.4b. The three nodes denoted by the larger circles are mutually rigid, but are only connected via nodes that are not rigid with respect to all three of them (Jacobs, 1998). This sort of structure leads to the breakdown of some theorems on which the pebble game rests. Hence, the natural extension of the pebble game to three-dimensional non-bond-bending networks results in an algorithm that is, in principle, only approximate and can make errors in counting the number of floppy modes and in the rigid cluster analysis. Nevertheless, it turns out that the extent of these errors is very small for certain types of network (Chubynsky and Thorpe, 2007). For example, the approximate pebble game algorithm has been found to perform well for bond-dilution networks, which consist of a regular pattern of sites and connections with a specified fraction of connections (rods) removed at random, and some disorder in constraint length introduced to avoid the possibility of perfectly parallel bonds. Hence, the efficiency of the pebble game makes it attractive even in cases where it is not guaranteed to be exact.

23.3 Nanorod Networks in Composite Materials

The theory of connectivity percolation in rods and rod-containing mixtures finds a direct application in the development of lightweight electrically conducting materials and antistatic films. Insulating materials such as plastics and ceramics can be made electrically conducting over macroscopic distances by embedding in them electrically conducting fibers that connect into a percolating network. As discussed in Section 23.2.1, highly elongated particles percolate at very low volume fractions (inversely proportional to their length), so that conducting composite materials can be obtained by adding only a small amount of conducting filler to an insulating matrix. The application of this idea using fibers with diameters on the micron scale goes back at least to the early 1980s, when stainless steel fibers or metalized glass fibers were employed as the filler (Crossman, 1985).

The characterization of CNTs in 1991 (Iijima, 1991) paved the way for a new generation of conducting composites using networks of nanoscale conductors, since individual CNTs can be excellent electrical conductors (Saito et al., 1998). One of the first

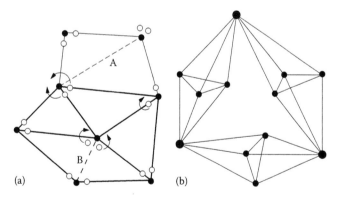

FIGURE 23.4 (a) Illustration of the pebble game for rigidity analysis of a two-dimensional network of rods. The rigid cluster is shown by thick lines and pebbles, representing constraints, by open circles. It is possible to free two pebbles at each end of the test constraint labeled A by the sequence of pebble movements indicated by the arrows. This is not the case for test constraint labeled B, which is therefore redundant. (b) A noncontiguous cluster in three dimensions. The nodes indicated by the larger circles are mutually rigid, but the bipyramidal structures joining them are free to rotate.

applications to exploit this property was the doping of luminescent polymers to control their conductivity (Curran et al., 1998). By making the polymer more conductive, it was possible to achieve electroluminescence (the emission of light upon application of a potential difference) at lower current densities. The conductivity of the composites varies over several orders of magnitude as the density of CNTs is increased, but shows a sharp increase at the percolation threshold (Coleman et al., 1998), effectively transforming the material from an insulator to a conductor.

In the vicinity of the percolation threshold, where macroscopically connected paths first appear, only a small fraction of CNTs contribute to the conductivity. Indeed, it has been argued that this regime provides an opportunity to access the intrinsic properties of individual nanotubes through macroscopic measurements (Benoit et al., 2002). Above the percolation threshold the "strength" of the network (the fraction of particles belonging to the infinite cluster) increases, and the conductivity continues to rise as $(\rho - \rho_c)^t$, where t is a universal critical exponent (Stauffer and Aharony, 1994). Calculations and experiments on a variety of three-dimensional conducting networks (including model resistor networks) indicate that $t \approx 2$ (Stauffer and Aharony, 1994). Although this approximate value has been confirmed in some experiments on CNT composites (Pötschke et al., 2003), considerably lower values (as low as $t \approx 1.4$) and even a temperature dependence of the exponent have also been reported (Barrau et al., 2003).

The experimental realization of theoretically predicted percolation behavior in CNT composites encounters a persistent obstacle. The nanotubes tend to bundle due to intermolecular attraction, including van der Waals forces, making it difficult to disperse them uniformly throughout the material. Incomplete dispersion leads to higher percolation thresholds than expected in the absence of bundling (Potschke et al., 2003, Sandler et al., 2005). There are both thermodynamic and kinetic aspects to achieving a proper dispersion; reducing the forces that lead to bundling makes the equilibrium state more dispersed, but it is still necessary to reach that state at a reasonable rate. Hence, the properties of the final composite can depend sensitively on the method of preparation and the conditions encountered in the process. Rapid stirring and sonication help to disperse the bundles, but a range of more invasive techniques have also been developed to assist the process. To improve the solubility of CNTs in organic solvents, one route is to modify the nanotubes chemically by attaching functional groups (Sun et al., 2002), while in mixtures of polymers and CNTs, the nature of the polymer plays an important role. CNT–polymer interactions can be strong (involving electrostatic attraction or even covalent attachment) or transient and weak, but in both cases, the interactions between polymer and CNT affect how the CNTs interact with each other (Szleifer and Yerushalmi-Rozen, 2005). For example, Du et al. have described a coagulation method, where precipitating polymers of poly(methyl methacrylate) entrap the nanotubes, preventing them from aggregating (Du et al., 2003). Naturally, these dispersion methods affect the individual and collective properties of CNTs, and the efficacy of the dispersion

must be balanced against retaining the desired properties of the resulting composite.

In addition to determining the extent of nanotube dispersion, composite preparation methods influence the overall distribution of nanotube orientations. When exposed to a flow, elongated objects tend to align themselves with the direction of movement. The resulting anisotropy of the orientational distribution leads to electrical conductivity that is also anisotropic, being greater in the direction of the flow (Haggenmueller et al., 2000). Alignment of CNTs in a stream can be exploited to produce macroscopic (sub-millimeter scale) ribbons and fibers of CNTs that are so flexible that they can be tied into knots (Vigolo et al., 2000). Recent work provides evidence that the percolation threshold does not change monotonically with the extent of CNT alignment. It is certainly the case that strong alignment of rod-like objects leads to a relatively high percolation density (Du et al., 2005), since the excluded volume for parallel rods scales only as the length of the rods rather than its square (see Figure 23.2). However, it seems that the opposite extreme of a uniformly random distribution does not produce the lowest possible percolation density. A slight average alignment of the rods within 70°–80° of some space-fixed axis is optimal (Li et al., 2008), a result that has been derived both from experimental investigation and from Monte Carlo simulations (Du et al., 2005). The anisotropic conductivity of composites with strongly aligned fibers may be advantageous in some applications. Indeed, anisotropic conducting adhesives are an important alternative to solder for connecting electronic components. However, the composites employed in such applications usually contain spherical conducting filler particles that become connected into chains that percolate in a particular direction by the application of pressure in that direction. Curing of the adhesive matrix then maintains the connected configuration (Lin and Zhong, 2008).

Although an isotropic dispersion of long rods percolates at a low density, there are ways of decreasing the percolation threshold even further. One approach is to induce weak attractions between the rods so as to increase the number of contacts between them but without inducing strong bundling or introducing anisotropy in the overall distribution of rod orientations. This goal can be achieved by entropic means through the depletion mechanism that was described in Section 23.2.1 and Figure 23.3. The depletion effect in the context of CNT suspensions has been studied experimentally using micelles as the depletion agent (Wang et al., 2004, Vigolo et al., 2005). In these studies, micelles were self-assembled out of surfactant molecules, which, once formed can be regarded as impenetrable but otherwise noninteracting spherical particles. The resulting two-component fluid of hard rods and spheres has been modeled in computer simulations (Schilling et al., 2007). The simulations are able to probe the correlations that arise from the hard core excluded volume of both rods and depletants in detail, and show that depletion enhances the local alignment of the rods relative to their close neighbors (see Figure 23.3) without making the overall distribution of single particle orientations anisotropic (which would lead to a nematic liquid crystalline phase). Although this mutual alignment is a weak form of bundling and alone would tend to raise the percolation density, the overall induced attraction between the rods leads to a greater number of contacts per rod, which overcomes the opposing effect of alignment and leads to a net reduction in the percolation threshold.

A rather different approach to lowering the volume fraction of conducting filler that is necessary for percolation involves not requiring the conducting network to fill the entire volume of the sample. In segregated composites, CNTs or other conductive particles are coated onto the surface of the particles of a polymer powder. The coated powder is then compressed to mold it into a continuous material. The compression forces the powder particles into polyhedral shapes with the conductive filler confined to the interfaces between them (Gao et al., 2008). A conducting network consisting of a skeleton running between the polyhedra emerges, allowing percolation to be achieved at volume fractions of well under a tenth of one percent (Mamunya et al., 2008). Although the network ultimately spans three-dimensional space, its structure is effectively two-dimensional at a microscopic level. The critical exponent, t, for conductivity has been measured and is indeed compatible with the accepted range of values for two-dimensional connectivity, which is around 1.3, substantially lower than the value of 2 in three dimensions (Gao et al., 2008).

A new range of applications of conducting CNT networks was opened up in 1999 by the discovery that the electrical and mechanical properties of CNTs are strongly coupled (Baughman et al., 1999). The injection of electrical charge into a CNT produces a mechanical deformation in the tube, primarily as a result of changes to its electronic structure rather than simple electrostatics. This electromechanical actuation can be used to extract mechanical work from CNT networks, as originally demonstrated in sheets of "nanotube paper"—entangled mats of nanotube bundles (Baughman et al., 1999). The reverse effect is also observed: the electrical properties of a CNT network are sensitive to mechanical deformation of the network by external forces. The resistivity of CNT films has been shown to increase almost linearly with applied strain, making such films attractive as strain sensors (Dharap et al., 2004).

By embedding CNT films within a material, it is possible not only to measure strain, but also to detect and quantify damage due to mechanical loading and unloading. Microcracks in a laminated material lead to the onset of delamination and step-like increases in the resistivity of the material due to loss of conductivity in the CNT network. The resistance of a damaged sample therefore has a qualitatively different relationship to its stress from the smooth increase observed in an undamaged sample (Thostenson and Chou, 2006). The shift in the stress–resistivity relationship on repeated loading–unloading cycles is an indication of irreversible damage. The fact that a percolating network of CNTs can be embedded in a composite material with the need for only a small volume fraction of nanotubes means that the ability for strain and damage sensing can be built into a material in a minimally invasive way (Li et al., 2008).

Calculations and simulations based on the vibrational modes of CNTs predict that they have a high thermal conductivity (Che et al., 2000) in addition to their electrical conductivity. This property of individual particles should mean that the thermal conductivity of a composite material can be greatly enhanced by the addition of a CNT network, and that this enhancement should be achievable at the low volume fractions needed for percolation (Foygel et al., 2005). Percolation behavior of the thermal conductivity has indeed been observed experimentally, with a sharp jump over several orders of magnitude at the percolation threshold (Biercuk et al., 2002) and an increase thereafter that is well described by the same kind of power law, $(\rho - \rho_c)^t$, as electrical conductivity (Foygel et al., 2005). Although vapor-grown carbon fibers with diameters in the micron range share and maybe even exceed the high thermal conductivity of CNTs, a much higher volume fraction of the fibers must be loaded into a matrix to enhance the conductivity, and the enhancement is weaker. The relatively poor performance of micron-diameter fibers has been attributed to their lower aspect ratio and the consequent difficulty of achieving a percolating network (Biercuk et al., 2002).

One further remarkable property of CNTs is their high mechanical strength, leading to the expectation that they can be used to make exceptionally strong materials. Unfortunately, the strength of individual nanotubes does not transfer directly to bulk composites, since the mechanical load must be transferred between the host matrix and the nanotubes and, in the case of multiwalled nanotubes, between the layers of the tubes themselves. Both interfaces are prone to slippage, though load transfer is better in compression than in tension (Schadler et al., 1998). However, this disappointing result can be turned to advantage by exploiting the slippage in applications where strong mechanical damping is needed. Addition of CNTs to a polymer matrix in characteristically low volume fractions can dramatically increase their viscous loss factor with minimal penalty from the added weight (Suhr et al., 2005). This application of CNTs demonstrates once again the scope for designing novel materials based not only on the remarkable properties of individual CNTs, but also on the generic ability of nanorods to form percolating networks at very low densities.

23.4 Networks of Biological Nanorods

A wide array of proteins can aggregate into elongated fibers, making nanoscale rod-like structures important in all biological matter. The formation of these structures sometimes takes place through a hierarchy of several levels. For example, collagen molecules, which are found abundantly in skin and bone, twist into helical ropes which aggregate into fibrils, which in turn bundle into fibers (Alberts et al., 2008). Other protein fibrils play a key role in disease when they form from incorrectly folded globular proteins that need to remain in solution in order to function properly (Chiti and Dobson, 2006).

Living cells owe their shape, mechanical response, and some of their transport properties to their cytoskeleton, a complex network of protein filaments. Three types of filament are usually found: actin filaments (two-stranded helices) and intermediate filaments (rope-like fibers) play an important role in cell shape and structure, while microtubules (hollow cylinders) are involved in intracellular organization. A host of other proteins are involved in linking the network and in transporting loads along the filaments. Furthermore, the network is dynamic, with individual filaments growing or shrinking at both ends simultaneously through association and dissociation processes (Alberts et al., 2008). The cytoskeleton performs many remarkable tasks in cells, but to understand how such a complex structure works it can be fruitful to isolate a limited number of its components and probe their behavior *in vitro*. By distilling the essential physics of these simplified networks into suitable models, it may be possible not only to gain insight into the properties of the cytoskeleton in living matter, but also to exploit this knowledge in the design of artificial network-based materials.

An important class of simplified system derived from the cytoskeleton consists of cross-linked networks of filamentous actin (F-actin). The mechanical behavior of actin networks depends both on the elastic properties of the individual filaments and on the architecture of the network. Actin filaments are semi-flexible, which means that there are both enthalpic and entropic contributions to their elastic modulus. A very stiff rod would have purely mechanical (enthalpic) contributions to its stretching and bending moduli, as illustrated schematically Figure 23.5b. In the opposite extreme, the stretching modulus of a completely flexible polymer is entirely entropic and arises from the fact that there is only one fully stretched conformation but many coiled ones with a shorter end-to-end distance. Hence, stretching a flexible polymer leads to a loss of entropy without an increase in the contour length of the chain, as illustrated Figure 23.5a. Actin filaments have an intermediate behavior; they are not flexible enough to form loops or coils, but an unconstrained filament may still experience significant thermal fluctuations. The stiffness of a chain can be quantified by its persistence length, i.e., the distance over which its tangent vector becomes uncorrelated. For actin, the persistence length is somewhat longer than the filament itself. In turn, the filament length is longer than the scale of the mesh in the network. Actin networks are therefore rather different (MacKintosh et al., 1995) from some other elastic materials like rubbers, which consist of permanently cross-linked networks of flexible polymers (Doi, 1995).

As might be expected, the density of cross-links in a suspension of actin filaments strongly influences the elastic response of the resulting network (Janmey et al., 1990). In the absence of cross-linking proteins, an actin suspension subjected to a constant shear stress responds by showing a large initial strain (deformation) which then continues to increase slowly. This creep is characteristic of viscoelastic fluids, as is the recovery of only part of the strain when the stress is released. In contrast, the addition of actin-binding proteins to the suspension results in a more elastic network, where the initial deformation upon shear strain is smaller, and is mostly recovered on release. Creep is practically eliminated by the mutual binding of the filaments. In the cross-linked network the elastic shear modulus, which measures the strain required to produce a given deformation, rises with increasing strain, i.e., the network is strain stiffening, becoming progressively harder to deform further. Abrupt decreases in the modulus are observed at sufficiently large strains, and have been attributed to rupture of the filaments rather than dissociation of the cross-links (Janmey et al., 1990).

Both connectivity and rigidity percolation phenomena are relevant to the properties of actin networks. The onset of a significant elastic response is associated with the connectivity percolation threshold, where the combination of rod density and linker density first produces a macroscopically connected network (Åström et al., 2008). The rigidity threshold then depends on the type of protein involved in the cross-linking. One can imagine two extremes: free rotation at cross-links or a constrained angle at the links (Wilhelm and Frey, 2003). The latter case is the bond-bending model described in Section 23.2.2, and in a two-dimensional network of this type, connectivity and rigidity percolation coincide. At the rigidity threshold the generic behavior, as described by the critical exponents, seems to be insensitive to the mechanism of rigidification at the links. For example, suppressing the bending of adjacent segments of any given rod with a certain probability leads to a rigidity transition in the same universality class as the standard rigidity percolation model (Latva-Kokko et al., 2001).

Simple models that capture the interplay between the bending and stretching of individual actin filaments are able to explain and predict regimes of qualitatively different behavior in the networks. For example, Wilhelm and Frey considered a random network of rods in two dimensions, linked at the points where they cross. Spring constants for the bending and stretching of segments were assigned according to the thickness of the rods and the length of the segments (Wilhelm and Frey, 2003). For thick rods and high densities (leading to many contacts along the length of a rod), it is harder to bend rods than to stretch or compress them. In this regime, the network behaves like a homogeneous elastic medium with shear modulus proportional to the filament compressional stiffness and the rod density. Local deformations of the network are affine, i.e., collinear sets of points remain collinear after the deformation and ratios of

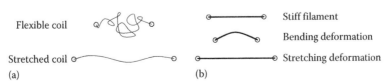

Flexible coil

Stretched coil

(a)

Stiff filament

Bending deformation

Stretching deformation

(b)

FIGURE 23.5 (a) The elasticity of flexible polymers arises from the loss of thermal fluctuations, leading to a decrease in entropy when the coil is stretched. (b) For stiff polymers, the penalty for bending or stretching is predominantly enthalpic.

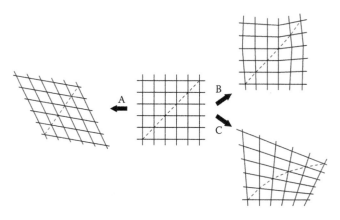

FIGURE 23.6 Example deformations of a square lattice. Deformation A is affine: it preserves collinearity and ratios of distances. Deformations B and C are non-affine due to explicit bending of the lines (B) or to loss of parallelism between the lines (C).

distances are preserved, as illustrated in Figure 23.6. In contrast, at lower densities or for thinner rods, bending deformations of individual filaments dominate and the shear modulus becomes independent of the aspect ratio of the filaments at a value determined by the density.

If thermal fluctuations are taken into account then it should be possible to distinguish two affine regimes (Head et al., 2003b). At the highest densities, the mechanical stiffness is still expected to dominate. However, at somewhat lower rod densities, the separation between cross-links may be sufficiently large that thermal fluctuations cause a nonnegligible contraction of the network (a scenario intermediate between the two extremes depicted in Figure 23.5). These fluctuations can be drawn out by an applied strain or shear without increasing the contour length of the filaments through stretching of the constituent subunits. This affine entropic regime is expected to be restricted to some maximum strain because of the limited size of the thermal fluctuations. The limit is reduced as the rod density increases, since the segments between cross-links become shorter and have less scope to fluctuate (Head et al., 2003b). The entropic contribution to the stiffness of a segment of length l scales as l^{-4} and is thus much more sensitive to the density of cross-links than the mechanical stiffness, which goes only as l^{-1}. This strong dependence also makes thermally fluctuating stiff networks highly sensitive to length polydispersity and to deviations from true randomness in the network architecture (Heussinger and Frey, 2007).

Thermal fluctuations in actin filaments are one of several characteristics that can explain the strain stiffening of networks, i.e., the increase in elastic modulus with the applied strain. The straightening of filaments accounts for the nonlinear elastic response at low strain until the point where this source of compliance has been exhausted (Storm et al., 2005). Beyond this point, enthalpic contributions must be considered to reproduce experimental modulus–strain curves. Nonlinearity in the enthalpic regime is influenced by the fact that, even at bulk mechanical equilibrium, individual segments are not at their equilibrium length due to the constraints imposed by the architecture of the

cross-linking network (Storm et al., 2005). A third contribution to strain stiffening comes from the cross-link proteins themselves, which may have both mechanical and thermal elasticity as well as a finite extensibility (Broedersz et al., 2008).

Strain-stiffening can account for the apparent discrepancy between the elastic modulus of actin networks measured *in vitro* and in cells, where the latter is found to be much higher (Gardel et al., 2006). The higher modulus is a direct consequence of strain stiffening if cells prestress their cytoskeletal frameworks with a constant background force. The frequency-dependent elastic response of actin networks cross-linked by flexible filamin A proteins with a constant stress offset has been characterized experimentally *in vitro* and shown to be much closer to the mechanics of living cells than the behavior about zero initial stress (Gardel et al., 2006).

In living cells, the architecture of actin networks is not static, since cross-linking proteins bind and unbind dynamically with characteristic rates. Transient cross-linking leads to a maximum and then a minimum in the viscous response of the network as a function of increasing frequency. The maximum coincides with a decrease in the elastic response at low frequencies. The effect of the unbinding rate on these properties has been studied experimentally in actin networks cross-linked by heavy meromyosin, whose unbinding rate can be increased by the addition of the nonhydrolyzable nucleotide analogue AMP-PNP (Lieleg et al., 2008). This work shows that the features in the viscous and elastic properties change in tandem, indicating a common origin of the effects. The unbinding of cross-links produces a local relaxation of the network that leads both to an increased viscous loss and to a decreased elastic response. The minimum in viscous response is a result of competition between stress release on unbinding and the friction induced by filament fluctuations. At sufficiently high frequencies of deformation, the unbinding rate is too slow to have an effect on the network properties.

Åström et al. have developed a model that incorporates the fact that there is a limit to how far an individual cross-link can be stressed before it ruptures. The model consists of a random three-dimensional network of mechanically deformable rods cross-linked by springs at locations of close proximity (Åström et al., 2008). If the cross-links are not allowed to break then strain stiffening arises directly when the non-affine regime is reached. However, if the cross-links rupture beyond a specified extension, there must be a transfer of the supported strain to nearby cross-links. These in turn may rupture due to their increased stress, resulting in correlated avalanches of ruptures. These ruptures reduce the network's elastic response, leading to a strain-softening regime. The model can be taken one step further by allowing new cross-links to form when segments are brought close together by the deformation. This dynamic cross-linking leads to bundling of the filaments under strain. The new cross-links can be strong enough to retain the bundling even when the external strain is released, resulting in pronounced hysteresis in the strain–stress relationship.

The survey of experiments and theoretical modeling on actin networks presented here has highlighted the rich properties of this particular example of biological nanorod network.

However, we have not even touched on the fact that in living cells, the cytoskeleton is constantly out of equilibrium. The constant consumption of energy makes the network into an "active gel." Molecular motors travel along the network's filaments, performing mechanical work, transporting loads, and enabling cell motility. Active gels constitute an important new area of biophysics, though one which begins to depart from the theme of this chapter. We recommend two recent reviews (Liverpool, 2006, Jülicher et al., 2007) for a survey of the current state of the field.

23.5 Summary

Explicit statistical modeling of composite materials or the cellular cytoskeleton with molecular detail would be a formidable task. Mastery of such systems at the molecular level can be important both for practical applications and for a deep understanding in specific cases. In this chapter, however, we have taken a highly coarse-grained approach to describing networks of rod-like particles. Retaining just the most essential detail has the advantage of highlighting unifying concepts that link some disparate systems. For example, we have emphasized the role of percolation theory, which lies at the heart of network formation and provides a general framework for analyzing physical properties of both finite and system-spanning clusters. The level of detail incorporated into this and other theories can be refined as needed, starting with ideal, fully penetrable rods, then proceeding in turn to particles that exclude volume through a hard core, that interact through longer-range forces, and that are flexible.

Many challenges remain in the understanding, control, and development of nanorod networks. We have seen that network structure is sensitive to intermolecular forces and to the nature of cross-links, necessitating innovative preparation methods to produce composite materials with the desired properties. The high aspect ratio of some nanorods means that computer simulations must employ large periodic cells and a correspondingly large number of particles, making the simulations computationally costly. Here, new simulation methods and coarse-graining schemes will expand the range of problems that can be tackled. Purely theoretical approaches offer insight and semi-quantitative predictions, but must be extended to cover increasingly complex nanorod systems. All these approaches have made important advances in recent years, and there is every reason to expect that the field will continue to progress rapidly.

References

Alberts, B., Johnson, A., Lewis, J., Raff, M., Roberts, K., and Walter, P. (2008). *Molecular Biology of the Cell*, 5th edn. Garland, New York.

Åström, J. A., Kumar, P. B. S., Vattulainen, I., and Karttunen, M. (2008). Strain hardening, avalanches, and strain softening in dense cross-linked actin networks. *Phys. Rev. E*, 77:051913.

Balberg, I., Anderson, C. H., Alexander, S., and Wagner, N. (1984). Excluded volume and its relation to the onset of percolation. *Phys. Rev. B*, 30:3933–3943.

Barrau, S., Demont, P., Peigney, A., Laurent, C., and Lacabanne, C. (2003). DC and AC conductivity of carbon nanotubes—Polyepoxy composites. *Macromolecules*, 36:5187–5194.

Baughman, R. H., Cui, C., Zakhidov, A. A., Iqbal, Z., Barisci, J. N., Spinks, G. M., Wallace, G. G., Mazzoldi, A., De Rossi, D., Rinzler, A. G., Jaschinski, O., Roth, S., and Kertesz, M. (1999). Carbon nanotube actuators. *Science*, 284:1340–1344.

Bawden, F. C., Pirie, N. W., Bernal, F. D., and Fankuchen, I. (1936). Liquid crystalline substances from virus-infected plants. *Nature*, 138:1051–1052.

Benoit, J. M., Corraze, B., and Chauvet, O. (2002). Localization, Coulomb interactions, and electrical heating in single-wall carbon nanotubes/polymer composites. *Phys. Rev. B*, 65:241405(R).

Berhan, L. and Sastry, A. M. (2007). Modeling percolation in high-aspect-ratio fiber systems. II. The effect of waviness on the percolation onset. *Phys. Rev. E*, 75:041121.

Biercuk, M., Llaguno, M., Radosavljevic, M., Hyun, J., Johnson, A., and Fischer, J. (2002). Carbon nanotube composites for thermal management. *Appl. Phys. Lett.*, 80:2767–2769.

Bollobás, B. and Riordan, O. (2006). *Percolation*. Cambridge University Press, Cambridge, U.K.

Broedersz, C. P., Storm, C., and MacKintosh, F. C. (2008). Nonlinear elasticity of composite networks of stiff biopolymers with flexible linkers. *J. Chem. Phys.*, 101:118103.

Bug, A. L. R., Safran, S. A., and Webman, I. (1985). Continuum percolation of rods. *Phys. Rev. Lett.*, 54:1412–1415.

Bug, A. L. R., Safran, S. A., and Webman, I. (1986). Continuum percolation of permeable objects. *Phys. Rev. B*, 33:4716–4724.

Chandler, D. and Andersen, H. (1972). Optimized cluster expansions for classical fluids. II. Theory of molecular liquids. *J. Chem. Phys.*, 57:1930–1937.

Chatterjee, A. P. (2000). Continuum percolation in macromolecular fluids. *J. Chem. Phys.*, 113:9310–9317.

Che, J. W., Cagin, T., and Goddard, W. A. (2000). Thermal conductivity of carbon nanotubes. *Nanotechnology*, 11:65–69.

Chiti, F. and Dobson, C. M. (2006). Protein misfolding, functional amyloid, and human disease. *Annu. Rev. Biochem.*, 75:333–366.

Chubynsky, M. V. and Thorpe, M. F. (2007). Algorithms for three-dimensional rigidity analysis and a first-order percolation transition. *Phys. Rev. E*, 76:041135.

Coleman, J. N., Curran, S., Dalton, A. B., Davey, A. P., McCarthy, B., Blau, W., and Barklie, R. C. (1998). Percolation-dominated conductivity in a conjugated-polymer-carbon-nanotube composite. *Phys. Rev. B*, 58:R7492–R7495.

Coniglio, A., de Angelis, U., and Forlani, A. (1977). Pair connectedness and cluster size. *J. Phys. A*, 10:1123–1139.

Crossman, R. (1985). Conductive composites past, present, and future. *Polym. Eng. Sci.*, 25:507–513.

Curran, S. A., Ajayan, P. M., Blau, W. J., Carroll, D. L., Coleman, J. N., Dalton, A. B., Davey, A. P., Drury, A., McCarthy, B., Maier, S., and Strevens, A. (1998). A composite from poly(*m*-phenylenevinylene-*co*-2, 5-dioctoxy-*p*-phenylenevinylene) and carbon nanotubes: A novel material for molecular optoelectronics. *Adv. Mater.*, 10:1091–1093.

DeSimone, T., Demoulini, S., and Stratt, R. M. (1986). A theory of percolation in liquids. *J. Chem. Phys.*, 85:391–400.

Dharap, P., Li, Z., Nagarajaiah, S., and Barrera, E. V. (2004). Nanotube film based on single-wall nanotubes for strain sensing. *Nanotechnology*, 15:379–382.

Doi, M. (1995). *Introduction to Polymer Physics*. Clarendon Press, Oxford, U.K.

Doi, M. and Edwards, S. (1986). *The Theory of Polymer Dynamics*. Clarendon Press, Oxford, U.K.

Du, F., Fischer, J., and Winey, K. (2003). Coagulation method for preparing single-walled carbon nanotube/poly(methyl methacrylate) composites and their modulus, electrical conductivity, and thermal stability. *J. Polym. Sci. B: Polym. Phys.*, 43:3333–3338.

Du, F., Fischer, J., and Winey, K. (2005). Effect of nanotube alignment on percolation conductivity in carbon nanotube/ polymer composites. *Phys. Rev. B*, 72:121404(R).

Foygel, M., Morris, R. D., Anez, D., French, S., and Sobolev, V. L. (2005). Theoretical and computational studies of carbon nanotube composites and suspensions: Electrical and thermal conductivity. *Phys. Rev. B*, 71:104201.

Frenkel, D. and Smit, B. (2002). *Understanding Molecular Simulation*, 2nd edn. Academic Press, San Diego, CA.

Gao, J.-F., Li, Z.-M., Meng, Q.-J., and Yang, Q. (2008). CNTs/ UHMWPE composites with a two-dimensional conductive network. *Mater. Lett.*, 62:3530–3532.

Gardel, M. L., Nakamura, F., Hartwig, J., Crocker, J. C., Stossel, T. P., and Weitz, D. A. (2006). Stress-dependent elasticity of composite actin networks as a model for cell behavior. *Phys. Rev. Lett.*, 96:088102.

Grimmett, G. (1999). *Percolation*, 2nd edn. Springer, Berlin, Germany.

Haggenmueller, R., Gommans, H., Rinzler, A., Fischer, J., and Winey, K. (2000). Aligned single-wall carbon nanotubes in composites by melt processing methods. *Chem. Phys. Lett.*, 330:219.

Hansen, J. and McDonald, I. (2006). *Theory of Simple Liquids*, 3rd edn. Academic Press, London, U.K.

Head, D. A., Levine, A. J., and MacKintosh, F. C. (2003a). Deformation of cross-linked semiflexible polymer networks. *Phys. Rev. Lett.*, 91:108102.

Head, D. A., Levine, A. J., and MacKintosh, F. C. (2003b). Distinct regimes of elastic response and deformation modes of cross-linked cytoskeletal and semiflexible polymer networks. *Phys. Rev. E*, 68:061907.

Heussinger, C. and Frey, C. (2007). Role of architecture in the elastic response of semiflexible polymer and fiber networks. *Phys. Rev. E*, 75:011917.

Hill, T. (1955). Molecular clusters in imperfect gases. *J. Chem. Phys.*, 23:617–622.

Iijima, S. (1991). Helical microtubules of graphitic carbon. *Nature*, 354:56–58.

Jacobs, M. F. (1998). Generic rigidity in three-dimensional bond-bending networks. *J. Phys. A*, 31:6653–6668.

Jacobs, D. J. and Thorpe, M. F. (1995). Generic rigidity percolation: The pebble game. *Phys. Rev. Lett.*, 75:4051–4054.

Janmey, P. A., Hvidt, S., Lamb, J., and Stossel, T. P. (1990). Resemblance of actin-binding protein/actin gels to covalently crosslinked networks. *Nature*, 345:89–92.

Jülicher, F., Kruse, K., Prost, J., and Joanny, J.-F. (2007). Active behavior of the cytoskeleton. *Phys. Rep.*, 449:3–28.

Kyrylyuk, A. and van der Schoot, P. (2008). Continuum percolation of carbon nanotubes in polymeric and colloidal media. *Proc. Natl. Acad. Sci. U.S.A.*, 105:8221–8226.

Latva-Kokko, M., Mäkinen, J., and Timonen, J. (2001). Rigidity transition in two-dimensional random fiber networks. *Phys. Rev. E*, 63:046113.

Leung, K. and Chandler, D. (1991). Theory of percolation in fluids of long molecules. *J. Stat. Phys.*, 63:837–856.

Li, C., Thostenson, E., and Chou, T.-W. (2008). Sensors and actuators based on carbon nanotubes and their composites: A review. *Composites Sci. Tech.*, 68:1227–1249.

Lieleg, O., Claessens, M. M. A. E., Luan, Y., and Bausch, A. R. (2008). Transient binding and dissipation in cross-linked actin networks. *Phys. Rev. Lett.*, 101:108101.

Lin, Y. and Zhong, J. (2008). A review of the influencing factors on anisotropic conductive adhesives joining technology in electrical applications. *J. Mater. Sci.*, 43:3072–3093.

Liverpool, T. B. (2006). Active gels: Where polymer physics meets cytoskeletal dynamics. *Philos. Trans. R. Soc. A*, 364:3335–3355.

MacKintosh, F. C., Käs, J., and Janmey, P. A. (1995). Elasticity of semiflexible biopolymer networks. *Phys. Rev. Lett.*, 75:4425–4428.

Mamunya, Y., Boudenne, A., Lebovska, N., Ibos, L., Candau, Y., and Lisunova, M. (2008). Electrical and thermophysical behaviour of PVC-MWCNT nanocomposites. *Composites Sci. Tech.*, 68:1981–1988.

Meester, R. and Roy, R. (1996). *Continuum Percolation*. Cambridge University Press, Cambridge, U.K.

Onsager, L. (1949). The effects of shape on the interaction of colloidal particles. *Ann. New York Acad. Sci.*, 51:627–659.

Ornstein, L. and Zernike, F. (1914). Accidental deviations of density and opalescence at the critical point in a single substance. *Proc. Acad. Sci. (Amsterdam)*, 17:793.

Pötschke, P., Dudkin, S. M., and Alig, I. (2003). Dielectric spectroscopy on melt processed polycarbonate–multiwalled carbon nanotube composites. *Polymer*, 44:5023–5030.

Saito, R., Dresselhaus, M., and Dresselhaus, G. (1998). *Physical Properties of Carbon Nanotubes*. Imperial College Press, London, U.K.

Sandler, J., Shaffer, M. S. P., Prasse, T., Bauhofer, W., Schulte, K., and Windle, A. H. (2005). Development of a dispersion process for carbon nanotubes in an epoxy matrix and the resulting electrical properties. *Polymer*, 40:5967–5971.

Schadler, L. S., Giannaris, S. C., and Ajayan, P. M. (1998). Load transfer in carbon nanotube epoxy composites. *Appl. Phys. Lett.*, 73:3842–3844.

Schilling, T., Jungblut, S., and Miller, M. A. (2007). Depletion-induced percolation in networks of nanorods. *Phys. Rev. Lett.*, 98:108303.

Škvor, J., Nezbeda, I., Brovchenko, I., and Oleinikova, A. (2007). Percolation transition in fluids: Scaling behaviour of the spanning probability functions. *Phys. Rev. Lett.*, 99:127801.

Stauffer, D. and Aharony, A. (1994). *Introduction to Percolation Theory*, 2nd edn. Taylor & Francis, London, U.K.

Storm, C., Pastore, J., MacKintosh, F., Lubensky, T., and Janmey, P. (2005). Nonlinear elasticity in biological gels. *Nature*, 435:191–194.

Suhr, J., Koratkar, N., Keblinski, P., and Ajayan, P. (2005). Viscoelasticity in carbon nanotube composites. *Nat. Mater.*, 4:134–137.

Sun, Y.-P., Fu, K., Lin, Y., and Huang, W. (2002). Functionalized carbon nanotubes: Properties and applications. *Acc. Chem. Res.*, 35:1096–1104.

Szleifer, I. and Yerushalmi-Rozen, R. (2005). Polymers and carbon nanotubes—Dimensionality, interactions and nanotechnology. *Polymer*, 46:7803–7818.

Thorpe, M. F. (1983). Continuous deformations in random networks. *J. Non-Cryst. Solids*, 57:355–370.

Thostenson, E. T. and Chou, T.-W. (2006). Carbon nanotube networks: Sensing of distributed strain and damage for life prediction and self healing. *Adv. Mater.*, 18:2837–2841.

Vigolo, B., Pénicaud, A., Coulon, C., Sauder, C., Pailler, R., Journet, C., Bernier, P., and Poulin, P. (2000). Macroscopic fibers and ribbons of oriented carbon nanotubes. *Science*, 290:1331–1334.

Vigolo, B., Coulon, C., Maugey, M., and Poulin, P. (2005). An experimental approach to the percolation of sticky nanotubes. *Science*, 309:920–923.

Wang, X. and Chatterjee, A. (2001). An integral equation study of percolation in systems of flexible and rigid macro-molecules. *J. Chem. Phys.*, 114:10544–10550.

Wang, X. and Chatterjee, A. (2002). Continuum percolation in athermal mixtures of flexible and rigid macromolecules. *J. Chem. Phys.*, 116:347–351.

Wang, X. and Chatterjee, A. (2003). Connectedness percolation in athermal mixtures of flexible and rigid macromolecules: Analytic theory. *J. Chem. Phys.*, 118:10787–10793.

Wang, H., Zhou, W., Ho, D., Winey, K., Fischer, J., Glinka, C., and Hobbie, E. (2004). Dispersing single-walled carbon nanotubes with surfactant: A small angle neutron scattering study. *Nano Lett.*, 4:1789–1793.

Wilhelm, J. and Frey, E. (2003). Elasticity of stiff polymer networks. *Phys. Rev. Lett.*, 91:108103.

V

Nanowire Properties

Mechanical Properties of GaN Nanowires

Zhiguo Wang
University of Electronic Science and Technology of China

and

Chinese Academy of Sciences

Fei Gao
Pacific Northwest National Laboratory

Xiaotao Zu
University of Electronic Science and Technology of China

Jingbo Li
Chinese Academy of Sciences

William J. Weber
Pacific Northwest National Laboratory

24.1 Introduction

The further development of the modern semiconductor industry is based on the continued miniaturization of silicon-based devices; however, existing materials and technologies are approaching their physical limits, which will soon prevent the industry from maintaining similar rates of improvement (Peercy 2000). Finding new materials and technologies adequate for the fabrication of highly integrated devices within the nanometer regime could potentially overcome such limitations. One-dimensional materials like nanowires that can efficiently transport electrical carriers and optical excitations can meet these needs. A variety of freestanding semiconductor nanowires with controlled electrical and optical properties, including the group IV, III–V, and II–VI semiconductor nanowires, can now be synthesized as a result of recent advances in crystal growth technology (Dai et al. 1995; Han et al. 1997; Morales and Lieber 1998; Lexholm et al. 2006; Tragardh et al. 2007). These one-dimensional structures often show novel physical properties that are different from their bulk (Kanemitsu et al. 1993; Zach et al. 2000; Makhlin et al. 2001). As such, the nanowires are considered ideal building blocks for highly integrated nanoscale electronics and optoelectronics.

Wide band-gap semiconductors are of particular interest since their large energy gap allows for the possibility of tuning optoelectronic devices to work from infrared to ultraviolet at high temperatures and high frequency. Among these semiconductors, gallium nitride (GaN) is especially of interest for its wide direct band gap. After the successful synthesis of blue light–emitting GaN-based materials, research interest on GaN has intensified due to a great desire for high-efficiency blue light–emitting diodes or laser diodes. The fabrication of nano-sized GaN materials has been a focused field of research, both due to the fundamental nanoscale mesoscopic physics and the developing nanoscale devices. Low-dimensional GaN nanostructures are expected to improve device characteristics and possess superior structural stability and mechanical properties at high temperatures; thus, they can be used as building blocks for nanodevices.

In this chapter, a brief introduction to the growth methods and the shape controlling of GaN nanowires is provided. The limitations in determining the mechanical behavior of GaN nanowires experimentally are then presented. In Section 24.3, recent progress is described on the modeling and simulation of the nanomechanical properties and behavior of GaN nanowires.

24.2 Growth Methods and Shape Controlling

Nanowires have small cross-sections, allowing them to accommodate much higher levels of strain without the formation of dislocations. The wires, furthermore, can rapidly exclude any dislocations that do form to the nearby sidewalls, where they terminate and cease to propagate through the crystal (Trampert et al. 2003). Transmission electron microscopy (TEM) performed on GaN nanowires detached from their substrate reveals that they are dislocation-free single crystals (Tchernycheva et al. 2007). This defect exclusion mechanism allows the growth of dislocation-free GaN nanowires on all types of substrates. In recent years, research interest in GaN nanowires has increased significantly because sufficiently small dimensions, such as those found in nanowires, promote quantum confinement effects that are expected to lead to novel or enhanced physical properties for potential use in future nanotechnology. GaN nanowires are typically synthesized through vapor phase methods, in which the initial starting reactants for the wire formation are gas phase species. Numerous techniques have been developed to prepare precursors into the gas phase for nanowire growth, including laser ablation, chemical vapor deposition, chemical vapor transport methods, molecular beam epitaxy, and sputtering. It should be noted that the concentrations of gaseous reactants must be carefully regulated for nanowire synthesis in order to allow the nanowire growth mechanism to predominate and suppress secondary nucleation events.

There are two approaches for vapor phase synthesis that have been used for GaN nanowire growth. The first one is based on the so-called vapor–liquid–solid (VLS) mechanism (Wagner and Ellis 1964), which uses a metallic catalyst (Ni, Fe, etc.) to promote one-dimensional growth (Duan and Lieber 2000; Zhong et al. 2003). GaN nanowires grown by the VLS mechanism have been successfully used to fabricate nano-devices such as lasers or modulators from single nanowires (Gradecak et al. 2005; Greytak et al. 2005). The second approach to GaN nanowire synthesis relies on a catalyst-free growth mode, which consists of a spontaneous transition to one-dimensional growth when nitrogen-rich conditions are used (Calleja et al. 1999). One advantage is that catalyst-free synthesis excludes the possible incorporation of catalyst metallic impurities in the nanowire material. For heterostructure formation, this growth mode also provides better control of the elemental composition, since it does not depend on a complex interaction between the constituents in the vapor phase, the metallic catalyst, and the semiconductor solid-state. Many groups report the fabrication of dense GaN nanowires ensembles (Cerutti et al. 2006; Meijers et al. 2006), as well as GaN/AlGaN and InGaN/GaN heterostructures in nanowires (Ristic et al. 2003; Kikuchi et al. 2004) by molecular beam epitaxy (MBE) using this procedure, but the exact growth mechanism remains an open question.

Due to its anisotropic and polar nature, GaN exhibits properties that depend on crystallographic orientation (Feng et al. 1999; Waltereit et al. 2000). Thus, controlling the growth direction of GaN nanowires is important for practical applications. Control of both the growth orientation, which has a strong effect on anisotropic properties, and the alignment are critical issues in nanowire growth. Control of the growth direction of GaN nanowires can be achieved through three methods demonstrated to date. The first method is to use heteroepitaxy on different single crystal templates, mediated by a catalyst cluster (Kuykendall et al. 2003). Epitaxial growth of wurtzite gallium nitride on Au- and Ni-coated sapphire substrates results in GaN nanowires with triangular cross-sections, with the wire axis oriented along the crystallographic [210] direction. However, the predominant nanowire growth direction is along the [110] direction instead of the [210] direction for nanowires grown on Fe-coated sapphire substrates. The second method to control the growth direction is choosing suitable substrates. For example, epitaxial growth of wurtzite gallium nitride on (100) γ-LiAlO$_2$ and (111) MgO single crystal substrates with Au as the initiator results in the selective growth of nanowires along the orthogonal [1$\bar{1}$0] and [001] directions, exhibiting triangular and hexagonal cross-sections, respectively (Kuykendall et al. 2004). Crystallographic alignment occurs presumably due to the close symmetry and lattice match between the substrates and observed nanowire growth directions. The third method to control the growth direction of nanowires is based on controlling the Ga flux during direct nitridation in dissociated ammonia on an amorphous substrate (Li et al. 2006). The nitridation of Ga droplets at high flux leads to GaN nanowire growth along the [001] direction, while nitridation with a low Ga flux leads to growth in the [100] direction.

The band gap of wurtzite GaN nanowires has been found to depend on the crystallographic orientation because of polarization in the [001] direction of GaN crystals (Waltereit et al. 2000). This is illustrated by the blue-shift in the bandgap of GaN nanowires grown in the [100] direction by about 100 meV, as compared with wires grown in the [001] direction at temperatures ranging from 0 to 300 K. Chin et al. (2007) also found the photoluminescence peak of [1$\bar{1}$0]-oriented nanowires is blue-shifted ~140 meV compared with that of the [001]-oriented nanowires. And they attributed the blue-shift to the surface states with act as traps of photoexcited carriers. The real mechanism of the blue shift of band gap needs further investigations.

Based on the particular structure characteristics and size effects, much progress has been made in nanodevice applications using these GaN nanowires. For example, GaN nanowires have been reported in numerous works to show promise as elemental building blocks for photonic, electronic, and optoelectronic nanodevices including logic gates, field-effect transistors (FETs), light-emitting diodes (LEDs), subwavelength photonics components, and so-walled "nanolasers" (Huang et al. 2001; Johnson et al. 2002; Zhong et al. 2003; Stern, et al. 2005; Pauzauskie et al. 2006; Qian et al. 2008). For integration into true nanodevices, the controllable assembly and precise location of fabricated nanowires must be established in device architectures. A better understanding of mechanical behavior is just one of many aspects that require detailed study.

24.3 Mechanical Behavior Studied Experimentally and Limitation

All of the promising future applications rely on the production of nanowires, in reasonable quantities, of controlled size, shape, and crystal structure. Ultimately, all applications will require that the nanowires be mechanically stable in the application environment. Measuring the mechanical properties of individual nanowires by conventional techniques is not trivial. Optical measurements used commonly in microelectromechanical systems are not readily applicable to nanowire resonators because the diameter is less than a visible wavelength.

There are several approaches to probe the mechanical properties of nanoscale specimens experimentally. The most widely used approach is the bending test, in which a nanowire suspended over a trench is indented by an atomic force microscopy (AFM) tip until failure occurs. Although this measurement is relatively easy to set up, the resulting force-displacement curve is difficult to analyze and interpret. The main difficulty comes from the unknown contact behavior between the AFM tip and the nanowire and the resulting high-stress gradient and possible contact damage. Chen et al. (2007) have investigated the mechanical elasticity of hexagonal wurtzite GaN nanowires using this method with a digital-pulsed force mode AFM. For these GaN nanowires, the stiffness and elastic modulus exhibit a dependence on diameter. With an increasing diameter, the elastic modulus decreases while the stiffness increases. Elastic moduli for the tested nanowires are in the range between 218.1 and 316.9 GPa. The second method is electromechanical resonance analysis in a TEM, which is based on applying an actuating signal between the nanostructure and a counter-electrode. The elastic beam theory can then be employed to relate the observed resonance frequency to Young's modulus (Nam et al. 2006). The modulus, *E*, obtained for an 84 nm nanowire was close to the theoretical bulk value; but the inferred *E* values decreased gradually for smaller diameters, which was difficult to resolve based on the present understanding of mechanics and materials at the nanoscale (Nam et al. 2006). Another approach is the tensile test, which measures the stress state throughout the entire specimen and thereby facilitates a direct comparison with theoretical models. The tensile test has been widely used to measure the mechanical properties of bulk materials. Unfortunately, tensile tests are difficult to perform at the nanoscale, mainly due to the challenge of manipulating the nanowire into the correct location and accurately applying and measuring the stress and strain. Recently, Hessman et al. (2007) have presented a stroboscopic detection method using an optical microscope that enables time-resolved imagining of the oscillating nanowire. This method may be a valuable tool for imaging and analyzing vibrating nanowires and needs further investigations.

The mechanical properties of GaN nanowires are not well established due to the complexities of mechanical testing presented at the nanometer size regime. Currently, with recent advances in computational power, atomistic simulations are the primary tools for investigating the mechanical deformation of nanowires and the associated mechanisms.

24.4 Mechanical Behaviors Studied by Molecular Dynamics

24.4.1 Molecular Dynamics

For atomistic or molecular simulations, the most popular methods include quantum mechanic (QM), molecular mechanic (MM), molecular dynamic (MD), coarse-grained (CG), and Monte Carlo (MC) simulations. Among them, QM methods can account explicitly for bond forming and breaking but are limited to system sizes of up to several hundred atoms. The others, however, can deal with systems of thousands, and even millions, of atoms and predict the static and dynamic behaviors of materials at the molecular level.

MD simulation is a technique for computing equilibrium and transport properties of a classical many-body system. It generates such information as atomic positions and velocities at the nanoscale level from which the macroscopic properties (e.g., pressure, energy, and heat capacities) can be derived by means of statistical mechanics. In MD, atoms move under the action of conservative forces that are additive, symmetric, and derived from intermolecular potential. The dynamic evolution of the system is governed by classical Newtonian mechanics, where for each atom *i*, the equation of motion is given by

$$F_i = M_i \frac{d^2 r_i}{dt^2} \qquad (24.1)$$

The atomic force F_i is obtained as the negative gradient of the effective potential. A physical simulation involves the proper selection of a numerical integration scheme, employment of appropriate boundary conditions, and stress and temperature control to mimic physically meaningful thermodynamic ensembles.

24.4.2 Stillinger–Weber Potential

The Stillinger–Weber (SW) potential (Stillinger and Weber 1985) is the most suitable potential for tetrahedral semiconductors like Si, and it is composed of both two- and three-atom contributions:

$$v_2(r_{ij}) = \varepsilon A \left[B \left(\frac{r_{ij}}{\sigma} \right)^{-4} - 1 \right] \exp \left\{ \left[\left(\frac{r_{ij}}{\sigma} \right) - \alpha \right]^{-1} \right\}, \quad \frac{r_{ij}}{\sigma} < \alpha \quad (24.2)$$

$$v_3(r_{ij}, r_{ik}, \theta_{jik}) = \varepsilon \lambda \exp \left[\gamma (r_{ij} - \alpha)^{-1} + \gamma (r_{ik} - \alpha)^{-1} \right] \times \left(\cos \theta_{jik} + \frac{1}{3} \right)^2 \quad (24.3)$$

where

ε and σ are the cohesive energy and length units, respectively

α represents the cut-off distance

θ_{ijk} is the angle formed by ji and the jk bonds

The original SW potential for GaN, which was developed by Aïchoune and coworkers (Aïchoune et al. 2000), takes into account the specificity of the different bonds, namely, Ga–N, Ga–Ga, and N–N bonds. The potential has been fitted to the lattice parameters, the experimental elastic constants, and the results of the *ab initio* calculations for an inversion domain boundary. Béré and Serra (2001) have modified the potential to stabilize an inversion domain boundary structure in GaN and to allow all interactions to smoothly decrease to zero just before the second-neighbor distance. Kioseoglou et al. (2003) modified the potential further to achieve a realistic description of the microscopic structure and the energetics of different planar defects as well as their interactions in the wurtzite GaN. Table 24.1 lists the parameters used to describe the interactions between atoms in the GaN nanowires. In addition, the potential can handle dangling bonds, wrong bonds, and excess bonds in bulk GaN very well. Therefore, this potential should be reliable to study the mechanical behavior of GaN nanowires.

24.4.3 Simulation Methodology

To investigate the atomic structures of single crystalline GaN nanowires observed in experiments as described in Section 24.2, the [001]-oriented GaN nanowires with hexagonal cross-sections and [1̄10]- and [110]-oriented GaN nanowires with triangular cross-sections are generated directly from bulk GaN by removing the atoms outside a hexagon or a triangle along the desired axial orientation and replacing them with vacant sites. The top views of these GaN nanowires are shown in Figure 24.1. The [001]-oriented nanowires are enclosed with {100} or {110} side planes, while the [1̄10]-oriented nanowires are enclosed with (112), (1̄1̄2), and (001) side planes between which the angles are 63.16°, 58.42°, and 58.42°, respectively. The [110]-oriented nanowires are enclosed with (11̄2), (1̄12), and (001) planes. The diameters and numbers of atoms of different oriented nanowires are summarized in Table 24.2.

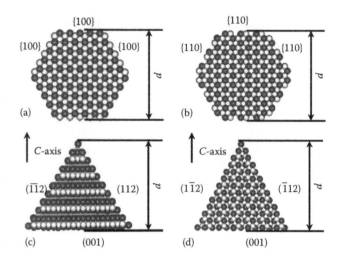

FIGURE 24.1 Cross-sectional views of the four types of nanowires used in this work. (a) [001]-oriented nanowires enclosed with {100} side planes; (b) [001]-oriented nanowires enclosed with {110} side planes; (c) [1̄10]-oriented nanowires enclosed with (112), (1̄1̄2), and (001) side planes; and (d) [110]-oriented nanowires enclosed with (11̄2), (1̄12), and (001) side planes. (From Wang, Z. et al., *Phys. Rev. B*, 76, 045310, 2007b. With permission.)

TABLE 24.2   The Diameters and Numbers of Atoms of Different Oriented Nanowires (NWs) Used

		Diameter (nm)	Number of Atoms
[001]-oriented NWs with {100} side planes	1	2.02	2304
	2	2.59	3600
	3	3.14	5184
[001]-oriented NWs with {110} side planes	4	1.92	2016
	5	2.88	4320
	6	3.20	5328
[1̄10]-oriented NWs	7	2.60	2652
	8	3.12	3756
	9	3.64	5052
[110]-oriented NWs	10	2.08	1620
	11	2.60	2420
	12	3.12	3380

Source: Wang, Z. et al., *Phys. Rev. B*, 76, 045310, 2007b. With permission.

A rigid boundary condition along the axial direction was applied to the axial direction with a length (L) of 6.12, 6.63, and 6.38 nm for the [001]-, [1̄10]-, and [110]-oriented nanowires, respectively. Strain–stress simulations were performed using the following procedure. The relative positions of atoms within the three atomic layers at the top and bottom of nanowires are fixed during simulations, forming two rigid borders. The initial structures of the nanowires were equilibrated for 100 ps at a given temperature, which allows the nanowires to have stable configurations. The strain was then applied along the axial direction to study the mechanical properties of the nanowires by imposing a displacement Δz. The atoms in the rigid borders were displaced by $\Delta z/2$, but the coordinates of the remaining atoms were scaled

TABLE 24.1   Parameters of the Stillinger–Weber Potential

	N–N	Ga–N	Ga–Ga
σ (a.u.)	2.9	3.203	2.72
ε (eV)	1.76	2.17	1.495
λ	26.76	32.5	26.76
α	1.8	1.8	1.8
A	7.917	7.917	7.917
B	0.720	0.720	0.720
γ	1.2	1.2	1.2
θ_0	1/3	1/3	0.184

by a factor $(L + \Delta z)/L$ along the z direction. This deformed tube was relaxed for 10 ps, and then the relaxed structure was used as an initial configuration for the next strain simulation. The procedure was repeated until each nanowire failed. The stress during each strain increment was computed by averaging over the final 2000 relaxation steps.

The axial stress is taken as the arithmetic mean of the local stresses on all atoms, as follows (Ju et al. 2004):

$$\sigma_z = \frac{1}{N} \sum_{i=1}^{N} \frac{1}{V_i} \left(m_i v_z^i v_z^i + \frac{1}{2} \sum_{\substack{j=1 \\ (j \neq i)}}^{N} F_z^{ij}(\varepsilon_z) r_z^{ij}(\varepsilon_z) \right) \quad (24.4)$$

where

- m_i is the mass of atom i
- v_z^i is the velocity along the axis direction
- F_z^{ij} refers to the component of the interatomic force along the axis direction between atoms i and j
- r_z^{ij} is the interatomic distance along the axis direction between atoms i and j
- V_i refers to the volume of atom i, which was assumed as a hard sphere in a closely packed crystal structure

The first term of the right-hand side of Equation 24.4 represents the kinetic effect associated with atomic motion, and will be affected by temperature. The second term is related to the interactive forces and distance between the atoms for the pair-wise interactions. The three-body forces can be similarly included as the second term of Equation 24.4, since each three-body term consists of a pair of interactions between a central vertex atom i and the neighboring atoms j and k.

24.4.4 Simulation Results

The relationship between tensile stress and strain for a [001]-oriented nanowire with {100} side planes and a diameter of 2.02 nm is shown in Figure 24.2 with a strain rate of 0.01% ps⁻¹ at various temperatures. For low strains, the stress–strain relationship follows Hooke's law, and the stress increases almost linearly with increasing strain up to a threshold value that is defined as the critical stress. It is of interest to note that the stress–strain curves show different behaviors at low and high temperatures when the strain is larger than the critical strain. For example, at 600 K, the stress increases with increasing strain up to 47.42 GPa (at a strain of 14.7%), but a further increase in strain leads to stress abruptly dropping to 0.0 GPa. At 1800 K, the stress–strain relationship displays a zigzag behavior after a critical stress level of 29.25 GPa (at 9.7% strain) and multiple maximums appear along the path. These multiple maximum stress levels suggest that the nanowires have multiple yield stresses before they completely fail. It can be observed that the stress required to produce the second decrease is much smaller than that required to produce the initial decrease. These results demonstrate that the nanowires fail in a brittle manner at low temperatures, but in a ductile manner

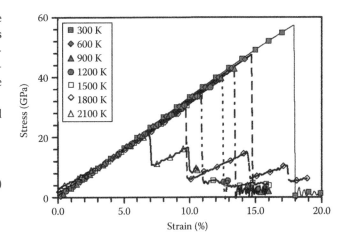

FIGURE 24.2 Tensile stress–strain curves for the [001]-oriented nanowires with {100} side planes and diameter of 2.02 nm with a strain rate of 0.001% ps⁻¹ at various temperatures. (From Wang, Z. et al., *Phys. Rev. B*, 76, 045310, 2007b. With permission.)

at higher temperatures. From Figure 24.2, it can be noted that the critical stress at low temperatures is larger than that at higher temperatures. As the temperature increases, a large number of atoms gain sufficient energy to overcome the activation energy barrier, and hence plastic deformation occurs, which may suggest that a thermally activated process plays a predominant role in the elongation of GaN nanowires.

Atomic configurations show that up to the strain of 14.7%, the nanowires increase uniformly and no structural defects appear at 600 K, as shown in Figure 24.3a. Upon reaching the critical strain of 14.7%, the crystal structure experiences an abrupt rupture with a clean cut, without any observed necking. Obviously, at 600 K, the tube shows brittle properties. At 1800 K, the extension of the nanowires also begins with an elastic deformation from its initial state to the critical strain of 9.7%, after which necking can be clearly seen in Figure 24.3b, and several atomic chains occur to link two separated GaN crystals before rupture. The above results show that the [001]-oriented nanowires exhibit ductility at high deformation temperatures and brittleness at lower temperatures. All the [001]-oriented nanowires studied here show the same characters, i.e., nanowires exhibit ductility at high deformation temperatures and brittleness at lower temperatures, except there exists some difference in the critical stress. The brittle to ductile transition (BDT) can be determined, and it lies in the temperature range between 1500 and 1800 K. At higher temperatures, the atomic structure has higher entropy, and its constituent atoms vibrate about their equilibrium positions at a much larger amplitude, as compared with atomic vibrational modes at lower temperatures, hence, plastic deformation occurs at higher temperatures.

The strain rate ($\dot{\varepsilon}$) also has an effect on the tensile behavior. Figure 24.4a presents the stress–strain curves of [001]-oriented nanowires with a diameter of 1.92 nm with the strain rates of 0.0005%, 0.00075%, and 0.001% ps⁻¹ at a simulation temperature of 1500 K. For all the strain rates applied, the stresses increase

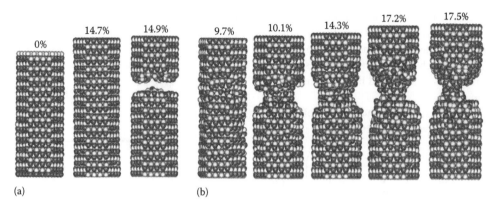

FIGURE 24.3 Atomic configurations of selected stages for the [001]-oriented nanowire with [100]-oriented lateral facets and a diameter of 2.02 nm at the temperature of (a) 600 K and (b) 1800 K. (From Wang, Z. et al., *Phys. Rev. B*, 76, 045310, 2007b. With permission.)

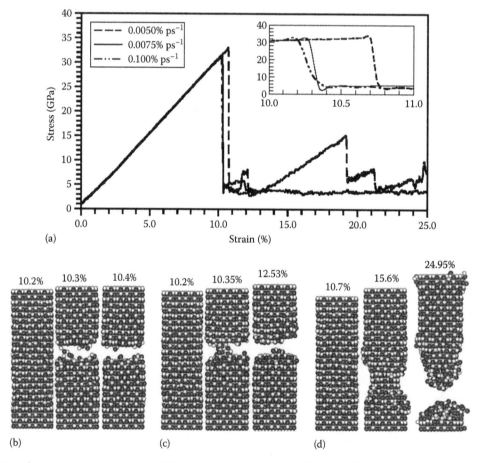

FIGURE 24.4 (a) Tensile stress–strain curves for the [001]-oriented nanowires with [110]-oriented lateral facets and a diameter of 1.92 nm, with a strain rate of 0.0005%, 0.00075% and 0.001% ps⁻¹ at 1500 K. The inset shows the enlarged parts of strains between 10% and 11%. Atomic configurations of selected stages during stretching with a strain rate of (b) 0.001%, (c) 0.00075%, and (d) 0.0005% ps⁻¹. (From Wang, Z. et al., *Phys. Rev. B*, 76, 045310, 2007b. With permission.)

linearly with strain up to 10.1%. Below this limit, the stress–strain curves for three strain rates almost completely overlap. This implies that the strain rate has no significant effect on the elastic properties of GaN nanowires. An abrupt decrease of the stress can be clearly seen at the critical strain for strain rates of

0.00075% and 0.001% ps⁻¹, whereas the stress–strain curve for a strain rate of 0.0005% ps⁻¹ exhibits a zigzag behavior. For a strain rate of 0.001% ps⁻¹, the critical stress is 31.43 GPa at a strain of 10.2%, while for strain rates of 0.00075% and 0.0005% ps⁻¹, the critical stress levels are 31.25 and 33.02 GPa at strains of 10.275%

and 10.7%, respectively. Obviously, the lower strain rates result in larger critical strains. The atomic configurations at selected stages during stretching of the nanowires, as shown in Figure 24.4b through d, reveal that the nanowire ruptures in a brittle manner at strain rates of 0.00075% and 0.001% ps⁻¹, but in a ductile manner at a strain rate of 0.0005% ps⁻¹. The results indicate that the BDT temperatures (T_{BDT}) shift to lower temperatures with a decrease in the strain rate. It should be noticed that the simulated strain rates are several orders of magnitude higher than those encountered in experiments. However, using the relationship between $\dot{\varepsilon}$ and T_{BDT}, the expected BDT temperature of nanowires can be predicted for a given experimental strain rate (Wang et al. 2006).

The stress–strain curves for [1$\bar{1}$0]-oriented nanowires with a diameter of 3.12 nm subject to uniaxial tensile loading with a strain rate of 0.01% ps⁻¹ at various temperatures are shown in Figure 24.5a. The nanowire deforms elastically until a critical strain is reached and then it begins to yield, during which bond breaking occurs and a significant decrease in stress occurs. The side views of the atomic configurations of the ruptured nanowire at 300, 500, 700, 900, and 1100 K are shown in Figure 24.5b. It can

be seen that the GaN nanowire ruptures with a clean cut, without any observed necking, which suggests that the nanowire grown along the [1$\bar{1}$0] direction fractures in a cleavage manner. A similar manner has been found for all the other nanowires considered.

Stress–strain curves of the [110]-oriented nanowires with a diameter of 2.60 nm during the deformation process with a strain rate of 0.01% ps⁻¹ at various temperatures are given in Figure 24.6a, whereas the side views of the atomic configurations at several stages at 500 K are shown in Figure 24.6b. All the nanowires show a linear stress–strain response for all applied strains up to a critical value, but a sudden decrease in stress is observed at the first yield. At this point, the nanowire slips in a {010} plane, which leads to stress relaxation, and a portion of the crystal structure is displaced. Upon full relaxation, the nanowire undergoes a second cycle of linear stress–strain increment, until a new slip is developed. This causes the whole cycle to repeat itself again. The cyclical stress–strain progression of such plastic deformation significantly weakens the nanowire structure, and thereby causes a progressively lower yield stress at each subsequent cycle. However, all the [110]-oriented nanowires studied have multiple yield stresses along their strain path for all the

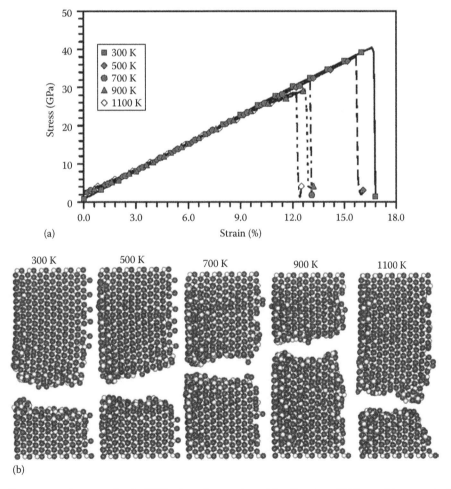

(a)

(b)

FIGURE 24.5 (a) Tensile stress–strain curves for the [1$\bar{1}$0]-oriented nanowires with a diameter of 3.12 nm subject to uniaxial tensile loading with a strain rate of 0.01% ps⁻¹ at various temperatures. (b) The side views of the atomic configurations of the ruptured nanowire at 300, 500, 700, 900 and, 1100 K. (From Wang, Z. et al., *Phys. Rev. B*, 76, 045310, 2007b. With permission.)

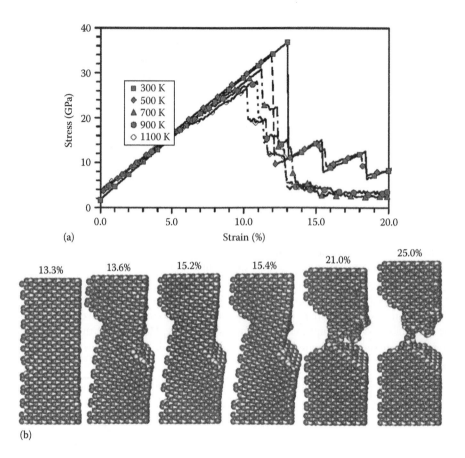

(a)

(b)

FIGURE 24.6 (a) Stress–strain curves of the [110]-oriented nanowires with a diameter of 2.60 nm during the deformation process with a strain rate of 0.01% ps⁻¹ at various temperatures. (From Wang, Z. et al., *Phys. Rev. B*, 76, 045310, 2007b. With permission.) (b) Side views of the atomic configurations at several stages at 500 K. (From Wang, Z. et al., *J. Mater. Sci.: Mater. Electron.*, 19, 863, 2008a. With permission.)

temperatures considered. These atomic configurations confirm that the slip occurred in the {010} planes.

The mechanism of ductile failure can be investigated by the MD simulations. The different atomic configurations of the cross-section of the neck at different stages, along with the stress–strain curve for the [001] oriented nanowire with a diameter of 2.02 nm at the simulation temperatures of 1800 K, are shown in Figure 24.7. As shown in the figure, the cross-sections show little change as the strain increases from 0% to 9.7%, but the bonds of the surface atoms rupture and the atoms within the neck almost randomized with increasing strain up to 9.8%, forming an amorphous-like structure. The atoms then relaxed and recovered to their normal crystal positions, and the atomic configurations reach another stable configuration. The bond rupture of the surface atoms and relaxation of other atoms induce the decrease in stress. The stress increases with the further increasing of strain to 14.3%, but little change in the cross-section is observed. The bond rupture of the surface atoms and the randomization of the central atoms within the neck occur again at strain of 14.6%, which induces the second drop of the stress. The above procedure repeats until the nanowire ruptures completely. The results indicate that the plastic deformation takes place through a phase transformation from crystal to amorphous structure,

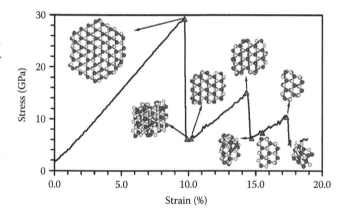

FIGURE 24.7 Atomic configurations of the cross-section within the neck, along with the stress–strain curve for the [001]-oriented nanowire with a diameter of 2.02 nm at the simulation temperatures of 1800 K. (From Wang, Z. et al., *Phys. Rev. B*, 76, 045310, 2007b. With permission.)

and generally starts from the surface of GaN nanowires at higher temperatures.

Figure 24.8 shows the stress–strain relationship of the [001]-oriented nanowires with {100} side facets with a length of 16.64 nm (*d* = 2.02 nm), [001]-oriented nanowires with {110}

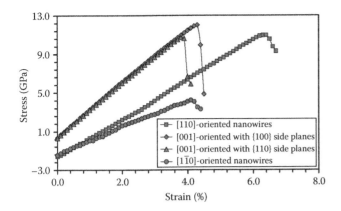

FIGURE 24.8 Compressive stress–strain relationship of the [001]-oriented nanowires with {100} side facets with length of 16.64 nm, [001]-oriented nanowires with {110} side facets with length of 16.64 nm, [1̄10]-oriented nanowires with length of 16.6 nm and [110]-oriented nanowires with length of 16.0 nm, simulated at 300 K, with strain rate of 0.001% ps⁻¹. (From Wang, Z. et al., *Phys. E: Low Dimens. Syst. Nanostruct.*, 40, 561, 2008b. With permission.)

side facets with a length of 16.64 nm ($d = 1.92$ nm), [1̄10]-oriented nanowires with a length of 16.6 nm ($d = 2.08$ nm), and [110]-oriented nanowires with a length of 16.0 nm ($d = 2.6$ nm), respectively simulated at 300 K, with strain rate of 0.001% ps⁻¹. Some representative atom configurations are shown in Figure 24.9. All four types of nanowires show that they retain their symmetry until the critical strain is reached, at which point they begin to buckle. The stress increases with increasing strain and decreases when the nanowire buckles. Figure 24.9a shows the [001]-oriented nanowire with {100} side facets with a length of 16.64 nm at zero strain, where there is no deformation. Up to a strain of 4.3%, the wire lengths decrease uniformly and no structural defects appear in the nanowires. Further compression causes the nanowire to buckle.

Figure 24.10 shows the effects of wire length on the buckling stress. It clearly shows that the buckling stress decreases with an increase in wire length. The critical stress decreases from 14.1 to 5.27 GPa as the wire length increases from 12.48 to 29.12 nm for the [001]-oriented nanowires with {100}-side planes. For a long

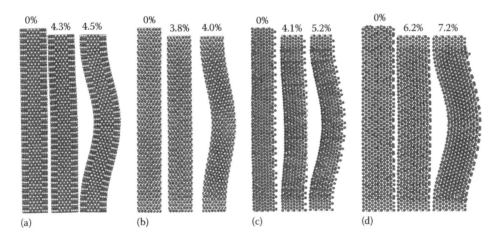

FIGURE 24.9 Snapshots of atomic configurations at various strain values of (a) [001]-oriented nanowires with {100} side facets with length of 16.64 nm, (b) [001]-oriented nanowires with {110} side facets with length of 16.64 nm, (c) [1̄10]-oriented nanowires with length of 16.6 nm and (d) [110]-oriented nanowires with length of 16.0 nm, simulated at 300 K, with strain rate of 0.001% ps⁻¹. (From Wang, Z. et al., *Phys. E: Low Dimens. Syst. Nanostruct.*, 40, 561, 2008b. With permission.)

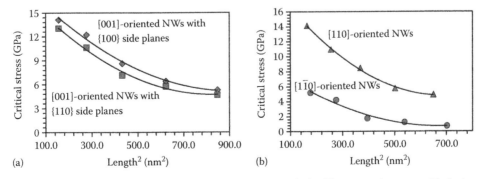

FIGURE 24.10 The wire length dependence of critical stress of GaN nanowires. The buckling stress decreases with the increase of wire length. (From Wang, Z. et al., *Phys. E: Low Dimens. Syst. Nanostruct.*, 40, 561, 2008b. With permission.)

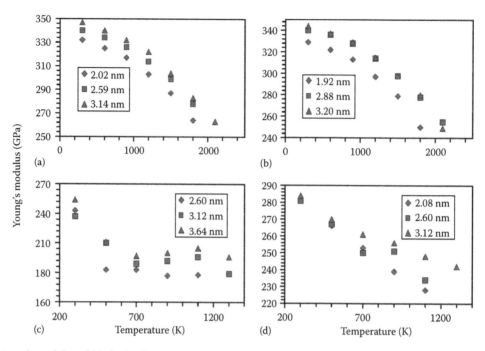

FIGURE 24.11 Young's modulus of (a) the [001]-oriented nanowires with {100} side planes; (b) the [001]-oriented nanowires with {110} side planes; (c) the [1$\bar{1}$0]- and (d) [110]-oriented nanowires at various simulation temperatures. (From Wang, Z. et al., *Phys. Rev. B*, 76, 045310, 2007b. With permission.)

column, the critical buckling load is predicted using the Euler theory (Timoshenko and Gere 1961):

$$P_{cr} = \frac{\pi^2 EI}{L_e^2} \tag{24.5}$$

where

E is the Young's modulus
L_e is the effective length of the wire
I is the moment of inertia

The behavior of an ideal column compressed by an axial load P may be summarized as follows: if $P < P_{cr}$, the column is in a stable equilibrium in a straight position; if $P = P_{cr}$, the column is in a neutral equilibrium in either a straight or slightly bent position; if $P > P_{cr}$, the column is in an unstable equilibrium in a straight position, which promotes buckling. Equation 24.5 shows that the critical load is inversely proportional to the square of the length. It can be clearly seen from Figure 24.10 that the longer the GaN nanowires, the smaller the critical stress (and corresponding critical strain) for buckling. The trend is in agreement with the Euler theory.

There are two approaches to obtain the Young's modulus, i.e., force approach and energy approach (Rafii-Tabar 2004). For the force approach, the Young's modulus can be directly obtained from the ratio of stress to strain, whereas the energy approach calculates the Young's modulus from the second derivative of strain energy with respect to the strain per unit volume. Using the force approach, Young's modulus is determined from the results of the tension tests for strains <2.5% using linear regression, and the calculated results for all the nanowires are shown

in Figure 24.11. The Young's modulus of the [001]-oriented nanowires is in the range between 347 and 249 GPa at temperatures between 300 and 2100 K, while it is in the range of 254–179 and 284–228 GPa for the [1$\bar{1}$0] and [110]-oriented nanowires, respectively at temperatures ranging between 300 and 1100 K. The calculated values generally agree well with the experimental results of 227–305 GPa of GaN nanowires (Nam et al. 2006). It is shown that Young's modulus decreases with increasing temperature, which suggests that the nanowires become softened at higher temperatures because of the higher kinetic energy of each atom on average, which results in easier slipping and elongation of the nanowires. From Figure 24.11, it can also be seen that Young's modulus decreases with the decreasing diameter of the nanowires, which is somewhat unexpected. This is in contrast to single-crystal materials, where Young's modulus is expected to increase with decreasing crystal size (Kulkarni et al. 2005; Chen et al. 2006). However, the results agree well with recent experimental results. The reason for this behavior may be due to a small atomic-coordination number and weak cohesion of the atoms near the surface, as compared with those in the bulk, and the increasing dominance of the surface would decrease the rigidity of the structure (Schmid et al. 1995).

24.4.5 Discussion

In general, a single-crystal subject to external tensile stress responds in one of two ways: either it yields plastically and deforms its shape or, alternatively, it fractures into two or more pieces. Plastic deformation occurs by shearing of the blocks of the crystal over slip planes along a slip direction—it is the shear component of the applied stress that is responsible for yielding. On the other hand, the fracture of a crystal occurs by the rupture of bonds holding the atoms together across the cleavage plane. In the latter case, the interatomic bonds can be ruptured by stretching them in a direction normal to the cleavage plane or by shearing the bonds in a direction normal or parallel to the crack front.

One way to represent the variations of the yield stress (τ_c) of a semiconductor with respect to temperature is to plot $\ln(\tau_c)$ versus $1/T$, as suggested by the kink-diffusion model (Hirth and Lothe 1982). In this model, the plastic strain rate can be described by Orowan's equation, which is proportional to the stress-independent activation enthalpy, H, for the glide of the dislocations corresponding to crystal deformation. Specifically,

$$\tau_c = A\dot{\varepsilon}^{1/n} \exp\left(\frac{\Delta H_\tau}{k_B T}\right) \qquad (24.6)$$

where

  A and n are constants

  k_B is the Boltzmann constant

  ΔH_τ is an energy parameter, such that $n\Delta H_\tau$ is approximately the activation enthalpy H (Rabier and George 1987)

The changes of $\ln(\tau_c)$ as a function of $1/T$ for the three types of nanowires are shown in Figure 24.12. In general, $\ln(\tau_c)$ increases linearly with increasing reciprocal temperature, but its slope abruptly changes at T_c for the [001]-oriented nanowire. The different slopes in the $\ln(\tau_c)$ curve characterize the transition, and thus correspond to two different activation enthalpies, H_t and H_l, for dislocation glide at high temperatures ($T > T_c$) and low temperatures ($T < T_c$), respectively. These features are qualitatively very similar to the experimental observations in InP (Suzuki et al. 1998), GaAs (Boivin et al. 1990), and SiC (Zhang et al. 2003). The BDT temperatures for the [001]-oriented nanowire deduced from Figure 24.12 are between 1500 and 1800 K, which is consistent with the atomic observations of rupture changes from a clean cut at low temperatures to necking at higher temperatures under tensile loading. Change in slopes for the [110]- and [1$\bar{1}$0]-oriented nanowires are not observed, which means that the nanowires fracture by a single mechanism.

The mechanical behavior of GaN nanowires shows a significant dependence on the crystallographic orientation. In particular, the [001], [1$\bar{1}$0], and [110] directions represent three orthogonal crystallographic orientations within the wurtzite GaN crystal structure—the first one is in a polar direction and the others are in nonpolar directions. It has been shown that the presence of spontaneous polarization in GaN has a drastic impact on electron-hole overlap, radiative lifetimes, and subsequent emission wavelength and quantum efficiencies for GaN (Waltereit et al. 2000). Spontaneous polarization can also have an effect on the mechanical behavior, as demonstrated in the present studies. The unique isosceles triangular cross-section of the [1$\bar{1}$0]- and [110]-oriented nanowires might also lead to the different tensile behavior compared with that of the [001]-oriented nanowires.

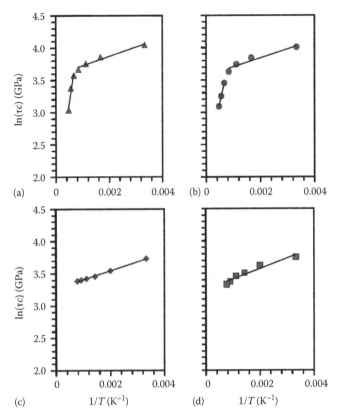

FIGURE 24.12 Evolution of $\ln(\tau_c)$ as a function of reciprocal temperature for the nanowires with growth direction along (a) [001]-crystal direction with {100} side planes, (b) [001]-crystal direction with {110} side planes, (c) [110]- and (d) [1$\bar{1}$0] crystal direction, respectively. (From Wang, Z. et al., *J. Mater. Sci.: Mater., Electron.* 19, 863, 2008a. With permission.)

The effect of orientation and cross-section shape on the properties of the nanostructures needs further investigations.

24.5 Summary

The fundamental deformation mechanisms in GaN nanowires with different orientations have been investigated by MD simulations. Due to its high crystal stability at low temperatures, the deformation behavior of [001]-oriented nanowires is characterized by brittle rupture. At higher temperatures, the crystal structure becomes less stable due to higher amplitudes of atomic vibrations around their equilibrium positions. The deformation of the nanowires changes from a brittle rupture to a ductile rupture with an increase in temperature. The mechanism of plastic deformation is through a phase transformation from a crystalline to an amorphous structure. The nanowires rupture in a brittle manner at a high strain rate and in a ductile manner at a low strain rate. Interestingly, the [110]-oriented nanowires slip in the {010} planes and there exists multiple yield stresses along the strain path, whereas the [1$\bar{1}$0]-oriented nanowires fracture in a cleavage manner under tensile loading.

It should be noticed that there are some drawbacks and advantages of this simulated method. The Stillinger–Weber potential is "short-range potential," which includes only the first nearest neighbor interactions. This may cause the interactions between the atoms to abruptly vanish outside a certain radius, and this affects the results somehow. For example, the melting temperature of bulk GaN using this method is determined to be 3000 K (Wang et al. 2007a,b), where the experimental value is about 2773 K. However, this short-range potential should not affect the main conclusions, and simulation results can provide a qualitative level. Characterizing the mechanical properties of individual one-dimensional nanostructures is a challenge to many existing testing and measuring techniques. The method provides a powerful tool to quantify the mechanical properties of individual nanowires.

24.6 Future Directions

Although classical MD simulations can predict some useful results, the interatomic potential employed is a "short-range potential" that only include the first nearest-neighbor interactions. This can cause the interactions between atoms to abruptly vanish outside a certain radius, which will affect the results to some degree. *Ab initio* or first principles molecular simulation should provide more accurate results, but these require extensive computational efforts. With the development of computer power, these will become useful methods to predict the properties of GaN nanowires.

Until now, calculations on nanowires only involved very small systems with diameters of only several nanometers. At the nanoscale, the behavior can change dramatically as the diameters increase. It is essential to quantify the behavior of nanowires with large diameters and increased lengths. Quantification and evolution description requires a fundamental approach for transitioning from a molecular description to continuum descriptions.

Acknowledgments

Z.G. Wang is grateful for the National Natural Science Foundation of China (10704014) and the Young Scientists Foundation of UESTC (JX0731). F. Gao and W. J. Weber were supported by the Division of Materials Sciences and Engineering, Office of Basic Energy Sciences, U.S. Department of Energy under Contract DE-AC05-76RL01830. J. Li gratefully acknowledges financial support from the "One-Hundred Talents Plan" of the Chinese Academy of Sciences. The authors also wish to thank the Molecular Science Computing Facility in the Environmental Molecular Sciences Laboratory at the Pacific Northwest National Laboratory for a grant of computer time.

References

Aïchoune, N., Potin, V., Ruterana, P. et al. 2000. An empirical potential for the calculation of the atomic structure of extended defects in wurtzite GaN. *Comput. Mater. Sci.* 17: 380–383.

Béré, A. and Serra, A. 2001. Atomic structure of dislocation cores in GaN. *Phys. Rev. B* 65: 205323.

Boivin, P., Rabier, J., and Garem, H. 1990. Plastic-deformation of GaAs single-crystals as a function of electronic doping. 1. Medium temperatures (150–650°C). *Philos. Mag. A* 61: 619–645.

Calleja, E., Sanchez-Garcia, M. A., Sanchez, F. J. et al. 1999. Growth of III-nitrides on Si(111) by molecular beam epitaxy doping, optical, and electrical properties. *J. Cryst. Growth* 201: 296–317.

Cerutti, L., Ristic, J., Fernandez-Garrido, S. et al. 2006. Wurtzite GaN nanocolumns grown on Si(001) by molecular beam epitaxy. *Appl. Phys. Lett.* 88: 213114.

Chen, C. Q., Shi, Y., Zhang, Y. S. et al. 2006. Size dependence of Young's modulus in ZnO nanowires. *Phys. Rev. Lett.* 96: 075505.

Chen, Y. X., Stevenson, I., Pouy, R. et al. 2007. Mechanical elasticity of vapour-liquid-solid grown GaN nanowires. *Nanotechnology* 18: 135708.

Chin, A. H., Ahn, T. S., Li, H. W. et al. 2007. Photoluminescence of GaN nanowires of different crystallographic orientations. *Nano Lett.* 7: 626–631.

Dai, H. J., Wong, E.W., Lu, Y. Z., Fan, S. S., and Lieber, C. M. 1995. Synthesis and characterization of carbide nanorods. *Nature* 375: 769–772.

Duan, X. F. and Lieber, C. M. 2000. Laser-assisted catalytic growth of single crystal GaN nanowires. *J. Am. Chem. Soc.* 122: 188–189.

Feng, D. P., Zhao, Y., and Zhang, G. Y. 1999. Anisotropy in electron mobility and microstructure of GaN grown by metalorganic vapor phase epitaxy. *Phys. Status Solidi A* 176: 1003–1008.

Gradecak, S., Qian, F., Li, Y., Park, H. G., and Lieber, C. M. 2005. GaN nanowire lasers with low lasing thresholds. *Appl. Phys. Lett.* 87: 173111.

Greytak, A. B., Barrelet, C. J., Li, Y., and Lieber, C. M. 2005. Semiconductor nanowire laser and nanowire waveguide electro-optic modulators. *Appl. Phys. Lett.* 87: 151103.

Han, W. Q., Fan, S. S., Li, Q. Q., and Hu, Y. D. 1997. Synthesis of gallium nitride nanorods through a carbon nanotube-confined reaction. *Science* 277: 1287–1289.

Hessman, D., Lexholm, M., and Dick, K. A. 2007. High-speed nanometer-scale imaging for studies of nanowire mechanics. *Small* 3: 1699–1702.

Hirth, J. P. and Lothe, J. 1982. *Theory of Dislocations*, Wiley, New York.

Huang, Y., Duan, X. F. Cui, Y. et al. 2001. Logic gates and computation from assembled nanowire building blocks. *Science* 294: 1313–1317.

Johnson, J. C., Choi, H. J., Knutsen, K. P. et al. 2002. Single gallium nitride nanowire lasers. *Nat. Mater.* 1: 106–110.

Ju, S. P., Lin, J. S., and Lee, W. J. 2004. A molecular dynamics study of the tensile behaviour of ultrathin gold nanowires. *Nanotechnology* 15: 1221–1225.

Kanemitsu, Y., Ogawa, T., Shiraishi, K., and Takeda, K. 1993. Visible photoluminescence from oxidized Si nanometer-sized spheres-exciton confinement on a spherical-shell. *Phys. Rev. B* 48: 4883–4886.

Kikuchi, A., Kawai, M., Tada, M., and Kishino, K. 2004. InGaN/ GaN multiple quantum disk nanocolumn light-emitting diodes grown on (111) Si substrate. *Jpn. J. Appl. Phys.* 43: L1524–L1526.

Kioseoglou, J., Polatoglou, H. M., Lymperakis, L. et al. 2003. A modified empirical potential for energetic calculations of planar defects in GaN. *Comput. Mater. Sci.* 27: 43–49.

Kulkarni, A. J., Zhou, M., and Ke, F. J. 2005. Orientation and size dependence of the elastic properties of zinc oxide nanobelts. *Nanotechnology* 16: 2749–2756.

Kuykendall, T., Pauzauskie, P., Lee, S. et al. 2003. Metalorganic chemical vapor deposition route to GaN nanowires with triangular cross sections. *Nano Lett.* 3: 1063–1066.

Kuykendall, T., Pauzauskie, P. J., Zhang, Y. F. et al. 2004. Crystallographic alignment of high-density gallium nitride nanowire arrays. *Nat. Mater.* 3: 524–528.

Lexholm, M., Hessman, D., and Samuelson, L. 2006. Optical interference from pairs and arrays of nanowires. *Nano Lett.* 6: 862–865.

Li, H. W., Chin, A. H., and Sunkara, M. K. 2006. Direction-dependent homoepitaxial growth of GaN nanowires. *Adv. Mater.* 18: 216–218.

Makhlin, Y., Schon, G., and Shnirman, A. 2001. Quantum-state engineering with Josephson-junction devices. *Rev. Mod. Phys.* 73: 357–400.

Meijers, R., Richter, T., Calarco, R. et al. 2006. GaN-nanowhiskers: MBE-growth conditions and optical properties. *J. Cryst. Growth* 289: 381–386.

Morales, A. M. and Lieber C. M. 1998. A laser ablation method for the synthesis of crystalline semiconductor nanowires. *Science* 279: 208–211.

Nam, C. Y., Jaroenapibal, P., Tham, D. et al. 2006. Diameter-dependent electromechanical properties of GaN nanowires. *Nano Lett.* 6: 153–158.

Pauzauskie, P. J., Sirbuly, D. J., and Yang, P. D. 2006. Semiconductor nanowire ring resonator laser. *Phys. Rev. Lett.* 96: 143903.

Peercy, P. S. 2000. The drive to miniaturization. *Nature (London)* 406: 1023–1026.

Qian, F., Li, Y., Gradecak, S. et al. 2008. Multi-quantum-well nanowire heterostructures for wavelength-controlled lasers. *Nat. Mater.* 7: 701–706.

Rabier, J. and George, A. 1987. Dislocations and plasticity in semi-conductors. 2. The relation between dislocation dynamics and plastic-deformation. *Rev. Phys. Appl.* 22: 1327–1351.

Rafii-Tabar, H. 2004. Computational modelling of thermomechanical and transport properties of carbon nanotubes. *Phys. Rep.* 390: 235–452.

Ristic, J., Calleja, E., Sanchez-Garcra, M. A. et al. 2003. Characterization of GaN quantum discs embedded in $Al_xGa_{1-x}N$ nanocolumns grown by molecular beam epitaxy. *Phys. Rev. B* 38: 125305.

Schmid, M., Hofer, W., Varga, P. et al. 1995. Surface stress, surface elasticity, and the size effect in surface segregation. *Phys. Rev. B* 51: 10937–10946.

Stern, E., Cheng, G., Cimpoiasu, E. et al. 2005. Electrical characterization of single GaN nanowires. *Nanotechnology* 16: 2941–2953.

Stillinger, F. H. and Weber, T. A. 1985. Computer-simulation of local order in condensed phases of silicon. *Phys. Rev. B* 31: 5262–5271.

Suzuki, T., Nishisako, T., Taru, T., and Yasutomi, T. 1998. Plastic deformation of InP at temperatures between 77 and 500 K. *Philos. Mag. Lett.* 77: 173–180.

Tchernycheva, M., Sartel, C., Cirlin, G. et al. 2007. Growth of GaN free-standing nanowires by plasma-assisted molecular beam epitaxy: Structural and optical characterization. *Nanotechnology* 18: 385306.

Timoshenko, S. P. and Gere, J. M. 1961. *Theory of Elastic Stability*. McGraw-Hill, New York.

Tragardh, J., Persson, A. I., and Wagner, J. B. 2007. Measurements of the band gap of wurtzite $InAs_{1-x}P_x$ nanowires using photocurrent spectroscopy. *J. Appl. Phys.* 101: 123701.

Trampert, A., Ristic, J., Jahn, U., Calleja, E., and Ploog, K. H. 2003. TEM study of (Ga,Al)N nanocolumns and embedded GaN nanodiscs. *Micros. Semicond. Mater.*, 180: 167–170.

Wagner, R. S. and Ellis, W. C. 1964. Vapor-liquid-solid mechanism of single crystal growth (new method growth catalysis from impurity whisker epitaxial + large crystals Si E). *Appl. Phys. Lett.* 4: 89–91.

Waltereit, P., Brandt, O., Trampert, A. et al. 2000. Nitride semiconductors free of electrostatic fields for efficient white light-emitting diodes. *Nature* 406: 865–868.

Wang, Z. G., Zu, X. T., Gao, F., and Weber, W. J. 2006. Brittle to ductile transition in GaN nanotubes. *Appl. Phys. Lett.* 89: 243123.

Wang, Z. G., Zu, X. T., Gao, F., and Weber, W. J. 2007a. Size dependence of melting of GaN nanowires with triangular cross-sections. *J. Appl. Phys.* 101: 043511.

Wang, Z., Zu, X., Yang, L., Gao, F., and Weber, W. J. 2007b. Atomistic simulations of the size, orientation, and temperature dependence of tensile behavior in GaN nanowires. *Phys. Rev. B* 76(4): 045310.

Wang, Z., Zu, X., Yang, L., Gao, F., and Weber, W. J. 2008a. Orientation and temperature dependence of the tensile behavior of GaN nanowires: An atomistic study. *J. Mater. Sci.: Mater. Electron.* 19(8–9): 863–867.

Wang, Z., Zu, X., Yang, L., Gao, F., and Weber, W. J. 2008b. Molecular dynamics simulation on the buckling behavior of GaN nanowires under uniaxial compression. *Phys. E: Low Dim. Syst. Nanostruct.* 40(3): 561–566.

Zach, M. P., Ng, K. H., and Penner, R. M. 2000. Molybdenum nanowires by electrodeposition. *Science* 290: 2120–2123.

Zhang, M., Hobgood, H. M., Demenet, J. L., and Pirouz, P. 2003. Transition from brittle fracture to ductile behavior in 4H-SiC. *J. Mater. Res.* 18: 1087–1095.

Zhong, Z. H., Qian, F., Wang, D. L., and Lieber, C. M. 2003. Synthesis of p-type gallium nitride nanowires for electronic and photonic nanodevices. *Nano Lett.* 3: 343–346.

Optical Properties of Anisotropic Metamaterial Nanowires

Wentao Trent Lu
Northeastern University

Srinivas Sridhar
Northeastern University

25.1 Introduction

The booming growth of nanoscience and nanotechnology is driven by our increasing capability of nanofabrication, new understanding of nanosized structures and materials, and new applications in material science and medicine. There are two methods for nanofabrication: top down and bottom up. The top-down method has been shrinking in size from the microscale to the nanoscale due to the use of shorter wavelengths in optical lithography. The bottom-up self-assembly method has been refined to fabricate more fancy and complicated structures.

The electrons and photons, the two fundamental information carriers of optoelectronics, have typical wavelengths of a few nanometers and hundreds of nanometers or a few microns, respectively. Thus a nanostructure with an internal length scale of tens or hundreds of nanometers gives rise to strong scattering and induces an interaction between electrons and photons, which leads to new optical properties of the nanostructure.

With Moore's law for shrinking the size of the silicon-based electronics expected to come to an end, completely different approaches to design computer chips such as using all-optical circuitry [13,18] and carbon-based electronics and photonics [1] are being seriously pursued. Along with the advance of nanoscience and nanotechnology, new concepts in physics such as negative refraction [57] and transformation optics [40] are being developed.

Nanowires are long objects whose diameters are of nanometer size. Due to the confinement of the motion of electrons, holes, photon, phonons, and other quasi-particles, many interesting features other than that of the bulk materials appear. For example for the quantum wires, the resistance will be quantized. Optical wave propagations in waveguides of nanometer size [49,50] have unique properties. Nanowires made of metal can be used to concentrate an electric field [48]. Some features can be advantageous and some can be adverse. Due to the shrinking size of metallic wires and the related overheating in computer chips, information exchange between different components of a chip is bottlenecked, preventing further increase in clock speeds.

The unique properties of nanowires depend not only on their sizes, they also strongly depend on the materials used. Metallic nanowires are used in chips for electron transportation. For carbon nanotubes used as quantum wires, all the electronic transport properties are determined by the graphene layer [12], which is a single layer of a honeycomb lattice of carbon atoms. Carbon nanotubes can be metallic or semiconducting depending on the geometric arrangement, such as the chirality and the number of layers. Nanowires made of silicon in silicon-on-insulator platform are used to guide optical waves [36].

Nanowires can be made of homogeneous and isotropic materials such as dielectrics or metals. It can also be made of inhomogeneous materials such as a grating. In this chapter, we focus on the optical properties of nanowires made of the so-called metamaterials.

25.2 Negative Refraction and Metamaterials

Metamaterials are a broadly defined class of materials that are artificially designed and fabricated, inhomogeneous in the microscopic scale, and posses certain desirable properties that are not available in natural materials and can be treated homogeneously in the scale of interest. These metamaterials in common have periodic or quasi-periodic structures in certain dimensions. Due to this periodicity or quasi-periodicity and the artificially designed resonant properties, new features of physical properties appear.

The concept of metamaterials is not new [5]. In radio frequency range, the frequency-selective surfaces can be viewed as metamaterials. Since the realization of negative refraction [56] in microwaves [41], there has been a renewed and intense interest in electromagnetic metamaterials. Negative refraction has added a new arena to physics, leading to new concepts such as perfect lens [22,33], superlens [11,21,33], and focusing by plano-concave lens [58,59]. Negative refraction has subsequently been achieved in microwaves [7,24,29,31,32], THz waves, and optical wavelengths [2,8,47], in metamaterials made of wire and split-ring resonators [45] or photonic crystals [14,25,26].

Ordinary materials have both positive permittivity and permeability, thus they possess positive refractive indices. Metamaterials can be double-negative with both permittivity (ϵ) and permeability (μ) being negative [39]. Double-negative metamaterials (DNM) allow negative refraction according to Snell's law, with a negative refractive index $n = \sqrt{\epsilon}\sqrt{\mu} < 0$ [43]. Metamaterials can be single-negative with either permittivity or permeability being negative. Single-negative materials such as ferrites and metals do not allow waves to propagate inside, and reflect waves back. Metamaterials can be periodic, such as photonic crystals [19]. They can also be non-periodic, such as the materials for cloaking [38].

The optical properties of metamaterials can be either isotropic or anisotropic. Among metamaterials, there is a subclass of materials called indefinite media [16,23,44,61] whose electromagnetic properties are extremely anisotropic. The permittivity and/or the permeability tensors are indefinite matrices, whose diagonal elements are not all positive if diagonalized. For ordinary anisotropic materials, the refractive index is given by an ellipsoid [4]. For an indefinite metamaterial (IM), the dispersion is hyperbolic for one polarization and elliptical for the other. Negative refraction, superlens imaging, and hyperlens focusing [20,46] can also be realized by using IMs. This broad range of properties opens infinite possibilities to use metamaterials in frequencies from microwave all the way up to the visible.

In this chapter, we further narrow our focus on nanowires made of IM [17]. To begin with, we consider a very general case in which the nanowire is made of a nonmagnetic material whose transverse permittivity component is different from the longitudinal one. In the case where the transverse permittivity is negative while the longitudinal one is positive ($\epsilon_\perp < 0$, $\epsilon_\parallel > 0$), these IM waveguides can support both forward- and backward-wave modes. A high effective phase index can be obtained for these modes. These waveguides can also support degenerate modes that can be used to slow down and trap light. Magnetic metamaterials and DNMs will not be considered, though our analysis can be easily extended to these cases. Due to the backward-wave nature of waves inside the waveguide, a nanowire waveguide made of DNM will have properties similar to that of a nanowire waveguide made of IM.

For easy understanding, we first consider a free-standing planar waveguide. Then we consider a cylindrical nanowire waveguide. Other shapes of nanowires, such as rectangular shapes, can also be considered and may be more suitable for optical integrated circuits. Though numerical solutions must be sought, these waveguides will have similar interesting properties and their designs can be guided by the available analytical solutions for the slab and the cylindrical nanowire waveguides.

25.3 Wave Propagation in an Anisotropic Planar Waveguide

We first consider a slab of planar waveguide made of anisotropic metamaterial in air. The wave propagation is along the z-direction with phase $e^{i(\beta z - \omega t)}$ and the transverse direction is in the x-direction. We consider the case of an indefinite medium with

$$\epsilon_z > 0, \quad \epsilon_x < 0. \tag{25.1}$$

For the transverse magnetic (TM) modes, the magnetic field is in the y-direction. The transverse wave vector wave inside the metamaterial is

$$k_x = \sqrt{\epsilon_z}\sqrt{k_0^2 - \beta^2/\epsilon_x}. \tag{25.2}$$

Here k_0 is the wave number in vacuum. Due to the negativity of ϵ_x, the dispersion is hyperbolic instead of elliptic.

Since the planar waveguide is symmetric, the magnetic field is

$$\begin{aligned} H_y(x) &= e^{\kappa_0(x+d/2)}, \quad x \le \frac{-d}{2}, \\ &= Ae^{ik_x x} + Be^{-ik_x x}, \quad \frac{-d}{2} \le x \le \frac{d}{2}, \\ &= e^{-\kappa_0(x-d/2)}, \quad x \ge \frac{d}{2}. \end{aligned} \tag{25.3}$$

where

d is the slab thickness

$\kappa_0 = \sqrt{\beta^2 - k_0^2}$ is the decay constant in the transverse direction in the air

The tangential electric field is

$$E_z(x) = \frac{i\kappa_0}{k_0}e^{\kappa_0(x+d/2)}, \quad x \leq \frac{-d}{2},$$

$$= -\frac{k_x}{\varepsilon_z k_0}\left(Ae^{ik_x x} - Be^{-ik_x x}\right), \quad \frac{-d}{2} \leq x \leq \frac{d}{2},$$

$$= -\frac{i\kappa_0}{k_0}e^{-\kappa_0(x-d/2)}, \quad x \geq \frac{d}{2}. \quad (25.4)$$

The matching of the tangential electric and magnetic fields at boundary $x = \pm d/2$ leads to the following eigen equation for the TM$_m$ modes

$$k_0 d = \frac{1}{\sqrt{\varepsilon_z}\sqrt{1 - n_p^2/\varepsilon_x}}\left(m\pi + 2\arctan\frac{\sqrt{\varepsilon_z}\sqrt{n_p^2 - 1}}{\sqrt{1 - n_p^2/\varepsilon_x}}\right). \quad (25.5)$$

Here $n_p \equiv \beta/k_0$ is the phase index of the guided mode.

At a fixed wavelength or frequency, ε_x and ε_z are constant. For easy plot of the phase index n_p for different waveguide thickness d, one can treat the thickness d as a function of n_p, which is chosen as a free parameter. The band structure for TM modes on a slab waveguide with $\varepsilon_x = -3$ and $\varepsilon_z = 2$ is plotted in Figure 25.1.

For waves inside the IM, one has

$$\beta = \sqrt{\varepsilon_x}\sqrt{k_0^2 - k_x^2/\varepsilon_z}. \quad (25.6)$$

So for $k_x > \sqrt{\varepsilon_z}k_0$, β will be real and negative if the imaginary part of the permittivity is ignored. So the waves inside the IM will be left-handed, $\beta S_z < 0$ with $S_z = E_x H_y$, similar to that inside a double-negative metamaterial. However, for waves confined in the transverse direction, S_z is no longer uniform. The total energy, P_z, is the sum of energy carried inside and outside the waveguide, $P_z = \int_{-\infty}^{\infty} S_z dx$. A guided wave is forward (backward) only if βP_z is positive (negative). For the TM$_m$ modes on the planar waveguide, we have evaluated the total energy as

$$P_z = \frac{n_p}{2}\left[\frac{1 + (-1)^m \cos k_x d}{\kappa_0} + \frac{d}{\varepsilon_x}\left(1 + (-1)^m \operatorname{sinc} k_x d\right)\right]. \quad (25.7)$$

The first term is the energy flow in air and the second term is that inside the slab. Due to the negative sign of ε_x, the energy flow inside the waveguide is negative and contra-directional to that in the air.

From the plotted example, a few salient features are evident. First unlike a conventional dielectric slab waveguide, most modes are backward-wave modes since $dn_p/dd < 0$. Only for modes near the light line where $n_p \sim 1$, they are forward-wave modes since $dn_p/dd > 0$. In Section 25.4, we will consider nanowire waveguide made of IMs. One will see that the modes on the cylindrical nanowire share most of the features of that on a slab waveguide. For the cylindrical nanowires, the determination of forward and backward wave will be discussed in details.

25.4 Wave Propagation on Anisotropic Nanowire Waveguides

We now consider wave propagation in a cylindrical waveguide. The axis of the waveguide is along the z-direction as shown in Figure 25.2. The waveguide is nonmagnetic and has an anisotropic optical property

$$\varepsilon_x = \varepsilon_y = \varepsilon_t \neq \varepsilon_z. \quad (25.8)$$

The waves propagate along the cylinder axis with

$$\mathbf{E} = \mathbf{E}_0 e^{i(\beta z - \omega t)}, \quad \mathbf{H} = \mathbf{H}_0 e^{i(\beta z - \omega t)}. \quad (25.9)$$

Here β is the propagation wave number along the nanowire waveguide. Though this waveguide still allows exact solutions for all the guided modes, the formulas are more involved and the features of the modes are not as apparent as that on a slab waveguide we have considered in Section 25.3. Nevertheless, we will arrive at the simplest expressions that can be easily generalized to more complex situations including magnetic waveguides. The optical properties

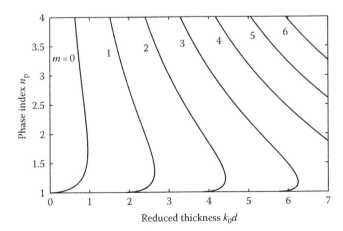

FIGURE 25.1 The phase index n_p of the guided TM$_m$ modes on a free-standing planar waveguide of thickness d with $\varepsilon_x = -3$ and $\varepsilon_z = 2$.

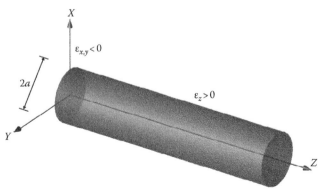

FIGURE 25.2 A nanowire waveguide made of an indefinite metamaterial. (From Huang, Y.J. et al., *Phys. Rev. A*, 77, 063836, 2008. With permission.)

of cylindrical nanowires will give guidelines for properties of other shapes of nanowires that may not allow exact solutions.

Due to the symmetry of the waveguide, all the field components can be expressed in terms of the longitudinal components E_z and H_z. In the polar coordinate system, one has for the fields inside the waveguide $r < a$ with a the radius,

$$
\begin{aligned}
E_r &= i\frac{1}{\varepsilon_t k_0^2 - \beta^2}\left(\beta\partial_r E_z + k_0\frac{1}{r}\partial_\phi H_z\right), \\
E_\phi &= i\frac{1}{\varepsilon_t k_0^2 - \beta^2}\left(\beta\frac{1}{r}\partial_\phi E_z - k_0\partial_r H_z\right), \\
H_r &= i\frac{1}{\varepsilon_t k_0^2 - \beta^2}\left(-\varepsilon_t k_0\frac{1}{r}\partial_\phi E_z + \beta\partial_r H_z\right), \\
H_\phi &= i\frac{1}{\varepsilon_t k_0^2 - \beta^2}\left(\varepsilon_t k_0\partial_r E_z + \beta\frac{1}{r}\partial_\phi H_z\right).
\end{aligned}
\tag{25.10}
$$

The wave equations for the longitudinal components inside the waveguide are

$$
\begin{aligned}
\left(\partial_x^2 + \partial_y^2\right)E_z + \varepsilon_z\left(k_0^2 - \frac{\beta^2}{\varepsilon_t}\right)E_z &= 0, \\
\left(\partial_x^2 + \partial_y^2\right)H_z + \left(\varepsilon_t k_0^2 - \beta^2\right)H_z &= 0.
\end{aligned}
\tag{25.11}
$$

The waveguide is free-standing in air, so the wave equations for $r > a$ are given by the above equations with the permittivity replaced by unity. The solutions are expressed in terms of the Bessel functions of various kinds

$$
\begin{aligned}
E_z &= AJ_n(Kr)e^{in\phi}, \quad r < a, \\
&= CK_n(\kappa_0 r)e^{in\phi}, \quad r > a
\end{aligned}
\tag{25.12}
$$

and

$$
\begin{aligned}
H_z &= BI_n(\kappa r)e^{in\phi}, \quad r < a, \\
&= DK_n(\kappa_0 r)e^{in\phi}, \quad r > a.
\end{aligned}
\tag{25.13}
$$

The coefficients will be determined by matching the boundary conditions. Here

$$
\begin{aligned}
K &= \sqrt{\varepsilon_z}\sqrt{k_0^2 - \frac{\beta^2}{\varepsilon_t}}, \\
\kappa &= \sqrt{\beta^2 - \varepsilon_t k_0^2}, \\
\kappa_0 &= \sqrt{\beta^2 - k_0^2}.
\end{aligned}
\tag{25.14}
$$

We only consider the extremely anisotropic case such that the longitudinal permittivity is positive while the transverse permittivity is negative

$$
\varepsilon_z > 0, \quad \varepsilon_t < 0.
\tag{25.15}
$$

One can see that due to the anisotropic nature of the waveguide, H_z and E_z inside the waveguide will have completely different behaviors.

The continuity of E_z and H_z at the interface $r = a$ gives

$$
\frac{C}{A} = \frac{J_n(Ka)}{K_n(\kappa_0 a)}, \quad \frac{D}{B} = \frac{I_n(\kappa a)}{K_n(\kappa_0 a)}.
\tag{25.16}
$$

The continuity of E_ϕ at the interface gives

$$
in\beta\frac{\kappa_0^2 - \kappa^2}{k_0 a^2 \kappa_0^2 \kappa^2} = \frac{B}{A}\frac{I_n(\kappa a)}{J_n(Ka)}\left[g_n(\kappa_0 a) + h_n(\kappa a)\right]
\tag{25.17}
$$

with the following defined functions

$$
\begin{aligned}
g_n(x) &= -\frac{K_n'(x)}{xK_n(x)} = \frac{K_{n-1}(x)}{xK_n(x)} + \frac{n}{x^2}, \\
h_n(x) &= \frac{I_n'(x)}{xI_n(x)} = \frac{I_{n-1}(x)}{xI_n(x)} - \frac{n}{x^2}.
\end{aligned}
\tag{25.18}
$$

The continuity of H_ϕ at the interface gives

$$
in\beta\frac{\kappa^2 - \kappa_0^2}{k_0 a^2 \kappa_0^2 \kappa^2} = \frac{A}{B}\frac{J_n(Ka)}{I_n(\kappa a)}\left[g_n(\kappa_0 a) - \varepsilon_z f_n(Ka)\right]
\tag{25.19}
$$

with

$$
f_n(x) = \frac{J_n'(x)}{xJ_n(x)} = \frac{J_{n-1}(x)}{xJ_n(x)} - \frac{n}{x^2}.
\tag{25.20}
$$

Thus we obtain the equation for all the modes

$$
\begin{aligned}
&\left[g_n(\kappa_0 a) - \varepsilon_z f_n(Ka)\right]\left[g_n(\kappa_0 a) + h_n(\kappa a)\right] \\
&= n^2\left[(\kappa_0 a)^{-2} - (\kappa a)^{-2}\right]\left[(\kappa_0 a)^{-2} - \varepsilon_t(\kappa a)^{-2}\right].
\end{aligned}
\tag{25.21}
$$

In the following, we discuss different modes in detail.

25.4.1 TE Modes

For the transverse electric (TE) modes, $E_z = 0$. The longitudinal magnetic field is given by Equation 25.13. One has

$$
\begin{aligned}
E_\phi &= i\frac{k_0}{\kappa^2}\partial_r H_z = i\frac{k_0}{\kappa}BI_1(\kappa r), \quad r < a, \\
&= i\frac{k_0}{\kappa_0^2}\partial_r H_z = -i\frac{k_0}{\kappa_0}DK_1(\kappa r), \quad r > a.
\end{aligned}
\tag{25.22}
$$

The continuity of E_ϕ at the interface requires that

$$
h_0(\kappa a) + g_0(\kappa_0 a) = 0.
\tag{25.23}
$$

For materials without loss, each term on the left side is positive, thus there is no solution. The waveguide does not support TE modes. This is exactly like that of a metallic wire, which does not support TE surface waves since current must flow along the waveguide. Only when $\varepsilon_t > 1$, the waveguide will support TE modes, like an ordinary dielectric fiber.

25.4.2 TM Modes

For TM modes, $H_z = 0$. The longitudinal electric field is given by Equation 25.12. One has

$$H_\phi = -i\varepsilon_z \frac{k_0}{K} A J_1(Kr), \quad r < a,$$

$$= i\frac{k_0}{\kappa_0} C K_1(\kappa_0 r), \quad r > a. \tag{25.24}$$

The continuity of H_ϕ leads to the equation

$$\varepsilon_z f_0(Ka) = g_0(\kappa_0 a). \tag{25.25}$$

The solutions to this equation give all the TM modes.

We first consider the solutions for fixed and real values of ε_z and ε_t. This is normally associated with a fixed k_0. It is convenient to consider a solution in the form of Ka or the reduced radius k_0a as a function of $\kappa_0 a$. The wave number along the waveguide can be obtained through $\beta = \left(k_0^2 + \kappa_0^2\right)^{1/2}$. Before we seek general solutions, it is better to consider the solutions in certain limits to reveal some important features of the TM modes on the anisotropic waveguide.

For the TM modes close to the light line, $\kappa_0 a \to 0$, one has

$$Ka \simeq x_{0,m} - \frac{\varepsilon_z}{x_{0,m}}(\kappa_0 a)^2 \left(\ln \frac{\kappa_0 a}{2} + \gamma\right).$$

Here, we have used $K_0(x) = -\ln(x/2) - \gamma$ for a small argument with γ the Euler constant. For complex ε_t with $\Re\varepsilon_t < 0$ and $\Im\varepsilon_t > 0$, the real and imaginary parts of β of the allowed modes will have the same signs. These modes are forward waves, similar to that of an ordinary optical fiber. We note that close to the light line,

the property of the TM modes of the anisotropic waveguide is similar to that of an isotropic fiber with $\varepsilon = 1 + \varepsilon_z(1 - \varepsilon_t^{-1})$.

In the limit of long wavelength or small waveguide radius, $k_0a \ll 1$, Equation 25.25 is reduced to

$$\varepsilon_z f_0\left(\frac{\kappa_0 a}{\eta}\right) = g_0(\kappa_0 a) \tag{25.26}$$

with $\eta = \sqrt{-\varepsilon_t/\varepsilon_z}$. This equation gives an infinite number of solutions $\kappa_0 a = \xi_{0,m}$, with $m = 1, 2, 3, \ldots$. This indicates that the anisotropic waveguide supports infinite number of propagating modes, no matter how thin the waveguide is. For $\kappa_0 a \to \infty$, since $g_0(\kappa_0 a) \simeq (\kappa_0 a)^{-1} \to 0$, one has $\xi_m \simeq \eta x_{1,m}$. Here $x_{n,m}$ is the mth zero of $J_n(x)$ away from the origin. For the mth TM band, one has $0 \le \kappa_0 a \le \xi_{0,m}$. The mth band starts with $k_0 a = x_{0,m}/\sqrt{\varepsilon_z - \varepsilon_z/\varepsilon_t}$ when $\kappa_0 a = 0$ and ends at $k_0 a = 0$ when $\kappa_0 a = \xi_{0,m}$. The modes with $\kappa_0 \gg k_0$ have $d(k_0 a)/d(\kappa_0 a) < 0$, and are backward waves. It will be obvious if we include a small imaginary part in ε_t with $\Im\varepsilon_t > 0$. The equation will give β with the real and imaginary parts having opposite signs. The energy flow is opposite to the phase velocity, which will be discussed later.

For arbitrary values of $\kappa_0 a$, the solution must be sought numerically. Since the right-hand side of Equation 25.25 is always positive, the solution requires that $J_1(Ka)$ and $J_0(Ka)$ have different signs. For the mth band, since $0 \le \kappa_0 a \le \xi_{0,m}$ with $\xi_{0,m}$ the solutions of Equation 25.26, one has $x_{0,m} \le Ka \le \xi_{0,m}/\eta < x_{1,m}$. For each $\kappa_0 a$ value, the Ka value can be searched within $[x_{0,m}, x_{1,m}]$ to satisfy Equation 25.25. Once the corresponding Ka is found, the reduced radius can be obtained as

$$k_0 a = \frac{\sqrt{(-\varepsilon_t/\varepsilon_z)(Ka)^2 - (\kappa_0 a)^2}}{\sqrt{1 - \varepsilon_t}}.$$

For the mth band, the corresponding longitudinal electric field E_z will have m nodes. The band structure and the phase index n_p of the TM modes on a nanowire waveguide with $\varepsilon_t = -3$ and $\varepsilon_z = 2$ is shown in Figure 25.3.

We next consider a waveguide of a fixed radius a with the following permittivity

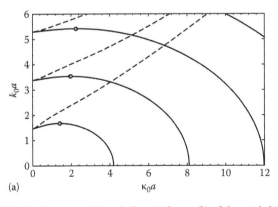

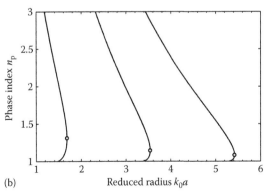

FIGURE 25.3 Band structure (a) and phase index n_p (b) of the guided TM modes in a nanowire waveguide of radius a with $\varepsilon_t = -3$ and $\varepsilon_z = 2$. Open circles denote the degenerate points of forward-wave and backward-wave modes. The dashed lines are for a dielectric waveguide with $\varepsilon = 1 + \varepsilon_z(1 - \varepsilon_t^{-1})$. (From Huang, Y.J. et al., *Phys. Rev. A*, 77, 063836, 2008. With permission.)

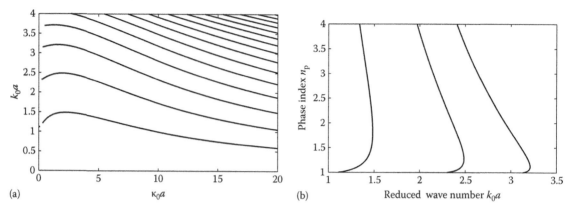

FIGURE 25.4 Band structure (a) and phase index n_p (b) of the guided TM modes on an anisotropic waveguide of radius a with ε_t and ε_z given by Equation 25.27 with $a = 10/k_p$ and $\varepsilon_a = 2.25$. (From Huang, Y.J. et al., *Phys. Rev. A*, 77, 063836, 2008. With permission.)

$$\varepsilon_t = \frac{1}{2}\left(1 + \varepsilon_a - \frac{k_p^2}{k_0^2}\right),$$

$$\varepsilon_z = 2\varepsilon_a \frac{k_0^2 - k_p^2}{k_0^2(1 + \varepsilon_a) - k_p^2}. \tag{25.27}$$

Here ε_a and k_p are positive constants. The realization of this property will be discussed later in Section 25.6. If $k_0 < k_p/(1 + \varepsilon_a)^{1/2}$, one has $\varepsilon_t < 0$ and $\varepsilon_z > 0$. The band structure of the TM modes on this waveguide is obtained by numeric means and plotted in Figure 25.4a with the corresponding phase index in Figure 25.4b. For this waveguide, there is no cutoff of $\kappa_0 a$ for each band. This is because that as $k_0 \to 0$, $\varepsilon_z \simeq 2\varepsilon_a$ and $\varepsilon_t \to -\infty$, thus $\eta = \sqrt{-\varepsilon_t/\varepsilon_z} \to \infty$. The cutoff $\xi_{0,m} \approx \eta x_{1,m} \to \infty$.

25.4.3 Hybrid Modes

The modes with both $E_z \neq 0$ and $H_z \neq 0$ are called hybrid modes. Their dispersions are contained in the solutions of Equation 25.21 with $n \neq 0$. We recast the equation in the following form:

$$\varepsilon_z f_n(Ka) = g_n(\xi) - \frac{n^2(\xi^{-2} - y^{-2})(\xi^{-2} - \varepsilon_t y^{-2})}{g_n(\xi) + h_n(y)}. \tag{25.28}$$

Here, we use the notation $\xi = \kappa_0 a$ and $y = \kappa a$. Note that $Ka = y/\eta$ with $\eta = \sqrt{-\varepsilon_t/\varepsilon_z}$.

At a fixed wavelength λ or wave number k_0, ε_t, ε_z are constant. Since $Ka = y/\eta$, if we use $\xi = \kappa_0 a$ as a free parameter, the eigen equation gives a single value of Ka or y for each ξ. Since

$$y = \sqrt{\xi^2 + (1 - \varepsilon_t)(k_0 a)^2},$$

so the eigen equation actually gives the reduced radius $k_0 a$ for each $\xi = \kappa_0 a$.

Close to the light line, $\xi \to 0$, the eigen equation can be simplified. Analytical solutions can be obtained [17]. These hybrid modes are all forward-wave modes.

In the limit of long wavelength or small waveguide radius, $k_0 a \to 0$, Equation 25.28 is reduced to

$$\varepsilon_z f_n\left(\frac{\xi}{\eta}\right) = g_n(\xi). \tag{25.29}$$

This will give a discrete set of solutions $\kappa_0 a = \xi_{n,m}$ for each n. The anisotropic waveguide supports infinite number of hybrid modes, no matter how thin the waveguide is.

For the allowed eigenmodes of the mth hybrid band, one has the range $0 \le \kappa_0 a < \xi_{n,m}$ with $\xi_{n,m}$ the solutions of Equation 25.29. The solutions near both ends of the above range can be obtained analytically as we have done. For arbitrary ξ within

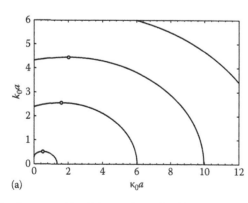
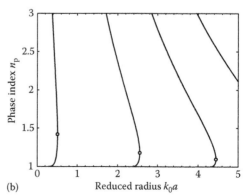

FIGURE 25.5 Band structure (a) and phase index n_p (b) of the guided hybrid modes with $n = 1$ on an anisotropic waveguide with $\varepsilon_t = -3$ and $\varepsilon_z = 2$. Open circles denote the degeneracy of forward-wave and backward-wave modes. (From Huang, Y.J. et al., *Phys. Rev. A*, 77, 063836, 2008. With permission.)

this range, the solution must be obtained numerically. However, only when $\varepsilon_t < -1$, one can have eigenmodes with $0 < Ka < x_{n,1}$. For the mth band, one has $x_{n,m} < Ka \leq \xi_{n,m}/\eta$. Otherwise, the solution for the first band requires $x_{n,1} < Ka \leq \xi_{n,1}/\eta$ with $\xi_{n,1} > \eta x_{n,1}$. The band structure and the phase index, n_p, of the hybrid modes on a nanowire waveguide with $\varepsilon_t = -3$ and $\varepsilon_z = 2$ are shown in Figure 25.5.

25.5 Energy Flow on Anisotropic Nanowires

For wave propagation on the cylindrical waveguide, the components of the Poynting vector are

$$S_z = \frac{1}{4\pi}(E_r H_\phi^* - E_\phi H_r^*),$$

$$S_r = -\frac{1}{4\pi}(E_z H_\phi^* - E_\phi H_z^*), \tag{25.30}$$

$$S_\phi = \frac{1}{4\pi}(E_z H_r^* - E_r H_z^*).$$

The physical Poynting vector is given by $\Re S$.

The total energy flow along the waveguide is the sum of energy flow inside and outside the waveguide:

$$P_z = P_z^{\text{in}} + P_z^{\text{out}} \tag{25.31}$$

with

$$P_z^{\text{in}} = 2\pi \int_0^a S_z r \, dr, \quad P_z^{\text{out}} = 2\pi \int_a^\infty S_z r \, dr. \tag{25.32}$$

Following Ref. [53], the total energy flow is normalized as

$$\langle P_z \rangle = \frac{P_z^{\text{in}} + P_z^{\text{out}}}{\left|P_z^{\text{in}}\right| + \left|P_z^{\text{out}}\right|}. \tag{25.33}$$

Thus one has $-1 < \langle P_z \rangle < 1$.

25.5.1 Energy Flow of TM Modes

Within the waveguide, one has $E_z = AJ_0(Kr)$, $E_r = -i(\varepsilon_z\beta/\varepsilon_t K) AJ_1(Kr)$, and $H_\phi = -i(\varepsilon_z k_0/K)AJ_1(Kr)$. So the Poynting vector component along the axis of the waveguide is

$$S_z = \frac{\beta}{\varepsilon_t k_0 a^2}\left|\frac{J_1(Kr)}{J_1(Ka)}\right|^2. \tag{25.34}$$

Here, we set the coefficient $A = K/[\varepsilon_z k_0 a J_1(Ka)]$. Since $\varepsilon_t < 0$, the energy flow inside the nanowire is always opposite to the phase velocity.

For the field in the air $r > a$, one has $E_z = CK_0(\kappa_0 r)$, $E_r = i(\beta/\kappa_0) CK_1(\kappa_0 r)$, and $H_\phi = i(k_0/\kappa_0)CK_1(\kappa_0 r)$, thus

$$S_z = \frac{\beta}{k_0 a^2}\left|\frac{K_1(\kappa_0 r)}{K_1(\kappa_0 a)}\right|^2. \tag{25.35}$$

Here the coefficient $C = -\kappa_0/[k_0 a K_1(\kappa_0 a)]$.

For the TM modes, one has

$$P_z^{\text{in}} = \frac{\beta}{2\varepsilon_t k_0 a^2}\int_0^a\left|\frac{J_1(Kr)}{J_1(Ka)}\right|^2 r \, dr,$$

$$\tag{25.36}$$

$$P_z^{\text{out}} = \frac{\beta}{2k_0 a^2}\int_a^\infty\left|\frac{K_1(\kappa_0 r)}{K_1(\kappa_0 a)}\right|^2 r \, dr.$$

The above integrals can be carried out and more compact expressions for the energy flow can be obtained as

$$P_z^{\text{in}} = \frac{\beta}{4\varepsilon_t k_0 (Ka)^2}\left[\frac{1}{f_0^2(Ka)} + \frac{2}{f_0(Ka)} + (Ka)^2\right]$$

$$= -\frac{\beta}{4k_0\varepsilon_t^2 f_0^2(Ka)}\frac{\varepsilon_z^2 f_0'(Ka)}{\varepsilon_t Ka},$$

$$\tag{25.37}$$

$$P_z^{\text{out}} = \frac{\beta}{4k_0(\kappa_0 a)^2}\left[\frac{1}{g_0^2(\kappa_0 a)} + \frac{2}{g_0(\kappa_0 a)} - (\kappa_0 a)^2\right]$$

$$= -\frac{\beta}{4k_0 g_0^2(\kappa_0 a)}\frac{g_0'(\kappa_0 a)}{\kappa_0 a}.$$

Here $g_0'(x)$ and $f_0'(x)$ are the derivatives of $g_0(x)$ and $f_0(x)$, respectively.

For convenience, we set $\beta > 0$ throughout the paper. Since $g_0'(x) < 0$ and $f_0'(x) < 0$, one has $P_z^{\text{in}} < 0$ and $P_z^{\text{out}} > 0$. In this convention, if $\langle P_z \rangle > 0$, this indicates that the energy flow and the phase propagation are in the same directions and the mode is a forward-wave mode. Otherwise, if $\langle P_z \rangle < 0$, the group velocity and the phase velocity are in the opposite direction and the mode is a backward-wave mode. The normalized energy flow for TM modes on a waveguide with $\varepsilon_t = -3$ and $\varepsilon_z = 2$ is shown in Figure 25.6. We note that for some portion of the bands the value of $\langle P_z \rangle$ is negative and thus these modes are backward waves.

25.5.2 Energy Flow of Hybrid Modes

The energy flow can also be evaluated for hybrid modes. The expression for S_z is much more complex than that of the TM

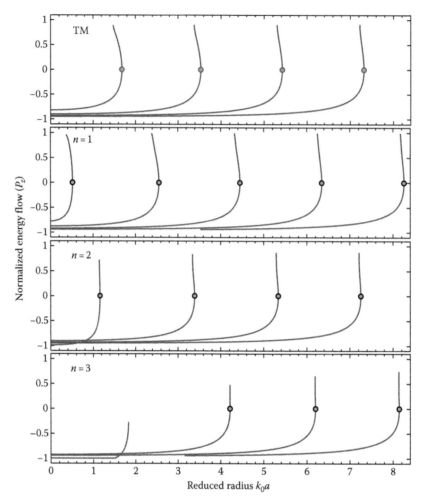

FIGURE 25.6 Normalized energy flow $\langle P_z \rangle$ for the first few bands of the TM modes and hybrid modes with $n = 1, 2, 3$ on an anisotropic waveguide with $\varepsilon_t = -3$ and $\varepsilon_z = -2$. Here we set $\beta > 0$. Open circles denote the degenerate points where $\langle P_z \rangle = 0$ and $v_g = 0$. (From Huang, Y.J. et al., *Phys. Rev. A*, 77, 063836, 2008. With permission.)

modes. However, the final expression for P_z is much simpler than expected, once the integrals are all carried out. One has

$$
P_z^{in} = \frac{\beta}{4k_0} \left\{ -\frac{\varepsilon_z^2}{\varepsilon_t Ka}(g_n + h_n)f_n' - \frac{1}{\kappa a}(g_n - \varepsilon_z f_n)h_n' \right.
$$

$$
\left. + \frac{2n^2}{(\kappa a)^4}\left[\frac{1+\varepsilon_t}{(\kappa_0 a)^2} - \frac{2\varepsilon_t}{(\kappa a)^2}\right] \right\},
$$

$$
(25.38)
$$

$$
P_z^{out} = \frac{\beta}{4k_0}\left\{ -\frac{1}{\kappa_0 a}(2g_n + h_n - \varepsilon_z f_n)g_n' \right.
$$

$$
\left. - \frac{2n^2}{(\kappa_0 a)^4}\left[\frac{2}{(\kappa_0 a)^2} - \frac{1+\varepsilon_t}{(\kappa a)^2}\right] \right\}.
$$

Here f_n', g_n', and h_n' are the derivatives of f_n, g_n, and h_n, respectively. In this derivation, we assume ε_t and ε_z are real, thus the arguments

of the Bessel functions are all real. We set $A = \sqrt{g_n + h_n}\,/(k_0 a J_n)$. Explicitly, one has

$$
f_n'(x) = -xf_n^2(x) - \frac{2}{x}f_n(x) + \frac{n^2}{x^3} - \frac{1}{x},
$$

$$
h_n'(x) = -xh_n^2(x) - \frac{2}{x}h_n(x) + \frac{n^2}{x^3} + \frac{1}{x},
\qquad (25.39)
$$

$$
g_n'(x) = xg_n^2(x) - \frac{2}{x}g_n(x) - \frac{n^2}{x^3} - \frac{1}{x}.
$$

We point out that the above expressions for P_z can be readily modified for dielectric or metallic cylindrical waveguide with the exchange of $f_n(ix) = -h_n(x)$ and $ixf_n'(ix) = -xh_n'(x)$.

The normalized energy flow on a waveguide with $\varepsilon_t = -3$ and $\varepsilon_z = 2$ for the hybrid modes with $n = 1, 2, 3$ is plotted in Figure 25.6.

25.5.3 Forward-Wave and Backward-Wave Modes

There are three ways to determine whether a guided mode is a forward-wave mode or backward-wave mode. One is through the sign of the derivative $d(k_0a)/d(\kappa_0a)$. From the band structure, we notice that for the modes near the light line $d(k_0a)/d(\kappa_0a) > 0$. These modes are forward waves. For large κ_0a or small k_0a, one has $d(k_0a)/d(\kappa_0a) < 0$, these modes are backward waves. From the band structure shown in Figures 25.3 and 25.5, the majority of the modes are backward waves.

The second way is through the sign of $\langle P_z \rangle$. For the TM modes when $\kappa_0a \to \infty$, one has $f_0(Ka) \to 0$ since $g_0(\kappa_0a) \to 0$. One thus has $Ka \to x_{0,m}$. This solution leads to the divergence of P_z^{in} which is negative, and the vanishing of P_z^{out} which is positive, subsequently $\langle P_z \rangle \to -1$, these modes are all backward waves. Correspondingly, one has $d(k_0a)/d(\kappa_0a) \to -\infty$ for $\kappa_0a \to \infty$. This is evident from the band structure shown in Figures 25.3 and 25.5.

One can prove that $d(k_0a)/d(\kappa_0a) \geq 0$ leads to $\langle P_z \rangle \geq 0$ and vice versa [17]. The degeneracy of forward- and backward-wave modes is located at $d(k_0a)/d(\kappa_0a) = 0$ or $\langle P_z \rangle = 0$.

The third way to determine whether a mode is a forward or backward wave is through the relative sign of the real and imaginary parts of β if dissipation is included. For example, we consider $\varepsilon_t = -3+0.05i$ and $\varepsilon_z = 2$. At $k_0a = 1.6$, the wave numbers of the first three eigenmodes are $\beta a = \pm(1.7112 + 0.0067i)$, $\pm(2.7250-0.0397i)$, $\pm(7.5756-0.0676i)$. Since the free space wave length is $\lambda = 3.927a$, this is a subwavelength waveguide. For the TM modes, except for the first mode, all the other modes are backward-wave modes since for those modes $\Re\beta$ and $\Im\beta$ have different signs. The normalized energy flow is $\langle P_z \rangle = 0.5151 -0.0020i$, $-0.4002 - 0.0058i, -0.8760 - 0.0078i$ for the above three modes, respectively. Here we set $\Re\beta > 0$. The field and Poynting vector profiles are plotted in Figure 25.7.

There is an interesting feature of the modes in the anisotropic waveguide. At a fixed k_0, for $a < a_m \equiv \sqrt{-\varepsilon_t/\varepsilon_z(1-\varepsilon_t)}x_{0,m}/k_0$, the mth band TM modes are backward waves. If the radius $a > a_m$, the waveguide supports two TM modes for the mth band, one forward and one backward. At $a = a_c$, these two modes become degenerate and the total energy flow is zero. This can be seen in Figures 25.3 and 25.6 where degenerate points are marked. Further increasing the radius, the waveguide will no longer support the mth band. The critical radius a_c is located such that $\langle P_z \rangle = 0$, $d(k_0a)/d(\kappa_0a) = 0$ or $dn_{eff}/da = \infty$. These degenerate modes can be used to slow down and even trap light. This will be discussed in Section 25.8.3.

25.6 Realization and Numerical Simulations

These extremely anisotropic media can be realized in a broad range of frequencies. For a multilayered structure of dielectric ε_a and metal ε_m, the effective permittivities can be obtained by using the effective medium theory [23,42],

$$\varepsilon_t = f\varepsilon_m + (1-f)\varepsilon_a,$$

$$\varepsilon_z = \frac{\varepsilon_a\varepsilon_m}{f\varepsilon_a + (1-f)\varepsilon_m}. \tag{25.40}$$

Here f is the filling ratio of the metal. For $f > f_{min} \equiv \varepsilon_a/(\varepsilon_a - \Re\varepsilon_m)$, one has $\Re\varepsilon_t < 0$. A realization of the anisotropic nanowire is shown in Figure 25.8.

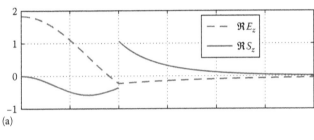
(a)

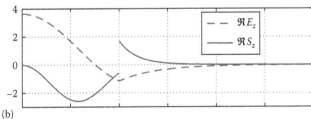
(b)

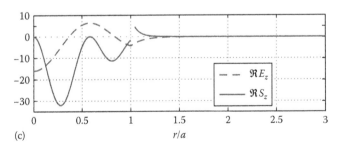
(c)

FIGURE 25.7 The longitudinal electric field and Poynting vector for the first three TM modes on an anisotropic waveguide of radius a with $\varepsilon_t = -3 + 0.05i$ and $\varepsilon_z = 2$ at $k_0a = 1.6$ with (a) $\beta a = 1.7112 + 0.0067i$, (b) $\beta a = 2.7250-0.0397i$, and (c) $\beta a = 7.5756-0.0676i$. The imaginary parts $\Im E_z$ and $\Im S_z$ are small and not plotted. (From Huang, Y.J. et al., *Phys. Rev. A*, 77, 063836, 2008. With permission.)

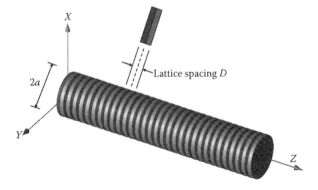

FIGURE 25.8 A sketch of the realization of a nanowire made of alternative disks of metal and dielectric. (From Huang, Y.J. et al., *Phys. Rev. A*, 77, 063836, 2008. With permission.)

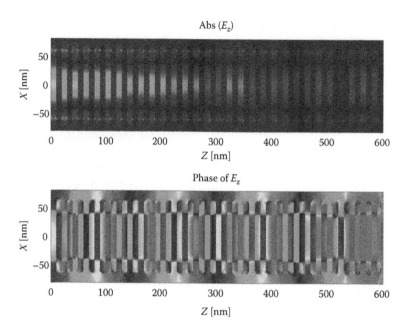

FIGURE 25.9 **(See color insert following page 20-16.)** FDTD simulation of the amplitude and phase propagation of the longitudinal electric field E_z along the nanowire with radius $a = 60\,\text{nm}$ at $\lambda = 488\,\text{nm}$. The metamaterial nanowire consists of alternative disks of silver and glass disks of thickness 10 nm. (From Huang, Y.J. et al., *Phys. Rev. A*, 77, 063836, 2008. With permission.)

We first consider a metamaterial waveguide at a fixed wavelength. For silver at $\lambda = 488\,\text{nm}$, one has $\varepsilon_m = -9.121 + 0.304i$ [28]. A nanowire made of alternative disks of silver and glass ($\varepsilon_a = 2.25$) of equal thickness will have $\varepsilon_t = -3.436 + 0.152i$ and $\varepsilon_z = 5.971 + 0.065i$ by using Equation 25.40. Here the disk thickness is 10 nm for both materials. For example, if one sets $a = 60\,\text{nm}$, one has $k_0a = 0.7725$. The first three TM modes will have $\beta a = 2.2525 - 0.0747i$, $4.9318-0.1419i$, $7.4031 - 0.2088i$. Thus one has $\lambda_\beta \sim 167\,\text{nm}$ and phase refractive index $n_{\text{eff}} = 2.92$ for the first TM mode. The decay length is 803, 423, and 287 nm, respectively. After traveling about 420 nm along the nanowire, only the first one will survive.

Finite-difference time-domain (FDTD) simulations [51] were performed to obtain the phase index n_p of modes on the metamaterial nanowire. The procedure is as follows. We illuminate the free-standing nanowire of finite length with a Gaussian beam, then get E_z after the termination of the simulation. The length of the waveguide is set to be larger than the decay length of the first TM mode. We get the phase from E_z, then determine β. Though the waveguide supports infinite number of modes including TM and hybrid modes, our method is legitimate due to the following two reasons: First, that the excitation of hybrid modes is small due to the profile of the incident Gaussian beam. So mainly the TM modes are excited. Second, that due to the dissipation in the metamaterial, after certain distance, only the first TM mode will survive. Thus the phase propagation is mainly due to the first TM mode. The amplitude and phase propagation of E_z along the above metamaterial nanowire is shown in Figure 25.9.

The relation between the phase index n_p and the nanowire radius a is shown in Figure 25.10. Very good agreement between FDTD simulations and analytical results has been obtained.

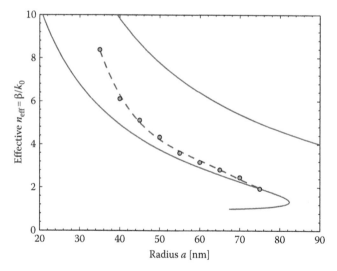

FIGURE 25.10 The phase index n_p of the first TM band on a nanowire with a different radius at $\lambda = 488\,\text{nm}$. The nanowire is made of alternative disks of silver and (see Figure 25.8). The disk thickness is 10 nm. Filled circle is obtained from FDTD simulations. The dashed line is the fitting of simulation data. The solid line is calculated from band equation with effective index $\varepsilon_t = -3.436 + 0.152i$ and $\varepsilon_z = 5.971 + 0.065i$. (From Huang, Y.J. et al., *Phys. Rev. A*, 77, 063836, 2008. With permission.)

However, for small radius, there is noticeable discrepancy. This is expected since when the radius is comparable with the lattice spacing of the multilayered metamaterial, the effective medium theory will fail. We have also performed FDTD simulations for the nanowire with smaller lattice spacing. Better agreement is indeed obtained.

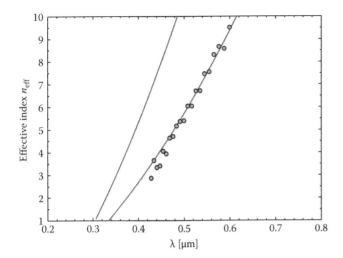

FIGURE 25.11 The phase index n_p for the TM modes on a nanowire with radius $a = 40\,$nm. The nanowire is a stack of equally thick alternative disks made of a Drude metal $\varepsilon_m = 1 - k_p^2 / k_0(k_0 + i\Gamma)$ and glass. Here $k_p a = 1.64$ and $\Gamma a = 0.0155$. The analytical curves (solid) are calculated by using real ε_m. (From Huang, Y.J. et al., *Phys. Rev. A*, 77, 063836, 2008. With permission.)

We also consider the band structure for different frequencies. The permittivity given by Equation 25.27 can be realized through the multilayered heterostructure with Drude metal with $\varepsilon_m = 1 - k_p^2 / k_0$ and dielectric ε_a. The nanowire is made of alternative disks of a Drude metal and a dielectric. The band structure and the phase index n_p of the TM modes are shown in Figure 25.4. One noticeable feature of these bands is the flatness of each band, which indicates small group velocity. We have performed the FDTD simulation for different frequencies for nanowire with a fixed radius. The results are shown in Figure 25.11. Again a good agreement between FDTD simulations and analytical results is achieved.

In the above example, we have shown only one realization of IM. Some other realizations may also be possible. For the metamaterials proposed and realized so far [40], most of them are anisotropic. Some of the metamaterials may have indefinite indices at some frequencies, such as the cloaking structures in Ref. [38].

A major barrier to the wide use of metamaterials is the absorptive loss. However, gain can be incorporated to reduce the loss. Active metamaterials have been pursued [3,6]. Since our focus is on the optical properties of nanowires, we will not discuss the details of active metamaterials here.

25.7 Light Coupling to Nanowire Waveguide

Due to its subwavelength size, coupling light into the nanowire waveguide is a technical challenge. Currently there are four ways to couple light to waveguides [36]: butt coupling, end-fire coupling, prism coupling, and grating coupling. The coupling efficiency depends on the coupling method and the optical properties of the nanowire waveguide.

The end-fire coupling is a butt coupling with a focal lens. For the butt coupling we have used in our FDTD simulations in the previous section, multiple modes will be excited. However, for nanowire waveguide made of realistic metamaterial, loss is unavoidable, thus only one or two modes will survive over certain distance and eventually only one mode will survive after a certain distance.

Though the nanowire waveguide supports infinite number of modes, selectively excitation of a single mode is possible. Thus, one can take the full advantage of the rich band structure provided by the nanowire waveguide. In order to excite a single mode, the prism coupling or grating coupling should be used. However, the phase-match condition must be satisfied for maximum energy transfer from the light source to the nanowire waveguide.

For a simple illustration, we use the prism coupling to excite the TM modes in a slab waveguide made of IM at $\lambda = 1.55/\mu$m. The metamaterial is formed by using alternative layers of silver and MgF_2. At this wavelength, one has $\varepsilon_m = -86.64 + 8.742i$ and $\varepsilon_a = 1.9$. Using Equation 25.40 with filling ratio $f = 5.6\%$, we have $\varepsilon_x = -3.0582 + 0.4896i$ and $\varepsilon_z = 2.0153 + 0.0003i$. For simplicity, we ignore the imaginary part of the permittivity. In the range $1.4 < n_p < 2.2$ and the slab thickness d between 225 and 240 nm, only the TM_0 modes will be excited as shown in Figure 25.12 (see Figure 25.1 for band structure of similar parameters and Figure 25.3 for that on a cylindrical nanowire). The critical thickness, such that the forward TM_0 and backward TM_0 will be merged into a single mode of zero group velocity, is $d_c = 236.9\,$nm with $n_p = 1.729$. At the thickness $d = 234\,$nm, two solutions are allowed, with $n_p = 1.553, 1.950$. The first one is a forward-wave mode while the second is a backward-wave mode.

We shine a Gaussian beam into a silicon prism of refractive index $n = 3.518$. At an incident angle $26.19°$ inside the prism, the forward-wave mode will be excited while at an incident angle $33.66°$, the backward-wave mode will be excited, which are shown in Figure 25.13.

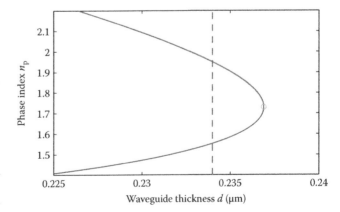

FIGURE 25.12 The phase index n_p of the guided TM_0 modes on a free-standing planar waveguide of thickness d with $\varepsilon_x = -3.0582$ and $\varepsilon_z = 2.0153$. The dashed line is for $d = 234\,$nm. The circle marks the location of zero group velocity at the critical thickness $d_c = 236.9\,$nm.

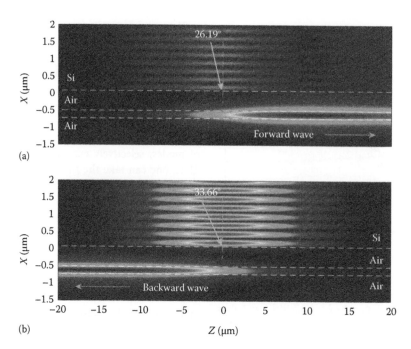

FIGURE 25.13 **(See color insert following page 20-16.)** Gaussian beam excitation through prism coupling of the forward-wave (a) and backward-wave mode (b) at incident angle 26.19° and 33.66°, respectively. The air gap between the prism and the waveguide ($d = 234$ nm) is 600 nm. Plotted is the absolute value of the magnetic field H_y.

Similarly, the prism coupling and grating coupling can be used to selectively excite the guided modes on the nanowire waveguide. We will not show examples here.

25.8 Applications of Nanowires Made of Indefinite Metamaterials

25.8.1 Phase Shifters

One salient feature of the modes on nanowire made of IMs is the large phase index. The high phase index is due to the hyperbolic dispersion in the metamaterial. The nanowire can be used for phase shifters with small footprint in optical integrated circuits.

The presence of loss in the metamaterials will restrict the use of long nanowire waveguides. However, for many applications other than the long-haul transportation, short waveguides have the advantage of small size and footprint. From our simulation, phase index $n_p \sim 8$ can be obtained. Thus to have a phase shift of π by a nanowire of radius 40 nm at $\lambda = 488$ nm, the length of the nanowire is about 41 nm. For excitation of such high phase index modes, grating coupling should be used.

25.8.2 Longitudinal Electric Field Enhancement

Besides the large phase index, modes with large longitudinal electric field can be excited on the nanowire waveguide. Recently there is a strong interest in large longitudinal electric fields [10]. Large longitudinal electric field can have a lot of applications, such as superfocusing to beat the diffraction limit [9,55] and trapping metallic nanoparticles in optical tweezer [62].

Due to the confinement in the transverse direction, there is a $\pi/2$ phase difference between E_z and E_r. Here, we consider the ratio

$$s = \frac{|E_z|_{max}}{|E_r|_{max}}. \tag{25.41}$$

As examples, we show the longitudinal and TE field of the first three TM mode on a nanowire of radius a with $\varepsilon_t = -3 + 0.05i$ and $\varepsilon_z = 2$ at $k_0 a = 1.6$ in Figure 25.14. One can see that $s \sim 200\%$, which is far more stronger than that on a silicon nanowire waveguide [10,50]. If one shrinks the radius, an even larger ratio would be expected. As we pointed out in Section 25.4.2, the band structure of the TM modes near the light line is similar to that of an ordinary optical fiber with a core index $\varepsilon = 1 + \varepsilon_z(1 - \varepsilon_t^{-1})$. In the above example, one has $\varepsilon = 1.9149$. For such a low contrast waveguide, very small longitudinal electric field will be obtained for the guided modes [10]. The fact that the nanowire waveguide made of IM will have large longitudinal electric field is due to the hyperbolic dispersion Equation 25.2, which allows much stronger confinement of light in the transverse direction, thus stronger longitudinal electric field.

25.8.3 Slow Light Waveguide

Recently Tsakmakidis et al. [54] proposed to trap light in a tapered waveguide with double-negative metamaterial as the core layer. The IM waveguide can also be used to slow down and trap light. These waveguides can thus be used as delay line in optical buffers [60]. The reason is that unlike the ordinary optical fiber, these waveguides support both forward and backward waves.

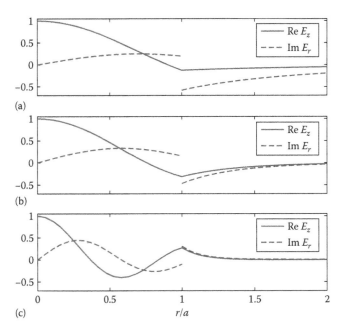

FIGURE 25.14 The longitudinal and transverse electric fields of the first three TM modes on a nanowire of radius a with $\varepsilon_t = -3 + 0.05i$ and $\varepsilon_z = 2$ at $k_0 a = 1.6$. One has $\beta a = 1.7112 + 0.0067i$, $2.7250 - 0.0397i$, and $7.5756 - 0.0676i$ for (a), (b), and (c), respectively.

For the anisotropic waveguide we have considered, $\varepsilon_t < 0$, one has $P_z^{in} < 0$ and $P_z^{out} > 0$ if one sets $\beta > 0$. If $P_z = P_z^{in} + P_z^{out} < 0$, the mode is a backward mode since the total energy flow is opposite to the phase velocity. Otherwise, the mode is a forward mode. At the critical radius a_c, the backward and forward modes become degenerate, and the energy flow inside the waveguide cancels out that in the air. One can prove that at the critical radius a_c where $P_z = 0$, the group velocity is indeed zero. One does not need to know the material dispersion to locate the zero group velocity point. This is due to the fact that for these waveguides, the dispersion due to geometric confinement dominates the material dispersion at and around the critical radius.

The unique properties of the modes on anisotropic waveguide can be used to slow down and even trap light. Even though the waveguide supports infinite number of both TM and hybrid modes at any fixed radius and frequency, with appropriate laser coupling, the excitation of the hybrid modes in the waveguide can be suppressed or even eliminated. Among the TM modes, the first TM mode will be more favorably excited. Furthermore, due to the material dissipation, the first TM mode will propagate the longest distance. The rest of the TM modes will all decay out at about half the decay length of the first TM mode. It is the first TM band that can be used for slow light application. Unlike the double negative waveguide [54], the anisotropic waveguide will slow down and trap light if one increases the radius to the critical radius. A sketch of a slow light waveguide is shown in Figure 25.15.

For the butt coupling we used in the FDTD simulations, multiple modes were excited. To excite a single mode, prism coupling or grating coupling can be used. By tapering the nanowire

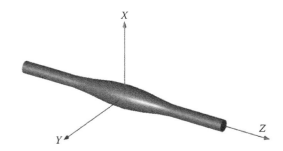

FIGURE 25.15 A sketch of the slow light waveguide made of an indefinite medium. (From Huang, Y.J. et al., *Phys. Rev. A*, 77, 063836, 2008. With permission.)

waveguide radius, any branches of the guided modes can be accessed in principle.

25.8.4 Light Wheel and Open Cavity Formation

Resonance is ubiquitous. They have many applications such as to store and confine energy in space, enhance the field concentration, and improve the defection accuracy. To have a resonance, a compact space such as cavities is required. Once they are made, cavities lack the translation symmetry in any direction. Examples are the quantum dots, microwave cavities, and photonic-crystal microcavities. Recently, negative-index metamaterials are also used to form open cavities [15,27,34,37], such as the checkerboard open resonators [35].

For optical integrated circuits, one of the most difficult tasks of nanofabrication is the alignment of different parts and devices, such as the alignment of active nanowires on photonic-crystal waveguide [30]. It would be more desirable to have an open cavity at any location.

Recently, a new concept, a light wheel, has been developed [52]. This is formed in a composite waveguide, which is made of an ordinary slab waveguide coupled with a properly designed slab waveguide made of DNM. If these two waveguides are separated infinitely away, the ordinary waveguide support a single forward-wave mode. The DNM waveguide supports a single backward-wave mode with the same phase index. Once these two waveguides are placed in the vicinity of each other, the composite waveguide no longer support propagating modes at the same wavelength. Instead it will support complex-conjugate decay modes. These two decay modes will form the so-called light wheel [52].

In order to have complex-conjugate decay modes, the waveguide should first be able to support degenerate propagating modes. In the example we considered in Section 25.7, the slab waveguide is made of a metamaterial with $\varepsilon_x = -3.0582$ and $\varepsilon_z = 2.0153$ at $\lambda = 1.55\,\mu m$. Below the critical thickness $d_c = 236.9\,nm$, the waveguide supports two modes of different phase indices (see Figure 25.12). One mode is a forward-wave TM_0 mode and the other is a backward-wave TM_0 mode. At the critical thickness (the effective thickness of the waveguide is zero due to the negative Goos–Hanchen lateral shift) a double light cone will be formed [54]. However, above the critical thickness, the waveguide supports no propagating modes. Instead, it supports complex-conjugate decay

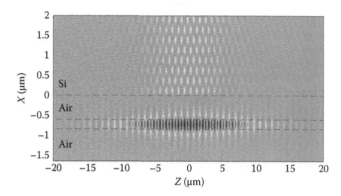

FIGURE 25.16 Light wheel formation through a prism coupling to a slab waveguide of thickness $d = 237$ nm with $\varepsilon_x = -3.0582$ and $\varepsilon_z = 2.0153$ at $\lambda = 1.55\,\mu$. The incident angle of the Gaussian beam in the silicon prism is 29.4°. The air gap between the prism and the waveguide is 600 nm. Plotted is the magnetic field H_y.

modes. For example at $d = 237$ nm, one has $n_p = 1.7286 \pm 0.0373i$. When an incident beam with $\beta = 1.7286k_0$ hits the waveguide, the two decay modes will be excited, one decays along and the other in the opposite direction of β, thus an open cavity will be formed. This cavity can be formed at any location along the waveguide, which is shown in Figure 25.16.

When dissipation is present, the decay modes no longer form a complex-conjugate pair. However, the imaginary part of their complex phase indices will still have opposite signs, thus light wheel formation is still allowed.

Light wheel and open cavity can also be formed on a nanowire waveguide made of IMs. If the radius of the nanowire is larger than the critical radius a_c, the nanowire will support complex-conjugate decay modes. Light confinement in all three dimensions can thus be realized on the nanowire at any location along the nanowire axis.

25.9 Conclusions

Indefinite metamaterials can be used to achieve negative refraction [16] and hyperlensing [20,46]. They can also be used as superlens [23]. In this chapter, we consider the wave propagation along a nanowire waveguide with an anisotropic optical constant. We have derived the eigenmodes equation and obtained the solutions for all the propagation modes. The field profiles and the energy flow on the waveguide are also analyzed. Closed-form expressions for the energy flow for all the modes are derived. For an extremely anisotropic cylinder, where the transverse component of the permittivity is negative and the longitudinal is positive ($\varepsilon_\perp < 0$, $\varepsilon_\parallel > 0$), the waveguide supports TM and hybrid modes but not the TE modes. Among the supported TM modes, at most only one mode can be forward wave. The rest of them are backward waves.

The possible realization of these extremely anisotropic nanowires is proposed by utilizing alternative layers of metal and dielectric. Extensive FDTD simulations have been performed and confirmed our analytical results.

Light couplings to the nanowire waveguide have also been discussed. To take full advantage of the rich band structure provided by the nanowire, prism coupling and grating coupling can be used to selectively excite the guided modes.

Four unique properties have been revealed for the modes on nanowire waveguides made of IMs. The first is that the backward-wave modes can have very large phase index. These nanowires can be used as phase shifters and filters in optics and telecommunications. The second is the large longitudinal electric field of the modes due to the hyperbolic dispersion of the metamaterial. The third is that the waveguide supports modes of zero group velocity. This is due to the fact that the waveguide can support both forward- and backward-wave modes at a fixed radius. If the nanowire waveguide is tapered, at certain critical radius, the two modes will be degenerate and carry zero net energy flow. At other radii, these waveguides support modes with small group velocity. These waveguides can thus be used as ultracompact delay lines in optical buffers [60]. The fourth is the formation of open cavity along the nanowire due to its support of complex-conjugate decay modes above the critical radius. The above features can lead to potential applications of these nanowires in optical integrated circuits.

Acknowledgments

This work was supported by the Air Force Research Laboratories, Hanscom through FA8718-06-C-0045 and the National Science Foundation through PHY-0457002.

References

1. Avouris, P. 2009. Carbon nanotube electronics and photonics. *Phys. Today* **64**(1): 34–40.

2. Berrier, A., M. Mulot, M. Swillo, M. Qiu, L. Thylén, A. Talneau, and S. Anand. 2004. Negative refraction at infrared wavelengths in a two-dimensional photonic crystal. *Phys. Rev. Lett.* **93**: 073902.

3. Boardman, A. D., Y. Rapoport, N. King, and V. N. Malnev. 2007. Creating stable gain in active metamaterials. *J. Opt. Soc. Am. B* **24**: A53–A61.

4. Born, M. and E. Wolf. 1999. *Principles of Optics: Electromagnetic Theory of Propagation, Interference and Diffraction of Light*, 7th ed., Cambridge University Press, Cambridge, U.K.

5. Caloz, C. and T. Itoh. 2006. *Electromagnetic Metamaterials: Transmission Line Theory and Microwave Applications*, John Wiley & Sons, Inc., Hoboken, NJ.

6. Chen, H.-T., W. J. Padilla1, J. M. O. Zide, A. C. Gossard, A. J. Taylor, and R. D. Averitt. 2006. Active terahertz metamaterial devices. *Nature* **444**: 597–600.

7. Cubukcu, E., K. Aydin, E. Ozbay, S. Foteinopoulou, and C. M. Soukoulis. 2003. Negative refraction by photonic crystals. *Nature* **423**: 604–605.

8. Dolling, G., C. Enrich, M. Wegener, C. M. Soukoulis, and S. Linden. 2006. Simultaneous negative phase and group velocity of light in a metamaterial. *Science* **312**: 892–894.

9. Dorn, R., S. Quabis, and G. Leuchs. 2003. Sharper focus for a radially polarized light beam. *Phys. Rev. Lett.* **91**: 233901.

10. Driscoll, J. B., X. Liu, S. Yasseri, I. Hsieh, J. I. Dadap, and R. M. Osgood. 2009. Large longitudinal electric fields (E_z) in silicon nanowire waveguides. *Opt. Express* **17**: 2797–2804.

11. Fang, N., H. Lee, C. Sun, and X. Zhang. 2005. Sub-diffraction-limited optical imaging with a silver superlens. *Science* **308**: 534–537.

12. Geim, A. K. and K. S. Novoselov. 2007. The rising of graphene. *Nat. Mater.* **6**: 183–191.

13. Gibbs, W. W. 2004. Computing at the speed of light. *Sci. Am.* **291**(11): 80–87.

14. Gralak, B., S. Enoch, and G. Tayeb. 2000. Anomalous refractive properties of photonic crystals. *J. Opt. Soc. Am. A* **17**: 1012–1020.

15. He, S., Y. Jin, Z. Ruan, and J. Huang. 2005. On subwavelength and open resonators involving meta-materials of negative refraction index. *New J. Phys.* **7**: 210.

16. Hoffman, A. J., L. Alekseyev, S. S. Howard, K. J. Franz, D. Wasserman, V. A. Podolskiy, E. E. Narimanov, D. L. Sivco, and C. Gmachl. 2007. Negative refraction in semiconductor metamaterials. *Nat. Mater.* **6**: 946–950.

17. Huang, Y. J., W. T. Lu, and S. Sridhar. 2008. Nanowire waveguide made from extremely anisotropic metamaterials. *Phys. Rev. A* **77**: 063836.

18. Jalali, B. 2007. Making silicon lase. *Sci. Am.* **296**(2): 58–65.

19. Joannopoulos, J. D., R. D. Meade, and J. N. Winn. 1995. *Photonic Crystals: Molding the Flow of Light*, Princeton University Press, Princeton, NJ.

20. Liu, Z., H. Lee, Y. Xiong, C. Sun, and X. Zhang. 2007. Far-field optical hyperlens magnifying sub-diffraction-limited objects. *Science* **315**: 1686.

21. Lu, W. T. and S. Sridhar. 2003. Near field imaging by negative permittivity media. *Microw. Opt. Technol. Lett.* **39**: 282–286.

22. Lu, W. T. and S. Sridhar. 2005. Flat lens without optical axis: Theory of imaging. *Opt. Express* **13**: 10673–10680.

23. Lu, W. T. and S. Sridhar. 2008. Superlens imaging theory for anisotropic nanostructured metamaterials with broadband all-angle negative refraction. *Phys. Rev. B* **77**: 233101.

24. Lu, Z., J. A. Murakowski, C. A. Schuetz, S. Shi, G. J. Schneider, and D. W. Prather. 2005. Three-dimensional subwavelength imaging by a photonic-crystal flat lens using negative refraction at microwave frequencies. *Phys. Rev. Lett.* **95**: 153901.

25. Luo, C., S. G. Johnson, J. D. Joannopoulos, and J. B. Pendry. 2002. All-angle negative refraction without negative effective index. *Phys. Rev. B* **65**: 201104.

26. Notomi, M. 2000. Theory of light propagation in strongly modulated photonic crystals: Refraction like behavior in the vicinity of the photonic band gap. *Phys. Rev. B* **62**: 10696.

27. Notomi, M. 2002. Negative refraction in photonic crystals. *Opt. Quantum Electron.* **34**: 133–143.

28. Palik, E. D. 1981. *Handbook of Optical Constants of Solids*, Academic Press, New York.

29. Parazzoli, C. G., R. B. Greegor, K. Li, B. E. C. Koltenbah, and M. Tanielian. 2003. Experimental verification and simulation of negative index of refraction using Snells law. *Phys. Rev. Lett.* **90**: 107401.

30. Park, H.-G., C. J. Barrelet, Y. Wu, B. Tian, F. Qian, and C. M. Lieber. 2008. A wavelength-selective photonic-crystal waveguide coupled to a nanowire light source. *Nat. Photonics* **2**: 622.

31. Parimi, P. V., W. T. Lu, P. Vodo, and S. Sridhar. 2003. Imaging by flat lens using negative refraction. *Nature* **426**: 404.

32. Parimi, P. V., W. T. Lu, P. Vodo, J. Sokoloff, J. S. Derov, and S. Sridhar. 2004. Negative refraction and left-handed electromagnetism in microwave photonic crystals. *Phys. Rev. Lett.* **92**: 127401.

33. Pendry, J. B. 2000. Negative refraction makes a perfect lens. *Phys. Rev. Lett.* **85**: 3966–3969.

34. Pendry, J. B. and S. A. Ramakrishna. 2003. Focusing light using negative refraction. *J. Phys.: Condens. Matter.* **15**: 6345–6364.

35. Ramakrishna, S. A., S. Guenneau, S. Enoch, G. Tayeb, and B. Gralak. 2007. Confining light with negative refraction in checkerboard metamaterials and photonic crystals. *Phys. Rev. A* **75**: 063830.

36. Reed, G. T. and A. P. Knights. 2004. *Silicon Photonics—An Introduction*, John Wiley & Sons Ltd., Chichester, U.K.

37. Ruan, Z. and S. He. 2005. Open cavity formed by a photonic crystal with negative effective index of refraction. *Opt. Lett.* **30**: 2308–2310.

38. Schurig, D., J. J. Mock, B. J. Justice, S. A. Cummer, J. B. Pendry, A. F. Starr, and D. R. Smith. 2006. Metamaterial electromagnetic cloak at microwave frequencies. *Science* **314**: 977–980.

39. Shalaev, V. M. 2007. Optical negative-index metamaterials. *Nat. Photonics* **1**: 41–48.

40. Shalaev, V. M. 2008. Transforming light. *Science* **322**: 384–386.

41. Shelby, R. A., D. R. Smith, and S. Schultz. 2001. Experimental verification of a negative index of refraction. *Science* **292**: 77–79.

42. Sihvola, A. 1999. *Electromagnetic Mixing Formulas and Applications*, The Institute of Electrical Engineers, London, U.K.

43. Smith, D. R. and N. Kroll. 2000. Negative refractive index in left-handed materials. *Phys. Rev. Lett.* **85**: 2933–2936.

44. Smith, D. R. and D. Schurig. 2003. Electromagnetic wave propagation in media with indefinite permittivity and permeability tensors. *Phys. Rev. Lett.* **90**: 077405.

45. Smith, D. R., J. B. Pendry, and M. C. Wiltshire. 2004. Metamaterials and negative refractive index. *Science* **305**: 788–792.

46. Smolyaninov, I. I., Y.-J. Hung, and C. C. Davis. 2007. Magnifying superlens in the visible frequency range. *Science* **315**: 1699–1701.

47. Soukoulis, C. M., S. Linden, and M. Wegener. 2007. Negative refractive index at optical wavelengths. *Science* **315**: 47–49.

48. Stockman, M. I. 2004. Nanofocusing of optical energy in tapered plasmonic waveguides. *Phys. Rev. Lett.* **93**: 137404.

49. Takahara, J., S. Yamagishi, H. Taki, A. Morimoto, and T. Kobayashi. 1997. Guiding of a one-dimensional optical beam with nanometer diameter. *Opt. Lett.* **22**: 475–477.

50. Tong, L., J. Lou, and E. Mazur. 2004. Single-mode guiding properties of subwavelength-diameter silica and silicon wire waveguides. *Opt. Express* **12**: 1025–1035.

51. Taflove, A. and S. C. Hagness. 2005. *Computational Electrodynamics: The Finite-Difference Time-Domain Method*, 3rd ed., Artech House Publishers, Norwood, MA.

52. Tichit, P. H., A. Moreau, and G. Granet. 2007. Localization of light in a lamellar structure with left-handed medium: The light wheel. *Opt. Express* **15**: 14961–14966.

53. Tsakmakidis, K. L., A. Klaedtke, D. A. Aryal, C. Jamois, and O. Hess. 2006. Single-mode operation in the slow-light regime using oscillatory waves in generalized left-handed heterostructures. *Appl. Phys. Lett.* **89**: 201103.

54. Tsakmakidis, K. L., A. D. Boardman, and O. Hess. 2007. 'Trapped rainbow' storage of light in meta-materials. *Nature* **450**: 397–401.

55. Urbach, H. P. and S. F. Pereira. 2008. Field in focus with a maximum longitudinal electric component. *Phys. Rev. Lett.* **100**: 123904.

56. Veselago, V. G. 1968. The electrodynamics of substances with simultaneously negative values of ε and μ. *Sov. Phys. USPEKHI* **10**: 509–514.

57. Veselago, V. G. and E. E. Narimanov. 2006. The left hand of brightness: past, present and future of negative index materials. *Nat. Mat.* **5**: 759–762.

58. Vodo, P., P. V. Parimi, W. T. Lu, and S. Sridhar. 2005. Focusing by plano-concave lens using negative refraction. *Appl. Phys. Lett.* **86**: 201108.

59. Vodo, P., W. T. Lu, Y. Huang, and S. Sridhar. 2006. Negative refraction and plano-concave lens focusing in one-dimensional photonic crystals. *Appl. Phys. Lett.* **89**: 084104.

60. Xia, F., L. Sekaric, and Y. Vlasov. 2007. Ultracompact optical buffers on a silicon chip. *Nat. Photonics* **1**: 65–71.

61. Yao, J., Z. Liu, Y. Liu, Y. Wang, C. Sun, G. Bartal, A. M. Stacy, and X. Zhang. 2008. Optical negative refraction in bulk metamaterials of nanowires. *Science* **321**: 930.

62. Zhan, Q. 2004. Trapping metallic Rayleigh particles with radial polarization. *Opt. Express* **12**: 3377–3382.

Thermal Transport in Semiconductor Nanowires

Padraig Murphy
California College of the Arts

and

University of California, Berkeley

Joel E. Moore
University of California, Berkeley

and

Lawrence Berkeley National Laboratory

26.1 Introduction and Motivation

Our motivation in this chapter is to understand the physics of thermal transport in semiconductor nanowires. These systems have the potential to revolutionize thermal management technology—a field which, if one looks over the entire course of human history, has been perhaps our most important manipulation of nature.

As is so often the case, it is materials with extreme properties that are the most sought after. An extremely high thermal conductivity material could be used for cooling in microelectronic systems [26,28]. Materials with a low thermal conductivity could of course be used for thermal insulation [21], but also in the resurgent field of thermoelectrics [8,16]. A "thermoelectric" is a short-hand term for a material that can be used to make a solid-state energy conversion device. Consider the case of a heat engine operating between hot and cold heat reservoirs. Instead of the usual gaseous working substance, we imagine placing a piece of solid between the two reservoirs. The temperature difference can then be used to drive the electrons in the solid from the hot to the cold reservoir, and hence around a circuit. The current in the circuit can then be used to do work.

Unfortunately, such a device is necessarily less efficient than a Carnot engine. The presence of the solid between the two reservoirs means that heat can leak from one reservoir to the other. Since this heat does no work, it reduces the overall efficiency of the engine. In the literature, the efficiency is parametrized by what is called the "figure of merit," ZT, given by

$$ZT = \frac{T\sigma S^2}{\kappa}. \tag{26.1}$$

Here

T is the mean temperature of the engine
σ is the electrical conductivity
S is the thermopower
κ is the thermal conductivity

The thermopower is a measure of the electric field generated by a temperature gradient in the material. If $\vec{E}$ is the electric field and ∇T the temperature gradient, then $\vec{E} = S\nabla T$. The efficiency, η, is monotonic in ZT, for $ZT = 0$; the efficiency $\eta = 0$, as $ZT \to \infty$; the efficiency approaches the Carnot efficiency, $\eta \to \eta_c$. (The precise form of η as a function of ZT can be found in [25].) Creating materials with low thermal conductivity (while maintaining a high electrical conductivity and thermopower) is thus a crucial part of the thermoelectrics research program.

Semiconductor nanowires are considered among the most promising materials currently under study [10]. For most semiconductors used in thermoelectric applications, the thermal conductivity is dominated by phonon transport, while the thermopower and electrical conductivity are dominated by electron transport. Hence, if the phonon mean free path (the distance between scattering events) can be reduced while electronic properties are unaffected, the figure of merit, ZT, will be increased. The versatility of semiconductors is an important part of their potential: doping can be used to manipulate thermopower and electrical conductivity, and various growth or nanostructuring processes can be used to modify electron and thermal transport via confinement and inhomogeneity effects. Nanowires can be grown of several semiconductor materials using a variety of techniques, and silicon in particular has been extensively explored, although InSb and other compound semiconductors might have better thermoelectric properties and can also be grown as nanowires. Beyond conventional semiconductors, some correlated oxide materials show useful thermoelectric properties in bulk [30], but relatively few oxides have successfully been synthesized as nanowires. Other one-dimensional systems, such as carbon nanotubes, may be of interest for their high intrinsic thermal conductance [3,11,19].

As well as being of technological interest, the fundamental physics of nanowires has also attracted much attention. One of the great advances of nanoscience in recent times was the observation of the quantum of thermal conductance, g_0 [23]. This is a thermal conductance constructed exclusively from the fundamental constants of nature:

$$g_0 = \frac{\pi^2 k_B^2 T}{3h}. \tag{26.2}$$

Here

    k_B is Boltzmann's constant
    h is Planck's constant
    T is the mean temperature of the system

The thermal conductance of a system, G, can be related to the thermal conductivity, κ, through $G = \kappa A/L$, where A is the cross-sectional area, and L the length of the system. For most systems, κ is independent of A and L, and so G has the dependence shown, i.e., $G \propto A/L$. However, since g_0 is constructed exclusively from fundamental constants, it gives a thermal conductance independent of both the cross-sectional area and the length of the system. In addition, note that g_0 is independent of the velocity of sound of the material, of the impurity density, etc.

The thermal conductance quantum has been observed at very low temperatures $T \lesssim 1\,\mathrm{K}$ for a phonon waveguide system [27]. The four suspended phonon waveguides are made from silicon nitride, and each has a width of 200 nm at the narrowest point. As in the electronic case, where the first measurement of the one-dimensional electrical conductance quantum e^2/h was made in quantum point contacts [29] rather than extended one-dimensional wires, observation of the thermal conductance quantum depends on quantum confinement in the transverse directions rather than in having an extended wire. The device is shown in Figure 26.1. The catenoidal shape is chosen to maximize coupling to the reservoirs. Figure 26.2 shows the thermal conductance divided by $16g_0$ as a function of temperature. Thus, per waveguide, the thermal conductance is $4g_0$ for $T \lesssim 1\,\mathrm{K}$. We will derive this result in detail in a later section.

At higher temperatures, the thermal conductivity of certain nanowires has an unexpected dependence on the temperature, T. One expects, for temperatures less than the Debye temperature (but not so small that one is in the quantum of thermal conductance limit), that the thermal conductivity, κ, should scale like $\kappa \sim T^3$. This can be understood by recalling the result for the thermal conductivity of particles: $\kappa = 1/3 c_v v l$, where c_v is the heat capacity, v is the velocity, and l is the mean free path.

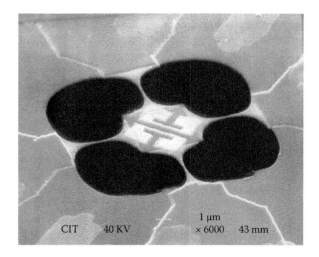

FIGURE 26.1 The experimental set up for Schwab et al. [27]. During the experiment, the phonon cavity (seen here at the center of the figure) is heated. This heat is then carried off by a heat current that flows through the four catenoidal phonon waveguides, that can be seen in the figure as thin slivers between the dark regions. The temperature of the phonon cavity can be measured, and so the thermal conductance of the phonon waveguides can be determined from the ratio of the heat current to the temperature increase.

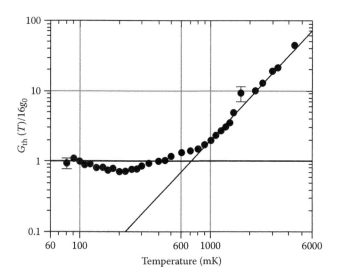

FIGURE 26.2 The thermal conductance as measured by Schwab et al. [27]. On the y-axis, the thermal conductance G divided by $16g_0$ is plotted. Clearly for $T \lesssim 1\,\mathrm{K}$, $G \simeq 16g_0$. Recall that the heat flux was through four separate phonon waveguides. The data thus determines that, at sufficiently low temperatures, the thermal conductance of a single phonon waveguide is approximately $4g_0$. (From Schwab, K. et al., *Nature*, 404, 974, 2000. With permission.)

The heat capacity scales like T^3 for a semiconductor or insulator, and so one expects $\kappa \sim T^3$. However, this has proven to not always be the case. For the thinnest silicon nanowires, with diameters of the order of 20 nm, the thermal conductivity has been shown experimentally to scale like $\kappa \sim T$ [6,14]; for wider wires, the T^3 dependence is obtained. In Figure 26.3, the points

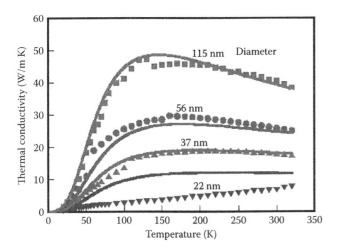

FIGURE 26.3 Here the thermal conductance is plotted as a function of temperature for wire diameters of 22, 37, 56, and 115 nm. The data is taken from Li et al. [14]. The solid lines are a fit to the data due to Mingo [18]. Note that the data and fit for the 22 nm wire do not agree. (From Murphy, P.G. and Moore, J.E., *Phys. Rev. B*, 76(15), 155313, 2007.)

are data, and the solid lines are from a model for the thermal conductivity that is adapted from the case of bulk silicon. (We will consider the model in more detail in a later section.) As one can see, the model gives a good fit to the data for larger diameters, but fails for the 22 nm diameter wire.

Figure 26.3 gives much insight into the source of thermal resistance in nanowires. While the other scattering mechanisms are independent of the diameter of the nanowire, surface scattering gives a mean free path of order its width, at least in the simplest models. Thus the reduction in thermal conductivity, and so mean free path, with the decreasing wire diameter implies that surface scattering is dominant. This is confirmed by Figure 26.4, which shows the scattering rates for different processes in silicon nanowires.

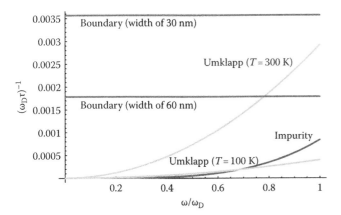

FIGURE 26.4 The scattering rate divided by the Debye frequency for various processes in nanowires. The boundary scattering rates (for two cases, diameters 30 and 60 nm) are given by the labeled horizontal lines. The umklapp rates are shown for two temperatures, $T = 300$ K and $T = 100$ K, as light lines. The rate for scattering from isotopic impurities is shown (dark line).

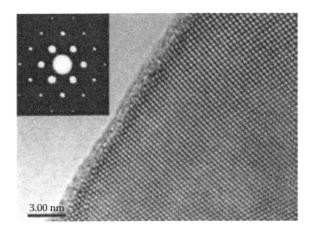

FIGURE 26.5 A TEM image of a 22 nm silicon nanowire. The inset is a selected area electron diffraction pattern of the nanowire. (From Li, D.Y. et al., *Appl. Phys. Lett.*, 83, 2934, 2003. With permission.)

It is also confirmed by transmission electron microscope (TEM) images of silicon nanowires, such as Figure 26.5 taken from Ref. [14]. These nanowires were grown using the vapor–liquid–solid (VLS) method, in which silicon is dissolved in a nanometer-scale gold cluster [15,31]. As the gold becomes saturated with silicon, the silicon is precipitated out. As more silicon is added, and so more is precipitated out, the nanowire grows. The length of the nanowires is typically of order a micron. The nanowires are clearly single crystals in the core, but with a rough boundary.

26.2 Phonons and Nanowires

In this chapter we will consider only semiconducting nanowires. In these systems the transport of heat is primarily through vibrations of the crystalline lattice, rather than through the motion of electrons, as would be the case in a metal. To fully understand these systems, it is important to treat the lattice vibrations quantum mechanically [12].

The first step in this procedure is to find the normal modes of the nanowire. These fall into four classes: longitudinal modes, torsional modes, and two kinds of flexural modes. These are illustrated in Figure 26.6. The longitudinal modes consist of successive expansions and contractions of the nanowire; in the limit of infinite wavelength, the longitudinal mode becomes a translation of the entire nanowire along its "long" direction. A torsional mode consists of successive rotations of a cross section of the nanowire; in the infinite wavelength limit, the torsional mode becomes a rotation of the entire nanowire around its long axis. The flexural modes arise as motion of a cross section of the nanowire in one of the directions perpendicular to the long direction, rather like the motion of a snake. There are two kinds of flexural modes, just as there are two choices for the direction perpendicular to the long direction of the nanowire.

Each mode has a particular wave vector, $\vec{k}$, associated with it; the equation of motion for the atoms in the crystal then gives an angular frequency for each mode, $\omega(\vec{k})$. In a bulk crystal,

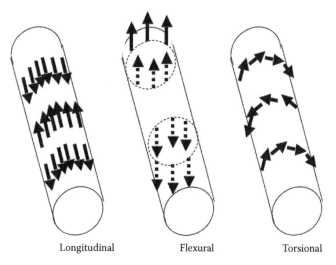

Longitudinal   Flexural   Torsional

FIGURE 26.6 The normal modes of a nanowire fall into four classes: longitudinal, which are analogous to the same modes found in bulk systems; flexural, which have a "snaking" motion and are analogous to transverse modes in a bulk solid; and torsional modes, which have no bulk analogue. There are two independent kinds of flexural modes, for the two directions perpendicular to the long axis of the rod into which the rod can move.

in the limit of small $|\vec{k}|$, one typically finds relations of the form $\omega(\vec{k}) = v|\vec{k}| = v\sqrt{k_x^2 + k_y^2 + k_z^2}$, where v is the velocity. In a nanowire, it is often convenient to isolate the long direction of the wire, which we will call k_z. For different choices of fixed k_x and k_y, we get different plots of $\omega(k_z)$. This is shown in Figure 26.7.

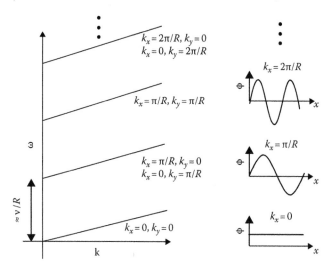

FIGURE 26.7 On the left is shown a schematic representation of the dispersion, i.e., plot of ω versus k, for say the longitudinal modes of the nanowire. The variable k represents the wave vector in the long direction of the nanowire. The graphs on the right represent what the normal modes look like in a cross section: the ungapped mode ($\omega = 0$ for $k = 0$) has zero transverse wave vector, whereas for the gapped modes k_x and/or k_y will be nonzero. Of course, if the shape of the cross section is not rectangular, the normal modes will not be labeled by k_x and k_y, but an analogous result will hold.

Note that for those modes with some transverse displacement, i.e., a nonzero k_x and k_y, then as $k_z \to 0$ one finds that ω tends to a constant greater than zero. We refer to those modes with a fixed k_x and k_y as a *subband*. Each k_x and k_y is of order n/R, where n is some integer, and R is the length scale of the width of the nanowire. Thus the angular frequency gap between successive subbands will be of order v/R, as shown in Figure 26.7.

Note also that due to the discrete nature of the lattice, there is a minimum allowed wavelength, and so a maximum allowed frequency ω_D, known as the Debye frequency. Thus, each subband has a minimum and maximum frequency.

For our purposes, the quantum mechanical nature of the system enters primarily through the Bose–Einstein formula for the number of phonons in a mode [12]:

$$n(\omega, T) = \frac{1}{e^{\hbar\omega/(k_B T)} - 1}. \qquad (26.3)$$

Here

ω is the frequency of the mode
$\hbar$ is defined as $\hbar = h/(2\pi)$, where h is Planck's constant
T is the temperature of the system

In addition, we note that the energy of a single phonon is given by $\hbar\omega$.

In practice, one can only find the normal modes of a "clean," harmonic system, i.e., one without impurities or interactions between the phonons. In the presence of such perturbations, a normal mode of the clean system evolves over time into other modes. This can be thought of as a phonon being transferred from one mode to another. In the case of transport, the most important consequence of this is that a phonon may be backscattered: a phonon incident on the nanowire from the left may be turned around by impurities or anharmonicity. Thus any phonon incident on the nanowire has a probability, not necessarily equal to 1, of being transmitted across it. We will denote the transmission probability of a phonon in the ith subband with frequency ω as $\mathcal{T}_i(\omega)$. If this probability of transmission is of order 1, and so independent of the length of the system, we refer to the phonon as *ballistic*; if the transmission probability is proportional to $1/L$, where L is the length of the system, we say the phonon is *ohmic* or *diffusive*; and finally if the transmission probability is proportional to $e^{-L/\xi}$, we say the phonon is *localized* with localization length ξ.

26.3 Landauer Formula for the Thermal Conductance

Let us consider a nanowire that is contacted at the left with a heat reservoir at a temperature T_h, and on the right with a heat reservoir at a temperature T_c, with $T_h > T_c$. The mean temperature, T, is defined as $T = (T_h + T_c)/2$, and the temperature difference, ΔT, is given by $\Delta T = T_h - T_c$. To find the total heat current, one takes the heat current that the left-hand side reservoir sends down the nanowire and subtracts from it the current being sent by the right-hand side reservoir. The heat current due to a single mode in subband i with frequency ω and velocity v is given by the energy transmitted per second by that subband, and so by $\hbar\omega n(\omega,T)\,(v/L)\,\mathcal{T}_i(\omega)$, where $\mathcal{T}_i(\omega)$ is the transmission probability. To find the total current one then sums over modes; this sum over modes can be expressed as [12]

$$\sum_n \approx \int dn = \int \frac{L}{2\pi}\,dk.$$

We can thus write the heat current from the left reservoir due to subband i as

$$\int \frac{L}{2\pi}\,dk\,\hbar\omega_i(k)\,n(\omega_i(k),T)\frac{v_i(k)}{L}\,\mathcal{T}_i(\omega_i(k))$$

$$= \int \frac{dk}{2\pi}\,v_i(k)\frac{\hbar\omega_i(k)}{e^{\hbar\omega_i(k)/(k_B T_L)}-1}\,\mathcal{T}_i(\omega_i(k)).$$

Then the total heat current from the left to the right is given by

$$I_Q = \sum_i \int \frac{dk}{2\pi}\,\hbar\omega_i(k)\,v_i(k)\left[\frac{1}{e^{\hbar\omega_i(k)/(k_B T_h)}-1}-\frac{1}{e^{\hbar\omega_i(k)/(k_B T_c)}-1}\right]$$

$$\times\,\mathcal{T}_i(\omega_i(k))$$

$$= \sum_i \frac{\hbar}{2\pi}\int_{\omega_i^{min}}^{\omega_i^{max}} d\omega\,\omega\,\mathcal{T}_i(\omega)\left[\frac{1}{e^{\hbar\omega/(k_B T_h)}-1}-\frac{1}{e^{\hbar\omega/(k_B T_c)}s-1}\right],$$

(26.4)

where we have used that $d\omega/dk = v$; as before, i labels the subband of the system, and $\mathcal{T}_i(\omega)$ is the transmission probability for a phonon of frequency ω in the ith subband. The index i runs over all possible subbands, which are labeled by the allowed wave vectors in the two directions transverse to the long direction of the wire and by the nature of the branch (i.e., longitudinal, etc.). We now Taylor expand the heat current to first order in ΔT; the thermal conductance is given by $G_{th} = I_Q/\Delta T$, and so can be expressed as

$$G_{th}(T) = \sum_i \int_{\omega_i^{min}}^{\omega_i^{max}} d\omega\,\frac{\hbar\omega}{2\pi}\,\frac{\hbar\omega}{k_B T^2}\,\frac{e^{\hbar\omega/(k_B T)}}{(e^{\hbar\omega/(k_B T)}-1)^2}\,\mathcal{T}_i(\omega).$$

(26.5)

If we define each of the functions $\mathcal{T}_i(\omega)$ to be zero for $\omega < \omega_i^{mir}$ and $\omega > \omega_i^{max}$, and we also define $\mathcal{T} = \sum_i \mathcal{T}_i$, which we will refer to as the transmission function, then

$$G_{th}(T) = \frac{k_B}{2\pi}\int d\omega\left(\frac{\hbar\omega}{k_B T}\right)^2\frac{e^{\hbar\omega/(k_B T)}}{(e^{\hbar\omega/(k_B T)}-1)^2}\,\mathcal{T}(\omega).$$

(26.6)

As in the electronic case [7], all of the details specific to a particular system relevant to transport (scattering rates, density of states, etc.) are now contained in the transmission function, $\mathcal{T}$. Once this function is specified, the thermal conductance can be found by numerically integrating $\mathcal{T}$ against the function shown in Equation 26.6.

In the ballistic regime, every phonon is transmitted across the nanowire with probability $\mathcal{T}_i = 1$. Let us consider a one-dimensional chain with longitudinal phonons only, in which case $\mathcal{T}(\omega) = 1$ for $0 \le \omega \le \omega_D$. It is convenient to change the integration variable to $x = \hbar\omega/(k_B T)$; then the heat current is

$$I_Q = \frac{k_B^2 T^2}{2\pi\hbar}\int_0^{\hbar\omega_{max}/(k_B T)} dx\,x\left[\frac{1}{e^{x\frac{1}{1+\Delta T/(2T)}}-1}-\frac{1}{e^{x\frac{1}{1-\Delta T/(2T)}}-1}\right].$$

(26.7)

At low temperatures (i.e., temperatures well below the Debye temperature for the chain), the upper limit of integration can be taken as infinity. It is straightforward to show that

$$\int_0^{\infty} dx\,\frac{x}{e^{x\frac{1}{1+\Delta T/(2T)}}-1} = \frac{\pi^2}{6}\left(1+\frac{\Delta T}{2T}\right)^2.$$

Therefore

$$I_Q = \frac{k_B^2 T^2}{h}\frac{\pi^2}{6}2\frac{\Delta T}{T} = \frac{\pi^2 k_B^2 T}{3h}\Delta T.$$

(26.8)

We have shown that the thermal conductance of the exactly one-dimensional chain—for temperatures well below the Debye temperature, and for ballistic transport—is given by

$$G_{th} = \frac{\pi^2 k_B^2 T}{3h} = g_0.$$

(26.9)

As mentioned before, it is independent of all material properties of the system. The Debye frequency was eliminated by choosing the low temperature limit, and the sound velocity vanished due to a cancellation between a velocity term in the current and an inverse velocity term in the density of modes. Of course, because of the assumption of ballistic transport, there was no dimensionful quantity associated with the scattering of the phonons.

Note also that we have a finite thermal resistance even though the phonons are not scattered inside the chain. This is the analogue of the "contact resistance" familiar from one-dimensional electronic systems.

It is interesting that the identical result for the thermal conductance can be derived for fermions, and for particles that satisfy fractional exclusion statistics [24]. This universality appears to be related to theorems limiting the information flow in a channel [4,22].

We are now in a position to understand the result seen by Schwab et al. [27]. For sufficiently low temperatures, $k_B T \ll \hbar v/R$, the gapped modes are unoccupied since

$$(e^{(\hbar v/R)/(k_B T)} - 1)^{-1} \simeq e^{-(\hbar v/R)/(k_B T)} \ll 1.$$

Only the four gapless subbands are then occupied, and the system is effectively one-dimensional: as the temperature changes within the prescribed range the number of occupied subbands is a constant. At low frequencies, phonons are weakly scattered by disorder, and since the temperature is low there is little scattering due to umklapp processes (see Section 26.4). Therefore, the mean free path exceeds the total length, making the system effectively ballistic. We can then apply the calculation of the thermal conductance for a one-dimensional ballistic system, with an additional factor of 4 for the four gapless subbands (one longitudinal, one torsional, and two flexural) of the rod. The thermal conductance of the system is then four thermal conductance quanta:

$$G_{th} = 4\frac{\pi^2 k_B^2 T}{3h} = 4g_0, \tag{26.10}$$

as was observed in the experiment.

26.4 Thermal Conductivity at Higher Temperatures

In the ballistic regime, the thermal conductance is independent of the length of the system, L. In the more familiar "diffusive" regime, one expects the thermal conductance to decrease with the length of the system—the "resistors in series" rule $(R_{1+2} = R_1 + R_2)$ means that the total resistance scales like L, and so the conductance like $1/L$. Thus the transmission function, $\mathcal{T}$, should also scale like $1/L$ in the diffusive regime. This can be derived formally using a transfer matrix method [2]. It can be derived informally using relatively simple arguments, familiar from the theory of random walks.

Consider a random walker who wishes to make his way home from the bar. The bar is located at $x = 0$, and his home is at the other end of the street, at $x = L$. He walks directly along each block, of length l, until he comes to the corner. There he becomes confused, and either continues in the same direction, with probability 1/2, or turns around and walks in the opposite direction, also with probability 1/2. We define $p(x)$ as the probability that starting at the point x, the man will reach home before he reaches the bar. Let us suppose that the man starts at the point x, in which case his probability of reaching home is $p(x)$. We allow him to walk one block, in which case his probability of getting home is $(p(x + l) + p(x - l))/2$. Since his probability of reaching home, having started at x, has not changed, we have

$$p(x) = \frac{p(x+l) + p(x-l)}{2}. \tag{26.11}$$

The boundary conditions are $p(0) = 0$, and $p(L) = 1$; the solution is obviously

$$p(x) = \frac{x/l}{L/l}. \tag{26.12}$$

We now think of each phonon as leaving the reservoir for the nanowire, and reaching the first scattering point at $x = l$. From the left edge, its probability of making it to the right edge of the wire is $(l/l)/(L/l) = l/L$. Thus if l is the mean free path of the phonon in a wire of length L, the transmission probability is l/L. Assuming that each phonon of frequency ω has the same mean free path, the total transmission function is

$$\mathcal{T}(\omega) = \frac{N(\omega) l(\omega)}{L}, \tag{26.13}$$

where $N(\omega)$ is the number of modes with frequency ω.

We must now consider the various processes that backscatter phonons: scattering from disorder; 3-phonon interaction processes, referred to as "umklapp"; and scattering from the rough surface of the nanowire. Given that the size of the disorder impurities is much smaller than the wavelength of the phonons, this scattering can be treated within the usual wave theory. The result for the mean free time is referred to as the Rayleigh–Klemens formula [13,17]:

$$\tau(\omega)^{-1} = \sigma_m^2 \omega^2 D(\omega), \tag{26.14}$$

where

σ_m^2 is the local variance of the mass divided by the mean mass squared

$D(\omega)$ is the density of modes

In d dimensions, and at low frequencies, the density of modes per atom is given by

$$D(\omega) \propto \frac{\omega^{d-1}}{\omega_D^d}. $$

Here ω_D is the Debye frequency of the material. The result is equivalent to a well-known result from electromagnetism: for light propagating in three spatial dimensions, and scattering elastically from impurities that are much smaller than the wavelength of the light, the mean free path scales like $l(\omega) \sim \omega^{-4}$.

As an example, we take the scattering of phonons due to the variation in the atomic mass of silicon (silicon has three stable isotopes). The result is [18]

$$(\omega_D \tau_i)^{-1} = (0.17\omega/\omega_D)^4. \tag{26.15}$$

We should point out that this equation for the scattering rate is not to be trusted too far for higher frequencies. There the density of modes deviates significantly from the simple ω^2 dependence that holds at low frequencies.

Another scattering process results from the nonlinearity of the lattice. The potential energy for an atom in the crystal is not exactly the simple harmonic oscillator potential proportional to the square of the displacement, u^2, but rather has higher-order terms proportional to u^3 and higher powers. The phonon modes considered earlier were eigenmodes of the harmonic system, and so the higher-order terms introduce scattering between the modes. When the displacement operator is written in terms of the operators that create and destroy phonon quanta, the term proportional to u^3 in the Hamiltonian will have terms with two destruction operators and a creation operator, and also terms with two creation operators and a destruction operator. This term then leads to processes where two phonons are destroyed and a third is created, and processes where two phonons are created from one. These processes must always satisfy energy conservation; momentum must always be conserved modulo a reciprocal lattice vector. If the process leads to the backscattering of a phonon, it is referred to as an *umklapp* process [1,12].

The exact mean free path associated with umklapp processes can be complicated to derive [33]. In silicon, the accepted form is [18]

$$\tau_u^{-1} = BT\omega^2 e^{-C/T}. \tag{26.16}$$

The constants B and C can be found by fitting to thermal conductivity data for bulk silicon. Expressing the result in terms of dimensionless parameters gives

$$\left(\omega_D \tau_u\right)^{-1} = \left(\frac{T}{T_D}\right)\left(\frac{0.1\omega}{\omega_D}\right)^2 e^{-0.21/(T/T_D)}. \tag{26.17}$$

The final process we will consider is scattering from the disordered boundary of the material. This, one imagines, is straightforward: the mean free path is expected to be of order the width of the system, giving a mean free time of $\tau_b^{-1} = \upsilon/(2R)$, where υ is approximately the sound velocity and R is the radius of the system (we should also include a dimensionless constant to specify the specularity of the surface; in practice, this appears to be of order unity [18]). In fact, surface scattering is much more complicated than is assumed in this simple picture, as we will see later. The simple picture appears to be adequate for nanowires with diameters of order 30 nm or greater.

To understand the relative magnitudes of these scattering rates, the reader is referred to Figure 26.4. For the boundary scattering rate we chose for υ the longitudinal acoustic phonon velocity along the 110 direction of silicon [9], which is 9180 m/s (see [9]). Note that for widths less that approximately 30 nm, boundary scattering is the dominant scattering mechanism for all phonon modes. The umklapp scattering rate is strongly temperature dependent, while the rate of scattering from isotopic impurities is negligible for all nanowires, when compared with the boundary scattering rate.

Given these three processes, we can find the total scattering rate by assuming them to be independent. Then the Matthiessen rule [1] gives

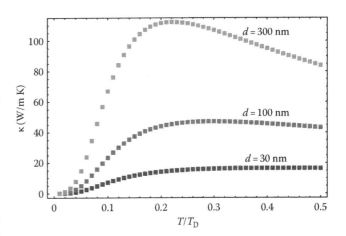

FIGURE 26.8 The thermal conductivity as a function of temperature for silicon wires of various diameters.

$$\tau_{tot}^{-1} = \tau_i^{-1} + \tau_u^{-1} + \tau_b^{-1}. \tag{26.18}$$

This total scattering rate can be used with the Landauer formula for the thermal conductivity derived earlier (see Equation 26.6). The result is shown in Figure 26.8. Note that if one uses the Debye approximation for the phonon density of modes, then at high temperatures the thermal conductivity is wrong by a factor of 4 or so. In the figure shown, we have used a cutoff in the density of modes at a frequency lower than the Debye frequency—in fact, at about $\omega_D/2$. As described in [18], this gives quantitative agreement with experiment. We stress though that the procedure is arbitrary; to predictively calculate the thermal conductivity at high temperatures one needs to find the phonon density of modes numerically from a microscopic model. In any case, the qualitative features we will now describe do not depend on introducing this cutoff.

The most obvious feature in the figure is that the thermal conductivity is strongly decreased by decreasing the diameter of the nanowire. This is not surprising: We know from Figure 26.4 that boundary scattering is dominant, and as the diameter of the wire is reduced the mean free path is correspondingly reduced. In addition, the downturn in thermal conductivity due to umklapp processes, familiar from thermal conductivity of bulk silicon, and visible in the figure for the 300 nm wire (and, to some extent, the 100 nm wire), is almost entirely absent for the thinnest wires. This can be understood if one recalls that for wires of order 30 nm, boundary scattering is the dominant scattering mechanism for all frequencies. In this case, we will show that the thermal conductivity has the same temperature dependence as the heat capacity. Note that the contribution of the phonons in subband i and with wave vector k to the heat capacity of the solid is

$$c_V^{(i)}(k) = k_B \left(\frac{\hbar\omega}{K_B T}\right)^2 \frac{e^{\hbar\omega_i(k)/(k_B T)}}{(e^{\hbar\omega_i(k)/(k_B T)} - 1)^2}. \tag{26.19}$$

In the resistive regime, with surface scattering dominant, the transmission function for a mode is given by $\mathcal{T}_i = 2R/L$. Thus, the thermal conductivity can be written as

$$\kappa \sim \frac{1}{A} \sum_i \int \frac{dk}{2\pi} c_V^{(i)}(k)\, v_i(k)\, 2R. \qquad (26.20)$$

If we assume that $v_i(k)$ is a constant independent of i and k, then $\kappa \propto c_V(T)$. If one recalls that the heat capacity is a constant at high temperatures (the Dulong–Petit law), then the thermal conductivity must also be a constant.

26.5 Localization

One of the most interesting aspects of quasi-one-dimensional transport is the phenomenon of localization. As the length of the system, L, becomes much longer than Nl, the conductance of the system becomes exponentially small in L. Localization is a wave effect, since it involves interference, and is strongest in low-dimensional systems [32]. There is a nonrigorous argument for why one expects waves to be localized where particles might not be. Suppose we have a system with time reversal invariance. Then, if A is the amplitude for a certain path and A^* is the amplitude for the time reversed path, then $A = A^*$ and so the return probability is given by $P_{\text{wave}} = |A + A^*|^2 = 4|A|^2$. By contrast, for particles we have that $P_{\text{particle}} = 2|A|^2$. Hence, return probabilities are greater in wave physics. It is important to note that in this analysis we are assuming that inelastic (3-phonon) scattering is not important—these processes will destroy the phase coherence and kill the interference effect. Another important aspect is that the phenomenon is very sensitive to the dimension of the space. All states are localized in one and two dimensions, but in three dimensions there is a mobility edge, an energy that separates localized and extended energy eigenfunctions.

While there are some theoretical suggestions that localization may be relevant to the physics of very thin nanowires [20], and may partly explain some experimental observations [5,10], unambigious experimental observation of phonon localization in nanowires is an important problem for the future.

26.6 Conclusions

We have reviewed the field thermal conductance of semiconductor nanowires. We have used a single formalism, known as the Landauer formalism, to understand the behavior over a broad range of temperatures. At low temperatures, one expects a thermal conductance given by $4g_0$, where g_0 is the quantum of thermal conductance. At higher temperatures, for systems of diameter of order 30 nm or greater, we have shown how scattering rates extracted from bulk can be used to calculate the thermal conductance, and match well to the data.

However, the field is still advancing. A recent work [6] suggests that one can understand the unusual scaling of the thermal conductivity with temperature, $\kappa \sim T$ rather than $\kappa \sim T^3$, by developing more elaborate models of surface scattering. Yet,

more recent data has shown that silicon nanowires that are created through electroless etching, with greater surface roughness than VLS wires, can have a thermal conductivity of the same order as amorphous silicon [10]. These and other results suggest that the field will continue to be very exciting in the future.

Acknowledgments

Our research in thermal transport was primarily supported by the Department of Energy via the Materials Sciences Division of the Lawrence Berkeley National Laboratory.

References

1. N. W. Ashcroft and N. D. Mermin, *Solid State Physics*, Holt, Rinehart and Winston, New York, 1976.
2. C. W. J. Beenakker, Random matrix theory of quantum transport, *Rev. Mod. Phys.* **69**, 731 (1997).
3. S. Berber, Y.-K. Kwon, and D. Tománek, Unusually high thermal conductivity of carbon nanotubes, *Phys. Rev. Lett.* **84**(20), 4613–4616 (2000).
4. M. P. Blencowe and V. Vitelli, Universal quantum limits on single-channel information, entropy, and heat flow, *Phys. Rev. A* **62**(5), 052104 (2000).
5. A. I. Boukai, Y. Bunimovich, J. Tahir-Kheli, J.-K. Yu, W. A. Goddard III, and J. R. Heath, Silicon nanowires as efficient thermoelectric materials, *Nature* **451**, 168 (2008).
6. R. Chen, A. Hochbaum, P. Murphy, J. Moore, P. Yang, and A. Majumdar, Thermal conductance of thin silicon nanowires, *Phys. Rev. Lett.* **101**, 105501 (2008).
7. S. Datta, *Electronic Transport in Mesoscopic Systems*, Cambridge University Press, Cambridge, U.K., 1997.
8. F. J. DiSalvo, Thermoelectric cooling and power generation, *Science* **285**(5428), 703–706 (1999).
9. H.-Y. Hao and H. J. Maris, Study of phonon dispersion in silicon and germanium at long wavelengths using picosecond ultrasonics, *Phys. Rev. Lett.* **84**(24), 5556–5559 (2000).
10. A. I. Hochbaum, R. Chen, R. D. Delgado, W. Liang, E. C. Garnett, M. Najar-ian, A. Majumdar, and P. Yang, Enhanced thermoelectric performance of rough silicon nanowires, *Nature* **451**, 163–167 (2008).
11. P. Kim, L. Shi, A. Majumdar, and P. L. McEuen, Thermal transport measurements of individual multiwalled nanotubes, *Phys. Rev. Lett.* **87**(21), 215502 (2001).
12. C. Kittel, *Introduction to Solid State Physics*, 8th edn., Wiley, New York (2004).
13. P. G. Klemens, The scattering of low-frequency lattice waves by static imperfections, *Proc. Phys. Soc. A* **68**, 1113–1128 (1955).
14. D. Y. Li, Y. Y. Wu, P. Kim, P. D. Yang, and A. Majumdar, Thermal conductivity of individual silicon nanowires, *Appl. Phys. Lett.* **83**, 2934 (2003).
15. W. Lu and C. M. Lieber, Topical review: Semiconductor nanowires, *J. Phys. D Appl. Phys.* **39**, 387 (2006).
16. G. Mahan, B. Sales, and J. Sharp, Thermoelectric materials: New approaches to an old problem, *Phys. Today* **50**, 42 (1997).

17. R. Maynard and E. Akkermans, Thermal conductance and giant fluctuations in one-dimensional disordered systems, *Phys. Rev. B* **32**, 5440 (1985).

18. N. Mingo, Calculation of Si nanowire thermal conductivity using complete phonon dispersion relations, *Phys. Rev. B* **68**(11), 113308 (2003).

19. N. Mingo and D. A. Broido, Carbon nanotube ballistic thermal conductance and its limits, *Phys. Rev. Lett.* **95**(9), 096105 (2005).

20. P. G. Murphy and J. E. Moore, Coherent phonon scattering effects on thermal transport in thin semiconductor nanowires, *Phys. Rev. B* **76**(15), 155313 (2007).

21. N. P. Padture, M. Gell, and E. H. Jordan, Thermal barrier coatings for gas-turbine engine applications, *Science* **296**, 280–284 (2002).

22. J. B. Pendry, Quantum limits to the flow of information and entropy, *J. Phys. A Math. Gen.* **16**, 2161–2171 (1983).

23. L. G. C. Rego and G. Kirczenow, Quantized thermal conductance of dielectric quantum wires, *Phys. Rev. Lett.* **81**(1), 232–235 (1998).

24. L. G. C. Rego and G. Kirczenow, Fractional exclusion statistics and the universal quantum of thermal conductance: A unifying approach, *Phys. Rev. B* **59**(20), 13080–13086 (1999).

25. D. M. Rowe, *CRC Handbook of Thermoelectrics*, CRC Press, Boca Raton, FL, 1995.

26. P. K. Schelling, L. Shi, and K. E. Goodson, Managing heat for electronics, *Mater. Today* **8**, 30–35 (2005).

27. K. Schwab, E. A. Henriksen, J. M. Worlock, and M. L. Roukes, Measurement of the quantum of thermal conductance, *Nature* **404**, 974–977 (2000).

28. A. Shakouri, Nanoscale thermal transport and microrefrigerators on a chip, *Proc. IEEE* **94**(8), 1613–1638 (2006).

29. B. J. van Wees, H. van Houten, C. W. J. Beenakker, J. G. Williamson, L. P. Kouwenhoven, D. Vandermarel, and C. T. Foxon, Quantized conductance of point contacts in a two-dimensional electron gas, *Phys. Rev. Lett.* **60**, 848–850 (1988).

30. Y. Wang, N. S. Rogado, R. J. Cava, and N. P. Ong, Spin entropy as the likely source of enhanced thermopower in NaxCo$_2$O$_4$, *Nature* **423**, 425–428 (2003).

31. Y. Xia, P. Yang, Y. Sun, Y. Wu, B. Mayers, B. Gates, Y. Yin, F. Kim, and H. Yan, One-dimensional nanostructures: Synthesis, characterization, and applications, *Adv. Mater.* **15**(5), 353 (2003).

32. J. M. Ziman, *Models of Disorder*, Cambridge University Press, Cambridge, U.K., 1979.

33. J. M. Ziman, *Electrons and Phonons*, Oxford University Press, Oxford, NY, 2001.

27

The Wigner Transition in Nanowires

David Hughes
Queen's University Belfast

Robinson Cortes-Huerto
Queen's University Belfast

Pietro Ballone
Queen's University Belfast

27.1 Introduction

Novel techniques to prepare and manipulate systems of nanometric size are being proposed and tested at an amazing and still quickening pace, holding the promise for a revolution in the technological basis of our economy and of our society (Roco et al. 2000). Nanotechnologies already are promoting major new steps in the miniaturization of electronic devices (Cui and Lieber 2001), introducing new concepts in medicine (Loo et al. 2004), changing well-established industrial processes (Zhou et al. 2003), and providing an ever-expanding variety of new ingredients for the next generation of nanostructured materials. In alliance with biotechnologies, nanotechnologies are set to be prominent players in the development of new energy sources and power generation techniques, as well as in the fight to control pollution of the natural environment.

From a purely intellectual perspective, nanotechnologies additionally hold the promise of a wide variety of fascinating new systems and exotic phenomena to explore and to understand. In this respect, nothing can be more intriguing than the electronic properties of nanomaterials, when low dimensionality, quantum mechanics, and drastic size effects combine together to produce unforeseen and counterintuitive effects. Examples of qualitatively new behaviors discovered in nanomaterials include the prominent role of magic sizes in the stability and electronic properties of nanoclusters (de Heer 1993), novel manifestations of superconductivity in nanostructured materials (Schuller et al. 2008), and the super-paramagnetic behavior observed in ferromagnetic nanoparticles (Billas et al. 1993).

Even within this collection of exotic systems and surprising phenomena, the properties of a special class of nanosystems, i.e., *nanowires*, stand out as particularly remarkable. In the last few years, nanometric conductors have been created by a variety of means, and investigated by experimental techniques ranging from photoemission spectroscopy (Zanolli et al. 2007) to transport measurements (Cao et al. 2005). Examples of the fabrication methods include deposition from the vapor phase and the selective doping of semiconducting nanostructures (Werner et al. 2006), the controlled doping of conducting polymers (Rahman et al. 2006, Rahman and Sanyal 2007), and the etching and mechanical thinning of metal wires. Finally, the fabrication of long and thin nanowires has reached macroscopic production rates with carbon nanotubes (CNTs), representing a wide research field by themselves (Charlier et al. 2007). Systems of this kind are bound to play an important role in every future application of nanosystems in electronic devices, since every active element will still need to be connected by a conducting nanowire to become part of a larger functional entity (Beckman et al. 2005).

For all these systems, the first and simplest characterization of their properties may be represented by the determination of their DC electric conductivity, sometimes carried out on individual wires by near-field (scanning tunneling microscopy [STM] or atomic force microscopy [AFM]) microscopy, or measured on mesoscopic samples consisting of many aligned elongated conductors. The electric conductivity of nanowires appears to be quantized in units of $G_0 = 2e^2/h$, that can be explained by scattering theory and by the Landauer theory of conductivity (Landauer 1970). Anomalous values of quantization, however, have been revealed in experiments, pointing to the spontaneous polarization of electrons in a near-1D confining potential (Thomas et al. 1996, Reilly et al. 2002, Bird and Ochiai 2004, Yoon et al. 2007, Danneau et al. 2008).

From a theoretical point of view, the interest in nanowires has been greatly enhanced by the early prediction of new exotic electron states in 1D metals, epitomized by the Tomonaga–Luttinger state (Tomonaga 1950, Luttinger 1963), more commonly referred to as the Luttinger-liquid (LL) phase. Recent experiments are increasingly providing support for the existence of these phases (Bockrath et al. 1999, Zaitsev-Zotov et al. 2000), especially for the thinnest and least defective nanowire samples.

In another manifestation of electron correlation and prominent quantum mechanical behavior, the transition from metal-like conduction by delocalized electrons to the insulating behavior of localized electrons has been revealed by experiments in lightly doped polymers (Rahman and Sanyal 2007). Needless to say, any new phenomena affecting conductivity have the potential for a major impact on the prospect of the future miniaturization of electronic components at the nanometric scale.

In 1D, as in higher dimensions, localization is primarily driven by a decrease of carrier density, which enhances the role of correlation with respect to kinetic energy. The original prediction of localization in the context of the homogeneous electron gas is due to Wigner (Wigner 1934). The only parameter relevant to describe the state of this model system is the electron density, ρ, or, equivalently, the Wigner–Seitz parameter, $r_s = (3/4\pi\rho)^{1/3}$ (in 3D), which measures the radius of the sphere that contains, on average, one electron. In the high-density limit, kinetic energy dominates over all other energy contributions (exchange and correlation), the ground state density is translationally invariant, with electrons behaving as independent particles whose wave function, in a constant potential, is a plane wave. Simple quantum mechanical considerations show that decreasing density changes the balance between potential energy (PE) and kinetic energy (KE) contributions, their ratio PE/KE being proportional to the Wigner–Seitz radius, r_s. At sufficiently low density (high r_s), the system reverts from the homogeneous to a broken-symmetry state in which electrons decrease their potential energy by localization, giving rise to a body-centered cubic lattice (bcc).

By and large, the qualitative picture outlined above for the 3D homogeneous electron gas is also valid in 1D, even though quantitative details need to be adjusted according to the space dimension. To be precise, the relative role of fluctuations increases with decreasing dimensionality and, in 1D, long-range periodicity is prevented by the divergence of the mean square fluctuation in the particles' position (Peierls instability). This last statement, however, is strictly valid only in the thermodynamic limit, i.e., for infinitely extended systems. Real systems, instead, are always finite, implying at the same time that the localization transition cannot be infinitely sharp, and that the role of fluctuations is mitigated with respect to the extended case. As a result, localized configurations persist over mesoscopic and even macroscopic times, and could be observed by experiments. Indeed, several groups have already measured phenomena that strongly point toward the formation of a Wigner crystal (WC) in a variety of quasi-1D (Q1D) systems (Hiraki and Kanoda 1998, Horsch et al. 2005, Rahman and Sanyal 2007). In all cases, the Q1D WC (Schulz 1993) is primarily the result of low density, with confinement playing a minor role.

This chapter is focused precisely on the localization (Wigner) transition taking place in Q1D metallic wires with decreasing carrier (electrons or holes) density. In what follows we provide, first of all, a brief overview of the phases believed to occur in Q1D systems. These include the "normal" regime, believed to be LL, which is contrasted to the regular Fermi liquid (FL); a magnetically ordered phase; a low-density, crystalline regime; and finally, other even more exotic phases identified by dynamical or excited-state properties. Moreover, we summarize a number of theoretical studies, ranging from the use of simple idealized models, to state-of-the-art correlated methods.

To emphasize the fact that the low-carrier-density nanowires we are interested in are not simply a theoretical curiosity, we provide a short summary of the experimental methods that have been used to fabricate wires of nanometric diameter over a wide range of carrier densities.

In the second half of the chapter, we introduce an idealized (jellium) model of metal wire, and we investigate the localization transition by a conceptually simple and computationally tractable approach based on density functional (DF) theory (Hohenberg and Kohn 1964, Kohn and Sham 1965). The approximation we use (local spin density, LSD) to treat exchange and correlation has been devised for electron systems in a low correlation regime, and the quantitative predictions of the method need to be taken with some caution. However, the method and the results it provides have a great didactical value, since they describe the system and its localization transition in simple and intuitive terms. According to the results of our DF computations, the localization transition is intrinsically related to the spontaneous spin polarization expected in low-density electron systems. For this reason, we also briefly cover this transition in our discussion.

27.2 The Phase Diagram of the Q1D Electron Gas

27.2.1 Fermi- versus Luttinger-Liquid Behavior

The FL picture (Giuliani and Vignale 2005) of two- and three-dimensional systems explains why the electron–electron (*ee*) interaction between particles close to the Fermi surface may be neglected, therefore justifying the apparent success of the independent particle, or free-electron gas description of such systems. Only at low charge carrier densities does the role of *ee* interaction begin to manifest itself, causing a breakdown in the quasi-particle description. Such systems are referred to as correlated and constitute "the exception rather than the rule." For 1D systems, the dimensional constraints give rise to enhanced correlation at all densities, rendering the quasi-particle formalism useless. In 1950, Tomonaga (1950) proposed a simplified model of the 1D electron gas, that was later generalized by Luttinger (1963) in 1963. It has subsequently been revised by several other authors (see Schulz et al. 2000, and references therein). These models provided a qualitatively different understanding of the behavior of 1D systems.

As it is also the case for the familiar FL model, Tomonaga and Luttinger considered the behavior of electrons whose energy is close to the chemical potential (or Fermi energy). Outside the range of energies considered, states are assumed not to be influenced as strongly by perturbations in the potential. The details of the Luttinger model are not presented here, though its consequences shall be discussed. A review article by Schulz et al. (2000) provides greater depth.

The solution of the Luttinger Hamiltonian is a nontrivial exercise and may not be achieved by perturbative methods. Rather, it is necessary to replace the field operators in the second quantized Hamiltonian (Inkson 1984, Negele and Orland 1998) with a new set of operators, which, upon inspection, are seen to fulfill the commutation relations of boson operators. It then becomes apparent that the low energy collective excitations of such systems behave as a set of massless bosons. Moreover, the Hamiltonian may be factorized into separate charge (C) and spin (S) components, $\hat{\mathcal{H}} = \hat{\mathcal{H}}_C + \hat{\mathcal{H}}_S$. This gives rise to a phenomenon known as spin–charge separation. Supposing an electron is injected into the system at a given time and position endowing it with a net spin and charge equivalent to one electron, at some later time the spin and charge will be spatially separated. This is not to say that the electron has in some way split, but rather, the system has entered an excited state, forming two collective excitations: a holon (charge quanta) and a spinon (spin quanta). A different variety of LL, also known as the spin-incoherent LL (Fiete 2007), has recently attracted considerable attention. This phase occurs when the energy scale associated with charge degrees of freedom is vastly larger than the energy scale of spin excitations.

Another interesting consequence of the Luttinger model is that the presence of even the smallest impurity will give rise to an insulating state at low temperatures. Linear response theory reveals that in the ground state the system is on the verge of forming a charge density wave (CDW) of wavelength $2k_F$. Introducing an impurity is sufficient to "pin" this wave in position, thereby reducing the conductivity to zero. This situation is often referred to as a dirty LL (Ogata and Fukuyama 1994). If the temperature of the system or the bias across the system is increased above a given threshold, the system enters the conducting phase again. Experimentalists look for specific, power law dependence of the conductivity on temperature $G \propto T^\alpha$, and of the differential conductance on the applied bias $dI/dV \propto T^{\alpha+1}$, as an indication of LL behavior (Giuliani and Vignale 2005). This property of 1D systems makes experimental investigation difficult, as only highly pure samples will exhibit true LL behavior. Moreover, experimentally distinguishing between the pinned CDW and WC formation presents difficulties (Horsch et al. 2005).

LL behavior is only observed in a small number of cases (Zaitsev-Zotov et al. 2000) though it is believed that it provides a more accurate description of Q1D systems than the FL formalism. Examples include, CNTs (Bockrath et al. 1999), and some less-common classes of organic (Biermann et al. 2002) and inorganic (Horsch et al. 2005) conducting polymers. Recently, it has been suggested that electrons at an armchair edge of an undoped 2D carbon plane (graphene) might exhibit LL properties (Fertig and Brey 2006).

27.2.2 Spontaneous Polarization of the Electron Gas in 3D, 2D, and 1D

At sufficiently low densities, the ground state of the 3D electron gas is known to become ferromagnetic (Ceperley and Alder 1980). Since the kinetic and exchange energies scale as r_s^{-2} and r_s^{-1}, respectively (Mahan 2000, Giuliani and Vignale 2005), it is clear that for large r_s, the exchange energy dominates and the system will polarize. Hartree–Fock calculations predict the polarization transition to occur at $r_s = 5.45$. However, the inclusion of correlation energy destabilizes the spin-polarized state, moving the transition to lower densities (Ceperley and Alder 1980). One recent study finds the polarization phase transition to occur over a continuous range of densities, $20 \pm 5 < r_s < 40$, where above $r_s = 40$ the system is fully polarized (Ortiz et al. 1999). The precise location of this phase transition, however, is still a matter of debate (Zong et al. 2002).

The effect of confining a system in two dimensions (or indeed one or zero dimensions) is to enhance the role of exchange and correlation energy (Zabala et al. 1998). It is therefore believed that the polarization state is stabilized, and the corresponding value of transition density is increased. Several groups have measured anomalous values of the conductance in quantum point contacts (QPC) (Thomas et al. 1996, Reilly et al. 2002, Bird and Ochiai 2004, Yoon et al. 2007, Danneau et al. 2008), which is believed to point toward the existence of a partially polarized ground state.

27.2.3 Wigner Crystallization in Q1D Systems

In 1934, Wigner (Wigner 1934) proposed that at sufficiently low densities, when *ee* interaction dominates over electron kinetic energy, electrons could attain a more stable state through localizing in space. Rather than being spread out in Bloch waves, electrons would reside in well-defined, localized regions, forming a bcc lattice. In doing so, a cost is paid in terms of kinetic energy; however, the resultant gain in correlation energy makes this phase preferable.

The WC was first observed experimentally in 1979 within a layer (2D) of liquid helium (Grimes and Adams 1979). It has subsequently been observed in several Q1D systems as we shall outline below (Hiraki and Kanoda 1998, Horsch et al. 2005, Rahman and Sanyal 2007).

The WC has been studied intensively by a variety of methods, including Hartree–Fock methods (Trail et al. 2003), which drastically overestimate its stability. Several DF computations, whose energy functionals are based on data gathered from QMC simulations, have been used to study this phase transition (Senatore and Pastore 1990, Ito and Teraoka 2004, 2006). Benchmark diffusion, fixed-node quantum Monte-Carlo simulation place the WC transition in the region $65 \pm 10 < r_s < 106$ (Ceperley and Alder 1980, Ortiz et al. 1999, Drummond et al. 2004).

Finally, the properties of a zigzag variety of Wigner crystal in Q1D systems have been reviewed by Meyer and Matveev (2009).

27.2.4 Other Phases

For completeness we briefly mention some additional phases observed in Q1D systems.

Mott insulators have been observed in organic nanowires, generated using nanoscale-electrocrystallization methods (Hasegawa et al. 2008), and also in single-crystal VO_2 nanowires, grown using physical vapor deposition methods (Baik et al. 2008). The transition to the Mott insulator has been investigated using many-body Green's function methods (Biermann et al. 2001).

The Bechgaard salts, a class of organic charge-transfer complexes, provide an example of a Q1D system that exhibits high temperature superconductivity (Biermann et al. 2002). Nickel et al. (2006) have used a renormalization group approach to investigate the phase diagram in Q1D systems, and have focused in detail on the superconducting phase.

27.3 Fabrication Techniques

A comprehensive summary of fabrication techniques for nanometric wires is well beyond the scope of this chapter. We provide here only a brief overview of the most common techniques for nanowire growth, focusing in particular on the methods that have been used to fabricate wires operating in the low-carrier-density regime, or high correlation limit.

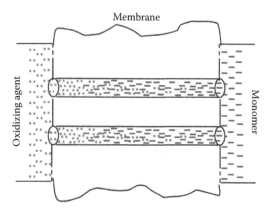

FIGURE 27.1 Schematic drawing of the experimental setup used to prepare conducting polypyrrole nanowires in porous membranes. (From Rahman, A. et al., *Phys. Rev. B*, 73, 125313, 2006. With permission.)

27.3.1 Conducting Polymer Chains

The channels of porous polycarbonate membranes have been used as a template for the growth of monodisperse, parallel bundles of polypyrrole nanowires (see Figure 27.1). The carrier concentration within such systems may be varied over a wide range by doping (Rahman and Sanyal 2007). In the procedure described in (Parthasarathy and Martin 1994), for instance, a polycarbonate membrane was immersed in a refrigerated solution containing the monomer pyrrole ring. Polymerization occurs in the pores of the membrane and over the surface. Surface polymers are removed by polishing techniques. If desired, it is possible to isolate individual wires by dissolving the membrane in dichloromethane; wires are then removed from solution by filtering. Electrical properties of the wires are measured by coating opposite surfaces (at either end of the wires) with gold, which has low electrical resistance. Both gold surfaces may then be connected to a current or voltage source and the electrical properties of many parallel wires measured and averaged over. Scanning electron microscopy (SEM) and tunneling electron microscopy (TEM) verify that polypyrrole wires are unbroken, and are not hollow, which would lead to erroneous measurements. Both of these failures plagued prior attempts to synthesize conducting polymer wires from polyaniline (Parthasarathy and Martin 1994).

Measurements of conductance properties, point to a strong temperature and bias dependence (Rahman and Sanyal 2007), findings that would be consistent with the formation of a WC.

27.3.2 Quantum Point Contacts

QPC represent arguably the simplest way to create nanowire structures. Several fabrication methods have been developed. These include wires created by contact between atomic force microscopes (AFM) and an underlying surface (Kuipers and Frenken 1993) (Figure 27.2). In such a case, the AFM is brought close to a surface, at which point it is speculated that van der Waals interactions give rise to a point contact.

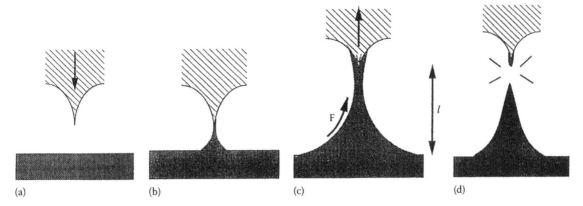

FIGURE 27.2 Schematic cycle of the formation and rupture of a surface-tip contact: (a) Approach of the tip. (b) Just after the jump to contact. (c) Growth of neck. (d) Breaking of the neck. (From Kuipers, L. and Frenken, J.W.M., *Phys. Rev. Lett.*, 70, 3907, 1993. With permission.)

Contacts created in this way have a low aspect ratio (radius/length), and translational invariance may not be assumed. Moreover, the range over which electron densities may be controlled, in either class of QPC, is very narrow, and is fundamentally limited by the material used.

A second type of QPC is commonly assembled using molecular beam epitaxy (MBE) (Werner et al. 2006), a form of thin film deposition (Fan et al. 2006). Current is free to flow through a GaAs 2D electron gas (2DEG) connecting two ohmic contacts (Figure 27.3). By increasing the negative bias to a set of gates, which are separated from the 2DEG by an insulating AlGaAs layer, the effective width of the channel through which electrons are free to move is progressively reduced. Recent experiments of Bird et al. (Bird and Ochiai 2004, Yoon et al. 2007) have measured anomalous conductances of a point contact fabricated using such techniques, pointing to spontaneous spin polarization in low density samples. This phenomenon, commonly referred to as the "0.7 anomaly", was first measured by Thomas et al. (1996), and its origin is still the subject of debate (Danneau et al. 2008). Other experiments have investigated the dependence of this anomaly on the electron density (Reilly et al. 2002).

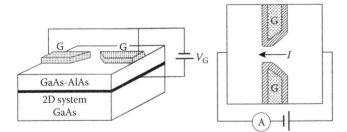

FIGURE 27.3 Schematic cross-sectional, and top view of a quantum point contact, defined in a high-mobility 2D electron gas at the interface of a GaAs/AlGaAs heterojunction. The point contact is formed when a negative voltage is applied to the gate electrodes on top of the AlGaAs layer. Transport measurements are made by employing contacts to the 2D electron gas at either side of the constriction. (From Kawabata, A., *Rep. Prog. Phys.*, 70, 219, 2007. With permission.)

27.3.3 Direct Measurements of Wave Function Localization

Auslaender et al. (Auslaender et al. 2002, 2005, Steinberg et al. 2006) have produced a novel, thin film based experimental setup, comprising two parallel wires embedded in a GaAs/AlGaAs heterostructure (Figure 27.4). This has been used to gain direct insight into the possible localization of the many-body wave function.

The device is formed using a method known as cleaved-edge overgrowth (Pfeiffer et al. 1990), a type of MBE. The setup uses a technique known as momentum-resolved tunneling spectroscopy to measure the tunneling current between two parallel quantum wires, and thereby build a picture of the many-body wave function of the upper wire (UW). It is possible to control the energy and density of the electrons in the UW by altering bias V_{SD} and V_G, respectively. A magnetic field B, applied in a direction perpendicular to the wires' plane, supplies the electrons with momentum $k_B = eBd/\hbar$ allowing tunneling to the longer of the two wires, or lower wire (LW). Knowledge of tunneling conductance $G(V_{SD},B)$ as a function of magnetic field strength, B, and bias V_{SD} is then related to the square of the many-body wave function in momentum space.

Not only has evidence of spin–charge separation been measured, pointing toward LL behavior (Auslaender et al. 2005), but when electron density in the UW has been depleted sufficiently, the wave function is seen clearly to localize, pointing toward the formation of a WC (Steinberg et al. 2006).

Mueller (2005) has modeled the system in three different regimes: the high-density regime, using a Thomas–Fermi description; the low-density regime, using semiclassical arguments; and in an attempt to understand the processes occurring during localization, a Hartree–Fock model was implemented. The models implemented are purely 1D, making it necessary to approximate the interelectron repulsion as follows: $U(r) = \min(|r|^{-1}, d^{-1})$, where r is the *ee* separation distance, and d is a parameter. Tunneling between wires is treated in lowest order perturbation theory, which the author identifies as a significant source of error. Moreover, the tunneling current is found

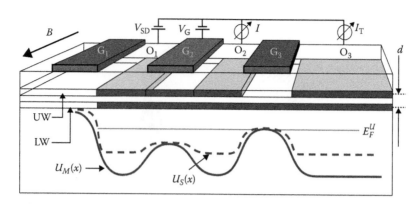

FIGURE 27.4 Schematic drawing of tunneling spectrometer. From top to bottom: Magnetic field, B, perpendicular to the plane of the wires. 2 μm wide gates (G_1, G_2, and G_3) and Ohmic contacts (O_1; source, and $O_{2,3}$; drains). Gate voltage, V_G, serves as an electron-density controller for the upper wire (UW), two-terminal current I, and I_T tunneling current. 20 nm thick UW at the edge of the 2D electron gas, 6 nm insulating AlGaAs barrier, and 30 nm thick lower wire (LW). Single mode (and multimode) potentials $U_S(x)$ ($U_M(x)$) experienced by UW. (From Steinberg, H. et al., *Phys. Rev. B*, 73, 113307, 2006. With permission.)

to be lowered by the use of the Hartree–Fock approximation, which additionally stabilizes the WC phase. Nevertheless, the method provides a qualitative view of the transition between a smooth density profile, which gradually develops Friedel oscillations, that become so marked that Wigner crystallization finally occurs.

Bird and collaborators have reported a similar experimental setup of two parallel wires, in which the effective width of one wire is varied. The conductance of the second wire is measured, which exhibits a resonant peak close to the pinch-off of the first wire (Morimoto et al. 2003). The origin of this effect remains unexplained, however, points to the existence of correlation between the two systems.

27.3.4 Carbon Nanotubes

CNTs consist of a graphene sheet rolled up to form a cylinder. CNTs can be generated with extremely large aspect ratios. Moreover, it is possible to form more exotic structures containing Y- and X-shaped junctions and budded structures. Also, nanotubes may be arranged coaxially (multi-walled CNTs) or in parallel bundles. It is possible to alter the chirality of the tube, specified by the graphene lattice vector: this affects electrical and optical properties. Dopant levels and electrical properties of CNTs may be controlled in such a way as to achieve metallic or semiconducting behaviors. For a review on most aspects of CNTs see Charlier et al. (2007). Moreover, by intercalating various types of nanotubes (for example, carbon, BN, MgO, Ga_2O_3) with other materials, very long wires, with highly varied electrical properties (Kim et al. 2007, Zanolli et al. 2007, Gong et al. 2008) have been created.

Cao et al. (2005) have successfully grown CNTs between tungsten/platinum electrodes suspended over a metal electrode (see Figure 27.5). Altering the potential of the electrode (gap voltage) enhances or reduces the active charge carrier concentration, modifying the electrical behavior of the nanowire. CNTs may be generated with an extremely low number of defects and impurities, a quality that is essential for the study of 1D systems.

Techniques for generating nanowires using CNTs are widely different, but chemical vapor deposition is one of the most popular (José-Yacamán et al. 1993).

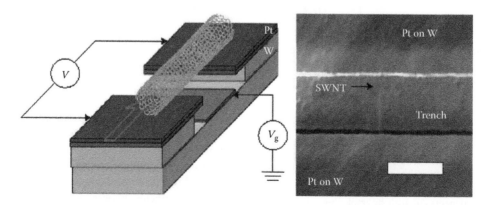

FIGURE 27.5 Left: Schematic drawing of a carbon nanotube (CNT) bridging two platinum electrodes. The CNT spans a trench containing a gate used to vary charge carrier concentration. Right: SEM image of the experimental setup. (From Cao, J. et al., *Nat. Mater.*, 4, 745, 2005. With permission.)

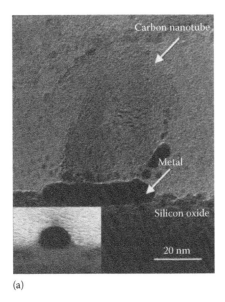

(a) (b)

FIGURE 27.6 Cross-sectional TEM images. (a) A 50 nm wide Au/Ti nanowire was formed underneath a CNT of about 60 nm in diameter. Inset: A Au/Ti nanowire of 12 nm in width is shown. (b) A metal nanowire of 40 nm in width was formed below a bundle of CNTs. White circle encloses one of the CNTs. (Reproduced from Yun, W.S. et al., *J. Vac. Sci. Technol. A*, 18, 1329, 2000, figure 4. With permission.)

In most cases, these structures have been seen to behave as FL; however, several studies have detected indications of LL behavior (Bockrath et al. 1999), and one more recent study points toward the formation of a 1D WC (Deshpande and Bockrath 2008).

In addition to acting as conducting wires themselves, CNTs have been used as masks for the etching of Si substrates (Yun et al. 2000) (Figure 27.6). With advances in CNT manipulation technologies (Hertel et al. 1998) the authors identify this as a realistic option for future, large-scale nanowire fabrication.

27.4 Overview of Theoretical Studies

The variety of theoretical methodologies implemented in the study of Q1D systems is truly extensive, ranging from the simplest fully-1D Thomas–Fermi (Tanatar et al. 1998) models, to sophisticated many-body techniques (Biermann et al. 2001). The study of transport properties in nanowires also constitutes a major portion of theoretical effort, this is due to a number of reasons, for example, the need to understand dynamical processes such as heating and power dissipation in structures intended for device applications (Montgomery et al. 2002).

We present a short summary of some of the main areas of recent theoretical focus concerning 1D and Q1D systems, drawing particular attention to the study of low-density (or highly correlated) systems.

27.4.1 Idealized Models of Nanowires

Idealized models play a powerful role in the study of quantum systems. By eliminating irrelevant degrees of freedom, for example, the nuclear degrees of freedom in the case of jellium models, one is free to explore the physics of correlated electronic motion, without added complications generated by interactions with a complex lattice of nuclei.

A number of studies have concentrated on truly 1D models (Tanatar et al. 1998, Mueller 2005, Casula et al. 2006, Shulenburger et al. 2008). In such cases, the Coulomb potential is modified to remove the singularity at zero separation. Typical approximations include $U(r) = \min(|r|^{-1}, d^{-1})$ or $U(r) = (d^2 + r^2)^{-1/2}$, where r is the interelectron separation, and d is a small, finite parameter.

As mentioned above, Mueller (2005) has investigated one such model. A combination of Thomas–Fermi, Hartree–Fock, and semiclassical techniques, in parallel with perturbation theory, have been used to gain insight into the observation of charge localization, and the possibility of Wigner crystallization in the tunneling spectroscopy experiments of Auslaender et al. (2002). This study provides a qualitative view of the transition of the electron density between regions of high charge density to low charge density. In doing so, crystallization is directly observed.

Tanatar et al. (1998) have used a 1D Thomas–Fermi approach to investigate the transition to the localized density regime for purely 1D wires. One additional level of approximation is introduced, in that the form of the density is assumed to have a cosine modulation. The results point to a Wigner transition at a density that appears to overestimate the one observed in experiment. The influence of introducing a second parallel wire has also been investigated: this additional interaction further stabilized the crystal phase.

Several authors have used more realistic Q1D models, often based on the jellium picture, which facilitate the treatment of much larger wire radii (Yannouleas and Landman 1997, Zabala et al. 1998). Several reviews (Brack 1993, de Heer 1993) highlight the success that the jellium model has enjoyed when applied to the valence electrons in simple metal clusters. One might therefore expect a comparable degree of success when applying a similar jellium model to nanowires. For brevity we do not elaborate on the specific details of these models, other than to say that the methods used vary greatly, from the simplest free-electron gas picture (Stafford et al. 1997), to modified Thomas–Fermi approaches (Yannouleas and Landman 1997, Yannouleas et al. 1998), and finally to DF-based simulations, in the local-density approximation (Perdew et al. 1990, Zabala et al. 1998, 1999). Also, the geometries used vary from infinitely extended, and having cylindrical cross-sections (Zabala et al. 1998, 1999), to finite systems (Yannouleas and Landman 1997, van Ruitenbeek et al. 1997), some of which have constrictions, or necks, introduced in the axial direction (Stafford et al. 1997, Yannouleas et al. 1998).

Such models (Yannouleas and Landman 1997, Stafford et al. 1997, van Ruitenbeek et al. 1997, Yannouleas et al. 1998) have been used to partially explain the experimentally observed oscillations in elongation forces, and conductivities, when a wire is subjected to strain. In addition to the abrupt changes in atomic position that occur when a wire is stretched, the mean radius of the wire will be reduced also (provided one is assuming a fixed volume). It is argued that for a given background density, parameterized by the Wigner–Seitz radius, r_s, certain critical values of the wire radius result in particularly stable electronic configurations, known as *magic radii*. The enhanced stability, produces peaks in the elongation force, defined as $F = -dE/dL$, where E is the energy of the wire of radius R_B and L is its length. Similar arguments reveal oscillations in the conductance as a function of radius.

Zabala et al. (1998, 1999) has used an infinitely extended, Q1D jellium model to investigate the existence of stable, partially polarized ground states in systems within metallic range of densities ($2 \leq r_s \leq 6$). The net polarization is small, only amounting to a fraction of a Bohr magneton, μ_B, per atomic unit of length, $a_0 = \hbar/m_e e^2$. An explanation, of such polarized states was provided using the Stoner criterion.

To the best of our knowledge, such Q1D models have not been used to investigate the crystalline or low-density phases of the confined electron gas.

27.4.2 Explicitly Correlated Methods

We have already emphasized that 1D systems experience enhanced *ee* correlation regardless of the charge carrier concentrations, resulting directly from the dimensional constraints. It is widely accepted that mean-field descriptions, for instance DF, are insufficiently accurate to provide a fully quantitative description of such correlated materials (Kotliar and Vollhardt 2004), and in several cases, even the qualitative picture provided by DF might be incorrect.

Dynamical mean-field theory (DMFT) was developed specifically for the treatment of correlated materials (Kotliar et al. 2006), based on a lattice representation of the system. In short, the theory maps any lattice to a single-site problem, which interacts with a self-consistent bath, or continuum of parameterized,

noninteracting excitations. Particles are free to enter and leave the site from the continuum, and vice-versa, which contains a number of occupied and unoccupied states. This hybridization with the continuum leads to the inclusion of dynamic or frequency-dependent properties. The resulting equations are typically solved by Monte-Carlo methods.

Chain-DMFT was developed specifically for the treatment of Q1D materials (Arrigoni 1999, 2000, Georges et al. 2000). In this case, particles are free to "hop" along individual chains, represented by a single site coupled to a self-consistent bath. However, it is possible to include perpendicular hopping between chains, in which case we recover DMFT. Biermann et al. (2001) have used chain-DMFT to investigate the de-confinement transition. That is, the transition between Mott-insulator (MI), to LL, to FL behavior. They have successfully identified the transition between LL and FL regimes as a function of lowering temperature. Moreover, the transition between MI and FL has been simulated by increasing the effective width of the chain, that is, by introducing perpendicular hopping. Deconfinement behavior has been realized experimentally in the Bechgaard salts (Biermann et al. 2002).

A renormalization group (Fisher 1998) approach has been taken by Nickel et al. (2006) in attempting to quantify the phase diagram of Q1D materials, placing particular focus on the superconducting phase. The formation mechanisms behind, and behavior of collective excitations, such as, CDW and spin density waves (SDW) has been explored in depth.

Casula et al. has performed benchmark, lattice-regularized, diffusion Monte-Carlo simulations of a Q1D electron gas, assuming a harmonic confining potential in the radial direction (Casula et al. 2006). The study has been used to parameterize a correlation functional, and has successfully observed the formation of a WC at low densities. Results are found to be consistent with the predictions of the Luttinger model.

Yang et al. have performed state-of-the-art Bethe–Salpeter and GW calculations of the optical spectra of Si nanowires of diameter in the range 1–2 nm (Yang et al. 2007). They find strong deviations in the spectra when compared to bulk silicon that are attributed to dimensional confinement effects. Several other GW-based investigations point to similar effects (Zhao et al. 2004, Yan et al. 2007).

27.4.3 Electron Excitations and Dynamics

Almost all experimental investigations into nanowire behavior and classification rely on measuring transmission currents, conductances, and differential conductances through the sample of interest (Bockrath et al. 1999, Steinberg et al. 2006, Rahman and Sanyal 2007, Deshpande and Bockrath 2008). Although, other techniques are commonly used, such as photoluminescence emission spectroscopy (Zanolli et al. 2007), transport measurements dominate. We therefore emphasize the importance of studying the dynamical and transport properties of nanowires. Moreover, the study of nanowires plays an essential role in the future of nanoelectronics and device miniaturization.

Landauer theory (Landauer 1970) predicts quantization of the conductance $G = (2e^2/h)TK$, where the interaction parameter $K = 1$ for noninteracting systems, $K < 1$ for repulsive interactions, and $K > 1$ for attractive interactions. Also, T is the transmission probability, often assumed to have value of unity.

As we saw above, recent experiments (Thomas et al. 1996, Reilly et al. 2002, Bird and Ochiai 2004, Yoon et al. 2007, Danneau et al. 2008) have observed intermediate plateaus in the conductance, in the range $0.5–0.7 \times 2e^2/h$. Such deviations from the traditional Landauer picture have been explained by the formation of spin moments, arising due to confinement.

A recent review paper by Kawabata (2007) summarizes many of the standard features of modern transport simulations. We also highlight the work of Todorov and his group, who have performed a number of time-dependent tight-binding simulations, focusing on the effects of heating and power dissipation due to current flow in nanometric wires (Montgomery et al. 2002). Indeed, transport computations constitute a large portion of all modern literature concerning nanowires.

As we have indicated previously, our primary area of interest is the low charge density regime. In this respect, we highlight the work of Lenac (2005), who has modeled the interaction of the phonons of a 2D WC with the electromagnetic field. This work has been used to model dispersion relations for polaritions and riplons, that might be observed experimentally, for example, on the surface of liquid He (Grimes and Adams 1979).

Gold et al. have used generalized, frequency-dependent linear response computations to model the excitation spectra of CDWs and SDWs in Q1D systems (Gold and Calmels 1998). The effects of varying the background density as well as the polarization of the system have been incorporated.

27.5 The Jellium Model of Metal Wires and the Density Functional Picture

A broad view of the different phases and properties relevant for nanometric conducting wires across a wide density range can be obtained by computational investigations based on the DF theory (Hughes and Ballone 2008). The jellium model and the DF methods used in this and in the following section are closely related to general models and computational techniques widely used in condensed matter physics, (Ashcroft and Mermin 1976, Brack 1993, de Heer 1993, Martin 2004) and are briefly described here mainly for completeness and for didactical purposes.

We consider a highly idealized model of nanowires consisting of a rod-like distribution of positive charge, and of N electrons moving in the electrostatic field due to all the charges in the system. The wire segment has length L, it is globally neutral, and is oriented along the z direction (see Figure 27.7). Cylindrical coordinates (r, ϕ, z) are used throughout the chapter, together with Hartree atomic units.

The simplest model is obtained by considering the following distribution of positive charge:

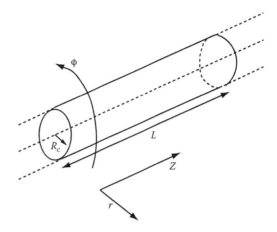

FIGURE 27.7 Schematic drawing of a cylindrical jellium nanowire.

$$\rho_+(\mathbf{r}) = \rho_+(r,z) = \begin{cases} \rho_b & \text{for } r \le R_c, 0 \le z < L \\ 0 & \text{for } r > R_c, 0 \le z < L \end{cases} \quad (27.1)$$

where R_c is the radius of the cylindrical charge density, and, in what follows, $R_c \ll L$. In other words, the positive density is constant and equal to ρ_b within the cylinder, and it vanishes outside. Following the standard practice of electron gas studies (Ashcroft and Mermin 1976), the background density, ρ_b, is measured by the radius, r_s, of the sphere that contains one unit charge, $(r_s = [3/4 \pi\rho_b]^{1/3})$.

According to the basic results of the *DF theory* (Hohenberg and Kohn 1964, Kohn and Sham 1965), the ground state energy and density of an A-electron system in an external potential, $V_{\text{ext}}(\mathbf{r})$, can be determined by minimizing a universal functional of the electron density:

$$E_{\text{KS}}\left[\{\psi_i\}; i = 1, ..., N\right]$$

$$= \sum_{i=1}^{N} \left\langle \psi_i \left| -\frac{1}{2}\nabla^2 + V_{\text{ext}}(\mathbf{r}) + \frac{1}{2}V_{\text{Ha}}(\mathbf{r}) + u_{\text{XC}}[\rho] \right| \psi_i \right\rangle \quad (27.2)$$

with respect to the N occupied Kohn–Sham (KS) orbitals $\{\psi_i\}$, that, in turn, determine the electron density according to the relation

$$\rho(\mathbf{r}) = \sum_{i=1}^{N} |\psi_i(\mathbf{r})|^2 \quad (27.3)$$

In Equation 27.2, V_{Ha} is the electrostatic potential of positive and negative charges, and therefore it includes the contribution of the *external* field due to the positive background as well as the Hartree potential due to the electron distribution. Moreover, $u_{\text{XC}}[\rho]$ is the exchange-correlation (XC) energy per electron, accounting for Fermi statistics (exchange) and other many-body effects (correlation). In our computations, the external field is represented by the Coulomb potential of the positive background, and therefore is included into V_{Ha}. Moreover, partly for the sake of simplicity, $u_{XC}[\rho]$ is given by the LSD approximation (Ceperley and Alder 1980).

Application of the Euler–Lagrange theorems to the minimization of the functional (27.2) shows that this problem is equivalent to the self-consistent solution of the Schroedinger-like equations:

$$\left\{-\frac{1}{2}\nabla^2 + V_{Ha}(\mathbf{r}) + \mu_{XC}[\rho]\right\}|\psi_i\rangle = \varepsilon_i|\psi_i\rangle \qquad (27.4)$$

for the N states of lowest eigenvalues. In the equation above, $\mu_{XC}[\rho] = u_{XC}[\rho] + \rho du_{XC}[\rho]/d\rho$ is the XC potential corresponding to the XC energy, $u_{XC}[\rho]$, of Equation 27.2.

Finite-size effects are minimized by periodically repeating the basic segment in the z direction, thus approaching the limit of a geometric wire extending to infinity along a single direction. The periodicity along z implicitly defines a 1D Brillouin zone (BZ) of width $2\pi/L$, and electron states can be labeled with a continuous wave vector, k_z, belonging to the BZ. Results do not depend on the choice of the periodicity L and of the corresponding number N of electrons in the simulation cell for systems whose density is translationally invariant along z.

Solving Equation 27.4, or minimizing directly the functional of Equation 27.2, forms the basis for two different computational approaches that are discussed in the following sections.

27.6 Axially Symmetric Solutions

At first, we assume that the electron charge distribution is rotationally and translationally invariant in ϕ and z degrees of freedom, respectively. This approximation is in the same spirit as the assumption of spherical symmetry used in the simulation of atomic systems. Moreover, as we shall demonstrate explicitly later, the cylindrical symmetry of the external potential is retained by the electron density at least for r_s in the $2 \leq r_s \leq 6$ range appropriate for the valence charge of simple metals.

Cylindrical symmetry ensures the wave function is separable and may be written as

$$\psi_{k_z}(\mathbf{r}) = \chi(r)e^{im\phi}e^{i((2\pi/L)l+k_z)z} \qquad (27.5)$$

where m and l are relative integers, and, following widely accepted conventions, k_z is selected in the interval $-(\pi/L) \leq k_z < (\pi/L)$. All the states corresponding to a given set of quantum numbers $\{nm\sigma\}$ are referred to as a sub-band.

The radial function, $\chi(r)$, is then determined by solving the differential equation

$$\frac{d^2\chi(r)}{dr^2} + \frac{1}{r}\frac{d\chi(r)}{dr} + \left[2\left(\varepsilon - V_{KS}(\mathbf{r})\right) - \left(\frac{2\pi l}{L} + k_z\right)^2 - \frac{m^2}{r^2}\right]\chi(r) = 0 \qquad (27.6)$$

subject to the appropriate boundary conditions, that for bound states read

$$\lim_{r \to 0}\chi(r) = r^m(a_0 + a_1r + a_2r^2) \qquad (27.7)$$

$$\lim_{r \to \infty}\chi(r) = \frac{\exp(-\sqrt{-\varepsilon}r)}{\sqrt{r}} \qquad (27.8)$$

These conditions can be satisfied for a discrete set of negative eigenvalues, that we indicate with $\varepsilon_{nm}(k_z + (2\pi l/L))$, where n is a positive integer analogous to the principal quantum number of atoms. The corresponding radial functions, $\chi(r)$, depend on the n, m quantum numbers, while they are independent of l, k_z, and in what follows they will be denoted by $\chi_{nm}(r)$. In addition to these bound states, the system has a continuum of scattering states whose eigenvalues are positive. The ground state density is given by

$$\rho(r) = \sum_{nml}\int_{BZ} f\left[\varepsilon_{nm}\left(k_z + \frac{2\pi l}{L}\right)\right]|\chi_{nm}(r)|^2 \, dk_z \qquad (27.9)$$

where $f[\varepsilon]$ is the occupation number. In the case of a cylindrical ground state, the z-momentum dependence of $\varepsilon_{nm}(k_z)$ is given by

$$\varepsilon_{nm}(k_z) = \varepsilon_{nm}(k_z = 0) + \frac{k_z^2}{2} \qquad (27.10)$$

and the integration over k_z can be performed analytically (Zabala et al. 1998). States are occupied up to an energy ε_F such that the number of states whose energy is less than ε_F is equal to the number of electrons in the system. The total number of distinct $\{nm\}$ combinations found for the occupied states is by definition the number of occupied sub-bands for the wire under investigation. As detailed below, this number is of the order of 10 for the sizes investigated in our study.

In our computation, a radial grid of 2000 points has been used, extending up to $R_{max} = 4R_c$ in the case of high density samples ($r_s < 10$), and up to $R_{max} = 2R_c$ for low-density systems. Each orbitals is integrated outward from $r = 0$, and inward from R_{max} using a predictor-corrector method. The corresponding eigenvalue is determined by the matching of the two solutions at R_c. Angular momenta up to $m = 20$ have been considered, and degenerate levels are equally populated at all stages of the calculation. A similar computational procedure is adopted for the determination of the Coulomb potential from the charge density.

Apart from the k_z label, and apart from obvious differences in the radial equations, the approach is completely analogous to methods routinely used to compute the electronic structure of atoms with the restriction to spherical symmetry, and high accuracy solutions can be obtained relatively easily with a limited computational effort. The method outlined above has been used several times to determine the ground state properties of jellium wires whose density has cylindrical symmetry (Stafford et al. 1997, Yannouleas and Landman 1997, Yannouleas et al. 1998).

27.6.1 Computations in the Radial-Cylindrical Approach

The results obtained by the radial code under the restriction of cylindrical symmetry are summarized in Figure 27.8. Our data agree with those of previous computations (Stafford et al. 1997, Zabala et al. 1998) whenever a comparison is possible.

As pointed out in Zabala et al. (1998), the enhancement of the exchange interaction due to confinement, together with the degeneracies of cylindrical wave functions in very thin wires, may gives rise to a progressive filling of sub-bands reminiscent of Hund's rule in atoms. In turn, this implies that partial polarization may arise at densities as high as those of bulk sodium ($r_s = 4$). The integral of the spin polarization density, $\rho_s(\mathbf{r}) = \rho_+(\mathbf{r}) - \rho_-(\mathbf{r})$, is small, amounting to a few electrons at most, and, therefore, the relative spin polarization is fairly low as soon as the wire radius exceeds monoatomic thickness. Moreover, as expected, $\rho_s(\mathbf{r}) = \rho_+(\mathbf{r}) - \rho_-(\mathbf{r})$ always peaks at the background edge in high density samples, and spin polarization might indeed represent the precursor of spin-polarized states quasi-localized at the surface of the semi-infinite jellium.

Extensive spin polarization in the ground state appears only at much lower density, far below those found in elemental metals, but achievable in artificial conductors obtained by doping semiconducting structures or polymeric chains. A plot of the spin polarization energy $\Delta E_S(\zeta) = E_{tot}(\zeta) - E_{tot}(0)$ for the wire and for the homogeneous electron gas shows that at low density and for the relatively thick wires considered in our study the stabilization of the spin polarization brought about by confinement and by inhomogeneity is fairly modest (see Figure 27.9). Nevertheless, computations in the cylindrical symmetry approximation show that the transition to a (partially) spin-polarized ground state

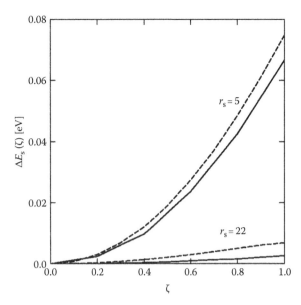

FIGURE 27.9 Spin-polarization energy $\Delta E_s(\zeta) = E_{tot}(\zeta) - E_{tot}(0)$ for the wire (full line) and for the homogeneous electron gas (dash line). Computations have been performed for a finite system with $N = 240$ electrons under the assumption of a cylindrically symmetric charge density. (From Hughes, D. and Ballone, P., *Phys. Rev. B*, 77, 245313, 2008. With permission.)

takes place at $r_s = 27$. At variance from what has been found in high-density wires, the spin polarization in low-density, low (nominal $\zeta \sim 0$) spin polarization samples tends to be localized in the central region of the wire, with only negligible contributions from regions beyond the background radius.

27.7 Iterative Minimization of the LSD Density Functional

The restriction to cylindrical symmetry for the electron density has to be abandoned in order to describe broken symmetry solutions such as the WC, expected to arise at low ρ_b densities, and also in the case of wires subject to an external perturbation not conserving the original symmetry. In those cases, a general solution may be obtained by expanding KS orbitals in plane waves:

$$\psi_k^{(i)}(\mathbf{r}) = \sum_G c_G^{(i)} e^{i\mathbf{G}\mathbf{r}} e^{i\mathbf{k}\mathbf{r}} \qquad (27.11)$$

In doing so we assume that the system is periodic in 3D, and we impose a fictitious periodicity of length (L_x, L_y) in the plane perpendicular to the z axis. Among other things, this periodicity implies that the label $\mathbf{k}$ is now a 3D vector. The plane wave expansion of Equation 27.11 is limited to those $\mathbf{G}$ vectors such that $|\mathbf{G} + \mathbf{k}|^2$ is less than a preselected cutoff E_{cut} which has the dimensions of an energy, and in what follows is measured in Rydbergs.

The functional $E_{KS}[\{\psi_i\}]$ is in fact a function of the multitude of $\{c_{G,k}^{(i)}\}$ coefficients, and it can be optimized by iterative minimization, using the information provided by the gradient:

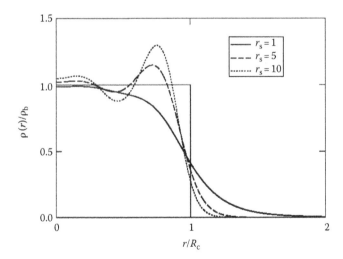

FIGURE 27.8 Electron density, $\rho(r)$, as a function of the radial coordinate, r, at three values of the electron gas parameter, r_s, computed using the 1D algorithm under the assumption of cylindrical symmetry. The positive density distribution has radius $R_c = 3.162 r_s$, and the ground state is paramagnetic for the three cases displayed here. (From Hughes, D. and Ballone, P., *Phys. Rev. B*, 77, 245312, 2008. With permission.)

$$\frac{\partial E_{KS}}{\partial c_{G,k}^{(i)}} = \sum_k \int \frac{\delta E_{KS}}{\delta \psi_k^{(i)}(\mathbf{r})} \frac{d\psi_k^{(i)}(\mathbf{r})}{dc_G^{(i)}} \, d\mathbf{r} = \sum_k \int \frac{\delta E_{KS}}{\delta \psi_k^{(i)}(\mathbf{r})} e^{i\mathbf{Gr}} \, d\mathbf{r}$$

$$= \sum_k \int \left\{ -\frac{1}{2}\nabla^2 + V_{Ha}(\mathbf{r}) + \mu_{XC}[\rho] \right\} \psi_k^{(i)}(\mathbf{r}) e^{i\mathbf{Gr}} \, d\mathbf{r} \quad (27.12)$$

The sum over **k** points is a discretized version of the integral over the BZ implied by the 3D periodicity.

The actual optimization may be achieved by a variety of methods (Marx and Hutter 2000), including steepest descent, conjugate gradient, etc. In all cases, the energy minimization is performed at fixed spin polarization $\zeta = (n_+ - n_-)/(n_+ + n_-)$, where n_+ and n_- are the spin-up and spin-down charges, respectively. The sum of n_+ and n_- is equal to N, although neither of the two partial charges needs to be an integer if the system is a metal.

As described below, for low r_s (high density) and up to at least $r_s = 20$ the ground state density turns out to be cylindrically symmetric, and the cutoff energy for the 3D-plane wave computation is tuned by the comparison with the results of the radial program using full cylindrical symmetry. Total energies computed by the two methods show that convergence in the plane wave expansion is already achieved at a cutoff energy of $E_{cut}(r_s) \sim 10/r_s^2$ Ry. The localized states found at low density ($r_s \geq 30$) are more difficult to represent in plane waves, and explicit tests have shown that a cutoff energy of $E_{cut} = 20/r_s^2$ Ry is required for a uniform convergence of ground state properties over the full density range $1 \leq r_s \leq 100$ explored in our study. Despite the relatively low cutoff, the large cell size implies that the number of variational degrees of freedom is of the order of 5×10^4 per state. Taking into account the high number of states included in our computations, it is not surprising to find that the minimization of the energy functional turns out to be relatively time consuming.

We emphasize that while the nominal (or net) spin polarization, $\zeta = (N_+ - N_-)/(N_+ + N_-)$, is an input parameter of our computations, the spin density, $\rho_s(\mathbf{r}) = \rho_+(\mathbf{r}) - \rho_-(\mathbf{r})$, is fully unconstrained both in the radial cylindrical and in the plane wave approach, and $I_s = \int |\rho_s(\mathbf{r})| d\mathbf{r}$ can be significantly larger than $|N_+ - N_-|$. In particular, at low background density ($r_s > 30$) local spin polarization is found also in nominally paramagnetic samples at $\zeta = 0$.

27.8 Broken-Symmetry Solutions

Computations have been performed for a series of wires of length $L = 32r_s$ periodically repeated along z, consisting of $N = 240$ electrons moving in the electrostatic potential of a cylindrical distribution of positive charge whose plasma parameter r_s spans the range $1 \leq r_s \leq 100$. At $r_s = 1$, therefore, the radius of the positive charge is $R_c = 3.162$ a.u., or 1.673 Å, and the wire segment explicitly included in the computation is 16.93 Å long. The radius reaches 167.3 Å at the lower density range ($r_s = 100$), and in such a case the periodicity along z is 1693 Å or 0.1693 μm. All wires have the same aspect ratio, $L/R_c = 10.12$.

Plane wave computations have been performed using a cubic simulation box of side L, and, therefore, the background density occupies ~3% of the simulation box.

Because of the fairly large size of our samples, and considering also their low average density, the sampling of the system BZ in the plane wave computations has been restricted to the Γ-point only. The role of k_z points is decreased by the fact that the systems we are primarily interested in (i.e., the low-electron-density wires) are in fact insulators.

The full range of spin polarizations, $0 \leq \zeta \leq 1$, has been explored by varying the relative number of spin-up (N_+) and spin-down (N_-) electrons, starting from the paramagnetic case ($N_+ = N_- = 120$), and progressively increasing (decreasing) $N_+(N_-)$ in steps of 10 electrons up to $N_+ = 240$ ($N_- = 0$). We use here N_+ and N_- instead of n_+ and n_- to indicate that the net spin imbalance, $N_{spin} = N_+ - N_-$, is restricted to integer values.

27.8.1 Plane Wave Computations

Electron localization is fully accounted for by the 3D-plane wave computations, whose results display the same magnetic transition at nearly the same density of the radial-cylindrical computations. More precisely, up to $r_s = 30$ the solutions found by the plane wave code also display translational invariance along z at all ζ, and, apart from occasional interchanges of nearly degenerate states, the corresponding sequence of KS orbitals agrees with that of the radial computations for full cylindrical symmetry. As anticipated in Section 27.5, the convergence of the plane wave expansion is confirmed by the good agreement of the total energy obtained by the two computational approaches.

Sizable differences between the solutions of the radial scheme and those of the plane wave computation first appear at $r_s = 30$. Inspection of the electron density found by the unconstrained plane wave minimization (see Figure 27.10) reveals that an apparent localization transition involving all coordinates (i.e., now including z) takes place in the samples of highest spin polarization ($\zeta \geq 0.5$). Localization can be described as *partial* because the overlap of different electron-density peaks is significant, and the density at local minima is still a sizable fraction of ρ_b (see the inset in Figure 27.10). We verified that the density modulation along z remains nearly unchanged when the sampling of the BZ is extended to more k_z points. Further analysis described below suggests that the electron configuration consists of a majority of delocalized states similar to those found at high density, coexisting with localized states whose energy is at the bottom and at the top of the occupied (valence) band.

We remark that up to $r_s = 40$ localization in the radial direction, giving rise to a sequence of well-defined electron-density shells, is more marked than localization within each of the radial shells. Moreover, localization is stronger in the inner region of the wire, and somewhat attenuated in the outer electron shell, as apparent in Figure 27.10.

Comparison of the different energy contributions for the z-invariant and for the localized states show that, as expected, localization is driven by a gain in correlation energy, only partly

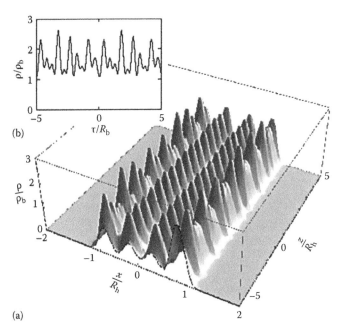

(b)

(a)

FIGURE 27.10 Electron density of a fully spin-polarized jellium wire obtained by the plane wave method. $N = 240$, $r_s = 30$. Panel (a): 2D plot of the density $\rho(r, z)$ on the axial plane $\phi = 0$. Panel (b): 1D plot of the electron density, $\rho(z)$, along the line parallel to the cylindrical axis of coordinates: $x - a = y = 0$, where $a = 2.25 \times r_s = 0.71R_c$. (From Hughes, D. and Ballone, P., *Phys. Rev. B*, 77, 245314, 2008. With permission.)

compensated by the kinetic energy term, which increases upon localization. This energy balance, in turn, explains why localization takes place at first in spin-polarized samples, since the kinetic energy (Ashcroft and Mermin 1976) of ferromagnetic states is higher than that of the paramagnetic state, while their correlation energy is lower. On the other hand, the kinetic energy of the WC is nearly independent of spin, and thus, the kinetic energy cost of localized states is less relevant for the ferromagnetic configuration, while the potential gain in correlation energy is comparatively larger. Both energy terms, therefore, point to high spin configurations as the first candidates for localization.

Despite the energy gain provided by localization, at $r_s = 30$ the ferromagnetic configuration is still slightly higher in energy than the paramagnetic, z-invariant state. The energy difference between the two, however, is very small, and, in fact, the ground state energy is almost constant over the entire $0 \leq \zeta \leq 1$ range, suggesting that in the vicinity of the localization transition spin glass features might arise from the near degeneracy of several different spin states. This same near degeneracy with respect to changes of ζ makes it difficult to provide an accurate determination of the net ground state spin polarization at densities close to the transition point (see the inset in Figure 27.11). Nevertheless, plane wave computations confirm the stability of partially polarized stated at r_s slightly higher than 30, as already anticipated by the radial-cylindrical computations. Of course, these observations imply that the transition to the spin-polarized state is at most weakly first order.

The fully ferromagnetic state becomes the state of lowest energy for $r_s \geq 35$, the shift in the transition point from $r_s = 27$ estimated in the radial-cylindrical method being due to the discretization

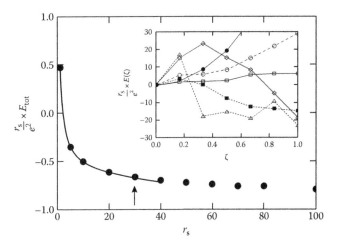

FIGURE 27.11 Total energy per electron of jellium wires computed by the radial-cylindrical method (full line) and by the 3D-plane wave method (full dots). At each r_s, the energy of the lowest energy spin configuration is reported. The arrow marks the transition between paramagnetic and partially spin-polarized configurations. The polarization energy per electron, $\Delta E_s(\zeta)$, at six background densities is shown in the inset. Full dots: $r_s = 1$; circles: $r_s = 20$; squares: $r_s = 30$; filled squares: $r_s = 40$; diamonds: $r_s = 70$; triangles: $r_s = 100$. (From Hughes, D. and Ballone, P., *Phys. Rev. B*, 77, 245315, 2008. With permission.)

of the DOS, and to the finite plane wave expansion of orbitals and electron density, which affects spin-polarized systems slightly more than the paramagnetic ones. In the low-density regime at $r_s \geq 40$, localization is apparent in all systems, irrespective of spin polarization, and becomes progressively more marked with increasing r_s. Plots of constant electron-density surfaces (isosurfaces, in what follows) provide a direct and intuitive view of

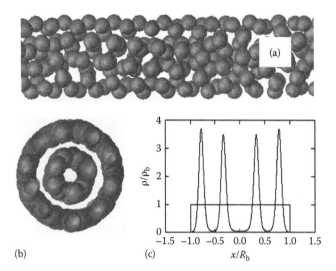

FIGURE 27.12 Electron-density contour plot for the ferromagnetic ground state of the $r_s = 70$, $N = 240$ wire. Panel (a): side view. Panel (b): transversal view. Panel (c): radial density profile obtained upon averaging the 3D density over ϕ and z. (From Hughes, D. and Ballone, P., *Phys. Rev. B*, 77, 245316, 2008. With permission.)

the localization extent in low-density systems, as apparent from Figure 27.12) showing the $\rho = 1.6\rho_b$ isosurface for the ferromagnetic ground state at $r_s = 70$. Shell effects are apparent from the transversal view of the density distribution (Figure 27.12b and c), and might be seen as the oversized and frozen-in version of the charge (Friedel) oscillations already present in the high density liquid phase (see Figure 27.8). The perspective view of the same isosurface (Figure 27.12a) clearly shows that the system consists of an assembly of well-defined charge droplets. In what follows, these droplets will sometimes be referred to as *charge blobs*, to account for their somewhat irregular shape.

The gradual organization of charge into shells with increasing r_s followed by the breakdown of shells into one-electron droplets is qualitatively similar to the two-stage freezing (or, equivalently, melting) observed in 2D circular quantum dots (Filinov et al. 2001), whose radial and orientational order are set in at different densities and/or temperatures. Only for sufficiently large systems, the two localization processes merge into a unique freezing transition.

The fairly regular pattern displayed by the droplets distribution in low-density samples suggests that a geometrical lattice, possibly closely related to the bcc structure of the extended WC, might underlay the ground state charge configuration. To identify this ideal geometry, the continuous density distribution provided by LSD is mapped onto a particles configuration by (1) first identifying connected regions whose density is higher than $\rho_{cut} = 2\rho_b$, and then (2) associating one particle to each of these domains, and locating it at the center of mass of the corresponding charge distribution. This procedure provides a fully unambiguous result only for systems of fairly high r_s ($r_s > 50$). For these low-density systems, the number of connected regions (and thus the number of associated particles) is always very close to the number N of electrons in the system, the difference

being at most a few units in all samples at $r_s \geq 70$, thus lending a reality flavor to the representation of the electron density by particles. The configuration obtained in this way closely resembles the low-temperature structure of classical particle models such as the one-component plasma (OCP), as obtained by slowly annealing liquid samples. In this respect, it is interesting to note that the radial distribution function, $g(r)$, of the representing particles belonging to the inner radial shells of the computed structures displays the same characteristic features found in the glassy state of the classical OCP (Tanaka and Ichimaru 1987), consisting in an asymmetric first peak and a split-second peak (see Figure 27.13).

The radial distribution function of particles representing charge blobs depends only weakly on spin polarization (see below) and on density for $r_s \geq 70$, apart from a trivial scaling of all distances. The dependence on sample size is also very weak up to the second peak of $g(r)$, while it becomes important at larger distances. These results are reflected in the weak density and size dependence of the running coordination number, $n_c(r)$, defined as

$$n_c(r) = 4\pi\rho_b \int_0^r r'^2 g(r')\,dr' \tag{27.13}$$

and displayed in the inset of Figure 27.13.

Despite the unambiguous mapping of charge blobs into particles, the identification of the ideal structure underlying the ground state charge distribution of low-density wires is made

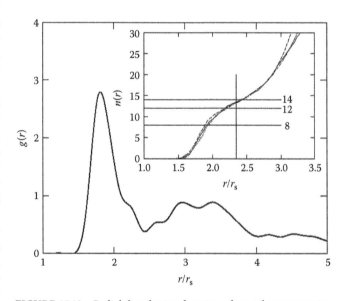

FIGURE 27.13 Radial distribution function of particles representing charge blobs (see text) for a wire of $N = 240$ electrons at $r_s = 70$ and $\zeta = 1$. Inset: running coordination number, $n_c(r)$, of particles representing charge blobs. Full line: $N = 240$ electrons, $r_s = 70$, $\zeta = 1$; dash line: $N = 480$ electrons, $r_s = 70$, $\zeta = 1$; dotted line: $N = 240$ electrons, $r_s = 100$, $\zeta = 1$. The horizontal lines correspond to full shells of neighbors in the bcc ($n_c = 8$ and $n_c = 14$) and in the fcc ($n_c = 12$) lattice. The vertical line corresponds to the minimum of $g(r)$, and defines the cutoff radius for the computation of the average coordination number. (From Hughes, D. and Ballone, P., *Phys. Rev. B*, 77, 245317, 2008. With permission.)

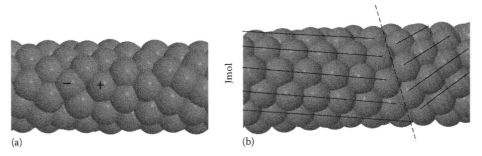

FIGURE 27.14 Defective configurations in the ground state distribution of charge droplets of fully spin-polarized wires at $r_s = 70$: (a) pair (+, −) of miscoordinated droplets in the middle shell of the $N = 480$ sample; (b) grain boundary on the inner shell of the $N = 240$ sample. (From Hughes, D. and Ballone, P., *Phys. Rev. B*, 77, 245318, 2008. With permission.)

difficult by the unavoidable distortions imposed by the finite sample size, and by the likely mismatch of the optimal lattice parameter with the other length scales entering the definition of our model, such as the background radius and the wire length. However, the major difficulty in characterizing the particles' geometry arises from a variety of point and extended defects such as dislocations and grain boundaries that are distributed in the structures produced by our energy optimization (see Figure 27.14) both for the $N = 480$ (Figure 27.14a) and for the $N = 240$ systems (Figure 27.14b). These defects are likely to result, at least to some extent, from limitations of our computational scheme, unable to reach the absolute minimum of the DF energy within an acceptable number of iterations. More importantly, the disorder in the electron droplets distribution certainly reflect a real and relevant property of these low energy systems, having a vast number of similar but different configurations of nearly equal energy. In turn, the positional disorder frozen into low energy configurations is likely to affect the properties of electron-density systems measured in experiments, and to give rise to glass-like features in the thermodynamics and real-time dynamics of low-density wires.

Despite the unavoidable uncertainties due to the intrinsic disorder of the structures resulting from our computations, information on the underlying ground state structure can be obtained from average quantities such as the radial distribution function, $g(r)$, and the coordination number, n_c, defined as the value of the running coordination number up to a distance corresponding to the first minimum of the radial distribution function ($r_{min} = 2.3 r_s$ for wires of $r_s \geq 50$). The similarity of the particles' $g(r)$ with those of the OCP already pointed out above clearly suggests a close relation with a bcc lattice. The coordination number, n_c, of the charge blobs residing in the inner shell of the computed structures is close to but nevertheless systematically lower than the $n_c = 14$ value that corresponds to the number of first and second neighbors in the bcc structure (see Figure 27.13). However, the absence of the shell closing at $n_c = 8$ also expected for bcc, once again prevents a fully unambiguous identification. A detailed analysis of the structures found for the $N = 240$ and $N = 480$ samples however suggest that it might be more appropriate to characterize the computed geometries as being intermediate between an fcc and a bcc lattice.

At the highest r_s's considered in our study ($r_s \geq 70$), the separation of the density peaks is so marked that it is possible and even easy to identify the spatial domain occupied by each blob. This allows us to verify that not only the number of droplets corresponds to the number N of electrons, but, in addition, the integral of the charge density for each blob is very close to one, the standard deviation amounting to only 5%. At low density, therefore, blobs can be identified with electrons, even though they should not be identified with KS orbitals, as it will be discussed later. The remarkable correspondence of density blobs and electrons arises from well-known anomalies in the response functions, that in reciprocal space identify the BZ of the WC, and in real space delimit the lattice unit cell, thus determining the size, charge, and spin of the basic building block.

Individual charge blobs always display partial (at $r_s < 70$) or full ($r_s \geq 70$) spin polarization for all systems in which localization is apparent, irrespective of the average polarization ζ and including nominally paramagnetic samples. Needless to say, this implies that the electronic structure of low-ζ systems includes a spin-compensating mechanism, such as antiferromagnetic ordering or a more general spin wave, bringing the net spin to the value imposed by the N_+ and N_- values. This is apparent in Figure 27.15, displaying the spin polarization of charge blobs for the wire $r_s = 70$, $\zeta = 0$ wire. The spin configuration of the outermost shell is fairly disordered, while the inner shell displays a regular helical pattern (not very clear in Figure 27.15, as in any 2D representation, but apparent in computer visualizations

FIGURE 27.15 Spin polarization of charge droplets for the $r_s = 70$, $\zeta = 0$ wire. Light droplets: spin up electrons. Dark droplets: spin down electrons. The inner (a) and outer (b) shells are shown separately. (From Hughes, D. and Ballone, P., *Phys. Rev. B*, 77, 245319, 2008. With permission.)

that allow one to rotate the isosurface) probably related to the enhancement of the spin-spin response function at $k_z = 2K_F$ and $k_z = 4K_F$ (Pouget and Ravy 1997).

The disordered spin distribution of the outer shell suggests that the spin–spin coupling constant is fairly small, as confirmed by the computation of the spin-resolved radial distribution functions $g_{++}(r)$ and $g_{+-}(r)$ that show only a slight predominance of antiferromagnetic coupling in the first coordination shell. This could be seen as the expected consequence of a nearly disjoint charge and spin blobs, reducing also the exchange interaction. It is important to realize, however, that localization concerns the density, not necessarily the KS orbitals. We verified, in fact, that even in samples displaying the most apparent charge localization, each KS orbital contributes to the density of several, widely spaced blobs, as can be seen in Figure 27.16. The qualitative information contained in this figure is confirmed by a quantitative measure of localization provided by the computation of the inverse participation ratio, that, apart from a few cases, points to a remarkably low localization for KS orbitals, as discussed below.

Charge and spin localization are nevertheless clearly reflected into basic properties of the KS orbitals, affecting, for instance, the system DOS. As can be seen in Figure 27.17, the density of states for the $r_s = 30$, $\zeta = 1$ sample shows a deep minimum (pseudogap) at the Fermi energy, pointing to an incipient metal–insulator

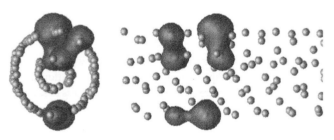

FIGURE 27.16 Density isosurface, $\rho_l = 0.05\,\rho_b$, for a KS orbital whose eigenvalue is close to the top of the occupied band. $r_s = 70$, $N = 240$, fully spin-polarized ground state. Solid particles mark the center-of-mass position of individual charge blobs. (From Hughes, D. and Ballone, P., *Phys. Rev. B*, 77, 245320, 2008. With permission.)

transition driven by localization. The $r_s = 40$, $\zeta = 1$ sample is clearly an insulator, and at $r_s = 70$, $\zeta = 1$ the energy gap separating occupied and unoccupied states is as wide as the total width of the occupied bands.

The localization of individual orbitals is measured by computing the inverse participation ratio, defined as follows. First of all, we compute the z-dependent planar average of each orbital, defined as

$$\psi_i^{\text{plane}}(z) = \frac{\int \psi_i(x,y,z)\,dx\,dy}{\pi R_c^2} \qquad (27.14)$$

then, the inverse participation ratio, p_i^{-1}, is computed according to

$$p_i^{-1} = L\,\frac{\int \left|\psi_i^{\text{plane}}(z)\right|^4 dz}{\left[\int \left|\psi_i^{\text{plane}}(z)\right|^2 dz\right]^2} \qquad (27.15)$$

The definition implies that localized states correspond to $p^{-1} \sim L$, and delocalized states to $p^{-1} \sim 1$. The results, reported in Figure 27.17, show that the central and major portion of the occupied band is made of fairly delocalized states, while the most localized orbitals are found at the low- and high-energy band edges. While the localization of these states could have been expected, the relative delocalization of all the other states is more surprising, given the apparent strong localization of the charge.

The different behavior of the density and of KS orbitals with respect to localization might be related to the invariance of LSD with respect to unitary transformation of the occupied states, that, by definition, leave the electron density and the kinetic energy unchanged. All the effects described in this section are likely to affect the transport properties of nearly 1D, low-carrier density conductors.

As a last observation, we would like to mention that at the lowest densities explored by our computations ($80 \le r_s \le 100$), the plane wave energy optimization gives rise to surprising new structures, especially for the low-spin samples, as shown in Figure 27.18 for the $N = 240$ wire of $r_s = 100$, $\zeta = 1$.

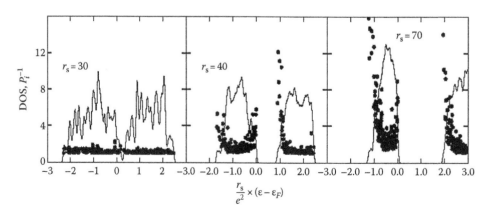

FIGURE 27.17 Density of states and inverse participation ratio, p_i^{-1}, for the fully spin-polarized wires at $r_s = 30$, 40, and 70, $N = 240$. (From Hughes, D. and Ballone, P., *Phys. Rev. B*, 77, 245321, 2008. With permission.)

FIGURE 27.18 Density isosurface, $\rho = 2\rho_b$, for the sample of $N = 240$ electrons, $r_s = 100$, $\zeta = 0$. (From Hughes, D. and Ballone, P., *Phys. Rev. B*, 77, 245322, 2008. With permission.)

FIGURE 27.20 Magnetization density close to a nanoconstriction, $m(\mathbf{r}) = 0.20\,\rho_b$ (darker surface), $m(\mathbf{r}) = -0.20\,\rho_b$ (lighter surface), background density $r_s = 25$, nominally paramagnetic sample of $N = 230$ electrons.

27.9 Nano-Constriction: Preliminary DF Results

The experiments of Bird et al. (Bird and Ochiai 2004, Yoon et al. 2007) point toward exotic magnetic behavior in QPC. By introducing a neck, or narrowing into the jellium background, we have been able to model such systems (Cortes-Huerto et al. 2009).

Once again, the background density is a constant within the wire and zero outside. The form of the neck is as follows:

$$\rho_N(\mathbf{r}) = \begin{cases} \rho_b & \text{if } \gamma(z) \leq r \leq R_b \text{ and } \left|z - z_0\right| \leq D \\ 0 & \text{otherwise} \end{cases} \quad (27.16)$$

where $D = 0.336\,R_b$. Moreover, $\gamma(z) = r_0 + \alpha(z - z_0)^2$, with r_0 being the minimum radius at the constriction, and $\alpha = (R_b - r_0)/D^2$. A plot of the corresponding background density is shown in Figure 27.19.

The corresponding Fourier transform is

$$\rho_N(\mathbf{G}) = \frac{2\pi\rho_b}{V} \int\limits_{z_0-D}^{z_0+D} dz \cos(G_z z) I(G_r, z) \quad (27.17)$$

where

$$I(G_r, z) = \begin{cases} \dfrac{1}{2}\left[R_b^2 - \gamma(z)^2\right] & \text{if } G_r = 0 \\ \dfrac{1}{G_r}\left[R_b J_1(G_r R_b) - \gamma(z) J_1(G_r \gamma(z))\right] & \text{otherwise} \end{cases}$$
$$\quad (27.18)$$

and $G_r^2 = G_x^2 + G_y^2$.

We shall not describe in detail the results of these computations that are still partially preliminary. However, we include a plot of the magnetization isosurfaces, where the position-dependent magnetization is defined by $m(\mathbf{r}) = n_\uparrow(\mathbf{r}) - n_\downarrow(\mathbf{r})$, in an $r_s = 20$ system (Figure 27.20). Our findings reveal the formation of complex magnetic structures, such as rings close to the

constriction. Work is in progress to assess the effect of these modulations of the spin density on properties such as DC and AC conductivity.

27.10 Discussion and Conclusive Remarks

Experimental and technological advances have raised great expectations on the potential applications of systems of nanometric size and on nanostructured materials. Among the systems of interest in this domain, nanowires represent a particularly important case, since every functional device will require the electrical connectivity of a vast amount of more elementary and equally nanometric components. Moreover, at the level of complexity achievable at this length scale, a network of conducting wires may acquire new functionalities (Yanushkevich and Steinbach 2008), going far beyond the passive role of traditional circuits.

Besides being important from the technological point of view, the conductivity of nanowires contains many features of interests from a more fundamental perspective. First of all, experimental measures for thin wires show that conductivity is quantized at values $2e^2/\hbar$ (van Wees et al. 1988). This behavior is explained by standard scattering theory (Landauer 1970), but violations to this rule have been observed (Bird and Ochiai 2004, Yoon et al. 2007), pointing to spontaneously spin-polarized electronic configurations in otherwise fully paramagnetic systems (Thomas et al. 1996). More generally, the Tomonaga–Luttinger (Tomonaga 1950, Luttinger 1963) theory predicts that the normal FL theory familiar from 3D metals (Giuliani and Vignale 2005) does not apply to conductors all the way down to atomically thin wires, and has to be replaced by a different picture in which the dynamics of charge and spin degrees of freedom becomes progressively more decoupled (Fiete 2007), opening the way to a wealth of new phenomena. Finally, and perhaps more importantly for this chapter, a metal–nonmetal transition is predicted to take place with decreasing density of the charge carriers, due to the localization of electrons and/or holes under the effect of their mutual Coulomb interactions (Schulz 1993). This transition, conceptually similar to the well-known Wigner transition predicted long ago for the 3D homogeneous electron gas (Wigner 1934), is comparably more relevant in the case of Q1D conductors, since confinement and low dimensionality enhance the role of electron correlation, and amplify the effect of localization, which can completely block conduction even when it takes place in a finite portion of the entire system.

FIGURE 27.19 Electron density for a wire with a narrow geometric constriction.

In this chapter, we have provided a brief description of the phases believed to occur in 1D and Q1D systems. These include the "normal" regime, believed to be LL, which is contrasted to the regular FL; a magnetically ordered phase; a low-density, crystalline regime; and finally, other even more exotic phases identified by dynamical or excited-state properties. Moreover, we have summarized a number of theoretical studies, ranging from the use of simple idealized models to state-of-the-art correlated methods.

A short summary of the experimental methods that have been used to fabricate wires of nanometric diameter over a wide carrier density range has been provided in Section 27.3, The summary aims at emphasizing that the low-density nanometric wires that are the subject of this chapter are not just theoretical or computational abstractions.

In the second half of our review we introduce an idealized (jellium) model of metal wire, and we investigate the localization transition by a conceptually simple and computational tractable approach based on density functional (DF) theory (Hohenberg and Kohn 1964, Kohn and Sham 1965). The LSD exchange-correlation approximation we use has been specifically devised for electron systems in a low correlation regime. Therefore, the quantitative predictions of the method for the highly correlated states found in low-density nanowires have to be taken with some caution. However, the method and the results it provides have a great didactical value, since they describe the system and its localization transition in simple and intuitive terms, especially taking advantage of the independent-electron picture underlying DF.

The computations described in Sections 27.5 through 27.9 offer a wealth of information on the ground state structure of the broken-symmetry phase. At sufficiently low density ($r_s \geq 40$), electron localization is complete and the ground state is a collection of charge droplets, each corresponding to one electron and 1/2 Bohr magneton. The broken-symmetry state is an insulator, and therefore the transition could be detected by measurements of the low-frequency electric conductivity. Other general spectroscopic properties such as the Raman spectrum might also be significantly affected by the transition that is also likely to change the frequency, strength, and dispersion relation of plasmon excitations.

The distribution of the charge droplets is fairly regular, and defines a lattice whose structure is intermediate between bcc and fcc. Nevertheless, several defects are distributed in the lowest energy structures found by our numerical minimization. Disorder is apparent also in the spin distribution, which shows only a weak preference for antiferromagnetic coupling. Either positional and spin disorder might give rise to nonlinear features in the conductivity and in the Raman excitation spectrum of low-density wires.

Both spontaneous spin polarization and electron localization found in our computations have a counterpart in the results of recent experiments (Rahman and Sanyal 2007, Deshpande and Bockrath 2008, Danneau et al. 2008). Noncollinear spin states, excluded by our simple LSD scheme, might instead appear as broken-symmetry solutions both in real systems and in more sophisticated determinations of the nanowire electronic structure. In this respect, the complex density distribution found by our energy minimization at $r_s = 100$, $\zeta = 0$ provides clear evidence that intriguing surprises may still be expected even from the simplest jellium model of nanowires.

References

E. Arrigoni (1999). Crossover from Luttinger- to Fermi-liquid behavior in strongly anisotropic systems in large dimensions. *Phys. Rev. Lett.* **83**:128.

E. Arrigoni (2000). Crossover to Fermi-liquid behavior for weakly coupled Luttinger liquids in the anisotropic large-dimension limit. *Phys. Rev. B* **61**:7909.

N. W. Ashcroft and N. D. Mermin (1976). *Solid State Physics.* Holt-Saunders, London, U.K.

O. M. Auslaender et al. (2002). Tunneling spectroscopy of the elementary excitations in a one-dimensional wire. *Science* **295**:825.

O. M. Auslaender et al. (2005). Spin-charge separation and localization in one dimension. *Science* **308**:88.

J. M. Baik et al. (2008). Nanostructure-dependent metal-insulator transitions in vanadium-oxide nanowires. *Phys. Chem. C Lett.* **112**:13328.

R. Beckman et al. (2005). Bridging dimensions: Demultiplexing ultrahigh-density nanowire circuits. *Science* **310**:465.

S. Biermann et al. (2001). Deconfinement transition and Luttinger liquid to Fermi liquid cross over in quasi-one-dimensional systems. *Phys. Rev. Lett.* **87**:276405.

S. Biermann et al. (2002). Quasi one-dimensional organic conductors: Dimensional crossover and some puzzles. In: *Strongly Correlated Fermions and Bosons in Low-Dimensional Disordered Systems.* Kluwer Academic Publishers, Dordrecht, the Netherlands.

I. M. L. Billas et al. (1993). Magnetic moments of iron clusters with 25 to 700 atoms and their dependence on temperature. *Phys. Rev. Lett.* **71**:4067.

J. P. Bird and Y. Ochiai (2004). Electron spin polarizations in nanoscale constrictions. *Nature* **303**:1621.

M. Bockrath et al. (1999). Luttinger-liquid behaviour in carbon nanotubes. *Nature* **397**:598.

M. Brack (1993). The physics of simple metal clusters: Self-consistent jellium model and semiclassical approaches. *Rev. Mod. Phys.* **65**:677.

J. Cao et al. (2005). Electron transport in very clean, As-grown suspended carbon nanotubes. *Nat. Mater.* **4**:745.

M. Casula et al. (2006). Ground state properties of the one-dimensional Coulomb gas using the lattice regularized diffusion Monte Carlo method. *Phys. Rev. B* **74**:245427.

D. M. Ceperley and B. J. Alder (1980). Ground state of the electron gas by a stochastic method. *Phys. Rev. Lett.* **45**:566.

J. C. Charlier et al. (2007). Electronic and transport properties of nanotubes. *Rev. Mod. Phys.* **79**:677.

R. Cortes-Huerto and P. Ballone (2009). *Phys. Rev. B*, in press.

Y. Cui and M. Lieber (2001). Functional nanoscale electronic devices assembled using silicon nanowire building blocks. *Science* **291**:851.

R. Danneau et al. (2008). 0.7 structure and zero bias anomaly in ballistic hole quantum wires. *Phys. Rev. Lett.* **100**:016403.

W. A. de Heer (1993). The physics of simple metal clusters: Experimental aspects and simple models. *Rev. Mod. Phys.* **65**:611.

V. V. Deshpande and M. Bockrath (2008). The one-dimensional Wigner crystal in carbon nanotubes. *Nat. Phys.* **4**:314.

N. D. Drummond et al. (2004). Diffusion quantum Monte-Carlo study of three-dimensional Wigner crystals. *Phys. Rev. B* **69**:085116.

H. J. Fan et al. (2006). Semiconductor nanowires: From self-organization to growth control. *Small* **2**:700.

H. A. Fertig and L. Brey (2006). Luttinger liquid at the edge of undoped graphene in a strong magnetic field. *Phys. Rev. Lett.* **97**:116805.

G. A. Fiete (2007). Colloquium: The spin-incoherent Luttinger liquid. *Rev. Mod. Phys.* **79**:801.

A. V. Filinov et al. (2001). Wigner crystallization in mesoscopic 2D electron systems. *Phys. Rev. Lett.* **86**:3851.

M. E. Fisher (1998). Renormalization group theory: Its basis and formulation in statistical physics. *Rev. Mod. Phys.* **70**:653.

A. Georges et al. (2000). Interchain conductivity of coupled Luttinger liquids and organic conductors. *Phys. Rev. B* **61**:16393.

G. F. Giuliani and G. Vignale (2005). *Quantum Theory of the Electron Liquid*, 1st edn. Cambridge University Press, Cambridge, U.K.

A. Gold and L. Calmels (1998). Excitation spectrum of the quasi-one-dimensional electron gas with long-range Coulomb interaction. *Phys. Rev. B* **58**:3497.

N. W. Gong et al. (2008). Au(Si)-filled β-Ga_2O_3 nanotubes as wide range high temperature nanothermometers. *Appl. Phys. Lett.* **92**:073101.

C. Grimes and G. Adams (1979). Evidence for a liquid-to-crystal phase transition in a classical, two-dimensional sheet of electrons. *Phys. Rev. Lett.* **42**:795.

H. Hasegawa et al. (2008). Organic Mott insulator-based nanowire formation by using the nanoscale-electrocrystallization. *Thin Solid Films* **516**:2491.

T. Hertel et al. (1998). Manipulation of individual carbon nanotubes and their interaction with surfaces. *J. Phys. Chem. B* **102**:910.

K. Hiraki and K. Kanoda (1998). Wigner crystal type of charge ordering in an organic conductor with a quarter-filled band: $(DI\text{-}DCNQI)_2Ag$. *Phys. Rev. Lett.* **80**:4737.

P. Hohenberg and W. Kohn (1964). Inhomogeneous electron gas. *Phys. Rev.* **136**:B864.

P. Horsch et al. (2005). Wigner crystallisation in $Na_3Cu_2O_4$ and $Na_8Cu_5O_{10}$ chain compounds. *Phys. Rev. Lett.* **94**:076403.

D. Hughes and P. Ballone (2008). Spontaneous spin polarization and electron localization in constrained geometries: The Wigner transition in nanowires. *Phys. Rev. B* **77**:245312–245322.

J. C. Inkson (1984). *Many Body Theory of Solids*. Plenum, New York.

Y. Ito and Y. Teraoka (2004). Magnetic phase diagram in a three-dimensional electron gas. *J. Magn. Magn. Mater.* **272–276S**:E273.

Y. Ito and Y. Teraoka (2006). Wigner film ground state in a three-dimensional electron gas. *J. Magn. Magn. Mater.* **310**:1073–1075.

M. José-Yacamán et al. (1993). Catalytic growth of carbon microtubules with fullerene structure. *Appl. Phys. Lett.* **62**:657.

A. Kawabata (2007). Electron conduction in one-dimension. *Rep. Prog. Phys.* **70**:219.

S. H. Kim et al. (2007). Cesium-filled single wall carbon nanotubes as conducting nanowires: Scanning tunneling spectroscopy study. *Phys. Rev. Lett.* **99**:256407.

W. Kohn and L. J. Sham (1965). Self-consistent equations including exchange and correlation. *Phys. Rev.* **140**:1133.

G. Kotliar and D. Vollhardt (2004). Strongly correlated materials: Insights from dynamical mean-field theory. *Phys. Today* **57**:53.

G. Kotliar et al. (2006). Electronic structure calculations with dynamical mean-field theory. *Rev. Mod. Phys.* **78**:865.

L. Kuipers and J. W. M. Frenken (1993). Jump to contact, neck formation, and surface melting in the scanning tunneling microscope. *Phys. Rev. Lett.* **70**:3907.

R. Landauer (1970). Electrical resistance of disordered one-dimensional lattices. *Philos. Mag.* **21**:863.

Z. Lenac (2005). Interaction of the electromagnetic field with a Wigner crystal. *Phys. Rev. B* **71**:035330.

C. Loo et al. (2004). Nanoshell-enabled photonics-based imaging and therapy of cancer. *Technol. Cancer Res. Treat.* **3**:33.

J. M. Luttinger (1963). An exactly soluble model of a many-fermion system. *J. Math. Phys.* **4**:1154.

G. D. Mahan (2000). *Many-Particle Physics*, 3rd edn. Kluwer Academic/Plenum Publishers, New York.

R. M. Martin (2004). *Electronic Structure. Basic Theory and Practical Methods*, 1st edn. Cambridge University Press, Cambridge, U.K.

D. Marx and J. Hutter (2000). *Ab Initio Molecular Dynamics: Theory and Implementation*, Vol. 1 of NIC Series. FZ Jülich, Berlin, Germany.

J. S. Meyer and K. A. Matveev (2009). Wigner crystal physics in quantum wires. *J. Phys.: Condens. Matter* **21**:023203.

M. J. Montgomery et al. (2002). Power dissipation in nanoscale conductors. *J. Phys.: Condens. Matter* **14**:5377.

T. Morimoto et al. (2003). Nonlocal resonant interaction between coupled quantum wires. *Appl. Phys. Lett.* **82**:3952.

E. J. Mueller (2005). Wigner crystallisation in inhomogeneous one-dimensional wires. *Phys. Rev. B* **72**:075322.

J. W. Negele and H. Orland (1998). *Quantum Many-Particle Systems*. Addison-Wesley, Reading, MA.

J. C. Nickel et al. (2006). Superconducting pairing and density-wave instabilities in quasi-one-dimensional conductors. *Phys. Rev. B* **73**:165126.

M. Ogata and H. Fukuyama (1994). Collapse of quantized conductance in a dirty Tomonaga-Luttinger liquid'. *Phys. Rev. Lett.* **73**:468.

G. Ortiz et al. (1999). Zero temperature phases of the electron gas. *Phys. Rev. Lett.* **82**:5317.

R. V. Parthasarathy and C. R. Martin (1994). Template-synthesised polyaniline microtubules. *Chem. Mater.* **6**:1627.

J. P. Perdew et al. (1990). Stabilized jellium: Structureless pseudo-potential model for the cohesive and surface properties of metals. *Phys. Rev. B* **42**:11627.

L. Pfeiffer et al. (1990). Formation of a high quality two-dimensional electron gas on cleaved GaAs. *Appl. Phys. Lett.* **56**:1697.

J. Pouget and S. Ravy (1997). X-ray evidence of charge density wave modulations in the magnetic phases of (TMTSF)(2) PF6 and (TMTTF)(2)Br. *Synth. Met.* **85**:1523.

A. Rahman and M. K. Sanyal (2007). Observation of charge density wave characteristics in conducting polymer nanowires: Possibility of Wigner crystallization. *Phys. Rev. B* **76**:045110.

A. Rahman et al. (2006). Evidence of a ratchet effect in nanowires of a conducting polymer. *Phys. Rev. B* **73**:125313.

D. J. Reilly et al. (2002). Density-dependent spin polarization in ultra-low-disorder quantum wires. *Phys. Rev. Lett.* **89**:246801.

M. C. Roco et al. (eds.) (2000). *Nanotechnology Research Directions: Vision for Nanotechnology Research and Development in the Next Decade.* Springer, Heidelberg, Germany.

K. I. Schuller et al. (2008). *Keynote Lecture at the "Trends in Nanotechnology" Conference.* Oviedo, Spain. 1–5 September, Abstract: www.tntconf.org/2008/Files/Abstracts/Keywords/TNT2008_Schuller.pdf.

H. J. Schulz (1993). Wigner crystal in one-dimension. *Phys. Rev. Lett.* **71**:1864.

H. J. Schulz et al. (2000). Fermi liquids and Luttinger liquids. *Field Theories for Low-Dimensional Condensed Matter Systems.* Springer, New York. (arXiv.org:cond-mat/9807366).

G. Senatore and G. Pastore (1990). Density-functional theory of freezing for quantum systems: The Wigner crystallization. *Phys. Rev. Lett.* **64**:303.

L. Shulenburger et al. (2008). Correlation effects in quasi-one-dimensional quantum wires. *Phys. Rev. B* **78**:165303.

C. A. Stafford et al. (1997). Jellium model of metallic nanocohesion. *Phys. Rev. Lett.* **79**:2863.

H. Steinberg et al. (2006). Localization transition in a ballistic quantum wire. *Phys. Rev. B* **73**.

S. Tanaka and S. Ichimaru (1987). Dynamic theory of correlations in strongly coupled, classical one-component plasmas: Glass transition in the generalized viscoelastic formalism. *Phys. Rev. A* **35**:4743.

B. Tanatar et al. (1998). Wigner crystallization in semiconductor quantum wires. *Phys. Rev. B* **58**:9886.

K. J. Thomas et al. (1996). Possible spin polarization in a one-dimensional electron gas. *Phys. Rev. Lett.* **77**:135.

S. Tomonaga (1950). Remarks on Bloch's method of sound waves applied to many-fermion problems. *Prog. Theor. Phys.* **5**:544.

J. R. Trail et al. (2003). Unrestricted Hartree-Fock theory of Wigner crystals. *Phys. Rev. B* **68**:045107.

J. M. van Ruitenbeek et al. (1997). Conductance quantization in metals: The influence of subband formation on the relative stability of specific contact diameters. *Phys. Rev. B* **56**:12566.

B. J. van Wees et al. (1988). Quantized conductance of point contacts in a two-dimensional electron gas. *Phys. Rev. Lett.* **60**:848.

P. Werner et al. (2006). On the formation of Si-nanowires by molecular beam epitaxy. *Int. J. Mater. Res.* **97**:1008.

E. Wigner (1934). On the interaction of electrons in metals. *Phys. Rev.* **46**:1002.

J. A. Yan et al. (2007). Size and orientation dependence in the electronic properties of silicon nanowires. *Phys. Rev. B* **76**:115319.

L. Yang et al. (2007). Enhanced electron-hole interaction and optical absorption in a silicon nanowire. *Phys. Rev. B.* **75**:201304.

C. Yannouleas and U. Landman (1997). On mesoscopic forces and quantised conductance in model metallic nanowires. *J. Phys. Chem. B* **101**:5780.

C. Yannouleas et al. (1998). Energetics, forces, and quantized conductance in jellium-modeled metallic nanowires. *Phys. Rev. B* **57**:4872.

S. N. Yanushkevich and B. Steinbach (2008). Spatial interconnect analysis for predictable nanotechnologies. *J. Comput. Theor. Nanosci.* **5**:56.

Y. Yoon et al. (2007). Probing the microscopic structure of bound states in quantum point contacts. *Phys. Rev. Lett.* **99**:136805.

W. S. Yun et al. (2000). Fabrication of metal nanowire using carbon nanotube as a mask. *J. Vac. Sci. Technol. A* **18**:1329.

N. Zabala et al. (1998). Spontaneous magnetization of simple metal nanowires. *Phys. Rev. Lett.* **80**:3336.

N. Zabala et al. (1999). Electronic structure of cylindrical simple-metal nanowires in the stabilized jellium model. *Phys. Rev. B* **59**:12652.

S. V. Zaitsev-Zotov et al. (2000). Lutinger-liquid-like transport in long InSb nanowires. *J. Phys.: Condens. Matter* **12**:L303.

Z. Zanolli et al. (2007). Quantum-confinement effects in InAs-InP core-shell nanowires. *J. Phys.: Condens. Matter* **19**:295219.

X. Zhao et al. (2004). Quantum confinement and electronic properties of Silicon nanowires. *Phys. Rev. Lett.* **92**:236805.

B. Zhou et al. (eds.) (2003). *Nanotechnology in Catalysis*, Vols. 1 and 2 (*Nanostructure Science and Technology*). Springer, New York.

F. H. Zong et al. (2002). Spin polarisation of the low-density three-dimensional electron gas. *Phys. Rev. E* **66**:036703.

28

Spin Relaxation in Quantum Wires

Paul Wenk
Jacobs University Bremen

Stefan Kettemann
Jacobs University Bremen

and

*Pohang University of
Science and Technology*

28.1 Introduction

The emerging technology of spintronics intends to use the manipulation of the spin degree of freedom of individual electrons for energy-efficient storage and transport of information.[72] In contrast to classical electronics, which relies on the steering of charge carriers through semiconductors, spintronics uses the spin carried by electrons, resembling tiny spinning tops. The difference to a classical top is that its angular momentum is quantized, it can only take two discrete values, up or down. To control the spin of electrons, a detailed understanding of the interaction between the spin and orbital degrees of freedom of electrons and other mechanisms that do not conserve its spin is necessary. These are typically weak perturbations, compared to the kinetic energy of conduction electrons, so that their spin relaxes slowly to the advantage of spintronic applications. The relaxation or depolarization of the electron spin can occur due to the randomization of the electron momentum by scattering from impurities, and dislocations in the material, and due to scattering with elementary excitations of the solid, such as phonons and other electrons, when it is transferred to the randomization of the electron spin due to the spin–orbit interaction. In addition, scattering from localized spins, such as nuclear spins and magnetic impurities, are sources of electron spin relaxation. The electron spin relaxation can be reduced by constraining the electrons in low-dimensional structures, quantum

wells (confined in one direction, free in two dimensions), quantum wires (confined in two directions, free in one direction), or quantum dots (confined in all three directions). Although spin relaxation is typically smallest in quantum dots due to their discrete energy level spectrum, the necessity to transfer the spin in spintronic devices recently led to intense research efforts to reduce the spin relaxation in quantum wires, where the energy spectrum is continuous. In the following, we review the theory of spin dynamics and relaxation in quantum wires, and compare it with recent experimental results. Section 28.2 provides a general introduction to spin dynamics. In Section 28.3, we discuss all relevant spin relaxation mechanisms and how they depend on dimension, temperature, mobility, charge carrier density, and magnetic field. In Section 28.4, we review recent results on spin relaxation in semiconducting quantum wires, and its influence on the quantum corrections to their conductance. These weak localization corrections are thereby a very sensitive measure of spin relaxation in quantum wires, in addition to optical methods, as we see in Section 28.5. We set $\hbar = 1$ in the following.

28.2 Spin Dynamics

Before we review the spin dynamics of conduction electrons and holes in semiconductors and metals, let us first reconsider the spin dynamics of a localized spin, as governed by the Bloch equations.

28.2.1 Dynamics of a Localized Spin

A localized spin $\hat{s}$, like a nuclear spin, or the spin of a magnetic impurity in a solid, precesses in an external magnetic field $\mathbf{B}$ due to the Zeeman interaction with Hamiltonian $H_z = -\gamma_g \hat{s}\mathbf{B}$, where γ_g is the corresponding gyromagnetic ratio of the nuclear spin or magnetic impurity spin, respectively, which we will set equal to one, unless needed explicitly. This spin dynamics is governed by the Bloch equation of a localized spin,

$$\partial_t \hat{s} = \gamma_g \hat{s} \times \mathbf{B}. \tag{28.1}$$

This equation is identical to the Heisenberg equation $\partial_t \hat{s} = -i[\hat{s}, H_z]$ for the quantum mechanical spin operator $\hat{s}$ of an $S = 1/2$-spin, interacting with the external magnetic field $\mathbf{B}$ due to the Zeeman interaction with Hamiltonian H_z. The solution of the Bloch equation for a magnetic field pointing in the z-direction is $\hat{s}_z(t) = \hat{s}_z(0)$, while the x- and y-components of the spin are precessing with frequency $\omega_0 = \gamma_g \mathbf{B}$ around the z-axis, $\hat{s}_x(t) = \hat{s}_x(0)\cos\omega_0 t + \hat{s}_y(0)\sin\omega_0 t$, $\hat{s}_y(t) = -\hat{s}_x(0)\sin\omega_0 t + \hat{s}_y(0)\cos\omega_0 t$. Since a localized spin interacts with its environment by exchange interaction and magnetic dipole interaction, the precession will dephase after a time τ_2, and the z-component of the spin relaxes to its equilibrium value s_{z0} within a relaxation time τ_1. This modifies the Bloch equations to the phenomenological equations:

$$\partial_t \hat{s}_x = \gamma_g(\hat{s}_y B_z - \hat{s}_z B_y) - \frac{1}{\tau_2}\hat{s}_x$$

$$\partial_t \hat{s}_y = \gamma_g(\hat{s}_z B_x - \hat{s}_x B_z) - \frac{1}{\tau_2}\hat{s}_y \tag{28.2}$$

$$\partial_t \hat{s}_z = \gamma_g(\hat{s}_x B_y - \hat{s}_y B_x) - \frac{1}{\tau_1}(\hat{s}_z - s_{z0}).$$

28.2.2 Spin Dynamics of Itinerant Electrons

28.2.2.1 Ballistic Spin Dynamics

The intrinsic degree of freedom spin is a direct consequence of the Lorentz invariant formulation of quantum mechanics. Expanding the relativistic Dirac equation in the ratio of the electron velocity and the constant velocity of light, c, one obtains in addition to the Zeeman term, a term that couples the spin $\mathbf{s}$ with the momentum $\mathbf{p}$ of the electrons, the spin–orbit coupling

$$H_{SO} = \frac{1}{2}-\frac{\mu_B}{mc^2}\hat{s}\mathbf{p}\times\mathbf{E} = -\hat{s}\mathbf{B}_{SO}(\mathbf{p}), \tag{28.3}$$

where we set the gyromagnetic ratio $\gamma_g = 1$. $\mathbf{E} = -\nabla V$, is an electrical field, and $\mathbf{B}_{SO}(\mathbf{p}) = (\mu_B/2mc^2)\mathbf{p}\times\mathbf{E}$. Substitution into the Heisenberg equation yields the Bloch equation in the presence of spin–orbit interaction:

$$\partial_t \hat{s} = \hat{s}\times\mathbf{B}_{SO}(\mathbf{p}), \tag{28.4}$$

so that the spin performs a precession around the momentum-dependent spin–orbit field $\mathbf{B}_{SO}(\mathbf{p})$. It is important to note that the spin–orbit field does not break the invariance under time reversal ($\hat{s}\to-\hat{s}$, $\mathbf{p}\to-\mathbf{p}$), in contrast to an external magnetic field $\mathbf{B}$. Therefore, averaging over all directions of momentum, there is no spin polarization of the conduction electrons. However, by injecting a spin-polarized electron with given momentum $\mathbf{p}$ into a translationally invariant wire, its spin precesses in the spin–orbit field as the electron moves through the wire. The spin will be oriented again in the initial direction after it has moved a length L_{SO}, which is the length of the spin precession. The precise magnitude of L_{SO} does not only depend on the strength of the spin–orbit interaction but may also depend on the direction of its movement in the crystal, as we will discuss below.

28.2.2.2 Spin Diffusion Equation

Translational invariance is broken by the presence of disorder due to impurities and lattice imperfections in the conductor. As the electrons scatter from the disorder potential elastically, their momentum changes in a stochastic way, resulting in diffusive motion. This results in a change of the local electron density $\rho(\mathbf{r},t)=\sum_{\alpha=\pm}|\psi_\alpha(\mathbf{r},t)|^2$, where $\alpha = \pm$ denotes the orientation of the electron spin, and $\psi_\alpha(\mathbf{r}, t)$ is the position- and time-dependent electron wave function amplitude. On length scales exceeding the elastic mean free path, l_e, this density is governed by the diffusion equation

$$\frac{\partial\rho}{\partial t}=D\nabla^2\rho \tag{28.5}$$

where the diffusion constant, D, is related to the elastic scattering time, τ, by $D=v_F^2\tau/d_D$, where v_F is the Fermi velocity, and d_D the diffusion dimension of the electron system. This diffusion constant is related to the mobility of the electrons, $\mu_e = e\tau/m$ by the Einstein relation $\mu_e\rho = e2\nu D$, where ν is the density of states per spin at the Fermi energy E_F. Injecting an electron at position $\mathbf{r}_0$ into a conductor with previously constant electron density ρ_0, the solution of the diffusion equation yields that the electron density spreads in space according to $\rho(\mathbf{r},t)=\rho_0+\exp(-(\mathbf{r}-\mathbf{r}_0)^2/4Dt)/(4\pi Dt)^{d_D/2}$, where d_D is the dimension of diffusion. This dimension is equal to the kinetic dimension d, $d_D = d$, if the elastic mean free path, l_e, is smaller than the size of the sample in all directions. If the elastic mean free path is larger than the sample size in one direction, the diffusion dimension accordingly reduces by one. Thus, on average, the variance of the distance the electron moves after time t is $\langle(\mathbf{r}-\mathbf{r}_0)^2\rangle = 2d_D Dt$. This introduces a new length scale, the diffusion length $L_D(t) = \sqrt{Dt}$. We can rewrite the density as $\rho = \langle\psi^\dagger(\mathbf{r}, t)\psi(\mathbf{r}, t)\rangle$, where $\psi^\dagger = (\psi_+^\dagger, \psi_-^\dagger)$ is the two-component vector of the up (+) and down (−) spin fermionic creation operators, and ψ the two-component vector of annihilation operators, respectively, $\langle\ldots\rangle$ denoting the expectation value. Accordingly, the spin density, $s(\mathbf{r}, t)$, is expected to satisfy a diffusion equation, as well. The spin density is defined by

$$s(\mathbf{r},t)=\frac{1}{2}\Big\langle \psi^{\dagger}(\mathbf{r},t)\sigma\psi(\mathbf{r},t)\Big\rangle, \qquad (28.6)$$

where

σ is the vector of Pauli matrices

$$\sigma_x=\begin{pmatrix}0 & 1\\ 1 & 0\end{pmatrix}, \quad \sigma_y=\begin{pmatrix}0 & -i\\ i & 0\end{pmatrix}, \quad \sigma_z=\begin{pmatrix}1 & 0\\ 0 & -1\end{pmatrix}.$$

Thus, the z-component of the spin density is half the difference between the density of the spin-up and -down electrons, $s_z = (\rho_+ - \rho_-)/2$, which is the local spin polarization of the electron system. Thus, we can directly infer the diffusion equation for s_z and, similarly, for the other components of the spin density, yielding, without magnetic field and spin–orbit interaction,[63]

$$\frac{\partial \mathbf{s}}{\partial t}=D\nabla^2\mathbf{s}-\frac{\mathbf{s}}{\hat{\tau}_s}. \qquad (28.7)$$

Here, in the spin relaxation term we introduced the tensor $\hat{\tau}_s$, which can have nondiagonal matrix elements. In case of a diagonal matrix, $\tau_{sxx} = \tau_{syy} = \tau_2$ is the spin dephasing time and $\tau_{szz} = \tau_1$ is the spin relaxation time. The spin diffusion equation can be written as a continuity equation for the spin density vector by defining the spin diffusion current of the spin components s_i,

$$\mathbf{J}_{s_i}=-D\nabla s_i. \qquad (28.8)$$

Thus, we get the continuity equation for the spin density components s_i,

$$\frac{\partial s_i}{\partial t}+\nabla\mathbf{J}_{s_i}=-\sum_j\frac{s_j}{\tau_{sij}}. \qquad (28.9)$$

28.2.2.3 Spin–Orbit Interaction in Semiconductors

While silicon and germanium have in their diamond structure an inversion symmetry around every midpoint on each line connecting nearest-neighbor atoms, this is not the case for III–V semiconductors like GaAs, InAs, InSb, or ZnS. These have a zinc-blende structure that can be obtained from a diamond structure with neighbored sites occupied by the two different elements. Therefore, the inversion symmetry is broken, which results in spin–orbit coupling. Similarly, this symmetry is broken in II–VI semiconductors. This bulk inversion asymmetry (BIA) coupling, often called the Dresselhaus coupling, is anisotropic, as given by[17]

$$H_D = \gamma_D\Big[\sigma_x k_x(k_y^2-k_z^2)+\sigma_y k_y(k_z^2-k_x^2)+\sigma_z k_z(k_x^2-k_y^2)\Big], \quad (28.10)$$

where γ_D is the Dresselhaus spin–orbit coefficient. Confinement in quantum wells with width a on the order of the Fermi wavelength λ_F accordingly yields a spin–orbit interaction where the momentum in growth direction is of the order of $1/a$. Because of the anisotropy of the Dresselhaus term, the spin–orbit interaction depends strongly on the growth direction of the quantum

well. Grown in [001]-direction, one gets, taking the expectation value of Equation 28.10 in the direction normal to the plane, noting that $\langle k_z\rangle = \langle k_z^3\rangle = 0$,[17]

$$H_{D[001]} = \alpha_1(-\sigma_x k_x+\sigma_y k_y)+\gamma_D(\sigma_x k_x k_y^2-\sigma_y k_y k_x^2) \quad (28.11)$$

where $\alpha_1 = \gamma_D\langle k_z^2\rangle$ is the linear Dresselhaus parameter. Thus, by inserting an electron with momentum along the x-direction, with its spin initially polarized in the z-direction, it will precess around the x-axis as it moves along. For narrow quantum wells, where $\langle k_z^2\rangle \sim 1/a^2 \ge k_F^2$, the linear term exceeds the cubic Dresselhaus terms.

A special situation arises for quantum wells grown in the [110]-direction, where the spin–orbit field points normal to the quantum well, as shown in Figure 28.1, so that an electron whose spin is initially polarized along the normal of the plane remains polarized as it moves in the quantum well.

In quantum wells with asymmetric electrical confinement, the inversion symmetry is broken as well. This structural inversion asymmetry (SIA) can be deliberately modified by changing the confinement potential by the application of a gate voltage. The resulting spin–orbit coupling, the SIA coupling, also called Rashba spin–orbit interaction[58] is given by

$$H_R = \alpha_2(\sigma_x k_y-\sigma_y k_x), \qquad (28.12)$$

where α_2 depends on the asymmetry of the confinement potential, $V(z)$, in the direction z, the growth direction of the quantum well, and can thus be deliberately changed by the application of a gate potential. At first sight, it looks as if the expectation value of the electrical field $\varepsilon_c = -\partial_z V(z)$ in the conduction band state vanishes, since the ground state of the quantum well must be symmetric in z. Taking into account the coupling to the valence band,[43,66] the discontinuities in the effective mass,[46] and corrections due to the coupling to odd excited states,[6] yields a sizable coupling parameter depending on the asymmetry of the confinement potential.[23,66]

This dependence allows one, in principle, to control the electron spin with a gate potential, which can therefore be used as the basis of a spin transistor.[14]

We can combine all spin–orbit couplings by introducing the spin–orbit field such that the Hamiltonian has the form of a Zeeman term

$$H_{SO} = -\mathbf{s}\mathbf{B}_{SO}(\mathbf{k}), \qquad (28.13)$$

where the spin vector is $\mathbf{s} = \sigma/2$. But we stress again that since $\mathbf{B}_{SO}(\mathbf{k}) \to \mathbf{B}_{SO}(-\mathbf{k}) = -\mathbf{B}_{SO}(\mathbf{k})$ under the time reversal operation, spin–orbit coupling does not break time reversal symmetry, since the time reversal operation also changes the sign of the spin, $\mathbf{s} \to -\mathbf{s}$. Only an external magnetic field $\mathbf{B}$ breaks the time reversal symmetry. Thus, the electron spin operator $\hat{\mathbf{s}}$ is for fixed electron momentum $\mathbf{k}$ governed by the Bloch equations with the spin–orbit field,

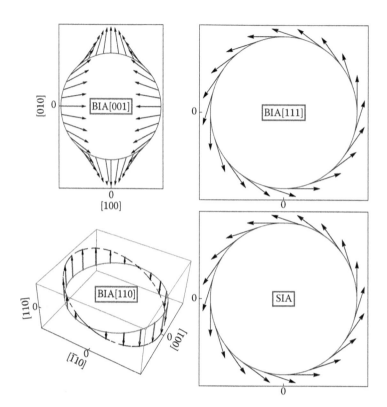

FIGURE 28.1 The spin–orbit vector fields for linear BIA spin–orbit coupling for quantum wells grown in [001]-, [110]-, and [111]-direction, and for linear-structure inversion asymmetry (Rashba) coupling, respectively.

$$\frac{\partial \hat{\mathbf{s}}}{\partial t} = \hat{\mathbf{s}} \times (\mathbf{B} + \mathbf{B}_{SO}(\mathbf{k})) - \frac{1}{\hat{\tau}_s} \hat{\mathbf{s}}. \qquad (28.14)$$

The spin relaxation tensor is no longer necessarily diagonal in the presence of spin–orbit interaction.

In narrow quantum wells where the cubic Dresselhaus coupling is weak compared to the linear Dresselhaus and Rashba couplings, the spin–orbit field is given by

$$\mathbf{B}_{SO}(\mathbf{k}) = -2 \begin{pmatrix} -\alpha_1 k_x + \alpha_2 k_y \\ \alpha_1 k_y - \alpha_2 k_x \\ 0 \end{pmatrix}, \qquad (28.15)$$

which changes both its direction and its amplitude, $|\mathbf{B}_{SO}(\mathbf{k})| = 2\sqrt{(\alpha_1^2 + \alpha_2^2)k^2 - 4\alpha_1\alpha_2 k_x k_y}$, as the direction of the momentum $\mathbf{k}$ is changed. Accordingly, the electron energy dispersion close to the Fermi energy is in general anisotropic, as given by

$$E_\pm = \frac{1}{2m^*} k^2 \pm \alpha k \sqrt{1 - 4\frac{\alpha_1\alpha_2}{\alpha^2} \cos\theta \sin\theta}, \qquad (28.16)$$

where

$k = |\mathbf{k}|$
$\alpha = \sqrt{\alpha_1^2 + \alpha_2^2}$
$k_x = k \cos\theta$

Thus, when an electron is injected with energy E, with momentum k along the [100]-direction, $k_x = k$, $k_y = 0$, its wave function is a superposition of plain waves with the positive momenta $k_\pm = \mp \alpha m^* + m^*(\alpha^2 + 2E/m^*)^{1/2}$. The momentum difference, $k_- - k_+ = 2m^*\alpha$, causes a rotation of the electron eigenstate in the spin subspace. When the electron spin was polarized up spin at $x = 0$ with the eigenvector

$$\psi(x = 0) = \begin{pmatrix} 1 \\ 0 \end{pmatrix}$$

and its momentum points in x-direction, at a distance x, it will have rotated the spin as described by the eigenvector

$$\psi(x) = \frac{1}{2} \begin{pmatrix} 1 \\ \frac{\alpha_1 + i\alpha_2}{\alpha} \end{pmatrix} e^{ik_+ x} + \frac{1}{2} \begin{pmatrix} 1 \\ -\frac{\alpha_1 + i\alpha_2}{\alpha} \end{pmatrix} e^{ik_- x}. \qquad (28.17)$$

In Figure 28.2, we plot the corresponding spin density as defined in Equation 28.6 for pure Rashba coupling, $\alpha_1 = 0$.

The spin will point again in the initial direction, when the phase difference between the two plain waves is 2π, which gives the condition for spin precession length as $2\pi = (k_- - k_+)L_{SO}$, yielding for linear Rashba and Dresselhaus coupling, and the electron moving in [100]- direction,

$$L_{SO} = \pi/m^*\alpha. \qquad (28.18)$$

We note that the period of spin precession changes with the direction of the electron momentum since the spin–orbit field, Equation 28.15, is anisotropic.

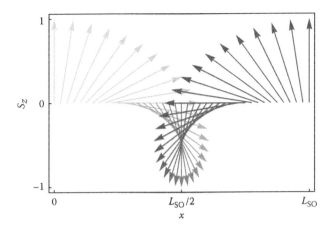

FIGURE 28.2 Precession of a spin injected at $x = 0$, polarized in z-direction, as it moves by one spin precession length $L_{SO} = \pi/m^{\star}\alpha$ through the wire with linear Rashba spin–orbit coupling, α_2.

28.2.2.4 Spin Diffusion in the Presence of Spin–Orbit Interaction

As the electrons are scattered by imperfections like impurities and dislocations, their momentum is changed randomly. Accordingly, the direction of the spin–orbit field, $\mathbf{B}_{SO}(\mathbf{k})$, changes randomly as the electron moves through the sample. This has two consequences: The electron spin direction becomes randomized thus dephasing the spin precession and relaxing the spin polarization. In addition, the spin precession term is modified as the momentum $\mathbf{k}$ changes randomly, and has no longer the form given in the ballistic Bloch-like equation, Equation 28.14. One can derive the diffusion equation for the expectation value of the spin, the spin density Equation 28.6, semiclassically[47,60] or by diagrammatic expansion.[65] In order to better understand this equation, we provide a simplified classical derivation in the following. The spin density at time $t + \Delta t$ can be related to the one at the earlier time t. Note that for ballistic times $\Delta t \leq \tau$, the distance the electron has moved with a probability $p_{\Delta \mathbf{x}}$, $\Delta \mathbf{x}$, is related to that time by the ballistic equation $\Delta \mathbf{x} = \mathbf{k}(t)\Delta t/m$ when the electron moves with the momentum $\mathbf{k}(t)$. On this timescale, the spin evolution is still governed by the ballistic Bloch equation (Equation 28.14). Thus, we can relate the spin density at the position $\mathbf{x}$ at the time $t + \Delta t$, to the one at the earlier time t at position $\mathbf{x} - \Delta \mathbf{x}$:

$$s(\mathbf{x}, t + \Delta t) = \sum_{\Delta \mathbf{x}} p_{\Delta \mathbf{x}}$$

$$\times \left(\left(1 - \frac{1}{\hat{\tau}_s}\Delta t \right) s(\mathbf{x} - \Delta \mathbf{x}, t) - \Delta t[\mathbf{B} + \mathbf{B}_{SO}(\mathbf{k}(t))] \times s(\mathbf{x} - \Delta \mathbf{x}, t) \right).$$

$$(28.19)$$

Now, we can expand in Δt to first order and in $\Delta \mathbf{x}$ to second order. Next, we average over the disorder potential, assuming that the electrons are scattered isotropically, and substitute $\sum_{\Delta \mathbf{x}} \Delta \mathbf{x}\, p_{\Delta \mathbf{x}} \cdots = \int (d\Omega/\Omega)\cdots$ where Ω is the total angle, and $\int d\Omega$ denotes the integral over all angles with $\int (d\Omega/\Omega) = 1$.

Also, we get $(s(\mathbf{x}, t + \Delta t) - s(\mathbf{x}, t))/\Delta t \rightarrow \partial_t s(\mathbf{x}, t)$ for $\Delta t \rightarrow 0$, and $\langle \Delta x_i^2 \rangle = 2D\Delta t$ where D is the diffusion constant. While the disorder average yields $\langle \Delta \mathbf{x} \rangle = 0$, and $\langle \mathbf{B}_{SO}(\mathbf{k}(t)) \rangle = 0$, separately, for isotropic impurity scattering, averaging their product yields a finite value, since $\Delta \mathbf{x}$ depends on the momentum at time t, $\mathbf{k}(t)$, yielding $\langle \Delta \mathbf{x} B_{SOi}(\mathbf{k}(t)) \rangle = 2\Delta t \langle \mathbf{v}_F B_{SOi}(\mathbf{k}(t)) \rangle$, where $\langle ... \rangle$ denotes the average over the Fermi surface. In this way, we can also evaluate the average of the spin–orbit term in Equation 28.19, expanded to first order in $\Delta \mathbf{x}$, and get, substituting $\Delta t \rightarrow \tau$ the spin diffusion equation,

$$\frac{\partial \mathbf{s}}{\partial t} = -\mathbf{B} \times \mathbf{s} + D\nabla^2 \mathbf{s} + 2\tau \left\langle (\nabla \mathbf{v}_F)\mathbf{B}_{SO}(\mathbf{p}) \right\rangle \times \mathbf{s} - \frac{1}{\tau_s}\mathbf{s}, \quad (28.20)$$

where $\langle ... \rangle$ denotes the average over the Fermi surface. Spin-polarized electrons injected into the sample spread diffusively, and their spin polarization, while spreading diffusively as well, decays in amplitude exponentially in time. As the spins precess around the spin–orbit fields between scattering events, one also expects an oscillation of the polarization amplitude in space. One can find the spatial distribution of the spin density, which is the solution of Equation 28.20 with the smallest decay rate Γ_s. As an example, the solution for linear Rashba coupling is[60]

$$s(\mathbf{x}, t) = (\hat{e}_q \cos \mathbf{q}\mathbf{x} + A\hat{e}_z \sin \mathbf{q}\mathbf{x})e^{-t/\tau_s}, \quad (28.21)$$

with $1/\tau_s = 7/16\tau_{s0}$, where $1/\tau_{s0} = 2\tau k_F^2 \alpha_2^2$ and where the amplitude of the momentum $\mathbf{q}$ is determined by $Dq^2 = 15/16\tau_{s0}$, and $A = 3/\sqrt{15}$, and $\hat{e}_q = \mathbf{q}/q$. This solution is plotted in Figure 28.3 for $\hat{e}_q = (1, 1, 0)/\sqrt{2}$. In Figure 28.4, we plot the linearly independent solution obtained by interchanging cos and sin in Equation 28.21, with the spin pointing in z-direction, initially. We choose $\hat{e}_q = \hat{e}_x$. Figure 28.4 shows that the period of precession is enhanced by the factor $4/\sqrt{15}$ in the diffusive wire, and that the amplitude of the spin density is modulated, changing from 1 to $A = 3/\sqrt{15}$ when compared with the ballistic precession of the spin.

Injecting a spin-polarized electron at one point, say $\mathbf{x} = 0$, its density spreads the same way it does without spin–orbit interaction, $\rho(\mathbf{r}, t) = \exp(-r^2/4Dt)/(4\pi Dt)^{d_D/2}$, where r is the distance to the injection point. However, the decay of the spin density is periodically modulated as a function of $2\pi\sqrt{15/16}\, r/L_{SO}$.[25]

The spin–orbit interaction together with the scattering from impurities is also a source of spin relaxation, as we discuss in the Section 28.3 together with other mechanisms of spin relaxation. We can find the classical spin diffusion current in the presence of spin–orbit interaction in a similar way as one can derive the classical diffusion current: The current at the position $\mathbf{r}$ is a sum over all currents in its vicinity, which are directed toward that position. Thus, $\mathbf{j}(\mathbf{r}, t) = \langle \mathbf{v}\rho(\mathbf{r} - \Delta \mathbf{x}) \rangle$ where an angular average over all possible directions of the velocity $\mathbf{v}$ is taken. Expanding in $\Delta \mathbf{x} = l_e \mathbf{v}/v$, and noting that $\langle \mathbf{v}\rho(\mathbf{r}) \rangle = 0$, one gets $\mathbf{j}(\mathbf{r}, t) = \langle \mathbf{v}(-\Delta \mathbf{x}) \nabla\rho(\mathbf{r}) \rangle = -(v_F l_e/2)\nabla\rho(\mathbf{r}) = -D\nabla\rho(\mathbf{r})$. For the classical spin diffusion

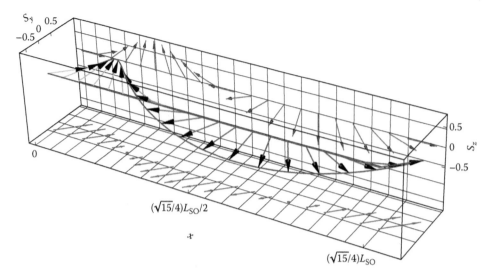

FIGURE 28.3 The spin density for linear Rashba coupling, which is a solution of the spin diffusion equation with the relaxation rate $7/16\tau_s$. The spin points initially in the x-y plane in the direction $(1,1,0)$.

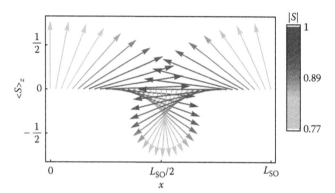

FIGURE 28.4 The spin density for linear Rashba coupling, which is a solution of the spin diffusion equation with the relaxation rate $1/\tau_s = 7/16\tau_{s0}$. Note that, compared to the ballistic spin density (Figure 28.2), the period is slightly enhanced by a factor $4/\sqrt{15}$. Also, the amplitude of the spin density changes with the position x, in contrast to the ballistic case. The color is changing in proportion to the spin density amplitude.

current of spin component S_i, as defined by $\mathbf{j}_{S_i}(\mathbf{r},t) = \mathbf{v}S_i(\mathbf{r},t)$, there is the complication that the spin keeps precessing as it moves from $\mathbf{r} - \Delta\mathbf{x}$ to $\mathbf{r}$, and that the spin–orbit field changes its direction with the direction of the electron velocity $\mathbf{v}$. Therefore, the 0th order term in the expansion in $\Delta\mathbf{x}$ does not vanish, rather, we get $\mathbf{j}_{S_i}(\mathbf{r},t) = \left\langle \mathbf{v}S_i^{\mathbf{k}}(\mathbf{r},t) \right\rangle - D_e\nabla S_i(\mathbf{r},t)$ where $S_i^{\mathbf{k}}$ is the part of the spin density that evolved from the spin density at $\mathbf{r} - \Delta\mathbf{x}$ moving with velocity $\mathbf{v}$ and momentum $\mathbf{k}$. Noting that the spin precession on ballistic scales $t \leq \tau$ is governed by the Bloch equation, Equation 28.14, we find by the integration of Equation 28.14 that $S_i^{\mathbf{k}} = -\tau(\mathbf{B}_{SO}(\mathbf{k})\times\mathbf{S})_i$, so that we can rewrite the first term yielding the total spin diffusion current as

$$\mathbf{j}_{S_i} = -\tau\left\langle \mathbf{v}_F(\mathbf{B}_{SO}(\mathbf{k})\times\mathbf{S})_i \right\rangle - D\nabla S_i. \qquad (28.22)$$

Thus, we can rewrite the spin diffusion equation in terms of this spin diffusion current and get the continuity equation

$$\frac{\partial \mathbf{s}_i}{\partial t} = -D\nabla\mathbf{j}_{S_i} + \tau\left\langle \nabla\mathbf{v}_F(\mathbf{B}_{SO}(\mathbf{k})\times\mathbf{S})_i \right\rangle - \frac{1}{\hat{\tau}_{sij}}\mathbf{s}_j. \qquad (28.23)$$

It is important to note that in contrast to the continuity equation for the density, there are two additional terms due to the spin–orbit interaction. The last one is the spin relaxation tensor, which will be considered in detail in the next section. The other term arises due to the fact that Equation 28.20 contains a factor 2 in front of the spin–orbit precession term, while the spin diffusion current, Equation 28.22, does not contain that factor. This has important physical consequences, resulting in the suppression of the spin relaxation rate in quantum wires and quantum dots as soon as their lateral extension is smaller than the spin precession length, L_{SO}, as we will see in the subsequent chapters.

28.3 Spin Relaxation Mechanisms

The intrinsic spin–orbit interaction itself causes the spin of the electrons to precess coherently, as the electrons move through a conductor, defining the spin precession length L_{SO}, Equation 28.18. Since impurities and dislocations in the conductor randomize the electron momentum, the impurity scattering is transferred into a randomization of the electron spin by the spin–orbit interaction, which thereby results in spin dephasing and spin relaxation. This results in a new length scale, the spin relaxation length, L_s, which is related to the spin relaxation rate $1/\tau_s$ by

$$L_s = \sqrt{D\tau_s}. \qquad (28.24)$$

28.3.1 D'yakonov–Perel Spin Relaxation

D'yakonov–Perel spin relaxation (DPS) can be understood qualitatively in the following way: The spin–orbit field $\mathbf{B}_{SO}(\mathbf{k})$ changes its direction randomly after each elastic scattering event from an

FIGURE 28.5 Elastic scattering from impurities changes the direction of the spin–orbit field around which the electron spin is precessing.

impurity, that is, after a time of the order of the elastic scattering time τ, when the momentum is changed randomly as sketched in Figure 28.5. Thus, the spin has the time τ to perform a precession around the present direction of the spin–orbit field, and can thus change its direction only by an angle of the order of $B_{SO}\tau$ by precession. After a time t with $N_t = t/\tau$ scattering events, the direction of the spin will therefore have changed by an angle of the order of $|B_{SO}|\tau\sqrt{N_t} = |B_{SO}|\sqrt{\tau t}$. Defining the spin relaxation time, τ_s, as the time by which the spin direction has changed by an angle of order one, we thus find that $1/\tau_s \sim \tau \langle \mathbf{B}_{SO}(\mathbf{k})^2 \rangle$, where the angular brackets denote integration over all angles. Remarkably, this spin relaxation rate becomes smaller the more scattering events take place, because the smaller the elastic scattering time τ is, the less time the spin has to change its direction by precession. Such a behavior is also well known as *motional* or *dynamic narrowing* of magnetic resonance lines.[9]

A more rigorous derivation for the kinetic equation of the spin density matrix yields additional interference terms, not taken into account in the above argument. It can be obtained by iterating the expansion of the spin density Equation 28.19 once in the spin precession term, which yields the term

$$\left\langle \mathbf{s}(\mathbf{x}, t) \times \int_0^{\Delta t} dt' \mathbf{B}_{SO}\big(\mathbf{k}(t')\big) \times \int_0^{\Delta t} dt'' \mathbf{B}_{SO}\big(\mathbf{k}(t'')\big) \right\rangle, \quad (28.25)$$

where $\langle ... \rangle$ denotes the average over all angles due to the scattering from impurities. Since the electrons move ballistically at times smaller than the elastic scattering time, the momenta are correlated only on timescales smaller than τ, yielding $\langle k_i(t') k_j(t'') \rangle = (1/2)k^2 \delta_{ij}\tau\delta(t' - t'')$.

Noting that $(\mathbf{A} \times \mathbf{B} \times \mathbf{C})_m = \varepsilon_{ijk}\varepsilon_{klm}A_i B_j C_l$ and $\sum \varepsilon_{ijk}\varepsilon_{klm} = \delta_{il}\delta_{jm} - \delta_{im}\delta_{jl}$, we find that Equation 28.25 simplifies to $-\sum_j(1/\tau_{sij})S_j$, where the matrix elements of the spin relaxation terms are given by[18]

$$\frac{1}{\tau_{sij}} = \tau\Big(\big\langle \mathbf{B}_{SO}(\mathbf{k})^2 \big\rangle \delta_{ij} - \big\langle \mathbf{B}_{SO}(\mathbf{k})_i \mathbf{B}_{SO}(\mathbf{k})_j \big\rangle \Big), \quad (28.26)$$

where $\langle ... \rangle$ denotes the average over the direction of the momentum $\mathbf{k}$. These nondiagonal terms can diminish the spin relaxation and even result in vanishing spin relaxation.

As an example, we consider a quantum well where the linear Dresselhaus coupling for quantum wells grown in the

[001]-direction, Equation 28.11, and linear Rashba coupling, Equation 28.12, are the dominant spin–orbit couplings. The energy dispersion is anisotropic, as given by Equation 28.16, and the spin–orbit field, $\mathbf{B}_{SO}(\mathbf{k})$, changes its direction and its amplitude with the direction of the momentum $\mathbf{k}$:

$$\mathbf{B}_{SO}(\mathbf{k}) = -2 \begin{pmatrix} -\alpha_1 k_x + \alpha_2 k_y \\ \alpha_1 k_y - \alpha_2 k_x \\ 0 \end{pmatrix}, \quad (28.27)$$

with $|\mathbf{B}_{SO}(\mathbf{k})| = 2\sqrt{(\alpha_1^2 + \alpha_2^2)k^2 - 4\alpha_1\alpha_2 k_x k_y}$. Thus we find the spin relaxation tensor as

$$\frac{1}{\hat{\tau}_s}(k) = 4\tau k^2 \begin{pmatrix} \dfrac{1}{2}\alpha^2 & -\alpha_1\alpha_2 & 0 \\ -\alpha_1\alpha_2 & \dfrac{1}{2}\alpha^2 & 0 \\ 0 & 0 & \alpha^2 \end{pmatrix}. \quad (28.28)$$

Diagonalizing this matrix, one finds the three eigenvalues $(1/\tau_s)(\alpha_1 \pm \alpha_2)^2/\alpha^2$ and $2/\tau_s$ where $\alpha^2 = \alpha_1^2 + \alpha_2^2$, and $1/\tau_s = 2k^2 \tau\alpha^2$. Note that one of these eigenvalues of the spin relaxation tensor vanishes when $\alpha_1 = \alpha_2 = \alpha_0$. In fact, this is a special case, when the spin–orbit field does not change its direction with the momentum:

$$\mathbf{B}_{SO}(\mathbf{k})\big|_{\alpha_1=\alpha_2=\alpha_0} = 2\alpha_0(k_x - k_y) \begin{pmatrix} 1 \\ 1 \\ 0 \end{pmatrix}. \quad (28.29)$$

In this case, the constant spin density given by

$$\mathbf{S} = S_0 \begin{pmatrix} 1 \\ 1 \\ 0 \end{pmatrix} \quad (28.30)$$

does not decay in time, since the spin density vector is parallel to the spin–orbit field $\mathbf{B}_{SO}(\mathbf{k})$, Equation 28.29, and cannot precess, as has been noted in Ref. [4]. It turns out, however, that there are two more modes that do not decay in time, whose spin relaxation rate vanishes for $\alpha_1 = \alpha_2$. These modes are not homogeneous in space, and correspond to precessing spin densities. They were found previously in a numerical Monte Carlo simulation and found not to decay in time, therefore being called *persistent spin helix*.[7,53] Recently, a long-living inhomogeneous spin density distribution has been detected experimentally in Ref. [64].

We can now get these *persistent spin helix modes* analytically by solving the full spin diffusion equation, Equation 28.20, with the spin relaxation tensor given by Equation 28.28. We can diagonalize that equation, noting that its eigenfunctions are plain waves $\mathbf{S}(\mathbf{x}) \sim \exp(i\mathbf{Q}\mathbf{x} - Et)$. Thereby one finds, first of all, the mode with eigenvalue $E_1 = DQ^2$, with the spin density

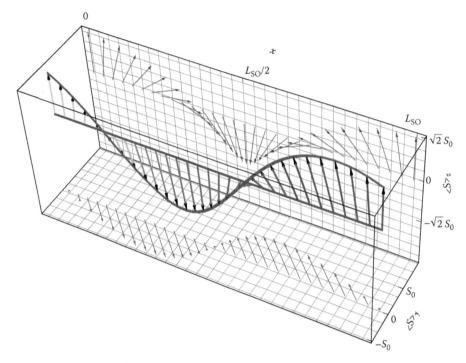

FIGURE 28.6 Persistent spin helix solution of the spin diffusion equation for equal magnitude of linear Rashba and linear Dresselhaus coupling (Equation 28.33).

$$\mathbf{S} = S_0 \begin{pmatrix} 1 \\ 1 \\ 0 \end{pmatrix} \exp(i\mathbf{Qx} - D\mathbf{Q}^2 t). \qquad (28.31)$$

Indeed for $\mathbf{Q} = 0$, the homogeneous solution, it does not decay in time, in agreement with the solution we found above, Equation 28.31. There are, however, two more modes with the eigenvalues

$$E_{\pm} = \frac{1}{\tau_s} \left(\tilde{\mathbf{Q}}^2 + 2 \pm 2 \left| \tilde{Q}_x - \tilde{Q}_y \right| \right), \qquad (28.32)$$

where $\tilde{Q} = L_{SO}Q/2\pi$. At $\tilde{Q}_x = -\tilde{Q}_y = \pm 1$, these modes do not decay in time. These two stationary solutions are

$$\mathbf{S} = S_0 \begin{pmatrix} 1 \\ -1 \\ 0 \end{pmatrix} \sin\left(\frac{2\pi}{L_{SO}}(x-y) \right) + S_0 \sqrt{2} \begin{pmatrix} 0 \\ 0 \\ 1 \end{pmatrix} \cos\left(\frac{2\pi}{L_{SO}}(x-y) \right),$$

$$(28.33)$$

and the linearly independent solution obtained by interchanging cos and sin in Equation 28.33. The spin precesses as the electrons diffuse along the quantum well with the period L_{SO}, the spin precession length, forming a *persistent spin helix*, as shown in Figure 28.6.

28.3.2 DP Spin Relaxation with Electron–Electron and Electron– Phonon Scattering

It has been noted, that the momentum scattering which limits the D'yakonov–Perel mechanism of spin relaxation is not restricted to impurity scattering, but can also be due to

electron–phonon or electron–electron interactions.[19,26,27,57] Thus, the scattering time τ is the total scattering time as defined, $1/\tau = 1/\tau_0 + 1/\tau_{ee} + 1/\tau_{ep}$, where $1/\tau_0$ is the elastic scattering rate due to scattering from impurities with potential V, given by $1/\tau_0 = 2\pi\nu n_i \int (d\theta/2\pi)(1 - \cos\theta) |V(\mathbf{k}, \mathbf{k}')|^2$, where ν is the density of states per spin at the Fermi energy, n_i is the concentration of impurities with potential V, and $\mathbf{kk}' = kk' \cos(\theta)$. In degenerate semiconductors and in metals, the electron–electron scattering rate is given by the Fermi liquid inelastic electron scattering rate $1/\tau_{ee} \sim T^2/\varepsilon_F$. The electron–phonon scattering time, $1/\tau_{ep} \sim T^5$, decays faster with temperature. Thus, at low temperatures, the DP spin relaxation is dominated by elastic impurity scattering τ_0.

28.3.3 Elliott–Yafet Spin Relaxation

Because of the spin–orbit interaction, the conduction electron wave functions are not eigenstates of the electron spin, but have an admixture of both spin-up and spin-down wave functions. Thus, a nonmagnetic impurity potential V can change the electron spin, by changing their momentum due to the spin–orbit coupling. This results in another source of spin relaxation which is stronger, the more often the electrons are scattered, and is thus proportional to the momentum scattering rate $1/\tau$.[21,68] For degenerate III–V semiconductors one finds[13,56]

$$\frac{1}{\tau_s} \sim \frac{\Delta_{SO}^2}{(E_G + \Delta_{SO})^2} \frac{E_k^2}{E_G^2} \frac{1}{\tau(k)}, \qquad (28.34)$$

where E_G is the gap between the valence and the conduction band of the semiconductor, and Δ_{SO} is the spin–orbit splitting of the valence band. Thus, the Elliott–Yafet spin relaxation (EYS) can be distinguished, being proportional to $1/\tau$, and thereby to the resistivity, in contrast to the DP spin scattering rate, Equation 28.26, which is proportional to the conductivity. Since the EYS decays in proportion to the inverse of the band gap, it is negligible in large band gap semiconductors like Si and GaAs. The scattering rate $1/\tau$ is again the sum of the impurity scattering rate,[21] the electron–phonon scattering rate,[29,68] and the electron–electron interaction,[10] so that all these scattering processes result in EY spin relaxation. In nondegenerate semiconductors, where the Fermi energy is below the conduction band edge, $1/\tau_s \sim \tau T^3 / E_G$ attains a stronger temperature dependence.

28.3.4 Spin Relaxation due to Spin–Orbit Interaction with Impurities

The spin–orbit interaction, as defined in Equation 28.3, arises whenever there is a gradient in an electrostatic potential. Thus, the impurity potential gives rise to the spin–orbit interaction

$$V_{SO} = \frac{1}{2m^2c^2} \nabla V \times \mathbf{k}\, \mathbf{s} \qquad (28.35)$$

Perturbation theory yields, then, directly the corresponding spin relaxation rate

$$\frac{1}{\tau_s} = \pi \nu n_i \sum_{\alpha,\beta} \int \frac{d\theta}{2\pi} (1 - \cos\theta) \left| V_{SO}(\mathbf{k},\mathbf{k}')_{\alpha\beta} \right|^2, \qquad (28.36)$$

proportional to the concentration of impurities, n_i. Here α, $\beta = \pm$ denotes the spin indices. Since the spin–orbit interaction increases with the atomic number Z of the impurity element, this spin relaxation increases as Z^2, being stronger for heavier element impurities.

28.3.5 Bir-Aronov–Pikus Spin Relaxation

The exchange interaction, J, between electrons and holes in p-doped semiconductors results in spin relaxation, as well.[8] Its strength is proportional to the density of holes p and depends on their itinerancy. If the holes are localized, they act like magnetic impurities. If they are itinerant, the spin of the conduction electrons is transferred by the exchange interaction to the holes, where the spin–orbit splitting of the valence bands results in fast spin relaxation of the hole spin due to the Elliott–Yafet, or the Dyakonov–Perel mechanism.

28.3.6 Magnetic Impurities

Magnetic impurities have a spin **S** that interacts with the spin of the conduction electrons by the exchange interaction J, resulting in a spatially and temporarily fluctuating local magnetic field

$$\mathbf{B}_{MI}(\mathbf{r}) = -\sum_i J \delta(\mathbf{r} - \mathbf{R}_i) \mathbf{S}, \qquad (28.37)$$

where the sum is over the position of the magnetic impurities $\mathbf{R}_i$. This gives rise to spin relaxation of the conduction electrons, with a rate given by

$$\frac{1}{\tau_{Ms}} = 2\pi n_M \nu J^2 S(S+1), \qquad (28.38)$$

where
  n_M is the density of magnetic impurities,
  ν is the density of states at the Fermi energy.

Here, S is the spin quantum number of the magnetic impurity, which can take the values $S = 1/2, 1, 3/2, 2....$

Antiferromagnetic exchange interaction between the magnetic impurity spin and the conduction electrons results in a competition between the conduction electrons to form a singlet with the impurity spin, which results in enhanced nonmagnetic and magnetic scattering. At low temperatures, the magnetic impurity spin is screened by the conduction electrons, resulting in a vanishing of the magnetic scattering rate. Thus, the spin scattering from magnetic impurities has a maximum at a temperature of the order of the Kondo temperature, $T_K \sim E_F \exp(-1/\nu J)$, where ν is the density of states at the Fermi energy.[50,52,71] In semiconductors, T_K is exponentially small due to the small effective mass and the resulting small density of states, ν. Therefore, the magnetic moments remain free at the experimentally achievable temperatures. At large concentration of magnetic impurities, the RKKY-exchange interaction between the magnetic impurities quenches however the spin quantum dynamics, so that $S(S+1)$ is replaced by its classical value S^2. In Mn-p-doped GaAs, the exchange interaction between the Mn dopants and the holes can result in compensation of the hole spins and therefore a suppression of the Bir-Aronov–Pikus (BAP) spin relaxation.[3]

28.3.7 Nuclear Spins

Nuclear spins interact by the hyperfine interaction with conduction electrons. The hyperfine interaction between nuclear spins, $\hat{\mathbf{I}}$, and the conduction electron spin, $\hat{s}$, results in a local Zeeman field given by[54]

$$\hat{\mathbf{B}}_N(\mathbf{r}) = -\frac{8\pi}{3} \frac{g_0 \mu_B}{\gamma_g} \sum_n \gamma_n \hat{\mathbf{I}}_n \delta(\mathbf{r} - \mathbf{R}_n), \qquad (28.39)$$

where γ_n is the gyromagnetic ratio of the nuclear spin. The spatial and temporal fluctuations of this hyperfine interaction field result in spin relaxation proportional to its variance, similar to the spin relaxation by magnetic impurities.

28.3.8 Magnetic Field Dependence of Spin Relaxation

The magnetic field changes the electron momentum due to the Lorentz force, resulting in a continuous change of the spin–orbit field, which similar to the momentum scattering results in motional narrowing and thereby a reduction of DPS:[12,34,56]

$$\frac{1}{\tau_s} \sim \frac{\tau}{1 + \omega_c^2 \tau^2}. \qquad (28.40)$$

Another source of a magnetic field dependence is the precession around the external magnetic field. In bulk semiconductors and for magnetic fields perpendicular to a quantum well, the orbital mechanism is dominating, however. This magnetic field dependence can be used to identify the spin relaxation mechanism, since the EYS does have only a weak magnetic field dependence due to the weak Pauli-paramagnetism.

28.3.9 Dimensional Reduction of Spin Relaxation

Electrostatic confinement of conduction electrons can reduce the effective dimension of their motion. In *quantum dots*, the electrons are confined in all three directions, and the energy spectrum consists of discrete levels like in atoms. Therefore, the energy conservation restricts relaxation processes severely, resulting in strongly enhanced spin relaxation times in quantum dots.[1,39] Then, spin relaxation can only occur due to the absorption or the emission of phonons, yielding spin relaxation rates proportional to the inelastic electron–phonon scattering rate.[39] Quantitative comparison of the various spin relaxation mechanisms in GaAs quantum dots resulted in the conclusion that the spin relaxation is dominated by the hyperfine interaction.[22,37,38]

A similar conclusion can be drawn from experiments on low temperature spin relaxation in low density n-type GaAs, where the localization of the electrons in the *impurity band* results in spin relaxation dominated by hyperfine interaction as well.[20,62]

For linear Rashba and linear Dresselhaus spin–orbit coupling, we can see from the spin diffusion equation (Equation 28.20) with the DP spin relaxation tensor (Equation 28.28) that the spin relaxation vanishes, when the spin current (Equation 28.22) vanishes, in which case the last two terms of Equation 28.20 cancel exactly. The vanishing of the spin current is imposed by hard wall boundary condition for which the spin diffusion current vanishes at the boundaries of the sample, $\mathbf{j}_{S_i}\mathbf{n}\big|_{\text{Boundary}} = 0$, where $\mathbf{n}$ is the normal to the boundary. When the quantum dot is smaller than the spin precession length L_{SO}, the lowest energy mode thus corresponds to a homogeneous solution with vanishing spin relaxation rate. Cubic spin–orbit coupling does not yield such a vanishing of the DP spin relaxation rate. Only in quantum dots whose size does not exceed the elastic mean free path, l_e, the DP spin relaxation from cubic spin relaxation becomes diminished.

In *quantum wires*, the electrons have a continuous spectrum of delocalized states. Still, transverse confinement can reduce the DP spin relaxation as we review in the next section.

28.4 Spin Dynamics in Quantum Wires

28.4.1 One-Dimensional Wires

In one-dimensional wires, whose width W is of the order of the Fermi wavelength λ_F, impurities can only reverse the momentum $p \rightarrow -p$. Therefore, the spin–orbit field can only change its sign, when a scattering from impurities occurs. $\mathbf{B}_{SO}(p) \rightarrow \mathbf{B}_{SO}(-p) = -\mathbf{B}_{SO}(p)$. Therefore, the precession axis and the amplitude of the spin–orbit field does not change, reversing only the

spin precession, so that the D'yakonov–Perel-spin relaxation is absent in one-dimensional wires.[49] In an external magnetic field, the precession around the magnetic field axis, due to the Zeeman interaction, is competing with the spin–orbit field, however. Then, as the electrons are scattered from impurities, both the precession axis and the amplitude of the total precession field is changing, since

$$|\mathbf{B} + \mathbf{B}_{SO}(-p)| = |\mathbf{B} - \mathbf{B}_{SO}(p)| \neq |\mathbf{B} + \mathbf{B}_{SO}(p)|,$$

resulting in spin dephasing and relaxation, as the sign of the momentum changes randomly.

28.4.2 Spin Diffusion in Quantum Wires

How does the spin relaxation rate depend on the wire width W when the quantum wire has more than one channel occupied, $W > \lambda_F$? Clearly, for large wire widths, the spin relaxation rate should converge to a finite value, while it vanishes for $W \rightarrow \lambda_F$. It is both of practical importance for spintronic applications and of fundamental interest to know on which length scales this crossover occurs. Basically, there are three intrinsic length scales characterizing the quantum wire relative to its width W. The Fermi wavelength λ_F, the elastic mean free path l_e, and the spin precession length L_{SO} (Equation 28.18). Suppression of spin relaxation for wire widths not exceeding the elastic mean free path, l_e, has been predicted and obtained numerically in Refs. [11,16,35,40,47,55]. Is the spin relaxation rate also suppressed in *diffusive wires* in which the elastic mean free path is smaller than the wire width as in the wire shown schematically in Figure 28.7? We will answer this question by means of an analytical derivation in the following.

The transversal confinement imposes that the spin current vanishes normal to the boundary, $\mathbf{j}_{S_i}\mathbf{n}\big|_{\text{Boundary}} = 0$. For a wire grown along the [010]-direction, $\mathbf{n} = \hat{e}_x$ is the unit vector in the x-direction. For wire widths W smaller than the spin precession length L_{SO}, the solutions with the lowest energy have thus a vanishing transverse spin current, and the spin diffusion equation (Equation 28.20) becomes

$$\frac{\partial s_i}{\partial t} = -D\partial_y j_{S_i,y} + \tau\left\langle\nabla_{VF}\left(\mathbf{B}_{SO}(\mathbf{k})\times\mathbf{S}\right)_i\right\rangle - \sum_j \frac{1}{\tau_{sij}} s_j \quad (28.41)$$

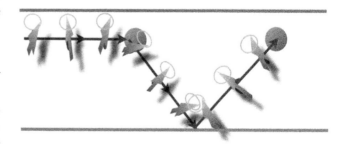

FIGURE 28.7 Elastic scatterings from impurities and from the boundary of the wire change the direction of the spin–orbit field around which the electron spin is precessing.

with

$$j_{S_i x}\Big|_{x=\pm W/2} = \left(-\tau\langle v_x(\mathbf{B}_{SO}(\mathbf{k})\times\mathbf{S})_i\rangle - D\partial_x S_i\right)\Big|_{x=\pm W/2} = 0, \tag{28.42}$$

where W is the width of the wire. One sees that this equation has a persistent solution, which does not decay in time and is homogeneous along the wire, $\partial_y S = 0$. In this special case, the spin diffusion equation simplifies to[60]

$$\partial_t \mathbf{S} = -\frac{1}{\tau_s\alpha^2}\begin{pmatrix} \alpha_1^2 & -\alpha_1\alpha_2 & 0 \\ -\alpha_1\alpha_2 & \alpha_2^2 & 0 \\ 0 & 0 & \alpha^2 \end{pmatrix}\mathbf{S}. \tag{28.43}$$

Indeed this has one persistent solution given by

$$\mathbf{S} = S_0\begin{pmatrix} \alpha_2 \\ \alpha_1 \\ 0 \end{pmatrix}. \tag{28.44}$$

Thus, we can conclude that the boundary conditions impose an effective alignment of all spin–orbit fields, in a direction identical to the one it would attain in a one-dimensional wire, along the [010]-direction, setting $k_x = 0$ in Equation 28.27,

$$\mathbf{B}_{SO}(\mathbf{k}) = -2k_y\begin{pmatrix} \alpha_2 \\ \alpha_1 \\ 0 \end{pmatrix}, \tag{28.45}$$

which therefore does not change its direction when the electrons are scattered. This is remarkable, since this alignment already occurs in wires with many channels, where the impurity scattering is two dimensional, and the transverse momentum, k_x, actually can be finite. Rather, the alignment of the spin–orbit field, accompanied by a suppression of the DP spin relaxation rate occurs due to the constraint on the spin dynamics imposed by the boundary conditions as soon as the wire width, W, is smaller than the length scale which governs the spin dynamics, namely, the spin precession length, L_{SO}.

It turns out that the spin diffusion equation (Equation 28.41) has also two long persisting spin helix solutions in narrow wires[36,65] that oscillate periodically with the period, $L_{SO} = \pi/m^*\alpha$. In contrast to the situation in 2D systems we reviewed in Section 28.3, in quantum wires of width $W < L_{SO}$ these solutions are long persisting even for $\alpha_1 \neq \alpha_2$. These two stationary solutions,

$$\mathbf{S} = S_0\begin{pmatrix} \dfrac{\alpha_1}{\alpha} \\ -\dfrac{\alpha_2}{\alpha} \\ 0 \end{pmatrix}\sin\left(\frac{2\pi}{L_{SO}}y\right) + S_0\begin{pmatrix} 0 \\ 0 \\ 1 \end{pmatrix}\cos\left(\frac{2\pi}{L_{SO}}y\right), \tag{28.46}$$

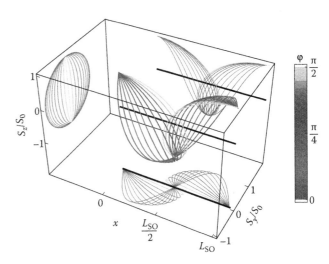

FIGURE 28.8 **(See color insert following page 20-16.)** Persistent spin helix solution of the spin diffusion equation in a quantum wire whose width, W, is smaller than the spin precession length, L_{SO}, for varying ratio of linear Rashba, $\alpha_2 = \alpha\sin\varphi$, and linear Dresselhaus coupling, $\alpha_1 = \alpha\cos\varphi$ (Equation 28.46) for fixed α and $L_{SO} = \pi/m^*\alpha$.

and the linearly independent solution, are obtained by interchanging cos and sin in Equation 28.46. The spin precesses as the electrons diffuse along the quantum wire with the period L_{SO}, the spin precession length, forming a *persistent spin helix*, whose x-component is proportional to the linear Dresselhaus coupling, α_x, while its y-component is proportional to the Rashba coupling, α_2, as seen in Figure 28.8.

A similar reduction of the spin relaxation rate is not effective for cubic spin–orbit coupling for wire widths exceeding the elastic mean free path, l_e.

One can derive the spin relaxation rate as a function of the wire width for diffusive wires, $l_e < W < L_{SO}$. The total spin relaxation rate, in the presence of both the linear Rashba spin–orbit coupling, α_2, and the linear and cubic Dresselhaus coupling α_1 and γ_D, respectively, is as function of wire width W given by[36]

$$\frac{1}{\tau_s}(W) = \frac{1}{12}\left(\frac{W}{L_{SO}}\right)^2\delta_{SO}^2\frac{1}{\tau_s} + D(m^{*2}\varepsilon_F\gamma_D)^2, \tag{28.47}$$

where $1/\tau_s = 2p_F^2\left(\alpha_2^2 + (\alpha_1 - m^*\gamma_D\varepsilon_F/2)^2\right)\tau$. We introduced the dimensionless factor $\delta_{SO} = (Q_R^2 - Q_D^2)/Q_{SO}^2$ with $Q_{SO}^2 = Q_D^2 + Q_R^2$, where Q_D depends on Dresselhaus spin–orbit coupling, $Q_D = m^*(2\alpha_1 - m^*\varepsilon_F\gamma)$. Q_R depends on Rashba coupling: $Q_R = 2m^*\alpha_2$. Thus, for negligible cubic Dresselhaus spin–orbit coupling, the spin relaxation length increases when decreasing the wire width W as

$$L_s(W) = \sqrt{D\tau_s(W)} \sim \frac{L_{SO}^2}{W}. \tag{28.48}$$

This can be understood as follows:[24,36,59] In a wire whose width exceeds the spin precession length L_{SO}, the area an electron covers by diffusion in time τ_s is WL_s. To achieve spin relaxation, this area should be equal to the corresponding 2D spin relaxation area $L_s(2D)^2$, where $L_s(2D) = L_{SO}/(2\pi)$. Thus, the smaller the wire width, the larger the spin relaxation length becomes, $L_s \sim (L_{SO})^2/W$ in agreement with Equation 28.48. For larger wire widths, the spin diffusion equation can be solved as well, and one finds that the spin relaxation rate does not increase monotonously to the 2D limiting value but shows oscillations on the scale L_{SO}, which can be understood in analogy to Fabry–Pérot resonances.[36] For pure linear Rashba coupling, that behavior can be derived analytically, in the approximation of a homogeneous spin density in transverse direction, yielding a relaxation rate given by

$$\frac{1}{\tau_s}(W) = \frac{D}{2} Q_{SO}^2 \left(1 - \frac{\sin(Q_{SO}W)}{Q_{SO}W} \right),   \quad (28.49)$$

where $Q_{SO} = 2\pi/L_{SO}$. Furthermore, taking into account the transverse modulation of the spin density by performing an exact diagonalization of the spin diffusion equation with the transverse boundary conditions (Equation 28.42), one finds for $W > L_{SO}$ modes which are localized at the boundaries and have a lower relaxation rate than the bulk modes.[60,65] For pure Rashba spin relaxation, we find that there is a spin helix solution located at the edge whose relaxation rate $1/\tau_s = 0.31/\tau_{s0}$ is smaller than the spin relaxation rate of bulk modes, $1/\tau_s = 7/16\tau_{s0}$.

28.4.3 Weak Localization Corrections

Quantum interference of electrons in low-dimensional, disordered conductors results in corrections to the electrical conductivity, $\Delta\sigma$. This quantum correction, the weak localization effect, is known to be a very sensitive tool to study dephasing and symmetry breaking mechanisms in conductors.[2] The entanglement of spin and charge by spin–orbit interaction reverses the effect of weak localization and thereby enhances the conductivity, the weak antilocalization effect.

The quantum correction to the conductivity, $\Delta\sigma$, arises from the fact that the quantum return probability to a given point $\mathbf{x}_0$ after a time t, $P(t)$, differs from the classical return probability, due to quantum interference. As the electrons scatter from impurities, there is a finite probability that they diffuse on closed paths, which

does increase the lower the dimension of the conductor. Since an electron can move on such a closed orbit clockwise or anticlockwise, as shown in gray and black in Figure 28.9, with equal probability, the probability amplitudes of both paths add coherently, if their length is smaller than the dephasing length, L_φ. In a magnetic field as indicated by the big arrow in the middle of Figure 28.9, the electrons acquire a magnetic flux phase. This phase depends on the direction in which the electron moves on the closed path. Thus, the quantum interference is diminished in an external magnetic field since the area of closed paths and thereby the flux phases are randomly distributed in a disordered wire, even though the magnetic field can be constant. Similarly, the scattering from magnetic impurities breaks the time reversal invariance between the two directions in which the closed path can be transversed. Therefore, magnetic impurities diminish the quantum corrections in proportion to the rate at which the electron spins scatter from them due to the exchange interaction, $1/\tau_{Ms}$ (Equation 28.38).

Thus, the quantum correction to the conductivity, $\Delta\sigma$, is proportional to the integral over all times smaller than the dephasing time, τ_φ, of the quantum mechanical return probability, $P(t) = \lambda_F^d \rho(\mathbf{x},t)$, where d is dimension of diffusion and ρ is the electron density.

In the presence of spin–orbit scattering, the sign of the quantum correction changes to weak antilocalization as was predicted by Hikami et al.[30] for conductors with impurities of heavy elements. As conduction electrons scatter from such impurities, the spin–orbit interaction randomizes their spin, Figure 28.10. The resulting spin relaxation suppresses interference of time-reversed paths in spin triplet configurations, while interference in singlet configuration remains unaffected, as indicated in Figure 28.10. Since singlet interference reduces the electron's return probability it enhances the conductivity, the weak antilocalization effect. Weak magnetic fields suppress also these singlet contributions, reducing the conductivity and resulting in negative magnetoconductivity. If the host lattice of the electrons provides spin–orbit interaction, the spin relaxation of DP or EY type does have the same effect of diminishing the quantum corrections in the triplet configuration.

When the dephasing length, L_φ, is smaller than the wire width W, the quantum corrections are determined by the interference of 2-dimensional closed diffusion paths, and as a result, the

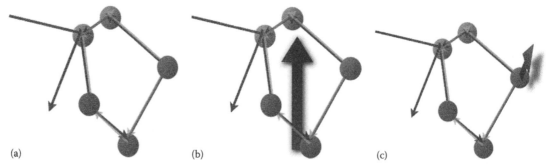

(a)   (b)   (c)

FIGURE 28.9   (a) Electrons can diffuse on closed paths, orbit clockwise or anticlockwise as indicated by the gray and black arrows, respectively. (b) Closed electron paths enclose a magnetic flux from an external magnetic field, indicated as the big arrow, breaking time reversal symmetry. (c) The scattering from a magnetic impurity spin breaks the time reversal symmetry between the clock- and anticlockwise electron paths.

SPIN up

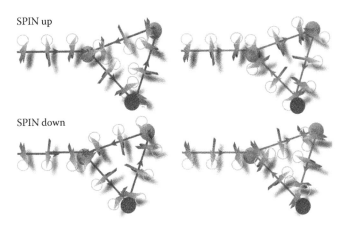

SPIN down

FIGURE 28.10 As electrons diffuse, their spin precesses around the spin–orbit field, which changes its orientation, when the electron is scattered. Electrons that enter closed paths with the same spin leave it therefore with a different spin if they choose the path in the opposite sense, as indicated by the light and dark blue arrows. However, electrons that enter the closed path with opposite spin, and move through the closed path in the opposite sense, attain the same quantum phase. This is a consequence of time reversal invariance.

conductivity increases logarithmically with L_φ, which increases itself as the temperature is lowered. At low temperatures, the electron–electron scattering is the dominating mechanism of spin dephasing, yielding $L_\varphi \sim T^{-1/2}$. One can derive the magnetic field dependence of that quantum correction nonperturbatively.[1,28,30,41,45,51] An approximate expression showing the logarithmic dependence explicitly is given by

$$\Delta\sigma = -\frac{1}{2\pi}\ln\frac{B + \frac{4}{3}H_{Ms} + H_\varphi}{H_\tau} + \frac{1}{2\pi}\ln\frac{B + H_\varphi + H_s + \frac{2}{3}H_{Ms}}{H_\tau}$$

$$+ \frac{1}{\pi}\ln\frac{B + H_\varphi + cH_s + \frac{2}{3}H_{Ms}}{H_\tau}, \qquad (28.50)$$

in units of e^2/h. All parameters are rescaled to dimensions of magnetic fields: $H_\varphi = 1/(4eD\tau_\varphi) = 1/(4eL_\varphi^2)$, $H_\tau = \hbar/(4eD\tau)$, the spin relaxation field due to spin–orbit relaxation, $H_s = \hbar/(4eD\tau_s)$,[41] and the spin relaxation field due to magnetic impurities $H_{Ms} = \hbar/(4eD\tau_{Ms})$. Here, $1/\tau_s$ is the DP relaxation rate in the 2D limit derived in the previous section.[33,41] The prefactor c depends on the particular spin–orbit interaction. For linear Rashba coupling, $c = 7/16$. Note that $7/16\tau_s$ is the smallest spin relaxation rate of an inhomogeneous spin density distribution[65] as derived in Section 28.2.2.4. $1/\tau_{Ms}$ is the magnetic scattering rate from magnetic impurities (Equation 28.38). Indeed, we see that the first term does not depend on the DP spin relaxation rate. This term originates from the interference of time-reversed paths, indicated in Figure 28.10, which contributes to the quantum conductance in the singlet state, $|S=0; m=0\rangle = \left(|\uparrow\downarrow\rangle - |\downarrow\uparrow\rangle\right)/\sqrt{2}$, where the minus sign in front of the second term is the origin of

the change in sign in the weak localization correction. The other three terms are suppressed by the spin relaxation rate, since they originate from interference in triplet states $|S=1; m=0\rangle = \left(|\uparrow\downarrow\rangle + |\downarrow\uparrow\rangle\right)/\sqrt{2}, |S=1; m=1\rangle, |S=1; m=-1\rangle$, which do not conserve the spin symmetry. Thus, at strong spin–orbit-induced spin relaxation, the last three terms are suppressed and the sign of the quantum correction switches to weak antilocalization.

In quasi-1-dimensional quantum wires that are coherent in transverse direction, $W < L_\varphi$, the weak localization correction is further enhanced, and increases linearly with the dephasing length, L_φ. Thus, for $WQ_{SO} \ll 1$ the weak localization correction is[36]

$$\Delta\sigma = \frac{\sqrt{H_W}}{\sqrt{H_\varphi + \frac{1}{4}B^\star(W) + \frac{2}{3}H_{Ms}}} - \frac{\sqrt{H_W}}{\sqrt{H_\varphi + \frac{1}{4}B^\star(W) + H_s(W) + \frac{2}{3}H_{Ms}}}$$

$$- 2\frac{\sqrt{H_W}}{\sqrt{H_\varphi + \frac{1}{4}B^\star(W) + \frac{1}{2}H_s(W) + \frac{4}{3}H_{Ms}}}, \qquad (28.51)$$

in units of e^2/h. We defined $H_W = \hbar/(4eW^2)$, and the effective external magnetic field,

$$\mathbf{B}^\star(W) = \left(1 - 1/\left(1 + \frac{W^2}{3l_B^2}\right)\right)\mathbf{B}. \qquad (28.52)$$

The spin relaxation field, $H_s(W)$, is for $W < L_{SO}$,

$$H_s(W) = \frac{1}{12}\left(\frac{W}{L_{SO}}\right)^2 \delta_{SO}^2 H_s, \qquad (28.53)$$

suppressed in proportion to $(W/L_{SO})^2$. In analogy to the effective magnetic field (Equation 28.52), the spin–orbit coupling acts in quantum wires like an effective magnetic vector potential.[24] One can expect that in ballistic wires, $l_e > W$, the spin relaxation rate is suppressed in analogy to the flux-cancellation effect, which yields the weaker rate, $1/\tau_s = (W/Cl_e)(DW^2/12L_{SO}^4)$ where $C = 10.8$.[5]

The dimensional crossover from weak antilocalization to weak localization, seen in Figure 28.11, as the wire width W is reduced, has recently been observed experimentally in quantum wires as we will review in Chapter 29.

28.5 Experimental Results on Spin Relaxation Rate in Semiconductor Quantum Wires

28.5.1 Optical Measurements

Optical time-resolved Faraday rotation (TRFR) spectroscopy[61] has been used to probe the spin dynamics in an array of n-doped InGaAs wires by Holleitner et al. in Refs. [31,32].

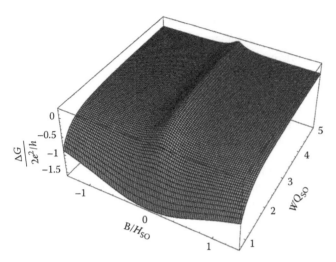

FIGURE 28.11 The quantum conductivity correction in units of $2e^2/h$ as function of magnetic field **B** (scaled with bulk relaxation field H_S) and the wire width W (scaled with $L_{SO}/2\pi$) for pure Rashba coupling, $\delta_{SO} = 1$.

The wires were dry etched from a quantum well grown in the [001]-direction with a distance of 1 μm between the wires. Spin-aligned charge carriers were created by the absorption of circularly polarized light. For normal incidence, the spins point then perpendicular to the quantum well plane, in the growth direction [001]. The time evolution of the spin polarization was then measured with a linearly polarized pulse (see inset of Fig. 1c of Ref. [31]). The time dependence fits well with an exponential decay, $\sim \exp(-\Delta t/\tau_s)$. As seen in Fig. 2a of Ref. [31], the thus measured lifetime, τ_s, at fixed temperature, $T = 5$ K, of the spin polarization is enhanced when the wire width, W, is reduced[31]: While for $W > 15$ μm it is $\tau_s = (12 \pm 1)$ ps, it increases for channels grown along the [100]-direction to almost $\tau_s = 30$ ps, and in the [110]-direction to about $\tau_s = 20$ ps. Thus, the experimental results show that the spin relaxation depends on the patterning direction of the wires: wires aligned along [100] and [010] show equivalent spin relaxation times, which are generally longer than the spin relaxation times of wires patterned along [110] and [$\bar{1}$10]. The dimensional reduction could be seen already for wire widths as wide as 10 μm, which is much wider than both the Fermi wavelength and the elastic mean free path, l_e, in the wires. This agrees well with the predicted reduction of the DP scattering rate (Equation 28.47) for wire widths smaller than the spin precession length, L_{SO}. From the measured 2D spin diffusion length, $L_s(2D) = (0.9 - 1.1)$ μm, and its relation to the spin precession length (Equation 28.18), $L_{SO} = 2\pi L_s(2D)$, we expect the crossover to occur on a scale of $L_{SO} = (5.7 - 6.9)$ μm as observed in Fig. 2a of Ref. [31]. From $L_{SO} = \pi/m^*\alpha$, we get, with $m^* = .064 m_e$, a spin–orbit coupling $\alpha = (5 - 6)$ meVÅ. According to $L_s = \sqrt{D\tau_s}$, the spin relaxation length increases by a factor of $\sqrt{30/12} = 1.6$ in the [100]-direction, and by $\sqrt{20/12} = 1.3$ in the [110]-direction.

The spin relaxation time has been found to attain a maximum, however, at about $W = 1$ μm $\approx L_s(2D)$, decaying

appreciably for smaller widths. While a saturation of τ_s could be expected according to Equation 28.47 for diffusive wires, due to cubic Dresselhaus coupling, a decrease is unexpected. Schwab et al., Ref. [60], noted that with wire boundary conditions that do not conserve the spin of the conduction electrons, one can obtain such a reduction. A mechanism for such spin-flip processes at the edges of the wire has not yet been identified, however.

The magnetic field dependence of the spin relaxation rate yields further confirmation that the dominant spin relaxation mechanism in these wires is DPS: It follows the predicted behavior Equation 28.40, as seen in Fig. 3a of Ref. [31], and the spin relaxation rate is enhanced to τ_s (**B** = 1 T) = 100 ps for all wire growth directions, at $T = 5$ K and wire widths of $W = 1.25$ μm.

28.5.2 Transport Measurements

A dimensional crossover from weak antilocalization to weak localization and a reduction of spin relaxation has recently been observed experimentally in n-doped InGaAs quantum wires,[48,67] in GaAs wires,[15] as well as in AlGaN/GaN wires.[44] The crossover indeed occurred in all experiments on the length scale of the spin precession length, L_{SO}. We summarize in the following the main results of these experiments.

Wirthmann et al. [67] measured the magnetoconductivity of inversion-doped InAs quantum wells with a density of $n = 9.7 \times 10^{11}$/cm², and a measured effective mass of $m^* = 0.04 m_e$. In the wide wires, the magnetoconductivity showed a pronounced weak antilocalization peak, which agreed well with the 2D theory,[33,41] with a spin–orbit-coupling parameter of $\alpha = 9.3$ meVÅ. They observed a diminishment of the antilocalization peak that occurred for wire widths $W < 0.6$ μm, at $T = 2$ K, indicating a dimensional reduction of the DP spin relaxation rate.

Schäpers et al. observed in $Ga_xIn_{1-x}As/InP$ quantum wires a complete crossover from weak antilocalization to weak localization for wire widths below $W = 500$ nm. Such a crossover has also been observed in GaAs quantum wires by Dinter et al.[15]

Very recently, Kunihashi et al.[42] observed the crossover from weak antilocalization to weak localization in gate-controlled InGaAs quantum wires. The asymmetric potential normal to the quantum well could be enhanced by application of a negative gate voltage, yielding an increase of the SIA-coupling parameter, α, with decreasing carrier density, as was obtained by fitting the magnetoconductivity of the quantum wells to the theory of 2D weak localization corrections of Iordanskii et al.[33] Thereby, the spin relaxation length, $L_s = L_{SO}/2\pi$, was found to decrease from 0.5 to 0.15 μm, which according to $L_{SO} = \pi/m^*\alpha$ corresponds to an increase of α from (20 ± 1) meVÅ at electron concentrations of $n = 1.4 \times 10^{12}$/cm² to $\alpha = (60 \pm 1)$ meVÅ at electron concentrations of $n = 0.3 \times 10^{12}$/cm². The magnetoconductivity of a sample with 95 quantum wires in parallel showed a clear crossover from weak antilocalization to localization. Fitting the data to Equation 28.51, a corresponding decrease of the spin relaxation rate was obtained, which was observable already at large widths of the order of the spin precession length, L_{SO}, in agreement with the theory Equation 28.47. However, a saturation as obtained theoretically in diffusive wires, due to cubic BIA coupling was not observed. This might be due to the limitation of Equation 28.47, to diffusive wire widths, $l_e < W$, while in ballistic wires a suppression also of the spin relaxation due to cubic BIA coupling can be expected, since it vanishes identically in 1D wires (see Section 28.4.1). Also, an increase of the spin scattering rate in narrower wires, $W < L_s(2D)$, was not observed in contrast to the results of the optical experiments (Ref. [31]) reviewed above.

The dimensional crossover has also been observed in the heterostructures of the wide-gap semiconductor GaN.[44] The magnetoconductivity of 160 AlGaN/GaN-quantum wires were measured. The effective mass is $m^* = .22m_e$, all wires were diffusive with $l_e <$ W. For electron densities of $n \approx 5 \times 10^{12}/cm^2$, an increase from $L_s(2D) \approx 550\,nm$ to $L_s(W \approx 130\,nm) > 1.8\,\mu m$, and for densities $n \approx 2 \times 10^{12}/cm^2$ an increase from $L_s(2D) \approx 500\,nm$ to $L_s(W \approx 120\,nm) > 1\,\mu m$ was observed. Using $L_s(2D) = 1/2m^*\alpha$, one obtains for both densities n, the spin–orbit coupling $\alpha \approx 5.8\,meV\text{\AA}$. A saturation of the spin relaxation rate could not be observed, suggesting that the cubic BIA coupling is negligible in these structures. We note, that an enhancement of the spin relaxation rate as in the optical experiments of narrow InGaAs quantum wires (Ref. [31]) was not observed in these AlGaN/GaN wires.

28.6 Critical Discussion and Future Perspective

The fact that optical and transport measurements seem to find opposite behavior, enhancement and suppression of the spin relaxation rate, respectively, in narrow wires, calls for an extension of the theory to describe the crossover to ballistic quantum wires. This can be done using the kinetic equation approach to the spin diffusion equation,[60] a semiclassical approach,[69,70] or an extension of the diagrammatic approach.[65] In particular, the dimensional crossover of DPS due to cubic Dresselhaus coupling, which we found not to be suppressed in diffusive wires, needs to be studied for ballistic wires, $l_e > W$, as many of the experimentally studied quantum wires are in this regime. Furthermore, using the spin diffusion equation, one can study the dependence on the growth direction of quantum wires, and find more information on the magnitude of the various spin–orbit coupling parameters, α_1, α_2, γ_D, by comparison with the directional dependence found in both the optical measurements[31] of the spin relaxation rate, as well as in recent gate-controlled transport experiments.[42] In narrow wires, corrections due to electron–electron interaction can become more important and influence especially the temperature dependence. Ref. [32] reports a strong temperature dependence of the spin relaxation rate in narrow quantum

wires. As shown in Ref. [57], the spin relaxation rates obtained from the spin diffusion equation and the quantum corrections to the magnetoconductivity can be different, when corrections due to electron–electron interaction become important. As the DPS becomes suppressed in quantum wires, other spin relaxation mechanisms like the EYS may become dominant, since it is expected that the dimensional dependence of EYS is less strong. In more narrow wires, disorder can also result in Anderson localization. Similar as in quantum dots,[38,39] this can yield enhanced spin relaxation due to hyperfine coupling (Equation 28.39). The spin relaxation in metal wires is believed to be dominated by the EYS mechanism, which is not expected to show such strong wire width dependence, although this needs to be explored in more detail. Even dilute concentrations of magnetic impurities of less than 1 ppm do yield measurable spin relaxation rates in metals and allow the study of the Kondo effect with unprecedented accuracy.[50,71]

28.7 Summary

The spin dynamics and spin relaxation of itinerant electrons in disordered quantum wires with spin–orbit coupling is governed by the spin diffusion equation (Equation 28.20). We have shown that it can be derived by using classical random walk arguments, in agreement with more elaborate derivations.[60,65] In semiconductor quantum wires, all available experiments show that the motional narrowing mechanism of spin relaxation, the DPS, is the dominant mechanism in quantum wires whose width exceeds the spin precession length, L_{SO}. The solution of the spin diffusion equation reveals the existence of persistent spin helix modes when the linear BIA- and the SIA-spin–orbit coupling are of equal magnitude. In quantum wires that are more narrow than the spin precession length L_{SO}, there is an effective alignment of the spin–orbit fields giving rise to long-living spin density modes for arbitrary ratio of the linear BIA- and the SIA-spin–orbit coupling. The resulting reduction in the spin relaxation rate results in a change in the sign of the quantum corrections to the conductivity. Recent experimental results confirm the increase of the spin relaxation rate in wires whose width is smaller than L_{SO}, both the direct optical measurement of the spin relaxation rate as well as transport measurements. These show a dimensional crossover from weak antilocalization to weak localization as the wire width is reduced. Open problems remain, in particular in narrower, ballistic wires, where optical and transport measurements seem to find opposite behavior of the spin relaxation rate: enhancement, suppression, respectively. The experimentally observed reduction of spin relaxation in quantum wires opens new perspectives for spintronic applications, since the spin–orbit coupling and therefore the spin precession length remains unaffected, allowing a better control of the itinerant electron spin. The observed directional dependence moreover can yield more detailed information about the spin–orbit coupling, enhancing the spin control for future spintronic devices further.

Symbols

τ_0 elastic scattering time

τ_{ee} scattering time due to electron–electron interaction

τ_{ep} scattering time due to electron–phonon interaction

τ total scattering time $1/\tau = 1/\tau_0 + 1/\tau_{ee} + 1/\tau_{ep}$

$\hat{\tau}_s$ spin relaxation tensor

D diffusion constant, $D = v_F^2 \tau / d_D$, where d_D is the dimension of diffusion

l_e elastic mean free path

L_{SO} spin precession length in 2D. The spin will be oriented again in the initial direction after it moved ballistically L_{SO}

Q_{SO} $= 2\pi / L_{SO}$

L_s spin relaxation length $L_s(W) = \sqrt{D\tau_s(W)}$ with $L_s(W)|_{w \to \infty} = L_s(2D) = L_{SO}/2\pi$

α_1 linear (bulk inversion asymmetry (BIA) = Dresselhaus)-parameter

α_2 linear (structural inversion asymmetry (SIA) = Bychkov-Rashba)-parameter

γ_D cubic (bulk inversion asymmetry (BIA) = Dresselhaus)-parameter

γ_g gyromagnetic ratio

Acknowledgments

We thank V. L. Fal'ko, F. E. Meijer, E. Mucciolo, I. Aleiner, C. Marcus, T. Ohtsuki, K. Slevin, J. Ohe, and A. Wirthmann for helpful discussions. This work was supported by SFB508 B9.

References

1. Aleiner, I. L. and V. I. Fal'ko. 2001. Spin-orbit coupling effects on quantum transport in lateral semiconductor dots. *Phys. Rev. Lett.* 87(25):256801.

2. Altshuler, B. L., A. G. Aronov, D. E. Khmelnitskii, and A. I. Larkin. 1982. *Quantum Theory of Solids*, Mir Publishers, Moscow, Russia.

3. Astakhov, G. V., R. I. Dzhioev, K. V. Kavokin, V. L. Korenev, M. V. Lazarev, M. N. Tkachuk, Yu. G. Kusrayev, T. Kiessling, W. Ossau, and L. W. Molenkamp. 2008. Suppression of electron spin relaxation in Mn-doped GaAs. *Phys. Rev. Lett.* 101(7):076602. http://link.aps.org/abstract/PRL/v101/e076602

4. Averkiev, N. S. and L. E. Golub. 1999. Giant spin relaxation anisotropy in zinc-blende heterostructures. *Phys. Rev. B* 60(23):15582–15584.

5. Beenakker, C. W. J. and H. van Houten. 1988. Flux-cancellation effect on narrow-channel magnetoresistance fluctuations. *Phys. Rev. B* 37(11):6544–6546.

6. Bernardes, E., J. Schliemann, M. Lee, J. C. Egues, and D. Loss. 2007. Spin-orbit interaction in symmetric wells with two subbands. *Phys. Rev. Lett.* 99(7):076603. http://link.aps.org/abstract/PRL/v99/e076603

7. Bernevig, B., Andrei, J., Orenstein, and S.-C. Zhang. 2006. Exact SU(2) symmetry and persistent spin helix in a spin-orbit coupled system. *Phys. Rev. Lett.* 97(23):236601. http://link.aps.org/abstract/PRL/v97/e236601

8. Bir, G. L., A. G. Aronov, and G. E. Pikus. 1976. Spin relaxation of electrons due to scattering by holes. *Sov. Phys. JETP* 42:705.

9. Bloembergen, N., E. M. Purcell, and R. V. Pound. 1948. Relaxation effects in nuclear magnetic resonance absorption. *Phys. Rev.* 73(7):679–712.

10. Boguslawski, P. 1980. Electron-electron spin-flip scattering and spin relaxation in III-V and II-VI semiconductors. *Solid State Commun.* 33:389.

11. Bournel, A., P. Dollfus, P. Bruno, and P. Hesto. 1998. Gate-induced spin precession in an In$_{0.53}$Ga$_{0.47}$As two dimensional electron gas. *Euro. Phys. J. AP.* 4(1):1–4. http://dx.doi.org/10.1051/epjap:1998238

12. Burkov, A. A. and L. Balents. 2004. Spin relaxation in a two-dimensional electron gas in a perpendicular magnetic field. *Phys. Rev. B* 69(24):245312.

13. Chazalviel, J. N. 1975. Spin relaxation of conduction electrons in *n*-type indium antimonide at low temperature. *Phys. Rev. B* 11(4):1555–1562.

14. Datta, S. and B. Das. 1990. Electronic analog of the electro-optic modulator. *Appl. Phys. Lett.* 56(7):665–667. http://link.aip.org/link/?APL/56/665/1

15. Dinter, R., S. Löhr, S. Schulz, Ch. Heyn, and W. Hansen. 2005. Unpublished.

16. Dragomirova, R. L. and B. K. Nikolic. 2007. Shot noise of spin-decohering transport in spin-orbit coupled nanostructures. *Phys. Rev. B* 75:085328.

17. Dresselhaus, G. 1955. Spin-orbit coupling effects in zinc blende structures. *Phys. Rev.* 100(2):580–586.

18. D'yakonov, M. I. and V. I. Perel. 1972. Spin relaxation of conduction electrons in noncen-trosymmetric semiconductors. *Sov. Phys. Solid State* 13:3023–3026.

19. Dyson, A. and B. K. Ridley. 2004. Spin relaxation in cubic III-V semiconductors via interaction with polar optical phonons. *Phys. Rev. B* 69(12):125211.

20. Dzhioev, R. I., K. V. Kavokin, V. L. Korenev, M. V. Lazarev, B. Ya. Meltser, M. N. Stepanova, B. P. Zakharchenya, D. Gammon, and D. S. Katzer. 2002. Low-temperature spin relaxation in n-type GaAs. *Phys. Rev. B* 66(24):245204.

21. Elliott, R. J. 1954. Theory of the effect of spin-orbit coupling on magnetic resonance in some semiconductors. *Phys. Rev.* 96(2):266–279.

22. Erlingsson, S. I., Y. V. Nazarov, and V. I. Fal'ko. 2001. Nucleus-mediated spin-flip transitions in GaAs quantum dots. *Phys. Rev. B* 64(19):195306.

23. Fabian, J., A. Matos-Abiague, C. Ertler, P. Stano, and I. Zutic. 2007. Semiconductor spintronics. *Acta Phys. Slovaca* 57:565.

24. Fal'ko, V. L. 2005. Private communication.

25. Froltsov, V. A. 2001. Diffusion of inhomogeneous spin distribution in a magnetic field parallel to interfaces of a III-V semiconductor quantum well. *Phys. Rev. B* 64(4):045311.

26. Glazov, M. M. and E. L. Ivchenko. 2002. Precession spin relaxation mechanism caused by frequent electron-electron collisions. *JETP Lett.* 75:403.

27. Glazov, M. M. and E. L. Ivchenko. 2004. Effect of electron-electron interaction on spin relaxation of charge carriers in semiconductors. *J. Exp. Theor. Phys.* 99:1279.

28. Golub, L. E. 2005. Weak antilocalization in high-mobility two-dimensional systems. *Phys. Rev. B* 71(23):235310. http://link.aps.org/abstract/PRB/v71/e235310

29. Grimaldi, C. and P. Fulde. 1997. Theory of screening of the phonon-modulated spin-orbit interaction in metals. *Phys. Rev. B* 55(23):15523–15530.

30. Hikami, S., A. I. Larkin, and Y. Nagaoka. 1980. Spin-orbit interaction and magnetoresistance in the two dimensional random system. *Prog. Theor. Phys.* 63(2):707–710. http://ptp.ipap.jp/link?PTP/63/707/

31. Holleitner, A. W., V. Sih, R. C. Myers, A. C. Gossard, and D. D. Awschalom. 2006. Suppression of spin relaxation in submicron InGaAs wires. *Phys. Rev. Lett.* 97(3):036805. http://link.aps.org/abstract/PRL/v97/e036805

32. Holleitner, A. W., V. Sih, R. C. Myers, A. C. Gossard, and D. D. Awschalom. 2007. Dimensionally constrained D'yakonov–Perel' spin relaxation in n-InGaAs channels: Transition from 2D to 1D. *New J. Phys.* 9:342–354.

33. Iordanskii, S. V., Y. B. Lyanda-Geller, and G. E. Pikus. 1994. Weak-localization in quantum-wells with spin-orbit interaction. *JETP Lett.* 60(3):206–211.

34. Ivchenko, E. L. 1973. Spin relaxation of free carriers in a noncentrosymmetric semiconductor in a longitudinal magnetic field. *Sov. Phys. Solid State* 15:1048.

35. Kaneko, T., M. Koshino, and T. Ando. 2008. Numerical study of spin relaxation in a quantum wire with spin-orbit interaction. *Phys. Rev. B* 78(24):245303. http://link.aps.org/abstract/PRB/v78/e245303

36. Kettemann, S. 2007. Dimensional control of antilocalization and spin relaxation in quantum wires. *Phys. Rev. Lett.* 98(17):176808. http://link.aps.org/abstract/PRL/v98/e176808

37. Khaetskii, A., D. Loss, and L. Glazman. 2003. Electron spin evolution induced by interaction with nuclei in a quantum dot. *Phys. Rev. B* 67(19):195329.

38. Khaetskii, A. V., D. Loss, and L. Glazman. 2002. Electron spin Decoherence in quantum dots due to interaction with nuclei. *Phys. Rev. Lett.* 88(18):186802.

39. Khaetskii, A. V. and Y. V. Nazarov. 2000. Spin relaxation in semiconductor quantum dots. *Phys. Rev. B* 61(19):12639–12642.

40. Kiselev, A. A. and K. W. Kim. 2000. Progressive suppression of spin relaxation in two-dimensional channels of finite width. *Phys. Rev. B* 61(19):13115–13120.

41. Knap, W., C. Skierbiszewski, A. Zduniak, E. Litwin-Staszewska, D. Bertho, F. Kobbi, J. L. Robert et al. 1996. Weak antilocalization and spin precession in quantum wells. *Phys. Rev. B* 53(7):3912–3924.

42. Kunihashi, Y., M. Kohda, and J. Nitta. 2009. Enhancement of spin lifetime in gate-fitted InGaAs narrow wires. *Phys. Rev. Lett.* 102(22):226601.

43. Lassnig, R. 1985. $k \to \cdot p \to$ theory, effective-mass approach, and spin splitting for two-dimensional electrons in GaAs-GaAlAs heterostructures. *Phys. Rev. B* 31(12):8076–8086.

44. Lehnen, P., T. Schäpers, N. Kaluza, N. Thillosen, and H. Hardtdegen. 2007. Enhanced spin-orbit scattering length in narrow $Al_xGa_{1-x}N$/GaN wires. *Phys. Rev. B* 76(20):205307. http://link.aps.org/abstract/PRB/v76/e205307

45. Lyanda-Geller, Y. 1998. Quantum interference and electron-electron interactions at strong spin-orbit coupling in disordered systems. *Phys. Rev. Lett.* 80(19):4273–4276.

46. Malcher, F., G. Lommer, and U. Rössler. 1986. Electron states in GaAs/$Ga_{1-x}Al_x$As heterostructures: Nonparabolicity and spin-splitting. *Superlatt. Microstruct.* 2(3):267–272. http://www.sciencedirect.com/science/article/B6WXB-4933G7S-1G/2/1b9e3dec0432f20bad957c6845e3db35

47. Mal'shukov, A. G. and K. A. Chao. 2000. Waveguide diffusion modes and slowdown of D'yakonov-Perel' spin relaxation in narrow two-dimensional semiconductor channels. *Phys. Rev. B* 61(4):R2413–R2416.

48. Meijer, F. E. 2005. Private communication.

49. Meyer, J. S., V. I. Fal'ko, and B. L. Altshuler. 2002. Vol. 72 of *NATO Science Series II*, Kluwer Academic Publishers, Dordrecht, the Netherlands, p. 117.

50. Micklitz, T., A. Altland, T. A. Costi, and A. Rosch. 2006. Universal dephasing rate due to diluted Kondo impurities. *Phys. Rev. Lett.* 96(22):226601. http://link.aps.org/abstract/PRL/v96/e226601

51. Miller, J. B., D. M. Zumbühl, C. M. Marcus, Y. B. Lyanda-Geller, D. Goldhaber-Gordon, K. Campman, and A. C. Gossard. 2003. Gate-controlled spin-orbit quantum interference effects in lateral transport. *Phys. Rev. Lett.* 90(7):076807.

52. Müller-Hartmann, E. and J. Zittartz. 1971. Kondo effect in superconductors. *Phys. Rev. Lett.* 26(8):428–432.

53. Ohno, Y., R. Terauchi, T. Adachi, F. Matsukura, and H. Ohno. 1999. Spin relaxation in GaAs(110) quantum wells. *Phys. Rev. Lett.* 83(20):4196–4199.

54. Overhauser, A. W. 1953. Paramagnetic relaxation in metals. *Phys. Rev.* 89(4):689–700.

55. Pareek, T. P. and P. Bruno. 2002. Spin coherence in a two-dimensional electron gas with Rashba spin-orbit interaction. *Phys. Rev. B* 65(24):241305.

56. Pikus, G. E. and A. N. Titkov. 1984. Spin relaxation under optical orientation in semiconductors. In *Optical Orientation*, F. Meier and B. P. Zakharchenya (eds.), Vol. 8 of *Modern Problems in Condensed Matter Sciences*, North-Holland, Amsterdam, the Netherlands, Chapter 3.

57. Punnoose, A. and A. M. Finkel'stein. 2006. Spin relaxation in the presence of electron-electron interactions. *Phys. Rev. Lett.* 96(5):057202. http://link.aps.org/abstract/PRL/v96/e057202

58. Rashba, EI. 1960. Properties of semiconductors with an extremum Loop.1. Cyclotron and combinational resonance in a magnetic field perpendicular to the plane of the loop. *Sov. Phys. Solid State* 2(6):1109–1122.

59. Schäpers, Th., V. A. Guzenko, M. G. Pala, U. Zülicke, M. Governale, J. Knobbe, and H. Hardtdegen. 2006. Suppression of weak antilocalization in $Ga_xIn_{1-x}As/InP$ narrow quantum wires. *Phys. Rev. B* 74(8):081301. http://link.aps.org/abstract/PRB/v74/e081301

60. Schwab, P., M. Dzierzawa, C. Gorini, and R. Raimondi. 2006. Spin relaxation in narrow wires of a two-dimensional electron gas. *Phys. Rev. B* 74(15):155316. http://link.aps.org/abstract/PRB/v74/e155316

61. Stich, D., J. H. Jiang, T. Korn, R. Schulz, D. Schuh, W. Wegscheider, M. W. Wu, and C. Schüller. 2007. Detection of large magnetoanisotropy of electron spin dephasing in a high-mobility two-dimensional electron system in a [001] $GaAs/Al_xGa_xAs$ quantum well. *Phys. Rev. B* 76(7):073309. http://link.aps.org/abstract/PRB/v76/e073309

62. Tamborenea, P. I., D. Weinmann, and R. A. Jalabert. 2007. Relaxation mechanism for electron spin in the impurity band of n-doped semiconductors. *Phys. Rev. B* 76(8):085209. http://link.aps.org/abstract/PRB/v76/e085209

63. Torrey, H. C. 1956. Bloch equations with diffusion terms. *Phys. Rev.* 104(3):563–565.

64. Weber, C. P., J. Orenstein, B. A. Bernevig, S.-C. Zhang, J. Stephens, and D. D. Awschalom. 2007. Nondiffusive spin dynamics in a two-dimensional electron gas. *Phys. Rev. Lett.* 98(7):076604. http://link.aps.org/abstract/PRL/v98/e076604

65. Wenk, P. and S. Kettemann. 2009. Dimensional dependence of weak localization corrections and spin relaxation in quantum wires with Rashba spin-orbit coupling. http://www.citebase.org/abstract? id=oai:arxiv.org:0907.1819

66. Winkler, R. 2003. *Spin-Orbit Coupling Effects in Two-Dimensional Electron and Hole Systems*, Vol. 191 of *Springer Tracts in Modern Physics*, Springer-Verlag, Berlin, Germany.

67. Wirthmann, A., Y. S. Gui, C. Zehnder, D. Heitmann, C.-M. Hu, and S. Kettemann. 2006. Weak antilocalization in InAs quantum wires. *Phys. E: Low-Dim. Syst. Nanostruct.* 34(1–2):493–496. *Proceedings of the 16th International Conference on Electronic Properties of Two-Dimensional Systems (EP2DS-16)*, Tokyo, Japan. http://www.sciencedirect.com/science/article/B6VMT-4JRVCNY-H/2/b34498d063a54a7949f4c00ed66a39c7

68. Yafet, Y. 1963. g-Factors and spin-lattice relaxation of conduction electrons. In *Solid State Physics*, Vol. 14, F. Seitz and D. Turnbull (eds.), Academic, New York.

69. Zaitsev, O., D. Frustaglia, and K. Richter. 2005a. The role of orbital dynamics in spin relaxation and weak antilocalization in quantum dots. *Phys. Rev. Lett.* 94:026809.

70. Zaitsev, O., D. Frustaglia, and K. Richter, 2005b. Semiclassical theory of weak antilocalization and spin relaxation in ballistic quantum dots. *Phys. Rev. B* 72:155325.

71. Zaránd, G., L. Borda, J. von Delft, and N. Andrei. 2004. Theory of inelastic scattering from magnetic impurities. *Phys. Rev. Lett.* 93(10):107204.

72. Zutic, I., J. Fabian, and S. Das Sarma. 2004. Spintronics: Fundamentals and applications. *Rev. Mod. Phys.* 76(2):323. http://arxiv.org/pdf/cond-mat/0405528v1

29

Quantum Magnetic Oscillations in Nanowires

A. Sasha Alexandrov
Loughborough University

Victor V. Kabanov
Josef Stefan Institute

Iorwerth O. Thomas
Loughborough University

29.1 Introduction

Periodic oscillations in the magnetization and magnetoresistance of metals in a magnetic field as the strength of the field is varied—known as the De Haas–van Alphen (dHvA) and the Shubnikov–De Haas (SdH) effects, respectively, and arising from the Fock–Landau quantization of the electron spectrum—have, since their discovery in the 1930s, become one of the standard techniques in the analysis of the structures of the Fermi surfaces (Shoenberg 1984a). A summary of the first 38 years of the field, focusing on the dHvA effect, may be found in Gold (1968), while more recent reviews relevant to our topic are Singleton (2000) and Kartsovnik (2004), which focus on magneto-oscillations in the two-dimensional (2D) and the quasi-two-dimensional (q2D) metals, such as the organic charge-transfer salts.

Both the dimensionality and the physical confinement of electrons may affect the quantum magneto-oscillations of a system, and it is therefore important to understand how these effects may manifest so that the structure of the Fermi surface may be properly understood. The attempt to define the effects of physical confinement (which is of paramount importance in the description of the quantum magnetic oscillations of nanowires) has a long history in this field, since a correct treatment of the boundary conditions is important even in bulk metals as the density of states of the system is finite (e.g., see the following classic papers: Fock 1928, Landau 1930, Darwin 1930, Dingle 1952b,c). Some attention, however, has also been paid to the case of electrons confined within a cylindrical wire exposed to a longitudinal magnetic field. Much of the theoretical work has focused on the weak field case—as in Dingle (1952c), Bogacheck and Gogadze (1973), Aronov and Sharvin (1987), and Gogadze (1984)—where there is only a small perturbation to the energy spectrum. In these cases, oscillations arise as a result of size quantization or the Aharonov–Bohm effect. In order to treat a system confined within a narrow wire at stronger fields, Alexandrov and Kabanov (2005a) have modeled the confinement using the Fock–Darwin model (Fock 1928, Darwin 1930), and in Alexandrov et al. (2007) this was extended to a stronger variety of confinements. In both these cases, a complex pattern of oscillations due to the interaction of size and Fock–Landau quantization was predicted.

In this chapter, we first introduce, in Section 29.2, the theory describing quantum magnetic oscillations in three-dimensional (3D) bulk metals. This is largely well understood, and functions as a foundation for the following discussion. Next, in Section 29.3, we describe some interesting properties of q2D metals. While this might seem irrelevant to the topic at hand, they serve as an important reminder that certain properties of 3D metals do not hold in systems of reduced dimensionality and are illustrative of some important new phenomena that might arise in such cases (the difference between the quantum magneto-oscillations of closed and open multiband systems in 2D and q2D metals does not exist in 3D bulk metals, for example). Finally, in Section 29.4, we review in detail the theory of quantum magnetic oscillations in nanowires.

29.2 Quantum Magnetic Oscillations in Bulk Metals

29.2.1 Fock–Landau Quantization

It has been known since the early twentieth century (Fock 1928, Landau 1930, Darwin 1930) that the spiral motion of electrons in a magnetic field is quantized, and that this has profound consequences on the behavior of metallic compounds that are exposed to such fields.

We begin with an overview of this Fock–Landau quantization, as discussed by Landau (Landau 1930). Accessible treatments of this topic may be found in textbooks such as Landau and Lifschitz (1977), Abrikosov (1988), and Hamguchi (2001), though in this section we mainly follow the account of Abrikosov (1988).

The presence of a magnetic field modifies the momentum operator $\hat{\mathbf{p}}$ to $\hat{\mathbf{p}}-e\mathbf{A}$, where $\mathbf{A}$ is the vector potential of the field. We may therefore write the Schrödinger equation for a free electron in a magnetic field parallel to the z-axis in the Landau gauge ($A_x = A_z = 0, A_y = Bx$):

$$-\frac{\hbar^2}{2m^*}\frac{\partial^2}{\partial x^2}\psi + \frac{1}{2m^*}\left(-i\hbar\frac{\partial}{\partial y}-eBx\right)^2\psi - \frac{\hbar^2}{2m^*}\frac{\partial^2}{\partial z^2}\psi = E\psi, \quad (29.1)$$

where

$\hbar = h/2\pi$ (h is Planck's constant)
m^* is the mass of the electron
B is the magnetic field
e is the charge of the electron
E is the energy

We solve this through the separation of variables: our solution has the form $\psi(x,y,z) = e^{ip_y y/\hbar}e^{ip_z z/\hbar}\psi(x)$ (p_y and p_z are the respective y and z components of $\mathbf{p}$). Substituting this into the above wave equation we acquire

$$-\frac{\hbar}{2m^*}\frac{\partial^2}{\partial x^2}\psi(x) + \frac{1}{2m^*}(p_y - eBx)^2\psi(x) + \frac{p_z^2}{2m^*}\psi(x) = E\psi(x). \quad (29.2)$$

This is similar to the equation of motion of a one-dimensional (1D) harmonic oscillator:

$$-\frac{\hbar^2}{2m^*}\frac{\partial^2}{\partial x^2}\psi - \frac{1}{2}Kx^2\psi = \epsilon\psi \quad (29.3)$$

where K is a quantity related to the oscillation frequency ω, so that $\omega = (K/m^*)^{1/2}$. This equation has the energy spectrum $\epsilon_n = \hbar\omega(n + 1/2)$, where n is an integer greater than or equal to 0. If we shift the center of the oscillation in Equation 29.2 by p_y/eB and set $\omega = eB/m^* \equiv \omega_c$ (we call ω_c the cyclotron frequency) and $\epsilon = E - p_z^2/2m^*$, we find that they are identical. Our spectrum is then

$$E_{p_z,n} = \frac{p_z^2}{2m^*} + \hbar\omega_c\left(n + \frac{1}{2}\right). \quad (29.4)$$

We now turn to the density of states. This requires some care, as we must take into account the boundary conditions of our system in order to avoid an infinite degeneracy of states for each Fock–Landau level n. Let us assume that we have a box whose sides are of different lengths L_x, L_y, L_z. We also assume that these lengths are very large, so that any effects due to the size quantization are negligible (see the discussion in Landau and Lifschitz (1977)); situations where this simplification cannot be applied, such as nanowires, are treated later on. Now, the position of the center of the oscillator is dependent on the value of p_y. In order to be contained within the box, it must therefore be within the interval

$$0 < p_y < eBL_x. \quad (29.5)$$

The number of states in the infinitesimal intervals dp_z and dp_y are

$$dn_y = \frac{dp_y L_y}{2\pi\hbar}, \quad dn_z = \frac{dp_z L_z}{2\pi\hbar}. \quad (29.6)$$

Overall, then, the total number of states in the interval p_y will be

$$n_y = \frac{L_x L_y eB}{2\pi\hbar}, \quad (29.7)$$

which when multiplied by dn_z gives us the number of states between p_z and $p_z + dp_z$:

$$eB\frac{Vdp_z}{(2\pi\hbar)^2}, \quad (29.8)$$

where $V = L_x L_y L_z$ is the sample volume.

The density of states of the system is hence given by

$$g(E) = \sum_{n,p_z,p_y}\delta(E - E_{p_z,n})$$

$$= eB\frac{V}{(2\pi\hbar)^2}\sum_n\int dp_z\,\delta(E - E_{p_z,n}), \quad (29.9)$$

where $\delta(E - E_{p_z,n})$ is a delta function.

Changing the variables so that we are integrating over $E_{p_z,n}$ gives us

$$g(E) = eB\frac{V\sqrt{2m^*}}{(2\pi\hbar)^2}\sum_n\int dE_{p_z,n}\frac{\delta(E - E_{p_z,n})}{\sqrt{E_{p_z,n} - \hbar\omega_c(n+1/2)}} \quad (29.10)$$

and performing this integration gives us the density of states

$$g(E) = eB\frac{V\sqrt{2m^*}}{(2\pi\hbar)^2}\sum_n\frac{1}{\sqrt{E - \hbar\omega_c(n+1/2)}}. \quad (29.11)$$

29.2.2 Quantum Magnetic Oscillations

This quantization has many intriguing consequences. The focus of this chapter is the oscillations it induces in the magnetization (de Haas–van Alphen or dHvA oscillations) and the conductivity (Shubnikov–de Haas or SdH oscillations) (Shoenberg 1984a), and how these manifest in low-dimensional systems. These oscillations, periodic in $1/B$, were first predicted by Landau (1930) (who did not think that they would be observable), and are now commonly used as a technique for examining the properties of the Fermi surface of metallic compounds.

Before we embark on an overview of the typical methods used to derive quantitative expressions for these effects, we shall first examine how they arise. At zero temperature, a 3D electron gas will fill all the energy levels up to the Fermi energy, which describes a spherical surface (the Fermi surface) in momentum space. Upon application of the magnetic field, the orbits of the electrons in the plane perpendicular to the field become quantized, and we have an allowed distribution of electrons in momentum space that resembles a set of concentric cylinders, whose axes are ordered in parallel with the direction of the field. The separation of these cylinders is governed by the strength of the magnetic field, and the highest occupied energy level is that closest to the Fermi energy. As the magnetic field is varied, the separation of the cylinders and the associated energy levels change. Eventually, the topmost occupied energy level will pass above the Fermi energy—and therefore all the electrons occupying it will fall to the next lowest energy level, resulting in a discontinuous change in the properties of the system that rely on the electron occupation. This repeats as the successive highest occupied energy levels pass above the Fermi energy, leading to oscillations in those quantities (Gold 1968, Shoenberg 1984a, Abrikosov 1988, Hamguchi 2001).

29.2.2.1 De Haas–Van Alphen Oscillations

Here, we outline the general procedure for calculating the dHvA oscillations of a free electron gas in a magnetic field. The same steps can be used in order to calculate the oscillations for a variety of systems, provided that one has an appropriate model dispersion relation for the system. (Here we follow the treatment in Abrikosov (1988), based on that of Lifshitz and Kosevich (1968); similar treatments may be found in Gold (1968) and Shoenberg (1984a).)

We begin with the formula for the energy of a free electron gas in a magnetic field B oriented along the z-axis,

$$E_{p_z,n} = \frac{p_z^2}{2m^\star} + \hbar\omega_c\left(n+\frac{1}{2}\right),$$ (29.12)

and the formula for the thermodynamic potential

$$\Omega = -k_B T \sum_\beta \ln(1+e^{(\mu-\epsilon_\beta)/T}),$$ (29.13)

where

the sum over β indicates summing over all states
T is the temperature
k_B is Boltzmann's constant
μ is the chemical potential
ϵ_β is the energy of the rth state

Summing over all identical states gives us

$$\Omega = -\frac{k_B T V e B}{(2\pi\hbar)^2}\sum_{n,\sigma}\int dp_z\,\ln(1+e^{(\mu-E(n,p_z))/k_B T}).$$ (29.14)

For reasons of clarity, we neglect spin splitting, and so the summation over the spin index σ merely results in the right-hand side of the above equation being multiplied by 2.

In order to proceed, we shall need to make use of Poisson's summation formula

$$\sum_{n_0}^\infty f(n) = \int_a^\infty dn\,f(n) + 2\Re\sum_{r=1}^\infty\int_a^\infty dn\,f(n)e^{i2\pi rn},$$ (29.15)

where

$f(n)$ is an arbitrary, real function, $n_0 - 1 < a < n_0$
r is the summation index
$\Re$ indicates that we take the real portion of the subsequent quantity

We are interested in the oscillatory portion of Ω, which after the application of Poisson's formula is given by

$$\tilde{\Omega} = -4\Re\sum_{r=1}^\infty I_r,$$ (29.16)

with I_r being given by

$$I_r = \frac{VeBk_B T}{(2\pi\hbar)^2}\int_a^\infty dn\int_{-\infty}^\infty dp_z\,\ln(1+e^{(\mu-E(n,p_z))/k_B T})e^{2\pi irn}.$$ (29.17)

We change the integration so that it is over E:

$$I_r = \frac{VeBk_B T}{(2\pi\hbar)^2}\int_0^\infty dE\int_{p_z^{min}}^{p_z^{max}} dp_z\left(\frac{\partial n(p_z,E)}{\partial E}\right)\ln(1+e^{(\mu-E)/T})e^{i2\pi rn(p_z,E)}.$$ (29.18)

The lower limit of the integral is set to 0 for convenience; since the dominant contribution to the phenomenon of interest is from states near the Fermi energy, its precise value is of little import. The values of p_z^{max} and p_z^{min} are the maximum and minimum possible values of p_z for a given E.

We now observe that $e^{i2\pi rn(p_z,E)}$ is a rapidly oscillating function. If $n(p_z,E)$ obtains an extremal value n_α at p_z^α, we can expand it as

$$n(p_z, E) = n_\alpha(E) + \frac{1}{2}\left(\frac{\partial^2 n}{\partial p_z^2}\right)_{p_z^\alpha}(p_z - p_z^m)^2. \qquad (29.19)$$

(There is no first-order term in the expansion since $\partial n/\partial p_z = 0$ at an extremum.) We can use this to integrate over p_z near the extrema, and the integrals over p_z can be well approximated by

$$\left(\frac{\partial n}{\partial E}\right)_m e^{i2\pi r n_m} \int_{-\infty}^{\infty} \exp\left(-i\pi r\left(\frac{\partial^2 n}{\partial p_z^2}\right)_m z^2\right) dz, \qquad (29.20)$$

which is a standard integral,

$$\int_{-\infty}^{\infty} \exp\left(-i\pi r\left(\frac{\partial^2 n}{\partial p_z^2}\right)_m z^2\right) dz = e^{\pm i\pi/4}\left(r\left|\frac{\partial^2 n}{\partial p_z^2}\right|_m\right)^{-1/2}. \qquad (29.21)$$

If there is more than one extremal surface, we must sum over them. For simplicity, however, we shall assume only the existence of one such surface and that it has a simple parabolic band structure given by Equation 29.12. (For more complicated structures, one may use the more general approach given in Abrikosov (1988).) Hence I_r following our integration is

$$I_r = \frac{V k_B T e B}{(2\pi\hbar)^2}\int_0^\infty dE \frac{\ln(1 + e^{(\mu-E)/k_B T})}{\hbar\omega_c} e^{i2\pi r n(E)\pm i\pi/4}\left(\frac{\hbar\omega_c m^*}{r}\right)^{1/2}, \qquad (29.22)$$

since $(\partial^2 n / \partial p_z^2) = 1 / \hbar\omega_c m^*$ and $(\partial n/\partial E) = 1/\hbar\omega_c$.

We now integrate this by parts and neglect the non-oscillatory term, obtaining

$$I_r = \frac{V e B \sqrt{\hbar\omega_c m}}{(2\pi\hbar)^2 c r^{3/2} 2\pi i} e^{\pm i\pi/4}\int_0^\infty dE \frac{e^{i2\pi r n(E)}}{e^{(E-\mu)/k_B T} + 1}. \qquad (29.23)$$

The dHvA effect is a phenomenon of the Fermi surface, so the dominant contribution should be from values of E close to μ. We need to account for the rapid oscillation of the exponential, however, so we expand $n(E)$ for $E \approx \mu$:

$$n(E) = n(\mu) + \left(\frac{\partial n}{\partial E}\right)_\mu (E - \mu). \qquad (29.24)$$

We then find that the rapidly varying part of the integral becomes

$$\int_0^\infty dE \frac{e^{i2\pi r n}}{e^{(E-\mu)/k_B T} + 1} = e^{i2\pi r n_m(\mu)}\int_{-\infty}^{\infty} dx \frac{\exp\left(\frac{i2\pi r x}{\hbar\omega_c}\right)}{e^{x/k_B T} + 1}, \qquad (29.25)$$

where $x = E - \mu$. We may integrate over this using the standard integral $\int_0^\infty dy \exp(i\alpha y)/(\exp y + 1) = -i\pi/\sinh(\alpha\pi)$, where α signifies a quantity of the correct form, and so acquire

$$I_r = -\frac{V e B k_B T \sqrt{\hbar\omega_c m^*}}{2(2\pi\hbar)^2 r^{3/2}} \frac{e^{i2\pi r n(\mu)}}{\sinh(2\pi^2 r k_B T/\hbar\omega_c)} \qquad (29.26)$$

Making use of the Onsager relation between $n(\mu)$ and the cross-sectional area of the Fermi surface S_F (Onsager 1952)

$$n_m \approx \frac{S_m}{2\pi e\hbar B}, \qquad (29.27)$$

and inserting our results into our overall expression for $\tilde{\Omega}$, we obtain the following expression:

$$\tilde{\Omega} = \frac{V\omega_c^{5/2}(m^*)^{3/2}}{4\pi^4\hbar^{1/2}} \sum_{r=1}^{\infty} \frac{R_T(r\lambda)}{r^{5/2}}\cos\left(r\frac{S_F}{e\hbar B} \pm \frac{\pi}{4}\right), \qquad (29.28)$$

where

$R_T(x) = x/\sinh(x)$ is the Lifshitz–Kosevich (LK) temperature damping factor

$\lambda = 2\pi^2 k_B T/\hbar\omega_c$

We may further derive expressions for the oscillatory magnetization $\tilde{M}$ and the magnetic susceptibility $\tilde{\chi}$ through the use of the relations $\tilde{M} = -\partial\tilde{\Omega}/\partial B$ and $\tilde{\chi} = \partial\tilde{M}/\partial B$, where we keep only the most rapidly oscillating parts.

It is also useful to consider the relationship when given in terms of the Fermi energy. In this case, $n(\mu) = (\mu/\hbar\omega_c) - 1/2$, and we find

$$\tilde{\Omega} = \frac{V\omega_c^{5/2}(m^*)^{3/2}}{4\pi^4\hbar^{1/2}} \sum_{r=1}^{\infty} \frac{(-1)^r R_T(r\lambda)}{r^{5/2}}\cos\left(r\frac{2\pi\mu}{\hbar\omega_c} \pm \frac{\pi}{4}\right). \qquad (29.29)$$

Thus far, we have ignored the effects of the splitting of energy levels by the spin quantum number. This is easy enough to repair. The energy spectrum incorporating the spin splitting is

$$E_\sigma = E(n, p_z) + \beta B\sigma, \qquad (29.30)$$

where

$\sigma = \pm 1$

$\beta = e\hbar/2m_e c$ is the Bohr magneton (m_e is the mass of the free electron, c the speed of light in a vacuum)

This functions as a straightforward shift of the chemical potential in the exponential term to $\mu_\sigma = \mu + \beta B\sigma$. Since $\beta B\sigma \ll \mu$, we may expand around $\mu_\sigma = \mu$ and insert this into the exponential factor. From that, we arrive at

$$2I_r \Rightarrow I_r \sum_\sigma \exp\left(i2\pi r \left(\frac{\partial n}{\partial E}\right)_\mu \beta B \sigma \right)$$

$$= 2I_r \cos\left(\frac{\pi r m^*}{m_e}\right). \tag{29.31}$$

The effect of spin splitting is therefore the introduction of an additional cosine factor that affects the amplitudes of the Fourier components of the magnetization. It depends on the ratio between the effective mass of the electron and the free mass of the electron.

Finally, we must also account for a further effect (Gold 1968, Abrikosov 1988) that arises due to the broadening of the Landau levels due to collisions, scattering off impurities, and crystal defects and similar phenomena that we have thus far ignored. Phenomenologically, these effects may be accounted for by the assumption that the broadening leads to a roughly Lorentzian line shape of half width Γ. Dingle (1952b) shows* that this leads to an additional multiplicative factor in each of the harmonic terms of the Fourier series with the form

$$R_{Dr} = (e^{-2\pi\Gamma/\hbar\omega_c})^r = e^{-2\pi^2 r k_B T_D/\hbar\omega_c}, \tag{29.32}$$

where

$T_D = \Gamma/\pi k_B = \hbar/2\pi\tau k_B$ is the Dingle "temperature"

τ is the relaxation time of the system

A more rigorous derivation of the Dingle damping factor for a particular form of scattering may be found in the calculation of the SdH oscillations from the Kubo formula in Section 29.2.2.2. Wasserman and Springford (1996) give more details on the influence of many-body interactions on the dHvA effect.

29.2.2.2 Shubnikov–De Haas Oscillations

A similar effect can be observed in the conductivity of an electron gas in an external field—the Shubnikov–De Haas effect. There are two methods of deriving these results: one is by making use of a semiclassical generalization of Boltzmann's formula (Abrikosov 1988, Hamguchi 2001), the other is by making use of a more rigorous derivation based on the Kubo formula of linear response theory (Kubo et al. 1965). For the longitudinal conductivity (i.e., the case where the electric field is oriented in the same direction as the magnetic field) in *three dimensions*, the approaches give similar results; however, this is not true in general. In q2D systems, for example, there may be some differences between them (Grigoriev 2002, 2003) since the linear response approach includes a number of small contributions that are negligible in three dimensions, but less so in the q2D case.

29.2.2.2.1 Linear Response Theory

We begin with the Kubo formula (Kubo et al. 1965), for the longitudinal conductivity, where we average over the locations of impurities:

$$\sigma_{zz} = \frac{\pi e^2}{V} \int dE \left[-\frac{\partial f(E)}{\partial E} \right] \overline{\mathrm{Tr}[\delta(E-\hat{H})\hat{v}_z \delta(E-\hat{H})\hat{v}_z]}, \tag{29.33}$$

where

$f(E)$ is the Fermi–Dirac distribution function

The bar indicates averaging over all impurity configurations

$\hat{v}_z$ is the velocity operator

$\hat{H}$ includes an electron interaction with impurities

We may rewrite the trace as follows:

$$\overline{\mathrm{Tr}[\delta(E-\hat{H})\hat{v}_z\delta(E-\hat{H})\hat{v}_z]}$$

$$= \sum_\beta \overline{\langle\beta|\delta(E-\hat{H})\hat{v}_z\,\delta(E-\hat{H})\hat{v}_z|\beta\rangle}$$

$$= \sum_{\beta,\beta',\beta'',\beta'''} \overline{\langle\beta|\,\delta(E-\hat{H})\,|\beta'\rangle\langle\beta'|\hat{v}_z|\beta''\rangle\langle\beta''|\,\delta(E-\hat{H})\,|\beta'''\rangle\langle\beta'''|\,\hat{v}_z|\beta\rangle}, \tag{29.34}$$

where

β denotes a given set of quantum numbers

Various primed β denote quantum numbers inserted through spectral decomposition.

$\hat{v}_z$ is diagonal in the Landau representation, so

$$\langle\alpha|\hat{v}_z|\alpha'\rangle = \delta_{\alpha\alpha'}v_{z\alpha}, \tag{29.35}$$

where $v_{z\alpha}$ is the velocity for a given set of quantum numbers α.

In addition, we assume that vertex corrections are negligible (Gerhardts and Hajdu 1971).

We then obtain

$$\overline{\mathrm{Tr}[\delta(E-\hat{H})\hat{v}_z\,\delta(E-\hat{H})\hat{v}_z]} = 2\sum_{\beta\beta'} v_{z\beta}v_{z\beta'}\left[\overline{\delta(E-\hat{H})_{\beta\beta'}}\right]^2, \tag{29.36}$$

where we have summed over the spin quantum number and ignored spin splitting (which can be reintroduced as per the dHvA effect). If the impurities are dilute and point-like, we may approximate the averaged delta functions as follows:

$\overline{\delta(E-\hat{H})_{\beta\beta'}} \approx (\delta_{\beta\beta'}/\pi)\Im G^R(\epsilon_\beta, E)$, where $G^R(\epsilon_\beta, E)$ is the retarded Green's function given by

$$G^R(\epsilon_\beta, E) = \frac{1}{E - \epsilon_\beta - \Sigma_\beta(E)}, \tag{29.37}$$

---

* While this calculation was performed for free electrons, it is valid for all Fermi surface shapes.

where

ϵ_β is the energy associated with the set of quantum numbers β

$\Sigma_\beta(E) = L_\beta(E) - i\Delta_\beta(E)$ is the retarded self-energy, with both $L_\beta(E)$ and $\Delta_\beta(E)$ being real functions

Typically, in 3D systems one may ignore $L_\beta(E)$ as it only shifts the value of E by a constant; however, one should be cautious, as this is not always the case (Grigoriev 2002, 2003).

For weakly disordered metals, where the scattering length l is sufficiently large, we may ignore any Feynman diagrams where the scattering lines are crossed in the self-energy, since these are suppressed by a factor of $1/k_F l$ (Bruus and Flensberg 2004). This allows us to make use of the self-consistent Born approximation (also known as the non-crossing approximation), where the self-energy is independent of the electron quantum numbers, and dependent only on the energy: $\Sigma_\beta(E) \approx \Sigma(E)$. This is a simplification that is needed to render the calculation analytically tractable.

Our final formula is then

$$\sigma_{zz} = \frac{2e^2\hbar}{\pi V} \int dE \left[-\frac{\partial f(E)}{\partial E} \right] \sum_\beta v_{z\beta}^2 \left[\Im G^R(\epsilon_\beta, E) \right]^2. \qquad (29.38)$$

Working with free electrons, we set $\beta = \{p_z, p_y, n\}$ and change the summation over p_z to an integration. We then change the integration over p_z to one over ϵ so that we arrive at

$$\sigma_{zz} = \frac{e^3 B}{\pi^3 \hbar} \int dE \left[-\frac{\partial f(E)}{\partial E} \right] \int d\epsilon \sum_n |v_z(\epsilon, n)| \left[\Im G^R(\epsilon, E) \right]^2. \qquad (29.39)$$

In order to proceed, we use the following generalization of the Poisson summation formula (Dingle 1952a, Hamguchi 2001):

$$\sum_n f\left(n + \frac{1}{2} \right) = \sum_{r=-\infty}^{\infty} (-1)^r \int_0^\infty dn f(n) e^{i2\pi rn}. \qquad (29.40)$$

It is easy to see that for a real function $f(n)$, this is equivalent to

$$\sum_n f\left(n + \frac{1}{2} \right) = \int_0^\infty dn f(n) + 2\Re \sum_{r=1}^{\infty} \int_0^\infty dn (-1)^r f(n) e^{i2\pi rn}. \qquad (29.41)$$

Since the Green's function in our approximation is no longer dependent on n, we may integrate the velocity alone, and in the 3D case we need only keep the contribution from the $r = 0$ term, since it is dominant over all the others. In general (such as in q2D systems), this will not be so. In addition, in the analytic structure of the Green's functions at $r > 0$ may give rise to an additional contribution from the second-order pole, which also cannot be ignored in such cases (Grigoriev 2002, 2003).

For the $r = 0$ term, we find

$$\frac{1}{2m^*\hbar\omega_c} \int_0^{2m^*\epsilon} dx \sqrt{2m^*\epsilon - x} = \frac{(2m^*\epsilon)^{3/2}}{2m^*\hbar eB}. \qquad (29.42)$$

Now we perform the integration over ϵ. We have

$$\int_0^\infty d\epsilon \epsilon^{3/2} \left[\Im G^R(\epsilon, E) \right]^2 = \int_0^\infty d\epsilon \frac{\Delta(E)^2 \epsilon^{3/2}}{\left[(E - \epsilon)^2 + \Delta(E)^2 \right]^2}$$

$$= -\int_E^\infty dA \frac{\Delta(E)^2 (E - A)^{3/2}}{\left[A^2 + \Delta(E)^2 \right]^2}, \qquad (29.43)$$

where $\Delta(E) = \Im \Sigma(E)$, $\Im$ signifying that we should use the imaginary portion of the following quantity, and we shift to a new variable of integration A in the second line of the above.

For $E \gg \Delta(E)$, the Green's functions are strongly peaked at $A = 0$. This implies that replacing $(E - A)^{3/2}$ with $E^{3/2}$ and extending the lower limit of integration to $-\infty$ is a good approximation (Altland and Simons 2006). We then close the contour of integration in the upper complex plane of A and through use of the residue theorem compute

$$-E^{3/2} \int_{-\infty}^\infty dA \frac{\Delta(E)^2}{[A^2 + \Delta(E)^2]^2} = -\frac{\pi E^{3/2}}{2\Delta(E)}. \qquad (29.44)$$

Gathering these results together, we have

$$\sigma_{zz} = \frac{e^2}{6\pi^2 m^* \hbar^2} \int dE \left[-\frac{\partial f(E)}{\partial E} \right] \frac{(2m^*E)^{3/2}}{\Delta(E)}. \qquad (29.45)$$

In this case, the dominant effect giving rise to the SdH oscillations is due to the effects of scattering on $\Delta(E)$. Our result is equivalent to that derived from the semiclassical generalization of the Boltzmann equation that is mentioned earlier. As noted before, this is not always true in materials of lower dimensionality or if the scattering effects are otherwise suppressed, such as in the presence of a large external reservoir of states.

How we next proceed depends on how we handle the behavior of the self-energy. As noted earlier, in 3D electron gases, the dominant form of scattering is normally due to impurities. We choose the self-consistent Born approximation (see Bruus and Flensberg (2004), for example) as described previously for what follows.

In this approximation and in three dimensions, as before, we may ignore the real portion of the self-energy in our calculations. The imaginary portion of the self-energy is proportional to the density of states multiplied by the square of the scattering amplitude W (whose value depends on the concentration of impurities and the strength of their interaction with the electrons, and into which we enfold the factor of 2 due to summation over spins) and may thus be written

$$\Delta(E) = \frac{W}{\pi} \sum_{\beta} \Im G^{R}(\epsilon_{\beta}, E). \qquad (29.46)$$

where β corresponds to the set of free electron quantum numbers defined earlier. Using Poisson's formula as above, we may calculate this to be

$$\Delta(E) = Wg(E)\left[1 - \left(\frac{\hbar\omega_c}{2E}\right)^{1/2} \sum_{r=1}^{\infty} \frac{(-1)^r}{r^{1/2}} R_D(r,E)\cos\left(\frac{2\pi r E}{\hbar\omega_c} - \frac{\pi}{4}\right)\right], \qquad (29.47)$$

where

$g(E) = \sqrt{E}\,(2m^*/\hbar^2)^{3/2}/2\pi^2$ is the 3D density of states for zero magnetic field (Kubo et al. 1965, Hamguchi 2001)

$R_D(r, E) = \exp(-2\pi|r|\Delta(E)/\hbar\omega_c)$ is similar to the Dingle damping factor

Since the average value of $\Delta(E)$ is directly proportional to the average Dingle temperature $T_D\pi$, we may then write

$$\Delta(E) = \pi k_B T_D\left[1 - \left(\frac{\hbar\omega_c}{2E}\right)^{1/2} \sum_{r=1}^{\infty} (-1)^r R_D(r,E)\cos\left(\frac{2\pi r E}{\hbar\omega_c} - \frac{\pi}{4}\right)\right]. \qquad (29.48)$$

We may then substitute this into the expression for the conductivity, obtaining (when $\mu/\hbar\omega_c \gg 1$)

$$\sigma_{zz} = \int dE\left[-\frac{\partial f(E)}{\partial E}\right]\sigma(E)\left[1 + \left(\frac{\hbar\omega_c}{2E}\right)^{1/2} \sum_{r=1}^{\infty} \frac{(-1)^r}{r^{1/2}} R_{Dr}\cos\left(\frac{2\pi r E}{\hbar\omega_c} - \frac{\pi}{4}\right)\right], \qquad (29.49)$$

where

$\sigma(E) = N(E)\hbar e^2/m^*\pi k_B T_D = N(E)\tau e^2/m^*$ is the zero-field conductivity

$N(E) = 2(2m^*E)^{3/2}/(6\pi^2\hbar^3)$ and we have made the approximation $R_D(r, E) \approx R_{Dr}$, R_{Dr} being given by Equation 29.32

Integrating over dE, we use the delta-function-like behavior of $-\partial f(E)/\partial E$ near the Fermi energy for $k_B T \ll \mu$ to obtain the constant term and expand around μ to obtain the LK damping factors for the oscillatory terms. Our final expression is

$$\sigma_{zz} = \sigma_0\left[1 + \left(\frac{\hbar\omega_c}{2\mu}\right)^{1/2} \sum_{r=1}^{\infty} \frac{(-1)^r}{r^{1/2}} R_{Dr} R_T(r\lambda)\cos\left(\frac{2\pi r \mu}{\hbar\omega_c} - \frac{\pi}{4}\right)\right], \qquad (29.50)$$

where

$\sigma_0 = \sigma(\mu)$

$N(\mu)$ is the number density of electrons in three dimensions

In Section 29.3.3 we turn to specific applications of these methods to low-dimensional systems.

29.3 Combination Frequencies in the Quantum Magnetic Oscillations of Multiband Quasi-2D Materials

29.3.1 Introduction

Quasi-two-dimensional (q2D) materials are those in which the ease of electron transport in the x–y plane is greater than in the z direction. Examples of compounds where this is the case include the organic charge-transfer salts* [extensively reviewed in Singleton (2000) and Kartsovnik (2004)] and cuprate superconductors—however, the phenomena typical to q2D systems are more easily observed in the former case than in the latter. Figure 29.1 shows the structure of the Fermi surface: a cylinder with periodic bulges.

A typical approach to such systems is to use a tight-binding model spectrum, which in the presence of a magnetic field is

$$E = \hbar\omega_c\left(n + \frac{1}{2}\right) - 2t\cos\left(\frac{p_z a}{\hbar}\right), \qquad (29.51)$$

where

$\omega_c = eB\cos(\Theta)/m$ is the cyclotron frequency

$t = t_0 J_0\,(k_F a \tan(\Theta))$ is the interlayer transfer integral

$J_0(x)$ is the zeroth-order Bessel function

a is the interlayer separation

Θ is the angle between the direction of the field and the normal to the x–y planes

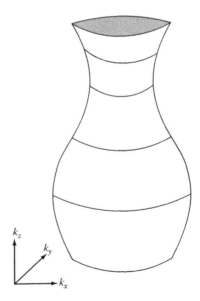

FIGURE 29.1 Diagram showing part of the Fermi surface of a q2D metal. The warping has been exaggerated so that the bulges are visible.

---

* These also exhibit a range of behaviors that are due to the existence of corrugated, open Fermi sheets, but these are beyond the scope of this chapter.

At certain values of Θ, $J_0(k_F a \tan(\Theta))$ is equal to zero. At these values, the electron group velocity perpendicular to the layers vanishes, causing the system to become effectively 2D (Yamaji 1989). This results in a peak in the resistivity; the oscillations in the resistance of the metal as Θ is varied and are called angular magnetoresistance oscillations (AMROs). For further information on these, see Singleton (2000) and Kartsovnik (2004)—in what follows, we shall take $\Theta = 0$, unless otherwise specified.

29.3.2 dHvA Oscillations

We begin with a multiband q2D metal with the dispersion

$$E_\alpha = \hbar\omega_{c,\alpha}\left(n+\frac{1}{2}\right) - 2t_\alpha \cos\left(\frac{p_z a}{\hbar}\right) + \Delta_{\alpha 0} + g_\alpha \sigma\mu_B B, \quad (29.52)$$

where

α labels the band number
$\Delta_{\alpha 0}$ is the band edge the absence of a magnetic field
μ_B is the Bohr magneton
$\omega_{c\alpha} = eB \cos(\Theta)/m_\alpha$ is the cyclotron frequency
m_α is the effective electron mass
$t_\alpha = t_{0\alpha} J_0(k_{F\alpha} a \tan(\Theta))$ is the transfer integral for band α
$\sigma = \pm 1/2$
g_α is the electron g-factor

It has been shown in Nakano (1999, 2000), Grigoriev (2001), Champel and Mineev (2001a), and Alexandrov and Bratkovsky (2002) (for the q2D case) that the oscillatory portion of the thermodynamic potential that gives rise to the dHvA effect has the following form:

$$\tilde{\Omega} = 4 \sum_\alpha^{\text{bands}} \sum_{r=1}^\infty (-1)^r A_\alpha^r \cos\left(\frac{2\pi r\tilde{\mu}}{\hbar\omega_{c\alpha}}\right), \quad (29.53)$$

where

$$A_\alpha^r = \frac{\rho_\alpha \hbar^2 \omega_{c,\alpha}^2}{4\pi r^2} R_T\left(\frac{2\pi^2 r k_B T}{\hbar\omega_{c,\alpha}}\right) J_0\left(\frac{2\pi r t_\alpha}{\hbar\omega_{c,\alpha}}\right) R_{D\,r\alpha} \cos\left(\pi r \frac{m^*}{m^e}\right). \quad (29.54)$$

Here, we have $\tilde{\mu} = \mu - \Delta_{\alpha 0}$, $\rho_\alpha = m_\alpha/2\pi\hbar^2 a$ being the density of states per unit volume for the energies of interest, R_T being the standard LK reduction factor, and $R_D = e^{-2\pi^2 r k_B T_{D\alpha}/\omega_{c\alpha}}$, where $T_{D\alpha}$ is the Dingle temperature for band α.

We observe the following features of these oscillations: First, the frequency of oscillation and the splitting of the amplitudes in the Fourier series are dependent on the orientation of the magnetic field through the angle Θ, which is not surprising, given that our Fermi surface is cylindrical, and hence is only rotationally symmetric in the x–y plane. Second, for large $t_\alpha/\hbar\omega_{c,\alpha}$, we will see that the Yamaji orientation factor $J_0(2\pi r t_\alpha/\hbar\omega_{c,\alpha})$, noted in Yamaji (1989), causes a further splitting of the amplitudes

into pairs, corresponding to the two extremal surfaces of the Fermi surface (the "neck" and the "throat" in the terminology of Shoenberg (1984a). This is a q2D phenomenon—when t_α is equal to zero, the Yamaji factor is equal to unity, and so we recover the expected 2D result of Alexandrov and Bratkovsky (2001). However, barring an improbable number of coincidences between various parameters of the system, this is unlikely to happen for all bands simultaneously (Alexandrov and Bratkovsky 2002).

Were this all, then the matter of dHvA oscillations in q2D systems would be relatively trivial. However, there are a number of complicating factors that give rise to some interesting phenomena that should be accounted for. One of these is the appearance of amplitudes in the Fourier spectrum that correspond to additional frequencies of oscillation whose periods are combinations of integral multiples of the frequencies of oscillation of the various bands. We discuss these in Section 29.3.2.1.

29.3.2.1 Combination Frequencies

One phenomenon that is often observed in q2D metals is the appearance of oscillations with frequencies that appear to be combinations of an expected dHvA frequency f_α and another frequency f_β. In the theory of the dHvA effect, there are several possible origins for these.

29.3.2.1.1 Magnetic Breakdown

This effect (Cohen and Falicov 1961, Pippard 1962, Falicov and Stachiowiak 1966, Singleton 2000, Kartsovnik 2004), observed for example in Sasaki et al. (1990), arises from the destruction of the zero-field semiclassical band structure at high values of the magnetic field. Above a given value of the field B_0, called the breakdown field, electrons may tunnel from one closed[*] band to a nearby open band, and so orbit around a "new" closed Fermi surface in addition to the original one (see Figure 29.2). The tunneling probability for this is $P = \exp(-B_0/B)$, with $B_0 \simeq m^* \Delta_g^2/e\hbar\mu$, where Δ_g is the interband gap at the junction of the two bands.

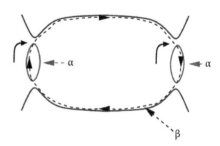

FIGURE 29.2 Schematic diagram of a possible magnetic breakdown network. α is the semiclassical path along the Fermi surface, and β is the new path allowed when magnetic breakdown occurs. The solid arrows show the direction of the cyclotron motion of the electrons. (Simplified from Kartsovnik, M.V., *Chem. Rev.*, 104, 5737, 2004, Figure 19.)

---

[*] In a closed band, the path taken along the Fermi-surface of the band by an electron eventually returns to its origin, where as with open bands, this is not the case.

As a result of this tunneling, there now exists a new frequency F_β, and also frequencies of oscillation that are combinations of it and integer multiples of the original semiclassical frequencies—that is, $F_\beta + nF_\alpha$, whose amplitudes resemble those of the standard LK theory multiplied by some path-dependent factor (Falicov and Stachiowiak 1966).

However, one occasionally encounters measurements with negative n. In order for these to be explained by magnetic breakdown, the electron orbit would be required to travel round the areas corresponding to F_α the *wrong way*, which indicates that these frequencies are "forbidden." How, then, do they arise? In magnetoresistance measurements, one possible source is the Shiba–Fukuyama–Stark effect (commonly known as the Stark effect), which is discussed below; however, this effect is irrelevant to thermodynamic phenomena such as the dHvA effect. This entails that there must be some other factor at play—perhaps the standard theory of magnetic breakdown fails in q2D systems (suggestions of this sort are assessed in Section 3.4.3 of Kartsovnik (2004), or there is some element of the physical system that has not been taken into account. We explore one such proposal next.

29.3.2.1.2 Chemical Potential Oscillations in Closed Multiband q2D Systems

One difference between 2D and 3D systems is the existence of significant oscillations in the chemical potential when the system is closed—that is, when the number of particles N is fixed. We shall describe in a qualitative fashion how this arises in purely 2D systems (following the discussion given in Shoenberg (1984b)) and then describe how this affects the behavior of multiband q2D metals.

Let us consider an ideal 2D metal with no scattering; hence $T_D = 0$ at zero temperature without spin splitting. In order to find the variation of μ that keeps N fixed, we should drop the assumption that all energy levels beneath μ are fully occupied. We postulate that μ is pinned or "stuck" to the highest occupied level n_F, so that its value is given by $\hbar\omega_c(n_F + 1/2)$. This is only partially occupied, with an occupation given by $N - n_F D$ (with $D = eB/\pi\hbar$ being the level degeneracy) ranging from 0 when $n_F D = N$ to N when $(n_F + 1)D = N$—that is, as D decreases from N/n_F to $N/(n_F + 1)$, μ jumps to the next highest level, $(n_F + 1)$, and is pinned to that. This gives rise to oscillations in μ as B is varied.

It was discovered in Alexandrov and Bratkovsky (1996) and subsequently numerically and analytically analyzed in Nakano (1999), Alexandrov and Bratkovsky (1997, 2002, 2004), Champel (2002, 2004), Kishigi and Hasegawa (2002, 2005), and Fortin et al. (2005) that this oscillation of the chemical potential gives rise to combination frequencies in closed 2D and q2D systems. These additional frequencies have been observed in several experimental situations (Ohmichi et al. 1999, Shepherd et al. 1999, Kartsovnik 2004). Here we shall discuss the q2D case as described in Alexandrov and Bratkovsky (2002), which should prove sufficient to demonstrate what is going on (though there has been some dispute over the region of the validity of the approximation (Champel 2002, 2004, Alexandrov and Bratkovsky 2004)).

To illustrate this effect, we consider the case of two bands with masses $m_1 = m_2 = m$ and equal Dingle factors R_D, but with different dHvA frequencies f_1 and f_2 (Alexandrov and Bratkovsky 2004). The relevant thermodynamic potential in the case of a closed system is the free energy $F = \Omega + \mu N$ where for a fixed number of electrons $N = -\partial\Omega/\partial\mu$. The overall chemical potential is given by $\mu = \mu_0 + \tilde\mu$, with

$$\tilde\mu = -\frac{\tilde N}{\rho} \equiv -\frac{\tilde N}{\rho_q + \rho_{bg}}, \qquad (29.55)$$

where

μ_0 is the non-oscillating part of μ
$\tilde\mu$ is the oscillating portion
$\tilde N = -\partial\tilde\Omega/\partial\mu$ is the oscillating part of the electron density
ρ_q is the density of states from quantized bands
ρ_{bg} is the density of the non-quantized background states

The derivative of $\tilde\Omega$ given by (29.53) with respect to μ and of the oscillatory components with respect to B at $T = 0$, and putting $z = 2\pi\tilde\mu/\hbar\omega_c$ give rise to

$$z = -\sum_{r=1}^{\infty} \frac{(-1)^r R_{Dr}}{r}\left[\sin\left(rz + \frac{f_1}{B}\right) + \sin\left(rz + \frac{f_2}{B}\right)\right], \qquad (29.56)$$

and

$$\tilde M = \frac{e^2}{4\pi^3 ma}\sum_{r=1}^{\infty} \frac{(-1)^r R_{Dr}}{r}\left[f_1\sin\left(rz + \frac{f_1}{B}\right) + f_2\sin\left(rz + \frac{f_2}{B}\right)\right], \qquad (29.57)$$

where $\tilde M$ is the oscillatory component of the magnetization. Taking z to be small and expanding $\tilde M$ in powers of z gives the following value for the first Fourier component of the $(f_1 \pm f_2)$ amplitudes:

$$M_{first}^{11} = \mp\frac{e^2 R_{D1}^2}{16\pi^3 ma}(f_1 \pm f_2). \qquad (29.58)$$

Another important feature of these harmonics (Alexandrov and Bratkovsky 2002) is that they display a strong dependence on the temperature and impurity as they contain damping factors from both of the original harmonics. In addition, the form of these factors is such that one cannot generally characterize the temperature dependence of the harmonics with a single effective mass except at high temperatures where $M_{tot}^* \approx rm_\alpha \pm r'm_{\alpha'}$ can be expected to hold, where r, r' and m_α, $m_{\alpha'}$ signify harmonics and effective masses from different bands. This is corroborated by numerical simulations.

All these features should ensure that it is relatively easy to distinguish combination frequencies that arise due to this effect from those due to others such as magnetic breakdown. They are also strongly suppressed as the 3D-limit is approached and chemical potential oscillations become negligible (Champel and Mineev 2001b).

29.3.2.1.3 Chemical Potential Oscillations in Closed Single-Band q2D Systems

Chemical potential oscillations may also give rise to oscillations in single-band q2D systems. Alexandrov and Kabanov (2007) have shown that the pair of extremal surfaces (the "neck" and the "belly" of the q2D Fermi surface) in the limit of intermediate field ($\hbar\omega_c \ll 4\pi t$—in this limit we use the approximation of the Bessel function $J_0(x) \approx \sqrt{2/\pi x}\cos(x - \pi/4)$, $x = 4\pi t/\hbar\omega_c$) may also give rise to additional oscillations in the magnetization of the form (at $T = 0$ K)

$$M_{mix} \approx \frac{e\mu\cos(\Theta)}{4\pi^2\hbar a}\frac{R_D^2 B}{2\pi B_\perp}\sin\left(\frac{2\pi F}{B}\right) \qquad (29.59)$$

where $B_\perp = 4\pi mt/e\hbar\cos(\Theta) \gg B$, $F = f_n + f_b = 2m\mu/e\hbar\cos(\Theta)$, and $f_{n,b} = m(\mu \mp 2t)/e\hbar\cos(\Theta)$ are the frequencies corresponding to the neck and the belly of the Fermi surface, respectively (see Figure 29.1). In principle, therefore, one should be able to observe in the Fourier Transforms of systems of this type a small amplitude between the two peaks corresponding to the second harmonics of the neck and belly frequencies. However, this amplitude is very small, and depending on the size of the magnetic field window used experimentally, it may prove difficult to detect.

29.3.3 SdH Oscillations

29.3.3.1 Phenomenological Differences from dHvA Oscillations

Shubnikov–de Haas oscillations in q2D systems exhibit a number of features that are quite unusual. These include both the presence of an unusual phase shift in the beating of the oscillations due to the warping of the Fermi surface and the appearance of additional slow oscillations, both of which are absent in the dHvA oscillations.

Phase shifts: From the asymptotic form of the Bessel function in the intermediate case, $J_0(x) \approx \sqrt{2/\pi x}\cos(x - \pi/4)$, it follows that one would typically expect that in that limit, the frequencies corresponding to the neck and belly oscillations would have a phase shift of $\pm\pi/4$, in accordance with the conventional 3D theory.

As it happens, this is true for the dHvA effect but not for the SdH effect. In q2D organic salts, it has been observed that the phase shift in the SdH effect has been as high as $\pm\pi/2$ (Weiss et al. 1999, Schiller et al. 2000), even in measurements strongly dominated by the first harmonic, where one would expect the conventional theory to apply very well.

Slow oscillations: The appearance of additional slow oscillations (Singleton 2000, Grigoriev 2002, Kartsovnik 2004, and the references therein) in measurements of the SdH effect has been attributed to the presence of small electron pockets in these materials. However, these pockets are not predicted by band structure calculations and, moreover, do not appear in the dHvA oscillations, whereas they are very pronounced in those of the SdH effect. This seems to argue against this interpretation.

Explanations: Grigoriev and collaborators (Grigoriev et al. 2002, Grigoriev 2002, 2003) have provided theoretical descriptions of the SdH effect in single-band q2D materials, using both the semiclassical Boltzmann formula and the linear response theory using the Born approximation. While there are some slight quantitative differences between the two accounts, they are qualitatively similar and provide the same explanations for the phenomena in question.

Phase shifts arise due to the importance of the oscillations in the mean free velocity, which becomes as important as the oscillations in the density of states or self-energy (depending on the approach taken) when $\hbar\omega_c$ is of the order of t (Harrison et al. 1996, Grigoriev 2002, Grigoriev et al. 2002). Whereas the effect of warping in the latter enters in a zeroth-order Bessel function $J_0(4\pi rt/\hbar\omega_c)$, it enters into the former as a first-order Bessel function $J_1(4\pi rt/\hbar\omega_c)$. In the limit where these can be replaced with the equivalent cosine or sine functions, we obtain (Grigoriev et al. 2002, Grigoriev 2002, 2003) (after integration over the energy)

$$\tilde{\sigma}_{zz} \propto \cos\left(\frac{2\mu\pi}{\hbar\omega_c}\right)\cos\left(\frac{4\pi t}{\hbar\omega_c} - \frac{\pi}{4} + \phi\right), \qquad (29.60)$$

with $\phi = \arctan(a)$ where $a = \hbar\omega_c/2\pi t$ if the Boltzmann approach is used (Grigoriev et al. 2002), or $a = \hbar\omega_c/2\pi t(1 + (2\pi^2 k_B T_D/\hbar\omega_c))$ (Grigoriev 2003) in the case of the linear response or Kubo formalism. While the latter gives closer agreement to experiment at magnetic fields of around 10 T, for higher fields it has been found that the measured maximum of the phase shift exceeds the theoretical maximum of $\pi/2$ (Kartsovnik et al. 2003).

Slow oscillations also arise due to the interference of fast quantum oscillations of the various quantities on which the conductivity depends, such as the mean free velocity and the various mechanisms implicitly included in τ. Taking the product of two such oscillations has the following result:

$(1 + A\cos x)(1 + B\cos x) = 1 + (A + B)\cos x + (AB/2)\cos 2x + AB/2$. The final $AB/2$ term corresponds in this case to the slow oscillations, which contains the slowly oscillating terms proportional to the Bessel functions. Taking $4\pi t \geq \hbar\omega_c$, so that we may use the sinusoidal asymptotics of the Bessel functions, we obtain (Grigoriev et al. 2002, Grigoriev 2002, 2003):

$$\sigma_{zz} = \sigma_0\left[1 + 2\sqrt{\frac{\hbar\omega_c}{2\pi^2 t}(1 + a^2)}\cos\left(\frac{2\pi\mu}{\hbar\omega_c}\right)\cos\left(\frac{4\pi t}{\hbar\omega_c} - \frac{\pi}{4} + \phi_b\right)R_T\left(\frac{2\pi^2 k_B T}{\hbar\omega_c}\right)R_D \right.$$
$$\left. + \frac{\hbar\omega_c}{2\pi^2 t}\sqrt{1 + a_s^2}\cos\left(2\left[\frac{4\pi t}{\hbar\omega_c} - \frac{\pi}{4} + \frac{\phi_s}{2}\right]\right)R_{D*}^2\right], \qquad (29.61)$$

where we have omitted the second harmonics and the constant terms, and $\sigma_0 = e^2 m^* t d^2/\pi^2 \hbar^2 k_B T_D$. $\phi_s = \arctan(a_s)$ and $\phi_b = \arctan(a)$, and R_{D*} is a Dingle factor that differs slightly from the expected factor R_D—see below. For derivations using both the Kubo formula (Grigoriev 2003) and Boltzmann semiclassical approximation (Grigoriev et al. 2002), $a_s = \hbar\omega_c/2\pi t$. For the Boltzmann derivation, $a = a_s$, whereas for the Kubo derivation

$a = \hbar\omega_c/2\pi t(1+(2\pi^2 k_B T_D/\hbar\omega_c))$, the difference between the two being due to the additional term corresponding to the second-order poles in the integrations over the Green's functions in the linear response calculation that cannot be neglected in a q2D system (Champel and Mineev 2002, Grigoriev 2003). The second term gives rise to the fast SdH oscillations, the third to the slow oscillations.

It should be apparent that the slow oscillations have no temperature damping, which follows from their lack of dependency on μ—they depend on the energy spectrum, not on the distribution function. This entails that they can be larger than the fast oscillations at $T \gtrsim T_D$. However, one would intuitively expect (and experiment shows (Kartsovnik et al. 2002)) that there would be some damping of these oscillations at reasonably high temperatures—this would likely be due to electron–electron and electron–phonon scattering, which has not been included in this calculation but can be expected to play a role. Grigoriev (2002) includes a brief account of how this might transpire, but to our knowledge a detailed calculation has yet to be carried out. It is likely that the temperature dependence of the oscillations would carry important information about the strength of the electron–electron and electron–phonon interaction in these materials.

In addition, the slow oscillations also have a different Dingle damping factor (R_{D^*}) from the fast oscillations (which have R_D). This is because the traditional Dingle factor includes all temperature-independent mechanisms of Fock–Landau level smearing, from microscopic causes such as scattering to more macroscopic inhomogeneities in the sample that can cause spatial variations in the electron energy equivalent to local shifts of the chemical potential μ. Since the measurements of the SdH oscillations are in effect an averaging over the entire sample, these variations normally result in increased damping of the oscillations. However, since the slow oscillations are not dependent on μ, they are sensitive only to the smearing introduced by microscopic mechanisms, and are thus damped less than they otherwise would be. One can estimate the degree to which the macroscopic inhomogeneities contribute to the damping of the SdH effect by taking the ratio of the two Dingle temperatures (Kartsovnik et al. 2002).

The above discussion is relevant to low and intermediate values of the magnetic field. At high values of the field where $\hbar\omega_c/t \gg 1$, we enter the 2D limit discussed in Section 3.5.2 of Kartsovnik et al. (2003). This is largely beyond the scope of this chapter, however. We should observe, though, that in this limit, it is likely that the model of point-like impurities assumed in the above calculations is no longer valid (Raikh and Shahbazyan 1993, Champel and Mineev 2006), and that various theoretical difficulties conspire so as to make the description of real materials in this limit nontrivial.

29.3.3.2 Scattering and Combination Frequencies

As in the dHvA effect, we may also observe combination SdH frequencies in multiband metals. In addition to the possible causes listed for the dHvA, all of which apply in this case, there are two additional causes: the Shiba–Fukuyama–Stark effect and interband scattering.

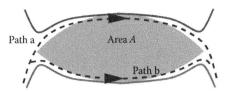

FIGURE 29.3 Schematic diagram showing the Shiba–Fukuyama–Stark effect. An electron can move along two possible paths, one of which encloses an orbit of area A. (Simplified from Singleton, J., *Rep. Prog. Phys.*, 63, 1111, 2000, Figure 10.)

Shiba–Fukuyama–Stark effect: This effect (Shiba and Fukuyama 1969, Stark and Friedberg 1974, Stark and Reifenberger 1977, Morrison and Stark 1981) produces subtractive combination frequencies in the conductivity that are not permitted by the traditional form of magnetic breakdown. In this case, breakdown occurs between two Fermi surfaces along which electrons are traveling in the same direction—they can go either by route a or b on the pathways described in Figure 29.3. As a result of this, we obtain oscillations in the resistivity that are periodic in B^{-1} with a frequency proportional to that of the area A (Singleton 2000). Since this effect requires no circulation of the electrons around the closed Fermi surface, the free energy of the system is not affected, and so this effect gives rise to no further oscillations in Ω, and hence no oscillations in the dHvA effect. In addition, since these effects do not arise from the Fock–Landau quantization of the system, they are not so strongly damped by the temperature as compared to the SdH oscillations.

Interband scattering: Thomas et al. (2008) have generalized Grigoriev's linear response theory (Grigoriev 2003) results to the case of multiple bands, examining the effects of chemical potential oscillations and scattering on these systems. In addition to the expected mixing effect that gives rise to combination frequencies, additional slow oscillations may be observed, as predicted in the single-band case by Grigoriev.

The most dramatic effect arises if impurities are permitted to scatter electrons between bands. In that case, the self-energy of the system must include oscillatory contributions from other bands, and so interference effects that give rise to combination frequencies arise.

Restricting ourselves to the case of two bands, and taking $R_D \ll 1$ so that the self-consistent equation is analytically solvable, and keeping terms up to the second order, we find (Thomas et al. 2008)

$$
\sigma_{zz} = \sigma_0 \left[1 + \sum_\alpha \left(D_1^\alpha \cos\left(\frac{2\pi\mu}{\hbar\omega_{c,\alpha}}\right) R_T\left(\frac{2\pi^2 k_B T}{\hbar\omega_{c,\alpha}}\right) \right. \right.
$$

$$
\left. + D_2^\alpha \cos\left(\frac{4\pi\mu}{\hbar\omega_{c,\alpha}}\right) R_T\left(\frac{4\pi^2 k_B T}{\hbar\omega_{c,\alpha}}\right) + D_s^\alpha \right)
$$

$$
\left. + D_{12}^+ \cos\left(\frac{2\pi\mu}{\hbar\omega_+}\right) R_T\left(\frac{2\pi^2 k_B T}{\hbar\omega_+}\right) + D_{12}^- \cos\left(\frac{2\pi\mu}{\hbar\omega_-}\right) R_T\left(\frac{2\pi^2 k_B T}{\hbar\omega_-}\right) \right],
$$

$$(29.62)$$

where

$$\frac{1}{\omega_\pm} = \frac{1}{\omega_{c2}} \pm \frac{1}{\omega_{c1}}. \qquad (29.63)$$

Here, the unmixed amplitudes are

$$D_1^\alpha = 2\frac{m_\alpha^{1/2}}{M} R_{D\alpha} \sqrt{\frac{\hbar eB}{2\pi^2 t}} (1 + (a_{1,\alpha})^2) \cos\left(\frac{4\pi t}{\hbar\omega_{c\alpha}} - \frac{\pi}{4} + \phi_{1,\alpha}\right) \qquad (29.64)$$

$$D_2^\alpha = \frac{(R_{D\alpha})^2 m_\alpha}{M}\left[\frac{4m_\alpha k_B T_D}{Mt}\sqrt{1 + (a_{1,\alpha})^2} \cos\left(\frac{4\pi t}{\hbar\omega_{c\alpha}} - \frac{\pi}{4}\right)\right.$$
$$\times \cos\left(\frac{4\pi t}{\hbar\omega_{c\alpha}} - \frac{\pi}{4} + \phi_{1,\alpha}\right) + 2\sqrt{\frac{\hbar\omega_{c,\alpha}}{4\pi^2 t}}(1 + (a_{2,\alpha})^2)$$
$$\times \cos\left(\frac{8\pi t}{\hbar\omega_\alpha} - \frac{\pi}{4} + \phi_{2,\alpha}\right)\bigg] + D_S^\alpha, \qquad (29.65)$$

and

$$D_S^\alpha = \frac{\hbar eB(R_{D\alpha})^2 m_\alpha}{2\pi^2 M^2 t}\left[\sqrt{1 + \left(\frac{\hbar\omega_{c,\alpha}}{2\pi t}\right)^2} \cos\left(2\left[\frac{4\pi t}{\hbar\omega_{c\alpha}} - \frac{\pi}{4} + \phi_{S\alpha}\right]\right) + 1\right], \qquad (29.66)$$

where

$$a_{r,\alpha} = \frac{1}{2}\left(\frac{\hbar\omega_{c,\alpha}}{r\pi t} + \frac{2\pi k_B T_D}{t}\right). \qquad (29.67)$$

The mixed frequencies are somewhat more complicated than those arising from chemical potential oscillations, with the amplitudes of the positive combinations $f_1 + f_2$ typically being more pronounced than those of the negative $f_1 - f_2$:

$$D_{12}^+ = \frac{\hbar eB R_{D1} R_{D2}\sqrt{m_1 m_2}}{M^2 t\pi^2}\left(1 + \frac{2\pi^2 T_D}{\hbar\omega_+}\right)$$
$$\times\left[\sqrt{1 + (y^-)^2} \cos\left(\frac{4\pi t}{\hbar\omega_-} + \phi_{y^-}\right)\right.$$
$$\left. + \sqrt{1 + (y^+)^2} \sin\left(\frac{4\pi t}{\hbar\omega_+} + \phi_{y^+}\right)\right] \qquad (29.68)$$

where

$$y^\pm = \left(\frac{\hbar(\omega_{c2} \pm \omega_{c1})}{4\pi t} + 2\pi^2 k_B T_D\left[\frac{a_{1,2}}{\hbar\omega_{c,2}} \pm \frac{a_{1,1}}{\hbar\omega_{c,1}}\right]\right)$$
$$\times\left(1 + \frac{2\pi^2 k_B T_D}{\hbar\omega_+}\right)^{-1}, \quad \phi_{y^\pm} = \arctan(y^\pm), \qquad (29.69)$$

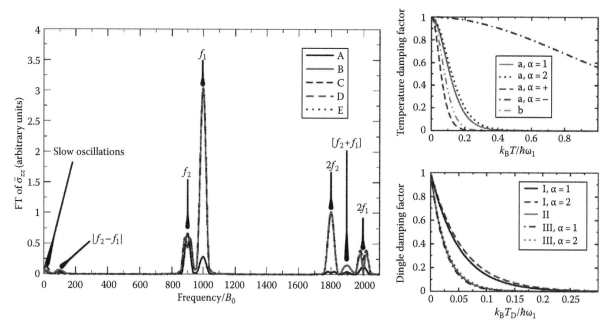

FIGURE 29.4 A graph showing the Fourier transform of the SdH oscillations from $0.9 \le B_0/B \le 0.95$ alongside plots of how the damping factors evolve as their respective temperatures are increased. Due to the small size of the field window, the slow oscillations are poorly resolved. The following parameters are used: $\mu = 500\,t$, $(\omega_1/\omega_2) = (m_2/m_1) = 0.9$, $k_B T_D = 0.026\,t$ and $k_B T = 0.00005\,t$, setting our unit of measurement to be $B_0 = m_1 t/2\hbar\,e$, which is around 41 Tesla if $t = 0.0/\text{eV}$ and $m_1 = m_e$. The legend in the Fourier transform plots should be interpreted as follows: A—closed system, fixed τ; B—open system, interband scattering; C—closed system, intra-band scattering; D—closed system, interband scattering; E—open system, intra-band scattering. The legend in the temperature damping plots should be interpreted as a—value of $R_T(2\pi^2 k_B T/\hbar\omega_{c,\alpha})$; b—value of $R_T(2\pi^2 k_B T/\hbar\omega_1) R_T(2\pi^2 k_B T/\hbar\omega_2)$. The legend in the Dingle damping plots should be interpreted as I—value of $R_{D\alpha}$; II—value of $R_{D1}R_{D2}$; III—value of $R_{D\alpha}^2$. (Adapted from Thomas, I.O. et al., *Phys. Rev. B*, 77, 075434, 2008, Figure 1.)

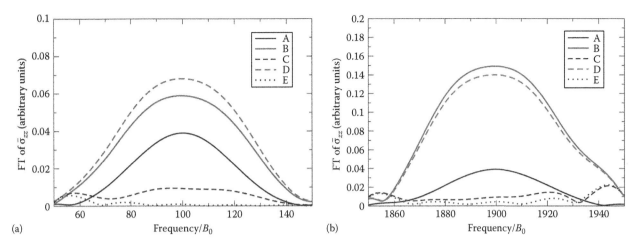

FIGURE 29.5   Figures showing the details of the (a) $|f_2 - f_1|$ amplitudes (From Thomas, I.O. et al., *Phys. Rev. B*, 77, 075434, 2008, Figure 5a. With permission.) and (b) $|f_2 + f_1|$ amplitudes (Adapted from Thomas, I.O. et al., *Phys. Rev. B*, 77, 075434, 2008, Figure 6a.) of Figure 29.4.

and

$$D_{12}^- = \frac{\hbar e B R_{D1} R_{D2} \sqrt{m_1 m_2}}{M^2 t \pi^2} \left[\sqrt{1 + (q^-)^2} \cos\left(\frac{4\pi t}{\hbar \omega_-} + \phi_{q^-} \right) \right.$$
$$\left. + \sqrt{1 + (q^+)^2} \sin\left(\frac{4\pi t}{\hbar \omega_+} + \phi_{q^+} \right) \right] \qquad (29.70)$$

where

$$q^\pm = \frac{\hbar(\omega_{c,2} \pm \omega_{c,1})}{4\pi t}, \quad \phi_{q^\pm} = \arctan(q^\pm), \qquad (29.71)$$

and $M = m_1 + m_2$. As can be seen from the equation, the damping of the amplitudes with respect to temperature and impurity scattering is different from the usual behavior. This can be combined with the behavior of the phase shift and similar dependencies to distinguish amplitudes that arise from this mechanism from those arising from others (see Figures 29.4 and 29.5).

In addition, it should also be noted that although this derivation was carried out under the assumption that $R_D \ll 1$, effects due to the mixing of bands by scattering should probably be observable at values of R_D closer to 1, whereas no such effects are likely to be visible in the dHvA effect. This is because the oscillations entering into the conductivity from scattering through $\tau \propto \Delta(E)^{-1}$ are enhanced as R_D nears unity, whereas in the dHvA effect they enter only through the Dingle factor and thus become negligible as T_D approaches zero (Wasserman and Springford 1996).

29.4 Quantum Magnetic Oscillations in Nanowires

29.4.1 Introduction

The question of how the physical confinement of a system affects the behavior of electrons within longitudinal magnetic fields (and hence the behavior of the quantum magnetic oscillations)

is one that has a venerable history in the field of condensed matter (see Fock 1928, Landau 1930, Darwin 1930, Dingle 1952a, for example) and it is one that is, as should be obvious, of importance when considering the theory of quantum magnetic oscillations in nanowires. A confined system within a magnetic field will exhibit size quantization due to the presence of physical boundaries in addition to the usual Fock–Landau quantization, and the behavior of the magneto-oscillations will be sensitive to the precise way that the two kinds of energy levels interact. Since there has been recent interest in the quantum magnetic oscillations of nanowires (see, for example, Brandt et al. 1987, Huber et al. 2004, Grozav and Condrea 2004, Nikolaeva et al. 2008) as well as studies of the effects of size on magnetoresistance (Cornelius et al. 2008), a theoretical account of these effects is of some importance.

It transpires that this is a nontrivial problem in many respects, and much depends on the various assumptions that one makes during one's calculations. Much work remains to be done, in particular regarding the appropriate approximation to choose for the energy spectrum.

29.4.2 Statement of the Problem

Let us assume that the system of interest is a long, clean metallic nanowire in a longitudinal magnetic field **B** parallel to the orientation of the wire (taken to be the z-axis of a cylindrical coordinate system). The electron mean free path, $l = v_F \tau$, is taken to be comparable or larger than the radius of the wire R, but smaller than the length L. In addition, we assume that the electron wavelength near the Fermi level is very small in these metallic nanowires, such that $L \gg l \gtrsim R \gg 2\pi\hbar/(m^* v_F)$; here and previously, v_F is the Fermi velocity and m^* is the band mass in the bulk metal. We also define a new frequency $\omega_s = \pi v_F / 2R$, which will account for some aspects of the confining effects later on.

The Schrödinger equation in polar coordinates for an electron moving in a longitudinal magnetic field B parallel to the

z-axis, where the vector potential has the form $A_\phi = B\rho/2$ and $A_z = A_\rho = 0$, is

$$-\frac{\hbar^2}{2m^*}\left[\frac{1}{\rho}\frac{\partial}{\partial\rho}\left(\rho\frac{\partial\psi}{\partial\rho}\right) + \frac{\partial^2\psi}{\partial z^2} + \frac{1}{\rho^2}\frac{\partial^2\psi}{\partial\phi^2}\right]$$

$$+ \frac{i\hbar\omega_c}{2}\frac{\partial\psi}{\partial\phi} + \frac{m^*\omega_c^2\rho^2}{8}\psi = E\psi. \qquad (29.72)$$

This is easy to solve (see Landau and Lifschitz 1977), for example, though one should be careful of conventions regarding e. The solution can be written in the form

$$\psi = \frac{1}{\sqrt{2\pi}}R(\rho)e^{im\phi}e^{ip_z z/\hbar}, \qquad (29.73)$$

where $R(\rho)$, following a redefinition in terms of $\xi = (m^*\omega_c/2\hbar)\rho^2$ is the radial function:

$$R(\xi) = e^{-\frac{\xi}{2}}\xi^{\frac{|m|}{2}}M\left(-\left(\beta - \frac{|m|}{2} - \frac{1}{2}\right), |m|+1, \xi\right) \qquad (29.74)$$

with $\beta = (\hbar\omega_c)^{-1}[E - (p_z^2/2m^*)] + m/2$. $M\left(-\left(\beta - (|m|/2) - (1/2)\right), |m|+1, \xi\right)$ is a confluent hypergeometric function. The zeroes of this are the eigenvalues of the Schrödinger equation for cylindrically confined electrons. If we write $\beta - (|m|/2) - (1/2)$ as $-a$ (here we follow the notation of Abramowitz and Stegun

(1965)), the energy levels of the particle are given by Landau and Lifschitz (1977)

$$E = \frac{p_z^2}{2m^*} + \hbar\omega_c\left(-a + \frac{|m|-m+1}{2}\right) \qquad (29.75)$$

If the wave function is finite everywhere, then $-a$ is an integer, and the eigenfunctions of the equation are the Laguerre polynomials with the Fock–Landau eigenvalues given by Equation 29.4. However, if it becomes zero at some finite radius R, then $-a$ is a positive, real number. This can lead to significant differences in the behavior of the system relative to the unconfined case (see Figure 29.6).

In order to proceed analytically in these cases, particularly where R is small, we must therefore obtain some form of approximate energy spectrum.

29.4.3 Weak Field Limit

29.4.3.1 Size Quantization in the Weak Field Limit

In the fourth part of his seminal series of papers on magnetization in metals, Dingle (1952c) treated the case of an electron gas confined within a thin cylindrical wire in a weak field. In this case, the influence of the field can be treated as a perturbation of the zero-field case, provided that for a given energy level of interest E_0, $\pi\omega_c^2/(8\omega_s^2) \ll E_0/\mu$, where ω_s is defined in the paragraph above Equation 29.72.

Perturbation of the zero-field Schrödinger equation gives us the following energy spectrum:

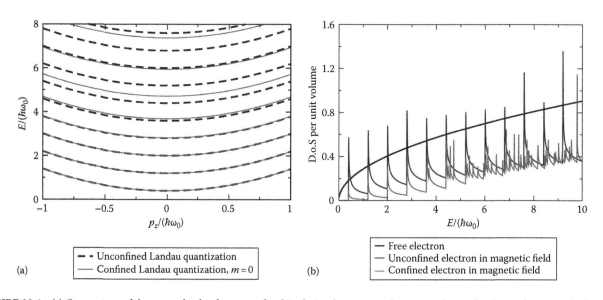

(a)

Unconfined Landau quantization

Confined Landau quantization, $m = 0$

(b)

Free electron

Unconfined electron in magnetic field

Confined electron in magnetic field

FIGURE 29.6 (a) Comparison of the energy levels of an unconfined Fock–Landau quantized system and a confined one, showing only the $m = 0$ energy levels from the latter. (b) Comparison of the density of states per unit volume for confined and unconfined systems. The required energy levels for the confined case are calculated numerically from the zeroes of the confluent hypergeometric function; the height of the peaks has been truncated in some cases so that the features of the graph are more readily observable. We have set $\mu = 10\hbar\omega_0$ and $\omega_c = 0.8\omega_0$, where $\omega_0 = \omega_s$ is a reference frequency.

$$E_{nm} = \frac{\hbar^2 \omega_s^2}{\pi^2 \mu} t_{nm}^2 \pm \frac{\hbar \omega_c m}{2} + \frac{\pi^2 \mu}{48} \frac{\omega_c^2}{\omega_s^2} \left[1 + \frac{2(m^2 - 1)}{t_{nm}^2} \right], \qquad (29.76)$$

where t_{nm} is the nth zero of a Bessel function that (in this case) can be found from the asymptotic

$$J_m(x) \sim \left(\frac{2}{\pi} \right)^{1/2} (x^2 - m^2)^{-1/2} \sin \left[(x^2 - m^2)^{1/2} - m \cos^{-1} \left(\frac{m}{x} \right) + \frac{\pi}{4} \right].$$

$$(29.77)$$

Using this approximation, Dingle calculated the following expression for the oscillatory portion of the magnetic susceptibility*:

$$\chi = -\frac{e^2 (2\mu)^{1/2}}{\hbar \pi^3 (m^*)^{1/2}} \frac{\omega_s}{\omega_c} \sum_{r=3}^{\infty} \sum_{r'=1}^{r/2} \frac{\Gamma}{r^2} \sin \left(\frac{2\pi r'}{r} \right) \sin \Lambda \sin \Pi, \qquad (29.78)$$

where r and r' are different summation indices and

$$\Gamma = \frac{2^{3/2} \pi^2 r k_B T}{\hbar \omega_s \sinh(2^{3/2} \pi^2 r k_B T / \hbar \omega_s)}, \qquad (29.79)$$

$$\Pi = \frac{2^{5/2} \pi \mu}{\hbar \omega_s} r \left[1 - \frac{7 \pi^2 \omega_c^2}{42 \omega_s^2} \sin^2 \left(\frac{\pi r'}{r} \right) \right] \sin \left(\frac{\pi r'}{r} \right) + \frac{\pi}{2} (r - 1), \quad (29.80)$$

and

$$\Lambda = \frac{\pi^2 \mu \omega_c}{4 \hbar \omega_s^2} r \sin \left(\frac{2\pi r'}{r} \right). \qquad (29.81)$$

As one might expect—given that this treats the field as a small perturbation—the dominant effect is due to the modification of the positions of the size-quantized energy levels by the magnetic field, which causes them to pass over the Fermi energy. One should also note that the temperature damping factor is controlled by the magnitude of ω_s, rather than ω_c as would be more usual for dHvA oscillations, which is most likely also due to the aforementioned dominance.

29.4.3.2 *"Grazing Orbits" and the Bohm–Aharonov Effect*

Bogacheck and Gogazde (BG; Bogacheck and Gogadzde 1973; see also the discussion in Aronov and Sharvin (1987) and further references in Gogadze (1984)) carried out a similar calculation to that of Dingle. Extending it out to the next order, they obtained a pair of oscillatory contributions to the density of

states ν_1^{osc} and ν_2^{osc}, the former corresponding to contributions from the states near the surface and the latter corresponding to Dingle-type oscillations due to the states toward the core of the wire. ν_2^{osc} is greater than ν_1^{osc} by a factor of $(E/E_0) \gg 1$, where $E_0 = \hbar^2 \omega_s^2 / \pi^2 \mu$.

One might then expect that the ν_2^{osc} oscillations would dominate, and in an ideal wire, that might well be true; however, BG note that in a physical wire, even the presence of weak diffusive scattering from the surface of a wire is sufficient to smooth out the spectrum of most of the electrons in the system to the extent that they have no oscillatory contribution. The exceptions to this are electrons whose semiclassical orbits "glance" off the surface, as their effective de Broglie wavelengths are large compared to the surface inhomogeneities. Their contribution is contained within ν_1^{osc}, which remains large within a physical wire, whereas ν_2^{osc} may be neglected.

From ν_1^{osc}, they derive the following expression for the oscillatory portion of the thermodynamic potential:

$$\tilde{\Omega} \approx \frac{2^{3/2} k_B T L (m^* \omega_s)^{1/2}}{\pi^{3/2} \hbar^{1/2}} \sum_{r=1}^{\infty} \frac{\cos(2\pi^2 r \mu / (\hbar \omega_s) - \pi/4)}{r^{3/2} \sinh(\pi^3 r k_B T / \hbar \omega_s)} \cos \left[2\pi r \frac{\Phi}{\Phi_0} \right],$$

$$(29.82)$$

where

Φ is the flux through the cylinder
$\Phi_0 = 2\pi \hbar / e$ is the flux quantum

These oscillations do not share the origin of the traditional dHvA effect. In this limit, since the magnetic field is weak, the cyclotron radius of the electrons is typically very large, much larger than the radius of the cylinder. Since that is the case, it follows that there is an upper limit on the permitted value of the magnetic quantum number for the electrons in the wire (at least in a semiclassical sense), since the upper and lower bounds of the semiclassical orbits associated with many of the possible values will lie outside of the wire. As the flux through the wire changes, the upper bounds of these orbits move in toward the boundary of the cylinder and pass through it, reconfiguring the states of the electrons in the metal as they do so, and thus causing the oscillations.

These oscillations have been observed in many experimental measurements—for example, Brandt et al.'s examination of the magnetoresistance of 150–1000 nm diameter bismuth wires (Brandt et al. 1987) was able to resolve both Dingle- and BG-type oscillations. It is interesting to note that in this case, both kinds of quantum size oscillation did not precisely match the predicted behavior extrapolated from the above theories; they persisted for fields beyond a hypothetical cutoff value determined by the radius of the wire, albeit sometimes with suppressed amplitudes. In the more recent experiment of Nikolaeva et al. (2008), it was found that the Aharonov–Bohm-type oscillations discussed by BG that dominate in single 55 nm Bi wires are unexpectedly suppressed in 30 nm arrays. Possible explanations for this include

---

* Here and wherever else has proven necessary, we have rewritten expressions so that they are given in the terms of our previously defined quantities.

the difficulty of measuring quantum interference effects of this sort in wire arrays, and the possible effects of the "surface" states required for the effect filling a very significant fraction of the wire at the field values of interest. In addition, matters such as spin–orbit coupling and factors unique to the structure of bismuth must be taken into account.

29.4.4 Beyond the Weak Field Limit

It is likely that at higher fields for narrow wires, we might expect additional effects of the interplay between size quantization and Landau quantization to become apparent.

There is a need for theoretical treatments of this region, and in what follows we discuss two recent approaches to the problem, which differ in how they handle the boundary conditions.

29.4.4.1 Confinement Model

The Fock–Darwin confinement model (Fock 1928, Darwin 1930) approximates the behavior of the physical confinement with that of a 2D parabolic confining potential; this is equivalent to imposing a "soft" form of boundary condition on the nanowire. It should approximate the true energy spectrum when $\mu/(\hbar\omega_s) \gg 1$, and has the following Hamiltonian:

$$\mathcal{H} = \frac{(\hat{p} - e\mathbf{A})^2}{2m^*} + \frac{m^*\omega_s^2}{2}(x^2 + y^2) \qquad (29.83)$$

and energy spectrum

$$E_\alpha = \frac{p_z^2}{2m^*} + 2\hbar\omega\left(n + \frac{|m| - m + 1}{2}\right) + m\hbar\omega^-, \qquad (29.84)$$

with $\omega^2 = \omega_s^2 + \omega_c^2/4$, $\omega^{\pm} = \omega \pm \omega_c/2$.

Alexandrov and Kabanov (AK; Alexandrov and Kabanov 2005a,b) have calculated the oscillatory portion of the thermodynamic potential of the confinement model in order to examine the behavior of the dHvA oscillations of a nanowire subject to these boundary conditions. We proceed as follows.

Replacing the negative values of m with $-m - 1$ and defining $\mu^+ = \mu - \hbar\omega$, $\mu^- = \mu - \hbar(\omega + \omega^+)$ and $\epsilon_{nm}^{\pm}(p_z) = 2\hbar n\omega + \hbar m\omega^{\pm} + p_z^2/(2m^*)$ gives us the following expression:

$$\Omega = k_B T L \sum_{\pm} \int \frac{dp_z}{2\pi\hbar} \sum_{n,m\geq 0} \ln\left[1 + \exp\left(\frac{\mu^{\pm} - \epsilon_{nm}^{\pm}(p_z)}{k_B T}\right)\right]. \qquad (29.85)$$

Using Poisson's formula, we may replace the summations over n and m with integrations over $x = 2\omega n + \omega^{\pm}m$ and $y = \omega n - \omega^{\pm}m/2$, as follows (this summary of the derivation of this result is due to Zerovnik (2006)):

$$\sum_{n,m=0}^{\infty} f(2\omega n + \omega^{\pm}m)$$

$$= \sum_{r,r'=-\infty}^{\infty} \iint_0^{\infty} dn\, dm\, f(2\omega n + \omega^{\pm}m) \exp(2\pi rn + 2\pi r'm)$$

$$= \sum_{r,r'=-\infty}^{\infty} \frac{1}{2\omega\omega^{\pm}} \int_0^{\infty} f(x)dx \int_{-x/2}^{x/2} e^{2\pi ix\left[(r/4\omega) + (r'/2\omega^{\pm})\right] + 2\pi iy\left[(r/2\omega) - (r'/\omega^{\pm})\right]} dy$$

$$= \sum_{r,r'=-\infty}^{\infty} \frac{1}{2\pi i(r\omega^{\pm} - 2\omega r')} \int_0^{\infty} \left(\exp\left[\frac{2\pi irx}{2\omega}\right] - \exp\left[\frac{2\pi ir'x}{\omega^{\pm}}\right]\right) f(x)dx. \qquad (29.86)$$

Oscillatory corrections have nonzero r or r'. Introducing a new variable $\xi = \hbar x + p_z^2/(2m^*) - \mu^{\pm}$, integrating by parts and extending the lower limit of ξ down to $-\infty$, and making use of the usual standard integrals and the summation formula $\sum_r (z - r)^{-1} = \pi\cot(\pi z)$ we finally acquire

$$\tilde{\Omega} = \sum_{r=1}^{\infty} \sum_{\pm} A_r(\omega, \omega^{\pm}) \sin\left(\frac{\pi r\mu}{\hbar\omega} - \frac{\pi r(\omega^+ \mp \omega^-)}{2\omega} - \frac{\pi}{4}\right)$$
$$+ A_r(\omega^{\pm}/2, 2\omega)\sin\left(\frac{2\pi r\mu}{\hbar\omega^{\pm}} \pm \frac{\pi r(\omega^+ \mp \omega^-)}{\omega^{\pm}} - \frac{\pi}{4}\right), \qquad (29.87)$$

with the amplitudes A_r given by

$$A_r(x, y) = \frac{Lk_B T(2xm^*/\hbar)^{1/2}}{2\pi r^{3/2}\sinh\left[\pi^2 k_B Tr/(\hbar x)\right]}\cot\left(\frac{\pi ry}{2x}\right). \qquad (29.88)$$

Similarly, assuming a field-independent relaxation time τ (this independence being due to the presence of a large reservoir of states, for example), AK found the following expression for the SdH oscillations of the conductance:

$$\tilde{\sigma} = \sum_{r=1}^{\infty} \sum_{\pm} B_r(\omega, \omega^{\pm}) \cos\left(\frac{\pi r\mu}{\hbar\omega} - \frac{\pi r(\omega^+ \mp \omega^-)}{2\omega} - \frac{\pi}{4}\right)$$
$$+ B_r(\omega^{\pm}/2, 2\omega)\sin\left(\frac{2\pi r\mu}{\omega^{\pm}} \pm \frac{\pi r(\omega^+ \mp \omega^-)}{\omega^{\pm}} - \frac{\pi}{4}\right), \qquad (29.89)$$

with the values of B_r being given by

$$B_r(x, y) \frac{e^2 \tau k_B T \exp\left[-\pi r/(2x\tau)\right]}{\hbar(2m^*\hbar x)^{1/2} r^{1/2}\sinh\left[\pi^2 k_B Tr/(\hbar x)\right]}\cot\left(\frac{\pi ry}{2x}\right). \qquad (29.90)$$

The effects of confinement are threefold. Firstly, we observe that the temperature and scattering dependences in (29.88) and

(29.90) indicate that as soon as ω_s is large enough ($\gg K_B T/\hbar$ or $1/\tau$), the damping of the amplitudes is more or less unaffected by the value of the field, and is thus limited by the width of the wire. This is consistent with the Dingle and BG results. Second, we observe the existence of three frequencies of oscillation, as opposed to the one expected in the bulk: 2ω, ω^+, and ω^-, where $\omega^2 = \omega_0^2 + \omega_c^2/4$ and $\omega^\pm = \omega \pm \omega_c/2$. If the conventional dHvA frequency F_0 is large ($\gg B$), we may write these novel frequencies as

$$F = \frac{F_0}{(1+\gamma)^{3/2}} \qquad (29.91)$$

and

$$F^\pm = \frac{2F_0}{\left|1+\gamma\pm(1+\gamma)^{1/2}\right|}, \qquad (29.92)$$

where we have $F_0 = S_F\hbar/(2\pi e)$, $\gamma = 4\omega_s^2/\omega_c^2 = 4\pi^2 S_F\hbar^2/(e^2 S B^2)$, $S = \pi R^2$ is the cross-sectional area of the wire. In low fields, where $\gamma \gg 1$, all the frequencies are much lower than F_0, with $F \approx F_0/\gamma^{3/2}$ and $F^\pm \approx 2F_0/\gamma$. In high fields, $\gamma \ll 1$ and we have $F \approx F^+ \approx F_0$ and $F^- \approx 4F_0/\gamma \gg F_0$.

Finally, we observe that the amplitudes of the oscillations contain cotangent functions, which become infinite for values of the field satisfying the condition $2\omega/(\omega^\pm) = (q+2)/r$, where q is an integer. A more careful analysis of the behavior of the amplitudes (Alexandrov et al. 2007) shows that this apparent infinity is analytically tractable, and in the region of the "magic" value in fact corresponds to an enhancement of the amplitude by a factor of the order of the Fermi energy $\mu/\hbar\omega_c \gg 1$. This is a significant and unexpected effect, arising due to the interaction of the boundary conditions with the Fock–Landau quantization (Alexandrov and Kabanov 2005a). In a bulk metal, Fock–Landau levels have a $SeB/(2\pi\hbar)$-fold degeneracy, but this degeneracy is lifted by the boundary conditions, reducing the density of states at each energy level and therefore resulting in smaller oscillatory amplitudes than those seen in the bulk. The magic resonance conditions partly restore the degeneracy of the spectrum. This can easily be seen if, following AK, we take the magic resonance at $\omega_c = \omega_s/\sqrt{2}$. Here, we have $2\omega = 3\omega_s/\sqrt{2}$ and $\omega^- = \omega_s/\sqrt{2}$ and the energy spectrum is given by

$$E_{nmk} = \frac{p_z^2}{2m^\star} + \frac{\hbar\omega_c}{2\sqrt{2}}(6n + 3|m| - m + 3). \qquad (29.93)$$

There exist combinations of different n and m which give the same values of $6n + 3|m| - m + 3$, thus indicating that the degeneracy has been partly restored, and that as a result the amplitude is greatly enhanced (Figure 29.7). The degree of the enhancement, however, would be further restricted in real nanowires due to the presence of anharmonic corrections that are neglected here, since the confinement model is a linear approximation to the true spectrum. A theoretical example of such a damping due to nonlinear corrections is presented in Section 29.4.4.2.

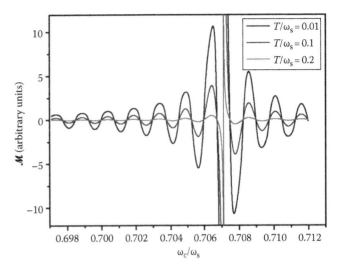

FIGURE 29.7 Oscillating part of the magnetization versus the magnetic field for relatively low fields and three temperatures. The resonance at $\omega_c = \omega_s/\sqrt{2}$ is due to a partial recovery of the energy-level degeneracy. $k_B = \hbar = 1$ here. (Taken from Alexandrov, A.S. and Kabanov, V.V., *Phys. Rev. Lett.*, 95, 076601, 2005a.)

29.4.4.2 Nonlinear Model

In order to further probe the effects of the boundary conditions on the system, Alexandrov, Kabanov, and Thomas (AKT) have been seeking a nonlinear approximation to the energy levels of the system when it has "hard" (infinite potential well) boundary conditions (Alexandrov et al. 2007). An approximation may be derived from the asymptotic description of the confluent hypergeometric function

$$M(a,b,\xi) = \Gamma(b)e^{\xi/2}\left(\frac{1}{2}b\xi - a\xi\right)^{(1-b)/2}$$
$$\times \pi^{-\frac{1}{2}}\cos\left(\sqrt{2b\xi - 4a\xi} - \frac{1}{2}b\pi + \frac{1}{4}\pi\right), \qquad (29.94)$$

where (as before) we take $a = (m - |m| - 1)/2 - (E - p_z^2/2m^\star)/\hbar\omega_c$ and $b = |m| + 1$. We impose our boundary conditions by requiring that $M(a,b,\xi)$ be equal to zero for a finite value of ξ, and after some algebra obtain

$$E_\alpha = \frac{\hbar\omega_s^2}{\mu}\left(n + \frac{|m|}{2} + \frac{3}{4}\right)^2 - \frac{\hbar\omega_c m}{2}. \qquad (29.95)$$

(A similar—up to a phase factor—result may be obtained through utilizing the Bohr–Sommerfeld quantization conditions.)

Let us set $\omega_c = 0$, and expand around the Fermi energy μ, defining $\epsilon = E - \mu$.

Our energy spectrum becomes

$$\epsilon \approx 2\hbar\omega_s\left(n + \frac{|m|}{2} + \frac{3}{4}\right) - 2\mu. \qquad (29.96)$$

When $\omega_c = 0$ and $E \approx \mu$, the spectrum of the parabolic confinement model is

$$\epsilon = E - \mu = 2\hbar\omega_s\left(n + \frac{|m|}{2} + \frac{1}{2}\right) - \mu. \qquad (29.97)$$

Therefore, up to an unimportant factor of 2 before the μ term of the nonlinear spectrum and the phase factors, the two are largely identical near the Fermi energy. This motivates a replacement of ω_s in (29.95) with ω, so as to better approximate the behavior of nonlinear spectra at large fields. That this replacement results in a fairly good approximation for small values of m is suggested by the numerical analysis of Alexandrov et al. (2007), though at large, positive values of m it begins to break down. This is because orbits with large, positive m correspond semiclassically to electron orbits that approach the boundaries of the wire; these correspond to a turning point due to the hard boundary conditions and since the model corresponds quite closely to a semiclassical approximation, we cannot expect it to hold accurately there. This is not as problematic as it might at first seem, since the hard boundary conditions that we have imposed are an unphysical idealization; the physical boundary conditions are likely to be intermediate between those of the confinement model and those of this nonlinear model.

The final, improved spectrum is given by

$$E = \frac{p_z^2}{2m^*} + \frac{\hbar\omega^2}{\mu}\left(n + \frac{|m|}{2} + \frac{3}{4}\right)^2 - \frac{\hbar\omega_c m}{2}. \qquad (29.98)$$

The calculation of Ω proceeds as usual; however, the details are somewhat nontrivial and we therefore only report the final result here. It should be noted that for some values of m, the energy becomes negative, but these contributions are an artifact of the approximation and can be safely neglected in the derivation. Eventually, we obtain

$$\tilde{\Omega} = \sum_{r=1}^{\infty}\sum_{\pm} C_r^{\pm} \sin\left[\frac{4\pi r\mu}{\hbar\tilde{\omega}^{\mp}} - \frac{\pi}{4}\right] + D_r^{\pm}\cos\left[\frac{4\pi r\mu}{\hbar\tilde{\omega}^{\mp}} - \frac{\pi}{4}\right], \qquad (29.99)$$

where

$$C_r^{\pm} = \frac{Lk_BT(m^*/\hbar)^{1/2}(\tilde{\omega}^+ + \tilde{\omega}^-)^{1/2}}{4\pi r^{3/2}\sinh\left[4\pi^2 k_B Tr/\hbar(\tilde{\omega}^+ + \tilde{\omega}^-)\right]}$$

$$\times \Re\cot\left[\frac{2\pi r\tilde{\omega}^{\pm}}{(\tilde{\omega}^+ + \tilde{\omega}^-)} \pm \frac{\beta(1-i)\omega^2(\hbar r/\mu)^{1/2}}{2(\tilde{\omega}^+ + \tilde{\omega}^-)^{3/2}}\right], \qquad (29.100)$$

and

$$D_r^{\pm} = \frac{Lk_BT(m^*/\hbar)^{1/2}(\tilde{\omega}^+ + \tilde{\omega}^-)^{1/2}}{4\pi r^{3/2}\sinh[4\pi^2 k_B Tr/\hbar(\tilde{\omega}^+ + \tilde{\omega}^-)]}$$

$$\times \Im\cot\left[\frac{2\pi r\tilde{\omega}^{\pm}}{(\tilde{\omega}^+ + \tilde{\omega}^-)} \pm \frac{\beta(1-i)\omega^2(\hbar r/\mu)^{1/2}}{2(\tilde{\omega}^+ + \tilde{\omega}^-)^{3/2}}\right], \qquad (29.101)$$

with $\omega^{\pm} = \omega \pm \omega_c/2$, $\tilde{\omega}^{\pm} = \omega\left(\sqrt{1 + \beta^2/4} \pm \beta/2\right)$ and $\beta = \omega_c/\omega$. Unlike the AK result, we have only *two* characteristic frequencies,

$$F^{\pm} = \frac{2F_0\left|\sqrt{2} \pm \sqrt{1 + \gamma/2}\right|}{\sqrt{1 + \gamma/2}\left[\sqrt{1 + \gamma/2} \pm (1/\sqrt{2})\right]^2} \qquad (29.102)$$

For low fields, $\gamma \gg 1$ and hence $F^{\pm} \approx 2^{5/2}F_0\left|\sqrt{2} \pm \sqrt{\gamma/2}\right|/\gamma^{3/2}$, so $F^+ > F^-$. For sufficiently high fields, $\gamma \ll 1$ and so $F^+ \approx 1.657F_0$ and $F^- \approx 9.657F_0$; in this case $F^- > F^+$. Evidence of this shift may be seen in Figures 29.8 and 29.9.

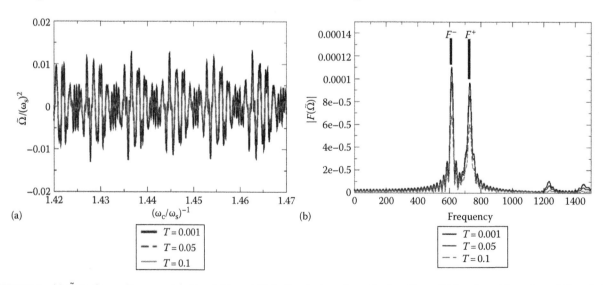

FIGURE 29.8 (a) $\tilde{\Omega}$ in the nonlinear model at low fields and (b) its Fourier transform. $\hbar = k_B = 1$ here. (Taken from Alexandrov, A.S. et al., *Phys. Rev. B*, 76, 155417, 2007.)

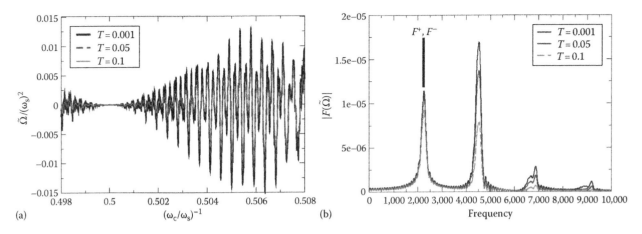

FIGURE 29.9 (a) $\tilde{\Omega}$ at high fields and (b) its Fourier transform. Note that the F^+ and F^- peaks in the lower Fourier transform are so close as to be indistinguishable; this may be why the second harmonic is larger than expected. $\hbar = k_B = 1$ here. (Taken from Alexandrov, A.S. et al., *Phys. Rev. B*, 76, 155417, 2007.)

In addition, while this result also displays magic resonances for particular ratios of ω_c/ω_s, their position has shifted and they are damped by the additional terms in the cotangent function, which confirms that nonlinear contributions to the energy spectrum suppress the magnitude of the enhancement to some degree (Figure 29.10).

It is apparent that there is a degree of model dependence in our results. Indeed, increasing the "hardness" of the boundary conditions can produce drastic changes in the nature of the expected oscillations. This could allow one to derive conclusions about the strength of the confinement within a nanowire, as well as the nature of its Fermi surface from experimental observations.

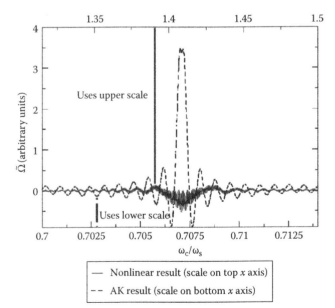

FIGURE 29.10 Graph comparing the first magic resonances of the parabolic confinement model with the nonlinear model. The temperature here is set at $k_B T = 0.025\hbar\omega_s$. (Taken from Alexandrov, A.S. et al., *Phys. Rev. B*, 76, 155417, 2007.)

However, observation of the novel oscillations is likely to be difficult, due to the small volume of the wires, though this could perhaps be ameliorated by studying arrays of wires. Examinations of the SdH (as in Huber et al. (2004) and Grozav and Condrea (2004)) and dHvA effects will require the gathering of data over a wider range of the magnetic field in order to allow accurate-enough Fourier analysis that the above theories can be tested. In addition, further theoretical work is needed in order to overcome the problems of the present accounts. This may well require the use of numerical techniques in order to escape the inaccuracies arising from the selection of a particular asymptotic approximation to the confluent hypergeometric function in order to derive the energy spectrum.

29.5 Conclusion

We have reviewed the theory of quantum magnetic oscillations in 3D bulk metals, and examined a number of ways in which this theory must be extended in order to account for effects that arise due to low dimensionality or due to confinement within narrow, nanoscale systems.

In multiband q2D systems, we have examined the phenomena that can give rise to additional combination frequencies in the SdH and dHvA effects, and have shown that when the effects of interband scattering are taken into account, one may see that such frequencies may be present in the SdH oscillations of open systems, but absent from their dHvA oscillations.

In nanowires, we have shown that the interaction of size quantization with Fock–Landau quantization results in drastic changes in the nature of the oscillations, with the precise effects depending on how the confinement is modeled. In general, the frequencies of the oscillations are renormalized by the interaction between the two, and the appearance of additional frequencies of oscillation can be observed. In addition, one also notes the appearance of "magic" resonances, which exist at values of the field where the degeneracy of the spectrum is partially restored. However, it appears that further work is required in order to overcome some of the problematic aspects of these approximate, analytic accounts, which will hopefully result in a complete description of quantum magnetic oscillations in cylindrical confined systems.

Acknowledgment

This work was funded by EPSRC grant No. EP/D035589.

References

Abramowitz, M. and I. A. Stegun. 1965. *Handbook of Mathematical Functions*. Dover, New York.

Abrikosov, A. A. 1988. *Fundamentals of the Theory of Metals*. North-Holland, Amsterdam, the Netherlands.

Alexandrov, A. S. and A. M. Bratkovsky. 1996. De Haas-van Alphen effect in canonical and grand canonical multiband fermi liquid. *Physical Review Letters* **76**: 1308.

Alexandrov, A. S. and A. M. Bratkovsky. 1997. New fundamental dHvA frequency in canonical low-dimensional Fermi liquids. *Physics Letters A* **234**: 53.

Alexandrov, A. S. and A. M. Bratkovsky. 2001. Semiclassical theory of magnetic quantum oscillations in a two-dimensional multiband canonical Fermi liquid. *Physical Review B* **63**: 033105.

Alexandrov, A. S. and A. M. Bratkovsky. 2002. Angular dependence of nonclassical magnetic quantum oscillations in a quasi-two dimensional fermi liquid with impurities. *Physical Review B* **65**: 035418.

Alexandrov, A. S. and A. M. Bratkovsky. 2004. Comment on "Origin of combination frequencies in quantum magnetic oscillations of two-dimensional multiband metals." *Physical Review B* **69**: 167401.

Alexandrov, A. S. and V. V. Kabanov. 2005a. Magnetic quantum oscillations in nanowires. *Physical Review Letters* **95**: 076601.

Alexandrov, A. S. and V. V. Kabanov. 2005b. Erratum: Magnetic quantum oscillations in nanowires. *Physical Review Letters* **95**: 169902(E).

Alexandrov, A. S. and V. V. Kabanov. 2007. Combination quantum oscillations in canonical single-band fermi liquids. *Physical Review B* **76**: 233101.

Alexandrov, A. S., V. V. Kabanov, and I. O. Thomas. 2007. Interplay of size and Landau quantizations in the de Haas-van Alphen oscillations of metallic nanowires. *Physical Review B* **76**: 155417.

Altland, A. and B. Simons. 2006. *Condensed Matter Field Theory*. Cambridge University Press, Cambridge, U.K.

Aronov, A. G. and Y. V. Sharvin. 1987. Magnetic flux effects in disordered conductors. *Reviews of Modern Physics* **59**: 755.

Bogacheck, E. N. and G. A. Gogadze. 1973. Oscillation effects of the "flux quantisation" type in normal metals. *Soviet Physics JETP* **36**: 973.

Brandt, N. B., D. B. Gitsu, V. A. Dolma, and Y. G. Ponomarev. 1987, Quantum size oscillations of the magnetoresistance of thin-filament bismuth single crystals. *Soviet Physics JETP* **65**: 515.

Bruus, H. and K. Flensberg. 2004. *Many-Body Theory in Condensed Matter Physics: An Introduction*. Oxford University Press, Oxford, NY.

Champel, T. 2002. Origin of combination frequencies in quantum magnetization oscillations of two-dimensional multiband metal. *Physical Review B* **65**: 153403.

Champel, T. 2004. Reply to 'Comment on "Origin of combination frequencies in quantum magnetic oscillations of two-dimensional multiband metals"'. *Physical Review B* **69**: 167402.

Champel, T. and V. P. Mineev. 2001a. Chemical potential oscillations and the de Haas-van Alphen effect. *Physical Review B* **64**: 054407.

Champel, T. and V. P. Mineev. 2001b. De Haas-van Alphen effect in two- and quasi-two-dimensional metals and superconductors. *Philosophical Magazine B* **81**: 55.

Champel, T. and V. P. Mineev. 2002. Magnetic quantum oscillations of the longitudinal conductivity σ_{zz} in quasi-two-dimensional metals. *Physical Review B* **66**: 195111.

Champel, T. and V. P. Mineev. 2006. Comment on "Magnetic quantum oscillations of the conductivity in layered conductors." *Physical Review B* **74**: 247101.

Cohen, M. H. and L. M. Falicov. 1961. Magnetic breakdown in crystals. *Physical Review Letters* **7**: 231.

Cornelius, T. W., M. E. Tomil-Molares, S. Karim, and R. Neumann. 2008. Oscillations of electrical conductivity in single bismuth nanowires. *Physical Review B* **77**: 125425.

Darwin, C. G. 1930. The diamagnetism of the free electron. *Proceedings of the Cambridge Philosophical Society* **27**: 86.

Dingle, R. B. 1952a. Some magnetic properties of metals. I. General introduction, and properties of large systems of electrons. *Proceedings of the Royal Society A (London)* **211**: 500.

Dingle, R. B. 1952b. Some magnetic properties of metals. II. The influence of collisions on the magnetic behaviour of large systems. *Proceedings of the Royal Society A (London)* **211**: 517.

Dingle, R. B. 1952c. Some magnetic properties of metals. IV. Properties of small systems of electrons. *Proceedings of the Royal Society A (London)* **212**: 47.

Falicov, L. M. and H. Stachiowiak. 1966. Theory of the dHvA effect in a system of coupled orbits. *Physical Review* **147**: 505.

Fock, V. 1928. Bermerkung zur quantelung des harmonischen oszillators im magnet-feld. *Zeitschrift für Physik* **47**: 446.

Fortin, J.-Y., E. Perez, and A. Audouard. 2005. Analytical treatment of the de Haas-van Alphen frequency combination due to chemical potential oscillations in an idealized two-band Fermi liquid. *Physical Review B* **71**: 155101.

Gerhardts, R. and J. Hadju. 1971. High field magnetoresistance at low temperatures. *Zeitschrift für Physik* **245**: 126.

Gogadze, G. A. 1984. Tunneling from normal cylyndrical conductor in weak field. *Solid State Communications* **49**: 277.

Gold, A. V. 1968. The de Haas-van Alphen Effect. In *Solid State Physics Volume I: Electrons in Metals*, eds. J. F. Cochran and R. R. Haering, pp. 39–127. Gordon and Breach, New York.

Grigoriev, P. 2001. The influence of the chemical potential oscillations on the dHvA effect in q2d compounds. *Soviet Physics JETP* **92**: 1090.

Grigoriev, P. 2002. *Magnetic quantum oscillations in quasi-two-dimensional metals*. PhD thesis. University of Konstanz, Konstanz, Germany.

Grigoriev, P. D. 2003. Theory of the Shubnikov-de Haas effect in quasi-two-dimensional metals. *Physical Review B* **67**: 144401.

Grigoriev, P. D., M. V. Kartsovnik, W. Biberacher, N. D. Kushch, and P. Wyder. 2002. Anomalous beating phase of the oscillating interlayer magnetoresistance in layered metals. *Physical Review B* **65**: 060403.

Grozav, A. D. and E. Condrea. 2004. Positive thermopower of single bismuth nanowires. *Journal of Physics: Condensed Matter* **16**: 6507.

Hamguchi, C. 2001. *Basic Semiconductor Physics*. Springer-Verlag, New York.

Harrison, N., R. Bogaerts, P. H. P. Reinders et al. 1996. Numerical model of quantum oscillations in quasi-two-dimensional organic metals in high magnetic fields. *Physical Review B* **54**: 9977.

Huber, T. E., A. Nikolaeva, D. Gitsu et al. 2004. Confinement effects and surface-induced charge carriers in Bi quantum wires. *Applied Physics Letters* **84**: 1326.

Kartsovnik, M. V. 2004. High magnetic fields: A tool for studying electronic properties of layered organic metals. *Chemical Reviews* **104**: 5737.

Kartsovnik, M. V., P. D. Grigoriev, W. Biberacher, N. D. Kushch, and P. Wyder. 2002. Slow oscillations of magnetoresistance in quasi-two-dimensional metals. *Physical Review Letters* **89**: 126802.

Kartsovnik, M., P. Grigoriev, A. G. Biberachera et al. 2003. Effects of low dimensionality on the classical and quantum parts of the magnetoresistance of layered metals with a coherent interlayer transport. *Synthetic Metals* **133–134**: 111.

Kishigi, K. and Y. Hasegawa. 2002. de Haas-van Alphen effect in two-dimensional and quasi-two-dimensional systems. *Physical Review B* **65**: 205405.

Kishigi, K. and Y. Hasegawa. 2005. Estimation of the cyclotron mass from the de Haas-van Alphen oscillation in the two-dimensional and quasi-two-dimensional multiband systems, *Physical Review B* **72**: 045410.

Kubo, R., S. J. Miyke, and N. Hashitsume. 1965. Quantum theory of galvanomagnetic effect at extremely strong magnetic fields. *Solid State Physics: Advances in Research and Applications* **17**: 269.

Landau, L. 1930. Diamagnetismus der metalle. *Zeitschrift für Physik* **64**: 629.

Landau, L. V. and E. M. Lifschitz. 1977. *Quantum Mechanics (Non-Relativistic Theory)*, 3rd edn. Pergamon Press, Oxford, U.K.

Lifshitz, I. M. and L. M. Kosevich. 1968. On the theory of the Shubnikov-de Haas effect, *Soviet Physics JETP* **6**: 67.

Morrison, D. and R. W. Stark. 1981. 2-Lifetime model-calculations of the quantum interference dominated transverse magnetoresistance of magnesium. *Journal of Low Temperature Physics* **45**: 531.

Nakano, M. 1999. Angle-dependent quantum-oscillations due to chemical potential oscillation in quasi-two-dimensional multiband Fermi liquids. *Journal of the Physical Society of Japan* **68**: 1801.

Nakano, M. 2000. Spin factor of de Haas-van Alphen oscillations in two-dimensional systems: Effect of background density of states and chemical-potential oscillations. *Physical Review B* **62**: 45.

Nikolaeva, A., D. Gitsu, L. Konopko, M. J. Graf, and T. E. Huber. 2008. Quantum interference of surface states in bismuth nanowires probed by the Aharonov-Bohm oscillatory behaviour of the magnetoresistance. *Physical Review B* **77**: 075332.

Ohmichi, E., Y. Maeno, and T. Ishiguro. 1999. Enhancement of the sum and difference frequencies in Shubnikov-de Haas oscillation at Yamaji angle in the layered perovskite sr_2ruo_4. *Journal of the Physical Society of Japan* **68**: 24.

Onsager, L. 1952. Interpretation of the de Haas-van Alphen effect. *Philosophical Magazine* **43**: 1006

Pippard, A. B. 1962. Quantization of coupled orbits in metals. *Proceedings of the Royal Society A (London)* **270**: 1.

Raikh, M. E. and T. V. Shahbazyan. 1993. High Landau levels in a smooth random potential for two-dimensional electrons. *Physical Review B* **47**: 1522.

Sasaki, T., H. Sato, and N. Toyota. 1990. Magnetic breakdown effect in organic superconductor kappa-(bedt-ttf)2cu(ncs)2. *Solid State Communications* **76**: 507.

Schiller, M., W. Schmidt, E. Balthes et al. 2000. Investigations of the Fermi surface of a new organic metal. *Europhysics Letters* **51**: 82.

Shepherd, R. A., M. Elliott, W. G. Herrenden-Harker et al. 1999. Experimental observation of the de Haas-van Alphen effect in a multiband quantum-well sample, *Physical Review B* **60**: R11277.

Shiba, H. and H. Fukuyama. 1969. A quantum theory of galvomagnetic effect in metals with magnetic breakdown I. *Journal of the Physical Society of Japan* **26**: 910.

Shoenberg, D. 1984a. *Magnetic Oscillations in Metals*. Cambridge University Press, Cambridge, U.K.

Shoenberg, D. 1984b. Magnetisation of a two-dimensional electron gas. *Journal of Low Temperature Physics* **56**: 417.

Singleton, J. 2000. Studies of quasi-two-dimensional organic conductors based on BEDT-TTF using high magnetic fields. *Reports on Progress in Physics* **63**: 1111.

Stark, R. W. and C. B. Friedberg. 1974. Interfering electron quantum states in ultra-pure magnesium. *Journal of Low Temperature Physics* **14**: 111.

Stark, R. W. and R. Reifenberger. 1977. Quantitative theory for the quantum interference effect in the transverse magnetisation of pure magnesium. *Journal of Low Temperature Physics* **26**: 763.

Thomas, I. O., V. V. Kabanov, and A. S. Alexandrov. 2008. Shubnikov-de Haas effect in multiband quasi-two-dimensional metals, *Physical Review B* **77**: 075434.

Wasserman, A. and M. Springford. 1996. The influence of many-body interactions on the de Haas-van Alphen effect. *Advances in Physics* **45**: 471.

Weiss, H., M. V. Kartsovnik, W. Biberacher et al. 1999. Angle-dependent magnetoquantum oscillations in M κ-(bedt-ttf)2cu[n(cn)2]br. *Physical Review B* **60**: R16259.

Yamaji, K. 1989. On the angle dependence of the magnetoresistance in quasi-2-dimensional organic superconductors. *Journal of the Physical Society of Japan* **58**: 1520.

Žerovnik, G. 2006. *Magnetne kvantne oscikcije v nanožicah.* Student essay. Josef Stefan Institute, Ljubljana, Slovenia.

Spin-Density Wave in a Quantum Wire

Oleg A. Starykh
University of Utah

30.1 Introduction

This chapter discusses possibility of realizing spin-density wave (SDW) in a single-channel one-dimensional quantum wire. It should be noted from the outset that purely one-dimensional system of electrons with spin-rotationally invariant interaction between particles (such as the usual Coulomb interaction) cannot have magnetically ordered ground state. At finite temperature, any such order is destroyed by thermal fluctuations, and the statement is known as the Mermin–Wagner theorem. At zero temperature, Goldstone modes associated with spontaneously broken *continuous* spin symmetry destroy the putative order parameter, making the original assumption of the broken symmetry incorrect. Thus, SDW states are not possible in one-dimensional wire. (Similarly, even one-dimensional superconductivity is not possible.)

This argument implies that small perturbations that reduce or break the rotational symmetry of the problem acquire a very important role. By reducing the symmetry from full spin-rotational invariance ($SU(2)$) to, for example, that of rotations about some particular axis ($U(1)$) or even to a smaller discrete symmetry (such as Z_2, for example), these symmetry-breaking perturbations open up ways to avoid restrictions of the Mermin–Wagner theorem. The crucial role of symmetry in low-dimensional systems can be readily illustrated by differences between the ground states of the $SU(2)$-invariant Heisenberg spin chain and the Z_2 invariant Ising one. The former has critical ground state and gapless excitations above it, while the latter demonstrates true zero-temperature long-range order with gapful excitations.

In the context of quantum wires and nanostructures, in general, such a symmetry-lowering perturbation is provided by the ever-present spin–orbital (SO) coupling.

Spin–orbital interaction originates from relativistic correction to electron's motion. Textbook examples of this consists in considering electrons moving with velocity $\vec{v}$ in an external electric field $\vec{E}_{s-o}$. In electron's reference frame that field produces magnetic field $\vec{B}_{s-o} = \vec{E}_{s-o} \times \vec{v}/c^2$. Then, spin–orbital coupling emerges as an interaction of electron's spin $\vec{S}$ with that magnetic field:

$$H_{s-o} = \frac{g\mu_B}{mc^2}\vec{S}\cdot\vec{E}_{s-o}\times\vec{p} \qquad (30.1)$$

where $\vec{p} = m\vec{v}$ is the momentum. In atoms, $\vec{E}_{s-o} \propto Ze/r^2$ is the electric field of the nuclei, while in fabricated nanostructures, $\vec{E}_{s-o} = -\vec{\nabla}V_{conf}(r)$ is associated with the structural asymmetry of the confinement potential. SO coupling due to structure inversion asymmetry (SIA) is known as the Rashba coupling (Bychkov and Rashba 1984). For the two-dimensional electron gas, where $\vec{E}_{s-o} \propto \hat{z}\partial_z V_{conf}(r)$ is related to the normal to the plane of motion gradient of the potential, it reads

$$H_R = \frac{\alpha_R}{\hbar}(\sigma_x p_y - \sigma_y p_x) \qquad (30.2)$$

The Rashba constant α_R is a phenomenological parameter that describes the magnitude of $\vec{E}_{s-o}$. The magnitude of asymmetry and hence the Rashba coupling strength can be controlled by the external gate voltage. In addition to the noted asymmetry of confining potentials (which include quantum-well potential that confines electrons to a two-dimensional layer as well as transverse [in-plane] potential that forms the one-dimensional channel (Moroz and Barnes 1999, 2000), spin–orbit interaction is inherent to semiconductors of either zinc-blende or wurtzite lattice structures lacking bulk inversion symmetry (Dresselhaus 1955).

Another very interesting system that motivates our investigation is provided by one-dimensional electron surface states on vicinal surface of gold (Mugarza et al. 2002) as well as by electron states of self-assembled gold chains on stepped Si(111) surface of silicon (Crain and Himpsel 2006). In both the systems, one-dimensional ballistic channels appear due to atomic reconstruction of surface layer of atoms (see also Mugarza and Ortega 2003, Ortega et al. 2005). The resultant surface electronic states lie within the bandgap of bulk states, and thus, to high accuracy, are decoupled from electrons in the bulk. Spin–orbit interaction is unexpectedly strong in these systems, with the spin–orbit energy splitting of the order of 100 meV. In fact, spin–split subbands of Rashba type have been observed in angular resolved photoemission spectroscopy (ARPES) in both two-dimensional (LaShell et al. 1996) and one-dimensional settings (Mugarza et al. 2002, Crain and Himpsel 2006). The very fact that the two (horizontally) spin–split parabolas are observed in ARPES speaks of the high quality and periodicity of the obtained surface channels. As we show below, the most interesting situation involves electrons subjected to both spin–orbital and magnetic fields. Although it is perhaps impractical to think of ARPES measurements in the presence of magnetic field, it is quite possible to imagine experiments on *magnetic* metal surfaces (Krupin et al. 2005, Dedkov et al. 2008).

With the current trends in modern state-of-the-art technology, the problem of interacting electrons subject to spin–orbit coupling in one-dimensional conducting channel (be it a quantum wire or a stepped surface channel) is quite a meaningful one. Recalling that the ordered state in one dimension is only possible when the spin rotational symmetry is sufficiently reduced, it makes sense to also include an external magnetic field, which couples to electron's spin via the standard Zeeman term (Sun et al. 2007, Gangadharaiah et al. 2008a). We shall see later that combined effect of non-commuting spin–orbital and magnetic fields will open up a way to selectively probe left- and right-moving excitations of the quantum wire.

We begin, in Section 30.2, with a "warm-up exercise" of a toy problem of two single-electron quantum dots coupled by strong Coulomb repulsion between electrons and subject to the spin–orbit interaction of Rashba type. The outcome of the simple analysis (Gangadharaiah et al. 2008b), described below, is easy to formulate: even in the absence of exchange interaction between the electrons in different dots, there is a (anisotropic) spin–orbit-mediated coupling between their spins. This coupling comes about from the simple fact that Coulomb interaction correlates *orbital* motion of the electrons. Simply put, the electrons move so as to maximize the distance between them. Since each of the electrons, in turn, has its spin correlated with its orbital motion by the Rashba term, it is perhaps not surprising that the spins of the two electrons end up being correlated as well. What is somewhat

surprising is the extreme similarity of this effect with the well-known van der Waals (vdW) interaction between neutral atoms.

Our main topic—spin-density wave in a quantum wire—is described in Section 30.3. Necessary technical details are outlined in the Appendix 30.A. The main findings of this chapter are summarized in Section 30.4.

30.2 Spin–Orbit-Mediated Interaction between Spins of Localized Electrons

Studies of exchange interaction between localized electrons constitutes one of the oldest topics in quantum mechanics. Strong current interest in the possibility to control and manipulate spin states of quantum dots has placed this topic at the center of spintronics and quantum computation research. As is known from the papers of Dzyaloshinskii (1958) and Moriya (1960a,b), in the presence of the spin–orbital interaction (SOI), the exchange is anisotropic in spin space.

Being a manifestation of quantum tunneling, the exchange is exponentially sensitive to the distance between electrons (Bernu et al. 2001). Here we show that when subjected to the spin-orbit interaction, as appropriate for the structure-asymmetric heterostructures and surfaces (Bychkov and Rashba 1984), interacting electrons acquire a novel *non-exchange* coupling between the spins. The mechanism of this coupling is very similar to that of the well-known vdW interaction between neutral atoms. This anisotropic interaction is of the *ferromagnetic Ising* type.

30.2.1 Calculation of the vdW Coupling

To illuminate the origin of the vdW coupling, we consider the toy problem of two single-electron quantum dots described by the double well potential (Burkard et al. 1999, Calderon et al. 2006), see Figure 30.1,

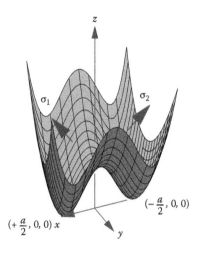

FIGURE 30.1 Two-dot potential (30.3). Blue (dark grey) arrows indicate electron's spins.

$$\tilde{V}(x_j, y_j) = \frac{m\omega_x^2}{2a^2}\left(x_j^2 - \frac{a^2}{4}\right)^2 + \frac{m\omega_y^2}{2}y_j^2 \tag{30.3}$$

where $\omega_{x/y}$ are confinement frequencies along x/y directions. The electrons, indexed by $j = 1, 2$, are subject to SOI of the Rashba type (Bychkov and Rashba 1984) with coupling α_R

$$H_{SO} = \sum_{j=1,2} \alpha_R \vec{p}_j \times \vec{\sigma}_j \cdot \hat{z} \tag{30.4}$$

where

$\vec{\sigma}$ are the Pauli matrices
$\hat{z}$ is normal to the plane of motion.

Finally, electrons experience mutual Coulomb repulsion so that the total Hamiltonian reads

$$H = \sum_{j=1,2}\left[\frac{\vec{p}_j^2}{2m} + \tilde{V}(x_j, y_j)\right] + \frac{e^2}{|\vec{r}_1 - \vec{r}_2|} + H_{SO} \tag{30.5}$$

At large separation between the two dots, the exchange is exponentially suppressed and the electrons can be treated as distinguishable particles. One then expects that Coulomb-induced correlations in the orbital motion of the electrons in two dots translate, via the spin–orbit interaction, into correlation between their spins. Consider the distance between the dots, a, much greater than the typical spread of the electron wave functions, $1/\sqrt{m\omega_x}$. In this limit, the electrons are centered about different wells, and the potential can be approximated as

$$V(\vec{r}_1, \vec{r}_2) \approx \frac{1}{2}m\omega_x^2((x_1 - a/2)^2 + (x_2 + a/2)^2) + \frac{1}{2}m\omega_y^2(y_1^2 + y_2^2) \tag{30.6}$$

At this stage, it is crucial to perform a unitary transformation (Shahbazyan and Raikh 1994, Aleiner and Fal'ko 2001), which removes the linear spin–orbit term from (30.5)

$$U = \exp[im\alpha_R\hat{z}\cdot(\vec{r}_1 \times \vec{\sigma}_1 + \vec{r}_2 \times \vec{\sigma}_2)] \tag{30.7}$$

Owing to the noncommutativity of Pauli spin matrices, SOI cannot be eliminated completely, resulting in higher order in the Rashba coupling α_R contributions, as given by $\tilde{H} = UHU^\dagger$ below

$$\tilde{H}_{SO} = \sum_{j=1,2}\left[-m\alpha_R^2\tilde{L}_j^z\tilde{\sigma}_j^z + \frac{4}{3}m^2\alpha_R^3(y_j\tilde{\sigma}_j^y + x_j\tilde{\sigma}_j^x)\tilde{L}_j^z\right.$$
$$\left.+\frac{2}{3}im^2\alpha_R^3(y_j\tilde{\sigma}_j^x - x_j\tilde{\sigma}_j^y)\right] + O(\alpha_R^4) \tag{30.8}$$

Here $\tilde{L}_j^z$ is the angular momentum of the jth electron, $\tilde{L}^z = x\tilde{p}_y - y\tilde{p}_x$, and *tilde* denotes unitarily rotated operators. The calculation is easiest when the confining energy is much greater than both the Coulomb energy e^2/a and the spin–orbit energy scale $\sqrt{m\omega}\alpha_R$. In terms of the new (primed) coordinates $\vec{r}_1' = \vec{r}_1 - \vec{a}/2$

and $\vec{r}_2' = \vec{r}_2 + \vec{a}/2$ centered about $(a/2, 0)$ and $(-a/2, 0)$, respectively, the interaction potential $e^2/|\vec{r}_1' - \vec{r}_2' + \vec{a}|$ is expanded in powers of $1/a$ keeping terms up to second order in the dimensionless relative distance $(\vec{r}_1' - \vec{r}_2')/a$. The linear term, $e^2(x_1' - x_2')/a^2$, slightly renormalizes the equilibrium distance between the electrons and can be dropped from further considerations. In terms of symmetric (S) and antisymmetric (A) coordinates,

$$x_{S/A} = \frac{x_1' \pm x_2'}{\sqrt{2}}; \quad y_{S/A} = \frac{y_1' \pm y_2'}{\sqrt{2}} \tag{30.9}$$

the quadratic term $e^2(2(x_1' - x_2')^2 - (y_1' - y_2')^2)/2a^3$ renormalizes the antisymmetric frequency $\omega_{Ax}^2 \to \omega_x^2 + 4e^2/(ma^3)$ and $\omega_{Ay}^2 = \omega_y^2 - 2e^2/(ma^3)$ while leaving the symmetric ones, $\omega_{Sx}^2 = \omega_x^2$ and $\omega_{Sy}^2 = \omega_y^2$, unmodified. Quite similar to the textbook calculation of the vdW force (Griffiths 2005), the resulting Hamiltonian $\tilde{H} = \tilde{H}_S + \tilde{H}_A + \tilde{H}_{SO}$ becomes that of harmonic oscillators

$$\tilde{H}_{S/A} = \frac{\vec{p}_{S/A}^2}{2m} + \frac{m}{2}(\omega_{xS/A}^2 x_{S/A}^2 + \omega_{yS/A}^2 y_{S/A}^2) \tag{30.10}$$

perturbed by $\tilde{H}_{SO} = \tilde{H}_{SO}^{(2)} + \delta\tilde{H}_{SO}^{(2)} + O(\alpha_R^3)$, where

$$\tilde{H}_{SO}^{(2)} = -\frac{m\alpha_R^2}{2}[(x_S\tilde{p}_{yS} - y_S\tilde{p}_{xS}) + S\leftrightarrow A](\tilde{\sigma}_1^z + \tilde{\sigma}_2^z)$$
$$-\frac{m\alpha_R^2}{2}[(x_S\tilde{p}_{yA} - y_A\tilde{p}_{xS}) + S\leftrightarrow A](\tilde{\sigma}_1^z + \tilde{\sigma}_2^z) \tag{30.11}$$

$$\delta\tilde{H}_{SO}^{(2)} = -\frac{m\alpha_R^2 a}{2\sqrt{2}}\left[\tilde{p}_{yS}(\tilde{\sigma}_1^z - \tilde{\sigma}_2^z) + \tilde{p}_{yA}(\tilde{\sigma}_1^z + \tilde{\sigma}_2^z)\right] \tag{30.12}$$

It is evident from Equations 30.11 and 30.12 that the leading corrections to the ground state energy is obtained either by the excitation of a single y-oscillator (through Equation 30.12) or by the simultaneous excitation of oscillators in both the x and y directions (through Equation 30.11),

$$\Delta E = \sum_{i,j=S,A} \frac{\left|\langle 0|\delta\tilde{H}_{SO}^{(2)}|1y_i\rangle\right|^2}{\omega_{iy}} + \frac{\left|\langle 0|\tilde{H}_{SO}^{(2)}|1x_i1y_j\rangle\right|^2}{\omega_{ix} + \omega_{jy}}$$

It is easy to see that the spin-dependent contributions from $\delta\tilde{H}_{SO}^{(2)}$ cancel exactly, while those originating from $\delta\tilde{H}_{SO}^{(2)}$ do not, resulting in the novel spin interaction

$$H_{vdW} = \frac{1}{8}m^2\alpha_R^4\tilde{\sigma}_1^z\tilde{\sigma}_2^z(\phi(\omega_{Sy}, \omega_{Sx}) + \phi(\omega_{Ay}, \omega_{Ax})$$
$$-\phi(\omega_{Ay}, \omega_{Sx}) - \phi(\omega_{Sy}, \omega_{Ax})) \tag{30.13}$$

where the function ϕ is given by a simple expression

$$\phi(x, y) = \frac{(x - y)^2}{xy(x + y)} \tag{30.14}$$

In case of cylindrically symmetric dots, $\omega_x = \omega_y$,

$$H_{vdW} = -\frac{\alpha_R^4 e^4}{4a^6 \omega_x^5} \tilde{\sigma}_1^z \tilde{\sigma}_2^z \qquad (30.15)$$

The physics of this novel interaction is straightforward: it comes from the interaction-induced correlation of the orbital motion of the two particles, which, in turn, induces correlations between their spins via the spin–orbit coupling. The net Ising interaction would have been zero if not for the shift in frequency of the antisymmetric mode due to the Coulomb interaction. Note that the coupling strength exhibits the same power-law decay with distance as the standard vdW interaction (Griffiths 2005).

From (30.13), it follows that in the extreme anisotropic limit of $\omega_y \to \infty$, or equivalently, the one-dimensional limit, there is no coupling between spins. This result is understood by noting that one-dimensional version of SOI, given by $\alpha_R \sum_j \sigma_j^y p_j^x$, can be gauged away to all orders in α_R by a unitary transformation $U_{1D} = \exp[im\alpha_R(x_1\sigma_1^y + x_2\sigma_2^y)]$. Hence there is an absence of the spin–spin coupling in this limit. However, either by including magnetic field (Zeeman interaction, see below) in a direction different from σ^y, or by increasing the dimensionality of the dots by reducing the anisotropy of the confining potential, the spin–orbital Hamiltonian acquires additional non-commuting spin operators. The presence of the mutually non-commuting spin operators (for example, σ^x and σ^y in (30.4)) makes it impossible to gauge the SOI completely, opening the possibility of fluctuation-generated coupling between distant spins, as in Equation 30.15.

30.2.2 Effect of the Magnetic Field

For simplicity, we neglect orbital effects and concentrate on the Zeeman coupling, $H_Z = -\Delta_z \sum_j \sigma_j^z/2$, where $\Delta_z = g\mu_B$. Unitary transformation (30.7) changes it to $H_Z - \Delta_z m\alpha_R a(\sigma_1^x - \sigma_2^x)/2 + \delta\tilde{H}_Z$. Here

$$\delta\tilde{H}_Z = -\sum_{j=1,2} m\alpha_R\Delta_z(x_j'\tilde{\sigma}_j^x + y_j'\tilde{\sigma}_j^y) \qquad (30.16)$$

describes the coupling between the Zeeman and Rashba terms. In the basis (30.9), it reduces to

$$\delta\tilde{H}_Z = -m\Delta_z\alpha_R \frac{y_S(\sigma_1^y + \sigma_2^y) + x_S(\sigma_1^x + \sigma_2^x)}{\sqrt{2}}$$
$$- m\Delta_z\alpha_R \frac{y_A(\sigma_1^y - \sigma_2^y) + x_A(\sigma_1^x - \sigma_2^x)}{\sqrt{2}} \qquad (30.17)$$

For sufficiently strong magnetic field, $\Delta_z \gg \sqrt{m\omega}\alpha_R$, $\tilde{H}_{SO}$ can be neglected in comparison with $\delta\tilde{H}_Z$. By calculating second-order

correction to the ground state energy of the two dots, represented as before by $\tilde{H}_S + \tilde{H}_A$, and extracting the spin-dependent contribution, we obtain

$$\Delta E_Z = -\Delta_z^2\alpha_R^2 \frac{e^2}{a^3}\left(2\frac{\sigma_1^x\sigma_2^x}{\omega_x^4} - \frac{\sigma_1^y\sigma_2^y}{\omega_y^4}\right) \qquad (30.18)$$

In the extreme anisotropic limit $\omega_y \to \infty$, the dots become 1D, and we recover the result of (Flindt et al. 2006). For the isotropic limit $\omega_x = \omega_y$, the coupling of spins acquires a magnetic dipolar structure identical to that found in (Trif et al. 2007).

It is interesting to note that for the *in-plane* orientation of the magnetic field,

$$H_Z = \frac{\Delta_z}{2}(\sigma^x\cos\theta + \sigma^y\sin\theta) \qquad (30.19)$$

the unitary transformation leads to

$$U^+H_ZU = H_Z + \frac{m\alpha_R\Delta_z}{\sqrt{2}}\left((x_S + x_A)\cos\theta + (y_S + y_A)\sin\theta\right) \qquad (30.20)$$

By calculating again the second-order correction to the ground state energy of the two dots to extract the spin-dependent contribution, we obtain an angle-dependent effective Hamiltonian:

$$\Delta E_Z = \Delta_z^2\alpha_R^2 e^2 \frac{3\cos^2\theta - 1}{2a^3}\sigma_1^z\sigma_2^z \qquad (30.21)$$

By changing the direction of the applied field, one can change the strength and type of the coupling between spins.

30.3 Magnetized Quantum Wire with Spin–Orbit Interaction

30.3.1 Noninteracting Electrons

We next consider significantly more challenging many-body problem of interacting electrons in a quantum wire with Rashba and Zeeman terms. We think of quantum wire as obtained by gating of two-dimensional electron gas. This is modeled by the addition of the transverse confining potential $V(x) = m\omega^2x^2/2$ to the noninteracting Hamiltonian H_0 of the system. Then,

$$H_0 = \frac{\hbar^2(p_x^2 + p_y^2)}{2m} + V(x) - g\mu_B\frac{\vec{\sigma}}{2}\cdot\vec{B} + H_R \qquad (30.22)$$

where g is the effective Bohr magneton, B is the magnetic field, σ_μ ($\mu = x, y, z$) are the Pauli matrices, and the SO Rashba interaction H_R is given by (30.2). When the confining potential is strong enough so that the width of the wire $\sqrt{\hbar/(2m\omega)}$ is of the order of the electron Fermi-wavelength, only the first subband of quantized transverse motion is occupied. The momentum of the transverse standing wave is then replaced by its expectation value, $p_x \to \langle p_x \rangle = 0$. Corrections to this approximation from the omitted term $\alpha_R \sigma_y p_x$ in (30.2) produce small spin-dependent variations of the velocities of right- and left-moving particles (Moroz et al. 2000). Then, the Hamiltonian (30.22) acquires an one-dimensional form

$$H_0 \approx \frac{k^2}{2m} + \alpha_R \sigma_x k - \frac{g\mu_B}{2} \vec{\sigma} \cdot \vec{B} \qquad (30.23)$$

Here and below k is electron's momentum along the axis of the wire, which we will denote as x-axis in the following for notational convenience. We also set $\hbar = 1$. Corrections to the one-dimensional form (30.23) are controlled by the smallness of the ratios $\Delta_{s-o}/E_F \ll 1$ and $\Delta_z/E_F \ll 1$, both of which are assumed in this work. Here $\Delta_{s-o} = \alpha_R k_F$ is the SO splitting, $\Delta_z = g\mu_B B$ is the Zeeman splitting, and $E_F = k_F^2/2m$ is the Fermi-energy, which is set by the two-dimensional reservoirs to which the wire is adiabatically connected.

It is easy to see that in the *absence* of magnetic field, SOI in (30.23) can be easily gauged away via the *spin-dependent* shift of the momentum, $H_0(B = 0) \to (k + m\alpha_R \sigma_x)^2/2m$. This shift describes the two orthogonal parabolic subbands, which are eigenstates of the matrix σ_x with eigenvalues $\pm 1/2$, centered about momenta $\mp m\alpha_R/2$. So far, this case does not contain new physics. With Coulomb interaction between electrons included, electrons form Luttinger liquid with somewhat modified critical exponents (Moroz et al. 2000, Iucci 2003), in comparison with the standard case of no SOI. The reason is that $H_0(B = 0)$ retains the continuous $U(1)$ symmetry, that of rotations about σ_x axis, which is enough in one dimension for the gapless ground state (Luttinger liquid state) to exist (Giamarchi 2004).

Most interesting situation arises when both SOI and Zeeman terms are present simultaneously and do not commute with each other as happens when spin–orbital axis [σ_x in (30.23)] is different from the magnetic field direction. Quite generally, we choose magnetic field to be in the $x - z$ plane, and make an angle $\pi/2 - \beta$ with the spin–orbital x-axis. Then, $\vec{B} = B(\sin \beta, 0, \cos \beta)$. Clearly, for as long as $\beta < \pi/2$, the two perturbations, the SO Rashba term and the Zeeman term, do not commute with each other because of the basic property of Pauli matrices $[\sigma_x, \sigma_z] = -2i\sigma_y \neq 0$. This implies the loss of the continuous spin-rotational symmetry and, by our arguments, the possibility of an interesting long-range magnetic order.

The energy eigenvalues of the resulting 2×2 matrix, representing Hamiltonian (30.23), is found as (Levitov and Rashba 2003, Pereira and Miranda 2005)

$$\epsilon_{\pm}(k) = \frac{k^2}{2m} \pm \sqrt{(\alpha_R k)^2 + \left(\frac{\Delta_z}{2}\right)^2 - \Delta_z \sin \beta \, \alpha_R k} \qquad (30.24)$$

Note that the two subbands, corresponding to $\pm$ signs in Equation 30.24, are separated by the gap Δ_z at the point $k = 0$.

The corresponding eigenstates are easy to find as well. Their most important property consists in the natural fact that they

are given by the momentum dependent spinors. For example, for the important case of mutually orthogonal SO and magnetic fields ($\beta = 0$), they are given by $|\chi \pm (k)\rangle$. Here

$$|\chi + (k)\rangle \equiv \begin{bmatrix} \sin\dfrac{\gamma(k)}{2} \\ \cos\dfrac{\gamma(k)}{2} \end{bmatrix}, \quad |\chi - (k)\rangle \equiv \begin{bmatrix} \cos\dfrac{\gamma(k)}{2} \\ -\sin\dfrac{\gamma(k)}{2} \end{bmatrix} \qquad (30.25),$$

and we introduced rotation angle $\gamma(k)$

$$\gamma(k) = \arctan\frac{2\alpha_R k}{\Delta_z} \qquad (30.26)$$

Observe that spin directions in each subband vary with the momentum. This, of course, is the direct consequence of the "momentum-dependent magnetic field" produced by the SO interaction H_R.

The obtained ($\pm$) subbands are filled up to the Fermi energy $E_F = k_F^2/2m$, which is parameterized by the Fermi-momentum k_F defined in the absence of both SO and magnetic field terms in (30.23). The Fermi-momenta $k_{\pm}^{R/L}$ of the subbands are easily found from Equation 30.24. Note that for $\beta \neq 0$, the Fermi-points are not symmetric about the origin, i.e., $k_{\pm}^R \neq -k_{\pm}^L$. Of crucial importance for the following is the observation that single-particle states at the Fermi-points of the two subbands have finite overlap. That is, $\langle \chi + (k_+^{R/L}) | \chi - (k_-^{R/L}) \rangle \neq 0$ as long as $\Delta_{s-o} \neq 0$. (They are, of course, always orthogonal at the same momentum: $\langle \chi + (k) | \chi - (k) \rangle = 0$.)

This fact allows for a qualitative discussion of the *intersubband interaction* effects in $|\chi\pm\rangle$ basis. Namely, the finite overlap between single-particle states of ($\pm$) bands guaranties that Coulomb interaction will open up *pair-tunneling processes* when a pair of electrons from the opposite Fermi-points of the one subband (say, k_+^R and k_+^L) tunnels into a similar pair of states in the other subband (at k_-^R and k_-^L). Such a process clearly conserves energy (both initial and final states are at E_F). For the case of orthogonal SO and magnetic field axes ($\beta = 0$ above), such a two-particle process also conserves momentum. This is because for $\beta = 0$ (and only in this case), the initial and final momenta of the pair are equal and are simply zero ($k_-^R + k_-^L = k_+^R + k_+^L = 0$), see Figure 30.2. The only exception to this is provided by the limit $\Delta_{s-o} = 0$, when $\gamma = 0$, see (30.26). In this limit, spinors $|\chi\pm\rangle$ become standard momentum-independent eigenstates of the Zeeman term, $|\chi\pm\rangle \to |\downarrow\rangle, |\uparrow\rangle$. Such Zeeman-split subbands are clearly orthogonal, $\langle \downarrow | \uparrow \rangle = 0$, and do not allow for the pair with, say, $S^z = -1$, to hop into the up-spin subband, and vice versa. Such a tunneling requires "conversion" of $S^z = -1$ pair into that with $S^z = +1$ and is possible only when S^z is not a good quantum number; that is, it requires spin nonconservation. This is exactly what the SO interaction does—it breaks the $U(1)$ symmetry about the axis of the field. Thus, as long as $\gamma(k_F) \neq 0$, the inter-subband pair-tunneling is possible. In view of the similarity with superconducting pair scattering, the inter-subband pair-tunneling process is often called Cooper scattering (Starykh et al. 2000).

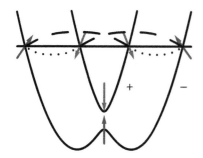

FIGURE 30.2 Occupied subbands $\epsilon_\pm$ of Equation 30.24 for $\beta = 0$. Arrows illustrate spin polarization in different subbands. Dashed (dotted) lines indicate exchange (direct) Cooper scattering processes. (Reprinted from Gangadharaiah, S. et al., *Phys. Rev. B*, 78, 054436, 2008a. With permission.)

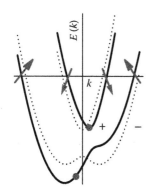

FIGURE 30.3 Occupied subbands $\epsilon_\pm$ for the case of nonorthogonal spin–orbital and magnetic field axes, $\beta \neq 0$. Arrows illustrate spin polarization, as in Figure 30.2. Filled dots indicate location of the center-of-mass for (+) (upper) and (–) (lower) subbands, which shift in opposite directions. Dashed lines show ($\pm$) subbands of Figure 30.2, corresponding to the orthogonal orientation, $\beta = 0$, for comparison. (Reprinted from Gangadharaiah, S. et al., *Phys. Rev. B*, 78, 054436, 2008a. With permission.)

Its existence represents qualitative difference between the wires with and without the spin–orbital terms. Quantitative consequences of the Cooper scattering are analyzed in details in the next section.

Observe that non-orthogonal orientation, $\beta > 0$, results in momentum mismatch between the pairs in the two subbands, see Figure 30.3. This mismatch suppresses inter-subband pair-tunneling, see (30.50) and the discussion after it.

30.3.2 Low-Energy Description

We now focus on the quantitative description of the problem. This is done by linearizing the *unperturbed* Hamiltonian (the one with $\alpha_R = 0$ and $B = 0$) in terms of low-energy fermions $R_s(L_s)$ that live near $+k_F(-k_F)$ Fermi-points. In terms of these, kinetic energy is simply

$$H_{\text{kin}} = \sum_{s=\uparrow,\downarrow} \int dx \{-iv_F R_s^+ \partial_x R_s + iv_F L_s^+ \partial_x L_s\} \tag{30.27}$$

where $v_F = k_F/m$ is the Fermi-velocity and s is the spin index. It is crucial at this stage to write the kinetic energy as a sum of commuting *charge* and *spin* parts (Sugawara formulation), $H_{\text{kin}} = H_{\text{charge}}^0 + H_{\text{spin}}^0$, where

$$H_{\text{charge}}^0 = \frac{\pi v_F}{2} \int dx \{J_R J_R + J_L J_L\} \tag{30.28}$$

$$H_{\text{spin}}^0 = \frac{2\pi v_F}{3} \sum_{a=x,y,z} \int dx \{J_R^a J_R^a + J_L^a J_L^a\} \tag{30.29}$$

and we introduced *charge*

$$J_R = \sum_s R_s^+ R_s, \quad J_L = \sum_s L_s^+ L_s \tag{30.30}$$

and *spin* currents ($a = x, y, z$)

$$J_R^a = \sum_{s,s'} R_s^+ \frac{\sigma_{ss'}^a}{2} R_{s'}, \quad J_L^a = \sum_{s,s'} L_s^+ \frac{\sigma_{ss'}^a}{2} L_{s'} \tag{30.31}$$

Magnetic field comes in via the Zeeman term

$$H_Z = -\Delta_z \int dx \{\cos\beta(J_R^z + J_L^z) + \sin\beta(J_R^x + J_L^x)\} \tag{30.32}$$

while the Rashba SO interaction reads

$$H_R = 2\Delta_{s-o} \int dx \{J_R^x - J_L^x\} \tag{30.33}$$

Note that both perturbations are written in terms of chiral (right and left) spin currents alone. The relative minus sign between the right and left spin currents in H_R follows from oddness of the Rashba term under spatial inversion ($x \to -x$), which interchanges right and left moving excitations. The coupling to magnetic field H_z of course is not sensitive to this operation.

The full noninteracting Hamiltonian (30.23) is given by the sum of the terms above, $H_0 = H_{\text{kin}} + H_Z + H_R$. The usefulness of the introduced spin-current formulation follows from the fact that it makes it explicitly clear that the two perturbations couple only to the spin sector of the theory.

This observation greatly simplifies subsequent calculations. We first notice that it is now easy to include *interactions* in the description. The interaction term can also be written as the sum of charge and spin contributions. Moreover, despite the fact that one-dimensional electrons interact with each other strongly (being constrained to a line, they simply cannot avoid collisions with each other), the interaction correction to the spin sector is small,

$$H_{\text{bs}} = -g \int dx \, \vec{J}_R \cdot \vec{J}_L \tag{30.34}$$

This interaction term is known as the *backscattering* correction, the amplitude of which is determined by the $2k_F$-component of the interaction potential: $g \sim U(2k_F) > 0$. For comparison, the interaction correction to the charge sector is controlled by the zero-momentum component of the potential, which is always much greater than the backscattering one, $U(0) \gg U(2k_F)$.

The backscattering interaction correction is not only numerically small but also *marginally irrelevant* in the renormalization group sense (to be clarified below). This means that the coupling constant g is in fact a function of energy (and/or momentum), which diminishes in magnitude as the energy approaches zero. This irrelevance implies that to the zeroth approximation, H_{bs} can be often (but not always, see below!) omitted altogether. When appropriate, such an approximation implies an *extended* symmetry of H_{spin}^0: being a sum of right and left spin current contributions, H_{spin}^0 is invariant under *independent* rotations of right and left spin currents.

Our solution to the problem exploits this extended symmetry. First, we observe that perturbations H_z and H_R add up to a single "vectorial" term

$$V = \int dx \{d_R J_R^x - d_L J_L^x - d_z(J_R^z + J_L^z)\} \qquad (30.35)$$

where $d_R = 2\Delta_{s-o} - \sin[\beta]\Delta_z$, $d_L = 2\Delta_{s-o} + \sin[\beta]\Delta_z$, and $d_z = \cos[\beta] \Delta_z$. Extended symmetry of H_{spin}^0 allows us to immediately conclude that right and left spin currents experience different total magnetic fields. The right spin current $\vec{J}_R$ couples to $\vec{h}_R$ with magnitude $h_R = \sqrt{d_R^2 + d_z^2}$, while the left current $\vec{J}_L$ experiences $\vec{h}_L$ of magnitude $h_L = \sqrt{d_L^2 + d_z^2}$. The two fields are different unless $\beta = 0$ when $h_R = h_L$.

Formal way to see this (Schnyder et al. 2008) is provided by rotation of the right (left) spin current $\vec{J}_R(\vec{J}_L)$ by angle θ_R (θ_L) about the $\hat{y}$-axis,

$$\vec{J}_R = \mathcal{R}(\theta_R)\vec{M}_R, \quad \vec{J}_L = \mathcal{R}(\theta_L)\vec{M}_L \qquad (30.36)$$

where $\mathcal{R}$ is the rotation matrix

$$\mathcal{R}(\theta) = \begin{pmatrix} \cos[\theta] & 0 & -\sin[\theta] \\ 0 & 1 & 0 \\ \sin[\theta] & 0 & \cos[\theta] \end{pmatrix} \qquad (30.37)$$

The rotation angles are given by $\tan[\theta_R] = d_R/d_z$ and $\tan[\theta_L] = -d_L/d_z$. These chiral rotations transform V into

$$V = -\int dx \{h_R M_R^z + h_L M_L^z\} \qquad (30.38)$$

It is worth pointing again that H_{kin}^0 is invariant under the rotations and is given by a simple substitution $\vec{J} \to \vec{M}$ in (30.29),

$$H_{spin}^0 = \frac{2\pi v_F}{3} \sum_{a=x,y,z} \int dx \{M_R^a M_R^a + M_L^a M_L^a\} \qquad (30.39)$$

Implications of Equation 30.38 are quite unusual: by varying the direction of the magnetic field, one can selectively couple to

right- and left-moving spin excitations. In particular, by lining $\vec{B}$ along the spin–orbital axis, which happens for $\beta = \pi/2$, one can adjust the magnitude $|\vec{B}|$ so that, for example, $h_R = 0$, while $h_L \neq 0$. It turns out that electron-spin-resonance (ESR) can, in effect, measure these chiral effective magnetic fields. More precisely, one finds that intensity I_{esr} of the ESR signal, in general, is given by a sum of two terms, representing two independent chiral contributions from Equation 30.38. Specifically, for a generic $h_R \neq h_L$ situation, one should observe two sharp lines (De Martino et al. 2002, 2004, Gangadharaiah et al. 2008a)

$$I_{esr}(\omega) = (\cos^2\theta_R + 1)h_R\delta(\omega - |h_R|) + (\cos^2\theta_L + 1)h_L\delta(\omega - |h_L|) \qquad (30.40)$$

Although certainly interesting, this result is based on complete neglect of the residual interaction between spin excitations H_{bs}. Little thinking shows that in this approximation, Equations 30.38 and 30.39 in fact represent a system of noninteracting fermions, R_s' and L_s', which are rotated version of R_s, L_s pair in (30.27). The relation between "old" and "new" fermions is simple

$$R_s = e^{i\theta_R \sigma_y/2}R_s', \quad L_s = e^{i\theta_L \sigma_y/2}L_s' \qquad (30.41)$$

where $\theta_{R/L}$ are the rotation angles from (30.36). That is, new (primed) fermions parameterize vector field $\vec{M}_{R/L}$ in the same way as the old (unprimed) ones parameterize fields $\vec{J}_{R/L}$. For example, under the right rotation; $\mathcal{R}(\theta_R)$

$$\vec{J}_R = \sum_{s,s'} R_s^+ \frac{\vec{\sigma}_{ss'}}{2} R_{s'} \rightarrow \vec{M}_R = \sum_{s,s'} R_s'^+ \frac{\vec{\sigma}_{ss'}}{2} R_{s'}' \qquad (30.42)$$

Using this relation, one can undo steps that led us from (30.27) through 30.29) (naturally, scalar charge fluctuations are not affected by the rotation at all) and find that (30.39) and (30.28) simply add up to a kinetic energy term (30.27) expressed via primed fermions R_s', L_s'. The extended symmetry discussed above is then simply the reflection of the symmetry of (30.27), which, for example, is invariant under independent rotations of right- and left-movers.

Now, it must also be clear that, irrespective of the basis used, noninteracting fermions cannot describe ordered state of any kind. This simple fact forces us to consider the fate of the two-particle backscattering interaction in greater detail. Under the rotations discussed, H_{bs} changes to

$$H_{bs} = -g\int dx \vec{M}_R \mathcal{R}^T(\theta_R)\mathcal{R}(\theta_L)\vec{M}_L$$

$$= -g\int dx \{\cos[\chi](M_R^x M_L^z + M_R^z M_L^z) + \sin[\chi](M_R^x M_L^z - M_R^z M_L^x)$$

$$+ M_R^y M_L^y\} \qquad (30.43)$$

where $\chi = \theta_R - \theta_L$ is the *relative* rotation angle.

Despite complicated appearance, (30.43) can be dealt with rather easily (Gangadharaiah et al. 2008a, Schnyder et al. 2008). The key to the subsequent progress is provided by switching to *abelian* bosonization technique, which is well suited for analyzing systems with $U(1)$ symmetry. Within this powerful technique, kinetic energy (30.39) turns into energy of free conjugated pair of bosons, φ_σ and θ_σ,

$$H^0_{\text{spin}} = \frac{\nu_F}{2}\int dx\{(\partial_x\varphi_\sigma)^2 + (\partial_x\theta_\sigma)^2\}. \tag{30.44}$$

At the same time, V transforms into linear in bosons form

$$V = -\int dx\left(\frac{\nu_F t_\varphi}{\sqrt{2\pi}}\partial_x\varphi_\sigma + \frac{\nu_F t_\theta}{\sqrt{2\pi}}\partial_x\theta_\sigma\right) \tag{30.45}$$

where

$$t_\varphi = (h_L + h_R)/2\nu_F, \quad t_\theta = (h_L - h_R)/2\nu_F \tag{30.46}$$

Observe that now V can be absorbed in H^0_{spin} with the help of simple shift of bosonic fields

$$\begin{aligned}\varphi_\sigma &\rightarrow \varphi_\sigma + \frac{t_\varphi x}{\sqrt{2\pi}},\\[6pt]\theta_\sigma &\rightarrow \theta_\sigma + \frac{t_\theta x}{\sqrt{2\pi}}\end{aligned} \tag{30.47}$$

The shift comes at a price—transverse components $M^{x,y}_{R/L}$ acquire oscillating position-dependent factors

$$M^+_R \rightarrow M^+_R e^{-i(t_\varphi - t_\theta)x}, \quad M^+_L \rightarrow M^+_L e^{i(t_\varphi + t_\theta)x} \tag{30.48}$$

The major consequence of this is that every term in H_{bs}, with a single exception of $M^z_R M^z_L$, acquires an oscillating prefactor

$$\begin{aligned}H_{\text{bs}} = -g\int dx\Bigg(&\cos[\chi]M^z_R M^z_L + \frac{\cos[\chi]-1}{4}(M^+_R M^+_L e^{i2t_\theta x} + \text{h.c.})\\[4pt]&+\frac{\cos[\chi]+1}{4}(M^+_R M^-_L e^{-i2t_\varphi x} + \text{h.c.})\\[4pt]&+\frac{\sin[\chi]}{2}(M^z_L M^+_R e^{-i(t_\varphi - t_\theta)x} - M^z_R M^+_L e^{i(t_\varphi + t_\theta)x} + \text{h.c.})\Bigg)\end{aligned} \tag{30.49}$$

The oscillating terms, which represent momentum-nonconserving two-particle scattering processes, average out to zero. This results in drastic simplification of the backscattering interaction, $H'_{\text{bs}} = -g\cos[\chi]\int dx\, M^z_R M^z_L$, which holds for as long as $t_\varphi \neq t_\theta \neq 0$. Using again bosonization rules, we see that the final Hamiltonian, $H^0_{\text{spin}} + H'_{\text{bs}}$, retains its harmonic form

$$H^0_{\text{spin}} + H'_{\text{bs}} = \frac{\nu_F}{2}\int dx\left\{\frac{1}{K}(\partial_x\varphi_\sigma)^2 + K(\partial_x\theta_\sigma)^2\right\} \tag{30.50}$$

where the dimensionless interaction parameter $K = 1 + g\cos[\chi]/(4\pi\nu_F) > 1$ is introduced. This Hamiltonian allows one to calculate any spin correlation function of interest, provided that one remains within the low-energy approximation assumed in the derivation. Such a situation corresponds to $\beta > 0$ and is illustrated in Figure 30.3: inter-subband pair tunneling is suppressed by the finite momentum mismatch.

30.3.3 Spin-Density Wave

We will not follow this well-understood route and instead focus on the case of $t_\theta = 0$, which is realized for $\beta = 0$. In this case, $\chi = 2\theta_R = -2\theta_L$. This special arrangement corresponds to the case of mutually orthogonal directions of the spin–orbital and magnetic fields, when the electron subbands are centered about $k = 0$ momentum. This is the configuration when inter-subband pair-tunneling is allowed by momentum conservation, see Figure 30.2. Focusing again on the nonoscillating terms alone, we observe that interaction between spin fluctuations is described by

$$H^\perp_{\text{bs}} = -g\int dx\left\{\cos[\chi]M^z_R M^z_L + \frac{\cos[\chi]-1}{4}(M^+_R M^+_L + M^-_R M^-_L)\right\} \tag{30.51}$$

Unlike the previously discussed H'_{bs}, the interaction term $H^\perp_{\text{bs}}$ does not reduce to simple quadratic form of φ and θ fields. In fact, it represents nontrivial interacting problem, the analysis of which requires renormalization group (RG) treatment. Omitting technical details, which can be found in modern textbooks (Gogolin et al. 1999, Giamarchi 2004) (a brief review is given in Appendix 30.A, see also (Giamarchi and Schulz 1988)), we cite the result

$$\frac{dg_z}{d\ell} = \frac{g^2_\perp}{2\pi\nu_F}, \quad \frac{dg_\perp}{d\ell} = \frac{g_\perp g_z}{2\pi\nu_F} \tag{30.52}$$

where g_z (correspondingly, $g_\perp/2$) is the coefficient of $M^z_R M^z_L$ (correspondingly, $M^+_R M^+_L$) term in $H^\perp_{\text{bs}}$. Here dimensionless RG parameter $\ell = \ln(a'/a)$ represents effect of integrating out fluctuations on the length scales between a' and a. The system of two coupled RG equations is nothing but famous Berezinskii–Kosterlitz–Thouless (BKT) renormalization group (Gogolin et al. 1999, Giamarchi 2004), flow diagram of which is shown in Figure 30.4. Its solution depends on the initial values of the couplings involved, $g_z(0) = g\cos[\chi]$ and $g_\perp(0) = g(\cos[\chi]-1)/2$, and is conveniently parameterized by the integral of motion $Y(\chi) = g^2_\perp - g^2_z$. It turns out that the solution is toward *strong coupling* (meaning that $|g_{z,\perp}(\ell)| \rightarrow \infty$ for sufficiently large ℓ) for all possible angles $\chi \in (0, \pi)$. This diverging

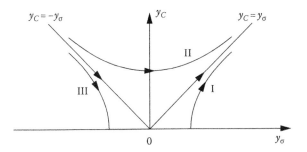

FIGURE 30.4 RG flow of Equation 30.52. Here $y_\sigma = g_z/(2\pi v_F)$, while $y_C = g_\perp/(2\pi v_F)$. Initial conditions on these couplings, as derived in (30.51), are such that region III is *not* accessible. All possible initial values lead to strong coupling, $y_C - y_\sigma \to \infty$. (Reprinted from Gangadharaiah, S. et al., *Phys. Rev. B*, 78, 054436, 2008a. With permission.)

solution implies instability toward a correlated spin state with finite energy gap between the ground and excited states. That gap can be estimated as $\Delta \sim v_F e^{-\ell_0}$, where ℓ_0 is the RG scale on which the couplings diverge. (The derivation of RG equations is perturbative in nature and relies on the smallness of the ratio $g_{z,\perp}/v_F$. Thus, solution with $g_{z,\perp}/v_F \geq 1$ cannot, strictly speaking, be trusted. But taking ℓ_0 for an *estimate* above remains sensible.) The *minimal* value of ℓ_0 occurs for $Y = 0$ (this corresponds to a line separating regions I and II in Figure 30.4), when $\chi_0 = 2\arccos\left[\sqrt{2/3}\right]$. The system then simplifies to a single equation,

$$\frac{dg_z}{d\ell} = \frac{g_z^2}{2\pi v_F} \tag{30.53}$$

(solution of which is: $g_z(\ell) = g_z(0)/(1 - g_z(0)\ell/(2\pi v_F))$). Thus, $\ell_0 = 2\pi v_F/g_z(0) = 3(2\pi v_F/g)$. The corresponding gap is exponentially small, $\Delta/v_F \sim \exp[-6\pi v_F/g] \ll 1$.

We now consider the physical meaning of the Cooper instability. This is best done by noting that the abelian form of the term that drives the instability, $(M_R^+ M_L^+ + \text{h.c.})$, is just $\cos\sqrt{8\pi}\theta_\sigma$. More specifically, abelian bosonization of (30.51) results in

$$H_{bs}^\perp = \int dx \left\{ -\frac{g_z}{8\pi}\left[(\partial_x\varphi_\sigma)^2 - (\partial_x\theta_\sigma)^2\right] + \frac{g_\perp}{4\pi^2 a_0^2}\cos\left[\sqrt{8\pi}\theta_\sigma\right] \right\} \tag{30.54}$$

As the coupling $g_\perp$ of that terms grows large (and positive) under the renormalization, the energy is minimized when θ_σ is at one of the semi-classical minima $\theta_\sigma^{cl} = (m+1/2)\sqrt{\pi/2}$ (m is integer). The energy cost of massive fluctuations $\delta\theta_\sigma$ near these minima represents the spin gap estimated above.

Physical meaning of these minima follows from the analysis of spin correlation functions. We consider spin density $S^a = \psi_s^\dagger \sigma_{s,s'}^a \psi_{s'}/2$, which is defined with respect to the standard spin basis, $s = \uparrow, \downarrow$. Consider first $2k_F$-components of the spin density. We find (Gangadharaiah et al. 2008a) (see Appendix 30.A, note that we use gauge, where $\eta_\uparrow \eta_\downarrow = i$)

$$\begin{pmatrix} S^x \\ S^y \\ S^z \end{pmatrix}_{2k_F} = -\frac{\cos\left[\sqrt{2\pi}\varphi_\rho + 2k_F x\right]}{\pi a_0}$$

$$\times \begin{pmatrix} -\sin\left[\sqrt{2\pi}\theta_\sigma\right] \\ \cos\left[\frac{\chi}{2}\right]\cos\left[\sqrt{2\pi}\theta_\sigma\right] + \sin\left[\frac{\chi}{2}\right]\cos\left[\sqrt{2\pi}\varphi_\sigma + t_\varphi x\right] \\ \sin\left[\sqrt{2\pi}\varphi_\sigma\right] \end{pmatrix}$$

$$\to -\frac{\cos\left[\sqrt{2\pi}\varphi_\rho + 2k_F x\right]}{\pi a_0}\begin{pmatrix} \pm 1 \\ 0 \\ 0 \end{pmatrix} \tag{30.55}$$

The last line of the above equation is somewhat symbolic, with zeros representing *exponentially* decaying correlations of the corresponding spin components, $S^{y,z}$. Here $\hat{z}$-component is disordered by strong quantum fluctuations of φ_σ field, which is *dual* to the ordered θ_σ one. The $\hat{y}$-component is absent because $\cos[\sqrt{2\pi}\theta_\sigma^d] = 0$. Thus, Cooper order found here, in fact, represents spin-density-wave (SDW) order at momentum $2k_F$ of the $\hat{x}$-component of spin density, as discussed previously in (Sun et al. 2007). Note that the ordered component, S^x, is oriented along the spin–orbital axis σ^x. Also, observe that S^x ordering is of *quasi-LRO* type as it involves free charge boson, φ_ρ. As a result, spin correlations do decay with time and distance, but very slowly $\langle S^x(x)S^x(0)\rangle \sim \cos[2k_F x]x^{-K_\rho}$. The rate of the spatial decay is determined by the interactions in the charge sector alone, via the parameter K_ρ.

The result (30.55) also hints a possibility of truly long-range-ordered spin correlations in the insulating state of the wire—Heisenberg spin chain. There the charge field φ_ρ is absent, which can be mimicked by setting $K_\rho \to 0$ in the spin correlation function above. Then, $\langle S^x(x)S^x(0)\rangle \sim \cos[2k_F x] = (-1)^x$ describes long-range ordered SDW state of the magnetic insulator (Gangadharaiah et al. 2008a) (with asymmetric Dzyaloshisnkii–Moriya interaction in place of the Rashba coupling considered here).

30.3.4 Transport Properties of the SDW State

SDW state manifests itself not in spin correlations alone. It turns out that its response to a weak potential scattering (impurity) is rather nontrivial. We consider here a weak delta-function impurity $V(x) = V_0\delta(x)$, with strength V_0, located at the origin. The condition $V_0 \ll \Delta$ means that impurity can be considered as a weak perturbation to the established SDW phase. (The limit of strong impurity, $V_0 \gg \Delta$, is rather standard: impurity destroys the SDW, and the wire flows into an insulator at low energies (Kane and Fisher 1992).)

The interaction of electrons with an impurity potential $V(x)$ is given by

$$\hat{V} = \int dx\, V(x) \sum_{s=\uparrow,\downarrow} \psi_s^\dagger(x)\psi_s(x) \qquad (30.56)$$

As usual, it is the *backscattering* (ρ_{2k_F}) part of the charge density that has to be considered, since it describes the process in which an electron with momentum $+k_F$ scatters back into $-k_F$ state, thereby losing large momentum $2k_F$: $\rho_{2k_F}(x=0) = \sum_s (R_s^\dagger L_s + \text{h.c.})$. Performing the rotation and bosonizing we find, see appendix and (30.A.33) in particular,

$$\rho_{2k_F}(x=0) = \frac{-2\sin\left[\sqrt{2\pi}\varphi_\rho\right]}{\pi a_0}$$

$$\left(\cos\left(\frac{\chi}{2}\right)\cos\left[\sqrt{2\pi}\varphi_\sigma\right] + \sin\left[\frac{\chi}{2}\right]\cos\left[\sqrt{2\pi}\theta_\sigma\right]\right) \qquad (30.57)$$

We now observe that setting $\theta_\sigma \to \theta_\sigma^{cl} = (m+1/2)\sqrt{\pi/2}$ as appropriate for the SDW phase *nullifies* the backscattering component of the density (30.57). The first term gets killed by diverging fluctuations of φ_σ, *dual* to θ_σ. Intriguingly, the second contribution is also zero because $\cos\left[\sqrt{2\pi}\theta_\sigma^{cl}\right] = \pm\cos\left[\pi/2\right] = 0$.

Physical explanation of this unusual result is grounded in the observation that x-components of electron's spins at $\pm k_F$ points are *antiparallel*, see (30.55). At the same time, scattering off potential impurity does not affect the spin of the scattered electron. Hence electron with, say, spin along $+\hat{x}$ axis and momentum $+k_F$ is scattered to a state with momentum $-k_F$ and the same spin orientation. However, the Fermi-point only supports states with spin along $-\hat{x}$, see Figure 30.1, for as long as one is interested in low-energy states (energy well below Δ). Thus, at such low energies, there is no electron backscattering, and potential impurity does not affect electron's propagation along the wire (Sun et al. 2007, Gangadharaiah et al. 2008a).

The situation is similar to that in recently proposed edge states in quantum spin Hall system (Kane and Mele 2005). Gapless spin-up and spin-down excitations propagate there in opposite directions along the edge, which forbids single-particle backscattering. Interacting electrons, however, can backscatter off the impurity in pairs (Xu and Moore 2006, Wu et al. 2006).

Pairwise scattering of electrons off the weak impurity can be analyzed as follows (Orignac and Giamarchi 1997, Egger and Gogolin 1998, Starykh et al. 2000). We parameterize $\theta_\sigma = \theta_\sigma^{cl} + \delta\theta$ and expand the relevant cosine term in Hamiltonian (30.54) to second order in fluctuations $\delta\theta$. One obtains a *massive* term $\propto \Delta(\delta\theta)^2$ in the Hamiltonian, which causes exponential decay in correlation functions of the dual φ_σ field. In particular, $\left\langle \cos\sqrt{2\pi}\varphi_\sigma(0,\tau)\cos\sqrt{2\pi}\varphi_\sigma(0,\tau')\right\rangle$ will decay as $\exp[-\Delta|\tau - \tau'|]$. Substituting $\theta_\sigma = \theta_\sigma^{cl} + \delta\theta$ in the second term in (30.57) converts it into $\sin[\chi/2]\sin\left[\sqrt{2\pi}\varphi_\rho\right]\sin\left[\sqrt{2\pi}\delta\theta\right]$. Correlations of $\sin\left[\sqrt{2\pi}\delta\theta\right]$ are also short-ranged

$$\left\langle \sin\left[\sqrt{2\pi}\delta\theta(\tau)\right]\sin\left[\sqrt{2\pi}\delta\theta(\tau')\right]\right\rangle \approx \frac{\Delta_c}{v_F k_F}\sinh\left[K_0\left(\Delta|\tau-\tau'|\right)\right]$$

$$(30.58)$$

where $K_0(x) \sim e^{-x}/\sqrt{x}$ (for $x \gg 1$) is the modified Bessel function (see Starykh et al. 2000). The two exponentially decaying contributions add up (in second order perturbation theory in impurity strength V_0) to produce an effective two-particle backscattering potential $\propto (V_0^2/\Delta)\cos\left[\sqrt{8\pi}\varphi_\rho(0,\tau)\right]$. This term describes interaction-generated *two-particle* backscattering. However, it is relevant only for strongly repulsive interactions, when $K_\rho < 1/2$. We are thus left with *irrelevant* impurity potential for both single- and two-particle scattering events. Hence, we conclude that as long as approximation $V_0 \ll \Delta$ is justified and for not too strong repulsion, when $1/2 \leq K_\rho \leq 1$, the SDW state is not sensitive to weak disorder.

This unusual prediction should manifest itself in strong sensitivity of the conductance of the quantum wire to the direction of the applied magnetic field. Weak potential impurity will remain irrelevant (and, thus, not observable) for as long as applied magnetic field is directed perpendicular to the spin–orbital (σ^x here) axis. By either turning the magnetic field off or simply rotating the sample (so that angle β between the magnetic field and the spin–orbital axis changes from 0), one should observe that conductance of the wire deteriorates. With $\beta \neq 0$, the wire will eventually turn insulating in the zero-temperature limit.

30.4 Conclusions

Spin–orbital interactions result in reduction of spin-rotational symmetry from $SU(2)$ to $U(1)$ in one-dimensional quantum wires and spin chains. This reduction, however, is not sufficient to change the critical (Luttinger liquid) nature of the one-dimensional interacting fermions. The situation changes dramatically once external magnetic field is applied. Most interesting situation occurs when the applied field is oriented along the axis orthogonal to the spin–orbital axis of the wire. The resulting combination of two non-commuting perturbations, taken together with electron–electron interactions, leads to a novel spin-density-wave order in the direction of the spin–orbital axis.

The physics of this order is elegantly described in terms of spin-nonconserving (Cooper) pair tunneling processes between Zeeman-split electron subbands. The tunneling matrix element is finite only due to the presence of the spin–orbit interaction, which allows for spin-up to spin-down (and vice versa) conversion.

The resulting SDW state affects both spin and charge properties of the wire. In particular, it suppresses the effect of (weak) potential impurity, resulting in the interesting phenomena of negative magnetoresistance in one-dimensional setting.

Even when the magnetic field and spin–orbital directions are not orthogonal, an arrangement when the critical Luttinger state survives down to the lowest temperature, the problem remains interesting. In this geometry, an ESR experiment should reveal two separate lines, which represent separate responses of right- and left-moving spin fluctuations in the system.

It is worth pointing out that unusual consequences of the interplay of spin–orbit and electron interactions are not restricted to one-dimensional systems alone. As shown in Section 30.2, Coulomb-coupled two-dimensional quantum dots acquire a novel vdW -like anisotropic interaction between spins of the localized electrons. The strength of this Ising interaction is determined by the forth power of the Rashba coupling α_R.

We hope that our work will stimulate experimental search and studies of strongly interacting quasi-one-dimensional systems with sizable spin–orbital interaction, particularly regarding their response to the (both magnitude and direction) applied magnetic field and/or magnetization. ESR studies of quantum wires and spin chain materials are very desirable as well.

Appendix 30.A: Bosonization Basics

Bosonization starts by expressing the fermionic operators $\psi_s(x)$

$$\psi_s = R_s e^{ik_Fx} + L_s e^{-ik_Fx} \tag{30.A.1}$$

in terms of right (R_s) and left (L_s) movers of unperturbed single-channel quantum wire. These represent electrons with momenta near the right, $+k_F$, and the left, $-k_F$, Fermi-momenta of the wire. They are then represented via exponentials of the chiral bosonic $\phi_{R/L,s}$ fields (Lin et al. 1997, Gogolin et al. 1999, Giamarchi 2004) as follows:

$$R_s = \frac{\eta_s}{\sqrt{2\pi a_0}} e^{i\sqrt{4\pi}\phi_{R,s}}, \quad L_s = \frac{\eta_s}{\sqrt{2\pi a_0}} e^{-i\sqrt{4\pi}\phi_{L,s}} \tag{30.A.2}$$

where $a_0 \sim k_F^{-1}$ is the short distance cutoff and η_s are the Klein factors, which are introduced to ensure the correct anticommutation relations for the fermions with different spin projection $s = \uparrow = +1$ and $s = \downarrow = -1$. The bosonic operators obey the following commutation relations:

$$\left[\phi_{R,s}, \phi_{L,s'}\right] = \frac{i}{4}\delta_{ss'} \tag{30.A.3}$$

$$\left[\phi_{R/L,s}(x), \phi_{R/L,s'}(y)\right] = \pm\frac{i}{4}\delta_{ss'}\,\text{sign}(x-y) \tag{30.A.4}$$

the first of which, (30.A.3), ensures anticommutation between right and left movers from the *same* subband, while the second is needed for the anticommutation between like species (i.e. right with right, left with left). Klein factors anticommute

$$\{\eta_s, \eta_{s'}\} = 2\delta_{ss'}, \quad \eta_s^\dagger = \eta_s \tag{30.A.5}$$

In the following, we choose the gauge where $\eta_{+1}\eta_{-1} = i$. The chiral $\phi_{R/L,s}$ are expressed in terms of ϕ_s and its dual θ_s as follows:

$$\phi_{R,s} = \frac{\phi_s - \theta_s}{2}; \quad \phi_{L,s} = \frac{\phi_s + \theta_s}{2}, \tag{30.A.6}$$

The bosonized form of the Hamiltonian is obtained by making use of Equations 30.A.2 through 30.A.6 as well as the following results for (chiral) densities:

$$R_s^\dagger R_s = \frac{\partial_x \phi_{R,s}}{\sqrt{\pi}} = \frac{\partial_x(\phi_s - \theta_s)}{\sqrt{4\pi}}$$
$$L_s^\dagger L_s = \frac{\partial_x \phi_{L,s}}{\sqrt{\pi}} = \frac{\partial_x(\phi_s + \theta_s)}{\sqrt{4\pi}}. \tag{30.A.7}$$

It then follows, from the above formalism, that kinetic energy of fermions with *linear* dispersion, see (30.27), transforms into kinetic energy of bosons

$$H_{kin} = \frac{v_F}{2}\sum_s \int dx\{(\partial_x\phi_s)^2 + (\partial_x\theta_s)^2\} \tag{30.A.8}$$

Next, we introduce charge and spin bosons via

$$\varphi_\rho = \frac{1}{\sqrt{2}}(\phi_\uparrow + \phi_\downarrow), \quad \varphi_\sigma = \frac{1}{\sqrt{2}}(\phi_\uparrow - \phi_\downarrow), \tag{30.A.9}$$

and similar for the $\theta_{\rho,\sigma}$. These independent modes allow us to represent kinetic Hamiltonian as a sum of two commuting, charge-only and spin-only, parts

$$H_{kin} = H_{kin,\rho} + H_{kin,\sigma} = \frac{v_F}{2}\sum_{v=\rho,\sigma} \int dx\{(\partial_x\phi_v)^2 + (\partial_x\theta_v)^2\} \tag{30.A.10}$$

The utility of this representation consists in the observation that *interaction* between electrons separates into charge and spin contributions as well. Interaction between electrons is described by

$$H_{int} = \frac{1}{2}\sum_{s,s'} \int dx dx' U(x-x')\psi_s^+(x)\psi_{s'}^+(x')\psi_{s'}(x')\psi_s(x) \tag{30.A.11}$$

where $U(x)$ is just (screened by gates) Coulomb interaction. Using (30.A.l), we observe that

$$\psi_s^+(x)\psi_s(x) = R_s^\dagger R_s + L_s^\dagger L_s + e^{-i2k_Fx}R_s^\dagger L_s + e^{i2k_Fx}L_s^\dagger R_s \tag{30.A.12}$$

Focusing on momentum-conserving terms and using identity (see 30.30 and 30.31)

$$\sum_{s,s'} R_s^\dagger R_{s'} L_s^\dagger L_s = 2\vec{J}_R \cdot \vec{J}_L + \frac{1}{2}J_RJ_L \tag{30.A.13}$$

we express $H_{int} = H_{int,\rho} + H_{int,\sigma}$ as a sum of charge and spin contributions

$$H_{int,\rho} = \int dx\left\{\frac{1}{2}U(0)(J_R^2 + J_L^2) + (U(0) - \frac{1}{2}U(2k_F))J_RJ_L\right\} \tag{30.A.14}$$

$$H_{int,\sigma} = \int dx\{-2U(2k_F)\vec{J}_R \cdot \vec{J}_L\} \tag{30.A.15}$$

The interaction is parameterized by two constants, $U(0)$ and $U(2k_F)$, which are zeroth and $2k_F$ components of the Fourier transform of $U(x-x')$: $U(q) = \int dx U(x)e^{iqx}$. Clearly $U(0) \gg U(2k_F)$

for a smoothly decaying potential. Observe that the spin sector involves only the weakest of the two interaction constants. This term, (30.A.15), is the backscattering correction (30.34) described in the main text.

The total charge Hamiltonian can be conveniently written in a quadratic boson form

$$H_\rho = \frac{v_\rho}{2} \int dx \{ K_\rho^{-1} (\partial_x \varphi_\rho)^2 + K_\rho (\partial_x \theta_\rho)^2 \} \qquad (30.A.16)$$

with charge velocity $v_\rho = v_F (1 + U(0)/(\pi v_F)) > v_F$ and dimensionless interaction parameter $K_\rho = 1 - (2U(0) - U(2k_F))/(2\pi v_F) < 1$. This important result is the starting point for many interesting features of strongly correlated one-dimensional electrons, as it allows for an essentially exact calculation of the low-energy properties of these electrons in terms of free bosons φ_ρ and θ_ρ.

The spin Hamiltonian does not have simple harmonic appearance because the product of spin currents in (30.A.15) includes very nonlinear $\cos[\sqrt{8\pi}\varphi_\sigma]$ term. Nevertheless, the progress is possible by attacking (30.A.15) using perturbative (in small $U(2k_F)/v_F$ ratio) RG, as described in the main text. The key idea is to exploit the charge-spin separation to the fullest: by its very derivation, the spin backscattering correction (30.A.15) involves only spin modes φ_σ and θ_σ. It is then allowed to disregard the charge part, $H_{\text{int},\rho}$, altogether and treat $H_{\text{int},\sigma}$ as the only perturbation to the free Hamiltonian (30.27). The charge part of kinetic energy $H_{\text{kin},\rho}$, which is contained in (30.27), is guaranteed (by the independence of charge and spin bosons) not to affect the result of such calculation in any way. The end result is that one can formulate perturbation theory in question in terms of weakly interacting fermions again! The bosonization is used here only to make the phenomenon of the charge-spin separation explicit on the level of operators. This observation provides for a convenient short-cut in deriving the operator-product relations between spin currents that are required for the perturbative RG (see Starykh et al. (2005) for more details).

In general, one is interested in RG equation for the anisotropic current–current interaction term,

$$H'_\sigma = - \sum_{a=x,y,z} \int dx \{ g_a J_R^a J_L^a \} \qquad (30.A.17)$$

For example, in (30.51), we have $g_z = g \cos[\chi]$, $g_x = -g_y = g(\cos[\chi] - 1)$ and $g = 2U(2k_F)$. We expand partition function $Z = \int e^S$, where the action $S = S_0 - \int d\tau H'_\sigma$, in powers of the perturbation H'_σ. The unperturbed action S_0 describes free spin excitations, which, by the logic outlined above, are faithfully represented by the free Hamiltonian (30.27). To second order in g, one has

$$Z = \int e^{S_0} \left\{ 1 + \sum_{a=x,y,z} \int dx \, d\tau \, g_a J_R^a(x,\tau) J_L^a(x,\tau) \right.$$
$$\left. + \frac{1}{2} \sum_{a,b} g_a g_b \int dx \, d\tau \, dx' \, d\tau' \, J_R^a(x,\tau) J_R^b(x',\tau') J_L^a(x,\tau) J_L^b(x',\tau') + \cdots \right\}$$
$$(30.A.18)$$

Renormalization is possible because the last term in the brackets contains terms that reduce to the second term, i.e., to H'_σ. This follows from the nontrivial property of the product of spin currents

$$J_R^a(x,\tau) J_R^b(x',\tau') = \frac{\delta^{ab}}{8\pi^2 (v_F \tau - ix)^2} + \frac{i\epsilon^{abc} J_R^c(x,\tau)}{2\pi (v_F \tau - ix)}$$
$$+ \text{ less singular terms.} \qquad (30.A.19)$$

This result follows from applying Wick's theorem (justified here because the unperturbed Hamiltonian is just H_{kin}) and making all possible fermion pairings. These are constrained by the fact that the only nonzero correlations are between like fermions (right with right, and left with left) of the same spin, as dictated by the structure of (30.27). Thus, the singular denominator in the equation above is just the Green's function of right fermions. Another ingredient of (30.A.19) is the well-known Pauli matrix property:

$$\sigma^a \sigma^b = \delta^{ab} + i \sum_c \epsilon^{abc} \sigma^c \qquad (30.A.20)$$

The product of left currents has similar expansion with the obvious replacement $v_F \tau - ix \to v_F \tau + ix$, as appropriate for the left-moving particles. With the help of these operator-product expansions, the last term (denoted V here) simplifies to

$$V = \text{const} - \frac{1}{2} \sum_{a,b,c,d} g_a g_b \int dx \, d\tau \, dx' \, d\tau' \frac{\epsilon^{abc} \epsilon^{abd} J_R^c(x,\tau) J_L^d(x,\tau)}{4\pi^2 (v_F \tau - ix)(v_F \tau + ix)}$$
$$(30.A.21)$$

where the constant is the contribution from the first term in (30.A.19). Switching to the center-of-mass and relative coordinates, we finally have

$$V = \frac{1}{2} \sum_c \delta g_c \int dx \, d\tau \, J_R^c(x,\tau) J_L^c(x,\tau) \qquad (30.A.22)$$

where

$$\delta g_c = -\frac{1}{2} \sum_{a,b \neq c} \frac{g_a g_b}{2\pi v_F} \int_a^{a'} \frac{dr}{r} = -\frac{1}{2} \sum_{a,b \neq c} \frac{g^a g^b}{2\pi v_F} \ln\left(\frac{a'}{a}\right) \qquad (30.A.23)$$

is given by the integral over relative coordinate. In terms of RG scale $\ell = \ln(a'/a)$, the differential change in the coupling g_c follows

$$\frac{dg_c}{d\ell} = - \sum_{a,b \neq c} \frac{g_a g_b}{4\pi v_F} \qquad (30.A.24)$$

This, of course, contains three equations

$$\frac{dg_x}{d\ell} = -\frac{g_y g_z}{2\pi v_F}$$
$$\frac{dg_y}{d\ell} = -\frac{g_x g_z}{2\pi v_F} \qquad (30.A.25)$$
$$\frac{dg_z}{d\ell} = -\frac{g_x g_y}{2\pi v_F}$$

RG equations in the main text follow from the ones above by simple rearrangements.

To obtain bosonic representation of spin in (30.55), one needs to start with the definition of $2k_F$ component of spin density,

$$\vec{S}_{2k_F}(x) = \frac{1}{2}\sum_{s,s'} R_s^+ \vec{\sigma}_{s,s'} L_{s'} e^{-i2k_F x} + L_s^+ \vec{\sigma}_{s,s'} R_{s'} e^{i2k_F x} \quad (30.A.26)$$

Next, we need to account for the rotation (30.41). This leads us to consider the following objects

$$e^{-i\theta_R \sigma_y/2} \sigma_x e^{i\theta_L \sigma_y/2} = e^{-i(\theta_R + \theta_L)\sigma_y/2} \sigma_x = \sigma_x \quad (30.A.27)$$

where we made use of (30.A.20) and the fact that in the current geometry, $\theta_R = -\theta_L$. Hence we find that S^x and S^z components of the spin density are not affected by the rotation, while S^y component does change. Indeed,

$$e^{-i\theta_R \sigma_y/2} \sigma_y e^{i\theta_L \sigma_y/2} = e^{-i(\theta_R - \theta_L)\sigma_y/2} \sigma_y = e^{-i\chi \sigma_y/2} \sigma_y \quad (30.A.28)$$

Hence,

$$S_{2k_F}^y = \cos\left[\frac{\chi}{2}\right] \frac{1}{2}\sum_{s,s'} \left\{ R_s'^+ \sigma_{s,s'}^y L_{s'}' e^{-i2k_F x} + \text{h.c.} \right\}$$

$$+ \sin\left[\frac{\chi}{2}\right] \frac{1}{2}\sum_{s,s'} \left\{ iR_s'^+ L_{s'}' e^{-i2k_F x} + \text{h.c.} \right\} \quad (30.A.29)$$

We now bosonize primed fermions using (30.A.2, 30.A.3, 30.A.6, and 30.A.9), as well as the Baker–Hausdorff formulae

$$e^A e^B = e^B e^A e^{[A,B]}, \quad e^A e^B = e^{A+B} e^{[A,B]/2} \quad (30.A.30)$$

In this way, we obtain

$$S_{2k_F}^y = \frac{1}{\pi a_0} \cos\left[\sqrt{2\pi}\varphi_\rho + 2k_F x\right]$$

$$\left(\cos\left[\frac{\chi}{2}\right] \cos\left[\sqrt{2\pi}\theta_\sigma\right] + \sin\left[\frac{\chi}{2}\right] \cos\left[\sqrt{2\pi}\varphi_\sigma + t_\varphi x\right] \right) \quad (30.A.31)$$

The appearance of the position-dependent phase $t_\varphi x$ in the last term in the brackets is due to the shift (30.47), where we need to use $t_\varphi = h_R/v_F = \sqrt{4\Delta_{s-o}^2 + \Delta_z^2}/v_F$ in order to account for the mutually orthogonal orientation of the spin–orbital and magnetic field axes. The other, unrotated, components of the spin density vector are obtained by similar calculations, resulting in (30.55) of the main text.

Calculation of the backscattering ($2k_F$) component of the density ρ_{2k_F} proceeds very similarly:

$$\rho_{2k_F} = \sum_s R_s^+ L_s e^{-i2k_F x} + L_s^+ R_s e^{i2k_F x}$$

$$\rightarrow \cos\left[\frac{\chi}{2}\right] \sum_s \left\{ R_s'^+ L_s' e^{-i2k_F x} + \text{h.c.} \right\}$$

$$+ \sin\left[\frac{\chi}{2}\right] \sum_{s,s'} \left\{ -iR_s'^+ \sigma_{s,s'}^y L_s' e^{-i2k_F x} + \text{h.c.} \right\} \quad (30.A.32)$$

Subsequent bosonization leads to

$$\rho_{2k_F} = -\frac{2}{\pi a_0} \sin\left[\sqrt{2\pi}\varphi_\rho + 2k_F x\right]$$

$$\left(\cos\left[\frac{\chi}{2}\right] \cos\left[\sqrt{2\pi}\varphi_\sigma + t_\varphi x\right] + \sin\left[\frac{\chi}{2}\right] \cos\left[\sqrt{2\pi}\theta_\sigma\right] \right)$$

$$(30.A.33)$$

Bosonized form of the nonlinear Cooper term $M_R^+ M_L^+$ follows from

$$M_R^+ = \frac{1}{2}\sum_{s,s'} R_s'^+ (\sigma_x + i\sigma_y)_{s,s'} R_{s'}' = \frac{i}{4\pi a_0} e^{-i\sqrt{2\pi}(\varphi_\sigma - \theta_\sigma)}$$

$$M_L^+ = \frac{1}{2}\sum_{s,s'} L_s'^+ (\sigma_x + i\sigma_y)_{s,s'} L_{s'}' = \frac{i}{4\pi a_0} e^{i\sqrt{2\pi}(\varphi_\sigma + \theta_\sigma)}$$

$$(30.A.34)$$

This result, together with (30.A.4), implies that $M_R^+ M_L^+ + \text{h.c.} \sim \cos\left[\sqrt{8\pi}\theta_\sigma\right]$. Note also that Equations 30.A.34 explain the effect of the shift (30.47) on transverse components $M_{R/L}^\pm$ of the magnetization, Equation 30.48.

Acknowledgments

I would like to thank my collaborators on this project, Jianmin Sun and Suhas Gangadharaiah, for their invaluable contributions to this topic. I also thank Suhas Gangadharaiah for careful reading of the manuscript. I am deeply grateful to Andreas Schnyder and Leon Balents for the collaboration on quantum kagome antiferromagnet where the trick of chiral rotations of spin currents has originated. I would like to thank I. Affleck, T. Giamarchi, K. Matveev, E. Mishchenko, M. Oshikawa, M. Raikh, and Y.-S. Wu for discussions and suggestions at various stages of this work. I thank Petroleum Research Fund of the American Chemical Society for the financial support of this research under the grant PRF 43219-AC10.

References

Aleiner, I.L. and V.I. Fal'ko, 2001. Spin-orbit coupling effects on quantum transport in lateral semiconductor dots. *Phys. Rev. Lett.* **87**: 256801.

Bernu, B., L. Candido, and D.M. Ceperley, 2001. Exchange frequencies in the 2D Wigner crystal. *Phys. Rev. Lett.* **86**: 870.

Burkard, G., D. Loss, and D.P. DiVincenzo, 1999. Coupled quantum dots as quantum gates. *Phys. Rev. B* **59**: 2070.

Bychkov, A. Yu and E.I. Rashba, 1984. Oscillatory effects and the magnetic susceptibility of carriers in inversion layers. *J. Phys. C* **17**: 6039.

Calderon, M.J., B. Koiller, and S. Das Sarma, 2006. Exchange coupling in semiconductor nanostructures: Validity and limitations of the Heitler-London approach. *Phys. Rev. B* **74**: 045310.

Crain, J.N. and F.J. Himpsel, 2006. Low-dimensional electronic states at silicon surfaces. *Appl. Phys. A* **82**: 431.

Dedkov, Yu. S., M. Fonin, U. Rüdiger, and C. Laubschat, 2008. Rashba effect in the graphene/Ni(111) system. *Phys. Rev. Lett.* **100**: 107602.

De Martino, A., R. Egger, K. Hallberg, and C.A. Balseiro, 2002. Spin-orbit coupling and electron spin resonance theory for carbon nanotubes. *Phys. Rev. Lett.* **88**: 206402.

De Martino, A., R. Egger, F. Murphy-Armando, and K. Hallberg, 2004. Spin-orbit coupling and electron spin resonance for interacting electrons in carbon nanotubes. *J. Phys.: Condens. Matter* **16**: S1437.

Dresselhaus, G. 1955. Spin-orbit coupling effects in zinc blende structures. *Phys. Rev.* **100**: 580.

Dzyaloshinskii, I.E. 1958. A thermodynamic theory of weak ferromagnetism of antiferro-magnetics. *J. Phys. Chem. Solids* **4**: 241.

Egger, R. and A.O. Gogolin, 1998. Correlated transport and non-fermi liquid behavior in single-wall carbon nanotubes. *Eur. Phys. J. B* **3**: 281.

Flindt, C., A.S. Sorensen, and K. Flensberg, 2006. Spin-orbit mediated control of spin qubits. *Phys. Rev. Lett.* **97**: 240501.

Gangadharaiah, S., J. Sun, and O.A. Starykh, 2008a. Spin-orbital effects in magnetized quantum wires and spin chains. *Phys. Rev. B* **78**: 054436.

Gangadharaiah, S., J. Sun, and O.A. Starykh, 2008b. Spin-orbit-mediated anisotropic spin interaction in interacting electron systems. *Phys. Rev. Lett.* **100**: 156402.

Giamarchi, T. 2004. *Quantum Physics in One Dimension*. Oxford University Press, Oxford, U.K.

Giamarchi, T. and H. J. Schulz, 1988. Theory of spin-anisotropic electron-electron interactions in quasi-one-dimensional metals. *J. Phys.* **49**: 819.

Gogolin, A.O., A.A. Nersesyan, and A.M. Tsvelik, 1999. *Bosonization and Strongly Correlated Systems*. Cambridge University Press, Cambridge, U.K.

Griffiths, D.J. 2005. *Introduction to Quantum Mechanics*, 2nd ed. Pearson, Upper Saddle River, NJ, p. 286.

Iucci, A. 2003. Correlation functions for one-dimensional interacting fermions with spin-orbit coupling. *Phys. Rev. B* **68**: 075107.

Kane, C.L. and M.P.A. Fisher, 1992. Transmission through barriers and resonant tunneling in an interacting one-dimensional electron gas. *Phys. Rev. B* **46**: 15233.

Kane, C.L. and E.J. Mele, 2005. Quantum spin hall effect in graphene. *Phys. Rev. Lett.* **95**: 226801.

Krupin, O., G. Bihlmayer, K. Starke et al., 2005. Rashba effect at magnetic metal surfaces. *Phys. Rev. B* **71**: 201403.

LaShell, S., B.A. McDougall, and E. Jensen, 1996. Spin splitting of an Au(111) surface state band observed with angle resolved photoelectron spectroscopy. *Phys. Rev. Lett.* **77**: 3419.

Levitov, L.S. and E.I. Rashba, 2003. Dynamical spin-electric coupling in a quantum dot. *Phys. Rev. B* **67**: 115324.

Lin, H.-H., L. Balents, and M.P.A. Fisher, 1997. N-chain Hubbard model in weak coupling. *Phys. Rev. B* **56**: 6569.

Moriya, T. 1960a. New mechanism of anisotropic superexchange interaction. *Phys. Rev. Lett.* **4**: 228.

Moriya, T. 1960b. Anisotropic superexchange interaction and weak ferromagnetism. *Phys. Rev.* **120**: 91.

Moroz, A.V. and C.H.W. Barnes, 1999. Effect of the spin-orbit interaction on the band structure and conductance of quasi-one-dimensional systems. *Phys. Rev. B* **60**: 14272.

Moroz, A.V. and C.H.W. Barnes, 2000. Spin-orbit interaction as a source of spectral and transport properties in quasi-one-dimensional systems. *Phys. Rev. B* **61**: R2464.

Moroz, A.V., K.V. Samokhin and C.H.W. Barnes, 2000. Theory of quasi-one-dimensional electron liquids with spin-orbit coupling. *Phys. Rev. B* **62**: 16900.

Mugarza, A. and J.E. Ortega, 2003. Electronic states at vicinal surfaces. *J. Phys.: Condens. Matter* **15**: S3281.

Mugarza, A., A. Mascaraque, V. Repain et al., 2002. Lateral quantum wells at vicinal Au(111) studied with angle-resolved photoemission. *Phys. Rev. B* **66**: 245419.

Orignac, E. and T. Giamarchi, 1997. Effects of disorder on two strongly correlated coupled chains. *Phys. Rev. B* **56**: 7167.

Ortega, J.E., M. Ruiz-Osés, J. Gordón et al., 2005. One-dimensional versus two-dimensional electronic states in vicinal surfaces. *New J. Phys.* **7**: 101.

Pereira, R.G. and E. Miranda, 2005. Magnetically controlled impurities in quantum wires with strong Rashba coupling. *Phys. Rev. B* **71**: 085318.

Schnyder, A.P., O.A. Starykh, and L. Balents, 2008. Spatially anisotropic Heisenberg kagome antiferromagnet. *Phys. Rev. B* **78**: 174420.

Shahbazyan, T.V. and M.E. Raikh, 1994. Low-field anomaly in 2D hopping magne-toresistance caused by spin-orbit term in the energy spectrum. *Phys. Rev. Lett.* **73**: 1408.

Starykh, O.A., D.L. Maslov, W. Häusler, and L.I. Glazman, 2000. Gapped phases of quantum wires. In *Low-Dimensional Systems: Interactions and Transport Properties*, ed. T. Brandes, pp. 37–78. *Lecture Notes in Physics*, Vol. 544, Springer, Berlin, Germany.

Starykh, O.A., A. Furusaki, and L. Balents, 2005. Anisotropic pyrochlores and the global phase diagram of the checkerboard antiferromagnet. *Phys. Rev. B* **72**: 094416.

Sun, J., S. Gangadharaiah, and O.A. Starykh, 2007. Spin-orbit-induced spin-density wave in a quantum wire. *Phys. Rev. Lett.* **98**: 126408.

Trif, M., V.N. Golovach, and D. Loss, 2007. Spin-spin coupling in electrostatically coupled quantum dots. *Phys. Rev. B* **75**: 085307.

Wu, C., B.A. Bernevig, and S.-C. Zhang, 2006. Helical liquid and the edge of quantum spin hall systems. *Phys. Rev. Lett.* **96**: 106401.

Xu, C. and J.E. Moore, 2006. Stability of the quantum spin hall effect: Effects of interactions, disorder, and Z_2 topology. *Phys. Rev. B* **73**: 045322.

Spin Waves in Ferromagnetic Nanowires and Nanotubes

Hock Siah Lim
National University of Singapore

Meng Hau Kuok
National University of Singapore

This chapter begins with a brief overview of the historical development of the theory of spin waves in magnetic nanostructures. State-of-the-art calculations for dipole–exchange spin waves in a ferromagnetic nanowire and a hollow ferromagnetic nanowire, both of cylindrical cross-sections, are presented. Additionally, the treatment of collective spin-wave modes in ordered or disordered nanowire arrays, within the multiple scattering framework, is discussed. Here we adopt a tutorial style approach, interspersed with examples, to illustrate how the calculations are performed. In particular, we show how the frequencies of spin waves and the corresponding eigenmode profiles are determined. The aim is to provide a solid foundation in concepts and techniques that will enable the reader to make his own independent calculations.

31.1 Introduction

In 1930, Bloch introduced the concept of a magnon in order to account for the reduction of spontaneous magnetization in a ferromagnet as a function of temperature T. At $T = 0\,\mathrm{K}$, all atomic spins point along the same direction so that the total energy is at a minimum. As the temperature increases, the spins begin to deviate randomly at an increasing rate from the common direction, thereby reducing the spontaneous magnetization. At sufficiently low temperatures, the low-lying energy states of spin systems have only a few misaligned spins and can be treated as a gas of quasiparticles called magnons or quantized spin waves. Because of exchange interactions, the disturbance arising from spin misalignments propagates through the spin system, as a collective motion of the atomic spins, as waves with discrete energies. Thus, spin waves can be viewed as the analog for magnetically ordered systems of lattice waves in solid systems. Based on this spin wave theory, Bloch derived the $T^{3/2}$ law for the

decrease of spontaneous magnetization. The quantitative theory of quantized spin waves was developed further by Holstein and Primakoff (1940) and Dyson (1956). For a comprehensive treatment of spin waves, the reader is referred to a book by Kittel (2004).

The theory of magnetostatic modes of an isotropic ferromagnetic slab, in the absence of exchange interactions, was first developed by Damon and Eshbach (1961). Within the long-wavelength approximation, a semi-classical continuum model (see, e.g., Cottam and Lockwood 1986) is applicable and the theory yields two types of spin waves: backward volume waves and nonreciprocal surface waves (commonly called "Damon–Eshbach" waves). As the mode frequency is much less than the corresponding electromagnetic frequency for a given wave number (and hence retardation effects can be ignored), these propagating modes are known as magnetostatic waves. When both dipolar and exchange interactions are important, the problem becomes more complicated. Kalinikos and Slavin (1986) have proposed a perturbation approach to treat this problem and have obtained analytical expressions for the dipole–exchange spin wave spectrum. Kalinikos et al. (1990) extended their theory to treat anisotropic ferromagnetic films, while Rado and Hicken (1988) and Hicken et al. (1995) considered surface as well as bulk anisotropies. Hillebrands (1990) investigated the magnetic field dependence of the spin mode frequencies for magnetic multilayers. The reader is referred to Demokritov and Tsymbal (1994) and Hillebrands (1999) for a review of spin waves in structured films.

In the 1990s, the quantization of spin waves was observed in Permalloy micron-sized stripes (Mathieu et al. 1998). The modes observed are identified as surface Damon–Eshbach waves, which are quantized due to lateral confinement. For a finite, nonellipsoidal, micron-sized magnetic thin film element, it was experimentally and theoretically demonstrated that the spin waves can exhibit strong spatial localization near the edge of the element due to the formation of a potential well for spin waves (Jorzick et al. 2002, Demokritov 2003). For a review of the theory and experiments on the propagation of linear and non-linear spin waves in magnetic films and arrays of micron-sized magnetic dots and wires, the reader is referred to a review article by Demokritov et al. (2001).

If two ferromagnetic stripes are brought sufficiently close together, a dynamic magnetic dipole field generated in the spatial region outside each stripe by the precession of the magnetization will couple with that generated by the other stripe, resulting in the formation of collective magnetostatic modes. This "cross-talk" between magnetic elements in densely packed arrays (e.g., in magnetic data storage) can be especially important as it can limit device performance based on these arrays. Kostylev et al. (2004) and Bayer et al. (2006) developed a method based on the Green's function approach to treat collective magnetostatic modes in a one-dimensional (1D) array of ferromagnetic stripes. Gubbiotti et al. (2007) extended the theory in order to interpret the observed frequency dispersion of collective spin modes in dense arrays of magnetic stripes.

In 2001, Arias and Mills (2001, 2002) formulated an exact theory within the long-wavelength approximation to treat dipole-exchange spin waves in a ferromagnetic cylindrical nanowire, where the magnetization is parallel to the axis of the wire. An approximate analytical expression based on this theory was used to interpret the quantization of spin waves in isolated nickel nanowires observed by Brillouin light scattering (BLS) spectroscopy (Wang et al. 2002). Arias and Mills (2003) extended the theory for a single isolated cylindrical nanowire to disordered and ordered arrays of these nanowires based on a real-space multiple scattering approach. Nguyen and Cottam (2006) used a microscopic theory to study spin waves in a single ferromagnetic nanotube, and Das and Cottam (2007) investigated the magnetostatic modes in an antiferromagnetic nanotube.

Arias et al. (2005) have developed an analytical theory for dipole–exchange spin waves in a ferromagnetic sphere and the response of a uniformly magnetized sphere to a spatially inhomogeneous microwave field. The theory was subsequently extended to treat spin-wave collective modes in a two-dimensional (2D) square lattice of ferromagnetic nanospheres (Arias and Mills 2004, Chu et al. 2006) and to calculate the BLS cross-section for the various spin-wave modes of a single metallic nanosphere (Chu and Mills 2007).

Wang et al. (2006) performed Brillouin studies of the magnetic field dependence of collective spin waves in hexagonally ordered 2D arrays of vertically oriented $Fe_{48}Co_{52}$ nanowires. The arrays comprised wires with fixed diameters of 20 nm and wire spacing-radius ratios ranging from 3 to 5.5. The calculated frequencies of the collective spin modes as a function of interwire separation based on the multiple scattering theory (Arias and Mills 2003) showed excellent agreement with the experimental BLS data. The influence of neighboring wires in the arrays is manifested as a depression of the frequency of the lowest-energy

collective spin-wave mode relative to that of the isolated wire. This frequency depression becomes progressively more pronounced with decreasing interwire spacing, which shows that interwire dipolar coupling plays an important role in high-density 2D arrays of nanomagnets.

The calculation of the spin-wave modes of small magnetic particles, in general, is a complicated problem when both dipolar and exchange contributions are taken into consideration. A numerical approach may be necessary and it often requires the solving of the Landau–Lifshitz–Gilbert equation (see for example, Aharoni 1996)

$$\frac{\mathrm{d}\boldsymbol{M}(\boldsymbol{r},t)}{\mathrm{d}t} = -\gamma \boldsymbol{M}(\boldsymbol{r},t) \times \boldsymbol{H}_{\mathrm{eff}}(\boldsymbol{r},t) + \frac{\alpha}{M_s}\boldsymbol{M}(\boldsymbol{r},t) \times \frac{\mathrm{d}\boldsymbol{M}(\boldsymbol{r},t)}{\mathrm{d}t}, \quad (31.1)$$

where

$\boldsymbol{H}_{\mathrm{eff}}$ is the effective magnetic field acting on the magnetization $\boldsymbol{M}(\boldsymbol{r},t)$

M_s is the saturation magnetization

γ is the gyromagnetic ratio

α is the Gilbert damping parameter

The damping corresponds to the energy dissipated during the precessional motion of the magnetization and allows $\boldsymbol{M}(\boldsymbol{r},t)$ to turn towards the effective field until both vectors are parallel in the static solution.

The energy minimum of a magnetic system can be obtained using the freeware OOMMF, which is available from the National Institute of Standards and Technology. To find the dynamic spin-wave modes, the damping is next reduced substantially and the system is given a small perturbation. The system is allowed to evolve in time and the time evolution of the average magnetization of the particle is then tracked. The Fourier transform of the time-dependent magnetization subsequently gives the spin-wave frequencies (Grimsditch et al. 2004a). Using this numerical procedure, Wang et al. (2005) studied the spin dynamics of a high-aspect-ratio nickel nanoring (Wang et al. 2005) in an external magnetic field. Grimsditch et al. (2004b) developed an alternative and powerful method for the calculation of magnetic normal modes of nanometer-sized particles based on the "dynamical matrix" approach. This has the advantage that a single calculation can yield the frequencies and eigenvectors of all the modes.

As the size of magnetic particles decreases, it is important to have access to experimental tools that can probe magnetic properties on the submicron scale. In this regard, BLS spectroscopy (Patton 1984, Demokritov and Tsymbal 1994, Carlotti and Gubbiotti 1999) provides an ideal tool to probe the spin-wave dynamics of magnetic nanostructures (Himpsel et al. 1998, Martín et al. 2003), such as dots, dot arrays, wires, and arrays of wires (Demokritov and Hillebrands 1999, Skomski 2003, Liu et al. 2005).

In this chapter, we shall discuss the low-frequency magnetic excitations that are accessible by BLS. A quantum mechanical

TABLE 31.1 Unit Conversion Table

Quantity	Symbol	CGS	SI
Magnetic field	$\boldsymbol{H}$	Oe	$\dfrac{10^3}{4\pi}$ A/m
Flux density	$\boldsymbol{B}$	G	10^{-4} T
Magnetization	$\boldsymbol{M}$	emu/cc	10^3 A/m
Surface anisotropy	K_s	erg/cm^2	10^{-3} J/m^2
Exchange stiffness	A	erg/cm	10^{-5} J/m

formalism is necessary at low temperatures or for films that are only a few atomic-layer thick. However, as the wavelengths of these spin waves are generally much larger than interatomic distances, a continuum description of dipole–exchange spin waves will therefore be adequate. In addition, we shall limit our discussion to infinitely long cylindrical nanowires and hollow cylindrical nanowires as well as arrays of these structures. Also presented are several original works that have not been previously published.

All equations in this chapter are expressed in the centimeter, gram, second (cgs) system of units in which the fundamental magnetic equation is $\boldsymbol{B} = \boldsymbol{H} + 4\pi\boldsymbol{M}$, where the flux density $\boldsymbol{B}$ is in gauss (G), the magnetic field strength $\boldsymbol{H}$ is in oersted (Oe), and the magnetization $\boldsymbol{M}$ is in emu/cc. Table 31.1 lists conversion factors for commonly used magnetic quantities. For further information on the various systems of units used in magnetism, the reader is referred to an article by Scholten (1995).

31.2 Isolated Cylindrical Nanowire

Let us consider an infinitely long ferromagnetic cylindrical wire of radius R, with saturation magnetization M_s, and an exchange stiffness constant A. The easy axis of magnetization and the externally applied magnetic field H_0 are aligned parallel to the axis of the wire. We shall assume that M_s is uniform throughout the wire and also that spin waves propagate without attenuation. Thus, ignoring the damping term, the Landau–Lifshitz equation of motion (see Equation 31.1) can then be written as

$$\frac{d\boldsymbol{M}(\boldsymbol{r},t)}{dt} = -\gamma\boldsymbol{M}(\boldsymbol{r},t) \times \boldsymbol{H}_{\text{eff}}(\boldsymbol{r},t). \tag{31.2}$$

By neglecting the magnetic volume anisotropy field, we can write $\boldsymbol{H}_{\text{eff}}$ as

$$\boldsymbol{H}_{\text{eff}}(\boldsymbol{r},t) = H_0\boldsymbol{e}_z + \boldsymbol{h}(\boldsymbol{r},t) + \frac{2A}{M_s^2}\nabla^2\boldsymbol{M}(\boldsymbol{r},t), \tag{31.3}$$

where
- $\boldsymbol{e}_z$ is a unit vector along the $\boldsymbol{z}$ direction (or symmetry axis of the wire)
- $\boldsymbol{h}(\boldsymbol{r},t)$ is the dynamic dipolar field, and the last term describes the exchange field

Assuming that the fluctuations in $\boldsymbol{M}$ associated with the spin waves are small compared with the static values, we may write $\boldsymbol{M}$ as

$$\boldsymbol{M}(\boldsymbol{r},t) = M_s\boldsymbol{e}_z + \boldsymbol{m}(\boldsymbol{r},t), \tag{31.4}$$

where $\boldsymbol{m}(\boldsymbol{r},t)$ is the dynamic component of the magnetization, and $|\boldsymbol{m}| = M_s$. We shall consider the time-harmonic solution, so that the time-dependence part of the spin wave is of the form $e^{-i\Omega t}$. This allows us to write $\boldsymbol{m}(\boldsymbol{r},t)$ and $\boldsymbol{h}(\boldsymbol{r},t)$ as a product of a spatial and a time-dependent part, namely,

$$\boldsymbol{m}(\boldsymbol{r},t) = \boldsymbol{m}(\boldsymbol{r})e^{-i\omega t}, \tag{31.5}$$

$$\boldsymbol{h}(\boldsymbol{r},t) = \boldsymbol{h}(\boldsymbol{r})e^{-i\omega t}. \tag{31.6}$$

Now, the magnetic dipolar field $\boldsymbol{h}(\boldsymbol{r})$ must satisfy the magnetostatic equations $\nabla \times \boldsymbol{h}(\boldsymbol{r}) = 0$ and $\nabla \cdot \boldsymbol{b} = \nabla \cdot [\boldsymbol{h}(\boldsymbol{r}) + 4\pi\boldsymbol{m}(\boldsymbol{r})] = 0$, where $\boldsymbol{b}$ is the magnetic field induction. Thus,

$$\boldsymbol{h}(\boldsymbol{r}) = -\nabla\Phi(\boldsymbol{r}), \tag{31.7}$$

where $\Phi(\boldsymbol{r})$ is a magnetostatic potential. Using Equations 31.3 through 31.7, we linearize Equation 31.2 by ignoring all the terms that are of quadratic and higher order in the components of $\boldsymbol{h}$ and $\boldsymbol{m}$. This leads to the following relations:

$$i\Omega m_x = (H_0 - D\nabla^2)m_y + M_s\frac{\partial\Phi}{\partial y}, \tag{31.8}$$

and

$$-i\Omega m_y = (H_0 - D\nabla^2)m_x + M_s\frac{\partial\Phi}{\partial x}, \tag{31.9}$$

where $\Omega = \Omega/\gamma$ and $D = 2A/M_s$. In this linear approximation, we have set $\boldsymbol{m}(\boldsymbol{r}) \cdot \boldsymbol{e}_z = 0$. In addition, within the material, the condition $\nabla \cdot \boldsymbol{b} = 0$ becomes

$$\nabla^2\Phi - 4\pi\left(\frac{\partial m_x}{\partial x} + \frac{\partial m_y}{\partial y}\right) = 0. \tag{31.10}$$

To derive a differential equation for Φ, we proceed as follows. We differentiate Equations 31.8 and 31.9 with respect to y and x to obtain

$$i\Omega\left(\frac{\partial m_x}{\partial y} - \frac{\partial m_y}{\partial x}\right) = (H_0 - D\nabla^2)\left(\frac{\partial m_x}{\partial x} + \frac{\partial m_y}{\partial y}\right) + M_s\nabla_\perp^2\Phi, \tag{31.11}$$

where $\nabla_\perp = (\partial^2/\partial x^2) + (\partial^2/\partial y^2)$. Applying the operator $(H_0 - D\nabla^2)$ to both sides of Equation 31.10, and using Equation 31.11, we arrive at

$$4\pi i\Omega\left(\frac{\partial m_x}{\partial y} - \frac{\partial m_y}{\partial x}\right) = (H_0 - D\nabla^2)\nabla^2\Phi + 4\pi M_s\nabla_\perp^2\Phi. \tag{31.12}$$

By differentiating Equations 31.8 and 31.9 with respect to x and y, and using Equation 31.12, we finally obtain the following homogeneous equation that the magnetic potential Φ must satisfy, namely,

$$\left[(D\nabla^2 - H_0)(D\nabla^2 - B_0) - \Omega^2\right]\nabla^2\Phi + 4\pi M_s(D\nabla^2 - H_0)\frac{\partial^2}{\partial z^2}\Phi = 0,$$

(31.13)

where $B_0 = H_0 + 4\pi M_s$. Arias and Mills (2001) derived Equation 31.13 using a different approach.

Using cylindrical polar coordinates, we seek solutions of Equation 31.13 for which the magnetic potential has the form

$$\Phi(\rho,\phi,z) = J_n(\kappa\rho)\exp(in\phi + ikz),$$

(31.14)

where

$J_n(\kappa\rho)$ is a Bessel function of order n

k is the wave vector of the mode in the direction parallel to the z-axis

Substitution of Equation 31.14 into Equation 31.13 yields

$$D^2(\kappa^2 + k^2)^3 + D(H_0 + B_0)(\kappa^2 + k^2)^2$$

$$+ (H_0 B_0 - \Omega^2 - 4\pi M_s Dk^2)(\kappa^2 + k^2) - 4\pi M_s H_0 k^2 = 0. \quad (31.15)$$

This equation is cubic in κ^2, so that for each choice of n and k, we have three linearly independent solutions of Equation 31.14. Thus, the magnetic potential within the cylinder can be written as

$$\Phi^{\text{wire}}(\rho,\phi,z) = \sum_{i=1}^{3} A_i J_n(\kappa_i\rho)e^{i(n\phi + kz)}.$$

(31.16)

Outside the cylindrical wire, the potential satisfies the Laplace equation, so that

$$\Phi^{\text{out}}(\rho,\phi,z) = CK_n(k\rho)e^{i(n\phi + kz)},$$

(31.17)

where

$K_n(k\rho)$ are the modified Bessel functions of the second kind

C is an arbitrary constant

To determine the four coefficients A_1, A_2, A_3, and C, we note that $m(r)$ and $h(r)$ have to fulfill the boundary conditions at the surface of the wire. Two of these boundary conditions are the continuity of Φ, namely,

$$\sum_{i=1}^{3} A_i J_n(\kappa_i R) = CK_n(kR),$$

(31.18)

and the continuity of the normal component of the magnetic field induction

$$-\frac{\partial \Phi^{\text{wire}}}{\partial \rho} + 4\pi m_\rho \bigg|_{\rho = R} = -\frac{\partial \Phi^{\text{out}}}{\partial \rho}.$$

(31.19)

Additionally, if we assume that the surface anisotropy contributes energy $-K_s(m_\rho/M_s)^2$, where K_s is in units of erg/cm^2, then the exchange boundary conditions require that $m(r)$ at the surface of the wire satisfies (Arias and Mills 2001)

$$\left(\frac{\partial m_\phi}{\partial \rho}\right)_{\rho = R} = 0,$$

(31.20)

and

$$\left(\frac{\partial m_\rho}{\partial \rho}\right)_{\rho = R} - b_{\text{pin}}m_\rho \big|_{\rho = R} = 0,$$

(31.21)

where $b_{\text{pin}} = 2K_s/AM_s$ is the pinning parameter. When $b_{\text{pin}} = 0$, there is no surface spin pinning. When b_{pin} is large, the spins are fully pinned.

To evaluate these boundary conditions, we need explicit expressions for the radial and azimuthal components of the dynamical magnetization. They are obtained as follows. From Equations 31.8 and 31.9, we have

$$(\Omega + H_0 - D\nabla^2)m_+ = M_s h_+,$$

(31.22)

where $m_+ = m_x + im_y$ and $h_+ = h_x + ih_y$. Let us assume that

$$m_+ = \sum_{i=1}^{3} Q_i J_{n+1}(\kappa_i\rho)e^{i(n\phi + kz)},$$

where Q_i are constants to be determined. Then

$$\nabla^2 m_+ = -\sum_{i=1}^{3} Q_i(\kappa_i^2 + k^2)J_{n+1}(\kappa_i\rho)e^{i((n+1)\phi + kz)},$$

so that

$$(\Omega + H_0 - D\nabla^2)m_+ = \sum_{i=1}^{3} Q_i\left[\Omega + H_0 + D(\kappa_i^2 + k^2)\right]J_{n+1}$$

$$(\kappa_i\rho)e^{i((n+1)\phi + kz)}.$$

(31.23)

Since

$$h_+ = -e^{i\phi}\left(\frac{\partial}{\partial \rho} + \frac{i}{\rho}\frac{\partial}{\partial \phi}\right)\Phi^{\text{wire}}$$

$$= \sum_{i=1}^{3} A_i\kappa_i e^{i((n+1)\phi + kz)}J_{n+1}(\kappa_i\rho),$$

(31.24)

substituting Equations 31.23 and 31.24 into Equation 31.22 allows us to determine Q_i, so that we can write

$$m_+ = M_s \sum_{i=1}^{3} \frac{A_i \kappa_i}{\Omega + H_0 + D(\kappa_i^2 + k^2)} J_{n+1}(\kappa_i \rho). \quad (31.25)$$

A similar procedure yields

$$m_- = -M_s \sum_{i=1}^{3} \frac{A_i \kappa_i}{-\Omega + H_0 + D(\kappa_i^2 + k^2)} J_{n-1}(\kappa_i \rho), \quad (31.26)$$

where $m_- = m_x - i m_y$.

Finally, the radial and azimuthal components of the dynamical magnetization are, respectively, given by

$$m_\rho(\rho, \phi, z) = \frac{1}{2} \left[e^{-i\phi} m_+ + e^{i\phi} m_- \right]$$

$$= \frac{1}{2} M_s \sum_{i=1}^{3} A_i \kappa_i \left[\frac{J_{n+1}(\kappa_i \rho)}{D(\kappa_i^2 + k^2) + H_0 + \Omega} \right.$$

$$\left. - \frac{J_{n-1}(\kappa_i \rho)}{D(\kappa_i^2 + k^2) + H_0 - \Omega} \right] e^{i(n\phi + kz)}$$

$$(31.27)$$

and

$$m_\phi(\rho, \phi, z) = \frac{i}{2} \left[-e^{-i\phi} m_+ + e^{i\phi} m_- \right]$$

$$= -\frac{i}{2} M_s \sum_{i=1}^{3} A_i \kappa_i \left[\frac{J_{n+1}(\kappa_i \rho)}{D(\kappa_i^2 + k^2) + H_0 + \Omega} \right.$$

$$\left. + \frac{J_{n-1}(\kappa_i \rho)}{D(\kappa_i^2 + k^2) + H_0 - \Omega} \right] e^{i(n\phi + kz)}.$$

$$(31.28)$$

These expressions allow us to write the boundary conditions as a set of four linear homogeneous equations, and the condition for the existence of nontrivial solutions can be written as

$$\det M(\Omega, \kappa_i) = 0, \quad (31.29)$$

where $M(\Omega, \kappa_i)$ is a 4×4-matrix. It should be noted that the determinant, called the boundary-value determinant, is a complex function (with real and imaginary parts) of two unknowns. In general, the algebraic complexity makes it impractical to derive an analytical expression for the spin wave frequency by solving Equations 31.15 and 31.29 simultaneously.

A better approach is to solve the problem numerically by developing a computer code based on the following procedure. As the determinant is complex, we search for simultaneous zero crossings of the real and imaginary parts of the determinant. This can be accomplished if we consider the *magnitude* of the determinant. Thus, for a given trial frequency Ω, the three linearly independent κ are calculated from Equation 31.15. Both Ω and κ are then substituted into the boundary-condition

determinant to determine if its *magnitude* "vanishes" within a preset accuracy. If it vanishes, a solution has been found; if not, a new trial frequency is attempted and the process repeated. Once the correct value of the frequency has been determined, the corresponding eigenvector (i.e., the four coefficients) can also be found, hence the corresponding mode profiles can be obtained. Thus, the spatial distribution of the dynamic potential can be calculated from Equations 31.16 and 31.17. The spatial profiles of the dynamical magnetization can be similarly obtained from Equations 31.27 and 31.28.

For illustration purposes, we shall assume that the magnetic material is Permalloy ($Ni_{80}Fe_{20}$) and use the following parameters in our calculations: saturation magnetization $M_s = 800 \times 10^3$ A/m, exchange constant $A = 10^{-11}$ J/m, and gyromagnetic ratio $\gamma = 190$ GHz/T. As Permalloy also exhibits negligible anisotropy, we may set the surface anisotropy energy $K_s = 0$, and hence $b_{pin} = 0$. Additionally, we shall also assume that this Permalloy material is amorphous or polycrystalline so that the uniaxial anisotropy constant $K_1 = 0$. However, for accurate studies, it is important to assign appropriate values to these parameters. They can be normally obtained from a fit to the experimental data, for instance, the dispersion relations of both Damon–Esbach spin waves and the standing bulk spin-wave modes of a reference thin Permalloy film.

We shall consider the case where the azimuthal quantum number is $n = 1$. In Figure 31.1, we show the magnitude of the determinant as a function of trial frequencies for a nanowire with a radius of 20 nm at an applied field of 0.3 T and $k = 10^7$ m^{-1}. Within the considered trial frequency range, four frequencies at 23.74, 49.57, 116.33, and 219.42 GHz satisfy the boundary conditions. The corresponding mode profile for the lowest eigenfrequency,

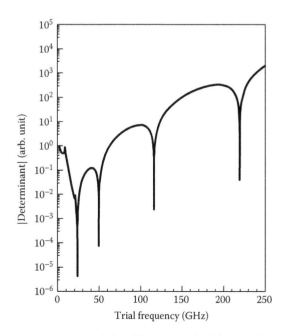

FIGURE 31.1 Computed plot of the magnitude of the boundary-value determinant versus trial frequency.

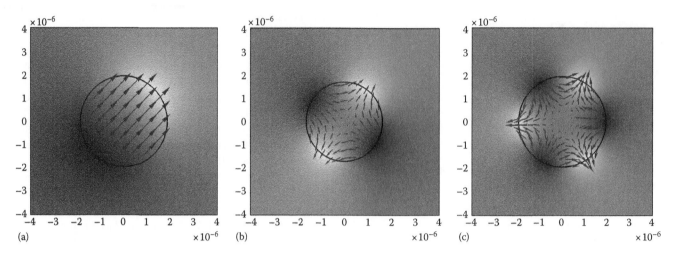

FIGURE 31.2 Mode profiles of an isolated Permalloy nanowire of radius 20 nm for an azimuthal quantum number (a) $n = 1$, (b) $n = 2$, and (c) $n = 3$. The arrows indicate the orientation of the dynamic magnetization $\boldsymbol{m}$ and the gray scale depicts the variation of the magnetic potential. The unit of the spatial dimension is centimeters.

23.74 GHz, is presented in Figure 31.2a. Similar calculated mode profiles for $n = 2$ and 3 of the lowest eigenfrequencies are also shown in Figure 31.2. These modes, as expected, exhibit the dipolar, quadrupolar, and sextupolar characteristics for $n = 1$, 2, and 3, respectively. The arrows indicate the orientation of the dynamical magnetization over the cross-section of the nanowire while the gray-scale depicts the variation of the magnetic scalar potential within and outside the nanowire.

The magnetic field dependence of the spin-wave frequencies for $n = 1$, 2, and 3 is shown in Figure 31.3, which reveals that the frequencies increase linearly with the magnetic field, with a slope given by $\gamma/2\pi$. This linear dependence can be understood as follows. While the full theory requires solving Equation 31.15 together with the required boundary conditions, an approximate analytical expression can be obtained that is applicable to

either small or large pinning. We note that Equation 31.15 can be factorized to give

$$(\kappa^2 + k^2)\left[D^2(\kappa^2 + k^2)^2 + D(H_0 + B_0)(\kappa^2 + k^2) \right.$$
$$\left. + (H_0 B_0 - \Omega^2 - 4\pi M_s D k^2) - \frac{4\pi M_s H_0 k^2}{\kappa^2 + k^2} \right] = 0. \quad (31.30)$$

The bulk standing modes of the nanowires correspond to real values of κ. If we make the assumption that $\kappa \gg k$, the spin-wave frequencies are then given by

$$f = \frac{\gamma}{2\pi}\left[H_0(H_0 + 4\pi M_s) + D\kappa^2(D\kappa^2 + 2H_0 + 4\pi M_s) \right]^{1/2}$$
$$= \frac{\gamma}{2\pi}\left[(H_0 + 2\pi M_s + D\kappa^2)^2 - (2\pi M_s)^2 \right]^{1/2}$$
$$= \frac{\gamma}{2\pi}\left[(H_0 + D\kappa^2)(H_0 + D\kappa^2 + 4\pi M_s) \right]^{1/2}, \quad (31.31)$$

where f is the frequency, in Hz, of the mode. This equation is similar to the following Herring–Kittel formula (Herring and Kittel 1951) for the dipole–exchange spin-wave spectrum in an infinite ferromagnetic medium

$$f = \frac{\gamma}{2\pi}\left[(H_0 + Dq^2)(H_0 + Dq^2 + 4\pi M_s \sin^2\theta_q) \right]^{1/2},$$

where θ_q is the angle between the direction of the wave vector $\boldsymbol{q}$ and that of the magnetization. Thus, Equation 31.31 represents the propagation of a spin wave, on the xy plane of the nanowire, with a wave vector κ whose values are quantized by the finite radius of the nanowire.

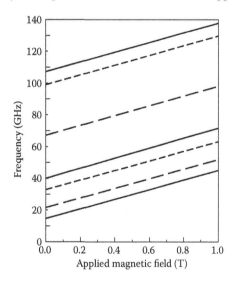

FIGURE 31.3 Magnetic field dependence of the spin wave frequencies for an isolated Permalloy nanowire of radius 20 nm and $k = 10^7$ nm^{-1}. Solid, long-dashed and short-dashed lines denote $n = 1$, 2 and 3, respectively.

For the applied field $H_0 = 0$, Equation 31.31 reduces to

$$f = \frac{\gamma}{2\pi} \left\{ D\kappa^2 [D\kappa^2 + 4\pi M_s] \right\}^{1/2}. \tag{31.32}$$

If the pinning is small, the radial function $J_n(\kappa\rho)$ will have antinodes (Cottam and Lockwood 1986, Wang et al. 2002) at $\rho = R$, such that the first three values of κ are $1.84/R$, $3.05/R$, and $4.20/R$ for $n = 1$, 2, and 3, respectively. It is obvious from Equation 31.31 that for H_0 to be sufficiently large, the frequency varies linearly with the field, with a slope of $\gamma/2\pi$.

31.3 Collective Spin-Wave Modes in an Array of Nanowires

If two infinitely long parallel ferromagnetic nanowires are brought close to each other, the dipolar fields generated, due to the spin precession within each nanowire, will couple so that collective spin-wave modes are formed. These modes are different from those of an isolated nanowire and cannot be described using a single azimuthal quantum number. In the following, we shall calculate the collective spin-wave modes of the array following the method described by Arias and Mills (2003). We employ the multiple scattering theory to account for the interwire interactions.

We now consider an array of infinitely long cylindrical wires, each of radius R and of saturation magnetization M_s, and with their symmetry axes parallel to the z axis. Adjacent wires are sufficiently far apart so that interwire exchange interactions can be neglected and the wires only interact through dipolar coupling. As the magnetization precesses in one wire, a dynamic magnetic dipole field h is generated outside this wire. Firstly, we need to derive an expression for the response of a single nanowire to an external driving field. Consider the wire to be at the origin of a coordinate system, with the wire axis parallel to the z axis. Then, this field in the xy plane, which must satisfy the Laplace equation in the spatial region outside the wire, can be described by a potential Φ of the form

$$\Phi(\rho, \phi) = \sum_{n=-\infty}^{\infty} A_n(0) I_n(k\rho) \exp(in\phi), \tag{31.33}$$

where $I_n(k\rho)$ is the modified Bessel function of the first kind. This external potential will excite the magnetization of the nanowire at the origin, resulting in the precession of the magnetization, which in turn will generate a dynamic dipole field. From Equation 31.17, the field generated outside the nanowire can be written as

$$\Phi^{(0)}(\rho, \phi) = \sum_{n=-\infty}^{\infty} B_n(0) K_n(k\rho) \exp(in\phi) \tag{31.34}$$

$$\Phi^{(0)}(\rho, \phi) = \sum_{n=-\infty}^{\infty} A_n(0) S_n K_n(k\rho) \exp(in\phi), \tag{31.35}$$

where we have defined $B_n(0) = A_n(0) S_n$ and S_n, in an analogy with the T-matrix of the scattering theory, describes the response of the nanowire to a driving field. S_n relates the incident amplitude of the driving field to the amplitude of the scattered field of a single nanowire. S_n can be determined as follows.

The potential outside the nanowire, consisting of the incident and scattered fields, can be written as

$$\Phi^{\text{total}} = \sum_{n=-\infty}^{\infty} \left[A_n(0) I_n(k\rho) + B_n(0) K_n(k\rho) \right] e^{in\phi}, \tag{31.36}$$

while the potential within the nanowire (summing over the azimuthal quantum number n) is

$$\Phi^{\text{wire}} = \sum_{n=-\infty}^{\infty} \sum_{i=1}^{3} A_i^n J_n(\kappa_i \rho) e^{in\phi}. \tag{31.37}$$

Continuity of the magnetic potential across the surface of the nanowire therefore gives

$$A_n(0) I_n(kR) + B_n(0) K_n(kR) = -\Gamma_n, \tag{31.38}$$

where

$$\Gamma_n = -\sum_{i=1}^{3} A_i^n J_n(\kappa_i R). \tag{31.39}$$

In addition, continuity of the normal component of the magnetic field induction yields

$$-\frac{\partial \Phi^{\text{total}}}{\partial \rho} = -\frac{\partial \Phi^{\text{wire}}}{\partial \rho} + 4\pi m_\rho. \tag{31.40}$$

By substituting Equations 31.27, 31.36, and 31.37 into Equation 31.40, we obtain

$$k A_n(0) I_n'(kR) + k B_n(0) K_n'(kR) = -G_n, \tag{31.41}$$

where the prime on the various Bessel functions indicates derivatives with respect to their respective arguments and

$$G_n = \sum_{i=1}^{3} \kappa_i A_i^n \left[2\pi M_s \left\{ \frac{J_{n+1}(\kappa_i R)}{D(\kappa_i^2 + k^2) + H_0 + \Omega} \right. \right.$$
$$\left. \left. - \frac{J_{n-1}(\kappa_i R)}{D(\kappa_i^2 + k^2) + H_0 - \Omega} \right\} - J_n'(\kappa_i R) \right]. \tag{31.42}$$

From Equations 31.38 and 31.41, we finally obtain the expression for S_n:

$$S_n = \frac{B_n(0)}{A_n(0)} = -\frac{G_n I_n(kR) - k\Gamma_n I_n'(kR)}{G_n K_n(kR) - k\Gamma_n K_n'(kR)}. \tag{31.43}$$

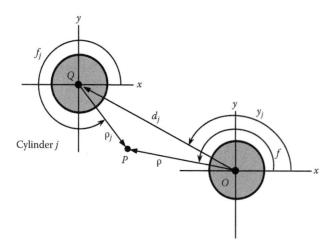

FIGURE 31.4 Coordinate system and quantities used in the multiple scattering theory.

The coefficients A_i^n appearing in the expressions for G_n and Γ_n are determined as follows. For a given n, we first set $A_i^n = 1$, and determine the two remaining coefficients by requiring that Equations 31.20 and 31.21 be satisfied. These two inhomogeneous equations will then yield the values of A_2^n and A_3^n.

We are now in a position to consider the collective spin waves in an array of nanowires. In Figure 31.4, we have a nanowire at the origin of a coordinate system and one of the nanowires in the array, wire j. The magnetization of the wire at the origin is driven by the field generated by all the other wires in the array, and can be calculated from the following magnetic potential

$$\tilde{\Phi}_M = \sum_{j \neq 0} \sum_{n=-\infty}^{\infty} B_n(j) K_n(k\rho_j) \exp(in\phi_j), \qquad (31.44)$$

where the quantities ρ_j and ϕ_j are defined in Figure 31.4.

Now, we want to express the incident wave, arising from the scattered waves of all the other nanowires, in the local cylindrical eigenfunction expansion. This can be accomplished by using the Graf addition theorem for Bessel functions given in the article by Abramowitz and Stegun (1964, Eq. 9.1.79). That is, this theorem allows us to relate the dynamic dipolar fields, of all but the nanowire at the origin, to the field expanded around this nanowire. By applying Graf's theorem to the triangle OPQ of Figure 31.4, we obtain

$$e^{i\nu\varsigma} K_\nu(k\rho_j) = \sum_{n=-\infty}^{\infty} K_{\nu+n}(kd_j) I_n(k\rho) e^{in\phi},$$

which is valid provided that $\rho < d_j$. ϕ is the angle between vectors $\mathbf{r}$ and $\mathbf{d}_j$ and ς is the angle opposite to $\mathbf{r}$. Now $\varsigma = \psi_j - (\Phi_j - \pi)$, so that

$$e^{-i\nu\phi_j} K_\nu(k\rho_j) = e^{-i\nu\psi_j - i\nu\pi} \sum_{n=-\infty}^{\infty} e^{-in\psi_j} K_{\nu+n}(kd_j) I_n(k\rho) e^{in\phi}.$$

By replacing ν by $-\nu$ in the above equation and noting that $K_{-\nu}(x) = K_\nu(x)$, we obtain

$$e^{i\nu\phi_j} K_\nu(k\rho_j) = (-1)^\nu \sum_{n=-\infty}^{\infty} e^{i\psi_j(\nu-n)} K_{\nu-n}(kd_j) I_n(k\rho) e^{in\phi}. \qquad (31.45)$$

When Equation 31.45 is combined with Equation 31.44, we have

$$\tilde{\Phi}_M = \sum_{n=-\infty}^{\infty} \tilde{A}_n(0) I_n(k\rho) e^{in\phi}, \qquad (31.46)$$

where

$$\tilde{A}_n(0) = \sum_{j \neq 0} \sum_{m=-\infty}^{\infty} (-1)^m e^{i(m-n)\psi_j} B_m(j) K_{m-n}(kd_j).$$

This driving potential $\tilde{\Phi}_M$ excites the magnetization in the nanowire at the origin, and using S_n, we can relate $B_n(0)$ to $\tilde{A}_n(0)$, i.e.,

$$B_n(0) = S_n \tilde{A}_n(0).$$

Thus, we have the result

$$S_n^{-1} B_n(0) = \sum_{j \neq 0} \sum_{m=-\infty}^{\infty} (-1)^m e^{i(m-n)\psi_j} K_{m-n}(kd_j) B_m(j). \qquad (31.47)$$

For clarity, and the fact that the method can easily incorporate variations in radii and magnetic properties from nanowire to nanowire, we rewrite Equation 31.47 as

$$S_n^{-1}(p) B_n(p) = \sum_{j \neq p} \sum_{m=-\infty}^{\infty} (-1)^m e^{i(m-n)\psi_j^p} K_{m-n}(kd_j^p) B_m(j), \qquad (31.48)$$

where $p = 1, 2, \ldots, L$, L is the total number of nanowires in the array, $S_n^{-1}(p)$ describes the scattering property of the nanowire p, and ψ_j^p and d_j^p are determined with respect to nanowire p. Thus, variations in the physical properties of the nanowires can be taken into account simply by using the appropriate expressions for S_n for each nanowire.

For numerical calculations, the infinite sums in Equation 31.48 have to be truncated. In practice, N is chosen to be sufficiently large so that a finite sum, taken from $-N$ to N, gives converged results. The frequencies of the various modes are determined by finding the zeros of the determinant of the coefficients formed from Equation 31.48. For a periodic array, the Bloch theorem requires that the coefficients $B_n(j)$ of nanowire j are related to those of the nanowire at the origin by $B_n(j) = B_n(0) \exp(i\mathbf{k}_\perp \cdot \mathbf{d}_j)$, where $\mathbf{k}_\perp$ is the Bloch wave vector in the xy plane, i.e., perpendicular to wires' axes. When an eigenfrequency has been calculated from Equation 31.48, the corresponding B_n coefficients of all the nanowires in the array can also

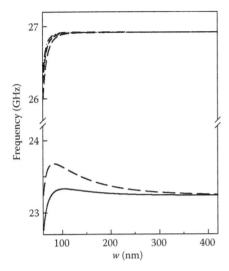

FIGURE 31.5 Frequencies of collective spin wave modes of an array of four parallel identical nanowires of radius 30 nm, placed at the corners of a square of side length w at $H_0 = 0.3$ T.

be evaluated. Hence, the magnetic potential in the spatial region outside the nanowires can be computed using Equation 31.36 with $A_n(0)$ replaced by $\tilde{A}_n(0)$, while inside the nanowire located at the origin, the potential is given by Equation 31.37. In addition, the dynamical magnetization m of this nanowire at the origin can be determined from Equations 31.27 and 31.28, and summing over n. The potential and dynamical magnetization inside the other nanowires in the array can also be similarly calculated.

As an illustration, we consider an array of four identical nanowires with a radius of 30 nm in an external field of 0.3 T and $k = 10^7$ m^{-1}. The wires are located at the corners of a square of side w, and the frequencies of the collective spin-wave modes of the array are calculated as a function of w. In Figure 31.5, we show the first few lowest eigenfrequencies. We note that, at

sufficiently large interwire separations, the eigenfrequencies of the lower branch correspond to the $n = 1$ mode of the dipolar character of an isolated nanowire. The higher branch, on the other hand, corresponds to the $n = 2$ mode of the quadrupolar character. As the interwire separation decreases, the branches begin to split, with the lower branch exhibiting the largest splitting. Additionally, the lower branch shows significant splitting over a wider range of separations, in comparison with the higher branch. This can be understood as follows. In the multiple scattering theory, modified Bessel functions $K_n(kd_j)$ appear in Equation 31.48. Because $K_n(x)$ diverges as x^{-n} as $x \to 0$, it is clear that the interaction associated with the $n = 1$ dipolar mode is comparatively longer in range, and the splitting, thus, decreases slowly with increasing interwire separation, compared to that of the $n = 2$ branch. Figure 31.6 displays the mode profiles of the lowest-frequency eigenmodes of the respective branches.

31.4 Hollow Cylindrical Nanowires

31.4.1 Isolated Hollow Cylindrical Nanowires

We can easily generalize the formalism outlined in Section 31.2 to the case of a *hollow* cylindrical nanowire. Let us consider an infinitely long hollow cylinder with an outer radius of R_o and an inner radius of R_i. The linearized equations, Equations 31.8 through 31.10, remain the same but we need to replace Equation 31.14 with

$$\Phi(\rho,\phi,z) = [AJ_n(\kappa\rho) + BY_n(\kappa\rho)]\exp(in\phi + ikz), \quad (31.49)$$

where

$Y_n(\kappa\rho)$ is a Bessel function of the second kind
A and B are arbitrary constants

For a solid cylinder, it is obvious that B is zero as $Y_n(\kappa\rho)$ diverges at the origin. Thus, the magnetic potential within the annulus of the hollow cylinder, i.e., for $R_i \le \rho \le R_o$, is

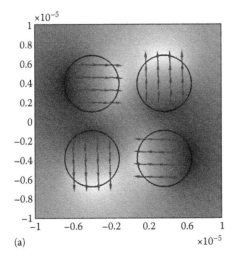

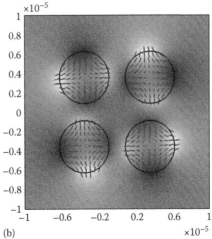

FIGURE 31.6 Mode profiles of an array of four identical nanowires of Figure 31.5. The arrows indicate the orientation of m and the gray scale depicts the variation of the magnetic potential for interwire separation w of 80 nm. Profiles of the lowest-frequency mode of the (a) $n = 1$ and (b) $n = 2$ branches. The unit of the spatial dimension is centimeters.

$$\Phi^{\text{hollow wire}}(\rho,\phi,z) = \sum_{i=1}^{3}\left[A_i J_n(\kappa_i\rho) + B_i Y_n(\kappa_i\rho)\right]e^{i(n\phi+kz)}. \quad (31.50)$$

Outside the wire and within its bore, the potential satisfies the Laplace equation, so that Equation 31.17 is replaced by

$$\Phi(\rho,\phi,z) = \begin{cases} C_1 I_n(k\rho)e^{i(n\phi+kz)} & \text{for } \rho < R_i, \\ C_2 K_n(k\rho)e^{i(n\phi+kz)} & \text{for } \rho > R_o. \end{cases} \quad (31.51)$$

Similarly, it can be shown that the respective expressions for the radial and azimuthal components of the dynamical magnetization are given by

$$m_\rho(\rho,\phi,z) = \frac{1}{2}M_s\sum_{i=1}^{3}\kappa_i\left\{A_i\left[\frac{J_{n+1}(\kappa_i\rho)}{D(\kappa_i^2+k^2)+H_0+\Omega}\right.\right.$$
$$\left.-\frac{J_{n-1}(\kappa_i\rho)}{D(\kappa_i^2+k^2)+H_0-\Omega}\right]$$
$$+B_i\left[\frac{Y_{n+1}(\kappa_i\rho)}{D(\kappa_i^2+k^2)+H_0+\Omega}\right.$$
$$\left.\left.-\frac{Y_{n-1}(\kappa_i\rho)}{D(\kappa_i^2+k^2)+H_0-\Omega}\right]\right\}e^{i(n\phi+kz)},$$
$$(31.52)$$

and

$$m_\phi(\rho,\phi,z) = -\frac{i}{2}M_s\sum_{i=1}^{3}\kappa_i\left\{A_i\left[\frac{J_{n+1}(\kappa_i\rho)}{D(\kappa_i^2+k^2)+H_0+\Omega}\right.\right.$$
$$\left.+\frac{J_{n-1}(\kappa_i\rho)}{D(\kappa_i^2+k^2)+H_0-\Omega}\right]$$
$$+B_i\left[\frac{Y_{n+1}(\kappa_i\rho)}{D(\kappa_i^2+k^2)+H_0+\Omega}\right.$$
$$\left.\left.+\frac{Y_{n-1}(\kappa_i\rho)}{D(\kappa_i^2+k^2)+H_0-\Omega}\right]\right\}e^{i(n\phi+kz)}.$$
$$(31.53)$$

The eight coefficients, A_1, A_2, A_3, B_1, B_2, B_3, C_1, and C_2, are determined by requiring that $m(r)$ and $h(r)$ satisfy the boundary conditions at the inner and outer surfaces of the hollow nanowire. Four of the boundary conditions are the continuity of Φ and the continuity of the normal component of the magnetic field induction at both these surfaces. The remaining four conditions require $m(r)$ at the inner and outer surfaces to satisfy Equations 31.20 and 31.21. The pinning parameters at these surfaces may differ, however. These boundary conditions lead to a set of eight linear homogeneous equations, and the condition for the existence of nontrivial solutions can be written as det $M(\Omega,\kappa) = 0$, where $M(\Omega,\kappa)$ is now a 8×8 matrix. The numerical procedure to solve for the eigenfrequencies of

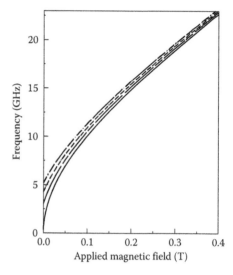

FIGURE 31.7 Magnetic field dependence of eigenfrequencies of spin wave modes in a ferromagnetic tube with inner radius 500 nm and thickness 10 nm.

the spin-wave modes is similar to that detailed in Section 31.2, and the corresponding mode profiles can be obtained when the eight coefficients are known.

As an illustration, we consider a tube with an inner radius of 500 nm with a thickness of 10 nm and set $k = 0$. In Figure 31.7, we show the first few calculated eigenfrequencies as a function of the external field, which is applied along the axis of the tube. The lowest-frequency eigenmode is nondegenerate, while higher-frequency modes appear to be two-fold degenerate in the plot as the difference in frequencies is less than 0.03 GHz. The two closely spaced eigenfrequencies correspond to spin waves propagating on the outer and inner surfaces of the nanotube. With a fixed thickness, and as $R_i \rightarrow \infty$, the two frequencies become identical.

For an understanding of the field dependence behavior of the eigenfrequencies, we may regard the cylindrical tube as locally equivalent to a planar film. This is reasonable as the thickness of the film is constant and small compared to the radius of the nanotube. According to Kalinikos and Slavin (1986), the frequencies of dipole–exchange Damon–Eshbach modes propagating on a ferromagnetic thin film are given by

$$f = \frac{\gamma}{2\pi M_s}\left[\left\{2Ak_p^2 + M_s(H_0 + 2\pi M_s k_p L)\right\}\right.$$
$$\left.\times\left\{2Ak_p^2 + M_s\left(H_0 - 2\pi M_s(-2 + k_p L)\right)\right\}\right]^{1/2}, \quad (31.54)$$

where

k_p is the magnitude of the wavevector perpendicular to the applied field

L is the film thickness

However, in the tube, the wavelength of the spin wave propagating around the circumference of the wire cannot assume any arbitrary value. Only an integral number of standing waves can fit into the circumference of the tube, and the following condition must therefore be satisfied

$$k_p = \frac{2\pi}{\lambda} = \frac{n}{R_0},$$

where $n = 0, 1, 2, \ldots$, is the azimuthal quantum number and $R_0 = 510\,\text{nm}$. If the discrete wavevector values are inserted into Equation 31.54, we reproduce exactly the same results shown in Figure 31.7. For $n = 0$, Equation 31.54 reduces to the well-known Kittel formula for the uniform precession mode in which all the moments precess together in phase with the same amplitude for an external field applied parallel to the surface of the thin film. Hence, we have identified the first few lowest eigenfrequencies as those of the uniform Kittel mode and the Damon–Eshbach surface spin-wave modes of the tube.

31.4.2 Array of Hollow Cylindrical Nanowires

The collective spin-wave modes in an array of hollow nanowires can also be calculated following the approach detailed in Section 31.3. The multiple scattering approach also yields Equation 31.48 with S_n (Equation 31.43) retaining the same form, except that Γ_n and G_n are now given respectively by

$$\Gamma_n = -\sum_{i=1}^{3}[A_i^n J_n(\kappa_i R_0) + B_i^n Y_n(\kappa_i R_0)], \qquad (31.55)$$

and

$$G_n = \sum_{i=1}^{3}\kappa_i A_i^n \left[2\pi M_s \left\{ \frac{J_{n+1}(\kappa_i R_0)}{D(\kappa_i^2 + k^2) + H_0 + \Omega} \right. \right.$$
$$\left. \left. - \frac{J_{n-1}(\kappa_i R_0)}{D(\kappa_i^2 + k^2) + H_0 - \Omega} \right\} - J_n'(\kappa_i R_0) \right]$$
$$+ \sum_{i=1}^{3}\kappa_i B_i^n \left[2\pi M_s \left\{ \frac{Y_{n+1}(\kappa_i R_0)}{D(\kappa_i^2 + k^2) + H_0 + \Omega} \right. \right.$$
$$\left. \left. - \frac{Y_{n-1}(\kappa_i R_0)}{D(\kappa_i^2 + k^2) + H_0 - \Omega} \right\} - Y_n'(\kappa_i R_0) \right].$$

The seven coefficients, A_1^n, A_2^n, A_3^n, B_1^n, B_2^n, B_3^n, and C_1, are obtained as follows. We set $A_1^n = 1$ and solve for the other six coefficients by requiring that six boundary conditions are satisfied: four from the exchange boundary conditions (at the inner and outer surfaces), and two from the continuity of both the magnetic potential and the normal component of the magnetic field induction at the *inner* surface of the nanowire. The frequencies of the collective spin-wave modes are then given by the solutions of Equation 31.48, and the corresponding mode profiles can be generated once the required coefficients are known.

As an illustration, we consider a 2D simple cubic array of parallel Permalloy nanotubes with an inner radius of 40 nm and a thickness of 5 nm. We confine our treatment to wave propagation

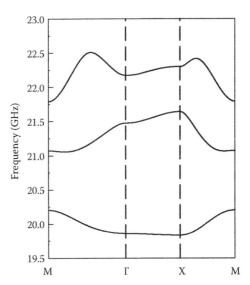

FIGURE 31.8 Dispersion curves of a 2D square array of Permalloy nanotubes, of inner radius 40 nm and thickness 5 nm, with a lattice constant of 94 nm for an applied field $H_0 = 0.3\,\text{T}$. The band structure is plotted in the three high-symmetry directions ΓXM of the first Brillouin zone.

perpendicular to the axis of the wires, and set $k = 0$, $H_0 = 0.3\,\text{T}$ and the interwire separation $a = 94\,\text{nm}$. Figure 31.8 shows the band structure of this square array of nanotubes, calculated along the path $\mathbf{M} = \pi/a(1,1) \rightarrow \mathbf{G} = \pi/a(0,0) \rightarrow \mathbf{X} = \pi/a(1,0) \rightarrow \mathbf{M} = \pi/a(1,1)$ in the irreducible part of the first Brillouin zone. A direct band gap of width 0.75 GHz exists between the first and second bands at the $\mathbf{M}$ point. The mode profile for the lowest-frequency collective spin excitation at the $\mathbf{X}$ point is presented in Figure 31.9.

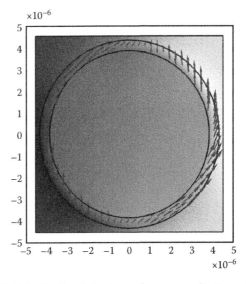

FIGURE 31.9 Profile of the lowest-frequency collective spin-wave mode at the high-symmetry $\mathbf{X}$ point. A unit cell of a 2D square lattice containing a nanotube is shown. The unit of the spatial dimension is centimeters.

Interestingly, the square array of ferromagnetic nanotubes may be regarded as a magnonic crystal, which is the magnetic analog of photonic crystals, with spin waves acting as information carriers. A magnonic crystal can have a complete band gap within which the propagation of spin waves is prohibited in any direction inside the crystal (Krawczyk and Puszkarski 2008). Recently, Wang et al. (2009) have observed frequency band gaps in a nanostructured magnonic crystal in the form of a 1D periodic array of alternating Permalloy and cobalt nanostripes. Dispersion relations of spin waves in the magnonic crystal have been mapped by Brillouin spectroscopy, and the results obtained accord with theoretical calculations.

31.5 Summary

In summary, we have discussed the dynamics of dipole–exchange spin waves in solid as well as hollow ferromagnetic nanowires of cylindrical cross-sections. By solving the linearized Landau–Lifshitz equation and the Maxwell equation, $\nabla \cdot \boldsymbol{b} = 0$ in a quasi-static approximation simultaneously, we obtain an important equation: Equation 31.15. In addition, by imposing the magnetostatic and the exchange boundary conditions at the surfaces of nanowires, we arrive at a system of homogeneous equations, whose determinant must vanish in order to yield the resonant frequencies of the spin-wave modes of an isolated nanowire.

In a densely packed array, provided the nanowires are sufficiently far apart that interwire exchange coupling is negligible, the dynamic dipolar coupling between nanowires is treated within the multiple scattering approach. The collective spin wave excitations can be calculated for any 2D array of infinitely long nanowires with different diameters and magnetic properties by using the expressions for the scattering matrix for each nanowire. Several examples have been provided to illustrate the various methods of calculations. We have also demonstrated, in a tutorial style approach, how frequencies of spin wave excitations and their profiles are computed.

References

Abramowitz, M. and Stegun, I., eds., 1964. *Handbook of Mathematical Functions*. Dover Inc., New York.

Aharoni, A. 1996. *Introduction to the Theory of Ferromagnetism*. Oxford University Press, New York.

Arias, R. and D. L. Mills. 2001. Theory of spin excitations and the microwave response of cylindrical ferromagnetic nanowires. *Physical Review B* **63**: 134439-1–134439-11.

Arias, R. and D. L. Mills. 2002. Erratum: Theory of spin excitations and the microwave response of cylindrical ferromagnetic nanowires. *Physical Review B* **66**: 149903-1.

Arias, R. and D. L. Mills. 2003. Theory of collective spin waves and microwave response of ferromagnetic nanowire arrays. *Physical Review B* **67**: 094423-1–094423-15.

Arias, R. and D. L. Mills. 2004. Theory of collective spin-wave modes of interacting ferromagnetic spheres. *Physical Review B* **70**: 104425-1–104425-7.

Arias, R., P. Chu, and D. L. Mills. 2005. Dipole exchange spin waves and microwave response of ferromagnetic spheres. *Physical Review B* **71**: 224410-1–224410-12.

Bayer, C., M. P. Kostylev, and B. Hillebrands. 2006. Spin-wave eigenmodes of an infinite thin film with periodically modulated exchange bias field. *Applied Physics Letters* **88**: 112504-1–112504-3.

Carlotti, G. and G. Gubbiotti. 1999. Brillouin scattering and magnetic excitations in layered structures. *Rivista Italiana del Nuovo Cimento* **22**: 1–60.

Chu, P. and D. L. Mills. 2007. Theory of Brillouin light scattering from ferromagnetic nanospheres. *Physical Review B* **73**: 054405-1–054405-8.

Chu, P., D. L. Mills, and R. Arias. 2006. Exchange/dipole collective spin-wave modes of ferromagnetic nanosphere arrays. *Physical Review B* **73**: 094405-1–094405-8.

Cottam, M. G. and D. J. Lockwood. 1986. *Light Scattering in Magnetic Solids*. Wiley, New York.

Damon, R. W. and J. R. Eshbach. 1961. Magnetostatic modes of a ferromagnet slab. *Journal of Physics and Chemistry of Solids* **19**: 308–320.

Das, T. K. and M. G. Cottam. 2007. Magnetostatic modes in nanometer-sized ferromagnetic and antiferromagnetic tubes. *Surface Review and Letters* **14**: 471–480.

Demokritov, S. O. 2003. Dynamic eigen-modes in magnetic stripes and dots. *Journal of Physics: Condensed Matter* **15**: S2575–S2598.

Demokritov, S. O. and E. Tsymbal. 1994. Light scattering from spin waves in thin films and layered systems. *Journal of Physics: Condensed Matter* **6**: 7145–7188.

Demokritov, S. O. and B. Hillebrands. 1999. Inelastic light scattering in magnetic dots and wires. *Journal of Magnetism and Magnetic Materials* **200**: 706–719.

Demokritov, S. O., B. Hillebrands, and A. N. Slavin. 2001. Brillouin light scattering studies of confined spin waves: Linear and nonlinear confinement. *Physics Reports* **348**: 441–489.

Dyson, F. J. 1956. General theory of spin-wave interactions. *Physical Review* **102**: 1217–1230.

Grimsditch, M., G. K. Leaf, H. G. Kaper, D. A. Karpeev, and R. E. Camley. 2004a. Normal modes of spin excitations in magnetic nanoparticles. *Physical Review B* **69**: 174428-1–174428-12.

Grimsditch, M., L. Giovannini, F. Montoncello, F. Nizzoli, G. K. Leaf, and H. G. Kaper. 2004b. Magnetic normal modes in ferromagnetic nanoparticles: A dynamical matrix approach. *Physical Review B* **70**: 054409-1–054409-7.

Gubbiotti, G., S. Tacchi, G. Carlotti et al. 2007. Collective spin modes in monodimensional magnonic crystals consisting of dipolarly coupled nanowires. *Applied Physics Letters* **90**: 092503-1–092503-3.

Herring, C. and C. Kittel. 1951. On the theory of spin waves in ferromagnetic media. *Physical Review* **81**: 869–880.

Hicken, R. J., D. E. P. Eley, M. Gester, S. J. Gray, C. Daboo, A. J. R. Ives, and J. A. C. Bland. 1995. Brillouin light scattering studies of magnetic anisotropy in epitaxial Fe/GaAs films. *Journal of Magnetism and Magnetic Materials* **145**: 278–292.

Hillebrands, B. 1990. Spin-wave calculations for multilayered structures. *Physical Review B* **41**: 530–540.

Hillebrands, B. 1999. In *Light Scattering in Solids VII*, M. Cardona and G. Güntherodt, eds., Springer, Berlin, Germany, pp. 174–269.

Himpsel, F. J., J. E. Ortega, G. J. Mankey, and R. F. Willis. 1998. Magnetic nanostructures. *Advances in Physics* **47**: 511–597.

Holstein, T. and H. Primakoff. 1940. Field dependence of the intrinsic domain magnetization of a ferromagnet. *Physical Review* **58**: 1098–1113.

Jorzick, J., S. O. Demokritov, B. Hillebrands, et al. 2002. Spin wave wells in nonellipsoidal micrometer size magnetic elements. *Physical Review Letters* **88**: 047204-1–047204-4.

Kalinikos, B. A. and A. N. Slavin. 1986. Theory of dipole-exchange spin wave spectrum for ferromagnetic films with mixed exchange boundary conditions. *Journal of Physics C: Solid State Physics* **19**: 7013–7033.

Kalinikos, B. A., M. P. Kostylevi, N. V. Kozhus, and A. N. Slavin. 1990. The dipole-exchange spin wave spectrum for anisotropic ferromagnetic films with mixed exchange boundary conditions. *Journal of Physics C: Condensed Matter* **2**: 9861–9877.

Kittel, C. 2004. *Introduction to Solid State Physics*, 8th edn. Wiley, Hoboken, NJ.

Kostylev, M. P., A. A. Stashkevich, and N. A. Sergeeva. 2004. Collective magnetostatic modes on a one-dimensional array of ferromagnetic stripes. *Physical Review B* **69**: 064408-1–064408-7.

Krawczyk, M. and H. Puszkarski. 2008. Plane-wave theory of three-dimensional magnonic crystals. *Physical Review B* **77**: 054437-1–054437-13.

Liu, H. Y., Z. K. Wang, H. S. Lim et al. 2005. Magnetic-field dependence of spin waves in ordered Permalloy nanowire arrays in two dimensions. *Journal of Applied Physics* **98**: 046103-1–046103-3.

Martín, J. I., J. Nogués, K. Liu, J. L. Vicente, and I. K. Schuller. 2003. Ordered magnetic nanostructures: Fabrication and properties. *Journal of Magnetism and Magnetic Materials* **256**: 449–501.

Mathieu, C., J. Jorzick, A. Frank et al. 1998. Lateral quantization of spin waves in micron size magnetic wires. *Physical Review Letters* **81**: 3968–3971.

Nguyen, T. M. and M. G. Cottam. 2006. Spin-wave excitations in ferromagnetic nanotubes. *Surface Science* **600**: 4151–4154.

OOMMF package is available at http://math.nist.gov/oommf.

Patton, C. E. 1984. Magnetic excitations in solids. *Physics Reports* **110**: 251–315.

Rado, G. T. and R. J. Hicken. 1988. Theory of magnetic surface anisotropy and exchange effects in the Brillouin scattering of light by magnetostatic spin waves. *Journal of Applied Physics* **63**: 3885–3889.

Scholten, P. C. 1995. Which SI? *Journal of Magnetism and Magnetic Materials* **149**: 57–59.

Skomski, R. 2003. Nanomagnetics. *Journal of Physics: Condensed Matter* **15**: R841–R896.

Wang, Z. K., M. H. Kuok, S. C. Ng et al. 2002. Spin-wave quantization in ferromagnetic nickel nanowires. *Physical Review Letters* **89**: 027201-1–027201-3.

Wang, Z. K., H. S. Lim, H. Y. Liu et al. 2005. Spinwaves in nickel nanorings of large aspect ratio. *Physical Review Letters* **94**: 137208-1–137208-4.

Wang, Z. K., H. S. Lim, L. Zhang et al. 2006. Collective spin waves in high-density two-dimensional arrays of FeCo nanowires. *Nano Letters* **6**: 1083–1086.

Wang, Z. K., V. L. Zhang, H. S. Lim et al. 2009. Observation of frequency band gaps in a one-dimensional nanostructured magnonic crystal. *Applied Physics Letters* **94**: 083112-1–083112-3.

Optical Antenna Effects in Semiconductor Nanowires

Jian Wu
The Pennsylvania State University

Peter C. Eklund
The Pennsylvania State University

32.1 Introduction

The past two decades have witnessed many rapid advances in the science and technology of submicron-scale solid state objects (Kittel 2005; Rahman 2008). It is now possible to synthesize many forms of electronic and electro-optic grade materials at the nanoscale, such as two-dimensional (2D) quantum wells (Yu and Cardona 2001) and graphene (Hashimoto et al. 2004; Novoselov et al. 2005; Zhang et al. 2005), one-dimensional (1D) single-walled carbon nanotubes (SWNTs) (Dresselhaus et al. 2001; O'Connell 2006; Popov et al. 2006) and semiconductor nanowires (SNWs) (Morales and Lieber 1998; Duan and Lieber 2000; Lu and Lieber 2007), and zero-dimensional (0D) quantum dots (Yoffe 1993, 2001, 2002). These nanostructures, to mention a few outstanding examples, provide ideal prototypes to explore the electrical, optical, thermal, and mechanical properties that originate from their dimensionality and scale.

A variety of elemental, binary, and ternary compositions have been synthesized in the nanowire (NW) form and, in some cases, with diameters, d, down to a few nanometers. NWs can be prepared with lengths of ~100's of μm to short segments of < 300 nm, the latter by cutting a long master NW using a focused ion beam (FIB) (Wu et al. 2009). SNWs have been demonstrated to be building blocks for future nanoscale electronics (Lieber and Wang 2007; Lu and Lieber 2007), thermoelectronics (Boukai et al. 2008; Hochbaum et al. 2008), photonics (Pauzauskie and Yang 2006; Lieber and Wang 2007), and biochemical sensors (Patolsky et al. 2006a–c).

Phonon confinement (Piscanec et al. 2003; Adu et al. 2005, 2006; Fukata et al. 2006) effects inside the waist of small diameter Si NWs ($\bar{d} < 10$ nm) have been demonstrated by several

groups via Raman scattering. Surface optical (SO) phonons that lie between the usual transverse optic (TO) and the higher frequency longitudinal optic (LO) phonons of binary semiconductors have been observed by Raman scattering in SNWs and proposed to be activated by quasiperiodic modulation of the surface potential via diameter modulations from a growth instability and from faceting (Gupta et al. 2003; Xiong et al. 2004). Indeed, the observation of a sharp SO peak in the Raman spectrum requires a *periodic* spatial perturbation of the surface potential of the semiconductor.

Photon confinement effects in NWs are the focus of this chapter. Considerable effort has been expended to understand photon confinement in spherical particles, e.g., their cavity resonances, the enhanced internal electromagnetic (EM) fields, the enhanced Raman scattering, and the photoluminescence emission have been the subject of extensive research in the 1990s (Chew and Wang 1982; Yamamoto and Slusher 1993; Slusher 1994; Chang and Campillo 1996). Photon confinement effects in spherical particles have also produced interesting nonlinear optical phenomena associated with the high quality factor (high Q) "whispering gallery modes" (Vahala 2003), where the photon circulates around the interior of the particle near to the surface.

Application of the classical EM theory to the long line antenna gives rise to the well-known dipolar intensity pattern, $I(\theta) \sim \cos^2\theta$, for both the absorbed and emitted intensity, where θ is the angle between the incoming or outgoing electric field and the antenna axis. The antenna problem was first published (for the infinitely long antenna) by Lord Rayleigh in 1918 using Bessel functions to solve the EM boundary value problem (Rayleigh 1918). The "nano" antenna effect was first predicted by Ajiki and Ando (1996) for SWNT. By calculating the electric dipole matrix elements for this molecular wire, they predicted that the optical absorption would be strongly suppressed if the incident electric field is perpendicular to the nanotube axis. This is a manifestation of the symmetry of the electronic wavefunctions of the small-diameter 1D nanotube. Four years later, the "Raman antenna effect" for SWNTs was reported by Duesberg et al. (2000). They observed that the Raman scattering intensity from both radial and tangential displacement phonons exhibits a classic $\cos^2\theta$ dependence, where θ is the angle between the incident laser electric field and the nanotube axis, i.e., a dipole pattern (*caution*: the functional form of the experimental polar intensity plot depends on whether an analyzer is used when observing the scattered radiation). Their antenna results were found to be in apparent contradiction with the expectations, which were based on group theory (Duesberg et al. 2000; Saito et al. 2001). However, since the SWNT Raman scattering is *resonant*, this contradiction is to be expected, and we will discuss this in more detail below.

We have reported strong antennae resonances at critical NW diameters associated with both Rayleigh (elastic) (Zhang et al. 2009) and Raman scattering (Chen et al. 2008). Strongly polarized resonant Raman backscattering has also been reported for 15–25 nm diameter WS_2 nanotubes (Rafailov et al. 2005); a detailed theory connecting the backscattering with the details of the nanotube antenna effect was not presented in this work. Finally, experimental studies of polarized Raman scattering from optical phonons in tapered Si NWs has been reported and analyzed in terms of the calculated fields internal to the wire (Cao et al. 2006). An array of multiwalled carbon nanotubes (MWNT) on a substrate has demonstrated the ability to absorb light as an antenna when the length of the MWNTs in the array matches the wavelength of the light (Wang et al. 2004).

Relatively little systematic work on the antenna effect from individual SNWs has been reported to date. To pin down the physical mechanisms behind the polarized scattering from these systems, studies should be made over an interesting range of materials as a function of their length, diameter, and variable photon wavelength. Indeed, the antenna effect is strongly connected with the details of the geometrical cavity resonances that can be excited by incident light with the proper wavelength. The geometric enhancement of the incident and the re-radiated fields from small particles has been known for some time (Chew and Wang 1982). It can be thought of in terms of classical EM fields (Bohren and Huffman 1983) or in terms of excitation and emission from quasi-stationary cavity modes of the nanoparticle or NW (Chang and Campillo 1996).

In this chapter, we review a systematic investigation of polarized elastic (Rayleigh) and inelastic (Raman) backscattering from individual GaP NWs (Chen et al. 2008; Zhang et al. 2009). The experiments and calculations are carried out for visible laser radiation on crystalline GaP NWs with diameters in the range $40 < d < 600$ nm. In the zinc-blende phase, GaP possesses a cubic lattice symmetry that greatly simplifies the analysis of the data. We first give a brief description of NW synthesis followed by a discussion of the experimental setup that we have used for polarized light scattering studies. Subsequently, we describe our theoretical model

for the polarization dependence of the elastic (Rayleigh) backscattered radiation from individual NWs via the polarizability tensor and examine the experimental results in light of this theory. Next, we proceed to the case of inelastic (Raman) backscattered light from individual wires via the Raman scattering tensor. Finally, we present concluding remarks and give some suggestions for directions for future research into NW antenna effects.

32.2 Nanowire Synthesis and Characterization

Extensive literature in the past 10 years have reported conditions for the synthesis of SNWs with diameters in the range $\sim 4 < d < 500$ nm and that grow via the vapor–liquid–solid (VLS) mechanism. The VLS mechanism was discovered in the context of micron-diameter single crystal "whisker" growth (Wagner and Ellis 1964; Givargizov 1975). In 1998, the VLS process was first used to create nano-sized whiskers or NWs (Morales and Lieber 1998). The VLS process generally involves exposing the *vapors* of a semiconductor (e.g., Si) to small *liquid* metal droplets (e.g., Au), whereby a *solid* NW is observed to grow from the surface of the droplet. The process, therefore, requires a sufficiently high temperature to create liquid metal or liquid metal-semiconductor alloy droplets. This liquid state is usually achieved by growing the NWs inside a quartz tube placed in a tube furnace capable of ~1200°C. In slightly more detail, the VLS process proceeds via a steady flux of semiconductor vapors striking the droplet surface. The atoms stick and rapidly diffuse, eventually saturating the droplet. Saturation then leads to the surface precipitation of a solid that seeds the growth of the SNW. The NW grows in the steady state as long as the supply of the semiconductor remains; the droplet surface is not poisoned and the temperature remains sufficiently high to maintain the liquid droplet state consistent with a high rate of atomic diffusion. Usually, one NW grows from each particle and the diameter of the NW is approximately the same as that of the droplet.

The SNWs can then be harvested from the quartz tube into lab air when the furnace is cooled. Scanning electron microscopy (SEM) and/or transmission electron microscopy (TEM) will then show that most NWs still have the "seed particle" attached to the NW. Two main approaches have been devised for the production of the semiconductor vapor (Xia et al. 2003): (1) pulsed laser vaporization (PLV) (Morales and Lieber 1998; Duan and Lieber 2000) of a semiconductor material via the ejected plume from a solid target and (2) the introduction of reactant vapors, i.e., via gases (e.g., SiH_4, GeH_4), or sublimation as practiced in sources used for the chemical vapor deposition (CVD) of thin films. For the same reason that pulsed laser deposition (PLD) has been found successful in growing a large variety of thin films, PLV has been similarly found successful in growing a large variety of NWs. This success exploits the general finding that the plume generated by the laser focused on a fresh target surface apparently has nearly the same stoichiometry as that of the target. PLV then provides a simple means to control the stoichiometry of the NW.

The GaP NWs studied here were grown by laser vaporization of a $(GaP)_{0.95}Au_{0.05}$ target in a flowing argon gas, as shown in the

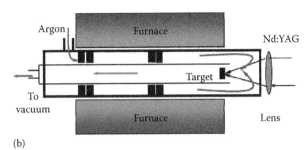

(a)   (b)

FIGURE 32.1 (a) Real image of our PLV apparatus showing the tube furnace, vacuum/gas handling components, Nd:YAG laser and optics for separating various harmonics housed inside the box on the upper right; (b) schematic of the PLV oven. Two quartz tubes are centered inside a furnace. A carrier gas (e.g., Ar) is introduced as shown, passing the preheated target down the central tube. For GaP NWs, the target containing GaP and Au was positioned just outside the furnace. The laser ablates the target to produce hot vapor that will condense into small clusters/nanoparticles. The furnace temperature is controlled to maintain the clusters in a liquid state.

schematic illustration and image of our PLV apparatus (Figure 32.1). In the PLV method, one can also mix metal powder (e.g., Au) with the semiconductor powder in the target. The success of an Au:GaP target suggests that there is an early formation of Au droplets in the vapor plume. In this case, the NWs grow from the droplets as they drift down the quartz tube entrained in Ar gas and the semiconductor vapor. Alternatively, metal nanoparticles can be deposited on a substrate that is positioned downstream from the target inside the growth furnace. At elevated temperatures, these particles melt and form droplets that then capture the semiconductor vapor and initiate the VLS growth. In our apparatus, the target is vaporized using focused pulses from an Nd:YAG laser (Spectron Laser Systems Inc., Model SL803) outputting pulsed radiation at either a wavelength of $1.064\,\mu m$ (fundamental, ~13 ns pulse width, 10 Hz repetition rate at 850 mJ/pulse) or at $0.532\,\mu m$ (second harmonic, 10 Hz repetition rate at 320 mJ/pulse). The focal spot is scanned over the surface of the target to continually present a fresh target surface to the beam.

GaP NWs grown in our PLV apparatus were harvested from the walls of the quartz tube and transferred to Si substrates or TEM grids (Ted Pella, Inc.) for study. For deposition on a Si substrate, the NWs were first dispersed in ethanol or isopropyl alcohol using mild ultrasound and then a small drop of the suspension was immediately placed on the substrate. The substrates often had lithographic markers to locate the NW position. These markers and the small diameter NWs could both be seen in a SEM or an atomic force microscopy (AFM) image. However, if the NWs were less than ~40 nm in diameter, they could not be seen in the spectrometer microscope image. However, the optical experiments could be carried out by superimposing the SEM and optical images of the large metal markers. NWs were usually deposited on TEM grids by gently rubbing the TEM grid over a substrate containing many NWs. AFM was used to measure the NW diameter when they were supported on substrates. The diameter of NWs on TEM grids could be obtained from calibrated TEM magnification and the crystallographic growth axis was determined using selected area electron diffraction (SAD) patterns. The same NW could then be studied optically as the TEM grid pattern was indexed.

As we shall see, the experimental results are expected to be somewhat sensitive to the local NW diameter, where the excitation beam falls. One, therefore, has to be careful to notice possible fluctuations in the NW diameter along its length. Real NWs can also have rough surfaces and/or be a tapered cylinder instead of a constant diameter cylinder. For most of our NWs, we estimate that the diameter we report in connection with the optical data is accurate to ±2 nm. Figure 32.2a shows an AFM topographic image of a GaP NW on an Si substrate with lithographic markers; Figure 32.2b shows a low magnification TEM image of an individual 80 nm GaP NW protruding over a hole in a TEM grid with a SAD pattern indicating that this particular NW grew along direction $\ll 01\bar{1} \gg$ (the NW growth axis will be indicated by double brackets in the chapter). High resolution TEM (HRTEM) images can also be used to indicate the uniformity of the NW. In this case, lattice images (Figure 32.2c) indicate where the wire is crystalline and if twinning (Xiong et al. 2006) has occurred. All of our GaP NWs examined by TEM were found to be crystalline. SAD patterns indicate that they have the cubic zinc-blende structure. Most have been observed by TEM to be single crystalline over lengths greater than several microns. We have observed several growth directions $\{\ll 001 \gg, \ll 011 \gg, \ll 111 \gg\}$ for PLV-grown GaP NWs. From TEM observations, the $\ll 011 \gg$ growth direction seem to be slightly preferred, followed closely in popularity by $\ll 111 \gg$, with $\ll 001 \gg$ being significantly less likely.

32.3 Optical Backscattering Experiments

Polarized Rayleigh (elastic) and Raman (inelastic) backscattering experiments were carried out using either a single-grating Renishaw in VIA micro-Raman spectrometer or a triple-grating Jobin-Yvon T64000 micro-Raman spectrometer. These spectrometers are referred to as "micro" because the incident and backscattered light pass through a microscope before entering the spectrometer. The microscope stage and objective lens carousel are shown in Figure 32.3a for the in VIA spectrometer. The focal

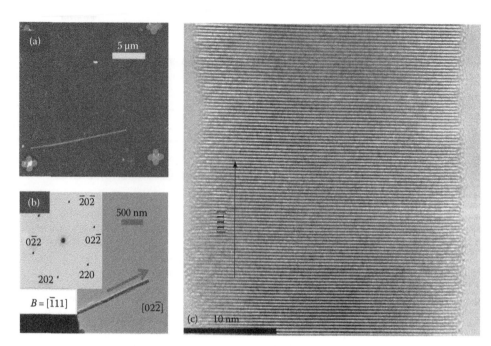

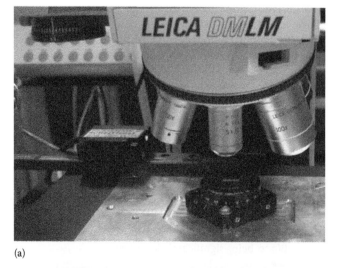

FIGURE 32.2 (a) AFM topographic image of a GaP NW on an Si substrate with Au/Cr markers to define NW positions. (b) TEM image of a GaP NW protruding out over a hole in a TEM grid. SAD pattern shows that this wire grew along ≪01Ī≫ direction. (c) HRTEM lattice image of a ≪111≫ GaP NW. Image shows that the crystallinity extends almost to the surface of the NW.

FIGURE 32.3 (a) Renishaw in VIA Raman microscope stage with objectives. A Sample rotator under the objective is used to align the NW to the $\hat{x}_2$ axis (Figure 32.4) and (b) polarization rotator (HWP) to rotate the laser polarization relative to the NW axis.

spot size at the sample is typically ~1 μm or less, depending on the objective magnification and the laser wavelength. A computer is used to control the translation of the microscope stage and thus the position of the NW in the horizontal focal plane of the instrument. We have added a manually operated rotary stage on top of the microscope translation stage (Figure 32.3a) to rotate the NW about a vertical axis perpendicular to the focal plane.

For Rayleigh scattering, we mount the NW on a TEM grid (over a hole in the grid) and place the grid over a small hole in the center of the rotary stage. The backscattered radiation then comes primarily from the NW. For Raman scattering experiments using a substrate to support the NW, the frequency of the backscattered light is characteristic of either the NW or the substrate. We therefore follow the intensity of the NW Raman peaks with the angle of the incident laser polarization.

The schematic backscattering geometry of the experiments is shown in Figure 32.4. The laser is directed along the $-\hat{x}_1$ direction of a lab $(\hat{x}_1,\hat{x}_2,\hat{x}_3)$ coordinate system and the beam focused by the microscope objective onto the NW lying in the $(\hat{x}_2,\hat{x}_3)$ plane. The laser polarization, $\vec{E}_0$, is initially polarized along the $\hat{x}_2$ axis. As it passes through a rotatable half wave plate (HWP), the laser polarization will be rotated by an angle, θ, in the horizontal plane (Duesberg et al. 2000). The backscattered light from the NW propagates along the $\hat{x}_1$-direction and is collected by the objective lens and then passes again through the polarization rotator where the electric vector is counter-rotated by the angle θ. The analyzer, positioned before the entrance slit, is fixed to only pass the scattered E-field component parallel to the NW that is aligned parallel to the entrance slit. Knowledge of the instrument functions for light polarized parallel and perpendicular to the slit is therefore

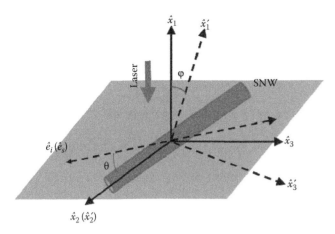

FIGURE 32.4 Coordinate systems defining the backscattering from a crystalline NW: primed orthogonal coordinate system refers to crystallographic directions <hkl> within the NW, as discussed in the text; unprimed coordinate system is the laboratory system. In the lab system, the NW lies in the $(\hat{x}_2,\hat{x}_3)$ plane and the laser is incident in the $-\hat{x}_1$ direction. The angle, θ, defines the polarization of incident (*i*)/scattered (*s*) photons. The rotational angle, φ, defines the orientation of NW crystal system relative to lab system.

not needed. The primed coordinate system shown in Figure 32.4 refers to an orthogonal set of axes of the NW. The relative orientation of the lab $(\hat{x}_1,\hat{x}_2,\hat{x}_3)$ and crystal $(\hat{x}_1',\hat{x}_2',\hat{x}_3')$ axes can be important for analyzing the light backscattering from the NWs.

We now give a simplified explanation of how the Rayleigh backscattered light from SNWs generates a $\cos^4\theta$ polar pattern in our apparatus. The $\cos^4\theta$ pattern is one signature of a strong NW antenna effect. As we will discuss, in the presence of a strong NW antenna effect, the induced EM field intensity inside the NW can be shown to stem only from the component of the time-dependent external field $(\vec{E}_0)$ parallel to the NW axis. The EM intensity internal to the NW that is driving the elastic scattering of the incident light should therefore be proportional to $\cos^2\theta$, where θ is the angle between the incident polarization and the NW axis. For Rayleigh scattering, the dipoles inside the NW then emit elastically with $\vec{E}$ parallel to the wire axis. As we have mentioned, only the

external emitted E-field with a component parallel to the polarization of the incident light will be collected by our spectrometer. This analyzer-based restriction therefore adds a second $\cos^2\theta$ factor to the emission process that results in an overall $\cos^4\theta$ polar pattern for the Rayleigh backscattering from the NW antenna. We present a complete derivation of the $\cos^4\theta$ factor below.

32.4 Classical Calculations of the Elastic Light Scattering from a Cylinder

The theory of elastic light scattering from an infinitely long wire ($L = \infty$) was first published in 1918 by Lord Rayleigh (Rayleigh 1918). Rayleigh solved the problem by expanding the incident, scattered, and internal EM fields in terms of Bessel functions of the first kind. Using the same approach, we have calculated results for infinite GaP wires (Bohren and Huffman 1983). Results for the field amplitude $|E|$ shown in Figure 32.5 are for a series of $L = \infty$ GaP NWs with d in the range $50 \Leftarrow d \Leftarrow 210\,nm$; the color scale is the same for all the E-field amplitude maps. Light was taken to be incident (blue arrow) normal to the NW axis and polarized either $\|$ (TM) or $\perp$ (TE) to the NW axis. The usual notation, TM or TE, refers to either transverse magnetic (TM) excitation or transverse electric (TE) excitation, i.e., transverse to the NW axis.

We now look at a second signature of the antenna effect. Antenna resonances can easily be observed by plotting the internal EM intensity ratio ($I_\| / I_\perp$) (inside the NW) vs. the diameter, d; the subscripts, $\|$ and $\perp$, refer to electric field polarization relative to the NW axis. Calculated results (via Bessel functions) for excitation with 514.5 or 488 nm light are shown in the Figure 32.6 for $L = \infty$. The calculations are made for the complex refractive index $n^\star = n + jk$ (GaP) (i.e., $3.6384 + j0.0000$ for 488 nm excitation and $3.5436 + j0.0000$ for 514.5 nm excitation). The imaginary part of $n^\star$ (or k) is taken to be approximately zero because these photons have frequency below the GaP direct gap at ~2.78 eV. A broad fundamental antenna resonance at small $d \sim 30\,nm$ can be seen in Figure 32.6, which is calculated together with relatively narrow resonances appearing at larger d. It can also be seen that the antenna resonances just shift

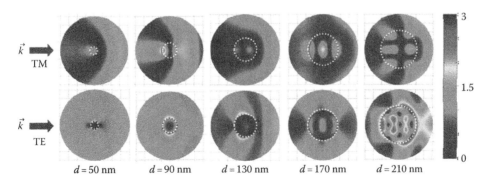

FIGURE 32.5 (See color insert following page 20-16.) Scattered internal electric field *amplitude* maps calculated for 488 nm waves incident from the left and for $L = \infty$ GaP NWs with diameters from 50 to 210 nm. The amplitude scale is color coded and appears to the right. TM and TE refer, respectively, to transverse magnetic ($\theta = 0°$) and transverse electric ($\theta = 90°$) excitation with the light incident at right angles to the NW axis. The dashed circles indicate the boundary of the NW. (From Chen, G. et al., *Nano Lett.*, 8, 1341, 2008. With permission.)

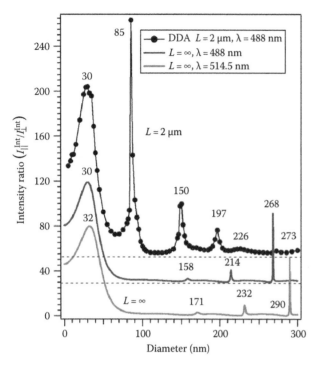

FIGURE 32.6 Calculated diameter (*d*)-dependence of the EM field intensity ratio inside the NW for the electric field polarization parallel and perpendicular to the NW axis. (Top) *L* = 2 μm NW with 488 nm excitation, (middle) *L* = ∞ NW with 488 nm excitation and (bottom) *L* = ∞ NW with 514.5 nm radiation (bottom). Numbers above the peaks refer to the diameter at the resonance in nm. (From Chen, G. et al., *Nano Lett.*, 8, 1341, 2008. With permission.)

position when the excitation wavelength is changed. This shift occurs because the physical variable in the classical calculation is *nd*/λ, where λ is the vacuum wavelength. Note that when the NW or dielectric cylinder is "off-resonance," the ratio $(I_{\parallel}/I_{\perp}) \sim 1$, i.e., it is equally easy to "squeeze" either EM polarization state into the NW.

Unfortunately, an exact analytical solution for a finite NW does not exist (Bohren and Huffman 1983). This difficulty can be circumvented by using standard (free) software available on the Internet (Draine and Flatau 2004) that approaches the scattering problem numerically and can be used to study particles of arbitrary shape. The software is based on the "discrete dipole approximation" (DDA) (Draine and Flatau 1994, 2004). In the DDA approximation, the particle is divided into numerous cubic cells, each containing a dipole whose local response is consistent with the refractive index, *n**, of the bulk material. As long as the NW dimensions are large enough, i.e., *d*, *L* > ~10 nm, quantum size effects that might change the electronic contribution to *n** can be neglected and it should be valid to input the bulk value for *n**. In Figure 32.6, we also display the results calculated (DDA) for the 488 nm antenna resonances for *L* = 2 μm GaP wires vs. diameter. It is interesting to compare the results for *L* = ∞ and *L* = 2 μm at λ = 488 nm. The effect of shortening the wire is (1) to cause a small downshift in the antenna resonances found for the infinite wire, and (2) to introduce a new resonance for the *L* = 2 μm wire, which should be identified with a length-related resonance where standing EM waves must also fit into the short length of the wire. Interestingly, the calculated lowest order 488 nm *L* = 2 μm antenna resonances (peak near *d* = 30 nm) appear nearly the same for the finite and infinite cylinders. In Figure 32.7, we display the TM and TE results for |*E*| calculated for an axial cross section of an *L* = 1 μm wire (DDA) for *d* =100, 150, 200 nm. The amplitude of the color scale in Figure 32.7 is variable from one diameter wire to the next to allow the |*E*| patterns to be more easily observed. Note that the finite length of the NW can bring axial structure into the problem that is absent in the case of an infinite wire.

32.5 Modeling the Backscattered Light from Semiconductor Nanowires

We have indicated that the elastic scattering of light from a particle or SNW can be calculated by solving the electromagnetic boundary value problem. The application to infinite NWs was discussed in Section 32.4. Here, we take an alternative approach to the scattering, similar to one commonly used to describe Raman scattering in the bulk (Hayes and Loudon 1978; Loudon

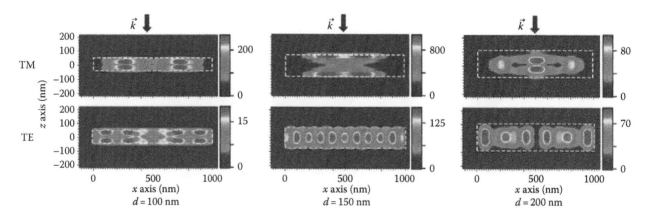

FIGURE 32.7 **(See color insert following page 20-16.)** Calculated internal electric field *amplitude* maps for the axial cross section. Results are from DDA simulation code using 488 nm excitation incident at right angles to the axis of *L* = 1 μm GaP NWs of diameter *d* = 100, 150, and 200 nm. Color-coded scales appear to the right correspond to differing maximum amplitude. The dashed lines indicate the boundary of the NW. (From Chen, G. et al., *Nano Lett.*, 8, 1341, 2008. With permission.)

2001; Yu and Cardona 2001). For bulk materials, the expression for the scattered Raman intensity is made with respect to the incident and scattered E-fields *inside* the material. In the case of nanoparticles or NWs, an explanation of the polarized scattering must also take into account how the incident and scattered fields enter and leave the scattering particle.

Consider the scattering geometry shown in Figure 32.4, i.e., a plane wave at wavelength λ is incident along $-\hat{x}_1$, where the NW lies in the $(\hat{x}_2, \hat{x}_3)$ plane with its axis along $\hat{x}_2$. The incident electric field from the laser external to the NW is given by

$$\vec{E}_0 = E_0 \hat{e}_i = \begin{pmatrix} 0 \\ E_0 \cos\theta \\ E_0 \sin\theta \end{pmatrix}, \qquad (32.1)$$

where θ is the angle between the polarized laser field direction, $\hat{e}_i$, and the NW axis which lies along $\hat{x}_2$. We denote the incident field *inside* the NW by $\vec{E}_0{}'$, which can be written as the product of a tensor, $\tilde{Q}$, and the incident field outside the wire,

$$\vec{E}_0{}' = \tilde{Q} \cdot \vec{E}_0. \qquad (32.2)$$

The form of $\tilde{Q}$ is taken to be diagonal

$$\tilde{Q} = \begin{pmatrix} 0 & 0 & 0 \\ 0 & Q_\| & 0 \\ 0 & 0 & Q_\perp \end{pmatrix}, \qquad (32.3)$$

where

$Q_\|$ and $Q_\perp$ refer to components parallel and perpendicular to the NW axis, respectively

$\tilde{Q}$ is introduced to describe the antenna property of the NW

When the NW is involved in an antenna resonance, $Q_\| \gg Q_\perp$. In general, $\tilde{Q}$ will also be a function of the position inside the NW or nanoparticle. It can be shown, however, that the volume average of the tensor components $(Q_\|, Q_\perp)$ is all that is required to compute the scattering. Values of $Q_\|$ and $Q_\perp$ could be obtained from the calculated polar plots of $|E(\theta)|$, as shown in Figure 32.5, or they can be obtained by experiments as shown below.

The scattered vector field, $\vec{E}_s{}'$, inside the NW is defined by the application of a scattering matrix, $\tilde{S}$, to the internal field, $\vec{E}_0{}'$. That is, we have

$$\vec{E}_s{}' = \tilde{S} \cdot \vec{E}_0{}'. \qquad (32.4)$$

The NW then emits an external field given by the product of an emission tensor, $\tilde{Q}_e$, and the internal scattered field,

$$\vec{E}_s = \tilde{Q}_e \cdot \vec{E}_s{}'. \qquad (32.5)$$

We also make the usual simplification, i.e., $\tilde{Q}_e \simeq \tilde{Q}$ (Chew and Wang 1982; Cao et al. 2006; Cardona and Merlin 2007). The polarization of the *detected* scattered electric field, $\hat{e}_s$, is parallel to the incident field, $\hat{e}_i$ (this follows from the experimental conditions; Section 32.3). Then, the external scattered intensity associated with this polarization can be written as

$$I^{NW} \propto \omega^4 \left| \hat{e}_s \cdot \vec{E}_s \right|^2 = \omega^4 \left| \hat{e}_s \cdot \tilde{Q} \cdot \tilde{S} \cdot \tilde{Q} \cdot \vec{E}_0 \right|^2 = \omega^4 \left| \hat{e}_s \cdot \tilde{Q} \cdot \tilde{S} \cdot \tilde{Q} \cdot \hat{e}_i \right|^2 I_0, \qquad (32.6)$$

where I_0 is the external incident light intensity.

For Rayleigh (elastic) scattering, the matrix, $\tilde{S}$, is the polarizability tensor, $\tilde{\alpha}$. For an isotropic material, e.g., GaP (zinc blende), $\tilde{\alpha}$ is just a scalar, i.e., $\tilde{S} = \alpha(\omega)\tilde{I}$ and $\tilde{I}$ is the unit matrix. In this case,

$$I^{NW}_{Rayleigh} \propto \omega^4 \left| \hat{e}_s \cdot \tilde{Q} \cdot \alpha \tilde{I} \cdot \tilde{Q} \cdot \hat{e}_i \right|^2 I_0 = \omega^4 \alpha^2 (Q_\|^2 \cos^2\theta + Q_\perp^2 \sin^2\theta)^2. \qquad (32.7)$$

We can use this expression to determine an experimental value for $Q_\|^2 / Q_\perp^2$ by fitting the backscattered Rayleigh polar patterns (Section 32.7).

32.6 Optical Phonons, Raman Scattering Matrices, and Manipulations

In the case where there are two atoms per primitive cell (Ga and P), we require six degrees of freedom to describe the atomic displacements of these atoms. Therefore, there will be three acoustic and three optic phonon branches (Yu and Cardona 2001). At zero phonon wave vector ($q = 0$), two acoustic phonon modes are transverse (TA) and one is longitudinal (LA). These phonon modes describe a rigid translation of the crystal lattice. The frequency of these modes is zero because there is no restoring force. Of the three $q = 0$ optic branches, two are transverse (TO) and one is longitudinal (LO). The eigenvectors of the $q = 0$ LO and TO modes for GaP are shown in Figure 32.8. These optic modes would be degenerate without the presence of a long range EM force that induces a small splitting between them (Yu and Cardona 2001). First-order Raman scattering can occur from optical phonons when they have the proper symmetry (Hayes and Loudon 1978; Loudon 2001; Yu and Cardona 2001). For GaP (zinc blende), both the LO and TO modes are Raman-active and can be observed for specific experimental conditions that prescribe the wave vectors, the polarization of the phonon and incident and scattered photons. The frequencies of the TO and LO phonons are obtained from the Raman spectrum of an individual GaP NW, as shown in Figure 32.8. The SO phonons appear as a low frequency shoulder on the higher frequency LO peak. The so-called selection rules that determine the scattering geometry where the TO and LO $q = 0$ phonons can be observed are summarized by the Raman tensor (Loudon 2001; Yu and Cardona 2001)

$$I^{TO, LO}_{Bulk\,Raman} \propto \omega^4 \left| \hat{e}_s \cdot \tilde{\mathfrak{R}} \cdot \hat{e}_i \right|^2 I_0, \qquad (32.8)$$

where

$\hat{e}_i$ and $\hat{e}_s$ are unit vectors for the incident (i) and scattered (s) fields *inside* the material

$\tilde{\mathfrak{R}}$ is the Raman tensor

Equation 32.8 ignores the transmission coefficients (Bohren and Huffman 1983) for the incident and scattered waves into and out

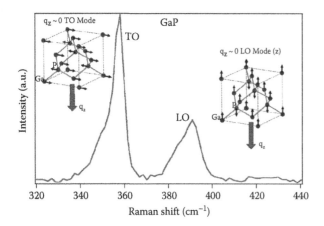

FIGURE 32.8 Typical Raman spectrum from an individual GaP NW showing TO and LO frequencies at 358 and 391 cm^{-1}, respectively. The insets present the vibrational eigenvectors for both TO and LO phonons. Ga and P atoms are indicated in the insets, respectively.

of the bulk sample. The form of the Raman tensor, $\tilde{\Re}$, is related to the symmetry of Raman-active phonons. For semiconductors with the zinc-blende structure, the Raman tensors for optic phonons with atomic displacements along the [100], [010], and [001] crystallographic axes are

$$\tilde{\Re}(\hat{e}_1) = \begin{pmatrix} 0 & 0 & 0 \\ 0 & 0 & a \\ 0 & a & 0 \end{pmatrix} \quad \tilde{\Re}(\hat{e}_2) = \begin{pmatrix} 0 & 0 & a \\ 0 & 0 & 0 \\ a & 0 & 0 \end{pmatrix} \quad \tilde{\Re}(\hat{e}_3) = \begin{pmatrix} 0 & a & 0 \\ a & 0 & 0 \\ 0 & 0 & 0 \end{pmatrix},$$

(32.9)

where the unit vectors, $\hat{e}_m$ ($m = 1,2,3$), represent the [100], [010], and [001] axes, respectively. We use the subscript, m, to distinguish them from the polarizations of incident and scattered light, $\hat{e}_i$ and $\hat{e}_s$. The constant tensor elements a take on different values for TO and LO phonons. The forms of the Raman tensors above apply only to the case of nonresonant scattering, i.e., the incident photon must not participate in strong electronic absorption. It is interesting to compare Equation 32.8 for bulk Raman scattering and Equation 32.6 for SNWs. Scattering from SNWs introduces the matrix, $\tilde{Q}$.

We are primarily interested in the change in the backscattered intensity as the incident polarization is rotated by θ about the incident photon wave vector, $\vec{k}_i$ (c.f., Figure 32.4). Depending on the orientation of the HWP, the polarization direction for the incident, $\hat{e}_i$, and backscattered radiation, $\hat{e}_s$, is

$$\hat{e}_i = \hat{e}_s = \begin{pmatrix} 0 \\ \cos\theta \\ \sin\theta \end{pmatrix} = \hat{e}.$$

(32.10)

The experimentally collected backscattered polarization component, $\hat{e}_s = \hat{e}_i$, is consistent with the discussion in Section 32.3 and involves the HWP polarization rotator and analyzer. The wave vectors ($k = 2\pi/\lambda$) for the incident and backscattered radiation,

$\hat{k}_i$ and $\hat{k}_s$, respectively, are both parallel to the $\hat{x}_1$ axis. To comply with the photon–phonon wave vector conservation rule (Yu and Cardona 2001), the wave vector of the optical phonon, $\hat{q}$, excited in the Raman process must also be parallel to the $\hat{x}_1$ axis (the phonon wave vector is $q = 2k_i$). We consider the case where $\hat{x}_j \parallel \hat{x}'_j$ ($j = 1,2,3$). Therefore, the displacement of the LO phonons must be along the $\hat{x}_1$ axis and the displacement of the TO phonons must be perpendicular to the $\hat{x}_1$ axis. Then from Equations 32.8 and 32.10, we have for scattering from the bulk that:

$$I_{LO}(\theta) \propto \left| \hat{e} \cdot \tilde{\Re}(\hat{x}_1) \cdot \hat{e} \right|^2,$$

(32.11a)

and

$$I_{TO}(\theta) \propto \left| \hat{e} \cdot \frac{1}{\sqrt{2}} \tilde{\Re}(\hat{x}_2) \cdot \hat{e} \right|^2 + \left| \hat{e} \cdot \frac{1}{\sqrt{2}} \tilde{\Re}(\hat{x}_3) \cdot \hat{e} \right|^2,$$

(32.11b)

where the two terms in Equation 32.11b can be identified with two orthogonal TO phonons. These results predict the polarization dependence of the bulk TO and LO scattering.

Next, we demonstrate how to calculate the Raman tensors of cubic GaP NWs with different growth directions and with a specific orientation, φ, about the NW axis (Figure 32.4). The scattering geometry in Figure 32.4 is defined by introducing two coordinate systems: the lab coordinate system ($\hat{x}_1, \hat{x}_2, \hat{x}_3$) and the crystal coordinate system ($\hat{x}'_1, \hat{x}'_2, \hat{x}'_3$) attached to the NW. $\hat{x}'_2(\hat{x}_2)$ is parallel to the NW growth axis; $\hat{x}'_1$ and $\hat{x}'_3$ are the other two crystallographic axes of the NW.

First, we suppose that

$$\hat{x}'_k = \sum_m \alpha_{km}\hat{e}_m, \quad \hat{x}_k = \sum_m \beta_{km}\hat{x}'_m,$$

(32.12)

where

$k,m = 1,2,3$

the $\hat{e}_m$ are defined as before

Then, we can define two transformation matrices, $\mathbf{T}'$ and $\mathbf{T}$ based, respectively, on the coefficients, α_{km} and β_{km}:

$$\mathbf{T}' = \begin{pmatrix} \alpha_{11} & \alpha_{12} & \alpha_{13} \\ \alpha_{21} & \alpha_{22} & \alpha_{23} \\ \alpha_{31} & \alpha_{32} & \alpha_{33} \end{pmatrix},$$

(32.12a)

$$\mathbf{T} = \begin{pmatrix} \beta_{11} & \beta_{12} & \beta_{13} \\ \beta_{21} & \beta_{22} & \beta_{23} \\ \beta_{31} & \beta_{32} & \beta_{33} \end{pmatrix} = \begin{pmatrix} \cos\varphi & -\sin\varphi & 0 \\ \sin\varphi & \cos\varphi & 0 \\ 0 & 0 & 1 \end{pmatrix}.$$

(32.12b)

The Raman tensors, $\Re(\hat{x}'_k)$, for optical phonons displaced along the directions, $\hat{x}'_k$, can then be written as

$$\tilde{\mathfrak{R}}(\hat{x}'_k) = \sum_m \alpha_{km} \mathfrak{R}(\hat{e}_m), \qquad (32.13)$$

where the Raman tensors, $\tilde{\mathfrak{R}}(\hat{e}_m)$, are given by Equation 32.9. Then the Raman tensors for phonon displacements in the directions, $\hat{x}'_k$, in the coordinate system $\{\hat{x}'_k\}$ are

$$\tilde{\mathfrak{R}}'(\hat{x}'_k) = \sum_m \alpha_{km} \mathbf{T}' \cdot \tilde{\mathfrak{R}}(\hat{e}_m) \cdot \mathbf{T}'^{-1}. \qquad (32.14)$$

Similarly, we have the Raman tensors for optical phonons displaced along the direction, $\hat{x}_k$, in the coordinate system, $\hat{x}'_k$ and $\hat{x}_k$, given by

$$\tilde{\mathfrak{R}}'(\hat{x}_k) = \sum_m \beta_{km} \tilde{\mathfrak{R}}'(\hat{x}'_k), \qquad (32.15a)$$

$$\tilde{\mathfrak{R}}(\hat{x}_k) = \sum_m \beta_{km} \mathbf{T} \cdot \tilde{\mathfrak{R}}'(\hat{x}_k) \cdot \mathbf{T}^{-1}. \qquad (32.15b)$$

Following the strategy above, we can calculate the Raman tensors, $\tilde{\mathfrak{R}}(\hat{x}_k)$, for NWs with growth axis along <<001>>, <<011>>, and <<111>> and for specific choices of the NW rotation angle φ. The bulk scattering TO and LO intensities vs. θ (polar plots) can be calculated by applying Equations 32.11 and 32.15b. This calculation leaves out the effects of the tensor, $\tilde{Q}$, as discussed above in Equation 32.6. For pedagogical reasons, we nevertheless present the calculated tensor-based polar backscattering patterns as the effects assignable to the Raman tensor alone.

The Raman tensor-based polar patterns are shown in Figures 32.9 through 32.11, respectively. These patterns correspond to bulk scattering for a specific orientation of the crystal axes

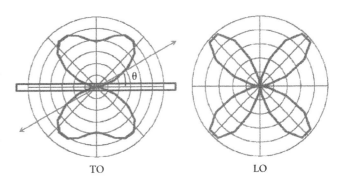

TO **LO**

FIGURE 32.9 Calculated θ-dependence of the contribution to the backscattered Raman intensity as seen at the detector for a <<100>> GaP NW oriented along the $\hat{x}_2$-axis and taking into account *only* the contribution from the Raman tensor. Results are calculated according to Equation 32.8 for angles φ between the <001> direction and the $\hat{x}_1$-axis. It is shown from the calculation that the shape of both TO and LO polar plot remains the same, independent of φ. Since the patterns do not include a contribution from $Q_\parallel$ and $Q_\perp$, the results do not depend on the diameter of the NW.

relative to the lab axes or they can be thought of as the result for $Q_\perp = Q_\parallel = 1$ (no antenna effect). They are calculated for $\hat{e}_s \parallel \hat{e}_i$. The NW is represented schematically in Figure 32.9; it is permanently oriented along $\theta = 0°$, where θ increases with counterclockwise rotation of the radial coordinate in the polar plot.

The tensor-based polar plots for GaP <<100>> NW are the simplest, as shown in Figure 32.9. The shapes of both TO and LO patterns stay the same, except for changes in intensity. For GaP <<111>> NWs shown in Figure 32.10, the polar patterns associated with LO phonon scattering do not change, while the TO patterns display some variety for $0° \le \varphi \le 30°$; the polar patterns just repeat themselves for $30° \le \varphi \le 60°$ and again in $60° \le \varphi \le 90°$. Figure 32.11 shows the TO and LO polar patterns of GaP <<110>>

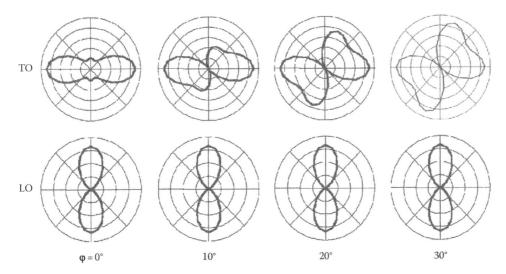

TO

LO

$\varphi = 0°$ $10°$ $20°$ $30°$

FIGURE 32.10 Calculated θ-dependence of the contribution to the backscattered Raman intensity as seen at the detector for a <<111>> GaP NW oriented along the $\hat{x}_2$-axis and taking into account *only* the contribution from the Raman tensor. Results are calculated according to Equation 32.8 for angles φ between the <11$\bar{2}$> direction and the $\hat{x}_1$-axis. It can be seen that the shape of the LO polar plot remains the same, independent of φ. The TO patterns change only from $0° < \varphi < 30°$, i.e., the patterns repeat every $30°$. Since the patterns do not include a contribution from $Q_\parallel$ and $Q_\perp$, the results do not depend on the diameter of the NW.

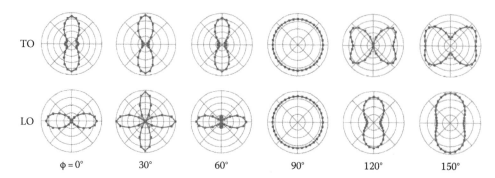

FIGURE 32.11 Calculated θ-dependence of the contribution to the backscattered Raman intensity as seen at the detector for a <<110>> GaP NW oriented along the $\hat{x}_2$-axis and taking into account *only* the contribution from the Raman tensor. Results are calculated according to Equation 32.8 for angles φ between the <1$\bar{1}$1> direction and the $\hat{x}_1$-axis. The shape of both the LO and TO polar plots depend on φ. Since the patterns do not include a contribution from $Q_\parallel$ and $Q_\perp$, the results do not depend on the diameter of the NW. (From Chen, G. et al., *Nano Lett.*, 8, 1341, 2008. With permission.)

NWs for $0° < \varphi < 180°$. The variety of the tensor-based polar patterns seen in the figure indicate the sensitivity of the scattering pattern to the NW orientation about the <<110>> growth axis.

32.7 Results and Discussion of Rayleigh Backscattering from Nanowires

A few papers have appeared recently on the Rayleigh spectroscopy of SWNTs. The backscattered intensity for fixed polarization, i.e., $I(\omega)$, was used to study the electronic transitions and optical properties of SWNTs (Heinz 2008). Rayleigh spectroscopy on

individual SWNTs demonstrated the capability to characterize the chirality of SWNTs (Sfeir et al. 2004, 2006) as well as study the interactions between two SWNTs (Wang et al. 2006). Also, Rayleigh spectroscopy was used in the case of SWNTs to directly measure the change of the electronic band structure with curvature (Huang et al. 2008).

Our experimental results for the Rayleigh backscattered polar plots, $I(\theta)$, for six different diameter GaP NWs are shown in Figure 32.12. The data are collected for $\hat{e}_s \parallel \hat{e}_i$. The squares represent the backscattered intensity data and the solid lines are the result of a least squares fit of Equation 32.7 to the data. This fit

GaP (cubic): Rayleigh scattering

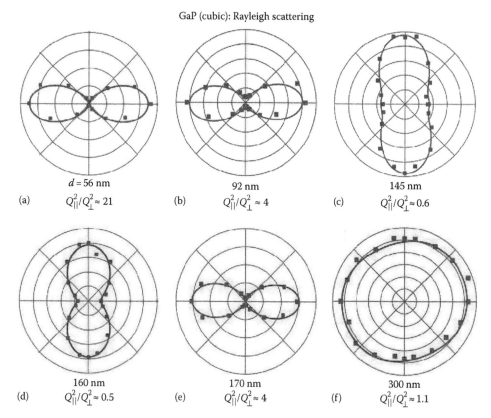

FIGURE 32.12 Backscattered Rayleigh polar plots for six GaP NWs with different diameter. The data (points) are fit using Equation 32.7, which results in experimental values for $Q_\parallel^2/Q_\perp^2$.

can yield an experimental value for $Q_\parallel^2 / Q_\perp^2$. In Figure 32.12a, we show the Rayleigh pattern for a $d = 56\,\text{nm}$ GaP NW. The strongest scattering is achieved when the incident laser and scattered light are polarized parallel to the wire axis. The data are well fit by $(Q_\parallel^2 / Q_\perp^2) \approx 21$. The polar pattern almost has the $\cos^4\theta$ form, indicative of a strong antenna effect. Indeed, we have found a ~$\cos^4\theta$ for all GaP NWs with $d \leq$ ~70 nm. As the wire diameter decreases, the intensity at $\theta = 90°$ becomes even smaller. For NWs with larger diameters (~70 nm $< d <$ ~100 nm), such as the case in Figure 32.12b, the center of the dipolar pattern opens up as the contribution from $Q_\perp$ becomes more important. For NWs with diameter ~100 nm $< d <$ ~150 nm, the strongest Rayleigh scattering is actually obtained when the incident and scattered light is polarized perpendicular to the wire, i.e., $Q_\perp > Q_\parallel$. When this occurs, the polar pattern appears as if the small d wire dipole pattern is rotated by 90°. Two of these patterns, i.e., for GaP $d = 145\,\text{nm}$ and $d = 160\,\text{nm}$, are illustrated in Figure 32.12c and d, respectively. At $d \sim 170\,\text{nm}$, as shown in Figure 32.12e, the pattern again closely resembles the $\cos^4\theta$ form, consistent with the excitation of a $d = 170\,\text{nm}$ antenna resonance predicted from the infinite long wire calculation shown in Figure 32.6. Finally, the pattern in Figure 32.12f for a $d = 300\,\text{nm}$ NW is seen to be close to that of a circle, i.e., $Q_\perp \approx Q_\parallel$. Figure 32.12a through f shows that the Rayleigh polar patterns are in good agreement with the theoretical approach developed in Section 32.5; the fits demonstrate the ability of elastic scattering to provide an experimental value for $Q_\parallel^2 / Q_\perp^2$.

In Figure 32.13, we compare our experimental Rayleigh data $(I_\parallel/I_\perp)$ to $(I_\parallel^{\text{int}} / I_\perp^{\text{int}})^2$ calculated for $L = \infty$ GaP NWs, as shown previously in Figure 32.6. The inset to Figure 32.13 is a magnified view that focuses on the diameter range from 70 to 200 nm. The experimental Rayleigh scattering results (+) from long ($L > 10\,\mu\text{m}$) GaP NWs illuminated by a ~1 μm laser spot behaves similar to the prediction for infinite wires. This agreement also suggests that we can also use the calculated results for $L = \infty$ $(I_\parallel^{\text{int}}/I_\perp^{\text{int}})$ to predict values for $(Q_\parallel^2 / Q_\perp^2)$.

32.8 Polarized Raman Backscattering from Semiconductor Nanowires

In Figure 32.14a and b, we display the measured integrated Raman band intensity ratio $(I_\parallel^X/I_\perp^X)$ for X = TO, LO phonons vs. d (Chen et al. 2008). Data in Figure 32.14a and b were obtained, respectively, using 514.5 or 488 nm excitation. For clarity, the TO data in each case have been displaced vertically. The data shown in these panels were collected from NWs lying either on an Si (100) substrate (with 200 nm oxide) or on a TEM grid. Notice that the smallest diameter wires studied, i.e., $d \sim 45\,\text{nm}$, exhibit a very large anisotropy in the scattering, i.e., $I_\parallel^X / I_\perp^X \sim 30$ for 488 nm excitation and even larger, i.e., $I_\parallel^X / I_\perp^X \sim 100$, for 514.5 nm excitation. These peaks are examples of strong *Raman* antenna effects coming from *photon* confinement in the waist of the NW. This antenna behavior is always observed for both LO and TO phonon Raman scattering in very small diameter NWs. As we shall see, this behavior is theoretically expected for TO scattering but *not* anticipated for LO scattering.

As we have shown in Section 32.6, the scattering matrix, $\tilde{S}$, appropriate for Raman scattering is the Raman tensor, $\tilde{\mathfrak{R}}$. Then the Raman intensity of NW including $\tilde{Q}$ can be written as

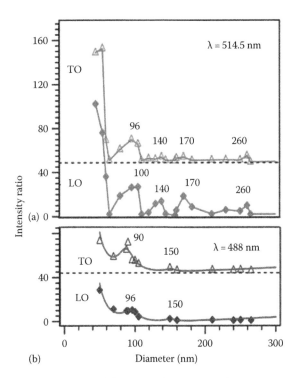

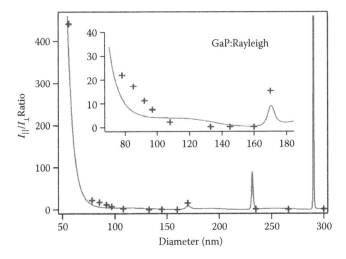

FIGURE 32.13 Ratio of the external backscattered Rayleigh radiation vs. GaP NW diameter d. Light incident at 514.5 nm perpendicular to the NW axis. Data are represented by (+) and the solid curve is the calculated result for an infinite GaP NW using the results of internal EM intensity (Figure 32.6 and Equation 32.7).

FIGURE 32.14 Ratio of the observed backscattered Raman intensity (LO and TO phonon backscattering) for incident polarization parallel and perpendicular to the NW axis: (a) with 514.5 nm excitation and (b) with 488 nm excitation. Peaks in the Raman intensity ratio are associated with the antenna effect (see text). (From Chen, G. et al., *Nano Lett.*, 8, 1341, 2008. With permission.)

$$I_{Raman}^{NW} \propto \omega^4 \left| \hat{e}_s \cdot \tilde{Q} \cdot \tilde{\mathfrak{R}} \cdot \tilde{Q} \cdot \hat{e}_i \right|^2 I_0. \qquad (32.16)$$

After transformations of **T** and **T′** in Equation 32.15, the Raman tensor has the general form

$$\tilde{\mathfrak{R}} = \begin{pmatrix} R_{11} & R_{12} & R_{13} \\ R_{21} & R_{22} & R_{23} \\ R_{31} & R_{32} & R_{33} \end{pmatrix}.$$

Using Equation 32.16, the scattered Raman intensity for SNWs then takes the form

$$I_{Raman}^{NW} \propto (R_{22}Q_{\parallel}^2 \cos^2\theta + R_{23}Q_{\parallel}Q_{\perp}\cos\theta\sin\theta$$

$$+ R_{32}Q_{\parallel}Q_{\perp}\cos\theta\sin\theta + R_{33}Q_{\perp}^2\sin^2\theta)^2 \qquad (32.17)$$

The values of R_{ij} depend on the "tipping" angle φ. Equation 32.17, together with the appropriate values for the transformed R_{ij} can be used to predict the backscattered Raman intensity pattern. In principle, values of $Q_{\parallel}$ and $Q_{\perp}$ can be obtained from the Rayleigh backscattering data and fitting to Equation 32.7, or the values can be obtained by calculation ($L = \infty$). If the Raman scattering is nonresonant, the R_{ij} are derived from the tensors, $\tilde{\mathfrak{R}}(\hat{e}_m)$ (Equation 32.9), via a suitable coordinate transformation by Equation 32.9 that depends on the growth direction $\langle\langle hkl \rangle\rangle$ and the particular axial rotation angle, φ. If the scattering is *resonant*, i.e., via the production of an exciton or via higher direct interband transitions, the form of the Raman matrix may be altogether different than the nonresonant forms found in Section 32.6. For example, a strong antenna effect was reported for WS_2 nanotube where strong resonant Raman scattering was expected. In this case, the Raman polar pattern was explained in terms of a Raman tensor that has the same form as the anisotropic polarizability tensor of the WS_2 nanotube (Rafailov et al. 2005).

As we have discussed above, SNWs at antenna resonance exhibit the property $Q_{\parallel} \gg Q_{\perp}$. In this case, Equation 32.17 can be simplified by neglecting terms involving $Q_{\perp}$. The Raman intensity for SNWs at strong antenna resonance is then

$$I_{NW\ Raman}^{TO,\ LO} \propto R_{22}^2 Q_{\parallel}^4 \cos^4\theta, \quad (Q_{\parallel} \gg Q_{\perp}; \text{any} \ll hkl \gg). \qquad (32.18)$$

This is a very important result. It tells us that for transformed LO or TO Raman tensors that exhibit a nonzero R_{22} element, the Raman intensity polar patterns for these phonons should both follow the $\sim\cos^4\theta$ form. In Figure 32.15, we show Raman results for two GaP NWs that exhibit a strong antenna effect for both TO and LO phonon scattering. The 50 nm wire (Figure 32.15a) was supported on an Si substrate. If this wire grows along the $\ll 110 \gg$ direction, the R_{22} element for both TO and LO are nonzero and the $\cos^4\theta$ fit agrees with Equation 32.18. In Figure 32.15b, we show results for a slightly larger diameter $d = 56$ nm GaP NW on TEM grid whose SAD pattern indicates a $\ll 111 \gg$

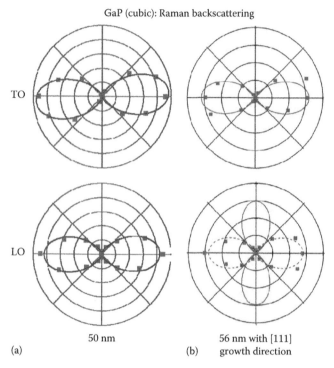

GaP (cubic): Raman backscattering

TO

LO

50 nm

(a)

56 nm with [111] growth direction

(b)

FIGURE 32.15 (a) Experimental data for the TO and LO polar plots for a 50 nm wire on Si substrate (unknown growth axis). Solid curves represent fits of a $\cos^4(\theta - \theta_0)$ to the data, where θ_0 is added as a parameter to account for a small rotation of the NW relative to the $\hat{x}_2$ axis. (b) TO and LO antenna polar plots for a 56 nm wire over a hole on a TEM grid; SAD indicates a $\ll 111 \gg$ growth axis. Data are indicated by the points, and solid curves for TO and LO are predictions according to Equation 32.16. The dashed curve in the LO polar plot is a $\cos^4\theta$ fitting to the data points and is not predicted by theory. For the NW in (b), the theory does not predict the observed LO pattern. This suggests that the form of the Raman tensor used to analyze the data for the 56 nm wire may be incorrect (see text).

growth direction. From the results calculated in Section 32.6, it can be shown that the transformed Raman tensor of a $\ll 111 \gg$ GaP NW has nonzero R_{22}^{TO} element for all tipping angles, φ. The $\cos^4\theta$ fit in Figure 32.15a agrees with Equation 32.18. However, the transformed LO Raman tensor elements of a GaP $\ll 111 \gg$ NW must satisfy $R_{22}^{LO} = R_{23}^{LO} = R_{32}^{LO} = 0$, and $R_{33}^{LO} \neq 0$. Therefore, from Equation 32.17, without any restrictions on $Q_{\parallel}$ and $Q_{\perp}$, we have for LO phonons

$$I_{NW\ Raman}^{LO} \propto R_{33}^2 Q_{\perp}^4 \sin^4\theta, \quad (\ll 111 \gg). \qquad (32.19)$$

In Figure 32.15b, the dashed curve is the $\cos^4\theta$ fitting to LO polar pattern, which is not the theoretical prediction. The rotated solid curve is the predicted LO polar pattern as indicated in Equation 32.19 for the GaP NW with $\ll 111 \gg$ growth axis.

In Figure 32.16, we display typical experimental Raman polar patterns, $I(\theta)$ (dots), for LO and TO backscattering collected from various individual GaP NWs using 514.5 nm excitation. Data in the figure (dots) are taken on wires either

GaP (cubic): Raman backscattering

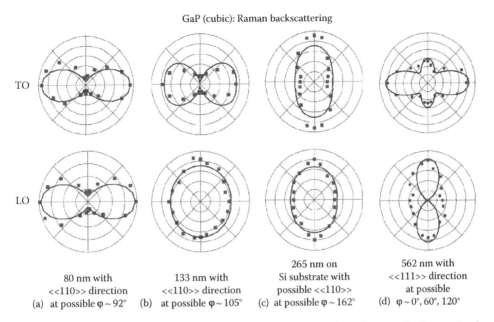

TO

LO

80 nm with <<110>> direction	133 nm with <<110>> direction	265 nm on Si substrate with possible <<110>>	562 nm with <<111>> direction at possible
(a) at possible φ ~ 92°	(b) at possible φ ~ 105°	(c) at possible φ ~ 162°	(d) φ ~ 0°, 60°, 120°

FIGURE 32.16 Experimental polar (θ) plots (dots) of the TO (upper) and LO (lower) polarized Raman backscattering from individual GaP NWs collected with 514.5 nm excitation (dots). Solid curves calculated according to Equation 32.16 for angles φ indicated. Column (a) $d = 80$ nm wire on TEM grid, column (b) $d = 133$ nm wire on a TEM grid, column (c) $d = 265$ nm on a Si substrate, column (d) $d = 562$ nm on TEM grid. The labels found below the diameter values indicate the growth axis and tipping angle φ measured between $\hat{x}_1$ and $\hat{x}_1'$. See text for method to obtain $Q_\parallel^2/Q_\perp^2$.

suspended over a hole in a TEM grid (c.f., $d = 80$, 133, and 562 nm) or supported on an Si substrate ($d = 265$ nm). As already discussed, for wires on a TEM grid, the TEM image and SAD pattern were used to determine the wire diameter and the growth axis; for NWs on Si substrates, an AFM z-scan was used to measure the diameter, and the growth direction, <<hkl>>, is not known. The theoretical curves are calculated on the basis of Equation 32.17 using either experimental values of $(Q_\parallel^2/Q_\perp^2)$ (from Rayleigh scattering) or results calculated from the infinite long wire $(I_\parallel^{int}/I_\perp^{int}) = (Q_\parallel^2/Q_\perp^2)$. For two NWs on TEM grids, we know the <<hkl>> from SAD. These NWs have $d = 80$ and 133 nm and they both grew along <<011>>. Both the experimental TO and LO patterns of a $d = 80$ nm wire (Figure 32.16a) can be described as an "open dipole" (i.e., the patterns are dipole-like and open at the origin); the solid curve is calculated from Equation 32.17 for a <<011>> wire with a $Q_\parallel/Q_\perp \sim 1.77$ (based on the infinite wire calculation) and φ = 92°. For the $d = 133$ nm <<011>> NW (Figure 32.16b), the experimental TO pattern is also an "open dipole," while the experimental LO pattern is nearly elliptical and oriented at 90° with respect to the TO pattern. Our best model calculation (solid curve) for a <<011>> wire uses $Q_\parallel/Q_\perp \sim 1.24$ (infinite long wire calculation) and φ = 105°. For the $d = 265$ nm NWs supported on Si, we observe nearly elliptical LO and TO patterns, oriented along θ = 90°; the solid curve is for a possible <<011>> wire with $Q_\parallel/Q_\perp \sim 0.89$ (value from Rayleigh scattering on a wire with same diameter) and φ = 162°. Finally, for the $d = 562$ nm <<111>> GaP NW, the solid curve is for $Q_\parallel/Q_\perp \sim 0.96$ (from Rayleigh scattering) and φ ~ 0°, 60°, or 120°.

32.9 Summary and Conclusions

Using GaP as a model NW system, we have presented results that indicate how polarized photons "squeeze" into individual NWs and how their antenna behavior can be observed. GaP (zinc blende) exhibits a cubic space lattice. This symmetry simplifies considerably the analysis of the data.

We have endeavored to show that when the wavelength of the incident light can drive a strong antenna resonance, the NW responds like a classical line antenna. For fixed wavelength and photons incident at right angles to the NW axis, a series of antenna resonances are predicted and observed. Rayleigh first predicted these resonances in 1918, but he probably was thinking about wires and wavelengths at a scale much larger than the tens to hundreds of nanometers applicable to the antennae we have considered here. During an antenna resonance, only the component of the incident and scattered E-field that is parallel to the NW determines the scattering from the wire. In our theoretical picture, this behavior is introduced via a diagonal tensor ($\tilde{Q}$) that describes the transmission of the EM fields into and out of the NW that are polarized parallel (∥) and perpendicular (⊥) to the NW axis. We have shown that Rayleigh scattering via the polarizability tensor can be used to provide an experimental value for the ratio ($Q_\parallel/Q_\perp$). This ratio can be obtained theoretically and provides a means of validating the model calculations. For conditions driving a strong antenna resonance, we have shown that $Q_\parallel \gg Q_\perp$; for conditions far away from resonance, we observe that $Q_\parallel \sim Q_\perp$. Large values of this ratio were then shown to distort the Raman backscattering polar pattern from what one observes in the bulk solid. However, this distortion is just due to $\tilde{Q}$ and the Raman selection rules that apply for the bulk still apply to the NW.

As matters now stand, the theoretical apparatus that we have used to explain the experimental results is doing a reasonably good job, except in the case of the smallest diameter NWs. There, we find trouble with the LO phonon backscattering pattern. Our best explanation (at this time) is that these NWs may have intragap optical adsorption that may invalidate the use of the nonresonant form of the Raman tensors. We are currently doing experiments to investigate this explanation.

Acknowledgments

This work was supported, in part, by the National Science Foundation grapheme NIRT program and the Materials Research Institute at Pennsylvania State University (PSU). The authors would like to thank Professor D. H. Werner, Dr. Gugang Chen, Dr. Qihua Xiong, Dr. M. E. Pellen, and J.S. Petko from PSU for their early, important contributions to this work. We are also indebted to Qiujie Lu and Duming Zhang for the helpful discussions as well as for having shared preliminary results of their work on Rayleigh scattering from NWs. We also gratefully acknowledge Dr. H. Gutierrez for his help with the NW growth and the TEM characterization of the GaP NWs.

References

Adu, K. W., H. R. Gutierrez, U. J. Kim, G. U. Sumanasekera, and P. C. Eklund 2005. Confined phonons in Si nanowires. *Nano Letters* **5**: 409–414.

Adu, K. W., Q. Xiong, H. R. Gutierrez, G. Chen, and P. C. Eklund 2006. Raman scattering as a probe of phonon confinement and surface optical modes in semiconducting nanowires. *Applied Physics A-Materials Science & Processing* **85**: 287–297.

Ajiki, H. and T. Ando 1996. Aharonov-Bohm effect on magnetic properties of carbon nanotubes. *Physica B* **216**: 358–361.

Bohren, C. F. and D. R. Huffman 1983. *Absorption and Scattering of Light by Small Particles*. New York, Wiley.

Boukai, A. I., Y. Bunimovich, J. Tahir-Kheli et al. 2008. Silicon nanowires as efficient thermoelectric materials. *Nature* **451**: 168–171.

Cao, L. Y., B. Nabet, and J. E. Spanier 2006. Enhanced Raman scattering from individual semiconductor nanocones and nanowires. *Physical Review Letters* **96**: 157402.

Cardona, M. and R. Merlin 2007. *Light Scattering in Solids. IX, Novel Materials and Techniques*. Berlin, Germany, Springer.

Chang, R. K. and A. J. Campillo 1996. *Optical Processes in Microcavities*. Singapore, World Scientific.

Chen, G., J. Wu, Q. J. Lu et al. 2008. Optical antenna effect in semiconducting nanowires. *Nano Letters* **8**: 1341–1346.

Chew, H. and D. S. Wang 1982. Double-resonance in fluorescent and Raman-scattering by molecules in small particles. *Physical Review Letters* **49**: 490–492.

Draine, B. T. and P. J. Flatau 1994. Discrete-dipole approximation for scattering calculations. *Journal of the Optical Society of America A-Optics Image Science and Vision* **11**: 1491–1499.

Draine, B. T. and P. J. Flatau 2004. *User Guide for the Discrete Dipole Approximation Code DDSCAT6.1*. http://arxiv.org/abs/astro-ph/0409262.

Dresselhaus, M. S., G. Dresselhaus, and P. Avouris 2001. *Carbon Nanotubes: Synthesis, Structure, Properties, and Applications*. Berlin, Germany, Springer.

Duan, X. F. and C. M. Lieber 2000. General synthesis of compound semiconductor nanowires. *Advanced Materials* **12**: 298–302.

Duesberg, G. S., I. Loa, M. Burghard, K. Syassen, and S. Roth 2000. Polarized Raman spectroscopy on isolated single-wall carbon nanotubes. *Physical Review Letters* **85**: 5436–5439.

Fukata, N., T. Oshima, N. Okada et al. 2006. Phonon confinement in silicon nanowires synthesized by laser ablation. *Physica B-Condensed Matter* **376**: 864–867.

Givargizov, E. I. 1975. Fundamental aspects of Vls growth. *Journal of Crystal Growth* **31**: 20–30.

Gupta, R., Q. Xiong, G. D. Mahan, and P. C. Eklund 2003. Surface optical phonons in gallium phosphide nanowires. *Nano Letters* **3**: 1745–1750.

Hashimoto, A., K. Suenaga, A. Gloter, K. Urita, and S. Iijima 2004. Direct evidence for atomic defects in graphene layers. *Nature* **430**: 870–873.

Hayes, W. and R. Loudon 1978. *Scattering of Light by Crystals*. New York, Wiley.

Heinz, T. F. 2008. Rayleigh scattering spectroscopy. *Carbon Nanotubes* **111**: 353–369.

Hochbaum, A. I., R. K. Chen, R. D. Delgado et al. 2008. Enhanced thermoelectric performance of rough silicon nanowires. *Nature* **451**: 163–167.

Huang, M. Y., Y. Wu, B. Chandra et al. 2008. Direct measurement of strain-induced changes in the band structure of carbon nanotubes. *Physical Review Letters* **100**: 136803.

Kittel, C. 2005. *Introduction to Solid State Physics*, Chapter 18, p. 515. Hoboken, NJ, Wiley.

Lieber, C. M. and Z. L. Wang 2007. Functional nanowires. *MRS Bulletin* **32**: 99–108.

Loudon, R. 2001. The Raman effect in crystals. *Advances in Physics* **50**: 813–864.

Lu, W. and C. M. Lieber 2007. Nanoelectronics from the bottom up. *Nature Materials* **6**: 841–850.

Morales, A. M. and C. M. Lieber 1998. A laser ablation method for the synthesis of crystalline semiconductor nanowires. *Science* **279**: 208–211.

Novoselov, K. S., A. K. Geim, S. V. Morozov et al. 2005. Two-dimensional gas of massless Dirac fermions in graphene. *Nature* **438**: 197–200.

O'Connell, M. 2006. *Carbon Nanotubes: Properties and Applications*. Boca Raton, FL, Taylor & Francis.

Patolsky, F., B. P. Timko, G. H. Yu et al. 2006a. Detection, stimulation, and inhibition of neuronal signals with high-density nanowire transistor arrays. *Science* **313**: 1100–1104.

Patolsky, F., G. F. Zheng, and C. M. Lieber 2006b. Fabrication of silicon nanowire devices for ultrasensitive, label-free, real-time detection of biological and chemical species. *Nature Protocols* 1: 1711–1724.

Patolsky, F., G. F. Zheng, and C. M. Lieber 2006c. Nanowire-based biosensors. *Analytical Chemistry* 78: 4260–4269.

Pauzauskie, P. J. and P. Yang 2006. Nanowire photonics. *Materials Today* 9: 36–45.

Piscanec, S., M. Cantoro, A. C. Ferrari et al. 2003. Raman spectroscopy of silicon nanowires. *Physical Review B* 68: 241312.

Popov, V. N., P. Lambin and North Atlantic Treaty Organization. 2006. *Carbon Nanotubes: From Basic Research to Nanotechnology*. Dordrecht, the Netherlands, Springer.

Rafailov, P. M., C. Thomsen, K. Gartsman, I. Kaplan-Ashiri, and R. Tenne 2005. Orientation dependence of the polarizability of an individual WS2 nanotube by resonant Raman spectroscopy. *Physical Review B* 72: 205436.

Rahman, F. 2008. *Nanostructures in Electronics and Photonics*. Singapore, Pan Stanford Publishing (distributed by World Scientific).

Rayleigh, L. 1918. On the dispersion of light by a dielectric cylinder. *Philosophical Magazine* 36: 365.

Saito, R., A. Jorio, J. H. Hafner et al. 2001. Chirality-dependent G-band Raman intensity of carbon nanotubes. *Physical Review B* 6408: art. no.-085312.

Sfeir, M. Y., F. Wang, L. M. Huang et al. 2004. Probing electronic transitions in individual carbon nanotubes by Rayleigh scattering. *Science* 306: 1540–1543.

Sfeir, M. Y., T. Beetz, F. Wang et al. 2006. Optical spectroscopy of individual single-walled carbon nanotubes of defined chiral structure. *Science* 312: 554–556.

Slusher, R. E. 1994. Optical processes in microcavities. *Semiconductor Science and Technology* 9: 2025–2030.

Vahala, K. J. 2003. Optical microcavities. *Nature* **424**: 839–846.

Wagner, R. S. and W. C. Ellis 1964. Vapor-liquid-solid mechanism of single crystal growth (new method growth catalysis from impurity whisker epitaxial + large crystals Si E). *Applied Physics Letters* 4: 89–90.

Wang, Y., K. Kempa, B. Kimball et al. 2004. Receiving and transmitting light-like radio waves: Antenna effect in arrays of aligned carbon nanotubes. *Applied Physics Letters* 85: 2607–2609.

Wang, F., M. Y. Sfeir, L. M. Huang et al. 2006. Interactions between individual carbon nanotubes studied by Rayleigh scattering spectroscopy. *Physical Review Letters* 96: 167401.

Wu, J., A. Gupta, H. R. Humberto, and P. C. Eklund 2009. Cavity-enhanced stimulated Raman scattering from short GaP nanowires, *Nano Letters* 9: 3252–3257.

Xia, Y. N., P. D. Yang, Y. G. Sun et al. 2003. One-dimensional nanostructures: Synthesis, characterization, and applications. *Advanced Materials* 15: 353–389.

Xiong, Q. H., J. G. Wang, O. Reese, L. Voon, and P. C. Eklund 2004. Raman scattering from surface phonons in rectangular cross-sectional w-ZnS nanowires. *Nano Letters* 4: 1991–1996.

Xiong, Q. H., J. Wang, and P. C. Eklund 2006. Coherent twinning phenomena: Towards twinning superlattices in III-V semiconducting nanowires. *Nano Letters* 6: 2736–2742.

Yamamoto, Y. and R. E. Slusher 1993. Optical processes in microcavities. *Physics Today* **46**: 66–73.

Yoffe, A. D. 1993. Low-dimensional systems: Quantum-size effects and electronic-properties of semiconductor microcrystallites (zero-dimensional systems) and some quasi-2-dimensional systems. *Advances in Physics* 42: 173–266.

Yoffe, A. D. 2001. Semiconductor quantum dots and related systems: Electronic, optical, luminescence and related properties of low dimensional systems. *Advances in Physics* 50: 1–208.

Yoffe, A. D. 2002. Low-dimensional systems: Quantum size effects and electronic properties of semiconductor microcrystallites (zero-dimensional systems) and some quasi-two-dimensional systems. *Advances in Physics* 51: 799–890.

Yu, P. Y. and M. Cardona 2001. *Fundamentals of Semiconductors: Physics and Materials Properties*. Berlin, Germany, Springer.

Zhang, Y. B., Y. W. Tan, H. L. Stormer, and P. Kim 2005. Experimental observation of the quantum Hall effect and Berry's phase in graphene. *Nature* **438**: 201–204.

Zhang, D. M., J. Wu, Q. J. Lu, H. R. Gutierrez, and P. C. Eklund 2009. Polarized Rayleigh back-scattering from individual GaP nanowires, currently under review.

33

Theory of Quantum Ballistic Transport in Nanowire Cross-Junctions

Kwok Sum Chan
City University of Hong Kong

33.1 Introduction

There is a widespread and intense research interest in the fabrication, characterization, and manipulation of semiconductor nanowires (Duan and Lieber 2000, Duan et al. 2000, Lee et al. 2001, Zhang et al. 2001, Bjork et al. 2002), with the aim to develop the novel "bottom up" approach to microelectronic circuit fabrication. In this approach, nanodevices and nanocircuits are assembled from the bottom up, using nano-size building blocks, such as carbon nanotubes (Rao et al. 1997, 2003, Tans et al. 1998, Liang et al. 2001) or semiconductor nanowires. This is a paradigm shift from the traditional "top down" approach, in which nano-size devices are etched into a bulk material using lithographic techniques (Kalliakos et al. 2007, Perez-Martinez et al. 2007). As it is now easy to produce nano-building blocks in large numbers with relatively inexpensive equipment, current research focus is also on developing techniques for manipulating the nano-building blocks (Whang et al. 2003a,b, Pauzauskie et al. 2006, Jamshidi et al. 2008), novel device structures (Cui and Lieber 2001, Duan et al. 2001, Huang et al. 2001b), and circuit architectures (Dehon 2003, Dehon et al. 2003, Zhong et al. 2003) that suit the "bottom up" approach.

Recently, fabrication of nanowire devices using the "bottom up" approach has been demonstrated experimentally (Cui and Lieber 2001, Duan et al. 2001, Huang et al. 2001b). In these experiments, nanowires were put on top of nanowires so that electrical contacts could be formed between some of the wires.

These contacts were then used to define or form the devices and circuits. The device structures of these novel nanodevices are different from the traditional ones fabricated using the "top down" approach as nano-size junctions play an important role in determining the device characteristics. In combination with some nanowire manipulation techniques, such as the solution-based Langmuir–Boldgett (LB) film technique (Whang et al. 2003b), the new device structure enables the fabrication of arrays or circuits of nanowire devices. For example, nanowires can be aligned to form a nanowire array, which is then rotated along a desirable direction and put on top of another nanowire array to form a nanowire grid structure (Huang et al. 2001a). It is possible to stack several layers of nanowires with this technique and construct 3D electronic circuits in the stack of nanowires.

In Figure 33.1, we show schematically how a p–n junction is constructed from two nanowires by putting an n-type wire on top of a p-type wire. Figure 33.2 shows the scanning electron micrograph of a silicon nanowire cross-junction and the *I–V* characteristics of p–n, p–p, and n–n nanowire junctions obtained by Lieber's group in Harvard University (Cui and Lieber 2001). The nanowire cross-junction was formed by a sequential deposition of solutions of n- and p-type nanowires, and contacts to the nanowires were defined by electron beam lithography. The rectifying behavior is clearly seen in the *I–V* characteristics, while approximately linear *I–V* behavior is found in p–p and n–n junctions. Similarly, nanowire p–n–p bipolar junction transistors can be made from three nanowires (2 p-type wires and 1 n-type wire)

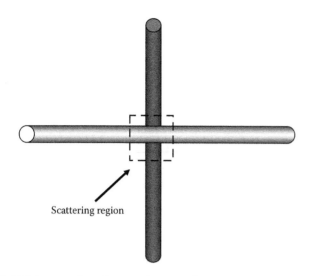

Scattering region

FIGURE 33.1 The schematic diagram of a nanowire p–n junction (cross-junction) formed by putting an n-type nanowire (light grey) on top of a p-type nanowire (dark grey). As an illustration, a dashed rectangle is also shown surrounding the scattering region of a quantum transport model. The wires outside the dashed rectangle are leads connected to the scattering region in the model.

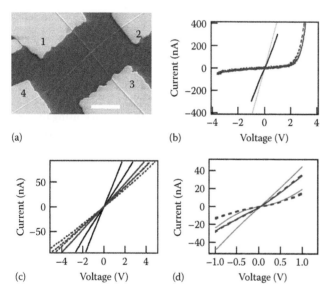

FIGURE 33.2 (a) shows the electron micrograph of a silicon nanowire cross-junction. (b), (c), and (d) show the *I–V* characteristics of silicon nanowire cross-junctions. One of the *I–V* characteristics shown in (b) shows the rectifying behavior of a p–n junction; others resemble Ohmic junctions. (Reprinted from Cui, Y. and Lieber, C.M., *Science*, 291, 851, 2001. With permission.)

as shown schematically in Figure 33.3. The experimental demonstration of nanowire bipolar junction transistors as well as of an inverter circuit using crossed silicon nanowires has also been reported (Cui and Lieber 2001).

Nanowire logic circuits and nanowire computation were also demonstrated by Lieber's research group (Huang et al. 2001b). Figure 33.4 shows the scanning electron micrographs of the three

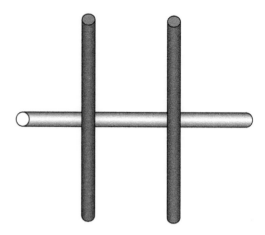

FIGURE 33.3 The schematic diagram of a nanowire p–n–p bipolar junction transistor formed from 2 p-type nanowires (dark grey) and one n-type nanowire (light grey).

logic gates, an OR gate, an AND gate, and a NOR gate, constructed from two nanowire p–n junctions. In the OR gate, three nanowires, two p-doped and one n-doped, are used to form two p–n junctions. In both the AND gate and the NOR gate, only two p–n junctions are formed from three of the four wires used in the circuit. The fourth wire is used as a gate to modify the resistivity of one of the nanowires to form a resistor in the circuit. Figure 33.5 shows a scanning electron micrograph of a nanowire field effect transistor decoder consisting of eight nanowires studied by Zhong et al. (2003). Four nanowires are used as inputs to control the other four output wires. The diagonal cross-junctions are specially modified so that the input wires I_1, I_2, I_3 and I_4 can be used to control the output wires O_1, O_2, O_3 and O_4.

From these experimental results, it is quite clear that nanowire cross-junctions play a key role in nanowire devices and circuits; an understanding of the transport properties of nanowire cross-junctions is crucial to the development of nanowire technology. There are a number of theoretical studies (Büttiker 1986, Ravenhall et al. 1989, Gaididei et al. 1992) of quantum wire (or quantum waveguide) cross-junctions in the literature, but most of these studies considered quantum wires lying on the same plane and cannot be applied to nanowire junctions with wires lying on different planes. The exceptions are the studies by Takagaki and Ploog (1993) and the author's group (Wei and Chan 2005, Chan and Wei 2007). Takagaki and Ploog (1993) used a continuum model and the wave-matching method to find the transmission functions between two square nanowires lying on different planes. It is difficult to apply this approach to circular wire junctions with different interwire coupling strengths. We have adopted a tight-binding model approach that can overcome all these shortcomings. It is, therefore, useful to present in this chapter a tutorial review of the theory and physics of quantum ballistic transport in nanowire cross-junctions for researchers with different backgrounds. We have included in some subsections background material that is not suitable for a journal article but useful for beginners in the field, such as the Green's function technique and the quantum transport theory;

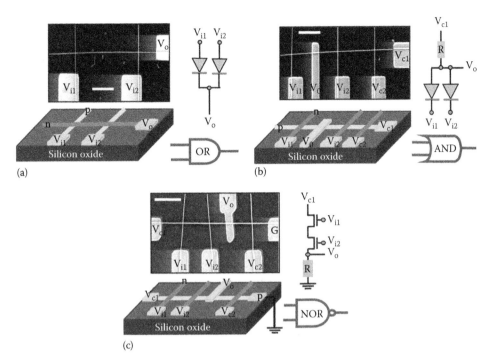

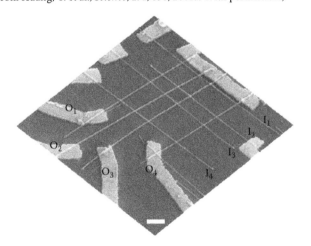

FIGURE 33.4 Electron micrographs and schematic diagrams of nanowire logic gate circuits. (a) OR gate. (b) AND gate. (c) NOR gate. (Reprinted from Huang, Y. et al., *Science*, 291, 630, 2001a. With permission.)

FIGURE 33.5 Electron micrograph of a nanowire decoder formed from eight nanowires. I_1, I_2, I_3, and I_4 are input wires. O_1, O_2, O_3, and O_4 are output wires. (Reprinted from Zhong, Z. et al., *Science*, 302, 1377, 2003. With permission.)

so, beginning researchers can readily learn how to model cross-junctions from this chapter. Experienced researchers, however, may want to skip these subsections and jump to those that deal directly with issues pertaining to nanowire cross-junctions.

33.2 Theory and Model

In this section, the background theory and the tight-binding models for determining the quantum ballistic transmission coefficients of an electron traveling through a nanowire cross-junction are discussed.

33.2.1 Transmission Function of a Cross-Junction

In nanowire devices or circuits built from cross-junctions, the transmission of an electron between wires is determined by the quantum scattering of the electron by the cross-junction as phase coherence is maintained when the electron travels through the cross-junction. The cross-junction is regarded as the scattering region (surrounded by the dotted rectangle in Figure 33.1) in a quantum transport model, which scatters electrons incident from one of the nanowires connected to the junction into the other nanowires. The nanowires connected to the cross-junction are usually referred to as the "terminals" or "leads" in the language of quantum transport theory. In the cross-junction shown in Figure 33.1, there are four terminals or leads connected to the scattering region in the center.

The electron transmission probability from one terminal into the other terminals, or back to the incoming terminal, is given by the scattering wave function. The scattering wave function in terminal, β, for an incident electron in mode, n, of terminal, α, is given by

$$\psi(x_\beta, y_\beta, z_\beta) = \sum_m S_{\beta,m;\alpha,n} \phi_m(x_\beta, y_\beta) \exp(ik_m z_\beta)$$
$$+ \delta_{\alpha\beta} \phi_n(x_\alpha, y_\alpha) \exp(-ik_n z_\alpha), \quad (33.1)$$

where

$(x_\beta, y_\beta, z_\beta)$ are local coordinates defined for terminal, β, as in Figure 33.6

$\phi_m(x_\beta, y_\beta)$ and $\phi_n(x_\alpha, y_\alpha)$ are wave functions of the transverse modes (subbands), m and n, of terminals, β and α, respectively

Here, different x–y–z coordinates are defined for the terminals and they are distinguished by the Greek letter subscripts.

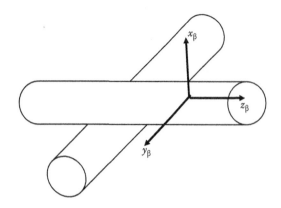

FIGURE 33.6 Schematic diagram of a nanowire cross-junction showing the coordinate axis used in the present chapter.

$S_{\beta,m;\alpha,n}(E)$ is the scattering matrix (in short, the S-matrix) relating the incident and outgoing wavefunctions. It depends on the incident energy, E, and is related to the transmission function, $T_{\beta\alpha}(E)$, by $T_{\beta\alpha}(E) = \sum_{m,n} \left| S_{\beta,m;\alpha,n}(E) \right|^2$. The transmission function can be used to calculate the current at lead, β, in a biased junction with different Fermi energies in the leads according to

$$I_\beta = \int \frac{2e}{h} T_{\beta\alpha}(E)[f_\beta(E) - f_\alpha(E)]dE, \qquad (33.2)$$

where $f_{\beta(\alpha)}(E) = 1/(\exp((E - E_F^{\beta(\alpha)})/kT)+1)$ is the Fermi–Dirac distribution for lead, $\beta(\alpha)$, with the Fermi energy, $E_F^{\beta(\alpha)}$. Here, we assume that an external bias voltage produces different Fermi energies in the leads without any modification of the potential profile of the scattering region. In the linear response regime, the bias voltage in the junction, which equals $-(E_F^\beta - E_F^\alpha)/e$, is small. The current can be approximated by

$$I_\beta = \frac{2e}{h} \int T_{\beta\alpha}(E)\left(-\frac{\partial f_\alpha}{\partial E}\right)dE \times (E_F^\beta - E_F^\alpha). \qquad (33.3)$$

At low temperatures, the Fermi–Dirac distribution resembles a step function, $f_\alpha(E) \approx \theta(E_F^\alpha - E)$; only electrons at the Fermi energy contribute to the current because the derivative of the Fermi–Dirac distribution sharply peaks at the Fermi energy, $-\frac{\partial f_\alpha}{\partial E} \approx \delta(E_F^\alpha - E)$. If the transmission function has a weak dependence on energy around the Fermi energy, the conductance of the junction between leads, α and β, is approximately given by $G_{\beta\alpha}(E_F^\alpha) = \frac{2e^2}{h} T_{\beta\alpha}(E_F^\alpha)$.

33.2.2 Green's Function

The S-matrix is related to the retarded Green's function of the junction by

$$S_{\beta,m;\alpha,n}(E) = -\delta_{\beta a} + i\hbar\sqrt{v_m v_n}$$
$$\times \iint_{A_\beta A_\alpha} \phi_m^*(\vec{\rho}_\beta) G_{\beta\alpha}^R(E,\vec{\rho}_\beta,z_\beta,\vec{\rho}_\alpha,z_\alpha)\phi_n(\vec{\rho}_\alpha)d^2\vec{\rho}_\beta d^2\vec{\rho}_\alpha,$$

$$(33.4)$$

where
  $v_{m(n)}$ is the electron velocity in mode, $m(n)$
  $\vec{\rho}_{\beta(\alpha)} = x_{\beta(\alpha)}\hat{i} + y_{\beta(\alpha)}\hat{j}$ is a position vector in the transverse plane of lead, $\beta(\alpha)$
  $A_{\beta(\alpha)}$ is the cross-sectional area of the lead

$G_{\beta\alpha}^R(E,\vec{\rho}_\beta,z_\beta,\vec{\rho}_\alpha,z_\alpha)$ is the retarded Green's function of the junction at energy, E, related to the junction's Hamiltonian, $\hat{H}(\vec{r})$, by

$$\left[E - \hat{H}(\vec{r})\right]G^R(E,\vec{r},\vec{r}') = \delta(\vec{r} - \vec{r}'). \qquad (33.5)$$

The retarded Green's function, $G^R(E,\vec{r},\vec{r}')$, is a Green's function of the time-independent Schrodinger equation, which can be used to find the stationary scattered wave solution of the time-independent Schrodinger equation (Schiff 1968). As a consequence, the S-matrix can be expressed in terms of the retarded Green's function as in Equation 33.4.

It is difficult to find a simple analytical expression for the Green's function of a structure with complex geometry. Nevertheless, the following approaches can be used to find the Green's function numerically. If the eigenfunctions, $\psi_{E'}(\vec{r})$, of the Hamiltonian are available, the Green's function can be obtained by using the expression, $G^R(E,\vec{r},\vec{r}') = \sum_{E'} (\psi_{E'}^*(\vec{r})\psi_{E'}(\vec{r}'))/(E - E' + i\gamma)$, where $i\gamma$ ($\gamma > 0$) is a small imaginary part. To obtain the advanced Green's function, $\gamma < 0$ is used. Both the retarded and advanced Green's functions are Green's functions of the time-independent Schrodinger equation, which can be used to find scattered wave solutions of the Schrodinger equation, but they correspond to imposing different boundary conditions on the solutions. The retarded Green's function corresponds to an outgoing wave solution, while the advanced Green's function corresponds to an incoming wave solution. For the present purpose, it is sufficient to consider the retarded Green's function only. For a system which has a perturbation, $\hat{V}$, the Green's function can be found by solving the Dyson equation

$$G^R(E,\vec{r},\vec{r}') = G_0^R(E,\vec{r},\vec{r}')$$
$$+ \iint G_0^R(E,\vec{r},\vec{r}'')\Sigma(E,\vec{r}'',\vec{r}''')G^R(E,\vec{r}''',\vec{r}')d^3\vec{r}''d^3\vec{r}'''.$$

$$(33.6)$$

where
  G_0^R is the Green's function of the unperturbed Hamiltonian, $\hat{H}_0 = \hat{H} - \hat{V}$
  $\Sigma(E,\vec{r}'',\vec{r}''')$ is the self energy due to the perturbation, $\hat{V}$

The wave functions or the Green's function, which can be used to determine the observable properties of a physical system, are continuous functions of the spatial coordinates in continuum models. However, to determine numerically the physical observables, knowing the wave function or the Green's function at discrete points in space is already sufficient, provided that the separations between the discrete points are small enough to have the desired accuracy. Moreover, the task of finding the Green's function can be simplified by discretizing the coordinates, which converts the Hamiltonian operator and the Green's function into matrices and operator equations into matrix equations. The matrix equation for the Green's function after discretization, which can be solved by matrix inversion, is

$$\sum_k \left[E - H(l,k) + i\gamma \right] G^R(E,k,j) = \delta_{l,j} \qquad (33.7)$$

or, in the matrix notation,

$$\left[(E+i\gamma)\tilde{I} - \tilde{H} \right] \tilde{G}^R(E) = \tilde{I}. \qquad (33.8)$$

Here, l, j, k are indexes used to denote the discrete points in space and $\tilde{I}$ represents the identity matrix. The small imaginary part, $i\gamma$ ($\gamma > 0$), is added to the matrix equation to obtain the retarded Green's function. Dyson's equation is also a matrix equation instead of an integral equation in discrete coordinates, $\tilde{G}^R(E) = \tilde{G}_0^R(E) + \tilde{G}_0^R(E)\tilde{\Sigma}(E)\tilde{G}^R(E)$.

For finite-size systems, the Green's functions are finite matrices; so they can be found from the inversion of the matrix, $(E + i\gamma)\tilde{I} - \tilde{H}$ as $\tilde{G}^R(E) = [(E + i\gamma)\tilde{I} - \tilde{H}]^{-1}$, or from the following equation derived from Dyson's equation: $\tilde{G}^R(E) = (\tilde{I} - \tilde{G}_0^R(E)\tilde{\Sigma}(E))^{-1}\tilde{G}_0^R(E)$. However, a nanowire cross-junction is an open system, which is infinite in size, and hence is not suitable for a straightforward application of these two equations. For an infinitely large structure, the problem of finding the Green's function is simplified by using a tight-binding Hamiltonian and dividing the infinite structure into a scattering region and leads connected to the scattering region. With this approach, a finite-size matrix equation for the Green's function, which is similar in structure to Equation 33.8 given above, can be derived from Dyson's equation as $[(E + i\gamma)\tilde{I} - \tilde{H}_s]\tilde{G}^R(E) = \tilde{I}$; here, $\tilde{H}_s$ is the modified Hamiltonian of the scattering region, which includes the effect of the leads as self-energies. The Green's function can then be easily found by inverting the finite matrix $(E + i\gamma)\tilde{I} - \tilde{H}_s$. In Section 33.2.4, we describe in detail how the modified Hamiltonian is derived using Dyson's equation for tight-binding models. When the dimensions of the wires are large, the number of tight-binding sites in the scattering region is large and the inversion of the matrix, $(E + i\gamma)\tilde{I} - \tilde{H}_s$, is computationally slow because the size of the matrix is large. For this situation, we have developed a modular approach based on Dyson's equation, which reduces the sizes of the matrices inverted in computation. In Section 33.2.4, this modular approach will be described in detail.

33.2.3 Tight-Binding Models

As pointed out above, the calculation of the Green's function of a finite region in a nanowire cross-junction is simplified by using a tight-binding Hamiltonian for the junction, because the Green's function for the finite region can be obtained by inverting a finite-size matrix as in a finite-size system. The effect of the infinite region outside the finite region can be included

as self-energy functions in the tight-binding formulation. Before we discuss in detail how the Green's function is calculated numerically in Section 33.2.4, it is necessary to describe the tight-binding models of nanowire junctions. Two tight-binding models are discussed; one is suitable for wires with polygonal cross section, and the other is suitable for wires with circular cross section.

The simple tight-binding (TB) models, which are widely used to approximate continuum Hamiltonians, consist of a number of tight-binding sites (or artificial atoms), each of which has a single electron s-orbital that can accommodate one electron if spin is ignored. Electrons in the s-orbitals can hop to the s-orbitals in neighboring TB sites. The spatial arrangement of the TB sites depends on the geometry or boundary of the system considered. For example, the TB sites of a circular wire are arranged in concentric circles, which is different from square or rectangular wires to approximate the circular boundary.

The simplest TB model has the TB sites occupying the lattice points of a simple cubic lattice (Figure 33.7), and all the hopping couplings between neighboring sites are identical. The TB sites are denoted by three integral indices, l, m, and n. The TB Hamiltonian is written in terms of the creation, $a_{l,m,n}^+$, and annihilation, $a_{l,m,n}$, operators of the s-orbital as

$$\hat{H} = \sum_{l,m,n} e a_{l,m,n}^+ a_{l,m,n} + \sum_{\substack{l,m,n \\ <l,m,n>}} t a_{<l,m,n>}^+ a_{l,m,n} \qquad (33.9)$$

where

$<l,m,n>$ denotes the neighboring TB sites of site (l,m,n)

e and t are respectively the orbital energy and the hopping matrix element between neighboring orbitals

This TB Hamiltonian can be obtained by replacing differential operators in the continuum kinetic energy operator, $\dfrac{-\hbar^2}{2m}((\partial^2/\partial x^2) + (\partial^2/\partial y^2) + (\partial^2/\partial z^2))$, by finite difference

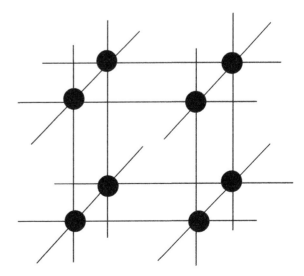

FIGURE 33.7 Tight-binding sites in a simple cubic tight-binding model.

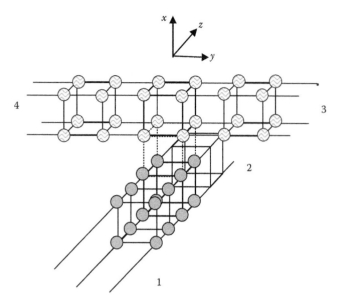

FIGURE 33.8 Tight-binding sites in a square nanowire cross-junction. Each square wire has a 2×2 cross section. 1, 2, 3, and 4 are labels of the leads.

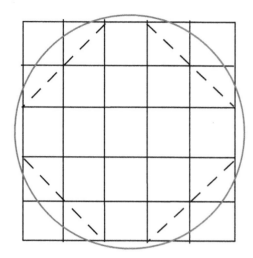

FIGURE 33.9 Schematic diagram showing how a simple cubic tight-binding model can be used to approximate a nanowire with circular cross section. Tight-binding sites within the grey circle have lower site energy, leading to electron confinement within the grey circle. As the number of the tight-binding sites is small, it is an approximation of a polygonal cross section.

approximations such as $\frac{\partial f}{\partial x} \approx (f_{l+1} - f_l)/\Delta$; here, Δ is the distance between two neighboring TB sites, $t = -\hbar^2/(2m\Delta^2)$ and $e = -6\hbar^2/(2m\Delta^2)$.

In Figure 33.8, a schematic diagram is shown of a cross-junction formed by 2 square wires (each has a 2×2 square cross section). The dotted lines joining the two nanowires represent the hopping coupling between the two wires. In reality, there is a thin layer of oxide separating the two wires in a junction and the probability of electron tunneling between the two nanowires is determined by the thickness of the oxide layer. Here, we use the simple model of hopping coupling between the two wires to represent the effect of the oxide layer. The strength of the hopping coupling can be varied to mimic the effect of oxide thickness. When the hopping coupling between two wires equals the hopping coupling between sites within the same wire, it is the strong coupling regime, in which there is no oxide layer hindering electron transfer between wires, and the hopping of electrons between the two wires is as easy as hopping within the same wire.

The total Hamiltonian of the nanowire junction is $\hat{H}_a + \hat{H}_c + \hat{H}_b$. $\hat{H}_a$ is the Hamiltonian of the upper wire with $a^+(a)$ representing the creation (annihilation) operator, and $\hat{H}_b$ is the Hamiltonian of the lower wire with $b^+(b)$ representing the creation (annihilation) operator. $\hat{H}_c$ is the coupling operator between two wires given by $\hat{H}_c = t_c \sum_{q=1}^{q=N_c} a_q^+ b_q + b_q^+ a_q$, where t_c is the interwire coupling constant, and $q(N_c)$ is the index (number) of the peripheral TB sites in the wires that participate in the interwire coupling.

The simple cubic TB model is not suitable for modeling a circular wire, although one can use a potential to confine the electron in a circular region in the center as in Figure 33.9, where sites inside the grey circle have lower potential energy than those

outside the circle. These central lattice points resemble a circular region only when the number of sites is very large. When the number of sites is small, the cross section defined by the grey circle resembles a polygon, which still has the fourfold symmetry of a square. In actual calculation, it is impractical to use a large number of TB sites to approximate a circular cross section, as computation time required is long.

To develop an accurate TB model with a small number of sites for circular wires, we need to use cylindrical coordinates and arrange the TB sites in concentric circular rings (a non-Cartesian lattice) as in Figure 33.10. To ensure the Hermitian property of the TB model defined on a non-Cartesian lattice, the TB Hamiltonian is derived by a variant approach, in which the energy of an electron in a circular wire defined on a circular grid is minimized. The total energy of an electron expressed in cylindrical coordinates is

$$E_{\text{total}} = \iiint \psi^*(\rho, \phi, z) \frac{-\hbar^2}{2m} \left[\frac{1}{\rho} \frac{\partial}{\partial \rho} \rho \frac{\partial}{\partial \rho} + \frac{1}{\rho^2} \frac{\partial^2}{\partial \phi^2} + \frac{\partial}{\partial z^2} \right]$$

$$\psi(\rho, \phi, z) \rho \, d\rho \, d\phi \, dz. \tag{33.10}$$

When the electron wave function, ψ, is confined to a wire of diameter, D, it satisfies the boundary condition, $\psi(\rho = D, \phi, z) = 0$. Before minimizing the total energy, the energy expression is rewritten in a more symmetrical form as

$$E_{\text{total}} = \frac{\hbar^2}{2m} \iiint \left\{ \left(\frac{\partial \psi^*}{\partial \rho} \right) \left(\frac{\partial \psi}{\partial \rho} \right) + \frac{1}{\rho^2} \left(\frac{\partial \psi^*}{\partial \phi} \right) \left(\frac{\partial \psi}{\partial \phi} \right) + \left(\frac{\partial \psi^*}{\partial z} \right) \left(\frac{\partial \psi}{\partial z} \right) \right\}$$

$$\rho \, d\rho \, d\phi \, dz. \tag{33.11}$$

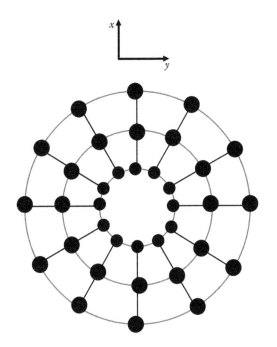

FIGURE 33.10 Tight-binding sites in the cross section for modeling a circular nanowire.

To avoid the singular term, $(\propto (1/\rho^2))$ in the Laplacian operator, the discrete coordinates are chosen to be $(\rho_i, \phi_j, z_k) = \left(\left[i + \dfrac{1}{2}\right]\Delta\rho, j\Delta\phi, k\Delta z\right)$. $i, j = 0, 1, 2, 3\ldots$, and $k = 0, \pm 1, \pm 2, \pm 3\ldots$ are the integral indices, and $\Delta\rho$, $\Delta\phi$, and Δz are the grid sizes. By replacing the differential operators by their finite difference expressions and integration by discrete summation, the approximate expression of the total energy is

$$E_{\text{total}} = \frac{\hbar^2}{2m}\Delta\rho\Delta\phi\Delta z$$

$$\left[\sum_{i,j,k}\rho_{i+1/2}\frac{(\psi^*(i+1,j,k) - \psi^*(i,j,k))(\psi(i+1,j,k) - \psi(i,j,k))}{\Delta\rho^2}\right.$$

$$+ \frac{1}{\rho_i}\frac{(\psi^*(i,j+1,k) - \psi^*(i,j,k))(\psi(i,j+1,k) - \psi(i,j,k))}{\Delta\phi^2}$$

$$+ \left.\frac{(\psi^*(i,j,k+1) - \psi^*(i,j,k))(\psi(i,j,k+1) - \psi(i,j,k))}{\Delta z^2}\right].$$

$$(33.12)$$

Note that the approximation of $\partial\psi/\partial\rho$ is taken at $\rho_{i+1/2} = (i + 1)$ $\Delta\rho$ in the expression to make the TB model Hermetian. The eigenvalues and the eigenfunctions of the Hamiltonian are given by minimizing the energy of the electron subjected to the normalization condition, $\iiint \rho\, d\rho\, d\phi\, dz|\psi|^2 = 1$, or, equivalently,

solving $(\partial E_{\text{total}}/\partial\psi^*(i,j,k)) - \zeta\rho_n\Delta\rho\Delta\phi\Delta z\psi(i,j,k) = 0$, where ζ is the Lagrange multiplier. The matrix eigenequation thus obtained is

$$\frac{\hbar^2}{2m}\left\{\frac{1}{\Delta\rho^2}\left[-\frac{\rho_{i+1/2}}{\rho_i}\psi(i+1,j,k) + 2\psi(i,j,k) - \frac{\rho_{i-1/2}}{\rho_i}\psi(i-1,j,k)\right]\right.$$

$$+ \frac{1}{\rho_i^2\Delta\phi^2}\left[-\psi(i,j+1,k) + 2\psi(i,j,k) - \psi(i,j-1,k)\right]$$

$$+ \left.\frac{1}{\Delta z^2}\left[-\psi(i,j,k+1) + 2\psi(i,j,k) - \psi(i,j,k-1)\right]\right\} = \zeta\psi(i,j,k).$$

$$(33.13)$$

Converting the matrix eigenequation into the TB format, the following TB model is obtained

$$\hat{H}_{\text{circular}} = \sum_{k}\sum_{i=1}^{N_\rho}\sum_{j=1}^{N_\phi}(e_k^z + e_i^\rho + e_j^\phi)a_{i,j,k}^+a_{i,j,k} + t_{i,i-1}^\rho a_{i,j,k}^+a_{i-1,j,k}$$

$$+ t_{i,i+1}^\rho a_{i,j,k}^+a_{i+1,j,k} + t_{j,j-1}^\phi a_{i,j,k}^+a_{i,j-1,k} + t_{j,j+1}^\phi a_{i,j,k}^+a_{i,j+1,k}$$

$$+ t^z(a_{i,j,k}^+a_{i,j,k-1} + a_{i,j,k}^+a_{i,j,k+1}).$$

$$(33.14)$$

where $a_{i,j,k}^+(a_{i,j,k})$ is the creation (annihilation) operator of the orbital at the TB site, i, j, k, of the circular grid. The TB parameters are given by

$$t^z = \frac{-\hbar^2}{2m\Delta z^2}, \quad e_k^z = -2t^z = \frac{\hbar^2}{m\Delta z^2}, \tag{33.15}$$

$$t_{i,i\pm 1}^\rho = \frac{-\hbar^2}{2m\Delta\rho^2}\frac{\rho_{i\pm 1/2}}{\sqrt{\rho_{i\pm 1}\rho_i}}, \quad e_i^\rho = \frac{\hbar^2}{m\Delta\rho^2}, \tag{33.16}$$

$$t_{j,j\pm 1}^\phi = -\frac{\hbar^2}{2m\rho_i^2\Delta\phi^2}, \quad e_j^\phi = \frac{\hbar^2}{m\rho_i^2\Delta\phi^2}. \tag{33.17}$$

As in square wire junctions, the total Hamiltonian of the cross-junction is the sum of three parts, $\hat{H}_L + \hat{H}_U + \hat{H}_T$, where $\hat{H}_L$, $\hat{H}_U$, and $\hat{H}_T$ are, respectively, the lower wire Hamiltonian, the upper wire Hamiltonian, and a hopping coupling between the two wires. The coupling term is similar to that in the square wire junction, $H_T = t_C\sum_{p=1}^{p=Q}(a_p^+b_p + b_p^+a_p)$, where a and b denote annihilation operators of the two wires of the junction, p denotes the indices of sites coupled by H_T, and Q is the number of sites in each wire participating in the interwire coupling. Figure 33.11 shows schematically how the TB sites are coupled through the interwire coupling. In the figure, four out of the sixteen TB sites in the circumference are coupled to another wire through $\hat{H}_T$.

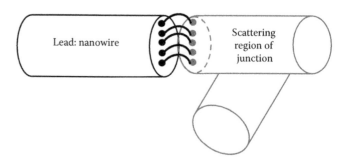

FIGURE 33.12 Schematic diagram showing how a lead is connected to the scattering region.

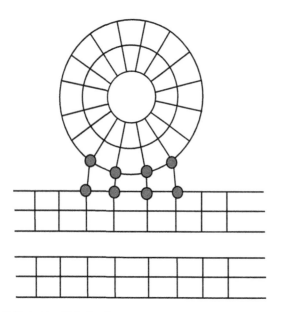

FIGURE 33.11 Tight-binding sites participate in interwire coupling in the cross section of a circular wire cross-junction.

33.2.4 Numerical Calculation of Green's Function

The nanowire junction is an open system with no boundary, which means the TB Hamiltonian is an infinite matrix. It is not possible to find the Green's function by simply inverting the matrix, $E\tilde{I} - \tilde{H}$, as no numerical packages can handle infinite matrices. The trick to circumvent this problem is to partition the system into a scattering region and the leads (the region outside the scattering region is the leads, which is also referred to as contacts), so that a finite-size, effective Hamiltonian matrix can be constructed for the scattering region, which can then be used to calculate the Green's function. The derivation of the effective Hamiltonian (Datta 1995) is as follows.

First, the Hamiltonian is partitioned into two parts: the Hamiltonian for the scattering region, $\hat{H}_S$, and the Hamiltonian for the leads, $\hat{H}_L$. The coupling between the two regions is $\hat{V}_C$. The equation for the Green function (Equation 33.8) is rewritten in terms of these sub-Hamiltonians or sub-matrices as

$$\begin{bmatrix} E + i\gamma - \tilde{H}_S & \tilde{V}_C \\ \tilde{V}_C^+ & E + i\gamma - \tilde{H}_L \end{bmatrix}\begin{bmatrix} \tilde{G}_S \\ \tilde{G}_C \end{bmatrix} = \begin{bmatrix} \tilde{I} & \tilde{0} \\ \tilde{0} & \tilde{I} \end{bmatrix}. \quad (33.18)$$

Eliminating $\tilde{G}_C$ from the equations, one can get

$$[E + i\gamma - \tilde{H}_S]\tilde{G}_S - \tilde{V}_C \frac{\tilde{I}}{(E + i\gamma - \tilde{H}_L)}\tilde{V}_C^+\tilde{G}_S = \tilde{I} \quad (33.19)$$

or

$$[E + i\gamma - \tilde{H}_S - \tilde{\Sigma}]\tilde{G}_S = \tilde{I} \quad (33.20)$$

with $\Sigma = \tilde{V}_C\,(\tilde{I}/(E + i\gamma - \tilde{H}_L))\tilde{V}_C^+ = \tilde{V}_C\tilde{G}_L\tilde{V}_C^+$ being the self-energy due to the coupling to the leads. $\tilde{G}_L = (\tilde{I}/(E + i\gamma - \tilde{H}_L))$ is the Green's function of the leads. Now, $\tilde{G}_S$ is given by the inverse of a finite-size matrix, $\tilde{G}_S = [E + i\gamma - \tilde{H}_S - \tilde{\Sigma}]^{-1}$. In a tight-binding model, the coupling $\tilde{V}_C$ couples the neighboring atomic sites on the surfaces of the lead and scattering region as shown schematically in Figure 33.12. So, we only need to know the Green's functions on the surfaces of the leads, i.e., the surface Green's functions of the leads. If the transverse and the longitudinal motions are separable in the leads, the surface Green's function of a lead can be written in terms of the transverse mode wave functions as $G_L(l, j) = \sum_m e^{ik_m a}\phi_m^*(l)\phi_m(j)$, where l and j denote the atomic sites on the surface of the lead, m denotes the transverse mode of the lead, a is the lattice spacing of the TB model, and k_m is the wave vector of mode, m. k_m is related to energy, E, by $E = E_m + 2t(1 - \cos k_m a)$, where E_m is the mth subband edge energy and t is the hopping strength along the longitudinal direction of the lead. To approximate the continuum, the number of TB sites in the model cannot be too small. Usually, a transverse dimension of 8–10 TB sites is sufficient for a good approximation of the continuum in an energy range covering the first six subbands.

If the size of the scattering region is very large, for example, in wires with large diameters or scattering regions with more than one junction, the calculation of the Green's function as the inverse of the effective Hamiltonian matrix is not computationally efficient. This problem can be circumvented using a modular approach that considers the junction as the coupling of two free-standing wires, and determines the junction's Green's function in terms of the free-standing wires' Green's function using Dyson's equation. For example, to find the scattering matrix between terminals, 1 and 3, shown in Figure 33.13, we need to find the Green's function between the two cross sections, 1 and 3, denoted as $G(1,3)$. This Green's function can be written in terms of the free-standing wire's Green's function by solving the following Dyson's equations:

$$G(1,3) = G^0(1,P)V_{PQ}G(Q,3), \quad (33.21)$$

$$G(Q,3) = G^0(Q,3) + G^0(Q,Q)V_{QP}G(P,3), \quad (33.22)$$

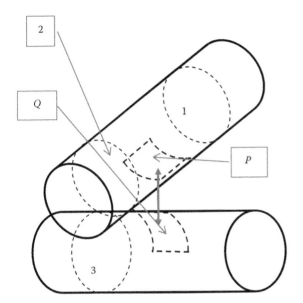

FIGURE 33.13 Schematic diagram showing the labeling of the cross sections and surfaces in a cross-junction used in Equations 33.21 through 33.28.

$$G(P,3) = G^0(P,P)V_{PQ}G(Q,3), \qquad (33.23)$$

where

G^0 is the Green's function of a free-standing wire
V_{PQ} is the tunneling coupling between the two wires through the surfaces, P and Q

Equation 33.23 can be substituted into Equation 33.22 to find $G(Q,3)$, which can then be substituted into Equation 33.21 to obtain $G(1,3)$ as

$$G(1,3) = G^0(1,P)V_{PQ}(1 - G^0(Q,Q)V_{QP}G^0(P,P)V_{PQ})^{-1}G^0(Q,3). \qquad (33.24)$$

To find the Green's function between cross sections 1 and 2, $G(1,2)$, we can use Dyson's equations shown below:

$$G(1,2) = G^0(1,2) + G^0(1,P)V_{PQ}G(Q,2), \qquad (33.25)$$

$$G(Q,2) = G^0(Q,Q)V_{QP}G(P,2), \qquad (33.26)$$

$$G(P,2) = G^0(P,2) + G^0(P,P)V_{PQ}G(Q,2). \qquad (33.27)$$

Substituting Equation 33.27 into Equation 33.26, we can obtain $G(Q,2)$, which can be substituted into Equation 33.25 to obtain $G(1,2)$ as

$$G(1,2) = G^0(1,2) + G^0(1,P)V_{PQ}[I - G^0(Q,Q)V_{QP}G^0(P,P)V_{PQ}]^{-1}$$
$$G^0(Q,Q)V_{QP}G^0(P,2). \qquad (33.28)$$

This approach is very efficient for calculating the Green's functions of scattering regions consisting of a large number of TB sites, because the sizes of the matrices used are the number of TB sites in either the cross sections or the interfaces between wires. Moreover, the Green's function of a free-standing wire is computationally easy to find (no inversion of large matrices is needed) because it can be expressed in terms of the transverse mode eigenfunctions and the Green's function of a one-dimensional TB chain as follows:

$$G^0(E,l_1,x_1,l_2,x_2) = \sum_\alpha \phi_\alpha^*(l_1)G_{1D}(E - E_\alpha,x_1,x_2)\phi_\alpha(l_2). \qquad (33.29)$$

G_{1D} can be obtained from the Green's function of a semi-infinite TB chain $G_{S1D}(E) = \left[-(1 - E/2t) \pm \sqrt{(1 - E/2t)^2 - 1} \right]$ by the following relation, which can be derived by using Dyson's equation:

$$G_{1D}(E,x_1,x_2) = \frac{G_{S1D}(-tG_{S1D})^{|x_1-x_2|}}{1 - t(G_{S1D})^2}. \qquad (33.30)$$

33.2.5 Transverse Modes

The transport properties of a nanowire junction depend on the coupling of the electron wave functions of the constituting wires. The transverse mode probability densities or mode profiles have a strong effect on this coupling. As bound and quasi-bound states, which lead to dips and peaks in the junction conductances, are formed at the junction by this coupling, knowledge of the transverse modes of a nanowire is crucial to the analysis of the features in the conductances. In this section, a discussion is given on the transverse modes, which forms the basis for the discussion of junction conductances in the following sections.

For a square or rectangular wire, the transverse x and y-direction wave functions are separable. A transverse mode is denoted by (α,β), where α is the quantum number for the y-direction wave function and β is that for the x-direction wave function, and the wave function is

$$\phi_{\alpha\beta}(l,j) = \frac{2\sin[l\beta\pi/(N+1)]\sin[j\alpha\pi/(M+1)]}{\sqrt{N+1}\sqrt{M+1}}, \qquad (33.31)$$

where

l and j are, respectively, the site indices along the x and y directions
N and M are the total number of sites along these two directions

Here, the x-direction is perpendicular to both wires in the junction, and the y-direction is parallel to one of the wires in the junction. The energies and the symmetry properties of the first six transverse modes are given in Table 33.1 for a rectangular wire. Note that the first and the fourth transverse modes are nondegenerate while the other transverse modes have degenerate modes in a square wire with $N = M$. In Figure 33.14, the transverse mode profiles of the first six modes of a square wire are shown schematically.

TABLE 33.1 The First Six Transverse Modes of an $N \times M$ Rectangular Wire and Their Subband Energies and Symmetry Properties

Mode	Subband Energy	Symmetry about X-Axis	Symmetry about Y-Axis
(1,1)	$E_1 = 4t - 2t \cos[\pi/(N+1)]$ $- 2t \cos[\pi/(M+1)]$	+	+
(1,2)	$E_2 = 4t - 2t \cos[2\pi/(N+1)]$ $- 2t \cos[\pi/(M+1)]$	+	−
(2,1)	$E_3 = 4t - 2t \cos[\pi/(N+1)]$ $- 2t \cos[2\pi/(M+1)]$	−	+
(2,2)	$E_4 = 4t - 2t \cos[2\pi/(N+1)]$ $- 2t \cos[2\pi/(M+1)]$	−	−
(3,1)	$E_5 = 4t - 2t \cos[\pi/(N+1)]$ $- 2t \cos[3\pi/(M+1)]$	+	+
(1,3)	$E_6 = 4t - 2t \cos[3\pi/(N+1)]$ $- 2t \cos[\pi/(M+1)]$	+	+

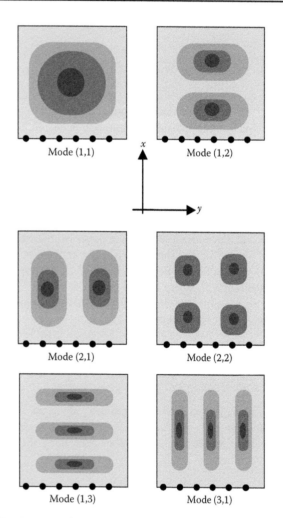

FIGURE 33.14 The transverse mode profiles of the lowest six subbands of a square nanowire.

The Schrodinger equation of a circular wire is separable into the longitudinal, angular, and radial parts, when expressed in cylindrical coordinates. Similarly, the TB eigenfunctions are also separable into a longitudinal part, Π_i, a radial part, Λ_j, and an angular part, Θ_k, as $\Phi_{i,j,k} = \Pi_i \Lambda_j \Theta_k / \sqrt{\rho}$, which satisfies respectively the following equations:

$$\frac{\hbar^2}{2m} \frac{[\Theta_{j-1} - 2\Theta_j + \Theta_{j+1}]}{\Delta\phi^2} = \omega\Theta_j, \tag{33.32}$$

$$\frac{\hbar^2}{2m\Delta\rho^2}\left(-\frac{\rho_{i-1/2}}{\sqrt{\rho_{i-1}\rho_i}}\Lambda_{[\omega],i-1} + 2\Lambda_{[\omega],i} - \frac{\rho_{i+1/2}}{\sqrt{\rho_i\rho_{i+1}}}\Lambda_{[\omega],i+1} \right) - \frac{\omega}{\rho_i^2}\Lambda_{[\omega],i}$$
$$= E(\omega)\Lambda_{[\omega],i}, \tag{33.33}$$

$$\frac{-\hbar^2}{2m\Delta z^2}(\Pi_{k-1} - 2\Pi_k + \Pi_{k+1}) = E\Pi_k. \tag{33.34}$$

ω is a constant introduced in the separation of the radial and angular parts of the discretized Schrodinger equation and is related to the angular momentum through $L_z = \sqrt{2m\omega}$. For a full circle with N_ϕ TB sites, the angular eigenfunction and eigenvalue for angular quantum number, n, are

$$\Theta_{[n],i} = \left[1/N_\phi\Delta\phi \right]^{1/2} \exp[in2\pi j/N_\phi], \tag{33.35}$$

$$\omega_n = \frac{\hbar^2\left[\cos\left(2n\pi/N_\phi\right) - 1 \right]}{(m\Delta\phi^2)}. \tag{33.36}$$

The radial function, Λ_i, has n_ρ nodes, where $n_\rho = 0,1,2,\dots$ (the position, $\rho = 0$, is not included in the counting of nodes), and the transverse modes are denoted by $[n_\rho, n]$. The two transverse modes, $[n_\rho, +n]$ and $[n_\rho, -n]$, are degenerate; so, they can both be denoted by $[n_\rho, |n|]$, where $|n|$ is the absolute value of n. To approximate the continuum, it is sufficient to use a model with $N_\rho = 4$ and $N_\phi = 20$, which has 8 TB sites along the cross-section diameter. The eigenenergies of the lowest six transverse modes of the wire are given in Table 33.2. The first and sixth subbands have rotational symmetry for $n_\rho = 0$ and, as a consequence, are symmetric with respect to reflection about the x–z plane shown. Any linear combination of the second and the third subbands, which are degenerate, is also an eigenfunction. So, the linear combination can be either symmetric or antisymmetric with respect to reflection about the x–z plane. This way of symmetrizing or antisymmetrizing the eigenfunctions can also be applied to the degenerate third and fourth subbands. In general, symmetrized wave functions have high electron density at sites involved in interwire coupling and therefore have stronger coupling to the other wire. Antisymmetrized modes have a node and thus a lower electron density near the coupling sites; so, they have weaker interwire coupling. In Figure 33.15, the transverse mode profiles of the first six modes of a circular wire are shown schematically.

33.2.6 Projected Green's Function

The coupling between two nanowires in a cross-junction lowers the electron energy and causes the formation of bound and quasi-bound states in the junction. The bound and quasi-bound state

TABLE 33.2 Subband Energies, Mode Indexes and Mode Labels of the Transverse Mode Wave Functions of the Lowest Six Subbands of a Circular Wire. Energy Is Expressed in Unit of t_z

Subband	Subband Energy (in Unit of t_z)	Number of Node along Radial Direction n_ρ	Angular Quantum Number n	Mode Label $[n_\rho, |n|]$
1st	0.2777	0	0	[0,0]
2nd	0.6985	0	±1	[0,1]
3rd	0.6985	0	±1	[0,1]
4th	1.2296	0	±2	[0,2]
5th	1.2296	0	±2	[0,2]
6th	1.3200	1	0	[1,0]

Source: From Chan, K.S. and Wei, J.H., *Phys. Rev. B*, 75, 125310, 2007. With permission.

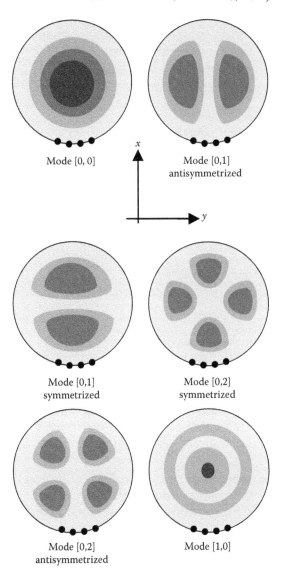

Mode [0, 0]

Mode [0,1] antisymmetrized

Mode [0,1] symmetrized

Mode [0,2] symmetrized

Mode [0,2] antisymmetrized

Mode [1,0]

FIGURE 33.15 The transverse mode profiles of the lowest six subbands of a circular nanowire.

wave functions are formed by mixing the electron wave functions of the two wires. When the two lowest subbands of the two wires are coupled together in a junction, they form a bound state. However, if two higher subbands are coupled in the junction, the bound state formed can be coupled to the unbound states of some lower subbands and consequently a quasi-bound state is formed. Prominent features, such as peaks and dips, in the conductances of a nanojunction can be explained in terms of the quasi-bound states; so, it is useful for analyzing the nanojunction conductance to have a means of finding out how subbands are coupled in a quasi-bound state. Chan and Wei (2007) have devised a mathematical tool called the projected Green's function for identifying the main subband components of a quasi-bound state. In this section, we discuss the definition of the projected Green's function and how it can be used to analyze quasi-bound states.

The projected Green's function for modes, m and n, of leads, β and α, at longitudinal positions, z_β and z_α, of a junction are defined as

$$G_P(m,n,z_\beta,z_\alpha) = \iint_{A_\beta A_\alpha} \phi_m^*(\vec{\rho}_\beta) G_{\beta\alpha}^R(E,\vec{\rho}_\beta,z_\beta,\vec{\rho}_\alpha,z_\alpha)\phi_n(\vec{\rho}_\alpha)d^2\vec{\rho}_\beta d^2\vec{\rho}_\alpha.$$

(33.37)

This definition applies to both cylindrical and square wires. Actually, the definition of the projected Green's function is related to the scattering matrix. The projected Green's function is obtained from the scattering matrix by removing the first term, $-\delta_{\alpha\beta}$, and the factor, $i\hbar\sqrt{v_m v_n}$, in the second term from Equation 33.4. If the subbands, ϕ_m and ϕ_n, are components of the quasi-bound state, rapidly changing features can be found near the quasi-bound state energy. Chan and Wei (2007) have given a detailed mathematical analysis of the projected Green's function. They proved that $G_P(m,n,z_\beta,z_\alpha) = \dfrac{c+id}{E-E_B+i\Gamma}$, where m and n are subbands with subband-edge energy higher than the quasi-bound state energy and $c+id = \langle\phi_m|\chi\rangle\langle\chi|\phi_n\rangle$, where χ is the bound state wave function formed from the m and n subbands by ignoring the coupling to the lower subbands. As χ is a bound state, it is localized around the junction. A quasi-bound state is formed from χ when χ is coupled to some lower subbands. The quasi-bound state formed is an unbound state, which gives rise to a resonance feature in the transmission or reflection coefficients. When the quasi-bound state has a long enough lifetime (or a small enough Γ) and ϕ_m and ϕ_n are important components of the bound state, χ, the complex number, $c+id$, is not small and the function, $\dfrac{c+id}{E-E_B+i\Gamma}$, has rapidly changing features in the real and imaginary parts at energy near E_B. These features can be readily identified in the graph and used to confirm that ϕ_m and ϕ_n, are important components of the quasi-bound state. Chan and Wei (2007) have shown several examples of how to use the projected Green's function to analyze the quasi-bound states in a cross-junction. The imaginary part of the projected Green's function, $(c\Gamma/((E - E_B)^2 + \Gamma^2)) + (d(E - E_B)/((E - E_B)^2 + \Gamma^2))$, is not a Lorentzian peak as it is not a diagonal matrix element of the Green's function. Nevertheless, the projected Green's function can be symmetrized so that the imaginary part is a Lorentzian peak according to

$$\frac{G_P(m,n,z_\beta,z_\alpha) + G_P(n,m,z_\alpha,z_\beta)}{2} = \frac{c}{E - E_B + i\Gamma}. \tag{33.38}$$

33.3 Conductances

In this section, the conductances of nano-junctions formed from square and circular wires are presented and discussed. In particular, the relationship between the features (dips and peaks) in the conductances and the quasi-bound states formed at the junction are analyzed in detail.

33.3.1 Conductances of Square Wire Junctions

In Figure 33.16, which is reprinted from Chan and Wei (2007), the interwire (G_{13}) and intrawire (G_{12}) conductances of an 8×8 square wire cross-junctions for several interwire coupling strengths are shown. In this figure, the energy is expressed in unit of $t_o = (\hbar^2/2ma^2)$, the hopping constant of a cubic lattice TB model. In this energy unit, the hopping constant of the model has the numerical value of 1 in the calculation. However, the results shown in the figure are obtained by using a numerical value of 0.67 for t. So, in SI unit, t equals $0.67t_o$ or $t = 0.67 (\hbar^2/2ma^2) = (\hbar^2/2m(1.22a)^2)$, which is equivalent to enlarging the lattice spacing to $a' = 1.22a$. For the 8×8 square wire considered, the transverse dimension is $9a' = 9 \times 1.22a \approx 11a$. Although the TB model has 8 TB sites along the transverse direction, the dimension is $11a$, which is equivalent to 10 TB sites along the transverse direction with a lattice spacing of a. In the adoption of TB models to approximate the continuum, there is always the question of how small the lattice spacing, a, should be for a desirable degree of accuracy. This question has been studied by Chan and Wei (2007) by comparing the conductances of models with different numbers of TB sites (including 4×4, 6×6, 8×8, and 10×10). It was found that an 8×8 model can already give good quantitative agreement with a 10×10 model for the lowest 6 subbands. So, although the number of sites in the transverse dimension is 8, the use of $t = 0.67$ allows the model to mimic a 10×10 square wire. The use of a smaller number of TB sites can reduce the computational time required for calculating the Green's function by matrix inversion.

When the interwire coupling is strong (here, strong coupling means the interwire hopping strength equals the intrawire hopping strength, $t_c = t = 0.67t_0$), the intrawire conductance (G_{12})

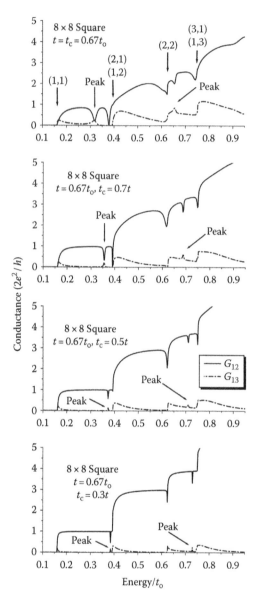

FIGURE 33.16 Intrawire (G_{12}) and interwire (G_{13}) conductances of an 8×8 square wire cross-junction. (Reprinted from Chan, K.S. and Wei, J.H., *Phys. Rev. B*, 75, 125310, 2007. With permission.)

near the subband edges (at $E \approx 0.16t_o$, $0.39t_o$, $0.63t_o$, $0.74t_o$) is suppressed. The intrawire conductance increases with electron incident energy measured from the subband edges. This means, near the subband edges, there is a strong scattering of the incident electron by the junction. There are dips in the intrawire conductance just below the subband edges, which are due to the quasi-bound states formed below the subband edges.

Bound states and quasi-bound states are formed at a junction between two wires, because the junction allows an electron to move from one wire to another wire and lowers its energy. Firstly, bound states, which are localized states, are formed from the states of the two wires by the interwire coupling. If not forbidden by symmetry, these bound states may have strong enough coupling to the unbound states of lower subbands (those with subband edges at lower energies) and form delocalized quasi-bound states. For the lowest subband (1,1), the bound states formed cannot form quasi-bound states because there is no lower subband providing any unbound states for coupling.

The dip and peak features in the intrawire and interwire conductances depend on the nature of the quasi-bound states. To explain the relationship, we first classify the intrawire conductance dips into two groups: Group A are dips in the intrawire conductance associated with corresponding peaks in the interwire conductance at the same energy, and Group B are dips in the intrawire conductance not associated with peaks in the interwire conductance. Quasi-bound states giving rise to Group A dips are delocalized in both wires; so, they can enhance interwire transmission and give rise to peaks in the interwire conductances at the same energy. The dip in the intrawire transmission is caused by this increase in interwire transmission due to the quasi-bound state. Quasi-bound states giving rise to Group B dips are localized in one of the two wires; they do not form transmission channels between wires and do not lead to features in the interwire conductances. Nevertheless, these quasi-bound states provide alternative channels for electron transmission between the two terminals of the same wire (here terminals 1 and 2). The electron waves going through the two transmission channels (one through the quasi-bound state and one not through the quasi-bound state) can interfere destructively, leading to transmission dips in intrawire transmission. Similar phenomena have been investigated by Shao et al. (1994) in quantum waveguides.

The characteristics of the quasi-bound states depend on the symmetry of the transverse modes involved. For example, below the second and third subbands (they are subbands with transverse modes, (1,2) and (2,1), hereafter we call them subbands (1,2) and (2,1)), the bound state formed from subbands (1,2) of both wires, has the same symmetry as the (1,1) transverse modes of both wires. So it can couple to the subbands (1,1) of the two wires and form quasi-bound states delocalized in both wires (Group A). On the other hand, the bound state formed from the coupling of subbands (1,2) to (2,1) can only couple to the subband (1,1) of one wire and the quasi-bound state formed is delocalized in one wire and localized in another wire (Group B). The Group A quasi-bound state from $(1,2) \times (1,2)$ forms a conduction channel and enhances electron transmission between two wires. So, there is a peak in G_{13} and a dip in G_{12} at energy around 0.32. This symmetry argument can be strictly applied to the second and third subbands because states of subband (1,1) are the only continuum states needed to be considered. For higher quasi-bound states, symmetry is not the only factor considered as continuum states of different symmetries are present; in these cases, coupling strength also plays an important role in determining the quasi-bound state characteristics. In Table 33.3, we give a summary of the conductance features and the related quasi-bound states for $t_c = t_o$, which can be explained in terms of the two groups of quasi-bound states discussed above.

As to the features at $E \approx 0.6$ and 0.74 in Table 33.3, there are some subtleties, which require further discussion. The dip around $E \approx 0.6$ actually consists of two dips from two quasi-bound states.

TABLE 33.3 Features in the Conductances of a Square Wire Cross-Junction ($t_c = t_o$)

Energy	G_{12}	G_{13}	Coupling of Subbands in the Quasi-Bound State
0.32	Dip	Peak	$(1,2) \times (1,2)$
0.38	Dip	No feature	$(1,2) \times (2,1)$
0.6	Dip (two close dips)	No feature	$(2,2) \times (2,2)$
			$(2,2) \times \{(1,1)+(1,2)\}$
0.65	Dip	Peak	$(1,3) \times (1,3)$
0.74	Dip	No feature	$(1,3) \times (3,1)$

The subband (2,2) of one wire can form a quasi-bound state with the subband (2,2) of another wire at energy around 0.627. This (2,2) subband could form another quasi-bound state with the (1,1) and (1,2) subbands of another wire at energy around 0.61. The $(2,2) \times (2,2)$ quasi-bound state is very close to the subband edge and the interwire transmission peak is probably very close to the subband edge and cannot be resolved from the subband edge. The second quasi-bound state could be delocalized in both wires as they can interact with continuum states with the appropriate symmetry. However, the interwire transmission peak is not very clear and this quasi-bound state has a Group B behavior.

The dip in G_{12} around energy 0.741 comes from the quasi-bound state formed from the mixing of (1,3) and (3,1) subbands. Symmetry does not preclude the formation of a quasi-bound state delocalized in both wires and a transmission peak between the two wires. However, no clear interwire transmission peak can be identified, indicating a strong Group B behavior.

The interwire conductance, G_{13}, is high when the electron incident energy is near a subband edge and G_{13} decreases with increasing energy. When t_c decreases, G_{13} at high incident energy decreases and the peaks shift toward the subband edges so that significant interwire transmission can only be found very close to the subband edges. For example, when $t_c = 0.3t$, the interwire conductance looks like an symmetric peak located at the subband edge.

33.3.2 Conductances of a Circular Wire Junction

In this section, we discuss the conductances of a circular wire cross-junction. As the conductance features are determined by the symmetry of the transverse mode profiles, there are some similar features between the square and circular wire junctions. So, a comparison with square wire junctions will be made to highlight the similarities. The conductances of a circular wire cross-junction are shown in Figure 33.17 with $t_c = 0.6, 0.8, 1.0$ (reprinted from Chan and Wei 2007). In general, the interwire conductance between two circular wires is weaker than that between two square wires because circular wires have smaller contact area in between. So, there is a weaker suppression of intrawire transmission near the subband edges. The transverse mode profiles of the degenerate second and third subbands (modes [0,1]) resemble the mode profiles of the (1,2) and (2,1) modes of the square wire. So, the nearby peak and dip structure

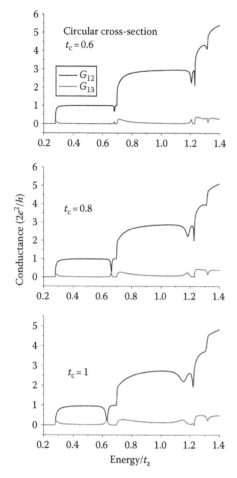

FIGURE 33.17 Intrawire (G_{12}) and interwire (G_{13}) conductances of a circular wire cross-junction. (Reprinted from Chan, K.S. and Wei, J.H., *Phys. Rev. B*, 75, 125310, 2007. With permission.)

should resemble that below the second and third subbands in the square wire junction. However, we can only find one dip and one peak, with the second dip missing. This difference is due to the fact that the second quasi-bound state has a smaller binding energy in the circular wire junction and thus the dip cannot be resolved from the subband edge. The mode profiles of the symmetrized mode of the fourth and fifth degenerate subbands (modes [0,2]) resemble the mode profile of (1,2) or (1,3) of a square wire because the TB sites participating in the interwire coupling concentrate in a small peripheral region in the wire. The electron probability distribution in this small region resembles that of mode (1,2) or (1,3) of a square wire. The antisymmetrized [0,2] mode resembles the (2,2) or (2,1) modes. As a result, the peak and dip structure just below the fourth and fifth subbands also resembles the peak and dip structure below the second and third subbands of the square wire junction.

For the subband with the edge at around $E = 1.3t_0$, the mode profile is symmetrical with respect to the x–z plane. So, the quasi-bound state formed resembles that of the symmetrized mode of [0,1] or the square wire (1,2) mode. A dip in G_{12} and a peak in G_{13} are expected. In Figure 33.17, a dip in G_{12} and a small peak in G_{13} are identified when $t_c = 0.6$ is used. For larger

t_c (0.8 and 1.0), the peak and dip are not clear because, probably, the interwire coupling is so strong that the dip and peak features are significantly distorted.

33.3.3 Energies and Probability Densities of Quasi-Bound States

To understand the characteristics of the conductance features of the junction, it is important to understand various factors that affect the quasi-bound state energy, which can be found from the local density of states obtained from the imaginary part of the Green's function. The local density of states at site J at energy E is defined as

$$\rho(J,E) = \text{Im}[G(J,J,E)] = \sum_k \psi_k^*(J)\psi_k(J)\delta(E - E_k).$$

$\rho(J,E)$ is the probability density of states with energy, E, at a particular site, J. Since the quasi-bound states of the junction have higher probability densities in the scattering region of the junction, we can use the total probability density in the scattering region (hereafter referred to as the scattering region probability density) to find the quasi-bound state energies. The scattering region probability density is $\rho_S(E) = \sum_J \rho(J,E)$, where J is a site in the scattering region. Here, the scattering region in a junction is defined as sections of the two wires that have some peripheral sites coupling to the other wires. For example, in an 8×8 square wire junction, the scattering region is an $8 \times 8 \times 16$ rectangular prism, consisting of 8 layers of TB sites of each wire. For a circular wire junction, when the number of peripheral sites coupling to the other wire is 4, the scattering region consists of 4 layers of TB sites from each of the two wires. In Figure 33.18, the layers of TB sites in the scattering region of a square wire junction and a circular wire junction are shown schematically. The square wire junction has 4 TB layers, so the transverse dimension is 4×4. For the circular wire junction, it is usually assumed that 1/5 of the peripheral sites participate in the interwire coupling; so, 4 sites couple to another wire in a ring of 20 sites. At the quasi-bound state energies we expect to find peaks of the scattering region probability density in a plot as a function of the incident energy. The quasi-bound state energies are usually defined as the energy positions of the peaks. In Figure 33.19, we show an example of the plot of total probability density versus incident energy. The peaks in the scattering region probability density are indicated by arrows.

In the work of Chan and Wei (2007), factors that determine the quasi-bound state energies were investigated. It was found that the quasi-bound state energy decreases with increase in the transverse dimension of the wire. This means that stronger confinement in the wire and larger separation between subbands lead to stronger electron binding in the junction region. This trend is similar to that found in the exciton-binding energy of quantum wells and wires: decrease in the confinement dimensions leads to increase in binding energy. The only exception

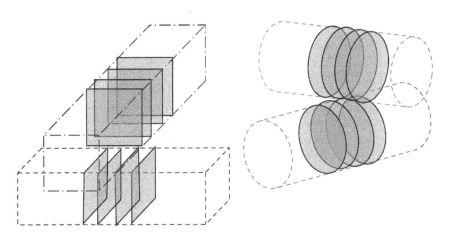

FIGURE 33.18 Layers of tight-binding sites in the scattering regions of a square wire junction and a circular wire junction. The square wire has a 4 × 4 cross section, so there are 4 layers in each wire. In the circular wire, 4 tight-binding sites participate in interwire coupling, so there are 4 layers of TB sites in each wire.

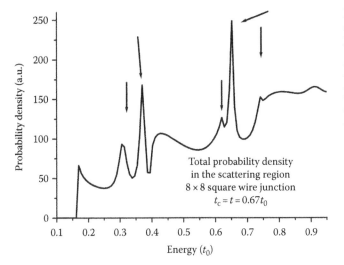

FIGURE 33.19 Total probability density in the scattering region plotted as a function of the electron incident energy.

is the quasi-bound state formed from subband [1,0] in a circular wire junction. In this case, the quasi-bound state energy is very small and is strongly affected by the coupling between the bound state of the [1,0] subbands and the unbound states of the subbands lying below. When the diameter of the circular wire is increased, the subband separation is reduced, resulting in an increase in the coupling between the bound state and unbound states. As a result, the quasi-bound state energy is increased, because the decrease in the quasi-bound state energy due to lower quantum confinement in larger wires is not large enough to cancel the effect.

The quasi-bound state energy also depends on the transverse mode profile along the x- and y-directions in square wire junctions (along the x- and y-directions in the upper wire, along the x- and z-directions in the lower wire). When the node number along the x-direction increases, the quasi-bound state energy increases. For example, the quasi-bound state, $(1,3) \times (1,3)$, has an energy about three times the energy of $(1,1) \times (1,1)$ and 20%

higher than that of $(1,2) \times (1,2)$. The reason is that the electron probability density at the coupling sites is higher in mode, $(1,3)$, than in other two modes. On the other hand, the quasi-bound state energy decreases with increase in the node number along the y-direction. For example, the quasi-bound state, $(2,2) \times (2,2)$, has an energy that is about a thousand times smaller than that of $(1,2) \times (1,2)$. The reason is that in subband, $(2,2)$, states near the subband edge with $k \approx 0$ have weak interwire coupling. Owing to the transverse mode profile, more effective coupling is to higher states with $k \approx 2\pi/L$. As a result, the quasi-bound state $(2,2) \times (2,2)$ energy is much smaller than that of $(1,2) \times (1,2)$. For the same reason, the quasi-bound state, $(2,1) \times (1,2)$, also has a smaller energy, which is about one-fifth of the energy of $(1,2) \times (1,2)$.

In circular wire junctions, angular momentum is an important factor that determines the quasi-bound state energy as higher angular momentum pushes the electron toward the boundary of the wire and enhances interwire coupling. For example, the quasi-bound state, $[0,0] \times [0,0]$, has an energy of about one-seventh of the quasi-bound state, $[0,1]_s \times [0,1]_s$. Nevertheless, the quasi-bound state, $[0,2]_s \times [0,2]_s$, has energy close to $[0,1]_s \times [0,1]_s$. The difference is only a few percent. This is the result of the transverse mode distribution in the $[0,2]_s$ mode, in which the electron distribution is divided into four sections, while there are two sections in the $[0,1]_s$ mode, which have higher probability density at sites coupled to the other wire in the junction.

The probability densities of some bound and quasi-bound states have also been obtained and studied by Chan and Wei (2007) using the Green's function. For quasi-bound states, the probability density is given by the local density of states (LDOS), $\rho(J,E) = \text{Im}[G(J,J,E)] = \sum_k \psi_k^*(J)\psi_k(J)\delta(E - E_k)$. However, for a bound state below the lowest subband edge, the imaginary part of the retarded Green's function is zero when the small imaginary number, δi, is set to be zero. To find the probability density of a bound state, the small imaginary part, δi, is not set to be zero so that a nonzero imaginary part is obtained. When the energy is close to the bound state energy, $E_B(|E - E_B| < \delta)$, and δ is small, the contribution to the Green's function from the unbound states can be ignored and the Green's function is approximately given by $G(J, J', E) = (f_B^*(J) f_B(J'))/(E - E_B + i\delta)$. The probability density of the bound state, $|f_B(J)|^2$, can be obtained from the Green's function as $|f_B(J)|^2 = i\delta G(J, J, E_B)$.

Chan and Wei (2007) have investigated the probability densities of a bound state and two quasi-bound states in a square-wire junction. They have plotted the electron probability densities on a cross section of a square wire junction (the cross section considered is shown schematically in Figure 33.20). The bound state studied is one formed from the two (1,1) subbands of the wires. When the coupling is strong ($t_c = 1$), it was found that a peak of electron density is found at the interface between the two wires. Significant density of electron is found around the interface region. When the interwire coupling is reduced, the peak in the probability density is split into two, with each peak located in one wire. In the quasi-bound state formed by two (1,2) subbands, there are three peaks of electron density in the cross section and they are of the same magnitude when the interwire coupling is strong. The middle density peak centers around the interface. Without the interwire coupling, there should be four peaks in the electron density in the cross section according to the mode density profile shown in Figure 33.14.

The LDOS cannot be used to determine the probability of all quasi-bound states because some quasi-bound states have degenerate states which cannot be distinguished by this approach. When a quasi-bound state is formed from two subbands with different transverse modes of the two wires, there is a degenerate quasi-bound state with the same energy. For example, a degenerate state, formed from $(1,2) \times (2,1)$ exists for the quasi-bound state, $(2,1) \times (1,2)$. In the LDOS approach, these two states are not distinguished and the probability density found is the sum of these two states. One can make use of the symmetry properties of the quasi-bound states to distinguish them. For example, one of the two states is symmetrical with respect to the x–z plane and antisymmetrical with respect

to the y–z plane, while the degenerate quasi-bound state has a different symmetry, antisymmetrical with respect to the x–z plane and symmetrical with respect to the y–z plane. It is possible to use a projection operator to distinguish these two states. The projection operator is an operator which can be used to project out wave functions with the desired symmetry as follows:

$$\hat{P}\psi = \psi, \quad \text{if } \psi \text{ has the symmetry of } \hat{P}$$

$$\hat{P}\psi = 0, \quad \text{if } \psi \text{ does not have the symmetry of } \hat{P}$$

For example, the projection operator for wave functions symmetric with respect to reflection about the x–z plane is defined as $\hat{P}_s\psi = (\psi + \psi')/2$, where ψ' is obtained from ψ by reflection about the x–z plane. The projection operator for antisymmetric wave functions is defined as $\hat{P}_a\psi = (\psi - \psi')/2$. We can then use the appropriate projection operator to project the desired wave functions in the Green's function as $G' = \hat{P}G\hat{P} = \sum_k (\hat{P}\psi_k^* \hat{P}\psi_k)/(E - E_k + i\delta)$. G' consists of only wave functions with the desired symmetry and so can be used to find the probability density of the quasi-bound state with the desired symmetry. Chan and Wei (2007) have used this approach to find the probability density of the quasi-bound state, $(2,1) \times (1,2)$, the probability density profile resembles the mode, $(2,1)$, in the upper wire, but with the density peaks shifted significantly toward the interface.

33.4 Summary and Future Perspective

This chapter gives a tutorial review of the theory of quantum ballistic transport in nanowire cross-junctions. The basic theory required for the determination of the charge conductance of cross nanowire junctions are presented and discussed. The features of the inter and intrawire conductances of cross-junctions are analyzed and explained in terms of the symmetry properties of the quasi-bound states formed at the junctions. This is also related to the symmetry of the transverse mode profiles involved. Theoretical tools, such as projected Green's function and local density of states, for the analysis of the characteristics of the quasi-bound states are also presented and discussed. This chapter provides the background for further study of the physics and device characteristics of nanowire cross-junction devices, which is a topic of current widespread research interest.

Despite the successful demonstration of nanowire cross-junction devices, our understanding of the physics and device characteristics of nanojunction devices are not complete, both theoretically and experimentally. There are still many important questions to be answered for the full development of these nanojunction devices. Topics to be explored in this area include the effects of electron–electron interaction, charge transport dynamics, contact effects, etc.

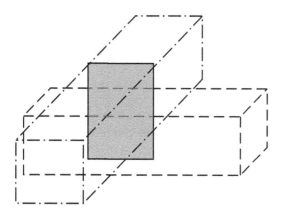

FIGURE 33.20 Cross section of a junction considered by Chan and Wei (2007) for plotting the electron probability density.

Acknowledgment

The work described in this chapter is fully supported by a grant of the Research Grants Council of Hong Kong SAR, China (Project No. CityU 100303/03P).

References

Björk, M. T., B. J. Ohlsson, T. Sass et al. 2002. One-dimensional heterostructures in semiconductor nanowhiskers. *Applied Physics Letters* **80**: 1058–1060.

Büttiker, M. 1986. Four-terminal phase-coherent conductance. *Physical Review Letters* **57**: 1761–1764.

Chan, K. S. and J. H. Wei. 2007. Quantum ballistic transport in nanowire junctions. *Physical Review B* **75**: 125310.

Cui, Y. and C. M. Lieber. 2001. Functional nanoscale electronic devices assembled using silicon nanowire building blocks. *Science* **291**: 851–853.

Datta, S. 1995. *Electronic Transport in Mesoscopic Systems*. Cambridge, U.K.: Cambridge University Press.

Dehon, A. 2003. Array-based architecture for FET-based nanoscale electronics. *IEEE Transactions on Nanotechnology* **2**: 23–32.

Dehon, A., P. Lincoln, and J. E. Savage. 2003. Stochastic assembly of sublithographic nanoscale interfaces. *IEEE Transactions on Nanotechnology* **2**: 165–174.

Duan, X. and C. M. Lieber. 2000. General synthesis of compound semiconductor nanowires. *Advance Materials* **12**: 298–302.

Duan, X., J. Wang, and C. M. Lieber. 2000. Synthesis and optical properties of gallium arsenide nanowires. *Applied Physics Letters* **76**: 1116–1118.

Duan, X., Y. Huang, Y. Cui et al. 2001. Indium phosphide nanowires as building blocks for nanoscale electronic and optoelectronic devices. *Nature* **409**: 66–69.

Gaididei, Yu. B., L. I. Malysheva, and A. I. Onipko. 1992. Electron scattering and bound-state energies in crossed N-chain wires. A comparative study of discrete and continuous models. *Journal of Physics: Condensed Matter* **4**: 7103–7114.

Huang, Y., X. Duan, Q. Wei et al. 2001a. Directed assembly of one-dimensional nanostructures into functional networks. *Science* **291**: 630–633.

Huang, Y., X. Duan, Y. Cui et al. 2001b. Logic gates and computation from assembled nanowire building blocks. *Science* **294**: 1313–1317.

Jamshidi, A., P. J. Pauzauskie, P. J. Schuck et al. 2008. *Nature Photonics* **2**: 86–89.

Kalliakos, S., C. P. Garcia, V. Pellegrini et al. 2007. Photoluminescence of individual doped GaAs/AlGaAs nanofabricated quantum dots. *Applied Physics Letters* **90**: 181902.

Lee, M. W., H. Z. Twu, C. C. Chen, and C. H. Chen. 2001. Optical characterization of wurtzite gallium nitride nanowires. *Applied Physics Letters* **79**: 3693–3695.

Liang, W., M. Bockrath, D. Bozovic et al. 2001. Fabry-Perot interference in a nanotube electron waveguide. *Nature* **411**: 665–669.

Pauzauskie, P. J., A. Radenovic, E. Trepagnier et al. 2006. Optical trapping and integration of semiconductor nanowire assemblies in water. *Nature Materials* **5**: 97–101.

Perez-Martinez, F., I. Farrer, D. Anderson et al. 2007. Demonstration of a quantum cellular automata cell in a GaAs/AlGaAs heterostructure. *Applied Physics Letters* **91**: 032102.

Rao, A. M., P. C. Eklund, S. Bandow et al. 1997. Evidence for charge transfer in doped carbon nanotube bundles from Raman scattering. *Nature* **388**: 257–259.

Rao, S. G., L. Huang, W. Setyawan et al. 2003. Large-scale assembly of carbon nanotubes. *Nature* **425**: 36–37.

Ravenhall, D. G., H. W. Wyld, and R. L. Schult. 1989. Quantum Hall effect at a four-terminal junction. *Physical Review Letters* **62**: 1780–1783.

Schiff, L. I. 1968. *Quantum Mechanics*. New York: McGraw-Hill.

Shao, Z., W. Porod, and C. S. Lent. 1994. Transmission resonances and zeros in quantum waveguide systems with attached resonators. *Physical Review B* **49**: 7453–7465.

Takagaki, Y. and K. Ploog. 1993. Quantum transmission in an out-of-plane crossed-wire junction. *Physical Review B* **48**: 11508–11511.

Tans, S. J., A. R. M. Verschueren, and C. Dekker. 1998. Room-temperature transistor based on a single carbon nanotube. *Nature* **393**: 49–52.

Wei, J. H. and K. S. Chan. 2005. Electron transport in crossed nano-wire junctions. *Physics Letters A* **341**: 224–250.

Whang, D., S. Jin, and C. M. Lieber. 2003a. Nanolithography using hierarchically assembled nanowire masks. *Nanoletters* **3**: 951–954.

Whang, D., S. Jin, Y. Wu et al. 2003b. Large-scale hierarchical organization of nanowire arrays for integrated nanosystems. *Nanoletters* **3**: 1255–1259.

Zhang, R. Q., T. S. Chu, H. F. Cheung et al. 2001. Mechanism of oxide-assisted nucleation and growth of silicon nanostructures. *Materials Science and Engineering C* **16**: 31–35.

Zhong, Z., D. Wang, Y. Cui et al. 2003. Nanowire crossbar array as address decoders for integrated nanosystems. *Science* **302**: 1377–1379.

VI

Atomic Wires and Point Contact

34

Atomic Wires

Nicolás Agraït
Universidad Autónoma de Madrid

and

*Instituto Madrileño de Estudios
Avanzados en Nanociencia*

34.1 Introduction

An atomic wire consisting of metal atoms forming a single file is certainly the thinnest wire that you could imagine. Atomic wires are classical textbook examples often used to illustrate many fundamental topics in condensed matter physics. The experimental realization of atomic wires has been possible recently (Ohnishi et al. 1998, Yanson et al. 1998). The electronic and mechanical properties of these exceptional systems are quite different to those of their macroscopic counterparts, directly reflecting the quantum nature of the atoms. For example, they can carry enormous current densities of up to 8×10^{14} A/m^2, and they are much stronger than bulk. These systems are very attractive because they are ideally suited for investigating in atomic detail the theories of electronic transport in the nanoscale and the mechanical behavior of matter with atomic detail. These topics are of fundamental interest in nanotechnology and have implications for the miniaturization of electronic components.

Here, we will only discuss atomic wires freely suspended between metallic electrodes, which make it possible to perform detailed transport experiments. Atomic wires supported on a substrate (Segovia et al. 1999, Nilius et al. 2002) can be considered to be a completely different system due to the strong coupling to the substrate. In addition, the lack of electrodes does not make it possible to perform transport experiments.

We will start by considering the basic experimental results in atomic contacts and atomic wires of different metals in Section 34.2. The conductance of these systems will be discussed theoretically in Section 34.3 starting with a simple free-electron model (Section 34.3.1), followed by the general approach of describing the conductance of a nanoscopic system (Section 34.3.2) and a review of the tight-binding method (Section 34.3.3), which will be used to shed light on the elastic scattering processes in atomic wires in Section 34.4. In Section 34.5, we will investigate the mechanical properties of atomic wires. The inelastic scattering of electrons in an atomic wire, which causes dissipation and heating but also makes it possible to obtain structural information will be considered in Section 34.6. Finally, we will discuss the theoretical calculations for the structure of the wires in Section 34.7. For an in-depth review of atomic contacts and wires, the interested reader can consult the article by Agrait et al. (2003).

34.2 Experiments on Atomic Wires and Atomic Contacts of Gold

Atomic wires form in the last stages of the breaking of a metallic contact of certain metals, particularly gold. The scanning tunneling microscope (STM) is a convenient tool to study a metallic contact because it allows for fine positional control between a metal tip and a metal substrate through the use of piezoceramics, and it has high mechanical stability. Using a gold tip and a gold substrate, it is possible to make a metallic contact on a given spot of the substrate by carefully touching the surface. If the contacting surfaces of the tip and the substrate are clean, the two pieces of metal will strongly adhere or *cold weld*. Now if the tip is retracted, the contact will first deform plastically forming a neck, and becoming thinner as its cross-sectional area decreases. Just before separation occurs, the contact will consist, in most cases, of a single atom and sometimes an atomic wire could form as shown schematically in Figure 34.1.

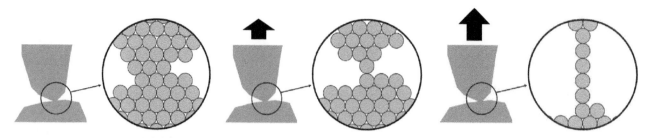

FIGURE 34.1 Metallic contact formation. As the metallic tip of an STM touches the metallic substrate, the atoms at the clean contacting surfaces become intimately joined forming a contact. Retraction of the tip results in a controlled thinning of the contact. Just before separation a single atom contact forms, which can eventually lead to the formation of an atomic wire.

The formation and evolution of the contact is monitored by measuring the current at a low fixed bias voltage, typically expressed as the conductance $G = I/V$, which is the inverse of the resistance. Figure 34.2a and b shows the evolution of the conductance for a gold contact as the tip retracts (black curve) and then advances to reform the contact (gray curve). Notice that in these and all subsequent figures, the conductance will be expressed in units of $G_0 = 2e^2/h$, where e is the charge of the electron and h is Planck's constant. This is the quantum unit of conductance and is the natural units conductance in the atomic scale as we will see in Section 34.3. The inverse of G_0 has units of resistance; its value is 12,906 Ω. In these conductance curves, two clearly different regimes can be observed: the contact regime and the tunneling regime. In the contact regime, the conductance is larger than G_0 and changes in steps. As we

will see when we discuss the mechanical properties, the contact deforms elastically, which results in a continuous variation in the conductance, until the accumulated stress is relaxed by sudden atomic rearrangement that is manifested as an abrupt jump in the conductance. A similar behavior is observed during contact formation. In the tunneling regime, the conductance is one or two orders of magnitude smaller and depends exponentially on the tip displacement, as evidenced by plotting the logarithm of the conductance. The conductance of the smallest contact of gold is given by the last conductance plateau before the transition to the tunneling regime. This smallest contact consists of just one atom and its conductance is one in units of G_0. The conductance of a one-atom contact of gold is very well defined and results in a sharp peak in the conductance histogram as shown in Figure 34.2c.

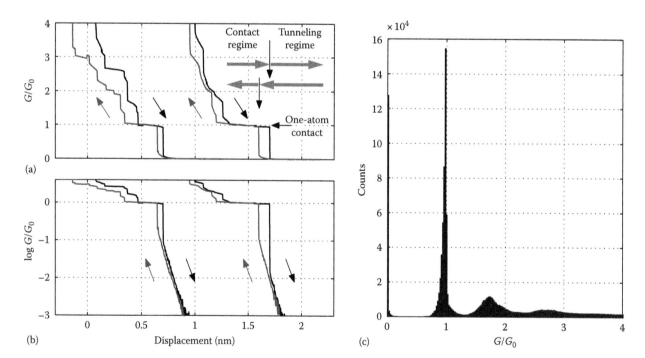

FIGURE 34.2 Conductance evolution during contact breaking (black curve) and contact formation (gray curve) for two different Au contacts, in linear (a) and logarithmic scales (b) at low temperature (4.2 K) using an STM. Note that both contacts show plateaus of conductance at $G_0 = 2e^2/h$, the quantum unit of conductance. (c) Histogram obtained from all the measured conductance data points (conductance histogram) for 6000 contacts.

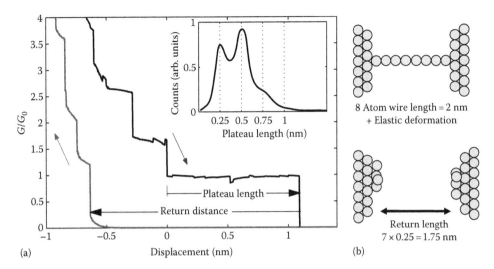

FIGURE 34.3 (a) A long plateau at a conductance of one G_0 indicates the formation of an atomic wire. (Inset) A histogram of plateau lengths shows peaks separated by an interatomic distance. (Adapted from Untiedt, C. et al., *Phys. Rev. B*, 66, 085418, 2002.) (b) The scheme shows how an 8-atom wire could give the observed return length taking into account that the atoms of the collapsed wire stay at the contact and the elastic deformation of the electrodes is relaxed after breaking.

Typically, the length of the last conductance plateau is less than 0.5 nm, however, sometimes it is possible to observe much longer plateaus, as the one shown in Figure 34.3a. These plateaus, which can be as long as 2.5 nm, correspond to the formation of a wire of single gold atoms. The probability of forming these atomic wires decreases rapidly with length, being less than 10^{-4} for plateaus longer than 2 nm (Yanson et al. 1998). A histogram of the length of the conductance plateaus, see the inset in Figure 34.3a, shows a preference for lengths that are multiples of 0.25 ± 0.2 nm, suggesting that that distance corresponds to the interatomic separation of the atoms in the atomic wire. This value is smaller than the bulk nearest neighbor interatomic distance of 0.288 nm, and in good agreement with the theoretical calculations, as we will see in Section 34.7. But how does the plateau length relate to the real length of the wire? We may argue that the *return length*, that is the tip displacement required to reform the contact, would give a better indication of the length of the wire because the plateau length would depend on the way the atomic wire forms. After breaking, the atoms of the collapsed wire will remain on the surface possibly forming two atomic layers (see the scheme in Figure 34.3) and consequently the return length would underestimate the real length of the 2 atomic diameters. However, we must also take into account that due to the elastic deformation of the electrodes at the point of maximum wire elongation, the return length will tend to give an overestimate of the real length. As we will see in Section 34.5, this elastic deformation will be typically 0.25 nm or 1 atomic diameter. Thus, we can conclude that the length of the atomic wire will be approximately equal to the return length plus an atomic diameter.

The experimental curves in Figures 34.2 and 34.3 were measured at cryogenic temperatures (4.2 K). At these temperatures, there is no atomic diffusion on the surface of the metal, the thermal drift of the STM is negligible, and the high energy resolution

is in spectroscopic measurements. In addition, as we mentioned above, the contact formation requires the contacting surfaces to be clean. In this respect, the low temperature environment is also convenient because it provides a cryogenic ultrahigh vacuum environment and residual adsorbates are effectively frozen. For many metals it suffices to prepare the contact in situ by making a number of large contacts, this effectively brings fresh metal to the surface that then remains clean. The cleanliness of the surface can be checked in the tunneling regime. From the theory of quantum tunneling, we should have $G \propto e^{-\alpha z}$ where z is the displacement of the tip, and $\alpha = 1.025\sqrt{\phi}$ where ϕ is the apparent tunneling barrier. From the slope of the log G versus z, we obtain $\phi \approx 5$ eV. This high value of the apparent tunneling barrier indicates that the contacting surfaces are clean, since in this case the apparent tunneling barrier should equal the work function, which for Au is 5.1 eV.

In Figure 34.4a and b, we show a typical conductance cycle at room temperature and in ambient conditions. The steps in the conductance are not as well defined as they are at low temperatures due to the much larger mobility of atoms. As a consequence, the peak in the conductance histogram at G_0 (Figure 34.4c) is not so sharp. In the tunneling regime, the tunneling barrier is much lower (~1 eV) and it is not uncommon to see some small steps due to the presence of adsorbates on the surface.

Another tool that has also served to gain a wealth of information on these systems is the mechanically controlled break-junction (MCBJ) (Agraït et al. 2003) in which the sample is a notched wire glued on a flexible substrate. Once the sample is cold and in a cryogenic vacuum, the notch is broken by bending the substrate with a piezo element resulting in the separation of the wire into two electrodes with fresh surfaces. This can be very advantageous for studying metals that oxidize in ambient conditions. As in STM, the separation of the electrodes can be controlled with

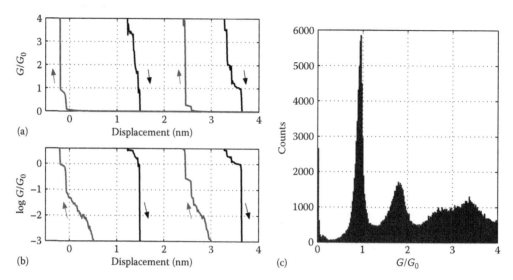

FIGURE 34.4 Conductance evolution during contact breaking (black curve) and contact formation (gray curve) for two different Au contacts, in linear (a) and logarithmic scales (b) at room temperature in ambient conditions using an STM. The right panel (c) shows a histogram of the measured conductance values or conductance histogram for 6000 contacts.

a resolution of the order of picometers, but the mechanical stability can be somewhat higher.

Many other metals have been investigated experimentally, in most cases using the MCBJ, as shown in Figure 34.5. The conductance traces and histograms are characteristic of each material. In most cases, the conductance of the last contact, the one-atom contact, is around G_0 but not so sharply defined as it is in gold. Atomic

wire formation, signaled by the appearance of long conductance plateaus, have been observed only in gold, platinum, and iridium. For other metals, the length of the one-atom conductance plateau is just of a few tenths of nanometer. In Section 34.7, we will discuss why atomic wires form only in certain metals.

34.3 Conductance of Atomic Contacts and Wires

As we saw in Section 34.2, in the case of gold, the conductance of a one-atom contact has a particular value of $2e^2/h$, which is independent of temperature, and this is also the value of the conductance of an atomic wire independently of length. This is quite surprising and shows that Ohm's law is not valid for atomic wires. Indeed Ohm's law states that the conductance G would be proportional to the transverse area S and the conductivity σ, a strongly temperature-dependent material property, and inversely proportional to the length L. The reason is that the dimensions of atomic wires are too small. Ohm's law results from a semiclassical description of the electron motion and is applicable to macroscopic conductors in which electrons scatter many times and lose all memory of their phase. In contrast, the dimensions of an atomic wire are just a few nanometers, and of the order of the electron wavelength λ_F. A full quantum description of electron transport is required.

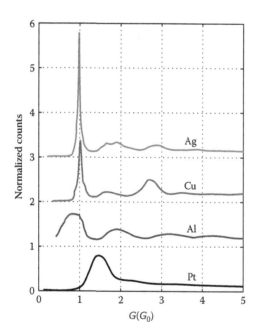

FIGURE 34.5 Normalized histograms for Cu, Ag, Pt, and K taken at low temperature using a MCBJ. The histograms for each metal are different and characteristic. (Adapted from Yanson, A., Atomic chains and electronic shells: Quantum mechanisms for the formation of nanowires, PhD thesis, Universiteit Leiden, Leiden, the Netherlands, 2001; Agraït, N. et al., *Phys. Rep.*, 377, 81, 2003.)

34.3.1 A Simple Free-Electron Model of an Atomic Wire

In quantum mechanics, the current through a conductor can be written as the probability of an electron to transmit through it. Consider a uniform cylindrical wire with a radius R and a length L with free and independent electrons connected to two bulk reservoirs (electrodes) (Agraït et al. 2003). If we neglect the effect of the reservoirs, it is easy to obtain the electronic states in the conductor solving Schrödinger's equation:

$$-\frac{\hbar^2}{2m^\star}\nabla^2\psi(r) = E\psi(r),$$

with the boundary condition $\psi[r=R]=0$. We find that the eigenstates are given by

$$\psi_{mnk}(r,\phi,z) = \psi_{mn}(r,\phi)e^{ikz} = \frac{J_m(\gamma_{mn}r/R)e^{im\phi}}{\sqrt{\pi L R}J_{m+1}(\gamma_{mn})}e^{ikz},$$

where

the z coordinate is taken along the cylinder axis
$m=0,\pm1,\pm2,\pm3,\ldots$ and $n=1,2,3,\ldots$ are the quantum numbers
γ_{mn} is the nth zero of the Bessel function of the order m, J_m

The energies of the eigenstates are

$$E_{mn}(k) = \frac{\hbar^2 k^2}{2m^\star} + \frac{\hbar^2}{2m^\star}\frac{\gamma_{mn}^2}{R^2} = \frac{\hbar^2 k^2}{2m^\star} + E_{mn}^c,$$

since $J_{-m}(r) = (-1)^m J_m(r)$, the states m and $-m$ are degenerate. The electron states are divided into a set of parabolic one-dimensional subbands with the bottom of each subband located at an energy E_{mn}^c as shown in Figure 34.6.

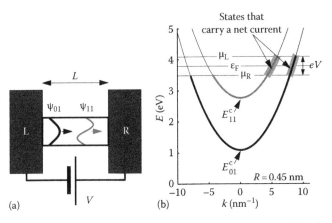

(a)

FIGURE 34.6 (a) Cylindrical wire of radius $R = 0.45\,\text{nm}$ and length L with an electronic charge density corresponding to Au, connected to two electrodes with an applied potential difference V. (b) For this small diameter two modes carry the current because only two of the one-dimensional electronic subbands are below the Fermi level ε_F. Subbands E_{01} is nondegenerate and E_{11} is doubly degenerate, and the Fermi level is determined by the condition that the wire remains charge neutral. The applied potential V unbalances the chemical potentials for right-going electrons μ_R and left-going electrons μ_L, resulting in a net current through the wire.

Each eigenstate can carry in the axial direction an amount of current I_{mnk} given by the integral over the transverse section of the conductor of the probability current density times the electron charge

$$I_{mnk} = \frac{e}{L}\frac{i\hbar}{2m^\star}\int\left(\psi_{mnk}\frac{\partial\psi_{mnk}^\star}{\partial z} - \psi_{mnk}^\star\frac{\partial\psi_{mnk}}{\partial z}\right)dS = \frac{e}{L}\frac{\hbar k}{m^\star} = \frac{e}{L}v_k.$$

Since there is a degenerate left-moving mode that carries the same current in the opposite direction for each right-moving mode, to have a net current flowing in the conductor there must be an imbalance between the population of the left-moving mode (fixed by the Fermi distribution on the left electrode, f_L) and of the right-moving mode (fixed by the Fermi distribution on the right electrode, f_R). Such an imbalance is provided by the voltage bias V applied between the electrodes. Each subband contributes to the current with

$$I_{mn} = \frac{e}{L}\sum_{k\sigma}v_k\left(f_L(\varepsilon_k) - f_R(\varepsilon_k)\right) = \frac{e}{\pi}\int dk v_k(f_L(\varepsilon_k) - f_R(\varepsilon_k)),$$

where

L is the length of the conductor
σ is the electron spin

The total current will be given by the sum of the contributions of all subbands that are populated, that is $E_{mn}^c < \varepsilon_F$, where ε_F is the Fermi level of the wire. For a long conductor, one can replace the sum over the allowed k values by an integral over k. As we are dealing with a one-dimensional system, the density of the states is $\rho(\varepsilon) = 1/v_k\hbar$ and then

$$I_{sub} = \frac{2e}{h}\int d\varepsilon(f_L(\varepsilon) - f_R(\varepsilon)).$$

At zero temperature, $f_L(\varepsilon)$ and $f_R(\varepsilon)$ are step functions, equal to 1 below their respective chemical potentials $\mu_L = \varepsilon_F + eV/2$ and $\mu_R = \varepsilon_F - eV/2$, and 0 above. Thus, the contribution to the conductance of each subband is identical $G_{sub} = I_{sub}/V = 2e^2/h$.

These results show that a perfect single mode conductor between two electrodes has a finite conductance, given by the universal quantity $2e^2/h$ (Datta 1997).

34.3.2 Landauer Theory of Conductance

A general approach for describing the conductance of a nanoscopic system like an atomic contact or an atomic wire is the scattering approach depicted schematically in Figure 34.7 (Datta 1997). The system is assumed to be connected to ideal electron reservoirs, the electrodes, by *perfect* leads. The basic idea is to relate the transport properties with the transmission and reflection probabilities for carriers incident on the system. The electrodes are assumed to

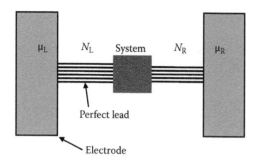

Perfect lead

Electrode

FIGURE 34.7 Scattering approach. The electrodes are in thermal equilibrium and have well-defined chemical potentials μ_L and μ_R. The perfect leads that connect the system to the left and right electrodes have N_L and N_R conducting modes, respectively.

be in a thermal equilibrium with a well-defined temperature and chemical potential. Phase coherence is assumed and inelastic scattering may occur only in the reservoirs. In these leads, the electrons propagate as plane waves along the longitudinal direction, while its transverse momentum is quantized due to the lateral confinement. This defines a number of conducting modes in these leads, say N_L and N_R for the left and right lead, respectively, in the two-terminal configuration. The use of perfect leads, an auxiliary construction that greatly simplifies the formalism, does not affect the results as long as the number of modes considered is sufficiently large.

In the scattering formalism, at a temperature of zero, the conductance of the system can be expressed as

$$G = \frac{2e^2}{h} \text{Tr}(\hat{t}^\dagger \hat{t}), \qquad (34.1)$$

where $\hat{t}$ is an $N_R \times N_L$ matrix whose elements t_{mn} give the probability amplitude for an electron in mode n on the left to be transmitted into mode m on the right. Equation 34.1 is known as the Landauer formula. The matrix $\hat{t}^\dagger \hat{t}$ is a hermitic $N_L \times N_L$ matrix. Consequently, it has real eigenvalues of τ_i, $i = 1, \ldots, N_L$ which can be shown to satisfy $0 \le \tau_i \le 1$. The eigenvectors of $\hat{t}^\dagger \hat{t}$ are called *eigenchannels* and correspond to particular linear combinations of the incoming modes that remain invariant upon transmission through the system. On the basis of the eigenchannels, the transport problem becomes a simple superposition of independent mode problems without any coupling, and the conductance can be written as

$$G = \frac{2e^2}{h} \sum_i \tau_i. \qquad (34.2)$$

The set of eigenvalues $\{\tau_i\}$ fully determines the transport properties of the junctions and is known as the PIN code of the junction. An experimental determination of these eigenvalues can be done in the superconducting state exploiting the nonlinearities of the current–voltage characteristics for contacts in the superconducting state (Scheer et al. 1997, 1998).

In the case of long wires, all nonzero eigenvalues can be close to unity and the value of the conductance will be a multiple of the quantum of the conductance, $G = N_c G_0$, where N_c is the number of

modes in the system. In a metallic wire of radius R, as discussed in Section 34.3.1, the number of modes is given by the number of subbands whose band bottom is lower than the Fermi level, that is, $E_{mn}^c < \varepsilon_F$, where ε_F. Taking into account that the number of zeros of the Bessel functions J_m below a certain value x is approximately equal to $x^2/4$, we find that $N_c \approx (\pi R/\lambda_F)^2$, where R is the wire radius.

This quantization effect was first observed in two-dimensional electron gases (2DEG) in the form of clear conductance plateaus at integer values of G_0. In metal contacts, the fact that λ_F is of the order of the size of the atom obscures the effect of conductance quantization. The conductance traces show plateaus at stable structural configurations that might not coincide with an integer value of the conductance in units of G_0 (see the histograms and conductance traces in Figure 34.5). As we have seen above, one-atom contacts of gold have a well-defined conductance of $1G_0$, which reflects the fact that one atom of gold provides a single completely open quantum channel and results in a sharp peak in the conductance histogram. This is also the case for silver and copper; in contrast, the one-atom contacts of other metals like aluminum and platinum are close to $1G_0$ but not so well defined, as evidenced in the broad peaks in the histogram. The explanation to this observation is that the atomic contacts of monovalent metals, like the noble metals and alkali metals, provide a single quantum channel for transmission, which is almost completely open (the corresponding eigenvalue is one), while in sp-metals and transition metals, which have more complicated electronic structures, several partially open eigenchannels contribute to the conductance. The transmissions of these partially open eigenchannels are quite sensitive to the detailed atomic arrangement.

34.3.3 Tight-Binding Models

As we saw in Section 34.3.2, calculating the conductance of an arbitrarily shaped conductance connected to electrodes requires computing its transmission matrix $\hat{t}$. In order to do this, it is necessary to solve Schrödinger's equation for the system composed of the conductor and the electrodes. A convenient way to do this is by using the tight-binding method in combination with Green's functions (Datta 1997).

In its simplest version, the tight-binding model uses an orthogonal basis $\{|i\rangle\}$ corresponding to a spherically symmetric local orbital at each atomic site in the system. Within this basis, the model Hamiltonian adopts the form

$$\hat{H} = \sum_i \varepsilon_i \, |i\rangle\langle i| + \sum_{i \ne j} t_{ij} \, |i\rangle\langle j|,$$

where

ε_i corresponds to the site energies

t_{ij} denotes the hopping elements between sites i and j, which are usually assumed to be nonzero only between the nearest neighbors

The retarded and advanced Green operators are defined as

$$\hat{G}^{\mathrm{r}}(E) = \lim_{\eta\to 0}\left[E + i\eta - \hat{H}\right]^{-1}, \quad \hat{G}^{\mathrm{a}}(E) = \lim_{\eta\to 0}\left[E - i\eta - \hat{H}\right]^{-1}.$$

The matrix elements of $\hat{G}^{\mathrm{r}}(E)$ are directly related to the local densities of states (LDOS) by

$$\rho_i(E) = \frac{1}{\pi}\mathrm{Im}\langle i\,|\,\hat{G}^{\mathrm{r}}(E)\,|\,i\rangle = -\frac{1}{\pi}\mathrm{Im}\langle i\,|\,\hat{G}^{\mathrm{a}}(E)\,|\,i\rangle. \quad (34.3)$$

In order to study the conductance of a finite system, like an atomic wire connected to semi-infinite electrodes, the total Hamiltonian can be decomposed as (see Figure 34.8)

$$\hat{H} = \begin{pmatrix} \hat{H}_{\mathrm{L}} & \hat{V}_{\mathrm{LC}} & 0 \\ \hat{V}_{\mathrm{CL}} & \hat{H}_{\mathrm{C}} & \hat{V}_{\mathrm{CR}} \\ 0 & \hat{V}_{\mathrm{RC}} & \hat{H}_{\mathrm{R}} \end{pmatrix},$$

where

$\hat{H}_{\mathrm{L}}$ and $\hat{H}_{\mathrm{R}}$ describe the electronic states in the uncoupled left and right leads, respectively

$\hat{H}_{\mathrm{C}}$ corresponds to the central region

$\hat{V}_{\mathrm{LC}}$, $\hat{V}_{\mathrm{CL}}$, $\hat{V}_{\mathrm{RC}}$, and $\hat{V}_{\mathrm{CR}}$ describe the coupling between the central region and the left lead and the central region and the right lead, respectively, and the corresponding retarded Green operator

$$\hat{G}^{\mathrm{r}}(E) = \begin{pmatrix} \hat{G}^{\mathrm{r}}_{\mathrm{L}} & \hat{G}^{\mathrm{r}}_{\mathrm{LC}} & \hat{G}^{\mathrm{r}}_{\mathrm{LR}} \\ \hat{G}^{\mathrm{r}}_{\mathrm{CL}} & \hat{G}^{\mathrm{r}}_{\mathrm{C}} & \hat{G}^{\mathrm{r}}_{\mathrm{CR}} \\ \hat{G}^{\mathrm{r}}_{\mathrm{RL}} & \hat{G}^{\mathrm{r}}_{\mathrm{RC}} & \hat{G}^{\mathrm{r}}_{\mathrm{R}} \end{pmatrix}$$

$$= \begin{pmatrix} (E+i\eta) - \hat{H}_{\mathrm{L}} & -\hat{V}_{\mathrm{LC}} & 0 \\ -\hat{V}_{\mathrm{CL}} & (E+i\eta) - \hat{H}_{\mathrm{C}} & -\hat{V}_{\mathrm{CR}} \\ 0 & -\hat{V}_{\mathrm{RC}} & (E+i\eta) - \hat{H}_{\mathrm{R}} \end{pmatrix}^{-1},$$

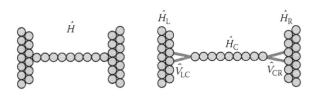

FIGURE 34.8 Decomposition of the Hamiltonian for an atomic wire connected to electrodes. $\hat{H}$ is the Hamiltonian of the whole system; $\hat{H}_{\mathrm{L}}$, $\hat{H}_{\mathrm{R}}$, and $\hat{H}_{\mathrm{C}}$ are the Hamiltonian for the uncoupled left and right electrodes, and atomic wire, respectively; and $\hat{V}_{\mathrm{LC}}$ and $\hat{V}_{\mathrm{CR}}$ describe the coupling of the atomic wire to the electrodes.

where we implicitly take the limit $\eta \to 0$. $\hat{G}^{\mathrm{r}}_{\mathrm{C}}$ is the Green operator for the central region modified due to the coupling with the leads. It can be written explicitly as

$$\hat{G}^{\mathrm{r}}_{\mathrm{C}} = (E - \hat{H}_{\mathrm{C}} - \hat{\Sigma}^{\mathrm{r}}_{\mathrm{L}} - \hat{\Sigma}^{\mathrm{r}}_{\mathrm{R}})^{-1}, \quad (34.4)$$

where $\hat{\Sigma}^{\mathrm{r}}_{\mathrm{L}} = \hat{V}_{\mathrm{CL}}\,\hat{g}^{\mathrm{r}}_{\mathrm{L}}\hat{V}_{\mathrm{LC}}$ and $\hat{\Sigma}^{\mathrm{r}}_{\mathrm{R}} = \hat{V}_{\mathrm{CR}}\,\hat{g}^{\mathrm{r}}_{\mathrm{R}}\hat{V}_{\mathrm{RC}}$ are self-energy operators introducing the effect of the coupling with the leads. We have introduced $\hat{g}^{\mathrm{r}}_{\mathrm{L}} = \lim_{\eta\to 0}(E + i\eta - \hat{H}_{\mathrm{L}})^{-1}$ and $\hat{g}^{\mathrm{r}}_{\mathrm{R}} = \lim_{\eta\to 0}(E + i\eta - \hat{H}_{\mathrm{R}})^{-1}$ as the Green operators of the uncoupled electrodes. The expressions for the advanced Green operators are obtained through the substitution $\eta \to -\eta$.

The zero-temperature linear conductance is given in terms of the Green operators as functions by the following expression (see Datta 1997, Todorov et al. 1993 for more details).

$$G = \frac{8e^2}{h}\mathrm{Tr}\left[\mathrm{Im}\hat{\Sigma}^{\mathrm{a}}_{\mathrm{L}}(E_{\mathrm{F}})\hat{G}^{\mathrm{a}}_{\mathrm{C}}(E_{\mathrm{F}})\mathrm{Im}\hat{\Sigma}^{\mathrm{r}}_{\mathrm{R}}(E_{\mathrm{F}})\hat{G}^{\mathrm{r}}_{\mathrm{C}}(E_{\mathrm{F}})\right], \quad (34.5)$$

are self-energy operators introducing the effects on the dynamics of the electrons in the central region due to the coupling with the leads, and $\hat{G}^{\mathrm{r}}_{\mathrm{C}}$ and $\hat{G}^{\mathrm{a}}_{\mathrm{C}}$ are the Green operators projected onto the central region.

The expression (34.5) can be written in the usual form

$$G = (2e^2/h)\mathrm{Tr}[\hat{t}(E_{\mathrm{F}})\hat{t}^{\dagger}(E_{\mathrm{F}})], \quad (34.6)$$

where

$$\hat{t}(E) = 2\left[\mathrm{Im}\hat{\Sigma}^{\mathrm{a}}_{\mathrm{L}}(E)\right]^{1/2} G^{\mathrm{r}}_{\mathrm{C}}(E)\left[\mathrm{Im}\hat{\Sigma}^{\mathrm{a}}_{\mathrm{R}}(E)\right]^{1/2}. \quad (34.7)$$

The knowledge of the $\hat{t}\hat{t}^{\dagger}$ matrix in terms of Green functions allows the determination of the conduction channels for a given contact geometry.

34.4 Elastic Scattering: Conductance Oscillations

As we have seen, a perfect atomic wire with a single quantum channel would have a conductance of $2e^2/h$. However, an imperfect coupling of the wire to the electrodes would result in reflections at the end of the wire leading to interference effects and a reduction of the conductance. The Au atomic wires in Figures 34.3, 34.11, and 34.13 indeed show a maximum conductance of $2e^2/h$ but lower values are also possible.

In order to investigate the effect of the coupling to the electrodes, we can apply the tight-binding scheme in conjunction with Green operators as detailed in Section 34.3.3. Consider the atoms with a single electronic state. For an infinite chain, we have a single subband with $E = \varepsilon_0 + 2t\cos ka$ where a is the separation between the atoms. The wave vectors extend from $(-\pi/a)$ to (π/a). If we assume that the each atom has a single electron, then the Fermi level is located at ε_0, the Fermi wave number is $k_F = \pi/2a$, and the Fermi velocity $v_F = 2ta/\hbar$. However, we are interested in the effect of coupling a finite wire to the electrodes.

First, we take the Hamiltonian of the isolated wire as

$$\hat{H}_C = \begin{pmatrix} \varepsilon_0 & -t & 0 & \cdots & 0 & 0 \\ -t & \varepsilon_0 & -t & \cdots & 0 & 0 \\ \vdots & \vdots & \vdots & \ddots & \vdots & \vdots \\ 0 & 0 & 0 & \cdots & -t & \varepsilon_0 \end{pmatrix}.$$

To determine the effect of the coupling on the conductance of the wire, we need to obtain the expression for the Green operators for the central region $\hat{G}_C^r$ and $\hat{G}_C^a$, given by Equation 34.4, and for the self-energy operators, $\hat{\Sigma}_R^r$, $\hat{\Sigma}_L^r$, $\hat{\Sigma}_R^a$, and $\hat{\Sigma}_L^a$, which in turn requires knowledge of the Green operator for the uncoupled electrodes $\hat{g}_R^r$, $\hat{g}_L^r$, $\hat{g}_R^a$, and $\hat{g}_L^a$, and the coupling matrices $\hat{V}_{RC}$, $\hat{V}_{CR}$, $\hat{V}_{LC}$, and $\hat{V}_{CL}$.

As a further simplification that still preserves the main features of the problem, we can describe the electrodes by semi-infinite one-dimensional wires with fixed hopping between the atoms t. In this case, the wire is connected by its first atom to only one atom of the electrode and by its last atom to only one atom of the right electrode. As a consequence, the coupling matrices have only one element different from zero, and we only need to know the Green function for the end element of the electrodes, which can be found analytically (see Todorov et al. 1993):

$$g_R(E) = g_L(E) = \frac{E + s\sqrt{E^2 - 4t^2}}{2t^2},$$

where $s = 1$ for $E > 0$ and $s = -1$ for $E < 0$ and the advanced is its complex conjugate, and we have taken $\varepsilon_0 = 0$. In Figure 34.9a through c, we show the resulting LDOS for a 5-atom wire calculated using Equation 34.3. The discrete levels of the isolated wire broaden progressively as the coupling increases. Note that for $t_1 = t_2 = t$, there will be no distinction between the wire and the electrodes and the LDOS is that corresponding to an infinite wire as shown in Figure 34.9c.

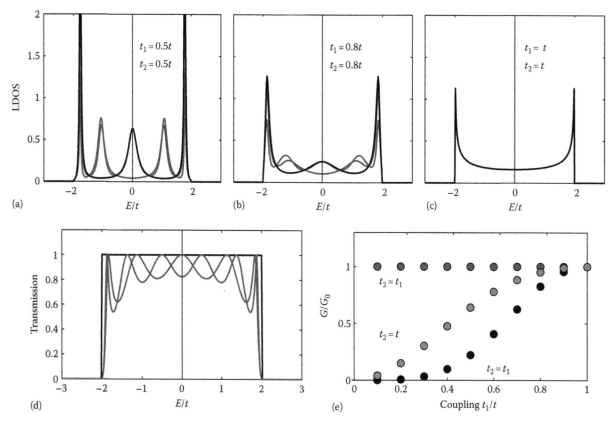

FIGURE 34.9 5-Atom wire coupled to semi-infinite 1d electrodes. The hopping within the wire and within the electrodes is t. (a and b) Local density of states (LDOS) for at the first and second (gray lines) and third or central (black line) atoms for a coupling to the electrodes of $0.5t$ (a) and $0.8t$ (b). Note that the peaks which correspond to the discrete states of the isolated wire broaden as a consequence of the coupling. (c) When the coupling to the electrodes is t the LDOS of all atoms is that of the infinite wire. (d) Transmission as a function of energy for a 5- and 6-atom wires, in both cases $t_1 = t_2 = 0.8t$. For a coupling $t_1 = t_2 = t$ the transmission equals one in all the band and is independent of the number of atoms (black curve). (e) Conductance as a function of coupling to the electrodes. For 5-atom wire the conductance is always G_0 if the coupling to the electrode is identical on both sides (upper curve). In contrast for a 6-atom wire the conductance decreases with decreasing coupling (lower curve). Note that for a 5-atom wire the conductance also decreases with decreasing coupling if the coupling is not symmetric (middle curve).

In Figure 34.9d, we show the transmission of the wire, that is $\mathrm{Tr}[\hat{t}(E)\hat{t}^{\dagger}(E)]$, where $\hat{t}(E)$ is given by Equation 34.7 for a 5-atom wire and the 6-atom wire for $t_1 = t_2 = 0.8t$ and for perfect coupling. For this system, the Fermi level is at $E = 0$, and consequently the low voltage conductance will be given by the transmission at $E = 0$ multiplied by G_0. It is important to realize that the conductance depends on the number of atoms; for an odd number of atoms, the conductance will be G_0 even for imperfect coupling as long as the coupling to the electrodes is symmetrical, while for an even number of atoms the conductance will decrease with decreasing coupling even in the case of symmetrical coupling, as illustrated in Figure 34.9e. In other words, the conductance is much more sensitive to the coupling to the electrodes if there is an even number of atoms in the wire as opposed to an odd number of atoms.

This even–odd effect is not directly observable in the conductance plateaus of atomic wires, however, after averaging many conductance traces, the oscillations become evident (Smit et al. 2003) as shown in Figure 34.10. In Au wires, the half-period of the oscillation is in agreement with the interatomic distance of the atoms in the wire as it is to be expected from the discussion above. For Pt and Ir, a similar periodicity is found but it is accompanied by a continuous decrease in conductance from $\sim 2.5G_0$ to $\sim 1G_0$ in Pt and from $\sim 2.2G_0$ to $\sim 1.8G_0$ in Ir. This behavior can be explained (de la Vega et al. 2004) by the existence in Pt and Ir of a partially filled $5d$ band. Indeed the presence of several channels in the wire gives rise to a more complex oscillatory pattern with a different wavelength for each channel.

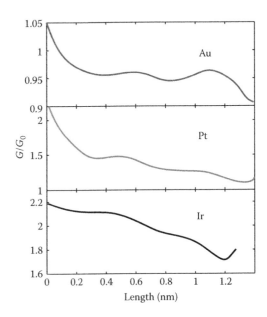

FIGURE 34.10 Averaged plateaus of conductance for atomic wires of Au, Pt, and Ir. Each of the curves is made by the average of many individual traces of conductance starting from the moment that an atomic contact is formed. The measurements were performed at low temperature using an STM in the case of Au and an MCBJ for Pt and Ir. (Adapted from Smit, R.H.M. et al., *Phys. Rev. Lett.*, 91, 076805, 2003.)

34.5 Mechanical Properties of Atomic Wires

How do the mechanical properties of atomic wires affect their transport properties? What is the force needed to break an atomic bond? Can we understand the results of the experimental conductance measurements without taking into consideration the mechanical processes taking place during the elongation of the contact? In this section, we will see how the experiments in which conductance and forces are measured simultaneously can provide answers to these questions and also to a more fundamental question: how do the mechanical properties of matter change as the size is reduced down to the atomic scale? We would like to remark that this question is not only academic, but has implications in nanotechnology and also in many technologically important problems like adhesion, friction, wear, and lubrication as all are related to small size contacts.

In Figure 34.11b, we can see the forces during the evolution of a gold contact. The forces and conductance were measured simultaneously using two STMs mounted in a series as sketched in Figure 34.11a. The top STM is used to make a metal contact as in the experiments described in the previous sections. The sample is a thin gold wire supported on one side. The forces exerted on the contact by the STM tip cause a deflection on this cantilever beam. An auxiliary STM is used to detect this deflection with picometer resolution. The experiments are performed at low temperatures, which ensure clean conditions and a negligible drift.

Let us follow the evolution of the contact shown in Figure 34.11. Consider the zero of the horizontal axis as the starting point. At this point, the contact consists of several atoms ($G \approx 6G_0$). In order to elongate the metallic contact, it is necessary to exert a tensile force, which in our plot is negative. In response to this force, the contact deforms elastically accumulating stress. This elastic deformation is naturally accompanied by a reduction in the cross-section, which results in a conductance plateau with a negative slope. Stress accumulation leads to the instability of the contact, which suddenly relaxes as the atoms in the contact rearrange to a new thinner and longer stable configuration. This force relaxation reflects as an abrupt decrease in the conductance. This process repeats several times leading to the formation of an atomic wire. During the elongation of this wire, the conductance variations are just a fraction of G_0 while the force shows the same sawtooth behavior observed for larger contacts. The *yield force*, that is the magnitude of the force right before relaxation, decreases as the size of the contact diminishes. During atomic wire elongation, the yield force remains approximately constant, being somewhat larger at the breaking point. After breaking, the atomic wire collapses and the tip has to cover the return distance to reform the contact. Surprisingly, forces at contact formation are also negative (that is tensile) while one would expect compressive forces. Only much larger contact forces would be compressive. In general, atomic wires are always in tensile stress. This can be understood as a surface tension effect.

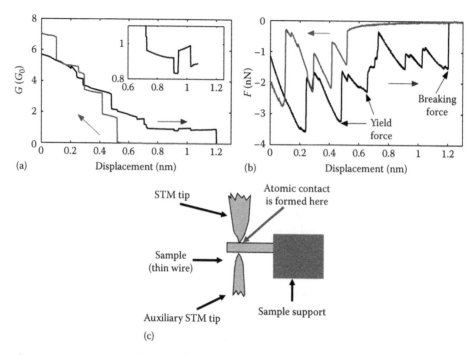

FIGURE 34.11 Simultaneous measurement of forces and conductance in gold. Notice the correlation between the sawtooth behavior of the force and the stepwise evolution of the conductance: the sudden force relaxations lead to the abrupt changes in conductance. During the elongation of the atomic wire the yield force are of the order of the breaking force, although smaller in magnitude. Force relaxations are always accompanied by an abrupt change in the conductance, of the order of G_0 for the contact and of only a fraction of G_0 for the atomic wire. The return length of this wire is 0.75 nm, about 3 atomic diameters, form the previous discussion this could correspond to 4 or 5 atoms. The measurements were performed at low temperatures using the setup sketched on the right. The force exerted on the contact by the (a) STM tip is obtained from the deflection of the sample, which is a thin gold wire, using the (b) auxiliary STM. We have used a similar procedure to that described in Rubio-Bollinger et al. (2001).

In a metallic contact of atomic dimensions, force relaxations reflect the atomic rearrangement processes that take place at the thinnest part of the contact, where the stress is largest. However, as the atomic wire starts forming, the atomic rearrangements do not occur in the atomic wire itself but in the electrode region close to the wire. As long as the force required to rearrange the atoms in the electrodes is smaller than the breaking force, the chain can grow in length by incorporating atoms from the electrodes.

The *breaking force* of gold atomic wires is 1.5 nN with a rather small dispersion (see Figure 34.12). This is in fact the force required to break an atomic bond. This force is found to be independent of

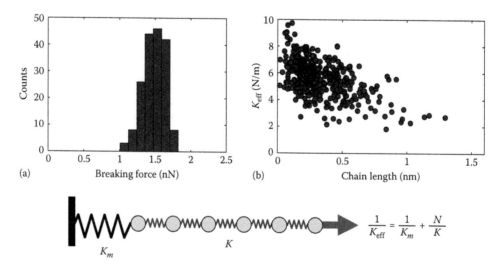

FIGURE 34.12 (a) Breaking force of gold atomic wires. (b) Effective stiffness, K_{eff}, measured in the experiments. The experimental system can be represented by a series of springs with stiffness K, representing the interatomic bonds, in series with a spring K_m, representing the electrodes.

the length of the wire and is considerably larger than the force required to break individual bonds in bulk gold, which is estimated at only 0.8–0.9 nN using density functional theory (DFT) calculations, as we will discuss below. This is direct experimental evidence that bonds of low-coordinated metal atoms are considerably stronger than bonds in the bulk (Rubio-Bollinger et al. 2001).

The slope of the force as a function of displacement defines the stiffness of the contact. However, to obtain the stiffness of an atomic wire, it is necessary to take into account the stiffness of the electrodes. In the bottom panel of Figure 34.12, the atomic wire and electrodes are shown schematically. The atomic wire is represented as a number of springs, each one having a stiffness K, connected in series, and the electrodes as another spring with a stiffness K_m. The effective compliance $1/K_{eff}$ of the system is the sum of the compliances of the different elements,

$$\frac{1}{K_{eff}} = \frac{1}{K_m} + \frac{N}{K}. \quad (34.8)$$

In Figure 34.12, the stiffness of atomic wires as a function of their length is plotted. Notice that the stiffness of the wires decreases with their length and that the electrodes (chain length zero) is typically of the order of 6 N/m. This plot shows that the atomic wire is unusually stiff: it takes a wire of three atoms in length to equal the compliance of the electrodes, giving a value of about 18 N/m for the stiffness of a single atomic bond. This low compliance of the electrodes implies that a large fraction of the elastic deformation takes place at the electrodes next to the wire. The breaking point will be of the order of 0.25 nm, which must be taken into account in the estimate of the length from the return distance.

34.6 Inelastic Scattering and Dissipation

As we have discussed in Section 34.3, atomic wires of gold are perfect one-dimensional conductors. The current is carried by a single quantum mode whose transmission probability deviates from unity only due to elastic scattering at points where the wire joins the electrodes. This description of transport through atomic wires is correct only if both temperature and bias voltage are low enough. Indeed, the thermal vibrational motion of the atoms in the wire alters the potential seen by the electrons and causes them to scatter inelastically. At room temperature, this is the mechanism that is responsible for most of the resistance in metals. In contrast, at low temperatures, the motion of the atoms in the wire is frozen and at low bias voltages, there is no inelastic scattering and consequently no dissipation. However, at higher voltages, the electrons will interact with the quantized vibrations or *phonons* of the atomic wire, which will result in a decrease in the conductance and heating of the atomic wire. The sensitivity to voltage of this electron–phonon interaction can be used for spectroscopy and to deduce structural information. This inelastic electron spectroscopy

(IES) usually receives different denominations depending on the transmission regime, being termed inelastic electron tunneling spectroscopy (IETS) in the low transmission regime and point contact spectroscopy (PCS) in the high transmission regime.

The differential conductance dI/dV as a function of the voltage measured in a gold atomic wire at low temperatures is shown in Figure 34.13b, and its derivative or *IES spectrum* is shown in Figure 34.13c and the inset of Figure 34.13a. We can see the conductance is close to unity at low voltages and drops by about 1% for voltages above a certain threshold in the range of 12–20 mV. This drop in the conductance, which is symmetric with respect to voltage, is due to the onset of the excitation of a particular vibrational mode in the wire with a well-defined energy $\hbar\omega = eV_p$, where V_p is the position of the peak in the IES spectrum. Stretching the atomic wire results in a shift of the peak to lower voltages, which reflects the softening of the vibrational modes due to the weakening of the interatomic bonds. In Figure 34.14a and b, we show the peak position V_p and amplitude A_p, respectively as the wire is stretched for different wires of different lengths. Figure 34.14c and d shows that the dependence of the peak amplitude with peak position and with length is proportional to L/ω^2. Note that the symmetry of the conductance drops in the conductance curve or antisymmetry of the peaks in the IES spectra is a signature of the phonon signal and serves in practice to differentiate them from the variations in the conductance due to elastic scattering in the electrodes or at the ends of the wire, which are typically asymmetric.

The experiments in Figure 34.13 show that for voltages below a certain threshold V_p the electrons cannot lose their energy through the interaction with the ions in the wire, that is, there is no dissipation in the wire. The dissipation of the excess energy of the electrons will take place inside the electrodes. Only for voltages above V_p will the electrons be able to give part of their energy to the ions and the atomic wire will start heating. Interestingly, as we will see in this section, the electrons traversing the wire both heat up and cool down the wire through phonon emission and absorption, respectively, resulting in a steady-state temperature as a function of voltage.

To gain insight into the fundamental mechanisms taking part in inelastic scattering processes in atomic wires, we will consider in detail the electron–phonon interaction in an ideal one-dimensional wire of length L at zero temperature. In an infinite atomic wire, the electron–phonon interaction is given by

$$\hat{H}_{ep} = \sum_{qk} q V_q \left(\frac{\hbar}{2MN\omega_q} \right)^{1/2} (\hat{a}_q \hat{c}^+_{k+q} \hat{c}_k + \hat{a}^+_q \hat{c}^+_{k-q} \hat{c}_k),$$

where

M is the mass of the ions
N is the number of ions
ω_q is the frequency of the phonon mode q
$\hat{c}_k$ and $\hat{c}^+_k$ are the annihilation and creation operators, respectively for the electrons in state k
$\hat{a}_q$ and $\hat{a}^+_q$ are the annihilation and creation operators, respectively for the phonons in state q

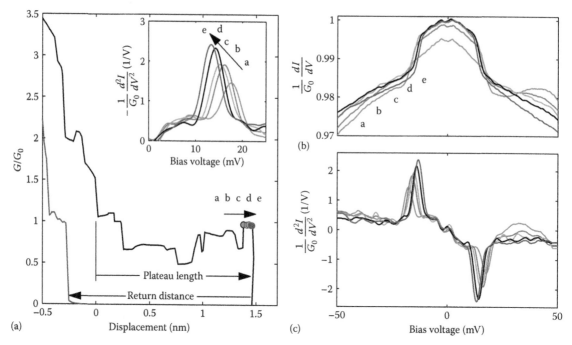

FIGURE 34.13 Inelastic scattering in Au atomic wires. (a) Shows the low bias conductance of a gold atomic contact. The conductance plateau indicates that an atomic wire of 5–6 atoms has formed. Note that the plateau presents quite a lot of structure, which is a result of elastic scattering possibly due to disorder at the point where the atoms are being extracted. The tip displacement was stopped before the wire broke and the bias voltage was ramped with a small superimposed ac modulation to obtain the differential conductance dI/dV curves shown in (b). These conductance curves present a marked symmetric drop in conductance for voltages of 10–20 mV. Between the different conductance curves the tip was displaced by a fraction of a nanometer stretching the atomic wire. In (c) the derivative of the differential conductance d^2I/dV^2 or IES spectrum is shown. In these IES spectra the conductance drop appears as an antisymmetric peak. Stretching results in a displacement of the peak in the spectra to lower frequencies as can be clearly observed in (c) and the inset in (a). These measurements were performed at 0.3 K using an STM and the procedure in Agraït et al. (2002a,b).

The term $\hat{a}_q^+\hat{c}_{k-q}^+\hat{c}_k$ represents the scattering of an electron from state k to state $k - q$ with the emission of a phonon q, and the term $\hat{a}_q\hat{c}_{k+q}^+\hat{c}_k$ corresponds to the scattering of an electron from state k to state $k + q$ with the absorption of a phonon q. We can write the probabilities per unit of time for an electron k to emit or absorb a phonon q, using Fermi's Golden Rule as

$$w_{kq}^{em} = \frac{2\pi}{\hbar}|\langle k-q, n_q+1|\hat{H}_{ep}|k, n_q\rangle|^2 \delta(\varepsilon_k - \hbar\omega_q - \varepsilon_{k-q})$$

$$= M_q(n_q+1)\delta(\varepsilon_k - \hbar\omega_q - \varepsilon_{k-q}),$$

$$w_{kq}^{ab} = \frac{2\pi}{\hbar}\left|\left\langle k+q, n_q-1\left|\hat{H}_{ep}\right|k, n_q\right\rangle\right|^2 \delta(\varepsilon_k + \hbar\omega_q - \varepsilon_{k+q})$$

$$= M_q n_q\delta(\varepsilon_k + \hbar\omega_q - \varepsilon_{k+q}),$$

where n_q is the occupation of the vibrational mode q and $M_q = \pi V_q^2 q^2/MN\omega_q$. To find the number of electrons that are inelastically scattered in the wire, we need to consider the scattering probabilities of all the states in the wire taking into account that the connection to the electrodes imposes a chemical potential μ_L for right-going electrons and μ_R for left-going electrons. In the following discussion, we will take $\mu_L > \mu_R$, which

implies a net flow of electrons from the left electrode to the right electrode. Then, taking the sum of all the allowed electron and phonon states we have

$$W^{em} = \sum_k\sum_q M_q(n_q+1)\delta(\varepsilon_k - \hbar\omega_q - \varepsilon_{k-q})f_L(\varepsilon_k)[1-f_R(\varepsilon_{k-q})],$$

$$W^{ab} = \sum_k\sum_q M_q n_q\delta(\varepsilon_k + \hbar\omega_q - \varepsilon_{k+q})f_L(\varepsilon_k)[1-f_R(\varepsilon_{k+q})].$$

These summations can be easily performed by taking into account the following considerations. The kinetic energy of phonons is typically more than two orders of magnitude smaller than the energy of electrons at the Fermi level, and since energy is conserved in the scattering process, as expressed by Dirac's delta function, this implies that $\varepsilon_k \approx \varepsilon_{k-q}$ and consequently $|k| \approx |k - q|$, that is, either $q \approx 0$ or $q \approx -2k$ (see Figure 34.15a). Both for phonon emission and for the absorption the final electronic states must be unoccupied. Emission is possible for electrons whose energy is $\varepsilon_k > \mu_R + \hbar\omega$ because this provides unoccupied left-going states (see Figure 34.15b). As all right-going states at lower energies are occupied, the electron must always backscatter in this process and consequently $q \approx -2k$. For low

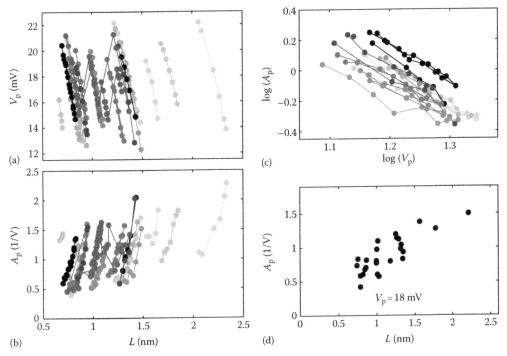

FIGURE 34.14 Frequency and amplitude of phonon scattering in Au wires. (a and b) Peak position V_p and amplitude A_p in the IES spectra as a function of length L for different atomic wires. Each color represents a different wire. As the wire is elastically stretched the peak position changes to lower values. In some of the wires, atomic rearrangements leading to a sudden increase of V_p and a sudden decrease in A_p are clearly observable. (c) Log-log plot of the peak amplitude versus peak position. The slope of the curve shows that $A_p \sim V_p^{-2}$. (d) Peak amplitude A_p for $V_p = 18$ mV versus atomic wire length L, showing $A_p \sim L$. These results are similar to those reported in Agraït et al. (2002a,b).

bias voltages, all electrons involved in transport will have $k = \pm k_F$, which for an atomic wire with one conduction electron per atom is $k_F = \pi/2a$ and consequently the wave number of the vibrational mode is $q = \pi/a$, which lies at the zone boundary. This vibrational mode $q_B = \pi/a$ has a wavelength equal to $2a$—it is the alternating bond length mode (see Figure 34.16). Then, for the phonon emission rate, we have

$$W^{em} = \begin{cases} 0, & \text{for } eV < \hbar\omega_B, \\ LC\left(\dfrac{eV - \hbar\omega_B}{\hbar\omega_B}\right)(n_B + 1) & \text{for } eV \geq \hbar\omega_B, \end{cases}$$

where

$$C = V_{q_B}^2 q_B a m_e / \pi M \hbar^2 v_F$$

m_e is the mass of the electron
v_F is the Fermi velocity of the electrons in the wire

Phonon absorption will be possible for electron $\varepsilon_k > \mu_R - \hbar\omega$ (see Figure 34.15b). Again in this case this implies backscattering and $q = \pi/a = q_B$ (note that q_B and $-q_B$ represent the same mode). In principle, there could be a contribution from electrons with energies very close to μ_L that could be forward

scattered by phonons with $q \approx 0$, but it is negligible. Note that for $eV < \hbar\omega_B$ it is possible to have phonon absorption by both right-going and left-going electrons, this process is possible even at $V = 0$ and provides a mechanism for phonon damping in the absence of a bias voltage (damping by electron-hole pair generation). The rate of phonon absorption in the atomic wire is then

$$W^{ab} = \begin{cases} LC\left(\dfrac{\hbar\omega_B + eV}{\hbar\omega_B} + \dfrac{\hbar\omega_B - eV}{\hbar\omega_B}\right)n_B, & \text{for } eV < \hbar\omega_B, \\ LC\left(\dfrac{eV + \hbar\omega_B}{\hbar\omega_B}\right)n_B, & \text{for } eV \geq \hbar\omega_B. \end{cases}$$

It is remarkable that due to the one-dimensional character of the one-dimensional wire, electrons can interact only with one phonon mode, the q_B mode of energy $\hbar\omega_B$. Thus, the passage of current through the wire creates and annihilates phonons in this mode, changing the occupation number n_B according to the emission and absorption rates W^{em} and W^{ab}

$$\frac{dn_B}{dt} = \begin{cases} -2LCn_B, & \text{for } eV < \hbar\omega_B, \\ LC\left(\dfrac{eV - \hbar\omega_B}{\hbar\omega_B}\right)(n_B + 1) - LC\dfrac{eV + \hbar\omega_B}{\hbar\omega_B}n_B, & \text{for } eV \geq \hbar\omega_B, \end{cases}$$

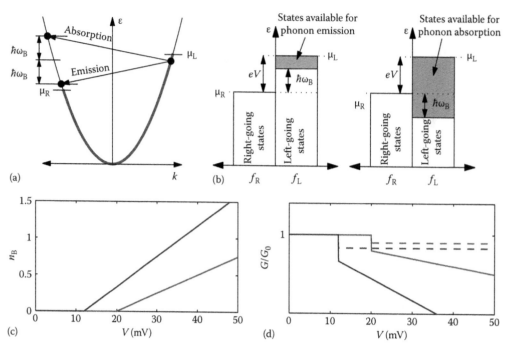

FIGURE 34.15 Inelastic effects in one-dimensional ideal wires. (a) Phonon emission and absorption processes in a one-dimensional conductor. The left electrode is at a chemical potential μ_L higher the chemical potential of the right electrode μ_R, such that $\mu_L - \mu_R = eV$. In the emission process the electron is backscattered and gives and energy $\hbar\omega_B$ to the emitted phonon. In the absorption process the electron is backscattered and receives and energy $\hbar\omega_B$ from the absorbed phonon. (b) The rightgoing-states available for phonon emission have energies between and $\mu_R + \hbar\omega_B$ and μ_L, while those available for absorption are more numerous and have energies between $\mu_R - \hbar\omega_B$ and μ_L. (c) Dependence of the occupation n_q of mode q_B (number of phonons) on voltage, in a wire in two different stretch states, with $\hbar\omega_B$ values of 12 and 20 meV. (d) Total conductance of the atomic wire taking into account inelastic effects for the same two stretch states considered in (c). The full line is for the *externally undamped limit* and the dashed line is for the *externally damped limit*.

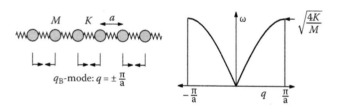

FIGURE 34.16 Simple mechanical model of an atomic wire. The interatomic interaction is modeled by a first-neighbor interaction with represented by spring with stiffness K, with a constant value for a given value of the interatomic separation a. M is the mass of the atoms. The dispersion relation is given by Equation 34.12. For a wire with a single half-filled band, the zone-boundary mode $q_B = \pm\pi/a$ will be the only mode capable of interacting with the electrons, and as a consequence the frequency detected in the IES spectra will be given by $\omega_B = \sqrt{4K/M}$. Note that the wavelength of the q_B-mode is precisely $2a$, which results in an alternating bond character.

whence n_B attains a steady-state value at any voltage (see Figure 34.15c) given by

$$n_B = \begin{cases} 0, & \text{for } eV < \hbar\omega_B, \\ \dfrac{eV - \hbar\omega_B}{2\hbar\omega_B}, & \text{for } eV \geq \hbar\omega_B. \end{cases} \quad (34.9)$$

Both phonon emission and absorption cause backscattering of the electrons, these backscattered electrons return to the electrode from which they originated and as a consequence they do not contribute to the current. The total backscattering current is given by

$$I_b = eW^{em} + eW^{ab} = \begin{cases} 0, & \text{for } eV < \hbar\omega_B, \\ eCL\left(\dfrac{e^2V^2 - \hbar^2\omega_B^2}{2\hbar^2\omega_B^2} + \dfrac{e^2V^2}{2\hbar^2\omega_B^2}\right), & \text{for } eV \geq \hbar\omega_B. \end{cases}$$

where we have used the steady-state value for n_B. The differential backscattering conductance is then

$$G_b = \frac{dI_b}{dV} = \begin{cases} 0, & \text{for } eV < \hbar\omega_B, \\ e^2CL\left(\dfrac{2eV}{\hbar^2\omega_B^2}\right) & \text{for } eV \geq \hbar\omega_B. \end{cases}$$

The drop in conductance occurs at $eV = \hbar\omega_B$ (see Figure 34.15) and has a magnitude of

$$\Delta G = -[G_b]_{eV=\hbar\omega_B} = \frac{2e^2CL}{\hbar\omega_B}. \tag{34.10}$$

Half of the drop is due to the electrons backscattered in the emission process and the other half is in the absorption.

So far we have considered that electrons once emitted could leak out of the wire, that is, the energy transferred from the electrons to the vibrations remains in the wire, this is the *externally undamped limit*. In general, the wire vibrations would couple to the electrodes to some degree. We can consider the limiting case in which the energy is instantaneously absorbed into an external heat bath, then the occupation would remain $n_B = 0$ for all voltages, this is the *externally damped limit*. The backscattered current will be due only to the emission process and the conductance will be

$$G_b = \frac{dI_b}{dV} = \begin{cases} 0, & \text{for } eV < \hbar\omega_B, \\ e^2CL\left(\dfrac{eV}{\hbar^2\omega_B^2}\right) & \text{for } eV \geq \hbar\omega_B. \end{cases}$$

and the drop in conductance at the threshold

$$\Delta G = -[G_b]_{eV=\hbar\omega_B} = \frac{e^2CL}{\hbar\omega_B}. \tag{34.11}$$

In the case of arbitrary coupling to the electrodes, the results would be between the externally damped and undamped limits.

According to Equations 34.10 and 34.11 for a given vibrational frequency, the magnitude of the conductance drop depends linearly on the length of the wire L, which agrees with the experimental observation (see Figure 34.14d). However, the observed experimental dependence (Figure 34.14c) is on $1/\omega_B^2$ rather than on $1/\omega_B$.

Now we can consider the effect of this exchange of energy between the electron and phonon systems on the temperature of the wire. According to Equation 34.9, for voltages above the threshold for the excitation of mode q_B, the onset of dissipation gives rise to a steady-state occupation of this mode. This occupation is completely out of equilibrium because modes of lower energies are unoccupied, and the temperature is not defined. However, we can get an equivalent temperature T_{eq} by equating the voltage-dependent occupation in Equation 34.9 to the equilibrium occupation

$$\frac{eV - \hbar\omega_B}{2\hbar\omega_B} = \frac{1}{\exp(\hbar\omega_B / k_B T_{eq}) + 1}, \quad \text{for } eV \geq \hbar\omega_B,$$

where k_B is the Boltzmann constant. We find that for $\hbar\omega_B = 20\,\text{meV}$ at $V = 60\,\text{mV}$, the occupation is one and $T_{eq} = 335\,\text{K}$; and for $\hbar\omega_B = 12\,\text{meV}$ at $V = 36\,\text{mV}$, the occupation is also one and $T_{eq} = 200\,\text{K}$. Note that these equivalent temperatures are independent of the wire length because in a longer wire there is

more phonon emission but this is compensated by more phonon absorption. Electrons both heat and cool the wire preventing a catastrophic increase in the number of phonons in the wire.

The coupling of the electronic current with the vibrational modes of the wire makes it possible to obtain structural information from the IES spectra. In order to illustrate how the mechanical and structural properties reflect on the electronic properties, we will consider the simplest model: an infinite monoatomic wire. If the atoms are separated a distance a, have a mass M, and their interaction is represented by a spring with spring constant K, as represented in Figure 34.16, the dispersion relation would be given by

$$\omega = 2\sqrt{\frac{K}{M}}\left|\sin\frac{1}{2}qa\right| \tag{34.12}$$

where q is the eigenmode wave number. The highest frequency $\omega_B = 2\sqrt{K/M}$ corresponds to the zone boundary mode $q_B = \pi/a = 2k_F$. In the case of a half-filled single band wire, this is the only mode that interacts with the electrons. From the position of the peak in the IES spectra that are in the range 10–22 meV, and taking into account the atomic mass of gold, we find values for K in the range of 18–90 N/m. These values should be considered just as a rough approximation, since this mechanical model is far too simple because we cannot approximate the metallic interaction by a simple first-neighbor interaction. A first-principles calculation as discussed in Section 34.7 will be necessary for a realistic quantitative comparison.

The preceding basic theoretical considerations for the inelastic scattering and dissipation in a one-dimensional wire and the coupling between the electrons and the atomic vibrations give a semiquantitative account of the experimental observations. For a more detailed treatment using a first-principles approach and with application to nanoscale devices, see the articles by Frederiksen et al. (2004, 2007).

34.7 Numerical Calculations of the Properties of Atomic Wires

A fully realistic computation of the properties of atomic wires taking into account an adequate number of atoms, dynamic structural arrangements, applied bias, temperature, and the effect of the electrodes is still beyond the capabilities of any theoretical model. Nevertheless, theoretical methods can be used to investigate the influence of different factors on the system (Agrait et al. 2003).

First principles or *ab initio* methods attempt a full quantum-mechanical treatment of both nuclear and electronic degrees of freedom. In order to obtain realistic configurations, the atoms are allowed to move in response to the forces they experience. These forces are obtained from the evaluation of the potential energy in a quantum-mechanical description of the system in which certain approximations must unavoidably be included.

Within this approach, the total energy of a system of ions and valence electrons can be written as

$$E_{\text{total}}(\{\mathbf{r}_I\},\{\dot{\mathbf{r}}_I\}) = \sum_I \frac{1}{2} m_I |\dot{\mathbf{r}}_I|^2 + \sum_{I>J} \frac{Z_I Z_J}{|\mathbf{r}_I - \mathbf{r}_J|} + E_{\text{elect}}(\{\mathbf{r}_I\}), \quad (34.13)$$

where

    $\mathbf{r}_I$, m_I, and Z_I are the position, mass, and charge of the *I*th ion, respectively

    $E_{\text{elect}}(\{\mathbf{r}_I\})$ is the ground-state energy of the valence electrons evaluated for the ionic configuration $\{\mathbf{r}_I\}$

The first two terms in this equation correspond to the ionic kinetic and interionic interaction energies, respectively. In this equation, we have considered that the electrons follow the instantaneous configuration of the ions (Born-Oppenheimer approximation). The major task in first principles or *ab initio* methods is to calculate the ground-state electronic energy, which is typically done via the Kohn–Sham (KS) formulation of the DFT of many-electron systems within the local density (LDA) or local-spin-density (LSD) approximation. The ground-state energy of the valence electrons, according to the Hohenberg–Kohn theorem, depends only on the electronic density. In the KS method, the many-body problem for the ground-state electronic density $n(\mathbf{r})$ of an inhomogeneous system of N electrons in a static external potential (due to the positive ions) is reduced to solving self-consistently the independent-particle Schödinger equation

$$\left[-\nabla^2 + v_{\text{eff}}(n)\right]\psi_j = \varepsilon_j \psi_j,$$

with the electronic density given by

$$n = \sum_j f_j |\psi_j|^2,$$

where f_j is the occupation of the *j*th (orthonormal) orbital. The KS effective potential v_{eff} is given by the functional derivative of the electronic energy and since it depends on the electronic density it must be obtained self-consistently. In *ab initio* methods, the computational requirements are so high that systems that can be studied this way are limited to a small number of atoms although this number is increasing due to the increasing power of computers and new schemes.

Ab initio methods have been used to investigate the strength of the atomic bonds in gold (Rubio-Bollinger et al. 2001). Interestingly, the bonds in the atomic wire were found to be much stronger than those in the bulk, which in fact makes the formation of the wires possible. The agreement with the experimentally measured breaking forces is quite good as mentioned in Section 34.5.

Although, as we have seen above, atomic wires have been found experimentally only for few metals (Au, Pt, and Ir); atomic wires of many different metals have been studied. Calculations have been performed for infinite wires using cells with one or several atoms and periodic boundary conditions, and for finite wires supported between pyramidal tips, which are generally considered to have a negligible effect on the structure and stability of the wires. These calculations show that the atomic wires are linear only if they are overstretched. As the available state per atom decreases for all metals, a zigzag structure appears (see Figure 34.17). Taking the *z*-axis along the axis of the chain and defining d_z as the projection of the interatomic distance over this axis, the energy per atom as a function of d_z, for calculations with two atoms per cell, has one or two minima depending on the metal. The absolute minimum corresponds to atomic wires where the interatomic bonds with the *z*-axis make an angle of about 60°. The 5*d* elements, in particular in Au, Pt, and Ir, show a second minimum corresponding to a stable zigzag configuration. This is rather surprising since it implies that a force will be needed to either stretch or contract the wire. The angle subtended with the *z*-axis for Au, Pt, and Ir are 20°, 25°, and 45°, respectively.

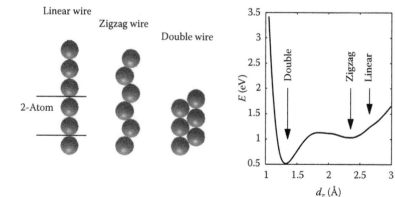

FIGURE 34.17 DFT calculated energy per atom of atomic wires of Au using a 2-atom cell (right) and corresponding atomic configurations. The absolute minimum corresponds to a double wire configuration, and the second minimum to a stable zigzag configuration. The atomic wire is linear only for bond lengths above 0.275 nm. (Adapted from Fernández-Suivane, L. et al., *Phys. Rev. B*, 75, 075415, 2007, and consistent with the results of Sánchez-Portal et al., 1997, 2001.)

The origin can be explained as a reduction in the transverse kinetic energy for the electrons due to the increased effective wire width. This mechanism is of the same nature as the shell structure observed for alkali metal nanowire. For Au, the bond lengths are consistent with those observed experimentally. The wire is stretched to a linear configuration only for bond lengths above about 0.275 nm, shortly before breaking. There is no experimental evidence of this zigzag structure, because as we saw in Section 34.5, the experimentally realized atomic wires are always under tensile stress.

The magnetic properties of atomic wires have also been investigated because the partially filled *d* orbitals might lead to magnetic states in a one-dimensional geometry. The calculations indicate that Au atomic wires would not show magnetism but Pt and Ir seem to have magnetic anisotropies (Fernández-Suivane et al. 2007). This has not been confirmed experimentally.

Calculating the conductance of an atomic wire using *ab initio* methods poses additional problems because it requires the inclusion of the electrodes in the calculations. In general, it is not enough to count the number of bands crossing the Fermi level in an infinite wire because the electrodes, which are strongly coupled to the wire, will modify the band structure and may also cause an interference effect. A possible approach is to use a combination of *ab initio* and Green's functions methods (de la Vega et al. 2004, García-Suárez et al. 2005, Jelínek et al. 2006).

In order to investigate the transport properties or the formation dynamics, it is necessary to include many more atoms to describe the electrodes. One possibility is to use the classical MD for the problems that require a large number of atoms and simulation time. Another possibility, in between first principles and empirical methods, is the tight-binding molecular dynamics (TBMD) method, which is more accurate than empirical potential methods because it explicitly includes the electronic structure and is much faster than the first principles methods. Conventional MD simulations use phenomenological inter-atomic potentials to model the energetics and dynamics of the system (Agrait et al. 2003). Pair potentials are not adequate for an accurate description of metallic systems that require potentials that include many-body interactions. These potentials contain the physics of the model systems and their functional form is selected on the basis of theoretical considerations and is typically fitted to a number of experimental or theoretically calculated data. The embedded atom method (EAM) and effective medium theory (EMT) potentials derived from the DFT are often used to model metallic systems. In these models, the potential energy of the system is written as a sum of a short-range pair-interaction repulsion and an embedding energy for placing an atom in the electron density of all the other atoms:

$$E_{\text{pot}} = \sum_i F_i[\rho_{h,i}] + \frac{1}{2} \sum_i \sum_{j \neq i} V_{ij}(r_{ij}), \qquad (34.14)$$

where

$V_{ij}(r_{ij})$ is a two-body potential that depends on the distance r_{ij} between atoms i and j

$F_i[\rho_{h,i}]$ is the *embedding energy* for placing an atom at position i, where the host electron density due to the rest of the atoms in the system is $\rho_{h,i}$

The latter is given by

$$\rho_{h,i} = \sum_{j \neq i} \rho(r_{ij}), \qquad (34.15)$$

where $\rho(r_{ij})$ is the "atomic density" function. The first term in Equation 34.14 represents, in an approximate manner, the many-body interactions in the system. These potentials provide a computationally efficient approximate description of bonding in metallic systems, and have been used with significant success in different studies. Closely related are the Finnis–Sinclair (FS) potentials, which have a particularly simple form

$$E_{\text{pot}}^{\text{FS}} = \varepsilon \left[\frac{1}{2} \sum_i \sum_{j \neq i} V(r_{ij}) - c \sum_i \sqrt{\rho_i} \right], \qquad (34.16)$$

with $V(r_{ij}) = (a/r_{ij})^n$ and $\rho_i = \sum_{j \neq i} (a/r_{ij}^m)$, where a is normally taken to be the equilibrium lattice constant, m and n are positive integers with $n > m$, and ε is a parameter with the dimensions of energy. For a particular metal, the potential is completely specified by the values of m and n, since the equilibrium lattice condition fixes the value of c.

It is important to remark that the predictions of molecular dynamics (MD) simulations using empirical potentials may be inaccurate in systems of atomic dimensions like atomic wires, since the usual potentials fitted to bulk properties do not describe adequately low-coordinated atoms. This situation may be improved by expanding the database used for the fitting to include a set of atomic configurations calculated by *ab initio* methods. Nevertheless, MD simulations present a picture of the events that lead to wire formation in gold contacts. Deformation involves a series of elastic stretching stages, each terminated by a change in the atomic arrangement, which involves the breaking of atomic bonds. Whether or not a chain is formed depends on the relative strength of the bonds for different atomic configurations. The way a surface atom gets incorporated into a chain is by keeping the bond with a low coordinated chain atom while breaking the bonds to more highly coordinated atoms in the shoulders of the contact. This shows that chain formation can occur if the bonds in the chain are much stronger than the bonds in the bulk so that it is harder to break the chain than to pull out an atom of the electrodes.

34.8 Summary

We have given a brief account of the basic transport and mechanical properties of atomic wires and atomic contacts. These systems due to their simplicity are ideally suited to investigate the fundamental properties of matter at the atomic scale, and serve as benchmark systems for theories that describe nano-scale systems. In addition, many of the techniques and concepts described in this chapter in relation to atomic wires can be applied to other nanosized systems.

Acknowledgments

This work was supported by MEC, Spain (MAT2004-03069); MICINN, Spain (MAT2008-01735 and CONSOLIDER-INGENIO-2010 CSD-2007-00010); and Comunidad de Madrid (Spain) through program Citecnomik (P-ESP-0337-0505).

References

Agrait, N., Untiedt, C., Rubio-Bollinger, G., and Vieira, S. 2002a, Electron transport and phonons in atomic wires, *Chem. Phys.* **281**, 231–234.

Agrait, N., Untiedt, C., Rubio-Bollinger, G., and Vieira, S. 2002b, Onset of dissipation in ballistic atomic wires, *Phys. Rev. Lett.* **88**, 216803.

Agrait, N., Yeyati, A., and van Ruitenbeek, J. 2003, Quantum properties of atomic-sized conductors, *Phys. Rep.* **377**, 81–279.

Datta, S. 1997, *Electronic Transport in Mesoscopic Systems*, Cambridge University Press, Cambridge, U.K.

de la Vega, L., Martín-Rodero, A., Yeyati, A. L., and Saúl, A. 2004, Different wavelength oscillations in the conductance of 5*d* metal atomic chains, *Phys. Rev. B* **70**(11), 113107.

Fernández-Suivane, L., García-Suárez, V. M., and Ferrer, J. 2007, Predictions for the formation of atomic chains in mechanically controllable break-junction experiments, *Phys. Rev. B* **75**, 075415.

Frederiksen, T., Brandbyge, M., Lorente, N., and Jauho, A. 2004, Inelastic scattering and local heating in atomic gold wires, *Phys. Rev. Lett.* **93**(25), 256601.

Frederiksen, T., Paulsson, M., Brandbyge, M., and Jauho, A.-P. 2007, Inelastic transport theory from first principles: Methodology and application to nanoscale devices, *Phys. Rev. B* **75**, 205413.

García-Suárez, V. M., Rocha, A. R., Bailey, S. W., Lambert, C. J., Sanvito, S., and Ferrer, J. 2005, Conductance oscillations in zigzag platinum chains, *Phys. Rev. Lett.* **95**, 256804.

Jelínek, P., Pérez, R., Ortega, J., and Flores, F. 2006, Hydrogen dissociation over au nanowires and the fractional conductance quantum, *Phys. Rev. Lett.* **96**, 046803.

Nilius, N., Wallis, T., and Ho, W. 2002, Development of one-dimensional band structure in artificial gold chains, *Science* **297**(5588), 1853–1856.

Ohnishi, H., Kondo, Y., and Takayanagi, K. 1998, Quantized conductance through individual rows of suspended gold atoms, *Nature* **395**, 780–785.

Rubio-Bollinger, G., Bahn, S., Agraït, N., Jacobsen, K., and Vieira, S. 2001, Mechanical properties and formation mechanisms of a wire of single gold atoms, *Phys. Rev. Lett.* **87**, 026101.

Sánchez-Portal, D., Artacho, E., Junquera, J., Garcia, A., and Soler, J. 2001, Zigzag equilibrium structure in monatomic wires, *Surf. Sci.* **482**, 1261–1265.

Sánchez-Portal, D., Untiedt, C., Soler, J., Sáenz, J., and Agraït, N. 1997, Nanocontacts: Probing electronic structure under extreme uniaxial strains, *Phys. Rev. Lett.* **79**, 4198–4201.

Scheer, E., Agraït, N., Cuevas, J., Levy Yeyati, A., Ludoph, B., Martn-Rodero, A., Rubio Bollinger, G., van Ruitenbeek, J., and Urbina, C. 1998, The signature of chemical valence in the electrical conduction through a single-atom contact, *Nature* **394**, 154–157.

Scheer, E., Joyez, P., Esteve, D., Urbina, C., and Devoret, M. 1997, Conduction channel transmissions of atomic-size aluminum contacts, *Phys. Rev. Lett.* **78**, 3535–3538.

Segovia, P., Purdie, D., Hengsberger, M., and Baer, Y. 1999, Observation of spin and charge collective modes in one-dimensional metallic chains, *Nature* **402**, 504–507.

Smit, R. H. M., Untiedt, C., Rubio-Bollinger, G., Segers, R. C., and van Ruitenbeek, J. M. 2003, Observation of a parity oscillation in the conductance of atomic wires, *Phys. Rev. Lett.* **91**, 076805.

Todorov, T., Briggs, G., and Sutton, A. 1993, Elastic quantum transport through small structures, *J. Phys.: Condens. Matter* **5**, 2389–2406.

Untiedt, C., Yanson, A. I., Grande, R., Rubio-Bollinger, G., Agraït, N., Vieira, S., and van Ruitenbeek, J. 2002, Calibration of the length of a chain of single gold atoms, *Phys. Rev. B* **66**, 085418.

Yanson, A. 2001, Atomic chains and electronic shells: Quantum mechanisms for the formation of nanowires, PhD thesis, Universiteit Leiden, Leiden, the Netherlands.

Yanson, A., Rubio Bollinger, G., van den Brom, H., Agraït, N., and van Ruitenbeek, J. 1998, Formation and manipulation of a metallic wire of single gold atoms, *Nature* **395**, 783–785.

Monatomic Chains

Roel H. M. Smit
Universiteit Leiden

Jan M. van Ruitenbeek
Universiteit Leiden

35.1 Introduction

35.1.1 Motivation

In the study of quantum mechanics, there are only a very limited number of realistic examples that can be used as illustrations, much to the frustration of undergraduate students. The complexity of solving Schrödinger's equation often leads to a necessary simplification to one dimension, and one-dimensional structures appear to be minimally realistic.

When one starts studying the behavior of metallic point contacts on the atomic scale (for a general introduction to metallic point contacts, see (Agraït et al., 2003)), one begins to explore computer simulations with realistic interatomic potentials. When these calculations indicate that gold (Sørensen et al., 1998) and platinum (Finbow et al., 1997) contacts form one-dimensional chains of single atoms upon stretching, the authors were implying that this could be an artifact of the model. It did not take long, however, before experiments confirmed the formation of chains. Experiments at both room temperature (Ohnishi et al., 1998) and liquid helium temperature (Yanson et al., 1998) showed the existence of strands of single atoms with lengths up to seven atoms. From this point on, many theory groups have used the metallic chains to fine-tune their theories. Having a reduced number of atoms and interactions, they form ideal test candidates, even for present-day models (Rocha et al., 2006).

A discussion of all models is beyond the scope of this chapter, but as we will see, the basic quantum mechanics of the plane-wave model is often enough to understand the physics of these metallic chains. These examples may reduce at least some of the frustrations of the students and allow the readers to enjoy the beauty of quantum mechanics.

35.1.2 Outline

Monatomic chains have not only been studied upon the breaking of metallic contacts, they have also been studied on surfaces. Two different techniques, namely, single atom manipulation with the use of the tip of a scanning tunneling microscope (Wallis et al., 2002) and the self-assembly of surface atoms (Segovia et al., 1999) have been used for this purpose. For a recent review on these studies, see Oncel (2008). This chapter only discusses freestanding monatomic chains suspended between two electrodes of the same material. In this case, the absence of interaction with a substrate simplifies the electron structure. This field of study has already led to many interesting physical discoveries and we will have to limit ourselves to only a few in this chapter. Following the chronology of the references, we begin in Section 35.2 with a description of the experimental setups in which they were produced. All experimental details necessary for the understanding of the remaining experiments of this chapter are provided in this section.

Once the basics are treated, Section 35.3 continues with a discussion of the atom-to-atom distance in the chain. Although this appears to be a minor detail at first, it has led to many imaginative proposals for alternative structures. Since it proved difficult to bridge the gap between experiment and theory, a significant number of papers has appeared on this problem. A small matter as the atom-to-atom distance provided the fuel for many other interesting studies on chains.

In Section 35.4, we discuss the reasons for the formation of chains and, especially, the reason for their forming only on certain metals. Although many models are able to reproduce the formation of chains by stretching metallic contacts, a fundamental comprehension was lacking for several years. As the key arguments are presented, we see that the relativistic corrections to the electron orbitals are vital in the formation of chains.

In Section 35.5, we address once again the conductance properties of chains. Within the framework of the plane-wave model, we see that it is possible to include interferences of the wave function. These interferences can be observed directly in the conductance of the chain, forming a powerful illustration of the wave character of electrons.

Many other interesting phenomena have been investigated for monatomic chains. Among others, it was found that the electron–phonon interaction inside chains reflects the one-dimensional character of the lattice (Agraït et al., 2002). There are studies of the influence of current and voltage on the stability of chains, where both electro migration forces (Todorov et al., 2001) and local heating (Frederiksen et al., 2007), due to the passing of electrons, lead to breakdown. In order to introduce the reader to monatomic chains, it was decided not to systematically discuss all the subjects, but to limit the discussion to a few key items, where we focus on the unique physics of these one-dimensional conductors.

35.1.3 One-Dimensional Conductance

Many of the conclusions about the properties of metallic chains have been drawn on the basis of their electrical conduction. Before starting our journey through this manifold of experimental and theoretical studies, we therefore need to have a basic understanding of conductance at the relevant scale. Atomic chains of gold have a diameter of the size of the Fermi wavelength. Macroscopic classical theories will therefore not suffice; we need to take the wave character of the electrons into account.

Let us therefore consider a two-dimensional wire, as shown in Figure 35.1. The wave function, f, of any electron with mass, m, inside the confining potential, (U), should obey Schrödinger's equation

$$-\frac{\hbar}{2m}\nabla^2\Psi(x,z)+U(x,z)\Psi(x,z)=E\Psi(x,z) \qquad (35.1)$$

where $\hbar$ is Planck's constant divided by $2n$. For smoothly shaped boundaries of the potential, the wave functions for the electrons can be separated into two components. For hardwall boundaries, the wave function should be zero at the boundary, resulting in

$$\Psi(x,z)=A\sin\frac{n\pi x}{W}\cdot\psi(z) \qquad (35.2)$$

where

    W represents the width of the structure
    n is the transverse quantum number

For the component parallel to the boundaries of the structure, the solution is a running wave e^{ikz}, such that the corresponding energy of every mode n is given by

$$E_n(k)=\frac{\hbar^2 k^2}{2m}+\frac{\hbar^2}{2m}\left(\frac{n\pi}{W}\right)^2 \qquad (35.3)$$

Let us take the constriction sufficiently narrow, such that only $n = 1$ for the transverse component of the wave function is of importance at the scale of the Fermi energy of the metal. When connecting the wire to macroscopic leads, as in Figure 35.2, matching of the wave functions gives rise to a partial reflection of the waves, which can be represented by a transmission probability amplitude t_1. The transmitted partial wave becomes

$$\psi_t(z)=\frac{1}{\sqrt{L}}t_1 e^{-ik_1 z} \qquad (35.4)$$

In general, t_1 depends on energy or wave vector, but for small energies we can ignore this dependence. The probability current for one electron wave to be transmitted is then given by

$$P(k_1)=\frac{\hbar}{2mi}\left(\psi^*(z,k_1)\frac{\partial\psi(z,k_1)}{\partial z}-\psi(z,k_1)\frac{\partial\psi^*(z,k_1)}{\partial z}\right)=\frac{\hbar k_1}{mL}t_1^* t_1 \qquad (35.5)$$

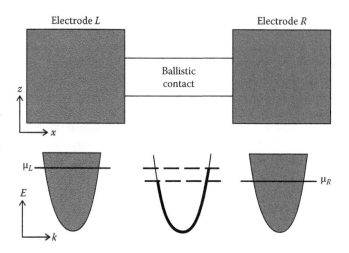

FIGURE 35.2 Graph showing the dispersion relation for the parallel wave vectors in the electrodes and the contact. The finite width of the contact only permits a finite number of bands. Both forward and backward moving states are occupied up to the chemical potential. At zero bias this leads to zero current.

FIGURE 35.1 Sketch of a two-dimensional wire of width W with parallel hard wall boundaries, which give rise to a set of transverse modes for the electron wave function.

The total electrical current through the constriction can now be given by a sum over all vectors k_1 multiplied by the charge carried per electron wave. Note that this value is $2e$ instead of e, assuming spin degeneracy of the wave function.

$$I_1 = 2e \sum_{k_1} P(k_1) = \frac{2e\hbar}{mL} T_1 \sum k_1 \qquad (35.6)$$

Here we define $T_1 \equiv t_1^* t_1$ as the transmittance of the contact for the transverse mode $n = 1$.

Figure 35.2 shows the dispersion relation of these k_1 values. As the contact is connected by macroscopic leads, the wave vectors form a continuum. The summation over k-vectors in Equation 35.6 can thus be replaced by an integral. This allows us to rewrite the current as

$$I_1 = \frac{2e}{h} T_1 \frac{L}{2\pi} \int k\, dk \qquad (35.7)$$

Substituting the dispersion relation $E = \hbar^2 k^2 / 2m$ leads to an integration over energies:

$$I_1 = \frac{2e}{h} T_1 \int_{\mu_L = E_F - eV/2}^{\mu_R = E_F + eV/2} dE = \frac{2e^2}{h} T_1 V \qquad (35.8)$$

For a realistic three-dimensional contact, the confining potential U will be more complex, resulting in a more complex transverse solution of the wave function. Let us, therefore, briefly look into the situation where more than one transverse wave function starts to play a role in the conduction. There will be a finite number of incoming waves from the left (right), amplitudes of which can be represented by a vector $\vec{i}_L (\vec{i}_R)$. Equivalently, $\vec{o}_R (\vec{o}_L)$ represents the outgoing amplitudes to the right (left). All possible scattering processes in the system are now described by the equation

$$\begin{pmatrix} \vec{o}_L \\ \vec{o}_R \end{pmatrix} = \begin{pmatrix} \widehat{s_{LL}} & \widehat{s_{LR}} \\ \widehat{s_{RL}} & \widehat{s_{RR}} \end{pmatrix} \cdot \begin{pmatrix} \vec{i}_L \\ \vec{i}_R \end{pmatrix} \qquad (35.9)$$

where the $\hat{s}_{ij}$ matrices describe the amplitude for scattering from the incoming waves from lead j to the outgoing waves in lead i.

While the general form of the matrix may be quite complicated, it can be simplified. As all wave functions are linearly independent and the scattering is linear, we can diagonalize the matrix. In other words, we can find a transformation to a new basis. In the new basis, the eigenstates are a property of the junction, such that

$$\hat{S}' = \begin{pmatrix} r_1 & \cdots & 0 & t_1 & \cdots & 0 \\ \vdots & \ddots & \vdots & \vdots & \ddots & \vdots \\ 0 & \cdots & r_N & 0 & \cdots & t_N \\ t_1 & \cdots & 0 & r_1 & \cdots & 0 \\ \vdots & \ddots & \vdots & \vdots & \ddots & \vdots \\ 0 & \cdots & t_N & 0 & \cdots & r_N \end{pmatrix} \equiv \begin{pmatrix} \hat{r} & \hat{t}' \\ \hat{t} & \hat{r}' \end{pmatrix} \qquad (35.10)$$

As electron waves can only be transmitted or reflected into the same eigenmodes, Equation 35.8 can be reduced to

$$G = \frac{1}{V} = \frac{2e^2}{h} \sum_{n=1}^{N} T_n \qquad (35.11)$$

where we define $T_n \equiv (\hat{t}^\dagger \hat{t})_{nn'}$, similarly as in Equation 35.8. Equation 35.11 is known as the Landauer expression.

The number of relevant channels ($T_n \neq 0$) is determined by the atoms forming the narrowest part of the junction. In the simplest case of a monovalent single atom chain there is only one channel with T_1 close to 1, as will be discussed in the following text.

35.2 The Initial Experiments

The one-dimensional structure of the monatomic chain was shown to be real by the experiments performed by Ohnishi et al. (1998) and Yanson et al. (1998). In the following paragraphs, we discuss these two experiments separately.

35.2.1 TEM Imaging

Of course, the structure is too small for most imaging techniques to be detected, but Ohnishi et al. (1998) showed that a transmission electron microscope (TEM) does meet the demands. They started with a very thin gold film, mounted in an ultra-high vacuum chamber ($p \approx 10^{-8}$ Pa) of a TEM. The film was sufficiently thin such that a high-intensity electron beam ($\sim 100\,\text{A/cm}^2$) was able to create holes in it. This gave them the opportunity to shape a freely hanging gold bridge by forming two holes in close proximity to each other. Further illumination of this bridge resulted in additional thinning such that a desired width of four atomic strands (≈ 1 nm) was easily obtained.

When the beam intensity was reduced, it appeared possible to image the final structure without influencing it any further. As such, the evolution of the structure over time could be studied. The structure continues thinning due to the diffusion of atoms at room temperature. In the final stages of this breaking process, the remaining structure frequently consists of a single strand of atoms, as shown in Figure 35.4d. This provides very appealing visual information on the existence of monatomic chains.

Within the same TEM, the authors also performed an experiment using an STM-like configuration, comparable to the one shown in Figure 35.3. Instead of a single gold film, they now started with a mechanically sharpened gold tip positioned close to an evaporated gold counter electrode. The gold tip was mounted on a shear-type piezoelectric transducer, allowing for movement in all three directions. The tip could therefore be dipped into and pulled from the counter electrode under computer control.

Although this configuration does not provide as much mechanical stability, hampering atomic resolution on the chains, it provides an opportunity to measure conductance simultaneously with the imaging. In Figure 35.4, we see how the

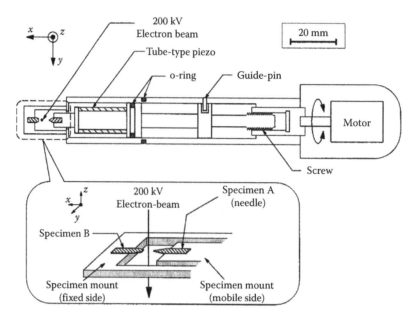

FIGURE 35.3 Schematic overview of a sample holder inside a TEM, including the piezoelectric element necessary for the contact breaking. (Reprinted from Kizuka, T. et al., *Phys. Rev. B*, 55(12), R7398, 1997. With permission.)

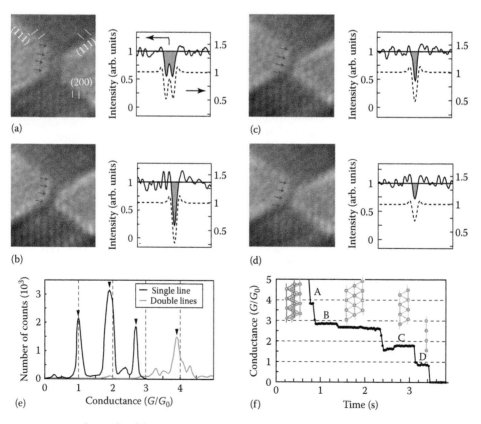

FIGURE 35.4 Figure summarizing the results of the TEM experiments with simultaneous conductance measurements. (a)–(d) Images taken by the TEM during the thinning, with an accompanying cross section of the image intensity. At the left bottom (e) the evolution of the conductance observed simultaneously with these images. At the right bottom (f) a histogram resulting from 50 of those evolutions. (Reprinted from Kurui, T. et al., *Phys. Rev. B*, 77(16), 161403R, 2008. With permission.)

conductance changes in steps of the order of $2e^2/h$ as the contact is broken. The size of these steps is in agreement with Equation 35.11 provided that the different modes close one by one. The images taken for the structures and their density profiles show how the number of strands is proportional to the conductance. Although the individual atoms cannot be identified in these images, this is convincing evidence that the conductance of a chain of gold atoms is close to $2e^2/h$.

35.2.2 Low Temperature Break Junctions

The second article in that same issue of *Nature* showed independent evidence of linear gold chains (Yanson et al., 1998). It showed that atomic chains are not limited to the parameters in the TEM experiment, but that they also appear under cryogenic conditions. Since surface diffusion at these temperatures is absent, the contacts need to be thinned down by stretching. The fact that chains also form under such circumstances is even more surprising. At room temperature, the contact starts as a bridge of several atomic strands in parallel, which break one by one. In this case, the final structure is already present at the start of the experiment. At low temperatures, however, the breaking of the contact is an irregular process during which surface diffusion plays no role.

At low temperatures, most of the experiments were performed either by the previously discussed STM-like technique or by mechanically controllable break junctions (MCBJ) presented in Figure 35.5. This second technique has the advantage of producing clean contacts by very simple means. One starts with a macroscopic wire (about 100 μm in diameter and 15 mm long) and cuts a notch halfway its length. Locally, the diameter is thus reduced to less than 50 μm, providing a weak spot. This wire is glued on a phosphor bronze bending beam ($20 \times 4 \times 1$ mm), having a Kapton foil as insulating layer. This gluing is done with two droplets on either side of the notch, in order to keep the two wire ends fixed with respect to each other.

After the insertion of the substrate in a vacuum chamber, it is cooled down to 4.2 K, providing a cryogenic vacuum. A piezoelectric element, situated at the middle of the bending beam, opposite to the metallic wire, can now be used to bend the substrate, resulting in the stretching of the wire. With an appropriate coarse movement (often provided by a mechanical drive) the incision can be stretched and broken down to a microscopic

size. Note that the surfaces are now freshly exposed. In combination with the cryogenic vacuum, this provides an extremely clean environment for the contact. Afterward, the piezoelectric element can be used for the fine control, resulting in an atom-by-atom thinning of the contact. During such a stretching of a gold wire, one obtains evolutions of the conductance similar to the one presented in Figure 35.6.

The information in this graph appears insufficient to prove the existence of monatomic chains, but one already notices the relatively long plateau situated around a conductance of $2e^2/h$. In Equation 35.11, we saw that this information agrees with the maximum transmission for a single channel, and for a monovalent atom such as gold, this represents a contact of one atom in diameter. To prove that this plateau is indeed caused by the formation of an atomic chain, the authors performed several tests. In one of these tests, they measured the distribution of the lengths of this last plateau, defined as the number of data points between 1.2 and $0.5G_0$, over many conductance-breaking traces. In Figure 35.7, this distribution shows a clear periodic structure. The chain thus periodically experiences an increased chance to break. As the chain consists of an integer number of atoms, the periodicity in this distribution directly reflects the atom-to-atom distance in the chain, confirming its existence. By now, many other experiments at low temperatures have been performed and the observed results are in accordance with the properties of a

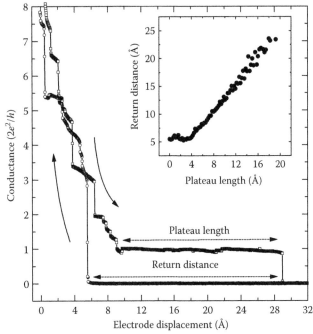

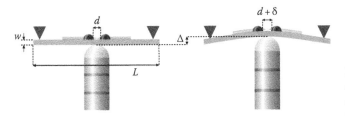

FIGURE 35.5 Schematic side view of the MCBJ. The left-hand image depicts the situation were the substrate is relaxed. At the right-hand side the piezo is pushed against the substrate resulting in bending of the substrate and stretching of the contact.

FIGURE 35.6 The conductance of a gold contact recorded while it is stretched and pushed back into contact. The conductance decreases stepwise, one step for every reconfiguration. Finally, the conductance reaches a value close to $2e^2/h$ before breaking. The length of this plateau reflects the one-dimensional character of the chain. After breaking a similar stepwise increase is found upon moving the electrodes back into contact. (Reprinted from Yanson, A.I. et al., *Nature*, 395(6704), 783, 1998. With permission.)

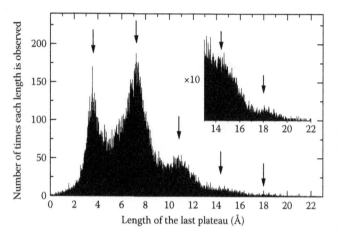

FIGURE 35.7 Distribution of lengths for the last conductance plateau before breaking. This result is obtained by collecting 10,000 breaking curves for Au contacts at 4.2 K. It shows a number of maxima at multiples of a fixed distance. Note that distance between peaks (and thus between atoms in the chain) was later corrected to 0.26 ± 0.02 nm. (Reprinted from Yanson, A.I. et al., *Nature*, 395(6704), 783, 1998. With permission.)

monatomic chain. It goes beyond the scope of this chapter to include all the evidence, but interested readers are encouraged to read reference (Yanson et al., 1998) and others mentioned in this chapter.

35.3 Distances beyond Expectation

35.3.1 The First Experimental Results

It came as a big surprise that breaking gold contacts spontaneously formed freely suspended, linear chains of a single atom wide. In hindsight, simulations proved to be more correct than they were expected to be. Interestingly, there were also many differences between theory and experiment. Even when looking at the atom-to-atom distance in the chain, there was already a disagreement.

The images made with TEM, such as Figure 35.4, leave little doubt about the Au-to-Au distance in the chain. The value can be directly extracted from the images and the first values represented in literature reported distances of 0.35–0.40 nm. The actual value can be slightly bigger when the chain is not perfectly perpendicularly oriented to the electron beam, but it can never be smaller. It is interesting to note that this interatomic distance varies between pairs of atoms. Even within one image, the atom-to-atom distance varies along the chain.

For the experiments performed at low temperatures, the atom-to-atom distance is not directly visible. Nevertheless, one can still extract its average value. The distribution of the plateau lengths depicted in Figure 35.7 shows a periodicity, which must coincide with the average atom-to-atom distance in the chain. Note that this distribution forms a statistical description over a large population of chains. Individual distances can deviate from the most frequently occurring value. The first value reported for

this distance was 0.36 ± 0.11 nm, which was later corrected to 0.26 ± 0.02 nm (Untiedt et al., 2002).

35.3.2 Stretching the Bond

Because of the small number of atoms in the chain and the reduced number of interactions, the chain forms a perfect litmus test for developing numerical models describing the physics of atomic contacts and their conductance. Almost all calculations predicted the atom-to-atom distance in the chains to be smaller than the nearest neighbor distance of 0.288 nm in the bulk (Sørensen et al., 1998). This value agrees with the corrected atom-to-atom distance at low temperatures, but is in strong disagreement with the TEM values at room temperature.

To bridge the gap to the 0.4 nm reported in experiment, several creative propositions were given. One was an imaginative idea by Sánchez-Portal et al. (1999). Their calculations indicated that the chain only becomes linear at higher tensile stresses. At lower tensile stress it has a zigzag structure, as this allows the electrons to delocalize more, reducing their energy. At a certain finite temperature (~40 K) thermal excitation induces a rotation of the chain around its axis, fixed by the anchoring points to the electrodes. For a chain with an odd number of atoms, all the odd-numbered atoms lie on the axis of rotation, such that only the even-numbered atoms will move. This movement results in a difference in contrast, which depends on the zigzag angle, but can be as large as a factor of 5, see Figure 35.8. The contrast for

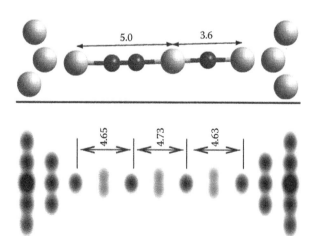

FIGURE 35.8 Two different proposed configurations to explain the interatomic distances observed in the room temperature TEM experiment. The top configuration of Legoas et al. has interstitial carbon atoms and carbon dimers inserted within the chemically reactive gold chains. Since the contrast of the electron beam is low on these light elements, atom-to-atom distances measured up to 0.5 nm represent the Au–Au distances, which can be explained. (Reprinted from Legoas, S.B. et al., *Phys. Rev. Lett.*, 88(7), 076105, 2002. With permission.) The bottom panel gives a configuration by Sánchez-Portal et al., which proposes the gold atoms to spin in a zigzag configuration such that the contrast for half of the atoms is smeared. Atom-to-atom distances up to 0.47 nm can then be explained. (Reprinted from Sánchez-Portal, D. et al., *Phys. Rev. Lett.*, 83(19), 3884, 1999. With permission.)

half of the atoms could thus be too low to be imaged in the TEM, resulting in almost double atom-to-atom distances.

Other theoretical studies showed that atoms of light elements such as carbon, oxygen, and sulfur could put themselves in interstitial positions between the atoms of a gold chain (Häkkinen et al., 1999; Legoas et al., 2002; Daniel and Astruc, 2004). As the contrast of an electron beam of a TEM goes as $Z^{2/3}$, with Z the atomic number, its sensitivity for these interstitial light-weight elements could be too low to detect them. The positioning of interstitial carbon atoms and dimers can explain distances up to the incidentally reported 0.5 nm, as shown in Figure 35.8. Although the TEM experiments were carefully performed in an ultra-high vacuum, these conditions could still not prove clean enough. While bulk gold is relatively inert, nano-sized gold particles have such strong interactions that they are widely explored as a catalyst (Daniel and Astruc, 2004). As also monatomic chains are shown to react easily with light elements (Häkkinen et al., 1999), a single atom diffusing over the electrodes toward the contact can be enough to contaminate the chain.

35.3.3 More Recent Experiments

A second step toward the verification of these models appeared quite a challenge. For the spinning zigzag geometry, one would have to freeze the structure below the activation energy of the spinning motion (~40 K). Obtaining such low temperatures inside the experimental chamber of the TEM is certainly not straightforward. But without performing the experiment, calculations on the contrast were already sufficient to indicate that the actual signal to noise ratio of commercial TEMs should be good enough to observe also the spinning atoms (Koizumi et al., 2001). As of yet, no experimental evidence for the zigzag configuration in Au has been presented. The hypothesis of this structure, however, shows how rich the properties of metals can be, even at these smallest scales. Although the structure might prove unrealistic in the end, the investigations on its plausibility have been very fruitful.

The interstitial atoms of light elements inside metallic chains have been the subject of intense debate as well. For the experimentalists the complication was that even the best cleanliness of an ultra-high vacuum could prove insufficient. The best opportunity was therefore to start from the cryogenic vacuum and deliberately contaminate the space with the desired molecular species. For both hydrogen (Csonka et al., 2006) and oxygen (Thijssen et al., 2006) it has been shown convincingly that atoms and even molecules can act as interstitial structures. At room temperature, however, the hydrogen is probably bound to loosely to stay bound during the atomic reconfigurations (Anglada et al., 2007). More recent experiments in TEM (Takai et al., 2001) show several examples of chain structures having interatomic distances of 0.29 nm at room temperature in agreement with calculations. It is thus very well possible that the first reports discussed contaminated structures.

Interestingly, Thijssen et al. were also able to induce the formation of chains in silver, by the admission of oxygen molecules.

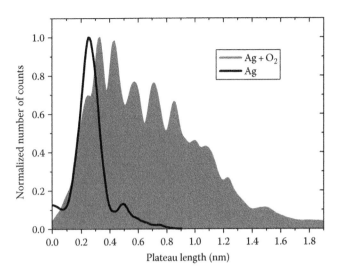

FIGURE 35.9 Length distribution for the last plateau in silver after the admission of oxygen gas into the vacuum. The gray line serves as a reference and indicates the length of the last plateau before oxygen admittance. (Reprinted from Thijssen, W.H.A. et al., *Phys. Rev. Lett.*, 96(2), 026806, 2006. With permission.)

First, they started from a clean Ag contact and obtained the length distribution for the last plateau depicted in Figure 35.11. In contrast to gold, these short plateaus indicate the absence of chains in silver. After adding 10 μmol of O_2 in the cryogenic vacuum, however, they repeated the experiment and found a spectacular change in behavior (Thijssen et al., 2006). The silver after exposure to oxygen, reproduced in Figure 35.9, shows plateaus up to 1.5 nm long, with many peaks in the distribution. As the diameters of O and Ag are different, it is understandable that these peaks are not equidistant anymore.

35.4 Why Do Chains Form? And for Which Metals?

35.4.1 The Problem

The fact that the numerical models predicted atomic chains (Finbow et al., 1997) does not imply that it was immediately obvious why some metals have the tendency to form these one-dimensional structures. The ingredients deposited into the models were approximate effective interaction potentials between atoms, which were optimized to reproduce bulk behavior, but not expected to be reliable at this scale. More elaborate density functional theory (DFT) calculations confirmed the stability of atomic chains for Au (Häkkinen et al., 1999; Portal et al., 1999; Sánchez-Portal et al., 1999; Bahn and Jacobsen, 2001; Palacios et al., 2002; da Silva et al., 2004). To be able to determine whether these chains also form during the breaking of a point contact, one needs to perform a dynamic simulation. This performance increases the computational demands greatly and only few studies have been performed within the DFT framework to reproduce the spontaneous chain formation (Krüger et al., 2002; Anglada et al., 2007).

With their one-dimensional structures, atomic chains account for a large surface area, making the whole structure energetically expensive. Each atom in a chain has only two nearest neighbors, while every atom on the surface has many more. In order to understand the stability of the chains we should actually look into the dynamics of stretching (Gall et al., 2004). Especially at low temperatures, the system does not have the internal energy to explore all possible configurations. It will simply be frozen to the first energetic minimum it encounters, irrespective of whether this minimum is more general or local. Let us start from a contact that has been pulled until it reaches a single atom at its narrowest constriction. For this atom to be able to pull additional atoms into a chain, we require that the bonds of the atoms with lower coordination are stronger than those with higher coordination. Although the structure would lower its energy more effectively by breaking of the chain, the barrier for this is too high to overcome.

Rubio-Bollinger et al. have succeeded in measuring the forces in gold contacts directly (Rubio-Bollinger et al., 2001). They replaced one of the two electrodes of the contact by a small gold cantilever, of which they determined the deflection with sub-nm resolution. Hooke's law states that this deflection is directly proportional with the force exerted on the contact. This initiative opens the possibility to follow the force in parallel with the conductance of the contact as shown in Figure 35.10. The authors estimate that the force required to break an individual bond in the bulk amounts to 0.8–0.9 nN. For the breaking force of the final contact, however, they found an average value of 1.5 nN. On average, the barrier against breaking the chain is thus indeed much higher.

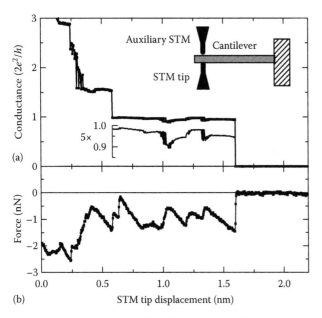

FIGURE 35.10 A combined plot of the evolution of the conductance of a gold contact and the force exerted on it during stretching. The discontinuities in force coincide with the breaking of metallic bonds. The final minimum in the force curve is on average 1.5 nN. (Reprinted from Rubio-Bollinger, G. et al., *Phys. Rev. Lett.*, 87(2), 026101, 2001. With permission.)

35.4.2 The Physics behind the Bond Strength

In order to understand the spontaneous formation of atomic chains in gold we need to understand why bonds between lowly coordinated atoms are stronger than bonds between highly coordinated atoms. To some extent, all metals share this property. For the transition metals with an almost full d-shell, however, it is especially large due to the interplay of d-electrons and sp-electrons at the Fermi energy. The d-electrons have a localized character and their interaction is thus more bond-like. As the d-band is gradually filled from 0 to 10 electrons per atom, this leads to a subsequent occupation of the bonding, nonbonding and antibonding states. The force of these bonds therefore has a parabolic dependence on the filling. The sp-electrons, on the contrary, have a spatially delocalized character and can be modeled as a Fermi gas. As the Fermi energy of this gas is dependent on the density of these sp-electrons, the electrons provide a large (Fermi-) pressure. For a free electron gas, this pressure can be deduced from the total energy of the electrons (Kittel, 2004):

$$U_0 = \frac{3}{5}N\frac{\hbar^2}{2m}\left(\frac{3\pi^2 N}{V}\right)^{2/3} \tag{35.12}$$

Here

N represents the number of electrons
V is the volume

The pressure of this gas can thus be obtained following

$$p = -\frac{\partial U}{\partial V} = \frac{2}{3}\frac{U_0}{V} = \frac{2}{5}n\frac{\hbar^2}{2m}(3\pi^2 n)^{2/3} \tag{35.13}$$

where n is the density of electrons. This amounts to a pressure of 2.1×10^{10} Pa for an s-electron density equal to that of the $6s$-electrons in gold.

At the surface of the metal, the equilibrium of these opposite forces is influenced by the boundary to vacuum. The s-electrons can spill out into the vacuum, releasing the pressure resulting in an increased bonding due to the d-electrons. In atomic chains, where a lot of vacuum is available for the spill-out of the s-electrons the bonding due to the d-electrons increases considerably.

However, as was mentioned already in Section 35.4.1, the spontaneous formation of chains is almost absent in pure silver (Rodrigues et al., 2002; Thijssen et al., 2006). Chains are seldom seen for this material and if they appear, they have short lifetimes. Where most chemical properties are shared between materials in the same column of the periodic table, this effect proved to be an exception.

When asked for the difference between gold and silver, most will answer as the color. Interestingly, that is exactly where we need to look for the answer. Gold has a higher atomic number and therefore has a larger relativistic correction to the mass of its electrons (Pyykkö, 1988). Due to this enhancement of the mass, m, the Bohr radius for the inner s-electrons, is given by

$$a_0 = \frac{4\pi\varepsilon_0\hbar^2}{me^2} \qquad (35.14)$$

with ε_0 the permittivity of vacuum, will be smaller for gold than it is for silver. This contraction will also appear for the outer s-electrons. Their orbitals must be orthogonal to the lower ones, and they feel the same relativistic correction directly as well. The consequence of this smaller radius is that the s-electrons come closer to the nucleus, lowering their energy.

The behavior of the d-electrons sharply contrasts to this. They are already orthogonal to the inner s-states by their angular momentum and their orbitals never come close to the nucleus. The s-band therefore shifts to lower energy with respect to the d-band bringing the d-band closer to the Fermi energy. This electronic effect causes the difference in colors for gold and silver. And if the d-band is more than half filled, as it is the case for gold, the charge transfer from d-states to s-states leads to a depopulation of antibonding states. The attractive force of the d-bonds in gold is thus further increased by the same relativistic

effects. In bulk gold, the stronger d-bond is compensated by a larger Fermi-pressure. At the surface, this s-pressure can be relieved and the stronger d-bond gives rise to enhanced binding, most prominently in atomic chains.

35.4.3 Relativistic Effects of Atomic Chains

In order to verify that chains are the result of relativistic corrections to the electronic density, one would like to switch off relativity and set the speed of light to infinity. But there are other tests, which are experimentally accessible, because when we look at the argumentation for chain formation, the arguments should equally apply for other heavy transition metals. As long as their d-band is more than half filled, depopulation will lead to a larger bonding, strengthening the bond.

Smit et al. produced length distributions for the $4d$-transition metals Rh, Pd, and Ag, as well as the $5d$-transition metals Ir, Pt, and Au (Smit et al., 2001). The results, obtained with a similar technique as explained in Section 35.2.2, are shown in Figure 35.11.

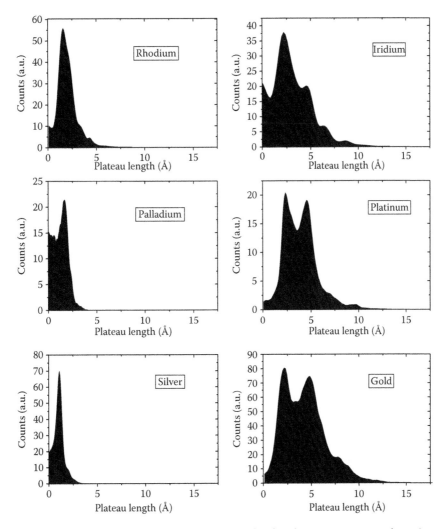

FIGURE 35.11 Length distributions measured for the $4d$ transition elements Rh, Pd, and Ag in comparison to their $5d$ counterparts (Ir, Pt, and Au). Where the $4d$-transition metals have the length of the final plateau limited to a single peak, attributed to a dimer forming the bridge, the $5d$ metals have multiple, equidistant peaks, indicating the formation of monatomic chains.

The clear difference between the length distributions for the $5d$-elements and the $4d$-elements shows that only the heavier transition metals show spontaneous chain formation. This chemical effect is not shared between elements of the same column in the periodic table, but by elements in the same row.

In parallel with this experimental paper, Bahn and Jacobsen (2001) published a DFT-based calculation, where they compare the bonding between bulk atoms and chain atoms. They agree that the stronger relativistic corrections in the $5d$-transition metals cause relatively stronger chain bonds, resulting in an increased possibility for chain formation.

35.5 The Conductance Revisited

35.5.1 Plane Wave Model for an Atomic Chain

In Section 35.1.3, we used elementary quantum mechanics to show that the conductance of a metallic contact at the scale of the Fermi wavelength is independent of its length. In this section, we show that with almost equally elementary quantum mechanics one can also find the first correction to this answer. Let us consider the model depicted in Figure 35.12. Here the different dimensionality of the chain and the electrodes is reflected by a different bandwidth for the conduction electrons. This inference leads to different values for the k-vectors. Although the Fermi energy is situated above the energy barrier, there still occurs scattering of the electronic plane waves at both ends of the chain, situated a distance L apart.

The wave functions in the three different regions of this potential are given by

$$
\begin{cases}
\Psi_{\mathrm{I}}(x,t) & = e^{i(k_1 x - \omega t)} \quad + A\, e^{-i(k_1 x + \omega t)} \\
\Psi_{\mathrm{II}}(x,t) & = B\, e^{i(k_2 x - \omega t)} \quad + C\, e^{-i(k_2 x + \omega t)} \\
\Psi_{\mathrm{III}}(x,t) & = D\, e^{i(k_3 x - \omega t)}
\end{cases}
\tag{35.15}
$$

In order to obtain a wave function describing the total system one needs to match the wave functions for the different regions, which results in restrictions for A, B, C, and D. The result for D is given by

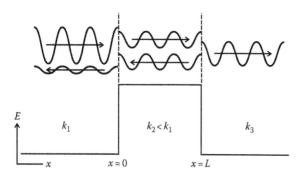

FIGURE 35.12 Elementary one-dimensional plane-wave model for the potential representing the contact of a chain of length L, to leads at both ends.

$$
D = \frac{4k_1 k_2}{(k_1 + k_2)(k_2 + k_3)e^{iL(k_3 - k_2)} + (k_1 - k_2)(k_2 - k_3)e^{iL(k_3 + k_2)}}
\tag{35.16}
$$

One then obtains for the transmission of the system

$$
T(k_1, k_2, k_3) = \frac{16 k_1^2 k_2^2}{F + G\cos 2k_2 L}
\tag{35.17}
$$

Where we used the substitution

$$
F = (k_1 + k_2)^2 (k_2 + k_3)^2 + (k_1 - k_2)^2 (k_2 - k_3)^2
\tag{35.18}
$$

$$
G = (k_1 + k_2)(k_2 + k_3)(k_1 - k_2)(k_2 - k_3)
\tag{35.19}
$$

Although the full result of this transmission looks rather cumbersome, the most important physics is dominated by the cosine term.

This model is the electronic analogue of the optical Fabry–Perot interferometer. In that case, a coherent beam of light is partially reflected by two parallel mirrors. The light is scattered backward and forward between these mirrors, leading to an interference pattern. This pattern can be destructive or constructive depending on the matching of the wavelength of the light with the length of the cavity. In case of the metallic chains, the interference in partial electron wave functions will result in a variation of the conductance as a function of length (Sim et al., 2001).

The extreme values of the transmission in Equation 35.17 are given by

$$
T_{\max} = \frac{4(k_2/k_1)(k_2/k_3)}{[(k_2/k_1) + (k_2/k_3)]^2}
\tag{35.20}
$$

in case of constructive interference ($k_2 = n \cdot \pi/L$) and

$$
T_{\min} = \frac{4(k_2/k_1)(k_2/k_3)}{[1 + (k_2/k_1)(k_2/k_3)]^2}
\tag{35.21}
$$

in case of destructive interference. If $k_1 = k_3$ in Equation 35.17, which is easily defensible when both electrodes are from the same wire, the maximum transmission in Equation 35.20 becomes $T_{\max} = 1$.

The cosine in Equation 35.17 depends only on the length of the chain and its k-vector at the Fermi energy. To calculate this Fermi k-vector we need to divide the volume inside the Fermi-sphere (or Fermi line in one dimension) by the volume per k-point, $\Delta k = 2\pi/L$, which will give us the number of occupied states,

$$
N = 2\frac{2k_{\mathrm{F}}}{(2\pi/L)} \quad \text{or} \quad k_{\mathrm{F}, L=\infty} = \frac{\pi N}{2L} = \frac{\pi}{2d}
\tag{35.22}
$$

Here, d represents the atom-to-atom distance in the chain. To the subscript of the Fermi k-vector we added $L = \infty$ to indicate that this expression is deduced for the limit of an infinite chain.

The result of Equation 35.22 has an elegant effect on the total transmission of the chain. For chains with an even number of atoms, the transmission in Equation 35.17 will be maximal, while for an odd number of atoms, it will be minimal. The conduction of a one-dimensional chain therefore shows a so-called parity effect. The effect is independent of the interatomic distance so that stretching will not be able to change it. As L increases due to the elasticity of the chain, the atom-to-atom distance d changes likewise. The absolute length is not of importance, only the length expressed in the number of atoms.

The interference pattern in the conductance was shown to remain valid also in more realistic numerical simulations (Sim et al., 2001; de la Vega et al., 2004; García-Suárez et al., 2005). The precise length, at which the maximum conductance occurs, however, depends on the precise geometry of the electrodes. This can be modeled effectively by including an additional phase shift at the scatter points at both end of the chain (Major et al., 2006).

35.5.2 Observation of Parity Effects in Real Chains

The evolution of conductance during the formation of a gold chain as the one presented in Figure 35.6 does not show a parity effect as predicted. The fact that individual contacts hardly ever present this behavior is explained by the irregular behavior of the contacts. In order to be able to extract the result of the parity effect on the conductance of chains one needs to average over atomic-scale configurations by obtaining enough statistics.

In order to have a statistical description on the length dependence of the conductance of a monatomic chain, Smit et al. (2003) repetitively broke a gold contact for over 10^5 times. The conductance of the data points, G_i of the last plateau of each curve were labeled by an index i, starting at 1 and increasing with successive digitized points for further stretching. Because the probability for the chain to survive decreases with stretching, the number of data points at each index, n_i, decreases monotonically with i. The average conductance $\langle G_i \rangle = (\Sigma_i G_i)/n_i$ then takes the shape of the top panel in Figure 35.13. After an initial decrease, due to tunneling through additional states at short distances, this curve indeed shows that the average conductance is oscillating as a function of length. Fitting to a sine function gives a periodicity of 0.47 ± 0.03 nm, which agrees within the error bar with twice the interatomic distance (0.50 ± 0.05 nm). Note that the background of this oscillation at larger lengths is indeed flat, agreeing with Equation 35.11.

It is important to note that the curve obtained for the average conductance in Figure 35.13 cannot be directly compared to Equation 35.17. In contrary to the model, the maxima of Figure 35.13 are not situated at 1 G_0. A realistic contact, namely, will be different in several aspects from our model. First of all, the contact will not be perfectly symmetric, leading to a decrease in the maximum conductance. Furthermore, there will be additional

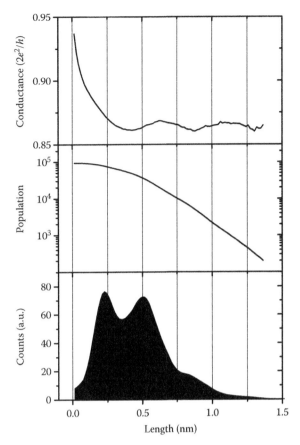

FIGURE 35.13 Three panels constructed from a series of conductance evolutions of a gold contact. The lower panel shows a length distribution similar as the one in Figure 35.7. The middle panel plots the integral of this length histogram, indicating how frequently a conductance value for a certain length was measured. The upper panel shows the average value of the conductance as a function of length, giving evidence for even–odd oscillations. Note that data points for short lengths have thus been calculated over a population of more than 10^5 data points, while this number drops toward the right. (From Smit, R.H.M. et al., *Phys. Rev. Lett.*, 91(7), 076805, 2003.)

scatterers close to the contact, which lead to a general decrease of the average conductance. And finally, a plateau of a given length will not always be representing a chain consisting of the same number of atoms, resulting from the irregular process of the chain formation. Although these effects are strong enough to suppress the parity oscillations in the conductance evolution of a single contact, the statistics over several thousands of contacts verify the existence of these oscillations.

35.5.3 Conductance Oscillations for Other Materials

Besides gold, parity oscillations were also seen in the conductance of Pt chains (Smit et al., 2003). The experimental result for this material shown in Figure 35.14 does not fit as well to a sine wave but shows a well-defined oscillation. Its periodicity of 0.50 ± 0.04 nm falls within the error bar of twice the atom-to-atom distance in the chain (0.46 ± 0.04 nm).

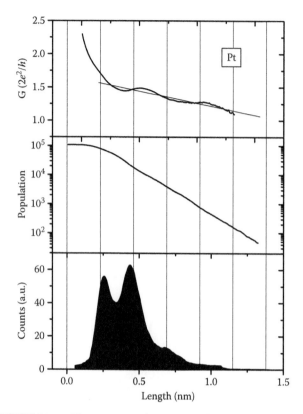

FIGURE 35.14 The average conductance, number of occurrences and length distribution as a function of length, presented as in Figure 35.13, but now for platinum contacts. (From Smit, R.H.M. From quantum point contacts to monatomic chains: Fabrication and characterization of the ultimate nanowire. PhD Thesis, Leiden University, Leiden, the Netherlands, 2003.)

This is surprising as the theory of Section 35.5.1 is only valid for monovalent metals. The pinning of the k-vector inside the chain leading to Equation 35.22 is not valid for a multivalent metal. For a multivalent metal, the electron energy is not simply given by a 1D parabolic dispersion relation. One needs to consider the k-vectors for all the electronic modes responsible for the conductance (Palacios et al., 2002) to be able to predict the interference pattern.

The conductance as a function of length for platinum chains is therefore much more involved. This can also be judged from the slope in the average conductance in Figure 35.14. A strongly decreasing conductance as a function of length is not expected based on Landauer's expression, Equation 35.11. Two interesting interpretations for this slope have been proposed. De la Vega et al. (2004) state that the slope is part of a very long wavelength period. In their calculations, one of the conduction bands passes the Fermi energy at a small k-value leading to such long wavelengths. García-Suárez et al. (2005), however, are of the opinion that this decrease is a fingerprint of the zigzag configuration of platinum chains. For longer chains, their calculations indicate a stronger zigzag configuration accompanied by a reduction in the conduction. Both these propositions remain to be verified experimentally, but show that the monatomic chains still provide us with interesting problems.

35.6 Conclusions and Outlook

The 5d-transition metals Ir, Pt, and Au show the spontaneous formation of chains of single atoms upon the thinning and stretching of contacts. This chapter presented the TEM and break junction experiments, which lead to their discovery. Considering the experimental data carefully and taking interstitial atoms into account, the atom-to-atom distance for pure metallic chains was proven to be slightly smaller than the bulk distance.

A combination of theoretical and experimental investigations leads to the conclusion that these chains are formed due to a transfer of charge from the *sp* to the *d*-bands. This transfer is larger for the heavier 5d metals than for their 4d counterparts, due to the stronger relativistic corrections. The result is a stronger interatomic bond at lower coordination, increasing the possibility of chain formation in comparison to an immediate rupture of the contact. The change in bonding strength for atoms with a lower coordination also explains why the bond distance is shorter inside the chain than it is for the bulk.

While the discovery of the chains stimulated the development of advanced computational methods, many of the observed phenomena in chains of Au can be explained by simple and elegant plane-wave models. It has been shown that the quantization of conductance in units of $2e^2/h$ follows from the transverse confinement of the wave function of the conduction electrons. If one also takes into account the partial reflection of wave functions at both ends of the chain, the resulting interferences lead to a small oscillation on top of this value. For monovalent metals, this leads to a parity effect, illustrating the wave character of the conduction electrons.

For the more complicated electronic structure of Pt and Ir, a plane-wave model is insufficient. Describing the same phenomena appropriately in these systems is a challenge even for the most advanced numerical models. There are, for instance, indications that chains of Pt should spontaneously show magnetic anisotropy (Delin et al., 2004; Smogunov et al., 2008). Experimental studies have not yet been able to verify the presence of magnetism in chains, as it is not directly reflected in the conductance of the contact (Untiedt et al., 2004).

A further challenge for the experimentalists is to provide longer chains. For longer chains, the electron–phonon and electron-electron interactions should play a more important role. This might lead to more exotic effects such as a Peierls transition or even the formation of a Luttinger liquid phenomena outside of the independent electron models presented here. Even though monatomic chains appear to be an elementary structure, they provide both experiment and theory with advanced and interesting problems.

References

Agraït, N., Untiedt, C., Rubio-Bollinger, G., and Vieira, S., 2002. Onset of energy dissipation in ballistic atomic wires. *Physical Review Letters*, 88(21), 216803.

Agraït, N., Yeyati, A.L., and van Ruitenbeek, J.M., 2003. Quantum properties of atomic-sized conductors. *Physics Reports*, 377(2–3), 81–279.

Anglada, E., Torres, J.A., Yndurain, F., and Soler, J.M., 2007. Formation of gold nanowires with impurities: A first-principles molecular dynamics simulation. *Physical Review Letters*, 98(9), 096102.

Bahn, S.R. and Jacobsen, K.W., 2001. Chain formation of metal atoms. *Physical Review Letters*, 87(26), 266101.

Csonka, S., Halbritter, A., and Mihály, G., 2006. Pulling gold nanowires with a hydrogen clamp: Strong interactions of hydrogen molecules with gold nanojunctions. *Physical Review B*, 73(7), 075405.

da Silva, E.Z., Novaes, F.D., da Silva, A.J.R., and Fazzio, A., 2004. Theoretical study of the formation, evolution, and breaking of gold nanowires. *Physical Review B*, 69(11), 115411.

Daniel, M.C. and Astruc, D., 2004. Gold nanoparticles: Assembly, supramolecular chemistry, quantum-size-related properties, and applications toward biology, catalysis, and nanotechnology. *Chemical Reviews*, 104(1), 293–346.

de la Vega, L., Martín-Rodero, A., Yeyati, A.L., and Saúl, A., 2004. Different wavelength oscillations in the conductance of 5d metal atomic chains. *Physical Review B*, 70(11), 113107.

Delin, A., Tosatti, E., and Weht, R., 2004. Magnetism in atomic-size palladium contacts and nanowires. *Physical Review Letters*, 92(5), 057201.

Finbow, G.M., Lynden-Bell, R.M., and McDonald, I.R., 1997. Atomistic simulation of the stretching of nanoscale metal wires. *Molecular physics*, 92(4), 705–714.

Frederiksen, T., Paulsson, M., Brandbyge, M., and Jauho, A.P., 2007. Inelastic transport theory from first principles: Methodology and application to nanoscale devices. *Physical Review B*, 75(20), 205413.

Gall, K., Diao, J.K., and Dunn, M.L., 2004. The strength of gold nanowires. *Nano Letters*, 4(12), 2431–2436.

García-Suárez, V.M. et al., 2005. Conductance oscillations in zigzag platinum chains. *Physical Review Letters*, 95(25), 256804.

Häkkinen, H., Barnett, R.N., and Landman, U., 1999. Gold nanowires and their chemical modifications. *Journal of Physical Chemistry B*, 103(42), 8814–8816.

Kittel, C., 2004. *Introduction to Solid State Physics*. John Wiley & Sons Ltd., New York.

Kizuka, T. et al., 1997. Cross-sectional time-resolved high-resolution transmission electron microscopy of atomic-scale contact and noncontact-type scannings on gold surfaces. *Physical Review B*, 55(12), R7398–R7401.

Koizumi, H., Oshima, Y., Kondo, Y., and Takayanagi, K., 2001. Quantitative high-resolution microscopy on a suspended chain of gold atoms. *Ultramicroscopy*, 88(1), 17–24.

Krüger, D. et al., 2002. Pulling monatomic gold wires with single molecules: An ab initio simulation. *Physical Review Letters*, 89(18), 186402.

Kurui, Y., Oshima, Y., Okamoto, M., and Takayanagi, K., 2008. Integer conductance quantization of gold atomic sheets. *Physical Review B*, 77(16), 161403R.

Legoas, S.B., Galvão, D.S., Rodrigues, V., and Ugarte, D., 2002. Origin of anomalously long interatomic distances in suspended gold chains. *Physical Review Letters*, 88(7), 076105.

Major, P. et al., 2006. Nonuniversal behavior of the parity effect in monovalent atomic wires. *Physical Review B*, 73(4), 045421.

Ohnishi, H., Kondo, Y., and Takayanagi, K., 1998. Quantized conductance through individual rows of suspended gold atoms. *Nature*, 395(6704), 780–783.

Oncel, N., 2008. Atomic chains on surfaces. *Journal of Physics: Condensed Matter*, 20, 393001.

Palacios, J.J. et al., 2002. First-principles approach to electrical transport in atomic-scale nanostructures. *Physical Review B*, 66(3), 035322.

Pyykkö, P., 1988. Relativistic effects in structural chemistry. *Chemical Reviews*, 88(3), 563–594.

Rocha, A.R. et al., 2006. Spin and molecular electronics in atomically generated orbital landscapes. *Physical Review B*, 73(8), 085414.

Rodrigues, V. et al., 2002. Quantum conductance in silver nanowires: Correlation between atomic structure and transport properties. *Physical Review B*, 65(15), 153402.

Rubio-Bollinger, G. et al., 2001. Mechanical properties and formation mechanisms of a wire of single gold atoms. *Physical Review Letters*, 87(2), 026101.

Sánchez-Portal, D. et al., 1999. Stiff monatomic gold wires with a spinning zigzag geometry. *Physical Review Letters*, 83(19), 3884–3887.

Segovia, P., Purdie, D., Hengsberger, M., and Baer, Y., 1999. Observation of spin and charge collective modes in one-dimensional metallic chains. *Nature*, 402(6761), 504–507.

Sim, H.S., Lee, H.W., and Chang, K.J., 2001. Even-odd behavior of conductance in monatomic sodium wires. *Physical Review Letters*, 87(9), 096803.

Smit, R.H.M., 2003. From quantum point contacts to monatomic chains: Fabrication and characterization of the ultimate nanowire. PhD Thesis, Leiden University, Leiden, the Netherlands.

Smit, R.H.M. et al., 2003. Observation of a parity oscillation in the conductance of atomic wires. *Physical Review Letters*, 91(7), 076805.

Smit, R.H.M., Untiedt, C., Yanson, A.I., and van Ruitenbeek, J.M., 2001. Common origin for surface reconstruction and the formation of chains of metal atoms. *Physical Review Letters*, 87(26), 266102.

Smogunov, A. et al., 2008. Colossal magnetic anisotropy of monatomic free and deposited platinum nanowires. *Nature Nanotechnology*, 3(1), 22–25.

Sørensen, M.R., Brandbyge, M., and Jacobsen, K.W., 1998. Mechanical deformation of atomic-scale metallic contacts: Structure and mechanisms. *Physical Review B*, 57(12), 3283–3294.

Takai, Y. et al., 2001. Dynamic observation of an atom-sized gold wire by phase electron microscopy. *Physical Review Letters*, 87(10), 106105.

Thijssen, W.H.A., Marjenburgh, D., Bremmer, R.H., and van Ruitenbeek, J.M., 2006. Oxygen-enhanced atomic chain formation. *Physical Review Letters*, 96(2), 026806.

Todorov, T.N., Hoekstra, J., and Sutton, A.P., 2001. Current-induced embrittlement of atomic wires. *Physical Review Letters*, 86(16), 3606–3609.

Untiedt, C. et al., 2002. Calibration of the length of a chain of single gold atoms. *Physical Review B*, 66(8), 085418.

Untiedt, C., Dekker, D.M.T., Djukic, D., and van Ruitenbeek, J.M., 2004. Absence of magnetically induced fractional quantization in atomic contacts. *Physical Review B*, 69(8), 081401.

Wallis, T.M., Nilius, N., and Ho, W., 2002. Electronic density oscillations in gold atomic chains assembled atom by atom. *Physical Review Letters*, 89(23), 236802.

Yanson, A.I. et al., 1998. Formation and manipulation of a metallic wire of single gold atoms. *Nature*, 395(6704), 783–785.

Ultrathin Gold Nanowires

Takeo Hoshi
Tottori University

and

*Japan Science and
Technology Agency*

Yusuke Iguchi
The University of Tokyo*

Takeo Fujiwara
The University of Tokyo

and

*Japan Science and
Technology Agency*

36.1 Introduction

Ultrathin nanowires of gold (Au) and other metals have been studied intensively, particularly from the 1990s, as a possible foundation of nano electronics (see Agraït et al. (2003) for a review). They are fabricated as nano-meter-scale contacts within two electrode parts and are composed of a couple of atoms in their wire length and/or diameter. Monoatomic chain, the thinnest wire, was also fabricated. The metal nanowires can show quantized conductance, even at room temperature, which is completely different from Ohm's law in macroscale samples. Structural and transport properties of nanowires were investigated by (1) fine experiments, such as high-resolution transmission electron microscopy (HRTEM), scanning tunneling microscope (STM), and atomic force microscopy (AFM) and (2) atomistic simulations with quantum-mechanical theory of electrons. Many experiments and simulations lead us to several common understandings among metal nanowires of various elements (Agraït et al. 2003).

This chapter focuses on the unique properties of Au nanowires. Common properties between Au and other nanowires are also discussed so as to figure out the uniqueness of Au nanowire from a general scientific viewpoint.

As a unique and fascinating property of Au nanowire, helical multishell structure was reported in 2000 (Kondo and Takayanagi 2000), as in carbon nanotube (Iijima 1991, Dresselhaus et al.

2001). A shell of helical Au nanowires consists of a folded hexagonal sheet, while carbon nanotube (Iijima 1991, Dresselhaus et al. 2001) consists of a folded graphene sheet. This chapter focuses particularly on the helical multishell nanowire of Au both for phenomena and for proposed formation mechanisms. The discussion is based on electronic structure and covers a wide range of nanostructures of Au and other metals.

This chapter is organized as follows: Section 36.2 is devoted to a tutorial for atomic structure and electronic wave functions of solid Au with other materials. Section 36.3 summarizes the structure and transport properties of Au and other metal nanowires. Section 36.4 focuses on the helical multishell Au nanowire, as the main topic of this chapter. Finally, in Section 36.5, a summary of this chapter is given and a future aspect is addressed for establishing the foundation of nano electronics. The appendix is devoted to quantum-mechanical molecular dynamics (QM-MD) simulation, which is important for understanding points of this chapter.

36.2 Basic Properties of Solid Gold

The basic properties of Au are summarized with those of other elements. Figure 36.1a shows periodic table up to period 6 elements and Figure 36.1b a part of periodic table that includes Au. Several elements in Figure 36.1b are discussed in this chapter. Among the elements, copper (Cu), silver (Ag), and Au are called

---

* He is currently affiliated with NEC Corporation, Kanagawa, Japan.

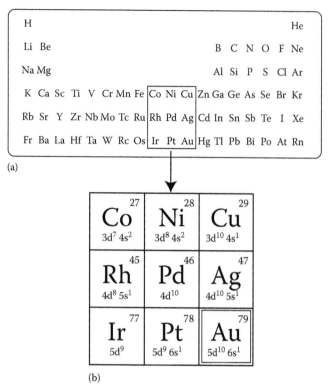

(a)

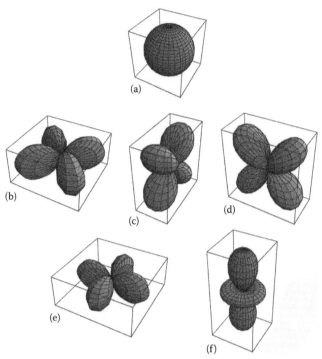

FIGURE 36.1 (a) Periodic Table up to Period 6 elements. (b) A part of Periodic Table that contains gold (Au). The atomic number is plotted at the upper right corner of each box and the valence electron configuration is plotted at the lower area of each box. All the elements shown in (b) form FCC structure in solid state, except cobalt (Co).

FIGURE 36.2 Schematic figure of the nonspherical forms $|Q(\theta, \phi)|$ of (a) s orbital and (b)–(f) five d orbitals; (b) xy orbital, (c) yz orbital, (d) zx orbital, (e) x^2-y^2 orbital, (f) $3z^2-r^2$ orbital. The xy orbital spreads mainly on the xy plane.

noble metals. They have 10 n d electrons and 1 (n + 1) s electron ($nd^{10}ns^1$) as valence electrons at each atom, where n = 3, 4, 5 for Cu, Ag, and Au, respectively. The labels of "s" and "d" indicate the character of atomic orbitals. In general, an atomic orbital is described as

$$R(r)Q(\theta,\phi) \qquad (36.1)$$

in polar coordinate. The nonspherical distribution comes from the part of $Q(\theta, \phi)$. The wave function of s electron is spherical ($Q(\theta, \phi)$ = const) and that of d electron is not. Figure 36.2 illustrates the (non-)spherical distribution $Q(\theta,\phi)$ of s and d wave functions by plotting the function of $r = r(\theta,\phi) = |Q(\theta, \phi)|$. For example, the xy orbital spreads mainly on the xy plane. Another type of nonspherical atomic orbital, "p" orbital, also appears later in this chapter. For details of the atomic orbitals, the interested reader can refer to elementary textbooks on quantum mechanics.

All the elements in Figure 36.1b form solid in face-centered-cubic (FCC) structure, shown in Figure 36.3a, except cobalt (Co) that forms close-packed hexagonal (HCP) structure. The lattice constant, the distance between the A and B atom in Figure 36.3a, is 3.61, 4.09, and 4.08 Å for Cu, Ag, and Au, respectively. In this chapter, as in many textbooks, x, y, z axes are defined and normalized so that the cubic box shown in

Figure 36.3a gives the cubic region of $0 \leq x, y, z \leq 1$. In Figure 36.3a, the eight corner atoms labeled by A, B, C, D, E, F, G, and H are described, for example, by (1,0,0) and (0,1,0) and the six face-center atoms, such as the atom I, are described, for example, (1/2,1/2,0), (0,1/2,1/2) in the normalized coordinate. FCC and HCP structures are ones in close packing (highest average density) as three-dimensional lattice and each atom has 12 nearest neighbor atoms. For example, the atom placed at (0,0,0) has the 12 nearest neighbor atoms that are placed at (±1/2, ±1/2, 0), (±1/2, 0, ±1/2), (0, ±1/2, ±1/2).

Several planes in the FCC structure are drawn in Figure 36.3b through d. Here the notations of planes are explained. A plane of $(x/l) + (y/m) + (z/n) = 1$ is indicated by an index of (l, m, n). The negative surface index is written with a bar. For example, the index of $(\bar{1}11)$ means the plane of $(x/(-1)) + (y/1) + (z/1) = 1$ or $-x + y + z = 1$. A direction vector is denoted as $[abc]$ and a "$[abc]$ nanowire" is a wire of which axis is in the $[abc]$ direction. Moreover, equivalent planes are called by "type" in this chapter. For example, the lattice structures on (111) and $(\bar{1}11)$ planes of FCC structure are equivalent to those shown in Figure 36.3d and these planes are called "(111)-type" planes. The (001)-type, (110)-type, and (111)-type planes are drawn in Figure 36.3b through d, respectively. A (111)-type plane has a hexagonal structure with triangular tiles, as shown in Figure 36.3d, and is one in close packing as two-dimensional lattice.

Electronic states among noble metals in FCC solid are described in Figures 36.4 and 36.5. The data are given by

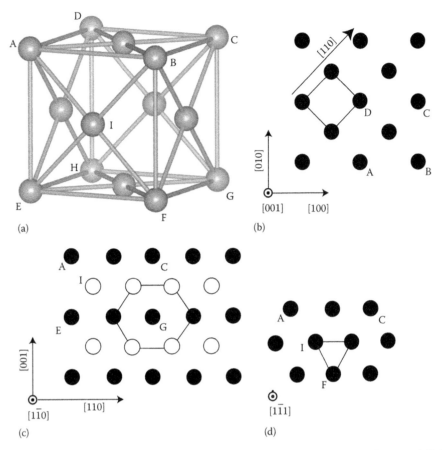

FIGURE 36.3 (a) FCC structure. "Bonds" are drawn between nearest neighbor atoms. (b) (001)-type, (c) (110)-type, and (d) (111)-type planes of FCC structure. The atoms marked as A, B, C, D, E, F, G, H, and I are common among (a)–(d). Shapes drawn by lines in (b), (c), and (d) will appear later in this chapter. In (c), two successive atomic layers are drawn and the atoms in the two atomic layers are depicted by open and filled circles, respectively.

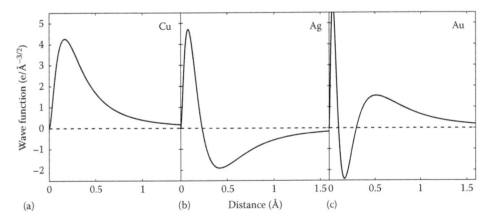

FIGURE 36.4 Radial wave functions $R(r)$ of noble metals, (a) copper, (b) silver, and (c) gold, in the bulk FCC structure. See text for details.

a modern quantum-mechanical calculation for electronic structure. The calculations are carried out by first-principles theory, the density-functional theory with the linear-muffin-tin-orbital method (Andersen and Jepsen 1984). See the textbook by Martin (2004) for an overview of modern electronic structure theories. Since Au is a heavy element, relativistic effect is included in these calculations, as scaler-relativistic formulation. See textbooks of quantum mechanics, such as Schiff (1968), for relativistic effect

and scaler-relativistic formulation. Figure 36.4 shows the radial wave functions $R(r)$ of the d orbitals. An (nd) orbital has (n–2) nodes in the radial wave function $R(r)$. The Cu wave function, a $3d$ wave function, does not have a node and the Ag and Au wave functions, $4d$ and $5d$ wave functions, have one and two node(s), respectively. Figure 36.5 shows the electronic density of states (DOS), or the energy spectrum of electronic states. The dashed line in Figure 36.5 indicates the Fermi level E_F, which means that

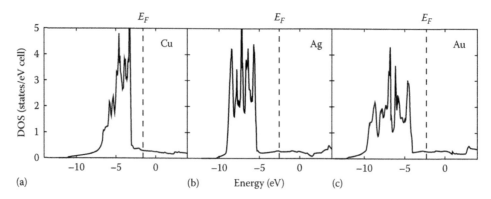

FIGURE 36.5 Density of states, or the energy spectrum of electronic states, in bulk FCC solids among noble metals, (a) copper, (b) silver, and (c) gold. See text for details.

the electrons occupy the energy region of $E < E_F$. As a common feature of the three elements, the d band is narrow and fully occupied, while the s band is broad and partially occupied. In the case of Au, for example, the d band lies in the narrow energy region of $-10\,eV \leq E \leq -4\,eV$, which is within the occupied energy region. The s band, on the other hand, lies in the region from $-13\,eV$ up to the right end of the graph. The above feature indicates that the 10 d electrons form a closed electronic shell and are nearly localized at atomic regions, while the 1 s electron is extended and can contribute to electrical current. Therefore, these solids are metallic and electrical current is observed with Ohm's law.

36.3 Metal Nanowire and Quantized Conductance

Nanowires or nanoscale contacts of metals are fabricated by deformation processes with, for example, STM tip or mechanically controllable break junction (Agraït et al. 2003). Figure 36.6 illustrates the fabrication process by a STM tip, in which a nanowire is formed between two electrode parts. Real-space image, like the one in Figure 36.6, is obtained by HRTEM. A formation process of a Au nanocontact was observed as successive HRTEM images (Kizuka et al. 1997, Kizuka 1998).

Quantized conductance of metal nanowires was reported, even at room temperature, and relatively clear quantized values were observed among noble metals and alkali metals (Agraït et al. 1993, Krans et al. 1993, Pascual et al. 1993, Olesen et al.

1994, Brandbyge et al. 1995, Krans et al. 1995, Muller et al. 1996, Rubio et al. 1996, Costa-Krämer et al. 1997, Hansen et al. 1997, Yanson et al. 1999, Yanson et al. 2000, Yanson et al. 2001, Agraït et al. 2003, Smit et al. 2003, Mares et al. 2004, Bettini et al. 2005, Mares and van Ruitenbeek 2005). The quantized conductance G is defined as

$$G = nG_0, \qquad (36.2)$$

where n is an integer and

$$G_0 \equiv \frac{2e^2}{h} \qquad (36.3)$$

is the conductance unit defined by the charge of electron e (>0) and Planck's constant h. Figure 36.7 is a schematic figure of conductance trace in thinning process, such as the process of Figure 36.6c and d. In Figure 36.7, conductance plateaus of $G \approx 2G_0$, $1G_0$ appear and the conductance reaches $G \approx 0$, when the wire is broken. In experimental research, the histogram of conductance values is constructed from many independent samples and a sharp peak at an integer value ($G = nG_0$) in the histogram is assigned to be a quantized conductance.

A simultaneous observation of structure and transport was realized by a combined experiment of HRTEM and STM (TEM-STM) (Ohnishi et al. 1998, Erts et al. 2000, Kizuka et al. 2001a), which is crucial for understanding nanowires, because the conductance value, in general, does not determine atomic structure

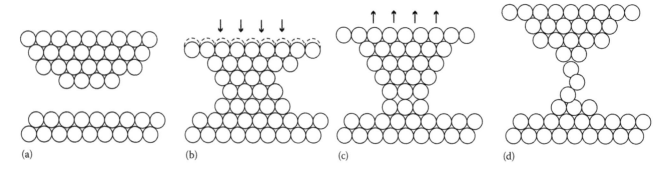

FIGURE 36.6 Schematic figures for the snapshots, (a)–(d), during the fabrication process of nanowire using a STM tip. Atoms are depicted as balls.

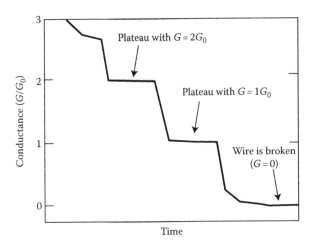

FIGURE 36.7 Schematic figure of conductance trace in thinning process.

uniquely. A TEM-STM experiment in 1998 (Ohnishi et al. 1998) obtained a direct image of a monoatomic Au chain with the length of several atoms and observed a quantized conductance in $G = G_0$.

Force measurement was realized by a combined experiment of AFM and STM (Rubio et al. 1996) and shows that a jump in conductance trace occurs with a jump in the force. The result implies that a jump in conductance trace is caused by the plastic deformation of nanowire. Stress–strain curve was obtained by the combined experiment of TEM and AFM (Kizuka et al. 2001b, Erts et al. 2002, Kizuka 2008) and the above statement was confirmed. In addition, the simultaneous measurement of stiffness and conductance was realized by mechanically controllable break junction with a force sensor and applied to Au and platinum nanowires (Rubio-Bollinger et al. 2004, Valkering et al. 2005, Shiota et al. 2008).

The quantum mechanical theory of electrical current has a rigorous foundation of nonequilibrium Green's function, which can be seen in the textbook by Datta (1995). Here only a resultant formulation is briefly explained. In quantum mechanics, electrons are described as "waves" with discretized energy levels and electrical current is given by the "wave modes" that are extended through nanowires. Such extended "wave modes" are usually called channels. The integer n in Equation 36.2 is the number of the channels. A general expression of the conductance is

$$G = G_0 \sum_i \tau_i, \tag{36.4}$$

where τ_i is the transmission rate of the ith "wave mode" ($0 \le \tau_i \le 1$) (Datta 1995). Equation 36.4 will be reduced to Equation 36.2, when all the channels are perfectly ballistic ($\tau_i = 1$).

The simulation of atomic structure is crucial for metal nanowires, since the transport properties are sensitive to atomic structure. A pioneering theoretical work in 1993 (Todorov and Sutton 1993) investigated the relation between the atomic structure and the conductance, in which the dynamics of atomic structure is determined by a classical potential. Nowadays, quantum

mechanical molecular dynamics (QM-MD) simulation is a standard simulation tool for nanowires and other nanostructures. In QM-MD simulations, an effective Schrödinger equation is solved at each time step and atomic structures are updated under the change of electronic states. See Appendix 36.A for a tutorial of QM-MD simulation. QM-MD simulation enables us to investigate (1) simultaneous discussion of structure and electronic properties in a single theoretical framework with quantum mechanics of electrons and (2) systematic research among various elements. Many simulations were carried out for monoatomic Au chain and related materials (Brandbyge et al. 1995, 1999, Sorensen et al. 1998, Okamoto and Takayanagi 1999, Sánchez-Portal et al. 1999, da Silva et al. 2001, Agraït et al. 2003, Fujimoto and Hirose 2003, da Silva et al. 2004). After intensive works of simulation and experiment, it is established that a monoatomic Au chain is formed and shows a quantized conductance of $G = 1G_0$, since only one s orbital contributes to the current.

"Thicker" nanowires, nanowires thicker than monoatomic chains, can show complex structural and/or transport properties. An interesting viewpoint is shell effect that enhances the stability of specific integer values in the conductance. The conductance histograms in alkali and noble metals were analyzed from the viewpoint of the shell effect (Yanson et al. 1999, 2000, 2001, Mares et al. 2004, Mares and van Ruitenbeek 2005). In the context of the shell effect of alkali and noble metals, the stability of elliptical metal nanowires was investigated quantum mechanically within a free-electron model in continuum media (Urban et al. 2004). Other theoretical investigations with free-electron models are found in the reference lists of the above papers or a review (Agraït et al. 2003). It is noteworthy that a difference of alkali and noble metals is the fact that the valence band of alkali metals consists of one s electron, while that of noble metals consists of $1s$ electron and $10d$ electrons, as shown in Figure 36.5. A TEM-STM experiment was carried out with thicker [110] Au nanowires for the direct relation between the structure (the shape of the cross section) and the conductance (Kurui et al. 2007). The appearance of helical multishell Au nanowire (Kondo and Takayanagi 2000) is a fascinating property of "thicker" nanowires and will be discussed in the rest of this chapter. As another systematic research of "thicker" nanowires, a TEM-STM experiment was carried out for Au nanocontacts parallel to the [001], [111], and [110] directions (Oshima et al. 2003a). The paper suggests that the variety in the crystalline orientation of the contacts can be the origin of the observed variety in the conductance trace among samples. The sliding of a twin boundary in a Au nanocontact was reported by direct HRTEM observation (Kizuka 2007). A recent TEM-STM experiment (Kurui et al. 2008) found (111)-type and (001)-type atomic sheets between electrode parts and they show quantized conductance.

To conclude this section, brief comments are made on metal nanowires in which p and d orbitals contribute to current. Such nanowires exhibit a complex transport behavior, even in monoatomic chains. For example, nanowires of aluminum has s and p valence orbitals and show that plateau structure in conductance trace is less regular but the plateaus-like structures are still observed nearly at integer conductance values (Krans et al. 1993, Cuevas et al. 1998, Yanson et al. 2008). In nanowires of nickel, a transition metal, the quantized conductance changes systematically under varying external parameters, the temperature (above and below the Curie temperature), and the applied magnetic field (Oshima and Miyano 1998).

36.4 Helical Multishell Structure

36.4.1 Overview

In 2000, helical multishell Au nanowires were synthesized and observed by HRTEM image (Kondo and Takayanagi 2000). They were fabricated by focusing an electron beam on a thin film (Kondo and Takayanagi 1997). The wire axis of the helical nanowires is the [110]-type direction of the original FCC geometry (see Figure 36.3b and c). The outermost shell is a folded (111)-type (hexagonal) atomic sheet (see Figure 36.3d) and helical around the nanowire axis. A single shell helical structure was synthesized later (Oshima et al. 2003b). Experiments also observed the thinning process (Oshima et al. 2003c) and the conductance (Oshima et al. 2006) of the multishell helical structures.

Hereafter multishell structures are denoted by the numbers of atoms in each shell. For example, a "14-7-1 nanowire" is a rod with three shells in which the outer, middle, and inner shells have fourteen, seven, and one atom(s) in the section view, respectively. The shape of the "6-1 nanowire" is depicted by lines in Figure 36.3c.

The observed multishell configurations are quite specific and are characterized by "magic numbers." The multishell structures in the 7-1, 11-4, 13-6, 14-7-1, 15-8-1 structures were experimentally observed. These numbers are called "magic numbers," since the difference of numbers between the outermost and the next outermost shells is seven, except the cases of the 7-1 structure. Such a rule of the observed multishell configurations implies that there is an intrinsic mechanism to form the specific multishell configurations.

The transport property of helical multishell Au nanowire was investigated both in theory (Ono and Hirose 2005) and in experiment (Oshima et al. 2006), which will be discussed in Section 36.5.

The appearance of helical metal wires was investigated by simulations with different methodologies. Before the experimental report of helical Au nanowire (in 1998), various "exotic" structures of nanowires were predicted with potentials of Al and Pb and one of them is helical structure (Gülseren et al. 1998). After the experimental report, several investigations, such as Bilalbegović (2003) and Lin et al. (2005), were carried out with classical potentials. Quantum mechanical simulations were also carried out (Tosatti et al. 2001, Senger et al. 2004, Yang 2004). A first principles calculation reported that the tension of nanowires gives the minimum values when the number of atoms of the lateral atomic row on the outermost shell is seven and the nanowire is helical (Tosatti et al. 2001). It is also showed that the tension does not have the minimum in model Ag nanowires. Another first principles calculation

was carried out for nanowires with atoms from three to five on the lateral atomic row on the outermost shell, and showed that helical nanowires are not the configuration of the minimum energy but of the minimum tension (Senger et al. 2004). Although these studies gave important progress on theory, they did not explain why the helical structures are formed in the specific multishell configurations with "magic numbers."

Pt nanowires were also synthesized with the same type of helicity (Oshima et al. 2002). Pt is placed at a neighbor of Au in Periodic Table (see Figure 36.1b) and its electronic configuration ($5d^9 6s^1$) is similar to that of Au ($5d^{10} 6s^1$). The above experiment implies that the mechanism of the formation of helical nanowires is generic between Au and Pt and may be inherent among some other elements.

In 2007, a theory (Iguchi et al. 2007) was proposed for the formation model of helical multishell nanowires. In this model, the transformation consists of two stages. At the first stage, the outermost shell is dissociated from the inner shell to move freely. At the second stage, an atom row slips on the wire surface and a (001)-type (square) face (Figure 36.3b) transforms into a folded (111)-type (hexagonal) one (Figure 36.3d) with helicity. The driving force for the helicity comes from the nature of nonspherical $5d$ electrons and a (111)-type surface structure is energetically favorable for $5d$ electrons. The theory contains the following points: (1) The theory explains the observed multishell configurations with "magic numbers" systematically. (2) The theory was validated by QM-MD simulations for Au and Cu with tight-binding form Hamiltonian (see Appendix 36.A). (3) The theory gives a general understanding among helical nanowire structures of Au and Pt and several reconstructed structures of equilibrium surfaces. After the proposal of the two-stage model (Iguchi et al. 2007), several related simulations were carried out for the formation of helical Au nanowire within tight-binding form Hamiltonian (Amorim and da Silva 2008, Fujiwara et al. 2008).

36.4.2 Two-Stage Formation Model

Hereafter, the multishell helical structures are systematically constructed, according to the proposed theory of two-stage formation model (Iguchi 2007; Iguchi et al. 2007).

First, a set of ideal [110] nanowires are prepared under two conditions: (a) there is no acute angle on the surface because of diminishing surface tension, and (b) there is no (001) side longer than any (111) side since the surface energy of a (001) surface is higher than that of (111). These structures are called "reference" structures in this chapter. Figure 36.8 shows the section views of the "reference" structures in the 6-1, 10-4, and 12-6 multishell configurations. Among them, the outermost shell has six more "bonds" on the lateral row than the inner shell, shown by bold lines in Figure 36.8.

The formation process of the helical 7-1 structure is depicted schematically in Figure 36.9. When an atom row appears at a (001)-type surface on the outermost shell of the reference 6-1 structure (Figure 36.9a), the structure turns into a 7-1 structure (Figure 36.9b)

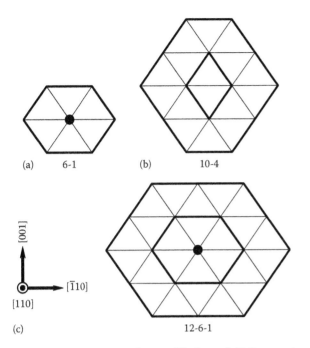

FIGURE 36.8 Section view of a set of "reference" [110] nanowires. The (a) 6-1, (b) 10-4, and (c) 12-6-1 structures are shown. See text for details. The shape of the 6-1 structure can be found in Figure 36.3c.

and is called "initial" structure in this chapter. The outermost shell has seven more atoms on a lateral atom row than the inner shell and the outermost shell can have room for atom row slip. This is the origin of the "magic numbers." Figure 36.9c shows an expanded lateral surface of the initial 7-1 structure. In Figure 36.9c, the surface region of B is (001)-type (square) lattice, while the surface regions of

A and C are (111)-type (hexagonal) ones. Dashed lines connect the same atoms at the right ends and the left ends. When the number of atoms on the lateral row in the outermost shell is odd, the surface reconstruction brings the helicity to the nanowire inevitably. A slip of atom row transforms the surface region B from (100)-type surface into a folded (111)-type one with helicity. As a result, the helical 7-1 structure is formed, as in Figure 36.9d.

36.4.3 Simulation of Formation Process

A QM-MD simulation was carried out, so as to confirm the proposed process model for forming helical structures. The simulation was carried out with a tight-binding form Hamiltonian that was used for several simulations of Au nanowire (da Silva et al. 2001, 2004). Details and theoretical foundations of the QM-MD simulations are described in Appendix 36.A. A Relaxation process with thermal fluctuation was simulated.

The samples in the simulations were finite [110] nanowires with the initial structures of the ideal 7-1, 10-4, 11-4, 12-6-1, 13-6-1, 15-7-1, and 15-8-1 structures. The simulation results are summarized in Figure 36.10 as section views, except those of the 10-4 and 12-6-1 structures. The initial structures appear in the left panels of Figure 36.10a through e. They are "reference" structures with one additional atom row on the outermost shell, as explained in Section 36.4.2. The additional atom rows are marked by arrows. The rod length is nine layers in the 7-1, 10-4, and 11-4 nanowires, seven layers in the 12-6-1 and 13-6-1 nanowires, or six layers in the 14-7-1, and 15-8-1 nanowires, respectively. Here the layer unit is defined as the periodic unit of the ideal FCC structure and is composed of two successive atomic layers shown in Figure 36.3c. The numbers of atoms in the simulations are from 76 to 156.

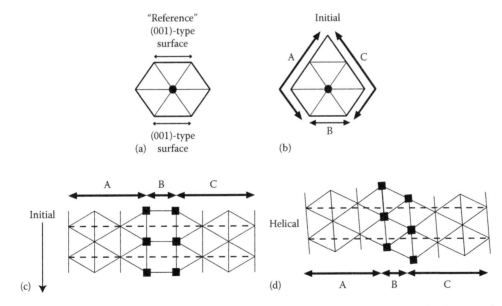

FIGURE 36.9 Schematic figure of the formation of the helical 7-1 structure: (a) "Reference" 6-1 structure. Two (001)-type surfaces appear and the other surfaces are (111)-type ones. (b) Initial 7-1 structure. The areas A and C are the (111)-type surfaces and the area B is the (001)-type surface. (c) Expanded lateral surface of the initial 7-1 structure of which the section view is given in (b). Dashed lines connect the same atoms at the right and the left ends. (d) Expanded lateral surface of the helical 7-1 structure transformed from the ideal non-helical structure of (c).

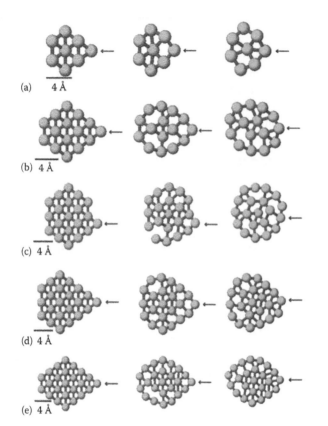

FIGURE 36.10 Section views in the relaxation process of the Au nanowires into helical structures; (a) 7-1 structure at the initial (left), 400-th (middle), and 5000-th iteration steps (right); (b) 11-4 structure at the initial (left), 400-th (middle), and 6000-th iteration steps (right); (c) 13-6-1 structure at the initial (left), 750-th (middle), and 9000-th iteration steps (right); (d) 14-7-1 structure at the initial (left), 400-th (middle), and 7000-th iteration steps (right); and (e) 15-8-1 structure at the initial (left), 400-th (middle), and 5000-th iteration steps (right). An atom row is marked by arrow for each system. See text for details.

The boundary condition is imposed by fixing the center of gravity of the top and bottom layers of the nanowires.

The finite-temperature dynamics is realized by the thermostat technique and the simulations were carried out at $T = 600$ and 900 K, which are lower than the melting temperature (1337 K). One iteration step corresponds to $\delta t = 1$ fs. Helical structures appear in all cases except the 12-6-1 nanowire at 600 K. The results shown in Figure 36.10 are those at $T = 600$ K for (a), (b), (c), and (e) and at $T = 900$ K for (d).

As a typical case, Figure 36.11a through d shows a set of side views in the case of the 11-4 structure at $T = 600$ K. Figure 36.11a through d indicates that the square tiles on the (001)-type surface at the initial structure are transformed into the hexagonal (111)-type surface, as in Figure 36.9c and d. The transformation propagates from the top and bottom of the nanowire. In the snapshot of Figure 36.11c, for example, the transformation has been completed except the region near the third lowest layer and the transformation. The energy of the nanowire is plotted in Figure 36.11e as the function of iteration step and decreases almost monotonically after the 1000-th iteration step.

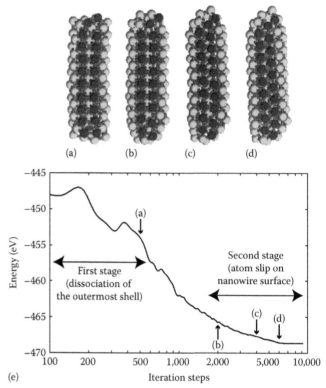

FIGURE 36.11 Formation of helical 11-4 Au nanowire in relaxation process. Side views at the (a) 500-th, (b) 2000-th, (c) 4000-th, and (d) 6000-th iteration step. Dark atoms are those that are placed initially on the (001)-type (square) surface and are transformed into a (111)-type (hexagonal) surface. (e) Change of the energy during the formation process. The iteration steps of (a), (b), (c), and (d) are indicated by arrows.

It should be noted that the present simulation is different from experiment in several points. For example, the time scale of the process is quite short, on the order of 10 ps, owing to the practical limit of computational resource. Therefore, the simulation result should be understood so that it captures an intrinsic energetical mechanism of the real process.

36.4.4 Analysis of Electronic Structure

The mechanism of the two-stage formation process is investigated by analyzing local electronic structure. A typical case of the 11-4 structure with $T = 600$ K is picked out for explanation.

Local density of states (LDOS) and "local energy" are used throughout the analysis. In general, LDOS is defined for each atom and can be decomposed into the contributions of each orbital. The profile of LDOS means the energy spectrum of electronic states at a local region near the atom. "Local energy" is defined for each atom or orbital by the energy integration of LDOS within the occupied energy region. A decrease of the local energy during the process means that the atom gains the energy by a binding mechanism.

Figure 36.12 shows the LDOS of specific atoms at specific iteration steps. The section views of Figure 36.12a are identical to those of Figure 36.10b. The LDOS for the atoms A, B, and C in

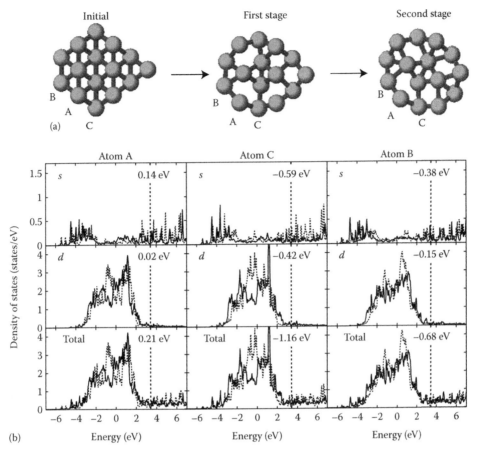

FIGURE 36.12 Analysis of the two-stage formation process in the 11-4 Au helical nanowire. (a) The section views with specifying the atoms A, B, and C. The snapshots are the same as those in Figure 36.10b. (b) Local density of states of the atoms A, C, and B at different iteration steps. The solid and broken lines in (b) are at the initial state and at the 500-th iteration step respectively for the atoms A and C, and at the 500-th and 5000-th steps for the atom B. The difference of the local energy between the two iteration steps is written at the right corner of each panel. The vertical broken line indicates the Fermi level. The upper, middle, and lower panels of (b) are the partial densities of states of *s* orbital, *d* orbital, and the total density of states, respectively. In each panel, the difference of the local energy between the two iteration steps is written at the upper right corner of each panel.

Figure 36.12a are plotted in Figure 36.12b. The solid and broken lines are the LDOS profile at the initial and 500-th steps for the atoms A and C, and are at the 500-th and 5000-th steps for the atom B, respectively. The vertical broken line indicates the Fermi level and the positions of the Fermi level are almost the same in the two iteration steps and indistinguishable in the graphs. The top panel of Figure 36.12b is the LDOS for the *s* orbital and the middle panel is the LDOS for the *d* orbitals. The *p* orbitals are also included in the simulation but their contribution in the energy range of Figure 36.12b is small. The bottom panel is the total LDOS value that is contributed by the *s*, *p*, and *d* orbitals. In each panel of Figure 36.12b, the difference of the local energy between the two iteration steps is written at the upper right corner. A negative or positive value of the difference means an energy gain or loss during the process, respectively.

The LDOS for each *d* orbital is also plotted Figure 36.13, which will be key for understanding the mechanism. The shape of the *d* orbitals is shown in Figure 36.2 Here a local coordinate system is defined for each atom as follows. The local *x*-axis is the nanowire axis, [110], the local *y*-axis is along the lateral direction and the

local *z*-axis is perpendicular to the surface. Therefore, the wire surface corresponds to the *xy* plane at each atom.

The first stage of the process can be explained by the change of LDOS of the atoms A and C. Figure 36.12 indicates that the dissociation of the atom A occurs with an energy loss by 0.21 eV, because of the reduction of its coordination number. It is remarkable, however, that the energy loss of the *d* electrons is quite small (0.02 eV), when it is compared with that of the *s* electron (0.14 eV). Here one should remember that the number of *d* electrons is larger, by about 10 times, than that of *s* electrons ($d^{10}s^1$). The left column of Figure 36.13, the data of the *d* orbitals at the atom A, indicates that the energy gain and loss among the *d* orbitals are on the order of 0.1 eV but they cancel with each other. The energy gain or loss mechanism of each orbital can be explained by the spatial spread of each orbital; The local energy of the *yz* orbital of the atom A increases, since the orbital extents perpendicularly to the wire surface and the nearest neighbor distance increases. The local energy of the *xy* orbital decreases, since the orbital can expand more to another (111)-type sheet through the

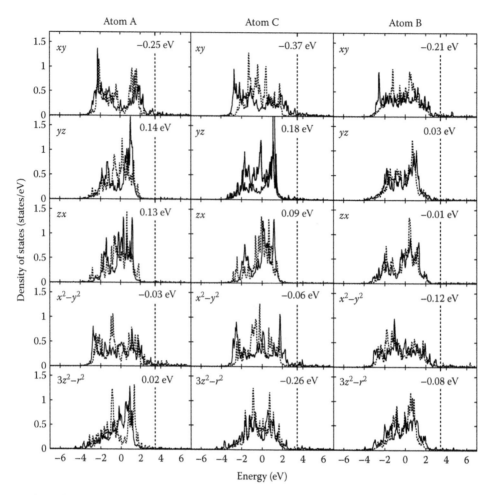

FIGURE 36.13 LDOS for each *d* orbital of the atoms A, C, and B at the two iteration steps in the relaxation process of the 11-4 Au nanowire. The definitions and notations are the same as in Figure 36.12. Local coordinate system for each atom is written in text.

atom C because of flattening two (111)-type sheets and the bandwidth becomes wider.

It is also remarkable in Figure 36.12 that the atom C has a large energy gain, by 1.16 eV, at the first stage. The middle column of Figure 36.13 indicates that the largest energy gain comes from that of the *xy* orbital (0.37 eV). Therefore, the local energy of the *d* orbitals decreases on the atom C. This energy gain of *d* orbitals can be attributed to the flattened surface structure around the atom C after the dissociation between the atom A and the inner shell. The energy of *s* orbital of the atom C also decreases appreciably but it may be not associated primarily with the dissociation, since the *s* orbital does not always favor the flatter atomic configurations.

The second stage is explained by the change of LDOS of the atom B. The local energy of the *s* orbital decreases by 0.38 eV (Figure 36.12b), since the coordination number of the atom B increases like an atom depicted by filled square in Figure 36.9c and d. The local energy of the *d* orbital of the atom B also decreases by 0.15 eV (Figure 36.12b). In particular, the *xy* orbital gives the largest energy gain among the *d* orbitals, by 0.21 eV (Figure 36.13), since the surface transforms to (111)-type and is flattened.

Here, the slip deformation is essential since it widens the area of the (111)-like hexagonal surface and the LDOS of the *xy* orbital transfers its weight from the antibonding region (the high energy region) to the bonding region (the low energy region) (Figure 36.12). Here one should recall that the atom A is connected with the atom B and can move relatively freely from the inner shell, after the dissociation at the first stage. Therefore, the atom B slips easily and introduces the helicity, since the atom B can trail other atoms without dissociating their bonds.

The above analysis is concluded that the first and second stages are governed by the energy gain mechanism of *d* orbitals among atoms on or near the (001)-type surface area. If a wire has a larger diameter than those in this section, the ratio of the energy gain to the total energy will be smaller and the transformation will not occur.

36.4.5 Discussions

Three points are discussed for the two-stage formation model.

First, the 10-4 and 12-6-1 nanowires are discussed. They are ones of the "reference" structures (see Figure 36.8) and

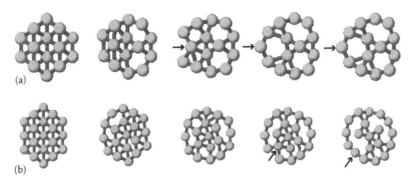

FIGURE 36.14 Relaxation process of the (a) 10-4 and (b) 12-6-1 Au nanowires into helical structures. The panels of (a), from right to left, are the snapshots at every 500 step from the initial one and the panels of (b) are those at every 2500 step from the initial one. An atom marked by arrow indicates the atom row that moves from an inner shell into the outermost shell.

do not have an additional atom row on the outermost shell unlike the nanowires in Figure 36.10. Figure 36.14 shows the relaxation process of the 10-4 nanowire at $T = 600$ K and the 12-6-1 nanowires at $T = 900$ K. These nanowires are transformed into helical ones. In the transformation process, an atom row indicated by arrows in Figure 36.14a or b moves from the inner shell into the outermost one, unlike the cases in Figure 36.10. The final snapshots in Figure 36.14a and b are the helical 11-3 and 13-5-1 (or 13-6) structures, respectively. Figure 36.15 shows the transformation process of the 12-6-1 case from a different viewpoint, so as to clarify the atom row movement.

The two-stage formation model holds also in these nanowires. The present results show that the inserted atom rows are supplied possibly from outer and inner regions. For example, in the 13-6-1 nanowire (see Figure 36.10c), an atom row is supplied from the outer region into the shell with 12 atoms, while, in the 12-6-1 nanowire (see Figure 36.14b), an atom row is supplied from the inner region into the shell with 12 atoms. Both nanowires show surface reconstruction into helical structure with the outermost shell of 13 atoms.

Second, QM-MD simulation for Cu nanowire was carried out for comparison with Au nanowires. The calculation of the Cu 11-4 nanowire was simulated and temperature was set to be 600 and 900 K, which is lower than the melting temperature (1358 K). As result, helical wire appeared at 900 K but did not at 600 K, since the surface atom did not dissociate from the inner shell at

600 K. These results lead us to the conclusion that Cu nanowire is more difficult than Au one to be transformed into helical structure, which is consistent to the fact that no helical Cu nanowire was observed in experiment.

Local electronic structure is analyzed in the 11-4 Cu nanowire at $T = 600$ K. Figure 36.16 shows the LDOS calculations for the atoms A, C at the initial structure and the 500-th iteration step. The definitions and notations are the same as in the Au case (Figure 36.12b). The two vertical broken lines are the values of the Fermi level at the two iterations and the higher value is that at the 500-th iteration step. Figure 36.17 shows LDOS for each d orbital of the atoms A, C at the two iteration steps. The definitions and notations are the same as in Figure 36.13. As common features of the LDOS of Cu nanowire (Figures 36.16 and 36.17) and Au nanowire (Figures 36.12b and 36.13), the energy gain of d orbitals at the atom C is seen and the maximum energy gain is given by the xy orbital among d orbitals.

The theory in the previous sections explains the above results; Cu, Ag, and Au have 10 d electrons ($d^{10}s^1$) but the d bandwidth of Au is wider than that of Cu and Ag (see Figure 36.5). Therefore, the energy gain mechanism for helical transformation is inherent in Cu but its effect is weaker than that in Au. Hence, helical structure appears in Au nanowire but not in Cu nanowire.

The above theory also explains why Pt helical nanowire can be formed, since Pt and Au has the 5d band (see Figure 36.1b), which is wider than the 3d and 4d band in the lighter elements of Figure 36.1b. The energy gain mechanism in the two-stage

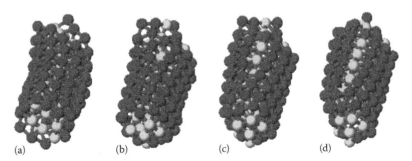

FIGURE 36.15 Relaxation process of 12-6-1 Au nanowire into helical structures: (a) 750-th, (b) 5000-th, (c) 9500-th, and (d) 14500-th iteration steps. The atoms in the outer and inner shells are shown as dark and light balls, respectively.

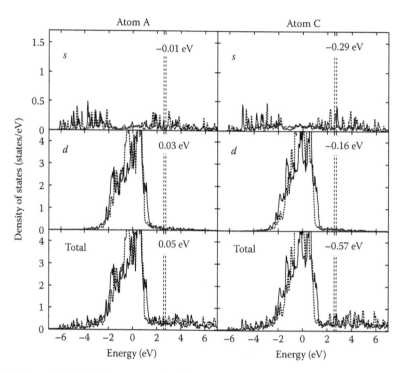

FIGURE 36.16 Local density of states in the 11-4 Cu nanowire for the atoms A and C. The solid and broken lines are at the initial state and at the 500-th iteration step, respectively. The difference of the local energy between the two iteration steps is written at the right corner of each panel. The definitions and notations are the same as in Figure 36.12b.

formation model is governed by the *d* bandwidth and the helical nanowires appear only among metals with a wider *d* band.

Third, the similarity in mechanism is discussed between helical nanowires and equilibrium surfaces. Since the present mechanism is based on the electronic structure of 5*d* band, the mechanism is inherent not only with nanowires but also with bulk surfaces of FCC 5*d* metals, Au, Pt, and iridium (Ir) (see Figure 36.1b). For example, the equilibrium (110) surfaces of these elements reconstruct into a 2 × 1 "missing row" structure (Fedak and Gjostein 1967, Binnig et al. 1983, Ho and Bohnen 1987), in which (111)-type hexagonal surfaces appear as successive nanofacets.

Figure 36.18a and b shows the top and side views of the missing row structure. The cubic cell of the ideal FCC structure, identical to Figure 36.3a, is also shown in Figure 36.18c for tutorial. The (110)-2 × 1 missing row structure appears, when one removes every other row in the [001] direction from the ideal (110) surface. The atom position of missing rows is drawn by the crosses in Figure 36.18b. In the ideal FCC structure of Figure 36.18c, the ideal (110) surface corresponds to the plane with the atoms R, P, S, and U. The atom positions of the "missing row" are those on the atom row that contains the atoms S and U. In Figure 36.18a and b, (111)-type facets appears, such as one that contains the atoms P, V, T, W, and R. One can confirm in Figure 18c that the atoms P, V, T, W, and R are on a (111)-type plane.

Moreover, the equilibrium (001) surfaces of these elements also show reconstruction in which a surface layer with (111)-type hexagonal regions is placed on the (001)-type square layers (Van Hove et al. 1981, Binnig et al. 1984, Abernathy et al. 1992, Jahns et al. 1999).

It should be noted that, an experimental paper of [110] Au nanowire (Kondo and Takayanagi 1997), earlier than the report of helical structure (Kondo and Takayanagi 2000), suggests that (001)-type regions on the nanowire surface reconstruct into (111)-type ones, as on equilibrium surfaces. The suggested reconstruction mechanism is consistent with the present one.

36.5 Summary and Future Aspect

This chapter gives a review of ultrathin Au nanowire and focuses two properties: (1) monoatomic chain with quantized conductance and (2) helical multishell structures with "magic numbers." The first property is general among different metal nanowires. The second property is unique for Au and Pt. The proposed theory with the two-stage formation process and the analysis of electronic structure explains how and why these helical structures appear. The appearance of helical multishell Au nanowire should be understood by the following two points: (1) This is a typical nanoscale effect, in which the numbers of atoms in surface (outermost) region and bulk (inner) region are comparable and the structure is determined by the energy gain mechanism of the surface region. (2) This is a typical quantum mechanical effect, in which the electronic structure, the nonspherical property of 5*d* band, governs the phenomena.

As a future aspect, structural and transport properties should be investigated further for "thicker" nanowires, so as to establish the foundation of nano electronics. The investigation of helical multishell nanowires is a typical one and other

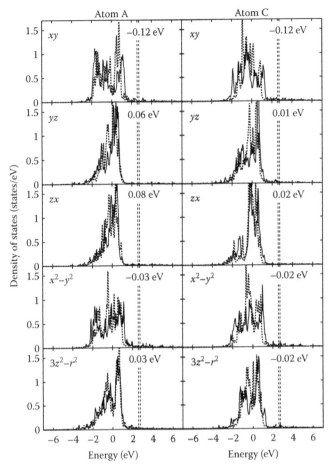

FIGURE 36.17 LDOS for each *d* orbital of the atoms A, C at the two iteration steps in the relaxation process of the 11-4 Cu nanowire. The solid and broken lines are at the initial state and at the 500-th iteration step, respectively. The definitions and notations are the same as in Figure 36.13.

investigations were explained in Section 36.3. The above two points should be of general importance among the investigations of "thicker" nanowires. Moreover, "thicker" nanowires can show complex transport property, since the interference of electron "wave" is influenced by the structures of nanowires and electrodes and the character of valence orbitals. Among such cases, the conductance value is, in general, not quantized and a more fundamental expression of Equation 36.4 should be considered. An example is found in theoretical (Ono and Hirose 2005) and experimental (Oshima et al. 2006) papers of helical multishell Au nanowires. These papers point out that the structure of electrode parts varies conductance value. For another example, transport behavior was investigated for model "thicker" nanowires (Shinaoka et al. 2008). The paper focuses on electrode effect and *d* orbital effect on conductance and local current.

The theoretical investigation of "thicker" nanowires requires quantum-mechanical simulation with a larger number of atoms. For an example, Figure 36.19 shows a simulation of a 15-8-1 nanowire with the wire length of 12 nm (Fujiwara et al. 2008). The number of atoms is 1020. The resultant nanowire contains multiple helical domains on wire surface with well-defined domain boundary, which cannot be obtained in smaller samples. More systematic investigations will be given in near future. When one would like to compare simulations with experiment, the above system size is still insufficient and electrode parts are missing in the simulation. A promising theoretical approach for larger quantum-mechanical simulations is "order-*N*" method, in which the computational time is "order-*N*" or proportional to the system size *N*. See articles cited in Hoshi and Fujiwara (2006) or a recent journal volume that includes Fujiwara et al. (2008) and focuses on the order-*N* methods.

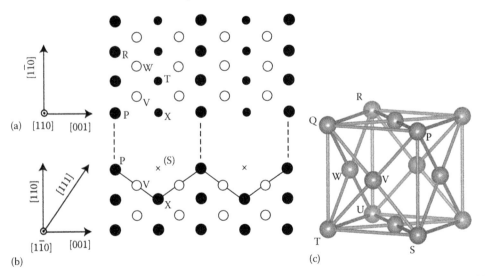

FIGURE 36.18 (a) Top and (b) side views of the "missing row" structure in FCC(110) surface. (c) Ideal FCC structure. In (a), three successive atomic layers are drawn. Atoms are distinguished by larger filled circles, open circles, or smaller filled circles, for the first, second, third surface layers, respectively. In (b), two successive atomic layers are drawn. Atoms are distinguished by filled and open circles. The crosses in (b) indicate the site of the "missing row." The lines in (b) indicate (111)-type surface regions as successive nanofacets. The atoms marked as P, R, R, S, T, U, V, and W appear among (a)–(c). The atom marked X appears only in (a) and (b).

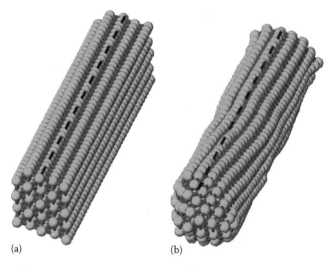

(a)　　　　　　　　　　　　(b)

FIGURE 36.19　Transformation of 15-8-1 helical structure of a Au nanowire with the length of 12 nm. Initial (left) and final (right) snapshots are shown. A dashed line is drawn for an eye guide commonly on the two snapshots.

Appendix 36.A: Note on Quantum-Mechanical Molecular Dynamics Simulation

This note is devoted to a tutorial of QM-MD simulation. See a textbook (Martin 2004) for general introduction of quantum mechanical (electronic structure) calculations. QM-MD simulation method is based on the quantum mechanical description of electrons and the classical description of atomic nuclei. Atomic structure is determined with the electronic wave functions. Since the mass of atomic nucleus is heavier by more than 1000 times than that of the electron, the motion of atomic nucleus is treated adiabatically. In many QM-MD simulations, only valence electrons are treated explicitly and nucleus and core electrons are treated as "ions."

Electronic wave functions are determined with the given position of the atomic nuclei or ions $\{R_I\}$ and the effective Schrödinger equation is obtained as

$$H\phi_i = \varepsilon_i\phi_i, \qquad (36.A.1)$$

where

　　H is an effective Hamiltonian
　　$\phi_i \equiv \phi_i(r)$ and ε_i are electronic wave function and its energy level, respectively

The motion of atomic nuclei or ions is described by their position $\{R_I\}$ and velocity $\{V_I\}$ and the Newton equation is derived

$$M_I\frac{dV_I}{dt} = F_I, \qquad (36.A.2)$$

where the force F_I is given by the derivative of the energy $E[\{R_I\}]$;

$$F_I = -\frac{\partial E}{\partial R_I}. \qquad (36.A.3)$$

The energy $E[\{R_I\}]$ and the force $\{F_I\}$ depend on the solutions $\{\phi_i(r)\}$ of Equation (36.A.1). A well-established method in QM-MD simulation is the first principles molecular dynamics that is realized by plane-wave bases and density functional theory (Car and Parrinello 1985, Payne et al. 1992, Martin 2004).

The procedure of QM-MD simulation within one time step is summarized as

$$\{R_I\} \Rightarrow \{\varepsilon_i, \phi_i(r)\} \Rightarrow E, \{F_I\} \Rightarrow \text{update}\{R_I\}; \qquad (36.A.4)$$

(1) With given positions $\{R_I\}$, the effective Schrödinger equation of Equation (36.A.1) is solved and the energy levels and wave functions for electrons $(\{\varepsilon_i, \phi_i(r)\})$ are obtained. (2) The total energy E and the forces $\{F_I\}$ are obtained. (3) The positions are updated $(R_I(t) \Rightarrow R_I(t + \delta t))$ by a numerical time evolution of Equation (36.A.2) with a tiny time interval δt. The time interval δt is usually on the order of femtosecond.

The QM-MD simulation shown in Section 36.4 was realized by the simulation code "ELSES" (=Extra-Large-Scale Electronic Structure calculations). See the web page (http://www.elses.jp/) or the papers (Hoshi and Fujiwara 2000, 2003, Geshi et al. 2003, Takayama et al. 2004, Hoshi et al. 2005, Hoshi and Fujiwara 2006, Takayama et al. 2006, Iguchi et al. 2007, Fujiwara et al. 2008, Hoshi and Fujiwara 2009). In the simulation, Slater–Koster or tight-binding form Hamiltonian are used. In general, a QM-MD simulation with tight-binding form Hamiltonians enable a much faster simulation than the first principles molecular dynamics, although it has not yet established to construct tight-binding form Hamiltonians among general materials from the first principles.

In the simulations of Au and Cu in Section 36.4, the energy $E[\{R_I\}]$ and the Hamiltonian H are written within the tight-binding form developed in Naval Research Laboratory (Mehl and Papaconstantopoulos 1996, Kirchhoff et al. 2001, Papaconstantopoulos and Mehl 2003, Haftel et al. 2004). The form contains several parameters and they are determined to represent electronic structures of bulk solids, surfaces, stacking faults, and point defects. Au nanowires were calculated by the present Hamiltonian (da Silva et al. 2001, 2004, Haftel and Gall 2006, Iguchi et al. 2007).

Figure 36.20 shows the energy band diagrams of FCC Au calculated by the present tight-binding form Hamiltonian and the first-principles Hamiltonian with linear muffin-tin orbital theory (Andersen and Jepsen 1984). In general, the energy band diagram is a standard method for visualizing electronic state in solids. The diagram describes the relation between energy e and wave vector k of electronic states $(e = e(k))$. Each point in the horizontal axis indicates a wave vector k. Several specific wave vectors are labeled, such as "Γ." See a textbook (Bradley and Cracknell 1972) for their definitions. In Figure 36.20, the energy band diagram by the tight-binding form Hamiltonian reproduces well that by the first-principles Hamiltonian, particularly in the occupied energy region $(E < E_F)$. The electronic structure in the occupied energy region contributes to the cohesive mechanism and determines atomic structures.

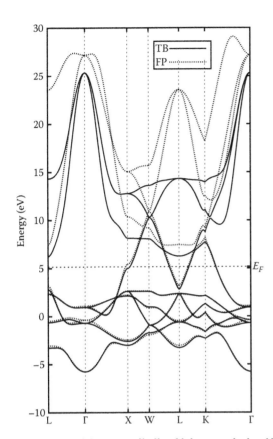

FIGURE 36.20 Band diagrams of bulk gold that are calculated by the tight-binding form Hamiltonian (TB) and the first-principles one (FP). See text for details.

Tight-binding form Hamiltonians for QM-MD simulation were developed also by many other groups, which can be found, for example, in a review paper (Goringe et al. 1997). Some of them are developed for specific elements, such as ones for C (Xu et al. 1992) and Si (Kwon et al. 1994) and some of them, Calzaferri et al. (1989), Nath and Anderson (1990), Calzaferri and Rytz (1996) for example, were developed for more general materials.

The analysis method of electronic wave function is also important for understanding phenomena. A bond between two atoms can be determined by electronic wave function through the theory of crystal orbital Hamiltonian population (COHP) (Dronskowski and Blöchl 1993), a well-defined energy spectrum of bond. In Section 36.4, the dissociation of two atoms was ascertained in two ways: (1) the interatomic distance increases by more than 20% and (2) the peak height of the COHP decreases down to 1/5 of that of the initial state.

References

Abernathy, D. L., S. G. J. Mochrie, D. M. Zehner, G. Grübel, and D. Gibbs. 1992. Orientational epitaxy and lateral structure of the hexagonally reconstructed Pt(001) and Au(001) surfaces. *Phys. Rev. B* **45**: 9272–9291.

Agraït, N., J. G. Rodrigo, and S. Vieira. 1993. Conductance steps and quantization in atomic-size contacts. *Phys. Rev. B* **47**: 12345–12348.

Agraït, N., A. L. Yeyati, and J. M. van Ruitenbeek. 2003. Quantum properties of atomic-sized conductors. *Phys. Rep.* **377**: 81–279.

Amorim, E. P. M. and E. Z. da Silva. 2008. Helical [110] gold nanowires make longer linear atomic chains. *Phys. Rev. Lett.* **101**: 125502 (4pp).

Andersen, O. K. and O. Jepsen. 1984. Explicit, first-principles tight-binding theory. *Phys. Rev. Lett.* **53**: 2571–2574.

Bettini, J., V. Rodrigues, J. C. González, and D. Ugarte. 2005. Real-time atomic resolution study of metal nanowires. *Appl. Phys. A* **81**: 1513–1518.

Bilalbegović, G. 2003. Gold nanotube: Structure and melting. *Vacuum* **71**: 165–169.

Binnig, G., H. Rohrer, Ch. Gerber, and E. Weibel. 1983. (111) facets as the origin of reconstructed Au(110) surfaces. *Surf. Sci.* **131**: L379–L384.

Binnig, G. K., H. Rohrer, Ch. Gerber, and E. Stoll. 1984. Real-space observation of the reconstruction of Au(100). *Surf. Sci.* **144**: 321–335.

Bradley, C. J. and A. P. Cracknell. 1972. *The Mathematical Theory of Symmetry in Solids*, Clarendon Press, Oxford, U.K.

Brandbyge, M., J. Schiøtz, M. R. Sørensen, P. Stoltze, K. W. Jacobsen, J. K. Nørskov, L. Olesen, E. Laegsgaard, I. Stensgaard, and F. Besenbacher. 1995. Quantized conductance in atom-sized wires between two metals. *Phys. Rev. B* **52**: 8499–8514.

Brandbyge, M., N. Kobayashi, and M. Tsukada. 1999. Conduction channels at finite bias in single-atom gold contacts. *Phys. Rev. B* **60**: 17064–17070.

Calzaferri, G. and R. Rytz. 1996. The band structure of diamond. *J. Phys. Chem.* **100**: 11122.

Calzaferri, G., L. Forss, and I. Kamber. 1989. Molecular geometries by the extended Hückel molecular orbital method. *J. Phys. Chem.* **93**: 5366.

Car, R. and M. Parrinello. 1985. Unified approach for molecular dynamics and density-functional theory. *Phys. Rev. Lett.* **55**: 2471–2474.

Costa-Krämer, J. L., N. García, and H. Olin. 1997. Conductance quantization histograms of gold nanowires at 4 K. *Phys. Rev. B* **55**: 12910–12913.

Cuevas, J. C., A. Levy Yeyati, and A. Martin-Rodero. 1998. Microscopic origin of conducting channels in metallic atomic-size contacts. *Phys. Rev. Lett.* **80**: 1066–1069.

da Silva, E. Z., A. J. R. da Silva, and A. Fazzio. 2001. How do gold wires break? *Phys. Rev. Lett.* **87**: 256102 (4pp).

da Silva, E. Z., F. D. Novaes, A. J. R. da Silva, and A. Fazzio. 2004. Theoretical study of the formation, evolution, and breaking of gold nanowires. *Phys. Rev. B* **69**: 115411 (11pp).

Datta, S. 1995. *Electronics Transport in Mesoscopic Systems*, Cambridge University Press, Cambridge, U.K.

Dresselhaus, M. S., G. Dresselhaus, and P. Avouris, Eds. 2001. *Carbon Nanotubes—Synthesis, Structure, Properties, and Applications*, Springer, Berlin, Germany.

Dronskowski, R. and P. E. Blöchl. 1993. Crystal orbital Hamilton populations (COHP). energy-resolved visualization of chemical bonding in solids based on density-functional calculations. *J. Phys. Chem.* **97**: 8617–8624.

Erts, D., H. Olin, L. Ryen, E. Olsson, and A. Thölén. 2000. Maxwell and Sharvin conductance in gold point contacts investigated using TEM-STM. *Phys. Rev. B* **61**: 12725–12727.

Erts, D., A. Lõhmus, R. Lõhmus, H. Olin, A. V. Pokropivny, L. Ryen, and K. Svensson. 2002. Force interactions and adhesion of gold contacts using a combined atomic force microscope and transmission electron microscope. *Appl. Surf. Sci.* **188**: 460–466.

Fedak, D. G. and N. A. Gjostein. 1967. A low energy electron diffraction study of the (100), (110) and (111) surfaces of gold. *Acta Metal.* **15**: 827–840.

Fujimoto, Y. and K. Hirose. 2003. First-principles treatments of electron transport properties for nanoscale junctions. *Phys. Rev. B* **67**: 195315 (12pp).

Fujiwara, T., T. Hoshi, and S. Yamamoto. 2008. Theory of large-scale matrix computation and applications to electronic structure calculation. *J. Phys.: Condens. Matter* **20**: 294202 (7pp).

Geshi, M., T. Hoshi, and T. Fujiwara. 2003. Million-atom molecular dynamics simulation by order-N electronic structure theory and parallel computation. *J. Phys. Soc. Jpn.* **72**: 2880–2885.

Goringe, C. M., D. R. Bowler, and E. Hernändez. 1997. Tight-binding modelling for materials. *Rep. Prog. Phys.* **60**: 1447–1512.

Gülseren, O., F. Ercolessi, and E. Tosatti. 1998. Noncrystalline structures of ultrathin unsupported nanowires. *Phys. Rev. Lett.* **80**: 3775–3778.

Haftel, M. I. and K. Gall. 2006. Density functional theory investigation of surface-stress-induced phase transformations in fcc metal nanowires. *Phys. Rev. B* **74**: 035420 (12pp).

Haftel, M. I., N. Bernstein, M. J. Mehl, and D. A. Papaconstantopoulos. 2004. Interlayer surface relaxations and energies of fcc metal surfaces by a tight-binding method. *Phys. Rev. B* **70**: 125419 (15pp).

Hansen, K., E. Laegsgaard, I. Stensgaard, and F. Besenbacher. 1997. Quantized conductance in relays. *Phys. Rev. B* **56**: 2208–2220.

Ho, K.-M. and K. P. Bohnen. 1987. Stability of the missing-row reconstruction on fcc (110) transition metal surfaces. *Phys. Rev. Lett.* **59**: 1833–1836.

Hoshi, T. and T. Fujiwara. 2000. Theory of composite-band Wannier states and order-N electronic-structure calculations. *J. Phys. Soc. Jpn.* **69**: 3773–3776.

Hoshi, T. and T. Fujiwara. 2003. Dynamical brittle fractures of nanocrystalline silicon using large-scale electronic structure calculations. *J. Phys. Soc. Jpn.* **72**: 2429–2432.

Hoshi, T. and T. Fujiwara. 2006. Large-scale electronic structure theory for simulating nanostructure processes. *J. Phys.: Condens. Matter* **18**: 10787–10802.

Hoshi, T. and T. Fujiwara. 2009. Development of the simulation package 'ELSES' for extra-large-scale electronic structure calculation. *J. Phys.: Condens. Matter* **21**: 064233 (4pp).

Hoshi, T., Y. Iguchi, and T. Fujiwara. 2005. Nanoscale structures formed in silicon cleavage studied with large-scale electronic structure calculations: Surface reconstruction, steps, and bending. *Phys. Rev. B* **72**: 075323 (10pp).

Iguchi, Y. 2007. Doctoral thesis (in Japanese), University of Tokyo, Tokyo, Japan.

Iguchi, Y., T. Hoshi, and T. Fujiwara. 2007. Two-stage formation model and helicity of gold nanowires. *Phys. Rev. Lett.* **99**: 125507 (4pp).

Iijima, S. 1991. Helical microtubules of graphitic carbon. *Nature* **354**: 56–58.

Jahns, V., D. M. Zehner, G. M. Watson, and D. Gibbs. 1999. Structure and phase behavior of the Ir(001) surface: X-ray scattering measurements. *Surf. Sci.* **430**: 55–66.

Kirchhoff, F., M. J. Mehl, N. I. Papanicolaou, D. A. Papaconstantopoulos, and F. S. Khan. 2001. Dynamical properties of Au from tight-binding molecular-dynamics simulations. *Phys. Rev. B* **63**: 195101 (7pp).

Kizuka, T. 1998. Atomic process of point contact in gold studied by time-resolved high-resolution transmission electron microscopy. *Phys. Rev. Lett.* **81**: 4448–4451.

Kizuka, T. 2007. Atomistic process of twin-boundary migration induced by shear deformation in gold. *Jpn. J. Appl. Phys.* **46**: 7396–7398.

Kizuka, T. 2008. Atomic configuration and mechanical and electrical properties of stable gold wires of single-atom width. *Phys. Rev. B* **77**: 155401 (11pp).

Kizuka, T., K. Yamada, S. Deguchi, M. Naruse, and N. Tanaka. 1997. Cross-sectional time-resolved high-resolution transmission electron microscopy of atomic-scale contact and noncontact-type scannings on gold surfaces. *Phys. Rev. B* **55**: R7398–R7401

Kizuka, T., S. Umehara, and S. Fujisawa. 2001a. Metal-insulator transition in stable one-dimensional arrangements of single gold atoms. *Jpn. J. Appl. Phys.* **40**: L71–L74.

Kizuka, T., H. Ohmi, T. Sumi, K. Kumazawa, S. Deguchi, M. Naruse, S. Fujisawa, S. Sasaki, A. Yabe, and Y. Enomoto. 2001b. Simultaneous observation of millisecond dynamics in atomistic structure, force and conductance on the basis of transmission electron microscopy. *Jpn. J. Appl. Phys.* **40**: L170–L173.

Kondo, Y. and K. Takayanagi. 1997. Gold nanobridge stabilized by surface structure. *Phys. Rev. Lett.* **79**: 3455–3458.

Kondo, Y. and K. Takayanagi. 2000. Synthesis and characterization of helical multi-shell gold nanowires. *Science* **289**: 606–608.

Krans, J. M., C. J. Muller, I. K. Yanson, Th. C. M. Govaert, R. Hesper, and J. M. van Ruitenbeek. 1993. One-atom point contacts. *Phys. Rev. B* **48**: 14721–14724.

Krans, J. M., J. M. van Ruitenbeek, V. V. Flsun, I. K. Yanson, and L. J. de Jongh. 1995. The signature of conductance quantization in metallic point contacts. *Nature* **375**: 767–769.

Kurui, Y., Y. Oshima, M. Okamoto, and K. Takayanagi. 2007. One-by-one evolution of conductance channel in gold [110] nanowires. *J. Phys. Soc. Jpn.* **76**: 123601 (4pp).

Kurui, Y., Y. Oshima, M. Okamoto, and K. Takayanagi. 2008. Integer conductance quantization of gold atomic sheets. *Phys. Rev. B* **77**: 161403(R) (4pp).

Kwon, I., R. Biswas, C. Z. Wang, K. M. Ho, and C. M. Soukoulis. 1994. Transferable tight-binding models for silicon. *Phys. Rev. B* **49**: 7242–7250.

Lin, J.-S., S.-P. Ju, and W.-J. Lee. 2005. Mechanical behavior of gold nanowires with a multishell helical structure. *Phys. Rev. B* **72**: 085448 (6pp).

Mares, A. I. and J. M. van Ruitenbeek. 2005. Observation of shell effects in nanowires for the noble metals Cu, Ag, and Au. *Phys. Rev. B* **72**: 205402 (7pp).

Mares, A. I., A. F. Otte, L. G. Soukiassian, R. H. M. Smit, and J. M. van Ruitenbeek. 2004. Observation of electronic and atomic shell effects in gold nanowires. *Phys. Rev. B* **70**: 073401 (4pp).

Martin, R. M. 2004. *Electronic Structure: Basic Theory and Practical Methods.* Cambridge University Press, Cambridge, U.K.

Mehl, M. J. and D. A. Papaconstantopoulos. 1996. Applications of a tight-binding total-energy method for transition and noble metals: Elastic constants, vacancies, and surfaces of monatomic metals. *Phys. Rev. B* **54**: 4519–4530.

Muller, C. J., J. M. Krans, T. N. Todorov, and M. A. Reed. 1996. Quantization effects in the conductance of metallic contacts at room temperature. *Phys. Rev. B* **53**: 1022–1025.

Nath, K. and A. B. Anderson. 1990. Atom-superposition and electron-delocalization tight-binding band theory. *Phys. Rev. B* **41**: 5652–5660.

Ohnishi, H., Y. Kondo, and K. Takayanagi. 1998. Quantized conductance through individual rows of suspended gold atoms. *Nature* **395**: 780–785.

Okamoto, M. and K. Takayanagi. 1999. Structure and conductance of a gold atomic chain. *Phys. Rev. B* **60**: 7808–7811.

Olesen, L., E. Lgsgaard, I. Stensgaard, and F. Besenbacher. 1994. Quantised conductance in an atom-sized point contact. *Phys. Rev. Lett.* **72**: 2251–2254.

Ono, T. and K. Hirose. 2005. First-principles study of electron-conduction properties of helical gold nanowires. *Phys. Rev. Lett.* **94**: 206806 (4pp).

Oshima, H. and K. Miyano. 1998. Spin-dependent conductance quantization in nickel point contacts. *Appl. Phys. Lett.* **73**: 2203–2205.

Oshima, Y., H. Koizumi, K. Mouri, H. Hirayama, K. Takayanagi, and Y. Kondo. 2002. Evidence of a single-wall platinum nanotube. *Phys. Rev. B* **65**: 121401(R).

Oshima, Y., K. Mouri, H. Hirayama, and K. Takayanagi. 2003a. Development of a miniature STM holder for study of electronic conductance of metal nanowires in UHV-TEM. *Surf. Sci.* **531**: 209–216.

Oshima, Y., A. Onga, and K. Takayanagi. 2003b. Helical gold nanotube synthesized at 150 K. *Phys. Rev. Lett.* **91**: 205503.

Oshima, Y., Y. Kondo, and K. Takayanagi. 2003c. High-resolution ultrahigh-vacuum electron microscopy of helical gold nanowires: Junction and thinning process. *J. Electron Microsc.* **52**: 49–55.

Oshima, Y., K. Mouri, H. Hirayama, and K. Takayanagi. 2006. Quantized electrical conductance of gold helical multishell nanowires. *J. Phys. Soc. Jpn.* **75**: 053705 (4pp).

Papaconstantopoulos, D. A. and M. J. Mehl. 2003. The Slater-Koster tight-binding method: A computationally efficient and accurate approach. *J. Phys.: Condens. Matter.* **15**: R413–R440.

Pascual, J. I., J. Mendez, J. Gómez-Herrero, A. M. Baró, and N. García. 1993. Quantum contact in gold nanostructures by scanning tunneling microscopy. *Phys. Rev. Lett.* **71**: 1852–1855.

Payne, M. C., M. P. Teter, D. C. Allan, T. A. Arias, and J. D. Joannopoulos. 1992. Iterative minimization techniques for ab initio total-energy calculations: Molecular dynamics and conjugate gradients. *Rev. Mod. Phys.* **64**: 1045–1097.

Rubio, G., N. Agraït, and S. Vieira. 1996. Atomic-sized metallic contacts: Mechanical properties and electronic transport. *Phys. Rev. Lett.* **76**: 2302–2305.

Rubio-Bollinger, G., P. Joyez, and N. Agraït. 2004. Metallic Ad-hesion in atomic-size junctions. *Phys. Rev. Lett.* **93**: 116803 (4pp).

Sánchez-Portal, D., E. Artacho, J. Junquera, P. Ordejón, A. García, and J. M. Soler. 1999. Stiff monoatomic gold wires with aspinning zigzag geometry. *Phys. Rev. Lett.* **83**: 3884–3887.

Schiff, L. I. 1968. *Quantum Mechanics*, 3rd ed., McGraw-Hill, New York.

Senger, R. T., S. Dag, and S. Ciraci. 2004. Chiral single-wall gold nanotubes. *Phys. Rev. Lett.* **93**: 196807 (4pp).

Shinaoka, H., T. Hoshi, and T. Fujiwara. 2008. Ill-contact effects of d-orbital channels in nanometer-scale conductor. *J. Phys. Soc. Jpn.* **77**: 114712 (7pp).

Shiota, T., A. I. Mares, A. M. C. Valkering, T. H. Oosterkamp, and J. M. van Ruitenbeek. 2008. Mechanical properties of Pt monatomic chains. *Phys. Rev. B* **77**: 125411 (5pp).

Smit, R. H. M., C. Untiedt, G. Rubio-Bollinger, R. C. Segers, and J. M. van Ruitenbeek. 2003. Observation of a parity oscillation in the conductance of atomic wires. *Phys. Rev. Lett.* **91**: 076805 (3pp).

Sorensen, M. R., M. Brandbyge, and K. W. Jacobsen. 1998. Mechanical deformation of atomic-scale metallic contacts: Structure and mechanisms. *Phys. Rev. B* **57**: 3283–3294.

Takayama, R., T. Hoshi, and T. Fujiwara. 2004. Krylov subspace method for molecular dynamics simulation based on large-scale electronic structure theory. *J. Phys. Soc. Jpn.* **73**: 1519–1524.

Takayama, R., T. Hoshi, T. Sogabe, S.-L. Zhang, and T. Fujiwara. 2006. Linear algebraic calculation of the Green's function for large-scale electronic structure theory. *Phys. Rev. B* **73**: 165108 (9pp).

Todorov, T. N. and A. P. Sutton. 1993. Jumps in electronic conductance due to mechanical instabilities. *Phys. Rev. Lett.* **70**: 2138–2141.

Tosatti, E., S. Prestipino, S. Kostlmeier, A. Dal Corso, and F. D. Di Tolla. 2001. String tension and stability of magic tip-suspended nanowires. *Science* **291**: 288.

Urban, D. F., J. Burki, C.-H. Zhang, C. A. Stafford, and H. Grabert. 2004. Jahn-Teller distortions and the supershell effect in metal nanowires. *Phys. Rev. Lett.* **93**: 186403 (4pp).

Valkering, A. M. C., A. I. Mares, C. Untiedt, K. Babaei Gavan, T. H. Oosterkamp, and J. M. van Ruitenbeek. 2005. A force sensor for atomic point contacts. *Rev. Sci. Instrum.* **76**: 103903 (5pp).

Van Hove, M. A., R. J. Koestner, P. C. Stair, J. P. Biberian, L. L. Kesmodel, I. Bartos, and G. A. Somorjai. 1981. The surface reconstructions of the (100) crystal faces of iridium, platinum and gold—I. Experimental observations and possible structural models. *Surf. Sci.* **103**: 189–217.

Xu, C. H., C. Z. Wang, C. T. Chan, and K. M. Ho. 1992. A transferable tight-binding potential for carbon. *J. Phys.: Condens. Matter* **4**: 6047–6054.

Yang, C.-K. 2004. Theoretical study of the single-walled gold (5,3) nanotube. *Appl. Phys. Lett.* **85**: 2923–2925.

Yanson, A. I., I. K. Yanson, and J. M. van Ruitenbeek. 1999. Observation of shell structure in sodium nanowires. *Nature* **400**: 144–146.

Yanson, A. I., I. K. Yanson, and J. M. van Ruitenbeek. 2000. Supershell structure in alkali metal nanowires. *Phys. Rev. Lett.* **84**: 5832–5835.

Yanson, A. I., I. K. Yanson, and J. M. van Ruitenbeek. 2001. Crossover from electronic to atomic shell structure in alkali metal nanowires. *Phys. Rev. B* **70**: 073401 (4pp).

Yanson, I. K., O. I. Shklyarevskii, J. M. van Ruitenbeek, and S. Speller. 2008. Aluminum nanowires: Influence of work hardening on conductance histograms. *Phys. Rev. B* **77**: 03411 (4pp).

Electronic Transport through Atomic-Size Point Contacts

Elke Scheer
University of Konstanz

37.1 Introduction

In this chapter, we will describe the electronic transport properties of a particular type of mesoscopic structure, namely, point contacts with lateral dimensions of a single or a few atoms. These structures, first realized in the early 1990s, are still attracting much interest because they represent test beds for important concepts of quantum mechanics, quantum chemistry, atomic physics, and solid-state physics. While the first years of research were devoted to the understanding of the intrinsic transport properties of atomic-size contacts (APCs) of normal metals, new fields of interest have now been entered, for example, the application of APCs to provide highly-spin polarized resistors or atomically sharp electrodes for contacting individual nano-objects in the context of molecular electronics.

Although manifold transport properties like noise, thermopower, and thermal conductance were studied and they revealed rich behavior, we will concentrate here on the understanding of the most basic transport property, namely, the linear electrical conductance, i.e., the conductance measured with a small voltage bias and the current-voltage characteristic as linear. A more comprehensive review, including the mechanical properties and the more complex transport quantities such as noise or thermopower has been given by Agraït, Levy Yeyati, and van Ruitenbeek (Agraït et al. 2003).

Furthermore, we exclude the discussion of the so-called atomic nanowires because they are the point of focus in several other chapters of this handbook. Atomic nanowires are constrictions with atomic size lateral dimensions and have been produced from manifold materials. Many properties of nanowires share the same concepts as those used to describe the physics of APCs. The term nanowire usually implies that the constriction has a length of nanometer scale and gives rise to a well pronounced one- or two-dimensional transport behavior. The term APC suggests that the constriction is short in the sense that any voltage applied at both ends drops locally at the position of the APC. Nevertheless, there is no sharp criterion distinguishing APCs from atomic nanowires. Finally, the term "quantum point contact" bears the risk of being mistaken with "quantum points," which is equivalent to "quantum dots." The latter terms denote structures that have finite size in all three dimensions. For this concept, manifold realizations exist as, among others, semiconductor quantum dots, clusters, or fullerenes. For example, a quantum point can be formed by two QPCs in a series. The physics of quantum points are addressed in several chapters of this handbook as well.

This chapter is organized as follows: In Section 37.2, we summarize the concept of conductance as a quantum mechanical wave scattering problem, we discuss its application to quantum point contacts (QPCs), and we point out the main differences between APCs and QPCs made from two-dimensional electron gases (2DEGs), which are described in detail in Chapter 38 (Zozoulenko and Ihnatsenka 2010). In Section 37.3, we first briefly describe the main fabrication schemes of APCs and then in Section 37.4 we comment on their electrical conductance. Section 37.5 gives a summary and an outlook of the field.

37.2 The Landauer Approach to Conductance

The theory of electron transport pioneered by Rolf Landauer (1957, 1970) treats an electronic transport experiment on a mesoscopic sample as an ordinary wave scattering experiment. Electrons are sent from a reservoir into the sample, which scatters them (Figure 37.1). A fraction of the electrons is reflected back and the rest is collected by another reservoir. Each reservoir plays the role of a source and a detector (for a review, see Zozoulenko and Ihnatsenka 2010, Beenakker and van Houten 1991, or Beenakker 1997). To establish a well defined set of electronic quantum states with which to describe the scattering experiment, the theory introduces "leads" connecting the mesoscopic sample to the reservoirs. In the leads, a finite number N of independent electronic ingoing and outgoing modes propagate. These modes constitute a finite and complete basis of states for the scattering problem. The scatterer connects in principle any incoming mode with all outgoing modes. In theory, this is represented by a transmission matrix t and a reflection matrix r. For example, the matrix element t_{nm} gives the probability amplitude for the left incoming mode m to be transmitted into the right outgoing mode n. These matrices carry all the information on the scatterer and, thus, determine all the transport properties of the mesoscopic sample under investigation. For example, the total conductance of a mesoscopic structure is given by the Landauer formula:

$$G = G_0 T$$

where

$T = \text{Tr}\{t^\dagger t\}$ is the total transmission of the structure
$G_0 = 2e^2/h$ is the conductance quantum

The factor 2 arises from the two possible spin orientations, assuming spin degeneracy. The quantity e^2/h also occurs in the context of the quantum Hall effect and describes the contribution of one (spin-split) electronic state to the Hall conductance. However, as the trace of a matrix remains invariant through any unitary basis transformation, there is no preferred basis to calculate the conductance. In other words, it is possible to chose a suitable basis in which the matrix t_{nm} has a simple form in order to calculate the eigenvalues τ_i of the matrix $t^\dagger t$ from which one can then deduce the total transmission

$$T = \sum (t^\dagger t)_{nn} = \sum_{i=1}^{N} \tau_i$$

with the transmission coefficients τ_i of the N individual transport channels. Conversely, a measurement of the conductance does not provide enough information to determine the ensemble $\{\tau_i\}$, unless the scatterer has only one mode. In fact, not even the total number N of transmission eigenmodes—nicknamed "channels"—of the scatterer can be extracted from the knowledge of the conductance. Methods for the determination of the channel ensemble will be described in Section 37.4.3.

Before going into this, we will briefly comment on the most important differences between the quantum point contacts of two-dimensional electron gases (2DEGQPC) and the atomic point contacts (APCs) (Figure 37.2). The transport through 2DGQPCs is easily interpreted in terms of the scattering theory of transport and is described in detail in Chapter 38 (Zozoulenko and Ihnatsenka 2009). A QPC, regardless of its realization, is a constriction between two metallic reservoirs of lateral size comparable to the Fermi wavelength λ_F of the electrons. A 2DEGQPC is fabricated on two-dimensional electron gases in epitaxially grown heterostructures of semiconductors. These gases are low electronic density (10^{12} cm^{-2}) metals in which the Fermi wavelength of the electrons can be as large as 30 nm. The constriction is defined and its width D is adjusted by means of nanofabricated electrostatic gates that deplete locally the electron layer. Experimentally, as the width is increased continuously, the conductance increases in steps of nearly G_0: this is the phenomenon of conductance quantization (van Wees et al. 1988, Wharam et al. 1988). The experiments are performed on very good realizations of ideal ballistic leads and QPCs. Moreover, the connection between the leads and the sample is almost perfectly adiabatic: the eigenmodes of both the leads and the QPC are of the same nature and they are perfectly matched. For these structures, the choice of a basis of laterally confined plane waves that are funneled through the constriction is quite natural. In this language, the conductance staircase has a very simple interpretation: if the width of the constriction is much smaller than λ_F, the conductance of the QPC is zero, i.e., all $\tau_i = 0$. Then, as the width is increased continuously, each time D reaches an integer multiple of $\lambda_F/2$, a new mode starts being transmitted ($\tau_i = 1$). Although, as explained before, a conductance measurement does not provide information on the individual eigenvalues,

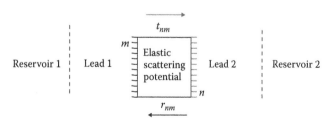

FIGURE 37.1 Illustration of the Landauer approach to electrical conductance.

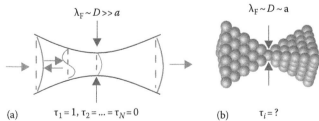

FIGURE 37.2 Comparison between a 2DEGQPC (a) and an APC (b). The dark gray lines in the left visualize the confining potential induced by the action of the gate.

the fact that in the experiments at least 15 conductance plateaus are observed in a reproducible and reversible fashion undoubtedly supports the view of channels opening one by one and reaching perfect transmission.

Moreover, measurements of the shot-noise (de Jong and Beenakker 1997, Blanter and Büttiker 2000) of the current going through a 2DEGQPC (Reznikov et al. 1995, Kumar et al. 1996) have shown that the noise power at the conductance plateaus is drastically reduced with respect to its full value, thus confirming the fact that the channel transmissions are either 0 or 1. The main reason for achieving perfect transmission is the long Fermi wavelength, which makes the motion of conduction electrons quite insensitive to atomic defects.

In contrast APCs, which nowadays can be fabricated using several techniques, provide contacts between two reservoirs of conventional three-dimensional metals with high electron density, like Au, Na, etc. In this case, λ_F is comparable to the interatomic distance and the conduction electrons are sensitive to all defects at this scale. The exact geometry is usually not known in the experiments, but obviously there is a region around an atomic contact in which the electrons experience boundary conditions that are rough at the atomic scale. There is a "central cluster" where many atoms are in fact surface atoms that cannot establish all their possible metallic bonds. This fact limits the number of the modes that can be transmitted through the constriction. In other words, the modes of the cluster do not connect adiabatically to the modes of the reservoirs (Brandbyge et al. 1997). An adequate language for the transport eigenmodes of the cluster is the one of linear combinations of the valence atomic orbitals. The set $\{\tau_i\}$ of the transmission eigenvalues can be constructed as superpositions from these orbitals as basis functions. It is thus determined both by the chemical properties of the atoms forming the contact and by their geometrical arrangement (Cuevas et al. 1998a,b, Scheer et al. 1998, Agraït et al. 2003).

37.3 Fabrication of Atomic-Size Contacts

37.3.1 Scanning Tunneling Microscope Technique

As introduced in the chapters by N. Agraït (2010) and J. Kröger (2010), the scanning tunneling microscope (STM) (for a review, see Wiesendanger 1994) is a suitable tool for the fabrication of atomic-size contacts and atomic chains and has been used for that purpose from the very beginning of its invention (Gimzewski and Möller 1987). While in the standard application of an STM a fine metallic tip is held at distance from a counter electrode (in general a metallic surface) by making use of the exponential distance dependence of the tunneling current, the tip can also be indented into the surface and carefully withdrawn until an atomic-size contact or short atomic wire forms (see Figure 37.3). For many metals it has been shown that the tip will be covered by several atomic layers of the metal of the counter electrode upon repeated indentation such that clean contacts may be formed

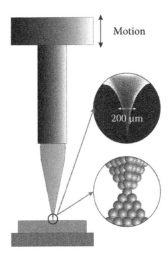

FIGURE 37.3 Working principle of the fabrication of APCs with an STM. The electron micrograph shows a W STM tip the width at half length is in the order of 100–200 μm. The lower inset gives an artist's view of the atomic arrangement of an APC. (Courtesy of C. Bacca.)

consisting of the same metal for both electrodes. The main advantages of the STM in this application are its speed and versatility to also form hetero-contacts, i.e., contacts between two different metals. When the electrodes forming the contacts are prepared in ultra high vacuum conditions, the STM allows us to gather information about the topography of the two electrodes on a somewhat larger scale than the atomic scale before or after the formation of the contact. Spectroscopic measurements on the scale of electron volts allow one to deduce information about the cleanliness and the electronic structure of the metal. The main drawbacks are its limited stability with respect to the change of external parameters such as the temperature or magnetic fields and the short lifetime of the contacts in general because of the sensibility of STM to vibrations.

37.3.2 TEM and Dangling Wires

Another interesting method for preparing and imaging atomic contacts are transient structures forming in a transmission electron microscope (TEM) when irradiating thin metal films on dewetting substrates (Kondo and Takayanagi 1997, Rodrigues and Ugarte 2001). The high energy impact caused by the intensive electron beam locally melts the metal film causing the formation of constrictions that eventually shrink down to the atomic size and finally pinch off building a vacuum tunnel gap. A typical system for these studies is Au on glassy carbon substrates. Several variations of this principle have been developed that allow one to contact both electrodes forming the contact. Because of the locally high temperatures, the typical lifetime of these contacts is very short, but they offer the unique possibility to simultaneously perform conductance measurements and imaging with atomic precision. Similar results have been obtained with variations of the STM inside a TEM (Kizuka 1998). This method enabled us to directly prove the existence of single-atom contacts and single-atom wide nanowires as well

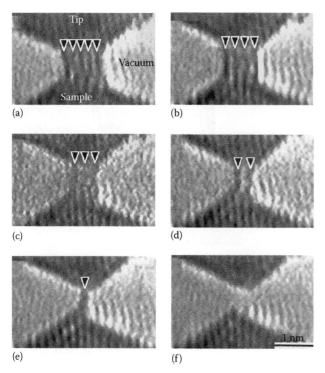

FIGURE 37.4 High resolution TEM images of short atomic wires fabricated with an STM inside the vacuum chamber of the TEM. The arrows indicate the number of atomic rows. In panel (f) the contact is broken to a tunnel contact. (Reprinted from Ohnishi, H. et al., *Nature*, 395, 780, 1998. With permission.)

as to establish a correlation between contact size and conductance. For Au and Ag contacts, it has been shown that preferably well ordered contacts in high symmetry growing directions are formed. The method is particularly fruitful for studying atomic nanowires (see Figure 37.4).

Transient nanowires and APCs with lifetimes in the millisecond range can also be fabricated in a table-top experiment first demonstrated by N. Garcia and coworkers (Costa-Krämer et al. 1995), which we call "dangling-wire contacts." Two metal wires in loose contact to each other are excited to mechanical vibrations so that the contact opens and closes repeatedly. One end of each wire is connected to the poles of a voltage source and the current is recorded with a fast oscilloscope. This method is in principle particularly versatile because it enables the formation of heterocontacts between various metals. However, in order to provide clean metallic contacts, a thorough cleaning of the wires would be required, similar to the tip and surface preparation in a STM. Another drawback is the lack of control of the distance of the electrodes. It is thus mostly used as a demonstration experiment in schools with Au-Au contacts.

37.3.3 Mechanically Controllable Break-Junctions

Already before the development of the first STM another technique enabling the fabrication of atomic-size contacts and tunable tunnel-contacts has been put forward. The first realizations

include the needle-anvil or wedge-wedge point contact technique pioneered by Yanson and coworkers (for a review, see Naidyuk and Yanson 2005) and the squeezable tunnel junction method described by Moreland and Hansma (1984) and Moreland and Ekin (1985) who used metal electrodes on two different substrates that have then carefully adjusted with respect to each other. The needle-anvil technique was mainly used to form contacts with diameters of typically several nanometers and thus having hundreds to thousands of atoms in the narrowest cross-section. These two techniques formed the starting point for the development of the mechanically controllable break-junctions (MCBJ) by Muller and coworkers (Muller et al. 1992), which nowadays is applied for the fabrication of APCs in various subforms, the most common of which are the so-called notched-wire (Krans et al. 1993) and thin-film MCBJs (van Ruitenbeek et al. 1996). The working principle is the same for both variations (Figure 37.5): A suspended metallic bridge is fixed on a flexible substrate, which itself is mounted in a three-point bending mechanism consisting of a pushing rod and two counter-supports. The position of the pushing rod relative to the counter supports is controlled by a motor or piezo drive or combinations of both. The electrodes on top of the substrate are elongated by increasing the bending of the substrate. The elongation can be reduced again by pulling back the pushing rod and thus reducing the curvature of the substrate. In order to break a junction to the tunneling regime, considerable displacements of the pushing rod and thus important bending of the substrate is required. Therefore, the most common substrates are metals with a relatively high elastic limit like spring steel or bronze. The substrates are covered by an electrically isolating material such as polyimide before the junction can be fixed on it.

The notched-wire MCBJ (Figure 37.6) uses a thin metallic wire (diameter 50–200 μm) which has a short, knife-cut constriction to a diameter of 20–50 μm. The wire is glued at both sides of the

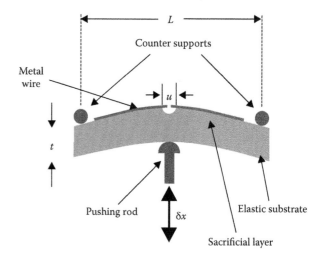

FIGURE 37.5 Working principle of the mechanically controllable break-junction (not to scale) with the metal wire, the elastic substrate, the insulating sacrificial layer, the pushing rod, the counter supports and the dimensions used for calculating the reduction ratio (see text).

FIGURE 37.6 Optical micrograph of a notched-wire MCBJ made of a 100 nm thin gold wire (top view). The wire is glued with epoxy resin (black) onto the substrate. The electrical contact is made by thin copper wires glued with silver paint. The inset shows a zoom into the notch region between the two black drops of epoxy resin. (Reprinted from Agraït, N. et al., *Phys. Rep.*, 377, 81, 2003. With permission.)

notch to the substrate and both ends. The distance between the glue drops is of the order of 50–200 μm.

The thin-film MCBJs are fabricated by the standard techniques of nanofabrication, i.e., electron beam lithography, metal deposition, and partial removal of a sacrificial layer underneath the metal film for suspending a bridge with the typical dimensions of 2 μm in length and 100×100 nm^2 at the narrowest part of the constriction (Figure 37.7).

Both versions of the techniques share the idea of enhanced stability due to the formation of the contact by breaking the very same piece of metal on a single substrate and by the transformation of the motion of the actuator into a much reduced motion of the electrodes perpendicular to it. The small dimensions of the freestanding bridge-arms give rise to high mechanical eigenfrequencies, much higher than the ground frequencies of the setup and thus reduced sensibility to perturbations by

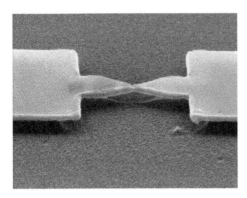

FIGURE 37.7 Electron micrograph of a thin-film MCBJ made of cobalt on polyimide taken under an inclination angle of 60° with respect to the normal. The distance between the rectangular shaped electrodes is 2 μm, the thickness of the thin film is 100 nm and the width of the constriction at its narrowest part is approximately 100 nm.

vibrations. Assuming homogeneous beam-bending of the substrates results in the so-called reduction ratio r between the length change of the bridge δu and the motion of the pushing rod δx (see Figure 37.5)

$$r = \frac{\delta u}{\delta x} = \frac{6tu}{L^2}$$

where
> t is the thickness of the substrate
> u is the length of the free-standing bridge arms
> L is the distance of the counter supports

This quantity thus denotes the factor with which any motion that acts on the pushing rod is reduced when it is transferred to the point contact. It has a typical value of 10^{-3} to 10^{-2} for the notched-wire MCBJs and 10^{-6} to 10^{-4} for the thin film MCBJs. The relatively weak reduction ratio of the notched-wire MCBJs usually requires the use of a piezo drive for controlling and stabilizing single-atom contacts, while the thin-film MCBJs can be controlled with purely mechanical drives, i.e., a dc motor with a combination of gear boxes. The typical motion speeds of the piezo drive lie between 10 nm/s and 10 μm/s, which results in 10 pm/s to 100 nm/s for the electrodes forming the atomic contacts. For thin-film MCBJs, these values are 10 nm/s to 1 μm/s for the pushing rod and 10 fm/s to 10 nm/s for the contact. Also, due to the in-built reduction, the piezo-driven setups are slower than STM systems.

On the other hand, the small r values require considerable absolute motion of the pushing rod and thus deformations of the substrate in order to achieve displacements of the electrodes. The high stability allows comprehensive studies on the very same atomic contact at various values of control parameters such as fields and temperature. MCBJ mechanisms have been developed for various environments including ambient conditions, vacuum, very low temperatures (Scheer et al. 1997), or liquid solutions (Grüter et al. 2005). The disadvantages as compared to STM techniques are the low speed, the fact that the surrounding area of the contact cannot easily be scanned, and clean contacts can only be guaranteed when working in good vacuum conditions. The sample preparation itself, however, does not require clean conditions because the atomic contacts are only formed during the measurement by breaking the bulk of the electrodes.

37.3.4 Electro-Migration Technique

A third method for the formation of atomic-size contacts is controlled burning of a wire by electro-migration caused by high currents. This technique has been optimized for the formation of nanometer-sized gaps for trapping individual molecules or other nano-objects (van der Zant et al. 2006, Wu 2007). Before the wire finally fails and the current drops drastically, atomic size contacts are formed for a rather short time span (Trouwborst et al. 2006, Hoffmann et al. 2008). During the electromigration

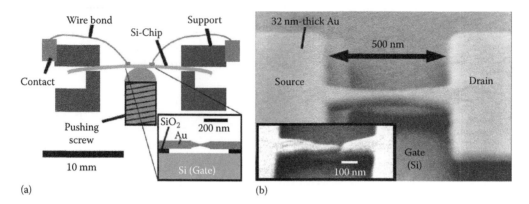

(a)   (b)

FIGURE 37.8 Working principle (a) and electron micrograph of an electromigrated MCBJ (b). (Reprinted from Champagne, A.R. et al., *Nano Lett.*, 5, 305, 2005. With permission.)

process, the transport changes from ohmic behavior, i.e., limited by scattering events to wave-like behavior, which can be described by the Landauer picture. The main drawback of this technique is the fact that it is a single-shot experiment; after burning through the wire, it cannot be closed again.

A combination of electromigration with the thin-film MCBJ overcomes this problem (Champagne et al. 2005): a thin-film MCBJ is thinned-out by electromigration to a narrow constriction with a cross-section of less than 10 nm (see Figure 37.8). The substrate is then bent carefully for completely breaking the wire or arranging single-atom contacts. This last step is reversible and repeatable for studying small contacts or trapped nano-objects. Because only the very last part of the breaking requires mechanical deformation of the substrate, it is rather fast and enables the use of more brittle substrates such as silicon.

37.3.5 Recent Developments

A new version of the MCBJ technique has been introduced by Waitz et al. (2008). It uses thin-film wires on silicon membranes with a thickness of a few hundred nanometers (Figure 37.9). The membrane is deformed by a fine tip on the rear side. At variance to the MCBJ techniques on bulk substrates, the elasticity of the membrane is used rather than the bending. Thus, the deformation is applied locally and it is possible to address particular positions while the rest of the circuit on the substrate remains unaffected.

37.4 Conductance of Atomic-Size Contacts

37.4.1 Opening and Closing Traces

The natural characterization method of atomic contacts that also enables us to roughly determine the size of the contacts is the recording of the so-called opening and closing curves, i.e., measurements of the conductance as a function of time when enhancing or reducing with constant speed the distance of the two electrodes forming the contact.

A typical opening and closing sequence measured consecutively on a gold thin-film MCBJ is shown in the top panel of Figure 37.10. Equivalent data for Al and Co is given in the central panel and the bottom panel.

Both the opening and the closing traces show regions in which the conductance is rather constant (so-called conductance plateaus)

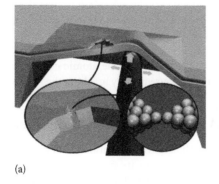

(a)

                                      (b)

FIGURE 37.9 MCBJ on silicon membranes. (a) Working principle of the MCBJ (not to scale). The insets illustrate the formation of a point contact as well as an artist's view of the atomic arrangement of single-atom contact. The thickness of the membrane is in the order of 300 nm, the lateral dimension of the membrane is typically 0.5 mm × 0.5 mm. The dimensions of the suspended metal film are comparable to the ones for thin film MCBJs on massive substrates but with smaller thickness of the sacrificial layer of typically 100 nm. (b) Optical micrograph of a membrane MCBJ with patterned Au electrodes.

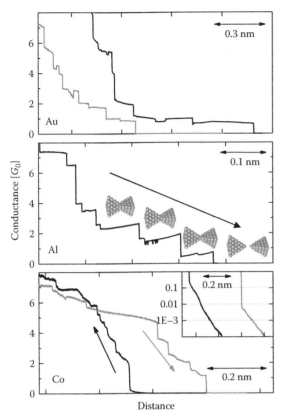

FIGURE 37.10 Conductance as a function of distance measured for thin-film MCBJs made of Au (top), Al (center), and Co (bottom) at low temperature $T < 10\,\mathrm{K}$. The distance axes have been calibrated using the exponential distance dependence in the tunneling regime. For Au two examples of opening traces, for Co one opening and one closing trace are shown. The inset displays the tunneling regime in logarithmic conductance scale. The atomic configurations are given to illustrate the typical size of the contacts on the plateaus and the transition to the tunneling regime where both electrodes are broken to form a vacuum gap.

interchanged with pronounced steps. The lengths of the plateaus and the step heights are of comparable size but not as regular as those observed for the 2DEGQPCs. Furthermore, the plateaus are not all horizontal but may have a fine structure such as small steps, curvature, or inclination (Sanchez-Portal 1997, Krans 1993, Agrait 2003). The typical shape depends on the metal under study and also on experimental details (see below).

At some point of the opening traces, the series of plateaus and steps comes to an end and the conductance decreases exponentially with the time, as can be seen in the inset of the bottom panel where the conductance is plotted on a logarithmic scale. This is the signature of vacuum tunneling: the two electrodes are broken to form a small tunnel gap. Because vacuum tunneling is a well-studied phenomenon, this region can be used to calibrate the setup and to transform the time axis into a distance axis. Assuming clean surfaces of the electrodes, the conductance G as a function of distance d is

$$G(d) \propto \exp\left(-\frac{2}{\hbar}\sqrt{2m^{*}\Phi}\cdot d\right)$$

where

Φ is the work function
m^{*} is the effective mass of the electrons

Using bulk values for Φ and m^{*} yields the distance scales given by the arrows in the individual panels of Figure 37.10. Since the work function of most metals is affected by contamination and surface geometry, the distance values determined with this method bear an error in the order of 20% as can be verified in an STM configuration with an independent calibration of the piezo. The plateau lengths are in the order of several tenths of nanometer, i.e., the typical atomic size. The interpretation of the opening and closing traces is that the plateaus correspond to a particular atomic configuration that can elastically be stretched until a reconfiguration occurs. This gives rise to a change of the minimal cross-section of the contact and the conductance. The relation between both of these quantities is not straightforward to deduce but requires a detailed consideration. A reasonable assumption is, however, that the smaller the cross-section. i.e., the smaller the number of atoms forming the contact, the smaller the conductance. Another reasonable assumption is that the last contact before breaking to the tunneling regime is given by a single-atom contact. Experimental support for this assumption will be given in Section 37.4.3. These considerations lead us to the artist's view of the opening trace given in the central panel of Figure 37.10. Starting from the right, it is well proven that a vacuum gap exists between the two electrodes, when the conductance depends exponentially on the distance. The last plateau before breaking is depicted as a contact formed by a single-atom that forms the apexes of the two pyramidal electrodes. The contacts with higher conductance are most likely formed by more than one atom; the number and arrangements are, however, not easy to guess from the bare conductance since several different arrangements may give rise to similar conductance values even for simple metals like gold (Dreher et al. 2005). We will come back to this point in Sections 37.4.2 and 37.4.3.

Additional important information that can be deduced from the examples shown in Figure 37.10 is the fact that the closing traces are not time reversed versions of the opening traces, although the adopted conductance values are similar. For most metals, the opening traces give rise to longer plateaus than the closing traces (Agraït et al. 1993, Böhler et al. 2004). From the area surrounded by the opening and closing trace, quantities like Young's module and deformation energy may be deduced (Rubio et al. 1996).

37.4.2 Conductance Histograms

An in-built problem of the study of atomic point contacts is the fact that each contact is unique in the sense that the exact atomic arrangement of the atoms in the narrowest part varies from contact to contact. This results in manifold transport behavior, such as varying conductance, varying noise properties, varying current-voltage characteristics, and conductance fluctuations. Nevertheless, experiments on a large ensemble of metallic contacts—realized with various techniques—have

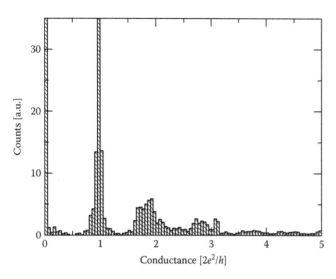

FIGURE 37.11 Conductance histogram of a Au thin-film MCBJ calculated from 66 opening traces recorded at $T = 10\,$K in ultra-high vacuum.

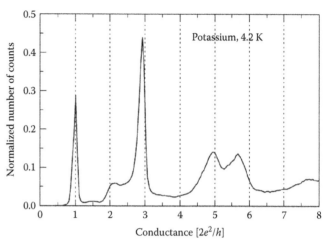

FIGURE 37.12 Conductance histogram of potassium measured at $4.2\,$K. The fact that the peaks at $G = 2$ and $G = 4$ are suppressed is typical for free electron systems. (Reprinted from Agraït, N. et al., *Phys. Rep.*, 377, 81, 2003. With permission.)

demonstrated the statistical tendency of atomic-size contacts to adopt configurations leading to some preferred values of conductance. The usual tool for deducing the statistical behavior is the recording of conductance histograms, i.e., a plot of the probability with which a particular conductance value is adopted as a function of the conductance. An example for Au recorded with a thin-film MCBJ at 10 K in ultrahigh vacuum conditions is given in Figure 37.11. Despite the relatively small number of opening traces ($M = 66$), it displays a pronounced and narrow maximum at $G = 1G_0$ and two smaller and wider ones at $G = 2G_0$, and $G = 3G_0$.

The actual preferred values do not only depend on the metal but also on the experimental conditions and on the way the histograms are built. Obviously, conductance histograms made from opening traces may differ from those obtained from closing curves, as discussed above. For most of the metals, the plateaus are not horizontal. It is thus crucial for the shape of the histogram whether the conductance values recorded all over the plateau or only the last ones–obtained just before break–are used. Both methods are applied, however. A more detailed discussion of this issue can be found in the review of Agraït et al. (2003). Other external parameters that determine the shape of the histograms are, among others, the temperature, the vacuum conditions, the applied voltage, the hardness of the electrode material (Yanson et al. 2005, 2008), and the number and the speed of the formation of the contacts. Luckily, most of these parameters mainly affect the height and the width of the histogram peaks, while the positions, i.e., the preferred conductance values, are rather robust.

For many metals, in particular monovalent ones (like Na, Au, etc.) that as bulk material fulfill the free-electron model, the smallest contacts have a conductance G close to G_0 (Olesen et al. 1994, Krans et al. 1995, Costa-Krämer 1997). Furthermore, many monovalent metals show several peaks in the histograms that presumably correspond to contacts with more than a single atom in the

smallest cross-section. Also, the higher-conductance peaks are often close to integer values of G_0, but not all integers are observed, a fact which can be explained with the help of electronic shell effects similar to magic numbers of clusters (Mares and van Ruitenbeek 2005). In particular, for the alkali metals, not all multiples of G_0 give rise to the same peak height and the pattern is strongly temperature dependent (Figure 37.12; Yanson et al. 1999).

Interestingly, there also exists a few multivalent metals such as Al (Yanson et al. 1997, 2008; see Figure 37.13) or Zn (Scheer et al. 2006) that show several rather well pronounced maxima, while the majority of the electronically more complex metals reveal either a single peak or wide bumps (for examples, please see Nb: Ludoph et al. 2000; Fe, Co, Ni: Untiedt et al. 2004, Sirvent et al. 1996; Figure 37.14) that do not have their maximum at integer multiples of G_0. The spacing between the neighboring maxima is in the *order* of G_0 as well but is only rarely exactly $1G_0$.

As shown in Section 37.4.4, the fact that the first peak in the conductance histograms often is close to $1G_0$ does not in general mean that the smallest contact (presumably a single-atom contact) corresponds to a single, perfectly transmitted Landauer channel with $\tau_i = 1$.

Summarizing, the origin of the peaks in the conductance histograms is not yet fully understood because it is an interplay of electronic and geometrical effects.

37.4.3 Determination of Conduction Channels

The total transmission $T = \Sigma \tau_i$ of the contacts can be easily deduced from their measured conductance using the Landauer formula $G = G_0 T$. As discussed in Section 37.2, the total ensemble of transmission coefficients $\{\tau_i\}$ cannot be determined solely by conductance measurements. However, transport quantities exist that do not depend linearly on the τ_i, which thus may serve to provide additional information on the transmission matrix.

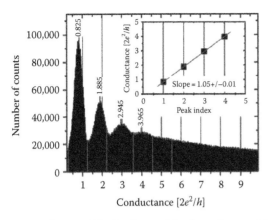
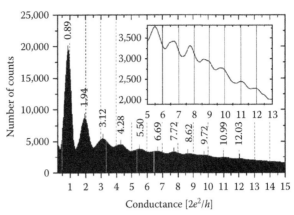

FIGURE 37.13 Influence of the hardness of the electrode material onto the conductance histogram of Al. Left (right): Histogram of annealed (work-hardened) Al measured with a notched-wire MCBJ at 4.2 K. The inset in the left panel displays the conductance values for the positions of the first four peaks. The inset in the right panel is an enlargement of the data for conductance values between 5 and 13G_0. (Reprinted from Yanson, I.K. et al., *Phys. Rev. B*, 77, 033411, 2008. With permission.)

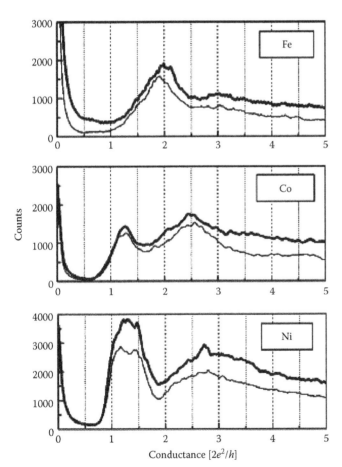

FIGURE 37.14 Conductance histograms of the ferromagnets Fe. Co, Ni measured with notched-wire MCBJs at 4 K with and without external magnetic field. (Reprinted from Untiedt, C. et al., *Phys. Rev. B*, 69, 081401, 2004. With permission.)

For example, the power of the shot-noise associated with the current at voltage V is a measure of the second-moment of the transmission distribution $\{\tau_i\}$ (Büttiker 1990) $S = 2eVG_0\Sigma\tau_i(1 - \tau_i)$.

A perfectly transmitted mode ($\tau_i =1$) does not contribute to the noise because of electron correlations arising from the Pauli principle. A measurement of this noise power provides a second relation between the matrix elements of $t^\dagger t$. However, this information is still not sufficient to determine the eigenvalues ensemble $\{\tau_i\}$ if the scatterer has more than two modes.

In principle, one needs to measure a number of moments equal to the total number N of channels in order to determine all the transmission coefficients, which is possible by analyzing measurements in the superconducting state of atomic-size contacts.

Instead of going through a complete opening trace, one can stop at any point and record a current-voltage characteristic (*IV*s) in the superconducting state. The first observation is that one finds different *IV*s for the same conductance i.e., the same total transmission. Examples for three Al contacts realized with a thin-film MCBJ on the same last plateau before breaking, all three with $G \approx 0.8G_0$, are given in Figure 37.15.

The origin of the well marked current increases at voltage values $V = 2\Delta/me$ (with the superconducting gap parameter Δ and m a natural number) are multiple Andreev reflections (MAR) (Averin and Kardas 1995, Cuevas et al. 1996). Each current onset corresponds to a charge transport with a varying number m of coherently transferred elementary charges. For example, the structure at $eV = 2\Delta$ is the well-known single-electron transfer also present in superconducting tunnel contacts. The height of the current onset depends nonlinearly on the transmission coefficients and thus allows one to determine this value. The set $j(V,\tau)$ of the current-voltage characteristics of single-channel superconducting QPCs with arbitrary transmission $0 < \tau \leq 1$ has been calculated exactly by several groups independently and applying different methods (Averin 1995, Cuevas 1996). Since the channels are independent of each other (see Section 37.2), the current contributions of N channels with an ensemble $\{\tau_i\}$ give just the total current

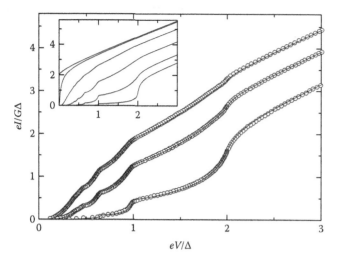

FIGURE 37.15 Current–voltage characteristics of three single-atom contacts of Al in the superconducting state measured at $T = 75\,mK$ with a thin-film MCBJ. The overall conductance of all three contacts is $G \approx 0.8G_0$. Symbols represent experimental data, lines are fit to the theory of MAR. Current and voltage are given normalized to the superconducting gap energy $e\Delta$. Inset: Calculated IV characteristics for a single channel superconducting QPC at $T = 0$ for various values of τ (from bottom to top: 0.1, 0.4, 0.7, 0.9, 0.99, 1). (After Cuevas, J.C. et al., *Phys. Rev. B*, 54, 7366, 1996; Averin, D. and Bardas, A., *Phys. Rev. Lett.*, 75, 1831, 1995.)

$$I(V) = \sum_{i}^{N} j(\tau_i, V)$$

This is used to fit the measured *IV*s and thus to decompose the total current into the contributions of the individual channels for contacts with more than one channel. It is worth mentioning that superconductivity itself does not alter the channel ensemble: it only serves as a tool for their determination as can be demonstrated for contacts with two channels, the ensemble of which can also be determined by analyzing the shot-noise or other methods that do not rely on superconductivity (van den Brom and van Ruitenbeek 1999).

The examples given in Figure 37.15 are each composed of the contributions of three channels with varying transmissions. Although the total conductance is smaller than $G = 1G_0$. In other words, these single-atom contacts that contribute to the first peak in the conductance histogram do accommodate three conductance channels instead of a single one. This observation in spite of the fact that all three contacts are part of the same conductance plateau, immediately yields to the conclusion that the transmissions do depend very sensitively on the precise atomic arrangement.

37.4.4 Atomic Contacts of Conventional Metals

The number $N = 3$ for a single atom contact is not at all unique. Experiments with lead ($N = 3$), niobium ($N = 5$), zinc ($N = 1$ or 2), and gold ($N = 2$, which has been rendered superconducting by the so-called proximity effect; Scheer et al. 2001) have demonstrated that not only the preferred conductance of single-atom

contacts, but also the number of channels is a function of the chemical valence (Scheer et al. 1998, Konrad et al. 2005, Scheer et al. 2006).

Several models have been put forward for explaining this material dependence. One of these is based on a free-electron model for the electrons neglecting the precise atom arrangement (Torres et al. 1994, Stafford et al. 1997; for a review see Grabert 2009). The other one uses a linear combination of atomic orbitals (LCAO) via tight-binding methods (Levy Yeyati et al. 1997, Cuevas et al. 1998; for a review see Agraït et al. 2003) or the so-called *ab initio* calculations (Bagrets et al. 2006). Tight-binding LCAO models are known to be very suitable for the description of the band structure of insulators and semiconductors because the wave functions are rather localized and only few of them overlap with the wave function of the given atom. This is why tight binding is rather unusual for calculating the electronic properties of metals since the wave functions of the conduction electrons are extended over many atom positions. Therefore, for bulk metals, the free-electron model seems to be more appropriate. However, when aiming at calculating the electronic situation in restricted geometries such as an APC in which only a few atomic waves overlap, the LCAO models have their advantages. The general rule is that the more wave functions overlap, the wider the resulting electronic band in the energy range.

According to the wave-guide model that successfully describes the behavior of 2DEGQPCs, the shape of the cross-section as well as the ratio between the size of the contact and the Fermi wave length determine the number of channels. Applying this model to APCs does not straightforwardly account for the observed variability of *N*. The LCAO model and the *ab initio* calculations, however, are able to describe the conductance behavior successfully. The latter ones are particularly useful for describing the influence of a finite measuring voltage, which is, however, out of the scope of this chapter.

For simplicity, we will describe here the main idea and findings of the LCAO approach. It is not necessary to go into the details of the models to understand how it addresses the most important observations (conductance of a single-atom contact, number of channels, transmissions of the channels, and sensitivity to atomic rearrangements). Due to the fabrication method of the contacts, the exact geometry of the contact on the atomic scale differs from contact to contact. Since the Fermi wavelength is of the same order as the diameter of the contacts, the electrons crossing the contact experience all the imperfections on this scale. Their wave functions will thus be influenced by these imperfections. The next important conclusion is that even for a good free-electron metal, the eigenfunctions of an APC will not be Bloch waves, like the ones of the electrodes. It is thus not taken for granted that the transmission coefficients achieve the value $\tau_i = 1$, even for a perfectly ordered cut-out of a single crystal of the metal. Rather, the eigenchannels are determined by the overlap of the wave functions of the atoms forming the central cluster with the wave functions of the electrodes and thus vary from contact to contact. The number of wave functions with overlap to the neighbors, i.e., the number of extended electronic states or in other words, the number of channels is at most the number of partially filled electronic orbitals, i.e., the number of valence orbitals. Since the overlap of the wave functions depends on the exact configuration of the atoms, it varies from contact to contact, and the transmission coefficients of the individual channels do so as well, giving rise to a broad variation of possible $\{\tau_i\}$ although the sum of the coefficients, i.e., the conductance, may be similar. Because the exact geometry of the contacts at the atomic scale is in general unknown, the channels are calculated for either perfectly ordered model geometries (Cuevas et al. 1998a) or the most likely configurations deduced from molecular dynamics calculations (Dreher et al. 2005). Interestingly, for most of the metals studied so far, it turned out that the number of channels with measurable transmission does not depend on the geometry of single-atom contacts, however, their transmissions vary strongly.

By comparing the results obtained for Al and Pb, Figure 37.17 shows how the chemical properties of the element enter into the conductance properties: Al and Pb are *sp*-like metals, i.e., because of their orbital structure single-atom contacts could give rise to up to four channels (1*s* orbital, 3*p* orbitals). The main differences between Al and Pb are that the level spacing between the *s* and the *p* orbitals in the isolated atom amounts to approximately 7 eV in the case of Al and 10 eV in the case of Pb. Furthermore, Al has three electrons on the outer shell while Pb has four (see the top of Figure 37.17). When combining these atomic wave functions in order to calculate the band structure of the bulk metal, one uses iterative processes varying the numerical parameters (Papaconstantopoulos 1986) in order to obtain good agreement with experimentally available data for the band structure. The results for Al and Pb are displayed in the two upper panels of Figure 37.17 named "bulk DOS". As a result of the mentioned element-specific ingredients, the Fermi energy of Al in the bulk metal lies in a region where both *s* and *p* electrons have a measurable density of states, while at the position of E_F of Pb the *p*-orbitals dominate by far. In the APC geometry of Figure 37.16, the central atom has only very few neighbors, which results in a weaker broadening of the electronic bands. As a consequence,

the local density of states (DOS) calculated for the central atom interpolates between the spiky energy levels of the isolated atom (see the top of Figure 37.17) and the bulk DOS. Furthermore, the broken isotropy makes the *p* orbital, which is directed along the transport direction (here: the p_z orbital), split-off in energy from the two remaining ones (here: p_x and p_y). The new eigenfunctions of the APC structure are two linear combinations of the p_z orbital with the s orbital and the two perpendicular p_x and p_y orbitals. Due to symmetry reasons, only the symmetric sp_z hybrid has a measurable weight while the antisymmetric one has negligible DOS throughout the whole energy range interesting for transport. This local DOS now enters into the calculation of the transmission functions, i.e., the energy dependent transmissions. Their values at E_F are called transmission coefficients and are entering the linear conductance via the Landauer formula (see Section 37.2), which can be written as

$$G = G_0 \operatorname{Tr}\!\left[\hat{t}g\left(E_F\right)\hat{t}^{\dagger}\left(E_F\right)\right]$$

where $\hat{t}(E) = 2\left[\operatorname{Im}\Sigma_L(E)\right]^{1/2}\hat{G}^r_{1N}(E)\left[\operatorname{Im}\Sigma_R(E)\right]^{1/2}$ are the transmission functions, $\Sigma_{L,R}$ are the so-called self-energies of the left and right lead, respectively, and $\hat{G}^r_{1N}(E)$ is the so-called Green's function of the central atom and corresponds to its local DOS. The self energies account for the coupling of the wave functions of the central atom to the leads. As a result, this coupling causes the difference between the energy dependence of the local DOS and the resulting transmission functions shown in the two lower panels. In the linear conductance, which is measured in the experiments when applying a small bias voltage, solely the value of $\hat{t}(E)$ at the Fermi enters. As can be deduced from Figure 37.17, in the case of Al, three channels contribute with a transmission varying between 0.1 and roughly 0.8, while for Pb three channels with an almost equal transmission of 0.8 are predicted.

For the geometry shown in Figure 37.16, i.e., two perfectly ordered pyramids grown in the (111) direction sharing one atom at the apex, the sum of the transmission amounts to roughly $1.2G_0$ for Al and $2.5G_0$ for Pb. For the same geometry but for the transition metal Nb (1*s* orbital, 5*d* orbitals), the calculation yields five channels with nonvanishing transmission, adding up to $G = 2.8G_0$.

The situation is more complex for the divalent metal Zn, which crystallizes in a hexagonal structure. Here, the number of channels does indeed depend on the growing direction of the contact (Konrad 2004, Häfner 2004, Scheer et al. 2006). Introducing disorder into the structure does, in general, reduce the transmission values and may lift degeneracies (Cuevas 1998a,b).

All multivalent metals have the following in common: even for perfectly ordered single-atom contacts, none of the τ_i arrives at the maximum possible value $\tau_i = 1$, reflecting that the eigenstates of the APC are not perfectly matched to the eigenstates of the electrodes. Thus, there is in general no transmission quantization going along with perfect conductance, i.e., perfect sums of transmissions. This is particularly striking for the element Al,

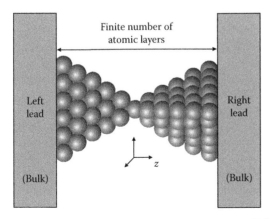

FIGURE 37.16 Schematics of a single-atom contact used for the calculations within the LCAO model. (Adapted from Cuevas, J.C. et al., *Phys. Rev. Lett.*, 80, 1066, 1998a.)

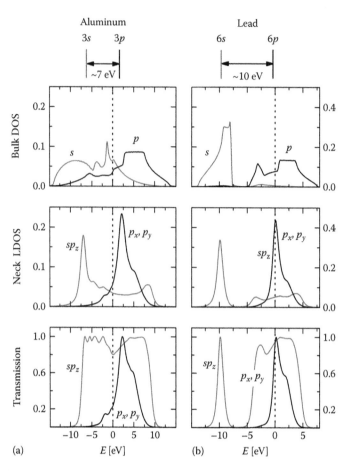

FIGURE 37.17 LCAO model for single-atom contacts of Al (a) and Pb (b) in the model geometry shown in Figure 37.16. Schematic development of the atomic valence levels into bulk conduction bands. The panels in the second row depict schematically the local density of states (LDOS) in eV^{-1} at the central atom of the model geometry. The global energy dependence of the transmission coefficients $\{\tau_i\}$ shown in the bottom most panels. The dotted lines indicate the position of the Fermi level. The sp_z mode is the best transmitted one for both materials. The p_x and p_y modes are degenerate due to the symmetry of the model geometry. The indicated energy differences between the atomic s and p levels are the ones used to fit the bulk band structure. (Data taken from Scheer, E. et al., *Nature (London)*, 394, 154, 1998.)

the histogram of which reveals several pronounced peaks. For instance, the first peak at $G = 0.8G_0$ is mainly caused by contacts accommodating three channels with varying distribution of transmissions (Scheer et al. 1998, Böhler et al. 2004), although its total conductance would be small enough to be provided by a single channel.

According to the LCAO model, single-atom contacts of monovalent metals are expected to accommodate only a single channel, which has been found by theory and by experiment for gold (van den Brom et al. 1999, Scheer et al. 2001) and several alkali metals (Yanson et al. 1999). Whether the transmission of this channel is equal to one that would give rise to conductance quantization (i.e., histogram peaks at multiples of G_0) depends on the geometry of the atomic arrangement and the local electronic band structure. This can be visualized in a simple model, assuming that the coupling of the single valence orbital of the central atom is given by a coupling constant to each of the two electrodes (Cuevas et al. 1998b). Two conditions have to be fulfilled for the resulting extended electronic mode to be perfectly matched: The coupling to both electrodes has to be equal, i.e., the central atom is symmetrically bonded to the

neighbor atoms on both sides. The second condition is that the energy of the valence orbital corresponds to the Fermi energy of the electrodes. This is fulfilled for monovalent free electron metals, since the valence band is half filled.

The origin of the existence of preferred conductance values in spite of the fact that the electronic wave functions are not quantized is still not completely settled. The situation is complex because, as mentioned earlier, several length scales that are important for conductance—the interatomic distance, the lateral dimension, the Fermi wavelength, momentum scattering length—do coincide. Thus, there is a competition between electronic and structural effects (Yanson et al. 2005, 2008). The electronic effects are closely connected with the physics of nanowires as mentioned in the introduction. We will therefore only briefly mention the most important facts.

The electronic effects include effects typical for free electron systems (Yanson et al. 1999, Grabert 2009) as well as interaction effects (Kirchner et al. 2003). For alkali metals, i.e., the best free-electron metals, the electronic effects dominate giving rise to shell effects that are equivalent to the existence of magic clusters consisting of a number of atoms for which the electronic shells are closed (see chapters by Grabert (2009) and Agraït (2009)). For contacts of alkali metals and gold, a transition from the electronic shell effects to the geometry shell effects for larger contacts have been found, which means that those contacts are the most stable and do give rise to pronounced peaks in the histograms, for which the structure is highly favorable, and electronic shells are closed (Yanson et al. 2001).

The investigation of atomic contacts with a TEM (see Section 37.3.2) suggests that highly symmetric and well-ordered atomic configurations are preferred. This could explain the appearance of particular conductance values. However, it would go along with preferred transmissions as well, a point that is much less studied so far. The few transmission histograms that are published so far suggest a wider distribution of transmission values (Böhler et al. 2004). It has been demonstrated that the height of the histogram peaks is markedly influenced by the hardness of the electrode material (Yanson et al. 2005, 2008). An outcome of these investigations is that the harder the metal, the more pronounced the shell effects.

37.4.5 Atomic Contacts of Semimetals

The coincidence of Fermi wavelength and contact size is lifted when studying APCs of semimetals. The term semimetal denotes materials for which the electronic bands cross the Fermi level in particular directions only. The two best studied elementary semimetals are Bi and Sb, which fail to be insulators because of a small lattice distortion with respect to a cubic symmetry. This slide distortion makes parts of the bands that would be unoccupied if in cubic symmetry to lie below the Fermi energy. Consequently, the density of conduction electrons is low giving rise to a rather large Fermi wavelength of 10–100 nm, comparable to the one in 2DEGs. It is thus instructive to study point contacts for the electronic properties that resemble 2DEGQPCs but the structure and size is that of APCs. The most obvious question is whether APCs of semimetals adopt preferred conductance values in the order of G_0 with steps in the same order. Such behavior would be expected for an electronic wave-guide similar to what was realized in the 2DEGQPCs. A single channel with perfect transmission would be expected for contacts with lateral sizes in the order of $\lambda_F/2$, much larger than an atom. If, however, atomic configurations dominate the transport behavior, much smaller steps both in height and in length should be observed.

When interpreting the data recoded on semimetal APCs, one has to keep in mind that because of the low crystal symmetry electrons and holes in different bands may have different effective mass. This makes the total electronic behavior complex and strongly dependent on external influences like external fields, disorder (Garcia-Mochales 1996, Garcia-Mochales and Serena 1997), and the fabrication method of contacts. It is thus necessary to fabricate the APCs with different methods that result in different contact structures.

At first glimpse, the experimental findings appear to some extent to be controversial. For Sb small plateaus with a typical height of $0.02G_0$ and plateau length in order of the lattice constant

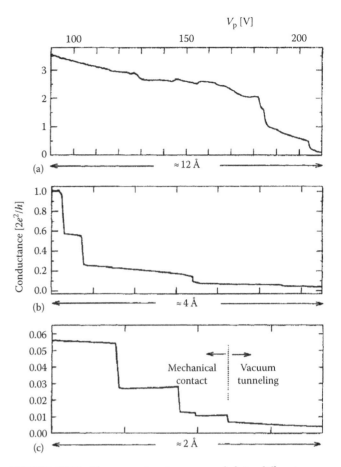

FIGURE 37.18 Three opening traces recorded in different conductance ranges of a Sb notched-wire MCBs recorded at 4.2 K. The horizontal axis is the voltage at the piezo which drives the pushing rod and is proportional to the distance Sb. The step sequence stops around $G \approx 0.01G_0$. (From Krans, J.M. and van Ruitenbeek, J.M., *Phys. Rev. B*, 50, 17659, 1994. With permission.)

have been found (Krans and van Ruitenbeek 1994). These findings are in accordance with the atomistic model (Figure 37.18). The value of conductance corresponds well to the ratio between the lattice constant and the Fermi wavelength $a_0/\lambda_F = 0.02$.

Several experiments have been performed for Bi with STMs (Costa-Krämer 1997, Rodrigo 2002) as well as with thin-film MCBJs. While the histogram in the article by Costa-Krämer does show rather wide bumps around multiples of G_0, individual opening traces depict extremely well marked and horizontal plateaus (Figure 37.19).

This seemingly contradictory behavior can partially be explained by the investigations of Rodrigo et al. (2002) who observed both the wave guide as well as the atomistic effects in the same experiment. Depending on the shape of the constriction on the length scale of λ_F, one or the other dominates, as depicted in Figure 37.20. For long necks, plateaus at multiples of G_0 are found. For short constrictions small steps are observed with a minimum conductance of $G_{Bi} \approx 0.2G_0$. This finding is in agreement with the results from the thin-film MCBJ measurements, from which $G_{Bi} \approx 0.15G_0$ for single-atom contacts was deduced.

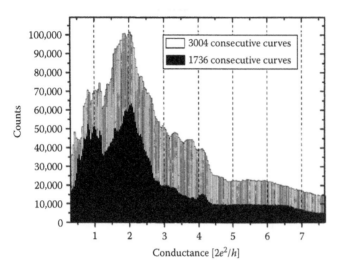

FIGURE 37.19 Conductance histogram of Bi measured with an STM at 4 K. (Reprinted from Costa-Krämer, J.L. et al., *Phys. Rev. Lett.*, 78, 4990, 1997. With permission.)

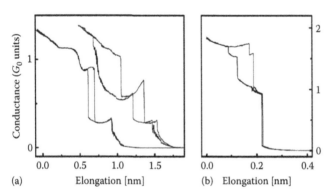

FIGURE 37.20 Opening and closing traces of Bi contacts realized with a STM at $T = 4.2$ K. (a) contacts revealing subquantum plateaus and (b) contact showing a last plateau with conductance close to G_0. (Reprinted from Rodrigo, J.G. et al., *Phys. Rev. Lett.*, 88, 246801, 2002. With permission.)

The histogram corresponding to the latter investigation reveals peaks at even multiples of G_{Bi}, supporting the atomistic model of conductance (Figure 37.21).

In summary, in APCs of semimetals, waveguide as well as atomic effects seem to be present. Proof of this interpretation could be delivered by measuring the number of conductance channels: According to the waveguide model contacts with $G_{Bi} < G \leq G_0$ would all have a single channel. Increasing the size would increase the transmission of this channel but not open new channels. In the atomistic model, however, these contacts would have more than one channel. This information could be deduced from shot noise measurements or with the method described in Section 37.4.3.

37.4.6 Atomic Contacts of Magnetic Metals

There is a particular interest in APCs of magnetic metals for several reasons. On the one hand, APCs of band magnets (Fe, Co, Ni) are proposed to act as sources of spin-polarized electrons. It has been predicted that the spin polarization, i.e., the relative

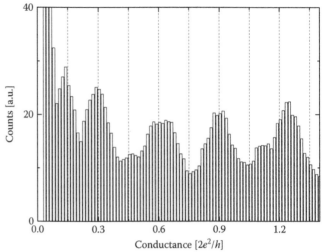

FIGURE 37.21 Histogram calculated from 80 opening traces of a Bi thin-film MCBJ measured at 30 mK.

preference of one of the two spin directions should strongly depend on the atomic configuration. For instance, in the tunnel limit of Ni, the spin polarization could achieve very high values close to 100% (Häfner et al. 2008). On the other hand, the interest was triggered by the observation of very large magnetoresistance values, i.e., changes of the resistance (or conductance) as a function of the applied magnetic field. The latter effect could be used for magnetic storage devices or magnetic field driven switches in nanoelectronic circuits. Although much research effort has been put into this field during the last decade, the mechanism behind the observations is still under debate.

According to the Jullière model, the spin polarization is given as

$$P = \frac{D^\uparrow - D^\downarrow}{D^\uparrow + D^\downarrow}$$

where $D^\uparrow$ and $D^\downarrow$ denote the electronic density of states at the Fermi energy for electrons with spin up or spin down, respectively. This quantity can, for example, be deduced from spin-resolved photo-emission measurements. In this definition, all electrons in all bands crossing the Fermi energy contribute to P. In the electronic conductance, however, not only the energy of the electronic states but also their wave functions are important. In ferromagnetic metals, the density of states at the Fermi energy is dominated by electrons in s-bands and in d-bands, both of which have very different wave functions. In general, the s-bands give rise to more extended wave functions than the d-bands and therefore a stronger contribution to the conductance. In APCs, the different wave function characters result in different values of the transmission coefficients for the channels with dominating s-characters or d-characters, respectively. Therefore, spin polarization values deduced from transport phenomena may differ from those determined from x-ray photoelectron spectroscopy (XPS) or other equilibrium electronic quantities. For the analysis of magneto-transport experiments, the definition of an effective transport spin polarization P_T is more appropriate:

$$P_T = \frac{G^\uparrow - G^\downarrow}{G^\uparrow + G^\downarrow}$$

Since in a transport measurement there is no direct access to the spin-orientation of the conducting electrons, yet another model has been established for the analysis of magneto-transport experiments, inspired by the observation of extremely strong magneto-resistance effects of ferromagnetic APCs.

Most of the magneto-resistance experiments are carried out in the following way: A single- or few-atom contact between two ferromagnetic electrodes is established using one of the methods described in Section 37.3. Opening and closing traces are recorded with and without applied magnetic fields. From these data histograms are calculated. In addition the conductance is measured as a function of the applied external magnetic field. In a third type of measurement, the conductance is recorded when rotating the sample in a constant external field.

MCBJ experiments on Co, Ni, and Fe do not reveal preferred conductance values at multiples of G_0 in accordance with findings at nonmagnetic multivalent metals (see Section 37.4.2). In most of the experiments, the histograms do not vary by much even when applying strong external fields of up to several Tesla (see Figure 37.14) (Untiedt et al. 2004).

However, it is possible only in very special realizations to prepare contacts with integer multiples of G_0 that split-up into half-integers in an external field (Ono et al. 1999).

The most common findings of the magneto-resistance investigations are the following: Depending on the field orientation, the starting conductance, the material under study, and the technique, the resistance undergoes pronounced changes. In the context of Giant Magneto-Resistance (GMR) or Tunnel Magneto-Resistance (TMR) used for read-write heads, the magneto-resistance ratio (MRR) is defined as

$$MRR = \frac{R_{\uparrow\downarrow} - R_{\uparrow\uparrow}}{R_{\downarrow\downarrow}}$$

where $R_{\uparrow\downarrow}$ and $R_{\uparrow\uparrow}$ are the resistance in antiparallel and parallel orientation of the magnetization of the two banks forming the GMR or TMR device. Usually the resistance is highest in antiparallel orientation and lowest in parallel orientation. Since in atomic devices the spin orientation is not always straightforward to deduce, and since additional effects may contribute to the MRR, the usual definition for the MRR of APCs is

$$MRR = \frac{R_{max} - R_{min}}{R_{min}}$$

These MRR values depending on the actual realization achieve the enormous values of 100.000% (Hua and Chopra 2003) (Figure 37.22). The size of the field that is necessary for switching between the two extremal resistance values usually is in the order of several

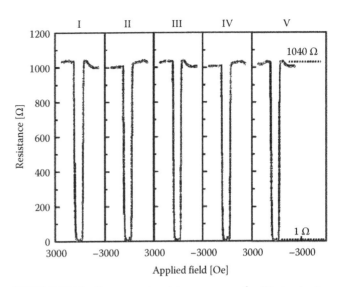

FIGURE 37.22 Five magnetoresistance curves of a Ni atomic-size contact as a function of applied magnetic field. The MRR value is in the order of 100.000%. (Reprinted from Hua, S.Z. and Chopra, H.D., *Phys. Rev. B*, 67, 060401, 2003. With permission.)

hundred millitesla but can arrive at several tesla for particular sample geometries and magneto-crystalline anisotropy (Egle et al. 2010). As mentioned in the introduction, the origin of the enormous MRR effects is still not clear because several microscopic processes may contribute to the resistance and its field dependence.

One possible mechanism is the so-called Ballistic Magneto-Resistance (BMR). If no spin-degeneracy is given, the Landauer formula has to be modified to distinguish the properties of the transport channels in the two spin orientations:

$$G = \frac{e^2}{h}\left(\sum_{i=1}^{N^\uparrow} \tau_i^\uparrow + \sum_{i=1}^{N^\downarrow} \tau_i^\downarrow\right)$$

Here the arrows indicate the spin-up and spin-down channels, respectively. Both the number and the transmissions of the individual channels may differ for spin-up and spin-down electrons. The magnetic field may now act in such a way that the transmissions of the channels in one spin direction would be suppressed, giving rise to strong MRR values depending on the distribution of the transmissions. So far, detailed calculations of the transmission coefficients of APCs only exist for a few model geometries of contacts of the band metals Fe, Co, and Ni (Häfner et al. 2008). The typical behavior is that the spin-up channel ensemble is very different from the spin-down channel ensemble. Both are strongly dependent on the exact geometry. For particular geometries, i.e., the tunnel geometry, only one spin direction would contribute to the transport giving rise to a fully spin-polarized current (Figure 37.23).

In other geometries, both spin orientations contribute to the transport with unequal weight. The application of a magnetic field could then act in such a way that it blocks the contribution of one of the spin directions. Depending on the size of its contribution, this could result in a considerable MRR. The calculation of the effect of the magnetic field onto the channels is

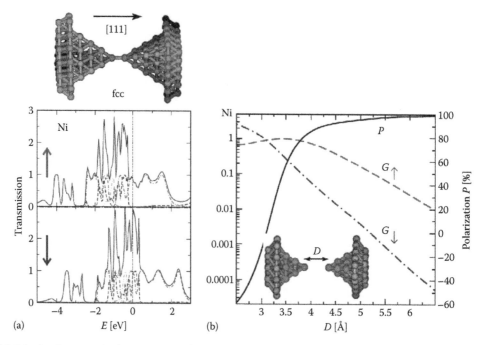

FIGURE 37.23 (a) Calculated spin resolved transmission of a single-atom contact of Ni in the geometry shown in the top as a function of energy. The distance between the two central atoms corresponds to the equilibrium distance in bulk. At the Fermi energy one almost perfectly transmitted channel $T_\uparrow = 0.86$ contributes in spin-up direction, while four channels with a total transmission $T_\downarrow = 2.66$ in spin down direction adding up to a total conductance of $1.8G_0$. (b) Calculated development of the spin-resolved conductance and the resulting spin polarization of a dimer atomic contact of Ni upon stretching to the tunnel regime. When the two apex atoms are in their bulk equilibrium distance the spin polarization is about −51% and arrives at +100% in the far tunnel regime. (Reprinted from Häfner, M. et al., *Phys. Rev. B*, 77, 104409, 2008. With permission.)

a tricky task because it requires the incorporation of spin-orbit coupling. First results exist for contacts of Ni, predicting that not only complete blocking of channels but also gradual changes of the conductance as a function of field would be expected. These gradual effects have also been observed when performing magneto-resistance measurements both in parallel and in perpendicular magnetic fields (Egle et al. 2009) and when changing the field orientation with respect to the current direction (Viret 2002, Gabureac et al. 2004, Viret et al. 2006, Bolotin 2006).

The latter are interpreted in terms of Anisotropic Magneto-Resistance (AMR), an effect that is well known from bulk materials and thin magnetic films (Barthelemy et al. 2002). It is caused by the spin orbit coupling and usually amounts to a few percent in bulk materials and wider contacts. The reason why this effect in general is very small is that the orbital angular momentum in extended metals is suppressed ("quenched") because of the interaction with the neighboring atoms. In atomic-size contacts, however, the quenching mechanism is less active due to the reduced number of neighboring atoms. The spin-orbit coupling and consequently the AMR can thus be strongly enhanced (Atomically-enhanced AMR [AAMR], Viret 2006).

Besides these intrinsic origins of magneto-resistance several extrinsic effects exist. These include domain walls in the constriction (Bruno 1999), shape anisotropies, and magneto-striction, i.e., a length change of the electrodes going along with the reorientation of the domains in the external field. This change in length may affect the contact size and therefore the resistance

of the contacts. From the theoretical side, promising tools have been developed that will hopefully lead to an increased understanding within the next years.

From the experimental point of view, the determination of the spin polarization P_T and the transport channels would be required in order to elucidate the complex behavior of the contacts. In superconductor-ferromagnet heterocontacts in the few nanometer range produced with the needle-anvil method mentioned in Section 37.3.1, P_T has been determined by analyzing the suppression of the Andreev reflection (i.e., the contribution of electron pairs $n = 2$) similarly to the determination of the transport channels of normal-conducting APCs (see Section 37.4.3; Soulen et al. 1998, Perez-Willard et al. 2004). The analysis is based on the assumption that all spin-up and spin-down electrons can be described by one effective transmission coefficient, each. Using the same method for APCs is therefore not straightforwardly possible because several channels with different transmissions contribute per spin direction. For the same reasons as detailed in the introduction, it is not possible to deduce the full channel set from only measuring the conductance, shot noise, or Andreev reflection if more than two quantities, i.e., spin-resolved transmission coefficients contribute.

37.5 Conclusions and Outlook

In summary, atomic-size contacts represent a powerful and versatile tool for the quantitative verification of the fully quantum mechanical description of electronic transport, i.e., the Landauer theory. Since their first realization approximately 20 years ago, much progress has been achieved in both fabrication methods and the understanding of their transport properties. On the experimental side, various experimental realizations have been developed that can be applied for different purposes, e.g., depending on whether highest flexibility or highest stability are required. Although several aspects of the understanding of their transport properties are still under debate, they now serve as test beds for advanced transport theories. As an example, we mention the issue of vibrational excitation and their relation to heat dissipation on the atomic level. Furthermore, the techniques used for the investigation of atomic-size contacts are now used to fabricate atomically sharp and well-characterized electrodes for contacting nano-objects such as individual molecules. The first results of these investigations are covered in the chapter by Devos (Devos 2009).

References

Agraït, N., Rodrigo, J. C., and Vieira, S. 1993. Conductance steps and quantization in atomic-size contacts. *Phys. Rev. B* 47: 12345–12348.

Agraït, N., Levy Yeyati, A., and van Ruitenbeek, J. M. 2003. Quantum properties of atomic-size conductors. *Phys. Rep.* 377: 81–279.

Agraït, N. 2010. Atomic wires. In: *Handbook of Nanophysics: Nanotubes and Nanowires*. Ed. K. Sattler. Boca Raton, FL: Taylor & Francis.

Averin, D. and Bardas, A. 1995. AC Josephson effect in a single quantum channel. *Phys. Rev. Lett.* 75: 1831–1834.

Bagrets, A., Papanikolaou, N., and Mertig, I. 2006. Ab initio approach to the ballistic transport through single atoms. *Phys. Rev. B* 73: 045428.

Barthelemy, A., Fert, A., Contour, J.-P. et al. 2002. Magnetoresistance and spin electronics. *J. Magn. Magn. Mater.* 242: 68–76.

Beenakker, C. W. J. 1997. Random-matrix theory of quantum transport. *Rev. Mod. Phys.* 69: 731–808.

Beenakker, C. W. J. and van Houten, H. 1991. Quantum transport in semiconductor nanostructures. *Solid State Phys.* 44: 1–111.

Blanter, Y. M. and Büttiker, M. 2000. Shot noise in mesoscopic conductors. *Phys. Rep.* 336, 2: 1–166.

Böhler, T., Grebing, J., Mayer-Gindner, A., Löhneysen, H. v., and Scheer, E. 2004. Mechanically controllable break junctions in use as electrodes in molecular electronics. *Nanotechnology* 15: 465–472.

Bolotin, K. I., Kuemmeth, F., and Ralph, D. C. 2006. Anisotropic magnetoresistance and anisotropic tunneling magnetoresistance due to quantum interference in ferromagnetic metal break junctions. *Phys Rev. Lett.* 97: 127202.

Brandbyge, M., Sørensen, M. R., and Jacobsen, K. W. 1997 Conductance eigenchannels in nanocontacts. *Phys. Rev. B* 56: 14956–14959.

Bruno, P. 1999. Geometrically constrained magnetic wall. *Phys. Rev. Lett.* 83: 2425–2428.

Büttiker, M. 1990. Scattering theory of thermal and excess noise in open conductors. *Phys. Rev. Lett.* 65: 2901–2904.

Champagne, A. R., Pasupathy, A. N., and Ralph, D. C. 2005. Mechanically adjustable and electrically gated single-molecule transistors. *Nano Lett.* 5: 305–308.

Costa-Krämer, J. L. 1997. Conductance quantization at room temperature in magnetic and nonmagnetic metallic nanowires. *Phys. Rev. B* 55: R4875–R4878.

Costa-Krämer, J. L., García-Mochales, N. P., and Serena, P. A. 1995. Nanowire formation in macroscopic metallic contacts: Quantum mechanical conductance tapping a table top. *Surf. Sci.* 342: L1144–L1149.

Costa-Krämer, J. L., Garcia, N., and Olin, H. 1997. Conductance quantization in bismuth nanowires at 4 K. *Phys. Rev. Lett.* 78: 4990–4993.

Cuevas, J. C., Martín-Rodero, A., and Levy Yeyati, A. 1996. Hamiltonian approach to the transport properties of superconducting quantum point contacts. *Phys. Rev. B* 54: 7366–7379.

Cuevas, J. C., Levy Yeyati, A., and Martín-Rodero, A. 1998a. Microscopic origin of conducting channels in metallic atomic-size contacts. *Phys. Rev. Lett.* 80: 1066–1069

Cuevas, J. C., Levy Yeyati, A., Martin-Rodero, A. et al. 1998b. Evolution of conducting channels in metallic atomic contacts under elastic deformation. *Phys. Rev. Lett.* 81: 2990–2993.

de Jong, M. J. M. and Beenakker, C. W. J. 1997. Shot noise in mesoscopic systems. In: *Mesoscopic Electron Transport*. Eds. L. L. Sohn, L. P. Kouwenhoven, and G. Schön, NATO-ASI Series E, *Appl. Sci.*, 345: 225–258. Dordrecht, the Netherlands: Kluwer Academic Publishers.

Devos, A. 2009. Phonons in nanoscale objects. In: *Handbook of Nanophysics*. Ed. K. Sattler. Boca Raton, FL: Taylor & Francis.

Dreher, M., Heurich, J., Cuevas, J. C., Nielaba, P., and Scheer, E. 2005. Theoretical analysis of the conductance histograms of Au atomic contacts. *Phys. Rev. B* 72: 075435.

Egle, S., Bacca, C., Pernau, H.-F., Hüfner, M., Hinzke, D., Nowak, U., and Scheer, E. 2010. Magneto-resistance of atomic-size contacts realized with mechanically controllable break-junctions. *Phys. Rev. B* (in press).

Gabureac, M., Viret, M., Ott, F., and Fermon, C. 2004. Magnetoresistance in nanocontacts induced by magnetostrictive effect. *Phys. Rev. B* 69: 100401 (R).

García-Mochales, P. and Serena, P. A. 1997. Disorder as origin of residual resistance in nanowires. *Phys. Rev. Lett.* 79: 2316–2319.

García-Mochales, P., Serena, P. A., Garcia N., and Costa-Krämer, J. L. 1996. Conductance in disordered nanowires: Forward and back-scattering. *Phys. Rev. B* 53: 10268–10280.

Gimzewski, J. K. and Möller, R. 1987. Transition from the tunnelling regime to point contact studied using scanning tunnelling microscopy. *Physica B* 36: 1284–1287.

Grabert, H. 2009. The nanoscale free-electron model. In: *Handbook of Nanophysics*. Ed. K. Sattler. Boca Raton, FL: Taylor & Francis.

Grüter, L., Gonzalez, M. T., Huber, R. et al. 2005. Electrical conductance of atomic contacts in liquid environments. *Small* 1: 1067–1070.

Häfner, M., Konrad, P., Pauly, F., Heurich, J., Cuevas, J. C., and Scheer, E. 2004. Conduction channels of one-atom zinc contacts. *Phys. Rev. B* 70: 241404(R).

Häfner, M., Viljas J. K., Frustraglia, D. et al. 2008. Theoretical study of the conductance of ferromagnetic atomic-sized contacts. *Phys. Rev. B* 77: 104409.

Hoffmann, R., Weissenberger, D., Hawecker, J. et al. 2008. Conductance of gold nanojunctions thinned by electromigration. *Appl. Phys. Lett.* 93: 043118.

Hua, S. Z. and Chopra, H. D. 2003. 100,000% ballistic magnetoresistance in stable Ni nanocontacts at room temperature. *Phys. Rev. B* 67: 060401(R).

Kirchner, S., Kroha, J., Wölfle, P., and Scheer, E. 2003. Conductance quasi quantization of quantum point contacts: Why tight binding models are insufficient. In: *Anderson Localization and Its Ramifications: Disorder, Phase Coherence, and Electron Correlations*. Eds. S. Kettemann and T. Brandes. Heidelberg, Germany: Springer.

Kizuka, T. 1998. Atomic process of point contact formation in gold studied by time-resolved high-resolution transmission electron microscopy. *Phys. Rev. Lett.* 81: 4448–4451.

Kondo, Y. and Takayanagi, K. 1997. Gold nanobridge stabilized by surface structure. *Phys. Rev. Lett.* 79: 3455–3458.

Konrad, P., Brenner, P. Bacca, C., Löhneysen, H. V., and Scheer, E. 2005. Stable single-atom contacts of zinc whiskers. *Appl. Phys. Lett.* 86: 213115.

Krans, J. M. and van Ruitenbeek, J. M. 1994. Subquantum conductance steps in atom-sized contacts of the semimetal Sb. *Phys. Rev. B* 50: 17659–17661.

Krans, J. M., Muller, C. J., Yanson, I. K., Govaert, Th. M., Hesper, R., and van Ruitenbeek, J. M. 1993. One-atom point contacts. *Phys. Rev. B* 48: 14721–14724.

Krans, J. M., van Ruitenbeek, J. M., Fisun, V. V., Yanson, I. K., and de Jongh, L. J. 1995. The signature of conductance quantization in metallic point contacts. *Nature* 375: 767–769.

Kröger, J. 2010. Contact experiments with a scanning tunneling microscope. In: *Handbook of Nanophysics: Principles and Methods*. Ed. K. Sattler. Boca Raton, FL: Taylor & Francis.

Kumar, A., Saminadayar, L., Glattli, D. C., Jin, Y., and Etienne, B. 1996. Experimental test of the quantum shot noise reduction theory. *Phys. Rev. Lett.* 76: 2778–2781.

Landauer, R. 1957. Spatial variation of currents and fields due to localized scatterers in metallic conduction. *IBM J. Res. Dev.* 1: 223–231.

Landauer, R. 1970. Electrical resistance of disordered one-dimensional lattices. *Philos. Mag.* 21: 863–867.

Levy Yeyati, A., Martín-Rodero, A., and Flores, F. 1997. Conductance quantization and electron resonances in sharp tips and atomic-size contacts. *Phys. Rev. B* 56, 10369–10372.

Ludoph, B, van der Post, N., Bratus', E. N. et al. 2000. Multiple Andreev reflection in single-atom niobium junctions. *Phys. Rev. B* 61: 8561–8569.

Mares, A. I. and van Ruitenbeek, J. M. 2005. Observation of shell effects in nanowires for the noble metals Cu, Ag, and Au. *Phys. Rev. B* 72: 205402.

Moreland, J. and Ekin, J. W. 1985. Electron tunneling experiments using Nb-Sn "break" junctions. *J. Appl. Phys.* 58: 3888–3895.

Moreland, J. and Hansma, P. K. 1984. Electromagnetic squeezer for compressing squeezable electron tunneling junctions. *Rev. Sci. Instrum.* 55: 399–403.

Muller, C. J., van Ruitenbeek, J. M., and de Jongh, L. J. 1992. Experimental observation of the transition from weak link to tunnel junction. *Physica C* 191: 485–504.

Naidyuk, Y. G. and Yanson, I. K. 2005. *Point-Contact Spectroscopy*. Berlin, Germany: Springer.

Ohnishi, H., Kondo, Y., and Takayanagi, K. 1998. Quantized conductance through individual rows of suspended gold atoms, *Nature* 395: 780–785.

Olesen, L., Lægsgaard, E., Stensgaard, I. et al. 1994. Quantized conductance in an atom-sized point contact. *Phys. Rev. Lett.* 72: 2251–2254.

Ono, T., Ooka, Y., Miyajima, H. et al. 1999. $2e^2/h$ to e^2/h switching of quantum conductance associated with a change in nanoscale ferromagnetic domain structure. *Appl. Phys. Lett.* 75: 1622–1624.

Papaconstantopoulos, D. A. 1986. *Handbook of the Band Structure of Elemental Solids*. New York : Plenum Press.

Pérez-Willard, F., Cuevas, J. C., Sürgers, C. et al. 2004. Determining the current polarization in Al/Co nanostructured point contacts. *Phys. Rev. B* 69: 140502 (R).

Reznikov, M., Heiblum, M., Shtrikmann, H., and Mahalu, D. 1995. Temporal correlation of electrons: Suppression of shot noise in a ballistic quantum point contact. *Phys. Rev. Lett.* 75: 3340–3343.

Rodrigo, J. G., García-Martín, A., Sáenz, J. J., and Vieira, S. 2002. Quantum conductance in semimetallic bismuth nanocontacts. *Phys. Rev. Lett.* 88: 246801.

Rodrigues, V. and Ugarte, D. 2001. Real-time imaging of atomistic process in one-atom-thick metal junctions. *Phys. Rev. B* 63: 073405.

Rubio, G., Agraït, N., and Vieira, S. 1996. Atomic-sized metallic contacts: Mechanical properties and electronic transport. *Phys. Rev. Lett.* 76: 2302–2305.

Sánchez-Portal, D., Untiedt, C., Soler, J. M., Sáenz, J. J., and Agraït, N. 1997. Nanocontacts: Probing electronic structure under extreme uniaxial strains. *Phys. Rev. Lett.* 79: 4198–4201.

Scheer, E., Joyez, P., Esteve, D., Urbina, C., and Devoret, M. H. 1997. Conduction channels transmissions of atomic-size aluminum contacts *Phys. Rev. Lett.* 78: 3535–3538.

Scheer, E., Agraït, N., Cuevas, J. C. et al. 1998. The signature of chemical valence in the electrical conduction through a single atom contact. *Nature (London)* 394: 154–157.

Scheer, E., Belzig, W., Naveh, Y., Devoret, M. H., Esteve, D., and Urbina, C. 2001. Proximity effect and multiple Andreev reflections in gold point contacts. *Phys. Rev. Lett.* 86: 284–287.

Scheer, E., Konrad, P., Bacca, C. et al. 2006. Correlation between transport properties and atomic configuration of atomic contacts of zinc, by low-temperature measurements. *Phys. Rev. B* 74: 205403.

Sirvent, C., Rodrigo, J. G., Vieira, S. et al. 1996. Conductance step for a single-atom contact in the scanning tunneling microscope: Noble and transition metals. *Phys. Rev. B* 53: 16086–16090.

Soulen, R. J., Byers, J. M., Osofsky, M. S. et al. 1998. Measuring the spin polarization of a metal with a superconducting point contact. *Science* 282: 85–88.

Stafford, C. A., Baeriswyl, D., and Bürki, J. 1997. Jellium model of metallic nanocohesion. *Phys. Rev. Lett.* 79: 2863–2866.

Torres, J. A., Pascual, J. I., and Sáenz, J. J. 1994. Theory of conduction through narrow constrictions in a three-dimensional electron gas. *Phys. Rev. B* 49: 1538–1541.

Trouwborst, M. L., van der Molen, S. J., and van Wees, B. J. 2006. The role of Joule heating in the formation of nanogaps by electromigration. *J. Appl. Phys.* 99: 114316.

Untiedt, C., Dekker, D. M. T., Djukic, D. et al. 2004. Absence of magnetically induced fractional quantization in atomic contacts. *Phys. Rev. B* 69: 081401.

van den Brom, H. and van Ruitenbeek, J. M. 1999. Quantum suppression of shot noise in atom-size metallic contacts. *Phys. Rev. Lett.* 82: 1526–1529.

van der Zant, H. S. J., Osorio, E. A., Poot, M. et al. 2006. Electromigrated molecular junctions. *Phys. Status Solidi B* 243: 3408–3412.

van Ruitenbeek, J. M., Alvarez, A., Piñeyro, I. et al. 1996. Adjustable nanofabricated atomic size contacts. *Rev. Sci. Instrum.* 67: 108–111.

van Wees, B. J., van Houten, H. H., Beenakker, C. W. J. et al. 1988. Quantised conductamce of point contacts in a two-dimensional electron gas. *Phys. Rev. Lett.* 60: 848–850.

Viret, M., Berger, S., Gabureac, M. et al. 2002. Magnetoresistance through a single nickel atom. *Phys. Rev. B* 66: 220401 (R).

Viret, M., Gabureac, M., Ott, F. et al. 2006. Giant anisotropic magneto-resistance in ferromagnetic atomic contacts. *Eur. Phys. J. B* 51: 1–4.

Waitz, R., Schecker, O., and Scheer, E. 2008. Nanofabricated adjustable multicontact devices on membranes. *Rev. Sci. Instrum.* 79: 093901.

Wharam, D. A., Thornton, T. J., Newbury, R. et al. 1988. One-dimensional transport and the quantisation of the ballistic resistance, *J. Phys. C* 21: L209–L214.

Wiesendanger, R. 1994. *Scanning Probe Microscopy and Spectroscopy.* Cambridge, U.K.: Cambridge University Press.

Wu, Z. M., Steinacher, M., Huber, R. et al. 2007. Feedback controlled electromigration in four-terminal nanojunctions. *Appl. Phys. Lett.* 91: 053118.

Yanson, A. I. and van Ruitenbeek, J. M. 1997. Do histograms constitute a proof for conductance quantization? *Phys. Rev. Lett.* 79: 2157.

Yanson, A. I., Yanson, I. K., and van Ruitenbeek, J. M. 1999. Observation of shell structure in sodium nanowires. *Nature* 400: 144–146.

Yanson, A. I., Yanson, I. K., and van Ruitenbeek, J. M. 2000. Supershell structure in alkali metal nanowires. *Phys. Rev. Lett.* 84: 5832–5835.

Yanson, A. I., Yanson, I. K., and van Ruitenbeek, J. M. 2001. Crossover from electronic to atomic shell structure in alkali metal nanowires. *Phys. Rev. Lett.* 87: 216805.

Yanson, I. K., Shklyarevskii, O. I., Csonka, S. et al. 2005. Atomic-size oscillations in conductance histograms for gold nanowires and the influence of work hardening. *Phys. Rev. Lett.* 95: 256806.

Yanson, I. K., Shklyarevskii, O. I., van Ruitenbeek, J. M., and Speller, S. 2008. Aluminum nanowires: Influence of work hardening on conductance histograms. *Phys. Rev. B* 77: 033411.

Zozoulenko, I. V. and Ihnatsenka, S. 2010. Quantum point contact in two-dimensional electron gas. In: *Handbook of Nanophysics: Nanotubes and Nanowires.* Ed. K. Sattler. Boca Raton, FL: Taylor & Francis.

38

Quantum Point Contact in Two-Dimensional Electron Gas

Igor V. Zozoulenko
Linköping University

Siarhei Ihnatsenka
Simon Fraser University

38.1 Introduction

Semiconductor nanostructures and mesoscopic electronic devices based on a two-dimensional electron gas (2DEG) have been a focus of attention for the semiconductor community during the past few decades. This is due to the new fundamental physics that these structures exhibit, and also due to possible applications in future electronic devices and devices for quantum information processing. A quantum point contact (QPC) represents the cornerstone of the mesoscopic physics. This is not only because the QPC is the simplest mesoscopic device, but also because most of the mesoscopic devices contain the QPC as their integral part. This has apparently motivated a strong interest in the various aspects of the electronic and transport properties of the QPC. In fact, one of the most important discoveries that gave a strong momentum to the whole field of mesoscopic physics was the discovery of the conductance quantization of the QPC in 1988 (van Wees et al., 1988; Wharam et al., 1988). Studies of the conductance quantization in the QPC provided valuable information not only on the fundamentals of the phase-coherent electron motion in low-dimensional structures, but also outlined important material aspects (such as the effect of impurities, potential confinement, etc.). The QPC has also proven to be the key system for studying the various aspects of the quantum Hall physics. This applies both to the pioneering studies in the beginning of the "mesoscopic era" in the early 1990s focusing on the basic aspects of the subband depopulation as well as to more recent studies of the fractional Hall regime revealing the exotic

features of interacting electrons in a high magnetic field such as fractional statistics and charge. It is also important to mention that apart from strong fundamental interest, a QPC found its important practical application as a noninvasive voltage probe and a single-electron charge detector. By now, many features of the electronic and transport properties of the QPC are well understood. However, even 20 years after the discovery of the conductance quantization, some of the important aspects of the QPC conductance still represent topics at the forefront of research and lively debates, where the emphasis is shifted to aspects of the electron interaction, spin, and nonequilibrium effects.

In this chapter, we present the basic physics of the QPC in a 2DEG. Section 38.2 outlines the major experimental results of the conductance quantization along with its essential theory. Section 38.3 discusses the transport in the QPC in the presence of a magnetic field. Section 38.4 addresses the many-body effects in the QPC with a particular focus on the so-called 0.7 anomaly in conductance. A brief conclusion and an outline are given in Section 38.5.

38.2 Conductance Quantization in the Quantum Point Contact

38.2.1 Electrostatic Confinement in Modulation-Doped Heterostructures

A QPC represents a constriction defined in a 2DEG formed in modulation-doped semiconductor heterostructures. Figure 38.1a

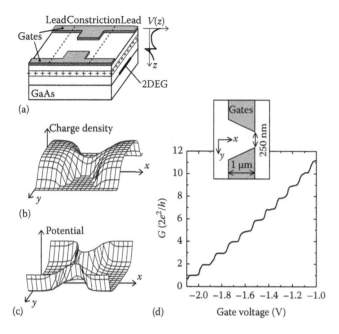

FIGURE 38.1 (a) A schematic geometry of a quantum point contact defined by split-gates in a GaAs/AlGaAs heterostructure. An inset to the right shows the band diagram. (b) The electron density and (c) the total self-consisted potential in the vicinity of the quantum point contact. (d) The conductance quantization of the QPC shown in the inset. (Adapted from van Wees, B.J. et al., *Phys. Rev. Lett.*, 60, 848, 1988.)

illustrates a typical GaAs/AlGaAs heterostructure. It has several layers including cap GaAs layers, an *n*-doped AlGaAs layer, a spacer AlGaAs layer, and a GaAs substrate. The calculated band diagram of the structure is shown in the inset to Figure 38.1a (for a detailed discussion of the 2DEG in modulation-doped heterostructures, see e.g., Davies, 1997). Due to the conduction band discontinuity between GaAs and AlGaAs, a triangular-shaped quantum well is formed at the interface between the spacer layer and the substrate, trapping electrons from the *n*-doped layer. Typically, only the lowest state in the well is occupied. The electrons trapped in the well form a high mobility two-dimensional electron gas. The electron mobility in the best samples reaches $\mu \sim 10^4$ m²/Vs, which corresponds to the mean free path of the order of 100 μm. This is to be contrasted with the mobilities $\mu \sim 1$ m²/Vs in Si MOSFET devices where 2DEG is trapped in an inversion layer at the interface between Si and SiO_2. There are two major factors that contribute to such high mobilities in the modulation-doped heterostructures. First, the dopants are spatially separated from the quantum well where the 2DEG resides, which eliminates the large-angle scattering due to direct collisions between electrons and donors. Second, the surface scattering on the interface between the spacer AlGaAs layer and the GaAs substrate is strongly suppressed because of an almost perfect match between the lattice constants of GaAs and AlGaAs. The electron phase coherence time in the GaAs/AlGaAs modulation-doped heterostructures at milliKelvin temperatures is of the order of $\tau_\phi \sim 0.1$–0.01 ns, which for typical sheet electron

densities ($n_s \sim 3 \times 10^{15}$ m⁻²) corresponds to the coherence length of the order of $l_\phi \sim 20$ μm.

One of the most common ways to define a lateral confinement in the 2DEG is to use the split-gate technique (Thornton et al., 1986). By applying a negative gate voltage to the gates, the electrons underneath the gates become depleted such that electron motion is confined to the regions defined by the lateral patterning of the AlGaAs heterostructure. Other popular techniques for the lateral patterning include an etching technique where depletion is achieved by the removal of a thin layer on top of the heterostructure. This brings the surface of the heterostructure (containing electrons trapped in the surface states) close to the 2DEG, which is sufficient to deplete the electrons below the etched regions. A closely related technique is the local oxidation method where, instead of etching, an oxide line is written with the help of an atomic force microscope (AFM; for a review of the experimental techniques, see e.g., the textbook by Heinzel, 2007).

The simplest lateral structure that can be defined in the 2DEG is a constriction, termed as a QPC, see Figure 38.1a. The constrictions defined in metals (known as Sharvin point contacts) have been intensively studied since the mid-1960s. Even thought these studies addressed the ballistic regime when the electron mean free path exceeded the characteristic size of the constriction, the electron transport regime in metals still remained classical because the Fermi wavelength ($\lambda \sim 0.5$ nm) was much smaller than any characteristic system size. In contrast, the semiconductor heterostructures offer a unique opportunity to study the quantum regime not only because of the exceptionally large electron mean free path, but primarily because the Fermi wavelength ($\lambda \sim 50$ nm) is comparable to a characteristic system size L, and, at the same time, the phase coherence length may significantly exceed it, $l_\phi \gg L$.

In 1988, two independent experimental groups made the remarkable discovery that the conductance through a narrow constriction in a 2DEG is quantized (van Wees et al., 1988; Wharam et al., 1988), see Figure 38.1b. The essential mechanism basic to this fact was immediately recognized: transverse quantization allows the propagation of a discrete set of modes through a QPC and, as follows from the Landauer–Buttiker formalisms, each such (spin degenerate) channel contributes $2e^2/h$ to the conductance. A more detailed understanding of the phenomenon, including the shape of the conductance steps, involves the precise geometry of the QPC and coupling to the reservoirs and, in addition, the influence of disorder and of temperature. In this section, we concentrate on the exact and approximate descriptions of transport through QPCs with various geometries including the effect of the impurities.

38.2.2 Basics of the Conductance Quantization

The central quantity in transport calculations is the conductance. In the linear response regime, it is given by the Landauer formula $G = \sum_\sigma G^\sigma$

$$G^{\sigma} = -\frac{e^2}{h} \int dE T^{\sigma}(E) \frac{\partial f(E - E_F)}{\partial E}, \qquad (38.1)$$

where

$T^{\sigma}(E)$ is the total transmission coefficient for the spin channel $\sigma = \uparrow, \downarrow$

$f(E - E_F)$ is the Fermi-Dirac distribution function

E_F is the Fermi energy (for the derivation of the Landauer formula, see e.g., the textbook by Davies, 1995)

In this section, we limit ourselves to the spin-degenerate electrons; the electron interaction and spin effects in the QPC will be discussed in the sections that follow. In the linear response regime for spin-degenerate electrons in the limit of zero temperature, the above expression for the conductance is reduced to

$$G = \frac{2e^2}{h} T(E_F). \qquad (38.2)$$

In order to calculate the transmission coefficient, let us define the scattering states of the system. For simplicity, let us first consider a hard-wall confining potential. Divide the QPC structure into three regions: two wide semi-infinite leads of the constant width w playing a role of electron reservoirs, and the central region including a constriction, see Figure 38.1a. Because the leads are of a constant width, in the leads regions one can separate the variables in the Schrödinger equation such that its general solution for the energy E reads

$$\psi_{lead}(x, y) = (A_{\alpha} e^{ik_{\alpha}x} + B_{\alpha} e^{-ik_{\alpha}x}) \phi_{\alpha}(y), \qquad (38.3)$$

where

$\phi_{\alpha}(y) = \sqrt{2/\pi} \sin(\pi\alpha y/w)$ represents the αth transverse mode in the leads

$E = (\hbar^2 k^2)/2m^*$, where $k^2 = (k_{\alpha})^2 + (k_{\alpha}^{\perp})^2$ with k_{α} and $k_{\alpha}^{\perp} = (\pi\alpha)/w$ being respectively the wave vectors of the longitudinal and transverse motions, m^* is the effective mass for GaAs, and w is the width of the leads

Note that the modes with $(k_{\alpha})^2 > 0$ are propagating, whereas those with an imaginary k-vector, $(k_{\alpha})^2 < 0$, are evanescent. Let us inject in the left lead electrons in the propagating transverse mode α. This state will be scattered by the constrictions and thus transmitted and reflected into all available modes of β (both propagating and evanescent) in the right and left leads with the corresponding transmission and reflections amplitudes $t_{\beta\alpha}$ and $r_{\beta\alpha}$. The total wave functions in the left and right leads corresponding to the incoming state α therefore reads

$$\psi_{left}(x, y) = e^{ik_{\alpha}x} + \sum_{\beta} r_{\beta\alpha} e^{-ik_{\beta}x} \phi_{\beta}(y),$$

$$\psi_{right}(x, y) = \sum_{\beta} t_{\beta\alpha} e^{ik_{\beta}x} \phi_{\beta}(y),$$

where the summation runs over all modes (both propagating and evanescent). The reflection and transmission coefficients from the propagating mode α to the propagating mode β are defined as the ratio of the fluxes associated with each mode:

$$T_{\beta\alpha} = \frac{k_{\beta}}{k_{\alpha}} |t_{\beta\alpha}|^2; \quad R_{\beta\alpha} = |r_{\beta\alpha}|^2.$$

Assuming that all incoming modes are independent (i.e., there is no interference between them), the total transmission and reflection coefficients are obtained by summing the contribution from all the transmitted and reflected states:

$$T = \sum_{\alpha,\beta} T_{\beta\alpha}; \quad R = \sum_{\alpha,\beta} R_{\beta\alpha}.$$

The transmission and reflection coefficients can be calculated exactly using well-established numerical methods such as the scattering matrix or Green's function techniques (Datta, 1997). In order to get a better insight into the problem at hand, we start with some approximate treatment. Let us expand the wave function for the QPC structure in terms of local transverse eigenfunctions, $\Psi(x, y) = \sum_{\alpha} \chi_{\alpha}(x) \varphi_{\alpha}(x, y)$ (note that in the lead regions, this form of the wave function reduces to Equation 38.3). Substituting this expansion into the Schrödinger equation, we obtain the following coupled differential equations for the mode coefficients $\chi_{\alpha}(x)$:

$$-\frac{\hbar^2}{2m^*} \frac{d^2\chi_{\alpha}(x)}{dx^2} + E_{\alpha}(x)\chi_{\alpha}(x)$$

$$= E\chi_{\alpha}(x) + \frac{\hbar^2}{2m^*} \sum_{\beta} \left[2A_{\alpha\beta}(x)\frac{d}{dx} + B_{\alpha\beta}(x) \right]\chi_{\beta}(x), \quad (38.4)$$

where the transverse eigenfunctions $\varphi_{\alpha}(x, y)$ satisfy the (local in x) transverse Schrödinger equation

$$-\frac{\hbar^2}{2m^*} \frac{\partial^2\varphi_{\alpha}(x, y)}{\partial x^2} + V(x, y)\varphi_{\alpha}(x, y) = E_{\alpha}(x)\varphi_{\alpha}(x, y),$$

with the transverse eigenenergy (local in x) $E_{\alpha}(x) = (\hbar^2 k_{\alpha}^{\perp 2}(x))/2m^*$ (with $k_{\alpha}^{\perp}(x) = (\pi\alpha)/w(x)$) and the coupling matrixes are given by the expressions

$$A_{\alpha\beta}(x) = \int \varphi_{\alpha}(x, y)\frac{\partial}{\partial x}\varphi_{\beta}(x, y)dy,$$

$$B_{\alpha\beta}(x) = \int \varphi_{\alpha}(x, y)\frac{\partial^2}{\partial x^2}\varphi_{\beta}(x, y)dy.$$

In a realistic QPC structure, the confining potential varies slowly along the x-direction and, therefore, one may expect an *adiabatic* transport regime when the mode mixing between different modes is strongly suppressed (Glazman et al., 1988; Yacoby, 1990; Castaño and Kirczenow, 1992; Maaø et al., 1994). This corresponds to the case when the last term on the right-hand side of Equation 38.4 (which gives rise to the mode mixing) is set to zero. Without the last term, Equation 38.4 has a transparent physical meaning, where each transverse mode a propagates through the constriction independently and the transverse eigenenergy $E_\alpha(x) = (\hbar^2 k_\alpha^{\perp 2})/2m^*$ (where $k_\alpha^\perp = (\pi\alpha)/w(x)$) plays the role of the effective potential energy. This is illustrated in Figure 38.2a, which shows the eigenenergies $E_\alpha(x)$ for different modes α along the constriction. As the width of the constriction $w(x)$ narrows, the effective potential $E_\alpha(x) \sim w(x)^{-2}$ raises. When it exceeds the energy of the incoming electrons E, the propagating mode α turns into an evanescent one and the corresponding transmission probability decreases drastically. Thus, the total transmission coefficient T is expected to be an integer number equal to the number of propagating modes in the narrowest part of the constriction. The integer plateaus in T are separated by the transitions regions corresponding to the tunneling of the evanescent states through the barrier in the narrowest part of the constriction. Figure 38.2b shows a comparison of the exact numerical calculation for the transmission coefficient with the adiabatic approximation for some representative QPC (shown in the inset). This comparison clearly shows that the adiabatic approximation neglecting the coupling of the modes is remarkably successful considering the abruptness of the constriction shown.

There was one aspect in the above discussion that has been silently brushed under the carpet. Namely, for any realistic constriction, the adiabatic approximation inevitably breaks down when the constriction becomes sufficiently wide. Nevertheless, as Figure 38.2b demonstrates, this approximation provides an excellent approximation to the exact conductance. In order to understand this, it is important to distinguish between the fundamental merits of the global approximation itself, which neglects intermode scattering altogether, and its qualities as a method for calculating the conductance. Reasons for the success

of the adiabatic approximation have been pointed out by Yacoby and Imry (1990). They calculated the leading corrections to the adiabatic approximation and demonstrated that the adiabaticity breaks down (i.e., the mode mixing becomes significant) at some distance from the constriction d when $w'(d) \sim 1/N_{prop}(d)$, with $N_{prop}(d) = \text{int}[kw(d)/\pi]$ being the local number of propagating modes (typically $N_{prop}(d) \gg 1$). At the same time, at this distance the correction to the reflection amplitudes is of the order of $|r_{\alpha\beta}(d)| \sim N_{prop}^{-2}(d) \ll 1$. This means that at the distance where the adiabaticity breaks down, the backscattering is negligible; thus, the total transmission coefficient is not affected (even though the current is redistributed between the modes because of the mode mixing).

It should be stressed that the adiabaticity is not the necessary condition for the observation of the conductance quantization of the QPC. The conductance of a QPC with abrupt corners has been studied by Szafer and Stone (1989). They started from the analysis of the impedance-matching and mode-conversion behaviors at the interfaces between the constriction and two-dimensional region (wide-narrow geometry, see Figure 38.3a). Based on the conservation of the transverse wave vectors in the narrow and wide regions, $q_W \approx q_N$, they provided a simple analytical expression for the conductance of the QPC that reproduces the corresponding exact numerical value for the transmission coefficient T, see Figure 38.3b.

In the discussions presented above, we used a model of a hard wall confinement. In a realistic QPC with electrostatically induced constrictions, the confinement is a smooth function forming a saddle in the narrowest part of the QPC. The confinement potential is well approximated by a parabolic function (see Figure 38.1c for illustration)

$$V(x,y) = V_0 - \frac{1}{2}m\omega_x^2 x^2 + \frac{1}{2}m\omega_y^2 y^2, \qquad (38.5)$$

where

V_0 is the electrostatic potential at the saddle
ω_x and ω_y give the curvature of the potential

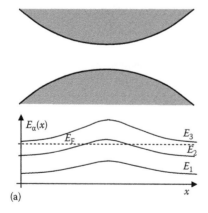
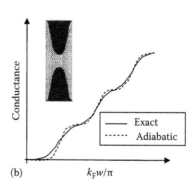

FIGURE 38.2 (a) Transverse eigenenergies $E_\alpha(x)$ playing the role of the effective potential for mode α in the adiabatic approximation. (b) A comparison between the adiabatic approximation and the numerically exact conductance of a QPC shown in the inset. (Adapted from Maaø, F.A. et al., *Phys. Rev. B*, 50, 17320, 1994.)

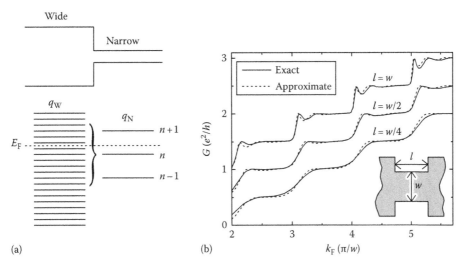

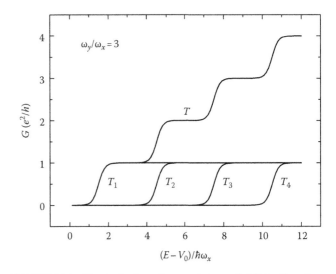

FIGURE 38.3 (a) A narrow-wide geometry and corresponding transfers wave vectors q_W and q_N in the wide and narrow regions. A brace illustrates the momentum conservation showing channels in the wide region that are coupled the mode n in the narrow region. (b) A comparison of the exact numerical conductance and the approximate analytical expression for the QPC with abrupt corners shown in the inset. (Adapted from Szafer, A. and Stone A.D., *Phys. Rev. Lett.*, 62, 300, 1989.)

The Hamiltonian is thus separable into a transverse part associated with energies $E_n = \hbar\omega_y(n + (1/2))$, $n = 1, 2, 3, \ldots$, and a part associated with the motion in the longitudinal x-direction in the effective potential $V_0 + E_n - (1/2)m\omega_x^2 x^2$. This effective potential can be viewed as the band bottom of the nth quantum channels (subband) near the saddle point (as discussed and illustrated in Figure 38.2a). Because of the separation of variables, there is no mode mixing between the different channels and the conductance is given by the sum of the corresponding transmission coefficient, $G = (2e^2/h)\sum_n T_n$, where the transmission probability for the nth channel is available in an analytic form (Buttiker, 1990a):

$$T_n = \frac{1}{1 + \exp\left(-2\pi((E - E_n - V_0)/\hbar\omega_x)\right)}. \quad (38.6)$$

Each T_n represents a step-like function where the transition region between $T = 0$ and 1 describes the tunneling through the saddle region, see Figure 38.4.

38.2.3 Breakdown of Quantized Conductance due to Impurities

A theoretical treatment of the conductance quantization in QPC presented above has been limited to ideal geometries and structures disregarding the effect of impurities. However, early experiments on QPC have already demonstrated that the step-like quantization of the conductance becomes quickly destroyed as the constriction is made longer. The breakdown of the quantized conductance is shown to be caused by the long-range impurity potential due to remote donors (note that this potential also represents a major factor limiting mobility in the unconfined 2DEG; Nixon et al., 1991). Figure 38.5 shows a typical pattern of the electron density in the

FIGURE 38.4 The single-channel transmission probabilities T_n, and the total transmission probability $T = \sum_n T_n$. Opening of successive quantum channels over narrow energy intervals leads to the quantization of the conductance. (Adapted from Buttiker, M., *Phys. Rev. B*, 41, 7906, 1990a.)

vicinity of the constriction in the heterostructure similar to the one shown in Figure 38.1a. The self-consistent potential on the depth of the 2DEG includes the contribution from the randomly positioned ionized donors in the doped layer of the heterostructure and from the gates that produce the guiding potential (the analytical expressions for the electrostatic potential from the gates of different shapes are provided by Davies et al., 1995). Surface states pin the chemical potential 0.8 eV below the conduction band on free surfaces of GaAs. The chemical potential is assumed to be flat throughout the structure, and the surface can therefore be treated as an equipotential with any gate voltages superposed. The total self-consisted electron density can be found from the solution of the Thomas-Fermi or Schrödinger equation. The conductance of

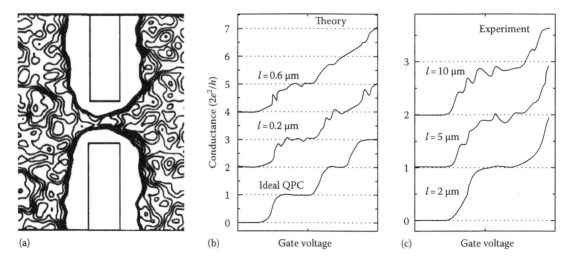

(a)   (b)   Gate voltage   (c)   Gate voltage

FIGURE 38.5 (a) A gate pattern on the surface and the density of electrons in 2DEG for point contacts with $l = 0.2\,\mu m$. (b) The conductance G as a function of the gate voltage V_g, for $l = 0.2$ and $0.6\,\mu m$ gate. The lowest curve shows the clean quantized steps found in a smooth system. (Adapted from Nixon, J.A. et al., *Phys. Rev. B*, 43, 12638, 1991.) (c) The conductance vs. the effective gate voltage at for QPCs of different lengths. (Adapted from Tarucha, S. et al., *Solid State Commun.*, 94, 413, 1995.) In (b) and (c) curves are offset for clarity. Note that in the shown example the experimental quantization survives in a QPC of the length of $l = 10\,\mu m$, whereas the calculated QPC the quantization is lost already for $l = 0.6\,\mu m$. This reflects a difference in parameters of the heterostructures.

electrons through the self-consistent potential is then found quantum mechanically. Perfect leads, in which no scattering occurs, are attached to the system by extending the potential profile at the left and right edges outward to infinity, and the conductance is calculated on the basis of the Landauer formula (Equation 38.1). The computed conductance shows that the quantization becomes gradually destroyed as the length of the QPC becomes larger than the correlation length of the impurity potential, see Figure 38.5. For a comparison, the experimental conductance exhibiting the same trend is also shown.

38.2.4 Visualizing the Electron Flow through QPC

Recent advances in scanning electron microscopy (SEM) made it possible to visualize the current flowing through a QPC (Crook et al., 2000; Topinka et al., 2000). Figure 38.6a shows the experimental technique used to image electron flow through the QPC. The heterostructure with a split-gate QPC is mounted in an AFM. Electron flow from the QPC is imaged by scanning a negatively charged AFM tip above the surface of the device and simultaneously measuring the position-dependent conductance. Capacitive coupling reduces the density of the 2DEG in a small spot directly beneath the tip, creating a depletion region that backscatters electron waves. When the tip is over areas of high electron flow, the conductance through the QPC is decreased; whereas when the tip is over areas of relatively low electron flow, the conductance through the QPC is unmodified. By scanning the tip over the sample, a two-dimensional image of electron flow can be obtained. Because the tip is capacitively coupled to the 2DEG, no current flows between the tip and the 2DEG.

Figure 38.6b and c shows the images of electron flow from the QPC corresponding to the second and third conduction plateaus. The current flow pattern on both sides of the QPC is independent of the current direction. As expected, the number of modes in the flow pattern equals the number of modes in the constriction. As each new mode is opened in the QPC, new angular lobes of electron flow appear, and the flow pattern widens. The widening of the angular pattern can be easily understood from a diagram illustrating the Fermi sphere in the k-space. The direction of the incoming/outgoing beam is given by the angle $\tan\alpha = k_y/k_x$, where k_y and k_x are the wave vectors of the transverse and longitudinal motion ($k_y^2 + k_x^2 = k_F^2$, with k_F being the Fermi wave vector). Because of the confinement, the transverse wave vector k_y takes upon quantized values, which results in a larger α for larger mode numbers.

Note also that the current flow pattern shows branching strands instead of smoothly spreading fans expected for an ideal system. These features in the current are due to the focusing of the electron paths by ripples in the background potential (Topinka et al., 2001).

38.2.5 QPC as a Charge Detector

One of the most important applications of a QPC is its utilization as a noninvasive voltage probe and charge detector (Field et al., 1993). When a QPC is biased in the tunneling regime, its conductance ($G \ll 2e^2/h$) is exponentially sensitive to minute variations of the electrical potential. As a result, even a change of the confining potential due to a nearby single electron can strongly affect the QPC conductance. Figure 38.7a shows a schematic diagram of a structure utilizing a QPC to detect a single electron charging in a nearby quantum dot (Field et al., 1993). The bar down the middle (G1) separates two electrical circuits

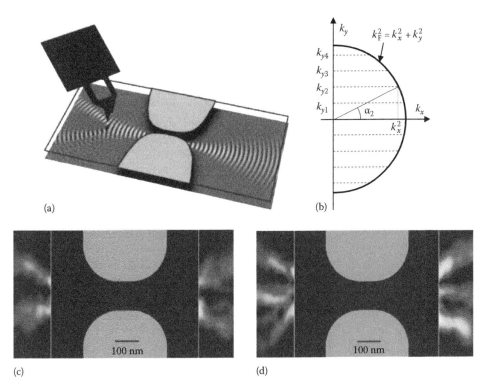

(a)

(b)

(c)

(d)

FIGURE 38.6 (a) An experimental setup showing an AFM cantilever scanning above the surface of a heterostructure. (After Topinka, M.A. et al., *Science*, 289, 2323, 2000.) (b) A schematic diagram of the Fermi sphere and the transverse wave vectors k_y,i. α_2 shows the direction of the electron beam in the second mode. (c) The angular pattern of the electron flow of in the (c) second and (d) third modes. The gray areas outline the position of electrostatic gates. (After Topinka, M.A. et al., *Nature*, 410, 183, 2001.)

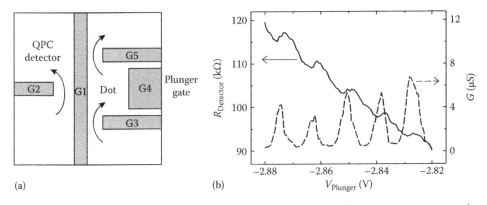

(a)

(b)

FIGURE 38.7 (a) A schematic geometry of a confined quantum dot in a close proximity to a detector QPC in a separate electrical circuit. (b) CB oscillations of the conductance vs. the gate voltage, together with the resistance of the QPC detector circuit. (Adapted from Field, M. et al., *Phys. Rev. Lett.*, 70, 1311, 1993.)

that interact via the electric fields present between the regions in close proximity. This gate is biased sufficiently negative that electrons cannot tunnel across. The gate G2 defines a QPC (detector) that operates in a regime close to the pinch-off. To the right of the bar is a quantum dot formed by gates G3–G5. An applied ac bias causes current to flow from top to bottom, passing through two constrictions formed by gates G3 and G5. These constrictions are nearly pinched off so that the electron number in the dot is nearly quantized and the dot is in the Coulomb blockade

(CB) regime. The extra gate G4 defining the right-hand edge of the dot acts as a "plunger." When the voltage on G4 is swept, CB oscillations are seen in the conductance of the quantum dot.

Figure 38.7b shows the conductance through the dot and the resistance of the split-gate detector as the plunger voltage is swept. The detector resistance has dips on a rising background. These dips directly correlate with the CB oscillations in the quantum dot, such that each dip signals a change in the electrostatic potential caused by charging due to just one single electron. Note that the sensitivity of the QPC detector is so high that it continues to work even after the conductance oscillations have become too small to be measured when gates G3 and G5 were practically pinched off.

Since its first demonstration as a noninvasive voltage probe, a QPC has been used extensively in various applications, including the detection of Coulomb charging around an antidot (Kataoka et al., 1999), a single-shot read-out of an individual electron spin (Elzerman et al., 2004), and a spin relaxation via the nuclei in quantum dots (Johnson et al., 2005), just to mention a few representative examples.

38.2.6 Shot Noise in QPC

In all electronic devices, the current fluctuates in time giving rise to noise. There are several types of noise phenomena of different origins. At low frequencies, the most pronounced is the so-called $1/f$-noise, which is related to the changes of the sample resistance with time (e.g., due to the activation processes leading to the ionization of impurities, etc.). At finite temperatures, a conductor can exhibit noise even in the absence of a current. This is a thermal noise with a "white" spectrum, i.e., frequency independent. By virtue of the fluctuation dissipation theorem, the thermal noise can be directly related to the conductance, hence it does not provide any additional information that can not be obtained from the resistance measurements. More interesting is the shot noise, a frequency independent noise whose origin is due to the fluctuation of electron charges passing through the conductor (to be compared with fluctuations of the number of water drops hitting a small puddle of water during a rainfall).

The noise is usually characterized by its power at a given frequency ω, defined as a Fourier-transform of the current–current correlation function, $P(\omega) = 2\int \langle \Delta I^2(t) \rangle \cos \omega t \, dt$, where $\langle \Delta I^2(t) \rangle \equiv \langle \Delta I(0) \Delta I(t) \rangle$, $\Delta I(t) = I(t) - \overline{I}$ with $\overline{I}$ being the dc average current. For completely uncorrelated events, the shot noise assumes its maximum value

$$P = 2e\overline{I} \tag{38.7}$$

(see e.g., Buttiker 1990b). In classical (macroscopic) conductors, the shot noise is completely suppressed because of a self-averaging character of electron transport where the sample size greatly exceeds the inelastic scattering length, $L \gg l_{in}$. However, in a mesoscopic transport regime when both the elastic and inelastic scattering lengths exceed the size of the structure (which is appropriate for a QPC), the shot noise exhibits new features due to the temporal correlation of electrons. In particular, Lesovik (1989) has shown that the shot noise is reduced because of the correlations that arise due to the Pauli principle that does not allow for electron occupation of the same state. In the regime of ballistic coherent transport, the shot noise power for the spin-degenerate electrons at zero temperature is given by the expression

$$P = 2e \frac{2e^2}{h} V \sum_{i=1}^{N} T_i (1 - T_i), \tag{38.8}$$

where

    V is the applied voltage
    T_i is the transmission coefficient for the ith channel

Since the current of the ith channel $I_i = (2e^2/h)VT_i$, Equation 38.8 predicts the suppression of the noise for each channel by a factor of $1 - T_i$ in comparison with its maximal (fully uncorrelated) value given by Equation 38.7.

The suppression of the shot noise in a QPC was experimentally studied by Reznikov et al. (1995). Figure 38.8 shows the noise power along with QPC conductance as a function of the gate voltage. The measured noise signal is in good agreement with the predicted dependence; Equation 38.8 shows the expected suppression of noise for fully transmitted channels, $T_i = 1$, and peak positions corresponding to $T_i = 1/2$. An investigation of the shot noise makes it possible to obtain information not easily accessible from usual resistance measurements. For example, measurements of quantum shot noise in a QPC in the fractional quantum Hall regime showed unambiguously that, in agreement with the theoretical predictions for the states with filling factors $\nu = q/p$, current is carried by fractional charges $e^* = e/p$, and for the state $\nu = 5/2$, the fractional charge is $e^* = 1/4e$ (see Dolev et al., 2008 and references therein).

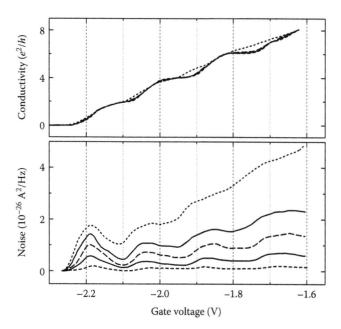

FIGURE 38.8 The noise power R and the conductance G of a QPC vs. the gate voltage V. The noise is measured for several source–drain voltages $V_{SD} = 0.5$, 1, 1.5, 2, and 3 mV (curves respectively from top to bottom). The conductance is shown for the source–drain voltage $V_{SD} = 0.5$ and 3 mV (solid and dashed lines, respectively). (Adapted from Reznikov, M. et al., *Phys. Rev. Lett.*, 75, 3340, 1995.)

38.3 Quantum Point Contact in Magnetic Field

38.3.1 Depopulation of Magnetosubbands: Basics

In a perpendicular magnetic field, a conductance of a QPC exhibits new features including the decrease of a number of quantized plateaus (at a given gate voltage), improved quantization, and the emergence of additional steps at odd values of e^2/h. Figure 38.9a shows the conductance of a QPC as a function of the gate voltage for different values of the magnetic field. The effect of a magnetic field is to reduce the number of plateaus in a given gate voltage interval (see also Figure 38.12d, which shows a decrease of the conductance at the fixed gate voltage as the magnetic field increases). Disregarding the effect of backscattering inside the constriction (which leads to the transition regions between the quantized steps), the conductance of the QPC according to Equation 38.2 can be expressed through the number of the occupied subband, $T \approx 2e^2/hN_{occ}(B)$ (note that Equations 38.1 and 38.2 hold irrespective of the presence/absence of the magnetic field). Thus, in order to understand the features in the QPC magnetoconductance, it is instrumental to inspect the nature of the propagating states and magnetosubband structure of the corresponding homogeneous quantum wire (Beenakker and van Houten, 1991; Davies et al., 1995).

We consider a case of a parabolic quantum wire for which an analytic solution is available. We start first from the case of homogeneous 2DEG in a perpendicular magnetic field. In the Landau gauge $\mathbf{A} = (-By, 0)$, the Schrödinger equation reads $H_0\psi = E\psi$, where H_0 represents the kinetic energy

$$H_0 = -\frac{\hbar^2}{2m^\star}\left\{\left(\frac{\partial}{\partial x} - \frac{ieBy}{\hbar}\right)^2 + \frac{\partial^2}{\partial y^2}\right\}.$$

By separating the variables and writing the solution in the form $\psi(x, y) = e^{ikx}u(y)$, we arrive at the equation for a one-dimensional harmonic oscillator

$$\left(-\frac{\hbar^2}{2m^\star}\frac{d^2}{dy^2} + \frac{1}{2}m\omega_c^2\left(y - y_k^0\right)^2\right)u(y) = Eu(y), \quad (38.9)$$

where $\omega_c = eB/m^\star$ is the cyclotron frequency

$$y_k^0 = -\frac{\hbar k}{eB}$$

The magnetic field effectively introduces a parabolic confinement with the vertex shifted by the distance y_k^0, which depends on the momentum $\hbar k$ in the x-direction. The solution of the above equation is well known and is represented by familiar Landau levels (LLs)

$$E_n = \left(n - \frac{1}{2}\right)\hbar\omega_c, \quad n = 1, 2, 3,\ldots \quad (38.10)$$

Let us now consider a parabolic quantum wire by adding to the Hamiltonian a harmonic confinement potential $V(y) = (1/2)\,m\omega_0^2 y^2$ in the transverse (y-direction), $H = H_0 + V(y)$. The translational invariance in the longitudinal (x-direction) dictates the Bloch form of the wave function, $\psi(x, y) = e^{ikx}u(y)$. By substituting $\psi(x, y)$ into the Schrödinger equation, we arrive at the equation for the transverse eigenfunction $u(y)$

$$\left(-\frac{\hbar^2}{2m}\frac{d^2}{dy^2} + \frac{1}{2}m\omega^2(y - y_k)^2 + \frac{\hbar^2 k^2}{2M_B}\right)u(y) = Eu(y), \quad (38.11)$$

where
$$\omega = (\omega_0^2 + \omega_c^2)^{1/2}$$
$$y_k = y_k^0\,(\omega_c/\omega)$$
$$M_B = m(\omega/\omega_0)^2$$

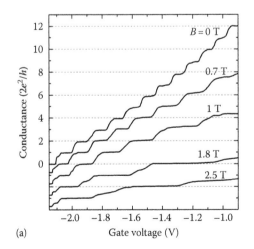

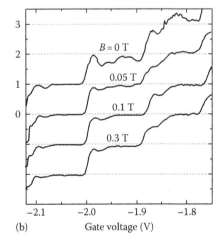

FIGURE 38.9 (a) Transition from quantization in zero field to quantization in high magnetic fields, obtained at several fixed values of the magnetic field at 0.6 K. (b) Improvement of the conductance quantization by the application of a magnetic field, measured at 40 mK. (Adapted from van Wees, B.J. et al., *Phys. Rev. B*, 43, 12431, 1991.)

This equation has the same form as Equation 38.9 where, however, the frequency ω of the parabolic potential and the position of the guiding center y_k are determined by the strength of both the magnetic ω_c and the electrostatic ω_0 confinements. The last term on the left-hand side of Equation 38.11 describes a free motion in the longitudinal direction with the effective mass M_B. This term lifts the degeneracy (in the k-vector) of the LLs, and the latter transforms into one-dimensional subbands of the quantum wire with the energy

$$E_{nk} = \left(n - \frac{1}{2}\right)\hbar\omega + \frac{\hbar^2 k^2}{2M_B}, \quad n = 1, 2, 3,\ldots \quad (38.12)$$

A structure of the energy levels is illustrated in Figure 38.10a showing a contribution to the total effective potential from the magnetic confinement (thick solid lines) and the electrostatic confinement (thin solid line) for three different values of the wave vector k. For a given k, the magnetic parabolas are shifted by the distance y_k from the origin such that the energy levels follow the parabolic electrostatic confinement. This is illustrated in the energy band diagram (Figure 38.10b) plotting the energy subbands E_{nk} against the guiding center y_k. Note that because the wave function for a given k is centered in the corresponding magnetic parabola, the energy band diagram also provides a position (i.e., the "center-of-the-mass") of the corresponding

wave function u_k. According to the Landauer formula (38.2), the conductance at zero temperature is given by the transmission coefficient at the Fermi energy E_F. Hence, the intersection of the magnetosubbands with the E_F gives the position of the propagating states that contribute to the conductance. These states are called the edge states because they follow the boundaries of the wire, see Figure 38.10e. Their classical counterparts are skipping orbits as illustrated in Figure 38.10b.

When the magnetic field is restricted to zero ($\omega_c = 0$), the distance between the subbands $\Delta E = E_{n+1,k} - E_{nk} = \hbar\omega$ is given by the strength of the electrostatic confinement ω_0. As the field grows, the subband spacing is increased and in the limit of a high field ($\omega_c \gg \omega_0$) it approaches the subband spacing $\hbar\omega_c$ of LLs of the homogeneous 2DEG. Due to the increase in the subband spacing, the number of occupied subbands N_{occ} below the Fermi level decreases. Because a number of conductance plateaus are equal to N_{opp}, the reduction of the plateau number directly reflects the effect of the magnetosubband depopulation.

As for the case of the zero magnetic field, the magnetoconductance of a realistic saddle-point parabolic QPC (described by Equation 38.5) can be written in a simple analytic form given by the expression similar to Equation 38.6 (for details, see Buttiker, 1990a,b). This is illustrated in Figure 38.10f for a QPC with a rather narrow constriction, $\omega_x = \omega_y$. Because for this case the tunneling barrier is thin, the zero-field quantization is poor with

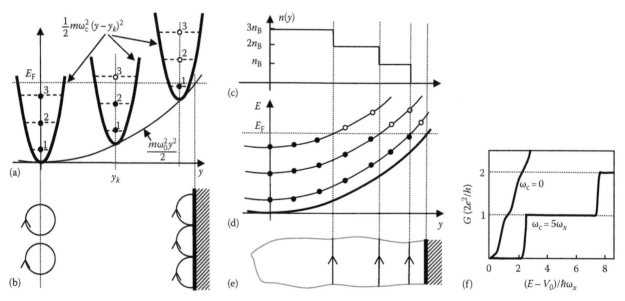

FIGURE 38.10 Magnetosubband structure and conductance of a parabolic quantum wire within a one-electron picture. (a) Schematic illustration of the structure of the energy levels for the case of a parabolic confinement. A thin parabola shows the electrostatic confinement; flat parabolas represent the magnetic confinement. Each parabola is shifted by the distance of the guiding center y_k that depends on the wave vector k of the longitudinal motion. For the given k the vertex of the magnetic parabola gives a position of the center-of-the-mass of the corresponding wave function. Dashed lines inside the parabolas (marked by 1, 2, 3) show positions of the three lowest states. Filled and empty circles indicate respectively fully occupied and empty states. (b) Classical cyclotron motion for the states inside the wire and classical skipping orbits near the wire boundary. (c) The electron density in a quantum wire for the case when three LLs are occupied in the center of the wire. (d) Adiabatic bending of the Landau levels by the external confinement as illustrated in (a). (e) Edge states. (After Chklovskii, D.B. et al., *Phys. Rev. B*, 47, 12605, 1993.) (f) Transition from the conductance quantization in a zero field to the conductance quantization in high magnetic fields for a quantum wire with $\omega_x = \omega_y$. For the case of magnetic field note wide quantization plateaus and transition regions of a negligible width. (After Buttiker, M., *Phys. Rev. B*, 41, 7906, 1990a.)

barely defined quantization steps. The effect of the magnetic field on the conductance is two-fold. First, the magnetic field reduces a number of conductance plateaus due to the magnetic subband depopulation as discussed above. Second, the quality of the quantization improves drastically. This is because of the edge-state character of propagating states in the magnetic field. Because left- and right-propagating edge states are spatially separated and localized in the vicinity of different boundaries, the coupling between them can be exponentially small. This, in turn, leads to a strongly suppressed backscattering and to almost perfect quantization. The improvement of the conductance quantization by the application of a magnetic field is illustrated in Figure 38.9b. At a magnetic field of zero, the conductance quantization of a QPC is deteriorated (presumably due to the effects of the impurities after several thermal cycling). A relatively small magnetic field (when the cyclotron radius is still smaller than the QPC width) is sufficient to improve the quantization. Note that the drastic improvement of the QPC quantization in the high magnetic field and the remarkable accuracy of the quantum Hall effect have the same origin related to the suppression of the backscattering in the edge-state transport regime.

38.3.2 Interacting Electrons in Quantum Point Contact

In the previous section, we presented the basics of the magnetoconductance in the QPC based on a one-electron picture of noninteracting electrons. It reproduces well the essential physics related to the depopulation of the magnetosubbands (Berggren et al., 1986). However, there are two important features in the experimental conductance that are not captured by the one-electron description. First, in the one-electron description, the calculated magnetoconductance exhibits wide quantized plateaus separated by transition regions of an essentially negligible width. The experiments, however, show that an extent of these transition regions can be comparable to the width of the plateaus (c.f. Figures 38.9a and 38.10f). Second, at a sufficiently high magnetic field, a spin degeneracy is lifted and odd plateaus are manifested in the experimental conductance (see Figure 38.9a). These two effects (that are due to enhanced screening and exchange interaction, respectively) will be discussed in this section.

Within a one-electron picture, the edge states have an essentially zero width and their spatial position is determined by the intersection of a corresponding subband with the Fermi energy as illustrated in Figure 38.10d and e. This naive picture does not account for screening that takes place at the Fermi level where the states are only partially filled; hence, the system has a metallic character. Because of the metallic behavior, the electron density can be easily redistributed (compressed) to screen the external potential in order to keep it constant. As a result, the edge states transform to the strips of the finite width called compressible strips. The compressible strips are separated by the incompressible regions where all the levels lie below E_F, hence they are completely filled, see Figure 38.11a through f. As a result, in these regions, the local electron density is constant

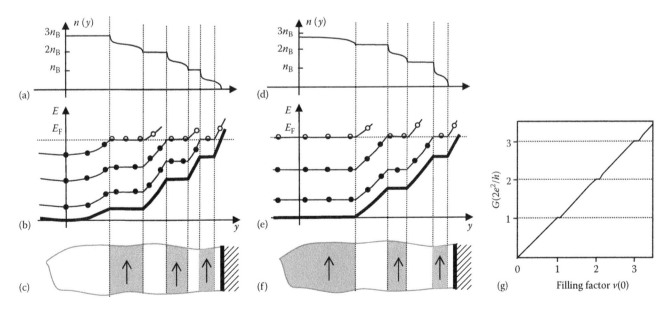

FIGURE 38.11 Magnetosubband structure and conductance of a parabolic quantum wire for spinless interacting electrons for the case of three occupied subbands in the center of the wire. (a–c) and (d–f) correspond to the cases of respectively incompressible and compressible strips in the middle of the wire. (a, d) The electron density in the quantum wire. In the region of the incompressible strips the electron density is quantized in units of n_B. (b, e) Magnetosubbands for interacting electrons. In the compressible regions the subbands are pinned to the Fermi energy. (c, f) The compressible and incompressible strips in the channel (shaded and unshaded regions, respectively). Filled and empty circles indicate respectively fully occupied and empty states. Half filled circles indicate partially filled states. (g) Magnetoconductance of a representative quantum point contact as a function of the filling factor in the center of the channel $v(0) = n(0)/n_B$ with $n(0)$ being the electron density in the absence of magnetic field. Note wide transition regions and quantization plateaus a negligible width. (Adapted from Chklovskii, D.B. et al., *Phys. Rev. B*, 47, 12605, 1993.)

(incompressible) and is equal to the number of allowed states in each LL, $n_B = eB/h$, times the number of occupied LLs.

A quantitative semi-classical treatment of the electrostatic of the edge states including the calculation of the positions and the widths of the compressible and incompressible strips for spinless electrons was proposed by Chklovskii et al. (1993). To calculate the magnetoconductance of the QPC, they used a conjecture that the QPC conductance is given by the filling factor at the saddle point of the electron-density distribution multiplied by $2e^2/h$. Hence, the plateaus in conductance are identified with the situation when there is an incompressible strip in the center of the channel (see Figure 38.11a through c), whereas the transition regions correspond to the central compressible strip (see Figure 38.11d through f). A conductance of a representative QPC calculated according to Chklovskii et al.'s conjecture is shown in Figure 38.11g. In contrast to the one-electron description, the magnetoconductance of interacting electron was shown to exhibit very narrow quantized plateaus separated by much broader rises where the conductance was not quantized. This conclusion (being opposite to the prediction of the one-electron picture, c.f. Figure 38.10f) is also in apparent disagreement with the experiments that typically show that an extent of the transition regions is smaller than the width of the plateaus. This indicates that even an accurate quantitative description of the magnetoconductance requires quantum mechanical treatment including electron interaction and spin effects.

A quantitative quantum mechanical description of the magnetoconductance of split-gate structures focusing on the formation and evolution of the even and odd (spin-resolved) conductance plateaus was given by Ihnatsenka and Zozoulenko (2008) on the basis of the spin density functional theory (DFT) in the local spin density approximation. The starting point is the Hamiltonian within the Kohn–Sham formalism

$$H(\mathbf{r}) = H_0 + V_{KS}(\mathbf{r}), \qquad (38.13)$$

where

   H_0 represents the kinetic energy
   $V_{KS}(\mathbf{r})$ is the mean-field Kohn–Sham potential (Giuliani and Vignale, 2005)

The central idea of the Kohn–Sham formalism is that the ground state density of the interacting system is represented by the ground state density of a *noninteracting* system in some local external potential $V_{KS}(\mathbf{r})$. The quantum mechanical exchange and correlation effects are included in this potential via the local exchange-correlation potential $V_{xc}(\mathbf{r})$. The exchange-correlation potential V_{xc} is typically calculated within a so-called local density approximation, when for a system with a spatially varying density, the exchange-correlation energy is locally approximated by that for the corresponding system at a constant density (for details, see e.g., Giuliani and Vignale, 2005). For the system at hand, the mean-field Kohn–Sham potential reads

$$V_{KS} = V_{conf} + V_H + V_{xc} + V_Z, \qquad (38.14)$$

where

   V_{conf} is the confining potential from metallic gates
   V_H is the Hartree potential
   V_Z is the Zeeman energy

The magnetosubband structure and the electron density are calculated self-consistently and the conductance is computed on the basis of the Landauer formula. Figure 38.12a shows the magnetoconductance of a representative wire calculated within the spin DFT and the Hartree approximations (note that the Hartree approximation corresponds to the case of spinless electrons when the exchange interaction is set to zero, $V_{xc} = 0$). The Hartree magnetoconductance shows the plateaus quantized in units of $2e^2/h$ separated by transition regions whose width grows as the magnetic field is increased. For large fields, the width of the transition regions is comparable or can even exceed the width of the neighboring plateaus. Analyses of the band structure (Figure 38.12b and c) demonstrate that the formation of the transition regions between the plateaus is related to the development of the compressible strip in the middle of the wire corresponding to the highest occupied subband. The transition between the conductance steps starts when the compressible strip reaches the center of the wire and it ends when the compressible strip disappears and two highest (spin-degenerate) magnetosubband are pushed above E_F.

Accounting for the exchange and correlation interactions within the spin DFT leads to the lifting of the spin degeneracy and the formation of the spin-resolved plateaus at odd values of e^2/h. The most striking feature of the magnetoconductance is that the width of the odd conductance steps in the spin DFT calculations is equal to the width of the transition intervals between the conductance steps in the Hartree calculations (Figure 38.12a). This is because the transition intervals in the Hartree magnetoconductance correspond to the formation of the compressible strip in the middle of the wire. At the same time, in the compressible strip in the center of the wire, the states are only partially occupied. As a result, the exchange interaction enhances the difference in the spin-up and spin-down population (triggered by the Zeeman interaction V_Z), which leads to the lifting of the subband spin degeneracy and the formation of the odd conductance plateaus, see Figure 38.12b and c. Another striking feature of the magnetoconductance is the effect of the collapse of the odd conductance plateaus for lower fields. This effect is attributed to the reduced screening efficiency in the confined (wire) geometry when the width of the compressible strip in the center becomes much smaller than the extent of the wave function (see the lower panels in Figure 38.12b and c).

A detailed comparison of the experimental data (see Figure 38.12d) demonstrates that the spin-DFT calculations reproduce not only qualitatively, but rather quantitatively all the features observed in the magnetoconductance of the QPC. This includes the dependence of the width of the odd and even plateaus on the magnetic field as well as the estimation of the subband index corresponding to the last resolved odd plateau in the magnetoconductance.

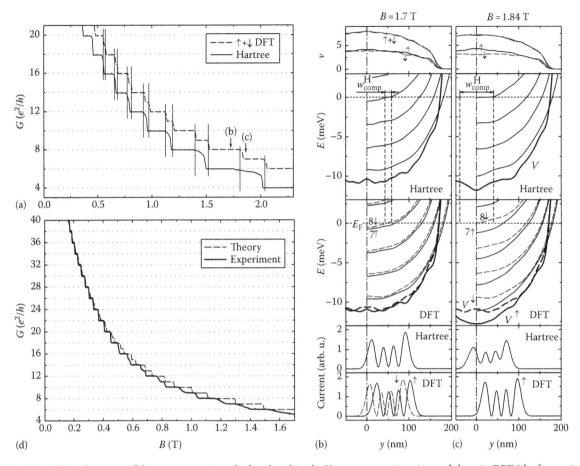

FIGURE 38.12 (a) Conductance of the quantum wire calculated within the Hartree approximation and the spin DFT (the former is shifted by $-2e^2/h$ for clarity). Note that the transition regions in the Hartree conductance correspond to the odd plateaus in the spin DFT conductance. (b, c) The Hartree and the spin DFT electron densities and the magnetosubband structure calculated for the magnetic fields marked by respectively (b) and (c) in (a) (corresponding to the cases of the incompressible (b) and the compressible (c) strips in the center of the wire). Lower panels show the current densities for the last two subbands ($N = 7, 8$). (d) Comparison of the calculated and experimental conductances (the effective width of a QPC is 700 nm). Experimental data is by Radu et al. (2008a). (After Ihnatsenka, S. and Zozoulenko, I.V., *Phys. Rev. B*, 78, 035340-1, 2008.)

38.3.3 Quantum Point Contact in the Quantum Hall Regime

In previous sections, we discussed the basics of the magnetoconductance of a QPC in a two terminal geometry. The majority of the magnetotransport measurements are done in the quantum Hall geometry including four or six voltage probes (see Figure 38.13). The resistance of a mesoscopic conductor in a multi-terminal geometry can be calculated on the basis of Landauer–Buttiker formalism (Buttiker, 1986). Consider a QPC in a multi-terminal geometry depicted in Figure 38.13. Each ith terminal (lead) is characterized by the chemical potential μ_i. Define the transmission probability from the lead i to the lead j as $T_{j \leftarrow i} \equiv T_{ji}$. (Note that T_{ji} can exceed 1 if there are more than one propagating modes in a lead j; $T_{ii} \equiv R_i$ is defined as a reflection coefficient in the lead i). The net current emitted by the lead i is then given by

$$I_i = \frac{2e^2}{h} \sum_i \left(T_{ji} V_i - T_{ij} V_j \right), \qquad (38.15)$$

where $V_i = \mu_i/e$ is the voltage in the terminal i. The current is drawn through the source and drain terminals, $I_s = -I_d = I$; the voltage probe terminals 1–4 draw no current. We define the multi-terminal resistance in a standard way, $R_{ij} = (V_i - V_j)/I$. Hence, the longitudinal resistance is $R_L = R_{12} = R_{34}$; the Hall resistance is $R_H = R_{13} = R_{24}$; and the two-terminal resistance is $R_{2t} = R_{sd}$. For the geometry of Figure 38.13, one can also define the two diagonal resistances R_{14} and R_{23}. Assume that in the bulk regions (i.e., in the contact regions), the system has a filling factor $v = N$ (which corresponds to N propagating edge states in the leads). The QPC is set to transmit only M states (i.e., $T_{43} = T_{12} = M$). By applying the Buttiker formula (38.15) to the geometry of Figure 38.13 and setting the drain voltage to $V_d = 0$, we obtain

$$R_L = \frac{h}{2e^2}\left(\frac{1}{M} - \frac{1}{N} \right), \quad R_H = \frac{h}{2e^2}\frac{1}{N}, \quad R_{2t} = \frac{h}{2e^2}\frac{1}{M},$$

$$R_{14} = \frac{h}{2e^2}\left(\frac{1}{M} - \frac{2}{N} \right), \quad R_{23} = \frac{h}{2e^2}\frac{1}{M}. \qquad (38.16)$$

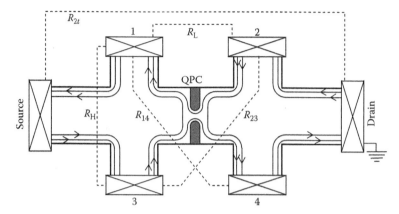

FIGURE 38.13 Multi-terminal measurements of the resistance of a QPC. A current is passed between the source and drain terminals, and the voltage is measured between terminals i and j as schematically indicated in the figure.

(Note that the diagonal resistances R_{14} and R_{23} are interchanged on field reversal.) The above relations predict "fractional" values of the conductance $G_L = R_L^{-1}$, as was confirmed by Kouwenhoven et al. (Beenakker and van Houten, 1991).

A concept of the edge channels utilized in Equation 38.15 can be generalized to the case of the fractional quantum Hall (FQH) effect (Beenakker, 1990). In the FQH regime, the energy of the homogeneous 2DEG has cusps at densities $n = \nu_p Be/h$ corresponding to certain fractional filling factors ν_p. As a result, the chemical potential has a discontinuity at ν_p such that in the vicinity of the boundary of the 2DEG the electron density decreases stepwise from its bulk value ν_{bulk} to zero with steps at ν_p. The strips of the constant filling factors ν_p can be identified as incompressible strips, whereas the compressible strips that separate them play a role of current-currying edge channels. This is illustrated in Figure 38.14, which shows alternating compressible and incompressible strips in the vicinity of a QPC in the FQH regime. A generalization of the Landauer–Buttiker formula for the FQH regime is given by equations similar to Equation 38.15 (Beenakker, 1990)

$$I_i = \frac{e^2}{h} \sum_j \left(T_{ji} V_i - T_{ij} V_j \right), \quad T_{ij} = \sum_{i=1}^{P_j} T_{p,ij} \Delta \nu_p, \quad (38.17)$$

where T_{ij} defines the transmission probability from lead j to lead i in terms of a sum over the generalized edge channels in lead j. The contribution from each edge channel $p = 1, 2, ..., P_j$ contains the weight factor $\Delta \nu_p = \nu_p - \nu_{p-1}$ and the fraction $T_{p,ij}$ of the current of the pth edge channel of lead j that reaches lead i. Note that Equation 38.17 describes the spin-resolved edge channels (hence the absence of a factor "2" on the right-hand side).

Consider now a QPC in the FQH regime corresponding to the filling factor ν_{bulk} in the bulk and ν_{QPC} in the narrowest part of the constriction, see Figure 38.14 for an illustration. In this case, the fractionally quantized resistance is given by expressions similar to Equation 38.16

$$R_L = \frac{h}{2e^2}\left(\frac{1}{\nu_{QPC}} - \frac{1}{\nu_{bulk}} \right), \quad R_H = \frac{h}{2e^2}\frac{1}{\nu_{bulk}}, \quad R_{2t} = \frac{h}{2e^2}\frac{1}{\nu_{QPC}},$$

$$(38.18)$$

$$R_{14} = \frac{h}{2e^2}\left(\frac{1}{\nu_{QPC}} - \frac{2}{\nu_{bulk}} \right), \quad R_{23} = \frac{h}{2e^2}\frac{1}{\nu_{QPC}}.$$

Quantization of the QPC resistance in the FQH regime according to Equation 38.18 was first confirmed by Kouwenhoven et al.

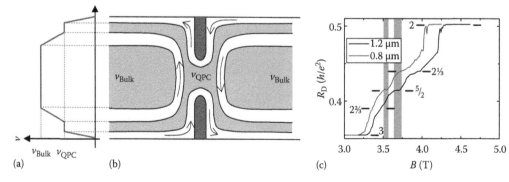

FIGURE 38.14 (a) Schematic drawing of the variation in the filling factor ν in a channel. (b) Schematic drawing of the incompressible bands of the fractional filling factor ν_{bulk} and ν_{QPC}, alternating with the edge channels (arrows indicate the direction of electron motion in each channel). (After Beenakker, C.W.J., *Phys. Rev. Lett.*, 64, 216, 1990.) (c) The diagonal resistance R_{23} from $\nu = 3$ to $\nu = 2$, measured concurrently in different QPCs of lithographic size 0.8 and 1.2 μm. Thin horizontal bars indicate filling factors for the corresponding plateaus. (Adapted from Miller, J.B. et al., *Nat. Phys.*, 3, 561, 2007.)

(1990). Figure 38.14c shows a representative example of a diagonal resistance QPC exhibiting plateaus corresponding to both fractional ($\nu_{QPC} = 5/2, 7/3, 8/3$) and integer ($\nu_{QPC} = 2, 3$) quantum Hall states.

The study of transport through a QPC in a FQH regime represents an active field of current research. These studies are motivated by e.g., a search for Luttinger liquid behavior (Roddaro et al., 2005), an interest in the non-Abelian quasiparticle statistics for certain filling factors such as $\nu = 5/2$, and the building of topologically protected gates for quantum computing by the manipulation of non-Abelian quasiparticles (Das Sarma et al., 2005; Miller et al., 2007; Dolev et al., 2008; Radu et al., 2008b).

38.4 0.7 Anomaly and Many-Body Effects in the Quantum Point Contact

38.4.1 Experimental Evidence of 0.7 Anomaly

In 1996, the Cambridge group pointed out that in addition to quantization in $2e^2/h$ steps, the zero-field conductance of a QPC exhibits a step-like feature at $G \sim 0.7 \times (2e^2/h)$ that was coined as a "0.7 anomaly" (Thomas et al., 1996). Since then this feature has attracted enormous experimental and theoretical attention and it is now widely believed that it is an intrinsic property of clean one-dimensional ballistic constrictions at low electron densities. The 0.7 anomaly has been seen in short and long QPCs

fabricated by the split-gated technique (Thomas et al., 1996, 1998; Reilly, 2001), quantum wires fabricated by shallow etching (Kristensen et al., 2000), GaAs wires patterned by focused ion beams (Tscheuschner and Wieck, 1996), and InP based quantum wires (Ramvall et al., 1997). The structure was observed with a different strength of confinement, different distances from the confining gates to one-dimensional electron gas, and different densities. In this section, we present the main experimental properties of the 0.7 anomaly, and in Section 38.4.2 we review the main theoretical models aimed at the explanation of this effect.

A hallmark of the 0.7 anomaly is its highly unusual temperature dependence, see Figure 38.15. At very low temperatures, the 0.7 anomaly is only weakly developed (and in some experiments is not seen at all). As the temperature is raised, the first quantization plateau at $2e^2/h$ becomes gradually thermally smeared, whereas the 0.7 plateau becomes even stronger. The 0.7 structure can be observable even at 4.2 K when all the quantized plateaus have disappeared. Kristensen et al. (2000) have demonstrated that the relative conductance suppression shows an activated Arrhenius-type behavior, $G(T)/G_0 = 1 - C \exp(-T_A/T)$ (where G_0 is the measured conductance value of the 0.7 feature, C is a constant, and T_A is the activation temperature that depends on the density).

The 0.7 feature strengthens as the two-dimensional carrier densities n_{2D} decreases, see Figure 38.15. As n_{2D} is decreased from 1.4 to 1.13×10^{15} m^{-2}, the 0.7 anomaly becomes strongly

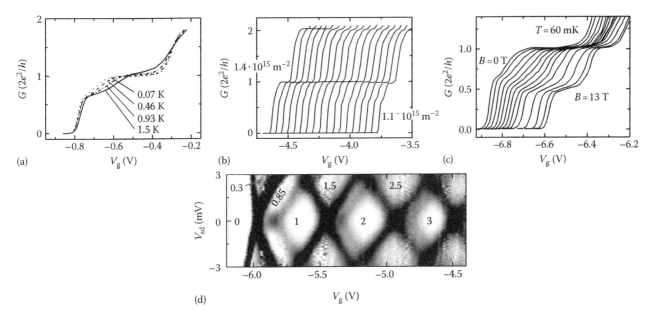

FIGURE 38.15 Experimental data for the temperature (a) (Adapted from Thomas, K.J. et al., *Phys. Rev. Lett.*, 77, 135, 1996.), electron density (b) (Adapted from Thomas, K.J. et al., *Phys. Rev. B*, 58, 4846, 1998.), magnetic field (c) (Adapted from Thomas, K.J. et al., *Phys. Rev. Lett.*, 77, 135, 1996.), and bias (d) (Adapted from Thomas, K.J. et al., *Phys. Rev. B*, 58, 4846, 1998.) dependence of the 0.7 anomaly. In (b), the electron density is subsequently reduced from 1.4×10^{15} to 1.1×10^{15} m^{-2} from left to right. (c) The evolution of the structure at $0.7(2e^2/h)$ into a step at e^2/h in a parallel magnetic field $B_{\parallel} = 0$–13 T, in steps of 1 T. For clarity successive traces have been horizontally offset by 0.015 V. (d) The grayscale plot of the zero-field transconductance dG/dV_g as a function of the side gate voltage V_g and the applied source-to-drain bias V_{sd} is shows for $T = 1.2$ K. The numbers indicate the plateau conductances in units of $2e^2/h$ and the 0.7 anomaly is the bright region at $V_{sd} = 0$ between $G = 0$ and $G = 2e^2/h$.

pronounced. At the highest density, shown in the left-hand trace, the 0.7 plateau is visible only as a weak knee.

A strong in-plane magnetic field $B_{\parallel}$ lifts the spin degeneracy of the one-dimensional subbands giving conductance plateaus quantized in units of e^2/h. Figure 38.15c shows that as $B_{\parallel}$ increases, the zero-field 0.7 structure evolves continuously into spin-split half-plateaus e^2/h.

The effect of a source–drain voltage V_{sd} on the conductance characteristics $G = G(V_g)$ is shown in Figure 38.15d. As V_{sd} is increased, half-plateaus appear at $(N + 1/2) \times 2e^2/h$ for $G > 2e^2/h$, whereas V_{sd}-induced structures appear at $0.85 \times 2e^2/h$ and $0.3 \times 2e^2/h$ for $G < 2e^2/h$. The gate voltage scale is a smooth measure of the one-dimensional confinement energy, so a grayscale plot of the transconductance dG/dV_{sd} allows one to follow the energy shifts of the subband features. The dark lines show transitions between plateaus and the white regions are the conductance plateaus (where the numbers denote the conductance in units of $2e^2/h$). Features moving to the right (left) with increasing V_{sd} do so as the electrochemical potential of the source (drain) crosses a subband edge, and if the subband energies were independent of their occupation, we would expect a linear evolution of the transconductance structures with V_{sd}. This is clearly not the case for the features associated with the 0.7 structure in the lowest subband, suggesting that the subband configuration is occupation-dependent, for which an interaction effect could be responsible.

It should be stressed that the clean quantized conductance plateaus and the absence of additional structures when the channel is moved from side to side by changing the gate voltage demonstrate the lack of potential fluctuations in and around the one-dimensional constriction, and hence rule out the potential fluctuations, resonant, or Coulomb blockade effects as possible origins of the 0.7 structure. Instead, the 0.7 anomaly was linked to the spontaneous lifting of spin degeneracy in the one-dimensional constriction driven by an electron–electron interaction effect related to the exchange interaction (Thomas et al., 1996, 1998). The strengthening of the 0.7 structure as n_{2D} is lowered and is consistent with

an exchange interaction mechanism. Further evidence that an exchange mechanism may be responsible for the 0.7 structure is provided by the source–drain measurements (see Figure 38.15d) where the features in the lowest subband are sensitive to the occupation statistics in the channel, as well as from the enhancement of the g factor for the last few occupied subbands.

Valuable independent information supporting the mechanism related to the spontaneous spin splitting is given by the shot noise measurements (Roche et al., 2004). The results for the measured shot noise power clearly indicate that spin-up and spin-down channels do not have the same transmission on the 0.7 structure. The evolution of the noise power with a parallel magnetic field B supports the picture of two channels with different spin orientations.

A direct measurement of the spin polarization of the 0.7 structure was performed by Rokhinson et al. (2006). In contrast to a majority of the studies utilizing 2DEG, Rokhinson et al. studied a QPC defined in a two-dimensional *hole gas* (2DHG). The 0.7 structure in *p-type* QPCs shows all the essential features reported for *n-type* QPCs, such as a gradual evolution into the $0.5(2e^2/h)$ plateau at high in-plane magnetic fields, survival at high temperatures, a gradual increase toward $(2e^2/h)$ plateau at low temperatures, and the zero bias anomaly, which is suppressed by either temperature increase or the application of a magnetic filed. The similarities between *p-type* and *n-type* QPCs suggest that the underlying physics responsible for the appearance of the 0.7 structure should be the same. The spin polarization was measured in a ballistic magnetic focusing geometry consisting of two QPC in parallel, see Figure 38.16. Due to the enhanced spin-orbit interaction in 2DHG, the carriers with opposite spin have different cyclotron orbits even in a small external perpendicular magnetic field $B_{\perp}$. By injecting current through the first QPC and monitoring the voltage across the detector QPC, the focusing peaks were clearly observed (Figure 38.16). The polarization of the injected current is related to the relative heights of the focusing peaks (similar peak heights correspond to the equal population of spin-up and spin-down

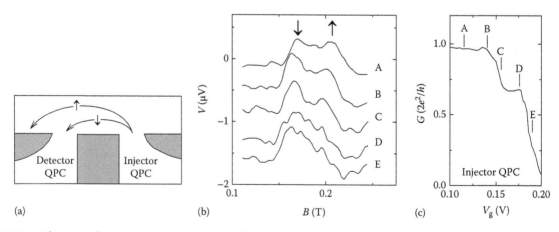

FIGURE 38.16 Polarization detection via magnetic focusing. (a) Magnetic focusing geometry and the schematic trajectories of the ballistic holes in a perpendicular magnetic field. (b) The first spin-split focusing peak is measured at different injector conductances. The curves are vertically offset relative to the top one. (c) The gate voltage characteristic of the injector QPC. Vertical arrows (A–E) mark the positions where the curves in (b) are taken. (Adapted from Rokhinson, L.P. et al., *Phys. Rev. Lett.*, 96, 156602-1, 2006.)

subbands). The main finding of Rokhinson et al. is a detection of a spin polarization below the first plateau, which is found to be as high as 40% in samples with a well-defined 0.7 structure.

In the first paper (Thomas et al., 1996), the origin of the 0.7 anomaly was already attributed to spontaneous spin polarization. Since then, many dozens (if not hundreds) of various studies have addressed this problem. Despite this, no consensus has been reached concerning the microscopic origin of the 0.7 anomaly. Different viewpoints with conflicting conclusions are reported in the literature, and in this section we briefly present some of these theories.

38.4.2 Theoretical Models for 0.7 Anomaly

There are several phenomenological models quantitatively reproducing the main features of the 0.7 anomaly including its unusual temperature dependence (Bruus et al., 2001; Reilly et al., 2002). While differing in details, these models essentially rely on the assumption of the spin gap that opens up between the quasi one-dimensional subbands in the constriction. Such approaches, while providing an important insight for an interpretation of the experiment, are not, however, able to uncover the microscopic origin of the observed effect.

The microscopic origin of the 0.7 anomaly was addressed in studies based on mean-field approaches. The spontaneous spin-splitting of the one-dimensional subband in quantum wires was studied by Wang and Berggren (1996) using a spin DFT. Starting with the Hamiltonian of the form of Equation 38.13 and solving the Schrödinger equation within the Kohn–Sham formalism, they demonstrated that the exchange interactions cause a large subband splitting whenever the Fermi energy passes the subband threshold energies and full spin polarization appears at low electron densities, see Figure 38.17. Numerous subsequent mean-field

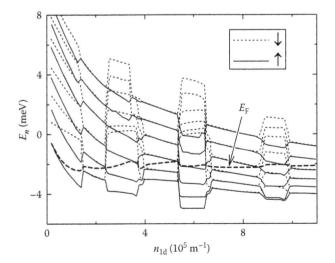

FIGURE 38.17 Sublevels in an infinite, straight quantum wire vs. the one-dimensional electron density n_{1d}. Solid and dotted lines correspond to the spin-up and spin-down electrons. The dashed line shows the position of the Fermi level E_F. (Adapted from Wang, C.-K. and Berggren, K.-F., *Phys. Rev. B*, 54, 14257, 1996.)

studies addressed spin polarization in a constricted geometry of a QPC. These studies include the spin-DFT approaches (Wang and Berggren, 1998; Berggren and Yakimenko, 2002; Hirose et al., 2003; Havu et al., 2004; Ihnatsenka and Zozoulenko, 2007), Hartree–Fock approaches (Sushkov, 2001), and Hubbard models (Cornaglia et al., 2005), just to name a few representative publications. Various mean-field approaches obviously treat the exchange interaction in different ways; the details of the modeling of a QPC (such as modeling of the confinement, calculation of the electron density, and the conductance.) differ in different works. This sometimes led to somehow different results and conclusions whose detailed review is far beyond the scope of the present chapter. Nevertheless, regardless of the approaches used (spin-DFT, Hartree–Fock, Hubbard) or the details of the modeling of the QPC, all mean-field calculations arrive to the same *qualitative* conclusion. Namely, as was already shown in 1998 by Wang and Berggren, the spontaneous spin polarization occurs locally in the region of the saddle point as the electron density is lowered. As a consequence, the effective potential barriers become different for spin-up and spin-down electrons. Transport associated with, say, spin-up electrons takes place via tunneling, while spin-down electrons still carry the current via propagating states in a normal way (Wang and Berggren, 1998). The difference of the effective potential might result in a formation of a localized quasi-bound state for one of the spin species (Hirose et al., 2003). The conductance of a representative QPC along with the effective potential and the local density of states for spin-up and spin-down electrons calculated within the spin-DFT approach is illustrated in Figure 38.18. Despite the prediction of the spontaneous spin polarization in the QPC consistent with the initial interpretation of the effect in the first paper of Thomas et al. (1996), the mean-field approaches fail to account for the key experimental features of the 0.7 anomaly. In particular, the mean-field approaches do not reproduce the unusual temperature dependence of the 0.7 anomaly. In contrast to the experimental data, the spin polarization predicted by the mean field theories is maximal for zero temperature and is gradually smeared as the temperature rises. Besides, the calculated conductance predicts a plateau value of 0.5 instead of 0.7. This is related to the above-mentioned fact that one of the spin channels remains blocked and the corresponding tunneling amplitude is exponentially small. Note that it has been argued recently that the failure to reproduce 0.7 anomaly within the spin-DFT approach might be related to the derivative discontinuity problem of the DFT, leading to spurious self-interaction errors not corrected in the standard local density approximation (Ihnatsenka and Zozoulenko, 2007).

There are several studies reported in the literature that did not attribute the 0.7 anomaly to spin effects. For example, Seelig and Matveev related the 0.7 anomaly to the enhanced backscattering by the acoustic phonons. For typical GaAs quantum wires, the effect of electron–phonon scattering on transport in quantum wires is strongly suppressed. On the other hand, the electron density in a QPC in the vicinity of the first conductance step is very low. As a result, a minimum energy of a phonon required to backscatter an electron inside a QPC is much

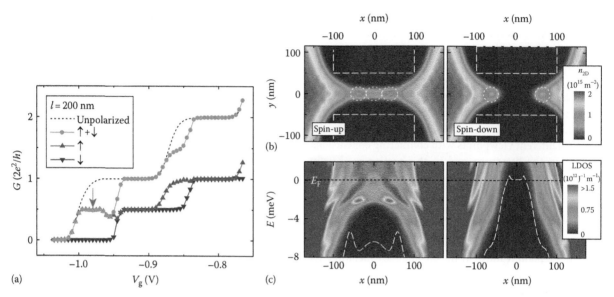

FIGURE 38.18 **(See color insert following page 20-16.)** Conductance of the quantum point contact calculated within the spin DFT approach as a function of the gate voltage V_g. The geometrical width of the constriction is $w = 100$ nm; the geometrical length is $l = 200$ nm. Dashed line corresponds to the spin-unpolarized solution. (b, c) Formation of quasibound states in the quantum point contact. (b) The charge density and (c) the local density of states are shown for the regime of one transmitted spin-up and totally blocked spin-down channel. The left and right columns correspond to the spin-up and spin-down electrons. White dashed lines in (c) indicate the self-consistent Kohn–Sham potential in the center of the QPC along the transport direction. The geometrical length and width of the QPC are $l = 200$ nm and $w = 100$ nm, respectively; the gate voltage $V_g = -0.98$ V. (Adapted from Ihnatsenka, S. and Zozoulenko, I.V., *Phys. Rev. B*, 76, 045338, 2007.)

smaller in comparison with a corresponding bulk value. Seelig and Matveev demonstrated that such a backscattering gives rise to the negative correction to the quantized value $2e^2/h$ of the conductance of a QPC with activated temperature dependence (consistent with the observation of Kristensen et al., 2000) and also gives rise to a zero-bias anomaly in conductance in agreement with the observation of Cronewett et al. (2002).

Sloggett et al. (2008) studied the effect of inelastic electron scattering on the conductance of a QPC. They demonstrated that at zero temperature, the approach results in the usual Landauer formula and the conductance does not show any structures. However, at nonzero temperature, the electron–electron interaction gives rise to a current of correlated electrons, which scales as squared temperature. The corresponding correction to conductance is negative and strongly enhanced in the region below the first conductance plateau. While the above studies accounting for the acoustic phonon scattering and the inelastic electron–electron scattering seem to reproduce the unusual temperature dependence of the 0.7 anomaly, it is not clear how to reconcile these theories with the experimental evidence of the spin polarization of the current below the first conductance plateau reported in the experiments of Roche et al. (2004) and Rokhinson et al. (2006).

38.4.3 0.7 Anomaly and Kondo Physics

A many-body origin of the 0.7 anomaly related to the Kondo physics was advocated in an experimental study by Cronewett et al. (2002; for an introduction to the Kondo physics see e.g.,

Hewson, 1997). They systematically measured the dependence of the conductance on temperature, magnetic field and applied source–drain voltage and found a number of similarities with the Kondo effect seen in quantum dots (Goldhaber-Gordon et al., 1998). In particular, (1) a narrow conductance peak forms at zero source–drain bias and low temperature, (2) conductance data collapses onto a single Kondo-like function with a single scaling parameter (designated as the Kondo temperature), (3) the width of the zero-bias peak is proportional to the Kondo scaling factor, and (4) the zero-bias peak splits in the magnetic field.

A theoretical model of the Kondo effect in a QPC was developed by Meir et al. (2002). Supported by the spin-DFT calculations predicting a formation of the localized state in a QPC, they used the Anderson impurity Hamiltonian, where the localized state inside the QPC (a "magnetic moment") plays a role of an unpaired spin. Within the Kondo model, this magnetic moment hybridizes with delocalized electrons in the leads forming a spin singlet state. According to the model of Meir et al., the 0.7 plateau is attributed to a high background conductance (giving rise to a 0.5 plateau) plus a Kondo enhancement. The Kondo enhancement of the conductance is suppressed with increasing temperature, which explains the unusual temperature dependence of the 0.7 anomaly that becomes more pronounced with the increase of the temperature. Because of the formation of a singlet state, the Kondo model predicts fully polarized transport below the first conductance plateau. This prediction, however, is in obvious disagreement with the experimental findings of Rokhinson et al. (2006) of the partially spin-polarized current for the 0.7 anomaly as well as with the measurements of the shot noise by Roche et al. (2004) indicating a presence of two channels with different spin orientations. Very recently, Sfigakis et al. (2008) reported a study of the Kondo effect in a QPC structure showing that the Kondo effect and the 0.7 structure are *separate and distinct effects*. Thus, the role of the Kondo correlations in the "0.7 anomaly" still remains an open question.

38.5 Conclusion and Outlook

The discovery of the conductance quantization in the QPC laid the foundation and provided a strong momentum for developments in many subfields of mesoscopic physics and the physics of low-dimensional semiconductor structures. Many aspects of the electron transport through a QPC are well understood by now. This includes, for example, the basics physics of the conductance quantization, effect of impurities, magnetosubband depopulation, and many others. Some features of the QPC have found their important practical application. For example, the QPC is routinely used as a standard tool for noninvasive voltage probing and charge detection. However, some important aspects of the electron transport in the QPC still remain highly controversial. This primarily concerns the 0.7 anomaly whose detailed microscopic origin is still unresolved and is under lively debates in the current literature. The current research is also focused on the many-body, nonlinear, and spin effects; study of a QPC in a two-dimensional *hole* gas; the search for the Luttinger liquid behavior; and the study of quasiparticles properties in the fractional quantum Hall regime. Despite the 20 years that have passed since the discovery of the conductance quantization, the electronic and transport properties of the QPC still remain an active field of research, and one can expect a number of new exciting discoveries in many years to come!

References

Beenakker, C. W. J. 1990. Edge channels for the fractional quantum Hall effect. *Phys. Rev. Lett.* 64: 216–219.

Beenakker, C. W. J. and van Houten, H. 1991. Quantum transport in semiconductor nanostructures. *Solid State Phys.* 44: 1–228.

Berggren, K.-F., Thornton, T. J., Newson, D. J., and Pepper, M. 1986. Magnetic depopulation of 1D subbands in a narrow 2D electron gas in a GaAs: AlGaAs heterojunction. *Phys. Rev. Lett.* 57: 1769–1772.

Berggren, K.-F. and Yakimenko, I. I. 2002. Effects of exchange and electron correlation on conductance and nanomagnetism in ballistic semiconductor quantum point contacts. *Phys. Rev. B* 66: 085323-1–085323-4.

Bruus, H., Cheianov, V. V., and Flensberg, K. 2001. The anomalous 0.5 and 0.7 conductance plateaus in quantum point contacts. *Physica E* 10: 97–100.

Buttiker, M. 1986. Four-terminal phase-coherent conductance. *Phys. Rev. Lett.* 57: 1761–1764.

Buttiker, M. 1990a. Quantized transmission of a saddle-point constriction. *Phys. Rev. B* 41: 7906–7909.

Buttiker, M. 1990b. Scattering theory of thermal and excess noise in open conductors. *Phys. Rev. Lett.* 65: 2901–2904.

Castaño, E. and Kirczenow, G. 1992. Case for nonadiabatic quantized conductance in smooth ballistic constrictions. *Phys. Rev. B* 45: 1514–1517.

Chklovskii, D. B., Matveev, K. A., and Shklovskii, B. I. 1993. Ballistic conductance of interacting electrons in the quantum Hall regime. *Phys. Rev. B* 47: 12605–12617.

Cornaglia, P. S., Balseiro, C. A., and Avignon, M. 2005. Magnetic moment formation in quantum point contacts. *Phys. Rev. B* 71: 024432-1–024432-7.

Cronenwett, S. M., Lynch, H. J., Goldhaber-Gordon, D., Kouwenhoven, L. P., Markus, C. M., Hirose, K., Wingreen, N. S., and Umansky, V. 2002. Low-temperature fate of the 0.7 structure in a point contact: A Kondo-like correlated state in an open system. *Phys. Rev. Lett.* 88: 226805-1–226805-4.

Crook, R., Smith, C. G., Barnes C. H. W., Simmons, M. Y., and Ritchie, D. A. 2000. Imaging diffraction-limited electronic collimation from a non-equilibrium one-dimensional ballistic constriction. *J. Phys.: Condens. Matter* 12: L167–L172.

Das Sarma, S., Freedman, M., and Nayak, C. 2005. Topologically protected qubits from a possible non-abelian fractional quantum Hall state. *Phys. Rev. Lett.* 94: 166802-1–166802-4.

Datta, S. 1997. *Electronic Transport in Mesoscopic Systems.* Cambridge, U.K.: Cambridge University Press.

Davies, J. P. 1997. *The Physics of Low-Dimensional Semiconductors.* Cambridge, U.K.: Cambridge University Press.

Davies, J. P., Larkin, I. A., and Sukhorukov, E. V. 1995. Modeling the patterned two-dimensional electron gas: Electrostatics. *J. Appl. Phys.* 77: 4504–4512.

Dolev, M., Heiblum, M., Umansky, V., Stern, and A., Mahalu, D. 2008. Observation of a quarter of an electron charge at the $\nu = 5/2$ quantum Hall state. *Nature* 452: 829–835.

Elzerman, J. M., Hanson, R., Willems van Beveren, L. H., Witkamp, B., Vandersypen L. M. K., and Kouwenhoven, L. P. 2004. Single-shot read-out of an individual electron spin in a quantum dot. *Nature* 430: 431–435.

Field, M., Smith, C. G., Pepper, M., Ritchie, D. A., Frost, J. E. F., Jones, G. A. C., and Hasko, D. G. 1993. Measurements of Coulomb blockade with a noninvasive voltage probe. *Phys. Rev. Lett.* 70: 1311–1314.

Giuliani, G. F. and Vignale, G. 2005. *Quantum Theory of the Electron Liquid.* Cambridge, U.K.: Cambridge University Press.

Glazman, L. I., Lesovik, G. B., Khmel'nitskii, D. E., and Shekter, R. I. 1988. Reflectionless quantum transport and fundamental steps in the ballistic conductance in microconstriction. *Pis'ma Zh. Eksp. Teor. Fiz.* 48: 218–220 [*JETP Lett.* 48: 238–240].

Goldhaber-Gordon, D., Shtrikman, H., Mahalu, D., Abusch-Magder, D., Meirav, U., and Kastner, M. A. 1998. Kondo effect in a single-electron transistor. *Nature* 391: 156–159.

Havu, P., Puska, M. J., Nieminen, R. M., and Puska, V. 2004. Electron transport through quantum wires and point contacts. *Phys. Rev. B* 70: 233308-1–233308-4.

Heinzel, T. 2007. *Mesoscopic Electronics in Solid State Nanostructures*. Weinheim, Germany: Wiley-VCH.

Hewson, A. C. 1997. *The Kondo Problem to Heavy Fermions*. Cambridge, U.K.: Cambridge University Press.

Hirose, K., Yigal Meir, Y., and Wingreen, N. S. 2003. Local moment formation in quantum point contacts. *Phys. Rev. Lett.* 90: 026804-1–026804-4.

Ihnatsenka, S. and Zozoulenko, I. V. 2007. Conductance of a quantum point contact based on spin-density-functional theory. *Phys. Rev. B* 76: 045338-1–045338-9.

Ihnatsenka, S. and Zozoulenko, I. V. 2008. Magnetoconductance of interacting electrons in quantum wires: Spin density functional theory study. *Phys. Rev. B* 78: 035340-1–035340-10.

Johnson, A. C., Petta, J. R., Taylor, J. M., Yacoby, A., Lukin, M. D., Marcus, C. M., Hanson, M. P., and Gossard, A. C. 2005. Triplet–singlet spin relaxation via nuclei in a double quantum dot. *Nature* 430: 431–435.

Kataoka, M., Ford, C. J. B., Faini, G., Mailly, D., Simmons, M. Y., Mace, D. R., Liang, C.-T., and Ritchie, D. A. 1999. Detection of Coulomb charging around an antidot in the quantum Hall regime. *Phys. Rev. Lett.* 80: 160–163.

Kouwenhoven, L. P., van Wees, B. J., van de Vaart, N. C., Harmans, C. J. P. M., Timmering, C. E., and Foxon, C. T. 1988. Selective population and detection of edge channels in the fractional quantum Hall regime. *Phys. Rev. Lett.* 64: 685–688.

Kristensen, A., Bruus, H., Hansen, A. E., Jensen, J. B., Lindelof, P. E., Marckmann, C. J., Nygard, J., Sorensen C. B., Beuscher, F., Forchel, A., and Michel, M. 2000. Bias and temperature dependence of the 0.7 conductance anomaly in quantum point contacts. *Phys. Rev. B* 62: 10950–10957.

Lesovik, G. B. 1989. Excess quantum noise in 2D ballistic point contacts. *Pis'ma Zh. Eksp. Teor. Fiz.* 49: 515–517. [*JETP Lett.* 49: 592–594.]

Maaø, F. A., Zozulenko, I. V., and Hauge, E. H. 1994. Quantum point contacts with smooth geometries: Exact versus approximate results. *Phys. Rev. B* 50: 17320–17327.

Meir, Y., Hirose, K., and Wingreen, N. S. 2002. Kondo model for the "0.7 anomaly" in transport through a quantum point contact. *Phys. Rev. Lett.* 89: 196802-1–196802-4.

Miller, J. B., Radu, I. P., Zumbuhl, D. M., Levenson-Falk, E., Kastner, M. A., Marcus, C. M., Pfeiffer, L. N., and West, K. W. 2007. Fractional quantum Hall effect in a quantum point contact at filling fraction 5/2. *Nat. Phys.* 3. 561–565.

Nixon, J. A., Davies, J. H., and Baranger, H. U. 1991. Breakdown of quantized conductance in point contacts calculated using realistic potentials. *Phys. Rev. B* 43: 12638–12641.

Radu, I. P., Miller, J. B., Amasha, S., Levenson-Falk, E., Zumbuhl, D. M., Kastner, M. A., Marcus, C. M., Pfeiffer, L. N., and West, K. W. 2008a. Suppression of spin-splitting in narrow channels of a high mobility electron gas, unpublished. (The layout and geometry of the devices and heterostructures are similar to those studied by Miller et al. 2007).

Radu, I. P., Miller, J. B., Marcus, C. M., Kastner, M. A., Pfeiffer, L. N., and West, K. W. 2008b. Quasi-particle properties from tunneling in the ν = 5/2 fractional quantum Hall state. *Science* 320: 899–902.

Ramvall, P., Carlsson, N., Maximov, I., Omling, P., Samuelson, L., Seifert, W., Wang, Q., and Lourdudoss, S. 1997. Quantized conductance in a heterostructurally defined $Ga_{0.25}In_{0.75}As/InP$ quantum wire. *Appl. Phys. Lett.* 71: 918–921.

Reilly, D. J., Facer, G. R., Dzurak, A. S., Kane, B. E., Clark, R. G., Stiles, P. J., Hamilton, A. R., O'Brien, J. L., Lumpkin, N. E., Pfeiffer, L. N., and West, K. W. 2001. Many-body spin-related phenomena in ultra low-disorder quantum wires. *Phys. Rev. B* 63: 121311-1–121311-4.

Reilly, D. J., Buehler, T. M., O'Brien, J. L., Hamilton, A. R., Dzurak, A. S., Clark, R. G., Kane, B. E., Pfeiffer, L. N., and West, K. W. 2002. Density-dependent spin polarization in ultra-low-disorder quantum wires. *Phys. Rev. Lett.* 89: 246801-1–246801-4.

Reznikov, M., Heiblum, M., Shtrikman, H., and Mahalu, D. 1995. Temporal correlation of electrons: Suppression of shot noise in a ballistic quantum point contact. *Phys. Rev. Lett.* 75: 3340–3343.

Roche, P., Segala, J., Glattli, D. C., Nicholls, J. T., Pepper, M., Graham, A. C., Thomas, K. J., Simmons, M. Y., and Ritchie, D. A. 2004. Fano factor reduction on the 0.7 conductance structure of a ballistic one-dimensional wire. *Phys. Rev. Lett.* 93: 116602-1–116602-4.

Roddaro, S., Pellegrini, V., Beltram, F., Pfeiffer, L. N., and West, K. W. 2005. Particle-hole symmetric Luttinger liquids in a quantum Hall circuit. *Phys. Rev. Lett.* 95: 156804-1–15680-4.

Rokhinson, L. P., Pfeiffer, L. N., and West, K.W. 2006. Spontaneous spin polarization in quantum point contacts. *Phys. Rev. Lett.* 96: 156602-1–156602-4.

Sfigakis, F., Ford, C. J. B., Pepper, M., Kataoka, M., Ritchie, D. A., and Simmons, M. Y. 2008. Kondo effect from a tunable bound state within a quantum wire. *Phys. Rev. Lett.* 62: 026807-1–026807-4.

Sloggett, C., Milstein, A. I., and Sushkov, O. P. 2008. Correlated electron current and temperature dependence of the conductance of a quantum point contact. *Eur. Phys. J. B* 61: 427–432.

Sushkov, O. P. 2001. Conductance anomalies in a one-dimensional quantum contact. *Phys. Rev. B* 64: 155319-1–155319-8.

Szafer, A. and Stone A. D. 1989. Theory of quantum conduction through a constriction. *Phys. Rev. Lett.* 62: 300–303.

Tarucha, S., Honda, T., and Saku, T. 1995. Reduction of quantized conductance at low temperatures observed in 2 to 10 μm-long quantum wires. *Solid State Commun.* 94: 413–418.

Thomas, K. J., Nicholls, J. T., Simmons, M. Y., Pepper, M., Mace, D. R., and Ritchie, D. A. 1996. Possible spin polarization in a one-dimensional electron gas. *Phys. Rev. Lett.* 77: 135–138.

Thomas, K. J., Nicholls, J. T., Appleyard, N. J., Simmons, M. Y., Pepper, M., Mace, D. R., Tribe, W. R., and Ritchie, D. A. 1998. Interaction effects in a one-dimensional constriction. *Phys. Rev. B* 58: 4846–4852.

Thornton, T. J., Pepper, M., Ahmed, H., Andrews, D., and Davies, G. J. 1986. One-dimensional conduction in the 2D electron gas of a GaAs: AlGaAs heterojunction. *Phys. Rev. Lett.* 56: 1198–1201.

Topinka, M. A., LeRoy, B. J., Shaw, S. E. J., Heller, E. J., Westervelt, R. M., Maranowski, K. D., and Gossard, A. C. 2000. Imaging coherent electron flow from a quantum point contact. *Science* 289: 2323–2326.

Topinka, M. A., LeRoy, B. J., Westervelt, R. M., Shaw, S. E. J., Fleischmann, R., Heller, E. J., Maranowski, K. D., and Gossard, A. C. 2001. Coherent branched flow in a two-dimensional electron gas. *Nature* 410: 183–186.

Tscheuschner, R. D. and Wieck, A. D. 1996. Quantum ballistic transport in in-plane-gate transistors showing onset of a novel ferromagnetic phase transition. *Superlattices Microstruct.* 20: 615–622.

van Wees, B. J., van Houten, H., Beenakker, C. W. J., Williamson, J. G., Kouwenhoven, L. P., van der Marel, D., and Foxon, C. T. 1988. Quantized conductance of point contacts in a two-dimensional electron gas. *Phys. Rev. Lett.* 60: 848–851.

van Wees, B. J., Kouwenhoven, L. P., Willems, E. M. M., Harmans, C. J. P. M., Mooij, J. E., van Houten, H., Beenakker, C. W. J., Williamson, J. G., and Foxon, C. T. 1991. Quantum ballistic and adiabatic electron transport studied with quantum point contacts. *Phys. Rev. B* 43: 12431–12453.

Wang, C.-K. and Berggren, K.-F. 1996. Spin splitting of subbands in quasi-one-dimensional electron quantum channels. *Phys. Rev. B* 54: 14257–14260.

Wang, C.-K. and Berggren, K.-F. 1998. Local spin polarization in ballistic quantum point contacts. *Phys. Rev. B* 54: 4552–4556.

Wharam, D. A., Thornton, T. J., Newbury, R., Pepper, M., Ahmed, H., Frost, J. E. F., Hasko, D. G., Peacockt, D. C., Ritchie D. A., and Jones G. A. C. 1988. One-dimensional transport and the quantisation of the ballistic resistance. *J. Phys. C: Solid State Phys.* 21: L209–L214.

Yacoby, A. and Imry, Y. 1990. Quantization of the conductance of ballistic point contacts beyond the adiabatic approximation. *Phys. Rev. B* 41: 5341–5350.

VII

Nanoscale Rings

39

Nanorings

Katla Sai Krishna
*Jawaharlal Nehru Centre for
Advanced Scientific Research*

Muthusamy
Eswaramoorthy
*Jawaharlal Nehru Centre for
Advanced Scientific Research*

39.1 Introduction

The exciting size- and shape-dependent properties associated with many materials at the nanometer scale evoked a great deal of interest in synthesizing materials with different lengths and shapes [1–9]. Spheres, wires, rods, and tubes are common shapes that are usually prevalent in material synthesis as against uncommon shapes like rings, bowls, and other complex morphologies. Nanorings, as the name implies, are nanoscale entities with ring-shaped geometry (i.e. the inner diameter of the ring is larger than the width and thickness of the rim) and with one of their dimensions (either the width or the thickness of the rim) in the nanometer range (Figure 39.1). Molecular nanorings are entities in which the size of the ring-shaped molecule is in the nanometer scale.

A nanoring can be a single entity with an end-to-end closed (a perfect tori structure) or coiled structure, or a superstructure formed by the self-assembly of smaller entities such as molecules and nanoparticles (Figure 39.2).

In the case of superstructures, the resulting symmetry is often determined by the shape of the smaller entities used to build the structure. Since the prevalent shapes of the smaller units are mostly spherical, disk/tile, and rod-like, it would be difficult to pack them in the form of a ring-like superstructure.

For example, opaline structures commonly result from the packing of (monodispersed) spherical entities (Figure 39.3). Stacked pancake structures (discotic phase) are obtained from the packing of disk-shaped entities. Nematic, semectic, or hexagonal phase superstructures are built from the packing of rod-like entities. These types of superstructures are usually observed in controlling the mesophases in colloidal suspensions and the superlattices in nanocrystal dispersions [10,11]. Nevertheless, the organization of sphere-, disk- and rod-shaped entities into ring-like superstructures [12] was also reported under certain conditions, which will be discussed later in this chapter.

39.2 Why Is Ring Shape Important?

Metal nanoparticles such as Ag, Au, and Cu are known to absorb and emit certain wavelengths of light depending upon their size and shape [13]. Among various shapes, ring-shaped nano objects of similar size exhibit several novel properties due to their unique structural features. For example, gold nanorings are shown to display tunable plasmon resonance in the near infrared, which is not possible with solid gold particles of similar size [14,15]. Further, the optical and electromagnetic properties of gold nanorings can be varied by varying the ratio of the ring radius to the wall thickness. When light falls on a gold nanoring, the electrons get excited and oscillate collectively as a wave (plasmon) that can be tuned by tuning the wavelength of light and the geometry of the ring. By tuning the wavelength of the incoming light, the pool of electrons in the ring can be made to resonate in the same wavelength. This resonance generates a strong and uniform electromagnetic field that will oscillate within the ring cavity. If the field inside the ring cavity is optimized to near infrared, it (cavity) can be used as a container for holding and probing smaller nanostructures in sensing and spectroscopy applications. Conducting the experiments inside the nanoring will give amplified infrared signals and better results that would be of interest to the drug industry and to biochemical researchers [16].

FIGURE 39.1 Illustration of a ring, a disk, and a cylinder.

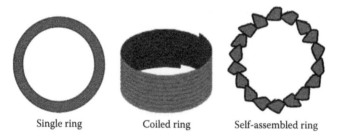

FIGURE 39.2 Schematic showing the single ring, the coiled ring, and the self-assembled ring.

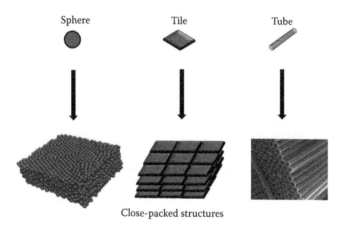

FIGURE 39.3 Scheme illustrating the packing of spheres, tiles, and tubes.

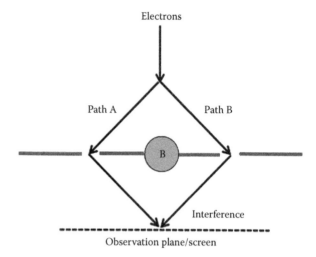

FIGURE 39.4 Schematic of double-slit experiment in which Aharonov–Bohm effect can be observed: electrons pass through two slits, interfering at an observation screen. The interference pattern will shift when a magnetic field, **B**, is turned on in the cylindrical solenoid.

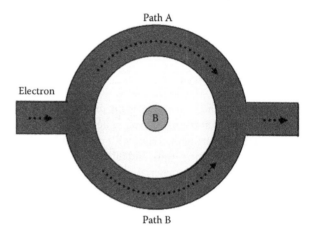

FIGURE 39.5 Schematic of Aharanov–Bohm effect in a nanoring.

Another interesting phenomenon observed in the metal nanorings is called the Aharanov–Bohm effect [17–19], which cannot be explained by classical mechanics. It is a quantum mechanical phenomenon. It predicts that if a coherent electron beam splits into two parts and the path encloses a finite magnetic flux (magnetic field **B** on the path of the electrons is zero), a phase difference will occur in the electron wave packets traveling along the two parts, which in turn would manifest in an interference pattern at the other end due to the nonzero magnetic vector potential A (Figure 39.4).

Similarly, the phase coherent electrons traveling around the magnetic field, **B**, through the arms of the metal nanorings (remember the magnetic field in the conductor regions, the arms of the rings, is zero) should show oscillations in the magnetoresistance because of the interference between the wave packets of electrons of the two arms (Figure 39.5). The phase shift of electrons occurs in both arms of the ring and can be tuned by changing the magnetic flux encircled by the ring.

Nanorings made up of metallic carbon nanotubes are expected to possess giant paramagnetic moments in the presence of an applied magnetic field (perpendicular to the plane of the ring), owing to the effective interplay between the ring geometry and the ballistic motion of π-electrons in the nanotube [20]. Theoretical studies predict that at 0 K and at applied magnetic fields of ~0.1 T, the rings of metallic carbon nanotubes can exhibit very large paramagnetic moments, which are three orders of magnitude higher than the diamagnetic moment of graphite.

Magnetic nanorings have been proposed for applications in high-density magnetic storage and vertical magnetic random access memory (V-MRAM) [21,22] because of their ability to retain the vortex states. Larger magnetic elements such as microdisks have rich domain structures determined by their geometry. On the other hand, very small nano entities generally have a single domain state, with all spins pointing to one direction (Figure 39.6). Further, the microdisks form a vortex state, whose magnetic moments form a closed structure from which

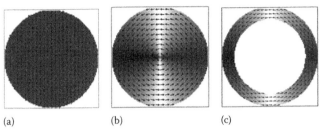

(a) (b) (c)

FIGURE 39.6 (a) Vortex state in single circular domain (all spins pointing upward), (b) vortex state in a circular disk (with a vortex core where spins are out of planarity and (c) vortex state of a nanoring (with a missing vortex core). In both the circular disk and nanoring, the left side spins are pointing downward and right side spins are pointing upward.

no stray field is leaked out. The center of the vortex, called vortex core, has the magnetic moments pointing out of the plane, either up or down.

As the size of the disks shrinks laterally, the vortex core becomes unstable and finally transforms to a single domain state below a certain critical size. The competition between the exchange energy (dominant at small scale) and the magneto static energy (dominant at large scale) basically decides such transitions. Since there is no vortex core, a stable vortex state can be achieved in the case of magnetic nanorings. All moments are completely confined within the plane and form a closure (Figure 39.6). The vortex chirality (clockwise and anticlockwise) in the magnetic nanorings can be used for information storage.

Nanorings made up of porphyrin entities through $\pi-\pi$ stacking and supramolecular self-assembly [23,24] are reported to show excellent florescence properties and are being explored for optical applications [25]. All these applications clearly demonstrate the significance of ring morphology at the nanoscale.

39.3 Synthetic Methods

As we discussed earlier, nanorings can be made either from a single entity or from the self-assembly of multiple entities like molecules, nanoparticles, etc. Various approaches are being adopted to make nanoring structures of different components. Among these, evaporation induced self-assembly, template-based synthesis, and chemical reactions and/or electrostatic interactions induced coiling are the most common approaches to make nanorings and are highlighted below.

39.3.1 Ring Formation by Evaporation-Induced Self-Assembly

In this approach, nanoring formation occurs when a solution containing nanocrystals/polymers/nanowires is allowed to evaporate on a substrate. Physical processes such as dewetting, surface tension, and solvent and solute dynamics play a major role in the ring formation. Two types of mechanism have been put forward in the literature to explain the formation of rings from the evaporating solutions.

39.3.1.1 Coffee-Stain Mechanism

This was introduced by Deegan et al. [26] to explain the formation of ring-like stains from the solution drop on evaporation. Enrichment of solute at the edges of the droplet will occur during the evaporation of the solvent (from the solution), if the contact line of the solution with the solid substrate is pinned. The solvent at the edge of a solution droplet evaporates faster than the bulk. This evaporation loss at the perimeter will be offset by an outward flow of fluid from the core, which draws the solute (dispersed material) from the interior to the edge of the drop and deposits it as a solid ring (Figure 39.7). If the solute transfer is not complete, a fraction of the material remains inside the resulting ring. Ring formation can occur over a wide variety of substrates, solutes, and solvents subject to the following conditions:

1. The solvent meets the surface at a nonzero contact angle
2. The contact line is pinned to its initial position
3. The solvent evaporates

Further, wetting of the substrate surface by the solvent is an important factor, which can be tuned by modifying the substitution pattern of the solute molecules used, the coating of the surface and the conditions under which the evaporation takes place. Several other factors such as solute concentration, solute–surface interaction, and surface tension gradients (Marangoni effect) also modify the ring pattern.

39.3.1.1.1 Marangoni Effect

During the process of evaporation, the solvent movement is determined by the forces of surface tension acting upon them. The phenomenon of liquid flowing along a gas–liquid or a liquid–liquid interface from areas having low surface tension to areas having higher surface tension is called Marangoni effect [27]. Marangoni convection is the flow caused by surface tension gradients originating from concentration gradients. The famous

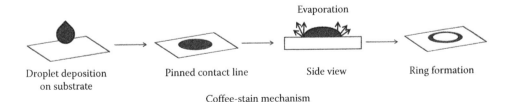

Evaporation

Droplet deposition on substrate   Pinned contact line   Side view   Ring formation

Coffee-stain mechanism

FIGURE 39.7 Schematic showing the coffee-stain mechanism. (Adapted from Lensen, M.C. et al., *Chem. Eur. J.*, 10, 831, 2004. With permission.)

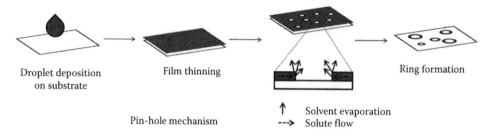

FIGURE 39.8 Schematic showing the pin-hole mechanism. (Adapted from Lensen, M.C. et al., *Chem. Eur. J.*, 10, 831, 2004. With permission.)

"tears of wine" is a result of this effect in which one can observe a ring of clear liquid, near the top of a glass of wine, from which droplets form and flow back into the wine. This effect becomes evident when an interface contains traces of surface-active substances. If an interface expands locally, these surface-active solutes are swept outward with the movement, creating a gradient in concentration of these surface-active substances. This concentration gradient implies a surface tension gradient, which acts opposite to the movement. The interfacial movement is therefore damped and this effect is labeled the Plateau-Marangoni–Gibbs effect.

39.3.1.2 Pin-Hole Mechanism

In the pin-hole mechanism, formation of holes in the liquid films during evaporation is responsible for ring-formation [23]. Holes nucleate when the evaporating solution film on a wetted substrate reaches a critical thickness, where there is a balance between thinning of liquid and wetting of the surface. The holes then open up and expand outward (in order to retain the optimum film thickness while evaporation continues) and push the solute particles toward the growing inner perimeter. As the evaporation continues, the solution around the rim becomes more and more concentrated. Finally, the resulting ring of particles gets stuck when the friction between the particles and the substrate can no longer be overcome by the force acting radially outward to thicken the film. Further, the evaporation rate at the edge of the pin-hole

is higher than in the bulk of the solution film, an inward flow of solute occurs to compensate this loss that will enrich the concentration of the solute at the inner edge of the ring. This process will continue till complete evaporation of solvent occurs, leaving the dispersed materials in the form of rings (Figure 39.8). In the case of wetting solvents, the holes open up when the film thickness reaches the nanometer level, so that the aggregation of dispersed material into rings is controlled by friction forces involving the substrate. For the nonwetting case, hole nucleation starts when the film thickness reaches micrometer level leading to the three-dimensional solution/precipitation induced aggregation.

39.3.2 Ring Formation Induced by Chemical Reaction

The functional groups present over the nanostructures would also induce the formation of ring-like superstructures around the water or organic droplet due to hydrophilic and hydrophobic interactions [28,29]. For example, the carbon nanofibers in CCl_4 derived from the carbonization of polymer followed by acid etching will fuse and roll around the water droplet present in carbon tetrachloride (Figure 39.9a). Sometimes, coiling can be facilitated by a chemical reaction. Carbon nanotubes functionalized at both the ends with acid groups curl to form the nanoring structures by an end-to-end chemical reaction using 1,3-dicyclohexylcarbodiimide (DCC) [30] (Figure 39.9b).

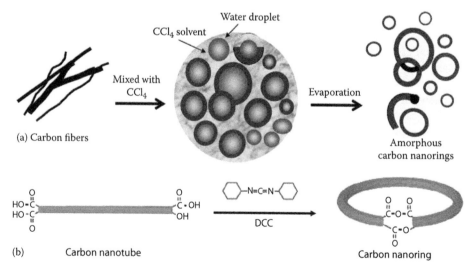

FIGURE 39.9 (a) Schematic of amorphous carbon nanoring formation. (b) Schematic showing the chemical fusion of carbon nanotube into nanoring using DCC. (Adapted from Sai Krishna, K. and Eswaramoorthy, M., *Chem. Phys. Lett.*, 431, 327, 2007; Sano, M. et al., *Science*, 293, 1299, 2001. With permission.)

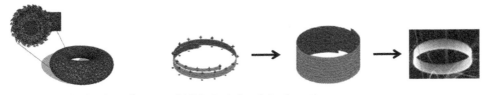

(a) Toroidal cylindrical micelles   (b) Polarity-induced ring formation

FIGURE 39.10 (a) Schematic of toroidal cylindrical micelles formed from triblock copolymers. (b) Schematic of polarity-induced ring formation through self-coiling ZnO nanobelt. (Reprinted from Pochan, D.J. et al., *Science*, 306, 94, 2004; Kong, X.Y. et al., *Science*, 303, 1348, 2004. With permission.)

39.3.3 Ring Formation Induced by Electrostatic Interactions

The electrostatic interaction between the charged surfaces of the nanostructures has been reported to give ring-shaped morphology in some cases. For example, the cylindrical micelles obtained with a negatively charged amphiphilic polymer undergo a shape change from a cylindrical to toroid shape on interaction with the positively charged divalent counterion [31] (Figure 39.10a). Rings of ZnO are known to form from polarity-induced self-coiling of ZnO nanobelts synthesized at high temperatures [32] (Figure 39.10b). Here again, electrostatic interaction of the charges over the nanobelt edges are responsible for the self-coiling process.

39.3.4 Ring Formation by Template-Based Approach

39.3.4.1 Colloidal Crystal Template

Colloidal crystal templating is a well-known method to produce ordered porous structures [33–35]. This method could also be tuned to obtain nanorings [21]. The process uses monodisperse polystyrene (PS) or silica spheres (of micron or submicron size) as templates. The deposition of nanostructure around the colloidal template can be carried out either through a physical or chemical method. In a physical method, a monolayer of PS nanospheres is first formed on a substrate (Figure 39.11a) by manipulating its surface chemistry. A thin film of a desired metal of required thickness is then sputter-deposited onto the substrate (Figure 39.11b). Finally, Ar+ ion-beam etching is used to remove all the deposited metal, except that under the nanospheres. This results in an array of metal nanorings as shown in Figure 39.11c. The PS nanospheres are then chemically removed using calcinations or a solvent extraction process. The size of the nanorings can be controlled by the size of the PS template, whereas the compositions and thickness of the nanorings can be controlled by the deposition process.

In the case of a chemical method, the packed colloidal spheres are functionalized before the desired metal oxide or organic precursors are allowed to deposit around the contact points of two spheres. Subsequent washing and chemical or thermal treatment removes the template, leaving the ring-like nanostructures of the desired materials [36,37] (Figure 39.12).

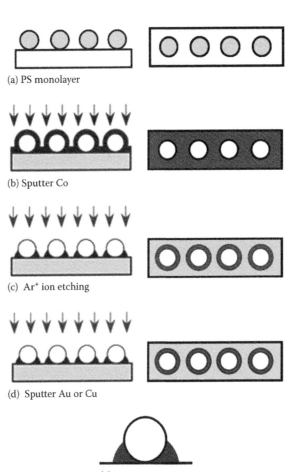

(a) PS monolayer

(b) Sputter Co

(c) Ar+ ion etching

(d) Sputter Au or Cu

(e)

FIGURE 39.11 Schematic of the fabricating process for arrays of nanorings shown from a side view (left column) and top view (right column). (a) A monolayer of polystyrene (PS) nanospheres is deposited onto the substrate. (b) A thin film of Co is sputter-deposited over the surface of PS nanospheres. (c) An Ar+ ion beam is used, in normal incidence, to etch away the sputtered Co film and thus leaving only the nanorings protected under the PS nanospheres. (d) A capping layer of Au or Cu is deposited over the entire surface to prevent oxidation. (e) Schematic of the tapered cross section of a nanoring. (Adapted from Zhu, F.Q. et al., *Adv. Mater.*, 16, 2155, 2004. With permission.)

In a different approach, the polymeric hollow array membranes were used as templates to make the nanorings [38]. The hollow scaffold can be obtained by polymerization of the organic components within the spaces of packed colloidal silica spheres,

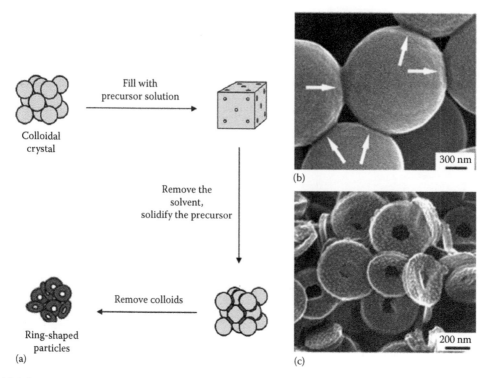

FIGURE 39.12 (a) Schematic outline of the formation procedure of rings. (b) SEM images of the positions of the rings. (c) Polystyrene rings obtained by the colloidal crystal template made from SiO_2 particles 1100 nm in diameter after the removal of the SiO_2 particles. (Reproduced from Yan, F. and Goedel, W.A., *Angew. Chem. Int. Ed.*, 44, 2084, 2005. With permission.)

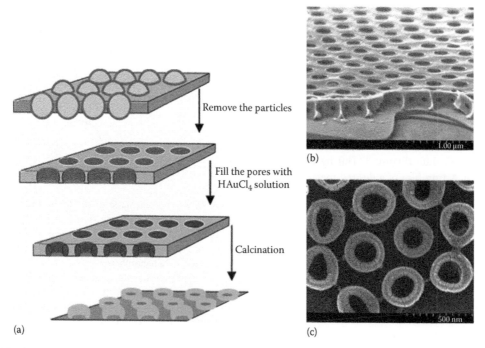

FIGURE 39.13 (a) Scheme of the preparation of gold rings on a mica sheet. (b) SEM images of a porous membrane prepared by using 330 nm silica particles as templates, transferred onto a mica sheet (partially removed by a scotch tape to allow a side view of the pores). (c) SEM images of gold rings on mica prepared using molds of porous membranes shown in (b). (Reprinted from Yan, F. and Goedel, W.A., *Nano Lett.*, 4, 1193, 2004. With permission.)

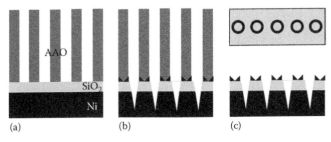

FIGURE 39.14 Schematic showing the (a) AAO template on 20 nm of SiO$_2$ on the desired ring material, (b) sputter redeposited material around the pore walls after sputter etching, and (c) the rings after AAO mask is removed (with a plan-view added to guide the eye). (Reprinted from Hobbs, K.L. et al., *Nano Lett.*, 4, 167, 2004. With permission.)

followed by the removal of silica. These polymeric porous scaffolds are partially filled with the metal or metal oxide precursors (e.g., HAuCl$_4$ for Au rings and Ti(OEt)$_4$ for TiO$_2$ rings). Subsequent reduction/calcination removes the organic template and retains the metal or metal oxide rings intact (Figure 39.13).

39.3.4.2 Porous Membrane Template

The extensive use of porous alumina membranes as templates to prepare an ordered array of nanotubes and nanorods is well known. On the contrary, the use of porous alumina membrane as the template for nanoring formation is very limited. Nevertheless, several metal nanorings like Au, Ni, Fe, and Si were prepared using this method. The process involves Ar$^+$ sputtering of the metal substrates through the pores of the anodic alumina membrane (AAO), which is used to mask the substrate [39,40]. Figure 39.14 shows the schematic diagram of the process. A porous alumina membrane is placed over a nickel substrate that is covered with a 20 nm thick SiO$_2$ layer. Ion etching of the nickel substrate through the holes of porous alumina membrane leaves the redeposition of Ni around the membrane wall as a ring. Subsequent removal of alumina membrane by chemical etching leaves the nickel nanorings on the SiO$_2$ layer. The purpose of the SiO$_2$ layer is to avoid the substrate interference while measuring the properties of resulting nanorings.

39.3.4.3 DNA Template

The ability of DNA to condense into well-defined toroids provides a unique opportunity to use them as templates to create toroidal nanostructures (nanorings) of controlled shape and dimensions. For example, the negatively charged condensed DNA rings will be an ideal surface for the silver ions to interact, which, on reduction, give silver nanorings [41] (Figure 39.15).

39.3.5 Ring Formation by Lithographic Technique

39.3.5.1 Soft Lithography

Soft-lithographic technique has been used to synthesize many novel nanostructures that are otherwise difficult to form. Single crystalline silicon nanorings of uniform diameter were

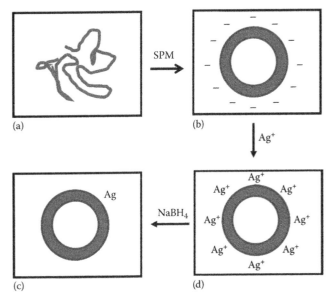

FIGURE 39.15 Schematic representation of the silver nanoring synthesis. (a) Unfolded DNA chain. (b) DNA toroidal condensate formed after the addition of a condensing agent (in this case, spermine, noted SPM^{4+}, a tetravalent polycation). (c) After the addition of AgNO$_3$, silver ions bound to the surface of the DNA toroid. (d) DNA-templated silver nanorings formed after Ag$^+$ reduced by NaBH4. (Reproduced from Zinchenko, A.A. et al., *Adv. Mater.*, 17, 2820, 2005. With permission.)

synthesized by Xia and coworkers [42] using this process (Figure 39.16a). The first step in this process involves the creation of nanorings in a thin film of photoresist (placed on a Si/SiO$_2$ substrate) by exposing it to a UV light source through a polydimethylsiloxane (PDMS) mask, patterned with ring-shaped features. These patterned features in the photoresist film were then subsequently transferred on to the underlying silicon layer (in the Si/SiO$_2$ substrate) by reactive ion etching (RIE) process. Further oxidation and wet-etching with hydrofluoric acid removes the underlying silica layer, leaving behind the individual silicon nanorings (Figure 39.16b).

39.3.5.2 Interference Lithography

Nanoring arrays of Au and Ni were fabricated based on interference lithography. It involves first the creation of Si$_3$N$_4$ nanohole arrays on silicon wafers by laser interference lithography (LIL), followed by selective electrochemical deposition on the step edge of periodic Si$_3$N$_4$ patterns. Large-scale fabrication of well-ordered metallic nanoring arrays is possible by this method [43].

39.4 Types of Nanorings

Depending on the composition, nanorings can be divided into different types namely

1. Carbon nanorings
   a. Molecular carbon nanorings
   b. Graphitic carbon nanorings
   c. Amorphous carbon nanorings

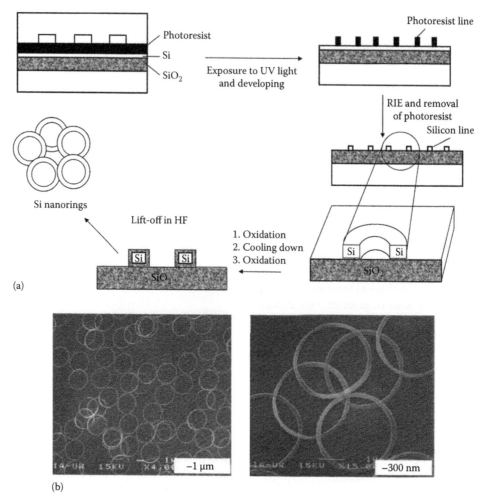

(a)

(b)

FIGURE 39.16 (a) Schematic of the soft lithography process. (b) Silicon nanorings formed through the lithographic route. (Reproduced from Yin, Y. et al., *Adv. Mater.*, 12, 1426, 2000. With permission.)

2. Metal nanorings
3. Metal oxide/sulfide nanorings
4. Organic nanorings
   a. Porphyrin nanorings
   b. DNA nanorings
   c. Protein nanorings
   d. Amphiphilic molecular nanorings

39.4.1 Carbon Nanorings

39.4.1.1 Molecular Carbon Nanorings

Molecular nanorings are rings that are formed from the molecules as the building blocks. For example, 6-cycloparaphenilacetylene molecule is a ring-shaped molecule that is formed from six repeating units of phenilacetylene in a cyclic fashion [44–46]. The ring size and the diameter can be varied by varying the number of phenilacetylene groups in the molecule. Figure 39.17 depicts a typical structure of a 6-cycloparaphenilacetylene molecule, where n denotes the number of repeating units of phenilacetylene in the molecule. By placing the n value with 1, 2, 3, 4,

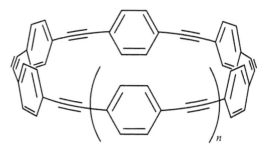

FIGURE 39.17 Molecular structure of a 6-cycloparaphenilacetylene molecule, where n = 1, 2, 3, 4. (Reproduced from Kawase, T. et al., *Angew. Chem. Int. Ed.*, 42, 1621, 2003. With permission.)

etc., different sizes of molecular nanorings can be obtained with specific ring diameters.

These rings are similar to the crown ether type of molecules, where the inside cavity can be utilized for inclusion of guest molecules [45]. Like crown ethers, these molecules also hold the guest molecules within the ring due to the intermolecular interactions of host–guest chemistry, and hence molecules such as

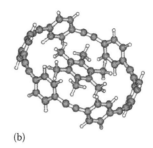
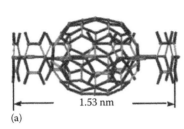

(a) 1.53 nm   (b)

FIGURE 39.18 Structure of the inclusion complex of (a) fullerene C_{70} molecule and (b) hexa methyl benzene in cycloparaphenilacetylene molecule. (Reproduced from Kawase, T. et al., *Angew. Chem. Int. Ed.*, 42, 5597, 2003; Cuesta, I.G. et al., *ChemPhysChem.*, 7, 2503, 2006. With permission.)

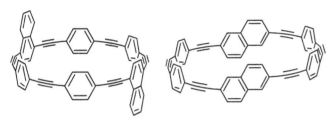

FIGURE 39.19 Naphthalene units substituted 6- CPPA molecule. (Reproduced from Kawase, T. et al., *Angew. Chem. Int. Ed.*, 42, 5597, 2003. With permission.)

fullerenes (Figure 39.18a) can be held inside a single molecule of 6-cycloparaphenilacetylene. Hexamethyl benzene (HMB) can form inclusion complexes with these molecular nanorings (Figure 39.18b) [46].

Different variants of these nanorings were also formed by replacing the phenilacetylene molecules in the ring with naphthalene units (Figure 39.19) [45]. These molecular nanorings find their application in the supramolecular chemistry wherein similar molecules like crown ethers, cryptands, cyclodextrin-type molecules play a major role.

39.4.1.2 Graphitic Carbon Nanorings

The carbon allotropes, graphite, diamond, fullerene, and nanotube in the form of sheets, spheres, rods, and tubes have been well known for a long time. However, the graphitic carbon in the form of nanoring was not known until Smalley and coworkers discovered it during the laser-assisted synthesis of single-walled carbon nanotubes (SWNTs) in the year 1997 [47]. The rings formed in the process have a perfect tori structure with no beginning or no end (Figure 39.20). To explain the formation of these toroidal carbon nanostructures, they proposed a Kekulean image of a growing nanotube eating its own tail. As the two ends of the bending nanotube come close together, the van der Waal's interaction makes them align and the bending strain makes them slide over one another. In that high-temperature process, the closed hemifullerene end of the nanotube often gets stuck to the metal nanoparticles (which is used as a catalyst to grow the nanotube) of the growing end of the nanotube, resulting in end-to-end fusion.

Avouris and coworkers have reported the formation of SWNT rings by the curling of nanotube ropes (no end-to-end fusion) [48]. Rings were produced by ultrasonicating a solution containing SWNTs in a mixture of concentrated sulfuric

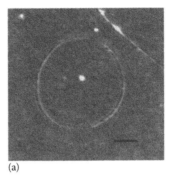
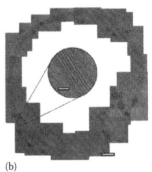

(a)   (b)

FIGURE 39.20 (a) Scanning force micrograph (SFM) of nanoring formed from SWNTs by laser-assisted growth of SWNT (reported by Smalley and coworkers). Scale bar: 100 nm. (b) TEM of a ring showing the higher magnification of the rim. The ring has an apparent height of 1.0–1.2 nm (actual height is closer to 1.5 nm) and width of 4–8 nm. Scale bars: 15 and 5 nm (inset). (Reprinted from Liu, J. et al., *Nature*, 385, 780, 1997. With permission.)

acid and hydrogen peroxide (Figure 39.21). This treatment not only disperses the nanotubes but also shortens the nanotube ropes. A high yield requires shortening the raw nanotubes to a length of 2–4 mm. The ring formation is a kinetic process, with the activation energy for the curling provided by the ultrasonic irradiation. Further, the strain energy caused by the increased curvature while coiling is compensated by the energy gained from the van der Waal's interactions. It was suggested that the hydrophobic nanotubes that act as nucleation centers for bubble formation bent mechanically at the bubble–liquid interface as a result of the bubbles collapsing during cavitation.

Carbon nanorings are also known to form from SWNTs by chemical modification and end-to-end functionalization [30] (Figure 39.22). Functionalization of SWNT were carried out by ultrasonicating them in a concentrated H_2SO_4/HNO_3 solution, followed by etching with H_2SO_4/H_2O_2 solution to obtain oxygen-containing groups, such as carboxylic acid groups at both ends. The end-to-end ring closure of the nanotubes was then achieved by using 1,3-dicyclohexylcarbodiimide (DCC) as the coupling agent (Figure 39.9b).

39.4.1.3 Amorphous Carbon Nanorings

Formation of amorphous carbon nanorings, different from the tubular and graphitic SWNT nanorings, has been reported by Eswaramoorthy and coworkers [28] through the fusion

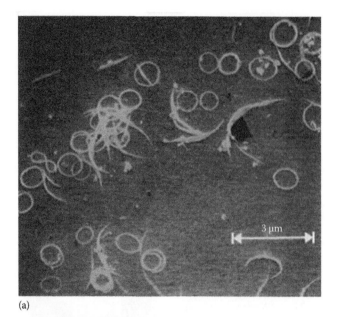

(a)

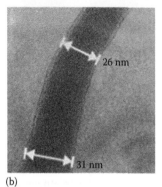

26 nm

31 nm

(b)

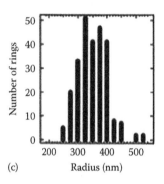

(c)

FIGURE 39.21 Nanoring formed by self-coiling of SWNTs prepared using laser ablation method followed by acid treatment and ultrasonication (Avouris and coworkers). (a) Scanning electron micrograph of a SWNT sample dispersed on a hydrogen-passivated silicon substrate, with rings clearly visible. (b) TEM image of a section of a ring wall. (c) Histogram showing the distribution of ring radii. (Reprinted from Martel, R. et al., *Nature*, 398, 299, 1999. With permission.)

and evaporation induced self-assembly of carbon nanofibers around water droplets dispersed in a carbon tetrachloride solution. Carbon nanofibers were generated within the pores of mesoporous silica (by carbonizing the polymer template used to synthesize mesoporous silica, SBA-15) prepared within the AAO membranes. These carbon fibers get functionalized during the acid etching treatment to remove the silica and alumina. The dried carbon fibers were then extracted with CCl_4. Drop casting of this solution on a copper grid followed by drying leaves a lot of amorphous carbon nanorings on the grid. The surface of the copper grid initially has a lot of adsorbed tiny water droplets. These droplets are replaced by the high density CCl_4 solvent containing the carbon nanofibers dispersed. During this process, the hydrophilic nature of the nanofibers self-assemble and fuse to form a ring around the upward moving water droplet (see the scheme, Figure 39.9a). These

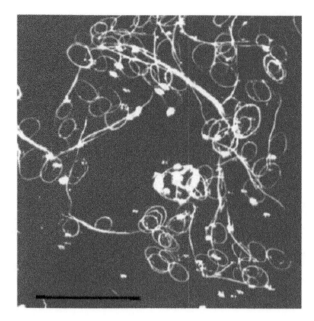

FIGURE 39.22 AFM images of carbon nanotube rings cast on mica. A reaction mixture after washing and filtering through 0.8 mm Teflon filter. Rings appear to stick to each other after these processes. Scale bar, 2 mm. (Reprinted from Sano, M. et al., *Science*, 293, 1299, 2001. With permission.)

nanorings are amorphous in nature and of diameter ranging from 0.5 to 20 μm with a rim thickness of about 50 nm. Figure 39.23 shows a distribution of rings with different sizes (with maximum of 2 μm diameter rings).

39.4.2 Metal Nanorings

Metal nanorings like Au, Ag, Ni, and Co were prepared following one of the techniques or multiple techniques discussed in the earlier section. For example, Au and Co nanorings were made through a template-based process using polystyrene spheres (PS) as the template [14,21]. The metals were deposited over the spheres by a sputtering technique followed by Ar^+ ion beam etching. Subsequent removal of the template resulted in the formation of the nanorings of these metals (Figure 39.24). Co nanorings were also prepared using evaporation induced, dipole-directed self-assembly [49]. The charged surface of DNA nanorings was used as a template to prepare the Ag nanorings by adsorption of silver ions by electrostatic interaction, followed by reduction [41]. Ni and Au nanorings were also reported by using laser interference lithography [43].

39.4.3 Metal Oxide/Sulfide Nanorings

Nanorings of metal oxides/nitrides/sulfides are unique as they show properties that are different from regular nanostructures. Wang and colleagues [32] have shown the formation of freestanding single-crystalline nanorings of ZnO by the epitaxial self-coiling of polar nanobelts synthesized at high temperatures. The synthesis was done by a solid-vapor technique, starting with powders of zinc oxide, indium oxide, and lithium carbonate in a

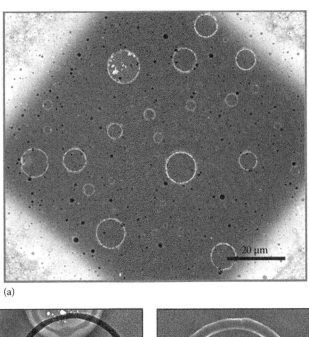

(a)

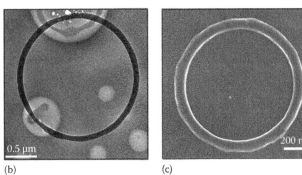

(b)   (c)

FIGURE 39.23 Amorphous carbon nanorings reported by Eswaramoorthy and coworkers. (a) FESEM image showing the carbon nanorings of different sizes. (b) TEM of a nanoring with nonuniform rim thickness. (c) FESEM image of a single amorphous carbon nanoring. (Reprinted from Sai Krishna, K. and Eswaramoorthy, M., *Chem. Phys. Lett.* 433, 327, 2007. With permission.)

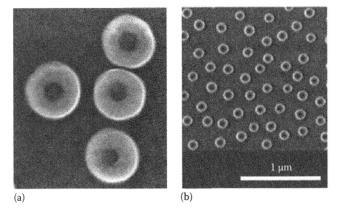

(a)   (b)

FIGURE 39.24 (a) SEM image of the Co nanorings made from PS spheres as template. (b) SEM image of gold nanorings synthesized using PS spheres as template. (Reproduced from Zhu, F.Q. et al., *Adv. Mater.*, 16, 2155, 2004; Reprinted from Aizpurua, J. et al., *Phys. Rev. Lett.*, 90, 057401, 2003. With permission.)

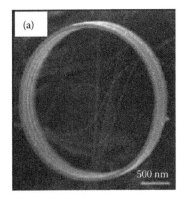

(a)

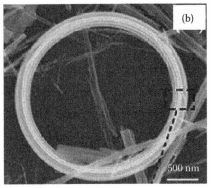

(b)

FIGURE 39.25 SEM images of nanorings formed through self-coiling of nanobelts: (a) $Ag_2V_4O_{11}$ and (b) $K_2Ti_6O_{13}$ nanorings. (Reprinted from Shen, G. and Chen, D., *J. Am. Chem. Soc.*, 128, 11762, 2006; Xu, C.Y. et al., *J. Phys. Chem. C*, 112, 7547, 2008. With permission.)

horizontal tube furnace. Heating the materials to 1400°C in argon causes ZnO material to deposit as rings (on a silicon substrate), with the ring diameter about 1–4 μm and shell thickness around 10–30 nm by a spontaneous self-coiling growth process induced by the polar surfaces of ZnO nanobelts. Nanorings of SnO_2 [50], $Ag_2V_4O_{11}$ (Figure 39.25a) [51], and $K_2Ti_6O_{13}$ (Figure 39.25b) [52] were also reported to form from the self-coiling of nanobelts. For example, Shen and coworkers [51] have reported the formation of $Ag_2V_4O_{11}$ nanorings through self-coiling under hydrothermal reaction conditions between $AgNO_3$ and V_2O_5 powders at 170°C. $Ag_2V_4O_{11}$ nanorings thus obtained have perfect circular shape and flat surfaces. Typical nanorings have diameters of 3–5 μm and thin and wide shells with a thickness of 30–50 nm.

Three dimensional macroporous scaffold obtained by sintering and selective dissolution of a colloidal crystal was used as a template to make crystalline titania nanorings [53] upon infiltration and calcination of titanium alkoxide precursor within the porous scaffold (Figure 39.26). The conventional infiltration step that introduces a polymer precursor prior to selective dissolution of colloidal crystal was not followed in this case. The resulting titania nanorings exhibit a robust, undisrupted rutile phase.

In a different approach, Li and coworkers [54] have synthesized α-Fe_2O_3 nanorings by a rapid microwave-assisted hydrothermal (MAH) method without the use of any template or organic surfactant. Their synthesis involves hydrolysis of an iron precursor ($FeCl_3$) in presence of $NH_4H_2PO_4$ under

FIGURE 39.26 SEM images of 2D interconnected titania rings fabricated on the honeycomb template. Scale bar is 1 μm. (Reprinted from Yi, D.K. and Kim, D.Y., *Nano Lett.*, 3, 207, 2003. With permission.)

MAH conditions. It is believed that MAH process facilitates the formation of single crystal nanorings through nucleation aggregation and coordination-assisted dissolution of α-Fe_2O_3 (by the ligand of phosphate ions) crystals. The superheating and nonthermal effects induced by the MAH process gradually generate a hole at the center of each primarily formed hematite nanodisk, resulting in the formation of nanorings (Figure 39.27). α-Fe_2O_3 nanorings were also prepared from the preferential dissolution along the long dimension of the elongated capsule-

shaped α-Fe_2O_3 nanoparticles in a double anion-assisted hydrothermal process [55].

Nanorings made up of semiconducting cadmium sulfide nanoparticles were prepared by the self-assembly of organically protected cadmium sulfide nanoparticles in aqueous phase [56] (Figure 39.28). The evolved structure was based on morphological controls of primary cadmium sulfide nanocrystals of intrinsic hexagonal symmetry. Since many semiconducting materials have similar hexagonal (or cubic) crystal symmetries, it is possible to extend this self-assembling processes to other semiconductor nanocrystals as well.

39.4.4 Organic Nanorings

39.4.4.1 Porphyrin Nanorings

Porphyrin molecular systems are well known to form ordered stacks due to π–π interactions between the porphyrin entities. Nolte and coworkers have done extensive studies over the formation of porphyrin nanorings and explored their unusual properties [23–25]. The nanorings of porphyrin derivatives formed through evaporation-induced self-assembly process is shown in Figure 39.29.

The ring formation is influenced by the properties of both the solvent and the substrate and together they determine the dewetting process. The ease by which the solution flow carries the porphyrin aggregates to the inner edge of the growing hole is determined by the specific interaction between solvent and substrate.

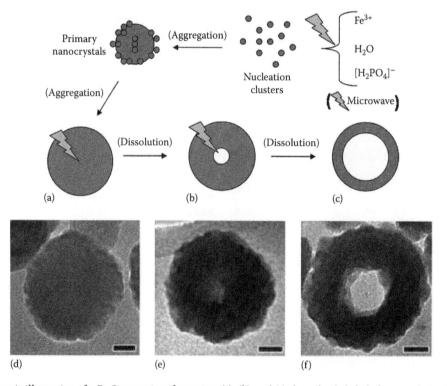

FIGURE 39.27 Schematic illustration of a-Fe_2O_3 nanorings formation. (a), (b), and (c) show the disk, hole form, and ring form, respectively. (d), (e), and (f) show the TEM images of the respective shapes in (a), (b), and (c). The TEM images of (d) are after 10 s, (e) after 50 s, and (f) after 25 min of synthesis time. Scale bar is 20 nm. (Reproduced from Hu, X. et al., *Adv. Mater.*, 19, 2324, 2007. With permission.)

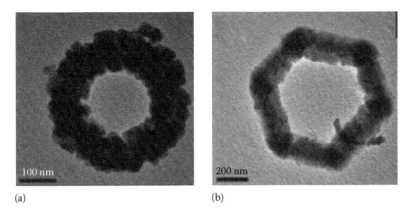

FIGURE 39.28 TEM images of (a) circular CdS ring and (b) hexagonal CdS ring. (Reprinted from Liu, B. and Zeng, H.C., *J. Am. Chem. Soc.*, 127, 18262, 2006. With permission.)

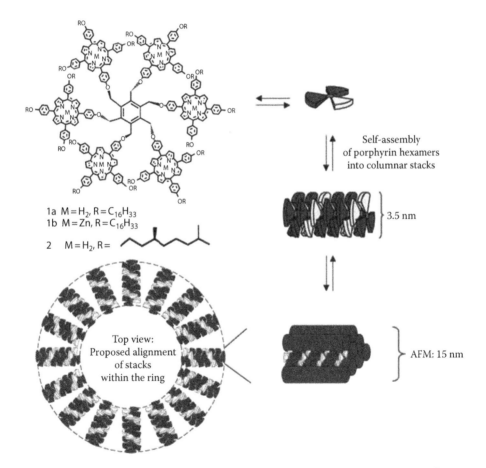

FIGURE 39.29 Chemical structure and schematic showing the stacking of individual porphyrin hexamer units to form a nanoring. (Reproduced from Lensen, M.C. et al., *Chem. Eur. J.*, 10, 831, 2004. With permission.)

The fluorescence microscopy of these rings using polarized excitation light and polarized detection of the emission reveals that only the small rings (up to 5 mm diameter) showed fluorescence anisotropy. The polarized fluorescence microscopy images of the rings on hydrophilic carbon-coated glass are depicted in the figure below (Figure 39.30a). When the excitation and the detection polarization are both oriented vertically, the left and right parts of the rings display higher fluorescence intensity than

the upper and lower parts, respectively. Turning both the excitation and the detection polarization 90° (Figure 39.30b) results in higher fluorescence intensity of the upper and lower parts of the rings.

The expression of polarization is directly related to the molecular ordering and these strong polarization effects indicate that the molecules within a ring are ordered. These rings can be designed and constructed with desired morphology and internal

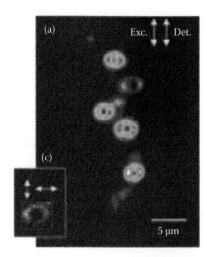
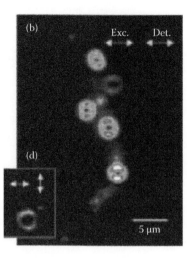

FIGURE 39.30 Polarized fluorescence microscopy images of porphyrin rings (a) excitation and the detection polarization are both oriented vertically (b) excitation and the detection polarization are both oriented horizontally. (Reproduced from Lensen, M.C. et al., *Chem. Eur. J.*, 10, 831, 2004. With permission.)

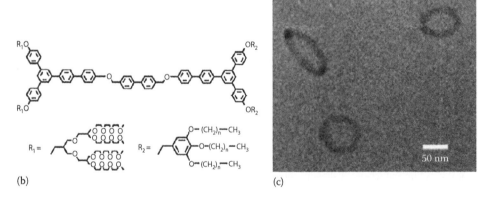

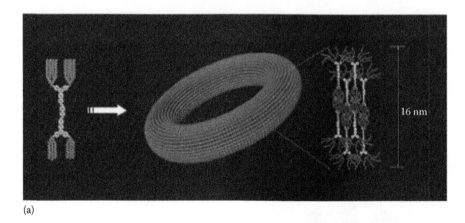

FIGURE 39.31 (a) Schematic representation of nanorings from amphiphilic molecular dumbbells. (b) Molecular structure of the dumbbell-shaped molecule. (c) Cryo-TEM image of the amphiphilic dumbbell-shaped molecules formed into toroidal structure. (Reprinted from Kim, J.K. et al., *J. Am. Chem. Soc.*, 128, 14022, 2006. With permission.)

molecular order, and with different porphyrin derivatives, such as porphyrins containing catalytic transition metals. Also, with respect to the orientation of the molecules within these rings, they are expected to show interesting properties that may find application in molecular electronics.

39.4.4.2 DNA Nanorings

The exploitation of biological macromolecules, such as nucleic acids, for the fabrication of advanced materials is a promising area of research. DNA nanorings are probably the simplest rigid objects with a nanometer size [57]. Small DNA circles were first prepared by designing two 21-mer DNA precursor sequences, which, upon hybridization and ligation, resulted in a statistical distribution of DNA nanorings containing 105, 126, 147, and 168, base pairs (bp). Atomic force microscopy (AFM) analysis of 168-bp nanorings confirmed their smooth circular structure without any ring deformation or supercoiling. These features predestine them as building blocks for the assembly of objects on the nanometer scale. However, a major drawback of DNA nanorings in the construction of higher ordered DNA architectures is their unbranched, continuously double stranded (ds) nature which prevents the guided aggregation of multiple rings. Furthermore, the statistical assembly of the short oligonucleotides that were applied in the known strategies for the synthesis of DNA nanorings prevents the controlled introduction of customized sequences into the circle that can serve as defined handles for the self-assembly of multiple rings.

39.4.4.3 Protein Nanorings

Greater variety of structural and functional uses can be envisioned for protein-based nanorings, given the functional specificity of proteins. In the presence of dimeric methotrexate (bisMTX), wild-type *Escherichia coli* dihydrofolate reductase (DHFR) molecules tethered together by a flexible peptide linker (ecDHFR2) are capable of spontaneously forming highly stable cyclic structures with diameters ranging from 8 to 20 nm [58]. The nanoring size is dependent on the length and composition of the peptide linker, on the affinity and conformational state of the dimerizer, and on induced protein–protein interactions. Control over protein assembly by chemical induction provides an avenue to the future design of protein-based materials and nanostructures.

39.4.4.4 Amphiphilic Molecular Nanorings

The exploitation of aromatic rod-like building blocks for the engineering of synthetic nanostructures is a promising area of research [59]. Incorporation of a rigid rod segment into amphiphilic molecular architectures leads to a number of well-defined nanostructures, including bundles, barrels, tubules, ribbons, and vesicles, in solution state [60]. Recently, researchers have demonstrated that dumbbell-shaped molecules consisting of an aromatic stem segment and hydrophilic dendritic branches, in aqueous solution, self-assemble into nanorings in which the rod segments stack on top of each other with mutual rotation (Figure 39.31). The primary driving force responsible for the helical arrangement of the aromatic rods was proposed to be the energy balance between repulsive interactions among the adjacent hydrophilic dendritic segments and attractive π–π stacking interactions. These results imply that incorporation of hydrophobic branches at one end of a molecular dumbbell further extends the supramolecular organization capabilities of stiff rod-like segments due to enhanced hydrophobic interactions.

Triblock copolymer poly(acrylic acid-*b*-methyl acrylate-*b*-styrene) (PAA_{99}-PMA_{73}-PS_{66}) nanorings [31] were reported to form by self-attraction of cylindrical micelles due to the interaction between negatively charged hydrophilic block of an amphiphilic triblock copolymer and a positively charged divalent organic counterion.

39.5 Conclusions

The search for existing and novel properties in nanomaterials has been an unending quest for researchers. The size- and shape-dependent properties associated with the materials at the nanoscale are fascinating. Nanorings are the new class of nanostructures that will play a prominent role in the future owing to their unusual properties. Their applications in VMRAM devices, nanocontainers for spectroscopy, and nanosized sensors have made nanorings an interesting area of research. This chapter provides an overview of the properties, applications, as well as various methods of fabrication of nanorings.

Acknowledgment

The authors thank Dr. N.S. Vidhyadhiraja for helpful discussions on Aharanov–Bohm effect in nanorings.

References

1. S. Mann and G. A. Ozin, *Nature*, **382**, 313–318, 1996.
2. S. Mann, *Angew. Chem.*, **112**, 3532–3548, 2000; *Angew. Chem. Int. Ed.*, **39**, 3392–3406, 2000.
3. Z. L. Wang, *J. Phys.: Condens. Mater*, **16**, R829–R858, 2004.
4. H. Yang, N. Coombs, and G. A. Ozin, *Nature*, **386**, 692–695, 1997.
5. Y. Xia, Y. Xiong, B. Lim, and S. E. Skrabalak, *Angew. Chem. Int. Ed.*, **48**, 60–103, 2009.
6. S. Mann, in *Biomineralization: Principles and Concepts in Bioinorganic Materials Chemistry*, Oxford University Press, Oxford, U.K., 2001.
7. K. S. Krishna, U. Mansoori, N. R. Selvi, and M. Eswaramoorthy, *Angew. Chem. Int. Ed.*, **46**, 5962–5965, 2007.
8. J. Dinesh, U. Mansoori, P. Mandal, A. Sundaresan, and M. Eswaramoorthy, *Angew. Chem. Int. Ed.*, **47**, 7685–7688, 2008.
9. C. Sanchez, H. Arribart, and M. G. Guille, *Nat. Mater.*, **4**, 277–287, 2005.
10. C. Murray, C. Kagan, and M. Bawendi, *Science*, **270**, 1335, 1995.
11. X. Peng, L. Manna, W. Yang et al., *Nature*, **404**, 59, 2000.
12. T. Vossmeyer, S.-W. Chung, W. M. Gelbart, and J. R. Heath, *Adv. Mater.*, **10**, 351, 1998.
13. G. Schmid, *Nanoparticles: From Theory to Applications*, Wiley-VCH Publications, Weinheim, Germany, 2004.
14. J. Aizpurua, P. Hanarp, D. S. Sutherland, M. Kall, G. W. Bryant, and F. J. García de Abajo, *Phys. Rev. Lett.*, **90**, 057401, 2003.
15. J. Aizpurua, L. Blanco, P. Hanarp et al., *J. Quant. Spectrox. Rad. Trans.*, **89**, 11, 2004.
16. Y. Rondelez, G. Tresset, K. V. Tabata et al., *Nat. Biotechol.*, **23**, 361–365, 2005.

17. Y. Aharonov and D. Bohm, *Phys. Rev.*, **115**, 485, 1959.

18. R. A. Webb, S. Washburn, C. P. Umbach, and R. B. Laibowitz, *Phys. Rev. Lett.*, **54**, 2696, 1985.

19. H. Hu, J.-L. Zhu, D.-J. Li, and J.-J. Xiong, *Phys. Rev. B*, **63**, 195307, 2001.

20. L. Liu, G. Y. Guo, C. S. Jayanthi, and S. Y. Wu, *Phys. Rev. Lett.*, **88**, 217206, 2002.

21. F. Q. Zhu, D. Fan, X. Zhu, J. G. Zhu, R. C. Cammarata, and C. L. Chien, *Adv. Mater.*, **16**, 2155–2159, 2004.

22. F. Q. Zhu, G. W. Chern, O. Tchernyshyov, X. C. Zhu, J. G. Zhu, and C. L. Chien, *Phys. Rev. Lett.*, **96**, 027205, 2006.

23. M. C. Lensen, K. Takazawa, J. A. A. W. Elemans et al., *Chem. Eur. J.*, **10**, 831, 2004.

24. H. A. M. Biemans, A. E. Rowan, A. Verhoeven et al., *J. Am. Chem. Soc.*, **120**, 11054–11060, 1998.

25. A. P. H. J. Schenning, F. B. G. Benneker, H. P. M. Geurts, X. Y. Liu, and R. J. M. Nolte, *J. Am. Chem. Soc.*, **118**, 8549, 1996.

26. R. D. Deegan, O. Bakajin, T. F. Dupont, G. Huber, S. R. Nagel, and T. A. Witten, *Nature (London)*, **389**, 827, 1997.

27. M. Maillard, L. Motte, A. T. Ngo, M. P. Pileni, *J. Phys. Chem. B*, **104**, 11871–11877, 2000.

28. K. S. Krishna and M. Eswaramoorthy, *Chem. Phys. Lett.*, **433**, 327–330, 2007.

29. B. P. Khanal and E. R. Zubarev, *Angew. Chem. Int. Ed.*, **46**, 2195, 2007.

30. M. Sano, A. Kamino, J. Okamura, and S. Shinkai, *Science*, **293**, 1299, 2001.

31. D. J. Pochan, Z. Chen, H. Cui, K. Hales, K. Qi, and K. L. Wooley, *Science*, **306**, 94–97, 2004.

32. X. Y. Kong, Y. Ding, R. Yang, and Z. L. Wang, *Science*, **303**, 1348–1351, 2004.

33. O. D. Velev and E. W. Kaler, *Adv. Mater.*, **12**, 531, 2000.

34. P. Jiang, J. Cizeron, J. F. Berton, and V. L. Colvin, *J. Am. Chem. Soc.*, **121**, 7957–7958, 1999.

35. O. D. Velev, P. M. Tessier, A. M. Lenhoff, and E. W. Kaler, *Nature*, **401**, 548, 1999.

36. H. Xu and W. A. Goedel, *Angew. Chem. Int. Ed.*, **42**, 4696, 2003.

37. F. Yan and W. A. Goedel, *Angew. Chem. Int. Ed.*, **44**, 2084, 2005.

38. F. Yan and W. A. Goedel, *Nano Lett.*, **4**, 1193, 2004.

39. K. L. Hobbs, P. R. Larson, G. D. Lian, J. C. Keay, and M. B. Johnson, *Nano Lett.*, **4**, 167–171, 2004.

40. S. Wang, G. J. Yu, J. L. Gong et al., *Nanotechnology*, **17**, 1594–1598, 2006.

41. A. A. Zinchenko, K. Yoshikawa, and D. Baigl, *Adv. Mater.*, **17**, 2820–2823, 2005.

42. Y. Yin, B. Gates, and Y. Xia, *Adv. Mater.*, **12**, 1426–1430, 2000.

43. R. Ji, W. Lee, R. Scholz, U. Gosele, and K. Nielsch, *Adv. Mater.*, **18**, 2593–2596, 2006.

44. T. Kawase, Y. Seirai, H. R. Darabi, M. Oda, Y. Sarakai, and K. Tashiro, *Angew. Chem. Int. Ed.*, **42**, 1621–1624, 2003.

45. T. Kawase, K. Tanaka, Y. Seirai, N. Shiono, and M. Oda, *Angew. Chem. Int. Ed.*, **42**, 5597–5600, 2003.

46. I. G. Cuesta, T. B. Pedersen, H. Koch, and A. S. de Meras, *ChemPhysChem.*, 7, 2503–2507, 2006.

47. J. Liu, H. Dai, J. H. Hafner et al., *Nature*, **385**, 780, 1997.

48. R. Martel, H. R. Shea, and Ph. Avouris, *Nature*, **398**, 299, 1999.

49. S. L. Tripp, S. V. Pusztay, A. E. Ribbe, and A. Wei, *J. Am. Chem. Soc.*, **124**, 7914–7915, 2002.

50. R. Yang and Z. L. Wang, *J. Am. Chem. Soc.*, **128**, 1466–1467, 2006.

51. G. Shen and D. Chen, *J. Am. Chem. Soc.*, **128**, 11762–11763, 2006.

52. C.-Y. Xu, Y.-Z. Liu, L. Zhen, and Z. L. Wang, *J. Phys. Chem. C*, **112**, 7547–7551, 2008.

53. D. K. Yi and D.-Y. Kim, *Nano Lett.*, **3**, 207–211, 2003.

54. X. Hu, J. C. Yu, J. Gong, Q. Li, and G. Li, *Adv. Mater.*, **19**, 2324–2329, 2007.

55. C.-J. Jia et al., *J. Am. Chem. Soc.*, **130**, 16968–16977, 2008.

56. B. Liu and H. C. Zeng, *J. Am. Chem. Soc.*, **127**, 18262, 2005.

57. G. Rasched, D. Ackermann, T. L. Schmidt, P. Broekmann, A. Heckel, and M. Famulok, *Angew. Chem. Int. Ed.*, **47**, 967–970, 2008.

58. J. C. T. Carlson, S. S. Jena, M. Flenniken, T. Chou, R. A. Siegel, and C. R. Wagner, *J. Am. Chem. Soc.*, **128**, 7630–7638, 2006.

59. J.-K. Kim, E. Lee, Z. Huang, and M. Lee, *J. Am. Chem. Soc.*, **128**, 14022–14023, 2006.

60. F. J. M. Hoeben, P. Jonkheijm, E. W. Meijer, and A. P. H. J. Schenning, *Chem. Rev.*, **105**, 1491–1546, 2005.

40

Superconducting Nanowires and Nanorings

Andrei D. Zaikin

Karlsruhe Institute of Technology

and

P.N. Lebedev Physics Institute

40.1 Introduction

The phenomenon of superconductivity was discovered (Kamerlingh Onnes 1911) as a sudden drop of resistance to an immeasurably small value. With the development of the topic it was realized that the superconducting phase transition is frequently not at all "sudden" and the measured dependence of the sample resistance, $R(T)$, in the vicinity of the critical temperature, T_c, may have a finite width. One possible reason for this behavior—and frequently the dominating factor—is the sample inhomogeneity, i.e., the sample might simply consist of regions with different local critical temperatures. However, with improving fabrication technologies it became clear that even for highly homogeneous samples the superconducting phase transition may remain broadened. This effect is usually very small in bulk samples and becomes more pronounced in systems with reduced dimensions. A fundamental physical reason behind such smearing of the transition is *superconducting fluctuations*.

The important role of fluctuations in reduced dimensions is well known. Above T_c, such fluctuations yield an enhanced conductivity of metallic systems (Larkin and Varlamov 2005). For instance, the so-called Aslamazov–Larkin fluctuation correction to conductivity, $\delta\sigma_{AL} \sim (T - T_c)^{-(2-D/2)}$, becomes large in the vicinity of T_c, and this effect increases with decreasing dimensionality D. Below T_c, according to the general Mermin–Wagner–Hohenberg theorem (Mermin and Wagner 1966, Hohenberg 1967), fluctuations should destroy the long-range order in low-dimensional superconductors. Thus, it could naively be concluded that low-dimensional conductors cannot exhibit superconducting properties because of strong phase fluctuation effects.

This conclusion, however, turns out to be premature. For instance, 2D structures undergo Berezinskii–Kosterlitz–Thouless phase transition (Berezinskii 1971, Kosterlitz and Thouless 1973, Kosterlitz 1974) as a result of which the decay of correlations in space changes from exponential at high enough T to power law at low temperatures. This result implies that at low T, long-range phase coherence essentially survives in samples of a finite size and, hence, 2D films can well exhibit superconducting properties.

Can superconductivity survive also in quasi-1D systems or do fluctuations suppress phase coherence, thus disrupting any supercurrent? The answer to this question would clearly be of both fundamental interest and practical importance. On one hand, investigations of this subject definitely help to encover novel physics and shed more light on the crucial role of superconducting fluctuations in 1D wires. On the other hand, rapidly progressing miniaturization of nanodevices opens new horizons for applications of superconducting nanocircuits and requires a better understanding of the fundamental limitations of the phenomenon of superconductivity in reduced dimensions. This chapter is devoted to a detailed discussion of the nontrivial interplay between superconductivity and the fluctuations in metallic nanowires and nanorings.

40.2 Background

It was first pointed out by Little (1967) that quasi-1D wires made of a superconducting material can acquire a finite resistance below T_c of a bulk material due to the mechanism of thermally activated phase slips (TAPS). Within the Ginzburg–Landau theory one can describe a superconducting wire by means of a complex order parameter $\Psi(x) = |\Psi(x)|e^{i\varphi(x)}$. Thermal fluctuations

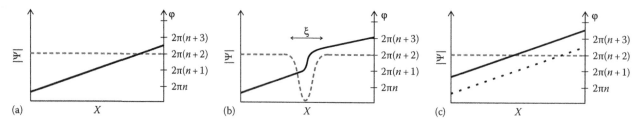

FIGURE 40.1 Schematics of the phase slip process. Spatial variation of the amplitude of the order parameter $|\Psi|$ (left axis, dashed line) and phase φ (right axis, solid line) at various moments of time: (a) before, (b) during, and (c) after the phase slippage. (From Arutyunov, K. Yu. et al., *Phys. Rep.*, 464, 1, 2008.)

cause deviations of both the modulus and the phase of this order parameter from their equilibrium values. A nontrivial fluctuation corresponds to a temporal suppression of $|\Psi(x)|$ down to zero at some point (e.g., $x = 0$) inside the wire, see Figure 40.1. As soon as the modulus of the order parameter $|\Psi(0)|$ vanishes, the phase $\varphi(0)$ becomes unrestricted and can jump by the value $2\pi n$, where n is any integer number. After this process the modulus $|\Psi(0)|$ gets restored, the phase becomes single valued again and the system returns to its initial state, accumulating the net phase shift $2\pi n$. Provided such phase slip events are sufficiently rare, one can restrict n by $n = \pm 1$ and totally disregard fluctuations with $|n| \geq 2$.

According to the Josephson relation $V = \hbar\dot{\varphi}/2e$, each such phase slip event causes a nonzero voltage drop V across the wire. In the absence of any bias current, the net average numbers of "positive" ($n = +1$) and "negative" ($n = -1$) phase slips are equal, thus the net voltage drop remains zero. Applying the current $I \propto |\Psi|^2 \nabla\varphi$ one creates a nonzero phase gradient along the wire and makes "positive" phase slips more likely than "negative" ones. Hence, the net voltage drop V due to TAPS differs from zero, i.e., thermal fluctuations cause a nonzero resistance $R = V/I$ of superconducting wires even below T_c. We would also like to emphasize that, in contrast to the so-called phase slip centers (Tidecks 1990, Kopnin 2001) produced by a large current above the critical one $I > I_c$, here we are dealing with *fluctuation-induced phase slips*, which can occur at arbitrarily small values of I.

A quantitative theory of the TAPS phenomenon was first proposed by Langer and Ambegaokar (1967) and then extended by McCumber and Halperin (1970). The LAMH theory predicts that the TAPS creation rate and, hence, the resistance of a superconducting wire R below T_c is determined by the activation exponent:

$$R(T) \propto \exp\left(\frac{-\delta U}{kT}\right), \quad \delta U \sim \frac{N_0 \Delta_0^2(T)}{2} s\xi(T), \quad (40.1)$$

where $\delta U(T)$ is the effective potential barrier for TAPS. This potential barrier is determined by the superconducting condensation energy per unit volume $N_0\Delta_0(T)^2/2$ (here and below N_0 is the metallic density of states at the Fermi energy and $\Delta_0(T)$ is the BCS order parameter) lost during a phase slip event in a part of the wire where the order parameter gets temporarily suppressed by thermal fluctuations. This volume is given by the wire cross

section s times the typical TAPS size which—according to the LAMH theory—is of the order of the superconducting coherence length $\xi(T)$. At temperatures very close to T_c, Equation 40.1 yields appreciable resistivity, which was indeed detected experimentally (Lukens et al. 1970, Newbower et al. 1972). Close to T_c, the experimental results fully confirm the activation behavior of $R(T)$ expected from Equation 40.1. However, as the temperature is lowered further below T_c, the number of TAPS inside the wire decreases exponentially and no measurable wire resistance is predicted by the LAMH theory except in the immediate vicinity of the critical temperature.

Experiments by Lukens et al. and by Newbower et al. have been performed with small whiskers of typical diameters $\sim 0.5\,\mu m$. Recent progress in the nanolithographic technique made it possible to fabricate samples with much smaller diameters down to— and even below—10 nm. In such systems, one can consider a possibility for phase slips to occur not only due to thermal, but also due to *quantum* fluctuations of the superconducting order parameter. The physical picture of quantum phase slippage is qualitatively similar to that of TAPS (see Figure 40.1), except the order parameter $|\Psi(x)|$ gets virtually suppressed due to the process of *quantum tunneling*.

Following the standard quantum mechanical arguments, one can expect that the probability of such a tunneling process should be controlled by the exponent $\sim\exp(-U/\hbar\omega_0)$, i.e., instead of temperature in the activation exponent (Equation 40.1) one should just substitute $\hbar\omega_0$, where ω_0 is an effective attempt frequency. This is because the order parameter field $\Psi(x)$ now tunnels under the barrier U rather than overcomes it by thermal activation. Since such a tunneling process should obviously persist down to $T = 0$, one arrives at the fundamentally important conclusion that *in nanowires superconductivity can be destroyed by quantum fluctuations at any temperature including* $T = 0$. Accordingly, such nanowires should demonstrate a nonvanishing resistivity down to zero temperature. *Assuming* that $\hbar\omega_0 \sim \Delta_0(T)$, one would expect that at $\Delta_0(T) \lesssim T < T_c$ the TAPS dependence (Equation 40.1) applies, while at lower $T \lesssim \Delta_0(T)$ quantum phase slips (QPS) take over, eventually leading to the saturation of the temperature dependence $R(T)$ to a nonzero value in the limit $T \ll \Delta_0$.

This behavior was indeed observed: experiments (Giordano 1988, 1994) clearly demonstrated a notable resistivity of ultrathin superconducting wires far below T_c. These observations could not be adequately interpreted within the TAPS theory and were attributed to QPS. Later other groups also reported noticeable deviations from the LAMH theory in thin (quasi-)1D wires. It should be noted, however, that the unambiguous interpretation of the observations by Giordano in terms of QPS was questioned by several authors because of possible granularity of the samples used in those experiments. If that was indeed the case, QPS could easily be created inside weak links connecting neighboring grains. Also in this case superconducting fluctuations play a very important role (Schön and Zaikin 1990); however, in contrast to the case of uniform wires, the superconducting order parameter *needs not to be destroyed* during a QPS event.

First attempts to theoretically analyze the QPS effects were based on the so-called time-dependent Ginzburg–Landau (TDGL) equations. Unfortunately the TDGL approach is by far insufficient for the problem in question for a number of reasons: (1) A trivial reason is that the Ginzburg–Landau (GL) expansion applies only at temperatures close to T_c, whereas in order to describe QPS one usually needs to go to lower temperatures down to $T \to 0$. (2) More importantly, also at $T_c - T \ll T_c$, the TDGL equation remains applicable only in a special limit of gapless superconductors, while it fails in a general situation considered here. (3) The TDGL approach does not account for dissipation effects due to quasiparticles inside the QPS core (in certain cases also outside this core) that are expected to reduce the probability of QPS events similarly to the standard problem of quantum tunneling with dissipation (Caldeira and Leggett 1983, Weiss 1999). (4) The TDGL approach is not fully adequate to properly describe the excitation of electromagnetic modes around the wire during a QPS event (this effect turns out to be particularly important for sufficiently long wires). For more details on this point we refer the reader to the papers by van Otterlo et al. (1999) and by Arutyunov et al. (2008). Thus, TDGL-based description of superconducting fluctuations simply cannot be trusted, and a much more elaborate theory is necessary in this situation.

A microscopic theory of QPS processes in superconducting nanowires was developed in a number of papers (Zaikin et al. 1997, 1998, Golubev and Zaikin 2001) with the aid of the imaginary time effective action technique (van Otterlo et al. 1999). This theory remains applicable down to $T = 0$ and properly accounts for the nonequilibrium, the dissipative, and the electromagnetic effects during a QPS event. One of the main conclusions of this theory is that in sufficiently dirty superconducting nanowires with diameters in the 10 nm range, QPS probability can already be large enough to yield experimentally observable phenomena. Also, further interesting effects including quantum phase transitions (QPT) caused by interactions between QPS were predicted.

An important parameter of this theory is the QPS fugacity:

$$y \sim S_{core} \exp(-S_{core}), \quad S_{core} \sim g_\xi,$$

where g_ξ is the dimensionless conductance of the wire segment of length ξ. Provided g_ξ is very large, typically $g_\xi \gtrsim 100$, the fugacity y remains vanishingly small, QPS events are very rare and in many cases can be totally neglected. In such case,

the standard BCS mean-field description should apply and a finite (though possibly sufficiently long) wire remains essentially superconducting outside an immediate vicinity of T_c. For smaller $g_\xi \lesssim 10 \div 20$ QPS effects already become important down to $T \to 0$. Finally, at even smaller $g_\xi \sim 1$ strong fluctuations should wipe out superconductivity everywhere in the wire. We also point out that, in the case of nanowires considered here, the parameter g_ξ is related to the well-known Ginzburg number Gi_{1D} (Larkin and Varlamov 2005) as $Gi_{1D} \sim 1/g_\xi^{2/3}$, i.e., the condition $g_\xi \sim 1$ also implies that the fluctuation region δT becomes of order T_c.

Another important parameter is the ratio between the (superconducting) quantum resistance unit, $R_q = \pi\hbar/2e^2 = 6.453\,k\Omega$, and the wire impedance, $Z_w = \sqrt{\tilde{L}/C}$:

$$\frac{R_q}{Z_w} = 2\mu,$$

where

C is the wire capacitance per unit length
$\tilde{L} = 4\pi\lambda_L^2/s$ is the wire kinetic inductance
λ_L is the London penetration depth

Provided this parameter becomes of order one, $\mu \sim 1$, superconductivity in sufficiently long wires gets fully suppressed due to intensive fluctuations of the phase φ of the superconducting order parameter. We note that both g_ξ and μ scale with the wire cross section s as $g_\xi \propto s$ and $\mu \propto \sqrt{s}$, respectively. *It follows immediately that with decreasing the cross section below a certain value the wire inevitably looses intrinsic superconducting properties due to strong fluctuation effects.* For generic parameters, both conditions $g_\xi \sim 1 \div 10$ and $\mu \sim 1$ are typically met for wire diameters in the range $\sqrt{s} \lesssim 10\,nm$.

A number of recent experimental observations are clearly consistent with the above theoretical conclusions. Perhaps, the first unambiguous evidence for QPS effects in quasi-1D wires was reported by Bezryadin, Lau, and Tinkham (Bezryadin et al. 2000) who fabricated sufficiently uniform superconducting wires with thicknesses down to $3 \div 5\,nm$, and observed that several samples showed no signs of superconductivity even at temperatures well below the bulk critical temperature. Those results were later confirmed and substantially extended by different experimental groups. At present there exists an overwhelming experimental evidence for QPS effects in superconducting nanowires fabricated to be sufficiently uniform and homogeneous.

Yet another interesting issue is related to persistent currents (PC) in superconducting nanorings. For instance, it was demonstrated that QPS effects can significantly modify PC in such systems and even lead to exponential suppression of supercurrent for sufficiently large ring perimeters (Matveev et al. 2002, Arutyunov et al. 2008). Another important factor that can substantially affect PC in isolated superconducting nanorings at low T is the electron parity number. Of particular interest is the behavior of rings with odd number of electrons that can develop *spontaneous* supercurrent in the ground state without any externally applied magnetic flux (Sharov and Zaikin 2005a,b). Experimental verification of these intriguing theoretical predictions still remains a challenge for future investigations.

40.3 Thermal and Quantum Fluctuations in Superconducting Nanowires

As we already emphasized above, the phenomenological TDGL approach is insufficient for an adequate analysis of superconducting fluctuations, in particular at temperatures below T_c. Here we will discuss a fully microscopic approach that allows for a complete description of fluctuation effects in superconductors. This approach is general, and as such it holds for any superconducting system. Therefore, the initial part of our consideration will be formulated in a rather general form. Then we will adopt our approach specifically to superconducting nanowires.

40.3.1 General Theory

The grand partition function $\mathcal{Z}$ for any quantum system described by the Hamiltonian $\hat{H}$ reads

$$\mathcal{Z} = \mathrm{Tr}\exp\left(\frac{-\hat{H}}{T}\right). \tag{40.2}$$

In our case, $\hat{H}$ is a standard Hamiltonian for electrons in a superconductor that includes a short-range attractive BCS and long-range repulsive Coulomb interactions as well as electromagnetic fields. The main idea of our method is to exactly integrate out the electronic degrees of freedom already on the level of the partition function, so that we are left with an effective theory in terms of collective fields (Popov 1987, Schön and Zaikin 1990). For this purpose, it is convenient to apply the imaginary time Matsubara technique and to express the partition function $\mathcal{Z}$ as a path integral over the anticommuting electronic fields $\bar{\psi}$, ψ, and the commuting gauge scalar and vector-potential fields V and $\mathbf{A}$, with Euclidean action:

$$S_E = \int dx \left(\bar{\psi}_\sigma \left[\partial_\tau - ieV + \xi\left(\nabla - \frac{ie}{c}\mathbf{A}\right) \right] \psi_\sigma \right.$$
$$\left. - \lambda\bar{\psi}_\uparrow\bar{\psi}_\downarrow\psi_\downarrow\psi_\uparrow + ienV + [\mathbf{E}^2 + \mathbf{B}^2]/8\pi \right). \tag{40.3}$$

Here $\xi(\nabla) \equiv -\nabla^2/2m - \mu + U(x)$ describes a single conduction band with quadratic dispersion, and also includes an arbitrary impurity potential, λ is the BCS coupling constant, $\sigma = \uparrow, \downarrow$ is the spin index, and en denotes the background charge density of the ions. In our notation dx denotes $d^3x\,d\tau$, where τ is the imaginary time. The field strengths are functions of the gauge fields through $\mathbf{E} = -\nabla V + (1/c)\partial_\tau\mathbf{A}$ and $\mathbf{B} = \nabla \times \mathbf{A}$ in the usual way for

the imaginary time formulation. Here and below, we will express our results in units in which both the Planck and the Boltzmann constants $\hbar$ and k, respectively, are set equal to unity. At certain points, however, we restore the explicit dependence on $\hbar$ for the sake of clarity.

We use a Hubbard–Stratonovich transformation to decouple the BCS interaction term and to introduce the superconducting order parameter field $\Delta = |\Delta|e^{i\varphi}$:

$$\exp\left(\lambda \int dx\, \bar{\psi}_\uparrow\bar{\psi}_\downarrow\psi_\downarrow\psi_\uparrow \right) = \left[\int \mathcal{D}^2\Delta\, e^{-\frac{1}{\lambda}\int dx\Delta^2} \right]^{-1}$$
$$\times \int \mathcal{D}^2\Delta\, e^{-\int dx\left(\frac{1}{\lambda}\Delta^2 + \Delta\bar{\psi}_\uparrow\bar{\psi}_\downarrow + \Delta^*\psi_\downarrow\psi_\uparrow\right)}, \tag{40.4}$$

where the first factor is for normalization and will not be important in the following. As a result, the partition function now reads

$$\mathcal{Z} = \int \mathcal{D}^2\Delta \int \mathcal{D}^3\mathbf{A} \int \mathcal{D}V\mathcal{D}^2\Psi\, e^{\left(-S_0 - \int dx\,\bar{\Psi}\mathcal{G}^{-1}\Psi\right)},$$
$$S_0[V, \mathbf{A}, \Delta] = \int dx \left(\frac{\mathbf{E}^2 + \mathbf{B}^2}{8\pi} + ienV + \frac{\Delta^2}{\lambda} \right), \tag{40.5}$$

where the Nambu spinor notation for the electronic fields and the matrix Green function in Nambu space

$$\Psi = \begin{pmatrix} \psi_\uparrow \\ \bar{\psi}_\downarrow \end{pmatrix}, \quad \bar{\Psi} = (\bar{\psi}_\uparrow \quad \psi_\downarrow);$$

$$\tilde{\mathcal{G}}^{-1} = \begin{pmatrix} \partial_\tau - ieV + \xi(\nabla - \frac{ie}{c}\mathbf{A}) & \Delta \\ \Delta^* & \partial_\tau + ieV - \xi(\nabla + \frac{ie}{c}\mathbf{A}) \end{pmatrix} \tag{40.6}$$

have been introduced. After the Gaussian integral over the electronic degrees of freedom, we obtain

$$\mathcal{Z} = \int \mathcal{D}^2\Delta\mathcal{D}V\mathcal{D}\mathbf{A}\, e^{-S_{\text{eff}}}, \tag{40.7}$$

where

$$S_{\text{eff}} = -\mathrm{Tr}\ln\tilde{\mathcal{G}}^{-1} + S_0[V, \mathbf{A}, \Delta] \tag{40.8}$$

is the effective action of our superconducting system. The gauge invariance of the theory enables us to rewrite the action (40.8) in a different form

$$S_{\text{eff}} = -\mathrm{Tr}\ln\mathcal{G}^{-1} + S_0[V, \mathbf{A}, \Delta], \tag{40.9}$$

where

$$\mathcal{G}^{-1} = \begin{pmatrix} \partial_\tau + \xi(\nabla) - ie\Phi + \frac{m\mathbf{v}_s^2}{2} - \frac{i}{2}\{\nabla, \mathbf{v}_s\} & |\Delta| \\ |\Delta| & \partial_\tau - \xi(\nabla) + ie\Phi - \frac{m\mathbf{v}_s^2}{2} - \frac{i}{2}\{\nabla, \mathbf{v}_s\} \end{pmatrix}, \tag{40.10}$$

and we have introduced the gauge invariant linear combinations of the electromagnetic potentials and the phase of the order parameter:

$$\Phi = V - \frac{\dot{\varphi}}{2e}, \quad \mathbf{v}_s = \frac{1}{2m}\left(\nabla_{\varphi} - \frac{2e}{c}\mathbf{A}\right). \quad (40.11)$$

The curly brackets $\{A, B\}$ denote an anticommutator.

The action (40.9) cannot be evaluated exactly. In order to bring our results to a tractable form we will have to implement several approximations. Here we will perform a perturbative expansion of the action up to the second order in Φ and $\mathbf{v}_s$. This perturbation theory is sufficient for nearly all practical purposes, because nonlinear electromagnetic effects (described by higher order terms) are usually very small in the systems under consideration. In addition, we assume that deviations of the amplitude of the order parameter field from its equilibrium value Δ_0 are relatively small. This assumption allows us to also expand the effective action in powers of $\delta\Delta(x,\tau) = \Delta(x,\tau) - \Delta_0$ up to the second order terms. The next step is to average over the random potential of impurities. After that the effective action becomes translationally invariant both in space and in time. All these steps are well documented in the literature (van Otterlo et al. 1999, Golubev and Zaikin 2001, Arutyunov et al. 2008); thus we avoid presenting the corresponding technical details here.

At this stage, we turn to the specific case of sufficiently long and very thin superconducting wires that will be of particular interest for us here. Let us define the coordinate along the wire x, the capacitance per unit length of the wire C, and the inductance times unit length L. Here and below, A will stand for the component of the vector potential parallel to the wire. For a cylindric wire with radius $r_0 \sim \sqrt{s}$ embedded in a dielectric environment with susceptibility ε, the capacitance C and inductance L are

$$C \approx \frac{\varepsilon}{2\ln(R_0/r_0)}, \quad L \approx 2\ln\left(\frac{R_0}{r_0}\right), \quad (40.12)$$

where R_0 is the distance from the center of the wire and the bulk metallic electrode.

Making use of the translational invariance of the (averaged over disorder) effective action, we perform the Fourier transformation with respect to both x and τ. Then we obtain (van Otterlo et al. 1999, Golubev and Zaikin 2001)

$$S = \frac{s}{2}\int\frac{d\omega dq}{(2\pi)^2}\left\{\frac{|A|^2}{Ls} + \frac{C|V|^2}{s} + \chi_{\mathrm{D}}\left|qV + \frac{\omega}{c}A\right|^2 + \chi_{\mathrm{J}}\left|V + \frac{i\omega}{2e}\varphi\right|^2\right.$$

$$\left. + \frac{\chi_{\mathrm{L}}}{4m^2}\left|iq\varphi + \frac{2e}{c}A\right|^2 + \chi_{\Delta}|\delta\Delta|^2\right\}. \quad (40.13)$$

Here the first two terms in the right-hand side account for the energy of fluctuating electromagnetic fields around the wire. The terms proportional to the functions χ_{D}, χ_{J}, χ_{L}, and χ_{Δ}, which depend both on the frequencies and the wave vectors,

account respectively for Drude, Josephson, London, and condensation energy contributions to the action. These functions are expressed in terms of the averaged products of the Matsubara Green functions (van Otterlo et al. 1999, Golubev and Zaikin 2001). The corresponding expressions are in general rather complicated and get simplified only in some special limits. For instance, in the limit $|\omega| \gg \Delta_0$ the Drude kernel χ_{D} reduces to the standard form

$$\chi_{\mathrm{D}} \simeq \frac{\sigma}{|\omega| + Dq^2},$$

where

- $\sigma = 2e^2 N_0 D$ is the wire Drude conductivity in the normal state
- N_0 is the electron density of state at the Fermi level
- $D = v_{\mathrm{F}}l/3$ is the diffusion coefficient
- l is the electron elastic mean free path

In the opposite limit $\omega, q \to 0$ and at low $T \ll T_{\mathrm{c}}$ we have

$$\chi_{\mathrm{D}} \simeq \frac{\pi\sigma}{8\Delta_0}\left[1 - \frac{3}{8}\left(\frac{\omega}{2\Delta_0}\right)^2 - \frac{8}{3\pi}\frac{Dq^2}{2\Delta_0}\right].$$

The Josephson kernel χ_{J} is proportional to the electron density of states N_0. In the limit $\omega, q \to 0$, it reads

$$\chi_{\mathrm{J}} = 2e^2 N_0\left[1 - \frac{2}{3}\left(\frac{\omega}{2\Delta_0}\right)^2 - \frac{\pi}{4}\frac{Dq^2}{2\Delta_0}\right].$$

For the London kernel χ_{L} at $T \to 0$ and in the limit of low frequencies and wave vectors, we obtain

$$\chi_{\mathrm{L}} = 2\pi N_0 Dm^2\Delta_0\left[1 - \frac{1}{4}\left(\frac{\omega}{2\Delta_0}\right)^2 - \frac{2}{\pi}\frac{Dq^2}{2\Delta_0}\right],$$

where m is the electron mass. Finally, in the same limit the function χ_{Δ} reads

$$\chi_{\Delta} \simeq 2N_0\left(1 + \frac{1}{3}\left(\frac{\omega}{2\Delta_0}\right)^2 + \frac{\pi}{4}\frac{Dq^2}{2\Delta_0}\right),$$

whereas at high temperatures $T \gg \Delta_0(T)$, we find

$$\chi_{\Delta} = 2N_0\left[\ln\frac{T}{T_{\mathrm{c}}} + \Psi\left(\frac{1}{2} + \frac{|\omega| + Dq^2}{4\pi T}\right) - \Psi\left(\frac{1}{2}\right)\right],$$

where $\Psi(x)$ is the digamma function.

As the action S (40.13) is quadratic both in the voltage V and the vector potential A, these variables can be integrated out exactly. Performing this integration one arrives at the effective action that only depends on φ and $\delta\Delta$. We obtain

$$S = \frac{s}{2}\int\frac{d\omega dq}{(2\pi)^2}\left\{\mathcal{F}(\omega,q)|\varphi|^2 + \chi_{\Delta}|\delta\Delta|^2\right\} \quad (40.14)$$

Since typically metallic nanowires (with the diameter of the order of superconducting coherence length $\xi = \sqrt{D/\Delta_0}$ or thinner) are quite strongly disordered, for generic values of the electron elastic mean free path l one has $1/L \gg \pi\sigma\Delta_0 s/c^2$ and $C \ll 2e^2 N_0 s$. The first condition allows to set $L = 0$ (which immediately yields $A = 0$), i.e., ignore fluctuations of the vector potential in our further analysis. In this limit, we obtain

$$\mathcal{F}(\omega,q) = \frac{\left(\frac{\chi_J}{4e^2}\omega^2 + \frac{\chi_L}{4m^2}q^2\right)\left(\frac{C}{s} + \chi_D q^2\right) + \frac{\chi_J \chi_L}{4m^2}q^2}{\frac{C}{s} + \chi_J + \chi_D q^2}. \quad (40.15)$$

This Gaussian integration also fixes the relation between the fluctuating voltage V and the superconducting phase φ:

$$V = \frac{\chi_J}{\frac{C}{s} + \chi_J + \chi_D q^2}\left(\frac{-i\omega}{2e}\varphi\right). \quad (40.16)$$

Note that Equation 40.16 allows to establish the applicability range of the Josephson relation $V = \dot{\varphi}/2e$ for the problem in question. This relation approximately holds only in the limit $\chi_J \gg C/s + \chi_D q^2$, which is obeyed in the limit of low frequencies and wave vectors $\omega/\Delta_0 \ll 1$ and $Dq^2/\Delta_0 \ll 1$. At even smaller wave vectors $Dq^2/2\Delta_0 \ll 2C/\pi e^2 N_0 s \ll 1$ the expression for the function $F(\omega, q)$ gets significantly simplified, and for the wire action we obtain

$$S = \frac{1}{2}\int\frac{d\omega dq}{(2\pi)^2}\left\{\left(C\omega^2 + \pi\sigma\Delta_0 sq^2\right)\left|\frac{\varphi}{2e}\right|^2 + s\chi_\Delta|\delta\Delta|^2\right\}. \quad (40.17)$$

Note that the phase-dependent part of this action describes the propagation of the plasmon Mooij–Schön mode (Mooij and Schön 1985) along the wire with dispersion

$$\omega = c_0 q, \quad (40.18)$$

where the velocity of this mode c_0 is

$$c_0 \simeq \frac{1}{\sqrt{\tilde{L}C}} = \sqrt{\frac{\pi\sigma\Delta_0 s}{C}} \quad (40.19)$$

and $\tilde{L} = 4\pi\lambda_L^2/s = 1/2\pi e^2 N_0\Delta_0 Ds$ is the kinetic inductance of a superconducting wire.

40.3.2 Gaussian Fluctuations

The effective action (40.14) allows to directly evaluate the fluctuation correction to the order parameter in superconducting nanowires. Performing Gaussian integration over both φ and $\delta\Delta$ we arrive at the wire free energy

$$F = F_{BCS} - \frac{T}{2}\sum_{\omega,q}\left[\ln\frac{\lambda\mathcal{F}(\omega,q)}{2N_0\Delta_0^2} + \ln\frac{\lambda\chi_\Delta(\omega,q)}{2N_0}\right], \quad (40.20)$$

where F_{BCS} is the standard BCS free energy (which is of no particular interest for us here) and two other terms account for the

fluctuation contributions from the phase and the modulus of the order parameter field, respectively. The equilibrium value of this field is defined by the saddle point equation

$$\partial F/\partial\Delta = 0$$

and can be written in the form $\Delta = \Delta_0 - \delta\Delta_0$, where $\Delta_0(T)$ is the solution of the BCS self-consistency equation $\partial F_{BCS}/\partial\Delta_0 = 0$, and the fluctuation correction $\delta\Delta_0(T)$ has the form

$$\delta\Delta_0 = -\frac{T}{2}\left(\frac{\partial^2 F_{BCS}}{\partial\Delta_0^2}\right)^{-1}\frac{\partial}{\partial\Delta_0}\sum_{\omega,q}\left[\ln\frac{\lambda\mathcal{F}(\omega,q)}{2N_0\Delta_0^2} + \ln\frac{\lambda\chi_\Delta(\omega,q)}{2N_0}\right]. \quad (40.21)$$

Making use of the above expressions for the functions $\mathcal{F}(\omega, q)$ and $\chi_\Delta(\omega, q)$ and employing the condition $\partial F/\partial\Delta_0 = 2N_0 sX$, at $T \to 0$, we obtain

$$\frac{\delta\Delta_0}{\Delta_0} \sim \frac{1}{g_\xi} \sim Gi_{1D}^{3/2}, \quad (40.22)$$

where $g_\xi = (2s/3\pi\lambda_F^2)(l/\xi)$ is the dimensionless conductance of the wire segment of length ξ, $\lambda_F = \hbar/p_F$ is the Fermi wavelength, and

$$Gi_{1D} = \frac{0.15}{(sN_0\sqrt{D\Delta_0})^{2/3}} \quad (40.23)$$

is the Ginzburg number for quasi-1D superconducting wires (Larkin and Varlamov 2005) defined as a value $(T_c - T)/T_c$ at which the fluctuation correction to the specific heat becomes equal to the specific heat jump at the phase transition point. In Equation 40.22, fluctuations of both the phase and the absolute value of the order parameter give contributions of the same order. The estimate (Equation 40.22) demonstrates that at low temperatures suppression of the order parameter in superconducting nanowires due to Gaussian fluctuations remains weak as long as $g_\xi \gg 1$ and it becomes important only for extremely thin wires with $Gi_{1D} \sim 1$. This conclusion is consistent with that obtained from the well-known Ginzburg–Levanyuk criterion, because for $Gi_{1D} \sim 1$ the width of the fluctuation region δT is comparable to T_c and, hence, the BCS mean-field approach becomes obsolete down to $T = 0$.

The condition $g_\xi \gg 1$ can be translated into that for the wire cross section:

$$\sqrt{s} \gg r_c \sim \lambda_F\sqrt{\frac{\xi}{l}}, \quad (40.24)$$

where $\xi \gg l$. This equation defines the critical wire radius r_c below which fluctuations completely wipe out superconductivity even at $T = 0$. For typical wire parameters we have $r_c \sim 1 \div 2$ nm. We conclude that wires with thicknesses in this range (or smaller) cannot become superconducting at any temperature down to zero.

Turning to higher temperatures, we observe that at $T \sim T_c$ it is necessary to retain only the contribution from zero Matsubara frequency in Equation 40.21. The terms originating from all nonzero frequencies are small in the parameter $\Delta_0(T)/T_c \ll 1$ and, hence, can be safely omitted. Integrating over the wave vector q, we get

$$\frac{\delta\Delta_0}{\Delta_0(T)} = \frac{T}{\delta F}, \tag{40.25}$$

where

$$\Delta_0(T) = \sqrt{\frac{8\pi^2 T(T_c - T)}{7\zeta(3)}} \tag{40.26}$$

is the BCS mean-field order parameter at $T \sim T_c$, $\zeta(3) \approx 1.2$ and

$$\delta F = \frac{16\pi^2}{21\zeta(3)} s N_0 \sqrt{\pi D}(T_c - T)^{3/2}. \tag{40.27}$$

We will see below that the latter expression for δF turns out to be exactly equal to the magnitude of the effective free energy barrier for TAPS in the LAMH theory in limit of small transport currents. Equations 40.25 and 40.27 demonstrate that at temperatures close to T_c Gaussian fluctuations of the superconducting order parameter in thin wires become more significant and effectively destroy superconductivity at $\delta F \lesssim T_c$, i.e., already in much thicker wires than in the case of low temperatures $T \ll T_c$. The corresponding condition for the wire cross section reads

$$\sqrt{s} \gg \frac{r_c}{(1 - T/T_c)^{3/4}}. \tag{40.28}$$

Only provided this condition is satisfied, Gaussian fluctuations of the order parameter remain insignificant at temperatures in the vicinity of T_c.

40.3.3 Thermally Activated Phase Slips

Let us remain at temperatures sufficiently close to T_c and restrict our attention to superconducting wires in which the condition (40.28) is well satisfied and, hence, the effect of Gaussian fluctuations on the order parameter $\Delta_0(T)$ can be safely neglected. This condition requires the wire to be not too thin and/or the temperature to be not too close to T_c, i.e., $(T_c - T)/T_c \gg Gi_{1D}$. At the same time we assume that the temperature is still not far from T_c, i.e., $T_c - T \ll T_c$ in which case the physics is dominated by TAPS. As we already discussed, sufficiently thin superconducting wires acquire nonzero resistance even below T_c due to TAPS, and this resistance is essentially determined by the TAPS rates, which will be analyzed below.

40.3.3.1 TAPS Rates

The TAPS rates corresponding to the phase slip by $\pm 2\pi$ can be represented by means of a standard activation dependence,

$$\Gamma_\pm = B_\pm e^{-\delta F_\pm / T}. \tag{40.29}$$

In order to evaluate the free energy barriers $\delta F\pm$ for TAPS, we will make use of a direct analogy to the problem of thermally activated escape of a particle from a metastable minimum of the potential $U(x)$. It is well known that with the exponential accuracy, this thermal escape rate has a simple form $\gamma \propto \exp(-\delta U/kT)$. Here the effective potential barrier $\delta U = U(x_0) - U(0)$ is determined as a difference between the local maximum and the local minimum values of the potential $U(x)$ reached respectively at $x = x_0$ and $x = 0$. These two extremum points of $U(x)$ are set by the standard saddle point condition $\partial U/\partial x = 0$.

Essentially the same situation occurs in our TAPS problem with the only difference that now we are dealing with thermally activated escape of the complex order parameter field, $\Delta(x)$, describing infinitely many degrees of freedom (two at each value of x). Following the same track, we find both the local minimum and maximum of the wire free energy, $F[\Delta]$, from the saddle point condition

$$\frac{\delta F[\Delta(x)]}{\delta\Delta(x)} = 0 \tag{40.30}$$

and form the difference between these two values that will define the TAPS free energy barriers $\delta F\pm$.

Let us employ the Ginzburg–Landau free energy functional for a quasi-1D superconducting wire of length X and cross section s:

$$F[\Delta(x)] = sN_0 \int_{-X/2}^{X/2} dx \left(\frac{\pi D}{8T} \left| \frac{\partial\Delta}{\partial x} \right|^2 + \frac{T - T_c}{T_c} |\Delta|^2 + \frac{7\zeta(3)}{16\pi^2 T^2} |\Delta|^4 \right)$$
$$- \frac{I}{2e}[\varphi(X/2) - \varphi(-X/2)]. \tag{40.31}$$

As before

$\varphi(x)$ is the phase of the order parameter $\Delta(x)$

I is the external current applied to the wire

The saddle point condition (40.30) for this functional yields the standard Ginzburg–Landau equation,

$$-\frac{\pi D}{8T}\frac{\partial^2\Delta}{\partial x^2} + \frac{T - T_c}{T_c}\Delta + \frac{7\zeta(3)}{8\pi^2 T^2}|\Delta|^2\Delta = 0. \tag{40.32}$$

For any given value of the bias current

$$I = \frac{\pi e N_0 D s}{2T}|\Delta|^2\nabla\varphi \tag{40.33}$$

this equation has a number of solutions. The TAPS free energy barrier δF_+ is determined by the two of them. The first one, $\Delta_m = |\Delta_m|\exp(i\varphi_m)$, corresponds to a metastable minimum of the free energy functional. This solution reads

$$|\Delta_m| = \Delta_0(T)\sqrt{\frac{1 + 2\cos\alpha}{3}}, \quad \varphi_m = \frac{2T}{\pi e N_0 D s}\frac{Ix}{|\Delta_m|^2}. \tag{40.34}$$

Here $\Delta_0(T)$ is defined in Equation 40.26 and the parameter

$$\alpha = \frac{\pi}{3}\theta\left(|I| - \frac{I_c}{\sqrt{2}}\right) + \frac{1}{3}\arctan\frac{2|I|\sqrt{1-(I/I_c)^2}}{I_c(1-2(I/I_c)^2)} \quad (40.35)$$

accounts for the external bias current I. The Ginzburg–Landau critical current, I_c, is given by the standard expression

$$I_c = \frac{16\sqrt{6}\pi^{5/2}}{63\zeta(3)}eN_0\sqrt{D}s(T_c - T)^{3/2}. \quad (40.36)$$

The second, saddle point, solution $\Delta_s(x) = |\Delta_s|\exp(i\varphi_s)$ of Equation 40.32 has the form

$$\frac{|\Delta_s|}{\Delta_0(T)} = \sqrt{\frac{1+2\cos\alpha}{3} - \frac{2\cos\alpha - 1}{\cosh^2\left[\sqrt{2\cos\alpha - 1}\,\frac{x}{\xi(T)}\right]}},$$

$$\varphi_s = \frac{2TI}{\pi eN_0 Ds}\int_0^x dx'\frac{dx'}{|\Delta_s(x')|^2}, \quad (40.37)$$

where $\xi(T) = \sqrt{\pi D/4(T_c - T)}$ is the superconducting coherence length in the vicinity of T_c.

As above, the free energy barrier, δF_+, is set by the difference $\delta F_+ = F[\Delta_s(x)] - F[\Delta_m(x)]$. Substituting the solutions (40.34) and (40.37) into the free energy functional (40.31) after some algebra, we obtain

$$\delta F_+ = \delta F\left[\sqrt{2\cos\alpha - 1} - \sqrt{\frac{2}{3}}\frac{I}{I_c}\arctan\left(\frac{\sqrt{3}}{2}\sqrt{\frac{2\cos\alpha - 1}{1-\cos\alpha}}\right)\right], \quad (40.38)$$

where δF is defined in Equation 40.27.

The free energy barrier δF_- for "negative" TAPS is determined analogously and is related to δF_+ as follows:

$$\delta F_- = \delta F_+ + \frac{\pi I}{e}. \quad (40.39)$$

This equation demonstrates that the free energy barrier for "negative" TAPS is higher than that for "positive" ones. This asymmetry is caused by a tilt of the potential energy by an external bias current I. In the limit $I \to 0$, these two free energy barriers become equal: $\delta F_- = \delta F_+ = \delta F$. In order to evaluate the pre-exponent $B_\pm$ in Equation 40.29 for the TAPS rate, it is necessary to account for fluctuations around both saddle point trajectories (40.34) and (40.37). As the order parameter Δ can fluctuate both in space and in time, this task requires going beyond the stationary GL functional (40.31) and employing a more general (time-dependent) theory. McCumber and Halperin (1970) derived the expression for the pre-exponent $B_\pm$ within the TDGL approach. Although their calculation is correct and sound, the TDGL approach itself is insufficient

for a number of reasons, as it was argued above. Hence, the pre-exponent $B_\pm$ needs to be recalculated within a fully microscopic theory that properly describes time-dependent fluctuations of the order parameter field along the wire. Only very recently this task was accomplished (Golubev and Zaikin 2008) with the result

$$B_\pm \sim T_c\frac{X}{\xi(T)}\sqrt{\frac{\delta F_\pm}{T}}, \quad (40.40)$$

where we omitted an unimportant numerical factor of order one. Together with Equations 40.29, 40.38, and 40.27, this result fully determines the TAPS rates in superconducting nanowires at temperatures not far from T_c.

It is interesting to compare the microscopic result (Equation 40.40) with that derived from the phenomenological TDGL analysis by McCumber and Halperin. We have

$$\frac{B_\pm}{B_{TDGL}} \sim \frac{1}{1-T/T_c}\frac{1}{(1-I/I_c)^{5/4}}, \quad (40.41)$$

i.e., within the whole applicability range of our calculation

$$Gi_{1D} \ll \frac{T_c - T}{T_c} \ll 1 \quad (40.42)$$

the microscopic expression for the TAPS rate turns out to be parametrically bigger than that evaluated within the TDGL-based approach. Although this difference affects only the pre-exponential factor $B_\pm$ it can nevertheless be significant at temperatures sufficiently close to T_c. Furthermore, in the vicinity of the critical current $I_c - I \ll I_c$ this difference gets further increased and—depending on T and I—can easily reach several orders of magnitude.

40.3.3.2 Temperature-Dependent Resistance and Noise

Let us now turn to the relation between the above TAPS rate and physical observables, such as, e.g., wire resistance and voltage noise. Every phase slip event implies changing of the superconducting phase in time in such a way that the total phase difference values along the wire before and after this event differ by $\pm 2\pi$. Since the average voltage is linked to the time derivative of the phase by means of the Josephson relation $V = \dot{\varphi}/2e$, for the net voltage drop across the wire we obtain

$$V = \frac{\pi}{e}[\Gamma_+(I) - \Gamma_-(I)], \quad (40.43)$$

where $\Gamma_\pm$ are given by Equations 40.40, 40.29, 40.38, and 40.27. In the absence of any bias current $I \to 0$ both rates are equal $\Gamma_\pm = \Gamma$ and the net voltage drop V vanishes. In the presence of small bias current $I \ll I_c$ from Equation 40.39 we obtain

$$\Gamma_\pm(I) = \Gamma e^{\pm\pi I/2eT}. \quad (40.44)$$

Thus, at such values of I and at temperatures slightly below T_c the I–V curve for quasi-1D superconducting wires takes a relatively simple form:

$$V = \frac{2\pi}{e}\Gamma\sinh\frac{\pi I}{2eT}. \qquad (40.45)$$

This important result of the LAMH theory implies that thermal fluctuations effectively destroy long-range phase coherence across the superconducting wire that acquires nonvanishing resistance even at temperatures below T_c. The zero bias resistance $R(T) = (\partial V/\partial I)_{I=0}$ demonstrates exponential dependence on temperature and the wire cross section:

$$\frac{R(T)}{R_q} \sim \frac{T_c}{T}\frac{X}{\xi(T)}\sqrt{\frac{\delta F}{T}}\exp\left[-\frac{\delta F}{T}\right]. \qquad (40.46)$$

This dependence was indeed reliably detected in a number of experiments, see Figure 40.2.

In addition to nonzero resistance (40.46) TAPS also cause the voltage noise below T_c. Treating TAPS as independent events one immediately arrives at the conclusion that they should obey Poissonian statistics. Hence, the voltage noise power

$$S_V = 2\int dt\,\langle\delta V(t)\delta V(0)\rangle$$

is defined as a sum of the contributions from "positive" and "negative" TAPS, i.e.,

$$S_V = \frac{2\pi^2}{e^2}[\Gamma_+(I)+\Gamma_-(I)]. \qquad (40.47)$$

At small currents $I \ll I_c$ this expression reduces to the following simple form:

$$S_V = \frac{4\pi^2}{e^2}\Gamma\cosh\frac{\pi I}{2eT}. \qquad (40.48)$$

Similarly to the wire resistance, the voltage noise rapidly decreases as one lowers the temperature below T_c. This TAPS noise remains appreciable and can be detected in experiments only in the vicinity of the critical temperature. The simultaneous measurement of both TAPS-induced resistance and noise appears to be an efficient way for quantitative experimental analysis of TAPS in superconducting nanowires.

40.3.4 Quantum Phase Slips

As temperature goes down thermal fluctuations decrease and, hence, TAPS become progressively less important and eventually die out in the limit $T \to 0$. At low enough temperatures *quantum fluctuations* of the order parameter field Δ take over and essentially determine the behavior of ultrathin superconducting wires. As we have already discussed, for not too thin wires with cross sections obeying Equation 40.24, the relevant quantum fluctuations are QPS. Each QPS event involves suppression of the order parameter in the phase slip core and a winding of the superconducting phase around this core. This configuration describes quantum tunneling of the order parameter field through an effective potential barrier and can be conveniently analyzed within the imaginary time formalism discussed above.

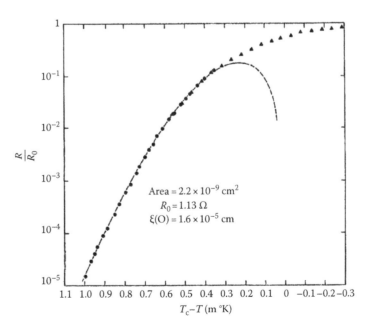

FIGURE 40.2 $R(T)$ dependence for a tin whisker. Solid symbols are the experimental data, dashed line represents the fit to the TAPS model. (From Lukens, J.E. et al., *Phys. Rev. Lett.*, 25, 1180, 1970.)

40.3.4.1 QPS Rate

In order to evaluate the rate γ_{QPS} with which QPS gets created inside the wire, it is necessary to find the action corresponding to a single phase slip. In principle, this is done analogously to the problem of TAPS. Technically, however, the QPS case is much more complicated since now the Ginzburg–Landau free energy functional cannot be applied anymore. Rather we have to deal with the microscopic effective action (40.13) that explicitly accounts for nonstationary and nonequilibrium effects.

Let us denote the typical size of the QPS core as x_0 and the typical (imaginary time) duration of the QPS event as τ_0. At this stage, both these parameters are not yet known. They remain to be determined later in the course of our consideration. We separate the total action of a single QPS S_{QPS} into a core part S_{core} around the phase slip center for which the condensation energy and dissipation by normal currents are important (scales $x \lesssim x_0$, $\tau \lesssim \tau_0$), and a hydrodynamic part outside the core S_{out} that depends on the hydrodynamics of the electromagnetic fields, i.e.,

$$S_{QPS} = S_{core} + S_{out}. \tag{40.49}$$

Let us first evaluate the hydrodynamic part S_{out}. This task is simplified by the fact that outside the core the absolute value of the order parameter field remains equal to its mean-field value Δ_0, and only its phase $\varphi(x, \tau)$ changes in space and time. Without loss of generality, we can assume that the absolute value of the order parameter is equal to zero at $\tau = 0$ and $x = 0$. For sufficiently long wires and outside the QPS core, the saddle point solution corresponding to a single QPS event should satisfy the identity

$$\partial_x \partial_\tau \tilde{\varphi} - \partial_\tau \partial_x \tilde{\varphi} = 2\pi \delta(\tau, x), \tag{40.50}$$

which follows from the fact that after a wind around the QPS center the phase should change by 2π. In a way QPS is just a vortex in space-time with the phase distribution $\varphi(x, \tau)$ described by the saddle point solution

$$\tilde{\varphi}(x, \tau) = -\arctan\left(\frac{x}{c_0 \tau}\right). \tag{40.51}$$

Substituting the solution (40.51) into the action (40.17), we obtain

$$S_{out} = \mu \ln\left[\frac{\min(c_0/T, X)}{\max(c_0 \tau_0, x_0)}\right], \tag{40.52}$$

where the parameter

$$\mu = \frac{\pi}{4e^2 c_0 (L + \tilde{L})} \simeq \frac{\pi}{4\alpha}\sqrt{\frac{sC}{4\pi\lambda_L^2}} \tag{40.53}$$

sets the scale for the hydrodynamic contribution to the QPS action. Here and below $\alpha = e^2/\hbar c \approx 1/137$ is the fine structure constant. We also note that at $T \to 0$, the contribution S_{out} (40.52)

diverges logarithmically for infinitely long wires, thus making *single* QPS events unlikely in this limit.

Let us now turn to the core contribution to the action of a single QPS. In order to exactly evaluate this contribution, it is necessary to explicitly find the QPS saddle point of the full nonlinear effective action (40.9). This is a formidable task which can hardly be accomplished in practice. On the other hand, this task is greatly simplified if one is aiming at estimating the term S_{core} up to a numerical prefactor of order one. Below, we will recover the full microscopic expression for the core contribution S_{core} leaving only this numerical prefactor undetermined. In this way we fully capture all the essential physics of QPS. The dimensionless prefactor can be regarded as a fit parameter that can be extracted, e.g., from the comparison with the available experimental data.

The above strategy allows us to approximate the complex order parameter field inside the QPS core by two simple functions that should satisfy several requirements. The absolute value of the order parameter $|\Delta(x, \tau)|$ should vanish at $x = 0$ and $\tau = 0$ and coincide with the mean-field value Δ_0 outside the QPS core. The phase $\varphi(x, \tau)$ should flip at $x = 0$ and $\tau = 0$ in a way to provide the change of the net phase difference across the wire by 2π. On top of that, in a short wire and outside the QPS core the phase φ should not depend on the spatial coordinate in the zero bias limit. All sufficiently smooth functions obeying these requirements can be used to estimate S_{core}. Let us, for instance, choose

$$|\delta\Delta(x, \tau)| = \Delta_0 \exp\left(\frac{-x^2}{2x_0^2} - \frac{\tau^2}{2\tau_0^2}\right) \tag{40.54}$$

for the amplitude of the order parameter field, and

$$\varphi(x, \tau) = -\frac{\pi}{2}\tanh\left(\frac{x\tau_0}{x_0\tau}\right) \tag{40.55}$$

for its phase. Substituting these functions into the action (40.13), minimizing the result with respect to x_0 and τ_0 and making use of the Drude expression for the normal conductivity of our wire $\sigma = 2e^2 N_0 D$, we obtain $x_0 \sim \xi$, $\tau_0 \sim 1/\Delta_0$ and

$$S_{core} = \pi A N_0 s \sqrt{D\Delta_0} = A\frac{R_q}{R_N}\frac{X}{\xi} = \frac{A}{4}g_\xi. \tag{40.56}$$

Here A is an unimportant numerical prefactor of order one, R_N is the total normal state wire resistance and, as before, $g_\xi = 4(R_q/R_N)$ (X/ξ) is the dimensionless normal conductance of a wire segment of length ξ.

Thus, we observe a clear separation between different fluctuation effects contributing to the QPS action: Fluctuations of the order parameter field and dissipative currents determine the core part (40.56) while electromagnetic fluctuations are responsible for the hydrodynamic term (40.52). In the interesting range of wire thicknesses, $\sqrt{s} \gtrsim 5 \div 10\,\text{nm}$, the core part S_{core} usually exceeds the hydrodynamic term $\sim\mu$. For example, for $C \sim 1$ we obtain $\mu \approx 30(r_0/\lambda_L)$. Setting $\sqrt{s} \sim 5 \div 10\,\text{nm}$ and estimating μ and S_{core} for typical system parameters $\lambda_F \sim 0.2\,nm$, $l \sim 7\,nm$, $\xi \sim 10\,nm$, $\lambda_L \sim 100\,nm$, $v_F = 10^6\,m/s$, and $\Delta_0 \sim 1 \div 10\,K$, we find $\mu \sim 1 \div 3$ and $S_{core} \gg \mu$. The latter inequality becomes even stronger for thicker wires. Note that the condition $S_{core} \gg \mu$ allows us to ignore the hydrodynamic part of the QPS action while minimizing the core part with respect to x_0 and τ_0. We also emphasize that both the "condensation energy" term (proportional to χ_Δ) as well as a dissipative contribution are crucially important for the above minimization procedure. Under the condition $C/e^2 N_0 s \ll 1$ (usually well satisfied in metallic nanowires) dissipation plays a dominant role during the phase slip event, and the correct QPS core action *cannot* be obtained without an adequate microscopic description of dissipative currents flowing inside the wire. Only employing the Drude formula $\sigma = 2e^2 N_0 D$ enables one to recover the correct result (40.56), whereas for some other models of dissipation different results for the core action would follow.

Now let us find the QPS rate γ_{QPS}. Provided the QPS action is sufficiently large, $S_{QPS} \gg 1$, this rate can be expressed in the form

$$\gamma_{QPS} = B_{QPS} \exp(-S_{QPS}). \qquad (40.57)$$

The results for the QPS action derived above allow to determine the rate γ_{QPS} with the exponential accuracy. By analyzing the contribution of fluctuations around the QPS saddle point trajectory one can also establish the pre-exponent B_{QPS} in Equation 40.57. This rather complicated task can be accomplished (Golubev and Zaikin 2001) with the result

$$B_{QPS} \sim \frac{S_{QPS} X}{\tau_0 x_0}, \qquad (40.58)$$

which determines the pre-exponent B_{QPS} up to an unimportant numerical factor of order one. The inequality $S_{core} \gg \mu$ allows to substitute S_{core} (40.56) instead of S_{QPS} in Equation 40.58. Substituting also $x_0 \sim \xi$ and $\tau_0 \sim 1/\Delta_0$, for the QPS rate we finally obtain

$$\gamma_{QPS} \sim \Delta_0 \frac{R_q}{R_N} \frac{X^2}{\xi^2} \exp(-S_{QPS}) \sim g_\xi \Delta_0 \frac{X}{\xi} \exp(-A g_\xi/4). \qquad (40.59)$$

40.3.4.2 Quantum Phase Transition and I–V Curve

Although typically the hydrodynamic part of the QPS action (40.52) is smaller than its core part (40.56), the former also plays an important role since it determines interactions between different QPS. Consider two such phase slips (two vortices in space-time) with the corresponding core coordinates (x_1, τ_1) and (x_2, τ_2). Provided these two cores do not overlap, i.e., provided $|x_2 - x_1| > x_0$ and $|\tau_2 - \tau_1| > \tau_0$, the core contributions are independent and simply add up. In order to evaluate the hydrodynamic part, we substitute the superposition of two solutions $\tilde{\varphi}(x - x_1, \tau - \tau_1) + \tilde{\varphi}(x - x_2, \tau - \tau_2)$ satisfying the identities

$$\partial_x \partial_\tau \tilde{\varphi} - \partial_\tau \partial_x \tilde{\varphi} = 2\pi v_{1,2} \delta(\tau - \tau_{1,2}, x - x_{1,2}) \qquad (40.60)$$

(where $v_{1,2} = \pm 1$ are topological charges of two QPS fixing the phase change after a wind around the QPS center to be $\pm 2\pi$) into the action (40.13) and obtain

$$S_{QPS}^{(2)} = 2S_{core} - \mu v_1 v_2 \ln\left[\frac{(x_1 - x_2)^2 + c_0^2(\tau_1 - \tau_2)^2}{\xi^2}\right], \qquad (40.61)$$

i.e., different QPS interact logarithmically in space-time. QPSs with opposite (equal) topological charges attract (repel) each other.

The next step is to consider a gas of n QPS. Again, assuming that the QPS cores do not overlap, we can substitute a simple superposition of the saddle point solutions for n QPS $\varphi = \sum_i^n \tilde{\varphi}(x - x_i, \tau - \tau_i)$ into the action and find

$$S_{QPS}^{(n)} = n S_{core} + S_{int}^{(n)}, \qquad (40.62)$$

where

$$S_{int}^{(n)} = -\mu \sum_{i \neq j} v_i v_j \ln\left(\frac{\rho_{ij}}{x_0}\right) + \frac{\Phi_0}{c} I \sum_i v_i \tau_i. \qquad (40.63)$$

Here $\rho_{ij} = \left(c_0^2(\tau_i - \tau_j)^2 + (x_i - x_j)^2\right)^{1/2}$ defines the distance between the ith and jth QPS in the (x, τ) plane, $v_i = \pm 1$ are the QPS topological charges, and $\Phi_0 = \pi \hbar c/e$ is the flux quantum. In Equation 40.63, we also included an additional term that keeps track of the applied current I flowing through the wire. This term trivially follows from the standard contribution to the action

$$\int d\tau \int dx (I/2e) \partial_x \varphi$$

The grand partition function of the wire is represented as a sum over all possible configurations of QPS (topological charges)

$$\mathcal{Z} = \sum_{n=0}^{\infty} \frac{1}{2n!} \left(\frac{y}{2}\right)^{2n} \int_{x_0}^{X} \frac{dx_1}{x_0} \cdots \int_{x_0}^{X} \frac{dx_{2n}}{x_0} \int_{\tau_0}^{1/T} \frac{d\tau_1}{\tau_0} \cdots \int_{\tau_0}^{1/T} \frac{d\tau_{2n}}{\tau_0}$$
$$\times \sum_{v_i = \pm 1} \exp(-S_{int}^{(2n)}) \qquad (40.64)$$

where an effective fugacity, y, of these charges is related to the QPS rate as

$$y = \frac{x_0 \tau_0 B_{QPS}}{X} \exp(-S_{core}) \sim S_{core} \exp(-S_{core}). \qquad (40.65)$$

We also note that only neutral QPS configurations with

$$\sum_i^n v_i = 0 \qquad (40.66)$$

(and hence n even) contribute to the partition function (40.64). This fact is a direct consequence of the boundary condition $\varphi(x, \tau) = \varphi(x, \tau + 1/T)$ in the path integral for the partition function.

It is easy to observe that for $I = 0$ Equations 40.63 and 40.64 define the standard model for a 2D gas of logarithmically interacting charges v_i. The only specific feature of our present model as compared to the standard situation is that here the space and time coordinates are not equivalent and one can consider different limiting cases of "long" and "short" wires.

Below we restrict our attention to the case of long wires and assume that $T \to 0$. We will follow the standard analysis of logarithmically interacting 2D Coulomb gas, which is based on the renormalization group (RG) equations both for the interaction parameter, μ, and the charge fugacity, y. Defining the scaling parameter, $\ell = \ln(\rho/\xi)$, we have (Kosterlitz 1974)

$$\partial_\ell \mu = -4\pi^2\mu^2 y^2, \quad \partial_\ell y = (2-\mu)y. \quad (40.67)$$

We observe that for large values of $\mu \propto \sqrt{s}$ the QPS fugacity decreases in the course of renormalization implying that QPS events are irrelevant for sufficiently thick wires. In contrast, for small enough μ the QPS fugacity, y, *increases* as one reaches longer and longer length scales. Hence, in the case of thinner wires QPS proliferate. Following the standard line of reasoning, we immediately conclude that a QPT for phase slips occurs in a long superconducting wire at $T \to 0$ and

$$\mu = \mu^* \equiv 2 + 4\pi y \approx 2. \quad (40.68)$$

This is essentially a Berezinskii–Kosterlitz–Thouless (BKT) phase transition for charges v_i in space-time. The difference from the standard BKT transition in 2D superconducting films is only that in our case the transition is driven by the wire thickness $\sqrt{s}$ (which enters into n) and not by temperature. In other words, for thicker wires with $\mu > \mu^*$ QPS with opposite topological charges are bound in pairs (dipols) and the *linear* resistance of a superconducting wire is strongly suppressed and T-dependent. This resistance tends to vanish in the limit $T \to 0$. Thus, we arrive at an important conclusion: *at $T = 0$ a long quasi-1D superconducting wire remains in a superconducting state, with vanishing linear resistance, provided its thickness is sufficiently large and, hence, the electromagnetic interaction between phase slips is sufficiently strong, i.e., $\mu > \mu^*$.*

On the other hand, for $\mu < \mu^*$ the density of free (unbound) QPS in the wire always remains finite; such fluctuations destroy the phase coherence (and, hence, superconductivity) and bring the wire into the normal state with nonvanishing resistance even at $T = 0$. Thus, another important conclusion is that superconductivity in sufficiently thin wires is *always* destroyed by quantum fluctuations.

Let us now turn to the calculation of the wire resistance in the presence of QPS. We again consider the limit of long wires. At any nonzero T such wires have a nonzero resistance $R(T, I)$ even in the "ordered" phase $\mu > \mu^*$. In order to evaluate $R(T)$ in this regime, we proceed perturbatively in the QPS fugacity, y. For $\mu > \mu^*$ QPS form closed pairs (dipols) and, hence, interactions between different dipols can be neglected. For this reason it suffices to evaluate the correction δF to the wire free energy due to one bound pair of QPS with opposite topological charges

(Schön and Zaikin 1990, Weiss 1999). Taking into account only logarithmic interactions within bound pairs of QPS we can easily sum up the series in Equation 40.64 and arrive at the result

$$\delta F = \frac{Xy^2}{x_0\tau_0} \int_{\tau_0}^{1/T} d\tau \int_{x_0}^{X} \frac{dx}{x_0} e^{(\Phi_0 I\tau/c)-2\mu\ln[\rho(\tau,x)/x_0]}, \quad (40.69)$$

where $\rho = (c_0^2\tau^2 + x^2)^{1/2}$. It is convenient to first integrate over the spatial coordinate x and take the wire length $X \to 0$. For nonzero I, the expression in Equation 40.69 is formally divergent for $T \to 0$ and acquires an imaginary part Im δF after analytic continuation of the integral over the imaginary time variable τ. This indicates a QPS-induced instability of the superconducting state of the wire: the state with a zero phase difference $\delta\varphi(X) = \varphi(X) - \varphi(0) = 0$ decays into a lower energy state with $\delta\varphi(X) = 2\pi$ with the decay rate γ_{QPS}. As before, the average voltage drop across the wire is given by the difference between the two rates:

$$V = \left(\frac{\Phi_0}{c}\right)[\gamma_{QPS}(I) - \gamma_{QPS}(-I)]. \quad (40.70)$$

Evaluating the imaginary part of the free energy Im δF, from Equations 40.69 and 40.70 we finally obtain

$$V = \frac{\Phi_0 Xy^2}{c\tau_0 x_0} \frac{\sqrt{\pi}\Gamma\left(\mu-\frac{1}{2}\right)}{\Gamma(\mu)\Gamma(2\mu-1)} \sinh\left(\frac{\Phi_0 I}{2cT}\right) \left|\Gamma\left(\mu-\frac{1}{2}+\frac{i}{\pi}\frac{\Phi_0 I}{2cT}\right)\right|^2$$

$$\times [2\pi\tau_0 T]^{2\mu-2}, \quad (40.71)$$

where $\Gamma(x)$ is the Euler gamma function. For the wire resistance $R(T, I) = V/I$ this expression yields

$$R \propto y^2 T^{2\mu-3} \quad (40.72)$$

for $T \gg \Phi_0 I$, and

$$R \propto y^2 I^{2\mu-3} \quad (40.73)$$

for $T \ll \Phi_0 I$. Thus, for sufficiently thick wires with $\mu > \mu^*$, we expect a strong temperature dependence of the linear resistance which eventually vanishes in the limit $T \to 0$ indicating the superconducting behavior of the wire.

For thinner wires with $\mu < \mu^*$ and at low temperatures, the above simple analysis becomes insufficient because of the presence of unbound QPS and the necessity to account for many-body effects in the gas of QPS. We expect that the temperature dependence of the wire resistivity should become linear at the transition to the non-superconducting phase.

40.3.4.3 Key Experiments

At present there exist numerous experimental observations confirming the existence of QPS effects in superconducting nanowires. Typically, the evidence for such effects is obtained by measuring the *I–V* curve of ultrathin superconducting wires at temperatures below T_c of the bulk material. Very close to T_c the I–V curve may follow the predictions of the TAPS model in which case an exponential decrease of the wire resistance $R(T)$ with temperature is observed in accordance with Equation 40.46. However, as the temperature goes down, one observes pronounced deviations from the dependence (40.46), such as "foot" or "knee" structures on the I–V curves, which can only be attributed to the effect of quantum fluctuations. For thinnest nanowires, one observes no resistance decrease below T_c at all, i.e., quantum fluctuations fully suppress superconductivity in such wires down to the lowest temperatures at which measurements can be performed. A detailed discussion of the present experimental situation in the field is provided in recent review papers (Arutyunov et al. 2008, Bezryadin 2008).

A clear manifestation of a crucial role played by quantum fluctuations in superconducting nanowires was provided in the experiments by Tinkham and coworkers (Bezryadin et al. 2000, Lau et al. 2001). These authors developed a novel technique that allowed them to fabricate sufficiently uniform superconducting wires considerably thinner than 10 nm with lengths ranging between ~100 nm and 1 μm. This was achieved by sputtering a superconducting alloy of amorphous $Mo_{79}Ge_{21}$ over a freestanding carbon nanotube or bundle of tubes laid down over a narrow and deep slit etched in the substrate.

Three out of eight samples studied in the experiments by Bezryadin et al. demonstrated no sign of superconductivity even well below the bulk critical temperature T_c. All three non-superconducting wires (i1, i2, and i3) had the normal state resistance above the quantum unit R_q, while R_N for the remaining five "superconducting" samples was lower than R_q. This observation allowed to conjecture that a dramatic difference in the behavior of these two groups of samples (otherwise having similar parameters) can be due to the dissipative QPT (Zaikin et al. 1998, Buchler et al. 2004) analogous to the Schmid phase transition observed earlier in Josephson junctions (Penttila et al. 1999). This interpretation, however, was not confirmed in the later experiments of the same group (Lau et al. 2001) who observed superconducting behavior in samples with normal resistances as high as $40 k\Omega \gg R_q$. Lau et al. concluded that "the relevant

parameter controlling the superconducting transition is not the ratio of R_q/R_N, but appears to be resistance per unit length, or equivalently, the cross-sectional area of a wire." This conclusion clearly favors interpretation of the data either in terms of a BKT-like QPT at $\mu \approx 2$ discussed above or just as a sharp crossover between the regimes of vanishingly small and sufficiently high QPS rates, γ_{QPS}. These two regimes correspond respectively to the superconducting and normal behavior of the nanowires. Both these QPT and crossover are expected to occur for wire diameters ~10 nm.

Further support for this interpretation comes from the data (Lau et al. 2001) obtained for more than 20 different nanowires. The resistance of these wires measured at $T = 1.5$ K is presented in Figure 40.3 versus the inverse normal resistance per unit length $X/R_N \propto s$. In Figure 40.3a, one observes a sharp crossover between normal and superconducting behavior at wire diameters $\sqrt{s} \sim 10$ nm. This crossover could be interpreted as an indication to the QPT controlled by the parameter $\mu \propto \sqrt{s}$. Note, however, that all wires studied by Lau et al. were quite short. Hence, this QPT should inevitably be significantly broadened by finite size effects. Replotting the same data on a semilog scale (Figure 40.3b), one indeed observes a rather broad distribution of measured resistances that—despite some scatter—can be fitted to the linear dependence on the wire cross section, s. This fit is highly suggestive of the crossover scenario although it cannot yet rule out the (broadened by size effects) QPT either.

In order to finally discriminate between these two options it is necessary to analyze the temperature dependence of the resistance $R(T)$. According to our theory, in the linear regime the wire resistance (caused by QPS) is defined by Equation 40.72. The dependence $R \propto y^2$ comes from pairs of QPS events where the fugacity (40.65) depends on temperature via $S_{core} \propto \Delta_0^{1/2}(T)$. An additional (weak) power-law dependence, $\propto T^{2\mu-3}$, enters because of interactions between QPS in space-time. In order to fit the data by Lau et al., we will ignore this power-law dependence and use the simplified formula

$$R(T) = b \frac{\Delta_0(T) S_{core}^2 X}{\xi(T)} \exp(-2S_{core}), \qquad (40.74)$$

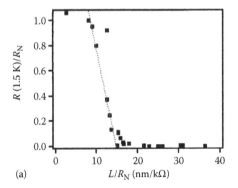

(a)

L/R_N (nm/kΩ)

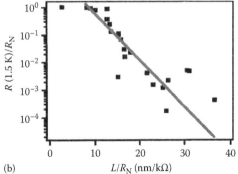

(b)

L/R_N (nm/kΩ)

FIGURE 40.3 Resistance at 1.5 K normalized to normal state resistance as a function of L/R_N. (a) Linear plot. The dotted line is a guide to the eye. (b) Semilog plot with an exponential fit. Slope of the fitted line is 0.39 kΩ/ nm. (From Lau, C.N. et al., *Phys. Rev. Lett.*, 87, 217003-1, 2001.)

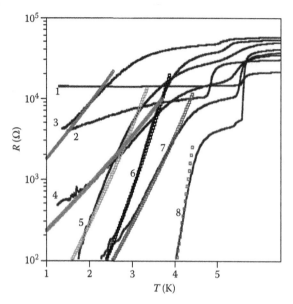

FIGURE 40.4 Superconducting transitions of "long" *MoGe* nanowires on top of an insulating carbon nanotube used as the substrate. Double-step shape of the $R(T)$ dependencies comes from the superconducting transition of the contact regions contributing to the two-probe measurement configuration. The samples' normal state resistances and lengths are 1: 14.8 kΩ, 135 nm; 2: 10.7 kΩ, 135 nm; 3: 47 kΩ, 745 nm; 4: 17.3 kΩ, 310 nm; 5: 32 kΩ, 730 nm; 6: 40 kΩ, 1050 nm; 7: 10 kΩ, 310 nm; 8: 4.5 kΩ, 165 nm. Symbols stand for calculations using Equation 40.74 with the single numerical coefficient $A = 0.7$. The critical temperature, T_c, and the dirty-limit coherence length, $\xi(0)$, used as fitting parameters for samples 3–8 are 3: 5.0 K, 8 nm; 4: 6.4 K, 8.5 nm; 5: 4.6 K, 8.9 nm; 6: 4.8 K, 8.9 nm; 7: 5.6 K, 11.9 nm; 8: 4.8 K, 8.5 nm. (From Bezryadin, A. et al., *Nature*, 404, 971, 2000; Lau, C.N. et al., *Phys. Rev. Lett.*, 87, 217003-1, 2001; Arutyunov, K.Yu. et al., *Phys. Rep.*, 464, 1, 2008.)

with S_{core} defined in (40.56) and b being an unimportant constant. The results of the fit to Equation 40.74 are presented in Figure 40.4. We observe a very good agreement between theory and experiment that is achieved with the same value $A \simeq 0.7$ for all six samples. Thus, the temperature dependence of the resistance of ultrathin *MoGe* wires studied by Bezryadin et al. and by Lau et al. is determined by QPS effects, and the observed sharp transition between normal and superconducting behavior is most likely a thickness-governed crossover between the regimes of respectively large and small QPS rates γ_{QPS}.

Another set of experiments (Zgirski et al. 2005, 2008) provides an independent experimental confirmation of the QPS theory described above. Aluminum was chosen by these authors for investigations of the properties of superconducting nanowires. It was demonstrated that low energy Ar⁺ ion sputtering can progressively and nondestructively reduce dimensions of such nanowires. In contrast to the experiments performed by the Tinkham's group (in which the dependence on the wire thickness was studied by fabricating many similar samples with different cross-section values), Zgirski et al. achieved progressive diameter reduction in *one and the same*

sample. The accuracy of the effective diameter determination was reported to be about ±2 nm. Only those samples that showed no obvious geometrical imperfections were used for further experiments.

After a sequence of sputterings (alternated with $R(T)$ measurements) the wire diameter was reduced from $\sqrt{s} \sim 100$ nm down to $\sqrt{s} \lesssim 10$ nm. Experiments were performed on several sets of aluminum nanowires with length X equal to 1, 5, and 10 μm. For larger diameters, $\sqrt{s} \gtrsim 20$ nm, the shape of the $R(T)$ dependence is rather "sharp" and can be qualitatively described by the TAPS mechanism. When the wire diameter is further reduced, deviations from the TAPS behavior become obvious (Figure 40.5). Fits to the TAPS model fail to provide any reasonable quantitative agreement with experiment for diameter values below $s^{1/2} \lesssim 20$ nm in which case the observations are attributed to the effect of quantum fluctuations.

A detailed comparison of the data with the QPS theory was performed and the results of this comparison are presented in Figure 40.5. We observe that as the sample diameter decreases the $R(T)$ curves become progressively broader, exactly as it is predicted by the QPS theory. Here the temperature dependence is merely determined by the QPS fugacity (40.65) via $S_{core} \propto \Delta_0^{1/2}(T)$. An additional weak power-law dependence enters because of inter-QPS interaction. The fits demonstrate a good agreement between theory and experiment, thus confirming the important role of QPS effects in the thinnest wires.

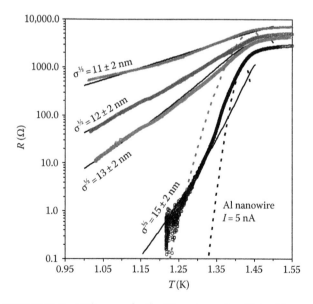

FIGURE 40.5 $R(T)$ curves for the thinnest samples obtained by progressive diameter reduction for the same aluminum nanowire with length $X = 10$ μm. The TAPS model fitting is shown with dashed lines for 11 and 15 nm samples with the best-fit mean free path $l = 3$ and 10 nm, correspondingly, $T_c = 1.46$ K, and the critical magnetic field $B_c(0) = 10$ mT. Fits to Equation 40.74 are shown by solid lines. For 11, 12, 13, and 15 nm wires the fit parameters are $A \simeq 0.1$; T_c: 1.64, 1.52, 1.47, and 1.47 K; mean free path l: 7.5, 8.2, 9.5, and 9.5 nm; normal state resistance R_N: 7200, 5300, 4200, and 2700 kΩ. (From Zgirski, M. et al., *Phys. Rev. B*, 77, 054508-1, 2008.)

For all aluminum nanowires the best-fit value for the parameter A was determined to be $A \simeq 0.1$. This value remains the same for all curves in Figure 40.5. This value of A is somewhat smaller than that extracted from the fits to the experimental data by Lau et al. This difference, however, is not important and it can easily be attributed to a different geometry and degree of inhomogeneity of samples used in these two experiments.

To summarize this part of our discussion we conclude that—although more experiments would be highly desirable—already existing experimental data unambiguously and overwhelmingly confirm our understanding of basic features of QPS physics in superconducting nanowires.

40.4 Persistent Currents in Superconducting Nanorings

40.4.1 Persistent Currents and Quantum Phase Slips

It is well known that superconducting rings pierced by external magnetic flux, Φ_x, develop circulating PC that never vanish. This phenomenon is a fundamental consequence of macroscopic phase coherence of Cooper pair wave functions. In the case of bulk metallic rings, fluctuations of the phase φ of the order parameter can be neglected, i.e., φ can be considered as a purely classical variable. In this case, the total phase difference, $\varphi(X) - \varphi(0)$, accumulated along the ring circumference, $X = 2\pi R$, is linked to the external flux, Φ_x, inside the ring by the well-known relation

$$\varphi(X) - \varphi(0) = 2\pi p + \phi_x, \qquad (40.75)$$

where

p is an integer number
$\phi_x = \Phi_x/\Phi_0$
as before, Φ_0 is the superconducting flux quantum

This relation implies the existence of a phase gradient along the ring which, in turn, means the presence of discrete set of current and energy states labeled by the number p. At sufficiently low temperatures, $T \ll \Delta_0$, quasiparticles are practically irrelevant and the grand partition function of the ring takes the form

$$\mathcal{Z}_{\phi_x} = \sum_{p=-\infty}^{\infty} \exp(-E_p(\phi_x)/T), \qquad (40.76)$$

where

$$E_p(\phi_x) = \frac{E_R}{2}(p + \phi_x)^2, \quad E_R = \frac{\pi^2 \hbar^2 N_0 D \Delta_0 s}{R} \qquad (40.77)$$

defines flux-dependent energy states of a superconducting ring with radius R and cross section s. The ground state energy, $E(\phi_x) = \min_p E_p(\phi_x)$, is a periodic function of the flux Φ_x with the

period Φ_0. The derivative of the ground state energy with respect to the flux defines the persistent current:

$$I(\Phi_x) = c\left(\frac{\partial E(\Phi_x)}{\partial \Phi_x}\right). \qquad (40.78)$$

Combining Equations 40.77 and 40.78, one finds

$$I = 2\pi e m v_s N_0 D \Delta_0 s, \qquad (40.79)$$

where m is the electron mass and

$$v_s(\phi_x) = \frac{\hbar}{2mR} \min_p \left(p + \frac{\Phi_x}{\Phi_0}\right) \qquad (40.80)$$

is the superconducting velocity. Both the current, I, and the velocity, v_s, are periodic functions of the magnetic flux, Φ_x, showing the familiar sawtooth behavior, see also Figure 40.6.

This picture remains applicable as long as fluctuations of the superconducting phase, φ, can be neglected, i.e., as long as the ring remains sufficiently thick. Upon decreasing the thickness of a superconducting wire $\sqrt{s}$ down to values $\sim 10\,\mathrm{nm}$ range, one eventually reaches the new regime in which quantum fluctuations of the phase gain importance and, as it will be demonstrated below, essentially modify the low temperature behavior of superconducting nanorings.

As we have already learned, at low T the most important fluctuations in superconducting nanowires are QPS. Each phase slip event implies transfer of one flux quantum, Φ_0, through the wire out of or into the ring and, hence, yields the change of the total

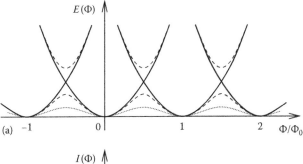

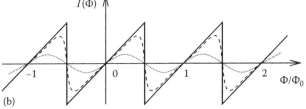

FIGURE 40.6 The energy (a) and persistent current (b) for a superconducting nanoring as functions of magnetic flux without fluctuations, with weak and with strong QPS effects shown respectively by solid, dashed, and dotted lines. (From Matveev, K.A. et al., *Phys. Rev. Lett.*, 89, 096802-1, 2002.)

flux inside the ring by exactly the same amount. In other words, each QPS yields a jump between two neighboring energy states E_p (Equation 40.77) (i.e., between two neighboring parabolas in Figure 40.6). As a result of such jumps the flux inside the ring fluctuates, its average value $\langle\Phi(\Phi_x)\rangle$ decreases and so does the persistent current I. One can also anticipate that the magnitude of this effect should increase with the ring perimeter X. This is because the QPS rate γ_{QPS} increases linearly with X, i.e., the bigger the ring perimeter the higher the probability for QPS to occur anywhere along the ring.

Before turning to technical details, it is instructive to point out a formal equivalence between the phenomenon discussed here and charging effects in ultrasmall Josephson junctions (Schön and Zaikin 1990). Indeed, the energy states of our ring (Equation 40.77) driven by the (normalized) external magnetic flux, ϕ_x, (see Figure 40.6) are fully analogous to such states of a capacitor in a Cooper pair box driven by the (normalized) gate charge, $Q_x/2e$. Furthermore, we will demonstrate that there exists a direct analogy between QPS events changing the flux inside the ring by Φ_0 and tunneling events of single Cooper pairs changing the capacitor charge by $2e$. This equivalence is reminiscent of the well-known duality between phase and charge representations (Schön and Zaikin 1990).

We start from the case of sufficiently thick superconducting rings where QPS effects can be neglected. In this case, by virtue of Poisson's resummation theorem one can identically transform the ring partition function (40.76) to the following expression:

$$Z_{\phi_x} = \sum_{k=-\infty}^{\infty} \exp(i2\pi k\phi_x)\int d\theta_0 \int_{\theta_0}^{\theta_0+2\pi k} \mathcal{D}\theta\exp(-S_0[\theta]), \qquad (40.81)$$

where

$$S_0[\theta] = \int_0^{1/T} d\tau \frac{1}{2E_R}\left(\frac{\partial\theta}{\partial\tau}\right)^2 \qquad (40.82)$$

and $1/E_R$ plays the role of a "mass" for a "particle" with the coordinate θ. This coordinate represents an effective angle. It is formally analogous to the Josephson phase variable in which case E_R just coincides with the charging energy.

Let us now consider thinner rings where QPS effects gain importance. In this case, we should include all transitions between different energy states (Equation 40.77) labeled by the number p. As we already discussed, these transitions are just QPS events with the rate γ_{QPS} defined in Equation 40.59. Let us fix $p=0$ and take into account virtual transitions to all other energy states and back. Performing summation over all such contributions, we arrive at the series in powers of the rate γ_{QPS} (or fugacity y) similar to Equation 40.64. Assuming that both the ring perimeter X and its thickness are not too large, (the latter condition restricts the parameter μ) one can neglect weak logarithmic interaction (40.61) between different phase slips. Then spatial QPS coordinates, x_n, can be trivially integrated out. As a result, we arrive at the following contribution to the partition function:

$$\tilde{Z}(\phi_x) = \int_{\phi_x}^{\phi_x} \mathcal{D}\phi(\tau) \sum_{n=0}^{\infty} \frac{1}{2n!}\left(\frac{\gamma_{QPS}}{2}\right)^{2n} \int_0^{1/T} d\tau_1 \cdots \int_0^{1/T} d\tau_{2n}$$

$$\times \sum_{v_i=\pm 1} \delta(\dot{\phi}(\tau) - \dot{\phi}_n(\tau))\exp\left(-\int_0^{1/T} d\tau \frac{E_R\phi^2(\tau)}{2}\right) \qquad (40.83)$$

where, as before, the summation is carried out over neutral charge configurations (40.66),

$$\phi_n(\tau) = \sum_{j=1}^{n} v_j\Theta(\tau - \tau_j), \qquad (40.84)$$

and $\Theta(\tau)$ is the theta function. We note that the term in the exponent describes virtual energy changes occurring due to QPS events. What remains is to add up similar contributions from all other parabolas with $p \neq 0$. In this way, we arrive at the final expression for the grand partition function of the nanoring:

$$Z_{\phi_x} = \sum_{p=-\infty}^{\infty} \tilde{Z}(p + \phi_x). \qquad (40.85)$$

Together with Equation 40.83 this result fully accounts for all QPS events in our system.

In the absence of QPS (i.e., for $\gamma_{QPS} \to 0$), Equations 40.85 and 40.83 obviously reduce to Equation 40.76. For nonzero γ_{QPS}, one can rewrite the partition function (40.83) and (40.85) in the equivalent form of the following path integral:

$$Z_{\phi_x} = \sum_{k=-\infty}^{\infty} \exp(i2\pi k\phi_x)\int d\theta_0 \int_{\theta_0}^{\theta_0+2\pi k} \mathcal{D}\theta\exp(-S[\theta]), \qquad (40.86)$$

where

$$S[\theta] = S_0[\theta] - \gamma_{QPS} \int_0^{1/T} d\tau \cos\theta(\tau). \qquad (40.87)$$

In order to demonstrate that the partition function (40.86) and (40.87) is identical to that defined in Equations 40.83 and 40.85, it suffices to perform the Hubbard–Stratonovich transformation of the kinetic term (40.82) (which amounts to introducing additional path integral over $\phi(\tau)$), to formally expand $\exp(-S[\theta])$ in powers of γ_{QPS} and then to integrate out the θ variable in all terms of these series. After these straightforward steps we arrive back at Equations 40.83 and 40.85.

Equations 40.86 and 40.87 define the grand partition function for a quantum particle in a cosine periodic potential which is, in turn, equivalent to that for a Josephson junction in the presence of charging effects (Schön and Zaikin 1990). This partition function can be identically rewritten as

$$Z_{\phi_x} = \sum_{p=-\infty}^{\infty} \exp\left(\frac{-\tilde{E}_p(\phi_x)}{T}\right), \qquad (40.88)$$

where $\tilde{E}_p(\phi_x)$ are the energy bands of the problem defined by the solutions of the well-known Mathieu equation. For $\gamma_{QPS} \ll E_R$ one has

$$\tilde{E}_0(\phi) = \frac{E_R}{2\pi^2} \arcsin^2\left[\left(1 - \frac{\pi^2}{2}\left(\frac{\gamma_{QPS}}{E_R}\right)^2\right)\sin(\pi\phi_x)\right], \quad (40.89)$$

i.e., the energy bands remain nearly parabolic, $\tilde{E}_p(\phi) \sim E_p(\phi)$, except in the vicinity of the crossing points where gaps open due to level repulsion (see Figure 40.6). In this limit, the value of the gap between the lowest and the first excited energy bands just coincides with the QPS rate, $\delta E_{01} = \gamma_{QPS}$. For larger γ_{QPS} the bandwidth shrinks while the gaps become bigger. In the limit $\gamma_{QPS} \gg E_R$ the lowest band coincides with

$$\tilde{E}_0(\phi) = \frac{8}{\sqrt{\pi}}\gamma_{QPS}^{3/4}\left(E_R\right)^{1/4} e^{-8\sqrt{\gamma_{QPS}/E_R}}(1 - \cos(2\pi\phi_x)), \quad (40.90)$$

The gap between the two lowest bands is $\delta E_{01} = \sqrt{\gamma_{QPS}E_R}$.

These results are sufficient to evaluate PC in superconducting nanorings in the presence of QPS. As before, taking the derivative of the ground state energy with respect to the flux Φ_x and making use of Equation 40.59 we find that for $\gamma_{QPS} \ll E_R$, and outside immediate vicinity of the points $\phi_x = 1/2 + p$ PC is again defined by Equations 40.79 and 40.80. In the opposite limit $\gamma_{QPS} \gg E_R$, we find

$$I = \tilde{I}_0 \sin(2\pi\phi_x), \quad \tilde{I}_0 = \frac{16e}{\sqrt{\pi\hbar}}\gamma_{QPS}^{3/4}\left(E_R\right)^{1/4} e^{-8\sqrt{\gamma_{QPS}/E_R}}. \quad (40.91)$$

We observe that in the latter limit PC is exponentially suppressed as

$$\tilde{I}_0 \propto \exp\left(\frac{-R}{R_c}\right), \quad (40.92)$$

where

$$R_c \sim \xi \exp(S_{core}/2). \quad (40.93)$$

This simple formula sets the size scale beyond which one would expect PC to be exponentially suppressed in superconducting nanorings with $\sqrt{s} \lesssim 10\,\text{nm}$. For example, for ξ of order $100\,\text{nm}$, $S_{core} \approx 10$ Equation 40.93 yields $R_c \sim 1 \div 2\,\mu\text{m}$.

The above results demonstrate practically the same qualitative features as those initially found within a different approach for the model of nanorings formed by Josephson chains (Matveev et al. 2002). In particular, both for granular and for homogeneous rings the dependence of PC on ϕ_x gradually changes from sawtooth to sinusoidal as the ring perimeter X increases. This change is accompanied by suppression of PC amplitude which eventually becomes an exponentially decaying function of the ring radius R (40.92) in the limit $R > R_c$. In other words, Equation 40.93 sets the length scale, R_c, beyond which PC in superconducting nanorings should be essentially washed out by quantum fluctuations. In some sense, the scale $2\pi R_c$ plays the role of a

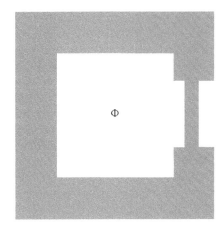

FIGURE 40.7 Superconducting ring with a "quantum phase slip junction."

dephasing length in our problem demonstrating that zero-point fluctuations can destroy macroscopic phase coherence down to $T \to 0$ even in superconducting systems.

Finally, we should mention that one can also consider a slightly modified situation of rings consisting of thicker and thinner parts, as it is shown in Figure 40.7. Assuming that QPS effects are negligible in a thicker part of the ring and they can only occur in its thinner part (of length d), we arrive at exactly the same results as above in which one should only replace the ring perimeter X by the length d. In particular, it follows from our analysis that such rings also exhibit the property of an exact duality to mesoscopic Josephson junctions if we identify E_R with the junction charging energy, E_c, and the QPS rate, γ_{QPS}, with the Josephson energy, E_J. For this reason, it was suggested (Mooij and Nazarov 2006) to call systems depicted in Figure 40.7 "quantum phase slip junctions" and argued that any known result on electron transport in circuits containing Josephson junctions can be exactly mapped onto a dual result for a QPS junction in a dual circuit. This observation can be used in metrology, e.g., for practical implementation of the electric current standard in the above structures. It was also proposed (Mooij and Harmans 2005) to use rings with QPS junctions for experimental realization of quantum phase slip flux qubits.

40.4.2 Parity Effect and Persistent Currents

In our previous analysis of QPS effects in superconducting nanorings, we followed the standard description developed for grand canonical ensembles, i.e., we implicitly assumed that the total number of electrons in the system N may fluctuate while the chemical potential is fixed. Obviously, this assumption is not correct for rings that are disconnected from any external circuit. In that case electrons cannot enter or leave the ring and, hence, the number N is strictly fixed, but the chemical potential, on the contrary, fluctuates. It turns out that novel effects emerge in this physical situation. These effects will be discussed below.

It is well known that the thermodynamic properties of isolated superconducting systems are sensitive to the parity of the total number of electrons (Averin and Nazarov 1992, Tuominen et al. 1992) even though this number is macroscopically large. This parity effect is a fundamental property of a superconducting ground state described by the condensate of Cooper pairs. The number of electrons in the condensate is necessarily even; hence, for odd N at least one electron remains unpaired having an extra energy equal to the superconducting energy gap, Δ_0. This effect makes thermodynamic properties of the ground state with even and odd N different. Clear evidence for such parity effect was demonstrated experimentally in small superconducting islands (Tuominen et al. 1992, Lafarge et al. 1993).

Consider now isolated superconducting rings pierced by the magnetic flux Φ_x. In accordance with the number–phase uncertainty relation the *global* superconducting phase of the ring fluctuates strongly in this case; however, these fluctuations are decoupled from the supercurrent and therefore cannot influence the latter. In this situation the parity effect may substantially modify PC in superconducting nanorings at sufficiently low temperatures. In particular, changing the electron parity number from even to odd results in spontaneous supercurrent in the ground state of such rings without any externally applied magnetic flux. In other words, a fundamental conclusion of our analysis will be that *BCS ground state of a canonical ensemble with odd number of electrons is the state with spontaneous supercurrent.*

Under what conditions is the parity effect important in superconducting nanorings?

Let us first consider homogeneous nanorings with cross section s and perimeter $X = 2\pi R$. As before, rings will be assumed sufficiently thin, $\sqrt{s} < \lambda_L$. On the other hand, below we will neglect QPS effects, i.e., describe superconducting properties of such rings within the mean-field BCS theory. As we have already learned, this description is justified provided the condition $g_\xi \gg 1$ is satisfied. Hence, the ring should not be too thin and the total number of conducting channels $\mathcal{N} = p_F^2 s/4\pi\hbar^2$ should remain large, $\mathcal{N} \gg 1$. In addition, the perimeter X should not exceed the scale $2\pi R_c$ (Equation 40.93). Finally, we will neglect a small difference between the mean-field values of the BCS order parameter for the even and odd ensembles (Golubev and Zaikin 1994, Janko et al. 1994). This is appropriate provided the ring volume is large enough, i.e., $\mathcal{V} = Xs \gg 1/N_0\Delta_0$.

Rigorous calculation of PC in canonical superconducting ensembles can be performed within the so-called parity projected formalism that was developed for superconducting grains (Golubev and Zaikin 1994, Janko et al. 1994) and then extended to superconducting nanorings (Zaikin 2004, Sharov and Zaikin 2005a,b). Here we avoid discussing rather complicated details of this formalism. Instead, we present simple physical arguments that allow to establish a transparent relation between the magnitudes of PC in canonical BCS ensembles with even and odd total number of electrons in the limit of low temperatures.

Let us set $T = 0$. As we already discussed, provided the total number of electrons N in the ring is even, they all form the condensate of Cooper pairs occupying all states with energies below

$\varepsilon_F - \Delta_0$, where ε_F is the Fermi energy. Exactly the same situation is realized for grand canonical BCS ensembles. Hence, at $T \to 0$ PC in the ring with even number of electrons I_e should exactly coincide with the standard grand canonical result ($I = en_s v_s s$) with $n_s \equiv n_e$. If, however, the number N is odd, one electron always remains unpaired and it occupies the lowest available energy state, ε_0. An external magnetic flux, Φ_x, applied to the ring can shift this energy level, i.e., we have $\varepsilon_0 = \varepsilon_0(\Phi_x)$. Accordingly, the "odd" electron occupying this level should provide an extra contribution to PC that should simply be added to I_e. In other words, at $T \to 0$ we obtain (Sharov and Zaikin 2005a,b, Kalenkov et al. 2006)

$$I_o(\Phi) = I_e(\Phi_x) + c\frac{\partial \varepsilon_0(\Phi_x)}{\partial \Phi_x}, \qquad (40.94)$$

where I_o represents PC in the BCS ensemble with odd number of electrons. Applying this relation to the case of homogeneous superconducting nanorings, one finds (Kang 2000, Kwon and Yakovenko 2002, Sharov and Zaikin 2005a,b)

$$I_o = en_o v_s s - e\frac{v_F}{X}\operatorname{sgn} v_s. \qquad (40.95)$$

Here we introduced the electron density in the case of even/odd total number of electrons, $n_{e/o} = N_{e/o}/\mathcal{V}$.

Hence, in the case of odd ensembles there exists an additional term that modifies the flux dependence of PC. As this term in Equation 40.95 is inversely proportional to the total number of conducting channels, $(I_e - I_o)/I_e \sim 1/\mathcal{N}$, the parity effect remains vanishingly small in generic uniform metallic rings with $\mathcal{N} \gtrsim 10^3$. On the other hand, for ultrathin nanorings with $\mathcal{N} \lesssim 10$, PC is essentially wiped out due to proliferation of QPS. Estimating $g_\xi \sim \mathcal{N}l/\xi$, we conclude that for $g_\xi \sim 1$ (i.e., when the QPS fugacity is already large and, hence, quantum suppression of PC becomes very strong) the number of conducting channels yet remains parametrically large, $\mathcal{N} \sim \xi/l \gg 1$. This estimate demonstrates that the parity effect on PC is never important in the case of *homogeneous* superconducting nanorings and appears to be practically unobservable in such systems.

Let us now slightly modify our system and consider a relatively thick superconducting ring with large $g_\xi \gg 1$ interrupted by a thin wire of length d with only few conducting channels

$$\mathcal{N}_n \sim 1, \qquad (40.96)$$

thus forming a weak link inside the superconducting ring, see Figure 40.8. Without loss of generality, this wire can be considered normal no matter if it is made of a normal or a superconducting material. In the latter case, quantum fluctuations would fully suppress the order parameter inside such a wire bringing it into the normal state. In contrast, QPS effects in superconducting parts of the ring can be neglected thus making the mean-field BCS description applicable. A clear advantage of these structures in comparison to homogeneous rings is that in the former case

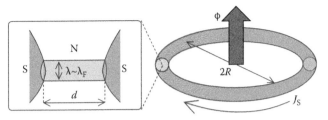

FIGURE 40.8 Superconducting ring with embedded SNS junction of length d. (From Sharov, S.V. and Zaikin, A.D., *Phys. Rev. B*, 71, 014518-1, 2005a.)

the effect of the electron parity number on PC can be large due to the condition (40.96).

Let us first consider the limit of short weak link $d \to 0$. In this case, the Josephson current can be expressed only via the contributions from discrete Andreev energy states inside this constriction $E_\pm = \pm\varepsilon(\varphi)$ as (Beenakker and van Houten 1991)

$$I(\varphi) = \frac{2e}{\hbar} \sum_{i=1}^{\mathcal{N}_n} \left[\frac{\partial E_{i-}}{\partial \varphi} f_-(E_{i-}) + \frac{\partial E_{i+}}{\partial \varphi} f_+(E_{i+}) \right], \quad (40.97)$$

where the index i labels conducting channels inside the constriction, $\varepsilon_i(\varphi) = \Delta_0\sqrt{1 - \mathcal{T}_i \sin^2(\varphi/2)}$, are Andreev energy levels in a quantum point contact (see Figure 40.9), $\mathcal{T}_i$ is the transmission of the ith channel, and φ is the superconducting phase across the contact linked to the magnetic flux Φ_x by means of a standard relation:

$$\varphi = \frac{2\pi\Phi_x}{\Phi_0} \quad (40.98)$$

Using the Fermi filling factors for these states, $f_\pm(E_{i\pm}) = [1 + \exp(\pm\varepsilon_i(\varphi)/T)]^{-1}$, one arrives at the standard grand canonical results. However, in the case of superconducting rings with fixed number of electrons N these filling factors should be modified. As before, let us set $T \to 0$. For even N, all electrons are paired occupying available states with energies below the Fermi level

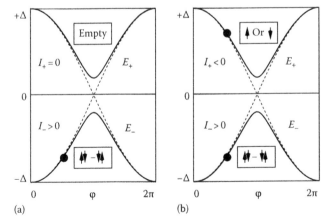

(a)   (b)

FIGURE 40.9 Andreev levels inside a quantum point contact and their occupation at $T = 0$ for even (a) and odd (b) ensembles. (From Sharov, S.V. and Zaikin, A.D., *Phys. Rev. B*, 71, 014518-1, 2005a.)

(see Figure 40.9a). In this case, one has $f_-(E_{i-}) = 1, f_+(E_{i+}) = 0$, the current is entirely determined by the contribution of all quasiparticle states E_{i-}, and Equation 40.97 yields the same result as one for the grand canonical ensemble:

$$I_e = -\frac{2e}{\hbar} \sum_{i=1}^{\mathcal{N}_n} \frac{\partial \varepsilon_i(\varphi)}{\partial \varphi}. \quad (40.99)$$

By contrast, in the case of odd N one electron always remains unpaired and occupies the lowest available energy state—in our case E_{k+} (where the index k corresponds to the conducting channel with the highest transmission value $\mathcal{T}_k$)—above the Fermi level. Hence, for odd N one has $f_-(E_{i-}) = 1, f_+(E_{k+}) = 1$, and $f_+(E_{i+}) = 0$ for all $i \neq k$, see Figure 40.9b. Then from Equation 40.97, we obtain

$$I_o = I_e + \frac{2e}{\hbar} \frac{\partial \varepsilon_k(\varphi)}{\partial \varphi} = -\frac{2e}{\hbar} \sum_{i=1, i\neq k}^{\mathcal{N}_n} \frac{\partial \varepsilon_i(\varphi)}{\partial \varphi}. \quad (40.100)$$

Comparing Equations 40.99 and 40.100, we observe that PC in even and odd ensembles differ because the contribution of the kth conducting channel is totally blocked in the odd case. This is because the contributions of the two Andreev states, E_{k-} and E_{k+}, in Equation 40.97 exactly cancel each other. For quantum point contacts with only one conducting channel, this effect implies *total blocking* of PC by the odd electron, i.e., in this case the supercurrent remains zero for any p or the magnetic flux Φ_x. We also stress that in quantum point contacts with several conducting channels the current through the most transparent channel will be blocked by the odd electron. Hence, though blocking of PC remains incomplete in this case, it may nevertheless be important also for quantum point contacts with $\mathcal{N}_n > 1$.

We now turn to another important limit of superconducting rings containing a normal wire of length $d > \xi_0 \sim \hbar v_F/\Delta_0$. In contrast to the case $d \to 0$ considered above, the Josephson current in SNS structures cannot anymore be attributed only to the discrete Andreev states inside a weak link, and an additional contribution from the states in the continuum should also be taken into account. Furthermore, for any nonzero d there are always more than two discrete Andreev levels in the system. Accordingly, significant modifications in the physical picture of the parity effect in such SNS rings can be expected.

The key difference can be understood already by comparing the typical structure of discrete Andreev levels in SNS junctions (Figure 40.10) with that of a quantum point contact (Figure 40.9). As before, in the limit $T \to 0$ all states below (above) the Fermi level are occupied (empty) provided the total number of electrons in the system is even (Figure 40.10a). If, on the other hand, this number is odd the lowest Andreev state above the Fermi energy is occupied as well (Figure 40.10b), thus providing an additional contribution to the Josephson current. This contribution, however, cancels only that of a symmetric Andreev level below the Fermi energy, while the contributions of all other occupied Andreev levels and of the continuum states remain

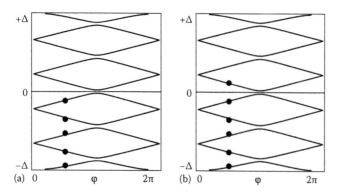

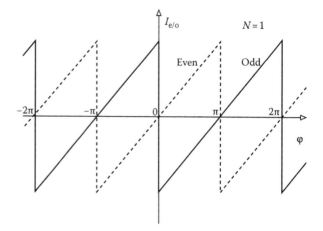

FIGURE 40.10 Andreev levels in a single mode SNS junction with $d = 6\hbar v_F/\Delta_0$ and their occupation at $T = 0$ for even (a) and odd (b) ensembles. (From Sharov, S.V. and Zaikin, A.D., *Phys. Rev. B*, 71, 014518-1, 2005a.)

FIGURE 40.11 The zero temperature current–phase dependence (40.101) for SNS rings with $\mathcal{N}_n = 1$: $I_e(\varphi)$ (dashed) line and $I_o(\varphi)$ (solid) line. (From Sharov, S.V. and Zaikin, A.D., *Phys. Rev. B*, 71, 014518-1, 2005a.)

uncompensated. Hence, unlike in the case of a short ($d \to 0$) single-channel weak link, in SNS rings one should not anymore expect the effect of PC blocking by the odd electron but rather some other nontrivial features of the parity effect.

The calculation of PC in SNS rings with $d \gg \hbar v_F/\Delta_0$ and $T \to 0$ is straightforward. In this case, the current I_e again coincides with that for the grand canonical ensembles (Ishii 1970) while the current I_o—according to Equation 40.94—also contains an additional contribution from the "odd" electron occupying the lowest available Andreev level $E_1(\varphi) = (\hbar v_F)/(2d)|\pi - \varphi|$ (where $0 < \varphi < 2\pi$), see also Figure 40.10b. As a result, from Equations 40.94 and 40.98 we obtain (Sharov and Zaikin 2005a,b)

$$I_e = \frac{e v_F \mathcal{N}_n}{\pi d} \varphi, \quad I_o = \frac{e v_F \mathcal{N}_n}{\pi d}\left(\varphi - \frac{\pi \operatorname{sgn}\varphi}{\mathcal{N}_n}\right). \quad (40.101)$$

This result applies for $-\pi < \varphi < \pi$ and should be 2π-periodically continued otherwise. We observe that in the case of odd ensembles the current–phase relation is shifted by the value $\pi/\mathcal{N}_n$ as compared to that in the even case. This shift is related to the odd electron contribution, $(2e/\hbar)\partial E_1/\partial \varphi$, from the lowest (above the Fermi level) Andreev state, $E_1(\varphi)$, inside the SNS junction. As we have expected, this contribution indeed does not compensate for the current from other electron states. Rather, it provides a possibility for a parity-induced so-called π-junction state in our system: According to Equation 40.101 for single mode SNS junctions the "sawtooth" current–phase relation will be shifted exactly by π, see Figure 40.11. More generally, we can talk about a novel π/N-junction state, because in the odd case the minimum Josephson energy (zero current) state is reached at $\varphi = \pm\pi/N$, see Figure 40.11. For any $\mathcal{N}_n > 1$, this is a twofold degenerate state within the interval $-\pi < \varphi < \pi$. In the particular case $\mathcal{N}_n = 2$, the current–phase relation $I_o(\varphi)$ turns π-periodic.

Let us recall that the π-junction state can be realized in SNS structures by driving the electron distribution function in the contact area out of equilibrium (Volkov 1995, Wilhelm et al. 1998, Yip 1998). Here, in contrast, the situation of a π- or

π/N-junction is achieved in thermodynamic equilibrium. Along with this important difference, there also exists a certain physical similarity between these two physical situations: In both cases the electron distribution function in the weak link deviates substantially from the Fermi function. It is this deviation that is responsible for the appearance of the π-junction state in both physical situations.

Perhaps the most spectacular physical consequence of the parity effect in SNS rings is the presence of *spontaneous* supercurrents *in the ground state* of such rings with odd number of electrons. Similarly to the case of standard π-junctions, such spontaneous supercurrents should flow even in the absence of an externally applied magnetic flux. Unlike in the usual case, however, here the spontaneous current state occurs for *any* inductance of the ring because of the non-sinusoidal dependence $I_o(\varphi)$ (see also Figure 40.12).

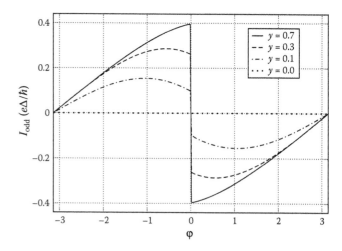

FIGURE 40.12 The zero temperature current–phase relation $I_o(\varphi)$ ($-\pi < \varphi < \pi$) for $\mathcal{N}_n = 1$ and different values of the parameter $y = d\Delta/\hbar v_F$. (From Sharov, S.V. and Zaikin, A.D., *Phys. Rev. B*, 71, 014518-1, 2005a.)

Consider, for instance, the limit $d \gg \hbar v_F / \Delta_0$. In the case of odd number of electrons, the ground state energy of an SNS ring can be written in a simple form

$$E = \frac{\Phi^2}{2c\mathcal{L}} + \frac{\pi \hbar v_F \mathcal{N}_n}{\Phi_0^2 d} \left(\Phi - \frac{\Phi_0 \operatorname{sgn} \Phi}{2\mathcal{N}_n} \right)^2, \quad (40.102)$$

where the first term is the magnetic energy of the ring ($\mathcal{L}$ is the ring inductance), while the second term represents the Josephson energy of an SNS junction. Minimizing (40.102) with respect to the flux Φ, one immediately concludes that the ground state of the ring is a twofold degenerate state with a nonvanishing spontaneous current,

$$I = \pm \frac{ev_F}{d} \left[1 + \frac{2ev_F \mathcal{N}_n}{d} \frac{\mathcal{L}}{\Phi_0} \right]^{-1}, \quad (40.103)$$

flowing either clockwise or counterclockwise. In the limit of small inductances $\mathcal{L} \to 0$ this current does not vanish, and its amplitude just reduces to that of the odd electron current at $\varphi \to 0$. From the above results, one finds

$$I_{sp} = e\Delta_0^2 d / \hbar^2 v_F, \quad d \ll \hbar v_F / \Delta_0, \quad (40.104)$$

$$I_{sp} = ev_F / \pi d, \quad d \gg \hbar v_F / \Delta_0. \quad (40.105)$$

At $d \sim \hbar v_F / \Delta_0$ the amplitude of the current I_{sp} can be evaluated numerically. In agreement with Equation 40.104, I_{sp} increases linearly with d at small d, reaches its maximum value $I_{max} \sim 0.4e\Delta_0 / \hbar$ at $d \sim \hbar v_F / \Delta_0$, and then decreases with further increase of d approaching the dependence (40.105) in the limit of large d. For generic BCS superconductors, the magnitude of this maximum current can be estimated as $I_{max} \sim 10 \div 100\,\text{nA}$. These values might be considered as surprisingly large ones having in mind that this current is associated with only one Andreev electron state.

The above analysis applies for sufficiently clean normal wires brought in a good electric contact with superconductors. Since both these assumptions can be violated in a realistic experiment it is important to discuss the corresponding modifications of our results. Here we restrict our discussion to the case of a disordered normal wire. As before, at $T = 0$ the current I_e identically coincides with one calculated for the grand canonical ensemble. An example for the current–phase relation, $I_e(\varphi)$, is displayed in Figure 40.13. PC in the odd ensemble I_o at $T = 0$ can be evaluated from Equation 40.94 having in mind that the lowest energy state available for an "odd" electron $\varepsilon_0(\varphi)$ in this case is defined by the proximity-induced minigap $\varepsilon_g(\varphi)$ in the density of states of the normal metal wire (Belzig et al. 1999).

Combining our results for $I_e(\varphi)$ with those for the minigap $\varepsilon_g(\varphi)$ (Kalenkov et al. 2006) we arrive at a typical dependence $I_o(\varphi)$ displayed in Figure 40.13. We observe that at sufficiently large values of $\varphi < \pi$ the absolute value of the odd electron contribution to PC $2e\partial \varepsilon_g / \partial \varphi$ exceeds the term $I_e(\varphi)$, and the total

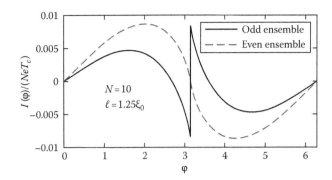

FIGURE 40.13 Phase dependence of the Josephson current at $T = 0$ for the odd and even number of electrons in the ring. (From Kalenkov, M.S. et al., *Phys. Rev. B*, 74, 184502-1, 2006.)

current I_o changes the sign. This nontrivial parity-affected current–phase relation is specific for SNS rings with disorder, and it substantially differs from the current–phase relations derived above for SNS rings with ballistic and resonant transmissions.

At the same time, as in the previous cases, in the odd ensemble there exists a possibility both for a π-junction state and for spontaneous currents in the ground state of the system without any externally applied magnetic flux.

Let us evaluate the ground state energy of the SNS junction by integrating Equation 40.94 with respect to the phase φ. One finds

$$E_o(\varphi) = E_e(\varphi) - \varepsilon_g(0) + \varepsilon_g(\varphi), \quad E_e(\varphi) = \frac{1}{2e} \int_0^\varphi I_e(\varphi) d\varphi,$$

$$(40.106)$$

where $E_{e/o}(\varphi)$ are the ground state energies of SNS junction for even and odd number of electrons in the ring. While the energy $E_e(\varphi)$ is always nonnegative and reaches its minimum at $\varphi = 0$, in the odd case the ground state energy, $E_0(\varphi)$, can become negative reaching its absolute minimum at $\varphi = \pi$. This physical situation of a π-junction is illustrated in Figure 40.14.

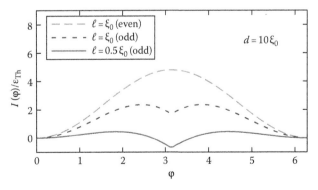

FIGURE 40.14 Josephson energy $E(\varphi)$ of an SNS ring as a function of the phase difference φ for the even and odd ensembles. The solid curve corresponds to a π-junction state. (From Kalenkov, M.S. et al., *Phys. Rev. B*, 74, 184502-1, 2006.)

It is easy to find out under which conditions the π-junction state becomes possible. For that purpose it is sufficient to observe that for any impurity concentration $E_e(\pi) = \gamma I_c/e$, where I_c is the grand canonical critical current at $T = 0$ and the prefactor $\gamma \sim 1$ depends on the form of the current–phase relation. The π-junction condition $E_0(\pi) < 0$ is equivalent to the inequality

$$\varepsilon_g(0) > \gamma I_c/e. \qquad (40.107)$$

It is obvious from Figure 40.14 that in the many channel limit the inequality (40.107) cannot be satisfied for sufficiently large l. On the other hand, for sufficiently short mean free paths $I_c \propto l^2$ decays faster with decreasing l as compared to the minigap $\varepsilon_g(0) \propto l$, and the π-junction state becomes possible. In particular, in the diffusive limit one finds (Dubos et al. 2001) $I_c \simeq 10.82\varepsilon_{Th}/eR_N = 1.53e\mathcal{N}_n v_F l^2/d^3$ and $\gamma \simeq 1.05$, where R_N is the Drude resistance of a normal metal.

Combining these results with the expression for the minigap $\varepsilon_g(0) \simeq 3.12\varepsilon_{Th}$ (Belzig et al. 1999), from the condition (40.107) we observe that in the odd case the π-junction state is realized provided the number of conducting channels in the junction $\mathcal{N}_n$ is smaller than

$$\mathcal{N}_n < 0.65d/l. \qquad (40.108)$$

This condition is not very restrictive and it can certainly be achieved in various experiments. For sufficiently dirty junctions it allows for a formation of the π-junction state even in the many channel limit. The condition (40.108) can also be rewritten as $g_N < 1.73$, where $g_N = 8\mathcal{N}_n l/3d$ is dimensionless conductance of a normal wire.

The condition for the presence of spontaneous currents in the ground state of SNS rings with an odd number of electrons is established analogously, one should only take into account an additional energy of the magnetic field produced by PC circulating inside the ring. The ground state with spontaneous currents is possible provided the total energy of the ring $E_{tot}(\pi)$ becomes negative, i.e.,

$$E_{tot}(\pi) = 1.8\varepsilon_{Th}[g_N - 1.73] + \frac{(\Phi_0/2)^2}{2\mathcal{L}} < 0. \qquad (40.109)$$

This condition is more stringent than that for the π-junction state, but can also be satisfied provided $\mathcal{L}$ exceeds a certain threshold value that can roughly be estimated as $\sim 0.1\ \Phi_0^2/\varepsilon_{Th}$.

We conclude that in the diffusive limit the current–phase relation in the odd case is entirely different from that in the ballistic case. Also, the restriction on the number of conducting channels $\mathcal{N}_n$ in the normal metal (40.108) is less stringent. This feature of diffusive SNS rings is rather advantageous for possible experimental observation of the effects discussed here.

Still, in practice it would be necessary to fabricate SNS rings with few conducting channels in the normal wire $\mathcal{N}_n \lesssim 10$. This condition can hardly be met for conventional normal metals. It appears, therefore, that the most promising candidates

for practical realization of such structures are junctions with N-layers formed by carbon nanotubes or organic molecules. In this respect it is important to point out that observations of clear signatures of dc Josephson effect in superconducting junctions with a weak link formed by carbon nanotubes were recently reported by several experimental groups (Jarillo-Herrero et al. 2006, Jorgensen et al. 2006, Cleuziou et al. 2006, Grove-Rasmussen et al. 2007, Tsuneta et al. 2007). SNS rings with carbon nanotubes appear to be most suitable objects in which it would be possible to observe the influence of the parity effect on PC discussed here.

40.5 Summary

It is well established that fluctuations play an important role in structures with reduced dimensionality. In this chapter we addressed fluctuation effects that occur in superconducting nanowires and nanorings at temperatures below the mean-field BCS critical temperature. Superconducting properties of such systems have been intensively studied—both theoretically and experimentally—during past years. The key conclusions of these investigations can be summarized as follows.

Thicker superconducting wires are characterized by very small Ginzburg numbers, $Gi_{1D} \lll 1$, and diameters typically $\gtrsim 100\,nm$. In such systems, the superconducting transition is broadened due to TAPS that cause nonzero resistance $R(T)$ at temperatures close enough to the critical temperature $T_c - T \ll T_c$. Upon decreasing temperature, TAPS events become less likely and quantum fluctuations of the order parameter take over. This is the regime of QPS that typically sets in at $T \lesssim \Delta_0(T)$. As long as the wire is sufficiently thick and the Ginzburg number remains very small, $Gi_{1D} \lll 1$, QPS events are rare and can hardly yield any measurable consequences. Hence, the behavior of thicker wires remains essentially superconducting outside an immediate vicinity of the critical temperature. Upon reduction of the wire diameter below $\sim50\,nm$ the QPS rate increases drastically. In this regime, the wire resistance $R(T)$ still decreases with temperature, but may remain well in the measurable range down to very low T. Provided the wire diameter is decreased further the dimensionless conductance of a wire segment of length ξ eventually becomes of order $g_\xi \sim 10$ or smaller. In such wires, QPS proliferate causing a sharp *crossover* from a superconducting to a normal behavior. For generic parameters this crossover is expected for wire diameters in the $10\,nm$ range. This crossover was indeed observed in a number of experiments in wires with thicknesses exactly in this range. Thus, *intrinsic superconductivity in wires with diameters $\lesssim 10\,nm$ is destroyed by quantum fluctuations of the order parameter at any temperature down to $T = 0$.*

Theoretical analysis of nanowires reveals further interesting effects, like the QPS-binding–unbinding QPT between superconducting and non-superconducting phases, which is predicted to occur as the impedance of a superconducting wire, $Z_w = \sqrt{\tilde{L}/C}$, becomes of order of the quantum resistance unit, $R_q \approx 6.5\,k\Omega$. For typical parameters, this condition is also achieved for wire diameters in the 10 nm range. To the best of our knowledge, no clear experimental evidence for this phase transition exists so far. This can be due to rather stringent requirements: QPT can only be observed in long nanowires at sufficiently low temperatures, ideally at $T \to 0$. Interesting effects may occur also in short nanowires forming weak links between superconducting electrodes. In many respects such systems can behave similarly to Josephson nanojunctions and weak links.

Novel effects are also expected in superconducting nanorings. While rings formed by thicker wires demonstrate the standard behavior familiar from the bulk samples, the situation changes drastically as soon as the wire thickness gets reduced down to ≈ 10 nm or below this value. In this case, QPS effects become important leading to strong fluctuations of the magnetic flux inside the ring. As a result, the amplitude of persistent current decreases and its flux dependence changes from the sawtooth-like to a smoother one. For such rings even at $T = 0$ persistent current gets exponentially suppressed by quantum fluctuations, provided the ring radius R exceeds the critical scale $R_c \approx \xi$ $\exp(S_{core}/2)$, where $S_{core} \sim g_\xi \sim Gi_{1D}^{-3/2}$ is the action of the QPS core. For typical wire parameters the length R_c can be of order one or few microns. This length constitutes another fundamental scale associated with QPS.

Yet another important factor that may influence PC in isolated superconducting nanorings is the electron parity number. This influence is particularly strong in nanorings containing a weak link with few conducting modes. Changing the electron parity number from even to odd may result in *spontaneous* supercurrent in the ground state of such rings without any externally applied magnetic flux. At $T = 0$ this current is produced by the only unpaired electron that occupies the lowest available Andreev state. Under certain conditions, this spontaneous supercurrent can reach remarkably large values up to $\sim e\Delta_0/\hbar \sim 10 \div 100$ nA, which can be reliably detected in modern experiments.

References

Arutyunov K.Yu., D.S. Golubev, and A.D. Zaikin. 2008. Superconductivity in one dimension. *Phys. Rep.* 464: 1–70.

Averin D.V. and Yu.V. Nazarov. 1992. Single-electron charging of a superconducting island. *Phys. Rev. Lett.* 69: 1993–1996.

Beenakker C.W.J. and H. van Houten. 1991. Josephson current through a superconducting quantum point contact shorter than the coherence length. *Phys. Rev. Lett.* 66: 3056–3059.

Belzig W., F.K. Wilhelm, C. Bruder, G. Schon, and A.D. Zaikin. 1999. Quasiclassical Green's function approach to mesoscopic superconductivity. *Superlatt. Microstruct.* 25: 1251–1288.

Berezinskii V.L. 2007. *Low Temperature Properties of Two-Dimensional Systems*. Moscow, Russia: Fizmatlit.

Bezryadin A. 2008. Quantum suppression of superconductivity in nanowires. *J. Phys.: Condens. Matter* 20: 043202-1–043202-19.

Bezryadin A., C.N. Lau, and M. Tinkham. 2000. Quantum suppression of superconductivity in ultrathin nanowires. *Nature* 404: 971–974.

Buchler H.P., V.B. Geshkenbein, and G. Blatter. 2004. Quantum fluctuations in thin superconducting wires of finite length. *Phys. Rev. Lett.* 92: 067007-1–067007-4.

Caldeira A.O. and A.J. Leggett. 1983. Quantum tunneling in a dissipative system. *Ann. Phys. (N.Y.)* 149: 374–456.

Cleuziou J.-P, W. Wernsdorfer, V. Bouchiat, T. Ondarcuhu, and M. Monthioux. 2006. Carbon nanotube superconducting quantum interference device. *Nat. Nanotechnol.* 1: 53–59.

Dubos P., H. Courtois, B. Pannetier, F.K. Wilhelm, A.D. Zaikin, and G. Schön. 2001. Josephson critical current in a long mesoscopic S-N-S junction. *Phys. Rev. B* 63: 064502-1–064502-5.

Giordano N. 1988. Evidence for macroscopic quantum tunneling in one-dimensional superconductors. *Phys. Rev. Lett.* 61: 2137–2140.

Giordano N. 1994. Superconducting fluctuations in one dimension. *Physica B* 203: 460–466.

Golubev D.S. and A.D. Zaikin. 1994. Parity effect and thermodynamics of canonical superconducting ensembles. *Phys. Lett. A* 195: 380–388.

Golubev D.S. and A.D. Zaikin. 2001. Quantum tunneling of the order parameter in superconducting nanowires. *Phys. Rev. B* 64: 014504-1–014504-14.

Golubev D.S. and A.D. Zaikin. 2008. Thermally activated phase slips in superconducting nanowires. *Phys. Rev. B* 78: 144502-1–144502-8.

Grove-Rasmussen K., H.I. Jorgensen, and P.E. Lindelof. 2007. Kondo resonance enhanced supercurrent in single wall carbon nanotube Josephson junctions. *New J. Phys.* 9: 124-1–124-10.

Hohenberg P.C. 1967. Existence of long-range order in one and two dimensions. *Phys. Rev.* 158: 383–386.

Ishii C. 1970. Josephson currents through junctions with normal metal barriers. *Progr. Theor. Phys.* 44: 1525–1547.

Janko B., A. Smith, and V. Ambegaokar. 1994. BCS superconductivity with fixed number parity. *Phys. Rev. B* 50: 1152–1161.

Jarillo-Herrero P., J.A. van Dam, and L.P. Kouwenhoven. 2006. Quantum supercurrent transistors in carbon nanotubes. *Nature* 439: 953–956.

Jorgensen H.I., K. Grove-Rasmussen, T. Novotny, K. Flensberg, and P.E. Lindelof. 2006. Electron transport in single-wall carbon nanotube weak links in the Fabry-Perot regime. *Phys. Rev. Lett.* 96: 207003-1–207003-4.

Kalenkov M.S., H. Kloos, and A.D. Zaikin. 2006. Minigap, parity effect, and persistent currents in SNS nanorings. *Phys. Rev. B* 74: 184502-1–184502-7.

Kamerlingh Onnes H. 1911. The superconductivity of mercury. *Comm. Phys. Lab. Univ. Leiden*, 122 and 124.

Kang K. 2000. Parity effect in a mesoscopic superconducting ring. *Europhys. Lett.* 51: 181–187.

Kopnin N. 2001. *Nonequilibrium Superconductivity*. New York: Oxford University Press.

Kosterlitz, J.M. 1974. The critical properties of the two-dimensional XY model. *J. Phys. C* 7: 1046–1060.

Kosterlitz J.M. and D.J. Thouless. 1973. Ordering, metastability and phase transitions in two-dimensional systems. *J. Phys. C* 6: 1181–1203.

Kwon H.-J. and V.M. Yakovenko. 2002. Spontaneous formation of a π soliton in a superconducting wire with an odd number of electrons. *Phys. Rev. Lett.* 89: 017002-1–017002-4.

Lafarge P., P. Joyez, D. Esteve, C. Urbina, and M.H. Devoret. 1993. Measurement of the even-odd free-energy difference of an isolated superconductor. *Phys. Rev. Lett.* 70: 994–997.

Langer J.S. and V. Ambegaokar. 1967. Intrinsic resistive transition in narrow superconducting channels. *Phys. Rev.* 164: 498–510.

Larkin A.I. and A. Varlamov. 2005. *Theory of Fluctuations in Superconductors*. Oxford, U.K.: Clarendon.

Lau C.N., N. Markovic, M. Bockrath, A. Bezryadin, and M. Tinkham. 2001. Quantum phase slips in superconducting nanowires. *Phys. Rev. Lett.* 87: 217003-1–217003-4.

Little W.A. 1967. Decay of persistent currents in small superconductors. *Phys. Rev.* 156: 396–403.

Lukens J.E., R.J. Warburton, and W.W. Webb. 1970. Onset of quantized thermal fluctuations in "one-dimensional" superconductors. *Phys. Rev. Lett.* 25: 1180–1184.

Matveev K.A., A.I. Larkin, and L.I. Glazman. 2002. Persistent current in superconducting nanorings. *Phys. Rev. Lett.* 89: 096802-1–096802-4.

McCumber D.E. and B.I. Halperin. 1970. Time scale of intrinsic resistive fluctuations in thin superconducting wires. *Phys. Rev. B* 1: 1054–1070.

Mermin N.D. and H. Wagner. 1966. Absence of ferromagnetism or antiferromagnetism in one-or two-dimensional isotropic Heisenberg models. *Phys. Rev. Lett.* 17: 1133–1136.

Mooij J.E. and C.J.P.M. Harmans. 2005. Phase-slip flux qubits. *New J. Phys.* 7: 219-1–219-7.

Mooij J.E. and Yu.V. Nazarov. 2006. Superconducting nanowires as quantum phase-slip junctions. *Nat. Phys.* 2: 169–172.

Mooij J.E. and G. Schön. 1985. Propagating plasma mode in thin superconducting filaments. *Phys. Rev. Lett.* 55: 114–117.

Newbower R.S., M.R. Beasley, and M. Tinkham. 1972. Fluctuation effects on the superconducting transition of tin whisker crystals. *Phys. Rev. B* 5: 864–868.

Penttila J.S., P.J. Hakonen, M.A. Paalanen, and E.B. Sonin. 1999. "Superconductor-insulator transition" in a single Josephson junction. *Phys. Rev. Lett.* 82: 1004–1007.

Popov V.N. 1987. *Functional Integrals and Collective Excitations*. Cambridge, U.K.: Cambridge University Press.

Schön G. and A.D. Zaikin. 1990. Quantum coherent effects, phase transitions, and the dissipative dynamics of ultra-small tunnel junctions. *Phys. Rep.* 198: 237–413.

Sharov S.V. and A.D. Zaikin. 2005a. Parity-affected persistent currents in superconducting nanorings. *Phys. Rev. B* 71: 014518-1–014518-7.

Sharov S.V. and A.D. Zaikin. 2005b. Parity effect and spontaneous currents in superconducting nanorings. *Physica E* 29: 360–368.

Tidecks R. 1990. *Current-Induced Nonequilibrium Phenomena in Quasi-One-Dimensional Superconductors*. New York: Springer.

Tsuneta T., L. Lechner, and P.J. Hakonen. 2007. Gate-controlled superconductivity in a diffusive multiwalled carbon nanotube. *Phys. Rev. Lett.* 98: 087002-1–087002-4.

Tuominen M.T., J.M. Hergenrother, T.S. Tighe, and M. Tinkham. 1992. Experimental evidence for parity-based 2e periodicity in a superconducting single-electron tunneling transistor. *Phys. Rev. Lett.* 69: 1997–2000.

van Otterlo A., D.S. Golubev, A.D. Zaikin, and G. Blatter. 1999. Dynamics and effective actions of BCS superconductors. *Eur. Phys. J. B* 10: 131–143.

Volkov A.F. 1995. New phenomena in Josephson SINIS junctions. *Phys. Rev. Lett.* 74: 4730–4733.

Weiss U. 1999. *Quantum Dissipative Systems*, 2nd edn. Singapore: World Scientific.

Wilhelm F.K., G. Schön, and A.D. Zaikin. 1998. The mesoscopic SNS transistor. *Phys. Rev. Lett.* 81: 1682–1685.

Yip S.K. 1998. Energy-resolved supercurrent between two superconductors. *Phys. Rev. B* 58: 5803–5807.

Zaikin A.D. 2004. Some novel effects in superconducting nanojunctions. *Fiz. Nizk. Temp. (Kharkov)* 30: 756–769 [*Low Temp. Phys.* 30: 568–581].

Zaikin A.D., D.S. Golubev, A. van Otterlo, and G.T. Zimanyi. 1997. Quantum fluctuations and resistivity of thin superconducting wires. *Phys. Rev. Lett.* 78: 1552–1555.

Zaikin A.D., D.S. Golubev, A. van Otterlo, and G.T. Zimanyi. 1998. Quantum fluctuations and dissipation in thin superconducting wires. *Usp. Fiz. Nauk* 168: 244–248 [*Physics Uspekhi* 42: 226–230].

Zgirski M., K.P. Riikonen, V. Tuboltsev, and K. Arutyunov. 2005. Size dependent breakdown of superconductivity in ultranarrow nanowires. *Nano Lett.* 5: 1029–1033.

Zgirski M., K.P. Riikonen, V. Tuboltsev, and K. Arutyunov. 2008. Quantum fluctuations in ultranarrow superconducting nanowires. *Phys. Rev. B* 77: 054508-1–054508-6.

41

Switching Mechanism in Ferromagnetic Nanorings

Wen Zhang
University of Southern California

Stephan Haas
University of Southern California

41.1 Introduction

During the last two decades, tremendous effort has been devoted to micron and submicron size magnetic elements, motivated both by fundamental scientific interest, such as novel magnetic states and switching processes [1–3], and by their potential for technological applications, such as magnetic random access memory (MRAM) [4,5] and magnetic sensors [6,7]. Characterizing the magnetic properties of nanostructures is a challenging task, as their shape significantly influences their physical response. Significant work has been invested in identifying geometries that offer the simplest, fastest, and the most reproducible switching mechanisms. Particular attention has been focused on magnetic structures with high-symmetry geometries, such as circular disks and squares, since spin configurations with high symmetry are expected for these elements, which, in turn, are believed to yield simple and reproducible memory states.

In analogy to the traditional approach to encode information in dipolar-like giant spins, quasi-uniform single domain states have been proposed and intensively studied. However, these typically suffer from three fundamental disadvantages: (1) they are sensitive to edge roughness so that the switching field typically has a broad distribution; (2) because of the long-ranged dipolar interactions between separate elements, high density arrays are hard to achieve; (3) they cannot be made too small due to the superparamagnetism effect. To overcome these problems, the use of the magnetic vortex state in disk geometries (see Figure 41.1a) has been suggested as it is insensitive to edge imperfections and entirely avoids the superparamagnetism effect. Moreover, the zero in-plane stray field opens the possibility of high density storage. However, the vortex state is stable only in disks of fairly

large sizes (diameter over about 100 nm) due to the existence of high energy penalty vortex core regions. What is worse is that the switching mechanisms for disks are complex and difficult to control. In seeking a solution, ring geometries (see Figure 41.1b) have been proposed and studied intensively in the last decade. In addition to all the advantages of the disk geometry, magnetic rings can hold completely stray field free and stable vortex states with diameters as small as 10 nm. Taking 10 nm spacing into account, ring arrays give an ultimate area storage density of about 0.25 Tbits/cm^2 (or 1.6 Tbits/in.^{2}), which is substantially higher than the traditional hard disk area storage density limit. Besides this promising application potential, ring geometries have proven to be a wonderful platform for the investigation of fundamental physical questions concerning domain walls [8].

For the study of magnetic properties of structures having micrometer and nanometer dimensions, micromagnetic simulations prove to be a very powerful tool [9]. To study dynamic properties, one has to resort to numerical integration of the Landau–Lifshitz–Gilbert (LLG) equation [10,11]. For (quasi) static properties, Monte Carlo simulations are also applied extensively [1]. Numerically evaluating the energy of such systems is highly nontrivial because of the long-range character of the dipolar interaction between magnetic moments that contributes to the magnetostatic energy. The complexity of a brute-force calculation is $\mathcal{O}(N^2)$. However, there are several approaches to improve this. The two most important techniques are the fast Fourier transform (FFT), which reduces the complexity to $\mathcal{O}(N \log(N))$, and the fast multipole method (FMM) [12,13], which reduces the complexity to $\mathcal{O}(N)$ but with fairly large coefficients. Based on these methods, several open

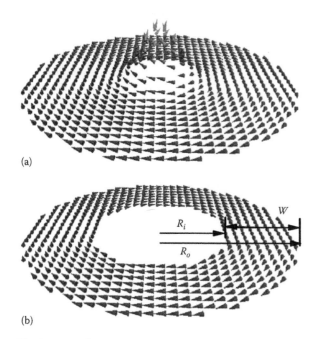

(a)

(b)

FIGURE 41.1 Schematic representation of a magnetic vortex state in a disk (a) and a ring (b). The core region where spin points out of plane is exaggerated and represented by the height. R_o and R_i are outer and inner radii of rings and W is their width.

software packages are available: OOMMF [14], magpar [15], Nmag [16], ψ-mag [17]. Of these, OOMMF is the most mature and popular, and has been extensively used worldwide.

This chapter is an overview of current research on magnetic rings, with a focus on clarifying their quasi-static properties. We will mainly refer to results in the literature and include some of our own recent findings. Considering the overwhelming amount of literature, we apologize for the inevitable omission of some important works. The chapter is organized as follows: in Section 41.2, we cover the basic background needed to study magnetic systems; in Section 41.3, we explain the equilibrium magnetic states (spin configuration) and metastable states in ring geometries; finally the mechanism of switching processes in rings is discussed in detail in Section 41.4, followed by a brief discussion of data storage applications in Section 41.5.

41.2 Background

In the quasi-classical approximation, the Hamiltonian ($\mathcal{H}$) of a magnetic nanoparticle in a magnetic field consists of four terms: exchange interaction, dipolar interaction, crystalline anisotropy, and Zeeman energy. The sum of the first three terms yields the internal energy. If each magnetic moment occupies a site of the underlying crystal lattice, $\mathcal{H}$ is given by

$$\mathcal{H} = -J\sum_{<i,j>}\vec{S}_i \cdot \vec{S}_j + D\sum_{i \cdot j}\frac{\vec{S}_i \cdot \vec{S}_j - 3(\vec{S}_i \cdot \hat{r}_{ij})(\vec{S}_j \cdot \hat{r}_{ij})}{r_{ij}^3} + U_k - \vec{H} \cdot \sum_i \vec{S}_i \tag{41.1}$$

where

  $J > 0$ is the ferromagnetic exchange constant (or the exchange integral, measured in units of energy), which is assumed to be nonzero only for the nearest neighbors
  D is the dipolar coupling parameter
  $\vec{r}_{ij}$ is the displacement vector between sites, i and j

The anisotropy term, U_k, can take various forms [18], among which the most common are the uniaxial anisotropy, $U_k = K\Sigma_i \sin^2\theta_i$, where θ_i is the angle $\vec{S}_i$ makes with the easy axis, and the cubic anisotropy, $U_k = K\sum_i \left[\alpha_i^2\beta_i^2 + \beta_i^2\gamma_i^2 + \alpha_i^2\gamma_i^2\right]$, where $\alpha_i, \beta_i, \gamma_i$ are the direction cosines of $\vec{S}_i$. Note that K is the single site anisotropy energy (not an energy density) and that $\vec{S}$ is a dimensionless unit spin vector with magnetic moments $\vec{\mu} = |\mu|\vec{S}$. Experimentally, the materials used are mostly Co, Fe, permalloy, and supermalloy. For these materials, the ratio, D/Ja^3, falls in the range of 10^{-3} to 10^{-5}, where $a \sim 0.3\,\text{nm}$ is lattice constant. For polycrystalline systems, U_k is usually omitted, since the crystalline anisotropy is very small. For epitaxial systems, however, crystalline anisotropy needs to be considered. Generally, it stabilizes certain directions and enhances the switching field. It also helps to resist the superparamagnetism effect. The Zeeman term above applies for uniform fields. If this is not the case, the term is adjusted to $\sum_i \vec{S}_i \cdot \vec{H}_i$. The main observables in magnetic systems are magnetization and susceptibility.

The competition between the various energy terms in $\mathcal{H}$ results in the interesting complexity of magnetic nanostructures. The exchange interaction tends to align spins in the same direction, whereas the dipolar interaction encourages spins to line up along the boundaries, which is the origin of the shape anisotropy. Thus, spins align in plane in a flat element while they point out of plane if the thickness is much larger than the lateral dimension. The former case attracts more attention as it exhibits rich new phenomena and is closely related to applications. Plenty of competing configurations exist, including flower, leaf, onion, buckle, and vortex states.

The study of micron and submicron magnetic elements has two major branches: quasi-static properties and dynamics. Here, we will focus on the former, in particular on the switching process, a first-order phase transition. To understand the switching process, we first discuss in detail all the possible states in magnetic rings. After that, the magnetization reversal mechanism (the hysteresis) is investigated.

Many magnetic phenomena involve vortices and domain walls. From a topological point of view, these are defects. The topological theory of defects [19–21] can help us to understand the complex creation and annihilation processes during switching. One of the most important principles is the conservation of topological charge. Topological charge is defined as the winding number, ω, i.e., a line integral around the defect center:

$$\omega = \frac{1}{2\pi}\oint \nabla\theta(\vec{r}) \cdot d\hat{r} \tag{41.2}$$

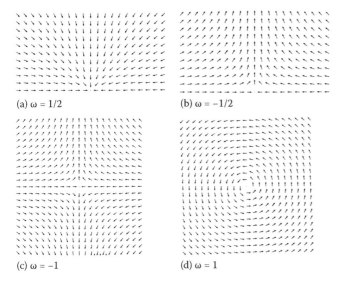

FIGURE 41.2 Elementary topological defects: (a) edge defect $\omega = 1/2$, (b) edge defect $\omega = -1/2$, (c) vortex $\omega = 1$, and (d) antivortex $\omega = -1$.

where θ is the angle between the local magnetic moment and the positive x axis. There are several elementary topological defects in magnetic systems. As is known, a vortex in the bulk has an integer winding number, $\omega = 1$, and an antivortex has $\omega = -1$. Furthermore, Tchernyshyov and Chern [20] identified two edge defects with fractional winding numbers, $\omega = \pm 1/2$. These four types of elementary topological defects are shown in Figure 41.2. All the intricate magnetic textures, including domain walls, are composite objects made of some of the above four elementary defects [20]. These defects have several properties: (1) in analogy to Coulomb interaction, two defects with the same winding number sign repel each other, while they attract each other if they have opposite winding number signs; (2) the vortex is repelled by the boundary since it has an image "charge" with the same sign; (3) edge defects are confined to the edge by an effective confining potential; (4) direct annihilation of two defects with the same sign is prohibited; (5) edge defects can change sign by introducing a bulk defect; (6) in sufficiently narrow rings, $\omega = \pm 1/2$ edge defects are degenerate, i.e., they have the same energy, whereas in thick rings the $\omega = +1/2$ will have a substantially higher energy than the $\omega = -1/2$ so that it can decay into a vortex $\omega = 1$ and an edge defect $\omega = -1/2$.

Concerning experimental fabrication techniques, interested readers can refer to Martin et al. [2] and Kläui et al. [22]. Experimentalists are able to produce two kinds of morphology: polycrystalline and epitaxial. In most cases, magnetic rings are made of polycrystalline permalloy, supermalloy, and Co, whose crystalline anisotropy is negligible. On the other hand, when one wants to study the effects of anisotropy and underlying lattice structure, epitaxial face centered cubic (fcc), face centered tetragonal (fct), and hexagonal close-packed (hcp) structure Co rings can be produced. The imaging techniques can be classified into two groups: (1) intrusive techniques such as magnetic force microscope (MFM): this method affects the magnetic state of the sample because of the external magnetic field of the tip; (2)

nonintrusive techniques such as photoemission electron microscopes (PEEM) and scanning electron microscopy with polarization analysis (SEMPA): the magnetic states remain the same after the image is scanned. The spatial resolution nowadays can be as high as 10 nm. Atomic resolution is also reported by SP-STM [23,24]. To measure the hysteresis curve, the most popular techniques used are the magneto-optic Kerr effect (MOKE) and SQUID. Other techniques include measurement of the Hall resistance [25] and of the spin-wave spectrum [26]. In most cases, the hysteresis is measured in arrays of nanorings, typically larger than 10×10. Thus, the transition field has a range that represents the transition field distribution among these nanoparticles. A few measurements have been reported for single particles where the transition is sharp, but the size of these particles is fairly big.

41.3 Magnetic States

A first step in the investigation of the magnetic properties of certain geometries is to identify the competing stable and metastable spin configurations at remanence. This information can be summarized in a geometric phase diagram.

Several works have been devoted to determining this phase diagram [27–29] for ring structures. Figure 41.3 shows an example with inner radius, R_i, fixed. Spins point out of plane in parallel when the element height, h, is much larger than width, w. The opposite situation is of more interest. When $h/w < 2$, there are two possible states: (1) the vortex state (V) (or the flux close state), characterized by the circulation of spins; (2) the onion state (O) (or the quasi-uniform state), characterized by two head-to-head domain walls. The vortex state is the only single domain state in ring structures. Because of the resemblance to strips, magnetic rings hold a lot of multidomain states, such as onion states and various twisted states discussed in detail below. The onion state has two domains, separated by two head-to-head domain walls.

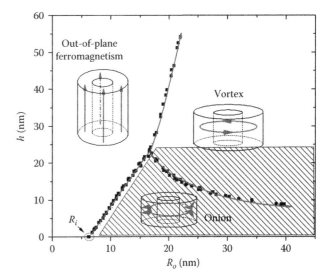

FIGURE 41.3 Schematic geometric magnetic phase diagram for nanorings with $R_i = 6$ nm. Shaded area is the region where the onion state could be (meta)stable, and thus it is of greatest interest.

The two domains are sometimes referred to as two arms. Since the domain walls in ring geometries are well confined, ring magnets serve as a perfect platform to study the motion of domain walls as well as the interaction between them [8,22].

The interesting behavior of magnetic systems stems from the fact that there is a great variety of metastable states. Several states may be stable at remanence depending on the history of how remanence is reached. One can see from Figure 41.3 that the dominant phase for thin films is the vortex state, but the onion state is actually a metastable state in a very large region. If the remanence state is obtained by relaxing from saturation, it is usually an onion state. A remanent state phase diagram is given for Co nanorings [30] where the onion state area is significantly increased. As mentioned above, onion states are characterized by two head-to-head domain walls [31] that have long been studied in magnetic strips. Two kinds of head-to-head domain walls exist: transverse domain walls in thin rings, see Figure 41.4a and vortex walls in wide rings, see Figure 41.4b. A geometric phase diagram (see Figure 41.5) for the two head-to-head domain walls was found by Laufenberg et al. [32].

Note that the phase diagram was obtained by relaxing the system from saturation, so it does not represent the ground state of the system at zero field and shows no vortex state at all. One can see from Figure 41.5b that analytical calculations tend to favor vortex walls while simulations tend to favor transverse walls compared with the experimental result. It results from the fact that there is an energy barrier between vortex walls and transverse walls. Analytical calculations give the lower energy state, while real systems can stay in the local minimum with transverse walls. Simulations are performed at zero temperature, so it is harder to form vortex walls than in the experimental situation, where thermal fluctuation can help the system overcome the barrier.

From a topological point of view, the transverse wall is composed of a $\omega = 1/2$ defect at the outer edge and a $\omega = -1/2$ defect at the inner edge [20]. The vortex wall is composed of two $\omega = -1/2$ defects at the outer and inner edge, respectively, together with a $\omega = +1$ vortex defect at the center of the rim [21].

Recently, a new category of metastable states has been discovered [33] in relatively small rings made of both Co and permalloy with $R_o \sim 180–520\,nm$, $w \sim 30–200\,nm$, $h \sim 10\,nm$: the twisted states (T) (or saddle state [34]). These states are characterized by 360° domain walls [35,36] (see Figure 41.6), which have previously been reported in narrow thin film strips [37,38]. They are shown to be stable within

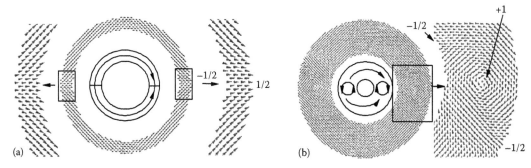

FIGURE 41.4 Head-to-head domain wall: (a) onion state with transverse domain wall and (b) onion state with vortex domain wall. The numbers in the figure indicate the type of the topological defect.

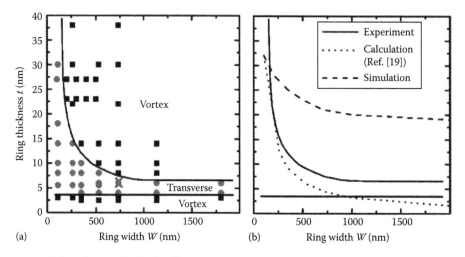

FIGURE 41.5 (a) Experimental phase diagram for head-to-head domain walls in NiFe rings at room temperature. Squares indicate vortex walls and circles transverse walls. The phase boundaries are shown as solid lines. (b) A comparison of the upper experimental phase boundary (solid line) with results from calculations (dotted line) and micromagnetic simulations (dashed line). Close to the phase boundaries, both wall types can be observed in nominally identical samples due to slight geometrical variations. The thermally activated wall transitions shown were observed for the ring geometry marked with a cross ($W = 730\,nm$, $t = 7\,nm$). (From Laufenberg, M. et al., *Appl. Phys. Lett.*, 88, 052507, 2006. With permission.)

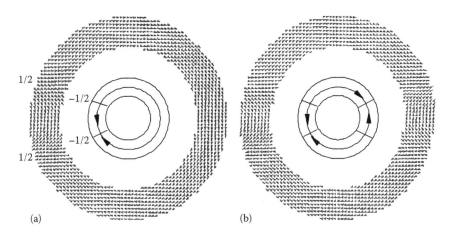

FIGURE 41.6 Twisted states: (a) with single 360° domain wall and (b) with two 360° domain walls. The numbers in the figure indicate the type of defects.

a field range of several hundred Oe [33]. Furthermore, they have low stray fields and are easy to control by current, which makes them attractive for data storage applications. There can be more than one 360° domain wall in rings, resulting in a multitwisted state. The 360° domain walls can be viewed as two transverse domain walls (Figure 41.6b). The attraction between the two transverse walls occurs because they have opposite senses of rotation. This tendency is balanced by the exchange energy in the region between the walls. The existence of this domain wall can also be understood fairly easily by topological arguments, referring to the properties (4) and (6) of topological defects in Section 41.2. When the width is small, the two transverse domain walls have the same defect on the same side, and they are stable. So these edge defects cannot annihilate, resulting in a 360° domain wall. When the width is larger, however, one of the $\omega = +1/2$ defects becomes unstable and transforms into a $\omega = -1/2$ defect by introducing a vortex into the center of the rim. Then the two transverse domain walls are able to annihilate each other, resulting in a vortex state. Therefore, only a portion of the region of stable transverse wall in the phase diagram (Figure 41.5) can possibly hold twisted states. Twisted states in large rings have almost zero remanence, so magnetization is incapable of distinguishing between vortex and twisted states and other quantities must be used. Toroidal moment [39] and winding number are supplementary order parameter candidates.

In the presence of an external magnetic field, several other states exist: wave states and shifted vortex core states. They usually exist in very thick rings, which is of less interest for practical reasons. Regarding the effect of crystalline anisotropy, it stabilizes quasi-uniform states (like onion states) and imposes some additional domain structures.

41.4 Switching Processes

For data storage applications, understanding switching processes is crucial. The main effort is directed toward identifying simple reproducible switching processes and reliable sensitive detection techniques. Meanwhile, in terms of fundamental physics,

the understanding of switching processes is just as important. As picosecond and nanometer time and spatial resolution detection techniques are still limited, the microscopic details of switching processes and other dynamic transitions are currently mainly investigated by micromagnetic simulations. One issue that is hard to quantify is the distribution of the switching field. Computationally, the transition in the hysteresis curve is very sharp, while the experimental curves have a broad transition region caused by edge roughness and defects.

The traditional way of studying switching processes is by applying a uniform magnetic field with changing magnitude. Motivated by the vortex configuration, circular fields are also studied in order to switch the two vortex states with opposite circulation. Here, we mainly discuss the uniform field switching case and say a few words about the circular field afterward.

41.4.1 Uniform Field

With a uniform field, three types of typical hysteresis curves are identified:

41.4.1.1 One-Step or Single Switching

This is a direct onion-to-onion state switching (O-O). Figure 41.7b) shows a typical hysteresis curve. In this process, an onion state is reversed to the opposite onion state directly. This process is usually only observed experimentally in small rings, especially in the thin film limit (small *h*). Theoretically speaking, it should happen in a much thicker system if the ring is perfectly symmetric, as it is observed in simulations. For real systems, however, there always exists some sort of asymmetry caused either by defects or by the environment. When asymmetry exists, the system tends to fall into its true ground vortex state, which will result in a double switching process.

It is believed that the switching mechanism is simply a coherent rotation, where two domain walls move in the same rotational direction. This can easily be observed computationally if one intentionally introduces some sort of asymmetry into the system.

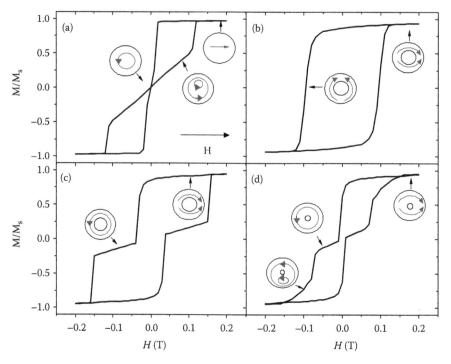

FIGURE 41.7 Schematic switching process and hysteresis: (a) hysteresis of magnetic disks, (b) one-step switching of rings, (c) double switching of rings, and (d) triple switching of rings.

Yoo et al. [40] offered experimental phase diagrams for Co rings with constant width and R_o (equivalent to $D/2$ in Figure 41.8) and varying height of 2–34 nm. They concluded that the main geometric factor is the film thickness. When $h < 6$ nm, the above one-step switching happens, while otherwise it is two-step switching. For small radii, this is not necessarily true, since Yoo et al. only study $R_o > 500$ nm. Rings with large radii resemble straight strips.

With shrinking radius, however, the story is different. Since magnetic processes including nucleation and annihilation depend strongly on the curvature, small rings can behave quite differently from their bigger counterparts.

41.4.1.2 Two-Step or Double Switching

The dominant switching process for magnetic rings is O-V-O switching or O-T-O switching (see Figure 41.7c), since the inevitable intrinsic asymmetry stimulates this process. A large regime in the geometric phase diagram falls into this category. In most cases, the remanence state starting from positive saturation is an onion state, and both transitions occur at negative field. When the element is very thick, the remanence state can be a vortex state, i.e., the first transition field is positive [25,30]. This behavior is found for elements near the boundary between double and triple switching in the switching phase diagrams. The mechanisms of the two steps are discussed separately below.

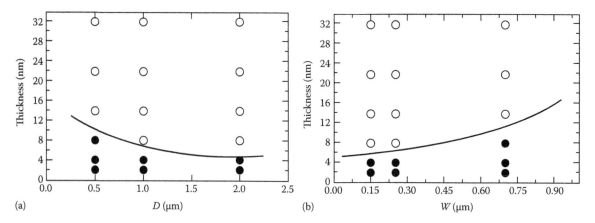

FIGURE 41.8 Phase diagrams of two-step switching (open circles) and single switching (full circles) as a function of the ring geometrical parameters: (a) for a ring width of 0.25 mm and (b) for a ring diameter of 2 mm. The solid lines define the boundary between the two different switching regimes. (From Yoo, Y.G. et al., *Appl. Phys. Lett.*, 82, 2470, 2003. With permission.)

For O-V or O-T, two possibilities exist, depending on whether it is initiated by domain wall motion or nucleation.

1. Nucleation and buckling: a strong buckling of the magnetization happens first in one arm of the onion state, where the magnetic moments are antiparallel to the external field, followed by a nucleation of a vortex passing through the arm [25,26]. This process is not typical in two-step transitions. Rather, it happens in some shifted inner circle rings or relatively wide rings. The transition field can be positive, resulting in a vortex state at remanence. It is similar to the triple switching-process, so it can also be regarded as a transient process.

2. Nucleation-free and domain wall motion: one wall is pinned more than the other due to some sort of asymmetry, so that the other wall moves toward it. When the two walls get close to each other, two things could happen: (a) The annihilation process—this is possible only if the ring is sufficiently wide (say larger than 10 nm). (b) 360° domain walls that have been found to exist for narrow and small rings recently [33]. The states formed are twisted states that have been discussed in the previous section. It is interesting and enlightening to understand this process from the topological point of view. As is shown in Section 41.3, each of the transverse domain walls consists of two half-vortices: one $\omega = -1/2$ defect on the outer edge and one $\omega = +1/2$ defect on the inner edge. The vortex state cannot form in arbitrarily thin rings since the half-vortices with the same winding number cannot annihilate directly and they cannot move to bulk either. Only when the vortex wall is also an energy minimum, one of the $\omega = +1/2$ edge defects can emit a $\omega = +1$ vortex into the bulk and transform into a $\omega = -1/2$ defect. The emitted vortex will travel to the other side and turn the $\omega = -1/2$ defect into a $\omega = +1/2$ defect [20,34]. Then the edge defects can annihilate. Overall, the choice of the above processes depends strongly on the ring width but less so on R_o and h (see Figure 41.8). These two processes happen for relatively small widths.

The V-O transition is sometimes called a first magnetization curve. In the half of the ring with magnetization antiparallel to magnetic field, $\vec{H}$, a vortex domain wall nucleates at the inner side of the rim and passes through it perpendicular to $\vec{H}$. In the meantime, two transverse head-to-head domain-wall-like structures appear next to the vortex and quickly propagate in opposite directions, forming the final onion state. This process is mainly shape dependent and has little to do with anisotropy because it depends on how easy it is to form a vortex wall on the edge, which can be assisted by the edge imperfection (or roughness). As expected, this process depends also on the local curvature of the ring. Since the local curvature and edge roughness affect small rings more, the distribution of the switching field will be larger in these systems. However, the magnitude of transition field, H_c, is insensitive to the radius. It increases with height, h [41], since thicker elements favor vortex structures and decreases

with width as the vortex state is more stable in thin rings. The nucleation requires a large twisting of the spins, which is harder to achieve in thin rings. Simulations tend to give higher H_c in the absence of defects, but defects are known to reduce H_c. This discrepancy is more severe for thin rings.

The field distribution of O-V is typically larger than in the case of V-O and can mainly be attributed to the variation of intrinsic defects among different rings. As temperature tends to wipe out the effects due to defects, i.e., thermal fluctuation makes it easier to overcome the local energy barrier induced by defects, the distribution becomes larger when the temperature goes down. This is quite surprising at first glance. However, this conclusion can be used to determine whether defects are responsible for the field distribution. If the distribution depends strongly on temperature, then defects are important. Generally speaking, transitions involving nucleation processes are less prone to defects and thermal fluctuation than processes involving domain wall or vortex core motion [42]. Since defects always exist, simulations at 0 K for perfect rings mimic more, in some sense, the experimental situation at high temperature.

41.4.1.3 Triple Switching

This process involves three intermediate steps: O-V-VC-O [25,42,43]. It is similar to O-V-O except that the vortex state does not deform into an onion state abruptly, but nucleates a vortex core in one arm, and the core then moves slowly to the outer rim. To some extent, it is similar to the switching process of magnetic disks, where the core in the vortex state is shifted by a magnetic field and finally exits at the boundary (see Figure 41.7a). To see this, Steiner and Nitta [25] have shown a sequence of hysteresis curves with various inner radii, R_i, from the disk limit ($R_i = 0$). The field distribution of the V-VC transition is affected little by temperature, so it is affected little by defects. On the other hand, the VC-O is a process affected strongly by temperature, so defects play an important role here as cores can be pinned by defects. The triple switching process only exists for thick and very wide rings. Since these rings lack the advantages of rings mentioned before, this switching process is less interesting and less studied.

Regarding the effects of anisotropy, it generally suppresses the field distribution and increases the switching field. One curious observation is that ring structures make the hard axis of the cubic anisotropy into the global easy axis [22].

41.4.2 Circular Field

Circular fields are interesting because they can be naturally generated by a perpendicular current through the disk center. This design is especially suitable for rings as there is a hole in the center to deposit a nanodot to conduct current. As a result, current-induced switching can be achieved, which is essential for data storage applications. At first, a circular field was proposed to switch the vortex state into opposite circulation [4], but it requires a fairly large current density and involves some fairly complicated transient states. Recently, after the twisted

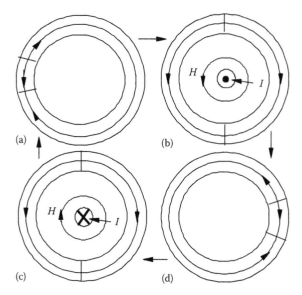

FIGURE 41.9 Schematic diagram of twisted state reversal by circular field.

states were discovered, a new scheme was proposed [35,36]. As is shown in Figure 41.9, switching is adjusted between two twisted states with the 360° domain wall located in the left and right rim (or any two positions along a diameter). This switching requires low current and has a very clean transition process that only involves a domain wall movement. In addition, it is very fast.

41.5 Applications

Nanomagnets feature in many applications. One of the most important application is in data storage devices. Three schemes have been proposed [44] for nanorings: (1) two onion states, similar to the traditional oblong memory element, except that the switching current is significantly less; however, the stray field is an obstacle for high density storage; (2) two vortex states with different circulation, studied most intensively but requiring a relatively high current density [4]; (3) two twisted states, most promising because of the low stray field and low current [36].

For the first choice, Kläui et al. [22] have performed a dynamics switching study. They used magnetic pulses to switch the magnetization. They found that switching is only possible when $H * \Delta t \approx 5 * 10^{-12} T * s$, keeping $H > 20\,mT$ and $\Delta t > 50\,ps$. The switching time is $T = 0.4\,ns$, so the possible switching rate is about $1.25\,GHz\ (=1/(2T))$.

Most attention has been focused on the second choice, for this one utilizes all the benefits of nanorings. Manmade asymmetry offers a way to control the vortex circulation. Several techniques are available: shifted inner circle [45,46], notches [40], and elongated shapes like ellipses [47,48]. Some works [45,49] claim that even when there exists asymmetry, the switching process is still a stochastic process due to the thermal fluctuation. Some rings in an array may follow O-O switching while other rings may follow O-V-O. Even for the same ring the circulation may not be deterministic, so the strength of the asymmetry needed is still an important open question.

Recently, increasing attention has been given to the third possibility [35,36,44], as it is suggested that the twisted state can be switched very easily by a small circular current and can achieve high speed. Experimental evidence is needed to confirm the idea.

Another obstacle to application is that ring arrays always show a wide distribution of switching fields. To make the application realistic, the width of the distribution has to be reduced. The distribution is attributed to both the interior and exterior defects. Though the influence is qualitatively known, quantitative analysis is still missing. Even though one would like to get rid of defects as much as possible in most cases, they can also be used to engineer the switching behavior [47,50].

41.6 Conclusions

We have reviewed the current results in the study of ferromagnetic rings, with emphasis on switching processes. Ring structures feature many advantages including being stable and insensitive to edge imperfection and completely stray field free. Ring arrays give an ultimate area storage density of about $0.25\,Tbits/cm^2$. We covered all possible states identified numerically and experimentally until now, among which vortex, onion, and twisted states are discussed in detail, since they are suggested to be candidates for data storage. There are three types of switching processes under uniform field, depending on the geometric parameters of the ring. The double switching, in particular, enjoys tremendous popularity. By introducing engineered asymmetry, the circulation direction of the vortex state can be easily controlled by a uniform field. The mechanisms of these switching processes were explained carefully in detail and topological argument was shown to be enlightening. On the other hand, circular field switching is easy to generate and is suitable to switch the twisted state. Finally, applications and limitations were briefly discussed.

Acknowledgments

We would like to thank Noah Jacobson and Yaqi Tao for useful discussions. Computing facilities were generously provided by the University of Southern California high-performance supercomputing center. We also acknowledge financial support by the Department of Energy under grant DE-FG02–05ER46240.

References

1. K. De'Bell, A. B. Maclsaac, and J. P. Whitehead, *Rev. Mod. Phys.* 72(1), 225 (2000).
2. J. I. Martin, J. Nogues, J. L. V. K. Liu, and I. K. Schuller, *J. Magn. Magn. Mater.* 256, 449 (2003).
3. R. P. Cowburn, *J. Phys. D: Appl. Phys.* 33, R1 (2001).
4. J. G. Zhu, Y. Zheng, and G. A. Prinz, *J. Appl. Phys.* (2000).
5. J. Akerman, *Science* 308, 508 (2005).
6. G. A. Prinz, *Science* 282, 1660 (1998).

7. M. M. Miller, G. A. Prinz, S. F. Cheng, and S. Bounnak, *Appl. Phys. Lett.* 81, 2211 (2002).

8. M. Kläui, *J. Phys.: Condens. Matter* 20, 313001 (2008).

9. J. Filer and T. Schrefl, *J. Phys. D: Appl. Phys.* 33, R135 (2000).

10. E. M. Lifshiftz and L. P. Pitaevskii, *Statistical Physics*, Part 2 Reed Educational and Professional Publishing Ltd., Oxford, U.K., 1980, pp. 286–288.

11. T. L. Gilbert, *Phys. Rev.* 100, 1243 (1955).

12. L. Greengard and V. Rokhlin, *J. Comput. Phys.* 73, 325 (1987).

13. W. Zhang and S. Haas, *arxiv:0810.0233* (2008).

14. M. J. Donahue and D. G. Porter, *OOMMF User's Guide* (NIST, Gaithersburg, MD) (1999).

15. W. Scholz, http://magnet.atp.tuwien.ac.at/scholz/magpar/ (2003).

16. T. Fischbacher, M. Franchin, G. Bordignon, and H. Fangohr, *IEEE Trans. Magn.* 43, 2896, (2007). http://nmag.soton.ac.uk/

17. G. Brown, T. C. Schulthess, D. M. Apalkov, and P. B. Visscher, *IEEE Trans. Magn.* 40, 2146 (2004).

18. C. Kittel, *Introduction to Solid State Physics*, 7th edn. (John Wiley & Sons, New York, 1996), pp. 565–566.

19. N. D. Mermin, *Rev. Mod. Phys.* 51, 591 (1979).

20. O. Tchernyshyov and G. W. Chern, *Phys. Rev. Lett.* 95, 197204 (2005).

21. H. Youk, G. W. Chern, K. Merit, B. Oppenheimer, and O. Tchernyshyov, *J. Appl. Phys.* 99, 08B101 (2005).

22. M. Kläui, C. A. F. Vaz, L. Lopez-Diaz, and J. A. C. Bland, *J. Phys.: Condens. Matter* 15, R985 (2003).

23. D. Wortmann, S. Heinze, and P. Kurz. et al., *Phys. Rev. Lett.* 86, 4132 (2001).

24. C. L. Gao, W. Wulfhekel, and J. Kirschner, *Phys. Rev. Lett.* 101, 267205 (2008).

25. M. Steiner and J. Nitta, *Appl. Phys. Lett.* 84, 939 (2004).

26. F. Giesen, J. Podbielski, B. Botters, and D. Grundler, *Phys. Rev. B* 75, 184428 (2007).

27. M. Beleggia, J. W. Lau, M. A. Schofield, Y. Zhu, S. Tandon, and M. DeGraef, *J. Magn. Magn. Mater.* 301, 131 (2006).

28. P. Landeros, J. Escrig, D. Altbir, M. Bahiana, and J. d'Albuquerque e Castro, *J. Appl. Phys.* 100, 044311 (2006).

29. W. Zhang, R. Singh, N. Bray-Ali, and S. Haas, *Phys. Rev. B* 77, 144428 (2008).

30. S. P. Li, D. Peyrade, M. Natali, A. Lebib, Y. Chen, U. Ebels, L. D. Buda, and K. Ounadjela, *Phys. Rev. Lett.* 86, 1102 (2001).

31. R. D. McMichael and M. J. Donahue, *IEEE Trans. Magn.* 33, 4167 (1997).

32. M. Laufenberg, D. Backes, W. Buhrer, D. Bedau, M. Kläui et al., *Appl. Phys. Lett.* 88, 052507 (2006).

33. F. J. Castano, C. A. Ross, C. Frandsen, A. Eilez, D. Gil, H. I. Smith, M. Redjdal, and F. B. Humphrey, *Phys. Rev. B* 67, 184425 (2003).

34. G. D. Chaves-O'Flynn, A. D. Kent, and D. L. Stein, *arxiv:0811.4440v1* (2008).

35. C. B. Muratov and V. V. Osipov, *J. Appl. Phys.* 104, 053908 (2008).

36. C. B. Muratov and V. V. Osipov, *arxiv:0811.4663v1* (2008).

37. Y. Zheng and J. G. Zhu, *IEEE Trans. Magn.* 33, 3286 (1997).

38. X. Portier and A. Petford-Long, *Appl. Phys. Lett.* (2000).

39. S. Prosandeev and L. Bellaiche, *Phys. Rev. B* 77, 060101(R) (2008).

40. Y. G. Yoo, M. Kläui, C. A. F. Vaz, L. J. Heyderman, and J. A. C. Bland, *Appl. Phys. Lett.* 82, 2470 (2003).

41. M. Kläui, L. Lopez-Diaz, J. Rothman, C. A. F. Vaz, J. A. C. Bland, and Z. Cui, *J. Magn. Magn. Mater.* 240, 7 (2002).

42. M. Kläui, C. A. F. Vaz, J. A. C. Bland et al., *Appl. Phys. Lett.* 84, 951 (2004).

43. M. Steiner, G. Meier, U. Merkt, and J. Nitta, *Phys. E* (2004).

44. C. L. Chien, F. Q. Zhu, and J. G. Zhu, *Phys. Today* (2007).

45. F. Q. Zhu, G. W. Chern, O. Tchernyshyov, X. C. Zhu, J. G. Zhu, and C. L. Chien, *Phys. Rev. Lett.* 96, 27205 (2006).

46. S. Prosandeev, I. Ponomareva, I. Kornev, and L. Bellaiche, *Phys. Rev. Lett.* 100, 047201 (2008).

47. N. Agarwal, D. J. Smith, and M. R. McCartney, *J. Appl. Phys.* 102, 023911 (2007).

48. N. Singh, S. Goolaup, W. Tan, A. O. Adeyeye, and N. Balasubramaniam, *Phys. Rev. B* 75, 104407 (2007).

49. T. J. Hayward, T. A. Moore, D. H. Y. Tse, J. A. C. Bland, F. J. Castano, and C. A. Ross, *Phys. Rev. B* 72, 184430 (2005).

50. X. S. Gao, A. O. Adeyeye, and C. A. Ross, *J. Appl. Phys.* 103, 063906 (2008).

Quantum Dot Nanorings

Ioan Bâldea
Universität Heidelberg

Lorenz S. Cederbaum
Universität Heidelberg

42.1 Introduction

Nowadays, a variety of methods (lithography, epitaxial growth, colloidal chemistry) allow the fabrication of very small islands of metals or semiconductors, wherein the electron motion is confined within nanometer ranges. These methods are usually called quantum dots (QDs), and were extensively investigated in the last few years (Kastner 1993, Kouwenhoven and Marcus 1998).

In the lithographic method (Kastner 1993), the confining potential along the growth direction is provided by a quantum well, whereas the lateral confinement is caused by an electrostatically induced potential barrier. Using *in situ* growth techniques (molecular beam epitaxy, metal–organic chemical–vapor deposition), an epitaxial layer with a small band gap is created by deposition on a substrate with a high band gap. This layer spontaneously breaks, which yields the formation of self-assembled QDs with a typical truncated pyramidal shape (Petroff et al. 2001). The confinement in all three directions is ensured by the high band gap of the surrounding material. Lithographically defined QDs have lateral sizes ~100 nm. The smallest epitaxially self-assembled QDs have heights ~2–3 nm and basal dimensions ~20 nm. Smaller QDs were prepared with routine techniques of colloidal chemistry; grains of CdSe with diameters $2R \simeq 5.5$ nm (Klein et al. 1997) and of silver with $2R \simeq 2.6$ nm (Collier et al. 1997) were fabricated in this way.

Because the electron motion is confined within nanometer ranges, the energy spectrum is quantized. The energy separation, Δ, between the lowest unoccupied and the highest occupied single-electron levels represents an important parameter for a QD. Another important parameter is the charging energy, U, that is, the energy that one has to pay to bring an electron on a dot where another electron is already present. Importantly, the QD properties can be widely controlled in all the methods of nanofabrication. In a spherical QD, Δ scales as $1/R^2$, and $U \simeq e^2/(4\pi\varepsilon_0\kappa_r R)$ (e is the elementary charge and κ_r is the dielectric constant).

Because of their discrete single-electron spectrum, QDs are often called "artificial atoms." However, the analogy between atoms and QDs should not be pushed too far. Besides the influence that surface or interface states may obviously have, the QD atom analogy is limited due to the role played by many-body (or correlation) effects. As is well known, electrons in atoms can be reasonably described within the single-particle picture, which assumes that an electron moves in a self-consistent field (SCF) accounting for the averaged influence of the other electrons. In contrast, in QDs, the electron–electron Coulomb interaction cannot be treated within the SCF. The Coulomb blockade peaks and Kondo plateaus observed in electric transport experiments through a QD (Goldhaber-Gordon et al. 1998a,b, van der Wiel et al. 2000) clearly demonstrate the important role of electron correlations.

The parameters Δ and U play a key role in single-electron nanodevices. The operation of such nanodevices requires sufficiently low temperatures. Effects of energy and charge quantization can be observed only for $\Delta > U > k_B T$ (k_B is Boltzmann's constant). For QDs fabricated by epitaxial growth, with typical charging energies $U \sim 1$ meV, experiments have to be carried out at very low temperatures. Ag-QDs with $2R = 2.6$ nm prepared by Heath's group (Collier et al. 1997) are characterized by a considerably larger value of $U = 0.34$ eV (Medeiros-Ribeiro et al. 1999). In general, metallic QDs have charging energies larger than semiconducting QDs of comparable sizes: the dielectric constant of the surrounding material in the case of semiconducting QDs ($\kappa_r \sim 10$–14) is larger than that in the case of metallic QDs ($\kappa_r \sim 2$–3) (Remacle 2001). The fact that the charging energy of Ag-QDs is larger by more than one order of magnitude than the thermal energy at room temperature (≈ 26 meV) suggests that single-electron nanodevices based on metallic QDs can operate at ambient temperatures (Andres et al. 1996, Markovich et al. 1997).

When two or more QDs become close enough to form an "artificial" molecule (Kouwenhoven 1995, Oosterkamp et al. 1998), the wave functions of the outermost (π, or "valence") electrons of adjacent QDs can significantly overlap. The result is that a small fraction of electrons can coherently tunnel from one QD to another and thus become delocalized over the whole nanostructure. Artificial nanostructures of assembled QDs are even more interesting than that of isolated QDs. Not only the properties of the building blocks (QDs) but also their coupling can be sensitively modified in a controlled manner by varying the interdot spacing (Collier et al. 1997, 1998, Markovich et al. 1999) or gate potentials (Goldhaber-Gordon et al. 1998a). Even the number of conducting electrons can be tuned by changing the dot energy (Goldhaber-Gordon et al. 1998a, Kawaharazuka et al. 2000).

The ability of preparing very narrow size distributions of QDs opened the way for fabricating regular extended one-, two-, and three-dimensional nanoarrays (Collier et al. 1997, 1998, Shiang et al. 1998, Markovich et al. 1999, Sun et al. 2000, Sun and Xia 2003, Liu and Levicky 2004a,b, Maksimenko and Slepyan 2004, Santhanam and Andres 2004, Behrens et al. 2006). What is impossible for the chemistry of the periodic table becomes conceivable for the QD chemistry: to design an "artificial" atomic solid ("architectonic QD solid," Markovich et al. 1999) with desired electronic, optical, and functional properties by tuning the electronegativity of the atoms, the strength of the covalent bond, and selecting the symmetry of the crystal structure.

42.2 Model of Quantum Dot Nanorings

In classical electrostatics, the energy ε of a collection of N charges Q_i can be expressed as follows (Landau and Lifshitz 1984):

$$\varepsilon = \frac{1}{2}\sum_{i=1}^{N} S_{ii}Q_i^2 + \sum_{i=1}^{N}\sum_{j>i}^{N} S_{ij}Q_iQ_j = \frac{1}{2}\sum_{i=1}^{N} U_i n_i^2 + \sum_{i=1}^{N}\sum_{j>i}^{N} V_{ij}n_in_j,$$

(42.1)

where S_{ii} and S_{ij} ($i \neq j$) are the self-elastances and mutual elastances, respectively; n_i is the number of excess electrons on

the ith QD; $U_i \equiv e^2 S_{ii}$; and $V_{i,j} \equiv e^2 S_{ij}$. (Let us remember that the elastance matrix $\mathbf{S} \equiv (S_{ij})$ represents the inverse of the more familiar capacity matrix $\mathbf{C} \equiv (C_{ij})$, $\mathbf{S} = \mathbf{C}^{-1}$.) The terms entering the first sum on the right-hand side of Equation 42.1, which represent the energies of the individual objects, correspond to the charging energies of the isolated QDs mentioned earlier. In addition, the QDs interact electrostatically among themselves, as expressed by the second sum on the right-hand side of Equation 42.1. In a nanoarray, the QDs are coupled not only electrostatically but also by electron tunneling. The interdot tunneling is characterized by the hopping (or resonance) integral t_{ij}. For small QDs, the energy separation, Δ, is large, and one can consider only one single-electron ("valence") level per dot. By further assuming that the interdot coupling is important only for adjacent QDs, one arrives at the following expression of the second quantized model Hamiltonian of a QD nanoring:

$$H = -t\sum_{l=1}^{N}\sum_{\sigma=\uparrow,\downarrow}(a_{l,\sigma}^{\dagger}a_{l+1,\sigma} + a_{l+1,\sigma}^{\dagger}a_{l,\sigma}) + \sum_{l=1}^{N}(\varepsilon_F\hat{n}_l + U\hat{n}_{l,\uparrow}\hat{n}_{l,\downarrow} + V\hat{n}_l\hat{n}_{l+1}),$$

(42.2)

where a ($a^{\dagger}$) denotes the creation (annihilation) operators for electrons, $\hat{n}_{l,\sigma} \equiv a_{l,\sigma}^{\dagger}a_{l,\sigma}$, and $\hat{n}_l \equiv \hat{n}_{l,\uparrow} + \hat{n}_{l,\downarrow}$. Note that, up to a constant ($\sum_l \hat{n}_l = N_e$ is the total number of electrons), the U term in Equation 42.2 represents the counterpart of the classical expression appearing in Equation 42.1, because of the operator identity $\hat{n}_l^2 = \hat{n}_l + 2\hat{n}_{l,\uparrow}\hat{n}_{l,\downarrow}$. In Equation 42.2, we have assumed that the QDs are identical (l-independent parameters ε_F, t, U, and V). This assumption is justified for reproducible QDs, like those prepared by Heath et al. (Collier et al. 1997, Heath et al. 1997, Markovich et al. 1998, Shiang et al. 1998, Medeiros-Ribeiro et al. 1999), characterized by very narrow size distributions (~2%–4%). In solid-state physics, the model of Equation 42.2 is known as the *extended Hubbard model*, which was introduced by Hubbard (1963) and later generalized to include nearest-neighbor Coulomb interaction (Beni and Pincus 1974). In molecular physics, it is called the Pariser–Parr–Pople (PPP) model (Parr 1963).

The *restricted Hubbard model* (Equation 42.2 with $V = 0$) was first used to study assembled Ag-QDs by Remacle and Levine (Remacle 2000, Remacle and Levine 2000a,b, Remacle 2001, Remacle and Levine 2001). For the on-site repulsion, they employed the value $U \simeq 0.34$ eV, directly measured for Ag-QDs of diameter $2R \simeq 2.6$ nm by scanning tunnel microscopy (STM) (Medeiros-Ribeiro et al. 1999). This value is in good agreement with the theoretical estimate $U = e^2/C$, using the capacity $C = 4\pi\varepsilon_0\kappa_r R$ for a spherical dot. The latter is justified, because fabricated Ag-QDs appear spherical even when imaged in high-resolution transmission electron microscopy (Heath et al. 1997). The hopping integral $t = t(d)$ was taken as a fit parameter [$d \equiv D/(2R)$]. It was determined by fitting (Remacle et al. 1998) the experimental data on second harmonic generation as a function of lattice compression (Collier et al. 1997). This yields $t(d) \approx t_0\exp[-5.5(d-1.2)]$, with $t_0 = 0.5$ eV. Further support

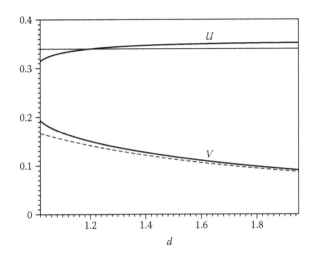

FIGURE 42.1 Hubbard parameters U and V (in eV) plotted versus interdot spacing d computed numerically for two identical spherical Ag-QDs (thick lines). They are reasonably approximated by a d-independent U and a naive Coulomb dependence $V = U/(2d)$ (thin lines).

for the value $t_0 = 0.5$ eV was found by analyzing the frequency-dependent dielectric response (Remacle and Levine 2000a). The exponential dependence on the interdot spacing d is similar to polyenes (Salem 1966).

In our studies on Ag-QD rings (Bâldea and Cederbaum 2002, Bâldea et al. 2004a,b, Bâldea and Cederbaum 2007, 2008a,b, Bâldea et al. 2009), we employed the *extended* Hubbard model of Equation 42.2 with $t(d)$ given by the aforementioned formula. To get the Hubbard parameters V and U, we computed numerically the mutual- and self-elastances for two identical spheres within classical electrostatics (Bâldea and Cederbaum 2002, and references therein). Figure 42.1 shows that these results are reasonably approximated by a constant U and $V \propto 1/d$. As visible in Figure 42.1, although smaller than U, the interdot repulsion V is still comparable with U and should be considered.

Unless otherwise specified, we present in the following text results obtained by exact (Lanczos) numerical diagonalization (see, e.g., Köppel et al. 1984) for QD nanorings described by the extended Hubbard model of Equation 42.2 with parameter values obtained as described earlier.

42.3 Tunable Electron Correlations

In the absence of the Coulomb interaction ($U = V = 0$), the single-electron energies range from $\varepsilon_F - 2t$ to $\varepsilon_F + 2t$. Intuitively, one expects that the single-particle picture is valid (weak correlation regime) when the bandwidth ($4t$) significantly exceeds the Hubbard terms $4t \gg U(>V)$. In the opposite situation ($4t \ll U$), this picture is expected to break down (strong correlation regime).

In Ag-QD nanoarrays, the interdot tunneling t can be precisely controlled by varying mechanically the interdot spacing d.

Using a Langmuir trough technique, monolayers of Ag-QDs can be reversibly compressed from $d \simeq 1.85$ to $d \simeq 1.1$ (Collier et al. 1997). Due to its sensitive (exponential) d-dependence, t can thus change by about two orders of magnitudes. Noteworthy, within the aforementioned range of d, the QD nanoarrays continuously change from a situation characterized by $4t \ll U$ to a situation where $4t \gg U$. This is most important, since this enables to continuously drive the nanosystem from a weak correlation regime to a strong correlation regime.

The crossover between weak and strong correlation regimes in the ground state G is illustrated in Figure 42.2. In the absence of correlations, the state of a many-electron system can be characterized by the single-particle states ("molecular" orbitals, MOs). For nanorings, the MO- (or Bloch-) operators $c_{k,\sigma}$ are defined in terms of the "atomic" orbitals operators $a_{l,\sigma}$ by $c_{p,\sigma} = 1/\sqrt{N} \sum_l a_{l,\sigma} \exp(-2\pi pli/N)$. The MO scheme of a 10-dot nanoring, characterized by the point group $D_{10,h}$, is presented in the inset of Figure 42.2. To facilitate understanding, it is useful to note that in this case the 10 MOs ordered by increasing energy are a nondegenerate a_{1g} ($p = 0$), two degenerate e_{1u} ($p = \pm 1$), two degenerate e_{2g} ($p = \pm 2$), two degenerate e_{3u} ($p = \pm 3$), two degenerate e_{4g} ($p = \pm 4$), and a nondegenerate b_{1u} ($p = 5$). For 10 electrons (half-filling case), the lower half of MOs is occupied, while the upper half is empty. This is indeed the case for almost touching QDs ($d \gtrsim 1$), as visible in Figure 42.2, where the calculated MO populations $n_p \equiv \sum_\sigma \langle G | c_{p,\sigma}^\dagger c_{p,\sigma} | G \rangle$ are plotted. With increasing d, the MO picture with fully occupied and empty MOs gradually deteriorates and eventually completely breaks down. At large d, the distinction between occupied and unoccupied MOs completely disappears: all MOs become democratically occupied by 0.5 electrons.

To conclude, these results demonstrate that by mechanical compression Ag-QD nanorings can be driven from weak to strong correlations (Bâldea and Cederbaum 2002, 2007, 2008a,b).

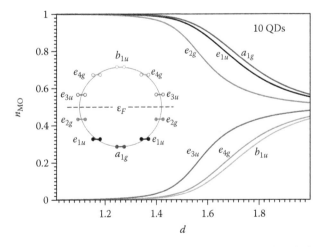

FIGURE 42.2 The MO scheme (symmetry group $D_{10,h}$) and the d-dependence of the MO populations n_{MO} for a half-filled 10-QD nanoring.

42.4 Optical Absorption

In this section, we shall discuss the optical absorption in the ground state. Within the linear response theory (Maldague 1977), the absorption coefficient of a nanoring shined by a light linearly polarized along j-direction ($j = x, y$) in the ring plane is expressed as a sum of contributions of various excited states $|\Psi_\lambda\rangle$:

$$\alpha_j(\omega) = \rho^{-2}\omega \sum_\gamma \sum_{\lambda \in \Gamma_\gamma} \left|\langle\Psi_\lambda | \mu_j | G\rangle\right|^2 \delta(\omega - \varepsilon_\lambda) \qquad (42.3)$$

where
  $\rho = D/2 \sin(\pi/N)$ is the ring radius
  ε_λ denotes excitation energies

The operator of electric dipole momentum $\boldsymbol{\mu}$ can be expressed as follows:

$$\boldsymbol{\mu} = -e\rho \sum_l \hat{n}_l \left[\hat{\mathbf{x}} \cos(2\pi l/N) + \hat{\mathbf{y}} \sin(2\pi l/N)\right]. \qquad (42.4)$$

Because of symmetry, $\alpha_x(\omega) = \alpha_y(\omega)$, and we can drop the subscript. The components (μ_x, μ_y) of the electric dipole operator $\boldsymbol{\mu}$ transform according to the two-dimensional irreducible representation E_{1u} of the point group D_{Nh}, which characterizes the symmetry of the nanorings with N QDs. Therefore, only eigenstates with Γ_γ-symmetry obeying

$$\Gamma_\gamma \otimes E_{1u} \otimes \Gamma_G \supset A_{1g} \qquad (42.5)$$

contribute to optical absorption. Here, Γ_G specifies the ground state symmetry. Equation 42.5 expresses the selection rule for the optical transitions allowed by the spatial symmetry. In addition, one should note that only excited states Ψ_λ that have the same total spin (S) and the same total spin projection (S_z) as the initial state (G) can contribute to optical absorption: both S and S_z are conserved in optical transitions.

The expression of $\boldsymbol{\mu}$ in terms of MO-operators $c_{k,\sigma}$ (defined in Section 42.3) is also useful:

$$\mu_x = -e\rho/2 \sum_{p,\sigma} (c^\dagger_{p+1,\sigma}c_{p,\sigma} + c^\dagger_{p,\sigma}c_{p+1,\sigma}),$$

$$\mu_y = ie\rho/2 \sum_{p,\sigma} (c^\dagger_{p+1,\sigma}c_{p,\sigma} - c^\dagger_{p,\sigma}c_{p+1,\sigma}), \qquad (42.6)$$

It is worth noting the occurrence of the consecutive indices p and $p + 1$ in Equation 42.6. This means that only optical transitions between *adjacent* MOs are allowed. For the sake of brevity, let us call this the selection rule of adjacent MOs.

In experiments, not only can the model parameters (t, ε_F, U, and V) be controlled, the number of electrons can also be modified within wide ranges by varying the voltage of a gate electrode.

Therefore, it makes sense to consider the physical properties of a nanoring with a given number N of QDs and a variable number of electrons N_e. Then, one should distinguish between the closed-shell and open-shell cases. From the detailed analysis of this problem recently published (Bâldea and Cederbaum 2007, 2008a,b), we shall consider one example for each case and emphasize the most relevant aspects.

Before starting to discuss the optical absorption for concrete cases of QD nanorings, we make the following remark. In order to calculate the optical absorption of infinite chains, a series of studies perform extrapolations to larger sizes ($N \to \infty$) of the results obtained in finite chains for the imaginary part of the optical conductivity $\sigma''(\omega)$ (Maldague 1977)

$$\sigma''(\omega) \propto \omega^{-1} \sum_{v \neq G} \left|\langle v | \hat{j}_{\parallel} | G\rangle\right|^2 \delta(\omega - \varepsilon_v), \qquad (42.7)$$

expressed in terms of the paramagnetic current $\hat{j}_{\parallel}$, and use for the latter the form

$$\hat{j}_{\parallel} = -it\frac{e}{\hbar} \sum_{l,\sigma} (a^\dagger_{l+1,\sigma}a_{l,\sigma} - a^\dagger_{l,\sigma}a_{l+1,\sigma}). \qquad (42.8)$$

If the expression of the electric dipole momentum

$$\hat{\mu}_{\parallel} = -e \sum_{l,\sigma} (l - 1/2)a^\dagger_{l,\sigma}a_{l,\sigma} \qquad (42.9)$$

is used in Equation 42.3, the quantities for $\alpha(\omega)$ and $\sigma''(\omega)$ are equivalent: the equivalence $\hat{j}_{\parallel} \Leftrightarrow \omega\hat{\mu}_{\parallel}$ emerges from the continuity equation. As far as the chains with open boundary conditions are concerned, $\hat{j}_{\parallel}$ and $\hat{\mu}_{\parallel}$ possess the same spatial symmetry: both are odd with respect to reflection to the middle of the chain. However, this is no longer the case for *periodic* boundary conditions. The current operator of Equation 42.9 transforms according to the irreducible representation A_{2g} of the group D_{Nh}, whereas, as already noted, the dipole operator $\boldsymbol{\mu}$, has E_{1u} symmetry. In the case of closed shell rings, the ground state possesses $\Gamma_G \equiv A_{1g}$ symmetry: the excited states that contribute to $\alpha(\omega)$ are of E_{1u} symmetry, while those contributing to $\sigma''(\omega)$ have A_{2g} symmetry. Therefore, the quantities of Equations 42.3 and 42.7 are completely different. In Figure 42.3a, we present the excitation energies of the lowest eigenstates contributing to $\sigma''(\omega)$ and $\alpha(\omega)$ in the case of a 10-dot nanoring at half filling. The curves are labeled by their spatial symmetries, A_{2g} and E_{1u}, respectively. The fact that, for larger d, these curves are close to each other is an illustration why, *coincidentally*, results for $\sigma''(\omega)$ and $\alpha(\omega)$ can be similar.

42.4.1 Case of Closed Shells

In the ground state of a closed-shell system, within the MO picture the MOs are fully occupied up to the highest occupied molecular orbital (HOMO). Above that, starting from the lowest unoccupied molecular orbital (LUMO), they are completely

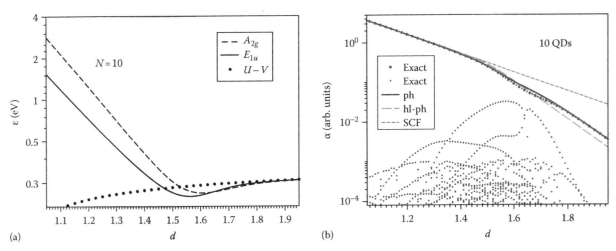

FIGURE 42.3 Results for half-filled 10-QD nanorings. (a) The lowest E_{1u} and A_{2g} excitation energies contributing to the spectrum of $\alpha(\omega)$ and $\sigma''(\omega)$, respectively. The dotted line represents the lowest excitation energy $U - V$ in the limit $t \to 0$. Note the logarithmic scale on the ordinate. (b) Absorption intensities α computed exactly and within several approximations discussed in the main text. Note the logarithmic scale on the ordinate.

empty. According to the selection rule of adjacent MOs, only the HOMO–LUMO transition is allowed. The excitation energy needed for this is called the HOMO–LUMO gap. The counterpart of the HOMO–LUMO gap in the presence of electron correlation is the optical gap, that is, the lowest frequency of absorption spectrum, which represents an important physical property. It is shown by the solid (E_{1u}) line of Figure 42.3a for a 10-dot nanoring at half-filling (10 electrons over 10 dots), a prototypical case of closed shells. This curve clearly reflects the crossover from the weak correlation regime (for small d) to the strong correlation regimes (at large d).

In the weak correlation regime, the optical gap is nothing but the HOMO–LUMO gap (in the nomenclature of solid state physics, the band gap) $2t \sin(\pi/N)$, scaling as the hopping integral t, which falls off exponentially with d (Figure 42.3a). In the strong correlation regime, the optical gap is the reminiscent of the Mott–Hubbard gap (charge gap), known from the study of the infinite restricted Hubbard model (Natan 1992). For larger d, t becomes small, and the interdot tunneling plays a reduced role. At half filling, the dots are singly occupied in the ground state, as the double occupancy is avoided because of its prohibitive energy cost $U(\gg t)$. The Mott–Hubbard gap is the energy needed to create a doubly occupied site. At large d, $U > V$ (see Figure 42.1), and to create a charge excitation requires an energy $U - V$. As visible in Figure 42.3a, the exact optical gap approaches just the limit $U - V$.

The results given earlier on MO populations and optical gap indicate the existence of strong electron correlations at larger d. For all d's, the exact ground state of the closed shell nanorings is a singlet state of A_{1g} symmetry. For larger d, its expansion contains many Slater determinants with significant contributions (Bâldea and Cederbaum 2008b). In view of the strong correlations and the selection rule of Equation 42.5, one expects that, out of all the ($\sim 10^4$) singlet E_{1u} states that can contribute to optical absorption, numerous states do significantly contribute. Strikingly, this is

not the case (Bâldea and Cederbaum 2007, 2008a,b). Despite the strong correlations, there is only a single excited state, which significantly contributes to optical absorption. Similar to the case of the MO spectrum, the exact absorption spectrum is *practically* monochromatic (Bâldea and Cederbaum 2007, 2008a,b). This behavior is illustrated by the optical intensities presented in Figure 42.3b for a 10-QD ring. By inspecting the exact intensity α, one can see that, out of the numerous contributions, the intensity of a single line is significant (note the logarithmic scale on the ordinate). This does not mean that the optical intensities are not affected by correlations. As seen in Figure 42.3b, at larger d the dominant line significantly deviates from the SCF curve.

42.4.2 Case of Open Shells

For open shell systems, within the MO picture there exists a partially occupied molecular orbital, lying between fully occupied MOs and completely unoccupied MOs. For nine electrons in a 10-dot nanoring, the case we shall consider for illustration in the following text, the e_{2g} MO is partially occupied: three out of the four available single-electron states are occupied. The e_{2g} MOs lie above the fully occupied e_{1u} MOs and below the completely empty e_{3u} MOs. According to the selection rule of adjacent orbitals, two optical transitions are allowed. The first absorption frequency corresponds to the excitation of one electron from the fully occupied e_{1u} MOs into the only empty e_{2g} MO (excitation energy t), and this yields an E_{1u} state. The second absorption frequency amounts to excite one of the three electrons from an e_{2g} MO into the empty e_{3u} MOs. The corresponding excitation energy is $1.236t$. This gives rise to a group of degenerate excited states, which comprises one E_{1u} state and three E_{3u} states.

Let us now discuss the exact results for this system. The ground state of this nanoring is a spin doublet E_{2g} state. By increasing the interdot spacing d, the MO populations change in a way analogous to that observed in Figure 42.2: the occupancies of the MOs that

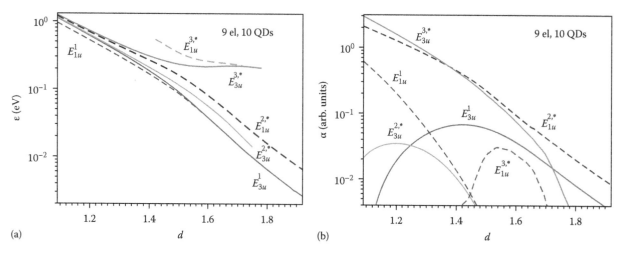

FIGURE 42.4 (a) Frequency ε and (b) intensity α of optical absorption for 9 electrons in a 10-QD ring. The symmetry E_{1u} and E_{3u} of the allowed final states is specified in the legend. The asterisk labels contributions of bright diabatic states. See the main text for details.

are occupied within the MO picture progressively decrease, the MOs that are unoccupied within the MO picture become gradually occupied. Similar to the case discussed in Section 42.4.1, unless the QDs are nearly touching, electron correlations are strong. Because of the selection rule of Equation 42.5 and because of the multiplication rule $E_{1u} \times E_{2g} = E_{1u} + E_{3u}$ of the $D_{10,h}$ symmetry group, only excited states, which are spin doublet of symmetries E_{1u} and E_{3u}, can contribute to optical absorption. In view of the strong correlations, one expects that many of the excited spin doublet states of symmetries E_{1u} and E_{3u} will contribute.

The comparison of the exact results with those of the MO picture reveals that, at smaller d, the main effect of electron correlations is to lift the degeneracy in the second group of optical excitations. The fact that this degeneracy is lifted by the Hubbard interactions U and V can be qualitatively understood even within the first order of the perturbation theory (Bâldea and Cederbaum 2008a). Quantitatively, the description based on the perturbation theory rapidly deteriorates at larger d. This is consistent to the fact that stronger correlations cannot be described perturbatively. As the dots become more and more distant, t ceases to be the dominant energy scale, and the electron hopping becomes more and more ineffective ($t \ll U, V$). As a result, there is an overall tendency that energy separations diminish. Groups of numerous states appear (the reminiscent of numerous states that are classically degenerate), whose energies agglomerate in narrow energy ranges of widths $\mathcal{O}(t^2)$ around 0, $U–2V$, U, etc. A consequence of this fact is the breakdown of the perturbation theory. Another important consequence is that, in a group of nearly degenerate states, there also exist two (or several) states with the same symmetry. This yields the phenomenon of avoided crossing, which will be discussed in Section 42.5. To facilitate understanding, instead of the results for the individual eigenstates calculated exactly, we present in Figure 42.4 the curves for the so-called bright diabatic states with significant contributions to the optical absorption (see Section 42.5 for details). One can anticipate that the results for the bright diabatic states

are just those needed for the pragmatic purpose of reproducing the experimental spectra.

If one inspects the results for the optical absorption of Figure 42.4, the most eye-catching fact is that for all d's the optical spectra are very scarce. For the present case of 9 electrons over 10 dots, out of 55,440 spin doublet eigenstates, there are ~10^4 eigenstates of symmetries E_{1u} and E_{3u}, which are allowed to contribute, but only very few do significantly contribute.

42.5 Avoided Crossings

Avoided crossing designates a situation where, as a certain parameter (in our case, d) approaches a certain critical value (d_c) from either direction, the energy difference of two states with the same symmetry reaches a small but nonvanishing minimum value. At d_c, the small energy splitting is due to the repulsion of the states of identical symmetry.

At the same time, the two states *rapidly* interchange their properties (in our case, spectral intensity α) around d_c. The bright state Ψ_1 below the critical point d_c becomes dark above it, whereas the mate state Ψ_2 is dark for $d < d_c$ but bright for $d > d_c$. The interchange is such that the properties of Ψ_1 for $d < d_c$ *smoothly* evolve into those of Ψ_2 for $d > d_c$, and vice versa, the properties of Ψ_1 for $d > d_c$ smoothly continue those of Ψ_2 from the range $d < d_c$. In such cases, instead of the (adiabatic) description in terms of the states Ψ_1 and Ψ_2, which are eigenstates of the electronic Hamiltonian, it is often more convenient to construct the *diabatic* states $\tilde{\Psi}_1$, which approximates Ψ_1 for $d < d_c$ and Ψ_2 for $d > d_c$, and $\tilde{\Psi}_2$, approximating Ψ_2 for $d < d_c$ and Ψ_1, for $d > d_c$ (Köppel et al. 1984). Unlike the adiabatic energies, the curves for the diabatic states do cross as a function of d and retain their electronic character.

To illustrate the behavior of diabatic states, we return to Figure 42.4. The curves labeled there by an asterisk correspond to bright diabatic states. As an example, we show in Figure 42.5 the results for all the (adiabatic) eigenstates, which exhibit avoided crossings pertaining to the bright diabatic state denoted by $E_{3u}^{3,*}$ in

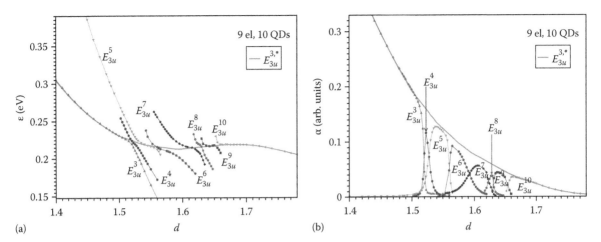

FIGURE 42.5 (a) Frequency ε and (b) intensity α of optical absorption for the bright diabatic state (labeled by asterisk, thick lines) and the relevant adiabatic excited eigenstates (3, 4, 5,...) of E_{3u} symmetry with significant contributions for nine electrons in a 10-QD ring. In contrast to the rapidly varying curves for the adiabatic states, the diabatic ones are smooth. Note the logarithmic scale.

Figure 42.4. The spectral intensities of the bright diabatic states have been obtained by summing the intensities of the individual states, $\alpha = \sum_i \alpha_i$, while for the energies, the averages weighted with the intensities, $\varepsilon = \sum_i \varepsilon_i \alpha_i / \sum_i \alpha_i$, have been employed. The rationale for considering the bright diabatic state is that in the region of avoided crossing, because of the small energy separation, one cannot measure individual signals separately, but rather the added intensities. As visible in Figures 42.4 and 42.5, the diabatic curves obtained in this manner are indeed smooth, in contrast to the true eigenstates (Figure 42.5), which exhibit very rapid variations in the region of avoided crossing.

42.6 Hidden Quasi-Symmetry

The fact encountered earlier that the optical spectrum is *practically* monochromatic for closed-shell QD nanorings and an unusual scarce for open shells is undoubtedly the most interesting issue in connection with optical absorption. It is hard to understand why, in spite of the strong electron correlations, out of numerous transitions allowed by spin conservation and spatial symmetry, only extremely few acquire substantial spectral weights. The optical spectra of QD nanorings with a variable number of electrons have been systematically analyzed (Bâldea and Cederbaum 2008a); the cases presented in Sections 42.4.1 and 42.4.2 are only two examples thereof. This analysis revealed that, apart from splitting the degenerate MO transitions, there are as many optical signals with significant intensity as predicted within the MO picture (Bâldea and Cederbaum 2008a). Therefore, it appears most natural to ascribe this behavior to the existence of a hidden quasi-symmetry (Bâldea and Cederbaum 2007, 2008a). It is not a true symmetry, but a *quasi*-symmetry. In all cases, in addition to the relevant optical signals, there are many other spectral lines, which, similar to those represented by points in Figure 42.3b, possess very small but definitely nonvanishing intensities. One might be tempted to ask whether

it will become a true symmetry in the case of the restricted Hubbard model. No, it remains a *quasi*-symmetry (Bâldea and Cederbaum 2007, 2008a, Bâldea et al. 2009). Still, it is a *hidden* quasi-symmetry: it is not directly related to the spatial ($D_{N,h}$), or other known symmetry. In spite of an involved in-depth analysis of the numerical data (I. Bâldea, unpublished), it was impossible to unravel the nature of this symmetry so far.

Although a physical explanation cannot yet be given, it was speculated that this behavior could be understood within a generalized Landau theory. As well known, Landau theory (see, e.g., Pines 1962) postulates a one-to-one map between the excitations near the Fermi level of interacting and noninteracting electron systems. One may attempt to describe the optical excitations by superpositions of particle–hole excitations $c_{p,\sigma}^\dagger c_{p,\sigma} |G_0\rangle$ in the noninteracting ground state $|G_0\rangle$, with electrons in the HOMO and holes in the LUMO. As expected in view of the strong correlations, this is a very poor approximation. To keep track of the strong correlations, one can try to use the interacting ground state $|G\rangle$ instead of $|G_0\rangle$. Qualitatively, this approximation is correct: it reproduces the number of the significant optical lines. Quantitatively, it is good only at half filling (see the curve labeled *hl-ph* in Figure 42.3b). Away from half filling, it is less satisfactory particularly at larger d, where correlations are stronger. This approximate method was generalized to include not only the HOMO–LUMO but all dressed particle–hole pairs $c_{p,\sigma}^\dagger c_{p\pm 1,\sigma} |G\rangle$ with E_{1u} symmetry (Bâldea and Cederbaum 2008a). The numerical diagonalization of Hamiltonian (Equation 42.2) in the Hilbert subspace spanned by these dressed particle–hole excitations turned out to provide eigenstates, which accurately reproduce the optical spectra both for closed and for open shells (Bâldea and Cederbaum 2008a). For illustration, in Figure 42.3b we show the curve denoted by *ph* obtained within this approach. Importantly, this method substantially reduces the numerical effort: the dimension of the *ph*-subspace is $\sim N$, much smaller than the dimension $\sim \binom{2N}{N_e}$ of the full Hilbert space.

42.7 Photoionization

In a typical ionization process, an electron (say, with spin up) is ejected from an MO p of a neutral system into the vacuum. An ionization process is characterized by an ionization potential $\omega_{p,i} = \langle \Psi_{p,i} | H | \Psi_{p,i} \rangle - \langle \Phi | H | \Phi \rangle$ and a spectroscopic factor $f_{p,i}$:

$$f_{p,i} = \left| \langle \Psi_{p,i} | c_{p,\uparrow} | G \rangle \right|^2. \qquad (42.10)$$

Here, $\Psi_{p,i}$ ($i = 1, 2, \ldots$, see the following text for notation) are eigenstates of the ionized nanoring. The spectroscopic factor $f_{p,i}$ is directly related to the partial-channel ionization cross section (Cederbaum et al. 1986). Here, we shall restrict ourselves to the case of a closed-shell neutral system in the ground state G (zero-temperature case). Because G is an A_{1g} state in this case, the symmetry of the final states $\Psi_{p,i}$ of the ionized nanoring coincides with the symmetry Γ_p of the corresponding MO. In the case where the MO picture holds, the removal of an electron from an occupied MO will leave the ionized system in a well-defined eigenstate. For that eigenstate, the spectroscopic factor will be

$$f_{SCF} = 1 \qquad (42.11)$$

and zero for all the other ionized eigenstates. This implies that within the MO picture the number of ionization signals is equal to the number of occupied MOs. This does not hold any more in the case of (strong) correlated systems. The ejection of an electron from the pth-MO will bring the ionized system in one of its eigenstates $\Psi_{p,1}, \Psi_{p,2}, \ldots$ By using Equation 42.10, one can easily show that the corresponding spectroscopic factors $f_{p,i}$ ($0 \le f_{p,i} \le 1$) obey the following sum rule:

$$\sum_i f_{p,i} = \langle G | c_{p,\uparrow}^\dagger c_{p,\uparrow} | G \rangle \equiv n_{p,\uparrow}. \qquad (42.12)$$

Equation 42.12 is important: it relates the integrated ionization weight to the MO occupancy.

To summarize, the breakdown of the MO picture of ionization manifests itself in spectroscopic factors substantially smaller than unity, and in the impossibility to single out a main line, which dominates the ionization spectra.

In Figure 42.6, we present exact and SCF results for the HOMO ionization, that is, corresponding to the lowest ionization energy. This process brings the ionized nanoring in its ground state, and an electron is ejected from the HOMO. The discrepancy between the exact and SCF results reveals the severe breakdown of MO picture: the important role played by electron correlations at larger d is clearly visible in Figure 42.6. The spectroscopic factor substantially differs from unity (cf. Equation 42.11), and the ionization energy approaches the value $-\varepsilon_F - 2V$, corresponding to the limit $t \to 0$. The limiting value is independent of N (Bâldea and Cederbaum 2008a,b), because for large d the dots are practically no longer coupled by tunneling. The breakdown of the MO picture of the HOMO ionization in QD nanorings (Bâldea and Cederbaum 2002) represents a remarkable difference from molecules. It is a well-established fact that the HOMO ionization of ordinary molecules can be accurately described within the SCF (Cederbaum et al. 1986). The breakdown of the MO picture of the HOMO ionization is an indication on the occurrence of a localization–delocalization transition (Bâldea and Cederbaum 2002). The latter is related to the metal–insulator transition, which was observed experimentally in Ag-QD arrays (Markovich et al. 1999). The fact that in Figure 42.6b the spectroscopic factor starts to deviate from 1 for $d \approx 1.2 - 1.4$ is notable, because this is just the d-range where the metal–insulator transition was experimentally observed.

Although, as noted earlier, the HOMO ionization behaves differently in assembled QDs and ordinary molecules, certain similarities exist. In both systems, the higher "occupied" MOs are less affected by correlations than the lower occupied MOs (Bâldea and

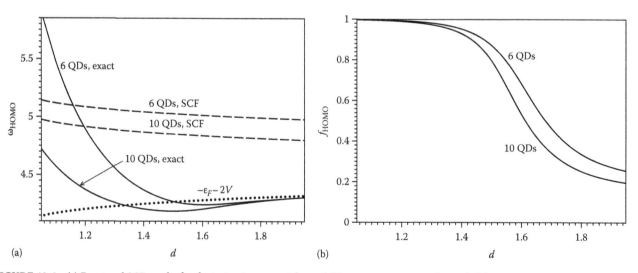

FIGURE 42.6 (a) Exact and SCF results for the ionization potential ω and (b) exact spectroscopic factor f of the HOMO ionization in half-filled 6- and 10-QD nanorings ($\varepsilon_F = -4.5\,\text{eV}$). The dotted line represents the HOMO ionization energy $-\varepsilon_F - 2V$ in the limit $t \to 0$. In SCF, the spectroscopic factor f is equal to 1.

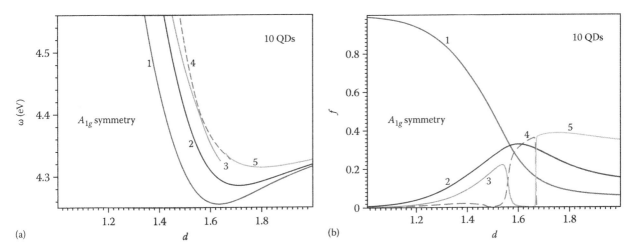

FIGURE 42.7 Exact results for the ionization potential ω and spectroscopic factor f of the A_{1g} ionization spectrum of 10-QD nanorings at half filling ($\varepsilon_F = -4.5\,\text{eV}$). See the main text for details. In both panels (a) and (b), the curves denoted by 1, 2, 3, 4, and 5 correspond to the eigenstates with significant ionization signal. Notice the avoided crossings $d \simeq 1.56$ and $d \simeq 1.67$ between the eigenstates 3 and 4, and 5 and 6, respectively.

Cederbaum 2008a). By "occupied" MOs, we mean here the MOs, which would be occupied if the MO picture were valid, since Figure 42.2 showed that all MOs are occupied at larger d. In Figure 42.7, we present results for the relevant ionization signals by ionization from the lowest "occupied" MO (Bâldea et al. 2009). At very small d, Figure 42.7b shows a dominant signal (1) that practically exhausts the whole A_{1g} ionization spectrum ($f \simeq 1$). By increasing d, it is gradually reduced, and concomitantly two other signals (2 and 3) acquire significant intensity. Two avoided crossings are visible for the signal with the highest ionization energy, one at $d \simeq 1.56$ (between 3 and 4) and another at $d \simeq 1.67$ (between 4 and 5). This example shows that not only states transforming according to a two-dimensional irreducible representation (like the states E_{3u} of Figure 42.5) but also those transforming according to a one-dimensional irreducible representation (like the present A_{1g} states) can be involved in avoided crossings.

In ordinary molecules, moderately strong correlations yield satellite lines of smaller intensities accompanying the main lines with dominant intensity. The latter are related to one-hole ($1h$) processes, the former to two-hole–one-particle ($2h1p$) or higher-order processes. Stronger correlations causing the breakdown of the MO picture lead to ionization spectra consisting of numerous lines with comparable intensities. Main lines cannot be identified in such molecules, because their properties cannot be tuned, and approaching the uncorrelated limit is impossible. By inspecting the results of Figure 42.7b, one can assign signal 1 as the "main" line of the A_{1g}-ionization spectrum; it traces back to the $1h$-process of the uncorrelated limit, valid for $d \approx 1$. However, it dominates the spectrum only up to $d \simeq 1.56$. Beyond this value, the intensities of the "satellite" lines (signals 2 and 3), which necessarily result from higher-order processes, exceed that of the "main" line.

While this is obviously a manifestation of the strong correlations, it is quite different from the case of ordinary molecules. Strong correlations in QD nanorings enhance a few "satellite" lines, which can even dominate over the "main" lines, but do not produce numerous lines with significant intensities. Similar to the

case of optical absorption (Section 42.4), the ionization spectra of QD nanorings for all MO symmetries comprise a few intense lines and a multitude of lines of very small but nonvanishing intensities (Bâldea et al. 2009), which are invisible in Figure 42.7. Despite the strong correlations, the ionization spectra are very scarce. This intriguing scarcity, which is characteristic for all symmetries (Bâldea et al. 2009), is similar to that for optical absorption (Section 42.4) and also pleads in favor of a hidden quasi-symmetry.

42.8 High Harmonic Generation by QD Nanorings

In this section, we briefly discuss the high harmonic generation (HHG) as an interesting potential application of QD nanorings (Bâldea et al. 2004a). The HHG is particularly interesting in the attempt of pushing the frequency of radiation sources to higher and higher values. Because multichromaticity is a common feature of the HHG-based sources, searching for systems and setups where most harmonics are forbidden, for example, for symmetry reasons, is particularly important. Employing a circularly polarized laser field incident on a system possessing a rotation axis of order N is particularly appealing. In this case, the invariance under dynamical symmetry yields unusual selection rules (Alon et al. 1998). According to them, the system only emits harmonics of orders $nN + 1$ circularly polarized as the incident field, and of orders $nN - 1$ polarized in opposite direction ($n = 1, 2, \ldots$).

In view of these selection rules, it is tempting to study the HHG spectrum of QD nanorings. Particularly encouraging to use the model of Equation 42.2 is that previous studies (Alon et al. 1998, Baer et al. 2003, Zdánská et al. 2003) found the HHG in benzene to proceed via bound-to-bound [and *not* via bound-to-continuum, which are not included in Equation 42.2] transitions. Calculations carried out for six-QD nanorings led to a series of interesting results (Bâldea et al. 2004a).

Most important for practical purposes, the HHG study of QD nanorings allowed the comparison with the HHG spectrum of

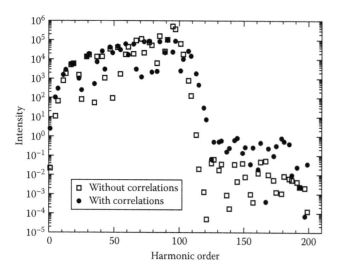

FIGURE 42.8 HHG spectrum for six Ag-QD nanorings at $d = 1.5$ with and without considering electron correlations. (Adapted from Bâldea, I. et al., *Phys. Rev. B*, 69, 245311, 2004a.)

a real molecule like benzene. Bâldea et al. (2004a) showed that a six-QD nanoring ("artificial" benzene) is considerably more efficient for HHG than the benzene molecule. For a given intensity of the incident laser field, harmonics of much higher orders are generated by artificial benzene than by ordinary benzene. An interesting feature of the HHG spectrum is the formation of a plateau, wherein the harmonic intensities remain roughly constant over many orders. For the highest available laser intensities, the HHG spectrum calculated for benzene consists of a single plateau only. Remarkably, the HHG spectrum of artificial benzene exhibits two plateaus (more precisely, even more, but only two are quantitatively significant).

Similar to the cases of optical absorption (Section 42.4) and photoionization (Section 42.7), the HHG spectra reveal the important part played by electron correlations. To illustrate this, we show in Figure 42.8 the HHG spectra computed with and without accounting for correlations. Interestingly, mainly affected are the harmonics belonging to the second plateau, which is missing in ordinary benzene.

42.9 Quantum Phase Transitions

By employing model parameters documented in the literature for QDs of silver, we have presented results in Sections 42.3, 42.4, and 42.7, which reveal the impact of electron correlations on various properties of Ag-QD nanorings. One may ask whether the correlation effects discussed earlier are general or depend on the particular parameters specific for a concrete example. It turns out that, more than the parameter values for a particular case, what is important for the behavior found earlier is the possibility to tune the ratio $U/4t$ from values substantially smaller than unity to values substantially larger that unity.

As concerns the value of V, extensive calculations (Bâldea, unpublished) show that the earlier statement holds as long as $V < U/2$. This is the case for the range $1.11 < d < 1.85$ of experimental

interest for Ag-QDs (see Figure 42.1). In this section, we shall briefly examine the general case of a QD nanoring described by the model of Equation 42.2 without imposing other restrictions on the parameters than general physical requirements. The only general constraint is that the quadratic form entering the right-hand side of Equation 42.1 be positively defined. By considering, consistently with the model of Equation 42.2, only nearest-neighbor couplings ($V_{ij} = V\delta_{j,i+1}$), this condition leads to two restrictions: $U > 0$ and $V < U$. No general restriction must be imposed on the hopping integral t, but if the bandwidth $4t$ exceeds the interaction terms $4t > U(>V)$ the nanosystem is essentially uncorrelated. Therefore, to reveal correlation effects one should consider a Hubbard strength U of at least $U = 4t$ (case of a moderately strong interaction) and monitor how the physical properties change when V is varied from 0 to U.

In doing this, we are offered the opportunity to discuss a *quantum phase transition* (Sachdev 2000) in QD nanorings. Quantum phase transitions are qualitatively different from conventional phase transitions. The latter (also called thermal phase transitions) occur by varying the temperature of a system and are due to thermal fluctuations. Quantum phase transitions occur by varying a physical parameter (like magnetic field, pressure, or interdot spacing) and are driven by quantum fluctuations. Therefore, a system can, in principle, undergo quantum phase transitions at zero temperature. Besides the well-studied superconducting-normal transition in thin films, a variety of other quantum phase transition are known (see, e.g., Sachdev 2000, Bâldea et al. 2000, 2001a).

Results on a quantum phase transition in a half-filled nanoring are presented in Figure 42.9, where the properties we monitor are the HOMO ionization and HOMO occupancy. As seen there, both the HOMO spectroscopic factor f_{HOMO} and occupancy n_{HOMO} exhibit maxima at $U \lesssim 2V$. This position coincides with

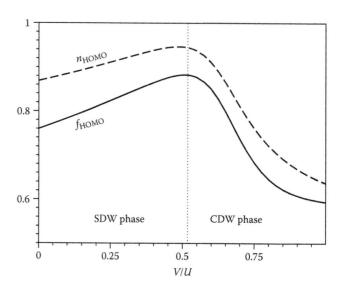

FIGURE 42.9 Exact results for the HOMO spectroscopic factor f and the occupancy n for six-QD rings at $U = 4$ and $t = 1$. Notice that these curves reach the maximum at the point $V/U \gtrsim 1/2$ of CDW–SDW transition.

the point of the so-called CDW–SDW quantum phase transition, known from the study of rings described by the extended Hubbard model (see, e.g., Waas et al. 1990, Bâldea et al. 2000, 2001a,b, and references therein). Depending on the model parameter values, several types of orderings arise in the ground state of the model of Equation 42.2 in order to minimize the total energy. For $U < 2V$, the ground state exhibits a charge density wave (CDW), which is characterized by an electron density modulation $n_l = \bar{n} + \delta n \cos(2k_F l)$. At half-filling, the Fermi wave vector is $k_F = \pi/2$, and the modulation amounts to a dimerization ($n_{2l} = \bar{n} + \delta n$, $n_{2l+1} = \bar{n} - \delta n$). In the opposite case $U > 2V$, the ground state is characterized by a spin density wave (SDW). There, the electron densities per spin directions are modulated out of phase $n_{l,\uparrow} \propto \bar{n} + \delta n \cos(2k_F l)$, $n_{l,\downarrow} \propto \bar{n} - \delta n \cos(2k_F l)$, while the total electron density remains homogeneous ($n_l = \bar{n}$).

CDW and SDW are due to correlations of the electrons close to the Fermi level. Therefore, one expects that the properties of the HOMO can be used to probe CDW and SDW orderings. The curves for f_{HOMO} and n_{HOMO} of Figure 42.9 shows that this is indeed the case. The strength of CDW and SDW electron correlations (measured by the deviations of f_{HOMO} and n_{HOMO} from unity) increases the more one moves deeper into either the CDW or the SDW ranges. As a result of the competition between CDW and SDW correlations, f_{HOMO} and n_{HOMO} reach their maximum at the CDW–SDW transition point.

42.10 Related Systems: Kondo Effect

In all of the earlier examples, we have examined nanorings described by the model of Equation 42.2 with homogeneous (site independent) parameters, suitable to describe the extreme case of identical QDs. In this section, we shall briefly consider another extreme situation, where all but one QDs are identical, and only retain the Coulomb on-site interaction for the nonidentical dot (in this section, referred to as *the* dot). This amounts to consider an inhomogeneous nanoring described by the Anderson impurity model (Anderson 1961):

$$H = H_D + H_{DE} + H_E,$$

$$H_E = \varepsilon_F \sum_{\sigma;l=2}^{N} a_{l,\sigma}^\dagger a_{l,\sigma} - t \sum_{\sigma;l=2}^{N-1} (a_{l,\sigma}^\dagger a_{l+1,\sigma} + h.c.),$$

(42.13)

$$H_D = \varepsilon_D \hat{n}_D + U \hat{n}_{D,\uparrow} \hat{n}_{D,\downarrow}, \quad H_{DE} = -t_D \sum_\sigma (a_{2,\sigma}^\dagger a_{D,\sigma} + a_{D,\sigma}^\dagger a_{N,\sigma} + h.c.).$$

(42.14)

Here, $a_{D,\sigma}(a_{D,\sigma}^\dagger)$ destroys (creates) electrons on the dot, $\hat{n}_{D,\sigma} \equiv a_{D,\sigma}^\dagger a_{D,\sigma}$ and $\hat{n}_D = \sum_\sigma \hat{n}_{D,\sigma}$.

Models similar to Equation 42.9 were often considered to describe single-electron transistors (SETs) consisting of a dot connected by tunneling junctions to metallic electrodes (Glazman and Raikh 1988, Ng and Lee 1988). Experimentally, t_D and ε_D can

be changed by varying gate potentials (Goldhaber-Gordon et al. 1998a). In the case of weak dot-electrode coupling at low temperatures and a sufficiently large charging energy U, the number n_D of electrons on the dot is a well-defined integer. Namely, n_D is either 0, 1, or 2, depending on whether the electrode Fermi energy is situated below the lower Hubbard "band" energy ($\varepsilon_F < \varepsilon_D$), between the upper and the lower Hubbard "band" energies ($\varepsilon_D < \varepsilon_F < \varepsilon_D + U$), or above the upper Hubbard "band" energy ($\varepsilon_D + U < \varepsilon_F$), respectively. Charge fluctuations (and thence electric conduction) can only occur within narrow ε_D ranges around the dot energies $\varepsilon_{D,u} = \varepsilon_F - U$ and $\varepsilon_{D,l} = \varepsilon_F$, where the dot states with different integral charges are nearly degenerate. The width $\Gamma \sim t_D^2/t$ of these ranges is determined by the broadening of the single electron levels in electrodes due to the finite dot-electrode coupling. This is the reason why, by varying the dot energy ε_D with the aid of a gate potential, the electrical conductance $\mathcal{G}$ exhibits sharp maxima (Coulomb blockade peaks) for dot energies $\sim \varepsilon_{D,u} \pm \Gamma$ and $\sim \varepsilon_{D,l} \pm \Gamma$, and vanishes otherwise. By lowering the temperature (starting from sufficiently low values, $k_B T \ll U$, Γ), the Coulomb blockade peaks become more and more pronounced, with the concomitant reduction of $\mathcal{G}$ in the region between them (Coulomb valley). By further decreasing T below a certain value T_K (Kondo temperature), the conductance in the Coulomb valley begins to gradually increase. The final stage is the formation of a plateau (Kondo plateau) ranging from $\varepsilon_D \approx \varepsilon_F - U$ to $\varepsilon_D \approx \varepsilon_F$ where the conductance reaches the ε_D-independent value $\mathcal{G}_0 = 2e^2/h$ of the ideal point contact (unitary limit). This effect was called the Kondo effect, in analogy to the effect known in solids with magnetic impurities (Hewson 1993).

The number n_D of electrons on the dot represents a key quantity for the electric transport through the SET. However, transport experiments do not allow to measure it directly. n_D can be only indirectly determined from the (zero bias) conductance $\mathcal{G}$ via the Friedel–Langreth sum rule (Langreth 1966, Meir and Wingreen 1992):

$$\mathcal{G} = \mathcal{G}_0 \sin^2(\pi n_D/2)$$

(42.15)

Importantly, it was recently proposed (Bâldea and Köppel 2009) to measure n_D directly, by means of photoionization, by shining the dot with photons of energy of the order of the work function ($h\nu \sim 1$ eV). In this case, the spectroscopic factor reads

$$f_k = \left| \left\langle \Psi_k \mid a_{D,\uparrow} \mid G \right\rangle \right|^2,$$

(42.16)

where Ψ_k is a final eigenstate of ionized system.[*] Equation 42.16 yields the sum rule

$$\sum_k f_k = \left\langle G \mid a_{D,\uparrow}^\dagger a_{D,\uparrow} \mid G \right\rangle = n_D/2.$$

(42.17)

---

[*] Again, we consider spin up for convenience: for the cases of interest, $n_{D,\uparrow} = n_{D,\downarrow} = n_D/2$. As far as photoionization is concerned, there is no significant difference between a dot attached to two electrodes and a dot embedded in a ring (the case we consider here). See Bâldea and Köppel (2009).

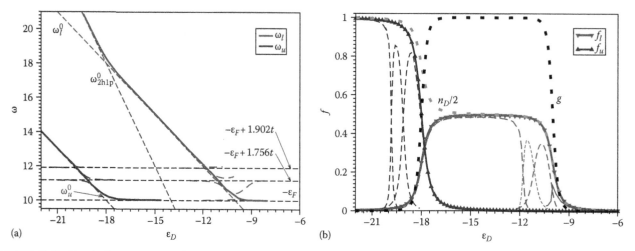

FIGURE 42.10 Results for energies ω and spectroscopic factors f of ionization of a QD embedded in a 10-site ring. The thin dashed lines of panel (a) represent ionization energies for $t_D \to 0$. In both panels (a) and (b), the thick dashed lines refer to the significant adiabatic ionized eigenstates, while the thick solid lines correspond to the two bright diabatic states l and u related to the ionization from the lower and upper Hubbard band, respectively. The dot occupancy per spin direction $n_D/2$ is related to the integrated ionization intensity and determines the normalized conductance $g \equiv \mathcal{G}/\mathcal{G}_0$ via Equation 42.15 ($t = 1$, $\varepsilon_F = -10$, $t_D = 0.2$, and $U = 8$).

So, the dot charge can be deduced from the integrated ionization weight via Equation 42.17.

The differential ionization can be used to obtain even more information. Let us analyze the limit $t_D \to 0$. Depending on whether the dot is doubly or singly occupied ($\varepsilon_D + U < \varepsilon_F$ or $\varepsilon_D < \varepsilon_F < \varepsilon_D + U$, respectively) the dot ionization energy is $\omega_u^0 = -\varepsilon_D - U$ or $\omega_l^0 = -\varepsilon_D$, respectively. The corresponding spectroscopic factors of the two ionization processes are $f_u^0 = 1$ and $f_u^0 = 1/2$. These ionization processes can become resonant with $1h-$, $2h1p-$, ... processes of removing an electron from the rest of the ring. The energies of the latter processes can be easily obtained by diagonalizing the bilinear Hamiltonian H_E of Equation 42.14. For the case considered in Figure 42.10 ($N = 10$), these energies are $-\varepsilon_F$, $-\varepsilon_F + 1.756t$, $-\varepsilon_F + 1.902t$,... In addition, the ω_l^0-process can be in resonance with a $2h1p$-process of energy $\omega_{2h1p}^0 = -2\varepsilon_D - U + \varepsilon_F$, which consists of creating two holes on the (doubly occupied) dot and one electron in the rest of the ring. These resonances are represented by the intersections of the thin dashed lines in Figure 42.10.

In the presence of a finite coupling $t_D \neq 0$, the earlier intersections evolve into avoiding crossings; true intersections cannot occur between eigenstates of identical symmetry. Each ionization process ω_u^0 and ω_l^0 is replaced by a juxtaposition of several signals, which represent the contributions of several ionized eigenstates involved in a succession of avoided crossings. Figure 42.10a and b shows three and four of such contributions, related to the ionization from the upper and lower Hubbard band, respectively. Besides these individual contributions, we also show the contributions of the two diabatic bright states (subscripts u and l in Figure 42.10). The latter are important, because only these can be observed experimentally. In particular, by extrapolating the portions of the curves ω_u and ω_l, which are linear and parallel among themselves, one can determine the charging energy U in a

direct way and not *indirectly*, as done in transport measurements (see Bâldea and Köppel 2009 for details).

A more detailed discussion of the ionization in SETs is beyond our present scope. Nevertheless, the aspects presented here demonstrate that (1) avoided crossings can frequently occur in QD nanosystems and (2) photoionization can be a powerful tool for their study.

42.11 Conclusion

Because QD nanoarrays possess properties tunable in broad ranges, they can be used to fabricate devices with designer-specified functional properties and to study fundamental quantum-mechanical processes. In this chapter, we have presented a variety of theoretical results for metallic QD nanorings. Particular emphasis has been laid on the manner in which the crossover from weak to strong correlations manifests itself in physical properties like optical absorption and photoionization. The impact of the strong electron correlations is very interesting. They are responsible for the occurrence of quantum phase transitions and numerous avoided crossings. Unlike in other cases studied earlier, the latter often involve more than two states. The most striking fact is that in spite of very strong correlations, optical absorption and photoionization in QD nanorings are very selective. It is most natural to ascribe this extreme scarcity of the optical and ionization spectra to the existence of a hidden quasi-symmetry of the QD nanorings described within the extended Hubbard model.

The study of assembled QDs remains a challenge for physicists and chemists to understand the properties of matter at nanoscale, without restrictions imposed on ordinary molecules and solids, for example, by the periodic table to the building units or by the fixed (optimized) geometry. Albeit not covered in this chapter, one should also mention that even standard experiments like optical absorption or photoionization in QD nanostructures can be conducted in a manner that is inconceivable in ordinary systems. The method of partial covering represents such an example (Bâldea and Cederbaum 2002, 2007, 2008b). By shining a molecule, photons interact with all the atoms it contains. In a QD nanoarray, one can fabricate a mask to cover (as a limiting case) all the nanostructures except for one QD. The information obtained by means of this technique can be richer than that in the case of irradiating the whole nanostructure (Bâldea and Cederbaum 2007; 2008b).

Some years ago, QDs were heavily advertised as promising building units for quantum computers. Because this expectation has not materialized so far, it is important to give other suggestions on possible applications using QDs. In this context, the theoretical predictions that QD nanorings are much more efficient for the higher harmonic generation than ordinary cyclic molecules like benzene are particularly valuable (Bâldea et al. 2004a).

The QDs with diameters of a few tens of nanometers employed for the nanorings (Petroff et al. 2001) and other one-dimensional arrays (Santhanam and Andres 2004) prepared so far are too large to play an important role in the phenomena discussed in this chapter. Fabricating QD nanorings using Ag-QDs of the type prepared by Heath's group appears to be most suitable to verify the theoretical predictions exposed here, in particular the predicted hidden quasi-symmetry. This is the reason why we have employed parameter values documented in the literature for this case. However, as already noted, these values are not critical for the findings discussed earlier. Another possibility would be to create equidistant barriers on semiconducting nanorings (Fuhrer et al. 2001, Granados and García 2003, Kong et al. 2004), such that the electron motion can be confined within a few nanometers. With these theoretical results, we hope to encourage experimentalists to fabricate and investigate nanorings consisting of small QDs, wherein electron correlations are important.

Acknowledgments

The authors are pleased to thank H. Köppel, J. Schirmer, and N. Moiseyev for numerous discussions over many years. Financial support for this work provided by the Deutsche Forschungsgemeinschaft (DFG) is gratefully acknowledged.

References

Alon, O. E., V. Averbukh, and N. Moiseyev. 1998. Selection rules for the high harmonic generation spectra. *Phys. Rev. Lett.* 80:3743–3746.

Anderson, P. W. 1961. Localized magnetic states in metals. *Phys. Rev.* 124:41–53.

Andres, R. P., T. Bein, M. Dorogi et al. 1996. "Coulomb staircase" at room temperature in a self-assembled molecular nanostructure. *Science* 272:1323–1325.

Baer, R., D. Neuhauser, P. R. Žďánská, and N. Moiseyev. 2003. Ionization and high-order harmonic generation in aligned benzene by a short intense circularly polarized laser pulse. *Phys. Rev. A* 68:043406.

Bâldea, I.. and L. S. Cederbaum. 2002. Orbital picture of ionization and its breakdown in nanoarrays of quantum dots. *Phys. Rev. Lett.* 89:133003.

Bâldea, I. and L. S. Cederbaum. 2007. Hidden quasisymmetry in the optical absorption of quantum dot nanorings. *Phys. Rev. B* 75:125323.

Bâldea, I. and L. S. Cederbaum. 2008a. Unusual scarcity in the optical absorption of metallic quantum-dot nanorings described by the extended Hubbard model. *Phys. Rev. B* 77:165339.

Bâldea, I. and L. S. Cederbaum. 2008b. Unusual features in optical absorption and photoionisation of quantum-dot nano-rings. In S. Wilson, P. J. Grout, G. Delgado-Barrio, J. Maruani, and P. Piecuch, editors, *Frontiers in Quantum Systems in Chemistry and Physics. Progress in Theoretical Chemistry and Physics*, Vol. 18, pp. 273–287. Springer Science + Business Media B. V., Heidelberg, Germany.

Bâldea, I. and H. Köppel. 2009. Studying the single–electron transistor by photoionization, arXiv:0811.2757. *Phys. Rev. B* 79:165317.

Bâldea, I., H. Köppel, and L. S. Cederbaum. 2000. Tunneling-driven quantum phase transitions in mesoscopic commensurate systems of strongly correlated electrons. *Solid State Commun.* 115:593–597.

Bâldea, I., H. Köppel, and L. S. Cederbaum. 2001a. Collective quantum tunneling of strongly correlated electrons in commensurate mesoscopic rings. *Eur. Phys. J. B* 20:289–299.

Bâldea, I., H. Köppel, and L. S. Cederbaum. 2001b. Symmetry-adapted BCS-type trial wave functions for mesoscopic rings. *Phys. Rev. B* 63:155308.

Bâldea, I., A. K. Gupta, L. S. Cederbaum, and N. Moiseyev. 2004a. High-harmonic generation by quantum-dot nanorings. *Phys. Rev. B* 69:245311.

Bâldea, I., H. Köppel, and L. S. Cederbaum. 2004b. Impact of phonons on quantum phase transitions in nanorings of coupled quantum dots. *Phys. Rev. B* 69:075307.

Bâldea, I., L. S. Cederbaum, and J. Schirmer. 2009. Intriguing electron correlation effects in the photoionization of metallic quantum-dot nanorings. arXiv:0807.3629. *Eur. Phys. J. B* 69:251–264.

Behrens, A. S., W. Habicht, K. Wagner, and E. Unger. 2006. Assembly of nanoparticle ring structures based on protein templates. *Adv. Mater.* 18:284–289.

Beni, G. and P. Pincus. 1974. Thermodynamics of an extended Hubbard model chain. i: Atomic limit for the half-filled band. *Phys. Rev. B* 9:2963–2970.

Cederbaum, L. S., W. Domcke, J. Schirmer, and W. von Niessen. 1986. Correlation effects in the ionization of molecules: Breakdown of the molecular orbital picture. *Adv. Chem. Phys.* 65:115–159.

Collier, C. P., R. J. Saykally, J. J. Shiang, S. E. Henrichs, and J. R. Heath. 1997. Reversible tuning of silver quantum dot monolayers through the metal-insulator transition. *Science* 277:1978–1981.

Collier, C. P., T. Vossmeyer, and J. R. Heath. 1998. Nanocrystal superlattices. *Ann. Rev. Phys. Chem.* 49:371–404.

Fuhrer, A., S. Luscher, T. Ihn et al. 2001. Energy spectra of quantum rings. *Nature* 413:822–825.

Glazman, L. I. and M. E. Raikh. 1988. Resonant Kondo transparency of a barrier with quasilocal impurity states. *JETP Lett.* 47:452–455.

Goldhaber-Gordon, D., J. Göres, M. A. Kastner, H. Shtrikman, D. Mahalu, and U. Meirav. 1998a. From the Kondo regime to the mixed-valence regime in a single-electron transistor. *Phys. Rev. Lett.* 81:5225–5228.

Goldhaber-Gordon, D., H. Shtrikman, D. Mahalu, D. Abusch-Magder, U. Meirav, and M. A. Kastner. 1998b. Kondo effect in a single-electron transistor. *Nature* 391:156–159.

Granados, D. and J. M. García. 2003. In(Ga)As self-assembled quantum ring formation by molecular beam epitaxy. *Appl. Phys. Lett.* 82:2401–2403.

Heath, J. R., C. M. Knobler, and D. V. Leff. 1997. Pressure/temperature phase diagrams and superlattices of organically functionalized metal nanocrystal monolayers: The influence of particle size, size distribution, and surface passivant. *J. Phys. Chem. B* 101:189–197.

Hewson, A. C. 1993. *The Kondo Problem to Heavy Fermions.* Cambridge University Press, Cambridge, U.K.

Hubbard, J. 1963. Electron correlations in narrow energy bands. *Proc. Roy. Soc. London A* 27:238–257.

Kastner, M. A.. 1993. Artificial atoms. *Phys. Today* 46:24–31.

Kawaharazuka, A., T. Saku, Y. Hirayama, and Y. Horikoshi. 2000. Formation of a two-dimensional electron gas in an inverted undoped heterostructure with a shallow channel depth. *J. Appl. Phys.* 87:952–954.

Klein, D. L., R. Roth, A. K. L. Lim, A. P. Alivisatos, and P. L. McEuen. 1997. A single-electron transistor made from a cadmium selenide nanocrystal. *Nature* 389:699–701.

Kong, X. Y., Y. Ding, R. Yang, and Z. L. Wang. 2004. Single-crystal nanorings formed by epitaxial self-coiling of polar nanobelts. *Science* 303:1348–1351.

Köppel, H., W. Domcke, and L. S. Cederbaum. 1984. Multimode molecular dynamics beyond the Born-Oppenheimer approximation. *Adv. Chem. Phys.* 57:59–246.

Kouwenhoven, L. 1995. Coupled quantum dots as artificial molecules. *Science* 268:1440–1441.

Kouwenhoven, L. and C. Marcus. 1998. Quantum dots. *Phys. World (UK)* 11:35–39.

Landau, L. D. and E. M. Lifshitz. 1984. Electrodynamics of continuous media. 2nd ed., *Landau and Lifshitz Course of Theoretical Physics*, Vol. 8. Pergamon Press, Oxford, NY.

Langreth, D. C. 1966. Friedel sum rule for Anderson's model of localized impurity states. *Phys. Rev.* 150:516–518.

Liu, Z. and R. Levicky. 2004a. Formation of nanoparticle rings on heterogeneous soft surfaces. *Nanotechnology* 15:1483–1488.

Liu, Z. and R. Levicky. 2004b. Ring structures from nanoparticles and other nanoscale building blocks. In J. A. Schwarz, C. I. Contescu, and K. Putyera, editors, *Dekker Encyclopedia of Nanoscience and Nanotechnology*, Vol. 3, pp. 3281–3288. CRC Press, Boca Raton, FL.

Maksimenko, S. A. and G. Y. Slepyan. 2004. Quantum dot arrays: Electromagnetic properties. In J. A. Schwarz, C. I. Contescu, and K. Putyera, editors, *Dekker Encyclopedia of Nanoscience and Nanotechnology*, Vol. 4, pp. 3097–3108. CRC Press, Boca Raton, FL.

Maldague, P. F. 1977. Optical spectrum of a Hubbard chain. *Phys. Rev. B* 16:2437–2446.

Markovich, G., D. V. Leff, S.-W. Chung, H. M. Soyez, B. Dunn, and J. R. Heath. 1997. Parallel fabrication and single-electron charging of devices based on ordered, two-dimensional phases of organically functionalized metal nanocrystals. *Appl. Phys. Lett.* 70:3107–3109.

Markovich, G., C. P. Collier, and J. R. Heath. 1998. Reversible metal-insulator transition in ordered metal nanocrystal monolayers observed by impedance spectroscopy. *Phys. Rev. Lett.* 80:3807–3810.

Markovich, G., C. P. Collier, S. E. Henrichs, F. Remacle, R. D. Levine, and J. R. Heath. 1999. Architectonic quantum dot solids. *Acc. Chem. Res.* 32:415–423.

Medeiros-Ribeiro, G., D. A. A. Ohlberg, R. S. Williams, and J. R. Heath. 1999. Rehybridization of electronic structure in compressed two-dimensional quantum dot superlattices. *Phys. Rev. B* 59:1633–1636.

Meir, Y. and N. S. Wingreen. 1992. Landauer formula for the current through an interacting electron region. *Phys. Rev. Lett.* 68:2512–2515.

Natan, A. 1992. Integrable models in condensed matter physics. In S. Lundquist, G. Morandi, and Y. Lu, editors, *Series on Modern Condensed Matter Physics, Lecture Notes of ICTP Summer Course*, Vol. 6, p. 458. World Scientific Singapore.

Ng, T. K. and P. A. Lee. 1988. On-site Coulomb repulsion and resonant tunneling. *Phys. Rev. Lett.* 61:1768–1771.

Oosterkamp, T. H., T. Fujisawa, W. G. van der Wiel et al. 1998. Microwave spectroscopy of a quantum-dot molecule. *Nature* 395:873–876.

Parr, R. G. 1963. *Quantum Theory of Molecular Electronic Structure*. W. A. Benjamin Inc., New York and Amsterdam, the Netherlands.

Petroff, P. M., A. Lorke, and A. Imamoglu. 2001. Epitaxially self-assembled quantum dots. *Phys. Today* 54:46–52.

Pines, D. 1962. The many-body problem. In *Frontier in Physics*. Benjamin, New York.

Remacle, F. 2000. On electronic properties of assemblies of quantum nanodots. *J. Phys. Chem. A* 104:4739–4747.

Remacle, F. 2001. Dissertation, Université de Liège. Liège, Belgium.

Remacle, F. and R. Levine. 2000a. Electronic response of assemblies of designer atoms: The metal-insulator transition and the role of disorder. *J. Am. Chem. Soc.* 122:4084–4091.

Remacle, F. and R. Levine. 2000b. Broken symmetry in the density of electronic states of an array of quantum dots as computed for scanning tunneling microscopy. *J. Phys. Chem. A* 104:10435–10441.

Remacle, F. and R. Levine. 2001. Superexchange, localized, and domain-localized charge states for intramolecular electron transfer in large molecules and in arrays of quantum dots. *J. Phys. Chem. B* 105:2153–2162.

Remacle, F., C. P. Collier, J. R. Heath, and R. D. Levine. 1998. The transition from localized to collective electronic states in a silver quantum dots monolayer examined by nonlinear optical response. *Chem. Phys. Lett.* 291:453–458.

Sachdev, S. 2000. *Quantum Phase Transitions*. Cambridge University Press, Cambridge, U.K.

Salem, L. 1966. *The Molecular Orbital Theory of Conjugated Systems*. Benjamin, New York.

Santhanam, V. and R. P. Andres. 2004. Metal nanoparticles and their self–assembly into electronic nanostructures. In J. A. Schwarz, C. I. Contescu, and K. Putyera, editors, *Dekker Encyclopedia of Nanoscience and Nanotechnology*, Vol. 3, pp. 1829–1840. CRC Press, Boca Raton, FL.

Shiang, J., J. Heath, C. Collier, and R. Saykally. 1998. Cooperative phenomena in artificial solids made from silver quantum dots: The importance of classical coupling. *J. Phys. Chem. B* 102:3425–3430.

Sun, S., C. B. Murray, D. Weller, L. Folks, and A. Moser. 2000. Monodisperse FePt nanoparticles and ferromagnetic FePt nanocrystal superlattices. *Science* 287:1989–1992.

Sun, Y. and Y. Xia. 2003. Triangular nanoplates of silver: Synthesis, characterization, and use as sacrificial templates for generating triangular nanorings of gold. *Adv. Mater.* 15:695–699.

van der Wiel, W. G., S. D. Franceschi, T. Fujisawa, J. M. Elzerman, S. Tarucha, and L. P. Kouwenhoven. 2000. The Kondo effect in the unitary limit. *Science* 289:2105–2108.

Waas, V., H. Büttner, and J. Voit. 1990. Finite-size studies of phases and dimerization in one-dimensional extended Peierls-Hubbard models. *Phys. Rev. B* 41:9366–9376.

Zdánská, P., V. Averbukh, and N. Moiseyev. 2003. High harmonic generation spectra of aligned benzene in circular polarized laser field. *J. Chem. Phys.* 118:8726–8738.

Index

Printed and bound by CPI Group (UK) Ltd, Croydon, CR0 4YY

23/10/2024

01778257-0010